AEROSOL MEASUREMENT

SECOND EDITION

AEROSOL MEASUREMENT

Principles, Techniques, and Applications

SECOND EDITION

Edited by

Paul A. Baron, Ph.D.
Physical Scientist
Centers for Disease Control and Prevention
National Institute for Occupational Safety and Health
Cincinnati, OH

Klaus Willeke, Ph.D.
Professor
Department of Environmental Health
University of Cincinnati
Cincinnati, OH

WILEY-INTERSCIENCE

A JOHN WILEY & SONS, INC., PUBLICATION

New York • Chichester • Weinheim • Brisbane • Singapore • Toronto

This book is printed on acid-free paper. ♾

Published simultaneously in Canada.

For ordering and customer service, call 1-800-CALL-WILEY.

Library of Congress Cataloging-in-Publication Data:

Aerosol measurement : principles, techniques, and applications / [edited] by Paul A. Baron and Klaus Willeke.—2nd ed.

p. cm.
Includes index.
ISBN 0-471-35636-0 (cloth)
1. Aerosols—Measurement. 2. Air—Pollution—Measurement. I. Baron, Paul A., 1944–
II. Willeke, Klaus.
TD884.5 .A33 2001
628.5′3′0287–dc21 2001017845

Printed in the United States of America.

10 9 8 7 6 5 4 3 2 1

CONTENTS

PREFACE

We dedicate this edition of *Aerosol Measurement* to two aerosol scientists, Professor David Swift and Professor Kvetoslav Spurny, both of whom have contributed to the current edition. Professor Swift, who provided many insights into the transport of aerosols within the respiratory system (Chapter 36), died in 1997 (obituary in *Aerosol Science and Technology* 1998, 28[4]). Professor Spurny died in 1999 (obituary in *Aerosol Science and Technology* 2000, 32[3]), only days after submitting a treatise on aerosol measurement history (Chapter 1). Both scientists made significant contributions to the aerosol community and will be greatly missed.

The measurement of aerosols has been practiced widely for several decades. Until the late 1980s, the development of new measurement methods was primarily motivated by the need to evaluate particulate pollution control devices and to find better means of monitoring indoor and outdoor aerosols. During the past several years, industry has become increasingly interested in modern aerosol measurement methods, not only to protect the health of their workers, as required by law, but also to increase productivity and, thereby, gain competitive advantage. For instance, in the production of semiconductor circuit boards a single submicrometer-sized particle may spoil the circuit if it adheres to the board where a circuit of submicrometer dimensions is being deposited. As a consequence, the number of undergraduate and graduate students taking courses in aerosol science and measurement has risen dramatically in recent years. The increased importance of this field is also evidenced by the creation and rapid growth of aerosol research associations, such as the American Association for Aerosol Research and several other national associations (a list is provided in Chapter 2).

In Part I of this book we present the fundamentals relevant for novices to this field, utilizing approaches developed in over 20 years of teaching university courses on aerosol science and measurement. Because we expect many readers to be air pollution regulators, industrial hygienists, and environmental scientists or engineers, we have applied our experience in teaching short courses to practitioners: The chapters in Part I stress the physics and give useful equations but avoid lengthy scientific derivations.

Almost all of the equations in the book have been incorporated into a spreadsheet program (freely available on the Internet) that is described in Chapter 2, allowing the reader to easily perform calculations and plot results. We believe that this can greatly aid in understanding aerosol mechanics and predicting behavior in experimental systems.

We have authored or co-authored several of the first chapters to provide models for the remaining chapter contributions in order to achieve a uniform style and a consistent structure in the book. Readers familiar with the principles of aerosol measurement can find details on specific instrumental techniques in Part II. Many of the chapters in Parts I and II offer sample calculations, thus making the book suitable for use as a teaching text. The practitioner concerned with the special requirements of his or her field, such as industrial hygiene or industrial aerosol processing, can find aerosol measurement applications in Part III. The bringing together of many applications fields by experts enables the reader to look into the practices of related fields so that technology transfer and adaptations may result.

Aerosol Measurement was first published in 1993. Since then, the original publisher, Van Nostrand Reinhold, was purchased and absorbed by John Wiley and Sons. This new edition of *Aerosol Measurement* contains new chapters and authors. Many of the original chapters have been significantly upgraded, reflecting the latest scientific and technological advances.

Several authors retired as aerosol practitioners or moved into other areas and thus did not want to contribute to this edition. We wish them all the best and hope that the new chapters fulfill the purpose of this book as well as the original ones did.

We thank all the contributors for generously providing their time and effort so that their expertise is available to the aerosol and associated communities. We also thank our wives Diane (P.A.B.) and Audrone (K.W.) for their support during the assembly of the two editions this book.

Paul A. Baron, Ph.D.
Cincinnati, Ohio
Klaus Willeke, Ph.D.
Cincinnati, Ohio

LIST OF PRINCIPAL SYMBOLS

ROMAN SYMBOLS

a	particle radius
amu	atomic mass units
A	area
	atomic mass
B	particle mobility (m/N·s); Eq. 4–14
c_m	mass concentration (g/m^3, mg/m^3, μg/m^3)
c_n	number concentration (particles/m^3)
c_q	particle charge concentration (C/m^3)
C	concentration of solute in solvent (m^3/m^3)
C_c	slip correction factor; Eqs. 4–8, 4–9
C_d	drag coefficient; Eqs. 4–19, 4–23, 4–24
d	diameter of an object, such as a particle (m, μm)
	characteristic dimension of an object (m)
	spatial dimension
d_a	aerodynamic diameter (m, μm); Eqs. 3–2, 3–3, 4–30
d_p	particle diameter (m, μm)
d_e	envelope equivalent diameter (m, μm)
d_m	mass equivalent diameter (m, μm); Eq. 4–21
$\overline{d}_p$	mean particle diameter
d_s	Stokes diameter (m, μm)
d_{50}	median particle diameter (m, μm)
D	diffusion coefficient of particle (m^2/s); Eq. 3–13
	Tube diameter
D_f	fractal dimension of an object
D_v	diffusion coefficient of vapor molecule (m^2/s); Eq. 4–4
e	charge on an electron (1.6×10^{-19} C)
f	frequency (Hz, s^{-1})
E	electric field (V/m)
	total efficiency of a filter; Eq. 9–1
	energy
F	force on particle (N)
g	gravitational constant (m/s^2)
G	gray level
h	height (m)
H	height of a chamber or duct (m)
	molecular accommodation coefficient; Eq. 4–51
I	number of charges (also n)
	intensity of light or radiation (J/s·m^2)
	electric current (amp)
J	flux of gas molecules or ions (number/s·m^2)

k	Boltzmann constant (1.38×10^{-23} J/K)
	thermal conductivity (W/m·K)
k_0	proportionality constant for radius of gyration; Eq. 23–9
K	coagulation coefficient (m^3/s); Eq. 5–9
K_r	wall loss rate; Eq. 33–8
L	length, light path length (m, μm)
m	refractive index
m_p	particle mass (g, mg, μg, ng)
M	gram molecular weight (g/cm^3)
N	number concentration (number/m^3)
N_p	particle concentration (number/m^3)
n	molecular concentration (number/m^3)
	number of unit charges (also i)
	number of particles
	number of measurements
p	partial pressure (N/m^2, Pa, atm)
	ion polarity
p_s	saturation vapor pressure (N/m^2, Pa, atm)
P	pressure (N/m^2, Pa, atm)
	penetration fraction
	scattered light flux
Pe	Peclet number; Eq. 4–16
q	charge on a particle (C)
	light-scattering vector
Q	flow rate (m^3/s)
Q_e	particle extinction efficiency
r	distance between two particles (m)
	radial distance (m)
	particle radius (m; μm)
R	specific gas constant (N·m/K·kg), Eq. 3–3
	ratio
	fractal particle perimeter radius
	resolution
R_g	radius of gyration; Eq. 23–12
R_u	universal gas constant (8.31×10^7 dyne cm/K mole); Eq. 4–3
Re_f	flow Reynolds number; Eqs. 4–1, 4–2
Re_p	particle Reynolds number; Eqs. 4–1, 4–2
Re_0	particle Reynolds number under initial conditions; Eq. 4–37
s	signal
S	stopping distance (m, μm); Eqs. 4–36, 4–37
	Sutherland constant (K), Eq. 4–10
	particle emission rate; Eq. 33–8
	signal
Sc	Schmidt number; Eq. 4–17
Sh	Sherwood number; Eq. 8–57
S_R	saturation ratio; Eq. 5–2
Stk	Stokes number; Eq. 4–39
$t_{1/2}$	half life or half time; Eq. 5–12
T	temperature (K, °C)
	transmittance
U	gas velocity (m/s)

	sampling velocity (m/s)
	output signal of a photometer (V)
U_0	ambient gas velocity (m/s)
v_a	sampled air volume (m^3)
v_p	particle volume (m^3)
V	velocity of particle relative to gas (m/s)
	potential (V)
$\overline{V}$	average molecular velocity (m/s); Eq. 4–5
V_p	particle velocity (m/s)
V_0	initial velocity of a particle (m/s)
V_{ts}	terminal settling velocity (m/s); Eq. 4–28
x	distance in x direction (m)
x_{rms}	root mean square Brownian motion in the x direction (m); Eq. 4–15
Z	electrical mobility (m^2/V·s); Eq. 4–45
	atomic number

Greek Symbols

α	coefficient in slip correction equation; Eq. 4–8
	thermal diffusivity (m^2/s)
	attachment coefficient
β	coefficient in slip correction equation; Eq. 4–8
	length to width ratio, aspect ratio
	frequency of ion attachment; Eq. 18–10
	flow rate ratio; Eq. 18–31
γ	coefficient in slip correction equation; Eq. 4–8
	surface tension (N/m)
	specific heat ratio
δ	limiting sphere radius for ion transport; Eq. 18–9
ε	dielectric constant
η	dynamic viscosity (P, N·m/s)
	efficiency
θ	angle (rad, °)
κ	relative permittivity
λ	mean free path (m, μm)
	wavelength (m, μm)
μ	mass absorption coefficient
ν	kinematic viscosity (m^2/s)
ρ_e	effective density that includes voids (kg/m^3)
ρ_f	fluid density (kg/m^3)
ρ_g	gas density (kg/m^3)
ρ_p	particle density (kg/m^3)
σ	standard deviation; Eq. 22–2
σ_g	geometric standard deviation
τ	relaxation time of particle; Eq. 4–34
ϕ	angle (rad, °)
χ	dynamic shape factor
ω	angular velocity (rad/s)
Ω	transfer function

Subscripts

a	air or gas
	aspiration
ac	alternating current
B	mobility equivalent
c	cylinder
d	droplet, droplet surface
	drag
dc	direct current
dep	deposition
diff	in a diffusiophoretic field
e	effective
elec	in an electric field
ev	equivalent volume
f	flow
	fluid
	fiber
	fractal
g	geometric
	gas
grav	in a gravity field
i	initial
	individual
j	jet
n	number
m	mass
mob	mobility
p	particle
r	reference to NTP
s	saturation condition
sonic	speed of sound
th	in a thermal gradient field
ts	terminal settling under influence of gravity
trans	transmission
0	initial condition
∞	far from particle surface

CONTRIBUTORS LIST

URS BALTENSPERGER, Paul Scherrer Institute, CH-5232 Villigen PSI, Switzerland

PAUL A. BARON, National Institute for Occupational Safety and Health, MS R-3, 4676 Columbia Parkway, Cincinnati, OH 45226

PRATIM BISWAS, Department of Chemical Engineering, Campus Box 1180, Washington University, St. Louis, MO 63130-4899

JOHN E. BROCKMANN, Sandia National Laboratories, Dept 9114-Mail Stop 0827, Albuquerque, NM 87185-0834

HEINZ BURTSCHER, Fachhochschule Aargau, University of Applied Science, CH-5210 Windisch, Switzerland

BRUCE K. CANTRELL, National Institute for Occupational Safety and Health, Pittsburgh Research Laboratory, Bldg 152, 626 Cohrans Mill Rd., P.O. Box 18070, Pittsburgh, PA 15236-0070

BEAN T. CHEN, National Institute for Occupational Safety and Health, MS 3030, 1095 Willowdale Rd., Morgantown, WV 26505-2845

YUNG SUNG-CHENG, Lovelace Respiratory Research Institute, P.O. Box 5890, Albuquerque, NM 87185

JUDITH C. CHOW, Desert Research Institute, University and Community College System of Nevada, 2215 Raggio Parkway, Reno, NV 89512-1095

BEVERLY S. COHEN, New York University School of Medicine, Nelson Institute of Environmental Medicine, 57 Old Forge Rd., Tuxedo, NY 10987

DOUGLAS W. COOPER, The Texwipe Company, LLC, 650 E. Crescent Ave., Upper Saddle River, NJ 07458-1827

CHATTEN COWHERD, JC., Midwest Research Institute, 425 Volker Blvd., Kansas City, MO 64110-2299

E. JAMES DAVIS, Department of Chemical Engineering, Box 351750, University of Washington, Seattle, WA 98915-1750

ROBERT P. DONOVAN, L&M Technologies, Inc., 4209 Balloon Park Rd., Albuquerque, NM 87109

RICHARD C. FLAGAN, California Institute of Technology, Chemical Engineering Dept., 1200 E. California Blvd, MC 210-41, Pasadena, CA 91125

ROBERT A. FLETCHER, National Institute of Standards and Technology, Chemistry A 113, Gaithersburg, MD 20899

Josef Gebhart, Kirchbornstrasse 13, D-63128 Dietzenbach, Germany

SERGEY GRINSHPUN, Department of Environmental Health, University of Cincinnati, P.O. Box 670056, Cincinnati, OH 45267-0056

WILLIAM A. HEITBRINK, National Institute for Occupational Safety and Health, MS R-5, 4676 Columbia Parkway, Cincinnati OH 45226

ANTHONY J. HICKEY, School of Pharmacy, University of North Carolina, Chapel Hill, NC 27599–7360

WILLIAM C. HINDS, School of Public Health, UCLA, 10833 LeContre Ave., Los Angeles, CA 90024-1772

MARK D. HOOVER, Lovelace Respiratory Research Institute, P.O. Box 5890, Albuquerque, NM 87185-5890

PAUL A. JENSEN, National Institute for Occupational Safety and Health, 1095 Willowdale Rd., Morgantown, VA 26505

WALTER JOHN, Particle Science, 195 Grover Lane, Walnut Creek, CA 94956

MURRAY V. JOHNSTON, Department of Chemistry and Biochemistry, University of Delaware, Newark, DE 19716

JORMA KESKINEN, Aerosol Physics Laboratory, Department of Physics, Tampere University of Technology, P.O. Box 692, FIN-33101 Tampere, Finland

TOIVO T. KODAS, Superior Powder Techniques, 3740 Hawkins NE, Albuquerque, NM 87109

MATTHEW LANDIS, U.S. Environmental Protection Agency, MD-47, Office of Research and Development, National Exposure Laboratory, Research Triangle Park, Durham, NC 27711

KEN W. LEE, Kwangju Institute of Science and Technology, Department of Environmental Science and Engineering, 572 Sangam-dong, Kwangsan-ku, Kwangju, 50-6303, South Korea

VIRGIL MARPLE, University of Minnesota, 142 Mechanical Engineering, 111 Church St, SE, Minneapolis, MN 55455

MALAY K. MAZUMDER, Department of Applied Science, University of Arkansas at Little Rock, 2801 S. University, Little Rock, AR 72204-1009

ANDREW D. MAYNARD, National Institute for Occupational Safety and Health, MS R-3, 4676 Columbia Parkway, Cincinnati, OH 45226

OWEN R. MOSS, Chemical Industry Institute of Toxicology, Research Triangle Park, Durham, NC 27709-2137

R. MUKUND, GE Power Systems, 111 Merchant Street, MD: S-30, Cincinnati, OH 45246

AINO NEVALAINEN, National Public Health Institute, Department of Environmental Biology, P.O. Box 95, FIN-70701 Kuopio, Finland

GEORGE J. NEWTON, 449 Graceland SE, Albuquerque, NM 87185-5890

GARY NORRIS, U.S. Environmental Protection Agency, MD-47, Office of Research and Development, Research Triangle Park, Durham, NC 27711

TIMOTHY J. O'HERN, Engineering Sciences Center, Sandia National Laboratories, MS 0834, Albuquerque, NM 87185-0834

BERNARD A. OLSON, University of Minnesota, 125 Mechanical Engineering Building, 111 Church St SE, Minneapolis, MN 55455

Sortiris E. Pratsinis, ETH Institut für Verfahrenstechnik, ETH Zentrum, ML F26, CH 8092 Zurich, Switzerland

Daniel J. Rader, Sandia National Laboratories, MS 9042, P.O. Box 969, Livermore, CA 94551-0969

Tiina Reponen, Department of Environmental Health, University of Cincinnati, P.O. Box 670056, Cincinnati, OH 45267-0056

Charles E. Rodes, Research Triangle Institute, P.O. Box 12194, Research Triangle Park, Durham, NC 27709-2194

Kenneth L. Rubow, Mott Corporation, Farmington, CT 06032

J. H. J. Scott, National Institute of Standards and Technology, Chemistry A 113, Gaithersburg, MD 20899

W. Russell Seebaugh, Department of Engineering, University of Denver, Denver, CO 80208-0177

Georgios Skillas, ETH Institut für Verfahrenstechnik, ETH Zentrum, ML F26, CH 8092 Zurich, Switzerland

Christopher M. Sorensen, Kansas State University, Department of Physics, Manhattan, KS 66506-2601

Paul A. Solomon, U.S. Environmental Protection Agency, 944 East Harmon Ave., Las Vegas, NV, 89119

John A. Small, National Institute of Standards and Technology, Chemistry A 113, Gaithersburg, MD 20899

Kvetoslav R. Spurny, Deceased. Grafschaft, Postfach 1260, D-57377 Schmallenberg, Germany

David Swift, Deceased. Division of Environmental Health Engineering, Johns Hopkins University, Baltimore, MD 21205

Michael Tolocka, U.S. Environmental Protection Agency, MD-46, Office of Research and Development National Exposure Laboratory, Research Triangle Park, Durham, NC 27711

Jon Volkwein, National Institute for Occupational Safety and Health, Pittsburgh Research Laboratory, Bldg 152, 626 Cochrans Mill Rd., Pittsburgh, PA 15236-0070

John G. Watson, University and Community College System of Nevada, Desert Research Institute, 2215 Raggio Parkway, Reno, NV 89512-1095

Ernest Weingartner, Paul Scherrer Institute, CH-5232 Villigen PSI, Switzerland

Anthony S. Wexler, Mechanical and Aeronautical Engineering, University of California, One Shields Avenue, Davis, CA 95616

Russell W. Wiener, National Exposure Assessment Research Laboratory, U.S. Environmental Protection Agency, MD-77, Research Triangle Park, NC 27711

Klaus Willeke, Department of Environmental Health, University of Cincinnati, PO Box 670056, Cincinnati, OH 45267-0056

James Charles Wilson, Department of Engineering, University of Denver, Denver, CO 80208

PART I
Principles

1 Historical Aspects of Aerosol Measurements

KVETOSLAV R. SPURNY*
Grafschaft, Schmallenberg, Germany

INTRODUCTION

Aerosols have been recognized as a specific topic of basic and applied science since World War II. As a discipline, aerosol science has its own history, which was created and positively influenced by renowned physicists, chemists, meteorologists, and so forth, as well as by political and economical events and technological development. The methodology of aerosol measurement is a substantial part of the history of aerosol science. Fortunately, several journals have shown interest in publishing papers dealing with the historical development of aerosol science. It was mainly the American Association for Aerosol Research's journal *Aerosol Science and Technology*, that recognized the importance of such publications for the further development of aerosol science in the third millennium (e.g., Davis, 1997; Kerker, 1997; Walton and Vincent, 1998; Spurny, 1998a, 2000b; Knutson, 1999; McMurry, 2000). The proceedings of the first Symposium on the History of Aerosol Science have been published (Preining and Davis, 2000).

THE EARLY DAYS

Aerosol science history is closely related to air pollution history. The existence of unpleasant and harmful particles in outdoor and indoor atmospheres was mentioned in very early literature. For example, the Romans complained of the foul air in ancient Rome. Serious particulate air pollution led to the prohibition of coal burning in London in 1273, followed by a Royal Proclamation by Edward I in 1306. In 1661, John Evelyn submitted the first major tract regarding particulate air pollution to Charles II. His *Fumifugium* contained a graphic description of pollution in the city of London (Lodge, 1969). However, the birth of aerosol science and aerosol measurement methodology did not occur until the second half of the nineteenth century.

PRECLASSICAL PERIOD OF AEROSOL MEASUREMENT

Research and development of measurement methodology in aerosol science before the 1900s are considered as events of the preclassical period. The first aerosol research efforts

* Deceased, November 3, 1999.

Aerosol Measurement: Principles, Techniques, and Applications, Second Edition, Edited by Paul A. Baron and Klaus Willeke.
ISBN 0-471-35636-0

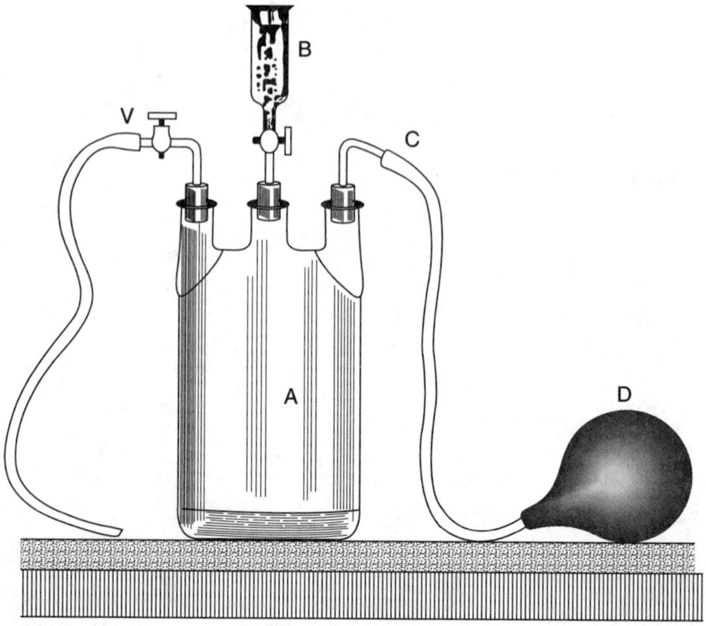

Fig. 1–1. Apparatus used by P. J. Coulier for detection of the condensation activity of dust particles. A, transparent flask; B, burette; C, connection; D, rubber ball; V, valve.

were closely associated with initial developments in colloid chemistry (Spurny, 1998a). Within the preclassical period, the first observations were made of fine particles dispersed in the atmosphere, and some early experiments were performed in the laboratory. According to McMurry's review (2000), as early as 1841, J. P. Espy built a "nephelescope" with which he was able to observe water cloud formation under laboratory conditions. He was not aware that condensation occurred on the particles. According to Podzimek's reviews (1985, 1989), in 1847 H. Becquerel hypothesized about the existence of fine particles in air, now called *condensation nuclei.* Their existence was confirmed about 30 years later by experiments of Coulier (1875). He was the first to publish work showing that when air is expanded adiabatically, condensation occurs more readily in unfiltered air than in filtered air. The apparatus of Coulier was in principle the first condensation nuclei detector (Fig. 1–1). It consisted of a transparent flask (A) with a rubber bulb (D) for compressing the air in the flask. By opening the valve (V) the air could expand and condensation on aerosol particles could occur, which was qualitatively determined by the cloudiness within the chamber.

John Aitken began his research on condensation in 1875 and published his results in 1880 and later (Aitken, 1880a,b, 1888–1889). His apparatus and experiments were very similar to those of Coulier. He recognized Coulier as the first to show the important part played by nuclei in the cloudy condensation of water (Spurny, 2000b). Aitken, who was born in Falkirk, Scotland, in 1839 and died in 1919, had developed and used the first portable instruments for counting dust particles in the atmosphere (Aitken, 1888–1889, 1890–1891, 1920). Figure 1–2 shows the original schematic of that counter. The air being tested was drawn through pipe A and passed into receiver R, where it was mixed with a certain quantity of dustless air and saturated with water. The air in R was then expanded by the pump, which produced a shower of rain. The number of drops that fell on a measured area were then counted. Already by the

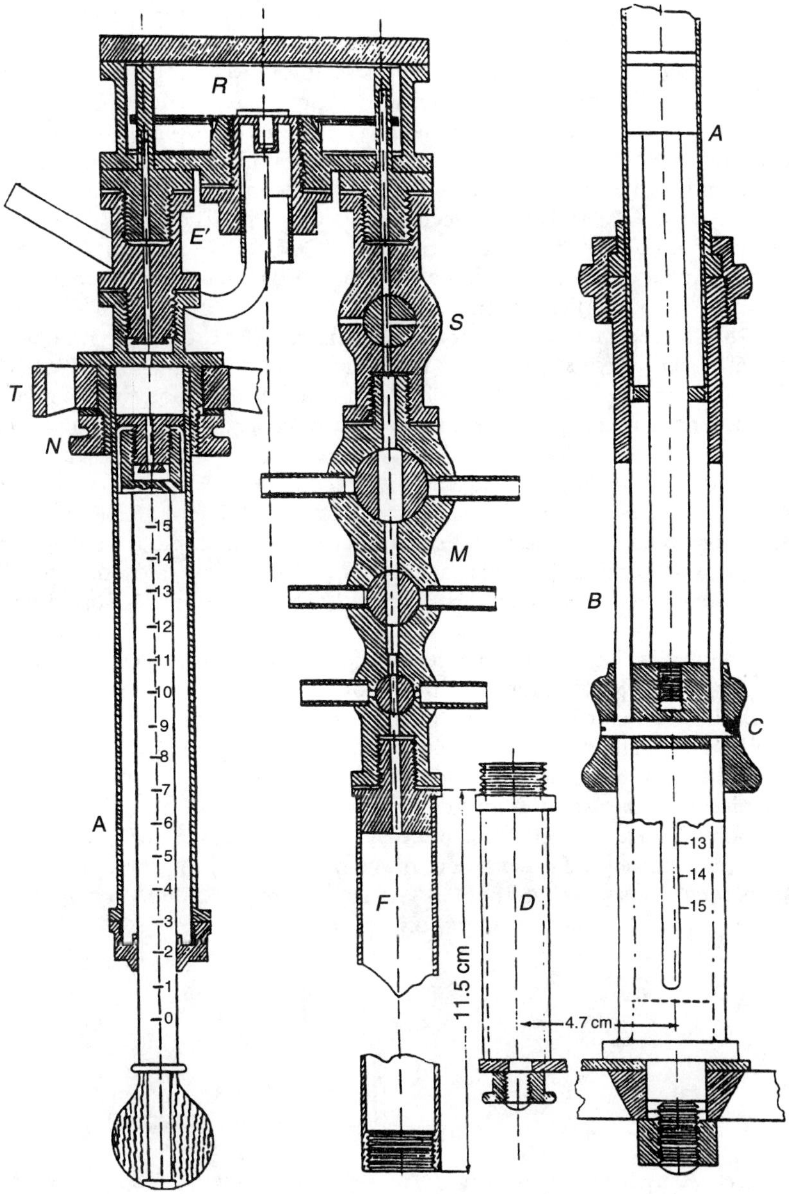

Fig. 1–2. The original schematic showing the design of a portable "dust counter." A, air-pump; R, receiver; S, stopcock; M, apparatus for measuring air to be tested; F, filter; T, "head" of tripod; N, nut; B, metal support; D, end of tube; E, part E.

end of the 1890s, C. T. R. Wilson had developed a refined expansion cloud chamber that he used to study homogeneous nucleation (Wilson, 1897).

The first recorded use of laboratory-generated aerosols was by Leonardo da Vinci (1452–1519) (Kerker, 1997). It was not until several centuries later that John Tyndall repeated da Vinci's experiments. John Tyndall was born in 1820 in Ireland and died in 1893 in England.

Tyndall studied in Marburg, Germany, with W. Bunsen and later worked with Michael Faraday, eventually becoming his successor. Tyndall's observations that dust and smoke in a room are easily detectable by the light that is scattered when a beam of sunlight enters the room was used in 1856 by Faraday to indicate the presence of colloidal particles in liquids. A decade later, Tyndall extended the method to detect aerosols and was the first to apply this method to the detection of particulate air pollution in indoor air (Tyndall, 1871; Gentry, 1997). At the end of the 1860s, he studied fine particles in the air by means of dark-field illumination (Tyndall, 1869a). He produced clouds of small particles by irradiation of mixtures of air with various vapors and gases, such as HCl, HBr, HI, and amyl nitrate (Tyndall, 1869b, 1870a). Tyndall was not only the inventor of the tyndallometer, nephelometer, ultramicroscope, and optical particle counter but also the indirect inventor of the thermal precipitator. In 1870 he reported the observation of a narrow dark region above a heated body in a dusty atmosphere. Several years later Lord Rayleigh (1882), Lodge (1883, 1884), Lodge and Clark (1885), and J. Aitken (1894) observed that a heated body was completely surrounded by a dark space. In principle, this was the discovery of the thermophoretic effect (Fuchs, 1971).

Another effect, which later led to the construction of electrostatic precipitators, was already observed in the early 1880s. Several investigators (Hohlfeld, 1824; Lodge and Clark, 1883, 1885; Lodge, 1886) reported that a small electric discharge into a smoke-laden atmosphere rapidly dissipated the smoke by coagulating the particles into agglomerates. This effect was probably already observed much earlier, in the 1770s, by Giovanni Battista Beccaria, who published similar observations (Beccaria, 1772).

The glass impingers also had their progenitor before the 1900s. A glass impinger was used for the sampling of bacterial aerosols in the laboratory of Robert Koch at the Institute of Hygiene in Berlin, Germany, in the 1880s. Koch's assistant, Michaelis, used this impinger (L) in his apparatus for the testing of dust respirators (Fig. 1–3) (Michaelis, 1890). In his application, the end of the glass tube was not placed under the water surface, but the bacterial aerosols were usually sampled into the water from above its surface. Early measurements of bacterial aerosols in room air were also reported before the 1900s (Singerson, 1870–1874).

In 1881, ultramicroscopical observations of the motion and deposition of smoke particles were described by Bodaszewsky (1881) in Germany. Somewhat later, Townsend (1900), in England, conceived of the first aerosol diffusion battery.

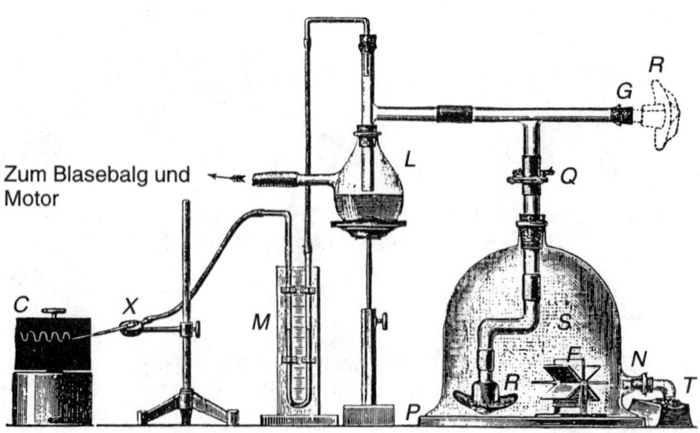

Fig. 1–3. Michaelis' apparatus for the testing of respirators. L is the glass impinger used at that time in the laboratory of Robert Koch in Berlin. Zum Blasebalg und Motor: To the bellows and motor. M, manometer; L, impinger; Q, pinch-cock.

CLASSICAL PERIOD OF AEROSOL MEASUREMENT

The period of classical aerosol physics (Spurny, 1993) was characterized by the use and exploitation of measurements and experimental techniques common during that time. The classical period of aerosol science research lasted approximately until the middle of the twentieth century and ended with the publication of *Mechanics of Aerosols* (Fuchs, 1955, 1964). No lasers, no computers, and no spectroscopic analytical tools were available during this period.

The term *aerosol* was first used at this time. It was coined by the physical chemist E. G. Donnan in about 1918 and introduced into the meteorological literature in 1920 by A. Schmauss, the director of the Meteorological Central Station in Munich, Germany (Schmauss, 1920a,b). He used the new term to compare the colloidal chemical processes with the processes in a cloudy atmosphere. He found important similarities. By analogy to the term *hydrosol*, he used the term *aerosol* for clouds of particles and droplets dispersed in air.

The broader development of aerosol measurement methods and equipment occurred after 1900, and primarily after 1920. During this period, the negative health effects of industrial aerosols and dusts were recognized (Sinclair, 1950; Davies, 1954; Drinker and Hatch, 1954). In general, aerosols, and especially industrial dusts, can be measured while particles are airborne or after the particles have been collected on a surface by physical or chemical means. In the early 1920s, as well as during the entire period before the 1960s, the latter collection method was preferred in the field of industrial hygiene.

Measurement Philosophy and Strategy

The first attempts to measure airborne dust were made at the beginning of the 1900s. The first approach was the simple and obvious one of drawing a known volume of dusty air through a filter and weighing the quantity collected. At first a cotton wool filter was used; by about 1906, this was superseded by the "sugar tube" in which dust was trapped on a bed of sugar granules. The sugar was dissolved in water, and the collected dust was weighed after deposition onto a filter (Walton, 1982). The total dust gravimetric method had several disadvantages. The most important one was that relatively small numbers of coarse particles constituted a major part of the measured mass. Microscopic analysis of silicotic lung tissues had shown that the dust retained in the lung consisted only of relatively small particles. More than 70% of the particles deposited in the lung were smaller than 1 μm (McCrae, 1913). This finding stimulated the development of a second generation of dust-sampling instruments designed to provide microscopic counts of the numbers of fine particles (smaller than 5 μm). The number of such particles in 1 ml of air was measured. This strategy continued until the introduction of respirable dust gravimetric samplers during the 1950s (Davies, 1952).

It is of interest that, according to the latest measurement strategy for atmospheric aerosols, the philosophy of particle number concentration measurement appears to be finding favor again (Spurny et al., 1969). It was recently found in epidemiological and toxicological studies that the concentration of particles smaller than 1 μm and their chemical composition were better correlated to human health effects than were those of larger sized particles. Within this particle size range of less than 1 μm, the number concentration of particles appears to be more informative than the particle mass. I began my aerosol measurement work at the end of the 1940s and can remember very well the philosophy of dust measurement at that time. The most important reason for dust measurement in the workplace was the high incidence of silicosis in both industry and in the mines. An important observation of the high mortality of hard-rock miners, accredited to Agricola (the Latinized name of Georg Bauer, who wrote "De Re Metallica"), first appeared in sixteenth century literature (Drinker and Hatch, 1954). It was in the late 1920s that silica (quartz dust) was recognized as producing the pulmonary diseases of pneumoconiosis and silicosis (Collis, 1926).

A broad need for the measurement of industrial dust in the workplace was therefore recognized before, but more so after, World War II. Generally speaking, knowledge already existed of several physical methods for dust sampling: inertial particle separation, filtration, thermophoresis, and electrostatic precipitation. However, very few sample analysis methods existed, although light microscopic methods were available for particle counting and sizing. Therefore, the procedures of choice were particle sampling on plain and smooth transparent surfaces and/or in liquids and particle counting and sizing by light microscope methods.

Aerosol Sampling Methods

Gravity is not a strong enough force to separate respirable dust and other aerosol particles from air samples in a reasonable time. Inertial, thermal, and electrostatic forces must be applied to speed up the particle deposition, or an efficient filtration system must be used. All of these methods are suitable for sampling aerosols to estimate the particle numbers or particle mass concentrations (Spurny et al., 1961).

Konimeters. The term *konimeter* was used to designate a one-stage impactor. Sir Robert Kotze in the Union of South Africa (Innes, 1919) developed the first successful konimeter in 1919. The dust particles were collected by impaction onto a glass plate covered with a thin film of petroleum or glycerin, which trapped and retained the dust. In the United Kingdom, the Owens Jet Dust Counter (Owens, 1922) was used for a long time (Fig. 1–4). It was very similar to the Kotze konimeter. It contained an entrance chamber in front of a rectangular nozzle. No adhesive substance was used on the impaction glass surface. Instead, the entrance chamber was lined with moistened blotting paper to ensure humidification of the sampled air volume. Later, in the 1930s, Behounek improved the Owens counter by using a vacuum reservoir combined with a hand pump (Fig. 1–5) (Behounek, 1939; Behounek et al., 1942) to avoid fluctuation of the flow rate through the nozzle. Commercial konimeters produced in the United Kingdom and in Germany have been available since about the 1930s.

An English konimeter, the Bausch and Lomb counter (Gurney et al., 1938), was an improved Owens konimeter. It could collect 12 samples on a circular glass plate. This instrument included a light microscope (Fig. 1–6) similar to the later Zeiss konimeter. The Zeiss konimeter (Lehmann et al., 1934; Löbner, 1935; Zeiss, 1950) could collect 30 dust samples on a single glass disk, which was rotated to permit the immediate examination of the spots under the built-in microscope. In England, Walton (1936) developed and introduced into practice a photoelectrical estimation of konimeter dust spots. Measurements with konimeters had many disadvantages. The sampling time was very short, in the range of seconds, and the repro-

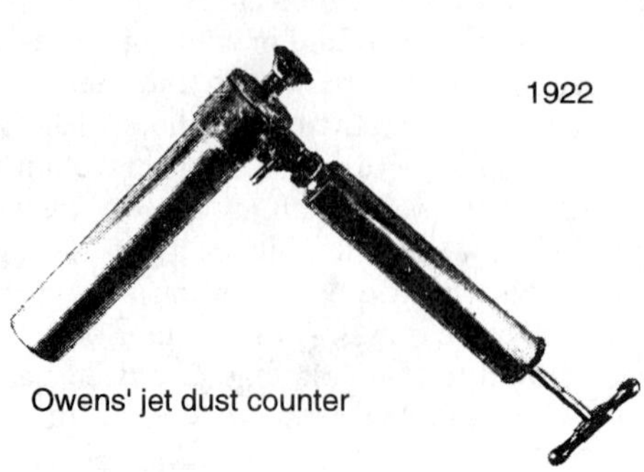

Fig. 1–4. The Owens Jet Dust Counter used in England in the 1920s.

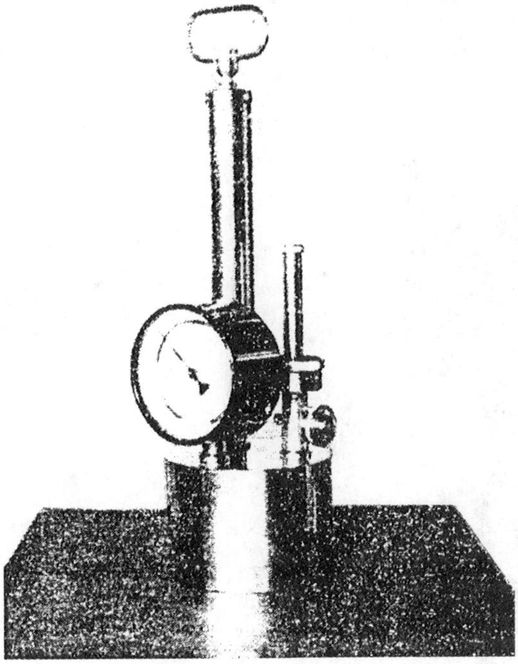

Konimetr Owens-Běhounek.

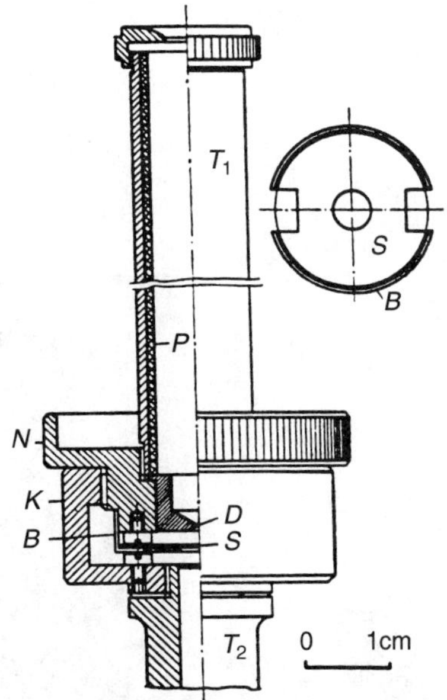

Bild 4. Einrichtung des Staubmessers. K = fester Teil, N = abnehmbarer Teil mit der Staubkammer, B = abnehmbarer Ansatz, S = Deckglas, D = Düse, T_1 = Einsaugrohr, T_2 = Verbindungsrohr zum Metallgefäß, P = Filterpapier.

Fig. 1–5. The original picture of the Owens' konimeter improved by Behounek. Arrangement of the dust measuring device. K = Solid part, N = detachable part with dust chamber, B = detachable lug, S = cover glass, D = nozzle, T1 = inlet tube, T2 = connecting tube to the metal vessel, P = filter paper.

ducibility and the results of interlaboratory comparisons were poor. There was no correlation between particle number and mass (Beadle, 1951). Ranz and Wong (1952) made the first important theoretical and experimental study of the collection of aerosol particles by inertial impaction.

Cascade Impactors. An important improvement in the field of dust sampling was achieved by an instrument consisting of four impaction stages developed in the 1940s by May in England (May, 1945). Four jets were arranged in a series, and the dust particles were collected on adhesive-coated microscope slides. The May cascade impactor was produced commercially in England by the Casella Company (Fig. 1–7). In 1946 Soskin improved such a cascade impactor for the sampling and sizing of aerosol particles below 1 μm in diameter (Soskin, 1946). Andersen introduced his cascade impactor, consisting of six stages, in 1958 (Andersen, 1958) (Fig. 1–8). The development of more sophisticated cascade impactors began after the 1960s (Mercer, 1973). See Chapter 10 for the discussion of current instrumentation.

Impingers. Impingers are very similar to konimeters. The only difference between the two is that, in an impinger, particle impaction onto a solid surface is combined with subsequent collision of dust particles with a liquid, such as water or alcohol. The impinger, shown schematically in Figure 1–9, is operated like a konimeter (impactor) except that the jet is immersed in the liquid. In operation, particles larger than about 1 μm are captured by inertia

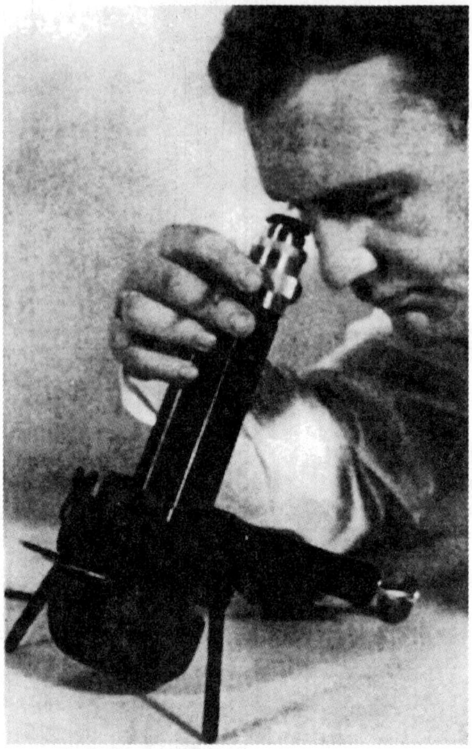

Bausch and Lomb dust counter

Fig. 1–6. Dust particle counting by means of the Bausch and Lomb konimeter.

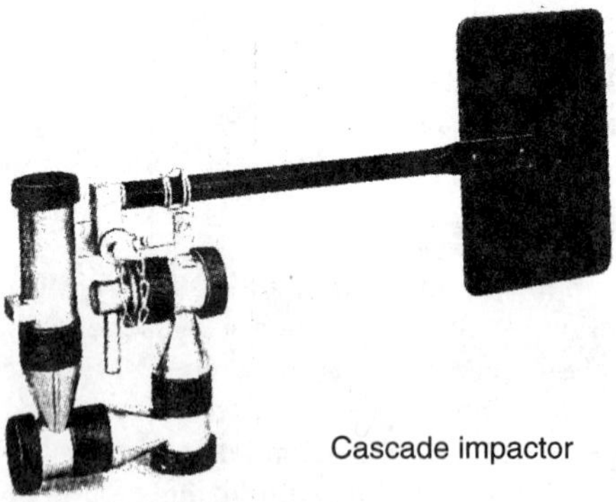

Cascade impactor

Fig. 1–7. The Casella cascade impactor for dust particle fractionation.

and end up suspended in the liquid. The collection efficiency drops off rapidly for particles less than 1 μm. After air of known volume is sampled, the dust concentration is evaluated by light microscopy in a counting glass cell.

The most important impinger developments and applications came after the 1920s (Greenburg and Smith, 1922; Emery, 1927; Greenburg, 1932; Dalla Valle, 1937; Holt, 1951). The Greenburg-Smith impinger and its smaller modification, the midget impinger, were used as standard equipment for a long time in the United States, as well as in several other coun-

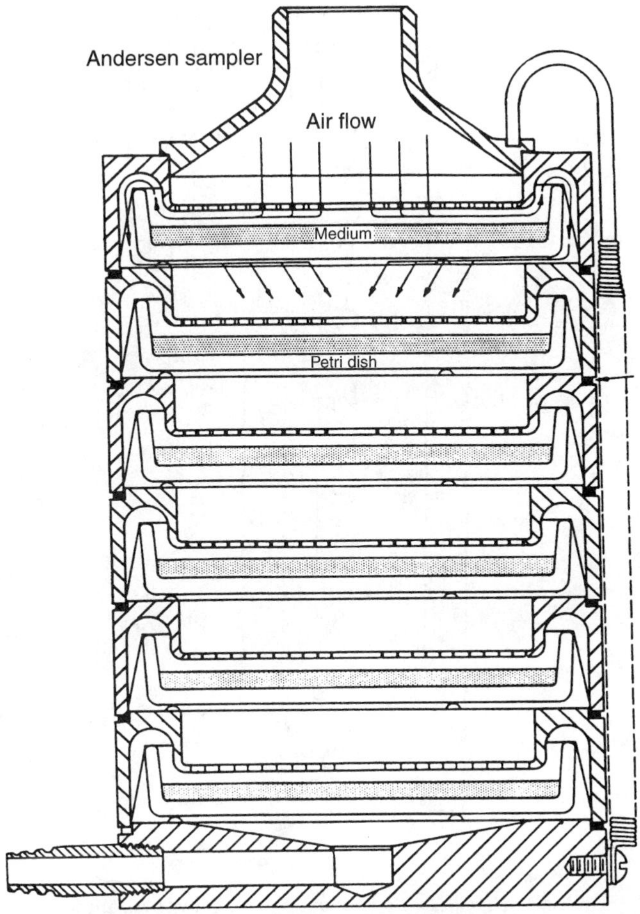

Fig. 1–8. Andersen's six-stage cascade impactor.

tries (Hatch et al., 1932). Bernz (1942) published a theoretical treatment of impingers. See Chapters 10, 24, 25, and 31 for discussions of current instrumentation.

Precipitators. Measurement devices for airborne dust measurement, in which the airborne dust particles were separated in thermal and electric fields, were called thermal and electrostatic precipitators, respectively.

Thermal Precipitators. Bancroft (1920) stated that thermophoresis plays an important role in dust separation and proposed the development of a thermal filter. Einstein (1924), Hettner (1924), and Epstein (1929) investigated radiometric forces and measurement methods. In the 1930s very important studies were conducted using thermophoretic forces to sample dust in workplace air. Impingers and konimeters were found unsuitable for basic scientific dust measurements. The thermal precipitator was found to be much more satisfactory (Green, 1934; Green and Watson, 1935). Miyake (1935) experimented with a heated platinum ribbon and separated dust particles from gas flow. Green and Watson (1935) and Watson (1936) conducted the most successful experiments. These studies resulted in the construction of a portable thermal precipitator.

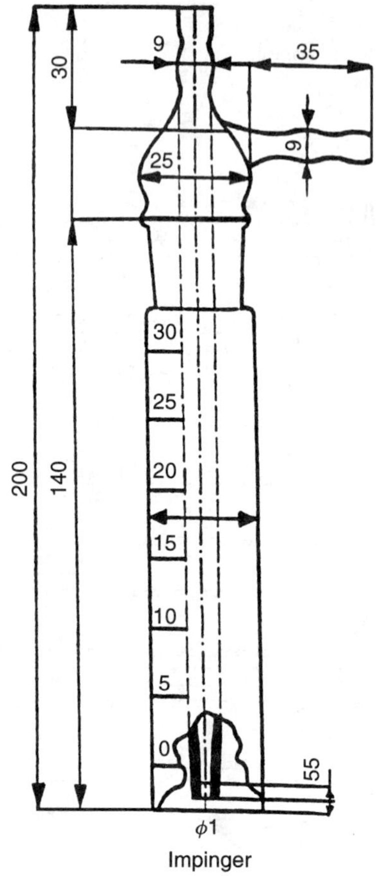

Fig. 1–9. Schematic of a dust impinger. Dimensions are in mm.

The Green-Watson thermal precipitator used a nichrome (NiCr) wire located across a slot. The wire was heated electrically to a temperature of about 100°C. Figure 1–10 shows a schematic of this instrument and its function. The walls of the slot are formed by two coverslips backed by blocks of brass. The dust deposit was evaluated by light microscopy.

The thermal precipitator was used for many years as a standard dust sampling instrument in several countries and underwent a number of modifications and improvements. Walton et al. (1947) built a modified thermal precipitator for the quantitative sampling of aerosols for electron microscopic evaluations. Laskin (1951) developed an oscillating thermal precipitator. Thermal precipitators with oscillatory or rotating collecting surfaces were designed to reduce the effect of particle overlap and to eliminate the problem of size segregation (Cember et al., 1953). Kathley et al. (1952) developed and used a thermal precipitator for sampling airborne bacteria. Hamilton (1952) designed a long-running thermal precipitator. Walkenhorst (1962) used a heated tungsten ribbon instead of a wire for improving the sampling conditions. Orr and Martin (1958) developed a thermal precipitator for continuous aerosol sampling. Bredl and Grieve (1951) constructed the first gravimetric thermal precipitator. A schematic of this instrument is shown in Figure 1–11. The dust was collected on an aluminum plate (AL), which could be weighed. The upper plate was heated electrically. See Chapter 10 for a discussion of current instrumentation.

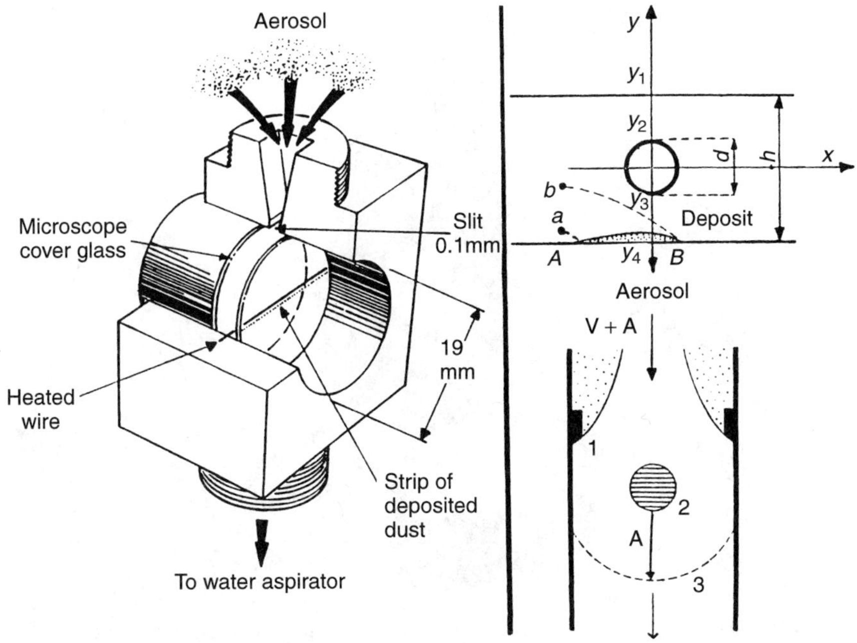

Fig. 1–10. Schematic of a standard thermal precipitator and its function. The limiting particle trajectories a and b result in a deposit with limits at points A and B; the heated wire has a diameter d, and the spacing between the two cover glasses is h.

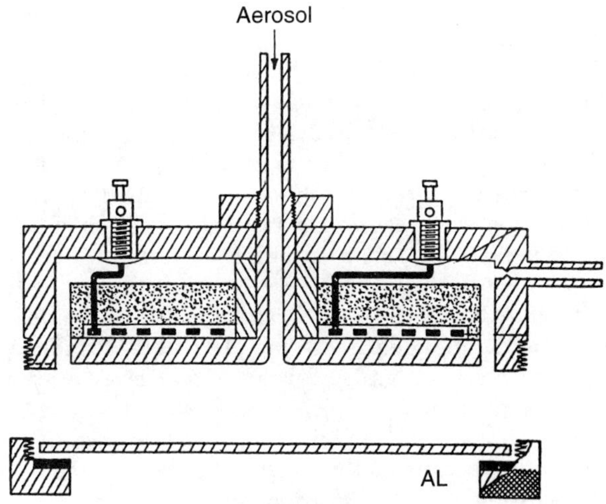

Fig. 1–11. Schematic of a thermal precipitator that collects dust samples large enough to weigh. AL, aluminum plate.

Electrostatic Precipitators. In 1824, Hohlfeld used electrostatic forces to remove airborne particles from the air when he, as mentioned above, applied high power to a wire suspended in a bottle filled with smoke and rapidly precipitated the smoke particles in the bottle (see also Mercer, 1973). Cottrell, in the United States (Cottrell, 1911), developed the

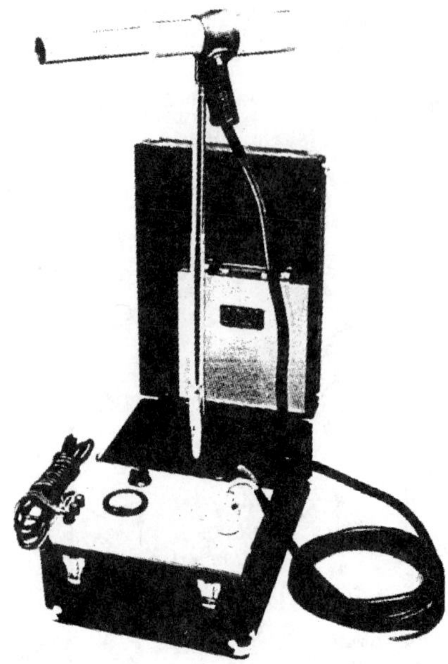

Fig. 1–12. A portable electrostatic precipitator for dust sampling in workplaces.

first electrofilter designed for air cleaning. Further development of portable electrostatic precipitators started after 1919. Tolman et al. (1919) built a small glass electrostatic precipitator and used it to collect smokes. Bill (1919) made the first application to industrial hygiene sampling. Lamb et al. (1919) used a modified electric precipitator for sampling smokes and bacteria. In 1924 the first electrostatic precipitator was used in Germany (Salmag, 1924). Subsequently, a number of electrostatic precipitators were described, for example, those by Drinker and Thompson (1925), Drinker (1932), and Barnes and Penny (1938). The instrument of Barnes and Penney was a portable apparatus (Fig. 1–12) and was later produced commercially. The electrically precharged dust particles were collected in a metal tube. The dust concentration was measured as mass/m^3. Luckiesh et al. (1946) described an electric precipitator suitable for sampling airborne bacteria on Petri dishes with cultural medium.

Pauthenier and Moreau-Hannot (1933) and Pauthenier and Chalande (1952) were the first to publish theoretical considerations and modeling. Davies (1952b) studied the collection efficiency of portable electric precipitators. He concluded that, given a reasonable length of gas path, it should be possible to collect completely particles larger than 0.5 μm in radius, very minute particles may also be removed, and the collection efficiency of particles in the region of 0.1 μm radius might be low because of the difficulty of charging and the low limiting charge. See Chapter 18 for discussion of current instrumentation.

Particle Counting and Sizing

As previously discussed, the measurement of particle number concentrations was the preferred measurement method during the classical period. According to microscope methods, the particles in dust samples collected by konimeters, impingers, and thermal and electrostatic precipitators were counted, and their sizes, mainly in the range 0.5 to 5 μm, were measured. In some cases, individual mineral particles were identified. The refractive index of transparent particles could be found by oil immersion methods, using a range of liquids whose indices embraced those of the particle. This procedure was applied, for example, to identify single SiO_2 (quartz) particles.

Quartz has two indices of refraction, 1.544 and 1.553. The minerals encountered in industrial dusts have higher or lower indices. When suitable liquids (such as mononitrobenzene, tetraline, and so forth) were used, the number of SiO_2 particles could be estimated. If a small portion of dust is well dispersed in a medium with a refractive index of 1.54 (a suitable oil) and examined microscopically, the quartz particles will exhibit central illumination when the objective of the light microscope is raised slightly above focus.

For the measurement of particle sizes, optical micrometers and standardized graticules were used. The eyepiece graticules consisted of a series of lines and circles of graduated size on a glass disk (Fig. 1–13). The sizes of irregular particles were described in terms of arbitrary dimensions, for example, as diameters measured in one arbitrarily defined dir-

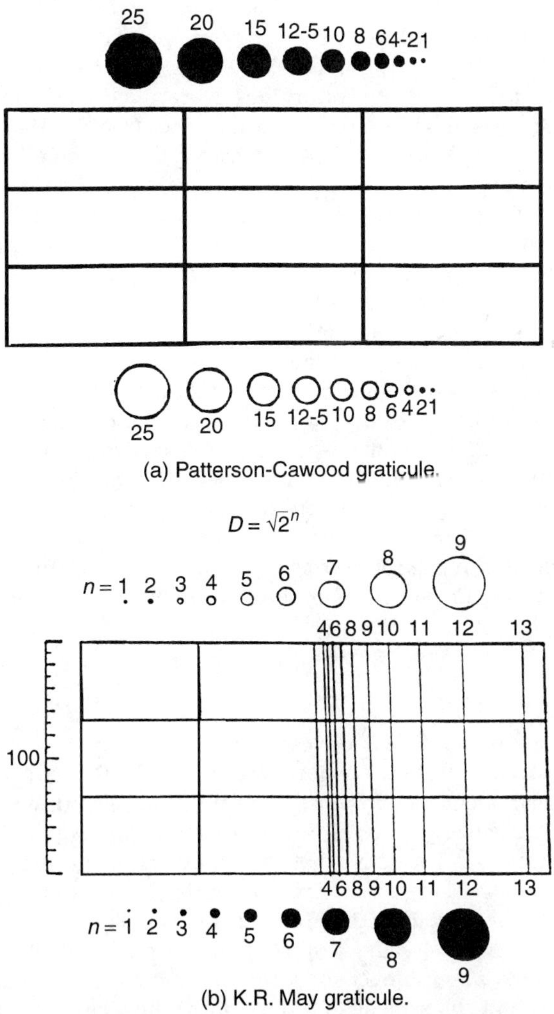

Fig. 1–13. Graticules for the estimation of particle sizes.

ection. When impingers were used, the dust particles were sampled in distilled water or ethyl alcohol. In the laboratory, the samples were first made up to a known volume, using, for example, a glass microcuvette. After the particles were counted, their sizes were measured. See Chapters 11, 12, 23, and 24 for discussions of current particle counting and sizing techniques.

Limitations of the "Classical" Methods

Many of the imperfections and the sampling and measurement errors of these methods were recognized when the methods were first used. All of the previously described instruments were plagued with problems of rebound, re-entrainment, and deagglomeration of particles during sampling. The sampling times of different instruments varied between seconds and several hours. The sampling and collection efficiencies of different instruments also varied substantially. For these reasons, a comparison of the concentrations measured by these instruments was practically impossible. The differences in the measured concentrations lay in the range of ±100%. Therefore, no single ratio or even approximate conversion factor was available that could be used to compare particle counts made by two different instruments.

The importance of isokinetic sampling conditions in the measurement of dusts and aerosols was not fully recognized before the 1960s. Walton had mentioned the possible errors due to nonisokinetic sampling of aerosols in 1954. Nevertheless, the first satisfactory theory for isokinetic sampling was described only after the 1960s (Davies, 1968).

Also, the thermal precipitator, which was considered the method of choice for a long period, was later found to be of little use. Several investigations during the 1960s and later (Mercer, 1973) showed important irregularities and a lack of homogeneity in particle deposits obtained during dust sampling. The theory of thermophoresis indicates that the thermophoretic force or thermophoretic velocity, at normal air pressure, depends on the particle size, that is, the smaller particles are deposited first. Therefore, in samples obtained in a thermal precipitator, the average particle size increases continuously, from the front edge (nearest the intake) to the back edge. Furthermore, nonuniform patterns of deposition exist with respect to both the number of particles per unit area and the size distribution. The collection efficiency of a thermal precipitator begins to decrease with an increase of particle sizes above 2 μm. Particles having large thermal conductivities are subjected to thermal forces many times greater than are particles having low thermal conductivities (Schadt and Cadle, 1957).

The general conclusion, based on theoretical and experimental investigations done during the 1950s, but primarily after the 1960s, is that the working principles of the "classical" methods and instruments remain useful and applicable to modern aerosol measurement techniques. However, the instruments themselves, in their original design and function, are of historical importance only.

Sampling by Filtration

Removing dust and aerosol particles from gases by filtering them through a suitable medium has provided a simple means of dust collection since the 1920s. The Soxhlet Filter was the standard filtration instrument used for dust sampling then, when dust concentrations in the workplace were in the range of several mg/m^3. Trostel and Frevert (1923) developed this instrument. The Soxhlet Filter used a Whatman paper extraction thimble filter, filled with fluffed-out cotton to reduce clogging. The gravimetric dust concentration was calculated from the change in weight of the dried thimble. Soluble dust sampling filters continued in use, too. The "sugar filter" of 1906 has already been mentioned. Holt (1951) used naphthalene filters for determining dust mass concentrations. After the sampling, the naphthalene was evaporated by heating. A soluble filter made of tetrachloronaphthalene crystals was used for dust sampling in France (Avy, 1956).

The next development in dust-sampling filters was the paper filter disk (Silverman and Ege, 1943). In the early 1950s, membrane filters (MFs) became the most important standard analytical and sampling filters for aerosols. These filters consisted of a porous membrane having a foam-like structure that was approximately 100 to 150 μm thick (Fig. 1–14). They were prepared from one or a mixture of several cellulose ester gels. The MFs had a pore volume of 75% to 80%. The manufacturing process controlled the pore size. The MFs

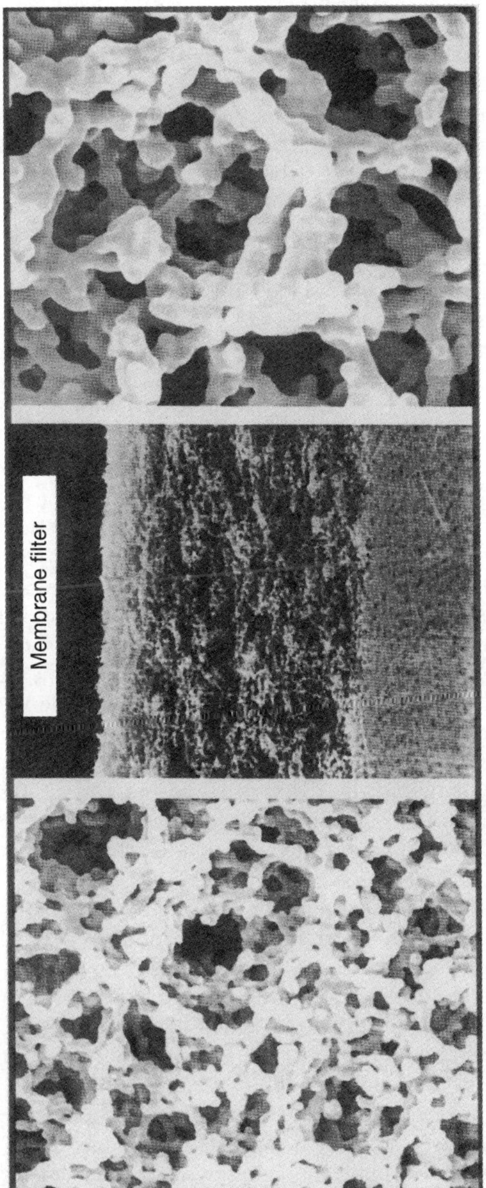

Fig. 1–14. The inner and surface structures of a membrane filter with a theoretical pore size of 0.8 μm. At three magnifications from top to bottom high, low, and intermediate.

were used both to weigh collected dust and to count particles with an optical microscope. Before the microscope method could be used, the MF containing the collected dust particles had to be treated with a few drops of immersion oil or an organic solvent to make it transparent.

The real history of the preparation and use of MFs began long before the 1950s, however. Cellulose ester MFs were produced commercially in Germany, beginning in 1927. The production was based primarily on the research of Zsigmondy and Bachmann (1916, 1918; see also Spurny 1965–1967). After World War II, the production of MFs began in the United States, Russia, England, and Czechoslovakia. The history of MF production procedures and applications to aerosol measurement has been well documented (Spurny, 1965–1967; Spurny and Gentry, 1979; Spurny, 1998b). The application of MFs in dust and aerosol measurement began approximately at the end of the 1940s, although Kruse (1952) had used MFs for measuring bioaerosols in Germany. A Russian publication was the first to discuss the use of MFs in gravimetric measurement of dust concentrations in the workplace (Reznik, 1951). In the United States, Alexander Goetz was the father of MF production and application. He used MFs for both aerosol (Goetz, 1953, 1956) and bioaerosol (Goetz and Tsuneishi, 1959) measurement. First and Silverman (1953), Fraser (1953), Burke (1953), and Kalmus (1954), also from the United States, did very important pioneering work on the application of MFs to dust measurement. In France, Le Bouffand (1954) and Le Bouffand and Davelu (1958) introduced MFs for dust measurements. At the same time, the MF became the standard method used for dust measurement in Czechoslovakia (Spurny and Vondracek, 1957).

Important progress in the field of dust and aerosol measurement by pore filters was achieved after the 1960s with the invention and application of polycarbonate filters, also called Nuclepore filters or NPFs (Spurny et al., 1969; Spurny, 1998b). See Chapter 9 for a discussion of modern filters.

Elutriators and Aerosol Centrifuges

In 1952, Walton had defined air elutriation as a process where particles are separated on the basis of size by contrasting their settling velocity to the velocity of the air current in which they move. Vertical elutriators were used mainly for size fractionation and/or measurements of the aerodynamic particle size distribution (Walton and Vincent, 1998). Concurrently, Timbrell (1952, 1972) designed a very useful elutriation spectrometer. It was a portable instrument consisting of a wedge-shaped sedimentation chamber. The aerosol was drawn by a laminar airflow through the sedimentation chamber. The relationship between the particle settling velocity and the distance along the particle deposit depended on the flow rate. Therefore size/distance relationships for several flow rates could be plotted. Timbrell also used his instrument to determine the shape factors and aerodynamic diameters of fibrous particles.

The deposition force for sampled particles could be substantially increased (e.g., up to a factor of 20,000) by using aerosol centrifuges. The operation of aerosol centrifuges was similar to the horizontal elutriators, having the force of gravity replaced by a centrifugal force (Mercer, 1973). The first aerosol centrifuges also were constructed in the 1950s.

Although Sawyer and Walton (1950) designed and produced the first conifuge in 1950, conifuge theory was formulated only after the 1960s (Stöber and Zessack, 1966). This instrument consisted of a metal cone mounted directly on the rotor of a high-speed electric motor and a conical metal cover that could be fastened rigidly to the cone, leaving a conical annular air space between the cone and the cover. When the unit was rotated, air was drawn into the opening at the top, pumped through the annulus, and exhausted through jet orifices at the bottom. Particles were deposited in narrow bands around the inner surface of the outer cone. The position of the center of each band was characteristic of the aerodynamic diameter of particles.

A second, already commercially available, aerosol centrifuge was the Goetz Aerosol Spectrometer, first described in 1957. This centrifuge consisted of an aluminum cone grooved and having two independent helical channels and covered with a close-fitting conical shell. The cone rotated at speeds up to 24,000 rpm. Aerosol particles moving through the channels were subjected to a constantly increasing centrifugal acceleration that deposited them on the channel floor. The floor consisted of a thin, removable foil that covered the inner surface of the outer cone. Particles deposited on this foil formed Archimedean spirals. The length of the spiral segment was correlated to the aerodynamic particle size. The theoretical description of particle separation as performed by this centrifuge and the evaluation procedures were published only after 1960 (Preining, 1962; Stöber and Zessack, 1966).

The next generation of aerosol centrifuges began in the 1960s, primarily incorporating the design and development of the spiral centrifuge aerosol spectrometer (Kast, 1961; Stöber and Flachsbart, 1969). The Stöber Aerosol Centrifuge was later developed as a very sophisticated and useful instrument (Stöber, 1972). A satisfactory theoretical description of the deposition of particles having different shapes has been published only recently (Asgharian and Godo, 1999). See Chapter 10 for a discussion of centrifuges; other aerodynamic sizing instruments are discussed in Chapters 13, 14, and 17.

Condensation Nuclei Counting and Measurement

The Aitken period of instruments for the detection and measurement of condensation nuclei has been mentioned. It has also been fully described in the review publications of Podzimek (1965, 1985, 1989). Important development of these instruments continued in Germany after 1900. Lüdeling (1903) and Scholz (1931) substantially improved the original Aitken instruments. Later Junge (1935) developed and used an expansion counter and replaced the microscope with a camera, which photographed the number of suspended droplets in a known volume.

Since the end of the 1930s, photoelectric expansion-type instruments have been used. Automated photoelectric condensation nuclei counters (CNCs) were first developed at General Electric's Research Laboratory beginning in the 1950s (Vonnegut, 1949; Rich, 1966; Skala, 1963). McMurry (2000) very recently published the full history of CNCs. The CNCs found broad applications in several fields of aerosol measurement, for example, to test aerosol filters by using diffusion battery measurements. The history of the latter technique is very well presented in the publication of Knutson (1999). See Chapter 19 for a discussion of current condensation-based instruments.

Ultramicroscopy Optical Particle Counters and Aerosol

Ultramicroscopy. Tyndall's phenomenon, which was studied first in 1869 on an aerodisperse system, is caused by light scattering from particles in the air. Tyndall used his methods to demonstrate that particles well below the visible limit can be observed, counted, and measured. This was the logical basis for the later invention of ultramicroscopes, nephelometers, tyndallometers, and optical particle counters (Gentry, 1997; Kerker, 1997).

The important ultramicroscopical observations of Bodaszewsky (1881) have been mentioned. The systematic development of the ultramicroscopes began with the introduction of the slit ultramicroscope by Siedentopf and Zsigmondy (1903, 1904; see also Fig. 1–15). Light scattered from particles was observed with an optical microscope. Zsigmondy (1909), awarded the Nobel Prize in 1926, used this apparatus, having dark-field illumination to count individual colloid particles within a defined volume. Further development and application of ultramicroscope methods were continued mainly in Vienna (Ehrenhaft, 1905, 1907).

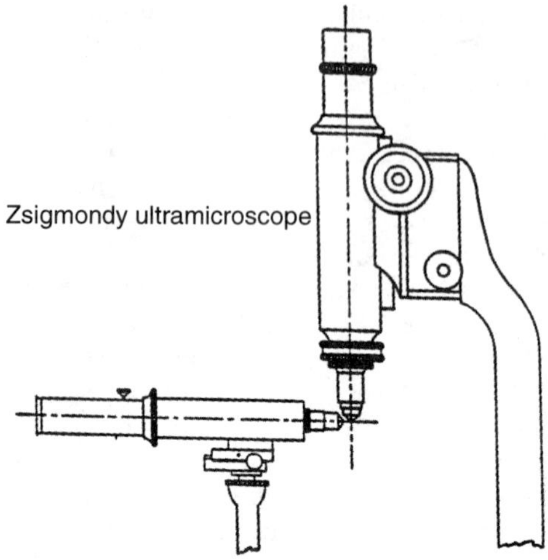

Fig. 1–15. The ultramicroscope developed by Siedentopf and Zsigmondy in 1903.

Vlasenko in Russia (Deryaghin and Vlasenko, 1953) developed the first aerosol ultramicroscope. It was a flow ultramicroscope. The aerosol was aspirated through a glass tube and through an illuminated zone in the center of this tube. Observations were made with a microscope in the direction of the flow axis. The particles passing through the illuminated zone appeared as flashes of light. A diaphragm situated in the microscope eyepiece defined the counting field. Numerical particle counting was obtained through this eyepiece. When I visited Vlasenko in Moscow in 1962, he demonstrated this new instrument for me with pride, stating: "It is very sensitive to small particles, because the human eye is a much more sensitive apparatus than any photocell." Perhaps he was right at that time. However, visual ultramicroscopy did not permit observation of several particles while simultaneously following the movement of the particles.

As early as 1919, Wells and Gerke used a photographic recording in combination with an ultramicroscope to measure the movement of charged aerosol particles. Fuchs subsequently improved this method for simultaneously measuring size and charge of individual aerosol particles (Fuchs and Petryanov, 1933). This method is shown in Figure 1–16. The movement of individual particles in a condenser with an alternating electric field was observed in a microscope and photographed. With the combination of gravity and electric force, the resulting motion of an individual particle showed a periodically changing path. By measuring the parameters L_g and L_e, an alternative evaluation of the particle size or the particle charge was possible (Kubie, 1965).

Tyndallometry. The measurement of the light scattered from all particles in a given volume occurred at a later time with the development of tyndallometers and nephelometers (Tolman and Vliet, 1919; Berek et al., 1936; Stuke, 1955). Tyndallometers measured the light scattering through an angle of 30°. In Germany, the E. Leitz company began commercial production of tyndallometers at the end of the 1930s (Fig. 1–17; Meldau, 1956). Hodkinson (1965) published a critical evaluation of light-scattering equipment existing before the 1960s. See Chapter 15 for discussion of current nephelometers.

Optical Particle Counters. A photoelectric optical particle counter (OPC), based on the theory of light scattering from individual particles, was initially developed in the second half of the 1940s (Gucker et al., 1947; Gucker and O'Konski, 1949; Gucker and Rose, 1954;

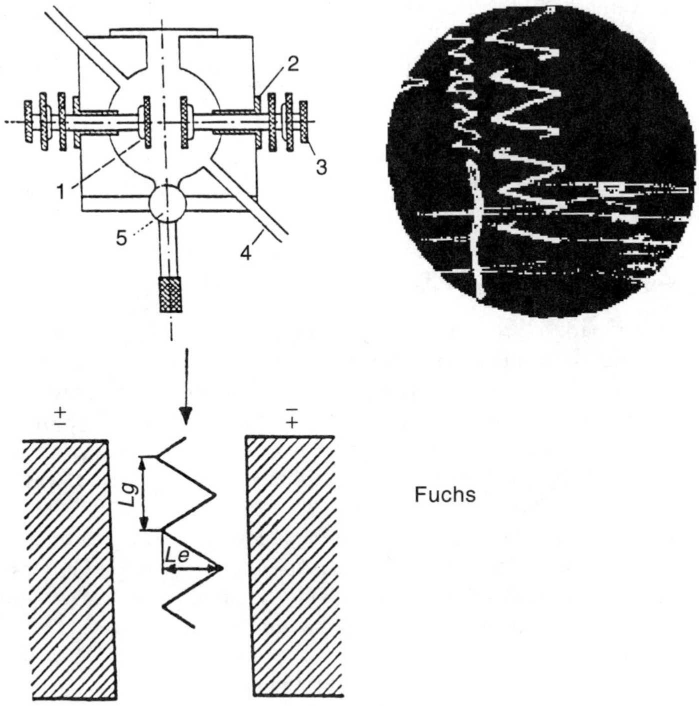

Fig. 1–16. Diagram of the condenser used to measure the electric charge of single aerosol particles.

Fig. 1–17. The German tyndallometer produced by the Leitz Company.

Gucker, 1947). The OPCs measured, on line and in situ, the number concentration of diluted aerosols having particles larger than 0.3 μm. The OPCs were refined later to record voltage pulses of different magnitudes. These voltage pulses were correlated with light pulses coming from aerosol particles of different sizes and used to determine particle size distributions (Gucker and Rose, 1954; Fisher at al., 1955). The commercial development of OPCs began in the 1960s, for example, by the Royco Company (Zinky, 1962). About 10 years later, the first photometer capable of simultaneous measurement of light-scattering signals from single particles in an aerosol stream over nearly 360° was built (Gucker et al., 1973). Kerker (1997) published the full history of light-scattering instrumentation for aerosol studies and measurements.

Single particle light scattering can be applied in a very broad manner in laboratory measurements by using the aerosol particle levitation techniques (Davis, 1997). See Chapter 20 for further discussion of this topic.

Mineralogical and Chemical Aerosol Analysis

As previously mentioned, quartz and other silicates, as well as heavy metals such as lead, were the mineral dust components of interest before 1960. The analytical procedures and instruments available at that time, such as gravimetry, titration, colorimetry, photometry, polarography, and X-ray diffraction, were used for further chemical dust and aerosol analyses. The microscope analysis of SiO_2 (quartz) was mentioned previously. The second, more quantitative method, standardized internationally at a later time, was the X-ray diffraction analysis method (Clark and Reynolds, 1936; Ballard et al., 1938; Klug et al., 1948). This method was sensitive to SiO_2 concentrations as low as 1%. Nevertheless, the method was not used until much later for routine analysis in industrial hygiene laboratories. One of the reasons for the delay that Drinker and Hatch (1954) mentioned was that "the equipment [X-ray diffractometer] was expensive and the technique was not simple." Dust samples were analyzed for heavy metals, as well as for SiO_2, lead, and cadmium, and these chemicals were considered the most important hazards from the viewpoint of industrial toxicology. After the sample dust was extracted and placed in a solution of strong inorganic acids, the concentration of both lead and cadmium chlorides was determined using the classical polarographic procedure.

MEASUREMENT OF FIBROUS AEROSOLS

In the history of fibrous aerosols, asbestos was the first source of fine fibrous particles that were dispersed into the workplace air, as well as into the general atmosphere. As early as the 1920s the medical profession began to suspect a causal effect between the presence of asbestos fibers in the air and fatality among asbestos workers who contracted lung disease. The peribronchial fibrosis was fully recognized in 1927 and was designated *asbestosis*. In 1935 the first cases of brochogenic carcinoma were reported. Mesothelioma had previously been a rare type of tumor. The association with asbestos dust was suspected from the 1930s and was established in the 1960s (Selikoff and Lee, 1978). Since this time, the carcinogenic potency of inhaled asbestos dust has become evident and widely known. The need for the measurement and control of airborne asbestos dust was officially recognized around the 1930s.

Asbestos fiber samples retained on the surface of a filter can be examined directly by a scanning electron microscope (SEM) after vacuum coating with gold to make the specimen electrically conducting. Fibers with diameters ≥0.1 µm are readily visible. The SEM procedure came into use in the early 1970s (Spurny et al., 1976, 1979). The transmission electron microscope (TEM) is the only instrument capable of giving a well-resolved image of the fine asbestos fibrils (Fig. 1–18). The crystal structure of individual fibers within the size range of about 0.7 to 0.03 µm diameter is readily visible. The first applications of TEM for counting and identifying asbestos fibers began also in the 1970s (Chatfield and Pullan, 1974; Chatfield, 1980).

Direct reading methods for the measurement of fibrous aerosols have been of interest since the middle 1970s. The tyndallometer was applied to the measurement of asbestos

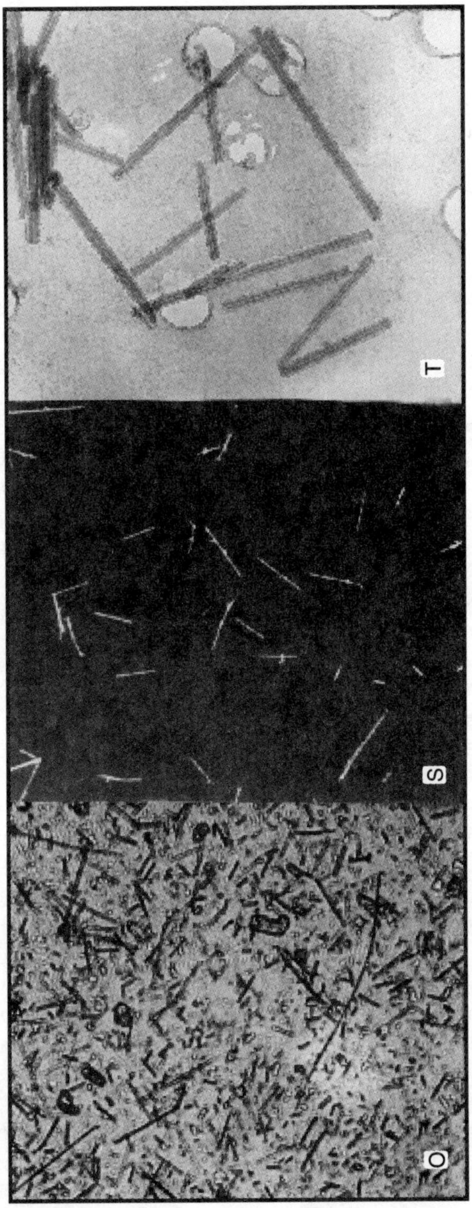

Fig. 1–18. Optical and electron micrographs of asbestos fibers sampled on filter surfaces. O, optical; S, scanning; T, transmission microscopy.

dust in Germany (Breuer and Robock, 1975). However, the method needed calibration in mass and was not specific for fibers. The first specific optical fiber counter, designated the FAM (fibrous aerosol monitor) was developed in the late 1970s (Baron, 1979). During the 1980s several important improvements in all aspects of the measurement of fibrous aerosols were realized, and, therefore, a satisfactory methodology has existed exist since 1990 (Spurny, 2000a). See Chapters 12, 17 and 23 for further discussions of current fiber analysis techniques.

CONCLUDING REMARKS

I intended this short historical review as a summary of my impressions of the approximately 100 years of development in the aerosol measurement field, although I concentrated primarily on the development between 1920 and 1960. An exhaustive description of activities during this time is available from the published papers of the period, as well as from recently published reviews, including the Proceedings of the first Vienna Conference on the History of Aerosol Science (Preining and Davis, 2000). I have tried to describe the atmosphere and the philosophy that existed during that period, hoping this could help younger aerosol scientists to better understand the methodological development and to evaluate the existing state of this field by comparison. Two important realities must be considered when comparing the aerosol measurement technology available at the beginning of the year 2000, which is presented in this book, with the technology before 1960.

First, development after 1960 exploited the knowledge of the classical period. The basic sampling and measurement principles and mechanisms were developed further. Second, development after 1960 profited from the rapid technical and instrumental progress in the fields of microelectronics, laser, and computer techniques, as well as modern physical methods in analytical chemistry and analytical electron microscopy. Most instruments used before 1960 were laboratory made and included no computer support and no automation. Much of the current methodology fully uses all the technical advancements made since 1960. Considering the most important successes in aerosol measurement techniques, I must mention, for example, the great improvements in cascade impactors (real and virtual, low pressure, and so forth) and in electric aerosol mobility analyzers, which are the logical successors of the earlier electrostatic precipitators.

Enormous improvements have been made in the optical particle counters and analyzers, facilitated by modern laser and spectroscopy techniques. The same is true for the very impressive improvement of photoelectric CNCs. New analytical aerosol filters (glass, polymer, carbon fiber, and especially the polycarbonate "Nuclepore" filters) have been developed and are commercially available. The developments in the field of chemical and mineralogical aerosol analysis have been very successful (Spurny, 2000a). Applications of very new and very sensitive methods of modern analytical chemistry (chromatography; mass spectrometry; plasma and laser spectroscopy; and radiation and nuclear beam methods, such as PIXE, NAA, and so forth) to aerosol analysis have made it possible to determine almost all inorganic and organic components in aerosol samples. Furthermore, the applications of some of these modern analytical techniques, which are fast and sensitive enough, have made possible the new and important field of real-time measurement and identification of single airborne particles (Spurny, 1986; see also Chapter 13). The dream of Sheldon Friedlander (1977)—the Single Particle Counter and Analyzer—has been practically realized.

And, how about the needs and perspectives for the twenty-first century? Research, as well as the applications of aerosol science and technology, will increase in several fields. Also, ultra-highly dispersed aerosols, consisting of particles as small as and much smaller than 1 μm, will assume great importance in the near future. These aerosols are of basic importance in the aerosol synthesis of nano-sized, high-technology materials and also in the field of human inhalation toxicology and its health effects. Very sensitive, on line, in situ and real-time aerosol measurement and analytical techniques will be necessary for further research in the areas of both "good" and "bad" highly dispersed aerosols.

REFERENCES

Aitken, J. 1880a. On dust, fogs, and clouds. *Proc. R. Soc. Edinb.* 9:14–18.

Aitken, J. 1880b. On dust, fogs, and clouds. *Nature* 14:384–385.

Aitken, J. 1890–1891. On a simple pocket dust counter. *Proc. R. Soc. Edinb.* 18:39–52.

Aitken, J. 1888. On the number of dust particles in the atmosphere. *Nature* 37:187–206.

Aitken, J. 1888–1889. An improvement in the apparatus for counting the dust particles in the atmosphere. *Proc. R. Soc. Edinb.* 16:135–172.

Aitken, J. 1894. On the formation of small clear spaces in dust air. *Trans. R. Soc. Edinb.* 32:239–272.

Aitken, J. 1920. *Collected Scientific Papers*, ed. G. G. Knott. London: College University, pp. 1–591.

Andersen, A. A. 1958. New sampler for the collection, sizing, and enumeration of viable airborne particles. *J. Bacteriol.* 76:471–484.

Asgharian, B. and M. N. Godo. 1999. Size separation of spherical particles and fibers in an aerosol centrifuge. *Aerosol Sci. Technol.* 30:383–400.

Avy, A. P. 1956. Les Aerosols. Paris: Dunod, pp. 1–292.

Ballard, J. W., H. I. Oshry, H. H. Schrenk, E. C. Barnes, and G. W. Penny. 1938. An electrostatic dust weight sampler. *J. Ind. Toxicol.* 20:259–266.

Bancroft, W. D. 1920. Thermal ficters. *J. Phys. Chem.* 24:421–433.

Barnes, E. C. and G. W. Penney. 1938. An electrostatic dust weight sampler. *J. Ind. Hyg. Toxicol.* 14: 259–265.

Baron, P. A. 1979. Recent development in fibrous aerosol measurement. *J. Environ. Pathol. Toxicol.* 12:335–343.

Beadle, D. G. 1951. Investigation of performance and limitations of konimeter. *J. Chem. Met. Mining Soc. S. Afr.* 51:265–283.

Beccaria, G. B. 1772. "Elettricimo Artificiale," *Nella Stamperia Realejorino.*

Behounek, F. 1939. The reality of dust counters. *Terrestrial Magnetism Atmospheric Electricity* 44:21–37.

Behounek, F. and E. Effenberger. 1942. Vergleichsmessungen mit dem Owens-Behounek Staubzähler und dem Konimeter. *Fa. Zeiss. Gerlands. Beiträge* 59:74–90.

Berek, M., K. Männchen, and W. Schäfer. 1936. Über tyndallometrische Messungen des Staubgehaltes der Luft und ein neues Staubmessgerät. *Z. Instrumentenkunde* 56:49–56.

Bernz, N. Z. 1942. Dust counts and their significance. *Heating, Piping, Air Conditioning* 14:173–175.

Bill, J. P. 1919. Electromethod of dust collection as applied to the sanitary analysis of air. *J. Ind. Hyg.* 1:323–342.

Bodaszewsky, L. J. 1881. Rauch und Dampf unter dem Mikroskop. *Dinglers Polytech. J.* 5:325.

Bredl, J. and T. W. Grieve. 1951. Thermal precipitator for gravimetric estimation of solid particles in flue gases. *J. Sci. Instrum.* 28:21–23.

Breuer, H. and K. Robock. 1975. Das Tyndallometer TM Digital zur unmittelbaren Bestimmung der Feinstaubkonzentration auf dem Gebiet der Staub- und Silikosebekämpfung in Steinkohlebau. *Verlag Glückauf*, Essen (Germany) 10:77–78.

Burke, W. C. 1953. Determination of silica particles collected on membrane filters. *Am. Ind. Hyg. Assoc. Q. J.* 14:299–302.

Cember, H., T. Hatch, and J. Watson. 1953. Rotating thermal precipitator. *Am. Hyg. Assoc. Q. J.* 14: 191–194.

Chatfield, E. J. 1980. *Measurement of Asbestos Fibers in Air and Water.* ISO/TC 147 /WG 18. Ontario, Canada: Ontario Research Foundation, pp. 1–20.

Chatfield, E. J. and H. Pullan. 1974. Measuring asbestos in the environment. *Can. Res. Dev.* December 23–27.

Clark, G. L. and D. H. Reynolds. 1936. Quantitative analysis of mine dust: An X-ray diffraction method. *Ind. Eng. Chem. Anal.* 8:36–42.

Collis, E. L. 1926. *Tuberculosis-Silicosis*, Brochure32. Geneva: Occupation and Health International. Labor Office.

Cottrell, F. G. 1911. The electrical precipitation of suspended particles. *J. Ind. Eng. Chem.* 3:242–250.

Coulier, P. J. 1875. Note sur une nouvelle propriete de I'air. *J. Pharm. Chim. Paris Ser. 4*, 22(l):165–172, 22(2):254–255.

DallaValle, J. M. 1937. Note on the comparative tests made with the Hatch and the Greenburg-Smith impingers (dust sampling device). *U.S. Pub. Health Repts.* 52:1114–1118.

Davies, C. N. 1952a. Dust sampling and lung disease. *Br. J. Ind. Med.* 9:120–126.

Davies, C. N. 1952b. The separation of airborne dust and particles. *Proc. Inst. Mech. Engrs. Lond.* pp. 199–213.

Davies, C. N. 1954. *Dust Is Dangerous.* London: Faber and Faber, Ltd.

Davies, C. N. 1968. The entry of aerosols into sampling tubes and heads. *Br. J. Appl. Phys. Ser. 2* 1:921–932.

Davis, E. J. 1997. A history of single aerosol particle levitation. *Aerosol Sci. Technol.* 26:212–254.

Deryaghin, B. V. and G. Y. Vlasenko. 1953. Ultramicroscopic measurement of particle concentration of aerosols. *Nature* (Moscow) 11.

Drinker, P. 1932. Alternating-current precipitators for sanitary air analysis. *J. Ind. Hyg. Toxicol.* 14:364–370.

Drinker, P. and T. Hatch. 1954. *Industrial Dust.* New York: McGraw-Hill.

Drinker, P. and R. M. Thompson. 1925. Determination of suspensoids by alternating-current precipitators. *J. Ind. Hyg.* 7:261–271.

Ehrenhaft, F. 1905. Die diffuse Zerstreuung des Lichtes an kleinen Kugeln: Ultramikroskopische Studie. *Wien. Anz.* 3:213–214.

Ehrenhaft, F. 1907. Über eine der Brownschen Molekularbewegung gleichartige Molekularbewegung in den Gasen und deren molekularkinetischen Erklärungsversch. *Wien. Anz.* 5:72, 331.

Einstein, A. 1924. Zur Theorie der Radiometerkräfte. *Z. Physik.* 27:1–6.

Emery, A. H. 1927. Instructions for sampling atmospheric dust by the impinger method. *Bureau Mines Rep. Invest.* No. 6048.

Epstein, P. 1929. Zur Theorie des Radiometers. *Z. Physik.* 54:537–563.

First, M. N. and L. Silverman. 1953. Air sampling with membrane filters. *AMA Arch. Ind. Health Occup. Med.* 7:1–14.

Fisher, M. A., S. Katz, A. Lieberman, and N. E. Alexander. 1955. The Aeroscope: An instrument of the automatic counting and sizing of aerosol particles. *Proceedings of the National Air Pollution Symposium*, Los Angeles, p. 112.

Fraser, D. A. 1953. Absolute method of sampling and measurement of solid airborne particulates. *AMA Arch. Ind. Hyg. Occup. Med.* 7:412–419.

Friedlander, S. K. 1977. *Smoke, Dust and Haze.* New York: J. Wiley & Sons.

Fuchs, N. A. 1955. *Mechanika Aerozolei.* Moscow: Academy of Science Press.

Fuchs, N. A. 1964. *The Mechanics of Aerosols.* Oxford: Pergamon Press.

Fuchs, N. A. 1971. *Collection of Aerosol Abstracts.* Moscow: Karpov Institute.

Fuchs, N. A. and J. Petryanov. 1933. Über die Bestimmung der Grösse und der Ladung der Nebeltröpfchen. *Kolloid Z.* 65:171–183.

Gentry, J. W. 1997. The legacy of John Tyndall in aerosol science. *J. Aerosol Sci.* 28:1365–1372.

Goetz, A. 1953. Application of molecular filter membranes to the analysis of aerosols. *Am. J. Public Health* 43:150–163.

Goetz, A. 1956. The application of the concentrometer principle for the determination of aerosols. *J. Air Pollut. Control Assoc.* 6:19–27.

Goetz, A. 1957. An instrument for the quantitative separation and size classification of airborne particulate matter down to 0.2 μm. *Geofis. Pura Appl.* 36:49–69.

Goetz, A. and N. Tsuneishi. 1959. A bacterial irritation analogue for aerosols. *AAL4 Ind. Health* 20:167–174.

Green, H. L. and H. H. Watson. 1935. Physical methods for estimation of the dust hazard in industry. *Medical Res. Council Spec. Report Ser.* No. 199.

Green, L. 1934. Recent development in methods of sampling dusts. *Bull. Inst. Mining* pp. 1–22.

Greenburg, L. 1932. The Greenburg-Smith impinger sampling apparatus for dusts, fumes, smoke, and gases. *Am. J. Public Health* 22:1077–1082.

Greenburg, L. and G. W. Smith. 1922. A new instrument for sampling aerial dust. *U.S. Bureau Mines Rep. Invest.* No. 2394.

Gucker, F. T. 1947. Sensitive photoelectric photometer. *Electronics* 20:106–110.

Gucker, F. T. and C. T. O'Konski. 1949. Electronic methods for counting aerosol particles. *Chem. Rev.* 44:373–388.

Gucker, F. T., H. B. Pickard, and C. T. O'Konski. 1947. A photoelectric instrument for comparing the concentrations of very dilute aerosols and measuring low light intensities. *J. Am. Chem. Soc.* 69:429–438.

Gucker, F. T. and D. G. Rose. 1954. A photoelectric instrument for counting and sizing aerosol particles. *Br. J. Appl. Phys.* (Suppl. 3) S138–S143.

Gucker, F. T., J. Tuma, M. H. Lin, C. M. Huang, S. C. Evans, and T. R. Marshall. 1973. 360 Degree scattering diagrams for individual aerosols in flowing streams. *J. Aerosol Sci.* 4:389–404.

Gurney, S. W., C. R. Williams, and R. P. Meigs. 1938. Investigation of characteristics of Bausch and Lomb dust counter. *J. Ind. Hyg. Toxicol.* 20:24–31.

Hamilton, R. J. 1952. Long-running thermal precipitator. *Natl. Coal Board Central Res. Establ. Rep.* No. 138.

Hatch, T., H. Warren, and P. Drinker. 1932. Modified form of the Greenburg-Smith impinger for field use with a study of its operating characteristics. *J. Ind. Hyg.* 14:301–311.

Hettner, G. 1924. Zur Theorie des Radiometers. *Z. Physik.* 27:12–22.

Hodkinson, J. R. 1965. The physical basis of dust measurement by light scattering. In *Aerosols—Physical Chemistry and Applications*, ed. K. Spurny. Prague: Press of the Academy of Science, pp. 181–194.

Hohlfeld, H. 1824. Das Niederschlagen des Rauchs durch Elektrizität. *Kastner Arch. Ges. Naturlehre* 2:205–206.

Holt, P. F. 1951. Determination of dust mass concentration by volatile filters. *Metallurgia* 44:52–54.

Innes, S. J. 1919. The estimation of injurious dust in mine air by the Kotze konimeter. *J. Chem. Met. Soc. S. Afr.* 19:132–145.

Junge, Chr. 1935. Supersaturation measurements of atmospheric condensation nuclei. *Gerlands Beitr. Geophys.* 46:108–129.

Kalmus, F. H. 1954 . Preparation of aerosols for electron microscopy. *J. Appl. Phys.* 25:87–89.

Kast, W. 1961. Neues Staubmessgerät zur Schnellbestimmung der Staubkonzentration und der Kornverteilung. *Staub* 21:215–223.

Kathley, T. W., M. T. Gordon, and O. Orr. 1952. A thermal precipitator for aerobiology. *Science* 116:468–469.

Kerker, M. 1997. Light scattering instrumentation for aerosol studies: An historical overview. *Aerosol Sci. Technol.* 27:522–540.

Klug, H. P., L. Alexander, and E. Kummer. 1948. X-ray diffraction analysis of crystalline dusts, *J. Ind. Hyg. Toxicol.* 30:166 –172

Knutson, E. O. 1999. History of diffusion batteries in aerosol measurements. *Aerosol Sci. Technol.* 31:83–128.

Kruse, H. 1952. Membrane filter. *Gas- u. Wasserfachz.* 93:712–715.

Kubie, G. 1965. On the ultramicroscopy of aerosols. In *Aerosols: Physical Chemistry and Applications*, ed. K. Spurny. Prague: Press of the Academy of Science, pp. 207–219.

Lamb, A. B., G. L. Wendt, and R. E. Wilson. 1919. Portable electrical filter for smokes and bacteria. *Trans. Am. Electrochem. Soc.* 35:357–369.

Laskin, S. 1951. Oscillating thermal precipitator. *Ind. Hyg. Dig.* 15:28.

Le Bouffand, L. 1954. Emploi des membranes a micropores pour l'analyse coniotique des atmospheres. *2eme Colloque Inter sur las Pousieres*, Strasbourg.

Le Bouffand, L. and M. Davelu. 1958. Membranfiltergerät zur Staubprobenahme. *Staub* 18:342–351.

Lehmann, H., F. Löwe, and K. A. Traenke. 1934. Das Zeissische Freiluftkonimeter. *Arch. Hyg.* 112:141–152.

Löbner, A. 1935. The Zeiss konimeter and its applications. *Phytopath. Z.* 8:633–636.

Lodge, J. P. 1969. *The Smoke of London.* New York: Maxwell Reprint Co.

Lodge, O. J. 1883. On Lord Rayleigh's dark plane. *Nature* 28:274–299.

Lodge, O. J. 1884. On the phenomena exhibited by dusty air in the neighborhood of strongly illuminated bodies. *Philos. Mag.* Ser. 5, 17:214–239.

Lodge, O. J. 1886. The electrical deposition of dust and smoke, with special reference to the collection of metallic fume, and to a possible purification of atmosphere. *J. Soc. Chem. Ind.* November 29, 572–576.

Lodge, O. J. and J. W. Clark. 1883. On the phenomenon exhibited by dusty air in the neighbourhood of strongly illuminated bodies. *Philos. Mag.* 17:214–239.

Lodge, O. J. and J. W. Clark. 1885. Dust. *Nature* 31:265–269.

Luckiesh, M., A. H. Taylor, and L. Holladay. 1946. Sampling device for airborne bacteria. *J. Bacteriol.* 52:55–66.

Lüdeling, G. 1903. Ein neuer Staubzähler. *Illustr. aeronautische Mitteilungen.* 7:321–335.

May, K. R. 1945. The cascade impactor: An instrument for sampling coarse aerosols. *J. Sci. Instrum.* 22:187–193.

McCrae, J. 1913. The ash of silicotic lungs. *S. Afr. Inst. Med. Res.* Johannesburg. 6 pages.

McMurry, P. H. 2000. The history of condensation nucleus counters. *Aerosol Sci. Technol.* 33:297–322.

Meldau, U. R. 1956. *Handbuch der Staubtechnik.* Düsseldorf, Germany: VDI Verlag.

Mercer, T. T. 1973. *Aerosol Technology in Hazard Evaluation.* New York: Academic Press.

Michaelis, H. 1890. Prüfung der Wirksamkeit von Staubrespiratoren. *Z. Hyg. Infektionskr.* 9:389–393.

Miyake, S. 1935. *Smokeless Zone Around Heated Platinum Ribbon*, Report 123. Tokyo: Aeronaut. Res. Inst., pp. 85–106.

Orr, C. and R. A. Martin. 1958. Thermal precipitator for continuous aerosol sampling. *Rev. Sci. Instrum.* 29:129–134.

Owens, J. S. 1922. Suspended impurities in the air. *Proc. R. Soc. Lond. A.* 101:18–25.

Pauthenier, M. and R. Chalande. 1952. Electron microscopical study of electrically precipitated deposits. *Contr. Rend.* 235:1113–1115.

Pauthenier, M. and M. Moreau-Hannot. 1933. Theoretical and experimental studies on electric dust precipitation. *Bull. Soc. Fr. Electr.* 3:411–432.

Podzimek, J. 1965. The Importance of Aerosols in Meteorology. In *Aerosols—Physical Chemistry and Applications*, ed. K. Spurny. Prague: Press of the Academy of Science, pp. 493–509.

Podzimek, J. 1985. 100 year evolution of Aitken nuclei counters, current and future problems. *J. Rech. Atmos.* 19:257–274.

Podzimek, J. 1989. John Aitken's contribution to atmospheric and aerosol sciences: On hundred years of condensation nuclei counting. *Bull. Am. Meteorol. Soc.* 70:1538–1545.

Preining, O. 1962. Das Goetzsche Aerosolspektrometer. *Staub* 22:129–133.

Preining, O. and E. J. Davis. (Eds.) 2000. *History of Aerosol Science*. Proceedings of the Conference in Vienna, 1999. Vienna: Austrian Academy of Science.

Ranz, W. E. and J. B. Wong. 1952. Impaction of dust and smoke particles on surface and body collectors. *Ind. Eng. Chem.* 44:1371–1381.

Rayleigh, Lord. 1882. On the dark plane which is formed over a heated wire in dusty air. *Proc. R. Soc.* 34:414–418.

Reznik, J. B. 1951. Dust measurement by means of membrane filters. *Gigiena Saniatatia* (Moscow) 10:28–41.

Rich, T. A. 1966. Apparatus and method for measuring the size of serosols. *J. Rech. Atmos.* 2:79–85.

Salmag, H. 1924. Analysis of smoke. *Z. Angew Chemie.* 27:97–98.

Sawyer, K. F. and W. H. Walton. 1950. The "Conifuge": A size-separating sampling device for airborne particles. *J. Sci. Instrum.* 27:272–276.

Schadt, C. F. and R. D. Cadle. 1957. Critical comparison of collection efficiencies of commonly used sampling devices. *Anal Chem.* 29:864–868.

Schmauss, A. 1920a. Kolloidchemie und Meteorologie. *Metorol. Z.* 37:1–81.

Schmauss, A. 1920b. Die Chemie des Nebels, der Wolken und des Regens. *Die Umschau* (Frankfurt/M., Germany) 24:61–63.

Scholz, J. 1931. Ein neuer Apparat zur Bestimmung der Zahl der geladenen und ungeladenen Kerne. *Z. Instrumentenkunde* 51:505–521.

Selikoff, I. J. and D. H. K. Lee. 1978. *Asbestos and Disease*. New York: Academic Press, 549 pp.

Siedentopf, H. and R. Zsigmondy. 1903. Über Sichtbarmachung und Grössenbestimmung ultramikroskopischer Teilchen, mit besonderer Anwendung auf Goldrubingläser. *Ann. Phys.* 10:1–39.

Siedentopf, H. and R. Zsigmondy. 1904. Ein neues Ultramikroskop. *Berl. Klin. Wochenschr.* 41:865–872.

Silverman, L. and J. F. Ege. 1943. A filter paper method for lead-fume collection. *J. Ind. Hyg. Toxicol.* 25:185–192.

Sinclair, D. 1950. *Handbook of Aerosols.* Washington, DC: AEC.

Singerson, G. 1870–1874. Micro atmospheric researches. *Proceedings of the Royal Irish Academy, Second Series, Science*, pp. 13–31.

Skala, G. F. 1963. An automatic Aitken nuclei counter. *Anal. Chem.* 35:702–706.

Soskin, L. S. 1946. A modified cascade impactor. A device for sampling and sizing aerosols of particles below one micron in diameter. *J. Ind. Hyg. Toxicol.* 28:269–272.

Spurny, K. 1965–1967. Membranfilter in der Aerosologie. *Zbl. Biol. Aerosol-Forschung.* 12:369–407; 13:44–101; 13:398–451.

Spurny, K. R., ed. 1986. *Physical and Chemical Characterization of Individual Airborne Particles.* Chichester, UK: Ellis Howood.

Spurny, K. R. 1993. Aerosol science of the early days. *J. Aerosol Sci.* (Suppl. 1) 24:SI–S2.

Spurny, K. R. 1998a. Methods of aerosol measurement before the 1960s. *Aerosol Sci. Technol.* 29:327–349.

Spurny, K. R., ed. 1998b. *Advances in Aerosol Filtration.* Boca Raton, FL: Lewis Publ., pp. 1–533.

Spurny, K. R., ed. 2000a. *Analytical Chemistry of Aerosols.* Boca Raton, FL: Lewis Publ., pp. 1–486.

Spurny, K. R. 2000b. Atmospheric condensation nuclei—P. J. Coulier 1875 and J. Aitken 1880. *Aerosol Sci. Technol.* 32:243–248.

Spurny, K. R., E. Ackerman, and K. Spurny, Jr. 1976. The sampling and electron microscopy of asbestos aerosol in ambient Air. *J. Air Pollut. Control Assoc.* 26:296–298.

Spurny, K. R. and J. W. Gentry. 1979. Aerosol fractionation by graded Nuclepore filters. *Powder Technol.* 24:129–142.

Spurny, K., C. Jech, B. Sedlacek, and O. Storch. 1961. *Aerosoly* [Aerosols]. Prague: Publ. Technical Literature.

Spurny, K. R., J. P. Lodge, E. R. Frank, and D. C. Sheesley. 1969. Aerosol sampling and measurement by means of Nuclepore filters. *Environ. Sci. Technol.* 3: 453–458.

Spurny, K. R., W. Stöber, H. Opiela, and G. Weiss. 1979. On the evaluation of fibrous particles in remote ambient air. *Sci. Total Environ.* 11:1–40.

Spurny, K., and V. Vondracek. 1957. Analytische Methoden zur Bestimmung von Aerosolen unter Verwendung der Membranfilter. *Collection Czechoslov. Chem. Commun.* 22:22–29.

Stöber, W. 1972. Dynamic of shape factors of nonspherical particles. In *Assessment of Airborne Particles*, eds. T. T. Mercer, P. E. Morrow, and W. Stöber. Springfield, IL: C. C. Thomas, pp. 249–289.

Stöber, W. and H. Flachsbart. 1969. Size-separating precipitation of aerosols in a spinning spiral duct. *Environ. Sci. Technol.* 3:1280–1296.

Stöber, W. and U. Zessack. 1966. Zur MessSung von Aerosol—Teichenspektren mit Hilfe von Zentrifugalabscheidern. *Z. Biol. Aerosol Forschung.* 13:263–281.

Stuke, J. 1955. Staubessungen mit dem Nephelometer. *Glückauf.* 91:1405–1406.

Timbrell, V. 1952. A method of measuring and grading microscopic spherical particles. *Nature* 170:318–320.

Timbrell, V. 1972. An aerosol spectrometer and its applications. In *Assessment of Airborne Particles*, eds. T. T. Mercer, P. E. Morrow, and W. Stöber. Springfield, IL: C. C. Thomas, pp. 290–330.

Tolman, R. C. and E. V. Vliet. 1919. Tyndallometer for the examination of disperse systems. *J. Am. Chem. Soc.* 41:297–312.

Tolman, R. C., L. H. Reyerson, A. P. Brooks, and D. H. Smyth. 1919. An electrical precipitator for analysing smokes. *J. Am. Chem. Soc.* 41:587–589.

Townsend, J. S. 1900. The diffusion of ions into gases. *Philos. Trans.* 193:129–158.

Trostel, L. S. and H. W. Frevert. 1923. Collection and examination of explosive dusts in air. *J. Ind. Eng. Chem.* 15:232–239.

Tyndall, J. 1869a. On a new series of chemical reactions produced by light. *Proc. R. Soc. Edinb.* 17:93–102.

Tyndall J. 1869b. On the blue color of sky, the polarisation of skylight and the polarisation of light by cloudy matter generally. *Proc. R. Soc. Edinb.* 17:223–233.

Tyndall, J. 1870a. On the action of rays of high refractibility upon gaseous matter. *Philos. Trans.* 160:333–336.

Tyndall, J. 1870b. On dust and disease. *Proc. R. Inst.* 6:1–14.

Tyndall, J. 1871. On dust and smoke. *The Royal Institute Library of Science* (TRILS-ps) 2:302–313.

Vonnegut, B. 1949. A continuous recording condensation nuclei matter. First National Air Pollution Symposium, Los Angeles, CA, pp. 36–44.

Walkenhorst, W. 1962. Ein neuer Thermalpräzipitator mit Heizband und seine Leistung. *Staub* 22:103–105.

Walton, W. H. 1936. The photoelectrical estimation of konimeter dust spots. *J. Ind. Hyg.* 18:689–698.

Walton, W. H. 1952. The modified thermal precipitator. *Br. J. Appl. Phys.* 5(Suppl. 3):S29–S37.

Walton, W. H. 1954. Theory of size classification of airborne dust clouds by elutriation. *Am. Ind. Hyg. Q.* 15:21–25.

Walton W. H. 1982 . The nature, hazard and assessment of occupational exposure to airborne asbestos dust. *Ann. Occup. Hyg.* 25:117–247.

Walton, W. H., R. G. Faust, and W. J. Harris. 1947. A modified thermal precipitator for the quantitative sampling of aerosols for electron microscopy. *Porton Techn. Paper Ser.* 83.

Walton, W. H. and J. H. Vincent. 1998. Aerosol instrumentation in occupational hygiene. An historical perspective. *Aerosol Sci. Technol.* 28:417–438.

Watson, H. H. 1936. Errors due to anisokinetic sampling. *Trans. Farad. Soc.* 32:1073–1078.

Wells, P. V. and R. H. Gerke. 1919. The dust free space surrounding hot bodies. *J. Am. Chem.* Soc. 41:312–321.

Wilson, C. T. R. 1897. Condensation of water vapour in the presence of dust-free air and other gases. *Philos. Trans. R. Soc. Lond. A.* 189:265–307.

Zeiss, C. 1950. *Konimeter.* Prospectus 650, Carl Zeiss, Inc., Oberkochen, Germany.

Zinky, W. R. 1962. A new tool for air pollution control: The aerosol particle counter. *Air Pollut. Control Assoc. J.* 12:578–583.

Zsigmondy, R. 1909. *Colloids and Ultramicroscope*. New York: J. Alexander.

Zsigmondy, R. and W. Bachmann. 1916. Cellulosenitrat-Ultrafilter. German Patents 329-060 (May 9) and 329–117 (August 22).

Zsigmondy, R. and W. Bachmann. 1918. Cellulosenitrat-Ultrafilter. *Z. Anorg. Chem.* 103:119–128. *J. Soc. Chem. Ind.* 37:453A–458A.

2 Bridging Science and Application in Aerosol Measurement: Accessing Available Tools

PAUL A. BARON

Centers for Disease Control and Prevention,* National Institute for Occupational Safety and Health, Cincinnati, OH

KLAUS WILLEKE

Department of Environmental Health, University of Cincinnati, Cincinnati, OH

INTRODUCTION

An aerosol is an assembly of liquid or solid particles suspended in a gaseous medium, for example, air, long enough to be observed and measured. The need to measure aerosols has increased dramatically in recent years. This need has arisen in various disciplines. Although much information is available in the published literature, the scientists and practitioners making aerosol measurements in one application may have little or no awareness of the knowledge and experience gained in other applications.

For instance, environmental engineers and industrial hygienists make aerosol measurements in order to ensure that the public and the industrial work force are not exposed to hazardous aerosols at undesirable concentration levels. Faced by increasingly complex and demanding regulations, the aerosol measurements are becoming more costly in equipment and in time to perform, and they may require more than elementary knowledge to perform and interpret. The results are critical as expensive control measures may have to be put in place as a result of such measurements.

In contrast, the scientists and engineers concerned with industrial materials are developing an ever-increasing number of manufacturing processes in which the material passes through an aerosol phase that needs to be monitored and controlled. For instance, powders and pigments may be produced by passing the feed materials into a flame, plasma, laser, or flow furnace where they evaporate. Upon cooling, a very high concentration of very small particles is formed. With time, the aerosols may agglomerate to a lower concentration of larger particles. At any time, the trajectories of these particles may be directed by external forces, for example, diffusion or an applied electric field, to deposit the particles in a prede-

*Mention of product or company name does not constitute endorsement by the Centers for Disease Control and Prevention.

Aerosol Measurement: Principles, Techniques, and Applications, Second Edition, Edited by Paul A. Baron and Klaus Willeke.
ISBN 0-471-35636-0

termined manner, thus forming products such as ceramics or optical fibers. While producing desirable materials in this manner, the aerosol may have to be measured not only to ensure an optimum amount of uniform product but also to avoid exposure of humans to processing materials that may be quite hazardous.

As a consequence, the novice and the experienced scientist or practitioner alike may have to become familiar with new principles, aerosol measurement techniques and applications. This book attempts to address all of these aspects by dealing with aerosol measurement in three parts. Part I is devoted to the basic concepts of aerosol mechanics (i.e., the behavior of particles suspended in air under the influence of various forces and conditions). This section ends with a chapter combining the concepts of aerosol mechanics with real world situations where measurements must take into account the aerosol-laden environment as well as which properties of the aerosol can be measured with the tools available. Part II expands the latter by devoting a chapter to each principle instrumental technique or group of techniques. Part III begins with a description of nonspherical properties and then discusses a wide range of applications. Each application requires a specific set of aerosol properties to be measured, thus dictating the type of measurement technique or group of associated techniques that can be used.

The book attempts to give the fundamental principles in sufficient detail so that scientists and practitioners may use them in deciding which aerosol properties to measure and how to interpret the results. The technique and application chapters attempt to guide them in performing the actual measurements. As such, the book bridges science and application in aerosol measurement.

There are a number of tools available to aid in understanding and measuring aerosols. The scientific literature provides a wealth of information to aid in selecting instrumentation and understanding aerosol behavior. Supplementary references are cited at the end of each chapter. A summary list of books and journals is given below (adapted from the Education Committee of the American Association for Aerosol Research, 1990, A Bibliography of Aerosol Science and Technology, *Aerosol Science and Technology*, 14:1–4). Perhaps a more complete list of currently available books on a host of aerosol and related topics can be found on the Internet at www.amazon.com or other bookseller sites by entering the appropriate search criteria. A new resource for the aerosol community is being developed at www.aerosolsonline.com. This site is supported by the International Aerosol Assembly and lists data available on members, instrument manufacturers, and other useful aerosol-related information.

ASSOCIATED FIELDS

Many aerosol studies grew historically from applications in the health- and environment-related areas. Quite often, publications featuring aerosol measurements are presented in the chemistry, physics, biology, optics, or engineering disciplines. Other areas involving overlapping particle sizes and similar particle dynamics arose in industrial applications and have retained their own societies and journals. Two such areas closely related to aerosol studies are the studies of powders and sprays. Sprays have been especially important in combustion technology, while both sprays and powders have been important techniques in material manufacturing. A relatively new area of research and development emphasis is in the small particle region of the aerosol range, approaching the molecular size range. This area has been given various names, including *nanotechnology*, referring to the approximate range of 1 to 100 nm usually addressed. Some of the journals and references in these fields are included in the References.

COMPUTER TECHNOLOGY

Computer technology has advanced significantly in recent years, closely following Moore's law of doubling in complexity or power every 18 to 24 months. This has opened up calculation opportunities in a number of areas that were previously very difficult or expensive. Easy-to-use programs are now available in a number of areas to aid in various aspects of aerosol research and applications. The Internet is now a rich source of information about aerosol instruments and measurement techniques, researchers, companies, and so forth. There are a variety of information sources with software reviews, lists of programs, and software companies. Two such sources are *Research and Development* magazine (www.rdmag.com) and *Scientific Computing and Instrumentation* magazine (www.scamag.com).

LANGUAGES

Traditional programming languages such as Basic, FORTRAN, Pascal, and C have generally become more powerful and integrated into the general computer environment. Higher level languages more suited to the scientific environment have been developed. The programs mentioned here are only provided as examples of the available computational tools and do not comprise a comprehensive list by any means. Many programs have multiple capabilities and include the functions for equation solving, statistical analysis, curve fitting, and graphing. Mathematica (***WOL***),* and MathCad (***MAS***) are examples of higher level programs (having their own "language") in which the user can enter equations directly and provide rapid calculation and graphing of complex sets of equations. Perhaps the most ubiquitous calculation tool currently in use is the spreadsheet. Originally developed as a business tool, current spreadsheet programs such as Excel (***MIC***) and Quattro Pro (***CRL***) have extensive libraries of functions that allow a user to perform complex calculations in a short period of time. Most of these programs can be linked to one another via special programming or other commercially available software. A collection of spreadsheets was prepared to allow the user to play "what if?" games with the formulas in this book to provide better understanding of concepts and to predict behavior of various aerosol systems. These spreadsheets are described below.

Aerosol behavior is largely dependent on air movement and properties. Air motion, turbulence, particle trajectories, and heat transfer in various systems can be calculated with computational fluid dynamics (CFD) programs such as Fluent and FIDAP (***FLT***). These calculations generally require a great deal of computing power, but simple problems can be solved on current personal computers. However, these are complex programs that require not only a background in fluid dynamics theory but also training in their use.

The chapter in the previous edition of this book on data acquisition was dropped because of the difficulty of addressing such a complex topic in the limited space available. The use of video recording was described as a powerful adjunct to aerosol measurements. This has been applied with great success in the industrial hygiene area, allowing correlation of recorded activities with aerosol and other measurements. The measurements can be recorded with data acquisition software. There are many data acquisition systems available, from stand-alone portable devices that can be integrated with hand-held instruments to more complex systems requiring computer support and extensive programming. This area has become much more amenable to the infrequent user, with powerful visually oriented languages such as Labview (***NAT***) and LabTech Notebook (***ADT***). These allow relative ease of integration of complex control and data acquisition functions with real-time computer screen visualization and control. Some training in these programs is helpful, but simple problems can generally be solved with the manuals available.

Many aerosol instruments come with automated data logging built in or can be readily interfaced with a computer to allow data viewing and analysis. Often, the manufacturers of these instruments also provide a means of interface with conventional programs, such as spreadsheets, so that a new language or operating system does not have to be learned for each instrument.

* See Appendix I for full manufacturer addresses referenced by the italicized three-letter codes.

Analysis of particle shape is often a useful adjunct to other types of particle measurement. Particle shape gives clues regarding the history of a particle and can be important for particle dynamics (see Chapter 23). Image analysis is a technique that can be used to quickly accumulate statistics on many particles. Image analysis can easily be integrated with a light microscope, although contrast and refraction effects limit the size of particle that can be accurately imaged, especially after collection on a filter. Electron microscopy generally allows better imaging for interface to image analysis systems. There are many commercial image analysis systems. An image analysis program developed by the National Institutes of Health (NIH Image) is freely available, and there is a list server where questions and discussion can be posted. Information and downloading instructions are available at rsb.info.nih.gov/nih-image/.

Light scattering from particles is commonly used as a detection or particle analysis technique. The prediction of light scattering from individual particles is discussed in Chapter 15 and can be accomplished with computer codes that implement the equations mentioned. Sources on the Internet for some of these codes have been compiled and listed at the web sites www.eee.metu.edu.tr/~ngencer/codes.htm and diogenes.iwt.uni-bremen.de/~wriedt/Mie_Type_Codes/body_mie_type_codes.html.

Data analysis and presentation have also been made easier, with many programs available for statistical analysis, curve fitting, and graphing. A program designed specifically for aerosol distributions is DistFit (***TSI***). Neural networks have also been used to analyze data. This technique uses a network of simulated neurons that "learn" relationships from a training set and can then be applied to new situations. However, to provide accurate results, the applications require that the relationship space is well-represented by the training set. This technique sometimes has an advantage in finding relationships in data more accurately than traditional statistical analysis. Once a network is trained, it can often provide more rapid analysis of new data than other types of analysis. A disadvantage is that a closed-form function describing the solution is not readily available, as the solution is stored in the neural network.

AEROSOL CALCULATOR

The Aerosol Calculator (Baron 1999) is a collection of spreadsheets (Excel, ***MIC***) that is available from several sources (www.tsi.com, www.bgiusa.com) and is keyed to the equations in this book as well as in the texts by Hinds (1982, 1999). The spreadsheets are updated as new additions and corrections are made. Some additional calculations are provided along with reference to the source literature. The principal spreadsheet (AeroCalc) contains a series of modules, each calculating a specific equation or set of equations. The modules can be copied into a new spreadsheet and, by creating arrays and linking appropriate modules, complex systems can be modeled and investigated. It is recommended that the calculations be performed on a copy of the spreadsheet, to prevent corruption of the original file by errors. The individual spreadsheets in the Aerosol Calculator collection are listed in Table 2–1. The following discussion assumes some familiarity with spreadsheet operation.

Each module in AeroCalc is self-contained and consists of a set of input parameters with indicated units followed by the calculated output parameter(s). Two separate sheets are provided, one in SI units and the other in cgs units, bridging the units used in the first (Willeke and Baron, 1993) and second editions of this book and by Hinds (1982, 1999). The calculations have been performed using relatively common functions in an attempt to make it easier for the user to understand and modify. No macros were used. However, some calculations require the iteration function to be turned on. These calculations will not work in spreadsheet programs that do not support the iteration function.

TABLE 2–1. Spreadsheet Programs Used to Perform Various Aerosol-Related Calculations[a]

AeroCalc.xl	Calculations keyed to specific equations in this text and several other references
sizedis.xl	Calculation of lognormal distributions with discrete size distribution intervals. Surface area and volume distributions are calculated assuming spherical particles. A simulation of random number count errors is included, as well as calculation of respirable, thoracic, inhalable, and PM10 dust fractions. See Chapters 25, 27, and 29 for definitions of these terms
size2d.xl	Calculation of a two-dimensional size distribution, e.g., for fibers with length and diameter. See Chapter 23
2Drect.xl	Calculation of Laplacian function in two-dimensional rectangular coordinate system. It can be used for electric fields or for potential flow in any two-dimensional field. For calculating potential flow, see the example in White, F. M., 1986, *Fluid Mechanics*, 2nd Ed., pp. 497–500. Each spreadsheet cell represents and x–y coordinate in the electric or flow field. The problem is defined by the boundary values, and the cells inside the boundary are filled with replicated calculation cells. Each internal cell performs an identical calculation based on nearest neighbor values. The array size that can be solved is limited by the memory and calculation speed of the computer. The array is solved by iteration, and the number of iterations increases with the number of cells in the array
2Daxial.xl	Similar to 2Drect.xl except that it is used for cylindrically symmetrical two-dimensional problems
Probit.xl	Calculation and plotting of probits for a size distribution measurement. This is useful for graphical presentation of size distribution data on a log probability plot. See Chapter 22

[a] Instructions and references for each module are in the corresponding spreadsheets.

The application of the equations provided in various references allows the user to better understand the limitations and usefulness of these equations through the use of the spreadsheet programs. Not only can the reader of the book perform calculations without having to do extensive programming, unit conversion, and error checking, but the software allows graphing of the results as a function of the variables, such as particle size, temperature, and flow rate, for better understanding of the text description. Some of the calculations are empirical and have limits based on the data set from which they were derived. Frequently, these limits have been indicated in AeroCalc, but the user must be aware of other limitations described in the text and in the original references.

The calculations possible with this program become much more meaningful and powerful if the user understands the equations and the assumptions behind the equations; errors can result if the user enters improper values or misinterprets the results. Thus, the program does not substitute for that understanding, though it certainly allows the user to "play" with the parameters in the equations for a better understanding of aerosol behavior. Some example applications of the AeroCalc program are indicated below.

AeroCalc

The general format of the AeroCalc spreadsheet is indicated in Figure 2–1. The modules are separated with a line of asterisks, and the input and calculated parameters are separated with a line of dashes. Column A contains the description of the module at the top, followed by descriptions of the input and calculated parameters. Column B contains the input and calculated parameters, while column C contains the units. Although not shown in this example, column C contains comments about specific parameters indicating value limits

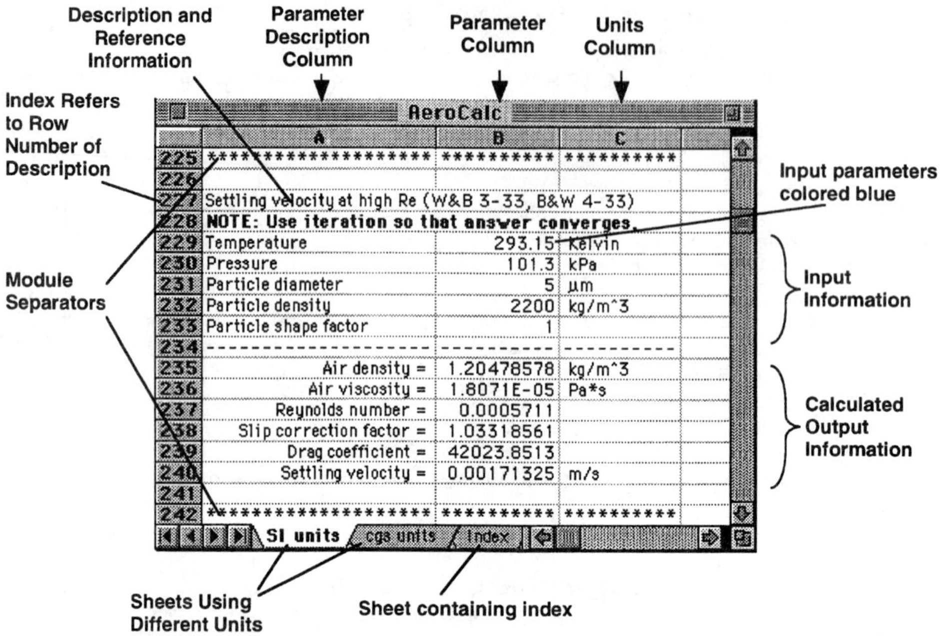

Fig. 2–1. Example module in AeroCalc spreadsheet indicating several typical features. Bold row 228 indicates that the iteration function (under menu Tools:Options:Calculation) must be turned on for the results to be calculated correctly. If iteration is not turned on, an error message regarding circular references appears on startup of the program. This message can be ignored if the modules being used do not require iteration.

or other warnings. Occasionally there is a table of useful values in columns C, D, and E. There is an index of all the calculations in columns I and J between rows 2 and approximately 105. There is also a table of conversion factors in columns I and J between rows 117 and 205.

In the example in Figure 2–1, iteration of the calculation is required to reach the correct answer. In the Excel menu bar, under Tools, Options, Calculation, there is an option to select Iteration with a maximum number of iterations and the degree of precision (maximum change). If this is not turned on, Excel will report on startup that "circular references" exist. This will not affect other calculations, and the warning can be turned off. Most of the calculations will reach a solution with fewer than 100 iterations at a maximum change of 0.0001, but some, as indicated in the module header, may take more.

In the example, the calculation of settling velocity depends on the drag coefficient when the velocity is outside the Stokes regime (Reynolds number, $Re < 0.1$ in Stokes regime). However, the drag coefficient depends on the final velocity, so iteration of the calculation is required to reach the correct solution. Any of the input parameters can be changed to observe the effect on the calculated values. This can provide useful information. However, it is easy to expand the usefulness by creating an array of calculations that are identical except for one changing parameter.

Figure 2–2 displays the calculation in Figure 2–1 with several additional columns inserted between columns B and C. Column B is then replicated to the right to fill the inserted columns. To observe the change in the calculated values of settling velocity as a function of particle diameter, each subsequent column can be changed by an increment. In Figure 2–2,

AeroCalc

	A	B	C	D	E	F	G	H
227	Settling velocity at high Re (W&B 3-33, B&W 4-33)							
228	**NOTE: Use iteration so that answer converges.**							
229	Temperature	293.15	293.15	293.15	293.15	293.15	293.15	Kelvin
230	Pressure	101.3	101.3	101.3	101.3	101.3	101.3	kPa
231	Particle diameter	5	10	20	40	80	160	µm
232	Particle density	2200	2200	2200	2200	2200	2200	kg/m^3
233	Particle shape factor	1	1	1	1	1	1	
234	-------------------	-------	-------	-------	-------	-------	-------	-------
235	Air density =	1.20479	1.20479	1.20479	1.20479	1.20479	1.20479	kg/m^3
236	Air viscosity =	1.8E-05	1.8E-05	1.8E-05	1.8E-05	1.8E-05	1.8E-05	Pa*s
237	Reynolds number =	0.00057	0.0045	0.03567	0.27715	1.92823	10.3567	
238	Slip correction factor =	1.03319	1.01659	1.0083	1.00415	1.00207	1.00104	
239	Drag coefficient =	42023.9	5338.72	672.831	88.7937	14.6451	4.05698	
240	Settling velocity =	0.00171	0.00674	0.02675	0.10393	0.36153	0.97091	m/s

SI units / cgs units / Index

Fig. 2–2. Particle diameter is changed, starting in column C, by multiplying the value in the previous column by a factor of two. The settling velocity can be plotted as a function of particle diameter.

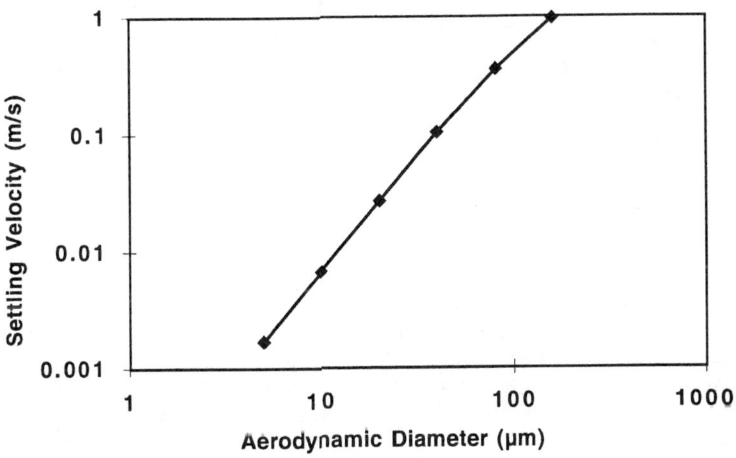

Fig. 2–3. Plot of particle setting velocity as a function of particle diameter on a long–log scale using the Excel chart capability.

[Col C, Row 231] is set = 2 × [Col B, Row 231]. This equation is replicated to the rest of Row 231. It is also convenient to set the other input parameter column data to equal the corresponding values in the column B. That way, only the values in column B need to be changed, and the entire array recalculates appropriately.

The values calculated in Figure 2–2 can immediately be plotted as indicated in Figure 2–3. Noncontiguous rows can be selected for plotting by replicating the rows elsewhere on the spreadsheet. Alternatively, they can be selected directly by holding down the keyboard control key (command key on the Macintosh) while selecting the second row. In this example, we have changed particle diameter in a geometric rather than a linear progression, and it is convenient to display the results on a log–log scale. The nonlinearity of the resulting curve is an indication of the change in the drag coefficient outside the Stokes regime. The number of columns can be increased, using smaller increments in particle diameter, and the data replotted to obtain a smoother curve.

It is often desirable to determine the input parameter that gives a target output value. This can be done by trial and error, replacing the input parameter with various values, until the target value is reached with acceptable precision. Alternatively, Excel has a feature called the "Solver" under the Tools menu that allows automatic calculation of the appropriate input value to give a selected target value. The Solver may not always work properly with modules that use iteration for their solution.

Several modules can be linked together to perform more complex calculations. For instance, a common situation in aerosol measurement is the desire to ensure that the sampling system conveying particles to a measurement instrument has minimal, or at least known, losses. There are AeroCalc modules that calculate the losses in various inlets, tubing, and bends under laminar and turbulent conditions. By assuming that each of these components act independently (not always a good assumption; see Chapter 8), we can estimate the overall loss in the inlet system. For instance, if we have an inlet, followed by a bend, a length of tubing, a second bend, and another section of tubing, the overall efficiency of aerosol reaching the sensor relative to the air concentration is, to first approximation, the product of each individual efficiency η

$$E = \eta_{inlet}\eta_{tube1}\eta_{bend1}\eta_{tube2}\eta_{bend2}\eta_{tube2} \qquad (2\text{–}1)$$

By linking spreadsheet modules that calculate each η and the overall efficiency, it is possible to optimize the overall efficiency by varying the flow rate and the lengths, diameters, and orientation of each component.

The calculations in the modules can also be linked to the size distribution spreadsheets indicated in Table 2–1. One example of such a calculation is given in Chen and Baron (1996) where the calculated fiber deposition efficiency in a tubular inlet was compared successfully with experimental deposition measurements. Another example is given in Baron (1996) where the aerodynamic diameters of fibers likely to deposit in the thoracic region of the respiratory system, represented by an aerodynamic selection system, were compared with current microscope measurement procedures.

The Aerosol Calculator provides a convenient set of tools that allow rapid calculation of a wide range of parameters in aerosol mechanics. It can be used to increase understanding of aerosol behavior, indicating the most important mechanisms operating in a given situation.

REFERENCES

Chapter References

Baron, P. A. 1996. Application of the thoracic sampling definition to fiber measurement. *Am. Ind. Hyg. Assoc. J.* 57:820–824.

Baron, P. A. 1999. Aerosol Calculator. Collection of computer spreadsheets available from *www.tsi.com* and *www.bgiusa.com*.

Chen, C.-C. and P. A. Baron. 1996. Aspiration efficiency and deposition in the fiber sampling cassette. *Am. Ind. Hyg. Assoc.* 52:142–152.

Hinds, W. 1982. *Aerosol Technology*. New York: John Wiley & Sons, Inc.

Hinds, W. 1999. *Aerosol Technology*. 2nd Ed. New York: John Wiley & Sons, Inc.

Willeke, K. and P. A. Baron. 1993. *Aerosol Measurement*. New York: Van Nostrand Reinhold.

General References

Abraham, F. F. 1974. *Homogeneous Nucleation Theory: The Pretransition Theory of Vapor Condensation, Supplement I: Advances in Theoretical Chemistry.* New York: Academic Press.

Bailey, A. G. 1988. *Electrostatic Spraying of Liquids.* New York: John Wiley & Sons, Inc.

Baron, P. A. 1996. Application of the thoracic sampling definition to fiber measurement. *Am. Ind. Hyg. Assoc. J.* 57:820–824.

Baron, P. A. 1999. Aerosol Calculator. Collection of computer spreadsheets available from *www.tsi.com* and *www.begiusa.com*.

Beddow, J. K. 1980. *Particulate Science and Technology.* New York: Chemical Publishing Co.

Bohren, C. F. and D. R. Huffman. 1983. *Absorption and Scattering of Light by Small Particles.* New York: Wiley & Sons, Inc.

Chen, C.-C. and P. A. Baron. 1996. Aspiration efficiency and deposition in the fiber sampling cassette. *Am. Ind. Hyg. Assoc.* 52:142–152.

Clift, R., J. R. Grace, and M. E. Weber. *Bubbles, Drops, and Particles*. Boston: Academic Press.

Colbeck, I., ed. 1997. *Physical and Chemical Properties of Aerosols.* Dordrecht: Kluwer Academic Publishers.

Davies, C. N., ed. 1966. *Aerosol Science.* New York: Wiley & Sons, Inc.

Dennis, R. 1976. *Handbook on Aerosols.* Publication TID-26608. Springfield, VA: National Technical Information Service, U.S. Dept. of Commerce.

Einstein, A. 1956. *Investigations on the Theory of Brownian Motion.* New York: Dover Publications, Inc.

Friedlander, S. K. 1977. *Smoke, Dust and Haze.* New York: Wiley & Sons, Inc.

Fuchs, N. A. 1989. *The Mechanics of Aerosols.* New York: Wiley & Sons, Inc.

Fuchs, N. A. and A. G. Sutugin. 1970. *Highly Dispersed Aerosols.* Ann Arbor, MI: Ann Arbor Science Publishers, Inc.

Green, H. L. and W. R. Lane. 1964. *Particulate Clouds, Dust, Smokes and Mists*, 2nd Ed. Princeton, NJ: Van-Nostrand Co.

Happel, J. and H. Brenner. 1973. *Low Reynolds Number Hydrodynamics with Special Applications to Particulate Media*, 2nd Rev. Ed. Leyden: Noordhoff International Publishing.

Hesketh, H. E. 1977. *Fine Particles in Gaseous Media.* Ann Arbor, MI: Ann Arbor Science Publishers, Inc.

Hidy, G. M. 1972. *Aerosols and Atmospheric Chemistry.* New York: Academic Press.

Hidy, G. M. and J. R. Brock. 1970. *The Dynamics of Aerocolloidal Systems.* New York: Pergamon Press.

Hidy, G. M. and J. R. Brock, eds. 1971. *Topics in Recent Aerosol Research.* New York: Pergamon Press.

Hidy, G. M. and J. R. Brock, eds. 1972. *Topics in Current Aerosol Research*, Part 2. New York: Pergamon Press.

Hinds, W. 1982. *Aerosol Technology*. New York: John Wiley & Sons, Inc.

Hinds, W. C. 1999. *Aerosol Technology.* New York: Wiley & Sons, Inc.

Hinds, W. 1999. *Aerosol Technology*. 2nd Ed. New York: John Wiley & Sons, Inc.

Kerker, M. 1969. *The Scattering of Light and Other Electromagnetic Radiation.* New York: Academic Press.

Lefebvre, A. H. 1989. *Atomization and Sprays.* New York: Hemisphere.

Liu, B. Y. H., ed. 1976. *Fine Particles.* New York: Academic Press.

Marlow, W. H., ed. 1982a. *Aerosol Microphysics I. Chemical Physics of Microparticles.* Berlin: Springer-Verlag.

Marlow, W. H., ed. 1982b. *Aerosol Microphysics II. Chemical Physics of Microparticles.* Berlin: Springer-Verlag.

Mason, B. J. 1971. *The Physics of Clouds.* Oxford: Clarendon Press.

McCrone, W. C., et al. 1980. *The Particle Atlas*, Vols. I–VII. Ann Arbor, MI: Ann Arbor Science Publishers.

Mednikov, E. P. 1980. *Turbulent Transport of Aerosols* [in Russian]. Science Publishers.

Orr, C., Jr. 1966. *Particulate Technology.* New York: Macmillan Co.

Reist, P. C. 1984. *Aerosol Science and Technology.* New York: McGraw-Hill.

Sanders, P. A. 1979. *Handbook of Aerosol Technology.* Melbourne, FL: Krieger Publishing.

Sedunov, Y. S. 1974. *Physics of Drop Formation in the Atmosphere* [translated from Russian]. New York: John Wiley & Sons, Inc.

Twomey, S. 1977. *Atmospheric Aerosols.* Amsterdam: Elsevier Science Publishers.

Van de Hulst, H. C. 1957. *Light Scattering by Small Particles.* New York: John Wiley & Sons, Inc. Republished (1981) unabridged and corrected. New York: Dover Publications, Inc.

Vohnsen, M. A. 1982. *Aerosol Handbook*, 2nd Ed. Mendham, NJ: Dorland Publishing Co.

Wen, C. S. 1996. *The Fundamentals of Aerosol Dynamics*. Singapore: World Scientific Publishing Co.

Whytlaw-Grey, R. W. and H. S. Patterson. 1932. *Smoke: A Study of Aerial Disperse Systems.* London: E. Arnold.

Willeke, K. and P. A. Baron. 1993. *Aerosol Measurement*. New York: Van Nostrand Reinhold.

Willeke, K., ed. 1980. *Generation of Aerosols and Facilities for Exposure Experiments.* Ann Arbor, MI: Ann Arbor Science Publishers.

Williams. M. M. R. and S. K. Loyalka. 1991. *Aerosol Science Theory and Practice: With Special Application to the Nuclear Industry.* Oxford: Pergamon Press.

Withers, R. S. 1979. *Transport of Charged Aerosols.* New York: Garland Publishers.

Yoshida, T., Y. Kousaka, and K. Okuyama. 1979. *Aerosol Science for Engineers.* Tokyo: Power Co., Ltd.

Zimon, A. D. 1976. *Adhesion of Dust and Powders*, 2nd Ed. [in Russian]. Moscow: Khimia. 1st Ed., 1969 [in English]. New York: Plenum Press.

Measurement Techniques

Allen, T. 1968. *Particle Size Measurement.* London: Chapman and Hall.

Allen, T. 1981. *Particle Size Measurement*, 3rd Ed. New York: Methuen, Inc.

Barth, H. G., ed. 1984. *Modern Methods of Particle Size Analysis.* New York: John Wiley & Sons, Inc.

Beddow, J. K. 1980. *Testing and Characterization of Powders and Fine Particles.* New York: John Wiley & Sons, Inc.

Beddow, J. K. 1984. *Particle Characterization in Technology.* Boca Raton, FL: CRC Press Inc.

Cadle, R. D. 1965. *Particle Size: Theory and Industrial Applications.* New York: Reinhold Publishing Corp.

Cadle, R. D. 1975. *The Measurement of Airborne Particles.* New York: John Wiley & Sons, Inc.

Cheremisinoff, P. N., ed. 1981. *Air Particulate Instrumentation and Analysis.* Ann Arbor, MI: Ann Arbor Science Publishers, Inc.

Dallavalle, J. M. 1948. *Micromeritics*, 2nd Ed., New York: Pitman Publishing Corp.

Dzubay, T. G. 1977. *X-Ray Fluorescence Analysis of Environmental Samples.* Ann Arbor, MI: Ann Arbor Science Publishers, Inc.

Herdan, G. 1953. *Small Particle Statistics.* New York: Elsevier Science Publishing Co., Inc.

Jelinek, Z. K. (translated by W. A. Bryce). 1974. *Particle Size Analysis.* New York: Halstead Press.

Lodge, J. P., Jr. and T. L. Chan, eds. 1986. *Cascade Impactor, Sampling and Data Analysis.* Akron, OH: American Industrial Hygiene Association.

Malissa, H., ed. 1978. *Analysis of Airborne Particles by Physical Methods.* Boca Raton, FL: CRC Press.

Nichols, A. L. 1999. *Aerosol Particle Size Analysis: Good Calibration Practices*. Cambridge: Royal Society of Chemistry.

Orr, C. and J. M. Dallavalle. 1959. *Fine Particle Measurement.* New York: Macmillan Co.

Rajhans, G. S. and J. Sullivan. 1981. *Asbestos Sampling and Analysis.* Ann Arbor, MI: Ann Arbor Science Publishers, Inc.

Silverman, L., C. Billings, and M. First. 1971. *Particle Size Analysis in Industrial Hygiene.* New York: Academic Press.

Stockham, J. D. and E. G. Fochtman. 1977. *Particle Size Analysis.* Ann Arbor, MI: Ann Arbor Science Publishers, Inc.

Vincent, J. H. 1989. *Aerosol Sampling: Science and Practice.* New York: John Wiley & Sons, Inc.

Gas Cleaning

Clayton, P. 1981. *The Filtration Efficiency of a Range of Filter Media for Submicrometer Aerosols.* New York: State Mutual Book and Periodical Service.

Davies, C. N. 1973. *Air Filtration.* London: Academic Press.

Dorman, R. G. 1974. *Dust Control and Air Cleaning.* New York: Pergamon Press.

Mednikov, E. P. 1965. *Acoustic Coagulation and Precipitation of Aerosols.* New York: Consultants Bureau.

Ogawa, A. 1984. *Separation of Particles From Air and Gases*, Vols. I and II. Boca Raton, FL: CRC Press.

Spurny, K. 1998. *Advances in Aerosol Filtration.* Boca Raton, FL: Lewis Publishers.

White, H. J. 1963. *Industrial Electrostatic Precipitation.* Reading, MA: Addison-Wesley Publishing Co., Inc.

Environmental Aerosols/Health Aspects

American Conference of Governmental Industrial Hygienists. *Air Sampling Instruments*, 8th Ed. 1995. Cincinnati, OH: American Conference of Governmental Industrial Hygienists.

Brenchly, D. L., C. D. Turley, and R. F. Yarmae. 1973. *Industrial Source Sampling.* Ann Arbor, MI: Ann Arbor Science Publishers, Inc.

Cadle, R. D. 1966. *Particles in the Atmosphere and Space.* New York: Reinhold Publishing Corp.

Cox, C. S. and C. M. Wathes. 1995. *Bioaerosols Handbook.* Boca Raton, FL: Lewis Publishing.

Drinker, P. and T. Hatch. 1954. *Industrial Dust.* New York: McGraw-Hill Book Co., Inc.

Flagan, R. C. and J. H. Seinfeld. 1988. *Fundamentals of Air Pollution Engineering.* New York: Prentice Hall.

Hickey, A. J. 1996. *Inhalation Aerosols.* New York: Marcel Dekker, Inc.

Hidy, G. M. 1972. *Aerosols and Atmospheric Chemistry.* New York: Academic Press.

Junge, C. 1963. *Air Chemistry and Radioactivity.* New York: Academic Press.

Lighthart, B. and A. J. Mohr. 1994. *Atmospheric Microbial Aerosols—Theory and Applications.* Chapman and Hall.

McCartney, E. J. 1976. *Optics of the Atmosphere.* New York: John Wiley & Sons, Inc.

Mercer, T. T. 1973. *Aerosol Technology in Hazard Evaluation.* New York: Academic Press.

Middleton, W. E. K. 1952. *Vision Through the Atmosphere.* Toronto: University of Toronto.

Moren, F., M. B. Dolovich, M. T. Newhouse, and S. P. Newman. 1993. *Aerosols in Medicine. Principles, Diagnosis, and Therapy.* Elsevier Science Ltd.

Muir, D. C. F., ed. 1972. *Clinical Aspects of Inhaled Particles.* London: Heinemann.

National Research Council, Subcommittee on Airborne Particles. 1979. *Airborne Particles.* Baltimore: University Park Press.

National Research Council. 1996. *A Plan for a Research Program on Aerosol Radiative Forcing and Climate Changes.* Washington DC: National Academy Press.

Perera, F. and A. K. Ahmen. 1979. *Respirable Particles: Impact of Airborne Fine Particles on Health and Environment.* Cambridge, MA: Ballinger Publishing.

Seinfeld, J. H. and S. Pandis. 1998. *Atmospheric Chemistry and Physics.* New York: John Wiley & Sons, Inc.

Spurny, K. 1999a. *Aerosol Chemical Processes in Polluted Atmospheres.* Boca Raton, FL: Lewis Publishers.

Spurny, K. 1999b. *Analytical Chemistry of Aerosols.* Boca Raton, FL: Lewis Publishers.

Vincent, J. H. 1995. *Aerosol Science for Industrial Hygienists.* Tarrytown NY: Elsevier.

Whitten, R. C., ed. 1982. *The Stratospheric Aerosol Layer.* Berlin: Springer-Verlag.

Industrial Applications and Processes

Andonyev, S. and O. Filipyev. 1977. *Dust and Fume Generation in the Iron and Steel Industry.* Chicago: Imported Publications.

Austin, P. R. and S. W. Timmerman. 1965. *Design and Operation of Clean Rooms.* Detroit: Business News Publishing Co.

Boothroyd, R. G. 1971. *Flowing Gas–Solids Suspensions.* London: Chapman and Hall.

Donnet, J. B. and A. Voet. 1976. *Carbon Black.* New York: Marcel Dekker Inc.

Kodas, T. T. and M. J. Hampden-Smith. 1999. *Aerosol Processing of Materials.* New York: John Wiley & Sons, Inc.

Marshall, W. R., Jr. 1954. *Atomization and Spray Drying.* New York: Chemical Engineering Progress Monograph Series, Vol. 50, No. 23, AIChE.

Proceedings of Meetings

Air and Waste Management Association. 1989. *Visibility and Fine Particles.* Proceedings of the 1989 EPA/A&WMA International Specialty Conference, Pittsburgh, Pennsylvania.

American Conference of Governmental Industrial Hygenists. *Advances in Air Sampling.* 1988. Papers from the American Conference of Governmental Industrial Hygienists Symposium. Ann Arbor, MI: Lewis Publishers, Inc.

ASTM Symposium on Particle Size Measurement. 1959. ASTM Special Technical Publication No. 234.

Barber, D. W. and R. K. Chang. 1988. *Optical Effects Associated with Small Particles.* Singapore: World Scientific Publishing Co.

Beard, M. E. and H. L. Rook, eds. 2000. *Advances in Environmental Measurement Methods for Asbestos.* STP 1342. Philadelphia: American Society for Testing Materials.

Beddow, J. K. and T. P. Meloy, eds. 1980. *Advanced Particulate Morphology.* Boca Raton, FL: CRC Press.

Davies, C. N. 1964. *Recent Advances in Aerosol Research.* New York: Macmillan Co.

Dodgson, J., R. I. McCallum, M. R. Bailey, and D. R. Fisher, eds. 1989. *Inhaled Particles VI.* Oxford: Pergamon Press.

Fedoseev, V. A. 1971. *Advances in Aerosol Physics* (translation of Fizika Aerodispersnykh Sistem). New York: Halsted Press.

Gerber, H. E. and E. E. Hindman, eds. 1982. *Light Absorption by Aerosol Particles.* Hampton, VA: Spectrum Press.

Hobbs, P. V. 1993. *Aerosol–Cloud–Climate Interactions.* New York: Academic Press.

Israel, G. 1986. *Aerosol Formation and Reactivity.* Proceedings of the Second International Aerosol Conference, September 22–26, 1986, Berlin (West). Oxford: Pergamon Press.

Kuhn, W. E., H. Lamprey, and C. Sheer, eds. 1963. *Ultrafine Particles.* New York: John Wiley & Sons, Inc.

Lee, S. D., T. Schneider, L. D. Grant, and P. J. Verkerk, eds. 1986. *Aerosols: Research, Risk Assessment and Control Strategies.* Proceedings of the Second U.S.–Dutch International Symposium, Williamsburg, Virginia May 19–25, 1985. Chelsea, MI: Lewis Publishers, Inc.

Liu, B. Y. H., D. Y. H. Pui, and H. J. Fissan. 1984. *Aerosols: Science, Technology and Industrial Applications of Airborne Particles.* 300 Extended Abstracts from the First International Aerosol Conference, Minneapolis, Minnesota, September 17–21, 1984. New York: Elsevier Science Publishing Co., Inc.

Lundgren, D. A., et al., eds. 1979. *Aerosol Measurement.* Gainesville, FL: University Presses of Florida.

Marple, V. A. and B. H. Y. Liu, eds. 1983. *Aerosols in the Mining and Industrial Work Environments.* Ann Arbor, MI: Ann Arbor Science Publishers, Inc.

Mercer, T. T., P. E. Morrow, and W. Stober, eds. 1972. *Assessment of Airborne Particles.* Springfield, IL: C.C. Thomas Publishers.

Mittal, K. L., ed. 1988. *Particles on Surfaces 1: Detection, Adhesion, and Removal.* Proceedings of a Symposium held at the Seventeenth Annual Meeting of the Fine Particle Society. July 28–August 2, 1986. New York: Plenum Publishing Corp.

Mittal, K. L., ed. 1990. *Particles on Surfaces 2: Detection, Adhesion, and Removal.* New York: Plenum Publishing Corp.

Preining, O. and E. J. Davis, eds. 2000. *History of Aerosol Science*. Proceedings of the History of Aerosol Science. August 31–September 2, 1999. Vienna: Austrian Academy of Science.

Richardson, E. G., ed. 1960. *Aerodynamic Capture of Particles.* New York: Pergamon Press.

Shaw, D. T., ed. 1978a. *Fundamentals of Aerosol Science.* New York: John Wiley & Sons, Inc.

Shaw, D. T., ed. 1978b. *Recent Developments in Aerosol Technology.* New York: John Wiley & Sons, Inc.

Siegla, P. C. and G. W. Smith, eds. 1981. *Particle Carbon: Formation During Combustion.* New York: Plenum Press.

Spurny, K. 1965. *Aerosols: Physical Chemistry and Applications.* Proceedings of the First National Conference on Aerosols. Prague: Publishing House of the Czechoslovak Academy of Sciences.

Walton, W. H., ed. 1971. *Inhaled Particles III.* Surrey: Unwin Brothers.

Walton, W. H., ed. 1977. *Inhaled Particles IV.* Oxford: Pergamon Press.

Walton, W. H., ed. 1982. *Inhaled Particles V.* Oxford: Pergamon Press.

Selected Journals on Aerosol Science and Applications

Aerosol Science and Technology
American Industrial Hygiene Association Journal
Annals of Occupational Hygiene
Atomization and Sprays
Atmospheric Environment
Environmental Science and Technology
International Journal of Multiphase Flow
Journal of Aerosols in Medicine
Journal of Aerosol Research, Japan
Journal of Aerosol Science
Journal of the Air and Waste Management Association (formerly *Journal of the Air Pollution Control Association*)
Journal of Colloid and Interface Science
Journal of Nanoparticle Research
Langmuir
Particle & Particle Systems Characterization
Particulate Science and Technology
Powder Technology
Staub Reinhaltung der Luft

3 Aerosol Fundamentals

PAUL A. BARON

Centers for Disease Control and Prevention, National Institute for Occupational Safety and Health Cincinnati, OH

KLAUS WILLEKE

Department of Environmental Health, University of Cincinnati, Cincinnati, OH

INTRODUCTION

The term *aerosol* refers to an assembly of liquid or solid particles suspended in a gaseous medium long enough to be observed or measured. The term originated as the gas phase equivalent to the term *hydrosol*, which refers to a suspension of particles in a liquid (from the Greek word combination "water particle"). Manufactured and naturally produced particles, found in ambient and industrial air environments or in industrial process gas streams, may have a great diversity in size, shape, density, and chemical composition. The diversity of microscopic particles, includes what we might think are ideal shapes, such as spheres (droplets of water or oil) or cylinders (glass fibers). However, it also includes more complex shapes such as crystalline particles, which have some regular and some fractured surfaces; asbestos fibers, which are often bundles of finer fibrils but may be matted clumps of fibrils; and carbon black particles, which often consist of an extended framework of very small spheroids. All these have a different chemical constitution. Even if all these particles have the same microscopically observed diameter, the mass, surface area, and other properties of each particle are likely to be quite different. Part III, the Applications Section of this book, discusses a wide range of particulate types in many different environmental and industrial settings.

A variety of techniques are available for obtaining useful information about these particles. One way to detect the particles is by scattering light from them. The scattering pattern produced by spheres is quite predictable, while that from glass fibers is predictable only if their orientation relative to the light beam is known. The light pattern from asbestos fibers might be similar to that from glass fibers in some cases, but may be much more complex when the fibers are splayed bundles or matted clumps. Much less light might be scattered by carbon black due to its light-absorbing properties.

To characterize the size of spherical droplets, a single parameter, such as the droplet diameter, may suffice. With fibers, two dimensions may be sufficient, although often two may not be enough for less regular shapes. Clearly, for more complex particles we have to choose the particle properties that are important for our purposes in order to make a meaningful measurement. We can characterize particles by shape parameters observed under a microscope; light-scattering properties using an optical particle counter or a nephelometer; elemental or

Aerosol Measurement: Principles, Techniques, and Applications, Second Edition, Edited by Paul A. Baron and Klaus Willeke.
ISBN 0-471-35636-0

chemical properties with X-ray fluorescence or infrared spectroscopy; surface properties with a pycnometer or by an adsorption measurement; or dynamic behavior from measurement of settling velocity or diffusion. Note that one characteristic (e.g., physical size) may not correlate well with another characteristic (e.g., chemical composition). However, any or all of these types of characterization may be important for the scientist. For example, to estimate the toxicity of an aerosol entering the lung, one needs to know the size-dependent diffusion, gravitational settling, impaction, and interception properties of the particles to determine the deposition rate within the lung. In addition, the chemistry, surface area, and fibrosity of the particles may indicate their interactions with the lung tissues once they are deposited.

Sand found on a beach also comes in many different sizes, but here the shapes tend to be closer to those of spheres because the material has been subjected to erosive forces of wind and water motion. When measuring these macroscopic materials as to their physical, chemical, and biological composition, we must use different methods and techniques. Similarly, when measuring microscopic materials, a diversity of methods and techniques is used. In this chapter, common aerosol characteristics are introduced in preparation for more detailed discussions of specific measurement techniques that are discussed in later chapters.

DESIRABLE VERSUS UNDESIRABLE AEROSOLS

The development of many aerosol sampling and analysis techniques has been stimulated by a variety of applications. Since approximately the 1950s, advances in aerosol measurement have been motivated by investigations into the health effects of radioactive aerosols and industrial aerosols in the workplace and the environment. More recently, a great deal of effort has gone into trying to understand the effect of various natural and manmade aerosols on global warming. The production of high-speed integrated circuits has required increasingly cleaner environments to reduce contamination by aerosol particles. These efforts not only have resulted in more refined and sensitive instruments but also have increased understanding of particle generation and transport mechanisms. All these efforts have largely been aimed at reducing contaminants. In contrast, a great deal of knowledge is gained today by researchers working with "desirable" aerosols used to produce high technology materials such as ceramic powders, superconducting materials, and optical fibers.

UNITS AND USE OF EQUATIONS

All equations and calculations in this book are in Système Internationale (SI) units. Whenever deemed appropriate, calculations in centimeter–gram–second (cgs) units are shown subsequently in brackets. Because aerosol particles range in diameter from about 10^{-9} m to about 10^{-4} m, the unit of *micrometer* (1 μm = 10^{-6} m) is generally used in discussions of particle dimensions. For instance, a particle most hazardous to the human respiratory system is on the order of 10^{-6} m in diameter and is conveniently described as a 1 μm particle. The term *micron* has been used in the past as a colloquial version of micrometer, but it is no longer accepted in technical writing. Researchers manufacturing aerosols through evaporation and subsequent condensation processes may deal with particles in the 0.01 to 0.1 μm range or even smaller and therefore prefer to express the particle sizes in nanometers (1 nm = 10^{-9} m). In calculations requiring the use of SI units, micrometer is converted to meter by multiplying by 10^{-6}. If calculations are performed in cgs units, micrometer is converted to centimeter by multiplying by 10^{-4}. In this book, *particle size* always refers to particle diameter. In some publications, particle radius is used instead.

The SI unit for aerosol mass concentration, that is, the mass of particulate matter in a unit volume of gas, is expressed in kg/m^3. Because the amount of aerosol mass is generally very low, the aerosol mass concentration is usually expressed in g/m^3, mg/m^3, μg/m^3, or ng/m^3. Particle velocity (e.g., under the influence of gravity or an electric field) is expressed in m/s [shown also in cm/s in brackets]. Volume is frequently indicated in liters (L = 10^{-3} m^3) because sampling volumes are often on the order of liters.

Aerosol number concentrations are expressed in number/m^3 [number/cm^3]. The older unit of millions of particles per cubic foot (mppcf) is given in parentheses when dealing with engineering systems and applications regulated by the U.S. Environmental Protection Agency. Tables in Appendix B give the conversion factors for the major units used by the practioner or researcher dealing with aerosols.

The SI unit for pressure is expressed in Pascal (1 Pa = 1 N/m^2). Atmospheric pressure (101 kPa = 1.01 × 10^6 dyne/cm^3) may also be referred to as 1 atm (= 14.7 psig = 760 mm Hg = 1040 cm H_2O = 408 inch H_2O). Gas and particle properties are listed at normal temperature and pressure (NTP), which refers to 101 kPa and 293 K [1 atm and 20°C = 68°F]. Many handbooks list values at 101 kPa and 273 K [1 atm and 0°C] (standard temperature and pressure = STP), which are less useful because most aerosol measurements in the environment are at temperatures close to 293 K [20°C].

EXAMPLE 3–1

A miner drilling into a rock face during his work shift hits a seam of quartz (a form of crystalline silica extremely hazardous to the lungs). The X-ray diffraction analysis on the personal sample taken over a period of 240 min at 1.7 L/min indicates 240 μg of respirable crystalline silica. Assume that this represents pure silica particles and that they are 2.8 μm diameter spherical particles with a density of 2660 kg/m^3 [2.66 g/cm^3]. What was the airborne exposure concentration in particles/m^3, particles/cm^3, mppcf (million particles per cubic foot), kg/m^3, g/m^3, mg/m^3, and μg/m^3?

Answer: The volume, v_p, of a single silica particle of physical diameter, d_p, is

$$v_p = \frac{\pi}{6} d_p^3 = \frac{\pi}{6}(2.80 \mu m)^3 = 11.5 \mu m^3$$
$$= 1.15 \times 10^{-17} m^3 \, [1.15 \times 10^{-11} cm^3]$$

The mass, m_p, of a particle with density, ρ_p, is

$$m_p = \rho_p v_p = (2.66 \times 10^3 \text{ kg/m}^3)(1.15 \times 10^{-17} \text{m}^3) = 3.06 \times 10^{-14} \text{kg} = 3.06 \times 10^{-11} \text{g}$$
$$= 3.06 \times 10^{-8} \text{mg} = 3.06 \times 10^{-5} \mu\text{g}$$

The number of particles, n_p, in 240 μg of silica mass is

$$n_p = \frac{\text{silica mass}}{\text{single particle mass}} = \frac{240 \mu\text{g}}{3.06 \times 10^{-5} \mu\text{g}} = 7.84 \times 10^6 \text{ particles}$$

At flow rate, Q, and sampling time, t, the volume of sampled air, v_a, is

$$v_a = Qt = \left(1.7 \frac{\text{L}}{\text{min}}\right)(240 \text{min}) = 408 \text{L}$$
$$= 0.408 \text{m}^3 \, [4.08 \times 10^{-5} \text{cm}^3]$$

The number concentration, c_n, of silica particles is, therefore,

Continued

$$c_n = \frac{n_p}{v_a} = \frac{7.84 \times 10^6 \text{particles}}{0.408 \text{m}^3} = 1.92 \times 10^7 \frac{\text{particles}}{\text{m}^3}$$

$$= 19.2 \frac{\text{particles}}{\text{cm}^3} = 0.545 \text{mppcf}$$

The particle mass concentration, c_m, is

$$c_m = m_p c_n = (3.06 \times 10^{-11} \text{g})\left(1.92 \times 10^7 \frac{\text{particles}}{\text{m}^3}\right)$$

$$= 5.88 \times 10^{-4} \frac{\text{g}}{\text{m}^3} = 0.588 \frac{\text{mg}}{\text{m}^3} = 588 \frac{\mu\text{g}}{\text{m}^3}$$

Silica is one of the most toxic dusts encountered in the workplace. The current exposure standard in the United States is 50 μg/m³. Thus, this measurement indicates excessive exposure. In comparison, the workplace standard for the least toxic materials is 10 mg/m³. The environmental air quality standard in the United States is 80 μg/m³. The units of mppcf are no longer in common use for air measurements. They were popular for the measurement of dust concentrations by light microscopy.

Calculations occasionally will also be performed in both these systems to facilitate conversion because each system has its advantages. Electrostatic calculations in the SI system has an advantage in that it uses the familiar units of volts and amperes. The elementary unit of charge, *e*, is equal to 1.6×10^{-19} coulomb. However, there is some convenience in using the cgs units because the proportionality constant in Coulomb's Law is unity. In this system, all electrical units are defined having the prefix "stat." The elementary unit of charge, *e*, is equal to 4.8×10^{-10} statcoulomb. The electric field is expressed in statvolts/cm. One statvolt equals 300 volts in SI units. Also, particle motions expressed in cm/s reflect convenient magnitudes of particle velocity in an electric field.

It is assumed that the reader has available a computer to perform calculations. The reader is encouraged to experiment with the equations given in this book by calculating the results with a variety of input parameters to gain a feeling for the resulting values and how they behave. Spreadsheet programs are particularly useful for this purpose, and most of the equations given in this book have been set up in the Aerosol Calculator spreadsheets described in Chapter 2.

COMMON TECHNICAL AND DESCRIPTIVE TERMS

Various names are used to describe airborne particulate matter. The name *particle* refers to a single unit of matter, generally having a density approaching the intrinsic density of the bulk material. Individual particles may be chemically homogeneous or contain a variety of chemical species as well as consist of solid or liquid materials or both. Particle shapes may be simple, as in spherical liquid droplets, or complex, as in fiber bundles or agglomerated smoke. Many of the following terms do not have strict scientific definitions but rather are in common use as merely descriptive terms, often indicating the appearance or source of the particles.

Aerosol: An assembly of liquid or solid particles suspended in a gaseous medium long enough to be observed or measured. Generally, the sizes of aerosol particles are in the range of 0.001 to 100 μm. If the particle concentration is large enough that the density of the aerosol is more than about 1% greater than the gas alone, the assembly is considered a cloud and has bulk properties that differ from a more dilute aerosol. Note: The technical use of the term *aerosol* is much broader than the every-day usage referring to droplets emitted from a spray can

Bioaerosol: An aerosol of biological origin, including airborne suspensions of viruses, pollen, bacteria, and fungal spores and their fragments

Cloud: A high-density suspension of particles in air, often with a well-defined boundary

Dust: Solid particles formed by crushing or other mechanical breakage of a parent material. These particles generally have irregular shapes and are larger than about 0.5 μm

Fog or Mist: Liquid particle aerosol. These can be formed by condensation of supersaturated vapors or by physical shearing of liquids, such as in nebulization, spraying, or bubbling

Fume: Particles that are usually the result of condensed vapor with subsequent agglomeration. Solid fume particles typically consist of complex chains of submicrometer-sized particles (usually < 0.05 μm) of similar dimension. Fumes are often the result of combustion and other high temperature processes. Note that the common usage of *fume* also refers to noxious vapor components

Haze: A visiblity-reducing aerosol

Nanoparticle: A particle in the size range of 1 to 100 nm

Particle: A small, discrete object

Particulate: An adjective indicating that the material in question has particle-like properties. Less properly used as a term for particle

Smog: An aerosol consisting of solid and liquid particles, created, at least in part, by the action of sunlight on vapors. The term smog is a combination of the words smoke and fog and often refers to the entire range of such pollutants, including the gaseous constituents

Smoke: A solid or liquid aerosol, the result of incomplete combustion or condensation of supersaturated vapor. Most smoke particles are submicrometer in size

Spray: Droplet aerosol formed by mechanical or electrostatic breakup of a liquid

A number of terms describe the shape and origin of particles in an aerosol. These include

Agglomerate: A group of particles held together by van der Waals forces or surface tension

Aggregate: A heterogeneous particle in which the various components are not easily broken apart. The term *heterogeneous* indicates that the individual components may differ from each other in size, shape, and chemical composition

Flocculate: A group of particles very loosely held together, usually by electrostatic forces. Flocculates can easily be broken apart by shear forces within the air

Primary particle: A particle introduced into the air in solid or liquid form. A primary particle is often contrasted to a secondary particle

Secondary particle: Usually a particle formed in the air by gas to particle conversion. This term is sometimes used to describe agglomerated or redispersed particles

Appendix A gives additional definitions of aerosol terms.

PARTICLE SIZE AND SHAPE

Particle size is important because it largely determines the behavior of the particle in gas suspension. Particles behave differently in different size ranges and are even governed by different physical laws. For example, on the earth's surface, particles only slightly larger than gas molecules are governed primarily by Brownian motion, while large, visible particles are affected primarily by gravitational and inertial forces.

Particle size and shape can be quite complex and are often defined only to the extent that one can measure or calculate them. Therefore, there are numerous definitions of particle size and shape that depend on the measurement technique or on the use to which the parameter will be put. For instance, an electron microscope is a common means for measuring the size and shape of a particle. To accomplish this type of measurement, a particle is collected on a substrate, a process that may place the particle on the surface in some preferred orientation. The analyst measures the particle by comparison with standard-sized objects within the observation area. Except for ideally shaped spherical particles, the analyst usually reduces a complex shape to one or two measured parameters, for example, width, or diameter and length. With an image analysis system, one may be able to extract more features of a particle's shape. The usual aim in collecting this type of information is to reduce the data collected from each particle to the fewest numbers that can adequately characterize the particle.

Size Parameters

A commonly used term in aerosol science and technology is that of ***equivalent diameter*.** This refers to a diameter that is a measurable index of a particle. When a particle is reported by a technique, the measurement usually corresponds to a specific physical property. Thus, an equivalent diameter is reported as the diameter of a sphere having the same value of a specific physical property as the irregularly shaped particle being measured (Fig. 3–1). When the motion of a particle is of concern, the *mobility equivalent diameter*, d_B, is the diameter of a sphere with the same mobility as the particle in question.

For instance, *aerodynamic (equivalent) diameter* (equivalent is sometimes left out or implied) is the diameter of a standard-density (1000 kg/m^3 or 1 g/cm^3) sphere having the same

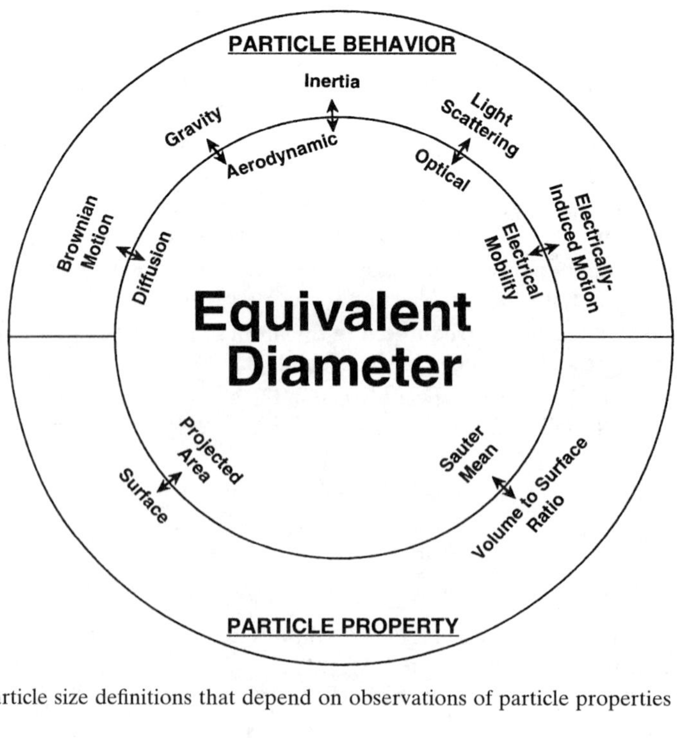

Fig. 3–1. Particle size definitions that depend on observations of particle properties or behavior.

gravitational settling velocity as the particle being measured. This definition is often used for characterizing particles that move primarily by settling as opposed to diffusion in still air (i.e., diameters larger than about 0.3 µm at normal atmospheric temperature and pressure). Reference to the aerodynamic diameter of a particle is useful for describing particle settling and inertial behavior in the respiratory tract, one of the body organs most at risk upon exposure to toxic aerosols. The behavior of particles in other devices such as filters, cyclones, and impactors is also often governed by the aerodynamic flows around the particles, and the sizes are therefore reported in terms of the particles' aerodynamic diameter. As will be seen in the next chapter, a solid, spherical particle's gravitational settling velocity is proportional to the *particle density*, ρ_p, the square of the *physical particle diameter*, d_p, and the *Cunningham slip correction factor*, C_c. The latter is introduced because the suspending gas is not a continuous fluid, but consists of discrete molecules. The slip correction factor, described in more detail in the next chapter, is a function of particle diameter, that is, $C_c = C_c(d_p) = C_{c,d_p}$. Thus

$$\rho_p C_{c,d_p} d_p^2 = \rho_0 C_{c,d_a} d_a^2 \tag{3–1}$$

One of the conditions chosen here is that of a sphere with standard particle density $\rho_0 = 1000\,\mathrm{kg/m^3}$ [$1\,\mathrm{g/cm^3}$], which defines the particle diameter as the aerodynamic diameter, d_a

$$d_a = d_p \left(\frac{C_{c,d_p}}{C_{c,d_a}} \right)^{1/2} \left(\frac{\rho_p}{\rho_0} \right)^{1/2} \tag{3–2}$$

When the particle density is close to standard density, C_{c,d_p} differs little from C_{c,d_a}, and the ratio of the two slip correction factors can be approximated by one. For particles above about 1 µm, the slip correction factor is close to unity, so this ratio can be approximated by one even for particle densities very different from unity. Therefore, for many applications, Eq. 3–2 reduces to

$$d_a = d_p \left(\frac{\rho_p}{\rho_0} \right)^{1/2} \tag{3–3}$$

EXAMPLE 3–2

What is the aerodynamic (equivalent) diameter of a spherical particle that is 3 µm in diameter and has a particle density of $4000\,\mathrm{kg/m^3}$ [$4\,\mathrm{g/cm^3}$]? Ignore the slip correction factors.
Answer: From Eq. 3–3:

$$d_a = d_p \left(\frac{\rho_p}{\rho_0} \right)^{1/2} = 3\,\mu\mathrm{m} \left(\frac{4000}{1000} \right)^{1/2} = 6\,\mu\mathrm{m}$$

This indicates that a 6 µm standard-density particle gravitationally settles at the same velocity as the 3 µm particle with the higher density.

A particle may be extremely complex in shape, such as an agglomerate. In this case, a significant part of the internal volume of the particle is made up of voids. When describing the properties or behavior of such a particle, two additional definitions are available: the *mass equivalent diameter*, for which the particle is compressed into a spherical particle without voids; and the *envelope equivalent diameter*, for which the particle voids are included in the sphere. The mass equivalent diameter is convenient because it uses the bulk density of the material, a parameter often available in the literature (e.g., Lide, 1994) or easily measured. Further discussions of nonspherical particles are presented in Chapters 4 and 23.

Microscopes and other particle imaging systems are often used to measure particles. For instance, observing a particle's silhouette and calculating the diameter of a circle that has the same area gives the *projected area (equivalent) diameter*. Collecting an aerosol particle for measurement in a microscope can cause a number of biases to occur in the assessment of the original aerosol. For instance, the collected particle may be oriented by the surface. A fiber usually settles onto a surface with its long axis parallel to the surface. An agglomerate may collapse onto the surface from gravity or from surface tension of adsorbed water and appear more spread out than in its original form. The continued air flow over the particle may desiccate it, thus reducing it in size and mass. The collected particle may also react with the collection substrate, which may change the particle's size and chemical composition. The analyst needs to consider these possibilities when using data from methods involving sample collection (see Chapter 12)

We note here a dichotomy in measurement technique, namely, that of collection of an aerosol particle for laboratory measurement versus direct, in situ measurement of the particle. Traditionally, collection followed by measurement was often the most readily available. This approach still has its advantages because it brings to bear the many powerful analytical techniques available in the laboratory. However, this approach has the disadvantages that the particles may be modified by the transport and collection processes and that the analytical result is not immediately available. In situ techniques, on the other hand, provide a more limited degree of particle characterization.

In situ techniques can be subdivided further into extractive and external sensing techniques. Extractive techniques require the aerosol to be brought to the instrument sensor, while external sensing techniques measure the aerosol in its undisturbed state. For example, Chapter 15 describes mainly instruments that detect light scattered from particles brought into an instrument,while Chapter 16 describes light-based systems that detect particles some distance from the instrument.

A common in situ technique is the measurement of light scattered from the particles. The amount of light scattered from individual particles is a complex function of particle parameters of size, shape, and refractive index as well as instrumental parameters such as the wavelength of light and the scattering angle. The usual approach is to define an optical equivalent diameter that is the diameter of a calibration particle that scatters as much light in a specific instrument as the particle being measured. For simple particle shapes, such as spheres, ellipsoids, and rods of known chemical composition, the amount of light scattered may be calculated exactly. For most particles with more complex shapes, the association between optical equivalent diameter and a physically useful property is often difficult to establish precisely. Despite this, light scattering as an instrumental technique has a number of distinct advantages. These advantages include rapid, continuous, and sensitive detection of particles, often at relatively low cost.

Spray aerosol droplets used as fuels in combustion processes burn or react at their surfaces. Therefore, a useful measurement parameter is the *Sauter mean diameter*, the diameter of a droplet whose surface-to-volume ratio is equal to that of all the droplets in the spray distribution.

Because submicrometer particles move primarily by Brownian diffusion, it is natural to define their size by a *diffusion equivalent diameter*, that is, the diameter of a standard-density spherical particle with the same rate of diffusion as the particle being measured. For compact particles, the diffusion equivalent diameter is very close to the physical diameter, as might be measured with an electron microscope.

The measurement of small particles by diffusion-based techniques is often relatively slow and has poor resolution (see Chapter 19). In an electric field, a particle of known charge moves along a predictable trajectory. Therefore, the electrical mobility of a charged particle in an electric field is the basis for defining the *electrical mobility equivalent diameter*. Particle motion in an electric field can yield high-resolution measurements as well as separation of desired particle sizes (see Chapter 18).

In addition to the various equivalent diameters mentioned above, any other physical property, such as mobility in a magnetic field, external surface area, radioactivity, and chemical or elemental concentration, can be used to determine an equivalent diameter.

Size Ranges

Although it is customary to discuss particulate clouds in terms of particle size, rarely is such cloud composed of single-diameter particles—only in the laboratory and then only with great care can single-sized aerosols be produced. Such single-sized particulate aerosols are referred to as *monodisperse*. These aerosols are useful for studying their size-dependent properties or for calibrating instruments. Whether dust, mist, or fume, virtually all naturally occurring aerosols are a mixture of a wide variety of particle sizes, that is, they are *polydisperse*.

A large airborne molecule can be considered a very small aerosol particle. Although air consists of nitrogen, oxygen, and other gases, air molecules can be considered for most calculations as having an average diameter of 0.37 nm (0.00037 μm). In comparison, aerosol particles are generally 1 nm (0.001 μm) in diameter or larger. Fume particles of this size can be seen only immediately upon condensation from the vapor state. A short time later, the high concentration of these very small particles causes coagulation into larger entities, ultimately reaching sizes near 1 μm.

Conversely, dust particles result from size reduction of larger materials. Generally, one considers particles less than 100 μm (0.1 mm) in diameter to stay airborne long enough to be observed and measured as aerosols. For example, human hairs range from about 50 to about 100 μm in diameter. If they were cut into small pieces and released into the air, they would be near the upper limit of the aerosol size range. Size reduction of bulk material by mechanical forces, be they natural or induced by human action, can occur only for sizes where the externally applied forces are greater than the internal cohesion forces. Particles smaller than about 0.5 μm are relatively rare in dust distributions for this reason.

PARTICLE SUSPENSIONS

Because an aerosol is a system of airborne particles suspended in a gas medium, one generally considers the gas properties and flow dynamics first and then evaluates how individual particles follow or deviate from the gas motion. The difference in trajectories between particles and gas molecules is the basis for many aerosol particle size measurement techniques. It is also the basis for many devices controlling aerosol contaminants and for techniques manipulating aerosol particles for manufacturing purposes. Changes in gas properties generally affect the particle trajectories.

As an example, one may appreciate the need for dealing with air flow characteristics first by asking how much aerosol deposition will occur 50 km from an aerosol-emitting power plant. The wind velocity determines the speed with which the aerosol is transported away from the power plant. Large particles gravitationally settling in a shorter time than is available for transport to the 50 km distant site will not be found at the receptor site. The mechanism of settling and dispersion is determined by the degree and mode of turbulence. Returning to aerosol measurement principles, a commonly used instrument, the horizontal elutriator, size-selectively removes particles in a horizontal flow channel. Here, the gas flow is generally "well behaved" by the careful avoidance of air turbulence.

INSTRUMENT CONSIDERATIONS

In general, one cannot obtain particle size information on the entire 5 decade size range of 0.001 to 100 μm with a single instrument. On a macroscopic scale, this would be equivalent to measuring a 1 mm distance with a small scale and then using the same scale for measuring a 1 km distance (which is six orders of magnitude larger than 1 mm). When sensing with optical techniques utilizing white light, the wavelength of visible light from about 0.4 to 0.7 μm limits the observation of particles to about this size range and larger. Inertial techniques become inefficient below about 0.5 μm at normal temperature and pressure. In an electron microscope, the observational tool is electromagnetic radiation (electrons) with a much smaller wavelength that can "see" much smaller particles. Therefore, one expects to apply different instrumental techniques, measuring different size parameters, for submicrometer- versus supermicrometer-sized aerosols.

Most aerosol sizing instruments effectively measure over a size range no larger than $1^1/_2$ orders of magnitude. Thus, the largest measurable size may be about 50 times the smallest measurable size for a given instrument. Because most of the size parameters measured relate to the particle surface, volume, or mass, this size range corresponds to a surface range of 2500 and a volume or mass range of 125,000. Instruments measuring a cumulative value (e.g., total mass or number) can cover a wider size range. Preferably, each aerosol sizing instrument should give a monotonically increasing response to increases in particle size. Unfortunately, some optical devices may detect the same amount of scattered light for more than one particle size, resulting in significant loss of size resolution.

When a single-source aerosol is measured by any of the above-mentioned size parameters, the representative particle size is usually quoted as the *mean size* (average of all sizes), *median size* (equal number of particles above and below this size), or the *mode* (size with the maximum number of particles). The spread of the particle size distribution is characterized by an *arithmetic* or *geometric* (logarithmic) *standard deviation*. Typically, the particle size distribution is *lognormal*, that is, the particle concentration versus particle size curve looks *normal* (also referred to as *bell-shaped* or *Gaussian*) when the particle size is plotted on a logarithmic scale (see Chapters 6 and 22).

The reason for the use of this logarithmic or geometric size scale can be conceptualized by breaking a piece of blackboard chalk. For example, a 64 mm long piece of chalk would break up into two pieces of 32 mm length each. Subsequent breakup yields pieces of 16, 8, 4, 2, and 1 mm, and so forth, length until the internal forces resist further breakage. The ratio of adjacent sizes is always two, thus appearing at the same linear distance on a logarithmic or geometric size scale. Because with each breakage step more and more particles are produced, the distribution is skewed so that there are many more small particles than large ones. This exercise of breaking up a piece of chalk mimics the way many natural and manmade forces generate aerosols. Generally, aerosol particle sizes therefore are plotted on a logarithmic size scale.

Many aerosols measured in ambient or industrial air environments or in industrial process streams are a mixture of aerosols, resulting in more than one particle mode and covering a wide size range. This may make the measurement and analysis of the aerosols considerably more complex. In general, one should attempt to first identify all aerosol sources and decide what information is needed and for what purpose. This decision will then point the way to the best available instrument to reach the desired objective.

Aerosol instruments not only differ by the size parameter that they measure, but each size parameter may "weigh" the particle size differently. A grocery store analogy can help elucidate this concept. If 10 large apples and 100 small raisins are purchased, the median size by *number count* is somewhat larger than the size of the raisins. The median is close to that of the raisins because the median size divides the "population" in two, but in this case most of the "population" consists of small raisins. If each piece of fruit is weighed on a scale, the weight of the apples dominates, and the median size "by mass" is considerably larger. Thus, aerosol measurement "by mass" results in a larger median size than aerosol measurement "by count," although the same particle size distribution is measured. Therefore, any size result should be accompanied by a description of the *weighing factor*, or the *weighting* as it is commonly called.

If many particles in an aerosol are measured and the particles are grouped into discrete, contiguous, size bins, the size distribution can be represented by plotting particle number versus size. The lower and upper particle diameter limits, d_l and d_u, of each size bin need to be chosen with some care in order to get a useful description of the overall size distribution. The number of particles in each bin will depend on the size of the bin, that is $d_u - d_l$. To remove this bin width dependence, the number of particles in each bin is usually normalized by dividing the number of particles in the bin by the bin width. Size distribution properties are discussed further in Chapters 6, 7, and 22.

PARTICLE SHAPE MEASUREMENT

Traditionally, particle shape has been acknowledged by including a *shape factor* in the particle motion equations. For a nonspherical particle, inclusion of this factor in the equation allows one to calculate the desired parameter while characterizing complex particle shapes by a single dimension. Although this provides an indication of the particle's behavior under certain conditions, it does not provide sufficient detail to fully characterize the particle. For instance, the particle's reactivity is a sensitive function of the particle's surface, and often the shape and texture provide clues as to the particle's formation and history. Because powerful computers are now available, image analysis methods will be described that characterize the shape more directly. It is generally difficult to measure the shape of all particles in an aerosol; therefore, careful measurements on a few particles are often assumed representative of the entire aerosol.

A variety of schemes have been developed to measure the outline shape and detailed texture of particles. Simple shapes, such as spheres (droplets) and rods (simple fibers), can be completely described by one or two dimensions, respectively. More complex shapes are difficult to characterize. For instance, measuring the distance of the particle perimeter from the particle centroid and analyzing this distance as a function of angle using Fourier analysis has been proposed as a classification scheme for particle shapes (Beddow et al., 1977). A number of similar shape description techniques have also been described (Kaye, 1981). However such techniques are generally limited to the outline profile of a particle and cannot characterize all the surface complexities and convolutions of many particles.

If a particle were a long, straight chain, it might be characterized as a one-dimensional object having primarily length. Complex branching, however, causes the particle to take up more space than a linear object while not meeting the criteria for a sheet-like (two-dimensional) or a spheroidal (three-dimensional) object. It can, however, be characterized by assigning it a fractional, or *fractal*, dimension. The term *fractal* was coined by Mandelbrot (1983) and has found use in a wide range of applications, including particle shape, turbulence, lung structure, and fibrous filter structure (Kaye, 1989). The principal characteristic of a fractal object is that a measure of complexity is similar on several measurement scales. For instance, if one looks at a fume particle at several different magnifications, the variation, or complexity, may appear very similar. This property of similar complexity at several scales, or scale invariance, is described by the mathematical concept of self-similarity. In Figure 3–2, an agglomerate is presented at several

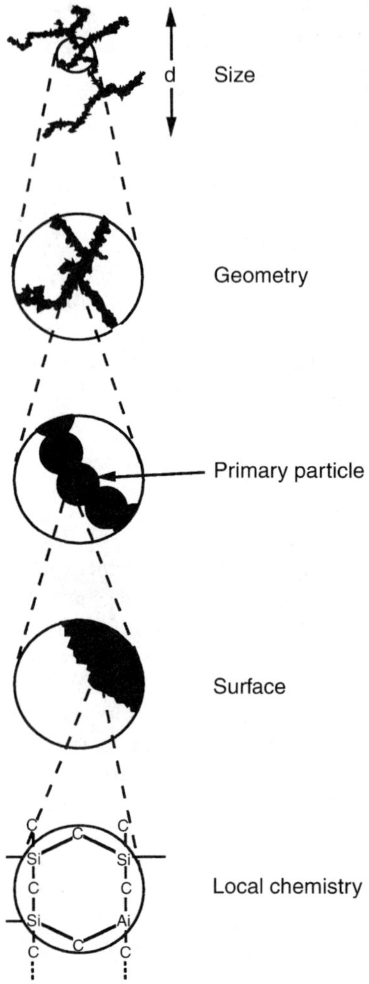

Fig. 3–2. Schematic structure of an agglomerate in two-dimensional space. Scanning down the figure corresponds to viewing the particle at ever higher magnification. (Adapted from Schaefer and Hurd, 1990.)

scales. Generally, self-similarity only occurs between the first and third scales from the top.

The observed shape of a particle is the result of its history. The fume particle, for instance, begins as a vapor condensing into spherules. The spherules, being very small, diffuse rapidly and coagulate into branched chains, as, for example, the zinc oxide fume particle in Figure 3–3. As the chains increase in size and the number of individual spherules in the neighborhood decreases, the chains may intercept one another and form larger agglomerates. Such an agglomerate might be observed at several magnifications, ranging from the structure of the entire particle down to the chemical structure of the surface. Thus, at lowest magnifications, the complex structure can be represented by a fractal dimension; at an intermediate magnification, the spherules have nearly integral dimension; at higher magnification still, the spherule surface may be rough and characterized by another fractal dimension. Fractal and other nonspherical particles are discussed further in Chapter 23.

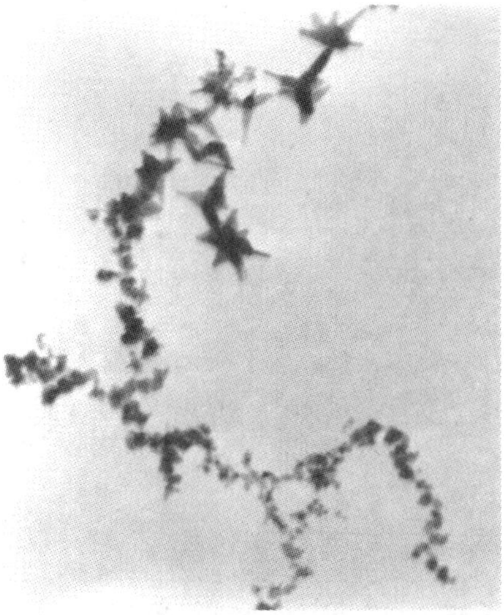

Fig. 3–3. A typical fume particle is an agglomerate of smaller particles formed from condensed vapor. This is a zinc oxide fume particle indicating condensation components with different levels of crystallization.

PARTICLE FORCES

The intra- and interparticle forces that hold particles together or to a surface, and the forces that detach particles from each other or from a surface, are difficult to quantitate for use by the practitioner. These forces may depend on particle bulk and surface parameters (size, shape, roughness, chemistry), the properties of the surrounding gas (temperature, humidity) and the mechanics of the contacting particles (relative particle velocity, contact time). These forces will, therefore, be described qualitatively.

When particles are subjected to an external force, such as gravity or an electrical force, the particles will move in the force field. The migration velocity in the force field is particle-size dependent, a fact that is exploited by most aerosol-sized spectrometers for particle size discrimination.

Adhesion Forces

In contrast to gas molecules, aerosol particles that contact one another generally adhere to each other and form agglomerates. If they contact a surface, such as a filter or any other particle collection device, they are assumed to adhere to the surface (i.e., particle adhesion is the working hypothesis of these devices).

The London–van der Waals forces, which are attractive in nature, act over very short distances relative to particle dimensions (Friedlander, 1977: 44). According to the theory of their origin, random motion of the electrons in an electrically neutral material creates instantaneous dipoles that may induce complementary dipoles in neighboring material and thus attract the surfaces to each other.

Most particles 0.1 μm or larger carry some small net charge that exerts an attractive force in the presence of a particle with an opposite charge (Hinds, 1999: 143). For two charged particles (point charges), this force is inversely proportional to the square of the separation distance. After two surfaces have made contact with each other by either or both of the above

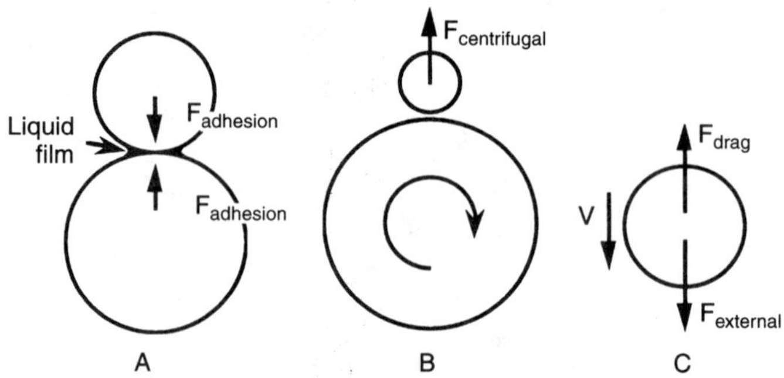

Fig. 3–4. Examples of particle forces. **A**, Adhesion due to liquid film. **B**, Detachment due to centrifugal force. **C**, Particle motion at velocity V due to balance between drag force and an external force.

forces, the surfaces may deform with time, thereby increasing the contact area and decreasing the separation distance and thus increasing the force of adhesion.

Figure 3–4A exemplifies how air humidity may affect particle adhesion. At high humidity, liquid molecules are adsorbed on the particle surface and fill the capillary spaces at and near the point of contact. The surface tension of this liquid layer increases the adhesion between the two surfaces.

Detachment Forces and Particle Bounce

Figure 3–4B exemplifies the detachment of a particle from a rotating body. The centrifugal force is proportional to the particle's mass or volume, that is, particle diameter cubed (d^3). Detachment by other types of motion, such as vibration, is similarly proportional to d^3, while detachment by air currents is proportional to the exposed surface area, that is, d^2. In contrast, most adhesion forces are linearly dependent on particle diameter. Thus, large particles are more readily detached than small ones. While individual particles less than 10 μm are not likely to be easily removed (e.g., by vibration), a thick layer of such particles may be easily dislodged in large (0.1 to 10 mm) chunks (Hinds 1999: 144). Re-entrainment of particles from a surface into an aerosol flow may, therefore, create measurement problems after a significant number of particles has been deposited from the aerosol.

If an aerosol flow is directed toward a surface (e.g., in filters and impactors), particles with sufficient inertia will deviate from the air stream lines and move toward the surface. Liquid and sticky small particles will deposit on the surface. Upon contact, a solid particle and the surface may deform. If the rebound energy is greater than the adhesion energy, a condition that may occur for sufficiently high-impact velocity, a solid particle will "bounce," that is, move away after contact with the surface. On contact with the surface, some or all of the particle's kinetic energy is converted to thermal energy, resulting in reduced kinetic energy on rebound or heating of the particle–surface interface on sticking, respectively. Grease or oil on the surface will generally increase the likelihood of adhesion, but, after a layer of particles has been deposited, the incoming particles may bounce from the top surface of the previously deposited particles. Particle adhesion on impact is an especially critical factor in inertial collection devices, as indicated in Chapter 10.

Externally Applied Forces

When an airborne particle is subjected to an externally applied force (e.g., gravity), it will be moved by that force. Opposing this external force is the aerodynamic drag force, as shown in Figure 3–4C. When the two forces are in equilibrium, which happens almost instantaneously (there is a very short relaxation time, having consequences that are discussed later), the particle moves in the force field with migration velocity V. Knowledge of the two opposing forces allows determination of this velocity. Particle velocity is important for estimating collection on surfaces as well as for separating particles by size. Quite often, aerosol measurements are designed to simulate some natural process, such as particle deposition in the respiratory system. Thus, it is important to understand the aerosol behavior in the original system as well as in the instrument in order to make accurate measurements. Instrumental techniques based on the forces involved in the original process (e.g., gravitational settling and diffusion as in particle deposition in the respiratory system) make the measurement more useful and relevant.

In space, astronauts must pay special attention to the dust generated by their clothing and the activities they engage in. Otherwise, their living space quickly becomes polluted with aerosols. On earth, gravity has a major cleaning effect on ambient and industrial aerosols. Larger particles tend to settle out more rapidly. Because the gravitational force is readily accessible for measurement applications, it is the basis for the definition of aerodynamic diameter.

We are familiar with the attraction of lint particles to clothing. This is due to charge differences between the lint and the clothing. Similarly, charged aerosol particles can be attracted to or repelled from charged surfaces or other particles. Few particles carry no charge, although the magnitude of charge can vary greatly. Particles that are freshly aerosolized tend to carry greater charge levels than particles that have been airborne for hours or longer. This aging effect is due to the attraction of oppositely charged airborne ions produced by natural radiation. For aerosol particles that are highly charged, the electric force may exceed the gravitational force by several orders of magnitude. This readily generated force can be used for air cleaning as well as particle separation and measurement, for example, with electrical mobility analyzers (see Chapter 18).

If there is a gradient in the number of particles present in the air, a diffusion force can be defined that moves the particles from the high concentration to the low concentration environment. It is often the dominant motive force for particles smaller than about 0.2 μm diameter. The diffusion battery, for example, is commonly used for measuring submicrometer particles. Diffusion is also important for understanding the particle and gas deposition properties of the human lung. If the suspending gas is a nonuniform mixture of gases, the particles also may be moved by diffusiophoretic forces caused by the concentration gradient of the gas components (see Chapter 19).

Inertial forces can be applied to particles by forcing the suspending air to change direction. Size-dependent inertial effects are used for particle separation, collection, and measurement in such devices as impactors, cyclones, and acceleration nozzles (see Chapters 10, 13, and 17). Impaction is an important mechanism for particle deposition in the respiratory system (see Chapter 25).

If there is a temperature gradient in the aerosol-containing space between two surfaces, the higher activity of the air molecules near the hot surface pushes the particles toward the colder surface (thermophoretic force). This property is exploited in the thermal precipitator, which is used to collect particles onto a desired surface (see Chapter 10). A special case of thermophoresis, but generally not very useful as a measurement tool, is produced by light. Illumination of a particle heats up one side of the particle as well as gas molecules nearby that push the particle toward the colder side. Illumination can also produce radiation pressure whereby the stream of photons exerts a force on the particle (photophoresis). A focused laser beam can been used as optical "tweezers" to move small particles (e.g., bacteria) in a liquid.

REFERENCES

Beddow, J. K., G. C. Philip, and A. F. Vetter. 1977. On relating some particle profile characteristics to the profile Fourier coefficients. *Powder Technol.* 18:19–25.

Friedlander, S. K. 1977. *Smoke, Dust and Haze*. New York: John Wiley & Sons.

Hinds, W. C. 1999. *Aerosol Technology*. New York: John Wiley & Sons.

Kaye, B. H. 1981. *Direct Characterization of Fine Particles.* New York: John Wiley & Sons.

Kaye, B. H. 1989. *A Random Walk Through Fractal Dimensions.* Weinheim, Federal Republic of Germany: VCH Verlagsgesellschaft mbH.

Lide, D. R., ed. 1994. *CRC Handbook of Chemistry and Physics.* Boca Raton: CRC.

Mandelbrot, B. B. 1983. *The Fractal Geometry of Nature*. New York: W. H. Freeman and Company.

Schaefer, D. W. and A. J. Hurd. 1990. Growth and structure of combustion aerosols. *Aerosol Sci. Technol.* 12:876–890.

4 Gas and Particle Motion

PAUL A. BARON

Centers for Disease Control and Prevention, National Institute for Occupational Safety and Health, Cincinnati, OH

KLAUS WILLEKE

Department of Environmental Health, University of Cincinnati, Cincinnati, OH

INTRODUCTION

Aerosols consist of two components: a gas or gas mixture, most commonly air; and the particles suspended in it. The behavior of the particles within the aerosol depends to a large extent on the motion and intrinsic properties of the suspending gas. Submicrometer-sized particles, especially those less than 0.1 μm diameter, are affected by the motion of individual gas molecules (the *free molecular regime*). Thus, the kinetic theory of gases is useful in understanding the behavior of these particles. Larger particles can be treated as being submersed in a continuous gaseous medium or, more broadly, a fluid (the *continuum regime*). The tools of gas or fluid dynamics are more useful for this size range. Intermediate-sized particles usually can be treated by adjustment of equations from the continuum regime. This intermediate range is termed the *transition* or *slip regime*. Whether considering a molecular ensemble or a continuous fluid, the motion of the gas will largely dictate the behavior of the suspended particles. In this chapter, concepts and parameters that affect gas and particle motion are discussed and quantified as migration and deposition parameters in specific force fields.

BULK GAS MOTION

Reynolds Number

When measuring an aerosol, elucidating the gas flow patterns is critical to understanding what happens to the aerosol in the environment, on its way into the sensor of the measuring instrument or in the actual sensor. While the aerosol particles follow the overall gas flow, their trajectories can deviate from the gas flow due to various external forces as well as changes in gas direction and velocity. Gas motion can be visualized by observing the streamlines, that is, tracing the motion of miniscule volumes of gas. Gas flowing around an aerosol particle can have the same flow streamline pattern as gas passing around a large object such as a basketball.

The equivalence of fluid motion for various-sized objects can be described in terms of the forces involved. The flow pattern, whether it is smooth or turbulent, is governed by the ratio

Aerosol Measurement: Principles, Techniques, and Applications, Second Edition, Edited by Paul A. Baron and Klaus Willeke.
ISBN 0-471-35636-0

of the inertial force of the gas to the friction force of the gas moving over the surface. This ratio is expressed by the Reynolds number, *Re*, an extremely useful parameter when dealing with aerosols.

$$Re = \frac{\rho_g V d}{\eta} = \frac{Vd}{\nu} \tag{4–1}$$

where V is the velocity of the gas, η is the dynamic gas viscosity, ν is the kinematic viscosity $(= \eta/\rho_g)$, and d is a characteristic dimension of the object, such as the diameter of a sphere. Because this dimensionless number characterizes the flow, it depends on gas density, ρ_g, not on the particle density. At normal temperature and pressure (NTP), that is, 293 K [20°C] and 101 kPa [1 atm], $\rho_g = 1.192\,\text{kg/m}^3$ [$1.192 \times 10^{-3}\,\text{g/cm}^3$] and $\eta = 1.833 \times 10^{-5}\,\text{Pa·s}$ [$1.833 \times 10^{-4}\,\text{dyne s/cm}^2$], which reduces Eq. 4–1 to

$$\text{Re} = 65{,}000\,Vd \quad \text{for } V \text{ in m/s and } d \text{ in m}$$
$$[\text{Re} = 6.\,Vd \qquad \text{for } V \text{ in cm/s and } d \text{ in cm}] \tag{4–2}$$

Distinction must be made between the *flow Reynolds number*, Re_f, and *the particle Reynolds number*, Re_p. Flow Reynolds number defines the gas flow in a tube or channel of cross-sectional dimension d. Particle Reynolds number defines the gas flow around a particle that may be found in this tube or channel flow. The characteristic dimension in the latter is particle diameter, d_p, and V expresses the relative velocity between the particle and the gas flow. Because the difference between these velocities is generally small and the particle's dimension is very small, the particle Reynolds number usually has a very small numerical value.

Common Gas Flows

When friction forces dominate the flow (i.e., at low Reynolds numbers), the flow is smooth, or *laminar*. Under laminar flow, no streamlines loop back on themselves. At higher Reynolds numbers, the inertial forces dominate, and loops appear in the streamlines until at still higher Reynolds numbers the flow becomes chaotic, or *turbulent*. The actual values of the Reynolds number depend on how the gas flow is bounded. For instance, laminar flow occurs in a circular duct when the flow Reynolds number is less than about 2000, while turbulent flow occurs for Reynolds numbers above 4000. In the intermediate range, the gas flow is sensitive to the previous history of the gas motion. For instance, if the gas velocity is increased into this intermediate range slowly, the flow may remain laminar. When a gas passes around a suspended object, such as a sphere, flow is laminar for particle Reynolds numbers below about 0.1.

Because often it is expensive and difficult to test collection and measurement systems at full scale and in situ, small-scale water (or other liquid) models operating at the same Reynolds number as the system being studied are a useful alternative. Dye injection into the flow stream allows visualization of the streamlines. Such models can operate on a smaller physical scale with a slower time response so that it is easy to observe the time evolution of flow patterns. The same technique can be used to model the behavior of particles.

Many gas-handling systems for instruments use cylindrical tubing to carry the aerosol from one place to the other. Understanding the flow patterns within the tubing is important for predicting the losses that occur within the tubing as well as predicting the distribution of particles within the tubing. If a gas begins to flow in a cylindrical tube, the friction at the wall slows the gas velocity relative to the motion in the center of the tube. At low Reynolds numbers, the dominating friction force produces a characteristic laminar parabolic velocity profile. The gas velocity in the center of the tube for this *Poiseuille flow* is twice that of the average velocity in the tube. Poiseuille flow does not become established immediately. A common rule of thumb is to assume that it takes 10 tube diameters for this equilibrium flow to be effectively established.

EXAMPLE 4–1

Silica dust of 10 μm diameter is removed by a 0.30 m diameter ventilation duct at 20 m/s (about 4000 fpm). An old rule of thumb in industrial hygiene is that silica dust of that size gravitationally settles at 1 cm/s (0.01 m/s). Calculate the flow and particle Reynolds numbers at 293 K [20°C].

Answer: The relevant parameters for the flow Reynolds number are the duct diameter and the gas flow velocity in the duct. From Eq. 4–2:

$$Re_f = 65{,}000\,Vd = 65{,}000\left(20\frac{\text{m}}{\text{s}}\right)0.30\text{m} = 3.90\times10^5$$

The relevant parameters for the particle Reynolds number are the particle diameter and the gravitational settling velocity perpendicular to the gas flow:

$$Re_p = 65{,}000\,Vd = 65{,}000\left(0.01\frac{\text{m}}{\text{s}}\right)10\times10^{-6}\text{m} = 6.5\times10^{-3}$$

The flow Reynolds number exceeds 4000, indicating turbulent flow in the ventilation duct. The particle Reynolds number is less than one, indicating that the flow around the particle can be laminar. However, it is not in this case because the gas flow is turbulent.

The region near a surface where the flow is dominated by the friction force is termed the *boundary layer*. When flow starts along a surface, in either time or space, the boundary layer consists only of the gas at the surface, where the relative velocity is zero. At low Reynolds numbers, the boundary layer grows until steady-state conditions are reached. For the cylinder flow example above, the boundary layer grows into a parabolic flow profile that fills the cylinder. At higher Reynolds numbers (in the turbulent regime) or during abrupt changes in flow conditions, the boundary layer can become separated from the surface. The development of the boundary layer and its relationship to the overall flow depends on the object immersed in the fluid. Such behavior has been described in many fluid mechanics texts (e.g., White, 1986) and especially in "Boundary-Layer Theory" (Schlichting, 1979).

There are a wide variety of flow situations for which empirical or experimentally verified theoretical solutions exist. For instance, when a gas passing through a cylindrical tube under laminar flow conditions negotiates a 90° bend, the cylindrical symmetry of the flow pattern in the tube is reduced to a plane of symmetry. Thus, the flow symmetry must also be reduced. Two circulation patterns, one sometimes described as secondary flow to differentiate it from the primary flow along the tube axis, are set up on either side of the plane of the bend, as shown in Figure 4–1. This secondary flow causes mixing of the gas as well as increased inertial forces on particles suspended in the gas (Tsai and Pui, 1990). In tubing used to transport aerosols, bends are generally undesirable because of increased particle loss.

For various reasons, there are often constrictions or expansions in a tube carrying a gas. A constriction will force the gas to increase in velocity and be focused in the center of the tubing, even more than the constriction in size. After this contraction region, or *vena contracta*, the gas flow eventually expands again to fill the tubing and re-establishes an equilibrium pattern. These disturbances will also cause increased particle deposition.

When a gas flows from an initial tube diameter into a suddenly expanded section or into free space, the flow pattern may persist for many initial tube diameters downstream. If the expansion of the tube is very slight, the flow does not separate from the walls, and the flow pattern can expand smoothly to fill the increased diameter of the tube. In general, the angle

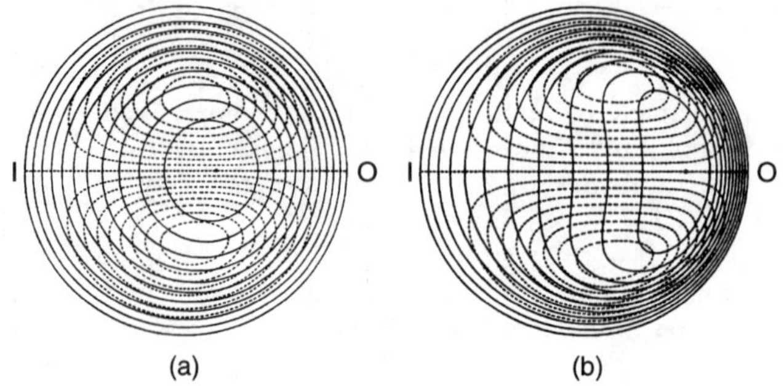

Fig. 4–1. Secondary flow streamlines (dotted lines) and primary flow velocity contours (solid lines) at a short distance downstream from the exit plane of a 90° bend in a tube. I and O refer to the inner and outer sides of the bend, respectively. The flows are calculated at two Dean numbers:

$$De = Re/\sqrt{\text{bend radius/tube radius}}$$

a, $De = 17$. **b**, $De = 107$. (From McConalogue and Srivastaval, 1968, with permission of the Royal Society.)

between the wall and the tube axis needs to be less than 7° to avoid flow separation from the tube wall.

Gas Density and Mach Number

The density of a gas, ρ_g, is related to its temperature, T, and pressure, P, through the *equation of state*:

$$P = \rho_g \frac{R_u}{M} T = \rho_g RT \tag{4–3}$$

where ρ_g is the gas density (1.192 kg/m^3 [1.192 × 10^{-3} g/cm^3] for air at NTP), T is the absolute gas temperature in K, M is the molecular weight in kg/mol, and R_u is the universal gas constant (= 8.31 Pa·m^3/mol·K [8.31 × 10^7 dyne·cm/mol·K]). In air, the effective molecular weight is 0.0289 kg/mol [28.9 g/mol]. Thus, the specific gas constant for air is R = 288 Pa·m^3/kg·K [2.88 × 10^6 dyne·cm/g·K]. One atmosphere equals 101 kPa, where 1 Pa = 1 N/m^2 = 10 dyne/cm^2.

When this gas moves at a high velocity relative to the acoustic velocity, U_g, in that gas, the gas becomes compressed. The degree of compression depends on the *Mach number*, *Ma*:

$$Ma = \frac{U}{U_{\text{sonic}}} \tag{4–4}$$

Here, the gas velocity is designated as U to distinguish it from particle velocity V. When $Ma \ll 1$, the gas flow is considered incompressible. This is true in most aerosol sampling situations. In air, the sonic or sound velocity at ambient temperature is about 340 m/s (1100 ft/s).

TRANSITION AND GAS MOLECULAR FLOW

Knudsen Number

Large aerosol particles are constantly bombarded from all directions by a great number of gas molecules. When a particle is small, less than 1 μm in size, its location in space may be affected by bombardment of individual gas molecules. Its motion is then no longer determined by continuum flow considerations, but by gas kinetics.

TABLE 4–1. Gas Properties for Several Gases at NTP (293.15 K and 101.3 kPa)

Gas	η (10^{-6} Pa·s)	S (K)	ρ_g (kg/m^3)	λ (μm)
Air	18.203	110.4	1.205	0.0665
Ar	22.292	141.4	1.662	0.0694
He	19.571	73.8	0.167	0.192
H_2	8.799	66.7	0.835	0.123
CH_4	10.977	173.7	0.668	0.0537
C_2H_6	9.249	223.2	1.264	0.0328
i-C_4H_{10}	7.433	255.0	2.431	0.0190
N_2O	14.646	241.0	1.837	0.0433
CO_2	14.673	220.5	1.842	0.0432

Source: Adapted from Rader (1990).

The *average velocity of a molecule*, $\overline{V}$, is a function of its molecular weight, M, and the gas temperature, T. In air (M_{air} = 0.0289 kg/mol) at normal temperature and pressure (NTP, 20°C, 1 atm), this molecular velocity is 463 m/s. Using these air reference values, the average velocity can be estimated for other gases and temperatures:

$$\overline{V} = \overline{V}_r \left(\frac{T}{T_r}\right)^{1/2} \left(\frac{M_r}{M}\right)^{1/2} \tag{4–5}$$

Mean free path, λ, is the mean distance a molecule travels before colliding with another molecule. In air at 293 K and atmospheric pressure, the mean free path, λ_r, is 0.0664 μm. The mean free path is an abstraction that is determined from a kinetic theory model that relates it to the coefficient of viscosity. Using these reference values, λ is determined for other pressures and temperatures (Willeke, 1976):

$$\lambda = \lambda_r \left(\frac{101}{P}\right)\left(\frac{T}{293}\right)\left(\frac{1+110/293}{1+110/T}\right) \tag{4–6}$$

where P is in kPa and T in K. If the unit of atmosphere is used for pressure, the factor of 101 used in Eq. 4–6 is substituted by one. The factor of 110 (K) is the Sutherland constant, and the value changes for different gases. The mean free path and the average molecular velocity are parameters that are frequently used to predict bulk properties of a gas, such as thermal conductivity, diffusion, and viscosity. Mean free paths for other gases are presented in Table 4–1.

The *Knudsen number*, *Kn*, relates the gas molecular mean free path to the physical dimension of the particle, usually the *particle radius*, *r*.

$$Kn = \frac{\lambda}{r} = \frac{2\lambda}{d_p} \tag{4–7}$$

where d_p is the physical diameter of the particle. The Knudsen number is somewhat counterintuitive as an indicator of particle size because it has an inverse size dependence. $Kn \ll 1$ indicates continuum flow, and $Kn \gg 1$ indicates free molecular flow. The intermediate range, approximately Kn = 0.4 to 20, is usually referred to as the transition or slip flow regime.

Slip Flow Regime and Correction Factor

If a particle is much smaller than the gas molecular mean free path ($Kn \gg 1$), it can travel past an obstacle at a very small distance from the object because no gas molecule may impede it. If the particle is very large ($Kn << 1$), many gas molecular collisions occur near the surface and the particle is decelerated. When the Knudsen number is the order of unity, the particle may slip by the obstacle. When the particle size is in this *slip flow regime*, it is convenient to assume that the particle is still moving in a continuum gas flow. To accommodate for the difference, a slip correction factor, C_c, also referred to as the *Cunningham slip correction factor*, is introduced into the equations. An empirical fit to air data for particles gives (Allen and Raabe, 1985)

$$C_c = 1 + Kn[\alpha + \beta \exp(-\gamma/Kn)] \tag{4–8}$$

Various values for α, β, and γ have been reported. However, it is important to use the mean free path with which these constants were determined. The value of λ_r used in Eq. 4–6 should also be consistent with the derivation of the slip coefficient constants. The following constants are consistent with $\lambda_r = 0.0664\,\mu m$ at NTP. For solid particles, $\alpha = 1.142$; $\beta = 0.558$; $\gamma = 0.999$ (Allen and Raabe, 1985). For oil droplets, $\alpha = 1.207$; $\beta = 0.440$; $\gamma = 0.596$ (Rader, 1990). C_c for other gases such as CO_2 and He are similar within a few percent. The slip correction and viscosity values are better determined than most other aerosol-related parameters and are therefore reported with a higher degree of precision.

For pressures other than atmospheric, the slip correction changes because of the pressure dependence of the mean free path in Kn, and the following may be used for solid particles:

$$C_c = 1 + \frac{1}{Pd_p}[15.60 + 7.00 \exp(-0.059 Pd_p)] \tag{4–9}$$

where P is the absolute pressure in kPa, and d_p is the particle diameter in μm (Hinds, 1999).

C_c is one in the continuum regime and becomes greater than one for decreasing particle diameter in the transition regime. For instance, $C_c = 1.02$ for 10 μm particles; 1.15 for 1 μm particles, and 2.9 for 0.1 μm particles. Note that the shape factor and the slip correction must be consistent with the type of equivalent diameter used in the same equation (Brockmann and Rader, 1990). For further discussion of shape factor, see Chapter 23.

Gas Viscosity

Gas viscosity is primarily due to the momentum transfer that occurs during molecular collisions. These frequent and rapid collisions tend to damp out differences in bulk gas motion as well as impede the net motion of particles relative to the gas. Thus, the mobility of a particle in a force field depends on the aerodynamic drag exerted on the particle through the gas viscosity. Fluid dynamic similitude, as expressed by Reynolds number, depends on gas viscosity, η. Therefore, knowledge of the gas viscosity is important when dealing with aerosol particle mechanics. The viscosity can be related to a reference viscosity η_r and a reference temperature, T_r, as follows:

$$\eta = \eta_r \left(\frac{T_r + S}{T + S}\right)\left(\frac{T}{T_r}\right)^{3/2} \tag{4–10}$$

where S is the Sutherland interpolation constant (Schlichting, 1979). Note that viscosity is independent of pressure.

In SI units, viscosity is expressed in Pa·s. In cgs units, viscosity is expressed in dyne s/cm^2, also referred to as *poise* or *P*. For air at 293 K, the viscosity is 1.833×10^{-5} Pa·s [183.3 μpoise] and S = 110.4 K. The interpolation formula is fitted to the data over the range 180 to 2000 K (Schlichting, 1979). Reference values of viscosity and Sutherland constants for other gases are presented in Table 4–1.

GAS AND PARTICLE DIFFUSION

The random movement of the gas molecules causes gas and particle diffusion if there is a concentration gradient. For instance, in a diffusion denuder, SO_2 gas molecules may diffuse to an absorbing surface due to their high diffusivity. Sulfate particles, which are larger and therefore have lower diffusivity, will mostly be transported through the device. Thus, the SO_2 gas molecules are separated from the sulfate particles.

Gas Diffusion

Diffusion always causes net movement from a higher concentration to a lower one. The *net flux* of gas molecules, J, is in the direction of lower concentration. Thus, in simple one-dimensional diffusion,

$$J = -D\frac{\partial N}{\partial x} \tag{4–11}$$

where x is the direction of diffusion, N is the concentration, and D is a proportionality constant referred to as the *diffusion coefficient*. The diffusion coefficient for a gas with molecular weight, M, is (Hinds, 1999:27)

$$D = \left(\frac{3\sqrt{2}\pi}{64Nd_{\text{molec}}^2}\right)\left(\frac{RT}{M}\right)^{1/2} \tag{4–12}$$

where N is the number of gas molecules/m^3 and d_{molec} is the molecular collision diameter (3.7×10^{-10} m for air). The diffusion coefficient of air molecules at 293 K is 1.8×10^{-5} m^2/s. This predicts a diffusion coefficient that is approximately 10% below the correct value (Hinds, 1999:27).

Particle Diffusion

Small particles can achieve significant diffusive motion in much the same fashion as des-cribed for gas molecules. The difference is only in the particle size and shape. Because of their increase in inertia with particle mass and the larger surface area over which the bombardment by the gas molecules is averaged, large particles will diffuse more slowly than small particles. For particles in a gas, the *diffusion coefficient* or *diffusivity*, D, can be computed by

$$D = \frac{kTC}{3\pi\eta d_{\text{p}}} = kTB \tag{4–13}$$

where k, the Boltzmann constant, is 1.38×10^{-23} N·m/K [1.38×10^{-16} dyne·cm/K] and the mechanical mobility, B in m/N·s [cm/dyne·s], is a convenient aerosol property that combines particle size with some of the properties of the suspending gas

$$B = \frac{C_{\text{c}}}{3\pi\eta d_{\text{p}}} \tag{4–14}$$

Particle diffusion, also referred to as *Brownian motion*, occurs because of the relatively high velocity of small particles, and it is sometimes useful to estimate how far, on the average, these particles move in a given time. The *root mean square* (rms) *distance*, x_{rms}, that the particles can travel in time, t, is

$$x_{\text{rms}} = \sqrt{2Dt} \tag{4–15}$$

EXAMPLE 4–2

Fume aerosols of 0.01 μm diameter are drawn into the deep lung regions of a worker whose alveoli can be approximated by 0.2 mm diameter spheres. We would like to estimate if these particles are likely to deposit in this area of the lungs during a breath-holding period of 4 seconds. Assume that body temperature is 330 K [37°C].

Answer: We note that calculation of x_{rms} (Eq. 4–15) requires knowledge of the diffusion coefficient, which in turn requires the slip correction factor and viscosity. To simplify the calculation, let us assume for the moment that the diffusion is taking place at room temperature. Thus, the air viscosity is 1.83×10^{-5} Pa·s and the mean free path is 0.0665 μm. (For a more exact estimate of these parameters at body temperature, use Eqs. 4–6 and 4–10, respectively.)

The slip correction factor can be determined from Eq. 3–8 using constants for solid particles:

$$C_c = 1 + Kn[1.142 + 0.558\exp(-0.999/Kn)]$$

$$C_c = 1 + \frac{2 \cdot 0.0665\,\mu\text{m}}{0.01\,\mu\text{m}}\left[1.142 + 0.558\exp\left(-0.999\frac{0.01\,\mu\text{m}}{2 \cdot 0.0665\,\mu\text{m}}\right)\right] = 23.1$$

We then estimate the diffusion coefficient, using Eq. 3–13:

$$D = \frac{kTC_c}{3\pi\eta d_p}$$

$$D = \frac{\left(1.38 \times 10^{-23}\,\frac{N{\cdot}m}{K}\right)(293\text{ K})(23.1)}{3 \cdot 3.14 \cdot (1.83 \times 10^{-5}\,\text{Pa·s})\left(0.01\,\mu\text{m} \times 10^{-6}\,\frac{\text{m}}{\mu\text{m}}\right)} = 5.42 \times 10^{-8}\,\frac{\text{m}^2}{\text{s}}$$

$$\left[D = \frac{\left(1.38 \times 10^{-16}\,\frac{\text{dyn·cm}}{\text{K}}\right)(293\text{ K})(23.1)}{3 \cdot 3.14 \cdot (1.83 \times 10^{-4}\,\text{poise})\left(0.01\mu\text{m} \times 10^{-4}\,\frac{\text{cm}}{\mu\text{m}}\right)} = 5.42 \times 10^{-4}\,\frac{\text{cm}^2}{\text{s}}\right]$$

Finally, using Eq. 3–15:

$$x_{rms} = \sqrt{2Dt} = \sqrt{2\left(5.42 \times 10^{-8}\,\frac{\text{m}^2}{\text{s}}\right)(4\text{ s})}$$
$$= 6.60 \times 10^{-4}\,\text{m}[0.0660\text{ cm}] = 0.660\text{ mm}$$

We find that, at room temperature, the rms displacement by diffusion is much larger than the alveolar size. At the elevated temperature in the lung (37°C), the particles are expected to move faster and diffuse further. If this air termperature is used in the calculation of the diffusion coefficient, x_{rms} is 0.675 mm. Thus, we know that most of these particles are likely to be collected in the alveolar space of the lung. A more exact analysis can be made by considering such factors as the spherical geometry of the alveoli, the location of the particles within the alveoli, and the air temperature.

By including the air temperature dependence also in the calculation of viscosity and mean free path, the rms displacement calculation results in 0.661 mm.

TABLE 4–2. Particle Parameters for Unit Density Particles at NTP

Particle Diameter, d_p (μm)	Slip Correction Factor, C_c	Settling Velocity, V_{grav} (m/s)	Relaxation Time, t (s)	Stopping Distance, S $V_0 = 10$ m/s (m)	Mobility, B (m/N·s)	Diffusion Coefficient, D (m²/s)	rms Brownian Displacement in 10 s (m)
0.00037[a]			2.6×10^{-10}	2.5×10^{-9}	9.7×10^{15}	1.8×10^{-5}[b]	2.8×10^{-2}
0.01	23.04	6.95×10^{-8}	7.1×10^{-9}	7.1×10^{-8}	1.4×10^{13}	5.5×10^{-8}	1.0×10^{-3}
0.1	2.866	8.65×10^{-7}	8.8×10^{-8}	8.8×10^{-7}	1.7×10^{11}	6.8×10^{-10}	1.2×10^{-4}
1	1.152	3.48×10^{-5}	3.5×10^{-6}	3.5×10^{-5}	6.8×10^{9}	2.7×10^{-11}	2.3×10^{-5}
10	1.015	3.06×10^{-3}	2.3×10^{-4}	2.3×10^{-3}	6.0×10^{8}	2.4×10^{-12}	7.0×10^{-6}
100	1.002	2.61×10^{-1}	1.3×10^{-2}	0.13	5.9×10^{7}	2.4×10^{-13}	2.2×10^{-6}

[a] Average diameter of a molecule in air.
[b] Calculated using Eq. 3–12.

Table 4–2 includes examples of x_{rms} for various-sized particles after a period of 10 s.

Peclet Number

The amount of convective transport of particles toward an object may be related to the diffusive transport through the dimensionless *Peclet number*, *Pe*,

$$Pe = \frac{Ud_c}{D} \tag{4–16}$$

where d_c is the significant dimension of the particle collecting surface and U is the upstream gas velocity toward the surface. The larger the value of *Pe*, the less important is the diffusional process (Licht, 1988:226). *Pe* is often used in the description of diffusional deposition on filters. For further discussions of diffusions, see Chapters 8, 9, and 19.

Schmidt Number

The ratio of the Peclet number, Eq. 4–16, to the Reynolds number, Eq. 4–1, is referred to as the *Schmidt number*, *Sc*. It expresses the ratio of kinematic viscosity to diffusion coefficient.

$$Sc = \frac{\eta_g}{\rho_g D} = \frac{\nu}{D} \tag{4–17}$$

As the Schmidt number increases, convective mass transfer increases relative to Brownian diffusion of particles. It has been used for describing diffusive transport in flowing fluids (convective diffusion), especially in the development of filtration theory (Friedlander, 1977). *Sc* is relatively independent of temperature and pressure near standard conditions.

AERODYNAMIC DRAG ON PARTICLES

Externally applied forces on an aerosol particle are opposed and rapidly balanced by the aerodynamic drag force. An example of this is a sky diver: The gravitational force pulling the sky diver toward the earth is eventually balanced by the air resistance, and the diver reaches a final falling speed of about 63 m/s (140 miles per hour).

A particle's drag force, F_{drag}, relates the resistive pressure of the gas to the velocity pressure given by the relative motion between the particle and the surrounding gas. When the particle dimensions are much larger than the distance between the gas molecules, the surrounding gas can be considered a continuous fluid (continuum regime). Under this condition, the *drag force* is given by

$$F_{drag} = \frac{\pi}{8} C_d \rho_g V^2 d_p^2 \tag{4–18}$$

Note that the aerodynamic drag is related to the gas density, ρ_g, not the particle density. The *drag coefficient*, C_d, relates the drag force to the velocity pressure. When the inertial force pushing the gas aside, due to the velocity difference between the gas and the particle, is much smaller than the viscous resistance force, the drag coefficient, C_d, is expressed in terms of gas flow parameters

$$C_d = \frac{24}{Re_p} \qquad Re_p < 0.1 \tag{4–19}$$

where Re_p is the particle Reynolds number. This relationship is accurate within 1% in the Re_p range indicated. If 10% accuracy is acceptable, Eq. 4–19 can be used up to $Re_p < 1.0$. Combining Eqs. 4–1, 4–18, and 4–19 results in

$$F_{drag} = 3\pi\eta V d_p \tag{4–20}$$

This equation is also known as *Stokes' law*. For Stokes' law flow of gas around the particle, the drag on the particle depends only on gas viscosity, η, particle velocity, V, and particle diameter, d_p. This assumes that the particle is spherical. The particle drag for shapes other than spheres is usually difficult to predict theoretically. Therefore, for particles of other shapes, a *dynamic shape factor* χ is introduced that relates the motion of the particle under consideration to that of a spherical particle

$$F_{drag} = 3\pi\eta V \chi d_m \tag{4–21}$$

where d_m is now the *mass equivalent diameter* defined as the diameter of a sphere composed of the particle bulk material with no voids that has the same mass as the particle in question. The shape factor is sometimes related to the equivalent volume or *volume equivalent diameter*, d_{ev}, defined as the diameter of a sphere of equivalent volume. This term may be ambiguous. When the equivalent volume is composed of particle bulk material with no void, $d_{ev} = d_m$. However, if the material includes voids, $d_{ev} > d_m$. If we determine the shape factors and equivalent diameters for particles that we wish to measure, the behavior of the particles can be predicted when they are influenced by various force fields (e.g., gravity or electrostatic).

We know that gases are not continuous fluids, as indicated above, but consist of discrete molecules. Therefore, when the particle size approaches the mean free path of the gas molecular motion (transition or slip regime), we can apply a correction that takes the "slip" between the particle and the gas into account. Thus, the Cunningham slip correction factor C_c is introduced into Eq. 4–21:

$$F_{drag} = \frac{3\pi\eta V \chi d_{ev}}{C_c} \tag{4–22}$$

Eq. 4–22 assumes that the flow around the particle is laminar. As particles move faster (i.e., have a larger Re_p), the above relationships must be modified further. As indicated above, the range over which Eq. 4–19 is accurate defines the Stokes regime. For larger Re_p, empirical relationships for C_d have been developed to extend Stokes' law. The drag coefficient, C_d, for a spherical particle is the ratio of the resistance pressure due to aerodynamic drag (drag force/cross-sectional area) to the velocity pressure of the flow toward the sphere based on the relative velocity between the particle and the suspending gas. Figure 4–2 shows the relationship of the drag coefficient to particle Reynolds number over a wide range of Reynolds numbers.

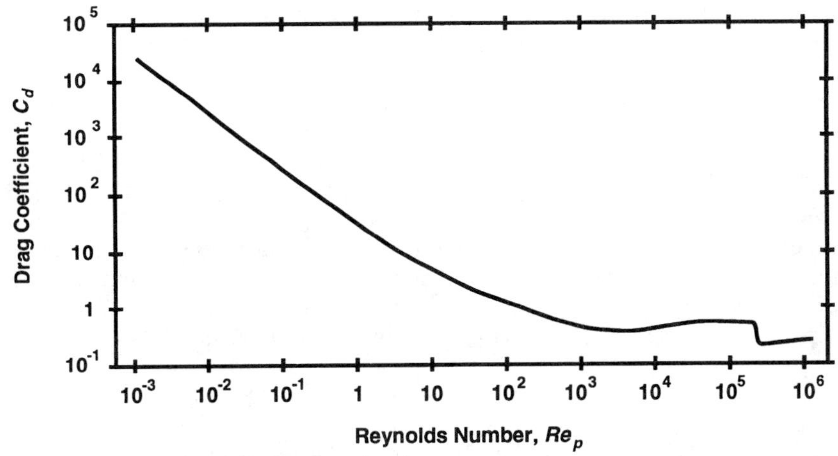

Fig. 4–2. Drag coefficient as a function of particle Reynolds number for spherical particles.

For Re_p above 0.1, Sartor and Abbott (1975) developed the following empirical relationship:

$$C_d = \frac{24}{Re_p}(1 + 0.0196 Re_p) \qquad 0.1 \le Re_p < 5 \tag{4–23}$$

Serafini (Friedlander, 1977:105) developed the following:

$$C_d = \frac{24}{Re_p}(1 + 0.158 Re_p^{2/3}) \qquad 5 \le Re_p < 1000 \tag{4–24}$$

Note that these relationships have been derived from data obtained with smooth spheres. Similar relationships have been derived and reviewed for particles such as droplets, solid spheroids, disks, and cylinders (Clift et al., 1978:142). Typically, Re_p is based on the equatorial diameter for disks and spheroids and on the cylinder diameter for cylinders, although other definitions can be used. Particles with extreme shapes may have a significantly different drag coefficient. For instance, C_d for fibers is up to four times lower than for spheres with $Re_p < 100$ when the fiber diameter is used as the significant dimension in the Reynolds number expression, Eq. 4–1.

Thus, by using the appropriate form of the drag coefficient (Eq. 4–19, 4–23, or 4–24) and including the shape factor and slip coefficient, the drag force can be calculated over a wide range of particles and conditions:

$$F_{drag} = \frac{\pi C_d \rho_g V^2 \chi^2 d_{ev}^2}{8 C_c} \tag{4–25}$$

For further discussion of shape factors and behavior of nonspherical particles, see Chapter 23.

PARTICLE MOTION DUE TO GRAVITY

The *gravitational force*, F_{grav}, is proportional to particle mass, m_p, and gravitational acceleration, g,

$$F_{grav} = m_p g = (\rho_p - \rho_g) v_p g \approx \rho_p v_p g \tag{4–26}$$

where ρ_g is the gas density. The gravitational pull depends on the difference between the density of the particle and that of the surrounding medium. For a particle in water, this buoyancy effect is significant. For a particle in air, the buoyancy effect can be neglected for compact particles because the particle density is generally much greater than the density of the gas. If the particle is spherical, particle volume v_p can be replaced with $\pi d_p^3/6$:

$$F_{grav} = \frac{\pi}{6} d_p^3 \rho_g g \quad (4\text{–}27)$$

The gravitational field of the earth was mentioned in the previous chapter. This field exerts a force pulling a particle down. As the particle begins to move, the gas surrounding the particle exerts an opposing drag force, which, after a short period of acceleration, equals the gravitational force, and the particle reaches its terminal settling velocity. By equating the drag force, Eq. 4–18 (with Cunningham slip correction factor C_c added), to the gravitational force, Eq. 4–27, and using Eq. 4–19 for C_d and Eq. 4–1 for Re_p, the following relationship is obtained for the spherical particle settling velocity, V_{grav}, in the Stokes regime (Eq. 4–19):

$$V_{grav} = V_{ts} = \frac{\rho_p d_p^2 g C_c}{18\eta} \qquad Re_p < 0.1 \quad (4\text{–}28)$$

To reflect the equilibrium between the two opposing forces, this velocity is also referred to as *terminal settling velocity*, V_{ts}. For spherical particles with negligible slip ($C_c = 1$) and $1 < d_p < 100$, this equation reduces to the following at NTP:

$$V_{grav} \cong 3 \times 10^{-8} \rho_p d_p^2$$
$$[V_{grav} \cong 0.003 \rho_p d_p^2] \text{ in cgs units} \quad (4\text{–}29)$$

where ρ_p is in kg/m^3 [g/cm^3] and d_p is in μm. Spherical particles (e.g., droplets), are common in nature, and their motion can be described mathematically. Therefore, behavior of nonspherical particles is often referenced to such particles through comparison of their behavior in a gravitational field.

EXAMPLE 4–3

An open-faced filter cassette samples at 2 L/min (3.33×10^{-5} m^3/s [33.3 cm^3/s]) over its inlet face of about 35 mm diameter. If the cassette is held facing downward, can a 25 μm diameter particle with a density of 3000 kg/m^3 [3 g/cm^3] be drawn upward onto the filter in calm air?

Answer: The cassette samples at a flow rate Q over a cross-sectional filter area A. The upward air velocity, U, is

$$U = \frac{Q}{A} = \frac{3.33 \times 10^{-5} \frac{\text{m}^3}{\text{s}}}{\pi \left(\frac{0.035 \text{ m}}{2} \right)^2} = 0.0346 \text{ m/s} [3.46 \text{ cm/s}]$$

The gravitational settling velocity of the 25 μm particle is, from Eq. 4–29,

$$V_{grav} = 3 \times 10^{-8} \rho_p d_p^2 = 3 \times 10^{-8} \left(3000 \frac{\text{kg}}{\text{m}^3} \right) \left(25 \,\mu\text{m} \cdot 10^{-6} \frac{\text{m}}{\mu\text{m}} \right)$$

$$\left[V_{grav} = 0.003 \rho_p d_p^2 = 0.003 \left(3 \frac{\text{g}}{\text{cm}^3} \right) (0.0025 \text{ cm})^2 \right]$$

$$= 0.0563 \text{ m/s} [5.63 \text{ cm/s}] > 0.0346 \text{ m/s} [3.46 \text{ cm/s}]$$

The particle cannot be drawn upward into the sampler.

This is also the principle of a vertical elutriator, which prevents particles above a certain size from passing through the device. However, in some implementations of this device (e.g., the cotton dust elutriator) inlet effects complicate the penetration efficiency.

The *aerodynamic diameter*, d_a, of a particle is the diameter of a standard-density sphere that has the same settling velocity as the particle in question, as shown in Eqs. 3–1 to 3–3. Another definition that is also commonly used is the *Stokes diameter*, d_s, which is the diameter of a spherical particle with the same density and settling velocity as the particle in question. The aerodynamic diameter can be related to the Stokes diameter through the settling velocity equation:

$$\rho_p d_s^2 = \rho_0 d_a^2 \tag{4–30}$$

A number of instruments, including the horizontal and the vertical elutriators (see Chapters 8, 10, 25, 26), use settling velocity to separate particles according to size. For instance, aerosol particles of a certain size, d_p, initially spread throughout a quiescent rectangular chamber or room of height H, will settle at a constant velocity, V_{grav}. After some time, t, the particle concentration in the chamber, $N(t)$, will be

$$N(t) = N_0\left(1 - \frac{V_{grav}t}{H}\right) \tag{4–31}$$

where N_0 is the initial particle concentration in the chamber. After time t, a vertical distance of $V_{grav} \cdot t$ is cleared of particles. The same relationship determines the concentration of particles in a rectangular channel with air flowing through it (a horizontal elutriator). At some distance downstream of the entrance to the channel (where the aerosol concentration is N_0), the concentration will be $N(t)$, where t is the time needed to reach that distance.

The above discussion of particle settling describes the behavior of particles in still air, a condition that is not often achieved in the environment or even in the laboratory. When the gas in a container undergoes continual and random motion, such as in a room with several randomly directed fans, the particles undergo *stirred settling*. The time-dependent concentration, $N(t)$, under these conditions is also expressed in terms of an initial particle concentration, N_0, the gravitational settling velocity in still air, V_{grav}, and the height of the container, H:

$$N(t) = N_0 \exp\left(\frac{-V_{grav}t}{H}\right) \tag{4–32}$$

This equation applies to any container shape with vertical walls and a horizontal bottom. This indicates that even under stirred or turbulent conditions, larger particles (higher settling velocities) will settle out more rapidly than smaller particles, even though some of the large particles may persist in the air for a long time because of the exponential decay. Note that the forms of Eqs. 4–31 and 4–32 are similar except for the exponential decay when stirring takes place during the settling. This similarity in form occurs for all such comparisons of uniform and stirred settling.

Gravitational Settling at Higher Reynolds Numbers

Particle settling velocity can be calculated accurately for $Re_p < 0.1$ using Eq. 4–28. At higher Reynolds numbers, the observed settling velocity is lower than predicted by that equation because the drag coefficient is higher than predicted by Eq. 4–19. For spherical particles ($c = 1$), the gravitational settling velocity can be expressed as a function of C_d by equating the drag force, Eq. 4–25, to the gravitational force Eq. 4–27:

$$V_{grav} = \left(\frac{4\rho_p C_c d_p g}{3\rho_g C_d}\right)^{1/2} \tag{4–33}$$

The drag coefficient has a complex dependence on the settling velocity, and Eq. 4–33 therefore cannot be solved in closed form. Graphical (Licht, 1988:160) and tabular (Hinds, 1999:56) determinations of the settling velocity at high Reynolds numbers have been used. Using Eq.

4–33 and the drag coefficient equation for the appropriate Reynolds number (e.g., Eq. 4–23 or 4–24), an iterative solution for the settling velocity can readily be obtained with a computer or calculator. A guess for C_d allows the calculation of an initial value for V_{grav}, which is then used to calculate a new value of C_d. The new value of C_d is then used in Eq. 4–33, and the iteration is continued until the values converge.

PARTICLE PARAMETERS

The gravitational force effectively removes large particles from the suspending gas. Particles of 1 μm or smaller take a long time to settle (see Table 4–2, below). To settle these, the removal force is increased by, for example, rotating the gas volume, as in a centrifuge. Other devices channel the gas flow in a circular fashion (e.g., cyclones) or through bends (e.g., impactors) to create an increased force field. The following parameters are useful for describing the inertial and settling behavior of particles.

Relaxation Time and Stopping Distance

With the Stokes settling velocity relationship (Eq. 4–28), several useful particle parameters can be defined. The first is the *particle relaxation time*:

$$\tau = \frac{\rho_p d_p^2 C_c}{18\eta} \tag{4–34}$$

This is the time a particle takes to reach 1/e of its final velocity when subjected to a gravitational field. The relaxation period is typically quite short, as indicated in Table 4–2, and can therefore be neglected for most practical applications. Use of this parameter simplifies the expression for *gravitational settling velocity* to

$$v_{grav} = \tau g \tag{4–35}$$

EXAMPLE 4–4

A grinding wheel dislodges many wheel and workpiece particles and projects them from the contact point toward the receiving hood of the ventilation system. A particle of a certain size and density is projected 10 mm away. How far will a particle twice this size be projected? Estimate the projected distance when the speed of the grinding wheel is doubled.

Answer: The projected distance is proportional to the stopping distance. From Eqs. 4–34 and 4–36,

$$S = V_0 \tau \propto V_0 \rho_p d_p^2$$

The stopping distance depends on the square of the particle diameter, so a two times larger particle will project four times the distance to 40 mm. At twice the grinding wheel speed, the particle will come off at approximately twice the initial velocity, resulting in a doubling of the distance to 20 mm.

The above stopping distance equation assumes that the particle is in the Stokes regime. If the particle diameter and velocity are such that Re_p is larger than 0.1, the stopping distance will be somewhat less than quadrupled for the larger particle. This is because the drag increases faster with diameter outside the Stokes regime (see Fig. 4–2). Similarly, increasing the initial velocity also increases Re_p and results in somewhat less than doubling of the distance.

Quite often, a particle, rather than starting from rest in a gravitational field, is injected into the air with an initial velocity, V_0. For instance, such a particle might be released from a rotating grinding wheel. The product of the relaxation time and the initial particle velocity is referred to as the *stopping distance*, S:

$$S = V_0 \tau \tag{4–36}$$

where S is in m. Values of S for an initial velocity of 10 m/s are given in Table 4–2. The concept of stopping distance is useful, for example, in impactors when evaluating how far a particle moves across the air streamlines when the flow makes a right angle bend.

Because Eq. 4–28 is accurate only in the Stokes regime, the following empirical relationship can be used at higher Re_p (Mercer, 1973:41):

$$S = \frac{\rho_p d_p}{\rho_g}\left(Re_0^{1/3} - \sqrt{6}\arctan\left(\frac{Re_0^{2/3}}{\sqrt{6}}\right)\right) \qquad 1 < Re_p < 400 \tag{4–37}$$

where Re_0 is the Reynolds number of the particle at the initial velocity.

Stokes Number

When gas flow conditions change suddenly, as at the particle collection surface of an impactor, the ratio of the stopping distance to a characteristic dimension, d, is defined as the *Stokes number*, *Stk*:

$$Stk = \frac{S}{d} \tag{4–38}$$

The characteristic dimension depends on the application, for example, in fibrous filtration it is the diameter of the fiber; in axisymmetrical impaction flows it is the radius or diameter of the impactor nozzle. For a given percent particle removal, the Stokes number value is therefore application specific. For example, the Stokes number of an impactor with one or several identical circular nozzles is

$$Stk = \frac{\rho_p d_p^2 V C_c}{9 \eta d_j} \tag{4–39}$$

where d_j is the impactor jet diameter in m and V is the particle velocity in the jet. V is assumed to be equal to the gas velocity in the jet. For further discussions of inertial devices, see Chapters 8, 9, 10, 14, and 17.

Shape Factor

As described above, particle aerodynamic diameter and Stokes diameter have been defined using ideal spherical particles. Apart from liquid droplets or particles produced from liquid droplets, few particles in nature are spheres. It is convenient to describe more complex shapes by a single diameter and have the additional flow resistance or drag represented by a factor. This *dynamic shape factor*, χ, is the ratio of the drag force of the particle in question (particle diameter, d_p) to that of a sphere of equivalent volume (equivalent volume diameter, d_{ev}). The expression for gravitational settling, Eq. 4–28, thus becomes

$$V_{grav} = \frac{\rho_p d_p^2 g C_{c.d_p}}{18\eta} = \frac{\rho_0 d_a^2 g C_{c.d_a}}{18\eta} = \frac{\rho_p d_{ev}^2 g C_{c.d_{ev}}}{18\eta\chi} \qquad Re_p < 0.1 \tag{4–40}$$

The value of the Cunningham slip factor C_c depends on the chosen diameter, d_p, d_a, or d_{ev}. The shape factor is always equal to or greater than one. Compact shapes typically have values between one and two, while more extreme shapes, such as fibers and high-volume agglomerates, may have larger values. Shape factors are useful for converting a readily measurable equivalent diameter to one that depends on particle behavior, such as aerodynamic diameter or diffusion equivalent diameter. Thus, shape factors have been defined in a variety of ways that have to do with the available means of measuring the physical and equivalent particle diameter as well as the means of measuring particle drag. Therefore, when applying published shape factors, it may be important to understand the experimental basis for their development.

Some particles have relatively regular shapes with volumes that can be calculated or compact shapes that can be measured with a microscope to determine an equivalent volume diameter. For such particles, the shape factor is, from Eq. 4–40,

$$\chi = \frac{\rho_p d_{ev}^2 C_{c.d_{ev}}}{\rho_0 d_a^2 C_{c.d_a}} \qquad Re_p < 0.1 \tag{4–41}$$

Three variables need to be measured: ρ_p, d_{ev}, and d_a. The equivalent volume diameter may be measured microscopically or determined from the mass (measured chemically or using radioactive tracers) and the number of particles (Barbe-le Borgne et al., 1986). The aerodynamic diameter can be measured in a settling chamber or centrifuge, and, if the particle contains no voids, the density is the bulk density of the particle material.

Shape factors have also been measured by settling macroscopic models of regularly shaped particles in liquids. For instance, this technique has been used to measure shape factors for cylinders and chains of spheres (Kasper et al., 1985); for rectangular prisms (Johnson et al., 1986); and for modified rectangular prisms (Sheaffer, 1987). These particles have two or three distinct symmetry axes and therefore may have two or three shape factors, depending on their orientation. Shape factors have also been derived for oblate and prolate spheroids (Fuchs, 1964:37). Table 4–3 exemplifies a few.

Porous particles, agglomerates, and fume particles may have an effective density (r_e, including internal voids) that is quite different from the *bulk material density*, ρ_p. In this case, a shape factor defined as a function of the *mass equivalent diameter* (d_m) may be more appro-

TABLE 4–3. Dynamic Shape Factors for Various Types of Compact Particles (No Internal Voids)

Shape	Dynamic Shape Factor (χ)
Sphere	1.00
Cluster of spheres	
2 sphere chain	1.12
3 sphere chain	1.27
4 sphere chain	1.32
Prolate spheroid (L/D = 5)[a]	
Axis horizontal	1.05
Axis vertical	1.39
Glass Fiber(L/D = 5)	1.71
Dusts	
Bituminous coal	1.05–1.11
High-ash soft coal	1.95
Quartz	1.36–1.82
Sand	1.57
Talc	2.04
UO_2	1.28
ThO_2	0.99

[a] Calculated values; all others are experimental.

Source: Adapted from Davies (1979).

priate (Brockmann and Rader, 1990), thus replacing d_{ev} with d_m in Eq. 4–41. The shape factor χ may be further broken down as the product of *envelope shape factor*, k, and second component d (defined as $[\rho_p/\rho_e]^{1/3}$), which is due to the porosity of the particle. Theoretically or empirically derived shape factors as described above can be used to match the approximate envelopes of the observed particles. For relatively compact particles, the porosity component of χ dominates, while for more sparse, branched chain agglomerates the envelope factor dominates. For additional discussion of complex particle shapes, see Chapter 23.

PARTICLE MOTION IN AN ELECTRIC FIELD

Application of electrostatic forces is particularly effective for submicrometer-sized particles for which gravity forces are weak because of the d_p^3 dependence (Eq. 4–27). On a large scale, removal of aerosols by electrostatic forces is practiced in electrostatic precipitators (also called *electrofilters*). In aerosol sampling and measuring instruments, electrostatic forces are applied to precipitate or redirect either all aerosol particles or those in a specific size range.

For a particle with a total charge equal to n times the elementary unit of charge, e, the *electrostatic force*, F_{elec}, in an *electric field* of intensity, E, is

$$F_{elec} = neE \tag{4–42}$$

EXAMPLE 4–5

A 0.5 µm diameter standard density particle has been diffusion charged with 18 elementary units of charge. Calculate the electrical force on the particle when it passes between two flat parallel plates (e.g., an electrostatic precipitator) that have 5 kV applied across a 0.02 m gap. Compare the electrical to the gravitational force.

Answer: The electric field between the plates is

$$E = \left(\frac{5000\ \text{V}}{0.02\ \text{m}}\right) = 2.5 \times 10^5\ \text{V/m}$$

$$\left[E = \left(\frac{5000\ \text{V}}{2\ \text{cm}}\right)\left(\frac{1\ \text{stV}}{300\ \text{V}}\right) = 8.33\ \text{stV/cm}\right]$$

Using Eq. 4–42,

$$F_{elec} = neE = 18(1.6 \times 10^{-19}\ \text{C})(2.5 \times 10^5\ \text{V/m}) = 7.2 \times 10^{-13}\ \text{N}$$

$$\left[F_{elec} = neE = 18(4.8 \times 10^{-10}\ \text{stC})(8.33\ \text{stV/cm}) = 7.2 \times 10^{-8}\ \text{dyn}\right]$$

One newton (N) in SI units equals 10^5 dyne in the cgs system of units. Using Eq. 3–27,

$$F_{grav} = \frac{\pi}{6} d_p^3 \rho_p g = \frac{\pi}{6}(0.5 \times 10^{-6}\ \text{m})(1000\ \text{kg/m}^3)(9.80\ \text{m/s}^2)$$

$$\left[F_{grav} = \frac{\pi}{6}(0.5 \times 10^{-4}\ \text{cm})^3 (1\ \text{g/cm}^3)(980\ \text{cm/s}^2)\right]$$

$$= 6.41 \times 10^{-16}\ \text{N}\left[6.41 \times 10^{-11}\ \text{dyn}\right]$$

Comparing the two forces,

$$\frac{F_{elec}}{F_{grav}} = \frac{7.2 \times 10^{-13}\ \text{N}}{6.41 \times 10^{-16}\ \text{N}} = 1120$$

The electric force exceeds the gravity force over 1000 times.

Electrostatic forces can affect particle motion and, to a certain extent, gas motion as well. These forces can be important during particle generation, transport, and measurement. Depending on the number of charges on a particle and the level of surrounding electric field, the force on that particle can range anywhere from zero to the largest of any force discussed here. If a particle is placed in an electric field described by Eq. 4–42, it will reach a terminal velocity, V_{elec}, when the field and drag forces are equal.

$$V_{\text{elec}} = \frac{neEC_c}{3\pi\eta d_p} \tag{4–43}$$

The electronic charge e is 1.602×10^{-19} coulombs (C) [4.803×10^{-10} statcoulombs (stC)]. This terminal or *drift velocity* can also be written in terms of the particle mobility, B:

$$V_{\text{elec}} = neEB \tag{4–44}$$

or, including the electric charge, the *particle electrical mobility*, $Z = neB$:

$$V_{\text{elec}} = ZE \tag{4–45}$$

where the electrical mobility, Z, has units of velocity/electric field or $\text{m}^2/\text{V·s}$ [$\text{cm}^2/\text{stV·s}$] (i.e., unit electrical mobility is a drift velocity of 1 m/s in a 1 V/m field). One statvolt (cgs unit) is equal to 300 volts (V).

EXAMPLE 4–6

Foundry fumes are sampled into an electrostatic precipitator for collection onto an electron microscope grid. A power supply is used to apply a potential of 5000 V across the condenser with a plate spacing, H, of 0.01 m. The aerosol flows through the condenser at a uniform velocity of 0.02 m/s. The particles of concern have an electrical mobility of $3.33 \times 10^{-9}\ \text{m}^2/\text{V·s}$. What is the minimum plate length, L, that will precipitate all of these particles?

Answer: For a potential of 5000 V, the precipitation time, t_e, in the electric field is

$$t_e = \frac{H}{V_{\text{elec}}} = \frac{H}{ZE} = \frac{0.01\text{m}}{\left(3.33 \times 10^{-9}\dfrac{\text{m}^2}{\text{V·s}}\right)\left(\dfrac{5000\ \text{V}}{0.01\ \text{m}}\right)} = 6\ \text{s}$$

In cgs units, the mobility is converted to $0.01\ \text{cm}^2/\text{stV·s}$, and the spacing is 1 cm.

$$\left[t_e = \frac{1\ \text{cm}}{\left(0.01\dfrac{\text{cm}^2}{\text{stV·s}}\right)\left(\dfrac{16.7\ \text{stV}}{1\ \text{cm}}\right)} = 6\ \text{s}\right]$$

The transit time t_t for the air flow at velocity U must equal or exceed this time.

$$t_t = \frac{L}{U} \ge t_e$$

$$L \ge 6\ \text{s}(0.02\text{m/s}) = 0.12\text{m}\ [12\ \text{cm}]$$

The simplest electric field is uniform, for example, between two large parallel plates:

$$E = \frac{\Delta V}{x} \tag{4–46}$$

where x(m) is the distance between the plates and ΔV is the difference in potential (volts). The field between two concentric tubes or between a tube and a concentric wire is also used for electrostatic precipitation. In this case the field depends on the distance, r, from the axis:

$$E = \frac{\Delta V}{r \ln(r_o/r_i)} \tag{4–47}$$

where ΔV is the difference in potential between the outer tube and inner tube (or wire) of radius r_o and r_i, respectively.

In the SI units, the force in newtons (N) on each of two particles with n_1 and n_2 unit charges on them is described by *Coulomb's law*:

$$F_{elec} = \frac{n_1 n_2 e^2}{4\pi\varepsilon_0 r^2} = K_E \frac{n_1 n_2 e^2}{r^2} \tag{4–48}$$

where r in m is the distance between the particles. The factor K_E, a proportionality constant that depends on the unit system, is 8.988×10^9 (SI units). This equation strictly applies only to point charges. However, it is a good approximation for the force between two particles or a particle at some distance from a charged object such as a sampler and indicates that the force drops off rapidly with distance. Aerosol particles, which typically carry a limited amount of charge because of their small surface area, are generally only affected electrically when they are quite close to another charged particle or close to a charged object.

In cgs units, Eq. 4–48 is converted to give the force in dynes (dyn):

$$F_{elec} = \frac{n_1 n_2 e^2}{r^2} \tag{4–49}$$

where r is in cm, the electronic charge is 4.80×10^{-10} stC, and the proportionality constant K_E is unity. For further discussions of charged particle dynamics, see Chapters 17, 18, and 20.

PARTICLE MOTION IN OTHER FORCE FIELDS

Particle motion is governed by a variety of other forces. Very small particles approach the behavior of the molecules of the surrounding gas (i.e., they diffuse readily and have little inertia); they can be affected by light pressure, acoustic pressure, and thermal pressure. In a similar fashion to gravitational and electrical forces, other forces can be used to cause particle motion and thus size-selective measurement. The same forces can also cause particles to be lost rapidly in the sampling inlet or on measurement instrument surfaces. Other forces not mentioned may have some effects but are generally much weaker than the ones mentioned here. For instance, magnetic forces are typically several orders of magnitude smaller than electrostatic forces, but have been used for fiber alignment (see Chapter 23).

Thermophoresis

Particles in a thermal gradient are bombarded more strongly by gas molecules on the hotter side and are therefore forced away from a heat source. Thus, heated surfaces tend to remain clean, while relatively cool surfaces tend to collect particles. This process is called *thermophoresis*, from the Greek "carried by heat." For particles smaller than the mean free path (λ), the thermophoretic velocity, V_{th}, is independent of particle size and is (Waldmann and Schmitt, 1966)

$$V_{th} = \frac{0.55\eta}{\rho_g} \nabla T \qquad d_p < \lambda \tag{4–50}$$

where ∇T is the thermal gradient in K/m. There is a slight increase (on the order of 3%) in the velocity of rough-surfaced particles versus spherical solids or droplets.

For particles larger than λ, the thermophoretic velocity depends on the ratio of the thermal conductivity of the gas to that of the particle and also on the particle size. For large conductive aerosol particles, the thermophoretic velocity may be about five times lower than for small, nonconductive ones. To calculate the thermophoretic velocity, the molecular accommodation coefficient (H) is needed:

$$H \cong \left(\frac{1}{1+6\lambda/d_{\mathrm{p}}}\right)\left(\frac{k_{\mathrm{g}}/k_{\mathrm{p}}+4.4\lambda/d_{\mathrm{p}}}{1+2k_{\mathrm{g}}/k_{\mathrm{p}}+8.8\lambda/d_{\mathrm{p}}}\right) \tag{4–51}$$

where k_{g} and k_{p} are the thermal conductivities of the gas and particle, respectively. The thermal conductivity of air is 0.026 W/m·K [5.6×10^{-5} cal/cm·s·K], while that for particles ranges from 66.9 W/m·K for a metal (iron) to 0.079 W/m·K for an insulator (asbestos) (Mercer, 1973:166). The thermally induced particle velocity is then (Waldmann and Schmidt, 1966)

$$V_{\mathrm{th}} = \frac{-3\eta C_{\mathrm{c}} H}{2\rho_{\mathrm{g}} T} \nabla T \qquad d_{\mathrm{p}} > \lambda \tag{4–52}$$

Thermophoresis is relatively independent of particle size over a wide range and has been used for collecting small samples, such as for electron microscope measurements, in thermal precipitators. The sampling rate of these instruments is low because of the difficulty of maintaining a thermal gradient, and thus thermal precipitators have not been scaled up for large-volume use (see Chapter 10).

Photophoresis

Photophoresis is similar to thermophoresis in that particle motion is caused by thermal gradients at the particle surface except that in this case the heating is caused by light absorption by the particle rather than by an external source. Light shining on a particle may be preferentially absorbed by the side near the light source or, under certain circumstances of weak absorption and focusing, by the far side of the particle. Thus, in the former case the particle will be repelled from the light source, while in the latter, called *reverse photophoresis*, it will be attracted.

Electromagnetic Radiation Pressure

Electromagnetic radiation can have a direct effect on particle motion by transferring momentum to the particle. Light impinging on a particle can be reflected, refracted, or absorbed. The fraction of momentum transfer from the light beam to the particle depends on the geometric cross section of the particle as well as the average direction of the scattered light. If a significant fraction of the light is absorbed by the particle, photophoresis, as described above, will be more important in deciding particle motion. Radiation pressure has been used to trap particles in focused laser beams and manipulate them for further study (see Chapter 20).

Acoustic Pressure

Acoustic waves, either stationary, as in a resonant box, or traveling in open space can be reflected, diffused, or absorbed by particles. Particle motion in an acoustic field includes oscillation in response to the gas motion, circulation in the acoustic field, net drift in some direction. Such waves have been used to increase particle coagulation or agglomeration and, in other cases, to enhance droplet evaporation or condensation (Hesketh, 1977:97). A resonant acoustic system has also been used to measure particle aerodynamic diameter by measuring a particle's ability to oscillate in response to the air motion (Mazumder et al., 1979; see also Chapter 17).

Diffusiophoresis and Stephan Flow

When the suspending gas differs in composition from one location to another, diffusion of the gas takes place. This gas diffusion results in suspended particles acquiring a net velocity as a function of the gas diffusion, that is, diffusiophoresis. The particles are pushed in the direction of the larger molecule flow. The force is a function of the molecular weight and diffusion coefficients of the diffusing gases and is largely independent of the particle size.

A special case of diffusiophoresis occurs near evaporating or condensing surfaces. A net flow of the gas–vapor mixture away from an evaporating surface is set up that creates a drag on particles. The converse situation holds for a condensing surface (i.e., gas and particles will flow toward the surface). This net motion of the gas–vapor mixture is called Stephan (also spelled Stefan) flow and can cause the motion of particles near these surfaces (Fuchs, 1964:67). Stephan flow can affect particle collection in industrial scrubbers and scavenging of the environment by growing cloud droplets. To increase particle collection by Stephan flow, the vapor must be supersaturated. Diffusiophoretic velocities are generally only significant for very small particles. For instance, diffusiophoresis of 0.005 to 0.05 μm diameter particles was found to have the following net *deposition velocity*, V_{diff}, toward surfaces condensing water vapor (Goldsmith and May, 1966);

$$V_{\text{diff}} = 1.9 \times 10^{-3} \frac{dP}{dx} \tag{4–53}$$

where the deposition velocity is in m/s and dP/dx is the pressure gradient of the diffusing vapor in Pa/m. Note that in condensing and evaporating droplets, thermophoretic effects can also be important.

REFERENCES

Allen, M. D. and O. G. Raabe. 1985. Slip correction measurements of spherical solid aerosol particles in an improved Millikan apparatus. *Aerosol Sci. Technol.* 4:269–286.

Barbe-le Borgne, M., D. Boulaud, G. Madelaine, and A. Renoux. 1986. Experimental determination of the dynamic shape factor of the primary sodium peroxide aerosol. *J. Aerosol Sci.* 17:79–86.

Brockmann, J. E. and D. J. Rader. 1990. APS response to nonspherical particles and experimental determination of dynamic shape factor. *Aerosol Sci. Technol.* 13:162–172.

Clift, R., J. R. Grace, and M. E. Weber. 1978. *Bubbles Drops and Particles.* New York: Academic Press.

Davies, C. N. 1979. Particle–fluid interaction. *J. Aerosol Sci.* 10:477–513.

Friedlander, S. K. 1977. *Smoke, Dust and Haze.* New York: John Wiley & Sons.

Fuchs, N. 1964. *The Mechanics of Aerosols.* Oxford: Pergamon Press. (Reprinted Mineola, NY: Dover Press, 1989.)

Goldsmith, P. and F. G. May. 1966. In *Aerosol Science*, ed. C. N. Davies. London: Academic Press.

Hesketh, H. E. 1977. *Fine Particles in Viscous Media.* Ann Arbor, MI: Ann Arbor Science Publishers.

Hinds, W. C. 1999. *Aerosol Technology.* New York: John Wiley & Sons.

Johnson, D. L., D. Leith, and P. C. Reist. 1987. Drag on non-spherical, orthotropic aerosol particles. *J. Aerosol Sci.* 18:87–97.

Kasper, G., T. Niida, and M. Yang. 1985. Measurements of viscous drag on cylinders and chains of spheres with aspect ratios between 2 and 50. *J. Aerosol Sci.* 16:535–556.

Licht, W. 1988. *Air Pollution Control Engineering: Basic Calculations for Particulate Collection.* New York: Marcel Dekker, Inc.

Mazumder, M. K., R. E. Ware, J. D. Wilson, R. G. Renninger, F. C. Hiller, P. C. McLeod, R. W. Raible, and M. K. Testerman. 1979. SPART analyzer: Its application to aerodynamic size measurement. *J. Aerosol Sci.* 10:561–569.

McConalogue, D. J. and R. S. Srivastava. 1968. Motion of a fluid in a curved tube. *Proc. R. Soc. A.* 307:37–53.

Mercer, T. T. 1973. *Aerosol Technology in Hazard Evaluation.* New York: Academic Press.

Rader, D. J. 1990. Momentum slip correction factor for small particles in nine common gases. *J. Aerosol Sci.* 21:161–168.

Sartor, J. D. and C. E. Abbott. 1975. Prediction and measurement of the accelerated motion of water drops in air. *J. Appl. Meteorol.* 14(2):232–239.

Schlichting, H. 1979. *Boundary-Layer Theory.* New York: McGraw Hill.

Sheaffer, A. W. 1987. Drag on modified rectangular prisms. *J. Aerosol Sci.* 18:11–16.

Tsai, C. J. and D. Y. H. Pui. 1990. Numerical study of particle deposition in bends of a circular cross-section–laminar flow regime. *Aerosol Sci. Technol.* 12:813–831.

Waldmann, L. and K. H. Schmitt. 1966. Thermophoresis and diffusiophoresis of aerosols. In *Aerosol Science*, ed. C. N. Davies. London: Academic Press.

White, F. M. 1986. *Fluid Mechanics.* New York: McGraw-Hill.

Willeke, K. 1976. Temperature dependence of particle slip in a gaseous medium. *J. Aerosol Sci.* 7:381–387.

5 Physical and Chemical Changes in the Particulate Phase

WILLIAM C. HINDS

Department of Environmental Health Sciences, UCLA School of Public Health, Los Angeles, CA

INTRODUCTION

Aerosols, by their nature, are somewhat unstable in the sense that concentration and particle properties change with time. These changes can be the result of external forces, such as the loss of larger particles by gravitational settling, or they may be the result of physical and chemical processes that serve to change the size or composition of the particles. This chapter addresses the latter category of processes. They all involve mass transfer to or from a particle. This transfer may be the result of molecular transfer between the particle and the surrounding gas, for example, condensation, evaporation, nucleation, adsorption, absorption, and chemical reaction, or it may result from interparticle mass transfer, such as by coagulation.

Processes that cause physical or chemical changes in the particulate phase influence the particle size distribution of nearly all aerosols. These processes contribute in an essential way to the earth's hydrological cycle. They are involved in the formation of photochemical smog and are key to shaping the atmospheric aerosol size distribution. These processes play a significant role in many occupational aerosol exposures and in the operation of the condensation nuclei counters, described in Chapter 19. They are central to industrial aerosol processing and for the generation of test aerosols.

Condensation, thermal coagulation, and adsorption are related processes that rely on the diffusion of molecules or particles to a particle's surface. Evaporation is the opposite of condensation and is governed by the same laws. Reactions may be nongrowth processes that change the composition or density of an aerosol particle with little or no change in particle size. Because the processes discussed in this chapter are related and may be occurring simultaneously, it is necessary to look at each process separately in order to obtain an accurate picture of the changes that occur. Furthermore, these processes depend on particle size in a complex way, so I adopt a single particle approach for much of the analysis that follows. I rely on the concepts of mean free path and diffusion coefficient, which are defined in Chapter 4.

Definitions

The *partial pressure* of a vapor is a way of expressing the concentration of that vapor in a volume of a gas. It is the pressure the vapor would exert if it were the only component present.

Aerosol Measurement: Principles, Techniques, and Applications, Second Edition, Edited by Paul A. Baron and Klaus Willeke.
ISBN 0-471-35636-0

This pressure, expressed as a fraction of the ambient pressure, is the fractional concentration of the vapor. Air at 293 K [20°C or 68°F] and 50% relative humidity has a partial pressure of water vapor of 1.17 kPa [8.8 mmHg], which means that the air–water vapor mixture is 1.17/101 [= 8.8/760] = 1.2% water vapor on a volume basis.

The *vapor pressure* or *saturation vapor pressure* is a unique property of any liquid at a given temperature. It represents the minimum partial pressure of that liquid's vapor that must be maintained at the gas–liquid interface to prevent evaporation. This is a condition required for mass equilibrium, no net transfer of molecules at the liquid surface, that is, no net condensation or evaporation. Vapor pressure as defined here is for a flat liquid surface, but, as will be explained below, a slightly greater partial pressure is required to maintain mass equilibrium around an aerosol particle. The partial pressure of vapor in a sealed chamber containing a liquid will eventually reach the vapor pressure of the liquid at the temperature of the container.

The vapor pressure of water in kPa and mmHg at a temperature T in K is given by

$$p_s = \exp\left(16.7 - \frac{4060}{T-37}\right)\text{kPa} = \exp\left(18.72 - \frac{4062}{T-37}\right)\text{mmHg} \qquad (5\text{–}1)$$

for T from 273 to 373 K.

For aerosol condensation and evaporation processes, it is the ratio of the partial pressure of vapor to the saturation vapor pressure that is important. This ratio is called the *saturation ratio*, S_R. When the saturation ratio is equal to one, the mixture is described as *saturated*; when it is greater than one, the mixture is *supersaturated*; and when less than one, it is *unsaturated.*

Nucleation or *nucleated condensation* refers to the process of initial formation of a particle from vapor. This process is usually facilitated by the presence of small particles, called *condensation nuclei*, that serve as sites for condensation.

EXAMPLE 5–1

Saturated air coming from the ocean at 293 K [20°C] is carried by air currents up the side of a mountain to an altitude of 1 km. Assuming this represents adiabatic expansion to a pressure of 89 kPa [670 mmHg], what would be the saturation ratio of this air mass if no condensation occurred?

Answer: The absolute temperature of the air mass after an adiabatic expansion of saturated air from p_1 to p_2 is given by

$$T_2 = T_1\left(\frac{p_2}{p_1}\right)^{0.28} = 293\left(\frac{89.0}{101}\right)^{0.28} = \left[293\left(\frac{670}{760}\right)^{0.28}\right] = 283\text{ K}$$

At 283 K the saturation vapor pressure for water is given by Eq. (5–1):

$$p_s = \exp\left(16.7 - \frac{4060}{283-37}\right) = 1.22\text{kPa} = [9.1\text{mmHg}]$$

The saturation ratio is the ratio of the actual partial pressure of vapor, 2.34 kPa [17.6 mmHg] (by Eq. 5–1 at 293 K), to the saturation vapor pressure for the ambient temperature, 1.22 kPa [9.1 mmHg]:

$$S_R = \frac{2.34}{1.22} = \left[\frac{17.6}{9.1}\right] = 1.92$$

Adsorption is the process whereby vapor molecules attach to solid surfaces. It is most important for porous solids, such as activated charcoal, that have large surface areas. *Absorption* refers to the process of vapor molecules transferring from the gas phase to the liquid phase.

For aerosol particles *condensation* occurs when more vapor molecules arrive at a particle's surface than leave. It results in a net growth of the particle. *Evaporation* is the reverse of condensation and results in a net loss of molecules and a shrinkage of the particle.

The Kelvin Effect

Vapor pressure has been defined as the partial pressure required for mass equilibrium (no net evaporation or condensation) for a flat liquid surface. Because liquid aerosol particles have a sharply curved surface, a greater partial pressure is required to maintain mass equilibrium for a droplet than for a flat liquid surface at a given temperature. This increase in partial pressure of vapor required for mass equilibrium increases with decreasing particle size. This effect is called the *Kelvin effect*. The saturation ratio required for mass equilibrium (no net condensation or evaporation) for a droplet of diameter d_p is given by the Kelvin equation:

$$S_R = \exp\left(\frac{4\gamma M}{\rho_p R T d_p}\right) \tag{5–2}$$

where γ, M, and ρ_p are the surface tension, molecular weight, and density of the liquid, respectively, and R is the gas constant. Thus, 0.1 and 0.01 µm diameter water droplets require an environment with a saturation ratio of least 1.022 and 1.24, respectively, to prevent evaporation. Evaporation will occur if the saturation ratio is less than that given by Eq. 5–2, even if the saturation ratio is greater than one. Likewise, if the saturation ratio is greater than that required by the Kelvin equation, then condensation and growth will occur. For a given supersaturation, the minimum droplet size required to prevent evaporation is given by Eq. 5–2 and is referred to as the *Kelvin diameter* for that condition. The Kelvin effect is illustrated by the line labeled "pure water" in Figure 5–1.

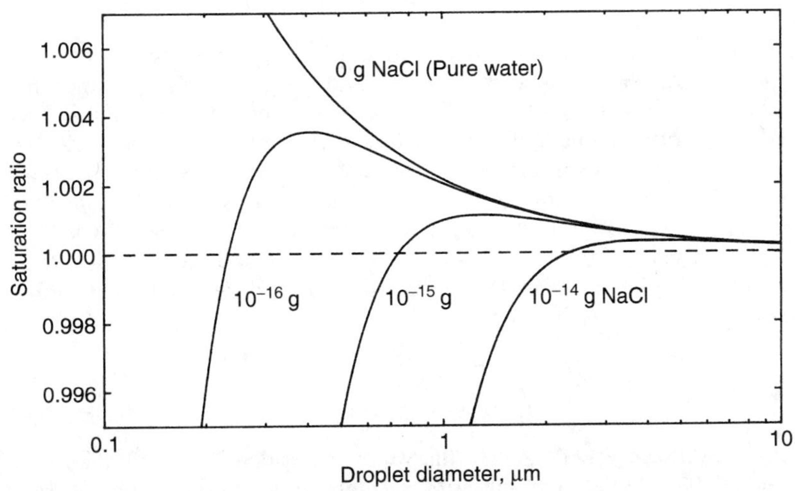

Fig. 5–1. Saturation ratio versus droplet size for pure water and droplets containing the indicated mass of sodium chloride at 293 K [20°C]. (Adapted from Hinds, 1999.)

EXAMPLE 5–2

What saturation ratio is required to prevent growth or evaporation of 0.05 μm pure water droplets?

Answer: Use the Kelvin equation, Eq. 5–2:

$$S_R = \exp\left(\frac{4\gamma M}{\rho_p R T d_p}\right)$$

where $\gamma = 0.0727\,\text{N/m}$; $M = 0.018\,\text{kg/mole}$; $\rho_p = 1000\,\text{kg/m}^3$; $R = 8.31\,\text{J/K·mole}$; $T = 293\,\text{K}$; and $d_p = 5 \times 10^{-8}\,\text{m}$.

Substituting in gives

$$S_R = \exp\left(\frac{4 \times 0.0727 \times 0.018}{1000 \times 8.31 \times 293 \times 5 \times 10^{-8}}\right) = \exp(0.043) = 1.044$$

CONDENSATION

Growth Rate

When a droplet of pure liquid is in a supersaturated environment that exceeds the requirement given by the Kelvin equation, the droplet grows by condensation of vapor on its surface. The rate of growth depends on the saturation ratio and the particle size. It is controlled by the rate of arrival of vapor molecules at the droplet surface. Initially the droplet will usually be less than the mean free path of the surrounding gas λ (0.066 μm at standard conditions; see Chapter 4) and the rate of arrival of vapor molecules is governed by the kinetic theory of gases. The growth rate, the rate of increase in droplet diameter, is given by Hinds (1999) as

$$\frac{d(d_p)}{dt} = \frac{2\alpha_c(p - p_d)}{\rho_p\sqrt{2\pi RT/M}} \quad \text{for } d_p < \lambda \qquad (5\text{–}3)$$

where α_c is the condensation coefficient, the fraction of arriving molecules that stick, approximately 0.04 (see Barrett and Clement, 1988); p is the partial pressure of vapor in the neighborhood of the droplet; and p_d is the partial pressure of vapor at the droplet surface as given by the Kelvin equation. In the application of Eq. 5–3 to obtain the growth rate in m/s [cm/s] requires that pressure be expressed in Pa[dyn/cm^2] (note that mmHg × 1330 = dyn/cm^2), density of the liquid is in kg/m^3 [g/cm^3], temperature is in K, and molecular weight is in kg/mole [g/mole]. The gas constant R is 8.31 J/K·mole [8.31×10^7 dyn·cm/K·mole].

Once a droplet's size is greater than the mean free path, the rate of arrival of vapor molecules is governed by the rate of molecular diffusion to the droplet surface. Under these conditions the rate of growth is given by

$$\frac{d(d_p)}{dt} = \frac{4D_v M}{\rho_p d_p R}\left(\frac{p_\infty}{T_\infty} - \frac{p_d}{T_d}\right)\phi \quad \text{for } d_p > \lambda \qquad (5\text{–}4)$$

where D_v is the diffusion coefficient of the vapor molecules, $2.4 \times 10^{-5}\,\text{m}^2/\text{s}$ [0.24 cm^2/s] for water vapor at 293 K [20°C]; the subscript ∞ refers to conditions removed from the particle; and the subscript d refers to conditions right at the particle surface. During rapid condensation ($S_R > 1.05$), the temperature of the droplet T_d will be greater than the surrounding air due to the release of heat of vaporization. The droplet temperature due to heating during condensation or cooling during evaporation can be estimated by (Hinds, 1999)

$$T_d = T_\infty + \frac{(6.65 + 0.345T_\infty + 0.0031T_\infty^2)(S_R - 1)}{1 + (0.082 + 0.00782T_\infty)S_R} \tag{5–5}$$

where T_∞ is in °C. The quantity p_d is evaluated at T_d by Eq. 5–1.

The last factor in Eq. 5–4 corrects for complications in the calculation of mass transfer by diffusion within one mean free path of the particle surface. This correction is known as the *Fuchs correction*. The Fuchs correction factor ϕ is given by Davies (1978) as

$$\phi = \frac{2\lambda + d_p}{d_p + 5.35(\lambda^2/d_p) + 3.42\lambda} \tag{5–6}$$

This factor can be omitted with little error for growing or evaporating droplets larger than about 2 µm.

Equations 5–3 to 5–5 apply only to pure materials, that is, single-component liquids without any dissolved salts or impurities. The growth rate for droplets less than the mean free path is independent of droplet size, but it is inversely proportional to droplet size for droplets larger than the mean free path.

EXAMPLE 5–3

What is the rate of growth by condensation for a 5 µm water droplet at a saturation ratio of 1.04 and a temperature of 293 K [20°C]?

Answer: Because 5 µm is greater than the mean free path (0.066 µm), we can use Eq. 5–4. We can neglect ϕ because $d_p > 2\,\mu m$.

$$\frac{d(d_p)}{dt} = \frac{4D_v M}{\rho_p R d_p}\left(\frac{P_\infty}{T_\infty} - \frac{P_d}{T_d}\right)$$

where $D_v = 2.4 \times 10^{-5}\,m^2/s$ [0.24 cm²/s]. Because $S_R < 1.05$, $T_d \approx T_\infty = 293\,K$ and $p_d \approx p_s$ at 293 K [20°C]. Saturation vapor pressure p_∞ is given by Eq. 5–1 for $T = 273 + 20 = 293\,K$:

$$p_s = \exp\left(16.7 - \frac{4060}{293 - 37}\right) = 2.318\,kPa = 2318\,Pa$$

$$p_\infty = 1.04 \times p_s = 1.04 \times 2.318 = 2.411\,kPa = 2411\,Pa$$

Substituting in gives

$$\frac{d(d_p)}{dt} = \frac{4(2.4 \times 10^{-5})0.018}{1000(8.31)5 \times 10^{-6}}\left(\frac{2410}{293} - \frac{2318}{293}\right)$$
$$= 4.159 \times 10^{-5}(0.317) = 1.32 \times 10^{-5}\,m/s = 13.2\,\mu m/s$$

Time Required for Growth

The time required for a droplet to grow from d_1 to d_2 can be obtained by integrating Eq. 5–4 over the size limits.

$$t = \frac{\rho_p R(d_2^2 - d_1^2)}{8D_v M\left(\frac{p_\infty}{T_\infty} - \frac{p_d}{T_d}\right)} \quad \text{for } d_1 >> \lambda \tag{5–7}$$

NUCLEATION

Homogeneous

The preceding section describes the growth process for pure materials once the droplets have been formed. The initial formation of the droplet from vapor is a more complicated process. Droplets can be formed in the absence of condensation nuclei, but this process, called *homogenous nucleation* or *self-nucleation*, requires large saturation ratios, usually in the range of 2 to 10, which normally occur only in special laboratory or chemical process situations. Pure water vapor at 293 K [20°C] and at a saturation ratio of 3.5 or greater spontaneously forms droplets by homogeneous nucleation. This corresponds to a Kelvin diameter of 0.0017 μm and suggests that molecular clusters of about 90 molecules are necessary for this process. A detailed description of homogenous nucleation is given by Seinfeld and Pandis (1998).

Heterogeneous

The more common formation mechanism is nucleated condensation or heterogenous nucleation. This process relies on existing submicrometer particles, called *condensation nuclei*, to serve as sites for condensation. Our natural atmosphere contains thousands of these nuclei in each cubic centimeter of air. To a first approximation, insoluble nuclei serve as passive sites on which condensation occurs for supersaturated conditions. Under supersaturated conditions, a solid nucleus with a wettable surface will have on its surface an adsorbed layer of vapor molecules. If the nucleus has a diameter greater than the Kelvin diameter for a particular condition of supersaturation, the nucleus "looks like" a droplet to surrounding vapor molecules and vapor will condense on its surface. Once condensation starts, droplet growth continues as described by Eqs. 5–3 and 5–4.

The situation with soluble nuclei is more complex and more important. Our normal atmosphere contains large numbers of soluble nuclei, formed as the solid residue left behind after the water has evaporated from a droplet containing dissolved material. Many are sodium chloride nuclei formed from droplets of sea water created by the action of waves and bubbles in the oceans. Because these soluble nuclei have a strong affinity for water, they facilitate the initial formation of droplets and enable their growth to occur at lower saturation ratios than would be the case for insoluble nuclei.

Because of the complex effect the presence of dissolved salt has on the rate of growth of a droplet, Eqs. 5–3 and 5–4 cannot be used to determine growth rates for such droplets. The stabilization time for droplets containing salt is described by Ferron and Soderholm (1990). In general dissolved salts increase the rate of growth and decrease the rate of evaporation compared with that for pure liquids. As a droplet grows by the addition of water vapor, the concentration of salt becomes more and more dilute. Consequently, it is convenient to characterize the amount of salt in a droplet not by its concentration but by the mass of salt in the droplet, a quantity that remains constant during condensation and evaporation processes. The mass of salt is also equal to the mass of the original nucleus upon which the droplet formed.

When a dissolved salt is present in a droplet there are two competing effects at work as the droplet evaporates or grows. As a droplet evaporates the concentration of salt increases, because only the water leaves. This enhances the affinity of the dissolved salt to hold water in the droplet. The other effect is the Kelvin effect that results in an increase in the equilibrium vapor pressure required for a droplet as its size decreases. The relationship between saturation ratio and particles size for droplets containing dissolved salts is illustrated in Figure 5–1 by the three lines, called *Kohler curves*, labeled with their mass of dissolved salt.

Equilibrium Conditions

As with pure materials, the region above a given curve in Figure 5–1 represents a growth region and below the curve represents an evaporation region. Thus, if the saturation ratio is greater than 1.002, any droplet (or nuclei) with more than 10^{-15} g of sodium chloride will grow to a large droplet, although its growth rate will slow as it gets larger, as predicted by Eq. 5–4. When environmental conditions and particle size give a location on Figure 5–1 that is *below and to the left* of the peak for a given curve, a droplet will either grow or evaporate until it reaches the curve. This portion of the line represents a true equilibrium region, and the droplet will remain at that size as long as environmental conditions stay constant. This is true even if the saturation ratio is less than 1.0. Thus, there are a large number of particles in the atmosphere that will experience an increase in their size with an increase in relative humidity and a decrease with a decrease in relative humidity. The line for pure water does not have this type of equilibrium region. It represents only a demarcation between the growth (above) and evaporation (below) regions. As droplets continue to grow the concentration of dissolved salts decreases, eventually reaching the point where the droplets behave the same as pure water, and their curves in Figure 5–1 merge with that for pure water.

EVAPORATION

Rate of Evaporation

The process of evaporation of a pure liquid droplet (no dissolved salts) is similar to the process of growth except that it proceeds in the opposite direction. Evaporation will occur when the ambient partial pressure of vapor is less than the saturated vapor pressure ($p < p_s$). The rate of particle shrinkage due to evaporation can be predicted by Eq. 5–4. During evaporation, the term in parentheses will be negative, giving a negative growth rate, which represents shrinkage due to evaporation.

For volatile particles such as water or alcohol, the quantity p_d must be evaluated at the cooler conditions prevailing at the droplet surface, T_d, which is given by Eq. 5–5.

For particles larger than about 50 μm an additional correction must be included to account for the disruption in the diffusion of vapor away from the droplet surface caused by the settling of the droplet. This effect increases the rate of evaporation of 50 and 100 μm droplets by 10% and 31%, respectively. More detailed information on this effect is given by Davies (1978) and Fuchs (1959).

Drying Time

Figure 5–2 gives droplet lifetimes or drying times, that is, the time for evaporation from an initial diameter to zero for pure water droplets at three conditions of relative humidity. The graph was obtained by numerical integration of Eq. 5–4 from the initial size to zero. For particles initially larger than about 2 μm at standard conditions, ϕ in Eq. 5–4 can be neglected and the equation integrated to give droplet lifetimes.

$$t = \frac{R\rho_p d^2}{8 D_v M\left(\dfrac{p_d}{T_d} - \dfrac{p_\infty}{T_\infty}\right)} \quad \text{for initial } d_p > 2\,\mu\text{m} \tag{5–8}$$

EXAMPLE 5–4

Water droplets 60 μm in diameter are sprayed into 50% relative humidity air at 293 K [20°C]. How long before they evaporate completely?

Answer: Use Eq. 5–8:

$$t = \frac{R\rho_p d^2}{8D_v M\left(\frac{p_d}{T_d} - \frac{p_\infty}{T_\infty}\right)}$$

where

$R = 8.31$ J/K·mole
$\rho_p = 1000\,\text{kg/m}^3$
$d = 6 \times 10^{-5}$ m
$D_v = 2.4 \times 10^{-5}\,\text{m}^2/\text{s}$
$T_\infty = 293$ K
$p_\infty = 0.5 \times 2.34$ kPa $= 1.17$ kPa $= 1170$ Pa
T_d is given by Eq. 5–5
p_d is given by Eq. 5–1 at 286.4 K.
Substituting into the original equation gives

$$t = \frac{8.31 \times 1000 \times (6 \times 10^{-5})^2}{8 \times (2.4 \times 10^{-5}) \times 0.018\left(\frac{1523}{286.4} - \frac{1170}{293}\right)} = 6.5\text{s}$$

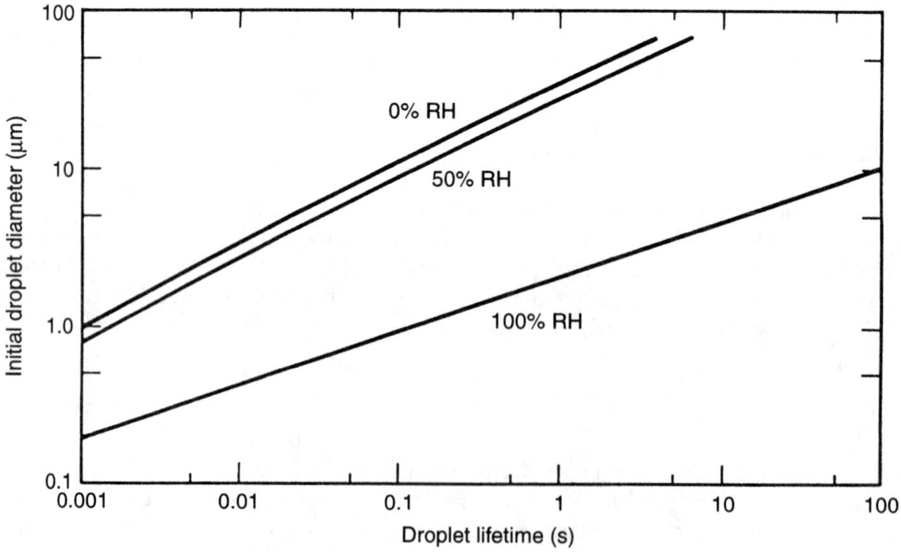

Fig. 5–2. Drying times for pure water droplets at 293 K [20°C]. RH, relative humidity. (From Hinds, 1999, reprinted with permission of John Wiley & Sons, Inc.)

TABLE 5–1. Droplet Lifetimes for Selected Materials[a]

Initial Droplet Diameter (μm)	Droplet lifetimes (s)			
	Ethyl Alcohol	Water	Mercury	Dioctyl Phthalate
0.01	4×10^{-7}	2×10^{-6}	0.005	1.8
0.1	9×10^{-6}	3×10^{-5}	0.3	740
1	3×10^{-4}	0.001	1.4	3×10^{4}
10	0.03	0.08	1200	2×10^{6}
40	0.4	1.3	2×10^{4}	4×10^{7}

[a] Calculated by Eq. 5–4 for vapor-free air at 293 K [20°C].

Source: Adapted from Hinds (1999).

Table 5–1 gives droplet lifetimes for four materials at standard conditions. It illustrates the wide range of droplet lifetimes for different materials. The effect of material properties on droplet lifetime is, to a first approximation, proportional to ρ_p/D_vM.

COAGULATION

Coagulation is an aerosol growth process that results from the collision of aerosol particles with each other. If the collisions are the result of Brownian motion, the process is called *thermal coagulation*; if they are the result of motion caused by external forces, the process is termed *kinematic coagulation*. Thermal coagulation is in some ways analogous to growth by condensation except that it is other particles diffusing to a particle's surface rather than molecules that causes the growth. It differs from condensation in that a supersaturation is not required, and it is a one-way process of growth with no equivalent process corresponding to evaporation. The result of many collisions between particles is an increase in particle size and a decrease in aerosol number concentration. In the absence of any loss or removal mechanisms there is no change in mass concentration as a result of coagulation.

To understand the process we look first at a simplified description of coagulation called *simple monodisperse coagulation* or *Smoluchowski coagulation*. The latter is named after the person who developed the theory in 1917. This approach illustrates the process well, is useful for analyzing many situations, and is the basis for further refinements.

Simple Monodisperse Coagulation

For simple monodisperse coagulation we make the simplifying assumptions that the particles are monodisperse, they will stick if they contact one another, and they grow slowly. The latter two are valid assumptions for most aerosol particles and situations. Aerosol particles exhibit Brownian motion and diffuse like gas molecules, but their diffusion occurs at a much slower pace; consequently, the diffusion coefficients for aerosol particles can be a million times smaller than those for gas molecules.

The derivation developed by Smoluchowski is based on the diffusion of other particles to the surface of each particle (see Hinds, 1999). It gives the rate of change (decrease) in aerosol number concentration as

$$\frac{dN}{dt} = -KN^2 \tag{5–9}$$

where N is particle number concentration and K is the coagulation coefficient. For particles larger than the gas mean free path, K is given by

TABLE 5–2. Coagulation Coefficients for Selected Particle Sizes at 293 K [20°C][a]

Particle Diameter (μm)	Coagulation Coefficient (m^3/s)
0.05	9.9×10^{-16}
0.1	7.2×10^{-16}
0.5	5.8×10^{-16}
1	3.4×10^{-16}
5	3.0×10^{-16}

[a] For coagulation coefficients in cm^3/s, multiply table values by 10^6.

Source: Adapted from Hinds (1999). Includes additional correction factors; see Hinds (1999).

$$K = 4\pi d_p D = \frac{2kTC_c}{3\eta} \quad \text{for } d_p > \lambda \tag{5–10}$$

where D is the particle diffusion coefficient m^2/s [cm^2/s], η is the gas viscosity in Pa·s [g/cm·s], and k is the Boltzmann constant, 1.38×10^{-23} J/K [1.38×10^{-16} dyn·cm/K]. The coagulation coefficient has units of m^3/s [cm^3/s] for number concentration expressed in particles/m^3 [particles/cm^3]. The coagulation coefficient is only slightly dependent on particle size being proportional to slip correction factor C_c. Table 5–2 gives coagulation coefficients for different size particles at standard conditions. In the usual situation, the extent of particle size increase is sufficiently limited that the coagulation coefficient can be considered a constant, and the rate of coagulation is proportional only to number concentration squared. Thus, coagulation is a rapid process at high number concentration and a slow one at low concentrations.

As a practical matter, the net effect of coagulation over some period of time is a more useful quantity than the rate of coagulation. The change in number concentration over a period of time t is obtained by integrating Eq. 5–9 to get

$$N(t) = \frac{N_0}{1 + N_0 K t} \tag{5–11}$$

where $N(t)$ is the number concentration at time t and N_0 is the initial number concentration. Number concentration must be expressed in particles/m^3 for K in m^3/s [particles/cm^3 for K in cm^3/s].

As number concentration decreases particle size increases, but, for a contained system with no losses, particle mass will remain constant. If number concentration decreases to one half of its original value, then the same mass (and volume) will be contained in half as many particles, so each particle will have twice its original mass (and volume). For liquid particles, particle size is proportional to the cube root of particle volume, and consequently it is also proportional to the inverse cube root of number concentration.

$$d(t) = d_0 \left(\frac{N_0}{N(t)} \right)^{1/3} \tag{5–12}$$

Thus an eightfold reduction in particle number concentration results in a doubling of particle size. Equations 5–11 and 5–12 can be combined to give a more direct expression for the change in particle size due to coagulation over a period of time t:

$$d(t) = d_0 (1 + N_0 K t)^{1/3} \tag{5–13}$$

Equations 5–12 and 5–13 are correct for liquid droplets and approximately correct for solid particles that form compact clusters. Table 5–3 gives the time required for various

TABLE 5–3. Time Required for Selected Coagulation Processes[a]

Initial Number Concentration (m^{-3})	Time for Number Concentration to Halve (s)	Time for Particle Size to Double (s)
10^{18}	0.002	0.014
10^{16}	0.2	1.4
10^{14}	20	140
10^{12}	2000 (33 min)	14,000 (4 h)
10^{10}	200,000 (55 h)	1,400,000 (16 d)

[a] Assumes simple monodisperse coagulation with K = $5 \times 10^{-16} m^3/s$ [$5 \times 10^{-10} cm^3/s$].

Source: Adapted from Hinds 1999.

initial concentrations to reach one half their number concentration and the time for particle size to double. It is apparent from Table 5–3 that whether or not coagulation can be neglected depends on the concentration and time scale under consideration. Thus, over a period of a few minutes coagulation is only important if particle number concentration exceeds $10^{12}/m^3$.

Polydisperse Coagulation

The previous description of coagulation is accurate enough for a wide variety of situations, but it requires the assumption of a monodisperse aerosol. In the real case we usually have a polydisperse aerosol, and the situation is more complicated. Because the coagulation process is governed by the rate of diffusion of particles to the surface of each particle, the process is enhanced when small particles with their high diffusion coefficients diffuse to a large particle with its large surface. A 10-fold difference in particle size produces a 3-fold increase in coagulation rate, and a 100-fold difference results in more than a 25-fold increase in coagulation rate. To use Eqs. 5–11 or 5–13 for polydisperse aerosols requires the use of numerical methods because the coagulation for every combination of particle sizes has a different value of K and has to be calculated separately (Zebel, 1966). For the case of coagulation of an aerosol with a lognormal size distribution having a count median diameter (CMD) and a geometric standard deviation σ_g, an equation derived by Lee and Chen (1984) can be used to calculate the average coagulation coefficient $\overline{K}$:

$$\overline{K} = \frac{2kT}{3\eta}\left\{1+\exp(\ln^2\sigma_g)+\left(\frac{2.49\lambda}{CMD}\right)\left[\exp(0.5\ln^2\sigma_g)+\exp(2.5\ln^2\sigma_g)\right]\right\} \tag{5–14}$$

This value of $\overline{K}$ can be used in place of K in Eq. 5–11 to predict the change in number concentration over a period of time t for which there is only a modest change in CMD. Equation 5–13 can be used with $\overline{K}$ to predict the increase in CMD over a time period for which $\overline{K}$ is approximately constant. For this type of calculation it is reasonable to assume that σ_g remains constant for modest changes in particle size. For large changes in particle size calculation can be done as a series of steps, each with a constant but different value of $\overline{K}$. Figure 5–3 shows the effect of polydisperse coagulation on number concentration and median size for the indicated initial condition.

Kinematic Coagulation

Kinematic coagulation is a coagulation process whereby the relative motion between particles is created by external forces rather than by Brownian motion. Brief descriptions of several such mechanisms are given below. In all cases the greater the particle number con-

EXAMPLE 5–5

A. An iron oxide fume has an initial number concentration of $10^{13}/m^3$. Assuming the aerosol is monodisperse with a diameter of 0.2 μm, what will be the number concentration and particle size after 2 minutes? Assume standard conditions.
B. Repeat the above example for a polydisperse aerosol having a CMD of 0.2 μm and σ_g of 2.0. Assume σ_g remains constant.

Answer: A. Use Eq. 5–11:

$$N(t)=\frac{N_0}{1+N_0Kt}$$

where:

$$N_0=10^{13}/\mathrm{m}^3$$

$$K=\frac{4kTC_c}{3\eta}=\frac{4(1.38\times10^{-23})293\times1.88}{3(1.81\times10^{-5})}=5.6\times10^{-16}\,\mathrm{m}^3/\mathrm{s}$$

$$t=120\mathrm{s}$$

substituting in

$$N(t)=\frac{10^{13}}{1+10^{13}(5.6\times10^{-16})120}=5.98\times10^{12}/\mathrm{m}^3$$

Use Eq. 5–12 or 5–13 to determine the change in diamete:

$$d(t)=d_0\left(\frac{N_0}{N(t)}\right)^{1/3}=0.2\left(\frac{10^{13}}{5.98\times10^{12}}\right)^{1/3}=0.24\mu\mathrm{m}$$

B. Use Eq. 5–14 to get $\overline{K}$

$$\overline{K}=\frac{2kT}{3\eta}\left[1+\exp(\ln^2\sigma_g)+\left(\frac{2.49\lambda}{CMD}\right)[\exp(0.5\ln^2\sigma_g)+\exp(2.5\ln^2\sigma_g)]\right]$$

$$\overline{K}=\frac{2(1.38\times10^{-23})\times293}{3(1.81\times10^{-5})}\times\left[1+\exp[\ln^2(2.0)]+\left(\frac{2.49\times0.066}{0.2}\right)(\exp[0.5\ln^2(2.0)]\right.$$

$$\left.+\exp[2.5\ln^2(2.0)])\right]=1.489\times10^{-16}[6.986]=1.04\times10^{-15}\mathrm{m}^3/\mathrm{s}$$

Substituting into Eq. 5–11 using $\overline{K}$ instead of K gives

$$N(t)=\frac{N_0}{1+N_0\overline{K}t}=\frac{10^{13}}{1+10^{13}(1.04\times10^{-15})120}=4.45\times10^{12}/\mathrm{m}^3$$

$$CMD_2=0.2\left(\frac{10^{13}}{4.45\times10^{12}}\right)^{1/3}=0.26\mu\mathrm{m}$$

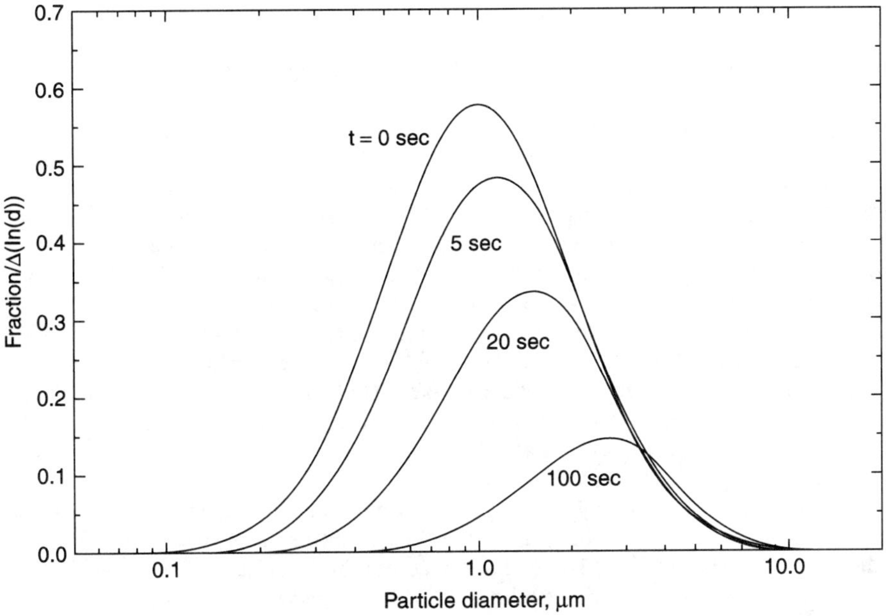

Fig. 5–3. Effect of coagulation on particle size distribution. $N_0 = 10^{14}/m^3$ $[10^8/cm^3]$, initial count median diameter = 1.0 μm, and initial geometric standard deviation = 2.0. (From Hinds, 1999, reprinted with the permission of John Wiley & Sons, Inc.)

centration, the greater the rate of coagulation. In general there are no simple equations that describe these process in a complete way. More detailed information is given by Fuchs (1964) and Hinds (1999).

Because particles of different sizes settle at different rates there is a relative motion between settling particles of different sizes. The aerodynamics of the collision process is complicated, and the collision efficiency is low except for the case of very large particles, such as raindrops, settling through micrometer-sized or larger aerosol particles.

A similar process occurs when particles are projected through an aerosol at high velocity. This is an important mechanism for the capture of particles by spray droplets in certain kinds of wet scrubbers used for gas cleaning.

Gradient or shear coagulation occurs for particles moving in a flow velocity gradient. Particles on slightly different streamlines in a velocity gradient travel at different velocities and faster particles eventually overtake the slower ones. If the particles are big enough, particle contact occurs by interception.

In turbulent flow, particles follow a complex path having strong velocity gradients. Relative motion between particles arises from these gradients and from inertial projection of the particles. The resulting coagulation is called *turbulent coagulation*. This mechanism is most effective when the turbulent eddy size is the same order of magnitude as the particle stopping distance. This mechanism is only important for particles larger than about 1 μm. Generally, the more intense the turbulence, the more coagulation that results from this mechanism.

Finally, there is acoustic coagulation where intense sound waves are used to create relative motion between particles. Depending on their size, particles respond to high intensity sound waves differently; large particle may be unaffected, whereas small particles oscillate with the sound waves. The relative motion that results leads to collisions, and the process is called *acoustic coagulation*. Generally sound pressure levels exceeding 120 dB are required to produce significant coagulation.

REACTIONS

Compared with bulk materials, aerosol particles have very high ratios of surface area to mass. For example, 1 g of standard density material (1000 kg/m^3) when divided into 0.1 μm particles has a surface area of 60 m^2. Because of their large specific surface, surface area per gram, aerosols participate actively in many kinds of interaction between gas molecules and liquid or solid particles. Particles can undergo three kinds of reactions: reactions between compounds within a particle, reactions between particles of different chemical composition, and reactions between the particle and one or more chemical species in the surrounding gas phase. In the first case reactions are governed by the usual chemical kinetics. The second case is most likely controlled by the rate of arrival of other particles, which is described by the coagulation process given above. Once dissimilar particles contact each other, reactions proceed by chemical kinetics. The third case may be controlled by the rate of arrival of the appropriate gas molecules at the particle surface. The rate of arrival of gas molecules is described by the condensation growth equations given in this chapter. Absorption and adsorption are related processes that also have as one of their necessary steps the arrival of gas molecules at the particle surface.

These processes can be thought of as having three mass transfer steps in series, any one of which may be the rate-controlling step. First there is diffusion of specific gas molecules to the surface of the particle. Next is the transfer across the interface or reaction at the interface, and finally there is diffusion into the solid or liquid particle.

Reaction

In the case of a chemical reaction between the suspending gas and a particle, any of the three steps given above may control the rate of reaction. For solid particles diffusion into the interior will be relatively slow even though the distances involved are small. Diffusion into the interior of liquid particles will be more rapid and may be augmented by internal circulation. If the reaction is controlled by the rate of arrival of gas molecules at the particle surface, then the maximum rate of reaction is given by

$$R_R = \frac{2\pi d_p D_v p}{kT} \quad \text{for } d_p > \lambda \tag{5–15}$$

where R_R is the rate of reaction in molecules/s. This is equivalent to a condensation process (Hinds, 1999) under uniform temperature conditions. This situation is called a *diffusion-controlled reaction*. The process can continue until all molecules of the particle have reacted.

Absorption

The process whereby gas molecules dissolve in a liquid droplet is called *absorption*. In this process the transfer at the interface is usually not controlling, but diffusion in either the gas phase or liquid phase may be. The process can continue until the limit of solubility of the gas in the liquid is reached. This limit may change with temperature or the presence of other dissolved components.

Adsorption

Adsorption is the transfer of gas molecules from the surrounding gas to a solid surface. There are two types of adsorption that can occur on the surface of a solid particle: physical adsorption, or physisorption; and chemical adsorption, or chemisorption. Physisorption is a physical process where gas molecules are held to a particle's surface by van der Waals forces. It occurs for all gases when the ambient temperature is below their critical temperature. It is a rapid and readily reversible process. Because the adsorption process is rapid, the diffusion of gas molecules to the particle surface is usually the rate-limiting step. The relationship between the amount of adsorbed gas and the partial pressure of the gas or vapor at a given temperature is called the *adsorption isotherm*. Physisorption is usually not significant if the saturation ratio is below 0.05, but can lead to an adsorbed layer several molecules thick when the saturation ratio is 0.8 or greater. For a particle in adsorption equilibrium, a reduction in the partial pressure of the vapor will lead to a transfer of adsorbed vapor molecules from the particle's surface to the gas.

The process of adsorption is similar to the process of condensation. Highly porous materials, like activated carbon, have enormous surface areas and contain numerous small pores and capillaries that facilitate condensation on their surface and inhibit evaporation. The isotherms for highly porous materials will differ significantly from those for smooth solids.

Chemisorption is similar except that chemical bonds are formed to hold the gas molecules on the particle's surface. It can occur above or below the critical temperature of the gas. In chemisorption only a monolayer can form, and, unlike physisorption, the process is not easily reversible because the chemical bonds are much stronger than van der Waals forces. Either the rate of gas phase diffusion or the rate of reaction can control the rate of this process. The rate of transfer slows as a complete monolayer is approached. In some cases molecules are first held to the surface by physisorption and then slowly react to attach by chemisorption. In other cases a physisorption layer may form on top of a chemisorption layer.

REFERENCES

Barrett, J. C. and C. F. Clement. 1988. Growth rates for liquid drops. *J. Aerosol Sci.* 9:223–242.

Davies, C. N. 1978. Evaporation of airborne droplets. In *Fundamental of Aerosol Science*, ed. T. Shaw. New York: John Wiley & Sons.

Ferron, G. A. and S. C. Soderholm. 1990. Estimation of the times for evaporation of pure water droplets and for stabilization of salt solution particles. *J. Aerosol Sci.* 21:415–429.

Fuchs, N. A. 1959. *Evaporation and Droplet Growth in Gaseous Media*. Oxford: Pergamon Press.

Fuchs, N. A. 1964. *The Mechanics of Aerosols*. Oxford: Pergamon Press.

Hinds, W. C. 1999. *Aerosol Technology*, 2nd Ed. New York: John Wiley & Sons.

Lee, K. W. and H. Chen. 1984. Coagulation rate of polydisperse particles. *Aerosol Sci. Technol.* 3:327–334.

Seinfeld, J. H. and S. N. Pandis. 1998. *Atmospheric Chemistry and Physics*. New York, Wiley.

Zebel, G. 1966. Coagulation of aerosols. In *Aerosol Science*, ed. C. N. Davies. London: Academic Press.

6 Size Distribution Characteristics of Aerosols

WALTER JOHN
Particle Science, Walnut Creek, CA

BASIC CONCEPTS OF PARTICLE SIZE AND SIZE DISTRIBUTIONS

Definitions of Particle Size

Size is probably the most fundamental parameter describing aerosol particles. To satisfy the definition of an aerosol, the particles must be suspended in a gas. This implies that the particles are small enough to be suspended for an appreciable time, an arbitrary criterion. Conventionally, the upper size limit is considered to be about 100 μm; such particles have a settling velocity of 0.25 m/s and experience a drag force deviating appreciably from that of Stoke's law. The lower size limit is likewise arbitrary, usually taken to be a few nanometers, the size of molecular clusters. These size limits encompass a size range of five decades. Over this enormous range, the properties and behavior of aerosols change greatly.

For a spherical particle of unit density, the size can be simply characterized by the geometric diameter. For particles of arbitrary shape and density, an equivalent diameter is used. The *aerodynamic diameter* is defined as that of a spherical particle of unit density having the same settling velocity as the particle in question. The aerodynamic diameter is useful for particles in the inertial size range (i.e., those larger than about 0.5 μm). Particles smaller than about 0.5 μm undergo Brownian diffusion and are characterized by the *diffusive diameter*, the diameter of a particle of unit density having the same rate of diffusion as the particle in question. The *Stokes diameter* is the diameter of a spherical particle having the same density and settling velocity as the particle in question. An *optical diameter* is defined as the diameter of a calibration particle having the same response in an instrument detecting particles by their interaction with light. There are a number of diameter definitions used to describe particles measured by microscopy.

The appropriate particle size definition depends primarily on the type of measurement made. For example, the aerodynamic diameter would be used to analyze the data from a cyclone, a cascade impactor, or an Aerodynamic Particle Sizer (APS; ***TSI***).* The diffusive diameter would be used for a diffusion battery measurement, the Stokes diameter for a

* See Appendix I for full manufacturer addresses referenced by the italicized three-letter codes.

Aerosol Measurement: Principles, Techniques, and Applications, Second Edition, Edited by Paul A. Baron and Klaus Willeke.
ISBN 0-471-35636-0

differential mobility analyzer, and, obviously, the optical diameter with an optical particle counter.

Size Distributions

The particles in an aerosol are seldom uniform in size. Such an aerosol would be said to be *monodisperse*. The most highly monodisperse aerosols are those generated in a laboratory, typically with a spread in particle diameter of a few percent. The spread is more precisely characterized by the geometric standard deviation, defined below. Conventionally, a distribution with a spread of less than about 10% to 20% is considered monodisperse. Aerosols that have a larger range in size are said to be *polydisperse*. Both monodisperse and polydisperse aerosols consist of particles with sizes distributed over a certain range. It is necessary to have a mathematical description of such size distributions in order to be able to analyze them quantitatively.

The simplest size distribution would be a histogram of the number of particles in successive size intervals. Data for such a histogram could be obtained by sampling an aerosol with a cascade impactor and counting the number of particles on each stage with the aid of a microscope. The size intervals would be determined from the known cutpoints of each stage. Finer size intervals would be afforded by the use of an instrument such as the APS. With sufficiently fine intervals, the distribution would become a *differential size distribution.* Because the dependent variable or the ordinate of the plot is the number of particles, such a distribution is called a *number distribution.* If $N(d_p)$ is the number of particles in the size interval from d_p to $d_p + dd_p$, where d_p is the particle diameter, the number distribution is

$$dN = N(d_p)dd_p \tag{6–1}$$

Because the particle diameter typically ranges over several orders of magnitude, it is convenient to use $d\ln d_p$ (or $d\log d_p$) for the size interval, and the number distribution becomes

$$dN = N(d_p)d\ln d_p \tag{6–2}$$

Similarly, let $S(d_p)$ be the total surface area of the particles in the same differential size interval. Then the surface area size distribution is

$$dS = S(d_p)d\ln d_p \tag{6–3}$$

Two additional size distributions are frequently used, the *volume distribution* and the mass distributions:

$$dV = V(d_p)d\ln d_p$$
$$dM = M(d_p)d\ln d_p \tag{6–4}$$

The data for one of the above distributions might be obtained directly by an appropriate particle sampler, for example, the mass distribution might be obtained by weighing the particle deposit on each of the stages of a cascade impactor. Alternatively, the number distribution might be obtained directly by an instrument such as an optical particle counter or an APS. Then the surface distribution could be obtained by transforming the number distribution, that is, by taking

$$S(d_p) = N(d_p)\cdot \pi d_p^2 \tag{6–5}$$

Likewise, the volume and *mass distributions* can be obtained from

$$V(d_p) = N(d_p)\cdot \frac{\pi d_p^3}{6}$$
$$M(d_p) = V(d_p)\cdot \rho \tag{6–6}$$

where ρ is the particle density.

While particle size distributions can be simply tabulated or plotted, it is convenient to fit the data to a function allowing the distribution to be characterized by only a few parameters. A variety of functions have been used for this purpose. Number distributions are frequently fitted to a power law. Mass distributions are commonly fitted to a lognormal function. The lognormal function is simply obtained from the normal function by using logarithmic variables. The lognormal has a peak, a peak width, and its most notable feature, a tail for large values of the independent variable, in this case, the particle diameter. Aerosol size distributions from many different sources have been found to fit the *lognormal distribution*. The lognormal number distribution is

$$dN = \frac{N}{\sqrt{2\pi}\ln\sigma_g}\exp\left[\frac{-(\ln d_p - \ln CMD)^2}{2(\ln\sigma_g)^2}\right]d\ln d_p \tag{6–7}$$

where N is the total number of particles, CMD is the count (number) median diameter (defined below), and σ_g is the geometric standard deviation, given by

$$\ln\sigma_g = \left[\frac{\int_0^\infty(\ln d_p - \ln d_g)^2 dn}{N-1}\right]^{1/2} \tag{6–8}$$

σ_g is a measure of the width of the peak; if $d_{84\%}$ and $d_{16\%}$ are the diameters that include 84% and 16% of all the particles with diameters from zero to the diameter in question, then

$$\sigma_g = \left(\frac{d_{84\%}}{d_{16\%}}\right)^{1/2} \tag{6–9}$$

In Eq. 6–8, d_g is the geometric mean diameter, defined by

$$\ln d_g = \frac{1}{N}\int_0^\infty(\ln d_p)dn \tag{6–10}$$

For a lognormal distribution, the count median diameter, CMD, is equal to d_g. The lognormal function has a number of remarkable features. If the particle number distribution is lognormal, then the surface and volume distributions are also lognormal and will be given by Eq. 6–7 by replacing N by S or V, the total surface and volume, respectively, and by replacing CMD by SMD or VMD, the surface median and volume median diameters, respectively. The median diameters are related:

$$\begin{aligned} SMD &= CMD\exp(2\ln^2\sigma_g) \\ VMD &= CMD\exp(3\ln^2\sigma_g) \end{aligned} \tag{6–11}$$

The diameter corresponding to the peak of the lognormal distribution is called the *mode* diameter and is given by

$$d_{mode} = CMD\exp(\ln^2\sigma_g) \tag{6–12}$$

If, for example, experimental data for an aerosol mass distribution is found to fit the lognormal distribution, the distribution can be completely characterized by the mode diameter (or by the mass median diameter), the geometric standard deviation, and the total mass (integral of the differential mass distribution or the area under the curve). It is not unusual for the size distribution to have more than one mode, especially when there is more than one source of the aerosol. Then the distribution may be fit by a sum of lognormals.

A measurement of the aerosol concentration as a function of particle size may also be analyzed as discussed above. If the aerosol is speciated (i.e., elemental or compound concentrations are obtained), the chemical concentrations versus particle size can be treated as size distributions. For more comprehensive treatments of the analysis of size distributions, the reader is referred to Chapter 22 and to Hinds (1999).

Use of Size Distribution Functions

The lognormal function is the most widely used for particle size distributions. However, there appears to be no justification for such use other than the fact that good fits are obtained to a variety of empirical data. In addition, the lognormal distribution has the very useful mathematical characteristics discussed above. The modified gamma distribution has also been used for atmospheric aerosols (Pruppacher and Klett, 1980). The Weibull distribution fits fragmentation aerosols somewhat better than the lognormal according to Brown and Wohletz (1995). The Rosin-Rammler (1933) distribution is related to the Weibull distribution. In the modeling of the evolution of atmospheric aerosol, complexities are encountered in condensation and coagulation processes requiring numerical rather than analytical treatment of the size distributions.

AMBIENT AEROSOLS

Introduction

Particles in the ambient atmosphere have diameters spanning the entire range within the definition of an aerosol. The particle sizes are determined by the formation processes and subsequent physical and chemical reactions in the atmosphere. Particle size is a key parameter in the transport and removal of the ambient aerosol. The principal effects of aerosol, including the respiratory health hazard, visibility reduction, and deposition to surfaces depend on particle size. Therefore, the measurement and interpretation of particle size distributions in the atmosphere are essential to the overall understanding of the origin and the effects of ambient aerosol.

An early and widely used representation of the ambient particle size distribution was that of Junge (1963), who fitted the plot of the logarithm of particle number concentration vs. the logarithm of particle radius with a simple power law. Later, Whitby (1978) showed that transforming atmospheric aerosol number distributions to volume distributions revealed three distinct size modes, which he labeled the nuclei, accumulation, and coarse modes (Fig. 6–1). Most importantly, he interpreted each mode in terms of a different formation process leading to different particle characteristics. This model provided a fundamental basis for the understanding of the properties of ambient aerosol. The characteristics of the atmospheric aerosol depend on location, meteorological conditions, time of day, the status of sources, and many other factors. Given this complexity, the Whitby trimodal model is a remarkable simplification that has proved to be very useful.

More recent studies have shown that ambient particle size distributions are considerably more complicated than the Whitby model. Whitby based his model mainly on number distributions taken with an electrical mobility analyzer from 0.01 to 0.4 μm and an optical particle counter from 0.4 to 6 μm. The transformation of number to volume distributions introduces uncertainty and provides limited size resolution; the data are weak near the 0.4 μm crossover size of the instruments, an important size range, as will be seen below. Limited chemical speciation data were available to Whitby, and the last stage of a typical cascade impactor at that time had a cutoff of about 0.5 μm. Since then, differential mobility

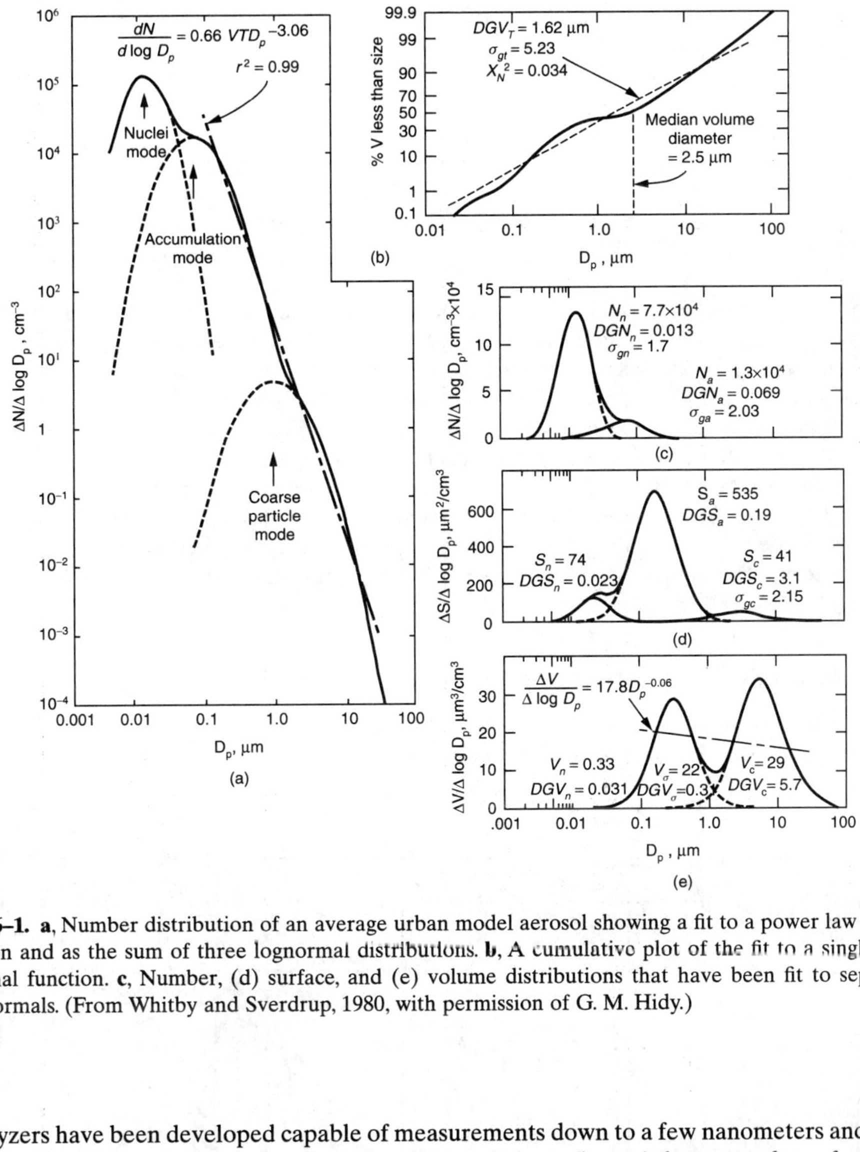

Fig. 6–1. **a**, Number distribution of an average urban model aerosol showing a fit to a power law distribution and as the sum of three lognormal distributions. **b**, A cumulative plot of the fit to a single lognormal function. **c**, Number, (d) surface, and (e) volume distributions that have been fit to separate lognormals. (From Whitby and Sverdrup, 1980, with permission of G. M. Hidy.)

analyzers have been developed capable of measurements down to a few nanometers and low-pressure impactors have stages extending down to about 5 nm. A large number of studies have been made with these instruments as well as with many others. The more recent data indicate that in the size range of each of Whitby's modes there can be multiple modes with particles of different origin and composition. These modes must be taken into account for proper assessment of aerosol effects.

In the following sections, the Whitby model will be discussed. Then each of the size ranges covered by the nuclei, accumulation, and coarse modes will be discussed, including the more recent findings. The intent is not to present a comprehensive review of the vast literature on atmospheric size distributions, but rather to convey a broad understanding of the main features within the constraints of current knowledge. Our knowledge of atmospheric size distributions is still incomplete despite decades of effort. The complexity of atmospheric processes presents difficult challenges to the measurement and modeling of ambient aerosol, which must be characterized by particle size and chemical composition and with temporal and spatial resolution. As more information is obtained, the picture presented here will no doubt require some revision.

The Whitby Model

Whitby (1978) described a trimodal distribution (Fig. 6–2) consisting of a nuclei mode in the size range of 0.005 to 0.1 µm, an accumulation mode from 0.1 to 2 µm, and a coarse mode of more than 2 µm. Each mode was fitted by a lognormal function. In Table 6–1, the modal parameters for eight different types of atmospheres are listed (from Whitby and Sverdrup, 1980).

Ambient particle size distributions typically have a minimum concentration between the accumulation and coarse modes (i.e., near 2 µm). Whitby divided the particles into two main fractions: fine particles, with diameters less than 2 µm; and coarse particles, with diameters of more than 2 µm. These two fractions have major differences both in origin and in physical and chemical characteristics (Fig. 6–3). The fine fraction derives mainly from combustion, whereas the coarse fraction is generated by mechanical processes. The fine fraction includes the nuclei mode, which are transient particles formed by condensation and coagulation. The nuclei rapidly grow into the accumulation mode. According to Whitby, the accumulation mode also contains droplets formed by the chemical conversion of gases to vapors that condense. The coarse particle fraction contains wind blown dust, sea spray, and plant particles.

Nuclei Mode, Size Range 0.005 to 0.1 µm

The smallest mode in the atmospheric aerosol, in terms of both particle size and mass concentration, is the nuclei mode, which, however, contains the highest number of particles. One

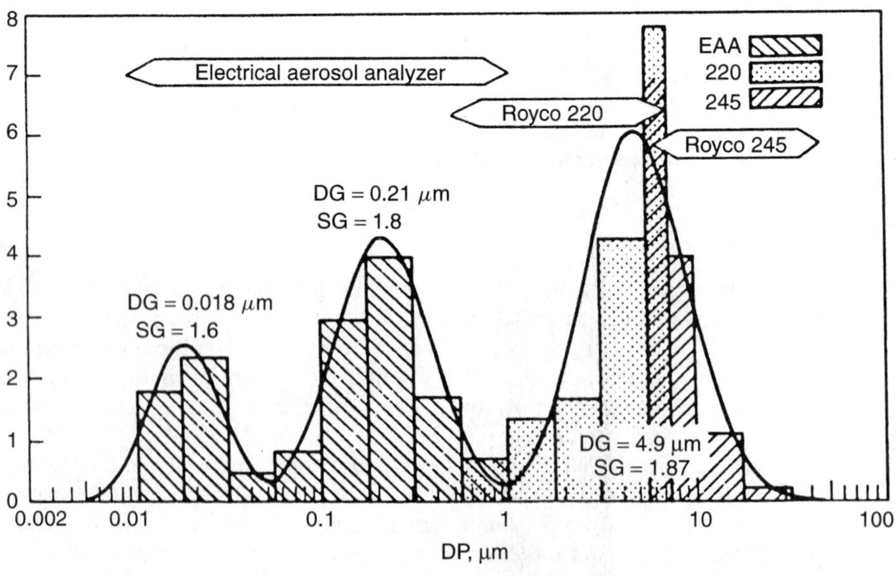

Fig. 6–2. Trimodal particle volume distribution measured at the General Motors Milford Proving Grounds, October 29, 1975. The size range measured with each instrument is indicated. The Roycos are optical particle counters. EAA, electrical aerosol analyzer; DG, the geometric mean particle diameter; SG, the geometric standard deviation. (From Whitby, 1978, with permission of Elsevier Science.)

TABLE 6–1. Lognormal Modal Parameters for Eight Typical Atmospheric Size Distributions

Atmospheric Distribution	Nuclei			Accumulation			Coarse		
	d_g (μm)	σ_g	V^a (μm/cm)^3	d_g (μm)	σ_g	V (μm/cm)^3	d_g (μm)	σ_g	V (μm)
Marine, surface	0.019	1.6[b]	0.0005	0.3	2[b]	0.10	12	2.7	12
Clean continental background	0.03	1.6	0.006	0.35	2.1	1.5	6	2.2	5
Average background	0.034	1.7	0.037	0.32	2.0	4.45	6.04	2.16	25.9
Background and aged urban plume	0.028	1.6	0.029	0.36	1.84	44	4.51	2.12	27.4
Background and local sources	0.021	1.7	0.62	0.25	2.11	3.02	5.6	2.09	39.1
Urban average	0.038	1.8	0.63	0.32	2.16	38.4	5.7	2.21	30.8
Urban and freeway	0.032	1.74	9.2	0.25	1.98	37.5	6.0	2.13	42.7
Labadie coal power plant	0.015	1.5	0.1		0.18	1.96	12	5.5	2.524
Mean	0.029	1.66	0.26[c]	0.29	2.02	21.5[d]	6.3	2.26	25.9
Standard deviation	± 0.007	± 0.1	± 0.33	± 0.06	± 0.1	± 20	± 2.3	± 0.22	± 13

[a] Volume of particles per volume of air.
[b] Assumed.
[c] Average omitting marine, urban and freeway, and Labadie.
[d] Average omitting marine.
Source: Whitby and Sverdrup (1980), with the permission of G. M. Hidy.

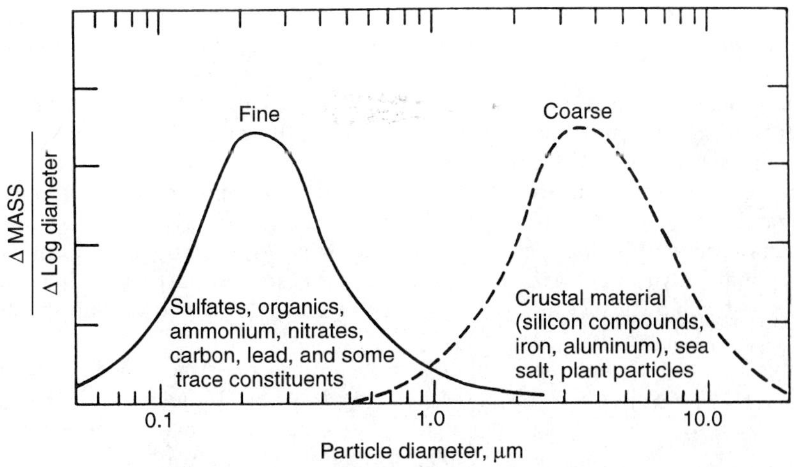

Fig. 6–3. Schematic of typical urban aerosol composition by particle size fraction. The chemical species are listed in approximate order of relative mass contribution. (From U.S. Environmental Protection Agency, 1982).

of the largest databases on the nuclei mode remains that of Whitby and his collaborators, who observed the geometric mean diameter by volume to range from 0.015 to 0.038 μm, averaging 0.029 μm, with a geometric standard deviation of 1.7, in a variety of locations (Table 6–1). The mode diameter increases with time by less than a factor of three because the nuclei mode particles coagulate more rapidly with particles in the condensation mode (defined below) than with other particles in the nuclei mode. Extensive measurements by Reischl and Winklmayr during the Southern California Air Quality Study (SCAQS) with an improved electrical mobility analyzer showed the nuclei mode to occur at a size consistent with the Whitby urban average value of 0.038 μm. Some recent data have shown two modes present in the size range of the nuclei mode.

Whitby (1978) has discussed the dominant role played by sulfur in the nucleation and growth of the nuclei mode. The nuclei mode is formed by photochemical reactions on gases in the atmosphere and by combustion. A striking demonstration of the photochemistry is afforded by the rapid appearance and growth of the nuclei mode at dawn. Because of its transient nature, the nuclei mode is significant only in the immediate vicinity of sources, for example, on a freeway.

There is considerable current interest in "ultrafine particles," loosely defined to be in the same size range as the nuclei mode, but with the emphasis near the lower end of the range. The interest stems from the possibility that particles this small might penetrate the tissue in the deep lung, leading to health effects.

Accumulation Mode, Size Range 0.1 to 2 μm

This size range contains most of the fine particle mass. The U.S. Environmental Protection Agency (EPA; 1997) has established a particle standard for PM-2.5 (i.e., for particles with aerodynamic diameters smaller than 2.5 μm) based on the minimum in the ambient particle size distribution near 2.5 μm and on the fact that the accumulation mode consists mainly of combustion products (i.e., anthropogenic emissions). Also, the EPA considered PM-2.5 to present a respiratory health hazard. We note that the cutpoint is slightly lower than the 4 μm cutpoint of the respirable aerosol sampling criterion of the American Conference of Governmental Hygienists (ACGIH; 1999), which would include some coarse mode particles. See Chapters 25 and 27 for further discussions of these sampling conventions.

The combustion of fossil fuels produces gases containing sulfur, nitrogen, and organic compounds. Complex reactions in the atmosphere result in the oxidation of the sulfur and nitrogen to produce particles in the accumulation mode containing inorganic compounds such as ammonium sulfate and ammonium nitrate. Organic and elemental carbon particles are also produced in the accumulation mode size range. Some of these chemicals are externally mixed (i.e., they are in separate particles) and some are internally mixed, being in the same particles. Because an internally mixed compound may have a concentration varying with particle size, the mode diameter of that compound may be different from that of the mode of the particles in question. Whether a compound is internally or externally mixed can sometimes be inferred indirectly from the size distributions, but is best determined directly by single particle analysis techniques such as microscopy or recently developed on-line spectrometers (Prather et al., 1994; see also Chapter 13).

Whitby described a single accumulation mode with a mass median diameter of about 0.3 μm. In a study of sulfur aerosols in the Los Angeles area, Hering and Friedlander (1982) found the size distributions in the accumulation mode size range on different days to fall into two different types, depending on atmospheric conditions, namely, whether the air was relatively clean and dry or polluted and humid. During SCAQS, John et al. (1990) (Fig. 6–4) found two modes in the particle size distributions of inorganic ions in this size range. One mode, designated the *condensation mode* by John et al., had an average aerodynamic diameter of 0.2 μm. The other mode, named the *droplet mode*, had an average aerodynamic diameter of 0.7 μm. Both modes contained sulfate, nitrate, and ammonium ion (Table 6–2). Size distributions measured with differential mobility analyzers and optical counters by Eldering and Cass (1994) showed similar modal structure. This was possible even though the instruments used do not speciate chemicals because the inorganic ions dominated the total particle concentration, as is typical for urban air.

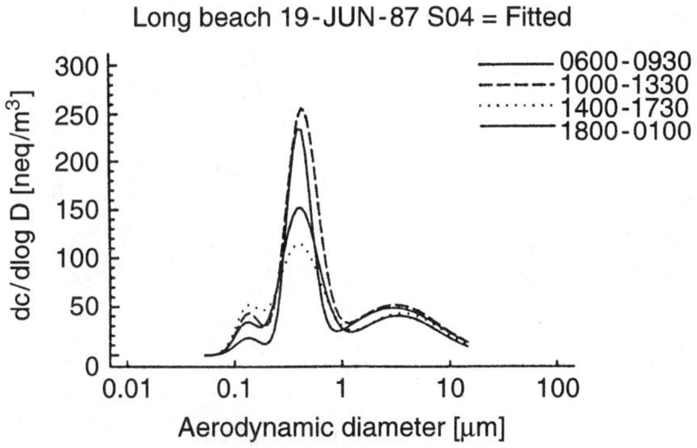

Fig. 6–4. Sulfate particle size distribution measured in ambient air with a Berner cascade impactor. The sulfate ion concentrations are given in units of nanoequivalents per cubic meter of air. (From John et al., 1990, with permission of Elsevier Science.)

TABLE 6–2. Average Mode Parameters for Atmospheric Inorganic Aerosol, Measured During the Southern California Air Quality Study of 1987

	Mode[a]		
	Condensation	Droplet	Coarse
Summer			
Aerodynamic Diameter (μm)	0.2 ± 0.1	0.7 ± 0.2	4.4 ± 1.2
σ_g	1.5 ± 0.2	1.7 ± 0.2	1.9 ± 0.3
Av. Concentration ($\mu g\,m^{-3}$)	5 ± 5	26 ± 21	13 ± 7
Fall			
Aerodynamic diameter (μm)	0.2 ± 0.1	0.7 ± 0.3	5.5 ± 0.7
σ_g	1.5 ± 0.2	1.9 ± 0.5	1.8 ± 0.4
Av. Concentration ($\mu g\,m^{-3}$)	9 ± 8	40 ± 29	5 ± 4

[a] Means and standard deviations are listed.

Source: John et al. (1990), with the permission of Elsevier Science.

The condensation mode was named to reflect its formation and growth by condensation of gases either directly or indirectly through coagulation with nuclei mode particles. The rate of growth of particles in the condensation mode decreases with increasing particle size. Therefore, in the time available in the ambient atmosphere, the condensation mode does not grow much beyond 0.2 μm.

The other fine particle mode observed by John et al. (1990) was named the droplet mode because particle deposits in that size range showed evidence of being wet. The droplet mode averaged 0.7 μm in diameter, but the diameter ranged from near the condensation mode diameter of 0.2 μm up to a maximum of 1 μm. The total ion mass in the droplet mode averaged 6.5 times that in the condensation mode. It was pointed out by McMurry and Wilson (1983) and by Hering and Friedlander (1982) that the formation of particles as large as those in the droplet mode requires aqueous phase reactions involving sulfur. Such reactions have rates that increase with increasing particle size. Meng and Seinfeld (1994) developed a model for the droplet mode involving the oxidation of sulfur dioxide in water droplets in fog or low clouds. Richards (1995) has discussed reactions in stratus clouds at night. Vertical mixing and droplet settling followed by evaporation is postulated to form large aerosol particles.

In addition to the condensation and droplet modes, John et al. (1990) also observed a coarse ion mode (discussed in the next section). Because the geometric standard deviations do not vary much (see Table 6–2), it is possible to characterize the modes by their mode diameters and concentrations. In Figure 6–5, a large data set has been summarized by plotting the relative mode concentrations versus the mode diameters. In Figure 6–5a, the sulfate data are seen to cluster in three modes that were relatively constant over the entire Los Angeles air basin and over the summer of 1987. In contrast, Figure 6–5b shows that the nitrate varied considerably. Prevailing westerly winds carry pollutants from the coastal sources toward the

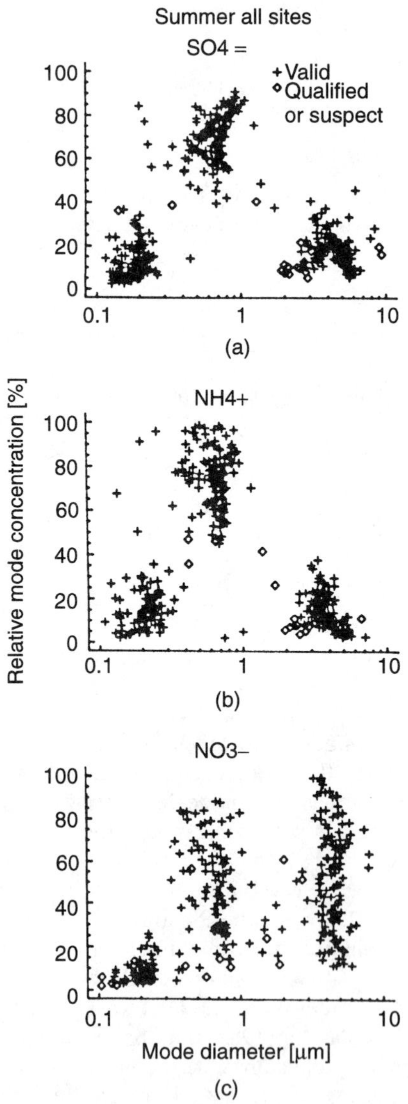

Fig. 6–5. Relative mode concentration versus mode diameter for all sampling sites and all sampling periods during the summer SCAQS. **a**, Sulfate. **b**, Ammonium. **c**, Nitrate. (From John et al., 1990, with permission of Elsevier Science.)

east, where ammonia converts the nitric acid to ammonium nitrate. As a result, nitrate concentrations are higher in the eastern end of the air basin.

In addition to a varying droplet mode diameter, there is appreciable overlap between the condensation and droplet modes. It is therefore misleading to quote a mass mean diameter over the fine particle range because that mixes the two modes. Even when the modes overlap, the particles are a mixture of two different populations with different origins and different compositions. The sulfate concentration in the droplet mode increases with increasing mode diameter, which is consistent with sulfur causing the formation of the mode. Others have observed sulfate size distributions peaking at 0.7 μm or larger. McMurry and Wilson (1983) reported sulfate particles as large as 3 μm in a power plant plume in Ohio. Georgi et al. (1986) observed sulfate in Hanover, Germany, peaking just above 1 μm and extending somewhat above 10 μm when the wind was from the east. Kasahara et al. (1994) reported sulfur distributions in Austria with mass mean diameters of 0.66 μm in Vienna and 0.65 μm in Marchegg. Koutrakis and Kelly (1993) found sulfate size distributions in Pennsylvania to peak at a geometric mean diameter of 0.75 μm. In Hungary, Meszaros et al. (1997) observed ammonium, nitrate, and sulfate modes in the range 0.5 to 1.0 μm, consistent with the droplet mode, but did not observe a condensation mode for these ions.

It appears that 0.7 μm is a typical mode diameter in many different locations, but the mode diameter varies considerably, depending on conditions. The mode diameter increases with residence time and can even exceed the large size limit of the accumulation mode. The size of the droplet mode has great significance for atmospheric visibility (John, 1993). A mode aerodynamic diameter of 0.7 μm corresponds to a geometric diameter of 0.57 μm, assuming a density of 1.5. This size is almost exactly on the peak of the light-scattering curve for sunlight. Thus, the droplet mode dominates visibility reduction; there is a smaller reduction due to extinction by particles of elemental carbon. Sloan et al. (1991) in a study of visibility in Denver observed two modes in the sulfate and nitrate size distributions with sizes consistent with condensation and droplet modes.

Koutrakis and Kelly (1993) found the size of sulfate particles in Pennsylvania to depend on relative humidity (RH) and acid content. Their data show the effect of RH to be most pronounced on ammonium bisulfate particles. Aerosols in Pennsylvania were found to contain little nitrate; the acidity of the aerosol results in most of the nitrate existing as gaseous nitric acid. In the SCAQS, nitrate and sulfate ions were closely balanced by ammonium ion; this is typical for California aerosols, which are nearly neutral. This is to be contrasted with typical aerosols in the eastern United States, which are acidic.

The accumulation mode size range also includes elemental and organic carbon particles. In general, carbon data are more uncertain than those of the inorganics because of experimental difficulties. Measurements by McMurry (1989) during SCAQS indicate a bimodal distribution in this size range, with one mode in the condensation mode but the other mode at a somewhat smaller diameter than the inorganic droplet mode. Venkataraman and Friedlander (1994) measured size distributions of polycyclic aromatic hydrocarbons (PAHs) and elemental carbon. Peaks were found at about 0.1 and 0.7 μm. Similar size distributions were found for aliphatic carbon, carbonyl carbon by Pickle et al. (1990), and by Mylonas et al. (1991) for organonitrates. Meszaros et al. (1997) found peaks for PAHs in the accumulation size range.

Milford and Davidson (1985) have summarized the size distributions of 38 particulate trace elements, mostly taken with cascade impactors. Most have a dominant peak in the accumulation mode size range with a smaller peak at about 3 to 5 μm.

Coarse Mode, Size Range Greater Than 2 μm

Whitby and Sverdrup (1980) reported an average coarse mode diameter of 6.3 ± 2.3 μm for extensive measurements with optical counters in a variety of locations. However, other modes have been observed, some considerably larger than 6 μm. The largest particles suspended in the atmosphere are in a size mode that will be referred to here as the *giant coarse mode* to distinguish it from smaller coarse modes to be discussed later. These particles can only be observed by in situ techniques or collected by special samplers such as the Noll Rotary Impactor or the Wide Range Aerosol Classifier, which has a very large inlet. Noll et al. (1985) measured coarse modes with mass median diameters ranging from 16 to 30 μm, with an average mode diameter of 20 μm and a standard deviation of 2.0 (Fig. 6–6). Measurements by Lundgren et al. (1984) are in general agreement. The particles consist of mineral particles derived from soil, biological particles, and sea salt.

The classic studies of Bagnold (1941), Gillette (1974), and Gillette et al. (1972, 1974) have explained how wind-blown dust is generated. Direct aerodynamic entrainment of soil particles is relatively insignificant. A process called *saltation* involves turbulent bursts, which eject particles approximately 100 μm in diameter from the ground (see Chapter 28). Subsequently, these particles impact the surface at a shallow angle, dislodging smaller particles that can then be entrained in the air. There has been little theoretical work on the prediction of the size distribution produced. Noll and Fang (1989) have proposed a model explaining the selective suspension of particles in the giant coarse mode size range by atmospheric turbulence. Particles that are too large fall out rapidly under gravity. Particles with too little inertia follow the eddies and do not acquire any net upward velocity from the wind. There is an intermediate size small enough to allow the particles to acquire upward velocity but with sufficient inertia to sustain an upward momentum. Corroborating evidence was obtained by Noll and Fang, who found that the particles deposited on the bottom surface of a collection plate had the same size distribution as those airborne.

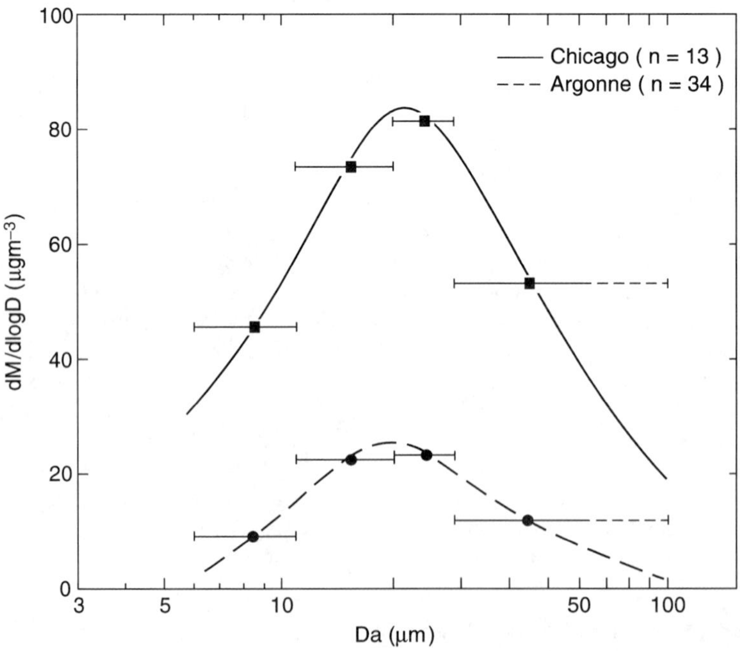

Fig. 6–6. Average mass size distributions for summer dry ground in Chicago and Argonne, measured with the Noll rotary impactor. (From Noll et al., 1985, with permission of Elsevier Science.)

Biological particles frequently consist of pollens, which are fairly monodisperse, and generally 20 μm or more in diameter. Large plant fragments are also present in the coarse mode. In urban areas, road dust generated by vehicles is found in the giant coarse mode. Particles of rubber containing mineral inclusions are seen. Coarse sea salt particles are found in coastal areas.

Measurements with cascade impactors find coarse modes with diameters typically ranging from 5 to 10 μm. These instruments are incapable of sampling the giant coarse mode, but the smaller modes that have been reported appear to be well within the capability of the samplers. Measurements by Lin et al. (1993) over the entire size range are revealing. They used an Andersen cascade impactor to collect particles smaller than 10 μm and the Noll rotary impactor for larger particles. Examination of their coarse mode shows a peak at 20 to 30 μm, depending on the sample. However, the size distribution is skewed to smaller sizes, and some of the distributions show evidence of a smaller peak at or slightly below 10 μm. Overall, available data suggest the existence of a coarse particle mode at a smaller size than the giant coarse mode.

One possibility is that the smaller coarse mode, hereafter referred to simply as the *coarse mode*, may be an aged version of the giant coarse particle mode. Fallout of the larger particles during transport from the originating location of the giant coarse mode would shift the mode diameter downward. The deposition velocity of a 25 μm particle is an order of magnitude greater than that of a 6 μm particle. At the same time, the mass concentration will decrease rapidly. However, the mass concentrations observed for the giant coarse mode are relatively high, ranging from about 30 μg/m^3 up to more than 100 μg/m^3. Also, the smaller the particle, the greater the distance traveled before deposition (i.e., the greater the extent of the area contributing). No calculation of the aging of a coarse distribution is available. It is interesting to note that, assuming that the giant coarse mode is lognormal, the aged mode will not be lognormal.

Gillette et al. (1974) found soil aerosol size distributions to be similar to the size distribution of the soil particles when they are separated in a water suspension. When the soil particles were separated in water containing detergent, there was an excess of particles smaller than 5 μm compared with the aerosol, implying that the forces producing the aerosol were not strong enough to completely deagglomerate the soil. They also observed that the shapes of the aerosol size distributions were insensitive to wind conditions, implying that aerodynamic suspension was not operating. By sampling at various heights above the ground, Gillette et al. (1972) measured aerosol size distributions during periods of vertical flux. This *excludes* the possibility that they observed aged giant coarse mode particles. By converting their number distributions to volume distributions, it can be estimated that the mass median diameter of the soil aerosol was about 9 μm. This is evidence of another source of a coarse mode smaller than the giant coarse mode. The distinction between the coarse mode and the giant coarse mode has an implication for aerosol composition because the ratio of clay to silt in soil varies with particle size.

It is well-known that in coastal areas nitrate is found in the coarse aerosol fraction as a result of the reaction of nitric acid with sea salt (Savoie and Prospero, 1982; Harrison and Pio, 1983; Bruynseels and Van Grieken, 1985; Wall et al., 1988). During the summer SCAQS study, the wind was primarily from the Pacific Ocean. John et al. (1993) observed a coarse ion mode containing nitrate, sulfate, chloride, sodium, ammonium, magnesium, and calcium. The mean mode diameter was 4.4 μm, and the geometric standard deviation was 1.9. This mode could be called the *coarse ion mode*. Wall et al. (1988) pointed out that small NaCl particles will be completely converted to $NaNO_3$, whereas large NaCl particles will be only partially converted. This leads to a nitrate mass distribution that has the same shape as that of the NaCl mass distribution for particles smaller than the NaCl mode diameter, but for larger particles the distribution is truncated relative to that of the NaCl. This is another case of a size distribution that is not lognormal.

Coarse nitrate is also seen in continental air, formed by the reaction of nitric acid with alkaline soil particles and, at night, possibly by the reaction of N_2O_2 with soil particles (Wolff, 1984). The nitrate on soil particles is not associated with ammonium.

Venkataraman et al. (1999) measured size distributions of PAHs in India, finding non-volatile PAHs to peak in the accumulation mode, but an average of 32% was in the coarse mode. Semivolatile PAHs were predominately in the coarse mode, averaging 60% in the coarse mode. They discuss the volatilization of the original particles in the nuclei or accumulation modes followed by adsorption of the organic compounds onto coarse mode particles. The details of the process are uncertain. The presence of PAHs and nitrates in the coarse mode exemplify the complexity of ambient aerosol composition because many of the substances in atmospheric aerosol are semivolatile.

INDOOR AEROSOLS

Indoor aerosol usually refers to aerosol in residences and offices as distinguished from that in industrial workplaces. Increasing emphasis is being placed on indoor aerosols, because people average 80% to 90% of their time indoors (Spengler and Sexton, 1983). Three major studies have been conducted in recent years: the Harvard 6-City Study (Spengler et al., 1981), the New York Energy Research and Development Authority (ERDA) Study (Sheldon et al., 1989), and the EPA particle total exposure assessment methodology (PTEAM) study (Pellizzari et al., 1992). Some of the indoor aerosol derives from infiltration of atmospheric aerosol. In the PTEAM study, the ratio of PM-2.5 to PM-10 indoors was about 0.5.

In homes with smokers, tobacco smoke is the largest component of the indoor aerosol. Tobacco smoke particles coagulate rapidly in the first few minutes after emission, causing the size distribution to shift toward larger diameters. The number distribution of tobacco smoke typically peaks in the accumulation mode size range. In Figure 6–7, the number distribution

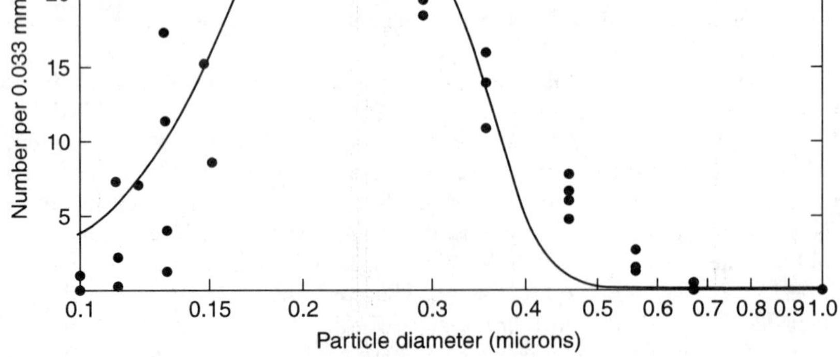

Fig. 6–7. Comparison of experimental size distribution data for tobacco smoke with prediction based on self-preserving size spectrum theory. (From Friedlander, 1977, with permission of S. K. Friedlander.)

TABLE 6–3. Size Distribution Parameters of Indoor Smoke Aerosols and the Soluble Fraction

Product	CMD (nm)	GSD	Soluble Fraction
Hollywood Peanut Oil	199.1	1.62	0.076 ± 0.026
Mazola Corn Oil	173.6	1.58	0.164 ± 0.008
Wesson Canola Oil	238.9	1.61	0.203 ± 0.045
Wesson Vegetable Oil	168.3	1.68	0.128 ± 0.033
Sweet Italian sausages	73.8	1.55	0.456 ± 0.087
Wood burning with flame	80.3	1.90	0.714 ± 0.022
Wood burning without flame	55.1	1.31	0.924 ± 0.098
Particles in living room (wood burning in wood stove)	96.7	1.81	

CMD, Count median diameter; GSD, geometric standard deviation.

Source: Aerosol Science & Technology: "Hygroscopic Growth of Assorted Indoor Aerosols." 24:151–160. Copyright 1996. Cincinnati, OH. Reprinted with permission of American Association for Aerosol Research.
Dua and Hopke (1996).

measured by Keith and Derrick (1960) is compared with a theoretical calculation based on self-preserving size spectrum theory (Friedlander and Hidy, 1969). Light-scattering measurements by Chung and Dunn-Rankin (1996) showed mainstream cigarette smoke to have a CMD of 0.14 μm for unfiltered cigarettes and 0.17 μm for filtered cigarettes, with GSDs of 2.1 and 2.0, respectively. The corresponding mass median diameters were 0.71 and 0.66 μm. Fresh sidestream smoke had a CMD of 0.27 μm. "Typical environmental tobacco smoke" measured with a Scanning Mobility Particle Sizer by Morawska and Jamriska (1997) gave a number distribution with a lognormal shape, peaking at about 0.12 μm. Cigarette smoke particles will undergo hygroscopic growth in the human respiratory tract (Robinson and Yu, 1998).

The next most significant source of indoor aerosol is from cooking. The size distribution parameters for smoke from cooking oils, sausages, and wood burning are listed in Table 6–3. The particles are in the accumulation size range. Table 6–3 also lists the soluble fraction of the smoke particles, which is low for the oils and sausages and high for wood smoke. Correspondingly, the oil and sausage smoke particles do not show hygroscopic growth, whereas the wood smoke particles do (Dua and Hopke, 1996).

Kerosene heaters, when present, contribute to indoor aerosol. The fine fraction (PM-2.5) of the indoor aerosol contains particles of soil, wood smoke, iron, steel, and particles from auto-related and sulfur-related sources (Spengler et al., 1981). Biological particles in indoor air include dander, fungi, bacteria, pollens, spores, and viruses. Walls, floors, and ceilings may release glass fibers, asbestos fibers, mineral wool, and metal particles. Aerosols are generated by consumer product spray cans. Paper products are a source of cellulose fibers, and clothing articles are sources of natural and synthetic organic fibers.

In homes where radon gas is present, radioactive aerosols may be formed by the attachment of radon daughters to suspended particles. Activity median diameters (AMD), from the size distribution of the radioactivity, are as small as a few nanometers. Morawski and Jamriska (1997) have discussed the difficulties in measuring radon progeny and reference the extensive literature.

INDUSTRIAL AEROSOLS

The characteristics of aerosols produced in industry are determined by the type of industry, the nature of the product, and the industrial operations. Detailed discussions of emissions from basic industries have been given (Stern, 1968). Basic industries include petroleum refineries, nonmetallic mineral product industries, ferrous metallurgical operations, nonferrous metallurgical operations, inorganic chemical industry, pulp and paper industry, and food and feed industries. Power plants and incinerators are examples of stationary combustion sources. The properties of the aerosols emitted to the atmosphere depend on the material burned, combustion conditions, and the type of controls on the stacks.

Within an industry, aerosols are generated by processing activities. Welding produces fumes, chain agglomerates of very fine particles. Mechanical operations such as grinding make coarse particles. Size distributions measured by Sioutas (1999) in an automotive machining facility show that welding and heat-treating operations produce fine particles as expected from the condensation of hot vapors, while machining and grinding produce relatively large particles by breakup of solid material. Spray painting produces liquid droplets in the aerosol size range. The transport and handling of powdered materials produces dust aerosols. Ore piles, coal piles, and tailings piles give rise to fugitive emissions.

Workplace operations can produce aerosols of very large particles that are difficult to monitor or sample, yet may pose a respiratory hazard. The ACGIH (1999) recommends sampling such coarse aerosols according to the inhalable particulate mass criteria (IPM), which specifies the desired sampling efficiency up to 100 µm (see Chapter 25). An example is wood dust aerosol, which can cause nasal cancer (Hinds, 1988).

Aerosols in mines have long been investigated because of their ability to cause respiratory diseases including pulmonary fibrosis, pneumoconiosis, and lung cancer. Indeed, work on mine aerosols has provided much of the existing aerosol sampling technology and the approach to sampling criteria for hazardous aerosols (Mercer, 1973). The mechanical operations in coal mines produce coarse aerosols as expected, with a peak in the mass distribution at about 7 µm. However, if the mine contains diesel machinery, the mass distribution is bimodal, with a second peak at about 0.2 µm (Cantrell and Rubow, 1990; see Chapter 26). Cavallo (1998) has published activity-weighted particle size distributions for two uranium mines with diesel engines; the AMDs are in the 0.1 to 0.2 µm range.

REFERENCES

American Conference of Governmental Industrial Hygienists (ACGIH). 1999. *TLVs and BEIs, appendix D: Particle size-selective sampling criteria for airborne particulate matter.* Cincinnati: ACGIH.

Bagnold, R. A. 1941. *The Physics of Blown Sand and Desert Dunes.* London: Methuen.

Brown, W. K. and K. H. Wohletz. 1995. Derivation of the Weibull distribution based on physical principles and its connection to the Rosin-Rammler and lognormal distributions. *J. Appl. Phys.* 78:2758–2763.

Bruynseels, F. and R. Van Grieken. 1985. Direct detection of sulfate and nitrate layers on sampled marine aerosol by laser microprobe mass analysis. *Atmos. Environ.* 19:1969–1970.

Cantrell, B. K. and K. L. Rubow. 1990. Mineral dust and diesel exhaust aerosol measurements in underground metal and nonmetal mines. In *Proc. VIIth International Pneumoconioses Conf. NIOSH Pub. No. 90-108*, pp. 651–655. NIOSH.

Cavallo, A. J. 1998. Reanalysis of 1973 activity-weighted particle size distribution measurements in active U.S. uranium mines. *Aerosol Sci. Technol.* 29:31–38.

Chung, I.-P. and D. Dunn-Rankin. 1996. In situ light scattering measurements of mainstream and sidestream cigarette smoke. *Aerosol Sci. Technol.* 24:85–101.

Dua, S. K. and P. K. Hopke. 1996. Hygroscopic growth of assorted indoor aerosols. *Aerosol Sci. Technol.* 24:151–160.

Eldering, A. and G. R. Cass. 1994. An air monitoring network using continuous particle size distribution monitors: Connecting pollutant properties to visibility via Mie scattering calculations. *Atmos. Environ.* 28:2733–2749.

Friedlander, S. K. 1977. *Smoke, Dust and Haze.* New York: John Wiley & Sons.

Friedlander, S. K. and G. M. Hidy. 1969. New concepts in aerosol size spectrum theory. In *Proceedings of the 7th International Conference on Condensation and Ice Nuclei*, ed. J. Pkodzimek. Prague: Academia.

Georgi, B., K. P. Giesen, and W. J. Muller. 1986. Measurements of airborne particles in Hannover. In *Aerosols, Formation and Reactivity, Proceedings*, Second International Aerosol Conference, September 22–26, 1986, Berlin (West). Oxford: Pergamon Press.

Gillette, D. A. 1974. On the production of wind erosion aerosols having the potential for long range transport. *J. Rech. Atmos*. 8:735–744.

Gillette, D. A., I. H. Blifford, Jr., and C. R. Fenster. 1972. Measurements of aerosol size distributions and vertical fluxes of aerosols on land subject to wind erosion. *J. Appl. Meteorol*. 11:977–987.

Gillette, D. A., I. H. Blifford, Jr., and D. W. Fryrear. 1974. The influence of wind velocity on the size distributions of aerosols generated by the wind erosion of soils. *J. Geophys. Res.* 79:4068–4075.

Harrison, R. M. and C. A. Pio. 1983. Size differentiated composition of inorganic atmospheric aerosols of both marine and polluted continental origin. *Atmos. Environ.* 17:1733–1738.

Hering, S. V. and S. K. Friedlander. 1982. Origins of aerosol sulfur size distributions in the Los Angeles basin. *Atmos. Environ.* 16:2647–2656.

Hinds, W. C. 1988. Basis for particle size-selective sampling for wood dust. *Appl. Ind. Hyg*. 3:67–72.

Hinds, W. C. 1999. Aerosol Technology. New York: John Wiley & Sons.

John, W. 1993. Multimodal size distributions of inorganic aerosol during SCAQS. In *Southern California Air Quality Study, Data Analysis*, Proceedings of an International Specialty Conference, Los Angeles, CA, July 1992, Air & Waste Management Association, Pittsburgh, PA, p. 167.

John, W., S. M. Wall, J. L. Ondo, and W. Winklmayr. 1990. Modes in the size distributions of atmospheric inorganic aerosol. *Atmos. Environ.* 24A:2349–2359.

Junge, C. E. 1963. Air chemistry and radioactivity. New York: Academic Press.

Kasahara, M., K. Takahashi, A. Berner, and O. Preining. 1994. Characteristics of Vienna aerosols sampled using rotating cascade impactor. *J. Aerosol Sci.* 25(S1):S53–S54.

Keith, C. H. and J. E. Derrick. 1960. Measurement of the particle size distribution and concentration of cigarette smoke by the "Conifuge". *J. Colloid Sci.* 15:340–356.

Koutrakis, P. and B. P. Kelly. 1993. Equilibrium size of atmospheric aerosol sulfates as a function of particle acidity and ambient relative humidity. *J. Geophys. Res.* 98:7141–7147.

Lin, J.-M., G.-C. Fang, T. M. Holsen, and K. E. Noll. 1993. A comparison of dry deposition modeled from size distribution data and measured with a smooth surface for total particle mass, lead and calcium in Chicago. *Atmos. Environ.* 27A:1131–1138.

Lundgren, D. A., B. J. Hausknecht, and R. M. Burton. 1984. Large particle size distribution in five U.S. cities and the effect on the new ambient particulate matter standard (PM_{10}). *Aerosol Sci. Technol.* 7:467–473.

McMurry, P. H. 1989. *Organic and elemental carbon size distributions of Los Angeles aerosols measured during SCAQS.* Final report to the Coordinating Research Council, project SCAQS-6–1, University of Minnesota Particle Technology Laboratory Report No. 713.

McMurry, P. H. and J. C. Wilson. 1983. Droplet phase (heterogeneous) and gas phase (homogeneous) contributions to secondary ambient aerosol formation as functions of relative humidity. *J. Geophys. Res*. 88:5101–5108.

Meng, Z. and J. H. Seinfeld. 1994. On the source of the submicrometer droplet mode of urban and regional aerosols. *Aerosol Sci. Technol.* 20:253–265.

Mercer, T. T. 1973. *Aerosol Technology in Hazard Evaluation.* New York: Academic Press.

Meszaros, E., T. Barcza, A. Gelencser, J. Hlavay, G. Kiss, Z. Krivacsy, A. Molnar, and K. Polyak. 1997. Size distributions of inorganic and organic species in the atmospheric aerosol in Hungary. *J. Aerosol Sci.* 28:1163–1175.

Milford, J. B. and C. I. Davidson. 1985. The sizes of particulate trace elements in the atmosphere—A review. *JAPCA* 33:1249–1260.

Morawska, L. and M. Jamriska. 1997. Determination of the activity size distribution of radon progeny. *Aerosol Sci. Technol.* 26:459–468.

Mylonas, D. T., D. T. Allen, S. H. Ehrman, and S. E. Pratsinis. 1991. The sources and size distributions of organonitrates in Los Angeles aerosol. *Atmos. Environ.* 25A:2855–2861.

Noll, K. E. and K. Y. P. Fang. 1989. Development of a dry deposition model for atmospheric coarse particles. *Atmos. Environ.* 23:585–594.

Noll, K. E., A. Pontius, R. Frey, and M. Gould. 1985. Comparison of atmospheric coarse particles at an urban and non-urban site. *Atmos. Environ.* 19:1931–1943.

Pellizzari, E. D., K. W. Thomas, C. A. Clayton, R. C. Whitmore, H. Shores, S. Zelon, and R. Peritt. 1992. Particle total exposure assessment methodology (PTEAM): *Riverside, California pilot study*, Vol. I (final report). Research Triangle Park, N.C. U.S. Environmental Protection Agency, Atmospheric Research and Exposure Assessment Laboratory, EPA Report No. EPA/600/SR-93/050. Available from NTIS, Springfield, VA, PB93-166957/XAB.

Pickle, T., D. T. Allen, and S. E. Pratsinis. 1990. The sources and size distributions of aliphatic and carbonyl carbon in Los Angeles aerosol. *Atmos. Environ.* 24A:2221–2228.

Prather, K. A., T. Nordmeyer, and K. Salt. 1994. *Anal. Chem.* 66:1403–1407.

Prupacher, H. R. and J. D. Klett. 1980. *Microphysics of Clouds and Precipitation*. Boston: Reidel.

Richards, L. W. 1995. Airborne chemical measurements in nighttime stratus clouds in the Los Angeles basin. *Atmos. Environ.* 29:27–46.

Robinson, R. J. and C. P. Yu. 1998. Theoretical analysis of hygroscopic growth rate of mainstream and sidestream cigarette smoke particles in the human respiratory tract. *Aerosol Sci. Technol.* 28:21–32.

Rosin, P. and E. Rammler. 1933. The laws governing the fineness of powdered coal. *J. Inst. Fuel* 7:29–36.

Savoie, D. L. and J. M. Prospero. 1982. Particle size distribution of nitrate and sulfate in the marine atmosphere. *Geophys. Res. Lett.* 9:1207–1210.

Sheldon, L. S., T. D. Hartwell, B. G. Cox, J. E. Sickles, II, E. D. Pellizari, M. L. Smith, R. L. Perritt, and S. M. Jones. 1989. An investigation of infiltration and indoor air quality, final report. New York State Energy Research and Development Authority, Albany, N.Y., Contract No. 736-CON-BCS-85.

Sioutas, C. 1999. A pilot study to characterize fine particles in the environment of an automotive machining facility. *Appl. Occup. Environ. Hyg.* 14:246–254.

Sloan, C. S., J. Watson, J. Chow, L. Pritchett, and L. W. Richards. 1991. Size-segregated fine particle measurements by chemical species and their impact on visibility impairment in Denver. *Atmos. Environ.* 25A:1013–1024.

Spengler, J. D., D. W. Dockery, W. A. Turner, J. M. Wolfson, and B. G. Ferris, Jr. 1981. Long-term measurements of sulfates and particles inside and outside homes. *Atmos. Environ.* 15:23–30.

Spengler, J. D. and K. Sexton. 1983. Indoor air pollution: A public health perspective. *Science* 221:9–17.

Stern, A. C., ed. 1968. *Air Pollution*, Vol. 111, *Sources of Air Pollution and Their Control*, 2nd Ed. New York: Academic Press.

U.S. Environmental Protection Agency. 1982. *Air quality criteria for particulate matter and sulfur.* EPA-600/882-029b, December 1982.

U.S. Environmental Protection Agency: 1997. National Ambient Air Quality Standards for Particulate Matter. *Fed. Reg.* 62 (138) July 18.

Venkataraman, C. and S. K. Friedlander. 1994. Size distributions of polycyclic aromatic hydrocarbons and elemental carbon. 2. Ambient measurements and effects of atmospheric processes. *Environ. Sci. Technol.* 28:563–572.

Venkataraman, C., S. Thomas, and P. Kulkarni. 1999. Size distributions of polycyclic aromatic hydrocarbons—Gas/particle partitioning to urban aerosols. *J. Aerosol Sci.* 30:759–770.

Wall, S. M., W. John, and J. L. Ondo. 1988. Measurement of aerosol size distributions for nitrate and major ionic species. *Atmos. Environ.* 22:1649–1656.

Whitby, K. T. 1978. The physical characteristics of sulfur aerosols. *Atmos. Environ.* 12:135–159.

Whitby, K. T. and G. M. Sverdrup. 1980. California aerosols: Their physical and chemical characteristics. In *The Character and Origins of Smog Aerosols*, eds. G. M. Hidy, et al., New York: John Wiley & Sons, p. 477.

Wolff, G. T. 1984. On the nature of nitrate in coarse continental aerosols. *Atmos. Environ.* 18:977–981.

7 An Approach to Performing Aerosol Measurements

PAUL A. BARON and WILLIAM A. HEITBRINK

Centers for Disease Control and Prevention*, National Institute for Occupational Safety and Health, Cincinnati, OH

INTRODUCTION

Today's scientists and engineers making aerosol measurements have available a diversity of aerosol monitors ranging from sample collection on a filter for later analysis to complex direct-reading instruments that detect the airborne particles in real time and display size distribution and chemical data. Instruments used for aerosol measurement frequently provide only an indirect measure of the desired information. For instance, commonly used optical particle counters measure an "optical size" that must then be converted to a physical or aerodynamic size using assumptions about particle properties. Most instruments also only operate over a limited particle size range, and often two or more instruments with different detection principles are used to cover a wider size range. Therefore, the aerosol practitioner must be able to assess the meaning and usefulness of data likely to be obtained with various instruments when selecting one or more for a specific purpose. Use of the simplest or most complex device may introduce errors in measurement and interpretation. While the data from the more complex sizing instruments may make errors evident and indicate the need for corrections, errors also occur in the less-sophisticated filter or inertial collectors. Lack of recognition of these errors may affect the interpretation of aerosol measurements. Approaching the aerosol measurement process with an appropriate plan will reduce the likelihood of major errors in the results.

QUALITY ASSURANCE: PLANNING A MEASUREMENT

In aerosol measurement, the difficulties of selecting an instrument often sidetrack the issue of applying quality assurance principles to obtain accurate and meaningful data. Many scientists prefer to address problems in a more investigative fashion. However, many of the principles of quality assurance can be integrated into this approach with relatively little effort and with large benefits in accuracy and efficiency. Quality assurance principles have been developed over many years, and programs applying these principles are present in most

*Mention of company or product name does not constitute endorsement by the Centers for Disease Control and Prevention.

Aerosol Measurement: Principles, Techniques, and Applications, Second Edition, Edited by Paul A. Baron and Klaus Willeke.
ISBN 0-471-35636-0

analytical laboratories around the world. It has been recognized that reliable data are much more likely to be produced by laboratories under such conditions. The following series of steps is one approach to quality assurance and was developed by the U.S. Environmental Protection Agency (EPA, 1994). It should be recognized that the process is usually a cyclic one: after one pass through the following steps, the steps are repeated until an optimal measurement approach has been achieved.

State the problem: Writing down the problem with a complete indication of the various parameters needed, along with the resources available, will help clarify the likely solutions to the measurement process. For instance, the emission of aerosol from a manufacturing process is contaminating a second process. The transfer of contaminant must be controlled.

Identify the decision: Specifying the decision with appropriate levels of confidence will indicate the complexity of measurement needed to carry out the measurement. In the example above, we need to determine the lowest cost-control measure that prevents the occurrence of cross contamination.

Identify the inputs to the decision: What data are needed to reach the decision? Do we need to measure the size distribution of aerosol and transmission paths throughout the plant or just identify when aerosol concentration at the second process has been sufficiently reduced?

Define the study boundaries: It is often easy to design elaborate experiments to measure parameters that do not contribute significantly to the final decision. Measuring the size distribution of the contaminating aerosol may help understand the process of transfer, but may not necessarily contribute to the final result.

Develop a decision rule: This is a statistical statement of the confidence expected in the data such that the decision can be made.

Specify limits on decision error: This presents a statistical "target" for the measurements to be made.

Optimize design: This is the step that uses experience gained in the initial data collection to refine and improve the previous steps. For example, initial experiments may indicate contamination from unexpected sources or pathways, or the results may be at the limit of the measurement instrument, suggesting that alternate measurement techniques may be better.

MEASUREMENT ACCURACY

If "measurement processes are to serve both the practical needs of humankind and excellence in the pursuit of new scientific knowledge, they must be endowed with an adequate level of accuracy. . . . Control, and acceptable bounds for imprecision and bias, are clearly prerequisites; but scientific conventions (communication) and scientific and technological means for approaching 'the truth' must also be considered" (Currie, 1992). Although nomenclature provides the basis for communication of accuracy of the measurements, the basis for developing the accuracy limits on measurements comes from experiments, assumptions, and scientific knowledge.

Although a formal quality assurance process may seem like overkill for each aerosol measurement process, understanding the principles of a good quality assurance program can highlight or alert the scientist to pitfalls in a proposed experimental approach. There are texts available on quality assurance principles. An example of detailed requirements of such a program applied to environmental measurements are available from the EPA web site (www.epa.gov/region10/www/offices/oea/qaindex.htm).

The discussion below presents problems that can occur when making aerosol measurements. These problems are presented to instill some caution into the practitioner when performing aerosol measurements. There are many measurement techniques presented in following chapters, and it may appear difficult to choose among these techniques. However, based on the desired aerosol property, time resolution, instrument size, resource constraints, and the accuracy required, the choices are often narrowed to one or two approaches. The chapters in Part II, Techniques, start out with techniques that involve collection of particles with subsequent analysis of the collected material (i.e., integral concentration measurements). Then, real-time instruments that collect particles and analyze them are discussed, followed by direct-reading instruments that present information about individual particles (usually size distribution). The final chapters in Part II present information that applies to all measurement techniques: sampling, data presentation, and instrument calibration.

SIZE RANGE

One of the first factors to consider in the selection of instrumentation for aerosol measurement is the size range of the aerosol. Chapter 6 presented size ranges encountered in several environments. Additional examples are given in this chapter and in Part III, Applications. At the small particle end of the spectrum, aerosol particles can form and grow from (photo)chemical reactions, condensational nucleation and growth, and coagulation or agglomeration. At the upper end of the spectrum particles are likely to be formed from mechanical action, such as abrasion and crushing, while droplets can be formed by spraying and bubbling. The typical dividing line between the small and large aerosol particles is about 1 μm, with the former types of aerosols rarely growing significantly above several micrometers and the latter aerosols rarely extending below about 0.5 μm. The type of system generating the aerosol can often give a clue to the particle size range likely to be produced. For example, hot processes such as smelting are likely to produce submicrometer fume particles, mechanical processes such as drilling tend to produce large particle dusts, while some processes such as welding and grinding may produce multimodal distributions covering a wide size range. A number of aerosol measurement instruments are mentioned here with only a very cursory description of their detection mechanism and capabilities. Further details are provided in the indicated chapters covering specific instruments.

The past 30 years of aerosol measurement research has been quite active, with explorations of different detection, classification, and analysis techniques. Some of these techniques have become successful commercial instruments, while others have languished for a variety of reasons. These include inaccuracy, insufficient sensitivity, lack of appropriate application, difficulty of use, high cost, or better competing techniques. There is a continual effort to build aerosol instruments that measure one or more useful aerosol properties over a wider size range. In most cases, the aerosol measurement process is a compromise, with the selection of the available instrumentation that measures an aerosol property closely related to that desired. In general, the smaller the corrections and the fewer the assumptions in conversion factors needed to provide the desired result, the better the information is likely to be. This makes the selection of instrumentation for a given application somewhat of an art.

Although there is not a strict separation between the two, there are two general approaches to aerosol measurement: *collection and analysis* and *direct-reading sensors*. The former is generally less expensive in capital investment, more time consuming, gives integral concentration measures, and allows qualitative and quantitative measurement of the aerosol. The latter approach requires much more expensive instrumentation, usually gives size distribution information, gives nearly instantaneous results, and allows many measurements to

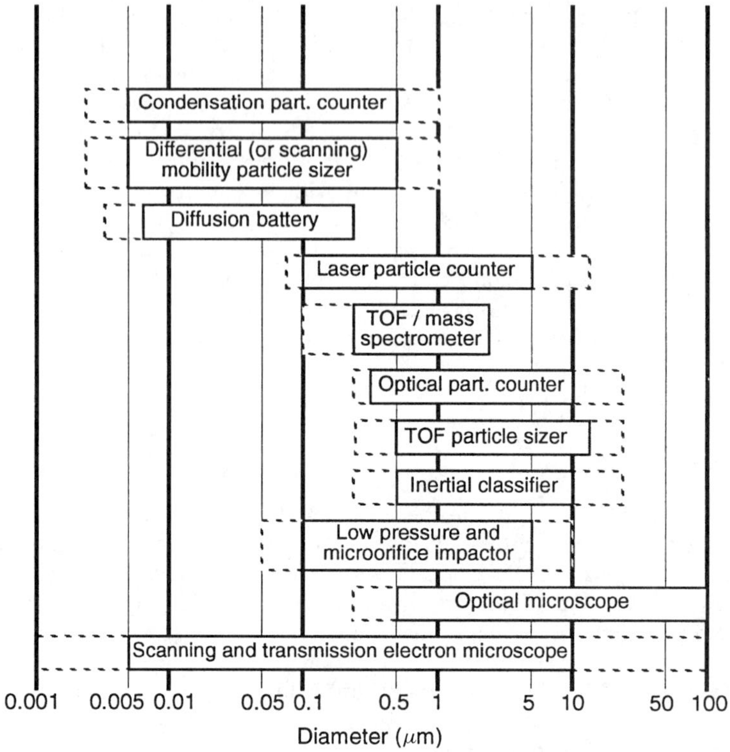

Fig. 7–1. Measurement size range of some principal aerosol sizing and analytical instruments. TOF, Time of flight. (Adapted from Pui, 1996.)

be made over time. An overview of the size range of several types of commonly used classes of instruments is presented in Figure 7–1, and a flow chart indicating the application of some of these and other instruments is presented in Figure 7–2.

COLLECTION AND ANALYSIS MEASUREMENTS

The most common collection technique involves the use of filters for collecting particles from the air. Most modern sampling filters are virtually 100% efficient for all particle sizes (see Chapter 9). The filter is placed in a holder that depends largely on the application. If the sampling device is intended to be a stand-alone device that collects particles from an environment, the enclosure and inlet of the device must be appropriately designed to give accurate, or at least known, sampling efficiency. The aspiration (or entry) efficiency and internal losses in various devices are discussed in Chapter 8. Sampling devices are frequently designed for specific applications, and some of these are discussed in Part III.

In addition to simply collecting all particles entering the sampling device, some instruments are designed to classify particles into two or more size fractions. Inertial separation devices, such as cyclones and impactors, are most commonly used for this purpose. A cyclone causes air to move in a swirling motion from which larger particles are deposited onto a

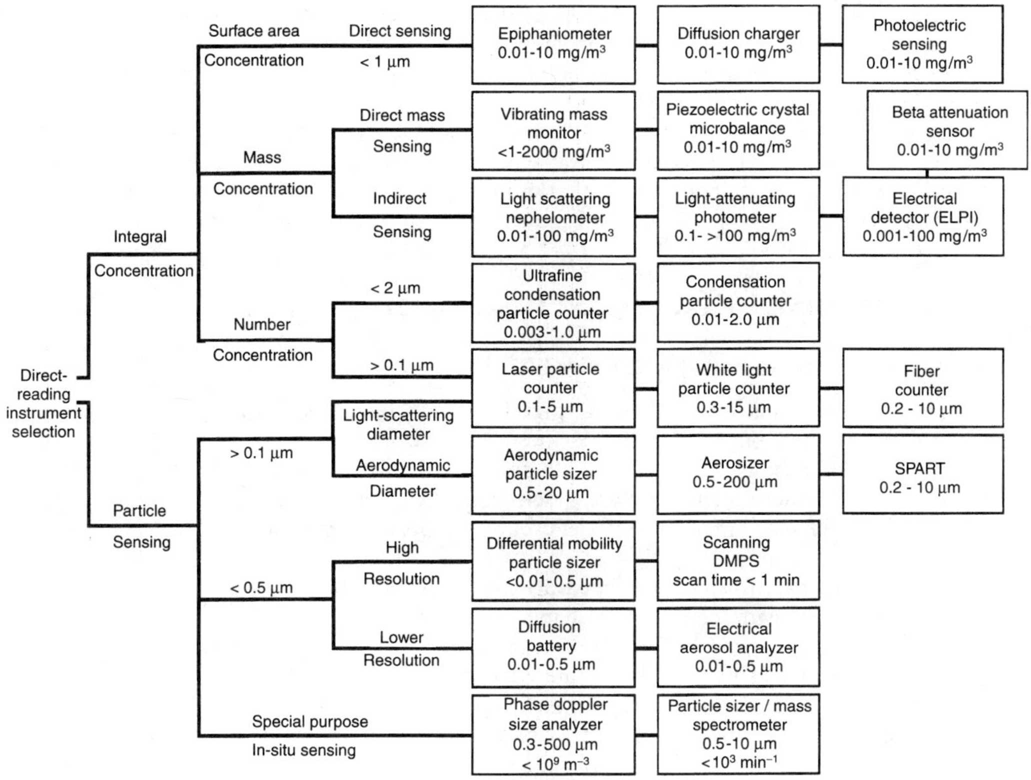

Fig. 7–2. Flow chart for selecting a direct-reading instrument for analyzing aerosol particles. (Adapted from Pui, 1996.)

surface by centrifugal action, while impactors cause a more abrupt change in airflow direction, also causing larger particles to be deposited onto a surface or substrate. For instance, a cyclone or impactor is often placed before the filter (as a "pre-classifier") to simulate the removal of particles by the upper respiratory system so that the material collected on the filter simulates particles reaching the gas exchange region of the lungs (see Chapters 25, 26, 27, and 29).

Particles collected on the filter can be analyzed in many different ways. The sample on the entire filter can be subjected to gravimetric, chemical, biological, or radioactive emission analysis (see Chapters 11, 24, and 34), or individual particles on the filter can be subjected to various forms of microscopy, spectroscopy, or shape analysis (see Chapters 12 and 23).

Classification or size distribution measurement of aerosols can be achieved by placing several classifiers in series as a "cascade." Typically, each stage collects larger particles than the subsequent stages. These devices have various names: cascade impactors, cascade centripeters, cascade cyclones, or diffusion batteries, depending on the separation mechanism. The first three are inertial separators, while the latter separates by a diffusional mechanism. Inertial separators remove larger particles from the air stream first, depositing them onto a clean or greased surface or a filter. The amounts collected on each stage of the cascade can then be analyzed to allow calculation of the size distribution (see Chapters 6 and 22).

Generally, the size classification is performed in a series of steps in which the size cuts decrease by a factor of about 1.5 to 2 from one stage to the next smaller one.

For classification of particles in the submicrometer range, diffusion batteries can be used. These devices consist of several screens or collimated hole structures that allow particles to diffuse to the surface (see Chapter 19). The material collected on the screens or structures can be analyzed (e.g., for radioactivity or chemical composition). The size resolution of these devices is generally much poorer than that of impactors (see Chapters 19 and 22) or electrical classifiers (see Chapter 18), but they are relatively inexpensive.

DIRECT-READING MEASUREMENT OF AEROSOLS

Direct-Reading Measurement of Collected Particles

A wide range of physical and chemical principles have been applied to the detection and analysis of collected particles. Some of these approaches have resulted in direct-reading instruments. For various reasons, few of these devices have survived as commercial instruments. Several of these techniques are described in Chapter 14. Radioactive aerosol monitors are described in Chapters 34 and 35. Perhaps the most common aerosol particle property measured is the mass.

The most direct approach to mass measurement is the deposition of particles onto a vibrating surface and measurement of the change in resonant frequency. Two distinct types of instrument use this approach. The first uses a piezoelectric crystal as the collection surface. This provides high sensitivity and accurate mass measurement, but only for relatively small and sticky particles and only in very limited regions of the crystal. Large particles (several micrometers) may not couple well to the vibrating surface and may be poorly detected. The crystal has vibrational nodes on its surface, and the particles must be precisely deposited on the appropriate nodes to achieve consistent response. For additional description, see Chapter 14 and Williams et al. (1993).

Another vibrational sensor is the Tapered Element Oscillating Microbalance (***R&P***).* The collection substrate, either a filter or an impaction surface, is placed at the end of a tapered vibrating tube. The amount of mass collected on the substrate is related to the decrease in the resonant vibrational frequency of the tube. This approach appears to have fewer artifacts, although variations in temperature, humidity, pressure, and external vibrations can sometimes affect the accuracy of the measurement (see Chapters 14 and 26).

Another approach to mass measurement is the use of β-radiation scattering from collected material. The sample detector places the sample between a β-radiation source and a detector. The radiation is scattered by the electron cloud around the atoms of the sample, attenuating the radiation reaching the detector. The amount of attenuation is approximately proportional to the mass of material, although materials with low atomic number (e.g., hydrogen) have reduced scattering efficiency and are thus underdetected. Hydrocarbon compounds thus require a different calibration than most other materials.

A more recent development is the use of electrical charge to measure the concentration of particles on the individual stages of a cascade impactor (Electrical Low Pressure Impactor [ELPI], ***TSI***; see Chapter 14). Particles are charged as they enter the instrument, and the amount of charge transferred to each impactor stage by the impacted particles is measured with an electrometer. Each stage has a calibration constant that is a function of the charging efficiency for particles collected on that stage and is determined by weighing the portion of a representative test aerosol collected there.

* See Appendix I for full manufacturer addresses referenced by the italicized three-letter codes.

Light scattering from a larger volume than in optical partical counters is used to measure an integral scattering function that averages the signal over particle type and size distribution. This technique provides a rapid readout device that can sometimes be calibrated to give particle mass if the aerosol does not change significantly with time. These devices are called *nephelometers* or *photometers* and some are used as hand-held aerosol indicators in workplace or indoor air, while other more sensitive devices are used for measuring visibility and level of light scattering in the outdoor environment (see Chapters 15 and 16).

Direct-Reading Surface Area Measurements

Several approaches have been developed to measure aerosol particle surface area. One approach is to expose the aerosol to an ion field and measure the net charge accumulated by the particles. In the ELPI, size segregation of particles is accomplished after charging to measure size-dependent surface area. Number and mass distributions are calculated from these data. Another instrument uses charging in a similar fashion, but without size segregation to give total surface area. The epiphaniometer exposes the aerosol to a radioactive gas that decays to radioactive metal atoms, which diffuse to the aerosol particles' surface. The gas is removed before detection, and the resulting radiation detected is indicative of the total particle surface area. A third approach to surface area measurement uses short wavelength light to produce electron emission from the surface of the aerosol particles. The charge detected is a function of surface area and may also be specific for certain chemical species. See Chapter 14 for a discussion of these techniques.

Direct-Reading Measurements of Individual Airborne Particles

Direct reading instruments in this class generally separate or classify particles according to size, but require a particle sensor that responds quickly and efficiently to each particle. The most widely used sensor is the *optical particle counter* (OPC). In an OPC, particles pass through a sensing zone that is illuminated by either a broadband (white light) or a monochromatic (laser or light-emitting diode) source. If the instrument uses a laser, it may be called *a laser particle counter* (LPC). The light scattered by each particle is detected over a range of angles and converted to an electronic pulse that is a complex, but generally increasing function of particle size. Light scattering provides a relatively inexpensive, nondestructive, high-speed technique for particle detection. An OPC can be used for obtaining information about individual particles or for determining total particle concentration, for example in clean rooms (see Chapters 15, 16, and 33). With the appropriate optics, OPCs can be designed so that the sensing volume is external to the instrument, thus allowing the measurement of particles in extreme environments, such as outside of aircraft in the atmosphere (see Chapter 30) or in high-temperature stacks or reactors (see Chapter 31).

Optical particle counters are used as stand-alone instruments to detect and size particles. However, the light scattered by each particle has a complex dependence on the light source, the range of detection angles, the particle size, particle shape, and the particle refractive index. It is usually difficult to predict or compensate for the latter two factors in real-world situations, and thus the sizing capability of OPCs is usually only approximate.

In the small particle size range, particle detection by light scattering loses sensitivity, with a lower limit of about 0.1 μm under optimum conditions. To detect particles smaller than 1 μm, the OPC is often aided by condensational growth of small particles in the condensation particle counter (CPC), also called a condensation nucleus counter (CNC). The CPC exposes particles to a supersaturated vapor that condenses onto particles. All the particles grow to approximately the same size (on the order of a micrometer) and can then be detected by light scattering. The CPC can detect particles down to several nanometers and, under certain conditions, even size particles in the 1 to 3 nm range (see Chapter 19).

Direct-Reading Particle Size Distribution Instruments

Small Particle Size Range. There are several forms of the electrical mobility classifier that allow size separation of submicrometer particles. These devices operate by providing a known fraction of the particles with one electrical charge each and subjecting the particles to an electric field. Particles that achieve a selected velocity in the electric field (i.e., a selected electrical mobility) pass through the classifier and can be detected, usually with a CPC. Several of these devices have been developed and commercialized, each optimized for a specific size range or application (see Chapter 18). These devices can provide high-resolution size information in the range of a few nanometers to about 0.5 μm within several minutes. Larger particles are not sized accurately because they are likely to retain multiple charges.

Diffusion batteries can also be used as direct-reading instruments by detecting the particles passing through the diffusive collecting elements with a CPC. However, because of the inherently lower resolution of the size separation elements, diffusion batteries have significantly lower resolution than the electrical mobility classifiers. The size distribution must be deconvoluted from the diffusion battery's raw penetration data, and the deconvoluted spectra are subject to significant errors (see Chapters 19 and 22). Although these devices are less costly than the electrical mobility instruments, they have largely lost favor as direct reading instruments and are used primarily as integral sampling devices, as noted above.

Large Particle Size Range. The most common instruments for particle sizing are the OPC and the LPC, which can operate over a relatively wide range of concentration and size. The LPC generally produces a higher intensity beam at the sensing volume, resulting in higher sensitivity to small particles. Solid-state lasers are available with shorter wavelength that can also be used to detect smaller particles. These instruments provide rapid readout and moderate size resolution. They are subject to complex sizing errors as a function of particle parameters, as noted above, but for many applications provide a lower cost solution. Some of the errors in sizing can be reduced by appropriate calibration with the aerosol being measured (see Chapters 15 and 21).

The time-of-flight particle sizers sample the aerosol through a nozzle, accelerating the particles so that their velocity is a function of particle aerodynamic diameter. The velocity of the particles is measured by the time of flight of the particles through the sensing zone. The high acceleration through the nozzle produces non-Stokesian effects in the sizing process, and corrections usually have to be applied to obtain true aerodynamic diameter. However, the corrections, especially for known density and gas viscosity, are predicted from theory and can be accurately applied. These spectrometers can provide high-resolution spectra in less than 1 minute and give reasonably accurate results. Because of the relative complexity of these instruments, the sizing and concentration errors, although usually not great, sometimes can be subtle and difficult to correct (see Chapter 17).

One of the ultimate goals of aerosol measurement is to provide a complete analysis (e.g., size and chemical analysis) of individual particles in real time. A new type of instrument approaches this goal, although in a very expensive and rather bulky, yet movable, package (see Chapter 13). The time-of-flight sizing principle has been coupled with a laser ablation system and a mass spectrometer to produce particle size–dependent chemical information about an aerosol. There are significant limits to the size range and concentration that this instrument can measure, but it can provide information virtually unattainable by other means.

AEROSOL MEASUREMENT ERRORS

Figure 7–3 summarizes some major sources of biases that may occur in an aerosol measurement. The original unsampled aerosol may range in particle size from about 0.001 to about 100 μm. Various portions of this range may be nondetectable with a given measurement tech-

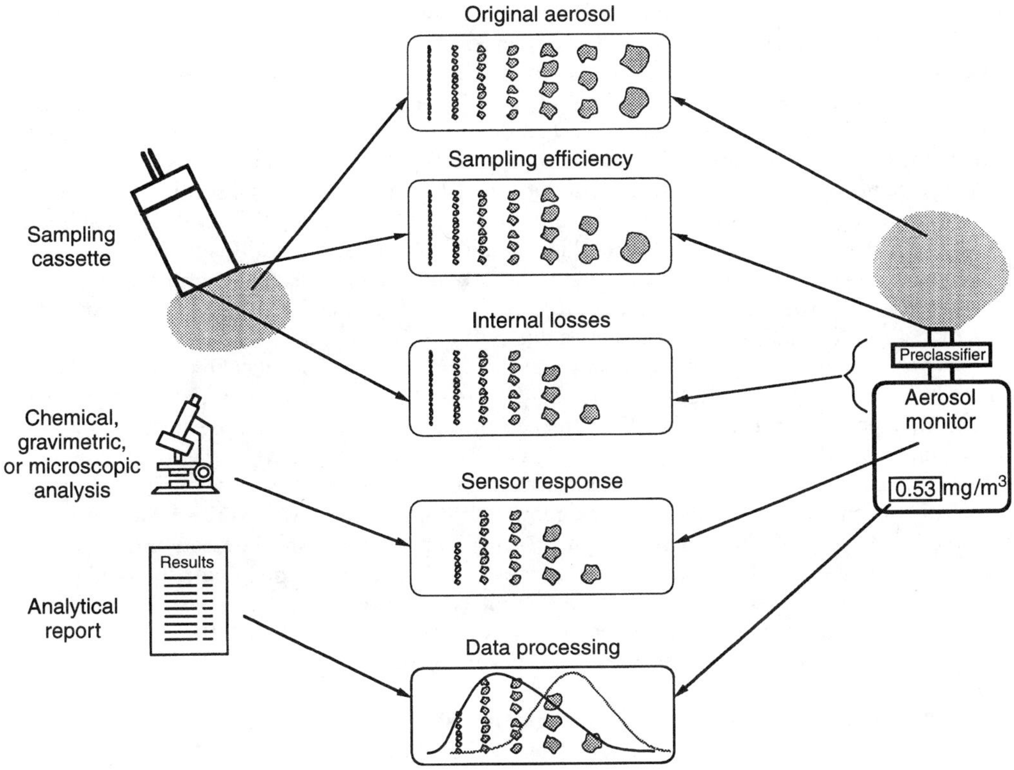

Fig. 7–3. Schematic representation of some important biases in aerosol monitoring. (Adapted from Willeke and Baron, 1990.)

nique. Particles smaller than about 20% of the wavelength of visible light (0.4 to 0.7 μm) are generally not detectable by optical means. Depending on the purpose of sampling and the type of aerosol present, different portions of this 0.001 to 100 μm size range may be of interest. For instance, the health scientist's concern has often focused on the 0.5 to 10 μm size range because the aerosol particle mass within this size range is likely to deposit in the biologically sensitive regions of the respiratory system. Measurement of such aerosols will be used as an example in some of the following discussion, parts of which have been adapted from Willeke and Baron (1990).

As the aerosol enters the sampling inlet of the aerosol measuring device, the ratio of ambient air velocity to sampling velocity, the air turbulence, as well as the size, shape, and orientation of the inlet, may affect the sampling efficiency of the inlet (Vincent, 1989; Okazaki et al., 1987a,b). Generally, the larger particles enter less efficiently, as illustrated in Figure 7–3, because of properties producing inertial losses and particle settling. Various particle size pre-classifiers, such as cyclones or elutriators, take advantage of these properties to impose size discrimination on sampled particles. Some of these devices are tailored to allow only a certain fraction of particles to pass through for detection. Aerosol particles reaching one specific region of health concern (i.e., the alveolar or gas exchange region of the lung) is defined as respirable dust. A cyclone is generally used to measure respirable dust as defined by the American Conference of Governmental Industrial Hygienists (ACGIH) definition of respirable dust, while a horizontal elutriator is used for the British Medical Research Council (BMRC) definition (ACGIH, 1984).

The section connecting the inlet to the collection device (e.g., filter) or sensor (e.g., detection region of photometer) is usually considered separately from the inlet or the point at which the aerosol enters the measurement device. For instance, in asbestos sampling a length of tubing equal in diameter to the filter collection area called a *cowl* (Baron, 1994) connects the sampling inlet to the collecting filter. In a direct-reading monitor, the aerosol is generally transported from the inlet to the sensor via a tube or channel. Particle losses may occur in these channels due to electrostatic attraction, impaction, or gravitational settling and further reduce the aerosol concentration, generally in the upper size range as illustrated in Figure 7–3. For devices with small inlets sampling submicrometer particles, diffusion may also contribute significantly to the losses. Thus, it is important to make this connection region out of conductive material to reduce electrostatically induced losses and, furthermore, to minimize the length of this region to reduce losses due to other forces.

When a filter sample is analyzed under a microscope, particles smaller than the wavelength of the illuminating light may not be detected efficiently. For an electron microscope, that wavelength is much smaller than for an optical microscope. Thus, the microscope and the human eye discriminate against detection of smaller particles. Other types of analysis also may have size-dependent biases introduced during sample preparation or analysis. The sensor of a direct-reading aerosol monitor has a lower threshold below which the smaller particles remain undetected, as illustrated in Figure 7–3. The upper size limit of detection is generally less of a problem for the sensor. However, most sensors do not respond equally to all particles of varying size and shape. This further modifies the measurement process. Sometimes, as with the electron microscope, the process of viewing particles can change the shape, state, or chemistry of the particle. Often these effects cannot be changed, but must be recognized during the analysis of the data.

A further bias can occur with instruments such as optical particle sizing instruments that depend on having only one particle at a time in the detector view volume. If more than one "coincide" in the view volume, the sensor only registers one particle, possibly of a larger size (Willeke and Liu, 1976). More complex instruments may produce more complex coincidence effects, modifying the observed size distributions in unusual ways (Heitbrink et al., 1990). These coincidence errors usually can be reduced by lowering the particle concentration (e.g., by inserting a dilutor before the sensor).

Data processing involves collection, storage, and analysis of the data. If too few particles have been sampled, the displayed particle size distribution may not reflect the true size distribution because of statistical considerations. If the particles are counted as a function of particle size under the optical microscope or in situ by an optical sensor, the volume or mass can be calculated for each particle, thus shifting the "weighting" from a "count" distribution to a "volume" or "mass" distribution. Various assumptions in this weighting procedure can bias the resulting distribution. The assumption of particle sphericity is usually an approximation, except for droplets. Because the particle volume depends on the cube of the particle size, a few large particles outweigh many small particles. Thus, presentation of the particle size by "count" for most naturally occurring aerosol size distributions focuses on a smaller size range than the size distribution weighted by "volume" or by "mass." The number of particles in the relevant size range, therefore, statistically limits the accuracy of the recorded aerosol concentration, indicating that a sufficient number of particles must be collected in the size range of interest.

The type of display, whether it is a histogram or a cumulative plot, emphasizes different aspects of the size distribution. Finally, the method of size calibration plays an important role in the accuracy of the results. For example, if a photometer or optical single particle counter is calibrated with particles that scatter but do not absorb light, an absorbing aerosol, such as coal dust, will be registered as having a smaller than actual particle size.

In the following sections, the aerosol size distributions used to demonstrate some of the above points were calculated using the Aerosol Calculator program (see Chapter 2). This type of program allowed the rapid calculation of lognormal size distributions (using Eq. 8–4) that can exist in sampled atmospheres, as well as how these distributions might be affected by biases and variability that occur with these measurements. The program used Eq. 7–1 for calculation of the number fraction (Δf) of particles in a size interval ranging from $\ln d$ to $\ln d + \Delta \ln d$ for a diameter d_i:

$$\Delta f = \frac{1}{\sqrt{2\pi}\ln\sigma_g}\exp\left[-\frac{(\ln CMD - \ln d_i)^2}{2(\ln\sigma_g)^2}\right]\Delta \ln d_p, \tag{7–1}$$

where CMD is the count median diameter of the lognormal distribution for which σ_g is the geometric standard deviation. Published sampling and measurement efficiency data were used to modify these lognormal distributions. Note that the number concentrations calculated were based on equal size increments on a log scale (i.e., $\Delta \log d_p$ = constant) so that the ordinate in each graph is $\Delta N/\Delta \log d_p$, where N is in the units of number of particles/m^3. Some curves were scaled to give a desired peak concentration.

The variability present in actual aerosol measurements of finite numbers of particles was simulated in some cases. Because aerosol particles arrive at a detector or sampler at random times, the count variability was described by a simulated Poisson distribution within each size increment. This variability was introduced by adding to each size increment a random number that was normally distributed (on the square root scale) about zero and had a variance equal to the particle count in that size increment.

In the following sections, some of the sources of bias and variability in measurement and interpretation are examined in more detail for some specific measurement situations. Note that this approach to calculating size distributions provides a convenient means of data analysis, both for planning experiments and for understanding published data.

Sampling and Transport

The measured size-dependent sampling efficiencies for the open and closed face 37 mm cassettes (Buchan et al., 1986), both widely used in industrial hygiene sampling with filters, have been multiplied by the corresponding values of an example lognormal size distribution with a median diameter d_{50} = 5.0 µm and a geometric standard deviation σ_g = 2.0 (Fig. 7–4). These samplers are used for a variety of dust measurements, and a smaller diameter version of the cassette is used for asbestos exposure measurement (Baron, 1994). Two sampling efficiency curves are calculated for an open and a closed face sampler hanging down with the inlet perpendicular to a horizontally moving wind stream of 100 cm/s; the third curve was calculated from measurements with the sampler on a mannequin facing the wind under the same wind conditions. The mannequin-mounted sampler curves were nearly identical for closed and open faced cassettes so a single average curve has been drawn for this case. It is apparent that the air flow conditions near the sampler inlet can significantly affect the collection efficiency of the sampler. The bluff mannequin body reduced the effect of wind speed on the sampler inlets. As pointed out in Chapter 6, the inlet efficiency is optimum when the air flow velocity and direction in the sampler and surrounding air are exactly or nearly matched. In Figure 7–4, there is a size-dependent reduction in particle concentration relative to the true concentration that varies with sampler placement.

Electrostatic attraction to the cassette inlet and its walls reduces the amount collected on the filter, especially if the cassette is constructed of nonconducting material (Baron and Deye, 1990). The loss increases with the number of electrical charges on the aerosol particles and on the sampler and decreases with increasing sampling rate. The number of charges on airborne particles depends on the process producing the particles, the air humidity or the amount of water on the particle surface during release, and the length of time the particles have been airborne (see Chapter 18). Direct-reading aerosol monitors may have similar sam-

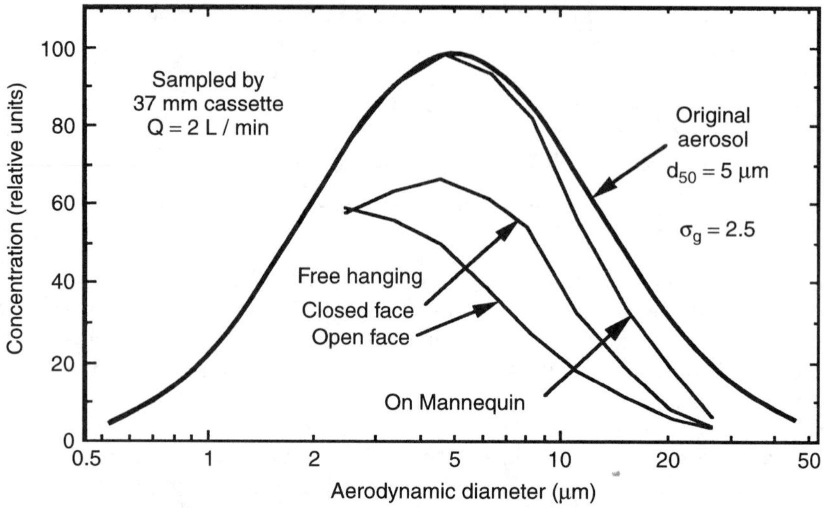

Fig. 7–4. Sampling and transport biases in several cassette configurations. Sampling efficiency data were taken from Buchan et al. (1986) and smoothed. Cassettes hung on a bluff body (a mannequin) appear to have smaller biases than free-hanging ones (Adapted from Willeke and Baron, 1990).

pling and transport losses, depending on the design of the inlet and the section leading to the sensor (Liu et al., 1985).

Sensor Sensitivity and Coincidence Effects

When the particles collected on a filter are analyzed by optical microscopy, many of the small particles are not detected by the microscopist, with none being counted below a certain size, say, 0.3 μm. The smaller particles of the original aerosol size distribution are thus not counted. If, in addition to inlet losses, the filter does not collect particles with 100% efficiency, the sample available for analysis may be further modified.

The combined effect of sensor response and inlet losses are illustrated for the Aerodynamic Particle Sizer (APS, ***TSI***), a time-of-flight aerosol spectrometer that uses light scattering to detect particles. To illustrate the effect of a sensor's size-dependent sensitivity, a lognormal size distribution with mass median aerodynamic diameter $d_{50} = 1\,\mu m$ and $\sigma_g = 2.5$ is calculated to simulate the measured aerosol (Fig. 7–5). Based on measured efficiency curves of Blackford et al. (1988), there is a modification of the "measured" size distribution at the low end due to a lack of detector response and at the high end due to a loss of particles at the instrument inlet. Note that neither of these losses changes rapidly with particle size and that the resulting distribution appears nearly lognormal. These modifications of the shape of the distribution may result in incorrect interpretation of the shape of the original aerosol distribution.

If the sensor is an optical device receiving a light-scattering signal each time a particle passes through the view volume, particle coincidence (i.e., simultaneous presence of two or more particles in the view volume) may result in the detection of a single larger particle, producing a slight shift to larger sizes and reducing the observed particle number over the entire size range. The importance of coincidence effects increases with particle number concentration.

In a time-of-flight device, such as the APS or the Aerosizer (***TSI***), the time of flight of a particle accelerated between two path-intersecting laser beams is a measure of its aerody-

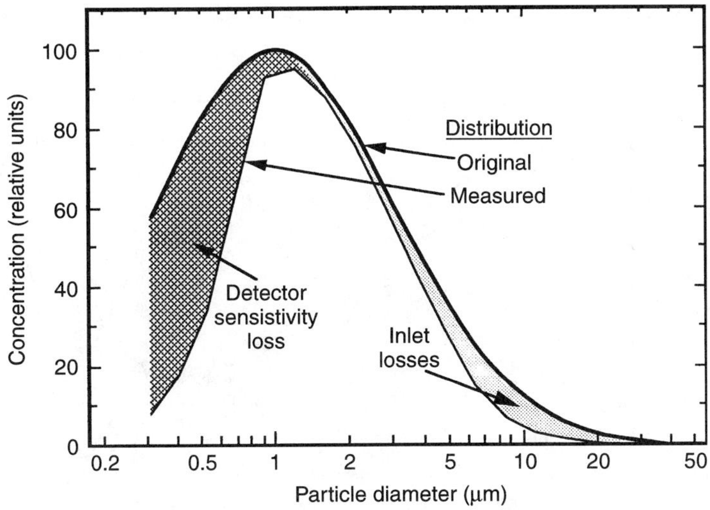

Fig. 7–5. Sensor bias data for the Aerodynamic Particle Sizer (APS3300) taken from Blackford et al. (1988). (Adapted from Willeke and Baron, 1990.)

namic liameter. These instruments can have coincidence losses like other optical particle counting instruments. In addition to a loss of particle counts, these instruments may produce a background of artifactual or phantom counts (due to more than one particle in the sensor at a time) at all particle sizes that may overshadow the fewer correctly detected particles at the tails of the distribution (Heitbrink et al., 1991). These phantom counts can be especially important if the distribution is converted to a mass distribution (a few large, phantom particles may outweigh the rest of the distribution) or if the data are used for comparison measurements (Wake, 1989), for example, the ratio of concentrations upstream and downstream of a cyclone.

Aerosol sensors of different types may be used to measure the same parameter, such as particle aerodynamic diameter (d_a). This can provide some estimate of the biases present in the measurements. However, when the readings from different instruments result in widely disparate readings, a detailed understanding of the detection and sampling processes can be used to estimate the "best" answer. A comparison of several measurement techniques used on a grinding wheel aerosol to measure aerodynamic diameter is shown in Figure 7–6 (O'Brien et al., 1986). Filter samples were analyzed by scanning electron microscope (SEM), and real-time measurements were made with two OPCs (Model CI-108, ***CLI***; Model ASAS-X, ***PMS***), a quartz crystal microbalance cascade impactor (Model PC-2 [instrument type discontinued], California Measurements, Berkeley), and an APS. The results from instruments not measuring aerodynamic diameter d_a (defined as the diameter of a unit density particle having the same gravitational settling speed as the particle in question) were converted to d_a. Such a conversion is generally made in health effect studies because gravitational settling and inertial impaction of particles in the human respiratory tract in the size range of about 0.5 μm and higher are directly dependent on aerodynamic particle diameter (Hinds, 1999).

One can make a best estimate of the aerosol d_a distribution based on knowledge of the size-dependent sensitivities of the instruments and the correction factors applied to each instrument's data. With the SEM and OPCs, relatively large correction factors based on assumed average particle shape, density, and refractive index may culminate in

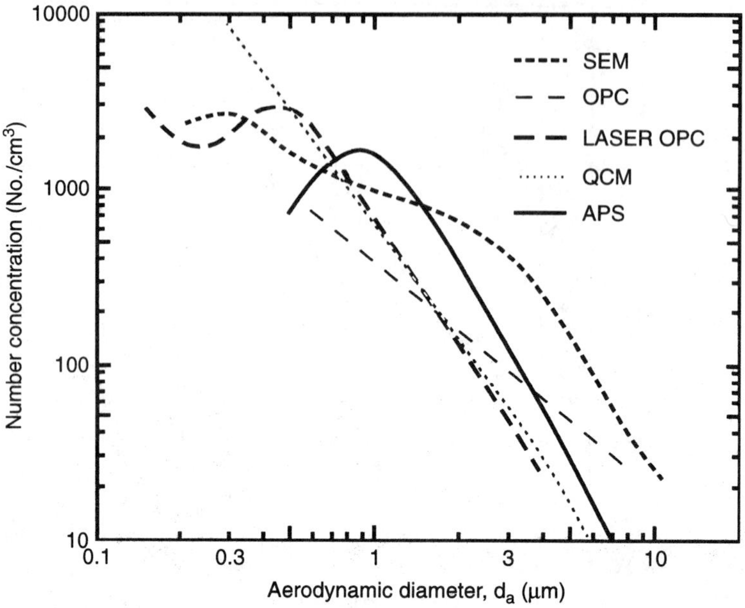

Fig. 7–6. Measurement of grinding wheel dust using six different measurement techniques, including a scanning electron micrograph (SEM), two optical particle counters (OPC), a quartz crystal microbalance cascade impactor (QCM), and an APS3300 (APS). (Adapted from O'Brien et al., 1986.)

relatively unsatisfactory results. The grinding aerosol was a difficult aerosol for such a comparison because of the presence of a number of materials with widely disparate properties.

Particle Statistics

Assuming that particles in an aerosol have been detected by a direct reading instrument, the distribution of particles can be simulated using the Aerosol Calculator spreadsheet indicated above. For a lognormal aerosol size distribution with a number median diameter of 2.5 μm and a geometric standard deviation of 2.0, the smooth number distribution curve calculated in Figure 7–7a results from a relatively large total count of 1 million particles distributed in 19 size increments over the size range 0.2 to 45 μm. Such a high particle count is realistic for dynamic sensors whose data acquisition systems permit multichannel analysis, but might overload a filter that must be analyzed by microscopy.

The surface area and the volume for each particle size may now be calculated. The surface area and volume of each size, multiplied by the number of particles in the respective size ranges, results in the distributions also shown in Figure 7–7a. The peak of each distribution is normalized to 100 relative units for illustration purposes. Inclusion of the particle density would allow conversion of the volume to a mass distribution. The representation of the aerosol size distribution by any of these weightings (count, surface, or volume) results in a smooth curve because a large number of particles was used.

When the total count is reduced to 1000 (Fig. 7–7b), the number distribution curve is still recognizable as approximately lognormal, although the additional variability due to a smaller count in each size increment is apparent. Example 7–1 indicates how to perform

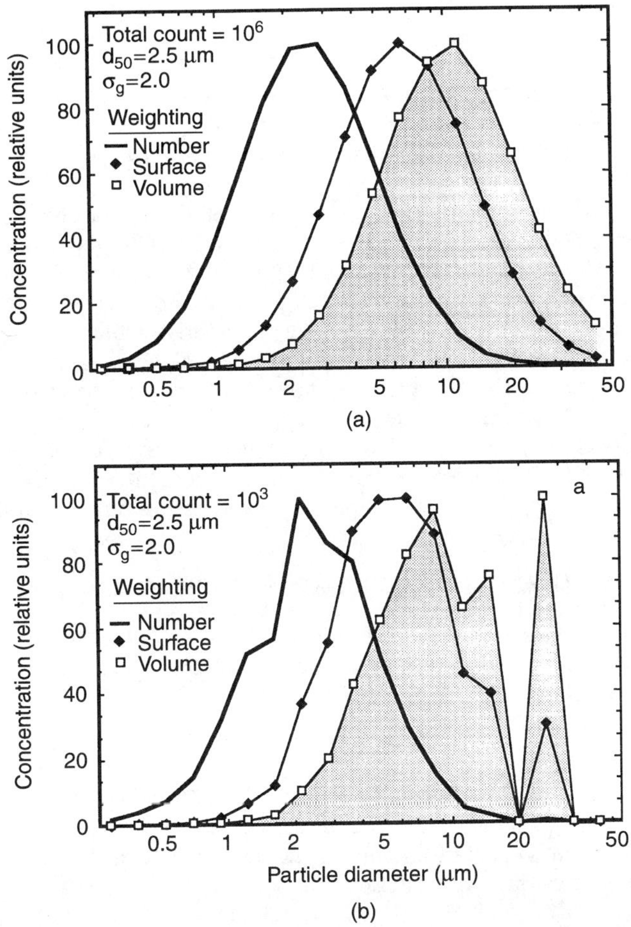

Fig. 7–7. Variability in volume, or mass, measurements. Surface area and volume distributions calculated from high (**a**) and low (**b**) particle count are given, and the curves are normalized to the same height. The volume variability at low count is due to statistical fluctuations, especially in the tail of the number distribution. (Adapted from Willeke and Baron, 1990.)

this calculation using the Aerosol Calculator. However, the surface distribution emphasizes the larger particles, of which there are fewer. The variability in particle count for these larger particles is greater. The volume or mass distribution (highlighted by shading) emphasizes even larger particles resulting in a poorly determined curve. Conversion of a count distribution to a volume or mass distribution by counting an insufficient number of particles, may, therefore, result in considerable imprecision. Figure 7–7b illustrates and emphasizes the need for measuring a large number of particles in the particle size range of interest. Several modern real-time aerosol monitors are computer based and offer easy conversion from one weighting to another. Such easy conversion may tempt the user to accept numbers that may have inherent biases and high variability. Note that the variability in mass due to a small number of large particles applies also to gravimetric measurements when small samples are obtained.

EXAMPLE 7–1

Calculate the number, surface and volume values for a lognormal distribution of spherical particles with a count median diameter of 5 μm and σ_g of 2. Simulate the variability as if the entire distribution contains approximately 1000 particles.

Answer: The following equations were developed in the spreadsheet program Excel (Microsoft Corp., Bellevue, WA) and were implemented in the Aerosol Calculator sizedis.xl module (see Chapter 2). The input values and constants are listed in the first four rows in the listing below. First we need to generate the diameters for which the lognormal distribution is produced. Column A has numbers starting at 0.25 with each following row multiplied by a constant factor, in this case 1.32, giving 19 size intervals or bins between 0.25 and 49 μm. The starting size and size interval can be changed to span the range of other size distributions if desired. The second column is the geometric mean of the upper and lower endpoints of each bin and is the size used to represent that bin. Thus, A6 to A7 is the first size interval and the geometric center of that interval is B6 $= \sqrt{A7 \times A6}$. C6 uses Eq. 7–1 to determine the concentration function in that bin or size interval.

$$C6 = ((A7 - A6)/(\$E\$1 * B6 * \$E\$3)) * EXP\left(-\left((LN(B6) - \$E\$2)\hat{}2\right)/\left(2 * (\$E\$3\hat{}2)\right)\right)$$

The "$" indicates that the reference does not change in the following rows, i.e., in C7, C8, etc. The concentration function is normalized to give the appropriate number of total counts, in this case 1000.

$$D6 = C6 * \$E\$4 / \$C\$25$$

C25 is the sum of all the values in column C. Next "noise" is added to the normalized particle density function to simulate the counting process. The following function produces a random number that is part of an approximately normal distribution, centered about zero with a standard deviation σ (Hansen, 1985)

$$\sigma = \sqrt{12n}\left[\left(\frac{\sum_{i=1}^{n} RAND_i}{n}\right) - 0.5\right]$$

where n is usually chosen to be a number 3 and $RAND_i$ is a random number between 0 and 1 that can be generated by the computer. The larger the value of n, the closer the resulting distribution will approximate a normal distribution, especially in the tails of the distribution. The Poisson distribution approaches a normal distribution for large particle counts, so this function provides a reasonable approximation to a Poisson distribution. Poisson statistics require that the variance of the particle counts be equal to the mean count. Thus the standard deviation of the count in each bin is equal to the square root of the value of the density function, i.e., the count in that bin. The function is rounded to integer values as would be produced by a counting instrument.

$$\begin{aligned} E6 = {} & ROUND(D6 + ((((RAND() + RAND() + RAND() \\ & + RAND() + RAND())/5) - 0.5) * SQRT(60) * SQRT(D6)), 0) \end{aligned}$$

where RAND() is a function that generates a random number between 0 and 1. E6 is the value for the number distribution. The surface and volume distributions are calculated from this distribution assuming spherical particles.

$$F6 = E6*PI()*B6\hat{}3$$

$$G6 = E6*PI()*B6\hat{}3/6$$

Finally, if one wishes to normalize the peak value of the distributions to the same value, e.g., 100, as in Figure 7–7, three more columns, H, I, J can be created that contain the normalized number, surface and volume distributions. These have not been included in the table below due to space considerations. H6, I6, and J6 would contain E6·100 / E25, F6·100 / F25, and G6·100 / G25, respectively. E25, F25 and G25 contain the maximum values in their respective columns. Note that the columns E, F and G (as well as H, I and J) will always appear somewhat different than indicated below since the random numbers will produce different results.

Further, sampling or detection efficiencies such as those indicated in Figure 7–4 and 7–5 can be calculated by multiplying the normalized density function (column C) by those efficiencies.

TABLE 7–1. Spreadsheet Size Distribution Calculation from Example 7–1 (Using the Aerosol Calculator Described in Chapter 2)

	A	B	C	D	E	F	G
1	d(50) =	5	SQRT (2*π) =		2.5066		
2	σ(g) =	2	LN (d(50)) =		1.6094		
3			LN (σ(g)) =		0.6931		
4	total	number of	particles =		1000		
5		diameter (μm)	Δf	Number	No. With Random Count	Surface	Volume
6	0.2500	0.2872	3E-05	0.03	0	0	0
7	0.3300	0.3791	0.0002	0.16	1	0.452	0.029
8	0.4356	0.5005	0.0006	0.65	1	0.787	0.066
9	0.5750	0.6606	0.0023	2.25	3	4.113	0.453
10	0.7590	0.8720	0.0067	6.69	4	9.556	1.389
11	1.0019	1.1511	0.0170	16.93	19	79.09	15.17
12	1.3225	1.5194	0.0366	36.50	30	217.6	55.10
13	1.7457	2.0056	0.0673	67.08	71	897.2	299.9
14	2.3043	2.6474	0.1052	104.96	110	2,422	1,069
15	3.0416	3.4946	0.1403	139.88	119	4,565	2,659
16	4.0149	4.6128	0.1592	158.79	154	10,294	7,914
17	5.2997	6.0889	0.1540	153.54	156	18,170	18,439
18	6.9956	8.0374	0.1268	126.45	122	24,759	33,167
19	9.2342	10.609	0.0890	88.71	81	28,643	50,646
20	12.189	14.004	0.0532	53.01	62	38,200	89,161
21	16.090	18.486	0.0271	26.98	19	20,397	62,843
22	21.238	24.401	0.0117	11.70	14	26,188	106,502
23	28.035	32.209	0.0043	4.32	5	16,296	87,482
24	37.006	42.517	0.0014	1.36	0	0	0
25	48.850		1.0027	1,000	156	38,200	106,502

Corrections for Density and Other Physical Properties

Curve A in Figure 7–8, shows a smooth, calculated representation of an aerosol size distribution with a median aerodynamic diameter of 5.0 μm and $\sigma_g = 2.0$. This aerodynamic particle diameter can be converted to physical particle diameter. The conversion is achieved by dividing the aerodynamic particle diameter by the square root of the particle density (see Chapter 3). A shape factor to account for nonspherical shapes also needs to be included in the conversion, but is not discussed here. For coal dust, which has a higher than unity particle density ($\rho_p \approx 1.45\,g/cm^3$), the physical diameter (curve B, Fig. 7–8) is smaller than the aerodynamic diameter. Thus, a particle of a given physical size settles in the same region of the respiratory tract as a physically larger, but less dense particle.

Curve A in Figure 7–8 represents the actual aerodynamic diameter distribution of a coal dust aerosol to be measured using an OPC. Typically, these counters, as well as photometers, are calibrated with spherical, nonabsorbing test aerosols such as dioctyl phthalate (DOP) or polystyrene latex spheres (PSL). For example, DOP has a refractive index $m = 1.49$ (no imaginary or absorptive component). All the light received by these test particles is scattered from the particles. However, if the particles are light absorbing, such as coal dust ($m = 1.54$ to $0.5i$, with 0.5 representing the absorptive component of the refractive index), a particle of a given size scatters less light. Therefore, a coal dust particle scatters much less light and appears much smaller than a similar-sized test particle; the particle size distribution is recorded to be in a smaller size range, as illustrated by curve C in Figure 7–8 (Liu et al., 1974). In addition, the size correction for the absorbing particle, such as coal dust, may be strongly particle size dependent, further distorting the measured size distribution. Although the distortion and shifting of the size distribution for coal dust is an extreme case, the assumptions involved (spherical particles, refractive index, and density values) illustrate some of the pitfalls of using optical sizing data to determine aerodynamic size.

An optical particle counter uses the optical properties of the individual particles for size discrimination. It therefore needs to be calibrated for the aerosol it measures. Other types of aerosol monitors use different physical properties for size discrimination. An electrical size

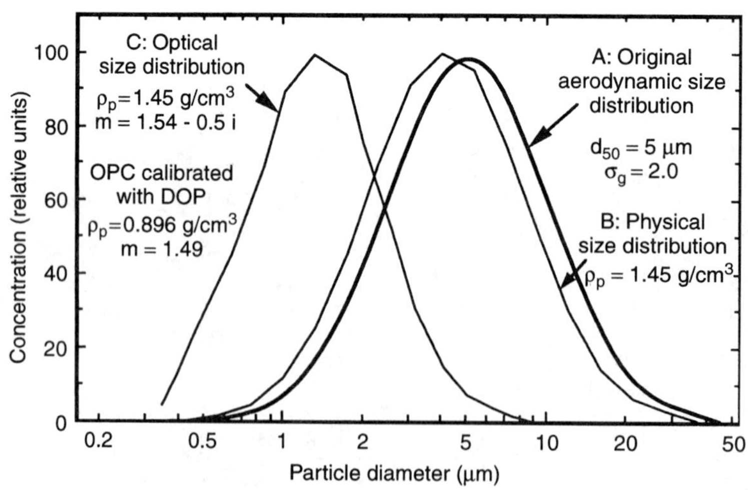

Fig. 7–8. Measurement of coal dust using various physical properties of the particles (density, ρ_p, and refractive index, m). Curve A represents the aerodynamic size distribution of a coal dust sample; curve B represents the physical size distribution of that dust (correcting for density); and curve C represents the measurement of the coal dust by an optical particle counter (OPC) calibrated with monodisperse dioctyl phthalate (DOP) particles (Liu et al., 1974). (Adapted from Willeke and Baron, 1990.)

classifier, for example, uses the electrical mobility of particles for size discrimination of submicrometer aerosols. Because the composition of the aerosol to be measured may be unknown, inadequate calibration may prevent the "size" obtained with one type of aerosol monitor from equaling the "size" obtained with another instrument.

Presentation of Size Distribution Data

There are several ways of presenting measured size distributions, each with advantages and disadvantages (see Chapter 22). Assume that two dusts are present in the air: dust 1 with a median diameter of 1.5 µm and dust 2 with a median diameter of 10 µm, both with a geometric standard deviation of 2.0. Measurement of the aerosol with a direct-reading aerosol size spectrometer is simulated using the Aerosol Calculator to give the bimodal size distribution shown in Figure 7–9A.

If this measurement is replotted on a cumulative plot where the value of the ordinate indicates the number of particles less than the given size, the wavy plot of Figure 7–9B results. Starting with the smaller particles, the curve increases with increasing particle size in an S-shaped manner. At sizes slightly larger than the median size of dust 1 the curve levels off and then increases in slope again as the median size of dust 2 is approached. This type of presentation is common for the results of low-resolution instruments such as a cascade impactor.

If one does not know that there are two dust modes present, one may be tempted to draw a straight line through the cumulative plot, as indicated by the heavy straight line in Figure 7–9B. This is frequently done and justified by attributing the deviation of the data from a straight line to experimental variability. The resulting graphically estimated or "measured" aerosol thus has a geometric median diameter of about 3.4 µm (corresponding to the minimum between the two dusts) and a geometric standard deviation of 3.5, indicating a single dust distribution much broader than each of the modes in the original bimodal distribution. Potentially valuable information is lost in this representation of the data because multiple modes usually indicate different sources of aerosol.

Some statistical tests may also indicate that the cumulative data in Figure 7–9B do not fit a single distribution. For instance, the Kolmogorov-Smirnov test (Gibson, 1971) would indicate whether the measured distribution fits a single mode distribution, and a plot of residuals (the differences between the measured and calculated values) qualitatively indicates whether adjacent measurements in the curve are correlated or whether the data fit the single lognormal distribution model.

Both types of representation have advantages and disadvantages. The differential plot gives a better presentation of the distribution shape: Modes show up directly, and any effect of bias is constrained to a narrow size range and is not propagated throughout the entire size distribution as in the cumulative plot. The cumulative plot provides a better estimate of the median diameter of the aerosol and allows easier presentation of data graphically without using a computer. Frequently, investigation of the data through several display techniques affords a more complete understanding of the physical meaning of the data.

Particle Size Selection

The type of aerosol monitor used depends on the purpose of sampling. The industrial hygienist generally samples from a health perspective. Because the physiological shape of the human respiratory system determines the region in which the particles will deposit, a pre-classifier is frequently mounted ahead of the sensor in order to intentionally limit the particle measurement to particles reaching the physiological region of concern. For example, a cyclone, impactor, or elutriator pre-classifier can separate the aerosol into respirable and nonrespirable fractions.

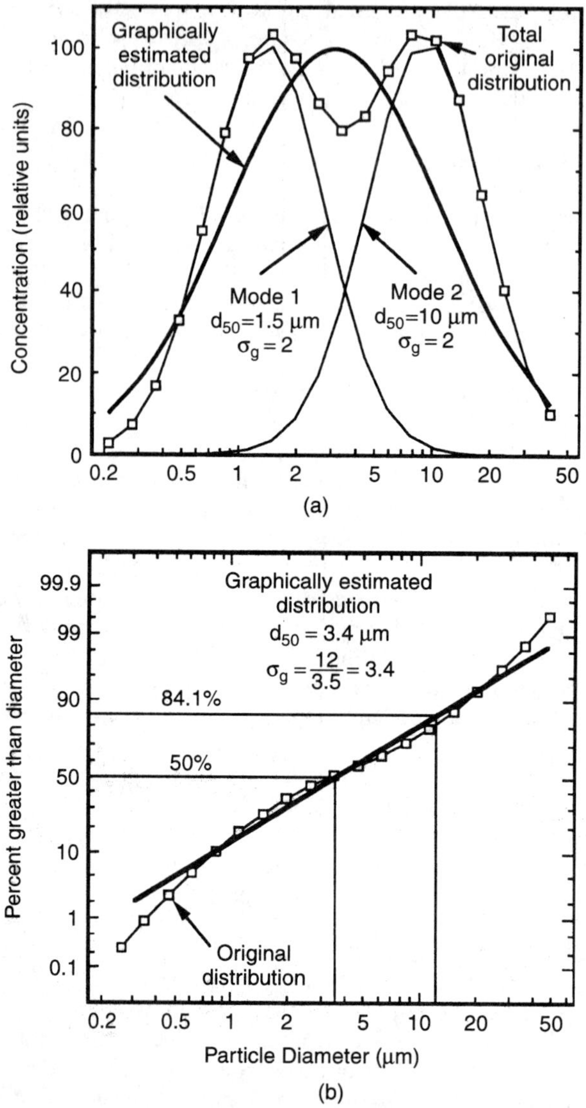

Fig. 7–9. Representations of a bimodal size distribution: histogram (**A**) versus log probability plot (**B**). It is possible to misinterpret the log probability plot of a bimodal distribution as being from a single mode. (Adapted from Willeke and Baron, 1990.)

The respirable mass fraction can also be obtained by sensor discrimination. Size distribution results can be weighted appropriately in each size range to give the respirable dust response (Baron and Willeke, 1986). Aerosol photometers are relatively inexpensive, direct-reading instruments that have a built-in size discrimination sometimes used for respirable dust measurements (Baron, 1994). Such light-scattering devices monitor the scatter of light from an aerosol cloud rather than from single particles. Figure 7–10 illustrates this for a specific photometer, the TM digital μP (***HUN***).

Figure 7–10 shows the calculated instrument response per unit mass concentration as a function of particle size for two kinds of aerosols with the same aerodynamic size distribu-

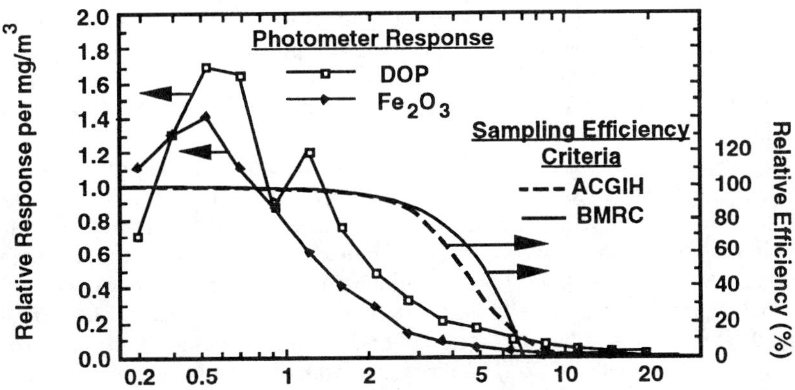

Fig. 7–10. Respirable mass response using a photometer for example size distributions of two materials with $d_p = 5\,\mu m$, $\sigma_g = 2.0$. Detection efficiency is for the TM Digital µP from the Hund Corp. Based on measurements by Armbruster (1987). Two definitions of respirable dust are also included. (Adapted from Willeke and Baron, 1990.)

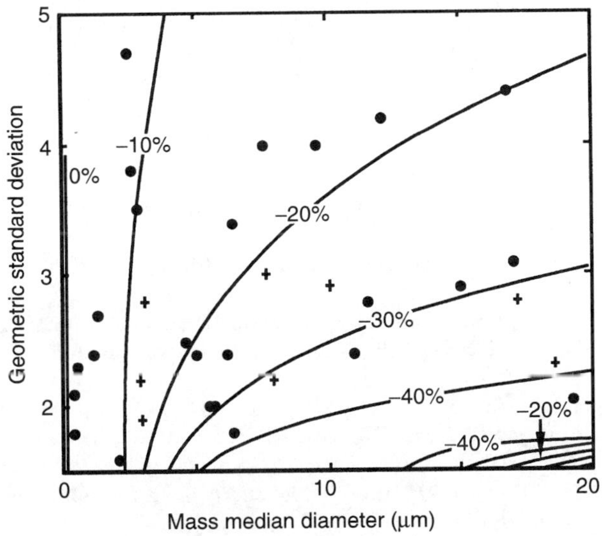

Fig. 7–11. Bias map comparing two defined respirable dust response curves for a range of lognormal size distributions. Data points represent distributions reported by (•) Hinds and Bellin (1988) and (+) Bowman et al. (1984).

tion ($d_{50} = 5\,\mu m$, $\sigma_g = 2.0$): non-light-absorbing DOP droplets and dense, light-absorbing iron oxide (Fe_2O_3) particles (Armbruster, 1987). The decline in response with increasing particle size above about $1\,\mu m$ is common to all photometers. This decrease approximately corresponds to the classification characteristics of the ACGIH and BMRC definitions for respirable dust (ACGIH, 1999a,b), also indicated on Figure 7–9. Complex interactions between the incident light and the particle result in similarly complex response curve patterns that differ from one type of aerosol to another.

A photometer calibrated with one type of aerosol will, therefore, generally be biased if used to measure another aerosol with different chemistry or size distribution. This bias can be adjusted for a specific aerosol by drawing the aerosol through a filter downstream of or parallel to the sensor and adjusting the sensor readout to equal the concentration measured using the filter. This procedure is valid as long as the type and size distribution of the aerosol remain unchanged. Instruments of this type can be used to make relative measurements, often providing useful real-time information, but should be used only with great care for situations requiring high accuracy.

One approach to evaluating the accuracy of a method over a wide range of aerosol size distributions is the use of a bias map (Caplan et al., 1977). This involves determining the range of size distributions over which the measurement is expected to occur. Hinds and Bellin (1988) reviewed aerosol distributions in more than 30 workplace operations and found size distributions with σ_g ranging from 1.5 to 5 and mass median aerodynamic diameters ranging from 0.1 to 20. One can examine the bias resulting from measurement of lognormal distributions throughout this range by comparison of one sampler versus a standard. As an example, calculation of the bias of the ACGIH definition versus the BMRC definition (Fig. 7–10) for each size distribution can be used to produce the bias map in Figure 7–11. This approach has been used to evaluate the optimum flow rate through a cyclone by comparing bias maps of the cyclone relative to a respirable dust definition at different flow rates (Bartley et al., 1994). Hinds and Bellin (1988) used their size-distribution data to estimate the effectiveness of respirators with measured size-dependent leakage.

REFERENCES

ACGIH. 1999a. *Particle Size-Selective Sampling of Particulate Air Contaminants*, ed. J. H. Vincent. Cincinnati, OH: American Conference of Governmental Industrial Hygienists.

ACGIH. 1999b. *Threshold Limit Values for Chemical Substances and Physical Agents*. Cincinnati, OH: American Conference of Governmental Industrial Hygienists.

Armbruster, L. 1987. A new generation of light-scattering instruments for respirable dust measurement. *Ann. Occup. Hyg.* 31:181–193.

Baron, P. A. 1984. Aerosol Photometers for respirable dust measurements. In *Manual of Analytical Methods*, 3rd Ed. (DHHS/NIOSH Pub. No. 84-100). Cincinnati, OH: National Institute for Occupational Safety and Health.

Baron, P. A. 1994. Asbestos and other fibers by PCM, Method 7400, Issue 2: 9/15/94. *NIOSH Manual of Analytical Methods, 3d eD.*, ed. P. M. Eller. (NIOSH) Pub. 84-100. Cincinnati, OH: National Institute for Occupational Safety and Health.

Baron, P. A. and G. J. Deye. 1990. Electrostatic effects in asbestos sampling I: Experimental measurements. *Am. Ind. Hyg. Assoc. J.* 51:51–62.

Baron, P. A. and K. Willeke. 1986. Respirable droplets from whirlpools: Measurements of size distribution and estimation of disease potential. *Environ. Res.* 39:8–18.

Bartley, D. L., C.-C. Chen, R. Song, and T. J. Fischback. 1994. Respirable aerosol sampler performance testing. *Am. Ind. Hyg. Assoc. J.* 55(11):1036–1046.

Blackford, D., A. E. Hansen, D. Y. H. Pui, P. Kinney, and G. P. Ananth. 1988. Details of recent work towards improving the performance of the TSI Aerodynamic Particle Sizer. In *Proceedings of the 2nd Annual Meeting of the Aerosol Society*. Bournemouth, U.K., March 22–24.

Bowman, J. D., D. L. Bartley, G. M. Breuer, L. J. Doemeny, D. J. Murdock. 1984. Accuracy criteria recommended for the certification of gravimetric coal mine dust samplers. Internal Report available from National Technical Information Service NTIS PB 85-222446. Cincinnati: National Institute for Occupational Safety and Health.

Buchan, R. M., S. C. Soderholm, and M. J. Tillery. 1986. Aerosol sampling efficiency of 37 mm filter cassettes. *Am. Ind. Hyg. Assoc. J.* 47:825–831.

Caplan, K. J., L. J. Doemeny, and S. D. Sorensen. 1977. Performance characteristics of the 10 mm cyclone respirable sampler: Part I—Monodisperse studies. *Am. Ind. Hyg. Assoc. J.* 38(2):83–95.

Currie, L. A. 1992. In pursuit of accuracy: Nomenclature, assumptions and standards. *Pure Appl. Chem.* 64:455–472.

Eisenhart, C., H. H. Ku, and R. Collé. 1990. Expression of the uncertainties of final measurement results: Reprints. In *Selected Publications for the EMAP Workshop.* NIST Internal Report 90-4272. Washington, DC: National Institute for Standards and Technology.

EPA. 1994. Guidance for the Data Quality Objectives Process, EPA QA/G4, in EPA/600/R-96/055. (Available at http://www.epa.gov/region10/www/offices/oea/qaindex.htm)

Gibson, J. D. 1971. *Nonparametric Statistical Inference.* New York: McGraw-Hill.

Hansen, A. G. 1985. Simulating the normal distribution. *BYTE* October:137–138.

Heitbrink, W. A., P. A. Baron, and K. Willeke. 1991. Coincidence in time-of-flight aerosol spectrometers: phantom particle creation. *Aerosol Sci. Technol.* 14:112–126.

Hinds, W. C. 1999. *Aerosol Technology.* New York: John Wiley & Sons.

Hinds, W. C. and P. Bellin. 1988. Effect of facial-seal leaks on protection provided by half-mask respirators. *Appl. Ind. Hyg.* 3:158–164.

Liu, B. Y. H., V. A. Marple, K. T. Whitby, and N. J. Barsic. 1974. Size distribution measurement of airborne coal dust by optical particle counters. *Am. Ind. Hyg. Assoc. J.* 8:443–451.

Liu, B. Y. H., D. Y. H. Pui, and W. Szymanski. 1985. Effects of electric charge on sampling and filtration of aerosols. *Ann. Occup. Hyg.* 29:251–269.

O'Brien, D. M., P. A. Baron, and K. Willeke. 1986. Size and concentration measurement of an industrial aerosol. *Am. Ind. Hyg. Assoc. J.* 47:386–392.

Okazaki, K., R. W. Wiener, and K. Willeke. 1987a. Isoaxial aerosol sampling: Non-dimensional representation of overall sampling efficiency. *Environ. Sci. Technol.* 21:178–182.

Okazaki, K., R. W. Wiener, and K. Willeke. 1987b. Non-isoaxial aerosol sampling: Mechanisms controlling the overall sampling efficiency. *Environ. Sci. Technol.* 21:183–187.

Pui, D. Y. H. 1996. Direct-reading instrumentation for workplace aerosol measurements—A review. *Analyst* 121:1215–1224.

Vincent, J. H. 1989. *Aerosol Sampling: Science and Practice.* New York: John Wiley & Sons.

Wake, D. 1989. Anomalous effects in filter penetration measurements using the aerodynamic particle sizer (APS 3300). *J. Aerosol Sci.* 20:1–7.

Willeke, K. and P. A. Baron. 1990. Sampling and interpretation errors in aerosol sampling. *Am. Ind. Hyg. Assoc. J.* 51:160–168.

Willeke, K. and B. Y. H. Liu. 1976. Single particle optical counter: Principles and application. In *Fine Particles: Aerosol Generation, Measurement, Sampling and Analysis*, ed. B. Y. H. Liu. New York: Academic Press.

Williams, K., C. Fairchild, and J. Jaklveic. 1993. Dynamic mass measurement techniques. In *Aerosol Measurement*, eds. K. Willeke and P. Baron. New York: Van Nostrand Reinhold.

PART II
Techniques

8 Sampling and Transport of Aerosols

JOHN E. BROCKMANN
Sandia National Laboratories*
Albuquerque, NM

INTRODUCTION

Aerosol measurement frequently requires that an aerosol sample be conveyed to a measurement device. This conveyance is accomplished by withdrawing a sample from its environment and transporting it through sample lines to the device. It is not uncommon for a sample to be transported to a chamber or bag for storage and subsequent measurement. An aerosol sampling system generally consists of

1. A sample inlet, where the aerosol sample is extracted from its ambient environment (the inlet shape and geometry may vary; although this variety is briefly discussed, this chapter focuses on sampling through thin-walled tubes)
2. A sample transport system consisting of the necessary plumbing to convey the aerosol sample to the measuring instrument or to a storage chamber (the components, or flow elements, consist of such items as tubes, elbows, and constrictions)
3. A sample storage volume (although this item is optional and its presence is determined by necessity rather than by choice) that will have an additional sample inlet and transport system to the measuring instrument (the storage volume is usually an inflatable bag that is filled with the aerosol sample over a time scale that is short compared with the time spent measuring the sample).

Figure 8–1 illustrates schematically a sampling system that withdraws an aerosol sample from the environment and transports it through a sample line to a measurement instrument. The inlet efficiency is defined as the fraction of aerosol particles in the environment that are aspirated (drawn) through the inlet plane of the inlet and transmitted through the inlet into the sampling line. Frequently a sample is drawn from a flowing gas stream. The free stream gas velocity is U_0 and the average velocity of gas flow in the inlet, the aspiration or sampling velocity, is U. Very small particles will follow the streamlines and will be aspirated through the inlet plane with nearly 100% efficiency. Large particles are influenced by inertia and are not as responsive to changes in the gas flow. In the limit, with coaxial free stream and sampling velocities, very large particles will approach the inlet plane at the free stream

* Sandia is a multiprogram laboratory operated by Sandia Corporation, a Lockheed Martin Company, for the United States Department of Energy under contract DE-AC04-94AL85000.

Aerosol Measurement: Principles, Techniques, and Applications, Second Edition, Edited by Paul A. Baron and Klaus Willeke.
ISBN 0-471-35636-0

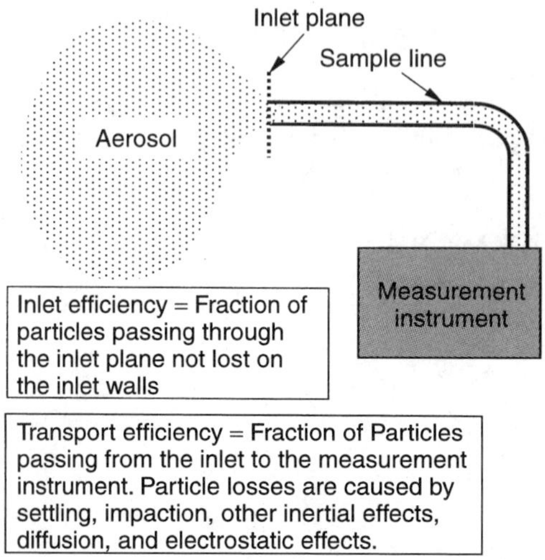

Fig. 8–1. Representative sampling of aerosols is critical to making accurate measurements. This schematic illustrates two important aspects of aerosol sampling efficiency: (1) the inlet efficiency, that is, the efficiency with which the aerosol is drawn from the environment into the transport line; and (2) the transport efficiency, that is, the efficiency with which that aerosol passes through the sampling line (possibly including a temporary storage volume) to the instrument.

velocity, and the efficiency with which these large particles are drawn through the inlet plane is U_0/U. This is defined as the aspiration efficiency. It is then advisable to keep this ratio of free stream gas velocity to sampling gas velocity close to 1 if we want representative sampling. Even if this aspiration efficiency is 100%, larger particles are lost in the inlet. Generally, the further from 100% the aspiration efficiency, the greater the losses of larger particles in the inlet.

It is desirable that the sample be representative of the aerosol in its original environment and not affected by the sampling process. Such characteristics as particle mass and number concentration and size distribution should remain unchanged between the point at which the aerosol is sampled and the instrument performing the measurement: This is representative sampling. It is, however, difficult to prevent changes from occurring during aerosol sampling and transport. Particles, because of their inertia, do not always enter the sampling inlet representatively. They can be lost from the sample flow by contact with the walls of the sampling system. Inertial, gravitational, and diffusional forces are among the mechanisms that can act to move the particles toward a wall. Any changes should be assessed quantitatively so that measurements may be corrected. Sampling practices that introduce uncharacterized changes should be avoided.

Many of the mechanisms that inhibit representative sampling depend on the aerosol particle size so that a given sampling system may exhibit representative sampling over some range of particle size but not for particles larger or smaller than that range. Generally speaking, larger particles are more strongly influenced by gravitational and inertial forces and are more difficult to sample representatively; smaller particles with higher diffusion coefficients are more easily lost to the walls of the sampling system by diffusion. If the sampled particles are charged, they can interact with electric fields near and inside the inlet. Use of conductive inlets, sampling lines, and storage containers can often minimize the effects of electrical charge on particles. Employing an aerosol sampling system that samples representatively for the particle size range of interest is of paramount importance. The potential factors that can cause changes in aerosol characteristics during the sampling process or can otherwise contribute to a nonrepresentative sample are

1. Aspiration efficiency and deposition in the sampling inlet during sample extraction
2. Deposition during transport through a sampling line or during storage
3. Extremes (high or low) or inhomogeneity in the ambient aerosol concentration
4. Agglomeration of particles during transport through the sampling line
5. Evaporation and/or condensation of aerosol material during transport through the sampling line
6. Re-entrainment of deposited aerosol material back into the sample flow
7. High local deposition causing flow restriction or plugging
8. Inhomogeneous particle concentrations in inlets and transport tubes

Each of these factors is addressed in the following sections of this chapter. The first two items are extensively dealt with in the aerosol extraction and aerosol transport sections, respectively. The third, fourth, and fifth items are addressed in the sample conditioning portion of the Other Sampling Issues section. The sixth, seventh, and eighth items are addressed in respective sections of the Other Sampling Issues section.

The representative extraction and transport of an aerosol sample is inhibited by loss and deposition mechanisms. The principal driving forces for these mechanisms are gravitational, inertial, and diffusive. The aerodynamic equivalent diameter (Eq. 4–40) is appropriate when gravitational or inertial forces drive deposition. Correlations describing this type of deposition are often functions of the particle terminal settling velocity, V_{ts}, the particle Stokes number, Stk, and the flow Reynolds number, Re_f:

$$V_{ts} = \tau g \tag{8–1}$$

$$Stk = \frac{\tau U}{d} \tag{8–2}$$

$$Re_f = \frac{\rho_g U d}{\eta} \tag{8–3}$$

where

τ = particle relaxation time (Eq. 4–34)
g = gravitational acceleration
U = characteristic gas velocity
d = characteristic system dimension
ρ_g = gas density
η = gas absolute viscosity

The mobility-equivalent, or diffusion-equivalent, diameter is appropriate when diffusive forces drive deposition. Correlations describing the diffusive deposition of particles are functions of the particle diffusion coefficient, D:

$$D = kTB \tag{8–4}$$

where

k = Boltzmann's constant
T = gas absolute temperature
B = particle dynamic mobility (Eq. 4–14)

The Stokes number is the ratio of the particle stopping distance (a measure of how quickly a particle can accommodate itself to a flowing gas) to the characteristic dimension of the flow geometry. The Stokes number characterizes the inertial behavior of particles. Particles with large stopping distances have high inertia and large Stokes numbers. Representative sample extraction and transport become more difficult for larger particles because their higher inertia makes them less susceptible to influence by the sample flow.

Calibration

Sampling systems should be calibrated for aerosol sampling and transport efficiency at the gas flows and over the size range of interest. Ideally, a sampling system should be calibrated fully assembled under the conditions in which it will be operating. Often, a calibration of component sections at operational conditions is adequate. Calibration at other than operational conditions may be sufficient if a defensible means, such as models and correlations from the literature, is employed to apply the calibration and predict the performance under operational conditions.

Under some circumstances, a user may not have the means for aerosol calibration. The use of a specified sampling protocol or the use of calibration data found in the literature for some of the commercial samplers or for components in the sampling system may be adequate to ensure that operation will be in a range with acceptable sampling efficiency for the user's application. In this case, at least a flow calibration should be performed.

A system's sampling efficiency can be estimated when the system is composed of components that have well-characterized efficiency data and models available in the literature. A number of these models are reviewed in this chapter. In this case, operation at an estimated sampling efficiency much different from 100% increases the uncertainty in the estimate of the actual sampling efficiency.

Sample Extraction

An aerosol sample is extracted from its environment into an inlet for transport to the measuring instrument. Drawing a representative aerosol sample into an inlet is not trivial. The velocity and direction of the gas from which the sample is being drawn, the orientation of the aerosol sampling probe, the size and geometry of the inlet, the velocity of the sample flow, and the particle size are important factors in how representative an extracted sample is. In extracting a sample, a particle must be sufficiently influenced by the sample gas flow to be drawn into the inlet. The particle must also be transported through the inlet without being deposited in the inlet. Particle inertia and gravitational settling are impediments to representative sample extraction, and representative sampling is more difficult with increasing aerodynamic particle size.

The aspiration efficiency, η_{asp}, of a given particle size is defined as the concentration of the particles of that size in the gas entering the inlet divided by their concentration in the ambient environment from which the sample is taken. The transmission efficiency, η_{trans}, of a given particle size is defined as the fraction of aspirated particles of that size that are transmitted through the inlet to the rest of the sampling system. The inlet efficiency, η_{inlet}, is the product of the inlet and transmission efficiencies and is the fraction of the ambient concentration that is delivered to the aerosol transport section of the sampling system by the inlet:

$$\eta_{\text{inlet}} = \eta_{\text{asp}}\eta_{\text{trans}} \tag{8–5}$$

Sample Transport

The transport of the aerosol sample through sample lines from the inlet may occur directly to the measurement instrument or into a temporary storage volume for subsequent transport to instruments via sample lines. These sample lines may contain bends, inclines, contractions, and other flow elements; flow may be laminar or turbulent. The deposition of particles during residence in a bag and during transport will alter the characteristics of the aerosol reaching the measurement instrument. Other phenomena that will change the characteristics of the aerosol in the sample flow are particle growth by agglomeration or condensation, particle evaporation, and re-entrainment of previously deposited material into the sample flow. These phenomena are discussed later; sample transport will address aerosol deposition.

A number of deposition mechanisms may be operating, and several can be operating in each flow element. Some are not well characterized, and conditions where these are encountered should be avoided wherever possible. Deposition mechanisms can be dependent on the flow regime (laminar or turbulent), the flow rate, the tube size and orientation, temperature gradients, vapor condensation onto the walls of the system, and particle size. Various deposition mechanisms depend on different particle equivalent diameters. Settling and inertial deposition depend on the particle's aerodynamic diameter, while diffusional deposition depends on the particle's diffusion, or mobility, diameter. The transport efficiency for a given particle size, through a given flow element, under the action of a given deposition mechanism, $\eta_{\text{flow element, mechanism}}$, is defined as the fraction of those particles entering the flow element that are not lost by that deposition mechanism during the transit of that flow element. It is defined as a function of particle size for specific deposition mechanisms that are operating in the flow element. The total transport efficiency, $\eta_{\text{transport}}$, for a given particle size is the product of the transport efficiencies for each mechanism in each flow element of the sample transport system for that particle size:

$$\eta_{\text{transport}} = \prod_{\substack{\text{flow} \\ \text{elements}}} \prod_{\text{mechanisms}} \eta_{\text{flow element, mechanism}} \tag{8–6}$$

The sampling efficiency, η_{sample}, is the product of the inlet and total transport efficiencies:

$$\eta_{\text{sample}} = \eta_{\text{inlet}}\eta_{\text{transport}} \tag{8–7}$$

This chapter presents correlations for the transport efficiencies for various mechanisms operating in various flow elements so that the reader can estimate the total transport efficiency for a sampling system.

Other Sampling Issues

There can be times when the sampled aerosol concentration (either mass or number) is too high for the sampling instrument. Under these circumstances, the sample must be diluted with clean gas to bring the concentration within the measurement range of the instrument. Uncertainty in the dilution and sample flows will produce uncertainty in the calculated concentration, which must be addressed. High number concentrations may drive the aerosol to undergo rapid coagulation, which alters the distribution; the number concentration decreases, and the mean particle size increases. Dilution of the sample will arrest the coagulation process so that a representative sample can be measured.

The sampled aerosol may be in a condensing or evaporating environment. Condensation or evaporation of material on or from aerosol particles will change the size of the particles and the total suspended mass of aerosol material. To obtain a representative sample from an environment in which material (such as water vapor) is condensing on the particles, the sample may have to be conditioned by dilution or heating. Obtaining a representative sample from an environment in which particle material is evaporating from the particle is more difficult and can be addressed by minimizing the time between sampling and measurement to keep evaporation to a minimum.

In sampling from the ambient atmosphere, from a room, or from a duct, one must be concerned with the homogeneity of the aerosol throughout the volume of gas. A representative sample requires sampling at a sufficient number of points to give an accurate picture of the aerosol throughout the volume of interest (Fissan and Schwientek, 1987). In the case of duct sampling, the American National Standards Institute (ANSI) standard N13.1 (ANSI, 1969) provides agreed-upon sampling locations to obtain a representative sample. In room sampling, convection in the room can cause considerable inhomogeneity in the aerosol. This is especially significant in situations corresponding to very low concentrations, such as in clean rooms, where long sampling times are required for meaningful particle counting statistics to be obtained (Fissan and Schwientek, 1987). Sampler placement in this situation may be made on the basis of flow modeling or by the use of tracer smokes or fogs. Sampler placement and sample inhomogeneity are further discussed in Chapter 29 (on indoor aerosols). Having made the reader aware of the pitfalls of inhomogeneity of the aerosol in attempting to obtain a representative sample, it is assumed that this problem has been addressed and, therefore, attention can now be focused on a single sampling point.

Summary

Correlations describing aspiration efficiency, transmission efficiencies, and transport efficiencies are given in subsequent sections. These correlations can be used to evaluate the performance of an existing sampling system or to aid in the design of a sampling system. Because these correlations are based on assumptions and experiments that are not always the same as the reader's application, they may not be applicable for calculated efficiencies much different from 1. Because the efficiencies are particle size dependent, the range of particle sizes over which sampling is representative (sampling efficiency close to 1) can be estimated with a fair degree of confidence using these correlations. In designing a sampling system, parameters such as flow, line size, orientation, and length can be adjusted using the correlations to estimate the efficiency for the particle size range of interest to achieve representative sampling. Of course, the sampling system should be experimentally evaluated whenever possible. At the end of the presentation for each type of efficiency correlation, a short qualitative discussion is given on how the efficiency changes with the dependent parameters. Sampling situations to be avoided and the methods of avoiding them are also discussed.

Although some of the phenomena discussed have been extensively investigated and characterized, others have not. It is the purpose of this chapter to provide the reader with some background information on how aerosol sampling can be accurately evaluated and appropriately designed and how sampling pitfalls may be avoided. For additional information on aerosol sampling, the reader is referred to the review paper on sampling of aerosols by Fuchs (1975), to the review on aerosol sampling and transport by Fissan and Schwientek (1987), and to the book by Vincent (1989) on aerosol sampling.

SAMPLE EXTRACTION

Aerosol sampling arises from a number of requirements. Some of these are

1. Monitoring the ambient air for pollution
2. Monitoring air in the workplace for hazardous materials
3. Monitoring of exhaust stacks and lines to monitor pollution control equipment
4. Monitoring clean rooms for particulate contamination

5. Monitoring from manufacturing or industrial processes
6. Monitoring in experimental research

In all these applications, the first step in obtaining a sample is sample extraction.

There are two basic situations in aerosol sampling:

1. Sampling of particles from a quiescent environment
2. Sampling from a gas flow that carries aerosol particles

Ambient air sampling must deal with both quiescent sampling and sampling from flowing gas. It often does so by the use of an inlet coupled with an inertial particle size fractionator. The inlet samples representatively for particles smaller than some specified diameter over a specified range of ambient wind velocities and the internal particle size fractionator passes 50% of the particles of that specified diameter. Ambient air sampling is usually performed with commonly available samplers that incorporate an inlet and size fractionator in conjunction with the measurement device, usually a filter or an impactor, located immediately after the size fractionator so that transport of the sampled aerosol and the attendant losses are minimized.

Liu and Pui (1981) and Armbruster and Zebel (1985) discuss inlet design and performance for ambient air samplers. These designs are tested in wind tunnels to determine their sampling efficiency as a function of particle size and wind speed. In 1987, the U.S. Environmental Protection Agency (EPA) set a standard for airborne particulate matter called PM-10 (U.S. Environmental Protection Agency, 1987). PM-10 required, among other things, samplers to sample 10 μm aerodynamic diameter particles with 50% efficiency. This can be accomplished by employing an inlet with a high inlet efficiency and a fractionator that allows 50% of the 10 μm aerodynamic diameter particles to pass into the rest of the sampler. The PM-10 regulations require that, for a sampler to be officially accepted as a PM-10 sampler, it must pass specified tests in a wind tunnel. The intent was to allow flexibility in sampler design while maintaining a consistency in sampler performance.

Although the inlets may perform as required, the internal processes of the samplers, specifically their size fractionators, may cause them to yield results different from those expected from their qualification testing. John et al. (1991) and John and Wang (1991) present a comparison between the Sierra-Andersen model 321A PM-10 sampler and the Wedding high-volume PM10 sampler. They show that loading of the samplers, and whether or not the fractionator was oiled, produced an effect on the sampling effectiveness. Deagglomeration and re-entrainment of collected material, caused by bombardment with sampled aerosol particles, was found to produce anomalous results in the Sierra-Andersen sampler. These difficulties with PM-10 samplers appear to be more in the area of instrument response, but they illustrate some of the pitfalls in sampler design.

PM-10 inlets may be too bulky for applications in which an aerosol sample must be extracted from a duct or in which room air must be sampled at a number of locations. Other inlets are required for these applications.

One type of inlet is the *blunt sampler*. This term encompasses a number of sampler inlets ranging from what could be called thick-walled nozzles to those in which the inlet is small compared with the overall sampler dimension. Vincent et al. (1982) describe a blunt sampler as one in which the sampler and inlet configuration present a large physical obstruction to the flow. This type of sampling nozzle may be configured as a flat disk with a small centrally located sampling orifice (e.g., Vincent et al., 1985). Alternatively, the sampling orifice may be in a spherical body or in some body of intermediate shape (Vincent, 1984; Vincent and Gibson, 1981). Particle deposition on the lip or face of a blunt sampler, particle bounce at the inlet, and re-entrainment of material into the inlet make blunt samplers difficult to characterize for larger particles. There may be difficulty in obtaining representative sampling of larger particles as well.

Another inlet type is the thin-walled nozzle. This nozzle is an idealized sampling nozzle that does not disturb ambient flow and that has no rebound of particles from the leading edge into the nozzle. Sampling with a thin-walled nozzle has received more extensive study than sampling with a blunt sampler or a thick-walled nozzle. For practical usage, a nozzle can be regarded as "thin walled" when the ratio of its external to internal diameter is less than 1.1 (Belyaev and Levin, 1972). This chapter deals with sample extraction employing thin-walled nozzles.

Sampling situations in which the flow velocities are varying present a problem. Generally, the sampling velocity is not a variable quantity. A variation in the sample flow rates introduces variable transmission efficiency through the sampling lines that can effectively result in nonrepresentative sampling. In fact, most instrumentation has a measurement response that is dependent on flow rate and, consequently, operates at a fixed flow rate. Exceptions are the instruments that perform a total integral collection, such as filters. These instruments can be positioned close to the inlet, minimizing the transport distances and losses, making them relatively independent of sample flow rate. This situation still requires some type of integral flow measurement should a variation of sample flow occur.

Still, sample flow rate may vary because a sampler is turned on and then off. If the sampling velocity is constant during the period of time in which the sampler is on, and the sampler is on for a long period of time compared with the sampled gas residence time in the sampler's inlet and sampling lines (at least a factor of 10 greater), then the dead volume of gas in the inlet and lines is cleared and is small compared with the total volume of gas sampled. The assumption of constant sampling velocity is valid.

Ambient free-stream gas velocity variations may be beyond the control of the user because of flow adjustments or conditions in the duct from which the sample is being drawn. This situation is commonly encountered. Under these conditions, one may sample at a constant sample flow rate over a range of free-stream flows and note the largest particle size for which representative sampling still occurs over this range. The measurements made with this sampling system would need to disregard particles larger than this noted size because their sampling efficiency would be effectively unknown. This is a similar approach to that used in ambient sampling but may not be optimized for large particles. One may develop an inlet along the lines of an ambient air sampler to optimize performance for larger particles over a wide range of free-stream velocities. This has been done by McFarland et al. (1988), in which they present a shrouded aerosol sampling probe that representatively samples approximately 10 μm and smaller particles from duct flow ranging from 2 to 4 m/s.

An alternative approach to the problem of varying free-stream flows is to vary the sample flow so that representative sampling is maintained over the range of free-stream flow variation. This entails variable sample flow and should only be used under conditions in which the particle loss in sampling lines and the instrument response do not depend strongly on sample flow rate. Null-type nozzles—in which pressure measurements responding to the flow inside and outside of the nozzle—are balanced at a null condition so that the sampling and free-stream velocities are matched (Paulus and Thron, 1976; Orr and Keng, 1976) can be used to obtain representative sampling. This is an active system in which the sample flow is adjusted to obtain the null condition. The null condition may, because of local fluctuation in the flow, not reflect equality in the sample and free-stream velocities. Kurz and Ramey (1988) suggest an active sampling nozzle that employs a flow sensor and flow controller to maintain representative sampling conditions over a range of duct velocities.

There are specific sampling protocols given by the EPA for source sampling of particles in stacks (U.S. Environmental Protection Agency, 1974) when the sampling data are required to verify or test compliance with rules. Sample trains and procedures are specified. The reader is referred to appropriate EPA documentation for this type of sampling.

This chapter concentrates on aerosol sampling through thin-walled nozzles and the transport of the sampled aerosol through sampling lines to the instrument.

Efficiency

Withdrawal of an aerosol sample from its environment into the sampling system requires making a particle enter the sampling inlet and conveying it to the transport portion of the system. The efficiency with which this is accomplished is called the *inlet efficiency*. There are two components to the inlet efficiency, η_{asp}, and η_{trans} of Eq. 8–5. These efficiencies are dependent on the ambient gas velocity, U_0, the inlet geometry, size, and position, the sampling gas velocity, U, and the particle's aerodynamic diameter, d_a.

In efficiently extracting a sample, the sampling gas velocity must be low enough so that the sampled particle can accommodate itself to the sampling gas flow within a distance comparable to the inlet diameter. This is an inertial condition. The sampling gas velocity must also be high enough so that the sampled particle does not settle appreciably in the time that sampling occurs. This is a gravitational settling condition (Davies, 1968).

In sampling from a flowing gas with a nozzle, it is implicitly assumed that the flow velocities are large compared with the settling velocity of the particles being sampled (i.e., that the gravitational-settling condition is met). Grinshpun et al. (1990) point out that in low-velocity sampling, the aspiration efficiency will depend on the ratio of the settling velocity to the ambient gas velocity. It is prudent to determine this ratio for the particle size of interest to ensure that the gravitational-settling condition is met.

A nozzle sampling from still air or from flowing gas may be used in various orientations with respect to gravity and the ambient gas stream flow direction. A nozzle is said to face in the direction opposite to the inlet sample flow direction. Thus, a nozzle facing upward draws the sample downward, and a nozzle facing the gas flow draws a sample in the same direction as the gas flow. A nozzle facing the gas flow where the direction of the sample flow is aligned with that of the gas flow is said to be sampling isoaxially. Anisoaxial, or nonisoaxial, sampling occurs when the gas flow and sample flow directions are not parallel.

Sampling is said to be isokinetic when it is isoaxial and the mean sample flow velocity through the face of the inlet is equal to the gas flow velocity. Strictly speaking, the term *isokinetic* applies only to laminar flow in the ambient free stream. The more general term, *iso-mean-velocity*, is applicable to both laminar and turbulent flow conditions in the free stream. Convention, however, applies the term *isokinetic* to both flow regimes. This chapter employs conventional terminology, but the reader should be aware of the distinction. Sampling with a sampling velocity not equal to the gas velocity is anisokinetic (*aniso-mean-velocity*) sampling. When the sampling velocity is higher than the gas velocity, the sampling is super-isokinetic (*super-iso-mean-velocity*), and when the sampling velocity is lower than the gas velocity, the sampling is sub-isokinetic (*subiso-mean-velocity*).

Figure 8–2 is a schematic of isoaxial sampling with a thin-walled nozzle for isokinetic ($U = U_0$), sub-isokinetic ($U < U_0$), and super-isokinetic ($U > U_0$) flow conditions. The limiting streamline represents the boundary between gas that enters the inlet and gas that does not. Gas is always sampled representatively, and particles that do not deviate from the gas streamlines will also be sampled representatively. Particles with sufficient inertia to deviate from the streamlines may not be sampled representatively. The figures are, strictly speaking for laminar flow in the ambient gas stream. This condition is not always encountered. Turbulent flow in the ambient gas stream introduces a lateral component to the gas velocity that in turn influences the particle motion. However, these figures are qualitatively correct in their depiction of flow and particle transport to and through the inlet for both laminar and turbulent flow conditions.

Figure 8–2a shows isokinetic sampling in which the limiting streamline flows directly into the nozzle without deviation. In this case, the aspiration efficiency is 1 (100%). Transmission

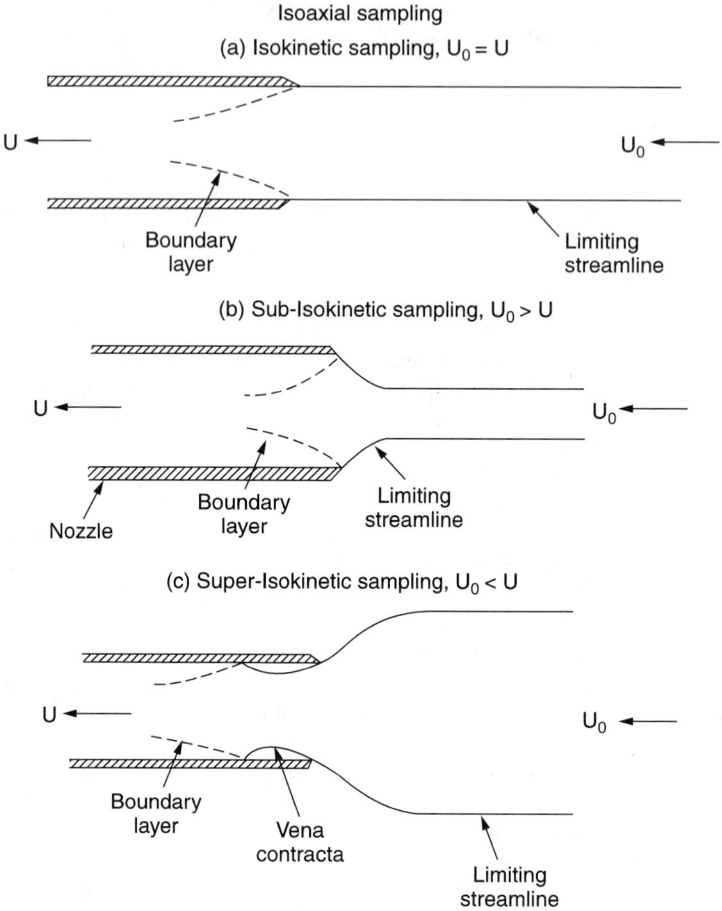

Fig. 8–2. Schematic diagram of isoaxial sampling with a thin-walled nozzle, with sample flow gas velocity U and free-stream ambient gas velocity U_0 under **a**, Isokinetic ($U = U_0$) sampling conditions; **b**, sub-isokinetic ($U < U_0$) sampling conditions; and **c**, super-isokinetic ($U > U_0$) sampling conditions.

losses arise from gravitational settling inside the nozzle (Okazaki et al., 1987b). Losses in the inlet can also be caused by free-stream turbulence (Wiener et al., 1988) in which the particles' lateral motion caused by turbulence causes them to impact the internal wall of the inlet.

Figure 8–2b shows sub-isokinetic sampling in which the limiting streamline must diverge from the ambient free-stream flow into the nozzle. Particles with sufficient inertia that lie outside the limiting streamline can cross the limiting streamline to be aspirated by the nozzle. In this case the aspiration efficiency is 1 or more for all particles, increasing from 1 to U_0/U for larger particles. Transmission losses arise from gravitational settling in the nozzle (Okazaki et al., 1987b) from free-stream turbulence effects (Wiener et al., 1988) and from inertial impaction on the inner wall of the nozzle by particles with velocity vectors toward the wall caused by the expanding streamlines (Liu et al., 1989).

Figure 8–2c shows super-isokinetic sampling in which the limiting streamline must converge from the ambient free-stream flow into the nozzle. Particles with sufficient inertia that lie within the limiting streamline can cross the limiting streamline and not be aspirated by the nozzle. In this case the aspiration efficiency is 1 or less for all particles, decreasing from 1 to a limit of U_0/U for larger particles. Transmission losses arise from gravitational settling

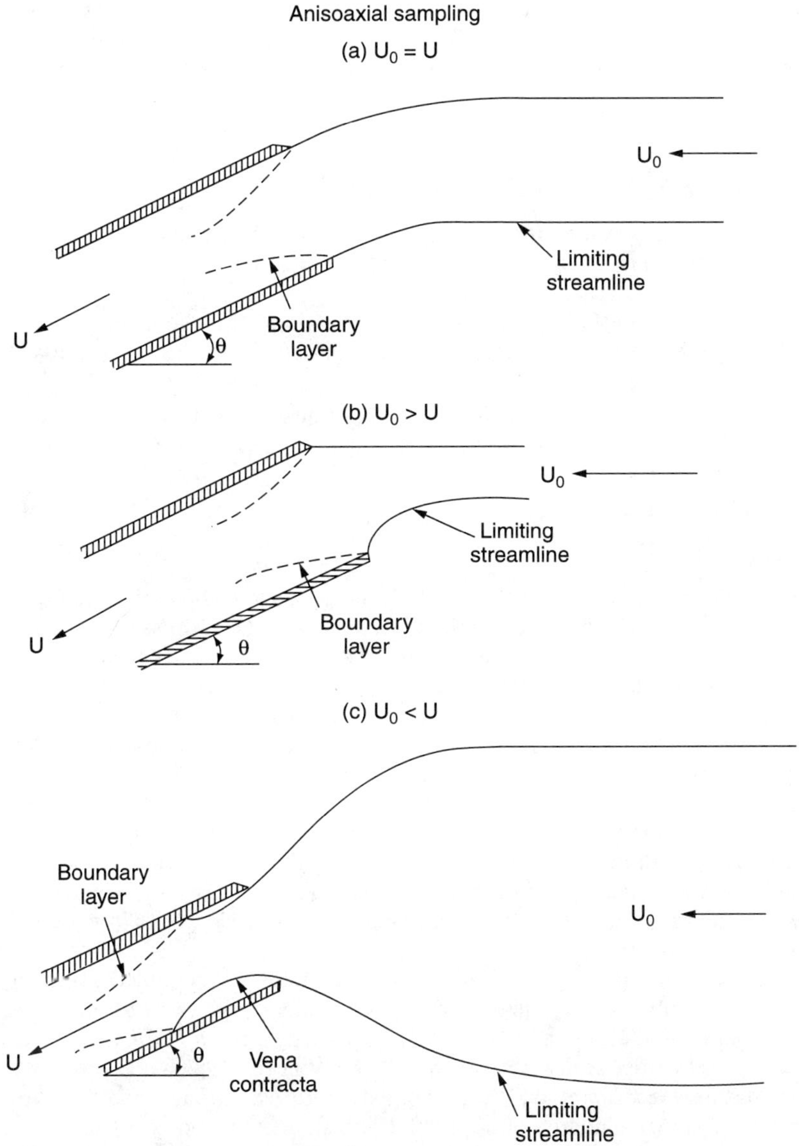

Fig. 8–3. Schematic diagram of anisoaxial sampling with a thin-walled nozzle, with sample flow gas velocity U, inclined at sampling angle θ to the direction of the free-stream ambient gas velocity U_0 under sampling conditions in which **a**, $U = U_0$; **b**, $U < U_0$; and **c**, $U > U_0$.

in the nozzle (Okazaki et al., 1987b) from free-stream turbulence effects (Wiener et al., 1988) and from turbulent deposition of particles in the vena contracta formed in super-isokinetic sampling (Hangal and Willeke, 1990b).

Figure 8–3 is a schematic diagram of anisokinetic sampling for flow conditions where $U_0 = U$, $U_0 > U$, and $U_0 < U$. The angle θ is the angle between the direction of the ambient free-stream gas velocity and the sampling gas velocity. Particles with sufficient inertia to cross the limiting streamlines will be aspirated with efficiencies different from 1. Transmission losses still arise from gravitational settling in the inlet, from free-stream turbulence effects, and from losses in the vena contracta (Hangal and Willeke, 1990b). An additional transmission loss arises from the impaction of particles on the inside lip of the nozzle facing the free-stream velocity (Hangal and Willeke, 1990b).

Sampling from Flowing Gas with a Thin-Walled Nozzle

The correlations for aspiration efficiency and transmission efficiency are listed below, along with their equation numbers and the conditions for which they apply:

- η_{asp} Aspiration efficiency for isoaxial sampling (Eq. 8–8), anisoaxial sampling between 0° and 60° (Eq. 8–20), and anisoaxial sampling between 45° and 90° (Eq. 8–22)
- $\eta_{trans,\ inert}$ Transport efficiency for inertial deposition in sub-isokinetic isoaxial sampling (Eq. 8–16), super-isokinetic isoaxial sampling (Eq. 8–18), and anisoaxial sampling (Eq. 8–25)
- $\eta_{trans,\ grav}$ Transport efficiency for gravitational settling in the inlet region of a nozzle (Eq. 8–23)

The inlet efficiency of a thin-walled nozzle is the product of the aspiration efficiency and the transmission efficiency. Isoaxial, isokinetic sampling is the ideal sampling configuration and will aspirate all particle sizes with nearly 100% efficiency. A departure from this ideal configuration into the regions of anisokinetic sampling and anisoaxial sampling results in non-representative sampling; the aspiration efficiency for large particles is different from 100%, and the larger the particles, the greater the difference. Transmission losses in isokinetic, isoaxial sampling arise principally from gravitational settling in horizontal flow and the effects of free-stream turbulence. If the flow is upward or downward with respect to gravity, the transmission losses from settling will be negligible. However, the sampling velocity needs to be large compared with the particle settling velocity. If the flow is neither isokinetic nor isoaxial, then losses in the inlet from inertial effects can occur; the flow can change direction in the course of entering the inlet, and the larger particles that do not follow the streamlines can be deposited on the walls.

Several researchers have theoretically and experimentally examined sampling from a flowing gas with thin-walled nozzles. They have examined isokinetic and anisokinetic sampling in isoaxial flow (Belyaev and Levin, 1972, 1974; Jayasekera and Davies, 1980; Davies and Subari, 1982; Lipatov et al., 1986; Stevens, 1986; Vincent, 1987; Okazaki et al., 1987a,b; Rader and Marple, 1988; Liu et al., 1989; Zhang and Liu, 1989; Hangal and Willeke, 1990a,b) and in anisoaxial flow (Lundgren et al., 1978; Durham and Lundgren, 1980; Okazaki et al., 1987c; Davies and Subari, 1982; Lipatov et al., 1986, 1988; Vincent et al., 1986; Grinshpun et al., 1990; Hangal and Willeke, 1990a,b). Rader and Marple (1988) give a concise summary of the work on isoaxial sampling. Hangal and Willeke (1990a,b) present a comprehensive summary of the correlations for thin-walled sampling nozzles under conditions of isoaxial and anisoaxial sampling. They further identify correlations that are applicable under each of the conditions.

It is implicit that the ambient free-stream gas velocity and the sample gas velocity remain constant over the period of sampling for the correlations to apply. The reader is cautioned that the following correlations apply only to conditions of constant gas velocities. In this section, the presented correlations are applicable to the case in which the sampling velocity and the wind velocity are large compared with the particle settling velocity (i.e., gravitational effects are negligible). In the case for which these gas velocities are comparable with particle settling velocity and gravitational effects are no longer negligible, the reader is referred to the sections on "Sampling in Calm Air" and "Sampling from Low-Velocity Flowing Gas."

Isoaxial Sampling. For isoaxial sampling where the ambient gas stream velocity is U_0 and the sampling velocity is U, the well-known correlation of Belyaev and Levin (1972, 1974) for aspiration efficiency, η_{asp}, has proved satisfactory with an accuracy to within 10%:

$$\eta_{asp} = 1 + \left[\frac{U_0}{U} - 1\right]\left[1 - \frac{1}{1 + k\,Stk}\right] \tag{8–8}$$

for $0.18 \le Stk \le 2.03$ and $0.17 \le U_0/U \le 5.6$, where

$$Stk = \frac{\tau U_0}{d} \tag{8–9}$$

$$k = 2 + 0.617\left[\frac{U_0}{U}\right]^{-1} \tag{8–10}$$

Stevens (1986) has reviewed the data of Belyaev and Levin (1972, 1974), Jayasekera and Davies (1980), and Davies and Subari (1982) and reported good agreement with the Belyaev and Levin correlation, extending the range of applicability down to a Stokes number of 0.05. At velocity ratios of $U_0/U < 0.2$ (super-isokinetic sampling), considerable discrepancy between the data of Davies and Subari (1982) and the correlation of Belyaev and Levin is seen. Lipatov et al. (1986) have reported experimental data for velocity ratios of U_0/U down to 0.029 (highly super-isokinetic sampling) that are in agreement with the Belyaev and Levin correlation. Lipatov et al. (1986, 1988) have concluded that the differences seen in these data are attributable to particle rebound and entrainment in the course of interaction with the outer surfaces of the sampling nozzle. Their data minimized the effects of bounce, while the data of Davies and Subari (1982) did not. These results indicate that for purely aspiration efficiency, the data of Belyaev and Levin (1974) are good for U_0/U values down to 0.029, but for $U_0/U < 0.2$ particle interactions with the walls of the sampling nozzle may begin to occur. The theoretical results of Rader and Marple (1988) support the use of the Belyaev and Levin (1974) correlation over the Stokes number range of $0.005 \le Stk \le 10$ and the velocity ratio range of $0.2 \le U_0/U \le 5$ with an accuracy to within 10%. Figure 8–4 gives the aspiration efficiency as a function of Stokes number, as calculated by the correlation of Belyaev and Levin (1974). At small Stokes numbers, the efficiency is close to 1 and at large Stokes numbers the efficiency is seen to approach the limiting value of U_0/U.

Rader and Marple (1988) give a correlation for isoaxial aspiration efficiency that includes interception by the nozzle lip:

$$\eta_{asp} = 1 + \left[\frac{U_0}{U} - 1\right]\left[1 - \frac{1}{1 + 3.77\,Stk^{0.883}}\right] \tag{8–11}$$

for $0.005 \le Stk \le 10$ and $0.2 \le U_0/U \le 5$.

Liu et al. (1989) and Zhang and Liu (1989) give a correlation for isoaxial aspiration efficiency based on numerical data:

$$\eta_{asp} = 1 + \frac{\left[\frac{U_0}{U} - 1\right]}{\left[1 + \frac{0.418}{Stk}\right]} \quad \text{for } \frac{U_0}{U} > 1$$

and

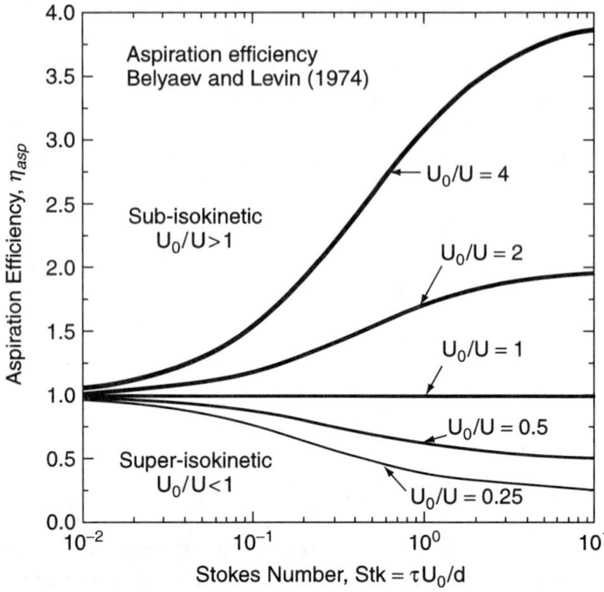

Fig. 8–4. Plot of the aspiration efficiency, η_{asp}, for a thin-walled nozzle as a function of the Stokes number (based on the free-stream ambient gas velocity, U_0, and the nozzle diameter, d) for various values of the free-stream to sampling gas velocities, U_0/U, as given by the correlation of Belyaev and Levin (1974).

$$\eta_{asp} = 1 + \frac{\left[\frac{U_0}{U} - 1\right]}{\left[1 + \frac{0.506\sqrt{U_0/U}}{Stk}\right]} \quad \text{for} \frac{U_0}{U} < 1 \tag{8–12}$$

for $0.01 \leq Stk \leq 100$ and $0.1 \leq U_0/U \leq 10$. These correlations are very close to the correlation of Belyaev and Levin (1974) and may be used interchangeably.

Particles are deposited in the nozzle inlet by gravitational settling and by inertial effects. The particle losses in the nozzle are accounted for by the transmission efficiency.

Okazaki et al. (1987a,b) assume that a particle that has penetrated into the boundary layer formed in the entrance region of the inlet will deposit by gravitational settling on the inside wall of the inlet. The boundary layer thickness is characterized by the inlet Reynolds number, $Re = Ud/\nu$ (Schlichting, 1968). The fraction of particles that penetrate into the boundary layer is characterized by the Stokes number based on the ambient gas velocity, $Stk = \tau U_0/d$. Gravitational deposition in the boundary layer of the inlet is characterized by the gravitational deposition parameter, Z:

$$Z = \frac{L}{d}\frac{V_{ts}}{U} \tag{8–13}$$

where L *is* the inlet region length.

The gravitational deposition parameter is the ratio of the particle-settling distance during transport in the inlet region, LV_{ts}/U, to the diameter of the inlet, d. Okazaki et al. (1987a,b) have performed experiments to measure the deposition inside an inlet for various particle sizes, velocity ratios, and nozzle diameters. They assume that deposition inside an inlet will be correlated to a combination of the quantities Z, Re, and Stk. The results are correlated as the gravitational-settling transmission efficiency, $\eta_{trans,\,grav}$, for horizontal isoaxial sampling by

$$\eta_{trans,grav} = \exp\left[-4.7K^{0.75}\right] \tag{8–14}$$

$$K = Z^{1/2}\, Stk^{1/2}\, Re^{-1/4} \tag{8–15}$$

The inlet diameters ranged from 3.2 to 15.9 mm, but the inlet length was 200 mm for all the test data. Yamano and Brockmann (1989) point out that the analysis is performed for a laminar boundary layer in the inlet tube and that it may not apply once turbulent flow in the tube has developed. Furthermore, if the flow remains laminar, once the laminar boundary layer fills the tube, deposition may no longer be part of the inlet effect but rather should be considered as deposition from tube flow. The single inlet length used in the data may have masked these effects and implicitly included in the correlation gravitational settling for this fixed inlet length. A comparison of the results of Eq. 8–14 to correlations of gravitational deposition in laminar and turbulent pipe flow shows that, generally, Eq. 8–14 gives a lower transmission efficiency through the inlet length. It should, therefore, provide a conservative estimate of the transmission efficiency for gravitational deposition in the inlet.

Inertial losses have been examined by Liu et al. (1989) and by Hangal and Willeke (1990b).

In the case where $U_0/U > 1$ (sub-isokinetic sampling), some particles with velocity vectors directed toward the nozzle walls are deposited and the transmission efficiency is less than 1 (Liu et al., 1989). Liu et al. (1989) give an inertial transmission efficiency, $\eta_{trans,\,inert}$, for sub-isokinetic isoaxial sampling of

$$\eta_{trans,inert} = \frac{1+\left[\dfrac{U_0}{U}-1\right]\Big/\left[1+\dfrac{2.66}{Stk^{2/3}}\right]}{1+\left[\dfrac{U_0}{U}-1\right]\Big/\left[1+\dfrac{0.418}{Stk}\right]} \tag{8–16}$$

for $0.01 \le Stk \le 100$ and $1 \le U_0/U \le 10$. Hangal and Willeke (1990b) assume no inertial transmission losses for sub-isokinetic isoaxial sampling.

Liu et al. (1989) maintain that for $U_0/U < 1$ (super-isokinetic sampling), particle velocities are not directed toward the walls and no particles are deposited. They give an inertial transmission efficiency for super-isokinetic sampling as

$$\eta_{trans,\,inert} = 1.0 \tag{8–17}$$

for $0.01 \le Stk \le 100$ and $0.01 \le U_0/U \le 1.0$.

Hangal and Willeke (1990b), however, maintain that in super-isokinetic sampling a vena contracta is formed in the nozzle inlet and that turbulence in the vena contracta will deposit particles contained in it. They give an inertial transmission efficiency for super-isokinetic sampling as

$$\eta_{trans,inert} = \exp\left[-75\,I_v^2\right] \tag{8–18}$$

$$I_v = 0.09\left[Stk\frac{U-U_0}{U_0}\right]^{0.3} \tag{8–19}$$

for $0.02 \le Stk \le 4$ and $0.25 \le U_0/U \le 1.0$, where I_v is the parameter describing inertial losses in the vena contracta.

Figure 8–5 shows the transmission efficiencies as a function of the Stokes number, as calculated by the Liu et al. (1989) correlation for sub-isokinetic sampling and by the Hangal and Willeke (1990) correlation for super-isokinetic sampling. Gravitational-settling loss is not included in Figure 8–5.

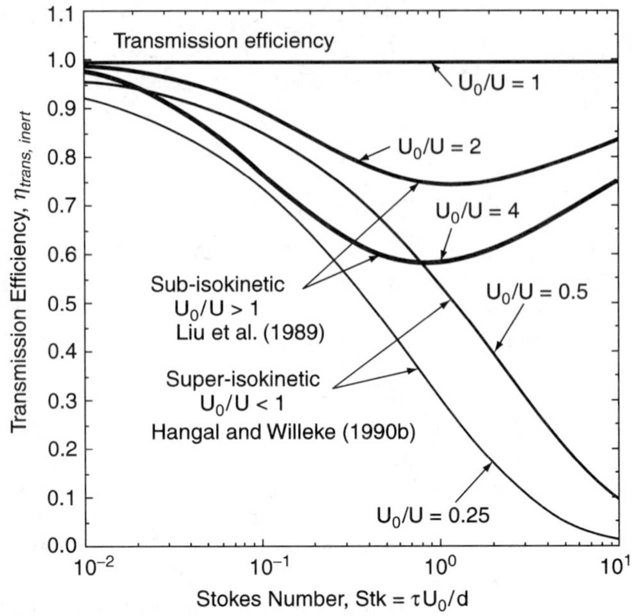

Fig. 8–5. Plot of the transmission efficiency, $\eta_{trans, inert}$, for a thin-walled nozzle as a function of the Stokes number (based on the free-stream ambient gas velocity, U_0, and the nozzle diameter, d) for various values of the free-stream to sampling gas velocities, as given by the inertial deposition correlation of Liu et al. (1989) for sub-isokinetic sampling and the inertial deposition correlation of Hangal and Willeke (1990b) for super-isokinetic sampling. Gravitational deposition is not included.

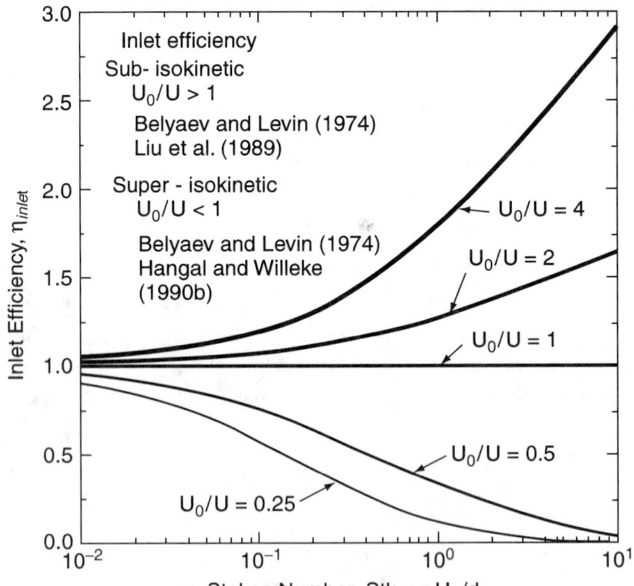

Fig. 8–6. Plot of the inlet efficiency, η_{inlet}, for a thin-walled nozzle as a function of the Stokes number (based on the free-stream ambient gas velocity, U_0, and the nozzle diameter, d) for various values of the free-stream to sampling gas velocities, U_0/U, as given by the correlation of Belyaev and Levin (1974) multiplied by the inertial deposition correlation of Liu et al. (1989) for sub-isokinetic sampling and the inertial deposition correlation of Hangal and Willeke (1990b) for super-isokinetic sampling. Gravitational deposition is not included.

The inlet efficiency for isoaxial sampling is the product of the aspiration efficiency and all applicable transmission efficiencies. Figure 8–6 plots the inlet efficiency as a function of the Stokes number. These results are calculated from the Belyaev and Levin (1974) correlation for aspiration efficiency and the transmission efficiencies given in Figure 8–5.

Anisoaxial Sampling. Hangal and Willeke (1990a,b) have surveyed the literature on anisoaxial sampling and have performed experiments to establish a database on anisoaxial sampling. They have identified correlations with ranges of applicability for anisoaxial sampling. Deposition of particles in the inlet occurs from gravitational settling and from vena contracta deposition, as discussed in the section on isoaxial sampling. Anisoaxial sampling has an additional deposition mechanism that Hangal and Willeke (1990b) refer to as *direct wall impaction*. This occurs on the inside nozzle wall facing the ambient free stream; particles with sufficient inertia cross streamlines and impact on the wall. This is similar to the inlet inertial deposition identified by Liu et al. (1989); however, their investigation dealt only with isoaxial sampling.

The anisoaxial data of Hangal and Willeke (1990a,b) were taken in horizontal free-stream flow, with the nozzle inclined upward or downward with respect to the horizontal. Their conventions indicate that a nozzle facing downward has a negative angle with respect to the horizontal, with the sample flow being directed upward; this case is referred to as *upward sampling*. A similar explanation is made for an upward-facing nozzle with a positive angle with respect to the horizontal and a downward sample flow; this case is referred to as *downward sampling*. In their correlations, Hangal and Willeke (1990a,b) use the magnitude of the sampling angle in degrees. The only correlation in which they differentiate between upward and downward sampling is for the impaction losses in the inlet lip.

Hangal and Willeke (1990b) found that the correlation for aspiration efficiency given by Durham and Lundgren (1980) fit their data for sampling angles from 0° to 60°. This expression is

$$\eta_{\text{asp}} = 1 + \left[\left(\frac{U_0}{U}\right)\cos\theta - 1\right]\frac{1-[1+[2+0.617(U/U_0)]Stk']^{-1}}{1-[1+2.617Stk']^{-1}}$$
$$\left[1-[1+0.55Stk'\exp(0.25Stk')]^{-1}\right] \tag{8–20}$$

$$Stk' = Stk\exp(0.022\theta) \tag{8–21}$$

for $0.02 \le Stk \le 4$ and $0.5 \le U_0/U \le 2$ and $0° \le \theta \le 60°$. They extended the correlation of Laktionov (1973), which was originally developed for 90° sampling, to angles between 45° and 90° and gave the correlation for aspiration efficiency as

$$\eta_{\text{asp}} = 1 + \left[\left(\frac{U_0}{U}\right)\cos\theta - 1\right]\left[3Stk^{\sqrt{\frac{U}{U_0}}}\right] \tag{8–22}$$

for $0.02 \le Stk \le 0.2$ and $0.5 \le U_0/U \le 2$ and $45° \le \theta \le 90°$.

The aspiration efficiency as a function of the Stokes number for sampling at 0°, 45°, and 90° for a range of U_0/U is shown in Figure 8–7. Equation 8–20 is used for the 0° and 45° sampling angles, and Eq. 8–22 is used for the 90° sampling angle. The 0° curves are essentially those of Belyaev and Levin (1974); one may see the rapid departure from representative sampling for anisoaxial sampling.

Hangal and Willeke (1990b) have modified the expression of Okazaki et al. (1987b) for gravitational settling in the nozzle inlet to account for the inclination of the nozzle. The transmission efficiency for gravitational settling, $\eta_{\text{trans, grav}}$, is

$$\eta_{\text{trans,grav}} = \exp[-4.7K_\theta^{0.75}] \tag{8–23}$$

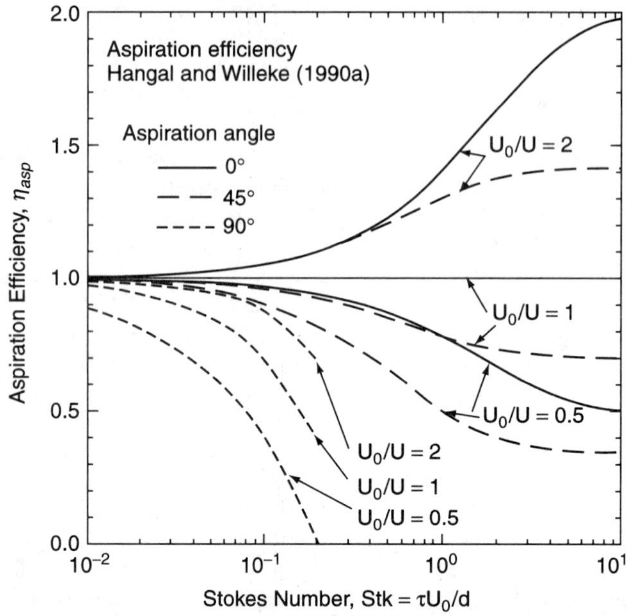

Fig. 8–7. Plot of the aspiration efficiency, η_{asp}, for a thin-walled nozzle at 0°, 45°, and 90° sampling angles as a function of the Stokes number (based on the free-stream ambient gas velocity, U_0, and the nozzle diameter, d) for various values of the free-stream to sampling gas velocities, U_0/U, as given by the correlations presented in Hangal and Willeke (1990a).

$$K_\theta = Z^{1/2}\, Stk^{1/2}\, Re^{-1/4} \cos\theta = K \cos\theta \tag{8–24}$$

It is apparent that for horizontal sampling with the free-stream velocity in the horizontal direction ($\theta = 0°$) K_θ is identical to K (Eq. 8–15) and Eq. 8–23 reduces to Eq. 8–14; for vertical sampling ($\theta = 90°$), $K_\theta = 0$ and there are no gravitational losses. The gravitational settling transmission efficiency depends only on the orientation of the sampling direction with respect to gravity and not on isoaxial or nonisoaxial sampling.

Hangal and Willeke (1990b) give the transmission efficiency for inertia that includes the losses in the vena contracta, accounted for by the parameter I_v, and the losses from direct impaction on the inner wall of the nozzle facing the ambient free-stream gas velocity, accounted for by the parameter I_w. The inertial losses from deposition in the vena contracta and from direct impaction on the nozzle inner wall are combined in the correlation for inertial transmission efficiency:

$$\eta_{\text{trans inert}} = \exp\left[-75(I_v + I_w)^2\right] \tag{8–25}$$

for $0.02 \le Stk \le 4$ and $0.25 \le U_0/U < 4$.

The vena contracta loss parameter is defined as

$$I_v = 0.09\left[Stk \cos\theta \frac{U - U_0}{U_0}\right]^{0.3} \tag{8–26}$$

for $0.25 \le U_0/U \le 1.0$ and $I_v = 0$ otherwise.

The losses from direct impaction are the only losses that depend strongly on whether or not the nozzle faces upward or downward. In downward sampling, the nozzle faces upward and gravitational settling acts to move the particles away from the wall, thus reducing impaction on the wall. This is accommodated in the correlation by subtracting a quantity, α, from the sampling angle, θ. Similarly, in upward sampling, the nozzle faces downward and gravitational settling acts to move particles toward the wall, increasing the impaction losses. In this case the quantity α is added to the sampling angle, θ, which is always taken as a positive quantity in these calculations. In this special case, the free-stream gas velocity is horizontal and the sampling angle θ is in the vertical plane. The direct impaction loss parameter is defined as

$$I_w = Stk\sqrt{U_0/U}\,\sin(\theta-\alpha)\sin\left(\frac{\theta-\alpha}{2}\right) \tag{8–27}$$

for downward sampling (nozzle faces upward) and as

$$I_w = Stk\sqrt{U_0/U}\,\sin(\theta+\alpha)\sin\left(\frac{\theta+\alpha}{2}\right) \tag{8–28}$$

for upward sampling (nozzle faces downward), where

$$\alpha = 12\left[\left(1-\frac{\theta}{90}\right)-\exp(-\theta)\right] \tag{8–29}$$

Figure 8–8 shows the transmission efficiency for inertial effects as a function of the Stokes number for sampling angles of 0°, 45°, and 90° at various values of U_0/U. The curve for $U_0/U = 2$ at $\theta = 0°$ is from Liu et al. (1989), and the remaining curves are calculated from Hangal and Willeke (1990b). Figure 8–9 shows the inlet efficiency for the same conditions shown in Figure 8–7 and 8–8. It is seen that anisoaxial sampling is less representative than isoaxial sampling.

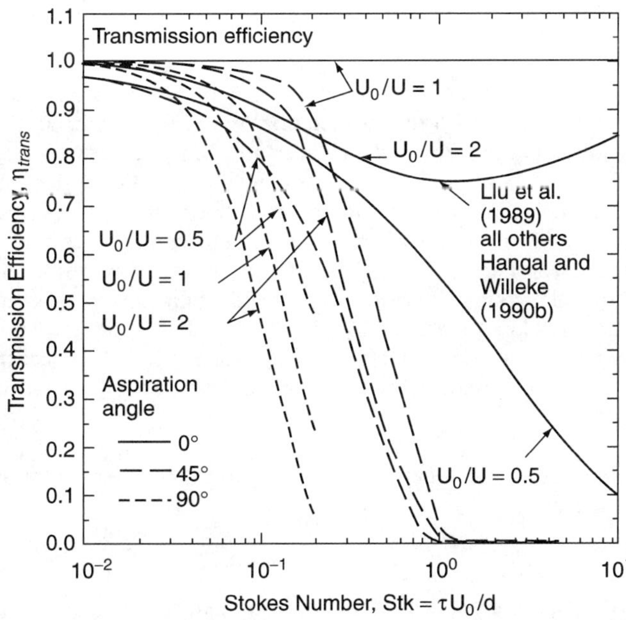

Fig. 8–8. Plot of the transmission efficiency, η_{trans}, for a thin-walled nozzle at 0°, 45°, and 90° sampling angles as a function of the Stokes number (based on the free-stream ambient gas velocity, U_0, and the nozzle diameter, d) for various values of the free-stream to sampling gas velocities, U_0/U, as given by the inertial deposition correlation of Liu et al. (1989) (for sub-isokinetic sampling at 0°) and the inertial deposition correlations presented in Hangal and Willeke (1990b). Gravitational deposition is not included.

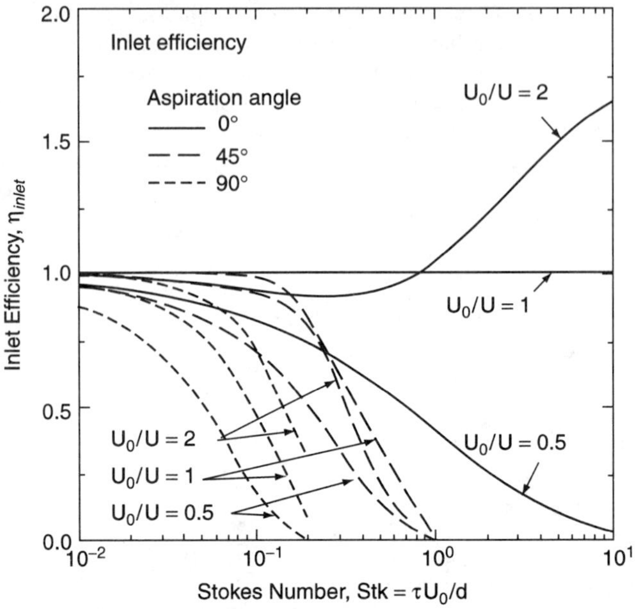

Fig. 8–9. Plot of the inlet efficiency, η_{inlet}, for a thin-walled nozzle at 0°, 45°, and 90° sampling angles as a function of the Stokes number (based on the free-stream ambient gas velocity, U_0, and the nozzle diameter, d) for various values of the free-stream to sampling gas velocities, U_0/U, as given by the aspiration efficiency correlations presented in Hangal and Willeke (1990b) multiplied by the inertial deposition correlation of Liu et al. (1989) (for sub-isokinetic sampling at 0°) and the inertial deposition correlations presented in Hangal and Willeke (1990b). Gravitational deposition is not included.

Free-Stream Turbulence Effects. The limited amount of research on the effects of free-stream turbulence in sampling with thin-walled nozzles seems to indicate that there is little effect on the isoaxial aspiration efficiency (Rader and Marple, 1988; Vincent et al., 1985). Wiener et al. (1988) note that although there does appear to be little effect on the aspiration efficiency, there is a measurable effect on the transmission efficiency that can increase or decrease the deposition in the nozzle inlet. Larger nozzle inlets (of the order of 1 cm in diameter) were less susceptible to these effects. They observed that for a Stokes number of less than 1 and a turbulence intensity of less than 7.5%, the spread in sampling efficiency caused by turbulence was less than 15%. This is of the order of the uncertainty in the sampling efficiency correlations. The effects of turbulence-induced concentration inhomogeneities are discussed below under "Inhomogeneous Particle Concentrations in Inlets and Transport Tubes."

Summary. The transmission efficiency, η_{trans}, is the product of the gravitational and inertial transmission efficiencies:

$$\eta_{\text{trans}} = \eta_{\text{trans, grav}}\ \eta_{\text{trans, inert}} \tag{8–30}$$

The inlet efficiency, η_{inlet}, is the product of the aspiration efficiency, η_{asp}, and the transmission efficiency, η_{trans}, as given in Eq. 8–5.

The inlet efficiency for sampling with the thin-walled nozzle depends on the Stokes number based on ambient gas velocity and the nozzle inlet diameter, the ratio of ambient gas velocity to sampling gas velocity, and the sampling angle. To obtain a representative sample, the sampling should be isoaxial and isokinetic (iso-mean-velocity) and the Stokes number ($\tau U_0/d$) should be kept small. The ambient free-stream and sampling gas velocities should be large compared with the particle settling velocity. Larger inlet diameters (of the order of 1 cm) are less susceptible to deposition caused by free-stream turbulence.

EXAMPLE 8–1

Particles of 15 μm aerodynamic diameter in air at 1.013×10^5 Pa (1 atm) and 293 K [20°C] are sampled sub-isokinetically (U_0 = 3.0 m/s, U = 1.5 m/s) from horizontal flow by an isoaxial thin-walled nozzle of diameter 0.0127 m [1.27 cm] and length 0.10 m [10 cm]. What are the aspiration, transmission, and inlet efficiencies for this particle size? If the nozzle is inclined 30° downward from the horizontal (upward sampling), what are the efficiencies?

Answer: For isoaxial sampling, the aspiration efficiency is calculated from the Belyaev and Levin correlation (Eq. 8–8), the transmission efficiency for loss from gravitational settling from Okazaki et al. (Eq. 8–14), and the transmission efficiency for loss from inertial deposition from Liu et al. (Eq. 8–16). The inlet efficiency is the product of these three efficiencies. For the above conditions, $\tau = 6.8 \times 10^{-4}$ s, $Stk = 0.161$, $Re = 1230$, $Z = 0.035$, and $U_0/U = 2$. This gives the following:

Aspiration efficiency = 1.27
Transmission efficiency for gravitational deposition = 0.84
Transmission efficiency for inertial deposition = 0.86
Inlet efficiency = 0.92

For anisoaxial sampling, the efficiencies are calculated from the correlations given in Hangal and Willeke, the aspiration efficiency is from Eq. 8–20, the transmission efficiency for loss from gravitational settling is from Eq. 8–23, and the transmission efficiency for loss from inertial deposition is from Eq. 8–25. The inlet efficiency is the product of these three efficiencies. For the above conditions, $\tau = 6.8 \times 10^{-4}$ s, $Stk = 0.161$, $Stk' = 0.31$, $Re = 1230$, $Z \cos\theta = 0.030$, $U_0/U = 2$, $\theta = 30°$, and $\alpha = 8°$. This gives the following:

Aspiration efficiency = 1.11
Transmission efficiency for gravitational deposition = 0.85
Transmission efficiency for inertial deposition = 0.86
Inlet efficiency = 0.81

Sampling from Flowing Gas with a Blunt Sampler

A type of inlet different from the thin-walled nozzle is the blunt sampler. This type encompasses a number of sampler inlets ranging from what could be called thick-walled nozzles to those in which the inlet is small compared with the overall sampler dimension. Vincent et al. (1982) describe a blunt sampler as one in which the sampler and inlet configuration present a large physical obstruction to the flow. An example of this type of sampling nozzle is given by Vincent et al. (1985) as a 40 mm diameter flat disk with a centrally located 4 mm diameter sampling orifice. The sampling orifice need not be in a flat plate. It can be in a spherical body or some intermediate shape (Vincent, 1984; Vincent and Gibson, 1981). Drawbacks to the blunt sampler and the thick-walled nozzle are particle deposition on the lip or face of the sampler and subsequent re-entrainment of material into the inlet, difficult-to-characterize particle bounce, and difficulty in obtaining representative sampling of larger particles. A definitive reference on theory and application for blunt body samplers is the book by Vincent (1989) and is recommended for the interested reader.

The theory for blunt sampler performance is not as developed as for thin-walled nozzles. Vincent (1989) points out that there are complicated aerosol mechanical and aerodynamic problems associated with blunt samplers that are not present in sampling with isokinetic, isoaxial thin-walled nozzles. Correlations for inlet performance for a disk-shaped blunt sampling probe are given below. Such a probe is described as a flat axisymmetrical disk of diameter D_S, with a centrally located hole of diameter d, through which the sample is drawn. The condition $d << D_S$ holds.

Isoaxial Sampling. Vincent (1989) gives a set of equations that can be used to model the aspiration performance of a blunt sampler consisting of a concentric hole in a larger flat disk facing into the wind. The model takes into account the more complex (compared with a thin-walled sampling probe) air flow coming into the sampler. The air flow can diverge as it slows when it approaches the sampler and then can converge as the sampler aspirates a portion of the flow. Inertial modeling breaks this into two regions, and particles can cross streamlines in both regions. An aspiration efficiency is formulated for each region, and the total aspiration efficiency is the product of the aspiration efficiencies for each region. This aspiration efficiency is expressed as

$$\eta_{asp} = \left[1+\alpha_1\left(\frac{1}{\psi}-1\right)\right]\left[1+\alpha_2\left(\frac{U_0}{U}\psi-1\right)\right] \tag{8–31}$$

where

$$\psi = \frac{1}{B^2}\left(\frac{d^2}{D_S^2}\frac{U}{U_0}\right)^{1/3} \tag{8–32}$$

$$\alpha_1 = 1-\frac{1}{1+G_1 Stk\sqrt{\psi\frac{U_0}{U}}} \tag{8–33}$$

$$\alpha_2 = 1-\frac{1}{1+G_2 Stk\,\psi} \tag{8–34}$$

Stk is the Stokes number defined in Eq. 8–9

B is a bluntness factor, approximately equal to 1 for the flat disk with $D_S >> d$

G_1 and G_2 are constants based on the sampler configuration, experimentally determined to be $G_1 = 0.25$ and $G_2 = 6.0$.

for $0.16 \le U_0/U \le 20$ and $Stk < 0.3$ for $U_0/U > 1$ and $Stk < \sim 5$ for $U_0/U < 1$ based on two data sets (Vincent et al., 1985; Chung and Ogden, 1986).

The aspiration efficiency expressed in Eq. 8–31 approaches U_0/U for large *Stk* as does the expression of Belyaev and Levin (1972) in Eq. 8–8. Vincent (1989) discusses other blunt sampler configurations and gives models of the same form as the blunt disk sampler. For d and D_S nearly equal in value as one would expect for a thin-walled probe, the results of the blunt disk sampler approach the results of Belyaev and Levin (1972). Figure 8–10 plots the correla-

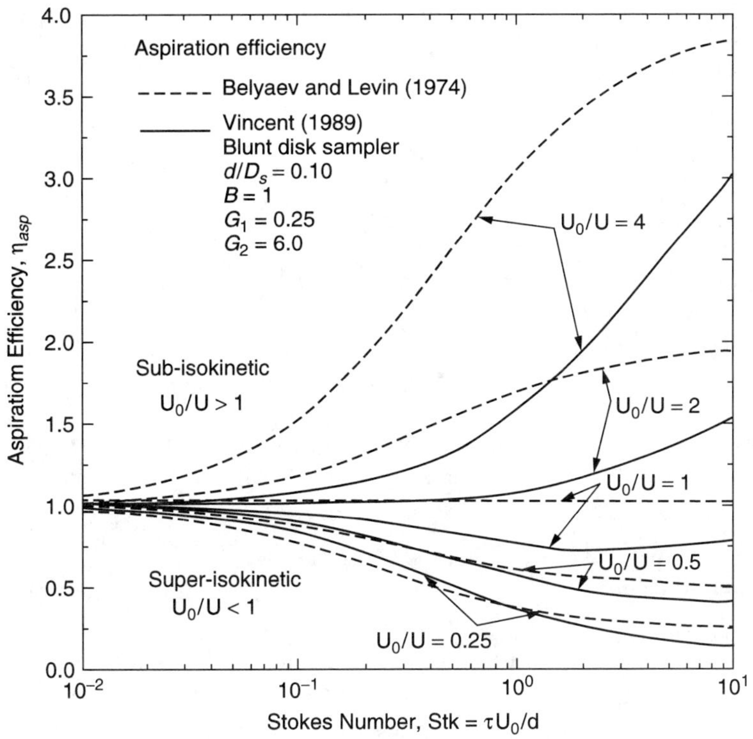

Fig. 8–10. Plot of the aspiration efficiency, η_{asp}, for a blunt disk sampler as a function of the Stokes number (based on the free-stream ambient gas velocity, U_0, and the sampling inlet diameter, d) for various values of the free-stream to sampling gas velocities, U_0/U, as given by the correlation of Vincent (1989). Also plotted for comparison are the results of the correlation of Belyaev and Levin (1974).

tion of Vincent (1989) for the aspiration efficiency of the blunt disk sampler discussed above and compares it to the correlation of Belyaev and Levin for the thin-walled nozzle.

Vincent (1989) also discusses the effects of the blunt sampler body on departure from the idealized aspiration efficiency. The blunt body sampler can collect particles from the free stream by impaction on the upwind surface. Adhesion forces act to keep the particles on the surface, and aerodynamic forces act to remove the particles. In the case of a blunt disk sampler, there are two regions described by concentric circles around the sampling orifice where the removal forces can be greater than the adhesion forces and the sampler surface is swept clean. These regions are the area immediately around the inlet and the annular area at the outer edge of the disk. The removal of the material in these regions is blow-off. Blow-off of material from the external surface of the sampler can affect the amount of material aspirated into the sampling orifice. It can be assumed that the material impacting the sampler face in the swept circular region immediately around the inlet orifice is drawn into the orifice and sampled. Experimental results (Vincent, 1989) indicate that the total material sampled can be as much as 1.5 times the initial aspiration efficiency.

Anisoaxial Sampling. Anisoaxial sampling with a thin-walled nozzle introduces a dependence on the aspiration angle θ. There is an expected dependence of the bluntness factor B on the aspiration angle, but this is not defined. The function $f(\theta)$ was proposed to address this dependence, but it is not well determined and is often assumed to be unity. G_1 and G_2 are still assumed to be constant. Inclusion of the angular dependence is reflected in the following minor changes to Eqs. 8–31, 8–32, and 8–33. Setting θ equal to zero yields the isoaxial expressions

$$\eta_{\text{asp}} = \left[1+\alpha_1\left(\frac{\cos\theta}{\psi}-1\right)\right]\left[1+\alpha_2\left(\frac{U_0}{U}\psi-1\right)\right] \tag{8–35}$$

where

$$\psi = \frac{1}{B^2 f(\theta)^2}\left(\frac{d^2}{D_S^2}\frac{U}{U_0}\right)^{1/3} \tag{8–36}$$

$$\alpha_1 = 1-\frac{1}{1+G_1 Stk\sqrt{\dfrac{U_0}{U}}\left(\sqrt{\psi}\cos\theta+4\sqrt{\sin\theta}\right)} \tag{8–37}$$

The aspiration efficiency expressed in Eq. 8–35 approaches $\cos\theta\, U_0/U$ for large Stk as do the expressions in Eqs. 8–20 (Durham and Lundgren, 1980) and 8–22 (Hangal and Willeke, 1990b).

Sampling in Calm Air

Davies (1968) points out that in sampling from calm air with a small tube at an arbitrary orientation, two conditions must be met for representative sampling. The first is an inertial condition to ensure that particles are drawn into the nozzle. This is expressed as

$$Stk_i \le 0.016 \tag{8–38}$$

where the Stokes number, Stk_i, is based on the average inlet sampling velocity, U, and the inlet diameter, d. The second is a particle settling velocity condition to ensure that the orientation of the nozzle has no influence on sampling. This is expressed in terms of the ratio of the settling velocity to the sampling velocity:

$$\frac{V_{ts}}{U} \le 0.04 \tag{8–39}$$

These two conditions constitute the Davies criterion for representative sampling through a tube in arbitrary orientation. This criterion has proved to be a sufficient condition for representative sampling.

Agarwal and Liu (1980) have established a somewhat more relaxed criterion than that of Davies. They have developed a theoretical prediction based on the solution of the Navier-Stokes equations for the flow field around an upward-facing inlet and a calculation of the particle trajectories and sampling efficiencies. Their prediction is supported by the experimental results of a number of researchers. The Agarwal and Liu criterion for accurate sampling (a sampling efficiency of 90% or higher) with an upward-facing nozzle is

$$Stk_i \frac{V_{ts}}{U} \le 0.5 \tag{8–40}$$

or

$$\frac{\tau V_{ts}}{d} \le 0.05 \tag{8–41}$$

This criterion depends only on particle relaxation time, τ, particle settling velocity, V_{ts}, and nozzle diameter, d; it does not depend on the sampling flow velocity. Agarwal and

Liu (1980) note that the experimental data indicate a dependence on the sampling gas velocity but that at higher sampling efficiencies, this dependence is reduced and the criterion is adequate. Grinshpun et al. (1990) have reviewed work on sampling from calm air. They present data for $V'_s = V_{ts}/U \geq 0.005$ and $Stk_i \geq 2.5$ that show lower efficiencies than the data of Agarwal and Liu (1980) indicate. Grinshpun et al. (1990) point out that although the Agarwal and Liu (1980) analysis is qualitatively correct, it is a first-order approximation. The supporting experimental data of Agarwal and Liu (1980) fall in a region in which V'_s is less than about 10^{-3} and the Stokes number is less than about 1000. These data are outside the Grinshpun et al. (1990) data range. This would suggest that the use of the Agarwal and Liu criterion might not apply for values of V'_s greater than 10^{-3} when the Stokes number is larger than about 1.

Grinshpun et al. (1993) give an empirical equation for the sampling efficiency of a sharp-edged round inlet with the inlet axis oriented at angle φ with respect to gravity ($\varphi = 0°$ is upward facing, $\varphi = 90°$ is horizontal)

$$\eta_{\text{asp, calm air}} = \frac{V_{ts}}{U} \cdot \cos(\varphi) + \exp\left[-\frac{4Stk_i^{1+\sqrt{\frac{V_{ts}}{U}}}}{1+2Stk_i}\right] \tag{8–42}$$

for $0° \leq \varphi \leq 90°$, $10^{-3} \leq V_{ts}/U \leq 1$ and $10^{-3} \leq Stk_i \leq 100$.

The first term on the right addresses the effect of gravitational settling of particles directly into the sampling nozzle. For vertical sampling, $\varphi = 0°$; it accounts for the enhancement of particles settling into the upward-facing nozzle. For horizontal sampling, $\varphi = 90°$; this enhancement is gone. The second term on the right-hand side of the equation addresses the other combined inertial and gravitational effects and is independent of orientation.

For vertical sampling, we can specify a criterion for greater than 95% sampling efficiency that relates the inlet size and flow

$$\frac{U^2}{g \cdot d} \leq 1 \tag{8–43}$$

or

$$\frac{4 \cdot Q}{\pi \cdot d^{2.5} \cdot \sqrt{g}} \leq 1 \tag{8–44}$$

where Q is the sample volumetric flow rate, d is the inlet diameter, and g is the gravitational acceleration.

Figure 8–11 is a plot of the criteria of Davies and of Agarwal and Liu showing the regions of representative sampling indicated by each along with the locus of Stk_i and V_{ts}/U for which Grinshpun et al. (1993) predict sampling efficiency of 95%. The regions below the curves are the regions for which sampling meets or exceeds the efficiency described by the relationship. The relative settling velocity, $V'_s = V_{ts}/U$, is on the horizontal axis, and the Stokes number, Stk, is on the vertical axis.

Summary. The Davies criterion is stringent and applies to any nozzle orientation. The Agarwal and Liu criterion is more relaxed and is supported for $V_{ts}/U \leq 10^{-3}$ and $Stk_i \leq$ ca. 1000. However, it applies to an upward-facing nozzle, and its use for other nozzle orientations is not recommended. The expression of Grinshpun et al. (1993) is based on data for $10^{-3} \leq V_{ts}/U \leq 1$ and $10^{-3} \leq Stk \leq 100$ and applied for a sampling inlet facing upward to facing horizontally. Their expression is more stringent than the Agarwal and Liu criterion, which is based on numerical calculations. The criterion set out in Eqs. 8–43 and 8–44 is recommended.

EXAMPLE 8–2

Particles of 15 μm aerodynamic diameter must be sampled representatively from still air at 1.013×10^5 Pa (1 atm) and 293 K [20°C] by an instrument with a sampling flow rate of 8.3×10^{-5} m^3/s [5 L/min]. What is the inlet diameter required to meet the Davies criterion, the Agarwal and Liu criterion, and the criterion in Eq. 8–44?

Answer: The Davies criterion consists of an inertial condition (Eq. 8–38) and a gravitational-settling condition (Eq. 8–39), which can be expressed in terms of volumetric flow and solved to give the following conditions for the inlet diameter:

$$d \geq \left(\frac{4Q\tau}{0.016\pi} \right)^{1/3} \quad \text{inertial condition}$$

$$d \leq \left(\frac{0.16Q}{\pi V_{ts}} \right)^{1/2} \quad \text{gravitational settling condition}$$

For the conditions given in the example, $\tau = 6.8 \times 10^{-4}$ s, $V_{ts} = 0.0067$ m/s [0.67 cm/s], and $Q = 8.3 \times 10^{-5}$ m^3/s [5 L/min]. This gives $0.0165\,\text{m} \leq d \leq 0.025\,\text{m}$ for the inlet size range that will meet the Davies criterion for representative sampling.

The Agarwal and Liu criterion is a single condition based on the inlet size, particle relaxation time, and settling velocity (Eq. 7–15). The condition for inlet diameter is

$$d \geq \frac{\tau V_{ts}}{0.05}$$

which gives $d \geq 9.0 \times 10^{-5}$ m [0.009 cm] for the inlet diameter to meet the Agarwal and Liu criterion for representative sampling.

For the specified flow of $Q = 8.3 \times 10^{-5}$ m^3/s [5 L/min], Eq. 8–44 can be used to directly determine the inlet diameter:

$$d \geq \left[\frac{4 \cdot Q}{\pi \cdot \sqrt{g}} \right]^{2/5}$$

which gives $d \geq 1.63 \times 10^{-2}$ m [1.63 cm] for the inlet diameter to meet the Eq. 8–44 criterion for representative sampling. Note that this is a reasonable inlet diameter, and the criterion is applicable to all particle sizes. Use Eq. 8–42 to calculate the actual sampling efficiency.

Sampling from Low-Velocity Gas Flow

The previous sections deal with sampling from a moving gas in which the gas flow velocities are large compared with the particle settling velocity and with sampling from calm air when the particle settling velocity and gravitational effects are important. This section deals with sampling from low-velocity gas flows in which particle settling velocity can influence the aspiration efficiency into a sharp-edged nozzle. Grinshpun et al. (1993, 1994) have examined the available literature and offer a correction factor for the effects of particle gravitational settling on aspiration efficiency that can be applied to the aspiration efficiency correlations given above. They also provide an interpolation method to employ the corrected aspiration effi-

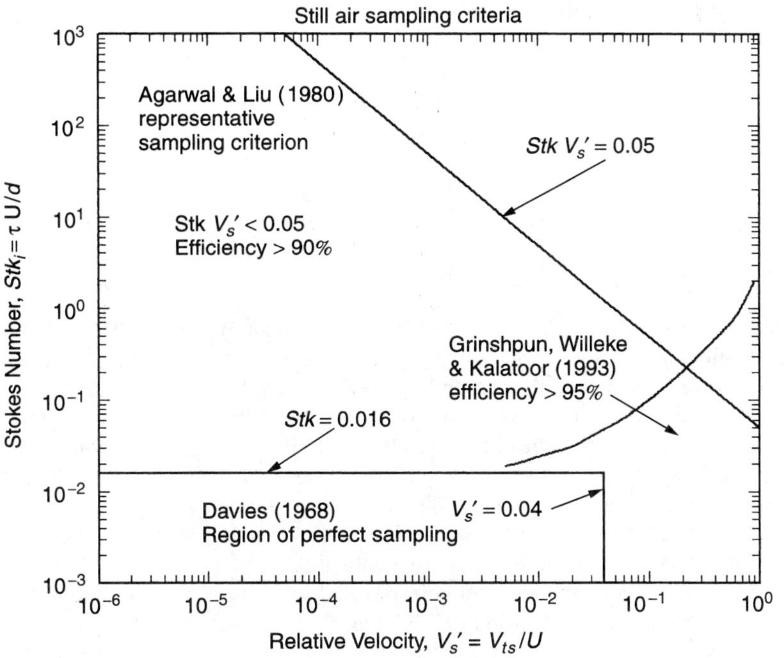

Fig. 8–11. Plot of the Stokes number (based on the inlet velocity, *U*, and the inlet diameter, *d*) with respect to the relative velocity, V'_s, (the ratio of particle-settling velocity, V_{ts}, to inlet velocity, *U*), showing the regions of representative sampling for a tube from still air as given by the sampling criteria of Davies (1968), Agarwal and Liu (1980), and Grinshpun et al. (1993).

ciencies at free-stream gas velocities down to calm air conditions. This has had the effect to extended the aspiration efficiency equations of Hangal and Willeke (1990a,b) for sharp-edged inlets to calm air sampling conditions. The overall aspiration efficiency is calculated as a combination of calm air and moving air efficiencies:

$$\eta_{\text{asp,overall}} = \eta_{\text{asp}} \cdot (1+\delta)^{1/2} \cdot f_{\text{moving}} + \eta_{\text{asp,calm air}} \cdot f_{\text{calm}} \tag{8–45}$$

where η_{asp} is the appropriate correlation from the above section "Sampling from Flowing Gas with a Thin-Walled Nozzle."

The correction factor for gravitational settling is given by $(1 + \delta)^{1/2}$ where δ is defined by

$$\delta = \frac{V_{\text{ts}}}{U_0}\left[\frac{V_{\text{ts}}}{U_0} + 2\cos(\theta + \varphi)\right] \tag{8–46}$$

where the angle θ is the aspiration angle defined as the angle from the sampling gas velocity vector to the free-stream velocity vector and the angle φ is the inclination from the vertical defined as the angle from the gravity vector to the sampling gas velocity vector. The sign convention used here for the angles is positive for the anticlockwise direction and negative for the clockwise direction when measured from and to the vectors as described. In this case, the angles are co-planar and are added in Eq. 8–46. The range of φ is $-90° \le \varphi \le 90°$, that is, sampling gas velocity vector vertically downward ($\varphi = 0°$) to horizontal ($\varphi = \pm 90°$). The range of θ is $-90° \le \theta \le 90°$, that is, sampling gas velocity and free-stream gas velocity vectors isoaxial ($\theta = 0°$) to perpendicular ($\theta = \pm 90°$).

The interpolation weighting factors are

$$f_{\text{moving}} = \exp\left(-\frac{V_s}{U_0}\right) \tag{8–47}$$

and

$$f_{\text{calm}} = 1 - \exp\left(-\frac{V_s}{U_0}\right) \tag{8–48}$$

Equation 8–42 is used for $\eta_{\text{asp, calm air}}$. Equation 8–8 (Belyaev and Levin, 1974, 1974) is used for isoaxial sampling, $\theta = 0°$. Equation 8–20 (Durham and Lundgren, 1980; Hangal and Willeke, 1990b) is used for $0° < \theta \leq 60°$. Equation 8–22 (Laktionov, 1973; Hangal and Willeke, 1990b) is used for $60° < \theta \leq 90°$. Values of parameters must be within the range of applicability for Stokes number and velocity ratios indicated for each correlation. These equations give good agreement with experimental data over a wide range of experimental results. The range of conditions for which these equations apply are generally given as $0° \leq \theta \leq 90°$; $0° \leq \varphi \leq 90°$; $0 \leq U_0/U \leq 10$; $10^{-3} \leq Stk_i \leq 10^2$; $10^{-3} \leq V_s/U \leq 1$, with the stipulation that each correlation used in Eq. 8–45 not be used outside its range of applicability.

Note that downward sampling is not considered, primarily because of a lack of data under calm air conditions. Chen and Baron (1996) found that aspiration efficiencies agreed well for the asbestos sampling inlet (25 mm diameter, 50 mm long) over most sampling conditions. However, at $\varphi = 60°$, $U_0/U = 15$ (outside the recommended range), these equations overestimated the aspiration efficiency more than fourfold. The Durham and Lundgren (1980) equations (Eq. 8–20) agreed well with the experimental data under this condition.

SAMPLE TRANSPORT

A sample transport system consists of the necessary plumbing to convey the aerosol sample to the measuring instrument or to a storage chamber. The components, or flow elements, consist of such items as tubes, elbows, and constrictions. The transport system should be designed to minimize the particle losses; this is often accomplished by minimizing the distance and the transit time between the inlet and the destination. A sample storage volume such as a bag or a chamber may be necessary in a number of applications. One example is the case of an airborne sampling system in which a sample must be taken over a short period of time, too short for the instruments to make a measurement. A sample is drawn into a bag over a short period of time, and the instrument samples from the bag to make the measurement. Another example is the case of testing high-efficiency filters when the aerosol concentration exiting the filter is low and the flow rate through the filter is high compared with the measurement instrument's sample flow rate. In this case, the low concentration requires a long sampling time, a length of time too long to be practical. A sample from downstream of the filter is drawn into a bag at a high flow rate for a short period of time, and the instrument samples from the bag.

During transport through the sampling line or residence time in a storage chamber, particles are lost by various deposition mechanisms. The more common ones that are discussed in this section are:

1. Gravitational settling
2. Diffusional deposition
3. Turbulent inertial deposition

4. Inertial deposition at a bend
5. Inertial deposition at flow constrictions
6. Electrostatic deposition
7. Thermophoretic deposition
8. Diffusiophoretic deposition

Correlations exist for many of the above mechanisms in laminar or turbulent flow; however, these correlations may not apply to transition flow regimes. It is prudent to avoid these transition flow regimes in the sampling lines. In the following review of particle deposition during transport, the sampling lines are assumed to be tubes of circular cross section. The overall transport efficiency, $\eta_{\text{transport}}$, is the product of the transport efficiencies in each flow element for each mechanism, $\eta_{\text{flow element, mechanism}}$, as given in Eq. 8–6.

The mechanisms and the flow elements in which they operate (the correlations are given in this section) are listed below along with the corresponding equation numbers:

$\eta_{\text{tube, turb inert}}$	Turbulent inertial deposition in sample lines for turbulent flow (Eq. 8–61)
$\eta_{\text{bend, inert}}$	Inertial deposition in sample line bends for laminar flow (Eq. 8–67) and turbulent flow (Eq. 8–68)
$\eta_{\text{cont, inert}}$	Inertial deposition in an abrupt contraction for laminar flow (Eq. 8–71)
$\eta_{\text{tube, th}}$	Thermophoretic deposition in sample lines for turbulent flow (Eq. 8–72)
$\eta_{\text{tube, dph}}$	Diffusiophoretic deposition in sample lines for turbulent flow (Eq. 8–77)
$\eta_{\text{bag, grav diff}}$	Combined gravitational and diffusional deposition in a storage chamber with a well-mixed volume (Eq. 8–78)

This is by no means an exhaustive list, but it covers conditions commonly encountered in aerosol transport.

Gravitational Settling in Sampling Lines

Particles settle due to gravitational force and deposit on the lower wall of nonvertical lines in a sampling system during transport. Correlations for gravitational settling and deposition under various tube orientations and flow conditions are available.

For laminar flow in a straight horizontal tube with circular cross section, Fuchs (1964) and Thomas (1958) independently solved the problem of gravitational settling by assuming a parabolic flow distribution. Their result for the transport efficiency for gravitational deposition, $\eta_{\text{tube, grav}}$, from laminar flow in a circular horizontal tube is

$$\eta_{\text{tube, grav}} = 1 - \frac{2}{\pi}\left[2\,\varepsilon\sqrt{1-\varepsilon^{2/3}} - \varepsilon^{1/3}\sqrt{1-\varepsilon^{2/3}} + \arcsin\left(\varepsilon^{1/3}\right)\right] \tag{8–49}$$

$$\varepsilon = \frac{3}{4}Z = \frac{3}{4}\frac{L}{d}\frac{V_{\text{ts}}}{U} \tag{8–50}$$

where

Z = gravitational deposition parameter
L = length of the tube
d = inside diameter of the tube

Pich (1972) has shown that Eq. 8–49 can be used for laminar flow in elliptical channels by substituting the length of the minor axis of the ellipse for d in the definition of ε in Eq. 8–50. This correlation is applicable for either the major or the minor axis horizontal.

In a circular tube that has an angle of inclination, θ, with respect to the horizontal, the component of the sedimentation velocity of a particle perpendicular to the wall of the pipe is $V_{ts}\cos\theta$. Heyder and Gebhart (1977) modified Eq. 8–51 and obtained the following equation for gravitational settling from laminar flow in a circular inclined pipe. This correlation is in good agreement with their experimental results and can be used as the general correlation for the transport efficiency, $\eta_{\text{tube, grav}}$, for gravitational deposition from laminar flow in a circular tube:

$$\eta_{\text{tube,grav}} = 1 - \frac{2}{\pi}\left[2\,\kappa\sqrt{1-\kappa^{2/3}} - \kappa^{1/3}\sqrt{1-\kappa^{2/3}} + \arcsin\left(\kappa^{1/3}\right)\right] \tag{8–51}$$

$$\kappa = \varepsilon\cos\theta = \frac{3}{4}\frac{L}{d}\frac{V_{ts}}{U}\cos\theta \tag{8–52}$$

for

$$\frac{V_{ts}\sin\theta}{U} << 1 \tag{8–53}$$

The criterion expressed by Eq. 8–53 states that the axial component of a particle's settling velocity should be small compared with the average gas velocity in the tube. For a horizontal flow ($\theta = 0°$), Eq. 8–51 reduces to Eq. 8–49.

Wang (1975) has derived a theoretical formula for gravitational settling from laminar flow in an inclined tube, which shows that the deposition rate depends on the flow direction in the tube; deposition from uphill flow is different from deposition from downhill flow. This observation is made when the criterion in Eq. 8–53 is not satisfied. Heyder and Gebhart (1977) show that Wang's formula reduces to Eq. 8–51 when the criterion is satisfied.

The transport efficiency for gravitational settling from laminar flow in a vertical tube is given by Eq. 8–51 as 1 (100%). There is no horizontal area on which particles will deposit by gravitation in a vertical tube. The criterion expressed in Eq. 8–53 must still be kept in mind to avoid sampling situations in which the sample gas flow is not high enough to convey the particles through a vertical sampling line.

Another consideration in laminar flow in a vertical tube is the Saffman force, which can drive particles toward the wall in downward flow or away from the wall in upward flow (Saffman, 1965, 1968). This force arises from the lift on a spherical particle produced by the velocity gradient in the boundary layer and leads to particle motion across streamlines. Lipatov et al. (1989, 1990) discuss this effect. It appears to be noticeable for particles of the order of 15 μm and increases with an increasing magnitude of velocity gradient and particle size.

In the case of turbulent flow in a horizontal tube, gravitational-settling loss is assumed to occur from a well-mixed volume through the boundary layer. The transport efficiency, $\eta_{\text{tube, grav}}$, is expressed as (Schwendiman et al., 1975)

$$\eta_{\text{tube,grav}} = \exp\left[-\frac{4Z}{\pi}\right] = \exp\left[-\frac{dLV_{ts}}{Q}\right] \tag{8–54}$$

where Q is the volumetric flow rate of gas through the tube.

In the case of an inclined or vertical tube, a similar modification to the one applied to gravitational settling from laminar flow may also be applied to turbulent flow. The transport efficiency for turbulent pipe flow for particles undergoing gravitational deposition is

$$\eta_{\text{tube,grav}} = \exp\left[-\frac{4\,Z\cos\theta}{\pi}\right] = \exp\left[-\frac{dLV_{ts}\cos\theta}{Q}\right] \tag{8–55}$$

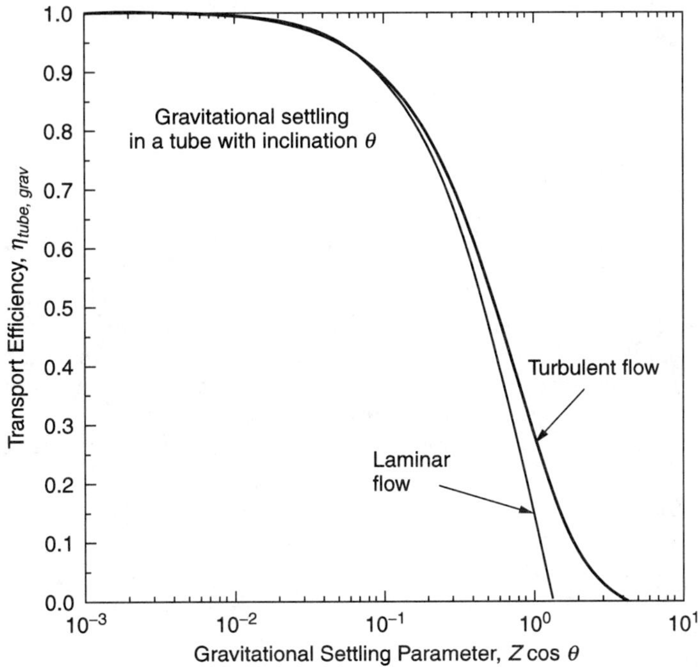

Fig. 8–12. Plot of the transport efficiency, $\eta_{\text{tube, grav}}$, of particles undergoing gravitational settling flowing through a tube inclined at an angle θ from the horizontal, as a function of the gravitational settling parameter Z. The correlations for laminar and turbulent flows are both given.

The criterion expressed by Eq. 8–53 is also applicable in this case as in the case of laminar flow, and the same caveats with respect to vertical flow apply.

Figure 8–12 shows the transport efficiency through a tube at incline θ with gravitational settling for both laminar and turbulent flow as calculated by Eqs. 8–51 and 8 55, respectively. The two flow conditions give essentially the same results for transport efficiencies greater than 0.5. An implication of this is that for a given tube diameter and length, the higher the flow, the lower the loss from gravitational settling. For both laminar and turbulent flow, gravitational deposition depends on the parameter $Z \cos\theta = (V_{ts}L)/(Ud)\cos\theta = (\pi d V_{ts} L)/(4Q)\cos\theta$. Decreasing Z yields higher transport efficiency; Z can be decreased by decreasing the transport length, L, by increasing the volumetric flow, Q, by decreasing the tube diameter for a given volumetric flow, and by increasing the inclination angle of the tube, θ.

Yamano and Brockmann (1989) show that for turbulent flow through a bending tube the transport efficiency for gravitational loss is the same as that for a horizontal tube with the same diameter and a length equal to the projected length of the bending tube onto the horizontal plane. If the flow in a bend is laminar, the velocity profile differs from that in a straight pipe, and the transport efficiency is not necessarily the same as that in a horizontal pipe whose length is the projected length of the bend onto the horizontal plane.

Diffusion in Sampling Lines

Small particles undergoing Brownian motion will diffuse from high particle concentrations to low particle concentrations. The wall of a tube acts as a sink for these diffusing particles, and the concentration at the wall is taken as zero; particles will diffuse toward a wall and be deposited there. In tube flow, the transport efficiency with diffusive particle loss, $\eta_{\text{tube, diff}}$, may, in general, be expressed as

$$\eta_{\text{tube, diff}} = \exp\left[-\frac{\pi\, d\, L V_{\text{diff}}}{Q}\right] = \exp[-\xi\, Sh] \tag{8–56}$$

where

V_{diff} = deposition velocity for particle diffusion loss to the wall
Sh = Sherwood number

The diffusive deposition velocity, V_{diff}, also called the *mass transfer coefficient*, can be determined from the available heat and mass transfer correlations. The Sherwood number ($Sh = V_{\text{diff}} d/D$) is a dimensionless mass transfer coefficient that contains the diffusive deposition velocity in the definition. The Sherwood number is correlated with the Reynolds number ($Re_{\text{f}} = \rho_{\text{f}} U d/\eta$) and the Schmidt number ($Sc = \eta/[\rho_{\text{f}} D]$) for both laminar and turbulent tube flow. For laminar flow, Holman (1972) gives

$$\begin{aligned} Sh &= 3.66 + \frac{0.0668 \dfrac{d}{L} Re_{\text{f}} Sc}{1 + 0.04\left[\dfrac{d}{L} Re_{\text{f}} Sc\right]^{2/3}} \\ &= 3.66 + \frac{0.2672}{\xi + 0.10079\xi^{1/3}} \end{aligned} \tag{8–57}$$

$$\xi = \frac{\pi\, D\, L}{Q} \tag{8–58}$$

where

D = particle diffusion coefficient
L = tube length
Q = volumetric flow rate through the tube

For turbulent flow, Friedlander (1977) gives

$$Sh = 0.0118\, Re_{\text{f}}^{7/8}\, Sc^{1/3} \tag{8–59}$$

The diffusional deposition velocity can be determined from the appropriate correlation above and used in Eq. 8–56 to compute the transport efficiency for tube flow in which particles are being deposited by diffusion.

This expression for the laminar flow case gives very good agreement with the analytic solution by Gormley and Kennedy (1949) for diffusional deposition from laminar tube flow. The well-known formulation of Gormley and Kennedy (1949) for the transport efficiency in laminar tube flow for particles undergoing diffusive deposition is

$$\eta_{\text{tube, diff}} = 1 - 2.56\xi^{2/3} + 1.2\xi + 0.177\xi^{4/3} \tag{8–60a}$$

for $\xi < 0.02$ and

$$\eta_{\text{tube, diff}} = 0.819\exp(-3.657\xi) + 0.097\exp(-22.3\xi) + 0.032\exp(-57\xi) \tag{8–60b}$$

for $\xi > 0.02$.

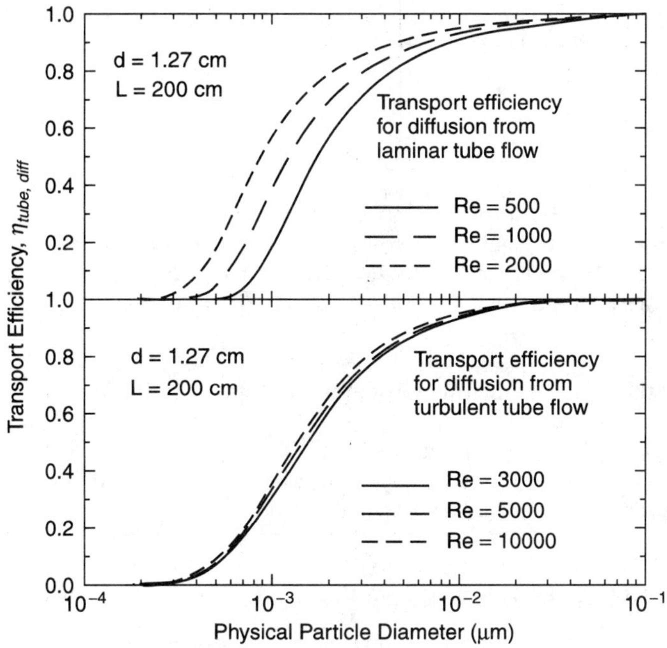

Fig. 8–13. Plots of the transport efficiency, $\eta_{\text{tube, diff}}$, of particles undergoing diffusional deposition flowing through a tube as a function of particle diameter in laminar tube flow (top) and turbulent tube flow (bottom) at 1 atm and 293 K in a 12.7 mm diameter, 2 m long tube.

Gormley and Kennedy (1949) also give an expression for the transport efficiency for diffusive deposition from laminar flow between two parallel plates.

Calculated results from Eqs. 8–57 and 8–56 and the Gormley and Kennedy solution in Eq. 8–60 differ by only a few percent over the range of interest.

In diffusive deposition from laminar flow, the transport efficiency is a function only of $\xi = \pi DL/Q$. There is no dependence on tube diameter. To increase transport efficiency, ξ should be kept small by keeping the transport distance, L, small or by increasing the flow rate, Q. Figure 8–13 shows the transport efficiency for particles deposited by diffusion as a function of particle diameter. The results for laminar flow (top portion) and for turbulent flow (bottom portion) are shown for conditions of air 1 atmosphere and 293 K in a 2 m long, 12.7 mm diameter tube. In the laminar flow portion of Figure 8–13, the increase in the Reynolds number is caused by an increase in Q, and one can see the transport efficiency increase at smaller sized particles for increasing Q.

The experimental work of Brockmann et al. (1982) supports the use of Eqs. 8–59 and 8–56 to calculate the diffusional deposition of particles from turbulent tube flow. An examination of the ξSh term for turbulent flow in Eq. 8–56 shows that transport efficiency in turbulent flow is not strongly dependent on flow rate, Q. This can be seen in Figure 8–13. Decreasing the transport distance, L, or increasing the tube diameter, d, will act to increase the transport efficiency.

Data reported by Lee and Gieseke (1994) in which they examine simultaneous turbulent diffusive and turbulent inertial deposition in sampling lines (turbulent inertial deposition is addressed in the next section) suggest an interaction of larger diffusing particles with the turbulence that enhances their deposition to greater values than could be explained by combining the diffusion correlation and turbulent inertial deposition correlation given in the next section. There seems to be a minimum deposition velocity based on Reynolds number. This phenomenon is discussed in the next section.

EXAMPLE 8–3

A diluter has been designed to measure the sample flow by use of a capillary flow meter through which the aerosol flows. The sample flow is 5×10^{-6} m^3/s [5 cm^3/s] through the 0.00181 m [0.181 cm] diameter 0.05 m [5 cm] long capillary and is drawn from the air stream at 1.013×10^5 Pa [1 atm] and 293 K. No particles larger than 0.1 μm are present. What is the transport efficiency for 0.1, 0.01, and 0.001 μm particles through the sampling capillary?

Answer: For the above conditions, the Reynolds number is 510 and the flow is laminar. The gravitational-settling parameter, Z, for a 0.1 μm particle is 3.9×10^{-6}, and gravitational deposition is negligible. The only losses come from diffusional deposition from laminar flow (Eqs. 8–56 and 8–57). The transport efficiency for 0.1 μm particles is 1.0, for 0.01 μm particles 0.96, and for 0.001 μm particles 0.45.

Turbulent Inertial Deposition in Sampling Lines

There are three regimes of deposition observed in turbulent tube flow (Guha, 1997). The first is the turbulent diffusion regime in which the turbulence influences central mixing, and Brownian diffusion is the mechanism whereby particles are transported through the laminar sublayer to the tube wall. This regime pertains to small particles and has been described in the previous section. The second is the turbulent diffusion–eddy impaction regime that applies to larger particles and in which particle inertia plays a role. In the second regime, particle deposition increases with increasing particle size. The third is the particle inertia–moderated regime that is applicable to still larger particles. In the third regime, the particle inertia is large so that the particle trajectories are less affected by turbulence, and turbulent inertial deposition actually decreases slightly with increasing particle size. It is the second two regimes that are referred to as *turbulent inertial deposition* and are addressed in this section.

Enhanced deposition of particles in turbulent flow has been observed experimentally (e.g., (Friedlander and Johnstone, 1957; Liu and Agarwal, 1974; Lee and Gieseke, 1994; Muyshondt et al., 1996a), to mention a few). There are several models of particle inertial deposition in turbulent flow. One of the original concepts is the free flight stopping distance. This concept can be described by the turbulence in the central region of the pipe flow propelling a particle into the laminar sublayer of the turbulent boundary layer. If the particle's inertia is sufficiently high, it will fully penetrate the sublayer and be collected on the wall. Other models are based on calculation of particle trajectories (e.g., Reynolds, 1999; Chen and Ahmadi, 1996) and can include lift forces (Ziskind et al., 1998). Guha (1997) developed an Eulerian approach including turbulent Brownian diffusion, thermophoresis, turbulent inertial deposition, electrostatics, gravity, and lift forces.

The transport efficiency, $\eta_{\text{tube, turb inert}}$, in a tube with turbulent inertial deposition of particles using the turbulent inertial deposition velocity, V_t, is expressed as

$$\eta_{\text{tube,turb inert}} = \exp\left[-\frac{\pi\, d\, L\, V_t}{Q}\right] \qquad (8\text{–}61)$$

Liu and Agarwal (1974) found that the dimensionless turbulent deposition velocity, V_+, increases rapidly with increasing dimensionless particle relaxation time, τ_+, reaching a peak of 0.14 at a τ_+ value of approximately 30. This is the turbulent diffusion–eddy impaction regime ($\sim 0.3 < \tau_+ < \sim 10$). Above $\tau_+ = 30$, V_+ shows only a moderate dependence on

τ_+, decreasing to 0.085 at $\tau_+ = 1000$. This is the particle inertia moderated regime ($\tau_+ > \sim 10$). In the region $\tau_+ < 12.9$, Liu and Agarwal (1974) found a correlation between V_+ and τ_+, which is given below with a slight modification based on the observations of Lee and Gieseke (1994).

Lee and Gieseke (1994) developed data for particle deposition in turbulent tube flow for particles undergoing simultaneous turbulent inertial deposition and Brownian diffusion. Their results are not predicted by a combination of correlations combining the two mechanisms. The observed dimensionless deposition velocities for these particles is higher than that predicted by the combination of the two mechanisms. The data suggest a minimum dimensionless deposition velocity based on Reynolds number that can be added to the dimensionless deposition velocity predicted by the turbulent inertial deposition correlation. The additional quantity is 2×10^{-8} *Re*. This has the effect of increasing the deposition velocity for the particles that are affected by turbulent inertial deposition and Brownian diffusion.

$$V_+ = 6 \cdot 10^{-4}\ \tau_+^2 + 2 \cdot 10^{-8}\ Re_f \qquad (8\text{–}62)$$

where

$$V_+ = 5.03 \frac{V_t}{U} Re_f^{1/8} \qquad (8\text{–}63)$$

$$\tau_+ = 0.0395\ Stk\ Re_f^{3/4} \qquad (8\text{–}64)$$

where

V_+ = dimensionless deposition velocity
τ_+ = dimensionless particle relaxation time
V_t = deposition velocity for turbulent inertial deposition
Stk = Stokes number formulated with the tube diameter and average gas velocity in the tube

Experimental results by Liu and Agarwal also show that V_+ may be regarded as constant for τ_+ larger than 12.9. That is,

$$V_+ = 0.1 \quad \text{for } \tau_+ > 12.9 \qquad (8\text{–}65)$$

Reeks and Skyrme (1976) offer an explanation for the decline in deposition velocity with particle size in the particle inertia–moderated regime, as reflected by the decline in V_+, based on a particle's declining susceptibility to turbulence transporting it to the boundary layer. Larger particles are less influenced by turbulence than smaller ones. They present a model that accounts for the decline in V_+ at higher values of τ_+. Im and Ahluwalia (1989) also present a model that accounts for this decline and is also applicable to rough as well as to smooth surfaces. Both of these models involve more than a few correlational equations and do not differ greatly from the Liu and Agarwal (1974) results.

Muyshondt et al. (1996a) report data for larger particles that show higher dimensionless deposition velocities than those given by the modified Liu and Agarwal (1974) correlation. Their data cover particle sizes larger than 1 μm in diameter, tube diameters between 12 and 100 mm [0.5 to 4 in.], Reynolds numbers between 2500 and 50,000, and dimensionless particle relaxation times between 0.5 and 30. Their correlation includes a Reynolds number dependence for these larger particles that they believe reflects turbulent diffusion effects not included in the formulation of the dimensionless relaxation time. Their correlation does not fit the Liu and Agarwal (1974) and Lee and Gieseke (1994) data as well as the modified Liu and Agarwal correlation given above. Their data do indicate for larger Reynolds numbers a higher deposition of particles than the data of Liu and Agarwal (1974).

Figure 8–14 shows the transport efficiency of particles undergoing turbulent inertial deposition as a function of particle aerodynamic diameter for various combinations of

EXAMPLE 8–4

Particles of 15 µm aerodynamic diameter in air at 1.013×10^5 Pa [1 atm] and 293 K are transported through a 0.0127 m [1.27 cm] diameter 1.0 m long tube inclined at 45° to the horizontal. Determine the total transport efficiency through the tube for flows of $8.3 \times 10^{-5}\,m^3/s$ [5 L/min] and $8.3 \times 10^{-4}\,m^3/s$ [50 L/min].

Answer: Particle deposition could be caused by gravitational settling, by diffusional deposition, and by turbulent inertial deposition. For the above conditions, the Reynolds number at $8.3 \times 10^{-5}\,m^3/s$ [5 L/min] is 540 (laminar flow) and at $8.3 \times 10^{-4}\,m^3/s$ [50 L/min] is 5400 (turbulent flow). The diffusion coefficient for a 15 µm diameter particle is $1.6 \times 10^{-12}\,m^2/s$ [$1.6 \times 10^{-8}\,cm^2/s$], and diffusional deposition is negligible in both cases.

For the $8.3 \times 10^{-5}\,m^3/s$ [5 L/min] case, the flow is laminar and the only deposition is from gravitational settling (Eq. 8–51). For the above conditions, $K = 0.42$, and the transmission efficiency is 0.42. For the $8.3 \times 10^{-4}\,m^3/s$ [50 L/min] case, the flow is turbulent and deposition is from gravitational settling (Eq. 8–55) and from turbulent inertial deposition (Eq. 8–61). For the above conditions, $Z \cos\theta = 0.056$, and the transport efficiency for gravitational settling is 0.93. The Stokes number is 0.35, and the transport efficiency for turbulent inertial deposition is 0.37. This gives a total transport efficiency of 0.34.

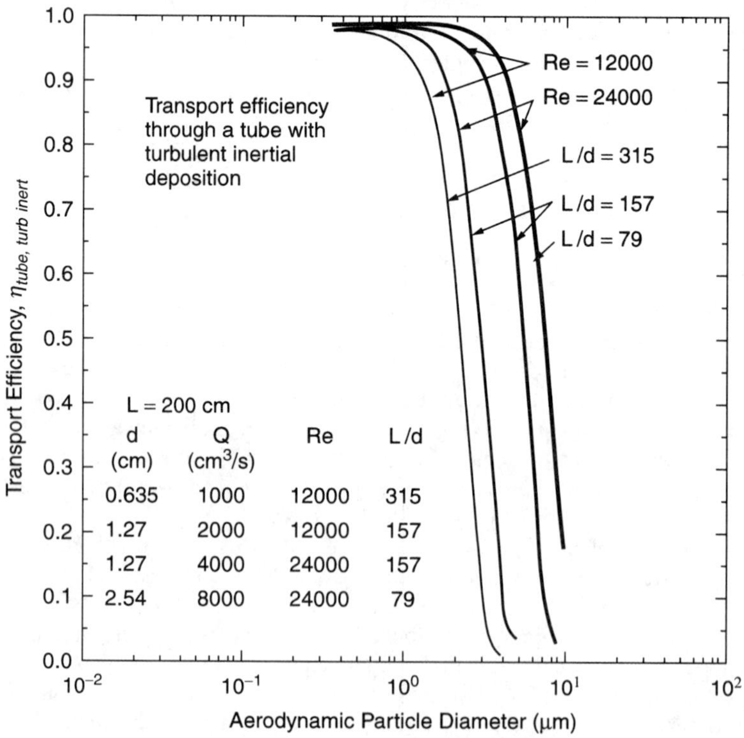

Fig. 8–14. Plot of the transport efficiency, $\eta_{tube,\,turb\;inert}$, of particles undergoing turbulent inertial deposition flowing through a 2 m long tube with diameters of 6.35, 12.7, and 25.4 mm at 1×10^{-3}, 2×10^{-3}, 2×10^{-3}, 4×10^{-3}, and $8 \times 10^{-3}\,m^3/s$, [1000, 2000, 4000, and 8000 cm^3/s] as a function of particle aerodynamic diameter at 1 atm and 293 K.

volumetric flow and tube diameter. Note that for the conditions in Figure 8–14, the modification to the Liu and Agarwal (1974) correlation suggested by Lee and Gieseke (1994) decreases the transport efficiency by no more than 2%. Transport efficiency is increased by increasing the tube diameter, d, by decreasing the volumetric flow, Q, or by decreasing the transport length, L.

Inertial Deposition in a Bend

When the direction of sampling gas flow is diverted in a bend, an aerosol particle may deviate from the gas flow due to its inertia and deposit on the wall of the bend. In laminar flow, secondary recirculation flow patterns develop that push the axial flow core to the outside of the bend and are responsible for particle deposition on both the inside and outside of the bend. These recirculation flows influence particle deposition and are, in turn, dependent on the flow Reynolds number and the radius of curvature of the bend. Laminar flow is more stable through a bend than through a straight tube; flow can remain laminar up to Reynolds numbers of 5000. The pertinent parameters in characterizing particle deposition from laminar flow through a bend are the curvature ratio, R_0, defined as the radius of the bend divided by the radius of the tube, the flow Reynolds number ($Re = \rho dU/\eta$), and the Stokes number formulated with the tube diameter and average gas velocity ($Stk = \tau U/d$).

The simple empirical correlation given by Crane and Evans (1977) for transport efficiency, $\eta_{\text{bend, inert}}$, of particles undergoing inertial deposition in a 90° bend is adequate for estimates of particle deposition from laminar flow in bends and can be extended to bends of other angles.

$$\eta_{\text{bend, inert}} = 1 - Stk\varphi \tag{8–66}$$

where φ is the angle of the bend in radians.

This correlation is an approximation of experimental data (Crane and Evans, 1977). It assumes zero deposition only at a Stokes number of zero. The numerically calculated deposition rates of particles from laminar flow in a circular bend (Cheng and Wang 1981) and the experimental work of Pui et al. (1987) indicate that for small Stokes numbers ($Stk < 0.05$ to 0.1, depending on the Reynolds number) no deposition from inertial effects in the bend occurs. If Stokes numbers in bends in laminar flow can be kept below this limit, then no inertial deposition is expected.

Cheng and Wang (1981) present only the results of their numerical calculations for Reynolds numbers of 100 and 1000. Pui et al. (1987) found that the Cheng and Wang (1981) numerical results for a Reynolds number of 1000 agreed well with their experimental data, whereas the calculations for a Reynolds number of 100 did not. This disagreement was attributed to differences in the assumed and actual flow fields at the lower Reynolds number. The experimental results support the use of the Cheng and Wang (1981) model for a Reynolds number of 1000 and curvature ratios, R_0, between 4 and 30. No correlation is given, but the transport efficiency for laminar flow through a 90° bend at $Re = 1000$ and $R_0 = 8$ is 1 at $Stk \simeq 0.05$, 0.5 at $Stk \simeq 0.16$, and 0.1 at $Stk \simeq 0.32$. These results may be used as a guide to estimate the transport efficiency.

Data for particle deposition in a bend from laminar flow are given by Pui et al. (1987) for a Reynolds number of 1000, curvature ratios of 5.6 and 5.7, and a tube internal diameter ranging from 4 to 8.5 mm. A fit to these data yields a correlation that may be more satisfying than the Crane and Evans (1977) relationship:

$$\eta_{\text{bend,inert}} = \left[1 + \left[\frac{Stk}{0.171}\right]^{0.452\frac{Stk}{0.171}+2.242}\right]^{-\frac{2}{\pi}\varphi} \tag{8–67}$$

Pui et al. (1987) also found that deposition from turbulent flow in a bend can be expressed independently of the Reynolds number. They present an analysis showing that particles are

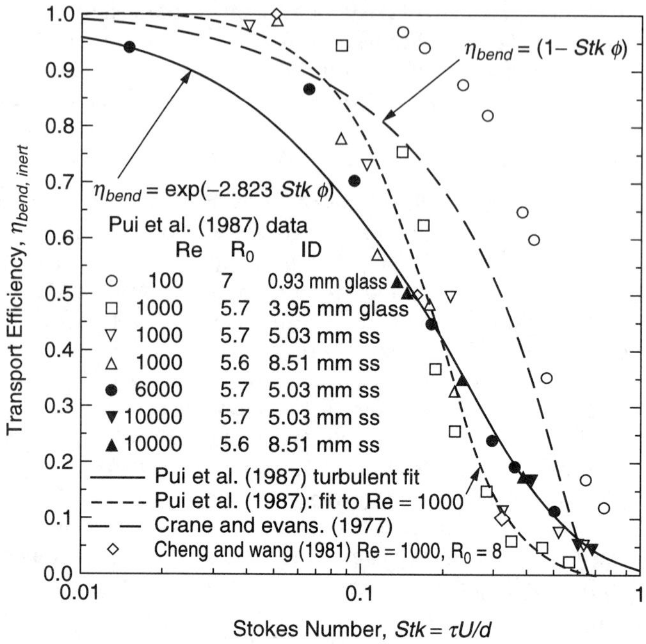

Fig. 8–15. Plot of the transport efficiency, $\eta_{\text{bend, inert}}$, of particles undergoing inertial deposition in a bend as a function of the Stokes number (based on the average gas velocity through the bend, U, and the tube diameter, d).

deposited through the boundary layer from a well-mixed core at a constant rate, with total deposition proportional to the angle through which the aerosol flows. This analysis is in good agreement with the data. A data-based correlation for the transport efficiency of particles through a bend in turbulent flow, $\eta_{\text{bend, inert}}$, is given as

$$\eta_{\text{bend, inert}} = \exp[-2.823\ Stk\ \varphi] \tag{8–68}$$

Figure 8–15 shows the transport efficiency of particles undergoing inertial deposition in a 90° bend in both laminar and turbulent flow as a function of the Stokes number. The correlations above are plotted along with data found in the literature. Data shown are the three values from Cheng and Wang (1981) mentioned above for $Re = 1000$ and $R_0 = 8$, as well as the Pui et al. (1987) data for laminar and turbulent flow.

When sampling lines must go through bends, the curvature ratio should be on the order of 4 or higher (i.e., abrupt turns should be avoided). To obtain a higher efficiency of transport through bends, the Stokes number should be kept small. If the flow is laminar and the Stokes number can be kept less than 0.05, losses from inertial impaction in the bend will be minimal. If the flow is turbulent, Eq. 8–68 is an adequate correlation for predicting the losses.

Inertial Deposition in Flow Constrictions in Sampling Lines

Flow constrictions in a sampling line should be avoided whenever possible; they produce losses that are difficult to characterize. Examples of constrictions are flow through a valve (although some ball valves present a little more disturbance to flow than the commonly used tube unions and connections), in-line orifices, abrupt changes in flow direction such as in the case of a tee or a cross or a right-angle bend with a curvature ratio close to 1, and changes from a large to small tube diameter either through an abrupt contraction or through a converging tube. If a sampling system having one or more of these features must be used, particle transport should be characterized experimentally over the range of applicable operating conditions. Similarly, sudden expansions in a flow path can produce eddies, from which particle deposition is difficult to characterize. These features should also be avoided.

Estimates for particle transport through flow constrictions can be made in some cases. For example, the transport efficiency correlations for a 90° bend from the previous section may be used to estimate the transport efficiency for flow elements such as a tee, a cross, or a sharp right-angle bend. In this case the highest velocity in the element should be used in the calculations.

Ye and Pui (1990) have developed a correlation for inertial particle deposition from laminar flow in a tube with an abrupt contraction. The correlation is based on numerical calculation and has been compared favorably with experimental data in the literature. The applicable geometry consists of coaxial circular tubes in series with a step reduction in cross section from the large to the small tube. The large-diameter tube (diameter d_i) is the inlet to the contraction, and the small-diameter tube (diameter d_o) is the outlet from the contraction. This type of geometry is found or approximated in common tube and pipe fittings. The transition from the larger diameter tube to the smaller diameter tube is characterized by the half angle of the contraction defined as the contraction angle. A blunt contraction would have a contraction angle of 90°, while a more gradual transition could have contraction angles less than 90°. Data have been reported by Muyshondt et al. (1996b) and by Chen and Pui (1995). We will relate these data below in the form with our definition of Stokes number and fractional penetration through the contraction. The Chen and Pui (1995) data are taken for a diameter ratio of 2 and at contraction angles of 75°, 45°, and 15°. The Reynolds number in the upstream section is nominally 1000. The Muyshondt et al. (1996b) data are taken for diameter ratios of 4.02, 2.15, and 1.323 and at contraction angles of 90°, 45° and 12°. The data are reported as penetration correlated with $Stk(1 - [d_o/d_i]^2)$. The Reynolds number for the Muyshondt et al. (1996b) data ranged into the turbulent regime.

Ye and Pui (1990) give a correlation of numerical data for the transport efficiency, $\eta_{\mathrm{cont,\,inert}}$, for particles undergoing inertial deposition in a 90° contraction. It is a function of the tube diameter ratio (d_o/d_i) and the Stokes number based on the small-diameter tube and the average velocity in the large tube, $Stk = \tau U/d_o$. Chen and Pui (1995) give a correlation of their data as functions of the same parameters. In laminar flow, the velocity profile gives a higher particle flux in the central portion of the tube and, taking this into account, we have the minimum penetration for large Stokes number in laminar flow as

$$\eta_{\mathrm{cont,inert,laminar\,min}} = 1 - \left[1 - \left(\frac{d_o}{d_i}\right)^2\right]^2 \tag{8–69}$$

The velocity profile in turbulent flow is assumed to be flat so that the particle flux through the tube is uniform over the cross section. This assumption gives the minimum penetration for large Stokes numbers in turbulent flow as

$$\eta_{\mathrm{cont,inert,turbulent\,min}} = \left(\frac{d_o}{d_i}\right)^2 \tag{8–70}$$

Muyshondt et al. (1996b) correlate their data with penetration as a function of the quantity $Stk(1 - [d_o/d_i]^2)$ given as

$$\eta_{\mathrm{cont,inert}} = 1 - \frac{1}{1 + \left\{\dfrac{2\,Stk\left[1 - \left(\dfrac{d_o}{d_i}\right)^2\right]}{3.14\exp(-0.0185\theta)}\right\}^{-1.24}} \tag{8–71}$$

where θ is the contraction angle in degrees.

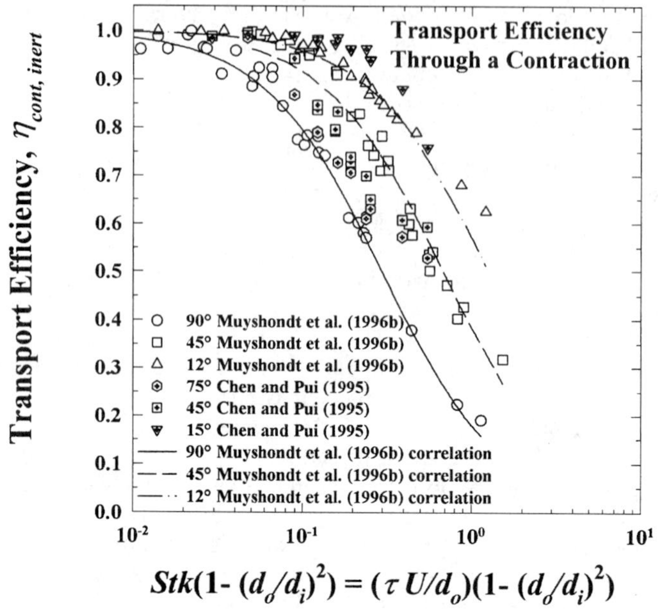

Fig. 8–16. Plot of the transport efficiency, $\eta_{cont, inert}$, of particles through a sudden contraction in a tube as a function of the Stokes number (based on the gas velocity in the large tube, U, and the diameter of the small tube, d_0) for contraction ratios of 0.9, 0.5, and 0.1.

This correlation fits the data well and is mathematically simpler than the correlations presented by Ye and Pui (1990) and Chen and Pui (1996).

Figure 8–16 plots the Chen and Pui (1995) and the Muyshondt et al. (1996b) data for the transport efficiency as a function of the quantity $Stk(1 - [d_o/d_i]^2)$ for particles flowing through abrupt contractions with contraction angles from 12° to 90°. The correlation of Muyshondt et al. (1996b) is also plotted for three contraction angles. Note that it fits all of the data well. The limits given above should be employed for penetration when the correlation yields lower values for penetration. Transport efficiency may be kept high by keeping the Stokes number small.

Electrostatic Deposition in Sampling Lines

The deposition of charged particles by electrostatic forces can occur in sampling lines during transport. Even in an aerosol that is, on average, neutral, some particles are charged. This charging occurs from diffusion of ions to the particles. Ions are constantly produced in our environment by cosmic rays. For a given ion concentration, there is an equilibrium charge distribution over the aerosol particles. Particles may also be charged by the mechanism that produces them. Static charge in the sampling line or an externally imposed electrical field in the line can produce particle deposition. Because it is not always possible to know the distribution of charge on aerosol particles or the electrical fields in a sampling line that is subject to static charge, electrostatic deposition of particles in sampling lines is most difficult to characterize. Note that electrostatic effects are negligible when the particles are uncharged or no electric field exists in the sampling line; on the other hand, these effects can be larger than those of any of the other forces discussed when transporting highly charged particles in non-conducting tubing.

Electrostatic deposition is largely avoidable. Metal or electrically conductive lines set up to avoid electrical fields by having no electrically isolated sections will obviate the problem of electrostatic deposition. If metal lines cannot be used, Tygon is an acceptable substitute (Liu et al., 1985). Materials to be avoided for aerosol transport are Teflon and Polyflo (Liu et al., 1985).

Thermophoretic Deposition in Sampling Lines

A temperature gradient in a gas will cause a suspended particle to move down the gradient from higher toward lower temperatures. This particle transport by a temperature gradient is called *thermophoresis*. The thermophoretic velocity, V_{th}, of a particle is the velocity that it achieves as a result of the balance between the thermophoretic force and the drag force. The thermophoretic velocity is dependent on the temperature gradient. It is independent of particle size for particles much larger than the gas molecule mean free path (continuum-regime particle) and for particles much smaller than the gas molecule mean free path (free-molecule regime). The transport efficiency in turbulent tube flow with thermophoretic deposition, $\eta_{tube,th}$, is expressed as

$$\eta_{tube,th} = \exp\left[-\frac{\pi\, d\, L\, V_{th}}{Q}\right] \tag{8–72}$$

In laminar tube flow, the flow and temperature gradient conditions become more involved, and no expression for the transport efficiency for laminar flow is given.

The thermophoretic velocity for particles in the continuum regime is expressed as (Friedlander, 1977)

$$V_{th} = \frac{2\left(\frac{k_g}{k_p}\right)k_g \nabla T}{5\,P\left[1+2\frac{k_g}{k_p}\right]} \tag{8–73}$$

For particles in the free-molecule regime, the thermophoretic velocity is expressed as (Friedlander, 1977)

$$V_{th} = \frac{3\nu\nabla T}{4\left[1+\pi\frac{0.9}{8}\right]T} \tag{8–74}$$

where

V_{th} = thermophoretic velocity
k_g = thermal conductivity of the gas
k_p = thermal conductivity of the particle
∇T = temperature gradient in the gas
P = pressure
T = gas temperature
ν = kinematic viscosity of the gas

Talbot et al. (1980) offer an expression for the magnitude of the thermophoretic velocity that covers particle sizes from the free molecule to the continuum regime and goes to the above limits:

$$V_{\text{th}} = \frac{2\,C_s v\left[\dfrac{k_g}{k_p} + C_t Kn\right]C(Kn)}{(1+3C_m Kn)\left[1+2\dfrac{k_g}{k_p}+2C_t Kn\right]}\frac{\nabla T}{T} \tag{8–75}$$

where

C_s = 1.13
C_t = 2.63
C_m = 1.14
Kn = particle Knudsen number defined as twice the gas molecular mean free path, λ, divided by the particle diameter, D_p
$C(Kn)$ = slip correction factor

The estimation of thermophoretic deposition in sample lines is not a straightforward process. The determination of the temperature gradients can be difficult. Because of the relatively low heat capacity of gases, the gas rapidly attains the tube wall temperature, causing a changing temperature gradient. The particle material thermal conductivity is not always known, and, in the case of an agglomerate particle, the particle thermal conductivity is not necessarily the same as the particle material thermal conductivity because of the included voids.

Thermophoretic deposition in sampling lines can be avoided by heating the lines to the gas temperature or cooling the gas to the line temperature by dilution.

Diffusiophoretic Deposition in Sampling Lines

A particle in a nonuniform gas mixture will be acted on by the diffusing gas molecules, where the molecular diffusion arises from the concentration gradients. The force that the particle experiences is called the *diffusiophoretic force*. This force arises from the unequal momentum transfer to the particle from the heavier molecules on the higher concentration side of the particle. The direction of this force is in the direction of diffusion of the heavier gas molecule. This force causes the particles to move with a velocity called the *diffusiophoretic velocity*, V_{dph}. This phenomenon should not be confused with Brownian diffusion of the particle.

Diffusiophoresis is caused by a concentration gradient in the gas molecules. Near a condensing or an evaporating surface, an additional effect influences the force acting on the particle. Near the surface, the condensing or evaporating vapor has associated with it a concentration gradient that drives the vapor's diffusion through the gas, effecting condensation or evaporation. To maintain a constant gas pressure, the gas must also have a concentration gradient equal to but in the opposite direction of the vapor's gradient. The concentration gradient in the gas produces diffusion of the gas toward the surface in the case of vapor evaporation and away from the surface in the case of vapor condensation. Because the surface is not a source or a sink for the gas, an aerodynamic flow called the *Stefan flow* is set up to transfer the gas away from an evaporating surface and toward a condensing surface to counteract the diffusive transfer of gas. The Stefan flow exerts a force on the particle in the direction of the flow, in addition to the diffusiophoretic force acting in the direction of the diffusion of the heavier molecule (Fuchs, 1964; Hinds, 1982). Waldman and Schmitt (1966) and Goldsmith and May (1966) give an expression for the diffusiophoretic velocity on a particle acted on by diffusiophoretic forces and the Stefan flow:

$$V_{\text{dph}} = \frac{1\sqrt{m_1}}{\gamma_1\sqrt{m_1} + \gamma_2\sqrt{m_2}} \frac{D}{\gamma_2} \nabla\gamma_1 \tag{8–76}$$

where

m_1 = the mass of a molecule of the diffusing species
m_2 = the mass of a molecule of the stagnant species
γ_1 = the mole fraction of the diffusing species
γ_2 = the mole fraction of the stagnant species
$\nabla\gamma_1$ = the gradient of the mole fraction of the diffusing species

In the case of water vapor and air, the velocity of a particle is away from an evaporating surface and toward a condensing surface. For diffusiophoresis arising from condensation in turbulent pipe flow, the transport efficiency, $\eta_{\text{tube, dph}}$, of particles can be estimated from

$$\eta_{\text{tube,dph}} = \exp\left[\frac{-\pi\, d\, L\, V_{\text{dph}}}{Q}\right] \tag{8–77}$$

Whitmore and Meisen (1978) produced data for diffusiophoretic deposition in turbulent flow for ammonia condensing from a number of gases. The results indicated that the particle transport efficiency was comparable with the mole fraction of the noncondensing gas. Thus, if 10% of the gas transporting particles condensed, about 10% of the particles would be deposited. These results of Whitmore and Meisen (1978) permit a rough estimate of the transport efficiency for particles undergoing diffusiophoretic deposition based on the noncondensable mole fraction of the gas.

Conditions under which diffusiophoretic deposition would occur should be avoided in the transport of aerosol samples. This can be accomplished by avoiding condensation in sampling lines by heating or dilution.

Deposition in Chambers and Bags

Situations arise in aerosol sampling in which a sampled aerosol is temporarily stored. This can happen when a sample must be taken at a rate faster than the instrument can accommodate. For example, an aircraft sampling from a plume will often have only a few seconds of time in the plume to acquire a sample, but the instrument cannot make a measurement in that short a time period. The sample is drawn into a sample chamber or bag during the short transit period, and the measurement is made on the stored sample over a longer period of time. Another such sampling application is the study of aerosol size distribution evolution from gas to particle conversion and coagulation. In all these applications, the deposition of particles during their residence time in the chamber can have a definite effect on the measurement. When the time to take an aerosol measurement or a series of measurements is comparable with the time for the aerosol distribution in the chamber to change appreciably from deposition, then the deposition must be accounted for.

Two deposition mechanisms are usually considered. They are gravitational settling and diffusion. Crump and Seinfeld (1981) and Crump et al. (1983) have developed a model for gravitational and diffusive deposition inside stirred vessels. Other less characterizable deposition mechanisms should be avoided. Diffusiophoresis and thermophoresis can be avoided by the sampling technique, that is, avoid temperature gradients at the chamber wall and avoid condensing conditions. Electrostatic deposition can be avoided by the selection of an electrically conductive chamber material, such as aluminized Mylar, or one that can be demonstrated not to be susceptible to electrostatic deposition. Teflon should not be used, for reasons outlined in the section on electrostatic deposition. McMurry and Rader (1985) and McMurry and Grosjean (1985) have reported experimental data in which an enhanced loss of charged particles inside Teflon film chambers has been observed. They have developed a model of electrostatic deposition in chambers with static charge that explains the data well.

A model for simultaneous gravitational and diffusional losses (Crump and Seinfeld, 1981) has been developed for a well-mixed vessel of arbitrary shape. For a spherical vessel, they give an expression for the wall loss coefficient, β, where the remaining fraction of particles of a given size after time t is

$$\eta_{\text{bag, grav diff}} = \exp\left[-\int_0^t \beta \, dt\right] \quad (8\text{–}78)$$

and

$$\beta = \frac{12k_e D}{\pi^2 R V_{ts}} \int_0^{\frac{\pi V_{ts}}{2\sqrt{k_e D}}} \frac{x}{e^x - 1} dx + \frac{3V_{ts}}{4R} \quad (8\text{–}79)$$

where

R = vessel radius
k_e = coefficient of eddy diffusion

The first term in Eq. 8–79 describes the losses from diffusion and includes the effect of declining eddy diffusivity (mixing level) approaching the wall. The second term in Eq. 8–79 is the contribution from gravitational settling. It is equal to the settling velocity times the ratio of the deposition area of the chamber to the volume of the chamber. The deposition area for gravitational settling is the projected area on the horizontal plane. This definition is applicable for stirred settling from any vessel shape.

The coefficient of eddy diffusion characterizes the level of mixing in the chamber. Higher values indicate a higher eddy diffusivity and, consequently, a higher loss by diffusion. A chamber with a constant throughput of gas will have a constant mixing level driven by the kinetic energy of the incoming gas, but a chamber that is initially filled and then closed off will have a mixing level that decays with time. Equation 8–78 contains the integral of β with respect to time because the coefficient of eddy diffusion, k_e, can change with time. This parameter is the only unknown in Eq. 8–79; it must be determined experimentally or estimated. Crump et al. (1983) indicate that k_e is proportional to the square root of the ratio of the turbulent energy dissipation rate to the kinematic viscosity of the fluid. This gives

$$k_e = \frac{CQ^{3/2}}{\sqrt{A^2 \nu V_c}} \quad (8\text{–}80)$$

where

C = proportionality constant
Q = flow rate into the bag
A = area of the flow inlet
V_c = volume of the chamber

They give for their experimental 118 L spherical vessel

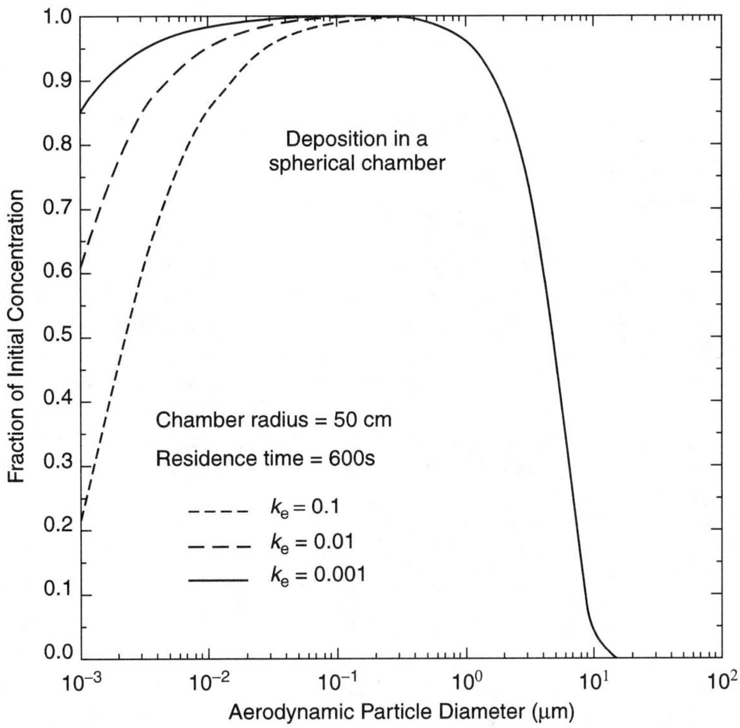

Fig. 8–17. Plot of the fraction of initial concentration, $\eta_{\text{bag, grav diff}}$, of particles undergoing diffusive and gravitational deposition in a 0.50 m radius spherical chamber, 600 s after filling, as a function of particle aerodynamic diameter at 1 atm and 293 K for various values of the coefficient of eddy diffusion, k_e.

$$k_e = 1.35 \cdot 10^5 \cdot Q^{3/2} \tag{8–81}$$

where k_e is in units of s^{-1} and Q is in units of m^3/s. Not enough information is given to estimate a proportionality constant for Eq. 8–80. The values of k_e given by Crump et al. (1983) range from 0.028 to 0.068 s^{-1}, and McMurry and Rader (1985) use values of 0.0064 and 0.12 s^{-1}. Conservative estimations can be performed with the higher values of k_e to assess the diffusion losses in a bag or chamber. The losses arising from settling are more straightforward.

Figure 8–17 shows the fraction of initial concentration of aerosol in a 0.50 m radius spherical chamber as a function of particle size, 600 s after filling. The conditions are for air at 1 atmosphere and 20°C. Calculations for three values of k_e that span the range mentioned above are shown, and k_e is assumed constant over the 600 s time period. Gravitational deposition depends only on the size of the chamber and can be reduced by using a larger chamber. The higher values of k_e produce a higher diffusion loss. A larger bag will result in a lower diffusion loss.

In filling a bag with flexible sides, particles on the bag walls can be resuspended in the gas volume as a result of the mechanical forces on the wall during filling. Generally, larger particles are more easily resuspended. The resuspension of particles from the bag should be considered in the selection of the bag and the decision to reuse bags. In some applications, single-use bags may be prudent.

OTHER SAMPLING ISSUES

Sample Conditioning by Dilution

When sampling aerosols it is often necessary to reduce concentrations by dilution. Dilution may be required to bring concentrations within the range of measuring instruments, to quench chemical reactions or coagulation, or to cool a hot gas. The factors to consider in the design or selection of a diluter include the time from sample extraction to dilution (in a coagulating or chemically reacting aerosol, this time should be as short as possible), required dilution rates, particle losses in the diluter, and flow rates. Brockmann et al. (1984) briefly discuss a variety of diluters and describe a diluter designed specifically for high dilution and transport of coagulating ultrafine aerosols. Diluters are also discussed in Chapters 30, 31, and 37.

An important consideration in selecting a diluter is the uncertainty with which the dilution rate is known. In diluting a sample, a dilution gas flow, Q_d, is mixed with a sample gas flow, Q_s. The total gas flow out of the diluter, Q_t, is the sum of the dilution and sample gas flows. The dilution rate, DR, is the total flow divided by the sample flow and can be expressed in terms of any of the two flows:

$$DR = \frac{Q_t}{Q_s} = \frac{Q_d + Q_s}{Q_s} = \frac{Q_t}{Q_t - Q_d} \tag{8–82}$$

Differences in uncertainty arise, from which two flows are measured to determine the dilution rate. When Q_s and Q_t are measured, the uncertainty in the dilution rate, $U(DR)$, is related to the uncertainties in the flows, $U(Q_s)$ and $U(Q_t)$, by the relation (see the discussion of propagation of errors in Chapter 22)

$$U(DR) = \left[U(Q_s)^2 + U(Q_t)^2\right]^{1/2} \tag{8–83}$$

The uncertainty is not much different when Q_s and Q_d are measured and approaches that for Q_s and Q_t for larger values of DR:

$$U(DR) = \left[1 - \frac{1}{DR}\right]\left[U(Q_s)^2 + U(Q_d)^2\right]^{1/2} \tag{8–84}$$

When the sample flow is one of the flows measured, the uncertainty in the dilution rate is only slightly larger than the uncertainties in the flow measurements. When Q_d and Q_t are measured, the sample flow rate is determined from the difference between the measured quantities. When this difference is of the order of the uncertainty in these measured flows, the uncertainty in the dilution rate becomes much larger than the uncertainty in either of the measured flows:

$$U(DR) = [DR - 1]\left[U(Q_t)^2 + U(Q_d)^2\right]^{1/2} \tag{8–85}$$

The uncertainty in dilution rate in this situation increases with dilution rate. If the uncertainty in flow measurement is ±3% and the dilution rate is 10, then the uncertainties in DR for the situations described by Eqs. 8–83, 8–84, and 8–85 are ±4.2%, ±3.8%, and ±38%, respectively. For higher dilutions, measurement of the sample flow is strongly recommended if the dilution rate must be known.

Measurement of the aerosol sample flow can be accomplished by drawing it through a capillary tube in laminar flow (Fuchs and Sutugin, 1965; Delattre and Friedlander, 1978; Brockmann et al., 1984); this allows the flow to be measured. This is a well-defined system in which the particle losses can be characterized. It lends itself to high dilution rates. One must be cautious, however, to avoid plugging of the capillary.

Dilution gas can be introduced into the flow stream through holes in the wall of the containing tube or through a porous wall (Ranade et al., 1976). It may also be introduced coaxially surrounding the sample gas flow. Care must be taken to avoid uneven flow, which would direct the undiluted sample flow against the walls of the sampling system that could promote uncharacterizable deposition.

The dilution gas must be well mixed with the sampled aerosol. This can be accomplished through turbulence either in the transport line or by passing the sample and dilution gas through a mixing nozzle (Brockmann et al., 1984). To avoid possible particle deposition in a flow constriction, care should be taken to sheath the sample with clean gas during flow in the converging section of the nozzle or to introduce the sample flow into the throat of the nozzle. Mixing will occur in the diverging portion of the nozzle.

Plugging of Sampling Lines and Inlets

Aerosol flowing through cracks and leaks in pressurized systems has been observed to deposit in the flow paths and plug them. This has also been observed when large amounts of aerosol have flowed through pipes (Morewitz, 1982). A simple model of aerosol plugging of flow paths has been developed (Vaughn, 1978; Morewitz, 1982) in which the integral mass of aerosol, m, that flows through a pipe up to the point of plugging is correlated to the diameter of the pipe, d, through which it flows:

$$m = \mathrm{K}d^3 \tag{8–86}$$

where

m = total mass in grams of aerosol that flowed through the pipe before plugging
K = proportionality constant that is between 10,000 and 50,000 kg/m^3
d = pipe diameter in m

Equation 8–86 has been referred to as the "Morewitz criterion for aerosol plugging" (Novick, 1994). It is a very general correlation that does not depend on aerosol concentration, particle size, or flow rate.

Novick (1994) has observed situations in which the Morewitz criterion was met but no plugging occurred. Novick (1994) has examined data, including those reviewed by Morewitz (1982), and has observed that Eq. 8–86 may be a necessary condition for plugging but is not a sufficient condition. He suggests that the data support a second condition being applied to the plugging criterion, that of the violation of the Davies inertial criterion for representative sampling given in Eq. 8–38 in which the Stokes number is based on the tube diameter and the average gas velocity in the tube. It should also be noted that the presence of obstructions to flow such as discussed in the section "Deposition in Flow Constrictions" can precipitate aerosol plugging.

These plugging criteria are not developed and are included to provide the reader with a means to identify situations in which plugging or high particle deposition may be encountered.

Re-entrainment of Deposited Particles

Fromentin (1989) discussed the current state of research and modeling of particle re-entrainment by flowing gas. Wen and Kasper (1989) point out that steady flow will not produce much re-entrainment, but that unsteady conditions will. These conditions are encountered in turbulent flow, with eddies impinging on the wall, and during the startup or shutdown of flows. Ziskind et al. (1998) model the effects of shear near a surface and its effects on particle re-entrainment. Re-entrainment is not a well-characterized process and should be avoided by using clean sample lines and bags, minimizing deposition, eliminating mechanical shock and vibration to the sampling system, and eliminating flow transients and shocks in the sampling system.

Inhomogeneous Particle Concentrations in Inlets and Transport Tubes

Much of the previous discussion presents aspiration efficiency and transport efficiency as a single number, assuming that at the chosen measurement point all the particles reaching that point will be detected. However, some instruments and techniques either measure a selected portion of the aerosol stream or otherwise assume that the aerosol stream is uniform. For instance, the Aerodynamic Particle Sizer (***TSI***) samples 20% of the air stream entering the instrument inlet; an asbestos sampler collects fibers onto a filter area having the same diameter as the inlet, while the final measurement involves microscope analysis of small portions of the filter. Various forces act on particles that enter and are transported through a sampling system. In addition, the air flow can be complex, with secondary flow vortices often occurring when the air flow changes direction. Electrostatic effects can cause attraction or repulsion of particles from the inlet and internal surfaces, especially if the walls are nonconductive. Gravitational settling of larger particles reduces the concentration on the top of horizontal transport ducts, especially under laminar flow conditions. Inertial movement of particles can produce nonuniform aerosol streams when the flow changes direction. All these forces interact with the flow field.

Pui, Romay-Novas, and Liu investigated flow fields in right angle tubing bends and found that two counter-rotating secondary flows produced mixing of the air stream and inertial deposition on the outside of the bend. Baron et al. (1994; Chen and Baron, 1996) observed particle deposits in asbestos sampling inlets and demonstrated nonuniform aerosol streams as a result of electrostatic, gravitational, and inertial forces. Figure 8–18 shows the filter deposits for samplers operating at 0.14 m/s [14 cm/s] inlet velocity and several wind velocities and an angle of 120° to the wind (vertically downward). In Figure 8–18, the dark areas represent the sampled methylene blue particle deposits. The white areas are regions that were cleared of particles by inertial deposition and gravitational settling on the outer wall of the inlet and were mixed by the vortices formed as the aerosol entered the inlet. Although the overall sampling efficiency might have been relatively high, measurement of the aerosol at specific points of the inlet cross section would have caused significantly biased results. Under turbulent conditions, inertial effects can cause localized stratification of the aerosol (Squires and Eaton, 1991).

SUMMARY AND CONCLUSIONS

Correlations for aspiration efficiency and transmission efficiency for thin-walled nozzles sampling from a flowing gas have been reported. The product of these efficiencies is the inlet efficiency. The inlet efficiency depends on the Stokes number, the ratio of ambient gas velocity to sampling gas velocity, and the sampling angle. To minimize loss and obtain a representative sample, the following suggestions are made:

1. Isokinetic (iso-mean-velocity) sampling (sampling velocity equal to ambient velocity and isoaxial sampling) should be used whenever possible.
2. Gas velocities should be large compared with the particle-settling velocity.
3. Inlet diameters should be of the order of 1 cm to reduce the deposition from free-stream turbulence.

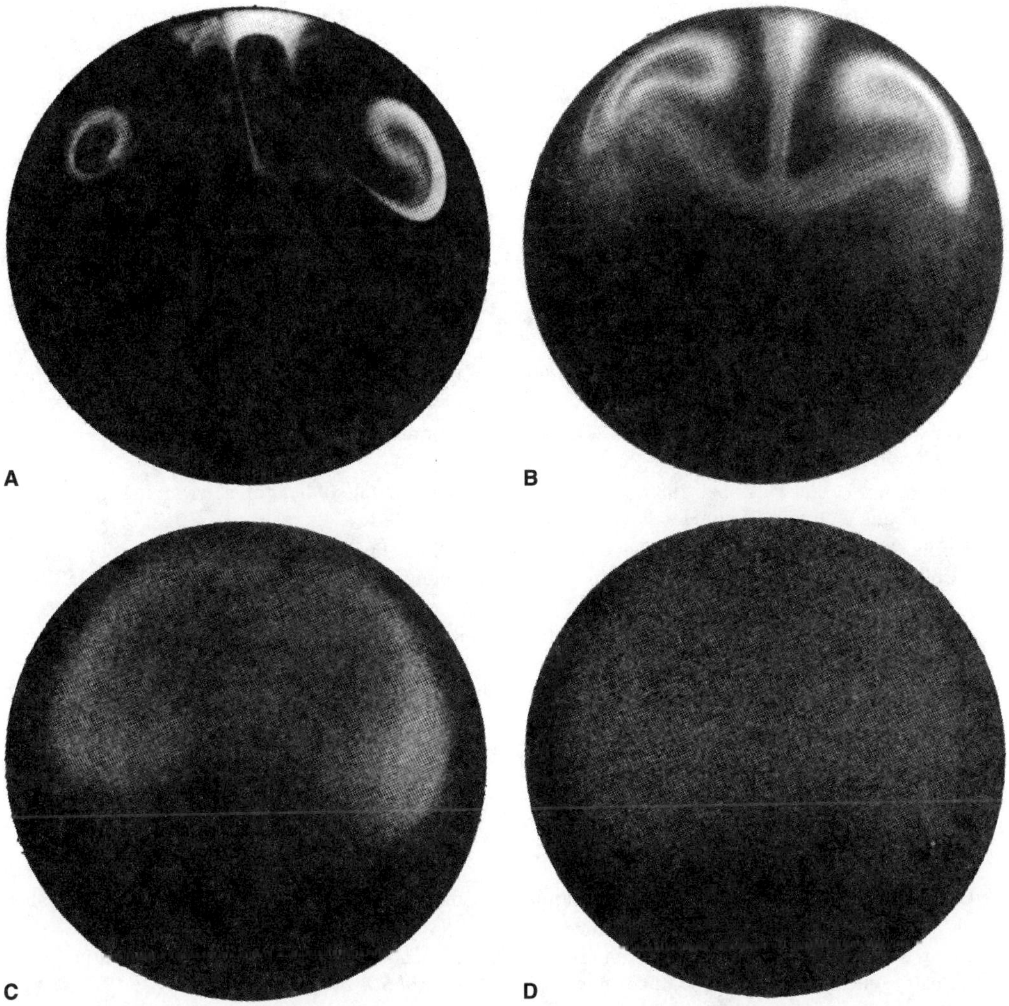

Fig. 8–18. Indications of anisoaxial sampling flow patterns. Methylene blue particles were deposited on a filter at the end of a 0.025 m (25 mm) diameter, 0.050 m (50 mm) long inlet oriented 120° to the air velocity while sampling at an inlet velocity of 0.32 m/s. Four samples were taken at wind velocities of 0.23 m/s (**A**), 0.30 m/s (**B**), 0.45 m/s (**C**), and 0.62 m/s (**D**). The white pattern on each filter is a result of a particle-free region created just outside the inlet and modified by the flow field inside the inlet.

4. Stokes numbers should be kept as low as possible.
5. Electrostatic effects should be minimized.

Criteria for representative sampling from calm air are also presented.

Transport efficiency through sample lines and for residence time in storage bags is discussed. The correlations reported in the literature for deposition from gravitational settling, diffusional deposition, turbulent inertial deposition, inertial deposition in a bend, inertial deposition in an abrupt contraction, and diffusional and settling deposition in a chamber have been presented. Their dependence on the parameters of the correlations is discussed. Deposition from electrostatic, thermophoretic, and diffusiophoretic forces has been discussed and suggested practices outlined to avoid these deposition mechanisms. Resuspension of deposited material has also been discussed. Dilution as a means of sample conditioning and the uncertainties entailed in diluting a sample have been presented. These correlations allow the reader to evaluate the performance of sampling systems and to optimize their design. Because these correlations cannot address every situation and have not exhaustively covered all possible deposition mechanisms, it is strongly recommended that their use to design or evaluate a sampling system be accompanied by validating experimental calibration of the sampling system.

REFERENCES

ANSI. 1969. *Guide to Sampling Airborne Radioactive Materials in Nuclear Facilities*. ANSI N13.1-1969. New York: American National Standards Institute, Inc.

Agarwal, J. and B. Y. H. Liu. 1980. A criterion for accurate aerosol sampling in calm air. *Am. Ind. Hyg. Assoc. J.* 41:191–197.

Armbruster, L. and G. Zebel. 1985. Theoretical and experimental studies for determining the aerosol sampling efficiency of annular slots. *J. Aerosol Sci.* 16:335–341.

Baron, P. A., C. C. Chen, D. R. Hemenway, and P. O'Shaughnessey. 1994. Non-uniform air flow in inlets: The effect on filter deposits in the fiber sampling cassette. *Am. Ind. Hyg. Assoc. J.* 55:722–732.

Belyaev, S. P. and L. M. Levin. 1972. Investigation of aerosol aspiration by photographing particle tracks under flash illumination. *J. Aerosol Sci.* 3:127–140.

Belyaev, S. P. and L. M. Levin. 1974. Techniques for collection of representative aerosol samples. *J. Aerosol Sci.* 5:325–338.

Brockmann, J. E., B. Y. H. Liu, and P. H. McMurry. 1984. A sample extraction diluter for ultrafine aerosol sampling. *Aerosol Sci. Technol.* 4:441–451.

Brockmann, J. E., P. H. McMurry, and B. Y. H. Liu. 1982. Experimental study of simultaneous coagulation and diffusional loss of free molecule aerosols in turbulent pipe flow. *J. Colloid Interf. Sci.* 88:522–529.

Chen, C.-C. and P. A. Baron. 1996. Aspiration efficiency and wall deposition in the fiber sampling cassette. *Am. Ind. Hyg. Assoc. J.* 52:142–152.

Chen, D.-R. and D. Y. H. Pui. 1995. Numerical and experimental studies of particle deposition in a tube with conical contraction—Laminar flow regime. *J. Aerosol Sci.* 26(4):563–574.

Chen, Q. and G. Ahmadi. 1997. Deposition of particles in turbulent pipe flow. *J. Aerosol Sci.* 28:789–796.

Cheng, Y. S. and C. S. Wang. 1981. Motion of particles in bends of circular pipes. *Atmos. Environ.* 15:301–306.

Chung, K. Y. K. and T. L. Ogden. 1986. Some entry efficiencies of disk-like samplers facing the wind. *Aerosol Sci. Technol.* 5:81–91.

Crane, R. L. and R. L. Evans. 1977. Inertial deposition of particles in a bent pipe. *J. Aerosol Sci.* 8:161–170.

Crump, J. G., R. C. Flagan, and J. H. Seinfeld. 1983. Particle wall loss rates in vessels. *Aerosol Sci. Technol.* 3:303–309.

Crump, J. G. and J. H. Seinfeld. 1981. Turbulent deposition and gravitational sedimentation of an aerosol in a vessel of arbitrary shape. *J. Aerosol Sci.* 12:405–415.

Davies, C. N. 1968. The entry of aerosols into sampling tubes and heads. *Br. J. Appl. Phys. (J. Phys. D) Ser.* 21:921–932.

Davies, C. N. and M. Subari. 1982. Aspiration above wind velocity of aerosols with thin-walled nozzles facing and at right angles to the wind direction. *J. Aerosol Sci.* 13:59–71.

Delattre, P. and S. K. Friedlander. 1978. Aerosol coagulation and diffusion in a turbulent jet. *Ind. Eng. Chem. Fund.* 17:189–194.

Durham, M. D. and D. A. Lundgren. 1980. Evaluation of aerosol aspiration efficiency as a function of Stokes number, velocity ratio and nozzle angle. *J. Aerosol Sci.* 11:179–188.

Fissan, H. and G. Schwientek. 1987. Sampling and transport of aerosols. *TSI J. Part. Instrum.* 2:3–10.

Friedlander, S. K. 1977. *Smoke, Dust, and Haze*. New York: John Wiley & Sons.

Friedlander, S. K. and H. F. Johnstone. 1957. Deposition of suspended particles from turbulent gas stream. *Ind. Eng. Chem.* 49:1151–1156.

Fromentin, A. 1989. *Particle Resuspension From a Multilayer Deposit by Turbulent Flow*. PSI Bericht No. 38. Paul Scherrer Institute, Switzerland.

Fuchs, N. A. 1964. *The Mechanics* of *Aerosols.* Oxford: Pergamon.

Fuchs, N. A. 1975. Sampling of aerosols. *Atmos. Environ.* 9:697–707.

Fuchs, N. A. and A. G. Sutugin. 1965. Coagulation rate of highly dispersed aerosols. *J. Colloid Sci.* 20:492–500.

Goldsmith, P. and F. G. May. 1966. Diffusiophoresis and thermophoresis in water vapor systems. In *Aerosol Science*, ed. C. N. Davies. New York: Academic Press, pp. 163–194.

Gormley, P. G. and M. Kennedy. 1949. Diffusion from a stream flowing through a cylindrical tube. *Proc. R. Irish Acad.* 52A:163–169.

Grinshpun, S. A., G. N. Lipatov, and A. G. Sutugin. 1990. Sampling errors in cylindrical nozzles. *Aerosol Sci. Technol.* 12:716–740.

Grinshpun, S., K. Willeke, and S. Kalatoors. 1993. A general equation for aerosol aspiration by thin-walled sampling probes in calm and moving air. *Atmos. Environ.* 27A:1459–1470.

Grinshpun, S., K. Willeke, and S. Kalatoors. 1994. Corrigendum: A general equation for aerosol aspiration by thin-walled sampling probes in calm and moving air. *Atmos. Environ.* 28:375.

Guha, A. 1997. A unified Eulerian theory of turbulent deposition to smooth and rough surfaces. *J. Aerosol Sci.* 28:1517–1537.

Hangal, S. and K. Willeke. 1990a. Aspiration efficiency: Unified model for all forward sampling angles. *Environ. Sci. Technol.* 24:688–691.

Hangal, S. and K. Willeke. 1990b. Overall efficiency of tubular inlets sampling at 0–90 degrees from horizontal aerosol flows. *Atmos. Environ.* 24A:2379–2386.

Heyder, J. and J. Gebhart. 1977. Gravitational deposition of particles from laminar aerosol flow through inclined circular tubes. *J. Aerosol Sci.* 8:289–295.

Hinds, W. C. 1982. *Aerosol Technology.* New York: John Wiley & Sons.

Holman, J. P. 1972. *Heat Transfer.* New York: McGrawHill.

Im, K. H. and R. K. Ahluwalia. 1989. Turbulent eddy deposition of particles on smooth and rough surfaces. *J. Aerosol Sci.* 20:431–436.

Jayasekera, P. N. and C. N. Davies. 1980. Aspiration below wind velocity of aerosols with sharp edged nozzles facing the wind. *J. Aerosol Sci.* 11:535–547.

John, W. and H. C. Wang. 1991. Laboratory testing methods for PM-10 samplers: Lowered effectiveness from particle loading. *Aerosol Sci. Technol.* 14:93–101.

John, W., W. Winklmayr, and H. C. Wang. 1991. Particle deagglomeration and re-entrainment in a PM-10 sampler. *Aerosol Sci. Technol.* 14:165–176.

Kurz, J. L. and T. C. Ramey. 1988. The development, performance, and application of an advanced isokinetic stack sampling system. In *Proc. 20th DOE/NRC Nuclear Air Cleaning Conf.* NUREG/CP0098, pp. 847–856.

Laktionov, A. B. 1973. Aspiration of an aerosol into a vertical tube from a flow transverse to it. *Fizika Aerozoley* 7:83–87. (Translation from Russian AD760 947, Foreign Technology Division, Wright-Patterson AFB, Dayton, OH.)

Lee, K. W. and J. A. Gieseke. 1994. Deposition of particles in turbulent pipe flows. *J. Aerosol Sci.* 25:699–709.

Lipatov, G. N., S. A. Grinshpun, and T. L. Semenyuk. 1989. Properties of crosswise migration of particles in ducts and inner aerosol deposition. *J. Aerosol Sci.* 20:935–938.

Lipatov, G. N., S. A. Grinshpun, T. L. Semenyuk, and A. G. Sutugin. 1988. Secondary aspiration of aerosol particles into thin-walled nozzles facing the wind. *Atmos. Environ.* 22:1724–1727.

Lipatov, G. N., S. A. Grinshpun, G. L. Shingaryov, and A. G. Sutugin. 1986. Aspiration of coarse aerosol by a thin-walled sampler. *J. Aerosol* Sci. 17:763–769.

Lipatov, G. N., T. L. Semenyuk, and S. A. Grinshpun. 1990. Aerosol migration in laminar and transition flows. *J. Aerosol* Sci. 21(Suppl. 1):S93–S96.

Liu, B. Y. H. and J. K. Agarwal. 1974. Experimental observation of aerosol deposition in turbulent flow. *J. Aerosol Sci.* 5:145–155.

Liu, B. Y. H. and D. Y. H. Pui. 1981. Aerosol sampling inlets and inhalable particles. *Atmos. Environ.* 15:589–600.

Liu, B. H. Y., D. Y. H. Pui, K. L. Rubow, and W. W. Szymanski. 1985. Electrostatic effects in aerosol sampling and filtration. *Ann. Occup. Hyg.* 29:251–269.

Liu, B. Y. H., Z. Q. Zhang, and T. H. Kuehn. 1989. A numerical study of inertial errors in anisokinetic sampling. *J. Aerosol Sci.* 20:367–380.

Lundgren, D. A., M. D. Durham, and K. W. Mason. 1978. Sampling of tangential flow streams. *Am. Ind. Hyg. Assoc. J.* 39:640–644.

McFarland, A. R., N. K. Anand, C. A. Ortiz, M. E. Moore, S. H. Kim, R. E. DeOtte, Jr., and S. Somasundaram. 1988. Continuous air sampling for radioactive aerosol. In *Proc. 20th DOEINRC Nuclear Air Cleaning Conf.* NUREG/CP-0098, pp. 834–846.

McMurry, P. H. and D. Grosjean. 1985. Gas and aerosol wall losses in Teflon film smog chambers. *Environ. Sci. Technol.* 19:1176–1182.

McMurry, P. H. and D. J. Rader. 1985. Aerosol wall losses in electrically charged chambers. *Aerosol Sci. Technol.* 4:249–268.

Morewitz, H. A. 1982. Leakage of aerosols from containment buildings. *Health Phys.* 42:195–207.

Muyshondt, A., N. K. Anand, and A. R. McFarland. 1996a. Turbulent deposition of aerosol particles in large transport tubes. *Aerosol Sci. Technol.* 24:107–116.

Muyshondt, A., A. R. McFarland, and N. K. Anand. 1996b. Deposition of aerosol particles in contraction fittings. *Aerosol Sci. Technol.* 24:205–216.

Novick, V. J. 1994. Plugging passages with particles: Refining the Morewitz criteria. *Aerosol Sci. Technol.* 21:219–222.

Okazaki, K., R. W. Wiener, and K. Willeke. 1987a. The combined effect of aspiration and transmission on aerosol sampling accuracy for horizontal isoaxial sampling. *Atmos. Environ.* 21:1181–1185.

Okazaki, K., R. W. Wiener, and K. Willeke. 1987b. Isoaxial aerosol sampling: Nondimensional representation of overall sampling efficiency. *Environ. Sci. Technol.* 21:178–182.

Okazaki, K., R. W. Wiener, and K. Willeke. 1987c. Non isoaxial aerosol sampling: Mechanisms controlling thc overall sampling efficiency. *Environ. Sci. Technol.* 21:183–187.

Orr, C., Jr. and E. Y. H. Keng. 1976. Sampling and, particle-size measurements. In *Handbook on Aerosols*, ed. R. Dennis. TID-26608, Technical Information Center, U.S. Department of Energy.

Paulus, H. J. and R. W. Thron. 1977. Stack sampling. In *Air Pollution, Vol. 3, Measuring, Monitoring, and Surveillance of Air Pollution*, 3rd eD., ed. A. C. Stern. New York: Academic Press, pp. 525–587.

Pich, J. 1972. Theory of gravitational deposition of particles from laminar flows in channels. *J. Aerosol Sci.* 3:351–361.

Pui, D. Y. H., F. Romay-Novas, and B. Y. H. Liu. 1987. Experimental study of particle deposition in bends of circular cross section. *Aerosol Sci. Technol.* 7:301–315.

Rader, D. J. and V. A. Marple. 1988. A study of the effect of anisokinetic sampling. *Aerosol Sci. Technol.* 8:283–299.

Ranade, M. B., D. K. Werle, and D. T. Wasan. 1976. Aerosol transport through a porous sampling probe with transpiration air flow. *J. Colloid Interf. Sci.* 56:42–52.

Reeks, M. W. and G. Skyrme. 1976. The dependence of particle deposition velocity on particle inertia in turbulent pipe flow. *J. Aerosol* Sci. 7:485–495.

Reynolds, A. M. 1999. A Lagrangian stochastic model for heavy particle deposition. *J. Colloid Interf. Sci.* 215:85–91.

Saffman, P. G. 1965. The lift on a small sphere in slow shear flow. *J. Fluid Mech.* 22:385–400.

Saffman, P. G. 1968. Corrigendum. *J. Fluid Mech.* 31:624.

Schlichting, H. 1968. *Boundary Layer Theory*, 6th Ed. New York: McGraw-Hill.

Schwendiman, L. C., G. E. Stegen, and J. A. Glissmeyer 1975. *Report BNWL-SA-5138*. Richland, WA: Battelle Pacific Northwest Laboratory.

Stevens, D. C. 1986. Review of aspiration coefficients of thin-walled sampling nozzle. *J. Aerosol Sci.* 17:729–743.

Squires, K. D. and J. K. Eaton 1991. Preferential concentration of particles by turbulence. *Physics Fluids A—Fluid Dynamics* 3:1169–1179.

Talbot, L., R. K. Cheng, R. W. Schefer, and D. R. Willis. 1980. Thermophoresis of particles in a heated boundary layer. *J. Fluid Mech.* 101:737–758.

Thomas, J. W. 1958. *J. Air Pollut. Control Assoc.* 8:32.

U.S. Environmental Protection Agency. 1974. *Administrative and Technical Aspects of Source Sampling for particulates*. EPA-450/3-74-047.

U.S. Environmental Protection Agency. 1987. *Fed. Reg.* 52:24634.

Vaughn, E. U. 1978. Simple model for plugging of duct, by aerosol deposits. *ANS Trans.* 28:507–508.

Vincent, J. H. 1984. A comparison between models for predicting the performances of blunt dust samplers. *Atmos. Environ.* 187:1033–1035.

Vincent, J. H. 1987. Recent advances in aspiration theory for thin-walled and blunt aerosol sampling probes. *J. Aerosol* Sci. 18:487–498.

Vincent, J. H. 1989. *Aerosol Sampling: Science and Practice.* New York: John Wiley & Sons.

Vincent, J. H., P. C. Emmett, and D. Mark. 1985. The effects of turbulence on the entry of airborne particles into a blunt dust sampler. *Aerosol Sci. Technol.* 4:17–29.

Vincent, J. H. and H. Gibson. 1981. Sampling errors in blunt dust samplers arising from external wall loss effects. *Atmos. Environ.* 15:703–712.

Vincent, J. H., D. Hutson, and D. Mark. 1982. The nature of air flow near the inlets of blunt dust sampling probes. *Atmos. Environ.* 16:1243–1249.

Vincent, J. H., D. C. Stevens, D. Mark, M. Marshall, and T. A. Smith. 1986. On the aspiration characteristics of large-diameter thin-walled aerosol sampling probes at yaw orientations with respect to the wind. *J. Aerosol Sci.* 17:211–224.

Waldmann, L. and K. H. Schmitt. 1966. Thermophoresis and diffusiophoresis of aerosols. In *Aerosol Science*, ed. C. N. Davies. New York: Academic Press, pp. 137–162.

Wang, C. S. 1975. Gravitational deposition of particles from laminar flows in inclined channels. *J. Aerosol Sci.* 6:191–204.

Wen, H. Y. and G. Kasper. 1989. On the kinetics of particle re-entrainment from surfaces. *J. Aerosol Sci.* 20:483–498.

Whitmore, P. J. and A. Meisen. 1978. Diffusiophoretic particle collection under turbulent conditions. *J. Aerosol Sci.* 9:135–145.

Wiener, R. K., K. Okazaki, and K. Willeke. 1988. Influence of turbulence on aerosol sampling efficiency. *Atmos. Environ.* 22:917–928.

Yamano, N. and J. E. Brockmann. 1989. *Aerosol Sampling and Transport Efficiency Calculation (ASTEC) and Application to Surtsey/DCH Aerosol Sampling System.* NUREG/CR-525. SAND88-1447. Albuquerque, NM: Sandia National Laboratories.

Ye, Y. and D. Y. H. Pui. 1990. Particle deposition in a tube with an abrupt contraction. *J. Aerosol Sci.* 21:29–40.

Zhang, Z. Q. and B. Y. H. Liu. 1989. On the empirical fitting equations for aspiration coefficients for thin-walled sampling probes. *J. Aerosol Sci.* 20:713–720.

Ziskind, G., M. Fichman, and C. Gutfinger. 1998. Effects of shear on particle motion near a surface—Application to resuspension. *J. Aerosol Sci.* 29:323–338.

9 Filter Collection

K. W. LEE

Kwangju Institute of Science and Technology, Department of Environmental Engineering, Kwangju, South Korea

R. MUKUND

GE Power Systems, Cincinnati, OH

INTRODUCTION

Filtration is probably the most widely utilized technique for aerosol measurement, owing to its flexibility, simplicity, and economy. The central concept in aerosol filtration is the collection, through removal from the gas phase, of a *representative* sample of the aerosol on a suitable porous medium or filter. The transfer of the aerosol from a dispersed state in the air to a compact sample on the filter then facilitates the storage, transport, and sample preparation requisite for gravimetric, microscopic, microchemical, or other analytical techniques. The term *air* is used interchangeably here with *gas* to refer to the gaseous phase in which the aerosol is dispersed.

The prime objective in selecting an aerosol filtration scheme appropriate to a study is to ensure that as representative a sample of the aerosol as possible is presented to the applied analytical technique. The various aerosol filtration topics discussed in this chapter are, thus, all designed to assist the reader in developing filtration methods that would best meet this goal in specific applications. In subsequent sections of this chapter, filter sampling principles are presented, followed by a review of the various types of available aerosol measurement filters, the theoretical principles for estimating filtration efficiency and pressure drop, and the salient features and requirements of common filter analysis methods that influence filter selection.

The reader should note that the reference literature on filtration for aerosol measurement is extensive, if somewhat scattered. Several excellent and comprehensive reviews can be found, including ones by Brown (1993), Lippmann (1995), and Spurny (1998), in addition to numerous reports in the aerosol literature on utilizing filter methods for aerosol measurement. Subsequent chapters in this book, particularly Chapters 11 and 12 and those in Part III, also provide practical examples of filter-based measurement techniques that have been perfected for various applications. Thus, the aerosol measurement practitioner would be well advised to survey the literature for references to filtration techniques utilized previously for similar applications before proceeding with filter system selection or design.

Aerosol Measurement: Principles, Techniques, and Applications, Second Edition, Edited by Paul A. Baron and Klaus Willeke.
ISBN 0-471-35636-0

GENERAL PRINCIPLES OF FILTER SAMPLING

The essential components of a filter sampling system for aerosol measurement are shown in Figure 9–1. Typically, aerosol-laden air is drawn through a sampling probe for isokinetic sampling from a flow stream into a filter holder containing an appropriately selected filter medium. Here, the aerosol is separated from the flow stream to the extent dictated by the characteristics of the filter, the air velocity through the filter, and other factors such as the particulate loading on the filter. The air drawn through the filter flows to an optional flow measurement device, such as a rotameter, mass flow meter, or dry test meter, and then into a flow-regulating mechanism, such as an orifice or valve coupled with an air-moving device or pump. Selecting the appropriate components and the optimum order for the flow progression through the system is crucial in achieving a representative sample of the aerosol on the filter. The sampling probe upstream of the filter holder is required when sampling from moving air streams. In such applications, the volumetric sampling rate and the cross-sectional area of the sampling probe nozzle determine the air velocity through the probe inlet. The combination of an air inlet velocity equal to that of the air stream in the vicinity of the inlet and the alignment of the probe parallel to the gas streamlines ensures isokinetic sampling of the aerosol. The importance of isokinetic sampling in ensuring that a representative aerosol sample is extracted from the air stream is discussed in greater detail in Chapter 8.

The effectiveness with which the sampling probe extracts the aerosol from a moving air stream is termed the *aspiration efficiency* of the inlet. An excellent treatment of the various factors affecting inlet aspiration efficiency is provided by Davies (1968). In addition, Vincent (1989:86–137) presents a detailed discussion of the aspiration efficiencies of thin-walled and blunt samplers in moving air streams. Chapter 8 presents an updated overview of these issues. Sampling from still air using an inlet probe also introduces possibilities of sampling biases, which are addressed by Davies (1968), Agarwal and Liu (1980), and Vincent (1989:144–164). In addition to the limitations imposed by aerosol dynamic behavior on inlet aspiration efficiency, particle losses due to temperature gradients between the probe inlet surfaces and the

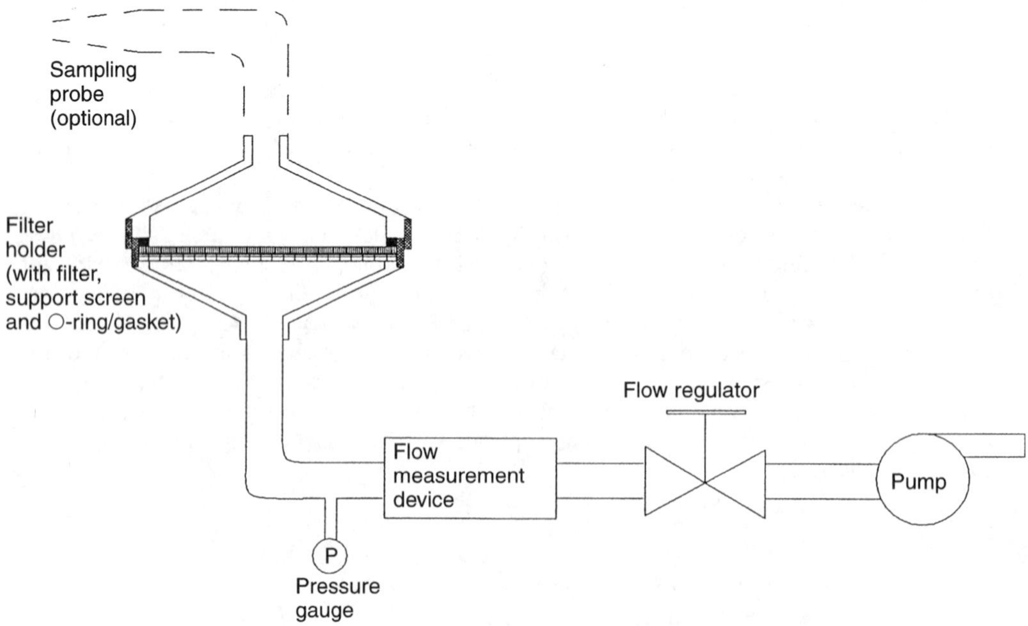

Fig. 9–1. Schematic example of a filter collection system typically used for aerosol measurement.

air stream, and electrostatic deposition losses due to charged probe inlet surfaces, particularly in the case of those made with plastic materials, must also be considered.

In many instances, the geometry of the sampling environment necessitates a length of connecting tubing between the inlet probe and the filter holder. The passage of the aerosol from the inlet to the filter holder is characterized by the *transport efficiency* and is influenced by a number of factors such as gravitational, diffusional, and inertial deposition onto the wall surfaces, as well as temperature gradient–related and electrostatic wall loss effects. These issues are discussed in Chapter 8 on the sampling and transport of aerosols. It is critical that the aspiration and transport efficiencies of a filter sampling system be determined in each application to adequately characterize any distortion in the aerosol size distribution or concentration obtained on the sampling filter. Maximizing these efficiencies is particularly critical due to their inherent particle size dependence, making corrections for sampling biases extremely difficult if the aerosol size distribution is unknown. In general, sampling losses are minimized by locating the filter holder as close as possible to the sampling inlet probe and, if possible, sampling with no inlet probe or connecting tubing upstream of the filter.

Filter holders provide the means by which the filtration media can be supported, typically on a coarse wire screen or backup medium, and held in a positive seal to constrain the air drawn by the sampling pump to pass through the filter. This seal is generally obtained through the use of an O-ring or a gasket of appropriate form and material that does not damage the filter material. Teflon gaskets are increasingly popular for this application due to their inherently low adhesivity to filter surfaces. Filter holders are available for all common filter sizes in the range 13 to 47 mm in diameter, as well as in 0.2×0.25 m [8×10 in] sizes for high-volume ambient air quality sampling applications. A compilation of the wide spectrum of commercially available filter holders is presented by Lippmann (1995).

Filter holders may be open-faced, in-line, or of a cassette variety, the latter commonly used in industrial hygiene sampling applications, as shown in Figure 9–2. In general, open-faced filter holders provide a greater assurance of uniform filter deposition and lower sampling losses in the inlet. However, closed-face or in-line filter holders are necessary for probe-based sampling applications and also protect filter media from possible mechanical damage and rupture during sampling. In-line filter holders generally use a gradual expansion from the inlet to the filter surface, as well as downstream to the outlet, so as to ensure uniform air velocities over the cross section of the filter. This enables the collection of a uniform aerosol deposit on the filter surface, which is critical for subsequent analyses that may utilize only a fraction of the filter.

The use of a filter holder for aerosol collection on a filter can introduce the potential for electrostatic and diffusional aerosol deposition losses onto filter holder inlet surfaces, as well

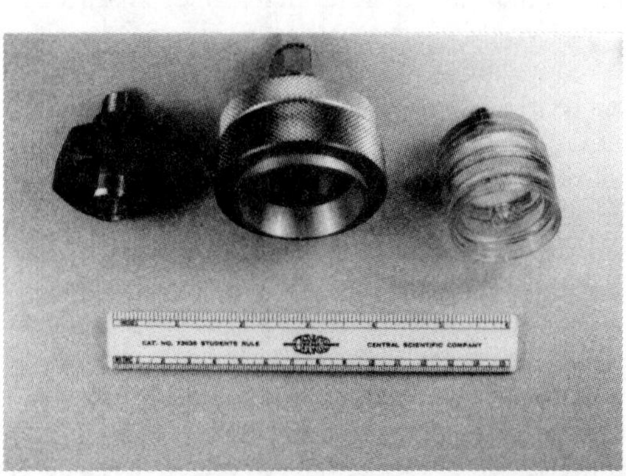

Fig. 9–2. Examples of filter holders available commercially for aerosol collection. From left: in-line, ***GEL***, 47 mm; open face, ***MIL***, 47 mm; and cassette style, ***GEL***, 25 mm (closed-face shown; can also be used as open-faced). (Photograph coutesy of H. T. Kim, Kwangju Institute of Science and Technology.)

as for aerosol transformations from condensation or evaporation induced by temperature gradients present between the air stream and the filter holder. These factors must be considered when designing aerosol measurement filtration systems for specific applications. The techniques utilized in the field to overcome these potential problems include the use of a controlled-temperature enclosure for housing the filter holder, thereby minimizing temperature gradients.

The most critical function of a filter holder—that of ensuring a positive seal around the circumference of the filter—is typically the most frequent source of errors in filter sampling. Such sampling errors result from failures in the outer seal, thus permitting air to flow around the filter. Seal integrity and leak testing procedures must be employed to ensure that the sampled air is drawn only through the filter in the filter holder. Among the techniques that may be applied for this purpose are (1) presampling positive pressurization tests conducted with the filter holder assembly capped-off appropriately; (2) the use of a pressure gauge to monitor the vacuum pressure in the sampling line connecting the filter holder to the pump during sampling—leaks and ruptures in the filter are then easily detected from any drop in the vacuum, with normal operation indicated by a steady or gradually increasing vacuum pressure; and (3) measurement of submicrometer aerosol (e.g., ambient aerosol) penetration through the filter assembly—if the filter collects these particles with high efficiency, any particles dectected downstream of the filter are an indication of bypass leakage around the filter.

A wide spectrum of filters is available commercially for aerosol measurement applications, providing the user with a selection of filter materials, pore sizes, and collection characteristics, as well as a variety of shapes such as disks and sheets sized to fit commonly available filter holders. The parameters of importance when selecting a particular type of filter for an application generally include collection efficiency of the filter for the aerosol size distribution to be sampled, pressure drop across the filter in relation to the air volumetric throughput required, compatibility of the filter with the sampling conditions, use and handling procedures, and analytical method to be employed. This includes issues of potential artifact formation on the filter surfaces from chemical reactions, interferences such as would occur from the use of hygroscopic filters in the gravimetric analysis of aerosols, and cost constraints relating to the size of the sampling effort and the number of filters required.

A more detailed discussion of the various types of filters and their collection efficiencies and pressure drop characteristics is presented subsequently. Also discussed are several examples of compatibility problems typically encountered during aerosol sampling and alternatives developed to circumvent these potential problems. A comprehensive catalog of commercially available filters is provided by Lippmann (1995), including a compilation of filter manufacturer trade names and corresponding filter characteristics such as pore size, thickness, pressure drop, ash content, weight per unit area, maximum operating temperature, tensile strength, and refractive index. In a subsequent section of this chapter, a compilation of commonly used filter types for various applications is presented, together with the relative advantages and limitations of each category of filters.

The use of appropriate filter handling procedures is important to ensure the collection of a representative aerosol sample. The procedures range from filter seal and surface integrity testing to minimizing interferences and artifacts from such factors as chemical reactions/transformations on the filter surface, electrostatic charging of filters, and moisture uptake in filter materials and hygroscopic aerosols. These and other procedures are discussed subsequently in connection with the requirements of filter analysis methods.

Finally, the measurement and control of the air flow rate through the filter is as important as the collection of a representative aerosol sample on the filter because the air flow rate or cumulative volume through the filter is necessary to calculate the aerosol concentration from the sample. A detailed discussion of the various alternatives available for use as flow regulating and/or measuring devices and as pumps or air movers is presented by Rubow and

Furtado (1989). In addition, established scientific procedures are available for the calibration of various types of flow measurement devices as outlined, for example, by Cheng (1995). A well-designed filter collection system for aerosol measurement thus involves the collection of a representative aerosol sample on a suitable filter, combined with an accurate knowledge of the air flow rate or cumulative air volume transported through the filter.

AEROSOL MEASUREMENT FILTERS

A logical way to classify aerosol measurement filters is by their characteristic structure. Accordingly, filters used for aerosol sampling may be classified as fibrous filters, porous-membrane filters, straight-through pore membrane filters, and granular-bed filters. The terms *porous-membrane* and *straight-through pore membrane* are very similar to those suggested by Hinds (1999), and discussions here closely follow his classification and the discussion provided by Lippmann (1995). A summary of the salient characteristics of each type of filter is also provided in Table 9–1, complementing the discussion that follows.

Fibrous Filters

Fibrous filters consist of a mat of individual fibers. Generally, filter porosity is relatively high, ranging from about 0.6 to 0.999. Porosities of less than 0.6 are not typically found in fibrous filters because of the difficulties in effectively compressing the component fibers into a smaller thin layer. Fiber sizes range from less than 1 μm to several hundred micrometers. The range of fiber diameters for a given filter is usually very diverse, although some types of fibrous filters may consist of fibers of a uniform size. Often, the filters are fabricated using a binder material to hold individual fibers together. The mass for the binder material can be as high as 10% of the filter material. Binder-free filters are generally selected for aerosol measurement because of the artifacts and interferences caused by the presence of the organic binder in the filter. The materials used for fibrous filters include cellulose, glass, quartz, and plastic fibers. Sometimes, mixed fibers of cellulose, asbestos, and glass are also used as filters for certain low-cost qualitative sampling applications. Figure 9–3 shows the microstructure of a glass fiber filter and reveals the fibrous, mat-like nature of the filter material.

Cellulose fiber (paper) filters were used once very widely for general-purpose air sampling. Whatman (***WHA***)* filters are one of the most representative filters in this category. The filters are inexpensive, come in various sizes, and have good mechanical strength and low pressure drop characteristics. Some of the critical limitations of cellulose fiber filters are their moisture sensitivity and relatively low filtration efficiency for submicrometer particles.

Glass fiber filters typically have a higher pressure drop than paper filters and often provide filtration efficiencies of greater than 99% for particles >0.3 μm. The filters are more expensive than paper filters. However, glass fiber filters are less affected by moisture than are cellulose fiber filters. Glass fiber filters are widely used as the standard filter media for high-volume air sampling. Teflon-coated glass fiber filters overcome some of the inherent inadequacies of glass fiber filters by being inert to catalyzing chemical transformations as well as by being less moisture sensitive.

Quartz fiber filters are commonly used in high-volume air sampling applications involving subsequent chemical analyses such as atomic absorption, ion chromatography, and carbon analysis due to their low trace contamination levels as well as to their relative inertness and ability to be baked at high temperatures to remove trace organic contaminants.

Polystyrene fiber filters have been used for sampling purposes to a limited extent. These filters have less mechanical strength than cellulose filters. However, their filtration efficiency

* See Appendix I for full manufacturer addresses referenced by the italicized three-letter codes.

TABLE 9–1. Summary of the Salient Characteristics of the Various Types of Filters Commonly Utilized for Aerosol Measurement

Filter Type	Characteristics
Fibrous filters	Mat/weave of fibers with diameters of 0.1–100 μm. Cellulose or wood (paper), glass, quartz, and polymer fiber filters are available Porosities of 60–99%, thicknesses of 0.15–0.5 mm Particle collection is throughout the depth of the filter from interception, impaction, and diffusion onto fibers High particle collection efficiencies require low air velocities Pressure drops are the lowest among all filters under comparable conditions
Porous-membrane filters	Microporous membranes with tortuous pores throughout the structure Polymer, sintered metal, and ceramic microporous filters available Pore sizes (determined from liquid filtration) in the range 0.02–10 μm Porosities of <85% and thicknesses of 0.05–0.2 mm Particle collection through attachment to microstructure elements High collection efficiencies, but highest pressure drop among all filters
Straight-through pore filters	Thin polycarbonate films (10 μm) with cylindrical pores perpendicular to film surface, with diameters in the range 0.1–8 μm Porosities are low, in the range of 5–10% Particle collection through impaction and interception near the pores and diffusion to tube walls of pores Collection efficiencies are intermediate between fibrous and microporous membrane Pressure drops are significantly higher than fibrous filters and comparable with or higher than microporous membrane filters for equivalent collection efficiency
Granular-bed filters	For special sampling, granules of specialty chemicals, sugar, naphthalene, sand, metal, and glass beads are used Samples are recovered by washing or volatilization Granular bead sizes range from 200 μm to a few millimeters Filtration is achieved by impaction, interception, diffusion, and gravitation Filter porosities of 40–60% for stationary beds Low collection efficiency due to large granule size. To enhance diffusion, low flow is used; bed depth is increased, or smaller granules are used
Porous-foam filters	Porosities of <97% Pore diameters from 10 to 50 μm are common Collection efficiencies are low

is comparable with that of glass fiber filters. Other plastic materials used in filters include polyvinyl chloride and dacron. For special applications involving high temperatures and corrosive environments, filters made of stainless steel fibers have also been recently introduced.

Porous-Membrane Filters

A variety of membrane filters made of cellulose esters, polyvinyl chloride, Teflon, and sintered metals are commercially available. Membrane filters are gels formed from a colloidal solution and have a very complicated and uniform microstructure providing a tortuous or irregular air flow path. Often, the complex filter structure consists of a series of layers formed by different processes, depending on the manufacturing technique. The pore sizes provided by manufacturers often do not match any of the physical filter pores or structural characteristics and are defined from liquid filtration. In general, the pressure drop and the particle

Fig. 9–3. Electron micrograph showing the typical microstructure of a glass fiber filter (***GEL*** Type A/E). Scale bar shown on micrograph.

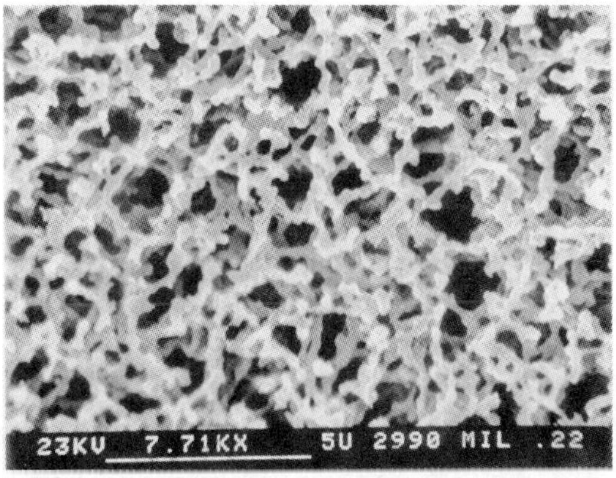

Fig. 9–4. Electron micrograph of a microporous-membrane filter (***MIL***, 0.22 μm pore size) showing the tortuous flow path and structural elements in the filter. Scale bar shown on micrograph.

collection efficiency are very high, even for particles significantly smaller than the characteristic pore size. Particles are captured by the surfaces provided by the filter structure, principally by Brownian motion and inertial impaction mechanisms. An example of a microporous membrane filter structure is shown in Figure 9–4.

Straight-Through Pore Membrane Filters

Straight-through pore membrane filters consist of a polycarbonate membrane with straight-through pores of a uniform size. They are very often called Nuclepore filters after their original manufacturer, Nuclepore Corp. (Pleasanton, CA), although they are also currently manufactured by Poretics Corp. (***POR***). The filters are manufactured by subjecting polycarbonate membranes to neutron bombardment, followed by an etching process that produces uniform-sized pores in the membrane. The number of pores is controlled by the bombardment time, and the pore size is determined by the etching time. Capillary pore membrane filters have a very different and simpler structure compared with porous-membrane filters.

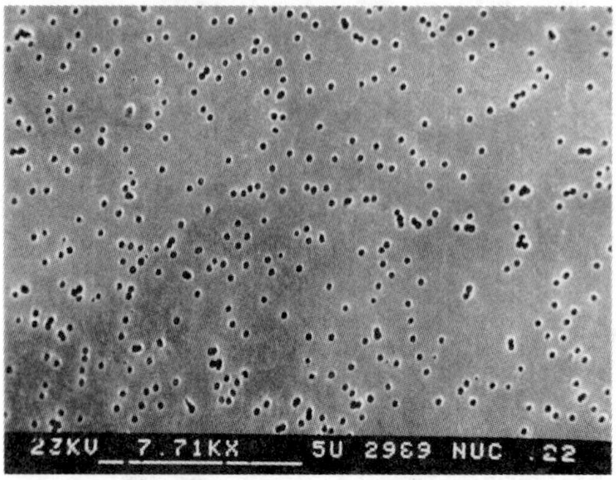

Fig. 9–5. Electron micrograph of a straight-through pore membrane filter (Nuclepore, 0.22 μm pore size) showing the smooth surface and uniform pores in the filter. Scale bar shown on micrograph.

They consist of a very smooth and translucent surface with straight-through capillary holes across the membrane structure. Figure 9–5 shows the microstructure of a Nuclepore filter and reveals the flat surface and uniform pores in the filter. Straight-through pore membrane filters are widely used for particle analysis using surface analytical techniques such as light and electron microscopy.

Mori et al. (1991) examined various membrane-type filters and proposed to classify both porous-membrane and straight-through pore filters into five new groups according to the appearance of the filter structure. The five groups proposed are (1) random directional fiber-shaped filters, (2) unidirectional fiber-shaped filters, (3) net-shaped filters, (4) agglomerate-shaped filters, and (5) pore-shaped filters. This scheme may have some utility when cataloguing the spectrum of membrane filters available.

Granular-Bed Filters

For specialized applications, aerosol sampling may be performed using a granular- or packed-bed filter. Filtration is accomplished by passing particle-laden air through a bed consisting of granules and recovering aerosols afterwards by extraction procedures. A major advantage of aerosol sampling by granular beds is that, by selecting the proper bed media, both particulate and gaseous pollutants can be simultaneously collected (Kogan et al., 1991). Possibilities for using this method at high temperature and pressures also make this filter type attractive. Granules of activated charcoal, XAD-2, Florisil, Tenax, sugar, naphthalene, glass, sand, quartz, and metal beads have been used. XAD-2 and Tenax are polymeric adsorbent resins made of spherical granules, and Florisil is a magnesia-silica material. Figure 9–6 shows two different types of granular aerosol collection media, Florisil and Tenax. Aerosols are usually recovered from the granular media for chemical analysis by washing, volatilization, or the use of solvents.

Porous-Foam Filters

Particle size–dependent penetration characteristics can be achieved by the use of porous-foam filters in an instrument that is very compact, simple, and cost effective. This type of filter is usually formed from reticulated polyurethane or polyethylene with a structure consisting of a matrix of bubbles that are pierced at their points of contact and create an open and three-dimensional lattice of connected short elements of approximately triangular cross

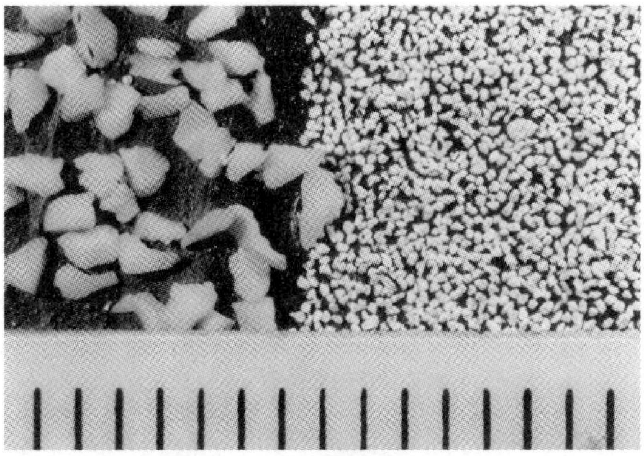

Fig. 9–6. Optical microscope photograph of two different types of granular-bed filtration material: Florisil (left) and Tenax (right). Millimeter scale also shown.

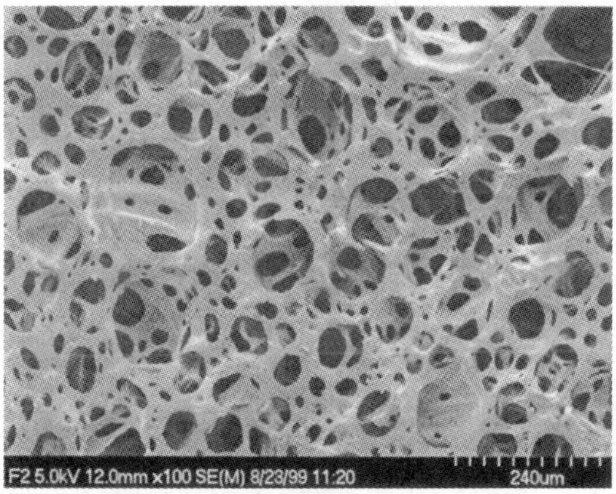

Fig. 9–7. Electron micrograph of a porous-foam filter. Scale bar shown on micrograph.

section. The foam structure is characterized in terms of the following geometrical parameter: The nominal porosity expressed conventionally by the foam manufacturers in terms of "pores per inch"; the volume fraction; and the effective diameter of fiber width (Vincent et al., 1993). Porosities up to 0.97 are possible, and pore diameters from 10 to 50 µm are common. Efficiencies are lower than the best fibrous filters. Figure 9–7 shows an electron micrograph of a typical foam filter that is made up of a large number of bubbles. The geometric arrangement of the bubbles is not clear in this figure.

A wide selection of fibrous and membrane filters are available commercially for use in aerosol measurement. Tables 9–2 to 9–4 list a number of different filter manufacturers for each type of filter discussed as well as a general price range indicating the relative differences in cost between the various filters.

FILTRATION THEORY

The filtration of particles by both fibrous and membrane filters has been the subject of numerous analytical, numerical, and experimental studies during recent years. As a result, the

TABLE 9–2. Commercial Sources and Typical Cost Ranges for Fibrous Aerosol Measurement Filters

Filter Material(s)	Variety of Characteristics Available	Commercial Sources[a]	Typical Cost Range[b]
Cellulose fiber	Fine, medium, coarse void sizes	***WHA*** ***MSI*** ***MFS*** ***S&S***	$10–20/100
Borosilicate glass fiber	With organic binder Without organic binder With Teflon coating With cellulose content	***WHA*** ***MSA*** ***GEL*** ***MIL*** ***H&V*** ***PAL*** ***NUC*** ***MFS, MSI, S&S***	$10–50/100 (Teflon-coated)
Quartz fiber	Pure With small percentage borosilicate glass	***WHA*** ***MFS*** ***GEL***	$50–100/100
Polymer fiber	Polypropylene, 0.45, 0.8, 1.2 μm uniform pore sizes	***MSI***	$40/100

[a] See Appendix I for full manufacturer addresses referenced to the italicized three-letter codes.
[b] Costs are for 47 mm diameter disks and are taken from manufacturers' 1998–1999 catalogs.

TABLE 9–3. Commercial Sources and Typical Cost Ranges for Microporous-Membrane Aerosol Measurement Filters

Filter Material(s)	Variety of Characteristics Available	Commercial Sources	Typical Cost Range[a]
Cellulose membrane	Cellulose nitrate Mixed ester Cellulose acetate, 0.02–8 μm pore size	***MIL*** ***GEL*** ***SAR*** ***NUC*** ***MFS, MSI*** ***S&S, WHA*** ***NAL***	$70–130/100
Teflon/PTFE membrane	Pure Polypropylene reinforced, 0.2–1.0 μm pore size	Same as above	$170–250/50
Polyester/polycarbonate/polypropylene membranes	0.02–15 μm pore sizes	Same as above	$50–200/100
Silver membrane	Pure metallic silver, 0.2–5 μm	***OSM*** ***MIL*** ***NUC***	$150–450/25
Nylon membrane	Pure nylon Laminated or impregnated with polypropylene for support, 0.1–20 μm pore size	***GEL*** ***MSI*** ***SAR***	$90–110/100
Polyvinyl chloride membrane	Pure PVC PVC with acrylonitrile, 0.45, 0.8, 5.0 μm pore sizes	***GEL*** ***NUC***	$100–150/100

[a] Costs are for 47 mm diameter disks and are taken from manufacturers' 1998–1999 catalogs.

TABLE 9–4. Commercial Sources and Typical Cost Ranges for Straight-Through Pore Membrane Aerosol Measurement Filters

Filter Material(s)	Variety of Characteristics Available	Commercial Sources	Typical Cost Range[a]
Polycarbonate	0.01–14 µm pore size (uniform with ±15%) 6–10 µm thickness 5–10% porosity	***NUC*** ***OSM***	\$75–185/100
Polyester	0.1–12.0 µm pore sizes	***NUC***	\$75–100/100

[a] Costs are for 47 mm diameter disks and are taken from manufacturers' 1998–1999 catalogs.

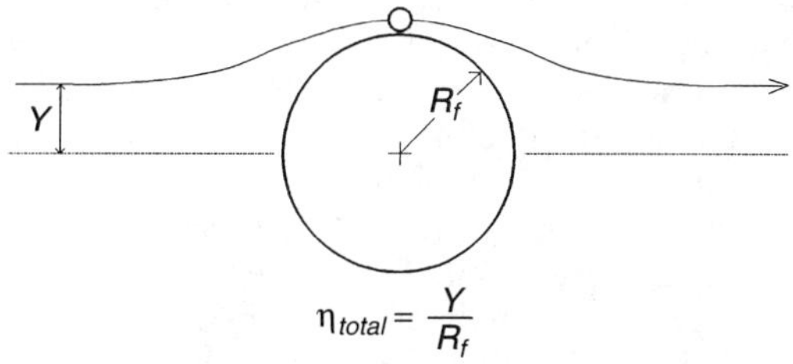

Fig. 9–8. Definition of single-fiber efficiency.

dependence of filtration efficiency on the particle size, the filter media characteristics, and the flow velocity is now well established, both qualitatively and quantitatively.

In this section, filtration mechanisms are discussed and some useful predictive equations for filter collection efficiency and the pressure drop across a filter are introduced for practical application. Filtration theory is discussed for fibrous filters, and its applicability to membrane filters is described subsequently.

The starting point in characterizing fiber filtration is to consider the capture of particles by a single fiber. The single-fiber efficiency, η, is defined as the ratio of the number of particles striking the fiber to the number that would strike if the streamlines were not diverted around the fiber. If a fiber of radius R_f removes all the particles contained in a layer of thickness Y as shown in Figure 9–8, the single fiber efficiency, η, is then defined as Y/R_f.

The overall efficiency, or the total efficiency, E, of a filter composed of many fibers in a mat can be related to the single-fiber efficiency as follows:

$$E = 1 - \exp\left[\frac{-4\eta\alpha L}{\pi d_f(1-\alpha)}\right] \tag{9–1}$$

where α is the solidity or packing density of the filter (1 – porosity), L the filter depth or thickness, and d_f the fiber diameter. Equation 9–1 relates the total efficiency of a filter to the single-fiber efficiency as previously defined. Usually, Eq. 9–1 is used to calculate the single-fiber efficiency from the total filter efficiency, E, which can be measured experimentally.

The advantage of using the single-fiber efficiency is that it is independent of the filter thickness, L. While it is not meaningful to compare the total efficiencies of two filters of two different thicknesses, different filters can be compared on the basis of their single-fiber efficiencies. This is an important point to consider when comparing the overall filtration efficiencies of filters because a filter with a lower single-fiber efficiency may have a higher total efficiency by virtue of being thicker.

Filtration Mechanisms

As air penetrates a filter, the trajectories of particles deviate from the streamlines due to several mechanisms. As a result, particles may collide with the fiber surface and become deposited on them. The important mechanisms causing particle deposition are diffusion, inertial impaction, interception, and gravitational settling. The single-fiber efficiency, η, can then be assumed in the first approximation to be composed of the arithmetic sum of the individual efficiencies from diffusion, η_{diff}, interception, η_{inter}, inertial impaction, η_{imp}, and gravitational-settling, η_{grav}, mechanisms. In addition, dendrite formation from particles collected on fibers can provide additional particle collection in filters.

Brownian Diffusion. Under normal conditions, aerosol particles undergo Brownian motion. Small particles generally do not follow the streamlines but continuously diffuse away from them. Once a particle is collected on a surface, it would adhere to it due to the van der Waals force. The particle concentration at the surface can, thus, be assumed to be zero, and the resulting concentration gradient normal to the media surface can be considered as the driving force for the diffusion of particles. Because Brownian motion generally increases with decreasing particle size, the diffusive deposition of particles is increased when the particle size is reduced. This phenomenon is illustrated in Figure 9–9. Similarly at low flow velocities, particles can spend more time in the vicinity of the fiber surfaces, thus enhancing diffusional collection. From the convective diffusion equation describing this process, a dimensionless parameter called the Peclet number can be defined as

$$P_e = \frac{d_c U}{D} \qquad (9\text{–}2)$$

where Pe is the Peclet number, d_c the characteristic length of collecting media, U the average air velocity inside the filter medium, and D the diffusion coefficient of the particle. For pure molecular diffusion, D can be written as

$$D = kTC_c/3\pi\eta d_p \qquad (9\text{–}3)$$

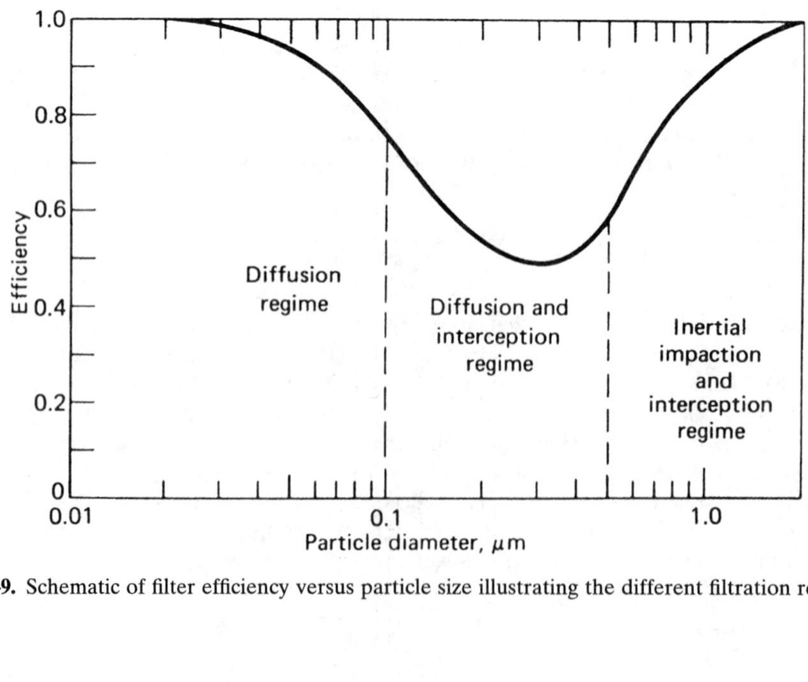

Fig. 9–9. Schematic of filter efficiency versus particle size illustrating the different filtration regimes.

where k is the Boltzmann constant, T the absolute temperature, η the air viscosity, d_p the particle diameter, and C_c, the Cunningham slip correction. From the above discussion, particle collection by diffusion is expected to decrease with increasing Peclet number. C_c is written as

$$C_c = 1 + 2.492\frac{\lambda}{d_p} + 0.84\frac{\lambda}{d_p}\exp\left(-0.435\frac{d_p}{\lambda}\right) \tag{9–4}$$

where λ is the mean free path of the gas molecules.

The approach taken to quantify the single-fiber diffusional collection efficiency, η_{diff}, by investigators such as Friedlander (1957), Natanson (1957), and Lee and Liu (1982a,b) is a boundary layer model that is commonly used in heat and mass transfer analysis. In early models by Friedlander and Natanson, a flow field that is an isolated cylinder model was utilized. Recent theories, such as that of Lee and Liu (1982), utilize a multiple-cylinder model that takes into account the flow interference effects of neighboring fibers. Thus, these models provide a better representation of the actual flow profile in the filters. The Lee and Liu (1982) theory yields

$$\eta_{diff} = 2.58\left(\frac{1-\alpha}{K}\right)^{1/3} Pe^{-2/3} \tag{9–5}$$

where K is the hydrodynamic factor,

$$K = -\frac{1}{2}\ln\alpha - \frac{3}{4} + \alpha - \frac{\alpha^2}{4} \tag{9–6}$$

Interception. Even if the trajectory of a particle does not depart from the streamline, a particle may still be collected if the streamline brings the particle center to within one particle radius from the fiber surface. The fact that particles with a finite size can be collected even in the absence of Brownian motion and inertial impaction indicates the importance of the interception effect. One would expect the interception to be relatively independent of flow velocity for a given fiber, and this characteristic can be contrasted to the flow-dependent characteristics of diffusion and inertial impaction. The dimensionless parameter describing the interception effect is the interception parameter, R, defined as the ratio of particle diameter to fiber diameter:

$$R = \frac{d_p}{d_f} \tag{9–7}$$

If the Kuwabara flow is used, one can obtain the following expression:

$$\eta_{inter} = \frac{1+R}{2K}\left[2\ln(1+R) - 1 + \alpha + \left(\frac{1}{1+R}\right)^2\left(1-\frac{\alpha}{2}\right) - \frac{\alpha}{2}(1+R)^2\right] \tag{9–8}$$

where η_{inter} is the single-fiber efficiency due to interception. Although Eq. 9–8 is a complete expression for the interception efficiency, the form of the equation is somewhat long, and it can be shown to be approximated into the following simpler form:

$$\eta_{inter} = \frac{1-\alpha}{K}\frac{R^2}{(1+R)} \tag{9–9}$$

Enhanced collection of particles by a fiber can also occur from the interception of diffusing particles and has been proposed by Stechkina et al. (1969). The magnitude of this additional efficiency term is of the same order as that of the errors involved in the approximation method used in the analysis. Spielman and Goren (1968) also indicate that the term is not theoretically consistent and is, consequently, not introduced here.

Inertial Impaction. The streamlines of a fluid around the fiber are curved. Particles with a finite mass and moving with the flow may not follow the streamlines exactly due to their inertia. If the curvature of a streamline is sufficiently large and the mass of a particle sufficiently high, the particle may deviate far enough from the streamline to collide with the media surface. The importance of this inertial impaction mechanism increases with increasing particle size and increasing air velocity, as shown in Figure 9–9. Therefore, the effect of increasing air velocity on the inertial impaction of particles is contrary to that for the diffusive deposition. The inertial impaction mechanism can be studied by the use of the dimensionless Stokes number, defined as

$$Stk = \frac{C d_p^2 \rho_p U}{18 \eta d_f} \qquad (9\text{–}10)$$

where ρ_p is the density of the particle. The Stokes number is the basic parameter describing the inertial impaction mechanism for particle collection in a filter. A large Stokes number implies a higher probability of collection by impaction, whereas a small Stokes number indicates a low probability of collection by impaction.

Stechkina et al. (1969) calculated the inertial impaction filtration efficiency for particles, using the Kuwabara flow field. Their expression for the filtration efficiency due to inertial impaction, η_{imp}, is as follows:

$$\eta_{imp} = \frac{1}{(2K)^2}\left[\left(29.6 - 28\alpha^{0.62}\right)R^2 - 27.5R^{2.8}\right]Stk \qquad (9\text{–}11)$$

Equation 9–11 has been used extensively for calculating the contribution by the inertial impaction mechanism.

Gravitational Settling. Particles will settle with a finite velocity in a gravitational force field. When the settling velocity is sufficiently large, the particles may deviate from the streamline. Under downward filtration conditions, this would cause an increased collection due to gravity. When flow is upward, this mechanism causes particles to move away from the collector, resulting in a negative contribution to filtration. This mechanism is important only for particles larger than at least a few micrometers in diameter and at low flow velocities. The dimensionless parameter governing the gravitational sedimentation mechanism is

$$Gr = \frac{V_g}{U} \qquad (9\text{–}12)$$

where U is the flow velocity and V_g is the settling velocity of the particle. It can be shown that the single-fiber filtration efficiency due to gravity, η_{grav}, can be approximated (Davies, 1973) as

$$\eta_{grav} = \frac{Gr}{1+Gr} \qquad (9\text{–}13)$$

In filtration theories, it is common to assume that the individual filtration mechanisms discussed above are independent of each other and additive. Therefore, η, the overall single-fiber collection efficiency in Eq. 9–1, can be written as the sum of individual single-filter efficiencies contributed by the different mechanisms. This approximation has been found to serve adequately for predicting the overall collection efficiencies of fibrous filters, owing to the different ranges in particle sizes and face velocities in which the different filtration mechanisms predominate, as illustrated in Figure 9–9.

EXAMPLE 9–1

Calculate the single-fiber efficiencies for a 0.5 μm diameter particle at 293 K [20°C] and 101.3 kPa [1 atm] due to (1) Brownian diffusion, (2) interception, (3) inertial impaction, and (4) gravitational settling for a fibrous filter having a fiber diameter of 5 μm and a solidity of 0.2 and operating at an air flow velocity of 0.15 m/s [15 cm/s]. Assume the particle density is 1000 kg/m³ [1 g/cm³].

Answer: (1) Brownian diffusion:

Using Eq. 9–3, the diffusion coefficient is calculated as

$$D = \frac{(1.38\times10^{-23}\,Nm/K)(293\,K)(1.33)}{3\pi(1.807\times10^{-5}\,\text{Pa s})(0.5\times10^{-6}\,\text{m})}\left[\frac{(1.38\times10^{-16}\,\text{g cm}^2/\text{s}^2/\text{K})(293\,\text{K})(1.33)}{3\pi(1.807\times10^{-4}\,\text{g/cm/s})(0.5\times10^{-4}\,\text{cm})}\right]$$

$$= 6.33\times10^{-11}\,\text{m}^2/\text{s}[6.33\times10^{-7}\ \text{cm}^2/\text{s}]$$

where C_c is calculated using Eq. 9–4:

$$C_c = 1 + 2.492\left(\frac{0.0653\,\mu\text{m}}{0.5\,\mu\text{m}}\right) + 0.84\left(\frac{0.0653\,\mu\text{m}}{0.5\,\mu\text{m}}\right)\exp\left(-0.435\frac{0.5\,\mu\text{m}}{0.0653\,\mu\text{m}}\right)$$

$$= 1.33$$

From Eq. 9–2, the Peclet number is

$$Pe = \frac{(5\times10^{-6}\,\text{m})(0.15\,\text{m/s})}{6.33\times10^{-11}\ \text{m}^2/\text{s}}\left[\frac{(5\times10^{-4}\,\text{cm})(15\,\text{cm/s})}{6.33\times10^{-7}\ \text{cm}^2/\text{s}}\right] = 1.18\times10^4$$

Using Eq. 9–6, the hydrodynamic factor is

$$K = -0.5\ln(0.2) - 0.75 + 0.2 - \frac{0.2^2}{4} = 0.245$$

Using Eq. 9–5, the single-fiber efficiency η_{diff} is

$$\eta_{\text{diff}} = 2.58\left(\frac{1-0.2}{0.245}\right)^{1/3}(1.18\times10^4)^{-2/3}$$

$$- 0.0074$$

(2) Interception (from Eq. 9–7):

$$R = \frac{d_p}{d_f} = \frac{0.5\,\mu\text{m}}{5\,\mu\text{m}} = 0.1$$

Therefore,

$$\eta_{\text{inter}} = \frac{1-0.2}{0.245}\frac{(0.1)^2}{(1+0.1)} = 0.02968$$

from Eq. 9–9.

(3) Inertial impaction:

The Stokes number is obtained using Eq. 9–10:

$$Stk = \frac{1.33\left[(5\times10^{-7})^2\text{m}^2\right](1000\,\text{kg/m}^3)(0.15\,\text{m/s})}{18(1.807\times10^{-5}\,\text{Pa s})(5\times10^{-6}\,\text{m})}$$

$$\left[\frac{1.33\left[(5\times10^{-5})^2\ \text{cm}^2\right](1\text{g/cm}^3)(15\ \text{cm/s})}{18(1.807\times10^{-4}\,\text{g/cm/s})(5\times10^{-4}\ \text{cm})}\right]$$

$$= 3.07\times10^{-2}$$

Continued

The single-fiber efficiency due to inertial impaction is obtained using Eq. 9–11:

$$\eta_{\text{imp}} = \frac{1}{(2\times 0.245)^2}\left[(29.6-28[0.2]^{0.62})0.01-27.5(0.1)^{2.8}\right]3.01\times 10^{-2}$$
$$= 0.01862$$

(4) Gravitational settling:
The settling velocity of a 0.5 μm particle is

$$V_g = \frac{\rho_p d_p^2 g C}{18\eta}$$
$$= \frac{1000\,\text{kg}/\text{m}^3\left[(5\times 10^{-7})^2\,\text{m}^2\right](9.807\,\text{m/s}^2)1.33}{18\times 1.807\times 10^{-5}\,\text{Pa s}}$$
$$= \left[\frac{1\,\text{g}/\text{cm}^3\left[(5\times 10^{-5})^2\,\text{cm}^2\right](980.7\,\text{cm/s}^2)1.33}{18\times 1.807\times 10^{-4}\,\text{g/cm/s}}\right]$$
$$= 1.01\times 10^{-5}\,\text{m/s}\quad\left[1.01\times 10^{-3}\,\text{cm/s}\right]$$

Then

$$Gr = \frac{V_g}{U} = \frac{1.01\times 10^{-5}\text{m/s}}{0.15\text{m/s}} = \left[\frac{1.01\times 10^{-3}\,\text{cm/s}}{15\text{cm/s}}\right] = 6.70\times 10^{-5}$$

Therefore,

$$\eta_{\text{grav}} = \frac{Gr}{1+Gr} = 6.70\times 10^{-5}$$

Thus, the collection efficiency for a 0.5 μm particle has the greatest contribution from inertial impaction, a bit less from diffusion, and little from settling.

Loading Effects. It is well recognized and observed that the particulate collection and the pressure drop increase if high particulate concentrations are filtered for an extended period of time. This takes place because particulates accumulate on filter media and the deposited particles provide additional surfaces for collecting incoming particles. This mechanism is inherently time dependent because the size, shape, and morphology of particle dendrites change continuously.

Payatakes (1976) treated this filtration mechanism by numerically solving sets of differential equations. Kanaoka (1989) has also developed a simple method for accounting for the dendrite filtration mechanism. According to his theory, the total particle collection efficiency, E_{loaded}, including the effects of dendrite formation, is

$$E_{\text{loaded}} = 1 - \frac{\exp(-\gamma\, AcUt)}{\exp(-\gamma\, AcUt)+\exp(AL)-1} \tag{9–14}$$

where

$$A = \frac{4\alpha}{\pi(1-\alpha)d_f}\eta_{\text{initial}} \tag{9–15}$$

γ is the efficiency increase factor, c the particulate mass concentration, U the air velocity, t the time, and L the filter mat thickness. His theory assumes that the filtration efficiency of a single fiber increases in the following manner:

$$\eta_{\text{loaded}}/\eta_{\text{initial}} = 1 + \gamma m \tag{9–16}$$

where η_{loaded} is the filtration efficiency for the fiber that accumulated the particle mass m, and η_{initial} is the filtration efficiency of the clean fiber. γ in Eq. 9–16 was recommended by Kanaoka to be determined from experimental data.

Most Penetrating Particle Size and Minimum Efficiency

As discussed, an increase in the particle size causes increased filtration by interception and inertial impaction mechanisms, whereas a decrease in particle size enhances collection by Brownian diffusion. As a consequence, there is an intermediate particle size region where two or more mechanisms are simultaneously operating, yet none is dominant. This is the region where the particle penetration through the filter is a maximum and the filter efficiency a minimum. This is schematically illustrated in Figure 9–9. The particle size at which the minimum efficiency occurs is termed the *most penetrating particle size*. This size was previously assumed to be ≈0.3 μm and is the basis for the so-called DOP (dioctyl phthalate) test method for high-efficiency particulate air (HEPA) filters.

As fibrous filtration theory has been improved in recent years, the most penetrating particle size and the corresponding minimum efficiency have been observed to vary with the type of filter and the filtration velocity. Lee and Liu (1980) derived the following equation for predicting the most penetrating particle diameter:

$$d_{\text{p,min}} = 0.885\left[\left(\frac{K}{1-\alpha}\right)\left(\frac{\sqrt{\lambda}kT}{\eta}\right)\left(\frac{d_{\text{f}}^2}{U}\right)\right]^{2/9} \tag{9–17}$$

Figure 9–10 is a comparison of Eq. 9–17 with experimental data. The most penetrating particle diameter decreases with increasing flow velocity and increasing filter solidity (1 – porosity). As the filter medium size increases, the most penetrating particle size increases. The corresponding minimum efficiency is given as

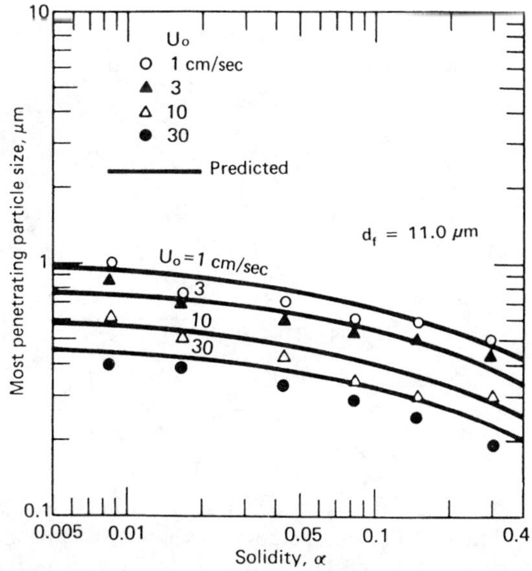

Fig. 9–10. Comparison of theory and experiment for the most penetrating particle diameter.

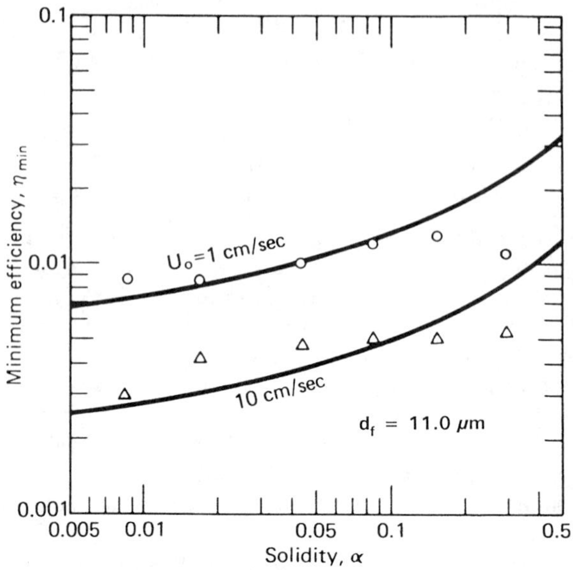

Fig. 9–11. Comparison of theory and experiment for the minimum single-fiber efficiency.

$$\eta_{min} = 1.44\left[\left(\frac{1-\alpha}{K}\right)^5\left(\frac{\sqrt{\lambda}kT}{\eta}\right)^4\left(\frac{1}{U^4 d_f^{10}}\right)\right]^{1/9} \tag{9–18}$$

A comparison of the equation with experimental data for minimum efficiency is shown in Figure 9–11.

EXAMPLE 9–2

What is the particle diameter that will give minimum efficiency for the filter given in Example 9–1?

Answer: From Eq. 9–17,

$$d_{p,min} = 0.885\left(\left(\frac{0.245}{1-0.2}\right)\left(\frac{\sqrt{6.65\times10^{-8}\,\text{m}}(1.38\times10^{-23}\,\text{kg m}^2/\text{s}^2/\text{K})293\text{K}}{1.807\times10^{-4}\,\text{Pa s}}\right)\frac{(5\times10^{-6})^2\,\text{m}^2}{0.15\text{m/s}}\right)^{2/9}$$

$$= 0.885\left(\left(\frac{0.245}{1-0.2}\right)\left(\frac{\sqrt{6.65\times10^{-6}\,\text{cm}}(1.38\times10^{-16}\,\text{g cm}^2/\text{s}^2/\text{K})293\text{K}}{1.807\times10^{-4}\,\text{g/cm/s}}\right)\frac{(5\times10^{-4})^2\,\text{cm}^2}{15\text{cm/s}}\right)^{2/9}$$

$$= 2.42\times10^{-7}\,\text{m} = [2.42\times10^{-5}\,\text{cm}] = 0.242\,\mu\text{m}$$

Membrane Filters

The predictive theory presented in this section was developed originally for fibrous filters. Studies of porous-membrane filters conducted by Rubow (1981) have indicated that the filtration mechanisms for porous-membrane filters are equivalent to those for fibrous filters. Furthermore, fibrous filtration theories were found to be applicable with the use of actual thickness and solidity, the only modification necessary being the use of an effective fiber diameter to represent the structural elements in the membrane. Similarly, a most penetrating particle size has also been shown to exist experimentally for porous-membrane filters. Thus, Eqs. 9–17 and 9–18 can also be expected to be applicable for such membrane filters with the appropriate choice of the effective fiber diameter value.

Particle collection in straight-through pore membrane filters can be estimated using of tube diffusion theory for diffusional collection in pore walls and impaction and interception theory for collection near pore inlet surfaces (see, e.g., Spurny et al., 1969). Typically, the efficiency of the filters is low for particle sizes less than the pore size and greater in the size range where diffusional collection is significant (i.e., $d_p > 0.1\,\mu m$). Particles greater than the pore size are collected with high efficiency. Due to their unique, somewhat "impactor-like" cutoff filtration characteristics, Nuclepore filters have been utilized as size-selective aerosol samplers in sequential collection stages using different pore size filters (e.g., Cahill et al., 1977; Parker et al., 1977). Heidam (1981) reviews aerosol fractionation by straight-through pore filters and concludes that particle bounce can be a potential problem in using such filters in these applications.

Pressure Drop of Filters

As air passes through filter media, the filter structure causes a resistance that is a measure of air permeability or the pressure drop. A consideration of the pressure drop across filter media is important in choosing a specific filter type in a particular application. The pressure drop is easily measurable and can be used as a check on the flow fields on which deposition mechanisms are based. More importantly, the measurement of the pressure drop across filter media plays a central role in the practical estimation of filtration efficiency. Ideally, filters that exhibit a high filtration efficiency at a low pressure drop are the most desirable ones. Due to the diversity of filter types and the complicated nature of filter structure, it is difficult to describe precisely the media geometry and the corresponding flow patterns. Therefore, a prediction of the pressure drop for real filters, such as porous-membrane filters, is not straightforward. In fact, a comparison of the calculated pressure drop based on an idealized flow model with the actual measured pressure drop is used conveniently as an indication of how uniformly the media structure elements, such as fibers and pores, are arranged. A factor utilizing this concept is called the *pressure drop factor*, β, and is written as

$$\beta = \frac{\Delta P_{ex}}{\Delta P_{th}} \tag{9–19}$$

where ΔP_{ex} and ΔP_{th} are the pressure drop measured experimentally and predicted by the model, respectively. This pressure drop factor is then applied to the theoretically calculated filtration efficiency value as

$$E_{ex} = \beta E_{th} \tag{9–20}$$

where E_{ex} and E_{th} are the filter efficiencies measured experimentally and predicted by the model, respectively (Davies, 1973; Lee et al., 1978; Schlichting, 1968).

The pressure drop for fibrous filters is given by the following theoretical equation:

$$\Delta P_{th} = \frac{16\eta\alpha UL}{Kd_f^2} \tag{9–21}$$

The measured pressure drop across a filter has been found to provide an adequate approximation, through Eq. 9–21, for the effective fiber diameter, d_f, for use in the filtration efficiency theory described previously. This is particularly useful for fibrous filters composed of a highly polydisperse range of fiber filters.

EXAMPLE 9–3

What is the filter efficiency for the filter given in Example 9–1 with a mat thickness of 0.1 cm? Suppose the measured pressure drop for this filter is 7.37 cm [= 2.9 in] of water. What is the expected filter efficiency, considering that the filter structure is not ideal?

Answer: Assuming the individual mechanisms are independent and additive, the single-fiber efficiency is

$$\eta = 0.0162 + 0.02968 + 0.01862 + 0.000067$$
$$= 0.0646$$

The filter efficiency is calculated using Eq. 9–1:

$$E = 1 - \exp\left(\frac{-4(0.0646)0.2(0.001\,\text{m})}{\pi(5\times10^{-6}\,\text{m})(1-0.2)}\right)$$
$$= 1 - \exp\left(\frac{-4(0.0646)0.2(0.1\,\text{cm})}{\pi(5\times10^{-4}\,\text{cm})(1-0.2)}\right)$$
$$= 0.984 = 98.4\%$$

From Eq. 9–21, the theoretical pressure drop becomes

$$\frac{16(1.807\times10^{-5}\,\text{Pa s})(0.2\times0.15\,\text{m/s})0.001\,\text{m}}{0.245\left[(5\times10^{-6})^2\,\text{m}^2\right]}\left[\frac{16(1.807\times10^{-4}\,\text{g/cm/s})(0.2\times0.15\,\text{cm/s})0.1\,\text{cm}}{0.245\left[(5\times10^{-4})^2\,\text{cm}^2\right]}\right]$$
$$= 1420\,\text{Pa} = \left[14{,}200\,\text{g/cm/s}^2 = 14.74\,\text{cm of water} = 5.8\,\text{in of water}\right]$$

The expected filter efficiency becomes

$$E_{\text{ex}} = E_{\text{th}}\frac{0.0737\,\text{m of water}}{0.147\,\text{m of water}}$$
$$= 0.9834\times0.5 = 0.492$$

Filter Testing Method

Filter testing is important for characterizing the operational characteristics of a filter medium. Traditionally, filters were tested using the conventional DOP test method. This method is described in the U.S. Army Military Standard (MIL-STD-282, replaced by ISO standard, see Chapter 33) and uses a 0.3 μm aerosol as test particles and a photometer. As discussed previously, however, particle collection efficiency depends on the particle size and the filtration velocity. Thus, a comprehensive particle collection efficiency test needs the ability to vary particle size. Furthermore, a number of highly efficient commercially available membrane filters exhibit considerable high pressure drops across the filter, and this needs to be accommodated when the efficiency testing is performed.

To address these requirements, an improved filter testing method was introduced by Liu and Lee (1976). Subsequently, the method was used extensively to test various types of filters (Liu et al., 1983). Figure 9–12 is a schematic of the experimental setup for measuring the filtration efficiency as a function of particle size and flow velocity. The system consists of an aerosol source that is capable of providing a series of monodisperse aerosols of known size, a filter testing section, and an aerosol detector.

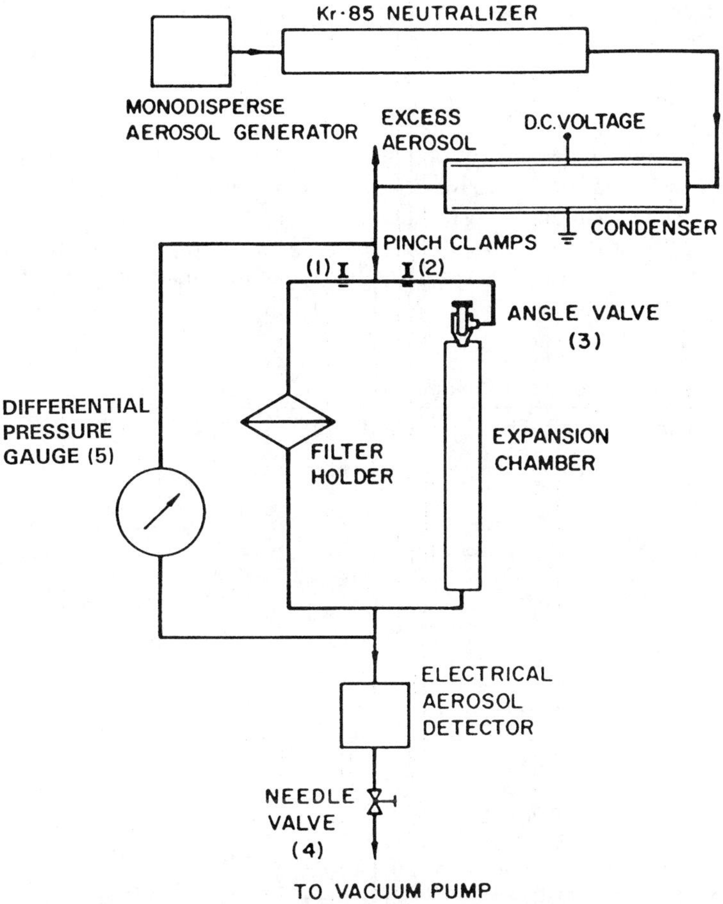

Fig. 9–12. An example of an experimental system for measuring filter efficiencies.

For the aerosol source, an atomization-condensation technique separately described by Liu and Lee (1975) is used. The technique initially atomizes a DOP solution dissolved in alcohol to produce initially polydisperse particles. Subsequently, the particles become monodisperse by the vaporization-condensation method. The particle size can be varied between about 0.03 and 1.3 μm in diameter this way. The geometric standard deviation of the test particles is found to be about 1.3. These test particles are allowed to pass through a Kr-85 electrical neutralizer to avoid possible electrostatic attraction effects on filter testing results. By passing the test aerosol through the filter to be tested and then through the reference line, the amounts of the particles penetrating the paths can be compared to compute the efficiency. For testing a filter of high pressure drop, it is necessary to expand the aerosol into a low pressure in the reference line using an expansion system. In this way, the flow stream condition of both the filter tested and the reference line can be maintained the same. This can be important to ensure that the performance of the particle-detecting instrument is not affected by the different pressure condition. Consideration must also be given to particle loss in the reference sampling line, and measurement for this purpose can be performed using the electrostatic charge characteristics of particles (Liu and Lee, 1976) or by calibration using chemical analysis. For measuring aerosol concentrations in the reference line and downstream of the filter, any aerosol instrument suitable for detecting the aerosols in the size range of interest can be used. In the aforementioned study, an electrical aerosol detector was used.

Liu et al. (1983) performed extensive filter testing for 75 different filter media and compiled the results for each filter; collection efficiencies were measured for 0.035, 0.01, 0.3, and 1.0 μm particles at four different pressure drops. Table 9–5 is an adaptation of the results of Liu et al. (1983) and subdivides the filters tested into various categories described previously. Note that filter manufacturing technology has improved since this study was performed, and some of the filter types reported in this reference currently have greater collection efficiencies. Manufacturers should be consulted for a better estimate of overall filter collection efficiencies.

TABLE 9–5. List of Filters Tested and Principal Results[a]

Type	Manufacturer	Filter	Material	Pore Size (μm)	Filter Permeability Velocity (cm/s) (ΔP = 1 cmHg)	Filter Efficiency Range (%)[b]
Fibrous filter	Whatman	No. 1	Cellulose fiber	NA	6.1	49–99.96
		No. 2			3.8	63–99.97
		No. 3			2.9	89.3–99.98
		No. 4			20.6	33–99.5
		No. 5			0.86	93.1–99.99
		No. 40			3.7	77–99.99
		No. 41			16.9	43–99.5
		No. 42			0.83	92.0–99.992
	Gelman	Type A	Glass fiber	NA	11.2	99.92 to >99.99
		Type A/E			15.5	99.6 to >99.99
		Spectrograde			15.8	99.6 to >99.99
		Microquartz			14.1	98.5 to >99.99
	MSA	1106B		NA	15.8	99.5 to >99.99
	Pallflex	2500 QAO	Quartz fiber	NA	41	84–99.9
		E70/2075W			36.5	84–99.95
		T60A20	Teflon-coated glass fiber		49.3	55–98.8
		Another lot			40.6	52–99.5
		T60A25			36.5	65–99.3
		TX40H 120			15.1	92.6–99.96
		Another lot			9.0	98.9 to >99.99
	Reeve Angel	934AH	Glass fiber	NA	12.5	98.9 to >99.99
		acid-treated			20	95.0–99.96
	Whatman	GF/A	Glass fiber		14.5	99.0 to >99.99
		GF/B			5.5	>99.99
		GF/C			12.8	99.6 to >99.99
		EPM 1000			13.9	99.0 to >99.99
	Delbag	Microsorban-98	Polystyrene	NA	13.4	98.2 to >99.99
Porous membrane filters	Millipore	MF-VS	Cellulose acetate/nitrate	0.025	0.028	99.999 to >99.999
		MF-VC		0.1	0.16	99.999 to >99.999
		MF-PH		0.3	0.86	99.999 to >99.999
		MF-HA		0.45	1.3	99.999 to >99.999
		MF-AA		0.8	4.2	99.999 to >99.999
		MF-RA		1.2	6.2	99.9 to >99.999

TABLE 9–5. *Continued*

Type	Manufacturer	Filter	Material	Pore Size (μm)	Filter Permeability Velocity (cm/s) (ΔP = 1 cmHg)	Filter Efficiency Range (%)[b]
		MF-SS		3.0	7.5	98.5 to >99.999
		MF-SM		5.0	10.0	98.1 to >99.9
		MF-SC		8.0	14.1	92.0 to >99.9
		Polyvic-BD	Polyvinyl chloride	0.6	0.86	99.94 to >99.99
		Polyvic-VS		2.0	5.07	88 to >99.99
		PVC-5		5.0	11	96.7 to >99.99
		Celotate-EG	Cellulose acetate	0.2	0.31	>99.95 to >99.99
		Celotate-EH		0.5	1.07	99.989 to >99.999
		Celolate-EA	Teflon	1.0	1.98	>99.99
		Mitex-LS		5.0	4.94	84 to >99.99
		Mitex-LC	PTFE-polyethylene reinforced	10.0	7.4	62 to >99.99
		Fluoropore		0.2	1.31	>99.90 to >99.99
		FG		0.5	2.32	>99.99
		FH		1.0	7.3	>99.99
		FA		3.0	23.5	98.2–99.98
		FS				
	Metricel	GM-6	Cellulose acetate/ nitrate	0.45	1.45	>99.8 to >99.99
		VM-1	Polyvinyl chloride	5.0	51.0	49–98.8
		DM-800	PVC/ acrylonitrile	0.8	2.7	>99.96 to >99.99
	Gelman	Gelman	Teflon	5.0	56.8	85–99.90
			Teflon			
	Ghia	S2 37PL 02	Teflon	1.0	12.9	>99.97 to >99.99
		S2 37PJ 02		2.0	23.4	99.89 to >99.99
		S2 37PK 02		3.0	24.2	92–99.98
	Zefluor	P5PJ03750	Teflon	2.0	32.5	94.6–99.96
		P5PI03750		3.0	31.6	88–99.9
	Chemplast	75-F	Teflon	1.5	3	83–99.99
		75-M		1.0	6.6	54 to >99.99
		75-C		1.0	32	26–999.8
	Selas Flotronics	FM0.45	Silver	0.45	1.8	93.6–99.98
		FM0.8		0.8	6.2	90–99.96
		FM 1.2		1.2	9.2	73–99.7
		FM5.0		5.0	19.0	25–99.2
Straight-through membrane filter	Nuclepore	N010	Polycarbonate	0.1	0.602	99.9 to >99.99
		N030		0.3	3.6	93.9 to >99.99
		N040		0.4	2.9	78 to >99.99
		N060		0.6	2.1	53–99.5
		N100		1.0	8.8	28–98.1
		N200		2.0	7.63	9–94.1
		N300		3.0	12	9–90.4
		N500		5.0	30.7	6–90.7
		N800		8.0	21.2	1–90.5
		N1000		10.0	95	1–46
		N1200		12.0	161	1–66

[a] Note that current versions of many of these filters have higher efficiency.

[b] Filter efficiency values generally correspond to a particle diameter of 0.035–1 μm, a pressure drop range of 1–30 cmHg, and a face velocity of 1–100 cm/s.

Source: Adapted from Liu et al. (1983).

FILTER SELECTION

The factors influencing the selection of a filter medium for a specific application can be numerous and varied. As mentioned earlier, the important considerations include particle collection efficiency, pressure drop through the filter at the flow required, compatibility with the analytical method to be employed, and cost constraints. In addition, constraints originating from the mechanical strength of the filter medium and compatibility with environmental sampling conditions such as temperature, pressure, humidity, and corrosiveness can also influence filter selection.

The nature and requirements of the analytical technique(s) employed to study the aerosol, following its collection on the filter, greatly influence the selection of the most appropriate filter medium. Filter analysis techniques for deposited aerosols can be divided into three general categories: gravimetric, microscopic, and microchemical. A review of these categories demonstrates a number of the filter selection factors involved, as well as the potential for errors or biases and possible corrective techniques.

Gravimetric Analysis

The measurement of the increase in weight of a filter following a well-defined sampling period is the most common way to determine aerosol mass concentration. The technique requires that the filter collect the aerosol with a high efficiency (close to 100%) and that the weight increase following sampling be entirely attributable to the collected aerosol, that is, filter weight must be independent of age and the temperature and humidity exposure conditions. Gravimetric filter analysis has been found to be most sensitive to the effects of moisture/relative humidity and static charge buildup on filter materials.

Moisture effects arise from the uptake of water vapor by the filter material and from the hygroscopicity of the aerosol sample. Filters of cellulose fibers are the most affected by water vapor uptake, with glass and cellulose quartz fiber filters being less susceptible. Water vapor uptake is lowest in Teflon membrane filters, closely followed by polycarbonate and some polyvinyl chloride (PVC) membrane filters (Mark, 1974; Demuynck, 1975; Charell and Hawley, 1981). Table 9–6 shows the weight stabilities and sensitivities to water vapor uptake of a selection of common filter media and is taken from the results of Lowry and Tillery (1979). A standard means of minimizing relative humidity interferences in gravimetric analysis involves equilibrating the filters at a constant temperature and humidity (e.g., 20, 50% RH) for 24 h before and after sampling. Overcoming the complicating effects of moisture uptake by hygroscopic aerosols collected on a filter is more difficult, with few approaches available other than to calibrate the weight gained under different humidity conditions using a control sample or to minimize the time lag between sampling and weight measurements.

The accumulation of static charges on a filter can result in handling difficulties, enhanced or diminished particle collection, and weighing errors in electrobalances (Engelbrecht et al., 1980). Depending on the filter material and the manufacturing process, certain types of filters such as polycarbonate and PVC membranes can become significantly charged and result in both sampling and measurement errors. A common approach to minimize these effects is to expose the filter to a source producing bipolar ions such as Po-210 or Am-241 before sampling and before gravimetric analysis. Sampling of fibers (e.g., asbestos) has also been found to be biased by charge accumulation on plastic, non-conducting filter holders—the use of conductive filter holders has been found to alleviate this problem.

TABLE 9–6. Weight Sensitivity and Stability of a Selection of Common Filter Media to Water Vapor Uptake

Filter Type	Average Weight 37 mm Diameter (mg)	Weight Stability Under Standard Conditions[a] (mg)	Average Weight Change Following 24 h in a Desiccator[b] (mg)	Average Weight Change Following 24 h at 80% RH[b] (mg)
Gelman glass filter type A/E (without organic binder)	86	0.14	−0.02	+0.01
MSA #457193; 5 μm pore size PVC membrane with fibrous backup filter	243	0.04	−0.04	0
Millipore AA; 0.8 μm pore size, cellulose ester membrane	53	0.13	0	+0.35
Millipore Teflon; 5 μm pore size, Teflon membrane	108	0.02	+0.01	0
Selas Flotronics FM-37; 0.8 μm pore size, silver membrane (no longer available)	460	0.03	0	+0.01

[a] 95% confidence interval based on three measurements a day for six filters of each type over 30 days.
[b] Filters were equilibrated under room conditions for 24 h, weighed, desiccated, or humidified for 24 h, equilibrated under room conditions for 24 h, and reweighed. RH, relative humidity.
Source: Lowry and Tillery (1979).

EXAMPLE 9–4

Ambient air is to be sampled to collect a particulate mass of at least 10 mg for gravimetric analysis. A flow velocity of 0.305 m/s [= 1 ft/s] will be adopted using an 0.20 m × 0.25 m [8 × 10 in] sheet quartz filter with an effective filtration area of 0.18 m × 0.230 m (= 0.0414 m^2). Calculate the minimum required sampling time assuming that the particle collection efficiency is 100%. The ambient average particle concentration is estimated to be ≈20 μg/m^3.

Answer: The flow rate through the filter is

$$0.0414\,\text{m}^2 \times 0.305\,\text{m/s} = 0.0126\,\text{m}^3/\text{s}$$

The particle mass collected per unit time at this flow rate is

$$(20\,\mu\text{g/m}^3)(0.0126\,\text{m}^3/\text{s}) = 0.253\,\mu\text{g/s}$$

The required sampling time for collecting 10 mg is

$$10{,}000\,\mu\text{g}/(0.253\,\mu\text{g/s}) = 39{,}500\,\text{s} = 11.0\,\text{h}$$

Microscopic Analysis

Particle analysis by light or electron microscopy is frequently used to obtain information on the size, morphology, and compositional characteristics of aerosol samples. Microscopic analysis requires that the collection of the particles occurs on, and as closely to, a flat filter surface as is the case with microporous and straight-through pore membrane filters.

Polycarbonate, straight-through pore membrane filters are particularly well suited to microscopic applications due to their smooth, flat surface and near complete surface collection characteristics (provided that pore sizes are selected appropriately).

Other surface analysis techniques that impose similar constraints on aerosol collection media include X-ray fluorescence (XRF), X-ray diffraction (XRD), and proton-induced X-ray emission (PIXE) analyses for elemental and chemical species concentration measurement and aerosol radioactivity measurement techniques. These techniques also benefit from the collection of the aerosol on or close to the surface of the filter and impose additional constraints of minimizing both the aerosol collection surface area and the background concentration or response of the blank filter material in the analysis.

Generally, microporous and straight-through pore membrane filters are both well suited to these other filter analysis techniques. Teflon membrane filters are most commonly used for XRF analysis of filter-sampled aerosols owing to their inertness and low blank value concentrations (e.g., Chow et al., 1990). Quartz and glass fiber filters can be useful for XRD analyses in cases of high particulate loading, while Teflon filters are superior for low loadings (Davis and Johnson, 1982). Silver membranes composed of metallic silver are useful for the analysis of crystalline silica using XRD techniques due to their extremely low interference in the quartz region of the diffraction spectrum. Aerosol radioactivity analysis using α- or β-radiation detectors usually requires a high flow rate, high collection efficiency, and the collection of particles as close to the surface as possible to minimize the absorption of radiation. Microporous cellulose ester membranes of pore sizes between 0.45 and 0.8 μm (e.g., Type AA, ***MIL***) are commonly employed to meet these objectives. Busigin et al. (1980) reviewed the collection characteristics of radon progeny radioactive aerosols by a variety of different filters.

The selection of filters for sampling airborne microorganisms (viruses, bacteria, and fungi) is also governed by the need to microscopically count the number of viable microorganisms or colony-forming units collected on the filter. In these applications, the loss of some viable microorganisms may result from the desiccation induced by their collection on the filter surface. Thus, collection on filter surfaces is limited to hardy microorganisms that can withstand desiccation and can be transferred to suitable growth media following collection. More detailed discussions of sampling airborne bioorganisms are presented in Chapter 24, and by Chatigny et al. (1989).

Microchemical Analysis

The chemical analysis of particles collected on filter media is becoming increasingly routine in applications such as air quality monitoring. The most important factors for consideration in selecting filter media for microchemical analysis are the quantity of particulate matter required for the analysis and the minimization of (1) interferences arising from the background response of the blank filters and (2) artifact formation from chemical transformations occurring on the filter during and after sampling. The magnitude and variability of the blank filter or background trace element/chemical species concentrations of different filters can be significant when determining the sensitivity and the limits of detection of the analytical technique. Maenhaut (1989) reviews a number of different analytical techniques for trace atmospheric elements and discusses the blank concentration ranges for different types of filters in various analyses. A number of other important aspects in aerosol sampling for microchemical analyses are also presented by Hopke (1985).

Most chemical analyses require the extraction of the particles collected on the filter following sampling in a manner suitable for input into analytical instrumentation. Filters of cellulose (Whatman-type paper, ***WHA***), glass, Teflon-coated glass, and quartz fibers are all commonly used in aerosol sampling for microchemical analysis due to their low pressure drop characteristics that permit high-volume sampling. Cellulose paper filters are conveniently processed for aerosol extraction using incineration, ashing, or digesting in acid solution. However, cellulose paper filters suffer from low particle collection efficiency characteristics at low particulate filter loadings (see Table 9–7).

Glass, Teflon-coated glass, and quartz fiber filters have significantly higher particle collection efficiencies but must be acid leached for recovery/extraction of the aerosols. Glass fiber filters suffer from a positive artifact mass in ambient air sampling due to their slight alkalinity that results in the in situ conversion of sulfur dioxide to sulfate (e.g., Coutant, 1977; Rodes and Evans, 1977; Stevens et al., 1978). Artifact particulate nitrate can also be formed on glass fiber filters depending on the gaseous nitric acid concentrations (Appel and Tokiwa, 1981). Quartz and paper fibers do not suffer from significant sulfate artifacts in ambient air sampling, as reported by Pierson et al. (1980).

Quartz fiber filters are particularly useful in aerosol sampling for microchemical analysis due to their low water vapor uptake characteristics and low background/blank elemental concentrations. Hence, they are commonly used in ion chromatographic analysis for species such as chloride, nitrate, sulfate, potassium, and ammonium ions (e.g., Chow et al., 1990). In

TABLE 9–7. Compilation of Common Applications, Advantages, and Disadvantages for Various Types of Fibrous Filters

Filter Type	Typical Applications	Advantages	Limitations
Fibrous filters (general)	Air quality sampling	Low pressure drop at high-volume sampling operation Low cost High particulate loading capacity	Lower collection efficiencies for submicrometer particles Particle collection occurs throughout the depth of the filter
Cellulose fiber	Typically used in limited/qualitative applications in air quality sampling	Inexpensive Convenient extraction of particulates	Highly moisture-sensitive Limited temperature range Low particle collection efficiency Low chemical resistance
Borosilicate glass fiber	Wide scope in air quality sampling. Used without organic binder	Temperature resistance to ≈500°C Chemically resistant to some extent	Sulfate artifact formation due to alkalinity of fibers Water vapor uptake can occur and must be equilibrated appropriately
Teflon-coated glass fiber	Wide range of air sampling applications—emissions analysis, gravimetric analysis, biological and mutogenic analysis	Low moisture uptake Minimizes chemical transformation artifacts	Artifact nitrate
Quartz fiber	Air sampling for chemical analysis of particulates—ion chromatography, atomic absorption, carbon analysis, PAH analysis, etc.	Low moisture uptake Stable to temperatures up to 800°C Low trace contaminant levels Can be baked to remove trace organics before sampling Low artifact formation	Friable Artifact nitrate formation has been observed

TABLE 9–8. Compilation of Common Applications, Advantages, and Disadvantages for Various Types of Microporous Membrane Filters

Filter Type	Typical Applications	Advantages	Limitations
Membrane filters (general) (apply to all below)	Used in air sampling for surface analytical techniques, submicrometer particle collection	High collection efficiency High mechanical strength	High pressure drop Low particulate loading capacity, rapid clogging Limited temperature range, typically
Cellulose Mixed esters Nitrate Acetate, etc. PVC membranes	Sampling of metals, cotton dust, asbestos, etc., in NIOSH standard methods	Inexpensive among membrane filters Low chemical resistance	Susceptible to water vapor uptake Operating temperature limited to 75°–130°C Electrostatic charge buildup observed in PVC membranes
Teflon membranes PTFE	Gravimetric analysis, neutron activities analysis, XRF, XRD (see text)	Inert to chemical transformations Extremely low moisture sensitivity Low trace/background concentrations Chemical resistant	Loss of nitrates observed Temperature range limited to ≈150°C for supported membranes and 260°C for pure membranes
Silver membranes (no longer available)	Organic particulate collection and analysis, e.g., benzo[a]pyrene, PAH, etc.; XRD analysis of silica	Chemical resistant Low background interferences High maximum operating temperature of 550°C	Most expensive among common membrane filters

TABLE 9–9. Compilation of Common Applications, Advantages, and Disadvantages for Straight-Through Pore Membrane Filters

Filter Type	Typical Applications	Advantages	Limitations
Polycarbonate membranes	Ideal for aerosol collection for subsequent surface analytical techniques, e.g., microscopy, PIXE (see text)	Flat, uniform surface Nonhygroscopic Low background/blank concentration Surface capture characteristics Semitransparent surface	High pressure drop Low particulate loading capacity Particle size range of low collection efficiency usually exists Susceptible to static charge buildup

addition, quartz fiber filters can be baked at high temperatures to lower blank organic concentration values, allowing them to be used for collecting particulate samples that can be extracted for organic species, polycyclic aromatic hydrocarbons, alklyating aspects, and mutagenic activity (e.g., Lioy and Daisey, 1983). Quartz fiber filters are also utilized in organic and elemental carbon analyses of particulate deposits on filters by a combination of flash and rapid heating of the filters and conversion of evolved carbon to CO_2 that can then be measured (e.g., Tanner et al., 1982).

Membrane filters can also be used for microchemical analysis, although they suffer from problems of limited particulate loading ability and the possibility of losing coarse particles during handling and transport following sampling (Dzubay and Barbour, 1983). Teflon membrane filters have the advantage of low blank chemical species concentrations, as well as chemical inertness that permits sampling with minimal sulfate artifact formation. However, ammonium nitrate (nitrate salts) and nitric acid can be lost from "inert" Teflon, as well as quartz, filters through volatilization (Rodes and Evans, 1977) and as a result of reactions with acidic species on the filter (Harker et al., 1977). Nylon filters have been utilized as an alternative for nitrate collection (Grosjean, 1982; Spicer et al., 1982), although they are susceptible to sulfate artifact formation (Chan et al., 1986).

Artifact formation in the sampling of organic aerosols on filters has also been reported in a number of studies. As discussed by Pitts and Pitts (1986), these include negative artifacts from volatilization of the more volatile particulate organics (Van Vaeck et al., 1984), positive artifacts from adsorption of gaseous organics on the filter (Stevens et al., 1980; Cadle et al., 1983), and transformations/reactions occurring with species such as O_3 and NO_2 sampled through the filter. De Raat et al. (1990) compared glass fiber, Teflon-coated glass fiber, and Teflon membrane filters for sampling of mutagens and polycyclic aromatic hydrocarbons (PAH) in ambient airborne particles and found: (1) a slightly greater mutagenicity when sampling with glass fiber filters, probably due to adsorption of gaseous PAH on the glass fibers followed by conversion to mutagens; (2) higher adsorption of volatile PAH on glass fiber filters and greater volatilization on the Teflon-coated glass fiber and Teflon membrane filters; and (3) lower concentrations of the more reactive PAH species on the glass fiber filters.

The preceding discussions indicate that the requirements of filter analysis techniques and the need to minimize interferences and artifacts are often inherently conflicting and can rarely be perfectly satisfied. However, an adequate filter medium can generally be selected to meet the needs of most applications through a careful consideration of the important issues. Tables 9–7 to 9–9 are compilations of the common applications, advantages, and limitations of the various types of filters discussed in previous sections. The information in the tables is derived from a review of the literature and illustrates the availability of an adequate selection of filter media for the wide spectrum of applications that utilize filter sampling.

REFERENCES

Agarwal, J. K. and B. Y. H. Liu. 1980. A criterion for accurate aerosol sampling in calm air. *Am. Ind. Hyg. Assoc. J.* 41:191–197.

Appel, B. R. and Y. Tokiwa. 1981. Atmospheric particulate nitrate sampling errors due to reactions with particulate and gaseous strong acids. *Atmos. Environ.* 15:1087.

Brown, R. C. 1993. *Air Filtration*. Oxford: Pergamon Press.

Busigin, A., A. W. van der Vooren, and C. R. Phillips. 1980. Collection of radon daughters on filter media. *Environ. Sci. Technol.* 14:533–536.

Cadle, S. H., P. J. Groblicki, and P. A. Mulaya. 1983. Problems in the sampling and analysis of carbon particulate. *Atmos. Environ.* 17:593.

Cahill, T. A., L. L. Ashbauch, and J. B. Barone. 1977. Analysis of respirable fractions of atmospheric particulates via sequential filtration. *J. Air Pollut. Control Assoc.* 27:675.

Chan, W. H., D. B. Orr, and D. H. S. Chung. 1986. An evaluation of artifact SO_4 formation on nylon filters under field conditions. *Atmos. Environ.* 20:2397.

Charell, P. R. and R. G. Hawley. 1981. Characteristics of water adsorption on air sampling filters. *Am. Ind. Hyg. Assoc. J.* 42:353–360.

Chatigny, M. A., J. M. Macher, H. A. Burge, and W. R. Soloman. 1989. Sampling airborne microorganisms and aeroallergens. In *Air Sampling Instruments for Evaluation of Atmospheric Contaminants*, 7th Ed. Cincinnati, OH: American Council of Governmental Industrial Hygienists, pp. 199–220.

Cheng, Y. S. 1995. Aerosol sampler calibration. In *Air Sampling Instruments for Evaluation of Atmospheric Contaminants*, 8th Ed. Cincinnati, OH: American Conference of Governmental Industrial Hygienists.

Chow, J. C., J. G. Watson, R. T. Egami, C. A. Frazier, Z. Lu, A. Goodrich, and A. Bird. 1990. Evaluation of regenerative-air vacuum street sweeping on geological contributions to PM_{10}. *J. Air Waste Manag. Assoc.* 40:1134.

Coutant, R. W. 1977. Effect of environmental variables on collection of atmospheric sulfate. *Environ. Sci. Technol.* 11:873.

Davies, C. N. 1968. The entry of aerosols into sampling tubes and heads. *Br. J. Appl. Phys. Ser.* 21:921.

Davies, C. N. 1973. *Air Filtration.* London: Academic Press.

Davis, B. L. and L. R. Johnson. 1982. On the use of various filter substrates for quantitative particulate analysis by X-ray diffraction. *Atmos. Environ.* 16:273.

Demuynck, M. 1975. Determination of irreversible absorption of water on air sampling filters. *Atmos. Environ.* 9:523–528.

de Raat, W. K., G. L. Bakker, and F. A. de Meijere. 1990. Comparison of filter materials used for sampling of mutagens and polycyclic aromatic hydrocarbons in ambient airborne particles. *Atm. Environ.* 24A:2875–2887.

Dzubay, T. G. and R. K. Baybour. 1983. A method to improve adhesion of aerosol particles on Teflon filters. *J. Air Pollut. Control Assoc.* 33:692.

Engelbrecht, D. R., T. A. Cahill, and P. J. Feeney. 1980. Electrostatic effects on gravimetric analysis of membrane filters. *J. Air Pollut. Control Assoc.* 30:391.

Friedlander, S. K. 1957. Mass and heat transfer to single spheres and cylinders at low Reynolds numbers. *A.I.Ch.E. J.* 3:43–48.

Grosjean, D. 1982. Quantitative collection of total inorganic atmospheric nitrate on nylon filters. *Anal. Lett.* 15(A9):785.

Harker, A., L. Richards, and W. Clark. 1977. Effect of atmospheric SO_2 photochemistry upon observed nitrate concentrations. *Atmos. Environ.* 11:87.

Heidam, N. Z. 1981. Review: Aerosol fractionation by sequential filtration with Nuclepore filters. *Atmos. Environ.* 15:891.

Hinds, W. D. 1999. *Aerosol Technology.* New York: John Wiley & Sons.

Hopke, P. K. 1985. *Receptor Modeling in Environmental Chemistry*. New York: John Wiley & Sons.

Kanaoka, C. 1989. Time dependency of air filter performance. *J. Aerosol Res. Jpn.* 4:256–264 [in Japanese].

Kogan, V., M. R. Kuhlman, R. W. Coutant, and R. G. Lewis. 1991. Aerosol filtration by sorbent beds. In JAWMA 43:1367–1373, press.

Lee, K. W., L. D. Reed, and J. A. Gieseke. 1978. Pressure drop across packed beds in the low Knudsen number regime. *J. Aerosol Sci.* 9:557–566.

Lee, K. W. and B. Y. H. Liu. 1980. On the minimum efficiency and the most penetrating particle size for fibrous filters. *J. Air Pollut. Control Assoc.* 30:377–381.

Lee, K. W. and B. Y. H. Liu. 1982a. Experimental study of aerosol filtration by fibrous filters. *Aerosol Sci. Technol.* 1:35–46.

Lee, K. W. and B. Y. H. Liu. 1982b. Theoretical study of aerosol filtration by fibrous filters. *Aerosol Sci. Technol.* 1:147–161.

Lioy, P. J. and J. M. Daisey. 1983. The New Jersey project on airborne toxic elements and organic substances (ATEOS): A summary of the 1981 summer and 1981 winter studies. *J. Air Pollut. Control Assoc.* 33:649.

Lippmann, M. 1995. Filters and filter holders. In *Air Sampling Instruments for Evaluation of Atmospheric Contaminants*, 8th Ed. Cincinnati, OH: American Conference of Governmental Industrial Hygienists.

Liu, B. Y. H. and K. W. Lee, 1975. An aerosol generator of high stability. *Am. Ind. Hyg. Assoc. J.* 36:861–865.

Liu, B. Y. H. and K. W. Lee. 1976. Efficiency of membrane and Nuclepore filters for submicrometer aerosols. *Environ. Sci. Technol.* 10:345–350.

Liu, B. Y. H., D. Y. H. Pui, and K. L. Rubow. 1983. Characteristics of air sampling filter media. In *Aerosols in the Mining and Industrial Work Environments*. Ann Arbor, MI: Ann Arbor Science.

Lowry, P. L. and M. 1. Tillery. 1979. *Filter Weight Stability Evaluation*. Los Alamos Scientific Laboratory, Report No. LA-8061-MS.

Maenhaut, W. 1989. Analytical techniques for atmospheric trace elements. In *Control and Fate of Atmospheric Trace Metals*, eds. J. M. Pacyna and B. Ottar. Dordrecht: Kluwer, pp. 259–301.

Mark, D. 1974. Problems associated with the use of membrane filters for dust sampling when compositional analysis is required. *Ann. Occup. Hyg.* 17:35.

Mori, J., H. Emi, and Y. Otani. 1991. Classification of membrane gas filters and their performance evaluation. *J. Aerosol Res. Jpn.* 6:149–156 [in Japanese].

Natanson, G. L. 1957. *Proc. Acad. Sci. USSR. Phys. Chem. Sec.* 112:21–25.

Parker, R. D., G. H. Buzzard, T. G. Dzubay, and J. P. Bell, 1977. A two-style respirable aerosol sampler using Nuclepore filters in series. *Atmos. Environ.* 11:617.

Payatakes, A. C. 1976. Model of transient aerosol particle deposition in fibrous media with dendritic pattern. *A.I.Ch.E. J.* 23:192–202.

Pierson, W. R., W. W. Brachaczek, T. J. Korniski, T. J. Truer, and J. W. Butler, 1980. Artifact formation of sulfate, nitrate and hydrogen ion on backup filters: Allegheny mountain experiment. *J. Air Pollut. Control Assoc.* 30:34.

Pitts, B. J. F. and J. R. Pitts, Jr. 1986. *Atmospheric Chemistry: Fundamentals and Experimental Techniques.* New York: John Wiley & Sons.

Rodes, C. E. and G. F. Evans. 1977. *Summary of LACS Integrated Measurements*. EPA-600/4-77-034. Research Triangle Park, NC: U.S. Environmental Protection Agency.

Rubow, K. L. 1981. *Submicrometer Aerosol Filtration Characteristics of Membrane Filters*. Ph.D. thesis, University of Minnesota, Minneapolis, Minnesota.

Rubow, K. L. and V. C. Furtado. 1989. Air movers and samplers. In *Air Sampling Instruments for Evaluation of Atmospheric Contaminants*, 7th Ed. Cincinnati, OH. American Conference of Governmental Industrial Hygienists.

Schlichting, H. 1968. *Boundary Layer Theory*, 6th ed. New York: McGraw Hill.

Spicer, C. W., J. E. Howes, T. A. Bishop, L. H. Arnold, and R. K. Stevens. 1982. Nitric acid measurement methods: an intercomparison. *Atm. Environ.* 16:1478–1500.

Spielman, L. and S. L. Goren. 1968. Model for predicting pressure drop and filtration efficiency in fibrous media. *Environ. Sci. Technol.* 2:279–287.

Spurny, K. R. 1998. *Advances in Aerosol Filtration.* Boca Raton, FL: Lewis Publishers.

Spurny, K. R., J. P. Lodge, Jr., E. R. Frank, and D. C. Sheesley. 1969. Aerosol filtration by means of Nuclepore filters: Structural and filtration properties. *Environ. Sci. Technol.* 3:453.

Stechkina, I. B., A. A. Kirsch, and N. A. Fuchs. 1969. Studies in fibrous aerosol filters—IV. Calculation of aerosol deposition in model filters in the range of maximum penetration. *Ann. Occup. Hyg.* 12:1–8.

Stevens, R. K., T. G. Dzubay, G. Russwurm, and D. Rickel. 1978. Sampling and analysis of atmospheric sulfates and related species. In *Sulfur in the Atmosphere, Proc. International Symposium*, United Nations, Dubrovnik, Yugoslavia, September 7–14, 1977. *Atmos. Environ.* 12:55.

Stevens, R. K., T. G. Dzubay, R. W. Shaw, Jr., W. A. McClenny, C. W. Lewis, and W. E. Silson. 1980. Characterization of the aerosol in the Great Smoky Mountains. *Environ. Sci. Technol.* 14:1491.

Tanner, R. L., T. S. Gaffney, and M. F. Phillips. 1982. Determination of organic and elemental carbon in atmospheric aerosol samples by thermal evolution. *Anal. Chem.* 54:1627.

Van Vaeck, L., K. Van Cauwenberghe, and J. Janssens. 1984. The gas-particle distribution of organic aerosol constituents: Measurement of the volatilization artifact in Hi-Vol cascade impactor sampling. *Atmos. Environ.* 17:900.

Vincent, J. H. 1989. *Aerosol Sampling Science and Practice.* New York: John Wiley & Sons.

Vincent, J. H., R. J. Aitken, and D. Mark. 1993. Porous plastic foam filtration media: Penetration characteristics and applications in particle size-selective sampling. *J. Aerosol Sci.* 24:929–944.

10 Inertial, Gravitational, Centrifugal, and Thermal Collection Techniques

VIRGIL A. MARPLE and BERNARD A. OLSON
Mechanical Engineering Department, University of Minnesota, Minneapolis, MN
KENNETH L. RUBOW
Mott Corporation, Farmington, CT

INTRODUCTION

Inertial classification, gravitational sedimentation, centrifugation, and thermal precipitation are techniques used to collect particles for subsequent analysis or for particle classification. The inertial classifiers, which include impactors, virtual impactors, and cyclones, are widely used in the sampling of particles. Settling chambers, which include centrifuges as well as gravitational settling devices, are less widely used. Thermal precipitators are seldom used.

Impactors, in particular, have been used extensively, studied both theoretically and experimentally, and are the instrument of choice for the determination of aerosol mass size distributions. Many versions of impactors are available commercially, and many more have been designed, built, and used in special studies.

Virtual impactors are a more recently developed inertial classifier and possess a feature not found in conventional inertial impactors; namely, the particles remain airborne after classification compared with the conventional impactor in which the particles are collected on a solid surface. The feature of keeping particles airborne is important for transporting the particles to other particle-analyzing instruments, or to filters, and in concentrating particles larger than the cut size in a small fraction of the total flow.

Cyclones are also widely used as particle samplers but have been much more difficult to analyze theoretically. In general, cyclone particle cut-off characteristics are not as sharp as for impactors but still sharp enough to be useful as classifiers. They differ from impactors primarily by their ability to collect much larger quantities of particles.

Settling chambers are devices that directly measure particle terminal settling velocities, but are not commonly used. However, centrifuges, which magnify settling forces by applying centrifugal forces to the particles, are more widely used.

Thermal precipitators are also used for the collection of particles but are much less popular than inertial classifiers or centrifuges because the particles are not classified by size. The thermal precipitator is most effective for collecting particles over a wide size range for subsequent sizing by a microscope. They do have the advantage that they collect small particles, have low pressure drops, and, therefore, require only small pumps.

These classifiers are discussed in the following sections. Because inertial classifiers (espe-

Aerosol Measurement: Principles, Techniques, and Applications, Second Edition, Edited by Paul A. Baron and Klaus Willeke.
ISBN 0-471-35636-0

cially impactors) are the most widely used of the classifiers, the major emphasis of this chapter is on these classifiers.

INERTIAL CLASSIFIERS

Numerous inertial classifiers have been designed and reported in the literature, with many of them being commercially available. Tables 10–1 to 10–3 list most of the commercially available impactors and cyclones by type and manufacturer. The following sections discuss the classifiers in general terms, providing information on specific devices only to illustrate specific points.

Principle of Inertial Classification

The principle of inertial classification of particles is quite simple in that the particles' inertia is used in their classification. Classification is achieved in these instruments by turning the gas flow and capturing the particles with sufficient inertia to cross gas streamlines and escape the flow. Particles with less inertia will remain in the gas flow.

The simplest type of inertial classifier is a body collector, which is a body (usually cylinder or ribbon) passing through particle-laden gas. As this body moves through the gas, the gas is deflected around the body. Large particles, however, due to their inertia, are not deflected as much as small particles and will strike the surface of the body.

An excellent example of a body collector is the automobile. As the automobile passes through the air, large particles in the air will impact on the automobile, and, as passengers, we are in an excellent position to observe impaction on the windshield. Probably the best observations are during a snowstorm when the path of the snowflakes can be observed. If the car is moving rather slowly, the snowflakes will approach the car and pass over the windshield without impaction. As the automobile increases in speed, the snowflakes will impact on the windshield. The two determining factors as to whether or not the snowflakes will impact on the windshield are the speed of the automobile and the size of each snowflake.

A similar observation is experienced when the automobile encounters flying insects. A large insect will have a trajectory undeflected by the airstream around the car and impact dramatically on the windshield. Smaller insects will follow the airstream and not impact. However, if the body is smaller, the smaller insects will be collected. This can be seen on smaller cross-sectional area bodies, such as the radio antenna. Inspection of the insects on the antenna will indicate that they are smaller than those collected on the automobile windshield.

The above example indicates that the velocity of the air, U, the size of the particle, d_p, and the size of the body, d_b, are three important parameters in determining whether or not a particle will be collected on the body. A dimensionless parameter, the Stokes number, defined as the ratio of the particle's stopping distance to the physical dimension of the body collector, is the governing relationship as to whether or not a particle will strike the body. If the Stokes number, Stk, defined as

$$Stk = \frac{\rho_p C_c d_p^2 U}{18 \eta d_b} \qquad (10\text{–}1)$$

where

ρ_p = particle density
C_c = slip correction
U = relative velocity of body to air (or gas)
d_p = particle diameter
η = air (or gas) viscosity
d_b = body diameter

is larger than approximately unity, the particle will impact on the body.

TABLE 10–1. Selected List of Commercially Available Impactors

Manufacturer[a]	Sampler Name	Flow Rate ($\times 10^{-5}$ m^3/s [L/min])	No. of Stages	Cut Points (Range, μm)	Comments[b]
Cascade Impactors for Ambient Air Sampling					
AND	Sierra/Marple Model 210	12 [7]	10	0.16–18	1
AND	Low Pressure Impactor	5 [3]	12	0.08–35	3
AND, GMW	One ACFM Ambient Impactor	47 [28]	8	0.4–10	
DEK	ELPI (Electrical Low Pressure Impactor)	17, 33, 50 [10, 20, 30]	13	0.03–10	4
DEK	Low Pressure Impactor	17, 50 [10, 30]	13	0.03–10	
DEK	PM-10 Impactor	17, 33, 50 [10, 20, 30]	3	10, 2.5, 1.0	
HAU	Berner Impactor	50 [30]	9	0.063–16.7	
INT	Mercer 7-Stage Impactor (02-100)	0.17 [0.1]	7	0.33–3.1	5
INT	Mercer 7-Stage Impactor (02-130)	1.7 [1]	7	0.32–4.5	
INT	Mercer 7-Stage Impactor (02-150)	3.3 [2]	7	0.25–5.0	
INT	Mercer 7-Stage Impactor (02-170)	8.3 [5]	7	0.5–5.0	
INT	Multijet CI (02-200, 02-220, 02-240)	17, 25, 33 [10, 15, 20]	7	0.5–8.0	
INT	Multijet CI (02-260)	47 [28]	7	0.5–9.0	
MSP	MOUDI (micro-orifice impactor)	50 [30]	10	0.056–10	6
MSP	Nano-MOUDI	17 [10]	3	0.01–0.032	6
MSP	Cleanroom Cascade Impactor	5 [3]	6	0.05–10	
MSP	Airborne Cascade Impactor	170 [100]	5	0.25–2.5	
CMI	QCM Real-Time Impactor, PC-2	0.42 [0.25]	10	0.05–25	7
CMI	QCM Real-Time Impactor, PC-2H	3.3 [2]	10	0.05–10	7
CMI	QCM Real-Time Impactor, PC-6H	3.3 [2]	6	0.05–6.0	7
Cascade Impactors for Pharmaceutical Applications					
AND, GMW	Eight Stage Non-Viable Cascade Impactor	47 [28]	8	0.4–10	
COP	"Andersen-Type" Cascade Impactor	47 [28]	8	0.4–10	
COP	Multistage Liquid Impinger (MSLI)	50 [30]	4	1.7–13	
MSP	Marple-Miller Pharmaceutical Impactor	8.2, 20, 50, 100 [4.9, 12, 30, or 60]	5	0.63–10	
MSP	Next Generation Pharmaceutical Impactor	50–170 [30–100]	7	0.23–11	

(continued)

TABLE 10–1. ***Continued***

Manufacturer[a]	Sampler Name	Flow Rate ($\times 10^{-5}$ m^3/s [L/min])	No. of Stages	Cut Points (Range, μm)	Comments[b]
Impactors for Ambient HiVol Samplers					
AND, GMW	HiVol Impactor, Series 230	1880 [1130]	4	0.49–7.2	9
AND, GMW	HiVol Impactor, Series 230	942 [565]	6	0.41–10	9
Single-Stage Impactors for Ambient Samplers					
ADE	MS&T Area Sampler	38 [23]	1	1.0	10
ADE	MS&T Area Sampler	6.7, 17, 33 [4, 10, or 20]	1	2.5	10
ADE	MS&T Area Sampler	6.7, 17, 33 [4, 10, or 20]	1	10	10
MSP	Micro-Environmental Monitor	17 [10]	1	2.5 or 10	
URG	Portable Size Selective Impactor	6.7 [4]	1	2.5	
Personal Samplers					
AND, GMW	Marple Personal Sampler (Model 290)	3.3 [2]	8	0.5–20	1
MSP	Personal Environmental Monitor	3.3, 6.7, 17 [2, 4, or 10]	1	2.5	
MSP	Personal Environmental Monitor	3.3, 6.7, 17 [2, 4, or 10]	1	10	
SKC	Spiral Sampler (respirable)	4.2 [2.5]	1	4.0	
SKC	Spiral Sampler (PM-2.5)	3.3 [2.0]	1	2.5	
URG	Personal PUF Pesticide Sampler	6.7 [4]	1	1, 2.5, or 10	
URG	Personal Impactor Filter Pack	6.7 [4]	1	1, 2.5, or 10	
URG	Personal Impactor Filter Pack	3.3 [2]	1	2.5	
Source Test Impactors					
AND	In-Stack Air Sampler, Series 220	12 [7]	9	0.16–18	
AND	Stack Sampling Head (Mark III, IV)	5–35 [3–21]	8	0.4–11	1
AND	Impactor Preseparator	1.7 [1]	21	10	
DEK	Automotive Mass Impactor	120 [70]	4	0.2–2.5	
INT	High Temp, High Pres. Impactor	16 [16]	7	0.62–8.8	
PCS	UW (Pilat) Mark V Cascade Impactor	47 [28]	13	0.2–20	
PCS	UW (Pilat) Mark III Cascade Impactor	4.7 [2.8]	7	0.2–20	
PCS	Mark 8 High Grain Loading Impactor	47 [28]	3	1.5–10.8	
PCS	UW (Pilat) Low Pressure Source Test Impactor	47 [28]	14	0.05–20	

PM Inlets					
AND, GMW	Hi-Volume PM-10 Inlet	1880 [1130]	1	10	
AND, GMW	Medium Flow PM-10 Inlet	187 [112]	1	10	
AND, GMW	Dichotomous Sampler Inlet	17.8 [16.7]	1	10	
AND, WED	Hi-Volume PM-10 Inlet	1880 [1130]	1	10	
BGI	Low Flow PM-10, PM-2.5 Inlet	17.8 [16.7]	1	2.5 or 10	
URG	Low Flow PM-10 Inlet	17.8 or 53 [16.7 or 32]	1	10	
Viable and Biological Impaction Samplers					
AND	Single Stage Bioaerosol Sampler	47 [28]	1	0.65	11
AND	Microbial Air Sampler	47 [28]	2	0.65, 3.5	11
AND	Particle Fractionating Viable Sampler	47 [28]	6	0.65–7	11
NBS	Slit-to-Agar Biological Sampler	92 [55]	1	Not stated	12
SKC	BioStage-1Bioaerosol Impactor	47 [28.3]	1	Not stated	
SKC	BioSampler	10.8 [12.5]	1	Not stated	
SSI	SAS Portable Sampler	150, 300 [90, 180]	1	Not stated	

[a] See Appendix I for full manufacturer addresses referenced to the italicized three-letter code.

[b] Comments:

1. Radial slot design.
2. Circular jets, interchangeable nozzles.
3. Four low-pressure stages.
4. Uses electrometers connected to each impactor stage for counting charges on particles for real-time measurement.
5. One round jet per stage.
6. Micro-orifice plates of 2000 jets on bottom stages.
7. Uses vibrating quartz crystal collection surfaces as mass-to-frequency transducers for real-time measurement.
8. Fits on HiVol, round jets.
9. Fits on HiVol, rectangular jets.
10. Also referred to as the "Harvard Impactor."
11. Collection directly onto agar plates.
12. Slot impactor with rotating turntable for agar plates.

TABLE 10–2. Selected List of Commercially Available Virtual Impactors

Manufacturer	Sampler Name	Flow Rate ($\times 10^{-5}$ m^3/s [L/min])	No. of Stages	Cut Points (Range, μm)
AND, GMW	Dichotomous Sampler	27.8 [16.7]	1	2.5
BGI, INT	Cascade Centripeter	50 [30]	3	1.2, 4, 14
INT	Virtual Impactor	1.7–8.3 [1–5]	2	0.5–10
MSP	High Volume Virtual Impactor	1880 [1130]	1	1.0, 2.5
MSP	Microcontaminant Particle Sampler	50 [30]	1	1.0
MSP	Universal Air Sampler	500 [300]	2	10, 1.0, or 2.5
MSP	Bioconcentrator	550 [330]	3	10, 2.0, 2.0
URG	VAPS	53 [32]	2	10, 2.5

TABLE 10–3. Selected List of Commercially Available Cyclones

Manufacturer	Cyclone Name	Flow Rate Range ($\times 10^{-5}$ m^3/s [L/min])	D_{50} Range (μm)
AND	Sharp Cut Cyclone	27.8 [16.7]	2.5
AND, INT	SRI V	12–47 [7–28]	0.3–2.0
AND, INT	SRI IV	12–47 [7–28]	0.5–3.0
AND, INT	SRI III	23–47 [14–28]	1.4–2.4
—	AIHL	13–45 [8–27]	2.0–7.0
AND, INT	SRI II	23–47 [14–28]	2.1–3.5
AND, INT	SRI I	23–47 [14–28]	5.4–8.4
BGI	Respirable	3.67, 7.00 [2.2, 4.2]	4.0
BGI	PM 1.0 Sharp Cut	27.8 [16.7]	1.0
BGI	PM 2.5 Sharp Cut	27.8 [16.7]	2.5
BGI	GK 2.05 (KTL)	6.67 [4.0]	2.5
BGI	Triplex (SCC 1.062)	5.83, 2.5, 1.75 [3.5, 1.5, 1.05]	1.0, 1.5, 4.0
BGI, GMW	Aerotec 2	583–833 [350–500]	2.5–4.0
—	Aerotec 3/4	37–92 [22–55]	1.0–5.0
DEK	Dekati Cyclone	17 [10]	10
MSA, SEN, SKC	10 mm Cyclone (also called Dorr-Oliver)	1.5–8.3 [0.9–5]	1.8–7.0
INT	STR	1.7–100 [1–60]	0.3–10
SEN	1/2″ HASL	13–17 [8–10]	2–5
SEN	1″ HASL	108–583 [65–350]	1.0–5.0
—	BK 76	670–1830 [400–1100]	1.0–3.0
—	BK-152	1920–4500 [1150–2700]	2.0–5.0
SKC	Aluminum	4.17 [2.5]	4.0
SKC	GS	4.58 [2.75]	4.0
URG	Sharp Cut-point	27.8 [16.7]	1.0
URG	Sharp Cut-point	5, 17, 27.8 [3, 10, 16.7]	2.5
URG	Sharp Cut-point	47.2 [28]	3.5
URG	Sharp Cut-point	27.8, 47.2 [16.7, 28.3]	10

Note that in the Stokes number, the three parameters discussed in the above example (U, d_p, and d_b) are included as well as properties of the gas and particles (η, C_c, and ρ_p). The Stokes number is important in all types of inertial collectors and not just body collectors. The Stokes number for a conventional impactor or a virtual impactor is the ratio of the stopping distance to the radius of a circular nozzle, or half width of a rectangular nozzle, and is defined as

$$Stk = \frac{\rho_p C_c d_p^2 U}{9\eta W} \tag{10–2}$$

where

U = average air (or gas) velocity at the nozzle exit = $Q/\pi(W/2)^2$ (round nozzle impactor) or = Q/LW (rectangular nozzle impactor)
W = nozzle diameter (circular impactor) or nozzle width (rectangular impactor)
Q = volumetric flow rate through the nozzle
L = rectangular nozzle length

The Stokes number is a dimensionless parameter that can be used to predict whether or not a particle will impact on a body, an impaction plate or in the collection probe of a virtual impactor, or will follow the air streamlines out of the impaction region and remain airborne. Actually, the square root of Stokes number, $\sqrt{Stk}$, is more commonly used because it is a dimensionless particle size. A critical value of $\sqrt{Stk}$, often used to characterize inertial classifiers, is $\sqrt{Stk_{50}}$. This is the value of $\sqrt{Stk}$ corresponding to d_{50}, the value of d_p at which particles are collected with 50% efficiency. Thus, if the value of $\sqrt{Stk_{50}}$ is known, the value of d_{50}, corresponding to the cut size of the impactor, can be found from

$$d_{50} = \sqrt{\frac{18\eta d_b}{\rho_p C_c U}}\sqrt{Stk_{50}} \tag{10–3}$$

for body impactors, and

$$d_{50} = \sqrt{\frac{9\eta W}{\rho_p C_c U}}\sqrt{Stk_{50}} \tag{10–4}$$

for conventional and virtual impactors.

General Description

Inertial classifiers have been widely used for the separation of particles by their aerodynamic diameters where the aerodynamic diameter is the diameter of a standard density (1000 kg/m^3 [1 g/cm^3]) sphere that has the same gravitational settling velocity as the particle in question. Four types of inertial classifiers in common use, as shown in Figure 10–1, are body impactors, conventional impactors, virtual impactors, and cyclones. The first of these, the body impactor, is the simplest in that it consists of only a body in a moving aerosol stream onto which particles impact. The latter three of the inertial classifiers all consist of a jet of gas impinging on a target.

A conventional impactor, in its simplest form, consists of a jet of particle-laden gas impinging on a flat plate with particles impacting on the plate. Variations of the impactor include the use of either round or rectangular nozzles, single or multiple nozzles, and flat or cylindrical impaction plates.

In a virtual impactor, the impaction plate is replaced by a collection probe slightly larger than the nozzle, with the classified particles penetrating into the collection probe. A small fraction of the gas passes through the collection probe to transport the classified particles out the lower end of the probe. The remainder of the gas, the major portion, reverses direction in the collection probe and escapes at the upper edge.

In the cyclone, the aerosol stream is drawn through the inlet and impinges tangentially on the inner surface of a cylinder, flows in a spiral pattern down the inside of the cylinder and cone walls, reverses direction, spirals upward around the cyclone axis, and exits through a centrally located tube at the upper end of the cylinder. Particles are collected on the

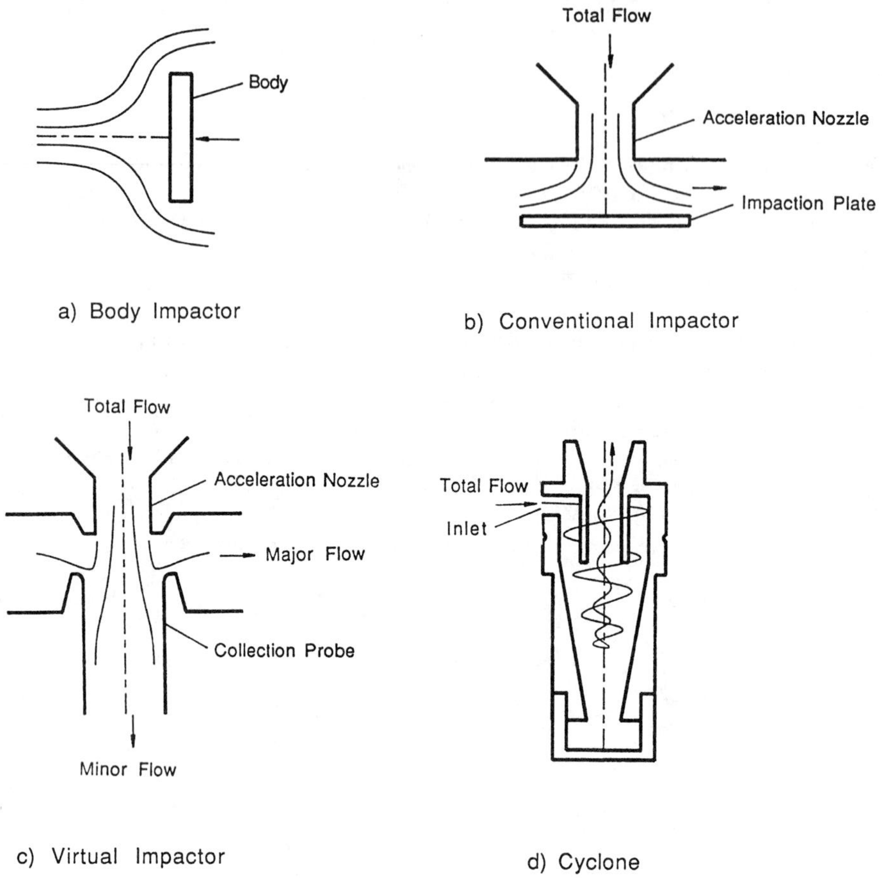

Fig. 10–1. Four types of inertial classifiers.

cylindrical and conical walls by inertial forces on the particles. Clumps of impacted particles that are knocked off or dropped from the cyclone walls tend to fall to the apex of the cone where they are collected in a cup, sometimes called a *dust cap* or *grit pot*.

Another inertial device, although not a particle classifier in the traditional sense, is the aerodynamic focusing lens, which consists of a series of small sharp-edged orifices on one axis (de Juan and Fernandez de la Mora, 1998; Liu et al., 1995a,b; Fernandez de la Mora and Riesco-Chueca, 1988). The focusing lens operates under the principle that the flow streamlines are radially inward as the air approaches the inlet to a sharp-edged orifice. The inertia of the particles will cause the particles to cross the streamlines and be closer to the center line of the flow when the flow exits the orifice than when it entered. By passing the flow through a series of these orifices of decreasing size, the particles will exit the lens with the particles focused near the center line. These lenses are described in detail in Chapter 13.

Conventional Impactors. The most common type of impactor consists of a single jet of particle-laden gas (aerosol) impinging on a flat plate, as shown in Figure 10–2. Particles larger than the cut size of the impactor will slip across the streamlines and impact on the plate, while smaller particles will follow the streamlines and not be collected. The most important

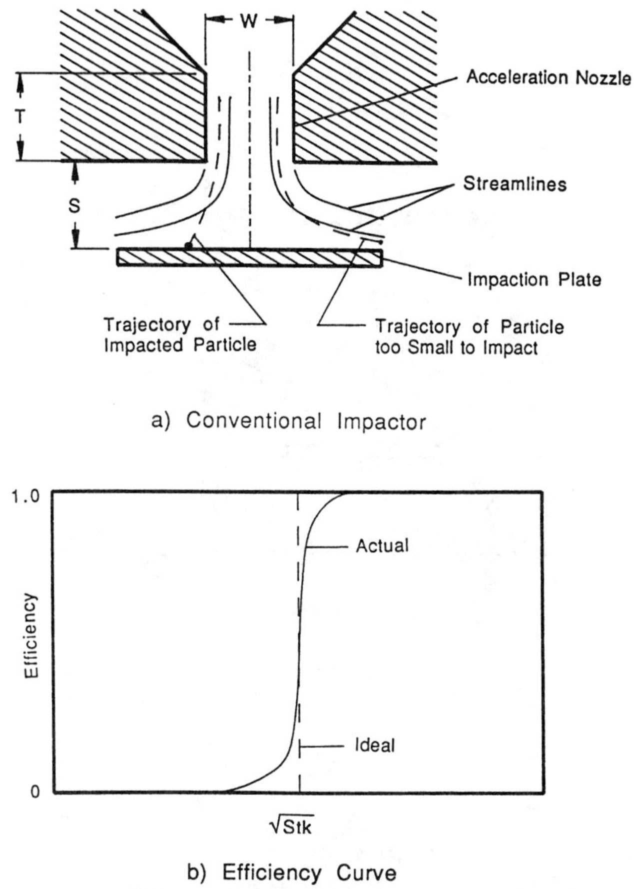

Fig. 10–2. Schematic diagram of a conventional impactor and corresponding particle collection efficiency curve.

impactor characteristic is the collection efficiency curve, also shown in Figure 10–2. The collection efficiency (as a function of particle size) is defined as the fraction of particles passing through the nozzle that are collected on the impaction plate. The ideal impactor has a perfectly sharp efficiency curve, that is, all particles larger than the cut size of the impactor are collected on the plate, while all smaller particles follow the gas flow out of the impaction region.

A nozzle and an impaction plate constitute a single-stage impactor that is useful when classifying particles into two size fractions. For example, this is the case for analyzing particles that are less than 10 μm (PM-10) or less than 2.5 μm (PM-2.5) (see Chapter 29). In these types of impactors, the particles larger than the cut size are removed from the air stream, while the smaller particles penetrate the impactor stage to be either collected on a filter, where they can be analyzed (e.g., for mass concentration or elemental composition), or passed into some other instrument for real-time mass or number concentration measurement.

Often, it is desirable to determine the entire size distribution of the aerosol and not just the quantity less than a certain size. In this case, a series of impactor stages are used in a cascaded fashion such that the gas passes from one stage to the next, as shown in Figure 10–3,

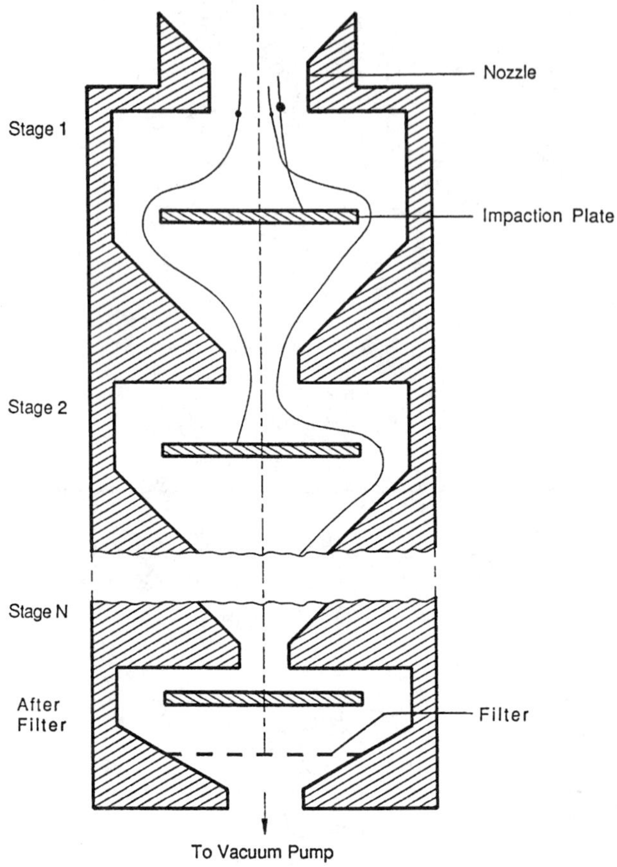

Fig. 10–3. Schematic diagram of a cascade impactor.

to remove particles in discrete size ranges (Lodge and Chan, 1986). This is known as a *cascade impactor* and is widely used for determining size distributions of aerosols.

A cascade impactor makes use of the fact that particle collection is governed by the Stokes number. The velocity of the particle-laden gas stream is increased in successive stages, resulting in the collection of successively smaller particles in the subsequent stages. For example, if a four-stage cascade impactor has cut-sizes of 10, 5, 2.5, and 1.25 μm, the first stage will collect particles larger than 10 μm, the second stage will collect particles between 5 and 10 μm, the third stage between 2.5 and 5 μm, and the fourth stage between 1.25 and 2.5 μm. Particles less than 1.25 μm penetrate the final stage of the impactor and can be collected on an after-filter.

The particle deposits on the impaction plates can be evaluated by a variety of methods. A few of the more common methods are (1) the particles are collected on glass plates, membrane filters, or foils and are inspected or counted under a microscope; (2) the particles are collected on foils and weighed to determine the mass of particles at each stage; (3) the particles are collected on quartz crystals, and the mass of particles is determined by the change in the natural frequency of the crystals (Fairchild and Wheat, 1984; see also Chapter 14); or (4) the particles are charged before passing through the impactor, and the current is measured at each impaction plate to determine the number of particles being collected

(Keskinen et al., 1992; see Chapter 14). The first two methods provide size distribution data integrated over time, while the latter two methods provide size distribution data in near real time.

The uncertainties in the particle size distribution are the size of the largest particles collected on the first stage and the size of the smallest particles collected on the after-filter. These sizes may be estimated, or, better, an impactor may be selected with a sufficient number of stages to span the entire size distribution of interest so that the mass collected on the first stage and the after-filter is minimized.

As stated above, many cascade impactor designs have been built and tested. The calibration curves for no two designs will be exactly alike due to differences in design parameters (e.g., nozzle diameters, number of nozzles, sampling flow rates) and to small influences of the boundary conditions on particle collection. Figure 10–4 shows a set of typical cascade impactor efficiency curves. Some impactors may have sharper cut-off characteristics and some not as sharp, but the general shape will be very similar to those shown.

Analysis of Eq. 10–4 reveals that the particle size range of an impactor can be lowered to very small sizes by either increasing the value of the slip correction, C_c, (i.e., by going to low pressures in the impactor), or by decreasing the nozzle diameter, W. Impactors utilizing these techniques are known as either *low-pressure impactors* or *micro-orifice impactors*. Both types have been developed and used successfully in sampling particles down to approximately 0.05 μm diameter (Berner et al., 1979; Hering et al., 1978, 1979; Hering and Marple, 1986; Hillamo and Kauppinen, 1991; Marple et al., 1991).

There are several differences that must be considered when selecting either low-pressure or micro-orifice impactors for a particle sampling program. In a low-pressure impactor, the particles are collected by increasing the value of the slip correction at highly reduced pres-

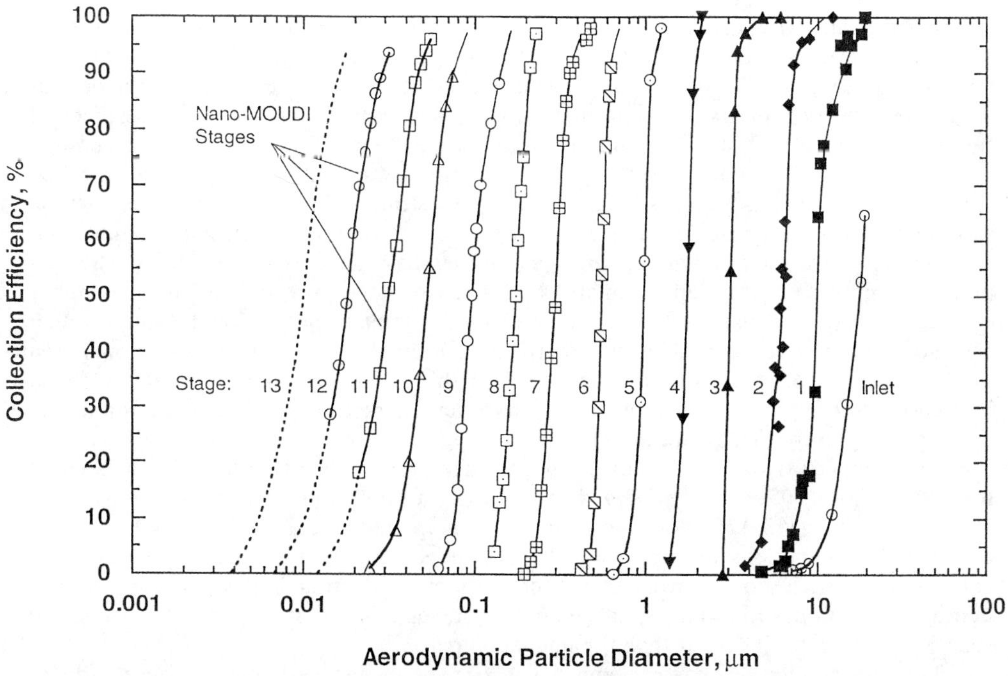

Fig. 10–4. Particle collection efficiency curves for the micro-orifice uniform deposit impactor (MOUDI), including nano-MOUDI stages.

sures down to approximately 3×10^3 Pa [0.03 atm]. This means that the vacuum pump drawing the flow through the impactor must be rather large or the flow rate rather small. In addition, particles that are sensitive to evaporation at low pressures experience a reduction in size during the collection process (Biswas et al., 1987). However, the low-pressure impactor is relatively simple to construct because the nozzle diameters are similar in size to those used in conventional impactors.

In a micro-orifice impactor, the pressures are substantially larger than in low-pressure impactors, and conventional vacuum pumps can be used to obtain moderate flow rates (Marple et al., 1991). Volatile particles can be more easily collected in this impactor because the pressure drop through the entire impactor is only about 4×10^4 Pa [0.4 atm] (Fang et al., 1991). The difficulty with the micro-orifice impactor is in its construction because the nozzle diameters are very small (approximately 50 µm for the final stages), and the number of nozzles is large to obtain an adequate flow rate (as many as 2000 for the final stages). However, these are manufacturing problems and not problems in the use of the impactor.

A special class of impactors is one that makes use of both low pressures and micro-orifices to achieve smaller cut sizes than can be obtained by either low pressures or micro-orifices alone. For example, by using micro-orifices, the lower limit of the cut size is about 0.05 µm. However, by using a combination of micro-orifices and low pressure, the cut size of the impactor can be pushed to even smaller sizes. The lower three efficiency curves shown in Figure 10–4 are from three such stages. Because the pressure is dropping at each stage of a cascade impactor, the volumetric flow rate increases at each stage. Therefore, at some point in the cascade impactor, it may be necessary to reduce the flow. In the example shown in Figure 10–4, the flow in the three lower stages, called *nanostages*, has been reduced to 1.7×10^{-4} m^3/s [10 L/min] from 5×10^{-4} m^3/s [30 L/min] in the upper stages (Marple and Olson, 1999). This is accomplished by bleeding 3.3×10^{-4} m^3/s [20 L/min] from the cascade impactor at the stage where the flow is reduced.

Impactors are normally designed to have sharp cut-off characteristics (steep efficiency curves). However, in some cases, it is desirable to have a cut-off curve (efficiency curve) that is not sharp but rather one that follows a particular retention curve. Such is the case in designing impactors with penetration curves matching the American Conference of Governmental Industrial Hygienists (ACGIH) or the British Medical Research Council (BMRC) respirable mass criteria curves (Lippman, 1989) (see Chapter 25). A special class of impactors, called *respirable impactors*, has been designed to emulate penetration characteristics similar to these respirable curves (Marple, 1978; Marple and McCormack, 1983). This is accomplished by multiple-nozzle, single-stage impactors with the nozzles having different diameters. The respirable penetration curve is approximated by a series of steps corresponding to the number of different diameters used in the impactor stage. The fraction of the flow passing through each set of nozzles of a particular diameter is proportional to the total cross-sectional area of the nozzles of that particular diameter. With this technology, it is possible to design impactors for nearly any flow rate, with penetration characteristics that approximate any monotonically decreasing penetration curve.

Inertial impactors can be used over a wide range of conditions and have been designed for cut sizes of 0.005 µm (Fernandez de la Mora et al., 1990) up to approximately 50 µm (Vanderpool et al., 1987) and flow rates from a few cm^3/min to several thousand m^3/min. There are, however, limitations to impactors when considering their use. Three major areas of concern are particle bounce from the collection surface, overloading of collected particle deposits on the impaction plates, and interstage losses (collection of particles on internal surfaces of the impactor other than the impaction plate).

Of the three limitations, particle bounce has received the most attention because particles bouncing from an impaction plate will be collected on a subsequent stage that has a smaller cut point or on the after-filter and, thus, bias the measured size distribution or become lodged in and clog smaller nozzles of subsequent stages. The most logical approach to solving a particle bounce problem, and the technique used by many researchers, is to provide a sticky surface on the impaction plate. The mass stability and chemical composition, purity, and stability of the sticky surface are all factors in selecting the appropriate substance. Numerous types of greases and oils, including petroleum jelly, Apiezon greases (***API***)* and silicone oils and sprays, have been used to coat the impaction plates; however, these coatings must be used correctly in order for the results to be satisfactory (Hering, 1995; Rao and Whitby, 1978a,b). For example, if the impaction plate has just a sticky surface, once a monolayer of collected particles forms on the surface, additional incoming particles will impact on the particles that have already been collected, and particle bounce is again possible. Therefore, if a sticky surface is to be used and a large quantity of particles collected, the sticky surface will not be adequate; a sticky substance must be used that will wick up through the particle deposit by capillary action and continually provide incoming particles with a sticky surface. As described in a later section, silicone oil works well.

The techniques that will be used for particle analysis must be considered when selecting an impaction surface. If the particles collected on the impaction plate are to be discarded, the impaction plate can be a porous material saturated with a light oil (Reischl and John, 1978). This oil will wick up through the deposit and continually provide the incoming particles with a sticky impaction surface. Because these particles are to be discarded, the weight stability of the oil is not important.

If the mass size distribution for solid particles is to be determined from the deposits collected on the stages of a cascade impactor, great care must be taken in providing a sticky impaction plate surface. Because the deposits must be weighed, particle collection substrates such as metal foils, plastic films, or filters are placed on the impaction plates so that they can be easily removed for weighing. Grease, oil, or other sticky substances are applied to the substrates. The sticky substance must have mass stability because treated substrates will be weighed on a balance before and after sampling. In most cases, the sticky substance will be dissolved into a solvent and applied in a light film to the substrates. The treated substrates are then baked in an oven to drive off the volatiles, thereby leaving weight-stable sticky substrates onto which the particles can be collected.

The problem of deposit overloading on an impaction plate is easier to control because the quantity of particles collected is a function of sampling time. It is normal practice to sample from several minutes to several hours to obtain an initial size distribution. The deposits on the impaction plates are then inspected, and, if any stage is overloaded, the next sampling period is shortened. Conversely, if the deposits are too light, the sampling period is extended. The major difficulty in determining the appropriate sampling time is dictated by the total quantity of particles that can be collected on an impaction plate (before overloading occurs). This quantity varies from impactor to impactor, and only experience or data from the manufacturer can provide this information.

The final area of concern is the collection of particles on surfaces other than the impaction plate. These particles are deposited on the interior surfaces of the impactor (interstage loss) and are a function of particle size, again distorting the measured size distribution. In the upper stages, where the collected particles are rather large, particles can be lost by impaction or turbulent deposition. However, because each stage removes particles larger than the cut size of that stage, the interstage losses rapidly decrease as particles penetrate the upper stages. In the lower stages, particles can be lost by diffusion (a problem only for very small particles). Most of the diffusional losses occur in the final stages, where the nozzles are small and particles can diffuse to the nozzle walls.

In a properly designed cascade impactor that covers a wide range of particle sizes, one would expect to find interstage losses in the initial stages and maybe in the final stages of the

* See Appendix I for full manufacturer addresses referenced to the italicized three-letter codes.

impactor (if there are cut sizes below 0.1 μm) with the losses being a minimum through the intermediate stages. Typically, particle loss varies but may be as high as 30% to 40% for large particles in the inlet and first stages.

If experiments have been performed to determine the interstage loss as a function of particle size, the particle size distribution data of the impactor can be corrected. One difficulty in estimating what this correction should be is that the losses will vary as a function of the nature of the particles. For example, if the particles are liquid or sticky, they will adhere to any surface with which they come in contact. A dry particle, however, may rebound when it hits a surface and remain airborne until it reaches the next stage. Thus, interstage losses are less for a particle that bounces easily than for a sticky particle.

In summary, the three major concerns for an impactor, that is, particle bounce, overloading, and interstage losses, are a function of both the particle properties and the impactor. If the particles are sticky, particle bounce is minimized, interstage losses are maximized, and overloading is of less concern. If the particles are solid and rebound easily from a surface, particle bounce is a problem, interstage losses are decreased, and overloading is of concern.

Normally a cascade impactor consisting of several stages is used to determine particle size distributions. The inertial spectrometer (INSPEC) and high-flow spectrometer (LASPEC) were developed to provide particle size distribution information on one stage (Belosi and Prodi, 1987; Prodi et al., 1979, 1983, 1984). The aerosol sample flow rates are up to $3.3 \times 10^{-6}\,m^3/s$ [0.2 L/min] and $1.7 \times 10^{-4}\,m^3/s$ [10 L/min] in the INSPEC and LASPEC, respectively. These devices employ a rectangular filter instead of an impaction plate and introduce the particles through a nozzle near one edge of the filter, as shown in Figure 10–5. The particles are introduced at the center of this nozzle with a sheath of clean air on both sides. The inertial properties of the particles distributes the particles in the sheath air at distances that are a function of the particles' aerodynamic diameters. Subsequently, as the particles are removed

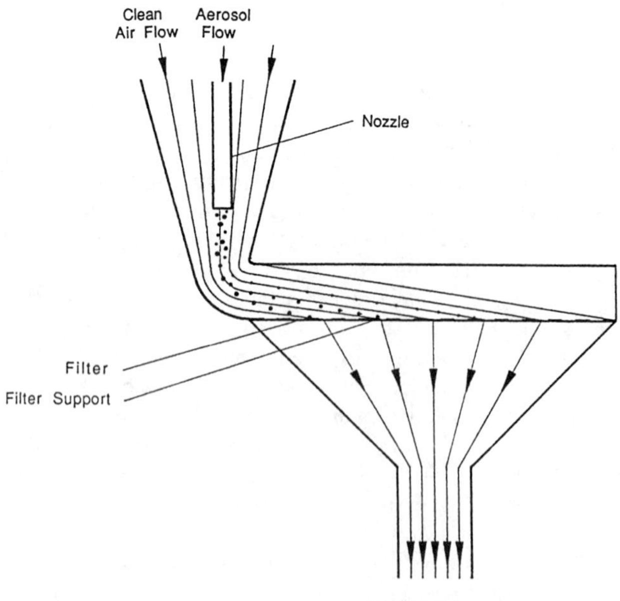

Fig. 10–5. Schematic diagram of an inertial spectrometer.

from the gas stream at the combination filter–impaction plate, the position of each particle on the filter is a function of the aerodynamic diameter of the particle. This inertial classifier provides the entire size distribution differentiated by size on the filter.

It is also possible to design a spectrometer with the nozzle being at the center of a circular filter. The radial deposition of particles on the filter is a function of particle size, with the larger particles being deposited closer to the center than the smaller particles. This has been developed in a personal sampler, called a *personal inertial spectrometer* (PERSPEC), operating at a flow rate of $3.3 \times 10^{-5}\,m^3/s$ [2 L/min] (Prodi et al., 1988).

Virtual Impactors. A virtual impactor is a particle inertial classification device that is very similar to the conventional inertial impactor, with the primary difference being that the impaction plate has been replaced by a collection probe as shown in Figure 10–6. The jet of particle-laden gas exiting the nozzle penetrates into the collection probe where classification occurs. Large particles penetrate further into the collection probe than do small particles. A small fraction of the flow, the minor flow, is allowed to penetrate through the collection probe carrying with it the particles larger than the cut size of the virtual impactor. Most of

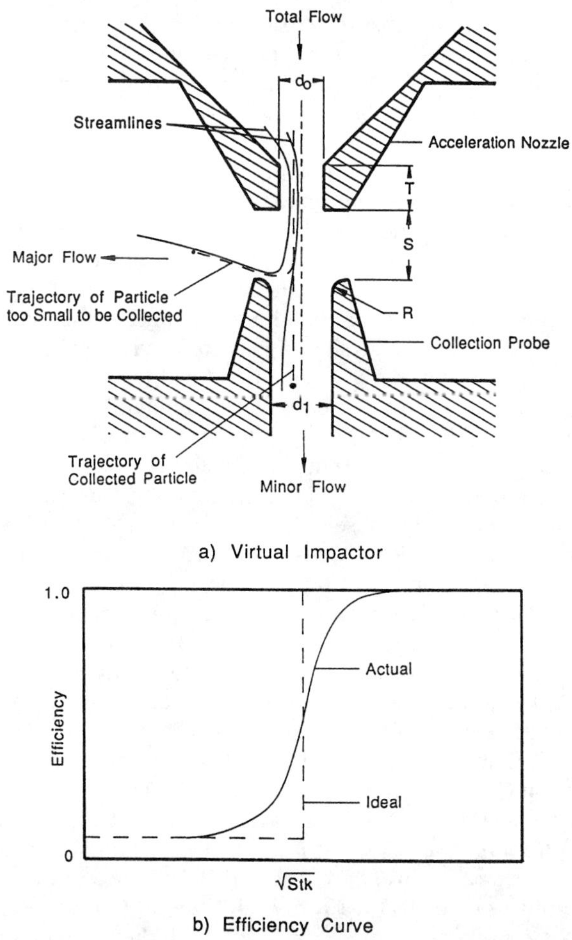

Fig. 10–6. Schematic diagram of a virtual impactor and corresponding particle collection efficiency curve.

the flow, the major flow, reverses in the collection probe and exits at the top of the collection probe, carrying with it the particles that are smaller than the cut size. Thus, both size fractions remain suspended in the gas. These two gas streams can then be directed to collection filters, into another inertial classification device, into another impactor stage, or into an instrument that automatically measures the concentration of particles in the gas streams. Conner (1966) described the first device to incorporate a virtual impactor with a controlled minor flow.

The particle collection efficiency curves for virtual impactors, as for conventional impactors, are quite sharp, as was shown in the theoretical study by Marple and Chien (1980). The major difference between the efficiency curves of virtual and conventional impactors is that particles less than the cut size of the virtual impactor remain in both the major and minor flows. Therefore, the collection efficiency curve of particles in the collection probe asymptotically approaches the percentage of total flow that is penetrating the collection probe. For example, if the minor flow is 10% of the total flow, 10% of the particles smaller than the cut size will remain in the minor flow and "contaminate" the large-particle fraction.

One method of eliminating the small particles in the large-particle fraction is to provide a central core of clean filtered air in the nozzle of the virtual impactor (Masuda et al., 1978; Chen et al., 1986). Because the air flow in the central portion of the nozzle is the air that constitutes the minor flow, the small particles will not be present in the minor flow. The major difficulty with this solution is that the virtual impactor becomes more complicated with the addition of the clean airflow. If some small particles in the large-particle fraction are not of concern, the contamination can be minimized by reducing the minor flow to the lowest value possible. In some cases this has been reduced to as low as 0.1% of the total flow (Xu, 1991).

Another method for determining the quantity of contamination by the small particles is their direct measurement. This is accomplished by extracting a sample from the major flow at the same flow rate as the minor flow (Marple and Olson, 1995). The minor flow sample is analyzed in the exact same manner as the minor flow sample and the results subtracted from the results of the minor flow sample. In this manner, contribution of the small-particle contamination in the minor flow can be quantified. For example, if the mass of particles in the minor flow is determined by filter analysis, the mass of particles taken from the major flow at the same flow rate and on the same filter material will indicate the mass of particles on the minor flow filter attributable to small-particle contamination.

Besides being an inertial classifier, a virtual impactor can also be considered a particle concentrator for particles larger than the cut size. This is very useful when sampling particles that are of low concentration or sampling particles into an instrument that requires concentrations greater than are normally present (Keskinen et al., 1987; Liebhaber et al., 1991; Marple et al., 1989; Wu et al., 1989). Because the particles larger than the cut size of the impactor are concentrated in the minor flow, the concentration factor is equal to the ratio of the total flow rate to the minor flow rate. For example, if the minor flow is 5% of the total flow, the concentrating factor is 20.

Virtual impactors have also been used to concentrate ambient particles for exposure studies. One such impactor uses rectangular nozzle and receiving tube geometry to concentrate particles in the size range of 0.15 to 2.5 μm by a factor of about 25 at a total inlet flow rate of 0.083 m^3/s [5000 L/min] (Sioutas et al., 1997). To design a virtual impactor that concentrates particles at an even smaller size, particles can be grown by condensing a vapor on the particles before passing them through the virtual impactor. After virtual impaction, the liquid is evaporated from the particles, restoring them to their original small sizes. This has been employed to concentrate ultrafine particles with a concentration factor of approximately 25 at an inlet flow rate of 1.77×10^{-3} m^3/s [106 L/min] (Sioutas et al., 1999).

Interstage particle loss is a major area of concern with the virtual impactor. These losses normally occur at the upper edge of the collection probe or on the backside of the nozzle plate. The losses are a maximum for particle sizes corresponding to the cut point of the impactor and can reach values as high as 60% if the virtual impactor is improperly designed or operated (Marple and Chien, 1980). Major factors that influence these losses include the ratio of the collection probe diameter to the nozzle diameter, the shape of the collection probe inlet, the alignment of the axes of the nozzle and collection probe, the shape of the nozzle protruding through the nozzle plate, the jet Reynolds number, and the minor flow percentage. Several investigators have studied these parameters, and the optimum values are still being refined (Jaenicke and Blifford, 1974; McFarland et al., 1978; Loo, 1981; Chen et al., 1985, 1986; Chen and Yeh, 1987; Loo and Cork, 1988; Xu, 1991). In general, however, the axes of the nozzle and collection probe must be aligned as close as possible, the collection probe diameter should be from 30% to 40% larger than the nozzle diameter, the inlet to the collection probe should be a smooth radius, the nozzle should protrude through the nozzle plate approximately two to three nozzle diameters, and the minor flow should be 5% to 15% of the total flow. Initially, nearly all virtual impactors use round nozzles and collection probes. However, attempts have been made to use rectangular configurations have been studied (Forney, 1976; Forney et al., 1982) and successfully applied to particle concentration of atmospheric particles (Sioutas et al., 1997).

Most virtual impactor designs have been single-stage units with one nozzle and one collection probe. However, some virtual impactor samplers have been designed with multiple nozzles and collection probes in a single stage. The purpose for this design feature is to reduce the size of the sampler and to reduce the pressure drop through the sampler (Marple et al., 1990; Szymanski and Liu, 1989).

Body Impactors. The major problem in designing a conventional impactor for classifying large particles is that of drawing large particles into a sampling inlet. However, large particles can be classified inertially by the use of body impactors that do not involve drawing particles through an inlet. As shown in Figure 10–7, the body impactor functions by simply sweeping an impaction surface (body) through the air or, conversely, drawing air past the body. In either case, the problems associated with the sampling of large particles is reduced.

The collection efficiency of the body impactor is defined as the fraction of particles that are impacted on the body from a volume of air swept out by the body (Fig. 10–7). Particle collection curves are presented in Figure 10–7 for particle deposition on ribbons, spheres, and cylinders. Golovin and Putnam (1962) and May and Clifford (1967) summarize particle impaction on bodies of various shapes. The cut size is a function of the body size (dimension normal to the flow direction) and the relative velocity between the air and body. The cut size is governed by the Stokes number equation (Eq. 10–1) where d_b is the body size and U is the velocity difference between the air and the body.

Examples of two body impactors that have been developed are the rotorod sampler and the Noll impactor. The rotorod sampler (***MUL***) is used for sampling particles larger than 15 μm, and the most common application is the sampling of pollen levels in outdoor air. The Noll impactor is designed for sampling coarse atmospheric particles. It has four rotating plates of various dimensions to collect particles from 6 to 29 μm (Noll, 1970; Noll et al., 1985).

Cyclones. In a cyclone, a jet of air impinges tangentially on the inner surface of a cylinder and then swirls downward in a cyclonic fashion inside the cylinder and into a conical section. In the conical section the air reverses direction and spirals upward around the cyclone axis to an exit tube at the upper end of the cylinder. Particles larger than the cut size are deposited on the inner surface of the cylinder and in the cone. The particles fall downward into the apex of the cone and into a dust collection cavity (grit pot).

Many cyclones have been designed in a variety of sizes for numerous applications and are very popular for collecting dust in industrial process lines. These units are normally large

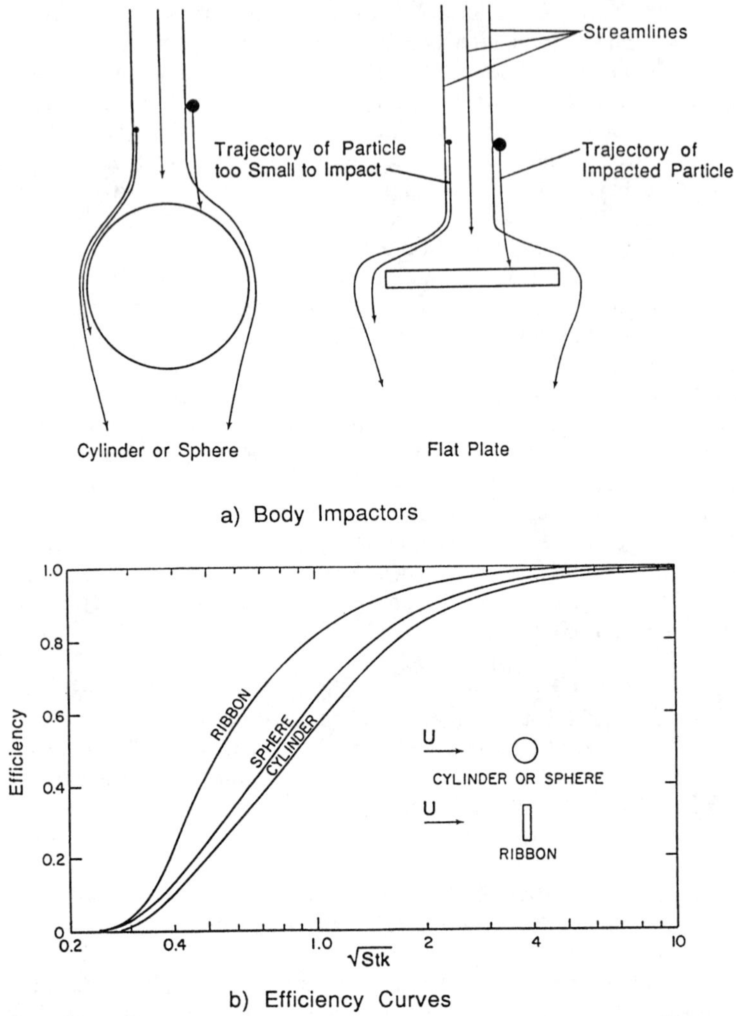

Fig. 10–7. Schematic diagrams of body impactors and corresponding particle collection efficiency curves from Golovin and Putnam (1962).

and have high flow rates. Cyclones are also popular for aerosol sampling. While a variety of cyclones have been developed for respirable dust samplers, the most popular cyclone used in the United States is the 10 mm nylon cyclone, which has penetration characteristics that simulate the ACGIH respirable mass criteria (Lippmann, 1995).

The cut size of the cyclone is governed by the flow rate, the sizes of the inlet and outlet tubes, and the size of the cylinder (e.g., Hering, 1995). A rigorous theoretical analysis of a cyclone is more difficult than that of an impactor because the flow is three-dimensional and must be analyzed using a three-dimensional numerical program. Although some numerical work has been performed by the authors and others, it is not common practice to apply this technique in the design of cyclones.

Most cyclones are used as a single stage; however, there have been cascaded versions developed (Smith et al., 1979; Liu and Rubow, 1984). The unit developed by Smith et al. (1979) for stack sampling consists of five stages with cut sizes ranging from 0.32 to 5.4 μm at a sampling flow rate of $4.7 \times 10^{-4}\,m^3/s$ [28.3 L/min].

Particle deposition in a cyclone is caused by the cyclonic action of the fluid in the cylinder. Most cyclones achieve this by the jet of gas impinging tangentially on the inner surface of a cylinder. However, it is possible to achieve the cyclonic spiral flow by turning vanes at the inlet of a tube. A cascade version of this type of cyclone has been developed for collecting dust particles in a size range from 1 to 12 μm (Liu and Rubow, 1984). Although the construction of this cyclone is more complicated than the conventional cyclone, it does lend itself to a compact cascade design. A simple model of axial cyclone performance was developed that approximately predicts the cyclone cutpoint (Maynard, 2000). The conventional cyclone configuration, where the air enters tangentially and exits along its axis, makes it difficult to design a compact cascade cyclone sampler.

Measurement Strategies

Before selecting an inertial classifier, one must first decide the purpose for which the particles are being collected. If the purpose is to determine the characteristics of aerosols as a function of size, such as obtaining the mass size distribution or the chemical composition of the aerosol at various particle sizes, it is most convenient to use a cascade sampler. If only the quantity of particles less than a specified size is desired, such as is often the case for compliance sampling, then a two-stage sampler consisting of an inertial classifier followed by a filtration stage is most convenient.

A decision must also be made as to whether or not the sample needs be time resolved or can be an integrated sample. Normally, when samples are taken with an inertial classifier, the sample is integrated over the time period. Some impactors provide time-resolved samples on rotating impaction plates, for example, the Lundgren cascade impactor (Lundgren, 1971) and the Davis Rotating-drum Universal-size-cut Monitoring (DRUM) impactor (Raabe et al., 1988).

Another important factor is the particle size range over which one needs to operate. Although inertial classifiers can classify particles in the size range of about 0.005 to 50 μm, a more precise size range must be known in order to select the device best suited for the test. In some cases, the size distribution of the aerosol may not be known until after the first sample has been taken. If little is known about the size distribution before the first test, a wide-range cascade sampler provides the most information.

Finally, the analysis technique to be employed for inspection of the deposits will influence the type of sampler used for particle collection. For example, if the mass size distribution is to be determined by gravimetric analysis of the collected samples, an impactor with substrates coated with a sticky surface can be used. However, the substrates and sticky surfaces must have good mass stability. If a particle sample is to be analyzed with scanning electron microscopy (SEM), then a sticky surface is not desirable. In this case, a method must be devised whereby the particles can be collected on a dry surface. When sticky coatings cannot be used to reduce particle bounce, it is desirable to have the cut sizes of a cascade impactor as close together as conveniently possible so that the inertia of particles impacting on the impaction plates are kept as small as possible. It has also been found that submicrometer particles do not appear to bounce as easily as supermicrometer particles. Therefore, if a sticky coating cannot be used, a greased 1 μm cut stage should follow an ungreased 1 μm stage to collect any supermicrometer particles that may have bounced to this point in the impactor. Thus, particles penetrating the 1 μm cut stage (the submicrometer particles) will not be contaminated with any supermicrometer particles that may have bounced through the upper stages.

Design Considerations

Of the various types of inertial classifiers discussed in the previous sections, impactors are most likely to be designed for specific applications. References have been provided in previous sections for the basic designs of other inertial samplers. In some cases the sampling criteria may be such that a suitable commercially available impactor does not exist. This may be for reasons of size, configuration, number of stages, or the particular cut sizes desired. For these situations it is recommended that a special impactor be designed and built for the study following a few simple guidelines (Marple and Rubow, 1986).

Inertial impactors have been studied extensively through theoretical analysis by numerical methods (Marple, 1970; Marple and Liu, 1974; Rader and Marple, 1985). From these studies, numerous particle collection efficiency curves have been calculated and reported in the literature. In most cases, the theories have been compared with experimental results with good agreement. For the most part, an impactor's efficiency curve can be determined with as much accuracy from theoretical analysis as it can from experimental calibration.

To operate correctly, two simple design guidelines must be followed. These guidelines involve two dimensionless parameters: jet-to-plate distance divided by the nozzle diameter (S/W) and jet Reynolds number (Re_j) (Marple, 1970; Marple and Liu, 1974; Marple and Willeke, 1976a,b; Rader and Marple, 1985). These guidelines were obtained from a theoretical, numerical analysis of the Navier-Stokes equations to determine the flow field and the subsequent numerical integration of particle trajectories. This process was used to determine guidelines necessary to obtain sharp collection efficiency curves. These theoretical guidelines have been used in the design of impactors and have been shown to result in impactors with collection efficiency curves that have sharp cut-off characteristics (Marple et al., 1988).

Figure 10–8 shows the theoretically determined efficiency curves as a function of jet-to-plate distance divided by the nozzle diameter (S/W) and jet Reynolds number (Re_j) of the flow. Both of these terms are important in that they will influence the efficiency curve and the 50% cut point of the impactor. For example, the position of the efficiency curve as a function of the S/W ratio is shown to be sensitive for small values of S/W. The values of Stk_{50} are relatively constant for S/W values larger than 0.5 and 1.0 for round and rectangular impactors, respectively. The S/W value should not be less than this value because small variations in S/W will then change the cut size of the impactor. To provide a margin of safety, it is recommended that the S/W value be greater than 1.0 for round impactors and 1.5 for rectangular impactors. The upper value of S/W that can be used in an impactor is not as well known. However, impactors have operated well with S/W values as high as 5 or 10. If values larger than these are to be used, a calibration is recommended to ensure that the jet has not dissipated before impinging on the plate.

The importance of the jet Reynolds number is more related to the sharpness of cut than to the cut size of the impactor. If the Reynolds number is low, the gas viscous forces will be large and the velocity profile will be parabolic at the nozzle exit and in the air jet as it approaches the impaction plate. This will enhance the collection of particles near the center line, where the smallest particles will be collected. The low velocity near the nozzle wall will require the particles in this portion of the flow to be larger in order for collection to occur. Therefore, the result of an impactor operating at a low Reynolds number is a less sharp collection efficiency curve. Theoretical analysis has shown, and experiments have verified, that the efficiency curve will be its sharpest if the Reynolds number is kept in the 500 to 3000 range for both round and rectangular impactors. The Re_j is expressed as

$$Re_j = \frac{\rho_g W U}{\eta} \text{ (round)} \qquad (10\text{–}5)$$

$$Re_j = \frac{2\rho_g W U}{\eta} \text{ (rectangular)} \qquad (10\text{–}6)$$

where ρ_g is the gas density. Example 10–1 shows the calculation of the nozzle diameter and Reynolds number for an impactor.

EXAMPLE 10–1

Calculate the nozzle diameter (W) and the jet Reynolds number (Re_j) for a 10 µm cut point impactor. The impactor has one round nozzle, and the flow rate is $5 \times 10^{-4} m^3/s$ [30 L/min]. Assume normal temperature and pressure.

Answer: The first step is to determine the required nozzle diameter for a cut point of 10 µm. Substituting the flow rate for the average jet velocity

$$U = \frac{4Q}{\pi W^2}$$

into the Stokes equation (Eq. 10–2) for a round jet impactor gives

$$Stk = \frac{4\rho_p C_c d_p^2 Q}{9\pi\eta W^3}$$

where

ρ_p = particle density (1000 kg/m³ [1 g/cm³])
d_p = cut point particle diameter (10×10^{-6} m [10×10^{-4} cm])
Q = volumetric flow rate (m³/s [cm³/s])
C_c = Cunningham's slip correction factor (approximately 1)

Solving the Stokes equation for the nozzle diameter gives

$$W = \sqrt[3]{\frac{4\rho_p C_c d_p^2 Q}{9\pi\eta Stk}}$$

The square root of the Stokes number ($\sqrt{Stk}$) corresponding to the 50% collection efficiency ($\sqrt{Stk_{50}}$) can be estimated from Figure 10–8, which shows the theoretical impactor efficiency curves for different values of the jet Reynolds number. Assume a jet Reynolds number of 3000; then the corresponding $\sqrt{Stk_{50}}$ is 0.47 and

$$W = \sqrt[3]{\frac{4 \times 1000 \times 1 \times (10 \times 10^{-6})^2 \times (5 \times 10^{-4})}{9\pi \times (1.81 \times 10^{-5}) \times (0.47)^2}} = 1.2 \times 10^{-2} m$$

$$\left[W = \sqrt[3]{\frac{4 \times 1 \times 1 \times (10 \times 10^{-4})^2 \times 500}{9\pi \times (1.81 \times 10^{-4})(0.47)^2}} = 1.2 cm \right]$$

Knowing the nozzle diameter, the jet Reynolds number is calculated from Eq. 10–5 by substituting the flow rate for the jet velocity:

$$Re_j = \frac{4\rho_g Q}{\pi W \eta} = \frac{4 \times (1.205) \times (5 \times 10^{-4})}{\pi \times 0.012 \times (1.81 \times 10^{-5})}$$

$$\left[= \frac{4 \times (1.205 \times 10^{-3}) \times 500}{\pi \times 1.2 \times (1.81 \times 10^{-4})} \right]$$

$$= 3500$$

The jet Reynolds number is outside of, but close to, the upper limit of the suggested range of 500 to 3000 and, therefore should provide sharp cut off characteristics.

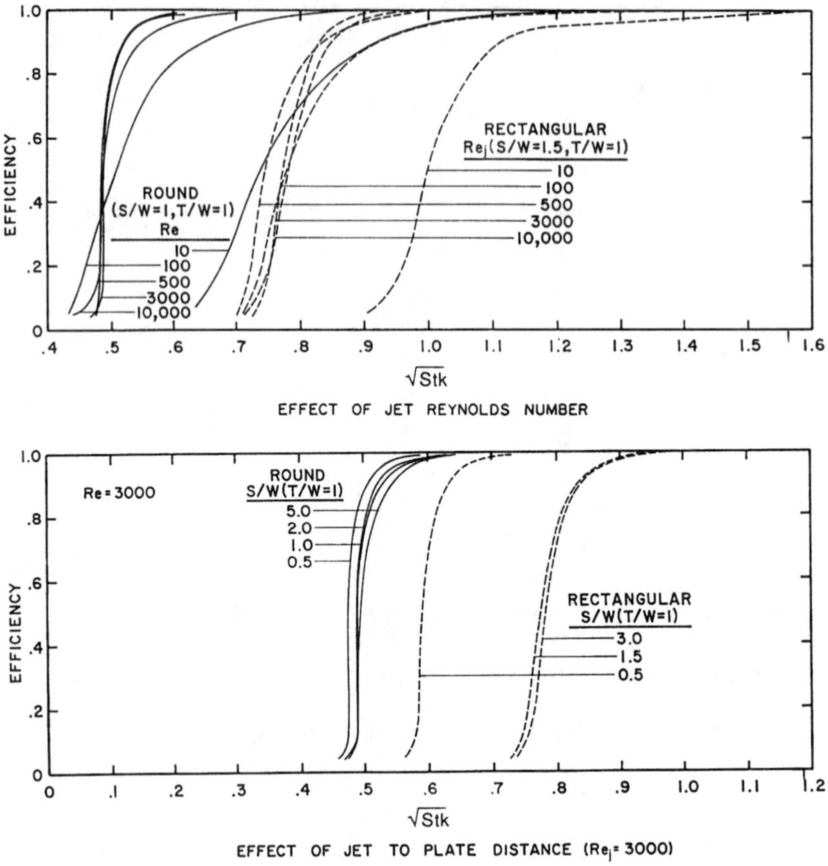

Fig. 10–8. Theoretical impactor efficiency curves for rectangular and round jet impactors showing the effect of jet-to-plate distance ratio (*S/W*) and jet Reynolds number (Re_j) (Rader and Marple, 1985). (Reproduced with permission of the Elsevier Sciences Publishing Co., Inc.)

Example of Impactor Application

Impactors have been used to sample a wide variety of aerosols in a variety of studies. Because it is impossible to discuss all of the situations, the next few sections describe the use of a generic impactor to solve a hypothetical problem. In each step of the process, remarks are made as to how the step applies to the general sampling of aerosols with impactors. Examples are given in these sections to demonstrate the type of calculations that must be made when using impactors.

The problem in this example was to determine the size distribution and mass concentration of dust particles in an industrial environment. In the initial phase of the program, it was suspected that the particle size distribution contained particles as small as 0.1 μm. The cut size of the upper stages of the impactor had to be large enough to cover the respirable range, with the first stage of the impactor having a cut point of at least 10 μm.

The impactor selected for this analysis had nine stages and an after-filter. The cut sizes of these stages are given. The flow rate through the impactor was $5 \times 10^{-4}\,m^3/s$ [30 L/min].

Substrate Preparation. Because the purpose of sampling the particles was to determine the mass size distribution, the substrate had to have a stable weight and, therefore, aluminum foil substrates were selected. Filter material could also have been used, although some of these materials have a tendency to change weight with different humidity conditions. In addition, because the dust particles were solid, the application of some type of sticky substance to the substrates to reduce particle bounce was necessary. Therefore, the material used for the substrate had to be impervious to the oil so that it did not migrate through the substrate and be lost on the support of the substrate. For this application it was decided to use a silicone oil spray. A mask was prepared by cutting a hole in a clear plastic sheet just large enough to accommodate the deposited particles and placed over the foil and oil applied. The substrates were then placed in an oven at a temperature of 65°C for 90 min to evaporate the volatiles. The substrates were weighed on a microbalance and then placed on the impaction plates for use in the sampling program.

Sampling Time Estimations. To determine the appropriate sampling period, an estimate of the aerosol mass concentration was made to prevent overloading and bounce on the substrates. The size distribution of the aerosol was not known, so it was assumed that the aerosol was spread uniformly over the number of stages used in the impactor. The regulatory limit for the particle mass concentration in this work environment was 2 mg/m^3. Thus, it was assumed that the mass concentration was no more than this value. The nine stages of the impactor covered a size range from 0.1 to 18 μm, and each stage was assumed to hold 1 mg of material. The sampling period was estimated to be 2.5 h at a sampling flow rate of 5×10^{-4} m^3/s [30 L/min].

After the samples were collected, the substrates were brought back to the laboratory for analysis. For this particular problem, it was necessary to determine the mass size distribution.

There were several other types of analysis that could have been performed with the substrates at this point, for example, X-ray diffraction analysis to determine the elemental composition, optical microscopy for inspection of the deposit, and SEM to investigate shape and elemental composition of individual particles. The analysis of these samples with SEM would have been difficult because of the presence of the sticky oil surface; an ungreased substrate on one of the stages would have allowed this type of analysis. If particle bounce had been too severe, this would not have resulted in good data. If grease coatings were necessary, the particles could have been washed from the substrates with a solvent and the particles separated from the solvent by filtration. These particles would then have been available for SEM analysis.

The mass concentration during one sampling period was 2.0 mg/m^3 (Example 10–2). The size distribution results are tabulated.

EXAMPLE 10–2

The nine-stage cascade impactor was used to sample an aerosol for 167 min operating at a flow rate of 5×10^{-4} m^3/s [30 L/min]. Table 10–4 lists the results of the gravimetric analysis. From Table 10–6, determine the total mass concentration. Plot a histogram of relative mass (i.e., $\Delta m/\Delta\log[d_a]$), versus aerodynamic diameter on semilog graph paper. Plot the cumulative mass versus aerodynamic diameter on lognormal probability graph paper. Determine the mass median diameter and the geometric standard deviation (σ_g).

continued

TABLE 10–4. Sample Size Distribution Data

Impactor Cutpoint (μm)	Δm (μg)	Cumulative Mass Percent Less Than Indicated Size	$\frac{\Delta M}{\Delta \log(d_a)}$
18	10	99.9	
			392
10	100	98.9	
			2,340
5.6	590	93.0	
			7,410
3.2	1,800	75.0	
			12,400
1.8	3,100	44.0	
			11,000
1.0	2,810	15.9	
			4,920
0.56	1,240	3.50	
			1,250
0.32	305	0.45	
			168
0.18	42	0.03	
Filter	3	0.00	

Answer: The total mass concentration is found by summing the Δm column and dividing that sum by the product of the flow rate and the sampling time. The total mass is 10.0 mg; therefore, total mass concentration

$$= \frac{10.0\text{mg}}{(5\times10^{-4}\text{m}^3/\text{s})\times(60\times167\text{s})\times(1\text{m}^3)}$$

$$= \left[\frac{10.0\text{mg}}{(30\text{L/min})\times(167\text{min})\times(1\text{m}^3/1000\text{L})}\right]$$

$$= 2.00\text{mg/m}^3$$

Figure 10–9 shows the relative mass histogram. The curve superimposed on the histogram shows that the mass distribution is lognormal. A plot of cumulative mass as a function of aerodynamic diameter is shown in Figure 10–10. From this plot the mass median diameter (MMD) and geometric standard deviation can be obtained. The mass median diameter is the diameter where 50% of the mass is collected and, from Figure 10–10, is 2.0 μm. The geometric standard deviation is given by

$$\sigma_g = \frac{d_{84.1\%}}{d_{50\%}} = \frac{4.0}{2.0} = 2.0$$

$$\sigma_g = \frac{d_{50\%}}{d_{15.9\%}} = \frac{2.0}{1.0} = 2.0$$

$$\sigma_g = \sqrt{\frac{d_{84.1\%}}{d_{15.9\%}}} = \sqrt{\frac{4.0}{1.0}} = 2.0$$

Size Distribution Data Analysis

Particle size distribution data are normally represented as number, surface area, or volume (mass) size distributions using histograms or cumulative graphs. Because an impactor collects particles that are weighed on a balance to provide the mass of the particles in a particular size range, the mass distribution is the most common method for presenting impactor data. The data from the previous example are shown as the mass of particles in each of the size ranges represented by the stages of the impactor. It is possible to estimate the surface area and number of particles in each of these size fractions by assuming the particles are spheres. To remove the bias of the width of the size fraction, the mass of particles is divided by the classification width. Furthermore, because the particle size is normally represented on a log scale, it is better to have the histogram height as $\Delta m/\Delta\log d_a$. When the histogram is plotted as shown in Figure 10–9, the area under the curve is representative of the percentage of mass in a particular size range. For this particular example, the mass size distribution is of interest, and Figure 10–9 shows the resulting mass size distribution.

A cumulative distribution can also be calculated. This has been performed, and the results are shown in Figure 10–10. Because the size distribution is made up of one distinct class of particles with a log-normal size distribution, the data lie along one straight line in the cumulative distribution.

In the above analysis it was assumed that the impactor had ideally sharp cuts at the 50% cut points of the stages. Because the true efficiency curve is probably S shaped but fairly sharp, this is a good assumption in that some particles are included from the size class above the stage and some are collected on the subsequent stage. In a rather broad distribution, these two errors tend to cancel each other, and the data are a fairly good representation of the actual size distribution. However, there are techniques by which the actual shape of the efficiency curve can be incorporated into the data analysis and a more accurate description of the particle size distribution obtained. In recent years a number of these data inversion techniques have been developed (e.g., Crump and Seinfeld, 1982; Markowski, 1987; Rader et al., 1991; Wolfenbarger and Seinfeld, 1990). Additional information on data analysis is presented in Chapter 8.

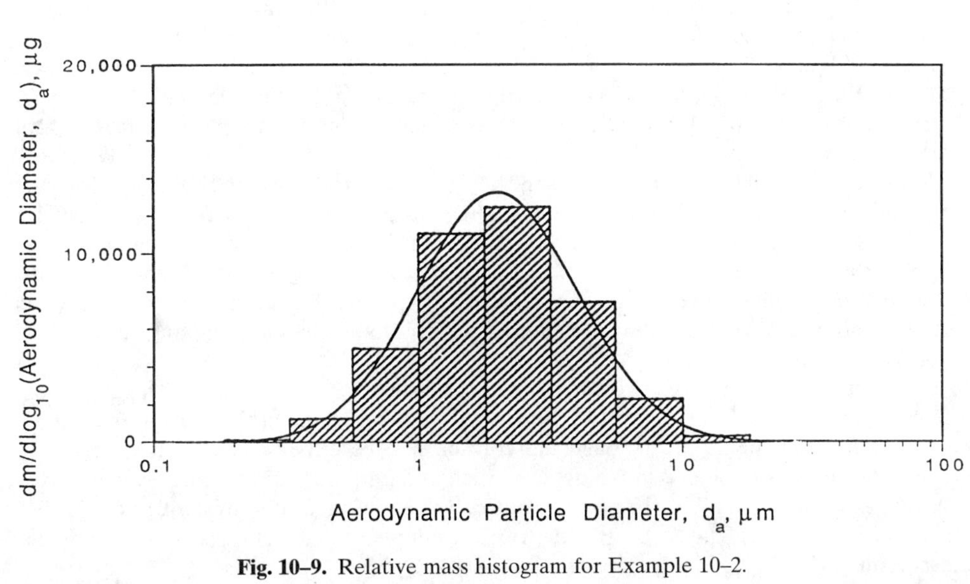

Fig. 10–9. Relative mass histogram for Example 10–2.

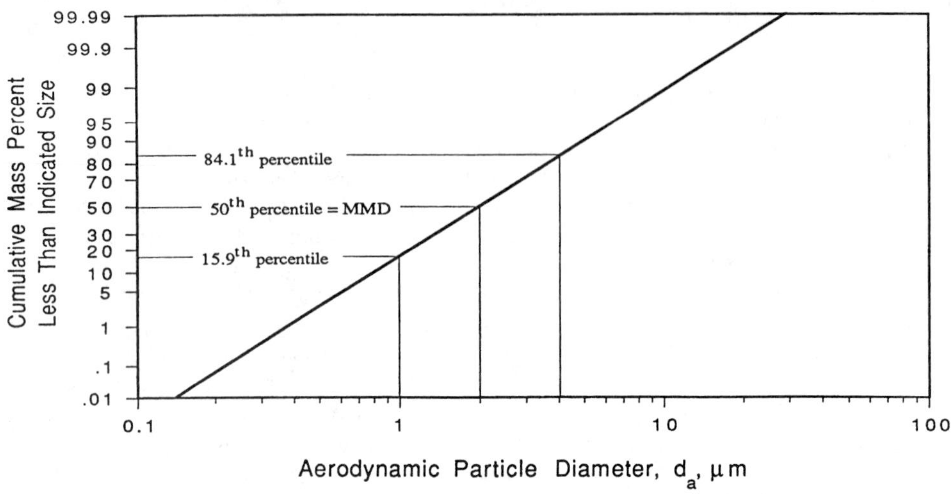

Fig. 10–10. Log-normal probability plot for Example 10–2.

SETTLING DEVICES AND CENTRIFUGES

The definition of a particle's aerodynamic diameter is the equivalent diameter of a unit density sphere that has the same gravitational settling velocity as the particle in question. Therefore, a device that measures the settling velocity directly, such as a settling chamber, is a natural selection for a device to make direct measurements of a particle's aerodynamic diameter.

Settling chambers do not operate well for small particles because of their low settling speeds (e.g., the settling velocity of a 1 μm particle is 0.035 mm/s). In addition, Brownian motion interferes with settling of small particles and sets a lower limit at about 0.6 μm (Orr and Keng, 1976). However, settling chambers can be used for larger particles (the settling speed of 10 μm particles is 3.05 mm/s). Furthermore, because of the low settling velocities, great care must be taken to eliminate any convective air currents in the chamber.

John and Wall (1983) constructed a device to measure the settling speeds of particles in the 10 to 20 μm size range. This was achieved by illuminating the particles with a laser beam directed up a tube in which the particles were settling and measuring the time taken to fall a predetermined distance.

A special form of a settling chamber is the horizontal elutriator where the aerosol is passed slowly along a horizontal channel. Particles settle onto the bottom of the flow channel at locations dependent on the particle size, particle density, gas velocity, and channel height. The larger aerodynamic diameter particles settle near the entrance while the smaller particles are deposited near the exit. Two devices that incorporate this technique for particle classifica-tion are the horizontal elutriator used for respirable dust sampling and the Timbrell aerosol spectrometer.

Horizontal elutriators are used in two-stage samplers for respirable dust measurement. The nonrespirable particles deposit on horizontal, parallel plates in the elutriator while the respirable particles penetrate through the elutriator to be either collected for subsequent gravimetric analysis or passed into a detector such as a photometer. The particle penetration characteristics of the ideal horizontal elutriator, by definition, are equivalent to the BMRC respirable dust criteria. The MRE gravimetric dust sampler was developed for respirable dust measurement (Wright, 1954; Dunmore et al., 1964).

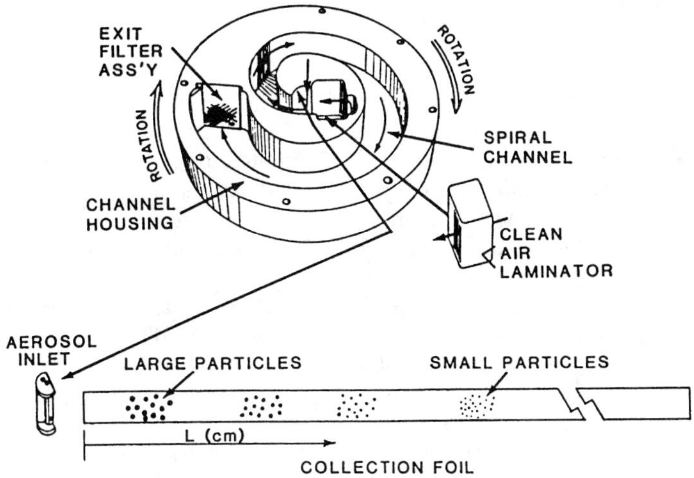

Fig. 10–11. Schematic diagram of a centrifuge (Cheng et al., 1988). (Reproduced with permission of the Elsevier Sciences Publishing Co., Inc.)

The physical description and operation of the Timbrell aerosol spectrometer are detailed by Timbrell (1954, 1972). This spectrometer achieves accurate particle size classification by winnowing sampled particles in a laminar stream of clean air where they settle along the bottom of a horizontal sedimentation chamber according to their aerodynamic diameter. The chamber is a wedge-shaped channel with microscope slides recessed into the horizontal floor. Particles ranging from 1.5 to 25 μm aerodynamic diameter can be classified at a winnowing air flow rate of $1.7 \times 10^{-6}\,m^3/s$ [0.1 L/min]. The instrument is calibrated in terms of aerodynamic particle diameter by means of spherical particles of known density. This spectrometer has been used to determine the aerodynamic size classification of spheres, particles of irregular shape, fibers, and aggregates (e.g., Griffiths and Vaughan, 1986).

The difficulty in using a settling chamber or horizontal elutriators is that the force on the particles is quite small. One method to greatly increase the forces and increase the settling speed is to use a centrifuge. In a centrifuge, the air and particles are rotated at a high rotational speed, and the centrifugal force is used to deposit the particles on the outer edge of an aerosol chamber.

Although several centrifuges have been developed, the one that has seen the greatest application and is currently in use today is the spiral centrifuge (Hoover et al., 1983; Stöber, 1976; Kotrappa and Light, 1972; Stöber and Flachsbart, 1969). One commercially available centrifuge is the Lovelace aerosol particle separator (LAPS, ***INT***). This classifier is actually a spectrometer in that the particles are introduced at the inner wall of a flow entering a spiral flow passage that is being rotated at a high rate of speed, as shown in Figure 10–11. At the entrance to the spiral, the particles are near the inner wall, and clean sheath air is adjacent to the outer wall. As the particles flow along the passage they are forced by centrifugal forces through the clean air and are deposited on the outer edge of the spiral. The forces on the larger particles are greater, and, therefore, they are deposited closer to the inlet of the spiral than are the smaller particles. Thus, the particles are deposited along the spiral passage as a function of decreasing particle size. Before sampling, a foil is placed in the outer wall of the spiral passage that is removed after sampling, and the particles are analyzed. The centrifuge can be calibrated with particles of known size and density such that the aerodynamic diameter as a function of distance from the inlet is known. Once calibrated, a centrifuge can be used to obtain the aerodynamic size distribution of irregular shaped particles, aggregate particles, and fibers (e.g., Martonen and Johnson, 1990).

THERMAL PRECIPITATORS

Thermal precipitators are a class of instruments that make use of thermophoretic forces on particles to collect particles onto a sampling surface. The thermal precipitator has been used for sampling respirable-sized particles in mines in Great Britain and South Africa.

The principle of operation is quite simple (Waldmann and Schmitt, 1966). When a particle passes through a temperature gradient in the air, the air molecules from the warmer side of the particle strike the particle with higher energy than the air molecules on the cold side. This provides a net force in the direction of the cold surface, and the particle migrates to the cold surface.

The results of this phenomenon can be experienced in everyday life. For example, where a hot water or steam radiator is adjacent to a wall, the wall behind the radiator becomes dirty. Also, in cold climates, the inside surfaces of windows in an automobile collect a film of contamination. Both of these phenomena are due to the particles migrating to a cold surface when a temperature gradient exists near the surface.

By making use of the thermophoretic forces, the thermal precipitator can be simple in design. All that is necessary is to place a hot wire or filament close to a cooler surface with particles passing between the filament and the surface. The particles then migrate to the cool surface, which is most often a microscope slide, and are collected. The particles can subsequently be analyzed microscopically.

Characteristics of thermal precipitators are that the flow rates are rather low (on the order of a few cm^3/min) and the collection efficiencies are good for particles in the submicrometer range (i.e., for particles down to 0.01 μm). The upper size limit of particles collected by this technique is on the order of 5 to 10 μm. With the thermal precipitator being efficient at collecting particles in this size range, it has been used most extensively for collecting respirable-sized particles in industrial atmospheres, especially in the mining industry. One configuration of the thermal precipitator is shown in Figure 10–12, including the heated filament and the cooler surface that contains the microscope slide.

The standard thermal precipitator was rather bulky and required a fair degree of expertise to operate, and samples had to be counted under a microscope, a labor intensive task. In the early 1950s, a modified thermal precipitator was developed that employed a photoelectric detector to replace the optical microscope counting of the particles (Kitto and Beadle,

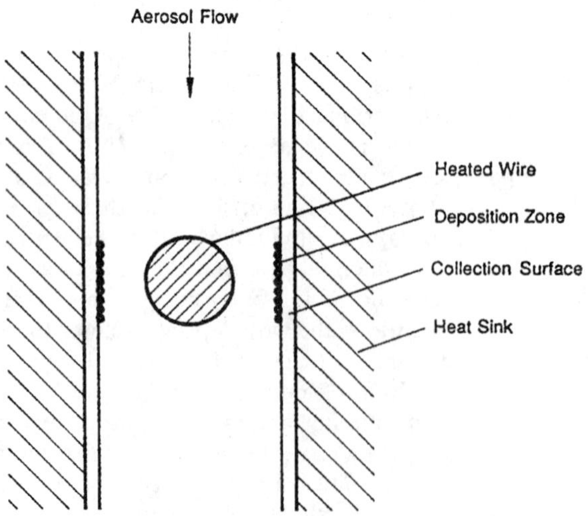

Fig. 10–12. Cross-sectional view of a heated wire and plate thermal precipitator.

1952; Beadle, 1954). The modified thermal precipitator operated at $1.7 \times 10^{-7}\,m^3/s$ [$10\,cm^3/min$] for sampling periods from 1 to 10 min. A horizontal elutriator was used at the inlet to remove particles that were nonrespirable and the particles deposited on a slide. The slide would be advanced between tests, and up to 11 samples were collected on one slide. The samples in the slide were then conditioned in a desiccator and placed in a photoelectric assessor that compared the light transmission through the deposit with the light transmission through the clean portion of the slide. The amounts of light transmitted through the deposit and slide were measured by photoelectric cells, and the numbers were quoted as dimensionless photoelectric readings that could be related to the mass of particles collected.

REFERENCES

Beadle, D. G. 1954. A photo-electric apparatus for assessing dust samples. *J. Chem. Met. Min. Soc. SA* 55:30–39.

Belosi, F. and V. Prodi. 1987. Particle deposition within the inertial spectrometer. *J. Aerosol Sci.* 18:37–42.

Berner, A., C. Lürzer, F. Pohl, O. Preining, and P. Wagner. 1979. The size distribution of the urban aerosol in Vienna. *Sci. Tot. Environ.* 13:245–261.

Biswas, P., C. L. Jones, and R. C. Flagan. 1987. Distortion of size distributions by condensation and evaporation in aerosol instruments. *Aerosol Sci. Technol.* 7:231–246.

Chen, B. T. and H. C. Yeh. 1987. An improved virtual impactor: Design and performance. *J. Aerosol Sci.* 18:203–214.

Chen, B. T., H. C. Yeh, and Y. S. Cheng. 1985. A novel virtual impactor: Calibration and use. *J. Aerosol Sci.* 16:343–354.

Chen, B. T., H. C. Yeh, and Y. S. Cheng. 1986. Performance of a modified virtual impactor. *Aerosol Sci. Technol.* 5:369–376.

Cheng, Y. S., H. C. Yeh, and M. D. Allen. 1988. Dynamic shape factor of a plate-like particle. *Aerosol Sci. Technol.* 8:109–123

Conner, W. D. 1966. An inertial-type particle separator for collecting large samples. *J. Air Pollut. Control Assoc.* 16:1:35.

Crump, J. G. and J. H. Seinfeld. 1982. Further results of inversion on aerosol size distribution data: Higher-order Sobolev spaces and constraints. *Aerosol Sci. Technol.* 1:363–369

de Juan, L. and J. Fernandez de la Mora. 1998. Sizing nanoparticles with a focusing impactor: Effect of the collector size. *J. Aerosol Sci.* 29:589–599.

Dunmore, J. H., R. J. Hamilton, and D. S. G. Smith. 1964. An instrument for the sampling of respirable dust for subsequent gravimetric assessment. *J. Sci. Instrum.* 41:669–672.

Fairchild, C. I. and L. D. Wheat. 1984. Calibration and evaluation of a real-time cascade impactor. *Am. Ind. Hyg. Assoc. J.* 45:205–211.

Fang, C. P., P. H. McMurry, V. A. Marple, and K. L. Rubow. 1991. Effect of flow-induced relative humidity changes on size cuts for sulfuric acid droplets in the MOUDI. *Aerosol Sci. Technol.* 14:266–277.

Fernandez de la Mora, J., S. V. Hering, N. Rao, and P. H. McMurry. 1990. Hypersonic impaction of ultrafine particles. *J. Aerosol Sci.* 21:169–187.

Fernandez de la Mora, J. and P. Riesco-Chueca. 1988. Aerodynamic focusing of particles in a carrier gas. *J. Fluid Mech.* 195:1–21.

Forney, L. J. 1976. Aerosol fractionator for large-scale sampling. *Rev. Sci. Instrum.* 47:1264–1269.

Forney, L. J., D. G. Ravenhall, and S. S. Lee. 1982. Experimental and theoretical study of a two-dimensional virtual impactor. *Environ. Sci. Technol.* 16:492–497.

Golovin, M. N. and A. A. Putnam. 1962. Inertial impaction on single elements. *Ind. Eng. Chem. Fund.* 1:264–273.

Griffiths, W. D. and N. P. Vaughan. 1986. The aerodynamic behaviour of cylindrical and spheroidal particles when settling under gravity. *J. Aerosol Sci.* 17:53–65.

Hering, S. 1995. Impactors, cyclones, and other inertial and gravitational collectors. In *Air Sampling Instruments*, 8th Ed., eds. B. S. Cohen and S. V. Hering. Cincinnati: American Conference of Governmental Industrial Hygienists, Inc., pp. 279–321.

Hering, S. V., R. C. Flagan, and S. K. Friedlander. 1978. Design and evaluation of a new low pressure impactor, 1. *Environ. Sci. Technol.* 12:667–673.

Hering, S. V., S. K. Friedlander, J. J. Collins, and L. W. Richards. 1979. Design and evaluation of a new low pressure impactor, 2. *Environ. Sci. Technol.* 13:184–188.

Hering, S. V. and V. A. Marple. 1986. Low pressure and micro-orifice impactors. In *Cascade Impactors: Sampling & Data Analysis*, eds. J. P. Lodge and T. L. Chan. Akron, OH: American Industrial Hygiene Association, pp. 103–127.

Hillamo, R. E. and E. I. Kauppinen. 1991. On the performance of the Berner low pressure impactor. *Aerosol Sci. Technol.* 14:33–47.

Hoover, M. D., G. Morawietz, and W. Stöber. 1983. Optimizing resolution and sampling rate in spinning duct aerosol centrifuges. *Am. Ind. Hyg. Assoc. J.* 44:131–134.

Jaenicki, R. and I. H. Blifford. 1974. Virtual impactors: A theoretical study. *Environ. Sci. Technol.* 8:648–654.

John, W. and S. M. Wall. 1983. Aerosol testing techniques for size-selective samplers. *J. Aerosol Sci.* 14:713–727.

Keskinen, J., K. Janka, and M. Lehtimäki. 1987. Virtual impactor as an accessory to optical particle counters. *Aerosol Sci. Technol.* 6:79–83.

Keskinen, J., K. Pietarinen, and M. Lehtimäki. 1992. Electrical low pressure impactor. *J. Aerosol Sci.* 23:353–360.

Kitto, P. H. and D. G. Beadle. 1952. A modified form of thermal precipitator. *J. Chem. Met. Min. Soc. SA* 52:284–306.

Kotrappa, P. and M. E. Light. 1972. Design and performance of the Lovelace aerosol particle separator. *Rev. Sci. Instrum.* 43:1106–1112.

Liebhaber, F. B., M. Lehtimäki, and K. Willeke. 1991. Low-cost virtual impactor for large-particle amplification in optical particle counters. *Aerosol Sci. Technol.* 15:208–213.

Lippman, M. 1995. Size-selective health hazard sampling. In *Air Sampling Instruments*, 8th Ed., eds. B. S. Cohen and S. V. Hering. Cincinnati: American Conference of Governmental Industrial Hygienists, Inc., pp. 81–119.

Liu, B. Y. H. and K. L. Rubow. 1984. A new axial flow cascade cyclone for size classification of airborne particulate matter. In *Aerosols: Science, Technology, and Industrial Applications of Airborne Particles*, eds. B. Y. H. Liu, D. Y. H. Pui, and H. Fissan. New York: Elsevier Science Publishing Co., pp. 115–118.

Liu, P., P. J. Ziemann, D. B. Kittelson, and P. H. McMurry. 1995a. Generating particle beams of controlled dimensions and divergence: I. Theory of particle motion in aerodynamic lenses and nozzle expansions. *Aerosol Sci. Technol.* 22:293–313.

Liu, P., P. J. Ziemann, D. B. Kittelson, and P. H. McMurry. 1995b. Generating particle beams of controlled dimensions and divergence: II. Experimental evaluation of particle motion in aerodynamic lenses and nozzle expansions. *Aerosol Sci. Technol.* 22:314–324.

Lodge, J. P. and T. L. Chan. 1986. *Cascade Impactors: Sampling and Data Analysis*. Akron, OH: American Industrial Hygiene Association.

Loo, B. W. 1981. *High Efficiency Virtual Impactor*. United States Patent No. 4,301,002.

Loo, B. W. and C. C. Cork. 1988. Development of high efficiency virtual impactors. *Aerosol Sci. Technol.* 9:167–176.

Lundgren, D. A. 1971. Determination of particle composition and size distribution changes with time. *Atmos. Environ.* 5:645–651.

Markowski, G. R. 1987. Improving Twomey's algorithm for inversion of aerosol measurement data. *Aerosol Sci. Technol.* 7:127–141.

Marple, V. A. 1970. *A Fundamental Study of Inertial Impactors.* Ph.D. Thesis, University of Minnesota, Minneapolis, MN.

Marple, V. A. 1978. Simulation of respirable penetration characteristics by inertial impaction. *J. Aerosol Sci.* 9:125–134.

Marple, V. A. and C. M. Chien. 1980. Virtual impactors: A theoretical study. *Environ. Sci. Technol.* 14:976–985.

Marple, V. A. and B. Y. H. Liu. 1974. Characteristics of laminar jet impactors. *Environ. Sci. Technol.* 8:648–654.

Marple, V. A., B. Y. H. Liu, and R. M. Burton. 1990. High-volume impactor for sampling fine and coarse particles. *J. Air Waste Manage* 40:762–767.

Marple, V. A., B. Y. H. Liu, and B. A. Olson. 1989. Evaluation of a cleanroom concentrating aerosol sampler. In *Proceedings of the 35th Annual Meeting of the Institute of Environmental Sciences*, pp. 360–363.

Marple, V. A. and J. E. McCormack. 1983. Personal sampling impactor with respirable aerosol penetration characteristics. *Am. Ind. Hyg. Assoc. J.* 44:916–922.

Marple, V. A. and B. A. Olson. 1995. A high volume PM10/2.5/1.0 trichotomous sampler. *Particulate Matter: Health and Regulatory Issues*. VIP-49. pp. 237–261. Pittsburgh, PA, Air and Waste Management Association International Specialty Conference.

Marple, V. A. and B. A. Olson. 1999. *A Micro-Orifice Impactor with Cut Sizes Down to 10 Nanometers for Diesel Exhaust Sampling*. Final Report. Minneapolis, MN: University of Minnesota Generic Technology Center for Respirable Dust.

Marple, V. A. and K. L. Rubow. 1986. Theory and design guidelines. In *Cascade Impactors: Sampling & Data Analysis*, eds. J. P. Lodge and T. L. Chan. Akron, OH: American Industrial Hygiene Association, pp. 79–101.

Marple, V. A., K. L. Rubow, and S. M. Behm. 1991. A micro-orifice uniform deposit impactor (MOUDI). *Aerosol Sci. Technol.* 14:434–446.

Marple, V. A., K. L. Rubow, W. Turner, and J. D. Spengler. 1988. Low flow rate sharp cut impactors for indoor air sampling: Design and calibration. *JAPCA* 37:1303–1307.

Marple, V. A. and K. Willeke. 1976a. Impactor design. *Atmos. Environ.* 12:891–896.

Marple, V. A. and K. Willeke. 1976b. Inertial impactors: Theory, design and use. In *Fine Particles*, ed. B. Y. H. Liu. New York: Academic Press, pp. 411–466.

Martonen, T. B. and D. L. Johnson. 1990. Aerodynamic classification of fibers with aerosol centrifuges. *Particulate Sci. Technol.* 8:37–63.

Masuda, H., D. Hochrainer, and W. Stöber. 1978. An improved virtual impactor for particle classification and generation of test aerosols with narrow size distributions. *J. Aerosol Sci.* 10:275–287.

May, K. R. and R. Clifford. 1967. The impaction of aerosol particles on cylinder, spheres, ribbons and discs. *Ann. Occup. Hyg.* 10:83–95.

Maynard, A. C. 2000. A simple model of axial flow cyclone performance under laminar flow conditions. *J. Aerosol Sci.* 31:151–167.

McFarland, A. R., C. A. Ortiz, and R. W. Bertch. 1978. Particle collection characteristics of a single-stage dichotomous sampler. *Environ. Sci Technol.* 12:679–682.

Noll, K. E. 1970. A rotary inertial impactor for sampling giant particles in the atmosphere. *Atmos. Environ.* 4:9–19.

Noll, K. E., A. Pontius, R. Frey, and M. Gould. 1985. Comparison of atmospheric coarse particles at an urban and non-urban site. *Atmos. Environ.* 19:1931–1943.

Orr, C. and E. Y. H. Keng. 1976. Sampling and particle-size measurement. In *Handbook on Aerosols*, ed. R. Dennis. U.S. Energy Research and Development Adm., NTIS publ. No. TID-26608, pp. 93–117.

Prodi, V., F. Belosi, and A. Mularoni. 1984. A high flow inertial spectrometer. In *Aerosols: Science, Technology, and Industrial Applications of Airborne Particles*, eds. B. Y. H. Liu, D. Y. H. Pui, and H. J. Fissan. New York: Elsevier Science Publishing Co., Inc., pp. 131–134.

Prodi, V., F. Belosi, A. Mularoni, and P. Lucialli. 1988. PERSPEC: A personal sampler with size characterization capabilities. *Am. Ind. Hyg. Assoc. J.* 49:75–80.

Prodi, V., T. De Zaiacomo, C. Melandri, G. Tarroni, M. Formignani, P. Olivieri, L. Barilli, and G. Oberdoerster. 1983. Description and application of the inertial spectrometer. In *Aerosols in Mining and Industrial Work Environments*, eds. V. A. Marple and B. Y. H. Liu. Ann Arbor: Ann Arbor Science Publications, Vol. 3, pp. 931–949.

Prodi, V., C. Melandri, G. Tarroni, T. De Zaiacomo, M. Formignani, and D. Hochrainer. 1979. An inertial spectrometer for aerosol particles. *J. Aerosol Sci.* 10:411–419.

Raabe, O. G., D. A. Braaten, R. L. Axelbaum, S. V. Teague, and T. A. Cahill. 1988. Calibration studies of the drum impactor. *J. Aerosol Sci.* 19:183–195.

Rader, D. J. and V. A. Marple. 1985. Effect of ultra-stokesian drag and particle interception on impaction characteristics. *Aerosol Sci. Technol.* 4:141–156.

Rader, D. J., L. A. Mondy, J. E. Brockmann, D. A. Lucero, and K. L. Rubow. 1991. Stage response calibration of the mark III and Marple personal cascade impactors. *Aerosol Sci. Technol.* 14:365–379.

Rao, A. K. and K. T. Whitby. 1978a. Non-ideal collection characteristics of inertial impactors—Single stage impactors and solid particles. *J. Aerosol Sci.* 9:77–86.

Rao, A. K. and K. T. Whitby. 1978b. Non-ideal collection characteristics of inertial impactors—Cascade impactors. *J. Aerosol Sci.* 9:87–100.

Reischl, G. P. and W. John. 1978. The collection efficiency of impaction surfaces: A new impaction surface. *Staub-Reinhalt. Luft* 38:55.

Smith, W. R., R. R. Wilson, and D. B. Harris. 1979. A five stage cyclone system for in situ sampling. *Environ. Sci. Technol.* 13:1387–1392.

Sioutas, C., S. Kim, and M. Chang. 1999. Development and evaluation of a prototype ultrafine particle concentrator. *J. Aerosol Sci.* 8:1001–1017.

Sioutas, C., P. Koutrakis, J. J. Godleski, S. T. Ferguson, C. S. Kim, and B. M. Burton. 1997. Fine particle concentrators for inhalation exposures—Effect of particle size and composition. *J. Aerosol Sci.* 6:1057–1071.

Stöber, W. 1976. Design performance and applications of spiral duct aerosol centrifuges. In *Fine Particles, Aerosol Generation, Measurement Sampling and Analysis*, ed. B. Y. H. Liu, New York: Academic Press, pp. 351–398.

Stöber, W. and H. Flachsbart. 1969. Size-separating precipitation of aerosols in a spinning spiral duct. *Environ. Sci. Technol.* 3:1280–1296.

Szymanski, W. S. and B. Y. H. Liu. 1989. An airborne particle sampler for the space shuttle. *J. Aerosol Sci.* 20:1569–1572.

Timbrell, V. 1954. The terminal velocity and size of airborne dust particles. *Br. J. Appl. Phys.* 5:S86.

Timbrell, V. 1972. An aerosol spectrometer and its application. In *Assessment of Airborne Particles*, eds. T. T. Mercer, P. E. Morrow, and W. Stöber. Springfield, IL: Charles C. Thomas, pp. 290–330.

Vanderpool, R. W., D. A. Lundgren, V. A. Marple, and K. L. Rubow. 1987. Cocalibration of four large-particle impactors. *Aerosol Sci. Technol.* 7:177–185.

Waldmann, L. and K. H. Schmitt. 1966. Thermophoresis and diffusiophoresis of aerosols. In *Aerosol Science*, ed. C. N. Davies. New York: Academic Press, pp. 137–162.

Wolfenbarger, J. K. and J. H. Seinfeld. 1990. Inversion of aerosol size distribution data. *J. Aerosol Sci.* 21:227–247.

Wright, B. M. 1954. A size-selecting sampler for airborne dust. *Br. J. Ind. Med.* 11:284.

Wu, J. J., D. W. Cooper, and R. J. Miller. 1989. Virtual impactor aerosol concentrator for cleanroom monitoring. *J. Environ. Sci.* 5:52–56.

Xu, X. 1991. *A Study of Virtual Impactor.* Ph.D. Thesis, University of Minnesota, Minneapolis, MN.

11 Chemical Analysis Methods for Atmospheric Aerosol Components

PAUL A. SOLOMON

U.S. Environmental Protection Agency, Office of Research and Development, National Exposure Research Laboratory, Las Vegas, NV

GARY NORRIS, MATTHEW LANDIS, and MICHAEL TOLOCKA

U.S. Environmental Protection Agency, MD-46, Office of Research and Development, National Exposure Research Laboratory, Research Triangle Park, NC

INTRODUCTION

The atmospheric aerosol is a complex mixture of liquid and solid particles suspended in a gas we call *air*. Atmospheric particulate matter (PM), the solid and liquid particles in air, has both biogenic (natural) and anthropogenic (man-made) sources and is either emitted directly into the air (as primary particles) or formed through chemical reactions during its lifetime in the atmosphere (as secondary particles). Primary particles are created for the most part through mechanical processes such as weathering or other grinding processes and, therefore, are typically observed in particles larger than 2.5 micrometers (μm) in aerodynamic diameter. In general, the major components of primary particles are soil-related material (e.g., Fe, Si, Ca, Mg) and organic material (e.g., pieces of pollens and spores). Precursors to secondary anthropogenic particles include nitrogen oxides, sulfur oxides, and organic compounds emitted mostly from combustion sources. Reaction products of these compounds include a wide variety of species, many of which exist in both the gas phase and condensed phase under normal atmospheric conditions (e.g., gas phase HNO_3 and NH_3 with solid NH_4NO_3 particles). Secondary particles are typically smaller than 2.5 μm in aerodynamic diameter and are composed mostly of nitrates, sulfates, organic carbon, and elemental carbon (soot). Historically, particulate matter has been collected on filters of various types, in some cases with specialized equipment to remove sampling artifacts, with subsequent chemical analysis in the laboratory, the latter being the focus of this chapter. Hence, particulate matter in the atmosphere is composed of a wide variety of compounds, the determination of which requires numerous techniques for collection and chemical analysis.

Understanding the sources of atmospheric particulate matter is important because ambient anthropogenic particles have been shown to have adverse effects on human health,

Aerosol Measurement: Principles, Techniques, and Applications, Second Edition, Edited by Paul A. Baron and Klaus Willeke.
ISBN 0-471-35636-0

cause visibility degradation, participate in acid deposition, result in material and crop damage, impact global climate change, and cause general soiling of property (U.S. EPA, 2001). The chemical components of atmospheric particulate matter provide clues to the sources of particles, as different sources can produce particles of different chemical composition. Source identification is then determined by advanced statistical techniques (e.g., receptor modeling, an advanced statistical analysis method for linking pollution observed at an ambient location back to its source) or by the use of advanced mathematical simulation models (e.g., grid based) that integrate source emissions, atmospheric chemistry, and meteorology to trace the fate of source emissions through the atmosphere.

Driven by relationships observed between ambient fine particulate matter (PM-2.5, i.e., particles less than 2.5 μm in diameter) concentrations and adverse health effects (U.S. EPA, 1996a,b, 2001; Albritton and Greenbaum 1998; U.S. EPA 1996a,b; U.S. EPA 2001), the U.S. Environmental Protection Agency (EPA) has established a series of ambient particulate matter monitoring networks to better characterize the spatial and temporal distributions of PM-2.5 mass and its composition throughout the United States. Data from these networks will help link particles observed at ambient receptor location species to their sources, resulting in a means for developing effective emissions control strategies for primary particles and secondary particle precursor species.

Three interlaced networks are being established by the EPA to provide different levels of information in space, time, and aerosol composition (U.S. EPA, 1998a,b). The base network consists of about 1100 monitors, distributed based on population, for measuring PM-2.5 mass using the Federal Reference Method (FRM; U.S. CFR 1997, Parts 50, 53, and 58).

The second network, the EPA National PM-2.5 Chemical Speciation Network (referred in this chapter as the Chemical Speciation Network), consists of up to 300 chemical speciation monitors, 54 of which are part of the U.S. EPA's long-term SLAMS (State and Local Air Monitoring Network) and NAMS (National Air Monitoring Network) networks for measuring trends in atmospheric pollutants (e.g., ozone, CO, SO_2, PM-10) (U.S. EPA, 1997). The chemical speciation monitors provide 24 h integrated chemical composition data, typically on a 1 in 3 or 1 in 6 day schedule. This network is not used for compliance, but to assist states with the development of equitable and efficient State Implementation Plans. The components measured include PM-2.5 mass, nitrate, sulfate, ammonium, chloride, sodium, and potassium ions; organic carbon (OC); elemental carbon (EC); and trace elements obtained by X-ray fluorescence (XRF). Size-selective inlets, particle fractionators, and denuder/filter-based methods are used in these samplers (Solomon et al., 2000a). Methods for collecting samples for monitoring networks in general are described in Chapter 27, and the methods used in the Chemical Speciation Network are described elsewhere (U.S. EPA, 1999).

The third network consists of eight Supersites projects that together comprise the EPA's Supersites Program (Albritton and Greenbaum, 1998; U.S. EPA, 1998a). Each project consists of one or more highly instrumented sites that are coordinated with active air quality monitoring and health effects–related studies. The Supersites Program is designed to provide (1) detailed information on the spatial and temporal nature of particulate matter and on atmospheric processes and thus to provide states with additional data for developing cost-effective emissions management strategies for lowering concentrations of PM-2.5 in ambient air; (2) support to health effects–related programs; and (3) evaluation of advanced monitoring methods for their potential transition to routine monitoring networks.

The Chemical Speciation Network measurements are obtained by collecting PM-2.5 on several filter types followed by chemical analysis in the laboratory (U.S. EPA, 1999; Solomon et al., 2000a,b). The primary chemical analysis methods discussed in this chapter relate to those used in the Chemical Speciation Network. These include gravimetric analysis for mass, ion chromatography for anions and cation species, thermal optical transmission for OC and EC, and energy dispersive X-ray fluorescence (EDXRF) for minor and trace elements.

Although these are well-established methods, they may not be suitable to all species desired for receptor modeling and understanding atmospheric processes and for obtaining improved relationships between particulate matter and its sources and adverse health effects (see Albritton and Greenbaum [1998] for 11 hypotheses for potential causal factors); and these methods may not be readily available to everyone. For example, detailed organic speciation of the collected aerosol is important for improved receptor modeling (Schauer, 1998; Schauer et al., 1996; Schauer and Cass, 2000), while acidity, soluble transition metals, and organic compounds are potential causal factors associated with adverse health effects (Albritton and Greenbaum, 1998). Finally, for example, laboratories may not have an XRF available to them, but may have other elemental measurement instrumentation, or they may prefer to analyze cation species by colorimetry (e.g., ammonium ion) or graphite furnace atomic absorption (GFAA; e.g., K^+, Na^+, Mg^{2+}). Therefore, other analysis methods are discussed below and summarized along with the primary methods in Table 11–1. More detailed descriptions of most of the methods described in this chapter can be found in Chow and Watson (1999) and Lodge (1989).

SCOPE AND OBJECTIVES

This chapter surveys the analytical techniques used to determine the concentrations of aerosol mass and its chemical components. The techniques surveyed are listed in Table 11–1. Table 11–2 provides information on the major components and trace inorganic components of the atmospheric particulate matter collected at several locations in the United States, including two urban and two rural location species. The urban location species represent periods of highest pollution in the east (Atlanta, GA), typically observed in the summer, and in the west (Rubidoux, CA), typically observed during the late fall and winter months. The rural sites represent an annual average during the last available year of sampling. The information presented is similar to that being collected in the Chemical Speciation Network (U.S. EPA, 1999). As can be seen, there are wide variations in the chemical compositions of the aerosols and their concentrations, thus, possibly requiring application of several analytical methods to obtain valid data.

Over the last decade, determination of organic aerosol species in ambient particulate matter has become feasible. Organic species marker compounds used in source apportionment modeling of particulate matter are given in Table 11–3 (Schauer, 1998). The availability of this type of detailed organic aerosol data has greatly enhanced the ability of receptor models to identify sources (Schauer et al., 1996; Schauer and Cass, 2000). Thus, analytical methods to determine organic aerosol species are described in this chapter.

Finally, semicontinuous species-specific methods are rapidly emerging that could eliminate the need to collect aerosols on filters with retrospective chemical analysis in the laboratory. These methods are becoming available for sulfate, nitrate, other anions and cation species, OC, EC, and trace elements. These methods are mentioned briefly at the end of this chapter.

The analysis method is only one aspect involved in determining the concentration of species in atmospheric particulate matter after collection of the sample on a filter or other substrate. Other issues include sample storage, where applicable, sample extraction from the filter, and quality control and quality assurance. Sample storage includes stabilizing the collected sample from the end of sampling through sample analysis. It also may include long-term storage to allow for reanalysis of the filter later. Finally, precision and accuracy of the measurements are needed to define uncertainty in data. These issues are also briefly mentioned in this chapter.

TABLE 11–1. Analysis Methods Suitable and Most Commonly Used for Determining the Chemical Compositions of Atmospheric Particulate Matter

Species	Method	Sample Preparation	Comments
PM-2.5 mass	Gravimetric[a]	Equilibrate at fixed relative humidity, temperature, and neutralized charge	This is the Federal Reference Method as described in 40 CFR, Part 50, Appendix J for PM-10 and in 40 CFR, Part 50, Appendix L for PM-2.5 (CFR, 1997)
	β-Attenuation semicontinuous method	None	Lillienfeld and Dulchinos (1972), Macias and Husar (1976)
	Inertial microbalance (TEOM)	Equilibrate at 50°C;	Standard commercial method, potential loss of volatile species
	Semicontinuous method (TEOM)	Equilibrate to 30°C with a Nafion dryer	Nonroutine in research mode, reduces volatilization of labile species (Patashnick and Rupprecht, 1991)
	CAMM semicontinuous method	None	Research method. Relates pressure drop across filter to mass loading (Koutrakis et al., 1995)
Anions and cation species	Ion chromatography[a]	Liquid extraction in water or other aqueous solvent	EPA method for PM-2.5 Chemical Speciation Network for anion and cation species (Mulik et al., 1978; Mulik et al. and Sawicki, 1979; U.S. EPA, 1999b)
	Ion selective electrode	Liquid extraction in water	Most often used for NH_4^+, but also for sulfate and nitrate (Appel et al., 1988; Lodge, 1989)
	Colorimetry	Liquid extraction in water	Most often used for NH_4^+ (Bolleter et al., 1961; Lodge, 1989)
	X-ray fluorescence	None	Only for sulfate as sulfur times 3 (Chow, 1995)
	Fourier transform infrared spectroscopy	None	Semiquantitative (Ingle and Crouch, 1988; McClenny et al., 1985; Allen et al., 1994)
	Steam condensation/ion chromatography	None	Semicontinuous for anion and cation species (Dasgupta, 1993; Ito and Thurston, 1996)
	Automated Nitrate Monitor	Condensation onto metal surface followed by flash volatilization	Semicontinuous for nitrate (sulfur and carbon under development) (Hering and Stolzenburg, 1998)
Particulate carbon	Thermal optical reflectance (TOR)	Bake quartz filters at temperatures greater than 773 K for several hours or as high as 1173 K for 10,800 s (3 h)	Provides data on OC, EC, CC, and TC. Method used to measure particulate carbon concentrations with the Interagency Monitoring of Protected Visual Environments (IMPROVE) protocol (Chow et al., 1993)
	Thermal optical transmittance[a] (TOT)	Bake quartz filters at temperatures greater than 773 K for several hours or as high as 1173 K for 10,800 s (3 h)	Provides data on OC, EC, CC, and TC. Method used to measure particulate carbon concentrations with the NIOSH Method 5040 protocol (Birch and Cary, 1996)

	Temperature-programmed volatilization (TPV)	Bake quartz filters at temperatures greater than 773 K for several hours or as high as 1173 K for 10,800 s (3 h)	Research method. Provides data on OC, EC, and TC on quartz filters and OC from carbon-impregnated filters (Eatough et al., 1995)
	Aethalometer	None	Provides semicontinuous measurement of optical absorption due to elemental or light absorbing carbon (Rosen et al., 1978; Hansen and Novakov, 1990)
	In situ thermal/optical carbon analyzer	None	Semicontinuous OC and EC by the TOT method (Turpin et al., 1990)
Organic aerosol speciation	GC-MS[a,b]	Bake quartz-fiber filters at temperatures greater than 773 K for several hours or as high as 1173 K for 10,800 s (3 h), followed by solvent extraction after sample collection	Provides a large number of nonpolar and a few polar organic compounds that are extremely useful as tracers for receptor modeling and understanding atmospheric chemical processing (Rogge et al., 1993; Schauer et al., 1996; Schauer and Cass, 2000)
Elements—nondestructive	X-ray fluorescence[a] (XRF)	None	Primarily provides data on crustal-related species (Si, Ca, Fe) and S, Zn, and Pb. Many other atmospheric species are usually below the limit of detection. See Watson et al. (1999)
	Proton-induced X-ray emission (PIXE)	None	Differs from XRF only in the method used to generate the fluorescence, i.e., high-energy protons versus X-rays (U.S. EPA Method IO-3.6) (Cahill et al., 1987)
	Instrumental neutron activation analysis (INAA)	Filters are folded or made into pellets and sealed in polypropylene bags or vials	With the exception of C, Si, Ni, Sn, and Pb, INAA provides quantitative results for many trace elements typically observed in ambient aerosols (Lodge, 1989)
Elements—destructive	Inductively coupled plasma mass spectrometry (ICP-MS)	Liquid extraction in water or acid solution	ICP-MS provides detection limits on the order of 1 to 100 parts per trillion for approximately 65 elements with a linear dynamic range in excess of eight orders of magnitude (Grohse, 1999)
	Semicontinuous metals monitor	Aerosols are collected by steam condensation and analyzed by GFAAS	A semicontinuous method that allows for determination of up to six elements simultaneously (Kidwell and Ondov, 2000)
	Single-particle mass spectrometry	None	In this mostly qualitative method, the chemical composition of single particles are measured in real time (Prather et al., 1994; Murphy and Thomson, 1995; Johnston and Wexler, 1995; Hinz et al., 1996; Noble and Prather, 1996)

[a] Indicates methods Employed in the EPA's National PM-2.5 Chemical Speciation Network.
[b] Limited number of analyses in PM-2.5 Chemical Speciation Network.

TABLE 11–2. PM-10 Mass and PM-2.5 Mass and Chemical Compositions at Selected Locations Throughout the United States

Species[a]	Atlanta, GA, August 1999[b] (n = 15)	Rubidoux, CA, Jan–Feb 1999[c] (n = 14)	Shenandoah National Park, VA, May 1998–May 1999[d] (n = 102)	Meadview, AZ, May 1998–May 1999[b] (n = 76)
	Concentrations (Average ± Standard Deviation) in μg/m³			
PM-10 mass	NA	38.4 ± 24.6	13.9 ± 9.2	9.2 ± 32.7
PM-2.5 mass	32.9 ± 9.0	26.4 ± 22.5	11.4 ± 8.1	3.1 ± 4.2
Sulfate	10.8 ± 4.1	1.6 ± 1.6	3.4 ± 3.6	0.7 ± 0.5
Nitrate[e]	0.61 ± 0.11	10.0 ± 9.6	0.3 ± 0.3	0.1 ± 0.1
Ammonium	3.7 ± 1.4	3.5 ± 3.7	NA	NA
OC	9.3 ± 2.4	6.6 ± 3.3	1.3 ± 0.9	0.5 ± 0.3
EC	0.86 ± 0.30	2.7 ± 1.7	0.8 ± 0.5	0.3 ± 0.1
	Concentrations (Average ± Standard Deviation) in ng/m³			
S	4100 ± 1500	640 ± 620	1500 ± 1300	20 ± 180
Si	180 ± 75	160 ± 120	120 ± 90	250 ± 1100
K	58 ± 15	86 ± 44	40 ± 20	37 ± 110
Ca	71 ± 42	160 ± 120	34 ± 25	54 ± 170
Mn	3.2 ± 1.9	3.7 ± 2.7	2.5 ± 10.2	1.9 ± 9.0
Fe	130 ± 42	180 ± 120	31 ± 20	42 ± 150
Cu	3.9 ± 3.2	5.0 ± 3.5	1.5 ± 10.1	1.3 ± 8.7
Zn	14.7 ± 8.9	60 ± 77	6.5 ± 10.0	2.2 ± 8.7
Pb	5.1 ± 3.3	11 ± 10	3.1 ± 10.0	1.6 ± 8.7
As	2.1 ± 0.8	0.4 ± 0.6	1.3 ± 10.1	1.2 ± 8.8

[a] This is a partial list of species that are usually observable above limits of detection (LOD). Other species often measured in ambient aerosols by filter methods are given with their LOD in Chow (1995).
[b] Solomon et al. (2000b).
[c] Tolocka et al. (2000).
[d] Calculated from publicly available IMPROVE data located at http://alta_vista.cira.colostate.edu/; go to FTP site, data, IMPROVE, aerosol, SHEN1, and MEVE1.
[e] Obtained using denuders and reactive filters. This value is typically higher than nitrate concentrations obtained from a Teflon filter.

MASS MEASUREMENTS

The U.S. Environmental Protection Agency has promulgated new primary and secondary National Ambient Air Quality Standards (NAAQS) for respirable particles or fine particulate matter (PM-2.5), defined as particles with an aerodynamic diameter less than 2.5 μm. The Code of Federal Regulations (40 CFR, Part 53, Sub-part E) specifies sampler design, performance characteristics, and operational requirements of the FRM for fine particulate matter. The new PM-2.5 standards, as well as the revised standards for inhalable particles (PM-10, defined as particles with an aerodynamic diameter of less than 10 μm), are based on gravimetric analysis of aerosols collected on a filter over a sampling period of 24 h (PM-10: 40 CFR, Part 50, Appendix J; PM-2.5: 40 CFR, Part 50, Appendix L) (U.S. CFR, 1997). Gravimetric analysis was selected because it has been used historically with past particle standards, that is, total suspended particulate (TSP) and PM-10 (TSP: 40 CFR, Part 50, Appendix B and J) (U.S. CFR, 1997). Also, epidemiological studies have found associations between PM mass determined gravimetrically on filters and mortality and morbidity (U.S. EPA, 1996a,b; U.S. EPA, 2001; 40 CFR, Part 50, Subpart II [U.S. CFR, 1997]; Dockery et al., 1993; Pope et al., 1995;

TABLE 11–3. Organic Species Molecular Markers and Associated Emissions Source Used for Source Apportionment of Airborne Particulate Matter

Molecular Marker	Major Urban Sources
n-Pentacosane	Gasoline vehicles, diesel vehicles
n-Hexacosane	Gasoline vehicles, diesel vehicles
n-Heptacosane	Gasoline vehicles, diesel vehicles
n-Octacosane	Gasoline vehicles, diesel vehicles
n-Nonacosane	Vegetative detritus
n-Triacontane	Motor vehicle
n-Hentriacontane	Vegetative detritus, cigarette smoke
n-Dotriacontane	Variety
n-Tritriacontane	Vegetative detritus, cigarette smoke
n-Tetratriacontane	Tire wear debris
n-Pentatriacontane	Tire wear debris
anteiso-Triacontane	Cigarette smoke
iso-Hentriacontane	Cigarette smoke
anteiso-Hentriacontane	Cigarette smoke
iso-Dotriacontane	Cigarette smoke
anteiso-Dotriacontane	Cigarette smoke
iso-Tritriacontane	Cigarette smoke
20S&R-5α(H),14β(H),17β(H)-Cholestanes	Gasoline vehicles, diesel vehicles
20R-5α(H),14α(H),17β(H)-Cholestanes	Gasoline vehicles, diesel vehicles
20S&R-5α(H),14β(H),17β(H)-Ergostanes	Gasoline vehicles, diesel vehicles
20S&R-5α(H),14β(H),17β(H)-Sitostanes	Gasoline vehicles, diesel vehicles
22,29,30-Trisnorneohopane	Gasoline vehicles, diesel vehicles
17α(H),21β(H)-29-Norhopane	Gasoline vehicles, diesel vehicles
17α(H),21β(H)-Hopane	Gasoline vehicles, diesel vehicles
22S-17α(H),21β(H)-30-Homohopane	Gasoline vehicles, diesel vehicles
22R-17α(H),21β(H)-30-Homohopane	Gasoline vehicles, diesel vehicles
22S-17α(H),21β(H)-30-Bishomohopane	Gasoline vehicles, diesel vehicles
22R-17α(H),21β(H)-30-Bishomohopane	Gasoline vehicles, diesel vehicles
Particle-phase nonanal	Meat cooking
n-9-Hexadecanoic acid	Meat cooking
Cholesterol	Meat cooking
Levoglucosan	Hardwood burning, softwood burning
8,15-Pimaradien-18-oic acid	Softwood burning
Pimaric acid	Softwood burning
Isopimaric acid	Softwood burning
Propionylsyringol	Hardwood burning
Butyrylsyringol	Hardwood burning
Benzo[k]fluoranthene	Gasoline vehicles, natural gas comb.
Benzo[b]fluoranthene	Gasoline vehicles, natural gas comb.
Benzo[e]pyrene	Gasoline vehicles, natural gas comb.
Indeno[1,2,3-cd]fluoranthene	Gasoline vehicles, natural gas comb.
Indeno[1,2,3-cd]pyrene	Gasoline vehicles, natural gas comb.
Benzo[ghi]perylene	Gasoline vehicles, natural gas comb.
Coronene	Noncatalyst cars
Elemental carbon	Diesel exhaust
Aluminum	Road dust
Silicon	Road dust

Source: Schauer (1998).

Schwartz et al., 1994, 1996; Thurston et al., 1994). Gravimetric mass also will be measured in the Chemical Speciation Network at up to 300 sites nationwide (U.S. EPA, 1996a,b; U.S. EPA, 2001).

Gravimetric analysis of aerosols collected on filters is a difference measurement, where filters are weighed before and after sampling under controlled conditions. The most widely used sampling media are Teflon, quartz fiber, Teflon-coated quartz fiber, or glass fiber. Properties of these filter media and others are described by Chow (1995) and in Chapters 9 and 27. A detailed protocol has been established for the gravimetric analysis of aerosols collected on filters (U.S. CFR 1997, Part 50 Appendices J and L). For PM-2.5, this protocol requires equilibration of filters within a narrow range of temperature (293 to 296 K [20° to 23°C] and not to vary by more than ±2 K [2°C] over 24 h) and relative humidity (30 to 40% RH and not to vary by more than ±5% RH over 24 h). These controls provide reproducible and standard conditions to account for possible variations in water associated with the filter and the collected aerosols during pre- and postweighing periods. The latter varies among location, species, seasons, and aerosol compositions and cannot be accounted for directly by other methods. Affinity to water or changes in relative humidity are the primary reason hygroscopic filter media (e.g., cellulose and cellulose ester membrane) are not recommended for gravimetric analysis of particulate matter in ambient air (Demuynck, 1978; Tierney and Conner, 1967; Chow, 1995).

Mass concentrations of aerosols in ambient air are obtained by weighing filters before and after sampling using an analytical balance with a sensitivity of ±100 μg for PM-10 collected on standard 0.20 by 0.25 m^2 [8 by 10 in^2] filters and ±1 μg for fine particles collected on 37 mm or 47 mm diameter Teflon filters. These analytical limits of detection (LOD) for the balance correspond to atmospheric limits of detection of 0.06 μg/m^3 for PM-10 and 0.04 μg/m^3 for PM-2.5. The LOD for PM-10 assumes the sample was collected over a 24 h sampling period, on a standard filter (0.18 by 0.23 m^2 [7 by 9 in^2] sampling area), and operated at a flow rate of 0.0189 m^3/s [40 ft^3/min]. The LOD for PM-2.5 assumes the sample was collected over a 24 h sampling period on a 47 mm filter and operated at a flow rate of 2.78×10^{-4} m^3/s [16.7 L/min or 1 m^3/h].

To achieve high accuracy and precision, the weighing area must be free of vibrations and the balance must be shielded from air currents. Static charge must be removed from the filters before weighing to minimize electrostatic attraction to balance surfaces (PM-10: Part 40, Part 50, Appendix J [U.S. CFR, 1997]; PM-2.5: Part 40, Part 50, Appendix L [U.S. CFR, 1997]; Lawless and Rodes, 1999). Potential errors due to electrostatic charge are especially significant with Teflon and other membrane-type filters used in low-volume samplers. Other factors influencing the accuracy and precision of aerosol mass measurements include sampling artifacts (see Chapter 27) and sampling, transport, and storage procedures. Precision of the mass measurement is usually obtained by reweighing filters and reporting their reproducibility or from collocated measurements. Precision values within 5% are typically obtained (40 CFR Part 58, Appendix A [U.S. CFR, 1997]; U.S. EPA, 1998).

WATER-EXTRACTABLE ANION AND CATION ANALYSIS METHODS

Overview

Ionic species can comprise the bulk of the mass of secondary atmospheric particulate matter (Tolocka et al., 2001; Solomon et al., 1988; Chow and Watson, 1999; U.S. EPA, 2001). Furthermore, the ionic content of the aerosol can be used to estimate source attributions at a receptor (Chow et al., 1992; Watson et al., 1994; Magliano et al., 1999). Ion chromatography is a widely used multispecies technique employed for anion and cation analysis and is the method of choice for the Chemical Speciation Network (U.S. EPA, 1999). Concentrations in water of both simple ions, such as chloride, potassium, and sodium, and polyatomic ions, including sulfate, nitrate, and organic acids, can be determined by ion chromatography. Colorimetry and ion-selective electrodes have been developed to measure a few ions, such as ammonium and hydrogen ions, in aqueous extracts of aerosol samples collected on filters. Finally, there is the possibility of sample loss and degradation during extraction, so methods have been developed or applied to measure a few species directly on the filters. These include, for example, Fourier transform infrared (FTIR) spectroscopy, an optical technique, and XRF, which have been used to determine ions such as sulfate or bisulfate directly in the aerosol sample collected on the filter media.

Aqueous Extraction

The extraction of atmospheric particulate matter from glass fiber and quartz fiber filters for the analysis of anions and cation species (e.g., NO_3^-, SO_4^{2-}, Cl^-, K^+, Na^+, and NH_4^+) presents few problems (Chow, 1995; Chow, and Watson, 1999). Filters that are hydrophilic, such as paper, glass fiber, or quartz fiber filters, typically have extraction efficiencies for water-soluble species approaching 100% (Appel et al., 1980b, 1981; Chow and Watson, 1999). Teflon filters, which are hydrophobic, however, have much lower extraction efficiencies and require special treatment before extracting water-soluble compounds with an aqueous solution. For example, based on the work of Derrick and Moyers (1981), many groups "wet" Teflon filters with 50 μl of ethanol just before extraction. The U.S. EPA (1989) suggests 100 μl of ethanol followed by extraction in 10^{-4} N $HClO_4$. After filters are wet, extraction usually is performed by shaking or using an ultrasonic bath for several hours to overnight (U.S. EPA, 1989; Chow and Watson, 1999). Maintaining the sample at reduced temperatures minimizes the potential loss of volatile species such as nitrate or ammonium nitrate during the extraction process.

Nylon filters are often used to efficiently collect aerosol nitrate and/or nitric acid. Nitrate bound to Nylon is not efficiently recovered as NO_3^- with water extraction (Sickles et al., 1986); however, an alkaline solvent (e.g., the dilute CO_3^{2-}–HCO_3^- buffer employed as eluent in ion chromatography) provides efficient extraction (Hering et al., 1988; Solomon et al., 1988), with the possible exception of very lightly loaded samples (Appel et al., 1988).

Total strong acidity, H^+ or H_3O^+, can be measured by extracting particles collected on a Teflon filter with water (Appel et al., 1980a; Appel, 1993; Koutrakis et al., 1992; Chow, 1995). However, analysis requires the elimination of dissolved CO_2 and a technique to eliminate or allow for the dissociation of weak acids. Strong acidity, as well as strong acid anions in particles, can be obtained using Teflon filters (47 mm) wet with 100 ± 5 μl of ethanol followed by ultrasonic extraction in 10^{-4} N $HClO_4$ (U.S. EPA, 1989). The use of a pH 4–extraction solvent represses the dissociation of weak acids and minimizes the concentration of dissolved CO_2. The hydrogen ion concentration is then measured using a combination pH meter equipped with a semimicroelectrode. Although the accuracy of hydrogen ion measurements (about 10%) is less than that of the related ionic species (e.g., sulfate and ammonium at about 5%), their accuracy is considerably better than measurements of carbonaceous aerosols and nitrate (about 15%) (Koutrakis et al., 1988).

Sampling Requirements

The filter material from which water-soluble ions will be extracted must be hydrophilic in nature, allowing the aqueous solution to penetrate the filter. As discussed above, Teflon filters must be "wet" before extraction, while a weak base must be used to efficiently extract Nylon filters for nitrate.

Ion Chromatography

Ion chromatography is a widely used method to rapidly determine the concentrations of anions and cation species in aqueous solutions (Mulik et al., 1978; Appel, 1993; Chow and Watson, 1999; U.S. EPA, 1999; U.S. EPA 2001). An excellent basic description of ion chromatography and related methods is given by Weiss and Johnson (1986). In ion chromatography, an aliquot of the extracted sample is injected into the ion chromatography sampling port. The eluent, in most cases, for strong acid anions is a dilute carbonate-bicarbonate solution; a dilute solution of boric acid is used for weak acid anions and a dilute solution of HCl for cation species.

Depending on the species to be analyzed, an anion or cation guard column is used to remove organic compounds and other interfering components from the sample. Anions are separated using a basic ion exchange resin, while cation species are separated using an acidic ion exchange resin (Weiss and Johnson, 1986; Chow and Watson, 1999). The ion exchange column is followed by the eluent suppressor column, which converts the ions in the solvent (e.g., carbonate or bicarbonate for anions) into less conductive forms (e.g., H_2CO_3) while converting the analytes into their acidic forms that are strongly conductive. This increases the signal-to-background ratio in the detector, allowing low, often sub-ppb concentrations, of the analyte to be detected (Mulik et al., 1978; Small, 1981; Colenutt and Trenchard, 1985; Lodge, 1989; Chow and Watson, 1999). Ions are detected using an electroconductivity detector. The determination of SO_4^{2-}, NO_3^-, and NH_4^+ can be made over a range of at least three orders of magnitude using this technique, called *chemically suppressed anion chromatography*. The instrument detection limit (IDL) values vary with the system configuration. Other species routinely determined by ion chromatography in atmospheric particles include Cl^-, F^-, formic acid, acetic acid, K^+, Na^+, NH_4^+, and many others (Weiss and Johnson, 1986; Mulik et al., 1978; Chow and Watson, 1999).

Ion Selective Electrodes

Ion selective electrodes (ISEs) have been used to determine concentrations of sulfate, nitrate, and ammonium ions in aqueous extracts of atmospheric particulate matter collected on filters (Clarke et al., 1999; Saleh et al., 1999; Appel et al., 1988). ISEs are membrane-type electrodes that produce a potential when immersed in a solution containing the appropriate ion. The potentiometric response is obtained by selectively extracting the ion of interest into the membrane by the appropriate counterion. The counterion is constrained within the membrane because the organic matrix is hydrophobic. The potential across the membrane is measured using a pH meter with an expanded scale in contact with a reference electrode on either side of the membrane. ISEs are easy to deploy and are inexpensive. However, they are prone to interferences from other ions in solution. The method for ammonium is described by Appel (1993). For ammonium, ISEs have an operating range from about 0.04 to 1700 μg/ml. Single laboratory precision values of 2% to 4% have been obtained for ammonium, with recoveries on the order of >90% from Teflon filters.

Colorimetry

Colorimetry, as the name implies, refers to measuring the color change of a solution by absorption spectroscopy. Today, several manufactures sell automated colorimetric systems for the determination of a variety of species. The most common ions measured are SO_4^{2-}, NO_3^-, Cl^-, and NH_4^+ (Butler et al., 1978; Mueller et al., 1979; Fung et al., 1979; Pyen and Fishman, 1979). A sensitive, selective, and rapid (up to 50 samples per hour) automated method is often employed to measure NH_4^+ based on the indophenol blue method (Bolleter et al., 1961). Sulfate is measured using the methylthymol-blue method, while nitrate is reacted with sulfanilamide to form a diazo compound. In each case, the ion in question is transformed quantitatively into a species that absorbs light at a particular wavelength. The amount of light absorbed is proportional to the amount of analyte in the sample extract. The Beer Lambert law describes the proportionality. Comparison studies between ion chromatography and automated colorimetry have found excellent agreement between the two methods for sulfate, nitrate, and ammonium (Butler et al., 1978; Fung et al., 1979).

X-Ray Fluorescence

XRF is used routinely to determine sulfur concentrations in atmospheric particles collected on Teflon filters as described in this chapter (see below for a description of the technique). Experience indicates that sulfur concentration times 3 (ratio of the molecular weights of S and SO_4^{2-}) is a reasonable estimate of the water-soluble sulfate concentration for most ambient aerosol samples (Dzubay and Stevens, 1975; Chow, 1995; Solomon et al., 2000a). Comparison of sulfur times 3 to sulfate also is used routinely as a quality assurance check of the data. Other alkali metals can be obtained by XRF; however, these values represent total metal content and not water-soluble content, which may be quite different. For example, water-soluble K^+ is a good tracer for wood smoke burning; however, it is not correlated with nonwater-soluble K, which derives from soil and sea salt emissions.

Fourier Transform Infrared Spectroscopy

Most polyatomic ions found in aerosols absorb infrared radiation. Nitrate, for example, has strong vibrational transitions that occur at $1400\,cm^{-1}$ ($\lambda = 7.14 \times 10^{-6}\,m$) while ammonium has medium-strength vibrational transitions in the same region. Infrared spectra are obtained using an FTIR spectrometer. A Michelson Interferometer, located inside the spectrophotometer, is used to measure and separate or deconvolute the composite signal (Ingle and Crouch, 1988). The interferometer splits the light beam from the sample in two, varying the path length taken by the separate beams. This introduces a phase difference in the signal between the two paths, and they recombine at the detector either destructively or constructively depending on the path length difference and the frequency of the absorption signal.

The resulting intensity observed at a detector is an interferogram, a time-dependent signal that contains the frequencies allowed to reach the detector (i.e., constructive recombination). This time-dependent signal is transformed into a frequency signal using a Fourier transform, a mathematical conversion. Species are detected by passing the infrared light through the filter substrate (McClenny et al., 1985) or impaction onto a crystal (Johnson et al., 1983). Background subtraction is used to remove absorption features due to the C–C stretching and bending modes found in Teflon filters.

FTIR can provide mass sensitivities in the high picogram to low nanogram range (Allen and Palen, 1989). However, the method is not directly quantitative and suffers from interferences. Concentration calibrations have been obtained relative to filter samples that have been analyzed by another method (Allen et al., 1994). Determination of sulfate, nitrate, ammonium, and organic species by functional groups on low-pressure impactor stages has resulted in unique aerosol size distribution information important for understanding aerosol chemistry (Allen et al., 1994).

Many other methods exist for determining the concentration of anion and cation species in aqueous solutions. Those presented here are some of the most common methods, including ion chromatography and the FTIR method, the latter of which is still under development. Ion chromatography is the method of choice for the Chemical Speciation Network. Other methods discussed here, such as colorimetry and XRF, are useful tools but report only a single water-soluble anion or cation of interest per analysis and are not necessarily quantitative. For example, ammonium ion is determined by colorimetry from aqueous extracts and sulfur by XRF, where the sulfate concentration is estimated from the XRF value by multiplying XRF sulfur concentration by three. The FTIR method is still under development for its application to atmospheric aerosols.

PARTICULATE CARBON

Particulate carbon is a major component of PM-2.5, typically accounting for 25% to over 50% of the mass for samples collected in the eastern and western United States (U.S. EPA, 1996a, 2001). Recent measurements in the eastern United States during the winter exceeded 70% OC (Tolocka et al., 2001). Particulate carbon is operationally classified into elemental or black carbon (EC), organic carbon (OC), and carbonate carbon (CC). Total carbon (TC) represents the sum of EC, OC, and CC. EC is emitted primarily from anthropogenic sources, such as diesel vehicle exhaust, especially in urban areas, due to incomplete combustion. OC is emitted from anthropogenic sources (primary aerosols) and is formed in the atmosphere from gaseous precursors (secondary aerosols). CC is typically associated with soil-related sources (Appel et al., 1983) and usually constitutes less than 5% of the TC (Chow et al., 1993).

A number of techniques have been used to measure particulate carbon collected on filters. Carbon analyses are destructive except for the indirect measurement of black carbon by laser transmission or adsorption (Novakov, 1982; Hansen and Novakov, 1990; Chow, 1995). Direct measurements of carbon are obtained by heating a filter containing collected aerosol and measuring the amount of carbon evolved from the particulate matter (thermal analysis). During the heating process, some of the particulate OC is pryrolized or converted to EC, which may or may not be accounted for by the different methods. Carbon analysis methods differ by the use of direct or indirect measurement of carbon, heating temperatures, length of analysis time at each temperature, rate of temperature increase, atmospheres used for oxidizing the organic compounds, and the method used to adjust for pyrolysis (i.e., light transmittance or reflectance) (Hering et al., 1990; Cadle and Groblicki, 1982; Birch, 1998; Chow et al., 1993b). Direct measurement methods for particulate carbon provide similar total carbon results (Chow, 1995); however, differences in the operational aspects of the methods result in variations between the reported EC and OC concentrations. Direct methods only provide estimates of EC, although multiwavelength light–absorbing methods are now commercially available and provide additional information on the type of carbon (functional groups) present in the sample.

Sampling Requirements

The filter medium used for carbon analysis by thermal methods cannot itself contain carbon, thus eliminating the use of Teflon filters. The most widely used filter media are quartz fiber and glass fiber. Quartz fiber filters provide a lower carbon blank value ($<1\,\mu g/cm^2$) because they can be pre-fired at temperatures up to 1173 K (900°C) (Chow et al., 1993b) compared with <723 K (450°C) for glass fiber filters. Heating the filter to reduce initial carbon blank levels is an important and essential step. Quartz filters are required for the commonly used thermal optical protocols because the highest temperature in the protocols is 1073 K (800°C) or higher. The most common quartz fiber filters used for carbon measurements include 2500 QAT- Ultra Pure (UT) type 7201 (***PAL***)* for 37 mm and 2500 QAT- Ultra Pure (UT) type 7202 (***PAL***) for 47 mm. After samples are collected, sample transport and storage should be at reduced temperatures (Chow, 1995). The effects of storage and shipping temperatures have not been investigated thoroughly; however, recommended temperatures range from

* See Appendix I for full manufacturer addresses referenced to the italicized three-letter codes.

refrigeration for OC and EC determination (<4°C; Chow, 1995) to temperatures from −25°C (Rogge et al., 1991) to −80°C for determination of organic species (Sverdrup et al., 1990). Sampling requirements needed for optical methods are discussed below.

Calibration, Precision, Limits of Detection, and Interferences

A carbon analyzer, using the thermal analysis approach, is calibrated by spiking a filter punch with a solid containing a known amount of carbon, usually sucrose, disodium salt of ethylenediaminetetraacetic acid (EDTA), or potassium hydrogen phthalate (KHP) (Chow et al., 1993b; Birch, 1998). These standards provide both OC and TC standards. An ambient particulate EC standard is presently not available from the National Institute of Standards and Technology (NIST), although EC standards have been prepared by collecting particles from a butane flame (Appel et al., 1983). Injections of CO_2 and CH_4 also can be used to calibrate the response from a nondispersive infrared (NDIR) detector (Cadle and Mulawa, 1990) or a flame ionization detector, respectively. Carbon dioxide also can be used to verify the operation of the methanator, which converts CO_2 to methane. Interferences are minimal in thermal analysis methods, expect for the pyrolysis correction, which has become part of the method itself. Precision and limits of detection are described below within the method descriptions.

The following sections describe several methods for determining OC and/or EC collected on filters. The first, thermal optical with correction for pyrolyzed carbon using light transmittance, is the method currently used in the Chemical Speciation Network for determining atmospheric concentrations of OC, EC, CC, and TC.

Thermal Optical and Thermal Techniques

The two most widely used thermal optical methods are thermal optical transmittance (TOT) (Birch, 1998; Birch and Cary, 1996) and thermal optical reflectance (TOR) (Huntzicker et al., 1982; Chow et al., 1993). The TOT and TOR methods are primarily used with the National Institute for Occupational Safety and Health (NIOSH) Method 5040 (Eller and Cassinelli, 1996; Birch, 1998) and the Interagency Monitoring of Protected Visual Environments (IMPROVE) (Chow et al., 1993a) protocols, respectively. A third, earlier method was developed by Novakov (1982) and is briefly described below. The three methods differ in the way they correct for the amount of OC pyrolyzed to EC during the He heating step, as well as the temperature ramp (temperature and length of time at each step) each employs.

The TOT and TOR methods evolve carbon in four temperature steps from a quartz fiber filter in an He atmosphere to determine OC and in an He/O_2 atmosphere in three thermal steps to evolve EC. Evolved carbon is catalytically first oxidized to CO_2, reduced to CH_4, and quantified with a flame ionization detector (Johnson et al., 1981; Stevens et al., 1982; Huntzicker et al., 1982; Grosjean, 1984; Hering et al., 1990). The split between OC and EC, and thus determination of the pyrolysis correction, in the TOT and TOR methods is based on either transmittance (Birch and Cary, 1996) or reflectance (Chow et al., 1993b). CC is determined by acidifying the sample (Chow et al., 1993). The acidification procedure requires thermal analysis of two identical portions of the sample, one before and one after acidification. The difference between the TC results obtained for the two punches gives an estimate of CC. Alternatively, the carbonate can be estimated by integrating the carbonate peak at 1093 K (820°C) as described below (Birch and Cary, 1996). The latter approach applies only when the carbonate is removed as a single peak during the fourth temperature step in He (e.g., calcium carbonate). The analytical limit of detection for TOR and TOT is about 0.2 $\mu g/cm^2$ carbon (Chow et al., 1993b; Birch and Cary, 1996). Assuming a sample collected at $2.78 \times 10^{-4} m^3/s$ [16.7 L/min or 1 m^3/h] for 24 h and a punch of 100 mm^2 (1 cm^2) from a 47 mm filter (deposit area = 1200 mm^2 [12.0 cm^2]), the atmospheric limit of detection is 0.1 $\mu g/m^3$. A semicontinuous TOT method has been evaluated using thermal optical transmission by

Turpin et al. (1990) and operated in Atlanta, GA, as part of an advanced methods intercomparison study during the summer of 1999 (Turpin et al., 2001).

Novakov (1982) measured TC by linearly heating a sample in an O_2 atmosphere and measuring the CO_2 with NDIR. The transmittance of a 633 nm HeNe laser is monitored as the temperature is increased at 10°C/min. The split between EC and TC is determined when the transmittance starts to decrease, indicating removal of EC (around 743 K [470°C]). A modification of this method, temperature-programmed volatilization (TPV) (Eatough et al., 1995), is well suited for determining OC from carbon-impregnated filters used as backup filters behind a quartz fiber filter to estimate carbon volatilized from the first filter (Tang et al., 1994). In the modified method, N_2 is used as the carrier gas to prevent oxidation of the EC and OC is determined by heating the filter to 623 K (350°C).

Thermal methods use a combination of temperature and an inert gas with oxidizing agents to distinguish between EC and OC. Separation of OC and EC has been obtained by heating a sample in an inert atmosphere with MnO_2 (Stevens et al., 1982; Mueller et al., 1982; Grosjean, 1984). OC was determined at 623 K (350°C) and then at temperatures greater than 1123 K (900°C) (Fung, 1990); the MnO_2 provides the O_2 required to oxidize the EC to CO_2. Carbon dioxide is then measured with an NDIR detector or an electrochemical cell (Countess, 1990). Other methods heat the samples in He to determine OC and use an He/O_2 atmosphere to measure EC. Evolved carbon is either measured as CO_2 or catalytically reduced to CH_4. Methane is quantified with a flame ionization detector. ***R&P*** manufacturers and sells a continuous carbon analyzer (Series 5400) that heats samples at 613 K (340°C) and 1023 K (900°C) to quantify OC and EC. No optical measurements are made to quantify pyrolysis. Samples are collected for 1 h at a flow rate of $2.78 \times 10^{-4}\,m^3/s$ [16.7 L/min or $1\,m^3/h$] and have an atmospheric limit of detection of about $0.2\,\mu g/m^3$.

Optical Techniques

Particulate matter collected on filters appears gray to black due to the presence of EC in the particulate matter (Rosen et al., 1978), with higher loadings producing darker colors. Organic material, on the other hand is typically transparent to visible light. Thus, optical methods only can be used to estimate concentrations of EC or black carbon in atmospheric particulate matter samples. Light absorption by particles collected on a filter is linearly related to the EC concentration in the collected particles by Beers Law (Novakov, 1982, 1984; Gundel et al., 1984). The relationship between the amount of light absorbed and the EC concentration (i.e., the amount on the filter)—the absorption coefficient—is determined experimentally and can vary depending on the source of EC and the collection media (Appel et al., 1983). Gundel et al. (1984) did not find differences depending on the source of EC and estimated a typical absorbance coefficient of $25.4\,m^2/g$ at 632 nm for particles collected on quartz fiber filters from a variety of combustion sources (Hansen and Novakov, 1990).

An aethalometer uses filter transmission to continuously measure the EC collected on a quartz fiber filter (Rosen et al., 1978; Hansen and Novakov, 1990). A commercial Aethalometer (***MAG***) uses a continuous quartz fiber filter tape. The instrument wavelength for a Model AE16 is 880 nm, and the absorbance coefficient is $16.7\,m^2/g$. The limit of detection for an older Aethalometer that used an incandescent lamp as its light source (AE- 9) was $100\,ng/m^3$ black carbon for a 1 h measurement (Allen et al., 1999). Recently, ***MAG*** has developed duel and multi wavelength aethalometers. These provide an improved estimate of EC concentrations as well as insight into other organic categories in the collected particulate matter.

Comparison of Methods

Cadle and Groblicki (1982) evaluated the measurement of EC on filters by comparing several methods: organic extractions, nitric acid digestion, vacuum stripping, thermal methods, and the integrating plate method. Systematic and correlated differences were found between the methods. Hering and colleagues (1990) compared seven variations of thermal carbon analysis methods. Similar results were found for OC and TC, but the EC concentrations varied between the methods. Cadle and Mulawa (1990) conducted a round-robin methods comparison study with 11 laboratories that used both thermal and thermal optical techniques. EC concentrations had the largest difference between methods, and differences were observed for OC, EC, and TC for diesel and automobile exhaust source samples. Birch (1998) conducted a study that compared TOC (NIOSH Method 5040) (Eller and Cassinelli, 1996; Birch, 1998), TOR (Chow et al., 1993), and thermal techniques with coulometric detection of CO_2. Similar TC results were found in the study, but the EC concentrations for the coulometric techniques were 2.5 to over 800 times higher than the TOT method, depending on the sample type. EC concentrations by TOR were 1.25 to 10 times higher than the EC concentrations obtained from TOT. Two recent comparison studies have indicated the primary difference between TOR (IMPROVE protocol) and TOT (NIOSH protocol) is due to their different temperature programs (Chow et al., 2001; Norris et al., 2001). EC typically varied by a factor of 2 (TOR > TOT) and OC by about 10% to 20% (TOT > TOR). The various carbon analysis methods showed more consistent agreement for TC; however, the measurement of EC is operationally defined, adding to its variability among the methods.

Organic Speciation Analysis of Aerosols

OC represents a large fraction of fine-particle mass (e.g., Gray et al., 1986; Solomon et al., 1989; Chow et al., 1993b, Tolocka et al., 2001) and is composed of many compounds, most of which partition between the gas and aerosol phases at ambient conditions and are referred to as *semivolatile organic compounds* (SVOC). As well, many of the compounds exist only as aerosols or just in the gas phase (volatile organic compounds [VOC]). It is the SVOC compounds that make filter sampling for organic aerosols a challenge, in a similar fashion to using filters to collect ammonium nitrate (Russell et al., 1983). In addition, gas phase organic compounds (VOC or SVOC) can be adsorbed onto the filter (known as a *positive artifact*), or SVOC compounds in aerosols collected on the filter can volatilize back into the gas phase during sampling (known as a *negative artifact*) (Tang et al., 1994; Eatough et al., 1990; Turpin et al., 1994). The debate on the magnitude of organic artifacts, positive and negative, is far from being concluded and appears to depend on many yet to be identified variables. Knowing more about the individual components of the organic aerosol in conjunction with the gas phase semivolatile species and VOC will help to resolve these issues.

Organic aerosol compound speciation also can provide a great deal of information regarding the sources and formation processes of carbonaceous particles (Schauer et al., 1996). A very limited number of samples collected in the Chemical Speciation Network are being analyzed for a series of individual organic compounds. Although speciation is desirable, it is not easy to perform because there is no single analytical method that can be used to analyze all classes of organic compounds. Usually, nonpolar organic compounds are analyzed using thermal desorption or solvent extraction followed by gas chromatography (GC) and mass spectrometry (MS) (Mazurek et al., 1987, 1993; Rogge, 1993; Rogge et al., 1993; Schauer et al., 1996). Analysis of polar organic compounds is more challenging because special sample preparation is required (e.g., derivatization) for the different compound classes (Blando et al., 1998; Hawthorne et al., 1992; J. O. Allen et al., 1997). Comprehensive determination of organic speciation is expensive and yields a large data set that may be time consuming to process and incorporate into air quality models. It may be more cost effective to develop inexpensive methods that allow the measurement of specific characteristic groups of organic compounds (Turpin, 1998). Similar lumping approaches for VOC species have been used successfully in ozone modeling.

The most common method for identifying and quantifying individual organic compounds found in ambient aerosol samples is gas chromatography coupled with a mass spectrometer as the detector (GC-MS) (Simoneit, 1986; Mazurek et al., 1987; Rogge et al., 1993; Rogge, 1993; Schauer et al., 1996). Samples are collected on preheated quartz fiber filters and extracted with various solvents depending on the species of interest. An aliquot of the extract is injected into the front end of the GC column. The organic compounds are then separated in the GC column based on the differential distribution of their components between the stationary and mobile phases as the sample passes through columns containing both phases. The distribution of the components, and thus the time they take to reach the end of the column or the detector, is determined by the chemical or physical attributes of the analyte and/or stationary phase. Peak areas are integrated to determine the concentration of that compound in the sample aliquot. The theory of separation by chromatography and discussion of important variables are found in Karger et al. (1973) and Grob (1995). After separation, the samples associated with individual peaks are injected into an MS for identification.

In the laboratory, samples are extracted via extraction protocols (e.g., Soxhlet, sonication, supercritical fluid) that are custom designed to optimize the removal of nonpolar and polar organic compounds from the quartz fiber filters (Paputa-Peck et al., 1983; Nielsen et al., 1984; Kamens et al., 1985; Bayona et al., 1988; Rogge et al., 1993; Rogge, 1993; Hildemann et al., 1994; Turpin et al., 1994; J. O. Allen et al., 1997). Typically, sample extracts can be concentrated using rotary evaporation. This is normally followed by further solvent evaporation with a high-purity stream of N_2 or other inert gas such as argon. The concentrated extract can be analyzed by chromatography using silica gel or alumina coated columns, divided into aliquots for derivitazation, or directly analyzed. Extracts from the fine carbon particle samples can be analyzed using liquid or gas chromatographic methods. Some compounds of interest used in receptor modeling (Schauer et al., 1996) are listed in detail in Table 11–3.

ELEMENTAL ANALYSIS BY NONDESTRUCTIVE TECHNIQUES

This section describes three multielemental analysis techniques that are nondestructive for the elements of interest; however, other species in the sample may be lost due to volatilization from sample heating or being placed in a vacuum (Solomon et al., 2000a). The Chemical Speciation Network uses XRF analysis for determining concentrations of elements in ambient aerosols collected on Teflon filters. The IMPROVE visibility network (Eldred et al., 1998) uses proton-induced X-ray emission (PIXE), and that method is described in this section as well. Finally, instrumental neutron activation analysis is described, a sensitive multielement method that complements XRF and PIXE.

X-Ray Fluorescence Analysis

XRF methods are common analytical techniques for quantifying a broad range of minor and trace elements in ambient aerosol collected on filters. Several comprehensive reviews of the theory and application of XRF methods have been published (Dzubay and Stevens, 1975; Lodge, 1989; Watson et al., 1999). XRF methods are convenient for elemental analysis because filters require no preparation, and the method is applicable to about 45 elements between Na and U (Stevens, 1984; Lodge, 1989; Chow, 1995; U.S. EPA, 2001; Chow and Watson, 1999). XRF analysis can be qualitative by comparing sample X-ray spectra to a library of known element spectra or quantitative by calculating spectral peak areas and relating them to calibration standards with corrections for X-ray attenuation (Dzubay and Stevens, 1991). Atmospheric concentrations are obtained by dividing the mass of the element calculated per filter by the volume of air sampled through the filter.

Summary of Technique. In general, XRF analysis involves irradiating a sample with a beam of X-ray radiation. The elements in the sample are excited by the X-rays and upon return to the ground state emit characteristic X-rays that are subsequently measured by an X-ray detector. XRF analysis methods can be broadly divided into two categories: wavelength dispersive (WDXRF) that measures elements sequentially using a crystal spectrometer (Wagman et al., 1977) and energy dispersive (EDXRF) that measures elements simultaneously using a solid-state silicon diode detector (Jaklevic et al., 1981; Watson et al., 1999). WDXRF features high spectral resolution but requires high power excitation to overcome a relatively low sensitivity for heavy elements (e.g., Pb) (Chow and Watson, 1999). WDXRF is rarely used for analysis of atmospheric samples because the high X-ray energy will destroy the filter and volatilize some elements (e.g., S and Cl) during the analysis. EDXRF features high sensitivity but requires a sophisticated multilinear spectral deconvolution procedure to eliminate interelement interferences (Stevens et al., 1978). However, techniques are well established to correct for these interferences.

XRF analysis techniques also differ by the method used to excite the elements in the sample, either direct or indirect excitation. In direct excitation, the X-ray beam is filtered and projected directly onto the sample, whereas in indirect excitation the initial X-ray beam is directed onto a secondary target and the X-rays emitted from that target are used to excite the sample with nearly mono-energetic and low-intensity radiation. High sensitivities for a wide range of elements are achieved using secondary targets consisting of titanium, molybdenum, and samarium and varying the X- ray tube voltages (Stevens et al., 1978). The use of secondary targets reduces spectral background, minimizes interelement interferences, and results in lower sample heating. Atmospheric detection limits for aerosols collected on Teflon filters range from 2 to 200 ng/m^3 for 44 of 49 elements (Lodge, 1989). However, at current pollution levels found in most U.S. cites, only about 12 to 15 elements, including S, Pb, and a number of crustal-related elements (e.g., Fe, Ca, Mg, Si, Ti), are observed above the detection limit for 24 h low-volume (16.7 L/min) samples. Additional information regarding atmospheric limits and filter media blank values can be found in Chow (1995), Appel (1993), Chow and Watson (1999), and Watson et al. (1999).

XRF is generally described as nondestructive; however, some lighter elements (e.g., S and halides) have the potential to volatilize from heat generated by direct radiation and from being in a vacuum generated to prevent air from attenuating X-rays. Other species, such as nitrate, also volatilize from the filter during analysis (Solomon et al., 2000a). Lower evaporative losses are expected during analysis using indirect radiation in an atmospheric pressure helium environment currently in use at the X-ray laboratory of the U.S. EPA (C. W. Lewis, U.S. EPA, NERL, Research Triangle Park, NC, 2000; Solomon et al., 2000a).

Sampling Requirements. XRF analysis of atmospheric aerosols collected on filters requires a lightly coated uniform deposit of particles on the surface of the filter. Heavily loaded filters require significant absorption corrections (Dzubay and Stevens, 1991), which depend on particle size and composition and are of particular concern for low atomic number elements, such as Si and Al. Separating fine particles (< PM-2.5) and coarse particles (< PM-2.5, often with an upper limit of PM-10 or PM-15) as done with the dichotomous sampler helps to minimize particle size absorption effects (Dzubay and Stevens, 1991).

Teflon and polycarbonate membrane filters are ideally suited for XRF analysis because they are relatively thin, have high particle collection efficiencies, and are free from significant contamination. Glass fiber and quartz fiber filters are not applicable for XRF analysis because they have relatively high blanks for many elements of interest and because particles are entrained in the filter matrix rather than on the surface. As a result, particle concentrations of light elements are underestimated because the filter medium attenuates soft X-rays.

Precision and Accuracy. The sources of uncertainty in XRF analysis include the stability of the instrument, accuracy of the calibration standards, uncertainties in the attenuation correction, uniformity of sample deposit, and variability in the counting statistics (Stevens et al., 1984; Appel, 1993; Watson et al., 1999). Because multielement standards that exactly duplicate the attenuation effects of all unknown samples do not exist, some laboratories rely on a series of calibration standards incorporated into a semiempirical model that derive calibration constants for each element. This methodology provides improved calibration for elements with accurate standards and accurate calibration for elements that have poor standards (Dzubay and Stevens, 1991). Precision values for atmospheric samples obtained from repeated measurement of the same sample range from 5% to 20%.

Proton-Induced X-Ray Emission Analysis

Proton-induced X-ray emission (PIXE) analysis differs from XRF only in the method used to generate the fluorescence (i.e., high-energy protons versus X-rays). Analyses of elements ranging in atomic number from Si to Pb are routinely obtained. A comparison of PIXE with other analysis methods is reported by Waetjen et al. (1983). The PIXE method has the added capability of analyzing small-diameter samples (2 to 3 mm), which permits analysis of single-orifice–type impactors and rotating drum samplers with high time resolution (U.S. EPA 1999b; Cahill et al., 1987).

Summary of Technique. In PIXE, a high-energy proton beam from a cyclotron or van de Graff accelerator is used to excite the elements in a sample. The excited elements give off X-rays, similar to XRF, and are detected using an X-ray energy-dispersive detector (e.g., silicon diode containing a small amount of Li; Nelson, 1977). The higher excitation energy of PIXE can volatilize light elements and damage membrane filters, creating problems with samples requiring long irradiation times (Watson et al., 1999). The precision and accuracy of the PIXE method are comparable with those of the EDXRF methods. However, when samples are collected for PIXE analysis, often the sample is concentrated onto a 2 to 3 mm diameter spot. Thus, the sample is often at a higher concentration per unit area than samples collected for EDXRF analysis. For example, in the EPA National Fine Particle Network, all samples are collected on 47 mm diameter Teflon filters, whereas in the IMPROVE Network samples collected for PIXE analysis are concentrated onto a (3 mm) diameter spot. The same interferences that affect EDXRF methods also affect PIXE methods.

Sampling Requirements. As with XRF, the best results for PIXE are obtained with thin high-purity filters (e.g., polycarbonate and Teflon) or impaction surfaces (e.g., Kapton and Mylar) (U.S. EPA Method IO-3.6). Polycarbonate filters have low density per unit area and are typically free from contamination. Particles collected on polycarbonate filters are retained on the surface rather than in a membrane web, as is the case with Teflon filters. This has some advantages when concentrations are not greater than 10 to 20 $\mu g/m^3$. Above those concentrations, particles may fall off the surface during transport to the laboratory for analysis. However, having the particles on the surface makes scanning electron microscopy analysis of individual particles feasible (see Chapter 12).

Instrumental Neutron Activation Analysis

Instrumental neutron activation analysis (INAA) is an excellent method for quantifying a broad range of trace elements in atmospheric aerosols because it is one of the most sensitive, selective, and reliable methods available (Lodge, 1989). The factors limiting widespread uses of INAA are its relatively high cost and the limited availability of experienced investigators with access to a nuclear reactor and associated detection equipment. With the exception of carbon, silicon, nickel, tin, and lead, INAA provides quantitative results for many trace elements typically observed in ambient aerosols. Several researchers have combined the results of XRF and INAA to construct comprehensive data sets of trace element observations for particulate matter source apportionment studies (Tuncel et al., 1985; Dzubay et al., 1988).

Summary of Technique. In general, INAA involves irradiating filter samples with thermal or epithermal neutrons in a nuclear reactor, converting some of the target nuclei into radioactive isotopes. As the radioactive isotopes decay, they emit γ-rays of characteristic energies and intensities proportional to the amount of the element in the sample. The γ-rays are detected using high-resolution spectrometers with lithium drifted or intrinsic germanium detectors (Dams et al., 1970; Ondov et al., 1975; Germani et al., 1980; Lodge, 1989; Landsberger, 1999). Quantification is obtained by comparing intensities (peak areas) of the sample to those of standards of known concentration.

INAA typically consists of two irradiation steps, the first for a few minutes to quantify short-lived radionuclides (<12 h). The filters are then allowed to "cool" for a few days while the radioactivity dissipates. The filters are subsequently repackaged and irradiated for several hours to quantify medium- and long-lived radionuclides (>12 h).

Although INAA is considered nondestructive, filters are physically processed in ways that make reanalysis by XRF, PIXE, or scanning electron microscopy virtually impossible. Filters prepared for INAA are either tightly folded or made into pellets to attain a symmetrical geometry before being sealed into polyethylene bags or vials. In the case of Teflon filters, the membrane is typically cut away from the polypropylene support rings and wadded up into a small ball.

Sampling Requirements. Unlike XRF and PIXE analysis, INAA utilizes the entire filter and sample, and has virtually no problem with filter matrix or particle self-absorption. As a result, investigators have much more flexibility in choosing samplers and filter media. Filters with nonuniform deposits (e.g., cascade impactors) or heavily loaded filters are routinely analyzed (Germani and Zoller, 1994; Ondov et al., 1990; Ondov and Divita, 1993). The sole considerations for filter selection are particulate matter collection efficiency and blank concentrations and variability.

Precision and Accuracy. Several factors affect the accuracy and precision of INAA, including changes in neutron flux profile, irradiation timing, inaccurate placement of samples in front of detectors, inaccurate correction for dead time losses and pulse pile-up, poor deconvolution of overlapping peaks in the spectra, sample blanks and vial contamination, and volatilization of elements during irradiation (Orvini and Speziali, 1998; Landsberger, 1999). All of the above factors can be minimized by implementation of a vigorous quality control program, including analysis of certified reference materials, re-irradiating and recounting, spectrum evaluation, analysis by different isotopes or γ-rays, and interlaboratory and multimethod comparison studies.

ELEMENTAL ANALYSIS BY DESTRUCTIVE TECHNIQUES

The U.S. EPA's National PM-2.5 Chemical Speciation Network uses XRF analysis for determining concentrations of elements in ambient aerosols collected on filters. This section describes inductively coupled plasma mass spectrometry, another popular technique used for multielemental analysis of particulate matter collected on filters. Other historically used methods, such as atomic absorption spectroscopy and inductively coupled plasma–atomic emissions spectroscopy, are described by Appel (1993).

Inductively Coupled Plasma Mass Spectroscopy

Inductively coupled plasma mass spectroscopy (ICP-MS) is gaining broad acceptance as a method for determining a wide range of elements in atmospheric particulate matter due to technological advances that continue to improve sensitivity and reduce interferences.

Montaser (1998) comprehensively reviews the fundamental principals of ICP-MS, while Grohse (1999) discusses ICP-MS for aerosol analysis. ICP-MS provides detection limits on the order of 1 to 100 parts per trillion for approximately 65 elements with a linear dynamic range in excess of eight orders of magnitude (Grohse, 1999). In addition, ICP-MS is capable of isotope determinations that are extremely useful for tracer and source apportionment application species.

Summary of the ICP-MS Technique

In ICP-MS the sample, usually a liquid, is aspirated through a nebulizer and injected as a droplet aerosol into an argon radio frequency plasma. In the plasma, the bombardment of the droplets by free electrons causes removal of the solvent, breakdown of molecules to atoms, and ionization of the atoms to give them a charge so that they can be identified by mass spectroscopy. The ions are extracted from the plasma through a differentially pumped vacuum interface using a series of electrostatic ion lenses that repel negative ions and direct positive ions into a quadrupole mass spectrometer. The ions are then sorted according to their mass-to-charge ratio, and individual ions are detected by a counting electron multiplier.

Higher resolution isotope analysis necessary for some source apportionment application species can be achieved using an ICP magnetic sector multiple collector mass spectrometer. This instrument is a double-focusing instrument that provides increased resolution and sensitivity by utilizing both a quadrupole system to electronically focus a specific ion profile according to its kinetic energy and a magnetic sector that focuses the ions according to their mass-to-charge ratio. The highly resolved ions are then quantified by up to nine Faraday detectors.

Sampling Requirements

Because ICP-MS offers instrumental detection limits orders of magnitude lower than standard atomic absorption spectroscopy and inductively coupled plasma–atomic emissions spectroscopy (ICP-AES), high-purity Teflon membrane filters are used in conjunction with ultraclean handling techniques to minimize the potential for contamination. Therefore, sources of contamination must be minimized during all aspects of sample handling, including sample collection, transport, storage, and analysis.

Precision and Accuracy

The main sources of uncertainty in ICP-MS are matrix effects, isobaric interferences, and polyatomic interferences (Montaser, 1998). Matrix effects are nonspectral interferences caused by nonlinear ion transmission through the mass spectrometer (Douglas and Tanner, 1998). Overlapping isotopes of different elements (e.g., ^{204}Hg and ^{204}Pb) causes isobaric interference. Most isobaric interferences are known and corrected automatically by contemporary instrument software. Polyatomic interference is the presence of molecular ions that have the same mass-to-charge ratio as atomic metal ions causing spectral interferences (e.g., $^{40}Ar^{16}O^+$ and $^{56}Fe^+$), which are not typically corrected by the instrument's software (Douglas and Tanner, 1998). A literature review of polyatomic interferences has been assembled by May and Wiedmeyer (1998). Polyatomic interferences are formed in the plasma from precursors in the sample matrix, reagents, atmospheric gases, and plasma gas.

Many instrument manufacturers are developing technologies to either minimize the formation of molecular ions (cool plasma) or reduce/eliminate molecular ions after being formed (collision or dynamic reaction cells). *Cool plasma* refers to a technique that allows for the determination of elements such as potassium and iron by operating the plasma torch at a lower power and increasing the sample injection flow rate to reduce Ar-based molecular ions. However, the cool plasma technique reduces the sensitivity for elements with high ionization potential such as arsenic and selenium and increases matrix effects (Vanhaecke and Moens, 1999). Collision and dynamic reaction cells use a collision gas such as helium to reduce interfering ions using charge transfer/collision-induced and chemical dissociation, respectively. These new techniques are quite promising and should improve ICP-MS sensitivity for potassium, iron, selenium, and arsenic, but additional research is required.

CONTINUOUS METHODS

As described above, the EPA currently uses filter-based measurement systems with retrospective analysis for the species of interest in the laboratory. However, use of filters in large networks, like the Chemical Speciation Network, is expensive and extremely resource intensive, especially if measurements are to occur more frequently then once every third day. As well, it is at least several months before it is known what the air pollution levels were at the time of sampling. Over the last few years, there has been a dramatic upsurge in the development of continuous or semicontinuous species-specific methods for aerosol constituents. These developments have been in part motivated by the new particulate matter standards, the need to better understand the chemical compositions of aerosols and their relationship to sources, and the need to relate the chemical compositions of the aerosols to adverse health effects, which require daily sampling or at least sampling every other day over long periods (years).

These semicontinuous or continuous methods essentially combine the collection and analysis method into one instrument. These approaches often also minimize sampling artifacts, because the sample is typically either not collected on a substrate or are collected for a short time period at low flow rates, both of which reduce the likelihood of sampling artifacts. Continuous or semicontinuous methods also will allow for near real-time monitoring of pollution levels for Air Quality Index warnings and projections. Therefore, a brief summary of continuous methods is given below, as these methods will be replacing filter-based sampling with subsequent chemical analysis in the laboratory, likely over the next 2 to 5 years. As well, the EPA is pushing development and evaluation of the continuous methods through its internal particulate matter methods research and through the EPA's PM Supersites Program (U.S. EPA, 1998a).

Mass

The Tapered Element Oscillating Microbalance (TEOM) (***R&P***) has been widely used as a continuous PM-10 mass measurement monitor (Patashnick and Rupprecht, 1991). By replacing the PM-10 inlet with a PM-2.5 inlet, the TEOM also has been used to measure continuously PM-2.5 mass concentrations. With this method, particles are collected on a filter mounted on the tip of a glass tube that oscillates in an applied electric field. The oscillation frequency of the glass tube decreases as the mass of the particles collected on the filter increases. The sensitivity of the method is about 1 to 2 $\mu g/m^3$, with a typical time resolution of 5 min. To maintain the necessary precision, the tapered element must be kept at a constant temperature to minimize the effects of thermal expansion and contraction and at an elevated temperature to remove adsorbed water to prevent artifacts due to variation in ambient humidity. However, heating the sample, both in the inlet line and at the tapered element, results in the loss of semivolatile species from the aerosol, and the determined mass is often lower than that obtained by the FRM (G. Allen et al., 1997). To help overcome the problems associated with heating the inlet, several groups are working on the design of systems that operate with controlled relative humidity (Rupprecht & Patashnick Co., Voorheesville, NY, personal communication, 1998; E. Edgerton, ARA, Inc., Durham, NC, personal communication, 1998). See also Chapters 14 and 26 for additional discussions of the TEOM technique.

The β-attenuation monitor determines aerosol mass concentrations by measuring the change in intensity of β-rays passing through a membrane filter as the particles are collected on the filter (Lillienfeld and Dulchinos, 1972; Macias and Husar, 1976). The intensity of the β-rays passing through the filter collection medium decreases proportionally with the particle mass loading. A continuous filter tape allows this to be an automated sampler, providing a "fresh" filter segment for each measurement. Initially, the β-attenuation is measured through an unexposed filter segment as a blank attenuation. The filter segment then can be exposed to ambient particles for a specified time for particle accumulation. The β-attenuation measurement is then repeated, and the mass concentration is converted after the blank correction (Wedding and Weigand, 1993). Although this method has the capability, in principle, of reporting hourly average measurements, 2 to 4h averaging periods are needed to obtain a sufficient deposit for accurate measurements at typical ambient particle mass concentrations (Chow, 1995). Also, the relationship between mass and energy absorption depends to some extent on particle composition. See Chapter 14 for further discussion of β-monitors.

Recently, a continuous fine-particle mass sampler that provides measurements with a time resolution of 20min has been developed (Koutrakis et al., 1995). This method, known as the Continuous Ambient Mass Monitor (CAMM), is based on the continuous measurement of pressure drop across a fibrous filter (Fluoropore) during particle sampling. The performance of this particle mass monitor has been validated through laboratory and preliminary field tests (Babich et al., 2000). A filter tape transportation system allows for unassisted particle sampling for several weeks. For each sampling period, a new segment of the filter tape is exposed. This method combines measurement at ambient temperatures, short sampling periods, and low face velocity, which together result in minimum volatilization or adsorption artifacts. In addition, because this technique requires a low flow rate ($5 \times 10^{-6} m^3/s$ [0.3 L/min]), the relative humidity of the air sample can be controlled to 40% or less by passing the air sample through a Nafion diffusion dryer before its collection. The CAMM in its current state of development has a limit of detection of less than $5 \mu g/m^3$. Accuracy or equivalency relative to the Harvard Impactor resulted in a slope of 1.04 and $r = 0.95$ and an average ratio of 1.07 ± 0.18.

Inorganic Ions

Sulfate and Other Ions. Continuous monitoring of particle sulfate or other ionic species (e.g., sulfate, nitrate, ammonium, and chloride) is possible using annular or wet denuders to remove interfering gases followed by a steam chamber to form condensation nuclei that are subsequently collected and analyzed in near real time by direct injection into an ion chromatograph. This method also can be used to measure other anions and cation species, as well as their gas phase counterparts (Dasgupta, 1993; Wyers et al., 1993; Simon and Dasgupta, 1995; Ito and Thurston, 1996; Weber et al., 2001). Ions in aerosols can be determined as often as about every 10min to about every 30min depending on which instrument is used. The sensitivities of these methods are typically less than $100 ng/m^3$. A modified version of the above method collects aerosols over short time periods on mini quartz fiber filters that are extracted in situ with small amounts of water (P. K. Dasgupta, Texas Tech University, Lubbock, TX, personal communication, 1999).

Nitrate. Besides the methods mentioned above, the Automated Nitrate Monitor (***R&P***) is a new method that provides automated and semicontinuous measurements of airborne particle nitrate concentrations with a time resolution of 10min (Hering and Stolzenburg, 1998;

Stolzenburg and Hering, 2000). The method uses an integrated collection and vaporization cell whereby particles are collected by a humidified impaction process and volatilized in place by flash vaporization. A chemiluminescence analyzer detects the NO_x generated during the heating process. Gas phase nitrates that may interfere with the method are removed before the sample is humidified with a base-coated honeycomb denuder. Operations are unattended apart from twice-weekly calibration checks. Preliminary results are available in near real time. The method has a detection limit of 0.2 μg/m^3 and a precision of 0.5 μg/m^3 (Hering and Stolzenburg, 1998; Stolzenburg and Hering, 2000). The approach is similar to the manual method that has been used for over 20 years to measure the size distribution of sulfate aerosols (Hering and Friedlander, 1982). In this case, however, the particle collection and analysis have been combined into a single cell, allowing the system to be automated. The sulfate method also has been automated by replacing the NO_x chemiluminescence detector with an SO_2 analyzer (S. V. Hering, Aerosol Dynamics, Inc., Berkely, CA, personal communication, 1999).

Carbonaceous Aerosol

An in situ thermal/optical carbon analyzer to determine concentrations of OC and EC at 1 h or better time resolution has been developed by Turpin et al. (1990). In this method, air is pulled through a PM-2.5 cut-point cyclone and split into two air streams. In the first air stream, the particles are collected directly on quartz fiber filters; two 0.015 m (1.5 cm) quartz-fiber filters are placed back to back. In the second air stream, particles are collected on a Teflon filter that also is followed by two back-to-back quartz fiber filters. The OCs measured on these quartz fiber filters are used to account for positive sampling artifacts (vapor adsorbed gases) because only interfering organic gases are allowed to reach the backup quartz fiber filters. After the sample is collected for 60 min, the air is purged from the system with oxygen-free helium.

Analysis for OC and EC is similar to the TOT thermal-optical method described earlier in this chapter, where OC is determined by heating the filter to about 973 K in oxygen-free He, followed by heating in 2% O_2 in He at temperatures up to 1073 K for determining the EC. In the final step of the analysis, a known amount of methane is automatically injected into the system as an internal standard. As with TOT, evolved carbon is converted to CO_2 and then to CH_4 and detected using a flame ionization detector. Correction for OC pyrolyzed during the first heating process in He is accomplished by monitoring the transmittance of an HeNe laser beam (633 nm wavelength) through the sampling filter during the analysis. The point where the transmittance regains its original value for the loaded filter is considered the split between organic and elemental carbon. Correction for vapor adsorption artifact is obtained by measuring OC adsorbed onto the quartz fiber filters located behind the Teflon filter. Precision from collocated sampling is 3.1%, with an accuracy of greater than 95% relative to a known laboratory carbon analyzer. The detection limit for a 1 h sample is about 0.2 μg/m^3.

Trace Elements

Current methods for determining elemental concentrations in ambient air typically require sampling times of 8 to 24 h, which are long compared with changes in meteorological conditions. The Semi-Continuous Metals Monitor is designed to sample at a high flow rate and collect PM-2.5 samples in 20 min, followed by offline multielemental analysis by graphite furnace atomic absorption spectroscopy (GFAAS) (Kidwell and Ondov, 2001). A virtual impactor inlet provides the 2.5 μm cut point at a flow rate of 2.83×10^{-3} m^3/s [170 L/min]. To facilitate separation of the particles from the air stream, growth by condensation using steam injection is used to grow particles as small as 0.05 μm diameter to a nearly uniform size of about 3 μm. The larger particles are concentrated into a $1.7 \times 10^{-4} m^3/s$ [10 L/min] flow using a custom-designed glass virtual impactor. A real impactor made of glass is then used to separate the particles from the air stream and collect them in a water slurry. After each 20 min sampling period, the collected slurry is pumped to a scintillation vial for offline analysis by GFAAS for Al, As, Cd, Cr, Cu, Fe, Mn, Ni, Pb, Sb, Se, and Zn. Computer control is used to gather meteorological data, maintain precise flow rates, adjust the steam injection rate, and operate solenoid valves for sample collection. All sensors are interrogated every 2 s, with 1 min averages stored in a data logger. Limits of detection for the above conditions assuming a 3 ml slurry in ng/m^3 are Mn (2.3), Ni (0.2), Cu (0.3), As (0.7), Zn (0.02), and Cd (0.03). The system has been installed in a 2.4 m × 7.3 m [8 × 24 foot] trailer.

Single-Particle Methods

One of the most significant advances in the last decade for characterizing atmospheric aerosols has been the development of methods to measure the chemical composition of single particles, in situ, and in real time (see Chapter 13 for more details). This is important for several reasons. First, many chemical/microphysical aerosol models need information about whether the complex chemical composition of particles is internally or externally mixed, which can be easily resolved observationally by detecting and chemically analyzing single atmospheric particles in situ (Murphy and Thomson, 1997a,b). Second, information on single-particle size and composition provides source-specific information that cannot be obtained by other means. Third, unique atmospheric chemistry can be observed in situ, such as the conversion of sea salt aerosol from NaCl to $NaNO_3$ as marine air passed over the Los Angeles coast (Gard et al., 1998) or detecting sea salt in sulfate particles aloft (Middlebrook et al., 1998). Such atmospheric detection has been demonstrated using laser vaporization of particles larger than about 0.1 to 10 μm diameters (Prather et al., 1994; Murphy and Thomson, 1995; Johnston and Wexler, 1995; Hinz et al., 1996; Noble and Prather, 1996). In some instruments, particles are detected by laser light scattering that triggers a high-powered ultraviolet pulsed laser that vaporizes the particle in a plasma creating molecular and atomic ions that are analyzed with a time-of-flight (TOF) mass spectrometer. Other groups have extended measurements to ultrafine particles by randomly firing the vaporization laser (Reents et al., 1995), including using a differential mobility analyzer to size-selected particles (Carson et al., 1997). Other instruments utilize thermal desorption and ionization sources (Jayne et al., 2000; Jimenez et al., 2000). The method is sufficiently sensitive to determine trace chemical composition in particles as small as 0.1 μm and below. However, atmospheric concentrations have not been obtained by most of the single-particle methods because sampling efficiencies are size dependent and not well characterized. Efforts to characterize sampling efficiencies for the aerosol TOF mass spectrometer are underway (K. A. Prather, University of Riverside, Riverside, CA, personal communication, 1999) as are efforts to quantify other single-particle methods (Jayne et al., 2000; Jimenez et al., 2000).

SUMMARY

This chapter summarizes a variety of chemical analysis methods for the determination of a number of major, minor, and trace species in atmospheric particulate matter. Most analyses consider that the sample has been collected on a filter (typically Teflon, quartz fiber, Nylon, or an impregnated quartz fiber or cellulose filter). The emphasis of this chapter was on methods used in the Chemical Speciation Network, which include mass by gravimetric analysis, anions (sulfate and nitrate) by ion chromatography, cation species (Na^+, K^+, NH_4^+) by ion chromatography, organic and elemental carbon by thermal-optical transmission, and trace elements by XRF. Other common methods for determining these species also are summarized. An advanced method for determining organic species in aerosols is described. Finally, there is a nationwide effort to develop near real-time, semicontinuous methods, in which the collection and analysis occur at nearly the same time and most often in the field, thus eliminating the need for retrospective chemical analysis in the laboratory. Current semicontinuous methods for the species of focus in this chapter are briefly described as well.

DISCLAIMER

The U.S. Environmental Protection Agency through its Office of Research and Development collaborated in the preparation of this review chapter. It has been subjected to Agency review and approved for publication. Mention of trade names or commercial products does not constitute an endorsement or recommendation for use.

REFERENCES

Albritton, D. and D. Greenbaum. 1998. *Atmospheric Observations: Helping Build the Scientific Basis for Decisions Related to Airborne Particulate Matter*. Report of the PM Measurements Research Workshop, Chapel Hill, North Carolina, 22–23 July. Prepared by the Health Effects Institute and the Aeronomy Laboratory of the National Oceanic and Atmospheric Administration. Health, Effects Institute, Cambridge, MA. http://www.al.noaa.gov/WWWHD/pubdocs/PMMRW.pdf.

Allen, D. T. and E. Palen. 1989. Recent advances in aerosol analysis by infrared spectroscopy. *J. Aerosol Sci.* 20:441–455.

Allen, D. T., E. Palen, J. Haimov, I. H. Mitchell, S. V. Hering, and J. R. Young. 1994. Fourier transform infrared spectroscopy of aerosol collected in a low pressure impactor (LPI/FTIR): Method development and field calibration. *Aerosol Sci. Technol.* 21:325–342.

Allen, G. A., J. Lawrence, and P. Koutrakis. 1999. Field validation of a semi-continuous method for aerosol black carbon (aethalometer) and temporal patterns of summertime hourly black carbon measurements in southwestern PA. *Atmos. Environ.* 33:817–823.

Allen, G., C. Sioutas, P. Koutrakis, R. Reiss, F. W. Lurmann, and P. T. Roberts. 1997. Evaluation of the TEOM method for measurement of ambient particulate mass in urban areas. *J. Air Waste Manage.* 47:682–689.

Allen, J. O., N. M. Dookeran, K. Taghizadeh, A. L. Lafleur, K. A. Smith, and A. F. Sarofim. 1997. Measurement of oxygenated polycyclic aromatic hydrocarbons associated with a size-segregated urban aerosol. *Environ. Sci. Technol.* 31:2064–2070.

Appel, B. R. 1993. Atmospheric sample analysis and sampling artifacts. In *Aerosol Measurement: Principles, Techniques and Application Species*, eds. K. Willeke and P. Baron. New York: Van Nostrand Reinhold.

Appel, B. R., E. M. Hoffer, Y. Tokiwa, M. Haik, and J. J. Wesolowski. 1980a. *Sampling and Analytical Problems in Air Pollution Monitoring.* Final Report to the U.S. EPA, Grant No. R806734-01-0.

Appel, B. R., Y. Tokiwa, and M. Haik. 1981. Sampling of nitrates in ambient air. *Atmos. Environ.* 15:283–289.

Appel, B. R., Y. Tokiwa, J. Hsu, E. L. Kothny, E. Hahn, and J. J. Wesolowski. 1983. *Visibility Reduction as Related to Aerosol Constituents.* Final Report to the California Air Resources Board, Agreement No. Al081-32. NTIS Report PB 84 243617.

Appel, B. R., Y. Tokiwa, E. L. Kothny, R. Wu, and V. Povard. 1988. Evaluation of procedures for measuring atmospheric nitric acid and ammonia. *Atmos. Environ.* 22:1565–1573.

Appel, B. R., S. M. Wall, M. Haik, E. L. Kothny, and Y. Tokiwa. 1980b. Evaluation of techniques for sulfuric acid and particulate strong acidity measurements in ambient air. *Atmos. Environ.* 14:559–1563.

Babich, P., P. Y. Wang, G. Allen, C. Sioutas, and P. Kourtrakis. 2000. Development and evaluation of a continuous ambient $PM_{2.5}$ mass monitor. *Aerosol Sci. Technol.* 32(4):309–324.

Bayona, J. M., K .E. Markides, and M. L. Lee. 1988. Characterization of polar polycyclic aromatic compounds in a heavy-duty diesel exhaust particulate by capillary column gas chromatography and high-resolution mass spectrometry. *Environ. Sci. Technol.* 22:1440–1447.

Birch, M. E. 1998. Analysis of carbonaceous aerosols: Interlaboratory comparison. *Analyst* 123:851–857.

Birch, M. E. and R. A. Cary. 1996. Environmental carbon-based method for monitoring occupational exposures to particulate diesel exhaust. *Aerosol Sci. Technol.* 25:221–241.

Blando, J. D., R. J. Porcja, T.-H. Li, D. Bowman, P. J. Lioy, and B. J. Turpin. 1998. Secondary formation and the Smoky Mountain organic aerosol: An examination of aerosol polarity and functional group composition during SEAVS. *Environ. Sci. Technol.* 32:604–613.

Bolleter, W. T., C. T. Bushman, and P. W. Tidwell. 1961. Spectrophotometric determinations of ammonium as indophenol. *Anal. Chem.* 33:592–594.

Butler, F. E., R. H. Jungers, L. F. Porter, A. E. Riley, and F. J. Toth. 1978. Analysis of air particulates by ion chromatography: Comparison with accepted methods. In *Ion Chromatographic Analysis of Environmental Pollutants*, eds. E. Sawicki, J. D. Mulik, and E. Wittgenstein. Ann Arbor, MI: Ann Arbor Science Publishers, Inc., pp. 65–76.

Cadle, S. H. and P. J. Groblicki. 1982. An evaluation of methods for the determination of organic and elemental carbon in particulate samples. In *Particulate Carbon: Atmospheric Life Cycle*, eds. G. T. Wolff and R. L. Klimisch. New York: Plenum Press, pp. 89–109.

Cadle, S. H. and P. A. Mulawa. 1990. Atmospheric carbonaceous species measurement methods comparison study: General Motors results. *Aerosol Sci. Technol.* 12:128–141.

Cahill, T. A., B. C. Gaelli, and U. P. Eldred. 1987. Size-time composition profile of aerosols using the drum sampler. *Nucl. Instrum. Meth. Phys. Res.* B22:344–351.

Carson, P. G., K. R. Neubauer, M. V. Johnston, and A. S. Wexler. 1997. On-line chemical analysis of aerosols by rapid single-particle mass spectrometry. *J. Aerosol Sci.* 26:535–545.

Chow, J. C. 1995. Measurement methods to determine compliance with ambient air quality standards for suspended particles. *J. Air Waste Manage. Assoc.* 45:320–382.

Chow, J. C., J. G. Watson, D. H. Lowenthal, P. A. Solomon, K. L. Magliano, S. D. Ziman, and L. W. Richards. 1992. PM_{10} source apportionment in California's San Joaquin Valley. *Atmos. Environ.* 26A:3335–3354.

Chow, J. C., J. G. Watson, D. H. Lowenthal, P. A. Solomon, K. L. Magliano, S. D. Ziman, and L. W. Richards. 1993a. PM_{10} and $PM_{2.5}$ compositions in California's San Joaquin Valley. *Aerosol Sci. Technol.* 18:105–128.

Chow, J. C. and J. G. Watson. 1999. *Guideline on Speciated Particulate Monitoring.* Draft 3, January. Prepared by the Desert Research Institute, Reno, NV, for U.S. EPA, OAQPS, Research Triangle Park, NC. Available at http://www.epa.gov/ttn/amtic/pmspec.html.

Chow, J. C., J. G. Watson, D. Crow, D. H. Lowenthal, and T. Merrifield. 2001. Comparison of IMPROVE and NIOSH Carbon Measurements. *Aerosol Sci. Technol.* 34:23–34.

Chow, J. C., J. G. Watson, L. C. Pritchett, W. R. Pierson, C. A. Frazier, and R. C. Purcell. 1993b. The DRI thermal/optical reflectance carbon analysis system: Description, evaluation and application species in U.S. air quality studies. *Atmos. Environ.* 8:1185–1201.

Clarke, A. G., G. A. Azadi-Boogar, and G. E. Andrews. 1999. Particle size and chemical composition of urban aerosols. *Sci. of Total Environ.* 235:15–24.

Colenutt, B. A. and P. Trenchard. 1985. Ion chromatography and its application to environmental analysis: A review. *Environ. Pollut.* (Series B) 10:77–96.

Countess, R. J. 1990. Interlaboratory analyses of carbonaceous aerosol samples. *Aerosol Sci. Technol.* 12:114–121.

Dams, R., J. A. Robbins, K. A. Rahn, and J. W. Winchester. 1970. Nondestructive neutron activation analysis of air pollution particulates. *Anal. Chem.* 42:861–867.

Dasgupta, P. K. 1993. Automated measurement of atmospheric trace gases: Diffusion-based collection and Analysis. In *Measurement Challenges in Atmospheric Chemistry*, ed. L. Newman. Washington, D.C.: American Chemical Society.

Demuynck, M. 1978. Determination of irreversible absorption of water by cellulose filters. *Atmos. Environ.* 9:523–528.

Derrick, M. R. and J. L. Moyers. 1981. Precise and sensitive water soluble ion extraction method for aerosol samplers collected on polytetrafluoroethylene filters. *Analyt. Lett.* 14:1637–1652.

Dockery, D. W., C. A. Pope III, X. Xu, J. D. Spengler, J. H. Ware, M. E. Fay, B. G. Ferris, and F. E. Speizer. 1993. An association between air pollution and mortality in six U.S. cities. *N. Engl. J. Med.* 329:1753–1759.

Douglas, D. J. and S. D. Tanner. 1998. Fundamental considerations in ICPMS. In *Inductively Coupled Plasma Mass Spectrometry*, ed. A. Montessori. New York: Wiley-VCH Inc., pp. 615–677.

Dzubay, T. G. and R. K. Stevens. 1975. Ambient air analysis with dichotomous sampler and X-ray fluorescence spectrometry. *Environ. Sci. Technol.* 9:663–667.

Dzubay, T. G. and R. K. Stevens. 1991. Sampling and analysis methods for ambient PM-10 aerosol. In *Receptor Modeling for Air Quality Management*, ed. P. Hopke. New York: Elsevier, pp. 11–44.

Dzubay, T. G., R. K. Stevens, G. E. Gordon, I. Olmez, A. E. Sheffield, and W. J. Courtney. 1988. A composite receptor method applied to Philadelphia aerosol. *Environ. Sci. Technol.* 22:46–52.

Eatough, D. J., N. Aghdair, M. Cottam, T. Gammon, L. D. Hansen, E. A. Lewis, and R. J. Farber. 1990. Loss of semi-volatile organic compounds from particles during sampling on filters. In *Transactions: Visibility and Fine Particles*, ed. C. V. Mathai. Pittsburgh: Air Waste Management Association, pp. 146–156.

Eatough, D. J., H. Tang, T. W. Cui, and J. Machir. 1995. Determination of the size distribution and chemical composition of fine particulate semi-volatile organic material in urban environments using diffusion denuder technology. *Inhal. Toxicol.* 7:691–710.

Eldred, R. A., P. J. Feeny, P. K. Wakabayashi, J. C. Chow, and E. Hardison. 1998. Methodology for chemical speciation measurements in the IMPROVE network. *Proceedings of an International Specialty Conference B $PM_{2.5}$: A Fine Particle Standard.* Long Beach, CA, January 28–30. Pittsburgh: Air Waste Management Association, pp. 352–364.

Eller, P. M. and M. E. Cassinelli. 1996. *Elemental Carbon (Diesel Particulate): Method 5040. NIOSH Manual of Analytical Methods*, 4th Ed. (1st Supplement). Cincinnati, OH: National Institute for Occupational Safety and Health, DHHS (NIOSH), Publication No. 96-135.

Fung, K. 1990. Particulate carbon speciation by MnO_2 oxidation. *Aerosol Sci. Technol.* 12:122–127.

Fung, K. K., S. L. Heisler, A. Price, B. V. Nuesca, and P. K. Mueller. 1979. Comparison of ion chromatography and automated wet chemical methods for analysis of sulfate and nitrate in ambient particulate filter samples. In *Ion Chromatographic Analysis of Environmental Pollutants*, eds. E. Sawicki, J. D. Mulik, and E. Wittgenstein. Ann Arbor, MI: Ann Arbor Science Publishers, Inc., pp. 203–209.

Gard, E. E., M. J. Kleeman, D. S. Gross, L. S. Hughes, J. O. Allen, B. D. Morrical, D. P. Fergenson, T. Dienes, M. E. Gälli, R. J. Johnson, G. R. Cass, and K. A. Prather. 1998. Direct observation of heterogeneous chemistry in the atmosphere. *Science* 279:1184–1187.

Germani, M. S., I. Gokmen, A. C. Sigleo, G. S. Kowalczyk, I. Olmez, A. M. Small, D. L. Anderson, M. P. Failey, M. C. Gulovali, C. E. Choquette, E. A. Lepel, G. E. Gordon, and W. H. Zoller. 1980. Concentrations of elements in the NBS bituminous and subbituminous coal standard reference materials. *Anal. Chem.* 52:240–245.

Germani, M. S. and W. H. Zoller. 1994. Solubilities of elements on in-stack suspended particles from a municipal incinerator. *Atmos. Environ.* 28:1393–1400.

Gray, H. A., G. R. Cass, J. J. Huntzicker, E. K. Heyerdahl, and J. A. Rau. 1986. Characteristics of atmospheric organic and elemental carbon particle concentrations in Los Angeles. *Environ. Sci. Technol.* 20:580–589.

Grob, R. L. 1995. *Modern Practice of Gas Chromatography*, 3rd Ed. New York: John Wiley & Sons.

Grohse, P. M. 1999. Trace element analysis of airborne particles by atomic absorption spectroscopy, inductively coupled plasma-atomic emission spectroscopy, and inductively coupled plasma-mass spectrometry. In *Environmental Analysis of Airborne Particles. Advances in Environmental, Industrial and Process Control Technologies*, vol. 1, eds. S. Landsberger and M. Creatchman. Amsterdam: the Netherlands: Gordon and Breach Science Publishers, pp. 1–65.

Grosjean, D. 1984. Particulate carbon in Los Angeles air. *Sci. Total Environ.* 32:133–145.

Gundel, L. A., R. L. Dod, H. Rosen, and T. Novakov. 1984. The relationship between optical attenuation and black carbon concentration for ambient and source particles. *Sci. Total Environ.* 36:197–202.

Hansen, A. D. A. and T. Novakov. 1990. Real-time measurement of aerosol black carbon during the carbonaceous species methods comparison study. *Aerosol Sci. Technol.* 122:194–199.

Hawthorne, S. B., D. J. Miller, J. J. Lagenfeld, and M. S. Krieger. 1992. PM_{10} High-volume collection and quantitation of semi- and non-volatile phenols, methoxylated phenols, alkanes, and polycyclic aromatic hydrocarbons from winter urban air and their relationship to wood smoke emissions. *Environ. Sci. Technol.* 26:2251–2262.

Hering, S. V., B. R. Appel, W. Cheng, F. Salaymeh, S. H. Cadle, P. A. Mulawa, T. A. Cahill, R. A. Eldred, M. Surovik, D. Fitz, J. E. Howes, K. T. Knapp, L. Stockburger, B. J. Turpin, J. J. Huntzicker, X. Zang, and P. H. McMurry. 1990. Comparison of sampling methods for carbonaceous aerosols in ambient air. *Aerosol Sci. Technol.* 12:200–213.

Hering, S. V. and S. K. Friedlander. 1982. Origins of sulfur size distributions in the Los Angeles basin. *Atmos. Environ.* 16:2647–2656.

Hering, S. V., D. R. Lawson, et al., 1988. The nitric acid shootout: Field comparison of measurement methods. *Atmos. Environ.* 22:1519–1539.

Hering, S. V. and M. R. Stolzenburg. 1998. A new method for the automated high-time resolution measurement of PM-2.5 nitrate. Presented at the AWMA Specialty Conference on Fine Particles, January 28–30, 1998, Long Beach, California.

Hildemann, L. M., M. A. Mazurek, G. R. Cass, B. R. T. Simoneit. 1994. Seasonal trends in Los Angeles ambient organic aerosol observed by high-resolution gas chromatography. *Aerosol Sci. Technol.* 20(4):303–317.

Hinz, K.-P., R. Kaufmann, and B. Spengler. 1996. Simultaneous detection of positive and negative ions from single airborne particles by real-time laser mass spectrometry. *Aerosol Sci. Technol.* 24:233–242.

Huntzicker, J. J., R. L. Johnson, J. J. Shah, and R. A. Cary. 1982. Analysis of organic and elemental carbon in ambient aerosols by a thermal-optical method. In *Particulate Carbon: Atmospheric Life Cycle*, eds. G. T. Wolff and R. L. Klimisch. New York: Plenum Press, pp. 79–88.

Ingle, J. D. and S. R. Crouch. 1988. *Spectrochemical Methods of Analysis*. Englewood Cliffs, NJ: Prentice Hall.

Ito, K. and G. D. Thurston. 1996. Daily PM_{10}/mortality associations: An investigation of at-risk subpopulations. *J. Exposure Analysis Environ. Epidemiol.* 6:79–95.

Jaklevic, J. M., R. C. Gatti, F. S. Goulding, B. W. Loo, and A. C. Thompson. 1981. Aerosol analysis for the regional air pollution study. Report No. EPA-600/4-81-006. Environmental Science Research Laboratory, U.S. EPA, Research Triangle Park, NC.

Jayne, J. T., D. C. Leard, X. Zhang, P. Davidovits, K. A. Smith, C. E. Kolb, and D. R. Worsnop. 2000. Development of an aerosol mass spectrometer for size and composition analysis of sub-micron particles. *Aerosol Sci. Technol.* 33:49–70.

Jimenez, J. L., J. T. Jayne, Q. Shi, D. R. Worsnop, C. E. Kolb, X. Zhang, K. A. Smith, I. Yourshaw, R. Flagan, J. H. Seinfeld, J. Morris, and P. Davidovits. 2000. Ambient aerosol sampling using an aerosol mass spectrometer. In *Proceedings of the 93rd Annual Meeting of the Air & Waste Management Association*, Salt Lake City, Utah. Pittsburgh: Air & Waste Management Association.

Johnson, R. L., J. J. Shaw, R. A. Cary, and J. J. Huntzicker. 1981. An automated thermal-optical method for the analysis of carbonaceous aerosol. In *ACS Symposium Series 167: Atmospheric Aerosol, Source/Air Quality Relationships*, eds. E. S. Macias and P. K. Hopke. New York: Plenum Press, pp. 223–233.

Johnson, S. A., R. Kumar, and P. T. Cunningham, 1983. Airborne detection of acidic sulfate aerosol using an ATR impactor. *Aerosol Sci. Technol.* 2:401–405.

Johnston, M. V. and A. S. Wexler. 1995. MS of individual aerosol particles. *Anal. Chem.* 67:721A–726A.

Kamens, R. M., D. Bell, A. Dietrich, J. Perry, R. Goodman, L. Claxton, and S. Tejada. 1985. Mutagenic transformations of dilute wood smoke systems in the presence of ozone and nitrogen dioxide. Analysis of selected high-pressure liquid chromatography fractions from wood smoke particle extracts. *Environ. Sci. Technol.* 19:63–69.

Karger, B. L., L. R. Snyder, and C. Horvath. 1973. *An Introduction to Separation Science.* New York: John Wiley & Sons.

Kidwell, C. B. and J. M. Ondov. 2001. Development and evaluation of a prototype system for collection of sub-hourly ambient aerosol for chemical analysis. *Aerosol Sci. Technol.* 35(1) In Press.

Koutrakis, P., K. M. Thompson, J. M. Woldson, J. D. Spengler, G. J. Keeler, and J. L. Slater. 1992. Determination of aerosol strong acidity losses due to interactions of collected particles: Results from laboratory and field studies. *Atmos. Environ.* 26(A)6:987–995.

Koutrakis, P., P. Y. Wang, C. Sioutas, and J. M. Wolfson. 1995. U.S. Patent Application, Serial No: 211/030.

Koutrakis, P., J. M. Wolfson, J. L. Slater, M. Brauer, J. D. Spengler, and R. K. Stevens. 1988. Evaluation of an annular denuder/filter pack system to collect acidic aerosols and gases. *Environ. Sci. Technol.* 22:1463–1468.

Landsberger, S. 1999. Trace element determination of airborne particles by neutron activation analysis. In *Elemental Analysis of Airborne Particles*, eds. S. Landsberger and M. Creatchman. Amsterdam, the Netherlands: Gordon and Breach Science Publishers, pp. 197–233.

Lawless, P. A. and C. E. Rodes. 1999. Maximizing data quality in the gravimetric analysis of personal exposure sample filters. *J. Air Waste Manage.* 49:1039.

Lillienfeld, P. and J. Dulchinos. 1972. Portable instantaneous mass monitoring for coal mine dust. *Am. Ind. Hyg. Assoc. J.* 33:136–145.

Lodge, J. P. Jr. 1989. *Methods of Air Sampling and Analysis*, 3rd Ed. Chelsea, MI: Lewis.

Macias, E. S. and R. B. Husar. 1976. A review of atmospheric particulate mass measurement via the beta attenuation technique. In *Fine Particles*, ed. B. Y. H. Liu. New York: Academic Press, Inc.

Magliano, K. L., V. Hughes, L. R. Chinkin, D. L. Coe, T. L. Haste, N. Kumar, and F. W. Lurmann. 1999. Spatial and temporal variations in PM_{10} and $PM_{2.5}$ source contributions and comparison to emissions during the 1995 integrated monitoring study. *Atmos. Environ.* 33(29):4757–4773.

May, T. W. and R. H. Wiedmeyer. 1998. A table of polyatomic interferences in ICP-MS. *Atom. Spectrosc.* 19(5):150–155.

Mazurek, M. A., L. M. Hildemann, G. R. Cass, B. R. T. Simoneit, and W. R. Rogge. 1993. Methods of analysis for complex organic aerosol mixtures from urban emission sources of particle carbon. In *Measurement of Airborne Compounds: Sampling, Analysis, and Data Interpretation*, eds. E. D. Winegarand and L. H. Keith. American Chemical Society Symposium Series. Boca Raton, FL: CRC Press.

Mazurek, M. A., B. R. T. Simoneit, G. R. Cass, and H. A. Gray. 1987. Quantitative high-resolution gas chromatography and high-resolution gas chromatography/mass spectroscopy analysis of carbonaceous fine aerosol particles. *Int. J. Environ. Anal. Chem.* 29:119–139.

McClenny, W. A., J. W. Childers, R. Rohl, and R. A. Palmer. 1985. FTIR transmission spectroscopy for the non destructive determination of ammonium and sulfate in ambient aerosols collected on Teflon filters. *Atmos. Environ.* 19:1891–1898.

Middlebrook, A. M., D. M. Murphy, and D. S. Thomson. 1998. Observations of organic material in individual marine particles at Cape Grim during the first Aerosol Characterization Experiment (ACE I). *J. Geophys. Res.* 103:16475–16483.

Montaser, A. 1998. *Inductively Coupled Plasma Mass Spectrometry.* New York: Wiley-VCH, Inc.

Mueller, P. K., K. K. Fung, S. I. Heisler, D. Grosjean, and G. M. Hidy. 1982. Atmospheric particulate carbon observations in urban and rural areas of the United States. In *Particulate Carbon: Atmospheric Life Cycle*, eds. G. T. Wolff and R. L. Klimisch. New York: Plenum Press, pp. 343–370.

Mueller, P. K., B. V. Mendoza, J. C. Collins, and E. S. Wilgus. 1979. Application of chromatography to the analysis of anions extracted from airborne particulate matter. In *Ion Chromatographic Analysis of Environmental Pollutants*, eds. E. Sawicki, J. D. Mulik, and E. Wittgenstein. Ann Arbor, MI: Ann Arbor Science Publishers, Inc., pp. 77–86.

Mulik, J. D., E. Estes, and E. Sawicki. 1978. Ion chromatographic analysis of ammonium ion in ambient aerosols. In *Ion Chromatographic Analysis of Environmental Pollutants*, Vol. I, eds. E. Sawicki, J. D. Mulik, and E. Wittgenstein, Ann Arbor, MI: Ann Arbor Science Publishers, Inc, pp. 41–51.

Mulik, J. and E. Sawicki, eds. 1979. *Ion Chromatographic Analysis of Environmental Pollutants*, Vols. I, II. Ann Arbor, MI: Ann Arbor Science Publishers, Inc.

Murphy, D. M. and D. S. Thomson. 1995. Laser ionization mass spectroscopy of single aerosol particles. *Aerosol Sci. Technol.* 22:237–249.

Murphy, D. M. and D. S. Thomson. 1997a. Chemical composition of single aerosol particles at Idaho Hill: Positive ion measurements. *J. Geophys. Res.* 102:6341–6352.

Murphy, D. M. and D. S. Thomson. 1997b. Chemical composition of single aerosol particles at Idaho Hill: Negative ion measurements. *J. Geophys. Res.* 102:6353–6368.

Nelson, J. W. 1977. Proton-induced aerosol analysis: Methods and samplers. In *X-Ray Fluorescence Analysis of Environmental Samples*, ed. T. G. Dzubay. Ann Arbor, MI: Ann Arbor Science, pp. 19–34.

Nielsen, T., T. Seitz, and T. Ramdahl. 1984. Occurrence of nitro-PAH in the atmosphere in a rural environment. *Atmos. Environ.* 18:2159–2165.

Noble, C. A. and K. A. Prather. 1996. Real-time measurement of correlated size and composition profiles of individual atmospheric aerosol particles. *Environ. Sci. Technol.* 30:2667.

Novakov, T. 1982. Soot in the atmosphere. In *Particulate Carbon: Atmospheric Life Cycle*, eds. G. T. Wolff and R. L. Klimisch. New York: Plenum Press, pp. 19–41.

Novakov, T. 1984. The role of soot and primary oxidants in atmospheric chemistry. *Sci. Total Environ.* 36:1–10.

Ondov, J. M. and F. Divita, Jr. 1993. Size spectra for trace elements in urban aerosol particles by instrumental neutron activation analysis. *J. Radioanal. Nucl. Chem.* 167:247–258.

Ondov, J. M., J. A. Dodd, and G. Tuncel. 1990. Nuclear analysis of trace elements in size-classified sub-micrometer aerosol particles from a rural air shed. *Aerosol Sci. Technol.* 13:249–263.

Ondov, J. M., W. H. Zoller, I. Olmez, N. Aras, G. Gordon, L. Rancitelli, K. Abel, R. Filby, K. Shah, and R. Ragaini. 1975. Elemental concentrations in the National Bureau of Standards' environmental coal and fly ash standards reference materials. *Anal. Chem.* 47:1102–1109.

Orvini, E. and M. Speziali. 1998. Applicability and limits of instrumental neutron activation analysis: State-of-the-art in the year 2000. *Micro-chem. J.* 59:160–172.

Paputa-Peck, M. C., R. S. Marano, D. Scheutzle, T. L. Riley, C. V. Hampton, T. J. Prater, L. M. Skewes, T. E. Jensen, P. H. Ruehle, L. C. Bosch, and W. P. Duncan. 1983. Determination of nitrated polynuclear aromatic hydrocarbons in particulate extracts by capillary column gas chromatography with nitrogen selective detection. *Anal. Chem.* 55:1946–1954.

Patashnick, H. and E. G. Rupprecht. 1991. Continuous PM-10 measurements using the Tapered Element Oscillating Microbalance. *J. Air Waste Manage.* 41:1079.

Pope, C. A. III, M. J. Thun, M. M. Namboori, D. W. Dockery, J. S. Evans, F. E. Speizer, and D. W. Heath, Jr. 1995. Particulate air pollution as a predictor of mortality in a prospective study of U.S. adults. *Am. J. Respir. Crit Care Med.* 151:669–674.

Prather, K. A., T. Nordmeyer, and K. Salt. 1994. Real-time characterization of individual aerosol particles using time-of-flight mass spectroscopy. *Anal. Chem.* 66:1403–1407.

Pyen, G. S. and M. J. Fishman. 1979. Determination of anions in pore waters from cores. In *Ion Chromatographic Analysis of Environmental Pollutants*, eds. E. Sawicki, J. D. Mulik, and E. Wittgenstein. Ann Arbor, MI: Ann Arbor Science Publishers, Inc., pp. 235–244.

Reents, W. D. S., S. W. Downey, et al. 1995. Single particle characterization by time-of-flight mass spectrometry. *Aerosol Sci. Technol.* 23:263–270.

Rogge, W. F. 1993. *Molecular Tracers for Sources of Atmospheric Carbon Particles: Measurements and Model Predictions*. Ph.D. Thesis. Pasadena, CA: California Institute of Technology.

Rogge, W. F., L. M. Hildemann, M. A. Mazurek, G. R. Cass, and B. R. T. Simoneit. 1991. Sources of fine organic aerosol. 1. Charbroilers and meat cooking operations. *Environ. Sci. Technol.* 25:1112–1125.

Rogge, W. F., M. A. Mazurek, L. M. Hildemann, G. R. Cass, and B. R. T. Simoneit. 1993. Quantification of urban organic aerosols at a molecular level: Identification, abundance and seasonal variation. *Atmos. Environ.* 27:1309–1330.

Rosen, H., A. D. A. Hansen, L. Gundel, and T. Novato. 1978. Identification of the optically absorbing component in urban aerosols. *Appl. Opt.* 17(24):3859–3861.

Russell, A. G., G. J. McRae, and G. R. Cass. 1983. Mathematical modeling of the formation and transport of ammonium nitrate aerosol. *Atmos. Environ.* 17:949–964.

Saleh, M. A., E. Ewane, and B. L. Wilson. 1999. Monitoring the Houston ship channel for inorganic pollutants by Ion Selective Electrodes, Ion Chromatography, and Inductively Coupled Mass Spectrometry. *Chemosphere*. 39:2357–2364.

Schauer, J. J. 1998. *Technical Report 1. Chemical Analysis for Molecular Marker Source Apportionment of Atmospheric Fine Particulate Matter.* Prepared for the U.S. EPA, Research Triangle Park, NC.

Schauer, J. J. and G. R. Cass. 2000. Source apportionment of wintertime gas-phase and particle-phase air pollutants using organic compounds as tracers. *Environ. Sci. Technol.* 34(9):1821–1832.

Schauer, J. J., W. F. Rogge, L. M. Hildemann, M. A. Mazurek, G. R. Cass, and B. R. T. Simoneit. 1996. Source apportionment of airborne particulate matter using organic compounds as tracers. *Atmos. Environ.* 30:3837–3855.

Schwartz, J., D. W. Dockery, and L. M. Neas. 1996. Is daily mortality associated specifically with fine particles? *J. Air Waste Manage.* 46:927–939.

Schwartz, J., D. W. Dockery, L. M. Neas, D. Wypij, J. H. Ware, J. D. Spengler, P. Koutrakis, F. E. Speizer, and B. G. Ferris, Jr. 1994. Acute effects of summer air pollution on respiratory symptom reporting in children. *Am. J. Respir. Crit. Care Med.* 150:1234–1242.

Sickles, J. E. II, C. Perrino, L. Allegrim, A. Febo, M. Possanzini, and R. J. Paur. 1986. *Measurement of Selected Inorganic Air Pollutants Near Los Angeles Using an Annular Denuder System.* Presented at the EPA/APCA Symposium on measurement of Toxic Air Pollutants, May 1986, Raleigh, NC.

Simon, P. K. and P. K. Dasgupta. 1995. Continuous automated measurement of the soluble fraction of atmospheric particulate matter. *Anal. Chem.* 67:71–78.

Simoneit, B. R. T. 1986. Characterization of organic constituents in aerosols in relation to their origins and transport: A review. *Int. J. Environ. Anal. Chem.* 23:207–237.

Small, H. 1981. Application species of Ion chromatography in trace analysis. In *Trace Analysis*, Vol. 1, ed. J. F. Lawrence. New York: Academic Press, pp. 267–322.

Solomon, P. A., T. Fall, L. Salmon, G. R. Cass, H. Gray, and A. Davidson. 1989. Chemical characteristics of PM_{10} aerosols collected in the Los Angeles area. *JAPCA* 39:154–163.

Solomon, P. A., S. M. Larson, T. Fall, and G. R. Cass. 1988. Basinwide nitric acid and related species concentrations observed during the Claremont nitrogen species comparison study. *Atmos. Environ.* 22:1587–1594.

Solomon, P. A., W. Mitchell, D. Gemmill, M. P. Tolocka, G. Norris, R. Wiener, S. Eberly, J. Rice, J. Homolya, R. Scheffe, R. Vanderpool, R. Murdoch, S. Natarajan, and E. Hardison. 2000a. *Evaluation of $PM_{2.5}$ Chemical Speciation Samplers for Use in the U.S. EPA National $PM_{2.5}$ Chemical Speciation Network.* Prepared for Office of Air Quality Planning and Standards by the Office of Research and Development, Research Triangle Park, NC: Report no. EPA 454/R-01-005.

Solomon, P. A., K. Baumann, E. Edgerton, R. Tanner, D. Eatough, H. Marin, D. Savoie, S. Natarajan, M. Meyers, and G. Norris. 2001. Comparison of integrated samplers for mass and composition during the 1999 Atlanta Supersites Project. *J. Geophys. Res., Atmospheres*, Special Issue for the Atlanta Supersites Project, Submitted for publication, July.

Stevens, R. K. 1984. *Sampling and Analysis of Atmospheric Aerosols.* U.S. EPA Report 600/D-84-283, NTIS No. PB85-128197.

Stevens, R. K., T. G. Dzubay, C. W. Lewis, and R. W. Shaw, Jr. 1984. Source apportionment methods applied to the determination of the origin of ambient aerosols that effect visibility in forested areas. *Atmos. Environ.* 18:261–272.

Stevens, R. K., T. G. Dzubay, G. Russwurm, and D. Rickel. 1978. Sampling and analysis of atmospheric sulfates and related species. *Atmos. Environ.* 12:55–68.

Stevens, R. K., W. A. McClenny, and T. G. Dzubay. 1982. Analytical method to measure the carbonaceous content of aerosols. In *Particulate Carbon: Atmospheric Life Cycle*, eds. G. T. Wolff and R. L. Klimisch. New York: Plenum Press, pp. 111–129.

Stolzenburg, M. R. and S. V. Hering. 2000. Method for the automated measurement of fine particle nitrate in the atmosphere. *Environ. Sci. Technol.* 34(5):907–914.

Sverdrup, G. M., B. E. Buxton, J. C. Chuang, and G. S. Casuccio. 1990. Determination of optimal storage conditions for particle samples. *Environ. Sci. Technol.* 24:1186–1195.

Tang, H., E. A., Lewis, D. J. Eatough, R. M. Burton, and R. J. Farber. 1994. Determination of the particle size distribution and chemical composition of semi-volatile organic compounds in atmospheric fine particles. *Atmos. Environ.* 28:939–947.

Thurston, G. D., K. Ito, C. G. Hayes, D. V. Bates, and M. Lippmann. 1994. Respiratory hospital admissions and summertime haze air pollution in Toronto, Ontario: Consideration of the role of acid aerosols. *Environ. Res.* 55:271–290.

Tierney, G. and W. Conner. 1967. Hygroscopic effects on weight determinations of particulates collected on glass-fiber filters. *J. Am. Ind. Hyg. Assoc.* July–August:363–365.

Tolocka, M. P., P. A. Solomon, W. Mitchell, G. Norris, D. Gemmill, R. Vanderpool, and J. Homolya. 2001. East versus West in the US: Chemical characteristics of $PM_{2.5}$ during the winter of 1999. *Aerosol Sci. Technol.* 34:88–96.

Tuncel, S. G., I. Olmez, J. R. Parrington, G. E. Gordon, and R. K. Stevens. 1985. Composition of fine particle regional sulfate component in Shenandoah Valley. *Environ. Sci. Technol.* 19:529–537.

Turpin, B. 1998. *Chemical Characterization of Atmospheric Organic Aerosols in Support of Health Studies: Current Options and Future Prospects—A Workshop Summary.* Prepared for EPRI, Palo Alto, CA.

Turpin, B. J., R. A. Cary, and J. J. Huntzicker. 1990. An in situ, time resolved analyzer for aerosol organic and elemental carbon. *Aerosol Sci. Technol.* 12:161–171.

Turpin, B. J., J. J. Huntzicker, and S. V. Hering. 1994. Investigation of organic aerosol sampling artifacts in the Los Angeles Basin. *Atmos. Environ.* 19:3061–3071.

Turpin, B. J., H.-J. Lim, E. Edgerton, S. Hering, G. Allen, H. Maring, P. Solomon. 2001. Semi-continuous aerosol carbon measurements: Comparison of Atlanta Supersites Measurements. *J. Geophys. Res., Atmospheres*, Special Issue for the Atlanta Supersites Project, Submitted for publication.

U.S. CFR. 1997. National Primary and Secondary Ambient Air Quality Standards, Final Rules. *Code of Federal Regulations, Title 40*, Parts 50–53 and 58. July 18, 1997.

U.S. EPA. 1989. Compendium Chapter IP-9. *Determination of Reactive Acidic and Basic Gases and Particulate Matter in Indoor Air.* Research Triangle Park, NC: Atmospheric Research and Exposure Assessment Laboratory, p. 21.

U.S. EPA. 1996a. *Air Quality Criteria for Particulate Matter.* National Center for Environmental Assessment, Office of Research and Development, Research Triangle Park, NC, April 12. http://www.epa.gov/ttn/amtic/codefed.html.

U.S. EPA. 1996b. *Review of the National Ambient Air Quality Standards for Particulate Matter: Policy Assessment of Scientific and Technical Information—Office of Air Quality Planning and Standards Staff Paper.* Office of Air Quality Planning and Standards, U.S. EPA, Research Triangle Park, NC, July. http://www.epa.gov/ttn/amtic/codefed.html.

U.S. EPA. 1997. *National $PM_{2.5}$ Speciation Program: Overview, Objectives, Requirements, and Approach.* Office of Air Quality Planning and Standards and the Office of Research and Development. U.S. EPA, Research Triangle Park, NC. http://www.epa.gov/ttn/amtic/pmspec.html.

U.S. EPA. 1998a. *Draft Supersites Conceptual Plan.* Prepared for the Technical Subcommittee on Fine Particle Monitoring of the Clean Air Scientific Advisory Committee, November 9, 1998. Office of Air Quality Planning and Standards and the Office of Research and Development, U.S. EPA, Research Triangle Park, NC. http://www.epa.gov/ttn/amtic/supsinfo.html.

U.S. EPA. 1998b. *Background Material on National Monitoring Network and Relationship to Supersites.* Office of Air Quality Planning and Standards and the Office of Research and Development, U.S. EPA, Research Triangle Park, NC. http://www.epa.gov/ttn/amtic/supsinfo.html.

U.S. EPA. 1999. *Particulate Matter ($PM_{2.5}$) Speciation Guidance.* Office of Air Quality Planning and Standards, U.S. EPA, Research Triangle Park, NC.

U.S. EPA. 1999b. PM—Metals in ambient by Proton X-Ray (PIXE). U.S. EPA Method IO-3.6, Report 625R-96-01a. Environmental Protection Agency, Washington, DC.

U.S. EPA. 2001. Air quality criteria for particulate matter [second external review draft]. Research Triangle Park, NC: Office of Research and Development, National Center for Environmental Assessment-RTP Offic: Report no. EPA 600/P-99/002a,b. 2v.

Vanhaecke, F. and L. Moens. 1999. Recent trends in trace element determination and speciation using inductively coupled plasma mass spectrometry. *Fresenius J. Anal. Chem.* 364:440–451.

Waetjen, U., E. Bombelka, F. W. Richter, and H. Ries. 1983. PIXE analysis of high-volume aerosol samples and intercomparison of results with different analytical methods. *J. Aerosol Sci.* 14:305–308.

Wagman, J., R. L. Bennett, and K. T. Knapp. 1977. Simultaneous multiwavelength spectrometer for rapid element analysis of particulate pollutants. In *X-Ray Fluorescence Analysis of Environmental Samples*, ed. T. G. Dzubay. Ann Arbor, MI: Ann Arbor Science Publishers Inc., pp. 35–55.

Wall, S. M., W. John, and J. L. Ondov. 1988. Measurement of aerosol size distributions for nitrate and major ionic species. *Atmos. Environ.* 22:1649–1656.

Watson, J. G., J. C. Chow, and C. A. Frazier. 1999. X-ray fluorescence analysis of ambient air samples. In *Elemental Analysis of Airborne Particles*, eds. S. Landsberger and M. Creatchman. Amsterdam, the Netherlands: Gordon and Breach Science Publishers, pp. 67–96.

Watson, J. G., J. C. Chow, Z. Lu, E. M. Fujita, D. H. Lowenthal, D. R. Lawson, and L. L. Ashbaugh. 1994. Chemical mass balance source apportionment of PM_{10} during the Southern California Air Quality Study. *Aerosol Sci. Technol.* 21:1–16.

Wedding, J. B. and M. A. Weigand. 1993. An automatic particle sampler with beta gauging. *J. Air Waste Manage.* 43:475–479.

Weber, R., D. Orsini, Y. Duam, K. Baumann, C. S. Kiang, W. Chameides, Y. N. Lee, F. Brechtel, P. Klotz, P. Jongejan, H. ten Brink, J. Slanina, P. Dasgupta, S. Hering, M. Stolzenburg, E. Edgerton, B. Harstell, P. A. Solomon, and R. Tanner. 2001. Intercomparison of near real-time monitors of PM2.5 nitrate and sulfate at the EPA Atlanta Supersite. *J. Geophys. Res.*, Special Issue for the Atlanta Supersites Project, Submitted for Publication, July.

Weiss, J. and E. L. Johnson, 1986. *Handbook of Ion Chromatography.* Sunnyvale, CA: Dionex Corporation.

Wyers, G. P., R. P. Otjes, and J. Slania. 1993. A continuous-flow denuder system for the measurement of ambient concentrations and surface-exchange fluxes of ammonia. *Atmos. Environ.* 27:2085–2090.

12 Analysis of Individual Collected Particles

R. A. FLETCHER, J. A. SMALL, and J. H. J. SCOTT
National Institute of Standards and Technology, Gaithersburg, MD

INTRODUCTION

Important information about an aerosol resides in the morphology and in the chemical composition of the individual particles. Chemical, morphological, phase, and crystallographic data are often crucial for purposes of tracing formation mechanisms and possible sources of airborne particulate matter. Applications of single-particle analysis include clean room technology, microelectronics, indoor and outdoor air pollution, forensics, and many problems related to microcontamination. Analytical instruments and techniques must be employed to obtain compositional information at the single-particle level. Digital image processing plays an increased role in providing new information about chemical composition and morphological properties of microstructures and particles.

Single-particle analysis is defined here to mean the analysis of collected, individual particles ranging from about 5 nm to several millimeters in lateral dimension. The information obtained includes the chemical constituents that make up the particle, the morphology (size, shape), and the physical and optical properties. Two requirements must be met for single-particle analysis: The analytical technique must have sufficient spatial resolution to differentiate a single particle from the background substrate material and adjacent particles and sufficient sensitivity to detect at least major components of the particle's composition (picogram detection levels for micrometer-sized unit density particles). Microanalysis, entailing the use of microscopes or microprobes, often fulfills these requirements. Microanalysis by means of microscopy originated with the advent of the light microscope, while the first microprobe work resulted from the invention of the electron microprobe (Castaing and Guinier, 1950). Microanalytical developments have been motivated by the need to obtain images, elemental and molecular (compound) information on a microscopic scale. In microanalysis, a beam of excitation radiation, which can be photons, electrons, protons, neutrons, or ions, is used to bombard the sample. The interaction of the beam with the sample results in emitted radiation that is separated by a spectrometer and collected by a detector, as shown in Figure 12–1. For microprobes, the excitation beam is focused on the particle with a spatial resolution ranging from nanometers to micrometers. To create an image, either the excitation beam is rastered across the particle or the mounted particle is translated under the beam. Microscopes usually flood the sample with excitation radiation, but form an image by focusing the secondary radiation. Spatially resolved analysis is achieved in the microscope either

Aerosol Measurement: Principles, Techniques, and Applications, Second Edition, Edited by Paul A. Baron and Klaus Willeke.
ISBN 0-471-35636-0

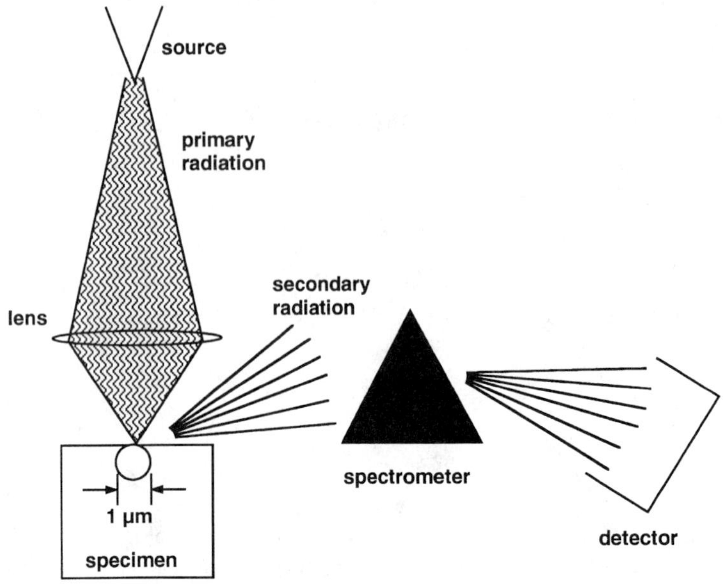

Fig. 12–1. Schematic representation of generic microanalysis showing excitation and emission radiation. The incident radiation can be electrons, photons, or ions. The emissions are photons in the infrared, ultraviolet, or visible; ions; or X-rays.

by masking a portion of the secondary radiation (aperturing) or by employing a spatially sensitive detector.

Some specific examples of instruments using electron beam excitation are electron probes, scanning electron microscopes, and analytical electron microscopes. The emitted radiation, backscattered electrons or X-rays, is used for single-particle analysis. Examples of instruments that use photon radiation are the Raman microprobe (or micro-Raman), where the detected beam is frequency-shifted photons; the laser microprobe, which detects ions in a time-of-flight mass spectrometer; and the Fourier transform infrared microscope, which is based on photon absorption. Examples of instruments that use ion beams are the ion microprobe and microscope based on secondary ion mass spectrometry (SIMS). In SIMS, an argon, oxygen, or cesium ion beam is used to bombard the sample, and either a time-of-flight mass spectrometer or more usually a magnetic sector or quadrupole mass spectrometer is used to detect the secondary ions generated in the interaction process.

Often in an analysis, complementary capabilities of two or more instruments can be used in sequence on an individual particle to provide more complete particle characterization (phase, chemical, and morphological information) about that particle (Steel et al., 1984; Fletcher et al., 1990).

This chapter describes microscopes and microprobes used for analysis of collected, individual particles. The instruments discussed are the light microscope, electron microscopes (both scanning and transmission), electron microprobes, laser, optical, scanning probe, and ion microprobes. The principles of operation and the instrumental capabilities are presented. The chapter also contains some basic information about sample preparation, useful for the aerosol scientist. Several excellent references that should be consulted for more detailed information include the book by Spurny (1986), the publication by Heinrich (1980), the article by Newbury (1990), and the review chapter by Grasserbauer (1978). For quick reference, Table 12–1 presents a comparison of the microanalytical techniques capable of single-particle analysis.

TABLE 12–1. Comparison of Typical Microbeam Analytical Instrument Capabilities

Analytical Method	SEM	TEM	EPMA	LMMS	SIMS	Micro-Raman	LM	FT-IR	AFM
Excitation	Electrons	Electrons	Electrons	Photons[a]	Ions	Photons	Photons	Photons	N.A.
Emission	Electrons, X-rays	Electrons, X-rays	Electrons X-rays	Ions	Ions	Photons	Photons	Photons	N.A.
Quantity measured	Electrons, X-rays	Electrons, X-rays	Electrons, X-rays	m/z	m/z	Frequency shift	Light intensity	Intensity at given wavelength	Distance, force, other interactions
Lateral resolution	5 nm	0.3 nm	0.05 μm	1 μm	1 μm, 0.1 μm (probe)	2 μm	0.5 μm	10–50 μm	
Detectable elements	>Carbon	>Carbon	>Carbon	All	All	N.A.	N.A.	N.A.	N.A.
Isotopic detection	No	No	No	Yes	Yes	N.A.	N.A.	N.A.	N.A.
Detectability of molecular or chemical compounds	None	Inorganic	None	Organic and inorganic	Organic and inorganic	Organic and inorganic	Inorganic	Organic and inorganic	N.A.
Absolute detection sensitivity	10^{-16} g	10^{-20} g	10^{-16} g	10^{-18} to 10^{-20} g	10^{-19} g	10^{-12} g	N.A.	5×10^{-12} g	N.A.
Relative sensitivity	0.1%	0.1%	0.1%	1–100 ppm	1–1000 ppm	1%	N.A.	1%	N.A.
Sample vacuum mounted	Yes	Yes	Yes	Yes	Yes	No	No	No	No
Destructive method	No	No	No	Yes	Yes	No	No	No	No
Surface sensitive	No	No	No	Unknown	Yes	No	No	No	Yes
Imaging capability	Yes	Yes	Yes	Demonstrated, not widely available	Yes	Yes	Yes	Yes	Yes
Automation demonstrated	Yes	Yes	Yes	No	No	Yes	Yes	Yes	Yes
Quantitative	Yes	Yes	Yes	No	No	No	No	Semi	N.A.

[a]Photons in the UV to IR range.

Source: Adapted from Wieser et al. (1980).

LIGHT MICROSCOPY

Underlying Principle

Optical microscopy or light microscopy (LM) is one of the more familiar microanalytical techniques. The operating principle of light microscopes is well known. In the simplest form, the light microscope utilizes light refraction via a lens system to form enlarged images of microscopic objects. The image is focused on the detector, which can be the human eye or a camera. An object must absorb approximately 0.3% of the incident light to be visible to the eye (Dovichi and Burgi, 1987).

Instrumentation

A schematic of a light microscope is presented in Figure 12–2. The important components of a light microscope are a light source, an objective lens, and an ocular. The light source can be diffuse or bright and serves to illuminate the sample. The objective lens collects light that passes through the sample or is reflected from the sample surface and projects an image near the ocular. The ocular magnifies the image that is projected by the objective for the eye. Normally, the virtual image seen resides below the sample plane. There are a number of optical accessories used in conjunction with the light microscope to characterize a sample physically. Some of these capabilities are discussed later.

Some considerations in LM are depth of field, referring to the distance beyond the plane of focus that the object remains in focus; magnification, quantifying the image enlargement; numerical aperture, relating the maximum light-gathering capability of the microscope objective; and resolving power, indicating the size of the smallest feature that can be discriminated. Table 12–2 presents characteristics of various common microscope objectives (Steel, 1980).

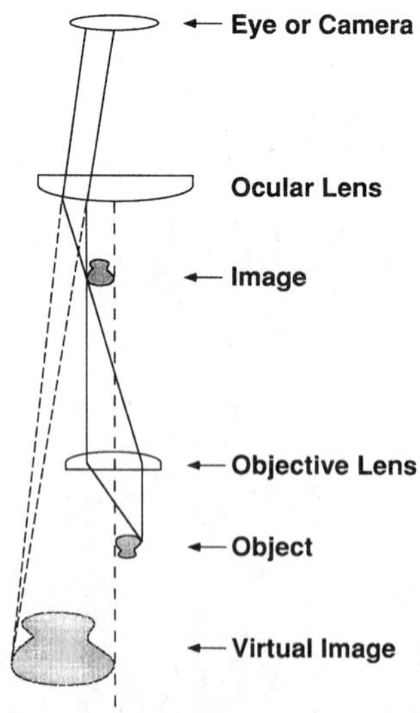

Fig. 12–2. Schematic of a light microscope. Light is transmitted through the sample and focused by an objective lens. The intermediate image is then enlarged and transmitted to the eye or the detector. (Adapted from McCrone and Delly, 1973.)

TABLE 12–2. Characteristics of Some Common Light Microscope Objectives

Magnification	Numerical Aperture	Depth of Field (μm)	Diameter[a] of Field (mm)	Resolving[b] Power (μm)
3	0.08	50	9	5
10	0.25	8	2	1.3
50	0.85	1	0.5	0.4
100	1.3	0.4	0.4	0.25

[a] Approximate value with ×10 oculars. Wide field is now common; values may be 1.5 to 2 times larger.
[b] Approximate value in green light.
Source: Steel (1980).

Capabilities and Applications

LM is often the first microanalytical technique used to examine a sample because it is a nondestructive approach. It is relatively easy to use as an imaging tool for many applications, but identifying a material through its optical properties can be difficult. A skilled microscopist can use the physical and optical properties of a particle (such as the size, shape, surface texture, color, refractive indices, crystallographic properties, and birefringence) to help identify a given particle and thus possibly its source (Grasserbauer, 1978). Two additional references containing information on analysis of particles by LM are Steel (1980) and Friedrichs (1986). A general detailed reference for LM is given by Chamot and Mason (1958). *The Particle Atlas* (McCrone and Delly, 1973) is regarded as a principal reference for identification of particles by LM.

Size and Shape Analysis. Determination of shape and size often represents the first step in single-particle analysis. Sometimes the shape will provide information about the particle type and thus the most probable formation mechanism of a particle. For example, fly ash normally appears under a light microscope as spheres, and dark, fractal-like (complex branched-chain) structures are usually formed from a combustion source. Fibers may be asbestos or glass or may come from a wide variety of other natural or man-made sources. *The Particle Atlas* (McCrone and Delly, 1973) can be used to compare the image of a sample particle to reference micrographs. Although particles can be resolved and thus observed at the 0.25 to 0.5 μm size level, shape and size determination is most reliably done on particles larger than several micrometers in diameter (Steel, 1980). Fibers are an exception to the above statement because even with a lateral dimension as small as 0.5 μm, size and shape can be determined.

Related to particle size, particle magnification in the light microscope is limited due to the wavelength of light used (diffraction limit). A good working rule of thumb is that the maximum magnification of a light microscope is 1000 times the numerical aperture of the objective (McCrone and Delly, 1973). The shape and size are useful for identifying particle origin, but some additional physical properties help to provide more definitive information about a particle's makeup.

Identification by Light Microscopy. Using the physical properties of particle shape and size and the optical properties of color, refractive index, and birefringence, the identity of an unknown particle can often be determined. Normally, optical characterization requires particles to be greater than 5 to 10 μm in lateral dimension (Steel, 1980).

Particles that have an index of refraction different from the substrate-mounting material are most easily viewed. It is the contrast between the background support and the particle that is most often important when trying to view an object by LM. The contrast can be

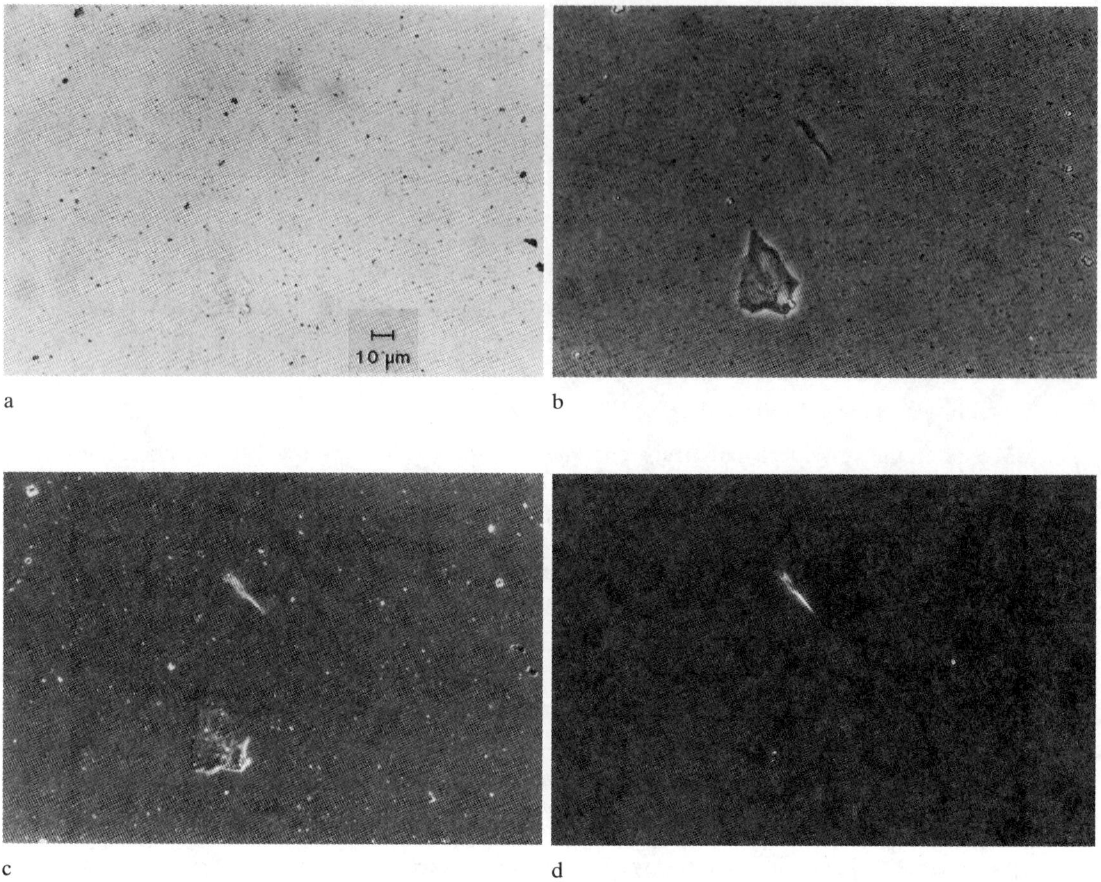

Fig. 12–3. Set of four light micrographs illustrating various LM techniques that help to increase particle visibility by increasing contrast and, in the last case (d), help to identify birefringent materials in the sample. **a**, Straight transmitted light; **b**, transmitted light using phase contrast; **c**, differential interference; **d**, effect of slightly uncrossed polarizers. The last frame (d) brings out the birefringent material present in the sample as the apparent luminous objects. (Courtesy of E. Steel, NIST.)

improved by a number of techniques. Figure 12–3 shows the same field of view of a collected airborne particle sample using different contrast enhancers.

The filter material is mixed esters of cellulose, and the sample is prepared by treating with acetone vapor to make the filter transparent (Baron and Pickford, 1986). The micrographs are taken in transmitted light. Figure 12–3a illustrates the problem with viewing the filter in direct (unaided) transmission. Two large objects in the central part of the micrograph are barely visible under these conditions. When phase contrast microscopy is applied (Fig. 12–3b), the particles become quite visible; the phase shift of the light transmitted through the particles is used to enhance the contrast. Another way to increase the visibility of the particle is by differential interference contrast, shown in Figure 12–3c. This is considered a complementary technique to phase contrast, but in this case the objects take on a three-dimensional appearance. Finally, Figure 12–3d shows the effect of slightly uncrossed polarizers. The advantage of uncrossed polarizers is that the particles are, for the most part, still visible in the field of view and the particles made of anisotropic material stand out as illuminated objects. Crossed polarizers, on the other hand, would make all of the isotropic materials (such as glass or amorphous plastics) "invisible" and show only those particles (as luminous objects) that rotate the polarized transmitted light. Detailed discussion of these techniques is not possible here, but is given in McCrone and Delly (1973).

Several additional techniques are useful in conjunction with LM. Some microscopes are equipped with monochromatic or near-monochromatic light sources that can be used to excite fluorescence, if present, in a particle. Often the observation of fluorescence, usually excited in the ultraviolet, provides information for identifying the particle's composition. Refractive index is another parameter that can be determined and used as a powerful tool for particle identification. The refractive index measurement, with accuracy to one part in one thousand (Grasserbauer, 1978), is accomplished by immersing the particle in a series of index matching fluids to find the matching refractive index that causes the particle to "disappear." Microchemical reactions can be used to help identify the particle composition (Seeley, 1952; Chamot and Mason, 1940). Analyses of these kinds require considerable experience to be useful for identifying particle composition.

Fiber Analysis. Baron (1993) presents an extensive discussion of microscopic techniques for fibers, and the following contributions relating to fibers have been extracted from his work. Phase contrast microscopy (PCM) is an interferometic technique that enables particles with low contrast (transparent particles) to be viewed. Light that is transmitted through the object is phase shifted relative to that transmitted through the substrate only. The phase shift is not detectable by the eye, but intensity differences are detected. The PCM transforms the phase difference to an intensity difference by forming an interference pattern using the phase shifted and unshifted light. PCM is primarily used for fiber counting to provide an index of asbestos fiber exposure in workplaces where asbestos is known to be present. It cannot be used to detect fibers thinner than about 0.25 μm. Although the morphology observed under PCM allows some discrimination between fiber types, it is not specific enough to allow positive identification of asbestos or other fibers. PCM also is used for measurement of man-made mineral fibers (MMMF), for example, mineral wool and fibrous glass (NIOSH, 1977). It can be used with other analytical techniques, such as polarized light microscopy (PLM) and scanning (SEM) and transmission (TEM) electron microscopy, for specific fiber types.

Measurement Accuracy of Fibers by PCM. The accuracy of various fiber-counting techniques is poor when compared with other analytical methods. For instance, most analytical methods in the NIOSH *Manual of Analytical Methods* state an overall uncertainty (combined variability and bias) of better than 10% (NIOSH, 1984). Under optimum analysis conditions (uniform sample deposit, no background dust interference, optimum loading) a relative standard deviation of 0.10 (or 10%) is expected. Thus, the accuracy (including bias and variability) can be no better than this level.

In fact, other sources of variability and bias can occur. Some of these are due to the small sample size observed as part of the measurement procedure. For instance, fibers may be nonuniformly distributed in the sample due to inertial, electrostatic, or other sampling influences. Because the analysis assumes a uniform distribution on the filter, taking a small portion for microscopic analysis can therefore result in significant variations in the reported concentration. In addition, one microscopist may introduce biases relative to other microscopists due to decisions about which particles to count as fibers. When comparisons are made between groups of microscopists, these biases may appear as increased variability in the overall results.

The use of established analytical procedures for fiber count analysis is extremely important. This is the only way that results from one laboratory can be compared reliably with those of other laboratories. Microscopist training, proper equipment, and established quality control procedures are all-important components of proper laboratory practice. To ensure uniformity of application of these analytical procedures, both within-laboratory and interlaboratory sample exchanges are necessary (Abell et al., 1989; Ogden et al., 1986). The usual technique for determining analytical biases, that is, the comparison with a reference method, does not work because an alternate fiber-counting technique that measures the "true" fiber concentration does not exist. Thus, the final test of fiber counting accuracy is the comparison of one's results with those of a group of competent laboratories.

Several formal programs of sample exchange have been established for PCM. These include the American Industrial Hygiene Association's (AIHA) Proficiency Analytical Testing (PAT) program (Groff et al., 1991), the U.K.'s Regular Inter-Laboratory Counting Exchange (RICE) program (Crawford and Cowie, 1984), and the International Asbestos Fibre Regular Interchange Counting Arrangement (AFRICA) program (Institute of Occupational Medicine, Edinburgh). For a laboratory performing PCM analyses to establish compliance with U.S. Occupational Safety and Health Administration (OSHA) asbestos fiber exposure level regulations, regular sample exchanges with other laboratories are required.

Measurement of fiber concentrations for comparative use within a single study may not need all the components of a complete quality assurance scheme. For instance, if a study is intended only to provide relative fiber concentrations to show differences with time or location, interlaboratory exchange of samples may not be necessary. However, the use of published counting and sample preparation procedures, as well as performance of blind repeat analyses, are important for establishing analytical confidence limits.

The Particle Atlas can be consulted to classify a particle on the basis of its physical and optical properties. This reference contains over 600 color photomicrographs of particles from various sources and of known composition (McCrone and Delly, 1973). In this reference, the types of particles are broken down into four categories: (1) wind-blown particles such as fibers and minerals; (2) industrial particles such as abrasives, polymers, fertilizers, and cleaners; (3) combustion particles such as auto, coal-fired, and oil-fired soots; and (4) miscellaneous particles. The authors provide a step-by-step characterization procedure for classifying the particles into one of the four categories. The same reference contains scanning electron micrographs of the same 600 particles shown in the photomicrographs.

Sample Preparation and Practical Applications. In a light microscope, either transmitted or reflected light can be used. For transmitted light, the particles are mounted on a glass or quartz slide. An index or immersion oil may be applied to the sample to improve viewing under transmitted light. An aliquot of the particles should be tested with the oil to ensure that no reaction or dissolution takes place. The oil will contaminate the particles prepared in this manner, so subsequent microanalysis on the particles is unlikely using other techniques. Reflected light can be used to view particles collected on an aerosol filter surface, but the particles usually must be >1 μm. Opaque particles and particles with large indices of refraction are most easily viewed in this manner. To overcome this size limitation, some filters can be made transparent or removed entirely to allow viewing with transmitted light. Friedrichs (1986) mentions three ways to transform the filter. Each has its own advantage and disadvantage. In the first method, filters can be treated with index matching fluid. Particles remain on the filter, but liquid particles or particles soluble in the fluid may be dissolved or possibly removed, and particles with a refractive index matching the immersion oil will be difficult to see. Some "filter clearing" agents are given in Table 12–3 (Friedrichs, 1986; LeGuen and Galvin, 1981; Baron and Pickford, 1986). A second method is to dissolve the filter using an appropriate solvent (e.g., polycarbonate filters dissolve in chloroform). In the third method, the filter is ashed, leaving the refractory particles behind. In the last two approaches, the particles no longer have a filter support and must be remounted on a transparent substrate.

The number of particles collected on a substrate can be determined and related to the particle concentration in an aerosol. The number of particles is normally determined per

TABLE 12–3. Membrane Filter Clearing Agents for Light Microscopy

Filter Type	Clearing Agent	Refractive Index
Mixed esters of cellulose	Acetone vapor/triacetin (AIA method)	1.44–1.48
	Dimethyl formamide/Euparal method	1.48
	Acetone vapor/Euparal method	1.48
Polycarbonate	Immersion oils	1.584 or 1.625
	Chloroform-dissolved materials	

Sources: Friedrichs (1986), Le Guen and Galvin (1981), and Baron and Pickford (1986).

viewing area, and, when a number of randomly selected viewing areas are taken together, an estimate can be made of the number of particles on the entire filter surface. This estimate can be related to the airborne particle number concentration (based on sample air volume) as in the case for asbestos number concentration (Asbestos International Association, 1979; Carter et al., 1989).

The relevance of depth of field, especially for particle counting, is illustrated by the set of micrographs in Figure 12–4. In these micrographs, certain particles are visible and in focus, while others are difficult to see. Figure 12–4 is an example of a typical LM application that might be employed by an aerosol scientist for examining a filter surface in reflectance.

The particles in Figure 12–4 were collected on a filter consisting of mixed esters of cellulose. The filter is slightly bowed in the center as a result of air flow through the filter cassette. This bowing causes a distorted planar surface for the microscopy, leading to poor particle detection due to limited depth of focus. In Figure 12–4a, the upper left of the field of view is in focus, but the lower right is out of focus. As the microscope focus is altered, the lower right regions shown in Figure 12–4b become clear images, and particles in the upper left gradually become fuzzy and indistinguishable. Clearly a shallow depth of field is problematic. The recent development of the confocal microscope provides a different approach by limiting the image formation strictly to those photons scattered within the depth of field. By changing the objective to specimen distance, a series of images can be obtained as "optical sections" of a three-dimensional object.

Depth of field usually encountered in LM is not as large as that found in the SEM. Figure 12–5 contains two micrographs of the same field of view of amosite asbestos. The top is a light micrograph and the bottom is an electron micrograph, both with approximately the same magnification. Note that only some of the asbestos fibers are in focus in the light micrograph while the entire electron-generated image is in focus.

ELECTRON BEAM ANALYSIS OF PARTICLES

Principle of Electron Beam Excitation

A schematic of a typical electron beam instrument and some of its analytical functions is shown in Figure 12–6. The source of the electron beam is an emitter such as a heated tungsten filament shaped into a fine tip. The electrons emitted from the filament are formed into a beam and focused by an ion lens system onto the specimen. The electron beam interacts with the atoms of the sample, resulting in the scattering of the beam electrons and the ejection of both electrons and X-ray photons from the specimen.

Capabilities

Electron Imaging. Electron imaging that provides the analyst with particle size and morphology is accomplished by two different methods. One of the imaging methods is used in

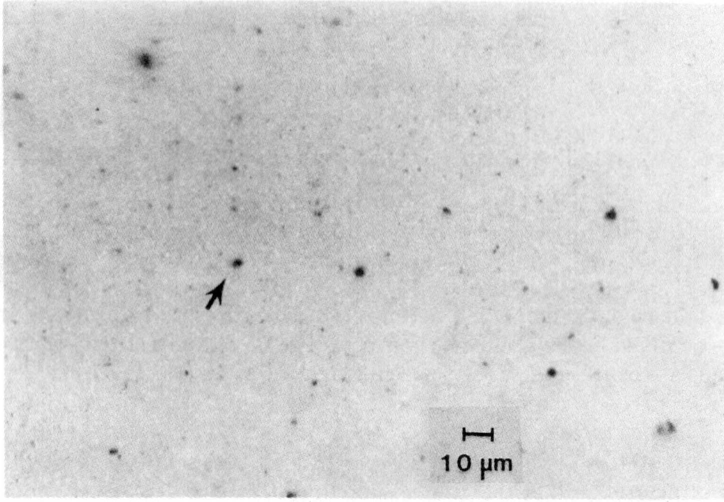

(a)

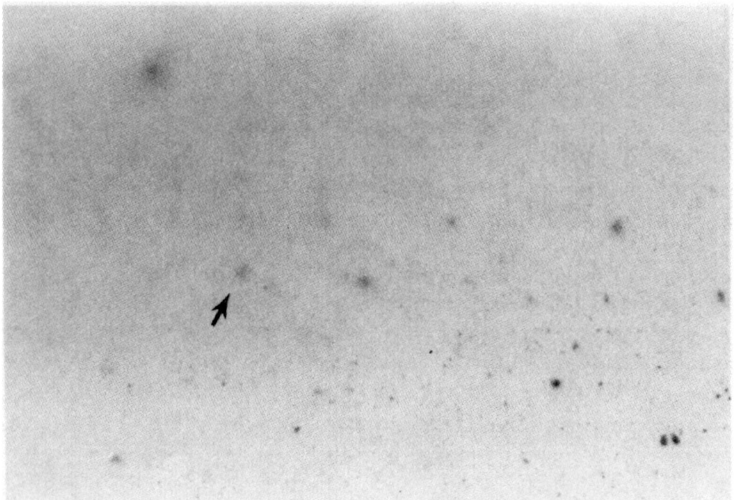

(b)

Fig. 12–4. Light micrographs of an untreated filter composed of mixed cellulose esters that contains collected airborne particles. The magnification of the presented image is ×620. The set of micrographs illustrates the impact of depth of field on a viewing volume by selectively bringing certain areas of the filter into focus while remaining areas containing particles are left out of focus. An arrow is used to indicate the same selected particle in both micrographs. The focused field is descending from the upper left corner in **a** down to the bottom right corner in **b**. (Courtesy of E. Steel, NIST.)

transmission electron microscopy (TEM). An example of a TEM image of asbestos is shown in Figure 12–7. The TEM image is observed on a phosphor screen or recorded on film and is similar to the image formed in the transmitted light microscope. The transmission microscope illuminates the sample with a static beam of high-energy (100 to 400 keV) electrons to image the specimen at magnifications of 1000 to about 1,000,000 times. Because the image is formed

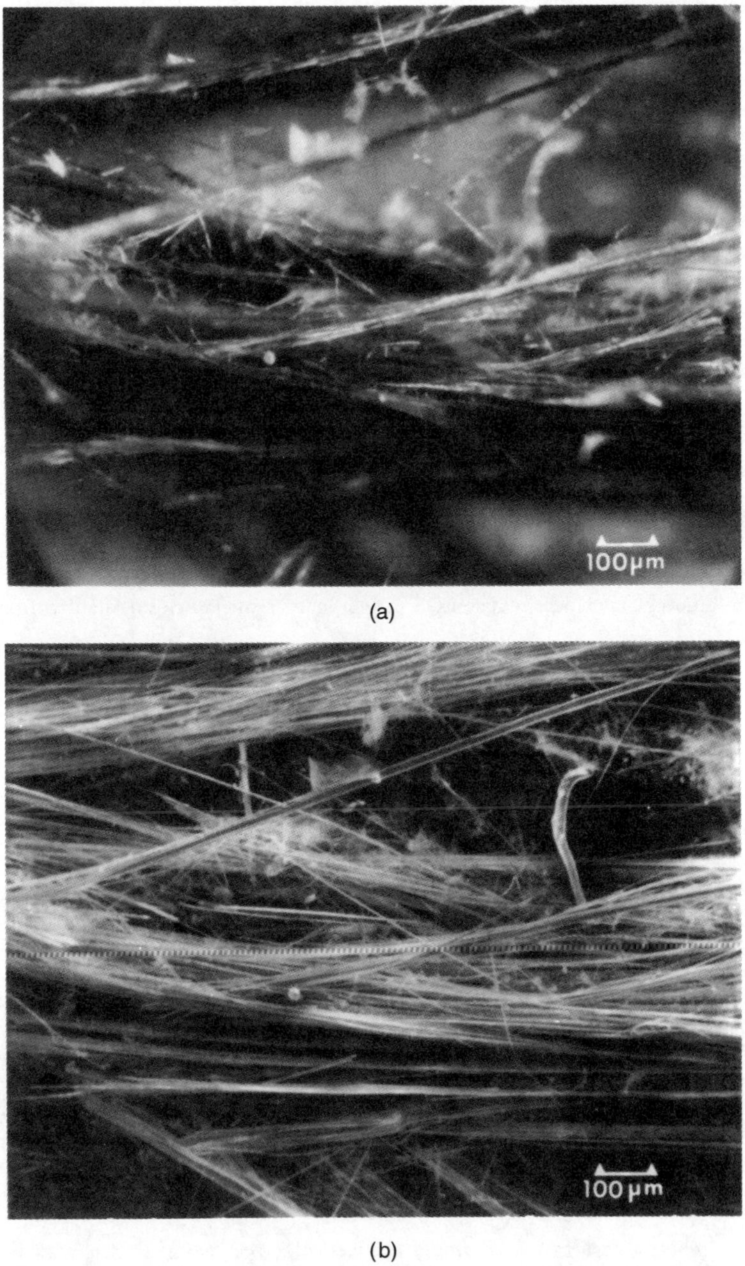

Fig. 12–5. Two micrographs of the same field of view taken of an amosite asbestos sample. The light micrograph **(a)** has limited depth of field. The depth of field in the electron micrograph **(b)**, taken with a scanning electron microscope, illustrates the advantage of using SEM in many cases.

by electrons that travel through the sample, TEM is best suited for imaging small particles less than 0.5 μm in diameter.

The other type of electron imaging, an example of which is shown in Figure 12–8, is used in conventional scanning electron microscopes and electron probe microanalysis (EPMA)

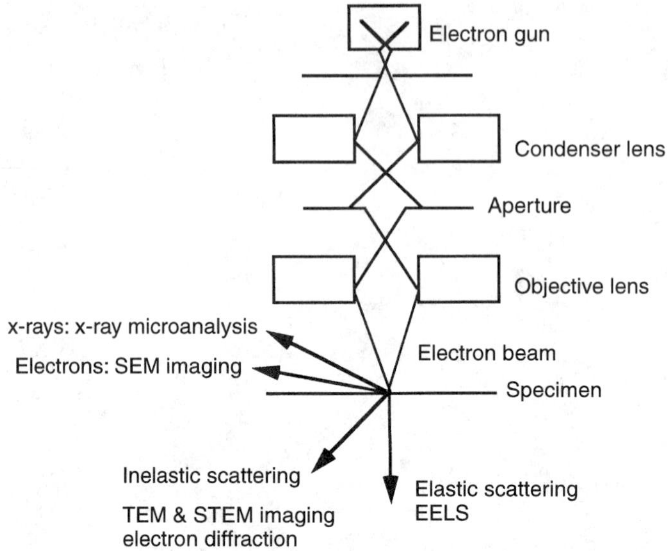

Fig. 12–6. A schematic of a typical electron beam instrument showing the electron source and several of the different analytical signals used in the analysis of particles. (Adapted from Goldstein et al., 1975.)

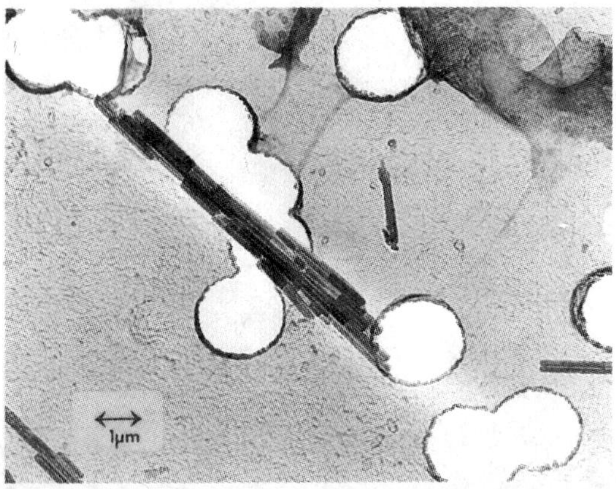

Fig. 12–7. An example of a transmission electron image of asbestos showing the "hollow-tube morphology" with dimensions of the center "tube" on the order of 20 nm. (Courtesy of S. Turner, NIST.)

systems. This imaging system uses a finely focused electron beam that is rastered over the sample area. The scattered electrons resulting from the elastic or inelastic interactions of the beam electrons with the sample are detected, and the amplified signal is used to modulate the brightness on a CRT monitor that has a scan pattern matched to the scan pattern of the beam. Unlike the TEM image, the image formed by the scanning instruments is an electronic image. In the SEM and the EPMA, the electron detector is generally positioned above the plane of the sample. In this configuration, the sample can be observed at magnifications of 10 to about 100,000 times, and the electron images provide topographical as well as morphological information. In TEM, the electron beam can also be rastered and an electron detector placed below the sample. This configuration is used for scanning transmission

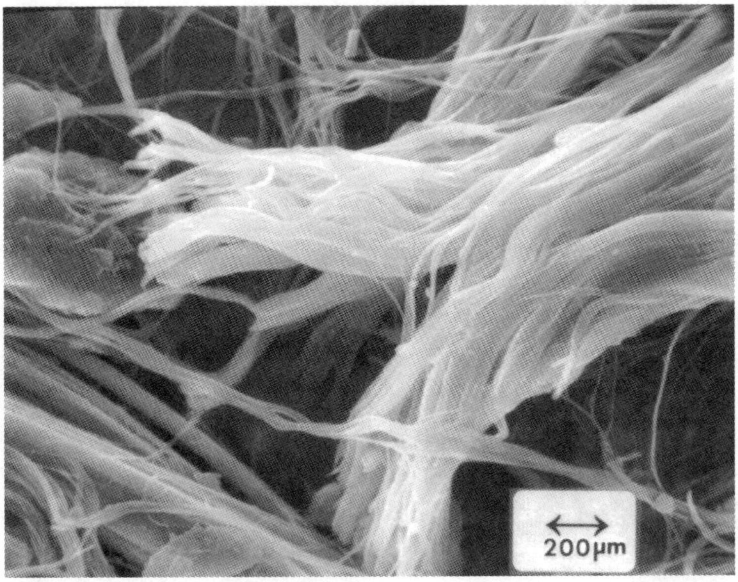

Fig. 12–8. An example of a secondary electron image of asbestos taken with a scanning electron microscope.

electron microscopy (STEM) at magnifications up to about 300,000 times. STEM imaging produces a shadow image of the particles that can be used to determine particle morphology but not topography.

Low Vacuum (LV) Scanning Electron Microscope. In conventional SEM and EPMA, the specimen chamber pressure is on the order of 1×10^{-4} Pa, which requires the coating of insulating samples with a conductive film such as carbon to prevent sample charging. In 1978, Danilatos developed the environmental scanning electron microscope (LV-SEM) in which the specimen chamber pressure operates in the pressure range from 1.3×10^{-4} to 1.3×10^{3} Pa. In the high-pressure regime above 100 Pa, the LV-SEM can be used to image a wide variety of samples including insulating materials without the conductive coating required in the conventional instruments. This ability is particularly advantageous for the analysis of particulate samples collected on insulating substrates such as membrane or cellulose filters. Figure 12–9 is a pair of images of mineral particles collected on a polycarbonate filter. The left side of the filter has been coated with carbon, providing electrical conductivity to the sample. Figure 12–9a was taken at high vacuum approximately 1×10^{-4} Pa in a conventional SEM, and Fig. 12–9b was taken on the LV-SEM at a chamber pressure of 800 Pa. The conventional image clearly shows significant sample charging on the right, uncoated side of the filter, making imaging impossible. In comparison, the LV-SEM image of the same area shows no sample charging.

Unlike conventional SEMs, the LV-SEM employs a series of pressure-limiting apertures that divides the LV-SEM vacuum system into five separate vacuum domains ranging from 1.3×10^{-5} Pa at the electron gun chamber to as high as 1.3×10^{3} Pa in the specimen chamber. In addition to the differential pumping system, the LV-SEM makes use of a unique electron-imaging detector. In this detector, shown in Figure 12–10, the gas molecules present in the high-pressure specimen chamber are used to amplify the secondary electron signal through a collision cascade resulting in creation of positive ions that migrate to the specimen surface and dissipate the charge buildup on the sample surface as a result of the beam impact.

(a)

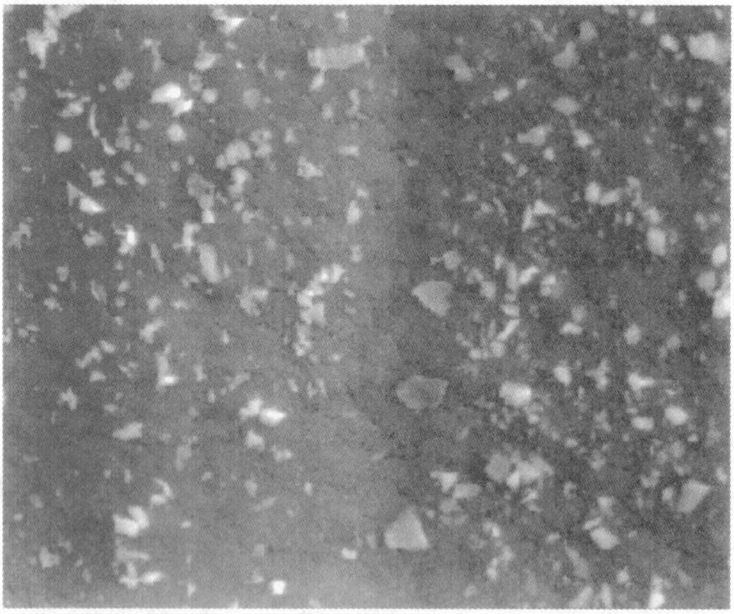

(b)

Fig. 12–9. Secondary electron image of mineral particles on a polycarbonate filter by **a**, conventional SEM that shows significant sample charging on the right side of the image and **b**, the same region of the filter by LV-SEM that shows no evidence of sample charging. (Courtesy S. Wight, NIST.)

X-Ray Microanalysis of Individual Particles. X-ray microanalysis is predicated on the ejection, by beam electrons, of an inner shell electron from an atom followed by the filling of these vacancies by orbital electrons and the subsequent emission of characteristic X-ray photons. Because the transitions of the electrons occur between sharply defined energy levels,

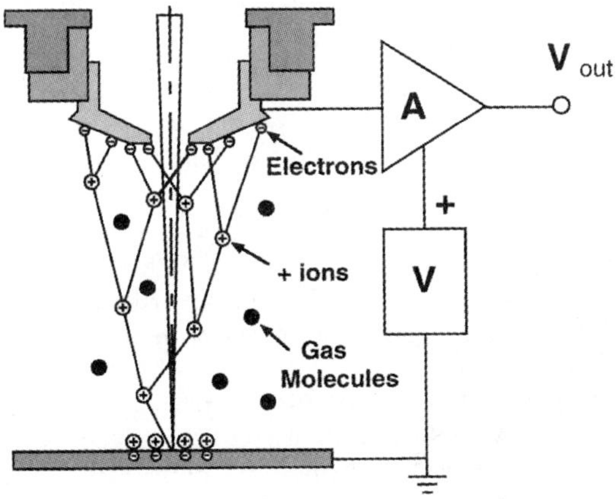

Fig. 12–10. Diagram of the environmental secondary detector and the charge elimination scheme. (From Philips Electron Optics, 1996.)

the X-rays are element specific and can be used for qualitative and quantitative analyses. In addition to X-ray emission, analysis of particles with diameters smaller than 0.1 μm can also be done in the analytical electron microscope with electron diffraction and electron energy-loss spectroscopy (Cowley, 1979; Joy, 1979). For most inorganic applications, X-ray microanalysis provides a nondestructive microanalytical technique with the sample remaining unchanged by the analysis. The analyst must keep in mind, however, that the sample is subjected to a vacuum and any volatile components will be lost. The concepts and principles of EPMA have been discussed in detail by several authors (Heinrich, 1981; Newbury et al., 1986).

Because the electron beam scatters when it interacts with the atoms of the sample and the absorption path lengths for the emitted X-rays are much longer than those for electrons, the spatial distribution of emitted X-rays and thus the resolution is always greater than the diameter of the electron probe. Generally, the spatial resolution of electron beam microanalysis techniques is in the range of 20 nm to approximately 2 μm, depending on the instrument selected for the analysis, the size of the particle, and other experimental conditions. The techniques can detect elements with atomic number 6 (carbon) and greater and have a lower detectability limit of about 0.1% by weight or approximately 10^{-16} g in the case of particles. Analytical results are generally reported in weight fractions or weight percent. The accuracy of X-ray microanalysis applied to conventional flat, polished specimens is quite good, about 2% by weight of the measured concentration, with a precision of about 0.1% by weight of the measured concentration (Newbury et al., 1986). The accuracy and precision associated with the analysis of particles are highly dependent on particle shape, size, and composition. In general, accuracies of approximately 10% to 20% by weight of the measured concentration with a precision of 5% can be expected for the quantitative analysis of particles.

The size of the particle to be analyzed often dictates the methods to be used for sample preparation as well as the instrument(s) and analytical scheme used for an analysis. For the purpose of discussing individual particle analysis, it is appropriate to divide particles into roughly three size categories (see Fig. 12–11). The size ranges shown are only a rough guide to the selection of a specific electron beam method. Small particles with diameters less than about 0.5 μm are best analyzed by analytical electron or scanning transmission microscopy with electron beam accelerating potentials in excess of 80 keV. Quantitative analysis by thin film methods is most accurate for the smallest particles and degrades as the particles become

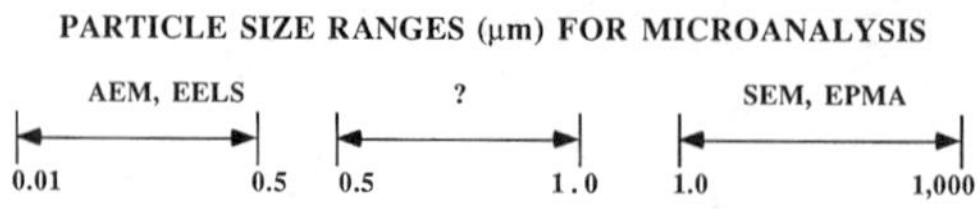

Fig. 12–11. Particle size ranges for analysis by electron beam instruments showing the three size ranges of particles used for analysis methods.

larger than about 0.1 μm in diameter. Above 0.1 μm, corrections for X-ray absorption, atomic number, and fluorescence become significant and must be incorporated into the "thin film" analytical schemes discussed below.

At the other end of the size spectrum, the larger particles in excess of 1 μm in diameter are best analyzed in an EPMA or SEM with accelerating potentials less than about 50 keV. Quantitative analysis by conventional electron probe methods is most accurate for the largest particles, greater than about 10 μm in diameter, because they approximate a bulk target at these dimensions. Below about 10 μm, quantitative analysis degrades as the particle diameter decreases, and corrections to compensate for differences between particles and bulk standards must be included in the analysis schemes.

Quantitative analysis of the particles in the intermediate size range 0.1 to 2 μm is difficult and can be approached in two ways. The analyst can use either analytical electron microscopy, in which case the particles are considered as increasing in size from a thin film, or EPMA/SEM, in which case the particles are considered as decreasing in size from a bulk sample. The remainder of this section is devoted to a discussion of the two different approaches to particle analysis.

Considerations Required for Applications

Small-Particle Analysis with Analytical Electron Microscopy Using X-Rays and Electrons

Chemical Analysis in the Analytical Electron Microscope. Chemical information about small particles is obtained in the analytical electron microscope (AEM) primarily through three analytical techniques: electron diffraction, X-ray microanalysis using energy-dispersive spectrometry (EDS), and electron energy-loss spectroscopy (EELS). Electron diffraction can be used to directly determine the crystallographic structure of small crystalline particles. Similar to the process used in X-ray diffraction, the symmetry elements and d-spacings revealed by the electron diffraction pattern can be used to determine the phase of the sample and, hence, its composition. EDS in the TEM is similar to EDS in the SEM or electron microprobe. Focused electrons from the microscope are used to excite atoms in the sample. The X-rays emitted by the sample during de-excitation are analyzed by an energy-dispersive detector almost identical to those used in SEMs. The resulting spectrum contains quantitative information about the elemental composition of the sample. EELS utilizes an electron spectrometer to analyze the energy lost by transmitted electrons during their interaction with the sample. Again, the energy spectrum can be exploited to extract compositional information from the specimen.

Each of these analytical tools has strengths and weaknesses. Fortunately the three techniques are largely complementary in the information that they provide, and in many cases all three can be brought to bear on a given sample during a single session on one microscope. Electron diffraction, the oldest and best established of the three techniques, has been used extensively to study the crystallography and atomic-scale defects of materials. Although it provides only indirect chemical information and is frequently used in conjunction with EDS or EELS, it is indispensable for distinguishing between chemically similar phases. EDS in the AEM is also a trusted, robust technique. Commercial EDS detectors for TEMs have been available for many years, and the technique is widely applied. Such widespread use has led to improved appreciation of the quantitative issues involved in chemical analysis and a practical understanding of the limits and capabilities of the technique. Quantitative EELS in the AEM is less widely used. This is unfortunate because it offers several advantages over EDS, including better spatial resolution, better collection efficiency, better sensitivity to low-Z elements, better trace detection limits, better energy resolution, and sensitivity to chemical bonding effects. Currently, the number of EELS-equipped AEMs with field emission electron sources capable of producing very small probes is increasing rapidly, and the widespread use of EELS as a chemical analysis tool is expected to follow.

Electron Diffraction. Electron diffraction is used to determine the structure of crystalline particles that have thicknesses less than about 500 nm. When coupled with qualitative chemical information, it provides an unambiguous identification of the material. As the electron beam interacts with a crystalline particle, the regular arrangement of the atoms in the specimen gives rise to coherent, elastic electron scattering specific to the spacing and orientation of the atomic planes in the crystal lattice. If a field-limiting aperture is placed over the particle to limit diffraction to only the area of interest, and a parallel electron beam is used to illuminate the sample, this coherent scattering forms a selected-area diffraction (SAD) pattern as shown in Figure 12–12. The spacing of the diffraction spots and the symmetry of their arrangement can be translated into crystalline lattice parameters. These parameters can be compared with those listed in a diffraction database, or the pattern can be compared directly to a known standard. An introduction to SAD in the AEM can be found in Williams and Carter (1996).

Alternatively, instead of using a field-limiting aperture to restrict diffraction from a broad beam, the beam can be focused to a small probe placed on the particle of interest. The resulting diffraction pattern is more complicated because the converged beam contains a range of incident angles, but this method can often be performed on particles too small for SAD. This technique, known as *convergent beam electron diffraction* (CBED), is more sensitive than SAD and can potentially provide more information about the particle, including thickness and subtle structural changes such as lattice distortions. Unfortunately it is also more

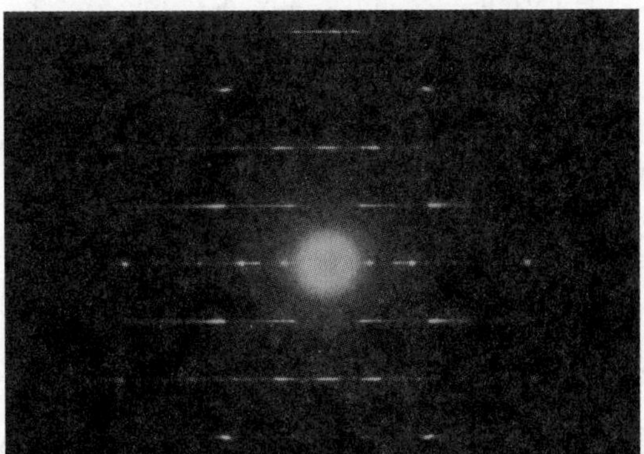

Fig. 12–12. An example of an electron diffraction pattern of chrysotile asbestos showing the intensity distribution resulting from scattering from the crystal lattice. (Courtesy of S. Turner, NIST.)

challenging experimentally, especially on small particles, and interpretation of the observed diffraction patterns is more complicated. The book by Spence and Zuo (1992) includes an extensive bibliography of CBED applications indexed by material.

Energy Dispersive Spectroscopy. Because the specimens examined in the AEM must be thinned to electron transparency, the complex partial-excitation (atomic number correction), X-ray absorption, and fluorescence corrections needed in quantitative X-ray microanalysis using EPMA are much smaller in the AEM case and often negligible. The brehmsstrahlung X-ray background is typically smaller as well, leading to more accurate measurements of net peak counts. These factors simplify the calculation of chemical concentrations from raw X-ray intensities.

When high-energy incident electrons are used for quantitative X-ray analysis of particles smaller than about 50nm, the effects of electron backscattering and electron energy loss can be neglected (Goldstein, 1986). In addition, if the particle can be approximated as an infinitely thin film, the effects of X-ray absorption and fluorescence can also be neglected. These conditions are referred to as the *thin-film approximation* and make it possible to express the relationship between the measured and generated X-ray intensities, for element a, as

$$I_a = I_a^* \varepsilon_a \qquad (12\text{–}1)$$

where I_a is the measured X-ray intensity from element a, I_a^* is the generated X-ray intensity, and ε_a is the experimentally determined efficiency of a particular EDS detector measuring X-rays from element a. Because the determination of particle thickness is difficult and the value of ε_a is not constant, most analysis schemes for small particles involve the measurement of ratios of elemental concentrations (expressed as weight fractions) according to the equation

$$C_a/C_b = k_{ab}\, I_a/I_b \qquad (12\text{–}2)$$

The term k_{ab} is a sensitivity factor often described in the literature as a Cliff-Lorimer factor. The k_{ab} factor is related to the generation and efficiency terms of element b ratioed to those for element a (Cliff and Lorimer, 1975). The analyst determines, from binary standards or first-principles calculations, a set of k_{ab} factors for the elements of interest. For a system containing unknown concentrations of elements a, b, c, . . . the concentrations C_a, C_b, C_c, . . . can be determined from Eqs. 12–3 to 12–5 and the knowledge that $\Sigma C_i = 1$. The concentrations for the various elements can be determined if the k_{ab} factors are known (as combinations of various binaries), as given by the following equations for a ternary system:

$$C_a/C_b = k_{ab}\, I_a/I_b \qquad (12\text{–}3)$$

$$C_c/C_b = k_{cb}\, I_c/I_b \qquad (12\text{–}4)$$

$$C_a + C_b + C_c = 1 \qquad (12\text{–}5)$$

The convention in most particle schemes is to express the k_{ab} factors relative to a common matrix element such as silicon (or iron), giving $k_{a\,Si}$ factors. It is important to note that these k factors are not constants, but sensitivity factors that vary with the accelerating potential, the particulars of the TEM/EDS system, and the data processing (e.g., background subtraction) procedures used to reduce the raw data to net intensities. These factors can be calculated from first principles or measured experimentally using standards. Because producing suitably homogeneous standards with well-characterized compositions can be difficult, it is often necessary to resort to calculation. Unfortunately, the relative uncertainties of calculated k factors are often as large as 10% or 20%, while the relative accuracies of properly measured experimental k factors are typically 2% to 3% (Williams and Carter, 1996, p. 605). As the thickness of the particle increases, the electron transparency decreases, eventually reaching a level where the sample no longer conforms to the "thin-film approximation." Under these circumstances, corrections for X-ray absorption and fluorescence must be included in the analysis scheme. According to Goldstein, the limits for the failure of the thin-film approximation are an X-ray absorption greater than 3% and/or a fluorescence larger than 5% of one element in the sample by another. For these larger particles, Eq. 12–2 is expanded to include X-ray absorption and fluorescence. The equations for the absorption and fluorescence require an estimate of the particle thickness and a calculation of the mass absorption coefficient for the elements a and b in the specimen (Goldstein et al., 1977).

Electron Energy-Loss Spectroscopy. Electron energy-loss spectroscopy (EELS) can be used for the quantitative elemental analysis of particles less than about 100 nm in diameter. In particles thicker than this, the electrons are more likely to undergo multiple scattering events than a single scattering event, complicating the quantitative analysis. This transition is gradual and depends on the composition of the sample and the energy of the electron beam used for analysis. Correction procedures exist for removing the effects of multiple scattering, but, as in X-ray microanalysis, the accuracy of the quantitative results is best in the regime of small corrections. Although EELS is capable of determining chemical bonding information and the nearest neighbor environment of atoms in the particle, this capability has not yet been widely used to analyze particles of unknown composition. The following is a very brief introduction to EELS, focusing on its ability to probe the elemental composition of individual particles. A more complete discussion of EELS and its capabilities, along with many references, is given by Egerton (1996).

In EELS, an electron spectrometer is used to analyze and display the energy distribution of the electrons transmitted through the sample. The resulting spectrum (Fig. 12–13) is generally divided into three energy-loss regions, roughly corresponding to the three types of electron–specimen interaction that occur during analysis. The first region, called the *elastic* region, contains an intense peak called the *zero-loss* peak. This peak contains unscattered and elastically scattered electrons. Despite its name, it also contains electrons that have lost a small amount of energy (a few meV) in the sample due to phonon scattering, thermal

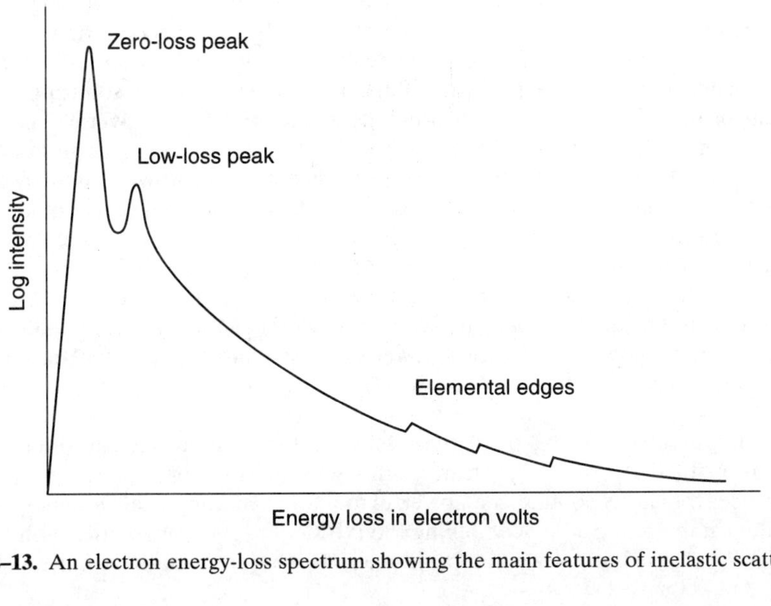

Fig. 12–13. An electron energy-loss spectrum showing the main features of inelastic scattering.

effects, and so forth. Its width is comparable with the energy resolution of the EELS system, typically about 0.3 to 3 eV. The zero-loss peak is easily the most intense feature in the spectrum of a thin sample, but it carries no elemental information. The second region, called the *low-loss* region, extends from the zero-loss peak to about 100 eV and arises from a variety of low-energy interactions such as the generation of plasmons, valence transitions, and solid-state effects. It can contain elemental information about the sample, but its sensitivity to valence and solid-state effects makes interpretation difficult for unknown materials. The third region, called the *core-loss* region, begins at about 100 eV and extends to the upper energy limit of the spectrometer, typically 2 to 3 keV. In this region the element-specific core-loss ionization edges occur, providing the most reliable data for elemental analysis of individual particles. As beam electrons interact inelastically with the atoms of the sample, they eject inner-shell atomic electrons and lose energy in the process. The amount of energy lost is related to the ionization energy of the specific shell from which an electron is ejected. The resulting energy-loss "edges" in the spectrum carry quantitative information about the presence and abundance of elements in the sample.

The spatial resolution of the analysis is determined by the diameter of the electron probe as well as by the thickness and elemental composition of the sample itself. Very thin samples can be probed with a spatial resolution comparable with the electron beam diameter. In dedicated STEM instruments this can be 1 nm or less. In thicker samples, especially those containing high-Z elements, the beam is broadened by scattering within the sample, and the lateral resolution is significantly degraded.

As mentioned earlier, EELS offers several advantages over EDS, including better spatial resolution, better collection efficiency, better sensitivity to low-Z elements, and better trace detection limits. Most of these benefits arise because EELS analyzes scattered primary electrons to characterize the sample, not secondary signals like characteristic X-ray photons. The scattering cross section of high-energy (100 to 400 keV) primary electrons in solids is strongly peaked in the forward direction. For a target of intermediate atomic number (copper) and a relatively low primary energy (100 keV), the most probable scattering angle is 2° to 3° (Newbury, 1986). This narrow solid angle of emergence allows most of the scattered electrons to be captured by the entrance aperture of the EELS spectrometer (located below the sample), resulting in high collection efficiencies.

In EDS the X-rays emitted by the de-exciting atoms are radiated isotropically over the full range of angles. A typical EDS detector area is about 30 mm^2, and the detector must be located at least several millimeters away from the sample due to geometric restrictions and interference with the objective lens pole piece. Because of these constraints an EDS spectrometer can only subtend a small fraction of the 4π steradian emission solid angle. This, combined with absorption losses in the detector window material, lead to overall EDS collection efficiencies of no more than a few percent, even for high-Z materials. The situation for low-Z atoms is more unfortunate. The characteristic X-ray emission pathway competes with other mechanisms for atomic relaxation, notably Auger electron emission. The likelihood of photon emission, called the *fluorescence yield*, increases rapidly with atomic number. While the fluorescence yield reaches about 90% at $Z = 54$, it is only 2% for $Z = 13$ (aluminum) (Cardona and Ley, 1978). The low-energy X-rays emitted by low-Z atom relaxations are also more strongly attenuated by the EDS detector window. EDS thus suffers a triple penalty compared with EELS: Fewer photons are emitted, fewer of those emitted travel toward the detector, and fewer that reach the detector are counted by the electronics.

Elemental Mapping. Up to this point the discussion has focused on acquisition of a single one-dimensional EELS or EDS spectrum. Although acquisition of point spectra using static, focused probes remains a popular and powerful tool for examining small volumes, much more information can be obtained by combining the AEM's imaging capabilities with the analytical modes. When used to extract compositional information from the sample, the result of

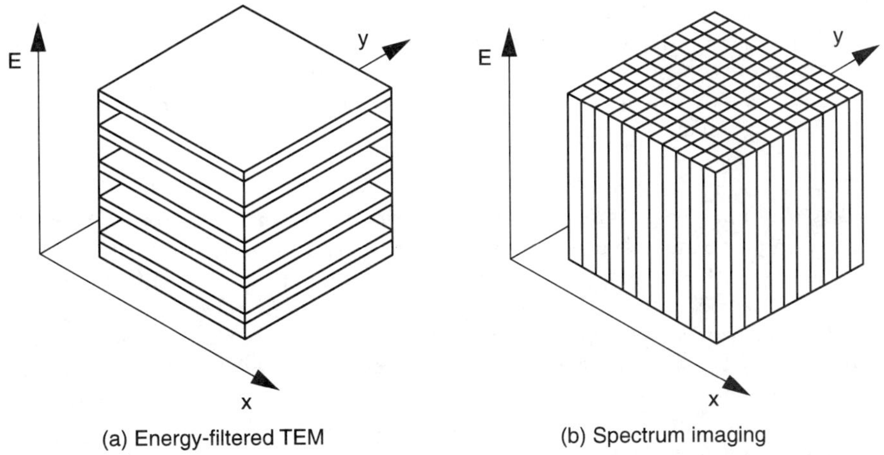

Fig. 12–14. Schematic showing how EFTEM **(a)** and spectrum imaging **(b)** build up an identical data cube using two different methods. E, energy loss.

this union is usually an elemental map—a two-dimensional image showing the spatial distribution of elements in the sample. Such images are very useful for visualizing chemical heterogeneities in a sample or for distinguishing between particles with different compositions. There are two techniques generally used to produce elemental maps in the AEM: energy-filtered transmission electron microscopy (EFTEM) and spectrum imaging. While EFTEM is performed exclusively with EELS, spectrum imaging can utilize either EELS or EDS as the analytical signal used to produce chemical contrast.

EFTEM is based on the acquisition of energy-selected images, that is, images formed from electrons whose energy loss falls within a narrow window (Krivanek et al., 1987). The microscope is operated in TEM mode, with a spread electron beam for illumination, not with a focused probe in STEM mode. Energy-selected images are acquired using an imaging energy filter, a device that converts a two-dimensional TEM image into an EELS spectrum, filters out all electrons except those that fall within the passband chosen by the analyst, and then reconverts the filtered spectrum back into a two-dimensional image. Because each energy-selected image only contains contributions from electrons of a single energy, if several such images are acquired at different energies they can be stacked to develop a data cube for the sample. Two dimensions of this cube are the familiar x and y spatial locations of the pixels on the sample. The third cube axis is energy loss, as shown schematically in Figure 12–14a. For each element to be mapped, typically three energy-selected images are acquired near the energy of the element's core-loss ionization edge. At least two energy slices are collected below the edge to allow an inverse power law background model to calculate, pixel by pixel, the background intensity at an energy loss above the ionization edge. This calculated image is subtracted digitally from the third observed energy slice, taken above the edge, to produce an elemental map where pixel intensity is proportional to the number of element-specific EELS counts above the background.

Spectrum imaging develops the same data cube for the sample, but in a different way (Jeanguillaume and Colliex, 1989; Hunt and Williams, 1991). In this technique, a focused STEM probe is used to acquire an entire EELS or EDS spectrum down the energy axis of the cube before the probe is rastered to the next pixel, as shown in Figure 12–14b. This is straightforward and is very similar to the equivalent process in the SEM (with EDS), except for the much higher spatial resolution that can be achieved. As in the SEM, multiple elemental maps may be acquired simultaneously because of the parallel detection nature of EDS. The sampling along the energy axis is typically much better for spectrum imaging than for EFTEM, but it takes much longer, requires that the sample be subjected to a larger electron dose, and is more susceptible to drift problems than EFTEM. Both techniques collect a large amount of information from the sample in a relatively short time and rely on computer post-processing to generate quantitative chemical distributions from the data set.

Large-Particle Analysis with the Electron Probe and Scanning Electron Microscope. In classic electron probe analysis procedures, both sample and standard must have flat surfaces and be infinitely thick with respect to the penetration of the electron beam. The corrections for the interaction of the electron beam with the sample and the subsequent X-ray emission can be calculated from simple geometric relationships. If the effects of particle geometry are not included in a given analysis scheme, the subsequent results will be qualitative and can be used for element identification only. Errors in composition for uncorrected analysis can be in the range of several hundred percent. It is therefore important to correct the X-ray data for the effects of particle shape and size.

In the quantitative analysis of particles, the shape and thickness of the specimen cannot be controlled or, in many cases, even measured. The difficulties in quantitative analysis of particles or rough surfaces result from three different effects that influence the generation and measurement of X-rays from these samples (Small, 1981).

The first effect is the result of the finite size (mass) of the sample. The mass effect is related to the elastic scattering of the electrons and is strongly affected by the average atomic number of the sample. This effect is important when the particle thickness is smaller than the range of the primary electron beam so that a fraction of the beam escapes the sample before exciting X-rays. As the size of the particle decreases, the mass of material from which X-rays are excited decreases, resulting in a reduction of X-ray intensities from the particle compared with a bulk sample of the identical composition. The mass effect can be demonstrated by comparing the X-ray emissions from a bulk target to the emissions from a particle of the same composition. This effect can be seen in Figure 12–15, which depicts the Ba Lα X-ray intensity from particles (normalized to the intensity from a bulk material of the same composition) plotted as a function of particle diameter. The energy of the Ba Lα X-rays is 4.47 keV, high enough that the absorption effects are minimal and the slope of the curve

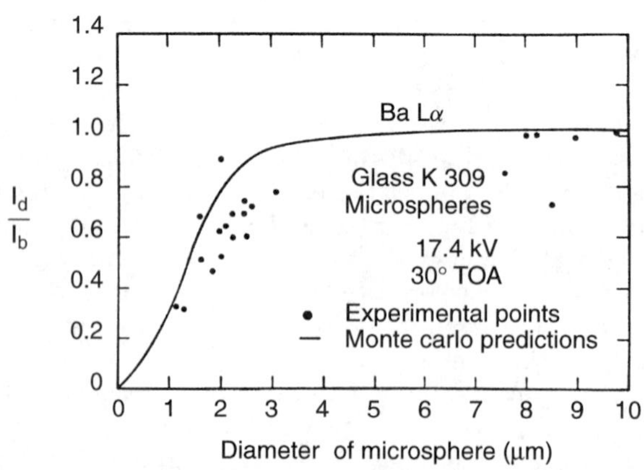

Fig. 12–15. Plot of the Ba Lα X-ray intensity from particles (I_d), normalized to the X-ray intensity from a bulk target (I_b), versus particle diameter. TOA = take off angle. This plot shows the decrease in the particle X-ray intensity compared with the bulk target for the smaller particles due to their decreased mass. (From Small et al., 1978.)

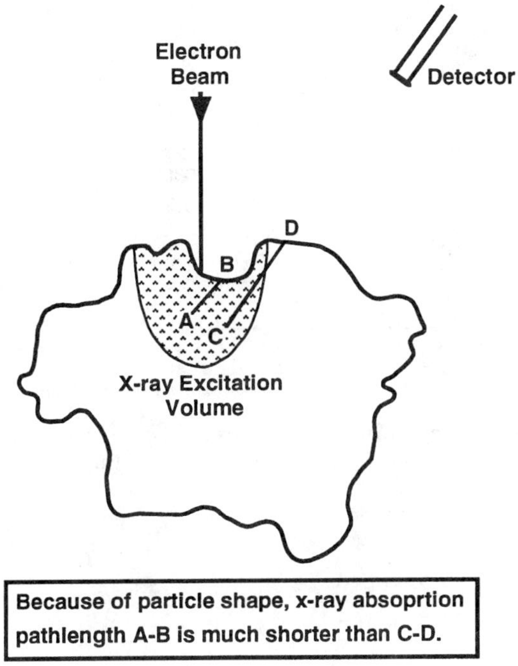

Fig. 12–16. Schematic showing the random X-ray absorption path lengths encountered in irregularly shaped particles.

reflects the mass effects. The mass effect is shown by the decrease in the Ba Lα X-ray intensity measured on particles less than 3 μm in diameter compared with the bulk. The net result of not correcting for the mass effect will be an underestimation of the composition for all elements analyzed in the particle (if the results are not normalized).

The second effect is the result of differences in X-ray absorption in particles compared with bulk standards. In the analysis of particles, the X-ray emergence angle (and therefore the absorption path length) cannot be as accurately predicted as it can for polished bulk specimens. For the particle shown in Figure 12–16, the X-ray intensity emitted from the particle along the A–B path will be higher than along the C–D path because the absorption path length is smaller and there is less mass for the X-ray to travel through before escaping the particle volume. The resulting elemental concentrations will therefore be higher than expected for X-rays measured from the shorter path length and lower than expected for X-rays measured from the longer path length. As expected, the magnitude of this effect is largest when there is high absorption, as is typically the case for soft X-rays from elements like Al or Si that have energies less than 2 keV. The difference between the absorption path length in a particle and a flat bulk sample can result in widely different values of emitted X-ray intensities. The result of the varying absorption path lengths is shown in Figure 12–17, which is the Si Kα X-ray intensities from particles, normalized to the intensity from a bulk material of the same composition, plotted as function of particle diameter.

The third effect is caused by the secondary excitation of characteristic X-rays of interest by electron-generated X-rays, either continuum or characteristic. Because X-ray absorption coefficients in solids are relatively small compared with electron attenuation, secondary X-ray fluorescence can occur over a much larger volume or range (tens of micrometers) than the range of primary electron excitation of X-rays (1 to 5 μm). In bulk samples and standards,

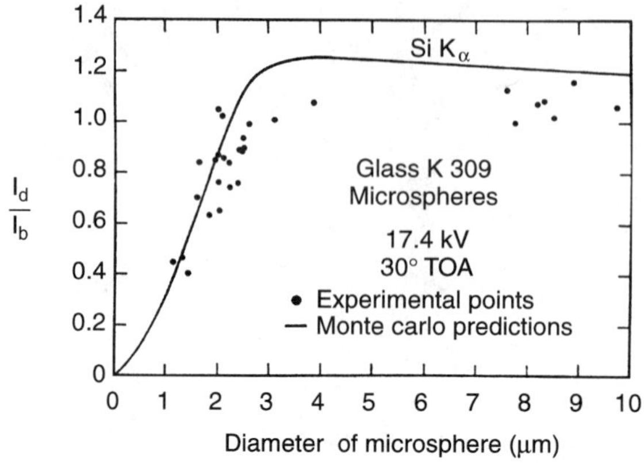

Fig. 12–17. Plot of the Si Kα X-ray intensities from particles (I_d), normalized to the X-ray intensity from a bulk target (I_b), versus particle diameter. TOA = take off angle. This plot shows the increase in X-ray intensity for the smaller particles compared with the bulk material due to the decreased X-ray absorption path length. (From Small et al., 1978.)

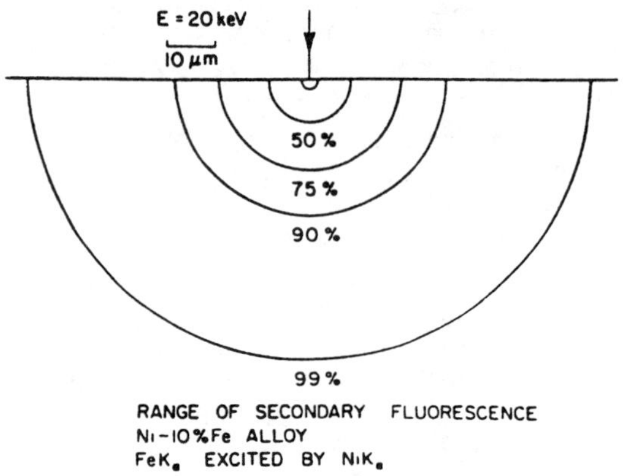

Fig. 12–18. Diagram showing the X-ray range in micrometers for Ni Kα X-rays fluorescing Fe Kα X-rays. This diagram indicates that Fe X-rays are fluoresced by Ni X-rays at a distance considerably greater than the dimensions of microscopic particles. (From Small, 1981.)

the electron-generated X-rays for the most part remain in the target because the samples are substantially larger than the X-ray ranges. In the case of particles, however, the particle volume may be only a small fraction of the exciting X-ray range. As a result, most of the electron-generated X-rays will exit the particle before exciting secondary X-rays. In those samples where the secondary fluorescence is important, the deviation of particle behavior relative to a bulk standard may be significant. This behavior is shown in Figure 12–18, which is a plot of the range for Ni Kα X-ray fluorescence of Fe Kα X-rays in an Ni-Fe alloy.

The net effects of not correcting for the low secondary excitation of X-rays in particle analysis schemes using bulk standards are

1. An underestimate of the concentration for elements that have a significant contribution to their characteristic line from excitation by other characteristic X-rays because the measured X-ray intensity from the particle will be reduced by the calculated value of the secondary X-rays

2. In the case of continuum fluorescence, an underestimate of the concentrations for all elements, particularly those with higher energy lines that are excited by the higher energy, longer range continuum

Normalization. One of the simplest methods for the quantitative analysis of particles is to normalize the results from a conventional quantitative procedure to 100% (Wright et al., 1963). The analyst, in selecting this method of correction, makes the assumption that the X-ray absorption and fluorescence are the same for the particle as for a bulk specimen and that the mass effect is the same for all elements. In practice, normalization of results is effective for the correction of the mass effect because the decrease in intensity as a function of particle size is nearly the same for all elements. Figure 12–19 shows that the different elemental curves merge together for particle diameters less than 2 μm. Because normalization does not compensate accurately for the absorption and fluorescence effects, the most accurate results will be obtained on particle systems that meet the following conditions:

1. Systems for which all the analytical lines for the elements are above 4 keV when the absorption effects are minimal
2. Systems for which the analytical lines are below about 4 keV in energy; the lines for all the elements should be as close together in energy as possible so that the matrix absorption is approximately the same in all cases
3. Systems for which there is no significant secondary fluorescence.

Table 12–4 lists the results from the analysis of lead silicate glass particles. The first set of results is taken from the analysis of the Pb Mα line at 2.3 keV that is close in energy to the Si Kα line at 1.74 keV. Because the absorption and mass corrections are similar for these two

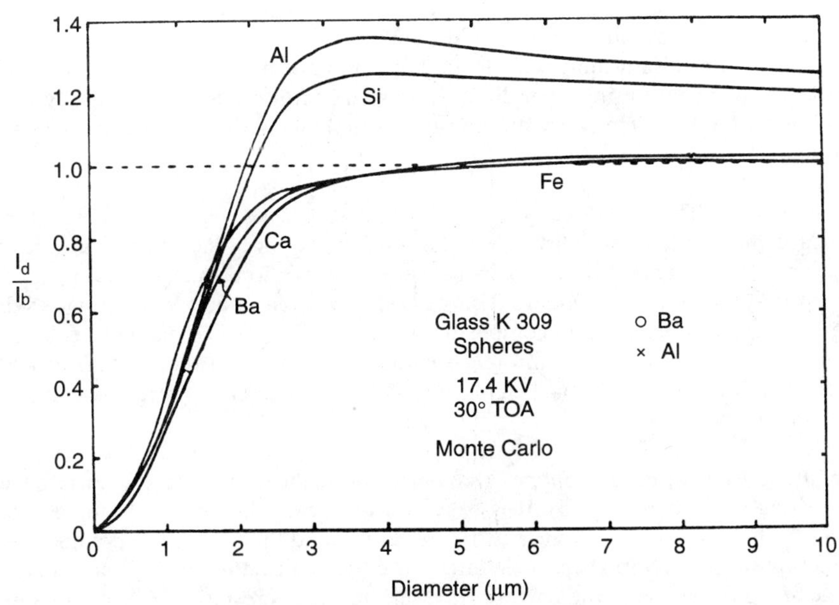

Fig. 12–19. Plot of calculated Monte Carlo curves for X-ray intensities of K-309 glass microspheres (I_d), normalized to the X-ray intensities from a bulk target (I_b), versus particle diameter. TOA = take off angle. The plots show that the different elemental curves merge for particle diameters smaller than approximately 2 μm.

TABLE 12–4. Analysis of Lead Silicate Glass K-229 by Normalization of Quantitative Results[a]

Analysis	Si wt.%	% Error	Pb wt.%	% Error
With Pb Mα lines[b]				
1	0.155	+10.7	0.620	−4.6
2	0.144	+2.9	0.643	−1.0
3	0.136	−2.7	0.658	+1.2
4	0.138	−1.1	0.653	+0.5
5	0.127	−9.0	0.675	+3.8
6	0.170	+22	0.588	−9.5
7	0.137	−2.5	0.657	+1.1
With Pb Lα line[c]				
1	0.134	−4.5	0.663	+1.9
2	0.177	+26.3	0.578	−11.0
3	0.159	+14.0	0.612	−5.8
4	0.166	+18.0	0.602	−7.4
5	0.017	+88.0	0.894	+37.0
6	0.100	−29.0	0.731	+12.4
7	0.157	−12.3	0.616	−5.2

[a] Nominal composition: Si = 0.140; Pb = 0.650.
[b] Meets conditions.
[c] Does not meet conditions.
Source: Small (1981).

lines, the lead and silicon concentrations are in good agreement with the true values. The second set of results were determined by analyzing the Pb Lα line at 10.6 keV. In this case, the two analytical lines have very different energies, and the particle absorption effect is not similar in magnitude. As expected, the errors associated with this analysis are considerably higher than those associated with the Pb Mα analysis.

One of the major disadvantages of the normalization of results from a bulk analysis procedure is that the analyst using a beryllium window detector cannot determine, by obtaining an analysis total of less than 100%, the presence of any unanalyzed elements such as those with atomic numbers less than 11.

Particle Standards. The analyst can use a conventional analysis scheme and substitute particle standards for the normal polished standards (White et al., 1966). In this procedure, the assumption is that the particle effects, particularly the absorption effect, will be approximately the same for the sample and standard. This assumption is reasonably valid provided that the sample and standard are close in composition and that the particle diameter is larger than about 2 μm. Below this size, any difference in size and shape between unknown and standards will be critical because a small change in effective diameter will result in a large change in X-ray intensity.

Geometric Modeling of Particle Shape. Geometric modeling of particle shape was developed by Armstrong and Buseck (1975). It is based on the determination of a simple geometric shape or combination of shapes such as square or pyramid that define the boundaries of the particle of interest. The various particle effects are then calculated for the chosen geometric shapes defining the shape of the particle. The various effects resulting in X-ray loss in particles are corrected as follows:

1. Electron transmission: The X-ray intensity lost as a result of electron transmission through the particle is calculated from a modified form of an expression developed by

TABLE 12–5. Results from the Analysis of Mineral Particles by the Geometric Modeling Method

	Actual composition (wt.%)	Determined (wt.%)
Anorthite oxide[a]		
CaO	18.9	19.0 ± 0.8
Al_2O_3	35.7	35.6 ± 1.2
SiO_2	44.3	44.4 ± 1.0
Rhodonite oxide[b]		
MnO	35.5	35.5 ± 1.1
FeO	13.0	13.1 ± 0.8
CaO	4.1	4.0 ± 0.2
SiO_2	46.9	46.8 ± 0.8
Pyrite element[c]		
FeS	46.6	46.4 ± 1.0
S	53.4	53.6 ± 1.0

[a] Average of 122 analyses.
[b] Average of 100 analyses.
[c] Average of 140 analyses.
Source: Adapted from Armstrong (1978).

Reuter (1972) to calculate relative X-ray production in thin films. In the modified form, the expression can be used to calculate X-ray production as a function of position within a particle.

2. Electron backscatter: The expression of Duncumb and Reed (1968) is used to calculate the loss of X-rays as a result of electron backscatter.
3. Electron sidescatter: The loss of primary X-rays from the sidescatter of electrons is minimized by rastering or defocusing the electron beam over the entire particle area by using an overvoltage less than 1.5 and by ratioing the concentrations.
4. X-ray absorption: The correction for X-ray absorption sets the integration limits for the absorption path length to the limits determined for the geometric shape or shapes that define the overall particle. The shapes that have been included in the program include rectangular prism, tetragonal prism, right triangular prism, square pyramid, hemisphere, and sphere.

The X-ray intensity from the secondary fluorescence of characteristic X-rays is calculated from the work of Armstrong and Buseck (1977) with the integration limits adjusted to calculate the probability of secondary fluorescence at a given location within a particle.

Results from this procedure are given in Table 12–5 for anorthite (calcium aluminum silicate), rhodonite (manganese silicate), and pyrite (iron sulfide) (Armstrong, 1978). In all cases, the analyses are in good agreement with the known compositions and have standard deviations less than 6% of the measured concentration. Careful attention must be paid to the sizes and shapes used to define the particle boundaries if accurate concentrations are needed.

Peak-to-Background Ratios. A fourth method for the quantitative analysis of particles was developed by Small et al. (1978) and Statham and Pawley (1978). This method, an extension of the method developed for biological specimens by Hall (1968), is based on the following observation: To a first approximation, the ratio of a characteristic X-ray intensity to the

TABLE 12–6. Analysis of Mineral Particles by the Peak-to Background Method (*FRAME P*); Conventional Analysis by *FRAME* C[a]

		Talc ($Mg_3[Si_4O_{10}](OH)_2$		
	Wt. % Mg	% Error	Wt. % Si	% Error
Nominal	19.3		29.8	
FRAME C	10.3	-47.0	15.8	-47.0
FRAME P	18.5	-4.0	29.0	-3.0
		FeS_2		
	Wt.% S	% Error	Wt. % Fe	% Error
Nominal	53.4		46.6	
FRAME C	39.9	-25.0	35.8	-23.0
FRAME P	52.9	-0.9	46.4	-0.4
		ZnS_2		
	Wt.% S	% Error	Wt.% Zn	% Error
Nominal	32.9		67.1	
FRAME C	30.3	-8.0	55.1	-18.0
FRAME P	36.0	+9.0	67.6	-0.7

[a] Results are the average of seven analyses.

Sources: Small et al. (1979b).

continuum intensity of the same energy for a flat, infinitely thick target is equivalent to the same ratio of intensity from a particle or rough surface to its continuum intensity of the same composition.

This observation can be expressed in the following equation:

$$(I/I_{\text{con}})_{\text{particle}} = (I/I_{\text{con}})_{\text{bulk}} \qquad (12\text{–}6)$$

where I is the background-corrected peak intensity and I_{con} is the continuum intensity for the same energy window as the peak. For this procedure it is assumed that the spatial distribution for characteristic X-ray excitation is identical to the distribution for continuum X-ray excitation. As a result, the effects of particle shape and size on measured X-ray intensity will be the same for the continuum and the characteristic X-rays. It therefore follows that by taking the ratio of the two intensities, the particle mass and absorption effects will cancel out.

Various types of mineral particles have been analyzed with the peak-to-background method using the algorithm *FRAME P* (Small et al., 1979b). The results of these analyses are reported in Table 12–6, along with the results from the conventional analysis algorithm, *FRAME C* (Myklebust et al., 1979). In all cases, the analyses with the peak-to-background routine are within 10% relative error of the stoichiometric values. In contrast, the errors with the conventional routine range from 8.0% for S in ZnS to 47% for Mg and Si in talc. In addition, the standard deviations for individual measurements are less for the peak-to-background routine than they are for the conventional atomic number absorption fluorescence (ZAF) algorithm (Henoc et al., 1973). ZAF is a quantitative analysis scheme that compares the X-ray intensities for the unknown with that for a standard.

At electron beam energies of 15 keV or greater, the assumption that the generation volumes for characteristic and continuum X-rays are identical is only valid for particles greater than about 1 μm in diameter. Below this size, the anisotropic generation of the continuum results in a significantly different excitation volume for the continuum compared with the isotropically generated characteristic X-rays. This effect is shown in Figure 12–20, where isotropic and anisotropic cross sections have been used to calculate peak-to-background

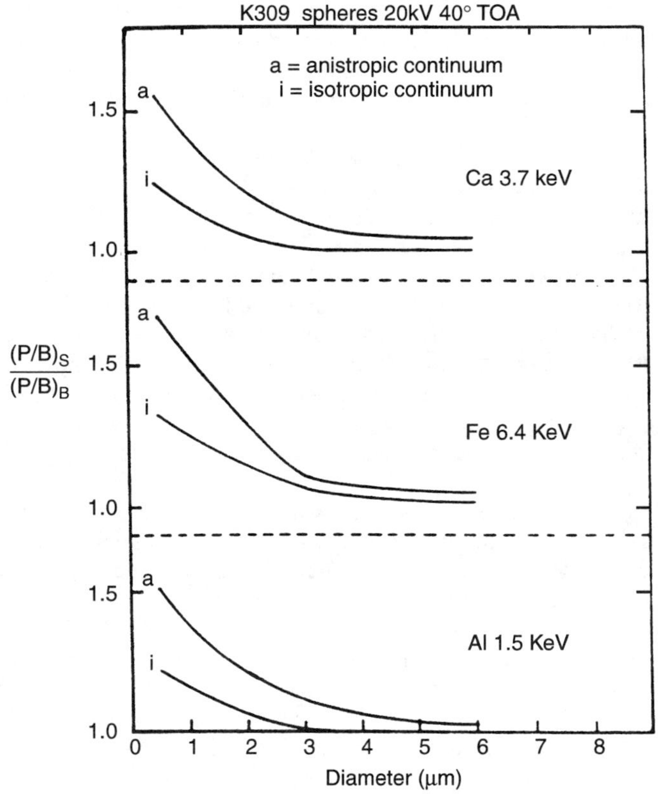

Fig. 12–20. Plot showing the effects of calculating peak-to-background ratios based on isotropic and anisotropic cross sections. (From Small et al., 1979a.)

ratios from K-309 glass particles normalized to bulk glass. The composition of K-309 is 7.9% Al, 18.7% Si, 10.7% Ca, 13.4% Ba, and 10.5% Fe. These plots show that the introduction of an anisotropic cross section for the continuum results in significantly higher peak-to-background ratios for the smaller particles. As a result of using peak-to-background procedures, it is necessary for quantitative analysis of particles less than about 1 μm in size to introduce into the analysis procedure a correction for anisotropic generation of the continuum (Small, 1981).

Particle Measurement Employing Low-Voltage Analysis. For conventional electron probe microanalysis, the accelerating voltage of the electron beam is generally in the range of 15 to 25 keV, which provides the necessary overvoltage to excite efficiently the K and L X-ray lines for elements with atomic numbers in the range of about 5 to 94. The same voltage criteria have been applied to the procedures developed for the X-ray microanalysis of individual particles. The main difference in analytical procedures for bulk samples and particles is that a series of particle corrections, described earlier in this chapter, must be applied to the particle k ratios or calculated compositions. These corrections compensate for the penetration or scattering of electrons out of the particle volume (particle mass effect) and the absorption of X-rays less than about 3 keV due to particle geometry (particle absorption effect; Small, 1993). In general, the analytical procedures that have been developed for quantitative particle analysis significantly improve the accuracy and reduce the relative uncertainty of analyses from approximately ±50% for uncorrected data to about ±10% for corrected data (Small and Armstrong, 1996). Although a significant improvement, a relative uncertainty of ±10% unfortunately is often not sufficient to distinguish between many materials with similar compositions.

In recent years, there have been a series of advancements in electron beam instruments and X-ray detectors that should make it possible to improve significantly the quality of results from the quantitative analysis of individual particles in the electron probe and scanning electron microscope. These advances include field-emission gun scanning electron microscopes (FEG-SEMs), which have high brightness electron guns with excellent performance at low electron accelerating voltages ($E_0 \leq 10$ keV), and high-resolution energy-dispersive X-ray spectrometers, like the NIST microcalorimeter detector, which provide high-resolution (<10 eV) and parallel X-ray collection (Newbury et al., 1999). The incorporation of these advances into analytical methodologies for particle analysis should make it possible to reduce the magnitude of analytical uncertainties associated with the X-ray microanalysis of single particles by employing low-energy electron beams to excite characteristic X-ray lines with energies <5 keV.

The rationale for low-voltage analysis of particles is to reduce the magnitude of the particle mass and absorption effects, thereby making the X-ray emission from a particle similar to that from a bulk polished target. If this can be accomplished, then the particle can be analyzed with conventional electron probe procedures. For the particle mass effect, lowering the accelerating voltage of the electron beam reduces the size of the electron beam interaction volume inside the particle and hence the number of electrons that scatter out of a given particle volume without generating X-rays. A comparison of the particle mass effect for high- and low-beam voltages is shown in Figure 12–21, which is a Monte Carlo plot of the electron trajectories for a 2 μm NIST Standard Reference Material (SRM) 2066 K-411 glass particle (see Table 12–7) excited with a 20 keV beam (Fig. 12–21a) and a 5 keV beam (Fig. 12–21b). In Figure 12–21a, a significant fraction of the 20 keV electrons penetrate through or scatter out the sides of the particle while for the 5 keV electron example, none of the electrons leave the particle (Fig. 12–21b). For particle analysis this means the analysis results from a 2 μm particle analyzed with a 20 keV beam will have a significant particle mass effect with low analytical totals for all elements. At 5 keV since essentially all of the electrons remain in the particle, the X-ray generation is essentially the same as that from a bulk material and the mass effect is minimal.

Lowering the beam voltage coupled with using the low energy X-ray lines, those below 5 keV in energy, reduces the X-ray emission volume and therefore the X-ray absorption pathlengths minimizing the effects of particle geometry. This is shown in Figure 12–22. Figure 12–22a is a Monte Carlo plot of the X-ray emission volume for the 1.25 keV Mg Kα X-rays in a 2 μm K-411 particle excited with a 20 keV electron beam. This plot shows that a significant fraction of the Mg X-rays originate several micrometers deep within the particle. As a result, the absorption of the Mg X-rays is significant and may be influenced by particle shape, as shown in Figure 12–16. Figure 12–22b is the Monte Carlo plot for Mg Kα X-rays excited in a 2 μm particle with a 5 keV beam. In this case, the Mg X-rays are generated much nearer the surface of the particle and the beam entry point, resulting in much lower absorption and minimizing the effects of particle geometry.

To study the effects of reducing the accelerating voltage on particle analysis, Small and Armstrong (1996) analyzed SRM 2066 microspheres manufactured from analytical glass K-411 (Marinenko, 1982). The analyzed spheres ranged in size from about 0.5 to 24 μm; they were analyzed at beam energies of 10 and 25 keV. The elemental X-ray intensities from the microspheres were compared with the intensities measured on the bulk K-411 glass sample.

The effect of reducing the acceleration voltage on the magnitude of the particle mass correction can be seen in Figure 12–23, which plots the ratio of the particle X-ray intensity to

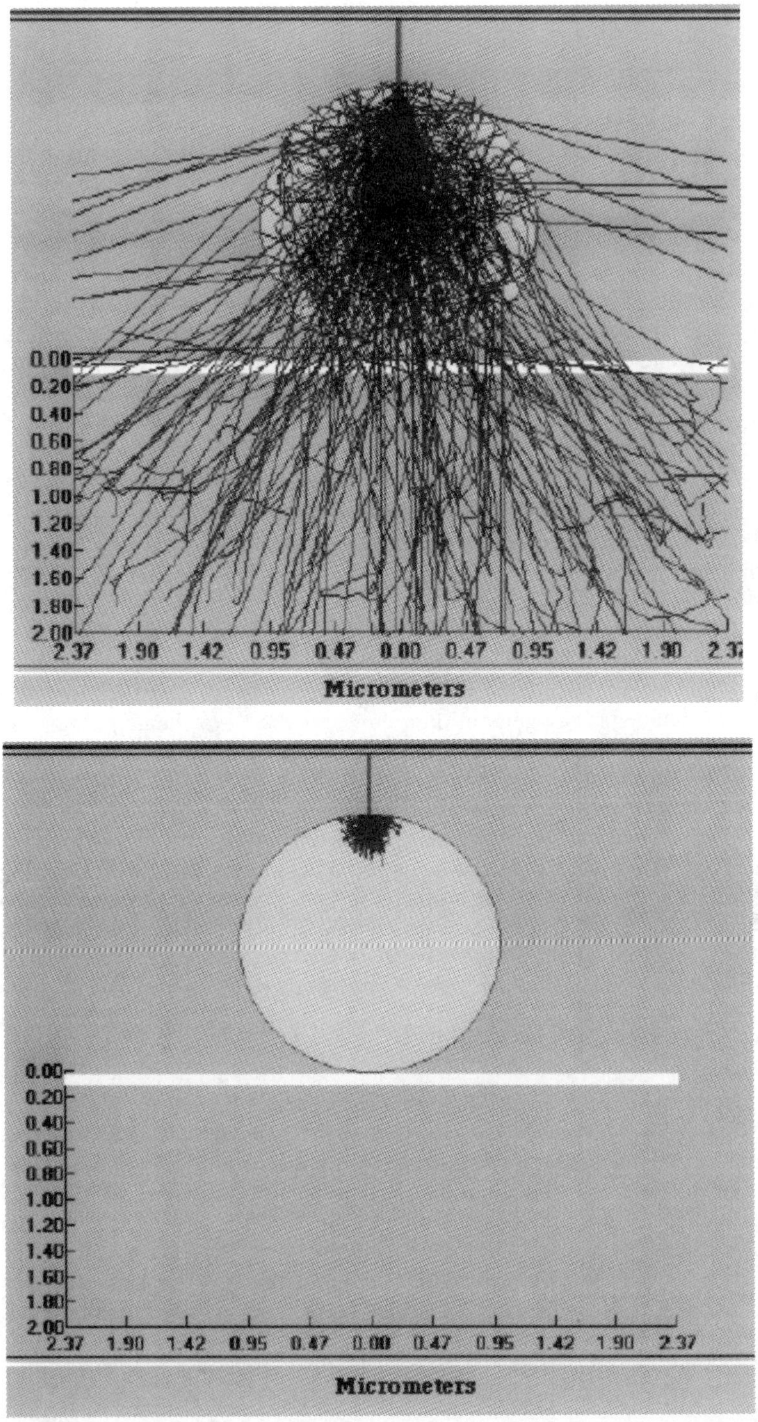

Fig. 12–21. Monte Carlo plot of the electron trajectories for a 2 μm K-411 glass particle excited with a 20 keV electron beam **(a)** and a 5 keV electron beam **(b)**. (Electron Flight Simulator, version 3.1E, copyright 1977 by Small World, based on Monte Carlo algorithms of Dr. David Joy, University of Tennessee.)

TABLE 12–7. Composition of K-411 Mineral Glass SRM 2066

Element	Certified Weight Fraction
Si	0.2538
Fe	0.1126
Mg	0.0884
Ca	0.1105
O	0.4238

the X-ray intensity from the bulk glass for the 3.7 keV Ca Kα and 6.4 keV Fe Kα X-ray lines versus particle diameter. These X-ray lines are energetic enough that the particle absorption correction is minimal and the shapes of the various plots are dominated by the particle mass correction. The 25 keV plot (Fig. 12–23a) shows that the ratios approach unity only for particles with diameters greater than about 5 μm. Below this diameter, the particle mass effect dominates due to electron scatter out the bottom and sides of the particles. This results in a 50% loss in X-ray intensity for a 2 μm sphere and a 75% loss for the smallest microspheres (~1 μm in diameter). Lowering the beam voltage to 10 keV significantly reduces the magnitude of the particle mass correction as shown in Figure 12–23b. The maximum loss of X-ray intensity for the 10 keV data is <12% for all of the particles measured. Unlike the 25 keV plots, the Fe and Ca ratios at 10 keV remain near unity for all of the particles in the size range analyzed (0.9 to 18 μm) with no detectable drop off even for the smaller diameter particles.

Lowering the beam energy will also reduce the magnitude of the particle absorption correction. The smaller X-ray generation volume, at the lower beam energies, reduces the X-ray absorption path length in both the particle and the standards and should minimize the effects of particle morphology on low-energy X-rays because the X-rays are generated closer to the sample surface. In addition to lowering the magnitude of the absorption correction over all particle sizes, reducing the beam voltage should also reduce the particle diameter at which a particle can be approximated as bulk material. In Figure 12–19 for a 17 keV beam, effects of particle morphology for Si and Al extend out about 10 μm before the particle X-ray intensities approximate those from a thick bulk sample, and the particle morphology effect on the particle/bulk (P/B) ratios for Si and Al are minimal. Reducing the beam voltage below 15 keV results in reduced excitation and generation volumes for X-rays, which means particles smaller than the 10 μm will approach the bulk sample situation.

The effect of reducing the acceleration voltage on the magnitude of the particle absorption effect can be seen in Figure 12–24, which shows the P/B plots for the Kα X-ray lines of Si (1.74 keV) and Mg (1.24 keV). The energies of these X-ray lines are low enough that X-ray absorption is significant, and the particle X-ray absorption effect dominates for microspheres larger than about 3 to 4 μm in diameter.

At 25 keV (Fig. 12–24a), the Mg P/B values are >1.0 for all particles larger than 3 μm, reaching a maximum of nearly 1.6 for a 5 μm sphere. The X-ray absorption effect is evident even for the larger diameter particles with ratios in excess of 1 for diameter approximately 10 μm. The Si P/B values are similar to the Mg values, reaching a maximum of 1.4 for a 6 μm sphere and decreasing to near 1 only for the largest spheres. As expected, the particle absorption effect is slightly smaller for Si than for Mg because the Si Kα X-rays are more energetic than the Mg Kα X-rays. For the smaller particles, diameters <3 μm, the particle mass effect still dominates as evidenced in both the Mg and Si P/B values being below 1.

Similar to the particle mass effect, the particle absorption effect is even further reduced at 10 keV (Fig. 12–24b). With the exception of a drop in the P/B ratio for the smallest particle at 0.93 μm, due to the particle mass effect, the P/B curves for both Si and Mg at 10 keV

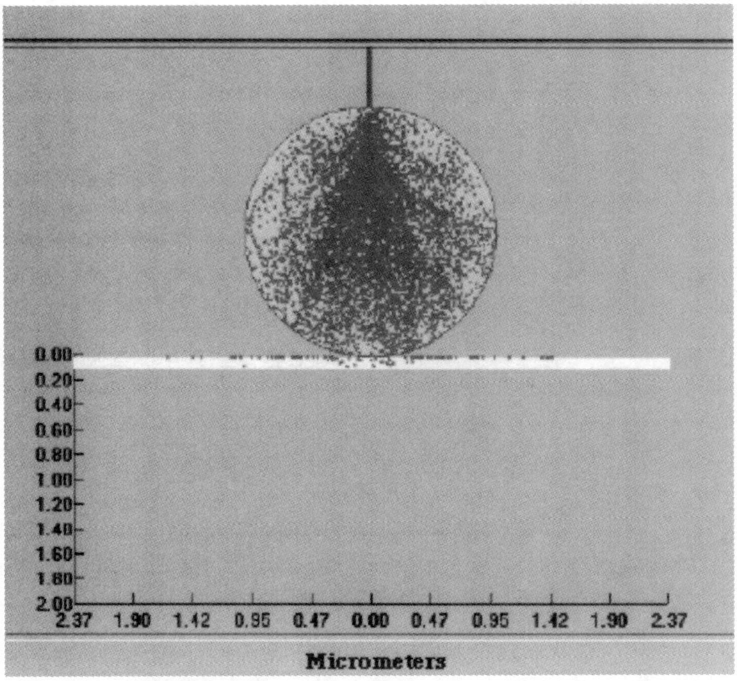

(a)

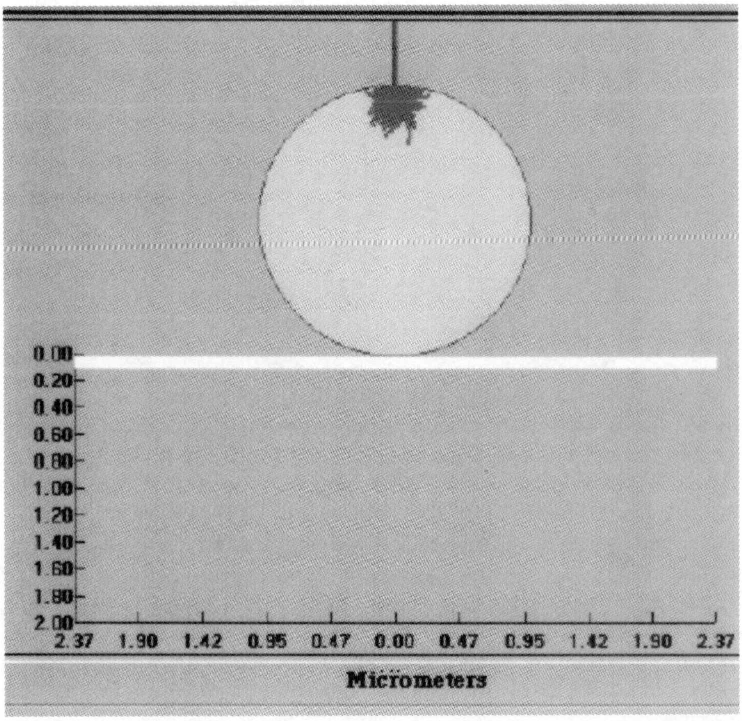

(b)

Fig. 12–22. Monte Carlo plot of the generated X-ray volume for the 1.25 keV MgK X-rays in a 2 μm K-411 glass particle excited with a 20 keV electron beam **(a)** and a 5 keV electron beam **(b)**. (Electron Flight Simulator, version 3.1E, copyright 1977 by Small World, based on Monte Carlo algorithms of Dr. David Joy, University of Tennessee.)

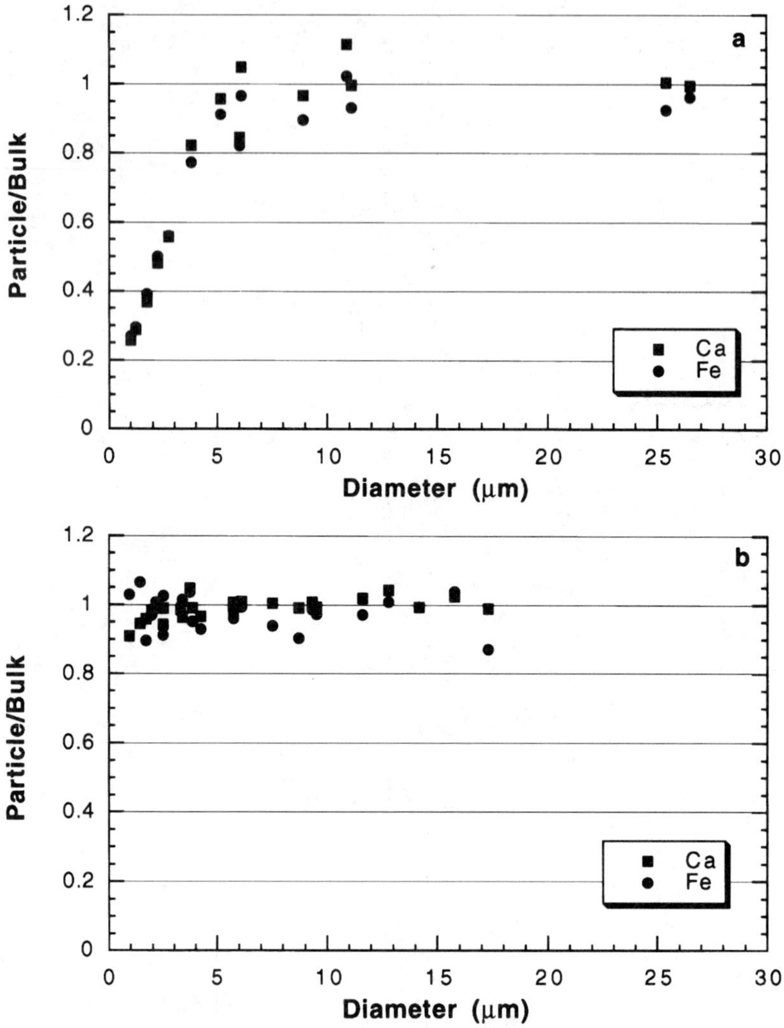

Fig. 12–23. Data for Ca and Fe particle/bulk ratios for 25 keV **(a)** and 10 keV **(b)**.

are relatively flat for all particle sizes compared with the 25 keV plots. In addition, the P/B values for the 10 keV plot do not show any significant trend above 1 for particles in the 2 to 10 μm size range, as was evident in the 25 keV plots. For the 10 keV data, the P/B values for Si and Mg are between 0.9 and 1.1 for all particles greater than 1.4 μm.

These results indicate that even a modest reduction, from 25 to 10 keV, in acceleration potential dramatically reduces both the particle mass and the absorption effects. Analysis at even lower accelerating voltages, 5 keV to 10 keV, should enable the development of a low-voltage particle analysis procedure that provides accuracies approaching those for bulk-polished samples, ±5% relative. With further reductions in beam voltage made possible by the shift of analyses to the transition-metal L lines, the particle mass effect may approach insignificant levels.

Automated Analysis of Particles with Electron Beam Instruments. An additional method of particle analysis that makes use of both the SEM and the EPMA is automated analysis

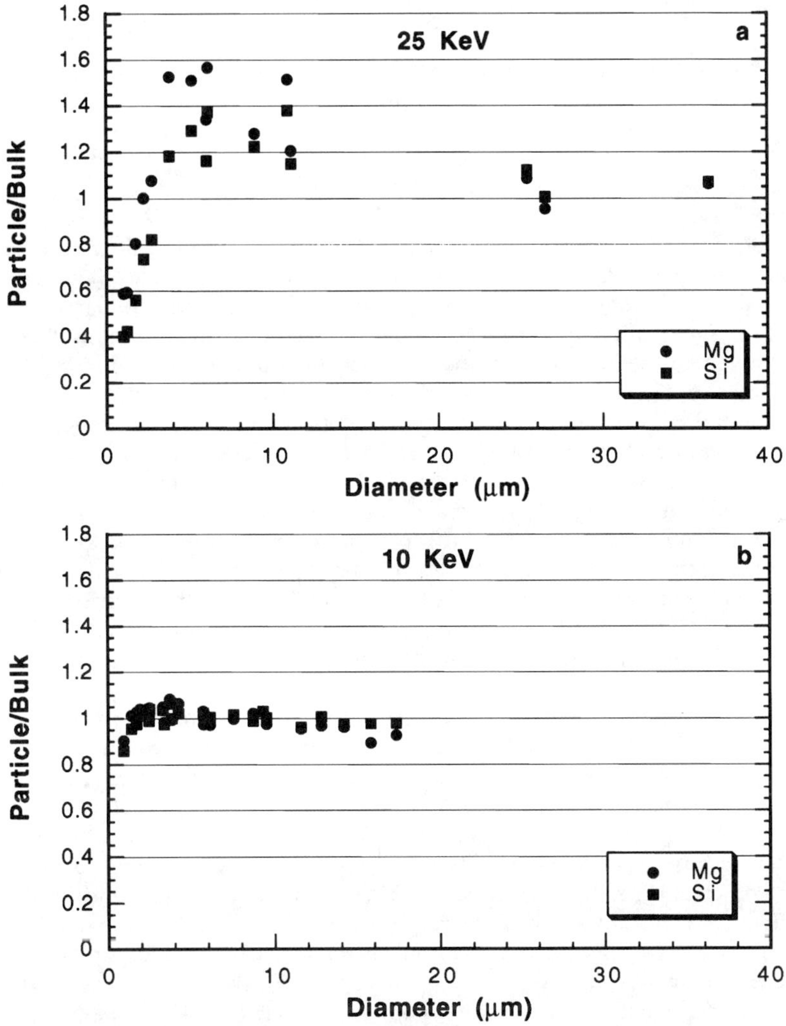

Fig. 12–24. Mg and Si particle/bulk ratios for 25keV **(a)** and 10keV **(b)**.

(Lee and Kelly, 1980; Germani and Buseck, 1991). In this procedure, the electron beam and specimen stage are controlled by computer, and a series of particles, usually numbering from a few hundred to a few thousand, are analyzed for elemental composition and morphology. The results from automated analysis are used to infer information about the total particulate population and the sampled aerosol. In this respect, automated particle analysis is different from the individual particle procedures described above. The elemental analyses are at best semiquantitative and, because of the short counting times, (often 10s or less) do not provide information on elements with concentrations lower than about 5 weight %.

Once data are collected, the different particles are divided into groups based on their composition and morphology. The groupings can be either preselected categories (discriminant analysis) or categories selected based on the data (cluster analysis) (Kaufman and Rousseeuw, 1990). After grouping the data, the analyst can then estimate various population characteristics, such as mass fraction of the total sample, resulting from a given particle type.

Analysis of Fibers by Electron Microscopy

Scanning Electron Microscopy of Fibers. The SEM provides analyses intermediate between those of the LM and the TEM, furnishing elemental analysis and higher resolution images than the LM. There is some disagreement about the effectiveness of *the conventional* SEM employing a tungsten filament electron source for fiber analysis. Under routine fiber-counting conditions (e.g., rapid scanning and direct visual inspection of the scanned image), fibers thinner than about 0.1 μm are not generally visible with the SEM, nor can the internal structures of fibers be seen. Additionally, definitive determination of fiber type by electron diffraction is not generally available with the SEM (Middleton, 1982; Steen et al., 1983). However, in the "photographic mode" (e.g., long scan times and photographic or digital recording), a high-quality SEM can be used to detect small fibers quite effectively. The photographs can be used to measure fiber size distributions (Platek et al., 1985).

The SEM has not been generally recommended for measuring asbestos, especially in environmental situations, because of the poor contrast (low signal-to-noise ratio) of chrysotile fibers (Small, 1987), inability to positively identify fibers by electron diffraction, and the lack of instrument standardization. However, the SEM does permit the microscopist (1) to obtain higher resolution and sensitivity than the LM, (2) to obtain qualitative information about fiber type from elemental analysis, and (3) to reduce sample preparation and analysis costs versus the TEM. In research studies of known fiber types, when analysis time is not important, photographic images from the SEM can be analyzed to obtain size distributions including all fiber sizes (Platek et al., 1985). The SEM field of view is larger than that of the TEM, and thus long fibers may be sized more accurately.

TEM Analysis of Fibers. The TEM allows the most definitive analysis of individual fibers: Particle shape can be observed and measured, elemental composition can be determined by X-ray EDS, and crystal structure can be deduced from electron diffraction patterns (Langer et al., 1974). TEM has sufficiently high resolution and sensitivity to allow observation of the smallest fiber sizes. However, TEM analysis is the most expensive one described in this chapter because of instrument cost and complexity, the requirement for high operator expertise, and complex sample preparation. A detailed study by Steel and Small (1985) of parameters affecting TEM analysis accuracy showed that instrument capabilities including mechanical stage, imaging and contrast, and diffraction can produce order-of-magnitude errors in asbestos analysis if they are not kept in control. Furthermore, quantitative fiber concentration measurements have relatively poor reproducibility (Environmental Protection Agency, 1985).

Two general approaches have been taken in preparing samples: direct transfer (the carbon film contains the particles in the same location and orientation as when they were originally deposited on the filter or substrate) and indirect transfer (collected particles are dispersed as a liquid suspension and redeposited on a filter for grid preparation) (Environmental Protection Agency, 1989). Many of the following procedures are described in the Environmental Protection Agency (1987) Asbestos Hazard Emergency Response Act (AHERA) method.

For asbestos fibril (single crystal of asbestos) detection, ×10,000 to ×20,000 magnification is used. In TEM analysis, particles scatter more of the electron beam out of the acceptance aperture than the supporting carbon film and appear as dark objects on a light background.

Particle shape is often an important indicator of type or origin of fiber. Larger asbestiform fibers can usually be distinguished by their fibrillar structure, especially at the ends where the fibrils tend to splay apart.

Besides observing particle shape, the TEM allows the use of electron diffraction, which indicates crystal structure, and X-ray EDS, which indicates elemental composition of particles. Diffraction and elemental composition patterns from reference materials are used to confirm the presence of specific types of fibers. It sometimes happens that, for a given fiber, the diffraction pattern and/or the elemental spectra may not be completely detected or may change over the length of the fiber. This may be due to fiber thickness, interfering materials, variations in fiber composition, or twisting of the crystallographic axis along the fiber length. Thus, full confirmation of all fibers' identities is often not possible.

Uncertainty in Fiber Analysis. A detailed study by Steel and Small (1985) of parameters affecting TEM analysis accuracy showed that instrument capabilities, including mechanical stage, imaging and contrast, and diffraction, can produce order-of-magnitude errors if they are not up to the demands of asbestos analysis. By careful comparison between several counters and microscopes using a technique called *verified analysis*, it was also found that a careful and experienced analyst could attain accuracies of 90% in finding all asbestos fibrils in a sample. In verified analysis, two analysts independently analyze a grid opening on a TEM grid recording a sketch drawing and dimensions of each fiber. Accuracy degraded significantly for fibrils shorter than 1 µm, and analysts had a 50% or less chance of finding those shorter than 0.5 µm. In an interlaboratory comparison on prepared grid samples, Turner and Steel (1991), Turner et al. (1991) found that the average result reported by the laboratories was 0.67 of the NIST-verified count. Forty percent of the fibers shorter than 1 µm were missed; twenty percent longer than 1 µm were missed.

Repeat counts by a single analyst of laboratory-generated amosite fiber samples gave a relative standard deviation of 0.27 (Taylor et al., 1984). The interlaboratory relative standard deviation for repeat counts on the same sample has been estimated to be approximately 1.0 (Environmental Protection Agency, 1985).

A laboratory accreditation program has been established by the National Institute for Standards and Technology (NIST) under the National Voluntary Laboratory Accreditation Program (NVLAP) (Turner et al., 1995a). Turner and Steel (1994) provided a standard test method for verifying the analysis of asbestos by TEM. Data regarding laboratory performance in this program have not yet been published, but it is possible to obtain reports (Turner et al., 1992, 1994a–c, 1995b). The NIST *Standard Reference Material 1876 and 1876b* (1992) is available for evaluating analytical methods and quality control (Turner et al., 1992; Small et al., 1985). Quality assurance guidelines for laboratories participating in this program are available (Berner et al., 1990; Turner et al., 1995a).

Phase Identification Analysis of Particles Employing Electron Backscatter Diffraction. Recently the technique of electron backscatter diffraction (EBSD) has been investigated as a method to provide a definitive identification of crystalline particles based on crystal structure rather than elemental concentrations (Small and Michael, 1999; Small and Michael, 2001). The EBSD patterns originate when beam electrons are elastically scattered from the near surface of a material and are subsequently diffracted by the crystal lattice of the sample as they exit. Once the electrons exit the sample, the diffracted electrons, in addition to a large number of "nondiffracted" or background electrons, can be imaged using an appropriate detector. Figure 12–25 shows the low-intensity EBSD pattern resulting from the angular modulation of the diffracted, backscattered electrons riding on the general angular distribution of backscattered electrons.

Backscattered electron diffraction patterns were first observed by Alam et al. in 1954. J. A. Venables and C. J. Harland (1973) made the initial observation of EBSD patterns in the SEM on bulk samples with a video-rate camera detection system. This system used a phosphor screen that was imaged by a low light–level video-rate camera. Goehner and Michael (1996) and Michael et al. (1997) developed an EBSD system that uses a 1024 × 1024 charge coupled device (CCD) camera directly coupled by a fiber optic reducer to a phosphor screen. Unlike previous EBSD systems that were designed primarily for determining grain orientation in bulk samples, this system was designed to identify a specific crystalline phase of an unknown.

Fig. 12–25. Uncorrected electron backscatter image containing both the low-intensity EBSD pattern and the nondiffracted backscattered electrons.

Fig. 12–26. EBSD pattern after background correction.

In their system, the raw EBSD images are corrected for the nondiffracting, background electrons, resulting in a very high-quality EBSD image as shown in Figure 12–26. Automated pattern analysis is carried out on the EBSD image; the crystallographic information is determined and is coupled with the elemental information from energy or wavelength dispersive X-ray spectrometry. Finally, the phase identification is made through a link to a database such

Fig. 12–27. a, Secondary electron image of a micrometer-sized U_3O_8 particle. **b**, EBSD image. **c**, Indexed image and overlay from the PIA analysis sequence.

as the Powder Diffraction Files published by the International Center for Diffraction Data (ICDD; Newton Square, PA).

Phase identification analysis (PIA) was first applied to conventional, bulk polished samples. Recent studies by Small and Michael (2001) demonstrate one of the first applications of EBSD to single-particle analysis. In their study they used PIA to analyze single particles with known compositions but unknown crystalline phase. Figures 12–27 and 12–28 show the PIA results from two representative particles, a 1 μm uranium oxide particle and a 0.5 μm lead oxide particle. Each figure contains a secondary electron image of the analyzed particle, a background-corrected EBSD pattern, and the synthesized pattern overlaid on the background-corrected EBSD pattern. The diffraction pattern obtained from the uranium oxide particle (Fig. 12–27b), has an excellent signal-to-noise ratio considering the small particle size. This particle was identified as orthorhombic U_3O_8. Figure 12–28 shows the results from the PIA of the 0.5 μm lead oxide particle that was identified as plattnerite, PbO_2. Similar

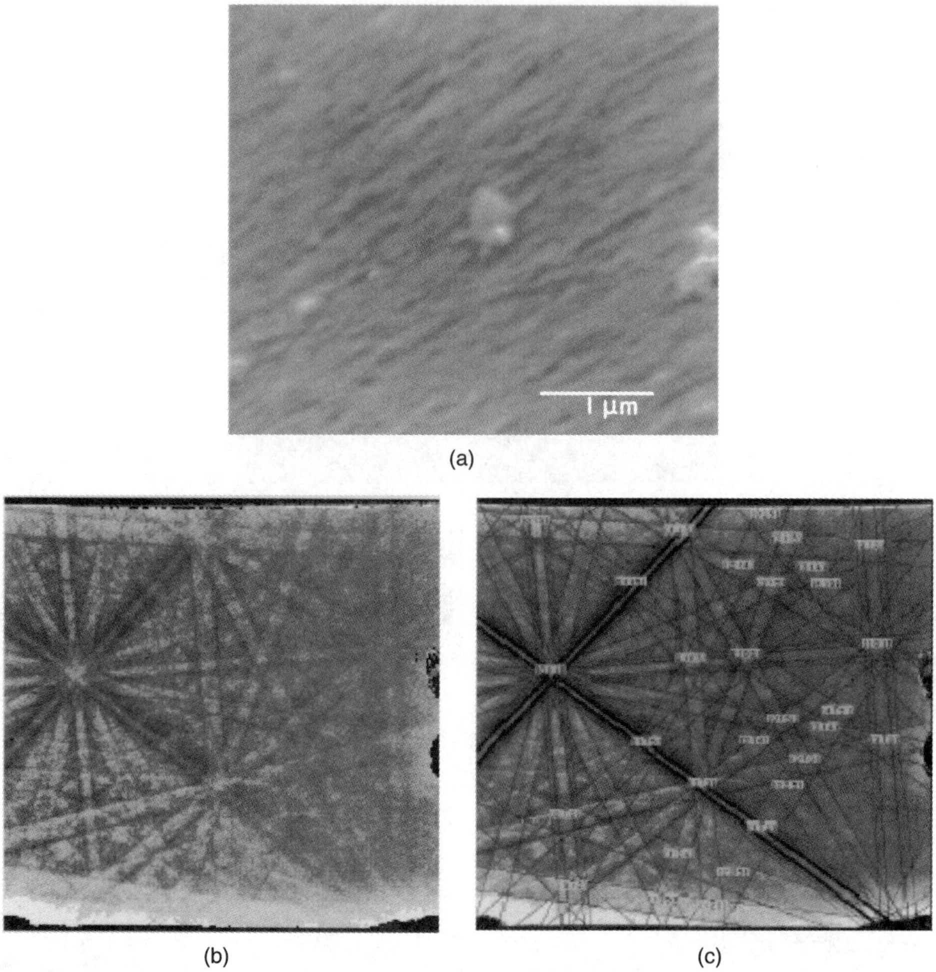

Fig. 12–28. **a**, Secondary electron image of a submicrometer PbO_2 particle. **b**, EBSD image. **c**, Indexed image and overlay from the PIA analysis sequence.

to the U_3O_8 particle, the quality of the diffraction image for the PbO_2 particle is excellent. The success of the PIA for these small particles implies that for high-Z and high-density materials such as these it may be possible to identify the crystalline phase of particles down to 100 nm or smaller.

The results of this study indicate that PIA is a valuable technique for single-particle analysis. EBSD when used in conjunction with an analytical SEM or EPMA provides the analyst with a very powerful and straightforward method for identifying the composition of submicrometer and larger crystalline particles.

LASER MICROPROBE MASS SPECTROMETRY

Underlying Principle

Laser microprobe mass spectrometry (LMMS) is a mass spectrometric technique that utilizes a focused, pulsed laser beam to generate ions by laser ablation/ionization. The process

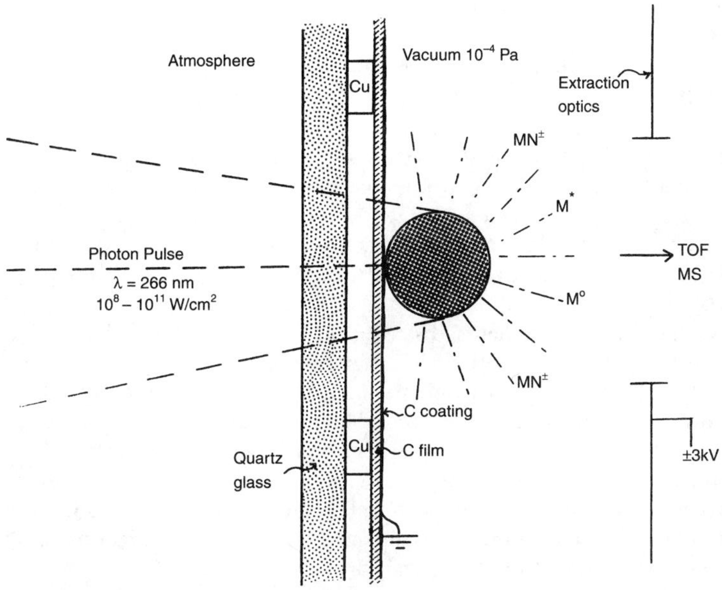

Fig. 12–29. Schematic of laser ablation in LMMS. A focused UV 10ns pulsed laser beam interacts with the particle, resulting in disruption of the solid and formation of ions, neutrals, and metastable species. (From Simons, 1984.)

is illustrated in Figure 12–29. The laser normally used has a short pulse length (tens of nanoseconds) and a focused high irradiance (10^6 to 10^9 W/cm^2). The absorption of the laser beam by the sample material produces heating over a short duration. The sample is volatilized and expands (usually) into vacuum. For the lower end of the irradiance range, 10^6 to 10^8 W/cm^2, material is removed from the sample in a "soft" manner frequently termed *laser desorption*. During high irradiance, intense heating can occur, resulting in plasma formation. The sample constituents are for the most part reduced to elements, small fragments, and clusters. In both energy regimes, nonlinear optical effects are thought to occur in addition to classic absorption.

Only ions are detected in the mass spectrometer. The ionization efficiency is low, on the order of one part in one thousand. Although there are ions present, the mechanism for ionization in laser desorption is not understood. In the high irradiance mode, ionization is thought to take place by thermal and collisional processes occurring in the plasma.

Extensive overviews of the fundamentals, the instruments, and the applications of lasers and mass spectrometry have been published (Vertes et al., 1993; Van Vaeck et al., 1994a,b). De Bock and Van Grieken (1999) provide a review of work done on environmental aerosol from 1990 to 1998.

Instrumentation

The laser microprobe is composed of a dual laser system coupled to a mass spectrometer through an optical microscope. The sample is viewed optically, and a particle is selected for analysis by aiming a co-aligned pilot laser beam spot (often from an HeNe laser) onto the particle. The ionizing laser is a high-power, pulsed laser, usually an Nd:YAG laser. The radiation from the Nd:YAG is frequency quadrupled to provide pulsed ultraviolet radiation (266 nm) and is focused on the sample by an optical microscope objective or other lens system. All ions generated are accelerated to uniform translational energy and mass separated by time-of-flight (TOF) mass spectrometry. In TOF mass spectrometry, flight time in the drift tube is proportional to the square root of the mass-to-charge ratio. The arrival of the ion packets at the end of the drift tube is determined by a secondary ion multiplier detector.

Two geometries, reflection and transmission, are commonly used in commercial instruments. *Transmission* refers to having the laser beam and the mass spectrometer in line, while *reflection* refers to the beam being at an acute angle with regard to the spectrometer. Figure 12–30 schematically presents only two possible examples of these two geometries. A complete description of LMMS is given by Denoyer et al. (1982), Hercules et al. (1982), and Simons (1988). Single-particle analysis is discussed by Kaufmann (1986), De Waele and Adams (1986), and Wieser and Wurster (1986). Early work in aerosol research applications is presented by Wieser et al. (1981).

In situ laser TOF-MS instruments, discussed in Chapter 13, often have advantages over laboratory-based systems, such as the laser microprobe mass analyzer, for analysis of environmental particles. With these instruments, sampling and mass analysis take place in one operation. However, there are cases where off-line measurements are needed because real-time measurements are not feasible. These may be in remote or inhospitable areas such as Antarctic aerosol (Hara et al., 1996) or in the case of particles that have been collected passively over great lengths of time such as particles found in ice cores (Biegalski et al., 1998). The laser microprobe preceded the laser aerosol TOF mass spectrometers, and much of the spectral information derived from both designs is closely related.

Ma et al. (1995) have developed a new laser microprobe using Schwarzschild optics that is capable of viewing particles and directing and focusing laser beams to ablate them. The

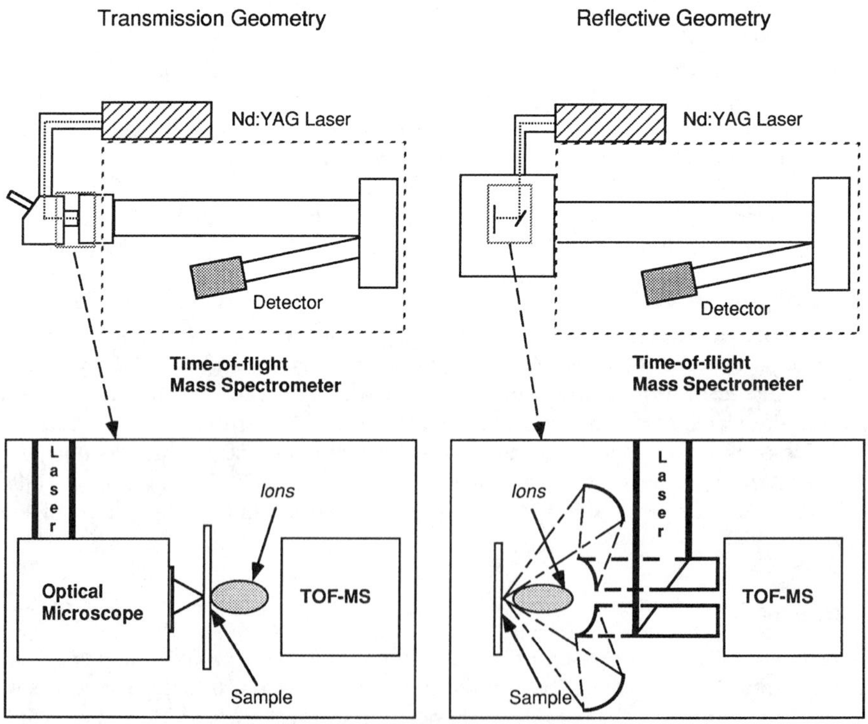

Fig. 12–30. Schematic showing transmission and reflective geometry of two existing LMMS instruments. The transmission geometry has higher spatial resolution and is normally considered the better design for particle analysis. (Courtesy of S. Hoeft.)

spatial resolution is an approximately 1 μm diameter spot size. Resonance ionization is accomplished using dye laser beams directed into the neutral ablation plume. Isotopic analysis was demonstrated for Ti on a nominal 2 μm particle taken from a meteorite. This instrument has the spatial resolution to analyze particles and to do chemical imaging much as the SIMS microprobe (Savina and Lykke, 1997).

Fourier-transform laser microprobe mass spectrometry (FT-LMMS) has been discussed and compared with TOF laser microprobe mass spectrometry (TOF-LMMS) (Van Vaeck et al., 1994a,b). Each is capable of single-particle analysis, but often give different information from the same compounds. FT-LMMS typically provides a 5 μm laser spot size, much higher mass resolving power (10^4 to 10^6), and mass accuracy to a nominal 1/1,000,000.

Capabilities

Some of the important aspects of the LMMS instruments are nominal micrometer spatial resolution, high detection sensitivity, and detection capability for either positive or negative ions produced in the laser ionization. In addition, it is usually possible to determine the elemental, isotopic, and, in many cases, even the molecular composition of a particle. The technique is complementary to electron microscopy (Steel et al., 1984) and micro-Raman spectroscopy (Fletcher et al., 1990) and should prove to be complementary to other spectroscopies. LMMS is generally not considered quantitative even for major constituents, except with regard to isotope abundance. Matrix effects in the sample play a dominant role during the laser beam–sample interaction, making quantitation difficult. The main strength of LMMS lies in providing useful qualitative information in a rapid manner.

Elemental Analysis. One important aspect of the LMMS is that all elements are usually ionized in a single laser pulse and are detected in a single mass scan. This single, full-mass scan capability inherent to TOF mass spectrometry makes the LMMS instrument well-suited for particle analysis. An example of an LMMS spectrum that displays 44 elements and their oxides present in NIST SRM 610 multielement glass is shown in Figure 12–31 (Simons, 1984). Most of these elements are present in the glass at only trace amounts (~500 μg/g level by weight).

Detection limits for the elements vary. In most sample matrices, the alkali metals are most easily detected and are reported to have a detection limit on the order of 10^{-19} g (Kaufmann, 1986). Most other elements have detection limits between 10^{-17} and 10^{-19} g, allowing detection of minor and in some cases trace elements from a single particle with micrometer dimensions and picogram mass. In terms of trace analysis on bulk samples, a relative concentration of 1 to 100 μg/g must be present for detection.

Isobaric interferences (same mass, different composition) resulting from the low mass resolution can lead to a problem in the interpretation of LMMS spectra, but can be overcome in many cases by observing the presence of isotopes or other characteristic peaks in the spectrum. For example, C_5^- and CO_3^- have the same m/z value of 60, and the two species would be difficult to distinguish if they were lone peaks in the mass spectrum. However, C_5^- rarely appears without associated carbon clusters, C_2^-, C_3^-, C_4^-. . . .

For qualitative analysis, the instrument capabilities include sensitivity, quickness, and simplicity of analysis. Unfortunately, one disadvantage is that different matrix materials ionize with different efficiencies, leading to problems for quantitative work (Kaufmann, 1986). One study that illustrates the problems involved with quantitative analysis was done by Musselman et al. (1988) for films, micrometer-sized spherical particles, and shards made from glass of known elemental composition. Relative sensitivity factors (*RSFs*) of Mg, Si, Ca, Cr, Fe, Zn, Ba, and Pb were compared for the film, particles, and shards made from the same glass. An *RSF* of element j

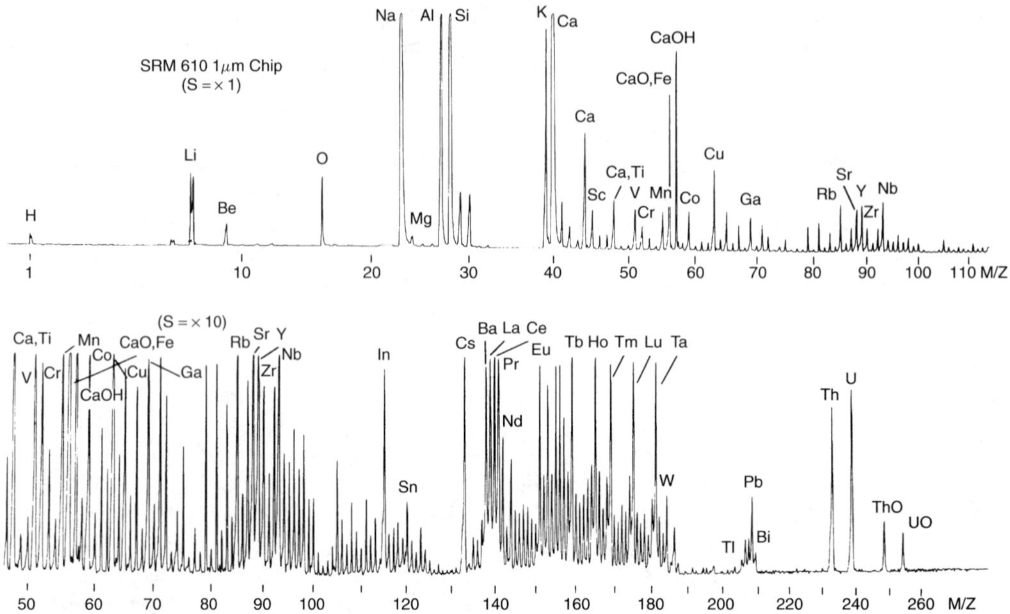

Fig. 12–31. A single LMMS spectrum of an NIST SRM 610 glass chip taken on the transmission instrument. Some 44 elements and their oxides are labeled as shown. Trace elements (e.g., U) are present at the 400 to 500 ppm level. It is characteristic of a TOF mass spectrum that all masses are shown in a single scan. (From Simons, 1984.)

$$RSF_j = \frac{I_j C_r f_r}{I_r C_j f_j} \qquad (12\text{–}7)$$

is defined here as a function of the concentration of the reference element, C_r, concentration of element j, C_j, and their respective measured ion intensities, I_r and I_j. In Eq. 12–7, f corresponds to the isotopic fraction. Using Ba as the reference element, ratios of the maximum to minimum *RSF* values for each of seven elements were determined. These ratios were found to vary from 1.63 for Ca to 5.17 for Si using all data for films, spheres, and shards. These results suggest that in the process of the laser beam–solid interaction, the elements are ionized with different efficiencies for the three geometries because the glass composition was the same in all three cases. When the authors restricted their data set to 1 μm spheres, shards, and films, the *RSF* maximum to minimum ratio was dramatically reduced, ranging from 1.17 for Ca to 2.57 for Fe. The authors concluded that laser focus plays a role in ionization efficiency for the planar and particle geometries. It is important to utilize standards of similar geometry, size, and composition before any meaningful quantitative work can be attempted.

Isotopic abundance information for all elements is one of the strengths of mass spectrometry and supplements nonisotope-specific elemental electron microprobe capabilities. Careful isotopic measurements provide quantitative information because the isotopes experience the same matrix effects and have the same ionization properties. High-accuracy (<4%) relative isotopic ratio measurements have been made by Lindner et al. (1986) on thin-film samples of rhenium/osmium to determine the half life of ^{187}Re. Details of the precautions necessary to obtain such precision and accuracy are described by Simons (1983). Other major isotopic investigations with LMMS have been conducted by Schroder and Fain (1984) regarding calcium release (using ^{44}Ca) from photoreceptors in the eye.

Molecular Analysis. Molecular characterization is accomplished using low irradiance (10^6 to 10^8 W/cm^2) directed onto the sample. Some frequently occurring anion molecular species, common to atmospheric particles, are often identified in LMMS spectra as NO_3^-, SO_4^-, HSO_4^-, PO_x^-, and CO_3^-. Usually all ions are singly charged regardless of the natural valence state. Organic compounds are often detected as molecular ions formed through the process of laser desorption or as adducts, that is, hydrogen or alkali metal (especially Na or K) attachment to the molecule. Characteristic fragment ions containing some information about the structure of the molecule are produced when sufficient laser irradiance is applied. It is interesting to note that the largest molecules detected by mass spectrometry, m/z 250,000, are measured by an extension of LMMS, matrix-assisted laser desorption/ionization (Hillenkamp and Karas, 1989). The mass spectra in LMMS, although unique to the technique, are similar to both electron impact (EI) and chemical ionization mass spectra. Certain types of compounds can be detected as molecular ions, such as the atmospherically important polycyclic aromatic hydrocarbons (PAHs) (Mauney, 1984). Conversely, organic compounds such as alkanes generate (as in EI) little or no molecular ion information whatsoever. Often the particle is not completely ablated away under low-irradiance conditions, and sometimes there is no visible (by LM) damage to the particle. Van Vaeck et al. (1990) present a discussion of organic compound analysis.

The mass spectra common to laser ionization mass spectrometry are often complicated. There are two approaches that can be taken to analyze the spectra, direct identification and fingerprinting. For specific compound identification, pure reference compounds are usually analyzed and compared with the unknown. It should be pointed out that for mixtures of organic compounds, as is often the case in atmospheric samples, direct identification of even major organic components may be difficult due to the complexity of the mass spectra. Thus fingerprinting is the only recourse. Fingerprinting is the comparison of mass peak patterns from sample particles to peaks obtained from particles of known source without specifically identifying the peaks. Fingerprinting can involve pattern recognition techniques, such as principal component analysis (Wieser and Wurster, 1986; Fletcher and Currie, 1987; Linder and Seydel, 1989; Ro et al., 1989; Odom et al., 1989).

Applications to Particle Analysis

LMMS is frequently used to provide information about environmental particles. Early work was done by Wieser et al. (1980) on both laboratory-generated and atmospheric aerosol particles. Mixtures and composite particles present difficult spectra to interpret, but the spectra nevertheless often contain information about origination or formation. For example, the coexistence of Pb and carbon in the same particle could suggest automobile exhaust. Kolaitis et al. (1989) could distinguish dimethanesulfonic acid particles collected on the lowest stages of a Battelle-type impactor from non-sea-generated particles. The authors report nitrates (derived from natural sources) forming around crystalline sea salt.

To obtain more information from a sample, the analytical electron microscope (AEM) can be teamed with LMMS; LMMS supplements the AEM analysis by identifying low atomic weight elements, trace elements, and molecular compounds. Sheridan and Musselman (1985) used an AEM to identify the major chemical composition and morphology and LMMS to identify low atomic weight (<Na) constituents of arctic haze aerosol. They found primarily sulfate particles. The AEM could detect the presence of sulfur in the particles, and the LMMS mass spectra revealed the characteristic sulfate ion at m/z 96 and bisulfate at m/z 97. Bruynseels et al. (1988) also combined electron microscopy and LMMS to identify the chemical composition of North Sea aerosol. The authors reported detection of nitrate particles, sulfates, sea salt, and PAHs desorbed from coal fly-ash particles. Denoyer et al. (1983) and Yokozawa et al. (1987) have characterized coal fly ash for elemental composition.

Surkyn et al. (1983) employed LMMS to look for elemental signatures for single-particle source identification.

Analysis of carbon soots (solid particles) and organic carbon is difficult on the single- or near-single-particle (~pg) basis. Yet in some cases the amount of carbon collected is so small that conventional bulk analysis techniques are not possible. LMMS has been used to try to distinguish soots from various combustion sources collected from controlled burns of wood, polyurethane, and heptane, as well as from atmospheric sources (Currie et al., 1987; Fletcher and Currie, 1988). Wood soot has a large amount of potassium associated with the carbon on an individual particle basis that makes it unique among the other sources studied. Efforts have been made to correlate the low mass/charge carbon cluster peak area ratios (C_4^-/C_2^-) to the amount of modern carbon present in the sample (Currie et al., 1989). Substrates and collection methods for carbon soots are described elsewhere (Fletcher, 1989; Cleary et al., 1992; Dobbins et al., 1998; Dobbins and Megaridis, 1987).

Hachimi et al. (1996) used two microprobe techniques, the LAMMA 500 and FTMS, as well as a bulk analysis approach, gas chromatography–tandem mass spectrometry, to characterize a nitro-PAH–containing aerosol collected from a steel foundry. The authors were successful in detecting PAHs and their derivatives by all three techniques. They propose the two-microprobe techniques as quick and efficient qualitative indicators of the presence or absence of PAH-like compounds followed by quantitative analysis by GC-MS.

LMMS has been utilized to identify particles collected by and residing in biological tissue (Hachimi et al., 1998). Tsugoshi et al. (1998) examined stone surfaces that were exposed to airborne pollutants. They performed semiquantitative analyses for sulfates and nitrates using an analysis scheme, frequency of detection (or occurrence), and found good agreement between LMMS and energy-dispersive X-ray analysis.

Clemett et al. (1993) analyzed 2 to 50 µm particles collected from the stratosphere by high-altitude sampling. Of particular interest was the determination of the chemical composition of certain interplanetary dust particles. For individual particles that were determined to be extraterrestrial by other microanalytical means, a microprobe instrument utilizing a two-laser system with mass spectrometry was applied. This instrument utilizes a pulsed infrared laser beam 40 µm in diameter to heat and volatilize compounds that are subsequently ionized from the gas phase by an ultraviolet laser beam. This is a sensitive technique for PAHs, and alkylated PAHs were found from single particles. The two-laser approach was further applied to 200 µm carbonaceous particles thought to be micrometeorites collected from the Antarctic (Clemett et al., 1998). The authors identified a number of PAH compounds in these relatively large particles.

SECONDARY ION MASS SPECTROMETRY

Underlying Principle

Secondary ion mass spectrometry (SIMS) is based on ion beam sputtering. When an ion beam interacts with a solid surface, a small fraction of the ions are reflected from the surface, but most of the ions implant themselves into the solid material, transferring some of their momentum into a collisional cascade. Some of the momentum from the cascade is imparted to surface atoms or molecules, providing them with sufficient kinetic energy to escape the surface layer. The process of removing material from the top layer using ion or neutral beam techniques is called *sputtering*. While the primary ions may penetrate several tens of nanometers below the sample surface, most of the sputtered ions and neutrals are liberated from the surface. Sputtered atoms can be neutral, charged, or in a metastable state. In addition to atomic and molecular sputtered products, electrons and photons are liberated (Behrisch, 1983).

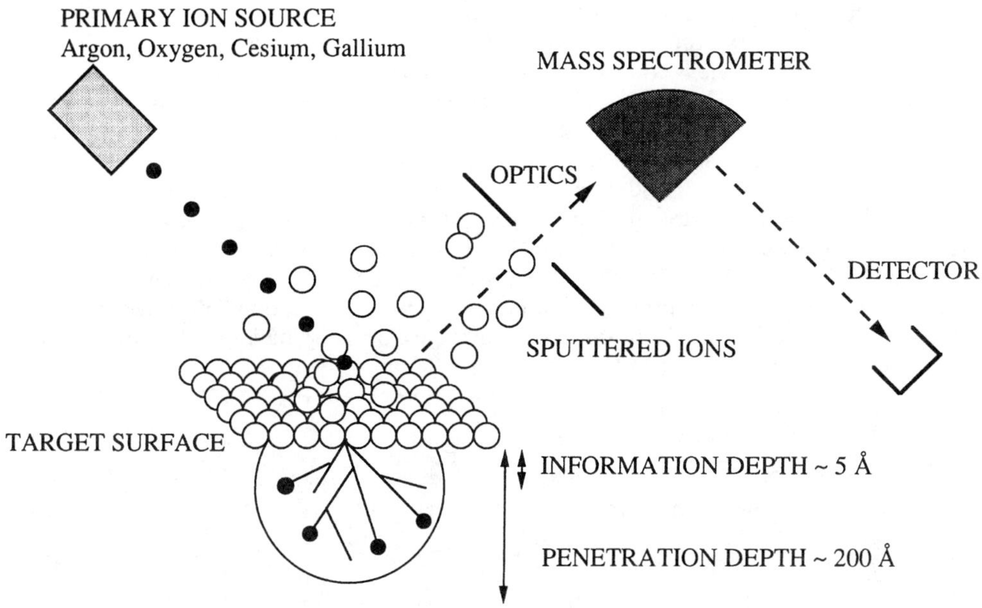

Fig. 12–32. Schematic of the ion beam sample bombardment in SIMS showing approximate depth of interaction between the ions and the sample surface and a representation of ion implantation. (Courtesy of G. Gillen, NIST.)

Figure 12–32 illustrates the interaction of a primary ion beam with a sample surface generating secondary ions in the process. The primary ion beam is usually composed of one of the following: oxygen, argon, cesium, or gallium ions. These secondary ions are mass separated on the basis of their mass/charge ratio and detected in a mass spectrometer. The surface of the sample is damaged in the process.

Instrumentation

There are two types of SIMS instruments: the ion microscope, which uses a defocused beam and specialized ion optics to form an image of the sample surface; and the ion microprobe, which uses a focused primary beam and produces ion images by sequentially rastering the beam over the sample. There are a number of commercial microprobe instruments using a highly focused liquid metal ion source. Advantages of the microprobe instrument over the ion microscope instrument are higher spatial resolution (on the order 0.1 μm), providing the opportunity of high spatial ion imaging and higher ion transmission through the use of a TOF mass spectrometer.

Different mass spectrometers are employed for ion detection. The most widely used are magnetic sector and quadrupole, but TOF is beginning to be used in ion microprobes.

Capabilities

The advantages of microanalysis SIMS include high sensitivity, high spatial resolution, elemental and isotopic analysis, molecular information, depth profiling, and ion imaging. All of the elements are detectable in both types of instruments. In the case of imaging, both elemental and molecular imaging are possible at the 1 μm size level. SIMS has a large dynamic range, making it applicable for trace analysis. As can be seen from Table 12–1, SIMS has detection limits comparable with or better than any technique presented. Gavrilovic (1984) estimates a yield on the order of 10^4 detectable ions from a particle with 0.1 μm diameter, assuming that the particle is completely consumed, an ionization efficiency of 1%, and a secondary ion collection rate of 10%. This is sufficient for analysis of major elemental composition. One shortcoming in SIMS is that the combined effect of sputtering and ion production efficiency of elements varies over five orders of magnitude (Storms et al., 1977). Quantitative analysis can be done provided calibration standards of like materials are used.

Similar to LMMS, there are two modes of operation: high primary beam current (10^{-3} A/cm^2), known as *dynamic* SIMS; and low primary beam current (10^{-9} A/cm^2), known as *static* SIMS. Dynamic SIMS is a high-damage regime and is used for elemental depth profiling, while static SIMS can provide the user with molecular compound information. In static SIMS, the ion dose is so low that probabilistically, each primary ion interacts with an undamaged area of surface.

Applications to Particle Analysis

Ion Microscope. Newbury (1980) reported an example of single-particle depth profiling on a 10 μm urban dust particle, as illustrated in Figure 12–33. The depth profile plot shows

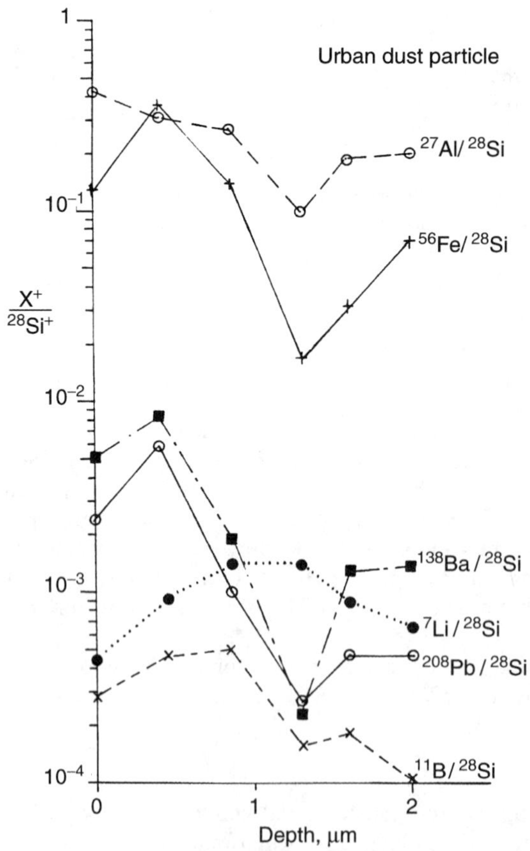

Fig. 12–33. Depth profiling by SIMS of a single 10 μm reference particle. Essentially the surface layer is successively sputtered away, generating ions that are separated in a mass spectrometer. Element ratio referenced to ^{28}Si. (From Newbury, 1980.)

various indicated elements ratioed to ^{28}Si as a function of distance into the particle. Linton et al. (1976) showed, using depth profiling, that elements such as As, Pb, and Cd are associated with the surface of fly-ash particles. This is not an example of single-particle analysis because the sample was composed of fly-ash particles compressed into a disk, but the example does show the unique information derived from the surface layer of the particles found using the technique. Work on atmospheric particles is generally qualitative rather than quantitative. Klaus (1984) used a SIMS instrument to analyze aerosol particles collected onto an aluminum foil in a low-pressure impactor. Although Klaus reports that Al foils are the best substrate due to low contaminants in the foils, gold offers a higher mass/charge element and can also be very pure. Oxides of various elements, certain anions such as sulfates, nitrates, and chlorides, and carbon in both graphitic and organic form were found by Klaus in some European continental aerosol plumes. A similar study was performed (Klaus, 1985) on automobile exhaust. Primarily elemental constituents were reported on the particle surface analysis; no organic molecular species were found. A chapter containing many references regarding particle analysis by SIMS is presented by Klaus (1986).

More recently, SIMS has been used in the ion image mode to identify composition in samples of collected particles and for doing isotopic analysis for actinide particles. The ability of SIMS to locate particles by ion imaging is shown in Figure 12–34, which indicates the relative location of fly-ash particles containing the specific elements shown. While this is an example of imaging the elemental composition, molecular compositional mapping is also

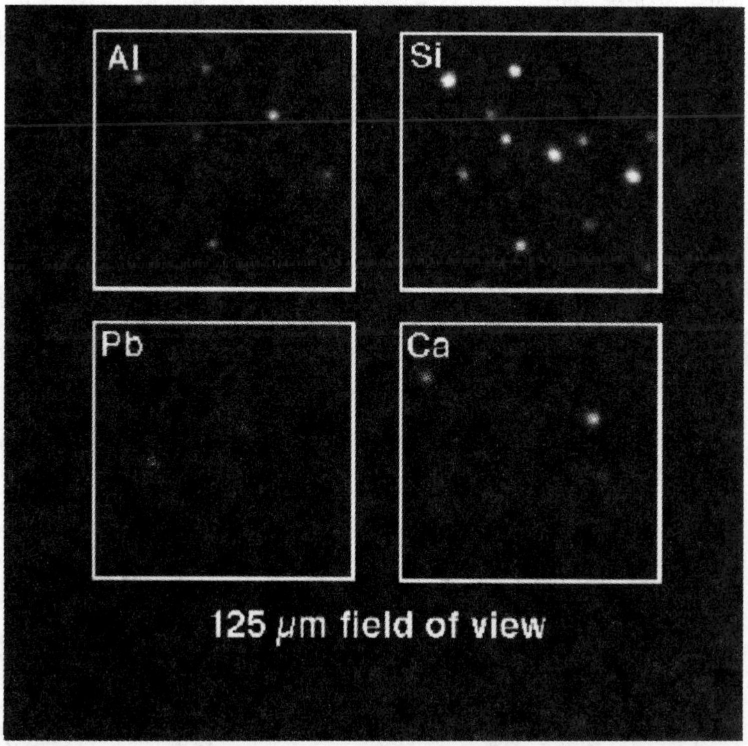

Fig. 12–34. SIMS ion images of the same set of fly-ash particles indicating the presence of the elements of Al, Si, Pb, and Ca found in the surface layer of the particles. (Courtesy of S. Wight and G. Gillen, NIST.)

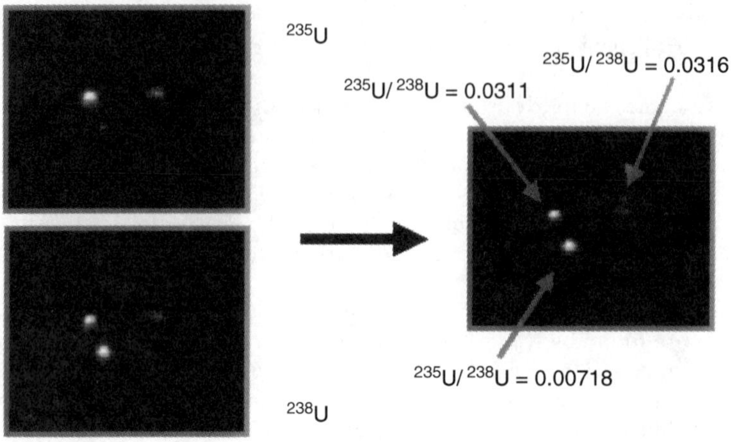

Fig. 12–35. SIMS images of uranium isotopic distributions found in three test particles. The images at the left show the isotope concentrations, and the frame at the right provides the 235/238 ratio found on these particles. (Courtesy of G. Gillen, NIST.)

possible. Imaging is a useful technique to identify a trace particle from a large group of particles if the desired particle(s) have a unique chemical composition.

Simons (1986) and Stoffel(s) et al. (1994) demonstrated isotopic analysis for uranium single particles and the isotopic analysis associated with verifying a new standard. Other particle standards have been reported (Simons, 1991). A method for automated SIMS analysis to obtain isotopic information from a set of randomly oriented particles, some containing uranium, has been demonstrated (Simons et al., 1997). In the latter case, 1000 ion images of micrometer-sized particles were obtained in an 8h period. The authors estimate that the system has the capacity to handle 10,000 particles (average loading of 10 per image field) per 8h. The usefulness of SIMS to survey many particles to find isotopic enrichments or to differentiate two particle populations was discussed. An application of single-particle analysis by SIMS for nuclear material was reported (Tamborini et al., 1998). Because the ionization efficiencies for the two isotopes are nearly equal, a measure of the relative amounts of each isotope (235/238) provides a way to identify isotopically enriched uranium from the naturally occurring element and thus serves as a forsenic analysis method (Betti et al., 1999). Detection limits were reported to be from ng/g to pg/g and the isotopic composition measured with a precision and accuracy of 0.5%. The authors report work on plutonium in the same paper. Figure 12–35 presents an ion image map of ^{238}U and ^{235}U on the set of particles using SIMS. The particles were control standards in the size range of 1 to 4 μm diameter. The images at the left provide the isotopic composition in a gray scale where white indicates the highest concentration. From the relative measurements of the two isotopes, isotopic ratios were determined and shown in the frame on the right.

Ion Microprobe. Bennett and Simons (1991) presented results for relative sensitivity factors determined for thin 20 nm glass films and micrometer glass shards made from the same glass composition. The sensitivity factors differed by a factor of 3 to 4 between film and shards.

Owari et al. (1989) presented a study using a submicrometer (on the order of 0.1 μm) liquid metal ion source that was highly focused for imaging small structures either by secondary electrons or by direct elemental secondary ion imaging. The authors demonstrated imaging on a mix of fly ash and NaCl particles fixed onto a metal substrate with conductive silver paint. Quantitative results were reported giving relative sensitivity factors derived from NBS SRM 309 glass microspheres. Sputtering with highly focused liquid metal ion beams appears to be a promising new research direction for particle analysis.

RAMAN MICROPROBE

Underlying Principle of Vibrational Spectroscopy

Raman scattering and infrared absorption result from vibrational and rotational processes that occur in molecules. Classical and quantum mechanical models describe the mechanism by which molecules inelastically scatter or absorb light. Classical models describe molecules as atoms (point masses) held together by chemical bonds (springs). The mass–spring systems have associated spring constants and resonant vibration frequencies. The fact that vibrational modes are distinct, exhibiting narrow bands in vibration spectra, is described quantum mechanically.

In Raman spectroscopy, the emitted radiation is inelastically scattered photons in the visible frequency range. If a beam of incident photons with energy, $h\nu$ (Planck's constant multiplied by the frequency), is directed upon a material composed of molecules that have a vibrational mode at ν_m, a certain number of the scattered photons will lose energy in the vibrational process and have a new energy, $h(\nu - \nu_m)$. Additionally, if the incident photons encounter vibrating molecules with the same vibrational mode as above, some photons will gain energy $[h(\nu + \nu_m)]$. The emitted radiation that has frequencies below the incident radiation is called Stokes shifted, and that shifted above the incident radiation is anti-Stokes shifted. Most of the photons are elastically scattered with no frequency shift at all (Rayleigh scattering). The number that are inelastically scattered with the above characteristic frequency shifts are small in number, on the order of 1 part in 10^6.

Molecules that have the property of allowing an induced instantaneous dipole to be established by the incident light beam are usually strong Raman scattering species. The feature normally used to characterize strong Raman active molecules is high molecular polarizability. Molecules with atomic members like iodine and sulfur exhibit Raman activity. Many factors that determine the scattering efficiency are discussed by Schrader (1986).

Instrumentation

An excellent source for both Raman and infrared spectroscopies is Coates (1998). Current instruments (Figure 12–36) are designed to take advantage of new technologies to eliminate the Rayleigh line, disperse, and detect the Raman signal. Modern Raman instruments can be classified into two categories: dispersive and interferometric. The dispersive-type instruments now use holographic notch filters to do the job of mechanical monochrometers, and the Raman signal is dispersed using a single fixed grating. Detection is by charge-coupled detector arrays (CCDs). These technological developments provide for a Raman microscope with no moving parts with respect to the frequency or wavelength measuring portions of the instrument, allowing for a stable and highly precise optical system (Coates, 1998).

Fourier-transform (FT) Raman instruments typically use an infrared excitation source and an interferometer to separate frequencies of the Raman-shifted signal. FT-Raman instruments overcome one major problem found in Raman spectroscopy, that is, broad-band fluorescence that can occur for a wide variety of compounds, especially organic compounds, and is pathogenic to the Raman measurement process by overwhelming the scattered signal.

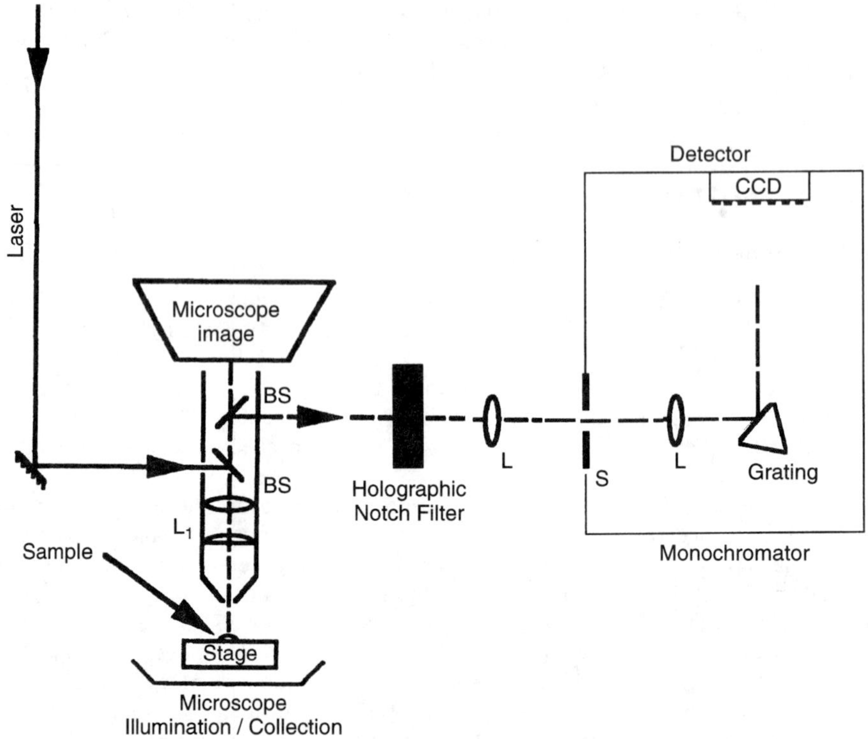

Fig. 12–36. Schematic of a micro-Raman instrument showing excitation radiation coming from an Ar ion laser, focused onto the sample through a microscope objective. The beam can be directed by normal microscope translation. The scattered Raman-shifted light from the sample is collected by the microscope objective and wavelength separated by the holographic notch filter. Raman-shifted photons enter a monochrometer and are detected by an intensified CCD array.

Fluorescence is reduced dramatically by using a near-infrared 1064 nm Nd:YAG laser for excitation, but scattering efficiency is lost due to the longer wavelength (Coates, 1998). In these instruments holographic notch filters are used and near-infrared single-channel detectors are employed. FT-Raman microscopes are frequently add-ons to FT-infrared instruments.

Capabilities

The capabilities and strengths of the micro-Raman technique are presented in some detail by Etz (1979) and Etz and Blaha (1980). The Raman microprobe offers direct molecular and structural analysis with picogram sensitivity at the single-particle level for particles 1 μm and larger in diameter. As in infrared spectroscopy, compound information may be readily obtained for crystalline materials, liquids, and, to a lesser extent, amorphous materials. Also, specific crystallographic phases can be determined by Raman spectroscopy. The technique is in principle nondestructive, but, in the case of radiation-sensitive compounds, high irradiance of the exciting, focused light can cause pyrolysis or photodecomposition of the sample. The technique is complementary to infrared spectroscopy, with sample fluorescence (intrinsic or from impurities) being a potential interference.

Schrader (1986) states that the intensity of a Raman shift is proportional to the concentration of the compound present and allows quantitation in many species. Quantitation is difficult from materials in the solid phase, but has been attempted for powders and single particles employing calibration standards of closely matching materials (Grynpas et al., 1982).

It has been demonstrated (Etz and Blaha, 1980) that reference spectra taken of bulk material of the same composition match those of the same materials examined as discrete micrometer-sized particles. Therefore, literature/library spectra of bulk materials can be utilized for compound identification of particles.

Delhaye and Dhamelincourt (1990) present a complete historical overview of the Raman microprobe, covering past and present technical developments. Other useful references are Rosasco et al. (1975), Delhaye and Dhamelincourt (1975), Rosasco (1980), and Adar (1988). Raman spectroscopy for chemical applications covering a wide range of topics including sample preparation, biological samples, and organic and inorganic compounds is presented by Grasselli et al. (1981).

Applications to Particle Analysis

Particle Collection and Preparation. The particles are held on a substrate having properties of low or non-interfering Raman bands and having relatively high thermal conductivity. Any material may be used as a substrate provided that it does not have interfering Raman-active bands. Substrates like Al_2O_3 are used to support the sample because they are thermally conductive and thus serve as a heat sink removing laser-induced heat from the particle. Particles can be collected directly onto Al_2O_3 disks or glass/quartz coverslips using a low-pressure cascade impactor to collect size-segregated atmospheric particles (Etz and Blaha, 1980). It is advantageous for micro-Raman spectroscopy not to use grease on the collection stages because it may become an interference in the spectrum. Similarly, membrane filters are usually not used directly because of interfering Raman bands that may mask the particle signal. Analysis from glass or quartz fiber filters is possible. LiF is another good substrate for Raman spectroscopy because it has no Raman-active vibrational modes, but it has one drawback—it is hygroscopic. Analysis is usually done at atmospheric pressure and ambient temperature. Thus, many particle samples can be examined by micro-Raman that would normally evaporate in a vacuum.

Particle Analysis. Some of the applications to particle analysis include indentification of airborne particles from stationary sources (Etz et al., 1978b), urban dust samples and particles from oil-fired power plants (Etz et al., 1978a), graphitic carbon determination on urban particles (Blaha et al., 1978), sulfate/sulfuric acid determination (Etz et al., 1977), and pure reference polyaromatic hydrocarbons (Etz et al., 1979). The information that can be derived from micro-Raman on a single-particle basis is illustrated by the example of a St. Louis urban dust particle mounted on an LiF substrate (Blaha et al., 1979). The Raman spectrum is shown in Figure 12–37. The particle was identified as $CaCO_3$ from the evidence of the Raman bands at 1432, 1088, 714, 283, and 156 cm^{-1}. $CaSO_4$ was present and identified by the weak band at 1018 cm^{-1}. The broad hump-like features at approximately 1350 and 1600 cm^{-1} indicated the presence of graphitic carbon, due to carbonaceous material in or on the particle surface. The analyst proposed that $CaSO_4$ resulted from the action of H_2SO_4 with the $CaCO_3$.

Sobanska et al. (1999) used Raman spectroscopy to analyze individual particles produced by a lead smelter chimney stack. The authors were able to identify the compound mixtures present in the individual particle spectra. The structural properties of carbon black due to heating and cooling were studied by Raman and micro-Raman (Zerda et al., 1998). Both analysis methods revealed the same spectral information associated with amorphous and graphitic carbon, documenting the crystalline transformation that the carbon black undergoes.

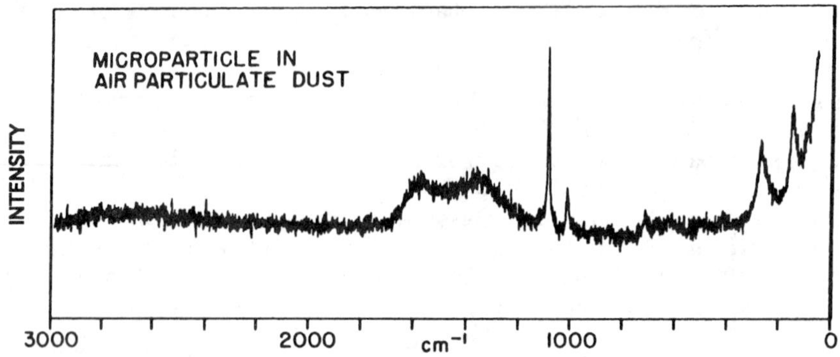

Fig. 12–37. Micro-Raman spectrum of an atmospheric 10 μm urban dust particle composed primarily of $CaCO_3$ as evidenced by the Raman active bands at 1432, 1088, 714, 283, and 156 cm^{-1}. The band at 1018 cm^{-1} shows the presence of $CaSO_4$, which is thought to result from the reaction of H_2SO_4 with $CaCO_3$ found in the particle. (From Blaha et al., 1979.)

INFRARED MICROSCOPY

Underlying Principle

Infrared (IR) spectroscopy is used to measure vibrational transitions in molecules occurring in the 10 to 10,000 cm^{-1} (1000 to 1 μm) range. The vibrational spectra are characteristic of functional groups in a given molecular compound (McMillan and Hofmeister, 1988).

The model describing vibrational processes involved in IR spectroscopy has been included in the above discussion of Raman microprobes. In IR spectroscopy, photons in the IR region of the spectrum are absorbed by molecular vibrational transitions. Molecules that have strong molecular dipoles are normally strong IR absorbers.

Instrumentation

The IR microscope contains an IR radiation source, optics to transfer the radiation, a spectrometer to separate light coming from or through the sample, and an IR detector. The IR microscope is an (optical) microscope that most commonly uses mid-range IR radiation, 4000 to 400 cm^{-1} (2.5 to 25 μm) to probe the sample. The incident radiation emanates from a black body radiation source, typically a silicon carbide "Globar" or filament. The beam-condensed incident radiation is directed at the sample using either lenses or reflecting optics. In modern IR microscopes, energy separation is achieved by an interferometer; the signal is Fourier transformed (Katon et al., 1990; McMillan and Hofmeister, 1988). Absorption bands, resulting from specific infrared-active molecular bonds, are usually seen in the IR spectrum as modulations in the signal intensity. One example of a typical IR detector used for mid-range wavelengths is the MCT (HgCdTe) liquid-nitrogen–cooled detector.

Capabilities

Fourier-transform infrared (FTIR) spectroscopy thus provides an analytical method for identifying or characterizing molecular compounds. A good general reference is by Messerschmidt and Harthcock (1988). Many of the reported applications of FTIR are not examples of single-particle analysis of micrometer-sized particles, but they do show that important molecular compositional information on microsamples of material is obtainable. The technique has demonstrated utility for qualitative analysis of organic functional groups (e.g., carbonyl or amide linkages), but direct chemical identification of complex mixtures found in atmospheric samples is often difficult. Dangler et al. (1987) point out that the technique offers fingerprint information for source apportioning an aerosol population.

One limitation in IR microscopy, as in visible light microscopy, is that the spatial resolution is determined by the diffraction limit for the wavelength of radiation used. To obtain a spectrum of a single particle, the particle must be approximately 50 μm away from neighboring particles or objects; otherwise, there will be signal mixing due to diffraction from the particles' edges (Katon et al., 1990). However, by aperturing in the image plane (a design common to most FTIR microscopes) in such a way as to define the sample area, spectra have been obtained of single objects as small as 5 μm (Messerschmidt and Reffner, 1988).

Transmission or reflectance measurements are done with FTIR microscopes depending on the sample presentation. Transmission measurements are possible from particulate samples mounted on IR-transparent materials such as ZnSe. If the particles are mounted on metal surfaces, reflectance IR is often the desired approach. An alternative to these two methods requires transferring representative quantities of collected particle material to a cell for reflectance measurements.

FTIR offers a number of attractive aspects. Sample collection and sample preparation are combined in one step; the particulate material is deposited onto a ZnSe disk and mounted directly into the microscope for a measurement in the transmission mode (Dangler et al., 1987). Another advantage of FTIR is that the sample is not mounted in vacuum so both high- and low-vapor pressure materials can be examined.

Given the breadth of IR reference spectral libraries and some of the new emerging multivariate analysis techniques being applied to FTIR, the technique provides chemical characterization for particulate samples, especially those containing organic compounds.

Applications to Particle Analysis

Application of FTIR to airborne particle material from vehicle emissions using a low-pressure impactor for particle collection has been reported (Dangler et al., 1987; Pickle et al., 1990; Brown et al., 1990). The authors report classes of organic and inorganic molecular constituents making up the sample. Applications of IR spectroscopy applied to particles have been reviewed by Allen and Palen (1989).

SCANNING PROBE MICROSCOPY

Principal of Operation

Scanning probe microscopy (SPM) includes a variety of techniques to characterize surfaces from the near atomic level to micrometer dimensions (Howland and Benatar, 1996). The most widely known probes are the scanning tunneling microscope (STM) and the atomic force microscope (AFM) and variations on this technique with several sensors—magnetic force microscope, electrostatic force microscope, scanning capacitance microscope, thermal scanning microscope, and the near-field scanning optical microscope (NSOM). Figure 12–38 shows a generic scanning probe microscope. Scanning probes utilize a small (1 to 100 nm) sensing probe that is brought near the sample surface. The sample is normally rastered under the probe by an x–y piezoelectric scanner and, in some cases (e.g., the AFM), the tip deflection or the force on the probe tip above the sample is measured. There is usually a feedback mechanism to reposition the probe tip in the z-direction (height). In other cases, such as the STM and certain modes of NSOM, tunneling electrons or photons are detected. For the AFM, the force applied to the probe tip is produced from atomic repulsion or attraction (van der Waals) forces between the atoms or molecules in the tip and the sample surface. When the tip is far

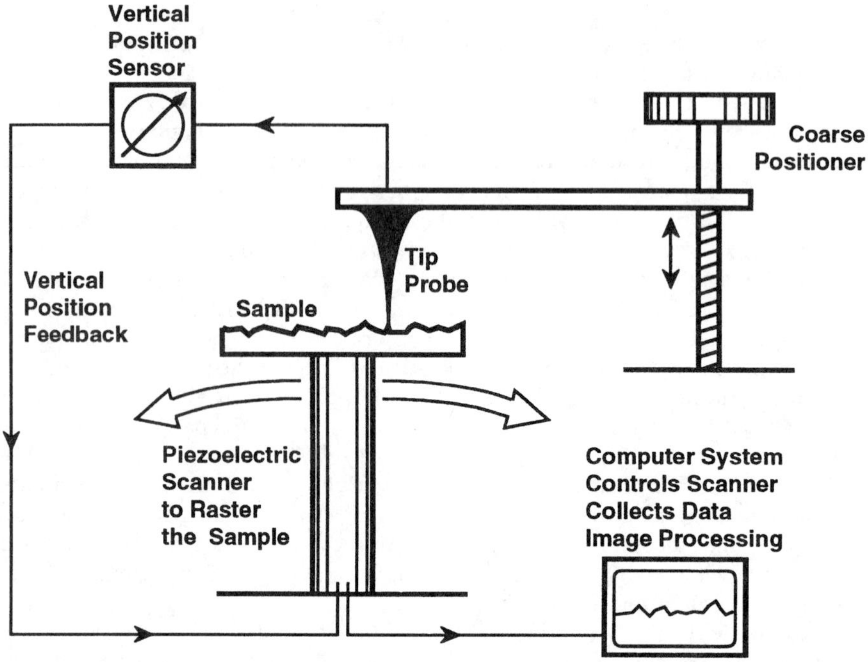

Fig. 12–38. Schematic of a scanning probe microscope. (From Howland and Benatar, 1996.)

away, attractive forces are exhibited, and as the tip nears the surface repulsive forces dominate. The probe tip is located on a small cantilever arm that bends or distorts with the applied force. Sensing of the probe position can be done by a reflection angle measurement from the back surface of the cantilever arm.

Capabilities

The STM is thought to have limited application to particle research because the material to be analyzed must be a conductor or semiconductor. STM images are convolutions of the electronic properties and the morphology of the sample and tip (McCarty and Weiss, 1999). The AFM is less discriminating, but has its own set of problems. The technique can be utilized either inside a vacuum or at atmospheric pressure conditions to measure nanometer-sized particles. The AFM provides a way to obtain the volume (three-dimensional information) of particles and microstructures because the probe height as related to the x–y position on the particle is measured. Particles can be analyzed that range from 10 nm to several micrometers. There are several modes of operation that place the probe tip either near or farther from the particle. In the near or "contact" mode, there is concern of touching and moving particles. Particles need to be firmly in contact with the underlying smooth substrate. Using image software, particle images and size distributions can be derived (Kollensperger et al., 1999). There are concerns with the sizing accuracy. The measurement result is a function of the "convolution" of the size and shape of the particle and the probe—both of which are unknowns (McCarty and Weiss, 1999). Other sources of uncertainty are water on the tip and surface, nonlinear motion and hysteresis of the scanner, and drifts due to thermal effects (Gonda et al., 1999). Uncertainties for some measurements of atmospheric particles were estimated to be at the 10% level (Kollensperger et al., 1997). Gonda et al. (1999) have built and tested an AFM utilizing an interferometer to accurately determine probe location in three dimensions.

The authors report standard deviations of 0.4 nm (n = 80) on a 180 nm height and 10 nm on a 3 μm pitch standard sample. These represent only type A uncertainties, those calculated from statistics methods.

Questions of measurement uncertainty associated with AFM are being addressed at the National Institute of Standards and Technology by Dixson et al. (1999, 2000) and Villarrubia et al. (1999). These researchers are using a modified commercial AFM for which the x, y, and z measurement directions are traceable to the International System of Units meter through optical wavelength interferometry. Work has been done to establish the uncertainty in the pitch measurement (length from object to object; e.g., line to line), the object width (or line width), and height. For particle analysis, object width and height are thought to be most relevant. Uncertainty arises in object width due primarily to the probe tip shape and the inability to detect (without distortion) the true edge of the object. Measurements and modeling of the tip shape have been done to improve the measurement capability. The expanded uncertainty ($k = 2$) (Taylor and Kuyatt, 1994), which includes contributions from multiple sources for object width, was found to be ±11 nm for a 449 nm linewidth (Villarrubia et al., 1999; Villarrubia, personal communication, 2000). In the same paper very favorable comparisons of measurement results for linewidth using SEM and electrical critical dimension are presented. The complete uncertainty statement in the height measurement includes the object height, the reproducibility per measurement (typically <1 nm), and certain motion errors. For the height of a 100 nm sphere the standard uncertainty would be approximately ±0.3 nm. This would apply for rigid objects. Soft objects like polymer beads can give erroneous results.

Certain modes of AFM operation provide a means for determining phase or compositional differences within the particle. The NSOM is currently in the research stage, but shows promise because of the potential to chemically characterize nanoscale particles. Vibrational spectroscopies, both IR and Raman, have been demonstrated in NSOM (Jordan et al., 1999; Stranick et al., 1998).

Applications

Characterization of nanoparticles by SPM is lacking because of the difficulties in interpreting images (McCarty and Weiss, 1999). Images were taken of atmospheric aerosol particles using an AFM (Friedbacher et al., 1995). Ambient aerosol was collected from an urban environment using an 11-stage low-pressure impactor. In the impactor, particles were deposited directly on relatively smooth substrates. Particle images obtained by AFM that are similar to those reported by the authors are shown in Figure 12–39. These are gray scale images in which the light shades correspond to particle heights of 20 nm and the darker gray levels represent zero height or the filter background. The height profiles over the substrate provide the particle image. Figure 12–39a shows particles collected on an impactor stage with a mean aerodynamic diameter of 21 nm, and Figure 12–39b shows particles from a different stage with a mean aerodynamic diameter of 170 nm.

Friedbacher et al. (1995) and Kollensperger et al. (1997) utilized software to extract particle size distributions from 20 to 125 nm. The root mean square roughness of the substrate was determined by the AFM and subtracted from the AFM images containing particles. Interestingly, it was reported that the particle height-to-width ratio was approximately 0.5, indicating that for the small, irregular-shaped particles, the particles reside in their most stable orientation. The authors estimate that there is approximately 10% uncertainty in the size measurements done in this manner; contributing factors are the particle size range and the probe tip to surface interaction.

Laboratory experiments were conducted using the AFM to determine size changes on ammonium sulfate and carbonaceous agglomerates exposed to controlled humidities and ozone concentrations (Kollensperger et al., 1999). The advantages of using the AFM for

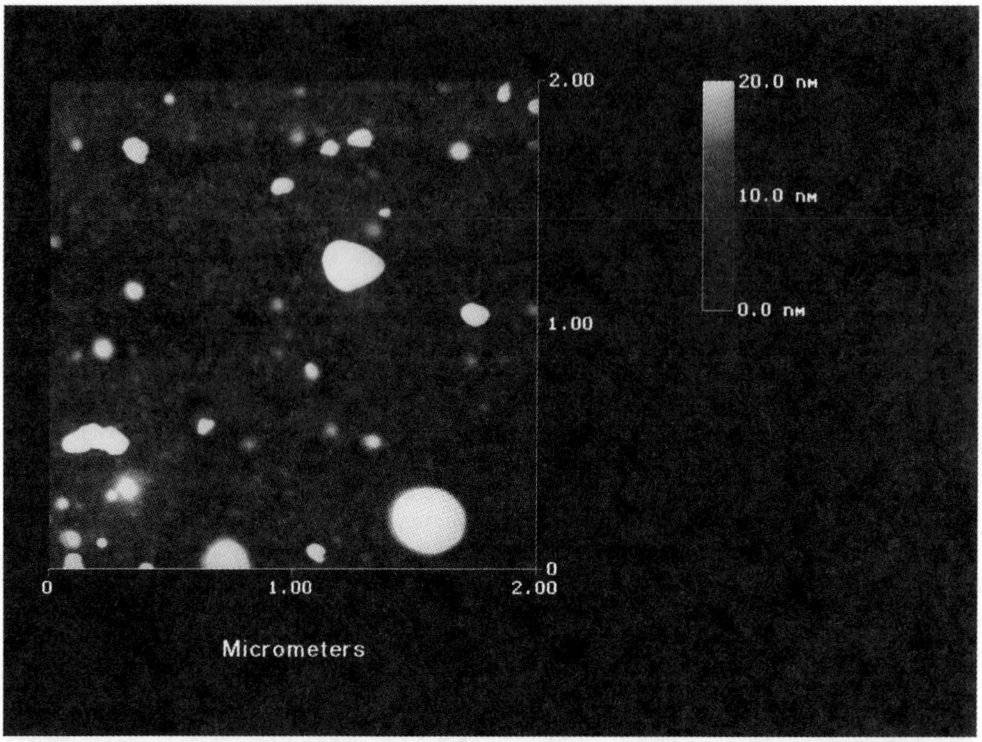

(a)

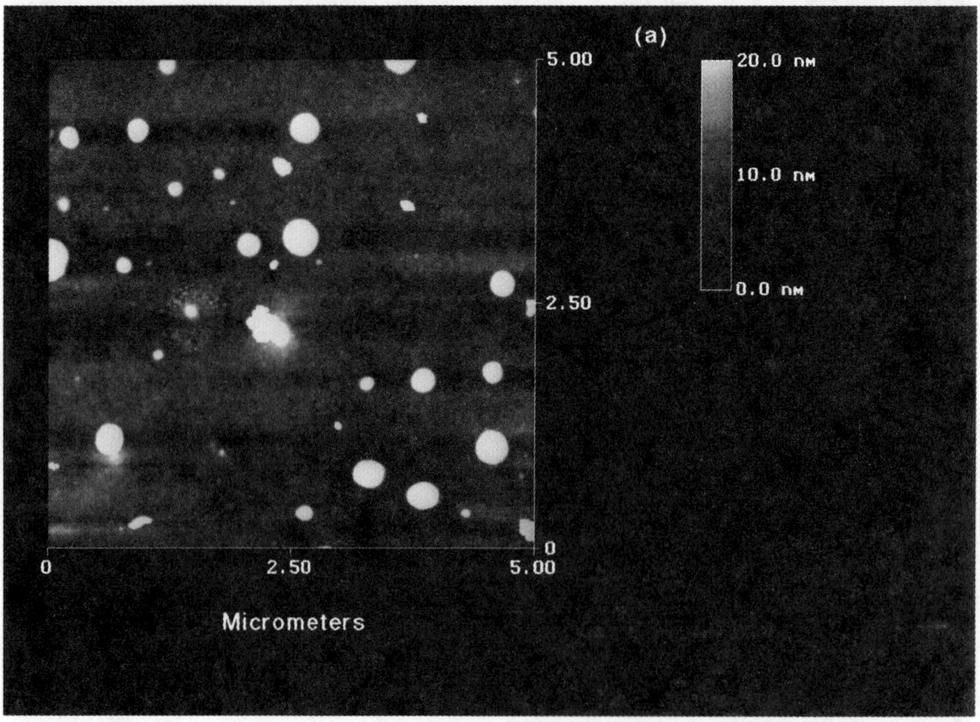

(b)

Fig. 12–39. Gray level images of atmospheric particles collected on impactor stages with mean aerodynamic diameters of 21 nm **(a)** and 170 nm **(b)**. Gray level range is from white to gray and corresponds to particle height above the substrate of 20 to 0 nm. (Unpublished images courtesy of G. Friedbacher.)

determining particle size distributions are presented by Kollensperger et al. (1999a,b). They report that it is a general technique that is not restricted by the conductance of the particles. The particle diameter is determined from a measure of the particle volume instead of the usual two-dimensional projection of the particle area found from electron microscopy. Measurements are made under ambient conditions; vacuum is required for many techniques, with the consequence that vacuum exposure can change the size and shape of the particles by removing volatile components.

COMPLEMENTARY CAPABILITIES OF MICROANALYTICAL INSTRUMENTATION

There is synergism that often results from using multiple instrument analysis; usually more information about a sample is gained by analysis with several instruments than by any single instrumental approach. The physical properties such as particle size and morphology (obtained by LM and electron microscopy characterization) are often augmented by information about the chemical composition. A complete chemical characterization may provide elemental composition (in terms of major, minor, and trace quantities in addition to isotopic abundance) and the chemical compounds that may be present. For example, elemental analysis from the electron instruments is complemented, many times, by isotopic abundance, trace analysis (mass spectrometry), and molecular composition (micro-Raman, FTIR, SIMS, and LMMS). SIMS and LMMS can further complement the elemental composition by extending the atomic number range to light elements (H through F).

A documented example of teaming two instruments together was given by Steel et al. (1984) where the capabilities of the AEM and LMMS were brought to bear on an analysis. In this work, the AEM was used to image and determine particle morphology, as well as major and minor elemental compositions of submicrometer particles of zircon on a thin-film carbon substrate. AEM provided the location on the carbon substrate of the submicrometer particles for subsequent analysis by LMMS, which in turn yielded trace and low-Z elemental composition along with isotopic abundances. The authors believed that this two-instrument approach provided a comprehensive characterization of small particles down to 0.1 μm in size.

Hachimi et al. (1998) demonstrated the usefulness of combining two techniques, micro-Raman and LMMS, for determining the oxidation state of chromium. De Bock and Van Grieken (1999) recognize that measurements made by microanalysis can and often do complement bulk chemical analysis on particle samples. An example is the work of Gieray et al. (1997) in which cloud particle chemistry is studied using both bulk chemical analysis and laser microprobe approaches.

Experimental designs applying different instruments in series or parallel depend on several factors. Obviously nondestructive analysis techniques should be attempted before analysis by methods that destroy some or all of the particle in the process of analysis (as is the case for mass spectrometry). This is a primary concern for samples with very limited numbers of particles or in the case of analysis on a single particle by several techniques.

Among the nondestructive techniques, the size of the particles in the sample dictates whether LM, SEM, or AEM can be used to obtain morphological, size, color, crystallographic properties, and surface features. LM often provides the first information about particles in a sample, and this information is often used to determine the course of subsequent analysis. The morphology, phase, and crystallographic nature of the particles can be an immediate clue as to which analysis technique should be pursued. A simple example is that droplets usually will not survive the vacuum environments common to many instruments, but could possibly be analyzed by one of the spectroscopies applicable at atmospheric pressures. Particles above a nominal 0.5 μm diameter can be viewed by LM and SEM, while particles smaller than 0.1 μm are best examined in the AEM.

There are also considerations with regard to conductive coatings that are usually applied for the electron microscopes, the amount of total material in the sample, the composition of the sample, and the information desired. For example, a sample can first be examined by light microscopy to yield size, shape, and various optical properties in the >1 μm size range. Then, if major elemental composition is required, SEM with EDS can be done for analysis, and an enlarged morphology can be imaged. Micro-Raman and FTIR provide molecular composition by relating the vibrational spectrum to chemical bond types or, in the case of organic molecules, functional groups. Information obtained from mass spectrometry augments the vibrational bond information by often providing the molecular weight (through molecular ion identification) and additional chemical/structural information through fragment ions also present in the mass spectrum. Trace elements at concentrations down to approximately 10 μg/g are often found by LMMS or SIMS in addition to the isotopic abundance. Trace elements and isotopic information may provide clues for determining particle origin.

Microanalysis utilizing multiple instruments provides a means for comprehensive physical and chemical characterization of a collected particles that is often relevant for the aerosol, environmental, and atmospheric scientist.

ACKNOWLEDGMENTS

The authors acknowledge the help of R. Dixson, E. Etz, G. Gillen, S. Hoeft, D. Newbury, D. Simons, E. Steel, S. Turner, J. Ver Koutern, M. Walker, and S. Wight for valuable contributions to this chapter. The authors thank Professor G. Friedbacher for providing unpublished AFM images.

REFERENCES

Abell, M. T., S. A. Shulman, and P. A. Baron. 1989. The quality of fiber count data. *Appl. Ind. Hyg.* 4:273–285.

Adar, F. 1988. Developments of the Raman microprobe—Instrument and applications. *Microchem. J.* 38:50–79.

Alam, M. N., M. Blackman, and D. W. Pashley. 1954. High angle Kikuchi patterns. *Proc. R. Soc. Lond. Ser.* 221A:224–242.

Allen, D. T. and E. Palen. 1989. Recent advances in aerosol analysis by infrared spectroscopy. *J. Aerosol Sci.* 20:441–455.

Armstrong, J. T. 1978. Methods of quantitative analysis of individual microparticles using electron instruments. *Scanning Electron Microsc.* I:455–467.

Armstrong, J. T. and P. R. Buseck. 1975. Methods of quantitative analysis of individual microparticles using electron microprobe: Theoretical. *Anal. Chem.* 47:2178–2192.

Armstrong, J. T. and P. R. Buseck. 1977. Development of a characteristic fluorescence correction for thin films and particles. *Proc. 12th Ann. Conf. Microbeam Anal.* Soc., Boston, pp. 42A–42F.

Asbestos International Association (AIA). 1979. Reference method for determination of airborne asbestos fibre concentration by light microscopy. Recommended Technical Method No.1 (RTM 1). AIA Health and Safety Publication. London: Asbestos International Association.

Baron, P. A. 1993. Measurements of asbestos and other fibers. In *Aerosol Measurement Principles Techniques and Applications*, eds. K. Willeke and P. A. Baron. New York: Van Nostrand Reinhold.

Baron, P. A. and G. C. Pickford. 1986. An asbestos sampling filter clearing procedure. *Appl. Ind. Hyg.* 4(1):169–171.

Behrisch, R. 1983. Introduction and overview. In *Sputtering by Particle Bombardment II.* Topics in Applied Physics, Vol. 52, ed. R. Behrisch. New York: Springer-Verlag, pp. 1–8.

Bennett, J. and D. Simons. 1991. Relative sensitivity factors and useful yields for microfocused Ga ion beam and TOF-SIMS system. *J. Am. Vac. Soc.* A9(3):1379–1384.

Berner, T., E. Chatfield, J. Chesson, and J. Rench. 1990. *Transmission electron microscopy asbestos laboratories: Quality assurance guidelines*. EPA/5-90-002.

Betti, M., G. Tamborini, and L. Koch 1999. Use of secondary ion mass spectrometry in nuclear forensic analysis for the charcterization of plutonium and highly enriched uranium particles. *Anal. Chem.* 71:2616–2622.

Biegalski, S. R., L. A. Currie, R. A. Fletcher, G. A. Klouda, and R. Weissenbok. 1998. AMS and microprobe analysis of combusted particles in ice and snow. *Radiocarbon* 40(1):3–10.

Blaha, J. J., E. S. Etz, and W. C. Cunningham. 1979. Molecular analysis of microscopic samples with a Raman microprobe: Applications to particle characterization. *Scanning Electron Microsc.* I:93–102.

Blaha, J. J., G. J. Rosaco, and E. S. Etz. 1978. Raman microprobe characterization of residual carbonaceous materials associated with urban airborne particulates. *Appl. Spectrosc.* 32:292–297.

Brown, S., M. C. Dangler, S. R. Burke, S. V. Hering, and D. T. Allen. 1990. Direct Fourier transform infrared analysis of size-segregated aerosols: Results from the carbonaceous species methods intercomparison study. *Aerosol Sci. Technol.* 12:172–181.

Bruynseels, F., H. Storm, R. Van Grieken, and L. V. Auwera. 1988. Characterization of North Sea aerosols by individual particle analysis. *Atmos. Environ.* 22:2593–2602.

Cardona, M. and L. Ley 1978. Introduction. In *Photoemission in Solids I, General Principles*, eds. M. Cardona and L. Ley. Berlin: Springer-Verlag, p. 78.

Carter, J., D. Taylor, and P. A. Baron. 1989. Fibers, Method 7400 revision #3: 5/15/89. *NIOSH Manual of Analytical Methods.* Cincinnati, OH: DHHS/NIOSH.

Castaing, R. and A. Guinier. 1950. *Proceedings of the 1st International Conference on Electron Microscopy*, Delft, 1949, p. 60.

Chamot, E. M. and C. W. Mason. 1940. *Handbook of Chemical Microscopy*, Vol. 2. New York: John Wiley & Sons.

Chamot, E. M. and C. W. Mason. 1958. *Handbook of Chemical Microscopy*, Vol. 1. New York: John Wiley & Sons.

Cleary, T. G., G. W. Mulholland, L. K. Ives, R. A. Fletcher, and J. W. Gentry. 1992. Ultrafine combustion aerosol generator. *Aerosol Sci. Technol.* 16(3):166.

Clemett, J. S., D. F. Chiller, S. Gillette, R. N. Zare, M. Marrette, C. Engrand, and G. Kurat. 1998. Observations of indigenous polycyclic aromatic hydrocarbons in "giant" carbonaceous Antarctic micrometeorites. *Origins Life Evol. Biosphere* 28:425–448.

Clemett, J. S., C. R. Maechling, N. Zare, P. D. Swan, and R. M. Walker. 1993. Identification of complex aromatic molecules in individual interplanetary dust particles. *Science* 262:721–725.

Cliff, G. and G. W. Lorimer. 1975. The quantitative analysis of thin specimens. *J. Microsc.* 103:203–207.

Coates, J. 1998.Vibrational spectroscopy: Instruments for infrared and Raman spectroscopy. *Appl. Spectrosc. Rev.* 33:267–425.

Cowley, J. 1979. Principles of image formation. In *Introduction to Analytical Electron Microscopy*, eds. J. J. Hren, J. I. Goldstein, and D. C. Joy. New York: Plenum Press, pp. 83–117.

Crawford, N. P. and A. J. Cowie. 1984. Quality control of asbestos fibre counts in the United Kingdom—The present position. *Ann. Occup. Hyg.* 28:391–398.

Currie, L. A., R. A. Fletcher, and G. A. Klouda. 1987. On the identification of carbonaceous aerosol via ^{14}C accelerator mass spectrometry and laser microprobe mass spectrometry. *Nucl. Instrum. Methods* B29:346–354.

Currie, L. A., R. A. Fletcher, and G. A. Klouda. 1989. Source apportionment of individual carbonaceous particles using ^{14}C and laser microprobe mass spectrometry. *Aerosol Sci. Technol.* 10:370–378.

Dangler, M., S. Burke, S. V. Hering, and D. T. Allen. 1987. A direct FTIR method for identifying functional groups, in size segregated atmospheric aerosols. *Atmos. Environ.* 21:1001–1004.

Danilatos, G. D. 1978. Foundations of environmental scanning electron microscopy. *Adv. Electrons Electron Phys.* 71:102–250.

De Bock, L. A. and R. E. Van Grieken. 1999. Single particle analysis techniques. In *Analytical Chemistry of Aerosols*, ed. K. R. Spurny. New York: Lewis, pp. 243–275.

Delhaye, M. and P. Dhamelincourt. 1975. Raman microprobe and microscope with laser excitation. *J. Raman Spectrosc.* 3:33–43.

Delhaye, M. and P. Dhamelincourt. 1990. A perspective of the historical developments and future trends in Raman microprobe spectroscopy. In *Microbeam Analysis—1990*, eds. J. R. Michael and P. Ingram, San Francisco: San Francisco Press, pp. 220–227.

Denoyer, E., D. F. S. Natusch, P. Surkyn, and F. Adams. 1983. Laser microprobe mass analysis (LAMMA) as a tool for particle characterization: A study of coal fly ash. *Environ. Sci. Technol.* 17:457–462.

Denoyer, E., R. Van Grieken, F. Adams, and D. F. S. Natusch. 1982. Laser microprobe mass spectrometry 1: Basic principles and performance characteristics. *Anal. Chem.* 54(1):26A–41A.

De Waele, J. K. E. and F. C. Adams. 1986. Laser-microprobe mass analysis of fibrous dusts. In *Physical and Chemical Characterization of Individual Airborne Particles*, ed. K. R. Spurny, pp. 271–297. New York: John Wiley & Sons.

Dixson, R., R. Koning, J. Fu, T. V. Vorburger, and B. Renegar. 2000. Dimensional metrology with atomic force microscopy. *SPIE Proc.* 3998:362–368.

Dixson, R., R. Koning, V. W. Tsai, J. Fu, and T. V. Vorburger. 1999. Dimensional metrology with the NIST calibrated atomic force microscope. *SPIE Conference on Metrology, Inspection and Process Control for Microlithography XIII*. SPIE 3677:20–34.

Dobbins, R. A., R. A. Fletcher, and H.-C. Chang. 1998. The evolution of soot precursor particles in a diffusion flame. *Comb. Flame* 115:285–298.

Dobbins, R. A. and C. M. Megaridis. 1987. Morphology of flame-generated soot as determined by thermophoretic sampling. *Langmuir* 3:254–259.

Dovichi, N. J. and D. S. Burgi. 1987. Photothermal microscope. In *Microbeam Analysis–1987*, ed. R. H. Geiss. San Francisco: San Francisco Press, pp. 155–157.

Duncumb, P. and S. J. B. Reed. 1968. The calculation of stopping power and backscatter effects in electron probe microanalysis. In *Quantitative Electron-Probe Microanalysis*, ed. K. F. J. Heinrich. NBS Special Pub. 298. Washington, DC: US Department of Commerce/National Institute of Standards and Technology, pp. 133–154.

Egerton, R. F. 1996. *Electron Energy Loss Spectroscopy in the Electron Microscope*. New York: Plenum Press.

Environmental Protection Agency. 1985. *Measuring Airborne Asbestos Following an Abatement Action*. EPA 600/4-85-049.

Environmental Protection Agency. 1987. Asbestos-containing materials in schools. Washington, DC: Government Printing Office, *Federal Register*, 40 CFR Part 763.

Environmental Protection Agency. 1989. *Comparison of Airborne Asbestos Levels Determined by Transmission Electron Microscopy (TEM) Using Direct and Indirect Transfer Techniques.* EPA 560/5-89-004.

Etz, E. S. 1979. Raman microprobe analysis: Principles and applications. *Scanning Electron Microscopy*, Vol. 1/1979. AMF O'Hare, IL: SEM Inc., pp. 67–92.

Etz, E. S. and J. Blaha. 1980. Scope and limitations of single particle analysis by Raman microprobe spectroscopy. In *Characterization of Particles*, ed. K. F. J. Heinrich. NBS Special Publication 553. Washington, DC: US Department of Commerce/National Institute of Standards and Technology, pp. 153–197.

Etz, E. S., G. J. Rosasco, and J. J. Blaha. 1978a. Observation of the Raman effect from small, single particles: Its use in chemical identification of airborne particulates. In *Environmental Pollutants*, eds. T. Y. Toribara, J. R. Coleman, B. E. Dahneke, I. Feldman. New York: Plenum Publishing Corp., pp. 413–456.

Etz, E. S., G. J. Rosasco, and W. C. Cunningham. 1977. The chemical identification of airborne particles by laser Raman spectroscopy. *Environmental Analysis*. London: Academic Press, Inc., pp. 295–340.

Etz, E. S., G. J. Rosasco, and K. F. J. Heinrich. 1978b. Chemical analysis of stationary source particulate pollutants by micro-Raman spectroscopy. EPA Report. EPA-600/2-78-193. Research Triangle Park, NC: U.S. EPA, pp. 1–37.

Etz, E., S. A. Wise, and K. F. J. Heinrich. 1979. Trace organic analysis: A new frontier in analytical chemistry. In *NBS Special Publication 519*. Washington, DC: US Department of Commerce/National Institute of Standards and Technology, pp. 723–729.

Fletcher, R. A. 1989. A new way to mount particulate material for laser microprobe mass analysis. *Anal. Chem.* 61:914–917.

Fletcher, R. A. and L. A. Currie. 1987. Observations derived from the application of principal-component analysis to laser microprobe mass spectrometry. In *Microbeam Analysis—1987*, ed. R. H. Geiss. San Francisco: San Francisco Press, pp. 369–371.

Fletcher, R. A. and L. A. Currie. 1988. Pattern differences in laser microprobe mass spectra of negative ion carbon clusters. In *Microbeam Analysis—1988*, ed. D. E. Newbury. San Francisco: San Francisco Press, pp. 367–370.

Fletcher, R. A., E. S. Etz, and S. Hoeft. 1990. Complementary molecular information on phthalocyanine compounds derived from laser microprobe mass spectrometry and micro-Raman spectroscopy. In *Microbeam Analysis—1990*, eds. J. R. Michael and P. Ingram. San Francisco: San Francisco Press, pp. 89–92.

Friedbacher, G., M. Grasserbauer, Y. Meslmani, N. Klaus, and M. Higatsberger. 1995. Investigation of environmental aerosol by atomic force microscopy. *Anal. Chem.* 67(10):1749–1754.

Friedrichs, K. H. 1986. Particle analysis by light microscopy. In *Physical and Chemical Characterization of Individual Airborne Particles*, ed. K. R. Spurny. New York: John Wiley & Sons, pp. 161–172.

Gavrilovic, J. 1984. Surface analysis of small individual particles by secondary ion mass spectroscopy. In *Particle Characterization in Technology, Volume 1, Applications and Microanalysis*, ed. J. K. Beddow. Boca Raton, FL: CRC Press, Inc., pp. 3–19.

Germani, M. S. and P. R. Buseck. 1991. Automated scanning electron microscopy for atmospheric particle analysis. *Anal. Chem.* 63:2232–2237.

Gieray, R., P. Wieser, T. Engelhardt, E. Swietlicki, H.-C. Hansson, B. Mentes, D. Orsini, B. Martinsson, B. Svenningsson, K. J. Noone, and J. Heintzenberg. 1997. Phase partitioning of aerosol constituents in cloud based on single-particle and bulk analysis. *Atmos. Environ.* 31:2491–2502.

Goehner, R. P., and J. R. Michael. 1996. Phase identification in a scanning electron microscope using backscattered electron Kikuchi patterns. *J. Res. Natl. Inst. Stand. Technol.* 101:301–308.

Goldstein, J. I. 1986. Principles of thin film X-ray microanalysis. In *Introduction to Analytical Electron Microscopy*, eds. J. J. Hren, J. I. Goldstein, and D. C. Joy. New York: Plenum Press, pp. 83–117.

Goldstein, J. I., J. L. Costley, G. W. Lorimer, and S. J. B. Reed 1977. Quantitative X-ray analysis in the electron microscope. *Scanning Electron Microscopy*, Vol. I, ed. O. Johari. Chicago: IITRI, p. 315.

Goldstein, J. I., H. Yakowitz, D. E. Newbury, J. W. Colby, and J. R. Coleman. 1975. *Practical Scanning Electron Microscopy*, New York: Plenum Press, p. 32.

Gonda, S., T. Doi, T. Kurosawa, Y. Tanimura, N. Hisata, T. Yamagishi, H. Fujimoto, and H. Yukawa 1999. Real-time interferometrically measuring atomic force microscope for direct calibration of standards. *Rev. Sci. Instrum.* 70:3362–3368.

Grasselli, J. G., M. K. Snavely, and B. J. Bulkin. 1981. *Chemical Applications of Raman Spectroscopy.* New York: John Wiley & Sons.

Grasserbauer, M. 1978. Characterization of individual airborne particles by light microscopy, electron and ion probe microanalysis, and electron microscopy. In *Analysis of Airborne Particles by Physical Methods*, ed. H. Malissa. West Palm Beach, FL: CRC Press, Inc., pp. 125–178.

Groff, J. H., P. C. Schlecht, and S. Shulman. 1991. *Laboratory Reports and Rating Criteria for the Proficiency Analytical Testing (PAT) Program.* DHHS (NIOSH) Publication No. 91–102. Cincinnati, OH: NIOSH.

Grynpas, M. D., E. S. Etz, and W. J. Landis. 1982. Studies of calcified tissues by Raman microprobe analysis. In *Microbeam Analysis—1982*, ed. K. F. J. Heinrich. San Francisco: San Francisco Press, Inc., pp. 333–337.

Hachimi, A., G. Krier, E. Poitevin, M. Schweigert, S. Peter, and J. Muller. 1996. Characterization of nitro-PAH adsorbed on environmental micro particles: A comparison of LAMMA, fourier transform mass spectrometry laser microprobes and gas chromatography-tandem mass spectrometry. *Int. J. Environ. Anal. Chem.* 62:219–230.

Hachimi, A., L. Van Vaeck, K. Poels, F. Adams, and J. F. Muller. 1998. Speciation of chromium, lead and nickel compounds by laser microprobe mass spectrometry and application to environmental and biological samples. *Spectrochim. Acta* Part B53:347–365.

Hall, T. A. 1968. Some aspects of the microprobe analysis of biological specimens. *Quantitative Electron-Probe Microanalysis*, ed. K. F. J. Heinrich. NBS Special Publication 298. Washington, DC: US Department of Commerce/National Institute of Standards and Technology, pp. 269–299.

Hara, K., T. Kikuchi, K. Furuya, M. Hayashi, and Y. Fujii. 1996. Characterization of Antarctic aerosol particles using laser microprobe mass spectrometry. *Environ. Sci. Technol.* 30:385–391.

Heinrich, K. F. J., ed. 1980. *Characterization of Particles*. NBS Special Publication 553. Washington, DC: U.S. Department of Commerce/National Institute of Standards and Technology.

Heinrich, K. F. J. 1981. *Electron Beam X-Ray Microanalysis*. New York: Van Nostrand Reinhold, Co.

Hercules, D. M., R. Day, K. Balasanmugam, T. A. Dang, and C. P. Li. 1982. Laser microprobe mass spectrometry 2. Applications to structural analysis. *Anal. Chem.* 54(2):208A–305A.

Hillenkamp, F. and M. Karas. 1989. *Proceedings of the 37th ASMS Conference on Mass Spectrometry and Allied Topics*. Miami Beach, FL, 1168 pp.

Henoc, J., K. F. J. Heinrich, and R. L. Myklebust. 1973. *A Rigorous Correction Procedure for Quantitative Electron Probe Microanalysis (COR 2)*. NBS Technical Note 769. Washington, DC: U.S. Department of Commerce/National Institute of Standards and Technology.

Howland, R. and L. Benatar. 1996. *A Practical Guide to Scanning Probe Microscopy.* Sunnyvale, CA.: Park Scientific Instruments.

Hunt, J. A. and D. B. Williams. 1991. Electron energy-loss spectrum-imaging. *Ultramicroscopy* 38:47.

Jeanguillaume, C. and C. Colliex. 1989. Spectrum-imaging—The next step in EELS digital acquisition and processing. *Ultramicroscopy* 28:252.

Jordan, C. J., S. J. Stranick, L. J. Rickter, R. R. Cavanagh, and D. B. Chase. 1999. Near-field scanning optical microscopy incorporating Raman scattering for vibrational mode contrast. *Surf. Sci.* 433–435:48–52.

Joy, D. 1979. The basic principles of electron energy loss spectroscopy. In *Introduction to Analytical Electron Microscopy,* eds. J. J. Hren, J. I. Goldstein, and D. C. Joy. New York: Plenum Press, pp. 83–117.

Katon, J. E., A. J. Sommer, and P. L. Lang. 1990. Infrared microspectroscopy. *Appl. Spectrosc. Rev.* 25:173–211.

Kaufman, L. and P. J. Rousseeuw. 1990. *Finding Groups in Data.* New York: Coiley & Sons, Inc.

Kaufmann, R. 1986. Laser-microprobe mass spectroscopy (LAMMA) of particulate matter. In *Physical and Chemical Characterization of Individual Airborne Particles*, ed. K. R. Spurny. New York: John Wiley & Sons, pp. 226–250.

Klaus, N. 1984. The effect of the substrate material on SIMS analysis of aerosols. *Sci. Total Environ.* 35:1–12.

Klaus, N. 1985. SIMS analysis of motor vehicle flue gas aerosols. *Sci. Total Environ.* 44:81–87.

Klaus, N. 1986. Aerosol analysis by secondary-ion mass-spectrometry. In *Physical and Chemical Characterization of Individual Airborne Particles*, ed. K. R. Spurny. New York: John Wiley & Sons, pp. 331–339.

Kolaitis, L. N., F. J. Bruynseels, R. E. Van Grieken, and M. O. Andreae. 1989. Determination of methanesulfonic acid and non-sea-salt in single marine aerosol particles. *Environ. Sci. Technol.* 23:236–240.

Kollensperger, G., G. Friedbacher, M. Grasserbauer, and L. Dorffner. 1997. Investigation of aerosol particles by atomic microscopy. *Fresenius J. Anal. Chem.* 358:268–273.

Kollensperger, G., G. Friedbacher, A. Krammer, and M. Grasserbauer. 1999a. Application of atomic force microscopy to particle sizing. *Fresenius J. Anal. Chem.* 363:323–332.

Kollensperger, G., G. Friedbacher, A. Krammer R. Kotzick, R. Niessner, and M. Grasserbauer. 1999b. Application of atomic force microscopy to particle sizing. *Fresenius J. Anal. Chem.* 364:294–304.

Krivanek, O. L., C. C. Ahn, and R. B. Keeney. 1987. Parallel detection electron spectrometer using quadrupole lenses. *Ultramicroscopy* 22:103.

Langer, A. M., A. D. Mackler, and F. D. Pooley. 1974. Electron microscopical investigation of asbestos fibers. *Environ. Health Perspect* 9:63–80.

Le Guen, J. M. M. and S. Galvin. 1981. Clearing and mounting techniques for the evaluation of asbestos fibres by membrane filter method. *Ann. Occup. Hyg.* 24:273–280.

Lee, R. J. and J. F. Kelly. 1980. Applications of SEM-based automatic image analysis. In *Microbeam Analysis—1980*, ed. D. B. Wittry. San Franscisco: San Francisco Press, Inc., pp. 13–16.

Linder, B. and U. Seydel. 1989. Pattern recognition as a complementary tool for the evaluation of complex LAMMS data. In *Microbeam Analysis—1989*, ed. P. E. Russell. San Francisco: San Francisco Press, pp. 286–292.

Lindner, M., D. A. Leich, R. J. Borg, G. P. Russ, J. M. Bazan, D. S. Simons, and A. R. Date. 1986. Direct laboratory determination of the ^{187}Re half-life. *Nature* 320:246–248.

Linton, R. W., A. Loh, D. F. S. Natusch, C. A. Evans, and P. Williams. 1976. Surface predominance of trace elements in airborne particles. *Science* 191:852–854.

Ma, Z., R. N. Thompson, K. R. Lykke, M. J. Pellin, and A. M. Davis. 1995. New instrument for microbeam analysis incorporating submicron imaging and resonance ionization mass spectrometry. *Rev. Sci. Instrum.* 66(5):3168–3176.

Marinenko, R. B. 1982. Preparation and characterization of K-411 and K-412 mineral glasses for microanalysis: SRM 470. *NBS Special Publication 260–274.*

Mauney, T. 1984. *Instrumental Effects in LAMMA, Ion Kinetic Energy Distributions, and Analysis of Soot Particles.* Ph.D. Thesis, Colorado State University. Fort Collins, Colorado, pp. 155–169.

McCarty, G. S. and P. S. Weiss. 1999. Scanning probe studies of single nanostructures. *Chem. Rev.* 99:1983–1990.

McCrone, W. C. and J. G. Delly. 1973. *The Particle Atlas*, Vols. 1–4, 2nd Ed. Ann Arbor, MI: Ann Arbor Science.

McMillan, P. F. and A. M. Hofmeister. 1988. Infrared and Raman spectroscopy. In *Reviews in Mineralogy*, Vol. 18, *Spectroscopic Methods in Mineralogy and Geology*, ed. F. C. Hawthorne. Chelsea, MI: BookCrafters, Inc., pp. 99–159.

Messerschmidt, R. and M. Harthcock, eds. 1988. *Infrared Microspectroscopy: Theory and Applications.* New York: Marcel Dekker, Inc.

Messerschmidt, R. and J. A. Reffner. 1988. FT-IR microscopy of biological samples: A new technique for probing cells. *Microbeam Analysis—1988*, ed. D. E. Newbury. San Francisco: San Francisco Press, pp. 215–218.

Michael, J. R., M. E. Schlienger, and R. P. Goehner. 1997. Electron backscatter diffraction in the SEM: Is electron diffraction in the TEM obsolete? *Microsc. Microanal.* 3(Suppl. 2):879–880.

Middleton, A. P. 1982. Visibility of fine fibres of asbestos during routine electron microscopical analysis. *Ann. Occup. Hyg.* 25:53–62.

Musselman, I. H., D. S. Simons, and R. W. Linton. 1988. Effects of sample geometry on interelement quantitation in laser microprobe mass spectrometry. In *Microbeam Analysis—1988*, ed. D. E. Newbury. San Francisco: San Francisco Press, pp. 356–366.

Myklebust, R. L., C. E. Fiori, and K. F. J. Heinrich. 1979. Frame C: A compact procedure for quantitative energy-dispersive electron probe x-ray analysis. *NBS Technical Note 1106*. Washington, DC: US Department of Commerce/National Institute of Standards and Technology, pp. 1–105.

Newbury, D. E. 1980. Secondary ion mass spectrometry for the analysis of single particles. In *Characterization of Particles*, ed. K. F. J. Heinrich. NBS Special Publication 553. Washington, DC: US Department of Commerce/National Institute of Standards and Technology, pp. 139–152.

Newbury, D. E. 1986. Electron beam-specimen interactions in the analytical electron microscope. In *Principles of Analytical Electron Microscopy*, eds. D. C. Joy, A. D. Romig, Jr., and J. Goldstein. New York: Plenum Press, pp. 1–27.

Newbury, D. E. 1990. Microanalysis to nanoanalysis: Measuring composition at high spatial resolution. *Nanotechnology* 1:103–130.

Newbury, D. E., D. C. Joy, P. Echlin, C. E. Fiori, and J. I. Goldstein. 1986. *Advanced Scanning Electron Microscopy and X-Ray Microanalysis.* New York: Plenum Press.

Newbury, D., D. Wollman, K. Irwin, G. Hilton, and J. Martinis. 1999. Lowering the limit of detection in high spatial resolution electron beam microanalysis with the microcalorimeter energy dispersive X-ray spectrometer. *Ultramicroscopy* 78:73–88.

NIOSH. 1977. *Criteria for a Recommended Standard: Occupational Exposure to Fibrous Glass.* DHEW (NIOSH) Publication No. 77-152. Washington, DC: U.S. Government Printing Office.

NIOSH. 1984. *NIOSH Manual of Analytical Methods*, Vol. 2. 3rd Ed. DHHS (NIOSH) Publication No. 84-100. Cincinnati: National Institute for Occupational Safety and Health.

NIST. 1992. *Certificate for Standard Reference Material 1876b*. Standard Reference Materials Program, National Institute of Standards and Technology, Gaithersburg, MD.

Odom, R. W., F. Radicati di Brozolo, P. B. Harrington, and K. J. Voorhees. 1989. LAMMS: Pattern recognition and cluster ions. *Microbeam Analysis—1989*, ed. P. E. Russell. San Francisco: San Francisco Press, pp. 283–285.

Ogden, T. L., T. Shenton-Taylor, J. W. Cherrie, N. P. Crawford, S. Moorcroft, M. J. Duggan, P. A. Jackson, and R. D. Treble. 1986. Within-laboratory quality control of asbestos counting. *Ann. Occup. Hyg.* 30:411–425.

Owari, M., H. Satoh, N. Hutigami, M. Kudo, and Y. Nihei. 1989. Secondary ion mass spectrometry using a liquid metal ion source. *J. Trace Microprobe Tech.* 7:59–85.

Philips Electron Optics Manual. 1996. *Environmental Scanning Electron Microscopy. An Introduction to ESEM.* Robert Johnson Associates, El Dorado, CA, p. 25.

Pickle, T., D. T. Allen, and S. E. Pratsinis. 1990. The source and size distributions of aliphatic and carbonyl carbon in Los Angeles aerosol. *Atmos. Environ.* 24A(8):2221–2228.

Platek, S. F., D. H. Groth, C. E. Ulrich, L. E. Stettler, M. S. Finnell, and M. Stoll. 1985. Chronic inhalation of short asbestos fibers. *Fund. Appl. Tox.* 5:327–340.

Reuter, W. 1972. The ionization function and its application to the electron probe analysis of thin films. In *Proceedings of the 6th International Conference on X-Ray Optics and Microanalysis*, eds. G. Shinoda, K. Kohra, and T. Ichinokawa. Tokyo: University of Tokyo Press, pp. 121–130.

Ro, C. U., I. Musselman, and R. Linton. 1989. Molecular speciation of micro-particle: Application of pattern-recognition techniques to laser microprobe mass spectrometry. In *Microbeam Analysis—1989*. ed. P. E. Russell. San Francisco: San Francisco Press, pp. 293–296.

Rosasco, G. J. 1980. Raman microprobe spectroscopy. In *Advances in Infrared and Raman Spectroscopy*, Vol. 7, eds. R. J. H. Clark and R. E. Hester. London: Heyden & Son Ltd., pp. 223–282.

Rosasco, G. J., E. S. Etz, and W. A. Cassatt. 1975. The analysis of discrete fine particles by Raman spectroscopy. *Appl. Spectrosc.* 29:396–404.

Savina, M. R. and K. R. Lykke 1997. Microscopic chemical imaging with laser desorption mass spectrometry. *Anal. Chem.* 69:3741–3746.

Schrader, B. 1986. Micro Raman, fluorescence and scattering spectroscopy of single particle. In *Physical and Chemical Characterization of Individual Airborne Particles*, ed. K. R. Spurny. New York: John Wiley & Sons, pp. 358–379.

Schroder, W. H. and G. L. Fain. 1984. Light-dependent calcium release from photoreceptors measured by laser micro-mass analysis. *Nature* 309(5965):268–270.

Seeley, B. K. 1952. Detection of micron and submicron chloride particles. *Anal. Chem.* 24(3):576–579.

Sheridan, P. and I. Musselman. 1985. Characterization of aircraft-collected particles present in arctic aerosol: Alaskan artic, spring 1983. *Atmos. Environ.* 19:2159–2166.

Simons, D. S. 1983. Isotopic analysis with the laser microprobe mass analyzer. *Int. J. Mass Spectrom.* 55:15–30.

Simons, D. S. 1984. Laser microprobe mass spectrometry. In *Secondary Ion Mass Spectrometry SIMS IV.* eds. A. Benninghoven, J. Okano, R. Shimizu, and H. W. Werner. Berlin: Springer-Verlag, p. 158.

Simons, D. S. 1986. Single particle standards for isotopic measurments of uranium by secondary ion mass spectrometry. *J. Trace Microprobe Techniques* 4:185–195.

Simons, D. S. 1988. Laser microprobe mass spectrometry: Description and selected applications. *Appl. Surf. Sci.* 31:103–117.

Simons, D. S. 1991. SIMS analysis of fused clay microspheres containing plutonium. *Proceedings of the Eighth International Conference on Secondary Ion Mass Spectrometry (SIMS VIII).* eds. A. Benninghoven, et al. New York: John Wiley & Sons, pp. 715–718.

Simons, D. S., G. Gillen, C. J. Zeissler, R. H. Fleming, and P. J. McNitt. 1997. Automated SIMS for determining isotopic distributions in particle populations. In *Proceedings of the Eleventh International Conference on Secondary Ion Mass Spectrometry (SIMS XI).* eds. G. Gillen, et al. New York: John Wiley & Sons.

Small, J. A. 1981. Particle analysis in electron beam instruments. In *Scanning Electron Microscopy*, Vol. I, pp. 447–461. O'hare, IL: SEM, Inc.

Small, J. A. 1987. Visibility of chrysotile asbestos in the scanning electron microscope. In *Proceedings of the Workshop on Abestos Fibre Measurements in Building Atmospheres.* Toronto, Canada, pp. 69–86.

Small, J. A. 1993. In *Handbook of X-Ray Spectrometry: Methods and Techniques*, eds. R. E. Van Grieken and A. A. Markowicz. New York: Marcel Dekker, pp. 621–634.

Small, J. A. and J. T. Armstrong. 1996. Quantitative particle analysis: Fact or fiction. In *Microscopy & Microanalysis*, eds. G. W. Bailey, J. M. Corbeth, R. Y. W. Dimlich, J. R. Michael, and N. J. Zaluzec. San Francisco: San Francisco Press Inc., pp. 496–497.

Small, J. A. and J. R. Michael. 1999. Phase identification of individual particles by electron backscatter diffraction (EBSD). *Microsc. Microanal.* 5(Suppl. 2):226–227.

Small, J. A. and J. R. Michael. 2001. Phase identification of individual crystalline particles by electron backscatter diffraction. *J. Microscopy* 201:59–69.

Small, J. A., K. F. J. Heinrich, C. E. Fiori, R. L. Myklebust, D. E. Newbury, and M. F. Dilmore. 1978. The production and characterization of glass fibers and spheres for microanalysis. In *Scanning Electron Microscopy*. Vol. I, pp. 445–454. O'hare, IL: SEM, Inc.

Small, J. A., K. F. J. Heinrich, D. E. Newbury, and R. L. Myklebust. 1979a. Progress in the development of the peak-to-background method for the quantitative analysis of single particles with the electron probe. In *Scanning Electron Microscopy*, Vol. 2, pp. 807–816.

Small, J. A., D. E. Newbury, and R. L. Myklebust. 1979b. Analysis of particles and rough samples by FRAME P, a ZAF method incorporating peak-to-background measurements. In *Microbeam Analysis—1979*. ed. D. E. Newbury. San Francisco: San Francisco Press, Inc., pp. 243–246.

Small, J. A., E. B. Steel, and P. J. Sheridan. 1985. Analytical standards for the analysis of chrysotile asbestos in ambient environments. *Anal. Chem.* 57:204–208.

Sobanska, S., N. R. Laboudigue, R. Guillermo, C. Bremard, J. Laureyns, J. C. Merlin, and J. P. Wignacourt. 1999. Microchemical investigations of dust emitted by a lead smelter. *Environ. Sci. Technol.* 33:1334–1339.

Spence, J. C. H. and J. M. Zuo. 1992. *Electron Microdiffraction*. New York: Plenum Press.

Spurny, K. R. 1986. *Physical and Chemical Characterization of Individual Airborne Particles*, ed. K. R. Spurny. New York: John Wiley & Sons.

Statham, P. J. and J. B. Pawley. 1978. A new method for particle X-ray micro-analysis based on peak to background measurements. *Scanning Electron Microscopy*, Vol. I, pp. 469–478. O'hare, IL: SEM, Inc.

Steel, E. B. 1980. Optical microscopy of particles. In *Characterization of Particles*, ed. K. F. J. Heinrich. NBS Special Publication 553. Washington, DC: US Department of Commerce/National Institute of Standards and Technology, pp. 5–11.

Steel, E. B., D. S. Simons, J. A. Small, and D. E. Newbury. 1984. Analysis of submicrometer particles by sequential AEM and LAMMA. In *Microbeam Analysis—1984*, eds. A. D. Romig, Jr. and J. I. Goldstein. San Francisco: San Francisco Press, pp. 27–29.

Steel, E. B. and J. A. Small. 1985. Accuracy of transmission electron microscopy for the analysis of asbestos in ambient environments. *Anal. Chem.* 57:209–213.

Steen, D., M. P. Guillemin, P. Buffat, and G. Litzistorf. 1983. Determination of asbestos fibres in air: Transmission electron microscopy as a reference method. *Atmos. Environ.* 17:2285–2297.

Stoffel(s), J. J., J. K. Briant, and D. S. Simons. 1994. A particle isotopic standard of uranium and plutonium in an aluminosilicate matrix. *J. Am. Soc. Mass Spectrom.* 5:852–858.

Storms, H. A., K. F. Brown, and J. D. Stein. 1977. Evaluation of cesium positive ion source for secondary ion mass spectrometry. *Anal. Chem.* 49:2023–2030.

Stranick, S. J., L. J. Rickter, and R. R. Cavanagh. 1998. High efficiency, dual collection mode near-field scanning optical microscope. *J. Vac. Sci. Technol. B* 16:1948–1952.

Surkyn, P., J. De Waele, and F. Adams. 1983. Laser microprobe mass analysis for source identification of air particulate matter. *Int. J. Environ. Anal. Chem.* 13:257–274.

Tamborini, G., M. Betti, V. Forcina, T. Hiernaut, B. Giovannone, and L. Koch. 1998. Application of secondary ion mass spectrometry of the identification of single particles of uranium and their isotopic measurement. *Spectrochim. Acta* Part B 53:1289–1302.

Taylor, B. N. and C. E. Kuyatt. 1994. *Guidelines for evaluating and expressing the uncertainty of NIST measurement results.* NIST Technical Note 1297. United States Department of Commerce, Technology Administration, National Institute of Standards and Technology.

Taylor, D. G., P. A. Baron, S. A. Shulman, and J. W. Carter. 1984. Identification and counting of asbestos fibers. *Am. Ind. Hyg. Assoc. J.* 45:84–88.

Tsugoshi, T., H. Chiba, T. Yokoyama, K. Furuya, and T. Kikuchi. 1998. Semi-quantitative analysis of heterogeneous samples using laser microprobe mass spectrometry, and application to environmental monitoring. *J. Trace Microprobe Techniques* 16:47–57.

Turner, S., S. S. Doorn, J. M. Phelps, E. S. Windsor, E. B. Steel and M. E. Beard. 1991. *Proceedings of the 49th Annual Meeting of the Electron Microscopy Society of America.* San Francisco: San Francisco Press, Inc., pp. 974–975.

Turner, S., S. S. Doorn, E. B. Steel, J. M. Phelps, E. S. Windsor, and K. K. Starner. 1994a. *Proficency Test for NIST Airborne Asbestos Program—1990.* NISTIR 5431. U.S. Department of Commerce, Technology Administration, National Institute of Standards and Technology.

Turner, S. and E. B. Steel. 1991. Accuracy of transmission electron microscopy analysis of asbestos on filters: Interlaboratory study. *Anal. Chem.* 63:868–872.

Turner, S. and E. B. Steel. 1994. *Airborne Asbestos Methods: Standard Test Methods for Verified Analysis of Asbestos by Transmission Electron Microscopy—Version. 2.0.* NISTIR 5351. U.S. Department of Commerce, Technology Administration, National Institute of Standards and Technology.

Turner, S., E. Steel, and D. Alderman. 1995a. *Airborne Asbestos Analysis.* NIST Handbook 150-13, U.S. Department of Commerce, Technology Administration, National Institute of Standards and Technology, C-4, C-21.

Turner, S., E. B. Steel, O. S. Crankshaw, S. Silberstein, and H. M. Richmond. 1995b. *Proficency Test for NIST Airborne Asbestos Program—1993.* NISTIR 5680. U.S. Department of Commerce, Technology Administration, National Institute of Standards and Technology.

Turner, S., E. B. Steel, S. S. Doorn, and S. B. Burris. 1994b. *Proficency Test for NIST Airborne Asbestos Program—1991.* NISTIR 5432. U.S. Department of Commerce, Technology Administration, National Institute of Standards and Technology.

Turner, S., E. B. Steel, S. S. Doorn, and S. B. Burris. 1994c. *Proficency Test for NIST Airborne Asbestos Program—1992.* NISTIR 5433. U.S. Department of Commerce, Technology Administration, National Institute of Standards and Technology.

Turner, S., E. B. Steel, J. M. Phelps, E. S. Windsor, S. S. Doorn, and S. D. Leigh. 1992. *Proceedings of the 50th Annual Meeting of the Electron Microscopy Society of America.* San Francisco: San Francisco Press, Inc., pp. 1706–1707.

Van Vaeck, L., J. Bennett, W. Lauwers, A. Vertes, and R. Gijbels. 1990. Laser microprobe mass spectrometry: Possibilities and limitations. *Mikrochim. Acta* III:283–303.

Van Vaeck, L., H. Struyf, W. Van Roy, and F. Adams. 1994a. Organic and inorganic analysis with laser microprobe mass spectrometry. Part I: Instrumentation and methodology. *Mass Spectrom. Rev.* 13:189–208.

Van Vaeck, L., H. Struyf, W. Van Roy, and F. Adams. 1994b. Organic and inorganic analysis with laser microprobe mass spectrometry. Part II: Applications. *Mass Spectrom. Rev.* 13:209–232.

Venables, J. A. and C. J. Harland. 1973. Electron backscattering patterns. A new technique for obtaining crystallographic information in the scanning electron microscope. *Philos. Mag.* 27:1193–1200.

Vertes, A., R. Gijbels, and F. Adams. 1993. *Laser Ionization Mass Analysis.* New York: John Wiley & Sons, Inc.

Villarrubia, J. S., R. Dixson, S. Jones, J. R. Lowney, M. T. Postek, R. A. Allen, and M. W. Cresswell. 1999. Intercomparison of SEM, AFM and electrical linewidths. *SPIE Conference on Metrology, Inspection and Process Control for Microlithography* XIII. SPIE 3677, pp. 587–598.

White, E. W., P. J. Denny, and S. M. Irving. 1966. *The Electron Microprobe*, eds. T. D. McKinley, K. F. J. Heinrich, and D. B. Wittry. New York: John Wiley & Sons, pp. 791–804.

Wieser, P. and R. Wurster. 1986. Application of laser-microprobe mass analysis to particle collections. In *Physical and Chemical Characterization of Individual Airborne Particles*, ed. K. R. Spurny. New York: John Wiley & Sons, pp. 251–270.

Wieser, P., R. Wurster, and U. Haas. 1981. Applications of LAMMA in aerosol research. *Fresenius Z. Anal. Chem.* 308:260–269.

Wieser, P., R. Wurster, and H. Seiler. 1980. Identification of airborne particles by laser induced mass spectroscopy. *Atmos. Environ.* 14:485–494.

Williams, D. B. and C. B. Carter. 1996. *Transmission Electron Microscopy: A Textbook for Materials Science.* New York: Plenum Press.

Wright, F. W., P. W. Hodge, and C. G. Langway. 1963. Studies of particles for extraterrestrial origin: 1. Chemical analysis of 118 particles. *J. Geophys. Res.* 68:5575–5587.

Yokozawa, H., T. Kikuchi, K. Furuya, S. Ando, and K. Hoshino. 1987. Characterization of coal fly-ash particles by laser microprobe mass spectrometry. *Anal. Chim. Acta.* 195:73–80.

Zerda, T. W., W. Xu, H. Yang, and M. Gerspacher. 1998. The effects of heating and cooling rates on the structure of carbon black. *Rubber Chem. Technol.* 71:26–37.

13 Real-Time Single-Particle Analysis

ANTHONY S. WEXLER
Department of Mechanical and Aeronautical Engineering, University of California, Davis, CA

MURRAY V. JOHNSTON
Department of Chemistry and Biochemistry, University of Delaware, Newark, DE

INTRODUCTION

The ideal aerosol instrument has been defined as one that counts each particle and reports its size, chemical composition (Friedlander, 1971; Flagan, 1993), and morphology rapidly. Chapter 12 discusses off-line single-particle analysis techniques. Off-line techniques involve deposition of particles on a substrate and later analysis of these particles. A new class of instruments is now available that sizes, counts, and analyzes individual particles on line and provides chemical analysis with high time resolution. Many of these instruments are in use or under construction around the world, and a commercial version is now available (Model 3800, ***TSI***).*

As discussed elsewhere in this text, a wide range of analytical techniques can be applied to particles, but a much narrower range can be applied to single-particle analysis. The more conventional techniques involve collecting a sample of many particles, size-selected or not, and then applying conventional chemical analysis to these samples. If the particles have a range of compositions, the mixing of the particles together obscures their original chemical distribution. Two somewhat confusing terms are used to describe aerosol mixing properties. A particle distribution is termed *externally mixed* if each particle is composed of only one chemical compound. A distribution is termed *internally mixed* if all the particles have the same composition, which is often a mixture of many compounds. It is a rare particle distribution that fits either of these extremes, and some confusion resides in the literature from loose application of these terms. The reality is often more complicated, especially in atmospheric situations. Aerosol particles often contain a number of chemical compounds, but within the aerosol distribution are groups of particles with similar composition mixes. Deciphering this mix is one of the strengths of single-particle analysis.

The field of on-line single-particle analysis is relatively young because as recently as 10 to 15 years ago a number of the necessary technologies were not yet available. Analog to digital converters and associated desktop computers have become significantly faster and cheaper, making time-of-flight mass spectrometry a reasonable option, as will be explained in more

* See Appendix I for full manufacturer addresses referenced to the italicized three-letter codes.

Aerosol Measurement: Principles, Techniques, and Applications, Second Edition, Edited by Paul A. Baron and Klaus Willeke.
ISBN 0-471-35636-0

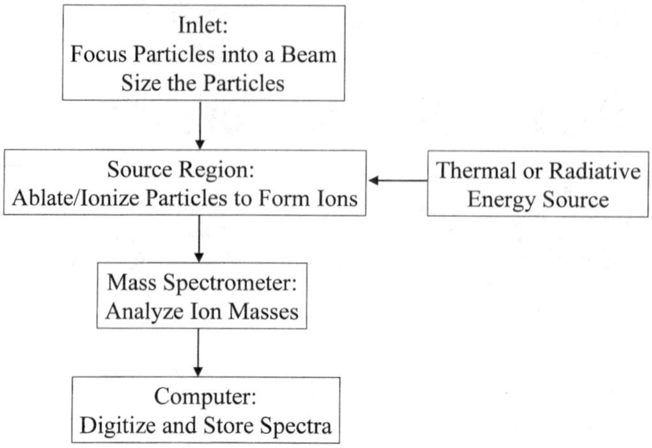

Fig. 13–1. Block diagram of the typical components in on-line single-particle analysis instruments.

detail below. Likewise, compact high-flux ultraviolet lasers are now available due to improved material quality in mirrors, lenses, crystals, and electrodes.

Most current on-line single-particle analyzers share a number of components (Fig. 13–1). First, the aerosol passes through an inlet where the gas is removed and a particle beam is formed. Second, the particles pass through one or more continuous laser beams where they are detected and sized by the light that they scatter. Third, one or more high-energy, pulsed lasers are fired in synchrony with a particle's arrival to ablate and ionize chemical constituents. Fourth, the ions are mass analyzed and the spectrum is stored. Several reviews of real-time single-particle mass spectrometers are available in the literature (Thomson and Murphy, 1994; Johnston and Wexler, 1995; Peter, 1996; Noble and Prather, 2000; Johnston, 2000). A special double issue on on-line single-particle analysis was published in *Aerosol Science and Technology* in July 2000 (Vol. 33[1–2]).

Particle Beam Formation

Inlets to on-line single-particle analyzers serve two purposes: They form a particle beam and drop the pressure from near-atmospheric to the nPa pressures required for operation of a mass spectrometer—a pressure drop of about eight orders of magnitude. The goal of a well-designed inlet is transmitting as many particles to the source region as possible, and the way to accomplish this is to process as much gas as possible. The amount of gas that can practically be processed is constrained by vacuum-pumping costs—the greater the gas load, the greater the energy consumption, and the larger and heavier the pumps that are required.

Most inlets are composed of a number of pressure reduction stages, as illustrated in Figure 13–2. Initially there is a primary orifice or capillary that restricts the flow, followed by one or usually more skimmer stages that reduce the pressure. Vacuum pumps remove gas and reduce the pressure between the initial orifice and the first skimmer and then between each subsequent skimmer. The orifices in each stage should be as small as possible to permit the minimum gas flow to the next stage. The orifice sizes cannot be too small, however, or they will block the passage of particles to the source region of the mass spectrometer. Also, the orifices must be precisely aligned to permit passage of the particle beam—the smaller the orifice, the more difficult this alignment becomes.

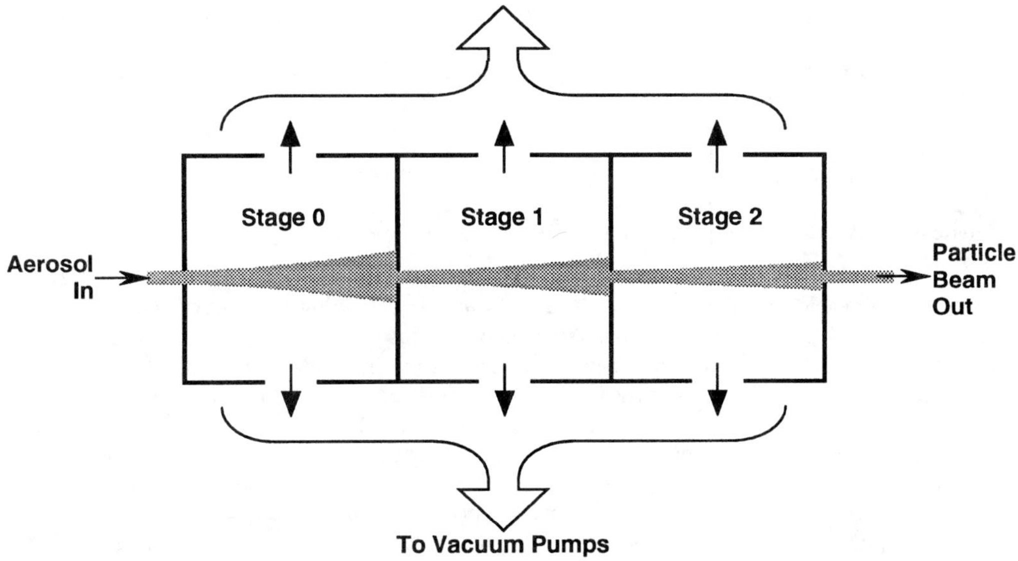

Fig. 13–2. Typical inlet configuration.

Fluid Mechanical Considerations in Inlet Design

Effective inlets minimize pumping costs while maximizing the particle transmission rate. The pumping costs are minimized by efficiently removing gas and reducing pressure at each skimmer stage. The transmission rate is maximized by directing particles into a beam that has low divergence and is fully contained within the ionization volume of the mass spectrometer (see "Particle Vaporization and Ionization," below). As mentioned earlier, smaller skimmer orifices permit less gas transmission to subsequent stages but also constrain the particle beam size and make alignment more difficult.

Let us now use these fluid mechanics principles to estimate the pressure in a skimmer stage. We will use subscript "0" to denote the previous stage, "1" to denote the stage we are analyzing, and "2" to denote the next stage (see Fig. 13–2). If we assume that the pressure drop from one stage to the next is large, two simplifications arise. First, the flow at the incoming orifice is choked so that the velocity is close to the speed of sound, U_{sonic}. In rapid contractions such as these, the gas undergoes a near-isentropic expansion resulting in a temperature and pressure at the orifice of $0.83T$ and $0.53p$, where T and p are the temperature and pressure upstream, respectively. Therefore, the mass flow rate into stage 1 is $A_0\rho(0.53p_0, 0.83T_0)U_{sonic}(0.83T_0)$, where A_0 is the area of the orifice and ρ is the density at a pressure of $0.53p_0$ and a temperature of $0.83T_0$. This flow enters stage 1 and expands to a much larger volume due to the substantially lower pressure.

Most of the air entering stage 1 leaves through the vacuum pumps drawing on this stage, which leads to the second assumption. In a well-designed inlet, the aerosol volume transmitted to the next stage should be negligible compared with that being pumped away. Let us assume that the pump draws a constant volume flow rate, V_{pump}, that the pressure change in the tubing between the pump and the stage is negligible, and that due to heat transfer through transfer tubing the air is at the ambient temperature, near 300 K, at the pump. Then the mass flow rate drawn by the pump is $\rho(p_1, T_{amb})V_{pump}$. Equating these two to conserve mass and employing the ideal gas equation of state gives an expression for the pressure ratio between stage 0 and 1:

$$p_1/p_0 = (0.53/0.83)(T_{amb}/T_0)[A_0 U_{sonic}(0.83T_0)/V_{pump}] \tag{13–1}$$

We desire a large pressure drop from stage 0 to 1, but the first two terms are on the order of 1 so do not help much. The last term is the ratio of the volume flow rate into stage 1 divided by the volume flow rate withdrawn. A small orifice and a large pump yield a large pressure decrease.

Particle Beam Transmission Considerations in Inlet Design

It is difficult to efficiently form and transmit a particle beam through inlets with large pressure drops. Particle transmission goals include minimizing clogging and accurately sizing particles, both of which must be accomplished while removing the carrier gas. Some inlet designs do not place much emphasis on aerodynamic particle sizing. This makes the design easier because there is one less consideration, but then less accurate sizing techniques must be employed, such as light-scattering intensity (see Chapter 15). The light scattered from particles is a function of the wavelength of the light, the size of the particle, and its shape and composition, but is limited to particles greater than about 200 nm. With conventional light intensity measurements, the morphology and composition confound accurate sizing. In single-particle analysis, the composition is obtained, so it may be possible to perform accurate sizing solely with light scattering if composition information is employed too. Currently, scattered light sizing is employed in some instruments to obtain an indication of particle size (e.g., Murphy and Thomson, 1995).

Aerodynamic sizing is used for more accurate particle sizing but imposes constraints on the inlet design. Two approaches have been taken to aerodynamic sizing in inlets: (1) transmitting a wide particle size range and imparting a size-dependent velocity to the particles and (2) transmitting a narrow particle size range that can be tuned so that a known particle size is transmitted.

Transmitting a Wide Particle Size Range—Capillaries and Aerodynamic Lens Arrays. Two techniques are used for transmitting a wide particle size range—capillaries and aerodynamic lens arrays. Capillaries are long, thin tubes that are effective at forming particle beams. Aerodynamic lenses are contractions in the flow that focus a range of particle sizes. An array of these lenses in series is able to focus a wide particle size range (Liu et al., 1995a,b). Capillaries were used in some of the earlier inlets but are becoming less common because they have two disadvantages relative to aerodynamic lens arrays—they clog, and they impart similar velocities to the particles transmitted. Let us examine these issues more carefully.

Consider a capillary inlet. Longer capillaries impose more friction on the flow, resulting in a lower volume flow rate through the inlet. For instance, the volume flow rate entering a capillary 160 μm in diameter and 12 mm long is about 100 ml/min, giving a velocity of about 80 m/s, whereas the velocity through a sharp orifice is sonic—340 m/s (Mallina et al., 1999). The lower velocity yields a lower volume flow rate into subsequent stages and consequently lower pumping costs and a lower hit rate. There are two Stokes numbers that are important to characterizing the performance of the capillary. The Stokes number at the entrance to the capillary determines the particle sizes that are deposited on the capillary walls, while the Stokes number based on the capillary length determines the capillary's aerodynamic sizing capabilities.

Table 13–1 contains the particle stokes numbers for a range of particle sizes. If we assume that the inlet is much wider than 160 μm before the capillary, then the gas velocity changes from near zero to 80 m/s. Likewise, the flow accelerates from a low value to 80 m/s over about the diameter of the capillary because the velocity is much lower just a few capillary diameters away from the capillary entrance. Using $Stk_{radial} = S/d = U\tau_p/d$, where

TABLE 13–1. Particle Radial and Axial Stokes Numbers for a Typical Capillary

d_p (µm)	Stk_{radial}	Stk_{axial}
0.1	0.05	0.002
1	1.8	0.076
10	120	5.0

$U = 80$ m/s and $d = 160$ µm, gives the Stokes number for radial acceleration as a function of diameter.

Stokes numbers much larger than one indicate that the particles travel in straight lines and are not focused into beams. As a result, particles larger than about 1 µm do not make it through the capillary and, more importantly, many hit the walls of the capillary, resulting in clogging (see Table 13–1).

The capillary entrance also focuses particles that have a Stokes number near 1. Simulations of this capillary show that particle diameters about 0.75 µm are optimally focused to a beam (Mallina et al., 1999). Particles with a Stokes number much larger deposit on the capillary walls. Particles with a Stokes number somewhat smaller are focused less well, while particles much smaller follow the fluid streamlines. At the exit of the capillary, the pressure drops dramatically, leading to much larger mean free paths for the gas, and the particles that are transmitted exit in straight-line trajectories. In fact, this is how capillaries are able to transmit a wide particle size range. Particles with a Stokes number near and below 1 at the entrance conditions flow down the capillary. Because it has a small diameter, the pressure drop along the capillary is significant. The lower exit pressure means less drag on the particles due to the pressure dependence in the Cunningham correction factor. Thus particles with a Stokes number greater than 1, evaluated at the capillary exit conditions, maintain a linear trajectory whereas particles whose exit Stokes numbers are less than 1 act like the gas and diverge.

The difference between the entrance and exit conditions of the capillary dictate which particle size range is focused into a beam. As particles traverse the capillary, they are also accelerated axially down the capillary, but now the flow accelerates over the whole length of the capillary, not just near the entrance, so the length scale is much larger than the radial acceleration at the entrance. Stokes numbers for axial acceleration in this capillary are given in Table 13–1. The goal of many inlets is to impart an aerodynamic diameter-dependent velocity to the particle so that the particle size can be inferred from this velocity. In capillaries, the particle accelerates over the length of the capillary, a relatively long time. Particles whose axial Stokes number is significantly less than 1 are accelerated to the speed of the gas. Particles larger than this limit have a size-dependent velocity. Note that very short capillaries, that is, orifices, are the best for imparting a size-dependent velocity over the largest particle size range, which is why this configuration is used in aerodynamic particle sizing instruments. This issue will be alluded to in the next section. As a result of clogging problems and limitations on aerodynamic sizing, capillaries are being used less frequently in favor of aerodynamic lens arrays.

Aerodynamic lenses consist of holes in plates and in many ways can be thought of as short capillaries. They focus particles with a Stokes number near 1. Lenses have the additional advantages that (1) they are relatively sharp compared with capillaries so that little deposition may occur, (2) they are usually relatively large so that deposition does not lead to significant clogging, and (3) their short length axially accelerates particles to different velocities.

Single lenses only focus a narrow range of particle sizes, so transmitting a wide range can be accomplished with a series of matched lenses, as illustrated in Figure 13–3. The lenses must

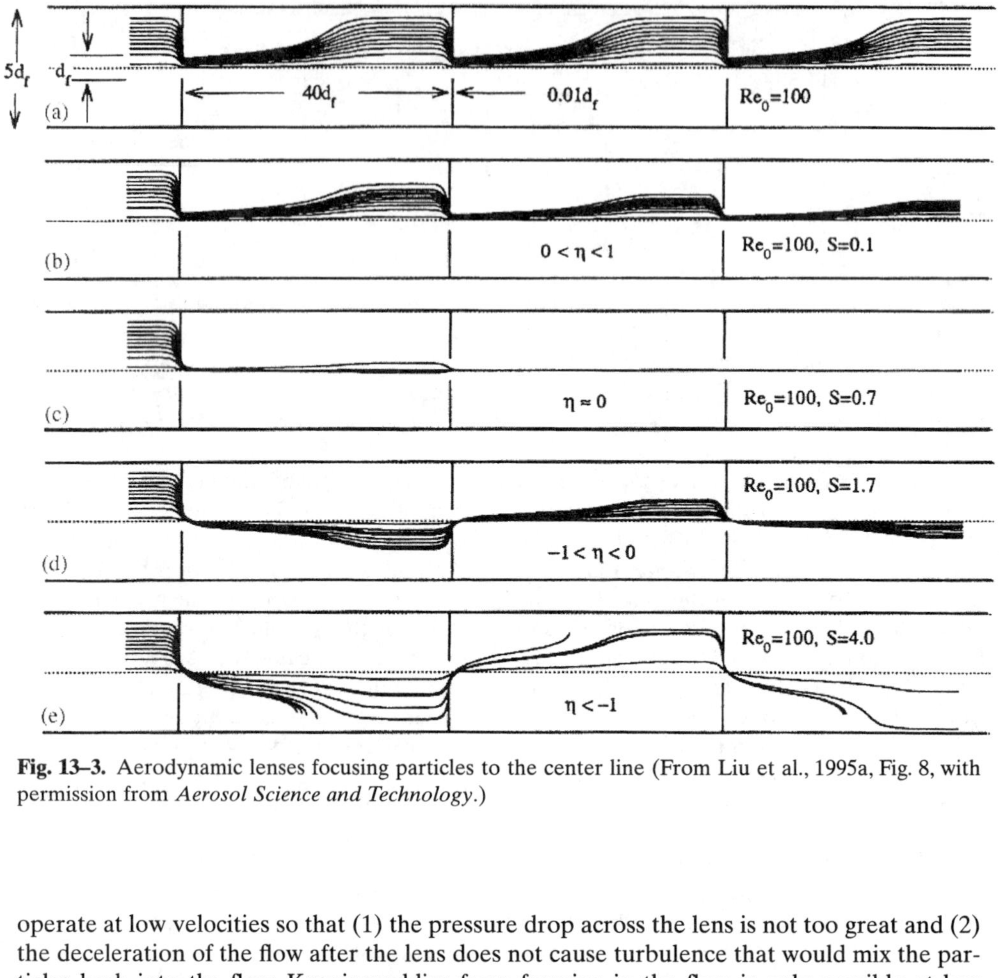

Fig. 13–3. Aerodynamic lenses focusing particles to the center line (From Liu et al., 1995a, Fig. 8, with permission from *Aerosol Science and Technology*.)

operate at low velocities so that (1) the pressure drop across the lens is not too great and (2) the deceleration of the flow after the lens does not cause turbulence that would mix the particles back into the flow. Keeping eddies from forming in the flow is only possible at low Reynolds numbers, but this requirement is in conflict with particle focusing because large velocities are needed to accelerate the particles to the center line. The solution to both problems lies in the pressure. The Reynolds number, $Re = Ud\rho/\mu$, is proportional to pressure, so low pressure gives a lower Reynolds number for the same velocity and orifice diameter. The viscosity, μ, is nearly independent of pressure. Lower pressure also enables smaller particles to be focused at the same velocity and orifice diameter because the Cunningham correction factor in the Stokes number is larger. That is, the lower pressure reduces the drag on the particles because the noncontinuum effect becomes more important. In the aerodynamic lens array of Liu and coworkers (1995a,b), a pressure of about 0.26 kPa (2 torr) was employed, which at standard temperature gives an air mean free path of about 20 μm.

By using a series of well-matched lenses, a wide range of particle sizes can be focused to the center line. The pressure, velocity, and size of the first lens focuses particles of a given size, say, d_{p1}, close to the center line, and a range of particles around this size are also moved closer to the center line. The next lens is somewhat smaller, focusing smaller particles, say, $d_{p2} < d_{p1}$. The larger particles that were brought very close to the center line by the first lens stay there. After a series of such lenses, carefully matched to each other, a relatively wide range of particle sizes can be focused to the center line (see, e.g., lens configuration *e* in Table 1 in Liu et al., 1995b).

Once the particles are focused to the center line, they pass through a choked orifice and the skimmer stages are traversed. Because the particles are close to the center line, they have a narrow divergence as they pass through the skimmers, but have a range of velocities due to their acceleration through the first choked orifice. Thus the particles have a range of velocities that can then be used to assess their size. This is discussed in more detail below.

Transmitting a Narrow Particle Size Range—Sharp Orifices. Whereas capillaries and lens arrays can be configured to transmit a range of particle sizes, single sharp orifices can be used to select only a narrow particle size range. As we have already discussed, passing an aerosol through a sharp orifice where the flow is choked focuses certain particles (Dahneke et al., 1982; Fernandez de la Mora and Riesco-Chueca, 1988; and references therein). The focused particles have a Stokes number around 1, the exact value depending on the nozzle geometry and the distance from the nozzle to the focal point. This principle can be employed to focus only a narrow range of particles, which has a few advantages and disadvantages relative to nozzles that transmit a wide size range. The primary disadvantage is that the instrument must then be operated in a scanning mode. That is, the operating conditions must be adjusted to focus particles of different sizes, one size at a time. This limits the particle sampling rate because only a narrow particle size range has been selected by the inlet. The primary advantage is that it enables particles to be sized that are quite small—10 nm or smaller.

The orifice focuses only a narrow range of particle sizes to the source region of the mass spectrometer, but this size range can be selected by adjusting the pressure upstream of the nozzle. Let us term the Stokes number that is focused Stk_f and approximate the Cunningham correction factor, $C_c = 1 + 1.66(2\lambda/d_p)$, to give

$$d_p = \left[(1.66\lambda)^2 + d_{p,max}^2\right]^{1/2} - 1.66\lambda \qquad (13\text{–}2)$$

where d_p is the particle diameter that is focused, $d_{p,max} = (18\mu d_n Stk_f/\rho_p U_{sonic})^{1/2}$ is the maximum particle size that the orifice can focus, λ is the gas mean free path, ρ_p is the particle density, the velocity through the orifice, U_{sonic}, is sonic since the flow is choked, μ is the viscosity of air, and d_n is the orifice diameter. The mean free path and the viscosity must be evaluated at the pressure and temperature in the orifice, which are $0.53p$ and $0.83T$, where p and T are the upstream pressure and temperature, respectively.

Therefore, the size that is focused can be adjusted by altering $d_{p,max}$ or λ. Most of the parameters that comprise $d_{p,max}$ are not adjustable, such as the viscosity and speed of sound in air and the particle density. The focused Stokes number, Stk_f, can be adjusted somewhat by changing the geometry of the orifice and the distance of the focus from the orifice, but substantial departures from $Stk_f = 1$ are difficult to achieve. The orifice diameter can be changed because a smaller orifice leads to a smaller $d_{p,max}$ and consequently a smaller d_p, but a smaller d_n also means that the volume of air brought through the orifice is lower. $d_{p,max}$ changes linearly with the square root of d_n, but the volume flow rate varies with its square. Thus, reducing the orifice diameter leads to sampling of smaller particles but at a much reduced volume flow rate. The only remaining parameter than can be adjusted is the gas mean free path, which may be varied over many orders of magnitude by suitably controlling the upstream pressure, p (see Chapter 4 for more on this dependence).

PARTICLE DETECTION

Particles transmitted through the inlet enter a mass spectrometer where they undergo chemical analysis. Because it is not known ahead of time when a particle will arrive, a particle detection step is normally required to achieve a high analysis rate. In principle, particle detection could be performed with a variety of techniques, which are discussed in Chapters 15 to 20. Light scattering is the most common method. A continuous laser beam intercepts the particle beam emerging from the inlet. The scattered radiation from a single particle indicates the arrival of that particle in the mass spectrometer. Particle detection by an independent method such as light scattering has two important advantages. First, the arrival of a particle can be synchronized with the chemical analysis sequence. Without synchronization, many or most particles may pass through the system without being analyzed. Second, the detection step provides an opportunity to size the particle before analysis (see "Particle Sizing," below). The main disadvantage of light scattering is that particles smaller than the wavelength of light are difficult to detect. For this reason, instruments that incorporate particle detection by light scattering (with a visible laser beam) are generally limited to particles greater than about 200 nm in diameter.

A fundamentally different detection strategy is to simply use the chemical analysis step as a means of particle detection. When a particle is vaporized and ionized, a burst of ions is produced over a short period of time. If the ion signal rises above a background or threshold level, the particle is detected because its mass spectrum is detected. The spectrum detection approach has the advantage that individual particles down to 10 nm diameter, and possibly smaller, can be analyzed (Reents et al., 1995; Carson et al., 1997a; Ge et al., 1998). However, this approach has the potential disadvantage that the probability of detecting very small particles may be composition dependent. That is, particles containing material that is easy to ionize may give larger ion signals and be more easily detected than particles containing material that is difficult to ionize (Kane and Johnston, 2000). In a particle stream containing a range of particle compositions, caution must be used when interpreting the number of particles detected by this approach. Depending on the ionization method used, this approach may also have the disadvantage of poor synchronization between the arrival of a particle and initiation of the chemical analysis sequence. For ionization methods that operate in a continuous mode (see "Particle Vaporization and Ionization," below), synchronization is not required. However, for pulsed ionization methods such as laser ablation, the laser is "on" for only a short fraction of the time. If there is no synchronization between the arrival of a particle and firing of the laser, most particles pass through the mass spectrometer while the laser is "off" and no spectrum is detected. In principle, it should be possible to overcome the synchronization problem by trapping one or more particles temporarily in an electrodynamic trap and injecting them into the mass spectrometer when the ablation laser fires (Frankevich et al., 1998).

PARTICLE SIZING

As with particle detection, sizing is often performed using similar techniques employed in other instruments. Particle sizing by the amplitude of the light scattered is discussed in Chapter 15. The amount of light scattered by a particle may depend on its shape and composition. In conventional instruments, this information is not available, which leads to substantial uncertainty in the sizing. In single-particle analysis instruments, the composition is available, and from the composition the morphology can often be estimated, so it may be possible to use the amplitude of the scattered light along with the composition to obtain a more accurate assessment of size. Such approaches have not yet been attempted.

Aerodynamic sizing instruments are discussed in Chapters 16 and 17. Some single-particle analysis instruments accelerate the particle in the inlet to velocities that are aerodynamic diameter dependent. Detection and sizing are done using one or two lasers. A number of design trade-offs must be chosen to employ this technique in single-particle analysis. If the particles are not accelerated rapidly, they may have similar velocities, so the size may be difficult to distinguish. Small differences in velocity can be ascertained by using two detection lasers spaced substantial distances apart from each other and then accurately timing the arrival of the two scatter signals. Unfortunately, particle beams invariably diverge so that a significant fraction of the detected particles may not be analyzed. The ***TSI*** Aerodynamic Particle Sizer and Aerosizer use a rapid acceleration to impart a wide range of velocities to the particles and then are able to position the two detection lasers close together. These instruments do not, however, need to form a well-focused particle beam because the detection occurs very close to the nozzle, something not possible in the inlets discussed here because of the intervening skimmer stages.

Sizing by selectively aerodynamically focusing particles of a known size to the source region of the mass spectrometer was discussed in "Introduction" under "Particle Beam Transmission Considerations in Inlet Design" (Mallina et al., 2000). Aerodynamic focusing is able to size particles over a wide size range, but is limited by the ability to detect, usually by spectrum detection, the particles that are focused. Aerodynamic particle selection also forces the instrument to be operated in a scanning mode whereby particle sizes must be analyzed in sequence.

Another sizing technique employed in single-particle analysis is particle beam chopping (Jayne et al., 2000). Here particles are aerodynamically accelerated to velocities that are size dependent, and a beam is formed. Then the particles pass through a chopper that only transmits a narrow packet of particles in time. The particles are then analyzed with a continuous ionization method, and the time delay between chopping of the beam and the detection of ions is used to infer the particle velocity and concomitant size. Similarly, two choppers could be used to transmit only particles of a given velocity for subsequent analysis.

PARTICLE VAPORIZATION AND IONIZATION

Chemical analysis is performed by vaporizing a particle and ionizing the atomic and molecular products. The vaporization and ionization processes can be performed simultaneously in a single step or sequentially in two steps, and in a continuous or pulsed manner.

Continuous Ionization Methods

Continuous methods involve directing the particle beam onto a heated surface (filament) where atomic and molecular species are vaporized. For single-particle work, the filament is heated to a high enough temperature to flash vaporize the particle in a short period of time (Stoffels and Allen, 1986). If the filament is set to a temperature that is too low, vaporization occurs over a long time scale, causing the signals from multiple particles to overlap. Ionization can be performed by either surface ionization or electron ionization. In the surface ionization mode, ions produced directly from particle vaporization at the filament are detected. The ionization efficiency is determined by the difference between the work function of the filament surface and the ionization energy of the atomic or molecular species vaporized from the particle. Typically, only atomic and molecular species having ionization energies less than about 8 eV are efficiently ionized. For this reason, most applications of surface ionization involve the detection of alkali metals in particles.

Alternatively, neutral species vaporized from a particle can be ionized with an electron beam. Electron ionization is more broadly applicable because most chemical components can be ionized with reasonable efficiency. For example, ammonium salts give ions such as NH_2^+ and NH_3^+; sulfate salts give ions such as S^+, SO^+, and so forth. Organic molecules give positively charged molecular ions (the original molecule minus an electron) and fragment ions. The molecular ion gives a direct measure of the molecular mass, while fragment ions indicate molecular structure. When many chemical components are present in the same particle, the mass spectra of the individual components overlap, making interpretation of the spectrum of the entire particle difficult. In these cases, it may be possible to replace electron ionization with a soft ionization method such as chemical ionization or photoionization. The soft ionization methods induce less fragmentation of the molecular ions, making the spectrum easier to interpret.

An advantage of the thermal vaporization electron ionization approach is that individual particles can be completely vaporized and analyzed. Thus, the absolute ion signal intensity can give a measure of particle mass, and both surface and interior chemical components can be detected. Surface ionization and thermal desorption electron ionization produce strong ion signals from materials that can be vaporized upon heating to several hundred degrees. Refractory compounds and other nonvolatile materials are vaporized only slightly or not at all and therefore give only weak ion signals. Extensive reviews of continuous ionization methods for particle analysis (Stoffels and Allen, 1986) and organic molecular characterization by electron ionization (McLafferty and Turecek, 1993) are available.

One-Step Laser Ablation and Ionization

Because the arrival of each particle in the mass spectrometer is a discrete event, pulsed ionization methods can be used to create much larger ion signals from individual particles than continuous methods. However, the pulsed methods must be synchronized to the arrival of a particle to achieve a high efficiency for particle analysis. Pulsed ionization can be performed in a single step where a single laser beam ablates the particle and ionizes vaporized material or in two steps where one laser beam ablates the particle and a second laser beam ionizes the vaporized material. Wavelengths from the infrared to the vacuum ultraviolet have been investigated for single-step ablation and ionization (Thomson and Murphy, 1993; Thomson et al., 1997). The threshold irradiance for ion production strongly depends on laser wavelength. With infrared radiation, the threshold irradiance is prohibitively high for analysis of individual micrometer and submicrometer sized particles. With ultraviolet radiation, intense ion signals can be obtained by ablating a single particle with a single laser pulse. The threshold irradiance for ion production depends on chemical composition. For example, aqueous particles generally have much higher threshold irradiances than dry particles (Neubauer et al., 1997, 1998).

The ions produced by one-step laser ablation and ionization correlate with chemical composition in a similar manner to those in a laser microprobe mass spectrometer (see Chapter 12; see also Kaufmann, 1986; Wieser and Wurster, 1986). Laser ablation is a robust ionization method. Strong ion signals can be obtained from most types of materials: refractory and semivolatile, organic and inorganic. The main difference between off-line and on-line laser ablation is that on-line analysis is able to detect semivolatile components such as particulate phase water (Neubauer et al., 1997) and methanesulfonic acid (Neubauer et al., 1996). In a laser microprobe experiment, semivolatile components evaporate from particles as they are mounted on a substrate and inserted into the mass spectrometer. In real-time mass spectrometry, particles experience a rapidly changing environment in both temperature and pressure as they pass through the inlet. The net effect is usually condensation induced by the gas expansion from the inlet (Mallina et al., 1997). Although re-evaporation could occur in the vacuum downstream of the inlet, the temperature decrease induced by the expansion and the short transit time to analysis inhibit evaporative losses.

Several decades of laser microprobe research have provided a general understanding of the characteristics of laser ablation mass spectra for real-time analysis. First, positive ion spectra primarily give information on cationic materials and organic compounds, while negative ion spectra primarily give information on anionic materials. Second, metals are generally detected with high sensitivity, while nonmetals are generally detected with low sensitivity. Third, absolute ion signal intensities are irreproducible and vary greatly from pulse to pulse.

TABLE 13–2. Common Ions in Laser Ablation Mass Spectra

	Positive Ion Spectrum (m/z)	Negative Ion Spectrum (m/z)
Elemental carbon	C^+ (12), C_2^+ (24), C_3^+ (36), C_4^+ (48), etc.	C^- (12), C_2^- (24), C_3^+ (36), C_4^- (48), etc.
Organic carbon	CH^+ (13), C_2H^+ (25), $C_2H_2^+$ (26), $C_2H_3^+$ (27), C_3H^+ (37), etc.	
Oxygenated carbon	$C_2H_3O^+$ (43)[a]	
NH_4^+	NH_3^+ (17), NH_4^+ (18)[a]	
H_2O	H_3O^+ (19)[b]	O^- (16), OH^- (17)
NaCl	Na^+ (23), Na_2Cl^+ (81, 83)	Cl^- (35, 37)
NO_3^-	NO^+ (30)	NO_2^- (46), NO_3^- (62)
SO_4^{-2}	S^+ (32), SO^+ (48)[b]	SO_2^- (64), SO_3^- (80), SO_4^- (96), HSO_4^- (97)
Metal oxides (Ca, Fe, etc.)	Fe^+ (56), FeO^+ (72), Ca^+ (40), CaO^+ (56), etc.	

[a] Low intensity.
[b] Very low intensity.

Finally, chemical components located at or near the particle surface tend to exhibit enhanced intensities in laser ablation mass spectra (Carson et al., 1997b). This enhancement is due to the fact that the particle core is incompletely ablated (Weiss et al., 1997).

Table 13–2 gives examples of common ions observed in laser ablation mass spectra of ambient particles. Two important principles are reflected in this table. First is the complementary nature of positive and negative ion spectra. Because different types of ions are prominent in each spectrum, the acquisition of both polarity spectra from a single particle gives a more complete representation of chemical composition, particularly for inorganic components, than either polarity alone. Whereas it is difficult or impossible to acquire both spectra from the same particle in a laser microprobe experiment, it is relatively easy to do with real-time mass spectrometry. Second is the difficulty of identifying organic components in particles. Elemental carbon produces carbon cluster ions ($C_n^{+/-}$) in both positive and negative ion spectra. In contrast, organic compounds can produce carbon–hydrogen clusters in positive ion spectra ($C_xH_y^+$). Thus, elemental and organic carbon can often be distinguished. However, the laser ablation process normally causes extensive bond breaking in organic molecules. Molecular ions are rarely observed, and the specific organic compounds present in the particle are difficult or impossible to identify. The only exceptions are certain aromatic molecules (particularly PAHs) that have low ionization energies and high fragmentation energy thresholds. Figure 13–4 shows representative positive and negative ion spectra patterns of different types of ambient particles (Hinz et al., 1999).

Two-Step Laser Ablation and Ionization

Some but not all of the drawbacks of one-step laser ablation and ionization can be overcome through the use of two separate lasers, one to vaporize and the other to ionize (Morrical et al., 1998; Zelenyuk et al., 1999). Vaporization is typically performed with an infrared laser having an irradiance below the threshold for ion formation. A second laser is fired approximately 1 µs later to photoionize vaporized species. The second laser operates typically in the ultraviolet region, and molecular species are ionized upon absorption of two or three photons. It is also possible to use coherent vacuum ultraviolet radiation to photoionize molecular species with a single photon (Van Bramer and Johnston, 1992).

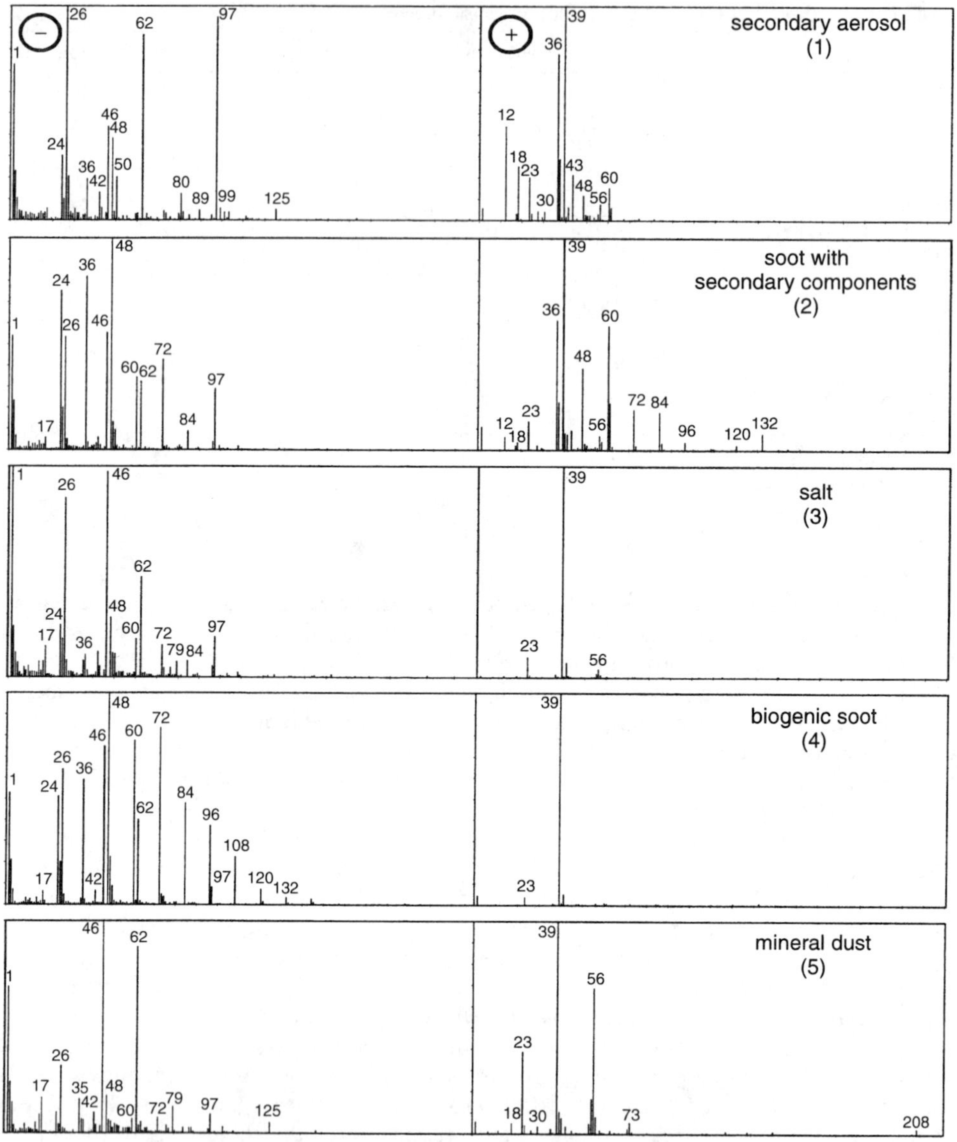

Fig. 13–4. Representative spectra patterns of individual ambient particles.

A direct comparison of one-step and two-step ionization of single particles shows that indeed aromatic molecular ions are enhanced by the latter method (Morrical et al., 1998). In addition, the large optical penetration depth of infrared radiation allows micrometer-sized particles to be completely ablated, suggesting that artifacts due to an inhomogeneous chemical composition will be reduced (Zelenyuk et al., 1999). These advantages are offset by the fact that the two-step method, under conditions optimized for aromatic molecular ion formation, does not produce strong ion signals from many inorganic species or aliphatic organic molecules. Maintaining optical alignment is also more difficult with the two step method.

MASS ANALYSIS

Four types of mass analyzers have been used to obtain single-particle mass spectra: magnetic sector, quadrupole, time of flight, and quadrupole ion trap. Magnetic sector instruments have found only limited use and are not discussed.

Quadrupole Mass Analyzers

Quadrupole instruments were used almost exclusively in early single-particle work. The analyzer consists of an entrance aperture, four hyperbolic rods, and an exit aperture (March, 1989). As the name suggests, a combination of time-varying radio frequency (rf) and constant potentials are applied to the rods to generate a quadrupole electric field. The rf amplitude and constant potential are selected to allow only one m/z to pass through the exit aperture at a given time. A mass spectrum is obtained by scanning the rf amplitude and constant potential to allow different m/z ions to sequentially pass through the exit aperture and strike the detector. Quadrupole mass analyzers are rugged, inexpensive, and easy to operate, making them ideal for many types of field measurements. When applied to real-time particle analysis, however, they are limited in that the spectrum cannot be scanned on the time scale that a burst of ions is produced from a single particle. If single-particle analysis is desired, then the analyzer must be tuned to transmit a specific m/z ion. Therefore, only one chemical component can be monitored from any given particle. Because of this limitation, other types of analyzers are more advantageous for single-particle analysis.

Time-Of-Flight Mass Analyzers

The most common analyzer used in current single-particle instruments is the time-of-flight mass analyzer because it is inexpensive, rugged, and easy to construct in-house (Cotter, 1997). Time-of-flight analyzers were not used in the early years of real-time single-particle analysis because of limitations in spectral acquisition. Recent advances in the performance of high-speed digitizers and computers has made this the analyzer of choice for many applications.

A schematic of a simple linear time-of-flight analyzer is shown in Figure 13–5. The analyzer consists of a source region, a secondary acceleration region, a field free drift region, and a microchannel plate detector. Ions produced at a specific point in time and space are extracted from the source region into the field free drift region where they are separated by time of flight. Because all ions are accelerated to the same nominal kinetic energy, different m/z ions will achieve different velocities and reach the detector at different times. Low m/z

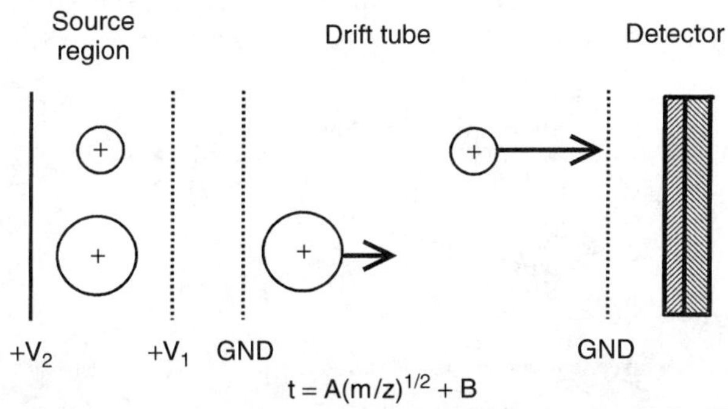

Fig. 13–5. Ion paths in a time-of-flight mass spectrometer.

ions will achieve the highest velocities and reach the detector first. High m/z ions will achieve the lowest velocities and reach the detector last. A mass spectrum is recorded by simply digitizing the ion signal from the detector as a function of time after the laser pulse.

In practice, all ions of a given m/z do not have the same kinetic energy and therefore reach the detector at slightly different times, causing peak broadening. The main contributors to peak broadening are spatial and velocity broadening. An effective approach to reduce broadening is to incorporate a reflecting field into the flight path (Fig. 13–6). In the reflector, ions are decelerated and turned back in the direction from which they came. Ions having a high kinetic energy penetrate further into the reflecting field than ions having a low kinetic energy. The additional time that high kinetic energy ions spend in the reflecting field compensates for the shorter time they spend in the field free regions. In this manner, all ions of a given m/z regardless of kinetic energy can be made to strike the detector at nearly the same time.

For real-time single-particle analysis, the particle beam traverses the source region perpendicular to the drift tube. Ionization is typically performed with a pulsed laser timed to coincide with arrival of the particle and triggering of the digitizer. This approach has three important advantages for single-particle analysis. First, the time-of-flight analyzer is relatively simple in design and easy to construct in-house for specialized instrumental configurations. Second, the time-of-flight analyzer is able to separate and detect all mass to charge ratios in the burst of ions produced from a single particle. Thus, all chemical components encompassed by the spectrum can be detected simultaneously. Third, positive and negative ions from a single particle can be simultaneously analyzed by putting dual drift tubes on either side of the source region. Positive ions are accelerated into one tube, while negative ions are accelerated into the opposite tube. This arrangement produces two separate spectra for each particle and allows a wider range of chemical components to be characterized.

The main difficulty with time-of-flight mass spectrometry is dynamic range. Digitizing boards and oscilloscopes are able to digitize the microchannel plate current at rates of 500 MHz and higher with nominal 8-bit resolution. However, only 6- or 7-bit accuracy is achieved in practice (1 part in 64 to 128). As a result, the digitization error for small signals can be a substantial fraction of the total. To partially alleviate this problem, logarithmic amplifiers can be interposed between the microchannel plate and digitizer. Logarithmic amplification can give a higher dynamic range and spread the digitization error evenly over that range (Murphy and Thomson, 1995). This is a significant advantage for ion formation by laser ablation because particle to particle variations of the absolute signal intensity can be an order of magnitude or more. However, these amplifiers have a 30 MHz bandwidth, which can result in a loss of mass resolution.

Resolution Enhancement of a Reflectron Time-of-Flight Mass Spectrometer

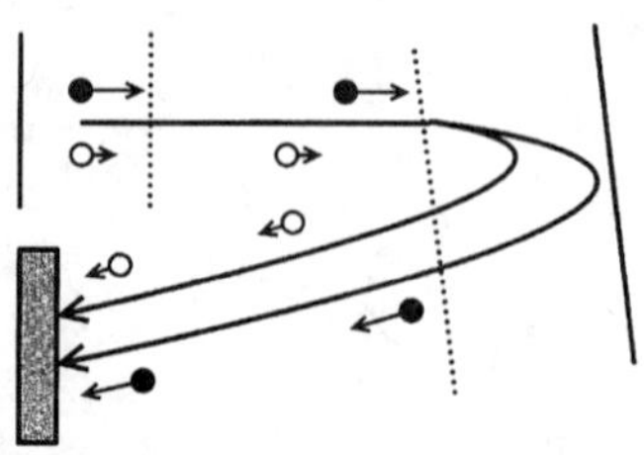

Δt minimized for same m/z ions having different initial velocities

Fig. 13–6. Ion paths in a reflecting-field time-of-flight mass spectrometer. ●, ○ = ions, all with the same m/z, having higher and lower initial velocity, respectively.

Ion Trap Mass Analyzers

Another method capable of obtaining a complete mass spectrum from a single particle is the quadrupole ion trap (March, 1997). Like the time-of-flight mass analyzer, the quadrupole ion trap can obtain a complete mass spectrum from the burst of ions from a single particle. The analyzer consists of three hyperbolic electrodes: two end caps and a ring electrode. Ions are trapped in a potential well created by applying a sinusoidal potential to the ring electrode. Ions are trapped across a broad m/z range, and each m/z is successively ejected from the trap to the detector by application of a second sinusoidal potential to the end caps. Quadrupole ion trap analyzers are compact, robust, and well suited for field measurements. They have not been as popular as time-of-flight analyzers because they are more difficult to construct in-house (commercial models are available), and it is more difficult to implement simultaneous positive and negative ion detection.

Quadrupole ion traps have a significant advantage over time-of-flight analyzers in their ability to perform tandem mass spectrometry. This capability is most useful for identifying organic molecules in complex samples. A specific m/z can be isolated in the ion trap by ejecting all other m/z ions by application of the various resonant frequencies. Once isolated, ions at the selected m/z can then be dissociated by applying a low-amplitude potential at the resonant frequency to the end caps. The amplitude is too low to eject the ion from the trap, but it is sufficient to increase the ion kinetic energy. Ions gain internal energy through collisions with background gas molecules in the trap and eventually dissociate. The dissociation products remain trapped in the potential well and are subsequently analyzed by resonant ejection. An example of the power of this approach is given by Gieray et al. (1997), who used laser ablation tandem mass spectrometry to characterize individual bacteria sampled in real time.

DATA HANDLING AND INTERPRETATION

The preceeding section discusses how chemical components in an aerosol particle can be determined from its mass spectrum. How can composition data for individual particles be transformed into useful information about the atmosphere? The simplest approach is to identify individual components or combinations of components that indicate the source of the particle and/or the chemical transformations it has undergone. For example, sea salt particles have large signal intensities of ions indicating sodium, potassium, and chloride; crustal matter has large signal intensities of ions indicating calcium and/or iron; secondary aerosols have large signal intensities of ions indicating ammonium, nitrate, and/or organic components; soot has large signal intensities of carbon cluster ions (C_n^+); and so forth. In early experiments, particles were grouped according to their chemical composition (and size) by manual inspection of the spectra. This can be a daunting task because tens of thousands of particle mass spectra might be obtained in a single experiment. Therefore, automated methods of particle classification are attractive.

Single-particle spectra can be classified into groups using methods such as principal components analysis, fuzzy cluster analysis, and neural network analysis. Indeed, similar approaches have been used for over a decade to interpret single-particle compositions obtained by electron and laser microprobe methods (see Chapter 12). Typically, these algorithms group particles without any a priori assumptions about their possible identities. Computers have become sufficiently fast that classification of individual particle spectra can be performed in near real time (Hinz et al., 1999; Song et al., 1999). Once a sufficient number of particles have been grouped, each group is interpreted by manual inspection of the centroid mass spectrum for that group. By counting the number of particles in each group, the relative loading of each type of particle in the aerosol can be determined. This loading can then be followed as a function of particle size, time of day, or location.

It should be emphasized that the chemical information derived from a single-particle mass spectrometer is fundamentally related to particle number rather than particle mass. In essence, single-particle mass spectrometers count particles of a given composition type and size. If the aerosol is externally mixed, then the mass of each chemical constituent in the aerosol can be estimated from the number and size distributions of particles in the group containing this constituent. If the aerosol is internally mixed, then mass composition cannot be obtained.

PUTTING IT ALL TOGETHER—SELECTED INSTRUMENTS

As discussed above, many possibilities exist for particle detection, sizing, vaporization, ionization, and mass analysis. The specific techniques that are incorporated into an instrumental design depend on which characteristics are most important for the intended application. For this reason, many types of instruments have been reported. Table 13–3 lists an instrument classification scheme based on how coupled size/composition measurements are made for single particles. For each measurement (size and chemical composition), particles can be analyzed in a sequential or parallel manner. Sequential analysis means that only one type of particle (i.e., one particular size and/or composition) can be analyzed at a time. In this case, the instrument must be "scanned" in some way to obtain information on all possible combinations of particle size and composition. Parallel analysis means that all particle sizes and/or compositions can be analyzed simultaneously. From a data throughput perspective, the least desirable instrument is one that must sequentially scan through both particle size and composition. It is not surprising that no such instrument has been reported. In principle, the most desirable instrument is one in which the full range of particle sizes and compositions can be determined in parallel. Indeed, several field instruments are based on this approach. In some cases, instruments that perform one measurement—either particle size or composition—sequentially and the other in parallel can equal or surpass the performance of a fully parallel instrument. If the sequential method allows particles to be transmitted and/or detected with greater efficiency than the parallel alternative, then the higher analysis rate can compensate for the inherent inefficiency of a sequential measurement. Table 13–3 gives examples of instruments in each category. Several of these instruments are discussed below.

TABLE 13–3. Classification of Real-Time Single-Particle Mass Spectrometers

Size	Chemical Composition	Examples
Parallel	Parallel	Murphy and Thomson (1995), Mie Scatter/LA-TOFMS Gard et al. (1997), laser velocimetry/LA-TOFMS Yang et al. (1996), laser velocimetry/LA-ITMS
Parallel	Sequential	Jayne et al. (2000), velocimetry/TD-EIMS
Sequential	Parallel	Mallina et al. (2000), dynamic focusing/LA-TOFMS Hinz et al. (1996), laser velocimetry/LA-TOFMS Weiss et al. (1997), laser velocimetry/LA-TOFMS
Sequential	Sequential	None

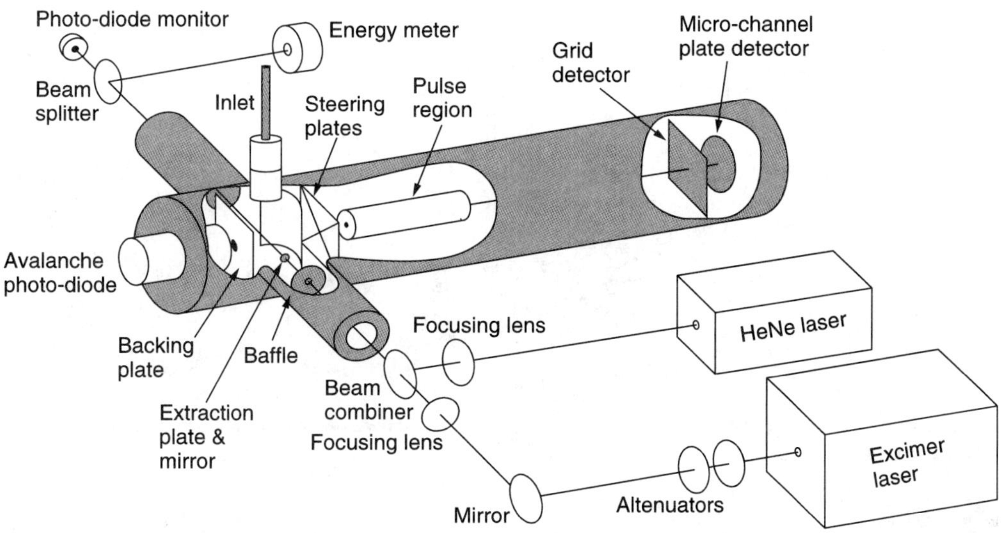

Fig. 13–7. The NOAA Aeronomy Laboratory mass spectrometer.

National Oceanic and Atmospheric Administration (NOAA) Aeronomy Laboratory

Figure 13–7 shows the first single-particle mass spectrometer used for ambient measurements (Murphy and Thomson, 1995). A similar design was used by the same group to perform airborne measurements (Murphy et al., 1998). Particles enter the mass spectrometer through a capillary inlet. In the source region, individual particles pass through a continuous wave helium-neon laser beam and are detected by light scattering. The sensitivity of particle detection is maximized with an elliptical mirror that focuses scattered radiation over a wide solid angle through a spatial filter (i.e., small aperture) onto an avalanche photodiode. The elliptical mirror also serves as the first acceleration plate of the time-of-flight mass spectrometer. The height of the scatter pulse provides a crude measure of particle size. The scatter pulse also triggers an excimer laser that fires promptly (within 1 μs) after receiving the trigger pulse. During this time period, the particle barely moves, so only a slight vertical offset between the helium-neon and excimer laser beams is needed for the excimer pulse to "hit" the particle. Ions produced by laser ablation of the particle are accelerated into a drift tube and detected with a grid detector/microchannel plate detector combination that provides a dynamic range of over four orders of magnitude in signal intensity. The advantage of the experimental configuration in Figure 13–7 is the compact inlet and optical design. Virtually all particles detected by light scattering are "hit" by the ablation laser. Both particle size and composition are determined in parallel, allowing for fast characterization of the aerosol. Even though less than 1 in 1000 particles entering the inlet reaches the source region, those particles traversing the laser beams are detected and ablated with very high efficiency. In an urban environment, many particles per second may be analyzed.

University of California, Riverside

Figure 13–8 shows an alternative design (Gard et al., 1997) that is commercially available (***TSI***). Particles enter the mass spectrometer through a conical inlet terminating in a short capillary. Particles along the center line of the inlet pass through two differentially pumped stages and intercept two continuous wave laser beams (Prather et al., 1994). A scatter pulse

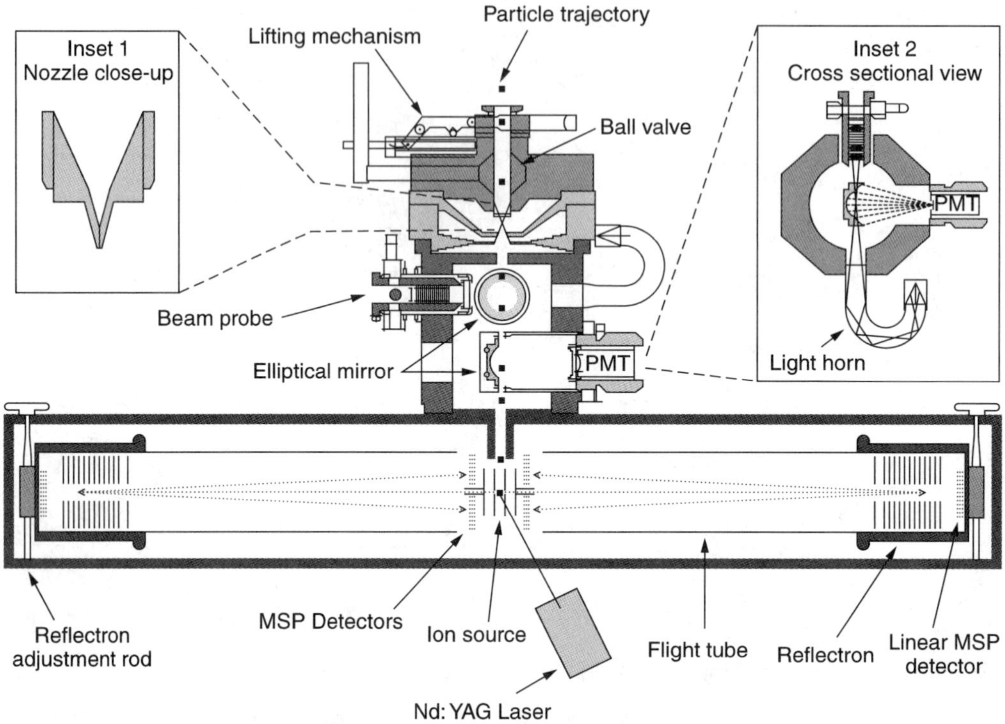

Fig. 13–8. The University of California, Riverside, mass spectrometer.

is detected as the particle passes through each laser beam. The time delay between the two pulses gives the velocity of the particle from which the aerodynamic diameter can be determined. After leaving the light-scattering region, the particle enters the source region of the mass spectrometer where it is "hit" by the ablation laser. The continuous laser beams are configured orthogonal to each other, and the time delay between the two scatter pulses is used to synchronize the arrival of the particle in the source region with the firing of the ablation laser. In this way, a majority of the particles detected by light scattering are ablated and analyzed. Dual time-of-flight mass analyzers simultaneously obtain positive and negative ion spectra for each particle that is ablated. The advantages of this configuration are precise aerodynamic sizing and greater opportunities for chemical characterization because both polarity mass spectra are obtained. As with the previous design, most particles entering the inlet are lost. However, the small fraction of particles that intercept the three laser beams are detected and analyzed with high efficiency, yielding hit rates of several per second in urban environments.

University of Delaware

Figure 13–9 shows a third type of instrument design (Mallina et al., 2000). Here, the aerosol is sampled through a dynamic focusing inlet that collimates only a narrow size range of particles for a given set of conditions. The size range selected passes through several stages of differential pumping into the mass spectrometer, where laser ablation occurs. The ablation laser is aligned collinear with the particle beam to maximize overlap. A continuous wave Nd:YAG laser beam intercepts the particle beam in the source region to detect "large" par-

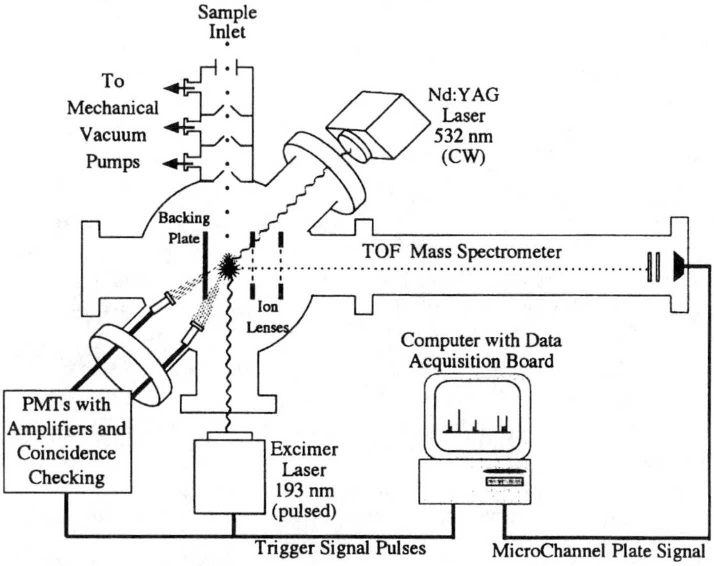

Fig. 13–9. The University of Delaware mass spectrometer.

ticles by light scattering. In this way, the ablation laser can be fired synchronously with the arrival of the particle. For small particles that cannot be detected by light scattering, the ablation laser is free-fired at a high repetition rate to maximize the probability that a particle is "hit." One advantage of this approach is the high transmission efficiency of the inlet. Particles having the proper aerodynamic diameter are transmitted with near unit efficiency into the mass spectrometer source region. The high transmission efficiency provides high data throughput even though different sized particles must be analyzed sequentially. Another advantage is the ability to select and analyze particle sizes below the limit to light scattering, roughly 200 nm. Ambient particles as small as 20 nm diameter have been detected.

Future Developments

The instruments described in this chapter have proven particularly useful for ambient measurements. Many other instrument designs are possible. Several are listed in Table 13–3; others have been developed and are in the testing stage; still others are in the conceptual stage. Given the dramatic growth in the field over the past few years, we expect that single-particle mass spectrometer design and operation will continue to evolve at a rapid pace.

REFERENCES

Carson, P. G., M. V. Johnston, and A. S. Wexler. 1997a. Laser desorption/ionization of ultrafine aerosol particles. *Rapid Commun. Mass Spectrom.* 11:993–996.

Carson, P. G., M. V. Johnston, and A. S. Wexler. 1997b. Real-time monitoring of the surface and total composition of aerosol particles. *Aerosol Sci. Technol.* 26:291–300.

Cotter, R. J. 1997. *Time-of-Flight Mass Spectrometry: Instrumentation and Applications in Biological Research.* Washington, DC: American Chemical Society.

Dahneke, B., J. Hoover, and Y. S. Cheng. 1982. Similarity theory for aerosol beams. *J. Colloid Interface Sci.* 87:167–179.

Fernandez de la Mora, J. and P. Riesco-Chueca. 1988. Aerodynamic focusing of particles in a carrier gas. *J. Fluid Mech.* 195:1–21.

Flagan, R. 1993. Probing the chemical dynamics of aerosols. In *Measurement Challenges in Atmospheric Chemistry*, ed. L. Newman. Washington, DC: American Chemical Society.

Frankevich, V., B. Oktem, and M. Johnston. 1998. Electrodynamic trapping of ultrafine particles for MS characterization. *Proceedings of the 46th American Society for Mass Spectrometry Conference on Mass Spectrometry and Allied Topics*, Orlando Fl, May 31–June 4. Santa Fe, NM: ASMS p. 519.

Friedlander, S. K. 1971. The characterization of aerosols distributed with respect to size and chemical composition—II. Classification and design of aerosol measuring devices. *J. Aerosol Sci.* 2:331–340.

Gard, E., J. E. Mayer, B. D. Morrical, T. Dienes, D. P. Fergenson, and K. A. Prather. 1997. Real-time analysis of individual atmospheric aerosol particles: Design and performance of a portable ATOFMS. *Anal. Chem.* 69:4083–4091.

Ge, Z, A. S. Wexler, and M. V. Johnston. 1998. Laser desorption/ionization of single ultrafine multicomponent aerosols. *Environ. Sci. Technol.* 32:3218–3223.

Gieray, R. A., P. T. A. Reilly, M. Yang, W. B. Whitten, and J. M. Ramsey. 1997. Real-time detection of individual airborne bacteria. *J. Microbiol. Meth.* 29:191–199.

Hinz, K.-P., M. Greweling, F. Drews, and B. Spengler. 1999. Data processing in on-line laser mass spectrometry of inorganic, organic or biological airborne particles. *J. Am. Soc. Mass Spectrom* 10:648–660.

Jayne, J. T., D. C. Leard, X. Zhang, P. Davidovits, K. A. Smith, C. E. Kolb, and D. R. Worsnop. 2000. Development of an aerosol mass spectrometer for size and composition analysis of submicron particles. In press. *Aerosol Sci. Technol.* 33:49–70.

Johnston, M. V. 2000. Sampling and analysis of individual particles by aerosol mass spectrometry. *J. Mass Spectrom.* 35:585–595.

Johnston, M. V. and A. S. Wexler. 1995. Mass spectrometry of individual aerosol particles. *Anal. Chem.* 67:721A–726A.

Kane, D. B. and M. V. Johnston. 2000. Size and composition biases on the detection of individual ultrafine particles by aerosol mass spectrometry. *Environ. Sci. Technol.* 34:4887–4893.

Kaufmann, R. L. 1986. Laser-microprobe mass spectroscopy of particulate matter. In *Physical and Chemical Characterization of Individual Airborne Particles*, ed. K. R. Spurny. New York: John Wiley & Sons.

Liu, P., P. J. Ziemann, D. B. Kittleson, and P. H. McMurry. 1995a. Generating particle beams of controlled dimensions and divergence: I. Theory of particle motion in aerodynamic lenses and nozzle expansions. *Aerosol Sci. Technol.* 22:293–313.

Liu, P., P. J. Ziemann, D. B. Kittleson, and P. H. McMurry. 1995b. Generating particle beams of controlled dimensions and divergence: I. Experimental evaluation of particle motion in aerodynamic lenses and nozzle expansions. *Aerosol Sci. Technol.* 22:314–324.

Mallina, R. V., A. S. Wexler, and M. V. Johnston. 1997. Particle growth in high-speed particle beam inlets. *J. Aerosol Sci.* 28:223–238.

Mallina, R. V., A. S. Wexler, and M. V. Johnston. 1999. High-speed particle beam generation: Simple focusing mechanisms. *J. Aerosol Sci.* 30:719–738.

Mallina, R. V., A. S., Wexler, K. P. Rhoads, and M. V. Johnston. 2001. High speed particle beam generation: A dynamic focusing mechanism for selecting ultrafine particles. *Aerosol Sci. Technol.* 33:87–104.

March, R. E. 1989. *Quadrupole Storage Mass Spectrometry.* New York: John Wiley & Sons.

March, R. E. 1997. An introduction to quadrupole ion trap mass spectrometry. *J. Mass Spectrom.* 32:351–369.

McLafferty, F. W. and F. Turecek. 1993. *Interpretation of Mass Spectra*, 4th Ed. Mill Valley, CA: University Science Books.

Morrical, B. D., D. P. Fergenson, and K. A. Prather. 1998. Coupling two-step laser desorption/ionization with aerosol time-of-flight mass spectrometry for the analysis of individual organic particles. *J. Am. Soc. Mass Spectrom.* 9:1068–1073.

Murphy, D. M. and D. S. Thomson. 1995. Laser ionization mass spectroscopy of single aerosol particles. *Aerosol Sci. Technol.* 22:237–249.

Murphy, D. M. and D. S. Thompson. 1997. Chemical composition of single aerosol particles at Iolaho Hills: Negative ion measurements. *J. Geophys. Res.* 120:6353–6368.

Murphy, D. M., D. S. Thomson, and M. J. Mahoney. 1998. In situ measurements of organics, meteoric material, mercury and other elements in aerosols at 5 to 19 kilometers. *Science* 282:1664–1669.

Neubauer, K. R., M. V. Johnston, and A. S. Wexler. 1997. On-line analysis of aqueous aerosols by laser desorption ionization. *Int. J. Mass Spectrom Ion Processes* 163:29–37.

Neubauer, K. R., M. V. Johnston, and A. S. Wexler. 1998. Humidity effects on the mass spectra of single aerosol particles. *Atmos. Environ.* 32:2521–2529.

Neubauer, K. R., S. T. Sum, M. V. Johnston, and A. S. Wexler. 1996. Sulfur speciation in individual aerosol particles. *J. Geophys. Res.* 101:18701–18707.

Noble, C. A. and K. A. Prather. 2000. Real time single particle mass spectrometry: A historical review of a quarter century of the chemical analysis of aerosols.. *Mass Spectrom. Rev.* 19:248–274.

Peter, T. 1996. Airborne particle analysis for climate studies. *Science* 273:1352–1353.

Prather, K. A., T. Nordmeyer, and K. Salt. 1994. Real-time characterization of individual aerosol particles using time-of-flight mass spectrometry. *Anal. Chem.* 66:1403–1407.

Reents, W. D. Jr., S. W. Downey, A. B. Emerson, A. M. Mujsce, A. J. Muller, D. J. Siconolfi, J. D. Sinclair, and A. G. Swanson. 1995. Single particle characterization by time-of-flight mass spectrometry. *Aerosol Sci. Technol.* 23:263–270.

Song, X.-H., P. K. Hopke, D. P. Fergenson, and K. A. Prather. 1999. Classification of single particles analyzed by ATOFMS using an artificial neural network, ART-2A. *Anal. Chem.* 71:860–865.

Stoffels, J. J. and J. Allen. 1986. Mass spectrometry of single particles in situ. In *Physical and Chemical Characterization of Individual Airborne Particles*, ed. K. R. Spurny. New York: John Wiley & Sons.

Thomson, D. S., A. M. Middlebrook, and D. M. Murphy. 1997. Threshold for laser-induced ion formation from aerosols in a vacuum using ultraviolet and vacuum-ultraviolet laser wavelengths. *Aerosol Sci. Technol.* 26:544–559.

Thomson, D. S. and D. M. Murphy. 1993. Laser-induced ion formation threshold of aerosol particles in a vacuum. *Appl. Optics* 32:6818–6826.

Thomson, D. S. and D. M. Murphy. 1994. Analyzing single aerosol particles in real time. *Chemtech* 24:30–35.

Van Bramer, S. E. and M. V. Johnston. 1992. Tunable, coherent vacuum ultraviolet radiation for photoionization mass spectrometry. *Appl. Spectrosc.* 46:255–261

Wieser, P. and R. Wurster. 1986. Application of laser-microprobe mass analysis to particle collections. In *Physical and Chemical Characterization of Individual Airborne Particles*, ed. K. R. Spurny. New York: John Wiley & Sons.

Weiss, M., P. J. T. Verheijen, J. C. M. Marijnissen, and B. Scarlett. 1997. On the performance of an on-line time-of-flight mass spectrometer for aerosols. *J. Aerosol Sci.* 28:159–171.

Yang, M., P. T. A. Reilly, K. B. Boraas, W. B. Whitten, and J. M. Ramsey. 1996. Real-time chemical analysis of aerosol particles using an ion trap mass spectrometer. *Rapid Commun. Mass Spectrom.* 10:347–351.

Zelenyuk, A., J. Cabalo, T. Baer, and R. E. Miller. 1999. Mass spectrometry of liquid aniline aerosol particles by IR/UV laser irradiation. *Anal. Chem.* 71:1802–1808.

14 Dynamic Mass and Surface Area Measurements

URS BALTENSPERGER, ERNEST WEINGARTNER
Paul Scherrer Institut, Villigen, Switzerland
HEINZ BURTSCHER
Fachhochschule Aargau, University of Applied Science, Windisch, Switzerland
JORMA KESKINEN
Tampere University of Technology, Institute of Physics, Tampere, Finland

INTRODUCTION

This chapter describes several techniques for aerosol measurement, each of which could be most appropriately described as a short-term batch process. Some approach the characteristics of instantaneous measurements such as optical or light-scattering techniques, but all can determine the concentration of aerosol samples in intervals convenient for many applications. The measured primary quantities of the instruments are either mass or aerosol surface area related.

MASS MEASUREMENT

The techniques presented in this section are the piezoelectric microbalance, the beta gauge and Tapered-Element Oscillating Microbalance (TEOM). The description of these instruments is based mainly on the chapter by Williams, Fairchild, and Jaklevic (1993), in the first edition of this book. The techniques have been used to measure total particulate mass levels or, in conjunction with an impactor or cyclone, to measure PM-10, PM-2.5, or PM-1 environmental particulate mass levels (see Chapter 27). A new technique, the Electrical Low-Pressure Impactor (ELPI), has been added. Further information about these techniques and their applications may be found in review articles by Chow (1995) and McMurry (2000).

PIEZOELECTRIC CRYSTAL MEASUREMENT METHOD

Measurement Principles

The quartz crystal microbalance technique, also covered in the first edition of this book, was a fairly common device in the 1970s and 1980s and continues to be used in some areas.

Aerosol Measurement: Principles, Techniques, and Applications, Second Edition, Edited by Paul A. Baron and Klaus Willeke.
ISBN 0-471-35636-0

In this type of instrument, particles are deposited by inertial impaction or electrostatic precipitation onto the surface of an oscillating piezoelectric quartz crystal disk. An electrode is attached to the center of both sides of the crystal, and the particles are deposited onto one of the electrodes. The natural resonant frequency of the crystal decreases as particle mass accumulates. The changing frequency of the sampling crystal is electronically compared with that of a clean reference crystal, generating a signal that is proportional to the collected mass. Quartz crystals have sensitivities of several hundred hertz per microgram. This sensitivity results in the ability to measure the mass concentration of about 10 μg/m^3 in less than one minute (Olin and Sem, 1971). However, the sensitivity over the crystal surface is nonuniform and depends on the excitation mode of the crystal. The highest sensitivity occurs near the center of the electrode. Each collection device is calibrated to compensate for the particle mass deposition pattern and the crystal sensitivity. A reference crystal is typically used to compensate for temperature and humidity deviations. The collection electrode surface is often coated with grease to improve particle coupling to the vibrating surface.

Potential Biases

The advantage of the technique is that it measures mass directly with high sensitivity and, under the proper conditions, high accuracy. The main drawback of this technique is that the relatively high frequency of the electrode surface (5 to 10 MHz) can result in poor coupling between the vibrating plate and collected particles. These coupling problems result in reduced sensitivity and appear to be present with agglomerates, with fibers, and with compact particles, increasing with particle size. Furthermore, saturation occurs at relatively low deposited mass, especially for nonsticky particles, a situation that requires frequent cleaning or re-greasing of the collection surface.

An investigation of a cascade impactor microbalance for drug inhaler aerosol size distributions indicated that the substrate crystals had to be cleaned and re-greased between each measurement to prevent modification of the measured size distribution by particle bounce (Tzou, 1999). However, the high sensitivity and high time resolution of quartz crystal sensors make this technique suitable for some applications.

Instruments

Three versions of a cascade impactor quartz crystal microbalance instrument are available from California Measurements (***CMI***).* A low flow version (PC-2, 4×10^{-6} m^3/s [0.24 L/min]) has been used for higher concentration aerosols, including pharmaceutical aerosols, and exposure monitoring; two higher flow versions have been used for lower concentration aerosols, such as indoor and ambient aerosol (PC-2H and PC-6H, 3.3×10^{-5} m^3/s [2 L/min]) (see Chapter 10 for cut points). An early version of this instrument was evaluated by Fairchild et al. (1980). Some improvements have been made to these instruments since the previous edition of this book (for details of the basic design, Williams, Fairchild, and Jaklevic, 1993, first edition), including lower dead volume in the sensing chambers for reduced internal losses and improved response time, and an isokinetic sampling inlet for more accurate high concentration measurements. An automated crystal cleaning system is in development that should alleviate some difficulties in the routine use of this instrument.

A portable self-contained Respirable Aerosol Mass Monitor (Model 3511, ***KAN***), capable of measurement after either a 24 s or a 120 s sampling period, has also been available since

* See to Appendix I for full manufacturer addresses referenced to the italicized three-letter codes.

the 1970s. A manually operated cleaning system in this instrument alleviates crystal loading problems.

A new version of a quartz microbalance monitor is being manufactured by Booker Systems Ltd. (***BKR***).

BETA GAUGE METHOD

Measurement Principles

The beta gauge method of mass determination depends on the near-exponential decrease in the number of beta particles transmitted through a thin sample as the deposit thickness is increased. The beta particles are emitted as a continuum energy distribution by a radioisotope source, and a suitable electron counter measures their intensity. The method has the advantages of instrumental simplicity and ease of automation for large-scale applications. The dynamic range of sensitivity is well matched to the mass range normally of interest in aerosol monitoring in which various filters are the substrates. However, a detailed understanding of the parameters that affect the measurements is necessary to ensure optimal instrumental implementation and correct interpretation of results.

Figure 14–1 is a schematic diagram of a beta gauge instrument using a two-beam compensating method. It consists basically of the radioactive source, detectors, and sample. The total flux in a continuous beta particle spectrum emitted by the radioisotopic source is determined in a reference section, as well as through the sample after transmission. This new development

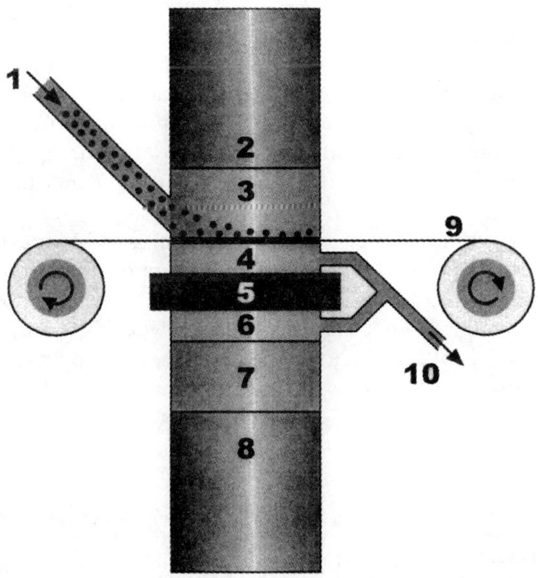

Fig. 14–1. Schematic diagram of a beta gauge. The aerosol enters the instrument at position 1 and is fed into a chamber (3) where the particles are deposited on a filter (9). The beta source (5) is surrounded by vacuum chambers (4, 6), which are connected to a flow controller and vacuum pump (10). In this setup, high measurement stability is achieved with two beta detectors (2, 8), which measure the signal (4, 3, 2) and the reference beam (6, 7, 8) to compensate for fluctuations in temperature and pressure.

increases the sensitivity of this instrument (***ESM***). Under proper experimental conditions, the transmitted flux (I) is related to the sample mass through the relationship (Evans, 1955)

$$I = I_0 e^{-\mu x} \tag{14–1}$$

where I_0 is the incident flux, μ is the mass absorption coefficient for β-radiation absorption (cm^2/g), and x is the mass thickness of the sample (g/cm^2). The mass absorption coefficient is normally determined through a calibration procedure involving the measurement of a series of known standards, which bracket the mass range of interest (Jaklevic et al., 1981). The incident flux, I_0, can be either derived during the same calibration procedure or made to cancel the value of I_0 by calculating the ratio between transmitted fluxes measured with and without the particle deposit, if the interval between successive measurements is short. The latter case applies to certain beta gauge designs where continuous particle deposition is monitored (Macias and Husar, 1974).

Instrument Design

The optimal choice of source and detector depends on many factors. The radioisotope source must have beta particle emission as the dominant mode of decay and exhibit a half-life sufficient for large decay corrections or frequent source replacements to be unnecessary. The source strength should provide adequate precision in counting statistics within the limitations of the detector rate-handling capabilities. Finally, as discussed in more detail below, the energy of the beta spectrum must be chosen to produce a mass absorption coefficient matched to the range of thicknesses to be measured. Table 14–1 lists some radioisotopes appropriate for aerosol beta gauge applications together with relevant parameters for each source (Lilienfeld, 1975).

The detector must be sensitive to the beta particles (i.e., electrons) in the energy range of interest and capable of operation at a counting rate sufficient to perform measurements in the required time interval. While other detectors have been used in the past, the most recent beta gauge systems employ semiconductor diode detectors and solid-state pulse processing electronics because of their simplicity and stability of operation. The following discussion is oriented toward this type of system, although most of the comments also apply to other types of detectors.

The source-detector geometry must be maintained in a stable mechanical configuration to minimize spurious counting-rate variations. Also, it is important that the spacing be as close as possible so that changes in atmospheric density within the gap are not interpreted as mass variations in the sample (Courtney et al., 1982). The lower limit to the spacing is normally determined by the thickness of sample holders and the associated handling mechanisms used with automated instrumentation.

TABLE 14–1. Commonly Available Sources Suitable for Beta Attenuation Measurements

Isotope	Half-Life (years)	E_{max} (MeV)	Range (mg/cm^2) in Carbon at E_{max}	Range (mg/cm^2) in Carbon at $E_{aver} = 0.4\ E_{max}$
^{64}Ni	92	0.067	7.7	1.6
^{14}C	5730	0.156	32	6.6
^{147}Pm	2.62	0.225	60	13
^{85}Kr	10.76	0.67	290	77
^{36}Cl	$3.1 \cdot 10^4$	0.712	320	84
^{204}Tl	3.8	0.765	340	94

Theoretical Considerations

Studies have shown that the functional dependence of beta particle transmission expressed in Eq. 14–1 is not valid for precise experimental measurements (Jaklevic et al., 1981; Heintzenberg and Winkler, 1984). This result is not unexpected because the exponential behavior is not a reflection of fundamental mechanisms associated with beta particle attenuation in matter. Electrons with kinetic energies less than 1 MeV lose energy primarily through collisions with atomic electrons present in the sample. As a consequence, an electron with a well-defined initial energy distribution will slow down through a series of discrete energy losses as it traverses the sample. An incident electron beam with a well-defined initial energy and direction will experience a gradual decrease in the average energy, accompanied by a spreading in the distribution of the beta ray's energy and angle of incidence on the target material.

The radioisotopic beta particle emission process results in a continuum of electron energies. Figure 14–2 is an idealized representation of a measured beta spectrum from a typical source. The energy distribution extends from a minimum energy, determined by the source window thickness, to a maximum endpoint energy, E_{max}, which is the total energy available for the radioactive decay process. An electronic discriminator level, E_{disc}, has been indicated above a low-energy electronic noise tail. As the mass between source and detector is increased, the counting rate observed above this discriminator level in the beta gauge detector represents those electrons in the original continuum spectrum that have not been totally stopped in the sample and are still incident on the detector. This rate reflects a complex energy loss process, which depends on several variables, including the average energy of the electrons in the beta spectrum and the amount of material traversed. When averaged over all these effects, the observed dependence of the counting rate on the thickness traversed is approximately exponential. Repeated measurements in a carefully defined experimental geometry using aluminum absorbers have established an approximate relationship between the mass attenuation coefficient in Eq. 14–1 and the beta spectrum endpoint energy (Gleason et al., 1951):

$$\mu(\text{cm}^2/\text{g}) = 0.017E_{\text{max}}^{-1.43} \qquad (14\text{–}2)$$

However, there is considerable variation among investigators regarding the values of the coefficients, which reflects the empirical nature of these parameters. Because the beta particle energy-loss process involves scattering from atomic electrons rather than nuclei, there is an additional dependence of the absorption coefficient on the average atomic number of the samples. Various authors have studied this effect, and an empirical relationship has been derived (Klein et al., 1984):

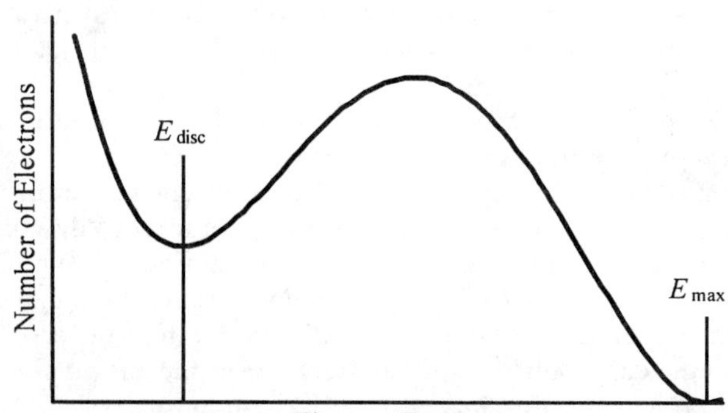

Fig. 14–2. Idealized beta particle spectrum emitted from a radioisotope source.

$$\mu(\text{cm}^2/\text{g}) = 0.016(Z/A)^{4/3} E_{\text{max}}^{-1.37} \tag{14–3}$$

where Z is the atomic number and A is the atomic weight. However, the validity of this relationship seems to depend on the specific geometry of the beta gauge. A less dramatic dependence of attenuation on Z/A has been observed by Jaklevic et al. (1981) and attributed to the relative importance of the particle angular distribution in the particular source-detector arrangement employed in that study. Regardless of the details of functional dependence, it should be noted that variations associated with this effect are normally tolerable because the range of Z/A is small for all elements, with the exception of hydrogen.

The practical consequences of these theoretical observations are that, although estimates of beta gauge sensitivity can be obtained using generalized expressions, precise mass measurements require that each specific instrument be calibrated using known gravimetric standards. It is also necessary to limit the dynamic range over which one attempts to apply a given calibration using a strictly exponential approximation. To the extent that the deposited mass is normally a small fraction of the tare weight of the filter medium, this is not a severe limitation in aerosol applications.

A more complete understanding of precision and accuracy requires a detailed error analysis. If one assumes that the value of μ has been carefully determined and that the counting interval is known with complete certainty, then the precision of the mass measurement is determined by experimental variations in the determination of I_0 and I. Using Eq. 14–1, the root mean square error, $\sigma^2(x)$, in the calculated concentration, x, can be derived as (Cooper, 1975)

$$\sigma^2(x) = \frac{1}{\mu^2}\left[\frac{\sigma^2(I_0)}{I_0^2} + \frac{\sigma^2(I)}{I^2}\right] \tag{14–4}$$

where I_0 and I are the incident and transmitted fluxes integrated for a fixed time interval. When the errors associated with I and I_0 are the result of Poisson counting statistics only, that is, $\sigma(I) = (I)^{1/2}$, and if the counting intervals of the two measurements are equal, the precision $\sigma(x)$ for a difference measurement varies as the inverse of the mass absorption coefficient and the inverse square root of measurement interval, as might be expected. The inverse dependence of precision on the mass absorption coefficient supports the intuitive observation that lower energy beta spectra will provide a more sensitive indicator of small mass changes in the sample. On the other hand, if the energy is too low, the exponential approximation is no longer valid because an increasing fraction of beta particles is totally stopped in the sample. Referring to Table 14–1, a useful rule of thumb is to select a beta spectral average energy corresponding to a range that is several times the maximum thickness to be measured. For this reason, most beta gauges designed for aerosol monitoring employ either ^{85}Kr, ^{14}C, or ^{147}Pm as reasonable choices for use with substrates in the range of 0.1 to 1 mg/mm^2 [10–100 mg/cm^2] and deposits in the range of 0.2 to 5 μg/mm^2 [20–500 μg/cm^2] (Klein et al., 1984).

The derivation of Eq. 14–4 assumed that the value of the mass absorption coefficient was known from a previous calibration procedure. However, the value of the mass absorption coefficient is normally calculated from transmission measurements performed on a series of mass standards that bracket the anticipated range of operation of the instrument. A more complete error analysis, including uncertainties in the fitting procedure, is discussed by Jaklevic et al. (1981). A general conclusion of that analysis is that, although the absolute accuracy of a mass measurement is affected by such calibration errors, the precision is dominated by the variability in the determinations of I and I_0. In principle, the precision of a given mass determination can be improved by increasing the counting interval to reduce the relative error associated with Poisson counting statistics. However, one eventually reaches a limit where the variability in the measured counting rate is dominated by other effects, which are not easily controlled. Sources of systematic errors can include fluctuations in atmospheric density; changes in laboratory relative humidity, which can affect the substrate mass for hygroscopic media; instabilities in the mechanical design; and variability in the placement of the sample in the instrument (Courtney et al., 1982). A major source of instrumental instability is the result of long- and short-term drift in the detector and analogue pulse processing electronics. Because of the difficulty in controlling such sources of errors, it should be requirements in all beta gauge measurement protocols that recalibration of the instrument be performed before each series of measurements and that replicate samples be repeatedly analyzed at the same time as the unknowns in order to monitor instrumental stability.

Potential Biases

The near-exponential behavior of the beta absorption process and the variations discussed above can result in several potential measurement artifacts that should be understood. Principal among these are particle size effects, substrate inhomogeneity, and atomic number dependence.

Particle size effects result from the fact that the beta gauge transmission measurements represent an average of the absorption experienced by an aggregate of particles deposited on the filter substrate. When this deposition is a homogeneous layer of small particles whose average diameter is much less than the layer thickness, the interpretation of the results in terms of exponential absorption by a uniform deposit is valid. On the other hand, one can imagine a deposit of equivalent mass, but consisting of only a few, very massive particles. In an extreme manifestation of this latter case, there could exist total absorption within a given particle. The transmission measurement in this limiting case would then reflect the fractional area covered by the particles, and the interpretation in terms of an exponential absorption by a uniform deposit would be invalid. A detailed discussion of this problem for the general case of exponential absorption is given by Cooper (1976). A similar treatment using a simplified model applied to the case of aerosol particles is given by Jaklevic et al. (1981). Their results indicate that one must either limit the size distribution to particles below 10 μm diameter or, if larger particles are to be analyzed, ensure that the average deposit thickness is sufficient for a statistically meaningful number of particles to be present. Similarly, for impaction samples, the deposit thickness must be uniform over the measurement area to a degree where the average of the exponential is equal to the exponential of the average.

An effect that can be explained using similar logic has to do with discrepancies caused by filter and source inhomogeneities. A microscopic examination of most membrane filter media shows that the substrate consists of a nonuniform distribution of fibers or flocculated material having relatively open spaces in between. Similarly, a radioisotope source is normally fabricated by methods that result in local inhomogeneities in radioactivity across the face of the source. Although a point source can, in principle, alleviate this problem, there is a practical limit to the specific activity that can be concentrated in a small volume. Variations in the apparent mass of the substrate can result from random alignments between the respective inhomogeneities causing spurious high or low mass readings. The problem is exacerbated by the large mass of the substrate relative to the deposit. Because neither the source nor the substrate can be made to be perfectly spatially uniform, it is important to constrain the measurement protocol to position the filter in the instrument identically for both the initial and final weight determinations. This is most easily implemented in the case of large-scale automated systems and is necessary for mass measurements, which aspire to achieve the limits of instrumental precision (Courtney et al., 1982).

The atomic number dependence of the mass absorption coefficient requires that certain precautions be taken regarding the choice of calibration standards and in the interpretation of results from discrete pollution sources. Table 14–2 shows the *Z/A* values compounds commonly observed in ambient aerosol sampling. The mass absorption coefficients are calculated

TABLE 14–2. Effect of Atomic Number Dependence on the Measured Mass of Several Compounds

Compound	Z/A	μ (cm^2/mg)[a]	μ (cm^2/mg)[b]
$(NH_4)_2SO_4$	0.530	0.153	0.166
NH_4HSO_4	0.521	0.152	0.163
$CaSO_4 \cdot 2H_2O$	0.511	0.152	0.159
SiO_2	0.499	0.154	0.154
$CaCO_3$	0.500	0.154	0.154
Carbon	0.500	0.154	0.154
Fe_2O_3	0.476	0.163	0.144
NaCl	0.478	0.172	0.145
$PbSO_4$	0.429	0.193	0.126
$PbCl_2$	0.417	0.204	0.121
PbBrCl	0.415	0.206	0.120

[a] From Jaklevic et al. (1981).
[b] Z/A dependence calculated from Eq. 14–3. Values normalized to 0.154 for carbon to account for instrumental differences.

using both the Z/A dependences observed by Jaklevic et al. (1981) and those calculated from Eq. 14–3. The mass absorption coefficient for a mixture of compounds would be the weighted sum of the respective coefficients. It is obvious that an inappropriate choice of calibration foils can affect the accuracy of the measurements if not corrected for in the data analysis. Similarly, measurements of a set of samples in which the relative contribution of diverging Z/A compounds varies widely will need special consideration in the interpretation. Although a complete correction requires that the sample composition be known, some estimate of the probable error can be obtained by observing the range of values of μ for the compounds listed in Table 14–2 and incorporating an error analysis based on Eq. 14–1. It should be noted that the errors associated with Z/A variations affect the accuracy of the measurements but not the precision and, as a consequence, have little effect on the lower limit of sensitivity of the beta gauge method.

Volatilization Losses

A general problem with particulate mass measurements is that the sampling temperature affects measured mass concentrations. An example of this problem is the water vapor content in the atmosphere, which affects the reading of a beta gauge instrument. At high relative humidity, the aerosol particles, as well as the filter substrate, will adsorb water, which may result in a significant increase in the indicated particle mass (Tsai and Cheng, 1996). In some beta gauge instruments the inlet system can be heated to reduce relative humidity in the sample, thus minimizing the contribution of water to the aerosol mass. However, this might result in substantial losses of semivolatile material and thus a negative bias of the particle mass (see also next section).

Results and Applications

Commercial sources and specifications of several beta attenuation instruments are given in Table 14–3. In recent years the monitoring of low aerosol concentration levels has been made possible through the use of the dual beam compensation method, which compensates for fluctuations in temperature, pressure, and supply voltages. According to the manufacturers, with these instruments, detection limits down to 3 and ~1 $\mu g/m^3$ can be achieved at a temporal resolution of 30 min and 24 h, respectively. There are a number of criteria other than precision

TABLE 14–3. List of Commercially Available Beta Gauges with Specifications

Producer	Device name	Source	Detection limit ($\mu g/m^3$)	Mode
ESM	FH 62 I-R	^{85}Kr	3 (1/2h)	Contin.
AND	GBAM-1020	^{14}C		Step
MET	BAM 1020	^{14}C		Step
KIM	SPM-611	^{147}Pm		Contin.
OPS	ADAM	^{14}C		Step
VER	F 701	^{14}C	10	Step
HOR	APDA-360	^{14}C	10 (1h)	Step

and individual designs that need to be evaluated in terms of specific applications. These include speed, convenience, cost, operating environment, and automated operation.

Beta gauge mass measurements have been incorporated into a number of studies. Hoek et al. (1997) compared different ambient mass measurement methods and found a good correlation between beta monitors and a PM-10 sampler. Speer et al. (1997) used the beta gauge method to measure the liquid water content of aerosol particles on a polytetrafluoroethylene (PTFE) membrane filter as a function of relative humidity. Other applications using continuous mass monitors have been described by Heintzenberg and Winkler (1984). A number of older applications are additionally found in the first edition of this book.

One such instrument, the Atmospheric Dust Automatic Monitor, ADAM (***OPS***) system, collects atmospheric particles on standard 47mm membrane filters and determines the amount of collected particle mass using the β-ray attenuation method. In addition, the residual beta activity on the sampled particles due to the presence of short-lived radon daughters (natural radioactivity) is measured and taken into account. The dual filter geometry permits the performance of the beta measurement on one filter while the next filter is sampled. The instrument automatically performs up to 40 unattended sequential measurements. The loaded filters are then available for the possible gravimetric determination of the collected mass and for subsequent chemical analysis of the particulate matter. The temperature of the airflow is kept as close as possible to ambient temperature (no heating) to minimize sampling artifacts due to the loss of volatile compounds.

TAPERED-ELEMENT OSCILLATING MICROBALANCE METHOD

Measurement Principles

In Tapered-Element Oscillating Microbalance (TEOM) devices, aerosol mass is collected on a vibrating collection substrate and measured through a change in oscillation frequency. Figure 14–3 shows a typical arrangement for a TEOM instrument. The active element of any TEOM system is a specially tapered hollow tube constructed of an elastic, glass-like material. The wide end of the tube is firmly mounted on a relatively massive base plate. The narrow end supports a replaceable collection medium, such as a filter or impaction plate, and is made to oscillate. Particle-laden gas streams are drawn through the collection medium, where particles are deposited. The filtered gas is then drawn through the hollow tube, typically controlled by an automatic mass flow controller.

An electronic feedback system initiates and maintains the oscillation of the tapered element. In 1983, the U.S. Bureau of Mines (BOM) and the National Institute of Occupational Safety and Health (NIOSH) funded the development of a prototype TEOM dust

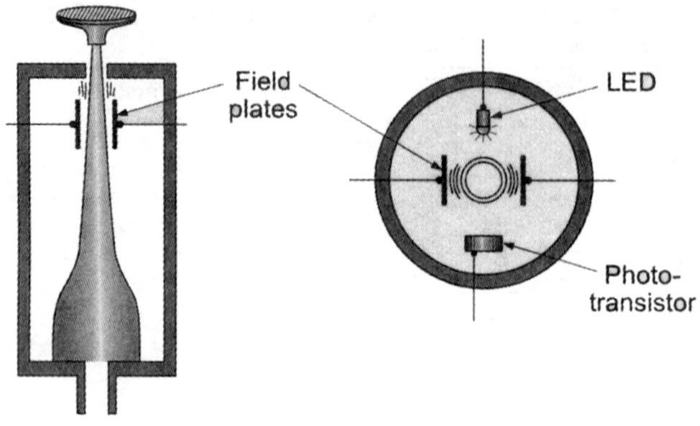

Fig. 14–3. Typical arrangement for the TEOM.

monitor for mining applications (Patashnick and Rupprecht, 1983). In that particular device, a light-emitting diode (LED)–phototransistor pair aligned perpendicular to the plane of oscillation of the tapered element detects the frequency of oscillation. The light-blocking effect of the oscillating element, positioned between the phototransistor and the LED, modulates the output signal of the phototransistor, which is then amplified. Part of the amplified signal is applied to a conductive coating on the outside of the tapered element. In the presence of constant electric field plates, this signal provides sufficient force to keep the tapered element in oscillation. In other words, part of the amplified signal from the LED-phototransistor pair is used in an electrical feedback loop to overcome any amplitude damping of the tapered element oscillation. The other part of the amplified signal from the LED-phototransistor pair is sent to a counter and data processing stage. Here, the oscillation frequency of the tapered element is calculated and stored in the memory. The manufacturer (***R&P***) has made several proprietary improvements to the feedback system since the early BOM/NIOSH prototype (Patashnick and Rupprecht, 1991).

The equation that describes the behavior of the TEOM system derives from the equations of motion for a simple harmonic oscillator:

$$\Delta m = K_0\left(\frac{1}{f_b^2} - \frac{1}{f_a^2}\right) \tag{14–5}$$

where Δm is the mass of the collected sample, f_b is the frequency of the oscillating element after sample collection, f_a is the frequency before sample collection, and K_0 is a constant (spring constant) unique to each tapered element. As the collection medium collects aerosol, the mass increases, thereby decreasing the frequency of oscillation. By measuring only the change in frequency, one can determine the gain in the aerosol mass on the collection medium. Although this expression for Δm is nonlinear, it is monotonic (single valued), independent of m, and depends only on the constant K_0. For subsequent measurements, f_b becomes f_a, a new initial frequency that reflects the total mass of the system. The new f_b after sampling will differ from f_a only because of the new mass uptake, Δm, collected during sampling.

Instrument Design

A TEOM instrument can be tailored for a particular application. To do so, the manufacturer must know the minimum mass concentration that the instrument must measure, how quickly each measurement must be made, and what sampling air flow rate will be used. To employ the highest sensitivity for a particular application, the manufacturer must consider the total mass of the tapered-element load. This mass, which in most applications is primarily the filter cartridge, must be held to a minimum to effect the maximum frequency change with a sample mass deposit. The reduction of the filter mass has practical limits that are related to both flow rate and filter life. A filter cartridge must have reasonable dimensions to sustain the desired flow before loading to a point where the flow drops to an unacceptable level. In addition to the filter mass, the tapered element has a certain amount of mass that also contributes to the total mass of the oscillating system. The limits on this mass depend on the dimensions of the tapered element. The element must have a sufficiently wide bore to allow the desired flow with minimum pressure drop and also have sufficient wall thickness to support the filter cartridge. Temperatures are maintained at a constant value, typically 303 or 323 K [30° or 50°C], to minimize thermal expansion of the tapered element and to reduce relative humidity.

Potential Biases

Calibration. Calibration of TEOM instruments is equivalent to determining the spring constant K_0. Because K_0 is determined by the physical characteristics of the tapered element, calibration is not likely to change over a period of time. The manufacturer provides a value of K_0, but the user can easily check the value of K_0 by adding a known mass to the tapered element, measuring the change in oscillation frequency, and using Eq. 14–5 to determine K_0. Using this method, Shore and Cuthbertson (1985) found that the manufacturer-supplied value for K_0 was correct within the experimental error. However, they also checked the calibration by injecting known masses of dioctyl phthalate (DOP) over the surface of the collection filter. This method suggested that the TEOM instrument measures particulate masses up to 10% lower (on an average) when the K_0 value supplied by the manufacturer was used. However, the masses of DOP used were much higher than those expected during particle collection. They also observed that limiting aerosol collection to the center of the filter influenced the measurements. These results suggest that if one elects to verify the manufacturer-supplied K_0, aerosol deposits similar to those expected during sampling should be used.

Particle Size Effects. Because TEOM devices typically use a filter collection medium, collection efficiency will not be significantly affected by particle size. Particles not collected by the filter medium would not represent significant mass. Particle size is likely to be more important to inlet bias of the sampling head; however, these problems are common to all sampling devices and not just to TEOM measurement technology.

Adherence Effects. Tapered elements typically oscillate at several hundred Hertz, and the collection medium is typically a filter. This reduces coupling problems between the filter and agglomerates, that could occur at higher frequencies. Theoretically, if sufficient mass were collected on the TEOM filter, particles would begin to flake off. In practice, however, the filters clog before collecting particle loadings become large enough to cause flaking. In fact, most TEOM instruments provide a warning to replace the filter at high loading.

Overloading. If the collection filter became sufficiently loaded, the added mass could conceivably damp the oscillations beyond the ability of the feedback system to sustain them. This situation, called *saturation*, could introduce serious error. The dynamic range of TEOM instruments, however, is several orders of magnitude. As discussed in the previous section, filters will clog before particle loadings become large enough to cause saturation.

Volatilization Losses. The volatilization problem mentioned for the beta gauges applies also to the TEOM. To reduce relative humidity, this instrument is often operated at 323 K [50°C], which is high enough to vaporize semivolatile aerosol material such as ammonium nitrate and certain organic compounds. Mignacca and Stubbs (1999) reported a 22% higher signal at an equilibration temperature of 303 K [30°C] compared with 323 K [50°C]. As a consequence, lower TEOM concentrations are generally measured when compared with reference methods, a situation especially encountered during the colder seasons (Allen et al., 1997; Salter and Parsons, 1999; Soutar et al., 1999; Muir, 2000; King et al., 2000). Brook et al. (1999) also compared PM-2.5 data from TEOM and dichotomous samplers and found a relatively small difference in the warmer months, while during the colder months TEOM values were lower by 23% on average. Ayers et al. (1999) compared 24 h PM-2.5 aerosol loadings determined by a TEOM and manual gravimetric samplers in Australian cities. They found systematically lower results from the TEOM by an average of >30%, indicating that aerosol material is lost from the heated sample filter employed in the TEOM. The problem may be minimized by using a diffusion dryer, which allows the temperature in the system to be reduced (Eatough et al., 1999). See Chapters 5, 9, and 27 for discussions of particle volatilization.

Damping. If the mass of the support structure for the element is sufficiently small, the element may induce vibrations in the structure itself. The oscillation of the tapered element will then be slightly damped. This phenomenon was observed when the prototype TEOM personal sampler was tested for miners (Williams and Vinson, 1986). Such damping will not be significant if the mass of the support structure is much larger than the mass of the element or if the support structure is properly clamped to a large mass. Tests of the prototype TEOM personal sampler for miners represent a special case; damping of this sort has not been reported for the commercial TEOM instruments.

Results and Applications

The TEOM technology had its beginnings at Dudley Observatory in the 1960s in conjunction with micrometeorite research (Patashnick and Hemenway, 1969). A microbalance, at that time consisting of a thin quartz fiber, was designed to measure particle masses in the range of 10^{-5} to 10^{-11} g. Since then, its inventors have made many improvements to the quartz fiber approach and formed Rupprecht & Patashnick Co. (***R&P***) to market the device. ***R&P*** is currently the only manufacturer and vendor of TEOM instruments. According to the manufacturer, the detection limit of these instruments is less than $2\,\mu g\,m^{-3}$ for a 24 h integrated sample.

A variety of applications utilizing the real-time mass-measuring capability of the TEOM have been investigated. Internal combustion engine emission measurements with the TEOM have allowed designers to identify conditions contributing to particulate emissions, and even transient vehicle-to-vehicle emissions have been detected (Shore and Cuthbertson, 1985). Egan and Litton (1986) monitored the mass concentrations of smoke from wood. Newman and Steciak (1987) used the TEOM instrument to help characterize smoke properties for fire modeling, fire detector evaluation, and assessment of the nuclear winter problem. In a field study, two TEOM instruments were used at the entrance and exit of a road tunnel to measure the emission factors of fine particles emitted from individual vehicles (Weingartner et al., 1997). Patterson and Eatough (2000) compared indoor and outdoor PM-2.5 concentrations measured with two TEOM monitors. Morawska et al. (1999) used a TEOM in conjunction with volume size distribution measurements to determine the average density of aerosol particles.

The U.S. Environmental Protection Agency (1989) has certified ***R&P*** models 1200 and 1400a for PM-10 measurements of ambient air quality. The TEOM instruments are often used to monitor urban PM-2.5 and PM-10 levels (Meyer et al., 1992; Deacon et al., 1997; Harrison et al., 1997, 1999; Keary et al., 1998; Olcese and Toselli, 1998). As already mentioned, care must be taken to properly account for volatile species.

ELECTRICAL LOW-PRESSURE IMPACTOR

Measurement Principles

The Electrical Low-Pressure Impactor (ELPI) was developed at Tampere University of Technology (TUT-ELPI) in the early 1990s (Keskinen, 1992; Keskinen et al., 1992). A modified version of the TUT-ELPI manufactured by Dekati, Ltd. (***DEK***) has been commercially available since 1995 (ELPI models 95 and 97). Unless otherwise specified, the following addresses the commercial models.

The instrument is based on the charging of particles, which enter a low-pressure cascade impactor, and subsequently measuring the electric current carried by the particles onto the impactor stages. Real-time operation is essentially achieved by measuring the current with a time constant of a few seconds.

Figure 14–4 shows a schematic diagram of the ELPI. It consists of three main components: a cascade impactor, a unipolar diode charger, and a multichannel electrometer. The charger is used to charge sampled particles to a well-defined charge level. The charged particles are then introduced into the cascade impactor that classifies the aerosols into 12 size fractions according to their inertia and consequently their aerodynamic diameter. The particles collected on a specific impactor stage produce an electrical current that is recorded, in real time, by the appropriate electrometer channel. A PC is required to run the operating software and record the measurement results.

Instrument Design

Impactor design is always a compromise. In the ELPI, the main design criteria were sharpness of cut, low losses of fine particles, and small stage and interstage volumes. Relatively

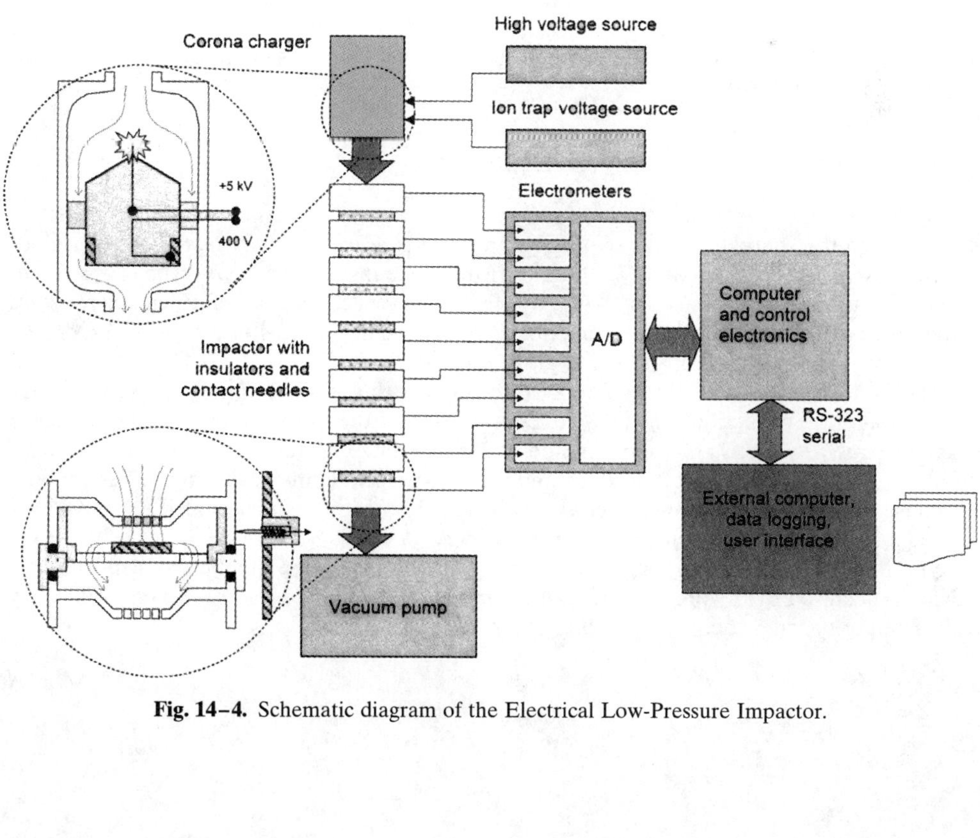

Fig. 14–4. Schematic diagram of the Electrical Low-Pressure Impactor.

high jet speeds and high-pressure ratios between stages are employed. To achieve an adequate size resolution, the impactor uses a relatively high number of stages. There are 12 stages equipped with current detection, plus an extra impactor stage providing the upper cut diameter of 10 µm. The lowest cut diameter (stage 1) of the instrument is 30 nm. Jet orifices are drilled symmetrically in rings around the center of each stage. Stage 1 serves as a critical orifice and controls the flow rate through the impactor. An external valve is used to obtain the correct operating pressure at the impactor outlet.

To charge particles, the ELPI uses a simple point-tube geometry corona charger. A high positive voltage (approximately 5 kV) is applied to a tungsten needle electrode in the center of a cylindrical tube. The particle stream is then introduced through the same region in the direction perpendicular to the applied electric field. Particles are thus exposed to a unipolar positive ion environment and electrically charged. After the charging zone, an electrical mobility analyzer of 0th order ("trap") is used to remove ions and those charged particles that have a smaller size than the measuring range of the instrument. This type of analyzer is characterized by the critical mobility (i.e., the particle mobility with zero penetration). The critical mobility of the trap depends on the physical dimensions, flow rate, and voltage. In the commercial unit, the cut size is approximately 20 nm. The advantages of this charger configuration are its simple construction and high charging efficiency. One disadvantage is the relatively high particle loss for the charger.

Theoretical Considerations

In the case of pure diffusion charging, the measured current can be used as a direct measure of the active surface area of the particles, as described in the surface area measurement section below (Eqs. 14–21 and 14–24). Therefore, the primary output of the instrument is related to the aerodynamic size distribution of the active surface area concentration. The desired output is, however, normally either particle number or mass distribution. To obtain these, the measured current signal is converted to a number concentration as a function of aerodynamic diameter, d_a, using the charger efficiency function $E_{ch}(d_a)$. The resulting response function relates the measured current to the number concentration at different particle sizes as follows:

$$N = \frac{I}{E_{ch}(d_p)} = \frac{I}{PneQ} \qquad (14\text{–}6)$$

where N is the number concentration (m^{-3}), I is the measured current (A), P is the penetration through the charger, n is the average number of charges per particle, e is the charge of an electron ($1.602 \cdot 10^{-19}$ C), and Q is the flow rate of the instrument (m^3/s). The charger efficiency function can be defined as $E_{ch}(d_p) = PneQ$. In practice, the charger efficiency function is determined experimentally. The function includes particle losses inside the charger, the fraction of particles charged, and the average particle charge.

The overall response function of an ELPI channel is the product of the charger efficiency function and the collection efficiency function of one of the impactor stages. The impactor stage response functions are approximated by ideal rectangular functions. They have a value of one between the cut size of the stage and the cut size of the one preceding it. The channel response function then equals the charger efficiency function evaluated at the stage i geometric midpoint diameter, which is defined as $(d_i\, d_k)^{1/2}$, where d_i is the cut point of the stage in question and d_k is the cut point of the next upstream stage. The number concentration at stage i is acquired by dividing the measured current of channel i with the charger efficiency calculated at the stage geometric midpoint d_i:

$$n_i = \frac{I_i}{E_{ch}(d_i)} \qquad (14\text{–}7)$$

In an electric field, particle acceleration is measured by electrical mobility, which is used to define an electrical mobility diameter d_i. The particles are classified in the impactor according to their aerodynamic diameter, d_{ai}. To evaluate the number concentration as a function, the effective density ρ_{eff} of the particles needs to be known or estimated (Moisio et al., 1997; Ahlvik et al., 1998). Using a standard density (1000 kg/m^3, ρ_0),

$$d_i = \sqrt{\frac{\rho_0 C_c(d_{ai})}{\rho_{eff} C_c(d_i)}} d_{ai} \tag{14–8}$$

where C_c is the cunningham slip correction factor.
The mass concentration m_i at channel i is calculated as

$$m_i = \rho_{eff} \frac{\pi}{6} d_i^3 n_i \tag{14–9}$$

Actual impactor collection efficiency curves differ from the ideal rectangular ones. Losses, primarily by diffusion and space charge forces, cause the curves to display a tail of increased collection on particle diameters much smaller than the actual impaction cut-off diameter (Juan et al., 1997; Keskinen et al., 1999). On gravimetric measurements made by cascade impactors, fine particle losses to the upper stages are not critical, as the mass introduced by diffusion deposition is insignificant when compared with the mass of particles actually impacting the stage. When electrical detection is used instead of the gravimetric method, the current carried by the fine particles can be significant compared with the signal caused by particles impacting the stage. To correct for the effect, a simple non-iterative calculation algorithm is used, solving the one-sided cross-correlation problem (Moisio, 1999).

Calibration

The concentration measurement is affected by the precision of the values describing charging efficiency, collection characteristics of the impactor, flow rate, and current measurement. The calibration of the flow rate of the instrument is straightforward and can be performed following the principles discussed in Chapter 21. The current measurement is affected by the whole amplification–A/D conversion train. This can be calibrated using a current standard.

For the calibration of charging efficiency and impactor cut values, aerosol techniques are needed. It is advantageous to employ the electrical current measurement system of the ELPI to calibrate the impactor. Indeed, the measurement principle can be used to calibrate any cascade impactor (Keskinen et al., 1999). Marjamäki et al. (2000) measured collection efficiency curves of all the impactor stages using monodisperse aerosols. The average value of the square root of the Stokes number at 50% collection efficiency was found to be 0.456, with a standard deviation of 0.017, so the charger efficiency can be determined using monodisperse aerosols. To approximate the charging efficiency, a power function is used to fit the experimental data in three different particle size ranges.

$$Pn = \begin{cases} 174.15 d_p^{1.9087}, & d_p < 0.095\,\mu m \\ 50.507 d_p^{1.3805}, & 0.095 < d_p < 1.8\,\mu m \\ 79.732 d_p^{0.5909}, & 1.8\,\mu m < d_p \end{cases} \tag{14–10}$$

It is impossible to produce absolutely identical cascade impactor units, as very small differences in the jet dimensions may cause changes in the cut values. In principle, therefore, every instrument should be individually calibrated. In practice, this is unnecessary as the procedure described by Hillamo and Kauppinen (1991) can be followed: First, the impactor jet sizes, flow rate, and operation pressures are measured. Second, the impactor cut values are calculated using Eq. 10–4, and the gas properties are evaluated for each stage at the upstream stagnation conditions (Flagan, 1982; Hering, 1987). For the Stokes number, the value mentioned above is used. Good agreement has been found between cut values calculated in this manner and the calibrated ones (Hillamo and Kauppinen, 1991). For minor changes in the flow rate, the *Pn* values of Eq. 14–10 can be used without correction.

Potential Biases

It is important to stress that the number and mass size distributions are obtained indirectly by calculation using the instrument software or a spreadsheet application. The relative importance of different nonideal phenomena depends on which of the two size distributions are being measured.

To obtain an accurate mass distribution, it is necessary to accurately determine the effective density of the measured aerosol particles. As described above, to obtain accurate results, the effective density of the particles should be known. This is possible by conducting simultaneous measurements with a DMA and an ELPI (e.g., Ahlvik et al., 1998). Without any a priori knowledge, there can be substantial error in the estimated particle density. This causes a potential source of bias. The size of bias in measured concentration depends on particle size through the size dependence of slip correction and charging efficiency. Moisio (1999) reported calculations of the effect on the TUT prototype ELPI. Rather against intuition, the effect of density on the instrument reading is higher in the number concentration measurement. In the mass measurement, the effect of particle density is small for coarse particles, but becomes quite significant for fine particles having a large slip correction factor.

The ELPI is based on impaction and, therefore, has some of the problems associated with impactor measurements, namely, particle bounce and blow-off (see Chapter 10). These effects are further complicated by possible contact charging during particle bounce. Contact charge transfer is difficult to predict without knowledge of the particle material (John, 1995; Horton et al., 1992). As in normal impactor use, preparation of the collection substrates is extremely important. Moisio (1999) reported severe distortion of the number size distribution of solid NaCl particles when ungreased aluminum foil substrates were used. Even careful greasing of the substrates did not completely solve the problem. Replacing the foil substrates with sintered metal plates covered with light oil solved the problem (Moisio, 1999). These substrates, however, also change the cut characteristics of the impactor, requiring a new calibration. Problems encountered trying to calibrate the instrument with solid polystyrene latex (PSL) particles (Maricq et al., 1999) can probably be attributed to particle bounce and contact charging.

The correction algorithm mentioned above assumes that all particles are collected. If there is a significant fraction of the aerosol distribution below the lowest cut size, this assumption is not met. This causes an overestimate in the measured concentration of particles with diameters that are much larger than the mode of the distribution. A similar bias is encountered if the calibrated loss values are too low, as described by Virtanen et al. (2001). They showed that for high-concentration measurements the space charge force has to be taken into account.

Results and Applications

The ELPI offers an indirect but rapid means of measuring particle size distribution. The most appealing applications are those where rapid response is necessary. As an example, Kymäläinen et al. (1996) used a TUT-ELPI to measure the time-dependent fume release rate during black liquor pyrolysis. In this application, the fuel is heated in a controlled atmosphere in a grid heater, where particle release takes place during short, 5 to 60 s periods. With real-

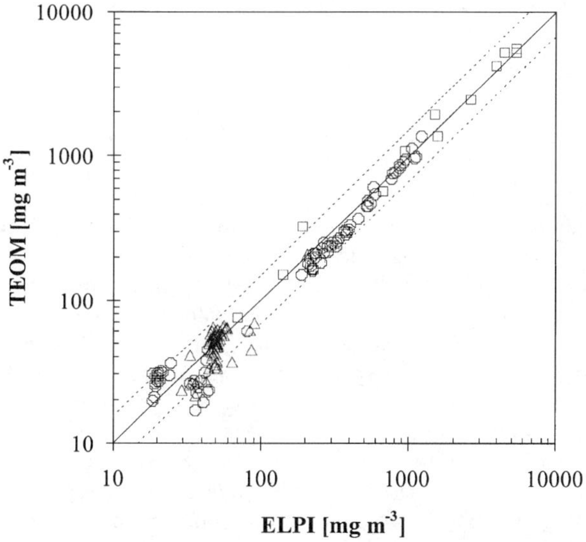

Fig. 14–5. Comparison of instantaneous mass concentration readings of ELPI and TEOM (in mg/m^3). Different markers indicate different power plants. Both instruments sampled through a 1/100 dilution system; the readings are the calculated stack concentrations (Moisio, 1999).

time measurement, it was possible to identify two successive particle production stages even during this short period. For the total particle mass integrated over the release period, good agreement was reported with gravimetric filter measurements.

The first application of the TUT-ELPI was a real-time measurement of combustion aerosol size distributions. Measurements were made at several full-scale power plants (Moisio et al., 1995; Moisio, 1999; Latva-Somppi et al., 1998). Among other things, instantaneous mass concentration values calculated over the measured size range were compared with TEOM results. These results are shown in Figure 14–5 (Moisio, 1999). There is clearly a very good correlation over a wide range of concentrations. The actual concentrations measured by the instruments were a factor of 100 lower than those shown in Figure 14–5 because both instruments sampled through a dilution system. Therefore, the large differences at low concentrations are probably influenced by instrumental noise.

The most important application of the ELPI is in measuring particles emitted by diesel and gasoline engines (Ahlvik et al., 1998; Klein et al., 1998; Pattas et al., 1998; Maricq et al., 1998, 1999). To evaluate emissions, the engine is typically operated according to a standardized test cycle, where the speed and load of the engine vary rapidly (for an overview of the cycles, see, e.g., http://www.dieselnet.com/). Ahlvik and co-workers (1998) used the ELPI to measure size distributions of diesel emissions. Light- and heavy-duty engines were tested for both steady-state and transient driving cycles. Relatively good agreement in total number concentration with the scanning mobility particle sizer (SMPS) was reported assuming standard density, and good agreement was found when using measured effective density values. The main limitation was found to be the lower size limit of 30 nm because a large number of particles had a smaller diameter. Maricq et al. (1999) used both the ELPI and the SMPS in measuring size-resolved particle emissions of gasoline vehicles during transient drive cycles. Good agreement was reported between ELPI and SMPS size distributions, except for the lowest ELPI channel, which tended to systematically overestimate the particle concentration.

SURFACE AREA MEASUREMENT

Introduction and Theoretical Background

The attachment of molecules or atoms to aerosols is a method that allows surface-related information to be obtained by on-line techniques. These techniques do not yield the geometric surface, through use of a Brunauer-Emmett-Teller (BET) analysis for example, but an "active surface," which is defined below. To detect attached atoms or molecules, they have to be labeled. Two techniques are therefore used: radioactive labeling or electrical charging. Instruments based on these two principles are described in the following sections.

Attachment is determined by two probabilities: the collision probability and the sticking probability or sticking coefficient. Only the first depends on geometry and is of importance here. As only the product of both can be measured, the sticking coefficient has to be unity. This is the case for ions, detected by an electrical measurement. Even if the "carrier molecule" does not stick, a charge transfer will occur. The condition is also fulfilled for large molecules or heavy atoms such as lead.

When a sticking coefficient of unity is assumed, the attachment probability equals the collision probability. For ions, image and Coulomb potentials may influence the collision probability in addition to geometry. To avoid Coulomb repulsion, the ion concentration has to be kept low enough to avoid multiple charging. The image force can be neglected if the particle diameter is larger than about 10 nm (Filippov et al., 1993). If these conditions are fulfilled, the integral collision cross section, or attachment cross section, can be measured. This is an important quantity by itself, as it determines adsorption kinetics and, thereby, has significant influence on chemical reactions between particles and the surrounding gas phase or gas phase reactions when the particles serve as a catalyst. It also determines particle growth by attachment of material from the gas phase. It could be described as the fraction of geometrical surface, which is directly accessible from outside. For this reason the term *active surface* is used. Another related designation, introduced in connection with the Epiphaniometer (Pandis et al., 1991), is the *Fuchs surface*.

In addition, it can be shown that for a certain particle mobility b, mobility and active surface A are inversely proportional within an uncertainty of some percent independent of particle size, shape, or material (Siegmann and Siegmann, 2000).

$$B \cdot A = const. \tag{14–11}$$

This is plausible because the drag force, defining B, is also due to collisions with carrier gas molecules. The collisions determine the momentum transfer and, thereby, also the drag force. This means that A will scale with the mobility diameter d^2 in the free molecular regime and with d in the continuum regime. In a more rigorous calculation, size and mass of the adsorbing species have to be considered, as is taken into account in the definition of the Fuchs surface by Pandis et al. (1991). In this discussion, it is assumed that $d_{molecule} << d_{particle}$. This makes the calculation more straightforward, and hence the resulting error is small compared with the experimental errors. Figure 14–6 (adapted from Gäggeler et al., 1989) shows the attachment rate as a function of the mobility diameter. Both regimes ($\propto d^2$ and $\propto d$) are clearly seen.

In the free molecular range, the relation between mobility B and mobility diameter d is given by (Fuchs, 1989)

$$B = \frac{\lambda}{\pi \delta \eta d^2} \tag{14–12}$$

where λ is the mean free path in the gas, η is the viscosity, and δ is a scattering parameter, depending on whether the scattering is diffuse or specular, which is close to unity. In this size range the active surface is defined by the geometric surface of a sphere

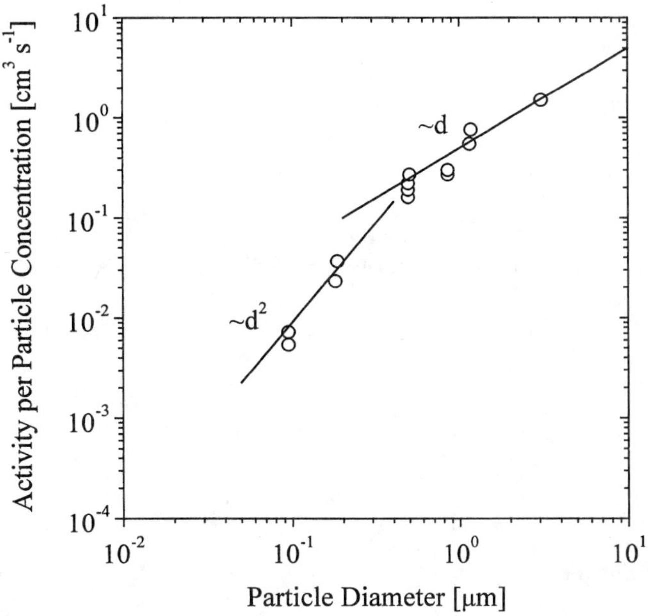

Fig. 14–6. Attachment rate as a function of the mobility diameter. (Adapted from Gäggeler et al., 1989.)

$$A = \pi d^2 \tag{14–13}$$

The relation between A and B then becomes

$$A = \frac{\lambda}{\delta \eta B} \tag{14–14}$$

As this relation is valid for the whole size range according to Eq. 14–11, it is used as a definition of the active surface of a particle of mobility b. The total active surface then becomes

$$A_{\text{tot}} = \int_{b=0}^{\infty} A(B)N(B)\text{d}B \tag{14–15}$$

What is really measured is the attachment rate dn/dt. For particles of a certain size, this is

$$\frac{\text{d}n}{\text{d}t} = -KNn \tag{14–16}$$

where n is the atom/ion concentration and dn/dt their decrease in concentration due to attachment to particles, N is the concentration of monodisperse particles of diameter d or mobility B, respectively, and K is the appropriate coefficient.

In the free molecular regime K is (Siegmann and Siegmann, 2000)

$$K = \overline{\upsilon}\frac{\pi}{4}d^2 \tag{14–17}$$

where $\overline{\upsilon}$ is the average velocity of the atom/ion and m its mass

$$\overline{\upsilon} = \sqrt{\frac{3kT}{m}} \tag{14–18}$$

Replacing d^2 by A according to Eq. 14–13 yields

$$K = \sqrt{\frac{3kT}{m}} A \tag{14–19}$$

The attachment rate then becomes

$$\frac{dn}{dt} = \sqrt{\frac{3kT}{m}} ANn \tag{14–20}$$

For polydisperse aerosols A has to be replaced by the total active surface A_{tot} (Eq. 14–15). A_{tot} can now be calculated from the measured dn/dt by

$$A_{tot} = \frac{1}{\sqrt{\frac{3kT}{m}} n} \frac{dn}{dt} \tag{14–21}$$

Again, no assumptions on particle shape or size are required.

EPIPHANIOMETER

Measurement Principles

The Epiphaniometer (EPI) was developed at the Paul Scherrer Institute (PSI), Switzerland, in the 1980s (Gäggeler et al., 1989). The EPI is based on the attachment of lead atoms (^{211}Pb), produced by the radioactive decay of a long-lived ^{227}Ac source:

$$^{227}\text{Ac}(T_{1/2} = 21.8\,\text{years})(\beta) \rightarrow {}^{227}\text{Th}(T_{1/2} = 18.7\,\text{days})(\alpha) \rightarrow {}^{223}\text{Ra}(T_{1/2} = 11.4\,\text{days})(\alpha) \rightarrow$$
$$^{219}\text{Rn}(T_{1/2} = 3.96\,\text{s})(\alpha) \rightarrow {}^{215}\text{Po}(T_{1/2} = 0.0018\,\text{s})(\alpha) \rightarrow {}^{211}\text{Pb}(T_{1/2} = 36.1\,\text{min})(\beta) \rightarrow$$
$$^{211}\text{Bi}(T_{1/2} = 2.17\,\text{min})(\alpha) \rightarrow {}^{207}\text{Tl}(T_{1/2} = 4.77\,\text{min})(\beta) \rightarrow {}^{207}\text{Pb}\,(\text{stable})$$

The number of attached ^{211}Pb lead atoms is then determined by counting the α-decay events of its daughter, ^{211}Bi.

Instrument Design

Figure 14–7 shows the instrumental configuration. The aerosol sample is first introduced into a 0.002 m^3 [2 L] exposure chamber. The actinium source at the bottom of this container provides a constant supply of ^{219}Rn gas. The radon decays with a half-life of 3.96 s to ^{211}Pb, which deposits on the particles. Next, the stream of particles and unattached lead flows through a 1 m tube. The unattached lead diffuses to the walls of the tube while the aerosol continues on to a total particle filter located directly beneath an α-detector. The attached ^{211}Pb concentration is measured via the α-decay of its daughter ^{211}Bi. The geometry of the detection unit is optimized to ensure correct measurement of α-particles. A Nuclepore filter collects particles on its smooth surface; thus the α-particles do not have to penetrate out of the filter material. The tubing is fed through the hole of the detector for optimized geometry. The α-detector is an annular PIPS-detector (Planar Ion-implanted Passivated Silicon, Canberra Semiconductor, NV) with a central 5 mm hole. This detector produces a signal proportional to the energy of the incident α-particle. The signals from the detector are classified into a 1024 channel histogram. This design offers the ability to simultaneously measure the activity concentration of ^{212}Pb, the daughter nuclide of the naturally occurring ^{220}Rn, and ^{211}Bi from the ^{227}Ac source. Due to overlapping α-energies, the measurement of the better-characterized ^{214}Pb (daughter nuclide of natural ^{222}Rn) is only possible if the ^{227}Ac source is eliminated from the system.

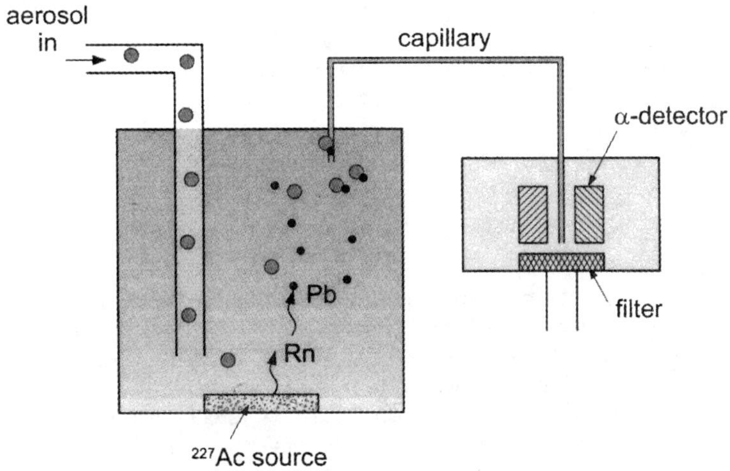

Fig. 14–7. Instrumental setup of the Epiphaniometer.

Flow control (typically $1.7 \times 10^{-5}\,m^3/s$ [1 L/min]) is achieved with a mass flow controller, which directly regulates the pump. Both an alternating current version and a battery-operated version are available. The battery version has a very low energy consumption of only 8 W. Improved versions of these instruments will soon be commercially available by Matter Engineering (***MAT***).

Principally, the detection limit of the EPI depends on the background of the detector signal, which is negligible. For low aerosol concentration the uncertainty of the measurement is given by counting statistics. An average Poisson error of 10% translates into a surface area concentration error of 0.2 and $0.003\,\mu m^2\,cm^{-3}$ for integration times of 30 min and 24 h, respectively.

Under steady-state conditions (i.e., a constant aerosol input) the decay rate of ^{211}Bi equals the rate of lead deposition on the filter. For nonsteady conditions, the instantaneous rate of ^{211}Pb deposition can be deconvoluted from the EPI signal using the known radioactive decay rates as follows. The EPI accumulates alpha counts for time intervals Δt, typically from 5 to 30 min. The total number of counts in this interval Y_i depends on the amount of lead deposited during the count interval and also on previously deposited lead. The average rate of lead deposition over the count interval $i + 1$ is, therefore, approximately (Rogak et al., 1991)

$$f_{i+1} = \frac{Y_{i+1} - Y_i \exp(-\Delta t \lambda)}{1 - \exp(-\Delta t \lambda)} \tag{14–22}$$

where $\lambda = \ln 2/36.1\,min^{-1}$ is the ^{211}Pb decay constant. This inversion procedure may be applied for time intervals of 10 min and higher. For shorter time intervals, a rigorous inversion giving a time resolution of a few minutes has been developed by Pandis et al. (1991). A higher time resolution is not advisable because of the inherent damping by the half-life of the ^{211}Pb.

Calibration of the instrument is best performed with a monodisperse aerosol obtained with a Differential Mobility Analyzer (DMA), where the number concentration is simultaneously determined with a Condensation Particle Counter.

Potential Biases

The EPI signal is proportional to the active surface area concentration only when the lead concentration in the chamber is unaffected by the aerosol. This is the case when most of the lead atoms are deposited on the walls of the instrument rather than on the particles. For large surface area concentrations, most of the lead produced in the chamber becomes attached to aerosol particles resulting in saturation. When saturated, the detector count rate reaches a maximum, s_{max}. It was experimentally shown that for count rates less than 45% of s_{max} the EPI signal is linearly related to the surface area (Rogak et al., 1991). For higher count rates the EPI response becomes significantly nonlinear, and a correction factor F was determined experimentally (Rogak et al., 1991) as

$$F = 1 + 0.144x - 0.0108x^2 - 4.88x^3 + 10.22x^4 \quad (14\text{–}23)$$

where $x = s/s_{max}$ is the measured activity normalized by the saturation value. This correction is not necessary as long as the measurements are not made too close to sources. Even though the attachment rate scales with the lead atom concentration in the container, this saturation limit may not be increased by decreasing the ^{227}Ac activity because it is determined by the relative change of the lead atom concentration. It can, in principle, be increased either by decreasing the container volume or increasing the flow rate, both resulting in a decreased residence time of the aerosol in the container. However, a complete decay of the gaseous ^{219}Rn has to be ensured, which otherwise penetrates to the filter and contributes to an increased background count rate. This takes at least 10 half-lives (40 s).

Another potential bias may result if the sticking coefficient is lower than unity because the attachment probability will then not equal the collision probability (see introduction to the discussion of surface area measurements, above). Tests using a variety of different aerosols, however, have confirmed that the sticking coefficient does not depend on the chemistry of the aerosol particles and is equal to unity in all cases (Rogak et al., 1991).

Results and Applications

Because of its low detection limit and robust design, the EPI is a suitable instrument to be operated at remote locations. Due to its low energy consumption, the instrument has been deployed on a glacier using a solar panel power source (Baltensperger et al., 1991; Lugauer et al., 1998). In 1988, EPI measurements were started at the high-alpine research station Jungfraujoch, 3454 m above sea level (Baltensperger et al., 1997). As an example, Figure 14–8 shows EPI measurements from the Jungfraujoch. Since July 1995 continuous measurements of other aerosol parameters have additionally been conducted, and the station has been incorporated into the European baseline station of the Global Atmosphere Watch Program (GAW) of the World Meteorological Organization (WMO), also comprising the Zugspitze (Germany) and the Sonnblick (Austria). The highest aerosol concentrations have been found from June to August and the lowest from December to February. A pronounced diurnal variation, having a maximum in the late afternoon, was observed in summer. This feature is due to thermal transport processes of air masses from the polluted planetary boundary layer into the free troposphere. A detailed meteorological analysis (Lugauer et al., 1998) has shown that seasonality and diurnal variations mainly depend on synoptic weather conditions, as does the vertical transport. This is most evident in the difference between anticyclonic and cyclonic weather conditions.

The EPI has also been deployed on a cruising ship in the Atlantic and southern oceans (Davison et al., 1996), in a tunnel study (Weingartner et al., 1997), and for measurements in an urban area (Harrison et al., 1999).

DIFFUSION CHARGER

Measurement Principles and Instrument Design

Diffusion charging has already been used for a long time mainly in connection with size analyzers. A unipolar diffusion charger is used, for example, in the ***TSI*** Electrical Aerosol

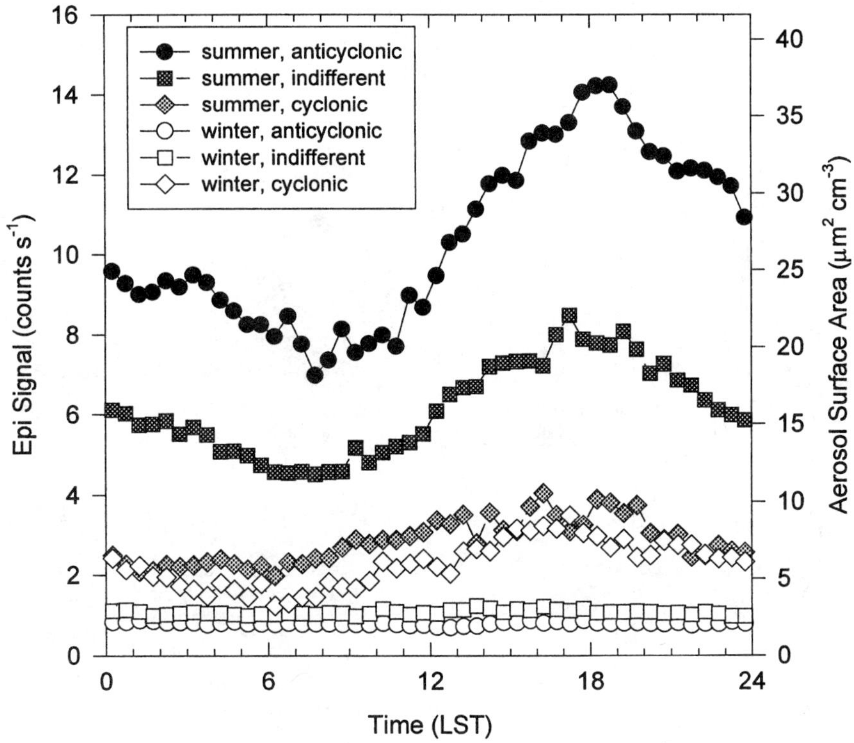

Fig. 14–8. Median diurnal variation of aerosol concentration at the high-alpine site Jungfraujoch (3454 masl) for anticyclonic, indifferent, and cyclonic conditions of the convective weather types in summer (June, July, August) and in winter (December, January, February). Data from 1991 to 1997 (Lugauer et al., 1998). Left axis: Epiphaniometer signal; right axis: aerosol surface area concentration, deduced from Epiphaniometer signal by assuming a constant shape of the number size distribution.

Analyzer (EAA) and in the ELPI, described above. Here, its use in combination with an "aerosol electrometer" (i.e., a device to measure the current corresponding to the flow of charged particles, termed *DC sensor*) will be discussed.

In a similar manner to the ELPI, the diffusion charger uses positive or negative air ions to measure the attachment coefficient of particles. Figure 14–9 illustrates the working principle of this instrument. Ions are created by an electrical corona discharge from a corona tip operated at several kilovolts. These ions may attach to the particles by Brownian diffusion. After the charging section, remaining ions are removed by an ion trap electrode to which a low voltage is applied. The particles are then precipitated onto an electrically insulated filter, and the filter current is measured using a sensitive current amplifier. This current I directly yields the ion attachment rate

$$\frac{dn}{dt} = \frac{I}{e} \tag{14–24}$$

which is required to calculate the active surface according to Eq. 14–21.

The simple design described above has some drawbacks. In particular, small particles are lost by precipitation in the high electric field in the vicinity of the corona electrode. A number of other charger designs have been developed. To our knowledge, only one commercial instru-

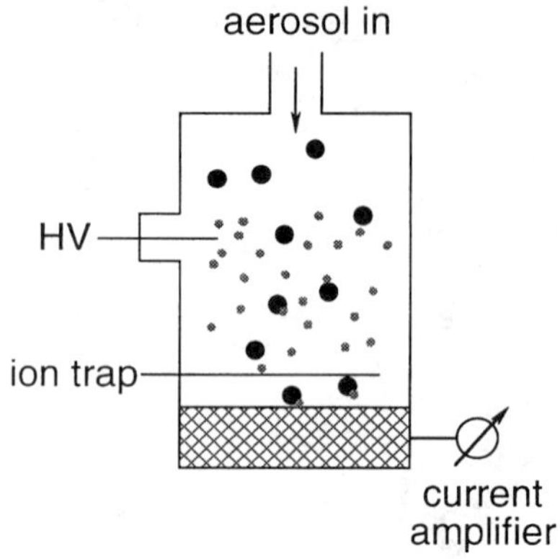

Fig. 14–9. Working principle of the diffusion charger. HV indicates the high voltage corona power supply.

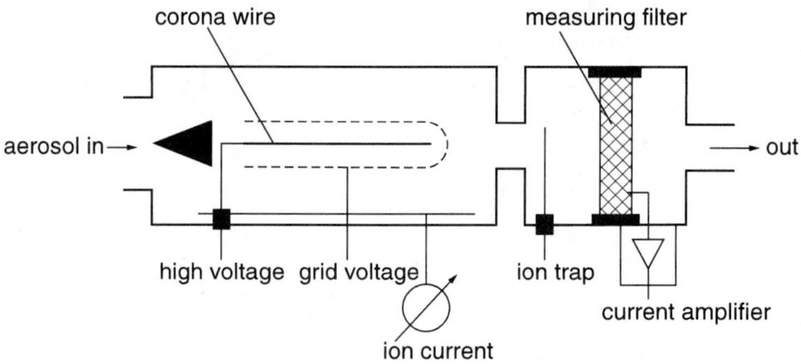

Fig. 14–10. Configuration of the LQ1-DC by Matter Engineering (***MAT***).

ment is available that allows charging plus current measurement in one unit. This is the LQ1-DC (***MAT***). It has the configuration shown in Figure 14–10, where the charging section is similar to the configuration used by ***TSI*** in the EAA. The corona discharge is not concentrated on a tip, as in Figure 14–9, but occurs along a thin Pt wire. This has the main advantage that it is less sensitive to contamination. To reduce particle losses due to the high-voltage field, the charging section and the corona discharge are separated by a screen. The electrical field is small in the charging section. Only several volts are applied between the screen and the outer tube. This field is required to make the ions penetrate the screen and enter into the charging section.

The ion concentration may be monitored by measuring the current between the outer tube and ground. This current can be used to keep the ion concentration constant by a feedback control, which controls the high-voltage supply. This feedback control is not yet included in the LQ1-DC, which uses a constant corona voltage. This design offers efficient charging and relatively low particle losses except for very small particles of only a few nanometers in diameter, which have an electrical mobility high enough to also be precipitated in the weak field. Again, ions have to be prevented from reaching the measuring filter through use of an ion trap, as described above. The positively charged particles are then deposited onto the measuring filter, which has a high insulation to ground. The filter is connected to the input of an ultralow current charge amplifier (resolution 1 fA = 10^{-15} A). The voltage signal from the charge amplifier is converted into an analog signal, calibrated to yield the active surface.

Other charger designs described in the literature are, for example, designed to minimize ultrafine particle losses. Büscher et al. (1994) described a charger using a square-wave AC field instead of a DC field in the charging section. This reduced small particle losses.

Potential Biases

As mentioned, several conditions have to be fulfilled to obtain a value that corresponds to the collision probability of *neutral* atoms or molecules onto the active surface, respectively.

1. The particle concentration or, more precisely, the total active surface has to be low enough to avoid depletion of ions due to attachment to particles; otherwise, saturation occurs. From this point of view, the time available for charging should be short.
2. Multiple charging should be avoided because Coulomb repulsion decreases the attachment probability. This condition is usually fulfilled for very small particles. For larger particles problems occur if the $n{\cdot}t$ product is high. This means that avoiding multiple charging requires a small charging efficiency or low $n{\cdot}t$ product.
3. For particles <10 nm the image potential will increase the effective collision cross section; the attachment rate of ions will be higher than that for neutral species.
4. The detection limit is determined by the noise and zero stability of the current amplifier. For the LQ1-DC, the detection limit is about 1 fA (10^{-15} A). To obtain a low detection limit in terms of the active surface with available amplifier characteristics, efficient charging is required. This can be obtained by means of higher $n{\cdot}t$ values, resulting in a measurable concentration range of 10 to 2000 $\mu m^2 cm^{-3}$ in the active surface area concentration.
5. Ion attachment is not material dependent as long as it is dominated by Brownian diffusion. As soon as electrical forces become important, material dependence has to be considered. This is the case for ultrafine particles smaller than 10 nm where, as mentioned above, the image potential becomes important. The image force depends on the polarizability, which is material dependent.

These considerations show that the requirements on the charging efficiency are contradictory. A compromise has to be found to obtain optimal performance. For maximum flexibility, the ion concentration needs to be adjustable, to be adapted to the specific measurements. The ion concentration should be high if a low detection limit is required and the particles are small and low if larger particles have to be measured.

Results and Applications

In principle, the EPI and DC sensor yield the same information, which is the active surface as defined above. However, when regarding possible applications, there are significant differences. The EPI is a very sensitive instrument, as it is based on counting single events (radioactive decays). It is, therefore, especially suited to measure very low concentrations, whereas at higher concentrations saturation occurs due to the long residence time. The diffusion charger, on the other hand, operates up to a much higher concentration, as the residence time is short. However, due to current amplifier limitations, its detection limit is significantly higher than that of the EPI. The response time of the diffusion charger is short; a time resolution of less than 1 s is possible. This makes this instrument suitable for monitoring transient processes such as engine acceleration. Examples are given after the next section.

PHOTOELECTRIC AEROSOL SENSOR

Measurement Principle

Photoemission has long been used as an analysis method in surface physics, mostly under vacuum conditions. Applied to aerosols, aerosol photoemission is a sensitive method for the characterization of fine particles. The method was developed at the ETH in Zürich (Schmidt-Ott and Federer, 1981), and the working principle of aerosol photoemission is described in detail by Burtscher (1992). If particles are irradiated with ultraviolet light of a photon energy higher than the ionization threshold of the particles, electrons are emitted from the particles. Depending on the carrier gas, these electrons will rapidly attach to a gas molecule and form an ion, diffusing in the vicinity of the particle. If the particle is small compared with the mean free path of the electron, the probability of back diffusion is negligibly small. For larger particles this probability increases, which means that only small particles can be charged efficiently by photoelectron emission. In practice, particles with diameters smaller than 1 μm can be charged. A quantitative treatment of this process was developed by Filippov et al. (1993).

As mentioned, the photon energy has to be higher than the threshold; on the other hand, it should be lower than the excitation energies of the carrier gas, and it has to be lower than the ionization threshold of the carrier gas. In air, this leaves a relatively small window in the range from 4 to 7 eV. At 7 eV, O_2 becomes excited; however, the efficiency is low enough to keep O_3 formation and other undesired photochemical reactions at a tolerable level.

Electron emission probability, also termed *photoelectric yield*, depends on light absorption leading to an excited electron and on the subsequent escape probability of this electron. Because the mean free path of electrons is on the order of several monolayers, only electrons excited close to the surface can be emitted. This method, therefore, is extremely sensitive to the state of the surface, for example, to material adsorbed on the particle surface.

A number of studies have shown that ambient particles, originating from incomplete combustion of hydrocarbons, exhibit the highest photoelectric yield and may, therefore, be selectively detected. A special role is ascribed to adsorbed polycyclic aromatic hydrocarbons (PAHs), which usually dominate the photoelectric yield in the case of urban ambient aerosols. On the other hand, diesel emissions may contain so few PAHs that photoemission is determined by elemental carbon, even though it has a significantly lower photoelectric yield.

Instrument Design

Figure 14–11 shows the basic components of a sensor to measure aerosol photoelectron emission. Sampled air is introduced into the system through a quartz glass tube where it is irradiated by an external ultraviolet light source. The resultant photoelectrons (or negative ions) are separated from the positively charged particles in an ion trap that follows the ionization chamber. The charged sample particles are finally collected from the sample stream in an electrically insulated filter that is placed in a Faraday cup. This filter is connected to a current amplifier where the current of positively charged particles is measured. The design of the photoelectric charger is similar to that of the diffusion charger. The major difference is in the way the particles are charged.

The EcoChem PAS2000 (***ECO***) uses an excimer lamp surrounding the sample air tube. This design feature allows a compact construction. In addition, excimer lamps have a long lifetime and a very short warm-up time. This means that the sensor can be operated in a

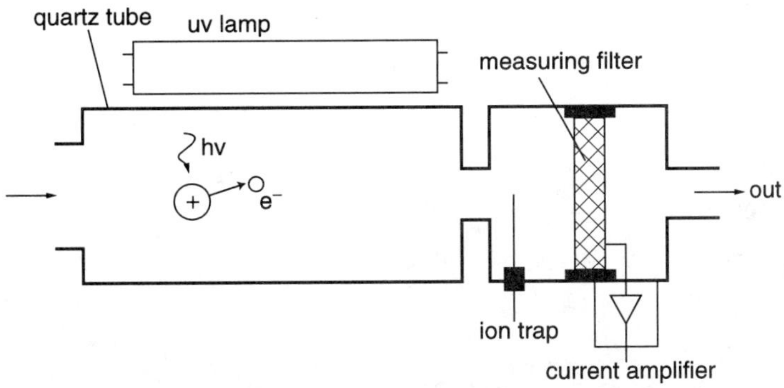

Fig. 14–11. Instrumental setup of a photoemission sensor.

pulsed mode by continuously switching the lamp on and off. This allows continuous offset compensation and increased sensitivity by measuring the difference signal between having the lamp on or off. Relatively small battery-operated units are available that can be used in personal monitors.

Potential Biases

The chamber wall, which is irradiated with ultraviolet light, should be made from a material with a high work function and that is transparent to ultraviolet light. However, such materials can readily achieve high static charge levels, and the resulting electric field can cause significant small particle losses. This is usually no problem for particles with $d > 10\,\text{nm}$. A solution for particles in the nanometer range is presented by Jung et al. (1988). The authors used two concentric metal grids placed on the inside of a chamber. Rosatzin and Burtscher (1988) presented a similar design, which can be used for high-temperature sample streams.

In the case of PAHs that are adsorbed on carbon particles, photoemission increases sharply for surface coverage less than a monolayer (Burtscher and Schmidt-Ott, 1986). Different PAHs have a different influence on the photoelectric yield. Some even cause a decrease compared with pure carbon particles (Niessner and Wilbrig, 1989). The major increase of photoelectric activity is observed for large planar PAHs. As long as the coating is thin, the increase in photoelectric activity is directly related to the degree of surface coating or amount of PAH on the particle. This relationship is no longer true for thick coatings. In this case photoemission only occurs at the top layer.

Results and Applications

Particles from incomplete combustion were found to have a high photoelectric yield. This was attributed to the activity of adsorbed PAH that partially covered the surface of the particles. The photoelectric yield of combustion particles has been shown to be linearly related to the total mass concentration of particle-bound PAHs (PPAH) in ambient air samples (McDow et al., 1990; Hart et al., 1993; Wilson and Barbour, 1994). For an urban aerosol, PPAH can be detected with these sensors in the concentration range of 1 to $5000\,\text{ng}\,\text{m}^{-3}$.

Experimental results and theoretical considerations show that the size dependence of photoelectric yield is very similar to that of atom attachment, as measured by an EPI or a DC sensor (Siegmann and Siegmann, 2000). Normalizing the signal of the Phototoelectric Aerosol Sensor to that of the EPI or the DC sensor therefore eliminates the influence of particle size and concentration. What remains is information on the chemical nature of the surface and

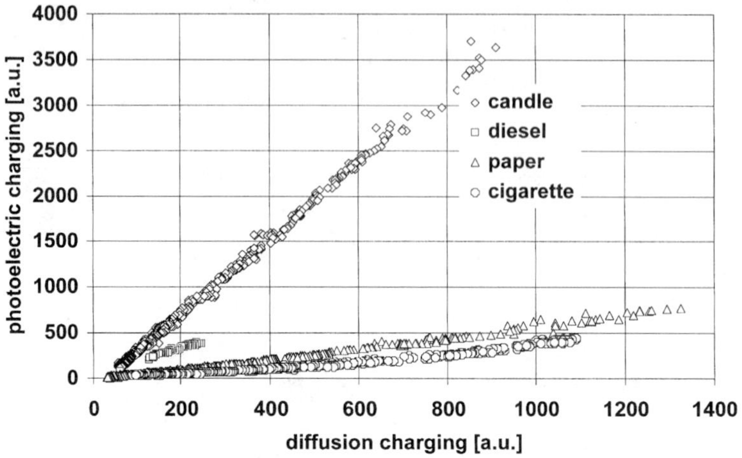

Fig. 14–12. Photoemission versus diffusion charging for particles from different types of combustion. (From Siegmann et al., 1999.)

the bulk of the particle. This method was used by Ammann et al. (1992, 1993) to characterize volcanic aerosol and by Matter et al. (1999) to investigate the source attribution of different combustion aerosols. As an example, Figure 14–12 shows the signal of the photoemission sensor versus the signal of the DC sensor for exhaust from different kinds of combustion. The slopes contain the above-mentioned chemical information, while the linear relation shows that the size and concentration do not influence the slope (= normalized signal). Photoelectric aerosol sensors are commercially available (***ECO*** and ***MAT***) and are often used for ambient air quality monitoring.

REFERENCES

Ahlvik, P., L. Ntziachristos, J. Keskinen, and A. Virtanen. 1998. Real time measurements of diesel particle size distribution with an electrical low pressure impactor. *SAE Technical Paper Series No 980410*. Warrendale, PA: Society of Automotive Engineers.

Allen, G., C. Sioutas, P. Koutrakis, R. Reiss, F. W. Lurmann, and P. T. Roberts. 1997. Evaluation of the TEOM method for measurement of ambient particulate mass in urban areas. *J. Air Waste Manage.* 47:682–689.

Ammann, M., R. Hauert, H. Burtscher, and H. C. Siegmann. 1993. Photoelectric charging of ultrafine volcanic aerosols: Detection of Cu(I) as a tracer of chlorides in magmatic gases. *J. Geophys. Res.* 98:551–556.

Ammann, M., L. Scherrer, W. Müller, H. Burtscher, and H. C. Siegmann. 1992. Continuous monitoring of ultrafine aerosol emissions at Mt. Etna. *Geophys. Res. Lett.* 19:1387–1390.

Ayers, G. P., M. D. Keywood, and J. L. Gras. 1999. TEOM vs. manual gravimetric methods for determination of PM2.5 aerosol mass concentrations. *Atmos. Environ.* 33:3717–3721.

Baltensperger, U., H. W. Gäggeler, D. T. Jost, M. Emmenegger, and W. Nägeli. 1991. Continuous background aerosol monitoring with the epiphaniometer. *Atmos. Environ.* 25A:629–634.

Baltensperger, U., H. W. Gäggeler, D. T. Jost, M. Lugauer, M. Schwikowski, E. Weingartner, and P. Seibert. 1997. Aerosol climatology at the high-alpine site Jungfraujoch, Switzerland. *J. Geophys. Res.* 102:19707–19715.

Brook, J. R., T. F. Dann, and Y. Bonvalot. 1999. Observations and interpretations from the Canadian fine particles program. *J. Air Waste Manage.* 49:35–44.

Burtscher, H. 1992. Measurements and characteristics of combustion aerosols with special consideration of photoelectric charging and charging by flame ions. *J. Aerosol Sci.* 23:549–595.

Burtscher, H. and A. Schmidt-Ott. 1986. In situ measurement and adsorption and condensation of a polyaromatic hydrocarbon on ultrafine particles by means of photoemission. *J. Aerosol Sci.* 17:699–703.

Büscher, P., A. Schmidt-Ott, and A. Wiedensohler. 1994. Performance of a unipolar "square wave" diffusion charger with variable nt-product. *J. Aerosol Sci.* 25:651–664.

Chow, J. C. 1995. Measurement methods to determine compliance with ambient air quality standards for suspended particles. *J. Air Waste Manage.* 45:320–382.

Cooper, D. W. 1975. Statistical errors in beta absorption measurements of particulate mass concentration. *J. Air Poll. Control Assoc.* 25:1154.

Cooper, D. W. 1976. Significant relationships concerning exponential transmission or penetration. *J. Air Poll. Control Assoc.* 26:366.

Courtney, W. J., R. W. Shaw, and T. C. Dzubay. 1982. Precision and accuracy of beta gauge for aerosol mass determinations. *Environ. Sci. Technol.* 16:236–239.

Davison, B., C. O'Dowd, C. N. Hewitt, M. H. Smith, R. M. Harrison, D. A. Peel, E. Wolf, R. Mulvaney, M. Schwikowski, and U. Baltensperger. 1996. Dimethyl sulfide and its oxidation products in the atmosphere of the Atlantic and southern oceans. *Atmos. Environ.* 30:1895–1906.

Deacon, A. R., R. G. Derwent, R. M. Harrison, D. R. Middleton, and S. Moorcroft. 1997. Analysis and interpretation of measurements of suspended particulate matter at urban background sites in the United Kingdom. *Sci. Tot. Environ.* 203:17–36.

Eatough, D. J., F. Obeidi, Y. B. Pang, Y. M. Ding, and W. E. Wilson. 1999. Integrated and real-time diffusion denuder sampler for PM2.5. *Atmos. Environ.* 33:2835–2844.

Egan, M. R. and C. D. Litton. 1986. Wood crib fires in a ventilated tunnel. *US Bureau of Mines RI 9045.*

Evans, R. D. 1955. *The Atomic Nucleus.* New York: McGraw-Hill.

Fairchild, C. I, M. I. Tillery, and H. J. Ettinger. 1980. *An Evaluation of Fast Response Aerosol Mass Monitors.* Los Alamos National Laboratory, Report LA-8220.

Filippov, A. V., A. Schmidt-Ott, and W. Fendel. 1993. Escape probability of photoelectrons from particles. *J. Aerosol Sci.* 24(Suppl. 1):S501–S502.

Flagan, R. C. 1982. Compressible flow inertial impactors. *J. Colloid Interface Sci.* 87:291–299.

Fuchs, N. A. 1989. *The Mechanics of Aerosols.* New York: Dover.

Gäggeler, H. W., U. Baltensperger, M. Emmenegger, D. T. Jost, A. Schmidt-Ott, P. Haller, and M. Hofmann. 1989. The epiphaniometer, a new device for continuous aerosol monitoring. *J. Aerosol Sci.* 20:557–564.

Gleason, G. I., J. D. Taylor, and P. L. Tabern. 1951. Absolute beta counting at defined geometries. *Nucleonics* 8:12–21.

Harrison, R. M., A. R. Deacon, M. R. Jones, and R. S. Appleby. 1997. Sources and processes affecting concentrations of PM10 and PM2.5 particulate matter in Birmingham (UK). *Atmos. Environ.* 31:4103–4117.

Harrison, R. M., M. Jones, and G. Collins. 1999. Measurements of the physical properties of particles in the urban atmosphere. *Atmos. Environ.* 33:309–321.

Hart, K. M., S. R. McDow, W. Giger, D. Steiner, and H. Burtscher. 1993. The correlation between in-situ, real-time aerosol photoemission intensity and particulate polycyclic aromatic hydrocarbons in combustion aerosols. *Water Air Soil Pollut.* 68:75–90.

Heintzenberg, J. and P. Winkler. 1984. Elemental carbon in the urban aerosol: Results of a seventeen month study in Hamburg, FRG. *Sci. Tot. Environ.* 36:27–38.

Hering, S. V. 1987. Calibration of the QCM impactor for stratospheric sampling. *Aerosol Sci. Technol.* 7:257–274.

Hillamo, R. E. and E. I. Kauppinen. 1991. On the performance of the Berner low pressure impactor. *Aerosol Sci. Technol.* 14:33–47.

Hoek, G., H. Welinder, E. Vaskovi, G. Ciacchini, N. Manalis, O. Royset, A. Reponen, J. Cyrys, and B. Brunekreef. 1997. Interlaboratory comparison of PM10 and black smoke measurements in the PEACE study. *Atmos. Environ.* 31:3341–3349.

Horton, K. D., M. H. E. Ball, and J. P. Mitchell. 1992. The calibration of a California Measurements PC-2 quartz crystal cascade impactor (QCM). *J. Aerosol Sci.* 23:505–524.

Jaklevic, J. M., R. C. Gatti, F. S. Goulding, and B. W. Loo. 1981. A beta gauge method applied to aerosol samples. *Environ. Sci. Tech.* 15:680–686.

Jaklevic, J. M., N. W. Madden, and C. E. Wiegand. 1983. A precision beta gauge using a plastic scintillator and photomultiplier detector. *Nucl. Intrum. Methods* 214:517–518.

John, W. 1995. Particle-surface interactions: Charge transfer, energy loss, resuspension and deagglomeration. *Aerosol Sci. Technol.* 23:2–24.

Juan, L., S. Brown, N. Serageldin, J. Rosell, J. Lazcano, and J. Fernández de la Mora. 1997. Electrostatic effects in inertial impactors. *J. Aerosol Sci.* 28:1029–1048.

Jung, T., H. Burtscher, and A. Schmidt-Ott. 1988. Multiple charging of ultrafine aerosol particles by aerosol photoemission. *J. Aerosol Sci.* 19:485–490.

Kasper, G. 1981. Electrostatic dispersion of homopolar charged aerosols. *J. Colloid Interf. Sci.* 81:32–40.

Keary, J., S. G. Jennings, T. C. O'Connor, B. McManus, and M. Lee. 1998. PM10 concentration measurements in Dublin City. *Environ. Monitor. Assess.* 52:3–18.

Keskinen, J. 1992. *Experimental Study of Real-Time Aerosol Measurement Techniques*. Tampere University of Technology Publications 94, Tampere.

Keskinen, J., M. Marjamäki, A. Virtanen, T. Mäkelä, and R. Hillamo. 1999. Electrical calibration method for cascade impactors. *J. Aerosol Sci.* 30:111–116.

Keskinen, J., K. Pietarinen, and M. Lehtimäki. 1992. Electrical low pressure impactor. *J. Aerosol Sci.* 23:353–360.

King, A. M., T. Pless-Mulloli, J. Merefield, and I. Stone. 2000. New directions: TEOMs and the volatility of UK non-urban PM_{10}: A regulatory dilemma? *Atmos. Environ.* 34:3211–3212.

Klein, H., E. Lox, T. Kreuzer, M. Kawanami, T. Ried, and K. Bächmann. 1998. *Diesel Particulate Emissions of Passenger Cars—New Insights into Structural Changes During the Process of Exhaust Aftertreatment Using Diesel Oxidation Catalyst.* SAE Technical Paper ser. No. 980196. Warrendale, PA: Society of Automotive Engineers.

Klein, F., C. Ranty, and L. Sowa. 1984. New examinations of the validity of the principle of beta radiation obsorption for determinations of ambient air dust concentrations. *J. Aerosol Sci.* 15:391–395.

Kymäläinen, M., K. Janka, J. Keskinen, M. Moisio, R. Backman, and M. Hupa. 1996. Measurement of time-dependent fume release rate during black liquor pyrolysis. *J. Pulp Paper Sci.* 22:17–23.

Latva-Somppi, J., M. Moiso, E. I. Kauppinen, T. Valmari, P. Ahonen, U. Tapper, and J. Keskinen. 1998. Ash formation during fluidized bed incineration of paper mill waste sludge. *J. Aerosol Sci.* 29:461–480.

Lilienfeld, P. 1975. Design and operation of dust measuring instrumentation based on the beta-radiation method. *Staub Reinhalt. Luft* 35:458–465.

Lugauer, M., U. Baltensperger, M. Furger, H. W. Gäggeler, D. T. Jost, M. Schwikowski, and H. Wanner. 1998. Aerosol transport to the high alpine sites Jungfraujoch (3454 m asl) and Colle Gnifetti (4454 m asl). *Tellus Ser. B* 50:76–92.

Macias, E. S. and R. B. Husar. 1974. High resolution online aerosol mass measurement by the beta attenuation technique. In *Proceedings of the 2nd International Conference on Nuclear Methods in Environmental Research*, eds. J. R. Vogt and W. Meyer. CONF-740701. 413 pp. U.S. Energy Research and Development Administration.

Maricq, M. M., R. E. Chase, D. H. Podsiadlik, W. O. Siegl, and E. W. Kaiser. 1998. *The Effect of Dimethoxy Methane Additive on Diesel Vehicle Particulate Emissions*. SAE Technical Paper Ser. No. 982572. Warrendale, PA: Society of Automotive Engineers.

Maricq, M. M., D. H. Podsiadlik, and R. E. Chase. 1999. Examination of the size-resolved and transient nature of motor vehicle particle emissions. *Environ. Sci. Technol.* 33:1618–1626.

Marjamäki, M., J. Keskinen, D.-R. Chen, and D. Y. H. Pui. 2000. Performance evaluation of the Electrical Low Pressure Impactor (ELPI). *J Aerosol Sci.* 31:249–261.

Matter, U., H. C. Siegmann, and H. Burtscher. 1999. Dynamic field measurements of submicron particles from diesel engines. *Environ. Sci. Technol.* 33:1946–1952.

McDow, S. R., W. Giger, H. Burtscher, A. Schmidt-Ott, and H. C. Siegmann. 1990. Polycyclic aromatic hydrocarbons and combustion aerosol photoemission. *Atmos. Environ.* 24A:2911–2916.

McMurry, P. 2000. A review of atmospheric aerosol measurements. *Atmos. Environ.* 34:1959–1999.

Meyer, M. B., J. Lijek, and D. Ono. 1992. Continuous PM_{10} measurements in a woodsmoke environment. In *Transitions: PM_{10} Standards and Nontraditional Particulate Source Controls*, eds. J. C. Chow and D. M. Ono. Pittsburgh: Air Waste Management Association, pp. 24–38.

Mignacca, D. and K. Stubbs. 1999. Effects of equilibration temperature on PM10 concentration from the TEOM method in the Lower Fraser Valley. *J. Air Waste Manage.* 49:1250–1254.

Moisio, M. 1999. *Real-Time Size Distribution Measurement of Combustion Aerosols*. PhD Thesis, Tampere University of Technology, Publications 279.

Moisio, M., J. Hautanen, M. Marjamäki, J. Keskinen, T. Valmari, and E. Kauppinen. 1995. Real-time measurement of size distribution and charge of flue gas particles. EPRI/DOE International Conference on Managing Hazardous and Particulate Air Pollutants, August 15–17, Toronto, Canada, Book 2.

Moisio, M., J. Hautanen, A. Virtanen, M. Marjamäki, and J. Keskinen. 1997. Electrical low pressure impactor data processing—Effect of particle density. *J. Aerosol Sci.* 28(Suppl. 1):S143–S144.

Morawska, L., G. Johnson, Z. D. Ristovski, and V. Agranovski. 1999. Relation between particle mass and number for submicrometer airborne particles. *Atmos. Environ.* 33:1983–1990.

Muir, D. 2000. New directions: The suitability of tapered element oscillating microbalances (TEOMs) for PM_{10} monitoring in Europe. The use of PM_{10} data as measured by TEOM for compliance with the European Air Quality Standard. *Atmos. Environ.* 34:3209–3211.

Newman, J. S. and J. Steciak. 1987. Characterization of particulates from diffusion flames. *Combust. Flame* 67:55–64.

Niessner, R. and P. Wilbrig. 1989. Ultrafine particles as trace catchers for polycyclic aromatic hydrocarbons: The photoelectric aerosol sensor as a tool for in situ sorption and desorption studies. *Anal. Chem.* 61:708–714.

Olcese, L. E. and B. M. Toselli. 1998. Statistical analysis of PM10 measurement in Cordoba City, Argentina. *Meteorol. Atmos. Phys.* 66:123–130.

Olin, J. G. and G. J. Sem. 1971. Piezoelectric microbalance for monitoring the mass concentration of suspended particles. *Atmos. Environ.* 5:653–668.

Pandis, S. N., U. Baltensperger, J. K. Wolfenbarger, and J. H. Seinfeld. 1991. Inversion of aerosol data from the epiphaniometer. *J. Aerosol Sci.* 22:417–428.

Patashnick, H. and C. L. Hemenway. 1969. Oscillating fiber microbalance. *Rev. Sci. Instrum.* 40:1008–1011

Patashnick, H. and G. Rupprecht. 1983. Personal dust exposure monitor based on the tapered element oscillating microbalance. *BuMines* OFR 56-84, NTIS PB 84-173749.

Patashnick, H. and E. G. Rupprecht. 1991. Continuous PM-10 measurements using the tapered element oscillating microbalance. *J. Air Waste Manage.* 41:1079–1083.

Pattas, K., N. Kyriakis, Z. Samaras, P. Pistikopoulos, and L. Ntziachristos. 1998. *Effect of DPF on Particulate Size Distribution Using an Electrical Low Pressure Impactor*. SAE Technical Paper Ser. No. 980544. Warrendale, PA: Society of Automotive Engineers.

Patterson, E. and D. J. Eatough. 2000. Indoor/outdoor relationships for ambient PM2.5 and associated pollutants: Epidemiological implications in Lindon, Utah. *J. Air Waste Manage.* 50:103–110.

Rogak, S. N., U. Baltensperger, and R. C. Flagan. 1991. Measurement of mass transfer to agglomerate aerosols. *Aerosol Sci. Technol.* 14:447–458.

Rosatzin, M. and H. Burtscher. 1988. Improved photoemission sensor for detection of particulate emission from combustion. *J. Aerosol Sci.* 19:633–637.

Salter, L. F. and B. Parsons. 1999. Field trials of the TEOM and Partisol for PM10 monitoring in the St. Austell China Clay area, Cornwall, UK. *Atmos. Environ.* 33:2111–2114.

Schmidt-Ott, A. and B. Federer. 1981. Photoelectron emission from small particles suspended in a gas. *Surface Sci.* 106:538–543.

Shore, P. R. and R. D. Cuthbertson. 1985. Application of a tapered element oscillating microbalance to continuous diesel particulate measurement. *Soc. Automotive Eng.* Report 850, 405. Warrendale, PA: Society of Automotive Engineers.

Siegmann, K., L. Scherrer, and H. C. Siegmann. 1999. Physical and chemical properties of airborne nanoscale particles and how to measure the impact on human health. *J. Mol. Struct.* (*THEOCHEM*) 458:191–201.

Siegmann, K. and H. C. Siegmann. 2000. *Fast and Reliable "In-Situ" Evaluation of Particles and Their Surface with Special Reference to Diesel Exhaust.* SAE Technical Paper Series 2000-01-1995. Warrendale, PA: Society of Automotive Engineers.

Soutar, A., M. Watt, J. W. Cherrie, and A. Seaton. 1999. Comparison between a personal PM10 sampling head and the tapered element oscillating microbalance (TEOM) system. *Atmos. Environ.* 33:4373–4377.

Speer, R. E., H. M. Barnes, and R. Brown. 1997. An instrument for measuring the liquid water content of aerosols. *Aerosol Sci. Technol.* 27:50–61.

Tsai, C.-J. and Y.-H. Cheng. 1996. Comparison of two ambient beta gauge PM10 samplers. *J. Air Waste Manage.* 46:142–147.

Tzou, T.-Z. 1999. Aerodynamic particle size of metered-dose inhalers determined by the quartz crystal microbalance and Andersen cascade impactor. *Int. J. Pharmacol.* 186:71–79.

U.S. Environmental Protection Agency. 1989. *Compendium Method IP-10B: Determination of Respirable Particulate Matter in Indoor Air Using a Continuous Particulate Monitor.* Washington, DC: U.S. Environmental Protection Agency.

Virtanen, A., M. Marjamäki, J. Ristimäki, and J. Keskinen. 2001. Fine particle losses in Electrical Low Pressure Impactor. *J. Aerosol Sci.* 32:389–401.

Weingartner, E., C. Keller, W. A. Stahel, H. Burtscher, and U. Baltensperger. 1997. Aerosol emission in a road tunnel. *Atmos. Environ.* 31:451–462.

Williams, K. L. and R. P. Vinson. 1986. Evaluation of the TEOM dust monitor. *U.S. Bureau of Mines*, IC 9119.

Wilson, N. K. and R. K. Barbour. 1994. Evaluation of a real-time monitor for fine particle-bound PAH in air. *Polycyclic Aromatic Comp.* 5:167–174.

Yu, C. P. 1977. Precipitation of unipolarly charged particles in cylindrical and spherical vessels. *J. Aerosol Sci.* 8:237–241.

15 Optical Direct-Reading Techniques: Light Intensity Systems

JOSEF GEBHART
Dietzenbach, Germany

INTRODUCTION

Light scattering and extinction by small particles suspended in gases are widely applied to obtain information on the concentration and size distribution of the particles. Devices based on this principle combine in situ measurements in the airborne state of the particles with a high degree of automation. This chapter presents measurement techniques that use the intensity of scattered light or the power of attenuated light to characterize particles and particle systems. For practical applications, two kinds of instruments exist: light scattering and extinction by single particles and light scattering and extinction by an assembly of particles. Single-particle techniques based on light scattering cover a size range from about 70 nm to more than 100 µm and are capable of measuring concentrations of less than 10^3 particles per m^3 (clean-room monitoring) to about 10^9 particles per m^3 (aerosol research). Ensemble scattering of light is applicable for concentration measurements of particulate matter in the atmosphere from a few $\mu g/m^3$ to several hundred mg/m^3.

In this section, light scattering and extinction by small particles are described by applying electromagnetic theory to spherical particles of known size and refractive index. The principle variables and basic assumptions of the theory are explained, and computer calculations of theoretical response functions are presented, giving a general survey of the characteristics of optical systems. For a better physical understanding of the phenomena of light scattering and extinction by small particles, two approximations are introduced: dipole or Rayleigh scattering for particles much smaller than the illuminating wavelength and classic optics, including diffraction, reflection, and refraction, for particles much larger than the wavelength. With respect to ensemble techniques (nephelometer, photometer) and their use for concentration measurements, general relationships between photometry and gravimetry are discussed. The main purpose of this theoretical section is to work out and present relations between basic variables of light scattering and extinction by small particles and quantities measurable with optical systems.

A section is devoted to single-particle techniques known as *optical particle counters* (OPCs). In an OPC, aerosol is drawn through a light beam, and light flashes scattered by single particles are received by a photodetector. The number concentration is determined from the count rate of the photoelectric pulses, and the size of the particles is estimated from the pulse height. OPCs are widely used in basic aerosol research, air pollution studies, and clean-room monitoring. Because these various applications require different specifications of

Aerosol Measurement: Principles, Techniques, and Applications, Second Edition, Edited by Paul A. Baron and Klaus Willeke.
ISBN 0-471-35636-0

an instrument, characteristic features of an OPC-like permissible range of number concentration, sensitivity, sampling flow rate, classification accuracy, and size resolution are discussed on the basis of experimental results.

Aerosol photometers are useful as real-time dust monitors in industrial hygiene, for concentration measurements of atmospheric aerosols, and in aerosol inhalation studies. The linearity range of light intensity versus aerosol concentration is limited at high concentrations by multiple scattering and at low concentrations by the stray-light background. The noise level of sensitive instruments is determined by Rayleigh scattering from the air molecules.

Whereas theoretical and experimental calibration curves of optical systems are usually based on particles of spherical shape, most real aerosol systems consist of particles of irregular shape. Therefore, the final section deals with the influence of particle shape on scattering and extinction of light.

LIGHT SCATTERING AND EXTINCTION BY A SINGLE SPHERE

Rigorous Electromagnetic Theory

Electromagnetic theory (Mie, 1908) yields the rigorous solution of light scattering and extinction by a spherical particle. The mathematical expressions have been published in extensive textbooks, such as those by van de Hulst (1957) and Kerker (1969). A list of programs available from several institutions to calculate these expressions is provided on the Internet at diogenes.iwt.uni-bremen.de/~wriedt/Mie_Type_Codes/body_mie_type_codes.html.

Light Scattering. Let a spherical particle of diameter d_p, made up of a material of refractive index m, be situated at the origin of a system of spherical coordinates and be illuminated from the negative z-direction by a plane, linearly polarized, monochromatic wave with electric vector in the x-direction (Fig. 15–1).

The spatial distribution of the light scattered by the particle per unit solid angle can then be described by a scattering function i that depends on the scattering angle θ, the polarization angle ϕ, the particle diameter d_p, the wavelength λ, and the refractive index $m = n - jk$ of the particle material (n, real part; k, imaginary part):

$$i = i(\theta, \phi, d_p, \lambda, m) \tag{15–1}$$

Particle diameter d_p and wavelength λ are connected by the dimensionless size parameter

$$\alpha = \pi d_p / \lambda \tag{15–2}$$

The power of light scattered by a particle per unit solid angle in direction θ is given by

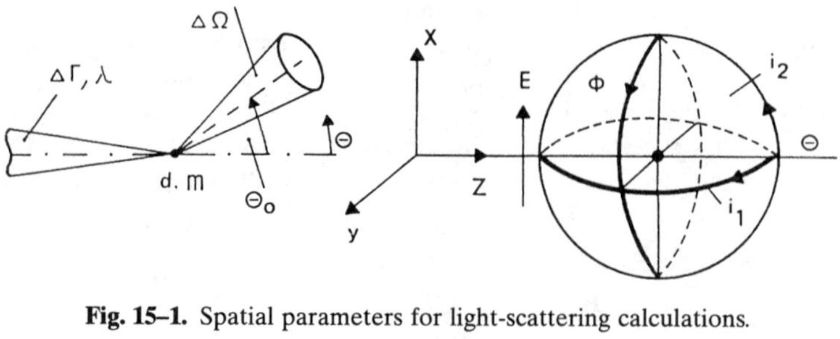

Fig. 15–1. Spatial parameters for light-scattering calculations.

$$S_\lambda(\theta,\phi,d_p,\lambda,m)=I_0\frac{\lambda^2}{4\pi^2}i(\theta,\phi,\alpha,m) \tag{15–3}$$

where I_0 is the illumination intensity (power of light per unit area).

Various optical arrangements used for making light-scattering measurements differ in the mean scattering angle θ_0, the receiver aperture $\Delta\Omega$, and the kind of illumination. Using linearly polarized light of a plane monochromatic wave for the illumination, the power of scattered light collected by a certain optical arrangement can be calculated by the integral

$$P_\lambda(d_p,\theta_0,\Delta\Omega,\lambda,m)=I_0\frac{\lambda^2}{4\pi^2}\iint_{\Delta\Omega} i(\theta,\phi,\alpha,m)\sin\theta d\theta d\phi \tag{15–4}$$

For the dependency of the scattering function i on the azimuthal (polarization) angle ϕ, one obtains

$$i=i_1\sin^2\phi+i_2\cos^2\phi \tag{15–5}$$

where i_1 (θ, α, m) and i_2 (θ, α, m) are the two components of the scattering function in the plane perpendicular and parallel to the polarization vector $\overline{E}$ of the incident light (Fig. 15–1),

For unpolarized light, one gets $i = 1/2(i_1 + i_2)$, and the scattering pattern is symmetrical in rotation with respect to the axis of the illuminating beam. White light illumination additionally requires an integration over the wavelength, whereby the spectral radiation density of the light source and the spectral sensitivity of the photodetector have to be taken into account (Heyder and Gebhart, 1979). If the particles are illuminated with noncollimated light, a further integration over the illumination aperture has to be carried out.

To characterize single-particle light scattering, two further quantities are commonly used: (1) partial scattering cross section:

$$C_s=P_\lambda/I_0 \tag{15–6}$$

and (2) angular scattering coefficient:

$$Q_S=\frac{S_\lambda}{I_0(\pi/4)d_p^2} \tag{15–7}$$

In this definition Q_S is the power of scattered light per unit solid angle in relation to the power $I_0(\pi/4)d_p^2$ striking the projected area of the particle.

Light Extinction. Light extinction is the attenuation of a parallel beam of light due to absorption and scattering by the particles. For a single sphere with size parameter α and refractive index m, light extinction can be described by its extinction coefficient $E_x(\alpha, m)$, which can be calculated by electromagnetic theory. The power of light $-\Delta P_\lambda$, which is removed by a particle from a parallel beam of light with intensity I_0, is then given by

$$-\Delta P_\lambda=I_0(\pi/4)d_p^2E_x(\alpha,m) \tag{15–8}$$

Approximations

The mathematical formalism of the rigorous electromagnetic theory does not allow a good physical understanding and interpretation of the phenomena of light scattering and extinction by small particles. This understanding can be improved using two approximations.

Light Scattering. When the particle size is much smaller than the wavelength ($\alpha << 1$), the particle is subjected to an almost uniform field. The particle then oscillates like a dipole with a polarization proportional to the electrical field of the incident wave. The scattering properties of such a particle can be expressed by its polarizability, p, which is generally a tensor and reduces to a scalar for a homogeneous sphere. In the dipole or Rayleigh approximation and for unpolarized monochromatic light, the flux of light scattered by a spherical particle per unit solid angle is given by

$$S_\lambda(\theta, d_p, \lambda, m) = I_0 \frac{\lambda^2}{8\pi^2}\left[\frac{\pi d_p}{\lambda}\right]^6 \left|\frac{m^2-1}{m^2+2}\right|^2 (1+\cos^2\theta) \tag{15–9}$$

Equation 15–9 contains the polarizability p_s of a sphere (van de Hulst, 1957) so that

$$S_\lambda(\theta, d_p, \lambda, m) = I_0 \frac{\pi^2}{2\lambda^4} p_s^2 (1+\cos^2\theta) \tag{15–10}$$

where $p_s = 3|(m^2 - 1)/(m^2 + 2)|V$ and V is the volume of the sphere.

In the limiting case of $\alpha >> 1$, the scattered light can be considered as consisting of three parts, which are due to the physical effects of diffraction, reflection, and refraction (Fig. 15–2).

For monochromatic light, the angular scattering coefficient Q_S can be expressed as the sum of its components according to

$$Q_s(\theta, \alpha, m) = Q_0(0, \alpha) + Q_1(\theta, m) + Q_2(\theta, m) \tag{15–11}$$

Higher order internal reflections Q_i, $i > 2$ (rainbows) are not contained in Eq. 15–11.

$Q_0(\phi, \alpha)$ is the diffracted part of scattered light. It is independent of the optical constants of the particle material. Its angular distribution, however, depends on the size parameter α.

Q_1 (θ, m) is the fraction of light scattered by reflection on the surface of the particle. Its angular distribution is independent of the particle diameter but is influenced by the optical constants of the material. $Q_2(\theta, m)$ is the component scattered by two refractions by the particle surface. Its angular distribution again depends on the optical constants but not on the particle diameter.

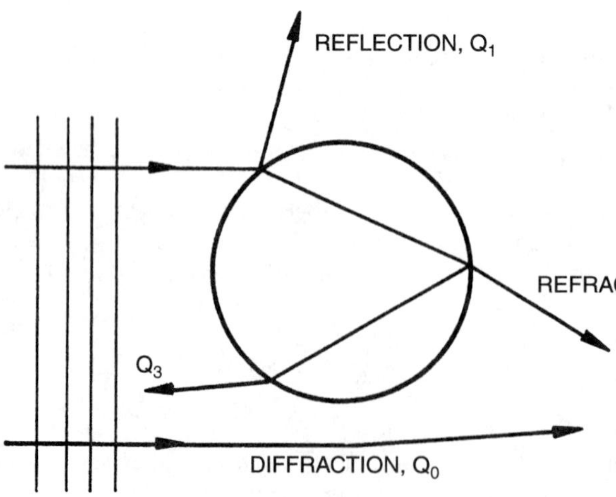

Fig. 15–2. Decomposition of scattered light into components according to ray optics.

For low scattering angles ($\theta < 10°$), the diffracted part of scattered light can be approximated by the Fraunhofer formula according to

$$Q_0(\alpha, \theta) = \frac{\alpha^2}{4\pi}\left[\frac{2I_1(\alpha \sin \theta)}{\alpha \sin \theta}\right]^2 \tag{15–12}$$

where I_1 ($\alpha \sin \theta$) is a first-order Bessel function.

Some angular distributions of the three fractions of scattered light are shown in Figure 15–3. The calculations are based on formulae published by Hodkinson (1962). For the reflected and the refracted light, only the components polarized perpendicular to the plane of observation are drawn in the diagram. Because the diffraction part changes with particle diameter, two α values have been considered. The forward lobe of diffraction is limited to the angular range $\theta < \theta_{min}$, where

$$\sin \theta_{min} = \frac{5\pi}{4\alpha} \tag{15–13}$$

While the reflected component Q_1 (θ, m) covers the whole angular range $0 < \theta < 180°$, the refracted part is confined to the range $\theta < \theta_{max}$, where

$$\cos \theta_{max}/2 = 1/m \tag{15–14}$$

Light Extinction. The extinction coefficient in the dipole or Rayleigh approximation can be expressed by two terms, according to

$$E_x = (\alpha, m) = 4\alpha\, Im\left[\frac{-(m^2-1)}{(m^2+2)}\right] + \frac{8}{3}\alpha^4\left[\frac{m^2-1}{m+2}\right]^2 \tag{15–15}$$

where *Im* is the imaginary part. The first term represents the absorption component of extinction and the second term the scattering component.

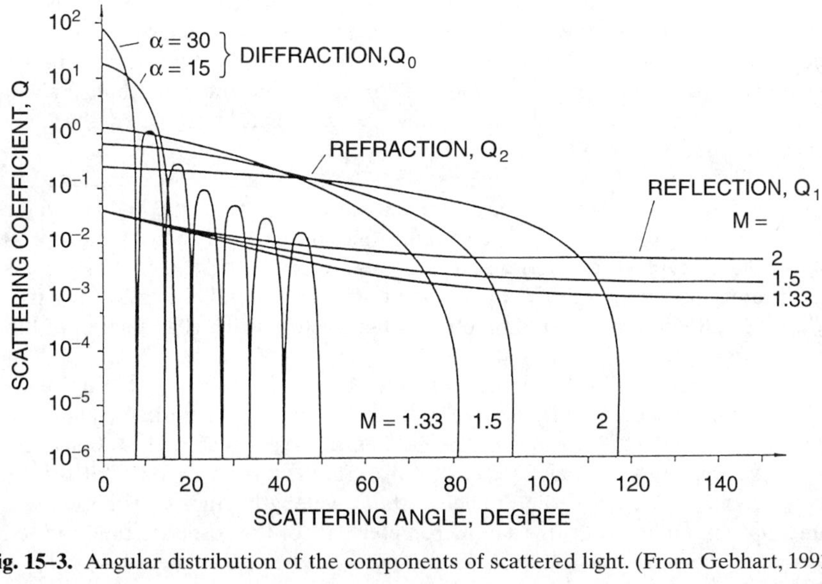

Fig. 15–3. Angular distribution of the components of scattered light. (From Gebhart, 1991.)

For classic optics ($\alpha >> 1$) the extinction coefficient has the value 2, including equal contributions of unity from light undergoing reflection, refraction, or absorption and from light being diffracted on the particle contour. In practice, the extinction coefficient is measured to be less than 2 because some of the scattered light is collected due to the finite aperture angle ϑ of the detection unit

$$E_x(\vartheta, \alpha, m) = 2 - 2\pi \int_0^{\vartheta} Q_s(\theta, \alpha, m) \sin\theta d\theta \tag{15–16}$$

Theoretical Response Functions

Although it is advisable to calibrate an optical instrument experimentally by means of test aerosols of known size and refractive index, theoretical response functions give a general survey of the characteristics of an optical system. With electromagnetic theory, the optical system response functions, which describe the power of light scattered through the collecting aperture, can be calculated as a function of the diameter of a spherical particle. The refractive index of the particle material is the principal parameter determining the behavior of these functions. Detailed calculations of this kind have been performed by Hodkinson and Greenfield (1965), Brossmann (1966), Gucker and Tuma (1968), Quenzel (1969), Oeseburg (1972), Cooke and Kerker (1975), and Heyder and Gebhart (1979). The response characteristics of coherent laser aerosol spectrometers have been calculated by Garvey and Pinnick (1983), Pettit and Peterson (1984), Soderholm and Salzman (1984), and Hinds and Kraske (1986).

Commercial optical systems can be divided roughly into instruments using low-angle scattering (diffraction lobe), those collecting scattered light in the forward direction ($\theta < 90°$), and those employing right-angle scattering. It is also important to distinguish between instruments using a polychromatic incandescent light source and those using a monochromatic laser light source.

The influence of monochromatic and polychromatic light on the response function is demonstrated in Figure 15–4, where the partial scattering cross section C_S is plotted against the particle diameter for monochromatic and white light and a mean scattering angle of $\theta_0 = 45°$. For particle diameters larger than the illuminating light wavelength, light scattering can be considered as a surface effect, and outside the diffraction lobe the mean flux of scattered light depends on the particle diameter squared. For monochromatic light, the scattering cross section curves show typical oscillations. These fluctuations can be smoothed by using white light and a large collecting aperture; however, the smoothing by white light by itself is more effective. In the diameter range smaller than the wavelength, the response functions are monotonic even for monochromatic light.

In Figure 15–5, the partial scattering cross section of submicrometer particles is shown for monochromatic light of an HeNe laser at a mean scattering angle of $\theta_0 = 40°$ and a receiver aperture of $\vartheta = 20°$. For $d_p < \lambda/2$, all curves are monotonic, and their slope indicates a d_p^6 dependency that is typical for dipole scattering, as expressed by Eq. 15–9. Response functions for different real parts of refractive index tend to run parallel in a log–log plot, which means that they differ by an approximately constant factor, which is a function of the refractive index.

For absorbing materials with a complex refractive index, $m = n - jk$, the influence of the k value is reduced considerably in the size range below the wavelength. A value of $k = 0.25$, for instance, represents the absorption properties of coal (Olaf and Robock, 1961). All response curves having the same real part, but different imaginary parts of the refractive index, cross over for particle diameters below the wavelength. High-sensitivity laser aerosol spectrometers can make use of this submicrometer part of the response function to achieve more accurate particle size analysis.

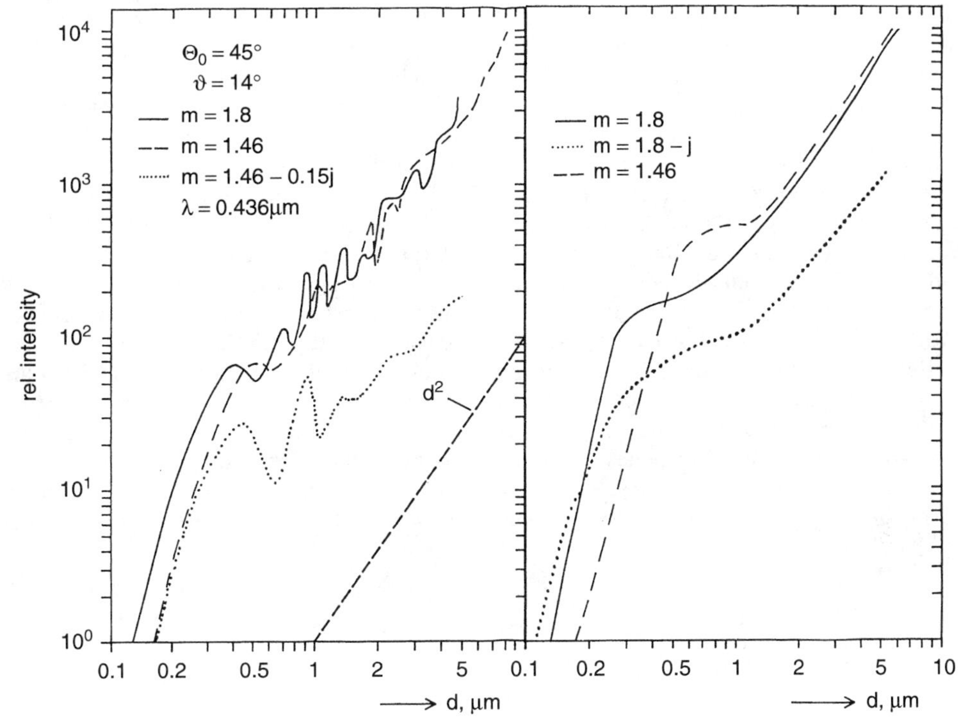

Fig. 15–4. Theoretical response functions of a forward-scattering system using monochromatic (**left**) and polychromatic (**right**) illumination. (Adapted from Gehhart et al., 1976.)

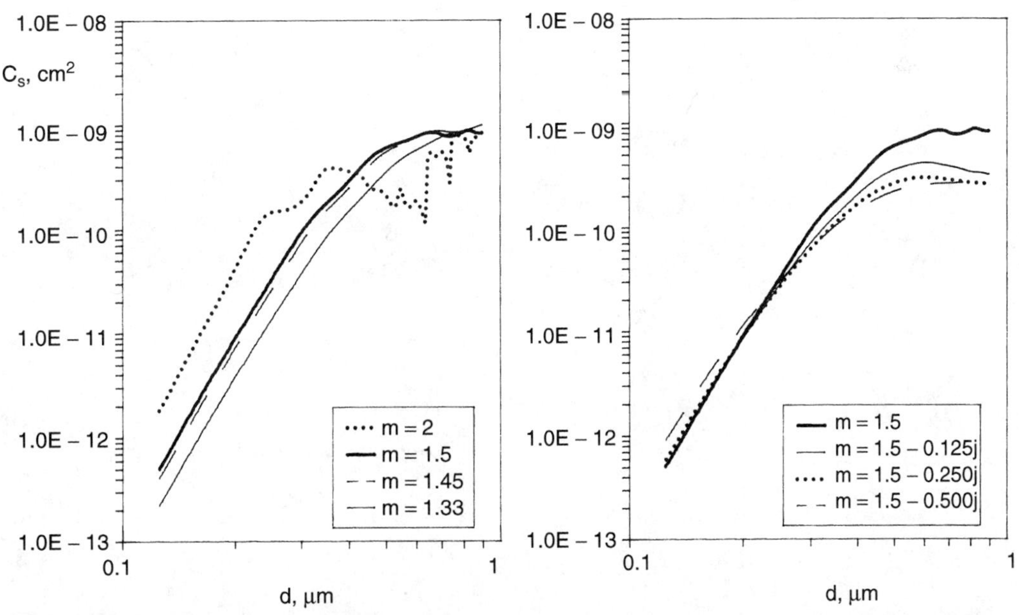

Fig. 15–5. Theoretical response functions in the submicrometer size range of a forward-scattering laser spectrometer. Wavelength $\lambda = 0.6328\,\mu m$.

According to classic optics, light entering the particle is scattered due to refraction. As can be seen from Figure 15–3, this component dominates in the angular range up to about 100° except within the forward diffraction lobe. For particles of opaque materials, the fraction of light entering the particle is absorbed. Instruments that mainly collect light scattered by refraction, therefore, show great differences in the response function between transparent and absorbing materials. Theoretical response functions, which include light refracted in the forward direction, are given in Figure 15–4. Experimental calibrations of optical particle counters with light-absorbing particles have been carried out by Whitby and Vomela (1967) with particles of India ink, by Willeke and Liu (1976) with coal particles, and by Szymanski and Liu (1986) with methylene blue. The experimental findings confirm the theoretical predictions of Figure 15–4 that opaque particles in the micrometer size range scatter a factor of 6 to 10 less light in the forward direction than the transparent particles of the same size.

Better conditions for particle size analysis can be realized if the light scattered within the forward lobe of diffraction ($\theta < 6°$) is collected. Within this angular range, the diffraction part $Q_0(\theta, \alpha)$ dominates, and the flux of scattered light is independent of the optical properties of the particles. Theoretical response curves for low-angle scattering using an HeNe laser are given in Figure 15–6. The response functions belonging to particles with different absorption now coincide, and those of transparent spheres show periodic oscillations rather than a sys-

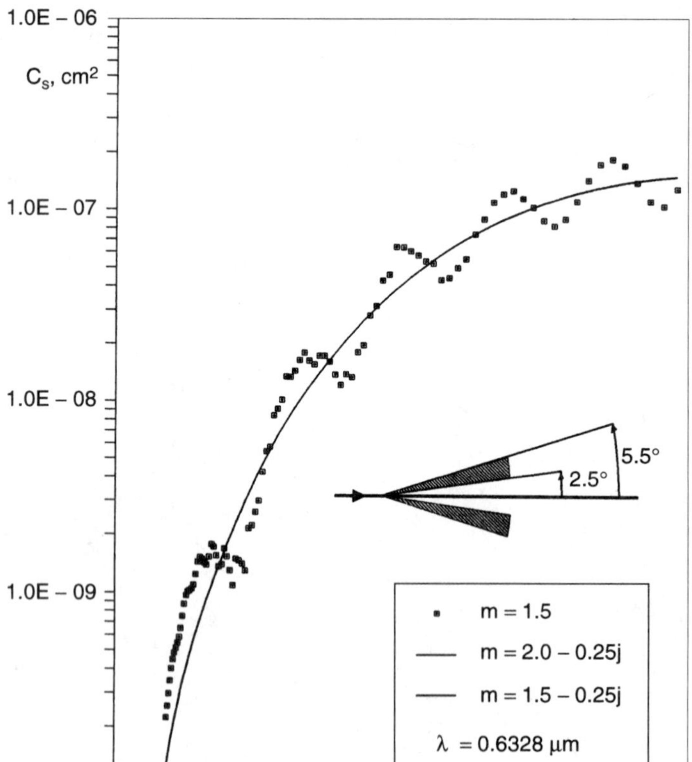

Fig. 15–6. Theoretical response functions of a low-angle scattering instrument. (Adapted from Gebhart et al., 1976.)

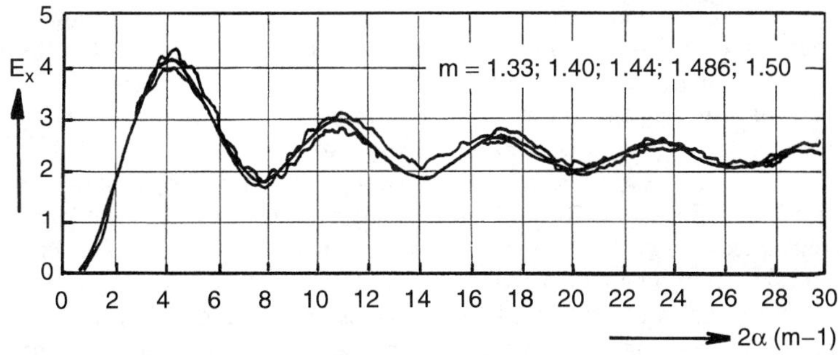

Fig. 15–7. Extinction coefficients of transparent spheres.

tematic deviation from the response calculated for absorbing materials. Because these oscillations originate from interference between diffracted and refracted light, they change with the real part of the refractive index and are smoothed by using polychromatic light or when detecting particles of irregular shape. For opaque materials, the refracted component is absorbed inside the particle, and no interference occurs.

A laboratory instrument that uses the polychromatic light from a mercury lamp and collects light scattered within the forward lobe of diffraction has been calibrated by Gebhart et al. (1976) with particles of different optical properties. The results confirm that light scattering within the forward lobe of diffraction is nearly independent of the optical properties of the particle material. Similar experimental findings have been reported by Szymanski and Liu (1986), who calibrated a low-angle laser counter (ASAS-300X, ***PMS***)* with particles of carbon black.

Light extinction of a particle can be best described by its extinction coefficient $E_x(\alpha, m)$, which is obtained from electromagnetic theory. In Figure 15–7, extinction coefficients of nonabsorbing spheres of different refractive indices are plotted versus the normalized size parameter $\alpha(m-1)$. For particle diameters of many wavelengths, the extinction coefficient is seen to approach a limiting value of 2. Another striking feature of the extinction curve is its oscillation, which is caused by constructive and destructive interference of diffracted and refracted light in the forward direction (see also Fig. 15–6). As the absorption inside the particle increases, the oscillations decline in importance and settle more quickly to a value of 2. The successive cycles in the extinction curve also depend on the high symmetry of a sphere and vanish for irregular particles.

LIGHT SCATTERING AND EXTINCTION BY AN ASSEMBLY OF PARTICLES

Ensemble Characteristics

Light Scattering. We define $P_\lambda(d_p, \lambda, m)$ as the flux of monochromatic light scattered by a single particle into the receiver aperture of an optical system. In the presence of many particles inside the sensing volume of the photometer the resulting light flux R collected by the detector is given by

$$R = c_n \int_0^\infty f(d_p) P_\lambda(d_p, \lambda, m) \mathrm{d}d_p \qquad (15\text{–}17)$$

* See Appendix I for full manufactuer addresses referenced to the italicized three-letter codes.

where c_n is the number concentration of the particles and $f(d_p)$ the probability density function of the particle size distribution. Equation 15–17 clearly demonstrates the problem of photometric measurements of aerosol concentrations: If the photometer response R varies, one cannot distinguish whether the number concentration c_n, the size distribution $f(d_p)$, or the optical properties m of the material have changed.

Using a test aerosol of a homogeneous chemical composition, the influence of the refractive index m can be eliminated so that only c_n and $f(d_p)$ remain as variables. In this case a linear relationship $R = const. \times c_n$ can be realized if either the function $f(d_p)$ can be replaced by one particle size (monodisperse aerosol) or the function $f(d_p)$ is constant during the measurements. In these cases the photometer response is linearly correlated with the number or mass concentration of an aerosol, and a simple gravimetric calibration procedure can be carried out. The simple proportionality of the scattered light flux to the number of particles inside the sensitive volume only holds if independent scattering by separate particles occurs and multiple scattering can be neglected.

Light Extinction. Let a parallel beam of monochromatic light with intensity I_0 and wavelength λ traverse a homogeneous mixture of particles with number concentration c_n and size distribution function $f(d_p)$. According to the Lambert-Beer law, the transmittance T of the light beam can then be expressed by

$$\begin{aligned} T &= I/I_0 \\ &= \exp\left\{-c_n s \frac{\pi}{4} \int_0^\infty d_p^2 E_x(\alpha, m) f(d_p) \mathrm{d}d_p \right\} \\ &= \exp\{-s\sigma\} \end{aligned} \tag{15–18}$$

where s is the extinction length according to Figure 15–8.

The term

$$\sigma = c_n \frac{\pi}{4} \int_0^\infty d_p^2 E_x(\alpha, m) f(d_p) \mathrm{d}d_p \tag{15–19}$$

is the total extinction coefficient (= scattering coefficient σ_s + absorption coefficient σ_a) of the particle assembly also known as *turbidity*. The expression $s\sigma = \tau$ represents the "optical depth." As can be seen from Eq. 15–19, the turbidity remains a unique function of the number concentration only if the size distribution $f(d_p)$ and the refractive index m of the particles do not change during the measurement.

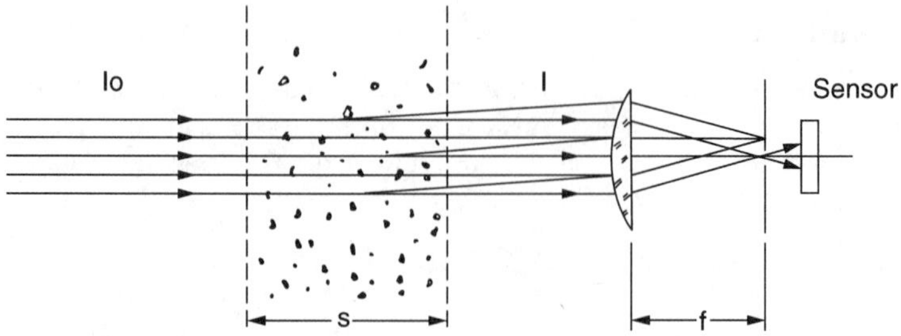

Fig. 15–8. Schematic of light extinction measurement system.

When the particle concentration becomes very high, the particles may no longer scatter independently and the coefficient $E_x(\alpha, m)$ of the single particle may change. Usually, this happens at concentrations much larger than those for which the Lambert-Beer law breaks down due to multiple scattering.

EXAMPLE 15–1

An aerosol generator based on controlled condensation of vapor produces uniform droplets of 3 μm diameter at a concentration of $2 \times 10^{12}\,m^{-3}$. The output of the generator passes through a glass tube of 30 mm inner diameter. Its constancy is monitored by a narrow-angle extinction technique that uses a collimated laser beam. Calculate the expected transmission of the measuring system.

Answer: The relevant parameters for the transmission T according to Eq. 15–18 are the total extinction coefficient σ of the particle assembly and the extinction length s. The extinction coefficient E_x of a 3 μm droplet in a parallel laser beam approaches 2.

$$\sigma = c_n \frac{\pi}{4} d_p^2 E_x = 2 \times 10^{12}\,\mathrm{m}^{-3} \left(\frac{\pi}{4}\right) (3 \times 10^{-6})^2 2 = 28.3\,\mathrm{m}^{-1}$$

$$T = \exp(-\sigma s) = \exp(-28.3\,\mathrm{m}^{-1} \times 3 \times 10^{-2}\,\mathrm{m}) = 0.428$$

The integral optical properties of atmospheric aerosols are usually summarized in terms of light interaction coefficients σ_i. If the number size distribution $f(d_p)$ of a chemically homogeneous aerosol consisting of spherical particles is known, the scattering coefficient σ_s of an assembly of particles can be calculated as

$$\sigma_s = c_n \int_0^\infty \frac{\pi d_p^2}{4} f(d_p) E_s(\alpha, m) \mathrm{d}d_p \tag{15–20}$$

where $E_s\,(\alpha, m)$ is the total scattering coefficient of a single particle, that is, the integral of the angular scattering coefficient Q_s over the whole solid angle. In this definition, σ_s, with the dimension [length^{-1}], describes the contribution of atmospheric aerosols to overall light scattering. The angular distribution of light scattered by an assembly of particles is usually described by a so-called phase function $\beta(\theta)$ defined as 4π times the ratio of the angular scattering cross section to the extinction cross section of the assembly. The phase function of an assembly of particles for unpolarized light is given by (Hodkinson, 1962)

$$\beta(\theta) = 4\pi \frac{\int_0^\infty f(d_p) d_p^2 Q_s(\theta, \alpha, m) \mathrm{d}d_p}{\int_0^\infty f(d_p) d_p^2 E_x(\alpha, m) \mathrm{d}d_p} \tag{15–21}$$

The main characteristic of $\beta(\theta)$ can be expressed by an asymmetry parameter g according to

$$g = \frac{\int_0^\pi \beta(\theta) \sin\theta \cos\theta \, \mathrm{d}\theta}{\int_0^\pi \beta(\theta) \sin\theta \, \mathrm{d}\theta} \tag{15–22}$$

For isotropic scattering (e.g., Rayleigh scattering) the asymmetry parameter is zero; for strongly forward scattering aerosols it approaches one.

In the case of nonabsorbing aerosols, light extinction occurs only due to light scattering, and the attenuation of a beam of light passing through such an aerosol is given by

$$I = I_0 \exp(-c_m b_s s) \tag{15–23}$$

where c_m is the aerosol mass concentration and $b_s = \sigma_s / c_m$ is the specific scattering coefficient (e.g., in m^2/g).

Effect of Multiple Scattering

Usually light scattering by an assembly of particles is evaluated by adding the intensities scattered by each particle as if it were present alone (Eq. 15–17). This kind of treatment of ensemble scattering, however, is only correct if the particles are randomly distributed in space (incoherent superposition of waves) and the aerosol is sufficiently dilute.

When the particles are less than a few wavelengths apart from each other, their fields interact and violate the boundary conditions used in Mie theory (van de Hulst, 1957). The scattering behavior changes relative to that for an isolated sphere. Even if the particles are several wavelengths apart so that mutual polarizations can be neglected, light scattered by one particle is likely to be incident on another one, which then scatters it again. This effect is known as *multiple scattering* and can be avoided only by diluting the aerosol. A criterion for independent and isolated scattering is a linear relationship between the flux of scattered light R and the number concentration c_n of the aerosol. In practice, the onset of multiple scattering is found by diluting a dense aerosol until scattering is a linear function of particle number concentration. Multiple scattering generally depends on number concentration c_n, particle diameter d_p, and sensing volume V_m. Hodkinson (1962) performed extinction measurements on suspensions of 1.8 μm polystyrene spheres in water and showed that within the transmittance (T) range $0.37 \le T \le 1$, the results obeyed the Lambert-Beer law. With narrow-angle extinction techniques, where only a negligible amount of scattered light can reach the detector, the validity range of the Lambert-Beer law may be extended even to $T < 0.1$. Comprehensive investigations of multiple scattering effects have been carried out by Szymanski (1992, 1996) using an apparatus described by Preining et al. (1981). With this experimental setup, light extinction, particle size, and number concentration of a monodisperse aerosol were measured simultaneously by independent methods. Transmittance through a medium undergoing light extinction in association with multiple scattering was described as function of the optical depth $\tau = s\,\sigma$ by (Szymanski, 1992):

$$T(s,\sigma) = \exp(-s\sigma)[1 + X(s,\sigma)] \tag{15–24}$$

The correction term X represents the radiation additionally reaching the detector unit due to multiple scattering. In the limit of diluted aerosols, the expression reduces to the Lambert-Beer law (Eq. 15–18).

Gravimetry and Photometry

Air quality standards for particulate matter in the environment are usually based on gravimetry. It is, therefore, of general interest to correlate photometer signals with the mass (volume) concentration of aerosols. The mass concentration c_m of a polydisperse aerosol is given by

$$c_m = c_n \rho_p \int_0^\infty f(d_p) \frac{\pi}{6} d_p^3 \, dd_p \tag{15–25}$$

where ρ_p is the density of the particle material.

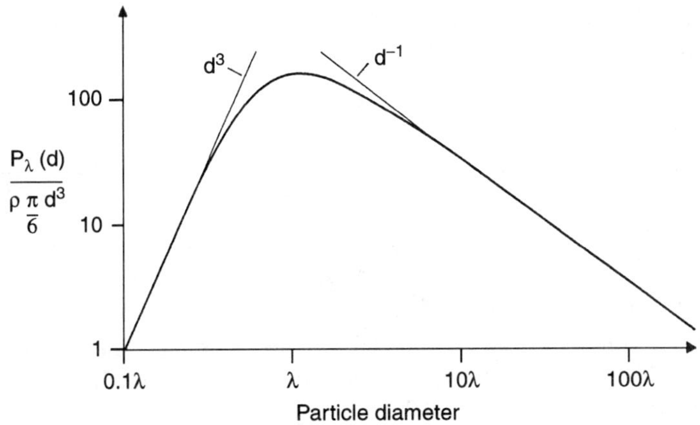

Fig. 15–9. Schematic diagram illustrating the relationship between photometry and gravimetry.

From the photometer response as expressed by Eq. 15–17, it follows that a linear relationship $R = const. \times c_m$ is only obtained if the flux of scattered light P_λ (d_p, m, λ) collected by an optical system is a function of the particle volume. Using the electromagnetic theory of light scattering on spheres, a specific scattering function $P_{\lambda,S}$ (d_p, m, λ) can be defined as

$$P_{\lambda,S}(d_p, m, \lambda) = \frac{P_\lambda(d_p, m, \lambda)}{\rho_p(\pi/6)d_p^3} \tag{15–26}$$

which is the flux of scattered light per unit mass (volume) concentration of aerosol. The general behavior of $P_{\lambda,S}$ for a fixed m and λ is illustrated in Figure 15–9.

In the micrometer size range, light scattering is a surface effect and $P_{\lambda,S}$ decreases proportional to d_p^{-1}. For particle diameters $d_p < 0.5\lambda$, light scattering can be approximated by Eq. 15–9 ($S_\lambda \sim d_p^6$), and the specific scattering function increases proportional to d_p^3. At about $d_p \sim \lambda$, the specific scattering function reaches a maximum, which, for dielectric spheres, has additional resonances superimposed. Only within this narrow plateau, light scattering is nearly proportional to particle volume. Armbruster et al. (1984) calibrated a respirable dust photometer that measures infrared light ($\lambda = 0.94\,\mu$m) scattered by airborne particles at a mean angle of 70°, with monodisperse test aerosols having different optical properties. Their results confirmed the general relationship of the specific scattering function as indicated in Figure 15–9. It is interesting to note that for atmospheric aerosols and visible light, particles within the accumulation mode (0.1 to 1 μm) nearly coincide with the maximum of the specific scattering function so that particles within this size range exhibit the greatest sensitivity when detected with optical means (Willeke and Brockmann, 1977).

Light extinction, expressed by the extinction cross section $(\pi/4)$ d_p^2 $E_x(\alpha, m)$ of a particle, generally shows characteristics similar to light scattering. Only for absorbing particles in the Rayleigh regime, where the first term of Eq. 15–15 dominates, does light extinction become proportional to the volume of the particles regardless of their size.

Light Scattering by Gas Molecules

For dielectric particles much smaller than the wavelength of light, dipole theory can be applied so that the power of scattered light per unit solid angle, S_λ $(\theta, d_p, \lambda, m)$, can be expressed by Eq. 15–10. For randomly oriented gas molecules, the mean polarizability, p_m, of a single molecule is connected with the macroscopic refractive index m_g of the gas according to Born (1965):

$$c_g p_m = m_g^2 - 1 \tag{15–27}$$

where c_g is the number concentration of the molecules. Substituting Eq. 15–27 into Eq. 15–10 (with $p_m \equiv p_s$) yields the contribution of a gas volume V_g to light scattering:

$$S_{\lambda,g} = V_g I_0 \left(\frac{\pi^2}{2\lambda^4 c_g}\right)(m_g^2 - 1)^2(1 + \cos^2\theta) \tag{15–28}$$

In Figure 15–10, Rayleigh scattering of air under normal pressure is compared with light scattering by single spherical particles of refractive index m = 1.5. Macroscopic refractive indices of some gases at λ = 0.546 µm and the corresponding fluxes of scattered light in relation to air are listed in Table 15–1.

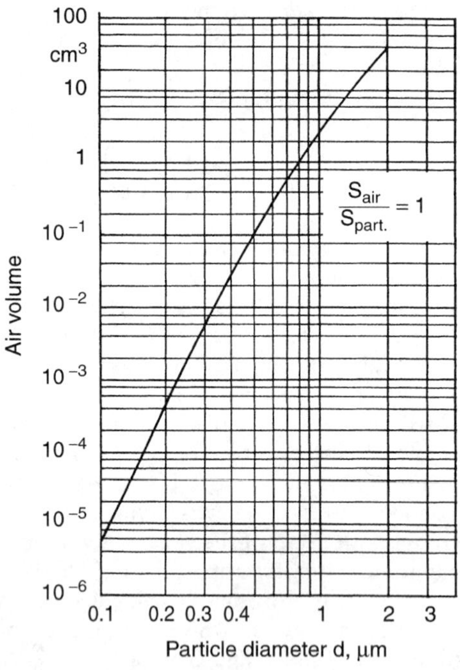

Fig. 15–10. Rayleigh scattering of air in relation to spheres with refractive index m = 1.5.

TABLE 15–1. Rayleigh Scattering of Different Gases in Relation to Air

	Air	He	Ne	H_2	H_2O	Ar	Kr	CO_2	Xe
$(m_g - 1) \times 10^6$	292	34.89	67.25	139.6	252.7	282.3	428.7	450.5	705.5
S_g/S_{air}	1	0.015	0.053	0.228	0.75	0.934	2.15	2.38	5.84

SINGLE-PARTICLE OPTICAL COUNTERS

General Remarks

An optical particle counter (OPC) measures the size and number concentration of aerosol particles in a limited size range by means of light scattering by single particles. For this purpose, a stream of aerosol is drawn through a condensed light beam. Light flashes scattered from single particles are received by a photodetector and converted into electrical pulses. From the count rate of the pulses, the number concentration, and the pulse height, the size of the particles is derived. The light power that an individual particle scatters is a function of its size, refractive index, and shape. Particle sizing based on this principle has been known for more than 50 years. Meanwhile, the technique has been steadily developed, and for about 40 years OPCs using white light illumination have been commercially available (Lieberman, 1986). The invention of the laser has allowed the successful replacement of the white light illumination by coherent and monochromatic laser light. Important characteristic features of an OPC are its permissible range of number concentration, its sampling flow rate, its sensitivity (lower detection limit), and its size measurement accuracy. OPCs are now widely used in basic aerosol research, air pollution studies, and clean-room monitoring. Because the requirements for an OPC change for these different kinds of applications, the specifications of an instrument have to be adjusted to the specific measurement problem. OPCs that cover these various kinds of applications are offered, for instance, by ***CLI***, ***PAC***, and ***PMS***. Trends in development are handheld, battery-operated counters with a laser diode as the light source.

Calibration Procedures

In recent decades, many theoretical response functions of commercial OPCs have been published. Meanwhile, menu-driven programs for PCs are available to carry out light-scattering calculations for spherical particles (Reist, 1990). Despite these theoretical calculations, OPCs are usually calibrated with monodisperse test aerosols of known size and refractive index. Instruments using incandescent lamps for particle illumination were calibrated by Whitby and Vomela (1967), Liu et al. (1974a,b), Willeke and Liu (1976), and Fissan et al. (1984). Experimental calibrations of white light as well as laser light counters with various test aerosols have been carried out by Gebhart et al. (1983), Chen et al. (1984), and Liu et al. (1985). More recent papers published by Liu and Szymanski (1986), Plomb et al. (1986), and Szymanski and Liu (1986) report experimental work on laser particle counters.

To produce monodisperse test aerosols, several generation techniques can be used. Aerosols with a high degree of monodispersity are obtained if aqueous suspensions of polystyrene latex (PSL) are nebulized, dried, and drawn into the OPC (Gebhart et al., 1980). Another technique is the vibrating-orifice generator, which allows the calculation of the particle diameter to an accuracy of about 1% from the operating parameters of the generator (Berglund and Liu, 1973). By varying the frequency of vibration, the liquid feed through the orifice, and the concentration of the aerosol material in the volatile solvent, monodisperse aerosols from about 0.3 to 30 μm in diameter can be produced for calibration purposes. For the generation of submicrometer aerosol standards, the principle of electrostatic classification can be used (Liu and Pui, 1974c). The aerosol material is either dissolved in a liquid or prepared as a colloidal suspension and then atomized through a jet nebulizer. After the mist is dried, a polydisperse aerosol consisting of solid particles or of low-volatility droplets remains. By passing the polydisperse aerosol through a differential-mobility analyzer, particles within a narrow size range are extracted according to their electrical mobility. Instrument calibration techniques are presented in Chapter 21 in greater detail.

When applying an experimentally calibrated OPC to an unknown aerosol, one has to keep in mind that light-scattering signals generally depend on the optical properties of the test aerosol. To overcome these problems, direct field calibration of OPCs has been proposed. A direct aerodynamic particle size calibration of OPCs by means of inertial impactors was performed by Marple and Rubow (1976). To obtain a single calibration point, two runs are made with the OPC. During one run, an impactor is attached to the inlet of the OPC, and during the second run the impactor is removed. An analysis of the data from the two runs yields a calibration point corresponding to the 50% cut point size of the impactor. By changing the size of the impactor nozzle, different calibration points can be obtained. Similar aerodynamic calibration procedures have been applied to a white-light 90° counter by Büttner (1983) using a sampling cyclone and by Friehmelt and Heidenreich (1999) using an aerodynamic particle sizer (***TSI***).

Optical Systems

In the following discussion, typical optical arrangements used in laboratory and commercial OPCs are presented.

The optical setup of a forward-scattering instrument with converging illumination from an incandescent light source and a coaxial collecting aperture is shown in Figure 15–11. This configuration was used, for instance, in the former models PC 215, PC 245, PC 247, and PC 2102 (***HIA***) and is typical for an optical arrangement using forward scattering. Light from an incandescent lamp is concentrated in the sensing volume by a system of condenser lenses forming an illumination cone of 5° semiangle. After passing the sensing volume, the primary light beam is absorbed in a concentric light trap with a half-angle of 16°. Light scattered by individual particles is collected through a coaxial aperture of 25° half-angle and directed onto the cathode of a photomultiplier. Considering the illumination cone, scattered light covering an angular range from about 10° to 30° is collected by the system.

Experimental response curves of the models PC 215 and PC 245 (***HIA***) measured with PSL spheres and droplets of dioctyl phthalate (DOP) were reported by Willeke and Liu (1976). A characteristic feature of this forward-scattering sensor is a multivalued region in the response curve between 0.7 and about 1.2 μm.

The counter model Cl-208 (***CLI***) shown in Figure 15–12 utilizes an elliptical mirror in its optical system. In the detector, the particle sensing zone is located at the primary focal point of the elliptical mirror. High-intensity light from a quartz halogen lamp is focused on the sensing zone, where it illuminates each traversing particle.

Light scattered from each particle is collected over an angular range from 15° to 105° and is directed to a photodetector located at the secondary focal point of the ellipsoid. The sample air is surrounded by a sheath of clean filtered air, allowing the precise placement of the particles at the primary focal point. At a sample flow rate of 1.2×10^{-4} m^3/s [7 L/min], a size sensitivity of 0.3 μm is quoted by the manufacturer. Experimental calibrations of the Climet instrument were carried out by Clark and Avol (1979), Mäkynen et al. (1982), and by Chen

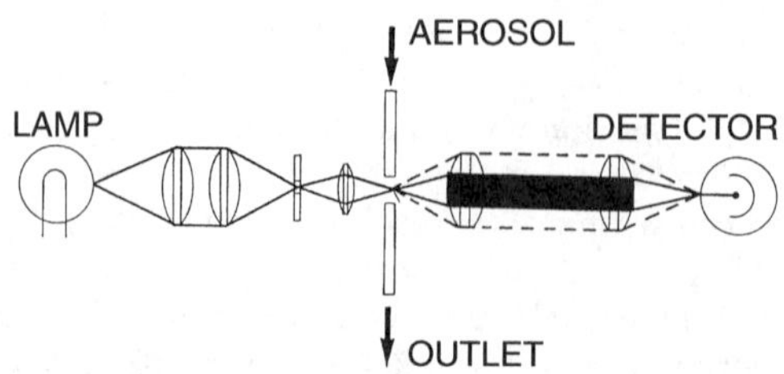

Fig. 15–11. Optical system of a forward-scattering instrument using an incandescent light source.

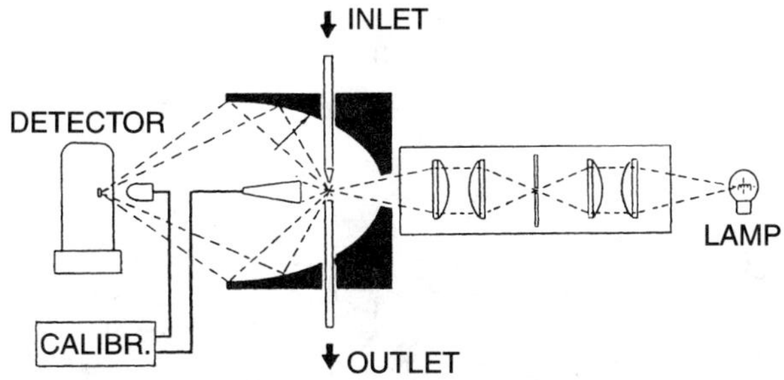

Fig. 15–12. Optical system of the particle counter CI-208 (***CLI***). (From Gebhart, 1991.)

et al. (1984). A monotonically increasing response function for PSL and DOP aerosols were found over the entire size range from 0.3 to 10 μm.

With laser light illumination, intensities can be realized in the sensing volume several orders of magnitude higher than those obtained with incandescent light sources. Furthermore, a collimated laser beam needs fewer optical elements, such as stops and lenses, reducing the stray-light background in the chamber considerably. Three kinds of particle illumination exist with laser light: (1) the output of a laser can be focused into the sensing volume like a common incandescent light source; (2) according to Knollenberg and Luehr (1976), high illumination intensities with a low-power laser can be achieved if the sensing volume is positioned inside the laser resonator (active scattering); and (3) in the passive cavity system, the original laser output is multiply reflected to create high-intensity illumination.

A high-resolution laser aerosol spectrometer for laboratory purposes was developed by Roth and Gebhart (1978) for the size range between 0.06 and 0.6 μm. An argon ion laser ($\lambda = 0.5145\,\mu m$) with an output of 2 W serves as the external light source. The laser light is focused by an astigmatic system of lenses into the sensing volume. The light scattered by the particles is collected by a microscope objective at a mean scattering angle of 40°.

An example of an active cavity sensor is shown in Figure 15–13. It is contained in models LAS-X, MS LAS, LPC 101, and MicroLPC0710 (***PMS***) and in the model 226/236 (***HIA***). This sensor employs a high-Q laser cavity to achieve a high illumination intensity (~ 500 W/cm^2). The primary collector of the scattered light is a parabolic reflector. Particles in the sample stream intersect the laser beam within the cavity at the focus of the paraboloid, which focuses the scattered light onto a 45° flat mirror. The light reflected from this mirror is refocused onto a photodetector by an aspheric lens. The whole system collects scattered light from 35° to 120°, providing a 2.2 steradian solid angle. The aerosol stream is aerodynamically focused. The theoretical response functions of this sensor have been published by Hinds and Kraske (1986), and experimental calibration data have been reported by Szymanski and Liu (1986).

The LASAIR series (***PMS***) are instruments with a passive cavity design. They are available at minimum sizing thresholds of 0.1, 0.2, 0.3, and 0.5 μm and with sample flow rates from 9.44×10^{-7} to 4.72×10^{-4} m^3/s (0.002 to 1 ft^3/min).

Most OPCs utilize an aerosol nozzle to transport the particles through the beam of illuminating light. In this case, the sensing volume is limited by the cross section of the aerosol stream and the width of the light beam. Using an aerosol nozzle has the advantage that, under stable flow conditions, the sensing volume and the sample flow rate are well defined. In most instruments, the aerosol stream is surrounded by a clean air sheath, which establishes a stable flow toward the outlet of the measuring chamber and thus avoids stray particles in the optical system. The clean air sheath can additionally be used to focus the aerosol stream aerodynamically down to a filament of less than 0.1 mm diameter. The problem that arises in con-

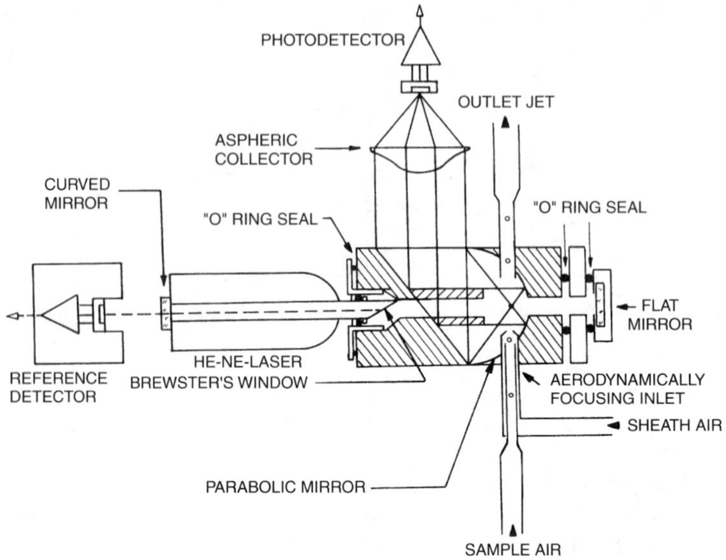

Fig. 15–13. Setup of an active scattering laser spectrometer used in the models LAS-X, HS-LAS, and LPC 101 (***PMS***) and in the models 226 and 236 (***HIA***). (From Gebhart, 1991.)

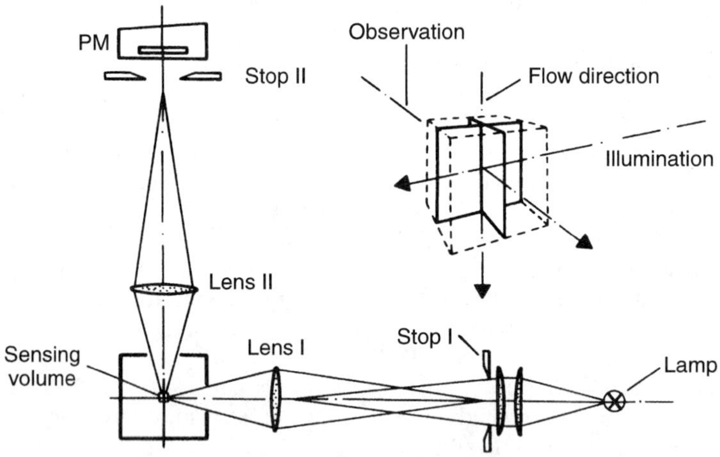

Fig. 15–14. Arrangement of Counter HC 15 (***PLY***), which forms the sensing volume by optical means only. (From Gebhart, 1991.)

nection with an aerosol nozzle is that a sample representative of all particle sizes has to be taken out of the sampled aerosol. Therefore, systems also have been developed that can count individual aerosol particles directly in the main stream of an aerosol.

Such an instrument is, for instance, the particle counter HC 15 (***PLY***), which forms its sensing volume by optical means only (Umhauer, 1980). The optical system, which is schematically drawn in Figure 15–14, consists of two optical pathways, one for illumination and one for the collection of scattered light.

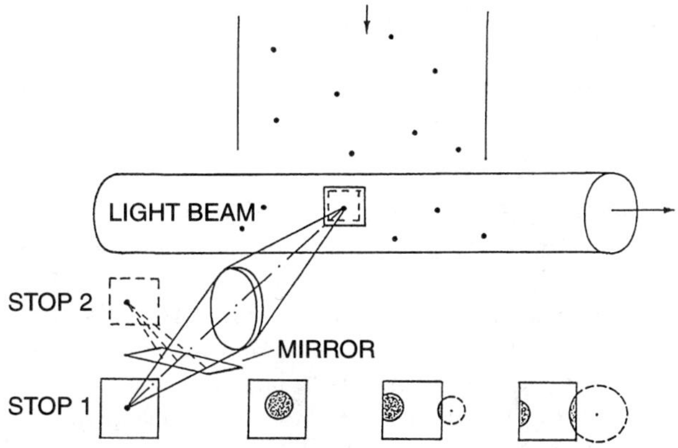

Fig. 15–15. Elimination of particles traversing the periphery of the sensing volume by means of a two-beam system and masked apertures.

The lenses I and II project miniaturized images of square masks (stops I and II) into the measuring chamber, where the two optical axes cross at right angles. The images of stops I and II, which coincide with the crossing point of the two axes, form the sensing volume. Particles passing this optically defined sensitive area are illuminated with incandescent light, and the light scattered by individual particles is collected at a mean scattering angle of 90°. Because an optically defined sensing volume can be made very small (about 100 μm across), high concentrations can be measured with such an instrument. The advantage of such a system is that all difficulties associated with aerosol sampling through inlets and small nozzles are avoided. There exist, however, problems in exactly defining the sensing volume and the sample flow rate. Whereas small particles are only counted when they pass the central region of the stop, larger particles can also give rise to countable pulses at the periphery of the imaging. Because these pulses are smaller than if the particle passed through the center of the detection volume, they are sensed as coming from smaller particles. Thus, the effective sample flow rate is particle size dependant. The bigger the particles, the more the peripheral detection contributes to the measured cumulative number distribution. This has been shown in an experimental analysis of the HC 15 counter by Helsper and Fissan (1980). The authors passed monodisperse 9 μm droplets through the sensing volume and found that about 50% of the count pulses were attenuated and simulated smaller droplets down to a detection limit of 0.3 μm.

A method to eliminate the effect of the border zone on the measured particle size distribution was reported by Knollenberg and Luehr (1976) and Umhauer (1983). Its principle of operation is illustrated in Figure 15–15.

A beam splitter behind the collecting optics directs the scattered light through masked apertures on two independent detectors. The signals from the two detectors, in conjunction with double pulse height analysis, lead to a system that determines the position of a particle in the light beam and that rejects all counts originating from particles in the border zone. The model HP-LPC (***PMS***) contains a sensor of this kind that can be directly installed in a conduit for gases. Although the sensing area of this model covers only a fraction of a percent of the conduit's cross section, a unique linear relationship between count rate and number concentration is obtained for all particle sizes above the detection limit. Umhauer (1983) constructed two independent rectangular collector and detector units for the model HC 15 (***PLY***), which overlapped in the sensing volume and which differed only in the size of the masks. The two different masks, in combination with double-pulse analysis, allowed identification and rejection of those particles that passed the sensing volume in the border zone. An improved version of this two-channel detector device was investigated by Sachweh et al. (1998). A modified type of this instrument is commercially available (PCS-2000; ***PAS***).

Pulse Processing

The light flashes originating from individual particles in the sensing volume are converted into photocurrent pulses. Due to the high time resolution of most photodetectors, the pulse shape follows the light intensity profile inside the sensing volume. The intensity profile is approximately a Gaussian one and has a pulse width from several to about 40 μs, depending upon the OPC used. The output pulses of the photodetector are usually fed to a current-to-voltage transducer, the main part of which consists of a high-frequency amplifier. Then a decision has to be made whether the pulse amplitude or its total charge (area) will be used as a measure of the scattered light. The charge integration method improves the resolution and the sensitivity of an OPC considerably, provided that the pulse duration is very constant for all particles. A clean air jacket in conjunction with aerodynamic focusing of the aerosol stream results in a constant pulse duration necessary for charge integration. In general, a multichannel analyzer requires a constant pulse height of a few microseconds duration in order to read the signal height. Therefore, the amplifier output signal—which is proportional to either the amplitude or the area of the original pulse—has to be stretched in a pulse converter. The pulse stretching introduces an electronic dead time during which no additional signals are processed. Additionally, the converter removes low-frequency components and dc offset from the incoming signal.

A further improvement of the signal evaluation is obtained, if the photomultiplier signals are digitized directly after the current-to-voltage transducer (Sachweh et al., 1989, 1998). By this procedure the raw signal containing electronic noise ripples is converted into nearly a square wave, the base line of which is given by the transit time of the particle through the sensing zone. This allows a well-defined determination of the mean pulse height.

Most OPCs have an analog output that can be connected directly to an oscilloscope. Observations of the pulse shape show that in some instruments many stray particles pass the sensing volume, producing pulses up to 10 times longer than the normal ones. Because pulse times of this length exceed the recovery time of the electronics, multiple counts can originate from a single stray particle that are classified as small particles (Gebhart et al., 1983). The existence of the stray particles themselves may be explained by unstable flow conditions that prevent particles from being completely removed from the measuring chamber after their initial passage through the light beam.

Range of Number Concentration

Number concentration c_0 of the aerosol, sample flow rate Q, and counting rate dN/dt of a particle counter are connected by the relationship

$$dN/dt = c_0 Q \tag{15–29}$$

Eq. 15–29 assumes that each particle contained in the sample flow produces a single count. In practice, however, counting losses due to coincidences occur. Less than 10% loss in particle counts approximately requires (van der Meulen et al., 1980)

$$c/c_0 = \exp(-c_{max} Q t_r) \geq 0.9 \tag{15–30}$$

where t_r is the recovery time of the electronics between successive count events, including the transit time of the particle through the light beam and the pulse processing time of the multichannel analyzer. From Eq. 15–30, it can be seen that for a given recovery time t_r, mea-

TABLE 15–2. Maximum Permissible Number Concentration in an OPC in Relation to the Volumetric Flow Rate[a]

Q (m³/s)	4.72×10^{-4}	4.72×10^{-5}	4.72×10^{-6}	4.72×10^{-7}	4.72×10^{-8}
c_{max} (m⁻³)	1.1×10^{7}	1.1×10^{8}	1.1×10^{9}	1.1×10^{10}	1.1×10^{11}

[a] Evaluated from Eq. 15–30 for an electronic recovery time of 20 μs.

surements at high number concentrations are only possible at the expense of the sample flow rate Q and vice versa. Some maximum permissible number concentrations derived from Eq. 15–30 for a typical recovery time of $t_r = 20\,\mu s$ are shown in Table 15–2. As indicated, c_{max} never falls below $3.5 \times 10^6/m^3$, which corresponds to ISO Class 8 (particles per m³ > 0.5 μm) as defined in ISO 14644-1 (see Table 33–1). Coincidence errors, however, are important if OPCs are applied to laboratory or environmental aerosols.

EXAMPLE 15–2

A filter with an expected penetration of 0.01 for 0.5 μm is checked with an OPC. The OPC samples aerosol at a flow rate of 47.16×10^{-6} m³/s [2.83 L/min] and has an electronic recovery time of 12 μs. Monodisperse 0.5 μm particles with a true number concentration of $c_0 = 4 \times 10^8\,m^{-3}$ serve as the test aerosol. Calculate the experimental error due to coincidence if no dilution steps are used.

Answer: The experimental penetration is c_2/c_1, where c_1 and c_2 are the measured concentrations upstream and downstream of the filter. Using Eq. 15–30, the count loss due to coincidence is

$$\frac{c_1}{c_0} = \exp(-c_0 Q t_r) = \exp[(-4 \times 10^8\,m^{-3}) \times (47.1 \times 10^{-6}\,m^3/s) \times (12 \times 10^{-6}\,s)] = 0.797$$

$$\frac{c_2}{c_0} = \exp(-0.01 c_0 Q t_r) = \exp[(-4 \times 10^6\,m^{-3}) \times (47.1 \times 10^{-6}\,m^3/s) \times (12 \times 10^{-6}\,s)] \approx 1$$

The error is:

$$\Delta\frac{c_2}{c_0} = \left[\frac{c_2}{c_1} - \frac{c_2}{c_0}\right] = \frac{c_2}{c_0}\left[\frac{1}{0.797} - 1\right] = 0.25\frac{c_2}{c_0}$$

The lower limit c_{min} of the detectable number concentration depends on the background noise according to

$$c_{min} > c_{ns} = [dN/dt]_{ns}/Q \qquad (15–31)$$

where $(dN/dt)_{ns}$ is the rate of noise pulses and c_{ns} the apparent number concentration produced by noise. Count noise can originate from internal particle sources, the electronics, ionizing radiation, and instabilities of the power supply. There is a general agreement that reliable concentration measurements with an OPC should start only at levels more than one order of magnitude above the background, that is, $c_{min} \geq 10\,c_{ns}$. For the determination of the noise level of an OPC, filtered air is drawn through the instrument for several hours. Measurements of this kind have been reported by Gebhart and Roth (1986), Wen and Kasper (1986), Liu and Szymanski (1987), and Gebhart (1989a). Some results of background noise are presented in Table 15–3.

TABLE 15–3. Specifications and Count Noise of a Selection of Commercial OPCs

Light Source	Model	Receiver Optics	Q (m^3/s)	Noise Counts per m^3	
				Pump On	Pump Off
Incandescent light	HIAC/Royco 4102	10°–30°	4.72×10^{-5}	530 > 21	Wash-out
	HIAC/Royco 5000	50° range	1.67×10^{-5}	84.8	~0
	CLIMET CI 8060	15°–150°	4.72×10^{-4}	0.95	~0
Laser light	PMS LAS-X	35°–120°	5.00×10^{-6}	10,840 > 210	Wash-out
	PMS LPC-110	35°–120°	4.72×10^{-5}	<17.6	~0
	HIAC/Royco 5100	60°–120°	4.72×10^{-4}	4236 > 28	Wash-out
	TSI 3755	15°–88°	4.72×10^{-5}	<35.3	
	CLIMET C1 6300	45°–135°	4.72×10^{-4}	<70.6	

Sources: Gebhart and Roth (1986) and Gebhart (1989a).

In most cases, particles originating from the internal flow system have been identified as being responsible for the count noise. This can be concluded from the wash-out effect of the instruments and their different behavior in the positions "pump on" and "pump off." Wash-out here means that the rate of noise counts decreases with time when the instrument is supplied with particle-free air. After an exposure to environmental aerosols, the instruments needed up to 15 h to obtain a stable count rate at a low level (Gebhart and Roth, 1986).

In general, OPCs with incandescent light sources have somewhat lower noise levels. After completion of the wash-out most instruments with a sensitivity of 0.5 μm are prepared to monitor ISO Class 4 or 3 (see Table 33–1). For the detection of ISO Class 3, for instance, instruments with a sensitivity of 0.5 μm must produce less than 3.5 noise counts per m^3, whereas laser counters with a sensitivity of 0.1 μm need a background level smaller than 100 counts per m^3 (see Chapter 33). In one model the count rate of noise in the first channel (0.1 to 0.3 μm) increased with time of operation, indicating that heating during the course of operation may have affected electronic noise. The high background in another instrument was caused by a leak in the housing and vanished when the instrument was put in a filtered laminar flow box (Gebhart and Roth, 1986). From these experimental findings it is obvious that a certain model does not necessarily guarantee a low noise level. It is rather the status of an individual device and its history (exposure to concentrated aerosols) that determines the background noise. Therefore, instruments used for monitoring clean room Class 100 or lower should be carefully checked before the measurements.

Sensitivity and Sample Flow Rate

Manufacturers usually express the sensitivity of an OPC as the smallest particle diameter d_{min} detectable with the instrument. According to a convention of the German standard setting organization (VDI 3491, number 3), however, sensitivity should be associated with counting efficiency and the particle size with 50% counting efficiency taken as the lower detection limit d_{min}. For the determination of the counting efficiency η near the lower detection limit, the measured concentration c has to be related to the concentration c_0 obtained by independent reference methods according to

$$\eta = \frac{c}{c_0} \tag{15–32}$$

During an experimental run, instruments to be checked simultaneously sample a monodisperse test aerosol from a common reservoir together with a reference instrument. With

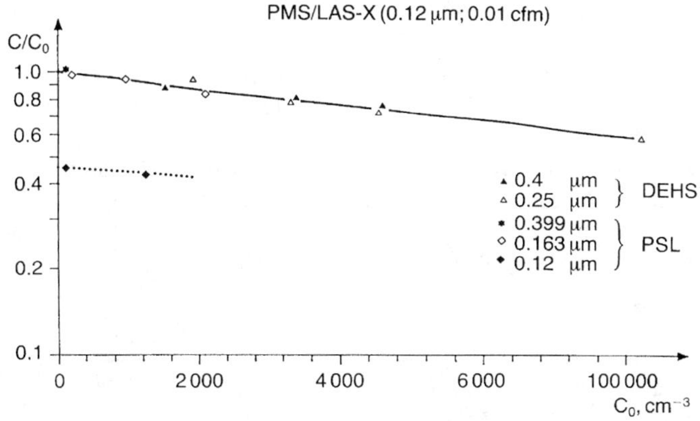

Fig. 15–16. Experimental coincidence curve of the laser spectrometer LAS-X (***PMS***).

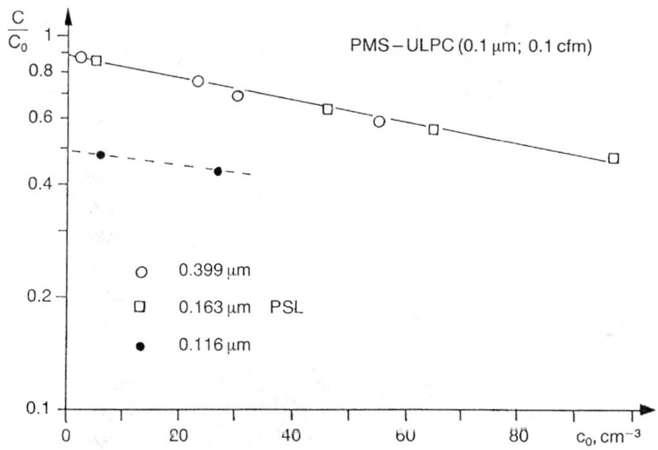

Fig. 15–17. Experimental coincidence curve of the laser counter ULPC (***PMS***).

this procedure, counting efficiencies of OPCs were measured by van der Meulen et al. (1980), Gebhart et al. (1983), Gebhart and Roth (1986), Wen and Kasper (1986), Liu and Szymanski (1987), and Gebhart (1989a).

A difference between c and c_0 can originate from (1) coincidence losses, (2) an incorrect sample flow rate Q, and (3) a decreasing sensitivity of the instrument. The effects of these parameters on counting efficiency can be separated by means of the coincidence curve of an instrument, where the ratio c/c_0 is plotted versus the reference concentration c_0. Coincidence curves of two instruments measured by Gebhart (1989a) are shown in Figures 15–16 and 15–17. Starting with test aerosols far above threshold, a single curve is obtained for all particle sizes in a semilogarithmic plot. If the extrapolation of this curve to zero concentration ($c_0 \rightarrow 0$) deviates from one, the sample flow rate of the instrument is not correct. Deviations from this curve for particle sizes approaching the lower detection limit indicate a decreasing sensitivity of the instrument. If counting efficiencies are generally evaluated by extrapolating the measured concentration ratio c/c_0 back to zero concentration, coincidence losses are eliminated and the counting efficiency reflects a specific property of an instrument.

Counting efficiencies of different OPCs obtained by this extrapolation technique are summarized in Table 15–4 (Gebhart, 1989a).

In the submicrometer size range, the counting efficiency decreases with decreasing particle size so that the lower detection limit is a gradual transition rather than a sharp step. The data confirm previous findings by van der Meulen et al. (1980) and Gebhart et al. (1983) that instruments using incandescent light count 0.32 μm particles with an efficiency of less than 20%. For most laser instruments, on the other hand, counting efficiencies near 100% are achieved down to particle diameters of 0.1 μm. With the highly sensitive model MS-LAS (***PMS***), even particles of 0.07 μm can be counted with an efficiency of about 30% at a reduced sample flow rate of 300 cm^3/min.

EXAMPLE 15–3

The sensitivity of a laser particle counter (LPC) has been determined with PSL spheres to be $d_{min} = 0.1\ \mu m$. The LPC shall now be applied to droplets of di-2-ethylhexyl sebacate (DEHS). Calculate the shift in the sensitivity of the LPC for DEHS droplets.

Answer: One has to calculate the diameter of a DEHS droplet that scatters the same flux of light as a PSL sphere of 0.1 μm in diameter. In the size range of 0.1 μm, the Rayleigh approximation of Eq. 15–9 can be applied, which for identical optical arrangements, reduces to

$$d_{PSL}^6 \left|\frac{m^2-1}{m^2+2}\right|_{PSL}^2 = d_{DEHS}^6 \left|\frac{m^2-1}{m^2+2}\right|_{DEHS}^2$$

PSL: m = 1.59, $d = 0.1\ \mu m$
DEHS: m = 1.45

$$d_{DEHS} = d_{PSL}\left[\frac{\left|\frac{m^2-1}{m^2+2}\right|_{PSL}^2}{\left|\frac{m^2-1}{m^2+2}\right|_{DEHS}^2}\right]^{1/6}$$

$$= d_{PSL}\left[\frac{\left|\frac{2.528-1}{2.528+2}\right|_{PSL}^2}{\left|\frac{2.103-1}{2.103+2}\right|_{DEHS}^2}\right]^{1/6}$$

$$= 0.108\mu m$$

In the supermicrometer size range, satisfactory results are obtained. However, for particle diameters above 2 μm, inlet losses in the sampling system may again affect counting efficiency, especially if instruments with lower sampling rates are used in connection with plastic tubes. The sampling efficiency of the particle counter Royco 245 was investigated by Willeke and Liu (1976). Comprehensive studies of inlet characteristics and their effect on the counting efficiency of an optical sensor have been carried out by Tufto and Willeke (1982) and by Okazaki and Willeke (1987). Fissan and Schwientek (1987) performed a review of aeerosol sampling and transport. Further information on sampling efficiency and aerosol transport is given in Chapter 8.

Table 15–4 indicates that the sensitivity of an OPC can be increased by using laser light illumination. On the other hand, it can be seen from this table that there exists a close

TABLE 15–4. Specifications and Counting Efficiencies of a Selection of Commercial OPCs

Incandescent Light Model	d_{min} (μm)	Q (m³/s)	Counting Efficiencies at Diameter d_p (μm)						
			0.163	0.22	0.32	0.47	0.72	0.95	2.02
HIAC/Royco 227	0.3	4.72×10^{-6}			0.12	0.92		1.05	
HIAC/Royco 4102	0.3	4.72×10^{-5}			0.18	1.01		0.98	
HIAC/Royco 5000	0.3	1.67×10^{-5}			0.18	0.89	0.91		
CLIMET CI 8060	0.3	4.72×10^{-4}			0.15	1.02	0.98		
Laser Light Model	d_{min} (μm)	Q (m³/s)	Counting Efficiencies at Diameter d_p (μm)						
			0.068	0.082	0.109	0.12	0.163	0.22	0.32
PMS LAS-X	0.12	5.00×10^{-6}				0.46	0.98		1.02
PMS LAS-X	0.9	1.00×10^{-6}			0.98	0.98	0.99		
PMS LPC-110	0.1	4.72×10^{-4}				0.47	0.98	1.0	
PMS MS-LAS	0.065	5.00×10^{-6}	0.27	1.03	0.97		1.04		
HIAC/Royco 5120	0.2	4.72×10^{-4}			0.24	0.94	0.99	1.0	
HIAC/Royco 236	0.12	5.00×10^{-6}					1.03	0.99	1.0

Sources: Gebhart and Roth (1986) and Gebhart (1989a).

connection between the sampling flow rate Q of a laser particle counter and its smallest detectable particle size d_{min}. The lower the sampling flow rate, the more sensitive is the instrument. This is because above a certain illumination intensity, light scattering by the air molecules within the sensing volume becomes the limiting factor for the sensitivity. Under these circumstances, the signal-to-noise ratio of an OPC can only be further improved by a reduction in the number of molecules that scatter light into the detector; in other words, an increase in sensitivity is only possible at the expense of the sensing volume and, in consequence, at the expense of the sampling flow rate. From Figure 15–10 it can be seen that 1 mm³ of air scatters about the same amount of light as a 0.22 μm particle with refractive index $m = 1.5$. This principal limitation of the sensitivity of an OPC exists regardless of the kind and intensity of illumination. The alternatives in clean-room technology, therefore, seem to be either to use an instrument with high sensitivity (0.1 μm diameter particles) but a lower sampling rate or a less sensitive device with a higher sampling rate (4.72×10^{-4} m³/s [28.3 L/min]).

In a more recent development by Particle Measuring System Inc. (***PMS***), successful attempts have been made to overcome the problem of Rayleigh scattering on gas molecules. In these models an elongated sensing volume is imaged onto a photodetector array consisting of independent elements. Owing to the facet structure of the detector unit, each element views only part of the gas molecules of the sampling flow but records the total flux of light scattered by a single traversing particle. With this segmentation technique, laser particle counters can be constructed that combine a high sensitivity (0.1 μm) with a high sampling rate of 28.3 L/min. For instance, the model LPC 110 (***PMS***) in Table 15–4 is an instrument based on this technique.

Sizing Accuracy and Resolving Power

The sizing accuracy of an OPC describes its ability to measure the geometrical diameter of aerosol particles by classifying their photoelectric pulse heights into channels of given size intervals. For the evaluation of the sizing accuracy, cumulative size distributions of well-characterized test aerosols are compared with those measured with the OPC (Gebhart et al., 1983; Liu et al., 1985; Szymanski and Liu, 1986). The count median diameter derived from

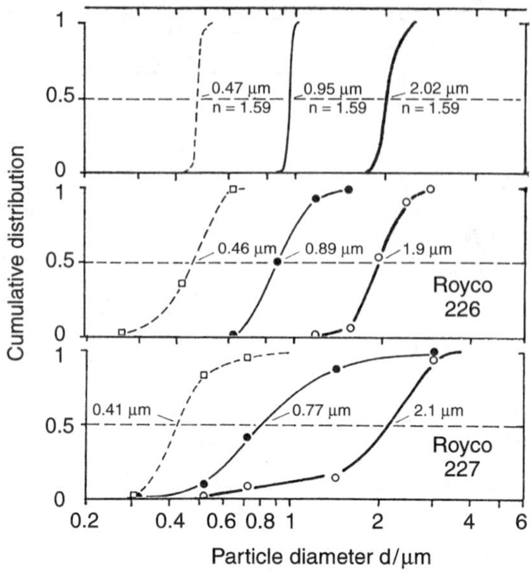

Fig. 15–18. Cumulative size distributions of different PSL test aerosols (above) compared with measurements with different OPCs. (Gebhart et al., 1983.)

these measured size spectra are compared with the true size of the test aerosol to provide a measure of the sizing accuracy. Examples of cumulative size spectra measured with different OPCs by Gebhart et al. (1983) are shown in Figure 15–18. The PSL test aerosols have been analyzed by electron microscopy.

Regarding the sizing accuracy, one has to distinguish between avoidable and unavoidable errors. Avoidable errors are systematic shifts on the size scale. They can be corrected by changing the linear amplification factor of the electronics. Unavoidable errors exist as long as the calibration curve has an ambiguous part or the calibration standard differs in refractive index from the aerosol to be investigated. Because of a multivalued part in the response function of the forward-scattering sensor shown in Figure 15–11, PSL spheres of 0.95 μm in diameter are classified as 0.77 μm particles (see Fig. 15–18). An extreme example of the effect of refractive index are measurements of opaque particles (coal dust, India ink) in an OPC that applies forward scattering and has been calibrated with transparent spheres (Whitby and Vomela, 1967; Willeke and Liu, 1976). In this case, particles of the same geometrical size differ by about one order of magnitude in the flux of scattered light, which reduces the measured size of the particles by about a factor of three.

The resolution power of an OPC describes its ability to distinguish two monodisperse aerosols of different mean particle size. It depends on the slope of the calibration curve and on the ability of an OPC to produce uniform pulses upon exposure to a monodisperse aerosol. The main factors influencing the resolution power are the homogeneity of illumination in the sensing volume and optoelectronic noise.

Because the homogeneity of illumination is affected by the alignment of the optics, instruments of the same model may not have the same resolution due to normal manufacturing tolerances. To measure the resolution power, a monodisperse aerosol of known size and standard deviation is sampled by the OPC, and the instrumental broadening of the size distribution is determined. This can be done directly either by recording differential spectra (Liu et al., 1974a; Roth and Gebhart, 1978) or by evaluating the slope of a cumulative size distribution (see Fig. 15–18).

A typical indication of nonuniform illumination of the particles is the recorded spectra having a sharp cutoff with respect to bigger particles and a gradual cutoff with respect to smaller sizes. Optoelectronic effects on the resolution power of an OPC become important only near the lower particle size detection threshold, where the noise ripples are comparable with the particle signals. The modulation of the particle signal by quantum-statistical processes in the photodetector reduces the resolution power near threshold considerably, but has no effect for bigger particles that contain enough photons in their scattered light pulses (Liu et al., 1974a). For instruments with an optically defined sensing volume, the resolution power is additionally diminished by the effect of the border zone (Helsper and Fissan, 1980).

With well-designed laser aerosol spectrometers, size resolutions comparable with that of electron microscopy can be achieved, especially in the size range below the wavelength of light (Roth and Gebhart, 1978; Knollenberg, 1989).

A Combined DMPS/LAS System

A differential mobility particle sizer (DMPS; see Chapter 18) and a laser aerosol spectrometer (model LAS-X, ***PMS***) were combined by Gebhart et al. (1990) for automated "in situ" evaluation of number concentration and size distribution of environmental aerosols in the diameter range between 0.01 and 3 µm.

In the particle size range of 0.01 to 0.4 µm, analysis was performed using the DMPS. The particles were classified according to their electrical mobility using a differential mobility analyzer (DMA; model 3071, ***TSI***) and counted using a condensation nucleus counter (CNC; model 3760, ***TSI***). To determine the mobility size of the original distribution, a numerical algorithm was applied to take into account the charge distribution of the particles and the transfer function of the DMA (Keady et al. 1983).

In the size range 0.09 to 3 µm, the LAS-X was used to size particles with laser illumination at a wavelength of 0.63 µm. To eliminate the effect of the refractive index on the classification accuracy of the LAS-X, the DMA was used to separate particle fractions of uniform electrical mobility from the unknown aerosol to be investigated. After sizing by the LAS-X, each of these fractions was split into several subfractions of uniform size but different charge. The peak of particles with one charge was suitable for a direct field calibration of the LAS-X in terms of mobility diameters up to about 1.2 µm. Using this procedure between 0.01 and about 1.2 µm, a uniform diameter calibration was obtained.

Another advantage of the combined system is that the numerical algorithm for the data inversion in the DMA can be validated using absolute measurements of the LAS-X particle number concentration in the range where both sensors overlapped (0.09 to 0.4 µm). This system was successfully applied by Brand et al. (1992) during a 1 month field study of environmental aerosols in the city of Frankfurt/Main.

Single-Particle Shape Analyzer

Illumination of a spherical particle with unpolarized or circularly polarized light results in a scattering pattern that is rotationally symmetrical relative to the axis of illumination. Deviations from spherical shape, on the other hand, lead to a variability in azimuthal light scattering. This principle can be used in an OPC to characterize the shape of a particle. An instrument that detects scattered light under a constant scattering angle of $\theta = 55°$ at eight azimuthal positions equally distributed between $0° \leq \Phi_0 \leq 360°$ was developed by Sachweh et al. (1995, 1999) to analyze the shape of various kinds of aerosol particles. For this purpose 1 minus the relative standard deviation of all eight detector responses was introduced as a sphericity index, *SPX*. The *SPX* is 1 for perfect spheres and decreases with increasing deviations from the spherical shape. The instrument had a lower detection limit of $d_{min} = 0.5\,\mu m$ and could clearly distinguish between PSL spheres and quartz particles. In a modified version, it could also identify fibers among an assembly of isometric particles (Sachweh et al., 1999). An apparatus based on the same principle is the Aspect Size and Shape Analyser (***BIR***). This instrument includes four detectors and covers the size range between 0.5 and 20 µm. A similar instrument, but mounted on an aircraft wing and using six detectors, was designed to discriminate between supercooled cloud droplets and ice crystals (Hirst et al., 2001).

MULTIPLE-PARTICLE OPTICAL TECHNIQUES

General Considerations

Multiple-particle detection instruments (photometers) based on scattered light intensity are useful for concentration measurements of aerosols if certain requirements are met. For the determination of concentration ratios or relative concentrations, the composition of the aerosol (particle size distribution, refractive index) must be constant during a series of experimental runs. For absolute measurements of mass concentration, the photometer must be calibrated with the aerosol to be investigated. In both cases, the instrument must be operated in its linear range, where the number of particles in the sensing volume is linearly correlated with the photometer signal. This range of linearity is limited at high concentrations by multiple scattering and at low concentrations by the stray-light background in the chamber.

Stray light originates from optical elements such as lenses, glass windows, and optical stops. Well-designed instruments are limited by Rayleigh scattering from the air molecules, resulting in nearly steady-state noise levels. By replacing air with a gas of known scattering properties, it is possible to calibrate a photometer in terms of scattering cross section. A measure of the stray-light background is the photometer response in the presence of particle-free air. Concentration ratios evaluated from photometer responses R_i are measured correctly only if the contribution of the background R_{ns} is either negligible or subtracted from the photometer reading:

$$\frac{c_1}{c_2} = \frac{R_1 - R_{ns}}{R_2 - R_{ns}} \tag{15–33}$$

In all cases where the analog signal is subjected to electronic data processing, the photometer should be adjusted to a zero response for particle-free air.

From the lower detection limit to onset of multiple scattering, the linearity range of a typical photometer can span at least three orders of magnitude in number concentration. Aerosol photometers in combination with inert aerosols are currently applied in filter testing and in aerosol medicine. They are also useful as real-time dust monitors in industrial hygiene (see Chapter 25) and for continuous recordings of aerosol concentrations in the atmosphere (see Chapter 27). In general, instruments based on light scattering are much more sensitive than light extinction systems.

Applications

Aerosol Inhalation Studies. Since the pioneering work of Altshuler et al. (1957), photometers in combination with monodisperse test aerosols have been used in aerosol research to measure the total deposition of aerosol particles in the human respiratory tract as well as to study gas mixing processes and deposition mechanisms inside the airways (Gebhart et al., 1988; Gebhart, 1989b). An integral photometer developed for aerosol inhalation studies is shown in Figure 15–19.

Its cylindrical aerosol channel can pass sample flow rates up to 1 m^3/min. An expanded parallel beam (3 mm diameter) of an HeNe laser traverses a system of diaphragms, crosses

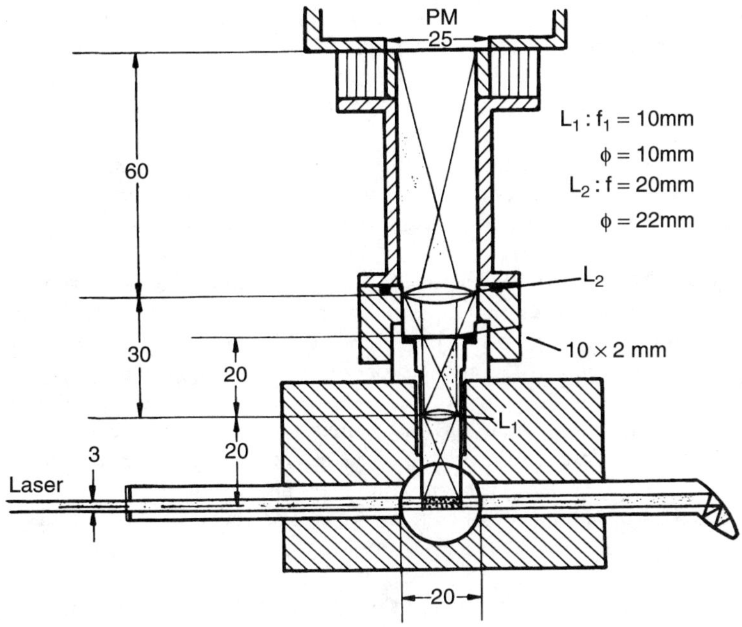

Fig. 15–19. Photometer for aerosol inhalation studies.

the aerosol channel, and is absorbed in a light trap. Light scattered by the particles within the angular range of 90° ± 14° is collected by a lens L_1, which images the illuminated particles onto a slit stop. The slit stop confines the sensitive volume to about $100\,mm^3$ and keeps stray light from reaching the photomultiplier. A second field lens L_2 images the aperture of lens L_1 onto the cathode of the photomultiplier. The instantaneous aerosol concentration is represented by the anode current of the photomultiplier, which, after amplification, can be used for data processing. The photometer technique permits continuous recording of the aerosol concentration close to the mouth during the entire course of a breathing cycle. From such records and simultaneous measurements of respiratory volumes and flow rates the amount of aerosol in successive fractions of inspired and expired air can be evaluated. Data of this kind provide new possibilities for lung function tests (Gebhart et al., 1988; Heyder, 1994).

An improved version of this inhalation measurement technique is the Respiratory Aerosol Probe (***PAR***). In this device, the analog signals of aerosol number concentration and of respiratory flow rate are measured simultaneously at the entrance of the respiratory tract and fed to a personal computer for automatic data collection and processing (Westenberger et al., 1992). In the photometer unit, the beam of a 3 mW laser diode ($\lambda = 0.78\,\mu m$) is collimated to illuminate about 70% of the cross section of an elliptically shaped respiratory channel. Light scattered by inspired particles within an angular range 90 ± 32° is collected and projected onto a miniaturized photomultiplier. The respiratory channel of the photometer is immediately connected with the mouthpiece and can be supplied with either clean air or aerosol by switching a set of pneumatic valves. The aerosol can be applied as a continuous flow or in boluses with volumes down to $2 \times 10^4\,mm^3$ [$20\,cm^3$]. The respiratory flow rate is measured with a pneumotachograph, which yields the inspired air volumes after integration.

Industrial and Environmental Hygiene. The TM digital μP (***HUN***) is an instantaneously reading respirable dust photometer constructed for concentration levels in workplaces. The portable, battery-supplied instrument measures the infrared light (0.94 μm) scattered by airborne dust particles at a mean scattering angle of 70°. Its open measuring chamber, which is insensitive to daylight, is filled by natural convection and includes coarse particles (total dust). The photometer reading, however, is calibrated against the mass concentration of respirable dust by means of a gravimetric respirable dust sampler (see Chapter 25). Experiments carried out by Armbruster et al. (1984) using monodisperse test aerosols and industrial dusts confirm the instrument's principle. The photometer values obtained for industrial dusts can be linearly converted into mass values of respirable dust as long as ultracoarse particles are excluded.

A respirable aerosol monitor (RAM; ***MIE***) allows real-time assessment of respirable dust by light scattering after aerodynamic separation of this fraction in a cyclone (Lilienfeld, 1983). A miniature real-time aerosol monitor (model MiniRAM, ***MIE***) incorporates a pulsed near-infrared light-emitting diode, a silicon detector, and various collimating and filtering optics, collecting light scattered over a range of 45° to 95°. Air surrounding the instrument passes freely through its open sensing chamber, and its measurement ranges cover mass concentrations from 0.01 to 100 mg/m^3. A further step of development of real-time aerosol monitoring is the personal/Data RAM (***MIE***). This instrument is also available in a model (pDR-1200, ***MIE***) using active sampling of aerosol through a size-selective device to allow monitoring of respirable, thoracic, and PM-2.5 fractions of environmental aerosols (see Chapters 25, 27, and 29).

Filter efficiencies can be checked within the linearity range of a photometer by measuring the concentrations of uniform test aerosols upstream and downstream of a filter by means of light scattering on an assembly of particles. In the case of high-efficiency filters, however, integral photometry requires a relatively high concentration upstream of the filter so that the use of an OPC or DMA is more reasonable.

Atmospheric Aerosol Measurement. Common mass concentration levels for respirable as well as total dusts in workplaces are in the range of 0.1 to 100 mg/m^3, whereas typical aerosol mass concentrations in the atmosphere cover the range from 10 to about 200 μg/m^3 (Charlson et al., 1967). Therefore, particulate pollution monitoring in the atmosphere requires nephelometers of high sensitivity. The integrating nephelometer was modified by Charlson et al. (1967) for air pollution and visibility studies. Its main component is a 0.15 m diameter pipe 2 m long containing all optical and sampling elements. The light source is a xenon flash lamp mounted on the pipe wall and powered at a rate of 1.2 flashes/s. An opal glass in front of the lamp provides a cosine characteristic of the light source. At one end of the pipe, a photomultiplier is mounted rigidly behind a series of collimating disks. The opposite end serves as a light trap, which is also constructed with collimating disks. Sample air is drawn through the central chamber at a flow rate of about 1.7×10^{-3} m^3/s [100 L/min]. The output from the phototube, a pulse of about 20 μs, is amplified and averaged over time. The device has sufficient sensitivity to distinguish gases on the basis of their Rayleigh scattering (see Table 15–1) so that different particle-free gases can be used to calibrate the instrument in terms of scattering cross sections. The results indicate that for aged atmospheric aerosols (accumulation mode), light scattering is proportional to the mass concentration of suspended particles (Charlson et al., 1967).

A newer generation of integrating nephelometers based on the same principle as described by Charlson et al. (1967) was developed by ***TSI***. Model 3551 is a high-sensitivity, one-wavelength (λ = 0.55 μm) instrument that determines the scattering coefficient σ_s of atmospheric aerosol particles (Eq. 15–20). Because of its high sensitivity, the nephelometer can measure scattering coefficients as low as 2×10^{-7} m^{-1} . In addition to model 3551, ***TSI*** offers annother model (model 3563) using three wavelengths and a backscatter option. It is designed for short- and long-term measurements of the light-scattering coefficient σ_s of atmospheric aerosols, including visibility and air quality studies.

LIGHT SCATTERING BY IRREGULAR PARTICLES

General Remarks

Theoretical and experimental calibration curves of optical techniques for particle characterization are usually confined to particles of spherical shape, whereas most aerosol systems consist mainly of particles having irregular shape. In this section, approximations valid for particle diameters $d_p << \alpha$ and $d_p >> \alpha$ are used to derive some general predictions about the influence of the particle shape on light scattering. The predictions derived from these approximations are compared with experimental results with optical particle counters using irregularly shaped particles (Gebhart, 1991). The experimental findings are correlated with angular scattering patterns of nonspherical particles obtained from microwave analog measurements (Zerull et al., 1977) and with light-scattering observations on single particles suspended in an electrodynamic balance (Bottlinger and Umhauer, 1989).

Particles Smaller than the Wavelength

When the particle size is much smaller than the wavelength, the particle is subjected to an almost uniform optical field. In this case, light scattering is mainly a volume effect, and the light-scattering diameter d_{sc} of such a small nonspherical particle approaches its volume-equivalent diameter, d_v. This can be shown theoretically by applying dipole theory (polarizability) to spheroids that are smaller than the wavelength (van de Hulst, 1957; Kerker, 1969). As has been shown by Gebhart (1991), a ratio of the semiaxes of $a/b = 5$ results in a variation of the scattering intensity by a factor of 1.27. Because of the d^6 dependence of the scattered light on the particle diameter, however, the light-scattering diameter d_{sc} of such a spheroid deviates only about 4% from that of a volume-equivalent sphere.

For experimental investigations, agglomerates of uniform polystyrene spheres are useful objects to study the response of an OPC to nonspherical particles. Let d_{scj} be the light-scattering diameter of an agglomerate formed by j uniform spheres with diameter d_1 and refractive index m. Then a relative light-scattering diameter, F_j, of the aggregates can be defined as

$$F_j = \frac{d_{sci}}{d_1} \qquad (15\text{–}34)$$

Relative light-scattering diameters F_j have been measured with a laser aerosol spectrometer by Gebhart et al. (1976) and Gebhart (1991). Owing to the high resolving power of the instrument, agglomerates consisting of up to six spheres could be distinguished. It was found that $F_j = j^{\frac{1}{3}}$ so that, within an experimental error of 3%, d_{scj} was identical to the light-scattering diameter of a sphere of equal volume. Chen and Cheng (1984) investigated agglomerates of PSL spheres with the Royco 236 laser counter (see Fig. 15–13) and also measured a volume-equivalent response for PSL particles below 0.5 µm. See Chapter 23 for further discussion of agglomerate measurements.

Particles Larger than the Wavelength

In the limiting case of $d_p >> \lambda$, the scattered light can be considered as consisting of three components, which are due to the physical effects of diffraction, reflection, and refraction. The light-scattering components are expressed by Eq. 15–11 and illustrated in Figure 15–3.

In accordance with the Fraunhofer diffraction formula, the light scattered within the forward diffraction lobe is a function of the projected area of a particle. Outside the diffraction lobe, however, light scattering is affected by reflection and refraction on the particle surface. Therefore, specular and internal reflections can produce angular distributions that deviate considerably from the scattering diagram of a sphere of equal projected area. Consequently, the effect of the particle shape is highest if light scattered outside the diffraction part is collected through a relatively small aperture.

For irregularly shaped particles with dimensions above the wavelength of light, the effect of particle shape on light scattering is lowest if the flux of scattered light is a function of the projected area of the particle, regardless of its shape. There are two possibilities for building instruments with a projected area response: either collect only diffracted light (low-angle scattering) or collect all light scattered by reflection and refraction (4π-geometry scattering). The response of a low-angle scattering instrument to agglomerates of polystyrene spheres was investigated by Gebhart et al. (1976) and Gebhart (1991). It was found that $F_j = j^{\frac{1}{2}}$ so that, within an experimental error of about 3%, d_{scj} was identical to the diameter of a sphere of equal projected area.

It should be mentioned in this connection that the response of a projected-area instrument to a nonspherical particle, of course, depends on particle orientation in the sensing volume. It can be shown, however, by statistical considerations that even for randomly oriented agglomerates of spheres the more frequent orientations are those that exhibit projected areas that are multiples of the cross section of a single sphere.

An instrument with a relatively large receiver aperture is the CI 208 counter (***CLI***) shown in Figure 15–12. The response of this instrument to doublet aggregates ($j = 2$) of polystyrene spheres was investigated by Chen and Cheng (1984). Their results demonstrated that for $d_1 > 0.6\,\mu m$, the light-scattering diameter d_{scj} of such doublet aggregates is identical to the diameter of a sphere that has the same projected area as the cluster.

Results obtained from microwave analog measurements (Zerull et al., 1977) showed that the scattering patterns of a nonspherical particle and of a sphere of equal projected area agreed quite well in the angular range where diffraction dominated. Outside the diffraction lobe, however, the intensity of light scattered by a nonspherical particle was, on average, higher than that predicted by Mie theory, and the intensity of light scattered in a fixed direction (small aperture) varies considerably with particle orientation. The experimental findings of Büttner (1983), who measured irregularly shaped particles of quartz and limestone with the right-angle scattering instrument shown in Figure 15–14, confirmed the microwave measurements. From Büttner's results, it is apparent that the particles consisting of quartz and limestone scatter much more light in the 90° range than spherical particles of polystyrene and glycerine having the same Stokes diameter, although the optical constants of the materials are comparable. Angular distributions of scattered laser light measured by Coletti (1984) on assemblies of isometric nonspherical particles support the experimental findings of Zerull et al. (1977) and Büttner (1983).

In a recent study, monodisperse NaCI crystals produced from a vibrating-orifice generator were classified in the counting mode of a two-mode laser aerosol photometer that collected light at 90° ± 11° (Gebhart et al., 1988). Whereas a low-angle scattering instrument indicated a monodisperse NaCI aerosol at the generator outlet, the 90° aerosol photometer classifies the NaCl crystals as polydisperse. Converting the same NaCl crystals into saturated NaCl-H_2O droplets, however, lead to a size distribution in the 90° aerosol photometer that could be characterized as monodisperse (Gebhart, 1991).

With a modified Millikan device using the electrodynamic balance (see Chapter 20), Bottlinger and Umhauer (1989) suspended single irregular particles in an optical sensor similar to that in Figure 15–14 and studied the signal variation for all possible orientations of the particle. Based on signal spectra reflecting all orientations of a particle, the authors developed a concept to eliminate the effect of particle shape on size analysis by single-particle light scattering.

When considering a random assembly of large irregular particles, the angular distribution of light scattered by external reflection should be approximately equal to that of large spheres because there exists, on average, a similar probability for angles of reflection. Considering light refraction by the surfaces of randomly oriented nonspherical particles, however, there is a greater probability of high internal reflection angles inside irregular particles, resulting in more total internal reflections at the expense of the refracted component Q_2 realized for spherical particles in the forward direction (see Fig. 15–3).

REFERENCES

Altshuler, B., L. Yarmus, E. D. Palmes, and N. Nelson. 1957. Aerosol deposition in the human respiratory tract. *A.M.A. Arch. Ind. Health* 15:293–303.

Armbruster, L., H. Breuer, J. Gebhart, and G. Neulinger. 1984. Photometric determination of respirable dust concentration without elutriation of coarse particles. *Part. Charact.* 1:96–101.

Berglund, R. N. and B. Y. H. Liu. 1973. Generation of monodisperse aerosol standards. *Environ. Sci. Technol.* 7:147–153.

Born, M. 1965. *Optik*. New York: Springer.

Bottlinger, M. and H. Umhauer. 1989. Single particle light scattering size analysis: Quantification and elimination of the effect of particle shape and structure. *Part. Syst. Charact.* 6:100–109.

Brand, P., K. Ruoss, and J. Gebhart. 1992. Performance of a mobile aerosol spectrometer for an in situ characterisation of environmental aerosols in Frankfurt City. *Atmos. Environ.* 26A:2451–2457.

Brossmann, R. 1966. *Die Lichtstreuung an kleinen Teilchen als Grundlage einer Teilchengrossenbestimmung*. Dissertation, Karlsruhe.

Büttner, H. 1983. Kalibrierung einer Streulichtmesseinrichtung zur Partikelgrossenanalyse mit Impaktoren. *Chemie Ing. Technol.* 55:65–76.

Charlson, R. J., H. Horvath, and R. F. Pueschel. 1967. The direct measurement of atmospheric light scattering coefficient for studies of visibility and pollution. *Atmos. Environ.* 1:469–478.

Chen, B. T. and Y. S. Cheng. 1984. Optical diameters of aggregate aerosols. *J. Aerosol Sci.* 16:615–623.

Chen, B. T., Y. S. Cheng, and H. C. Yeh. 1984. Experimental response of two optical particle counters. *J. Aerosol Sci.* 15:457–464.

Clark, W. E. and E. L. Avol. 1979. An evaluation of the Climet 208 and Royco 220 light-scattering optical particle counters. In *Aerosol Measurement*, eds. D. A. Lundgren, F. S. Harris, W. H. Marlow, M. Lippmann, W. E. Clark, and M. D. Durham. Gainesville: University Press of Florida, p. 219.

Coletti, A. 1984. Light scattering by nonspherical particles: A laboratory study. *Aerosol Sci. Technol.* 3:39–52.

Cooke, D. D. and M. Kerker. 1975. Response calculations for light scattering aerosol particle counters. *Appl. Optics* 14:734–739.

Fissan, H., C. Helsper, and W. Kasper. 1984. Calibration of optical particle counters with respect to particle size. *Part. Charact.* 1:108–111.

Fissan, H. and G. Schwientek. 1987. Sampling and transport of aerosols. *TSI J. Part. Instrum.* 2(2): 3–10.

Friehmelt, R. and S. Heidenreich. 1999. Calibration of a white light/90^0 optical particle counter for aerodynamic size measurements. *J. Aerosol Sci.* 30:1271–1279.

Garvey, D. M. and R. G. Pinnick. 1983. Response characteristics of the particle measuring systems active aerosol spectrometer probe (ASASP-X). *Aerosol Sci. Technol.* 2:477–488.

Gebhart, J. 1989a. Funktionsweise und Eigenschaften Optischer Partikelzahler. *Tech. Mess.* 56:192–203.

Gebhart, J. 1989b. Dosimetry of inhaled particles by means of light scattering. In *Extrapolation of Dosimetric Relationships for Inhaled Particles and Gases*, eds. J. Crapo, F. J. Miller, J. A. Graham, and A. W. Hayes. New York: Academic Press, pp. 235–245.

Gebhart, J. 1991. Response of single-particle optical counters to particles of irregular shape. *Part. Syst. Charact.* 8:40–47.

Gebhart, J., P. Blankenberg, S. Bormann, and C. Roth. 1983. Vergleichsmessungen an optischen Partikelzahlern. *Staub Reinhalt Luft.* 43:439–447.

Gebhart, J., P. Brand, and C. Roth. 1990. A combined DMPS-LAS-SYSTEM for the characterization of fine particles in gases. In *Proc. 2nd World Congress Particle Technology*, September 19–22. Kyoto, Japan, pp. 353–361.

Gebhart, J., G. Heigwer, J. Heyder, C. Roth, and W. Stahlhofen. 1988. The use of light scattering photometry in aerosol medicine. *J. Aerosol Med.* 1:89–112.

Gebhart, J., J. Heyder, C. Roth, and W. Stahlhofen. 1976. Optical size spectrometry below and above the wavelength of light; a comparison. In *Fine Particles*, ed. B. Y. H. Liu. London: Academic Press, pp. 794–815.

Gebhart, J., J. Heyder, C. Roth, and W. Stahlhofen. 1980. Herstellung und Eigenschaften von Latex-Aerosolen. *Staub Reinhalt. Luft.* 40:1–8.

Gebhart, J. and C. Roth. 1986. Background noise and counting efficiency of single optical particle counters. In *Aerosols, Formation and Reactivity, Proc. 2nd Internat. Aerosol Conf.*, September 22–26. New York: Pergamon.

Gucker, F. T. and J. Tuma. 1968. Influence of collecting lens aperture on the light-scattering diagrams from single aerosol particles. J. *Colloid Interface Sci.* 27:402–411.

Helsper, C. and H. J. Fissan. 1980. Response characteristics of a Polytec HC-15 optical particle counter. 8th Annual Meeting Gesellschaft f. Aerosolforschung Schmallenberg, Germany.

Heyder, J. 1994. Biomedical aerosol research. In *Proceedings of the 4th International Aerosol Conference* Aug. 29 to September 2, Los Angeles, California.

Heyder, J. and J. Gebhart. 1979. Optimization of response functions of light scattering instruments for size evaluation of aerosol particles. *Appl. Opt.* 18:705–711.

Hinds, W. C. and G. Kraske. 1986. Performance of PMS model LAS-X optical particle counter. J. *Aerosol Sci.* 17:67–72.

Hirst, E., P. H. Kaye, R. S. Greenaway, P. Field, and D. W. Johnson. 2001. discrimination of micrometre-sized ice and super-cooled droplets in mixed-phase cloud. *Atm. Environ.* 35:33–47.

Hodkinson, J. R. 1962. *Dust Measurement by Light Scattering and Absorption*. Ph.D. Thesis, School of Hygiene and Tropical Medicine, London.

Hodkinson, J. R. and J. R. Greenfield. 1965. Response calculations for light-scattering aerosol counters and photometers. *Appl. Opt.* 4:1463–1474.

Keady, P. B., F. R. Quant, and G. J. Sem. 1983. Differential mobility particle sizer. *TSI Q. Rep.* IX:3.

Kerker, M. 1969. *The Scattering of Light and Other Electromagnetic Radiation.* London: Academic Press.

Knollenberg, R. G. 1989. The measurement of latex particle sizes using scattering ratios in the Rayleigh scattering size range. *J. Aerosol Sci.* 20:331–345.

Knollenberg, R. G. and R. Luehr. 1976. Open cavity laser active scattering particle spectrometry from 0.05 to 5 microns. In *Fine Particles*, ed. B. Y. H. Liu. London: Academic Press, pp. 669–696.

Lieberman, A. 1986. Evolution of optical airborne particle counters in the U.S.A. In *Aerosols: Formation and Reactivity.* Proceedings of the 2nd Internional Aerosol Conference, September 22–26. Berlin, pp. 590–593.

Lilienfeld, P. 1983. Current mine dust monitoring developments. In *Aerosols in the Mining and Industrial Work Environments*, eds. V. A. Marple and B. Y. H. Liu. Ann Arbor, MI: Ann Arbor Science Pub., pp. 733–757.

Liu, B. Y. H., R. N. Berglund, and J. K. Agarwal. 1974a. Experimental studies of optical particle counters. *Atmos. Environ.* 8:717–732.

Liu, B. Y. H., V. A. Marple, K. T. Whitby, and N. J. Barsic. 1974b. Size distribution measurement of airborne coal dust by optical particle counters. *Am. Ind. Hyg. Assoc.* J. 8:443–451.

Liu, B. Y. H. and D. Y. H. Pui. 1974c. A submicron aerosol standard and the primary absolute calibration of the condensation nucleus counter. *J. Colloid Interface Sci.* 47:155–171.

Liu, B. Y. H. and W. W. Szymanski. 1986. On sizing accuracy, counting efficiency and noise level of optical particle counters. In *Aerosols, Formation and Reactivity.* Proceedings of the 2nd International Aerosol Conference, September 22–26. Berlin, pp. 603–606.

Liu, B. Y. H. and W. W. Szymanski. 1987. Counting efficiency, lower detection limit and noise level of optical particle counters. In *Proc. 33rd Annual Meeting of IES*, San Jose, CA.

Liu, B. Y. H., W. W. Szymanski, and K. H. Ahn. 1985. On aerosol size distribution measurements by laser and white light optical particle counters. *J. Environ. Sci.* 28:19–24.

Mäkynen, J., J. Hakulinen, T. Kivistö, and M. Lektimäki. 1982. Optical particle counters: Response, resolution and counting efficiency. *J. Aerosol Sci.* 13:529–535.

Marple, V. A. and K. L. Rubow. 1976. Aerodynamic particle size calibration of optical particle counters. *J. Aerosol Sci.* 7:425–433.

Mie, G. 1908. Beugung an leitenden Kugeln. *Ann. Physik.* 25:377.

Oeseburg, F. 1972. The influence of the aperture of the optical system of aerosol particle counters on the response curve. *J. Aerosol Sci.* 3:307–311.

Okazaki, K. and K. Willeke. 1987. Transmission and deposition behaviour of aerosols in sampling inlets. *Aerosol Sci. Technol.* 7:275–283.

Olaf, J. and K. Robock. 1961. Zur Theorie der Lichtstreuung an Kohle- und Bergepartikeln. *Staub* 21:495–499.

Pettit, D. R. and T. W. Peterson. 1984. Theoretical response characteristics of the coherent optical particle spectrometer. *Aerosol Sci. Technol.* 3:305–315.

Plomb, A., P. T. Alderliesten, and F. W. Galjee. 1986. On calibration and performance of laser optical particle counters. In *Aerosols, Formation and Reactivity, Proceedings of the 2nd International Aerosol Conference*, September 22–26, Berlin, pp. 594–598. Oxford: Pergamon Press.

Preining, O., P. E. Wagner, F. G. Pohl, and W. W. Szymanski 1981. *Aerosol Research at* the *Institute of Experimental Physics of the University of Vienna*, ed. O. Preining. University of Vienna.

Quenzel, H. 1969. Influence of refractive index on the accuracy of size determination of aerosol particles with light-scattering aerosol counters. *Appl. Opt.* 8:165–169.

Reist, P. 1990. *Mie theory calculations for your PC.* R. Enterprises, Chapel Hill, NC.

Roth, C. and J. Gebhart. 1978. Rapid particle size analysis with an ultra-microscope. *Microsc. Acta* 81:119–129.

Sachweh, B., H. Barthel, R. Polke, H. Umhauer, and H. Büttner. 1999. Particle shape and structure analysis from the spatial intensity pattern of scattered light using different measuring devices. *J. Aerosol Sci.* 30:1257–1270.

Sachweh, B., H. Büttner, and F. Ebert. 1989. Improvement of the resolution and the counting accuracy of an optical particle counter by fast digital signal recording. *J. Aerosol Sci.* 20:1541–1544.

Sachweh, B. A., W. D. Dick, and P. H. McMurry. 1995. Distinguishing between spherical and nonspherical particles by measuring the variability in atimuthal light scattering. *Aerosol Sci. Technol.* 23:373–391.

Sachweh, B., H. Umhauer, F. Ebert, H. Büttner, and R. Friehmelt. 1998. In situ optical particle counter with improved coincidence error correction for number concentrations up to 10^7 particles cm^{-3}. *J. Aerosol Sci.* 29:1075–1086.

Soderholm, S. C. and G. C. Salzman. 1984. Laser spectrometer: Theory and practice. In *Aerosols: Science, Technology, and Industrial Applications*, eds. B. Y. H. Liu, D. Y. H. Pui, and H. J. Fissan. New York: Elsevier, pp. 11–14.

Szymanski, W. W. 1992. Multiple scattering enhancement of transmitted radiation in water droplet aerosols. *J. Aerosol Sci.* 23:425–435.

Szymanski, W. W. 1996. Experimental determination of light diffusion coefficients in optically dense aerosols. *J. Aerosol Sci., Suppl. 1* pp. 529–530.

Szymanski, W. W. and B. Y. H. Liu. 1986. On the sizing accuracy of laser optical particle counters. *Part. Charact.* 3:1–7.

Tufto, P. A. and K. Willeke. 1982. Dependence of particulate sampling efficiency on inlet orientation and flow velocities. *Am. Ind. Hyg. Assoc. J.* 43:436–443.

Umhauer, H. 1980. Partikelgrossenbestimmung in Suspensionen mit Hilfe eines Streulichtzahlverfahrens. *Chem. Ing. Technol.* 52:55–63.

Umhauer, H. 1983. Particle size distribution analysis by scattered light measurements using an optically defined measuring volume. *J. Aerosol Sci.* 14:765–770.

van de Hulst, H. C. 1957. *Light Scattering by Small Particles*. New York: Wiley.

van der Meulen, A., A. Plomp, F. Oeseburg, E. Buringh, R. M. van Aalst, and W. Hoevers. 1980. Intercomparison of optical particle counters under conditions of normal operation. *Atmos. Environ.* 14: 495–499.

VDI-3489, Blatt 3. Messen von Partikeln; Methoden zur Characterisierung und Uberwachung von Prüfaerosolen; Optischer partikelzahler. VDI, Dusseldorf, FRG.

Wen, H. Y. and G. Kasper. 1986. Counting efficiencies of six commercial particle counters. *J. Aerosol Sci.* 17:947–961.

Westenberger, S., J. Gebhart, S. Jaser, M. Koch, and R. Köstler. 1992. A novel device for the generation and recording of aerosol micro-pulses in lung diagnostics. *J. Aerosol Sci.* 23(Suppl. 1)449–452.

Whitby, K. T. and R. A. Vomela. 1967. Response of single particle optical counters to nonideal particles. *Environ. Sci. Technol.* 1:801–814.

Willeke, K. and J. E. Brockmann. 1977. Extinction coefficients for multimodal atmospheric particle size distributions. *Atmos. Environ.* 11:995–999.

Willeke, K. and B. Y. H. Liu. 1976. Single particle optical counter: Principle and application. In *Fine Particles*, ed. B. Y. H. Liu. London: Academic Press, pp. 697–729.

Zerull, R. H., R. H. Giese, and K. Weiss. 1977. Scattering functions of non-spherical particles vs. Mie theory. *Appl. Opt.* 16:777–779.

16 Optical Direct-Reading Techniques: In Situ Sensing

DANIEL J. RADER AND TIMOTHY J. O'HERN
Engineering Sciences Center, Sandia National Laboratories, Albuquerque, NM

INTRODUCTION

The characterization of airborne particles and droplets is critical in the study of a wide range of fields, including combustion, environment, industrial processing, clean-room monitoring, and cloud and fog formation. The distributions of particle size, shape, structure, charge, and chemical composition are each important in some context. In the broadest sense, the instruments available for characterizing particles can be divided into two classes: extractive and in situ. In extractive sampling, a particle-laden volume of gas is removed from its environment and transported to a separate location where the particle measurement is made. Many of the most common aerosol measurement techniques operate in this mode (see Chapter 15), as it allows the careful control of the conditions under which the measurement is made. The success of extractive techniques, however, relies on the ability to sample and transport particles without biasing the properties of interest. This condition is sometimes difficult to meet. Inlet inefficiencies, wall losses, and rapid aerosol dynamics (evaporation, condensation, coagulation) are examples of physical processes that can alter the particle size distribution. Extractive techniques can also fail when measurements need to be made in hostile environments: extremes in pressure or temperature, reactive or corrosive environments, and so forth. In situ (noninvasive) measurement techniques can overcome many of these limitations, allowing particle characterization under conditions where extractive techniques are not suitable.

Most of the in situ aerosol measurement techniques currently available are optically based, as the combination of optical imaging and laser illumination has provided a wealth of measurement possibilities. In fact, such a wide array of in situ optical techniques are available that it is necessary to limit the scope of the present review in three ways. First, we restrict discussion to instruments that are commercially available. This is a fairly strict limitation, as the laboratory development of new noninvasive particle characterization techniques is an area of active research. This decision was motivated by the desire to promote techniques that are (1) readily available to the reader (without undue investment in hardware or time) and (2) reasonably portable. Even using commercial techniques, the researcher will find that these systems are generally expensive and require significant investments in time for training. Portability is desirable as many aerosol measurements must be made in field environments.

Aerosol Measurement: Principles, Techniques, and Applications, Second Edition, Edited by Paul A. Baron and Klaus Willeke.
ISBN 0-471-35636-0

As a second limitation, we generally consider only real-time or direct-reading equipment. This is not a serious restriction, as most in situ techniques readily lend themselves to on-line analysis and presentation. It should be noted, of course, that subsequent (and more sophisticated) data analysis is often essential to ensure data integrity. One exception is that holography is considered. Although holograms must be first developed and later reconstructed for analysis, automated techniques are beginning to appear, and so this powerful technique has been included. Finally, we restrict discussion to techniques that provide size distribution measurements; excluded by this restriction are techniques that provide only limited information about the particle cloud. One example is a transmissometer, which is capable of measuring either particle volume concentration or Sauter mean diameter (Holve and Self, 1980).

In short, this chapter reviews commercially available, direct-reading, optically based, in situ aerosol measurement instruments. Among them, capabilities exist for measurement of individual particle sizes from about 0.1 to above 1000 μm, concentrations as high as $10^6/cm^3$, and speeds in the kilometer per second range. Ensemble techniques can measure mean diameters as low as 1 nm. While in situ instruments overcome many of the limitations encountered with extractive methods, they do suffer (as a class and individually) from a wide range of new limitations. To describe these limitations, the next section provides an overview of in situ optical particle-sizing systems. A brief section introducing light scattering theory follows. With this background in place, the chapter continues with a review of the instruments that are currently commercially available to the researcher. The individual reviews are by necessity short, but sufficient references are provided to help the reader to explore each method further. Although every effort has been made to include all of the available equipment, some manufacturers may have been overlooked. The chapter concludes with a section on the crucial topic of performance verification, including the issues of standards, calibration generators, and instrument comparisons.

OVERVIEW

The in situ measurement of particles by optical methods has been an area of active research. Thus, many excellent reviews are available on the topic (Black et al., 1996; Koo and Hirleman, 1996; Tayali and Bates, 1990; Lefebvre, 1989; Hirleman, 1983, 1984, 1988a; Hovenac, 1987). Several sets of proceedings contain current applications and discussions of in situ techniques (Kuo, 1996; Hirleman et al., 1990; Hirleman, 1990; Gouesbet and Grehan, 1988).

It is helpful to divide optical in situ techniques into two general classes, based on whether they analyze single-particle events or aggregate cloud properties. Single-particle counters (SPC) generally make a size determination on one particle at a time by analyzing its scattering behavior while it passes through a well-defined (usually small) volume of high-intensity (usually laser) light. Intensity, phase, or image information in the scattered light have all been used for particle sizing. A size distribution is obtained by sizing a number of particles sufficient to ensure statistical accuracy. This class of instruments is similar in principle to the extractive, optical techniques covered in the previous chapter except that the measurement volume is now located external to the instrument. SPCs generally provide a wealth of information on the counted particles, providing correlations among particle properties such as size, velocity, and time of arrival and allowing spatial characterization of the particle field. At high number concentrations, however, single-particle counting techniques suffer from coincidence errors, which occur when more than one particle occupies the sensing volume at the same instant.

The second class of in situ systems, collectively called *ensemble techniques*, generally operates by illuminating a volume containing a large number of particles and analyzing the collective scattering. An illustrative example of an ensemble technique would be a photographic snapshot (or a hologram in three dimensions), which captures the state of a particle distribution at one instant in time. A drawback of photographic systems is that it is difficult to obtain real-time readout of results. Real-time ensemble techniques are available that remedy this limitation, but require a mathematical inversion of the data to determine the size distribution. Ensemble techniques are well suited for measurements at high particle concentration, but become ineffective at low concentration. Generally, ensemble techniques do not provide as detailed information as SPCs because individual particle information is lost in the averaging. Real-time ensemble techniques provide only limited spatial resolution of the particle field.

Generally, ensemble techniques measure particle concentration (number/volume), while SPC systems measure particle flux (number/area/time) (Hirleman, 1988a). That is, ensemble techniques report the number (and sizes) of particles present in the sampling volume over the measurement time (spatial averaging), whereas SPCs report the number (and sizes) of particles *passing through* the sampling volume during the measurement time (temporal averaging). To obtain aerosol concentration, SPCs require additional particle velocity information. As Hirleman (1988a) points out, the distributions measured by concentration- or flux-based techniques will differ if a systematic correlation exists between particle size and velocity, as demonstrated in Example 16–1.

As each SPC or ensemble particle sizing technique offers distinct strengths and weaknesses, an ideal instrument can only be defined in terms of measuring a specific set of properties for a specific aerosol in a specific environment. In this vein, Hovenac (1987) and Hirleman (1988a) outline an approach to in situ optical sizing in terms of instrument operating envelopes. The central idea is that the choice of instrument must be a two-step process: First, identify the particle properties that need to be measured and the conditions under which the measurement must be made, and, second, establish that these conditions fall within the instrument's operating envelope. The final step is critical. As Hirleman (1988a) points out, many instruments will continue to "merrily report erroneous data and not notify the user." An instrument operating envelope will be defined by the ability of the instrument to measure the desired property over an appropriate range to an acceptable accuracy. Hirleman (1988a) groups the parameters that comprise the operating envelope into three domains: particle, instrument, and environmental properties. Based on Hirleman's scheme, a general overview of the operating envelopes of in situ methods follows.

Particle Properties

A variety of particle properties can be of interest, including size, shape, concentration, velocity, and index of refraction. Each of these properties can be distributed among a population of particles, and the problem becomes one of measuring the related distributions. With nonspherical particles, an ambiguity arises in selecting a dimension to characterize particle size. Moreover, most measurement techniques actually infer particle size indirectly from an observation of some particle behavior (settling speed, light scattering intensity, and so forth). Thus, size distribution measurements must be reported in terms of equivalent diameters: Optical, aerodynamic, hydrodynamic, or electric mobility equivalent diameters are commonly reported. Care must be taken even when comparing among optical techniques, as the scattering behavior of the same particle will depend greatly on the details of the measurement technique used to observe it. Besides size, the particle concentration (number, surface area, or mass of particles per unit volume of gas) is also frequently of interest. A further complication arises as all particle properties can show spatial or temporal variation.

Measurement of particle size distributions demands that both particle sizing and counting be accomplished with great accuracy. High spectral resolution is required when the size distribution is itself of fundamental importance. For example, an exact description of the size

EXAMPLE 16–1

A fluid containing a bimodal droplet distribution is rapidly accelerated before a measurement access window. The smaller mode consists of monodisperse droplets with diameter $d_{p1} = 10\,\mu m$ and concentration $c_1 = 100$ particles/cm^3 that are moving with the fluid at velocity $V_1 = 10$ m/s. The larger mode consists of monodisperse droplets ($d_{p2} = 100\,\mu m$, $c_2 = 20$ particles/cm^3) that are lagging the flow at $V_2 = 2$ m/s. This aerosol is measured simultaneously by an ensemble diffraction technique (sensing volume $v = 1$ cm^3) and an SPC (sensing area $A = 1$ cm^2 normal to the flow, sample time $\Delta t = 1$ s). What is the number mean diameter measured as (1) a spatial average by the ensemble diffraction system, (2) a temporal average by the SPC (no correction for particle velocity), and (3) a spatial average by the SPC? Assume perfect measurement by both techniques, that is, neglect trajectory ambiguity, edge effects, and so forth.

Answer:

1. The ensemble technique responds to the number of particles of each size in the measurement volume. Thus, it would give a number mean of

$$\begin{aligned}\bar{d}_p &= \frac{vc_1d_{p1} + vc_2d_{p2}}{vc_1 + vc_2} \\ &= \frac{(1)(100)(10\mu m) + (1)(20)(100\mu m)}{(1)(100+20)} \\ &= 25\mu m\end{aligned}$$

2. If no correction is made for the discrepancy in droplet velocities, the SPC would weight the mean according to the number of counts during the sample time, $n_i = A\ \Delta t\ V_i\ c_i$. During the sample time, the SPC would record $n_1 = 100{,}000$ counts and $n_2 = 4000$ counts. To find the mean,

$$\begin{aligned}\bar{d}_p &= \frac{n_1d_{p1} + n_2d_{p2}}{n_1 + n_2} \\ &= \frac{(100{,}000)(10\mu m) + (4000)(100\mu m)}{(100{,}000 + 4000)} \\ &= 13.5\mu m\end{aligned}$$

3. An SPC can be used to measure a spatial average concentration if the particle velocities are measured. To do this, the observed number of counts, n_i, is divided by an effective sample volume given by $A\ \Delta t\ V_i$. This calculation gives back the true concentrations ($c_1 = 100$/cm^3 and $c_2 = 20$/cm^3) given in the problem statement and therefore gives the same mean diameter as in case 1.

distribution can be essential in understanding or predicting physical processes or in identifying origin or formation mechanisms. In some settings, however, accuracy may be less important than reproducibility. Typically, the mean size, the spread, and the shape of the distribution are all of interest. Ideally, the selected instrument's sizing range should suitably span the actual particle size range. This can complicate the characterization of wide distributions, as particle sizing over more than one order of magnitude in size is difficult to cover with one instrument in one configuration. The distribution's behavior at its tails can be important, particularly when transforming from a frequency to a mass weighted distribution.

A second property of interest is particle concentration: Aerosol mass, area, and number per volume of gas are each of interest in some context. The variety of concentrations that are encountered in particle measurement is impressive, ranging from a few particles per cubic meter in ultraclean areas to trillions of particles per cubic meter in some industrial settings. Obviously, one technique cannot be expected to cover this entire range. In most situations, it is impractical (or impossible) to characterize every particle present; thus, it becomes necessary to infer the true aerosol properties from a measurement of some subset. Difficulties arise when the particles are present only in small numbers, as is typical at larger sizes or in clean environments. Instrument noise (phantom counts) can become important under this condition, and an effort must be made to ensure statistical significance. Instrument limitations also become evident at high concentrations, as is discussed below. Concentration measurement errors can be amplified when extrapolating volume or mass distributions from measured frequency distributions.

The particle velocity distribution can be important in understanding dispersal, transport, or flux. In some applications, the correlation between particle size and velocity is desired. Even when particle velocity is not of interest itself, it may be a limiting factor in system performance. For example, particles moving at high speeds can pose signal-processing and response-time difficulties in SPCs. If the electronics are not fast enough, the signal from a high-speed particle will broaden and its intensity peak will diminish; the result is that the particle is undersized. Size–velocity correlations can also adversely influence system performance. Generally, SPCs provide particle velocity information while ensemble systems do not (a notable exception being pulsed photography or holography). For SPCs, a velocity measurement is typically required to infer particle concentration from the measured distribution of particle size: Otherwise, faster particles will be counted preferentially in a flux-based measurement. Lower limits for particle velocity in SPCs are discussed by Hovenac (1987); upper limits are typically about 300 m/s, but some systems can operate in the kilometer per second range.

Particle shape and index of refraction are less commonly of interest to the researcher, but are always important through their role in determining a particle's scattering characteristics. Note that some of the imaging systems discussed below are capable of recording particle shape.

Instrument Properties

An accurate determination of a particle size distribution requires that the instrument both size and count particles accurately. Hovenac (1987) describes factors that adversely affect SPC sizing and counting performance. Although both size and count sensitivity are crucial for ensemble techniques as well, the discussion is complicated by the averaging nature of the measurement. The discussion below focuses on the measurement limits imposed by instrument features.

Perhaps the most difficult aspect of making an accurate in situ measurement is in defining the sample volume, as particle velocities and trajectories cannot be controlled as in sampling-type instruments (Holve, 1980). This difficulty applies to both ensemble and SPC techniques and can lead to both sizing and counting errors. For most in situ systems, the sample volume is determined by the intensity profile of the illuminating beam and by the geometry and characteristics of the receiving optics (apertures, stops, lenses, filters, and so forth). Laser beam intensity nonuniformities within the sampling volume (in either the axial or radial direction) result in trajectory-dependent scattered intensity profiles for even monodisperse particles. For the common case of a laser beam with a Gaussian intensity profile, a particle passing through the axis of a laser beam will scatter more light than if it passed through the edge of the beam. Thus, a small particle passing through the beam axis and a large particle passing through the beam edge could give comparable scattering amplitudes ("trajectory ambiguity"; Grehan and Gouesbet, 1986). For intensity-based SPC techniques, such multivalued response degrades instrument accuracy. Moreover, the combination of a nonuniform beam profile and photodetector sensitivity creates the situation where the effective sample volume becomes size dependent; for example, small particles are detected only by passing through the central portion of the beam, whereas large particles are detected over a much larger cross section. Both ensemble and SPC in situ techniques can suffer this counting bias, and all SPCs require some form of sample volume correction (e.g., Holve and Self, 1979a,b; Holve, 1980).

One of the key parameters of interest is particle size. Several issues arise with regard to particle sizing with in situ techniques: precision (repeatability), accuracy (resolution), sensitivity (lowest detectable size), and dynamic range. One requirement for sizing accuracy is a monotonic response curve (intensity or phase versus size); unfortunately, light-scattering techniques are frequently multivalued due to Lorenz-Mie scattering effects (see below). Variations in particle shape and refractive index effects can dramatically affect the shape of the response curve and will limit system accuracy unless calibrations or calculations are performed with similar particles. Many in situ optical systems are based on near-forward scattering techniques, which minimize (but do not eliminate) shape and refractive index effects. The trajectory ambiguity discussed above also degrades accuracy for intensity-based techniques. All optical in situ techniques require that the laser beam waist be four to five times the size of the largest particle to ensure nominally uniform illumination over the particle's surface (Holve, 1980). For example, Hovenac (1987) has shown for SPCs that a single large particle in a small beam could be counted as two smaller ones. Making the linear dimensions of the measurement volume much larger than the largest particles also reduces the fraction of particles that suffer edge effects (Holve and Self, 1979a). Note that enlarging the measurement volume can increase coincidence errors, and so trade-offs must be made.

Lens imperfections, misalignment, electronic and photodetector nonlinearities, and other nonidealities can significantly degrade all aspects of system performance (Holve and Davis, 1985). Beam intensity fluctuations and system misalignment transients can impair both instrument precision and accuracy. As a rule of thumb, optical and signal-processing limitations generally limit the dynamic size range that can be measured (with one instrument at one setting) to about a factor of 30. Instrument noise is frequently a limiting factor in determining dynamic range and can also influence precision, accuracy, and sensitivity.

There is always a desire for improved instrument sensitivity. For in situ SPCs, a lower detection limit of about 0.3 μm is typical, although sampling-type SPCs can currently detect particles to about 0.05 μm. Knollenberg (1985) describes theoretical detection limits for SPCs and shows that the limit is dominated by background scattering from stray light or gas molecules present in the sampling volume. Interestingly, operating SPCs in vacuum can improve instrument sensitivity by a factor of two to six (Knollenberg, 1985). In summary, Knollenberg places the theoretical limit for SPCs operating in air or vacuum at around 0.02 μm. Some ensemble techniques offer much lower detection limits, for example, 0.01 μm for dynamic light scattering.

High particle concentrations can also limit system performance. For example, in SPCs this can lead to coincidence, dead time, and intensity attenuation errors. Coincidence occurs when two particles occupy the measuring volume at the same time, which may be counted as a single large particle, resulting in both sizing and counting errors and consequently skewing the size distribution to larger sizes. Coincidence places an upper limit on the number concentration that can be measured without significant interference for a given system configuration. This upper limit has been shown to be proportional to the probability of interference and inversely proportional to the effective measurement volume (Holve, 1980). Dead time occurs when the electronics are not ready when an event occurs because a previous event is still being analyzed; dead time effects can reduce or skew the measured size distribution.

High particle concentrations between the sample volume and the receiving optics can reduce the intensity of light scattered by the particle. The resulting error in intensity-based techniques would be to undersize all particles. In ensemble systems, multiple particle scattering occurs at high concentrations. In this case, measurements of the size distribution become concentration dependent. Corrections for coincidence, dead time, or multiple scattering can often be made using either hardware or software.

All of the techniques discussed in this review require sophisticated data analysis, and most require a full inversion or deconvolution to finally resolve the desired size and number distributions. Real-time ensemble instruments demonstrate the classic case of inverting a finite set of measured responses to infer an unknown distribution (Hirleman, 1988a). For intensity-based SPC techniques, Holve (1980) has discussed the need to deconvolve the resulting intensity histograms to account for trajectory ambiguity. Although beam intensity variations have a minimal effect on particle sizing with phase-Doppler techniques, corrections still need to be made to account for size-velocity correlations and size-dependent sample volumes when concentration is required. The importance of proper data analysis or inversion cannot be overemphasized. Some of the problems and pitfalls of data analysis and inversion are presented in Chapter 22.

The use of exposed laser illumination by most of these systems poses eye safety concerns. Most of the systems use low power lasers, but focusing can generate dangerous intensity levels. Proper laser safety operations are essential and should be considered in instrument selection, location, and operation.

Environmental Properties

Refractive index gradients along the optical path can cause beam steering, with a resulting change in optical collection angles. The length of the optical path and medium temperature and pressure gradients determine the extent of beam steering. Gas conditions (temperature, pressure, composition) also affect the gas refractive index. Laser systems are readily adaptable to high temperature environments, as they can mitigate the influence of high thermal radiation background. There are also practical issues like optical access and window contamination that must be considered. Also, application of optical techniques in environments with high ambient light levels can lead to spurious measurements unless suitably filtered.

LIGHT SCATTERING

The field of light scattering by particles is very broad and dynamic, and a thorough presentation is well beyond the scope of this brief introduction. Instead, this section provides a limited introduction to key concepts in light scattering that are used in this chapter. For further background, the reader is referred to van de Hulst (1981), Kerker (1969, 1988), and Bohren and Huffman (1983). Recent reviews of light scattering for particle characterization by Jones (1998) and Black et al. (1996) are of particular interest (see also Chapter 15).

Particle Light-Scattering Properties

When an electromagnetic beam is projected through a particle field, some portion is transmitted through the field, while some is absorbed and some is scattered by particles in the field. Figure 16–1 is a schematic representation of light scattering from a single particle. When the scattered light has the same wavelength as the incident beam the scattering is called *elastic scattering*, while *inelastic scattering* describes scattering with a shift in wavelength, as is seen

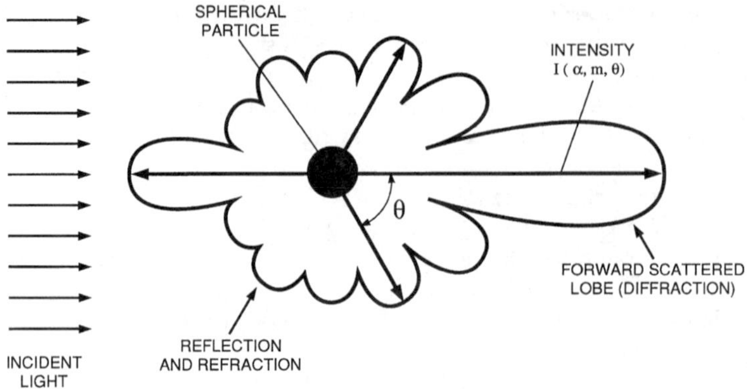

Fig. 16–1. Single-particle light scattering.

in Raman scattering or Doppler-shifted scattering. In general, the distribution of light scattered by a particle is a function of the particle size and shape, the incident wavelength, and the refractive indices of the particle and the surrounding medium. Both scattering and absorption characteristics of a particle can be included by describing the particle's optical properties with a complex refractive index $m = n - in'$, where the real part n describes scattered light characteristics and the imaginary part n' describes absorption. Therefore, the complex refractive index for relatively transparent particles has a very small imaginary part (absorption coefficient), while for strongly absorbing particles the imaginary part is larger, for example, $n' < 10^{-8}$ for water (Bohren and Huffman, 1983) and $n' \approx 1$ for metals in the visible range (van de Hulst, 1981). van de Hulst presents characteristic complex refractive indices for a number of different materials. Most handbook values of refractive index provide only the real part n, which is sufficient to calculate purely refractive properties.

Solution of the complete electromagnetic theory describing light-scattering processes (Lorenz-Mie theory) is sometimes either difficult or invalid, so a number of approximations have been developed that are valid within certain ranges. These ranges are determined using a dimensionless size parameter, which for a sphere is given as $\alpha = (\pi d_p)/\lambda$, where d_p is the diameter and λ is the incident wavelength. Note that the symbol x is also commonly used to represent the size parameter.

Rayleigh Scattering

Rayleigh scattering is typically used to describe light scattering characteristics from particles with very small values of the size parameter (diameters much smaller than the incident wavelength). Rayleigh theory is typically valid for $\alpha << 1$, for example, for $\lambda = 0.6328\,\mu m$ (HeNe laser), $d_p << 0.2\,\mu m$. In the Rayleigh scattering regime, the oscillating electric field of the light wave induces an oscillating dipole in the particle, causing symmetrical scattering (in the forward and backward directions) about the particle. The intensity for Rayleigh scattering is proportional to the sixth power of the particle diameter ($I \propto d_p^6$), placing severe requirements on the dynamic range of photodetectors used to measure Rayleigh scattered light from particles of various sizes. In addition, the absolute intensity of scattered light is small, often requiring intense sources to achieve measurable Rayleigh scatter, risking particle vaporization (Hovenac, 1987). As the particle size increases or as the difference between particle and medium refractive indices increases, the phase shift of light passing through particles becomes significant, the scattering becomes asymmetrical, and the Rayleigh-Gans approximation must be applied to describe the scattering polar distribution.

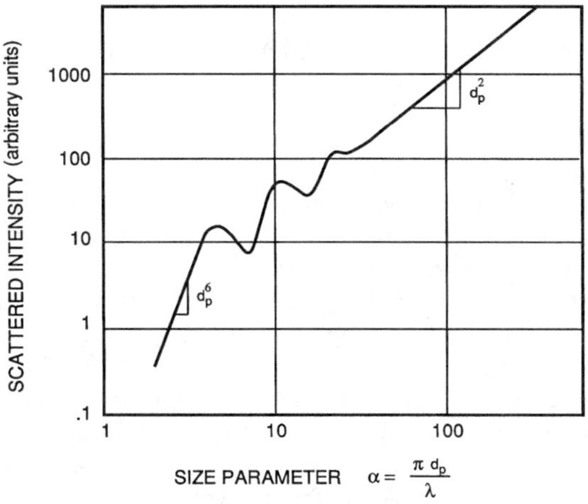

Fig. 16–2. Lorenz-Mie scattering response curve.

Lorenz-Mie Theory

Particles with diameters of the same order as the incident light (say, 0.1 to 1 μm) are too large for Rayleigh scattering because the local electric field is no longer nearly uniform at any instant. Thus, different parts of the particle scatter portions of the incident beam with different properties. In this size range, the appropriate light-scattering theory for homogeneous spheres illuminated by a plane wave is the more complex Lorenz-Mie scattering (van de Hulst, 1981), which in fact provides an exact solution to Maxwell's equations of electromagnetic propagation. In this regime there is a strong interaction between the particle and the incident beam. There is no simple relation between scattered intensity I and particle diameter d_p; the plot of I as a function of d_p (Fig. 16–2) is multivalued, so particles of several sizes can scatter with the same intensity. Strong dependence of scattered intensity on refractive index further complicates measurements in this regime. A number of Lorenz-Mie scattering computer programs are available (e.g., Bohren and Huffman, 1983). Lorenz-Mie theory breaks down when the particle size becomes of the same order as the focused beam diameter or for nonspherical or inhomogenous particles. To account for the effects of arbitrary incident illuminating beams on scattering by spheres, the Generalized Lorenz-Mie Theory was introduced (e.g., Gouesbet, 1994). Light scattering by nonspherical, inhomogeneous particles is discussed below.

Geometric Optics Approximation

Scattering from large particles, where $\alpha >> 1$, can be treated using simple ray optics by considering the light wave incident on the particle to be made up of a collection of individual rays. The rays hitting the particle lead to reflection, refraction, and absorption, while those passing along the edge of the particle give rise to diffraction. The effects of each of these phenomena can be evaluated individually for particle sizes larger than the incident wavelength, unless the particle refractive index is very close to one (van de Hulst, 1981).

The scattered intensity in this regime is proportional to the particle cross-sectional area ($I \propto d_p^2$) and not strongly dependent on shape or composition for aspect ratios close to one. Most geometrical optics approaches assume that the particle is illuminated by plane wave of uniform intensity and so are only valid for particle diameters much smaller than the beam diameter. As the particle size approaches or exceeds the incident beam diameter, corrections to standard geometrical scattering must be applied.

Diffraction (Fraunhofer and Fresnel)

For $d_p > \lambda$, diffraction begins to dominate refraction. For $d_p > 4$ to 5λ, Lorenz-Mie theory reduces to Fraunhofer diffraction theory when limited to the near-forward direction. Diffraction from a single point particle will, in general, produce a Fresnel diffraction pattern. However, when the far-field condition is satisfied (i.e., $z >> d_p^2/\lambda$, where z is the object to observation plane distance), the Fraunhofer diffraction pattern will be seen at the observation plane. The intensity of Fraunhofer diffraction is independent of particle refractive index, a fact that is often useful in particle-sizing applications. Diffraction is dominant at near-forward scattering angles, with an intensity proportional to the particle cross-sectional area. The angular distribution of scattered light intensity is determined by the particle shape and is inversely proportional to the diameter for spherical particles. Fraunhofer diffraction can be described theoretically using the Airy function, with the amplitude of the scattered light described by a Bessel function of the first kind.

Refraction and Reflection

Refraction in a particle can be described by Snell's law, similar to refraction at a plane refractive index interface. Reflection from a particle can be described using standard laws of reflection, with specular and diffuse reflection acting the same as for plane surfaces. The angular distribution of reflected and refracted light is independent of particle size, but depends on complex refractive index, with a higher refractive index spreading the light more strongly, and can depend strongly on particle shape (Gebhart and Anselm, 1988).

Laser Optics

Coherent light interferes, while incoherent light does not. Most light sources (except lasers) generate incoherent or partially coherent light. Incoherent light is sometimes preferred for optical systems in order to prevent formation of interference fringes in the image of interest, usually caused by edge diffraction. However, most optical particle-sizing techniques use coherent laser light because of its intensity, monochromaticity, and coherence, needed to provide the interference key to such techniques as laser Doppler velocimetry, phase-Doppler particle sizing, and holography.

Most systems use lasers with a Gaussian intensity profile (TEM_{oo}). Propagation of a Gaussian beam can be described using diffraction theory (Dickson, 1970), including the size and location of the focused waist, and the intensity distribution along the optical axis. When a Gaussian beam is used for light-scattering measurements, the illumination field is not uniform, and, depending on the relative size of the beam and particle, the incident light intensity may vary over the particle's surface (Jones, 1999; Black et al., 1996). Generalized Lorenz-Mie Theory was developed to account for light scattering by a single spherical particle in an arbitrary incident illuminating beam (Gouesbet, 1994). The theory also can accommodate scattering by particles located off beam center line.

Multiple Scattering

The Lorenz-Mie scattering theory, and the approximations used for light-scattering predictions in different size regimes, assume (among other things) single-particle light scattering. In a concentrated aerosol sample or over long path lengths, light scattered by a single particle can be rescattered by other particles in the illuminated field. The scattered light intensity measured by a photodetector will be reduced by multiple scattering. Clearly, multiple scattering must be taken into consideration in absolute intensity-based measurements and must also be considered for diffraction-based measurements.

Nonspherical Particles

Many particles of interest in aerosol measurements are not spherical, so the detailed light-scattering models must be modified to account for the particle shape. The light-scattering behavior of nonspherical, inhomogeneous particles remains an active area of research (e.g., Damaschke et al., 1998b; Jones, 1999). For sufficiently large particles, where the geometric optics approximation is valid, near-forward diffraction measurements are useful because they are independent of particle shape, as mentioned above. Gebhart and Anselm (1988) discuss approximations to light-scattering theory to include scattering by nonspherical particles for $d_p << \lambda$ and $d_p >> \lambda$ and provide experimental comparisons using agglomerate particles and NaCl crystals. They conclude that the scattered light intensity from small nonspherical particles is within about 4% of that of an equivalent volume sphere. The light-scattering characteristics of large nonspherical particles are similar to equivalent projected area spheres only when collected in the forward diffraction lobe or when essentially all of the reflected and refracted light is collected. Killinger and Zerull (1988) discuss scattering from nonspherical particles in the intermediate size range. They show that scattering from a 3 µm diameter rough sphere is quite close to that predicted by Lorenz-Mie theory but that scattering from a rough, slightly nonspherical 50 µm diameter glass bead varies significantly from scattering by an equivalent sphere. They also found up to one order of magnitude variation in scattered intensity depending on nonspherical particle orientation. See also Chapter 23 for discussion of nonspherical particle properties.

Optical Imaging

In situ systems use optical configurations consisting of lenses, mirrors, filters, and apertures. All components have practical limits. An ideal lens would transform a collimated input beam to a point at the focal distance. Real lenses can cause image aberrations due to chromatic effects, astigmatism, and other effects. Physical optics shows that diffraction places limits on image resolution so that the image of a point is a small Airy disk rather than a point. The highest achievable resolution of an optical imaging system is therefore the diffraction-limited resolution δ given by $\delta = 1.22\lambda l/D$, where l is the image distance and D is the limiting aperture of the system. Two points lying within δ of each other will be indistinguishable in the image, as they will lie within the same Airy disk. Resolution typically limits imaging systems to examination of particles with diameters of the order 5 µm and above. Note that the resolution decreases (poorer resolving ability) linearly with measurement distance from the object. The depth of focus Δf is given by $\Delta f = 2\delta^2/\lambda$, so high resolution (small δ) leads to short depth of field, making it difficult to define the boundaries of the sampled volume, that is, to determine whether an observed particle lies inside or outside the sample volume. Because the depth of field is proportional to the square of the resolution, large particles remain in focus over a much greater distance than do smaller particles. Thompson (1984) reviews imaging techniques for particle sizing, including both coherent and incoherent light, and one- and two-lens systems, along with applications in droplet measurements.

SINGLE-PARTICLE COUNTERS: INTENSITY BASED

The first class of instrument sizes and counts individual particles as they pass through an illuminated sample volume. As the particles pass through this region, they scatter light that is collected over some solid angle by the receiving optics and focused onto a photodetector

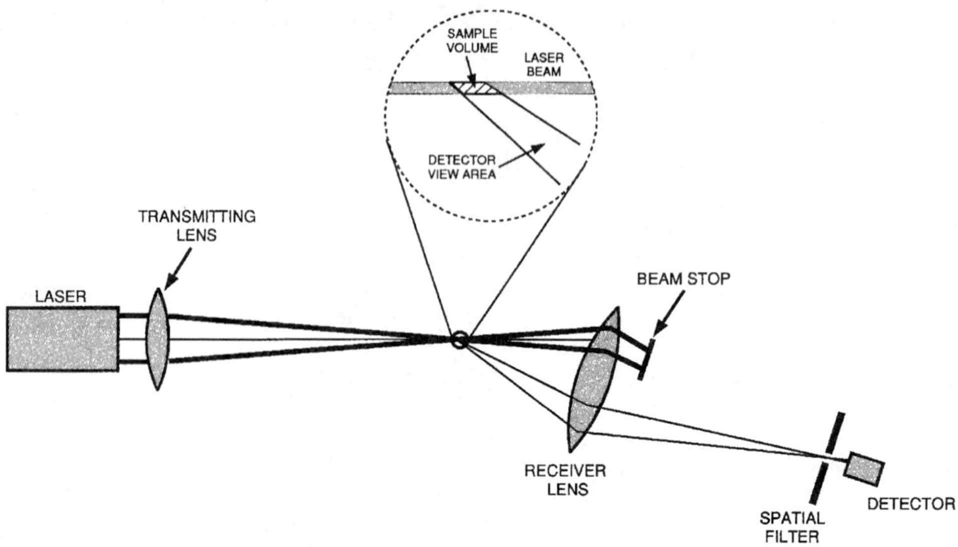

Fig. 16–3. Example of a single-particle light-scattering system.

(Fig. 16–3). The particle size is determined by the peak intensity of the scattered light. A variety of such techniques are now available, and many reviews of the topic are available (Holve et al., 1981; Knollenberg, 1979, 1981; Hovenac, 1987; Tayali and Bates, 1990; Koo and Hirleman, 1996; Black et al., 1996). All of the limitations and concerns reported for SPCs in the "Overview" apply to this class of techniques, including counting statistics at low concentrations and coincidence and dead time effects at high concentrations. In particular, nonuniformities in the illuminating beam can result in both sizing and counting errors for this class of equipment, and some form of correction (either hardware or analytic deconvolution) is required. Intensity-based techniques are particularly sensitive to environmental features that alter either illuminating beam or scattered light intensities, such as window contamination or high particle densities between the sample volume and collection optics.

Forward-Scattering Spectrometer Probe

The forward-scattering spectrometer probe (FSSP) models (***PMS***)* are ground-based or aircraft-mountable probes that size particles based on the intensity of forward-scattered light as they pass through a laser-illuminated sensing volume. The newer model FSSP-300 provides better sensitivity (down to 0.3 μm) and higher resolution (31 channels) over its range (0.3 to 20 μm) than the mechanically identical FSSP-100 (15 channels over several size ranges, such as 0.5 to 8.0 and 5.0 to 95 μm). The velocity operating range for these instruments is from about 10 to 125 m/s. In the typical configuration, a particle velocity distribution is not measured by the FSSP. However, an option is available that converts the running mean particle transit time (measured to correct for edge effects) into a mean velocity. The system has been used extensively in atmospheric aerosol studies (e.g., Knollenberg, 1981).

The operating principles for the FSSP-100 have been described by Knollenberg (1981) and for the FSSP-300 by Baumgardner et al. (1992). A patented dual-detector arrangement is used to size only those particles passing through a prescribed sampling volume. Briefly, the

* See Appendix I for full manufacturer addresses referenced to the italicized three-letter codes.

sampling volume between two probe tips is illuminated by an HeNe laser from one of the tips. When a particle enters the volume, it scatters light that is collected by optics located in the other probe tip. While a dump spot blocks the main beam, the forward-scattered light enters a beam-splitting prism and is focused onto two photodetectors. The signal photodetector is unmasked and reports an intensity maximum used to size the particle, while the annulus detector is masked to eliminate light from in-focus, centered particles. A comparison between the two signals for each particle is used as an acceptance criterion: particles passing far from the focal plane scatter a larger proportion of light into the annular detector and are rejected. A transit time test is also performed to eliminate particles traversing the beam near an edge. This test can bias the size distribution, particularly for broad distributions with size-dependent particle velocities (Baumgardner et al., 1990).

Several authors have reviewed the optical and electronic limitations of the FSSP technique (Baumgardner et al., 1990, 1992; Wendisch et al., 1996). Issues addressed include sample volume, sizing, and counting uncertainties. Concentration measurement uncertainties can be quite large (>50%) (e.g., Knollenberg, 1981), although correction algorithms can be applied to improve accuracy (Baumgardner et al., 1990). Size calibration studies using latex and glass spheres (Pinnick et al., 1981) and water droplets (Wendisch et al., 1996) have been reported. Later, Hovenac and Hirleman (1991) developed a rotating pinhole calibrator for use with the FSSP. Investigators have shown that experimental FSSP calibration curves are in reasonably good agreement with Mie calculations if particle refractive index and shape (Jaenicke and Hanusch, 1993) and beam nonuniformities (Hovenac and Lock, 1993) are considered.

90° White-Light-Scattering Analyzers

A description of an in situ SPC based on white-light illumination and detection at 90° scattering angle was given by Umhauer (1983). The choice of white-light illumination is intended to maximize monotonicity of the scattering intensity versus diameter response curve and to reduce (though not eliminate) index of refraction effects. These white-light systems are well suited to filter efficiency testing, especially at high pressures or temperatures, and have also been used widely in pharmaceutical spray sizing. The velocity operating range is typically from 0.1 to 10 m/s, although particle velocities are not measured. Commercial systems based on the Umhauer design have been marketed as the HC series particle sizers by ***PLY*** (now discontinued) and as the Model PCS-2000 series by ***PAS***. Both manufacturers offered several models, differing in optical geometry and, hence, in nominal size and concentration ranges.

The discontinued HC series ***PLY*** provided configurations with particle size ranges from 0.4 to 22 μm (model HC-2015) to 1.5 to 100 μm (model HC-2470). The larger measurement volume required for the latter size range made the HC-2470 more susceptible to coincidence errors, but the system was less susceptible to edge errors (Borho, 1970). Maximum concentrations achievable with the HC series is in the range of 10^5 particles/cm^3. Performance of the HC series has been investigated in several calibration studies: Mitchell et al. (1989) (according to British Standard BS3406; British Standards Institution, 1988), Fissan and Helsper (1981), Sachweh et al. (1998), and Friehmelt and Heidenreich (1999) (nonspherical particles). These authors generally report smooth, monotonic instrument response functions in good agreement with Lorenz-Mie calculations for spheres, although a strong index of refraction effect was observed. Sachweh et al. (1998) modified an Umhauer-type system to achieve a smaller optical measurement volume, allowing measurements at concentrations up to ~10^7 particles/cm^3 with a lower size detection limit of about 0.2 μm.

Depending on the configuration of the PCS-2000 system (***PAL***), the particle size range can be chosen within 0.15 and 100 μm for particle concentrations up to 10^6 particles/cm^3. PCS-2000 systems offer two independent photomultipliers to minimize border zone errors (i.e., trajectory ambiguity). Only a few studies of the PCS-2000 system have appeared in the literature. Stier and Quinten (1998) report on an index of refraction correction, and Maus and Umhauer (1996) used a PCS-like instrument in a filter study. Performance of PCS-2000 systems is expected to be generally similar to the HC series (***PLY***), based on their shared ancestry (Umhauer, 1983).

Particle Counter Sizer Velocimeter

The particle counter sizer velocimeter (PCSV) system (***PRM***) is an SPC that measures particle size based on the peak intensity of HeNe laser light scattered in the near-forward direction (Holve and Self, 1979a,b) (see Fig. 16–3). Using near-forward-scattered (predominately diffracted) light helps reduce particle shape and refractive index effects; thus, instrument response is mainly dependent on particle cross-sectional area. Particle speed is determined from the widths of the scattered light pulses (Holve, 1982). Only average speed is reported, although the entire speed distribution is recorded in the data set. Depending on the specific system configuration, the manufacturer gives the instrument's operating envelope as particle size between about 0.2 and 200 μm, concentration up to 10^7 particles/cm^3 for submicrometer and up to 100 ppm by volume for supermicrometer particles, and particle velocity between 0.1 and 400 m/s. A maximum particle count rate of 500 kHz is claimed for the system. To cover the wide range of sizes, two separate laser beams are used to form two independent measurement volumes; the narrower beam (nominal diameter of 20 μm) is used for sizing smaller particles, while the wider beam (nominal diameter of 200 μm) is used for sizing larger particles. Process Metrix (***PRM***) claims an accuracy of ±5% and a precision of ±3% of the indicated size. The system is available in both bench-top (PCSV-E) and probe (PCSV-P) fiber-coupled models. The water-cooled probe version is designed for operation in hostile environments (temperatures to 1400°C and pressures from vacuum to 10 atm) and provides gas purging to minimize window contamination. The bench-top version consists of two towers supported by a common bridge about 1 m apart, with the measurement volume centrally located. Both systems contain in situ alignment systems to correct for beam steering in hostile environments (Holve and Annen, 1984). Alignment sensitivity was explored analytically by Holve and Davis (1985).

A major feature of the PCSV system is the use of a deconvolution of the measured scattered intensity histogram to infer the size distribution (Holve and Self, 1979a,b; Holve and Annen, 1984; Holve and Davis, 1985). The deconvolution is required due to the trajectory ambiguity. Generalized Lorenz-Mie scattering theory is used to predict the scattering response function (scattered intensity versus particle size) for the desired geometry and has been experimentally confirmed (Holve and Self, 1979b). Although early work characterized the intensity profile of the sample volume experimentally (Holve, 1980), the current technique relies on an analytic description (Holve and Davis, 1985) to generate the profile and a single-point calibration to bring predicted and observed instrument responses into agreement. The validity of the sample volume analytic model was experimentally checked with monodisperse latex spheres (Holve and Davis, 1985), and the accuracy of the deconvolution algorithm was established using a mixture of monodisperse oleic acid droplets (Holve and Self, 1979b). ***PRM*** provides a rotating chrome on glass reference reticle for instrument calibration in the 2 to 80 μm size range.

PCSV systems have been used to measure particle size distributions in a variety of combustion environments (Bonin and Queiroz, 1991, 1996; Holve and Annen, 1984; Holve, 1980). Measurements of soda-lime glass beads in both cold and hot flows (Holve and Self, 1979b) gave self-consistent results, showing the PCSV's ability to size particles in flames at temperatures of 1600 K. More recently, the PCSV has been used to study plating processes (Bonin et al., 1995).

Dual-Beam and Top-Hat Beam Systems

Several techniques have been developed that attempt to provide intensity-based particle sizing that does not depend on the particle trajectory through the beam, thus avoiding deconvolution algorithms. To reduce trajectory ambiguity, investigators have either used two-beam systems or have tried to modify the intensity profile of a single incident beam.

Dual-beam, two-color, or pointer-beam techniques typically employ one beam that has a relatively wide diameter and a second, coincident beam focused to a smaller waist coaxial with the larger beam. The light-scattering signal from the smaller (pointer) beam is used to trigger an intensity measurement from the larger beam, which is used for particle sizing. For example, Hess (1984) describes and demonstrates measurements made with a two-color system using a small pointing beam inside a larger Gaussian laser beam of a different wavelength. The pointing beam defines a relatively uniform portion of the large Gaussian beam. Particles are only measured when detected by the pointing beam, indicating that they are in the uniform portion of the outer beam, so the scattered pedestal intensity can be recorded and the size calculated free of trajectory ambiguity. The MetroLaser PAS-100 and PAS-200 (***MEL***) instruments use a similar technique to provide a uniform measurement region for combined LDV, particle sizing, and concentration measurements. The system can be configured to collect diffracted light in cases where the particles are of unknown shape or refractive index and is designed to operate with no dead time by collecting groups of particle-scattering events before processing the raw data. The manufacturer-stated operating envelope includes a 0.4 to 6000 µm diameter range, with 2% typical resolution and a 30:1 dynamic range and a velocity range up to 1000 m/s. Data rates up to 3×10^6 particles/s can be measured.

Grehan and Gouesbet (1986) describe a laboratory system in which a particle-sizing laser beam was modified to a uniform (top-hat) profile using a Gaussian absorption filter. The top-hat concept reduces trajectory ambiguity by creating a beam of nearly constant laser intensity across most of the beam width. Of course, the realized intensity distribution of a top-hat beam is not uniform across the entire beam width, but is characterized by a constant intensity region in the center surrounded by Gaussian-like (intensity decaying) tails toward the beam edge. Thus, some trajectory ambiguity is still present because of the tails. Black et al. (1996) provide additional discussion of the top-hat beam technique. Note that the present authors are not aware of a commercially available top-hat beam system.

SINGLE-PARTICLE COUNTERS: LDV VISIBILITY BASED

Laser Doppler velocimetry (LDV) is a well-established and documented technique for noninvasive measurement of particle velocities, made by measuring the Doppler-shifted frequency of light scattered by individual particles passing through a laser beam-defined measurement volume (Durst et al., 1981). The most common LDV configuration uses crossed laser beams to define a measurement volume with typical dimensions on the order 1 mm or less. Particles passing through the measurement volume scatter light with a Doppler shift proportional to the particle speed. Speeds as high as several hundred meters/s can be measured using conventional detection electronics. The scattered light intensity signal from each particle passage ("Doppler burst") consists of a high-frequency Doppler component superimposed on a low-frequency "pedestal" component due to the Gaussian intensity distribution of the illuminating beams. After filtering out the low-frequency pedestal, the remaining component (Fig. 16–4) is the Doppler frequency, directly proportional to the particle velocity. The extent of modulation of the Doppler signal (ratio of the Doppler signal to the pedestal intensity) is called the signal *visibility*.

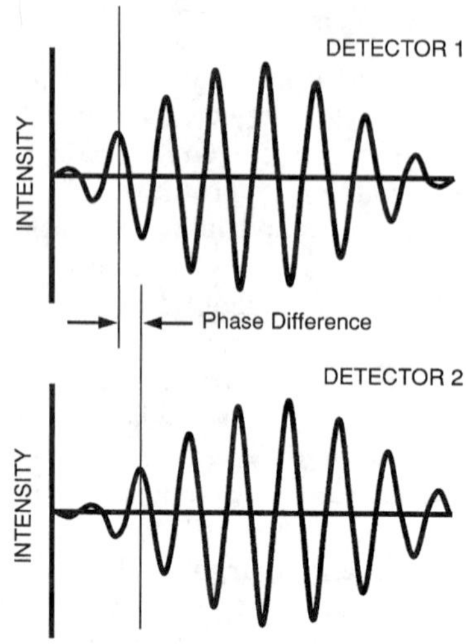

Fig. 16–4. Characteristic Doppler burst signals.

A number of particle measurement techniques have been developed based on the pedestal intensity or visibility of LDV signals. Jackson (1990) reviews interferometry-based droplet sizing techniques, including on-axis LDV visibility measurements (diffraction dominated), off-axis LDV visibility measurements (refraction dominated), and phase Doppler techniques. Such techniques are attractive because they can provide simultaneous measurement of single-particle size and velocity. Ideally, the peak intensity of the Doppler burst would be directly related to particle size. However, the Gaussian nature of the illuminating beams complicates matters due to trajectory ambiguity.

Particle size determination based on signal visibility was theoretically established by Farmer (1972), and the technique was attractive as a particle's size could be measured independently of its trajectory through the beam crossing. Recent reviews of the method are available (Jones, 1999; Black et al., 1996). One drawback of the technique is that the particle size–visibility relationship becomes oscillatory for larger sizes, thus limiting the available dynamic range (see Bachalo, 1980; Durst et al., 1981; Takeo and Hattori, 1990). For example, Jackson and Samuelsen (1987) compared phase Doppler and visibility-based interferometers in a spray system and found that the phase-based systems offered broader size and velocity ranges. Consequently, phase-based interferometers have found wider application than visibility-based systems, and no commercially available visibility systems are known to the authors.

SINGLE-PARTICLE COUNTERS: PHASE BASED

The phase Doppler technique is an LDV-based method for simultaneous measurement of single-particle size and velocity (see Hirleman, 1996, for a historical review). This technique is not intensity dependent like the previous group of SPC techniques and can therefore offer superior performance by minimizing effects such as beam attenuation or window fouling. A phase Doppler system measures the spatial and temporal frequencies of the Doppler-shifted light scattered by individual particles passing through a laser beam-crossing measurement volume. Phase Doppler systems use multiple photodetectors to sample slightly different spatial portions of the light scattered by individual particles. Figure 16–4 demonstrates high-

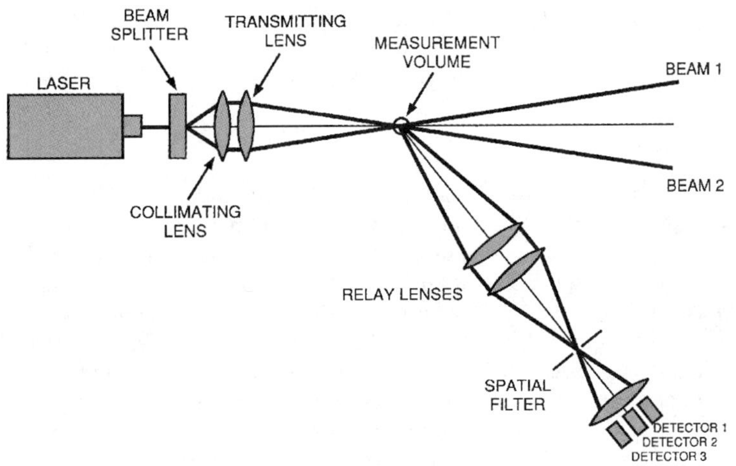

Fig. 16–5. Layout of a phase Doppler system.

pass–filtered Doppler bursts measured by two such detectors. The phase shift between the two signals is a measure of the scattered light spatial frequency, which can be directly related to the particle diameter, refractive index, and receiver geometry. The linear relationship between the spatial frequency of the scattered light and the particle diameter for fixed refractive index and receiver geometry can be shown using Lorenz-Mie scattering theory (Saffman et al., 1984), geometrical optics (Bachalo and Houser, 1984), or Generalized Lorenz-Mie Theory (Grehan et al., 1994). Particle velocity is related to the temporal frequency in the same manner as in conventional LDV. Figure 16–5 is a schematic layout of a generic phase Doppler system.

Particle sphericity is required because the phase shift is calculated for either rays refracted through spherical particles of known, constant refractive index or reflected off the surface of reflective particles. Preliminary work on measurement of nonspherical particles is presented below.

Most theoretical analyses of the Doppler phase shift have been performed using the geometric optics approximation and neglecting diffraction because scattered light is collected at sufficiently large off-axis angles (typically 30° or more). One of the key assumptions in phase Doppler analysis is that a single component of scattering, either refraction or reflection, dominates. Many authors have examined the effect of this assumption when it is violated by trajectory ambiguity, that is, when scattering is first dominated by reflection when a particle's external surface enters the laser beam and then refraction as it passes through the beam center (e.g., Qiu et al., 2000; Strakey et al., 1998; Hardalupas and Liu, 1997); these investigators and others have proposed and tested several postprocessing trajectory correction routines.

Phase Doppler Particle Analyzer

The fundamentals of the phase Doppler particle analyzer (PDPA) are described by Bachalo and Houser (1984). The PDPA (***AER***, ***TSI***) consists of a laser, transmitting optics, receiver optics package, signal processors, and data collection and analysis software, with all operations, data collection, and analysis computer controlled. The ***TSI*** system can be supplied to measure one, two, or three velocity components in addition to particle size. The manufacturer-stated operating envelope includes a 0.5 to 10,000 μm diameter range, with 1% typical accuracy and a 50:1 dynamic range, and a velocity range to over 500 m/s, with 0.2%

typical accuracy. This system also calculates number density based on the number of particles passing through a calculated size-dependent measurement volume (to correct for trajectory ambiguity effects). The maximum measurable number density is $10^6/cm^3$.

Particle Dynamics Analyzer

The Particle Dynamics Analyzer (PDA; ***DAN***) is described theoretically by Saffman (1987) and with applications by Saffman et al. (1984). This system is similar to the TSI/Aerometrics PDPA except that signal phase is measured using a cross-correlation technique instead of a Fourier transform technique. The manufacturer-stated operating envelope for the PDA system includes a 0.5 to 10,000 μm size range, with 1% typical accuracy and a 40:1 dynamic range, and a velocity range to greater than 500 m/s, with 1% typical accuracy. The number density is so highly dependent on the specifics of the application that customers are advised to refer to the open literature for specific concentration and mass flux limits for their application. ***DAN*** offers a dual PDA system that makes two simultaneous particle measurements in perpendicular planes in order to minimize trajectory ambiguity and slit effect (Xu and Tropea, 1994) errors (see next section).

Phase Doppler Performance and Applications

There has been an abundance of recent work with these instruments in both instrument performance characterization and applications in industrial and research settings. Note that many of the performance issues pertain to older instruments, many of which are still routinely used in industry and research; the newer generations of commercial instruments have been improved to address such concerns (e.g., frequency shift technique, photomultiplier gain optimization). Jackson (1990) reviews the older Aerometrics PDPA instrument and discusses the effect of the rotating diffraction grating (formerly used as a combined beam splitter and frequency shifter, replaced on later models with a beam splitter and Bragg cell frequency shifter) on PDPA concentration and velocity measurements. Dodge et al. (1987) performed liquid droplet measurements in sprays with the PDPA and noted that the PDPA was very sensitive to optical alignment. McDonell and Samuelsen (1990) evaluated the sensitivity of PDPA measurements to operator input parameters, specifically the photomultiplier tube (PMT) gain voltage and the frequency shift level. They found that mean velocity values were insensitive to PMT voltage and frequency shift but that fluctuating velocities were strongly dependent on PMT voltage and somewhat dependent on frequency shift because of wobble of the rotating diffraction grating. Volume flux measurement errors were caused by inconsistencies in determination of probe volume size and by the strong influence of PMT voltage and the weak influence of frequency shift on particle sizing. They conclude that PDPA operations must be made with a carefully chosen standard operating procedure and that a need exists for a reference calibration standard to guide operator selections of setup parameters for the PDPA instrument. Ceman (1990) and Bever and Hovenac (1999) have developed two different embodiments of such a standard calibration device (see "Performance Verification," below).

Dressler and Kraemer (1990) calibrated an older PDPA using a multijet droplet generator. They found PDPA measurement accuracy of about 5%. In addition, up to 5% measurement error could be caused by lens aberrations. They concluded that the PDPA system should be refocused for each different beam spacing to minimize lens spherical aberration errors. They also found that measurements were quite sensitive to the PMT gain voltage in that a 50 V difference in gain voltage could lead to spurious data validations and artificially broadened distributions. They concluded that the lowest usable gain setting should be employed for measurements. Determination of this value can be difficult in measurements of unknown, transient particle distributions.

Ceman (1990) and Ceman et al. (1993) report detailed investigations into PDPA oleic acid droplet sizing performance in the 2.3 to 25 µm range using a vibrating orifice aerosol generator and several different PDPA receiver geometries. They found that there were nonlinearities in PDPA measurements of particles below some critical diameter, in agreement with previous experiments (O'Hern et al., 1989; Ceman et al., 1990) and calculations (Al-Chalabi et al., 1988) using both Lorenz-Mie theory and geometrical optics. Saffman et al. (1984) found that a large receiver lens collection solid angle tended to damp these oscillations. Sankar et al. (1990) indicate that reflections from the surface of the droplet may also contribute to the oscillations in the phase versus diameter curve in the smaller diameter regime. Their work suggests that reflections can be minimized and linearity attained by collecting light at an angle close to the droplet Brewster angle; however, Ceman's results show that droplet sizing oscillations can be substantially reduced by using a larger collection solid angle (shorter focal length receiver lens) but that working near the droplet Brewster angle provided little or no additional improvement. Gobel et al. (1998) examined the accuracy of phase Doppler particle sizing at 1 µm and below and concluded that ambiguities exist but can be accounted for when sufficient information is available on the particles.

Alexander et al. (1985) used the PDPA for measurement of nonspherical methanol droplets produced by a vibrating orifice aerosol generator. The mean droplet diameters were 98 µm, with aspect ratios ranging from 0.7 to 1.4. The PDPA diameter measurements were very sensitive to aspect ratio, with the greatest error (45% oversizing) for elongated ellipsoids oriented perpendicular to the horizontal fringes in the PDPA measurement volume. Breña de la Rosa et al. (1989) have shown that the PDPA can be used to examine the shape of large nonspherical bubbles in water (1200 to 1800 µm diameter) by making separate measurements of the bubble major and minor diameters. This technique could possibly be extended to small particle measurements, although such an extension does not appear to be straightforward, especially for particles with rough or angular surfaces. Damaschke et al. (1998a) experimentally and numerically analyzed the effects of nonspherical particles on phase measurement by the phase Doppler technique, demonstrating that nonspherical droplets can be a significant error source, especially for the standard three-detector systems. They suggested that alternative detector arrangements and larger measurement volumes may help minimize such errors.

The PDPA has been employed in a large number of applications. Strakey et al. (1998) examined use of the PDPA for measuring sprays from high-pressure injectors. Their work included a careful analysis of probe volume dimensions smaller than the droplet diameters and showed that such a device could perform well even in dense sprays provided that phase and intensity validation and probe volume correction algorithms were implemented. Van Den Moortel et al. (1997) applied PDPA to examine turbulent gas–solid flow in a circulating fluidized bed and developed a postprocessing routine to correct for artificially high count rates measured from noisy Doppler bursts. Hardalupas et al. (1988) used both Aerometrics and custom-made phase Doppler devices to measure solid particles transported in a gas–solid dusty jet flow and found that asphericity of the glass beads used as seed particles in the flow limited the measurement accuracy. Bachalo et al. (1990) presented details of PDPA operations including data analysis routines and applied the PDPA to measurements of liquid spray droplets in both cold and combusting regimes. Droplet velocity time records showed significant spray fluctuations, leading to droplet cluster formation. Several other recent PDPA applications are discussed below in "Performance Verification."

SINGLE-PARTICLE COUNTERS: IMAGING

Determining a particle's properties by direct imaging is among the earliest techniques used in particle measurement: Consider the optical (and subsequently the electron) microscope.

A significant advantage is that the shape and index of refraction issues that complicate single-particle light-scattering measurements are avoided. In fact, imaging techniques provide one of the few avenues for investigating particle shape. The accuracy of single-particle imaging systems is limited by Fresnel diffraction and depth of field effects (Hovenac, 1987). Fresnel diffraction blurs image edges and complicates sizing. The depth of field effect arises from its dependence on particle size, with the result that large particles remain in focus over a greater axial distance than smaller ones.

Knollenberg (1970, 1979, 1981) designed an automated, in situ, single-particle, optical imaging system that is commercially available (***PMS***) as the Optical Array Imaging Probe (OAP). In this family of probes, a collimated laser beam defines a measurement volume located between two sensing tips that extend forward from the main body of the system. Receiving optics direct the beam to illuminate a linear array of photodiodes. A particle passing through the measurement volume casts a shadow on the array, resulting in a decreased signal from the individual elements that lie in the shadow. Three methods are used to analyze the resulting data from the array: one-dimensional, two-dimensional, and grey-level processing. In a one-dimensional OAP system, the array elements are read and latched during the particle transit in a way that only provides particle size information. In a standard two-dimensional OAP system, the entire two-dimensional image of the particle is stored in high-speed memory as a series of "snapshots" of the particle during its transit. For each image, the status of each of the 32 elements in the photodiode array is recorded by one bit, indicating if the element is being shadowed or not. The great advantage of acquiring two-dimensional particle images becomes apparent when measuring nonspherical particles. In the Grey Probe two-dimensional OAP system, a 64-element array is used where each element reports one of four shadow levels. The increased sophistication of the Grey Probe provides twice the resolution (twice as many elements) of the standard two-dimensional system, as well as depth-of-field information.

For all of these imaging systems, both instrument resolution and size range depend on physical spacing of the array elements, magnification, and particle velocity. Consequently, the user must identify the expected velocity range in order to properly configure the system. OAP probes also suffer from depth-of-field limitations, leading to erroneous interpretation of out-of-focus (blurred) particle images. To characterize OAP performance, Hovenac and Hirleman (1991) and Reuter and Bakan (1998) used a rotating reticle calibrator; the latter authors proposed a depth-of-field correction that allowed accurate sizing (rms uncertainty less than 6%) of particles from 50 to 500 μm. In addition, OAP probes reject particles (resulting in counting bias) that shadow elements at either edge of the array, as the fraction of the particle falling outside the array cannot be determined and thus precludes correct sizing.

Rugged, pylon-mountable airborne versions of the one-dimensional, two-dimensional, or grey-probe OAP configurations are available for either cloud droplet or precipitation measurements. Particle sizing ranges and resolution depend on the particular models, which differ in the number of array elements and optical configuration used. The cloud-droplet models are suited to sizing smaller particles, with ranges such as 10 to 620 μm (with 10 μm resolution) or 200 to 6000 μm (with 200 μm resolution). The precipitation models are suited to sizing larger particles, with ranges such as 50 to 3100 μm (with 50 μm resolution) or 150 to 9300 μm (with 150 μm resolution). The resolution limits given above assume instrument operation at aircraft speeds (100 m/s); significant improvement in instrument resolution can be achieved at lower velocities. The lower limit for OAP sizing (somewhere between 1 and 10 μm according to Knollenberg [1979]) results from the vanishingly small depth of field at these sizes. Particle velocities are not measured explicitly, but could be recovered by later analysis of the image sequences using the known imaging frequency. A ground-based precipitation OAP also is available. The system is contained in a weatherized package, with the two sensing tips extending above the top. The distance between probe sensing tips is 50 cm, providing a large sampling area for measuring the droplets in freefall.

ENSEMBLE TECHNIQUES: PARTICLE FIELD IMAGING

Imaging systems are useful because they can "freeze" the motion of a particle-laden flow, allowing later analysis and providing a permanent record of transient events. In addition, imaging techniques make no assumption of sphericity and can be used to allow visual examination of the shape of individual particles, along with statistical information on the particle-size distribution, concentration, velocity distribution, and so forth. Imaging measurements provide a primary measurement standard because individual particle sizes are measured directly. In addition, an archival record of the particle field is formed, allowing subsequent examination by various means. However, field imaging techniques do not provide real-time analysis, and imaging resolution is often degraded by dense particle fields, window effects, and other common conditions. A number of imaging techniques are available for measurement of particle fields, including photography and holography (often combined with image processing). Discussions of photographic and holographic techniques are included here because of their widespread use and unique capabilities. Several firms can provide custom design and assembly of aerosol-sizing equipment using these techniques.

Photographic investigations are attractive because the technique is well established and reasonably high resolution is possible. With recent advances in digital imaging, immediate feedback is often possible. Photography of small particles places special restrictions on the imaging process, often requiring very short exposure times to avoid blurring of moving particles and high resolution to detect the size range of interest. The achievable resolution of photographic systems is often sufficient for particle measurements; however, the depth of field suffers. Image processing techniques promise to improve the speed and accuracy of photographic studies and have been applied in numerous studies (e.g., Oberdier, 1984; Lavergne et al., 1988).

Holography has become a fairly common technique for the study of particles with diameters larger than about 5 μm. Holographic particle imaging has undergone a minor resurgence over the past few years as holographic particle image velocimetry has been developed and applied for three-dimensional velocity field measurement (e.g., Chan and Li, 1998; Zhang et al., 1997). A hologram is an interference pattern formed by the mixing of two coherent wave components: a reference wave and a subject wave reflected or scattered from the object or field of interest. Holographic reconstruction creates three-dimensional images of the original illuminated volume, which can then be examined in detail for particle size and shape, as well as velocity and acceleration in systems with multiple-pulse illumination. The three-dimensional aspect allows simultaneous matching of the requirements for high resolution and good depth of field, unlike the photographic process. In addition, image premagnification (before recording) can allow examination of smaller particles at the cost of reducing the sampled volume. A complete examination of the holographic process can be found in Collier et al. (1971). Aerosol particle examination was historically one of the first practical applications of holography (Thompson and Ward, 1966). Reviews of the use of holography for particle field measurements are given in Thompson (1974), Trolinger (1975, 1980), and Tyler and Thompson (1976). Holographic images of particle fields contain information on the size, shape, and three-dimensional spatial position of each individual particle comprising the field. Trolinger (1980) demonstrates that the great depth of field of a hologram gives it several orders of magnitude greater capacity than a photograph for storing volumetric information. Santangelo and Sojka (1993) have recently reviewed the field of holographic particle diagnostics.

Both standard dual-beam and in-line or Fraunhofer holographic techniques have been applied to aerosol studies. The Fraunhofer holographic technique uses a single laser beam as both reference and subject beam during hologram recording. This is accomplished by illuminating the sample volume with a collimated coherent light beam, part of which is scattered by objects in the sample volume, while the remainder passes through the volume undiffracted,

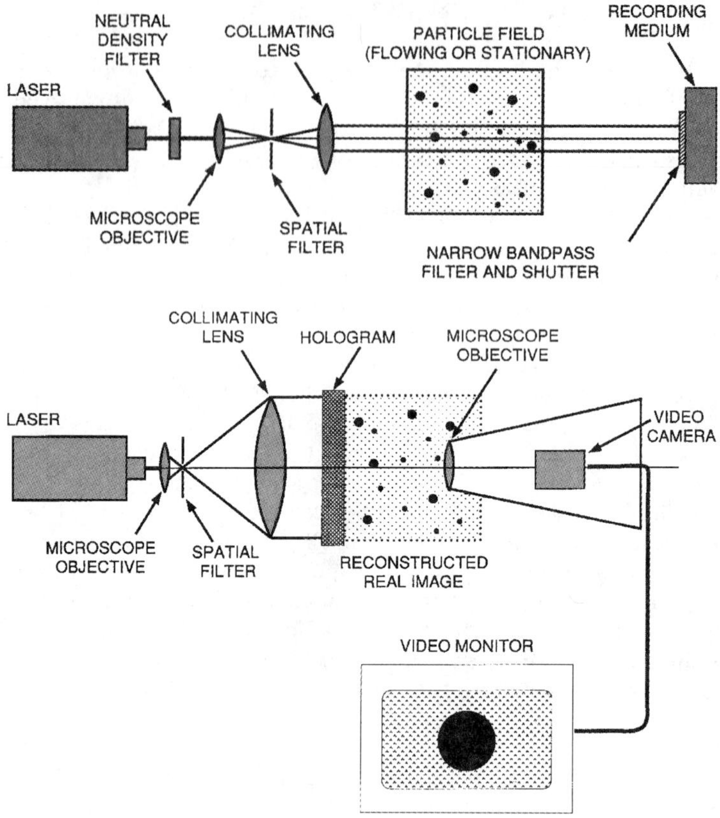

Fig. 16–6. Layout for Fraunhofer holography: **a**, recording system; **b**, reconstruction system.

acting as a collinear reference beam. The hologram is formed by recording the interference of these two beam components on a high resolution film. Figure 16–6a is a schematic diagram of a generic Fraunhofer holographic recording system. The in-line holographic technique is simpler and requires considerably fewer optical components than the off-axis or dual-beam method and minimizes recording material resolution requirements. However, in-line holographic real images generally contain a high background noise component caused by the background presence of the virtual image and out-of-focus objects. The Fraunhofer holographic technique is limited to cases in which the sample is composed of a dilute concentration of small particles in order to maintain a sufficiently strong undiffracted collinear reference beam.

Some advantages of the off-axis holographic technique are that it allows the use of diffuse illumination of the sample volume; permits the use of a perfectly filtered and well-controlled reference beam; is not subject to the far-field restrictions on object size and distance from the film plane; and is not limited to dilute particle concentrations. The use of separate subject and reference beams does, however, place additional coherence requirements on the light source.

Sufficient mechanical stability is required to maintain all optical components stationary to within $\lambda/10$ during exposure, where λ is the wavelength of the illuminating beam, so pulsed lasers are typically used for examination of moving aerosols. Holographic examination of moving objects imposes a restriction on the exposure time, with the general rule of thumb for image clarity being that the object move less than 1/10 of its diameter during the exposure; so, for example, the exposure time for a 10 μm diameter object moving at 10 m/s should be less than 100 ns. High resolution generally requires that the sample volume be located close to the film plane, often requiring the use of a lens system to relay the volume image onto the film plane. The recording medium, usually film, must have sufficient resolution to be able to record accurately the fringe pattern formed by beam interference. Spatial filtering by a pinhole aperture is generally required to eliminate beam impurities due to diffraction caused by dust particles, impurities of optical components, or inherent optical noise of the laser beam.

After recording, the hologram is mounted in a reconstruction system, where it is illuminated with another coherent light source, acting as the conjugate of the original reference beam. The hologram acts as a diffraction grating to form three-dimensional real and virtual images of the original sample volume. Use of translating stages allows detailed examination of the reconstructed image. Figure 16–6b is a schematic of an in-line reconstruction system. Unlike the in-line technique, in the off-axis reconstruction configuration the axis of the holographic images does not coincide with the reconstruction laser beam, so the viewing system is not continuously exposed to direct illumination by the reconstruction beam and the noise due to the reconstruction light source can be eliminated.

Holographic reconstruction and detailed data acquisition are very time consuming, often requiring on the order of several "man-days" for examination of the several hundred particle images needed for statistical significance. Examinations are typically performed by visual observation of the reconstructed images, although automated analysis techniques are a topic of current interest (e.g., Christnacher and Royer, 1994; Haussmann and Lauterborn, 1980; Schäfer and Umhauer, 1987; Chávez and Mayinger, 1990). Ewan et al. (1984) and Hess and Trolinger (1985) describe unique applications of the Malvern diffractometer (***MAL***) (see below) for evaluation of reconstructed holographic images. Instead of measuring the size distribution of a particle field, the Malvern is set up to examine a particle field hologram, yielding ensemble size distributions for each region probed by the Malvern laser beam. Use of the Malvern device could be a very useful step toward automated reconstruction (at least for ensemble properties).

While no off-the-shelf holographic aerosol measuring systems are currently available (to the knowledge of the authors), several companies (e.g., ***MEL***) can design and install custom holographic systems.

ENSEMBLE TECHNIQUES: FRAUNHOFER DIFFRACTION

Some of the first laser-based, in situ particle measurement techniques used Fraunhofer diffraction to characterize droplet sprays (Cornillault, 1972; Wertheimer and Wilcock, 1976; Swithenbank et al., 1977). Simply, the technique determines a particle distribution from a measurement of the ensemble diffraction pattern that results from the illumination of a particle cloud by a collimated laser beam. The technique has since been developed into a variety of commercial systems that have been extensively characterized, calibrated, and used in a wide range of particle studies. Excellent reviews of Fraunhofer diffraction theory and measurement techniques are available (Jones, 1998; Koo and Hirleman, 1996; Azzopardi, 1992; Swithenbank et al., 1991; Felton, 1990; Lefebvre, 1989; Meyer and Chigier, 1986; Hirleman, 1984).

In typical ensemble diffraction techniques (Fig. 16–7), a laser beam is expanded and then collimated into a beam several millimeters in diameter that passes through the particle cloud. Particles in the beam scatter light in all directions, although with particular efficiency in the near-forward direction for particles much larger than the wavelength. A receiving lens is used

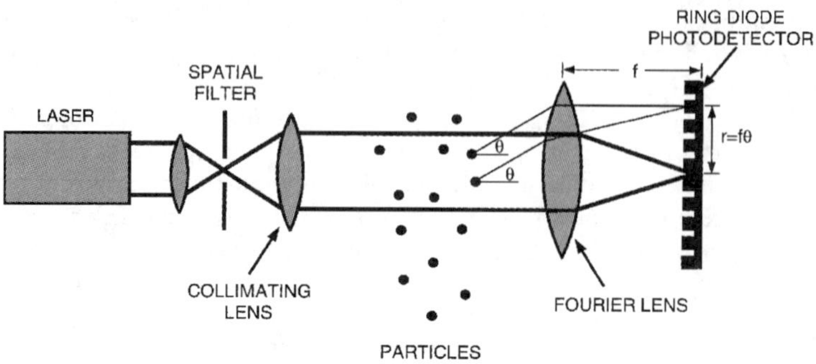

Fig. 16–7. Layout of an ensemble diffraction system.

to focus both the transmitted beam and forward scattered (predominately diffracted) light onto a detector located at the back focal plane of the lens. The transmitted light is focused to a point on the optical axis, while the diffracted light forms a series of concentric rings (Fraunhofer diffraction pattern). As the receiving lens performs a Fourier transform on the scattered light, light scattered at a given angle θ by a particle located anywhere in the illuminated sample volume will be focused at the same radial position in the transform plane. The resulting pattern is unaffected by particle location or motion. No measurement of particle velocity is made by diffraction techniques.

Commercial techniques vary in the method of detecting the diffraction signature formed at the transform plane. Swithenbank et al. (1977) showed that by using a multi-element set of annular, concentric photodetectors, the radial light energy distribution of the scattered light could be used to infer the size distribution; Swithenbank's work led to the Malvern family of diffractometers (***MAL***). As the scattered light is most intense near the axis, the ring detectors increase in area (usually logarithmically) with increasing radial distance. The theory of the annular scattering profile from diffraction is discussed elsewhere (Swithenbank et al., 1977; Felton, 1990). These techniques have been used to detect the diffraction signature (Cornillault, 1972; Wertheimer and Wilcock, 1976), but have not found as widespread commercial application. As a practical concern, the entire particle cloud should be within a maximum allowable distance of the receiver lens (Dodge, 1984b). If particles are further away than this, they may scatter light beyond the receiving lens; this effect is called *vignetting* (see Lefebvre, 1989). Vignetting tends to bias the distribution toward larger sizes, as the small particle (large scattering angle) signal is preferentially lost.

The ensemble diffraction technique results in a classic inversion problem wherein a continuous size distribution is sought that provides best agreement with a finite set of experimental measurements (the scattered energies measured by each annular detector). A description of this inverse scattering problem is given by Koo and Hirleman (1996), Felton (1990), and Hirleman (1987). Because this class of inversion problems is ill-posed, the search for an optimal solution method continues to be an area of active research. Two classes of mathematical solutions to the inverse scattering problem have been pursued in the literature (Koo and Hirleman, 1996): (1) indirect numerical quadrature (e.g., Hirleman, 1987; Swithenbank et al., 1977) and (2) direct integral transform solutions (e.g., Koo and Hirleman, 1992; Knight et al., 1991; Bayvel et al., 1987). Most prevalent are quadrature-based inversions, which use an iterative solution to minimize the sum of squared errors between theoretically predicted and experimentally measured detector responses. This method offers several advantages (ensures positivity of the size distribution and guarantees stability), but requires an assumption for the size distribution (e.g., Malvern systems offer a choice among a lognormal, Rosin-Rammler, or proprietary model-independent distribution). The choice of inversion algorithm is a critical step in the data analysis process. For example, several experimental and numerical simulation studies have found that the proprietary Malvern inversion software has difficulty in accurately inverting multimodal size distributions (Annapragada and Adjei, 1996; Knight et al., 1991; Meyer and Chigier, 1986). Although integral transform inversion methods seem to fare better, these methods also can lead to erroneous results under certain conditions (Koo and Hirleman, 1992). Clearly, caution is advised when analyzing ensemble diffraction data. Once the size distribution is determined, the particle concentration is determined by a direct measurement of the beam attenuation through the cloud (transmitted beam intensity at the center of the detector measured before and after the aerosol is introduced).

Implicit in the standard analysis of ensemble diffraction techniques is the assumption that the particle diameter is much larger than the wavelength of the illuminating light. Thus, the lower limit for an ensemble diffraction instrument (for visible light) is about 1 μm. As an upper limit, diffraction from large particles is concentrated at very small angles near the axis of the illuminating beam which makes measurement difficult.

A second common assumption in the standard analysis is that each photon undergoes at most one scattering event while passing through the cloud. For high concentrations or long path lengths, the probability that a photon undergoes two or more particle interactions before reaching the detector increases. Multiple scattering spreads light over a wider angle with the result that the size distribution appears wider and is skewed toward smaller sizes. Both experimental (Dodge, 1984a; Felton, 1990) and theoretical (Felton, 1990; Hirleman, 1988b, 1991) studies of multiple scattering effects have been reported. Most studies agree that multiple scattering effects can be safely ignored for obscurations (fraction of the beam that is scattered by the particle cloud) below about 50% (e.g., Felton, 1990). For higher obscurations, some authors introduce correction factors (i.e., to distribution parameters such as mean diameter or standard deviation) and present empirical correlations to account for multiple scattering (Felton, 1990; Dodge, 1984a). Note that these correlations must be used carefully, as they depend on both the size distribution model and the obscuration. A general analytic approach to multiple scattering has been developed by Hirleman (1988b, 1991) that requires no assumption of the form of the particle-size distribution; Hirleman (1988b) validated the method against experimental data. Building on the Hirleman multiple-scattering model, Harvill and Holve (1997) patented a computationally fast data inversion method for computing a particle size distribution from ensemble diffraction measurements with obscurations as high as ~95%. Harvill et al. (1995) published experimental verification of their algorithm, which is now incorporated into the data reduction software of the Malvern family (***MAL***) of diffractometers.

Several other general limitations of ensemble diffraction techniques should be discussed. First, the technique implicitly assumes a uniform light intensity, while the true radial intensity profile is Gaussian. As a result, the measurement volume for small particles will be smaller than for larger particles, thus biasing the measurement toward larger sizes (Meyer and Chigier, 1986). Second, beam steering can result when measurements are made through regions characterized by high (particularly time-varying) gradients in index of refraction. The result is similar to system misalignment, where the beam begins to illuminate the inner rings, resulting in a false large particle response. Miles et al. (1990) offer an approach to making measurements in systems with a time-varying index of refraction gradients. Finally, the ensemble technique gives a line-of-sight measurement with little or no spatial resolution. Techniques for deconvolving line of sight data to obtain radial profiles for axially symmetrical sprays (tomography) have been discussed by Hammond (1981) and Zhu et al. (1987).

EXAMPLE 16–2

An ensemble diffraction-based instrument measures a Rosin-Rammler droplet distribution with parameters d_{RR} = 20.0 μm (where d_{RR} is the drop diameter such that 63.2% of the spray volume is in drops of smaller diameter) and exponent q = 2.0 (where q is a measure of the spread in drop sizes) at an obscuration of 80% (OB = 0.8). What are the true parameter values ($d_{RR,0}$ and q_0) after correcting for multiple scattering effects?

Answer: For Rosin-Rammler distributions, Felton (1990) gives correlations for correcting ensemble diffraction data for multiple scattering effects. The correlations apply for obscurations between 65% and 98% and for values of the exponent between 1.2 and 3.8 with an uncertainty of ±2% and so may be used in this example. To correct for d_{RR}:

$$\frac{d_{RR,0}}{d_{RR}} = 1.0 + \left(0.036 + \left(0.4947(OB)^{8.998}\right)\right)q^{[1.9-3.437(OB)]}$$

$$= 1.0 + (0.102)(2.0)^{-0.85} = 1.057$$

To correct the exponent:

$$\frac{q_0}{q} = 1.0 + \left(0.035 + 0.1099(OB)^{8.65}\right)q^{[0.35+1.45(OB)]}$$

$$= 1.0 + (0.0509)(2.0)^{1.510} = 1.145$$

Thus, the final corrected values for the parameters are $d_{RR,0}$ = 21.1 μm and q_0 = 2.29. Note that the observed distribution is slightly shifted to smaller sizes and is somewhat broader (smaller q) than the true distribution.

Due to the relatively long history of ensemble diffraction techniques, a large literature has evolved on its performance and on various applications. Calibration work has included the use of both standard reference materials (Felton, 1990) and a calibration reticle developed by Hirleman (Hirleman et al., 1984; Dodge, 1984b; Hirleman and Dodge, 1985). Dodge (1984b) describes the importance of a calibration to correct for variations in ring detector sensitivities. Using this correction, instrument-to-instrument reproducibilities were found to be better than 3% using a reticle (Hirleman and Dodge, 1985). Many comparisons have been made among ensemble diffraction instruments (e.g., Azzopardi, 1992; Dodge, 1987; Dodge et al., 1987) and between ensemble diffraction and other particle measurement techniques (see below). The quality of agreement depends on the techniques and on the cloud property being compared. An in-line diffractometer for process monitoring (Ensemble Particle Concentration and Size; EPCS) has been described by Harvill et al. (1995).

Of the many ensemble diffraction systems available, only the Spraytec (***MAL***), HELOS (***SYM***), and PSA (***CIL***) systems allow true in situ aerosol measurements. The Spraytec is a bench-top system with large clearances between the transmitter and receiver modules and a claimed size range of 0.5 to 850 μm (depending on configuration) at data speeds up to 2500 Hz. Malvern also offers the Insitec EPCS for in-process particle sizing (0.5 to 1000 μm), featuring a rugged optical bench and continuously purged windows. The HELOS and PSA systems claim size ranges of 0.1 to 8750 μm and 0.04 to 2500 μm, respectively. Although not suited to true in situ particle sizing, ensemble diffraction systems for powder and suspension sizing are offered by several companies (***BEC***, ***SHI***, ***HOR***, ***MCR***).

ENSEMBLE TECHNIQUES: DYNAMIC LIGHT SCATTERING

Few techniques are available for noninvasive measurement of submicron diameter particles. Dynamic light scattering (DLS), also known as photon correlation spectroscopy (PCS), diffusion-broadening spectroscopy (DBS), or quasielastic light scattering (QELS), is an ensemble laser scattering technique for particle sizing in the 1 nm to 1 μm diameter range. The technique allows measurement of the mean particle size of an aerosol or suspension without knowledge of particle properties (density, refractive index), and no calibration is required. In general, dynamic light scattering works by measuring the degree of spectral broadening induced in an incident laser beam by Brownian motion of particles in an illuminated sample. The mean particle diffusion coefficient is measured directly, which can then be used to determine particle sizes. PCS measures the dynamic behavior (time-dependent intensity) of the scattered light signal caused by this particle motion. Large particles produce low-frequency oscillations in the scattered signal, and small particles produce high-frequency oscillations. DLS techniques require enough particles in the measurement volume for an ensemble measurement. Note that particle concentration is not directly measured, only ensemble light scattering. However, recent work has been done in the area of determining concentration directly from the PCS signals (Weber and Schweiger, 1999). Multiple scattering can lead to errors, although a two-color cross-correlation system can minimize these errors (Drewel et al., 1990). In addition, limited shape information is provided on the particles, and the data analysis deconvolution is difficult for broad particle distributions.

As shown in Figure 16–8, the optical setup of a DLS system is quite simple; a vertically polarized laser beam is focused into the sample of interest, and a photomultiplier is used to

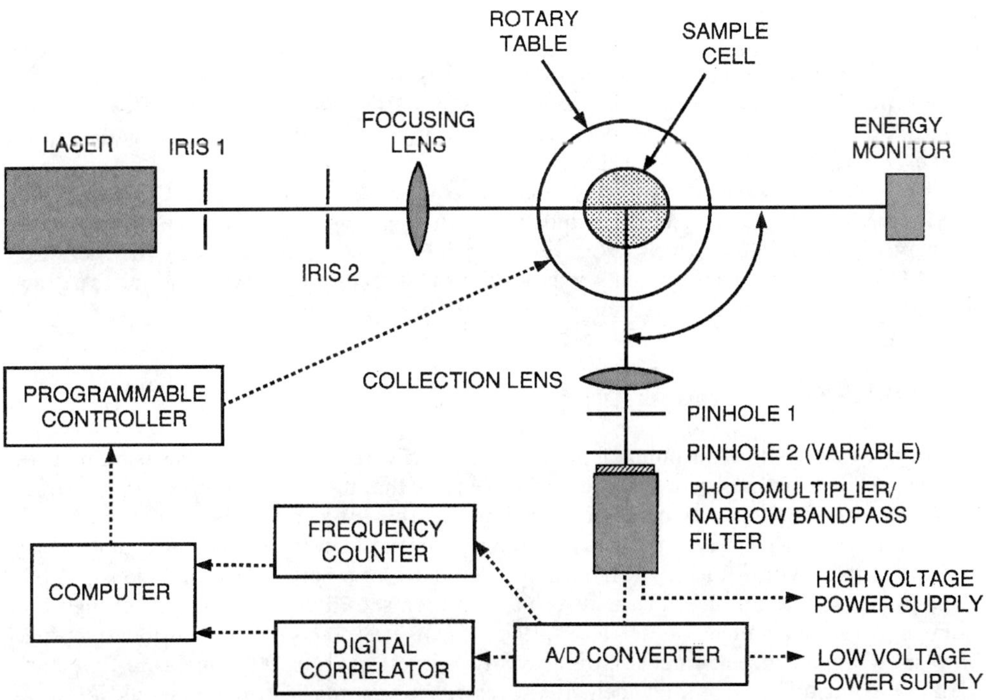

Fig. 16–8. Photon correlation spectroscopy layout.

collect the time-dependent light scattered at some angle by all of the particles in the beam (Weiner, 1984). The signal is then processed by either a spectrum analyzer (QELS or DBS techniques) or a digital correlator (PCS). In the PCS technique, a time autocorrelation function is performed on the dynamic scattered light signal and the characteristic decay rate of the exponential autocorrelation is used to determine the diffusion coefficient D. Particle effective hydrodynamic diameter is determined using the relations between D and d_p for the Knudsen number of interest, for example, Stokes-Einstein for continuum flow, Epstein diffusion for free-molecular flow, or Stokes-Einstein with Cunningham empirical slip correction for all Knudsen numbers (Bernard, 1988).

Analysis of the single exponential curve produced by PCS of a monodisperse sample is easily achieved, but difficulties arise in polydisperse cases, where the autocorrelation is a sum of the exponentials produced by particles in various size ranges. Extraction of polydispersity information by Laplace transform inversion is discussed by Bertero et al. (1988). Successful application of PCS to a system containing an unknown distribution of particle sizes frequently requires measurements at two or more scattering angles (Bott, 1987). This can be accomplished by mounting the photomultiplier on an optical rail that can rotate about the sample cell while maintaining a fixed geometry between the input laser beam and sample.

The PCS technique has theoretical limitations on the size of the largest particles that can be measured with high precision (Martin, 1987). This limit appears to be on the order of several hundred nanometers, with particle sizes above this limit achievable but with lower precision.

While commonly used for measurements of colloidal suspensions at rest, dynamic light-scattering techniques have been applied to a number of aerosol measurements, often in harsh environments. Flower and Hurd (1987) used dynamic light scattering to measure silica particles in premixed flames. They measured hydrodynamic particle diameters in the 0.04 to 0.15 µm range and noted that this technique is particularly useful for measurements of particles with purely real refractive indices (such as silica) where absorption techniques are invalid. Lhuissier et al. (1988) applied PCS for the measurement of diesel exhaust particles on the order 0.4 µm diameter in a wind tunnel flowing at 0.8 m/s. Taylor and Sorensen (1986) present the theoretical effects of Gaussian beam properties and flow of the diffusing particles in PCS measurements and verify experimentally using a flowing dioctyl phthalate aerosol with 0.09 and 0.9 µm diameter particles.

There are several commercial manufacturers of PCS devices (***MAL***, ***BEC***, ***BRK***, ***WYA***, ***PSS***). Most commercial systems include a multi-angle measurement capability and a temperature-controlled test cell to prevent convective flow from disturbing the measurements. However, use of these systems in their standard configuration typically requires extractive sampling. True in situ application requires system modification.

PERFORMANCE VERIFICATION

Although in situ optical techniques provide a powerful resource for the measurement of particle size distributions, care must be taken to ensure that they are used within their proper operating envelope. Thus, a means of verifying instrument performance—a standard—is essential in order to characterize instrument limits and to identify inherent instrument bias and inaccuracies. Unfortunately, establishing a standard for a particular technique is complicated by the well-established fact that results are often sensitive to instrument settings: For example, McDonell and Samuelsen (1990, 1996) discuss the effects of user-controlled settings on phase Doppler measurements. Consequently, Scarlett et al. (1990) emphasize that standards must include both a document that prescribes how the instrument should be used (written test method) and a standard reference material (SRM) made up of precisely characterized particles. McDonell and Samuelsen (1996) point out that a standard must reflect the complexity of the unknown system under study; for example, they show that a successful calibration of a phase Doppler interferometer with dilute monodisperse drops does not guarantee accurate measurements in a dense, swirling spray.

Examples of standard protocols and standard materials available for instrument calibration are discussed immediately below. In addition, a technique that allows an aerodynamic equivalent diameter based calibration of optical techniques is mentioned. Finally, a brief review of several instrument comparisons is provided as another method of verifying instrument performance.

Calibration

In calibration, a well-characterized aerosol or simulant (poly- or monodisperse) is presented to the instrument under prescribed conditions, and the instrument response is compared with the known result. For sampling instruments, it usually suffices to know the aerosol size distribution, shape, and composition (refractive index) for instrument calibration (see Chapter 21). For in situ techniques, however, concerns over trajectory ambiguity and coincidence typically require that particle trajectory (direction and velocity) and concentration also be controlled. Previous efforts toward developing calibration standards can be grouped into four general areas: (1) standard reference materials (SRMs), (2) monodisperse generators, (3) reticles, and (4) standardized sprays. Calibration techniques for in situ systems have been reviewed by Koo and Hirleman (1996), McDonell and Samuelsen (1996), and Hemsley et al. (1988).

SRMs include suspensions or powders that have been thoroughly characterized. Polystyrene latex (PSL) microspheres (with specified mean size and standard deviation) are available as dry powders or aqueous suspensions. PSL preparations are available with either National Institute of Standards and Technology (NIST) or Community Bureau of Reference (BCR) size certification. A suitable airborne dispersion is typically made by spraying a dilute suspension and drying (Mitchell, 1986). Various certified powders (quartz, metals, and so forth) are also available. Of particular interest are glass (standard or high refractive index) and silica microspheres, which come in a variety of sizes and can be NIST traceable. Analytic reference particles are commercially available (***DUK, BAN, SRD***). Although these techniques provide a very accurate size standard, it is difficult to quantify the generated concentration.

A second approach to calibration is to use a monodisperse droplet generator (***TSI***). See Chapter 21 for a description of this device. Droplet volume and, hence, diameter can be accurately calculated from the liquid flow rate and disturbance frequency (Berglund and Liu, 1973). Both single- and multiple-jet droplet generators have been used to characterize the performance of the phase Doppler particle analyzer (McDonell and Samuelsen, 1996; Ceman et al. 1993; Dressler and Kraemer, 1990). McDonell and Samuelsen (1996) have shown the value of using monodisperse droplets to assess instrument response variations due to field effects, such as beam steering in reacting environments.

A third approach to calibration, reticles have been used extensively in the calibration of in situ particle counters. Typically, a calibration reticle consists of a glass or plastic disk with either circular patterns of various sizes etched on its surface or with precision spheres embedded in its matrix. The disks are typically fixed for calibrating ensemble systems and rotated for calibrating SPCs. Circular-pattern reticles are typically used to calibrate diffraction-based instruments, as filled circles, spherical particles, or circular apertures of the same diameter give equivalent scattering patterns for near-forward scattering. Hirleman et al. (1984)

describe a calibration reticle for ensemble diffraction instruments that consists of a pattern of opaque chrome-filled circles photoetched onto a glass substrate; the circles are randomly positioned, and their diameters follow a Rosin-Rammler size distribution. These reticles have been used extensively in round-robin comparisons of ensemble diffraction instruments (Hirleman et al. 1984; Hirleman and Dodge, 1985; Dodge, 1987). Reticles are commercially available (***MAL***).

Reticles also have been used for calibrating SPCs. Hovenac and Hirleman (1991) reported the use of a rotating reticle for calibrating an Optical Array Probe (***PMS***) and a rotating pinhole calibrator for a PMS Forward-Scattering Spectrometer Probe. A reticle is available for calibrating the PCSV (***PRM***) near-forward SPC systems. In this case, several tracks of monodisperse images (circle diameters from 2 to 80 μm) are etched at discrete radii from the center of the disk. A precision DC motor is used to rotate the reticle to provide a sequence of monodisperse "particles" to the sample volume. Hemsley et al. (1988) discussed the use of both precision pinholes and a circular-pattern reticle (chrome disks sputtered on glass plates) for instrument calibration.

Two types of reticles have been proposed for calibrating phase Doppler interferometers. Ceman (1990) has developed a standard calibration device consisting of a transparent epoxy disk in which a known distribution of well-characterized glass beads is embedded. Ceman (1990) used the reticle to provide a known velocity, size distribution, and number density calibration particle field to characterize the performance of a PDPA (***AER***). Several changes must be made to the standard PDPA configuration to accommodate the disk calibrations: adjustment to the transmitter angle to account for refraction through the epoxy while maintaining a fixed collection angle (see Example 16–3); correction of the effective particle refractive index used in the PDPA software to refer it to the local medium; and correction of the effective receiver lens focal length to account for changes in optical path length caused by refraction through the epoxy. His preliminary investigations show that the PDPA is able to accurately measure calibration size distributions as well as individual particle diameters in the epoxy disk. Bever and Hovenac (1999) more recently reported on a microlens reticle (MLR) for PDPA calibration. The MLR consists of a single polymer 60 μm microlens affixed to a glass substrate; the microlens can be held either stationary in or rotated through the measurement volume. Bever and Hovenac (1999) demonstrated that after first calibrating a PDPA using a droplet generator, the correct operation of the PDPA can be subsequently verified using only the MLR.

Standardized sprays offer a fourth calibration possibility. As noted by McDonell and Samuelsen (1996), the use of a standardized research article (spray generator) is particularly desirable for calibrating in situ instruments intended for studies of complex (dense, reacting, swirling) flows. Dodge (1987) and Dodge and co-workers (1987) used pressure-swirl–type atomizers to compare the drop sizing performance of a variety of in situ instruments. McDonell et al. (1994) later reported on an interlaboratory comparison of phase Doppler measurements using a standardized research atomizer. More recently, McDonell and Samuelsen (1996) published a nice review of the use of standardized sprays to characterize the performance of phase Doppler interferometers.

Finally, a method for an aerodynamic particle size calibration of optical particle counters (OPCs) has been reported by Marple and Rubow (1976). A major advantage of this technique is providing a bridge between aerodynamic diameter (typically of interest in flowing systems) and the optical equivalent diameter reported by the systems of interest here. In their original method, an initial OPC size distribution is compared with one taken with a single-stage inertial impactor placed on the OPC inlet. By knowing the aerodynamic cutoff of the impactor, the two size distributions can be used to establish a single calibration point for the OPC. A complete OPC calibration is obtained by using impactors with different cutoff diameters. Although this technique was originally intended for calibrating sampling-type instruments, it can also be applied to the calibration of noninvasive systems. For example, Mitchell,

EXAMPLE 16–3

Aerosol measurements must often be made in a region physically separated from the measuring instrument by a window. For example, windows can be installed to provide optical access to an industrial flow of particle-laden gas. Consider measurements made in air (n_1 = 1.0), with the measurement beam passing through a 1 cm thick glass window (n_2 = 1.4). The beam has a 30° incident angle on the window, as shown in Figure 16–9. Find the angle of the beam (relative to its initial direction) and the apparent offset of the beam emerging from the window.

Answer: The change of angle in the window is simply given by application of Snell's law at the air–window interface:

$$n_1\ \alpha_1 = n_2\ \alpha_2$$

So, for the sample case, a 30° incident beam leads to a 20.92° beam angle inside the window. The angle inside the window is clearly independent of the window thickness. A second application of Snell's law at the inner window–air interface shows that the beam transmitted through the window recovers its original 30° angle. However, the window effectively shifts the location of any point on the beam in the measurement region (e.g., the beam crossing location for a dual beam LDV). The extent of shift is given by

$$\Delta y = t\left[1 - \frac{\cos\alpha_1}{\sqrt{(n_2/n_1)^2 - \sin^2\alpha_1}}\right]$$

where t is the window thickness (Durst et al., 1981). Using the given values in the formula shows that the beam is shifted by $\Delta y = 0.34$ cm by passing through the window. The reader is referred to Durst et al. (1981) for an expression for Δy in the configuration where regions of differing refractive index exist on either side of the window (e.g., measurements in heated air or in gases other than air).

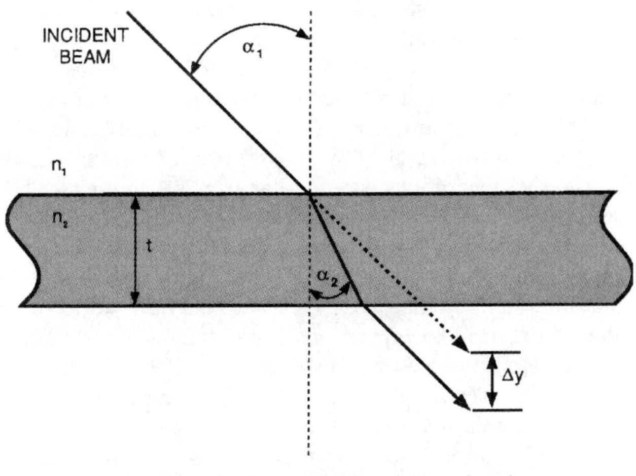

Fig. 16–9. Measurement in an enclosed region.

Nichols, and van Santen (1989) and Fissan and Helsper (1983) have calibrated a Polytec HC-15/1 with water droplets using the single-stage impactor technique.

Comparisons

One means of evaluating instrument performance is by comparison of several instruments measuring the same aerosol, which is often simpler than providing and measuring a primary standard aerosol or simulated aerosol. However, such comparisons are not straightforward for a number of reasons: differences in instrument principles, measured statistics, data reduction, definition of particle diameter, etc. Dodge (1987) and Jackson and Samuelsen (1987) discuss comparison of instruments for in situ measurement of spray droplets, including effects of spatial versus temporal sampling, and converting between point and ensemble measurements. With these limitations in mind, a number of researchers have compared performance of various particle sizing instruments.

In a classic study, Dodge (1987) reports on an inter-laboratory comparison of seventeen instruments of six different types, including Malvern (***MAL***) diffractometers, the Aerometrics (***AER***) PDPA, the ***PMS*** models FSSP-100, OAP-260X, and OAP-2D-GA1, and a Bete droplet analyzer (video imaging system) (***GRE***). In general the Malvern, Aerometrics, and Bete results were found to be in rough agreement for mean particle diameter, with the PMS instruments indicating larger diameters.

Presser et al. (1986) examined a spray with both the PDPA (***AER***) and an intensity deconvolution technique similar to the PCSV (***PRM***). They showed that the PDPA-measured mean particle sizes were considerably larger than those measured with the intensity deconvolution technique, and that both instruments gave results which showed similar trends as a function of position in the spray. Several authors have compared phase Doppler interferometer performance, including interlaboratory comparison of two different instruments from the same manufacturer (e.g., McDonell et al., 1994) or intralaboratory comparison of instruments from different manufacturers (Domnick et al., 1994). Good agreement is typically found for measurements of mean diameters, although volume flux measurements show more scatter. As mentioned above, care must be taken in setting user-controlled options, as these can significantly effect the measurements.

Several studies have compared the particle sizing performance of phase Doppler interferometers and diffraction-based instruments. Such comparisons are appealing, as the two techniques rely on fundamentally different measurement strategies and agreement would reinforce the integrity of both measurements. As the phase Doppler instruments provide temporal averaged point measurements (flux based) whereas the diffraction instruments provide spatial averaged line-of-sight measurements (spatial), data analysis is required to provide proper comparisons (e.g., Dodge et al., 1987; Cossali and Hardalupas, 1992). Qualitative agreement between the two techniques (i.e., similar trends) is typically observed, although excellent agreement has not been consistently reported (McDonell and Samuelsen, 1996; Jackson and Samuelsen, 1987). The degree of agreement depends on the complexity of the system under consideration, with a tendency toward better agreement for dilute sprays (Corcoran et al., 2000). In general, the phase Doppler systems tend to report a larger mean particle diameter than diffraction-based systems (e.g., McDonell and Samuelsen 1996; Zaidi et al., 1998; Cossali and Hardalupas, 1992; Dodge et al., 1987). McDonell and Samuelsen (1996) suggest that the accuracy of phase Doppler measurements tends to improve with increasingly sophisticated signal analyzers and probe volume correction algorithms. They report that newer, frequency domain–based processors provide more accurate results than older, counter-based processors, particularly at small sizes.

Young and Bachalo (1988) compare the PDPA (***AER***), the 2600 (***MAL***), and the OAP-two-dimensional–GA2 (***PMS***) performance by simultaneous measurement of falling glass beads and a droplet spray with all three instruments and discuss calibration techniques and comparison difficulties for these instruments. In general, mean glass bead diameters measured by all three instruments were in good agreement. For spray droplet measurements, where the droplet size–velocity correlation became significant, the 2600 gave a considerably smaller mean diameter than the other two instruments. However, the PDPA data were shown to agree fairly well with those of the 2600 when corrected from a temporal to a spatial basis (like that of the 2600).

On some occasions, the particle size distribution of interest will be too broad for one technique or instrument to measure. Combining different techniques can be difficult, as one is in fact performing a comparison in the overlap region (if one exists) with all the limitations discussed above. As one example, Picot et al. (1990) combined an FSSP-100 (***PMS***), an OAP-260X (***PMS***), and a flash photography system to characterize an aerial spray nozzle and discussed a method for incorporating data from the three systems.

CONCLUSIONS

From the extent of the previous discussion, it is easy to see that there are a wide range of optical in situ techniques available for measuring particle size distributions. Although this variety makes it likely that a researcher will find at least one instrument that will perform the required measurement, it is essential that a careful selection process be pursued. Particular attention is required in determining the instrument operating envelope relative to the measurement requirements in order to find the technique best suited to the application. Performance verification should be made as needed to ensure data accuracy and instrument reliability. The various commercial techniques are listed in Table 16–1. We have tried to include all vendors and equipment in these lists, but realize that the field is extensive and that considerable development efforts are always proceeding. We apologize for any omissions.

TABLE 16–1. Listing of Selected Optical In Situ Particle Size Measurement Systems and Their Characteristics

Method	Commercial Source	Measure Velocity	Comments
SPC intensity	***MEL***	LDV	PAS
SPC intensity	***PAL***	N	PCS-2000
SPC intensity	***PMS***	Y	FSSP, mean velocity option
SPC intensity	***PRM***	Y	PCSV (probe and bench)
SPC phase	***TSI***	LDV	PDPA
SPC phase	***DAN***	LDV	PDA
SPC imaging	***PMS***	Y	OAP
Ensemble imaging	***MEL***	Y	Custom holography
Ensemble diffraction	***CIL***	N	PSA (Particle Size Analyzer)
Ensemble diffraction	***MAL***	N	Spraytec and EPCS
Ensemble diffraction	***MCR***	N	Microtrac SRA, X-100
Ensemble diffraction	***SYM***	N	HELOS
Ensemble PCS	***BRK***	N	
Ensemble PCS	***BEC***	N	
Ensemble PCS	***MAL***	N	
Ensemble PCS	***PSS***	N	
Ensemble PCS	***WYA***	N	
Calibration	***BAN***	—	Standard reference particles
Calibration	***DUK***	—	Standard reference particles
Calibration	***SRD***	—	Standard reference particles
Calibration	***TSI***	—	Single-jet droplet generator

ACKNOWLEDGMENTS

Sandia is a multiprogram laboratory operated by Sandia Corporation, a Lockheed Martin Company, for the United States Department of Energy under Contract DE-AC04-94AL85000.

REFERENCES

Al-Chalabi, S. A. M., Y. Hardalupas, A. R. Jones, and A. M. K. P. Taylor. 1988. Calculation of calibration curves for the phase doppler technique: Comparison between Mie theory and geometrical optics. In *Optical Particle Sizing*, eds. G. Gouesbet and G. Grehan. New York: Plenum Press, pp. 107–120.

Alexander D. R., K. J. Wiles, S. A. Schaub, and M. P. Seeman. 1985. Effects of non-spherical drops on a phase Doppler spray analyzer. In *Particle Sizing and Spray Analysis*, SPIE Vol. 573, eds. N. Chigier and G. W. Stewart. Bellingham, WA: SPIE, pp. 67–72.

Annapragada, A. and A. Adjei. 1996. An analysis of the Fraunhofer diffraction method for particle size distribution analysis and its application to aerosolized sprays. *Int. J. Pharmacol.* 127:219–227.

Azzopardi, B. J. 1992. Instrumentation for particle size analysis by far field diffraction: Accuracy, limitations, and future. In *Particle Size Analysis*, eds. N. G. Stanley-Wood and R. W. Lines. Cambridge: Royal Society of Chemistry, pp. 108–132.

Bachalo, W. D. 1980. Method for measuring the size and velocity of spheres by dual-beam light-scatter interferometry. *Appl. Opt.* 19(3):363–370.

Bachalo, W. D. and M. J. Houser. 1984. Phase Doppler spray analyzer for simultaneous measurements of drop size and velocity distributions. *Opt. Eng.* 23:583–590.

Bachalo, W. D., R. C. Rudoff, and S. V. Sankar. 1990. Time-resolved measurements of spray drop size and velocity. In *Liquid Particle Size Measurement Techniques*, 2nd Ed., ASTM STP 1083, eds. E. D. Hirleman, W. D. Bachalo, and P. G. Felton. Philadelphia: American Society for Testing and Materials, pp. 209–224.

Baumgardner, D., W. A. Cooper, and J. E. Dye. 1990. Optical and electronic limitations of the forward-scattering spectrometer probe. In *Liquid Particle Size Measurement Techniques*, ASTM STP 1083, eds. E. D. Hirleman, W. D. Bachalo, and P. G. Felton. Philadelphia: American Society for Testing and Materials, pp. 115–127.

Baumgardner, D., J. E. Dye, B. W. Gandrud, and R. G. Knollenberg. 1992. Interpretation of measurements made by the forward scattering spectrometer probe (FSSP-300) during the airborne arctic stratospheric expedition. *J. Geophys. Res.* 97(D8):8035–8046.

Bayvel, L. P., J. Knight, and G. Robertson. 1987. Alternative model-independent inversion programme for Malvern particle sizer. *Part. Charact.* 4:49–53.

Berglund, R. N. and B. Y. H. Liu. 1973. Generation of monodisperse aerosol standards. *Environ. Sci. Technol.* 7:147–153.

Bernard, J. M. 1988. Particle sizing in combustion systems using scattered laser light. *J. Quant. Spectrosc. Radiat. Transfer* 40:321–330.

Bertero, M., P. Boccacci, C. De Mol, and E. R. Pike. 1988. Extraction of polydispersity information in photon correlation spectroscopy. In *Optical Particle Sizing*, eds. G. Gouesbet and G. Grehan. New York: Plenum Press, pp. 89–98.

Bever, S. J. and E. A. Hovenac. 1999. Reticle for verification of the correct operation of the phase Doppler particle analyzer. *Opt. Eng.* 38(10):1730–1734.

Black, D. L., M. Q. McQuay, and M. P. Bonin. 1996. Laser-based techniques for particle-size measurement: A review of sizing methods and their industrial applications. *Prog. Energy Combust. Sci.* 22: 267–306.

Bohren, C. F. and D. R. Huffman. 1983. *Absorption and Scattering of Light by Small Particles*. New York: John Wiley & Sons.

Bonin, M. P., W. L. Flower, R. F. Renzi, and L. W. Peng. 1995. Size and concentration measurements of particles produced in commercial chromium plating processes. *J. Air Waste Manage.* 45:902–907.

Bonin, M. P. and M. Queiroz, 1991. Local particle velocity, size, and concentration measurements in an industrial-scale pulverized coal-fired boiler. *Combust. Flame* 85:121–133.

Bonin, M. P. and M. Queiroz. 1996. A parametric evaluation of particle-phase dynamics in an industrial pulverized-coal-fired boiler. *Fuel* 75(2):195–206.

Borho, K. 1970. A scattered-light measuring instrument for high dust concentrations. *Staub Reinhalt. Luft.* 30:45–49.

Bott, S. E. 1987. Submicrometer particle sizing by photon correlation spectroscopy: Use of multiple-angle detection. In *Particle Size Distribution: Assessment and Characterization*, ed. T. Provder. Washington, DC: American Chemical Society, pp. 74–88.

Breña de la Rosa, A., S. V. Sankar, B. J. Weber, G. Wang, and W. D. Bachalo. 1989. A theoretical and experimental study of the characterization of bubbles using light scattering interferometry. In *ASME International Symposium on Cavitation Inception—1989*, eds. W. B. Morgan and B. R. Parkin. New York: American Society of Mechanical Engineers, pp. 63–72.

British Standards Institution. 1988. *British Standard Methods for Determination of Particle Size Distribution—Recommendations for Single Particle Light Interaction Methods*. London: British Standards Institution, BS3406: Part 7.

Ceman, D. L. 1990. *Phase Doppler Technique: Factors Affecting Instrument Response and Novel Calibration System. Master of Science Thesis*. The University of New Mexico, Albuquerque.

Ceman, D. L., T. J. O'Hern, and D. J. Rader. 1990. Phase Doppler droplet sizing—Scattering angle effects. In *ASME Fluid Measurements and Instrumentation Forum*, eds. E. P. Rood and C. J. Blechinger. New York: American Society of Mechanical Engineers, pp. 61–63.

Ceman, D. L., D. J. Rader, and T. J. O'Hern. 1993. Calibration of the phase Doppler particle analyzer with monodisperse droplets. *Aerosol Sci. Technol.* 18:346–358.

Chan, K. T. and Y. J. Li. 1998. Pipe flow measurement by using a side-scattering holographic particle imaging technique. *Opt. Laser Technol.* 30(1):7–14.

Chávez, A. and F. Mayinger. 1990. Evaluation of pulsed laser holograms of spray droplets using digital image processing. In *Proceedings of the Second International Congress on Optical Particle Sizing*, ed. E. D. Hirleman. Phoenix AZ: Arizona State University Printing Services, pp. 462–471.

Christnacher, F. and H. Royer. 1994. Automatic sizing of particle holographic images via optical correlation and edge extraction. *Opt. Commun.* 108:17–23.

Collier, R. J., C. B. Burkhardt, and L. H. Lin. 1971. *Optical Holography*, New York: Academic Press.

Corcoran, T. E., R. Hitron, W. Humphrey, and N. Chigier. 2000. Optical measurement of nebulizer sprays: A quantitative comparison of diffraction, phase Doppler interferometry, and time of flight techniques. *J. Aerosol Sci.* 31(1):35–50.

Cornillault, J. 1972. Particle size analyzer. *Appl. Opt.* 11:265–268.

Cossali, E. and Y. Hardalupas. 1992. Comparison between laser diffraction and phase Doppler velocimeter techniques in high turbidity, small diameter sprays. *Exp. Fluids* 13:414–422.

Damaschke, N., G. Gouesbet, G. Grehan, H. Mignon, and C. Tropea. 1998a. Response of phase Doppler anemometer systems to nonspherical droplets. *Appl. Opt.* 37(10):1752–1761.

Damaschke, N., G. Gouesbet, G. Grehan, and C. Tropea. 1998b. Optical techniques for the characterization of non-spherical and non-homogeneous particles. *Meas. Sci. Technol.* 9:134–140.

Dickson, L. D. 1970. Characteristics of a propagating Gaussian beam. *Appl. Opt.* 9:1854–1861.

Dodge, L. G. 1984a. Change of calibration of diffraction-based particle sizers in dense sprays. *Opt. Eng.* 23:626–630.

Dodge, L. G. 1984b. Calibration of the Malvern particle sizer. *Appl. Opt.* 23:2415–2419.

Dodge, L. G. 1987. Comparison of performance of drop-sizing instruments. *Appl. Opt.* 26:1328–1341.

Dodge, L. G., D. J. Rhodes, and R. D. Reitz. 1987. Drop-size measurement techniques for sprays: Comparison of Malvern laser-diffraction and Aerometrics phase/Doppler. *Appl. Opt.* 26:2144–2154.

Domnick, J., V. Dorfner, F. Durst, and M. Yamashita. 1994. Performance evaluation of different phase-Doppler systems. *Part. Part. Syst. Charact.* 11:91–100.

Dressler, J. L. and G. O. Kraemer. 1990. A multiple drop-size drop generator for calibration of a phase-Doppler particle analyzer. In *Liquid Particle Size Measurement Techniques*, 2nd Ed., ASTM STP 1083, eds. E. D. Hirleman, W. D. Bachalo, and P. G. Felton. Philadelphia: American Society for Testing and Materials, pp. 30–44.

Drewel, M., J. Ahrens, and K. Schatzel. 1990. Suppression of multiple scattering errors in particle sizing by dynamic light scattering. In *Proceedings of the Second International Congress on Optical Particle Sizing*, ed. E. D. Hirleman. Phoenix, AZ: Arizona State University Printing Services, pp. 130–138.

Durst, F., A. Melling, and J. H. Whitelaw. 1981. *Principles and Practice of Laser-Doppler Anemometry*. London: Academic Press.

Ewan, B. C. R., J. Swithenbank, and C. Sorusbay. 1984. Measurement of transient spray size distributions. *Opt. Eng.* 23:620–625.

Farmer, W. M. 1972. Measurement of particle size, number density, and velocity using a laser interferometer. *Appl. Opt.* 11:2603–2612.

Felton, P. G. 1990. A review of the Fraunhofer diffraction particle-sizing technique. In *Liquid Particle Size Measurement Techniques, ASTM STP 1083*, eds. E. D. Hirleman, W. D. Bachalo, and P. G. Felton. Philadelphia: American Society for Testing and Materials, pp. 47–59.

Fissan, H. J. and C. Helsper. 1981. Calibration of the Polytec CH-15 and HC-70 optical particle counters. In *Aerosols in the Mining and Industrial Work Environments*, Vol. 3, eds. V. A. Marple and B. Y. H. Liu. Ann Arbor, MI: Ann Arbor Science, pp. 825–831.

Flower, W. L. and A. J. Hurd. 1987. In situ measurement of flame-formed silica particles using dynamic light scattering. *Appl. Opt.* 26:2236–2239.

Friehmelt, R. and S. Heidenreich. 1999. Calibration of a white-light/90° optical particle counter for "aerodynamic" size measurements—Experiments and calculations for spherical particles and quartz dust. *J. Aerosol Sci.* 30(10):1271–1279.

Gebhart, J. and A. Anselm. 1988. Effect of particle shape on the response of single particle optical counters. In *Optical Particle Sizing*, eds. G. Gouesbet and G. Grehan. New York: Plenum Press, pp. 393–409.

Gobel, G., T. Wriedt, and K. Bauckhage. 1998. Micron and sub-micron aerosol sizing with a standard phase-Doppler anemometer. *J. Aerosol Sci.* 29(9):1063–1073.

Gouesbet, G. 1994. Generalized Lorenz-Mie theory and applications. *Part. Part. Syst. Charact.* 11:22–34.

Gouesbet, G. and G. Grehan, eds. 1988. *Optical Particle Sizing*. New York: Plenum Press.

Grehan, G. and G. Gouesbet. 1986. Simultaneous measurements of velocities and sizes of particles in flows using a combined system incorporating a top-hat beam technique. *Appl. Opt.* 25:3527–3538.

Grehan, G., G. Gouesbet, A. Naqwi, and F. Durst. 1994. Trajectory ambiguities in phase Doppler systems: Study of a near-forward and a near-backward geometry. *Part. Part. Syst. Charact.* 11(12):133–144.

Hammond, D. C. 1981. Deconvolution technique for line-of-sight optical scattering measurements in axisymmetric sprays. *Appl. Opt.* 20:493–499.

Hardalupas, Y. and C. H. Liu. 1997. Implications of the Gaussian intensity distribution of laser beams on the performance of the phase Doppler technique sizing uncertainties. *Prog. Energy Combust. Sci.* 23:41–63.

Hardalupas, Y., A. M. K. P. Taylor, and J. H. Whitelaw. 1988. Measurements in heavily-laden dusty jets with phase-Doppler anemometry. In *Transport Phenomena in Turbulent Flows: Theory, Experiment, and Numerical Simulation*, eds. M. Hirata and N. Kasagi. New York: Hemisphere, pp. 821–835.

Harvill, T. L. and D. J. Holve. 1997. Method for measuring particle size in the presence of multiple scattering. United States Patent 5,619,324.

Harvill, T. L., J. H. Hoog, and D. J. Holve. 1995. In-process particle size distribution measurements and control. *Part. Part. Syst. Charact.* 12:309–313.

Haussmann, G. and W. Lauterborn. 1980. Determination of the size and position of fast moving gas bubbles in liquids by digital 3-D image processing of hologram reconstructions. *Appl. Opt.* 19: 3529–3535.

Hemsley, D. J., M. L. Yeoman, C. J. Bates, and O. Hadded. 1988. The use of calibration techniques for the development and application of optical particle sizing instruments. In *Optical Particle Sizing*, eds. G. Gouesbet and G. Grehan. New York: Plenum Press, pp. 585–601.

Hess, C. F. 1984. Nonintrusive optical single-particle counter for measuring the size and velocity of droplets in a spray. *Appl. Opt.* 23:4375–4382.

Hess, C. F. and J. D. Trolinger. 1985. Particle field holography data reduction by Fourier transform analysis. *Opt. Eng.* 24:470–474.

Hirleman, E. D. 1983. Nonintrusive laser-based particle diagnostics. *Prog. Astron. Aeron.* 92:177–207.

Hirleman, E. D. 1984. Particle sizing by optical, nonimaging techniques. In *Liquid Particle Size Measurement Techniques, ASTM STP 848*, eds. J. M. Tishkoff, R. D. Ingebo, and J. B. Kennedy. Philadelphia: American Society for Testing and Materials, pp. 35–60.

Hirleman, E. D. 1987. Optimal scaling of the inverse Fraunhofer diffraction particle sizing problem: The linear system produced by quadrature. *Part. Charact.* 4:128–133.

Hirleman, E. D. 1988a. Optical techniques for particle size analysis. *Laser Topics* 10:7–10.

Hirleman, E. D. 1988b. Modeling of multiple scattering effects in Fraunhofer diffraction particle size analysis. *Part. Part. Syst. Charact.* 5:57–65.

Hirleman, E. D., ed. 1990. *Proceedings of the Second International Congress on Optical Particle Sizing*. Tempe: Arizona State University Printing Services.

Hirleman, E. D. 1991. General solution to the inverse near-forward-scattering particle-sizing problem in multiple-scattering environments: Theory. *Appl. Opt.* 30(33):4832–4838.

Hirleman, E. D. 1996. History of development of the phase-Doppler particle-sizing velocimeter. *Part. Part. Syst. Charact.* 13:59–67.

Hirleman, E. D., W. D. Bachalo, and P. G. Felton. eds. 1990. *Liquid Particle Size Measurement Techniques*: 2nd Ed., ASTM STP 1083. Philadelphia: American Society for Testing and Materials.

Hirleman, E. D. and L. G. Dodge. 1985. Performance comparison of Malvern instruments laser diffraction drop size analyzers. In *Proceedings, ICLASS-85, Third International Conference on Liquid Atomisation and Spray Systems*, Vol. 2. London: Institute of Energy, pp. IVA/3/1–14.

Hirleman, E. D., V. Oechsle, and N. A. Chigier. 1984. Response characteristics of laser diffraction particle size analyzers: Optical sample volume extent and lens effects. *Opt. Eng.* 23:610–619.

Holve, D. J. 1980. In situ optical particle sizing technique. *J. Energy* 4(4):176–183.

Holve, D. J. 1982. Transit timing velocimetry (TTV) for two-phase reacting flows. *Combust. Flame* 48:105–108.

Holve, D. J. and K. D. Annen. 1984. Optical particle counting, sizing, and velocimetry using intensity deconvolution. *Opt. Eng.* 23:591–603.

Holve, D. J. and G. W. Davis. 1985. Sample volume and alignment analysis for an optical particle counter sizer, and other applications. *Appl. Opt.* 24:998–1005.

Holve, D. J. and S. A. Self. 1979a. Optical particle sizing for in situ measurements Part 1. *Appl. Opt.* 18:1632–1645.

Holve, D. J. and S. A. Self. 1979b. Optical particle sizing for in situ measurements Part 2. *Appl. Opt.* 18:1646–1652.

Holve, D. and S. Self. 1980. Optical measurements of mean particle size in coal-fired MHD flows. *Combust. Flame* 37:211–214.

Holve, D. J., D. Tichenor, J. C. F. Wang, and D. R. Hardesty. 1981. Design criteria and recent developments of optical single particle counters for fossil fuel systems. *Opt. Eng.* 20:529–539.

Hovenac, E. A. 1987. *Performance and Operating Envelope of Imaging and Scattering Particle Sizing Instruments*. NASA CR-180859, National Aeronautics and Space Administration. Cleveland, OH: Lewis Research Center.

Hovenac, E. A. and E. D. Hirleman. 1991. Use of rotating pinholes and reticles for calibration of cloud droplet instrumentation. *J. Atmos. Oceanic Technol.* 8:166–171.

Hovenac, E. A. and J. A. Lock. 1993. Calibration of the FSSP: Modeling scattering from a multimode laser beam. *J. Atmos. Oceanic Technol.* 10:518–525.

Jackson, T. A. 1990. Droplet sizing interferometry. In *Liquid Particle Size Measurement Techniques, ASTM STP 1083*, 2nd Ed., eds. E. D. Hirleman, W. D. Bachalo, and P. G. Felton. Philadelphia: American Society for Testing and Materials, pp. 151–169.

Jackson, T. A. and G. S. Samuelsen. 1987. Droplet sizing interferometry: A comparison of the visibility and phase/Doppler techniques. *Appl. Opt.* 26:2137–2143.

Jaenicke, R. and T. Hanusch. 1993. Simulation of the optical particle counter forward scattering spectrometer probe 100 (FSSP-100). *Aerosol Sci. Technol.* 18:309–322.

Jones, A. R. 1999. Light scattering for particle characterization. *Prog. Energy Combust. Sci.* 25:1–53.

Kerker, M. 1969. *The Scattering of Light and Other Electromagnetic Radiation*. New York: Academic Press.

Kerker, M., ed. 1988. *Selected Papers on Light Scattering, SPIE*, Vol. 951. Bellingham, Washington: SPIE.

Killinger, R. T. and R. H. Zerull. 1988. Effects of shape and orientation to be considered for optical particle sizing. In *Optical Particle Sizing*, eds. G. Gouesbet and G. Grehan. New York: Plenum Press, pp. 419–429.

Knight, J. C., D. Ball, and G. N. Robertson. 1991. Analytical inversion for laser diffraction spectrometry giving improved resolution and accuracy in size distribution. *Appl. Opt.* 30(33):4795–4799.

Knollenberg, R. G. 1970. The optical array: An alternative to scattering or extinction for airborne particle size determination. *J. Appl. Meteor.* 9:86–103.

Knollenberg, R. G. 1979. Single particle light scattering spectrometers. In *Aerosol Measurement*, eds. D. A. Lundgren, F. S. Harris, W. H. Marlow, M. Lippmann, W. E. Clark, and M. D. Durham. Gainseville: University Press of Florida, pp. 271–293.

Knollenberg, R. G. 1981. Techniques for probing cloud microstructure. In *Clouds: Their Formation, Optical Properties, and Effects*, eds. P. V. Hobbs and A. Deepak. New York: Academic Press, pp. 15–89.

Knollenberg, R. G. 1985. The measurement of particle sizes below 0.1 micrometers. *J. Environ. Sci.* 28: 32–47.

Koo, J. H. and E. D. Hirleman. 1992. Synthesis of integral transform solutions for the reconstruction of particle-size distributions from forward-scattered light. *Appl. Opt.* 31(12):2130–2140.

Koo, J. H. and E. D. Hirleman. 1996. Review of principles of optical techniques for particle size measurements. In *Recent Advances in Spray Combustion: Spray Atomization and Drop Burning Phenomena*, Vol. 1, ed. K. K. Kuo. Reston, VA: AIAA, pp. 3–32.

Kuo, K. K. (ed.) 1996. *Recent Advances in Spray Combustion: Spray Atomization and Drop Burning Phenomena*, Vol. 1. Reston, VA: AIAA.

Lavergne, G., Y. Biscos, F. Bismes, and P. Hebrard. 1988. Optical particle sizing: Digital video image processing application. In *Optical Particle Sizing*, eds. G. Gouesbet and G. Grehan. New York: Plenum Press, pp. 471–482.

Lefebvre, A. H. 1989. *Atomization and Sprays*. New York: Hemisphere, pp. 367–409.

Lhuissier, N., A. Nazih, and M. E. Weill. 1988. Measurement of diesel exhaust particles in a dilution tunnel by photon correlation spectroscopy. In *Photon Correlation Techniques and Applications, Proceedings of the OSA Topical Meeting*, Vol. 1, eds. J. B. Abbiss and A. E. Smart, pp. 84–89. Washington, DC: Optical Society of America.

Marple, V. A. and K. L. Rubow. 1976. Aerodynamic particle size calibration of optical particle counters. *J. Aerosol Sci.* 7:425–433.

Martin, J. E. 1987. Slow aggregation of colloidal silica. *Phys. Rev. A* 36:3415–3426.

Maus, R. and H. Umhauer. 1996. Determination of the fractional efficiencies of fibrous filter media by optical in situ measurements. *Aerosol Sci. Technol.* 24:161–173.

McDonell, V. G. and G. S. Samuelsen. 1990. Sensitivity assessment of a phase-Doppler interferometer to user-controlled settings. In *Liquid Particle Size Measurement Techniques, ASTM STP 1083*, 2nd Ed., eds. E. D. Hirleman, W. D. Bachalo, and P. G. Felton. Philadelphia: American Society for Testing and Materials, pp. 170–189.

McDonell, V. G. and G. S. Samuelsen. 1996. Intra- and interlaboratory experiments to assess performance of phase Doppler interferometry. In *Recent Advances in Spray Combustion: Spray Atomization and Drop Burning Phenomena*, Vol. 1, ed. K. K. Kuo. Reston, VA: AIAA, pp. 57–106.

McDonell, V. G., G. S. Samuelsen, M. R. Wang, C. H. Hong, and W. H. Lai. 1994. Interlaboratory comparison of phase Doppler measurements in a research simplex atomizer spray. *J. Propul. Power* 10:402–409.

Meyer, P. and N. Chigier. 1986. Dropsize measurements using a Malvern 2200 particle sizer. *Atomisation Spray Technol.* 2:261–298.

Miles, B. H., P. E. Sojka, and G. B. King. 1990. Malvern particle size measurements in media with time varying index of refraction gradients. *Appl. Opt.* 29:4563–4573.

Mitchell, J. P. 1986. *Aerosol Generation for Instrument Calibration*. UK Atomic Energy Report AEEW-R 2092. US. Atomic Energy Commission.

Mitchell, J. P., A. L. Nichols, and A. van Santen. 1989. The characterization of water-droplet aerosols by Polytec optical particle analysers. *Part. Part. Syst. Charact.* 6:119–123.

Oberdier, L. M. 1984. An instrumentation system to automate the analysis of fuel-spray images using computer vision. In *Liquid Particle Size Measurement Techniques, ASTM STP 848*, eds. J. M. Tishkoff, R. D. Ingebo, and J. B. Kennedy. Philadelphia: American Society for Testing and Materials, pp. 123–136.

O'Hern, T. J., D. J. Rader, and D. L. Ceman. 1989. Droplet sizing calibration of the phase Doppler particle analyzer. In *ASME Fluid Measurements and Instrumentation Forum*, eds. E. P. Rood and C. J. Blechinger. New York: American Society of Mechanical Engineers, pp. 49–51.

Pinnick, R. G., D. M. Garvey, and L. D. Duncan. 1981. Calibration of Knollenberg FSSP light-scattering counters for measurement of cloud droplets. *J. Appl. Meteor.* 20:1049–1057.

Picot, J. C., M. W. van Vliet, N. J. Payne, and D. D. Kristmanson. 1990. Characterization of aerial spray nozzles with laser light-scattering and imaging probes and flash photography. In *Liquid Particle Size Measurement Techniques, ASTM STP 1083*, eds. E. D. Hirleman, W. D. Bachalo, and P. G. Felton. Philadelphia: American Society for Testing and Materials, pp. 142–150.

Presser, C., A. K. Gupta, R. J. Santoro, and H. G. Semerjian. 1986. Laser diagnostics for characterization of fuel sprays. In *ICALEO '86 Proceedings*, Vol. 58, *Flow and Particle Diagnostics*. Toledo, OH: Laser Institute of America, pp. 160–167.

Qiu, H.-H., W. Jia, C. T. Hsu, and M. Sommerfeld. 2000. High accuracy optical particle sizing in phase-Doppler anemometry. *Meas. Sci. Technol.* 11:142–151.

Reuter, A. and S. Bakan. 1998. Improvements of cloud particle sizing with a 2D-grey probe. *J. Atmos. Oceanic Technol.* 15:1196–1203.

Sachweh, B., H. Umhauer, F. Ebert, H. Buttner, and R. Friehmelt. 1998. In situ optical particle counter with improved coincidence error correction for number concentrations up to 10^7 particles/cm^3. *J Aerosol Sci.* 29(9):1075–1086.

Saffman, M. 1987. Optical particle sizing using the phase of LDA signals. *Dantec Inform. Measure. Anal.* 5:8–13.

Saffman, M., P. Buchave, and H. Tanger. 1984. Simultaneous measurement of size, concentration and velocity of spherical particles by a laser Doppler method. In *Second International Symposium on Applications of Laser Anemometry to Fluid Mechanics*, eds. R. J. Adrian, D. F. G. Durão, F. Durst, H. Mishina, and J. H. Whitelaw. Lisbon: Ladoan, pp. 85–103.

Sankar, S. V., B. J. Weber, D. Y. Kamemoto, and W. D. Bachalo. 1990. Sizing fine particles with the phase Doppler interferometric technique. In *Proceedings of the Second International Congress on Optical Particle Sizing*, ed. E. D. Hirleman. Phoenix, AZ: Arizona State University Printing Services, pp. 277–287.

Santangelo, P. J. and P. E. Sojka. 1993. Holographic particle diagnostics. *Prog. Energy Combust.* 19:587–603.

Scarlett, B., H. G. Merkus, and G. M. H. Meesters. 1990. European progress on calibration and standardization for particle sizing. In *Liquid Particle Size Measurement Techniques, ASTM STP 1083*, 2nd Ed., eds. E. D. Hirleman, W. D. Bachalo, and P. G. Felton. Philadelphia: American Society for Testing and Materials, pp. 9–18.

Schäfer, M. and H. Umhauer. 1987. Realization of a concept for the complete evaluation of double pulse holograms of particulate phases in flows. *Part. Charact.* 4:166–174.

Stier, J. and M. Quinten. 1998. Simple refractive index correction for the optical particle counter PCS 2000 by Palas. *J. Aerosol Sci.* 29:223–225.

Strakey, P. A., D. G. Talley, S. V. Sankar, and W. D. Bachalo. 1998. The use of small probe volumes with phase Doppler interferometry. In *Proc. ILASS Americas '98*, Sacramento, CA, May 17–20.

Swithenbank, J., J. M. Beer, D. S. Taylor, D. Abbot, and G. C. McCreath. 1977. A laser diagnostic technique for the measurement of droplet and particle size distribution. In *Experimental Diagnostics in*

Gas Phase Combustion Systems, eds. B. T. Zinn and C. T. Bowman. Vol. 53, Progress in Astronautics and Aeronautics. New York: AIAA, pp. 421–447.

Swithenbank, J., J. Cao, and A. A. Hamidi. 1991. Spray diagnostics by laser diffraction. In *Combustion Measurements*, ed. N. Chigier. Washington, DC: Hemisphere, pp. 179–227.

Takeo, T. and H. Hattori. 1990. Visibility analysis of laser Doppler anemometry for spherical particles smaller than several light wavelengths. *Jpn. J. Appl. Phys.* 29:419–426.

Tayali, N. E. and C. J. Bates. 1990. Particle sizing techniques in multiphase flows: A review. *Flow Measure Instrum.* 1:77–105.

Taylor, T. W. and C. M. Sorensen. 1986. Gaussian beam effects on the photon correlation spectrum from a flowing Brownian motion system. *Appl. Opt.* 25:2421–2426.

Thompson, B. J. 1974. Holographic particle sizing techniques. *J. Phys. E Sci. Inst.* 7:781–788.

Thompson, B. J. 1984. Droplet characteristics with conventional and holographic imaging techniques. In *Liquid Particle Size Measurement Techniques, ASTM STP 848*, eds. J. M. Tishkoff, R. D. Ingebo, and J. B. Kennedy. Philadelphia: American Society for Testing and Materials, pp. 111–122.

Thompson, B. J. and J. H. Ward. 1966. Particle sizing—The first direct use of holography. *Sci. Res.* 1:37–40.

Trolinger, J. D. 1975. Particle field holography. *Opt. Eng.* 14:383–392.

Trolinger, J. D. 1980. Analysis of holographic diagnostic systems. *Opt. Eng.* 19:722–726.

Tyler, G. A. and B. J. Thompson. 1976. Fraunhofer holography applied to particle size analysis—A reassessment. *Opt. Acta* 23:685–700.

Umhauer, H. 1983. Particle size distribution analysis by scattered light measurements using an optically defined measuring volume. *J. Aerosol Sci.* 14:765–770.

van de Hulst, H. C. 1981. *Light Scattering by Small Particles*. New York: Dover Publications, Inc.

Van Den Moortel, T., R. Santini, L. Tadrist, and J. Pantaloni. 1997. Experimental study of the particle flow in a circulating fluidized bed using a phase Doppler particle analyser: A new post-processing data algorithm. *Int. J. Multiphase Flow* 23(6):1189–1209.

Weber, R. and G. Schweiger. 1999. Simultaneous in situ measurement of local particle size, particle concentration, and velocity of aerosols. *J. Colloid Interface Sci.* 210:86–96.

Weiner, B. B. 1984. Particle sizing using photon correlation spectroscopy. In *Modern Methods of Particle Size Analysis*, ed. H. G. Barth. New York: John Wiley & Sons, pp. 93–116.

Wendisch, M., A. Keil, and A. V. Korolev. 1996. FSSP characterization with monodisperse water droplets. *J. Atmos. Oceanic Technol.* 13:1152–1165.

Wertheimer, A. L. and W. L. Wilcock. 1976. Light scattering measurements of particle distributions. *Appl. Opt.* 15:1616–1620.

Xu, T.-H. and C. Tropea. 1994. Improving the performance of two-component phase Doppler anemometers. *Measure Sci. Technol.* 5:969–975.

Young, B. W. and W. D. Bachalo. 1988. The direct comparison of three "in-flight" droplet sizing techniques for pesticide spray research. In *Optical Particle Sizing*, eds. G. Gouesbet and G. Grehan. New York: Plenum Press, pp. 483–497.

Zaidi, S. H., A. Altunbas, and B. J. Azzopardi. 1998. A comparative study of phase Doppler and laser diffraction techniques to investigate drop sizes in annular two-phase flow. *Chem. Eng. J.* 71:135–143.

Zhang, J., B. Tao, and J. Katz. 1997. Turbulent flow measurement in a square duct with hybrid holographic PIV. *Exp. Fluids* 23:373–381.

Zhu, H. M., T. Y. Sun, and N. Chigier. 1987. Tomographical transformation of Malvern spray measurements. *Atom. Spray Technol.* 3:89–105.

17 Direct-Reading Techniques Using Particle Motion and Optical Detection

PAUL A. BARON

Centers for Disease Control and Prevention*, National Institute for Occupational Safety and Health, Cincinnati, OH

MALAY K. MAZUMDER

Department of Applied Science, University of Arkansas at Little Rock, Little Rock, AR

YUNG-SUNG CHENG

Lovelace Respiratory Research Institute, Albuquerque, NM

INTRODUCTION

Optical direct-reading particle-counting techniques have the advantage of rapid, continuous, nondestructive particle detection. However, the amount of light scattered may not be directly related to the property that one wishes to measure. By combining the advantages of optical detection techniques with the manipulation of particle motion, several instruments have been developed that detect more specific properties of aerosol particles. The aerodynamic size of particles is used to describe the behavior of particles in gravitational settling, filtration, respiratory deposition, sampling systems, and so forth.

Measurement of aerodynamic size at one time could only be achieved by manually observing the settling velocity of individual particles. Subsequently, impactors allowed the measurement of size distributions on a routine basis, although gravimetric and/or chemical analysis still had to be carried out in the laboratory. With the advent of new technology (e.g., lasers and microcomputers), real-time measurements became possible. Several instruments were developed to measure aerodynamic size as rapidly and accurately as possible. These included the Electric–Single Particle Aerodynamic Relaxation Time (E-SPART; ***HOS***)** analyzer, the Aerodynamic Particle Sizer (APS; ***TSI***) and the Aerosizer (***TSI***). While the latter two instruments allow rapid determination of size distributions, they measure particle behavior largely outside the Stokes regime, and the recorded size must be corrected to give an accurate aerodynamic size of individual particles.

*Mention of company or product names does not constitute endorsement by the Centers for Disease Control and Prevention.

** See Appendix I for full manufacturer addresses referenced to the italicized three-letter codes.

Aerosol Measurement: Principles, Techniques, and Applications, Second Edition, Edited by Paul A. Baron and Klaus Willeke.

ISBN 0-471-35636-0

While the aerodynamic diameter describes the inertial properties, the electrostatic charge influences the electrodynamic behavior of the particle in transport processes. Both aerodynamic diameter and electrostatic charge measurements on individual particles are needed in many electrodynamic processes; some examples are electrophotography and laser printing, electrostatic powder coating, electrostatic precipitation, electrostatically enhanced fabric filtration, and electrostatic beneficiation of minerals and coal. The E-SPART is capable of measuring particle charge as well as aerodynamic diameter.

Airborne asbestos fiber measurements went through a similar progression over time in that, originally, relatively crude measurements of concentration were made by collection with midget impingers and microscope counting of all large particles. Filter collection with microscopic analysis was developed so that only fibers were detected. Finally, the development of the Fibrous Aerosol Monitor (model FAM-1, ***MIE***) allowed continuous, real-time detection of airborne fibers. The FAM-1 was designed to give results close to those of the phase contrast light microscopic method (see Chapters 12 and 26).

These sophisticated instruments provide more specific data about aerosols; however, because of the complexity of their detection and analysis systems, they may also have various limitations and subtle problems associated with the interpretation of the data. The following sections present a discussion of these instruments.

ELECTRIC–SINGLE PARTICLE AERODYNAMIC RELAXATION TIME ANALYZER

Measurement Principles

The Electric Single Particle Aerodynamic Relaxation Time (E-SPART) analyzer can be operated in several modes, the first of which is the original SPART mode (Mazumder and Kirsch, 1977; Mazumder et al., 1979). The SPART analyzer determines aerodynamic diameter by subjecting particles to an acoustic field of frequency f and measuring the response of these particles to the acoustic excitation. In its typical sampling configuration, the aerosol is sampled in a laminar air flow moving vertically downward through the SPART's sensing volume (Fig. 17–1). The acoustic field induces an oscillatory velocity component to the particle motion in the horizontal direction. The inertia of each particle causes a phase lag ϕ in the particle's periodic motion with respect to the acoustic field. This phase lag ϕ is related to the relaxation time of the particle τ_p, which is a function of the aerodynamic diameter d_a of the particle. The SPART analyzer employs a differential laser Doppler velocimeter (LDV) to measure the oscillatory velocity component (in the horizontal direction) of the particle and a microphone to measure the acoustic field. The phase lag of the particle motion with respect to the acoustic field driving the particle is converted to aerodynamic diameter using a microcomputer. The microcomputer stores the aerodynamic size data for the particles sampled and provides the measured size distribution. Although the response of the SPART can be calculated theoretically, it is calibrated with monodisperse latex particles because some instrumental parameters are difficult to measure.

In the E-SPART analyzer, an electrical particle acceleration field is used. There are two configurations: (1) a dc electric field superimposed on the acoustic field (Mazumder et al., 1983) and (2) an ac electric field replacing the acoustic field (Renninger et al., 1981). In the first configuration (i.e., an acoustic E-SPART analyzer), the horizontal motion of a charged particle is caused by the superposition of two fields: (1) the acoustic field forcing the particle in an oscillatory motion and (2) a dc electric field inducing a migration velocity component that depends on the polarity and magnitude of the electrical charge q of the particle and the field strength. In this configuration, d_a is measured for either electrically charged or uncharged particles by determining the phase lag ϕ. Electrical charge q is determined from

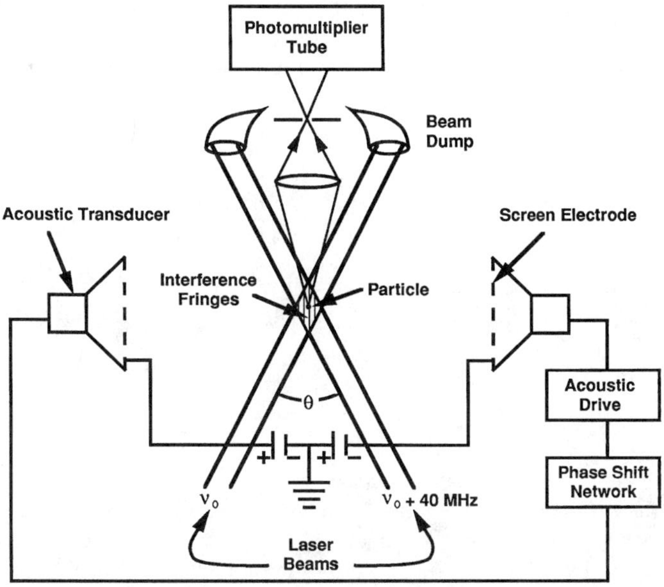

Fig. 17–1. Aerosol relaxation chamber showing the geometrical configuration of acoustic transducers and electrodes for applying acoustic and electric fields in the Electric–Single Particle Aerodynamic Relaxation Time (E-SPART) analyzer. One of the illuminating laser beams has been shifted by 40 MHz.

the measured electrical migration velocity V_e and the aerodynamic diameter d_a. The direction of V_e, which is also measured, provides the polarity of q.

In the second E-SPART configuration, no acoustic field is used, and the particles are subjected to an ac electric field. Therefore, the measurement process is applicable only to electrically charged particles. If a particle is charged, it experiences an oscillatory motion caused by the applied ac electric field. This oscillatory velocity component of the particle will have a phase lag ϕ with respect to the applied electric field, and measurement of ϕ again allows determination of d_a. The amplitude of the oscillatory velocity component of the particle is directly proportional to the electric charge on the particle. Thus, from the measurement of the velocity amplitude V_p and the phase lag ϕ, the electrostatic charge for individual particles can be calculated. There is a phase shift of 180° for particles of opposite polarity, and this 180° phase shift is detected to determine the polarity of the electrical charge. Both the phase lag and the amplitude information are obtained by the LDV and the associated signal processing electronics. More recently, applications of acoustic and electric drives to the measurement of particle size, charge, and density have been described (Cole, 1999; Cole and Tennal, 1993).

Particle Motion: External Oscillating Force

The oscillation of a particle experiencing a sinusoidal force F_e in a gaseous medium was derived by Stokes and can be represented in the Stokes regime by the following equation (Fuchs, 1964:80):

$$m_r \frac{dv_p}{dt} + \frac{v_p}{B_r} - F_e = 0 \tag{17–1}$$

where m_r is the effective mass of the particle, B_r is the effective mobility of the particle, v_p is the time-dependent (instantaneous) particle velocity, and F_e is the external force. Because particle velocity is not constant, Stokes showed that m_r and B_r can be replaced by

$$m_r = m_p + \frac{9m'\xi}{4} \tag{17–2}$$

$$B_r = \left(\frac{3\pi\eta d_p}{C_c} + \frac{9m'\omega\xi}{4}\right)^{-1} \tag{17–3}$$

where m_p is the particle mass, η is the gas viscosity, C_c is the Cunningham slip correction factor, m' is the mass of air displaced by the particle, ω is the angular frequency of oscillation ($2\pi f$), and

$$\xi = \frac{2}{d_p\sqrt{2\nu/\omega}} \tag{17–4}$$

where ν is the kinematic viscosity (η/ρ_g), and ρ_g is the gas density. Equation 17–1 is a simplified version of the original Stokes equation for $m_p >> m'$. When the frequency approaches zero, that is, the velocity approaches a constant, ξ approaches zero and the effective particle mass can be replaced by m_p, B_r is reduced to the mobility B, and

$$F_r = -3\pi\eta d_p(V_p - U_g)/C_c \tag{17–5}$$

where V_p is the particle velocity and U_g is the steady-state gas velocity. This is the more familiar form of the Stokes equation.

Particle Motion: Acoustic Field. Equation 17–1 describes the motion of a particle under the influence of an external field, for example, a charged particle in an ac electric field. When there is no external field, but the medium itself is oscillating, for example, the particle experiences an acoustic excitation, the time-dependent gas velocity u_g can be expressed as

$$u_g = U_g \sin(\omega t) \tag{17–6}$$

where U_g is the maximum gas velocity and t is time. The parameters measured by the instrument are the velocity amplitude ratio

$$\frac{V_p}{U_g} = \sqrt{\frac{1 + 3\xi + \frac{9}{2}\xi^2 + \frac{9}{2}\xi^3 + \frac{9}{4}\xi^4}{\alpha^2 + 3\alpha\xi + \frac{9}{2}\xi^2 + \frac{9}{2}\xi^3 + \frac{9}{4}\xi^4}} \tag{17–7}$$

and the phase lag of the particle behind the air motion, given by

$$(\phi - \theta) = \tan^{-1}\left(\frac{3(\alpha - 1)\xi(\xi + 1)}{2\left(\alpha + \frac{3}{2}\alpha\xi + \frac{3}{2}\xi + \frac{9}{2}\xi^2 + \frac{9}{2}\xi^3 + \frac{9}{4}\xi^4\right)}\right) \tag{17–8}$$

where

$$\theta = \tan^{-1}\left(\frac{\frac{2}{3} + \xi}{\xi(\xi + 1)}\right) \tag{17–9}$$

and

$$\alpha = \frac{2\rho_p}{3\rho_g} \tag{17–10}$$

The right-hand sides of Eqs. 17–7 and 17–8 are described by the terms α, ξ, and ν, which are functions of the particle and gas properties. Figures 17–2 and 17–3 show $(\phi - \theta)$ and V_p/U_g, respectively, plotted as a function of aerodynamic diameter for several acoustic drive frequencies.

If the inertial terms caused by the acceleration of the particle in the medium are neglected, then particle motion can be represented using the particle relaxation time τ_p

$$\tau_p = \frac{\rho_o d_a^2 C_c}{18\eta} \tag{17–11}$$

where ρ_o is standard density (1000 kg/m^3 [1 g/cm^3]). The phase lag ϕ of the particle behind the air motion in the acoustic field is given by

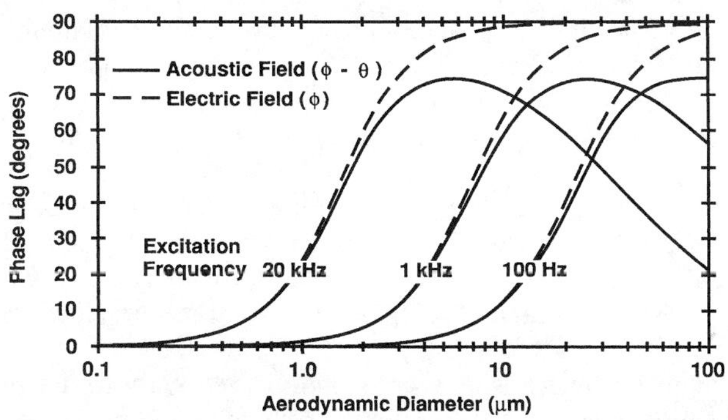

Fig. 17–2. Phase lag of the particle motion with respect to an acoustic excitation field (Eq. 17–8) and an electrostatic excitation field (Eq. 17–12) plotted as a function of aerodynamic diameter for several drive frequencies.

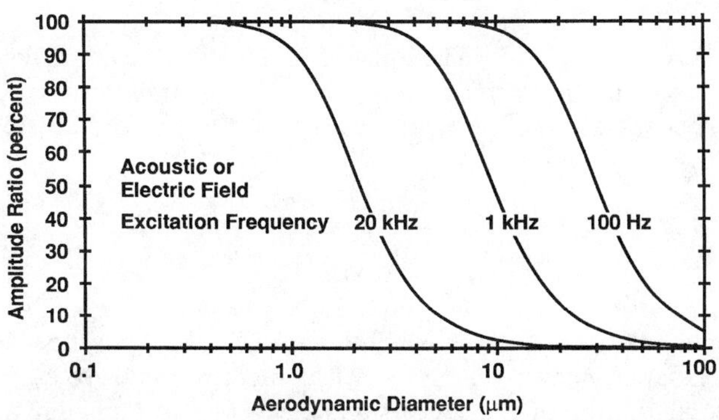

Fig. 17–3. Amplitude ratio of the particle motion with respect to an electrostatic field drive plotted as a function of aerodynamic diameter for several drive frequencies for the E-SPART (Eq. 17–7). The simpler equation (Eq. 17–13) gives similar curves throughout the entire range of sizes.

$$\phi = \tan^{-1}(\omega\tau_p) \tag{17–12}$$

where $\omega/2\pi$ is the acoustic frequency. The phase lag calculated using this equation is also plotted in Figure 17–2. Under this condition, the ratio of the amplitude of particle velocity V_p to the amplitude of the gas motion U_g due to the acoustic field is

$$\frac{V_p}{U_g} = \frac{1}{\sqrt{1+\omega^2\tau_p^2}} \tag{17–13}$$

Equations 17–11 and 17–12 can also be applied when the particle motion is induced by an external field, such as with a charged particle in an ac electric field. These equations, as well as Eqs. 17–7 and 17–8, indicate that the measurement of either the phase lag of the particle motion relative to the gas motion or the velocity amplitude ratio of the particle in the acoustic field is sufficient to determine τ_p or d_a.

In the case of acoustic excitation, there are two major forces acting on the particle besides gravitational field: (1) the viscous drag force and (2) the force caused by the pressure gradient in the medium. The first one is caused by the fluid resistance due to the viscosity of the medium, and the second is due to the inertial resistance. The effective fluid resistance will depend on the product $\omega\tau_p$. For small values of this product, the fluid resistance is primarily viscous, and for large values it is inertial. For sizing aerodynamic diameter, the product $\omega\tau_p$ can vary from 0.01 to 100. In the range 0.01 to 2, the resistance can be approximated by the viscous drag (Eq. 17–12), and, when $\omega\tau_p > 2$, both viscous and inertial resistance need to be considered (Eq. 17–8). Note that for $d_a \leq 100\,\mu m$, particle Reynolds number Re_p is less than 0.1.

The value of $\omega\tau_p = 2$ corresponds to $\phi = 63.5°$. For $\phi \leq 63.5°$, Eq. 17–12 gives results for an acoustic excitation within 15% of Eq. 17–8. For larger values of phase lag, the two curves are quite different (Fig. 17–2). However, the amplitude ratio curves given by Eqs. 17–13 and 17–7 are nearly identical (within less than 1%, Fig. 17–3), and, therefore, the simplified Eq. 17–13 holds over the entire range of $\omega\tau_p$.

To operate the analyzer over a wide size range, it is necessary to use two frequencies of excitation either in tandem inside a single relaxation chamber or simultaneously in two relaxation chambers connected in series. For example, a prototype E-SPART analyzer has been operated at two frequencies, 24 kHz (for 0.3 to 4.0 µm) and 1.0 kHz (for 2.0 to 20.0 µm), using two chambers connected in series. Integration of the experimental data can be accomplished with the appropriate software. The aerosol sampling system with only one relaxation chamber used in the E-SPART analyzer, is shown in Figure 17–4.

Particle Motion: dc Electric Field. When placed in a constant electric field E, the electrostatic force on a charged particle can be expressed as $F_e = qE$, where q is the particle charge. A particle of diameter d_a with n elementary charges will move with an electrical migration velocity V_e given by (see Chapter 3)

$$V_e = \frac{neEC_c}{3\pi\eta d_a} \tag{17–14}$$

where e is the elementary charge. As shown in Figure 17–3, the acoustic velocity component $V_p \sin(\omega t - \phi + \theta)$, is superimposed on this electrical migration velocity V_e. Figure 17–1 shows the geometrical configuration of transducers and electrodes for applying acoustic and electric fields. The field E is calculated from the voltage applied across the electrodes divided by the distance between them. Hence, a measurement of V_e can be used to calculate n, the number of elementary charges, once the aerodynamic diameter of the particle has been deter-

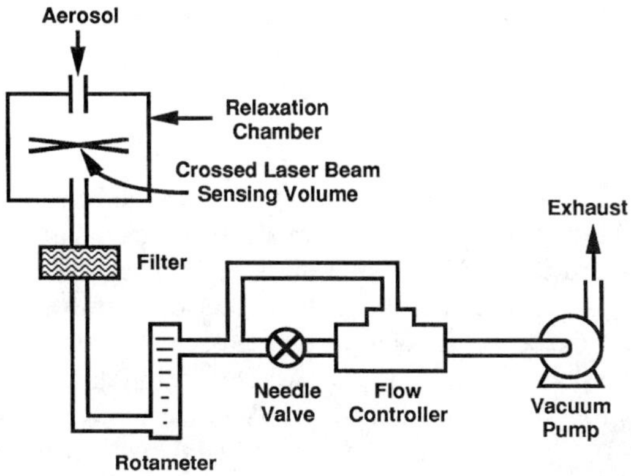

Fig. 17–4. A schematic of the aerosol sampling system used in the E-SPART analyzer.

mined. The software performing the charge measurement reads the voltage (which is adjustable) applied across the electrodes and computes the field E for determining the magnitude of the charge $q(ne)$ for each particle. The analyzer also recognizes the direction of V_e, which depends on the polarity of the charge q of the particle. Thus, from the direction and magnitude of V_e, the computer can record both polarity and magnitude of particle charge.

Particle Motion: ac Electric Field. The phase lag measurement technique can also be used on a charged particle in an ac E-SPART analyzer with the same electrode configuration shown in Figure 17–1 with no acoustic field applied. An electrical sinusoidal voltage $V_o \sin(\omega t)$ is applied across the two electrodes. In this process, it is necessary that the particles be electrically charged in order to make size and charge measurements. When a charged particle transits the LDV sensing volume, the particle will experience an oscillatory electric field, $E_o \sin(\omega t)$, and a zero gas velocity. When the time $t >> \tau_p$, ϕ has the same expression as Eq. 17–12, and the amplitude ratio is

$$\frac{V_p}{E_o} = \frac{qC_c}{3\pi\eta d_a\sqrt{1+\omega^2\tau_p^2}} \qquad (17\text{–}15)$$

Equations 17–12 and 17–15 indicate that, for a charged particle, d_a can be determined from the measured value of the phase lag or the amplitude ratio as with the acoustical SPART analyzer.

In the case of electric excitation, there are two major forces acting on the particle besides the gravitational field: (1) the coulombic force and (2) the viscous drag force. For $Re_p < 0.1$, which is the case for many practical applications, the inertial resistance of the fluid can be neglected, and the phase lag is given by Eq. 17–12. Unlike acoustic excitation, there is no pressure gradient force, and hence there is no foldover in the phase lag relationship (Fig. 17–2). The amplitude ratio, in the case of either electric or acoustic excitation, is given by Eq. 17–13 without significant error. When $Re_p > 1$, appropriate correction will be needed to compute the viscous and inertial resistance forces acting on the particle for accurate size and charge measurements.

In the E-SPART analyzer, the measurement of d_a is independent of the driving field amplitude E_o and the magnitude of the particle charge q. Once d_a is determined from ϕ, the analyzer then calculates the electrical mobility $(q \cdot \mathrm{B})$ or the electrostatic charge q of the particle from Eq. 17–15. The phase lag measurement technique is independent of the amplitude of the driving force as long as the particle amplitude is sufficiently large for accurate measurement of ϕ.

E-SPART Analyzer

The E-SPART analyzer consists of four components: (1) a dual-beam, frequency-biased laser Doppler velocimeter; (2) a relaxation cell; (3) an electronic signal and data processing system; and (4) a personal computer. The LDV measures the particle velocity. The sensing volume of the LDV is formed by the intersection of the two laser beams and is located between the electrodes as shown in Figure 17–1. As a particle passes through the sensing volume in the direction normal to the plane containing the two converging laser beams, the particle experiences an acoustic and/or an electric field. The LDV detects only the horizontal velocity component of the particle. It does not detect the vertically downward sampling velocity. However, the duration of the LDV signal burst is the residence time of the particle within the sensing volume, and it is inversely proportional to the sampling velocity. The residence time must be long enough to measure $(\phi - \theta)$ and V_e or ϕ and V_p/E_o. The residence time is discussed further in "Aerosol Sampling."

A helium-neon laser (632 nm) or an argon-ion laser (488 nm) is used as the monochromatic light source for the LDV. The choice of laser (HeNe or Ar+) and the output power (10 mW for HeNe or 50 to 500 mW for Ar+) depends on the application. Two output laser beams are derived by passing the laser beam through an acousto-optic cell (Bragg cell) modulator. The output beams have nearly equal power, but one of the beams is shifted in frequency by 40 MHz by the modulator (Figure 17–1). The two beams intersect at the focal volume within the relaxation cell. Light scattered from aerosol particles passing through the sensing volume is collected by the receiving lenses and focused to a pinhole directly in front of a photomultiplier tube. The output of the photomultiplier is an electrical signal that represents the Doppler burst containing the particle motion information.

Aerosol Sampling

Figure 17–4 shows a schematic of the flow control system used for the E-SPART analyzer. The aerosol sample is drawn into the relaxation chamber by using a vacuum pump. A differential flow controller is used to maintain a constant rate of sampling flow approximately 0.5 L/min through the relaxation chamber. The sampling rate through the LDV sensing volume is a few milliliters per minute, depending on the specific optical configuration and the diameter of the particle. If the length of the sensing volume is L and the sampling velocity of the aerosol particles is V_z, then the maximum residence time of the particle in the sensing volume will be L/V_z. This residence time is set equal to $N \cdot t$, where N is the number of acoustic cycles and t is the time period of the acoustic or electric excitation. Under this condition, an aerosol particle passing vertically downward through the center of the sensing volume will undergo periodic motion for N excitation cycles. Phase lag ϕ measurements on each individual cycle for this particle can be performed by the E-SPART analyzer, and the average value of ϕ over N cycles is used to determine d_a.

The choice of N depends on three factors: (1) the size resolution desired, (2) the response time of the signal-processing electronics, and (3) the particle concentration. Typically, N is set between 3 and 8 by adjusting the sampling velocity V_z. Because the time period t depends on the frequency f of the acoustic or electric excitation drive, the residence time for particles is varied depending on f and the size measurements. All the particles may not pass through the center of the sensing volume, resulting in a shorter sensing time; however, the particle must stay in the volume for at least one cycle to be measured.

It is essential that a laminar flow field is maintained as the aerosol sample passes through and around the LDV sensing volume. Velocity components of the particle in the x direction in the absence of any acoustic and electrical excitation should be less than the particle's Brownian motion.

Signal and Data Processing Electronics

The E-SPART signal-processing electronics are organized into five functional sections: (1) a receiver containing the RF amplifier, mixer, and demodulator; (2) the signal conditioning circuitry; (3) the size and charge measurement circuitry; (4) a direct memory access board for interfacing with the computer; and (5) a personal computer.

The instantaneous Doppler signal frequency generated by a particle traversing the sensing volume in the absence of any excitation is given by f_o, which is the LDV bias frequency (40 MHz) as determined by the Bragg cell. When a particle experiences an acoustic excitation of frequency f and a dc electric field, the Doppler frequency f_D is

$$f_D = f_o + \Delta f + \frac{2}{\lambda} V_p \sin(\omega t - \phi)\sin(\Theta/2) \tag{17–16}$$

where Θ is the intersection angle of the two laser beams, λ is the laser radiation wavelength, and ω is $2\pi f$. It is assumed ϕ is <63.5° in Eq. 17–16. A similar equation can be used when ac electric field excitation is used. The carrier frequency shift Δf is

$$\Delta f = \frac{2neEC_c}{3\pi\eta d_a \lambda}\sin(\theta/2) \tag{17–17}$$

where n is the number of electronic charges on the particles and E is the dc electric field.

In an acoustic E-SPART, the phase lag ϕ of the particle motion is determined from the time interval between the zero crossings of the acoustic field and the resultant particle motion (Fig. 17–5). The d_a for the particle is computed from f as follows:

$$d_a = \sqrt{\frac{9\eta \tan(\phi)}{\pi f \rho_o C_c}} \tag{17–18}$$

and from Eq. 17–17,

$$ne = q = \frac{3\pi \Delta f \eta d_a \lambda}{2EC_c \sin(\Theta/2)} \tag{17–19}$$

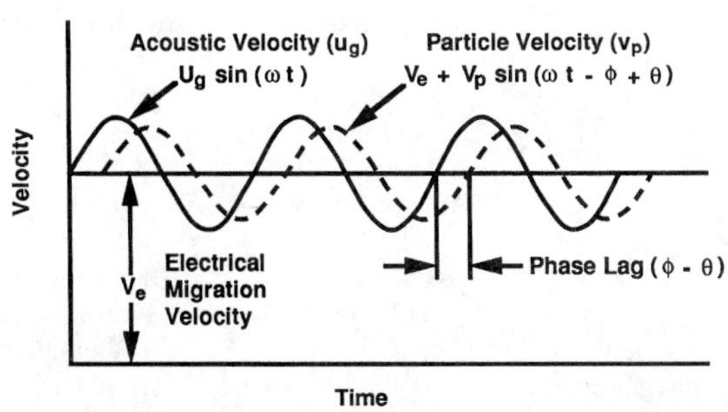

Fig. 17–5. Wave forms of the motion of charged particles within the sensing volume of the E-SPART analyzer.

In the case of the ac E-SPART, an ac electric field $E_o \sin(\omega t)$ replaces the dc electric field and the acoustic excitation. d_a is determined from ϕ, and q is calculated from the ratio of the amplitude of the particle motion (V_p) to the amplitude of the ac electric field E_o as shown in Eq. 17–15. The maximum value of the frequency deviation is

$$\Delta f_o = 2V_p \sin(\Theta/2)/\lambda \qquad (17\text{–}20)$$

and

$$ne = q = \frac{3\pi|\Delta f_o|\eta d_a \lambda\sqrt{1+\omega^2\tau_p^2}}{2EC_c \sin(\Theta/2)} \qquad (17\text{–}21)$$

The frequency deviation Δf_o, with respect to f_o, can be either positive or negative depending on the polarity of the charge. The magnitude of the charge is determined from $|\Delta f_o|$.

Table 17–1 shows a comparative analysis of two drive systems for the E-SPART analyzer: (1) Acoustic and DC Electric Drives and (2) AC Electric Drive. For such powder applica-

TABLE 17–1. Advantages and Disadvantages of Different Excitation Methods in the E-SPART Analyzer

Operational Features	Acoustic and DC Drive		AC Drive	
	Measurement of d_a	Measurement of q	Measurement of d_a	Measurement of q
Range of operation	From ϕ measurement in the range 0° to 70° (d_a can be measured from amplitude ratio [V_p/U_g])	From V_{TE} measurement in the range 0 to $\pm\, q_{max}$	From ϕ measurement in the range 0° to 90°	From V_p/E_o measurement in the range 0 to $\pm\, q_{max}$
Need for corrections	Stokes law does not remain valid when $\phi > 70°$ (no corrections are needed if amplitude ratio measurements are used)	Particle Reynolds number (*Re*) may exceed 1 for highly charged large particles	Stokes law can be applied without significant error	Particle Reynolds number (*Re*) does not exceed 1, even for the highly charged large particles
Counting efficiency and sampling error	Applicable to both charged and uncharged particles	Highly charged particles may be deflected away from the sensing volume	Applicable only to charged particles	No sampling loss caused by excitation
Change of size range/noise immunity	Change of acoustic drive frequency to change size range may need adjustments in electrode spacing	Flow turbulence and acoustically generated flow field affect q/m measurement	Range of operation can be changed continuously by changing frequency of the ac drive	No acoustically generated flow field noise. Measurement of q is insensitive to flow turbulence

tions as toners and powder coatings, the AC Drive System is more convenient for operating the instrument over a wide particle size range (Mazumder et al., 1999).

Instrument Operation

A data summary can be obtained that will provide the total number of particles counted with the average charge tabulated for a given size channel for both positive and negative charged particles. The software can provide plots of the size distribution of the aerosol in terms of number (Fig. 17–6), cumulative number, volume, cumulative volume, as well as statistics including count median diameter, mass median diameter, and geometric standard deviation.

For each particle, the aerodynamic diameter (d_a) and the charge (q) are determined in the E-SPART analyzer and the average value of the charge-to-mass ratio computed. For a spherical particle of diameter d_p and specific gravity ρ_p, we can write an approximate relationship:

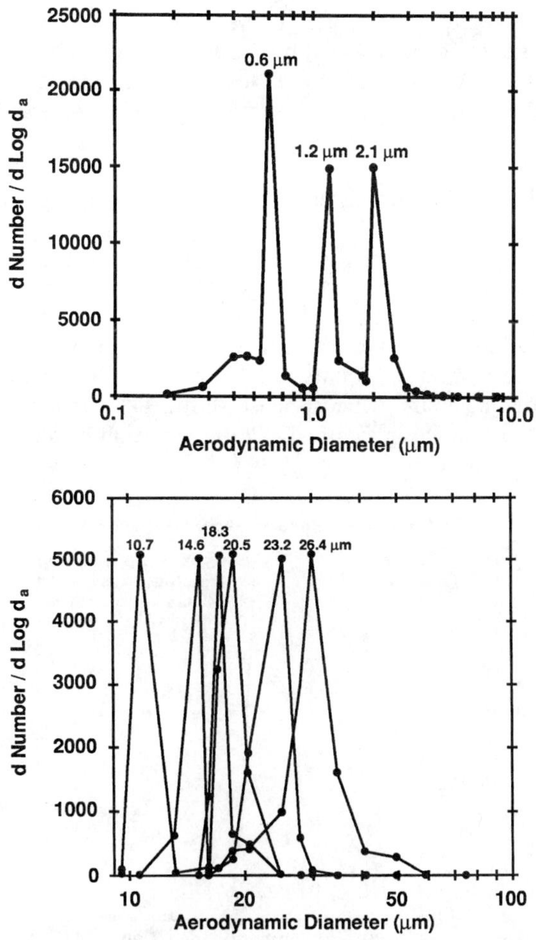

Fig. 17–6. a, Size frequency distribution $d\,N/d\log d_a$ for a mixture of polystyrene latex (PSL) spheres with diameters of 0.6, 1.2, and 2.1 µm (acoustic E-SPART). **b**, Size frequency distribution d $N/\text{d}\log d_a$ for monodisperse bis-ethyl hexyl sebacate droplet aerosols of different sizes (ac E-SPART). The data were obtained to give equal peak concentrations.

$$d_p^2 \rho_p = d_a^2 \rho_o \tag{17–22}$$

The mass m_p of the particle can be computed from the measured value of d_a, if ρ_o is known. Thus,

$$m_p = \pi \rho_o^{3/2} d_a^3 / (6 \rho_p^{1/2}) \tag{17–23}$$

For each size channel $(d_a)_i$, from $i = 1$ to $i = 32$, the particle count is stored as n_i; m_p for each channel is approximately $\pi n_i \rho_o^{3/2} d_a^3 / (6 \rho_p^{1/2})$. The total mass of the particle sample is given by summing over i channels:

$$m = \sum (m_p)_i = \frac{\pi \rho_0^{3/2}}{6 \rho_p^{1/2}} \sum n_i (d_a)_i^3 \tag{17–24}$$

For each size channel, $(d_a)_i$, the total count n_i is also stored in the charge channels. The sums are performed over all 32 channels. The number of charged particles, n_i, is equal to $n^o + n_i^+ + n_i^-$ where n^o, n^+, and n^- represent the total number of particles with zero, positive, and negative charges with diameter $(d_a)_i$, respectively. The software provides computations and plots of n_i^+, n^o, n_i^-, versus charge-to-mass (q/m) ratio for any channel $(d_a)_i$, as shown in Figure 17–7. A three-dimensional plot of number versus charge-to-mass ratio and d_a is also available (Fig. 17–8).

Size Resolution. Particle size is measured by determining the phase lag ϕ between the particle motion and the driving force (acoustic or electric). In practice, a time interval Δt is measured, and the relationship between the phase lag ϕ and Δt is given by $\Delta t = \phi/\omega$. The signal-processing electronics determines Δt by generating a phase comparator pulse with the duration Δt and then counting the time period of that pulse using a counter of frequency f_c. To obtain good resolution, the counter frequency f_c is made many times larger than the excitation frequency f. The number of counts n_c for a given time interval Δt can be written as $n_c = \Delta t\, f_c$. Because the maximum value of phase shift is 90°, the maximum count will occur for a 90° phase shift. As shown in Figure 17–2, the variation of phase shift with respect to

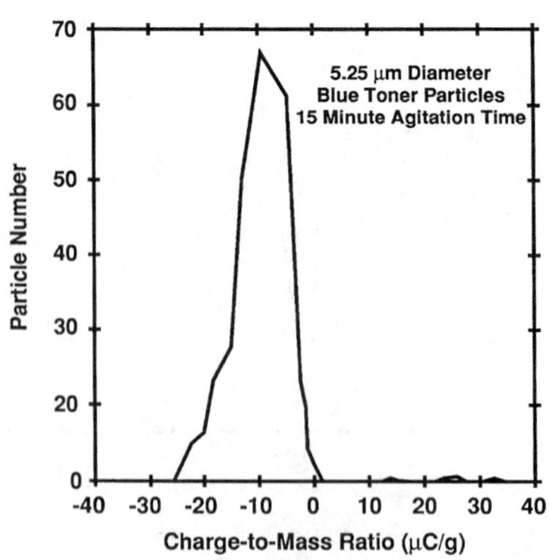

Fig. 17–7. Electrostatic charge distribution of a sample of blue toner showing the variation of charge density for a selected aerodynamic diameter.

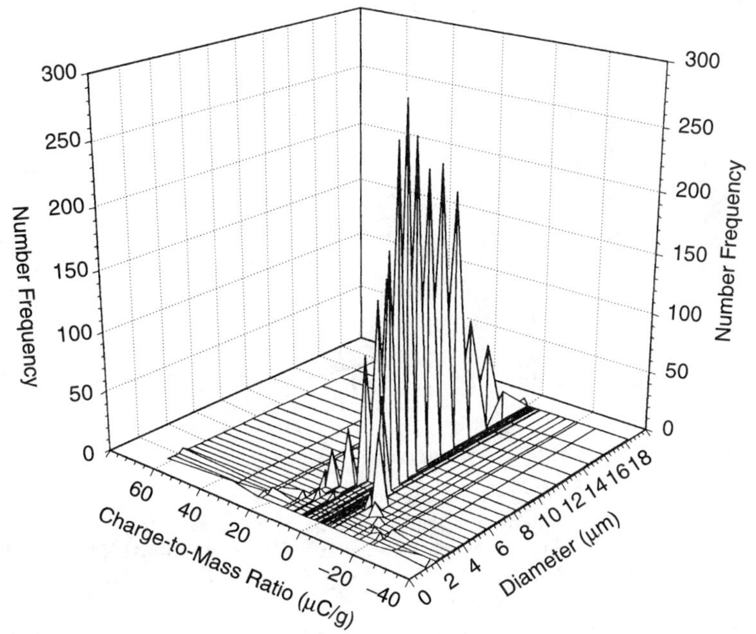

Fig. 17–8. A three-dimensional representation of the variation of particle count and charge density as a function of aerodynamic diameter.

aerodynamic diameter $d\phi/d(d_a)$ depends on the frequency of operation. For particles with the Cunningham correction factor equal to 1, that is, $d_a > 2\,\mu m$, maximum resolution (Renninger et al., 1981) is obtained when $d\phi/d(d_a)$ is maximum, which gives the phase angle $\phi = 30°$ or $\omega\tau_p = 1/\sqrt{3}$ for maximum resolution. In terms of channel number, the greatest resolution occurs when n_c is approximately equal to one third of the maximum count n_m. For example, it is possible to easily distinguish particle size differences between singlets, doublets, and triplets of 1.01 μm PSL particles.

Charge Resolution. The resolution in measuring particle charge also depends on the parameters of the instrument and the magnitude of the electric field. High-resolution charge measurement is needed to study the Boltzmann distribution of the particle charge. In such applications, a high electric field (dc) was used in the acoustic E-SPART, and the charge distribution was resolved within ±2.0 electronic charge, with the average charge varying from 0 to 50 electronic charges for 0.8 μm diameter monodisperse particles. For charged polydispersed particles, q/m measured using a Faraday cage agreed well with the q/m measured using the E-SPART. The resolution of the instrument is difficult to determine because there is no readily available process to control q accurately or any other instrument to measure the exact value of charge based on a single particle.

Range of Operation. The E-SPART analyzer can be operated in a range of 0.3 to 75 μm in aerodynamic diameter. The commercial instrument is operated at a single frequency. However, to cover this entire range, it is necessary to modify the analyzer to operate with at least two different frequencies of excitation, 25 and 0.5 kHz, in tandem or simultaneously. For high-resolution sizing of particles, it is necessary to operate the analyzer at three frequencies: 25, 2.0, and 0.5 kHz. The maximum particle count rate (10 particles/s to 2000 particles/s) depends on the frequency of operation.

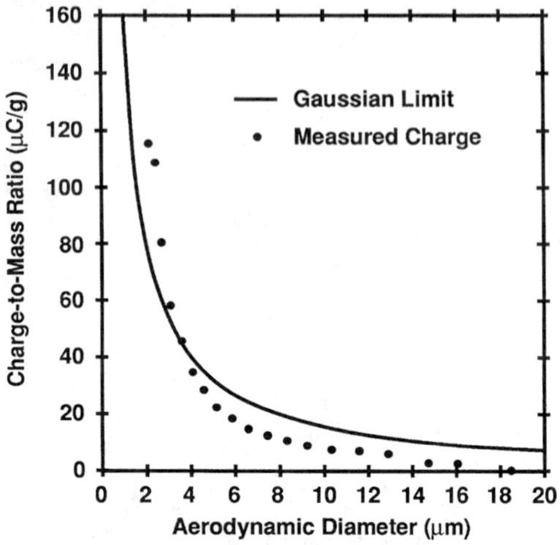

Fig. 17–9. The charge distribution of a tribo-charged toner sample as measured by an E-SPART analyzer. The solid line shows the saturation charge calculated from the Gaussian limit. The experimental data show that the E-SPART analyzer can measure particle charge near the saturation limit.

The desired range of measurement of electrostatic charge on each particle is from zero charge to its saturation value with positive or negative polarity. Figure 17–9 shows the average values of the charge-to-mass ratio of a toner sample experimentally measured as a function of d_a and the calculated Gaussian limit for maximum q/m as a function of diameter (Mazumder et al., 1991). The particles were tribo-charged. The saturation charge-to-mass ratio for tribo-charged, dielectric solid particles varies inversely with particle diameter. The data show that the analyzer can measure particles with saturation charge. For highly charged particles, the AC drive should be used to minimize sampling losses.

Precision and Accuracy. The basic principle applied in the E-SPART analyzer can provide absolute measurements of particle size and electric charge if the physical parameters involved are accurately known. For example, aerodynamic diameter depends on the viscosity of the gas in which the particles are suspended. Because viscosity is independent of pressure, the size measurement can be performed at different ambient pressures. However, if the temperature or the constituents of the gas change, the viscosity will change, and, therefore, the measured value of the aerodynamic diameter will be related to the properties of the gas in which the particles are suspended. This is an advantage of the E-SPART analyzer for in situ measurements. However, if there are uncertain variations in the ambient conditions from sample to sample, such as changes in temperature, the instrument's operation will be affected adversely. When the instrument is operated at a relatively high acoustic frequency (25 kHz), it is important to maintain a constant temperature in the relaxation cell so that the phase offset value does not change. This constraint is less severe when the acoustic E-SPART is operated at a frequency of 1 kHz or lower or is absent when an ac excitation is used.

AERODYNAMIC PARTICLE SIZER

The development of the Aerodynamic Particle Sizer (APS; ***TSI***) was based on a particle acceleration nozzle and laser Doppler detection system constructed by Wilson and Liu (1980). In this study particles were introduced into the center of an accelerating nozzle. Small particles followed the motion of the air closely while larger particles lagged behind, causing an increase in relative velocity between air and particle. This increase in relative velocity is analogous to the increase in settling velocity with particle d_a. Wilson and Liu (1980) indicated that the particle velocity is a function of d_a as long as Re_p stays small (within the Stokes regime, $Re_p < 0.1$). As Re_p increases, apparent particle size becomes a function of particle density and shape as well as d_a. In addition, there is a trade-off in size resolution and nozzle velocity. At high nozzle velocities, particle motion is more non-Stokesian (less accurate aerodynamic sizing) but particle sizing is more rapid. Artifacts in the observed aerodynamic distribution may also occur because the light scattering used for detection of particles results in incomplete detection or particle coincidence.

Based on similar principles, ***TSI*** developed the APS with support from the National Institute for Occupational Safety and Health (NIOSH) (Agarwal and Remiarz, 1981). The APS sizes particles by measuring their velocity relative to the air velocity within an acceleration nozzle. This velocity is compared with a calibration curve established using monodisperse spheres. Several commercial models have been available over the past two decades:

Model 3300	Based on prototype design; used an Apple II computer for data analysis. HeNe laser light source (commercially discontinued)
Model 3310	Updated version of 3300 using an IBM PC-compatible computer for data analysis. Improved data display software. (commercially discontinued)
Model 3320	Particle acceleration system identical to other APS models, but using a solid-state laser and redesigned sensor package with integrated particle spectrum readout. Further data analysis and logging possible using external computer. Lower size resolution and better coincidence rejection than 3310. Light-scattering intensity data available
Model 3312	Particle sizing capability identical to Model 3320 (UV-APS), but with ultraviolet fluorescence and optical scattering signal information also available. Designed specifically for biologically based aerosols (also termed Fluorescence APS or FLAPS)

Instrument Description

The APS 3310 consists of a sensor unit containing the sampling system, detector, preliminary processing electronics and internal flow rate indicator, and a computer. The computer receives the data from the sensor unit about once per second and updates the calculated aerodynamic size distribution. One version of the software that collects and displays the data comes with the instrument; more sophisticated software providing near-real-time display is provided. While the entire unit is sufficiently portable and rugged that it can be used for field measurements (Baron, 1986; Baron and Willeke, 1986; Szewczyk et al., 1992), it is generally more suited to laboratory environments.

The APS 3320 sensing unit is more compact and can be used as a stand-alone unit, having a direct-reading display of the size distribution. The more compact sensor unit size and more stable flow system make it more amenable to field measurements than the 3310. For additional data analysis and recording of the data, a computer must be attached.

The inlet system and nozzle in all four APS models are identical. Aerosol is introduced to the inlet at a flow rate of 5 L/min. Four liters per minute of this flow is removed, passed through a filter, and reintroduced upstream of the acceleration nozzle as sheath air. The remaining 1 L/min aerosol flow is fed through a focusing nozzle, recombined with the sheath air, and accelerated through the final nozzle (Fig. 17–10). The pressure below the nozzle is approximately 100 mm Hg (Chen et al., 1985). The sheath and total flow are controlled by valves and monitored with thermal mass flow meters (3300 and 3310). The 3320 has microprocessor volumetric flow controllers for total and sheath flows.

At the exit of the acceleration nozzle, each particle passes through two light beams. The light comes from an HeNe laser in the 3310, while a laser diode (LD) provides illumination in the 3312 and 3320. The light scattered from the particle causes two pulses to be detected by a photomultiplier, and the time lag between the two pulses, representing time of flight (TOF) of the particle between the two beams, is recorded. Because larger particles have not accelerated to the air velocity in the sensing zone, they are represented by larger time lags. The TOF data are stored in an accumulator in bins representing equal time intervals.

In the 3310, two sets of data are stored: one by the small particle processor (SPP) in 4 ns bin intervals and the other by the large particle processor (LPP) in bin intervals of 66.7 ns. The software gives the option of using just the SPP for particle distributions in the range 0.5 to 15.9 μm, while the LPP can be used to extend that range up to 30 μm. The LPP has anti-coincidence circuitry that virtually eliminates excess counts due to coincidence. When both

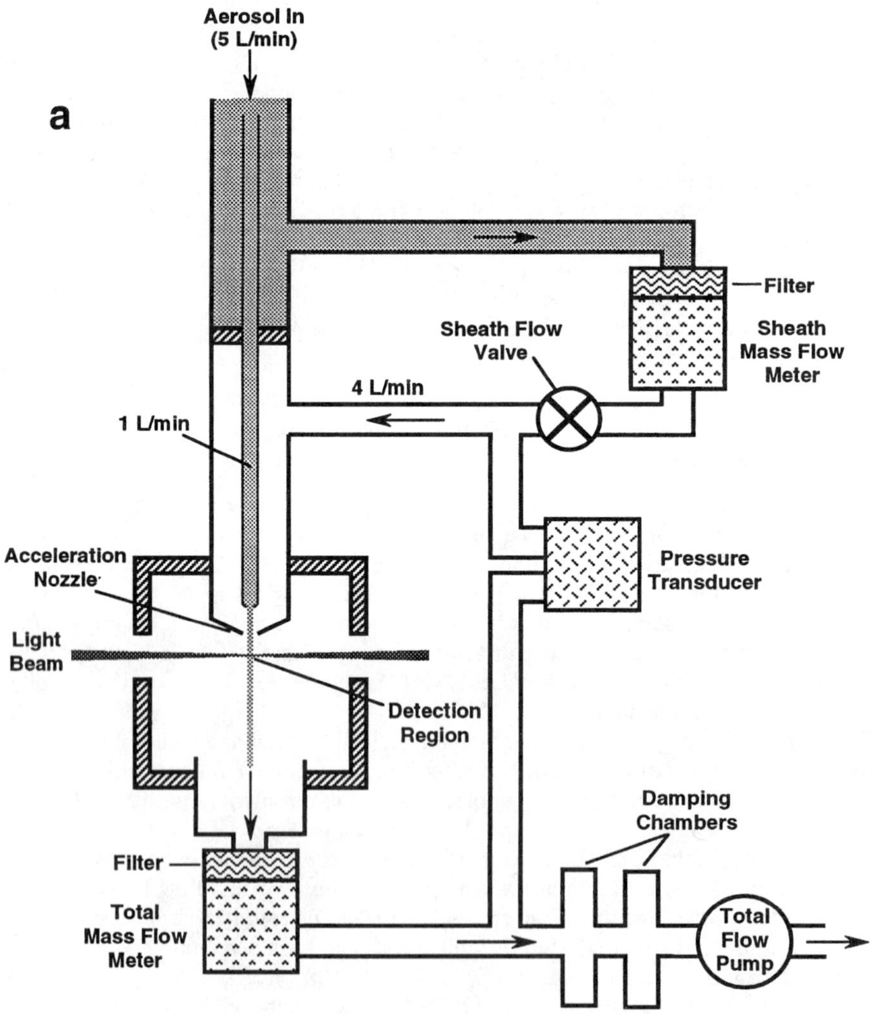

Fig. 17–10. Schematic of Aerodynamic Particle Sizer nozzle and laser velocimeter for models APS 3300 and 3310 (**a**) and for models APS 3320 and 3312 (**b**).

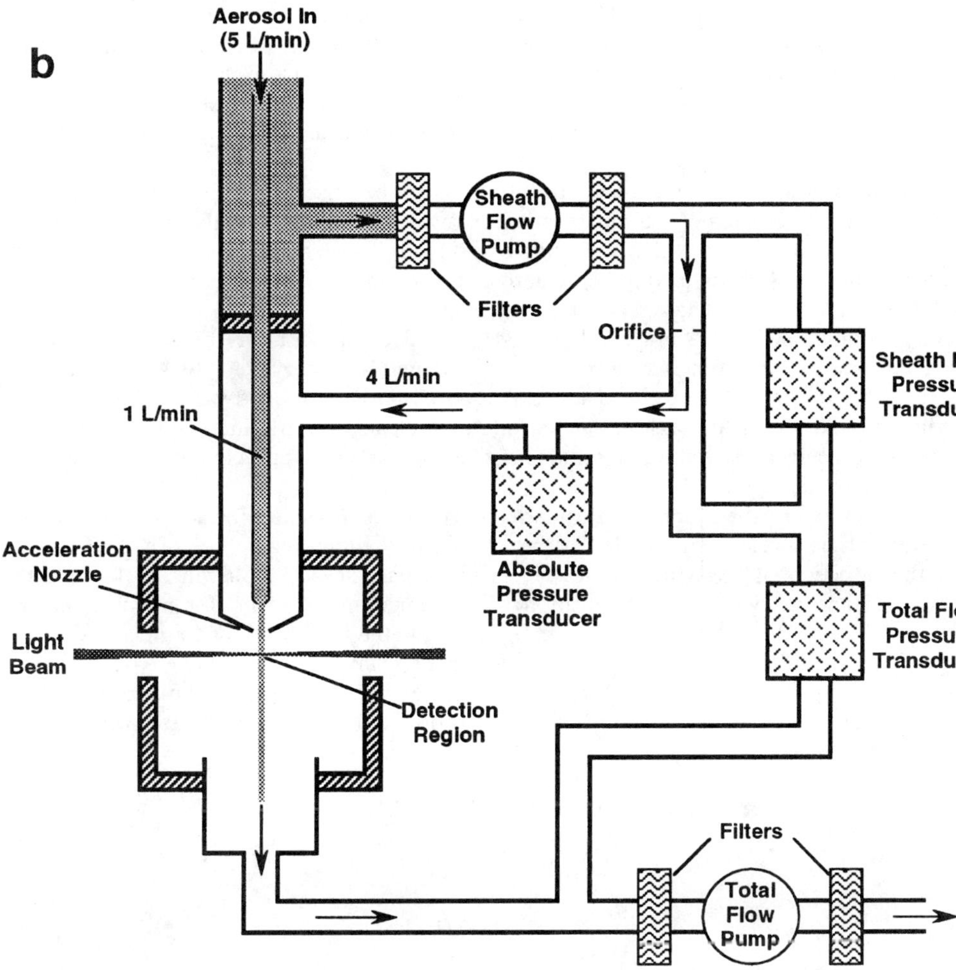

Fig. 17–10. *Continued*

processors are used, the two sets of data are linearly combined in the range from 5.7 to 15.9 μm.

In the 3320 and 3312, the LD illumination beams are relatively broad, and the light pulses overlap for each particle. The pulse shape is differentiated, and the inflection point of each peak defines the center point or mode of that peak. The light-scattering from each peak pair is measured and can be correlated to the TOF, allowing post process rejection of peak pairs whose heights do not correspond to the measured TOF. The light-scattering data are also available as a separate spectrum in the recorded data.

The final aerodynamic size is determined from a calibration of the accumulator spectrum using monodisperse spherical standard density (1000 kg/m^3) reference particles.

Sample Inlet

The 20 mm diameter inlet of the APS is located at the top of the instrument and is not conveniently located for sampling moving air directly. Thus, aerosols are typically ducted to the

APS with external tubing. Particle losses and aerosol nonuniformity within this tubing must be determined separately. Within the inlet, the air is split between the inner inlet (measured aerosol flow) and the sheath flow as indicated in Figure 17–10. The gas velocity at the measured flow inlet is higher than the velocity at the APS inlet (i.e., superisokinetic). This sampling arrangement produces some oversampling of larger particles to compensate for losses within the inner nozzle tube.

Aerosol entering the inlet is assumed to be uniform because only 20% of the aerosol is taken from the inlet stream and measured. Upstream manipulation of the aerosol stream, such as inertial stratification due to bends or cyclones, can bias the concentration at the measurement inlet. Careful mixing of the aerosol upstream of the measurement inlet may be necessary to reduce these effects with minimal losses.

Sample dilution systems are available as optional equipment for the APS to reduce problems with particle coincidence in the sensor. Penetration curves for these dilutors are measured by the manufacturer and are provided as part of the software to correct size distributions. At 15 μm the loss within the dilutor is near 50% and increases rapidly with increasing d_a; corrections of this magnitude indicate that larger particle channels provide data of questionable accuracy.

At the bottom of the inner inlet, a nozzle constricts the flow and focuses the aerosol in the center of the acceleration nozzle. The inner walls of the focusing nozzle form a 60° angle with the direction of flow. Impaction may produce particle accumulation on this nozzle surface, further restricting the penetration of the inner inlet to 50% for about 8 μm oil droplets (Kinney, 1990). Kinney et al. (1989) also evaluated modifications to this nozzle and found that a smaller nozzle angle (2° or 8°) produces less internal loss but decreases the resolution of the APS. The amount of particle loss for the 60° inlet nozzle can be approximated using an equation developed for particle deposition efficiency η in a tube with a 90° contraction.

$$\eta = [1 - \exp(1.721 - 8.557x + 2.227x^2)]^2 \tag{17–25}$$

where

$$x = \sqrt{Stk} / (D_i/D_n)^{0.31} \tag{17–26}$$

Stk is the Stokes number in the inlet tube, D_i is the diameter of the inlet tube, and D_n is the diameter of the nozzle (Ye and Pui, 1990). Impaction of liquid particles may follow the deposition efficiency described by Eq. 17–25, while solid particles may bounce and exhibit lower deposition.

Laser Velocimeter Sensor

Aerosol passing through the inner nozzle is combined with the sheath flow and focused into the center of the acceleration nozzle. The airflow conditions in the nozzle region have been modeled and agree well with experimental measurements (Ananth and Wilson, 1988). The air velocity reaches approximately 150 m/s at the exit of the acceleration nozzle. The flow conditions affecting the particle acceleration depend on the nozzle dimensions as well as the spacing of the nozzles. In the 3300 and 3310, the laser beam is split into two parallel, flattened beams that intersect the particle path 200 to 500 μm from the acceleration nozzle. The distance of these beams from the acceleration nozzle also affects the measured particle velocity. These various dimensions are difficult to control precisely during instrument manufacture. Thus, the calibration of each of these APS instruments is slightly different.

In the APS 3320 and 3312, the distance of the LD beam from the nozzle is more tightly controlled during manufacture and can be reset without extensive recalibration. The sensor region is thus more accurately positioned, and the calibration from instrument to instrument should be more consistent.

Detection and Data Analysis

As each particle passes through the two laser beams, the pulses are detected by a photomultiplier and the TOF is recorded. Because the light scattered from particles changes rapidly with particle size, two high-speed data accumulator systems are used in the detector module of the APS 3300 and 3310: a small particle processor (SPP) and a large particle processor (LPP). The SPP collects the TOF data in an accumulator in increments of 4 ns, and the LPP collects data in increments of 66.7 ns. These data are passed from the detection module to a personal computer for transformation to size distribution according to a stored calibration curve, as well as for any further manipulation or storage as desired. The SPP covers the aerodynamic diameter range of 0.5 to 15.9 μm, and the LPP covers 5 to 30 μm. In the overlap range, the data from the two processors are blended together, proportionately increasing the LPP contribution with increasing size. The treatment of the overlap range is discussed further below in the section on coincidence.

The APS 3320 uses a single processor for particle detection. It can also correlate the particle velocity and light-scattering signal for post process rejection of particles whose velocities do not indicate particle sizes that match the observed light-scattering signal. For example, a particle with a low velocity (indicating a large particle) that produced a small scattering signal could be rejected as not being physically reasonable. This correlation changes for particles with different refractive indices. Once a number distribution has been measured, various other differential and cumulative distributions can be calculated in a similar fashion to those described for the E-SPART in the previous section.

Calibration

Monodisperse latex spheres are typically used for calibration of the full size range of the APS if it is to be used for measuring solid particles. Latex spheres smaller than about 5 μm can readily be generated by nebulizing a water suspension of the spheres. Note that while isopropyl alcohol suspensions of latex spheres may be easier to generate and dry, the alcohol slowly dissolves in the spheres and will cause a slight increase in size after a period of time. Larger calibration particles can be generated dry from a surface by suction, as with the Small Scale Powder Disperser (Model 3433, ***TSI***) or by gently brushing the calibration particles from a clean surface, such as a glass slide. Because latex particles are only available in specific sizes, the calibration curve is completed using a spline or polynomial function to fit the calibration points.

The calibration of each APS 3300 and 3310 instrument is unique due to variations in the nozzle sizes, spacing, and laser beam locations. However, once the calibration of the APS has been completed in air at ambient pressure, calibration for other gas viscosities and pressures can be achieved as described by Rader et al. (1990). The gas velocity U_g in the nozzle can be calculated from the Bernoulli equation for compressible flow:

$$U_g = \left[\frac{2RT}{M}\ln\left(\frac{P}{P-\Delta P}\right)\right]^{1/2} \qquad (17\text{–}27)$$

where R is the universal gas constant, T is the absolute temperature, M is the gas molecular weight, P is the ambient pressure, and ΔP is the pressure drop across the nozzle. ΔP is measured by the flow transducer in the APS. The particle velocity V_p can be determined from

$$V_p = \frac{U_g t_{min}}{t} \qquad (17\text{–}28)$$

where t is the transit time of the particle between the laser beams and t_{min} is the minimum transit time for small particles observable in the APS accumulator. Plotting the ratio V_p/U_g as a function of Stokes number results in a universal response curve. This means that the check on the APS size response under the same or new pressure or viscosity conditions can be achieved always setting U_g to the same value.

The design of the APS 3320 detection system is more compact and precise, resulting in reproducible spacing of the particle acceleration and illumination components. These components can be replaced by the user without the extensive factory recalibration procedure required with the earlier models. The APS also has a second pump that helps regulate the sheath flow. This results in less sensitivity to external pressure changes.

The flow system in the APS 3300 and 3310 is not as carefully controlled as in the APS 3320. Slight changes in pressure at the inlet of the earlier APSs could cause significant shifts in the apparent size distribution of submicrometer particles. The pressure drop change produced a slight shift in the flow through the acceleration nozzle, resulting in a slight change in the calibration curve as noted above. The channels in the submicrometer range were especially sensitive to slight changes in the calibration. The increased pressure drop shifted the calibration, decreasing the measured concentration of the smallest particles. This would become especially apparent when comparing size distributions upstream and downstream of a classifier (e.g., cyclone or impactor) that had small but measurable pressure drops. The ratio of the downstream to upstream concentrations could drop as low as 50% in the submicrometer range. The improved flow control in the APS 3320 sheath flow reduces the likelihood of such a calibration shift with small inlet pressure changes.

Other monodisperse particles, such as those generated from the vibrating orifice monodisperse aerosol generator (model VOMAG, ***TSI***) can also be used for calibration. However, it was found that oil droplets generated in this fashion distorted into oblate spheroids due to the high acceleration (see below) and therefore exhibited a smaller aerodynamic diameter than predicted for a spherical shape (Baron, 1986). Unless the calibration is used for measuring droplets of the same oil, only solid particles should be used for the particle size calibration of the APS.

Non-Stokesian Corrections

The acceleration in the nozzle produces Reynolds numbers outside the Stokes regime, as indicated in Table 17–2 for particles in the APS size range. Thus, the measured size depends on

TABLE 17–2. Particle Properties in the APS Nozzle

Particle Diameter (μm)	Relative Velocity (cm/s)	Particle Reynolds Number	Weber Number (Oil Droplets)[a]
0.5	40	0.013	2.9×10^{-6}
1.0	1,750	1.16	0.0113
3.0	6,490	12.9	0.468
10.0	10,600	69.6	4.13
15.0	11,500	114.0	7.36
20.0	12,300	163.0	11.2

[a] These represent either oleic acid or di-octyl phthalate, both of which have a surface tension of about 0.033 N/m [33 dyne/cm].

Source: Baron (1986).

other factors besides the aerodynamic size, including gas density, viscosity, particle density, and particle shape. Using the approach of Wang and John (1987), correction factors for the measured size of compact particles can be calculated if the particle density, gas density, and gas viscosity are known. These calculations have been validated by the Navier-Stokes calculation of the flow field in the APS nozzle by Ananth and Wilson (1988). The following equations (Rader et al., 1990) are iterated until no further significant change occurs.

$$\sqrt{Stk_2} = \sqrt{Stk_1}\left(\frac{6+R_2^{2/3}}{6+R_1^{2/3}}\right)^{1/2} \qquad (17\text{–}29)$$

$$R_1 = \xi_i^{3/2}\sqrt{Stk_i}\,|U_g - V_p| \qquad (17\text{–}30)$$

$$\xi_i = \left(\frac{18\rho_{gi}^2 S}{\rho_{pi}\mu_i U_g}\right)^{1/3} \qquad (17\text{–}31)$$

where subscript 1 refers to calibration conditions with unit density spheres, subscript 2 refers to measurement conditions, *Stk* is the Stokes number, and S ($= U_g t_{min}$) is the distance between the laser beams. Measurements were made in argon and N_2O to confirm that this approach improved the accuracy of aerodynamic size measurement (Lee et al., 1990; Rader et al., 1990). The largest error in d_a (12%) was noted when these corrections were applied to large (30 μm) particles in argon. The slip correction factor must also be modified in the above equation because of the reduced pressure in the nozzle, and computer code is available to perform these corrections (Wang and John, 1989).

The high acceleration in the APS nozzle may also cause inaccuracies in measuring the d_a of nonspherical particles. Cheng et al. (1990) found that the measured size decreased with increasing shape factor. The above iterative correction was further modified to include shape factor. For more extreme shapes such as fibers, this approach may not be adequate. Identical-diameter fibers with different lengths gave the same measured size. Fibers, as well as other nonspherical particles, tend to orient themselves with their maximum cross section oriented perpendicular to the flow (Clift et al., 1978:142). However, larger fibers (on the order of 10 μm diameter) may not have sufficient time to orient in the flow field and may produce a measured size intermediate between the perpendicular and parallel orientation. Thus, the initial conditions of the particle (e.g., orientation, location in the flow field) during acceleration can affect the measured aerodynamic size.

The APS allows the rapid, precise measurement of aerodynamic size of most particles. Due to non-Stokesian flow in the acceleration nozzle, various factors bias that measurement. As described above, the biases caused by particle density, particle shape factor, gas viscosity, and gas density are sufficiently well understood that corrections to measured size can be made. The size of these biases is often on the order of 25% or less. Thus for many purposes, an estimated value of the particle density can yield sufficient accuracy in the corrected aerodynamic size. An exception may occur when the dynamic shape factor of the particle is large, as with fibers.

Chen and co-workers (1990) suggested that the correction factor for density is sufficiently well characterized that it can be used to provide estimates of aerosol particle density. Brockmann and Rader (1990) also used the APS response to measure shape factors for several types of particles.

Droplet Deformation

As indicated above, the high acceleration field in the APS will distort droplets into oblate spheroids with the maximum cross section perpendicular to the direction of motion, increasing the drag and causing them to be recorded as smaller particles. Figure 17–11 shows two

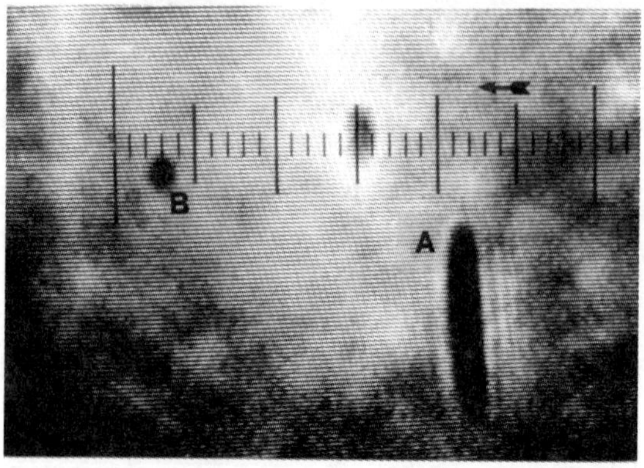

Fig. 17–11. Picture of droplets in the high-velocity air jet just beyond the APS nozzle taken with a high-speed laser-imaging system showing a droplet flattened by the drag force. The nozzle tip is about 200 μm to the right of droplet A, with the air and the droplets moving to the left. The scale markers are approximately 5 μm apart. The larger droplet A has an extreme 10 by 60 μm spheroidal shape, while the smaller droplet B is about 8 by 10 μm and is only slightly flattened.

dioctyl phthalate droplets detected with a laser imaging system (***LAS***) just past the tip of the acceleration nozzle. Air motion is directed from right to left, causing the deformation of the larger droplet into an extreme oblate shape and increasing the droplet drag. The flattened droplets in Figure 17–11 indicate that an oblate spheroidal shape is produced. The shape is not a true ellipsoid because the surface tension limits the curvature at the rim of the distorted droplet.

The distortion of a droplet in the sensing zone will depend on droplet size, the liquid surface tension, and viscosity. The Weber number, $We = u^2 \rho d_p / \gamma$, where u is the particle velocity relative to the air and γ is the droplet surface tension, represents the ratio of the air pressure force to the surface tension force. Droplets will eventually break up when experiencing Weber numbers between 12 and 20. Distortion increases with droplet size because the force on the droplet increases with size. While the Weber number indicates the maximum distortion that the droplet can undergo, the droplet viscosity determines the rate at which the droplet distorts. The rapid acceleration in the APS nozzle usually precludes droplet break up before reaching the sensing zone. Because the acceleration is high, viscosity is the controlling factor for distortion of many liquids (Griffiths et al., 1986). The degree of distortion has been calculated and agrees well with experimental measurements of the droplet undersizing for several oils with different viscosities (Bartley et al., 2000). Water droplets, which have a low viscosity but relatively high surface tension, distort less in the acceleration field (Baron, 1986; Bartley et al., 2000). In addition, the degree of distortion depends on the precise acceleration history and therefore can vary from instrument to instrument.

Coincidence Effects (APS 3310)

Accurate detection of a particle in the TOF detection system of the APS requires full detection of two pulses from the same particle. The coincidence effects resulting from such a system are more complex than those of a standard optical particle counter, where two coincident particles produce one somewhat larger measured particle. For the APS 3300 and 3310, several coincidence scenarios are presented in Figure 17–12. Coincidence between two particles can result in two smaller particles (Fig. 17–12a), one smaller particle, one randomly sized particle (Fig. 17–12b), or no particles (detected). The relative frequency of these possible

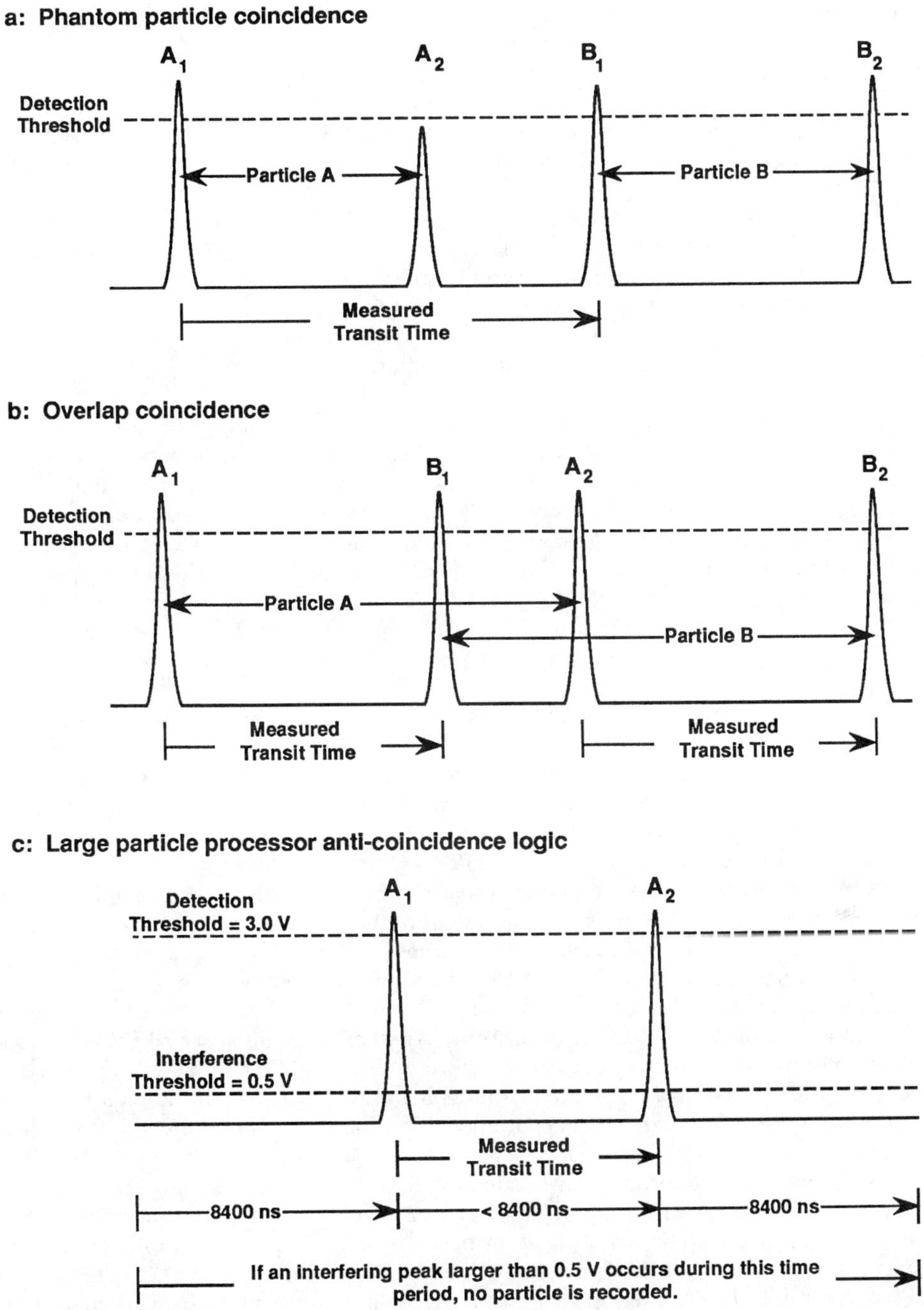

Fig. 17–12. Coincidence scenarios in the small particle processor (SPP) and the large particle processor (LPP) in the APS. **a**, A single detected pulse triggers the timer, while a pulse from a second particle produces a measured transit time that may indicate a particle of random size. **b**, Overlap of the pulses from two coincident particles produces two smaller detected particles. **c**, A particle detected by the LPP must have pulses larger than 3 V. The LPP anti-coincidence circuitry prevents detection of a particle when an interfering pulse occurs within 8400 ns before or after the evaluated pulse pair. (Adapted from Heitbrink et al., 1991.)

coincidence results and the effect on the measured distribution depends on the shape of the size distribution, the particle concentration, and which of the two signal processors (SPP or LPP) is used to detect the particles. The LPP is designed to completely eliminate coincidence counts (Fig. 17–12c) in the large particle range where coincidence can produce relatively large changes in apparent concentration. The coincidence effects in the APS 3310 have been modeled mathematically as well as with a Monte Carlo calculation (Heitbrink et al., 1991).

Particle-counting systems are prone to detection problems when more than one particle is present, or coincident, in the detection volume at the same time. The number of coincidence events in a measured distribution can be estimated from the difference between the actual concentration C_a and the measured concentration C_m:

$$C_a - C_m = C_a[1 - \exp(-C_a Q t)] \qquad (17\text{–}33)$$

where Q is the flow rate through the detection volume and t is the residence time of the particle in the detection volume (Willeke and Liu, 1976). This equation can be used to estimate the APS coincidence loss to the peak of a monodisperse size distribution. Obtaining an accurate coincidence level for most size distributions is more difficult because it depends on particle size. For instance, the concentration giving 1% coincidence in the SPP for 0.8, 3, and 10 μm particles is approximately 560, 390, and 230 particles/cm^3, respectively. For the LPP, a 1% coincidence level is predicted for 10, 20, and 29 μm particles at concentrations of 55, 48, and 43 particles/cm^3 (TSI Inc., 1987). By using Eq. 17–33 with several particle sizes, the upper limit to the number of coincidence events can be estimated for broader size distributions.

C_a may also be difficult to estimate for many distributions where many of the particles detected by the SPP are smaller than 1 μm. These particles may be only partially detected, resulting in single detected pulses that contribute to coincidence events (Fig. 17–12a), but not to the observed small particle concentration. These coincidence events result in a randomly sized "phantom" particle. The result is a nearly constant background of these coincidence-induced phantom particles (Heitbrink et al., 1991).

The phantom particle background produced by the SPP is therefore dependent on the number of particles near the pulse detection limit of the sensor as well as the concentration of fully detected particles. This background becomes important in size regions where relatively few real particles are detected. Thus, when particle number distributions are converted to mass distributions, a few large phantom particles can bias the calculated mass and unrealistically skew the distribution (Baron, 1986). Another situation arises when two particle distributions are being compared, such as before and after a filter to measure penetration efficiency (Wake, 1989). In the size range where phantom particle concentration is more than a few percent of the real particle concentration, the ratio of the upstream and downstream distributions will be inaccurate.

The SPP thus tends to produce overestimates of particle concentration near the tail of a distribution, such as often occurs at large particle sizes. On the other hand, the LPP is designed to completely eliminate phantom particles (Fig. 17–12c). Therefore, coincidence results solely in a loss of LPP-detected particles. The difference between the LPP and the SPP concentration in the overlap range can give a hint of the magnitude of coincidence effects. The SPP coincidence can sometimes be reduced by lowering the photomultiplier gain. For size distributions skewed to small particle sizes, this reduces the number of small particles detected, thus reducing the phantom particle creation. If a region of the size spectrum is known not to contain any real particles, the detected particles can be assumed to be phantom particles created by coincidence. The average detected particle number per channel in this region can be subtracted from the entire distribution to obtain a more accurate distribution.

Coincidence Effects (APS 3312, 3320)

In the APS 3312 and 3320, the detection circuitry rejects all particles that may have experienced coincidence, much the same as with the APS 3310 LPP described above. This causes a reduction in observed particle concentrations that is likely to increase with particle size, because the detection time for larger particles is increased. The number of coincidence events is recorded and provided to the user. The number of single pulses is also recorded and used to indicate the bin below the lowest fully measured particles. These reported values should aid in estimating the importance of coincidence events. Further research is needed to allow a quantitative estimate of the change in the measured size distribution due to particle coincidence.

Resolution and Accuracy

As noted above, the APS measures particles largely outside the Stokes regime and requires corrections to the data to provide an accurate d_a. As with any complex instrument, frequent size calibration provides additional confidence in the accuracy of the results (see Chapter 21). Measurement of spherical particles can be corrected largely by taking into account particle density and, if necessary, changes in sampled gas density and viscosity. Liquid particles can also be accurately measured if calibrated with the same liquid. The resolution for spherical particles is high. For example, Remiarz et al. (1983) found geometric standard deviations in the range of 1.0058 (6.8 μm oil) to 1.025 (0.8 μm latex) for monodisperse particles using the APS prototype instrument. The particle size resolution of the APS 3320 is lower than in the previous models due to the poorer velocity measurement precision with the relatively broad LD light beams.

Due to non-Stokesian behavior and variations in the acceleration flow field experienced by particles approaching the detection region, resolution and accuracy may be diminished for nonspherical particles. If these are not corrected, the measurement accuracy will decrease. Marshall et al. (1991) found that the d_a of particles with a shape factor of 1.19 was underestimated by 25% in the APS.

Applications

The APS 3300, 3310, and 3320 can be used to measure size distributions in a variety of applications. These instruments have been combined with an electrical sizing instrument (see Chapter 18) to obtain size distribution measurements over a wide size range (0.02 to 30 μm). Sioutas et al. (1999) found reasonably good agreement between calculations of mass concentration from APS spectra and direct mass measurements in the size range of 0.5 to 9.2 μm. Peters et al. (1993) found good agreement between APS size distribution measurements and low pressure impactor measurements (see Chapter 10).

The APS has been used for bioaerosol measurement (Baron and Willeke, 1986). Specific detection of biologically based particles has been enhanced using the APS 3312 where the fluorescence signal can be combined with the aerodynamic size to obtain characteristic size distributions of bioaerosols (Ho et al., 1999). Biological particle growth has been observed (Madelin and Johnson, 1992). Several biological species were measured using the APS 3312 to evaluate the response (Brosseau et al., 2000). Differences in the fluorescence signals of biological particles versus non-fluorescent control particles were seen, but no relationship to culturability of the bacteria or fungi was found, and the differences in fluorescence signal could not be related to known bacterical species.

The APS has been applied to the measurement of penetration curves of several types of aerodynamic classification devices, such as impactors (Baron, 1983), cyclones (Kenny and Gussman, 1997; Chen et al., 1999; Maynard, 1999), and open-pore foams (Fabriés et al., 1998;

Chen et al., 1999). Systems optimized for measurement of penetration curves have been developed (John and Kreisberg, 1999; Maynard et al., 1999). The challenge aerosol for this type of measurement preferably consists of spherical particles with a density of approximately 1000 kg/m^3 [1 g/cm^3] and with a distribution centered near the 50% cutpoint of the classification device. There should be an adequate concentration of particles throughout the size range of measurement, with a minimal concentration in the 0.5 μm range, to reduce the effects of coincidence. The challenge aerosol concentration is generally optimal in the 2×10^{-5} to 1×10^{-4} particles/m^3 (20 to 100 particles/cm^3) range.

AEROSIZER

Principles of Measurement

The Aerosizer (***TSI***) is based on the acceleration of particles and TOF principles, although at higher particle acceleration than in either the E-SPART or APS. The idea of accelerating particles in a sonic expansion flow and measuring the terminal velocity was first proposed and demonstrated in laboratory prototype instruments by Dahneke and co-workers

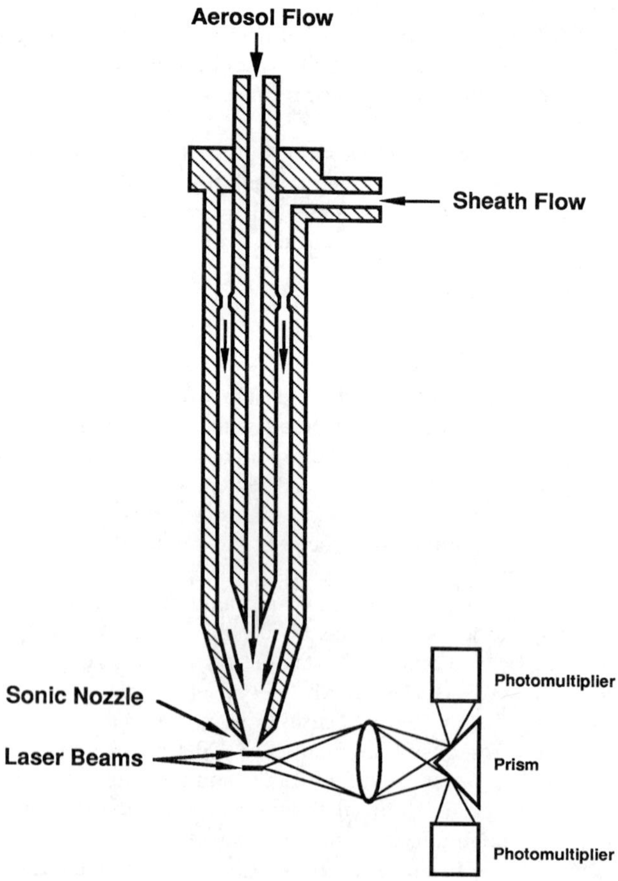

Fig. 17–13. Schematic of the detection system of the Aerosizer. The two laser beams are perpendicular to both the aerosol flow and detection direction.

(Dahneke, 1973; Dahneke and Padliya, 1977; Cheng and Dahneke, 1979; Dahneke and Cheng, 1979). The Aerosizer is the commercial product of this aerosol-beam research. Figure 17–13 depicts the particle-detection section of the device. The aerosol enters the inner capillary tube, surrounded by particle-free sheath air in the outer tube. Air and particles are accelerated through a convergent nozzle (0.75 mm diameter) with a 15° half angle and are delivered into a partially evacuated chamber. The ratio of pressure in the chamber and that in the ambient air is much smaller than 0.53; therefore, the air velocity at the nozzle exit, V_g, attains sonic velocity:

$$V_g = \sqrt{\frac{\gamma RT}{M}} \qquad (17\text{–}34)$$

where γ is the ratio of specific heat capacities (1.4 for air), R is the gas constant, T is the temperature in K, and M is the molecular weight (28.96 for air). The air continues to expand in the chamber with a supersonic free-jet flow. The flow field in the convergent nozzle and the supersonic expansion have been described (Dahneke and Cheng, 1979). From numerical calculations, it has been shown that the particle attains a terminal velocity soon after exiting the nozzle (Dahneke and Cheng, 1979). At a distance of five nozzle diameters downstream of the nozzle, the calculated axial velocity is within about 2% of that at a distance of 50 diameters downstream.

The TOF of a particle is measured with two laser beams located close to the nozzle. As particles pass through the laser beams, the light scattered from the particles is detected and converted into electronic signals by two photomultiplier tubes (PMTs). The distance between the two laser beams is about 1 mm with about a 20 to 30 μm spread of the individual beams. One PMT detects light as the particle passes through the first beam, while the other PMT detects the light from the second beam. The time between these two events (TOF) is measured and recorded to an accuracy of ±25 ns. Each PMT converts the scattered light into an electrical pulse that is sent to the data-acquisition system. A low PMT threshold senses the weak pulses generated by small particles. However, at the low PMT threshold larger particles can cause the PMT to produce an electrical oscillation that generates multiple detected pulses from a single true pulse, a behavior called *ringing* (Thornburg et al., 1999). Therefore, the Aerosizer uses several PMT thresholds that are selected based on the initial pulse height so that spurious triggering on these subsequent oscillations does not occur.

Several models of the Aerosizer have been produced by Amherst Process Instruments, including the original Aerosizer, the Aerosizer LD, and the Aerosizer DSP. Amherst Process Instruments was purchased by ***TSI*** in 1999, and the DSP version was designated Model 3220 (***TSI***). The principal improvements differentiating the models were that the Aerosizer LD had the HeNe laser replaced by a LD light source and the DSP model had an improved signal processor to reduce spurious triggering due to noise and ringing in large particle signals.

Calibration with Spherical Particles

The terminal velocity and TOF are functions of particle diameter, density, shape, and ambient pressure. Calibration curves (Fig. 17–14) provided by the manufacturer are based on theoretical calculations and experimental data from solid spherical particles. The experimental validation of the calibration curves, also provided by the manufacturer, uses a limited number of monodisperse spherical particles of polystyrene latex (PSL) and glass. A more extensive calibration of the instrument requires both PSL (density = 1050 kg/m^3 [1.05 g/cm^3]) and glass (density = 2460 kg/m^3 [2.46 g/cm^3]) particles with a size range of 0.5 to 150 μm obtained at an ambient pressure of 0.82 and 1 atm. This work is shown in Figure 17–15 (Cheng et al., 1993). Calibration curves for particles of these densities were calculated from the Aerosizer calibration table using an interpolation method and are plotted along with the measured data

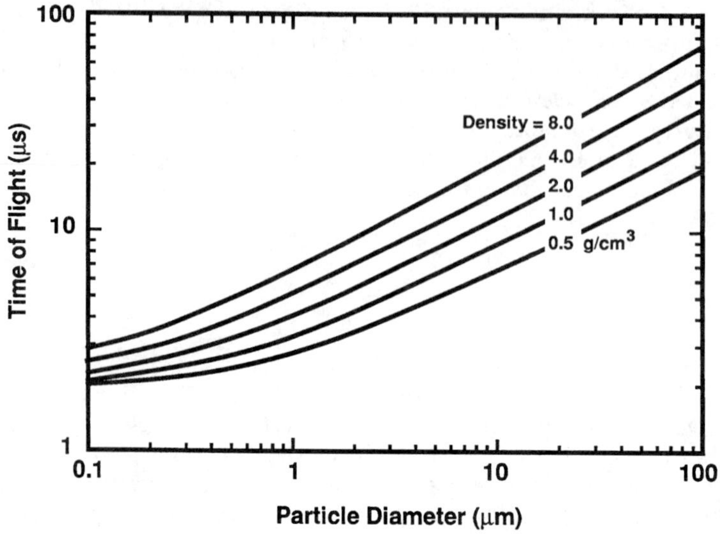

Fig. 17–14. Calculated Aerosizer calibration curves for spherical particles of different densities. (Courtesy of Amherst Process Instruments, Inc.)

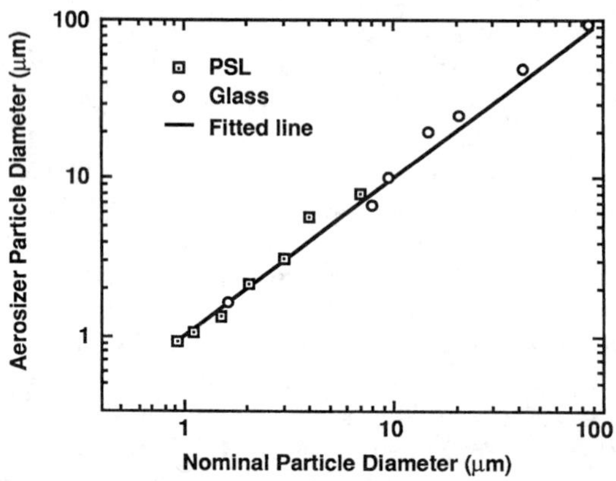

Fig. 17–15. Aerosizer calibration data for polystyrene latex (PSL) particles (density = 1050 kg/m^3 [1.05 g/cm^3]) and glass beads (density = 2450 kg/m^3 [2.45 g/cm^3]).

in Figure 17–15. Excellent agreement was found between the TOF of the PSL particles in the size range of 2 to 10 μm at 101.3 kPa [1 atm]. At a reduced pressure of 0.82 atm, the TOFs obtained for the PSL and glass particles were slightly higher than predicted by the calibration curves for the size range from 0.4 to 150 μm geometric diameter. However, the calibration tables were developed for normal ambient conditions, and theory predicts higher TOFs for operation under reduced ambient pressure (Dahneke and Cheng, 1979; Oskouie et al., 1998; Tsai et al., 1998); thus, the disagreement between the measured data and calibration curves was expected. The magnitude of the overestimation of particle size by the Aerosizer is expressed in terms of the ratio of the geometric diameter measured by the Aerosizer to the true geometric diameter from microscopy (d_{API}/d_g). This ratio took values between 1.08 and 1.27 for the PSL particles and glass beads.

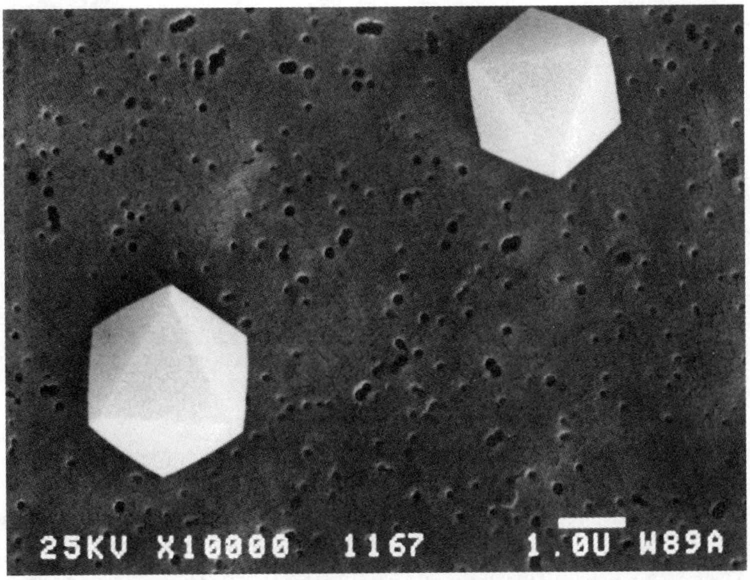

Fig. 17–16. Natrojarosite particles. (From Cheng et al., 1993, with permission.)

Tsai et al. (1998) calculated the particle trajectory assuming compressible flow in the Aerosizer nozzle, with a corresponding drag coefficient formula suitable for a range of Mach numbers and Reynolds numbers and appropriate for the Aerosizer operating conditions. The calculated TOF is in good agreement with experimental data obtained at 83.1 kPa [0.82 atm] pressure (Fig. 17–15). Compared with the calibration curve in Figure 17–14, the calculated results show deviation for particles greater than 10 µm. This deviation was due to the fact that Figure 17–14 was calculated assuming a one-dimensional flow field and a different formula for drag coefficient.

Instrument Responses with Nonspherical Particles and Droplets

Responses from the Aerosizer for nonspherical natrojarosite particles ($Na\ Fe_3[SO_4]_2\ [OH]_6$) were measured (Cheng et al., 1993). Preparation and characterization of the uniform-sized particles formed as single, symmetrical truncated cubes (Fig. 17–16) with a density of 3.11 g/cm^3 have been described (Marshall et al., 1991). Particles having sizes between 7.3 and 18.8 µm aerodynamic diameter were prepared and classified by aerodynamic size in a Timbrell spectrometer (see Chapter 10). Figure 17–17 presents plots of the Aerosizer-measured aerodynamic diameter against mean aerodynamic diameter. In all cases, the Aerosizer significantly undersized the natrojarosite particles, and the degree of size reduction was size dependent. Thus, the largest particles with true aerodynamic diameters close to 18.8 µm were undersized by as much as 51%, whereas the smallest particles analyzed with true aerodynamic diameters of 7.3 µm were undersized by only 21%.

Liquid droplets also behave differently from solid spherical particles in the Aerosizer. The Aerosizer underestimates the true aerodynamic diameter of oleic acid droplets (Tsai et al., 1998), similar to results shown by Baron (1986) and Bartley et al. (2000) from the APS. The deformation is a function of the Weber number (Baron, 1986; Lefebvre, 1989). A Weber number greater than the range of 12 to 20 suggest that the droplets broke up, whereas Weber numbers within that range indicate that the droplets may have deformed (Lefebvre, 1989). Calculations of Weber numbers for 1 and 5 µm oleic acid showed that particles larger than

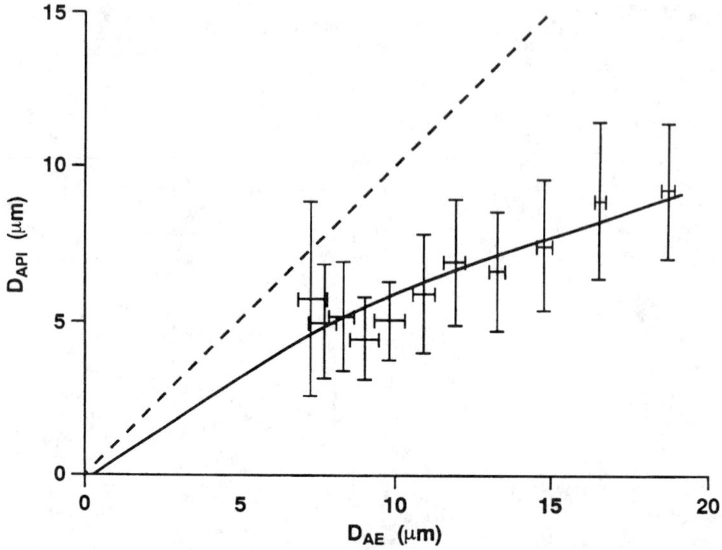

Fig. 17–17. Comparison between the Timbrell Spectrometer and the Aerosizer-measured aerodynamic diameters of natrojarosite at normal ambient pressure (100 kPa [750 mm]). (From Cheng et al., 1993, with permission.)

5 μm have Weber numbers in the 12 to 20 range and may deform in the measurement region (Thornburg et al., 1999). These results agree with experimental data showing particle deformation for oleic acid particles greater than 7 μm (Baron et al., 1996; Tsai et al., 1998). The Aerosizer is thus likely to exhibit a greater degree of droplet deformation than the APS due to its higher particle acceleration.

Transmission Efficiency

There are several reasons that some particles are not counted by the Aerosizer. The first is the transmission efficiency; the aerosol must penetrate the inlet nozzle and pass through the laser beams. Large particles may deposit in the inlet nozzle by impaction. After particles pass through the nozzle exit, small particles may follow the divergent gas streamline and fail to intercept the laser beams. Both scenarios result in lowered counting efficiency. In a numerical simulation, aerosol transmission efficiency in the Aerosizer was estimated by considering impaction in the nozzle, rebounding of solid particles, and passage through the sensing zone (Tsai et al., 1999). Theoretical calculations showed that the transmission efficiency is a complex function of particle size, beam diameter, and the solid or liquid state of the particle. For liquid particles at a typical laser beam diameter of 300 μm, the counting efficiency was about 4% for 0.1 μm particles, increasing to a maximum of about 27% for 2.5 to 3 μm particles. The counting efficiency decreased for larger particles due to deposition in the nozzle. For solid particles, the counting efficiency was essentially the same as for liquid particles in the size range of 1 to 10 μm. For particles greater than 10 μm, the counting efficiency increased further due to particle bounce in the inlet. The counting efficiencies were predicted to be higher for 600 and 800 μm diameter laser beams, suggesting that redesign of the detection optics would improve the instrument response.

The counting efficiency of the Aerosizer LD was determined experimentally using polydisperse oleic liquid droplets (Thornburg et al., 1999). The determination included effects of transmission efficiency and factors such as signal processing, where signals lower than the threshold were not considered. The counting efficiency of the Aerosizer for 0.3 to 10 µm particles varied substantially with particle size, PMT voltage, and count rate. The counting efficiency was less than 1% for particles smaller than 0.5 µm, whereas efficiencies for particles greater than 7 µm occasionally exceeded 100%. The low efficiency for small particles was due to low light scattering intensity, and the artificially high counting efficiency for larger particles was due to the ringing effects of multiple signals from each pulse. The strong dependence of counting efficiency on particle size increased the median diameter of the size distribution. The Aerosizer DSP signal processing system may give substantially improved results (see below).

Coincidence Errors

At high concentrations, the counting efficiency may decrease due to several causes associated with the measurement process that result in undetected particles (Mitchell et al., 1999). First, the transit time of a particle through each laser beam along with the electronic counting dead time for each pulse determines the upper limit of particle detection. When more than one particle is present in the same beam during the first particle's presence, information about one particle is lost, decreasing the measured concentration.

Second, the operation of the particle-detection electronics may be as important as beam-residence coincidence in determining an upper concentration limit associated with an arbitrarily defined, acceptable level of accuracy. At least two potential sources of particle loss from this cause are known. The amount of dead time associated with the two PMTs that detect light scattered by particles as they interact with the "start" and "stop" laser beams can cause particles to be lost if a second particle arrives during the period of detector insensitivity. Dead times are 400 and 800 ns for the low-sensitivity (large particle) and high-sensitivity (small particle) detection systems, respectively. Based on these dead times, it follows that limiting particle concentrations are 1.5×10^{12} and 7.5×10^{11} particles/m^3 for the fine and coarse particle-detection systems, respectively. These limits are at least an order of magnitude larger than values of concentrations measured previously using the Aerosizer with some pharmaceutical aerosols (Mitchell and Nagel, 1996). Dead time coincidence is likely to contribute only a small error under these conditions.

The operation of the timers for the particle detection system may introduce a more significant bias in the accurate determination of concentration, as the duration of insensitivity to incoming particles under overloaded conditions can be close to two orders of magnitude longer. As each particle passes through each laser beam, the scattered light is detected and converted into either a low- or high-amplitude electrical signal by the "start" and "stop" PMTs. The low signal threshold is set at high sensitivity (high-gain region) to measure scattered light from small particles, and conversely the high-signal threshold is set at low sensitivity (low-gain region) to measure light scattered by large particles. Thus, the two PMTs allow a differentiation between the entry and exit pulses associated with each particle. However, instead of linking two adjacent "start" and "stop" pulses with each other to define the TOF for a given particle, the Aerosizer detection system considers all the pulses over a defined time period and correlates pulses of equal intensity to define the TOF of individual particles. Signals having unequal intensity are randomly combined and averaged into a background signal. These background "particles" are equivalent to the phantom particles produced by coincidence in the APS, but are routinely subtracted from the final spectrum.

In more detail, each sensitivity region of the Aerosizer LD has four timers, each of which remains in the ON condition for a maximum of 52 µs after being triggered by a detected particle before being automatically reset. A timer starts when a particle is detected by the "start" PMT. If additional particles arrive at the "start" laser beam before the first particle arrives at the "stop" beam, each particle starts an additional timer until all four timers are operating. When a particle reaches the "stop" beam, its pulse detected by the "stop" PMT is correlated with as many pulses from the other timers that have been triggered. It follows that if a particle arrives after all four timers have been triggered and before any have been reset, its pulse is lost to the system and a real particle count has not been detected. This so-called timer unavailable coincidence becomes progressively worse as the incoming particle concentration increases.

The manufacturer has estimated particle detection losses from all causes in a simulation that is based on the behavior of perfectly monodisperse small (0.5 μm aerodynamic diameter) and large (5.0 μm aerodynamic diameter) particles delivered to the measurement zone at progressively increasing feed rates. If a 10% loss is arbitrarily deemed as the limit for acceptable bias in the measurement, it follows that the maximum aerosol concentration should not exceed approximately 1×10^{10} particles/m^3.

Signal Processing. There are two versions of the Aerosizer processing system. The significant difference between the signal processing in the older Aerosizer systems (e.g., Aerosizer LD) and the Aerosizer DSP systems is the introduction of an adaptive threshold system. Older Aerosizer LD systems used two detection thresholds ("high sensitivity" and "low sensitivity"), which were effectively fixed during a run (the PMT voltage, and thus the effective threshold, could be varied on a run-by-run basis). This usually resulted in "phantom particles" from multiple triggers caused by effects such as PMT afterpulsing and laser-beam fringes on large particles and also made it difficult to select a sensitivity setting appropriate for the sample(s) being run (especially for mixtures of large and small particles). Because these "phantom particles" were correlated with real particles, they showed up in the distribution as additional particles of approximately the same size as the real particles. These problems often produced significant errors in measured distributions (Heitbrink et al., 1991).

The Aerosizer DSP digitizes the PMT outputs and uses a digital delay line to allow the detection threshold to be dynamically set based on the behavior of the input signal. This delay line provides the ability to look "into the future" and thus set the thresholds based on how the signal will behave in the near term. These dynamic thresholds significantly improve the dynamic range of the detection process, providing the correlators with much better data from which to derive the TOF statistics.

Comparison Between the Aerosizer and the APS

Both the Aerosizer and the APS are based on the acceleration of particles in an aerosol beam and TOF measurement for size determination. However, the details of design and operation of these two instruments differ substantially. Whereas the APS is operated under subsonic flow with about 13.3 kPa [100 mm Hg] pressure differential between the ambient air and sensing volume (Chen et al., 1985), the Aerosizer is operated under supersonic conditions with a much higher pressure drop (greater than 100 kPa [750 torr] under 101.3 kPa [1 atm] ambient pressure). Therefore, a more powerful vacuum pump is needed for the Aerosizer, whereas the APS uses much smaller pumps placed inside the instrument. Because the nozzle in the Aerosizer is operated as a critical orifice, it controls the flow by itself. However, the APS requires a more elaborate flow-control system to keep a constant sampling flow rate. The Aerosizer has higher gas and particle velocities than the APS, for example, a 10 μm PSL particle attains about 100 m/s velocity in the Aerosizer versus about 38 m/s in the APS. With a higher velocity and therefore a higher Reynolds number in the Aerosizer, the drag force experienced by a particle will be higher than that in the Stokes regime ($Re < 0.1$). Therefore, the effects of particle density and shape factor on the instrument response will be greater in the Aerosizer than in the APS.

Another major difference between the Aerosizer and the APS is the method for measuring the TOF. The Aerosizer uses four timers for the TOF measurement. Only one clock is used in the APS to determine the TOF. This difference may be reflected in the noise level.

The noise of both instruments arises when the PMT produces a signal that is not associated with a particle event, and this signal is detected as a particle passing through the sensing volume. The calculated particle size from a false signal depends on the true particle concentration and the exact time sequence of this false signal. With a single clock system, the noise signals tend to produce more signals for small particles, whereas for the Aerosizer the noise level appears to be uniform across the size range. Both instruments have an internal mechanism to reduce or correct for the background noise level. However, it is not completely removed, and false signals affect the accuracy of the size distribution. Both the Aerosizer and the APS measure the size number distribution, but they also calculate the surface area and volume distributions, assuming that the particles are spherical. With noise corrections, false signals do not significantly affect the number distribution. However, when the same size distribution is transformed into a volume distribution, false signals in large particles are amplified. Therefore, the noise tends to skew the size distribution toward the large particles, especially in the Aerosizer.

The third feature of the Aerosizer that is different from the APS is the data presentation. Both instruments calculate the measured size distribution in terms of number, surface area, and volume distributions. The APS data are presented in the absolute units of particle number/cm^3, surface area/cm^3, and volume/cm^3, whereas the Aerosizer data are normalized with respect to the peak concentration. The Aerosizer distribution can be adjusted to give an estimated true concentration, but may be in error due to the way the background is subtracted from the TOF spectrum.

FIBROUS AEROSOL MONITOR

The measurement of fibers is useful for controlling the health risk due to exposure to airborne asbestos fibers and other similar materials. Currently, the most common method for estimating such exposure is personal filter cassette sampling followed by phase contrast light microscope counting of fibers. This method, exemplified by NIOSH Method 7400 (National Institute for Occupational Safety and Health, 1994), often requires sampling times of hours and analysis times of 5 to 20 min (see Chapters 12 and 26). The Fibrous Aerosol Monitor (originally model FAM-1, currently model FM-7400, ***MIE***) was developed under joint sponsorship of NIOSH, the Bureau of Mines, and the Environmental Protection Agency (Lilienfeld et al., 1979) to supplement this technique with real-time indication of fiber concentration.

The FAM preferentially detects fibers by aligning the fibers in an oscillating electric field, illuminating the fibers with a laser beam and detecting the resulting pulses of light with photomultiplier. Compact particles do not align in the oscillating field and thus do not produce pulses synchronous with the oscillating electric field.

Fiber Alignment

Particles in the sensing zone are subjected to a constant 300 V/mm electric field perpendicular to both the laser beam and the detector axis. A smaller ac field is applied at 45° to this field to oscillate the electric field in the ϕ direction (Fig. 17–18). With this field rocking motion, fibers aligned parallel to the electric field vector spend more time scattering light into the detector, thus improving sensitivity, than if the field exhibited a full 360° rotation (used in the prototype version).

Conductive fibers are predicted to align in the FAM with less than a 0.1° lag behind the rotating field (Lilienfeld, 1985). Nonconductive fibers are expected to align very poorly, if at all. Conductivity on the scale of fibers that are a few micrometers or tens of micrometers in length depends on surface conductivity as well as bulk conductivity. For instance, glass fibers have a very low bulk conductivity, but can align in the electric field of the FAM at a relative

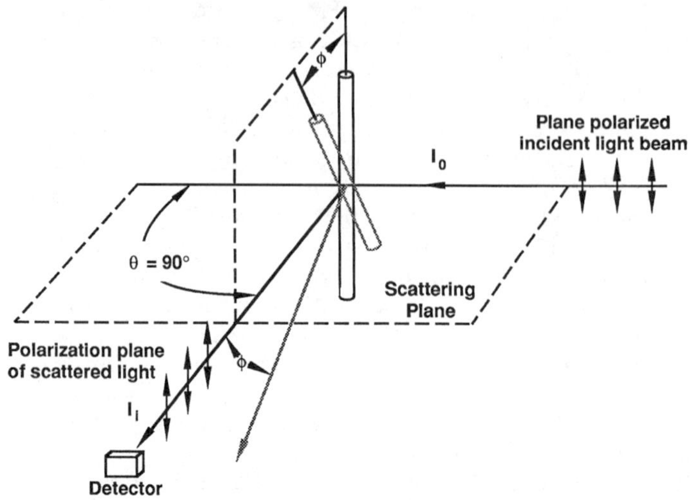

Fig. 17–18. Geometry of light scattering from a fiber in the Fibrous Aerosol Monitor (model FAM-1, MIE, Inc.). With the fiber perpendicular to the laser beam, the light scattered from the fiber occurs primarily in the plane perpendicular to both the fiber and the laser beam and results in a pulse of light reaching the detector as the fiber rotates. (Adapted from Lilienfeld, 1987.)

humidity above about RH 30%. Thus adsorbed water can provide enough surface conductivity to allow complete alignment. Asbestos fibers adsorb water more readily and align at much lower humidity levels.

Brownian motion can cause fibers to be randomly displaced from a fully aligned condition. At room temperature, the average angular deflection from alignment θ_B can be calculated from

$$\sin\theta_B \cong \frac{1}{\pi E}\sqrt{\frac{kT(\sin[2\beta]-1)}{2\varepsilon_o L^3}} \tag{17–35}$$

where E is the electric field, k is the Boltzmann constant, T is the absolute temperature, β is the fiber length-to-width (aspect) ratio, ε_o is the permittivity of free space, and L is the fiber length. The deflection is a strong function of fiber length, and, while fibers longer than 5 μm are predicted to align well in the FAM, shorter fibers may not hold alignment sufficiently well to be detected.

Sensing System

The FAM uses two fiber properties to exclude other types of particles: The alignment of conductive fibers with their long axis parallel to an electric field and the specific light-scattering pattern produced by fibers. A fiber in the FAM sensing zone is oriented by an electric field such that its fiber's major axis is always perpendicular to the incident light beam (Fig. 17–18). Under these conditions, the light scatters primarily into a plane that is perpendicular to the fiber axis and contains the light beam. The detector is situated to receive the maximum scattered light when the fiber is also perpendicular to the detector fiber axis. With very long fibers, this plane of scattering is quite thin. With a smaller length to diameter ratio, or aspect ratio, the scattering plane becomes broader, eventually approaching a uniform pattern as a function of ϕ for spherical particles. As with compact particles, more light is scattered in the

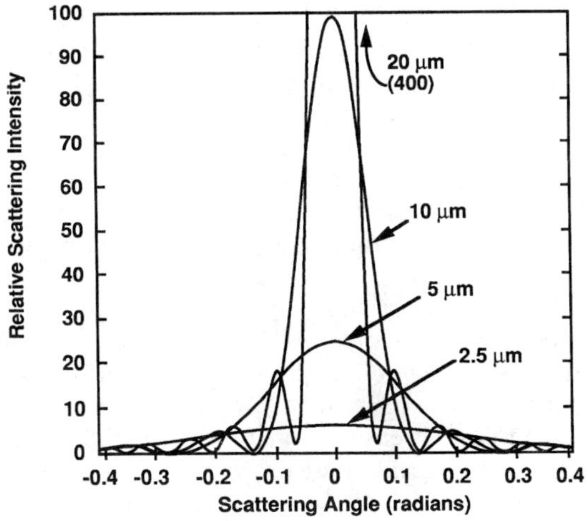

Fig. 17–19. Light-scattering pulses calculated for several fiber lengths for a single fiber diameter (Eq. 17–36). (Adapted from Lilienfeld, 1987.)

forward direction (small θ) than in the backward direction, and this proportion increases with fiber diameter. As with scattering from compact particles, the scattering intensity and angle dependence is a function of refractive index. The 90° detection angle in the FAM allows multiple detection of the same particle as it moves down the detection volume, improving sensitivity and specificity. In addition, for small-diameter fibers the scattered light tends to be plane-polarized parallel to the fiber axis. To increase the amount of scattered light from these fibers relative to that from nonaligned compact particles, a polarizing filter is placed in front of the photomultiplier detector.

A detector is placed at a right angle to the incident beam in the scattering plane as indicated in Figure 17–18. As a fiber rotates in the ϕ direction, a pulse of light is detected. The intensity profile of this pulse for small values of ϕ is indicated by

$$I \propto L^2 \left(\frac{\sin(\pi L \phi / \lambda)}{\pi L \phi / \lambda} \right)^2 \qquad (17\text{–}36)$$

where I is the scattering intensity, L is the fiber length, and λ is the light wavelength (0.6328 μm) (Lilienfeld, 1987). The sharpness of the scattering pattern indicated by this function depends on the fiber length. Pulse sharpness is used as a discrimination in the FAM to limit the detection to fibers longer than 5 μm. The curves in Figure 17–19 indicate the relative scattering intensity profiles as calculated by Eq. 17–36 for fibers of several lengths. A calculation of the sensitivity of the FAM indicates that it is capable of detecting 0.075 μm diameter fibers that are longer than 5 μm (Lilienfeld, 1987).

Nonideal Fiber Behavior

The theoretical considerations that are the basis for the fiber alignment and detection system in the FAM assume that ideal cylindrical fibers are being detected. However, there are a number of factors that can make fiber detection difficult to characterize theoretically. For instance, fibers may not have the ideal cylindrical or even an ellipsoidal shape. Asbestos fibers are typically bundles of fibrils that may have splayed ends, noncylindrical cross section, curvature, attached particles, and so forth. Fibers may exist as clumps of multiple fibers floating in the air. Different fiber types have different refractive indices and hence different scattering efficiencies. Fibers with curvature or with other particles attached may not align at 90° to the detector axis to give optimum detection geometry.

The fibers of one of the most common types of asbestos, chrysotile, often exhibit curvature. This curvature was found to result in lower sensitivity in the FAM prototype (Lilienfeld et al., 1979). However, by applying a sufficiently high voltage field in the detection volume the fibers can be straightened so that they produce responses similar to more ideal fibers (Lilienfeld, 1985).

The laser beam does not uniformly illuminate the detection volume. The profile of the beam is Gaussian, indicating that fibers at the edge of the beam will scatter less light than fibers in the center of the beam. Large-diameter fibers may settle enough to be incompletely detected. In addition, charged fibers may drift away from the beam due to the high electric field. All these factors will affect the sensitivity and response of the instrument. Thus, as with other light-scattering instruments, a calibration procedure is required to make the instrument respond in a manner equivalent to light microscopic detection.

Instrument Characteristics

Sampling System. The airflow in the FAM is controlled by a diaphragm pump to $3.33 \times 10^{-5}\,m^3/s$ [2 L/min]. A 25 mm cellulose ester filter collects fibers that have been detected by the FAM; this filter can be used to check the FAM calibration. An adjustment to the pump voltage allows calibration of the flow rate. The aerosol enters a 12.7 mm diameter tube inlet and passes through two right-angle bends. This arrangement is required by laser safety regulations to prevent direct viewing of the laser beam. The aerosol flow continues down the 10 mm diameter sensing tube to the sensing volume and then out of the other end of the sensing tube (Fig. 17–20). A minor amount of clean air flow from the pump through the beam

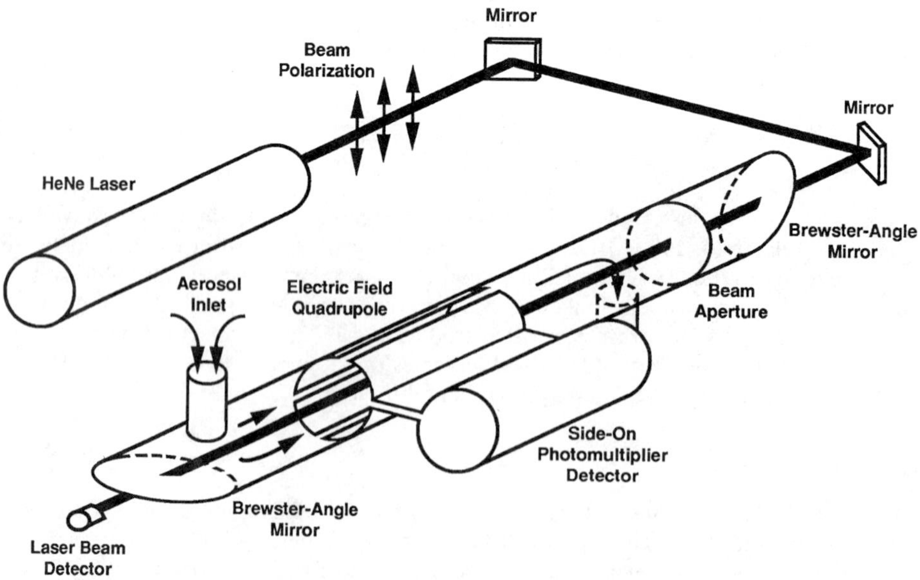

Fig. 17–20. Sensing system of the FAM-1, including the light-scattering system, fiber alignment system, and air flow pattern. (Adapted from Lilienfeld, 1987.)

orifice keeps the Brewster angle window (used to prevent backscatter of light) and orifice clean.

An additional option is available with the FAM that uses a virtual impactor to remove particles larger than 3 μm aerodynamic diameter from the sampled air stream. The impactor is driven entirely by the flow into the FAM inlet. The jet pressure in the impactor forces 10% of the flow with the large particles into the minor flow orifice, through a glass fiber filter, and through a rotameter. This 10% is thus impacted, filtered, and returned to the major flow entering the FAM. This arrangement has the advantage that an additional pump to control the minor flow is not required. With 10% of the small particles removed as well as the large particles, the final fiber count must be increased to adjust for this 10% loss.

The sensing volume is far enough downstream from the inlet that laminar flow is established. Under these conditions, the fibers are expected to align parallel to the flow until they reach the sensing zone. As indicated in Figure 17–20, the laser beam shines down the center of the sensing tube so only those fibers in the center, illuminated by the beam, are detected. In addition, fibers must stay in the beam for a sufficient time to be recorded as a fiber. Thus, any fibers that settle or otherwise deviate more than a fraction of the beam radius (about 0.7 mm) in a straight trajectory in the 0.1 s it takes the fiber to traverse the sensing zone will not be recorded as a fiber.

As with other particle counters, the FAM exhibits coincidence errors at high fiber concentrations. Coincidence of two fibers in the detection volume results in only one fiber being counted. The coincidence level can be estimated using Eq. 17–33 with a detection volume ($Q \cdot t$) of 8.3 mm^3. This indicates a coincidence loss of about 15% at a measured concentration of 15×10^6 fibers/m^3 [15 fibers/cm^3].

The FAM is a small suitcase-sized instrument with a mass about 12 kg and is normally operated from 115 V ac power, although a battery pack option is available that allows up to 3 h of operation. The readout is by liquid crystal display in fibers/cm^3. The FM-7400 has several outputs for data recording and calculates several statistics to indicate when a selected threshold has been reached with reasonable confidence. An alarm feature also indicates when the concentration exceeds a selected value. The pulse width of the signal is analyzed to estimate fiber length according Eq. 17–36 so that fiber length distributions in the range of approximately 2 to 300 μm can be measured.

Calibration

The original aim in the development of the FAM was to provide a real-time direct-reading monitor capable of giving fiber concentrations equivalent to those provided by the reference method using filter sample collection and phase contrast light microscope analysis (e.g., NIOSH Method 7400, as described in Chapters 12 and 23). Because the instrument response cannot be predicted theoretically, it must be calibrated by a side-by-side comparison with the reference method. Each instrument is calibrated in this fashion by the manufacturer with an amosite aerosol produced from a vibrating fluidized bed generator. During calibration, the instrument and filter inlets are placed close together. Most asbestos fibers from the generator have small enough Stokes numbers not to be significantly affected by anisokinetic sampling.

The instrument detection thresholds (for pulse height and sharpness) are set such that the measurement rate is effectively 167 mm^3/s [10 cm^3/min]. This is the flow rate through the laser beam in the sensing region. The FAM was originally designed to measure concentrations at levels near the U.S. asbestos standard in 1978, namely, 2×10^6 fiber/m^3 [2 fibers/cm^3]. The permissible exposure limit in workplaces is currently 1×10^5 fiber/m^3 [0.1 fibers/cm^3]. This sensor flow rate requires relatively long sample times to obtain good confidence in the results at low concentrations, for example, at a concentration of 10^5 fiber/m^3 [0.1 fibers/cm^3], 100 fibers are counted in 100 min.

FAM Evaluations and Application

A number of studies have been performed on the FAM, although most of these have not been in-depth evaluations. In general, these studies indicated that the FAM provides comparable results to the reference method. An evaluation was carried out by Iles and Shenton-Taylor (1986) that indicated good correlation of laboratory measurements with the reference phase contrast method, but field measurements gave much poorer agreement. The FAM manufacturer indicated that this study may have suffered from poor reliability of the instrument. It was found that instruments constructed in the first several years of production were notoriously unreliable, primarily because of the ease of laser misalignment. Improvements in the instrument ruggedness largely eliminated these and other stability problems. However, no further detailed studies of the FAM have been carried out since the mid-1980s.

The FAM was used as a monitor for asbestos removal operations. The FAM provided real-time indication of fiber concentrations inside and outside these sites to ensure the integrity of the asbestos aerosol containment system. Because the filter sampling typically takes an hour or more with filter analysis adding at least another 30 min, the FAM provided much more rapid feedback to prevent unnecessary exposure to potentially high concentrations of asbestos aerosols.

REFERENCES

Agarwal, J. K. and R. J. Remiarz. 1981. *Development of an Aerodynamic Particle Size Analyzer*. USDHEW-NIOSH Contract Report No. 210-80-0800. Cincinnati, OH: NIOSH.

Ananth, G. and J. C. Wilson. 1988. Theoretical analysis of the performance of the TSI aerodynamic particle sizer. *J. Aerosol Sci.* 9:189–199.

Baron, P. 1983. *Sampler Evaluation with an Aerodynamic Particle Sizer*. International Symposium on Aerosols in the Mining and Industrial Work Environment. Minneapolis, MN: Ann Arbor Press.

Baron, P. A. 1986. Calibration and use of the aerodynamic particle sizer (APS 3300). *Aerosol Sci. Technol.* 5(1):55–67.

Baron, P. A. and K. Willeke. 1986. Respirable droplets from whirlpools: Measurements of size distribution and estimation of disease potential. *Environ. Res.* 39:8–18.

Baron, P. A., J. M. Yacher, and W. A. Heitbrink. 1996. *Some Observations on the Response of the Aerosizer to Droplets in the 4–18 μm Range*. Presented at the American Association for Aerosol Research Annual Conference, Orlando, FL.

Bartley, D. L., P. A. Baron, A. B. Martinez, D. R. Secker, and E. Hirsch. 2000. Droplet distortion in accelerating flow. *J. Aerosol Sci.* 31:1447–1460.

Brockmann, J. E. and D. J. Rader. 1990. APS response to nonspherical particles and experimental determination of dynamic shape factor. *Aerosol Sci. Tech.* 13:162–172.

Brosseau, L. M., D. Vesley, N. Rice, K. Goodell, M. Nellis, and P. Hairston. 2000. Differences in detected fluorescence among several bacterial species measured with a direct-reading particle sizer and fluorescence detector. *Aerosol Sci. Technol.* 32(6):545–558.

Chen, B. T., Y. S. Cheng, and H. C. Yeh. 1985. Performance of a TSI aerodynamic particle sizer. *Aerosol Sci. Technol.* 4:89–97.

Chen, B. T., Y. S. Cheng, and H. C. Yeh. 1990. A study of density effect and droplet deformation in the TSI aerodynamic particle sizer. *Aerosol Sci. Technol.* 12:278–285.

Chen, C.-C., C.-Y. Lai, T.-S. Sheng, and J.-S. Hwang. 1999. Laboratory performance comparison of respirable samplers. *Am. Ind. Hyg. Assoc. J.* 60(5):601–611.

Cheng, Y. S., E. B. Barr, I. A. Marshall, and J. P Mitchell. 1993. Calibration and performance of an API Aerosizer. *J. Aerosol Sci.* 24:501–514

Cheng, Y. S., B. T. Chen, and H. C. Yeh. 1990. Behavior of isometric nonspherical aerosol particles in the aerodynamic particle sizer. *J. Aerosol Sci.* 21(5):701–710.

Cheng, Y. S. and B. E. Dahneke. 1979. Properties of continuum source particle beams. II. Beams generated in capillary expansions. *J. Aerosol Sci.* 10:363–368.

Clift, R., J. R. Grace, and M. E. Weber. 1978. *Bubbles, Drops and Particles*. New York: Academic Press.

Cole, K. R. 1999. *High Precision Micromotion Aerosol Spectrometry*. Ph.D. Dissertation. University of Arkansas, Little Rock.

Cole, K. R. and K. B. Tennal. 1993. Acoustic measurement of aerosol particles. *Aerosol Sci. Tech.* 19: 339–350.

Dahneke, B. 1973. Aerosol beam spectrometry. *Nature Physical Sci.* 244:54–55.

Dahneke, B. E. and Y. S. Cheng. 1979. Properties of continuum source particle beams I. Calculation methods and results. *J. Aerosol Sci.* 10:257–274.

Dahneke, B. and D. Padliya. 1977. Nozzle-inlet design for aerosol beam. *Instruments in Rarefied Gas Dynamics*, 51, Part II, pp. 1163–1172.

Fabriés, J. F., P. Görner, E. Kauffer, R. Wrobel, and J. C. Vigneron 1998. Personal thoracic CIP10-T sampler and the static version CATHIA-T. *Ann. Occup. Hyg.* 42(7):453–466.

Fuchs, N. A. 1964. *Mechanics of Aerosols.* New York: Pergamon Press.

Griffiths, W. D., P. J. Iles, and N. P. Vaughan. 1986. The behaviour of liquid droplet aerosols in an APS3300. *J. Aerosol Sci.* 17(6):921–930.

Heitbrink, W. A., P. A. Baron, and K. Willeke. 1991. Coincidence in time-of-flight spectrometers: Phantom particle creation. *Aerosol Sci. Technol.* 14:112–126.

Ho, J., M. Spence, and P. Hairston. 1999. Measurement of biological aerosol with a fluorescent aerodynamic particle sizer (FLAPS): Correlation of optical data with biological data. *Aerobiologia.* 15:281–291.

Iles, P. J. and T. Shenton-Taylor. 1986. Comparison of a fibrous aerosol monitor (FAM) with the membrane filter method for measuring airborne asbestos concentrations. *Ann. Occup. Hyg.* 30(1):77–87.

John, W. and N. Kreisberg. 1999. Calibration and testing of samplers with dry, polydisperse latex. *Aerosol Sci. Technol.* 31(2–3):221–225.

Kenny, L. C. and R. A. Gussman. 1997. Characterization and modelling of a family of cyclone aerosol preseparators. *J. Aerosol Sci.* 28(4):677–688.

Kinney, P. D. 1990. *Inlet efficiency study for the TSI Aerodynamic Particle Sizer*. M.S. thesis. Department of Mechanical Engineering, University of Minnesota.

Kinney, P. D., D. Y. H. Pui, and D. B. Blackford. 1989. *Particle sizing resolution and transmission efficiency of several inlet designs for the TSI Aerodynamic Particle Sizer*. American Association for Aerosol Research Annual Meeting, Reno, Available from TSI.

Lee, K. W., J. C. Kim, and D. S. Han. 1990. Effects of gas density and viscosity on response of Aerodynamic Particle Sizer. *Aerosol Sci. Technol.* 13:203–212.

Lefebvre, A. H. 1989. *Atomization and Sprays*. Bristol, PA: Taylor and Francis.

Lilienfeld, P. 1985. Rotational electrodynamics of airborne fibers. *J. Aerosol Sci.* 16(4):315–322.

Lilienfeld, P. 1987. Light scattering from oscillating fibers at normal incidence. *J. Aerosol Sci.* 18(4):389.

Lilienfeld, P., P. Elterman, and P. Baron. 1979. Development of a prototype fibrous aerosol monitor. *Am. Ind. Hyg. Assoc. J.* 40(4):270–282.

Madelin, T. M. and H. E. Johnson. 1992. Fungal and *Actinomycete* spore aerosols measured at different humidities with an aerodynamic particle sizer. *J. Appl. Bacteriol.* 72(5):400–409.

Marshall, I. A., J. P. Mitchell, and W. D. Griffiths. 1991. The behaviour of regular-shaped nonspherical particles in a TSI aerodynamic particle sizer. *J. Aerosol Sci.* 22(1):73–89.

Maynard, A. D. 1999. Measurement of aerosol penetration through six personal thoracic samplers under calm air conditions. *J. Aerosol Sci.* 30(9):1227–1242.

Maynard, A. D., L. C. Kenny, and P. E. J. Baldwin. 1999. Development of a system to rapidly measure sampler penetration up to 20 µm aerodynamic diameter in calm air using the Aerodynamic Particle Sizer. *J. Aerosol Sci.* 30(9):1215–1226.

Mazumder, M. K., N. Grable, Y. Tang, S. O'Connor, and R. A. Sims. 1999. Real-time particle size and electrostatic charge distirubtion analysis and its applications to electrostatic processes. *Inst. Phys. Conf. No. 163*, pp. 335–347.

Mazumder, M. K. and K. J. Kirsch. 1977. Single particle aerodynamic relaxation time analyzer. *Rev. Sci. Instrum.* 48(4):622.

Mazumder, M. K., R. E. Ware, and W. G. Hood. 1983. Simultaneous measurements of aerodynamic diameter and electrostatic charge on single-particle basis. In *Measurements of Suspended Particles by Quasi-Elastic Light Scattering*, ed. B. Dahneke. New York: John Wiley & Sons.

Mazumder, M. K., R. E. Ware, J. D. Wilson, R. G. Renninger, F. C. Hiller, P. C. McLeod, R. W. Raible, and M. K. Testerman. 1979. SPART analyzer: Its application to aerodynamic size distribution measurements. *J. Aerosol Sci.* 10:561–569.

Mazumder, M. K., R. E. Ware, T. Yokoyama, B. J. Rubin, and D. Kamp. 1991. Measurement of particle size and electrostatic charge distributions on toners using E-SPART analyzer. *IEEE Trans. Ind. Appl.* 27(4):611–619.

Mitchell, J. P. and M. W. Nagel. 1996. An assessment of the API Aerosizer for the real time measurement of medical aerosols from the pressurized metered-dose inhaler system. *Aerosol Sci. Technol.* 25: 411–423.

Mitchell, J. P., M. W. Nagel, and Y. S. Cheng. 1999. Use of the Aerosizer aerodynamic particle size analysis to characterize aerosols from pressurized metered-dose inhalers (pMDIs) for medication delivery. *J. Aerosol Sci.* 30:467–477.

National Institute for Occupational Safety and Health. 1994. Fibers, Method 7400 Issue No. 2 (8/15/94). *NIOSH Manual of Analytical Methods*. Cincinnati, OH: DHHS (NIOSH). Pub. No. 84–100.

Oskouie, A. K., H.-C. Wang, R. Mavliev, and K. E. Noll. 1998. Calculated calibration curves for particle size determination based on time-of-flight (TOF). *Aerosol Sci. Technol.* 29(5):433–441.

Peters, T. M., H. M. Chein, D. A. Lundgren, and P. B. Keady. 1993. Comparison and combination of aerosol size distributions measured with a low pressure impactor, differential mobility particle sizer, electrical aerosol analyzer, and aerodynamic particle sizer. *Aerosol Sci. Technol.* 19(3):396–405.

Rader, D. J., J. E. Brockmann, D. L. Ceman, and D. A. Lucero. 1990. A method to employ the APS factory calibration under different operating conditions. *Aerosol Sci. Technol.* 13(4):514–521.

Remiarz, R. J., J. K. Agarwal, F. R. Quant, and G. J. Sem. 1983. Real-time aerodynamic particle size analyzer. In *Aerosols in the Mining and Industrial Work Environments*, Vol. 3, eds. V. A. Marple and B. Y. H. Liu. Ann Arbor, MI: Ann Arbor Science.

Renninger, R. G., M. K. Mazumder, and M. K. Testerman. 1981. Particle sizing by electrical single particle aerodynamic relaxation time analyzer. *Rev. Sci. Instrum.* 52(2):242.

Sioutas, C., E. Abt, J. M. Wolfson, and P. Koutrakis. 1999. Evaluation of measurement performance of the scanning mobility particle sizer and aerodynamic particle sizer. *Aerosol Sci. Technol.* 30(1):84–92.

Szewczyk, K., M. Lehtimaki, K. Willeke, D. E. Elliot, and W. L. Muth 1992. Measurement of the change in aerosol size distribution with bacterial growth in a pilot scale fermentor. *Biotech. Bioeng.* 39: 243–245.

Thornburg, S., J. Cooper, and D. Leith. 1999. Counting efficiency of the API Aerosizer. *J. Aerosol Sci.* 30:479–488.

Tsai, C. J., H. M. Chein, S. T. Chang, and J. Y. Kuo. 1998. Performance evaluation of an API Aerosizer. *J. Aerosol Sci.* 29:839–853.

Tsai, C. J., W. S. Yang, W. W. Szymanski, and H. M. Chein. 1999. Particle transmission efficiency through the nozzle of the API Aerosizer. *J. Aerosol Sci.* 30:1019–1028.

TSI, Inc. 1987. *Model APS33B System Aerodynamic Particle Sizer Instruction Manual.* TSI, Inc., St. Paul, MN.

Wake, D. 1989. Anomalous effects in filter penetration measurements using the aerodynamic particle sizer (APS 3300). *J. Aerosol Sci.* 20(1):13–17.

Wang, H. and W. John. 1989. A simple iteration procedure to correct for the density effect in the Aerodynamic Particle Sizer. *Aerosol Sci. Technol.* 10:501–505.

Wang, J. C. and W. John. 1987. Particle density correction for the Aerodynamic Particle Sizer. *Aerosol Sci. Technol.* 6:191–198.

Willeke, K. and B. Y. H. Liu. 1976. Single particle optical counter: Principles and application. In *Fine Particles: Aerosol Generation. Measurement, Sampling and Analysis*, ed. B. Liu. New York: Academic Press.

Wilson, J. C. and B. Y. H. Liu. 1980. Aerodynamic particle size measurement by laser-Doppler velocimetry. *J. Aerosol Sci.* 11(2):139–150.

Ye, Y. and D. Y. H. Pui. 1990. Particle deposition in a tube with an abrupt contraction. *J. Aerosol Sci.* 21(1):29–40.

18 Electrical Techniques

RICHARD C. FLAGAN*
California Institute of Technology, Pasadena, CA

INTRODUCTION

Ultrafine aerosol particles, those below about 0.1 μm diameter, are difficult to measure optically due to the weak light scattering from particles much smaller than the wavelength of light. Inertial or gravitational separations of such particles are possible only at pressures well below atmospheric. Despite these difficulties in separation and direct detection of ultrafine particles, they can be sized with higher resolution and counted with higher efficiency than are generally possible for larger particles. The principal method for sizing ultrafine particles is based on the ease with which a known electrostatic force can be applied to such particles.

All sizes of aerosol particles are readily charged during generation by a variety of charge emission mechanisms or by the attachment of gas ions. Large particles frequently acquire multiple charges, but the probability of charging decreases with decreasing particle size. For sufficiently small particles, most charged particles will carry only one elementary unit of charge so that the charge on that particle is $q = \pm e$. If a singly charged particle is placed in an electric field of known strength, it will migrate with a velocity that depends only on the particle size and structure. As a result, the particle size can readily be determined. Moreover, even though such particles have relatively high diffusivities, the forces imparted on them can be sufficiently large to overwhelm diffusional effects, making high-resolution sizing possible.

To measure the particle size distribution by mobility techniques, one must know the distribution of charges on particles with respect to size and among particles of a given size. Typically, few ultrafine particles acquire a charge, so the particles with a given velocity will account for only a small fraction of the total particles of a given size. Aerosol charging systems for which the charge distributions are well known make it possible, however, to determine the size distribution of aerosol particles from the numbers of charged particles migrating at different velocities. A variety of techniques have been developed that allow determination of the particle size distribution based on the distribution of particles with respect to their electrophoretic migration velocities. Because of the increasing probability of a particle acquiring multiple charges with increasing size, the migration velocities of large particles can vary from particle to particle, so migration-based particle size analysis is generally limited to particles smaller than about 1 μm in diameter. Even at that size, the finite probability of multiple charging complicates data analysis.

Aerosol Measurement: Principles, Techniques, and Applications, Second Edition, Edited by Paul A. Baron and Klaus Willeke.
ISBN 0-471-35636-0

Electrostatic forces are also used in sampling systems, in other technologies that are employed for aerosol particle characterization, and in particle levitation apparatuses that facilitate extended study of a single charged particle without interferences created by a substrate. The latter systems are discussed in Chapter 20. This chapter addresses ways that electrostatic forces are employed for sampling and measurement of aerosol particles.

BEHAVIOR OF CHARGED PARTICLES

A particle with charge q in an electric field of strength E experiences a force of strength $F = qE$.* Because of the low ion densities and slow charge transfer kinetics in ambient temperature gases, aerosol particles usually carry only a small number of elementary charges. The charge is, therefore, conveniently represented as $q = ne$, where $e = 1.609 \times 10^{-19}$ is the magnitude of the elementary unit of charge and n is the number of charges that a particle carries ($n < 0$ for negatively charged particles, and $n > 0$ for positively charged particles). For times that are long compared with the aerodynamic relaxation time, $\tau_a = m_p B$, or frequencies sufficiently low compared with τ_a^{-1}, a charged particle will migrate at a steady-state migration velocity relative to the gas motion of

$$\boldsymbol{v}_e = \boldsymbol{v} - \boldsymbol{u} = neB\boldsymbol{E} = Z_p\boldsymbol{E} \tag{18–1}$$

where B and Z_p are the mechanical and electrical mobilities of the particle, respectively. For motion of spherical particles with diameter d_p in the Stokes regime, this mobility is

$$Z_p = \frac{neC_c(Kn)}{3\pi\eta d_p} \tag{18–2}$$

where η is the gas viscosity, C_c is the slip correction factor that accounts for noncontinuum effects when the particle size becomes comparable with or smaller than the mean free path of the gas molecules, λ, or in dimensionless terms, when the Knudsen number,

$$Kn = \frac{2\lambda}{d_p} \tag{18–3}$$

becomes large (see Chapter 4). The slip correction factor is given by

$$C_c = 1 + Kn\left(\alpha + \beta\exp\left[-\frac{\gamma}{Kn}\right]\right) \tag{18–4}$$

where the empirically determined coefficients are $\alpha = 1.142$, $\beta = 0.558$, and $\gamma = 0.999$ (Allen and Raabe, 1985).

Figure 18–1 shows the variation of the electrical mobility with particle size under normal ambient atmospheric conditions for particles carrying different numbers of elementary units of charge. For particles with a given charge in the continuum size regime, $Z_p \propto d_p^{-1}$, while in the free molecular regime, $Z_p \propto d_p^{-2}$. Because of this strong dependence of the mobility on particle size, the range of electric field strengths required to classify particles throughout the

* Caution must be exercised to avoid unit conversion problems in calculations of the forces and energies where electrostatics are involved. The force produced by a charge of 1C by an electric field of strength $E = 1\ \mathrm{V\,m^{-1}}$ is 1N. The force between two charges q_1 and q_2 separated by a distance r is $F = q_1q_2/4\pi\varepsilon_0 r^2\varepsilon$, where $\varepsilon_0 = 8.542 \times 10^{-12}\ \mathrm{C^2\,N^{-1}\,m^{-2}}$, and ε is the dielectric constant of the medium ($\varepsilon_{air} = 1.00054$).

EXAMPLE 18–1: CALCULATION OF MIGRATION VELOCITY OF A CHARGED PARTICLE IN AN ELECTRIC FIELD

Consider a singly charged 1 μm diameter particle. What is its migration velocity in an electric field of strength $E = 10^6\,\mathrm{V\,m^{-1}}$?

Answer: To evaluate the mobility of the particle using Eq. 18–2, the slip correction must be known. The particle Knudsen number is

$$Kn = \frac{2\lambda}{d_p} = \frac{1 \times 6.5 \times 10^{-8}\,\mathrm{m}}{1 \times 10^{-8}\,\mathrm{m}} = 0.130$$

The corresponding slip correction factor is

$$C_c = 1 + Kn\left[1.142 + 0.558\exp\left(-\frac{0.999}{Kn}\right)\right] = 1.16$$

The viscosity of air at ambient conditions is $m = 1.81 \times 10^{-5}\,\mathrm{kg\,m^{-1}\,s^{-1}}$, and the charge on the electron is $e = 1.61 \times 10^{-19}\,\mathrm{C}$, so the particle mobility becomes

$$Z_p = \frac{1 \times 1.61 \times 10^{-19}\,\mathrm{C} \times 1.16}{3\pi \times 1.81 \times 10^{-5}\,\mathrm{kg\ m^{-1}s^{-1}} \times 10^{-6}\,\mathrm{m}} = 1.09 \times 10^{-9}\,\mathrm{m^2V^{-1}s^{-1}}$$

The migration velocity is the product of the mobility and the electric field strength, that is,

$$v_{mig} = 1.09 \times 10^{-9}\,\mathrm{m^2V^{-1}s^{-1}} \times 10^6\,\mathrm{Vm^{-1}} = 1.09 \times 10^{-3}\,\mathrm{ms^{-1}}$$

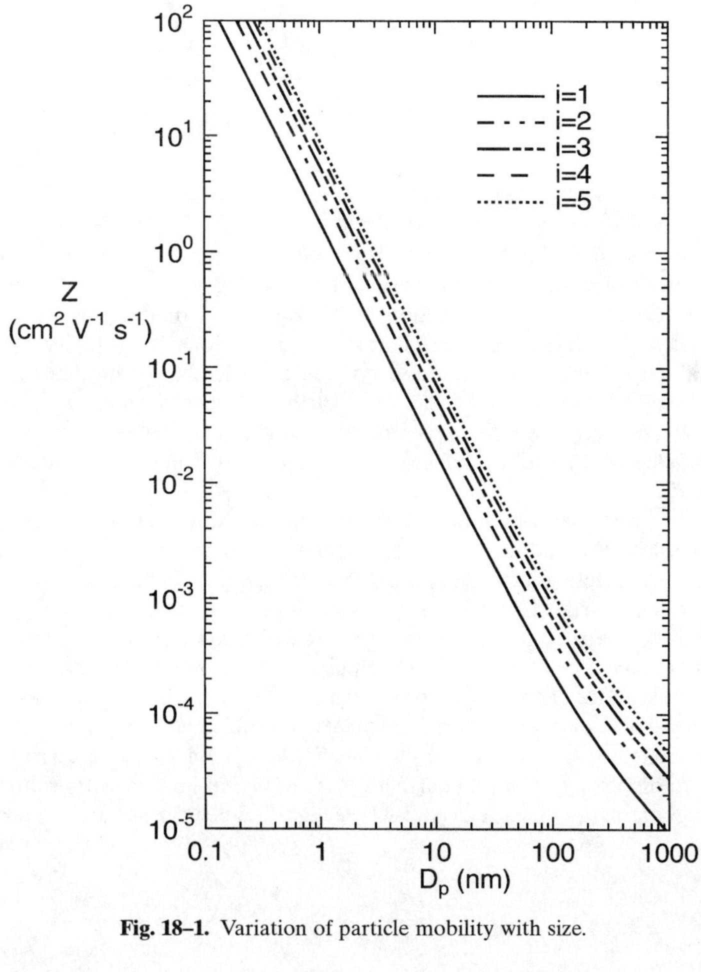

Fig. 18–1. Variation of particle mobility with size.

submicrometer size range exceeds the capabilities of a single instrument operating at fixed flow rates.

RELATIONSHIP BETWEEN MIGRATION AND DIFFUSION

Mobility techniques are generally applied to small particles that migrate at relatively high velocities in attainable electric fields. These same particles diffuse rapidly. To compare these two effects, consider particles that migrate a distance b under the action of an applied electric field, E, produced by applying a voltage difference V over the distance b. The time required to migrate that distance is $\tau = b/Z_pE$. The particle diffusivity is related to the mobility by the Einstein relation $D = BkT$, where k is the Boltzmann constant and T is the temperature. The ratio of the migration distance to the root mean square displacement due to Brownian diffusion during the migration time, $\langle x^2\rangle^{\frac{1}{2}} = \sqrt{2Dt}$ (Eq. 4–15), is simply

$$\frac{b}{\langle x^2\rangle^{\frac{1}{2}}} = \sqrt{\frac{neV}{2kT}} \qquad (18\text{–}5)$$

This measure of the resolution of mobility-based sizing techniques is, thus, proportional to the square root of the ratio of the electrostatic potential energy of the charged particle in the applied electric field to its thermal energy. For large voltages, migration will dominate, while for small applied voltages diffusion will distort the response of any mobility-based particle measurement. At ambient temperatures, $kT/e \approx 26\,\text{mV}$. The voltage that is required for a given measurement depends on the desired mobility or size resolution. Electrostatic breakdown within the apparatus limits the maximum operating voltage and, therefore, the highest attainable resolution.

AEROSOL CHARGE CONDITIONING

Aerosols generally include substantial numbers of charged particles. The charged accumulation-mode aerosol particles were initially called *large ions* by Langevin (1903). Mobility characterization of aerosols requires a known charge on those particles that are classified—ideally a single elementary charge in most applications. Known charge on the classified particle allows accurate determination of the particle size from measurements of the electrical mobility. Measurement of the size distribution also requires that the fraction of particles of a given size that carry a given number of charges be accurately known. Hence, a critical facet of electrical mobility characterization of an aerosol is the production of a known charge distribution.

A number of processes contribute to the charging of aerosol particles: static electrification, photoemission, thermionic emission, charging by small gas ions, and self-charging of radioactive aerosols. Static electrification involves transfer of charge to a particle as it is separated from bulk material. This mechanism causes particles to acquire charge as they are generated from the bulk material and can contribute to charging when particles strike and rebound from surfaces. Several physical mechanisms contribute to this electrolytic charging. These include (1) contact charging resulting from differences in the electrochemical potential between a particle and a contacting surface; (2) induced charging resulting from charge transfer when a particle separates from a surface in the presence of an electric field; (3) spray charging due to disruption of liquid surfaces (often in combination with induced charging); and (4) electrolytic charging that occurs when highly dielectric liquids are separated from solid surfaces. The particle charge acquired during static electrification is difficult to predict, so it is rarely employed in size distribution measurements.

When surfaces are heated to sufficiently high temperature, they may undergo thermionic emission. This mechanism causes particles processed in flames or other high-temperature systems to acquire similar charge, that is, all positive at high temperature, and may lead to differences in polarity of charge in different regions of the flame. Gas ions are generally present in high concentrations in such environments, so ion–particle reactions can lead to rapid charge redistribution as the particles leave the hot region.

Radioactive particles may become charged due to loss of α or β particles. Self-charging may also occur due to ejection of valence electrons or the release of charged fragments during α, β, or γ radiation (Yeh, 1976; Yeh et al., 1978).

Only rarely can the charge distribution of the sampled aerosol be used directly for mobility measurements. Because the initial charge state of the aerosol is rarely known with any precision, the aerosol must be conditioned to produce a known charge state. In an ideal system, one might seek to produce an equilibrium charge distribution because that distribution can be determined from well-established thermodynamic considerations. The process of attaining that distribution by charge transfer reactions with an electrically neutral cloud of positive and negative ions is called *bipolar diffusion charging* when it is used to produce the charge state for aerosol classification and aerosol neutralization when applied to dissipating a nonequilibrium charge state on the aerosol. Although, as will be shown below, the method can produce a well-characterized charge distribution, at the small end of the electrical mobility sizing range only a tiny fraction of particles acquire any change. Moreover, the charge distribution generally differs from that at thermodynamic equilibrium. Numerous studies have attempted to resolve the charge state of the aerosol produced in aerosol neutralizers, while others report on attempts to design new aerosol chargers that produce a known charge distribution with much higher probabilities of charging the ultrafine particles.

Most approaches to bringing the charge on aerosol particles to a known distribution involve reactions with small gas ions. Depending on whether the gas contains only ions of a single polarity or contains a mix of positive and negative ions, the aerosol may undergo unipolar or bipolar charging, respectively. The migration of the charges to the particle may be by Brownian diffusion alone, in which case the aerosol is said to undergo *diffusion charging*, or it may be augmented by an imposed, macroscopic electric field, in which case the aerosol undergoes *field charging*. The electric field strengths required for field charging to be significant are large enough that the particles undergo significant migration during the charging process. This introduces considerable uncertainty in mobility analysis of particle properties and leads to substantial deposition. Because of these complications to mobility analysis, field charging is rarely employed for aerosol measurements, although it is often important in collection of aerosol samples by electrostatic precipitation, as it is in gas cleaning. A number of chargers do, however, impose weak electric fields to manipulate the gas ions and achieve an additional degree of control over the diffusion charging process. Most mobility analyses rely on some form of diffusional ion–particle reactions, taking advantage of the high degree of control of the ultimate charge state of the aerosol that diffusion charging enables.

One additional approach to particle charging is used in some special cases, that is, photoelectron emission charging of particles due to illumination with short wavelength radiation. Photoelectron emission is highly dependent on the composition of the surface being illuminated, so it affords an opportunity to infer particle composition from its charging behavior.

Equilibrium Charge Distribution

Although the minimum charge on an aerosol particle is, in principle, zero, this condition is rarely achieved because ions present in the gas attach to some particles. Pairs of positive and negative ions are produced by collisions of cosmic rays and energetic nuclear particles produced by radioactive decay with the gas molecules. Ambient ion concentrations typically amount to about $10^3\,cm^{-3}$, although this number can vary widely as gas ions are lost to the surfaces of aerosol particles and undergo ion–ion recombination reactions.

It has often been hypothesized that, after long exposure to such bipolar ion mixtures, frequent ion–particle collisions will bring the particles to a state of charge equilibrium with the ionic atmosphere. In this equilibrium state, the fraction of particles of diameter d_p that carry n charges is described by the Boltzmann charge distribution, that is,

$$f_n = \frac{\exp\left(-\frac{n^2}{2\sigma^2}\right)}{\sum_{k=-\infty}^{\infty}\exp\left(-\frac{k^2}{2\sigma^2}\right)} \tag{18–6}$$

$$\sigma^2 = \frac{d_p kT}{2e^2} \tag{18–7}$$

While the simplicity of this distribution is attractive, and some investigators report that measured charge distributions have been found to be in reasonable agreement with it (Liu and Pui, 1974), its use would only be rigorously justified if charge detachment were to occur at appreciable rates (Fuchs, 1963). The forces between an ion and a particle to which it has become attached are, however, so strong that such ion emissions do not occur spontaneously at ambient temperatures unless the total charge on the particle is extremely high.

Nonetheless, measured charge distributions are often similar in form to the Boltzmann distribution. These similarities result from reactions of particles with positive and negative ions in a bipolar ion mixture. Reactions of ions with particles carrying opposite charges are favored over reactions when the charges have the same polarity. After exposure to a bipolar ion mixture, the aerosol charge distributions asymptotically approaches a steady state. While similar to the equilibrium distribution, the steady-state charge distribution is generally asymmetrical due to differences in the properties of positive and negative ions. When the concentrations of positive and negative ions differ, this asymmetry increases and no true steady-state exists. The extreme limit of unipolar charging occurs when the aerosol is exposed only to ions of one polarity. To understand real charge distributions anywhere in this continuum of diffusional charging, we must examine the kinetics of aerosol particle charging.

Diffusion Charging: Charge Transfer Kinetics

The rate of ion attachment to particles in the absence of an external electric field is determined by the coupled rates of diffusion of the ions and ion migration in the electric field that is induced by the interactions between their respective charges. Coulombic forces lead to attraction or repulsion at long distances; image forces lead to attractive forces at small separations. For particles large compared with the ionic mean free path,

$$\lambda_i = \frac{4 m_g \bar{c}_i |Z_i|}{3e\left(1+\frac{m_g}{m_i}\right)^{\frac{1}{2}}} \tag{18–8}$$

The rate of ion arrival at the surface of the particle is given by the convective diffusion equation, taking into account the ion migration induced by the electric field created by the charge already present on the particle and Brownian diffusion (m_i and m_g are the molecular masses of the ion and background gas, respectively, and $\bar{c}_i = \sqrt{8k_BT/\pi m_i}$ is the mean thermal speed of the ions).

For smaller particles, Fuchs (1963) developed the so-called limiting sphere approach: Outside a region of radius δ_F, the ion transport is described by the continuum convective diffusion equation; inside that radius a kinetic (free-molecular) transport model is applied. δ_F can be calculated in terms of the particle ion Knudsen number, $Kn_i = 2\lambda_i/d_p$, that is,

$$\delta_F = \frac{d_p}{2}\left(\frac{2}{Kn}\right)^2\left[\frac{1}{5}(1+Kn_i)^5 - \frac{1}{3}(1+Kn_i^2)(1+Kn_i)^3 + \frac{2}{15}(1+Kn_i^2)^{\frac{5}{2}}\right] \tag{18–9}$$

The number of ions striking the particle per unit time is the product of the number that would strike the particle in the absence of electrostatic forces, $J_0 = \pi\delta_F^2 n(\delta_F)\bar{c}_i(d_p/2\delta_F)^2$, and an attachment coefficient α. Here, $n(\delta_F)$ is the ion concentration at the limiting sphere, $4\pi\delta_F^2 n(\delta_F)\bar{c}_i/4$ is the number of ions emerging from the limiting sphere, and $(d_p/2\delta_F)^2$ is the fraction of those ions that reaches the particle. The attachment coefficient is the fraction of ions that enters the limiting sphere and ultimately collides with the particle; $\alpha > (d_p/2\delta_F)^2$ when the forces are attractive, and $\alpha < (d_p/2\delta_F)^2$ for repulsive forces.

In the presence of electrical forces, the fraction α is determined from the radius of closest approach, called the *apsoidal distance*, r_a. Without image forces, the ion would only collide with the particle when $r_a < d_p/2$. The short-range image forces can produce particle trajectories that have no apse and spiral in toward the particle. The minimum apse divides particle trajectories that escape capture from those that spiral into the particle (Hoppel and Frick, 1986). Hoppel and Frick (1986) note that, in the original development of this model, Fuchs (1963) incorrectly assumed that $\alpha = 1$ for all attractive encounters. Hoppel and Frick (1986) corrected this error and also accounted for three-body trapping to produce the ion attachment coefficients shown in Figure 18–2. With these attachment coefficients, it is now possible to evaluate the charge distributions resulting from various aerosol charging systems.

Given α, the frequency of ion attachment to particles carrying charge k is

$$\beta_{p,k} = 4\pi\delta_F^2\frac{n_p(\delta_F)\bar{c}_i}{4}\alpha \quad (s^{-1}) \tag{18–10}$$

where $p = \pm$ denotes the polarity of the ion. Using the attachment coefficients, the charge state of the aerosol can be determined by solving a system of balance equations for the particle concentration N_k at each charge level k. The rates of loss of positive and negative gas ions, concentrations n_+ and n_-, are

$$\frac{dn_+}{dt} = \rho_i - \beta_\pm n_+ n_- - n_+\sum_{k=-\infty}^{\infty}\beta_{+,k}N_k \tag{18–11}$$

$$\frac{dn_-}{dt} = \rho_i - \beta_\pm n_+ n_- - n_-\sum_{k=-\infty}^{\infty}\beta_{-,k}N_k \tag{18–12}$$

where ρ_i is the rate of production of ion pairs and $\beta_\pm$ is the ion–ion recombination rate coefficient. The rate equation for N_k is

$$\frac{dN_k}{dt} = \beta_{+,(k-1)}n_+N_{k-1} + \beta_{-,(k+1)}n_-N_{k+1} - \beta_{+,k}n_+N_k - \beta_{-,k}n_-N_k : -\infty < k < \infty \tag{18–13}$$

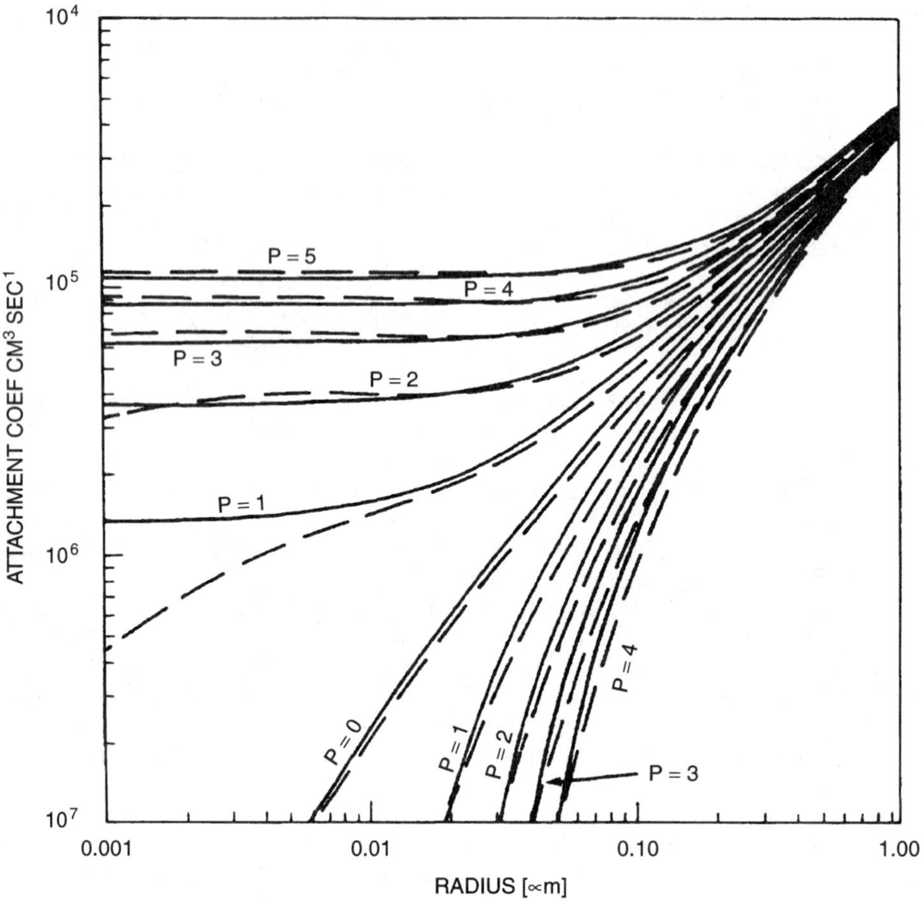

Fig. 18–2. Ion attachment coefficients calculated using the modified Fuchs (1963) limiting sphere approach of Hoppel and Frick (1986). The dashed lines show the attachment coefficients that result when the Fuchs α coefficients are not corrected. (From Hoppel and Frick, 1986, with permission.)

The total particle concentration

$$N_{tot} = \sum_{k=-\infty}^{\infty} N_k \tag{18–14}$$

is constant, leading to a redundancy that is usually eliminated by replacing the equation Eq. 18–13 for $k = 0$ with Eq. 18–14.

Bipolar Diffusion Charging

When a mixture of positive and negative ions is present, the aerosol is said to undergo bipolar diffusion charging. This is a process in which the magnitude of the charge on the particle may increase due to the attachment of charges of like polarity or may decrease when charges of opposite polarity attach to the particle. If the ion populations are initially balanced, that is, $n_+(t = 0) = n_-(t = 0)$, the charge state of the aerosol will eventually reach a quasi steady state at which the rate of attachment of positive ions exactly balances the rate of attachment of negative ions. Assuming that the ratio of the concentrations of positive and negative gas ions is fixed, the aerosol balance equations decouple from the ion balance equations. Detailed balancing then requires that in equilibrium

$$\beta_{+,(k-1)} n_+ N_{k-1} = \beta_{-,k} n_- N_k \tag{18–15}$$

and the ratio of the number of particles carrying k charges to the number of uncharged particles, N_0, becomes

$$\frac{N_k}{N_{tot}} = \left(\frac{n_+}{n_-}\right)^k \prod_{m=1}^{|k|} \frac{\beta_{p,p(m-1)}}{\beta_{-p,pm}} \tag{18–16}$$

where $p = k/|k|$ accounts for the polarity of the charge on the particle. The fraction of particles carrying k charges thus becomes

$$f_k = \frac{N_k/N_0}{N_{tot}/N_0} \tag{18–17}$$

and

$$\frac{N_{tot}}{N_0} = 1 + \sum_{k=1}^{\infty} \left(\frac{n_+}{n_-}\right)^k \prod_{m=1}^{k} \frac{\beta_{+,(m-1)}}{\beta_{-,m}} + \sum_{k=1}^{\infty} \left(\frac{n_-}{n_+}\right)^k \prod_{m=1}^{k} \frac{\beta_{-,-(m-1)}}{\beta_{+,-m}} \tag{18–18}$$

Gas ions have different mobilities due to differences in their molecular weights and, in atmospheric measurements, due to differences in the propensity to form water clusters around the different types of ions. Hoppel and Frick (1986) estimated that $M_+ = 150\,amu$, $M_- = 90\,amu$, $Z_{i+} = 1.20\,\text{cm}^2\,\text{C}^{-1}\,\text{s}^{-1}$, and $Z_{i-} = 1.35\,\text{cm}^2\,\text{V}^{-1}\,\text{s}^{-1}$. The resulting charge state of the aerosol has a distribution of positive and negative charges that is not perfectly balanced due to differences in the ion attachment rates due to differences in the ion mobilities. Figure 18–3 shows the variation of the fraction of particles carrying a given charge with particle size. Calculations are presented for the Boltzmann charge distribution (dots), the uncorrected Fuchs model (dashed lines), and the Hoppel and Frick model including three-body stabilization of ion attachment reactions (solid lines). Several features are apparent in this plot: (1) At the ultrafine end of the size spectrum, the charging probability, while larger than that predicted at equilibrium, is small; (2) for particles larger than about 100 nm diameter, the probability that a particle will acquire more than one charge becomes appreciable; and (3) for particles larger than about 200 nm, several charge states will be present. The low charging probability for ultrafine particles often leads to poor counting statistics in mobility measurements. Multiple charging of large particles causes several sizes of particles to have the same mobility, degrading instrumental resolution and complicating data analysis.

Calculation of the steady-state bipolar charge distribution using the corrected Fuchs model is quite involved. Wiedensohler (1988) developed an approximate representation of the corrected Fuchs charge distribution that is more readily applied. The fraction of particles carrying k charges is estimated to be

$$\frac{N_k}{N_{tot}} = 10^{\left[\sum_{i=0}^{5} a_{i,k} (\log_{10} d_{nm})\right]} \tag{18–19}$$

where d_{nm} is the particle diameter in nanometers; the coefficients $a_{i,k}$ are given in Table 18–1.

While the charge distribution deviates from the equilibrium Boltzmann distribution and is asymmetrical, the stability of the distribution and insensitivity to such parameters as flow

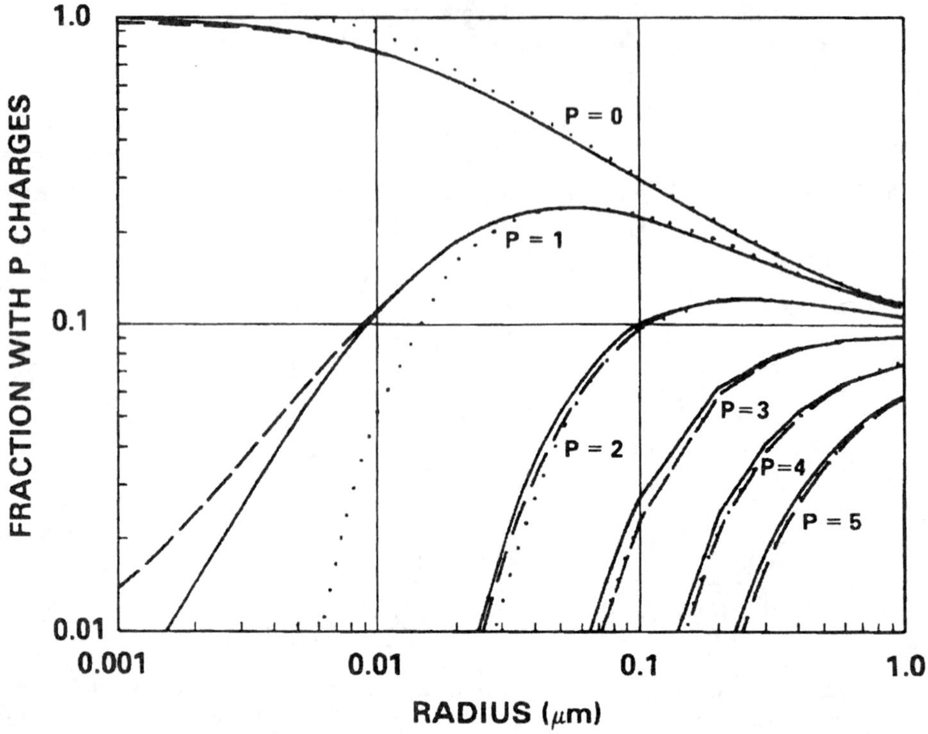

Fig. 18–3. Variation of the steady-state charge fraction with particle size (Hoppel and Frick, 1986). Solid lines are the predictions using the corrected Fuchs model. The dashed lines are the predictions using the Fuchs α coefficients. The dots show the boltzman charge distribution. (From Hoppel and Frick, 1986, with permission.)

TABLE 18–1. Coefficients $\alpha_{i,k}$ for the Wiedensohler (1988) Approximation to the Corrected Fuchs Charge Distribution

k	−2	−1	0	1	2
$\alpha_{0,k}$	−26.3328	−2.3197	−0.0003	−2.3484	−44.4756
$\alpha_{1,k}$	35.9044	0.6175	−0.1014	0.6044	79.3772
$\alpha_{2,k}$	−21.4608	0.6201	0.3073	0.4800	−62.8900
$\alpha_{3,k}$	7.0867	−0.1105	−0.3372	0.0013	26.4492
$\alpha_{4,k}$	−1.3088	−0.1260	0.1023	−0.1553	−5.7480
$\alpha_{5,k}$	0.1051	0.0297	−0.0105	0.0320	0.5049

Note that coefficients $\alpha_{4,1}$ and $\alpha_{5,2}$ differ from those originally published.

rates makes it useful for mobility-based measurements of the particle size distribution. Unfortunately, not all bipolar diffusion chargers attain this ideal distribution. Hoppel and Frick (1990) demonstrated that the imbalance in ion concentrations due to the unequal diffusion of ions to aerosol particles and apparatus walls is magnified in the ionic ratio because ionic recombination rapidly depletes both polarities equally. Hence, the charge distribution changes from the steady state that may (or may not) be present in the region where ion generation takes place in response to the changing environment.

Finally, it is important to note that the large molecular weight of the gas ions, and the correspondingly low mobilities, reflect ions comprising molecular clusters surrounding a core molecular ion. In atmospheric measurements, the primary clustering species is water vapor due to its abundance and high polarity. Observations of such clustering date back to the earliest measurements of gas ions (Nolan, 1926), yet the relationship between the ion properties and atmospheric parameters such as relative humidity remain poorly understood. Most investigations employ published values of the ionic properties and do not consider the possibility that they may vary.

Unipolar Diffusion Charging

Unipolar diffusion charging occurs when only ions of one polarity are present. As charge accumulates on the particle, the repulsive forces cause the rate of charging to slow. Because the accumulated charge is not neutralized by attachment of ions of opposite polarity, higher charge levels on large particles and higher charging probabilities for small ones can be attained than with bipolar diffusion charging. Unipolar diffusion charging was employed in the electrical aerosol analyzer, the first commercial mobility sizer for aerosol particles, to measure the concentrations of transmitted particles as small as 50 nm diameter with an electrometer.

Many theoretical and experimental studies of unipolar diffusion charging have been reported (e.g., Arendt and Kallmann, 1926; Fuchs, 1947; Bricard, 1948; White, 1951; Liu and Pui, 1977; Marlow and Brock, 1975; Davidson and Gentry, 1985; Pui and Liu, 1988). Pui and Liu (1988) showed that the theory of Marlow and Brock (1975) best describes charging of ultrafine particles ($Kn_i = 2\lambda_i/d_p << 1$), while Fuchs and Bricard's theories are best for much larger particles. The kinetic treatment of Fuchs (1963) as extended by Hoppel and Frick (1986) is readily extended to unipolar charging, although closed-form solutions are not obtained and numerical integration is needed. An estimate of the order of magnitude of the mean number of charges, k, acquired by a particle of diameter d_p can be obtained using White's equation (1951), that is,

$$\langle k \rangle = \frac{2\pi\varepsilon_0 d_p kT}{e^2} \ln\left[1 + \frac{\bar{c}_i d_p e^2 N_i t}{8\pi\varepsilon_0 kT}\right] \qquad (18\text{–}20)$$

where t is the charging time in seconds. It should be noted that White's equation only describes the mean charge state; there will be a distribution of charge levels for any particle size for any value of the $N_i t$ product. For large $N_i t$, the charge level varies with particle size as $<k> \propto d_p^2$. In the continuum limit, the electrical mobility of the particle varies as $Z_p = <k>e/3\pi\mu d_p \propto d_p$. Hence, in the large particle limit at high $N_i t$, the particle mobility increases linearly with particle diameter (assuming the particles are spherical). Mirme (1994) has taken advantage of this variation to use a DMA to measure size distributions well beyond the limits of single charging imposed by bipolar diffusion charging. Even in unipolar diffusion charging, charged particles below some size will carry only one charge, so the mobility varies with size as $Z_p \propto d_p^{-2}$. Thus, there is a minimum in the mobility of those particles that do acquire charge. To correct for the contribution of small particles to his large particle size distribution measurements, Mirme simultaneously measured the size distribution of the small particles with a DMA equipped with a bipolar diffusion charger. He then subtracted the signal contributed by the small particles from the large particle measurements.

Field Charging

When a dielectric particle carrying charge q is placed in a uniform electric field E_∞, the distortion of the electric field by the particle causes ions migrating along field lines in the ambient electric field to be deflected toward the particle, causing its charge to increase as

$$q = \frac{q_s e Z_i N_{i\infty} t}{e Z_i N_{i\infty} t + 4\varepsilon_0} \tag{18–21}$$

where q_s is the saturation charge level for a particle of relative permittivity κ

$$q_s = \pi\varepsilon_0 \left(\frac{3\kappa}{\kappa+2}\right) E_\infty d_p^2 \tag{18–22}$$

The permittivity ranges from 1 for a vacuum to ∞ for a conductor. The time required to attain the saturation charge is often quite short, so the saturation charge can be used as a reasonable estimate of the charge level attained under field-charging conditions. When both field and diffusion charging take place, the saturation charge provides an initial charge state for the diffusion-charging process.

Ion Generation

Controlled particle charging is essential to electrical mobility size distribution measurements. From the discussion of charging mechanisms above, it is apparent that production of a well-defined charge distribution can be attained only if the gas contains ions at known, or at least consistent, concentrations. Gas ions can be produced at ambient temperatures by high-voltage corona discharge (White, 1951), radioactive decay (Liu and Pui, 1974), photoelectron emission (Schmidt-Ott et al., 1980), and droplet formation in the presence of an electric field. An extreme form of droplet charging, the electrospray has been used to generate specific ions from solute molecules in atomized solutions and coupled with differential mobility analysis for the characterization of large molecules in solution (Gamero-Castano and de la Mora, 2000). All of the other methods have been used to charge aerosol particles for mobility analysis.

A gas usually contains a few free electrons and a comparable number of positive ions. At sufficiently high electric fields, the free electrons can be accelerated to sufficiently high velocities that their collisions with gas molecules lead to the ejection of additional electrons. A cascade of such events, called an *electron avalanche*, creates the corona and generates large numbers of positive ions and free electrons in the gas. The corona may be operated with the discharge electrode (a needle or wire) held at a high positive potential, in which case the free electrons are attracted toward the wire, or with the discharge electrode at a negative potential to repel the electrons. The negative corona is generally more stable than the positive corona, but it requires that the gas contain a species that can absorb free electrons to be effective. Moreover, it produces O_3. The positive corona does not require an electron-absorbing gas. It is frequently used to charge aerosol particles for measurement and was used in numerous electrostatic precipitators and the original electrical aerosol analyzer charger, which is shown in Figure 18–4. In that charger, the corona is operated with a potential between a central wire and a coaxial screen. The aerosol passes between that screen, which is maintained at a small positive voltage, and a coaxial electrode so that charging takes place in a region of weak electric field and charged particle losses due to migration to the wall are minimized. By monitoring and controlling the ion current arriving at the outer electrode, the $N_i t$ product can be determined. In an earlier version of that charger, Hewitt (1957) introduced an alternating potential in the outer region to reverse the migration of charged particles and further reduce their deposition. This approach has been employed more recently in efforts to improve the efficiency of unipolar charging of particles in the low nanometer size range (Buscher et al., 1994)

The original chargers used in the differential mobility analyzer and in aerosol neutralizers were based on ion generation by radioactive decay. In a typical aerosol device, the gas

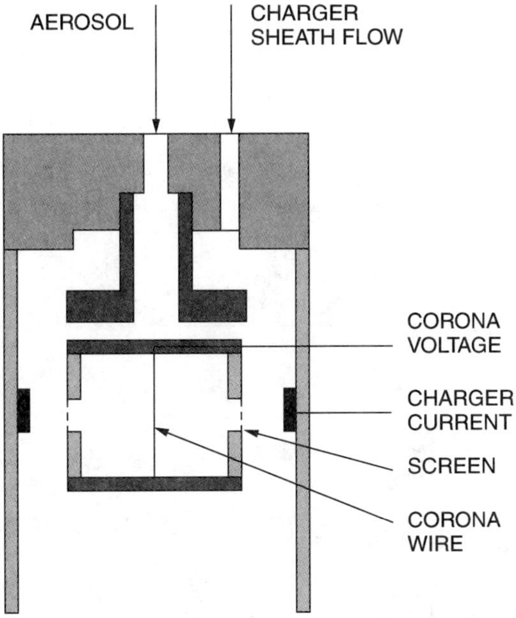

Fig. 18–4. Unipolar diffusion charger used in the ***TSI*** electrical aerosol analyzer.

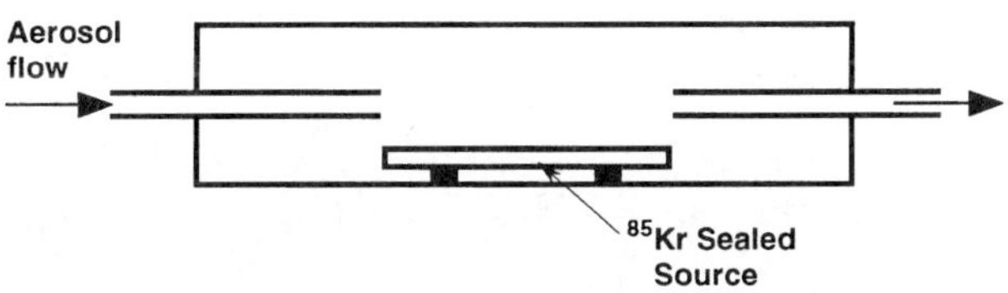

Fig. 18–5. Bipolar diffusion charger based on a sealed ^{85}Kr β-particle source.

flows through a chamber that contains a small radioactive source, such as that shown in Figure 18–5. The design of the charger is a compromise between safety, that is, providing adequate shielding and preventing line-of-sight access to the radioactive material, and factors that determine the charging efficiency, notably the residence time and flow geometry. A number of different isotopes have been used in aerosol neutralizers and chargers. Table 18–2 summarizes some of their properties.

Cooper and Reist (1973) provide a detailed discussion of the issues involved in charge generation by radioactive decay. Heavy α particles (He nucleus) produce short straight tracks with a range in air $R_\alpha = 0.56E$ (cm), where E is the energy of the emitted particle in MeV. Ions are generated at a rate of about 1 ion pair/35.5 eV, with a slight increase in ionization occurring at the end of the trajectory as the ion slows down. The light β particle (electron) produces a long irregular path. These particles are emitted with a continuous spectrum. The energy flux decays with distance as

$$I(x) = I_0 e^{-\mu x} (\text{MeV cm}^{-2}\text{s}^{-1}) \qquad (18\text{–}23)$$

TABLE 18–2. Properties of Radionuclides That Have Been Used in Gas Ion Generation

Source	$\tau_{\frac{1}{2}}$	Radiation Type	Energy (MeV)	R_{air} (cm)	μ (cm^{-1})	k (ion pairs/cm)
^{63}Ni	100 yr	β	0.065		0.449	
^{90}Sr	27.7 yr	β	0.546 (max)		0.044	230
^{90}Y	2.7 d	β	2.18 (max)		0.0089	196
^{85}Kr	10.76 yr	β	0.67 (max)		0.035	220
		γ	0.514 (0.41%)			
^{241}Am	458.6 yr	α	5.49 (85%)	4.0		
			5.44 (13%)	4.0		
^{210}Po	138.4 d	α	5.30 (100%)	3.8		

where the decay rate is $\mu = 17(\text{cm}^4\,\text{g}^{-1})\rho E_{\text{max}}^{-1.14}$. β Particles generate ion pairs at a rate of about 1 ion pair/34 eV. Cooper and Reist address the geometry of ion generation, providing guidance in the design of aerosol chargers and neutralizers that will produce well-known ion densities, thereby enabling reproducible particle charging.

Photoelectron Emission Charging

Photoelectron emission is a well-known method for probing the basic properties of condensed matter. Schmidt-Ott et al. (1980) demonstrated that aerosol particles emit photoelectrons when illuminated with photons with energies above the intrinsic photoemission threshold of the material, ϕ_i. The threshold for electron emission is enhanced by the attractive Coulombic force that will lead to recapture of the emitted electron if the photon energy, $h\nu$, is less than

$$\phi = \phi_i + \frac{e^2(k+1)}{2\pi\varepsilon_0 d_p} \quad k \geq -1 \tag{18–24}$$

where k is the charge on the particle. For a given photon energy, $h\nu$, the maximum number of charges that a particle can acquire by photoemission is, therefore,

$$k_{\text{max}} = (h\nu - \phi_i)\frac{2\pi\varepsilon_0 d_p}{e^2} \tag{18–25}$$

Because photoemission depends on material properties, photoemission charging provides information about the composition of the particles being probed, at least at their surfaces. Particles charged by photoemission may be measured either by mobility analysis or directly. Matter et al. (1995) have demonstrated that the negative charges produced by photoemission can be removed by a weak electric field, enabling the current carried by the positively charged particles to be measured with a Faraday cup aerosol electrometer. See Chapter 14 for use of this technique to measure particle surface area.

Nanoparticle Charging

In the steady-state charge distribution produced by bipolar diffusion charging, only a very small fraction of particles in the low nanometer size range acquire any charge. Combined with high diffusive losses, this leads to low sensitivity of mobility analyzers to nanoparticles. Development of charging systems for this important size range is ongoing. Unipolar diffusion chargers, such as that used in the early electrical aerosol analyzer, produce higher charge levels, although losses due to charged particle migration can be substantial for ultrafine particles. However, as the charging efficiency for ultrafine particles is increased, the size of particles that acquire multiple charges decreases. Modulation of the electric field in the charging region can reduce migration-induced losses (Hewitt, 1957; Buscher et al., 1994). If, however, an electric field is applied in the flow direction, charged particles can be induced to migrate in the direction of the gas flow (Yun et al., 1997; Fomichev et al., 1997; Chen and Pui, 1999). This reduces the losses of the charged particles and decreases the probability that the particle will acquire multiple charges.

PARTICLE SAMPLING

The drift velocity of charged aerosol particles in strong electric fields can be high enough to make the electrostatic precipitator much more efficient than gravity or inertia for particle separations, particularly for ultrafine particles. Moreover, carefully designed systems based on electrophoretic migration can be used to measure the charge and size of aerosol particles.

Electrostatic Precipitation

Electrostatic precipitators (ESPs) have long been used to collect samples of aerosol particles (Tolman et al., 1919). Two basic steps are involved in electrostatic precipitation: (1) charging the particles and (2) subjecting them to a strong electric field to cause them to migrate toward and deposit on a substrate that is suitable for subsequent analysis. Tolman et al. (1919) collected aerosol samples to determine the mass and chemical composition of smokes, an application that has generally been replaced by filters today. More commonly, electrostatic precipitation is now used to collect samples for analysis by microscopy. Field charging is usually employed to charge the particles quickly. The high unipolar ion concentrations produced by corona discharge and the high field strengths lead to rapid charge saturation.

The point-to-plane ESP is widely used to deposit aerosol particles onto electron microscope grids or platens. Figure 18–6 illustrates a point-to-plane ESP (Morrow and Mercer, 1964). The corona and electric field are formed between the sharp point of a corona needle and a flat deposition substrate that is located a few millimeters from the needle. A voltage of several kilovolts is applied to the needle, with the plane grounded. Typical ion currents are a few microamperes. Aerosol particles passing between the needle and the substrate are charged and then deposited on the substrate.

Because the mobility of a particle is a function of particle size, the efficiency of collection in the short transit time of the collector illustrated in Figure 18–6 may be a function of size and samples may be biased (Cheng et al., 1983). Liu et al. (1967) developed a two-stage, pulsed electrostatic precipitator for aerosol sampling to minimize this bias. The particles are first charged and then introduced into the collection region where the electric field is pulsed for 1.5 s with 3 s intervals during which the electric field is turned off. Because only those particles that are in the region between electrodes at the beginning of a voltage pulse are collected, the deposition is relatively uniform.

PARTICLE SIZE DISTRIBUTION MEASUREMENT

Measurements of the distribution of particles with respect to their electrical mobilities have been made since 1902, when Langevin used a simple coaxial cylinder ion condenser in his

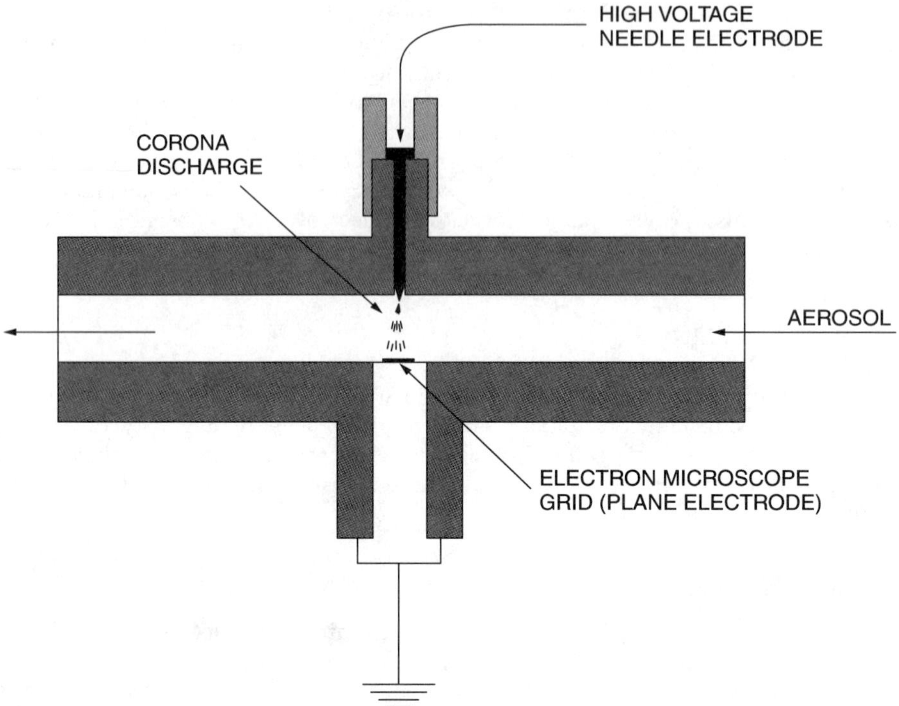

Fig. 18–6. Point-to-plane electrostatic precipitator for depositing aerosol particles onto electron microscope grids.

discovery of "large ions" in the atmosphere. He observed that these ions had mobilities that were 3000 times smaller than gas ions, placing them in the accumulation mode of the atmospheric aerosol. His measurement involved drawing a sample of gas containing ions and charged particles through the annular region between two electrodes. The current collected on the central electrode when a voltage was applied to the outer one provided a measure of charged particles with mobilities greater than a threshold value. This technique was used by many researchers over the following decades (see Flagan, 1998, for a detailed survey of these pioneering investigations). Efforts to improve the resolution of measurements of the distributions of ion and particle mobilities predated Langevin's discovery with the introduction of a segmented central electrode and culminated with the development of a differential electrical mobility classifier for gas ion measurements by Erikson in 1921 and for particles by Rohrmann in 1923. Although it saw little use for several decades, this instrument reappeared in a paper by Hewitt (1957) that introduced a number of innovations that would finally enter common usage years later.

Electrical mobility methods did not see much use until commercial instruments were introduced by ***TSI**** in the 1960s and 1970s. The first, called the Whitby Aerosol Analyzer (WAA), modified the traditional condenser analyzer by introducing the aerosol near the outer electrode and particle-free sheath air in the gap between the electrodes (Whitby and Clark, 1966). In the WAA the potential was applied to the inner electrode, and the electrical current carried in the flow exiting the downstream end of the condenser was measured. Dif-

* See Appendix I for full manufacturer addresses referenced to the italicized three-letter codes.

ferences in that current conveyed as the applied voltage was varied were used to infer the particle size distribution. A more compact and robust version of that instrument, known as the *electrical aerosol analyzer* (EAA), was introduced by Liu and Pui in 1974 and commercialized by ***TSI***, making the technique widely available to researchers around the world. That instrument has, however, been largely displaced by a commercial version of the differential electrical mobility classifier of Erikson (1921), Rohrmann (1923), and Hewitt (1957) that was introduced at the same time as the EAA.

Differential Mobility Analyzer

The differential mobility classifier, more commonly called the *differential mobility analyzer* (DMA), was introduced by Knutson and Whitby (1975) as a source of monodisperse submicrometer particles for use in calibrating other instruments. Although Knutson and Whitby understood the potential of their instrument for measurement of aerosol size distributions, it had to await the development of suitable detectors before that application became practical.

The cylindrical differential mobility classifier, illustrated schematically in Figure 18–7a, is a coaxial flow condenser. Aerosol particles enter the condenser through a narrow slot in the outer electrode. A particle-free sheath flow separates them from the high-voltage inner electrode. Downstream from the aerosol injection slot, a classified aerosol sample is extracted through a port in the inner electrode. Particles that migrate across the gap between the electrodes during the time that they are carried that length through the classifier are extracted. Particles with higher mobilities migrate across the annulus, depositing on the central electrode upstream of the sample extraction port. Particles with too low mobility deposit farther downstream or are discharged with the excess sheath flow. Thus, only particles having mobilities within a narrow range are transmitted.

To determine the size of particles transmitted through the DMA, we consider a simplistic model of the DMA that yields the same results as more complex and rigorous models. We first limit the discussion to classification in the absence of particle diffusion. The electric field in the common cylindrical DMA (CDMA) is

$$E = \frac{V}{r \ln \frac{R_2}{R_1}} \tag{18–26}$$

where V is the potential applied to the inner electrode, and the radii of the inner and outer electrodes are R_1 and R_2, respectively. The particles migrate in the radial direction at velocity, $v(r) = ZV/(r \ln R_2/R_1)$, while they are advected in the axial direction at the gas velocity, $u(r, x)$. Assuming that the gas flow is parallel to the electrodes throughout the volume, that is, $u = u(r)$, the particle trajectory is described by

$$\frac{\mathrm{d}r}{\mathrm{d}x} = \frac{\mathrm{d}x/\mathrm{d}t}{\mathrm{d}x/\mathrm{d}t} = \frac{ZV}{ru(r) \ln \frac{R_2}{R_1}} \tag{18–27}$$

Rearranging, multiplying each side by 2π, and integrating, yields

$$\int_{R_{in}}^{R_{out}} 2\pi r u(r) dr = Q_{out} - Q_{in} = \frac{2\pi ZVR_2}{\ln \frac{R_2}{R_1}} \tag{18–28}$$

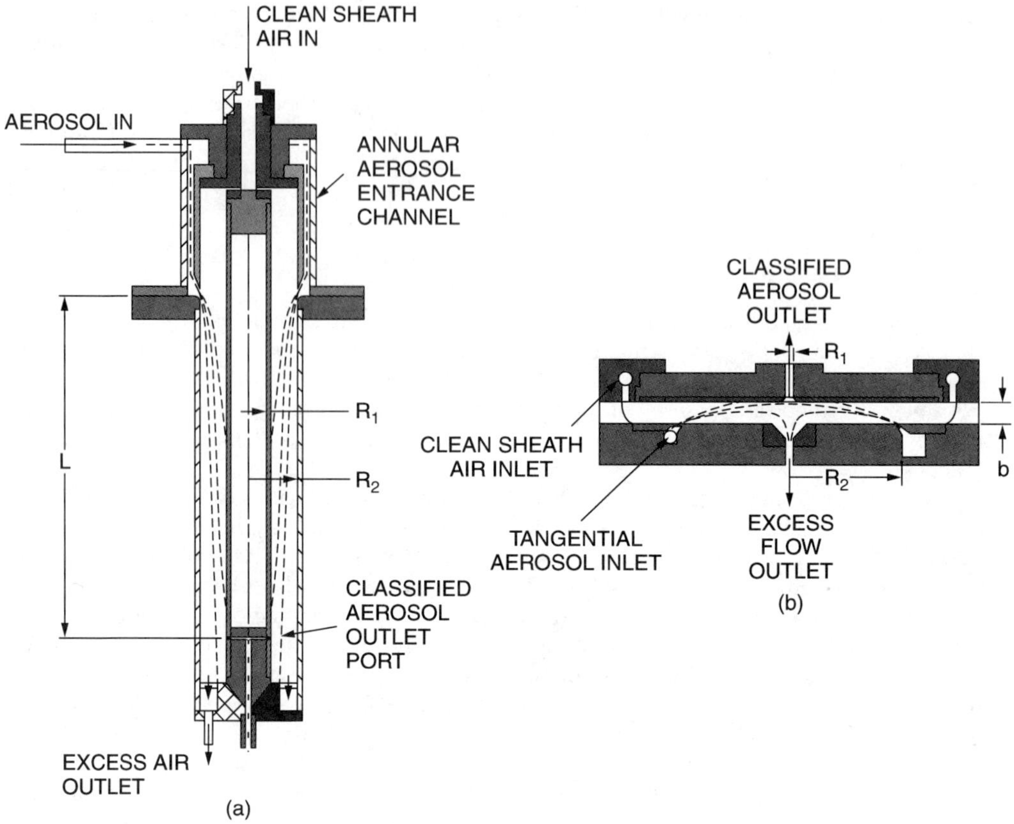

Fig. 18–7. Schematic diagram of a cylindrical DMA designed by Knutson and Whitby (1975) (a) and the radial DMA of Zhang et al. (1995) (b) Trajectories for particles with mobilities $Z \approx Z^*$ and for particles with higher and lower mobilities are illustrated.

The characteristic mobility of the particles classified at given flow rates and applied voltage, Z^*, is that corresponding to particles entering at the middle of the incoming aerosol flow rate, that is, $Q_{in} = Q_{sh} + Q_a/2 = (Q_e + Q_{sh} + Q_s)/2$, and exits at the middle of the classified aerosol outlet flow rate, $Q_{out} = Q_s/2$. The characteristic mobility is, thus,

$$Z^* = \frac{(Q_{sh} + Q_e)\ln \dfrac{R_2}{R_1}}{4\pi VL} \tag{18–29}$$

By a similar derivation, the characteristic mobility is

$$Z^* = \frac{Q_{sh} + Q_e}{2\pi \dfrac{V}{b}\left(R_2^2 - R_1^2\right)} \tag{18–30}$$

for a radial-flow DMA (RDMA; Fig. 18–7b) with a gap between the two electrodes of b and in which the aerosol enters at a radial position R_2 and flows inward to a radial position R_1 where the classified aerosol is discharged. If the particles are large enough (i.e., the classifi-

EXAMPLE 18–2: DMA OPERATION

The ***TSI*** long column cylindrical DMA (based on the Knutson and Whitby design) is to be used to measure particles as large as 1 μm diameter at a flow rate ratio of $\beta = (Q_a + Q_s)/(Q_{sh} = Q_e) = 0.1$, and balanced flows, that is, $Q_a = Q_s$. Figure 18–1 shows that the resolution of the DMA falls to about one-half of its ideal value of $\beta^{-1} = 10$ at about 10 V. What flow rates are needed, and what is the smallest particle size that can be measured with reasonable resolution? (The dimensions of the DMA are $R_1 = 9.38\,\text{mm}$, $R_2 = 19.58\,\text{mm}$, and $L = 444\,\text{mm}$.)

Answer:

1. Calculate the flow rates needed to classify 1 μm particles at 10,000 V. From Example 18–1, the mobility of a singly charged, 1 μm particle is $Z_p = 1.09 \times 10^{-9}\,\text{m}^2\,\text{V}^{-1}\,\text{s}^{-1}$. Note that $Q_{sh} + Q_a = Q_e + Q_s$, and balanced flows implies $Q_e = Q_{sh}$. From Eq. 18–30,

$$Q_{sh} = \frac{2\pi Z^* VL}{\ln\dfrac{R_2}{R_1}} = \frac{2\pi \times 10^4 \times 1.09 \times 10^{-9} \times 0.444}{\ln\dfrac{0.01958}{0.00937}} = 4.12 \times 10^{-5}\,\text{m}^3\text{s}^{-1}\left[= 2.48\,\text{L}\,\text{min}^{-1}\right]$$

2. The maximum mobility classified with reasonable resolution is that transmitted at a classification voltage of about 10 V, that is,

$$Z^*_{max} = Z^*_{min}\frac{V_{max}}{V_{min}} 1.09 \times 10^{-9}\,\text{m}^2\text{V}^{-1}\text{s}^{-1}\frac{10{,}000\text{V}}{10\text{V}} = 1.09 \times 10^{-6}\,\text{m}^2\text{V}^{-1}\,\text{s}^{-1}$$

From Figure 18–1 it is apparent that this corresponds to approximately 15 nm diameter. Thus, the size range that can be classified at given flow rates extends over a factor of about 65 in diameter.

3. Smaller particles can be classified if the flow rate is increased, although the maximum particle size is then reduced as well. Examination of Figure 18–1 reveals that the size range that can be classified with reasonable resolution narrows for smaller particles. Thus, a DMA designed to classify particles as small as 3 nm diameter is limited to particles smaller than about 120 nm by the breakdown voltage, covering only a factor of about 40 in diameter.

4. A given DMA can cover a wider sizing range by changing flow rates or by accepting lower resolution at the high mobility (small particle) end of the distribution. Collins et al. (2000b) have demonstrated that, by scanning the DMA flows simultaneously with the voltage scan, the sizing range can be increased by a factor of 4 or more while maximizing the particle size resolution over the entire size range.

cation voltage is high enough) that diffusion is unimportant, particles included in the classified aerosol flow will span a range of mobilities of $Z_p^* - \Delta Z \le Z_p \le Z_p^* + \Delta Z$, where

$$\Delta Z = Z_p^* \beta \tag{18–31}$$

and $\beta = (Q_a + Q_c)/(Q_{sh} + Q_e)$.

The probability that particles with mobility Z that enter the classification region of the DMA will be included in the classified aerosol flow when the flow rates and applied voltage are for a characteristic mobility of Z_p^* is $\Omega(Z_p, Z_p^*)$, which is called the *classifier transfer function*. Knutson and Whitby (1975) derived an expression for the transfer function for particle classification when diffusion is unimportant. Most uses of the differential mobility classifier extend to sufficiently small particles that diffusion cannot be neglected. A number of investigators have modeled the effect of diffusion on the transfer function. Stolzenburg (1988) derived a diffusional transfer function of the cylindrical DMA that can be expressed as

$$\Omega(\tilde{Z}) = \frac{\tilde{\sigma}}{\sqrt{2}(1-\delta)}\left[\varepsilon\left(\frac{\tilde{Z}-[1+\beta]}{\sqrt{2}\tilde{\sigma}}\right)+\varepsilon\left(\frac{\tilde{Z}-[1-\beta]}{\sqrt{2}\tilde{\sigma}}\right)-\varepsilon\left(\frac{\tilde{Z}-[1+\delta\beta]}{\sqrt{2}\tilde{\sigma}}\right)-\varepsilon\left(\frac{\tilde{Z}-[1-\delta\beta]}{\sqrt{2}\tilde{\sigma}}\right)\right] \tag{18–32}$$

where

$$\varepsilon(y) = y\,\mathrm{erf}(y) + \frac{1}{\sqrt{\pi}} e^{y^2} \tag{18–33}$$

erf (y) is the error function, and

$$\delta = \frac{Q_s - Q_a}{Q_s + Q_a} \tag{18–34}$$

$$\tilde{\sigma}^2 = \frac{G}{Pe_{\mathrm{mig}}} \frac{Z_p}{Z_p^*} \tag{18–35}$$

where the Peclet number for particle migration across the gap between the electrodes is

$$Pe_{\mathrm{mig}} = \frac{Z_p E b}{D} = \frac{qV}{kT} f(\mathrm{geometry}) \tag{18–36}$$

and G is a parameter that depends weakly on the DMA geometry and the flow rate ratio, β. This result applies to both radial and cylindrical DMAs (Zhang and Flagan, 1996). The factor f(geometry) accounts for nonuniformities in the electric field along the migration path. Because the electric field in the RDMA is uniform, $f_{\mathrm{RDMA}} = 1$. For the CDMA, $f_{\mathrm{CDMA}} = (R_2 - R_1)/(R_2 \ln R_2/R_1)$.

The resolution of a differential mobility classifier can be expressed conveniently as the ratio of Z_p^* to the observed range of mobilities of the transmitted particles. Following the approach taken in a wide range of spectroscopies, we use those mobilities corresponding to the full width of the DMA transfer function evaluated where the transmission efficiency is one-half the maximum value, ΔZ_{fwhm}. The resolution is bounded by a value that is determined by the ratio of flow rates, β, in the nondiffusive limit for balanced flows, that is, $Q_a = Q_s$; the value is (Flagan, 1999)

$$R_{\mathrm{nd}} = \frac{Z_p^*}{\Delta Z_{\mathrm{fwhm}}} = \beta^{-1} \tag{18–37}$$

At low particle migration Peclet numbers, diffusion dominates and the resolution becomes

$$R_{\mathrm{diff}} = 0.425\left(\frac{Pe_{\mathrm{mig}}^*}{G}\right)^{\frac{1}{2}} = 2.66\left(\frac{fV}{G}\right)^{\frac{1}{2}} \tag{18–38}$$

Figure 18–8 shows the variation of the resolution of several DMA designs for a flow rate ratio of $\beta = 0.1$ and a commercial cylindrical DMA (***TSI*** long column) for several values of

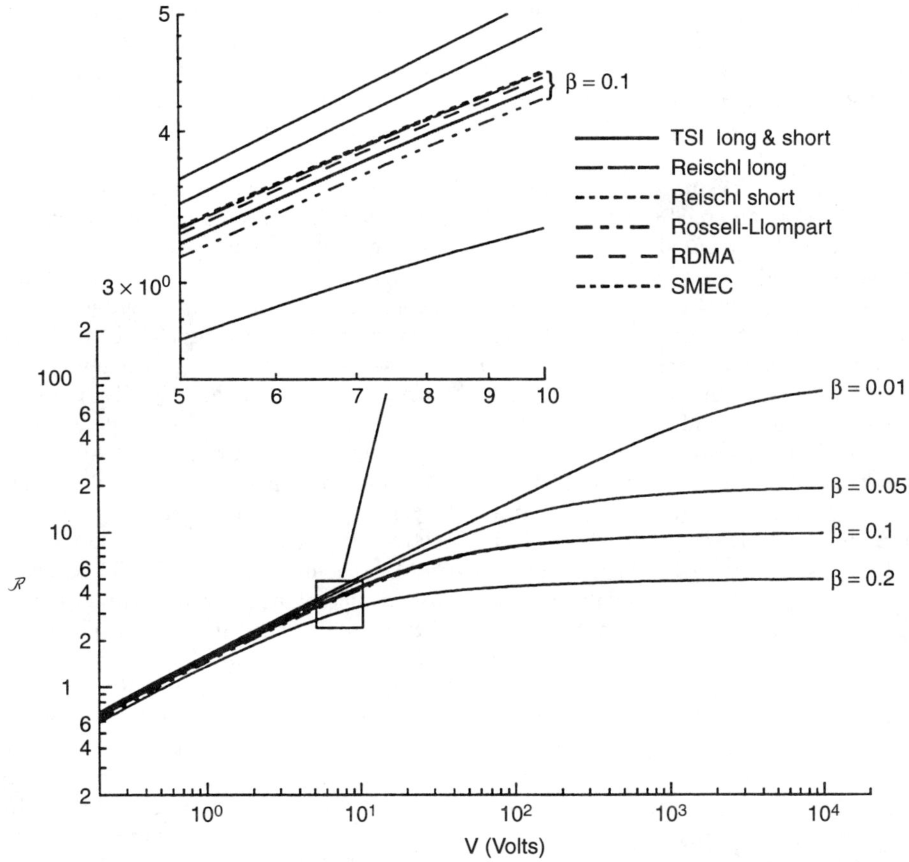

Fig. 18–8. Ideal resolution of several DMA designs as a function of applied voltage and flow rate ratio (Flagan, 1999).

β. For balanced flows within the DMA, that is, $\delta = 0$, the two limiting expressions intersect when

$$V = V_{\text{diff}} = \frac{0.141G}{\beta^2 f} \tag{18–39}$$

providing an estimate of the transition from nearly ideal operation where diffusion plays only a minor role to the low-voltage region where resolution decreases as $V^{1/2}$. Because the maximum voltage of any DMA design is limited to about 10 kV by electrostatic breakdown (arcing) for dry gas at atmospheric pressure, the range of particle mobilities that can be analyzed with resolution near that determined by the flow rate ratio decreases with decreasing β. Lower pressures or high humidity may lower the breakdown potential and further constrain the operating range.

Differential Mobility Analyzer Designs

Most of the measurements made with differential mobility analyzers before the mid-1990s were obtained either with the commercial version of the Knutson and Whitby (1975) cylindrical DMA (KW-CDMA) that was produced by TSI (St. Paul, MN) or with similar instruments. That device is illustrated in Figure 18–7. The coaxial cylinder instrument performed well for particles between about 10 nm and 1 μm diameter. A shortened version of that instrument was used to measure smaller particles, but diffusive losses of ultrafine aerosol particles are high due to the narrow annular entrance region seen at the top of the instrument in Figure 18–7a.

Reischl and coworkers at the University of Vienna (see Winklmayr et al., 1991) developed a new DMA, commercially produced by ***HAU***, that was better suited to measurement of nanoparticles. A tangential aerosol inlet greatly reduced the particle losses. The instrument was initially used with a built-in high-sensitivity electrometer to measure size distributions extending down to about 2.5 nm diameter. That design has been the basis of a number of new designs. Rossell-Llompart et al. (1996) constructed a version of the Vienna DMA with a very short analysis region to optimize its performance in the low nanometer size regime.

DMA development continues at too rapid a pace to allow an exhaustive listing of all of the developments, so only a few key innovations are summarized below. Zhang et al. (1995) and Pourprix and co-workers (see Fissan et al., 1996) applied the tangential inlet to the development of a new type of DMA in which the classification takes place in an inward radial flow between parallel disk electrodes (illustrated in Fig. 18–7b). These instruments have been called the Radial DMA (RDMA) and the Spectrometre de Mobilite Elitrique Circulaire (SMEC), respectively. The design of the KW-CDMA was modified by Pui and co-workers (see Chen et al., 1998) to extend its operation into the nanometer size range. This instrument, called the nanoDMA, passes a large aerosol flow through the annular entrance region of the KW design to reduce losses. A small portion of the flow conveys aerosol into the classification region of the DMA, with the remainder being discharged. The aerosol flow rate into the classification region, Q_a, is determined as the difference between the much larger flows that enter and are discharged from the annulus.

The ideal performance of all of these depends on the DMA geometry. Table 18–3 summarizes these geometry parameters for several DMAs. As seen in the inset in Figure 18–8, the theoretical performance does not vary significantly with the geometry of the DMA. Real instruments vary in how closely they approach the ideal. Degraded performance may result from imperfections in the construction of the instrument, maldistribution of the aerosol or sheath flows around the axis of the classifier, geometrical distortions of the electric field, turbulence, or imprecise control of the flow rates and voltages. Measurement of the transfer function in the high-voltage limit where diffusion plays only a minor role provides a clear test for these nonidealities.

DMA Operation

DMA classification of an aerosol requires precise control of four flow rates: the incoming aerosol and sheath flows, Q_a and Q_{sh}, and the outgoing classified aerosol and exhaust flows, Q_s and Q_e, as illustrated schematically in Figure 18–9. Typically, the sheath and exhaust flow rates are much larger than either the aerosol or classified sample flow rates. These flows combine in the classification region, which, from a flow-balancing point of view, can be treated as a plenum with no pressure drops. The metering and precision control of the four flows is particularly challenging because of the need to process the two aerosol flows with minimal particle losses and, furthermore, because the flows must be measured and controlled with very low pressure drops. The large particle limit of the resolution of the DMA is equal to β^{-1}. The differences in the resolving power of different DMA designs (assuming no imperfections that will degrade their performance) are small as illustrated by the magnified inset.

The DMA can be operated in underpressure (illustrated in Fig. 18–9) or overpressure mode. In the underpressure operation, the flows are pulled through the DMA by pumps. If the system being sampled is at higher pressure than ambient, that pressure can be used to

TABLE 18–3. Geometry Parameters for Various DMAs for $\delta = 0.0$

DMAs	R_1 (mm)	R_2 (mm)	L (mm)	β	G	f
Cylindrical						
TSI long	9.37	19.58	444.4	0.01	1.87	0.707
TSI long	9.37	19.58	444.4	0.05	1.99	0.707
TSI long	9.37	19.58	444.4	0.1	2.14	0.707
TSI long	9.37	19.58	444.4	0.2	2.42	0.707
TSI short	9.37	19.58	111.1	0.1	2.15	0.707
Vienna short	25.0	33.0	110	0.1	2.48	0.837
Vienna long	29.0	37.0	600	0.1	2.51	0.867
Rossell-Llompart	25.0	33.0	16	0.1	2.80	0.867
Radial	R_1 (mm)	R_2 (mm)	L (mm)	β	G	f
Caltech RDMA	24	50.4	10	0.1	2.92	1.0
SMEC	~5	65	4	0.1	2.81	1.0

The University of Vienna DMA (produced by Hauke) and the short column nanoparticle version of it are described by Winklmayr et al. (1991), and Rossell-Llompart et al. (1996), respectively. The RDMA and the Spectromètre de Mobilite Electriqué Circulaire (SMEC) are described by Zhang et al. (1995) and Fissan et al. (1996), respectively. Geometry parameters are calculated assuming fully developed laminar flow within the classification regions of the instruments.

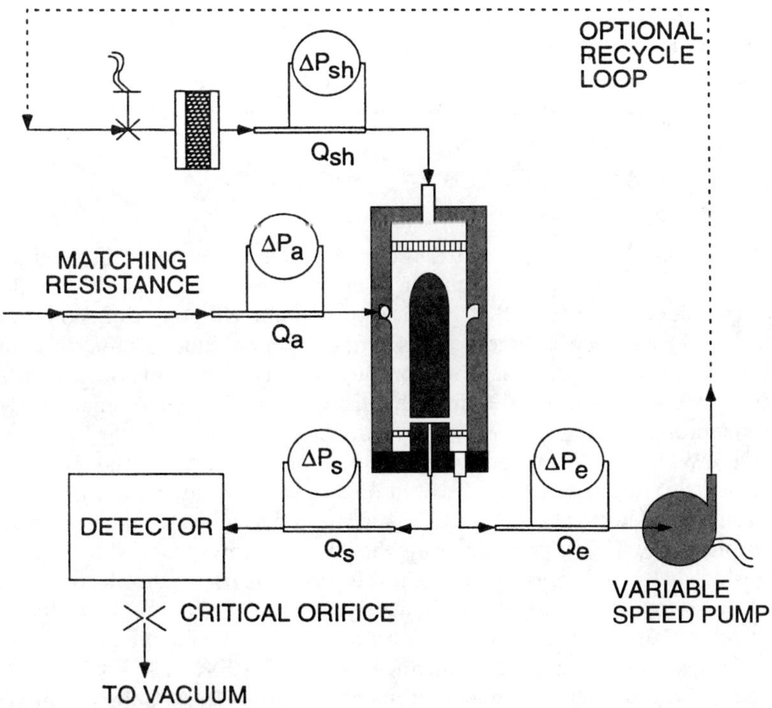

Fig. 18–9. Schematic diagram of flow control system for differential mobility analyzer operation. The system shown includes laminar flow metering of all four DMA flows. Using a variable-speed pump, the excess flow can either be discharged or recycled as indicated by the dashed lines. The DMA illustrated is the design of Winklmayr et al. (1991).

force the flow through the DMA, producing the overpressure mode of operation. In either case, balancing of the flows is critical. Because the pressure drop through the filter that removes particles from the sheath flow is often much larger than the pressure drops in the sample flows, an additional pressure drop in the form of small-bore tubing of appropriate diameter is sometimes added to the sample flow line to aid in balancing the flows. If the pressure drops are not balanced, it can be quite difficult to maintain stable flows through the DMA, leading to uncertainty in the sizes of particles that are classified.

The first generation of commercial DMAs used mass flow meters and manual control valves on the aerosol, sheath flows, and excess flows. Passage of either of the aerosol flows through a conventional valve is undesirable because particle losses within the valve may vary with valve setting. One approach to improving precision in the control of the flow rate ratio is to filter and recycle the exhaust flow (Rogak et al., 1991), allowing the aerosol and classified sample flows to be matched easily and to be metered downstream of the particle detection system as illustrated by the dashed line in Figure 18–9. It should be noted, however, that small leaks in the recycle system can lead to substantial errors in the much smaller aerosol flow rates. Diaphragm pumps reduce, but do not eliminate, the likelihood of such leaks. Moreover, they introduce pressure pulsations that may affect the measurement. These are generally minimized by the addition of large plenums on the flows entering and leaving the recirculation pump. Rotary blowers or pumps can also be used, but extreme care is required to ensure that leaks do not compromise the measurement. Recycle systems also heat the sheath gas, introducing the possibility that particles may decrease in size due to evaporation as they migrate through the DMA.

Ideally, each of the flows would be measured and controlled. Replacing the mass flow meters with laminar pressure drop elements and high-sensitivity, electronic differential pressure transducers facilitates design of flow meters with low particle losses. Such meters make it possible to monitor all four flows as illustrated in Figure 18–9. Combined with active control of the sheath and exhaust flows, highly stable flow control can be achieved (Russell et al., 1996). The active control of the critical flow rate ratios enables robust and precise operation of the DMA even in variable-pressure environments such as those encountered in airborne measurements.

Differential Mobility Particle Sizer

Routine mobility-based aerosol size distribution measurements were facilitated by the EAA, which employed a Faraday-cup electrometer to measure the current carried by those particles that escaped collection within the classification column. Unipolar diffusion charging was employed in that instrument to increase the charging probabilities to sufficiently high levels that ambient particle concentrations would produce electrical currents of a small fraction of a picoammeter. Multiple charging became significant at particle sizes as small as 50 nm diameter in that operating mode.

Knutson and Whitby (1975) recognized the potential of the differential mobility classifier for measurements of the particle size distribution, but the technique had to await the development of a suitable detector before it saw significant use as a measurement instrument. The high resolution of the DMA resulted from the low charging probabilities achieved with bipolar diffusion charging. The lower charge levels made the use of the electrometer difficult at ambient concentrations, although Winklmayr et al. (1991) eventually produced an electrometer-based size distribution measurement system. The introduction of the continuous-flow, single-particle-counting condensation nucleus counter (CNC) (Agarwal and Sem, 1978) first enabled routine size distribution measurements with the DMA. Coupled with computer-controlled stepping of the DMA voltage, the ***TSI*** 3020 CNC made it possible to take full advantage of the resolution of the DMA in making size distribution measurements (Helsper et al., 1982; TenBrink et al., 1983). This computerized size distribution measurement system was later commercialized by ***TSI*** as the Differential Mobility Particle Sizer (DMPS). This automated DMA system soon displaced the EAA as the primary system for mobility-based size distribution measurements because the newer technique determined the concentration of particles within a narrow mobility interval instead of measuring all particles with mobilities less than a threshold value. In the DMPS measurement, the classification voltage is stepped and, after a delay to allow the aerosol signal to reach steady state, the particle concentration is measured. The voltage is then stepped again, and the process is repeated. Computer control allowed the instrument to step through the many voltages required to determine the size distribution relatively rapidly, although the measurement time of 10 min or more was too long for the DMPS to be used in the study of rapidly varying aerosols.

Scanning Mobility Particle Sizer

While the DMA afforded much greater resolution than its predecessor, the measurement was too slow for many applications. Wang and Flagan (1990) observed that the critical aspect of mobility measurements was the dependence of the particle migration velocity on the strength of the electric field, but noted that separation could be achieved if the electric field were ramped continuously and particle counts were continuously recorded in time bins. Moreover, they showed that, in the nondiffusive limit, the same transfer function would be achieved in the system they called the Scanning Electrical Mobility Spectrometer (SEMS) as in the DMPS provided that the mean field strength during the particle migration time were used in the data analysis. Differential mobility analysis of particle size distributions over the full DMA classification range was reduced to less than 1 min with the SEMS. This scanning mode of DMA operation was later commercialized by ***TSI*** as the Scanning Mobility Particle Sizer (SMPS).

To determine the relationship between counts in a time bin and particle mobility, one must take delays in the transmission of particles from the DMA outlet to the point of measurement in the detector into account. Wang and Flagan (1990) took into account a time offset due to the residence time of the particles in the plumbing and detector. Russell et al. (1995) later showed that this fixed offset model was oversimplified. The SEMS/SMPS systems require a time response for which the available detectors (either CNCs or electrometers) were not designed. Mixing within the flow passages of the CNC produced a distribution of delay times that smear the SEMS/SMPS response function by allowing particle counts to appear in later time bins than expected. Russell et al. (1995) took this residence and distortion into account by deriving a transfer function based on a model of the flow between the DMA outlet as a plug flow region in series with a perfectly mixed volume. A much simpler approach was later used by Collins et al. (2000), who developed a deconvolution algorithm that was sufficiently fast to allow mobility spectra to be "desmeared" immediately upon measurement.

Data Analysis/Inversion

Efforts to deduce particle size distributions from DMA data predated the development of the automated instruments. The particle concentration signal obtained during steady-state operation of the DMA is the integral over particles transmitted over the entire range of mobilities sampled, that is,

$$S(V_j) = S_j = Q_c \int_0^{\infty} \sum_{i=-\infty}^{\infty} n(d_p)\eta_{trans}(d_p)s_{count}(d_p,i)\phi(d_p,i)\Omega[Z(d_p,i),Z_j^*]dd_p \tag{18–40}$$

where $\eta_{trans}(d_p)$ is the transmission probability for particles of diameter d_p and charge i, and $s_{count}(d_p)$ is the mean signal (counts in the case of a CNC detector or electrical current in the case of an electrometer) produced by the detector in response to a particle of diameter d_p

and charge i. The transfer function, Ω for particles of diameter d_p and charge i depends on the mobility of that particle compared with the nominal classification mobility for channel j, $Z_j^*(V,\beta,\delta)$. In the limit of singly charged particles, this reduces to

$$S(V_j) = S_j = Q_c \int_0^\infty n(d_p)\eta_{trans}(d_p)s_{count}(d_p,1)\phi(d_p,1)\Omega[Z(d_p,1)Z_j^*]dd_p \tag{18–41}$$

Assuming that the size distribution, $n(d_p)$, varies little over the interval of transmitted mobilities, that is, $n(d_p) \approx n(d_{p,j})$, and that the transmission efficiency, detector response function, and charging probability are slowly varying functions of d_p, the size distribution becomes

$$n(d_{p,j}) = \frac{S_j \left.\frac{dZ_p}{dd_p}\right|_{D_p=D_{p,j}}}{Q_c\eta_{trans}(d_{p,j})s_{count}(d_{p,j})\phi(d_{p,j},1)\int_0^\infty \Omega(Z_p,Z_p^*)dZ_p} \tag{18–42}$$

When evaluated using the nondiffusive transfer function for the DMA, Eq. 18–42 represents the zeroth order approximation to the particle size distribution. In that limit, and for balanced flows ($\delta = 0$), $\int_0^\infty \Omega(Z_p,Z_j^*)dZ_p = \Delta Z_p = \beta Z_j^*(d_p,1)$. More generally, the diffusive transfer function must be used in the data inversion.

A number of higher order data inversion algorithms have been applied to inference of particle size distributions, for example, non-negative least squares matrix inversion (Lawson and Hansen, 1974) and constrained regularization (Wolfenbarger and Seinfeld, 1990). One of the most widely used inversion algorithms is the very simple one proposed by Twomey (1975); see also Markowski (1987). In this approach, the general Fredholm integral equation for the instrument response, Eq. 18–40, is approximated by a summation,

$$S_j = \sum_{k=1}^{m} K_{j,k}n(d_k)\Delta d_k = \sum_{k=1}^{m} k_{j,k}N_k \tag{18–43}$$

where the summation is taken over the m sizes of particles used to represent the particle size distribution, $n(d_k)\Delta d_k = N_k$, and

$$K_{j,k} = \int_0^\infty \sum_{i=-\infty}^{\infty} \eta_{trans}(d_p)s_{count}(d_p,i)\Omega[Z(d_p,i),Z_j^*]dd_p \tag{18–44}$$

Starting with a reasonable guess of the particle size distribution, N_k^0, $k = 1, 2, \ldots, m$, the solution is calculated by iteration. At any iteration, l, a new trial solution is estimated as

$$N^{l+1}(d_k) = [1 + (X_{j,k}^l - 1)K_{j,k}]N^l(d_k) \tag{18–45}$$

where $X_{j,k}^l = S_j/\Sigma_{i=1}^m K_{j,k}N_k^l$ is the ratio of the actual instrument response to the calculated instrument response. $X_{j,k}^l$ may be calculated once for each channel j in a given iteration l. The iteration, Eq. 18–45, is then done for each measurement channel j and particle size k, producing a cycle of $m \times n$ iterations. This process is repeated until adequate convergence is achieved.

The entire data inversion can be performed in a single operation, although partial inversion to assign counts to the appropriate mobility bins often precedes inversion to account for the classifier response function. Such a stepwise inversion might involve a time deconvolution to correct for the mixing-induced distribution of delays in counting particles, followed by an iterative correction for multiple charging. Approaches for carrying out these partial inversions are described below.

Residence Time Deconvolution

Particle counts appear in a range of channels when the DMA voltage is scanned or stepped rapidly, distorting the inferred particle size distribution function. Collins et al. (2001) developed a simple and efficient deconvolution algorithm that can correct for this distortion. To a first approximation, the fraction of particles that exit the DMA at time t' that are detected at later time t can be expressed as the sum of a fixed "plumbing" delay, τ_p, and an exponentially decaying function, that is,

$$E(t-t') = \frac{1}{\tau_s} e^{-\frac{t-\tau_p-t'}{\tau_s}} \tag{18–46}$$

where τ_s is the mixing-induced *smearing* time. Note that an electrometer produces similar response smearing due to the inherent electronic time constant of the electrometer circuit. The counts appearing in time bin m due to particles that exited the DMA at times corresponding to earlier counting bins, n, can be expressed as a summation,

$$C_m = \sum_{n=0}^{m} R_n \theta_{(m-n)} \tag{18–47}$$

For the exponential distribution of delay times, the estimated instrument response in the absence of smearing can be computed recursively,

$$R_n = \frac{C_n - F_n}{1 - \dfrac{u-1}{u \ln u}} \tag{18–48}$$

where $u = e^{tc/\tau}$, $F_n = R_{n-1} \dfrac{(\mu - 1)^2}{u \ln u} + F_{n-1}$. The value of F_{n-1} will have been calculated for the previous bin. It equals

$$F_{n-1} = \sum_{k=-\infty}^{n-2} R_k \frac{(u-1)^2}{u \ln u} \frac{1}{u^{n-k}} \tag{18–49}$$

For a continuous stream of data, Eq. 18–48 can be used directly after a brief start-up transient. This recursive relationship enables real-time display of DMA size distributions that have been corrected for any smearing that may occur. This calculation attributes particle counts to the proper time bin. It should be performed before the correction for multiple charging and inversion.

Multiple Charging Correction

Some particles at the upper end of the DMA sizing range ($D_p < 0.1\,\mu m$) acquire multiple charges in bipolar diffusion charging. Multiple charging can occur for even smaller particles when unipolar chargers are employed. To transform the measured counts versus nominal mobility to a particle size distribution, the presence of large, multiply-charged particles in a measurement channel (counting bin) that nominally corresponds to smaller particle sizes must be taken into account. Obviously, knowledge of the charge distribution as a function of particle size is essential for this part of the data inversion, but this is not enough. The channel corresponding to the highest classification voltage may include larger particles that are outside the nominal size range of the DMA, but that have sufficiently high mobilities for classification due to their charge states. To avoid this problem, the DMA is often used with a precut impactor that removes particles larger than that transmitted when singly charged at the highest operating voltage. A recursive procedure can then be used in conjunction with the known charge distribution to assign counts to the mobility channels where they would have appeared if singly charged. Because no multiply charged particles are present at the largest classified particle size, the number of particles corresponding to that mobility is directly measured. Any particles of that size that are multiply charged will appear in bins with smaller, more mobile particles. Using the known charge distribution, the number counts in the respective bins are reduced accordingly. By sequentially making these corrections to the succession of channels from the lowest to highest mobility, detected particles are attributed to the counting bins for mobilities corresponding to single charging. A similar procedure can be employed when no large particle precut is used provided that the distribution of larger particles is known, either through auxiliary data from optical particle counters, aerodynamic particle sizers, or related instruments, or because no particles are present in the lowest mobility channels.

Tandem Differential Mobility Analyzer

The DMA can be used to study changes in aerosol particles due to condensation, evaporation, or other processes that will change the particle size or structure. One DMA is used to classify the aerosol particles at fixed voltage, selecting particles from a narrow interval of particle sizes. The classified aerosol is then processed to change the particle size, for example, by drying or humidification. The size distribution of the modified particles is then measured using a second DMA that is stepped or scanned through voltages. The size shift observed in this *tandem* DMA (TDMA; McMurry et al., 1983) can be used to infer such properties as vapor pressures or the influence of humidity on aerosol optical properties.

Time-of-Flight Mobility Analyzers

Mobility analysis can also be accomplished by measuring the time of flight of charged particles in an electric field. This method, called *ion mobility spectrometry* (Karasek, 1970; Eiceman and Karpas, 1994), is ideally suited for measurements of very high mobility particles or ions. Figure 18–10 illustrates a typical ion mobility spectrometer. A pulse of charged particles is injected through a shutter grid into a drift region. The current carried to a collector electrode at the opposite end of the drift region determines the number of particles transmitted within intervals of transit time. Typical drift times are 10 to 50 ms, and drift region lengths are several centimeters. The mobility resolution of the ion mobility spectrometer is determined by the ratio of the transit time to the time that the shutter is open. Brownian diffusion within the

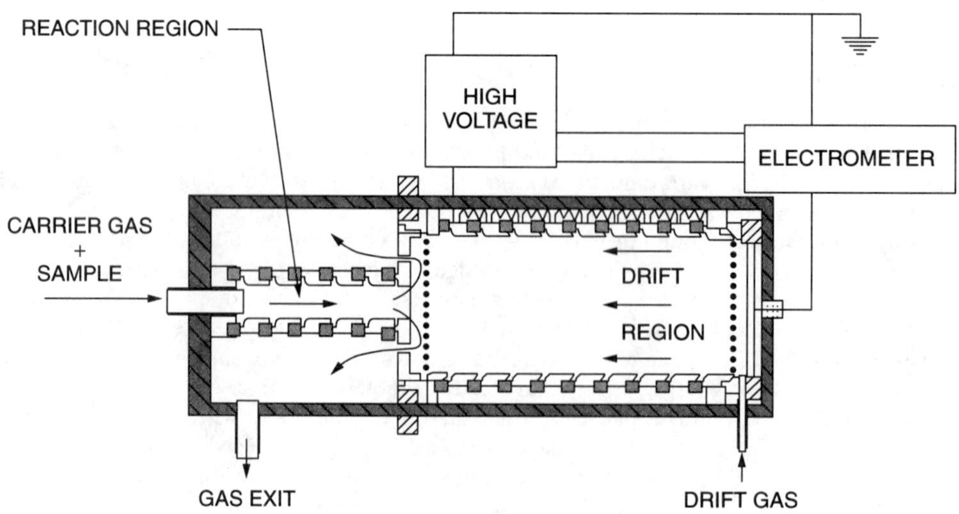

Fig. 18–10. Ion mobility spectrometer.

ion mobility spectrometer is generally relatively minor because of the high migration Peclet number, $Pe_{mig} = eV/k_BT$.

Nonspherical Particles

The discussion above has assumed that the particles being analyzed are spherical, but solid aerosols frequently contain nonspherical particles. The particle shape can influence mobility analysis in two ways: (1) the drag on the particle will differ from that of a sphere of equal volume; and (2) the charging probability may also differ. In the absence of detailed knowledge about the structures of particles, mobility analyzer data can be reported as mobility equivalent diameter, that is, the size of the spherical particle that has the same electrical mobility as the particle being measured. Studies of particle shape effects have appeared in the literature for small clusters of spheres (Kousaka et al., 1996), aggregate particles (Schmidt-Ott, 1988; Rogak and Flagan, 1992; Rogak et al., 1993), and fibers (Han et al., 1994; Baron et al., 1994). Schmidt-Ott (1988) and Rogak et al. (1993) found that the mobilities of agglomerate particles scale with their projected areas. When the fractal dimension of the particle is small so that little of the internal surface is shadowed by outer portions of the agglomerate, then the mobility is proportional to the total area of the particle. Similarly, Rogak and Flagan (1992) found that bipolar diffusion charging of low fractal dimension aggregates was similar to that for spheres of the same *surface area*. See Chapter 23 for further discussion of nonspherical particles.

Fibers

Mobility analysis of fiber particles has been the subject of a number of investigations. Laframboise and Chang (1977) developed a diffusion charging model for particles of arbitrary geometry. Han and Gentry (1993a,b) modeled unipolar diffusion charging of fibers in the free molecular regime and combined diffusion and field charging in the continuum regime. Their calculations and subsequent experiments showed that fiber separation on the basis of length can be achieved by unipolar charging followed by electrostatic precipitation. The high rate of charging of long, thin fibers produced sufficient differences in mobility to enable effective length separations.

Lipowicz and Yeh (1989) demonstrated that macroscopic (0.5 to 5 mm long) neutral aluminum wires can be separated by an electric field gradient in a liquid. This separation, known as *dielectrophoresis*, results from charge separation within the particles that is induced by the electric field gradient. The terminal dielectrophoretic migration velocity is (Baron et al., 1994)

$$v = \frac{\kappa_m \varepsilon_0}{36\mu}\left\{g(\beta)\left(\frac{\alpha}{\alpha-1} - f(\beta)\right)\right\}^{-1} d_f L_f \nabla(E^2) \tag{18–50}$$

where κ_m is the dielectric constant of the medium, α is the ratio of the dielectric constant of the fiber to that of the medium, and $\beta = L_f/d_f$ is the fiber aspect ratio, the ratio of the fiber length to its diameter. The functions of the aspect ratio, $f(\beta)$ and $g(\beta)$, are given by

$$f(\beta) = \frac{\beta}{\sqrt{\beta^2-1}}\left[\left(1 - \frac{\beta}{\beta^2-1}\right)\coth^{-1}\left(\frac{\beta}{\sqrt{\beta^2-1}}\right) + \frac{\beta}{\sqrt{\beta^2-1}}\right] \tag{18–51}$$

$$g(\beta) = \frac{8}{3}\left[-\frac{2\beta}{\beta^2-1} + \frac{2\beta^2-1}{(\beta^2-1)^{\frac{3}{2}}}\ln\left(\frac{\beta+\sqrt{\beta^2-1}}{\beta-\sqrt{\beta^2-1}}\right)\right]^{-1} \tag{18–52}$$

For high-aspect-ratio ($\beta < 5$) conductive fibers, the velocity is a function of L_f^2, but is only a weak function of the fiber diameter, so the length of the fiber can be determined with high resolution. High resolution of fiber length is also possible for low conductivity fibers, but the migration velocity of fibers of intermediate conductivity depends on both length and diameter. However, small fibers can be considered to be conductive because the distances that charges must migrate are small. Baron et al. (1994) constructed a coaxial flow fiber classifier in which the classifier surfaces were coated with a high dielectric strength material to prevent sparks or corona discharge. With a 500 Hz ac potential of 12 to 14 kV peak to peak, asbestos fibers were effectively classified, and size distributions were determined. Glass fibers have lower conductivity and are, therefore, more difficult to classify, but they can be made more conductive by raising the humidity before classification.

REFERENCES

Agarwal, J. K. and G. J. Sem. 1978. Generating submicron monodisperse aerosol for instrument calibration. *TSI Quarterly* IV(2):3–8.

Allen, M. D. and O. G. Raabe. 1985. Slip correction measurements of spherical solid aerosol particles in an Millikan apparatus. *Aerosol Sci. Technol.* 4:269–286.

Arendt, P. and H. Kallmann. 1926. The mechanism of charging dust particles. *Z. Physik*. 35:421–441.

Baron, P. A., G. J. Deye, and J. Fernback. 1994. Length separation of fibers. *Aerosol Sci. Technol.* 21:179–192.

Bricard, J. 1948. Sur l'equilibre ionique de la basse atmosphere. *C. R. Acad. Sci. (Paris)* 226:1536–1538.

Buscher, P., A. Schmidt-Ott, and A. Wiedensohler. 1994. Performance of a unipolar square-wave diffusion charger with variable Nt product. *J. Aerosol Sci.* 25:651–663.

Chen, D. R., D. Y. H. Pui, D. Hummes, H. Fissan, F. R. Quant, and G. J. Sem. 1998. Design and evaluation of a nanometer aerosol differential mobility analyzer (Nano-DMA). *J. Aerosol Sci.* 29:497–509.

Chen, D. R., and D. Y. H. Pui. 1999. A high-efficiency, high-throughput unipolar aerosol charger for nanoparticles. *J. Nanoparticle Res*. 1:115–126.

Cheng, Y. S., H. C. Yeh, and G. M. Kanapilly. 1983. Collection efficiencies of a point-to-plane electrostatic precipitator. *AIHAJ* 42:605–610.

Collins, D. R., R. C. Flagan, and J. H. Seinfeld. 2001. Improved inversion of scanning DMA data. *Aerosol Sci. Technol.* (in press).

Collins, D. R., R. C. Flagan, and J. H. Seinfeld. 2000. Scanning flow DMA. *J. Aerosol Sci.* 31:1129–1144.

Cooper, D. W. and P. C. Reist. 1973. Neutralizing charged aerosols with radioactive sources. *J. Colloid Interface Sci.* 45:17–26.

Davidson, S. W. and J. W. Gentry. 1985. Differences in diffusion charging of dielectric and conducting ultrafine aerosols. *Aerosol Sci. Technol.* 4:157–163.

Eiceman, G. A. and Z. Karpas. 1994. *Ion mobility spectrometry*. Boca Raton, FL: CRC Press.

Erikson, H. A. 1921. The change of mobility of the positive ions in air with age. *Phys. Rev.* 18:100–101.

Fissan, H., D. Hummes, F. Stratmann, P. Buscher, S. Neumann, D. Y. H. Pui, and D. Chen. 1996. Experimental comparison of four differential mobility analyzers for nanometer aerosol measurements. *Aerosol Sci. Technol.* 24:1–13.

Flagan, R. C. 1998. History of Electrical Aerosol Measurements. *Aerosol Sci. Technol.* 28:301–380.

Flagan, R. C. 1999. On differential mobility analyzer resolution. *Aerosol Sci. Technol.* 30:556–570.

Fomichev, S. V., N. M. Trotsenko, and A. V. Zagnitko. 1997. Aerosol chargers using ionizing radiation and electric field collinear to flow: Simulation and experiment for fine particle charging in electronegative air and electropositive nitrogen. *Aerosol Sci. Technol.* 26:21–42.

Fuchs, N. A. 1947. Izvestiya AN SSSR (Ser. Geophysicheskaya) 2:341.

Fuchs, N. A. 1963. On the stationary charge distribution on aerosol particles in a bipolar ionic atmosphere. *Geofis. Pura Appl.* 56:185–193.

Gamero-Castano, M. and J. F. de la Mora. 2000. Kinetics of small ion evaporation from the charge and mass distribution of multiply charged clusters in electrosprays. *J. Mass Spectrom.* 35:790–803.

Han, R. J. and J. W. Gentry. 1993a. Field and combined diffusional and field charging of fibrous aerosols. *Aerosol Sci. Technol.* 18:165–179.

Han, R. J. and J. W. Gentry. 1993b. Unipolar diffusional charging of fibrous aerosols—Theory and experiment. *J. Aerosol Sci.* 24:211–226.

Han, R. J., O. R. Moss, and B. A. Wong. 1994. Airborne fiber separation by electrophoresis and dielectrophoresis—Theory and design considerations. *Aerosol Sci. Technol.* 21:241–258.

Helsper, C., H. Fissan, A. Kapadia, and B. Y. H. Liu. 1982. Data inversion by simplex minimization for the electrical aerosol analyzer. *Aerosol Sci. Technol.* 1:135–146.

Hewitt, G. W. 1957. The charging of small particles for electrostatic precipitation. *Trans. Am. Inst. Elect. Engr.* 76:300–306.

Hoppel, W. A. and G. M. Frick. 1986. Ion attachment coefficients and the steady state charge distribution on aerosols in a bipolar ion environment. *Aerosol Sci. Technol.* 5:1–21.

Hoppel, W. A. and G. M. Frick. 1990. The nonequilibrium character of the aerosol charge distributions produced by neutralizers. *Aerosol Sci. Technol.* 12:471–496.

Karasek, F. W. 1970. The plasma chromatograph. Res. Dev. 21:34.

Knutson, E. O. and K. T. Whitby. 1975. Aerosol Classification by electric mobility: Apparatus, theory, and applications. *J. Aerosol Sci.* 6:443–451.

Kousaka, Y., Y. Endo, H. Ichitsubo, and M. Alonso. 1996. Orientation specific dynamic shape factors for doublets and triplets of spheres in the transition regime. *Aerosol Sci. Technol.* 24:36–44.

Laframboise, J. G. and J.-S. Chang. 1977. Theory of charge deposition on charged aerosol particles of arbitrary shape. *J. Aerosol Sci.* 8:331–338.

Langevin, P. 1903. L'Ionization des gaz. *Ann. Chim. Phys.* 28:289–384.

Lawson, C. L. and R. S. Hansen. 1974. *Solving Least-Squares Problems.* Englewood Cliffs, NJ: Prentice-Hall.

Lipowicz, P. J. and H. C. Yeh. 1989. Fiber dielectrophoresis. *Aerosol Sci. Technol.* 11:206–212.

Liu, B. Y. H. and D. Y. H. Pui. 1974. Equilibrium bipolar charge distribution of aerosols. *J. Colloid Interface Sci.* 49:305–312.

Liu, B. Y. H. and D. Y. H. Pui. 1977. On unipolar diffusion charging of aerosols in the continuum regime. *J. Colloid Interface Sci.* 58:142–149.

Liu, B. Y. H., K. T. Whitby, and H. S. Yu. 1967. Electrostatic sampler for light and electron microscopy. *Rev. Sci. Instrum.* 38.100–102.

Markowski, G. R. 1987. Improving Twomey algorithm for inversion of aerosol measurement data. *Aerosol Sci. Technol.* 7:127–141.

Matter, D., M. Mohr, W. Fendel, A. Schmidt-Ott, and H. Burtscher. 1995. Multiple wavelength aerosol photoemission by excimer lamps. *J. Aerosol Sci.* 26:1101–1115.

McMurry, P. H., H. Takano, and G. R. Anderson. 1983. Study of the ammonia (gas)–sulfuric acid (aerosol) reaction rate. *Environ. Sci. Technol.* 17:347.

Mirme, A. 1994. *Electrical Aerosol Spectrometry.* Ph.D. thesis, Universitatis Tartuensis, Tartu, Estonia.

Morrow, P. E. and T. T. Mercer. 1964. A point-to-plane electrostatic precipitator for particle size sampling. *AIHAJ* 25:8–14.

Nolan, J. J. 1926. The characteristics of the ionization produced by spraying water. *Phil. Mag.* 1:417–428.

Pui, D. Y. H. and B. Y. H. Liu. 1988. Advances in instrumentation for atmospheric aerosol measurement. *Physica Scripta* 37:252–269.

Rogak, S. N., U. Baltensperger, and R. C. Flagan. 1991. Measurement of mass transfer to agglomerate aerosols. *Aerosol Sci. Technol.* 14:447–458.

Rogak, S. N. and R. C. Flagan. 1992. Bipolar diffusion charging of spheres and agglomerate aerosol particles. *J. Aerosol Sci.* 23:693–710.

Rogak, S. N., R. C. Flagan, and H. V. Nguyen. 1993. The mobility and structure of aerosol agglomerates. *Aerosol Sci. Technol.* 18:25–47.

Rohrmann, H. 1923. Methode sur Messung der Grösse von Schwebelteilchen. *Z. Phys.* 17:253–265.

Rossell-Llompart, J., I. G. Loscertales, D. Bingham, and J. F. de la Mora. 1996. Sizing nanoparticles and ions with a short differential mobility analyzer. *J. Aerosol Sci.* 27:695–719.

Russell, L. M., R. C. Flagan, and J. H. Seinfeld. 1995. Assymmetric instrument response resulting from mixing effects in accelerated DMA-CPC measurements. *Aerosol Sci. Technol.* 23:491–509.

Russell, L. M., M. R. Stolzenburg, S. H. Zhang, R. Caldow, R. C. Flagan, and J. H. Seinfeld. 1996. Radially classified aerosol detector for aircraft based submicron aerosol measurements. *J. Atmos. Oceanic Technol.* 13:598–609.

Schmidt-Ott, A. 1988. In situ measurements of the fractal dimensionality of ultrafine aerosol particles. *Appl. Phys. Lett.* 52:954–956.

Schmidt-Ott, A., P. Schurtenberger, and H. C. Siegmann. 1980. Enormous yield of photoelectrons from small particles. *Phys. Rev. Lett.* 45:1284–1287.

Stolzenburg, M. R. 1988. *An ultrafine aerosol size distribution measuring system.* Ph.D. Thesis. University of Minnesota, Minneapolis MN.

TenBrink, H. M., A. Plomp, H. Spoelstra, and J. F. van de Vate. 1983. A high resolution electrical mobility aerosol spectrometer (MAS). *J. Aerosol Sci.* 14:589–597.

Tolman, R. C., L. H. Reyerson, A. P. Brooks, and H. D. Smith. 1919. An electrical precipitator for analyzing smokes. *J. Am. Chem. Soc.* 41:587–589.

Twomey, S. 1975. Comparison of constrained linear inversion and an iterative nonlinear algorithm applied to the indirect estimation of particle size distributions. *J. Comput. Phys.* 18:188–200.

Wang, S. C. and R. C. Flagan. 1990. Scanning electrical mobility spectrometer. *Aerosol Sci. Technol.* 13:230–240.

Whitby, K. T. and W. E. Clark. 1966. Electrical aerosol particle counting and size distribution measuring system for the 0.015 to 1 μ size range. *Tellus* 18:573–586.

White, H. J. 1951. *AIEE Trans.* 70:1186–1191.

Wiedensohler, A. 1988. An approximation of the bipolar charge distribution for particles in the submicron size range. *J. Aerosol Sci.* 19:387–389.

Winklmayr, W., G. P. Reischl, A. O. Lindner, and A. Berner. 1991. A new electromobility spectrometer for the measurement of aerosol size distributions in the size range from 1 to 1000 nm. *J. Aerosol Sci.* 22:289–296.

Wolfenbarger, J. K. and J. H. Seinfeld. 1990. Inversion of aerosol size distribution data. *J. Aerosol Sci.* 21:227–247.

Yeh, H. C. 1976. A theoretical study of electrical discharging of self-charging aerosols. *J. Aerosol Sci.* 7:343–349.

Yeh, H. C., G. J. Newton, and S. V. Teague. 1978. Charge distribution on plutonium-containing aerosols produced in mixed-oxide reactor fuel fabrication and the laboratory. *Health Phys.* 35:500–503.

Yun, C. M., Y. Otani, and H. Emi. 1997. Development of unipolar ion generator—Separation of ions in axial direction of flow. *Aerosol Sci. Technol.* 26:389–397.

Zhang, S. H., Y. Akutsu, L. M. Russell, and R. C. Flagan. 1995. Radial differential mobility analyzer. *Aerosol Sci. Technol.* 23:357–372.

Zhang, S. H. and R. C. Flagan. 1996. Resolution of the radial differential mobility analyzer for ultrafine particles. *J. Aerosol Sci.* 27:1179–1200.

19 Condensation Detection and Diffusion Size Separation Techniques

YUNG-SUNG CHENG
Lovelace Respiratory Research Institute, Albuquerque, NM

INTRODUCTION

The diffusion and condensation techniques described in this chapter are used primarily for measurement of ultrafine particles (<0.2 μm in diameter). Condensation techniques have been used in the study of atmospheric aerosols since the time of John Aitken (1888). The sizes of atmospheric nuclei that he measured were between 0.02 and 0.2 μm (Aitken nuclei). The Aitken nuclei and smaller particles could not be seen under an optical microscope, so he constructed an instrument to increase the particle size by saturating the atmosphere with water vapors, and the subsequent cooling of atmospheres caused condensation. The particles grew to larger than 1.0 μm and then could be counted in an optical microscope. From this early technique have evolved two types of instruments: (1) condensation nuclei counters (CNCs)—also called condensation particle counters (CPCs)—for determination of the number concentration of ultrafine particles, and (2) cloud chambers for measurement of high-energy charged particles at atomic and subatomic levels. This chapter focuses on the CNC. Improvements to the CNC design have continued over the last century, and now several types have been developed. They vary in the methods by which supersaturation is induced and particles are detected.

The diffusion technique is used to collect ultrafine particles and vapors and to determine the size distribution of ultrafine particles. The technique was first conceived following the observation that losses of atmospheric nuclei in tubes were related to their diffusion coefficients (Nolan and Guerrini, 1935). Mathematical equations for diffusion losses in rectangular or circular tubes were subsequently derived (Nolan and Nolan, 1938; Gormley and Kennedy, 1949). Diffusion coefficients and submicrometer particle sizes could then be determined from measurement of particle penetration through these tubes.

Diffusion samplers separate particles or vapors by differential diffusion mobilities. Two types of diffusion samplers are often used in air sampling. A diffusion battery can be used to measure the size distribution of submicrometer particles, and a diffusion denuder is designed to separate and collect gases or vapors from airborne particles. Diffusion battery samplers can consist of tubes of different shapes or stacks of fine-mesh screens of well-defined characteristics. Some diffusion denuders can also be used to determine the diffusion coefficient and therefore the size of gas or vapor molecules.

Aerosol Measurement: Principles, Techniques, and Applications, Second Edition, Edited by Paul A. Baron and Klaus Willeke.
ISBN 0-471-35636-0

Both diffusion and condensation devices are used for the measurement of ultrafine aerosol particles. A diffusion battery is often used with a CNC in a sampling train to determine both the concentration and particle size distribution. This chapter describes the operating principle, theory, design and applications, and data analyses of the CNCs, diffusion batteries, and diffusion denuders.

CONDENSATION THEORY

A CNC is an instrument for detecting ultrafine particles. The principle upon which the condensation technique is based involves three processes: (1) supersaturation of water or other working fluids, (2) growth of particles by condensation of vapors, and (3) detection of particles.

Supersaturation

Ultrafine particles including Aitken nuclei are too small to be detected by an optical microscope, which has a detection limit of about 0.1 µm. In a CNC, the aerosol is first saturated with a vapor of water or alcohol and subsequently cooled to induce a supersaturated condition, that is, the vapor pressure is higher than the saturation pressure at a given temperature (see Chapter 5). Vapor molecules condense on particles to form liquid droplets of larger sizes. For bulk liquids, the vapor pressure at the saturation or equilibrium condition (p_s in Pa) as a function of temperature (T, in degrees Kelvin) is shown by the following equation:

$$\log_{10} p_s = a - \frac{b}{T - c} \tag{19–1}$$

For some compounds, a slightly different equation gives the best fit to the data:

$$\log_{10} p_s = \frac{-52.3b}{T} + c \tag{19–2}$$

Table 19–1 lists the constants (a, b, and c) for some working fluids used in the CNCs, including water and alcohols. The saturation vapor pressure (p_s) is defined as the equilibrium partial vapor pressure for a flat liquid surface. For liquid droplets in an aerosol system, the partial pressure required to maintain the equilibrium is greater than that for a flat surface. This is called the *Kelvin effect*. The relationship between the saturation vapor pressure (p_d) on the droplet surface and particle diameter (d_p) can be expressed as the following function (see, for example, Hinds, 1999):

$$\frac{p_d}{p_s} = \exp\left(\frac{4v\gamma}{RTd_p}\right) \tag{19–3}$$

TABLE 19–1. Vapor Pressures of CNC Working Fluids

Fluid	Equation	a	b	c
Water	19–1	10.23	1750	38
Methanol	19–1	10.00	1233	45
Ethanol	19–1	10.16	1554	50.2
N-butanol	19–2		46.78	11.26
Dibutyl phthalate	19–1	18.39	5099	109

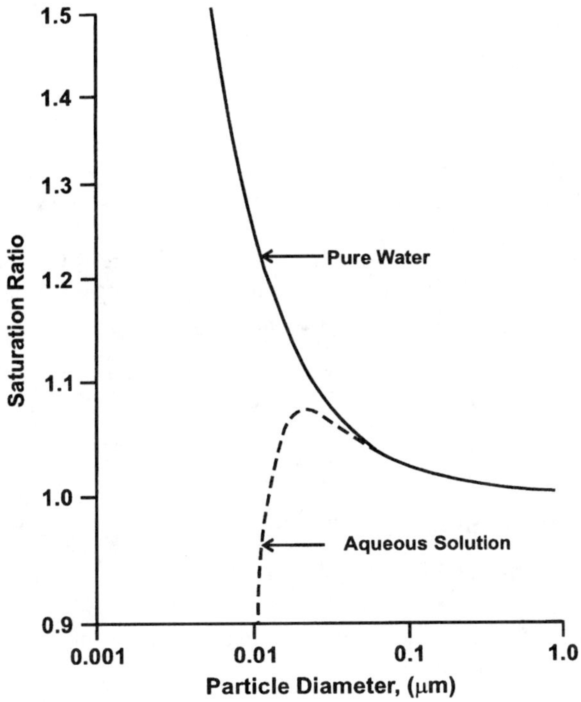

Fig. 19–1. Saturated ratio of water as a function of droplet size at 293 K [20°C]. The solid line is the pure water droplet, and the dashed line is the solution particle.

where γ is the surface tension, v is the molar volume of the liquid, and R is the gas constant. Figure 19–1 plots the saturation ratio at equilibrium ($S_R = p_d/p_s$) as a function of the particle size for a water droplet. As the diameter becomes smaller, the saturation ratio needed to induce condensation becomes larger. At a given level of saturation ratio in a CNC (S_R between 1.5 and 3), the Kelvin diameter, d^*, as defined by Eq. 19–3 by setting $d^* = d_p$, is the minimum size to initiate the condensation. Particles larger than the Kelvin diameter will grow, while those smaller than d^* will be too small for vapor condensation. Note that Eq. 19–3 is derived for vapor condensed on liquid droplets of the same material or on insoluble particles with wettable surface properties for the working fluid.

For aerosol systems consisting of nonvolatile solute particles and a volatile solvent, such as hygroscopic particles in existence with water vapor, the saturation vapor pressure is lowered because there are fewer solvent molecules in the surface layers than in the case of a pure solvent. For an ideal solution, the reduction of vapor pressure should be proportional to concentration (Raoult's law). In a solution droplet, the vapor pressure can be expressed in a form similar to the Kelvin equation (Eq. 19–3) (Tang, 1976; Friedlander, 1977):

$$\frac{p_d}{p_s} = \delta m_f \exp\left(\frac{4\gamma v_l}{RTd_p}\right) \quad (19\text{–}4)$$

where δ, v_l, and m_f are the activity coefficients, partial molar volume of the solvent, and mole fraction of the solute, respectively. Thus, there are two competing effects for solution droplets. The Kelvin effect tends to increase the vapor pressure, whereas the solute tends to reduce it. For an ideal solution ($\delta = 1$), Eq. 19–4 can be approximated in the following form (Friedlander, 1977):

$$\frac{p_d}{p_s} = \exp\left(\frac{4\gamma v_l}{RTd_p} - \frac{6n_2 v_l}{\pi d_{p^3}}\right) \tag{19–5}$$

where n_2 is the number of moles of the solute. Figure 19–1 compares vapor pressure for a pure water droplet to that of a solution particle, showing that the latter may be stable even at vapor pressures below saturation. In a CNC using water as the working fluid, hygroscopic particles, such as sodium chlorides and other salts, may have lower detection limits because they are soluble in water.

If a particle carries an electrostatic charge, the vapor pressure at the surface is also reduced, as well as the saturation ratio required for the condensation (Scheibel and Porstendörfer, 1986):

$$\frac{p_d}{p_s} = \exp\left\{\frac{4v}{RT}\left[\frac{\sigma}{d_p} - \frac{q^2\left(1-\frac{1}{\varepsilon}\right)}{2\pi d_{p^4}}\right]\right\} \tag{19–6}$$

where q is the electrostatic charge and ε is the dielectric constant of the droplet.

Droplet Growth

The droplet growth by condensation can be calculated by the following equation (Sutugin and Fuchs, 1965; Zhang and Liu, 1990):

$$d_p \frac{d}{dt}(d_p) = \frac{4D_a v}{R}\left(\frac{p}{T} - \frac{p_d}{T_d}\right) f(Kn) \tag{19–7}$$

where D_a is the diffusion coefficient of the condensing vapor; p and T are the vapor pressure and temperature in the surrounding gas far away from the particle, respectively; and v is the molar volume. The Fuchs correction, *f(Kn)*, is important for particles smaller than 0.1 µm:

$$f(Kn) = \frac{1 + Kn}{1 + 1.71Kn + 1.333Kn^2} \tag{19–8}$$

where Kn ($=2\lambda/d_p$) is the Knudsen number and λ is the mean free path of the gas medium. Furthermore, p_d and T_d are the vapor pressure and temperature at the surface, respectively. For the Kelvin effect, p_d can be calculated from Eq. 19–3. The temperature on the droplet surface, T_d, includes the effect of the droplet temperature increase due to the latent heat of condensation and can be estimated as follows (Hinds, 1999):

$$T_d = T + \frac{LMD}{k_v R}\left(\frac{p}{T} - \frac{p_d}{T_d}\right) \tag{19–9}$$

where L is the latent heat of the working fluid and k_v is the thermal conductivity of the carrier gas. Equation 19–7 describes the growth of a homogeneous liquid droplet or the growth of insoluble particles with wettable surfaces. Friedlander (1977) has described the growth of soluble particles.

Droplet Detection

In a CNC, particles grow to a near uniform size between 5 and 15 µm, large enough to be detected by optical means. Aitken (1888) used an optical microscope to count particles collected on slides for determining the concentration. Later, Pollak and others (Pollak and Daly, 1958; Jaenicke and Kanter, 1976) mounted a camera on the CNC to photograph the particles. The Aitken and photographic CNCs are sometimes referred to as the *absolute CNCs* because the number concentration is determined by direct counting. However, these CNCs are manually operated, and they have been used primarily as calibration references for newer types of CNCs in which the particles are detected by photoelectrical means. Early photoelectric CNCs, such as the Pollak counter (Pollak and Metnieks, 1958), use the light-extinction method, whereas recent models use light-scattering techniques. The detection system consists of a light source and a photodetector. In a light-extinction device, the transmission of light is detected; whereas in newer CNCs, the scattered light from the aerosol is detected in the photodetector. The signals from individual particles can be identified and counted, or the intensity of the scattered light is used as an indication of the particle concentration. The response curve of a photoelectric CNC must be calibrated. Early calibrations were performed with the photographic or Aitken CNCs. In recent studies, positively charged, monodisperse aerosol particles generated from an electrostatic classifier have been used; the concentration of a charged aerosol is determined with an electrometer. The aerosol concentration of the CNC is number concentration, conventionally expressed as particles/cm^3.

CONDENSATION NUCLEI COUNTERS

CNCs can be classified by the technique used to activate condensation and growth of particles or by the aerosol detection method. Three techniques have been used to induce supersaturation conditions: (1) adiabatic expansion, (2) conductive cooling, and (3) mixing of hot and cold air streams. The aerosol detection systems include photography, light extinction, and light scattering as described in the section on theory. Both photographic and light-extinction methods are associated exclusively with CNCs having expansion chambers, whereas the light-scattering detection method is incorporated into more recent CNCs—with condensation provided by thermal systems. This section describes the CNCs according to the supersaturation technique used, with emphasis on the commercially available instruments (also listed in Table 19–2).

TABLE 19–2. Commercial CNCs and Diffusion Batteries

Company	Model	Type	Type of Detection	Detection Limit (nm)
CNCs				
BGI	Pollak Counter[a]	Expansion/ overpressure	Light extinction	2.8
Gardner		Expansion/ overpressure	Light scattering	3.8
Environment One	Rich 200[a]	Expansion/ underpressure	Light scattering	
TSI	3007[b]	Conductive	Light scattering	10
TSI	3010	Conductive	Light scattering	10
TSI	3022A	Conductive	Light scattering	7
TSI	3025A	Conductive	Light scattering	3
TSI	3760A/3762	Conductive	Light scattering	10
Diffusion batteries				
INT	02-2000	In-line flow	—	—
INT	02-1900	Parallel flow	—	—

[a] No longer commercially available.
[b] Portable instrument.

Description of Condensation Nuclei Counter

Three types of CNCs based on the technique used to induce supersaturation conditions are described here.

Expansion-Type CNCs. An expansion-type CNC consists of a humidifier, an expansion chamber, and a detector. Water is the working fluid for this type of CNC. The aerosol stream is first humidified to reach the saturation of water vapor at room temperature, and then the aerosol in an expansion chamber is suddenly cooled by volume expansion or pressure release. At lower temperatures the chamber becomes supersaturated with water vapor, which then condenses on particles. The expansion-type CNC was first designed by Aitken (1888) and modified later by others. Several commercial instruments are based on these modifications dating back to the 1950s and 1960s, including the Pollak counter (Pollak and Metnieks, 1958) and the Rich counters (Rich, 1955, 1961), the Environment One and Gardner counters (Hogan and Gardner, 1968), and the GE counters (Skala, 1963; Haberl, 1979).

EXAMPLE 19–1

An expansion-type CNC with water as the working fluid has a pressure expansion ratio of 1.21. If the CNC is operated at a room temperature of 293 K [20°C] and 98.6 kPa pressure, what is the minimum particle diameter that can grow in the CNC (Kelvin diameter)?

Answer: The following are steps to solve the problem:

1. Estimate the temperature after expansion, T_f, from Eq. 19–13:

$$T_f = T_i\left(\frac{P_f}{P_i}\right)^{\frac{\gamma-1}{\gamma}} = 293.2\cdot\left(\frac{1}{1.21}\right)^{\frac{0.4}{1.4}} = 277.7\text{K}[=4.5\,^\circ\text{C}]$$

2. Calculate the saturated water pressure, $p_s(T_i)$ and $p_s(T_f)$, from Eq. 19–1:

$$p_s(T_i) = 10^{\left(10.23-\frac{1750}{293.1-38}\right)} = 2.36\text{ kPa}$$

Similarly, $p_s(T_f) = 850$ Pa.

3. Estimate the supersaturation ratio using Eq. 19–25:

$$S_R = \frac{2360}{850}\times\frac{1}{1.21} = 2.30$$

4. Calculate the Kelvin diameter using Eq. 19–3 and surface tension of 75 dyne/cm for water:

$$d^* = \frac{4\times 0.075\,\text{N/m}\cdot 1.8\times10^{-5}\,\text{m}^3/\text{gmol}}{8.31\,\text{N/m/gmol/K}\cdot 277.7\text{ K}\times\ln 2.30}$$
$$= 2.81\times10^{-9}\,\text{m} = 2.81\text{nm}$$

In the original Aitken counter, the air volume was expanded by using a piston; in other models, the pressure inside the expansion chamber was reduced to cause the expansion. In an overpressure system, the chamber was pressurized by pumping air into the chamber using a hand pump (Pollak counter; Pollak and Metnieks, 1958) or bellows (Gardner counter;

Hogan and Gardner, 1968), and then a valve was opened to release the chamber pressure to the ambient value. In the Environment One Model 200 CNC (Rich, 1961; Skala, 1963), underpressure systems were used in which air in the chamber under ambient pressure was released to an evacuated chamber. The volume or pressure expansion ratios as defined in the following equations were determined by the geometry of the expansion chamber:

$$\text{pressure expansion ratio} = \frac{P_i}{P_f} \tag{19–10}$$

$$\text{volume expansion ratio} = \frac{v_f}{v_i} \tag{19–11}$$

where P_i and P_f are pressures at the beginning and the end stage of expansion and v_i and v_f denote volumes at the beginning and end stage of the expansion. The expansion ratios in a CNC determine the supersaturation ratio of vapors. Usually the dry adiabatic expansion of ideal gases is assumed for air and water vapor because, in the short time of expansion, there is little time for heat transfer. Under these assumptions, the two expansion ratios are related:

$$\frac{P_i}{P_f} = \left(\frac{v_f}{v_i}\right)^{\gamma} \tag{19–12}$$

where γ is the specific heat ratio and is 1.4 for air. Because both the expansion ratios are greater than 1, Eq. 19–12 shows that the pressure expansion ratio is larger than the volume expansion ratio. The pressure expansion ratio of CNCs ranges from 1.1 to 1.5 (Miller and Bodhaine, 1982).

After expansion, the chamber temperature at the onset of condensation (T_f) can be related to the initial temperature (T_i) before expansion and the expansion ratios (Miller and Bodhaine, 1982):

$$\frac{T_f}{T_i} = \left(\frac{P_f}{P_i}\right)^{\frac{\gamma-1}{\gamma}} = \left(\frac{V_i}{V_f}\right)^{\gamma-1} \tag{19–13}$$

At this point, the water vapor in the system becomes supersaturated, and the saturation ratio is defined as the ratio of the partial pressure of water, $p(T_f)$, to the saturated water pressure, $p_s(T_f)$:

$$S_R = \frac{p(T_f)}{p_s(T_f)} \tag{19–14}$$

Assuming that the air is initially saturated with water, the partial pressure of water after expansion can then be calculated similar to Eq. 19–12:

$$\frac{p(T_f)}{p_s(T_i)} = \left(\frac{v_i}{v_f}\right)^{\lambda} = \left(\frac{P_f}{P_i}\right) \tag{19–15}$$

Substituting Eq. 19–15 into Eq. 19–14, we obtain the saturation ratio in terms of saturated vapor pressures and the pressure expansion ratio:

$$S_R = \frac{p_s(T_i)}{p_s(T_f)}\left(\frac{P_f}{P_i}\right) \tag{19–16}$$

An expansion-type CNC is normally operated in a cyclic fashion. The chamber is first filled with aerosol and vapor, the expansion occurs, the aerosol particles grow, and the concentration is determined. In photographic CNCs (Pollak and Daly, 1958; Jaenicke and Kanter, 1976; Scheibel and Porstendörfer, 1986), additional steps are necessary to develop photographs and to count the particles for concentration measurements. Some of the early CNCs, including the Pollak counter and other photographic CNCs, used an overpressure system with slow manual operation. They were primary standards for other CNCs because the counting method was considered a direct measurement of particle concentration. The Environment One and Gardner CNCs (Skala, 1963) use underpressure systems, and the expansion cycles are controlled by rotary valves. The sampling rates are 1 to 5 cycles per second, providing near continuous operation.

Figure 19–2 shows a schematic diagram of a Pollak counter (Pollak and Metnieks, 1958). The counter includes a vertical expansion chamber or fog-tube (0.025 m inner diameter and 0.60 m long) with a water-saturated ceramic lining, a light source at the top, a photodetector at the bottom, and a hand pump. Thus, the fog-tube combines the functions of humidification and expansion and also provides the light path for photoelectrical detection. The aerosol is first allowed to flush through the chamber, and then the inlet and the exit to the chamber are closed. Next, the fog-tube is pressurized to 21.3 kPa [160 mm Hg] over the ambient pressure by operating the hand pump. After a delay to allow the air to become saturated, the initial

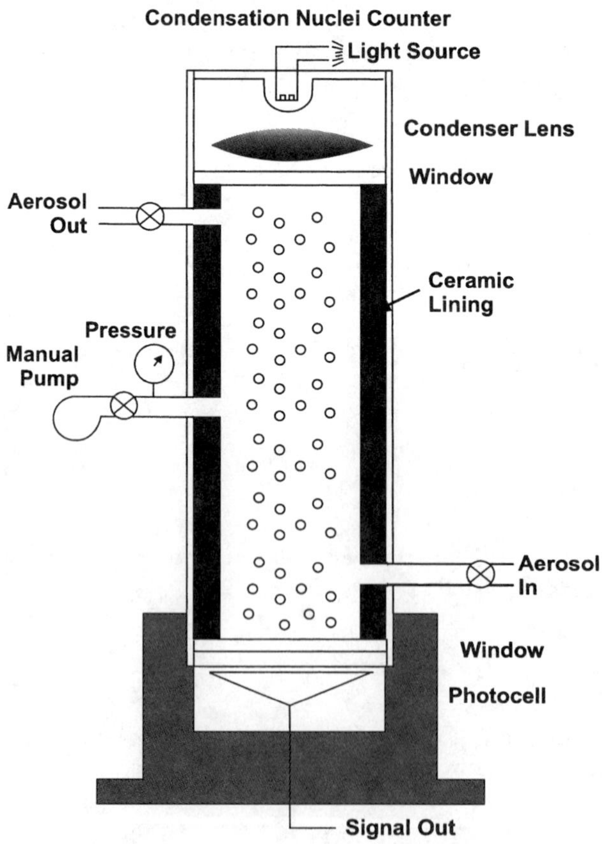

Fig. 19–2. Schematic of the Pollak counter.

light intensity is taken. Then the expansion valve is opened for air to expand to the ambient pressure. A second reading is taken, and the ratio of the final reading to the initial reading is the light transmission. It is calibrated against the aerosol concentration measured simultaneously from a photographic CNC as listed in the tables of Pollak and Metnieks (1958). These tables have been checked more recently using a monodisperse aerosol generated from electrical aerosol analyzers (Liu et al., 1975; Sinclair, 1984).

Conductive Cooling-Type CNCs. A major disadvantage of the expansion-type CNC is that the flow is cyclic and therefore incompatible with the requirement of steady-state flow when the CNC is used to measure concentration from a diffusion battery or electrical mobility analyzer. Continuous-flow CNCs are based on the principle of thermal cooling to induce the supersaturation of a working fluid with a steady flow in the system. Sinclair and Hoopes (1975b), Bricard et al. (1976), and Agarwal and Sem (1980) first designed the conductive cooling-type CNC. As shown in Figure 19–3, the conductive cooling CNC consists of a saturator, condenser, and particle detector. The aerosol passes through an alcohol reservoir kept at an elevated temperature. The residence time is such that the aerosol will be saturated with the working fluid at the set temperature. Then the aerosol enters a condenser tube, which is kept at a lower temperature by cooling the wall. In the condenser, gas cooling takes place by conduction and convection, which leads to supersaturation in the cooled aerosol stream. Particles then grow by condensation to form droplets. The concentrations of alcohol vapor and temperature are functions of the location in the condenser tube and are determined by numerically solving the heat and mass transfer equations (Ahn and Liu, 1990; Zhang and Liu, 1990).

Figure 19–3 shows a continuous-flow CNC (Model 3022, ***TSI****) described by Agarwal and Sem (1980). The saturator containing butanol is kept at 35°C, and the condenser tube is maintained at 10°C. The aerosol flow rate is $5 \times 10^{-6}\,m^3 s^{-1}$ [300 mL/min]. Under normal operating conditions at ambient pressure, the saturation ratio is calculated as a function of location as shown in Figure 19–3 (Zhang and Liu, 1990). The highest saturation ratio is at the center of the tube mainly because of the high vapor concentration. The vapor concentration and saturation ratio decrease near the tube wall. Because of the distribution of the vapor concentration, not all particles are activated for growth, resulting in a decreased counting efficiency for particles smaller than 10 nm (Su et al., 1990; McDermott et al., 1991; Kesten et al., 1991). Particles that grow to a size of about 12 µm are detected with a light-scattering system. The photodetector is operated as a single-particle counter for any concentration lower than 10^9 particles/m^3 [1000 particles/cm^3] and as a photometer for higher concentrations. Decreasing counting efficiency for nanometer-sized particles can be attributed to diffusional particle loss in the flow passage and incomplete activation due to inhomogeneous vapor concentration distribution in the condenser (Egilmez and Davies, 1984). To increase the counting efficiency of nanometer-sized particles, Wilson et al. (1983) modified the condenser tube so that for stratospheric aerosol measurements the aerosol is introduced in the center surrounded with clean sheath air saturated with alcohol for a low-pressure CNC. Based on similar concepts, Stolzenburg and McMurry (1991) designed continuous-flow CNCs ***(TSI)*** with lower detection limits (over 70% at a particle size of 3 nm) as shown in Figure 19–4.

Mixing-Type CNCs. The mixing-type CNC combines an aerosol stream at room temperature and a hot air stream with a saturated vapor of dibutyphthalate or dioctyl sebacate with a high boiling point. The CNC consists of a saturator with a reservoir of working fluid, a mixing chamber, and a particle detector. The aerosol flow is mixed rapidly with the hot stream in a nozzle assembly. The resultant mixed air is cooled down adiabatically. The temperature and vapor concentration can then be estimated from the initial temperatures and flow rates

* See Appendix I for full manufacturer addresses referenced to the italicized three-letter codes.

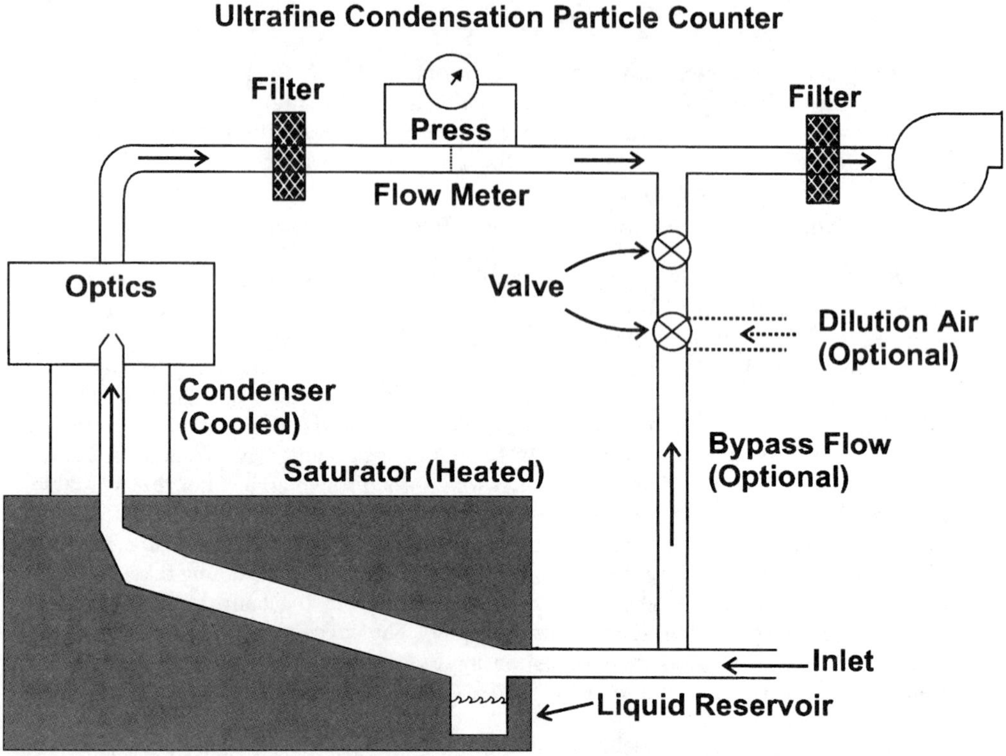

Fig. 19–3. Schematic of a TSI model 3022 CNC.

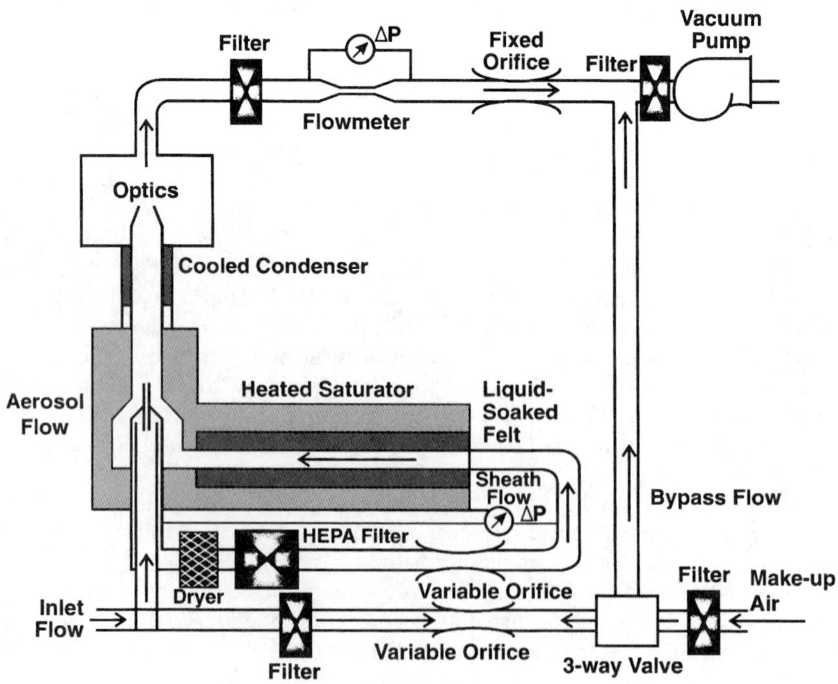

Fig. 19–4. Schematic of a TSI model 3025 CNC.

of the two air streams. Vapor condenses on particles at a steady-state continuous flow. The particle size can be estimated from the operating conditions and the number concentration (Okuyama et al., 1984). This mixing device was first used in Russia (Sutugin and Fuchs, 1965) primarily as an ultrafine particle generator called a *particle size magnifier*. It has been used as a CNC by incorporating a particle detection system (Okuyama et al., 1984). This type of CNC has two advantages over the thermal-cooling CNC: (1) higher aerosol flow rates (0.83 to $3.33 \times 10^{-5}\,m^3\,s^{-1}$ [0.5 to 2 L/min]) can be used, and (2) minimum diffusional loss of aerosol particles occurs because of the short aerosol delivery distance since the aerosol stream does not pass through the saturator.

THEORIES OF THE DIFFUSION MEASUREMENT TECHNIQUE

These mathematical expressions relate collection or penetration of vapors and particles through cylindrical and rectangular tubes and screens. These expressions can be used to calculate diffusion coefficients or particle sizes from experimental measurements through diffusion samplers.

These mathematical expressions were derived from the convective diffusion equation describing the concentration (c) in various geometries and flow profiles:

$$\frac{D}{r}\frac{\partial}{\partial r}\left(r\frac{\partial c}{\partial r}\right) = u(r)\frac{\partial c}{\partial z} \tag{19–17}$$

where D is the particle diffusion coefficient, r is the radial direction, z is the axial direction, and $u(r)$ is the velocity profile in the axial direction. Several assumptions were made in the derivation of Eq. 19–17: (1) the concentration is in a steady-state condition; (2) the flow field in the device is a fully developed laminar flow; (3) the effect of diffusion in the direction of flow is neglected; (4) no production or reaction of the gas or aerosol occurs in the device; and (5) the sticking coefficient of the gas or particle is 100% on the collection surface (walls or screens). Diffusion devices can be classified as tube (channel) type or screen type, and each type has different flow profiles. Solutions to Eq. 19–17 for different types of diffusion samplers are summarized in the following section.

Tube (Channel) Type

Penetration (P) of particles or gases due to the diffusion mechanism has been derived for channels of different geometries, including cylindrical, rectangular, disk, and annular shapes. The general solution of Eq. 19–15 can be expressed as a series of exponential functions:

$$P = \sum_{n=1}^{\infty} A_n \exp(-\beta_n \mu) \tag{19–18}$$

where μ is the dimensionless argument relating the diffusion coefficient, channel length, and flow rate, and β_n is the eigenvalue. Convergence of Eq. 19–18 depends on the magnitude of μ. For larger values of μ (low penetration), few terms are needed for convergence, whereas at smaller values of μ (high penetration), many terms are required. For low penetration, alternative equations have been derived. Specific equations for each channel type are described below.

Cylindrical Tubes. Penetration through a circular tube (Fig. 19–5) at a flow rate, Q, for particles with a diffusion coefficient of D has been derived by several investigators as a function of the parameter μ (Gormley and Kennedy, 1949; Sideman et al., 1965; Davis and Parkins, 1970; Tan and Hsu, 1971; Lekhtmakher, 1971; Bowen et al., 1976).

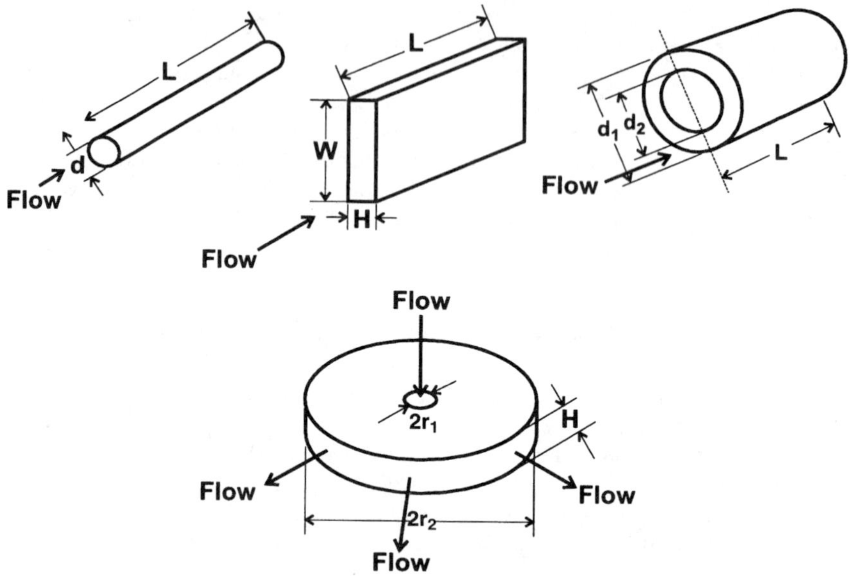

Fig. 19–5. Schematics of different shapes of tubes.

$$\mu = \frac{\pi D L}{Q} \tag{19–19}$$

The numerical solution obtained by Bowen et al. (1976) for μ between 1×10^{-7} and 1 is the most accurate. Results obtained by Davis and Parkins (1970), Tan and Hsu (1971), Sideman et al. (1965), and Lekhtmakher (1971) agree substantially with those of Bowen et al. (1976). By comparison of various expressions, the following analytical solutions have the accuracy of four significant figures as compared to Bowen's result in the entire range of μ:

$$\begin{aligned} P = \; & 0.81905\exp(-3.6568\mu) + 0.09753\exp(-22.305\mu) \\ & + 0.0325\exp(-56.961\mu) + 0.01544\exp(-107.62\mu) \\ & \text{for} \quad \mu > 0.02 \end{aligned} \tag{19–20}$$

$$P = 1.0 - 2.5638\mu^{2/3} + 1.2\mu + 0.1767\mu^{4/3} \quad \text{for} \quad \mu \leq 0.02 \tag{19–21}$$

The formula for the small values of μ is taken from Gormley and Kennedy (1949), Newman (1969), and Ingham (1975).

Rectangular Channels and Parallel Circular Plates. Particle penetration through a parallel narrow rectangular tube (Fig. 19–5) of width W and separation H, where $H << W$, has been derived as a function of μ defined as $8DLW/3QH$ (Gormley, cited by Nolan and Nolan, 1938; DeMarcus and Thomas, 1952; Sideman et al., 1965; Mercer and Mercer, 1970; Tan and Thomas, 1972; Bowen et al., 1976). The same equation can be used to calculate penetration through parallel circular plates (Fig. 19–5), where the diffusion parameter μ is defined as $8\pi D(r_2^2 - r_1^2)/3QH$, where r_2 and r_1 are outer and inner radii of the disks (Mercer and Mercer, 1970). Tan and Thomas (1972) and Bowen et al. (1976) gave the most accurate solution; other investigators agreed substantially with their results. Using the first four terms of the solution given by Tan and Thomas, the penetration can be calculated with the accuracy of four significant numbers as compared with the numerical solution of Bowen et al. (1976):

$$\begin{aligned} P = {} & 0.9104\exp(-2.8278\mu) + 0.0531\exp(-32.147\mu) \\ & + 0.01528\exp(-93.475\mu) + 0.00681\exp(-186.805\mu) \\ & \text{for} \quad \mu > 0.05 \end{aligned} \tag{19–22}$$

$$P = 1 - 1.526\mu^{2/3} + 0.15\mu + 0.0342\mu^{4/3} \quad \text{for} \quad \mu \le 0.05 \tag{19–23}$$

Ingham (1976) gave the formula for the small values of μ. Kennedy (quoted by Nolan and Kennedy, 1953) derived a similar formula with different coefficients, but the results are different by only 1%.

Annular Tubes. The theoretical calculation of diffusional losses through an annular tube (Fig. 19–6) has been derived by solving Eq. 19–17 using a fully developed flow in the annular tube (Winiwarter, 1989). The penetration is then a function of the inner to outer radius of the denuder, k, and the diffusional parameter, $\mu = \pi DL(d_2 + d_1)/2Q(d_2 - d_1)]$:

$$\begin{aligned} P = {} & A_0(k)\exp[-\beta_0^2(k)\mu] + A_1(k)\exp[-\beta_1^2(k)\mu] + A_2(k)\exp(-\beta_2^2(k)\mu) \\ & + A_3(k)\exp(-\beta_3^2(k)\mu) \end{aligned} \tag{19–24}$$

where d_2 and d_1 are outer and inner diameters, respectively. Values of $A_n(k)$ and $\beta_n(k)$ for selected values of k are given in Table 19–3. For $k = 0$ (cylindrical tube) and $k = 1$ (parallel plate), Eqs. 19–20 and 19–22 are the asymptotic solutions (Kerouanton et al., 1996). A numerical solution of the diffusional equation in an annular tube has been determined by solving the velocity profile instead of using the fully developed flow (Fan et al., 1996). From the numerical solution, the fitted equation is

$$P = 1 - \exp\left[-0.03711(1-k)^{1.317} Pe^{0.678}\right] \tag{19–25}$$

where Pe (= $u_m d_2^2/4DL$) is the Peclet number and u_m is the mean velocity. These equations are in general agreement with experimental data (Possanzini et al., 1983).

Screen Type

The aerosol penetration through a stack of fine-mesh screens with circular fibers of uniform diameter and arrangement has been derived (Cheng and Yeh, 1980; Cheng et al., 1980; Yeh

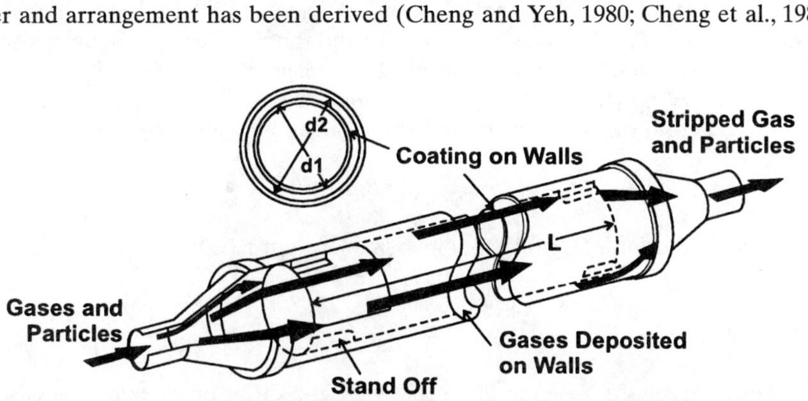

Fig. 19–6. Schematic diagram of an annular denuder. (From Stevens [1986] with permission of Lewis Publishers.)

TABLE 19–3. A_n and β_n for Selected Values of k in Eq. 19–24

				k			
	0.05	0.1	0.25	0.5	0.75	0.85	0.95
β_0	3.493	3.610	3.765	3.851	3.878	3.882	3.884
β_1	7.906	8.086	8.315	8.436	8.473	8.479	8.481
β_2	12.31	12.56	12.87	13.03	13.08	13.09	13.09
β_3	16.71	17.02	17.42	17.63	17.70	17.71	17.71
A_0	0.879	0.889	0.900	0.908	0.910	0.911	0.911
A_1	0.036	0.026	0.012	0.003	0.001	0.000	0.000
A_2	0.044	0.046	0.05	0.052	0.053	0.053	0.053
A_3	0.006	0.005	0.002	0.001	0.000	0.000	0.000

et al., 1982; Cheng et al., 1991). A stack of fine-mesh screens simulates a fan model filter (Kirsch and Fuchs, 1968; Kirsch and Stechkina, 1978) in terms of flow resistance and aerosol deposition characteristics (Cheng et al., 1985). The theoretical penetration was derived based on the aerosol filtration in the fan model filter:

$$P = \exp\left[-\beta_f n\left(2.7Pe^{-2/3} + \frac{1}{\kappa}R^2 + \frac{1.24}{\kappa^{1/2}}Pe^{-1/2}R^{2/3}\right)\right] \tag{19–26}$$

$$B_f = \frac{4\alpha h}{\pi(1-\alpha)d_f} \tag{19–27}$$

where n is the number of screens, d_f is the fiber diameter, h is the thickness of a single screen, α is the solid volume fraction of the screen, and k is the hydrodynamic factor of the screen:

$$k = -0.5\ln(2\alpha/\pi) + (2\alpha/\pi) - 0.75 - 0.25(2\alpha/\pi)^2 \tag{19–28}$$

where $R = d_p/d_f$ is the interception parameter and Pe is the Peclet number:

$$Pe = Ud_f/D \tag{19–29}$$

In addition, U is the velocity entering the screen. Equation 19–26 includes the diffusional and interceptional losses of aerosol on screens and is valid for particles up to 1 μm in size (Cheng et al., 1985). For particles larger than 1 μm, inertial impaction becomes an important mechanism, and Eq. 19–26 may not be adequate. For smaller particles ($d_p < 0.01$ μm), diffusional deposition is the dominant mechanism, and Eq. 19–26 is simplified to

$$P = \exp(-2.7B_f nPe^{-2/3}) \tag{19–30}$$

This equation is valid for a Reynolds number smaller than 1 (Cheng et al., 1992).

DIFFUSION DENUDERS

Gas or vapor molecules diffuse rapidly to the wall of a diffusion sampler and adsorb onto the wall coated with material suitable for collecting the gas. Diffusion tubes have been used to measure diffusion coefficients of several gases in the air (Thomas, 1955; Fish and Durham,

1971; Ferm, 1979; Durham et al., 1987). Since 1980, diffusion denuders followed by a filter pack have been developed to sample atmospheric nitric acid vapors and nitrate particulate aerosols. Using this sampling technique, called the *denuder difference method*, one can separate gaseous species, such as HNO_3 and NH_3, from particulate nitrates and thus minimize sampling artifacts due to the presence of these gases (Stevens et al., 1978; Appel et al., 1981; Shaw et al., 1982; Forrest et al., 1982; Ferm, 1986; Stevens, 1986). Diffusion denuders are also used to monitor vapors, such as formaldehyde, chlorinated organics, and tetra alkyl lead, in the ambient air or work environments (Johnson et al., 1985; Cecchini et al., 1985; Febo et al., 1986). Some personal samplers have also been developed for industrial hygiene use (DeSantis and Perrino, 1986; Gunderson and Anderson, 1987).

Description of Diffusion Denuders

Two types of diffusion denuders have been designed: the cylindrical tube and annular tube.

Cylindrical Denuders. In cylindrical denuders, a single cylindrical glass or Teflon tube is often used for collecting gases or vapors. The diameter and length of the tube and the sampling flow rate are designed to have greater than 99% collection efficiency. For example, a glass tube of 3 mm inner diameter (ID) and 0.35 m [35 cm] long would have over 99% efficiency for ammonia ($D_a = 2.47 \times 10^{-5}\,m^2\,s^{-1}$) at $5 \times 10^{-5}\,m^3\,s^{-1}$ [3 L/min] (Ferm, 1979). For higher sampling flow rates, parallel tube assemblies have been designed (Stevens et al., 1978; Forrest et al., 1982), consisting of 16 glass tubes 5 mm ID and 0.30 m [30 cm] long. The sampling flow rate was $8.33 \times 10^{-4}\,m^3\,s^{-1}$ [50 L/min], and the collection efficiency for ammonia was over 99% (Stevens et al., 1978). A two-stage diffusion denuder consisting of 212 glass honeycomb tubes, each with a height of 25.4 mm [2.54 cm] and an ID of 2 mm [0.2 cm] has been designed to removes gases for collection of ambient particles (Sioutas et al., 1994).

Penetration through the tube-type denuders can be estimated by taking the first term of Eq. 19–20 only:

$$P = 0.819 \exp(-3.66\mu) \tag{19–31}$$

This simplified equation is accurate at higher values of μ (>0.4) and at a lower penetration ($P < 0.190$). The error of the estimated penetration from Eq. 19–31 increases with the decreasing value of m (0.25% error for $\mu = 0.2$ and $P = 0.395$, and –2.7% for $\mu = 0.1$ and $P = 0.628$). Equation 19–31 is applicable for the fully developed laminar flow region in the tube. The flow Reynolds number in the tube should be less than 2300 for laminar flow:

$$Re_f = \frac{4\rho_f Q}{\eta \pi d} < 2300 \tag{19–32}$$

In the entrance of the tube, the flow is in a transition region from plug flow to developed flow. The length of entrance, L_e, is defined by the following equation and should be minimized:

$$L_e = 0.035\, d\, Re_f \tag{19–33}$$

Annular Denuders. Higher sampling flow rates are desirable, especially for sampling trains consisting of denuders and filters or dichotomous samplers (Shaw et al., 1982). An annular tube denuder has been designed for this purpose (Possanzini et al., 1983). This denuder consists of two coaxial cylinders with the inner one sealed at both ends so that air is forced to pass through the annular space (Fig. 19–6). The collection efficiency of the annular tube can be estimated from Eq. 19–24 for lower Reynolds numbers ($Re < 2300$) defined as:

$$Re_f = \frac{4\rho_f Q}{\eta\pi(d_1 + d_2)} \tag{19–34}$$

Comparing the performance of the cylindrical and annular denuders in removing a gas from an air stream, the flow rate of a typical annular denuder ($d_2 = 33$ mm and $d_1 = 30$ mm) can be determined by equating Eqs. 19–24 and 19–31. It can be shown that

$$\frac{Q}{L}\bigg|_{\text{annlar}} = 31.5\frac{Q}{L}\bigg|_{\text{cylinder}} \tag{19–35}$$

This relationship shows that for equal sampling time and tube length, the annular denuder can operate at 30 times the flow rate of the cylindrical denuder and still have the same removal efficiency. Also, the Reynolds number still indicates laminar flow conditions for the annular tube system. A multichannel annular diffusion denuder has been tested and used in ambient air sampling (Johnson et al., 1985).

Compact Coil Denuder. Pui et al. (1990) designed a compact coil denuder consisting of a 10 mm [1.0 cm] ID and 0.95 m [95 cm] long (L) glass tube bent into a three-turn helical coil with a 0.1 m [10 cm] diameter (Fig. 19–7). The heat and mass transfer rates to the tube wall in a curved tube are much higher than those in a straight tube operated at the same conditions (Mori and Nakayama, 1967a,b). This denuder is operated at $1.67 \times 10^{-4}\,\text{m s}^{-1}$ [10 L/min] (Q) with a Reynolds number of 1400. The penetration through the denuder can be expressed as (Pui et al., 1990)

$$P = 0.82\exp\left(-\frac{\pi L D}{Q}Sh\right) \tag{19–36}$$

$$Sh = \frac{0.864}{\delta}De^{-1/2}\left(1 + 2.35De^{-1/2}\right) \tag{19–37}$$

where Sh is the Sherwood number; De is the Dean number, $De = (Re_f/[\text{coil diameter/tube diameter}]^{1/2})$, the flow Reynolds number divided by the square root of the curvature; and δ

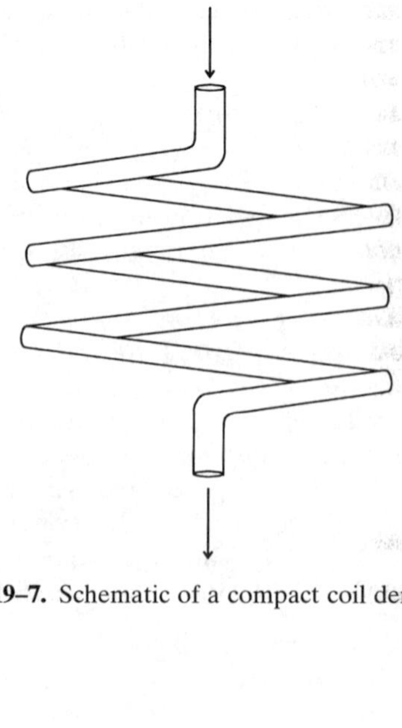

Fig. 19–7. Schematic of a compact coil denuder.

is the thickness ratio of the concentration boundary layer to the momentum boundary layer. This unit has 99.3% collection efficiency for SO_2 with less than 6% particle losses for particles between 0.015 to 2.5 μm in diameter. It is compact and easy to operate.

Transition-Flow Denuder. Both cylindrical and annular denuders are operated under laminar flow conditions, and they are designed to remove all gases of interest from the aerosol stream. In passing through such denuders, particles may evaporate or decompose and increase the concentration of some gases, especially in the case of the decomposition of NH_4NO_3 into HNO_3 and NH_3 gases. Biases in sampling due to evaporation of particles can be avoided by collecting only a known fraction of gases in the denuder and then calculating the gas concentration.

Durham et al. (1986) designed a transition-flow denuder, which permits a higher sampling flow. The cylindrical denuder has an ID of 9.5 mm with a 60 mm distance of the first active surface to allow for development of a stable flow profile. The denuder section is lined with a 32 mm long nylon sheet. Assuming a complete mixing in the active section, the penetration can be expressed as

$$P = \exp\left(-\frac{2\pi\alpha L}{Q}\right) \tag{19–38}$$

$$\alpha = \frac{rD_a}{\delta} \tag{19–39}$$

where L is the length of the active surface; Q is the flow rate; r is radius of the pipe; and δ is the boundary thickness, which is a function of the flow Reynolds number. The penetration must be determined empirically. Operating the denuder at $2.68 \times 10^{-4}\,m^3 s^{-1}$ [16.1 L/min] ($Re = 2500$), Durham et al. (1986) obtained a penetration of 0.911 for HNO_3.

Scrubber-Type Diffusion Denuders. Most diffusion denuders have a solid coating on the wall to collect gaseous species, and the coating substrates are washed after sampling for analysis. For continuous analysis of gas species, diffusion scrubbers are used, where the absorbent or solvent in the liquid form is continuously flowing along the tube wall, and the analyte can be analyzed in real time (Dasgupta, 1984; Dasgupta et al., 1988). A tubular scrubber is made by inserting a membrane tube into the glass tube to form a jacket between the glass wall and the membrane (Fig. 19–8). Porous membranes, such as polytetrafluoroethylene and polypropylene, allow gases but not particles to permeate and dissolve in a solution that flows continuously through the jacket. The collection characteristics of this type of diffusion scrubber should be similar to the tube diffusion denuder. Another type of diffusion denuder consists of a tube with a membrane tube at the center in which air flows through the annular space while the solvent passes through the membrane tube.

Coating Substrates

Absorbent material can be coated onto the tube wall of a denuder to collect the gas of interest from the air stream. Table 19–4 lists substrates for removal of some gases as reported in the literature. Some materials absorb more than one gaseous species. For example, sodium carbonate can absorb acidic gaseous species found in the ambient air, including HCl, HNO_2, HNO_3, and SO_2. How the material is applied to the tube wall depends largely on the nature of the material. Most materials are first dissolved and then applied to the tube wall. Solvents are allowed to evaporate, leaving the absorbent on the glass tube wall. In some cases, the glass denuder wall has been etched by sand blasting the surface to increase the capacity of walls to support the denuding chemical substrate (Possanzini et al., 1983). Absorbent paper

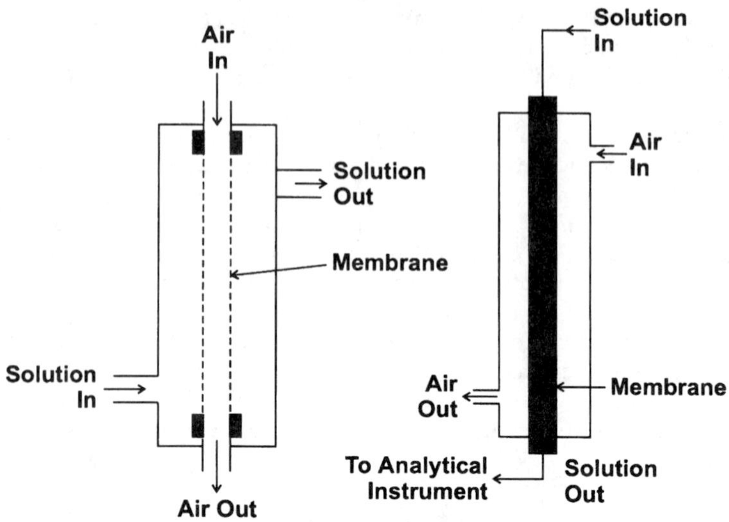

Fig. 19–8. Schematic of diffusion scrubbers.

TABLE 19–4. Materials for Absorbing Gases in the Diffusion Denuder

Coating Material	Gas Absorbed	Reference
Oxalic acid	NH_3, aniline	Ferm (1979), DeSantis and Perrino (1986)
Oleic acid	SO_3	Thomas (1955)
uH_3PO_3	NH_3	Stevens et al. (1978)
K_2CO_3	SO_2, H_2S	Durham et al. (1978)
Na_2CO_3	SO_2, HCI, HNO_3, HNO_2	Forrest et al. (1982)
$CuSO_4$	NH_3	Thomas (1955)
PbO_2	SO_2, H_2S	Durham et al. (1978)
WO_3	NH_3, HNO_3	Braman et al. (1982)
MgO	HNO_3	Stevens et al. (1978)
NaF	HNO_3	Slanina et al. (1981)
NaOH and guaiacol	NO_2	Buttini et al. (1987)
Bisulfite-triethanolamine	Formaldehyde	Cecchini et al. (1985)
Nylon sheet	SO_2, HNO_3	Durham et al. (1986)
Tenax powder	Chlorinated organics	Johnson et al. (1985)
Silica gel	Aniline	Gunderson and Anderson (1987)
ICI	Tetra alkyl lead	Febo et al. (1986)

impregnated with liquid or solution substrate, such as oleic acid, has been used to line the inside of the denuder wall (Thomas, 1955). A nylon sheet has also been used as a liner (Durham et al., 1986). Anodized aluminum surfaces have been found to be a good absorbing surface for nitric acid. Annular denuders made of anodized aluminum do not need a coating (W. John, personal communication, 1987). Tenax or silica gel powder is more difficult to apply; however, these materials adhere to the glass wall coated with silicon grease (Johnson et al., 1985; Gunderson and Anderson, 1987).

Sampling Trains

When sampling ambient or working atmospheres, it is sometimes necessary to collect gas species and particulate materials separately. In this case, a sampling train consisting of diffu-

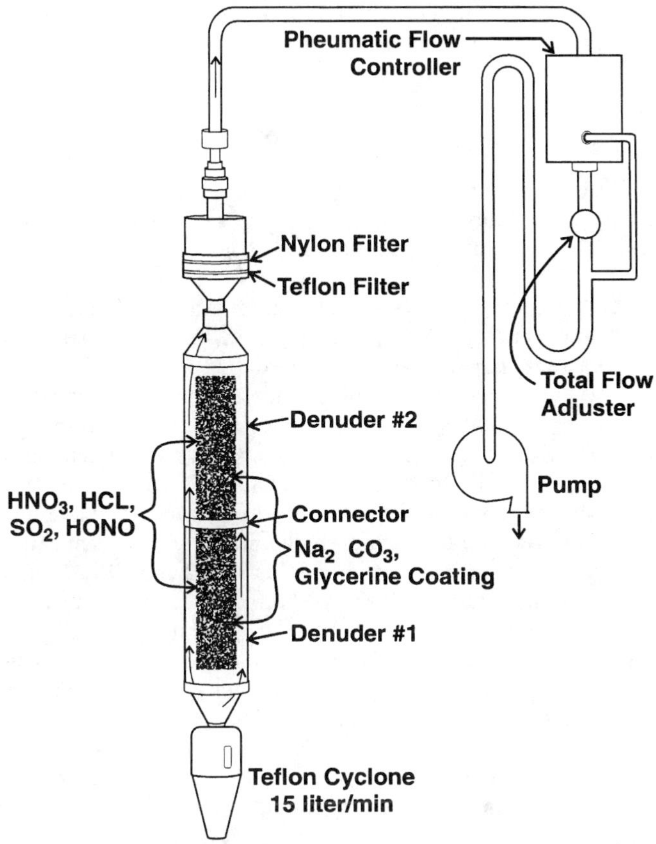

Fig. 19–9. An ambient acidic aerosol sampler consisting of a precutter, two annular denuders, and a filter pack. (From Stevens [1986] with permission of Lewis Publishers.)

sion denuders and a filter pack have been used. A more complex system (Fig. 19–9) includes a *cyclone preclassifier*, two Na_2CO_3-coated annular denuders, and a filter pack with a Teflon and a nylon filter. This system has been used to collect acidic gases (HNO_3, HNO_2, SO_2, and HCl) separately from nitrate and sulfate particles (Stevens, 1986). The first denuder removes gases quantitatively, whereas the second accounts for the interference from particulate material deposited on the wall assuming that particle deposition on each denuder is the same (Febo et al., 1986). The denuders are placed vertically to avoid particle deposition on the walls by sedimentation. A diffusion scrubber can be connected to an ion chromatograph or other analytical instruments for real-time analysis of gaseous species (Lindgren and Dasgupta, 1988; Dasgupta et al., 1988).

DIFFUSION BATTERIES

Diffusion batteries were originally developed to measure the diffusion coefficient of particles less than 0.1 μm diameter. They have since been used for determination of particle size distributions by converting the diffusion coefficient to the particle size. Diffusion batteries are one type of only a few instruments that are applicable in measuring ultrafine particles between 0.1 μm and about 0.8 nm, corresponding to the size of molecular clusters. In this section, various designs of the instrument, detection of particles, and methods of data analysis are discussed.

Description of Diffusion Batteries

Several types of diffusion batteries have been designed. Those based on rectangular channels and parallel circular plates are single-stage diffusion batteries. Cylindrical-tube and screen-type diffusion batteries usually have several stages.

Rectangular Channel. Rectangular channel diffusion batteries usually consist of many rectangular plates forming parallel channels of equal width. These plates are separated by spacers and glued to a container with an airtight seal. For example, Thomas (1955) designed a diffusion battery (0.1 mm wide, 0.127 m high, and 0.473 m long) consisting of 19 parallel channels made from graphite plates for a $1.67 \times 10^{-5}\,m^3\,s^{-1}$ [1 L/min] sampling flow rate. Other instruments have been made from aluminum or glass plates with similar construction (Nolan and Nolan, 1938; Nolan and Doherty, 1950; Pollak et al., 1956; Megaw and Wiffen, 1963; Rich, 1966). Each channel should be parallel and have the same width. Deviation of channel width results in a nonuniform flow rate through each channel, which in turn causes the deviation of penetration from the theoretical prediction of Eqs. 19–22 and 19–23. Pollak and Metnieks (1958) designed a diffusion battery of 10 single channels, each separately housed in a box.

A single-stage diffusion battery can be used to measure the diffusion coefficient of monodisperse aerosols at one flow rate. When it is used to measure polydisperse aerosols, such as those found in ambient air, several measurements must be taken at different flow rates to determine the distribution of diffusion coefficients.

Parallel Disks. Kotrappa et al. (1975) designed a diffusion sampler. It is based on the diffusional losses of particles from a fluid flowing radially inward between two coaxial, parallel, or circular plates as originally proposed by Mercer and Mercer (1970). Stainless steel plates (37.7 mm diameter) with a central hole of 2 mm diameter in the upper plate are the collecting substrate. Separation between the plates is 2.25 mm. An absolute filter is used to collect material penetrating the device. This sampler has been used to determine the diffusion coefficient of radon decay products, which have diffusion coefficients on the order of $5 \times 10^{-6}\,m^2\,s^{-1}$ [$0.05\,cm^2/s$]. The amount of radioactivity collected at the plates and absolute filter was determined and the diffusion coefficient calculated from Eq. 19–22, simplified to contain only the first term:

$$P = 0.9104 \exp\left(-2.8278 \frac{8\pi D(r_2^2 - r_1^2)}{3QH}\right) \tag{19–40}$$

Cylindrical Tubes. Tube-type diffusion batteries made of cylindrical tubes usually consist of a cluster of thin-walled tubes with IDs less than 1 mm. Large equivalent length (actual length × number of tubes) is required for measurement of particle size, because particles have much smaller diffusion coefficients than do gas molecules. Sinclair (1969), Breslin et al. (1971), and Scheibel and Porstendörfer (1984) have designed several cluster tube diffusion batteries. Figure 19–10 shows the schematic of a tube-type diffusion battery (Scheibel and Porstendörfer, 1984). Three diffusion batteries with 100, 484, and 1000 single tubes were used with lengths of 0.05, 0.093, and 0.3903 m, respectively. Tube-type diffusion batteries use materials that are commercially available and are also easier to construct than the parallel-plate diffusion battery. A lightweight material such as aluminum is often used; however, this type of diffusion battery is still heavy, bulky, and expensive. Most cluster tube diffusion batteries

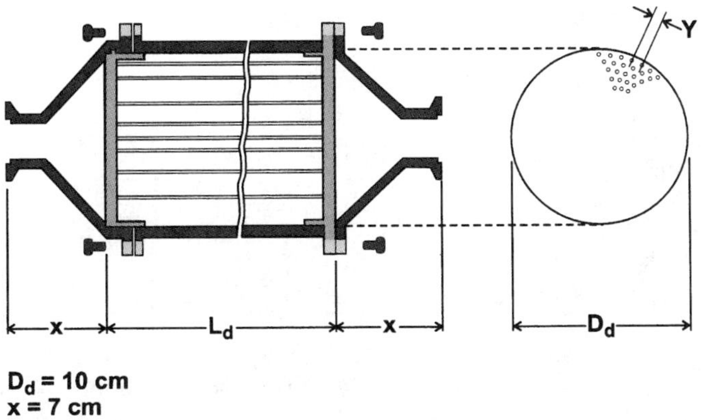

Fig. 19–10. Schematic of a cluster-tube diffusion battery (From H. G. Scheibel and J. Porstendörfer *J. Aerosol Sci.* 15:673–682, 1984, with permission.)

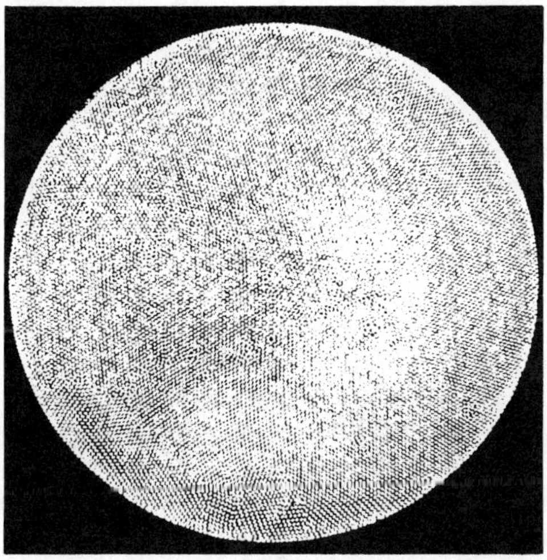

Fig. 19–11. A stainless steel collimated hole structure disk.

consist of one to three stages (Breslin et al., 1971; Scheibel and Porstendörfer, 1984), although an eight-stage diffusion battery was constructed (Sinclair, 1969).

Compact diffusion batteries with many stages have been designed by using collimated hole structures (CHSs). The CHSs are disks containing a large number of near circular holes. Figure 19–11 shows a 44.5 mm diameter CHS disk made from stainless steel containing 14,500 holes 0.23 mm in diameter (Brunswick Co., Chicago, IL). With a thickness of 3.2 to 25.4 mm, the equivalent length ranged from 46 to 369 m. A portable 11-stage diffusion battery was designed with CHS elements (Sinclair, 1972). The total length is 0.60 m, and the equivalent length is 5094 m. Figure 19–12 shows the schematic of a five-stage diffusion battery made from CHS elements. A multiple-stage diffusion battery is required to measure the size distribution of a polydisperse aerosol. The development of a multiple-stage CHS diffusion battery opens up the possibility of routine measurements of submicrometer aerosols. Other CHS disks made from glass capillary tubes of 25 or 50 mm in diameter and a thickness of 0.5 to 2.0 mm are also available (***GLL***). A six-stage CHS diffusion battery made from glass has also been designed (Brown et al., 1984).

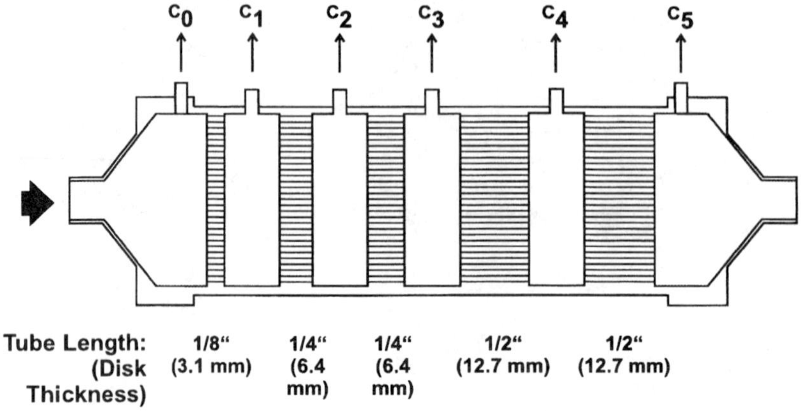

Fig. 19–12. Schematic of a five-stage diffusion battery consisting of a stainless steel collimated hole structure.

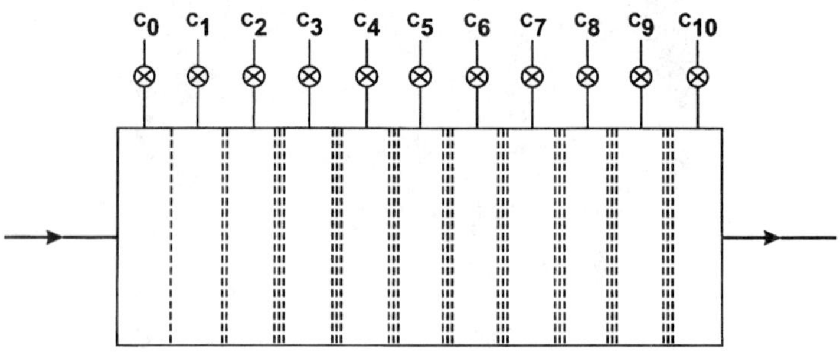

Fig. 19–13. Schematic of a 10-stage screen-type diffusion battery.

Screen Type. Sinclair and Hinchliffe (1972) and Twomey and Zalabsky (1981) have used diffusion batteries with stacks of filters as the cell material. This material is lightweight and inexpensive to build. However, commercial fiber or membrane filters are not ideal materials because of the nonuniform fiber diameter and packing. Aerosol penetration through the filter may not be consistent and could not be predicted accurately by filtration theory. Sinclair and Hoopes (1975a) designed a 10-stage unit using stainless steel mesh 635 screens of uniform diameter, opening, and thickness (Fig. 19–13). The designed flow rate ranged from 6.66 to $10 \times 10^{-5}\,m^3 s^{-1}$ [4 to 6 L/min]. Stacks of these well-defined screens simulate a fan model in both geometry and flow resistance (Cheng et al., 1985). Penetration through screens can be predicted by the fan model filtration theory (Eq. 19–25) (Cheng and Yeh, 1980; Cheng et al., 1980). Subsequently, this unit has become commercially available (Model 3040, ***TSI***). Other types of screens have also been tested and found useful (Yeh et al., 1982; Cheng et al., 1985). Table 19–5 lists characteristics of the different screens shown in Figure 19–14. Screen-type diffusion batteries are compact in size and simple to construct. Screens can be cleaned and replaced easily when contaminated or worn out.

Most multistage diffusion batteries described here are arranged in a series so that the aerosol concentration decreases continuously through the cells. Aerosol penetration is usually detected by a CNC. Based on parallel flow and mass collection principles, Cheng et al. (1984) designed a parallel flow diffusion battery (PFDB). This unit measures the penetration by mass or radioactivity without a particle-detecting unit. It is also more useful to detect unstable

TABLE 19–5. Characteristic Dimensions and Constants for Various Types of Screens in Screen-Type Diffusion Batteries

Weave	Square 145	Square 200	Square 400	Twill 400	Twill 635
Screen diameter (mm)	55.9	40.6	25.4	25.4	20.0
Screen thickness (mm)	122	96.3	57.1	63.5	50.0
Solid volume fraction	0.244	0.230	0.292	0.313	0.345
B_f	0.8969	0.9021	0.1180	1.450	1.677
k	0.330	0.352	0.269	0.246	0.216

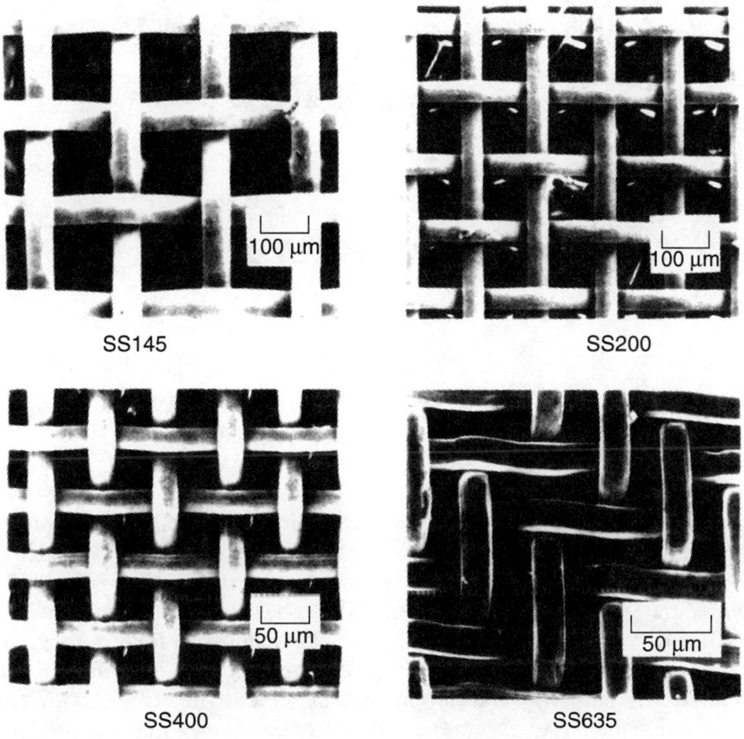

Fig. 19–14. Photomicrographs of stainless mesh screens.

aerosols with fluctuating size and concentration. A schematic diagram of the PFDB is shown in Figure 19–15. It consists of a conical cap and a collection section containing seven cells. Each diffusion cell contains a different number of stainless steel 200-mesh screens followed by a 25 mm Zefluor filter (***GEL***). The seven cells typically contain 0 through 35 screens. Critical orifices provide a $3.33 \times 10^{-5}\,m^3 s^{-1}$ [2 L/min] flow rate through each cell, resulting in a total flow rate of $2.33 \times 10^{-4}\,m^3 s^{-1}$ [14 L/min]. Gravimetric determination of collected filter samples from each cell provides the direct mass penetration as a function of screen number for the determination of aerosol size distribution, thereby eliminating the sometimes-inaccurate conversion of number to mass.

Other Diffusion Batteries. Screen diffusion batteries have been used routinely to determine the activity size distribution of radon progeny. A single screen and a filter have been used to

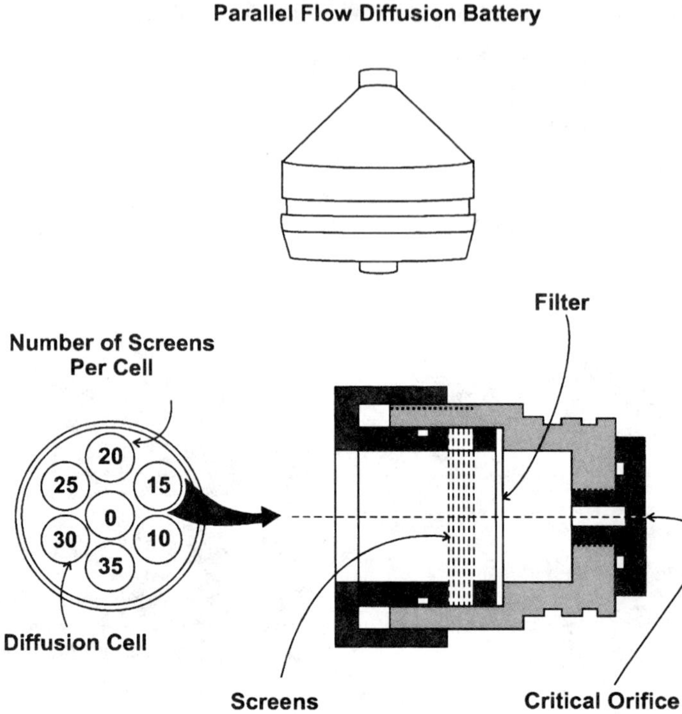

Fig. 19–15. Schematic of a parallel flow diffusion battery.

estimate the unattached fraction. The screen and flow rate with a 50% penetration in the screen of about 4 to 10 nm are usually used. The radioactivity collected on both the screen and the filter are counted, and the amount of activity collected on the screen is assumed to be the unattached radon progeny. The activity size distribution of the radon progeny can be measured with either graded screen diffusion batteries (Holub et al., 1988; Cheng et al., 1992; Knutson et al., 1997) or PFDBs (Reineking and Porstendörfer, 1986; Strong, 1988; Ramamurthi and Hopke, 1991; Wasiolek and Cheng, 1995). A graded diffusion battery (GDB) consists of several stages, each with a different type of screen. The screens with a lower mesh number are used to collect particles in the nanometer size range, whereas screens of a larger mesh number are used to collect larger particles. Figure 19–16 shows a schematic of a GDB including five stages of screens and a backup filter (Cheng et al., 1992). After a brief sampling time (5 to 10 min), the screens and filters are counted directly for radioactivity. Because of the geometry of the screen, the efficiency of the alpha counting is not 100%, and corrections should be made to allow for a lower counting efficiency on screens (Solomon and Ren, 1992). PFDBs with much higher flow rates (over $5.0 \times 10^{-4}\,m^3\,s^{-1}$ [30 L/min]) are used to collect indoor radon progeny, which often have very low concentrations. Usually, only the backup filters for each stage of the PFDB are counted for radioactivity with a detector mounted near each back filter. Thus, this method eliminates possible errors due to the attenuation of α rays by the screen and also provides near-real-time measurement of the activity size distribution.

Wire screen-type diffusion batteries are most commonly used for the activity size distribution, but other types have also been reported. Holub and Knutson (1987) designed a parallel-wire GDB consisting of four screens of parallel wires having uniform diameters and spaces and a filter. A theoretical equation developed for the model filter of parallel staggered array cylinders has been used for data analysis (Cheng et al., 2000). Sinclair (1986) designed a PFDB made of carbonized open-pore foam called *reticulated vitreous carbon*.

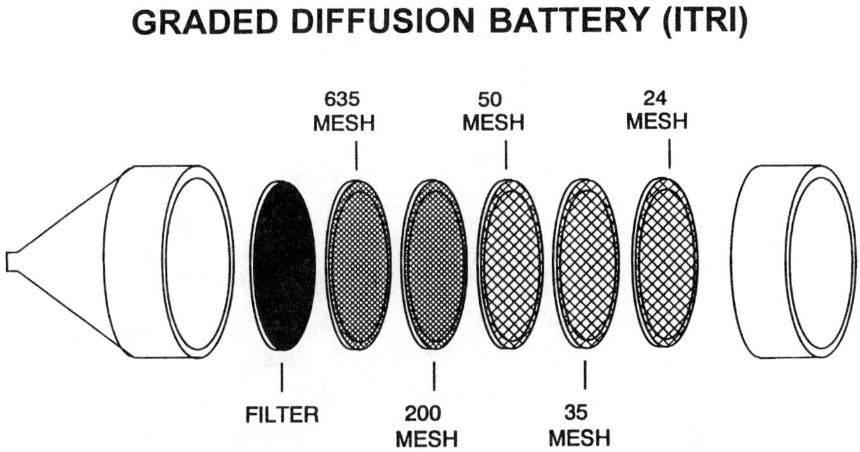

Fig. 19–16. Schematic of a five-stage graded diffusion battery.

Boulaud and Diouri (1988) report a five-stage PFDB with glass beads of different sizes. A working equation for the configuration has been developed. Recently, a new diffusion sampler consisting of an annular tube with a solid-state alpha-track detector placed in the inner cylinder tube has been designed for indoor radon measurement (Tymen et al., 1999). The activity size distribution of ^{218}Po can be determined after exposure to radon progeny for a relatively long sampling period (a few hours compared with 5 to 10 min for GDBs).

Use and Data Analysis

Aerosol penetration through a diffusion battery provides data for the determination of particle size distribution. Aerosol penetration through a diffusion cell is obtained by measuring the number, mass, or activity concentrations at the inlet and outlet of each cell. A CNC is used to measure the number concentration. Figure 19–17 shows the schematic diagram of a system including a diffusion battery, automatic switching valve, and a CNC. With the automatic sampling system, it takes 3 min to complete an 11-channel measurement.

For radioactive aerosols, penetration based on activity can be obtained by collecting samples at the diffusion cell and a backup filter at the end of the diffusion battery. The single-stage parallel disk diffusion sampler (Kotrappa et al., 1975) and screen diffusion batteries have been used for this purpose. Screens can be counted directly for radioactivity (Reineking and Porstendörfer, 1986). Penetration based on the mass can be obtained by using a PFDB.

Data Analysis

Monodisperse Aerosol. Particle size distributions are calculated from penetration data obtained from the diffusion battery measurement. For a monodisperse aerosol, the diffusion coefficient, D_p, can be calculated directly from the corresponding Eqs. 19–20 to 19–25. The particle size is then calculated from the following relationship:

$$D_p = \frac{kTC_c(d_p)}{3\pi\eta d_p} \tag{19–41}$$

$$C_c(d_p) = 1 + \frac{2\lambda}{d_p}\left[1.142 + 0.558\exp\left(-0.999\frac{d_p}{2\lambda}\right)\right] \tag{19–42}$$

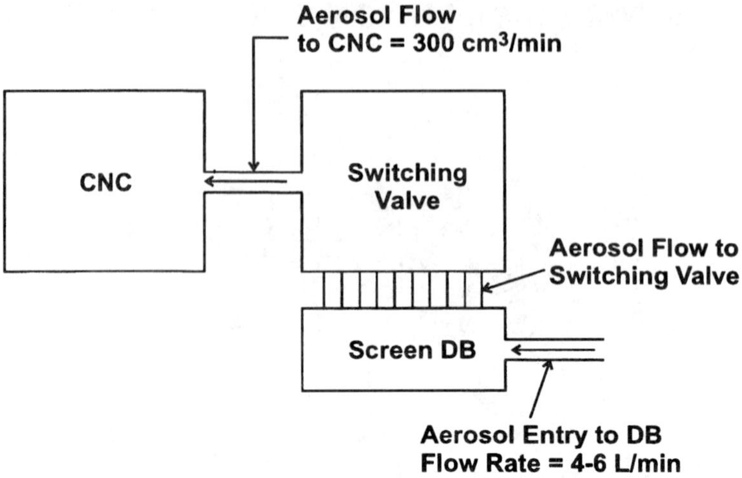

Fig. 19–17. Schematic of a diffusion battery, automatic switch valve, and condensation nuclei counter.

EXAMPLE 19–2

What is the aerosol penetration through a single-stage tube diffusion battery (0.10 m long, 100 tubes, and $6.67 \times 10^{-5}\,\mathrm{m^3\,s^{-1}}$ flow) for a 0.01 μm particle diameter, under an ambient pressure of 101.3 kPa and 23°C?

Answer:

1. Calculate the diffusion coefficient under the ambient condition using Eq. 19–41 by first calculating the slip correction factor (C_c [Eq. 19–42]):

$$C_c = 1 + \frac{2 \cdot 0.0673}{0.01}\left[1.142 + 0.558\exp\left(-0.999 \cdot \frac{0.01}{2 \times 0.0673}\right)\right] = 23.3$$

$$D_p = \frac{1.38 \cdot 10^{-23} \cdot 296.2 \times 23.3}{3 \times 3.14 \cdot 1.81 \cdot 10^{-5} \cdot 0.01 \cdot 10^{-6}} = 5.58 \times 10^{-8}\,\mathrm{m^2\,s^{-1}}$$

2. Calculate the penetration through a circular tube using Eqs. 9–19 and 9–20:

$$\mu = \frac{3.14 \cdot 5.58 \cdot 10^{-8}\,\mathrm{m^2\,s^{-1}} \cdot 0.1\,\mathrm{m}}{\frac{4}{100} \cdot 6.67 \times 10^{-5}\,\mathrm{m^3\,s^{-1}}} = 0.0263$$

$$P = 0.819\exp(-3.65 \times 0.0263) + 0.0975\exp(-22.3 \times 0.0263) + 0.0325\exp(-57.0 \times 0.0263) + 0.01544\exp(-107 \times 0.0263) = 0.806$$

where k is the Boltzmann constant (1.38×10^{-23} J K^{-1}), T is the absolute temperature (K), C is the Cunningham slip correction factor, and λ is the mean free path of air (0.0673 μm at 293 K [23°C] and 101.3 kPa). With monodisperse particles, measurements from a single-stage diffusion battery are sufficient, and measurements from multiple-stage devices should improve the accuracy.

Polydisperse Aerosols. Most aerosols in ambient environments and workplaces have polydisperse size distributions, and the method described in the previous section does not apply. Three penetration data points are the minimum required, but more will improve the accuracy of the size determination. Both graphical and numerical inversion methods have been developed for the size determination from penetration data.

Fuchs et al. (1962) generated a family of penetration curves for rectangular channel diffusion batteries assuming that the aerosol size distribution is lognormally distributed. Mercer and Greene (1974) provided curves representing the penetration of aerosols in both cylindrical and rectangular channels as functions of the diffusion parameter, μ, and the geometrical standard deviation from 1 to 5. Once the data are properly aligned with one of the curves, this method gives a rough estimate of the mean and geometric standard deviation of the diffusion coefficient. Similar curves have been derived for screen-type diffusion batteries (Lee et al., 1981). This method does not apply to aerosols that do not follow lognormal size distributions.

Sinclair (1969) used a graphical "stripping" method to estimate the particle size distribution from penetration data through a multistage cylindrical-type diffusion battery. A family of penetration curves has been calculated for monodisperse particles over a range of equivalent length. The experimental penetration data are plotted on a different paper of the same scale. The experimental curve is matched against the theoretical curves, and the one having the best fit at the right hand of the experimental curves (i.e., where penetration is least) is subtracted, leaving a new experimental curve. The process is repeated until the original experimental curve is entirely eliminated. Particle size and fractions of each size in the original aerosol are indicated by the matched theoretical curves and their intercepts with the ordinate of the graph. A similar method has been applied to the screen-type diffusion battery (Sinclair et al., 1979). This method does not assume a certain size distribution and thus is more useful. However, results for both graphical methods depend on judgment in matching curves.

More consistent results can be obtained by using numerical inversion methods. In a diffusion battery, aerosol penetration through stage i, P_i, can be expressed mathematically as the integration of the aerosol penetration equation for monodisperse aerosol $P_i(x)$:

$$P_i = \int_0^{\infty} P_i(x) f(x) \mathrm{d}x \tag{19–43}$$

where $f(x)$ is the size distribution and $P_i(x)$ is the aerosol penetration of size x in stage i. Each observed penetration for stage $i = 1$ through n can be expressed in the form of Eq. 19–43. Several numerical inversion methods have been developed to obtain the aerosol size distribution, $f(x)$. Raabe (1978) developed a nonlinear least-square regression to solve Eq. 19–43 under the assumption of a lognormal distribution for $f(x)$. Soderholm (1979) used a similar method for diffusion battery data analysis. Twomey (1975) proposed a nonlinear iterative method that Knutson and Sinclair (1979) applied to diffusion batteries. A modification of Twomey's method is used for data analysis of screen-type diffusion batteries (Kapadia, 1980; Cheng and Yeh, 1984). Other commonly used data inversion techniques include the simplex method (Nelder and Mead, 1965) and an expectation-maximization algorithm that have been developed for the screen-type diffusion battery (Maher and Laird, 1985). Knutson (1999) has summarized additional techniques. Based on tests performed by Wu et al. (1989), both simplex and expectation-maximization techniques performed satisfactorily in the data analysis of diffusion batteries. See Chapter 22 for further discussion of data inversion techniques.

CONCLUSIONS

Diffusion batteries and CNCs are useful instruments in measuring the aerosol concentration and size distribution of ultrafine particles. The lower limit of detection in a CNC is about 2 to 5 nm, thus limiting the usefulness of a CNC to large molecular clusters. However, diffusion denuders and diffusion batteries can still be used to collect and separate particles and gases smaller than a few nanometers. Other techniques, including counting radioactivity and analytical chemical techniques, can be used to identify and quantitate materials collected on diffusion devices. These techniques are important in studying particles between the size range of gases and aerosols, a region where interesting phenomena can occur.

ACKNOWLEDGMENTS

This work was supported by the Office of Health and Biological Research of the U.S. Department of Energy under Cooperative Agreement number DE-FC04-96AL76406.

REFERENCES

Agarwal, J. K. and G. J. Sem. 1980. Continuous flow, single-particle-counting condensation nucleus counter. *J. Aerosol Sci.* 11:343–357.

Ahn, K. H. and B. Y. H. Liu. 1990. Particle activation and droplet growth processes in condensation nucleus counter—I. Theoretical background. *J. Aerosol Sci.* 21:263–276.

Aitken, J. 1888. On the number of dust particles in the atmosphere. *Proc. R. Soc. Edinb.* 35.

Appel, B. R., Y. Tokiwa, and M. Haik. 1981. Sampling of nitrates in ambient air. *Atmos. Environ.* 15:283–289.

Boulaud, D. and M. Diouri. 1988. A new inertial and diffusional device (SDI 2000). *J. Aerosol Sci.* 19:927–930.

Bowen, B. D., S. Levine, and N. Epstein. 1976. Fine particle deposition in laminar flow through parallel-plate and cylindrical channels. *J. Colloid Interface Sci.* 54:375–390.

Braman, R. S., T. Shelley, and W. A. McClenny. 1982. Tungstic acid for preconcentration and determination of gaseous and particulate ammonia and nitric acid in ambient air. *Anal. Chem.* 54:358–364.

Breslin, A. J., S. F. Guggenheim, and A. C. George. 1971. Compact high efficiency diffusion batteries. *Staub-Rein. Luft* 31(8):1–5.

Bricard, J., P. Delattre, G. Madelaine, and M. Pourprix. 1976. Detection of ultra-fine particles by means of a continuous flux condensation nuclei counter. In *Fine Particles*, ed. B. Y. H. Liu. New York: Academic Press, pp. 566–580.

Brown, K. E., J. Beyer, and J. W. Gentry. 1984. Calibration and design of diffusion batteries for ultrafine aerosols. *J. Aerosol Sci.* 15:133–145.

Buttini, P., V. Di Palo, and M. Possanzini. 1987. Coupling of denuder and ion chromatographic techniques for NO_2 trace level determination in air. *Sci. Total Environ.* 61:59–72.

Cecchini, F., A. Febo, and M. Possanzini. 1985. High efficiency annular denuder for formaldehyde monitoring. *Anal. Lett.* 18:681–693.

Cheng, Y. S., T. R. Chen, J. Bigu, R. F. Holub, K. W. Tu, E. O. Knutson, and R. Falk. 2000. Intercomparison of activity size distribution of thoron progeny and a mixture of radon and thoron progeny. *J. Environ. Radioact.* 51:59–78.

Cheng, Y. S., J. A. Keating, and G. M. Kanapilly. 1980. Theory and calibration of a screen-type diffusion battery. *J. Aerosol Sci.* 11:549–556.

Cheng, Y. S., Y. F. Su, G. J. Newton, and H. C. Yeh. 1992. Use of a graded diffusion battery in measuring the radon activity size distribution. *J. Aerosol Sci.* 23:361–372.

Cheng, Y. S., Y. Yamada, and H. C. Yeh. 1991. Diffusion deposition on model fibrous filters with intermediate porosity. *Aerosol Sci. Technol.* 12:286–299.

Cheng, Y. S. and H. C. Yeh. 1980. Theory of a screen-type diffusion battery. *J. Aerosol Sci.* 11:313–320.

Cheng, Y. S. and H. C. Yeh. 1984. Analysis of screen diffusion battery data. *Am. Ind. Hyg. Assoc. J.* 45:556–561.

Cheng, Y. S., H. C. Yeh, and K. J. Brinsko. 1985. Use of wire screens as a fan model filter. *Aerosol Sci. Technol.* 4:165–174.

Cheng, Y. S., H. C. Yeh, J. L. Mauderly, and B. V. Mokler. 1984. Characterization of diesel exhaust in a chronic inhalation study. *Am. Ind. Hyg. Assoc. J.* 45:547–555.

Dasgupta, P. K. 1984. A diffusion scrubber for the collection of atmospheric gases. *Atmos. Environ.* 18:1593–1599.

Dasgupta, P. K., S. Dong, H. Hwang, H. C. Yang, and G. Zhang. 1988. Continuous liquid-phase fluorometry coupled to a diffusion scrubber for the real-time determination of atmospheric formaldehyde hydrogren peroxide and sulfur dioxide. *Atmos. Environ.* 22:946–963.

Davis, H. R. and G. V. Parkins. 1970. Mass transfer from small capillaries with wall resistance in the laminar flow regime. *Appl. Sci. Res.* 22:20–30.

DeMarcus, W. and J. W. Thomas. 1952. *Theory of a Diffusion Battery*. Oak Ridge National Laboratory Report ORNL-1413. Oak Ridge, TN: ORNL.

DeSantis, F. and C. Perrino. 1986. Personal sampling of aniline in working site by using high efficiency annular denuders. *Ann. Chim.* 76:355–364.

Durham, J. L., T. G. Ellestad, L. Stockburger, K. T. Knapp, and L. L. Spiller. 1986. A transition-flow reactor tube for measuring trace gas concentrations. *JAPCA* 36:1228–1232.

Durham, J. L., L. L. Spiller, and T. G. Ellestad. 1987. Nitric acid–nitrate aerosol measurements by a diffusion denuder, a performance evaluation. *Atmos. Environ.* 21:589–598.

Durham, J. L., W. E. Wilson, and E. B. Bailey. 1978. Application of an SO_2 denuder for continuous measurement of sulfur in submicrometric aerosols. *Atmos. Environ.* 12:883–886.

Egilmez, N. and C. N. Davies. 1984. An investigation into the loss of particles by diffusion in the Nolan-Pollack condensation nucleus counter. *J. Aerosol Sci.* 15:177–181.

Fan, B. J., Y. S. Cheng, and H. C. Yeh. 1996. Gas collection efficiency and entrance flow effect of an annular diffusion denuder. *Aerosol Sci. Technol.* 25:113–120.

Febo, A., V. DiPalo, and M. Possanzini. 1986. The determination of tetraalkyl lead air by a denuder diffusion technique. *Sci. Total Environ.* 48:187–194.

Ferm, M. 1979. Method for determination of atmospheric ammonia. *Atmos. Environ.* 13:1385–1393.

Ferm, M. 1986. A Na_2CO_3-coated denuder and filter for determination of gaseous HNO_3 and particulate NO in the atmosphere. *Atmos. Environ.* 20:1193–1201.

Fish, B. R. and J. L. Durham. 1971. Diffusion coefficient of SO_2 in air. *Environ. Lett.* 2:13–21.

Forrest, J., D. J. Spandau, R. L. Tanner, and L. Newman. 1982. Determination of atmospheric nitrate and nitric acid employing a diffusion denuder with a filter pack. *Atmos. Environ.* 16:1473–1485.

Friedlander, S. K. 1977. *Smoke, Dust and Haze*. New York: John Wiley & Sons.

Fuchs, N. A., I. B. Stechkina, and V. I. Starosselskii. 1962. On the determination of particle size distribution in polydisperse aerosols by the diffusion method. *Br. J. Appl. Phys.* 13:280–281.

Gormley, P. G. and M. Kennedy. 1949. Diffusion from a stream flowing through a cylindrical tube. *Proc. R. Ir. Acad.* A52:163–169.

Gunderson, E. C. and C. C. Anderson. 1987. Collection device for separating airborne vapor and particulates. *Am. Ind. Hyg. Assoc. J.* 48:634–638.

Haberl, J. B. 1979. General Electric condensation nuclei counters. In *Aerosol Measurement*, eds. D. A. Lundgren, F. S. Harris, W. H. Marlow, M. Lippmann, W. E. Clark, and M. D. Durham. Gainesville: University Press of Florida, pp. 568–573.

Hinds, W. C. 1999. *Aerosol Technology: Properties, Behavior, and Measurement of Airborne Particles*. New York: John Wiley & Sons.

Hogan, A. W. and G. Gardner. 1968. A nucleus counter of increased sensitivity. *J. Rech. Atmos.* 3:59–61.

Holub, R. F. and E. O. Knutson. 1987. Measuring polonium-218 diffusion coefficient spectra using multiple wire screens. In *Radon and Its Decay Products*, ed. P. K. Hopke. Washington, DC: American Chemical Society, pp. 357–364.

Holub, R. F., E. O. Knutson, and S. Solomon. 1988. Tests of the graded wire screen technique for measuring the amount and size distribution of unattached radon progeny. *Radiat. Prot. Dosim.* 24:265–268.

Ingham, D. B. 1975. Diffusion of disintegration products of radioactive gases in circular and flat channels. *J. Aerosol Sci.* 6:395–402.

Ingham, D. B. 1976. Simultaneous diffusion and sedimentation of aerosol particles in rectangular tubes. *J. Aerosol Sci.* 7:373–380.

Jaenicke, R. K. and H. J. Kanter. 1976. Direct condensation nuclei counter with automatic photographic recording, and general problems of absolute counters. *J. Appl. Meteorol.* 15:620–632.

Johnson, N. D., S. C. Barton, G. H. S. Thomas, D. A. Lane, and W. H. Schroeder. 1985. *Development of Gas/Particle Fractionating Sampler of Chlorinated Organics*. 78th Annual Meeting of Air Pollution Control Association, Detroit, MI.

Kapadia, A. 1980. *Data Reduction Techniques for Aerosol Size Distribution Measurement Instruments*. Ph.D. Thesis, University of Minnesota.

Kerouanton, D., G. Tymen, and D. Boulaud. 1996. Small particle diffusion penetration of an annular duct compared to other geometries. *J. Aerosol Sci.* 27:345–349.

Kesten, J., A. Reineking, and J. Porstendörfer. 1991. Calibration of a TSI model 3025 ultrafine condensation nucleus counter. *Aerosol. Sci. Technol.* 15:107–111.

Kirsch, A. A. and N. A. Fuchs. 1968. Studies of fibrous aerosol filters-III. Diffusional deposition of aerosols in fibrous filters. *Ann. Occup. Hyg.* 11:299–304.

Kirsch, A. A. and I. B. Stechkina. 1978. The theory of aerosol filtration with fibrous filter. In *Fundamentals of Aerosol Science*. ed. D. T. Shaw. New York: John Wiley & Sons, pp. 156–165.

Knutson, E. O. 1999. Diffusion battery history. *Aerosol Sci. Technol.* 31:83–128.

Knutson, E. O., A. C. George, and K. W. Tu. 1997. The graded screen technique for measuring the diffusion coefficient of radon decay products. *Aerosol Sci. Technol.* 27:604–624.

Knutson, E. O. and D. Sinclair. 1979. Experience in sampling urban aerosols with the Sinclair diffusion battery and nucleus counter. In *Proceedings of Advances in Particle Sampling and Measurement*, ed. W. B. Smith. EPA 600/7-79-065, pp. 98–120.

Kotrappa, K., D. P. Bhanti, and R. Dhandayutham. 1975. Diffusion sampler useful for measuring diffusion coefficients and unattached fraction of radon and thoron decay products. *Health Phys.* 29:155–162.

Lee, K. W., P. A. Connick, and J. A. Gieseke. 1981. Extension of the screen-type diffusion battery. *J. Aerosol Sci.* 12:385–386.

Lekhtmakher, S. O. 1971. Effect of Peclet number on the precipitation of particles from a laminar flow. *J. Eng. Phys.* 20:400–402.

Lindgren, P. F. and P. K. Dasgupta. 1988. Measurement of atmospheric sulfur dioxide by diffusion scrubber coupled ion chromatography. *Anal. Chem.* 61:19–24.

Liu, B. Y. H., D. Y. H. Pui, A. W. Hogan, and T. A. Rich. 1975. Calibration of the Pollak counter with monodisperse aerosols. *J. Appl. Meteorol.* 14:46–51.

Maher, E. F. and N. M. Laird. 1985. EM algorithm reconstruction of particle size distributions from diffusion battery data. *J. Aerosol Sci.* 16:557–570.

McDermott, W. T., R. C. Ockovic, and M. R. Stolzenburg. 1991. Counting efficiency of an improved 30-A condensation nucleus counter. *Aerosol. Sci. Technol.* 14:278–287.

Megaw, W. J. and R. D. Wiffen. 1963. Measurement of the diffusion coefficient of homogeneous and other nuclei. *J. Rech. Atmos.* 1:113–125.

Mercer, T. T. and T. D. Greene. 1974. Interpretation of diffusion battery data. *J. Aerosol Sci.* 5:251–255.

Mercer, T. T. and R. L. Mercer. 1970. Diffusional deposition from a fluid flowing radially between concentric, parallel, circular plates. *J. Aerosol Sci.* 1:279–285.

Miller, S. W. and B. A. Bodhaine. 1982. Supersaturation and expansion ratios in condensation nuclei counters: An historical perspective. *J. Aerosol Sci.* 13:481–490.

Mori, Y. and W. Nakayama. 1967a. Study on forced convective heat transfer in curved pipes. *Int. J. Heat Mass Transfer*. 10:37–59.

Mori, Y. and W. Nakayama. 1967b. Study on forced convective heat transfer in curved pipes. *Int. J. Heat Mass Transfer*. 10:681–695.

Nelder, J. A. and R. Mead. 1965. Simplex method for function minimization. *Comput. J.* 7:308–313.

Newman, J. 1969. Extension of the Levesque solution. *J. Heat Transfer*. 91:177–178.

Nolan, J. J. and V. H. Guerrini. 1935. The diffusion coefficient of condensation nuclei and velocity of fall in air of atmospheric nuclei. *Proc. R. Ir. Acad.* 43:5–24.

Nolan, J. J. and P. J. Nolan. 1938. Diffusion and fall of atmospheric condensation nuclei. *Proc. R. Ir. Acad.* A45:47–63.

Nolan, P. J. and D. J. Doherty. 1950. Size and charge distribution of atmospheric condensation nuclei. *Proc. R. Ir. Acad.* 53A:163–179.

Nolan, P. J. and P. J. Kennedy. 1953. Anomalous loss of condensation nuclei in rubber tubing. *J. Atmos. Terrestrial Phys.* 3:181–185.

Okuyama, K., Y. Kousaka, and T. Motouchi. 1984. Condensational growth of ultrafine aerosol particles in a new particle size magnifier. *Aerosol Sci. Technol.* 3:353–366.

Pollak, L. M. and J. Daly. 1958. An improved model of the condensation nucleus counter with stereo-photomicrographic recording. *Geofis. Pura Appl.* 41:211–216.

Pollak, L. W., T. C. O'Conner, and A. L. Metnieks. 1956. On the determination of the diffusion coefficient of condensation nuclei using the static and dynamic methods. *Geofis. Pura Appl.* 34:177–194.

Pollak, L. W. and A. L. Metnieks. 1958. New calibration of photo-electric nucleus counters. *Geofis. Pura Appl.* 41:201–210.

Possanzini, M., A. Febo, and A. Aliberti. 1983. New design of a high-performance denuder for the sampling of atmospheric pollutants. *Atmos. Environ.* 17:2605–2610.

Pui, D. Y. H., C. W. Lewis, C. J. Tsai, and B. Y. H. Liu. 1990. A compact coiled denuder for atmospheric sampling. *Environ. Sci. Technol.* 24:307–312.

Raabe, O. G. 1978. A general method for fitting size distributions to multi-component aerosol data using weighted least-squares. *Environ. Sci. Technol.* 12:1162–1167.

Ramamurthi, M. and P. K. Hopke. 1991. An automated, semicontinuous system for measuring indoor radon progeny activity-weighted size distributions, dp: 0.5–500 nm. *Aerosol Sci. Technol.* 14:82–92.

Reineking, A. and J. Porstendörfer. 1986. High-volume screen diffusion batteries and α-spectroscopy for measurement of the radon daughter activity size distributions in the environment. *J. Aerosol Sci.* 17:873–880.

Rich, T. A. 1955. A photo-electric nucleus counter with size discrimination. *Geofis. Pura Appl.* 31:60–65.

Rich, T. A. 1961. A continuous recorder for condensation nuclei. *Geofis. Pura Appl.* 50:46–52.

Rich, T. A. 1966. Apparatus and method for measuring the size of aerosols. *J. Rech. Atmos.* 2:79–85.

Scheibel, H. G. and J. Porstendörfer. 1984. Penetration measurements for tube and screen-type diffusion batteries in the ultrafine particle size range. *J. Aerosol Sci.* 15:673–682.

Scheibel, H. G. and J. Porstendörfer. 1986. Counting efficiency and detection limit of condensation nuclei counters for submicrometer aerosols. I. Theoretical evaluation of the influence of heterogeneous nucleation and wall losses. *J. Colloid Interface Sci.* 109:261–274.

Shaw, R. W., R. K. Stevens, J. Bowermaster, J. W. Tesch, and E. Tew. 1982. Measurements of atmospheric nitrate and nitric acid; the denuder difference experiment. *Atmos. Environ.* 16:845–853.

Sideman, S., D. Luss, and R. E. Peck. 1965. Heat transfer in laminar flow in circular and flat conduits with (constant) surface resistance. *Appl. Sci. Res.* A14:157–171.

Sinclair, D. 1969. Measurement and production of submicron aerosols. In *Proceedings of the 7th Conference on Condensation and Ice Nuclei*, ed. J. Podzimek. New York, NY: U S Atomic Energy Commission. pp. 132–137.

Sinclair, D. 1972. A portable diffusion battery. *Am. Ind. Hyg. Assoc. J.* 33:729–735.

Sinclair, D. 1984. Intrinsic calibration of the Pollak Counter—A revision. *Aerosol Sci. Technol.* 3:125–134.

Sinclair, D. 1986. Measurement of nanometer aerosols. *Aerosol Sci. Technol.* 5:187–204.

Sinclair, D., R. J. Countess, B. Y. H. Liu, and D. Y. H. Pui. 1979. Automatic analysis of submicron aerosols. In *Aerosol Measurement*, eds. W. E. Clark and M. D. Durham. Gainesville: University Press of Florida, pp. 544–563.

Sinclair, D. and L. Hinchliffe. 1972. Production and measurement of submicron aerosols. In *Assessment of Airborne Particles*, ed. T. T. Mercer et al. Springfield, IL: Charles C. Thomas, pp. 182–199.

Sinclair, D. and G. S. Hoopes. 1975a. A novel form of diffusion battery. *Am. Ind. Hyg. Assoc. J.* 36:39–42.

Sinclair, D. and G. S. Hoopes. 1975b. A continuous flow condensation nucleus counter. *J. Aerosol Sci.* 6:1–7.

Sioutas, C., P. Koutrakis, and J. M. Wolfson. 1994. Particle losses in glass honeycomb denuder sampler. *Aerosol Sci. Technol.* 21:137–148.

Skala, G. F. 1963. A new instrument for the continuous measurement of condensation nuclei. *Anal. Chem.* 35:702–706.

Slanina, J., L. V. Lamoen-Doornebal, W. A. Lingerak, and W. Meilof. 1981. Application of a thermo-denuder analyzer to the determination of H_2SO_4, HNO_3 and NH_3 in air. *Int. J. Environ. Anal. Chem.* 9:59–70.

Soderholm, S. C. 1979. Analysis of diffusion battery data. *J. Aerosol Sci.* 10:163–175.

Solomon, S. B. and T. Ren. 1992. Counting efficiencies for alpha particles emitted from wire screens. *Aerosol Sci. Technol.* 17:69–83.

Stevens, R. K. 1986. Modern methods to measure air pollutants. In *Aerosols: Research, Risk Assessment and Control Strategies*, eds. S. D. Lee, T. Schneider, L. D. Grant, and P. J. Verkerk. Ann Arbor, MI: Lewis, pp. 69–95.

Stevens, R. K., T. G. Dzubay, G. Russwurm, and D. Rickel. 1978. Sampling and analysis of atmospheric sulfates and related species. *Atmos. Environ.* 12:55–68.

Stolzenburg, M. R. and P. H. McMurry. 1991. An ultrafine aerosol condensation nucleus counter. *Aerosol Sci. Technol.* 14:48–65.

Strong, J. C. 1988. The size of attached and unattached radon daughters in room air. *J. Aerosol Sci.* 19:1327–1330.

Su, Y. F., Y. S. Cheng, G. J. Newton, and H. C. Yeh. 1990. Counting efficiency of the TSI model 3020 condensation nucleus counter. *Aerosol Sci. Technol.* 12:1050–1054.

Sutugin, A. G. and N. A. Fuchs. 1965. Coagulation rate of highly dispersed aerosols. *J. Colloid Sci.* 20: 492–500.

Tan, C. W. and C. J. Hsu. 1971. Diffusion of aerosols in laminar flow in a cylindrical tube. *J. Aerosol Sci.* 2:117–124.

Tan, C. W. and J. W. Thomas. 1972. Aerosol penetration through a parallel-plate diffusion battery. *J. Aerosol Sci.* 3:39–43.

Tang, I. K. 1976. Phase transformation and growth of aerosol particles composed of mixed salts. *J. Aerosol Sci.* 7:361–372.

Thomas, J. W. 1955. The diffusion battery method for aerosol particle size determination. *J. Colloid Sci.* 10:246–255.

Twomey, S. 1975. Comparison of constrained linear inversion and an alternative nonlinear algorithm applied to the indirect estimation of particle size distribution. *J. Comput. Phys.* 18:188–200.

Twomey, S. A. and R. A. Zalabsky. 1981. Multifilter technique for examination of the size distribution of the natural aerosol in the submicrometer size range. *Environ. Sci. Technol.* 15:177–184.

Tymen, G., D. Kerouanton, C. Huet, and D. Boulaud. 1999. An annular diffusion channel equipped with a track detector film for long-term measurements of activity concentration and size distribution of nanometer ^{218}Po particles. *J. Aerosol Sci.* 30:205–216.

Wasiolek, P. T. and Y. S. Cheng. 1995. Measurement of the activity-weight size distributions of radon decay products outdoors in central New Mexico with parallel and serial screen diffusion batteries. *Aerosol Sci. Technol.* 23:401–410.

Wilson, J. C., J. H. Hyun, and E. D. Blackshear. 1983. The function and response of an improved stratospheric condensation nucleus counter. *J. Geophys. Res.* 88:6781–6785.

Winiwarter, W. 1989. A calculation procedure for the determination of the efficiency in annular denuders. *Atmos. Environ.* 23:1997–2002.

Wu, J. J., D. W. Cooper, and R. J. Miller. 1989. Evaluation of aerosol deconvolution algorithms for determining submicron particle size distributions with diffusion battery and condensation nucleus counter. *J. Aerosol Sci.* 20:477–482.

Yeh, H. C., Y. S. Cheng, and M. M. Orman. 1982. Evaluation of various types of wire screens as diffusion battery cells. *J. Colloid Interface Sci.* 86:12–16.

Zhang, Z. Q. and B. Y. H. Liu. 1990. Dependence of the performance of TSI 3020 condensation nucleus counter on pressure, flow rate and temperature. *Aerosol Sci. Technol.* 13:493–504.

20 Electrodynamic Levitation of Particles

E. JAMES DAVIS

Department of Chemical Engineering, University of Washington, Seattle, WA

INTRODUCTION

Numerous characteristics and properties of single aerosol particles can be measured by levitating the particle in one or more laser beams using ac and dc electrical fields to fix the particle in space. The apparatus, which consists of ac and dc electrodes mounted in an enclosure, is called the *electrodynamic balance* (EDB). Gas–particle chemical reactions can also be explored by flowing a reactive gas through the chamber. For spherical microparticles and liquid droplets, light-scattering measurements can then be used to determine the size, refractive index, and chemical characteristics of the small mass. The optical properties are determined by comparing the measured intensity of the elastically scattered light with Mie theory, and chemical information is obtained using a spectrometer to record Raman and fluorescence spectra. Computer codes for Mie theory calculations are available in the appendix of the monograph by Bohren and Huffman (1983) and other sources, so we need not provide details here. By *elastic scattering* we mean that the scattered light has the same wavelength (or frequency) as the incident light. The interaction of laser light with the chemical bonds in a particle can produce a shift in the wavelength (frequency), and this so-called inelastic scattering or Raman scattering is the basis for chemical analysis. In addition, droplet evaporation measurements provide a method for determining vapor pressures (Ray et al., 1979), evaporation coefficients (Tang and Munkelwitz, 1991), and gas phase diffusion coefficients (Davis and Ray, 1977).

Because a droplet can be isolated from the walls of the container, electrodynamic levitation provides a method for studying the thermodynamics of highly concentrated salt solutions, for heterogeneous nucleation, and the resulting precipitation can be avoided. Water activities for a variety of concentrated salt solutions have been measured using electrodynamic levitation by Richardson and Spann (1984), Kurtz and Richardson (1984), Richardson and Kurtz (1984), Tang et al. (1986), Cohen et al. (1987a,b), Chan et al. (1992, 1997), Kim et al. (1994), Tang and Munkelwitz (1994), and Liang and Chan (1997).

Electrodynamic levitation is an outgrowth of the classic Millikan (1911) oil drop experiment in which a charged droplet is held and moved in a dc electric field, but there is an important difference between Millikan's electrostatic suspension and the use of an additional ac field to trap the particle electrodynamically. This difference is illustrated in Figure 20–1, which compares the Millikan condenser with a simplified EDB introduced by Straubel (1959). Straubel added a ring (or disk) electrode midway between the dc electrodes of a Millikan

Aerosol Measurement: Principles, Techniques, and Applications, Second Edition, Edited by Paul A. Baron and Klaus Willeke.
ISBN 0-471-35636-0

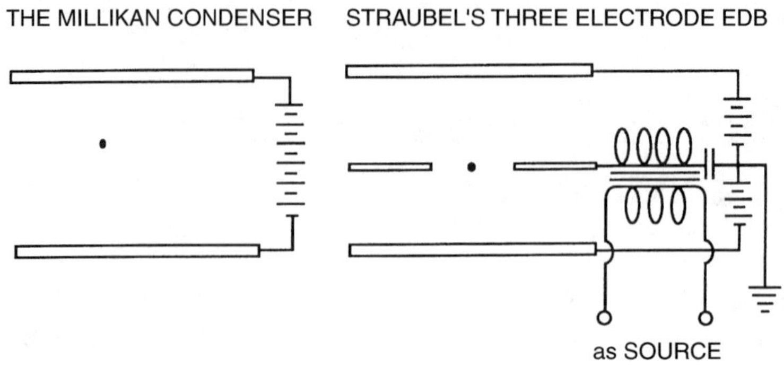

Fig. 20–1. The Millikan condenser and Straubel's (1959) three electrode EDB.

cell, and an ac potential was applied to the center electrode. The ac field exerts periodic radial and vertical forces that, near the center of the balance, are proportional to the distance from the nullpoint (center) and vanish at the nullpoint. These oscillatory forces trap the particle. When the dc field balances the resultant of the gravitational force and any other vertical forces on the particle, the ac field exerts no time-average force on the particle. If the only vertical forces are the electrostatic force and gravity, the EDB can be used as an analytical balance, for in this case force balance yields

$$C_0 q \frac{V_{dc,0}}{z_0} = -mg \tag{20–1}$$

where C_0 is a geometrical constant that is unity for a Millikan condenser with infinitely large electodes, q is the coulombic charge on the particle of mass m, $V_{dc,0}$ is the potential on the upper electrode when the lower electrode is at potential $-V_{dc,0}$, $2z_0$ is the distance between

EXAMPLE 20–1

An electrodynamic balance with $C_0 = 0.8$ and $z_0 = 5\,\text{mm}$ is used to trap a 1.0 μm diameter water droplet. The dc potential on the top electrode is 8.0 V. What is the number of elementary charges on the droplet?

Answer: If the droplet density, ρ_p, is $1000\,\text{kg/m}^3$ and d_p is the diameter, the droplet mass is given by

$$m = \pi d_p^3 \rho_p/6 = \pi(1.0 \times 10^{-6}\ \text{m})^3 (1000\ \text{kg/m}^3)/6 = 5.236 \times 10^{-16}\ \text{kg} = 5.236 \times 10^{-13}\ \text{g}$$

Solving for q using Eq. 20–1, one obtains

$$q = -mgz_0/C_0V_{dc} = -(5.236 \times 10^{-16}\ \text{kg})(9.81\ \text{m/s}^2)(5 \times 10^{-3}\ \text{m})/(0.8)(8.0)$$
$$= -4.013 \times 10^{-18}\,\text{C}$$

Now the unit charge is 1.6022×10^{-19} C, so the number of charges is

$$Z = |-4.013 \times 10^{-18}\ \text{C}|/(1.602 \times 10^{-19}\ \text{C/electron}) = 25\ \text{elementary charges}$$

the dc electrodes, and g is the acceleration of gravity. For finite electrodes and when an ac electrode is in place, $C_0 < 1.0$.

The uniform field of the Millikan condenser makes it impossible to position a particle in the center of the chamber when perturbations in position occur unless radial focusing and some type of vertical feedback control are applied. Arnold (1979) adapted a Millikan condenser to measure particle mass and charge by providing a radial electric field and an electro-optic feedback control system.

Many different electrode configurations have been used for electrodynamic trapping, and Hartung and Avedisian (1992) discussed nine of the dozen or more (Davis, 1997) that have been proposed. Three of the most frequently encountered configurations are presented in Figure 20–2. Figure 20–2a shows the classic bihyperboloidal EDB introduced by Wuerker et al. (1959), Figure 20–2b shows the double ring balance that is based on superposition of the ac and dc potentials on the ring electrodes, and Figure 20–2c shows the spherical void EDB of Arnold and Folan (1987). The latter device was developed to maximize fluorescence signals by coating the interior surface with a reflective material.

LEVITATION PRINCIPLES

Many applications of the EDB are based on the stability characteristics of a trapped particle, so it is appropriate here to outline the theory needed to interpret experimental measurements. It is sufficient to consider only the vertical motion of a particle, for the vertical component of the ac field is larger than the radial component for all of the electrode configurations used. In the absence of a vertical gas flow through the chamber, the equation of vertical motion for the bihyperboloidal EDB is

$$m\frac{d^2z}{dt^2} = -3\pi\kappa d_p\mu\frac{dz}{dt} + F_z - mg + qC_0\frac{V_{dc}}{z_0} - qC_1\frac{V_{ac}}{z_0}z\cos\omega t \tag{20–2}$$

in which z is the vertical distance from the midplane of the balance, d_p is the "effective" particle diameter if the particle is nonspherical, and κ is a correction factor for Stokes's law for a nonspherical particle. The gas viscosity is μ, F_z is a vertical force such as the radiation pressure force of a laser beam or the thermophoretic force, V_{ac} is the amplitude of the ac potential, and $\omega = 2\pi f$, in which f is the ac frequency. The dc and ac balance constants, C_0 and C_1, depend on the electrode geometry. We shall take $F_z = 0$ in the analysis of particle stability.

The equation of motion can be written in a more convenient form by introducing dimensionless variables defined by $Z = z/b$ and $\tau = \omega t/2$ in which $b = z_0C_0/C_1$. The parameter b can be obtained by a calibration procedure outlined below. The nondimensional equation of motion becomes

$$\frac{d^2Z}{d\tau^2} + \delta\frac{dZ}{d\tau} - 2\beta Z\cos 2\tau = \sigma \tag{20–3}$$

where δ is the drag parameter, β is the ac field strength parameter, and σ is a dc offset parameter defined by

$$\delta = \frac{6\pi\mu d_p\kappa}{m\omega},\ \beta = \frac{2g}{\omega^2 b}\left(\frac{V_{ac}}{V_{dc,0}}\right),\ \sigma = \frac{4g}{\omega^2 b}\left(\frac{V_{dc}}{V_{dc,0}} - 1\right) \tag{20–4}$$

Here $V_{dc,0}$ satisfies Eq. 20–1 in the absence of force F_z. The three dimensionless parameters δ, β, and σ govern the particle stability.

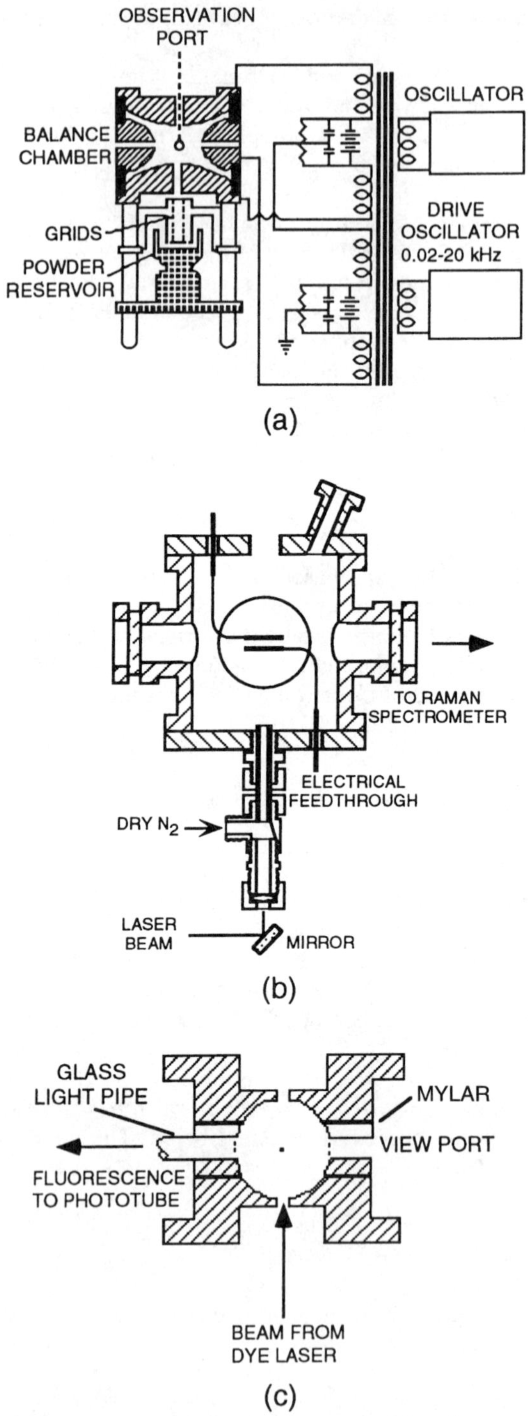

Fig. 20–2. Three frequently used EDB electrode configurations.

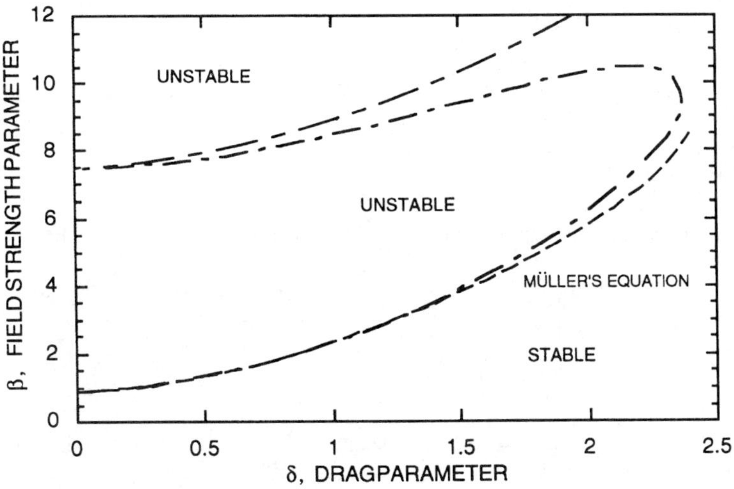

Fig. 20–3. The lower stability envelopes determined by solution of the equation of motion compared with Müller's approximation for the lowest springpoint.

When the dc potential is adjusted to balance the gravitational force, $V_{dc} = V_{dc,0}$ and $\sigma = 0$. In this case the particle will be stably balanced at the nullpoint provided that β is not too large. As β is increased by increasing V_{ac} or by decreasing ω, a point is reached at which a violent oscillation will ensue. Above this so-called "spring point" the large-amplitude oscillation can cause the particle to be lost. If β is maintained below the spring point and if $\sigma \neq 0$, the particle oscillates at the drive frequency with an amplitude that depends on the magnitude of σ and with a midpoint of oscillation that is shifted from the midplane of the balance.

The solution of the equation of motion examined in considerable detail by Frickel et al. (1978), Davis (1985), and others shows that there are multiple regions of instability, and two of these regions are shown in Figure 20–3 as a graph of β versus δ for $\sigma = 0$. The "marginal stability curve" or spring point of the lowest unstable region is particularly important, and Müller (1960) developed an approximate solution for that curve, which is

$$\beta_{crit}^2 = \frac{1}{2}(99 + 12\delta^2) - \sqrt{\frac{1}{4}(99 + 12\delta^2)^2 - (1 + 4\delta^2)(81 + 36\delta^2)} \qquad (20\text{–}5)$$

This equation is also plotted on Figure 20–3 and is seen to agree with the more rigorous analysis for $\delta < 1.5$.

Spring point measurements can be used to size particles of a known density provided that the parameter b in β is determined. This geometrical parameter can be obtained by measuring the spring point of a sphere whose size and density are known. The size, in turn, can be measured by light-scattering techniques or other methods. The calibration involves charging and trapping the sphere in the stable region of the β–δ stability map, then driving it to instability either by increasing V_{ac} or decreasing ω. At the spring point, V_{dc}, V_{ac}, and $f = \omega/2\pi$ are measured. Drag parameter δ can then be calculated from knowledge of the gas viscosity, the particle diameter, the particle mass, and the ac frequency. The spring point or critical value of β corresponding to this value of δ is then determined from the stability map or from Eq. 20–5 if it is appropriate. The only unknown in the definition of β is b, so it can be calculated from β_{crit}.

EXAMPLE 20–2

A 20 μm diameter glass sphere having a density of 2200 kg/m^3 (2.20 g/cm^3) is levitated in air at 300 K ($\mu = 1.846 \times 10^{-5}$ kg/m – s) using a dc voltage of 23 V. The spring point is reached by increasing the ac voltage to 1500 V at a frequency of 100 Hz. Determine the geometrical constant, b, for this EDB.

Answer: Using the definitions of β and δ and the spring point data, we obtain

$$\beta_{\text{crit}} = \frac{2gV_{\text{ac}}}{(2\pi f)^2 V_{\text{dc},0} b} = \frac{2(9.807\,\text{m/s}^2)(1500\,\text{V})}{(2\pi \times 100\,\text{s}^{-1})^2 (23\,\text{V}) b} = 2.025 \times 10^{-4}/b$$

and

$$\delta = \frac{6\pi\mu d_p}{m\omega} = \frac{6\pi\mu d_p}{(\rho\pi d_p^3/6)(2\pi f)} = \frac{18\mu}{\rho d_p^2 \pi f} = \frac{(18)(1.846 \times 10^{-5}\,\text{kg/m}-\text{s})}{(2200\,\text{kg/m}^3)(20 \times 10^{-6}\,\text{m})^2 \pi (100\,\text{s}^{-1})} = 1.202$$

Because $\delta < 1.5$, we can use Müller's equation to relate β_{crit} and δ. Thus, with Eq. 20–5, the result is

$$\beta_{\text{crit}}^2 = \frac{1}{2}\left[99 + 12(1.202)^2\right] - \sqrt{\left[\frac{99 + 12\,(1.202)^2}{2}\right]^2 - \left[1 + 4(1.202)^2\right]\left[81 + 36\,(1.202)^2\right]} = 8.350$$

Then $\beta_{\text{crit}} = 2.890$. Equating this with β_{crit} calculated above from the spring point data, the geometrical constant is $b = 0.8 \times 10^{-4}$. Once b is known, spring point measurements can be used to size a microsphere of known density, for the drag parameter corresponding to β_{crit} yields information about the diameter.

PARTICLE SIZING

The determination of the particle size by spring point measurements can be complemented by several other techniques. These include a variety of methods based on particle motion in an EDB. For a sphere, light-scattering techniques provide a highly accurate determination of the size, but we shall examine sizing procedures based on the particle dynamics first.

Particle Dynamics

Equation 20–3 is the basis not only for spring point measurements but also for three other approaches related to the particle dynamics. When the dc field does not balance the other vertical forces, that is, when $\sigma \neq 0$, the particle oscillates at the ac frequency and is, in general, out of phase with the ac source. Göbel et al. (1997) used phase lag measurements to size droplets in a four-ring EDB. Frickel and his co-workers recognized that it should be possible to determine the size of a particle by measuring either the amplitude of its oscillation or the shift of the center of the oscillation from the nullpoint. Zheng and Davis (1999) and Zheng et al. (2000) explored these methods with a double-ring EDB of the type shown in Figure 20–2b. They mounted a two-dimensional CD array on one port of the chamber and recorded the image of the oscillating particle with a PC equipped with Labview software to process the data. Figure 20–4 shows the line scan image and the resulting particle trajectory from which the amplitude, shift, and phase lag were determined.

By comparing the solution of Eq. 20–3 with the measured amplitude, shift, or phase lag, the particle size can be determined. Figure 20–5 shows a comparison between theory and

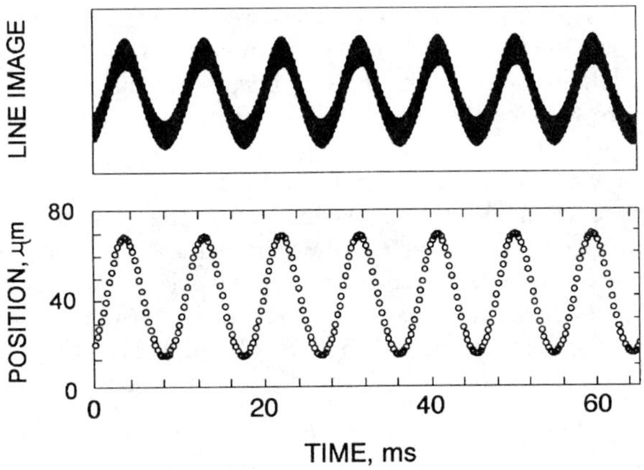

Fig. 20–4. A line scan image and particle trajectory for a GeO_2 particle. (*Source*: Zheng et al., 2000).

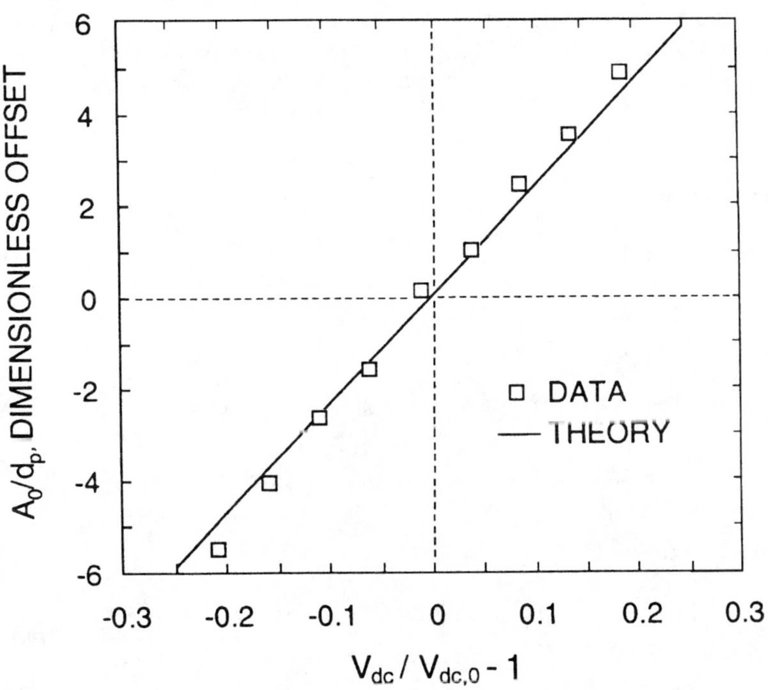

Fig. 20–5. A comparison between the measured dc offset and stability theory for a 10.6 μm GeO_2 particle. (*Source*: Zheng and Davis, 1999).

experiment as a function of $(V_{dc}/V_{dc,0} - 1)$ for a GeO_2 particle with an effective diameter of 10.6 μm based on spring point measurements. Zheng and Davis found the oscillation methods to be within 3% of the light-scattering size for glass microspheres and within 4% for polystyrene latex (PSL) microspheres.

Simpler methods of sizing a particle in an EDB were demonstrated by Davis and Periasamy (1985) and by Sageev et al. (1986). Both methods are applications of Stokes's law for the drag force on a sphere. Davis and Periasamy levitated a particle in a laminar jet passing upward through the balance chamber and measured the force on the particle as a function of the center line velocity of the jet, which was determined by calibration. When no dc field was required to balance the gravitational force, it was balanced by the aerodynamic drag.

Sageev and his co-workers momentarily turned off the ac electric field in an EDB and changed the dc voltage from $V_{dc,0}$ to V_{dc}, where $V_{dc,0}$ is the levitation voltage required to balance the gravitational force. The particle then moved under the influence of gravity and the dc field. They measured the time required for a particle to fall a given distance in their EDB chamber. They assumed that the particle fell at its terminal velocity, justifying this assumption by letting the particle fall 100 to 200 ms before measuring the velocity. The distance of fall was 0.9 mm. When both the dc field and gravity act on the particle, the size is given by

$$d_p = \sqrt{\frac{18\mu U}{(\rho_p - \rho_g)g}\left(\frac{V_{dc,0}}{V_{dc,0} - V_{dc}}\right)} \qquad (20\text{–}6)$$

in which U is the velocity, ρ_p is the particle density, ρ_g is the gas density, and the Cunningham correction factor has been neglected. Sageev and colleagues (1986) also determined the particle size of their PSL spheres using the spring point method and found that it gave diameters 2% to 3% larger than the terminal velocity measurement method. Equation 20–6 applies to the experiments of Davis and Periasamy (1985) with $V_{dc} = 0$.

EXAMPLE 20–3

A polystyrene latex (PSL) sphere having a density of 1055 kg/m³ (1.055 g/cm³) is found to fall 1.0 mm in 0.85 s in air at 300 K and atmospheric pressure ($\mu = 1.846 \times 10^{-5}$ kg/m–s and ρ_g = 1.1774 kg/m³) when no dc field is applied. What is the diameter of a PSL particle?

Answer: In this case, $V_{dc} = 0$ in Eq. 20–6, and the particle velocity is

$$U = \frac{\Delta y}{\Delta t} = \frac{1.0 \times 10^{-3}\,\text{m}}{0.85\,\text{s}} = 1.18 \times 10^{-3}\,\text{m/s} = 0.118\,\text{cm/s}$$

The diameter is given by

$$d_p = \sqrt{\frac{18(1.846 \times 10^{-5}\,\text{kg/m}-\text{s})(1.18 \times 10^{-3}\,\text{m/s})}{(1055 - 1.1774\,\text{kg/m}^3)(9.807\,\text{m/s}^2)}} = 6.15 \times 10^{-6}\,\text{m} = 6.15\,\mu\text{m}$$

Note: The Cunningham correction for slip has been neglected in this calculation.

Light Scattering

In the nineteenth century, Lord Rayleigh derived expressions for the intensity of light scattered by spheres much smaller than the wavelength of light, and at the beginning of the twentieth century Mie (1908) applied electromagnetic theory to analyze light scattering from spheres of any size. Mie theory predicts that the intensity of the scattered light depends on the polarization of the incident light (usually from a laser beam now), the size of the sphere, the position of the observer (the detector), and the refractive index ratio, m, of the sphere

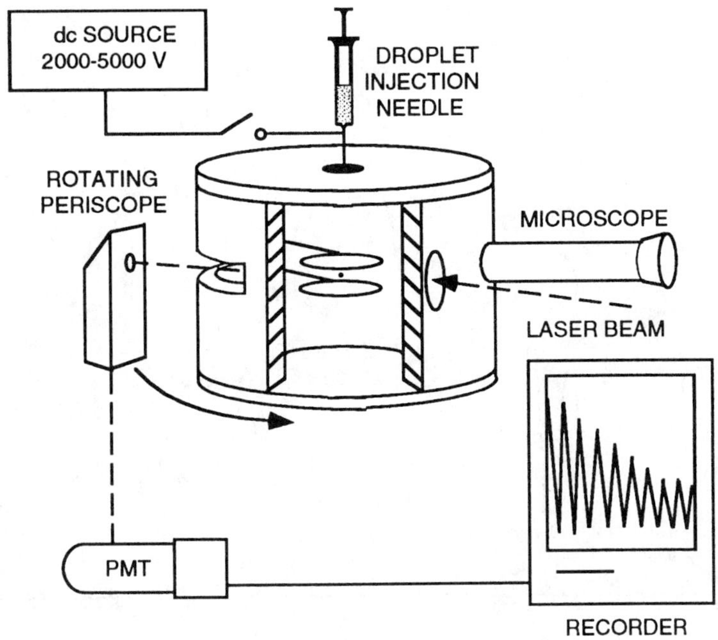

Fig. 20–6. A double-ring EDB equipped with light scattering peripherals for phase function measurements.

and surrounding medium. Measurements of the intensity of the scattered light can be compared with Mie theory to determine the size and/or refractive index of a sphere to approximately 2 parts in 10^5 (Ray et al., 1991). The light-scattering size is usually written in dimensionless form as $x = \pi d_p/\lambda$, where λ is the wavelength of the incident light. A graph of the intensity versus scattering angle, θ, is often called a *phase function*.

One measurement technique for obtaining phase functions is illustrated in Figure 20–6. A rotating periscope/photomultiplier tube (PMT) driven by a stepper motor detects light scattered from the particle through a window in the EDB. This type of device was used by Ray et al. (1991) to obtain the results shown in Figure 20–7 for a levitated droplet of dioctyl phthalate (DOP). Also shown in Figure 20–7 is the "best fit" of Mie theory. The excellent agreement between theory and experiment yields a light-scattering size of $x = 95.13 \pm 0.02$.

An alternate approach for measuring angular light scattering, which is used in the author's laboratory (Davis, 1987), consists of an EDB with a linear photodiode array mounted on the chamber wall. This permits rapid acquisition of phase functions and is particularly well-suited to study droplet evaporation and condensation processes. Figure 20–8 presents a set of phase functions obtained by Davis et al. (1987) for an evaporating droplet of hexadecane compared with calculated phase functions. Two features are particularly relevant; the number of peaks in a given range of scattering angles decreases as the size decreases, and the angle-average intensity of the scattered light decreases as the size decreases. Although the fine structure of the phase functions depends on the refractive index of the particle and its size, the number of fringes (peaks) in a given range of angles can be used to obtain an estimate of the light-scattering size. Thus, if N_p is the number of peaks per degree, a rough approximation based on Mie theory yields

$$x = 192N_p \tag{20–9}$$

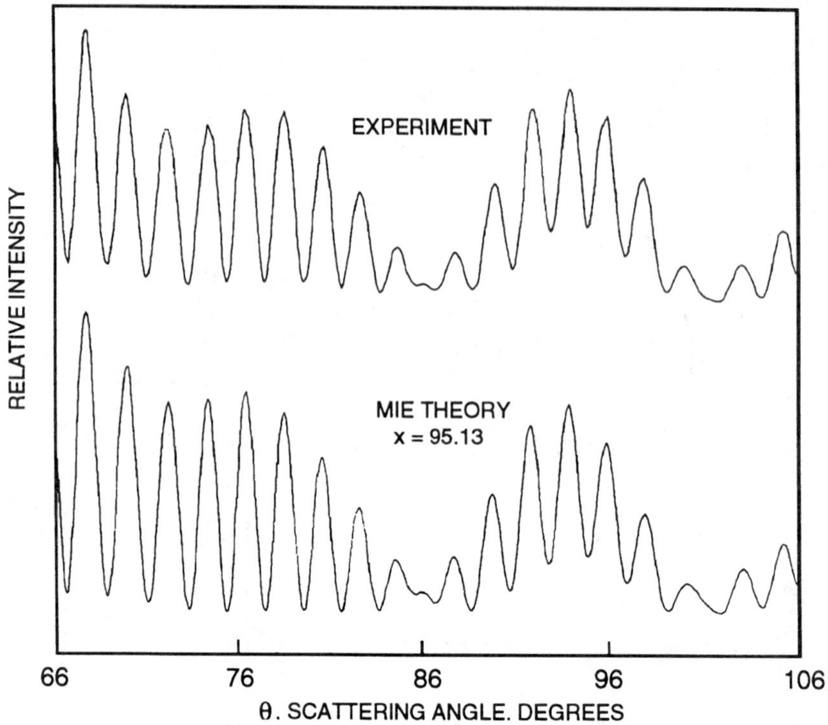

Fig. 20–7. A comparison between experimental and calculated phase functions for a droplet of DOP. (*Source*: Ray et al., 1991 with permission from Optical Society of America).

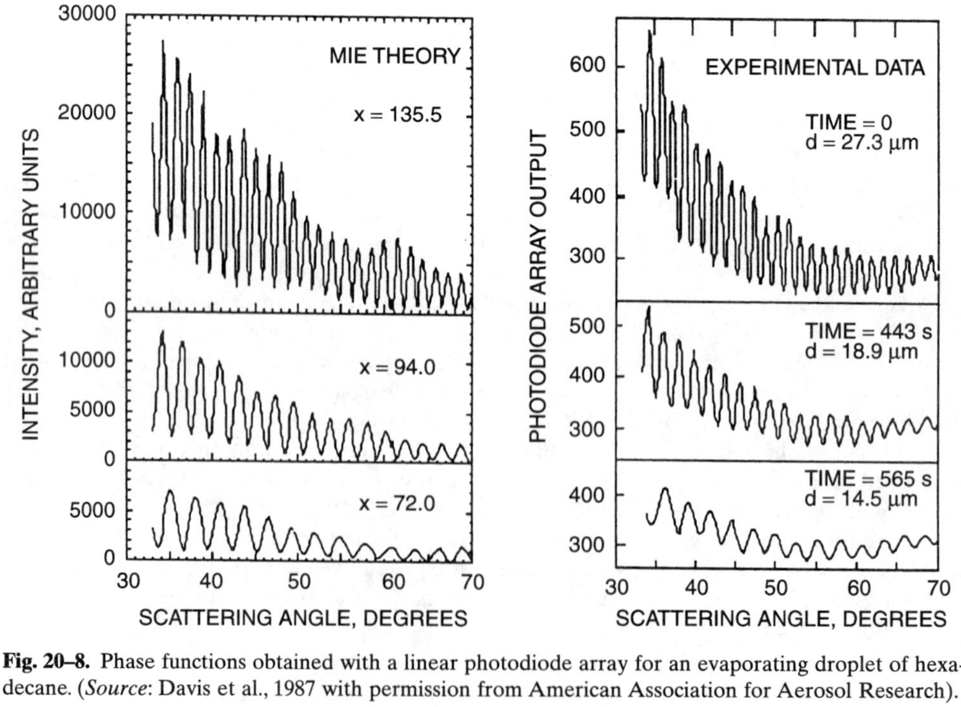

Fig. 20–8. Phase functions obtained with a linear photodiode array for an evaporating droplet of hexadecane. (*Source*: Davis et al., 1987 with permission from American Association for Aerosol Research).

Ragucci et al. (1990) verified the linear relationship between x and N_p using a calibrated Berglund and Liu (1973) droplet generator to produce water droplets of uniform size, and they used a vertically polarized pulsed Nd-YAG laser with $\lambda = 532\,\text{nm}$ to obtain phase functions. Their results agree with Eq. 20–9 if the constant is 197. They also showed that the number of fringes per degree is the same for horizontal and vertical polarized incident light.

EXAMPLE 20–4

Estimate the size of the DOP droplet whose phase function is shown in Figure 20–7 by peak counting.

Answer: The angle range presented in Figure 20–7 is $\Delta\theta = 40°$. If we include the small peak at 86° and consider the fringe at the far left as one-half peak, we obtain 19.5 peaks. Using Eq. 20–9, the dimensionless size is

$$x = 192\frac{19.5}{40} = 93.6$$

This is within 2% of the value obtained by direct comparison of the measured phase function with Mie theory ($x = 95.13$).

If the laser light source was an HeNe with a wavelength of 632.8 nm, the droplet diameter is

$$d_p = (0.6328\,\mu\text{m})(93.6)/\pi = 18.9\,\mu\text{m}$$

Spherical particles and spheroids have unique light-scattering features that can be used to size them. These are morphology-dependent resonances (MDRs). They have also been called whispering gallery modes because of their similarity to the modes associated with acoustic waves in domed structures (e.g., the dome of the cathedral of Florence). For certain combinations of microsphere diameter, particle refractive index, and wavelength of light, the scattered intensity is very large compared with off-resonance conditions. The MDRs are so sensitive to size that they can be used to determine the size with high precision. Barber and Chang (1988) edited a book that deals with such optical effects in considerable detail.

If the size (or refractive index or wavelength) is changed in a systematic way and if the intensity of the scattered light is measured at a single angle, typically at 90° to the incident laser beam, the graph of intensity versus time has been called a *resonance spectrum*. Figure 20–9 shows a comparison between a resonance spectrum obtained by Richardson et al. (1986) for an evaporating droplet of sulfuric acid and a calculated spectrum based on Mie theory. Note the sharp peaks corresponding to MDRs. The measured spectrum is seen to be in very good agreement with the theoretical result that predicts resonance peaks at $x = 22.88, 21.38, 19.89$, and 18.37. The droplet was levitated in vacuum in an EDB and was illuminated with an HeNe laser ($\lambda = 632.8\,\text{nm}$).

FORCE MEASUREMENT

The forces that can be exerted on an aerosol particle include the radiation pressure force, thermophoretic and diffusiophoretic forces, photophoretic forces, electrical forces, and, possibly, magnetic forces, in addition to aerodynamic drag and gravity. The thermophoretic force arises when a particle is placed in a surrounding gas having a nonuniform temperature.

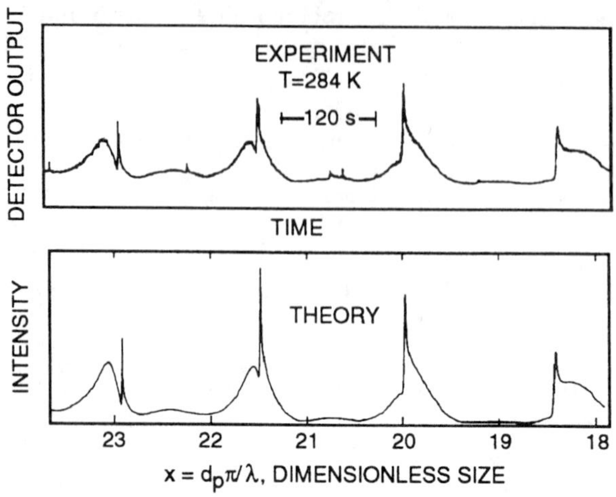

Fig. 20–9. Experimental and theoretical resonance spectra for an evaporating droplet of sulfuric acid. (*Source*: Richardson et al., 1986 with permission from American Association for Aerosol Research).

Molecules colliding with the particle from hotter regions transfer more momentum to the particle than those coming from colder regions. The somewhat related phenomenon of photophoresis is the result of a temperature gradient generated within the particle by electromagnetic heating. Diffusiophoresis results when a particle exists in a gas–vapor mixture having a concentration gradient of vapor.

If F_z represents one of these vertical forces, the dc potential required to balance gravity and force F_z is given by

$$C_0 q \frac{V_{dc}}{z_0} = -mg + F_z \tag{20–10}$$

Using Eq. 20–1 to eliminate $C_0 q / z_0$, the ratio of force F_z to the gravitational force is given by

$$\frac{F_z}{mg} = 1 - \frac{V_{dc}}{V_{dc,0}} \tag{20–11}$$

Thus, the ratio F_z/mg is obtained from two voltage measurements, $V_{dc,0}$ in the absence of the force and V_{dc} with the force exerted.

Li and Davis (1995) used this principle to measure the thermophoretic force on particles in the Knudsen regime. They mounted two heat exchangers inside an EDB to establish a temperature gradient in the gas phase, and they measured the dc potential required for levitation with and without the temperature gradient. The balance chamber could be evacuated to vary the mean free path, λ, of the gas molecules, thereby changing the Knudsen number, $Kn = 2\lambda/d$, in which d is the particle diameter. Figure 20–10 presents representative results for a PSL microsphere in air obtained by Li (1995). The thermophoretic force is seen to increase with Knudsen number until a maximum is reached that corresponds to the free-molecule regime. The force decreases for $Kn > 4$ because of edge effects associated with the isothermal chamber walls.

In a similar way, Lin and Campillo (1985) measured the photophoretic force on crystalline ammonium sulfate particles, and Allen et al. (1990) measured the radiation pressure force on microspheres.

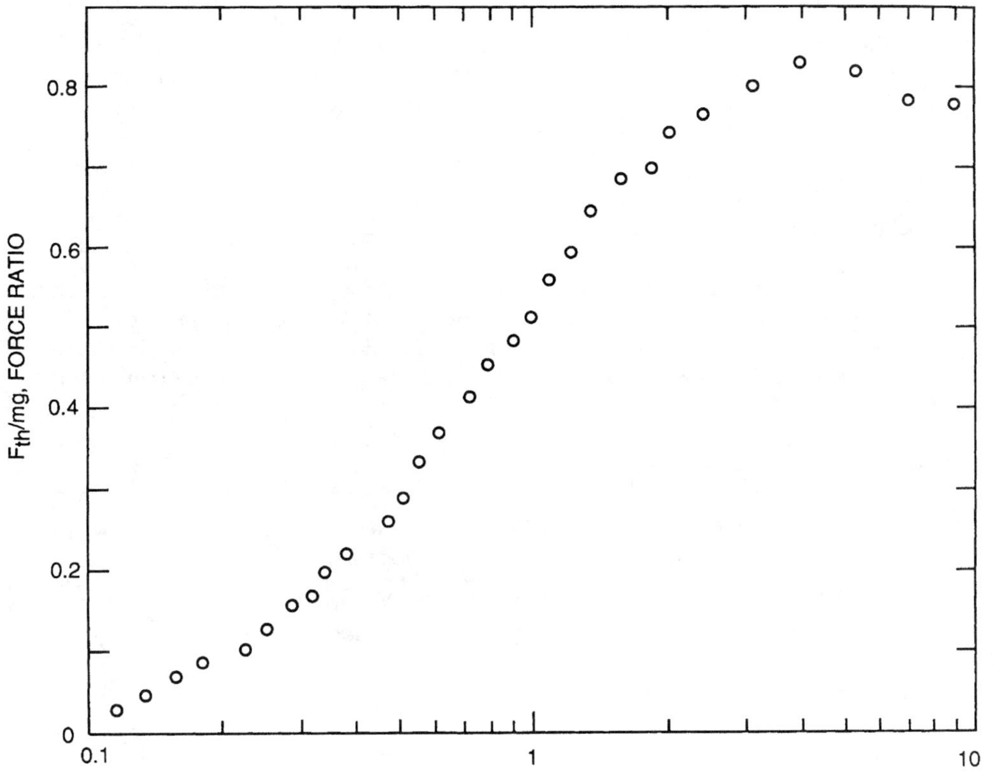

Fig. 20–10. The ratio of the thermophoretic force to the particle weight for a PSL sphere in air. (*Source*: Li, 1995).

MASS AND CHARGE MEASUREMENT

Although the mass of a spherical particle of known density can be determined by measuring its diameter using the methods described above, an alternative method must be applied for particles of irregular shape and/or unknown density. Arnold (1979) proposed such an alternative in which he changed the charge on a particle of constant mass by photoemission of electrons. Charge loss can occur due to ionizing radiation (Ward and Davis, 1989), photoemission due to ultraviolet (UV) radiation (Arnold and Hessel, 1985), and thermionic emission at elevated temperatures (Bar-Ziv et al., 1989).

Arnold's method involves changing the charge by one electron at a time. Equation 20–1 relates the dc voltage and the charge q. Let V_n be the levitation voltage when the charge is given by $q_n = ne$, where n is the number of elementary charges on the particle and e the charge on the electron ($e = 1.6022 \times 10^{-19}$ C). If the charge is q_{n-1} after one electron is ejected, then the dc levitation voltages before and after the loss of a unit charge are related by

$$q_n - q_{n-1} = e = \frac{mgz_0}{C_0}\left(\frac{1}{V_n} - \frac{1}{V_{n-1}}\right) = \frac{mgz_0}{C_0}\frac{(V_{n-1} - V_n)}{V_n V_{n-1}} \qquad (20\text{–}12)$$

Consequently, the absolute mass can be obtained from two voltage measurements. Arnold applied this method repeatedly to yield a series of steps when plotted as levitation voltage versus time of exposure to UV light. A step involving more than one electron emission can be clearly distinguished from a single emission in this manner. Furthermore, once the mass has been determined, the charge on the particle is obtained using Eq. 20–1.

EXAMPLE 20–5

An electrodynamic balance with $z_0 = 6\,\text{mm}$ and $C_0 = 0.80$ is used to measure the absolute mass by the one-electron emission method. If the dc voltage before illumination with an ultraviolet source is 9.850 V and after a short period of time the voltage must be changed to 9.796 V, what is the mass of the particle, its initial charge, and the initial number of elementary charges?

Answer: Using Eq. 20–12, one obtains

$$
\begin{aligned}
m &= C_0 e V_{n-1} V_n / g z_0 (V_n - V_{n-1}) \\
&= \frac{(0.80)(-1.6022\times10^{-19}\,\text{C})(9.796\,\text{V})(9.850\,\text{V})}{(9.807\,\text{m/s}^2)(0.006\,\text{m})(9.850\,\text{V} - 9.796\,\text{V})} \\
&= 3.89\times10^{-15}\,\text{kg} = 3.89\times10^{-12}\,\text{g} = 3.89\,\text{pg}
\end{aligned}
$$

From Eq. 20–1 the initial charge is

$$q_n = -mgz_0/V_n C_0 = -(3.89\times10^{-5}\,\text{kg})(9.807\,\text{m/s}^2)(0.006\,\text{m})/(9.850\,\text{V})(0.80) = -2.88\times10^{-17}\,\text{C}$$

The number of elementary charges is

$$n = |q_n|/e = (2.88\times10^{-17}\,\text{C})/(1.6022\times10^{-19}\,\text{C/electron}) = 180$$

The charge on a liquid droplet can be determined by levitating it in a flowing gas. Taflin et al. (1989) developed the method to determine the Rayleigh limit of charge. When the surface tension of a droplet is balanced by the repulsive force associated with the surface charge, a droplet explodes. Taflin and his coworkers let a drop evaporate until the surface charge density reached the point of explosion. For upward flow in the Stokes regime, Eq. 20–10 becomes

$$C_0 q \frac{V_{dc}}{z_0} = \frac{\pi}{6} d_p^3 \rho_p g + 3\pi d_p \mu U \tag{20–13}$$

This equation can be rearranged to give

$$d_p^2 = -\frac{6}{\pi}\frac{C_0 q}{\rho_p g z_0}\frac{V_{dc}}{d_p} + 18\frac{\mu U}{\rho_p g} \tag{20–14}$$

Thus, if the droplet diameter is measured by light scattering, a graph of d_p^2 versus V_{dc}/d_p should yield a straight line with slope $-6C_0 q/\pi\rho_p g z_0$. Taflin et al. (1989) and Bridges (1990) used this method to determine the droplet charge, and Figure 20–11 shows evaporation rate data obtained by Bridges (1990) for a dodecanol drop evaporating in air at room temperature.

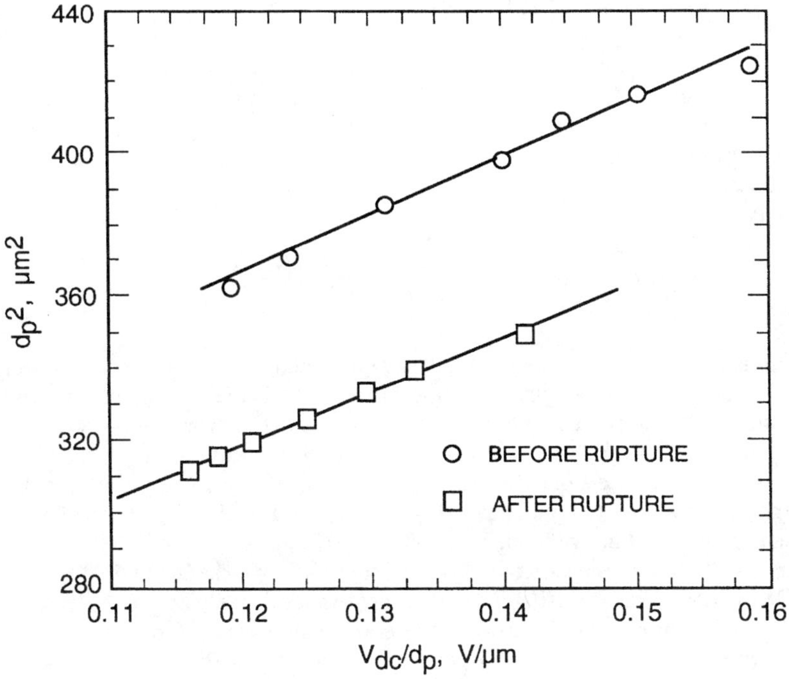

Fig. 20–11. Evaporation data for a dodecanol droplet before and after explosion upon reaching the Rayleigh limit. (*Source*: Bridges, 1990).

EXAMPLE 20–6

For the EDB used by Bridges $z_0 = 15.4$ mm and $C_0 = 0.64$. The density of dodecanol at room temperature is 820 kg/m³. Based on the data in Figure 20–11 after droplet explosion, what was the coulombic charge on the droplet? How many elementary charges correspond to that coulombic charge?

Answer: The slope of the lower line in Figure 20–11 is estimated to be $S = 1700\,\mu\text{m}^3/\text{V}$. Based on Eq. 20–14 the slope is

$$S = -6C_0 q/\pi\rho_p g z_0$$

Solving for q, one obtains

$$q = \frac{\pi S \rho_p g z_0}{6C_0} = \frac{\pi(1700\times10^{-18}\,\text{m}^3/\text{V})(820\,\text{kg/m}^3)(9.807\,\text{m/s}^2)(15.4\times10^{-3}\,\text{m})}{(6)(0.64)}$$
$$= -1.72\times10^{-13}\,\text{C}$$

The number of elementary charges is

$$n = |q|/e = (1.72\times10^{-13}\,\text{C})/(1.6022\times10^{-19}\,\text{C/electron}) = 1.08\times10^{6}\text{ elementary charges}$$

EVAPORATION/CONDENSATION

Davis and Ray (1977) and Ray et al. (1979) showed that gas phase diffusion coefficients and vapor pressures can be determined from droplet evaporation experiments in the continuum regime where the evaporation is diffusion controlled. For sufficiently slow rates of single-component droplet evaporation into a gas containing no vapor, the droplet diameter, d, satisfies the equation

$$d^2 - d_0^2 = -8\frac{M_i D_{ij} p_i^0(T_a)}{\rho_p R T_a}(t - t_0) \qquad (20\text{–}15)$$

in which d_0 is the diameter at time t_0, M_i is the molecular weight of the evaporating species, $p_i^0(T_a)$ is the vapor pressure at surface temperature T_a, D_{ij} is the diffusion coefficient of vapor i in surrounding gas j, and R is the ideal gas constant. If the evaporation rate is sufficiently low, the surface temperature is very close to the surrounding gas temperature, but rapid evaporation lowers the surface temperature because the heat of vaporization must be supplied by heat transfer from the gas phase. Note that, based on Eq. 20–15, a graph of a^2 versus time should yield a straight line with slope $S = -8M_i D_{ij} p_i^0(T_a)/\rho_p R T_a$.

Figure 20–12 shows some of the data of Ray and coworkers (1979) for dibutyl sebacate (DBS) evaporating in nitrogen at various temperatures. The slopes of the d^2 versus time plots become more negative as the temperature increases due primarily to the increase in vapor pressure and to a lesser extent to the increase in the diffusion coefficient with temperature. If the vapor pressure is known, the diffusion coefficient can be determined from the slope. Conversely, if the diffusion coefficient is known, the vapor pressure can be determined.

EXAMPLE 20–7

Ravindran et al. (1979) reported the diffusion coefficient of DBS in N_2 at 307 K to be $3.07 \times 10^{-6}\,\mathrm{m^2\,s^{-1}}$ ($0.0307\,\mathrm{cm^2/s}$). Based on the data in Figure 20–12, determine the vapor pressure of DBS at 307 K. The molecular weight of DBS is 314.47, and its density is 940.5 $\mathrm{kg/m^3}$.

Answer: Using Figure 20–12, the slope of the line for 307 K is estimated to be

$$S = -(1.40\,\mu\mathrm{m}^2)/(1280\,\mathrm{s}) = -0.00109\,\mu\mathrm{m}^2/\mathrm{s} = -1.09 \times 10^{-15}\,\mathrm{m}^2/\mathrm{s}$$

Using the definition of S, one obtains

$$S = -1.09 \times 10^{-15}\,\mathrm{m}^2/\mathrm{s} = -2\frac{(314.47\,\mathrm{kg/kmol})(3.07 \times 10^{-6}\,\mathrm{m}^2/\mathrm{s})}{(940.5\,\mathrm{kg/m}^3)(8314.4\,\mathrm{m^3 Pa/kmol\,K})(307\,\mathrm{K})} p_i^0$$

and, solving for the vapor pressure,

$$p_i^0 = 1.36 \times 10^{-3}\,\mathrm{Pa} = 1.02 \times 10^{-5}\,\mathrm{mmHg}$$

This result is in reasonable agreement with the vapor pressure equation proposed by Ray et al. (1979), which yields p_i^0 (307 K) = 1.19×10^{-5} mmHg.

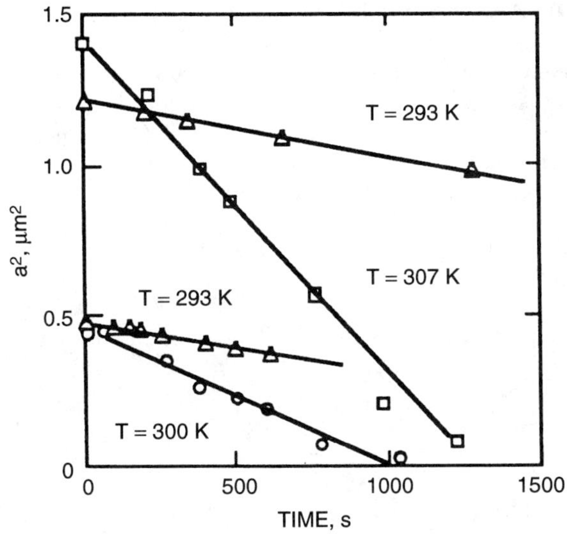

Fig. 20–12. Evaporation rate data for droplets of DBS evaporating in N_2 at various temperatures. (*Source*: Ray et al., 1979 with permission from American Institute of Physics).

EXAMPLE 20–8

Richardson et al. (1986) fitted their vapor pressure results to the equation

$$\ln P_i^0 = 20.70 - 9360/T\,(p_i^0 \text{ in Torr})$$

Using data extracted from Figure 20–9, estimate the evaporation coefficient for sulfuric acid at 284 K.

Answer: The time between the resonance peak at $x_0 = 22.88$ and the peak at $x_1 = 18.37$ is estimated to be 560 s. The corresponding droplet radii are $d_0 = (22.88)\ (0.6328\,\mu\text{m})/\pi = 4.608\,\mu\text{m}$ and $d_1 = (18.37)\ (0.6328\,\mu\text{m})/\pi = 3.700\,\mu\text{m}$. The molecular weight of sulfuric acid is 98.08, and the density is $1834\,\text{kg/m}^3$. The mean molecular speed is

$$c_i = \sqrt{\frac{(8)(8314.4\,\text{Pa/m}^3\,\text{kmolK})(284\text{K})}{(98.08\text{kg/kmol})\pi}} = 247.6\text{m/s} = 24760\,\text{cm/s}$$

The vapor pressure at 284 K is given by

$$p_i^0 = \exp(20.70 - 9360/T) = \exp(20.70 - 9360/284) = 4.748\times10^{-6}\ \text{Torr} = 6.33\times10^{-4}\,\text{Pa}$$

Based on Eq. 20–16

$$\varepsilon = -\frac{(d_1 - d_0)2\rho_p R T_a}{(t_1 - t_0)c_i M_i p_i^0}$$

$$= -\frac{[(3.700 - 4.608)(10^{-6}\,\text{m})](4)(1834\,\text{kg/m}^3)(8314.4\ \text{m}^3\text{Pa/kmol K})(284\,\text{K})}{(560\,\text{s})(247.6\,\text{m/s})(98.08\,\text{kg/kmol})(6.33\times10^{-4}\ \text{Pa})}$$

$$= 0.914$$

Thus, considering the accuracy of the vapor pressure equation, the evaporation coefficient is of order unity.

Tang and Munkelwitz (1991) performed evaporation experiments in the continuum regime as well as in the free-molecule regime ($Kn >> 1$) to determine vapor pressures, and the sulfuric acid evaporation rate data of Richardson et al. (1987) shown in Figure 20–9 were obtained by operating in the free-molecule regime. For isothermal evaporation in the free-molecule regime, the time-dependent radius satisfies

$$d - d_0 = -\varepsilon \frac{c_i}{2} \frac{M_i p_i^0(T_a)}{\rho_p R T_a}(t - t_0) \tag{20–16}$$

where ε is the evaporation coefficient and c_i is the molecular speed given by

$$c_i = \sqrt{\frac{8RT_a}{\pi M_i}} \tag{20–17}$$

CHEMICAL REACTIONS

The electrodynamic balance can serve as a platform for studying chemical reactions. Rubel and Gentry (1984) and Straubel and Straubel (1984) were among the first to use electrodynamic levitation to follow chemical reactions gravimetrically (using the dc levitation voltage). The former explored the reaction between a phosphoric acid droplet and ammonia:

$$NH_3 + H_3PO_4 \rightarrow NH_4H_2PO_4 \tag{20–18}$$

Rubel and Gentry reported the extent of the reaction as a function of time for various droplet sizes and NH_3 gas pressure. They observed an initially rapid change in the extent of the reaction followed by a slower diffusion-controlled process.

Straubel and Straubel made gravimetric measurements for HCl reactions with crystals of $ZnCO_3 \cdot 3Zn(OH)_2$, with $Na_2S_2O_3 \cdot 5H_2O$, and with $CaCO_3$ in the presence of water, but no attempts were made to interpret the data to obtain reaction velocity constants or other kinetics information. The Straubels photographed the diffraction patterns from particles levitated in a three-plate balance shown in Figure 20–1 and illuminated from above by a laser beam. The photographic film was placed below the lower electrode, which had a window in it to permit observation of the diffraction rings. They presented the dc voltage as a function of time for the reaction between $CaCO_3$ and HCl vapor, which is

$$CaCO_3 + 2HCl \rightarrow CaCl_2 + 2H_2CO_3 \tag{20–19}$$

The levitation voltage changed from ~150 V at the outset of the reaction to ~600 V at the end, which corresponds to a mass increase of 300%, assuming no charge loss. From the stoichiometry of the reaction ($M_{CaCO_3} = 100.09$, $M_{CaCl_2} = 110.99$, $M_{H_2CO_3} = 62.03$) one would expect the mass increase to be 135%. It appears that there was significant water uptake. This is consistent with the observation that the final state was a droplet.

Gravimetric analysis is not sufficient to understand the chemistry involved with reactions such as described by Eq. 20–19 because of the uptake of water and the possible decomposition of H_2CO_3. To chemically characterize such reactions Davis and his co-workers (Davis and Buehler, 1990; Buehler et al., 1991; Rassat and Davis, 1992; Aardahl and Davis, 1996; Aardahl et al., 1998) coupled the EDB with a Raman spectrometer, and Figure 20–13 shows one configuration of the system used. In this case, Rassat and Davis (1994) used an Argon-ion laser to illuminate the particle from below, and a second CO_2 laser was used to heat the particle (infrared (IR) heating). The incident beam for Raman measurements typically had

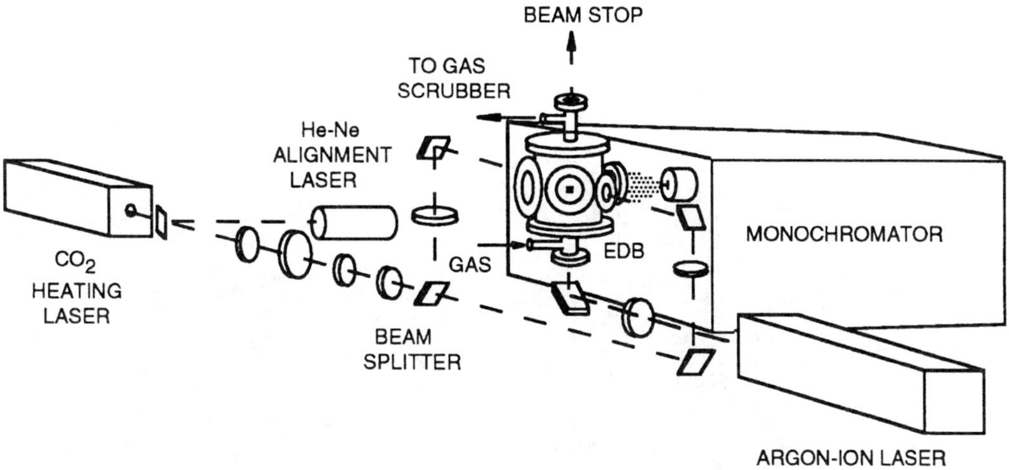

Fig. 20–13. The EDB/Raman spectrometer system. (*Source*: Rassat and Davis, 1994 with permission from Society of Applied Spectroscopy.)

a power of 0.75 W (an irradiance of 2300 W/cm^2) at a wavelength of 514.8 nm, and at this power the radiation pressure force on the particle was significant. The IR beam was split to heat the particle from two sides, thereby balancing the radiation pressure force in the horizontal direction. The inelastically scattered light was collected at 90° to the incident beam and was focused on the slit of a monochromator. An optical channel analyzer and a single channel detector mounted on the monochromator (not shown in Figure 20–13) were used to record the Raman spectra.

Rassat and Davis (1992) used the EDB/Raman system in Figure 20–13 to study the chemical reaction between a sorbent particle, CaO, and a humidified stream of SO_2 in air. This reaction constitutes a "dry scrubbing" process used for stack gas desulfurization. At elevated temperatures ($T > 800$ K) the product of the reaction is $CaSO_4$, but at low temperatures the overall reaction is

$$CaO + SO_2 + \frac{1}{2}H_2O \rightarrow CaSO_3 \cdot \frac{1}{2}H_2O \tag{20–20}$$

Rassat and Davis levitated a CaO particle and then introduced a polluted gas stream into the balance chamber and followed the reaction by recording Raman spectra during the course of the reaction. Figure 20–14 shows a sequence of Raman spectra reported by Davis et al. (1992) for the reaction at room temperature. The sequence shows peaks associated with the gases SO_2 (at 1152 cm^{-1}) and N_2 (at 2328 cm^{-1}) as indicated in Figure 20–14. The peak at approximately 3620 cm^{-1} corresponds to the OH stretching mode of $Ca(OH)_2$, and the peak at approximately 1000 cm^{-1} is associated with the SO bond in the product. The broad peak around 3380 cm^{-1} is identified as that for the water of hydration in the calcium sulfite product.

The presence of $Ca(OH)_2$ in the spectra indicates that the reaction

$$CaO + H_2O \rightarrow Ca(OH)_2 \tag{20–21}$$

occurred either simultaneously with the reaction in Eq. 20–20 or that the reaction in Eq. 20–21 was followed by the reaction

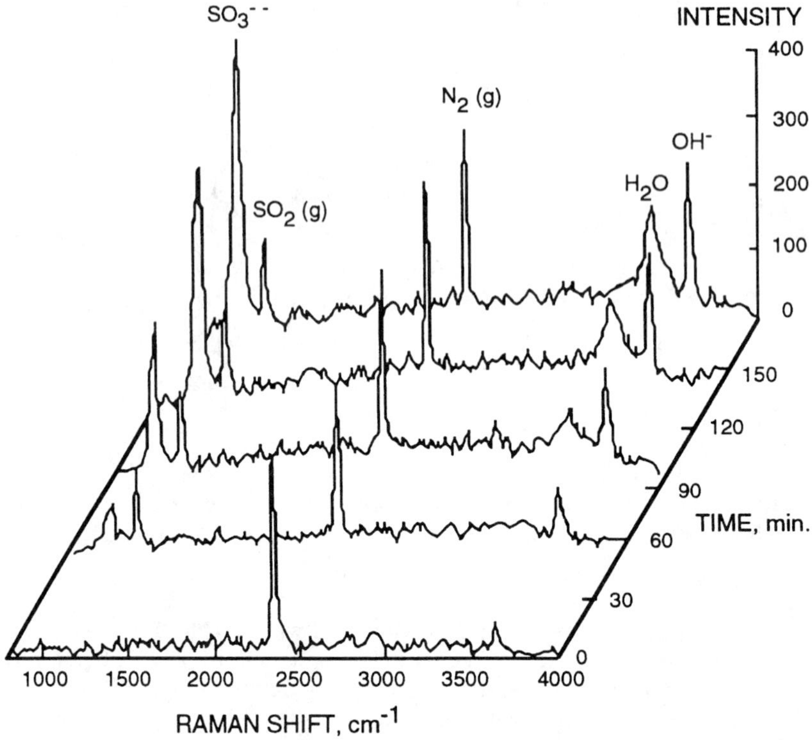

Fig. 20–14. A sequence of Raman spectra for the reaction between a CaO particle and SO_2 in a humid sir stream. (*Source*: Davis et al., 1992 with permission from Elsevier Science).

$$Ca(OH)_2 + SO_2 \rightarrow CaSO_3 \cdot \frac{1}{2}H_2O + \frac{1}{2}H_2O \tag{20–22}$$

Gravimetric analysis cannot help elucidate the mechanism.

The overall reaction did not go to completion, for it generally ceased at less than 40% conversion. This can be attributed to either pore-plugging or coating of the reacting particle by the product.

CONCLUDING REMARKS

The applications considered here represent only a few of the many applications of electrodynamic levitation of aerosol particles that have been investigated, and it is likely that new uses will continue to be found. The simplicity of the measurements makes the technique attractive, and the coupling of the EDB to elastic and inelastic light-scattering equipment greatly extends the capabilities.

EXAMPLE 20–9

Table 20–1 presents the dc levitation voltage data for two different CaO particles suspended in a humid air stream (without SO_2). Because there is no SO_2 present the reaction should be represented by Eq. 20–21. What can you conclude about the extent of the reaction from the data of Figure 20–15?

TABLE 20–1. Levitation Voltage Data for CaO Particles Reacting with Water Vapor

Particle 1		Particle 2	
Time (min)	Voltage	Time (min)	Voltage
0	20.7	0	15.8
95	21.3	110	16.9
150	22.3	160	18.5
250	24.1	200	19.6
380	26.5	300	23.1
440	28.1	340	23.7
540	29.7	400	23.7
650	29.7		

Answer: The molecular weights of CaO, H_2O, and $Ca(OH)_2$ are 56.08, 18.01, and 74.09, respectively. If complete conversion to $Ca(OH)_2$ occurred, the ratio of the final voltage to the initial voltage would be 74.09/56.08 = 1.321, which is also the mass ratio. However, for Particle 1 the ratio is 29.7/20.7 = 1.435, and for Particle 2 the ratio is 23.7/15.8 = 1.500. In both cases the final mass is too large for the particle to be only the hydroxide. It is likely that there was an uptake of water from the humid air in both experiments. For Particle 1 the excess mass is (1.435 – 1.321)/(1.321) = 0.0863 or 8.63% of the mass of pure $Ca(OH)_2$, and for Particle 2 the excess mass is (1.500 – 1.321)/(1.321) = 0.136 or 13.6% of the mass of pure $Ca(OH)_2$.

REFERENCES

Aardahl, C. L. and E. J. Davis. 1996. Gas/aerosol chemical reactions in the NaOH–SO_2–H_2O system. *Appl. Spectrosc.* 50:71–77.

Aardahl, C. L., J. F. Widmann, and E. J. Davis. 1998. Raman analysis of chemical reactions resulting from the collision of micrometer-size particles. *Appl. Spectrosc.* 52:47–53.

Allen, T. M., M. F. Buehler, and E. J. Davis. 1990. Radiometric effects on absorbing microspheres. *J. Colloid Interface Sci.* 145:343–356.

Arnold, S. 1979. Determination of particle mass and charge by one electron differentials. *J. Aerosol Sci.* 10:49–53.

Arnold, S. and L. M. Folan. 1987. A spherical void electrodynamic levitator. *Rev. Sci. Instrum.* 58:1732–1736.

Arnold, S. and N. Hessel. 1985. Photoemission from single electrodynamically levitated microparticles. *Rev. Sci. Instrum.* 56:2066–2069.

Barber, P. W. and R. K. Chang. 1988. *Optical Effects Associated with Small Particles*. Singpore: World Scientific.

Bar-Ziv, E., D. B. Jones, R. E. Spjut, D. R. Dudek, A. F. Sarofim, and J. P. Longwell. 1989. Measurement of combustion kinetics of a single char particle in an electrodynamic thermogravimetric analyzer. *Combustion Flame* 75:81–106.

Berglund R. N. and B. Y. H. Liu. 1973. Generation of monodisperse aerosol standards. *Environ. Sci. Technol.* 7:147–153.

Bohren, C. F. and D. R. Huffman. 1983. *Absorption and Scattering of Light by Small Particles.* New York: Wiley.

Bridges, M. A. 1990. *Measurement of Surface Tension Using an Electrodynamic Balance*. M.S. thesis, University of Washington.

Buehler, M. F., T. M. Allen, and E. J. Davis. 1991. Microparticle Raman spectroscopy of multicomponent aerosols. *J. Colloid Interface Sci.* 146:79–89.

Chan, C. K., R. C. Flagan, and J. H. Seinfeld. 1992. Water activities of $NH_4NO_3/(NH_4)_2SO_4$ solutions. *Atmos. Environ.* 26A:1661–1673.

Chan, C. K., Z. Liang, J. Zheng, S. L. Clegg, and P. Brimblecombe. 1997. Thermodynamic properties of aqueous aerosols to high supersaturation: I—Measurements of water activity of the system Na^+–Cl^-–NO_3^-–SO_4^{2-}–H_2O at ~298.15 K. *Aerosol Sci. Technol.* 27:324–344.

Cohen, M. D., R. C. Flagan, and J. H. Seinfeld. 1987a. Studies of concentrated electrolyte solutions using the electrodynamic balance. 1. Water activities for single-electrolyte solutions. *J. Phys. Chem.* 91:4563–4574.

Cohen, M. D., R. C. Flagan, and J. H. Seinfeld. 1987b. Studies of concentrated electrolyte solutions using the electrodynamic balance. 2. Water activities for mixed-electrolyte solutions. *J. Phys. Chem.* 91:4575–4582.

Davis, E. J. 1985. Electrodynamic balance stability characteristics and applications to the study of aerocolloidal particles. *Langmuir* 1:379–387.

Davis, E. J. 1987. The picobalance for single microparticle measurements. *ISA Trans.* 26:1–5.

Davis, E. J. 1997. A history of single aerosol particle levitation. *Aerosol Sci. Technol.* 26:212–254.

Davis, E. J. and M. F. Buehler. 1990. Chemical reactions with single microparticles. *MRS Bull.* 15:26–33.

Davis, E. J. and R. Periasamy. 1985. Light scattering and aerodynamic size measurements for homogeneous and inhomogeneous microspheres. *Langmuir* 1:373–379.

Davis, E. J., S. D. Rassat, and W. Foss. 1992. Measurement of aerosol/gas reaction rates by microparticle Raman spectroscopy. *J. Aerosol Sci.* 24:S429–S432.

Davis, E. J. and A. K. Ray. 1977. Determination of diffusion coefficients by submicron droplet evaporation. *J. Chem. Phys.* 67:414–419.

Davis, E. J., S. H. Zhang, J. H. Fulton, and R. Periasamy. 1987. Measurement of the aerodynamic drag force on single aerosol particles. *Aerosol Sci. Technol.* 6:273–287.

Frickel, R. H., R. E. Shaffer, and J. B. Stamatoff. 1978. Report No. ARCSL-TR-77041, Chemical Systems Laboratory, Aberdeen Proving Ground, MD.

Göbel, G., T. Wriedt, and K. Bauckhage. 1997. Periodic drag force and particle size measurement in a double ring electrodynamic trap. *Rev. Sci. Instrum.* 68:3046–3052.

Hartung, W. H. and C. T. Avedisian. 1992. On the electrodynamic balance. *Proc. R. Soc. Lond. A* 437:237–266.

Kim Y. P., B. Pun, C. K. Chan, R. C. Flagan, and J. H. Seinfeld. 1994. Determination of water activity in ammonium sulfate and sulfuric acid mixtures using levitated single particles. *Aerosol Sci. Technol.* 20:275–284.

Kurtz, C. A. and C. B. Richardson. 1984. Measurement of phase changes in a microscopic lithium iodide particle levitated in water vapor. *Chem. Phys. Lett.* 109:190–194.

Li, W. 1995. *Experimental Study of the Thermophoretic Force and Evaporation Rates for Single Microparticles in the Knudsen Regime*. Ph.D. Dissertation, University of Washington.

Li, W. and E. J. Davis. 1995. Measurement of the thermophoretic force by electrodynamic levitation: Microspheres in air. *J. Aerosol Sci.* 26:1063–1083.

Liang, Z. and C. K. Chan. 1997. A fast technique for measuring water activity of atmospheric aerosols. *Aerosol Sci. Technol.* 26:255–268.

Lin, H.-B. and A. J. Campillo. 1985. Photothermal aerosol absorption spectroscopy. *Appl. Opt.* 24:422–433.

Mie, G. 1908. Beiträge zur Optik trüber Medien, speziell kolloidaler Metallösungen. *Ann. Physik.* 25:377.

Millikan, R. A. 1911. The isolation of an ion, a precision measurement of its charge, and the correction of Stokes's law. *Phys. Rev.* 32:349–397.

Müller, A. 1960. Theoretische untersuchungen über das verhalten geladener teilchen in sattelpunkten electrischer wechselfelder. *Ann. Phys.* 6:206–220.

Ragucci, R., A. Cavaliere, and P. Massoli. 1990. Drop sizing by laser light scattering exploiting intensity angular oscillation in the Mie regime. *Part. Part. Syst. Charact.* 7:221–225.

Rassat, S. D. and E. J. Davis. 1992. Chemical reaction between sulfur dioxide and a calcium oxide aerosol particle. *J. Aerosol Sci.* 23:765–780.

Rassat, S. D. and E. J. Davis. 1994. Temperature measurement of single levitated microparticles using Stokes/anti-Stokes Raman intensity ratios. *Appl. Spectrosc.* 48:1498–1505.

Ravindran, P., E. J. Davis, and A. K. Ray. 1979. Diffusivities of low-volatility species in light gases. *AIChE J.* 25:966–975.

Ray, A. K., E. J. Davis, and P. Ravindran. 1979. Determination of ultra-low vapor pressures by submicron droplet evaporation. *J. Chem. Phys.* 71:582–587.

Ray, A. K., A. Souryi, E. J. Davis, and T. M. Allen. 1991. The precision of light scattering techniques for measuring optical parameters of microspheres. *Appl. Opt.* 30:3974–3983.

Richardson, C. B., R. L. Hightower, and A. L. Pigg. 1986. Optical measurement of the evaporation of sulfuric acid droplets. *Appl. Opt.* 25:1226–1229.

Richardson, C. B. and C. A. Kurtz. 1984. A novel isopiestic measurement of water activity in concentrated and supersaturated lithium halide solutions. *J. Am. Chem. Soc.* 106:6615–6618.

Richardson, C. B. and J. F. Spann. 1984. Measurement of the water cycle in a levitated ammonium sulfate particle. *J. Aerosol Sci.* 15:563–571.

Rubel G. O. and J. W. Gentry. 1984. Investigation of the reaction between single aerosol acid droplets and ammonia Gas. *J. Aerosol Sci.* 15:661–671.

Sageev, G., J. H. Seinfeld, and R. C. Flagan. 1986. Particle sizing in the electrodynamic balance. *Rev. Sci. Instrum.* 57:933–936.

Straubel, H. 1959. Verdampfunsgeschwindigkeit und Ladungsänderung von Flüssigkeitstropfen. *DECHEMA Monogr.* 32:153–159.

Straubel, E. and H. Straubel. 1984. Investigations of chemical reactions on aerosols. *J. Aerosol Sci.* 15:301–305.

Taflin, D. C., T. L. Ward, and E. J. Davis. 1989. Electrified droplet fission and the Rayleigh limit. *Langmuir* 5:376–384.

Tang, I. N. and H. R. Munkelwitz. 1991. Determination of vapor pressure from droplet evaporation kinetics. *J. Colloid Interface Sci.* 141:109–118.

Tang, I. N. and H. R. Munkelwitz. 1994. Water activities, densities, and refractive indices of aqueous sulphates and sodium nitrate droplets of atmospheric importance. *J. Geophys. Res.* 99:18801–18808.

Tang, I. N., H. R. Munkelwitz, and N. Wang. 1986. Water activity measurements with single suspended droplets: The NaCl-H_2O and KCl-H_2O systems. *J. Colloid Interface Sci.* 114:409–415.

Ward, T. L. and E. J. Davis. 1989. Electrodynamic radioactivity detector for microparticles. *Rev. Sci. Instrum.* 60:414–421.

Wuerker, R. F., H. Shelton, and R. V. Langmuir. 1959. Electrodynamic containment of charged particles. *J. Appl. Phys.* 30:342–349.

Zheng, F. and E. J. Davis. 1999. Size determination of single crystalline particles by electrodynamic oscillation measurements. Paper 6C5, AAAR 1999, Tacoma, WA, October 13, 1999.

Zheng, F., M. L. Laucks, and E. J. Davis. 2000. Aerodynamic particle size measurement by electrodynamic oscillation techniques. *J. Aerosol Sci.* 31:1173–1185.

21 Instrument Calibration

BEAN T. CHEN

National Institute for Occupational Safety and Health, Center for Disease Control and Prevention, Public Health Service, Department of Health and Human Services, Morgantown, WV

WALTER JOHN

Particle Science, Walnut Creek, CA

INTRODUCTION

Most knowledge concerning aerosol properties has been obtained by experimental means using aerosol instruments. These instruments can be categorized as (1) collection devices such as cascade impactors, Aitken-type condensation nuclei counters, or filter samplers, which are designed to remove particles from gas streams to obtain samples for analysis; and (2) real-time, direct-reading instruments, such as an optical particle counter, photoelectric condensation nuclei counter, or a photometer. Ideally, instrument response can be theoretically computed based on equations and procedures described in the previous chapters. However, practical considerations, such as compactness, portability of the instrument, and convenience of operation, may influence the design of the instrument. Theoretical prediction of instrument response based on ideal conditions may not be fulfilled. For example, although 50% effective cutoff diameters and collection efficiencies for an impactor stage can be computed, the phenomena of particle bounce, re-entrainment, electrostatic charge effects, and wall losses can modify performance (Rao and Whitby, 1978; Cheng and Yeh, 1979). Therefore, experimental calibration is essential.

Instruments are usually calibrated and evaluated by the manufacturer or the inventor before being used by others. For an instrument intended to collect an aerosol for analysis, collection efficiency and wall loss are generally determined in the calibration. For a real-time, direct-reading instrument, calibration establishes the relationship between an instrument's response (e.g., electronic signal or channel number) and the value of the property (e.g., particle size, number concentration, or mass concentration) being measured. However, the operating conditions and the parameters used during the original calibration can vary from those under which the eventual user operates. As a result, the original calibration data may not apply, and the user must recalibrate the instrument to operate it with confidence. In general, a reliable and accurate calibration requires (1) sufficient knowledge of the capabilities and limitations of the instrument, (2) adequate information on the environment where the instrument will be used, (3) appropriate test facilities, (4) proper selection of a desired test aerosol, (5) a thorough investigation of relevant parameters, and (6) a quality assurance program that is followed throughout the test.

Aerosol Measurement: Principles, Techniques, and Applications, Second Edition, Edited by Paul A. Baron and Klaus Willeke.
ISBN 0-471-35636-0

In the last two decades, developments in aerosol generation and classification, progress in electron microscopy and imaging analysis, and improvement of test facilities have made instrument calibration easier and the results more reproducible. This chapter reviews calibration techniques relevant to aerosol measurement devices, such as sizing instruments, condensation nuclei counters, and mass monitors. The generation methods for test aerosols and important parameters in instrument calibration are emphasized. Also reviewed are the calibration and use of flow monitoring devices, which play an integral role in aerosol sampling and instrument calibration.

MEASUREMENT METHODS AND CALIBRATION STANDARDS

Aerosol instruments can be categorized according to the particle properties characterized (inertial, gravitational, optical, diffusional, thermal, or electrical) or the measuring techniques (real time or sample collection, personal or area, passive or active). The measured parameters are usually particle size, number concentration, or mass concentration and distributions of these parameters. The calibration of an aerosol instrument implies that the instrument response will be related to a particle standard; for example, for particle sizing, a suitable standard may be latex particles that have been independently sized by methods traceable to the usual laboratory standards. For concentration, the sampled volume is needed, necessitating calibration of the instrument's flow rate. This can be accomplished by the use of various flowmeters that have been calibrated by traceable standards.

In practice, high accuracy is rarely needed in aerosol measurements. For example, the Environmental Protection Agency requires that the cut point of a PM-10 (particulate matter smaller than a 10 μm cut size) sampler be determined to within ±0.5 μm, or ±5%, and the mass concentration must be within 10% of that of an ideal sampler (Federal Register, 1987). Usually the nature of the aerosol source and the conditions under which the sampling is done result in values of precision that do not justify high accuracy. This does not mean that care is not needed in the calibration procedure, but that the National Institute of Standards and Technology (NIST) standards are not normally needed. The calibration may be done by comparison to an instrument calibrated by the manufacturer, for example, an Aerodynamic Particle Sizer (APS). It is advisable, however, to check the sizing of such a device for a few particle sizes using commercial samples of calibrated latex particles and to check the flow rate of the APS with a calibrated flowmeter.

Some aerosol parameters can be measured absolutely in the laboratory, that is, by determining the parameter through combined measurements of length, mass, and time (LMT). For example, the aerodynamic diameter of large particles can be determined by measuring the time required to fall a given distance (Wall et al., 1985). The particle diameter calculated from the operating parameters of the vibrating-orifice generator can be considered absolute because all of the involved quantities can be reduced to LMT (see discussion of the vibrating-orifice generator, below). Geometric particle diameters can be measured in an electron microscope. However, attention must be paid to possible effects of beam heating and exposure to vacuum. Liquid particles are sometimes sized by deposition on a plate and optical microscopy. This requires a correction for droplet distortion (Liu et al., 1982b; John and Wall, 1983; Cheng et al., 1986). Manual measurements by microscopy have limited accuracy because of the small sample size. This can be overcome by using image analysis under computer control. Monodisperse, submicrometer particles of known size can be obtained by using an electrostatic classifier where the particles are selected by their electrical mobility (Mulholland et al., 1999).

For the calibration of filter samplers, the collected aerosol mass concentration can be obtained directly by weighing the filters on a microbalance. Other mass samplers can be calibrated by comparison to a filter sampler. The air volume is calculated from the flow rate and the sampling duration. Flow rates can be calibrated with a variety of flowmeters. Some afford an absolute measurement as in the case of a bubble meter, where the volume swept out by the bubble is measured for a given time.

Because instrument calibration is time consuming, there is a trend toward using well-calibrated, real-time laboratory instruments with direct readouts to characterize the test aerosol and to measure the aerosol penetrating the instrument under test. It is important that the laboratory instrument be operated according to the manufacturer's specifications and to perform checks on its functioning.

GENERAL CONSIDERATIONS

Instrument calibration is essential to a successful measurement of aerosol properties in a sampling environment. However, before engaging in a rigorous process of instrument calibration, one should decide why the calibration is needed, where the instrument is to be used, which important parameters are to be measured, what levels of efforts are to be made, and how the task is to be conducted and the data be processed. All these issues are considered and discussed here. Scientific knowledge and technical experience play an important role in making the right decisions, facilitating the calibration process, and obtaining defendable results.

Rationale for Instrument Calibration

Aerosol sampling is often employed within the context of a general survey, investigating a specific complaint, regulatory compliance, or simply for scientific research purposes. The measurement data are used to characterize emission sources, to assess human exposures, as well as to evaluate control devices. To obtain reliable data, the sampling instrument must be calibrated. For some applications, the instrument must perform according to performance criteria recommended by different organizations, such as the National Institute for Occupational Safety and Health (NIOSH), American Industrial Hygiene Association (AIHA), American Conference of Governmental Industrial Hygienists (ACGIH), American National Standards Institute (ANSI), International Standards Organization (ISO), and the European Standards Commission (CEN), or regulatory standards established by government agencies, such as Occupational Safety and Health Administration (OSHA), Mine Safety and Health Administration (MSHA), and the Environmental Protection Agency (EPA). Table 21–1 summarizes some existing performance criteria and regulatory standards for aerosol sampling instruments. As an example, according to 40 CFR Part 53, *Ambient Air Monitoring*

TABLE 21–1. Performance Criteria for Aerosol Sampling Instrumentation

Type of Instrumentation	Performance Criteria/Guidelines	Regulatory Agency or Organization
PM-10 inlet sampler	Sampling effectiveness, 50% cut point, precision, flow-rate stability	EPA
Workplace sampling instrument	Inhalable, thoracic, and respirable fractions; Sampling efficiency; 50% cut point; sampling precision	ACGIH, ISO, CEN
Portable field instrument	Portability, reliability, calibration, interference, etc.	NIOSH
Respirable dust sampler	Entry, penetration, and sampling efficiencies	ISO, CEN
Personal sampling pump	Interferences	OSHA, ANSI

Source: Kenoyer and Leong (1995).

Reference and Equivalent Methods (Federal Register, 1987), a PM-10 sampler is to be calibrated in a wind tunnel using 10 different sizes of solid or liquid particles ranging from 3 to 25 μm at wind speeds of 2, 8, and 24 km/h (0.56, 2.2, and 6.7 m/s, respectively).

Environment to be Surveyed

The type of aerosol instrument selected and the manner in which it is calibrated may strongly depend on the environment in which the aerosol is to be sampled. In general, one should first attempt to identify the aerosol sources in the environment and decide what information is needed and for what purpose before selecting the parameters and the test aerosol for calibration. Depending on the wind speed in the environment, the sampling can be classified as still (or calm) air sampling or sampling in a moving air stream (Vincent, 1989, 1995; Hinds, 1999; see also Chapter 8). Still-air sampling generally refers to a wind speed less than 0.5 m/s and applies to indoor environments, including residential homes, offices, schools, and factories. Moving air stream sampling refers to environments with higher wind speed, such as ambient atmosphere or inside ventilation ducts and stacks. Settling chambers with uniform, low flow rates are suitable for testing instruments under still-air conditions (Kenny et al., 1999), while wind tunnels are more appropriate for testing instruments under moving air stream conditions.

Parameters to be Investigated

It is necessary to select a set of parameters to be investigated during the instrument calibration. These parameters should be chosen depending on the type of instrument as well as the aerosol properties of interest in the sampling environment. For example, volumetric flow rate, pressure drop, and light source intensity are operating parameters, while particle size and composition and the nature of the suspending gas medium are aerosol parameters. The parameters selected can be different between two instruments. To calibrate an aerodynamic sizing device, the effects of particle density, velocity, and ambient pressure on the instrument response are important while in an optical particle counter (OPC) the particle refractive index, wavelength of light source, and collection angles of scattered light are the important parameters.

Design of the Calibration Program

The level of effort to be undertaken in the calibration should be considered. A full-scale calibration examining the instrument response over its full operational range requires elaborate test facilities and extensive effort in terms of time and labor. However, when a calibration curve with the full-range response is available, it may be sufficient to perform a single or two-point calibration. It is not unusual for different components of the same instrument to be calibrated separately. For example, when one is interested in the performance of the sampling inlet (inlet efficiency), transport line (transport loss), and detection (counting efficiency) or collection section (collection efficiency) of a high-volume aerosol sampling system, a series of calibrations can be made, one for each individual component (Chen et al., 1999). Different parameters might be selected for investigation in the different components, and, consequently, different test facilities with different test aerosols are often employed during the component calibrations. In contrast, several instruments are sometimes arranged serially and calibrated as an integrated unit. For example, a two-stage virtual impactor and an electrical classifier were combined in series to investigate their integrated performance in fiber classification (Chen et al., 1996).

Selection of Test Aerosols

Proper selection of test aerosols is essential to instrument calibration. Because most instruments have responses strongly dependent on the physical and chemical properties of the aerosol particles, the calibration curve of an instrument is strictly valid only for the test aerosol. For an aerosol whose physical and chemical properties are significantly different from those of the test aerosol, data interpretation based on the calibration could be misleading (Willeke and Baron, 1990). For example, one would underestimate the size distribution of a carbon black aerosol by using the calibration curve of an OPC obtained from polystyrene latex spheres because the carbon particles cannot scatter as much light as the polystyrene latex spheres. Ideally, an aerosol that has similar physical and chemical properties (e.g., size, shape, density, refractive index, dielectric constant, and thermal conductivity) to the aerosol to be measured should be selected as the test aerosol during calibration.

Sampler performance can be investigated by use of selected test aerosols. Liquid particles can be used to simulate sticky particles that suffer wall losses. Solid, bouncy particles can be used to test for particle bounce and/or re-entrainment.

Data Analysis

A calibration curve contains the relationship between the instrument response and the values of a certain aerosol property to be measured. In the case of a direct-reading instrument, the calibration provides an adjustment (or a correction factor) to the indicated value. In addition, the resolution and sensitivity of the instrument should be examined and analyzed. After collecting calibration data, it is desirable to express that data in a generalized mathematical equation, relating the instrument response to a single parameter (Chen et al., 1985; Zhang and Liu, 1990).

For data analysis, instrument manufacturers sometimes provide a built-in algorithm whose properties, accuracy, and limitations are often unknown to the user. It may be desirable to base the analysis only on the raw calibration data without manipulation by the built-in algorithm.

Safety Precautions

When generating aerosols, it should always be borne in mind that a respiratory health hazard may be created. A primary consideration is containment of the aerosol. A chemistry hood is a good location for an aerosol generator. Even if the exhaust from the generator is vented, there are usually times when the apparatus is open or there may be leaks. A walk-in hood is especially convenient to accommodate an auxiliary apparatus. If a hood cannot be used, the exhaust should be vented or filtered. Hazardous substances require more stringent containment measures.

Care should be exercised in the choice of aerosol materials. For example, in the past dioctyl phthalate (DOP) was commonly used as a test aerosol because it has nearly unit density and is an oil with low volatility. However, animal tests have implicated DOP as a possible carcinogen. A good substitute is oleic acid, also a nonvolatile oil, which is available in food grade. A side benefit is that uranine, which is frequently added as a fluorescent tracer, is soluble in oleic acid, whereas it is insoluble in DOP. This means that the uranine is uniformly dispersed in the oleic acid droplets. Uranine is commonly used to trace waterways and is presumably harmless. Of course, even when the aerosol material is believed to be safe, it is prudent to avoid exposure.

Another hazard is associated with the use of radioactive sources to "neutralize" the electrical charges on aerosols resulting from the generation process. ^{85}Kr, a β (high-energy electron) emitter, is commonly used in source strengths up to 10 mCi. Unfortunately, ^{85}Kr also emits γ rays. Whereas the β rays are absorbed by the walls of the container, the γ rays penetrate. It is recommended that a qualified health physicist check the radiation level to evaluate the adequacy of the shielding. α particle sources, such as ^{210}Po, represent a hazard when ingested and must be handled with care.

CALIBRATION APPARATUS AND PROCEDURES

Figure 21–1a is a schematic of a typical calibration apparatus for aerosol instruments. It includes an aerosol generator, aerosol conditioning devices (e.g., diffusion dryer, charge neutralizer, aerosol classifier, aerosol concentrator, and dilution air supply), a mixing chamber, pressure and air flow monitoring equipment, the instrument to be calibrated, and a calibration standard. The aerosol produced from the generator can be monodisperse or polydisperse, solid or liquid, wet or dry, charged or uncharged, or spherical or nonspherical (described later). Generally, this aerosol requires several steps of conditioning before use. For an aerosol containing volatile vapors or water droplets, a diffusion dryer with desiccant and/or charcoal is commonly used to produce a dry aerosol. In some cases, a heat treatment using a high temperature furnace is required for the production of a test aerosol (Kanapilly et al., 1970; Chen et al., 1990). The heat treatment involves either sintering or fusing the particles to reach the desired particle morphology and chemical form or initiating particle evaporation and subsequent condensation to produce monodisperse particles. Because aerosol particles are usually charged by static electrification during formation, a neutralizer containing a bipolar ion source (e.g., ^{63}Ni, ^{85}Kr, and ^{241}Am) is often used in the aerosol treatment. This reduces the number of charges on particles and results in an aerosol with charge equilibrium (John, 1980). In addition, a size-classifying device is often used in the aerosol treatment to segregate particles of similar size or of desired size fraction (Liu and Pui, 1974; Chen et al., 1988; Romay-Novas and Pui, 1988). In addition, a concentrator or a dilutor is often used to adjust the aerosol concentration (Barr et al., 1983; Yeh and Carpenter, 1983).

The desired test aerosol can be used to calibrate instruments in several ways. The simplest way, as shown in Figure 21–1a, is to introduce the test aerosol into a mixing chamber in which the aerosol is uniformly distributed and sampled by both the instrument to be calibrated and the calibration standard. Pressure in the chamber and flow rate through the instrument are monitored. A sampling device, such as a filter sampler or an electrostatic precipitator, is often used to collect reference samples for the calibration standard. To ensure that both the calibration device and the instrument to be calibrated have comparable aerosol samples, an aerosol divider is used as a common sampling port for calibrating a mass monitor (Marple and Rubow, 1978). In the aerosol divider, the flow is split isokinetically into two streams: One passes directly into the instrument to be calibrated, and the other flows through the calibration standard (Fig. 21–1a). This mixing chamber setup is inexpensive, easy to use, and does not require a large working area. It is most widely used for instrument calibrations that require particle sizes less than 5 μm, such as those for obtaining the response curve of an OPC or the collection efficiency of an impactor. For calibrations requiring particles larger than 5 or 10 μm, it is relatively difficult to provide a stable aerosol with sufficiently high concentration. Because the instrument is placed outside the mixing chamber, the setup is not adequate for testing the aspiration efficiency of the instrument inlet.

Another way of calibrating an instrument is to introduce the aerosol particles into an aerosol test chamber (Fig. 21–1b) that contains the subject instrument and the test standard (Marple and Rubow, 1983; Chen et al., 1999). This chamber usually has a large test section to provide a quiescent atmosphere in which the entire instrument can be exposed to the aerosol as in the real sampling environment. The test aerosol is introduced at the top of the

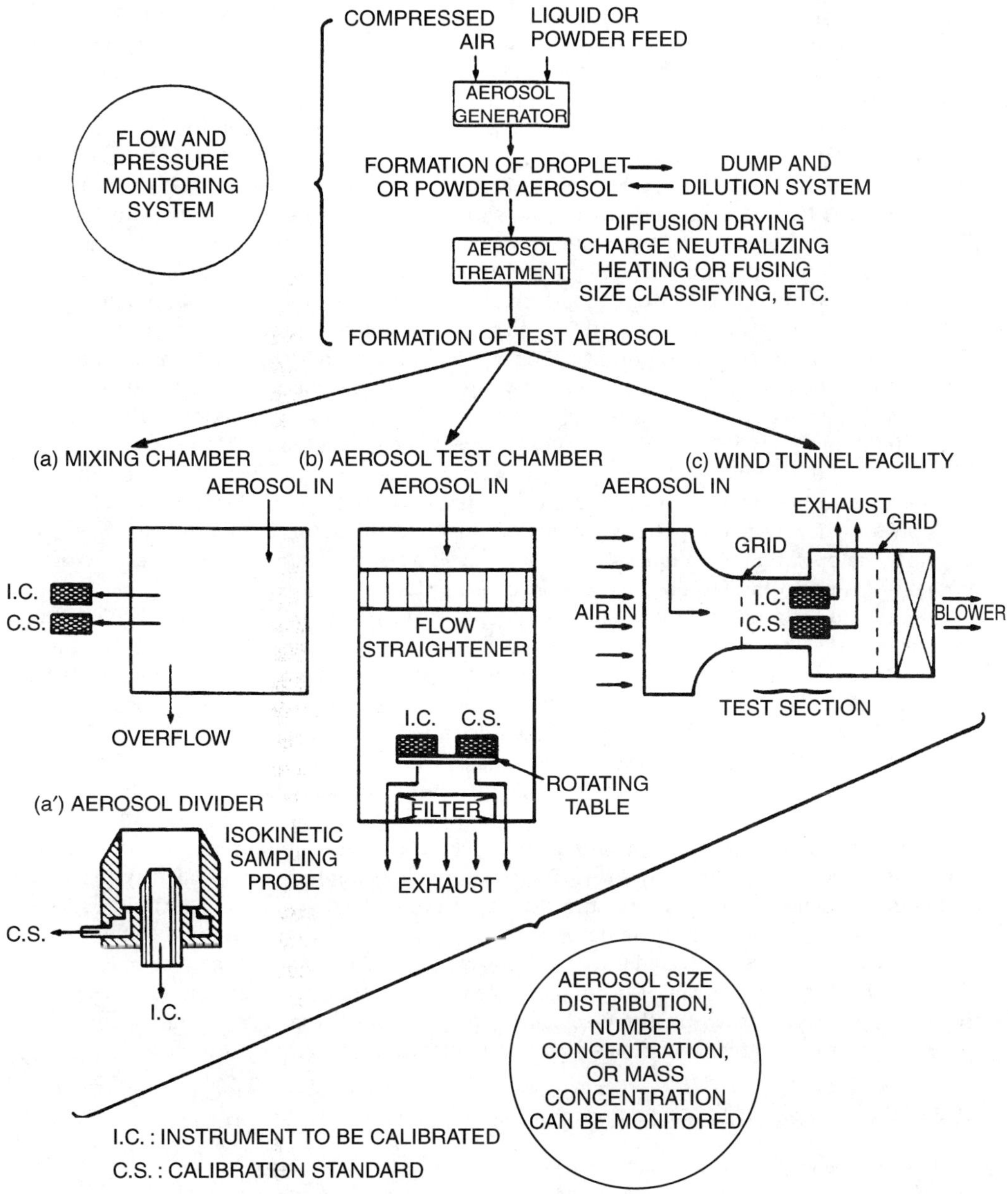

Fig. 21–1. Schematic diagrams of a setup for instrument calibration using (a) a mixing chamber, (á) an aerosol divider, (b) a test aerosol chamber, and (c) a wind tunnel facility.

chamber and uniformly distributed in the section where the instrument is set on a rotating table. Rotation provides a means to reduce any effects due to possible temporal and spatial variations in aerosol concentration. This test chamber setup can provide uniform concentrations of aerosol particles as large as 90 μm for instrument calibration (Maynard and Kenny, 1995; Aitken et al., 1999; Kenny et al., 1999). Several instruments can be placed inside the chamber for side-by-side comparison, including the sampling inlet. The flow rate and turbulence intensity in the chamber are low, simulating still-air sampling conditions. It should be noted that the air is exhausted from the chamber (i.e., the air is not completely static). A test chamber with static air is not recommended because it is very difficult to avoid convection currents that can affect the measurements.

Both the mixing chamber and the aerosol test chamber are used when the instrument to be calibrated is operated in a low or zero ambient wind velocity. To evaluate a sampler that will be operated in moving air, a wind tunnel facility (Fig. 21–1c) is needed (Prandtl, 1952). The sampler is located inside the tunnel and should not occupy more than 10% to 15% of the cross-sectional area of the tunnel's test section to avoid blockage effects. Personal samplers are often mounted on the upper torso of a full-sized manikin placed within the test section of the wind tunnel (Vincent and Mark, 1982; Kenny et al., 1997). The wind tunnel provides a wide range of wind speeds (0.5 to 10 m/s) to simulate different atmospheric conditions. The wind velocity, flow uniformity, and turbulence are monitored using flow-monitoring devices (described later). During calibration of the test sampler, an isokinetic sampler is generally used to collect reference samples. Two types of wind tunnels are commonly used: an open circuit tunnel (Vincent and Mark, 1982) and a closed circuit tunnel (Ranade et al., 1990). The open circuit tunnel operates by drawing filtered ambient air into the system and exhausting the air into the ambient downstream of the test section; the closed circuit tunnel circulates the air in a continuous path. Each type of tunnel has advantages and disadvantages. For example, the open tunnel occupies a smaller space with less installation cost, while the closed tunnel is less noisy and requires less energy consumption.

Before any calibration, a standard operating protocol should be prepared. First, the manual for the instrument to be calibrated should be read carefully to learn as much as possible about the operating principles of the instrument, the construction, and the recommended operating procedures. However, the manual may not cover all aspects relevant to the application. For example, a laser-operated OPC tends to produce oscillatory responses when the particles are larger than the wave length of the laser beam; however, the calibration curve provided by the manufacturer seldom shows this phenomenon (Chen et al., 1984).

The condition of the instrument should be checked before calibration. The integrity of the flow system can be quantified by a series of pressure measurements on a sealed system that has initially been brought to a pressure slightly above or below ambient pressure (Mokler and White, 1983). Any leakage can be discovered by various methods. The simplest one is to pressurize the system slightly and then put soapy water on the surface of the system to detect the leakage. A tracer gas can also be injected into the system to detect the leak location. It may be advisable to check the electronics by observing the signals on an oscilloscope, especially if the instrument outputs a signal to be processed by other instruments.

General Approaches to Instrument Calibration

Calibration methods can be characterized by whether the test aerosol is monodisperse or polydisperse and how the aerosol is measured. Monodisperse aerosol is either produced directly by a generator or classified after generation by auxiliary apparatus. The use of polydisperse test aerosol implies that particle sizing will be done on the airborne particles or by analysis of collected samples. In one approach to calibration, the particles collected or deposited within the instrument or sampler is analyzed (e.g., gravimetrically or by analysis of tracers). An alternative approach is to measure the aerosol entering and leaving the sampler. This may involve testing the sampler without a final filter.

The choice of the calibration method is made in consideration of the type of instrument to be calibrated, the kind of information needed, and available resources. The various methods have distinct advantages and disadvantages. The use of monodisperse particles necessitates many repetitions with different particle sizes. However, it may yield unambiguous information, for example, on whether 20 μm particles penetrate a size selector for PM-10. Similarly, the measurement of deposited particles may involve tedious extractions and quantitations, but may determine where wall losses occur. The use of polydisperse aerosol with a sizing instrument that produces a real-time size distribution makes possible rapid calibration measurements. This can be important if a parameter such as sampler flow rate is to be varied between calibrations or variations in the sampler's configuration are to be explored (John and Kreisberg, 1999).

TEST AEROSOL GENERATION

Test aerosols contain either monodisperse or polydisperse, spherical or nonspherical, and solid or liquid particles (Mercer, 1973; Raabe, 1976; Hinds, 1999; Cheng and Chen, 1995). The characteristics of an ideal generator are a constant and reproducible output of stable aerosol particles whose size and concentration can be easily controlled. For general instrument calibration, the test aerosol often contains monodisperse, spherical particles. Table 21–2 lists the test aerosols frequently used for instrument calibration. Monodisperse aerosols containing spherical particles are frequently used. Particles with nonspherical shapes are sometimes used in calibration to study the possible effect of shape on the instrument response. Polydisperse dust particles have also been used in calibrating dust monitors. This is important because most real aerosols contain nonspherical particles of different sizes.

The size distribution and concentration of a test aerosol depend on the characteristics of both the generator and the feed material. The information given in this section is intended as a guide for selection of appropriate generation techniques. The actual size distribution in each application should always be measured directly with the appropriate instruments.

Monodisperse Aerosols with Spherical Particles

The methods for producing monodisperse aerosols with spherical particles have been reviewed by Fuchs and Sutugin (1966), Mercer (1973), and Raabe (1976). These methods include the atomization of a suspension of monodisperse particles, the formation of uniform droplets by dispersion of liquid jets with periodic vibration or a spinning disk, and the growth of uniform particles or droplets by controlled condensation.

Atomization of Suspensions of Monodisperse Particles. A common way of producing monodisperse aerosols is by nebulizing a dilute liquid suspension containing monodisperse polystyrene (PSL) or polyvinyltoluene (PVT) latex spheres. These spheres are commercially available in sizes from 0.01 to over 100 μm (***BAN, DUK, DYN, JSR, MMM, POL, SER***).* PSL particles of different sizes have also been concurrently produced in an aerosol to obtain more than one data point per experimental run. Monodisperse latex particles containing fluorescent dye or radiolabeled isotopes are also used in calibrations when quantitative measurements by fluorometric or radiometric techniques are needed (Newton et al., 1980; Chen et al., 1999).

Two problems arise in the generation of these latex particles: formation of agglomerates and existence of residual particles. Agglomerates are formed when more than one latex particle is in a nebulized droplet. The percentage of agglomerates can be reduced by diluting the suspension. Assuming that the probability of the number of particles in an atomized droplet can be described by Poisson statistics and that the droplet-size distribution can be approximated by a lognormal distribution, Raabe (1968) derived the following equation to calculate the latex dilution factor, Y, necessary to give a desired singlet ratio, R, which is the number of droplets containing single particles relative to the total number of droplets containing particles:

* See Appendix I for full manufacturer addresses referenced to the italicized three-letter codes.

TABLE 21–2. Test Aerosols and Generation Methods Used for Instrument Calibration

Test Aerosol[a]	Particle Morphology	Size Range[b]		Density (kg/m^3)	Refractive Index	Generation Method	Aerosol Output ($particles/m^3$)
		VMD (μm)	σ_g				
PSL (PVT)	Spherical, solid	0.01 to >100	≤1.02[c]	1,050 (1,027)	1.59	Nebulization	$<10^{10}$
Fluorescent PSL	Spherical, solid	6 to >100[d]	1.08–1.17	1,050	1.59	Dry powder dispersion	—[e]
Soda lime glass (borosilicate glass)	Spherical, solid	1.1 to >100	1.07–1.3	2,460 (2,500–2,550)	1.51 (1.56)	Dry powder dispersion	—[e]
Oleic acid	Spherical, liquid	0.5–40	≤1.1	890	1.46	Vibrating-orifice atomization	$<10^{11}$
Ammonium fluorescein	Spherical, solid	0.5–50	≤1.1	1,350	—	Vibrating-orifice atomization	$<10^{11}$
Fused ferric oxide	Spherical, solid	0.2–10	≤1.1	2,300	—	Spinning-disk atomization	$<10^{13}$
Fused aluminosilicate	Spherical, solid	0.2–10	≤1.1	3,500	—	Spinning-disk atomization	$<10^{13}$
Fused cerium oxide	Spherical, solid	0.2–10	≤1.1	4,330	—	Spinning-disk atomization	$<10^{13}$
Sodium chloride	Irregular, solid	0.002–0.3	≤1.2	2,170	1.54	Evaporation/ condensation	$<10^{12}$
Silver	Irregular, solid	0.002–0.3	≤1.2	10,500	0.54	Evaporation/ condensation	$<10^{12}$
Coal dust	Irregular, solid	~3.3	~3.2	1,450	1.54–0.5i	Dry powder dispersion	<30 mg/m^3
Arizona road dust	Irregular, solid	~3.8	~3.0	2,610	—	Dry powder dispersion	<30 mg/m^3

[a] Standard particles, such as the PSL, fluorescent PSL, glass spheres, and Arizona road dust, are commercially available from companies such as ***BAN***, ***DUK***, ***IDC***, and ***POL***.

[b] Aerosol treatment of drying, charge neutralization, and size classification is generally used.

[c] For VMD less than 0.1 μm, σ_g is between 1.03 and 1.14.

[d] This size range is for fluorescent particles in dry powder form; particles of submicrometer sizes are available in suspension.

[e] The aerosol output for these dry particles depends on the particle size, bulk concentration, and generation parameters. Normally, particles of a larger size have a smaller concentration.

$$Y = F(\text{VMD})^3 \exp(4.5\ln^2 \sigma_g)[1 - 0.5\exp(\ln^2 \sigma_g)]/(1 - R)d_p^3 \tag{21–1}$$

EXAMPLE 21–1

A bottle of 1 μm PSL suspension containing a 10% solid is being used to produce a test aerosol containing at least 95% singlets. What is the dilution factor required in this suspension if the Retec X-70/N nebulizer is used and operated at 20 psig?

Answer: Using eq. 21–1:

$$Y = F(\text{VMD})^3 \exp(4.5\ln^2 \sigma_g)[1 - 0.5\exp(\ln^2 \sigma_g)]/(1 - R)d_p^3$$

$$F = 10\% = 0.1$$

$$R = 95\% = 0.95$$

$$d_p = 1\,\mu\text{m}$$

Based on size distribution data given in Table 21–3:

$$\text{VMD} = 5.7\,\mu\text{m},\ \sigma_g = 1.8$$

$$\begin{aligned} Y &= (0.1)(5.7)^3 \exp[4.5\ln^2(1.8)](1 - 0.5\exp[\ln^2(1.8)])/(1 - 0.95)(1)^3 \\ &= (0.1)(5.7)^3 \exp(1.555)[1 - 0.5\exp(0.345)]/(0.05) \\ &= (0.1)(5.7)^3(4.734)(0.294)/(0.05) \\ &= 514.9 \end{aligned}$$

A dilution factor of at least 515 is needed to produce an aerosol containing 95% of singlet PSL particles.

where F is the volumetric fraction of individual particles of diameter d_p in the original latex suspension, and VMD and σ_g are the volume median diameter and the geometric standard deviation of the droplet size distribution, respectively. The values of VMD and σ_g of commonly used air-blast atomizers are listed in Table 21–3. This equation is limited to values of $\sigma_g < 2.1$ and $R > 0.9$.

The second problem arises when nonlatex residual particles are present in the aerosol as a result of the surfactant usually present in the liquid suspension to prevent coagulation. Because most of the atomized droplets contain no latex particles, the nonlatex particles form a large background of small particles. If this background interferes with the measurements, the surfactant may be removed from the suspension before use by diluting, centrifuging, and discarding the supernate. In recent years, latex suspensions containing no surfactant have become available. These suspensions are stabilized by surface coatings of functional groups (***IDC***).

Vibrating-Orifice and Spinning-Disk Aerosol Generators. The vibrating-orifice aerosol generator can produce highly monodisperse aerosols in the approximate size range from 0.5 to 50 μm (Fulwyler et al., 1969; Raabe and Newton, 1970; Berglund and Liu, 1973).

TABLE 21–3. Operating Parameters of Air-Blast and Ultrasonic Nebulizers

Nebulizer	Operating Conditions				Aerosol Output (μL/L)	Droplet Size Distribution		Commercial Source
	Orifice Diameter (mm)	Air Pressure (kPa [psig])	Frequency (mHz)	Flow[a] Rate ($\times 10^{-5}$ m^3/s [L/min])		VMD (μm)	σ_g	
Air-blast type								
Collison	0.35	100 [15]		3.3 [2.0]	8.8	2.5–3.0	3.0	***BGI***
		170 [25]		4.5 [2.7]	7.7	1.9–2.0	2.0	
DeVilbiss[b]	0.84	100 [15]		20.7 [12.4]	15.5	4.2	1.8	***DEV***
D-40		200 [30]		34.8 [20.9]	12.1	2.8	1.9	
DeVilbiss	0.76	100 [15]		15.7 [9.4]	23.2	4.0	—	***DEV***
D-45		200 [30]		24.2 [14.5]	22.9	3.4	—	
Lovelace	0.26	140 [20]		2.5 [1.5]	40	5.8	1.8	***INT***
		350 [50]		3.8 [2.3]	27	2.6	2.3	
Retec	0.46	140 [20]		8.3 [5.0]	46	5.7	1.8	***INT***
X-70/N		350 [50]		16.2 [9.7]	47	3.2	2.2	
Ultrasonic type								
DeVilbiss	(2)[c]		1.35	68.3 [41.0]	54	5.7	1.5	***DEV***
880	(4)[c]		1.35	68.3 [41.0]	150	6.9	1.6	
Sono-Tek			0.025–0.12	10^{-6}–0.73 [10^{-6}–0.44]	—	18–80	—	***SON***

[a] Output per orifice.
[b] Vent closed.
[c] Power settings.

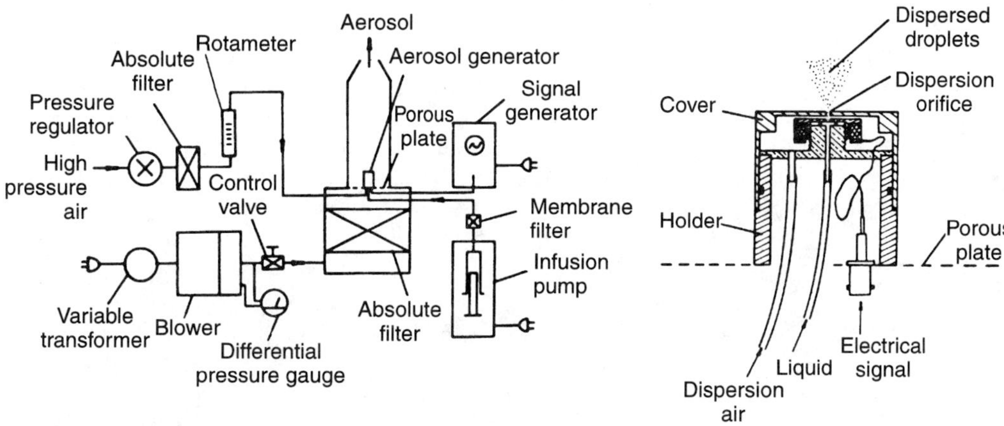

Fig. 21–2. Diagram of a Vibrating-Orifice Aerosol Generator of the Berglund-Liu Design. (Reprinted from Liu (1974) with the Permission of Air and Waste Management Assoc.).

The particle diameter can be calculated from the generator's operating conditions so that the aerosol can be considered a primary particle size standard. Also, the aerosol concentration is inherently stable. In the vibrating-orifice generator (Fig. 21–2), a liquid is forced through an orifice. The resulting liquid jet is made to break up into uniform droplets by subjecting the jet to a mechanical disturbance of constant frequency. The droplet diameter, d_d, is then given by

$$d_d = 10^6 (6Q_L/\pi f)^{1/3} \tag{21–2}$$

where Q_L is the liquid feed rate in m^3/s and f is the vibrating (disturbing) frequency in Hz. The droplet diameter is typically some tens of micrometers. To generate smaller particles, a nonvolatile solute can be dissolved in a volatile solvent. After evaporation of the solvent, the particle diameter is related to the volumetric concentration of the solute, C_v, by

$$d_p = C_v^{1/3} d_d \tag{21–3}$$

Liquid particles can be produced, for example, from a solution of oleic acid in isopropyl alcohol, or solid particles of sodium chloride can be produced from an aqueous solution. The minimum particle size attainable, in practice, depends on the purity of the solvent. The maximum practicable particle size is not well defined. Particles larger than about 20 μm diameter become more difficult to generate as the diameter is increased. It also becomes more difficult to avoid particle losses in transport. Therefore, the generation of particles larger than about 50 μm requires special effort.

Referring to Figure 21–2, the solution is forced through the orifice by a syringe pump. An alternating voltage from a signal generator is applied to a piezoelectric crystal, which then vibrates the assembly holding the orifice plate. A turbulent jet of air issuing from the hole in the cover above the orifice disperses the droplets before they can coagulate. Filtered, dry dilution air is introduced to dry the droplets and transport the aerosol from the generator.

There is a range of about a factor of two in the frequency producing uniform droplets. Within this range, certain frequencies may produce satellite droplets (i.e., droplets much smaller than the main drops that are being produced). They can be eliminated by adjustments of the vibrating frequency. Another undesirable characteristic is the production of multiplets

EXAMPLE 21–2

Oleic acid aerosol is produced by a vibrating-orifice atomizer from a solution of oleic acid in isopropyl alcohol, with a volume concentration of 1.48×10^{-2}. The liquid feed rate is $3 \times 10^{-9}\,m^3/s$ [$0.18\,cm^3/min$], and the vibrating frequency is 5.5×10^4 Hz. The dilution air flow rate is $3.33 \times 10^{-4}\,m^3/s$ [20 L/min]. What are the diameters of the droplets and the oleic acid particles? What is the particle number concentration?

Answer: Using Eqs. 21–2 and 21–3:

$$d_d = 10^6 (6Q_L/\pi f)^{1/3}$$

$$d_p = C^{1/3}\, d_d$$

substituting Q_L, f, and C, we find the droplet diameter $d_d = 10^6\,[6\,(3 \times 10^{-9})/55{,}000\pi]^{1/3} =$ 47.1 µm and the particle diameter $d_p = (1.48 \times 10^{-2})^{1/3}(47.1) = 11.6\,\mu m$

The rate of particle production is the same as the vibrating frequency, $55{,}000\,s^{-1}$. Dividing this by the flow rate of the dilution air, we find that the particle number concentration is $1.7 \times 10^8\,m^{-3}$.

because coagulation of the droplets cannot be completely suppressed. In practice, the operating conditions are adjusted to minimize the multiplets, and corrections are applied to the data taken with the aerosol, if necessary.

Particles from the vibrating orifice typically carry several thousand elementary charges. Because the presence of a high electrical charge may affect subsequent processes involving the aerosol, it is common practice to "neutralize" the charge by exposure of the aerosol to a radioactive source. The source creates charged ions in the gas that are attracted to charges of the opposite sign on the particles, reducing the particle charge distribution to a Boltzmann equilibrium. The radioactive sources commonly used include β emitters such as ^{85}Kr or tritium and α emitters such as ^{210}Po or ^{241}Am.

In general, solid particles dry with the formation of voids, resulting in a density less than that of the bulk material. To some extent, the drying process can be controlled by varying the volatility of the solvent, for example, by varying the proportions of water and alcohol and by controlling the amount of dilution air. If the drying is too rapid, the particles tend to have more voids. The average density of the particles, including voids, can be determined by the following method:

$$\rho_{p,av} = (d_d/d_g)^3 C_m \tag{21–4}$$

where $\rho_{p,av}$ is the average density of the particles (in kg/m^3), d_g is the particle geometric diameter (in µm) determined with a microscope, and C_m is the mass concentration (in g/L) of the solute (John and Wall, 1983).

One example of a solid particle aerosol, namely, ammonium fluorescein, deserves special mention because of its useful particle properties and the requirement for special generation procedures. Ammonium fluorescein particles are very smooth and have essentially bulk density ($1350\,kg/m^3$ [$1.35\,g/cm^3$]). The material has low hygroscopicity, and the fluorescence can be used for detection with high sensitivity. The particles are useful for sampler calibration and for checking for particle bounce. The solution is prepared by dissolving fluorescein in ammonium hydroxide. A reaction takes place with an ammonium group replacing a hydrogen atom on the fluorescein molecule. As a result, the molecular weight increases by 5%; thus, 12.8 g fluorescein per liter of solution is a 1% volume concentration. When particles of ammonium fluorescein larger than about 10 μm are generated, the droplets tend to dry too fast. To produce smooth particles, it is necessary to humidify the dilution air to slow down the drying. By this method, particles as large as 70 μm have been generated (Vanderpool and Rubow, 1988).

The vibrating-orifice aerosol generator can be used to produce uniformly charged, monodisperse aerosols (Reischl et al., 1977). The technique involves insulating the cap over the orifice assembly and applying a dc voltage to the cap. Charges are induced onto the top of the jet and are trapped on the droplets when they separate from the jet. The monodispersity of the droplets leads to uniform charging. The application of modest voltages, that is, up to ±10 V, produces particles with $\pm 10^4$ elementary charges. Reischl et al. (1977) present the theory and data demonstrating the method, which is useful for experimentation with charged aerosols.

Another method of producing monodisperse droplets is by the spinning disk, in which a liquid jet is fed at a constant rate onto the center of a rotating disk. The liquid spreads over the disk's surface in a thin film, accumulating at the rim until the centrifugal force exceeds the capillary force acting to hold it together, and a droplet is thrown off. Droplet size d_d depends on disk diameter, d_s (μm), and rotating speed, ω_s (rpm), as follows:

$$d_d = \left(W \gamma_L / \rho_L \omega_s^2 d_s\right)^{1/2} \tag{21–5}$$

where γ_L and ρ_L are the surface tension and the density of the liquid and W is a constant. Spinning disks have been investigated by Walton and Prewett (1949) and May (1949) using an air-driven spinning top and by Whitby et al. (1965) and Lippmann and Albert (1967) using a motor-driven spinning disk. Unlike a vibrating-orifice atomizer, aqueous suspensions as well as solutions can be used. The spinning disk produces an order of magnitude higher aerosol concentrations compared with the vibrating orifice. However, the monodispersity is not as high; the σ_g values are approximately 1.02 and 1.1 for the vibrating orifices and spinning disks, respectively. A disadvantage of the spinning disk is that undesired satellite droplets are formed and must be removed from the useful aerosol. In addition, the constant W (Eq. 21–5) varies with the instrument and the feed material used so that the particle size cannot be easily calculated as for the vibrating-orifice atomizer.

Controlled Condensation Techniques. Condensation is also a method that produces monodisperse aerosols for calibration purposes. In this method, the heated vapor of a substance is mixed with nuclei on which it subsequently condenses when it passes in laminar flow through a cooling zone. If the condensation process is diffusion controlled, the surface area of the growing droplet will increase at a constant rate, producing a particle having a diameter, d_t, at time t related to the initial diameter, d_o, of the nucleus, by

$$d_t^2 = d_o^2 + bt \tag{21–6}$$

where b is a constant related to the concentration, the diffusivity of the vapor, and the temperature. If bt is the same for all particles and much larger than d_o^2, an aerosol containing monodisperse particles is produced. In practice, a uniform temperature profile, sufficient vapor concentration, and sufficient residence time in the condensation region are the key controls, and a constant nuclei concentration provides a stable aerosol concentration (Sinclair and LaMer, 1949; Rapaport and Weinstock, 1955; Prodi, 1972; Liu and Lee, 1975; Tu, 1982). Particle sizes from 0.03 to greater than 2 μm with σ_g of 1.2 to 1.3 can be produced this way. The number concentration can be as high as 10^{13} particles/m^3 [10^7 particles/cm^3]. An example of a condensation aerosol generator is shown in Figure 21–3.

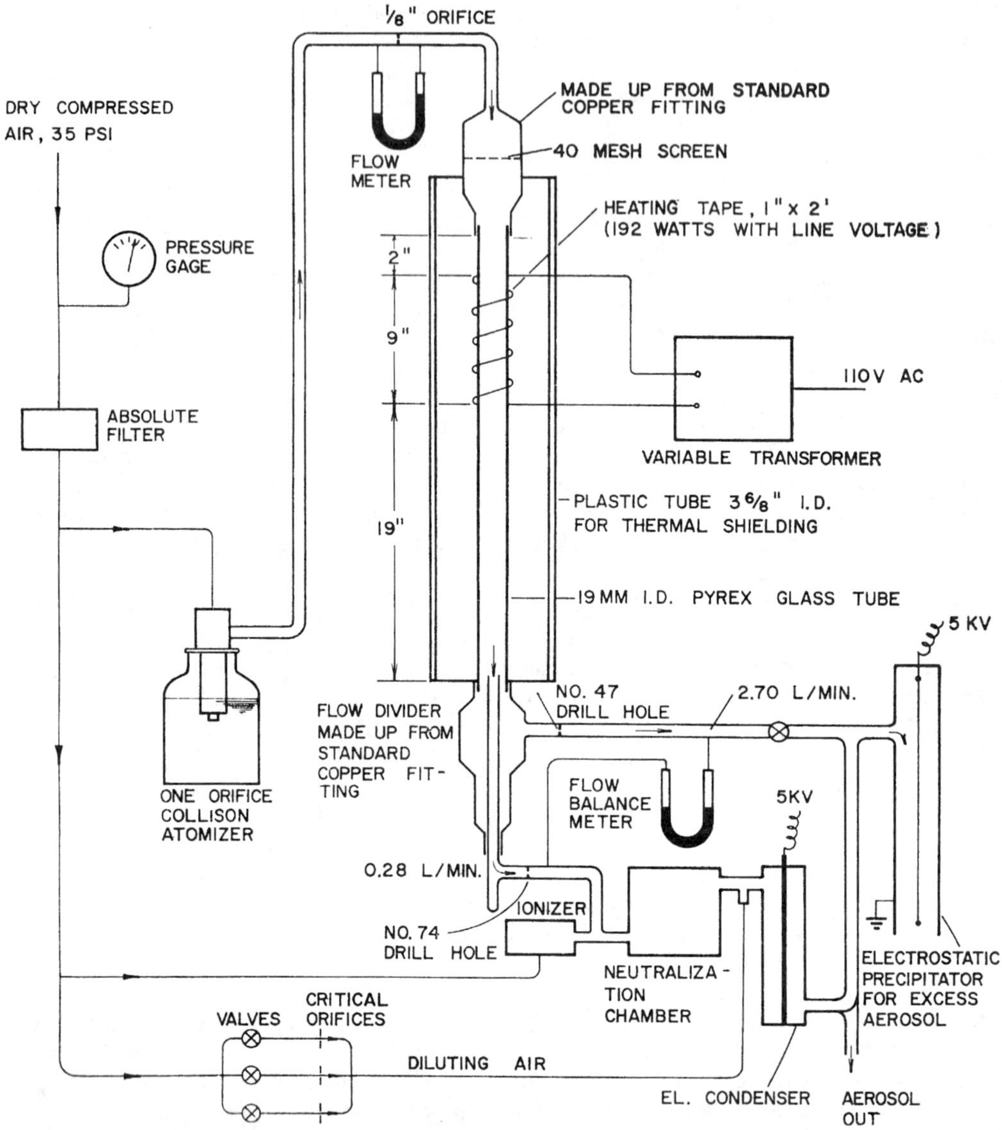

Fig. 21–3. Condensation-Type Monodisperse Aerosol Generator. (Reprinted from Tomaides, Liu, and Whitby (1971) with the Permission of Pergamon Press, Inc.).

Electrospray Techniques (Generation of Monodisperse Nanoparticles). Another method for producing monodisperse aerosols is to use an electrostatic atomizer or electrospray device (Hayati et al., 1987a,b; Fernandez de la Mora et al., 1990; Meesters et al., 1992; Grace and Marijnissen, 1994). Electrospraying refers to the generation of liquid droplets by feeding semiconducting fluid through a capillary tube and applying an electric field. There are several modes by which the liquid can break up into droplets, depending on flow rate, field strength, and other parameters. For certain conditions, when the field is strong enough, the liquid meniscus at the capillary outlet forms a cone from whose tip a very thin liquid jet emerges in the cone-jet mode (Cloupeau and Prunet-Foch, 1989, 1994). The microjet breaks by varicose wave instabilities into a stream of charged droplets, having diameters roughly twice as large as the jet diameter but much smaller than the capillary diameter (Rosell-Llompart and Fernandez de la Mora, 1994; Tang and Gomez, 1994). A system using this technique can generate very small droplets without the clogging problems associated with a very small orifice. The mean droplet size is usually in the range of 0.3 to 50 µm, but can be as small as 10 nm. The size is a function of the nozzle diameter, liquid feed rate, field strength, and properties of the liquid, including surface tension, electrical conductivity, and viscosity (Smith, 1986). The droplets are charged up to an appreciable fraction of the Rayleigh limit and initially repel each other until they are neutralized.

Tang and Gomez (1994) have demonstrated a generation system producing monodisperse droplets in the size range from 2 to 12 µm with a σ_g of 1.15 for small droplets and 1.05 for large droplets. Monodisperse droplets of 0.3 to 4 µm have been produced with a σ_g of 1.1 (Rosell-Llompart and Fernandez de la Mora, 1994). By using volatile solvents, particles down to nanometer diameters can be produced. More recently, a practical system has been developed using the electrospray technique to produce monodisperse particles with a mean diameter from 4 nm to 1.8 µm and a σ_g of 1.1. The operating ranges of the important parameters such as liquid feed rate, electrical conductivity, and concentration of the solution were determined (Chen et al., 1995). Unfortunately, the main group of particles is accompanied by a second group having diameters approximately eight times smaller. For some applications it will be necessary to remove the second group by using a size-selective device.

The electrospray technique produces supermicrometer aerosol less monodisperse and with lower concentration than that from a vibrating-orifice generator. However, for submicrometer particles and especially for nanoparticles, electrosprays offer unique advantages.

Monodisperse Aerosols with Nonspherical Particles

The effects of particle shape on instrument response are important, especially for sizing instruments. The effects of shape on instrument response can be investigated by using monodisperse aerosols of nonspherical particles during calibration. One way of generating these aerosols is to nebulize a liquid suspension containing monodisperse nonspherical particles. Various techniques have been used to produce monodisperse particles of highly uniform particle size and shape. Matijevic (1985) produced inorganic and polymer colloid particles of cubic, spindle, and rhombohedral shapes by chemical reactions. Fiber-like particles of a narrow size range were also produced using different methods (Esman et al., 1980; Loo et al., 1982; Vaughan, 1990; Hoover et al., 1990; Baron et al., 1994; Chen et al., 1996; Deye et al., 1999). The vibrating-orifice and spinning-disk aerosol generators described above can also be used to generate nonspherical particles, such as crystalline sodium chloride particles. Although the generators produce spherical droplets, the crystal form of the solid particles determines the shape of the final aerosol after drying the liquid.

In addition, naturally occurring materials, such as fungal spores, pollens, and bacteria or multiplets of spheres, have been used as test aerosols of nonspherical particles (Corn and Esmen, 1976; Adams et al., 1985). The aerosols of fungal spores, bacteria, and pollens are commonly generated using either the wet dispersion or the dry powder dispersion technique (described later). Depending on the needs when using these test particles of biological origin, the concerns of viability and culturability might need to be considered (Henningson and Ahlberg, 1994; Griffiths et al., 1996; Reponen et al., 1996; Ulevicius et al., 1997). The details of instrument calibration such as the generation, collection, and assay methods for microorganisms and other bioaerosols are described in Chapter 24.

Size Classification of Polydisperse Aerosols

An aerosol with a narrow size range can be produced from a polydisperse aerosol by passing the aerosol through a size classifier. For particles smaller than 0.1 µm, Liu and Pui (1974) developed a differential electrical mobility analyzer to classify aerosol particles of the same electrical mobility. Because most of the classified particles are singly charged, most of the aerosol produced is monodisperse, but there is a smaller amount of doubly charged particles with the same electrical mobility but different particle size. This classification technique has been used to produce a submicrometer aerosol for calibrating CNCs and diffusion batteries and for determining particle deposition in human nasal and oral casts (Liu et al., 1975; Scheibel and Porstendörfer, 1984; Cheng et al., 1990). For particles greater than 1 μm, a size-classifying technique based on particle inertia is generally used. Two virtual impactors can be placed in a series to segregate the desired fraction of the input aerosol for use in instrument calibration (Chen et al., 1988; Pilacinski et al., 1990). To classify aerosols in the 0.1 to 1.0 μm range, a technique involving both the mobility analyzer and a single-stage micro-orifice impactor has been used (Romay-Novas and Pui, 1988). The above techniques are also used for reducing undesired particles, such as PSL agglomerates from an air nebulizer or satellite particles from a spinning disk generator.

All the devices and techniques described above classify aerosol particles while the particles are airborne. Other instruments, such as elutriators, spectrometers, cascade impactors, and cascade cyclones, can be used to classify particles by collecting size-classified particles on a substrate that can then be resuspended. For example, a spiral centrifuge can collect aerodynamically classified particles on aluminum foil; resuspension of the particles caught on a narrow segment of the foil can be used to produce monodisperse aerosols (Kotrappa and Moss, 1971). The disadvantage of most size-classifying techniques is that only a small quantity of particles is produced.

Polydisperse Aerosols

Polydisperse aerosols can be used as test aerosols to calibrate instruments and samplers when used with an auxiliary sizing instrument such as the APS. Because the entire size distribution typically can be obtained in a minute, this method has significant advantages. Some polydisperse aerosols, such as aluminum oxide, coal dust, and Arizona road dust, are used in calibrating dust monitors, including samplers for respirable dust. There are two common ways to generate polydisperse aerosols: wet droplet dispersion and dry powder dispersion.

Wet Dispersion. The simplest way to disperse a droplet aerosol is by nebulization. Two types of nebulizers are commonly used to produce aerosols. Air-blast nebulizers (Mercer et al., 1968) use compressed air (15 to 50 psig; 1 psig = 6.87×10^4 dyne/cm^2) to draw bulk liquid from a reservoir as a result of Bernoulli effect (Fig. 21–4). The high-velocity air breaks up the liquid into droplets and then suspends the droplets to form an aerosol. Droplets produced from this method have a VMD of 1 to 10 μm and σ_g of 1.4 to 2.5 (Table 21–3). The particle size distribution can be modified by varying the pressure of the compressed air or the dilution ratio in the solution. A problem arises when the bulk liquid contains a volatile solvent that evaporates rapidly after droplet formation. The continuous loss of solvent increases the solute concentration in the reservoir and causes particle size to increase gradually with time. This problem can be circumvented by circulating the solution through a large reservoir (DeFord et al., 1981), delivering the solution at a constant rate (Liu and Lee, 1975), and presaturating the supply air and cooling the nebulizer.

In the ultrasonic nebulizer, the mechanical energy necessary to atomize a liquid comes from a piezoelectric crystal vibrating under the influence of an alternating electric field. The vibrations are transmitted through a coupling fluid to a cup containing the solution to be aerosolized. At a certain frequency (1.3–1.7 MHz), a heavy mist appears above the liquid surface of the cup. The diameter of the droplets making up the mist is related to the wavelength of the capillary waves, which decreases with increasing frequency of the ultrasonic vibrations. Normally the VMD is 5 to 10 μm, with a σ_g of 1.4 to 2.0 (Table 21–3).

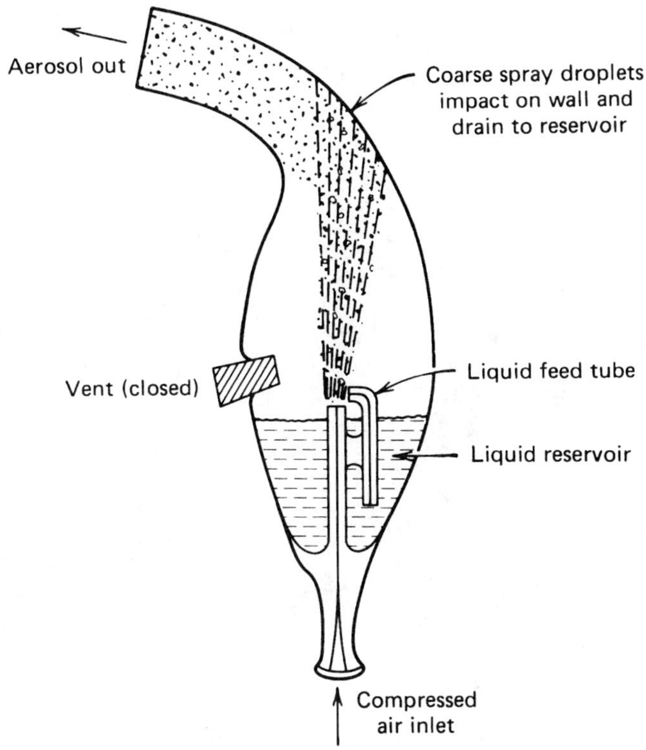

Fig. 21–4. Drawing of a DeVilbiss Model 40 Glass Nebulizer. (Reprinted from Hinds (1999) with the permission of John Wiley & Sons, Inc.).

Aerosol particles with chemical properties different from those of the liquid feed material can be produced through wet dispersion by using suitable gas phase reactions, such as polymerization or oxidation. Production of spherical particles of insoluble oxides and aluminosilicate particles with entrapped radionuclides has been described by Kanapilly et al. (1970) and Newton et al. (1980).

Dry Dispersion. The dry dispersion of powders can produce aerosols that have physical and chemical characteristics that are the same as or similar to those that will be sampled by the instrument under calibration. Numerous techniques have been described by Hinds (1980) for dispersing dust or fiber particles (Table 21–4). Basically, the techniques consist of two steps: (1) a means of delivering powder into the dispenser at a constant rate and (2) a means of dispersing the powder to form an aerosol. However, the dispersibility of a powder depends on the powder material, particle size, particle shape, and moisture content. Two common methods for generating aerosols from powders are the Wright Dust Feed (Fig. 21–5) and the fluidized bed (Fig. 21–6).

Aerosol generators using fluidized beds as the dispersing mechanism have the ability to thoroughly deagglomerate powdered samples. When equipped with a suitable dust feed mechanism, fluidized beds can operate stably over long periods of time. Fluidized beds can be scaled over a wide range of sizes, from very small to extremely large, producing aerosol concentrations from milligrams to tens of grams per cubic meter.

A fluidized bed consists of relatively large bed particles, typically on the order of 100 μm diameter, in a cylindrical container. The floor of the bed is of a porous material, such as a fine

TABLE 21–4. Operating Parameters of Dry Powder Dispersers

	Wright Dust Feed	Fluidized Bed	NBS II Dust Generator	Small Scale Powder Disperser	Jet-O-Mizer Model 00
Type of operation	Scraping the packed plug and dispersing it with air	Feeding the powder to the bed on a conveyor and air fluidizing it	Using metering gear to deliver the powder and air dispersing it	Using rotating plate to deliver the powder and dispersing it with Venturi suction	Using Venturi suction to feed the powder into a fluid energy mill in which centrifugal force and air velocity are used to break up the agglomerate and to disperse the powder
Air flow rate ($\times 10^{-5}$ m^3/s [L/min])	14–67 [8.5–40]	8–33 [5–20]	80–140 [50–85]	20–35 [12–21]	23–188 [14–113]
Feed flow rate (mm^3/min)	0.24–210	1.2–36	1,200–50,000	0.9–2.5	2,000–30,000
Output mass concentration, g/m^3 ($\rho = 1000$ kg/m^3)	0.012–11.5	0.13–4.0	15–100	0.0003–0.04	10–1,500
Recommended size range (μm)	0.2–100	0.5–100	1–100	1–50	0.2–30
Source	***BGI***	***TSI***	***BGI***	***TSI***	***FLU***

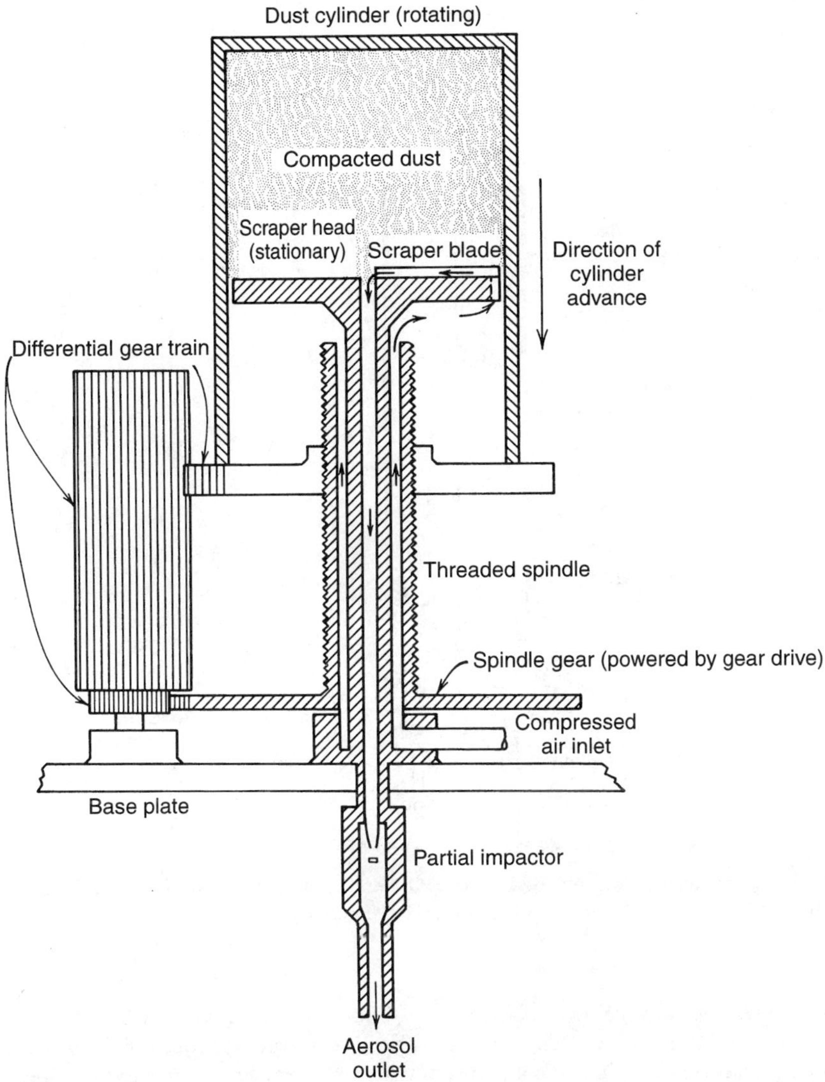

Fig. 21–5. Wright Dust Feed. (Reprinted from Hinds (1999) with the Permission of John Wiley & Sons, Inc.).

screen or a filter, which can retain the bed particles but allow an upward flow of air. In operation, a fluidized bed resembles a boiling liquid, but the best indicator of fluidization is the pressure drop across the bed. As the upward airflow is increased from zero, the pressure drop initially rises linearly with flow rate. Eventually, a condition is reached where the air drag on the particles is equal to the weight of the bed. The pressure drop curve then levels off. The flow velocity at the break in the curve is called the *minimum fluidization velocity* (MFV) (Carpenter and Yerkes, 1980). The MFV is correlated with the bed particle Reynolds number. For a typical fluidized-bed aerosol generator, the MFV is on the order of 10 cm/s.

In the two-component fluidized bed initially described by Guichard (1976), a relatively fine powder of the material to be aerosolized is added to a fluidized bed containing larger

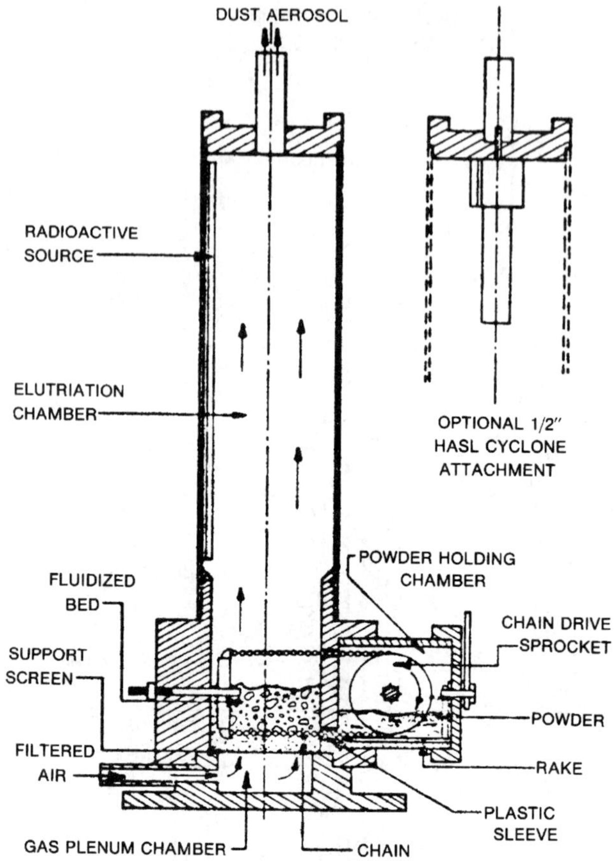

Fig. 21–6. Two-Component Fluidized-Bed Aerosol Generator. (Reprinted from Marple, Liu, and Rubow (1978) with the Permission of American Industrial Hygiene Assoc.).

bed particles. An airflow velocity sufficient to fluidize the bed will exceed the elutriation velocity of the powder particles. The details of the aerosolization process are not known, but a plausible scenario is that in the collisions between bed particles, adhering powder or dust particles are knocked off the bed particles. The constant action of the bed promotes uniform coating of the bed particles with a layer of dust particles, accounting for the deagglomeration that is observed. Fresh surfaces are constantly generated by the grinding action in the bed. Therefore, the use of dry air is recommended because moisture promotes oxidation of particle surfaces. Strong electrical charging is usually produced from contact and triboelectrification. This indicates that the aerosol should be neutralized before use. Some fluidized-bed generators incorporate strong sources of sound or vibration. Vibration can improve the performance in several ways. There is a tendency for channeling to occur in the bed, whereby the air flows at a higher rate in localized areas. Vibration can suppress the channeling, promoting a more uniform flow. Dust may collect on the walls above the bed. Periodically, the accumulation breaks loose, producing a burst in concentration. Vibration inhibits such an accumulation of dust. Vibration also improves the feeding of dust into the bed from a screw or chain.

An example of a two-component fluidized bed aerosol generator (Marple et al., 1978) is shown in Figure 21–6. The fluidized bed has a 1.4 cm thick layer of 100 μm brass beads (stainless steel is also frequently used) in a 5.1 cm diameter chamber. This type of generator is commercially available (Model 3400, ***TSI***). While such a generator is useful for many purposes, there are some nonideal aspects that require attention. When the generator is turned on with fresh metal bed particles, an aerosol is produced from the bed particles themselves. Initially, the source of particles is the fraction of small particles present in the bed particle sample. These small particles can be rapidly cleared from the bed by operating first at a higher than normal flow velocity. Even after the small particles are removed, a fine aerosol persists for a long period of time due to the grinding action that removes asperities from the bed particles. This type of background is not observed with glass beads, which are smoother and have less violent collisions because of their lower mass. After operation with a dust sample, a fluidized bed cannot be cleaned up effectively. If the same type of aerosol is needed again, the bed can be emptied and the material stored for future use.

The properties of some dusts may be altered in a generator having large metal bed particles. For example, aluminum particles can be flattened, and aluminum oxide particles can be broken up in the bed. Such alterations of particle properties could be significant for the subsequent use of the aerosol. The problem can be alleviated by using smaller bed particles or bed particles of lower density.

John and Wall (1983) developed a sonic fluidized bed, which avoids some of the problems of the large beds and which is useful for some applications. The bed's main feature is its small size, 25 mm diameter at the base, requiring less than 1 g of bed particles. The bed is funnel shaped so that the fluidizing velocity is higher than the exit velocity, favoring control of the elutriation velocity. For bed particles, 200 μm glass beads of the type used in gas chromatographic columns can be used. Such beads are highly uniform and clean. Because of the small amount of glass beads required, they can be discarded when dust samples are changed. A refinement is the addition of sonic energy to the bed. The sonic fluidized bed (Fig. 21–7) is vibrated by inexpensive piezoelectric crystals driven by an electronic oscillator at approximately 9 kHz. Because the bed lacks a feed system, it can only be used in a batch mode. It has been used successfully to generate aerosols of glass beads, A/C test dust, and soil for the testing of aerosol samplers. The soil was simply passed through a coarse screen and placed in the bed without bed particles, the coarsest soil particles functioning as bed particles.

A feed system for a fluidized bed generator was designed by Sussman et al. (1985) that allowed more constant and controllable output over time. The powder and bed beads were mixed, placed in a hopper, and pneumatically fed in small amounts at selected time increments into the fluidized bed. The overflow from the fluidized bed was allowed to fall into an overflow chamber to keep the bed height constant.

Spurny et al. (1975) developed a fluidized bed for the generation of aerosols of asbestos fibers. A special feature of this generator is a mechanical vibrator with adjustable amplitude and frequency. The effect of these vibration parameters on the aerosol concentration, fiber diameter, and fiber length was explored for several varieties of asbestos. It was found that the aerosol characteristics can be controlled to some extent by adjusting the vibration parameters. The generator was found capable of producing useful asbestos aerosols. In a similar approach, Weyel et al. (1984) used a low-frequency sonically fluidized bed to generate cotton fibers.

Besides the Wright Dust Feed and the fluidized bed, several dry powder generators are commercially available (Table 21–4). The ***TSI*** small-scale powder disperser is used to produce a small quantity of powder aerosols primarily for laboratory testing (B. T. Chen et al., 1995), and the Jet-O-Mizer is able to produce a large quantity of powder aerosols for inhalation studies (Cheng et al., 1985).

A simple technique for generating brief bursts of latex particles is to place a small quantity of the suspension on a glass slide or other clean surface, allow the suspension to dry, and gently brush the deposit off the surface toward the inlet of the instrument to be calibrated. In the approximate range 2 to 20 μm and larger, this approach is useful for size calibration

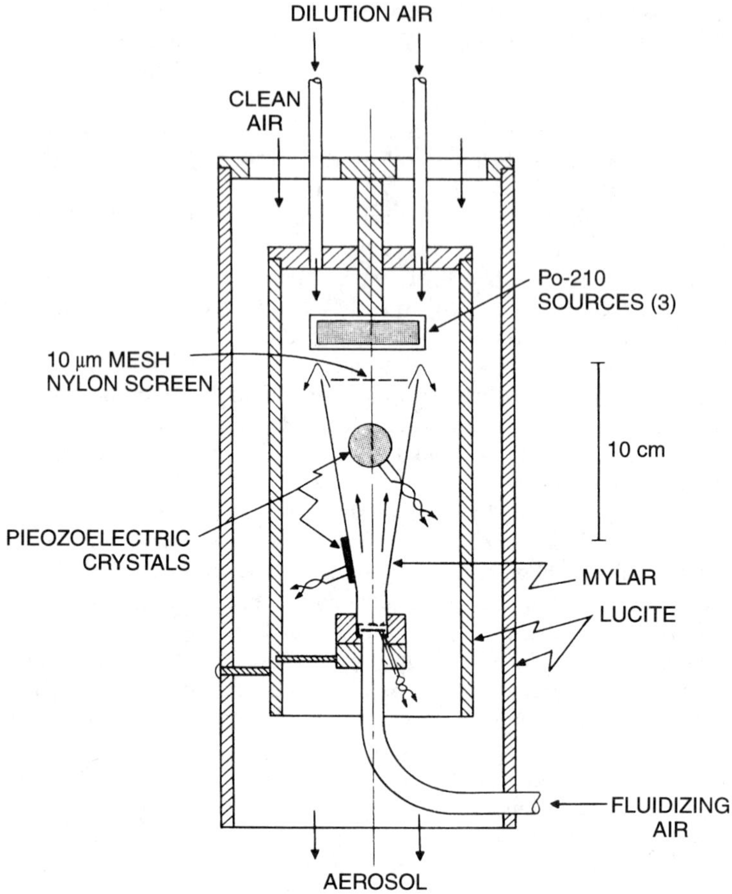

Fig. 21–7. Sonic Fluidized-Bed Aerosol Generator. (Reprinted from John and Wall (1983) with the Permission of Pergamon Press, Inc.)

of high-resolution instruments. When generated in this fashion, particles are more likely to be agglomerated, especially at the low end of the indicated size range.

Some generation techniques for fibrous aerosols are described in Chapter 23.

Test Aerosols with Tagging Materials

For some applications, particle detection is facilitated by incorporating fluorescent dye or radioisotope tags in the particles during their production. A fluorescent tagging material such as fluorescein can be analyzed in solution with nanogram sensitivity. Colored substances such as methylene blue can be analyzed with microgram sensitivity. The tagged aerosol may be extracted from a filter or a surface, enabling the quantitation of collection efficiency and wall losses within a sampler. Radiolabeling techniques have been used with the capability of detection of extremely low concentrations (Newton et al., 1980).

CALIBRATION OF FLOW, PRESSURE, AND VELOCITY

Accurate measurement of gas flow rate, pressure, and velocity is an integral part of instrument calibration (Mercer, 1973; Lippmann, 1995; Hinds, 1999). Various instruments

TABLE 21–5. Instruments that Measure Flow Rate, Volume, Pressure, and Velocity of Gases[a]

Quantity Measured	Instrument	Range	Standard
Volume	Spirometer	1 L to 1 m^3	Primary standard for flow rate calibration
	Bubble flowmeter	1 cm^3 to 10 L	
	Piston-displacement meter[b]	1 cm^3 to 12 L	
	Dry gas meter	Unlimited[c]	Secondary standard for flow rate calibration
	Wet gas meter	Unlimited[d]	
Volumetric flow rate	Venturi meter[e]	0.001–100 m^3/s	
	Orifice meter[e]	10^{-6}–100 m^3/s	
	Rotameter[e]	10^{-8}–0.05 m^3/s	
Pressure differential	Manometer	0–2 atm	Calibration standard
	Pressure gauge[e]	0–20 atm	
	Pressure transducer[e]	0–220 atm	
Velocity	Pitot tube	>5 m/s	Calibration standard
	Hot wire anemometer[e]	5 cm/s to 40 m/s	

[a] Commercial sources of these instruments can be found in Lippmann (1995).
[b] Mercury sealed.
[c] Range for flow rate calibration is 5 to 150 L/min.
[d] Range for flow rate calibration is 0.5 to 230 L/min.
[e] Frequent calibration against a standard is needed.

(Table 21–5) and techniques involved in the measurement of these parameters are discussed in this section.

Flow Rate Measurement

Depending on the flow rate and other practical considerations, various types of flowmeters can be used in an aerosol system to measure flow rate: variable (pressure) head meters such as orifice or venturi meters, variable area meters such as rotameters, bubble meters, and others (Table 21–5). Normally, the meters have a flow restriction that causes an increase in gas velocity and kinetic energy with a corresponding decrease in potential energy (i.e., static pressure). The flow rate can be calculated by knowing the pressure drop, the cross-sectional area upstream and at the constriction, the density of the gas, and the discharge coefficient. Taking into account the flow constriction and frictional effects, the discharge coefficient is the ratio of actual flow rate to ideal flow rate and depends on the design of the flow restriction.

The variable head meter determines the average flow rate by measuring the pressure differential across a calibrated resistance in the flow stream. The venturi meter has a streamlined constriction throat in the flow stream to minimize energy loss, and the discharge coefficient for this meter is slightly less than the ideal value of unity. A simpler form of a variable head meter is the orifice meter, in which a thin plate with a sharp-edged circular orifice is inserted at the center of the flow. Although a large energy loss takes place in the orifice meter, the meter is widely used because of its ease of installation and low cost. The discharge coefficient for an orifice meter depends on the orifice design and is generally much less than unity (~0.61; see Mercer, 1973). For constant flow control of filter sampling, a type of orifice meter called a *critical orifice meter* can be used downstream of the filter. The orifice is small enough to provide a downstream pressure less than 0.53 of the upstream pressure, under which conditions the velocity in the constriction reaches the speed of sound; a further reduction in the downstream pressure does not increase the velocity through the system. For a critical orifice, the flow rate Q (m^3/s [cm^3/s]) is proportional to the air pressure P_1 (Pa [dyne/cm^2]) and air density ρ_1 (kg/m^3 [g/cm^3]) at the upstream and is expressed as

EXAMPLE 21–3

A critical orifice with 4×10^{-3} m [0.4 mm] diameter is fabricated for air sampling purposes and used downstream of a filter. The flow rate is measured to be $1.67 \times 10^{-5}\,m^3/s$ [1 L/min] when the upstream pressure is close to the ambient (1.01×10^5 Pa [760 mm Hg], 20°C).

1. What size of orifice must be fabricated if the sampling flow rate is $3.33 \times 10^{-5}\,m^3/s$ [2 L/min] (assuming that the downstream pressure is still less than 0.53 of the upstream pressure)?
2. What is the sampling flow rate when this orifice is used in Albuquerque, NM (8.33×10^4 Pa [625 mm Hg])?
3. When the filter is loaded and the pressure gauge at the upstream of the orifice reads –980 Pa [–10 cm H_2O], what is the sampling flow rate?

Answer: Using Eq. 21–7: $Q \propto A_o(\rho_1 P_1)^{1/2}/\rho_a$:

1. $Q \propto A_o$. If the sampling flow rate is doubled, then the orifice diameter needs to be $(2)^{1/2}$ times the original size, that is, $(4 \times 10^{-3}) \times (2)^{1/2} = 5.66 \times 10^{-3}$ m [0.57 mm].
2. $Q \propto (\rho_1 P_1)^{1/2}/\rho_a$. Because $\rho_1 = \rho_a$ and $P_1 \propto \rho_1$, the sampling flow rate remains unchanged ($1.67 \times 10^{-5}\,m^3/s$ [1 L/min]).
3. $Q \propto (\rho_1 P_1)^{1/2} \propto P_1$ (because $\rho_1 \propto P_1$). The sampling flow rate is 1.67×10^{-5} $(1.01 \times 10^5 - 980)/1.01 \times 10^5 = 1.65 \times 10^{-5}\,m^3/s$ [0.99 L/min].

$$Q = 0.58 k \gamma^{1/2} (A_o/\rho_a)(\rho_1 P_1)^{1/2} \tag{21–7}$$

where k is the discharge coefficient, γ is the ratio of specific heats (=1.4 for air), A_o (m^2 [cm^2]) is the area of the orifice, and ρ_a (kg/m^3 [g/cm^3]) is the ambient air density. It is therefore essential during sampling practice not to overload the filter to cause significant reduction in pressure P_1 unless the pressure is monitored and appropriate correction of the flow rate can be made afterward. Kotrappa et al. (1977) reported a mean discharge coefficient of 0.70 when using 2 mm long hypodermic needles as critical orifices.

Different from the variable head meter, the variable area meter changes the orifice area with flow to maintain a constant pressure drop. The most common type of the variable area meter is the rotameter (Fig. 21–8a). It consists of a "float" that moves up and down within a vertical tapered tube that is larger at the top than the bottom. The gas flows upward, causing the float to rise until the pressure drop across the annular area between the float and the tube wall is just sufficient to support the float. The height of the float indicates the flow rate. For a rotameter calibrated at an ambient pressure, $P_{a,c}$, and used at a different ambient pressure, $P_{a,i}$, the true volumetric flow rate Q_i at a fixed float position is given by

$$Q_i = Q_c (P_{a,c}/P_{a,i})^{1/2} \tag{21–8}$$

where Q_c is the flow rate indicated by the rotameter during calibration (Mercer, 1973). For a rotameter calibrated and used at the same ambient pressure, the measured flow rate depends on the gas density ρ_g (pressure, P_g) in the tube. For example, if the rotameter is operated at a different gas density in the tube than that used during calibration (e.g., when the rotameter is located downstream of a filter or an impactor during aerosol sampling), then the actual volumetric flow rate at a fixed float position is given by

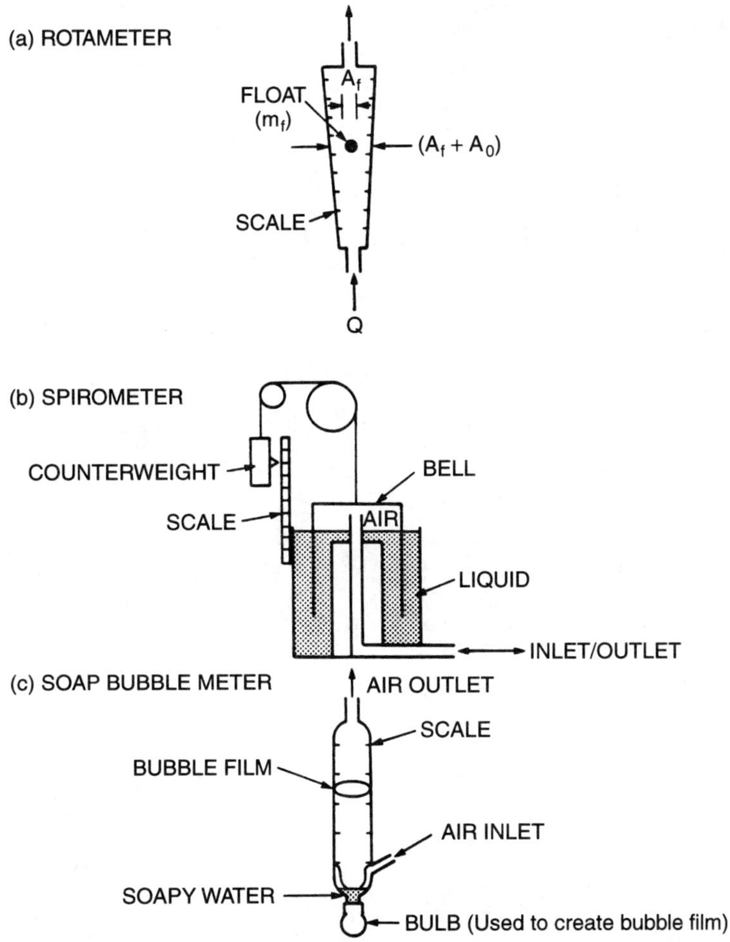

Fig. 21–8. Schematic diagrams of flow-measuring instruments: (a) a rotameter, (b) a spirometer, and (c) a soap bubble meter.

$$Q_i = Q_c(\rho_{g,i}/\rho_{g,c})^{1/2} = Q_c(P_{g,i}/P_{g,c})^{1/2} \tag{21–9}$$

where the subscripts i and c refer to the actual condition in the rotameter and the condition during calibration, respectively. Normally, $P_{g,c}$ is the ambient pressure, P_a, and $P_{g,i}$ is $(P_a - \Delta P)$, where ΔP is the gauge pressure downstream of the sampler.

A rotameter is generally calibrated using a calibrated dry gas meter, bubble flowmeter, or spirometer. A spirometer (Fig. 21–8b) measures the volume of oil displaced in a container when the air under measurement is introduced; the volume can be as high as $1\,m^3$. For a smaller volume, a soap bubble meter (Fig. 21–8c) is a widely used primary standard. A bubble film is created in a tube from a reservoir of soapy water and acts as a nearly frictionless piston as the air passes through the tube. The distances of the bubble displaced along the tube and the cross-sectional area of the tube are used to determine the volume of the air entering the bubble meter. Several automated bubble meters that incorporate bubble detecting sensors, automatic timing, and readout of flow rate are commercially available (***BUC***, ***GIL***) and widely used as calibration standards for flow rates up to 40 L/min.

EXAMPLE 21–4

A rotameter is calibrated by the manufacturer at sea level (1.01×10^5 Pa [760 mm Hg]) and used in Albuquerque, NM (8.33×10^4 Pa [625 mm Hg]). What percentage of error in flow rate will result if the rotameter is not recalibrated? Assuming that this rotameter is recalibrated and then used downstream of a filter to measure the sampling flow rate, what will be the true flow rate if the flow rate indicated by the recalibrated rotameter is 8.33×10^{-5} m^3/s [5 L/min] and the pressure drop is 2.49×10^3 Pa [10 in H_2O]?

Answer: Using Eq. 21–8:

$$Q_i = Q_c (P_{a,c}/P_{a,i})^{1/2}$$

$$P_{a,c} = 1.01 \times 10^5 \text{ Pa}, \; P_{a,i} = 8.33 \times 10^4 \text{ Pa}$$

$$Q_c/Q_i = (8.33/10.1)^{1/2} = 0.907$$

$$\% \text{ error} = [(Q_i - Q_c)/Q_i]100 = (1 - Q_c/Q_i)100 = 9.3\%$$

Using Eq. 21–9:

$$Q_i = Q_c (P_{g,i}/P_{g,c})^{1/2}$$

$$P_{g,i} = P_a - \Delta P = 8.33 \times 10^4 - 2.49 \times 10^3 = 8.08 \times 10^4 \text{ Pa} \, [606 \text{mm Hg}]$$

$$P_{g,c} = P_a = 8.33 \times 10^4 \text{ Pa} \, [625 \text{mm Hg}]$$

$$Q_c = 8.33 \times 10^{-5} \text{ m}^3/\text{s}$$

and the true flow rate

$$Q_i = 8.33 \times 10^{-5} (8.08/8.33)^{1/2} = 8.20 \times 10^{-5} \text{ m}^3/\text{s} \; [4.92 \text{L/min}]$$

In addition, a flow calibrator using a solid graphite piston (***BII***), instead of the manually created soap films, has become popular as a flow calibration device because it eliminates some problems associated with consistent bubble generation. However, one should be aware that air leakage or unnecessary friction could occur between the solid piston and the internal wall surface, and regular checkup is strongly recommended. Because of its valving system, this type of calibrator can also cause errors of a few percent when measuring flow from mass flow controllers.

A dry gas meter is also often used to calibrate rotameters and orifice meters, although it must be calibrated against a primary standard, such as spirometer. The dry gas meter contains two bellows that are alternately filled and emptied by the metered gas. Movement of the bellows controls the action of mechanical valves that direct flow and operates a cycle-counting device that registers the total volume of gas passing through the meter. In operation, one inlet of the meter is always open to ambient pressure because the housing of the device cannot support a very large pressure drop. In addition, at least 10 revolutions are recommended for each measurement to avoid the effect of nonlinear strokes. In a wet test meter, gas flows into a rotating system of chambers that connects to a revolution counter. The water level seals the chambers and acts as a valve to direct the flow to the proper chamber. The measured volume from this meter must be corrected for water vapor content. There is, however, no limit on the total gas volume measured.

Pressure Measurement

Three types of pressure-sensing devices are commonly used to determine the pressure differential between two points in a system or the gauge pressure: manometers, mechanical gauges, and pressure transducers (Table 21–5). The liquid-filled manometer (Fig. 21–9a) consists of a glass or plastic tube sized to allow the height of the fluid level to balance against the incoming pressure. The manometer expresses the pressure differential by the difference in liquid column height. There are three types of manometers: U tube, well type, and inclined type. Normally, a manometer does not require any calibration and can be used as a pressure standard provided that the specific gravity of the liquid is known.

The mechanical pressure gauge (Fig. 21–9b) is widely used as a pressure sensor of an aerosol system in both the laboratory and the field. It normally consists of a metal or plastic housing that contains a diaphragm assembly. The diaphragm movement due to the pressure differential is transferred to the dial indicator mechanism. The pressure gauge provides an accurate reading as a percentage of the full-scale range of the device. The most common pressure gauge is the Magnehelic (***DWY***), with a range from 0.01 inch of water to 100 psig. In this device, the diaphragm transmits the effect of pressure to an indicator by means of magnetic linkage without direct physical contact to ensure the accuracy and sensitivity of the instrument.

Pressure transducers are available to convert pressure to an electrical voltage. A transducer can be used to determine the pressure differential across an orifice meter and provide a real-time digital reading of pressure.

Velocity Measurement

Measurement of the local gas velocity in a duct is needed for proper isokinetic sampling and for calibrating flow-measuring devices. The most widely used device for measuring velocity is the pitot tube that directly measures the velocity pressure in a moving gas flow. The pitot tube is normally considered the calibration standard for gas velocity measurement.

A hot wire anemometer measures gas velocity by sensing the convective heat loss on a hot wire when the gas flows across it. The wire is heated electrically; heat loss to the air changes the wire's temperature, and the resulting change in resistance is sensed electronically and converted to velocity. Note that this device measures the mass flow rate of the gas. Temperature and pressure are needed to obtain the velocity, and periodic calibration is needed to provide laboratory-quality accuracy (Chen, 1993).

The velocity of particles in the air can be measured using a laser Doppler anemometer (or a velocimeter [LDV] can be used) (Fig. 21–9c). This allows either measurement of air velocity using small particles (on the order of 1 μm) or the velocity of larger particles that move at different velocities due to inertial or gravitational effects. This device does not require insertion of a sensing probe. It uses two laser beams to form an interference pattern with fringes. As a particle travels through the fringe pattern, its scattered light intensity provides a shift in the detected frequency, a phenomenon known as the *Doppler effect* (Chen, 1993). Another device, the particle imaging velocimeter (PIV), uses two laser-illuminated images of particles separated by a measured time interval to measure velocity. The velocity can be measured in near-real time over an area defined by the intersection of the laser

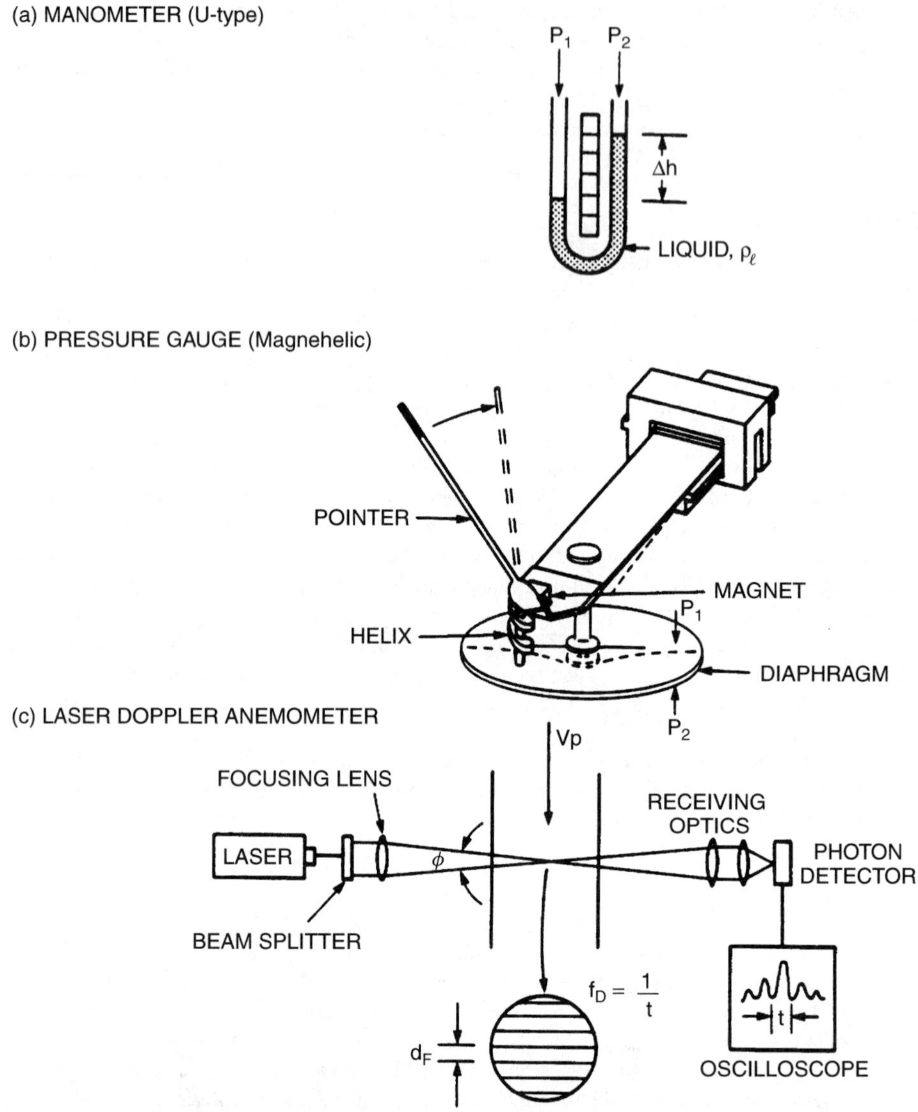

Fig. 21–9. Schematic diagrams of instruments: (a) a manometer, (b) a pressure gauge [for gas pressure measurement], and (c) a laser Doppler anemometer [for particle velocity measurement].

illumination plane and the detector viewing area. The LDV measures velocity at a point, while the PIV measures velocity over a selected area. Several companies produce these instruments (***DAN***, ***TSI***, ***OXF***).

INSTRUMENT CALIBRATION

In this section, several aerosol instruments that determine particle size, number concentration, and mass concentration are briefly described. Their calibration standards and important parameters are summarized in Table 21–6.

Particle Sizing Instruments

Most instruments used in particle size analysis actually measure some physical property of the particle rather than a simple linear dimension. Particle size is then related to the diameter of a spherical reference particle that possesses the same physical property in the same amount. This can be an aerodynamic, optical, electrical, or diffusional property of the particle (Chen et al., 1989).

Aerodynamic sizing instruments, including collection-and-analysis devices (cascade impactors, elutriators, aerosol centrifuges, and cascade cyclones) and real-time analyzers are generally designed to measure particle sizes between 0.2 and 25 µm. The collection-and-analysis devices measure masses of particles that have been separated according to their aerodynamic properties in different force fields (see Chapter 10). Real-time analyzers measure particle velocity as they pass through a sensing zone that can be a Doppler interference fringe pattern or a two-laser beam arrangement (see Chapters 16 and 17). Parameters such as particle size, flow rate (velocity), density, and intrinsic gas properties can affect collection efficiency or instrument response (Stöber, 1976; Marple and Willeke, 1976; Chen et al., 1985; Baron, 1986; Hering, 1995). In addition, loading capacities and wall losses in each instrument should be fully examined to avoid incorrect data interpretation.

The optical particle counter (OPC) is a real-time instrument that uses a single-particle light-scattering technique to measure aerosol size distribution (0.1 to 20 µm) and number concentration (see Chapter 15). Important parameters affecting the response include the size, shape, orientation, and refractive index of the aerosol particles, as well as the wavelength of the light source, the range of scattering angles, and the sensitivity of the photodetector. Particle size and refractive index are the two most important variables in OPC calibrations. Mie scattering equations are used to predict theoretically the response of an OPC; however, an OPC is usually calibrated with test aerosol (Hodkinson, 1966; Willeke and Liu, 1976; Gebhart et al., 1976; Chen et al., 1989). The particle concentration in an OPC must be limited to minimize the error due to multiple particles in the sensing zone; the coincidence error can be calculated from Poisson statistics.

Electrical mobility analyzers (see Chapter 18) and diffusion batteries (see Chapter 19) are sizing instruments based on the electrical and diffusional properties of submicrometer particles (<0.5 µm). Important parameters for the electrical mobility analyzer are the geometric diameter and dielectric constant of the particle and the flow rate and charging mechanism in the instrument (Liu and Pui, 1974; Pui and Liu, 1979; Yeh et al., 1983). The flow rate, temperature, particle size, as well as the geometric dimensions of the diffusion surfaces (e.g., screen wire diameter and tube length) are important for diffusion batteries (Cheng, 1995).

Condensation Nuclei Counters

Particle number concentration can be determined by sampling particles through a high-efficiency membrane filter counting the particles with an optical or electron microscope. This is, however, time consuming. For submicrometer aerosol particles, the number concentration is generally determined by using condensation nuclei counters (CNCs; see Chapter 19). In a CNC, supersaturation conditions are used to initiate water or alcohol vapor condensation on particle surfaces. Droplets grow to micrometer diameters regardless of their initial size and are then detected by microscopic, photographic, or photoelectric methods. The Pollak counter and photographic-type counters have been the standards against which CNCs are normally calibrated (Liu et al., 1975; Jaenicke and Kanter, 1976; Winters et al., 1977; Podzimek et al., 1982; Sinclair, 1984). Submicrometer aerosols produced from a differential mobility analyzer have also been used as a standard (Liu and Pui, 1974). Parameters important in CNC calibration include particle size, number concentration, hydrophilic and hydrophobic properties

TABLE 21–6. Summary of Direct Measurements and Primary Standards of Aerosol Instruments and

Instrument	Operating Principle	Measured Quantity	Important Aerosol Parameters	Important Instrument Parameters
Size Measurement				
Cascade Impactor	Particle inertial impaction	Mass	Size, shape, density	Flow rate, gas medium, physical dimensions
Aerodynamic Particle Sizing Instrument	Time of flight during deceleration	Velocity	Size, shape, density, rigidity	Flow rate, pressure, gas medium
Optical Particle Counter	Interaction between particle and light	Scattered light intensity from Single particles	Size, shape, refractive index	Wavelength of the light source, range of scattering angles, sensitivity of detector
Electrical Mobility Analyzer	Size classification based on electric mobility	Electric charge or particle count	Size, shape, dielectric constant, humidity	Flow rate, charging mechanism, electric field strength
Diffusion Battery	Particle diffusional deposition	Particle count or mass	Size, shape, number concentration	Flow rate, temperature, deposition surface
Number Concentration Measurement				
Condensation Nuclei Counter	Vapor condenses on particles and makes them detectable	Particle count	Size, number concentration, hygroscopicity	Flow rate, saturation ratio, temperature gradient
Mass Concentration Measurement				
Photometer	Interaction between particle and light	Total light scattering from all particles in sensing volume	Size, shape, refractive index, density	Wavelength of the light source, range of scattering angles, sensitivity of detector
β-attenuation Monitor	Mass dependent absorption of β-radiation	Mass	Size, elemental composition	Uniformity of particle deposit
Quartz Crystal Mass Balance	Mass dependent resonant frequency of the crystal	Mass	Size	Sensitivity of the sensor, mass loading

[a] For particle size smaller than 0.01 Φm, electrically classified monodisperse aerosols are used as a calibration standard.

Important Parameters to be Considered in Instrument Calibration

Particle Size Range (μm)	Direct Calibration Standard	Primary Calibration Standard	Main Advantage	Main Disadvantage
0.05–30	Monodisperse spherical particles with a known size and density	—	Aerodynamic size distribution based on mass concentration	Internal loss, particle bounce and re-entrainment
0.5–20	Monodisperse spherical particles with a known size, shape, and density	—	Real-time instrument with good sensitivity and resolution	Coincidence, density, and shape effects
0.3–15	Monodisperse, spherical particles with a known size and refractive index	—	Noninvasive, real-time, *in situ* measurement; also good for number concentration measurement	Calibration changes with the material
0.001–0.1	Monodisperse,[a] spherical particles with a known size and dielectric constant	—	Suitable for particles smaller than 0.1 μm	Multiple charges on the particle
0.001–0.1	Monodisperse,[a] spherical particles with a known size	—	Suitable for particles smaller than 0.1 μm	Unsuitable for large particles or particles with a large aspect ratio
0.001–0.5	Aitken counter with a microscopic measurement	Pollak counter, photographic counter, or electrically classified monodisperse aerosol	Suitable for concentration measurement of submicrometer particles	Size dependent counting efficiency
0.3–1.5	Gravimetric measurement of filter samples	—	Real-time, continuous readout	Calibration changes with material type
1–15	Gravimetric (equivalent) measurement of filter samples	—	Real-time measurement	Low sensitivity
0.02–10	Gravimetric measurement of filter samples	—	Real-time measurement	Frequent sensor cleaning

of the aerosol, vapor saturation ratio, temperature gradient, and flow rate. Cross calibrations among different instruments and techniques have indicated that the results generally agree among the different instruments and techniques involved (Liu et al., 1982a).

Mass Concentration Monitors

The most common way to determine aerosol mass concentration is to determine the mass collected on a filter and the gas volume sampled (see Chapter 9). This direct gravimetric approach is best achieved with a filter of high collection efficiency, for example, a glass fiber filter or a membrane filter (Liu et al., 1983). Several real-time monitors have been developed to determine aerosol mass concentration: β-attenuation mass monitors, the tapered element oscillating microbalance (TEOM; described in Chapter 14), and photometers (see Chapters 15 and 26). These monitors can provide the total mass concentration or only a desired mass fraction such as the respirable, thoracic, PM-2.5 or PM-10 mass concentrations by preceding them with a suitable particle size-selective device.

The β-attenuation monitor determines aerosol concentration on a filter or other substrate by measuring the attenuation of β particles from a radioactive source. In the TEOM, particles are collected on a filter supported by an oscillating element. Increase in mass causes a shift in frequency. Besides these three collection-and-analysis instruments, real-time photometers have been used for determining mass concentration of aerosols. In a photometer, integral light scattering (or extinction) signals due to many particles in a sensing volume are related to mass concentrations, and instrument calibration is always required. Several calibrations and comparisons among the above instruments have been done by Kuusisto (1983), Marple and Rubow (1984), Smith et al. (1987), and Baron (1988). Filter samples taken in parallel to real-time mass monitors can be used to calibrate the integrated mass response; however, it is difficult to calibrate real-time monitors over the short time intervals for which they are designed.

SUMMARY OF CALIBRATION PROCEDURES

Because the accuracy of aerosol data depends directly on the measurements, care must be taken to properly calibrate aerosol instruments and samplers. The following comments summarize calibration procedures and recommendations:

1. Aerosol instruments and samplers should be checked periodically to ensure that they are in proper operating condition. Before use, the calibration should be checked.
2. A device should be calibrated after it has been changed or repaired by the manufacturer, mishandled or damaged, or at any time when there is a question as to its accuracy.
3. A newly acquired instrument should be calibrated. Data supplied by the manufacturer may not be directly applicable to the user if the ambient pressure (altitude), temperature, and wind velocity in the sampling environment are very different from those used during the manufacturer's calibration.
4. Before calibration of a new instrument, the operating principles and the construction should be studied. The instrument's manual should be read, and the manufacturer's recommended operating procedures should be noted. It should be verified that the instrument is in proper operating condition.
5. Test aerosols that have physical and chemical properties similar to the aerosol to be measured should be selected for instrument calibration. The test aerosol should be monitored during the calibration to ensure its consistency in particle size and concentration.

6. Care must be taken in designing the setup to ensure that both the calibration device and the instrument to be calibrated will receive comparable aerosol samples. If one of the instruments needs dilution or augmentation of its flow, for example, by addition of particle-free air, this must be accomplished without alteration of the particle size distribution. This point concerns one of the most difficult aspects of instrument calibration.
7. Sufficient time should be allowed for the instruments to warm up, flow equilibrium to be established, and conditions to stabilize.
8. Flow rates should be checked before and after calibration runs.
9. Enough data should be obtained to give confidence in the calibration curve for a given parameter. Each calibration point should be made up of a sufficient number of readings to ensure statistical confidence in the measurement. The entire calibration should be repeated in separate runs.
10. A permanent record of all procedures, data, and results should be maintained.
11. Calibration curves and factors should be properly identified, including the date and conditions of calibration, the instrument involved, and who performed the procedure. It is useful to attach a tag to the instrument indicating where the original data are filed.

REFERENCES

Adams, A. J., D. E. Wennerstrom, and M. K. Mazumder. 1985. Use of bacteria as model nonspherical aerosol particles. *J. Aerosol Sci.* 16:193–200.

Aitken, R. J., P. E. J. Baldwin, G. C. Beaumont, L. C. Kenny, and A. D. Maynard. 1999. Aerosol inhalability in low air movement environments. *J. Aerosol Sci.* 30:613–626.

Baron, P. A. 1986. Calibration and use of the aerodynamic particle sizer (APS 3300). *Aerosol Sci. Technol.* 5:55–67.

Baron, P. A. 1988. Modern real-time aerosol samplers. *Appl. Ind. Hyg.* 3:97–103.

Baron, P. A., G. J. Deye, and J. Fernback. 1994. Length separation of fibers. *Aerosol Sci. Technol.* 21:179–192.

Barr, E. B., M. D. Hoover, G. M. Kanapilly, H. C. Yeh, and S. J. Rothenberg. 1983. Aerosol concentrator: Design, construction, calibration, and use. *Aerosol Sci. Technol.* 2:437–442.

Berglund, R. N. and B. Y. H. Liu. 1973. Generation of monodisperse aerosol standards. *Environ. Sci. Technol.* 7:147–153.

Carpenter, R. L. and K. Yerkes. 1980. Relationship between fluid bed aerosol generator operation and the aerosol produced. *Am. Ind. Hyg. Assoc. J.* 41:888–894.

Chen, B. T. 1993. Instrument calibration. In *Aerosol Management: Principles, Techniques and Applications,* eds. K. Willeke and P. A. Baron. New York: Van Nostrand Reinhold, pp. 493–520.

Chen, B. T., Y. S. Cheng, and H. C. Yeh. 1984. Experimental responses of two optical particle counters. *J. Aerosol Sci.* 15:457–464.

Chen, B. T., Y. S. Cheng, and H. C. Yeh. 1985. Performance of a TSI aerodynamic particle sizer. *Aerosol Sci. Technol.* 4:89–97.

Chen, B. T., Y. S. Cheng, and H. C. Yeh. 1990. A study of density effect and droplet deformation in the TSI aerodynamic particle sizer. *Aerosol Sci. Technol.* 12:278–285.

Chen, B. T., M. D. Hoover, G. J. Newton, S. J. Montano, and D. S. Gregory. 1999. Performance evaluation of the sampling head and annular kinetic impactor in the Savannah River Site alpha continuous air monitor. *Aerosol Sci. Technol.* 31:24–38.

Chen, B. T., H. C. Yeh, Y. S. Cheng, and G. J. Newton. 1989. Particle size analyzer for air quality studies. In *Encyclopedia of Environmental Control Technology*, Vol. II, ed. P. N. Cheremisinoff. Houston: Gulf Publ., pp. 453–514.

Chen, B. T., H. C. Yeh, and B. J. Fan. 1995. Evaluation of the TSI small-scale powder disperser. *J. Aerosol Sci.* 26:1303–1313.

Chen, B. T., H. C. Yeh, and N. F. Johnson. 1996. Design and use of a virtual impactor and an electrical classifier for generation of test fiber aerosols with narrow size. *J. Aerosol Sci.* 27:83–94.

Chen, B. T., H. C. Yeh, and M. A. Rivero. 1988. Use of two virtual impactors in series as an aerosol generator. *J. Aerosol Sci.* 19:137–146.

Chen, D. R., D. Y. H. Pui, and S. L. Kaufman. 1995. Electrospraying of conducting liquids for monodisperse aerosol generation in the 4 nm to 1.8 μm diameter range. *J. Aerosol Sci.* 26:963–977.

Cheng, Y. S. 1995. Denuder systems and diffusion batteries. In *Air Sampling Instruments*, eds. B. S. Cohen and S. V. Hering. Cincinnati: ACGIH, Inc., pp. 511–525.

Cheng, Y. S. and B. T. Chen. 1995. Aerosol sampler calibration. In *Air Sampling Instruments.* eds. B. S. Cohen and S. V. Hering. Cincinnati: ACGIH, Inc., pp. 165–186.

Cheng, Y. S., B. T. Chen, and H. C. Yeh. 1986. Size measurement of liquid aerosols. *J. Aerosol Sci.* 17:803–809.

Cheng, Y. S., T. C. Marshall, R. F. Henderson, and G. J. Newton. 1985. Use of a jet mill for dispensing dry powder for inhalation studies. *Am. Ind. Hyg. Assoc. J.* 46:449–454.

Cheng, Y. S., Y. Yamada, H. C. Yeh, and D. L. Swift. 1990. Deposition of ultrafine aerosols in a human oral cast. *Aerosol Sci. Technol.* 12:1075–1081.

Cheng, Y. S. and H. C. Yeh. 1979. Particle bounce in cascade impactors. *Environ. Sci. Technol.* 13: 1392–1396.

Cloupeau, M. and B. Prunet-Foch. 1989. Electrostatic spraying of liquid in cone-jet mode. *J. Electrostatics* 22:135–159.

Cloupeau, M. and B. Prunet-Foch. 1994. Electrohydrodynamic spraying functioning modes: A critical review. *J. Aerosol Sci.* 25:1021–1036.

Corn, M. and N. A. Esmen. 1976. Aerosol generation. In *Handbook on Aerosols*, ed. R. Dennis. Springfield, VA: National Technical Information Service, U. S. Dept. of Commerce, Publ. TID-26608, pp. 9–39.

DeFord, H. S., M. L. Clark, and O. R. Moss. 1981. A stabilized aerosol generator. *Am. Ind. Hyg. Assoc. J.* 42:602–604.

Deye, G. J., P. Gao, P. A. Baron, and J. Fernback. 1999. Performance evaluation of a fiber length classifier. *Aerosol Sci. Technol.* 30:420–437.

Esmen, N. A., R. A. Kahn, D. LaPietra, and E. D. McGovern. 1980. Generation of monodisperse fibrous glass aerosols. *Am. Ind. Hyg. Assoc. J.* 41:175–179.

Federal Register. 1987. Ambient air monitoring reference and equivalent methods: 40 CFR Part 53. *Federal Register* 52:24724.

Fernandez de la Mora, J., J. Navascues, F. Fernandez, and J. Rosell-Llompart. 1990. Generation of submicron monodisperse aerosols in electrosprays. *J. Aerosol Sci.* 21:S673–S674.

Fuchs, N. A. and A. G. Sutugin. 1966. Generation and use of monodisperse aerosols. In *Aerosol Science*, ed. C. N. Davies. New York: Academic Press, pp. 1–30.

Fulwyler, M. J., R. B. Glascock, and R. D. Hiebert. 1969. Device which separates minute particles according to electronically sensed volume. *Rev. Sci. Instrum.* 40:42–48.

Gebhart, J., J. Heyder, C. Roth, and W. Stahlhofen. 1976. Optical aerosol size spectrometry below and above the wavelength of light—a comparison. In *Fine Particles: Aerosol Generation, Measurement, Sampling, and Analysis*, ed. B. Y. H. Liu. New York: Academic Press, pp. 793–815.

Grace, J. M. and J. C. M. Marijnissen. 1994. A review of liquid atomization by electrical means. *J. Aerosol Sci.* 25:1005–1019.

Griffiths, W. D., I. W. Stewart, A. R. Reading, and S. J. Futter. 1996. Effect of aerosolization, growth phase and residence time in spray and collection fluids on the culturability of cells and spores. *J. Aerosol Sci.* 27:803–820.

Guichard, J. C. 1976. Aerosol generation using fluidized beds. In *Fine Particles*, ed. B. Y. H. Liu. New York: Academic Press, pp. 173–193.

Hayati, I., A. Bailey, and T. F. Tadros. 1987a. Investigations into the mechanism of electrohydrodynamic spraying of liquids, I. *J. Colloid Interface Sci.* 117:205.

Hayati, I., A. Bailey, and T. F. Tadros. 1987b. Investigations into the mechanism of electrohydrodynamic spraying of liquids, II. *J. Colloid Interface Sci.* 117:222.

Henningson, E. W. and M. S. Ahlberg. 1994. Evaluation of microbiological aerosol samplers: A review. *J. Aerosol Sci.* 25:1459–1492.

Hering, S. 1995. Impactors, cyclones, and other inertial and gravitational collectors. In *Air Sampling Instruments*, eds. B. S. Cohen and S. V. Hering. Cincinnati: ACGIH, Inc., pp. 279–321.

Hinds, W. 1980. Dry dispersion aerosol generator. In *Generation of Aerosols and Facilities for Exposure Experiments*, ed. K. Willeke. Ann Arbor, MI: Ann Arbor Science Publ., pp. 171–188.

Hinds, W. 1999. *Aerosol Technology*. New York: John Wiley & Sons.

Hodkinson, J. R. 1966. The optical measurement of aerosols. In *Aerosol Science*, ed. C. N. Davies. New York: Academic Press, pp. 287–357.

Hoover, M. D., S. A. Casalnuovo, P. J. Lipowicz, H. C. Yeh, R. W. Hanson, and A. J. Hurd. 1990. A method for producing non-spherical monodisperse particles using integrated circuit fabrication techniques. *J. Aerosol Sci.* 21:569–575.

Jaenicke, R. and H. J. Kanter. 1976. Direct condensation nuclei counter with automatic photographic recording, and general problems of absolute counters. *J. Appl. Meteorol.* 15:620–632.

John, W. 1980. Particle charge effects. In *Generation of Aerosols and Facilities for Exposure Experiments*, ed. K. Willeke. Ann Arbor, MI: Ann Arbor Science Publ., pp. 141–151.

John, W. and N. Kreisberg. 1999. Calibration and testing of samplers with dry, polydisperse latex. *Aerosol Sci. Technol.* 31:221–225.

John, W. and S. M. Wall. 1983. Aerosol testing techniques for size-selective samplers. *J. Aerosol Sci.* 14:713–727.

Kanapilly, G. M., O. G. Raabe, and G. J. Newton. 1970. A new method for the generation of aerosols of insoluble particles. *J. Aerosol Sci.* 1:313–323.

Kenny, L. C., R. J. Aitken, P. E. J. Baldwin, G. C. Beaumont, and A. D. Maynard. 1999. The sampling efficiency of personal inhalable aerosol samplers in low air movement environments. *J. Aerosol Sci.* 30:627–638.

Kenny, L. C., R. J. Aitken, C. Chalmers, J. F. Fabries, E. Gonzalez-Fernandez, H. Kromhout, G. Liden, D. Mark, G. Riediger, and V. Prodi. 1997. A collaborative European study of personal inhalable aerosol sampler performance. *Ann. Occup. Hyg.* 41:135–153.

Kenoyer, J. and D. Leong. 1995. Performance testing criteria for air sampling instrumentation. In *Air Sampling Instruments*, eds. B. S. Cohen and S. V. Hering. Cincinnati: ACGIH, Inc., pp. 195–201.

Kotrappa, P. and O. R. Moss. 1971. Production of relatively monodisperse aerosols for inhalation experiments by aerosol centrifugation. *Health Phys.* 21:531–535.

Kotrappa, P., N. S. Pimpale, P. S. S. Subrahmanyam, and P. P. Joshi. 1977. Evaluation of critical orifices made from sections of hypodermic needles. *Ann. Occup. Hyg.* 20:189–194.

Kuusisto, P. 1983. Evaluation of the direct reading instruments for the measurement of aerosols. *Am. Ind. Hyg. Assoc. J.* 44:863–874.

Lippmann, M. 1995. Airflow calibration. In *Air Sampling Instruments*, eds. B. S. Cohen and S. V. Hering. Cincinnati: ACGIH, Inc., pp. 139–150.

Lippmann, M. and R. E. Albert. 1967. A compact electric-motor driven spinning disc aerosol generator. *Am. Ind. Hyg. Assoc. J.* 28:501–506.

Liu, B. Y. H. 1974. Laboratory generation of particulates with emphasis on submicron aerosols. *JAPCA* 24:1170–1172.

Liu, B. Y. H. and K. W. Lee. 1975. An aerosol generator of high stability. *Am. Ind. Hyg. Assoc. J.* 36:861–865.

Liu, B. Y. H. and D. Y. H. Pui. 1974. A submicron aerosol standard and the primary, absolute calibration of the condensation nuclei counter. *J. Colloid Interface Sci.* 47:155–171.

Liu, B. Y. H., D. Y. H. Pui, A. W. Hogan, and T. A. Rich. 1975. Calibration of the Pollak counter with monodisperse aerosols. *J. Appl. Meteorol.* 14:46–51.

Liu, B. Y. H., D. Y. H. Pui, R. L. McKenzie, J. K. Agarwal, R. Jaenicke, F. G. Pohl, O. Prening, G. Reischl, W. Szymanski, and P. E. Wagner. 1982a. Intercomparison of different absolute instruments for measurement of aerosol number concentration. *J. Aerosol Sci.* 13:429–450.

Liu, B. Y. H., D. Y. H. Pui, and K. L. Rubow. 1983. Characteristics of air sampling filter media. In *Aerosols in the Mining and Industrial Work Environments*, eds. B. Y. H. Liu and V. A. Marple. Ann Arbor, MI: Ann Arbor Science, pp. 989–1038.

Liu, B. Y. H., D. Y. H. Pui, and X. Q. Wang. 1982b. Drop size measurement of liquid aerosols. *Atmos. Environ.* 16:563–567.

Loo, B. W., C. P. Cork, and N. W. Madden. 1982. A laser-based monodisperse carbon fiber generator. *J. Aerosol Sci.* 13:241–248.

Marple, V. A., B. Y. H. Liu, and K. L. Rubow. 1978. A dust generator for laboratory use. *Am. Ind. Hyg. Assoc. J.* 39:26–32.

Marple, V. A. and K. L. Rubow. 1978. An evaluation of the GCA respirable dust monitor 101-1. *Am. Ind. Hyg. Assoc. J.* 39:17–25.

Marple, V. A. and K. L. Rubow. 1983. An aerosol chamber for instrument evaluation and calibration. *Am. Ind. Hyg. Assoc. J.* 44:361–367.

Marple, V. A. and K. L. Rubow. 1984. *Respirable Dust Measurement.* A Mining Research Contract Report. Bureau of Mines, U.S. Dept. of the Interior.

Marple, V. A. and K. Willeke. 1976. Inertial impactors: Theory, design, and use. In *Fine Particles: Aerosol Generation, Measurement, Sampling, and Analysis*, ed. B. Y. H. Liu. New York: Academic Press, pp. 411–446.

Matijevic, E. 1985. Production of monodisperse colloidal particles. *Annu. Rev. Mater. Sci.* 15:483–516.

May, K. R. 1949. An improved spinning top homogeneous spray apparatus. *J. Appl. Phys.* 20:932–938.

Maynard, A. D. and L. C. Kenny. 1995. Sampling efficiency determination for three models of personal cyclones. *J. Aersosol Sci.* 26:671–684.

Meesters, G. M. H., P. H. W. Versoulen, J. C. M. Marijnissen, and B. Scarlett. 1992. Generation of micron-sized droplets from the Taylor cone. *J. Aerosol Sci.* 23:37–49.

Mercer, T. T. 1973. *Aerosol Technology in Hazard Evaluation*. New York: Academic Press.

Mercer, T. T., M. I. Tillery, and H. Y. Chow. 1968. Operating characteristics of some compressed-air nebulizers. *Am. Ind. Hyg. Assoc. J.* 29:66–78.

Mokler, B. V. and R. K. White. 1983. Quantitative standard for exposure chamber integrity. *Am. Ind. Hyg. Assoc. J.* 44:292–295.

Mulholland, G. W., N. P. Bryner, and C. Croarkin. 1999. Measurement of the 100 nm NIST SRM 1963 by differential mobility analysis. *J. Aerosol Sci.* 31:39–55.

Newton, G. J., G. M. Kanapilly, B. B. Boecker, and O. G. Raabe. 1980. Radioactive labeling of aerosols: Generation methods and characteristics. In *Generation of Aerosols and Facilities for Exposure Experiments*, ed. K. Willeke. Ann Arbor, MI: Ann Arbor Sci. Publ., pp. 399–425.

Pilacinski, W., J. Ruuskanen, C. C. Chen, M. J. Pan, and K. Willeke. 1990. Size-fractionating aerosol generator. *Aerosol Sci. Technol.* 13:450–458.

Podzimek, J., J. C. Carstens, and P. C. Yue. 1982. Comparison of several Aitken nuclei counters. *Atmos. Environ.* 16:1–11.

Prandtl, L. 1952. *Essentials of Fluid Dynamics*. London: Blackie & Son, Ltd., pp. 247–249, 306–311.

Prodi, V. 1972. A condensation aerosol generator for solid monodisperse particles. In *Assessment of Airborne Particles*, eds. T. T. Mercer, P. E. Morrow, and W. Stöber. Springfield, IL: C. C. Thomas Publisher, pp. 169–181.

Pui, D. Y. H. and B. Y. H. Liu. 1979. Electrical aerosol analyzer: Calibration and performance. In *Aerosol Measurement*, eds. D. A. Lundgren et al. Gainesville: University Presses Florida, pp. 384–399.

Raabe, O. G. 1968. The dilution of monodisperse suspensions for aerosolization. *Am. Ind. Hyg. Assoc. J.* 29:439–443.

Raabe, O. G. 1976. The generation of fine particles. In *Fine Particles: Aerosol Generation, Measurement, Sampling, and Analysis.* ed. B. Y. H. Liu. New York: Academic Press, pp. 57–110.

Raabe, O. G. and G. L. Newton. 1970. Development of techniques for generating monodisperse aerosols with the Fulwyler droplet generator. In *Fission Product Inhalation Program Annual Report (LF-43)*. pp. 13–17. Albuquerque: Lovelace Foundation for Medical Education and Research.

Ranade, M. B., M. C. Wood, F. L. Chen, L. J. Purdue, and K. A. Rehme. 1990. Wind tunnel evaluation of PM_{10} samplers. *Aerosol Sci. Technol.* 13:54–71.

Rao, A. K. and K. T. Whitby. 1978. Non-ideal collection characteristics of inertial impactors—II. Cascade impactors. *J. Aerosol Sci.* 9:87–100.

Rapaport, E. and S. E. Weinstock. 1955. A generator for homogeneous aerosols. *Experientia* 11:363–364.

Reischl, G., W. John, and W. Devor. 1977. Uniform electrical charging of monodisperse aerosols. *J. Aerosol Sci.* 8:55–65.

Reponen, T., K. Willeke, V. Ulevicius, and S. A. Grinshpun. 1996. Effect of relative humidity on the aerodynamic diameter and respiratory deposition of fungal spores. *Atmos. Environ.* 30:3967–3974.

Romay-Novas, F. J. and D. Y. H. Pui. 1988. Generation of monodisperse aerosols in the 0.1–1.0 μm diameter range using a mobility classification–inertial impaction technique. *Aerosol Sci. Technol.* 9:123–131.

Rosell-Llompart, J. and J. Fernandez de la Mora. 1994. Generation of monodisperse droplets 0.3 to 4 μm in diameter from electrified cone-jets of highly conducting and viscous liquid. *J. Aerosol Sci.* 25: 1093–1119.

Scheibel, H. G. and J. Porstendörfer. 1984. Penetration measurements for tube and screen-type diffusion batteries in the ultrafine particle size range. *J. Aerosol Sci.* 15:673–682.

Sinclair, D. 1984. Intrinsic calibration of the Pollak counter—A revision. *Aerosol Sci. Technol.* 3:125–134.

Sinclair, D. and V. K. LaMer. 1949. Light scattering as a measure of particle size in aerosols. *Chem. Rev.* 44:245–267.

Smith, D. P. H. 1986. The electrohydrodynamic atomization of liquids. *IEEE Trans. Ind. Appl.* 1A-22:527–535.

Smith, J. P., P. A. Baron, and D. J. Murdock. 1987. Response characteristics of scattered light aerosol sensors used for control monitoring. *Am. Ind. Hyg. Assoc. J.* 48:219–229.

Spurny, K., C. Boose, and D. Hochrainer. 1975. On the pulverization of asbestos fibers in a fluidized-bed aerosol generator. *Staub Reinhalt. Luft* 35:440–445.

Stöber, W. 1976. Design, performance and applications of spiral duct aerosol centrifuges. In *Fine Particles: Aerosol Generation, Measurement, Sampling, and Analysis.* ed. B. Y. H. Liu. New York: Academic Press, pp. 351–397.

Sussman, R. G., J. M. Gearhart, and M. Lippmann. 1985. A Variable Feed Rate Mechanism for Fluidized Bed Asbestos Generators. *Am. Ind. Hyg. Assoc. J.* 46:24–27.

Tang, K. and A. Gomez. 1994. Generation by electrospray of mondisperse water droplets for targeted drug delivery by inhalation. *J. Aerosol Sci.* 25:1237–1249.

Tomaides, M., B. Y. H. Liu, and K. T. Whitby. 1971. Evaluation of the condensation aerosol generator for producing monodisperse aerosols. *J. Aerosol Sci.* 2:39–46.

Tu, K. W. 1982. A condensation aerosol generator system for monodisperse aerosols of different physic ochemical properties. *J. Aerosol Sci.* 13:363–371.

Ulevicius, V., K. Willeke, S. A. Grinshpun, J. Donnelly, X. Lin, and G. Mainelis. 1997. Aerosolization of particles from a bubbling liquid: Characteristics and generator development. *Aerosol Sci. Technol.* 26:175–190.

Vanderpool, R., W. and K. L. Rubow. 1988. Generation of large, solid monodisperse calibration aerosols. *Aerosol Sci. Technol.* 9:65–69.

Vaughan, N. P. 1990. The generation of monodisperse fibers of caffeine. *J. Aerosol Sci.* 21:453–462.

Vincent, J. H. 1989. *Aerosol Sampling: Science and Practice.* New York: John Wiley & Sons.

Vincent, J. H. 1995. *Aerosol Science for Industrial Hygienists*, pp. 238–303. New York: Elsevier Science Inc.

Vincent, J. H. and D. Mark. 1982. Applications of blunt sampler theory to the definition and mesurement of inhalable dust. *Ann. Occup. Hyg.* 26:3–19.

Wall, S., W. John, and D. Rodgers. 1985. Laser settling velocimeter: Aerodynamic size measurements on large particles. *Aerosol Sci. Technol.* 4:81–87.

Walton, W. H. and W. C. Prewett. 1949. The production of sprays and mists of uniform drop size by means of spinning disc type sprayers. *Proc. Phys. Soc.* B62:341–350.

Weyel, D. A., M. Ellakani, Y. Alari, and M. Karol. 1984. An aerosol generator for the resuspension of cotton dust. *Toxic. Appl. Pharmacol.* 76:544–547.

Whitby, K. T., D. A. Lundgren, and C. M. Peterson. 1965. Homogeneous aerosol generator. *Int. J. Air Water Pollut.* 9:263–277.

Willeke, K. and P. A. Baron. 1990. Sampling and interpretation errors in aerosol monitoring. *Am. Ind. Hyg. Assoc. J.* 51:160–168.

Willeke, K. and B. Y. H. Liu. 1976. Single particle optical counter: Principle and application. In *Fine Particles: Aerosol Generation, Measurement, Sampling, and Analysis*, ed. B. Y. H. Liu. New York: Academic Press, pp. 697–729.

Winters, W., S. Barnard, and A. Hogan. 1977. A portable, photo-recording Aitken counter. *J. Appl. Meteorol.* 16:992–996.

Yeh, H. C. and R. L. Carpenter. 1983. Evaluation of an in-line dilutor for submicron aerosols. *Am. Ind. Hyg. Assoc. J.* 44:358–360.

Yeh, H. C., Y. S. Cheng, and G. M. Kanapilly. 1983. Use of the electrical aerosol analyzer at reduced pressure. In *Aerosols in the Mining and Industrial Work Environments*, eds. V. A. Marple and B. Y. H. Liu. Ann Arbor, MI: Ann Arbor Science Publ., pp. 1117–1133.

Zhang, Z. Q. and B. Y. H. Liu. 1990. Dependence of the performance of TSI 3020 condensation nucleus counter on pressure, flow rate, and temperature. *Aerosol Sci. Technol.* 13:493–504.

22 Methods of Size Distribution Data Analysis and Presentation

DOUGLAS W. COOPER
The Texwipe Company LLC, Upper Saddle River, NJ

INTRODUCTION

The behavior of an aerosol particle depends strongly on its size. Various sources produce aerosols that range from nearly molecular dimensions to supermicrometer sizes. Transport and deposition due to gravity, diffusion, inertia, and electrostatic mechanisms depend on particle size. Effects such as light scattering and impact on health vary with particle size. The choice of equipment and materials to reduce particle concentrations in air and other gases also depends in part on the size of the particles of concern. The equipment appropriate for sampling various aerosols, devices such as impactors and impingers and precipitators and filters, must be selected with consideration given to particle size. Even the beneficial uses of aerosols (materials fabrication, therapeutic aerosols) are more or less effective depending on the sizes of the particles.

Determining the particle size distribution requires isolating a sample, discriminating among the particles on the basis of particle size, and quantifying (e.g., counting, weighing) the particles in each sizing interval. Four methods of aerosol size distribution measurement are particularly common: (1) Particles are sampled and then illuminated and detected by an optical particle counter, with the number of particles in each of many optical equivalent diameter intervals being (see Chapters 15, 16) recorded. (2) Particles are sampled and then deposited by filtration and are sized and counted with techniques such as optical or electron microscopy (see Chapter 12). (3) Particles are sampled by an impactor, and the mass of particles in each of the many aerodynamic equivalent diameter intervals is recorded (see Chapter 10). (4) Particles are accelerated through a sensing zone and their sizes determined by behavior in that zone (see Chapter 17). Optical particle counter and surface analysis measurements are quite useful in contamination control work, as the damage done by particles is often related to their number concentration and size. The impactor measurements are quite useful in health-related assessments, as the risk posed by particles is often related to their masses and aerodynamic diameters.

Most aerosol texts deal with particle sizing and size distributions because they are such important topics, and there is an introduction available in Chapters 2, 6, and 7 of this book. Another fine introduction is given by Knutson and Lioy (1995).

Aerosol Measurement: Principles, Techniques, and Applications, Second Edition, Edited by Paul A. Baron and Klaus Willeke.
ISBN 0-471-35636-0

Types of Particle Size

There are different kinds of geometric particle size used to characterize nonspherical particles. A nonspherical particle's size might be its longest dimension, shortest dimension, or some combination of the two. One might use the diameter of a sphere with the same volume or the same surface area. For spherical and nonspherical particles measured with a technique that does not give geometric size, other "equivalent" diameters are often used: aerodynamic, optical, diffusion, electrical mobility, and so forth. These are described in various chapters throughout this book.

Particle Shape

Particle shape is also very important, and different shapes can produce rather different behaviors for particles of the same "equivalent size." There is a substantial body of literature concerning characterizing particle shape, much of which has been contributed by J. K. Beddow and co-workers. A valuable starting place for accessing this information is Beddow's excellent book, *Particulate Science and Technology* (1980), and more information is available in a second text of his (Beddow, 1997). With a narrower focus, a fine review of spray–pyrolysis particle morphology is presented by Jain et al. (1997). Further information about particle shape as it relates to aerosol measurement is described in Chapter 23.

PARTICLE SIZE DISTRIBUTIONS

Having decided on an appropriate measure of the particle size, which here we label d, one is still faced with not a single size, but a wide range of sizes, in most aerosols. Although summary statistics, such as the mean and standard deviation of the particle diameter, can be used to advantage, one may want to give a more detailed, albeit approximate, description of particle sizes. To do so, we often use either a frequency distribution or a cumulative distribution.

Cumulative Versus Frequency Size Distributions

A cumulative distribution gives the fraction of particles (by some measure) smaller than the particle size (by some measure). Table 22–1 shows a cumulative distribution $F(d)$, in this case a normal (also called *Gaussian*) distribution, with a mean of $\mu = 10$ and a standard deviation of $\sigma = 1$. The first column shows the particle size at the upper end of each sizing interval. The next column shows the fraction of the number of particles smaller than the size indicated in the column. This is a cumulative distribution.

Figure 22–1 shows the full cumulative size distribution, abbreviated in Table 22–1, plotted on linear axes. The distribution is S shaped, "sigmoidal."

TABLE 22–1. Normal (Gaussian) Distribution, with a Mean of $\mu = 10\,\mu m$ and a Standard Deviation of $\sigma = 1\,\mu m$

Particle Size (μm)	Percentile (Percent Smaller)
$7 = \mu - 3\sigma$	0.135
$8 = \mu - 2\sigma$	2.27
$9 = \mu - 1\sigma$	15.9
$10 = \mu$	50.0
$11 = \mu + 1\sigma$	84.1
$12 = \mu + 2\sigma$	97.73
$13 = \mu + 3\sigma$	99.865

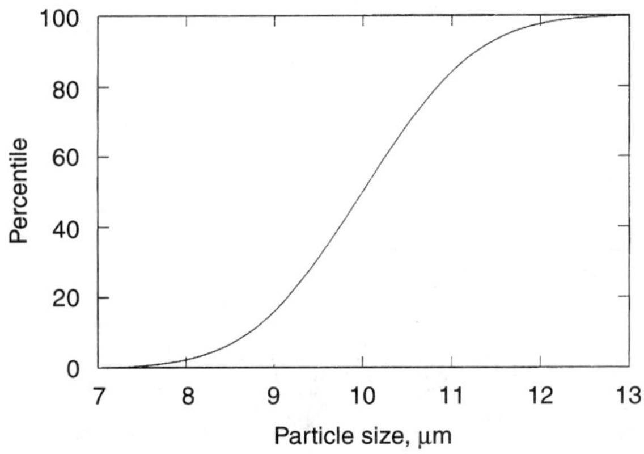

Fig. 22–1. Cumulative distribution, particle size: Gaussian (normal) distribution on linear axes. (Mean is $\mu = 10$, and standard deviation is $\sigma = 1$.)

The frequency distribution is determined from the change in the cumulative distribution divided by the change in particle size. The data in Table 22–1 can be analyzed to create an approximate frequency distribution, the fraction of the aerosol in various size intervals. The most common approach is to take the cumulative distribution value at one diameter, $F(d_2)$, and subtract from it the cumulative distribution value at a smaller diameter, $F(d_1)$, and divide by the difference between the two diameters:

$$f(d) \approx [F(d_2) - F(d_1)]/(d_2 - d_1) \tag{22–1}$$

The value of d to which $f(d)$ corresponds is usually chosen to be the mean of d_1 and d_2:

$$d = (d_2 + d_1)/2 \tag{22–2}$$

Such an approximate frequency distribution might best be plotted as a histogram, a bar graph with the bars filling each size interval and having a height representing the frequency. The choice of the size intervals will somewhat affect the shape of the distribution, however.

This process of finding the frequency distribution numerically is an approximation of the definition of the relationship between the frequency and cumulative distributions:

$$f(d) = \mathrm{d}F(d)/\mathrm{d}d \tag{22–3}$$

The term dd is a bit strange in appearance, but it denotes an infinitesimal change in a variable, in this case d. The frequency distribution is the derivative of the cumulative distribution. The frequency distribution and the cumulative distribution contain the same information, in different forms.

Figure 22–2 shows the frequency distribution that would be calculated in the limit of infinitesimal size intervals. The frequency distribution is the Gaussian (or normal) distribution, discussed next.

The Normal (Gaussian) Distribution

Most scientists and engineers are familiar with the Gaussian (normal) frequency distribution, the "bell-shaped curve" that describes many phenomena. For particles with a Gaussian size

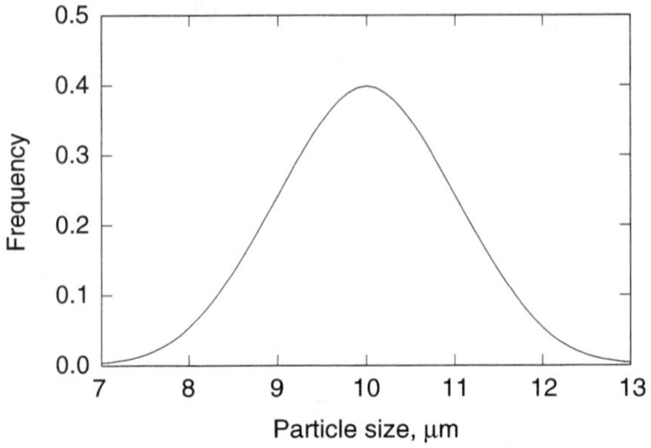

Fig. 22–2. Frequency distribution, particle size: Gaussian (normal) distribution on linear axes. (Mean is $\mu = 10$, and standard deviation is $\sigma = 1$.)

distribution, the fraction of the total (by count) within an infinitesimal size interval, d*d*, centered on size *d* is

$$f(d)\mathrm{d}d = (1/\sigma\sqrt{2\pi}) \times \exp\left[-(\mathrm{d}-\mu)^2/2\sigma^2\right]\mathrm{d}d \tag{22–4}$$

in which σ is the population standard deviation and μ is the population mean. The mean is a measure of the central tendency of the sizes, their sum divided by the number measured. The standard deviation is a measure of their spread, the square root of the mean squared difference between the sizes and the mean size.

Normal distributions tend to arise when a multitude of small additive (±) factors influence a variable that would otherwise have a single value. The normal distribution is found in aerosol science in cases where the particles are all nearly of one size, nearly monodisperse, such as polystyrene latex spheres used for calibration of particle measurement equipment or the droplets produced by the vibrating orifice or spinning disk generators.

The Lognormal Distribution

Lognormal particle size distributions are more common than normal distributions in aerosol science and technology. Particle size distributions often have standard deviations that are large in comparison with their mean sizes, which cannot occur for the inherently nonnegative normal distribution. Such distributions are often described mathematically by the lognormal distribution. Lognormal distributions arise when a multitude of multiplicative factors (both greater and less than 1) act on a variable that would otherwise have a single magnitude.

The lognormal distribution (Aitchison and Brown, 1957; Fuchs, 1964; Hinds, 1982; Crow and Shimizu, 1988; Heintzenberg, 1994) is the distribution that results when the distribution of log(x) is Gaussian ("normal"). Lognormal distributions can be shown to result from the proportional breakup of large objects into smaller ones (Kolmogorov, 1941; Epstein, 1947) or from certain types of agglomeration of small objects into larger ones (Friedlander, 1977). The lognormal distribution will result from the growth or breakup process $\mathrm{d}x(i)/\mathrm{d}t = k(i)x(i)$ when the growth constants $k(i)$ are normally distributed. The $x(i)$ are variables having to do with the sizes of the species i, typically the volumes of the species. Many aerosol particle size distributions have been well approximated by the lognormal distribution. The estimation of its parameters from truncated measurements was analyzed by Gentry (1977).

TABLE 22–2. Percentile Values for the Lognormal Distribution[a]

Size	Percentile
d_{50}/σ_g^3	0.135
d_{50}/σ_g^2	2.27
d_{50}/σ_g	15.9
d_{50}	50.0
$d_{50}\ \sigma_g$	84.1
$d_{50}\ \sigma_g^2$	97.73
$d_{50}\ \sigma_g^3$	99.865

[a] The median is d_{50}, and the geometric standard deviation is σ_g. Note the similarity to the results in Table 22–1.

Fiber length distributions and length-weighted diameter distributions have also been found to be approximately lognormal (Fogel et al., 1999). Some counter-examples were presented by Christensen et al. (1993). Using the assumption of lognormal length distribution and Monte Carlo simulations of fiber interactions with a rectangular filter grid (see also Cooper et al., 1978), Myojo (1999) showed that penetration through the grid depended on the ratio of the length of the fibers to the spacing of the grid elements; he also gave an extensive list of references on fibrous aerosols.

The lognormal distribution is actually a normal distribution of logarithms of particle size, $\log(d/d_o)$. The reference size, d_o, is usually 1 µm. The lognormal distribution's measure of central tendency is the geometric mean for a lognormal distribution, and the measure of spread is the geometric standard deviation, labeled σ_g. The geometric mean is the nth root of the product of n values; it is the antilog of the mean of the logarithm of the n values. The median is the 50th percentile particle size and equals the geometric mean for the lognormal. Common medians are the number median, area median, activity (radioactivity) median, volume median, and mass median. The geometric standard deviation (σ_g) for a lognormal distribution is the ratio of the 50th percentile size to the 16th percentile size, which for a lognormal distribution is also equal to the ratio of the 84th percentile size to the 50th percentile size. The geometric standard deviation is related to the standard deviation of the logarithms of size. As the geometric standard deviation approaches $\sigma_g = 1.0$, the lognormal distribution approaches the normal distribution having a relative standard deviation of $\sigma = \sigma_g - 1$.

Table 22–2 shows the percentiles for the lognormal distribution in terms of the median and the factors of the geometric standard deviation, σ_g, from the median.

Figure 22–3 shows a cumulative lognormal distribution. The median is $\mu = 1$, and the geometric standard deviation is $\sigma_g = 2$. Later in this chapter information is presented on transformations that allow plotting this as a straight line.

The Power-Law Distribution

The power-law frequency distribution is given by

$$f(d) = ad^b \tag{22–5}$$

where b is typically a power less than 0. The cumulative power-law distribution is

$$F(d) = [a/(b+1)]d^{(b+1)} \tag{22–6}$$

In the atmospheric sciences, the power-law distribution is often called the "Junge" distribution after C. Junge, an important contributor to the field. The cumulative size distribution is proportional to the reciprocal of the particle size approximately cubed, thus $b = -4$.

In the contamination field, a power-law distribution is used in Federal Standard 209E (U.S. General Services Administration, 1992) shown in Figure 22–4 to characterize the typical particle size distribution in a clean room. The cumulative count size distribution is proportional to $d^{-2.2}$. The cleanroom class in U.S. customary units is determined by the number of particles larger than 0.5 µm in the distribution per cubic foot of air. "Class 100" describes a clean room that has been shown to have fewer than 100 particles 0.5 µm or larger in diameter per cubic foot. The metric version, uses concentrations in terms of per cubic meter and steps of powers (M) of ten and the square root of ten, $10^{0.5} = 3.16$.

The power-law distribution is just a useful approximation, allowing one to predict easily the approximate behavior of the entire aerosol rather than just of individual particles. For example, if the deposition velocity of the aerosol is approximated by a power law, then the flux (deposition velocity times concentration) can be readily determined as a power-law function, too.

A power-law distribution can be created by a process that subdivides material with breakage rates proportional to the size of the material. This mechanism may explain why certain aerosols from wear processes are well described by this distribution. Power-law distributions

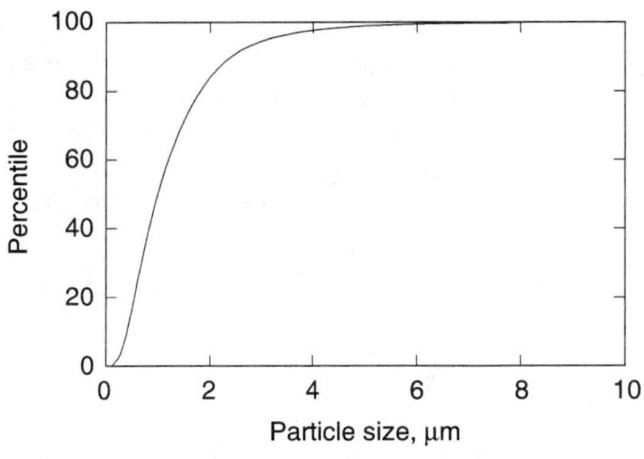

Fig. 22–3. Cumulative particle size distribution: lognormal on linear axes. (Median size is $d_{50} = 1$, and geometric standard deviation is $\sigma_g = 2$.)

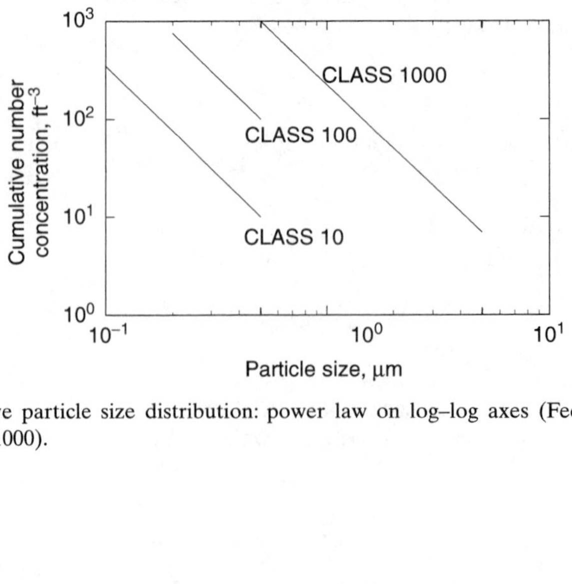

Fig. 22–4. Cumulative particle size distribution: power law on log–log axes (Federal Standard 209E Classes 10, 100, and 1000).

also arise in the study of fractal materials, some of which are formed by agglomeration. Here again, the agglomeration rate is proportional to a power of the particle size. A fine survey of fractal geometries is presented by Mandelbrot (1977). In Kaye (1989) are many illustrative examples of fractals in particle contexts, and there is even a section devoted to "fractal geometry and aerosol physics." Kaye also contributed two valuable chapters on sizing and on shape characterization in the Fayed and Otten (1997) handbook on powders.

One, two, or three different fractal dimensions were identified by Xie et al. (1994) from images of different types of particles, using "multifractal" analysis, essentially fitting straight lines to "Richardson" plots, log[P (perimeter)] versus log[L (step size)]. The effects of repetitive condensation/evaporation on agglomerate shapes were studied by Ramachandran and Reist (1995) using fractal descriptive techniques; the size at the slope transition in the Richardson plots was a good indicator of the size of the primary particles in the agglomerates. The surface fractal dimension was used to characterize powders of industrial significance, following the measurement of their nitrogen adsorption isotherms, and found to be a useful approach, though one of relatively low resolution (Wu, 1996). Multifractal analysis has also been used on the time series data from particle concentration measurements to get fractal time "lengths." See Chapter 23 for further discussion of fractal particles.

Other Distributions

Other distributions have been useful in various contexts, such as the Rosin-Rammler for coarse dusts and sprays (Lefebvre, 1989), the exponential for powders, the Nukiyama-Tanasawa for sprays, the exponential for some powders, and the Khrgian-Mazin for cloud droplets. One should consult texts such as those by Fuchs (1964) and by Hinds (1999) for more information. For light scattering from atmospheric aerosols ("haze"), McCartney (1976) described in detail the modified γ function, the product of the size to a power times the negative exponential of the size to a separate power, with four adjustable constants. In some cases, there may be value in using the four-parameter hyperbolic distribution described by Christiansen and Hartmann in Syvitski (1991). This distribution includes the normal, lognormal, Laplace, and exponential distributions as special cases; its four adjustable parameters allow very flexible fitting to data. The logβ distribution was used to characterize aerosol size distributions by Bunz et al. (1987). The distributions listed here are all unimodal distributions, but multimodal distributions can be formed from the addition of unimodal distributions.

CONCENTRATION DISTRIBUTIONS

Aerosol concentrations are rarely uniform and constant. Instead, one finds distributions of concentration values. Depending on how the data are accumulated over time, such changes may add uncertainty to the determination of the particle size distribution. Concentration distribution information can often be summarized with the same methods as those used for particle size distributions. For example, the sample mean and the sample standard deviation are obtained for the particle count just as for the particle size.

Three such distributions are discussed next: normal, lognormal, and Poisson.

The Normal Distribution

Normal distributions of concentrations are unusual. As noted above, the normal distribution arises from a large number of (positive and negative) additive effects. Sometimes, instruments that are themselves variable will be used to measure a relatively constant aerosol concentration, and the resulting data will be approximately normally distributed.

The Lognormal Distribution

Lognormal distributions of concentrations are not unusual. The lognormal distribution arises from a large number of multiplicative effects. It can arise from rate equations that have variable rates, as well. The geometric mean concentration and the geometric standard deviation of the concentration values fully describe such lognormal concentration distributions.

The Poisson Distribution

The Poisson distribution, discussed next, is a very useful approximation. It describes the particle count data that one would obtain from sampling an aerosol that has a constant mean number concentration. Even though the concentration is constant, the entrance of particles into the sampler is probabilistic, leading to some variation, the smallest amount of variation encountered in practice.

Denote the particle count in the ith time interval as $n(i)$. The mean count in an interval is μ, and the standard deviation of the count is σ. For the Poisson distribution the mean equals the variance, $\mu = \sigma^2$, so that the count standard deviation can be obtained from the square root of the mean count.

If one has accumulated counts over many intervals to form one long-duration sample, then the estimated standard deviation for this long-duration sample is the square root of the total count. The total count is the best estimate of the mean total count. The ratio of the standard deviation to the mean will be inversely proportional to the square root of the total count, indicating that the percentage uncertainty in the total count decreases inversely with the square root of the total count.

For the Poisson distribution, the probability of getting n counts in an interval from a population that has a mean μ in the interval is

$$P(n|\mu) = \mu^n \exp(-\mu)/n! \tag{22–7}$$

This is the distribution one would expect when using an optical particle counter to obtain particle counts over many time intervals from a constant concentration aerosol or when inspecting a large number of equal surface areas that have been exposed to a constant concentration aerosol for the same duration.

A simple application of the Poisson distribution is the estimation of an upper limit for the true count when the interval that was sampled gave no counts. One uses

$$P(0|\mu) = \mu^0 \exp(-\mu)/0! \tag{22–8}$$

$$P(0|\mu) = \exp(-\mu) \tag{22–9}$$

and notes that if the concentration had been as high as $\mu = 2.3$, there is only a 10% probability, $P(0|2.3) = 0.1$, that $n = 0$ particles would have been found. The hypothesis that $\mu = 4.6$ yields a 1% probability that $n = 0$ would be found, and so forth. One has 90% confidence that the true mean count for the interval is 2.3 or less and 99% confidence that it is 4.6 or less, and so forth.

The Poisson distribution can be used as the expected or hypothetical distribution when testing the hypothesis that the counts came from a constant, uniform count concentration distribution, using the chi-square analysis described below, for example.

Another use of the Poisson distribution is for modeling coincidence, the arrival of two or more particles (or their signals) in an interval (of space or time), discussed below.

For means much larger than 10, the Poisson distribution is very similar to the normal distribution (evaluated only at integer values). The use of the square root of the count as the variable transforms a Poisson distribution into a more nearly normal distribution (at integer count values) (See Box et al., 1978.)

Transforming count data by taking the square root not only stabilizes the standard deviation (to 0.5) for a Poisson distribution but has the general advantage of allowing convenient graphing of counts that have a large range that includes 0, which would make use of the logarithmic transformation awkward. With the square root transformation, the range of 0 to 10,000 would use the same extent as would 0 to 100 without the transformation.

SUMMARIZING DATA WITH A FEW PARAMETERS

Data for Individual Particle Sizes

One kind of data to summarize with a few parameters is particle size data obtained by sizing and counting individual particles. Assume that each particle has a measured diameter, $d(i)$, and that the number of particles measured is N. The sample mean particle diameter is then the sum of all the diameters divided by the number measured:

$$M = \sum d(i)/N \tag{22–10}$$

in which we have used M to denote the sample mean.

The sample standard deviation, s, of the particle diameter is the square root of the sum of the squared differences between the particle diameters and the mean, divided by $N - 1$, one less than the number measured:

$$s = \sqrt{\frac{\sum [d(i) - M]^2}{N - 1}} \tag{22–11}$$

The sample variance is s^2, the square of the sample standard deviation.

The sample mean, M, is an estimate of the population mean, μ, and the sample standard deviation, s, is an estimate of the population standard deviation, σ. The estimates become increasingly accurate as the number N in the sample increases. If the population has a Gaussian (normal) distribution, then it is completely described by these two parameters.

Data for the Fraction of the Aerosol in Each Interval

If the data are such that each datum represents the fraction of the aerosol (e.g., by count or by mass), $f(d[i])$, in each size interval, $\Delta d(i)$, then these equations for the sample mean diameter and the sample standard deviation become

$$M = \sum f[d(i)]d(i)\Delta d(i) \tag{22–12}$$

$$s = \sqrt{\sum f[d(i)][d(i) - M]^2 \Delta d(i)} \tag{22–13}$$

Elsewhere in this book is an example of hypothetical count data for some size intervals and the $f(d[i])$ calculated from these data.

Functional Form of Distribution

From the functional form of the distribution, $f(d)$, one can calculate the population mean from

$$\mu = \int f(d)(d)\mathrm{d}d \tag{22–14}$$

and the population standard deviation from

$$\sigma = \sqrt{\int f(d)(d-\mu)^2 \mathrm{d}d} \tag{22–15}$$

Often, the expected value notation is used for population means:

$$E[g(d)] = \int f(d)g(d)\mathrm{d}d \tag{22–16}$$

where $E(g[d])$ *is* the expected value (= population mean) of $g(d)$, $f(d)$ is the frequency distribution, and $g(d)$ *is* a function of diameter, such as area, volume, and so forth.

If $f(d)$ is based on count, then $E(g[d])$ would be the count mean of $g(d)$. If $f(d)$ is based on mass, then $E(g[d])$ would be the mass mean of $g(d)$. Other possibilities exist, such as the surface mean of $g(d)$ or the mass mean of $g(d)$, and so forth. The most common means are the count (number) mean diameter and the mass mean diameter, which use $g(d) = d$ and use the count or the mass frequency distributions for $f(d)$.

Parameters for the Normal Distribution

As can be seen from Eq. 22–4, the normal distribution is completely determined once the mean and standard deviation are specified.

Parameters for the Lognormal Distribution

The two parameters that totally describe the lognormal distribution are the geometric mean (which, for the lognormal, equals the median) and the geometric standard deviation. These are usually obtained from count or mass distributions. The median diameters are then labeled "count median diameter" or "mass median diameter." Sometimes, one uses the surface median or activity median diameter. In all these cases, the definition of the median diameter is that half (50%) of the count or mass or surface or activity, as appropriate, is contained in particles that are smaller than the median diameter. Of course, other percentiles (such as the 90th) could be used as the basis of a descriptive diameter, and occasionally are.

The geometric standard deviation for a lognormal distribution is independent of the power of the diameter being measured. Thus, σ_g will be the same whether particles are counted or weighed.

The various diameters for a lognormal distribution are related to each other (see Example 22–3). These "Hatch-Choate" relationships for various means and medians for the lognormal distribution are presented in the text by Hinds (1999), from which we extracted the following equations, in which MMD is the mass median diameter, SMD is the surface median, LMD is the length median diameter, and CMD is the count median diameter:

$$\mathrm{MMD} = \mathrm{CMD}\exp\left[3(\ln \sigma_g)^2\right] \tag{22–17a}$$

$$\mathrm{SMD} = \mathrm{CMD}\exp\left[2(\ln \sigma_g)^2\right] \tag{22–17b}$$

$$\mathrm{LMD} = \mathrm{CMD}\exp\left[1(\ln \sigma_g)^2\right] \tag{22–17c}$$

$$\text{modal diameter} = \mathrm{CMD}\exp\left[-1(\ln \sigma_g)^2\right] \tag{22–17d}$$

$$\text{mass mean diameter} = \mathrm{CMD}\exp\left[3.5(\ln \sigma_g)^2\right] \tag{22–17e}$$

Often, MMD is much larger than CMD. For example, if $\sigma_g = 2$, MMD/CMD = $\exp[3\ln(2)^2]$ = 4.23. $\sigma_g = 2$ may be a typical geometric standard deviation for a poydisperse aerosol. If $\sigma_g = 3$, MMD/CMD would be 37.4, and so forth.

TABLE 22–3. Various Means and Medians for a Lognormal Distribution having a Count Median of 1 and a Geometric Standard Deviation of 2

Count mode diameter (frequency peak)	0.62
Count median diameter (CMD)	1.00
Count mean diameter	1.27
Diameter of particle of average area	1.62
Diameter of particle of average mass	2.06
Area median diameter	2.61
Area mean diameter	3.32
Mass (volume) median diameter (MMD)	4.23
Mass (volume) mean diameter	5.37

Based on Reist (1984).

Another good reference on the Hatch-Choate relations for lognormal distributions is the book by Reist (1984). Heintzenberg (1994) gave not only the familiar relationships for the lognormal distribution but also equations for integrals involving the lognormal function.

Table 22–3 shows the various weighted means and medians for a lognormal frequency distribution (CMD = 1, σ_g = 2) based on the values given by Reist (1984). To get the area mean diameter, for example, one must transform the count or mass distribution to an area distribution and then integrate the area frequency distribution function times the particle diameter. For lognormal distributions, the Hatch-Choate relationships include such transformations.

Unfortunately, it is rare for any aerosol to be so nearly lognormal that measurements by count of the count median diameter and the geometric standard deviation can be used to estimate accurately the mass median diameter, or vice versa, except for cases of nearly monodisperse aerosols ($\sigma_g = 1$). Thus, impactor data, generally based on mass, cannot be relied on for accurate count median diameter estimates, and optical particle counter data cannot be relied on for accurate mass median diameter estimates except for nearly monodisperse aerosols. This is true whether the estimates are made from the Hatch-Choate relationships or from the numerical methods in the early part of this chapter.

Another advantage of the lognormal distribution is that it can be used to simplify the analysis of the size distribution that results when a lognormal aerosol penetrates a device with a sigmoidal penetration curve that can be approximated by a cumulative lognormal function. The size distribution of the penetrating fraction is itself approximately lognormally distributed, with parameters that can be readily derived from those of the inlet distribution and the penetration function (see Cooper, 1982). The fraction penetrating was also readily determined.

Linear Regression, Primarily for Power-Law Distributions

A linear regression equation is the least-squares best fit of the linear equation $y = mx + b$ to sets of $x(i)$, $y(i)$ pairs. The slope (dy/dx) of the line is m, and the intercept (y at $x = 0$) is b. The least-squares best fit produces the estimates for m and b that minimize $\Sigma(y[i] - mx[i] - b)^2$. The equations for estimating m and b from the data are given in most statistics texts, such as Hays and Winkler (1970) or Draper and Smith (1981), and form the basis for many different computer programs, which is how virtually all regression analyses are done now.

The computer programs generally give not only the best estimates of m and b but also the uncertainty in the estimates, the standard errors of m and of b, SE(m) and SE(b). These standard errors are analogous to the standard errors of the mean for a variable x. If the data were to match the assumptions of linear regression, then in 68% of the instances the confidence interval of $m \pm 1.0\mathrm{SE}(m)$ would include the true slope, and in 95% of the instances the confidence interval of $m \pm 1.96\mathrm{SE}(m)$ would include the true slope. The same statements can be made about $b \pm \mathrm{SE}(b)$ and $b \pm 1.96\mathrm{SE}(b)$ and the true intercept.

The more data, and the smaller the error in the data, the narrower will be the confidence limits for the slope and the intercept. Note that the assumptions underlying linear regression are rarely fulfilled exactly. These assumptions are (1) the relationship between y and x is truly linear; (2) the measurements of the independent variable, $x(i)$, contain no errors; (3) the measurements of the dependent variable, $y(i)$, contain only an additive error that is Gaussian with a mean of zero and a standard deviation that does not change. Because these assumptions are rarely fulfilled exactly, the probabilities associated with the confidence intervals are at best approximations. If the measurements of the independent variable, x, do contain an error, then the slope will be erroneously biased toward zero. The intercept will be biased in this case, too.

For a power-law distribution, the relationship becomes linear if logarithms are used:

$$f(d) = ad^b \tag{22–19a}$$

$$\log f(d) = \log a + b \log d \tag{22–19b}$$

$$F(d) = [a/(b+1)]d^{b+1} \tag{22–20a}$$

$$\log F(d) = \log[a/(b+1)] + (b+1)\log d \tag{22–20b}$$

It is easy to use a linear regression program on the logarithm of the cumulative size distribution versus the logarithm of the size. One can also simply plot $\log F(d)$ versus $\log d$, or plot $F(d)$ versus d on log–log axes, and draw the apparent best-fit line through the data. The use of a linear regression package is preferable, not so much for improved accuracy (recall that the regression assumptions are rarely met) as for being objective rather than subjective and, thus, being wholly reproducible by others.

Figure 22–5 shows hypothetical data and the best-fit straight line obtained from linear regression. Note that the assumption that the standard deviation of y is independent of x is clearly violated because the spread in the values of y gets bigger as y gets smaller. It would be advisable to transform y to $\log(y)$ or to $\sqrt{y}$ and try regression again. $\log(y)$ versus $\log(x)$ should also be tried.

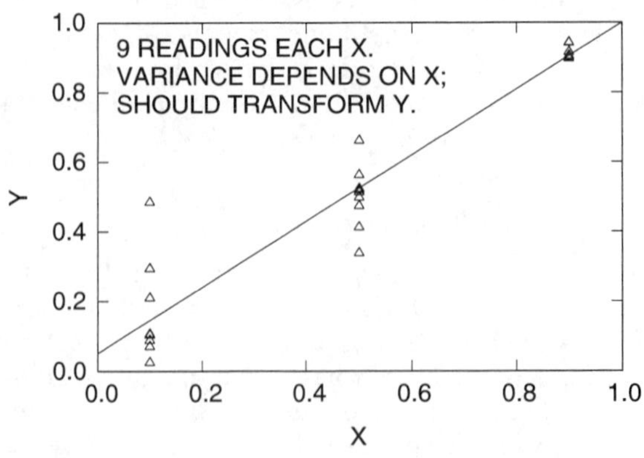

Fig. 22–5. Example of data and a least-squares linear regression line of best fit.

Often the correlation coefficient ($-1 < r < 1$) is used as a measure of the degree to which y is linearly dependent on x. This can be misleading. The smaller the errors in y in comparison to the change in y across the range of x, the better is the correlation coefficient and the better is the fit of the line to the data. Note that data taken with a very large range in y can have large errors (in y) and still give impressive correlations ($|r|$ close to 1). Also, seemingly high correlation values can be obtained from small numbers of samples even if there is no underlying relationship between the variables.

The reader is encouraged to consult statisticians or statistics texts about such fine points as hypothesis testing with correlation coefficients, carrying out regressions when there are errors in the independent variable (x), and accounting for variation in the error of y over the range of x by using regression with weights adjusted to the estimated errors in y (Draper and Smith, 1981).

SUMMARIZING SIZE DISTRIBUTIONS GRAPHICALLY

Although equations or tables are more useful for further analysis of the data, graphs can quickly convey the essential features of particle size distributions.

General Advice

An excellent book on the topic of graphing data is the work by Cleveland (1985). Some of his valuable suggestions include the following:

1. Make the legend describing the graph complete and succinct.
2. Choose symbols for the data that are easily distinguished.
3. Check carefully for errors: labels, tick marks, captions, data.
4. Make sure that the elements are large enough and clear enough for reproduction and any planned size reduction.
5. Make the data prominent.
6. Keep necessary elements only.
7. Surround the region with the scale lines, putting the tick marks outside the rectangular "data region."
8. Keep labels, captions, and so forth, outside the data region if possible.
9. Explain any error bars (e.g., standard deviation, standard error).
10. Choose axis scales that help minimize empty space. Zero need not always be shown, for example.
11. Choose axis scales that help readers draw conclusions (e.g., log–log where a power law, $y = ax^b$, is being investigated).

We add that when two or more graphs are to be compared, their axes should be as similar as possible.

Cumulative Versus Frequency Representations

The familiar bell-shaped curve of the normal frequency distribution shows readily the location and width of that distribution. This frequency presentation is also convenient and clear when the distribution is made up of the sum of several distributions, for example, in multimodal distributions, having more than one local maximum. A frequency distribution having more than one component may become a cumulative distribution with a shape that is not so readily interpreted by eye. To determine the fraction of the distribution that lies between two sizes, for example, the frequency distribution format is awkward because one must integrate the area under the curve. Shifting the interval widths or changing from linear to logarithmic progression in the independent variable can produce rather dramatic changes in the appearance of a frequency distribution.

The cumulative distribution format contains the same information as the frequency distribution. The central location can be estimated from the median, the 50th percentile, and the spread can be estimated from the difference between two percentiles, such as the 16th to the 84th percentiles or the 25th to the 75th, called the *interquartile range*. The percentage between two sizes is readily obtained by subtraction using the cumulative distribution rather than by the more difficult process of integration (estimation of the area) using the frequency distribution. There are transformations that convert some cumulative distributions into easily fitted straight lines.

Thus, it seems generally better to use the cumulative size distribution rather than the frequency size distribution.

Plotting Data and a Fitted Curve

For plotting one or a few distributions, the presentation of both the data and the best-fit curve(s) on the same graph has the advantages of completeness and clarity. Each curve helps to summarize, and the data are shown to allow others to decide on how good a summary each curve provides or to help them take the analysis further. The data should be made available in a tabulated form, too. The easiest way to plot the best-fit curve is to use axes that transform the cumulative distribution relationship it into a straight line. This is discussed next.

Plotting with Linear Axes

The size data usually start as counts (or mass) in particle size intervals. By dividing by the total count (or mass), the data can be converted to segments $dF(d[i])$ of the cumulative distribution in the intervals around $d(i)$. By accumulating the segments dF, one can create the useful cumulative distribution function, $F(d)$.

If $F(d)$ plotted against d using linear axes produces a rather symmetrical S-shaped curve, then it is likely that the distribution is normal or nearly so. If the curve is somewhat S shaped but with the large-diameter part much more extended than the small-diameter part, then the distribution may well be lognormal or nearly so. If the $F(d)$ curve is a straight line or if it has only one curved region rather than two, then the power-law distribution is a candidate. Plotting each of these is discussed next.

Plotting with Transformed Axes

A cumulative normal size distribution will plot as a straight line on axes that have their percentile values spaced in equal intervals of probits as the ordinate scale and micrometers (or other units of length) as the abscissa scale. Probits indicate how many standard deviations from the mean the percentile value represents for a truly normal distribution. The probit value for the 50th percentile is 0, and the probit values for the 16th and 84th percentile values are –1 and 1, respectively, and so forth. Seventeen probit–percentile pairs are given in Table 22–4, with more values readily obtained in most statistical texts and from some pocket calculators or from computer programs.

Other probit values can be obtained in this range from the approximation formula

$$\text{probits} = 4.9\left[F^{0.14} - (1 - F)^{0.14}\right] \tag{22–21}$$

Plotting data from a cumulative normal distribution on a probit axis versus the size on a linear axis will produce a straight line. However, the data will rarely be perfectly normal, so

TABLE 22–4. Percentile Values and Corresponding Probit Values

Percentile	Probit Value
00.003	−4.0
00.023	−3.5
00.135	−3.0
00.621	−2.5
02.275	−2.0
06.681	−1.5
15.866	−1.0
30.854	−0.5
50.000	+0.0
69.146	+0.5
84.134	+1.0
93.319	+1.5
97.725	+2.0
99.379	+2.5
99.865	+3.0
99.977	+3.5
99.997	+4.0

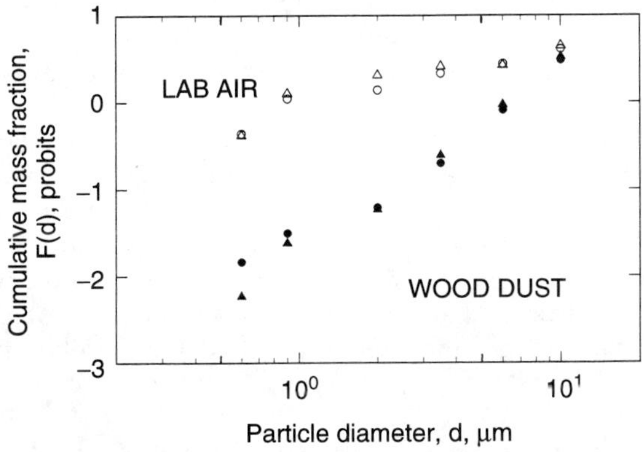

Fig. 22–6. Cumulative size distribution data (Shen and Ring, 1986) plotted as mass fraction measured in probits versus the logarithm of the apparent diameter. Open symbols indicate laboratory air; closed symbols indicate wood dust. Impactor 296A data are denoted with triangles, 296B data with circles. Lognormal distributions would plot as straight lines.

some deviation from a straight line can be expected. In plotting the best-fit straight line to data that are approximately normal, it is better to calculate the mean (M) and the standard deviation (s) directly from the formulas given above than to use linear regression to obtain the slope and the intercept of the curve. The probit value for points on the cumulative normal probability curve will be

$$\text{probits} = (d - M)/s \qquad (22\text{–}22)$$

Figure 22–6 shows an example of cumulative distributions plotted on a probit scale after the particle size d has been transformed by $\log(d/d_0)$, a transformation that creates a normal distribution in $\log d$ from a lognormal distribution in d. The data were taken from impactor measurements reported by Shen and Ring (1986), discussed more fully below.

The lognormal distribution independent variable is generally referred to as the logarithm of the particle size. Because logarithms take dimensionless numbers for their arguments, the variable is actually $\log(d/d_0)$, where d_0 is often in micrometers.

Finally, the cumulative distribution for a power-law aerosol is readily plotted either as $\log F$ versus $\log(d/d_0)$ on linear axes or F versus d/d_0 on logarithmic axes.

CONFIDENCE INTERVALS AND ERROR ANALYSIS

Having made measurements, one would like an estimate of their precision, an estimate of where future measurements of the same type would fall if they were made. One can also calculate a mean from the measurements, and one would like to know how precise an estimate the sample mean is of the true mean (unknown). The measured values may be used in other equations to obtain derived values, and one would like to know how precise these derived values are. These aspects are discussed in this section.

Confidence Intervals

The confidence intervals described here are of the following type: If the data have only random errors that are nearly normal in their distribution with a standard deviation of σ, then one can estimate σ by using the sample standard deviation, s, and can estimate a confidence interval in which P% of future measurements of the same kind would fall. One can also estimate a confidence interval for the measured mean, M, which would include the real mean, μ, in P% of the instances if repeated sets of the same kind of measurement were made.

Confidence Intervals for Single Readings. If one has a large number of estimates of d, $N > 30$, from a distribution that is approximately normal, then one can calculate the sample mean and the sample standard deviation and estimate that the readings come from a normal distribution with that mean and that standard deviation. Thus, a 95% confidence interval would be the measured mean plus or minus two measured standard deviations: $M \pm 2s$. The prediction that the next measured value would be within $M \pm 2s$ would be correct in about 95% of the cases. Looking at the tabulation of percentiles and probit values in Table 22–4, it is clear that $M \pm 3s$ (–3 probits to +3 probits) would cover 99.865 – 0.135 = 99.73% of the instances, and so forth.

With a smaller number of readings, the estimate of the mean and the estimate of the standard deviation would have more uncertainty, and the ability to predict future readings would be less.

Figure 22–7 shows 100 simulated observations, taken at random from a normal distribution with mean = 10 and standard deviation = 1. Lines have been drawn at ±3 standard deviations from the mean. About 68% of the readings are within 1 standard deviation, about 95% within 2 standard deviations, and none are outside 3 standard deviations. This kind of chart is a "control chart" and can be used to identify when a change in a process has occurred by noting an unusual number of values lying outside the three standard-deviation limits. More often, the means of many readings are plotted rather than individual readings, and their limits are in terms of standard errors.

When long-term records are not available to set control chart limits, the counts in one interval can be compared with those in another by the use of chi-square methods that assume Poisson statistics, a test of the constancy of the concentration (Box et al., 1978). The statistic

$$u = (x - y)/\sqrt{x + y} \tag{22–23}$$

for counts from one Poisson distribution will itself be normal with a mean of $E(u) = 0$ and a standard deviation of 1. This can be used to test whether the count, x, from one counting

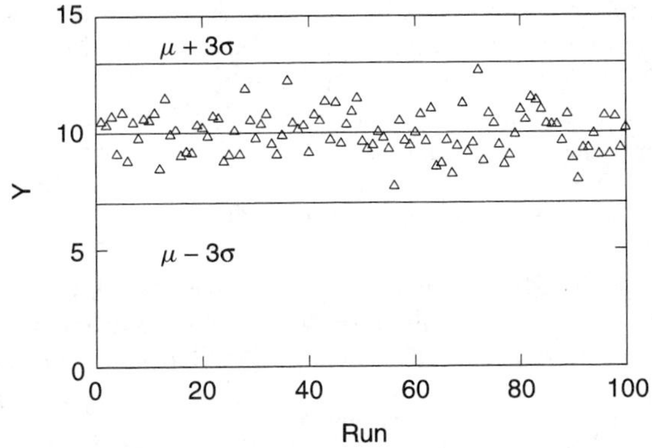

Fig. 22–7. "Data" drawn from a Gaussian (normal) distribution; $\mu = 10$, and $\sigma = 1$, with lines drawn for 3 standard deviations from the mean, $\mu \pm 3\sigma$.

interval is likely to have come from the same concentration as the count, y, from another counting interval (Cooper, 1991a). A value $|u| > 2$ provides strong evidence (95% confidence) that the concentrations are not the same.

Confidence Intervals for Means. If the measurements follow a normal distribution, then the means of the measurements will have a distribution that will be normal and will have a standard deviation equal to $1/\sqrt{N}$ times the standard deviation of the individual readings. N is the number of readings used to calculate an individual mean. Even if the measurements are not quite normal, the means will be given approximately by the t distribution (often called *Student's t* after the pseudonym of its discoverer).

For samples drawn from a normal distribution, the confidence interval for the true mean, μ, is given by the inequalities

$$M - t(N-1,\alpha/2)s/\sqrt{N} < \mu \le M + t(N-1,\alpha/2)s/\sqrt{N} \qquad (22\text{–}24)$$

where $t(N-1, \alpha/2)$ is obtained from tables of the Student's t distribution, available in statistics texts. $N-1$ is the number of "degrees of freedom." The value of $\alpha/2$ is half of the fraction of instances that will *not* occur within the confidence interval. If the confidence interval is a 90% confidence interval, then $\alpha/2$ is 5%. Asymmetrical confidence intervals can also be obtained, such as

$$\mu \le M - t(N-1, 0.05)s/\sqrt{N} \qquad (22\text{–}25)$$

which would cover 95% of the instances. This formula was used in the clean-room classification methodology used for Federal Standard 209E (U.S. General Services Administration, 1992).

Although the t distribution is more complicated than the normal distribution, its values are tabulated widely. Furthermore, as N becomes much larger than 10, t becomes distributed as nearly a normal distribution, and the t values become the corresponding probit values. For example, for such large N values,

$$M - 2.0s/\sqrt{N} < \mu \le M + 2.0s/\sqrt{N} \qquad (22\text{–}26)$$

would be a 95% confidence interval for the mean.

The value of multiple measurements (large N) in decreasing the uncertainty (improving the precision) of the mean is evident. Regardless of the distribution that the readings come from, the means of N readings will have a distribution that has a standard deviation ("standard error of the mean") that is $1/\sqrt{N}$ times the standard deviation of the individual readings. The $1/\sqrt{N}$ dependence implies that halving the uncertainty in the mean can be achieved by going from $N = 1$ to $N = 4$ or from $N = 4$ to $N = 16$ or from $N = 16$ to $N = 64$, giving diminishing improvements as N increases by one.

Confidence Intervals for Standard Deviations and Variances. A very similar approach, but using the chi-square distribution rather than the Student's t distribution, can be made for generating confidence intervals for the standard deviation or its square, the variance (Crow et al., 1960).

Error Analysis: Propagation of Error

"Propagation of error" concerns how the errors in measurements affect calculations based on these measurements. To carry out the analysis, it is more convenient to work with the variance, s^2, rather than the standard deviation.

The general formula for the propagation of error makes use of partial derivatives: the variance of $g(x[i])$, a function of one or more variables $x(i)$, is the sum of the products of the squares of the partial derivatives of g with respect to $x(i)$ and the variances of $x(i)$. Often, one uses just the particle diameter d as a variable (so $x[i] = d$) and calculates the variance of a function $g(d)$.

The variance in $g(d)$, $s(g)^2$, is obtained from the variance in d, $s(d)^2$, as follows:

$$s(g)^2 = [\mathrm{d}g(d)/\mathrm{d}d]^2 s(d)^2 \tag{22–27}$$

If $g(d) = ad$, then $s(g)^2 = a^2s(d)^2$, a simple result. If $g(d) = ad^b$, then the manipulation becomes somewhat more complicated, but can be simplified by using

$$s(ad^b)^2/(ad^b)^2 \approx b^2 s(d)^2/d^2 \tag{22–28a}$$

$$s(ad^b)/(ad^b) \approx bs(d)/d \tag{22–28b}$$

where the value of d to be used is the mean. If the volume is predicted from the diameter, then the relative standard deviation of the volume (approx. $s[d]/M[d]$) is $b = 3$ times the relative standard deviation of the diameter. If the standard error of the mean volume is desired, then the standard error of the diameter, rather than the standard deviation, should be used in this equation. Clearly, going from the diameter to the volume, where $b = 3$, increases the uncertainty, in contrast to going from the volume to the diameter, where $b = 1/3$, which decreases the uncertainty.

An example of the use of propagation of error in analyzing experimental determination of mass concentration was presented in detail by Evans et al. (1984).

TESTING HYPOTHESES WITH SIZE DISTRIBUTION DATA

Often, it is desired to compare one measured aerosol size distribution against another or against a theoretical distribution. One wants objective statistical measures to discriminate between two distributions in order to answer many types of questions: Are the aerosols likely to be from different sources? Has something changed during the period between the measurements of the two distributions? Do two instruments measuring the same aerosol give virtually the same results? Did different conditions produce different size distributions?

Analyzing Data from a Normal Distribution: Student's *t*

If a particle counter has been used, then the mean particle size, M_1, from one distribution can be compared with a hypothetical mean, M, by the Student's t test, the statistic

$$t = (M_1 - M)/(s_1/\sqrt{N_1}) \tag{22–29}$$

in which s_1 is the sample standard deviation and N_1 is the number of particles sized. This value of t is to be compared against the critical values of t at very low or very high percentiles, such as 1%, 5%, 10%, 90%, 95% and 99%. What is being sought is the probability that a t value of this size would have occurred if there were no difference between the means. For example, if one is testing whether $M_1 = M$, then an unusually negative t value (e.g., $t << -1$) or an unusually positive t value (e.g., $t >> +1$) would be a strong evidence against the hypothesis. Tables of the critical values of t for various N are available in most statistics texts.

When $N >> 10$, the t statistic can be evaluated from the normal distribution, with $t < 0$ expected in 50% of the instances, $t < 1$ expected in 84% of the instances, $t < 2$ expected in 97.7% of the instances, and so forth. Usually, $|t| > 2$ would be taken as strong evidence that M was not equal to M (this is the 95% significance level).

If the first distribution of mean M_1 is to be compared not against a theoretical mean, μ, but against a measured mean, M_2, then there are other formulas for Student's t test available in the literature (e.g., Hays and Winkler, 1970). One must decide whether to assume that the true underlying population distributions have equal standard deviations or not. If, as is often done, one assumes that the population standard deviations are equal (variances are equal) for the N_1 and N_2 measurements, one has

$$t = (M_1 - M_2)/\sqrt{s^2/N_1 + s^2/N_2} \tag{22–30}$$

$$s^2 = \left[(N_1 - 1)s_1^{\,2} + (N_2 - 1)s_2^{\,2}\right]/(N_1 + N_2 - 2) \tag{22–31}$$

the latter being the "pooled estimate of variance" (Meyer, 1975). The denominator of the expression for s^2, $N_1 + N_2 - 2$, is the number of degrees of freedom. If the denominator is much larger than 10, t can be obtained from the cumulative normal distribution, as above. If the denominator is not much larger than 10, then tables of Student's t distribution should be used with the appropriate number of degrees of freedom.

If the observations under two different conditions to be compared can be analyzed as pairs, then the differences, $z_i = x_i - y_i$, can be used as the data to be analyzed. The mean of z can be tested against zero to show some effect or against another value to show an effect larger than that value. In most instances, paired comparisons are preferred, as each member serves as the "control" for the other.

Correlation and Regression

There are three commonly used measures of correlation: (1) the Pearson product-moment correlation coefficient, which is the most common; (2) the Spearman rank correlation coefficient; and (3) the Kendall tau coefficient. Each has a somewhat different set of assumptions (see, e.g., Hays and Winkler, 1970).

Analysis of Variance

Analysis of variance (ANOVA) has some dependence on the form of the underlying data distribution, but it is rather robust (Hays and Winkler, 1970). The data need to come from a balanced statistical experimental design, which can be awkward to obtain (see standard statistics texts, e.g., Hays and Winkler, 1970.) Transformations can be used to make the data more nearly normal. Examples are the use of the logarithm of the particle size or the square root of the count. Multiple measurements of the distribution allow a statistical comparison because the variability can be estimated from the replications.

Distinguishing Among Count Distributions: Chi-Square

Comparing count distributions is simpler than comparing mass distributions because the uncertainty in the count, its standard deviation, can often be assumed to be the square root of the count (Poisson statistics). By using as the dependent variable the square root of the count, a Poisson distribution can be transformed to be more nearly normal. Furthermore, this square root transformation makes the variance for Poisson statistics approximately constant (0.25), which helps satisfy the assumptions in linear regression (Box et al., 1978).

Without making assumptions about the size distribution, one can use a chi-square (χ^2) analysis, after apportioning the counts into matching size intervals. Chi-square is the sum of squares of the differences between the counts, $C(i)$, in the ith interval or category, and the expected counts, $E(i)$, divided by the expected counts:

$$\chi^2 = \sum [C(i) - E(i)]^2 / E(i) \tag{22–32}$$

The values for the expected counts come from the hypothesis being tested.

The χ^2 value is to be compared with percentile values in statistical tables of χ^2 distributions for the appropriate number of degrees of freedom. Details are available in standard statistical texts.

An example of the application of χ^2 to compare size distributions is a test to see whether one size distribution of particles contaminating a surface is different from another. The data (sample A and sample B) are count data, subdivided into particle counts for diameters larger or smaller than 5 μm:

	Sample A	Sample B	*Totals*
$d < 5\,\mu m$	160	105	265
$d > 5\,\mu m$	75	105	180
Totals	235	210	445

The expected values are obtained by using the subtotals to divide the total particle count proportionally into the four cells so that the upper-left-hand cell would have an expected count of $E(i) = (445)\ (265/445)\ (235/445) = 140.0$, and so forth. The value of χ^2 here computes to 15, a value expected in much less than 1% of the cases if the two samples were drawn from the same population. The fraction of particles larger than 5 μm in sample A is larger than that of sample B, and the difference is statistically significant.

Chi-square analysis is also often used in "goodness of fit" comparisons with hypothesized distributions, where the expected values come from the hypothesized distribution.

Distinguishing Count or Mass Distributions: Kolmogorov-Smirnov

If the data can be put into a large number of intervals ($N >> 10$ generally) that have the same boundaries, then the Kolmogorov-Smirnov (K-S) test can be quite effective. Although many aerosol sizing instruments do not have a large number of sizing channels, Heitbrink et al. (1991) noted that the Aerodynamic Particle Sizer (***TSI***)* does have many such intervals. They

* See Appendix I for full manufacturer addresses referenced by the italicized three-letter codes.

used the K-S test successfully to analyze particle count data from this instrument. In other cases, the exact sizing of a large number of particles ($N >> 10$) could be used; in that case, one could select the size intervals to analyze, with the interval end points for the two size distributions chosen so as to provide a large number of common boundaries and comparisons.

The K-S test involves comparing the cumulative distributions at the common end points of each interval and calculating the maximum absolute difference between them. The maximum difference is compared with the tabulated values for various levels of statistical significance and various numbers of intervals tested. The maximum difference values needed for statistical significance of 90% or more are differences on the order of magnitude of $1/\sqrt{N}$. For more information, see standard statistical texts, such as Hays and Winkler (1970).

Detailed Example of Size Distribution Comparison

This example demonstrates some methods for comparing size distributions using data from the literature. It assumes that the reader has substantial familiarity with statistical techniques.

Table 22–5 shows the weight data obtained from an article by Shen and Ring (1986). "WA" is a wood dust aerosol sampled by Sierra impactor 296A; "LB" is a laboratory air sampled by Sierra impactor 296B, and so forth. The weights were in milligrams and had estimated standard deviations of about 0.1 mg for the wood aerosol and 0.01 mg for the laboratory air, according to the authors.

Figure 22–6 is plotted from the data. The axes are similar to log-probability axes, with cumulative mass fraction measured in probits. The log of the apparent diameter (effective cut diameter for the impactor) is the other axis. A cumulative lognormal distribution would plot as a straight line on these axes. The data certainly seem to indicate (1) the two distributions for the wood dust (WA and WB) are similar to each other, (2) the two distributions for the laboratory air (LA and LB) are similar to each other, and (3) the distributions for the wood dust are different from the distributions for the laboratory air. For both aerosols, sampler 296B reported a somewhat larger mass median diameter and a somewhat larger geometric standard deviation than did sampler 296A, which may indicate a problem of bias.

We analyzed the data using ANOVA. The factors tested for were the effects of the samplers (A and B) and the effects of the aerosols (wood dust and laboratory air). Analysis of variance tests whether assuming there is a contribution to the mean that is associated with a factor reduces the variance by a statistically significant degree. The values that Shen and Ring inferred for the mass median diameters were used as inputs. First, the logarithms of the mass medians were used (shown in the top part of Table 22–6). This transformation was made to

TABLE 22–5. Weights (mg) in each Impactor Size Interval for Impactor A Sampling Laboratory Air (LA), Impactor B Sampling Wood Dust Aerosol (WB), and so Forth

Diameter (μm)	WA	WB	LA	LB
>10	0.68	0.86	0.066	0.068
6–10	0.47	0.60	0.021	0.015
3.5–6	0.49	0.60	0.001	0.010
2.0–3.5	0.36	0.35	0.009	0.018
0.22–2.0	0.13	0.13	0.021	0.010
0.6–0.9	0.09	0.09	0.049	0.040
0.0–0.6	0.03	0.09	0.090	0.090

Source: Shen and Ring (1986).

TABLE 22–6. Data Used in Analysis of Variance

Mass Median Diameter (MMD)			
Sampler	Dust	MMD	ln(MMD)
A	Wood	6.16	1.818
B	Wood	6.65	1.895
A	Lab	1.22	0.199
B	Lab	1.33	0.285
Geometric Standard Deviation (σ_g)			
Sampler	Dust	σ_g	$\ln(\sigma_g)$
A	Wood	2.87	1.054
B	Wood	3.53	1.261
A	Lab	16.4	2.797
B	Lab	21.5	3.068

have values more closely approximating what would come from a normal distribution. A similar analysis on the logarithms of the inferred geometric standard deviations, shown in the bottom part of Table 22–6, was also performed.

The ANOVA procedure was from the statistical library provided by SAS, Inc. (***SAS***). There were two readings at each of two levels for each of two factors: aerosol, sampler. This is a balanced design. As applied to the logarithm of the mass median diameter, the procedure indicated that there was a difference between the two dusts at a better than 99.9% level of confidence and a difference between the two samplers at a better than 95% level of confidence. As applied to the log of the mass geometric standard deviation, the ANOVA procedure indicated there was a difference between the two dusts at more than a 98% level of confidence and a difference between the two samplers at a better than 90% level of confidence.

When ANOVA was applied to the untransformed mass median diameters and to the geometric standard deviations, there was no statistical significance to the differences between the samplers. Thus, using the logarithmic transformation, which better matched the assumptions of ANOVA, produced a more sensitive test for discriminating between the samplers. ANOVA applied to the untransformed mass median diameters and geometric standard deviations indicated that the dusts were different at the 98% confidence level (median diameters) and at the 90% confidence level (geometric standard deviations).

We performed the Student's t test on the mass median diameters, the geometric standard deviations, and their logarithms (using $N_1 = N_2 = 2$). The degrees of freedom are calculated not from the number of sizing intervals, but from the number 2, for the two samples being compared. Multiple determinations of the distribution with a mass sampler do allow statistical comparison, because the variability can be estimated from the replications. The estimates of the mass median diameter can be used as the data being analyzed. The logarithms of the mass median diameters give the mean logarithms, and these means can be compared: 1.818 and 1.895 for the wood dust; 0.199 and 0.285 for the laboratory air. The difference between the two estimates for the mean logarithm for the same aerosol is an estimate of the standard deviation of the estimate of the mean logarithm. The differences are 0.08 for the wood dust and 0.09 for the laboratory aerosol, suggesting that we can pool the variances. Using Eqs. 22–30 and 22–31 for comparing means by the t test, with $N_1 = 2$, $N_2 = 2$ for the number of samples for each mean, we obtained $t = 27.9$, significant at a better than 99% level. The t test is appropriate because the logarithms were distributed almost normally. It is not so clear that the t test is appropriate to analyze the logarithms of the geometric standard deviations, but the test is often not very sensitive to the form of the underlying distribution. Applying it to the logarithms of the geometric standard deviations gave us $t = 10.4$, also significant at a 99% level. The t test applied to the untransformed median diameters gave $t = 20.2$, a difference between the wood dust and the laboratory air at greater than the 99% confidence level. Applied to the untransformed geometric standard deviations, the t test gave $t = 6.1$, significant at between 95% and 99%, thus somewhat less definitive.

Our statistical analyses confirmed what "common sense" indicates when looking at the graphs of the cumulative size distributions: The wood dust and the laboratory aerosol size distribution were different in median size and in slope.

COINCIDENCE ERRORS

When two or more particles are present simultaneously in the sensing zone of an instrument, they may be counted as 1 particle (type 1 coincidence), perhaps of a different size, or as 0 particle (type 0 coincidence), depending on the instrument. Coincidence is similar to "saturation," where the pulses from the sensing zone come too rapidly for the electronics to separate. In general, coincidence causes a loss of particles, underestimating concentration, and a shift in the particle size distribution to larger sizes.

Coincidence has been treated by many authors, including Jaenicke (1972); Bader et al. (1972); Julanov et al. (1984, 1986); Raasch and Umhauer (1984), and Cooper and Miller (1987). A series of useful articles on the topic are those of Knapp and Abramson (1994, 1996) and Knapp et al. (1994). When the particles arrive in the zone at random, the probability of having n particles in the zone can be described by the Poisson distribution. The probability that the zone is empty is $\exp(-cV)$, in which c is the concentration (number per extent) and V is the zone extent (volume, area, time, for example). The mean number will be cV. The apparent concentration (assuming coincidence type zero) is $(c)\exp(-cV)$, which is nearly c for $cV < 1$, and reaches a maximum at $cV = 1$; the concentration will appear to decrease if cV becomes > 1. If the coincidence causes a count in a larger size interval, then this type 1 coincidence causes less loss of count but does create spurious counts of larger particles, further distorting the size distribution (see, e.g., Raasch and Umhauer, 1984). For type 1 coincidence, the counting efficiency becomes $[1 - \exp(-cV)]/cV$. These formulas can be adapted to correct the counts for coincidence losses.

The statistics of coincidence and time of arrival was explored in a pair of papers by Julanov et al. (1984, 1986), where methods for obtaining concentration estimates using the behavior of multiple nearly simultaneous arrivals were presented.

Counting errors of various types for surface scanners were analyzed by Pecen et al. (1987) for scanning-beam types and by Cooper and Miller (1987) and Cooper and Rottmann (1988) for vidicon-based (pixel) types. Special precautions with design and data analysis have to be taken for scanning instruments not to overcount particles on surfaces due to multiple intersections of the scan with the same particle (Galbraith and Neukermans, 1987). Such surface monitors can be calibrated with patterned surfaces, with surfaces having a known number of particles, or—by repeated counting—with surfaces having an unknown number of particles of the same size (Cooper and Neukermans, 1991).

More complicated particle-counting instruments, such as the Aerodynamic Particle Sizer and the Aerosizer, have more complicated coincidence effects (Heitbrink et al., 1991; see Chapter 17 for more information).

Measurement of particles in liquids and on surfaces has been covered extensively in a handbook edited by Knapp et al. (1996). Dahneke (1983) covered the use of the variation over time of light scattered from particles in fluids ("quasi-elastic light scattering") to infer particle size from correlations.

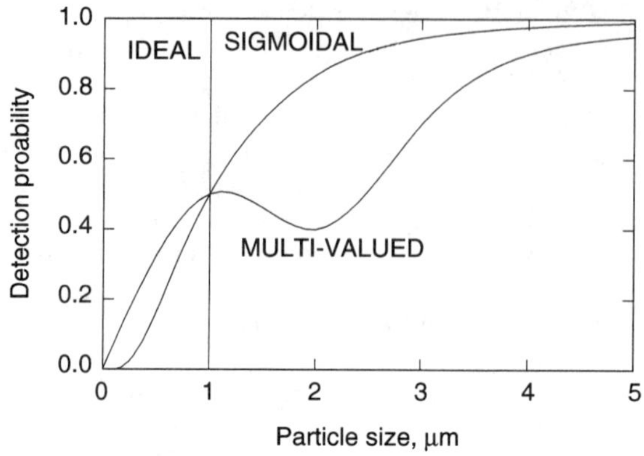

Fig. 22–8. Some hypothetical response functions for a particle-sizing instrument: step function, sigmoidal function, and multivalued function.

CHOOSING SIZING INTERVAL DEMARCATIONS

Particle-sizing techniques that classify particles into size intervals have boundaries between the size intervals that are not sharp step functions, but are generally sigmoidal (Fig. 22–8). For example, the instrument may go from having zero probability of including a particle in the interval to having a probability of 1 of including the particle (absent other sizing stages), as the particle size increases. The common approach to ascribing a characteristic size ("cut size") to one end of the interval is to use the size corresponding to a 0.5 probability of detection. A somewhat more precise choice is the size that divides the response function into equal regions of underestimation and overestimation (Cooper and Guttrich, 1981).

Calibration of multistage classifiers that takes into account the interaction among the response functions for various stages has been modeled by Lu et al. (1993) and demonstrated experimentally by Lu et al. (1995). References to some earlier work are available therein.

If the precision obtained from such cut-diameter approximations is inadequate, one will be tempted to improve it by techniques of data inversion (sometimes called *deconvolution* or *unfolding*), as discussed next.

DATA INVERSION

If one cannot obtain exact measurements, then perhaps one can use knowledge of the behavior of the measurement instrument to correct the data. This is the goal of various data inversion methods that have been presented in the literature of aerosol science and technology and in the literature of many other fields as well (some of the early examples are cited by Cooper and Davis, 1972; Cooper, 1976; Yu et al., 1983; Yu and Gentry, 1984). There is even a journal (*Inverse Problems*) dedicated to this topic.

Although there are instances—such as computerized axial tomography—where data inversion works very well, there have been many alternatives published in the scientific literature, even just in the literature on aerosol science, criticizing previous methods and presenting a new method. There has yet to emerge a clear-cut best-choice method for analyzing particle size distribution data, which should be a warning to those who hope to salvage ambiguous data using mathematical magic.

An excellent review article with more than 50 references on the subject is that of Kandlikar and Ramachandran (1999). They describe the problem, many contexts, and various approaches. They provide a useful tabulation of the various approaches on the basis of technique, references, constraints, or a priori information needed, relative computational effort, and comments. The techniques were subdivided into (A) Linear: 1, least squares; 2, constrained least squares (Philips, 1962; Twomey, 1963, 1977); 3, Tikhonov regularization (Tikhonov and Arsenin, 1977); 4, synthesis of basis functions (Twomey, 1963). (B) Nonlinear: 1, the Chahine (1968) method; 2, the Twomey (1975) method. (C) Extreme value estimation (Paatero, 1991). (D) Bayesian (Ramachandran and Kandlikar, 1996). Kandlikar and Ramachandran (1999) concluded that "there is no single algorithm that can be considered to be universally superior to the others. . . ."

Typically, the "exact" solution to the input that created the output is unrealistic due to small errors in the data, in the instrument description, or in the problem definition. Various treatments of the data, the instrument description, or the description of the unknown are available to obtain more realistic results (e.g., smoother or non-negative) at the cost of reduced agreement with the measurements.

An ideal instrument would have a probability of 1 of responding to particles within the nominal size boundaries of the sizing channel and a probability of 0 of responding to particles outside the nominal size interval boundaries. Figure 22–8 shows some hypothetical response functions for one particle size boundary of one channel of an instrument. An ideal instrument would have a step-function response. Typically, one has a sigmoidal boundary. In some cases, the boundary is multivalued rather than monotonic, complicating the picture still further. The instrument would have a set of such response functions to make up the channel responses. An ideal instrument would not need to have its data subjected to data inversion. Typical instruments may profitably have their data inverted, depending on (1) the size of the channel widths, (2) the size interval over which the response goes from nearly 0 to nearly 1, and (3) the size interval over which the particle size distribution changes appreciably. Instruments that have multivalued responses are not good candidates for data inversion unless the sizing intervals are chosen to segregate completely the regions of multivaluedness within distinct intervals.

This section outlines the topic of data inversion. As noted, much has been written on the topic. Tikhonov and Arsenin (1977), as cited by Wolfenbarger and Seinfeld (1991), summarize the difficulties of these often ill-posed problems: "Solutions will become unstable when the number of data is large. Solutions are not unique. Exact solutions often do not exist." Ill-posed, or "ill-conditioned," problems typically show large changes in the inferred solutions due to small changes, or errors, in the data.

Before discussing data inversion below, we emphasize the advice from Noble (1969): The best way to deal with ill-conditioning, the sensitivity of the results to small changes in problem formulation or in the data, is to avoid it through the choice of measuring instrument and conditions of measurement. This advice was echoed by Cooper and Wu (1990) some 20 years later. More details are given below.

What follows in this section is intended for readers strong in mathematics.

The Problem: An Integral Equation

Assume that a multichannel instrument produces a set of data that gives the fraction of the aerosol (by count or mass, typically) in each of n channels. The response $F(i)'$ can be described by the integral equation

$$F(i)' = \int K(i,d) f(d) \mathrm{d}d + e(i) \qquad (22\text{–}33)$$

where $f(d)$ is the particle size frequency distribution and $K(i, d)$ is the probability that a particle of size d will be counted (or weighed) in the ith channel of the instrument. The data will usually have at least some error, $e(i)$, as well, error being the degree to which $F(i)'$ does not exactly match the integral.

Although $K(i, d)$ is known, generally through calibration and perhaps through theory, $f(d)$ is unknown. The n-channel instrument, $i = 1$ to n, gives only n data points from which one tries to infer $f(d)$, which is a continuous function. The best we can expect to do is to get n elements of information on $f(d)$. "Elements" is deliberately ambiguous. One kind of element is the value of $f(d)$ at n values of $d(i)$ throughout the range of the instrument. Another kind of element is the first n moments of the $f(d)$ distribution: the mean diameter, the mean squared diameter, the mean cubed diameter, or the mean, the variance, the kurtosis, and so forth. Another kind of element might be the $k(i)$ coefficients of a polynomial expansion, $f(d) = k(0) + k(1)d^1 + k(2)d^2 + \ldots$. Yet another kind of element would be the Fourier expansion coefficients. The list seems boundless. The most common choice is to obtain estimates of $f(d)$ or $F(d)$ at n or fewer values of d. With only n measurements, at most only n values of $f(d)$ (or other parameters) can be determined, and an infinite number of different curves can be drawn through a finite number of points, so the unknown function is never fully determined by the data alone.

Solving by Converting to a Set of Linear Equations

A simple method to convert the integral equation into a set of linear equations in coefficients $f[d(j)] = f(j)$ is to evaluate the integral with numerical quadrature, which means to evaluate the integral by evaluating the function at a set of d values. For example, the midpoint quadrature evaluates $f(d)$ at the midpoint of the sizing intervals 1 to m and produces the approximation

$$F(i)' = K[i, d(1)]f[d(1)]\Delta d(1) + \ldots + K[i, d(m)]f(m)\Delta d(m) + e(i) \tag{22–34}$$

for $j = 1, \ldots, m$. $\Delta d(j)$ is the width of the size interval for which $f(d(j))$ is at the midpoint in this formulation. The values of $K(i, d)$ are known, so the quadrature produces a set of n equations in m unknowns.

These equations can be written compactly:

$$F(i)' \cong \sum K(i, j)f(j)\Delta d(j) + e(i) \tag{22–35}$$

The "approximately equals" sign is used to emphasize that not only is there error in the data but also there is error in going from the integral equation to the set of linear equations. For n equations in m unknowns: (1) we may have a solution if the number of equations equals the number of unknowns, $m = n$; (2) we do not have a solution if the number of equations is less than the number of unknowns, $n < m$; and (3) we may have many possible solutions if the number of equations is greater than the number of unknowns, $n > m$. Where $n > m$, more data (equations) than unknowns, we need to choose one of the possible solutions, generally by requiring that the chosen solution satisfy another constraint, such as being the smoothest of the alternative solutions or being the least-squares fit to the data.

The unknown values are often labeled as $x(j)$, and the known coefficients on the right-hand side are labeled $A(i, j)$. This avoids making the decision whether the user is solving for $f(j)$ or $f(j)\ \Delta d(j)$. The known values are often labeled as $b(i)$, and they may be part of a cumulative distribution description or part of a frequency distribution description. Regardless of the details of the description, we want to solve the set of linear equations

$$b(i) \cong \sum A(i, j)x(j) + e(i) \tag{22–36}$$

which in vector-matrix notation becomes

$$b \cong Ax + e \tag{22–37}$$

Some definitions are needed: $A(i, j)^T$ = transpose of A; $A(i, j)^T = A(j, i)$. $1/A$ = inverse of A; $(A)(1/A) = I$, the identity matrix. $I(i, j) = 1$ for $i = j$; $I(i, j) = 0$ otherwise.

Solving the Equations when n Equations Equal m Unknowns

If we have n equations in m unknowns and if $n = m$ and if all the equations are linearly independent of each other (i.e., none is simply equal to a constant times another), then the solution is straightforward:

$$x \cong (1/A)b - (1/A)e \tag{22–38}$$

the x vector (our solution) is the inverse matrix operating on the data vector. In fact, we know b, but not e, so that we calculate an $x = (1/A)b$ that is in error by about $(1/A)e$.

An example from Cooper and Wu (1990) shows a simple two-by-two matrix from a hypothetical instrument with rather good resolution, which one can see by noting that the matrix is nearly the identity matrix (which would need no correction).

$$A(i,j) = \begin{bmatrix} 0.95 & 0.05 \\ 0.05 & 0.95 \end{bmatrix} \tag{22–39}$$

The inverse matrix is

$$1/A(i,j) = \begin{bmatrix} 1.056 & -0.056 \\ -0.056 & 1.056 \end{bmatrix} \tag{22–40}$$

The solutions from data inversion become

$$x(1) = 1.056\, b(1) - 0.056\, b(2)$$
$$x(2) = -0.056\, b(1) + 1.056\, b(2)$$

The corrections are of the same magnitude as the off-diagonal elements of $A(i, j)$. These are rather small corrections and do not change $x(i)$ from the data $b(i)$ by very much nor would they multiply the error term by any large factor. Examples where much larger correction factors need be applied, having greater error magnification, were given in the same article.

Solving the Equations when n Equations Exceed m Unknowns

Where we have more equations than unknowns, we need an added constraint to select one solution from the possible solutions. Generally, the constraint chosen is to minimize the sum of the squared differences between what is calculated from Ax and what was measured, b. The least-squares criterion can be shown to lead to the following equation:

$$x \cong (1/A^T A)(A^T b) - (1/A^T A)(A^T e) \tag{22–41}$$

where, again, the error term cannot be calculated because e is unknown. The approximate solution for x is obtained by finding the matrix that is the product of A^T and A, and then taking its inverse and using this inverse to operate on the vector $A^T b$.

Perhaps surprisingly, this is a relatively convenient approach because it comes down to using the readily available programs for multiple linear regression to solve the equations

$Ax = b$. Simply use the multiple regression programs, giving the data, $b(i)$, and the $A(i, j)$ coefficients. The solution will be the least-squares fit, $x(j)$.

The difficulties arise when the answers are unrealistic: negative values for some of the size distribution components or components that oscillate as particle size increases. To get around this problem, one can try to solve for fewer unknowns by reformulating the equations. For example, $x(k)$ and $x(k + 1)$ can be combined to form a new $x(k')$, for two or more of the adjacent intervals and the set of equations solved again.

Inversion as Regression

Once it is recognized that the solution of the linear equations is simply linear regression, widely studied by statisticians and mathematicians, then a deep and broad literature opens up. There are many books on regression, and various computer programs are available, such as those from ***SAS***. An excellent source is the book by Draper and Smith (1981). Many candidate methods are revealed; for example, (1) one can choose to select only those $x(i)$ values that are statistically significantly different from zero (Student's t test on the regression coefficients); (2) one can carry out stepwise regression, starting by keeping only the most statistically significant component or starting by dropping the least significant component; and (3) to the regression equations can also be added confidence intervals on the true means for the coefficients (the $x[i]$) and on individual values. For example, Liley (1992) used a generalized linear model (GLM) on optical particle counter data along with assumed size distributions of (1) the lognormal, (2) the power-law (Junge), and (3) the modified Gamma forms to obtain fits to the particle count data. Liley also showed the advantage of the square root transformation of the counts, as it made the variance almost independent of particle size, better matching the conditions under which such regression is valid.

A measure of the sensitivity of the inferred solutions to the errors in measurement and definition is the condition number (Forsythe and Moler, 1967). In unfortunate cases, this number multiplies the error e. One definition of the condition number, cond(A), shows that it can be as large as the largest matrix element of A divided by the smallest matrix element of ($1/A$). The condition number can be orders of magnitude larger than 1. Most multiple linear regression programs will not indicate what cond(A) is. To estimate the error magnification, simply make a small change in the data $b(j)$ for each channel, one after another, and look at the change in the computed results for $x(i)$.

Some Guidelines on Reducing Sensitivity to Errors

Designing an experiment to reduce problems with data inversion can be done by calculating the condition number for various options (Cooper, 1974; Jochum et al., 1981; Yu, 1983; Farzanah et al., 1985; Kaplan and Gentry, 1987) or inspecting the inversion matrix, $(1/A^TA)A^T$, for particularly large coefficients (Cooper and Wu, 1990).

Even without calculating an inverse, one can examine the original matrix. The best matrices to invert are those that are most nearly like the identity matrix, $I(i, j)$, which has 1 on its upper left to lower right diagonal and 0 elsewhere.

Methods that Change the Response Function: Smoothing, Regularization

Wolfenbarger and Seinfeld (1990, 1991) favored the technique of "regularization," associated with Tikhonov (1963). Related books are those by Tikhonov and Arsenin (1977) and Groetsch (1984). Regularization solves a family of problems that are almost the same as the original problem, with the degree of difference specified by the investigator to reflect the uncertainty in the data. The modified problems are chosen to have stable inverses. Phillips (1962) and Twomey (1963) applied this approach to problems in which the solution was not constrained to be non-negative but was constrained in other ways, such as the smoothest solution fitting the data within a specified degree of precision. Wolfenbarger and Seinfeld applied regularization and the constraint that the solutions be nonnegative. Wolfenbarger and Seinfeld noted, as others have, that the results can be quite misleading if the device has poor sensitivity or poor resolution in size ranges where the size distribution is still significant. This is one advantage for combining the results of more than one type of instrument (Wu et al., 1989).

Defining a level of smoothness to rule out very unsmooth (highly oscillatory) solutions has its complications. Wahba (1985) discussed this in the context of the generalized cross-validation (GCV) inversion method. Wahba (1977) presented suggestions for handling the difficulties caused by noisy data. The Phillips-Twomey smoothing approach was used by Amato et al. (1995) on atmospheric optical depths measured with a grating spectrometer; a chi-squared criterion was used to select the optimal value of the smoothing parameter.

Methods that Change the Data

As with any signal, the data can be decomposed by Fourier analysis into a sample from a weighted sum of sine waves, $\sin(2\pi fd)$, where f is a frequency in units of reciprocal length, such as μm^{-1}. High-frequency components are those that vary greatly with small changes in particle diameter. High-frequency components in the data become high-frequency components in the results. By averaging the data over several runs and then carrying out the inversion on the averages, some reduction in these components can be achieved. Fitting curves to the data and then inverting the values interpolated by the curves is a variation on this theme. There should be some improvement, and one is not requiring that the answer look plausible so that if the data are still poor an implausible answer can result and warn the user of the method. Markowski (1987) used smoothing the data to improve the Twomey (1975) nonlinear method.

Methods that Constrain the Answer

One approach to preventing unrealistic solutions is to constrain the solutions. A simple constraint is normalization: The sum of the size distribution contributions must be the total number or mass or add to 1.0, and so forth. Another constraint is non-negativity, not allowing a negative solution. These two were applied, using nonlinear programming (simplex minimization; after Nelder and Mead, 1965) by Cooper and Spielman (1976) for the variable-slit impactor and by Kapadia (1980) and Helsper et al. (1982) for the electrical aerosol analyzer. Removing the high-frequency solution components can be used to produce smoother solutions: Twomey (1965) set lower limits on eigenvalues and used these to drop eigenvectors that contributed to the solution vector, a method related to smoothing to avoid highly oscillatory solutions that are clearly nonphysical. Similar work was presented by Baker et al. (1964), but neither they nor Twomey seems to have been aware of each other's publications. Twomey (1975) achieved non-negativity in the solutions with his nonlinear method of inversion (see the criticism by Cooper and Wu, 1990, however). Crump and Seinfeld (1982) demonstrated shortcomings in Twomey's approach and presented their own, involving constraints and "crossvalidation," later found to work in some cases and not in others. Wolfenbarger and Seinfeld pointed out the weaknesses in other approaches and incorporated non-negativity and smoothness in their regularization methods.

Constraining the solution helps to reject implausible solutions, but may also hide the fact that the solutions are still wrong (e.g., good and bad solutions in Fig. 1 in Wolfenbarger and Seinfeld, 1991). Ramachandran and Leith (1992) used a deconvolution technique on their multispectral light extinction data. Instead of using a set of orthogonal functions for the unknown function, they used such a set for the second derivative of the unknown function, using cross-validation to determine how much smoothing to use (the smallest eigenvalues to retain). Their approach was robust in the presence of as much as 10% noise. They studied the effects of increasing the number of measurements and found that more measurements did not necessarily provide better results. Subsequently Ramachandran and Vincent (1997) compared this method with one constraining the answer to be lognormal and found that the latter approach was somewhat better for the impactor data they studied (see following paragraph, also).

A particularly constrained method is to choose to fit a preselected functional form to the data, such as a set of normal distributions (Jaenicke, 1972) or one or more lognormal distributions (Kubie, 1971; Puttock, 1981). Such approaches give plausible results, sometimes with answers extending beyond the range of the data, which can be misleading. Chang et al. (1995) assumed lognormality for the components and employed light scattering and extinction data to determine size distributions in multicomponent aerosols. The assumption of lognormality was used by Morawska and Jamriska (1997) to obtain the parameters of radon progeny as measured with wire screen diffusion batteries; they reported that this procedure performed better than did the Twomey (1975) deconvolution algorithm.

Another form of constrained solution is the "apparent diameter" method, as used by Yu et al. (1983) and Yu and Gentry (1984). The apparent diameter is that which gives the measured response when substituted into the kernel. By assuming a power-law form for the kernel (collection efficiency here) and a lognormal form for the aerosol distribution, they extended this to obtain the geometric standard deviation, the second parameter for the lognormal distribution.

Extreme value estimation (EVE) tries to put bounds on the upper and lower values of the size distribution elements that are consistent with the data. Some success has been had with this approach (Paatero and Raunemaa, 1989), originally set forth by Replogle, Holcomb, and Burrus at a time when the computations were much less convenient than they are now, which may explain why it was not widely used immediately thereafter.

Other Methods

The expectation-maximization (EM) algorithm has had some success (Maher and Laird, 1985) in inverting diffusion battery data, a challenging task. The method takes advantage of relations among columns as well as rows in the matrices to be inverted and of the Poisson nature of particle counts, with the errors proportional to the square root of the counts. Used to infer particle size distribution values from diffusion experiments involving graded-screen-array diffusion batteries, the EM algorithm also did well (Knutson et al., 1997), although some arbitrary decisions regarding smoothing and the number of iterations to use still had to be made.

An application of deconvolution was the extraction of size distribution values from photon correlation data (Bertero et al., 1989), the response of particles undergoing Brownian motion to illumination with coherent (laser) light. The inversion is equivalent to the difficult problem of inverting the Laplace transform numerically, known to be highly sensitive to errors in the data.

Some authors have focused on rewriting the equations to estimate the various integral moments of the size distributions, the mean, variance, kurtosis, and so forth. Unfortunately, as shown by White (1990), some distribution types cannot be recovered from these moments. Rather different distributions can have the same values of the mean, variance, and kurtosis, for example.

For diffusion battery inversion, Yu and Gentry (1984) compared their apparent size method with nonlinear inversion and linear inversion with cross-validation. The apparent size method depends on tractable choices of the particle size distribution and the kernel of the integral equation.

REFERENCES

Aitchison, J. and J. A. C. Brown. 1957. *The Lognormal Distribution*. Cambridge: Cambridge University Press.

Amato, U., F. Esposito, C. Serio, G. Pavese, and F. Romano. 1995. Inverting high spectral resolution aerosol optical depth to determine the size distribution of atmospheric aerosol. *Aerosol Sci. Technol.* 23: 591–602.

Bader, H., M. R. Gordon, and O. B. Brown. 1972. Theory of coincidence and simple practical methods of coincidence count correction for optical and resistive pulse particle counters. *Rev. Sci. Instr.* 43: 1407–1412.

Baker, C. T. H., L. Fox, D. F. Mayers, and K. Wright. 1964. Numerical solution of Fredholm integral equations of the first kind. *Comput. J.* 7:141–148.

Beddow, J. K. 1980. *Particulate Science and Technology*. New York: Chemical Publishing Company.

Beddow, J. K. 1997. *Image Analysis Sourcebook*. Santa Barbara, CA: American Universities Science and Technology Press.

Bertero, M., P. Boccacci, C. De Mol, and E. R. Pike. 1989. Extraction of polydispersity information from photon correlation data. *J. Aerosol Sci.* 20:91–99.

Box, G. E. P., W. G. Hunter, and J. S. Hunter. 1978. *Statistics for Experimenters*. New York: John Wiley & Sons.

Bunz, H., W. Schock, M. Koyro, J. W. Gentry, S. Plunkett, M. Runyan, C. Pearson, and C. Wang. 1987. Application of the log beta distribution to aerosol size distributions. *J. Aerosol Sci.* 18:663–666.

Chahine, M. T. 1968. Determination of the temperature profile in an atmosphere from its outgoing radiance. *J. Opt. Soc. Am.* 58:1634–1637.

Chang, H., W. Y. Lin, and P. Biswas. 1995. An inversion technique to determine the aerosol size distribution in multicomponent systems from in situ light scattering measurements. *Aerosol Sci. Technol.* 22: 24–32.

Christensen V. R., W. Eastes, R. D. Hamilton, and A. W. Struss. 1993. Fiber diameter distributions in typical MMVF wool insulation products. *Am. Ind. Hyg. Assoc. J.* 54:232–238.

Cleveland, W. S. 1985. *The Elements of Graphing Data*. Monterey, CA: Wadsworth Advanced Books.

Cooper, D. W. 1974. *The Variable-Slit Impactor and Aerosol Size Distribution Analysis*. Ph.D. Dissertation, Harvard Division of Engineering and Applied Physics, Cambridge, MA.

Cooper, D. W. 1976. Data inversion method and error estimate for cascade centripeters. *Am. Ind. Hyg. Assoc. J.* 37:622–627.

Cooper, D. W. 1982. On the products of lognormal and cumulative lognormal particle size distributions. *J. Aerosol Sci.* 13:111–120.

Cooper, D. W. 1991a. Applying a simple statistical test to compare two different particle counts. *J. Aerosol Sci.* 22:773–777.

Cooper, D. W. 1991b. Comparing three environmental particle size distributions: Power law (FED-STD-209D), lognormal, approximate lognormal (MIL-STD-1246B). *J. Instr. Environ. Sci.* 5:21–24.

Cooper, D. W. and J. W. Davis. 1972. Cascade impactors for aerosols: Improved data analysis. *Am. Ind. Hyg. Assoc. J.* 33:79–89.

Cooper, D. W., H. A. Feldman, and G. R. Chase. 1978. Fiber counting: A source of error corrected. *Am. Ind. Hyg. Assoc. J.* 39:362–367.

Cooper, D. W. and G. L. Guttrich. 1981. A study of the cut diameter concept for interpreting particle sizing data. *Atmos. Environ.* 15:1699–1707.

Cooper, D. W. and R. J. Miller. 1987. Analysis of coincidence losses for a monitor of particle contamination on surfaces. *J. Electrochem. Soc.* 134:2871–2875.

Cooper, D. W. and A. Neukermans. 1991. Estimating an instrument's counting efficiency by repeated counts on one sample. *J. Colloid Interface Sci.* 147:98–102.

Cooper, D. W. and H. R. Rottmann. 1988. Particle sizing from disk images by counting contiguous grid squares or vidicon pixels. *J. Colloid Interface Sci.* 126:251–259.

Cooper, D. W. and L. A. Spielman. 1976. Data inversion using nonlinear programming with physical constraints: Aerosol size distribution measurement by impactors. *Atmos. Environ.* 10:723–729.

Cooper, D. W. and J. J. Wu. 1990. The inversion matrix and error estimation in data inversion: Application to diffusion battery measurements. *J. Aerosol Sci.* 21:217–226.

Crow, E. L., F. A. Davis, and M. W. Maxfield. 1990. *Statistics Manual.* New York: Dover.

Crow, E. L. and K. Shimizu. 1988. *Lognormal Distributions.* New York: Marcel Dekker.

Crump, J. G. and J. H. Seinfeld. 1982. A new algorithm for inversion of aerosol size distribution data. *Aerosol Sci. Technol.* 1:15–34.

Dahneke, B. 1983. *Measurement of Suspended Particles by Quasi-Elastic Light Scattering.* New York: John Wiley & Sons.

Draper, N. R. and H. Smith. 1981. *Applied Regression Analysis*, 2nd Ed. New York: John Wiley & Sons.

Epstein, B. 1947. The mathematical description of certain breakage mechanisms leading to the logarithmic-normal distribution. *J. Franklin Inst.* 244:471–477.

Evans, J. S., D. W. Cooper, and P. Kinney. 1984. On the propagation of error in air pollution measurements. *Environ. Mon. Assess.* 4:1322–1353.

Farzanah, F. F., C. R. Kaplan, P. Y. Yu, J. Hong, and J. W. Gentry. 1985. Condition numbers as criteria for evaluation of atmospheric aerosol measurement techniques. *Environ. Sci. Technol.* 19:121–126.

Fayed, M. E. and L. Otten, eds. 1997. *Handbook of Powder Science & Technology.* New York: Chapman & Hall.

Fogel, P., D. Y. Hanton, A. DeMeringo, and C. Morscheidt. 1999. The reliability of the dimensional measurements of man-made vitreous fibers used for biopersistence assays: A statistical approach. *Aerosol Sci. Technol.* 30:571–581.

Forsythe, G. E. and C. B. Moler. 1967. *Computer Solution of Linear Algebraic Systems.* Englewood Cliffs, NJ: Prentice-Hall.

Friedlander, S. K. 1977. *Smoke, Dust and Haze.* New York: John Wiley & Sons.

Fuchs, N. A. 1964. *Mechanics of Aerosols.* New York: Pergamon.

Galbraith, L. and A. Neukermans. 1987. Contamination detection on semiconductor wafers. In *Lasers in Microlithography*. Proceedings of SPIE. eds. J. S. Batchelder, D. J. Ehrlich, and J. Y. Tsao. 774: 32–40.

Gentry, J. W. 1977. Estimation of parameters for a log-normal distribution with truncated measurements. *J. Powder Technol.* 18:225–229.

Groetsch, C. W. 1984. *The Theory of Tikhonov Regularization for Fredholm Equations of the First Kind.* London: Pitman.

Hays, W. L. and R. L. Winkler. 1970. *Statistics.* New York: Holt, Rinehart and Winston.

Heintzenberg, J. 1994. Properties of the log-normal particle size distribution. *Aerosol Sci. Technol.* 21: 46–48.

Heitbrink, W., P. A. Baron, and K. Willeke. 1991. Coincidence in time-of-flight aerosol spectrometers: Phantom particle creation. *Aerosol Sci. Technol.* 14:112–126.

Helsper, C., H. Fissan, A. Kapadia, and B. Y. H. Liu. 1982. Data inversion by simplex minimization for the electrical aerosol analyzer. *Aerosol Sci. Technol.* 1:135–146.

Hinds, W. C. 1999. *Aerosol Technology*. 2nd ed. New York: John Wiley & Sons.

Jaenicke, R. 1972. The optical particle counter: Cross-sensitivity and coincidence. *J. Aerosol Sci.* 3:95–111.

Jain, S., D. J. Skamser, and T. T. Kodas. 1997. Morphology of single-component particles produced by spray pyrolysis. *Aerosol Sci. Technol.* 27:575–590.

Jochum, C., P. Jochum, and B. R. Kowalski. 1981. Error propagation and optimal performance in multicomponent analysis. *Anal. Chem.* 53:85–92.

Julanov, Y. V., A. A. Lushnikov, and I. A. Nevskii. 1984. Statistics of multiple counting in aerosol counters. *J. Aerosol Sci.* 15:69–79.

Julanov, Y. V., A. A. Lushnikov, and I. A. Nevskii. 1986. Statistics of multiple counting in aerosol counters—II. *J. Aerosol Sci.* 17:87–93.

Kandlikar, M. and G. Ramachandran. 1999. Inverse methods for analyzing aerosol spectrometer measurements: A critical review. *J. Aerosol Sci.* 30:413–437.

Kapadia, A. 1980. *Data Reduction Techniques for Aerosol Size Distribution Measuring Instruments*. Ph.D. Thesis, Department of Mechanical Engineering, University of Minnesota, Minneapolis, MN.

Kaplan, C. and J. W. Gentry. 1987. Use of condition numbers for short-cut experimental design. *Am. Inst. Chem. Eng. J.* 33:681–685.

Kaye, B. H. 1989. *A Random Walk Through Fractal Dimensions*. New York: VCH.

Knapp, J. Z. and L. R. Abramson. 1994. A new coincidence model for single particle counters, part I: Theory and experimental verification. *J. Pharm. Sci. Technol.* 48:110–134.

Knapp, J. Z. and L. R. Abramson. 1996. A new coincidence model for single particle counters, part III: Realization of single particle counting accuracy. *PDA J. Pharm. Sci. Technol.* 50:99–122.

Knapp, J. Z., T. A. Barber, A. Lieberman, eds. 1996. *Liquid and Surface Borne Particle Measurement Handbook*. New York: Marcel Dekker.

Knapp, J. Z., A. Lieberman, and L. R. Abramson. 1994. A new coincidence model for single particle counters, part II: Advances and applications. *PDA J. Pharm. Sci. Technol.* 48:255–292.

Knutson, E. O., A. C. George, and K. W. Tu. 1997. The graded screen technique for measuring the diffusion coefficient of radon decay products. *Aerosol Sci. Technol.* 27:604–624.

Knutson, E. O. and P. J. Lioy. 1995. Measurement and presentation of aerosol size distributions. In *Air Sampling Instruments*. Cincinnati, OH: American Conference of Governmental Industrial Hygienists. B. S. Cohen and S. V. Hering. eds.

Kolmogorov, A. N. 1941. Uber das logarithmisch normale Verteilungsgesetz der Dimensionen der Teilchen bei Zerstuckelung. *C.R. Acad. Sci.* (*Doklady*) *URSS* XXXI:99–101.

Kubie, G. 1971. A note on the treatment of impactor data for some aerosols. *J. Aerosol Sci.* 2:23–30.

Lefebvre, A. H. 1989. *Atomization and Sprays*. Bristol, PA: Taylor & Francis.

Liley, J. B. 1992. Fitting size distributions to optical particle counter data. *Aerosol Sci. Technol.* 17:84–92.

Lu, Z., A. Li, and D. W. Cooper. 1993. Correcting particle size distribution data for the differences between single-stage and multi-stage collection or counting efficiencies. *Part. Sci. Technol.* 11:49–55.

Lu, Z., A. Li, and C. Fu. 1995. Experimental verification of using cumulative efficiency curves for determining stage constants of a cascade impactor. *Aerosol Sci. Technol.* 23:253–256.

Maher, E. F. and N. M. Laird. 1985. EM algorithm reconstruction of particle size distributions from diffusion battery data. *J. Aerosol Sci.* 16:557–570.

Mandelbrot, B. 1977. *The Fractal Geometry of Nature*. New York: Freeman.

Markowski, G. R. 1987. Improving Twomey's algorithm for inversion of aerosol measurement data. *Aerosol Sci. Technol.* 7:127–141.

McCartney, E. J. 1976. *Optics of the Atmosphere*. New York: John Wiley & Sons.

Meyer, S. L. 1975. *Data Analysis for Scientists and Engineers*. New York: John Wiley & Sons.

Morawska, L. and M. Jamriska. 1997. Determination of the activity size distribution of radon progeny. *Aerosol Sci. Technol.* 26:459–468.

Myojo, T. 1999. A simple method to determine the length distribution of fibrous aerosols. *Aerosol Sci. Technol.* 30:30–39.

Nelder, A. and R. Mead. 1965. A simplex method for function minimization. *Comput. J.* 7:308–312.

Noble, B. 1969. *Applied Linear Algebra*. Englewood Cliffs, NJ: Prentice-Hall.

Paatero, P. 1991. Extreme value estimation, a method for regularizing ill-posed inversion problems. In *Ill-Posed Problems in Natural Sciences. Proc. of Intl. Conf., Moscow*. Utrecht, The Netherlands: VSP.

Paatero, P. and T. Raunemaa. 1989. Analysis of CO_2 thermograms by the new extreme-value estimation (EVE) deconvolution principle. *Aerosol Sci. Technol.* 10:365–369.

Pecen, J., A. Neukermans, G. Kren, and L. Galbraith. 1987. Counting errors in particulate detection on unpatterned wafers. *Solid State Technol.* 30:1422–1454.

Phillips, D. L. 1962. A technique for the numerical solution of certain integral equations of the first kind. *J. Assoc. Comput. Mach.* 9:84–97.

Puttock, J. S. 1981. Data inversion for cascade impactors: Fitting sums of lognormal distributions. *Atmos. Environ.* 15:1710–1716.

Raasch, J. and H. Umhauer. 1984. Error in the determination of particle size distributions caused by coincidences in optical particle counters. *Part. Charact.* 1:53–58.

Ramachandran, G. and M. Kandlikar. 1996. Bayesian analysis for inversion of aerosol size distribution data. *J. Aerosol Sci.* 27:1099–1112.

Ramachandran, G. and D. Leith. 1992. Extraction of aerosol-size distributions from multispectral light extinction data. *Aerosol Sci. Technol.* 17:303–325.

Ramachandran, G. and P. C. Reist. 1995. Characterization of morphological changes in agglomerates subject to condensation and evaporation using multiple fractal dimensions. *Aerosol Sci. Technol.* 23:431–442.

Ramachandran, G. and J. H. Vincent. 1997. Evaluation of two inversion techniques for retrieving health-related aerosol fractions from personal cascade impactor measurements. *Am. Ind. Hyg. Assoc. J.* 58:15–22.

Reist, P. C. 1984. *Introduction to Aerosol Science*. New York: Macmillan.

Shen, A.-T. and T. A. Ring. 1986. Distinguishing between two aerosol size distributions. *Aerosol Sci. Technol.* 5:477–482.

Syvitski, J. P. M., ed. 1991. *Principles, Methods and Application of Particle Size Analysis*. Cambridge: Cambridge University.

Tikhonov, A. N. 1963. On the solution of incorrectly posed problems and the method of regularization. *Soviet Math.* 4:1035–1038.

Tikhonov, A. N. and V. Y. Arsenin. 1977. *Solutions of Ill Posed Problems*. Washington, DC: V. H. Winston and Sons.

Twomey, S. 1963. On the numerical solution of Fredholm integral equations of the first kind by the inversion of the linear system produced by quadrature. *J. Assoc. Comput. Mach.* 10:97–101.

Twomey, S. 1965. The application of numerical filtering to the solution of integral equations encountered in indirect sensing measurements. *J. Franklin Inst.* 279:95–109.

Twomey, S. 1975. Comparison of constrained linear inversion and an iterative nonlinear algorithm applied to the indirect estimation of particle size distributions. *J. Comput. Phys.* 18:188–200.

Twomey, S. 1977. *Introduction to the Mathematics of Inversion in Remote Sensing and Indirect Measurements*. New York: Elsevier.

U.S. General Services Administration. 1992. Federal Standard 209E, airborne particulate cleanliness classes in cleanrooms and clean zones. *J. Environ. Sci.* 35(5):27–32.

Wahba, G. 1977. Practical approximate solutions to linear operator equations when the data are noisy. *SIAM J. Numer. Anal.* 14:645–667.

Wahba, G. 1985. A comparison of GCV and GML for choosing the smoothing parameter in the generalized spline smoothing problem. *Ann. Stat.* 13:1378–1402.

White, W. H. 1990. Particle size distributions that cannot be distinguished by their integral moments. *J. Colloid Interface Sci.* 135:297–299.

Wolfenbarger, J. K. and J. H. Seinfeld. 1990. Inversion of aerosol size distribution data. *J. Aerosol Sci.* 21:227–247.

Wolfenbarger, J. K. and J. H. Seinfeld. 1991. Estimating the variance in solutions to the aerosol data inversion problem. *Aerosol Sci. Technol.* 14:348–357.

Wu, J. J., D. W. Cooper, and R. J. Miller. 1989. Evaluation of aerosol deconvolution algorithms for determining submicron particle size distributions. *J. Aerosol Sci.* 20:477–482.

Wu, M. K. 1996. The roughness of aerosol particles: Surface fractal dimension measured using nitrogen adsorption. *Aerosol Sci. Technol.* 25:392–398.

Xie, Y. P. K. Hopke, G. Casuccio, and B. Henderson. 1994. Use of multiple fractal dimensions to quantify airborne particle shape. *Aerosol Sci. Technol.* 20:161–168.

Yu, P. Y. 1983. *The Simulation and Experimental Determination of Size and Charge Distributions of Fibrous Particles from Penetration Measurements*. Ph.D. Dissertation, University of Maryland, College Park, MD.

Yu, P. Y. and J. W. Gentry. 1984. A critical comparison of three size distribution analysis methods: Apparent size, non-linear inversion, and linear inversion with cross validation. *J. Aerosol Sci.* 15: 407–411.

Yu, P. Y., J. San, and J. W. Gentry. 1983. Application of the apparent diameter method for inversion of penetration: Isometric particles, eds. V. A. Marple and B. Y. H. Liu. *Aerosols in the Mining and Industrial Work Environments, Vol. 1, Fundamentals and Status*. Ann Arbor: Ann Arbor Science.

PART III
Applications

23 Nonspherical Particle Measurements: Shape Factors, Fractals, and Fibers

PAUL A. BARON

Centers for Disease Control and Prevention,* National Institute for Occupational Safety and Health, Cincinnati, OH

CHRISTOPHER M. SORENSEN

Department of Physics and Program for Complex Fluid Flows, Kansas State University, Manhattan, KS

JOHN E. BROCKMANN

Sandia National Laboratories,† Albuquerque, NM

INTRODUCTION

Many aerosol measurement techniques are based on the behavior of ideal aerosol particles, that is, spherical particles with a density close to $1000\,kg/m^3$ [$1\,g/cm^3$]. In dealing with most real-world particles, allowances must be made for nonspherical particle behavior and other characteristics. In many situations, nonideal particle behavior can be considered as a modification of ideal particle behavior by using a correction factor, commonly called the *shape factor*. Two types of particles, agglomerates (or clusters of particles) and fibers, have been dealt with extensively in the literature and are discussed in more detail in this chapter. As well as being important for predicting particle behavior, particle shape can provide clues to the particle formation mechanisms.

SHAPE FACTOR

The dynamic shape factor relates the drag on an irregular particle to the drag on a sphere with the same particle mass and composed of the same material. The dynamic shape factor facilitates the transformation between the mass (or volume) equivalent diameter and the aerodynamic equivalent diameter of nonspherical and agglomerated particles (Brockmann and Rader, 1990). This transformation is necessary in aerosol dynamics codes (Gelbard, 1982), which describe the aerodynamic behavior of aerosol particle distributions while conserving mass. Helton et al. (1986) have shown that the results of aerosol behavior calculations can

* Mention of product or company name does not constitute endorsement by the Centers for Disease Control and Prevention.

† Sandia is a multiprogram laboratory operated by Sandia Corporation, a Lockheed Martin Company, for the United States Department of Energy under contract DE-AC04-94AL85000.

Aerosol Measurement: Principles, Techniques, and Applications, Second Edition, Edited by Paul A. Baron and Klaus Willeke.

ISBN 0-471-35636-0

be highly sensitive to the dynamic shape factor. Brockmann (1985, 1987) reviewed the wide range of dynamic shape factors reported in the literature.

The dynamic shape factor, χ, is defined as the ratio of the drag force on a particle, F_D, to the drag force on that particle's mass equivalent sphere, F_{DM}, at the same velocity, V.

$$\chi = F_D / F_{DM} \tag{23–1}$$

The dynamic shape factor has been developed as a correction to Stokes law drag for nonspherical particles (Hinds, 1999), that is, a Stokes law drag ratio (Clift et al., 1978). This definition expresses the drag on a nonspherical particle in the Stokes law regime as

$$F_D = 3\pi\mu\chi d_m V / C(d_m) \tag{23–2}$$

where μ is gas absolute viscosity, V is gas velocity past the particle, d_m is mass equivalent diameter of the particle, and $C(d_m)$ is slip correction factor for the mass equivalent diameter.

The choice of particle diameter used in the expression for slip correction has been a subject of concern (Cheng et al., 1988). Here, the use of the same equivalent diameter in the argument of the slip correction as is used in the expression for drag simplifies the presentation. This treatment will introduce a slight particle size dependence on dynamic shape factor in the transition regime. In the case of spherical nonporous particles, this treatment reduces to the correct usage.

Several equivalent particle diameters are defined in the literature. The mass equivalent diameter, d_m, of a particle is the diameter of a nonporous sphere composed of the bulk particle material that has the same mass as the particle. The envelope equivalent diameter, d_e, of a particle is the diameter of a sphere, which is composed of the bulk particle material and included void spaces (or voids), that has the same mass as the particle. When there are no included voids in the particle, these two diameters are identical. When there are included voids in the particle, these two diameters are different and $d_e > d_m$. The mobility equivalent diameter, d_B, of a particle is the diameter of a sphere with the same dynamic mobility as the particle. The aerodynamic equivalent diameter, or aerodynamic diameter, d_{ae}, of a particle is the diameter of a sphere of unit specific gravity that settles at the same terminal velocity as the particle.

These equivalent diameters are related by the dynamic shape factor χ, the particle material density ρ_p, and the unit specific gravity density $\rho_0 = 1000\,\text{g/m}^3$ [$1\,\text{gm/cm}^3$], and by the effective particle density ρ_e of a particle containing voids.

$$d_{ae}^2 \rho_0 C(d_{ae}) = d_m^2 \rho_p C(d_m) / \chi \tag{23–3}$$

$$d_B / C(d_B) = d_m \chi / C(d_m) \tag{23–4}$$

$$\rho_e d_e^3 = \rho_p d_m^3 \tag{23–5}$$

When the dynamic shape factor is defined with the particle material bulk density, it is comprised of two components:

$$\chi = \kappa \delta C(d_m) / C(\delta d_m) \tag{23–6}$$

The first component κ is due solely to the shape of the particle envelope (Fuchs, 1964). For oblate and prolate spheroids (Fuchs, 1964) and rectangular prisms (Johnson et al., 1987), κ may be calculated. These calculated values compare well with experimentally determined values (Johnson et al., 1987). The second component δ is due to particle porosity. This second component is defined in terms of the particle bulk material density and the particle effective density as

$$\delta = (\rho_p/\rho_e)^{1/3} \tag{23–7}$$

The void fraction of a particle can be defined as $(1 - 1/\delta^3)$, and the envelope and mass equivalent diameters are related as

$$d_e = \delta d_m \tag{23–8}$$

Both κ and δ are potentially important for aggregate or agglomerate particles. Agglomerate particles with two basic morphologies are seen in the data of Kops et al. (1975), van de Vate et al. (1980), Allen and Briant (1978), Allen et al. (1978, 1979), Kasper and Shaw (1983), and Stöber et al. (1970). These morphologies are (1) a branched chain structure and (2) a compact aggregate with a more discernible enveloping shape. The dynamic shape factors of each of these two morphologies are also distinct. For the branched chain aggregates, the dynamic shape factor increases with the addition of primary spheres to the aggregate and is directly proportional to the cube root of the total number of primary particles (Stöber, 1971; Kops et al., 1975; Allen and Briant, 1978; Allen et al., 1979). This behavior is consistent with particles' shape determining their dynamic shape factor.

In compact agglomerates, the porosity component dominates the dynamic shape factor. Gieseke et al. (1977) define this component as the dynamic shape factor and employ it to characterize sodium oxide aerosols. The porosity component is also employed by Wegrzyn and Shaw (1979). For compact agglomerates, the shape factor is relatively independent of the number of primary particles but depends rather on the packing density of the primary particles in the aggregate (Kops et al., 1975; van de Vate et al., 1980). This behavior is consistent with the porosity of the agglomerate determining its dynamic shape factor.

FRACTAL PARTICLES

Introduction

Fractals are scale invariant objects. This means that they have dilation symmetry, that is, they look the same on all scales. Mathematical examples include the Koch curve, which has bumps upon bumps upon bumps, or the Sierpinski gasket, which has a descending series of holes, or the "dog chow" symbol shown in Figure 23–1, which has some similarity to the aggregates to be described here. Mathematical fractals are scale invariant over all scales, yet Nature provides many examples of objects that are fractals over a finite range of scales. Examples include coastlines that display inlets and peninsulas on a large scale (e.g., the western coastline of the island of Great Britain) and similar bumps and wiggles at much finer scale (e.g., a local map of a few miles of coastline). This scale invariance is not present in simple geometric objects. Thus, changing the scale of a circle makes it flatter or rounder; similarly, a square's corners get farther or closer apart. Fractal and geometric objects differ in another very important manner and that is in their dimensionality, a quantifiable parameter. Geometric objects have integer dimensions, while fractals have noninteger, fractal dimensions, D_f. An excellent introductory description of fractals is given by Family (1991).

The mathematical background of fractals and their relevance to natural objects put forward by Mandelbrot (1977, 1983) allowed Forrest and Witten (1979) to experimentally establish that random aggregates of metal smoke particles over a finite range of scales were fractals with a noninteger dimension. This seminal work caused an explosion of subsequent interest involving both simulation of aggregation mechanisms and experimental work on

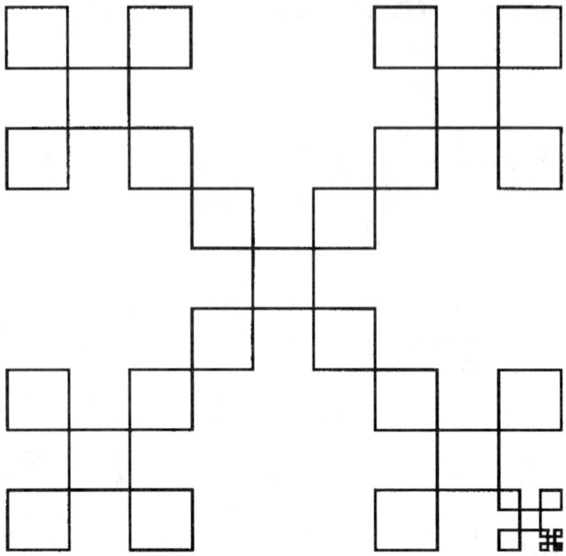

Fig. 23–1. "Dog chow" fractal with $D_f = \ln 5/\ln 3 = 1.465\ldots$. The pattern of five boxes in a larger box with three times the side length continues to all scales as implied by the box in the lower right.

aggregation in colloids and aerosols that continues today. The purpose of this chapter is to describe methods whereby the size and morphological parameters of fractal aggregates (i.e., aggregates composed of simple particles that display a fractal morphology) can be determined.

Fractal Aggregates

A fractal aggregate is an aggregate or cluster of particles that displays fractal scaling over the length scales from the primary or monomer particle size to the overall size of the aggregate. Fractal aggregates occur in aerosols and colloids as a result of random aggregation. Examples of both soot and titania fractal aggregates are given in Figures 23–2 and 23–3. The motion can be diffusive, ballistic (straight line), or in the crossover between these, and the resulting aggregate will be a fractal within the scale limits monomer to overall aggregate size. Both computer simulation and experiments have been important in establishing our knowledge of fractal aggregates (Family and Landau, 1984; Meakin, 1988; Viscek, 1992) because aggregation can be readily simulated on the computer. An example of a computer-generated fractal aggregate is given in Figure 23–4, and the similarity to the real-world examples given in Figures 23–2 and 23–3 is apparent. We now know that the aggregates can be classified into two major categories:

1. Particle-cluster aggregation, or diffusion-limited aggregation (DLA), which occurs when single monomers diffuse to and stick to a stationary, growing cluster (Witten and Sander, 1981). In three dimensions the clusters that result have a fractal dimension of $D_f \simeq 2.5$. It has recently been shown that those aggregates are not fractal over the entire range of their length scales (Oh and Sorensen, 1998).
2. Cluster–cluster aggregation, or diffusion-limited cluster aggregation (DLCA or DLCCA), which occurs when all clusters diffuse and then stick when they randomly touch (Kolb et al., 1983; Meakin, 1983). In three dimensions the clusters that result have a fractal dimension of $D_f \approx 1.75$ when the cluster motion is diffusive. Ballistic motion

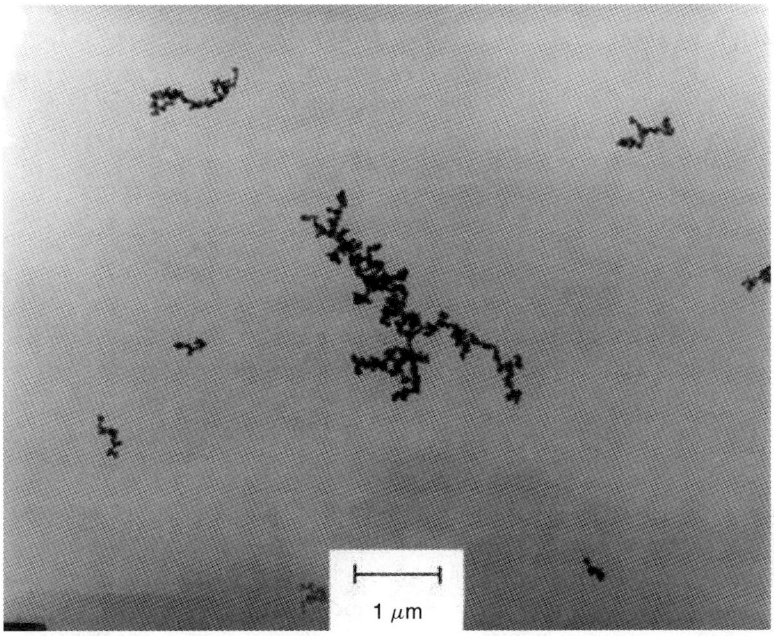

Fig. 23–2. Soot fractal aggregates from a premixed methane flame. The fractal dimension is $D_f \approx 1.8$.

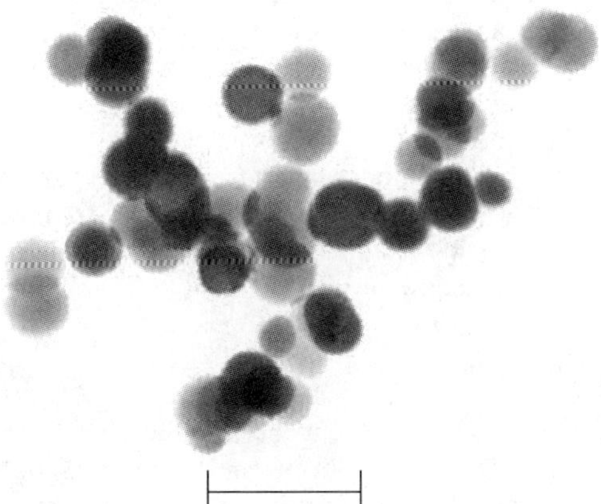

Fig. 23–3. TEM picture of titania (TiO_2) fractal aggregates with $D_f \approx 1.8$ produced by pyrolysis of titanium isopropoxide.

causes $D_f \approx 1.9$ (Meakin, 1984). If the sticking probability is significantly less than one, the reaction-limited cluster aggregation (RLCA) regime is entered for which $D_f \approx 2.15$, a situation important for many colloids (Lin et al., 1990). It is now known that only cluster–cluster aggregates occur in aerosols and colloids, with the DLA morphology finding application in other areas.

The meaning of the fractal dimension is in the relation between linear and volumetric size, the latter of which is linearly related to the mass or number of primary particles per aggre-

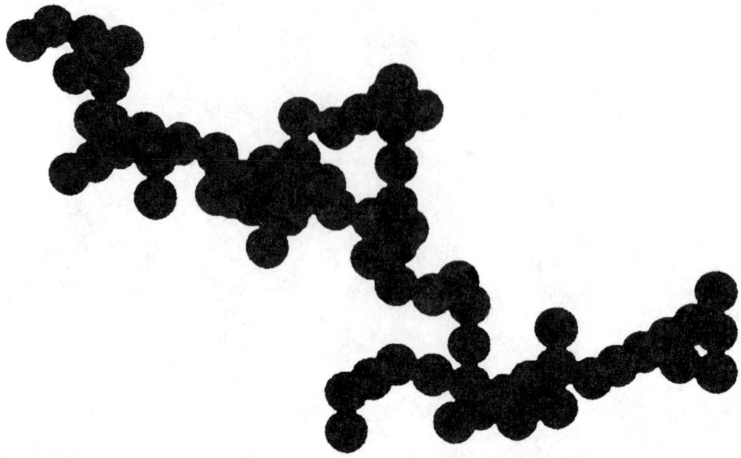

Fig. 23–4. A simulated DLCA aggregate with $D_f = 1.79$.

gate N. If R is a linear size of the aggregate, then $N \propto R^{D_f}$. We say that "the mass" N scales with linear size to the fractal dimension D_f. Geometric objects scale as well; for example, the volume of a sphere increases by a factor of eight when the diameter is doubled because its dimensionality is three, but fractals have a noninteger dimension D_f less than the embedding spatial dimension d, that is, $D_f < d$.

The beauty of the fractal dimension is that it allows a quantitative description of the degree of openness, or ramification, of the random aggregate. The smaller the D_f relative to the spatial dimension d, the more quickly the aggregate fills space as R increases. The fractal dimension plays a part in determining the aggregate density, the optical properties, the way in which it diffuses, and the kinetics of its further growth.

The relation between "mass" and linear size for a fractal aggregate can be quantified in three ways, all of which will form a basis for measurement of D_f to be describe below. We let a be the monomer (primary particle) radius and R_g be the radius of gyration (a root mean square radius; see below) of the aggregate. Then

$$N = k_o (R_g/a)^{D_f} \tag{23–9}$$

In Eq. 23–9 k_o is a constant of order unity (Wu and Friedlander, 1993), perhaps best described by $k_o \approx 1.3$ to 1.4 (Cai et al., 1995a; Oh and Sorensen, 1997; Sorensen, 1997) but see also Köylü and Faeth (1992). This is perhaps the most important defining relation for a practical description of a fractal aggregate. Another important relation describes the spatial correlation function of the density $g(r)$ given by

$$g(r) \approx r^{D_f - d} h(r/\xi) \tag{23–10}$$

where $g(r)$ is a conditional probability relating on average the density at two points separated by a distance r. The function $h(x)$ is a cut-off function for the power law, $h(x < 1) \approx 1$, but $h(x)$ decreases more rapidly than any power law for $x > 1$. This implies that the length ξ is on the order of the size of the aggregate. For DLCA the Gaussian $h(x) \approx \exp(-x^2)$ is fairly accurate (Sorensen et al., 1992b; Cai et al., 1995a; Sorensen and Wang, 1999). A third valuable relation involves the amount of material within regions of side length s centered on the aggregates. For a fractal particle

$$N \approx s^{D_f} \tag{23–11}$$

All three of these equations will be used for the particle analysis described below.

Real Space Analysis

Collection Methods

Thermophoretic Collection. Often aerosols are hot, for example, soot in a flame or metal oxides from a reactor, and insertion of a colder probe will cause thermophoresis of the hot aerosol particles down the thermal gradient to the colder probe. The great advantage of thermophoresis is that all particle sizes move at the same rate; hence the sampling is unbiased (Rosner et al., 1991) (at least for submicrometer particles; see Sorensen and Feke, 1996).

A "frog tongue" probe device designed after Dobbins and Megaridis (1987) has been used to sample flame soot and TiO_2 and SiO_2 aerosols with apparent success (Cai et al., 1993). This device was built from a modified disk hard drive and a carbon arrow shaft. The essential quality is the ability to inject quickly a probe into the aerosol, hold it there for a residence time determined by the operator, and then quickly remove the probe. The device can move the probe 5 cm in 3 ms and has a selectable residence time of 15 to 150 ms.

The "frog tongue" part of the probe is a thin "knife-blade" of metal. Transmission electron microscope (TEM) copper grids are mounted on the blade with polystyrene cement. This cement is easily broken so that the grid can be removed. The grids consist of copper mesh, having either a carbon or a Formvar coating. The blade is inserted with its face parallel to the aerosol flow to avoid perturbation of the flow and impaction of particles.

Impaction Collection. Placing a probe with its face perpendicular to the aerosol flow will allow impaction of aggregates onto the probe. This approach was used to collect soot particles 1 μm sized or larger (Sorensen and Feke, 1996; Sorensen et al., 1998). Impaction relies on inertia, so there is a bias toward collecting a larger fraction of the bigger aggregates. The Stokes number quantifies the likelihood of impaction and is discussed in Chapters 4 and 10.

Collection by Settling. Often the aerosol settles out to form a powder at the bottom or on the sides of the chamber, container, room, or region of aerosol study. This powder can be carefully collected and redispersed to create samples for microscopic examination. Collection should proceed with some care because the fractals may be fragile, as implied by their tenuous nature. Little is known about the fragility of fractal aggregates. Experiments exist in which aggregates have been stretched and then snapped back (Friedlander et al., 1998). These imply considerable resiliency. Despite this, any reasonable caution when handling would not be wasted.

Redispersion has been accomplished using volatile liquids, perhaps with surfactants, and then drying. Water, ethanol, and acetone have been used to study carbonaceous soot and TiO_2. Water has a large surface tension, and this tends to crush the aggregate as it dries. Aerosol methods have been used for redispersion of soot (Prenni et al., 2000), though it has not been established that individual particles retain their original structure. Each type of material has its own peculiarities, so experimentation is warranted.

In all cases collection densities on the microscope substrate (e.g., TEM grid) should not be great, the aggregates occupying 10% of the area or less (Cai et al., 1993). This is necessary to avoid significant "artificial" aggregations on the substrate. If two aggregates overlap on the substrate, there is no way to distinguish the resulting cluster from one that formed in the aerosol phase.

Analysis of Projected Images

Visualization. An electron microscope is necessary to study nearly all conceivable fractal aggregates because the primary particles (or monomers) are smaller than optical wavelengths ("big" particles do not stick together readily). Standard carbon- or Formvar-coated copper grids are sufficient to hold the sample. The magnification should be large enough to allow an accurate measurement of the monomer. For example, soot is usually composed of aggregates of monomers with radii $a \approx 10\,\text{nm}$. To magnify these to 1 mm images requires 100,000 magnification.

Photographs can be digitized with a digital scanner for computer analysis. The image data will be in the form of a two-dimensional array with a magnitude representing the gray level at a given pixel. A computer analysis routine can be written to identify individual clusters in the digitized array, or the operator can look at the image on the monitor, pick out the images visually, and store them as separate gray level arrays, one for each cluster. Often it is useful to subtract a background from this cluster gray level. The background is the average gray level of the pixels near the circumference of cluster. With this subtraction, all the nonaggregate pixels are set to zero ("white"). Once a gray level array for each cluster is achieved, analysis for morphological parameters can begin.

A major problem in the analysis of the cluster morphology is that the three-dimensional structures are viewed as two-dimensional projections as a consequence of the microphotography. One way to overcome this is to view the clusters in at least two different projections and with this stereo technique regenerate the true three-dimensional structure. This has been done in the past (Sampson et al., 1987; Köylü et al., 1995) but the method is laborious and, as will be shown, largely unnecessary. If the analysis is limited to one projection, and if the density of this projection can give accurate information regarding the total mass along a given projection through the cluster, then a viable analysis of the three-dimensional morphology can be obtained. Such a mass-preserving image is difficult to achieve, however, because the attenuation of the electrons producing the projected image is not linearly related to the total mass of material through which the electrons passed. Furthermore, the response of the photographic film that captures the image is linear only over a small range before it saturates and becomes insensitive to the mass of the cluster above it. In previous work involving small soot clusters (Cai et al., 1993), some success was achieved with mass-preserving projections largely because the clusters were small enough to keep the gray level–to–projected mass relationship approximately linear. In general, however, mass-preserving projection is uncertain, so we are left with projection of the cluster onto the two-dimensional plane in a binary format, that is, a shadow, in which any part of the cluster is the same degree of gray (black) as any other and the background is white. The advantage of this method is that it eliminates the response of the detector. The *conversion to a binary format appears not only easier to apply, but also more reliable and accurate.* What is needed is a quantitative method to convert two-dimensional information into three-dimensional information, and such a method is presented below. The bulk of the discussion applies to DLCA aggregates with $D_f \approx 1.75$, which are very typical. When $D_f > 2$, the problem of determining the size parameters of aggregates is much less explored, but a possible direction will be suggested.

Binary Projection Analysis

THE RADIUS OF GYRATION. We first consider the radius of gyration R_g of a three-dimensional body as given by

$$R_g^2 = \frac{\int r^2 \rho(r) \mathrm{d}^3 r}{\int \rho(r) \mathrm{d}^3 r} \tag{23–12}$$

where $\rho(r)$ is the density. A reasonable assumption is made that an ensemble of clusters on a TEM grid when viewed from one direction will yield an average spherical symmetry. Then, because $r^2 = x^2 + y^2 + z^2$, and because a projection onto a plane eliminates one of the dimensions, it follows from Eq. 23–12 that (Cai et al., 1993)

$$R_{g,3} = \sqrt{3/2}\, R_{g,proj} \tag{23–13}$$

In Eq. 23–13, $R_{g,3}$ is the true, three-dimensional radius of gyration of the cluster and $R_{g,proj}$ is that observed for the mass-preserving projected image. The factor 3/2 results from the elimination of one of the three dimensions. Furthermore, Eq. 23–13 applies to a mass-preserving projection. Equation 23–13 is verified by the computer simulations of Köylü et al. (1995), who found the empirical factor relating the two radii to be 1.24 ± 0.01, in good agreement with $\sqrt{3/2} = 1.225$.

The previous discussion shows the difficulty in achieving an accurate mass-preserving projection, so Eq. 23–13, while informative, is of questionable utility. The purpose here is to show that a better measurement can be obtained with a two-dimensional binary representation of the cluster. It is well established that the fractal dimension of $d = 3$ DLCA clusters is less than 2; typically, D_f is in the range 1.7 to 1.8. Thus, it might be expected that the image of a cluster projected onto a plane in a binary format would be mass preserving (i.e., no significant screening or occultation between monomers would occur). This would imply that the number of monomers in the aggregate would be proportional to the projected area of the cluster (i.e., $N \propto A_c$). It must be stressed that this expectation is only correct for asymptotically large ($N \to \infty$) clusters. For finite size clusters, however, screening occurs, and it is found empirically that

$$N = A_c^{\alpha} \tag{23–14}$$

where $\alpha = 1.1$ (for references and a complete discussion, see "Determination of N," below). Thus, the effective fractal dimension in the two-dimensional plane of the binary projection should be different from the fractal dimension of the real, three-dimensional cluster. In the immediately following argument, we will call these fractal dimensions $D_{f,2}$ and $D_{f,3}$, respectively.

Consider how the three-dimensional cluster is projected onto the two-dimensional plane. With spherical or circular symmetry, we assume the density profile of either the three-dimensional fractal cluster or its projection is given by

$$\rho(r) \propto r^{D_f - d} \quad \text{for } r \le R \tag{23–15a}$$

$$= 0 \quad \text{for } r > R \tag{23–15b}$$

where R is the perimeter radius and $D_f = D_{f,2}$ or $D_{f,3}$, depending on the spatial dimension of $d = 2$ or 3 for the binary projected or real cluster, respectively. Then Eq. 23–12 yields

$$R_{g,3}^2 = \frac{D_{f,3}}{D_{f,3} + 2} R^2 \tag{23–16}$$

$$R_{g,binary}^2 = \frac{D_{f,2}}{D_{f,2} + 2} R^2 \tag{23–17}$$

Thus, a relation between the true radius of gyration $R_{g,3}$ and the measured, binary projection radius of gyration $R_{g,binary}$ can be determined if we have a relation between $D_{f,3}$ and $D_{f,2}$.

To determine this latter relation, consider the empirical fact of Eq. 23–14 that $N_3 \approx A_c^\alpha$, where we now label the number of monomers per cluster with a subscript of three to designate that this is the number in three-dimensional space (i.e., the true number). We also have by Eq. 23–9 the relation $N_3 \approx R_{g,3}^{D_{f,3}}$. The binary projection has analogous relations such that $N_2 \approx R^{D_{f,2}}_{g,binary}$, which defines $D_{f,2}$, but, and here is the key, $N_2 \approx A_c$. Furthermore, by Eqs. 23–16 and 23–17, $R_{g,3} \approx R_{g,binary}$. All these proportionalities together yield

$$D_{f,2} = D_{f,3}/\alpha \tag{23–18}$$

This result is consistent with past work that has measured the fractal dimension of clusters in terms of both three-dimensional quantities and projectional quantities, where it was found that the projectional dimension is typically 10% less than that determined with the three-dimensional quantities (Sampson et al., 1987; Zhang et al., 1988; Cai et al., 1993). It is also consistent with simulations by Jullien et al. (1994), who also found the projectional fractal dimension to be about 10% less than the fractal dimension of the unprojected clusters. Because $\alpha \approx 1.1$, we believe Eq. 23–18 explains these past observations. Finally, we use Eqs. 23–16 to 23–18 to find

$$R_{g,3} = \left(\frac{D_{f,3} + 2\alpha}{D_{f,3} + 2} \right)^{1/2} R_{g,binary} \tag{23–19}$$

Now recall that $R_g \equiv R_{g,3}$ and $D_f \equiv D_{f,3}$. Then for typical values of $D_f = 1.8$ and $\alpha \approx 1.1$, this correction factor is 1.026. Thus, as anticipated and qualitatively explained earlier (Cai et al., 1993), the binary projection yields a remarkably accurate measure of the true, three-dimensional radius of gyration.

Computer analysis of the clusters begins with the total gray level defined as

$$G_{tot} = \sum_{x,y} G(x, y) \tag{23–20}$$

where $G(x, y) = 0$ or 1 is the gray level of the pixel at position (x, y). Because $G(x, y)$ is binary, G_{tot} is the total number of pixels in a cluster. To determine the radius of gyration R_g of a cluster, first calculate the cluster center of mass:

$$\vec{r} = G_{tot}^{-1} \sum_{x,y} G(x, y) \vec{r}(x, y) \tag{23–21}$$

and then the radius of gyration

$$R_{g,binary}^2 = G_{tot}^{-1} \sum_{x,y} G(x, y) [\vec{r}(x, y) - \vec{r}_{cm}]^2 \tag{23–22}$$

Correction using Eq. 23–19 to obtain $R_g/R_{g,3}$ could now be made, but because the correction is only about 2%, this is hardly warranted given that other experimental errors are most likely larger.

Another useful, and fairly simple, method to determine R_g is using the maximum, projected length of the aggregate image. Computer simulations indicate that half this length (hence a radius) R_2 is a constant ratio to R_g independent of N. This is shown in Figure 23–5. The result is that

$$R_g/R_2 = 0.69 \pm 0.03 \tag{23–23}$$

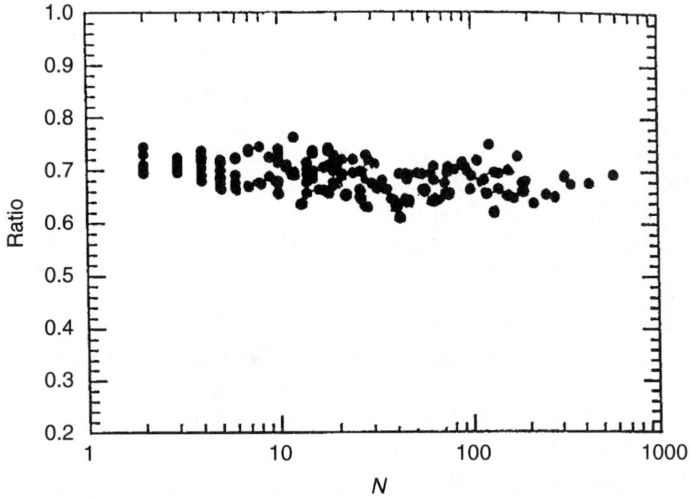

Fig. 23–5. Ratio of the aggregate radius of gyration R_g (in $d = 3$ space) to half the longest length R_2 of the clusters projected onto a $d = 2$ plane as a function of N for simulated DLCA aggregates with $D_f = 1.79$.

This method is simpler than calculation of R_g from the gray level, Eq. 23–22, but relies on the clusters being DLCA with $D_f \approx 1.79$ (i.e., equivalent to the simulation that produced Fig. 23–5).

DETERMINATION OF N. Determination of the number of monomers per aggregate N from the projected area of a cluster has a long and well established history for DLCA ($D_f \approx 1.75$) (Medalia, 1967; Medalia and Heckman, 1969; Mandelbrot, 1977; Sampson et al., 1987; Megaridis and Dobbins, 1990; Köylü and Faeth, 1992; Cai et al., 1993; Köylü et al., 1995). In general it is found that

$$N = k_a (A_c / A_p)^{\alpha} \tag{23–24}$$

where k_a and α are constants near unity and A_c and A_p are the projected areas of the cluster (aggregate) and primary particle (monomer), respectively. Medalia and Heckman (Medalia, 1967; Medalia and Heckman, 1969) first used this form and found empirically $k_a = 1.0$ and $\alpha = 1.1$. This has subsequently been corroborated by a number of workers with α varying by a few hundredths. Sorensen and co-workers (Oh and Sorensen, 1997) found that fractal soot clusters with $D_f \approx 1.75$ obeyed Eq. 23–24 with $\alpha = 1.09$. Köylü et al. (1995) analyzed both computer-simulated and real soot clusters and found $k_a = 1.15 - 1.16$ and $\alpha = 1.09 - 1.10$. Note that with these results the limit as $N \rightarrow 1$ is not preserved because k_a is not unity. In another simulation, Meakin et al. (1989) created DLCA clusters with $D_f = 1.8$ and N up to $N = 10^4$, larger than any in any other work that has compared N to the projected area. They fit their data with

$$A_c / A_p = 0.4784N + 0.5218N^{0.7689} \tag{23–25}$$

This result is equivalent to Eq. 23–24 with $k_a = 1.00$ and $\alpha = 1.10$ over the range of $N = 1$ to 100, $k_a = 1.00$ and $\alpha = 1.084$ for $N = 1$ to 1000, $k_a = 1.075$ and $\alpha = 1.083$ for $N = 10$ to 100, and $k_a = 1.106$ and $\alpha = 1.069$ for $N = 10$ to 1000. The slope of a log N versus log (A_c/A_p) graph is α, and Eq. 23–25 yields a slowly decreasing α with increasing N. This is consistent with the notion that, for clusters with $D_f < 2$, as $N \to \infty$, N should be linear with A_c, that is, α asymptotically approaches 1.00 because the cluster dimension is less than the dimension of the plane onto which it is projected.

In summary, all these results agree fairly well. Given this and because we desire to conclude with a recommendation for all plausible situations, we use the computer results of Meakin et al. (1989), Eq. 23–25, which are the most extensive (up to $N = 10^4$). A good procedure is to estimate the range of N values for the clusters to be analyzed and then fit Eq. 23–24 to Eq. 23–25 for this range to determine α and k_a. Then use these values to determine N from A_c and A_p. For example, if the clusters lie in the approximate range $N = 10$ to 1000, use $k_a = 1.106$ and $\alpha = 1.069$. Most likely other sources of experimental error will be larger than the uncertainty incurred by use of k_a and α.

THE FRACTAL DIMENSION. There are three useful ways to determine the fractal dimension D_f. Two methods analyze a single cluster; they are the method of nested circles or squares and the analysis of the density correlation function. The third method requires an ensemble of aggregates and compares N to R_g (or any measure of the aggregate's linear size) via Eq. 23–9. We discuss these methods sequentially below.

The Method of Nested Circles or Squares. In this method circles or squares of increasing radius or side length s are computer drawn on the aggregate centered on the aggregate's center of mass. The total binary black area within the circle or square is calculated and plotted versus s. Then from Eq. 23–11 the following scaling relation holds:

$$D \approx s^{D_{f2}} \tag{23–26}$$

Equation 23–26 is best used as a log–log plot of G versus s, which will have a slope of D_{f2}. D_{f2} must be converted to the three-dimensional fractal dimension with Eq. 23–18 and α.

Often a given cluster will yield a nonlinear, strangely shaped plot of N versus s. One must remember that "fractal" is a statistical concept, and not all clusters are alike. Examination of an ensemble of clusters is therefore highly recommended. While most clusters will exhibit fractal behavior, that is, have a linear $\log G$ versus $\log s$ plot, occasional "odd" clusters will occur. An example of this cluster-to-cluster variation is given by Zhang et al. (1988) and reproduced in Figure 23–6. There it is found that the sum of the black areas for three clusters is better described by Eq. 23–26 than for any individual cluster. On the other hand, if there are more "odd" clusters than "fractals," then the system is not a fractal system.

The Density Correlation Method. The density correlation function is a conditional probability that, given material at one point, defines the probability that there will be material at another. It is represented by $g(r)$ and is expressed by

$$g(r) = [\rho(R+r)\rho(R)] \tag{23–27}$$

In Eq. 23–27 $\rho(R)$ is the density at point R, and the brackets mean an average over all positions R.

The digitized, binary images can be used to calculate $g(r)$. Because the density is proportional to the gray level of the image, Eq. 23–20, we can write

$$g(r) = \sum_{u}^{N_u}\sum_{v}^{N_v} D(u+x)D(v+y)/N_u N_v \tag{23–28}$$

where

$$r = \sqrt{x^2 + y^2} \tag{23–29}$$

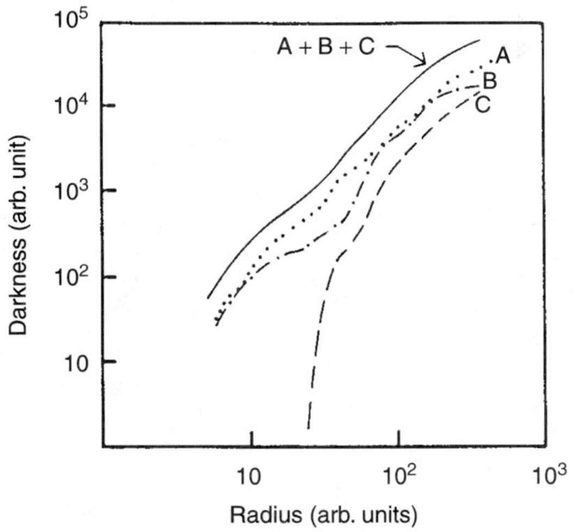

Fig. 23–6. The total gray level (G_{tot}) within a series of nested circles centered on the cluster center of mass for three different soot clusters A, B, and C determined from their binary projected images. The curve $A + B + C$ is their sum, which is fairly linear to imply an average $D_f = 1.72 \pm 0.10$.

In calculating with Eq. 23–28, u and v should be constrained to points (pixels) within the aggregate. This is not necessary, but otherwise $G = 0$ and unnecessary computation results. Note that Eq. 23–28 could also be restricted to averages over u or v separately to yield $g(x)$ and $g(y)$. If these differed, anisotropy would be implied. Many different x and y values yield the same r, so these could be averaged over a range r to $r + dr$ to calculate a final $g(r)$.

Once the density correlation function is calculated, it can also be plotted on a log–log plot to display its power-law nature, as written in Eq. 23–10. If the projected image has been stored in the computer in a binary format, the projected image has pixels with gray level $G = 0$ or 1 only, and the effective spatial dimension is $d = 2$ to be used in Eq. 23–10. Also for the binary format, the fractal dimension is $D_{f,2} = D_{f,3}/\alpha$ (i.e., the correction of Eq. 23–18 must be made). Here again cluster-to-cluster variation is expected, but typically not as much as in the "nested" method above. Figure 23–7 gives an example for a soot fractal aggregate.

The Ensemble Method. Both N and R_g can be extracted from the projected, binary images of the fractal aggregates. Given these parameters, Eq. 23–9 suggests that a log–log plot for a polydisperse ensemble of aggregates will yield D_f from the slope and k_o from the intercept. This fact has been used many times in the literature, and Figure 23–8 gives an example. Note that this analysis yields $D_f \equiv D_{f,3}$.

If k_o is not needed, this analysis can be simplified by using the total gray level of the binary projected image of the aggregate versus any aggregate length (e.g., the longest, 2*d* projected length). The slope of such a graph would yield $D_{f,2}$. This method usually yields excellent results.

Aggregates with $D_f > 2$. Jullien et al. (1994) modeled $d = 3$ random fractal aggregates on a computer with $1 \le D_f \le 2.5$ and then studied their projection onto a plane. Figure 23–9 is a reproduction of their Figure 2a, which is particularly useful. It shows the binary projected fractal dimension $D_{f,2}$ versus the true $d = 3$ fractal dimension D_f of the aggregate for a variety of aggregate sizes ranging from $N = 16$ to 8192. A line representing Eq. 23–18 with $\alpha = 1.1$ is included. The data for $D_f < 2$ support the line, giving yet more credence to the analysis above. For $D_f > 2$, this line fails, more so with increasing D_f. However, larger values of α in Eq. 23–18

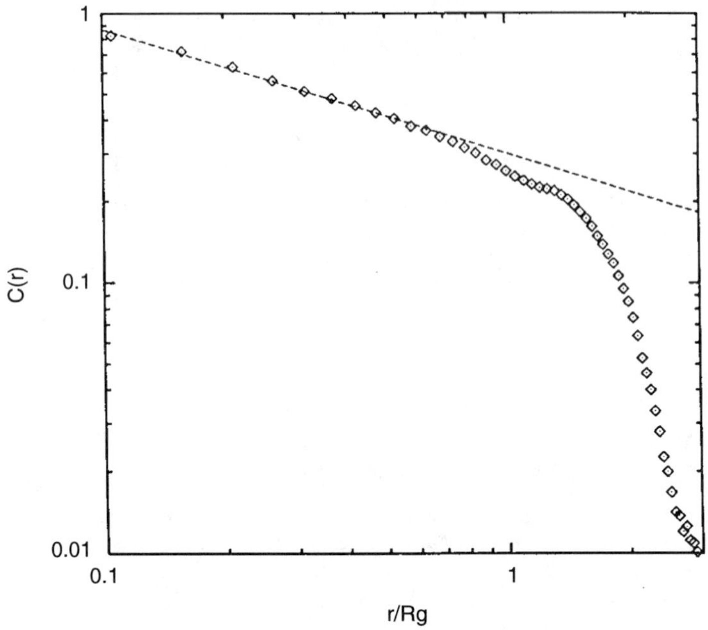

Fig. 23–7. Density correlation function for a soot cluster obtained from a premixed CH_4/O_2 flame. Dashed line is a fit to the first 13 points (thereafter the cut-off function $h[x]$ takes over) of the power law $g(r) \approx r^{D_{f,2}-d}$.

could describe the relationship between D_{f2} and $D_f > 2$. Given this, if one need analyze an aggregate with $D_f > 2$, α can be extracted from Figure 23–9, and then the analysis technique described above using Eqs. 23–18, 23–19, and 23–24 can be applied.

Optical Characterization of Fractal Aggregates

Light scattering is an excellent method for in situ determination of fractal aggregate size and morphology (Sorensen, 1997, 2001; Oh and Sorensen, 1999). The method involves measuring the scattered intensity I as a function of the scattering wave vector q, where

$$q = 4\pi\lambda^{-1}\sin(\theta/2) \tag{23–30}$$

λ is the optical wavelength, and θ is the scattering angle. Although θ is the experimental parameter, it is important to both think and work in terms of q because inverse q is the length scale of the scattering experiment. Plots of I versus q will have changes of slope whenever q^{-1} passes through a length scale change of the system. I versus q is called the *optical structure factor*. Figure 23–10 is a schematic diagram of the information that can be obtained from an optical structure factor measurement. At small q the scattering is constant and proportional to the cluster number density n and the number of monomers per cluster N squared. This is termed the *Rayleigh regime*. Note that in an aggegating system $nN = n_m$, the monomer number density, which is ideally a conserved quantity. Then $I \approx nN^2 = n_mN$, which will increase as the system aggregates because N increases. This increase in scattering (in the Rayleigh regime) as the system coarsens is the Tyndall effect (see Kerker, 1969) for a fractal aggregate system.

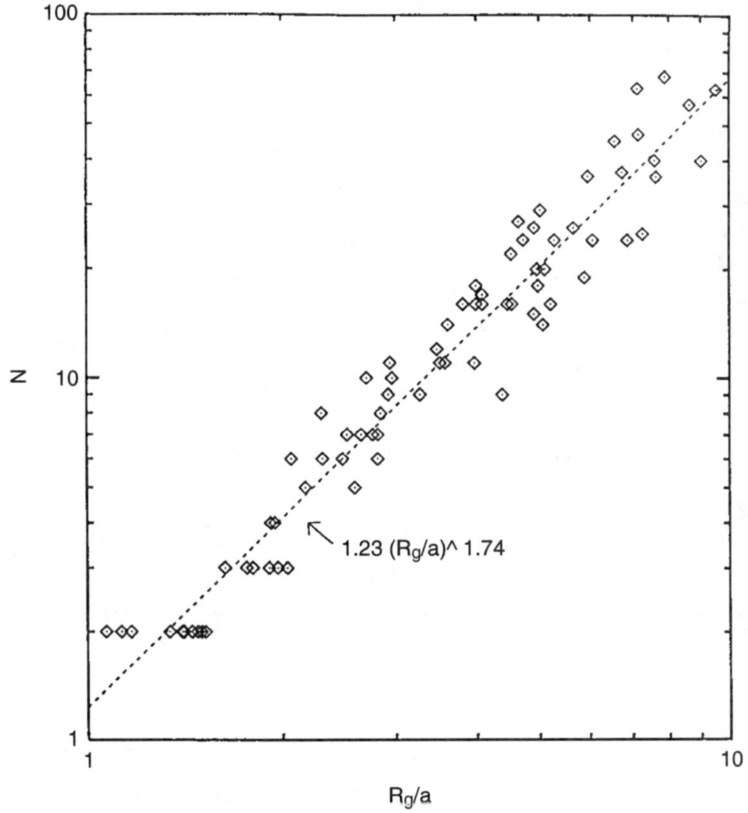

Fig. 23–8. Total number of monomers per aggregate versus radius of gyration divided by the monomer radius for an ensemble of soot clusters collected from a premixed methane/oxygen flame. The slope of this log–log plot demonstrates the power-law dependency of Eq. 23–1 and yields a fractal dimension of $D_f = 1.74 \pm 0.04$. The intercept yields $k_o = 1.23 \pm 0.07$.

As q increases from the Rayleigh regime, the slope eventually changes, indicating an aggregate length scale. Qualitatively one need only measure q where this first bend in I versus q occurs, invert it to q^{-1}, and identify this length as the overall average size of the aggregates. This can be quantified because this first bend is the Guinier regime where (Guinier et al., 1955)

$$I(q) = I(0)\left(1 - q^2 R_{g,z}^2/3\right) \tag{23–31}$$

In Eq. 23–31 $R_{g,z}$ is the z-average radius of gyration

$$R_{g,z}^2 = \int R_g^2(N) N^2 n(N)\,dN \Big/ \int n(N)\,dN \tag{23–32}$$

$n(N)$ is the number of aggregates of size N per unit volume (i.e., the aggregate size distribution). After the Guinier regime comes a regime of constant slope (on this log–log plot), the so-called power-law regime where (Sorensen and Wang, 1999)

$$I(q) = I(0) C_p C (q R_{g,z})^{-D_f} \tag{23–33}$$

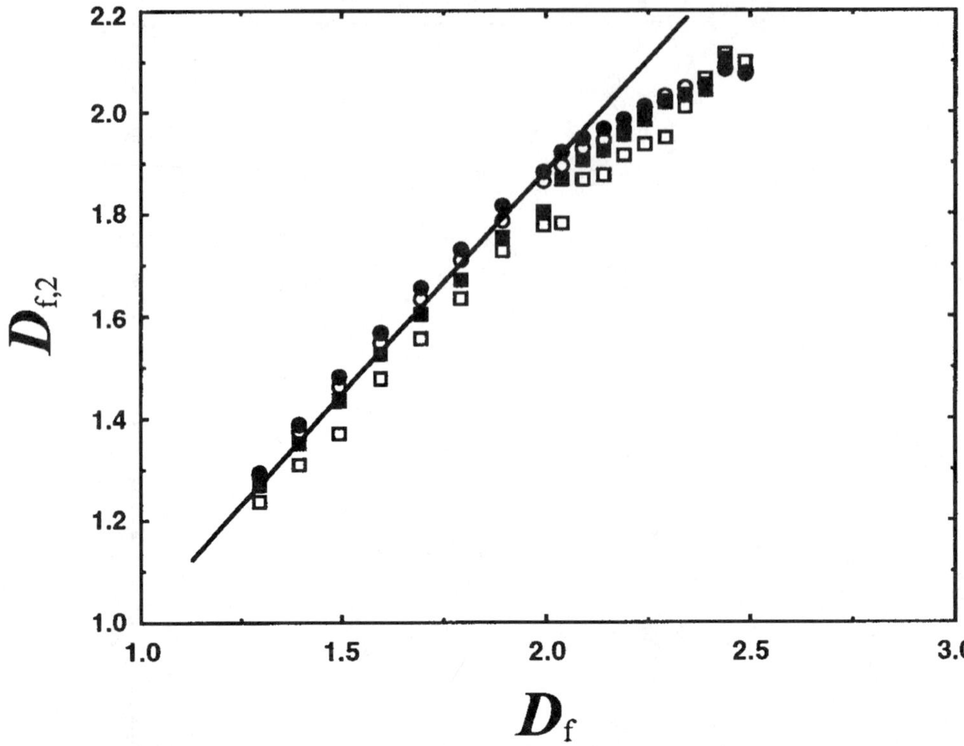

Fig. 23–9. Fractal dimension $D_{f,2}$ of the projection of a fractal aggregate versus its d = 3 space fractal dimension, D_f, for N = 16 (open circles), 128 (filled circles), 1024 (open squares), 8192 (filled circles) for computer-simulated aggregates after Jullien et al. (1994). Line is Eq. 23–10 with $\alpha = 1.1$.

In Eq. 23–33 C_p and C are constants to be described below. The important dependency is $I \propto q^{-D_f}$, which means that the slope of the double-log plot of I versus q is the negative fractal dimension; hence, D_f can be measured.

Finally, at highest q, another bend occurs indicating the monomer radius a. Thereafter, the monomer regime is entered with slope –4 (a Porod regime) if the monomers have a smooth, nonfractal surface.

Optical Structure Factor Measurements—Quantitative Description. With the qualitative description of the optical structure above, we now describe how to obtain accurate values of fractal aggregate R_g and D_f from light-scattering measurements.

Measurement of R_g. An example of an optical structure factor from some of our early work (Gangopadhyay et al., 1991) is given in Figure 23–11 for a soot aerosol in a premixed methane/oxygen flame. Aggregation increases with increasing height above the burner. Inspection of Figure 23–11 shows that with increasing height, the bend in the optical structure factor (i.e., where the slope of I versus q goes from zero to negative) progresses to smaller q. The cardinal rule is that a change in slope implies a length scale. In this case the length scale is the overall aggregate size, and because $R \approx q^{-1}$, this is a direct observation, albeit qualitative, of the aggregate size increasing with time. Notice the essentially isotropic scattering at h = 8 mm to indicate very small particles. Figure 23–12 presents a more recent example of scattering from a titania aerosol. Note the scale in q is an order of magnitude smaller than in Figure 23–11, and hence the clusters of titania are an order of magnitude larger. In Figure 23–12 a significant power-law regime is seen with a slope implying $D_f \approx 1.7$.

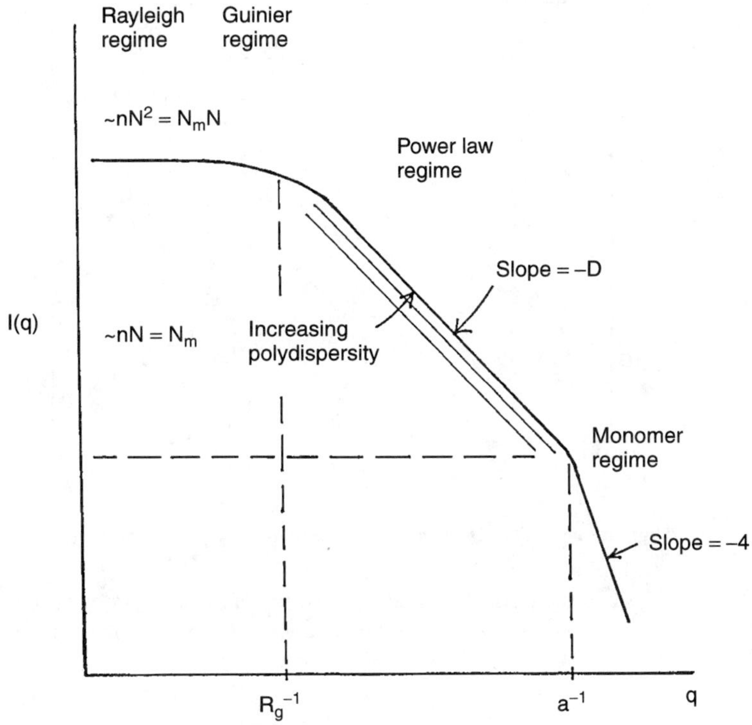

Fig. 23–10. Schematic representation of the scattered light intensity $I(q)$ versus $q = 4\pi\lambda^{-1}\sin\theta/2$, where θ is the scattering angle, from an ensemble of fractal aggregates of dimension D_f with N monomers per aggregate on a log–log plot. In the ensemble n is the number density of clusters and N_m is the total number of monomers, $nN = N_m$.

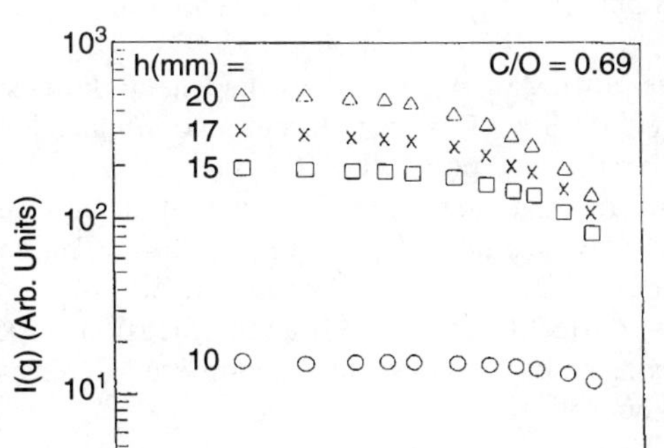

Fig. 23–11. Scattered light intensity $I(q)$ as a function of the scattering wave vector q for five different heights above burner h for a premixed methane/oxygen flame.

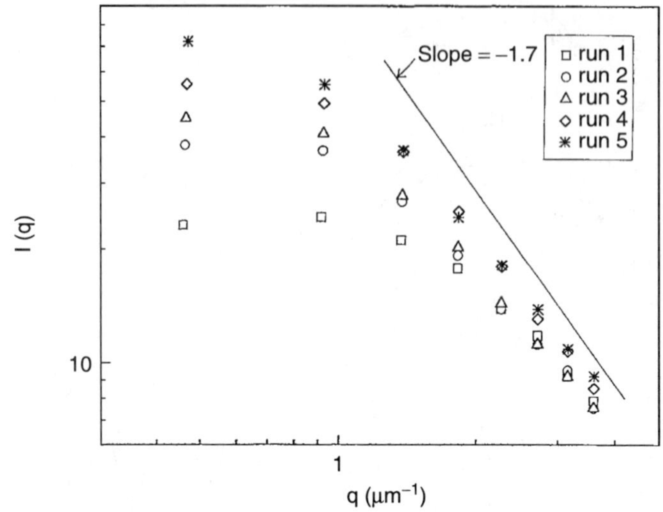

Fig. 23–12. Scattered light intensity $I(q)$ as a function of the scattering wave vector q for a titania aerosol.

Quantitative analysis of the optical structure factor proceeds in two steps (Sorensen et al., 1992a; Sorensen, 2001). First, the Guinier regime is analyzed to yield the aggregate radius of gyration. The Guinier equation, Eq. 23–31, may be inverted to yield

$$I(0)/I(q) \approx 1 + \frac{1}{3} R_{g,z}^2 q^2 \qquad (23\text{–}34)$$

This implies that a graph of $I(0)/I(q)$ versus q^2 should be linear with a slope of 1/3 $R_{g,z}$. This plot is similar to the Zimm plot of biophysics (Zimm, 1948; Tanford, 1961; Kerker, 1969). The data of Figure 23–11 are so plotted in Figure 23–13. Figure 23–13 is a proper Guinier analysis in that, for the most part, for the data used, $qR_g < 1$, which is equivalent to $I(0)/I(q) < 4/3$. Precisely speaking, this Guinier analysis should be limited to $qR_g \leq 1$. Often, however, the data are not plentiful and precise enough within these bounds to yield an accurate R_g. It has been found (Cai et al., 1995b) through experience with both real data and numerical calculations to create simulated data, that $I(0)/I(q)$ versus q^2 remains linear well beyond these limits, and data up to about $I(0)/I(q) \approx 2$ can be trusted to yield accurate R_g values.

Measurement of D_f. The power-law regime yields the fractal dimension D_f by its slope. One would like to have a good decade of linearity to get a good measure of this slope but this is rarely the case. Moreover, the true power-law character does not really show until $qR_g \geq 5$. Hence one must beware when data in this regime are limited. An example of structure factor data with an ample power-law regime is given in Figure 23–12. The power-law regime is quite linear in this log–log plot with a slope that yields $D_f = 1.75$.

Polydispersity does not affect the negative fractal dimension slope of the structure factor for $qR_g \geq 5$ except in extreme cases, as yet not encountered in aerosols. Martin and Ackerson (1985) discuss this theoretically.

Measurement of Polydispersity. A well-endowed power-law regime, that is, one that extends out to $qR_g \approx 10$ or more, has another advantage besides easy extraction of an accurate fractal dimension. A measure of the polydispersity can be obtained using the coefficient C_p in Eq. 23–33. In a recent study Sorensen and Wang (1999) proposed the large $qR_{g,z}$ form in Eq. 23–33 to differentiate between single-cluster and polydispersity effects on this regime. The constant C depends on the cut-off function $h(x)$ of the density correlation function, Eq. 23–10. It was

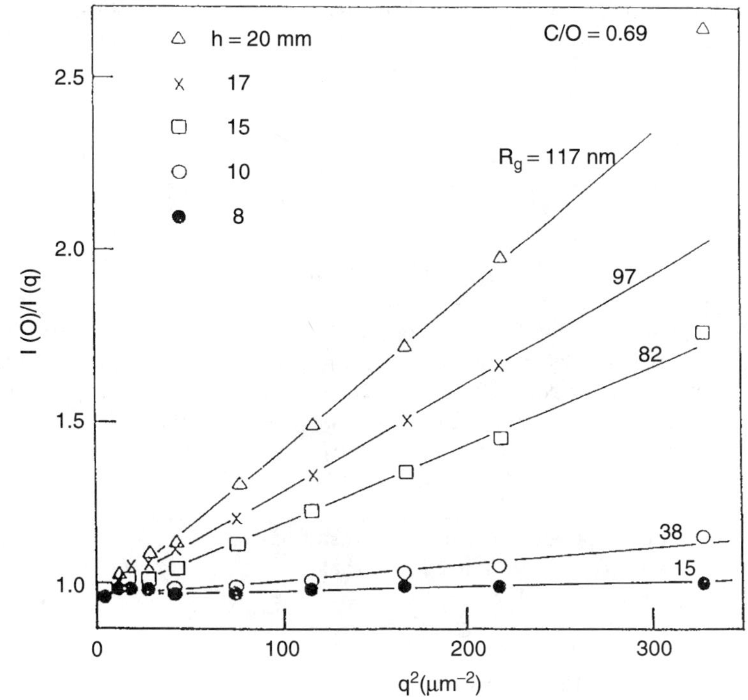

Fig. 23–13. Flame soot aerosol data of Figure 23–11 plotted for a Guinier analysis.

concluded that $C = 1.0 \pm 0.1$ best describes cluster–cluster fractal aggregates. The constant C_p depends solely on the polydispersity of the cluster system and can be used to measure the polydispersity. For further information on this topic, see Sorensen and Wang (1999).

In summary, the measurement of optical structure factor is a very useful method capable of yielding particle R_g, D_f, and the polydispersity. The particle refractive index need not be known. If data beyond $qR_g < 5$ are not available, the D_f measurement can only be considered qualitative. An excellent procedure is to determine R_g with the Guinier analysis using small q data and then D_f with a log–log plot (or fit) using large q data. See Sorensen (2001) for an extensive review of fractal aggregate light scattering.

Shape Factor. Sorensen and co-workers (Cai and Sorensen, 1994; Wang and Sorensen, 1999) have used light scattering to study fractal aggregate diffusivity. This is relevant to the dynamic shape factor because the diffusion coefficient D and the coefficient of the drag force f in $F = fV$ are related by the Einstein relation

$$D = kT/f \tag{23–35}$$

where k is the Boltzmann constant and T is the temperature. Their method used static light scattering to measure the aggregate radius of gyration, R_g, and the fractal dimension D_f and dynamic light scattering to measure the diffusion coefficient. The fractal aggregates were soot in a premixed methane/oxygen flame and TiO_2 aggregates in air at room temperature and pressures from 1/15 to 1 atmosphere. Wang and Sorensen (1999) also reinterpreted aerosol mobility data of Schmidt-Ott (1988) and Rogak et al. (1990) and the colloidal data of Wiltzius (1987) and combined all these with their data to achieve a general picture of the mobility of fractal aggregates at all values of the Knudsen number.

Wang and Sorensen (1999) gave their results in terms of the ratio of the mobility radius, R_{mob}, to the radius of gyration, R_g. To recast their results into the dynamic shape factor, recognize that Eq. 23–1 implies

$$\chi = R_{mob}/R_m \tag{23–36}$$

where R_m is the mass equivalent radius. The mobility radius was defined by the modified Stokes relation

$$F = 6\pi\mu R_{mob}/C(R_{mob}) \tag{23–37}$$

Wang and Sorensen's results (1999) pertain to the common situation of diffusion-limited cluster aggregates, which have a fractal dimension $D_f \approx 1.75$. Under these conditions two regimes of functionality for R_{mob} were found depending on the number of monomers or primary particles (N) per aggregate. When N is small (i.e., ≤60), then

$$R_{mob} = aN^{0.44} \tag{23–38}$$

where a is the monomer radius. This empirical result has the correct $N \to 1$ limit. Because $R_m^3 = a^3N$, it follows from Eqs. 23–36 and 23–37 that

$$\chi = N^{0.11}, \quad N \le 60 \tag{23–39}$$

For large N it was found empirically that

$$R_{mob} = 0.7\ R_g \tag{23–40}$$

R_g is related to N via

$$N = k_o(R_g/a)^{D_f} \tag{23–41}$$

Where $k_o \approx 1.3$. It then follows that

$$\chi = 0.6\ N^{0.24}, \quad N \ge 60 \tag{23–42}$$

Before we discuss these results, we consider two other functional relationships. First, Chan and Dahneke (1981) calculated R_{mob} for straight chains of N individual spheres in the free molecular limit. From this we find

$$\chi = \frac{\sqrt{0.802(N-1)+1}}{N^{1/3}} \tag{23–43}$$

Second, a limiting case, often called the *free draining limit* in polymer science, would be when the drag on an aggregate is the sum of the drags due to the N individual spheres. Then it follows that

$$\chi = N^{2/3} \tag{23–44}$$

All these results, Eqs. 23–39, 23–42, 23–43, and 23–44, are plotted in Figure 23–14. We see that the independent spheres approximation works very poorly. This implies that there is considerable interaction of the flow fields around each monomer. On the other had, the linear chain result agrees fairly well with the fractal aggregate when N is small, especially when $N \le 10$.

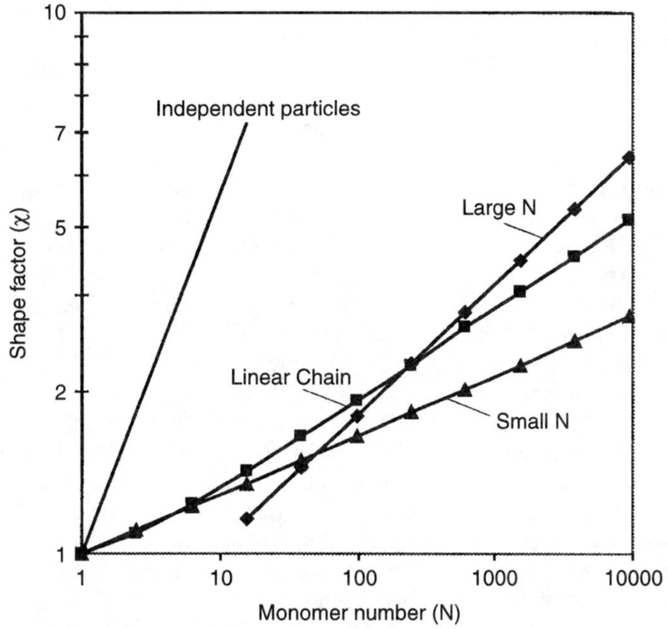

Fig. 23–14. Calculated values of shape factor for several particle agglomerate configurations.

This implies that the geometry of the arrangement of the monomers is not important until N is larger than 10.

Figure 23–14 may be used to determine the dynamic shape factor for fractal aggregates for all values of the Knudsen number. When $Kn \approx 0$, the continuum regime, χ, is given by the greater of the two solid lines, representing Eqs. 23–39 and 23–42, in Figure 23–14. For large Kn, the behavior of χ is less certain. A description for this regime consistent with the conclusions of Wang and Sorensen (1999) is that χ is given by the small N curve whenever $Kn > 5$. For $N > 60$ and $1 < Kn < 5$, we expect χ crosses over smoothly between the small and large N curves as $Kn \to 0$.

FIBERS

Introduction

The term *fiber* has been applied to a wide variety of particles having an elongated shape (i.e., one particle dimension significantly greater than the other two). Because of this elongation, fibers can have aerodynamic and other properties quite different from more compact particles. Certain fibers have several unique properties that make them not only useful from a commercial standpoint but also important from a health standpoint. Asbestos, for instance, includes six commercial fibrous minerals that have high tensile strength, chemical resistance, and excellent thermal and acoustic insulation characteristics. These properties have made asbestos useful in a variety of products, including friction materials, high-temperature insulating materials, acoustic insulation, fire-proof cloth and rope, and floor tiles. While the bulk materials in these products may consist primarily of macroscopically sized fibers, many of them can release long, thin fibers into the air.

A variety of materials can be considered fibers from an aerosol behavior standpoint. Besides asbestos, other mineral fibers exist in nature. Several materials, including glass and mineral slags, have been melted and spun into fibers. Ceramic materials have similarly been spun into fibers, as well as grown by chemical and vapor crystallization. Carbon and graphite fibers are produced commercially for high-strength products. Organic fibers, such as cotton, wood, and other cellulosic materials, are widely present in the environment, from both commercially produced materials and natural sources. Carbon "nanotubes" are a tubular form of carbon "buckyballs" (C_{60}), a molecular structure of pure carbon with very high strength and high conductivity (Ren et al., 1998). These nanotubes are being commercially developed for a variety of applications. Besides cylindrical particles that have relatively high strength, chains of particles also may behave as fibers and can serve as models for some aerodynamic properties of fibers. Some organic materials can be crystallized into well-defined fibrous shapes and can be used to test theories of fiber aerodynamic behavior.

Asbestos fiber aerosols have been closely associated with several diseases, such as asbestosis (a fibrosis of the lung), mesothelioma (cancer of the lining around the lung), and lung cancer (National Institute for Occupational Safety and Health, 1976). Thus, the fibers that can enter the respiratory system are of greatest concern. The seriousness of the diseases has driven measurement technology to provide maximum sensitivity and accuracy for measuring asbestos aerosols. Other airborne fibers may have one or more of the same physical and chemical properties as asbestos. In some cases, human exposures and/or animal studies suggest the disease potential of these fibers. Thus, there is concern regarding the health effects of fibers other than asbestos. While the commercial properties of mineral fibers have created a store of knowledge about physical and geologic properties, it is the health concerns that have largely driven the technology for detecting and quantifying airborne concentrations of fibers. Thus, much of the fiber aerosol research and measurement capability relates to the ability of microscopically sized fibers to enter, deposit, and remain in the human respiratory system.

The dimensions of fibers in aerosols can cover a wide range. Asbestos consists of material composed of individual fibrils as small as 0.025 μm (Langer et al., 1974). Airborne fibers can be single fibrils or various-sized bundles with a range of diameters, while lengths can be less than 0.5 μm to several hundred micrometers. The dimension distributions depend on the fiber type as well as on how the fibers were comminuted from the bulk material. The magnitude of disparity between length and diameter often makes it difficult to make accurate size distribution measurements. Several protocols, using various types of microscopes, have been developed to deal with fiber distribution measurement. Other types of instruments, primarily using light-scattering properties, have been developed to characterize fibers. However, these instruments usually provide only an approximate indication of fiber dimensions.

Note that the following discussion, except where otherwise indicated, deals largely with measurement of aerosolized fibers, generally visible only with a microscope, and not with the macroscopic or bulk properties of the fibrous material. Because asbestos has been the most intensely studied type of fiber, many comments will relate to this material. Many issues regarding asbestos mineralogy, health effects, and measurement techniques are discussed in a review by Walton (1982); further reviews in Environmental Health Perspectives are introduced by Langer (1974) and Dement (1990). Additional topics are presented in books by Selikoff and Hammond (1979), Rajhans and Sullivan (1981), Michaels and Chissick (1979), Chissick and Derricott (1983), and Holt (1987). Similar reviews have been carried out for manmade fibers (National Institute for Occupational Safety and Health, 1976, 1977; IARC, 1988).

Fiber Shape

The behavior of fibers suspended in a gas is a function of the fiber dimensions. Assuming either a cylindrical or prolate spheroidal shape, these dimensions can be defined by two para-

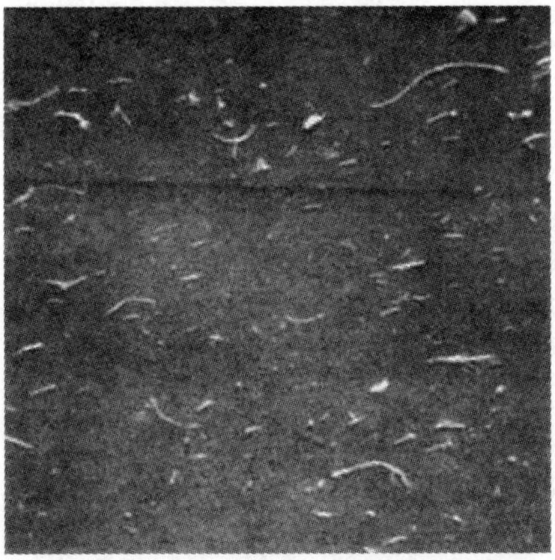

Fig. 23–15. Scanning electron micrograph (×1500) of an aqueous sample of magnetically aligned UICC Canadian chrysotile collected on a 0.1 μm pore size filter. (From Timbrell, 1973.)

meters, length and diameter. A third parameter β is often invoked to indicate the fibrosity or aspect ratio (i.e., the ratio of the length to the diameter). However, real fibers frequently do not meet the either the ideal cylindrical or the prolate spheroid shape assumption. Glass or mineral fibers are often nearly cylindrical, but even these fibers frequently display curvature along their length as well as bulbous or jagged ends. Asbestos fibers are formed from a unique crystal habitat in which the bulk mineral has slip planes in two directions, but only rarely in the third. This results in a propensity to produce particles that can split longitudinally to produce thinner and thinner fibers, ultimately resulting in fibrils about 0.02 to 0.05 μm diameter. Thus, while some asbestos fibers exhibit a nearly ideal cylindrical shape, others may have various combinations and degrees of splayed ends, curvature, splitting, noncircular circumference, and so forth. For instance, the magnetically aligned chrysotile fibers in Figure 23–15 show many of these characteristics. Despite these possible variations in shape, fibers are still most often characterized simply by length and diameter.

Distributions of natural fibers are rarely monodisperse in diameter and even more rarely in length. This has made it difficult to provide adequate calibration for instruments that attempt to measure fibers as well as to perform measurements of fiber toxicity as a function of fiber dimension. Distributions of fibers can often be described by a two-dimensional (length L and diameter W) lognormal distribution (Schneider et al., 1983; Cheng, 1986), that is, $\ln L$ and $\ln W$ are each distributed normally. The probability density function is given by

$$f(L,W) = \frac{1}{2\pi\sigma_w\sigma_L\sqrt{1-\tau^2}LW}\exp\left(-\frac{A^2+B^2-2\tau AB}{2(1-\tau^2)}\right) \tag{23–45}$$

where $A = (\ln W - \mu_W)/\sigma_W$ and $B = (\ln L - \mu_L)/\sigma_L$.

μ_i and σ_i^2 are the mean and variance of the natural logarithm of L and W, respectively, and τ is the correlation between $\ln L$ and $\ln W, \tau$ ranges from –1 to 1. The five parameters μ_L, μ_W, σ_L, σ_W, and τ are needed to define completely a two-dimensional size distribution. The two-dimensional lognormal size distribution has the properties that the marginal and the conditional distributions are lognormal (Holst and Schneider, 1985). The former property indicates that the length and diameter distributions are each separately lognormal. The latter indicates that functions of length and diameter of the form kW^pL^q, where k, p, and q are constants, are also lognormal. Such functions include aspect ratio, surface area, volume, and aero-

TABLE 23–1. Examples of Measured Fiber Size Distributions

Material	Diameter (μm)	σ_g	Length (μm)	σ_g	MMAD (μm)	σ_g	Measurement Technique
Chromoglycic acid[1]	0.205	1.58	2.09	1.83			SEM
					0.65	1.88	Cascade Impactor
Sugar cane silicate[2]	0.3–1.5*		3.5–65*				TEM
Caffeine[3]	1.13	1.08	5.55	1.12			SEM
					2.1	1.1	Sedimentation
Ceramic fibers[4]							
sample a	0.5		10.1				TEM
sample b	0.66		8.3				TEM
sample c	0.98		22.8				TEM
Chrysotile[5]							
Preform ring	0.13	2.15	1.6	2.7			TEM
Yarn dressing	0.08	1.92	1.0	2.4			TEM
Cure press	0.13	1.94	1.5	2.2			TEM
Crocidolite[6]							
Mine/Mill**	0.08–0.10	1.86–2.08	0.98–1.25	2.30–2.55			TEM
Manufacturing	0.04	1.58	0.54	2.32			TEM
Fibrous glass[7]							
Code 100	0.12	1.8†	2.7	2.2†			TEM
Code 110	1.8	1.7†	26	2.0†			TEM
Iron Oxide Chains[8]							
	0.059	1.1	1††	2.0			TEM
					0.32	1.11	Centrifuge

* These values represent the range of particle sizes rather than the median diameters.
** These values represent the range of several measurements that produced similar results.
† Estimated from data in reference.
†† Estimated from mean chain length of 22 primary particles.
[1] Chan and Gonda (1989).
[2] Boeniger et al. (1988).
[3] Vaughan (1990).
[4] Rood (1988).
[5] Rood and •• (1989); Pinkerton et al. (1983).
[6] Hwang and Gibbs (1981).
[7] Timbrell (1974).
[8] Kaspar and Shaw (1983).

dynamic diameter. Deviations from lognormality can sometimes be attributed to artifacts in sampling or analysis or to multiple aerosol generation sources.

Many fiber distributions reported in the literature include the length and diameter means and variances, but unfortunately do not include τ. However, if the original data are reported in a table as a function of both length and diameter, the correlation term can be estimated (Cheng, 1986). Most fiber distributions have positive τ, suggesting that diameter often increases with length.

There have been measurements of a variety of fibrous aerosols. Table 23–1 lists the results of some examples. Some of these materials have been generated for toxicity studies, some have been measured in environmental studies, while others have been generated as calibration materials.

Fiber Behavior

Translational Motion. As with other aerosol particles, fiber dimensions can cover a relatively wide range: The smaller fibers are affected primarily by diffusional forces, while the larger ones are primarily affected by flow shear and inertial and gravitational forces. Fiber

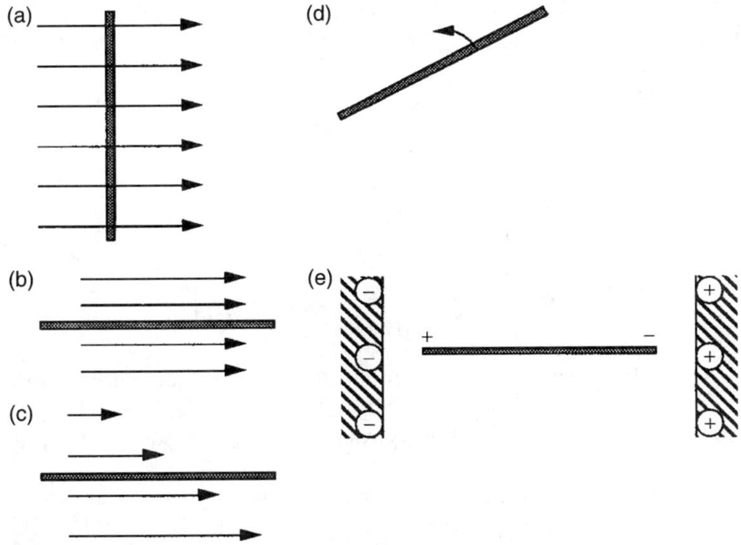

Fig. 23–16. Fiber alignment in various force fields. **a**, Fiber aligned perpendicular to relative gas motion. This is the preferred orientation during gravitational settling and acceleration at 0.01 < Re < 100 in the absence of other forces. **b**, Fiber parallel to relative gas motion. Fiber motion is often treated as a combination of cases a and b. **c**, Fiber is readily oriented parallel to, or at some small angle to, the direction of shear flow in the suspending gas medium. **d**, Small fibers governed by diffusional forces may exhibit completely random orientation. **e**, Conductive fibers are aligned parallel to an electric field. Many fibers are also aligned in a magnetic field, usually parallel to the field lines, though they may be aligned perpendicular to the field or, for some materials, at some intermediate angle.

behavior has been observed and theoretically calculated for several fibrous shapes, including prolate ellipsoids, cylinders, and chains of spheres. The motions of these various shapes generally differ only slightly.

Fiber behavior differs depending on whether the major axis is oriented parallel or perpendicular to the direction of the motion relative to the surrounding gas (Fig. 23–16a,b). The drag on a fiber is greatest when it is oriented perpendicular to the flow of the surrounding gas. Fiber behavior is often described in terms of a combination of the two orientations. While the difference in drag between the two orientations is typically about 15% to 30%, it can be difficult to determine the contribution of each orientation in experimental systems. At low Reynolds number (Re_p), fiber orientation will be stable (discounting Brownian rotation) and not change due to translational motion, for example, during gravitational settling (Gallily, 1971). In addition, fibers settling in still air will not settle exactly in the direction of the gravitational force, but will drift somewhat due to orientation (Weiss et al., 1978). Larger fibers, with Re_p greater than about 0.01 will settle with their major axis oriented perpendicular to the direction of motion (Fig. 23–16a). With increasing Re_p ($Re_p > 100$), longer fibers ($\beta > 20$) are still stable in the perpendicular orientation, but there is an increasing trend toward instability (Clift et al., 1978:154).

The aerodynamic diameter d_a of a prolate spheroid has been calculated

$$d_a = d_f\sqrt{\frac{\rho_f\beta}{\rho_0\chi}} \tag{23–46}$$

by using the numerical shape factor χ of a prolate ellipsoid of revolution (Fuchs, 1964) d_f is the physical fiber diameter, ρ_f is the fiber density, and ρ_0 is standard density. A cylinder with the same diameter and length as a prolate ellipsoid has 3/2 greater volume and mass. Therefore, for cylinders with the same axial dimensions, the right-hand side of Eq. 23–2 must be multiplied by $(3/2)^{1/3}$ or $(3/2)^{1/2}$ to obtain the equivalent volume diameter or equivalent weight diameter, respectively (Griffiths and Vaughan, 1986). For motion parallel and perpendicular to the fiber major axis, the respective shape factors $\chi_{\parallel}^{x}$ and $\chi_{\perp}^{x}$ are

$$\chi_{\parallel} = \frac{4(\beta^2 - 1)}{3} \bigg/ \left\{ \frac{2\beta^2 - 1}{\sqrt{\beta^2 - 1}} \ln\left(\beta + \sqrt{\beta^2 - 1}\right) - \beta \right\} \tag{23–47}$$

$$\chi_{\perp} = \frac{8(\beta^2 - 1)}{3} \bigg/ \left\{ \frac{2\beta^2 - 3}{\sqrt{\beta^2 - 1}} \ln\left(\beta + \sqrt{\beta^2 - 1}\right) + \beta \right\} \tag{23–48}$$

(see Fig. 23–16a,b). Note that a dynamic shape factor χ_d is also defined that is related to the numerical shape factor for prolate spheroids by $\chi = \chi_d \beta^{-1/3}$ (Stöber, 1972; Kasper, 1982). The dynamic shape factor is applied when the equivalent volume diameter of the particle is used rather than the physical diameter.

An alternate approach for directly calculating the aerodynamic diameter of cylinders (Cox, 1970) gives similar results:

$$d_{a,\parallel} = d_f \sqrt{\frac{9\rho_f}{4\rho_0}[\ln(2\beta) - 0.807]} \tag{23–49}$$

and

$$d_{a,\perp} = d_f \sqrt{\frac{9\rho_f}{8\rho_0}[\ln(2\beta) + 0.193]} \tag{23–50}$$

Others have also provided formulas for prolate ellipsoids and cylinders (Gonda and Khalik, 1985).

If fibers are not preferentially oriented by a drag force or other alignment force, the orientation may be completely random. Then, a single average shape factor $\bar{\chi}$ that is a function of the two noted above may be used

$$\bar{\chi} = \beta^{1/3}\left(\frac{1}{3}\chi_{\parallel} + \frac{2}{3}\chi_{\perp}\right) \tag{23–51}$$

In the presence of air gradients, the fiber will experience a torque until the fiber is oriented parallel to the direction of shear force (Fig. 23–16c). Thus, a fiber settling in a horizontal laminar flow will tend to be oriented horizontally (parallel to the shear). However, the fiber will experience a periodic instability and perform a "flip." This instability is a function of fiber dimensions as well as the flow gradients. Under such conditions, the aerodynamic diameter is not strictly an inherent property of the particle and depends on the experimental conditions of measurement (Gallily and Eisner, 1979).

Inertial separation is commonly used for particle separation and sizing (e.g., in impactors and cyclones). In such systems where flow conditions are rapidly changing, the fiber mechanics are governed by initial orientation and flow relaxation time besides the usual parameters observed for spherical particles (Gallily et al., 1986). For instance, fibers with large rotational inertia (especially long fibers) may not orient completely or may over-rotate in passing through a nozzle. Fiber behavior under such a situation may be only approximately defined by Stokes number or other nondimensionalized parameters.

Experimental measurement of fiber deposition has been carried out in horizontal elutriators (Gallily and Eisner, 1979; Griffiths and Vaughan, 1986; Iles, 1990), centrifuges (Stöber et al., 1970; Martonen and Johnson, 1990; Asgharian and Godo, 1999), impactors (Burke and

Esmen, 1978; Prodi et al., 1982; Asgharian et al., 1997), and cyclones (Fairchild et al., 1976; Iles, 1990) for a variety of fiber types.

The extended shape also means that interception during translational motion plays a larger role in fiber deposition than for compact particles. However, the alignment of fibers by shear flow often reduces the effect of length on interception.

Rotational Motion. The rotational mobility B_r of a high aspect ratio ellipsoid can be approximated by (Lilienfeld, 1985)

$$B_r = \frac{3[2\ln(2\beta - 1)]}{2\pi\eta L^3} \tag{23–52}$$

where η is the viscosity of the gas. Note that the rotational mobility is a strong inverse function of fiber length. Similarly the rotational diffusion coefficient D_r for fibers is also a strong function of fiber length

$$D_r = \frac{3kT}{\pi\eta\beta L^3}(\ln 2\beta - \delta) \tag{23–53}$$

where k is the Boltzmann constant, T is temperature, and δ is 1.4 for aspect ratios β larger than 10. Rotational mobility can be estimated by measuring the rate of relaxation after removal of an electrostatic alignment force (Cheng et al., 1991).

Behavior in the Transition Regime. Under molecular bombardment, fibers can exhibit both rotational diffusion and translational diffusion. Such fibers are likely to be randomly oriented (Fig. 23–16d). As for fibers in the Stokes regime, it is often convenient to separate the translational motion of fibers into motion in which the major axis is parallel to the direction of motion and another in which the major axis is perpendicular to the translational motion. Diffusion of fibers is described by the diffusion coefficient D_f (m^2/s)

$$D_f = BkT = \frac{kT}{f} = \frac{kTC_f}{f^0} \tag{23–54}$$

where B is the fiber mobility (dyne-cm/s), f^0 is the drag per unit velocity of the fiber in the continuum regime, and f is the drag per unit velocity of the fiber corrected for slip by the fiber slip correction factor C_f. A theory for the slip correction factors for nonspherical particles is described by Dahneke (1973a–c, 1982).

Fiber diffusional behavior is usually treated as a modification of spherical particle diffusion using particle shape factors (Asgharian and Yu, 1988). This approach has agreed well with experimental diffusion coefficient measurement of fibers with mean diameters between 0.24 and 0.38 µm (Gentry et al., 1983). Diffusional coefficients of much smaller fibers have also been measured (Gentry et al., 1988) that show higher diffusion coefficients than expected.

As with the stagnant flow conditions for fibers in the continuum regime, fibers are expected to be randomly oriented unless affected by shear flow or other forces. Again, the longer the fiber, the more likely it is to be oriented by such forces.

Several studies have estimated effects of various deposition mechanisms (diffusion, impaction, interception) to determine overall particle deposition in filters (Fu et al., 1990) and lung airways (Asgharian and Yu, 1989; Balásházy et al., 1990; Asgharian et al., 1997).

Charging. Theories for unipolar diffusion charging (Laframboise and Chang, 1977) and bipolar diffusion charging (Wen et al., 1984) of fibers have been developed. Unipolar charging of fibers causes the charge of long, thin fibers to increase dramatically, though the electrical mobility of such fibers changes more slowly with aspect ratio (Yu et al., 1987). Such a variation of mobility with fiber aspect ratio may allow separation of fibers of different lengths.

Electric Field Effects. A fiber may be aligned in an electric field by an induced dipole in the fiber. This requires that charges in the fiber be separated so that the polarity is opposite to that of the surrounding electric field as shown in Figure 23–16e. The charge separation from conduction is usually greater than from polarization of the material. For charge separation to occur, the fiber must be sufficiently conductive so that the charges can migrate the length of the fiber in a reasonable time. Aerosol particles, even those consisting of a normally nonconducting material, can often be considered conductive because of their low capacitance and small dimensions (Fuchs, 1964; Lilienfeld, 1985). Surface impurities can also contribute to a particle's conductivity. In addition, water adsorbed on the fiber surface (e.g., glass fibers at approx. 50% relative humidity) can increase the conductivity and allow fiber alignment in an electric field. Thus, an electric field of sufficient strength (1×10^5 to 5×10^5 V/m) can overcome diffusional randomization and flow shear forces to align most types of fibers, including relatively nonconductive ones. For instance, electrostatically aligned zinc oxide fibers were used to modulate microwave radiation (Tolles et al., 1974).

When fibers and compact particles of the same aerodynamic diameter are charged under the same conditions, the fibers may have higher mobility than compact particles. Field studies of work environments suggested that fibers carried a charge proportional to fiber length (Johnston et al., 1985). Other studies indicated that unipolar, charged particles can be separated according to aspect ratio (Griffiths et al., 1985; Yu et al., 1987).

Electrostatic enhancement of fiber deposition in lungs (conductive tubing) has been observed (Jones et al., 1983). Calculations support such enhancement of sedimenting charged fibers (Chen and Yu, 1990).

Dielectrophoresis. Dielectrophoresis was investigated for separating aluminum wires of different lengths in liquid suspension (Lipowicz and Yeh, 1989). The velocity of conductive fibers with an aspect ratio >3 in a gradient electric field was given by Eq. 18–50, indicating that the fiber velocity was proportional to the electric field squared and the fiber length squared. Baron et al. (1994) demonstrated that fiber length classification could be achieved for fibers as short as 4 μm in a 0.76 m long classifier. This technique has been used for production of small quantities of classified fibers for in vitro cell assay studies and for fiber size measurement (Baron et al., 1998; Ye et al., 1999). For further discussion of this technique, see Chapter 18.

Magnetic Field Effects. If a suspension of fibers in a liquid or gas is subjected to a magnetic field, fibers with sufficient magnetic susceptibility will align at some angle to the field. Usually this angle is either 0° or 90°; some amphibole asbestos samples have fibers aligned at both angles. Timbrell (1975) developed a technique for preparing permanently aligned samples by allowing a suspension of fibers in 0.5% celloidin/amyl acetate to dry in a 5000 to 10,000 gauss magnetic field. Several fiber types have been aligned by Timbrell (1972, 1973), including carbon fibers and the various types of asbestos. Fibers of glass, silicon carbide, silicon nitride, and tungsten-cored boron did not align in similar fields.

Figures 23–17, 23–18, and 23–19 contain images of fibers magnetically aligned on a slide surface with light-scattering patterns from magnetically aligned liquid suspensions of the same types of fibers. The direction of the magnetic field is shown in the figures. The scattering pattern has the main laser beam in the center, with the plane of scattering radiating in opposite directions. In Figure 23–17, monodisperse diameter carbon fibers all are aligned parallel to the field so that a well-defined scattering pattern perpendicular to the field is

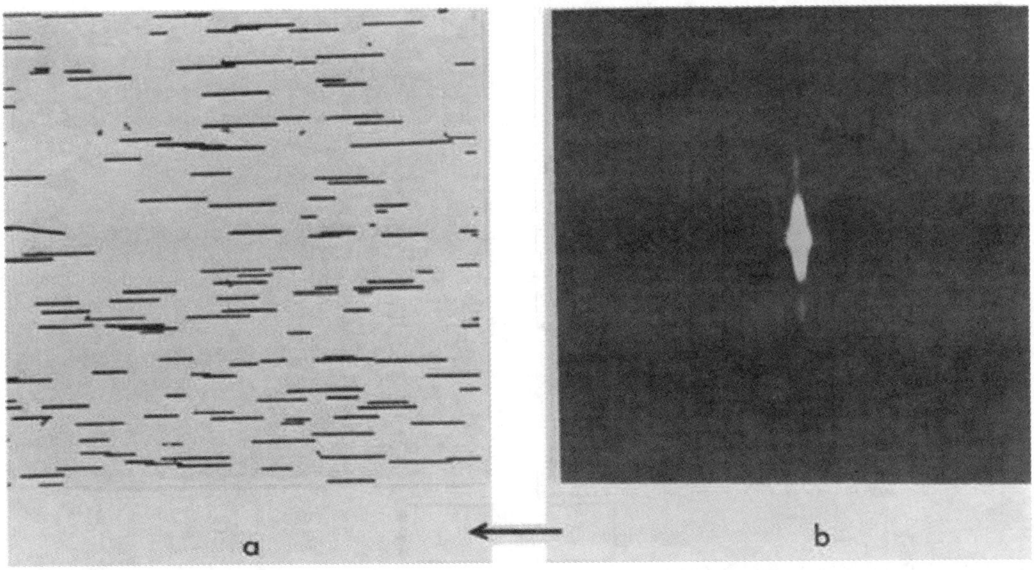

Fig. 23–17. **a**, Phase contrast microscope (PCM) image of magnetically aligned carbon fibers suspended in celloidin on a glass slide. **b**, Light-scattering pattern from the same fibers in aqueous suspension. The direction of the magnetic field is indicated by the arrow. Note the monodisperse diameter of the fibers, reflected in the sharply defined scattering pattern. (From Timbrell, 1973.)

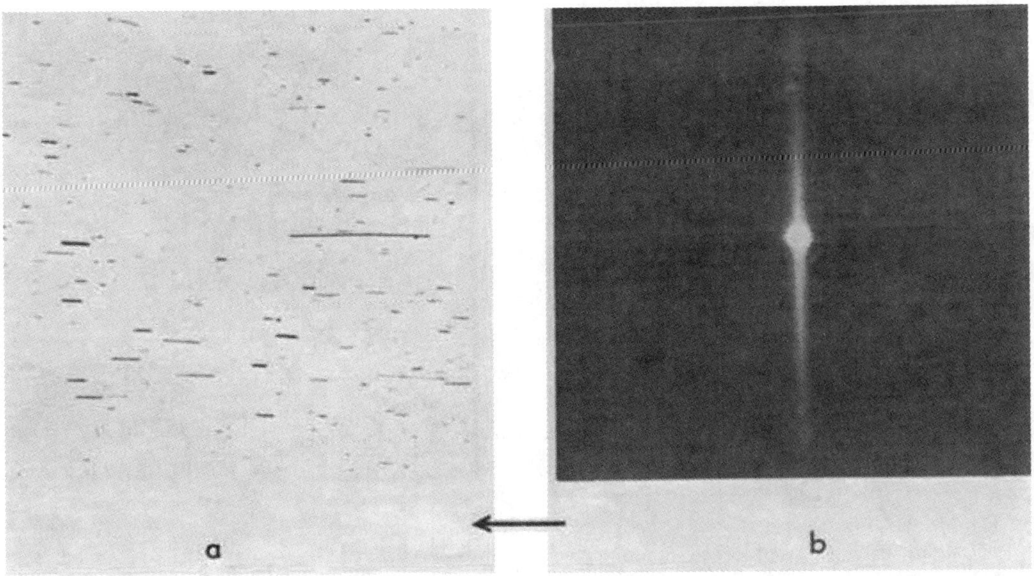

Fig. 23–18. **a**, PCM image of magnetically aligned UICC crocidolite fibers suspended in celloidin on a glass slide. **b**, Light-scattering pattern from the same fibers in aqueous suspension. The direction of the magnetic field is indicated by the arrow. (From Timbrell, 1973.)

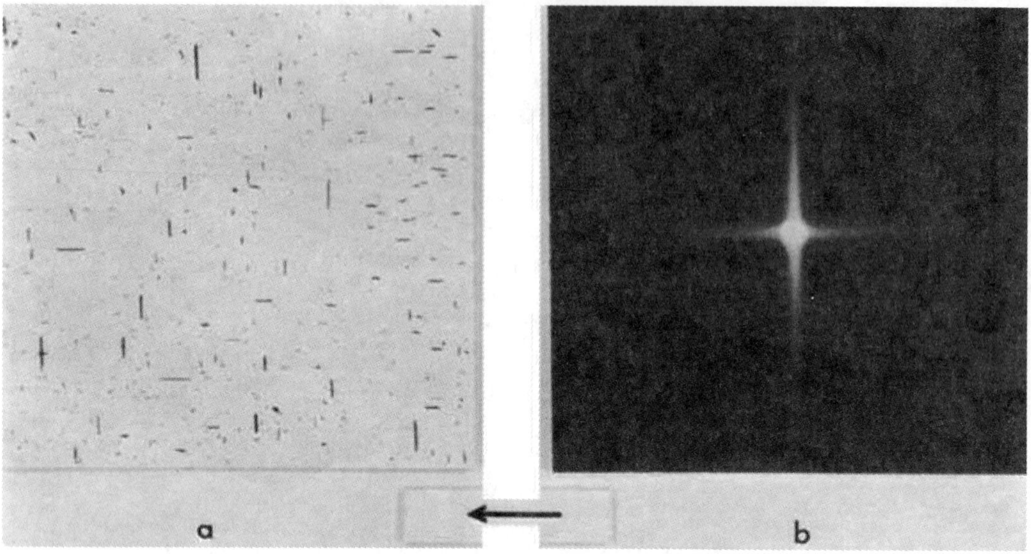

Fig. 23–19. **a**, PCM image of magnetically aligned UICC amosite fibers suspended in celloidin on a glass slide. **b**, Light-scattering pattern from the same fibers in aqueous suspension. The direction of the magnetic field is indicated by the arrow. (From Timbrell, 1973.)

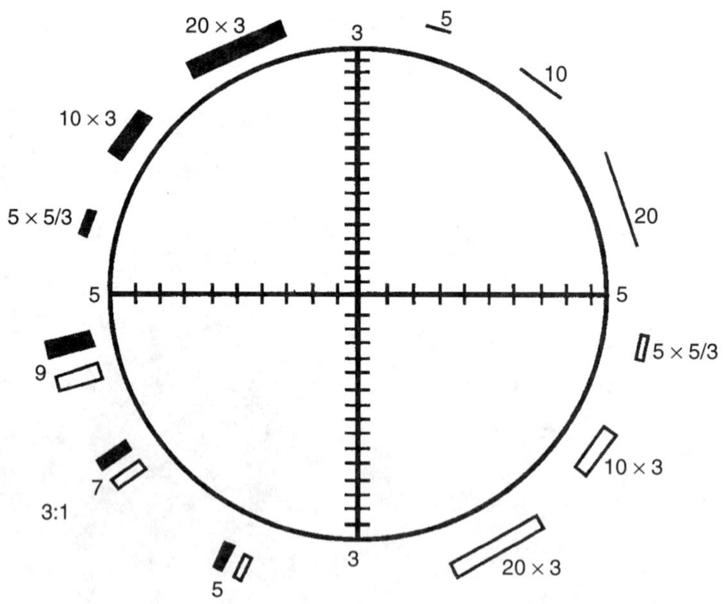

Fig. 23–20. Graticule used to define counting field in a phase contrast microscope. The defined objects outside the circle are used to help determine aspect ratios and sizes of fibers that should be counted.

produced. The crocidolite fibers in Figure 23–18 are aligned the same way, but are not monodisperse. In other cases, the fibers are aligned perpendicular or both perpendicular and parallel to the magnetic field (e.g., Fig. 23–20), the latter resulting in two planes of scattering. A synthetic fluoroamphibole was observed to align at ±65° to the magnetic field direction.

The degree and direction of alignment have not been adequately explained; however, they appear to be more a function of the mineralogical source of the material than of the primary crystal structure. Thus, Ugandan tremolite was observed to align perpendicular to the magnetic field, while Zululand tremolite aligned parallel to the field (Timbrell, 1973).

Light Scattering. With a flashlight beam shining on a glass rod such that the light beam was perpendicular to the rod axis, one would expect the refracted and scattered light to be dispersed into a plane perpendicular to the rod's axis. Light scattered and refracted from microscopic fibers produces a similarly unique pattern. Such scattering patterns from magnetically aligned fibers in an aqueous suspension are given in Figures 23–17, 23–18, and 23–19. If the fiber is not perpendicular to the light beam, the light-scattering pattern is not as conveniently constrained to a plane, becoming a cone of light. The details of light scattering from infinitely long cylinders, ellipsoids, and several other regular elongated shapes can be described with Mie theory (Van de Hulst, 1957; Kerker, 1969). As with scattering from spherical and compact particles, the scattering from fibers with diameters larger than the light wavelength is concentrated in the forward direction. The scattering from smaller diameter fibers is less in magnitude, but more uniform in all directions around the fiber axis. In addition, the scattered light tends to be polarized in the direction parallel to the fiber axis.

The unique, planar scattering pattern for right-angle illumination has been the basis of several useful fiber detection techniques. As noted above, the fiber must be held perpendicular to the light beam axis to obtain the characteristic fiber pattern. A fiber collected on a surface such as a glass slide will generally be parallel to the slide surface, thus allowing a beam perpendicular to the slide surface to produce the narrow, planar scattering pattern from a single fiber. To obtain a characteristic scattering pattern from a group of fibers, they must be aligned using some force, such as magnetic, electric, or shear flow.

Figures 23–17, 23–18, and 23–19 show the scattering patterns for several types of fibers. Note the well-defined scattering pattern for monodisperse carbon fibers (Fig. 23–17), while the broad distribution of diameters for the other fiber types produces a more diffuse pattern.

Laboratory Fiber Generation

Fibers are more difficult to generate than compact particles because of their tendency to intertwine when in contact with each other. This tendency is the basis for some commercial properties of fibers (e.g., the formation of rope and felt). Various types of fibers can be generated in various concentration ranges for instrument calibration, analytical method validation and quality assurance, and toxicology studies. Various generation mechanisms have been used to produce well-dispersed fibers.

Sample preparation of fibers for generation is an important step in producing useful aerosols. Some fibers may be available in sufficiently comminuted form for generation. Others need to be ground, chopped, or otherwise reduced in size. Blenders and grinders have been used. Air or water suspensions can be elutriated to reduce primarily the diameter range of the fibers. One technique for glass fibers used compression to produce fiber lengths with a mode in the 5 to 20 μm range, reducing the relative number of short fragments typically produced during grinding (Hanton et al., 1998).

Nebulization of liquid suspensions has been used for generating relatively short fibers at low concentrations. Fibers larger than the nebulized droplet diameter may not be generated efficiently, and, if the concentration is too high, more than one fiber may be present in a droplet, resulting in fiber agglomerates.

Several researchers have used high-speed chopping of a packed fiber plug. Timbrell et al. (1968b) developed a version of this approach using a household coffee grinder. A specially ground reference material, Union Internationale Contre le Cancer (UICC) asbestos

(Timbrell et al., 1968a), was packed uniformly in a syringe body and pushed slowly into the rotating blades of the grinder. The consumption rate of asbestos was 0.6 to 1.0 g/h, though not all this was generated as an aerosol. The dispersion appeared to be largely single fibers with relatively few clumps or flocculates, though no size distribution was reported. A generator with similar operating principle and dimensions was constructed using more durable materials, including tungsten carbide blades and a stainless steel body (Fairchild et al., 1976). This device was used for both chrysotile and fibrous glass. Although there were relatively few clumps of chrysotile by number, the mass in these clumps accounted for the largest fraction of the mass distribution. Airborne concentrations of 6 to 8 mg/m^3 were achieved, though the feed rate could be lowered to one fourth or increased to 10 times the value used for this measurement. Fibrous glass aerosols were more successfully generated with this device (Fairchild et al., 1978).

Fluidized bed generators have also been used for creating fibrous aerosols. Fluidized beds usually consist of two phases: a powder phase, containing one or more components, and an air phase passing through the powder. The powder is made to act like a boiling fluid by passing a sufficietly high air flow through it or by applying vibrational or acoustic energy. The powder may consist solely of the dust to be generated, or it may consist of the dust plus particles that are too large to be carried away by the airflow. These larger particles separate the dust particles from each other and tend to break apart agglomerates.

A two-component fluidized bed was used for inhalation exposure experiments with fibrous glass (Carpenter et al., 1981) and crocidolite (Griffis et al., 1983). The bed consisted of a stainless steel powder mixed with the fibers as a slurry and then dried. Air passing through the bed fluidized the bed and released the fibers, initially at a high rate and then decreasing exponentially. A similar air-fluidized bed with bronze powder as the fluidizing powder was used for generating multiple filter samples of chrysotile for a quality assurance program (Baron and Deye, 1987).

Charge must be reduced on generated aerosols to produce uniform air concentrations and consistent measurements. This can be accomplished by charge neutralizers (see Chapter 19) or by modifying the generation conditions. It was found that fluidized beds produced highly charged fibers when operated with dry air. The charge level dropped about 10-fold when the relative humidity of the air was increased to about 15% (Baron and Deye, 1990).

A two-component fluidized bed generator with a screw feed system to continually refresh the bed with premixed powder was found to produce a constant output concentration (Tanaka and Akiyama, 1984). This generator, using glass beads as the large particle fluidizing component, was found to produce a constant 6 mg/m^3 fibrous glass (from a glass fiber filter) for 1 week (Tanaka and Akiyama, 1987). A similar system using stainless steel beads and asbestos fibers was developed by Sussman et al. (1985).

An acoustically fluidized bed was developed to generate cellulose, ceramic, and glass fibers (Weyel et al., 1984; Frazer et al., 1986; Craig et al., 1991; Blake et al., 1997). This design allowed introduction of bulk fibrous powder into the generator with relatively little processing and produced reasonably constant output concentrations over long periods of time.

A one-component fluidized bed developed by Spurny (1980; Spurny et al., 1975) used vibrational energy to assist bed fluidization. The output of the bed for several types of asbestos was constant with time, and the fiber size was somewhat controlled by the vibration frequency and amplitude. A fluidized bed generator of this type using mechanical vibration is commercially available (***PAL***).

Chains of iron oxide particles have been generated using a laminar carbon monoxide flame (Kasper et al., 1980). This generation system created a high concentration of relatively monodisperse ($\sigma_g \approx 1.4$), nearly spherical particles that, under appropriate conditions, agglomerated into chains with little branching, perhaps due to alignment of magnetic dipoles in the agglomerating particles. Similar chains have been observed in other flame and arc systems.

An example particle chain size distribution is noted in Table 23–1. These chains can be used for investigating fiber diffusion, alignment, and aerodynamic size, especially in the transition regime (Kasper, 1983).

Fiber Health Effects

While asbestos fibers have many useful commercial properties, there has been much concern regarding their ability to cause disease. There are three primary diseases that have been attributed to asbestos fiber exposure: asbestosis (a fibrosis or scarring of the lung tissue), mesothelioma (a cancer of the pleura or peritoneum), and lung cancer. Other cancers (e.g., of the gastrointestinal tract) also have been attributed to asbestos exposure. Despite an abundance of research into disease mechanisms of asbestos fibers, the etiologies of these diseases are still not well understood. Fiber shape appears to play a major role, although other properties, such as fiber chemistry and solubility in body fluids, clearly are also important.

Fiber diameter must play a role in disease because the aerodynamic properties resulting in respiratory system deposition are strongly dependent on diameter (Timbrell, 1982). In general, mineral fibers must be smaller than about 2 to 3 μm diameter to reach the thoracic region and thinner still to reach the air exchange regions of the respiratory system (Stöber et al., 1970). It has been hypothesized that short fibers have much less disease potential because macrophages in the lung can engulf these particles and remove them from the lung with relative ease. Longer fibers cannot be completely engulfed by these cells and therefore tend to remain in the lung much longer (Holt, 1987). Timbrell (1982) found that clearance occurred for fibers up to 17 μm long, approximately equal to the human macrophage diameter. Use of the dielectrophoretic fiber length classifier to produce small quantities of length-classified fibers in combination with macrophage assays has confirmed that macrophages are damaged and killed by fibers slightly longer than the macrophage diameter. During macrophage death, an "oxidative burst" occurs, and the macrophage produces a series of factors that are linked to lung inflammation and ultimately fibrosis (Blake et al., 1997; Ye et al., 1999).

Besides shape, the chemical properties of fibers also appear to play a role. Some fibers, especially glass, have been found to dissolve in lung tissue over an extended period, reducing their potential for disease (Johnson et al., 1984; Law et al., 1990). Chrysotile fibers longer than 5 μm were found to increase in number in lung tissue, apparently due to longitudinal splitting of the fibers (Bellmann et al., 1986). In vitro studies have suggested that surface properties of fibers also can affect cell toxicity (Light and Wei, 1977).

Extrapolation of these properties indicates that long (especially those >17 μm long), thin (<3 μm diameter) insoluble fibers may have significant disease potential; for example, several researchers have postulated more specific size ranges as causing the various diseases (Pott, 1978; Lippmann, 1988; Timbrell, 1989).

Fiber Regulations

The potentially severe health effects of asbestos fiber exposure have prompted several regulatory and health research organizations to publish regulations and guidelines for controlling airborne concentrations of asbestos fibers. Because the health-based data indicate that disease at current exposure concentrations is primarily related to fiber number, most regulatory air concentration measurements are based on asbestos fiber number concentration rather than on mass concentration. In the United States, for example, the Occupational Safety and Health Administration (OSHA) provides regulations for exposure to hazardous agents in industrial and other workplace settings. The OSHA regulations require that workers are not to be exposed to more than 1×10^5 fibers/m^3 [0.1 asbestos fibers/cc] averaged over an 8h period or more than 1×10^6 fibers/m^3 [1.0 fiber/cc] over 30min as measured using the filter collection/phase contrast microscope (PCM) method (Occupational Safety and Health Administration, 1986). The Mine Safety and Health Administration (MSHA) regulates exposures in mines and mills and limits miner exposure to 2×10^6 fibers/m^3 [2.0 fibers/cc] for an 8h average and 1×10^7 fibers/m^3 [10 fibers/cc] for a 15min period (Mine Safety and Health Administration, 1988).

The Environmental Protection Agency (EPA) regulates environmental levels of pollutants. Apart from prohibiting visible emissions, the EPA has not implemented limits for environmental concentrations of asbestos. However, to protect children from being exposed to asbestos in schools, the EPA has mandated procedures for removing and measuring asbestos in schools (Environmental Protection Agency, 1987). The EPA has defined asbestos-containing material (ACM) as material containing more than 1% asbestos, to be measured using polarized light microscopy. After removal of ACM in schools, the EPA requires that the airborne asbestos concentration in the cleaned area be no greater than that outside the area. Measurement is conducted using five air samples inside the area and five outside for the comparison. Analysis by transmission electron microscope (TEM) is required for monitoring the completion of all asbestos removal operations, except that the PCM can be used when removing small amounts of asbestos. Guidance documents describing methods for controlling asbestos in buildings (Environmental Protection Agency, 1985, 1990) and for measurement of asbestos after removal (Environmental Protection Agency, 1985) have been provided.

Because ceramic fibers have become an important commercial material, the EPA promulgated regulations on fiber exposure as negotiated with the ceramic fiber industry to keep exposure below 1.0×10^6 fibers/m^3 [1.0 fibers/cc].

The Consumer Product Safety Commission (CPSC) provides guidance to manufacturers regarding the material content and potential hazards of commercial products. One such product to be targeted was hairdryers, prompting measurements of their emissions (Geraci et al., 1979). Individual state agencies set regulations that are often more stringent than those of the national agencies (Abbott, 1990).

The National Institute for Occupational Safety and Health (NIOSH) recommends health standards to OSHA and MSHA. NIOSH urged elimination, or reduction to the lowest possible levels, of exposures to asbestos fibers and recommended an exposure guideline of 0.1 fiber/cc based on practical limitations of PCM measurements (National Institute for Occupational Safety and Health, 1990).

Regulation of other fibers (e.g., fibrous glass and mineral wool) has generally dealt with these materials as nuisance dust. However, this may change because fibers other than asbestos have been demonstrated to have disease potential in humans and animals. For instance, erionite (a fibrous zeolite) has been associated with human mesotheliomas (Baris, 1980), and several manmade fibers have produced disease in animal exposure studies (Pott et al., 1987; Smith et al., 1987). Refractory ceramic fibers have produced mesothelioma in animals, and nylon fibers have produced acute inflammatory reactions in the lungs of rats (Porter et al., 1999) and humans (Jones et al., 1998).

Asbestos Terminology

Asbestos is a term applied to several commercial minerals exploited for their useful properties, largely due to their tendency to produce long, thin fibers. The development of asbestos crystal structure occurs primarily along one crystal axis. This results in a structure consisting of fibrils (the smallest diameter fibers, 0.025 to 0.05 μm diameter) bundled together. When subjected to comminution, asbestos normally breaks down into particles (fibers) that have high aspect ratios on both macroscopic and microscopic scales; the mineralogical term for this condition is *asbestiform* (American Society for Testing Materials, 1982). Other minerals with the same chemistry and crystal structure but without the unequal crystal development tend to produce more regular particles (termed *cleavage fragments*), though some of these particles also may be elongated. Although these cleavage fragments have a length/width distribution different from the asbestiform fibers (Virta et al., 1983), individual elongated cleavage fragments are often indistinguishable from asbestiform fibers when measured with commonly available techniques.

The asbestiform minerals that have been regulated include the following six: chrysotile (serpentine), amosite (cummingtonite-grunerite), crocidolite (glaucophane-riebeckite), tremolite asbestos, actinolite asbestos, and anthophyllite asbestos. The nonasbestiform mineral names for the first three materials are provided in parentheses. The latter three have the term *asbestos* attached because nonasbestiform varieties of these minerals have the same name. The latter five types of asbestos are classified as amphiboles. There are other asbestiform or fibrous minerals, but these are relatively rare and, except for attapulgite and wollastonite, have not been exploited commercially (Zumwalde and Dement, 1977). Richterite, or soda-tremolite, is similar to tremolite except that calcium in the crystal structure is replaced with sodium. It has been found as a contaminant of some vermiculite.

Health-related regulations, based on available microscope measurement methods, have specified asbestos fibers as particles with the elemental composition (from X-ray analysis) and crystal structure (from electron diffraction) appropriate to asbestos and with a length greater than 5 μm and an aspect ratio greater than 3:1 (Occupational Safety and Health Administration, 1986; National Institute for Occupational Safety and Health, 1990). Thus, the health-related regulatory definitions of asbestos fibers are based on measurements that may include particles that are cleavage fragments and not necessarily asbestiform. This definition is in contrast to the mineralogical definition relating to the asbestiform crystal structure, which often produces particle distributions with higher mean aspect ratios (Wylie, 1979; Kelse and Thompson, 1989).

The term *asbestos structure* is used for reporting asbestos air concentration under the Asbestos Hazard Emergency Relief Act (AHERA) regulations (Environmental Protection Agency, 1987). An asbestos structure is any particle (fiber, bundle, cluster, or matrix) consisting of or containing an observed fiber segment. This fiber segment must be longer than 0.5 μm with an aspect ratio of 5:1 or greater and be identified as asbestos by elemental analysis and electron diffraction. This definition is used to provide a sensitive method for assessing cleanliness after asbestos removal has been completed.

Measurement Techniques

Sampling and Measurement. There are two main classes of measurement techniques for fibers: microscopic observation of individual fibers and light-scattering–based instruments. Other instruments described in this book also can detect fiber aerosols, but, because they are not specific for fibers, are not considered here. Microscopic techniques require the collection of samples, most often filter samples, that are returned to the laboratory for preparation and analysis by a microscopist. Sample collection is most often carried out using a 25 mm diameter conductive plastic cassette with a 50 mm long inlet (cowl) to prevent direct deposition or contamination of the filter surface. The sampler provides some measure of size selection of collected fibers that varies with sampling rate and environmental conditions (Chen and Baron, 1996). Fiber sampling with a classifying inlet has been proposed to remove fibers and other particles not likely to deposit in the lungs (Baron, 1996).

Four principle types of microscopes are used for fiber detection and analysis: the phase contrast light microscope (PCM), the polarized light microscope (PLM), the scanning electron microscope (SEM), and the transmission electron microscope (TEM). A review of microscope techniques for workplace and environmental asbestos measurements is given by

Chatfield (1986). Descriptions of various light and electron microscopic techniques as well as pictures of many different types of fibers and particles are given in the seven volume *Particle Atlas* (McCrone and Delly, 1973). Other techniques described in Chapter 12 have also been used to analyze fibers.

Sample analysis can take place at various levels of complexity, for example, counting fibers having dimensions within prescribed size limits; determining the fiber size distribution; and measurement of the complete size distribution as well as qualitative analysis of individual fibers. The first is usually applied to establish compliance with regulations and may require some qualitative analysis as well as simple counting of fibers. The latter types of analysis are usually reserved for research studies or environmental assessments where fiber sources are unknown.

Sample preparation is extremely important for microscopic analysis because the view of the fibers and other particles is largely two-dimensional and only a small fraction of the sample is actually observed. Thus, the particles must be uniformly distributed at optimum concentration or loading over the sample surface. If the loading is too high, fibers and particles will overlap and be difficult to analyze; if the loading is too low, it will take too long to find a useful number of fibers (Peck et al., 1986; IARC, 1988). Low filter loading may also cause an overestimate of fiber concentration in some cases because fibers stand out more readily in a clean background (Cherrie et al., 1986).

Phase contrast microscopy is perhaps the most often used technique for monitoring airborne concentrations of asbestos and other fibers in workplace and other environments because it can be accomplished relatively quickly and inexpensively. Cellulose-based filters are used to collect the fibers, and then the filter is chemically cleared on a glass slide. The fibers are observed in a phase contrast microscope at approximately ×450 magnification and counted within observation fields defined by a graticule such as indicated in Figure 23–20. Between 20 and 100 of such fields are normally counted to obtain a statistically meaningful number of fibers (National Institute for Occupational Safety and Health, 1994).

The Environmental Protection Agency (1989) has evaluated asbestos sample preparation techniques. Most analyses of fibrous aerosol samples are performed on filter samples that are prepared without disturbing the location of the fibers on the filter surface (direct transfer sample preparation). This approach has been taken because asbestos fibers can break up when resuspended in a liquid (especially when ultrasonicated), increasing the fiber number. When sampling other fiber types that do not break apart or when sampling problems dictate dilution of the collected sample, a liquid suspension and redeposition of the fibers may be performed (indirect transfer sample preparation). Indirect sample preparation has the advantages of allowing removal of some interfering particles as well as providing a more uniformly deposited sample.

Bulk sample analysis by polarized light microscopy is often used in conjunction with air sampling to find potential sources of airborne fibers. X-ray diffraction (Abell, 1984) and other bulk analysis techniques have also been used for asbestos and other fibers; however, these techniques are not specific for fibers and often are not sufficiently sensitive for aerosol sample analysis. Chapter 12 provides details of these and other analytical techniques that may be applied to asbestos or other fibers.

There have been several attempts to improve the accuracy and speed and to reduce the subjectivity of microscope analysis by using automated image analysis of asbestos fiber samples (Whisnant, 1975). Automated image analysis involves taking the image from a microscope, digitizing the image, and using a computer to evaluate the number and size of objects present in the image. For fiber analysis, this may involve counting the number of fibers present or obtaining a size distribution of the fibers.

An image analysis system was coupled with PCM analysis of asbestos fibers. The Manchester Asbestos Program (MAP) was developed at the Manchester University (U.K.) with support from the Health and Safety Executive (Kenny, 1984, 1988). The MAP operated in a semiautomated mode, with the analyst selecting fields and focusing the microscope. The MAP was used for a time in various quality assurance programs but did not exhibit sufficient equivalency to human counting for continued use (Baron and Shulman, 1987).

Many image analysis systems are available with simple software for counting larger, well-defined fibers. For instance, they work reasonably well for fibrous glass insulation or synthetic organic textile fibers. Such programs typically have difficulty with asbestos fiber images because of the difficulty of detecting fibers of various diameters and curvatures, some barely visible, some overlapping, in the presence of a noisy image background containing particles of different shapes and sizes. However, with increased computer power and improved image analysis techniques, more accurate automated fiber counting may be possible in the future. A recently developed asbestos fiber counting program shows promise (Inoue et al., 1998).

Although not a complete image analysis system, Lundgren and co-workers (1995) used a computer display of a PCM microscope image to aid the analyst by recording the fiber count and location of each fiber during the analysis. A similar computer-based display system was coupled to a polarizing light microscope and used for quantitative measurement of tremolite fibers in dolomite (Lundgren et al., 1996).

Direct-Reading Fiber Measurement. An optical particle counter has been used for detection of chrysotile fibers in a textile plant where asbestos was the primary aerosol contaminant (Rickards, 1978). Other direct-reading monitors also may be used when the fibers are the major constituent in the aerosol, as in laboratory studies.

A fibrous aerosol monitor was developed for more specific measurement of asbestos fibers in the presence of compact particles (Lilienfeld et al., 1979). This instrument detects fibers via a combined electrostatic alignment/optical scattering technique. A commercial version of the Fibrous Aerosol Monitor (FAM-1, ***MIE***) was produced and has undergone continuous improvement, primarily in ruggedness for field use, over the past 10 years. A more recent version having a data logger and computer control was produced (FM-7400, ***MIE***). See Chapter 18 for more discussion of the FAM-1 and FM-7400 operation principles and evaluations.

Fiber length measurement using light scattering in the FAM was investigated (Cluff and Patitsas, 1992) and indicated good correlation between lengths measured using the FAM and by SEM analysis. A dielectrophoresis-based fiber length classifier was combined with an aerodynamic sizing instrument to obtain a fiber length and diameter distribution in near real time (Baron et al., 2000).

Other direct-reading instruments combining fiber alignment and light scattering have been developed for asbestos monitoring. Rood et al. (1992) developed a technique using fiber alignment by shear forces in an inlet tube, deposition by electrostatic precipitation, and detection by differential light scattering. Another device, the Fibrecheck (***CAS***), measured scattering at multiple angles to differentiate between fibers and compact particles. No field trials of these instruments have been published. All of these devices suffer from similar problems in accurately relating observed fiber counts to the standard approach (i.e., the filter collection/PCM counting method).

A general-purpose light-scattering instrument was developed that uses photomultipliers to detect light scattering in the forward direction and at three azimuthal angles, each gathering light between 27° and 140° from the forward direction. The four scattering values for each particle allowed classification of particles into shape-related groups (Kaye et al., 1996). A commercial version of this instrument is available (***BIR***). Further development of this concept used a detector array to give more complete scattering patterns (Kaye et al., 1992). An extension of the multiangle fiber measurement approach was developed by Sachweh and co-workers (1999). See Chapter 15 for further information about light-scattering instruments.

REFERENCES

Abbott, S. H. 1990. "State regulatory watch." *Asbestos Issues* December:16–25.

Abell, M. T. 1984. Chrysotile asbestos method 9000. In *NIOSH Manual of Analytical Methods*. P. Eller, ed. Cincinnati: National Institute for Occupational Safety and Health.

Allen, M. D. and J. K. Briant. 1978. Characterization of LMFBR fuel sodium aerosols. *Health Phys.* 35(2):237–254.

Allen, M. D., J. K. Briant, and O. R. Moss. 1979. Comparison of the aerodynamic size distribution of chain-like aggregates measured with a cascade impactor and a spiral centrifuge. *Am. Ind. Hyg. Assoc. J.* 40:474–481.

Allen, M. D., O. R. Moss, and J. K. Briant. 1978. Dynamic shape factors for LMFBR mixed-oxide fuel aggregates. *J. Aerosol Sci.* 10:43–48.

American Society for Testing Materials. 1982. Safety and health requirements relating to occupational exposure to asbestos. E 849-82. American Society for Testing Materials.

Asgharian, B. and M. N. Godo. 1999. Size separation of spherical particles and fibers in an aerosol centrifuge. *Aerosol Sci. Technol.* 30(4):383–400.

Asgharian, B. and C. P. Yu. 1988. Deposition of inhaled fibrous particles in the human lung. *J. Aerosol Med.* 1(1):37–50.

Asgharian, B. and C. P. Yu. 1989. A simplified model of interceptional deposition of fibers at airway bifurcations. *Aerosol Sci. Technol.* 11:80–88.

Asgharian, B., L. Zhang, and C. P. Fang. 1997. Theoretical calculations of the collection efficiency of spherical particles and fibers in an impactor. *Aerosol Sci. Technol.* 28:277–287.

Balásházy, I., T. B. Martonen, and W. Hofmann. 1990. Fiber deposition in airway bifurcations. *J. Aerosol Med.* 3:243–260.

Baris, Y. 1980. The clinical and radiological aspects of 185 cases of malignant pleural mesothelioma, ed. J. C. Wagner. *Biological Effects of Mineral Fibres*, Vol. 2. Lyon: International Agency for Research on Cancer, pp. 937–947.

Baron, P. A. 1996. Application of the thoracic sampling definition to fiber measurement. *Am. Ind. Hyg. Assoc. J.* 57:820–824.

Baron, P. A. and G. J. Deye. 1987. Generation of replicate asbestos aerosol samples for quality assurance. *Appl. Ind. Hyg.* 2:114–118.

Baron, P. A. and G. J. Deye. 1990. Electrostatic effects in asbestos sampling I: Experimental measurements. *Am. Ind. Hyg. Assoc. J.* 51:51–62.

Baron, P. A., G. J. Deye, and J. Fernback. 1994. Length separation of fibers. *Aerosol Sci. Technol.* 21: 179–192.

Baron, P. A., G. J. Deye, J. E. Fernback, and W. G. Jones. 2000. Direct-reading measurement of fiber length/diameter distributions, eds. M. E. Beard and H. L. Rook. *Advances in Environmental Measurement Methods for Asbestos*. Boulder, CO: American Society for Testing Materials, STP 1342.

Baron, P. A., P. Gao, G. J. Deye, and A. C. Maynard. 1998. Performance of a fiber length classifier. *J. Aerosol Sci.* 29(Suppl. 1):S11–S12.

Baron, P. A. and S. A. Shulman. 1987. Evaluation of the Magiscan image analyzer for asbestos fiber counting. *Am. Ind. Hyg. Assoc. J.* 48:39–46.

Bellmann, B., H. König, H. Muhle, and F. Pott. 1986. Chemical durability of asbestos and of man-made mineral fibres in vivo. *J. Aerosol Sci.* 17:341–345.

Blake, T., V. Castranova, D. Schwegler-Berry, G. J. Deye, P. Baron, C. Li, and W. Jones. 1997. Effect of fiber length on glass microfiber cytotoxicity. *J. Tox. Environ. Health* 54A:243–259.

Brockmann, J. E. 1985. Appendix F. *Uncertainty in Radionuclide Release Under Specific LWR Accident Conditions*. Albuquerque, NM: Sandia National Laboratories, SAND84-0410, Vol. 2.

Brockmann, J. E. 1987. Ex-vessel releases: Aerosol source terms in reactor accidents. *Prog. Nucl. Energy* 19:7–68.

Brockmann, J. E. and D. J. Rader. 1990. APS response to nonspherical particles and experimental determination of dynamic shape factor. *Aerosol Sci. Technol.* 13:162–172.

Burke, W. A. and N. A. Esmen. 1978. The inertial behavior of fibers. *Am. Ind. Hyg. Assoc. J.* 39:400–405.

Cai, J., N. Lu, and C. M. Sorensen. 1993. Comparison of size and morphology of soot aggregates as determined by light scattering and electron microscope analysis. *Langmuir* 9:2861–2068.

Cai, J., N. Lu, and C. M. Sorensen. 1995a. Analysis of fractal cluster morphology parameters: Structural coefficient and density autocorrelation function cutoff. *J. Colloid Interface Sci.* 171:470–473.

Cai, J., N. Lu, and C. M. Sorensen. 1995b. Fractal cluster size distribution measurement using static light scattering. *J. Colloid Interface Sci.* 174:456–460.

Cai, J. and C. M. Sorensen. 1994. Diffusion of fractal aggregates in the free molecular regime. *Phys. Rev. E.* 50:3397–3401.

Carpenter, R. L., J. A. Pickrell, B. V. Mokler, H. C. Yeh, and P. B. DeNee. 1981. Generation of respirable glass fiber aerosols using a fluidized bed aerosol generator. *Am. Ind. Hyg. Assoc. J.* 42:777–784.

Chan, P. and B. Dahneke. 1981. Free-molecular drag on straight chains of uniform spheres. *J. Appl. Phys.* 52:3106.

Chatfield, E. J. 1986. Asbestos measurements in workplaces and ambient atmospheres, eds. S. Basu and J. R. Millette. *Electron Microscopy in Forensic, Occupational and Environmental Health Sciences*. New York: Plenum Publishing Corp.

Chen, C.-C. and P. A. Baron. 1996. Aspiration efficiency and wall deposition in the fiber sampling cassette. *Am. Ind. Hyg. Assoc. J.* 52(2):142–152.

Chen, Y. K. and C. P. Yu. 1990. Sedimentation of charged fibers from a two-dimensional channel flow. *Aerosol Sci. Technol.* 12:786–792.

Cheng, M. T., G. W. Xie, M. Yang, and D. T. Shaw. 1991. Experimental characterization of chain-aggregate aerosol by electrooptic scattering. *Aerosol Sci. Technol.* 14(1):74–81.

Cheng, Y.-S. 1986. Bivariate lognormal distribution for characterizing asbestos fiber aerosols. *Aerosol Sci. Technol.* 5(3):359–368.

Cheng, Y.-S., M. D. Allen, D. P. Gallegos, and H.-C. Yeh. 1988. Drag force and slip correction of aggregate aerosols. *Aerosol Sci. Technol.* 8:199–214.

Cherrie, J., A. D. Jones, and A. M. Johnston. 1986. The influence of fiber density on the assessment of fiber concentration using the membrane filter method. *Am. Ind. Hyg. Assoc. J.* 47:465–474.

Chissick, S. S. and R. Derricot. 1983. *Asbestos, Properties, and Applications and Hazards*. Chichester: John Wiley & Sons.

Clift, R., J. R. Grace, and M. E. Weber. 1978. *Bubbles, Drops and Particles*. New York: Academic Press.

Cluff, D. L. and A. J. Patitsas. 1992. Size characterization of asbestos fibers by means of electrostatic alignment and light-scattering techniques. *Aerosol Sci. Technol.* 17(3):186–198.

Cox, R. G. 1970. The motion of long slender bodies in a viscous fluid I: General theory. *J. Fluid Mech.* 44:791.

Craig, D. K., C. A. Lapin, and G. E. Butterfield. 1991. The generation and characterization of silicon carbide whiskers (fibers) for inhalation toxicology studies. *Am. Ind. Hyg. Assoc. J.* 52(8):315–319.

Dahneke, B. 1982. Viscous resistance of straight-chain aggregates of uniform spheres. *Aerosol Sci. Technol.* 1:179–185.

Dahneke, B. E. 1973a. Slip correction factors for nonspherical bodies—I. Introduction and continuum flow. *J. Aerosol Sci.* 4:139–145.

Dahneke, B. E. 1973b. Slip correction factors for nonspherical bodies—II. Free molecule flow. *J. Aerosol Sci.* 4:147–161.

Dahneke, B. E. 1973c. Slip correction factors for nonspherical bodies—III. The form of the general law. *J. Aerosol Sci.* 4:163–170.

Dement, J. M. 1990. Overview: Workshop on fiber toxicology research needs. *Environ. Health Perspect.* 88:261–268.

Dobbins, R. A. and C. M. Magaridis. 1987. Morphology of flame-generated soot as determined by thermophoretic sampling. *Langmuir* 3:254–259.

Environmental Protection Agency. 1985. *Measuring Airborne Asbestos Following an Abatement Action*. EPA 600/4-85-049. EPA 600/4-85-049. Quality Assurance Division, Environmental Protection Agency.

Environmental Protection Agency. 1987. Asbestos-containing materials in schools. *Federal Register 40 CFR Part 763*. Washington, DC: Government Printing Office, April 30, 1987, 40 CFR Part 763.

Environmental Protection Agency. 1989. *Comparison of Airborne Asbestos Levels Determined by Transmission Electron Microscopy (TEM) Using Direct and Indirect Transfer Techniques*. EPA 560/5-89-004. EPA 560/5-89-004. Environmental Protection Agency.

Environmental Protection Agency. 1990. *Managing Asbestos in Place: A Building Owners Guide to Operations and Maintenance Programs for Asbestos Containing Materials*. EPA 20T-2003. EPA 20T-2003. Environmental Protection Agency.

Fairchild, C. I., L. W. Ortiz, H. J. Ettinger, and M. I. Tillery. 1976. *Aerosol Research and Development Related to Health Hazard Analysis*. LA-6277-PR. Los Alamos Scientific Laboratory.

Fairchild, C. I., L. W. Ortiz, M. I. Tillery, and H. J. Ettinger. 1978. *Aerosol Research and Development Related to Health Hazard Analysis*. LA-7380-PR. Los Alamos Scientific Laboratory.

Family, F. 1991. Fractals, eds. R. G. Lerner and G. L. Trigg. *Encyclopedia of Physics*, 2nd ed. New York: VCH Publishers, Inc., p. 414.

Family, F. and D. P. Landau. 1984. *Kinetics of Aggregation and Gelation*. Amsterdam: North-Holland.

Forrest, S. R. and T. A. Witten, Jr. 1979. Long-range correlations in smoke-particle aggregates. *J. Phys. A Math. Gen.* 12:L109–L117.

Frazer, D. G., V. Robinson, K. Jayaraman, K. C. Weber, D. S. DeLong, and C. Glance. 1986. *Improved Operating Parameters for the Pitt 3 Aerosol Generator During Resuspension of Respirable Cotton Dust*. The Tenth Cotton Dust Research Conference, National Cotton Council, Memphis, TN.

Friedlander, S. K., H. D. Jang, and K. H. Ryu. 1998. Elastic behavior of nanoparticle chain aggregates. *Appl. Phys. Lett.* 72:173–175.

Fu, T.-H., M.-T. Cheng, and D. Shaw. 1990. Filtration of chain aggregate aerosols by model screen filter. *Aerosol Sci. Technol.* 13(2):151–161.

Fuchs, N. A. 1964. *The Mechanics of Aerosols*. Oxford: Pergamon Press.

Gallily, I. 1971. On the drag experienced by a spheroidal, small particle in a gravitational and electrostatic field. *J. Colloid Interface Sci.* 36(3):325–339.

Gallily, I. and A. D. Eisner. 1979. On the orderly nature of the motion of nonspherical aerosol particles I. Deposition from a laminar flow. *J. Colloid Interface Sci.* 68:320–337.

Gallily, I., D. Schiby, A. H. Cohen, W. Holländer, D. Schless, and W. Stöber. 1986. On the inertial separation of nonspherical aerosol particles from laminar flows. I. The cylindrical case. *Aerosol Sci. Technol.* 5(2):267–286.

Gangopadhyay, S., I. Elminyawi, and C. M. Sorensen. 1991. Optical structure factor measurements of soot particles in a premixed flame. *Appl. Opt.* 30(4859–4864).

Gelbard, F. 1982. *MAEROS User Manual*. NUREG/CR-1391 SAND80-0822. Sandia National Laboratories.

Gentry, J. W., K. R. Spurny, J. Schörmann, and H. Opiela. 1983. Measurement of the diffusion coefficient of asbestos fibers, eds. V. A. Marple and B. Y. H. Liu. *Aerosols in the Mining and Industrial Work Environments*. Ann Arbor: Ann Arbor Science Publishers. 2:597–612.

Gentry, J. W., K. R. Spurny, S. A. Soulen, and J. Schörmann. 1988. Measurements of the diffusion coefficients of ultrafine asbestos fibers. *J. Aerosol Sci.* 19(7):1041–1044.

Geraci, C., P. A. Baron, J. W. Carter, and D. L. Smith. 1979. *Testing of Hair Dryer Emissions*, IA-79-29. National Institute for Occupational Safety and Health.

Gieseke, J. A., L. D. Reed, H. Jordan, and K. W. Lee. 1977. *Characteristics of Agglomerates of Sodium Oxide Aerosols*. BMI-NUREG-1977 NRC-7. Columbus, OH: Battelle Columbus Laboratories.

Gonda, I. and A. F. A. E. Khalik. 1985. On the calculation of aerodynamic diameters of fibers. *Aerosol Sci. Technol.* 4:233–238.

Griffis, L. C., J. A. Pickrell, R. L. Carpenter, R. K. Wolff, S. J. McAllen, and K. L. Yerkes. 1983. Deposition of crocidolite asbestos and glass microfibers inhaled by the beagle dog. *Am. Ind. Hyg. Assoc. J.* 44(3):216–222.

Griffiths, W. D., L. C. Kenny, and S. T. Chase. 1985. The electrostatic separation of fibres and compact particles. *Ann. Occup. Hyg.* 16(3):229–243.

Griffiths, W. D. and N. P. Vaughan. 1986. The aerodynamic behaviour of cylindrical and spheroidal particles when settling under gravity. *J. Aerosol Sci.* 17(1):53–65.

Guinier, A., G. Fournet, C. B. Walker, and K. L. Yudowitch. 1955. *Small Angle Scattering of X-Rays*. New York: John Wiley & Sons.

Hanton, D. Y., H. Furtak, and H. G. Grimm. 1998. Preparation and handling conditions of MMVF for invivo experiments. *Aerosol Sci. Technol.* 29:449–456.

Helton, J. C., R. L. Iman, J. D. Johnson, and C. D. Leigh. 1986. *Uncertainty and Sensitivity Analysis of a Dry Containment Test Problem for the MAEROS Aerosol Model*. SAND85-2795. Sandia National Laboratories.

Hinds, W. C. 1999. *Aerosol Technology*. New York: John Wiley & Sons.

Holst, E. and T. Schneider. 1985. Fibre size characterization and size analysis using general and bivariate log-normal distributions. *J. Aerosol Sci.* 5:407–413.

Holt, P. F. 1987. *Dust and Disease*. New York: John Wiley & Sons.

IARC. 1988. Man-Made Mineral Fibers and Radon. *IARC Monographs of the Evaluation of the Carcionogenic Risk of Chemicals in Humans*. Lyon, France: International Agency for Research on Cancer, pp. 33–171.

Iles, P. J. 1990. Size selection of fibres by cyclone and horizontal elutriator. *J. Aerosol Sci.* 21(6):745–760.

Inoue, Y., A. Kaga, K. Yamaguchi, and S. Kamoi. 1998. Development of an automatic system for counting asbestos fibers using image processing. *Part. Sci. Technol.* 16:263–279.

Johnson, D. L., D. Leith, and P. C. Reist. 1987. Drag on non-spherical orthotropic aerosol particles. *J. Aerosol Sci.* 18(1):87–97.

Johnson, N. F., D. M. Griffiths, and R. J. Hill. 1984. Size distribution following long term inhalation of MMMF. *Biological Effects of Man-Made Mineral Fibers*, Vol. 2. Copenhagen: World Health Organization, pp. 102–125.

Johnston, A. M., J. H. Vincent, and A. D. Jones. 1985. Measurements of electric charge for workplace aerosols. *Ann. Occup. Hyg.* 29(2):271–284.

Jones, A. D., A. M. Johnston, and J. H. Vincent. 1983. Static electrification of airborne asbestos dust, eds. V. A. Marple and B. Y. H. Liu. *Aerosols in the Mining and Industrial Work Environment*, Vol. 2. Ann Arbor, MI: Ann Arbor Science, pp. 613–632.

Jones, W. G., C. Piacitelli, D. Schwegler-Berry, and J. Burkhart. 1998. *Environmental Study of Nylon Flocking Process*. NIOSH.

Jullien, R., R. Thouy, and F. Ehrburger-Doll. 1994. Numerical investigation of two-dimensional projections of random fractal aggregates. *Phys. Rev.* E50:3878–3882.

Kasper, G. 1982. Dynamics and measurement of smokes. II. The aerodynamic diameter of chain aggregates in the transition regime. *Aerosol Sci. Technol.* 1(2):201–216.

Kasper, G. 1983. Note on the slip coefficient of doublets of spheres. *J. Aerosol Sci.* 14(6):753–754.

Kasper, G. and D. T. Shaw. 1983. Comparative size distribution measurements on chain aggregates. *Aerosol Sci. Technol.* 2:369–381.

Kasper, G., S. N. Shon, and D. T. Shaw. 1980. Controlled formation of chain aggregates from very small metal oxide particles. *Am. Ind. Hyg. Assoc. J.* 41:288–296.

Kaye, P. H., K. Alexander-Buckley, E. Hirst, et al. 1996. A real-time monitorng system for airborne particle shape and size analysis. *J. Geophys. Res.* 101:19215–19221.

Kaye, P. H., E. Hirst, J. M. Clark, and F. Micheli. 1992. Airborne particle shape and size classification from spatial light scattering profiles. *J. Aerosol Sci.* 23(6):597–612.

Kelse, J. W. and C. S. Thompson. 1989. The regulatory and mineralogical definitions of asbestos and their impact on amphibole dust analysis. *Am. Ind. Hyg. Assoc. J.* 50:613–622.

Kenny, L. C. 1984. Asbestos fibre counting by image analysis—The performance of the Manchester Asbestos Program on Magiscan. *Ann. Occup. Hyg.* 28:401–415.

Kenny, L. C. 1988. Automated analysis of asbestos clearance samples. *Ann. Occup. Hyg.* 32(1):115–128.

Kerker, M. 1969. *The Scattering of Light and Other Electromagnetic Radiation*. New York: Academic Press.

Kolb, M., R. Botet, and R. Jullien. 1983. Scaling of kinetically growing clusters. *Phys. Rev. Lett.* 51: 1123–1126.

Kops, J., G. Dibbets, L. Hermans, and J. F. Van der Vate. 1975. The aerodynamic diameter of branched chain-like aggregates. *J. Aerosol Sci.* 6(5):329–334.

Köylü, U. O. and G. M. Faeth. 1992. Structure of overfire soot in buoyant turbulent diffusion flames at long residence times. *Combust. Flame* 89:140–156.

Köylü, U. O., G. M. Faeth, T. L. Farias, and M. G. Carvalho. 1995. Fractal and projected structure properties of soot aggregates. *Combust. Flame* 100:621–633.

Laframboise, J. G. and J.-S. Chang. 1977. Theory of charge deposition on charged aerosol particles of arbitrary shape. *J. Aerosol Sci.* 8:331–338.

Langer, A. 1974. The subject of continuous vigilance. *Environ. Health Perspect.* 9:53–56.

Langer, A. M., A. D. Mackler, and F. D. Pooley. 1974. Electron microscopical investigation of asbestos fibers. *Environ. Health Perspect.* 9:63–80.

Law, B. D., W. B. Bunn, and T. W. Hesterburg. 1990. Solubility of polymeric organic fibers and manmade vitreous fibers in gambles solution. *Inhal. Toxicol.* 2:321–339.

Light, W. G. and E. T. Wei. 1977. Surface charge and asbestos toxicity. *Science* 265:537–539.

Lilienfeld, P. 1985. Rotational electrodynamics of airborne fibers. *J. Aerosol Sci.* 16(4):315–322.

Lilienfeld, P., P. Elterman, and P. Baron. 1979. Development of a prototype fibrous aerosol monitor. *Am. Ind. Hyg. Assoc. J.* 40(4):270–282.

Lin, M. Y., H. M. Lindsay, D. A. Weitz, R. C. Ball, R. Klein, and P. Meakin. 1990. Universal reaction-limited colloid aggregation. *Phys. Rev. A* 41:2005–2020.

Lipowicz, P. J. and H. C. Yeh. 1989. Fiber dielectrophoresis. *Aerosol Sci. Technol.* 11:206–212.

Lippmann, M. 1988. Asbestos exposure indices. *Environ. Res.* 46:86–106.

Lundgren, L., S. Lundström, I. Laszlo, and B. Westling. 1995. Modern fibre counting—A technique with the phase-contrast microscope on-line to a Macintosh computer. *Ann. Occup. Hyg.* 39(4):455–467.

Lundgren, L., S. Lundström, G. Sundström, G. Bergman, and S. Krantz. 1996. A quantitative method using a light microscope on-line to a Macintosh computer for the analysis of tremolite fibres in dolomite. *Ann. Occup. Hyg.* 40:197–209.

Mandelbrot, B. B. 1977. *Fractals: Form, Chance and Dimension*. San Francisco: W. H. Freeman.

Mandelbrot, B. B. 1983. *The Fractal Geometry of Nature*. New York: W. H. Freeman and Company.

Martin, J. E. and B. J. Ackerson. 1985. *Phys. Rev. A* 31:1180–1182.

Martonen, T. B. and D. L. Johnson. 1990. Aerodynamic classification of fibers with aerosol centrifuges. *Part. Sci. Technol.* 8:37–53.

McCrone, W. C. and J. G. Delly. 1973. *The Particle Atlas*. Ann Arbor, MI: Ann Arbor Science Publishers, Inc.

Meakin, P. 1983. Formation of fractal clusters and networks by irreversible diffusion-limited aggregation. *Phys. Rev. Lett.* 51:1119–1122.

Meakin, P. 1984. Effects of cluster trajectories on cluster-cluster aggregations: A comparison of linear and brownian trajectories in two- and three-dimensional simulations. *Phys. Rev. A* 29:997–999.

Meakin, P. 1988. Fractal aggregates. *Adv. Colloid Interface Sci.* 28:249–331.

Meakin, P., B. Donn, and G. W. Mulholland. 1989. Collisions between point masses and fractal aggregates. *Langmuir* 5:510–518.

Medalia, A. I. 1967. Morphology of aggregates. *J. Colloid Interface Sci.* 24:393–404.

Medalia, A. I. and F. A. Heckman. 1969. Morphology of aggregates—II. Size and shape factors of carbon black aggregates from electron microscopy. *Carbon* 7:567–582.

Megaridis, C. M. and R. A. Dobbins. 1990. Morphological description of flame generated materials. *Combust. Sci. Technol.* 71:95–109.

Michaels, L. and S. S. Chissick. 1979. *Asbestos: Properties, Applications and Hazards*. Chichester: John Wiley & Sons.

Mine Safety and Health Administration. 1988. *Exposure Limits for Airborne Contaminants*. CFR Part 56.5001. Washington, DC: U.S. Department of Commerce.

National Institute for Occupational Safety and Health. 1976. *Revised Recommended Asbestos Standard*. Publ. (NIOSH) 77–169. Cincinnati: U.S. Department of Health Education and Welfare.

National Institute for Occupational Safety and Health. 1977. *Occupational Exposure to Fibrous Glass*. DHEW (NIOSH) Publication No. 77–152. Cincinnati: National Institute for Occupational Safety and Health.

National Institute for Occupational Safety and Health. 1990. *Testimony of NIOSH on Occupational Exposure to Asbestos, Tremolite, Anthophyllite an Actinolite*. 29CFR Parts 1910 and 1926. Cincinnati: National Institute of Occupational Safety and Health.

National Institute for Occupational Safety and Health. 1994. *Asbestos and Other Fibers by PCM*. Method 7400. *NIOSH Manual of Analytical Methods*, 4th ed. Cincinnati: National Institute for Occupational Safety and Health.

Occupational Safety and Health Administration. 1986. *Occupational Exposure to Asbestos, Tremolite, Anthophyllite, and Actinolite Asbestos; Final Rules*. 29 CFR Part 1910.1001 and 1926. Washington, DC: Occupational Safety and Health Administration.

Oh, C. and C. M. Sorensen. 1997. The effect of monomer overlap on the morphology of fractal aggregates. *J. Colloid. Interface Sci.* 193:17–25.

Oh, C. and C. M. Sorensen. 1998. Structure factor of diffusion-limited aggregation clusters: Local structure and non-self-similarity. *Phys. Rev. E* 57:784–790.

Oh, C. and C. M. Sorensen. 1999. Scaling approach for the structure factor of a generalized system of scatterers. *J. Nanopart. Res.* 1:369–377.

Peck, A. S., J. J. Serocki, and L. C. Dicker. 1986. Sample density and the quantitative capabilities of PCM analysis for the measurement of airborne asbestos. *Am. Ind. Hyg. Assoc. J.* 47:A230–A234.

Porter, D. W., V. Castranova, R. A. Robinson, A. F. Hubbs, R. R. Mercer, J. Scabilloni, T. Goldsmith, D. Schwegler-Berry, L. Battelli, et al. 1999. Acute inflammatory reaction in rats after intratracheal instillation of material collected from a nylon flocking plant. *J. Toxicol. Environ. Health*. Part A 57:25–45.

Pott, F. 1978. Some aspects on the dosimetry of the carcinogenic potency of asbestos and other fibrous dusts. *Staub Rein. Luft* 38:486.

Pott, F., U. Ziem, F. J. Reiffer, F. Huth, H. Ernst, and U. Mohr. 1987. Carcinogenicity studies of fibers, metal compounds, and some other dusts in rats. *Exp. Pathol.* 32:129–152.

Prenni, A. J., R. L. Siefert, T. B. Onasch, and M. A. Tolbert. 2000. Design and characterization of a fluidized bed aerosol generator: A source for dry, submicrometer aerosol. *Aerosol Sci. Technol.* 32(5):465–481.

Prodi, V., T. De Zaiocomo, D. Hochrainer, and K. Spurny. 1982. Fibre collection and measurement with the inertial spectrometer. *J. Aerosol Sci.* 13:49–58.

Rajhans, G. S. and J. L. Sullivan. 1981. *Asbestos Sampling and Analysis*. Ann Arbor, MI: Ann Arbor Science Publishers.

Ren, Z. F., Z. P. Huang, J. W. Xu, J. H. Wang, P. Bush, M. P. Siegal, and P. N. Provencio. 1998. Synthesis of large arrays of well-aligned carbon nanotubes on glass. *Science* 282:1105–1107.

Rickards, A. L. 1978. The routine monitoring of airborne asbestos in an occupational environment. *Ann. Occup. Hyg.* 21:315–322.

Rogak, S. N., U. Baltensperger, and R. C. Flagan. 1990. Direct measurement of mass transfer to agglomerates in the transition regime. *J. Aerosol Sci.* 21(Suppl. 1):S51–S55.

Rood, A. P., E. J. Walker, and D. Moore. 1992. Construction of a portable fiber monitor measuring the differential light scattering from aligned fibres. *Aerosol Sci. Technol.* 17(1):1–8.

Rosner, D. E., D. W. Mackowski, and P. Garcia-Ybana. 1991. Size- and structure-insensitivity of the thermophoretic transport of aggregated soot particles in gases. *Combust. Sci. Tech.* 80:87–101.

Sachweh, B., H. Barthel, R. Polke, H. Umhauer, and H. Buttner. 1999. Particle shape and structure analysis from the spatial intensity pattern of scattered light using different measuring devices. *J. Aerosol Sci.* 30(10):1257–1270.

Sampson, R. J., G. W. Mulholland, and J. W. Gentry. 1987. Structural analysis of soot agglomerates. *Langmuir* 3:272–281.

Schmidt-Ott, A. 1988. In situ measurement of the fractal dimensionality of ultrafine aerosol particles. *Appl. Phys. Lett.* 52:954.

Schneider, T., E. Holst, and J. Skotte. 1983. Size distribution of airborne fibres generated from man-made mineral fiber products. *Ann. Occup. Hyg.* 27(2):157–171.

Selikoff, I. J. and E. C. Hammond. 1979. *Health Hazards of Asbestos Exposure*. New York: New York Academy of Sciences.

Smith, D. M., L. W. Ortiz, and R. F. Archuleta. 1987. Long-term health effects in hamsters and rats exposed to man-made vitreous fibers. *Ann. Occup. Hyg.* 31(4B):731–754.

Sorensen, C. M. 1997. Scattering and absorption of light by particles and aggregates, ed. K. S. Birdi. *Handbook of Surface and Colloid Chemistry*. Boca Raton, FL: CRC Press, pp. 533–558.

Sorensen, C. M. 2001. Light scattering from fractal aggregates. A review. *Aerosol Sci. Technol.* In Press.

Sorensen, C. M., J. Cai, and N. Lu. 1992a. Light-scattering measurements of monomer size, monomers per aggregate, and fractal dimension for soot aggregates in flames. *Appl. Opt.* 31:6547–6557.

Sorensen, C. M., J. Cai, and N. Lu. 1992b. Test of static structure factors for describing light scattering from fractal soot aggregates. *Langmuir* 8:2064–2069.

Sorensen, C. M. and G. D. Feke. 1996. The morphology of macroscopic soot. *Aerosol Sci. Technol.* 25(3):328–337.

Sorensen, C. M., W. B. Hageman, T. J. Rush, H. Huang, and C. Oh. 1998. Aerogelation in a flame soot aerosol. *Phys. Rev. Lett.* 80:1782–1785.

Sorensen, C. M. and G. C. Roberts. 1997. The prefactor of fractal aggregates. *J. Colloid Interface Sci.* 186:447–452.

Sorensen, C. M. and G. M. Wang. 1999. Size distribution effect on the power law regime of the structure factor of fractal aggregates. *Phys. Rev. E* Part B 60:7143–7148.

Spurny, K., C. Boose, and D. Hochrainer. 1975. Zur zerstäubung von asbestfasern in einem fliessbett-aerosolgenerator. *Staub Rein. Luft* 35(12):440–445.

Spurny, K. R. 1980. Fiber generation and length classification, ed. K. Willeke. *Generation of Aerosols and Facilities for Exposure Experiments*. Ann Arbor: Ann Arbor Science Publishers, pp. 257–298.

Stöber, W. 1971. A note on the aerodynamic diameter and the mobility of non-spherical aerosol particles. *J. Aerosol Sci.* 2(4):453–456.

Stöber, W. 1972. Dynamic shape factors of nonspherical aerosol particles, eds. T. T. Mercer, P. E. Morrow, and W. Stöber. *Assessment of Airborne Particles*. Springfield, IL: Charles C. Thomas, pp. 249–289.

Stöber, W., H. Flachsbart, and D. Hochrainer. 1970. The aerodynamic diameter of latex aggregates and asbestos fibers. *Staub Rein. Luft* 30(7):1–12.

Sussman, R. G., J. M. Gearhart, and M. Lippmann. 1985. A variable feed rate mechanism for fluidized bed asbestos generators. *Am. Ind. Hyg. Assoc. J.* 46(1):24–27.

Tanaka, I. and T. Akiyama. 1984. A new dust generator for inhalation toxicity studies. *Ann. Occup. Hyg.* 28(2):157–162.

Tanaka, I. and T. Akiyama. 1987. Fibrous particles generator for inhalation toxicity studies. *Ann. Occup. Hyg.* 31:401–403.

Tanford, C. 1961. *Physical Chemistry of Macromolecules*. New York: John Wiley & Sons.

Timbrell, V. 1972. Alignment of carbon and other man-made fibers by magnetic fields. *J. Appl. Phys.* 43:4839–4840.

Timbrell, V. 1973. *Desired Characteristics of Fibres for Biological Experiments*. Montreal, Canada: Institute of Occupational and Environmental Health.

Timbrell, V. 1975. Alignment of respirable asbestos fibres by magnetic fields. *Ann. Occup. Hyg.* 18: 299–311.

Timbrell, V. 1982. Deposition and retention of fibres in the human lung. *Ann. Occup. Hyg.* 26:347–369.

Timbrell, V. 1989. Review of the significance of fibre size in fibre-related lung disease: A centrifuge cell for preparing accurate microscope-evaluation specimens from slurries used in inoculation studies. *Ann. Occup. Hyg.* 33:483–505.

Timbrell, V., J. C. Gilson, and I. Webster. 1968a. UICC standard reference samples of asbestos. *Int. J. Cancer* 3:406–408.

Timbrell, V, A. W. Hyatt, and J. W. Skidmore. 1968b. A simple dispenser for generating dust clouds from standard reference samples of asbestos. *Ann. Occup. Hyg.* 11:273–281.

Tolles, W. M., R. A. Sanders, and G. W. Fritz. 1974. Dielectric response of anisotropic polarized particles observed with microwaves: A new method for characterizing the properties of nonspherical particles in suspension. *J. Appl. Phys.* 45(9):3777–3783.

Van de Hulst, H. C. 1957. *Light Scattering by Small Particles*. New York: John Wiley & Sons.

van de Vate, J. F., W. F. van Leeuwen, A. Plomp, and H. C. D. Smit. 1980. Dynamic shape factors: Measurement techniques and results on aggregates of solid primaries. *J. Aerosol Sci.* 11:67–75.

Virta, R. L., K. B. Shedd, A. G. Wylie, and J. G. Snyder. 1983. Size and shape characteristics of amphibole asbestos (amosite) and amphibole cleavage fragments (actinolite, cummingtonite) collected on occupational air monitoring filters. In *Aerosols in the Mining and Industrial Work Environments*, V. A. Marple and B. Y. H. Liu, eds. Vol. 2. Ann Arbor: Ann Arbor Science.

Viscek, T. 1992. *Fractal Growth Phenomena*. San Francisco: World Scientific.

Walton, W. H. 1982. The nature, hazards, and assessment of occupational exposure to airborne asbestos dust: A review. *Ann. Occup. Hyg.* 25:115–247.

Wang, G. M. and C. M. Sorensen. 1999. Diffusive mobility of fractal aggregates over the entire Knudsen number range. *Phys. Rev. E* 60:3036.

Wegrzyn, J. and D. T. Shaw. 1979. *NUREG/CR-0799*. Nuclear Regulatory Commission.

Weiss, M. A., A.-H. Cohen, and I. Gallily. 1978. On the stochastic nature of the motion of nonspherical aerosol particles. II. The overall drift angle in sedimentation. *J. Aerosol Sci.* 9:527–541.

Wen, H. Y., G. P. Reischl, and G. Kasper. 1984. Bipolar diffusion charging of fibrous aerosol particles—I. Charging theory. *J. Aerosol Sci.* 15(2):89–101.

Weyel, D. A., M. Ellakkani, Y. Alarie, and M. Karol. 1984. An aerosol generator for the resuspension of cotton dust. *Toxicol. Appl. Pharmacol.* 76:544–547.

Whisnant, R. A. 1975. *Evaluation of Image Analysis Equipment Applied to Asbestos Fiber Counting*. 210-75-0080/5. Cincinnati: National Institute for Occupational Safety and Health.

Wiltzius, P. 1987. Hydrodynamic behavior of fractal aggregates. *Phys. Rev. Lett.* 58:710.

Witten, T. A., Jr. and L. M. Sander. 1981. Diffusion-limited aggregation, a kinetic critical phenomena. *Phys. Rev. Lett.* 47:1400–1403.

Wu, M. K. and S. K. Friedlander. 1993. Note on the powerlaw equation for fractal-like aerosol agglomerates. *J. Colloid Interface Sci.* 159:246–247.

Wylie, A. G. 1979. Fiber length and aspect ratio of some selected asbestos samples. *Ann. NY Acad. Sci* 330:605–610.

Ye, J., X. Shi, W. Jones, Y. Rojanasakul, N. Cheng, D. Schwegler-Berry, P. A. Baron, G. J. Deye, C. Li, et al. 1999. Critical role of glass fiber length in TNF-α production and transcription factor activation in macrophages. *Am. J. Physiol. Lung C.* 276(March [3 Pat. 1]):L426–L434.

Yu, P. Y., C. C. Wang, and J. W. Gentry. 1987. Experimental measurement of the rate of unipolar charging of actinolite fibers. *J. Aerosol Sci.* 18(1):73–85.

Zhang, H. X., C. M. Sorensen, E. R. Ramer, B. J. Olivier, and J. F. Merklin. 1988. In situ optical structure factor measurements of an aggregating soot aerosol. *Langmuir* 4:867–871.

Zimm, B. H. 1948. Apparatus and methods for measurement and interpretation of the angular variation of light-scattering, preliminary results on polystyrene solutions. *J. Chem. Phys.* 16:1099–1116.

Zumwalde, R. D. and J. M. Dement. 1977. *Review and Evaluation of Analytical Methods for Environmental Studies of Fibrous Particulate Exposures*. DHEW (NIOSH) Publication No. 77-204. Cincinnati: National Institute for Occupational Safety and Health.

24 Biological Particle Sampling

TIINA REPONEN, KLAUS WILLEKE, and
SERGEY GRINSHPUN
Department of Environmental Health, University of Cincinnati,
Cincinnati, OH
AINO NEVALAINEN
National Public Health Institute, Kuopio, Finland

INTRODUCTION

Bioaerosols are particles of variable biological origin, for example, pollen, fungal spores or fragments of fungal mycelium, bacterial cells and spores, viruses, protozoa, excreta or fragments of insects, skin scales or hair of mammals, or other components, residues, or products of organisms, such as bacterial lipopolysaccharides, that is, endotoxins, or fungal mycotoxins. Bioaerosols are ubiquitous in both indoor and outdoor air. Among all particles larger than 0.2 μm in outdoor air, 30% appear to be of biological origin (Matthias-Maser and Jaenicke, 1995). Generally, the collection of bioaerosol particles utilizes the same sampling principles as those used for nonbiological aerosols. However, ensuring the survival or biological activity of the bioaerosol particles during and after collection is an important concern that differs from physical aerosol particle sampling. Furthermore, sample handling and storage, as well as the analysis of the collected biological particles, are considerably different from the procedures used in general particle sampling.

Bioaerosols have specific characteristics that influence their sampling. From the sampling perspective, bioaerosol particles may be

1. Single spores, pollen grains, bacterial cells, or viruses
2. Aggregates of several spores or cells or other biological material (e.g., mammalian allergens)
3. Products or fragments of spores and cells or other biological material
4. Biological material carried by other, nonbiological particles

Many bioaerosol particles, such as fungal spores and pollen, are designed by nature for transmission to other areas and to stay alive during the transmission. This type of hardy bioaerosol is resistant to environmental stresses, such as ultraviolet light, cold, heat, dryness, and toxic gases, and to sampling stresses, such as shear forces acting on them during sampling. In contrast, vegetative cells of bacteria are easily damaged, and their viability may be

Aerosol Measurement: Principles, Techniques, and Applications, Second Edition, Edited by Paul A. Baron and Klaus Willeke.
ISBN 0-471-35636-0

compromised by both environmental and sampling stresses (Cox, 1987). Some bioaerosol particles are bits and pieces of biological origin that are shed by an ecosystem. Bacteria emitted from aerated wastewater are residues of evaporated droplets. Many of the bacteria shed by humans, for example, are residues of the skin. They probably stay viable while airborne because they are adapted to the dry conditions of their environment and protected by their original substrate (i.e., the skin scale). Otherwise, most vegetative cells of bacteria are prone to damage as a result of becoming airborne.

Microbial cells in an aerosol may be viable or nonviable. The definition of the viability of a cell is not explicit (Roszak and Colwell, 1987), but, generally, viable cells are able to reproduce or they have metabolic activity. Nonviable organisms are dead or unable to reproduce. The nutritional requirements of many environmental microorganisms are not known; also, not all microorganisms can be grown on laboratory media. From a natural soil or water sample, for example, it is typically possible to culture less than 1% of the microorganisms present (Atlas, 1988), and this may also be true for airborne microorganisms. Bioaerosols that are not whole cells, such as endotoxins, mycotoxins, or various allergens, may also be present. Therefore, depending on the detection method, the results are commonly expressed as colony-forming units (cfu) when viable organisms are determined or as cells, spores, or pollen grains per unit volume of air when the viability of the organisms is not determined. The results of chemical analyses (e.g., for endotoxins) are expressed in ng/m^3 or mg/m^3.

The density of microbial cells is somewhat variable, depending on the degree of cell hydration, the reserve materials, and the lipid content of the cell (Doetsch and Cook, 1973). Microbial cells consist mainly of water, about 70% of their weight. The rest of the cell material consists of macromolecules, such as nucleic acids, proteins, lipids, carbohydrates, and their combinations. The density of a single microbial cell has been reported to be 1070 to 1090 kg/m^3 [1.07 to 1.09 g/cm^3] (Hamer, 1985), 900 to 1300 kg/m^3 [0.9 to 1.3 g/cm^3] (Orr and Gordon, 1956), 1090 to 1240 kg/m^3 [1.09 to 1.24 g/cm^3] (Bratbak and Dundas, 1984), or 1500 kg/m^3 [1.5 g/cm^3] (May, 1966).

Bioaerosol particles cover a wide size range. Viruses are the smallest potentially living particles, about 0.02 to 0.3 μm in length. Bacteria and fungal spores cover a size range from about 0.3 μm up to about 100 μm. Pollen, algae, protozoa, and dander are several tens to hundreds of micrometers in diameter. Fungi and plants are unique in their ability to produce monodisperse spore or pollen aerosol in large quantities. When microbial cells or spores are carried by other materials or when they are present as aggregates, their migration and deposition depend on the overall size of the ensemble.

Although most of the bioaerosols are harmless constituents of normal environments, some bioaerosol particles may be infectious agents or allergens or may carry toxic or irritant components or metabolites. To be infectious, an organism must be viable, but to cause allergenic or toxic effects viability is not a prerequisite. Thus, dead cells as well as cell residues may affect human health (Hirvonen et al., 1997). Host factors, including the genetic and environmental factors affecting the individual immunological response, also play an important role.

Biological characteristics may be utilized in controlling bioaerosol sources. Control measures can be targeted on environmental factors that regulate the growth of microorganisms (e.g., temperature and moisture).

BIOAEROSOL TYPES

Bacteria

Bacteria are generally found in any soils, waters, plants, and animals. In air, bacteria may occur as vegetative cells or endospores. They may be carried by other particles, such as water droplet residues, plant materials, or the skin fragments of animals. Bacteria are single-cell microorganisms that range in size from 0.5 to 30 μm. The aerodynamic size of most indoor air bacteria in relatively clean environments has been reported to be 1 to 3 μm (Nevalainen, 1989; Górny et al., 1999). In homes with high concentrations of other aerosols, such as cigarette smoke, the particle size distribution has been found to have a mean size in the 0.5 to 10 μm range, which suggests that the bacteria are aggregated with other particles (Górny et al., 1999). The shape of bacteria varies from spherical to rod shaped, spiral or filamentous. Many spherical bacteria occur as pairs, tetrads, or clusters (e.g., *Micrococcus* and *Staphylococcus*) or as chains (e.g., *Streptococcus*). The rod-shaped bacteria may occur as single cells or in chains (e.g., *Lactobacillus*). Bacteria can be divided into two groups based on the ability of the cell wall to retain crystal violet dye. Gram-positive bacteria, such as *Bacillus* and *Staphylococcus*, retain the dye, while gram-negative bacteria, such as *Pseudomonas* and *Legionella*, do not retain it. Gram-positive cell wall consists mainly of peptidoglycan, whereas gram-negative cell wall contains relatively little peptidoglycan but contains an outer membrane composed of lipopolysaccharides (endotoxin), lipoprotein, and other complex macromolecules. Both peptidoglycans and lipopolysaccharides can cause adverse health effects, as described below.

Bacterial cells that actively metabolize and divide into new cells are called *vegetative* cells. Endospores are formed within the vegetative cells of certain bacterial genera (e.g., *Bacillus* and *Thermoactinomyces*). Bacterial endospores are dormant forms of cells and are very resistant to cold, heat, radiation, and other environmental stresses. The sizes of bacterial spores range from 0.5 to 3 μm. They are easily carried away by air currents because gravitational settling is fairly insignificant for particles of this size range.

Pathogenic bacteria are known to cause disease in humans, animals, and plants. Pathogens are often specific and cause disease only to a certain species of animal or plant. Most animal and plant pathogens are different from human pathogens. Knowledge about human pathogens is the major focus of clinical microbiology.

Environmental or saprophytic (i.e., living on dead or decaying organic matter) bacteria are found everywhere, and their nutritional and temperature requirements vary. Only a fraction of environmental bacteria appear to be identified and characterized thus far. Some of the environmental bacteria are opportunistic pathogens (i.e., they may attack an individual having weakened immunological response). For instance, some species in the well-known genus *Legionella* are opportunistic pathogens. For reasons presently not known, some *Legionella* species find an ecological niche in manmade warm-water systems, where they may multiply and, if aerosolized, cause serious disease in exposed people (Keleti and Shapiro, 1987).

Actinomycetes are a group of soil bacteria that can both grow in a yeast form and produce mycelium and spores like filamentous fungi. According to the modern taxonomy based on genetic sequencing, they belong to the larger class of *Actinobacteria* (Stackebrandt et al., 1997). Their spores may contribute significantly to occupational exposure in agricultural work situations (Lacey and Dutkiewicz, 1994). In occupational and environmental hygiene, important genera of actinomycetes are *Streptomyces*, which gives soil its characteristic odor, and the thermophilic genera *Thermoactinomyces* and *Faenia*. These organisms cause hypersensitivity pneumonitis. Actinomycete spores may be found in office or residential buildings that have excessive microbial growth due to moisture accumulation within the structure or inside the heating, ventilation, and air conditioning (HVAC) system.

In air, bacteria may occur alone or may be carried by other particles. Bacteria tend to grow in colonies in their natural habitats, such as water and soil, and as biofilms on different surfaces. Therefore, whenever they become aerosolized, bacterial particles often occur as aggregates or microcolonies attached to other materials (Eduard et al., 1990). For example, the skin scales of mammals, which are abundant in indoor air, usually contain colonies of bacteria, such as *Micrococcus* and *Staphylococcus* (Lundholm, 1982; Noble, 1975).

Bacterial endotoxins are lipopolysaccharides that are specific to the cell wall of gram-negative bacteria. Endotoxins are heat resistant and chemically stable. They maintain their biological activity after the bacterial cell is no longer viable. Endotoxins cause acute toxic effects, including fever, malaise, and decrements in pulmonary function. They can be abundant in agricultural environments, some industries, and in humidified indoor air and may contribute to symptoms of humidifier fever (Rylander and Haglind, 1984; Rylander and Jacobs, 1994).

Bacterial cell wall also contains peptidoglycan; it comprises 90% of the cell wall of gram-positive bacteria and 10% of the cell wall of gram-negative bacteria. The health effects of peptidoglycan are not as well known as those of endotoxin, but it has been suggested that it has biological activities similar to endotoxin (AIHA, 1996; ACGIH, 1999).

Fungi

Fungi are also omnipresent microorganisms. They are responsible for most of the aerobic decay of natural organic material (Kendrick, 1985). Fungi include yeasts, molds, and mildews, which are often called *microfungi*, as well as large mushrooms, puffballs, and bracket fungi (Burge, 1995). The terms *mold* and *mildew* refer to visible fungal growth on surfaces. Fungi can be unicellular (e.g., yeasts) but are usually multicellular, forming long chains of cells called *hyphae*, which, as an ensemble, are called *mycelium*. Fungi are classified into groups based on their methods of spore production. The primary form of dissemination for this organism is by the release of fungal spores, which are well adapted to airborne transport. The size range of fungal spores, 1.5 to 30 μm or sometimes even larger, allows their transport by winds to long distances. Most of the fungal spores in indoor environments have been reported to be 2 to 4 μm in aerodynamic diameter (Reponen et al., 1994; Górny et al., 1999). Fungal spores are often resistant to various environmental stresses such as dryness, cold, heat, and ultraviolet radiation.

Some fungi can utilize living plant materials and cause crop diseases of great economic importance. A few fungi (e.g., *Histoplasma*, *Blastomyces*, *Coccidioides*, and *Candida*) readily invade living animal tissue and may cause infectious disease. Others are opportunistic pathogens (e.g., *Aspergillus* and *Cryptococcus*), which mainly cause infection only in people with impaired immunity. However, most fungi are saprophytic, provided adequate moisture is present. In buildings, moisture is the main controlling factor for mold growth. The relative humidity of the air (RH) affects the water content of materials in the room. However, it is the available moisture in the substrate, not the RH of the room air, that limits the growth of microorganisms in or on the materials. Virtually no microorganisms found in indoor air are able to grow if the equilibrium RH of the material is below 65% (Flannigan and Morey, 1996).

Most fungal aerosols can cause allergic reactions and diseases, such as asthma, allergic rhinitis, or, in some cases, hypersensitivity pneumonitis. Studies of these allergens usually focus on fungal spores, although metabolites and fragments of mycelium can also become airborne. Although a few genera, such as *Cladosporium*, *Alternaria*, basidiospores, and ascospores, dominate the outdoor aerosols in most of the world, there is some geographical variation. In areas of seasonal variation, the levels of fungal spores are highest in summer and fall and lowest in winter. Massive exposure to fungal spores may occur in farming and food handling occupations as well as in some industries (Rylander and Jacobs, 1994). In office and residential indoor environments, the outdoor air is an important source of fungal spores. The detection of airborne spores, resulting from growth on indoor substrates, can be difficult in the presence of normal background levels of outdoor bioaerosols.

In the cell wall of plants and fungi one finds glucans. The major source of β-(1 $\rightarrow$ 3)-*D*-glucans is the cell wall of fungi (Rylander and Jacobs, 1994). Glucans may play a role in various respiratory diseases and symptoms related to fungi, by themselves or by interacting with endotoxin or other environmental agents.

Some fungi may produce mycotoxins, which are toxic secondary metobolites. Several species of the genus *Aspergillus* produce carcinogenic toxins. These toxins have been studied primarily with respect to ingestion. There is some indication that aerosol exposure is a hazard as well (Sorenson et al., 1987). Aerosol exposure has been related to trichothecenes, which are mycotoxins produced by, for example, *Fusarium* and *Stachybotrys*. These can cause serious acute effects, including headaches, dizziness, immunosuppression, and pulmonary hemorrhage (Sorenson et al., 1987; Etzel et al., 1998).

Viruses

Viruses differ from other microorganisms because they can reproduce only inside a host cell. Therefore, viruses are intracellular parasites and never grow on nonliving substrates. They may infect bacteria, plants, animals, or humans and are usually targeted toward a specific subgroup. Those viruses that replicate only within bacterial cells are called *bacteriophages*. The smallest of all microorganisms, 0.02 to 0.3 μm, they consist of only one type of nucleic acid, either RNA or DNA, and are, therefore, not able to generate genetic information without a complete host cell. Viruses are surrounded by a protein layer called a *capsid*.

Viruses can be transmitted through air in the absence of the host cell (Gerone et al., 1966). Although individually small, they usually travel in air carried by other materials such as droplets of respiratory secretions. The size of a virus-containing particle may vary greatly, depending on many factors, such as RH of the surrounding air.

Viruses can cause infectious diseases (e.g., influenza, measles, and chicken pox), and the viral sources are almost always other infected humans. The viral agent for hantavirus pulmonary syndrome spreads from infected rodent droppings or urine. There is epidemiological evidence for the transmission of viral infections inside and between buildings (Riley et al., 1978; Donaldson, 1978). There is no evidence of airborne transmission of hepatitis or human immunodeficiency viruses, but there is some concern about blood and tissue aerosols generated during general and dental surgery that might contribute to the transmission of these agents.

Pollen

Pollen grains are produced by plants to transmit the "male" genetic material to "female" flower structures. Many ornamental plants produce pollen grains that are transported through the air by insects. However, many trees, grasses, and weeds produce pollen grains that are dispersed in the air by wind forces without the need of a carrier. To be successfully transmitted, this kind of pollen is usually produced in large amounts. For example, one shoot of hemp may produce over 10^9 pollen grains (Faegri and Iversen, 1989). Airborne pollen grains are usually resistant to environmental stresses such as desiccation, temperature, and light; hence, they tend to resist sampling stresses. Pollen grains from different plants vary in size, surface structure, and, to some extent, shape. The aerodynamic diameters of pollen grains are not well known, but the physical size range is approximately 10 to 100 μm, with many types of pollen grain being between 25 and 50 μm. Thus, pollen grains are not in the respirable size fraction. However, many kinds of pollen contain important allergens, which may be present in the air in smaller fragments (Rantio-Lehtimäki, 1995). Patterns of prevalence for airborne pollen vary with geography and climate. For example, ragweed pollen is considered to be one of the most important allergens in North America, while birch pollen is considered the most important in Northern Europe.

Cat, Dog, House Dust Mite, and Cockroach Allergens

A wide variety of materials derived from mammals contain potentially allergenic material. Cats and dogs are the most common household pets. Cat allergens are found in saliva and skin dander and are carried in small particles, less than 2.5 μm. Dog allergens are found in saliva, skin dander, and urine. Dog allergens are associated mainly with large particles, >9 μm (Custovic et al., 1997). Cat and dog allergens have been detected in schools and offices where these animals are not kept (ACGIH, 1999), which suggests that the allergens are carried on clothing.

House dust mites (e.g., *Dermatophagoides pteronyssinus* and *D. farinae*) are common allergens in residential environments. Dust mites are most prevalent in mattresses, carpets, and upholstered furniture where human skin scales collect and serve as food. Mites also need high humidity, over 50%, for survival and reproduction. Mite allergens are carried on mite fecal particles and dried body fragments, which are 10 to 20 μm in size (Burge, 1995).

The two most common cockroaches inside homes in temperate climate areas are *Periplaneta americana* and *Blatella germanica*. The source for cockroach allergens is not completely clear: Potential sources include cast skins, whole bodies, egg shells, fecal particles, and saliva. Cockroach allergens are found on particles over 5 μm in size (ACGIH, 1999).

Although the exposure route for allergens is assumed to be through air, most of the allergens are difficult to detect in air samples due to their relatively large size. Therefore, allergen samples are usually collected with a vacuum cleaner directly from the reservoir (e.g., from mattresses and carpets). Immunochemical methods are widely used for allergen analyses. The amount of dust mite allergens in a dust sample can also be determined indirectly through colorimetric indication of the amount of guanine, which is a component of dust mite excreta (von Bischoff, 1989). Specialized knowledge is needed to visually identify mites in samples.

SOURCES OF BIOAEROSOLS

Many indoor bioaerosols originate outdoors. The surfaces of living and dead plants are probably the most important sources of airborne fungal spores and bacteria. Therefore, the mechanical movement of plants or soil (e.g., through farming activities or construction) generates bioaerosols together with nonbiological dust. Actinomycetes may become airborne from soil (Atlas and Bartha, 1987). All natural waters as well as anthropogenic waters, such as sewage lagoons and cooling water systems, contain large numbers of microorganisms. For example, gram-negative bacteria, actinomycetes, and algae are common constituents of water ecosystems. Therefore, droplets resulting from rain, splashes, or bubbling processes may contain bioaerosols, which may remain airborne after evaporation of the liquid.

Strong sources of bioaerosols may be present in various occupational environments when organic material is handled, such as plants, hay, straw, wood chips, cereal grains, tobacco, cotton, organic waste, wastewater, or metalworking fluids. In agricultural and horticultural environments, exposure to fungal and actinomycete spores may be severe (Kotimaa, 1990; Lacey and Dutkiewicz, 1994; Rylander and Jacobs, 1994). Avian and rodent droppings can be a source for viral and fungal agents.

In industrial and nonindustrial situations, specific bioaerosol sources may develop due to microbial growth in a building's HVAC systems or in the building's structure itself. Generally, the prerequisite for microbial growth is excessive and accessible moisture. Standing water is always a good reservoir for microbial growth and therefore a potential source of microbial aerosols when disturbed (Keleti and Shapiro, 1987; ACGIH, 1999). In addition to water, microorganisms only need minute amounts of nutrients, which are available in the water or in building materials, such as cellulose, wood, or concrete. Therefore, a source of spores or other bioaerosol material may develop wherever water is leaking or condensing inside a building.

In nonindustrial indoor environments, the most important source of airborne bacteria is usually the presence of humans. Air with a high concentration of human bacteria is not necessarily a health hazard, but may indicate the presence of humans and their physical activities. It may also signal inefficient ventilation. Although pathogenic bacteria or viruses are often present in these air environments, their presence may be difficult to verify by conventional bioaerosol sampling techniques. Therefore, the concentration of normal human skin bacteria in air is sometimes used as an indicator of indoor air quality.

Recently, the threat of biological warfare and terrorism has gained increased attention. Agents of concern include pathogenic microorganisms, such as *Bacillus anthracis* and smallpox virus, as well as microbial toxins such as botulinum toxin.

GENERAL SAMPLING CONSIDERATIONS

The purpose of bioaerosol sampling is usually to verify and quantify the presence of bioaerosols for exposure assessment, to identify their sources for control, or to monitor the effectiveness of control measures. Dose–response relationships are poorly known, and therefore exposure guidelines have not been established for acceptable healthful levels of any bioaerosol (ACGIH, 1999). Bioaerosol concentrations have time-dependent variations ranging over several orders of magnitude. For example, bacteria and fungal spore concentrations on the order of 10 to 10^3 cfu/m^3 can be found in residential homes or occupational environments with moderate sources (e.g., tobacco processing, sanitary landfills, or biotechnical industries; Macher et al., 1991; Martinez et al., 1988; Rahkonen et al., 1990; Verhoeff et al., 1990; Reponen et al., 1992). Lower concentrations of $\leq 10^2$ cfu/m^3 can be found in well-ventilated facilities without significant bioaerosol sources, such as offices, laboratories, clean rooms, and operating rooms in hospitals. High concentrations of bacteria and fungal spores, with peak concentrations from 10^4 to as high as 10^{10} cfu/m^3, can be found in environments such as textile mills, sawmills, some agricultural exposure situations, and in seriously contaminated homes and offices (Kotimaa, 1990; Eduard et al., 1990; Lacey and Dutkiewicz, 1994). In most of these environments, the bioaerosol concentrations vary considerably in time and space. This is partly because bioaerosol sources do not necessarily generate particles continuously. For example, spore production and spore release from fungal mycelium may occur in bursts under certain air humidity and velocity conditions.

Sampling Strategy

No single sampling method can collect, identify, and quantify all of the bioaerosol components existing in a particular environment. Therefore, a source inventory is important and useful. It may include a preliminary microbiological analysis of the water reservoirs and a contact sample from a surface with assumed fungal growth. In industrial exposure situations, the type and location of sources are frequently evident. In nonindustrial environments, the sources are often less obvious, which necessitates a complex sampling strategy.

Sampling Efficiency of Bioaerosol Samplers

The overall sampling efficiency of a bioaerosol sampler can be divided into three components:

1. The inlet sampling efficiency is a function of the sampler inlet's ability to extract particles from the ambient air environment; it usually depends on the size, shape, and aerodynamic behavior of the particles being sampled.
2. The removal efficiency is determined by the sampler's ability to remove the particles from the air stream of the sampling inlet and deposit them into or onto the collection medium.

3. The biological aspect of sampling efficiency depends on the sampling and removal of the biological particles without altering their viability or biological activity and on the conditions for the organisms to form colonies or to be otherwise detected as biological particles.

The physical (inlet sampling and removal) and the biological factors must be separated in order to quantify their effects on the overall sampling performance.

None of the presently available samplers for culturable bioaerosols can be considered as a reference method, although the all-glass liquid impinger (AGI-30; ***AGI***, ***HAM***, ***MIL****) and the six-stage Andersen impactor (***AND***) have been suggested for that purpose (Brachman et al., 1964). Few of the currently available samplers have been adequately characterized as to their sampling performance. The results of reported field comparison studies are not easily comparable to each other, partly because the samplers, sampling times, and sampled volumes have varied within and between studies and partly due to the different operational principles and parameters of each of the samplers. Eduard and Heederik (1998) have summarized the results of several field comparison studies performed with different bioaerosol samplers. They found that some samplers are more efficient collectors of viable bioaerosol particles than others.

Bioaerosol particles must be collected from the ambient air with as little bias as possible. For minimal bias over a broad particle size range, aspiration into the inlet should occur under isokinetic conditions, that is, at the inlet air velocity equal to the ambient air velocity and the inlet is oriented facing the ambient air flow. Long sampling lines after the inlet may cause significant wall losses of large particles. Dependence of the inlet efficiency on the particle size as well as on wind and sampling conditions can cause significant over- or underestimation of the concentration. For example, theoretical calculations have shown that the Burkard Portable Air Sampler (***BUR***) overestimates the aerosol concentration of 10 μm particles by 2.5 times when positioned horizontally parallel to the wind direction at wind speeds of 0.5 m/s. If the sampler is placed vertically with all other conditions the same, particles larger than 5 μm will not be able to enter the sampler at all, while the concentration of particles of about 2 to 5 μm is considerably underestimated (Grinshpun et al., 1994). Nonisokinetic sampling is discussed in detail in Chapter 8.

The physics of particle motion in air and removal from it is common to all airborne particles. Therefore, physical principles can be applied to the sampling of bioaerosols. These principles determine the amount of sample collected and the sampling time needed for proper analysis.

A bioaerosol sampler has a high biological efficiency if it retains the biological properties that are used for analyzing the biological particles. For example, the sampling should not change the culturability of microorganisms if culture analysis is subsequently used. In the case of microscopic analysis, the sampling should not change the key morphological details of the particles. Table 24–1 shows that in the selected bioaerosol samplers (utilizing impaction or impingement principles) the impact velocity U_o ranges from about 1 to about 265 m/s. High impact velocity can result in metabolic and structural injuries of the collected microorganisms. Survival of bacteria has been found to decrease when the impaction velocity increases; bacterial survival also depends on the degree of embedding of bacteria into collection media (particularly for impactors) and the sampling time (Stewart et al., 1995; Terzieva et al., 1996; Lin et al., 1999a). The effect of mechanical impaction stress is more pronounced for the higher impaction velocity; the desiccation stress increases with the sampling time increase and when embedding is insufficient. The survival of bioaerosol particles depends also on their internal characteristics. Pollen grains and microbial spores are generally more protected against environmental stresses than vegetative cells (Cox and Wathes, 1995).

* See Appendix I for full manufacturer addresses referenced to the italicized three-letter codes.

TABLE 24–1. Sampling Parameters of Selected Bioaerosol Samplers

Bioaerosol Sampler[a]	Collection Medium	Q (10^{-4} m^3/s [L/min])	U_o (m/s)	Nozzle					d_{50} (μm)	
				Shape	W (mm)	L (mm)	A (mm^2)[b]	Number	Calculated[c]	Reported
AGI-30	Liquid	2.1 [12.5]	265.2	Round	1.0		0.79	1	0.31	<0.3[d]
Air-O-Cell	Adhesive	2.5 [15.0]	16.5	Rectangle	1.055	14.4	15.2	1	2.3	2.6[e]
Andersen, first stage	Nutrient	4.71 [28.3]	1.08	Round	1.18		1.09	400	6.61	7.0[f]
Andersen, sixth stage	Nutrient	4.71 [28.3]	24.02	Round	0.25		0.05	400	0.57	0.65[f]
Burkard Personal Sampler	Adhesive	1.67 [10.0]	11.90	Rectangle	1.0	14	14	1	2.52	2.3–2.4[g]
SAS	Nutrient	30 [180]	17.34	Round	1.0		0.79	219	1.45	1.9[h]
MK-II	Nutrient	5 [30.0]	51.42	Rectangle	0.35	28	9.8	1	0.67	

[a] AGI, All-Glass-Impinger; SAS, Surface Air Sampler; MK-II, Casella Airborne Bacteria Sampler.
[b] Area per nozzle.
[c] Equation 10–4, $T_{air} = 293$ K [20°C], $P_{air} = 1$ atm, $\eta_{air} = 1.81 \times 10^{-4}$ g/cm.s [1.81×10^{-4} poise], $\rho_p = 1000$ kg/m^3 [1.0 g/cm^3].
[d] Lin et al. (1997).
[e] Trunov et al. (2000).
[f] Andersen (1958).
[g] Aizenberg et al. (2000).
[h] Lach (1985).

Furthermore, aggregation of bacteria has been shown to increase their survival rate (Lighthart and Shaffer, 1997).

PRINCIPLES OF BIOAEROSOL COLLECTION

Bioaerosol sampling involves separating the particle trajectory from the air streamline trajectory. To achieve this, different physical forces are applied as illustrated in Figure 24–1.

In Figure 24–1A, the inertia of the particle forces its impaction onto a solid or semisolid impaction surface, usually either a culture medium or an adhesive surface that can be examined microscopically. This principle is used in single-stage impactors (e.g., the Surface Air Sampler; ***PBI***, ***SPI***) and in cascade impactors with two or more impaction stages (e.g., the Andersen cascade impactor). If the impaction stage consists of one or more slits instead of one or more circular holes, the impactor is frequently referred to as a *slit sampler* (e.g., Burkard spore traps and Casella [***CAS***], New Brunswick [***NBS***], and Mattson-Garvin [***BAS***]

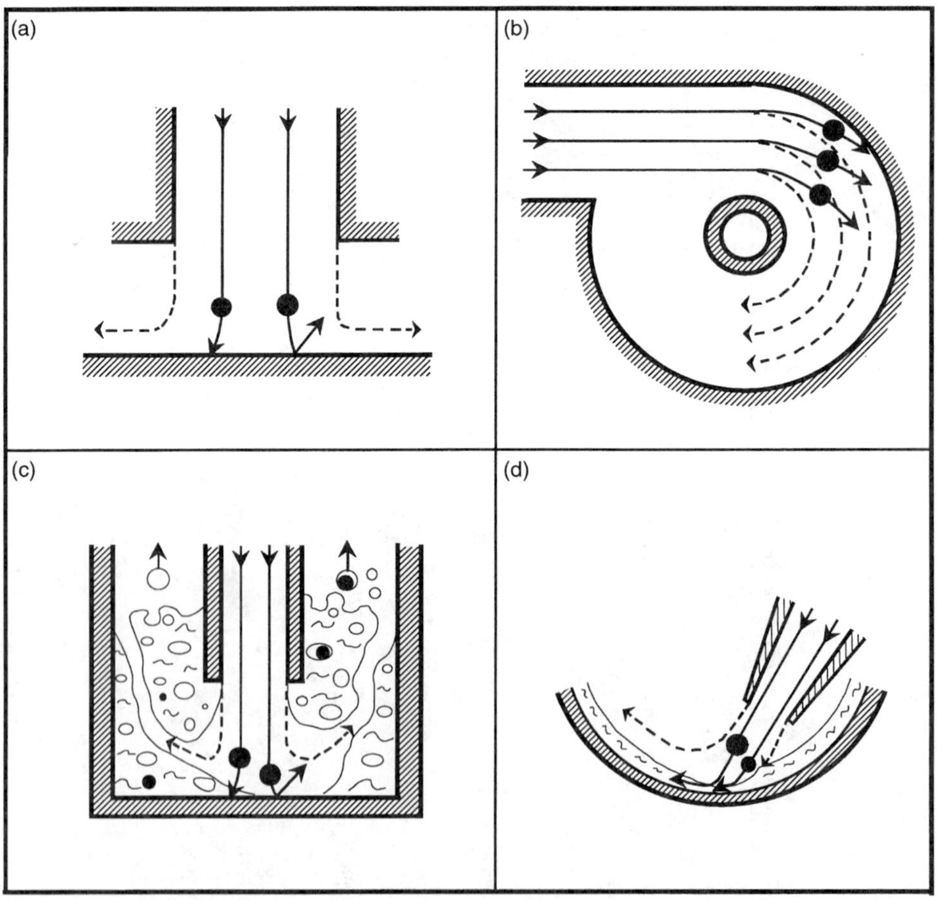

Fig. 24–1. Comparison of collection mechanisms for different inertial aerosol samplers. **A**, Inertial impaction. **B**, Centrifugal impaction. **C**, Liquid impingement. **D**, Tangential impingement. (From Willeke et al., 1998, reproduced by permission of American Association for Aerosol Research.)

culture plate samplers; more information is given below). Inertial gravitational and centrifugal particle collection techniques are described in Chapter 10.

The principle of particle separation by centrifugal force is shown in Figure 24–1B. It also uses the inertial behavior of the particle, but in a radial geometry. The Reuter centrifugal sampler (***BIO***) is of this type.

When filtration is the collection mechanism, inertial forces are also responsible for separating particles from the air stream. However, other mechanisms, such as interception, diffusion, and electrostatic attraction, also contribute to the deposition of particles onto the filter material.

In liquid impingement (Fig. 24–1C), the particles are collected primarily by inertial impaction into a liquid, but also by particle diffusion within the bubbles. Several versions of liquid impingers are currently available (e.g., the AGI-4 and AGI-30 impingers). In a multistage impinger, several impingement stages remove successively smaller particles (May, 1966). Figure 24–1D illustrates the sampling principle of a tangential impinger, which collects particles by inertial impaction, and centrifugation. This sampling principle is used in the BioSampler (***SKC***) (Willeke et al., 1998).

Particles may also be removed from the air stream by externally applied forces, such as electrical forces on charged particles, or thermal forces on an aerosol flow, which has a thermal gradient perpendicular to its flow. An electrostatic sampler has been used for virus sampling (Gerone et al., 1966) and has also been explored for bacterial sampling (Mainelis et al., 1999).

Aspiration of the airborne organisms into an inlet and subsequent collection is the preferred method for bioaerosol sampling. An exception to this is the rotating arm device, for example, the Rotorod sampler (***SAM***) that is commonly used to sample pollen. Settling plates rely on gravity to deposit airborne microorganisms onto a culture plate. The gravitational settling of particles is highly particle size dependent and is strongly influenced by the air motion in the room. Therefore, the colony-forming units counted on a settling plate after incubation or the spores counted on a settling slide cannot be related to the concentration of biological particles in the air. At the same time, gravitational settling methods are suitable to identify the larger microorganisms present in the air environment because they all settle during relatively short time (regardless of their size and air motion).

Particles can bounce when they strike the impaction surface or other previously collected particles. The latter often occurs when the sample is overloaded. Bouncing of larger particles can lead to undersampling in single-stage impactors when the particles bounce away from the collection substrate. Particle bounce can affect the size distribution obtained with cascade impactors when large particles bounce to the next collection stage. The collection efficiency, especially for larger particles, gradually decreases as the sampling surface becomes loaded. In a bioaerosol impactor, the collection efficiency may also decrease because the initially soft agar surface dries up with time, encouraging particle bounce (Juozaitis et al., 1994). This effect has not been accurately examined for commercially available bioaerosol samplers. However, it is a common practice to apply a thin film of grease to the impaction surface to reduce the bounce effect, hence, to increase the collection efficiency. The bouncing phenomenon is further discussed in Chapter 10.

In all of the sampling methods mentioned above, the bioaerosol particles are collected and analyzed in two steps: First the particles are removed from the air, and then they are analyzed and identified. Direct-reading aerosol instruments are used to study aerosols without separating them from the air. These can be used for studying bioaerosol particles in the laboratory, when other particles are eliminated (e.g., Qian et al., 1995). There is increased interest nowadays—particularly among military and antiterrorist agencies—in developing direct-reading instruments that can distinguish the particles of biological origin from other particles. One commercially available, but very expensive, device, the Ultraviolet Aerodynamic Particle Size Spectrometer (UV-APS; ***TSI***), simultaneously measures the aerodynamic particle size, the light-scattering intensity, and the fluorescence intensity of the particles passing through the instrument. See Chapter 17. The fluorescence is measured between 400 and 580 nm upon exitation of the particles by a UV-laser beam. This fluorescence is related to the presence of life-indicating biomolecules, such as NAD(P)H (nicotinamide adenine dinucleotide phosphate) and riboflavin (Hairston et al., 1997).

Particle Removal by Inertial Impaction

Inertial impaction is the most widely used mechanism for particle removal in bioaerosol samplers. In an impactor, the particle-laden airflow is pulled or pushed through a nozzle, which accelerates the air and the particles to high speed. The air jet emerging from the nozzle is deflected by the impaction surface and makes a sharp turn. Particles with sufficient inertia continue their forward movement toward the collection surface and impact. The impaction process depends on the particles' inertial properties, such as size, density, and velocity, and on the impactor's physical parameters, such as the nozzle dimensions and airflow paths.

The non-dimensional Stokes' number, Stk, is used to relate the particle trajectories to the air streamlines in an impactor (see Eqs. 4–38 and 4–39). The Stk is defined as a ratio of a particle's stopping distance (see Eq. 4–36) to the dimension of the nozzle. Stk can thus be used to predict whether a particle will impact on the collection surface. Stk_{50} defines the numerical value of Stk for which 50% of the particles are collected and 50% pass through the impaction stage. For a round nozzle, Stk_{50} is approximately one fourth, and for a rectangular nozzle it is approximately one half.

Similarly, the "cut-off size", d_{50}, designates the particle diameter for 50% removal. Because most impaction stages have sharp cut-off characteristics, almost all the particles larger than that size are collected. Therefore, d_{50} is generally assumed to be the size above which all particles larger than that size are collected. The cut-off size d_{50} is an important characteristic of any bioaerosol sampler. To evaluate the sampling performances of different samplers, the cut-off size can be calculated (see Eq. 10–4 and Example 24–1).

Numerical Estimation of Cut-Off Sizes

The design and performance characteristics of six commonly used bioaerosol samplers are compared in Table 24–1. These samplers have a single impactor nozzle or several in parallel. The range of their volumetric flow rate, Q, is from $1.67 \times 10^{-4}\,m^3/s$ to $3 \times 10^{-3}\,m^3/s$ [10 to 180 L/min]. For the six-stage Andersen impactor, data are given for the first and sixth stages. As seen, particle cut-off diameters (d_{50}) are close to the experimentally determined values. The small differences between calculated and experimental values are due to the dependence of Stk_{50} on the impaction nozzle geometry of each sampler and the air flow pattern in the impaction stage that does not exactly match the pattern used in the theoretical calculations.

The d_{50} value of an impactor can be used to determine which microorganism species are likely to be collected in the impactor if the mean aerodynamic sizes of the bioaerosol particles are known. The aerodynamic diameter of a particle equals the diameter of a spherical particle with density $1000\,kg/m^3$ [$1\,g/cm^3$] (= density of water) that has the same gravitational settling velocity as the particle in question (see also Chapter 3 and Example 3–2). For bioaerosol particles with a density close to that of water and a shape close to spherical, the physical and aerodynamic diameters are approximately equal to each other. However, for rod-shaped particles, or for particles with density other than that of water, the aerodynamic diameter predicts the particle behavior more accurately than the bioaerosol particle's geometrical dimensions. The physical sizes (length and width as measured under a microscope) of bioaerosol particles are better known than their aerodynamic sizes. Table 24–2 summa-

Example 24–1

An Andersen N-6 sampler (i.e., the sixth stage of a six-stage cascade impactor) is operated at a volumetric flow rate of 20 L/min instead of the required 28.3 L/min (1 cfm). This is a single-stage sampler with 400 nozzles of 0.25 mm diameter each, that is, 400 jets of aerosol flow are directed toward the nutrient medium below them. (1) Calculate the cut-off diameter d_{50} for the reduced flow. What is the smallest particle collected? (2) The N-6 is operated at the required flow rate of 28.3 L/min, but 40 of its 400 nozzles are plugged. What is the smallest particle collected?

Answer: (1) The air velocity, U_0, and, therefore, the approximate particle velocity in each of the n jets of cross-sectional area A_j are

$$U_o = \frac{Q}{nA_j} = \frac{Q}{n\left(\frac{\pi W^2}{4}\right)} = \frac{\left(20\times 10^3\,\frac{\text{cm}^3}{\text{min}}\right)\left(\frac{\text{min}}{60\text{s}}\right)}{400\left[\frac{\pi(0.025\text{cm})^2}{4}\right]} = 1697\,\frac{\text{cm}}{\text{s}}$$

To determine the cut-off diameter, d_{50}, use Eq. 10–4. At normal pressure and air temperature (1 atm and 20°C), the viscosity of air, η, equals 1.81×10^{-4} poise (1.81×10^{-4} g/cm s). The density of bioaerosol particles ranges from 900 to 1500 kg/m^3 [0.9 – 1.5 g/cm^3] and is assumed to equal 1000 kg/m^3 [1 g/cm^3] for this example. For the round nozzles, the value of Stk_{50} is approximately one fourth.

$$d_{50} = \sqrt{\frac{9\eta W Stk_{50}}{\rho_p U_o C_c}} = \sqrt{\frac{9\times 1.81\times(10^{-4}\,\text{g/cms})(0.025\text{cm})\times 0.25}{(1\,\text{g/cm}^3)(1697\,\text{cms})\times 1}} = 7.74\times 10^{-5} cm = 0.77\mu m$$

Therefore, a 0.77 µm particle is approximately the smallest one collected. In the above calculation, the Cunningham correction factor, C_c, is assumed to be 1. It is actually somewhat larger than 1 for the calculated d_{50} (see Eqs. 4–23 and 4–24, Table 4–2, and Appendix F). Through an iteration procedure, the d_{50} is therefore calculated to be somewhat smaller than indicated. At the recommended flow rate of 28.3 L/min, d_{50} is 0.57 µm (see Table 24–1).

(2) If 40 of the nozzles are plugged, the air velocity through the 360 remaining nozzles at a flow rate of 28.3 L/min is 2668 cm/s. For this velocity and $C_c = 1$, the new cut size is $d_{50} = 0.62$ µm. Because $C_c > 1$ for this particle size, the actual cut size is somewhat smaller.

rizes data on the measured aerodynamic sizes of several fungal and bacterial species. It also shows the aspect ratios for some of the species. Aspect ratio is defined as the ratio of length to cross-sectional area diameter of a particle and can influence the collection efficiency of particles sampled onto a filter.

The aerodynamic particle sizes shown in Table 24–2 can now be compared with the particle cut-off sizes in Table 24–1. As seen in Table 24–1, the cut-off sizes for the various samplers or sampler stages range from less than 0.5 µm to over 5 µm. A particle is collected by an impactor if its aerodynamic size is higher than the cut-off size of the impactor. For example, the aerodynamic size of *Aspergillus fumigatus* spores has been reported to be 2.0 to 2.1 µm (see Table 24–2). Thus, the Burkard sampler and the first stage of the Andersen sampler do not effectively collect these fungal spores. The differences in the cut-off sizes of the various samplers partly explain the differences between sampler performances that have been reported in the literature (e.g., Eduard and Heederik, 1998).

TABLE 24–2. Aerodynamic Sizes and Aspect Ratios of Microorganisms as Measured in Laboratory Experiments

Microorganism	Aerodynamic Size (μm)	Aspect Ratio
Fungal spores		
Aspergillus flavus	3.6[a]	
Aspergillus fumigatus	2.0[b], 2.1[a,c]	
Aspergillus versicolor	2.4[c]	
Cladosporium cladosporoides	1.8[c,d], 2.4[a]	1.9[e]
Paecilomyces variotii	2.6[a]	
Penicillium brevicompactum	2.1[c], 2.3[d]	1.3[e]
Penicillium chrysogenum	2.8[a]	
Penicillium melinii	2.7[c], 3.1[d]	1.1[e]
Penicillium minioluteum	1.7[a]	
Scopulariopsis brevicaulis	5.3[a]	
Bacterial spores		
Bacillus subtilis var. *niger*	0.9[d]	
Faenia rectivirgula	1.1[a]	
Saccharomonospora viridis	1.3[a]	
Streptomyces albus	0.8[d], 0.9[c], 1.2[a]	1.2[e]
Thermoactinomyces vulgaris	0.6[f]	1.2[e]
Bacterial vegetative cells		
Pseudomonas fluorescens	0.8[g]	3.0–3.1[g]
Micrococcus luteus	1.0[h]	
Bacillus subtilis	0.8[i]	
Bacillus megatherium	1.0[i]	2.6[g]
Mycobacterium smegmatis	1.2[e]	
Mycobacterium bovis	0.9[e], 0.8–1.0[k]	3–8[j]

[a] Madelin and Johnson (1992).
[b] Pasanen et al. (1991).
[c] Reponen et al. (1996).
[d] Aizenberg et al. (2000).
[e] Reponen, Willeke, Grinshpun (unpublished data).
[f] Reponen et al. (1998).
[g] Willeke et al. (1996).
[h] Stewart et al. (1995).
[i] Qian et al. (1997).
[j] Schafer et al. (1998).

COLLECTION TIME

An essential part of the sampling strategy is to define the sample collection time. Bioaerosol concentrations vary greatly with time, which is graphically shown in Part I of Figure 24–2. Typically, periods of low concentrations (e.g., t_1 to t_2) are followed by periods of high concentrations (e.g., t_3 to t_4) and vice versa. These large fluctuations are best represented on a logarithmic concentration scale, as shown. In this time plot, the average concentration, c_a, is 1000 bioaerosol particles/m^3. Bioaerosol concentrations of this order of magnitude are common in outdoor air and in many indoor situations. The ambient concentrations rarely remain stable within a narrow concentration range unless the time period is relatively short (i.e., minutes) or the atmosphere is undisturbed, such as in an unventilated, nonpopulated, closed room.

Sampling during periods of changing concentration must be sufficiently long, or many short samples must be combined to properly represent the average environmental concentration. Part II of Figure 24–2 reflects how the concentration varies in air volume, v, being

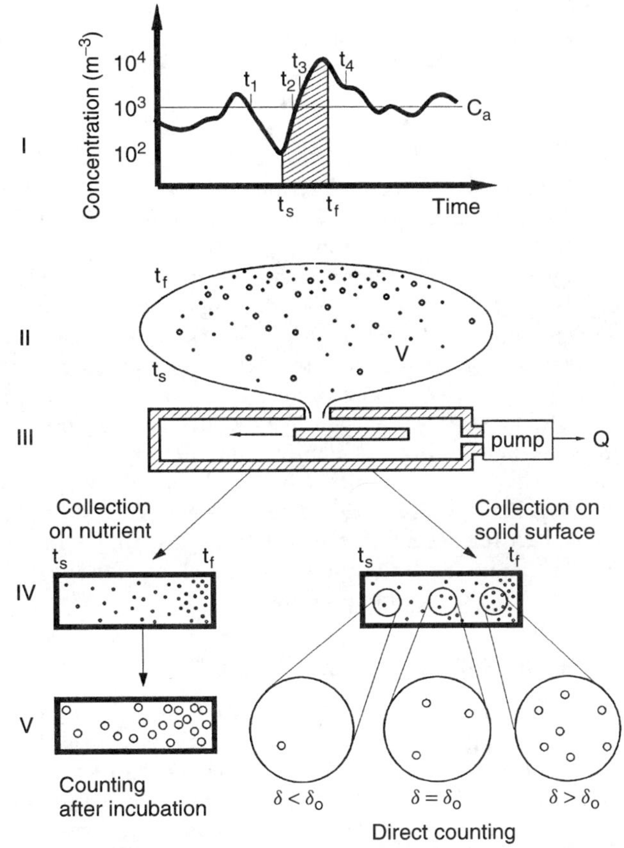

Fig. 24–2. The process of bioaerosol sampling. (From Nevalainen et al., 1992, with permission of Elsevier Science.)

sampled during the sampling period from starting time, t_s, through the finish, t_f. The volume of air equals the product of sampling flow rate, Q, and sampling time, t:

$$v = Qt \tag{24–1}$$

The sampling process actually integrates the instantaneous concentrations within the sampled period with respect to time. The number of particles, N, collected on the impaction substrate, Part III of Figure 24–2, equals the product of the average particle concentration, c_a, and the sampled air volume, v (Eq. 24–1):

$$N = c_a Qt \tag{24–2}$$

Surface Density of Collected Particles

Part IV of the sampling process in Figure 24–2 shows the bioaerosol particles being collected on a nutrient or solid surface that moves to the left during sampling, resulting in varying numbers of particles per unit area according to the changing bioaerosol particle concentration in the sampled air volume. The number of objects on the surface per viewing area, A—

microbial colonies on a Petri dish or microscopic particles on an adhesive surface—is referred to as the *surface density* of a sample. This surface density, δ, is

$$\delta = \frac{N}{A} = \frac{c_a Q}{A} t \tag{24–3}$$

Part V of Figure 24–2 presents the postcollection phase when the collected material is analyzed. For example, this can occur either immediately upon collection by viewing the collected particles directly with a microscope or after an incubation period when the colonies are sufficiently developed to be viewed and identified by visual inspection. Viewing, counting, and identifying the particles in a sample—either optically or otherwise–is facilitated if the surface density is optimal, δ_o. If the sample surface density is very low, $\delta \ll \delta_o$, the sampling and counting errors may be high, and the calculated aerosol concentration may not accurately reflect the true airborne concentration. If the sample surface density on a microscopic slide is very high, $\delta \gg \delta_o$, the particles may be located too closely to each other and pose resolution problems in counting and identification. In addition, the bioaerosol particles may be covered and masked by dust particles. If $\delta \gg \delta_o$ on a nutrient surface, the collected organisms may grow together or inhibit each other's growth. This is especially important for fungal spores, which often release substances that inhibit germination of adjacent spores. Nonbiological particles that impact on nutrient surfaces along with the microorganisms may not cause viewing problems, but may inhibit growth.

Optimal Sampling Time for Solid Surface Samplers

As seen in Eq. 24–3, the surface density of microorganisms collected on a substrate is linearly proportional to sampling time. The other parameters in Eq. 24–3—c_a, Q, and A—are normally beyond the control of the investigator. Thus, insufficiently loaded samples ($\delta \ll \delta_o$) and overloaded samples ($\delta \gg \delta_o$) can be avoided only by adjusting the sampling period.

By inverting Eq. 24–3, the sampling time can be calculated for each sampler by using the desired value of surface density and assuming a value for the airborne bioaerosol concentration:

$$t = \frac{\delta}{c_a} \frac{A}{Q} \tag{24–4}$$

The optimal sampling time for a given bioaerosol concentration is different for each sampler, depending on the sampler's flow rate and collection surface area. For purposes of this discussion, we assume an ambient concentration of 10^3 bioaerosol particles/m^3 and a sample collection efficiency of 100%. We also assume that $\delta_{macro} = 1$ particle/cm^2 is the ideal surface density for colony counting on a culture plate and that $\delta_{micro} = 10^4$ particles/cm^2 is the ideal surface density for microscopic particle counting on a sample slide. The latter equals three particles per microscopic field of 200 μm diameter. Using these assumptions and Eq. 24–4, we have calculated the optimal sampling times for several of the bioaerosol impactors/impaction stages listed in Table 24–1. The surface density δ was determined as the density of countable objects on the sampling surface. The definition of the sampling surface area A is specific for each sampler design. For a slit sampler with bioaerosol particles impacting onto an adhesive surface (e.g., the Burkard sampler), A is assumed equal to the area of the slit nozzle. For slit-to-agar impactors and sieve-type impactors, such as the Andersen impactor with 400 jets or the Surface Air Sampler (SAS) with 219 jets, the sampling surface is postulated to be equal to the sum of all nozzle cross sections. In culture plate samplers, the sampled particles are allowed to develop into colonies before viewing, and, therefore, the area of the sampling surface A is assumed equal to the area of the nutrient plate. For the

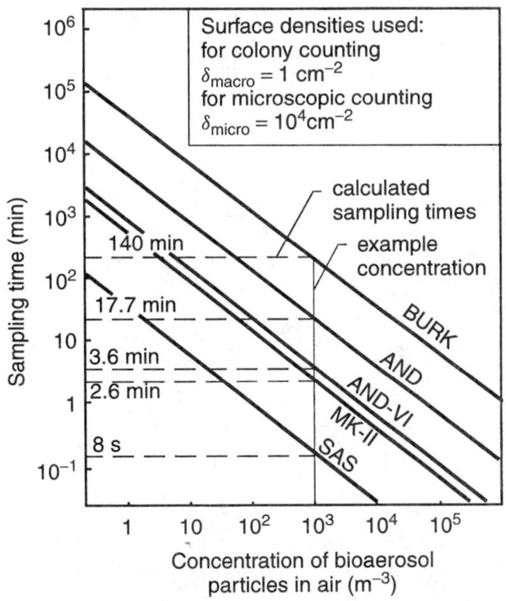

Fig. 24–3. Collection times for selected bioaerosol samplers. (From Nevalainen et al., 1992, with permission of Elsevier Science.)

Andersen six-stage impactor (referred to as AND below), it is assumed that bioaerosol particles impact onto each of the six stages. Therefore, the total area of all six stages is used in calculating the sampling time. This assumption is valid when the particle size distribution produces an equal number of particles deposited on each of the six stages. When the sixth stage is used as a separate sampler, annotated here as the AND-VI and often referred to as the N-6, only the surface area of one Petri dish is used.

It can be seen in Figure 24–3 that for $c_a = 10^3$ particles/m^3 and optimal surface density, the sampling time for the SAS (utilizing colony counting) is 8 s, while the Burkard personal air sampler has a sampling time of 140 min (utilizing microscopic counting). This represents a difference of three orders of magnitude. Lengthening the sampling time of a culture-plate sampler may result in a more representative air sample, but is likely to cause overloading, leading to counting errors and inhibition effects. Therefore, many culture-plate samplers are best suited for environments where a very low bioaerosol concentration is expected (e.g., $\leq 10^2$ cfu/m^3). In the latter case, the sampling periods can be longer, or several short-term samples can be taken over a long period of time.

As seen in Figure 24–3, the calculated sampling time for the Burkard personal sampler is 140 min for an expected average bioaerosol concentration of 10^3 bioaerosol particles/m^3. However, in most environments this would result in a sample that is grossly overloaded with nonbiological particles. Therefore, the assumed or expected values of c and δ can be modified in order to obtain a sampling time more in line with realistic expectations. For instance, for a higher bioaerosol concentration of 10^4/m^3, the sampling time would be as short as 14 min.

Optimal Sampling Time for Impingers

Impinger samples are not sensitive to overloading or undersampling because the liquid sample can be either diluted or concentrated, depending on the concentration of collected

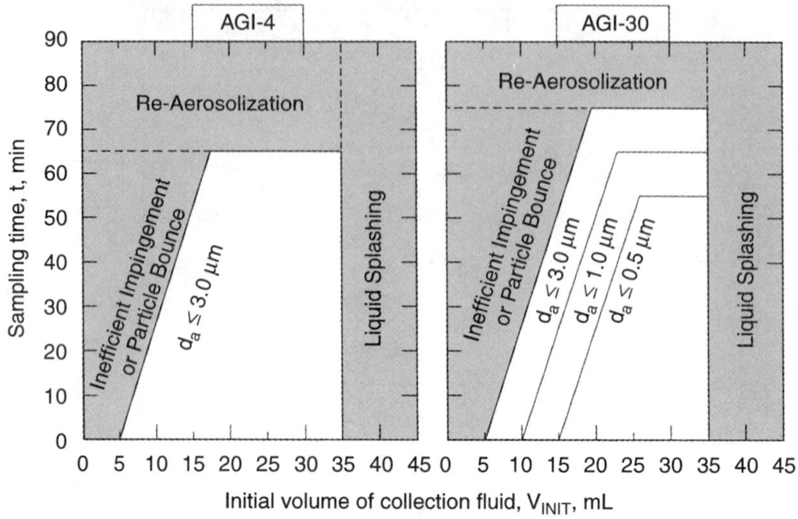

Fig. 24–4. Permissible sampling parameter ranges for less than 10% change in collection efficicency with the AGI-4 and AGI-30 impingers when operated at a sampling flow rate of 12.5 L/min. (Copyright American Industrial Hygiene Association, from Lin et al., 1997.)

bioaerosol particles in the liquid. However, evaporation of the sampling liquid and reaerosolization of already collected particles limit the sampling time in most impingers (Lin et al., 1997). Figure 24–4 shows the permissible sampling times and initial collection fluid volumes for sampling with AGI-4 and AGI-30 impingers when less than about 10% change

EXAMPLE 24–2

The Surface Air System (SAS) sampler is used to collect bioaerosols at an expected concentration level of 500 particles/m^3. The microbiologist who will analyze the results would like to see an average of 50 colonies on each 55 mm diameter Rodac plate sample (nutrient dish). The SAS sampler has 219 sampling jets and a flow rate of 180 L/min. Calculate the optimal sampling time.

Answer: The area for the 55 mm diameter Rodac plate equals 23.8 cm^2. Thus, the surface density for colony counting (Eq. 24–3) is

$$\delta = \frac{N}{A} = \frac{50\ \text{particles}}{23.8\ \text{cm}^2} = 2.1\,\frac{\text{particles}}{\text{cm}^2}$$

To determine the optimal sampling time, use Eq. 24–4:

$$t = \frac{\delta A}{c_a Q} = \frac{(2.1\ \text{particles/cm}^2)(23.8\ \text{cm}^2)}{(500\ \text{particles/m}^3)(180 \times 10^{-3}\ \text{m}^3/\text{min})} = 0.55\ \text{min} = 35\ \text{s}$$

This sampling time may be too short for an environment with variable concentrations of bioaerosols. Several consecutive samples should be taken, or a different sampler, which allows a longer sampling period, should be chosen (see Fig. 24–3).

in the collection efficiency is permitted. The AGI-4 is an All-Glass-Impinger with its impingement nozzle 4 mm above the bottom of the liquid; in the AGI-30, this distance is 30 mm, while the liquid height is usually about 20 mm, that is, the top surface of the liquid—when there is no airflow—is 10 mm below the impingement nozzle. If the volume of the collection fluid is over 35 mL, a significant amount of liquid splashes out of the impinger. On the other hand, a minimum volume of collection fluid is required for each sampling time so that the collection efficiency does not change by more than 10% during the sampling. The upper particle size limit of 3 μm is due to decreasing inlet efficiency when increasing particle size in the AGI-4 and the AGI-30 (Grinshpun et al., 1994). When the sampling time exceeds 65 min with the AGI-4 and 75 min with the AGI-30, a significant amount of already collected particles is reaerosolized.

EXAMPLE 24–3

In planning bioaerosol sampling in a hospital you are told that cleaning of the postoperative patient room will take 20 min. Previous reports indicate an average bacterial aerosol concentration of about 10^2 cfu/m^3. Which sampler would you choose to sample for 20 min in this patient room?

Answer: In Figure 24–3, the concentration of 10^2 bioaerosol particles/m^3 can be located on the x axis and followed up to the 20 min level on the y axis. At this intersection, the MK-II sampler and the Andersen N-6 sampler are indicated as being suitable for surface density $\delta_{macro} = 1\ cm^{-2}$ when the collected bioaerosol particles are counted as colonies after incubation.

SELECTION OF SAMPLER

The selection of a suitable bioaerosol sampler depends on the types and levels of bioaerosols of interest in the environment to be sampled. For example, when evaluating allergens, total rather than viable counts are important. Also, when selecting a cultural sampler one must consider the vulnerability of the bioaerosol to the sampling forces because some samplers affect viability more than others. The most commonly used instruments are discussed briefly in order to provide the user with the basis for deciding which sampler to use and how it operates. The names and manufacturers of some of the commonly used instruments are given in Table 24–3. More sampler information is given by Willeke and Macher (1999).

The Andersen six-stage sampler has been used in many studies (e.g., Reponen et al., 1992; DeKoster and Thorne, 1995; Górny et al., 1999). It allows characterization of bioaerosols in specific size ranges. Two-stage and one-stage versions are also commercially available. One usually expects that the ambient bioaerosol concentration is low enough so that by the end of the sampling period bioaerosol particles have passed through only about 10% of the 400 nozzles in each impaction stage. When the bioaerosol concentration is high, several bioaerosol particles may deposit below each impaction nozzle. Only one of the organisms may grow, preventing the development of others into visible colonies. In some cases, two or more organisms may grow together into a single colony. When a high percentage of the impaction spots below the nozzles of a multiple-nozzle impaction stage develops colonies, the actual concentration of bioaerosol particles can be statistically calculated by the positive-hole conversion method described by Andersen (1958) and Macher (1989). The masking effect has been studied theoretically and experimentally by Chang et al. (1995a,b).

TABLE 24–3. Company Information about Bioaerosol Samplers

Sampler	Manufacturer/ Supplier[a]
Inertial impactors	
Air-O-Cell	***ZAA/SKC***
Allergenco Air Sampler (MK-3)	***ALL***
Andersen Sampler, 1-, 2- or 6-stage	***AND***
Burkard Sampler	***BUR***
Casella Airborne Bacteria Sampler (MK-II)	***CAS***
Mattson-Garvin Slit Sampler	***BAR***
Rotorod	***SAM***
Surface Air Sampler (SAS)	***PBI/SPI***
Centrifugal impactors	
BioSampler[b]	***SKC***
Multistage Liquid Impinger	***BUR***
Reuter Centrifugal Sampler (RCS)	***BIO***
Impingers	
AGI-4, AGI-30	***AGI/HAM/MIL***
BioSampler[b]	***SKC***
Filter samplers	
37-mm Cassette	***CCO/MIL/SKC***
Button Sampler	***SKC***

[a] See key to company abbreviations in Appendix I.
[b] the BioSampler utilizes several sampling principles, including centrifugal forces and tangential impingement.

The SAS samplers are portable one-stage multiple-hole impactors that are commercially available in three models. These samplers utilize 55 mm diameter contact plates filled with nutrient medium and also require statistical adjustments for multiple impactions.

Several models are available among the slit impactors. Some rotate a Petri dish below the inlet nozzle (e.g., the Mattson-Garvin slit sampler); others impact particles on a microscope slide or tape (e.g., the Air-O-Cell and the Burkard samplers). The rotating culture plate impactors are especially useful for determining temporal changes in viable bioaerosol concentrations. This feature can be used to implicate a specific source of bioaerosol when used before, during, and after an emission episode. The Burkard recording spore traps yield time discrimination for sampling periods from 1 hour to 1 week.

The Reuter centrifugal sampler (RCS, ***BIO***) is portable and inconspicuous and, therefore, causes minimal disturbance to a room's occupants. However, the results obtained with the earlier version, the Standard RCS, must be interpreted carefully because the d_{50} is about 3.8 μm (i.e., higher than for the other inertial impactors), and the device cannot be easily calibrated. The newer version, the RCS Plus, has a lower cut-off size, 0.82 μm (Buttner et al., 1997), and a built-in calibration system.

Filtration is an easy-to-use method to sample robust or microscopically identifiable bioaerosols in heavily contaminated environments (Palmgren et al., 1986; Eduard et al., 1990). High physical collection efficiency can be achieved when the filter sampler is used with an appropriate filter. For example, porous membrane filters with a pore size of 5 μm collect 0.3 μm particles with an efficiency of 95% or higher, whereas capillary pore membrane filters must have a pore size of 0.6 μm to reach the same efficiency (Eduard and Heederik, 1998). However, due to desiccation of bioaerosol particles by the air flow through the filter after collection, filtration is not a suitable method for evaluating the levels of vegetative cells if used in combination with culture analysis (Näsman et al., 1999; Wang et al., 2000). Gelatin filters (Li et al., 1999) and wetted porous foams (Kenny et al., 1999) have been used to decrease the desiccation effect. Filter samplers are usually small and can mostly be used with personal sampling pumps, thus allowing the collection of personal samples. Recently, filter samplers have been adapted for the sampling of inhalable bioaerosols (Kenny et al., 1999; Aizenberg et al., 2000).

The liquid of an impinger-type sampler can be used for serial dilutions and subsequent cultivation or microscopic analysis. Collection in a liquid can also be used for endotoxin determinations (Milton et al., 1990), as well as for immunological, genetic, and viral analyses. Conventional impingers, such as the AGI-30 and the AGI-4, can only be used with water-based collection fluids. These fluids evaporate quickly and are not efficient for the collection of hydrophobic particles, such as fungal and bacterial spores, due to the particle bounce and the reaerosolization of already collected particles (Grinshpun et al., 1997; Lin et al., 2000; Lin and Li, 1999). The BioSampler can be used with nonevaporative liquids, such as glycerol or mineral oil, which do not evaporate and thus permit long sampling times. The viscosities of glycerol and mineral oil are up to three decades higher than that of water, thus decreasing the potential for the reaerosolization of hydrophobic particles (Lin et al., 1999a,b). Mineral oil maintains the viability of the collected microorganisms and thus can be used with culture analysis (Lin et al., 1999; Lin et al., 2000).

CALIBRATION

No matter how efficient a sampler is, unless the air flow through the device is properly calibrated, reliable bioaerosol quantification will not be possible. Chapter 21 on instrument calibration discusses various flow calibration procedures. In addition to precisely knowing the volume of air sampled for the quantification of the sampled bioaerosol concentration, accurately calibrated air flow rates are especially critical for impactors because they are designed to operate at specific air flow rates. Improperly adjusted air flow rates will alter an impactor's cut-off size and, thus, its ability to collect an air sample that represents the ambient bioaerosol size distribution. When sampling with impactors, the distance from the inlet to the surface of the collection medium must also be correct. If the impactor is equipped with a movable stage, as in slit-to-agar impactors, the adjustment of the nozzle-to-medium distance is fairly easy. If the sampler is not equipped with an alignment device, the nutrient medium must be poured into the Petri dish to a predetermined depth.

The performance characteristics of bioaerosol samplers can be evaluated in the laboratory using aerosolized test particles. Inert test particles simulating the aerodynamic size of the bioaerosol particles in question can be used to evaluate the physical sampling efficiency of bioaerosol samplers, whereas testing of the biological sampling efficiency requires the aerosolization of the specific biological particles of interest. Bioaerosols can be aerosolized in the laboratory using various wet or dry dispersion methods (Griffiths et al., 1996; Reponen et al., 1997). Ideally, the aerosolization method simulates the natural release of the bioaerosol particles. For example, fungal spores are released from moldy building materials as dry particles by air currents.

CONTAMINATION

To avoid contaminating and, therefore, compromising bioaerosol samples, the principles of aseptic techniques must be followed. Aseptic technique means that a systematic practice is maintained to prevent undesired microorganisms and spores from contaminating the sample. In the case of ambient air sampling, for example, both the microorganisms on the human skin and within the respiratory system are among the nondesirable particles.

All surfaces, including washed hands, contain bacteria and spores unless they are specifically sterilized. Sterilization means total destruction of both cells and spores. This can be done by autoclaving or flaming the objects or materials in question. For example, all microbial culture media are sterilized before use. Not all objects can be sterilized, and therefore disinfection is used to remove the majority of microorganisms from the surfaces and materials. Disinfection treatment with an oxidizing chemical or alcohol destroys pathogenic organisms and most other vegetative cells. Although it does not destroy all spores, disinfecting sampling equipment is usually enough to prevent significant contamination of the air sample.

Most samplers are reusable and should be thoroughly cleaned before each use and decontaminated either by autoclaving or by disinfection via a chemical soak or wipe-down. Special care should be given to samplers that have convoluted inlets and pathways leading to the collection media; contaminating organisms and debris can accumulate within the sampler, making it difficult to disinfect. This creates the potential for compromised successive samples. When mounting the nutrient medium dishes or slides into samplers, touching of the nutrient surfaces by hand or other nonsterile objects should be avoided. This includes falling droplets and settling dust (e.g., from inside a ventilation system). A set of control nutrient dishes should be present during the same sampling period and be incubated with the samples in order to confirm the sterility of the medium. Once used, the Petri dishes should be sealed and transported gently with the sampling surface facing down.

SAMPLE ANALYSIS

The sample analysis method should be selected as part of the sampling plan. There are different ways to detect and quantify the collected bioaerosol particles. Traditional methods include microscopic counting and cultivation analysis. However, limitations of these traditional methods have led to the development of other methods, such as biochemical, immunological, and molecular biological assays.

Microscopy

Bioaerosol particles can be enumerated under the objective lens of the microscope after collection on a glass slide, tape, or appropriate filter. This method does not distinguish between culturable and nonculturable bioaerosol particles. Large bioaerosol particles, such as pollen grains and fungal spores, are readily enumerated under the light microscope by an experienced microscopist. Pollen grains can be identified based on their morphology, but identification of fungal species is limited with the microscopic techniques. Smaller bioaerosol particles, such as bacterial cells, are easily masked by other more numerous particles. Also, bacterial cells are not visible with a light microscope unless stained. Particles containing biological material can be detected by staining their nucleic acids with a fluorescent stain (e.g., acridine orange) and counting them with an epifluorescence microscope. Labeled antibody stains can be used to identify selected microorganisms, for example, *Legionella* (AIHA, 1996). Various microscopic analysis methods have been discussed in detail by Morris (1995).

Culture Methods

Culture-based assays are used for bacteria and fungi by collecting them directly onto a nutrient agar or transferring them onto an agar from a liquid or a filter sample. When microorganisms are cultured into countable colonies, the incubation conditions and the medium should be suitable for the organisms of interest, ideally for all viable microorganisms. In most cases, however, one cannot culture all viable microorganisms on the same medium because of the great differences in growing needs for the different organisms present. With this type of sample analysis, one cannot determine the absolute number of bacterial cells or fungal spores in the sampled air because the bioaerosol particles usually contain aggregates of two or more cells or spores (Eduard et al., 1990). The results are given in colony-forming units per cubic meter of air (cfu/m^3).

For air sampling purposes, a nonselective medium is most often used. Many media provide the basic nutrients needed for the growth of the most common environmental microorganisms. These nutrients usually include carbon, nitrogen, phosphate, sulfate, iron, magnesium, sodium, potassium, and chloride ions. Many broad-spectrum media are commercially available (e.g., tryptone-glucose-yeast agar, nutrient agar, or casein-soy-peptone agar for bacteria and malt-extract agar for fungi). By varying the composition of the medium or using very specific nutrients, selected microorganisms can be cultivated or excluded. Selective media are available for many types of microorganisms (e.g., eosin-methylene blue agar for gram-negative bacteria and dichloran glycerol agar for xerophilic fungi, which grow on media with low water content). Fungi and bacteria may prevent each other's growth. Therefore, it is sometimes beneficial to use a fungicide in the bacterial medium and an antibiotic against bacteria in a fungal medium. A commonly used fungicide for this purpose is cycloheximide (~ 500 mg/L). Streptomycin or chlortetracycline (~40 mg/L) are used as antibiotics.

Bioaerosol samples for culturable counts are usually incubated at room temperature (21° to 25°C) or at 28°C or 35°C in an incubator. These temperatures may favor the growth of some microorganisms more than others, but generally most environmental strains grow well at temperatures below 30°C. A temperature of 35°C is generally used when monitoring food and water. As a general rule, it is important to incubate all the samples, including the reference samples, at the same temperature. Exceptions to this are selective media, for which the incubation temperature is usually given by the manufacturer. If thermophilic organisms are of interest, they are usually differentiated from other species by using a much higher incubation temperature (e.g., 55°C).

Identification of fungal colonies is based on the morphology of the colonies, spores, and hyphae as well as on different physiological tests. Although a number of handbooks are available that allow identification of fungi to the genus level (e.g., Kendrick, 1985), this remains an elaborate task that requires specific expertise in environmental mycology.

Identification of bacterial colonies is based mainly on the morphology and staining properties of the cells and different physiological and biochemical tests. Easy-to-use test kits are available that identify the bacteria with a certain probability when supported by morphological characteristics. Identification of environmental bacteria requires special experience because these organisms differ from the clinically significant species. The principles of identification and classification are presented in bacteriological handbooks, such as Bergey's *Manual of Systematic Bacteriology* (Holt, 1984) and the handbook of Truper and Kramer (1981). In most cases, general classification of the colony types or genus identification provides enough information to draw conclusions from the sampling results. For instance, the colony type may be a gram-negative rod versus a gram-positive coccus, while the genus identification differentiates between *Pseudomonas* and *Staphylococcus*. In some cases, however, species identification is needed. For example, it may be important to know whether pathogenic *Pseudomonas aeruginosa* or *Staphylococcus aureus* is present.

Other Analysis Methods

Biochemical methods measure certain biological molecules in bioaerosol particles, such as endotoxin, mycotoxins, β-glucans, and fatty acids. Depending on the agent, the analysis requires gas chromatography, mass spectrometry, high-performance liquid chromatography, or spectrophotometry. In an immunoassay, antibodies are bound to a specific target antigen, such as dust mite, cockroach, animal allergen, or fungal allergen. The major limitation of immonuassay is that specific antigens for microorganisms are difficult to define and standardize (Buttner et al., 1997). Polemerase chain reaction (PCR) is a molecular biological assay method in which selected nucleic acid sequences contained in bioaerosol particles are replicated and detected. It is especially used for the rapid detection of organisms that are difficult or impossible to culture in the laboratory, such as *Mycobacterium tuberculosis* and *Histoplasma capsulatum* (Schafer et al., 1998).

These analysis methods are usually applied to filter or liquid samples. Although many of the new methods are promising, most are still research techniques. If one of these techniques is to be used, arrangements must be made with an appropriate research laboratory. Various analysis methods are discussed in detail elsewhere (AIHA, 1996).

Data Analysis and Interpretation

There are no guidelines for acceptable or harmful levels of bioaerosols. Therefore, it is necessary to decide in advance the criteria that will be used to determine whether or not an environment is contaminated. For this purpose, reference data on the range of bioaerosol concentrations and the airborne microbial flora in outdoor air and in indoor air in nonproblem environments should be collected with the sampler and the analysis method to be used. For example, for fungal spores in nonindustrial indoor environments, the levels should be lower than those outdoors except during the periods of snow cover, when the outdoor concentrations are close to zero. Furthermore, the composition of the fungal species in the environment under study can be compared with that in control environments. The data interpretation is discussed in more detail elsewhere (ACGIH, 1999).

REFERENCES

ACGIH. 1999. *Bioaerosols: Assessment and Control*, ed. J. M. Macher. Cincinnati, OH: American Conference of Governmental Industrial Hygienists.

AIHA. 1996. *Field Guide for the Determination of Biological Contaminants in Environmental Samples*, eds. H. K. Dillon, P. A. Heinsohn, and J. D. Miller. Fairfax, VA: American Industrial Hygiene Association.

Aizenberg, V., T. Reponen, S. Grinshpun, and K. Willeke. 2000. Performance of Air-O-Cell, Burkard, and Button samplers for total enumeration of airborne spores. *Am. Ind. Hyg. Assoc. J.* 61:855–864.

Andersen, A. A. 1958. New sampler for the collection, sizing and enumeration of viable airborne particles. *J. Bacteriol.* 76:471–484.

Atlas, R. M. 1988. *Microbiology: Fundamentals and Applications*. New York: Macmillan.

Atlas, R. M. and R. Bartha. 1987. *Microbial Ecology: Fundamentals and Applications*. Reading, MA: Addison-Wesley.

Brachman, P. S., R. Ehrlich, H. F. Eichenwald, V. J. Gabelli, T. W. Kethley, S. H. Madin, J. R. Maltman, G. Middlebrook, J. D. Morton, I. H. Silver, and E. K. Wolfe. 1964. Standard sampler for assay of airborne microorganisms. *Science* 144:1295.

Bratbak, G. and I. Dundas. 1984. Bacterial dry matter content and biomass estimations. *Appl. Environ. Microbiol.* 48:755–757.

Burge, H. A. 1995. *Bioaerosols*. Boca Raton, FL: CRC Press.

Buttner, M. P., K. Willeke, and S. A. Grinshpun. 1997. Sampling and analysis of airborne microorganisms. In *Manual of Environmental Microbiology*, ed. C. J. Hurst. Washington, DC: ASM Press, pp. 629–640.

Chang, C.-W., S. A. Grinshpun, K. Willeke, J. A. Macher, J. Donnelly, S. Clark, and A. Juozaitis. 1995a. Factors affecting colony enumeration in bioaerosol sampling and analysis. *Am. Ind. Hyg. Assoc. J.* 56:979–986.

Chang, C.-W., Y.-H. Hwang, S. A. Grinshpun, J. A. Macher, and K. Willeke. 1995b. Evaluation of counting error due to colony masking in bioaerosol sampling. *Appl. Environ. Microbiol.* 60:3732–3738.

Cox, C. S. 1987. *The Aerobiological Pathway of Microorganisms*. Chichester: John Wiley & Sons.

Cox, C. S. and C. M. Wathes, eds. 1995. *Bioaerosol Handbook*. Boca Raton, FL: Lewis Publishers.

Custovic, A., R. Green, A. Fletcher, A. Smith, C. A. C. Pickering, M. D. Chapman, and A. Woodcock. 1997. Aerodynamic properties of the major dog allergen Can FL: Distribution in homes, concentration, and particle size of allergen in the air. *Am. J. Respir. Crit. Care Med.* 155:94–98.

DeKoster, J. A. and P. S. Thorne. 1995. Bioaerosol concentrations in noncompliant, complaint, and intervention homes in the Midwest. *Am. Ind. Hyg. Assoc. J.* 56:573–580.

Doetsch, R. N. and T. M. Cook. 1973. *Introduction to Bacteria and Their Ecobiology*. Baltimore: University Park Press.

Donaldson, A. I. 1978. Factors influencing the dispersal, survival and deposition of airborne pathogens of farm animals. *Vet. Bull.* 48:83–94.

Eduard, W. and D. Heederik. 1998. Methods for quantitative assessment of airborne levels of noninfectious microorganisms in highly contaminated work environments. *Am. Ind. Hyg. Assoc. J.* 59:113–127.

Eduard, W., J. Lacey, K. Karlsson, U. Palmgren, G. Ström, and G. Blomquist. 1990. Evaluation of methods for enumerating microorganisms in filter samples from highly contaminated occupational environments. *Am. Ind. Hyg. Assoc. J.* 51:427–436.

Etzel, R., E. Montana, W. Sorenson, G. Kullman, T. Allan, and D. Dearborn. 1998. Acute pulmonary hemorrhage in infants associated with exposure to *Stachybotrys atra* and other fungi. *Arch. Pediatr. Adolesc. Med.* 152:757–762.

Faegri, K. and J. Iversen. 1989. *Textbook of Pollen Analysis*, IV Ed., eds. K. Faegri, P. E. Kaland, and K. Krzywinski. Chichester: John Wiley & Sons.

Flannigan, B. and P. R. Morey. 1996. *Control of Moisture Problems Affecting Biological Indoor Air Quality*. (ISIAQ) International Society of Indoor Air Quality and Climate Guideline TFI-1996, Ottawa, Canada.

Gerone, P. J., R. B. Couch, G. V. Keefer, R. G. Douglas, E. B. Derrenbacher, and V. Knight. 1966. Assessment of experimental and natural viral aerosols. *Bacteriol. Rev.* 30:576–584.

Górny, R., J. Dutkiewicz, and E. Krysinska-Traczyk. 1999. The size distribution of bacterial and fungal bioaerosols in the indoor air. *Ann. Agric. Environ. Med.* 6:105–113.

Griffiths, W. D., I. W. Stewart, A. R. Reading, and S. J. Futter. 1996. Effect of aerosolization, growth phase and residence time in spray and collection fluids on the culturability of cells and spores. *J. Aerosol Sci.* 27:803–820.

Grinshpun, S. A., C. W. Chang, A. Nevalainen, and K. Willeke. 1994. Inlet characteristics of bioaerosol samplers. *J. Aerosol Sci.* 25:1503–1522.

Grinshpun, S. A., K. Willeke, V. Ulevicius, A. Juozaitis, S. Terzieva, J. Donnelly, G. N. Stelma, and K. P. Brenner. 1997. Effect of impaction, bounce and reaerosolization on the collection efficiency of impingers. *Aerosol Sci. Technol.* 26:326–342.

Hairston, P. P., J. Ho, and F. R. Quant. 1997. Design of an instrument for real-time detection of bioaerosols using simultaneous measurement of particle aerodynamic size and intrinsic fluorescence. *J. Aerosol Sci.* 28:471–482.

Hamer, G. 1985. Chemical engineering and biotechnology. In *Biotechnology, Principles and Applications*, eds. I. J. Higgins, D. J. Best, and J. Jones. Oxford: Blackwell.

Hirvonen, M. R., M. Ruotsalainen, K. Savolainen, and A. Nevalainen. 1997. Effect of viability of *Actinomycete* spores and their ability to stimulate production of nitric oxide and reactive oxygen species in RAW264.7 macrophages. *Toxicology* 124:105–114.

Holt, J. G. 1984. *Bergey's Manual of Systematic Bacteriology*, vols 1–4. Baltimore: Williams & Wilkins.

Juozaitis, A., K. Willeke, S. Grinshpun, and J. Donnelly. 1994. Impaction onto a glass slide or agar versus impingement into a liquid for the collection and recovery of airborne microorganisms. *Appl. Environ. Microbiol.* 60:861–870.

Keleti, G. and M. A. Shapiro. 1987. *Legionella* and the environment. *CRC Crit. Rev. Environ. Control* 17:133–185.

Kendrick, B. 1985. *The Fifth Kingdom*. Waterloo, Ontario, Canada: Mycologue Publications.

Kenny, L. C., A. Bowry, B. Crook, and J. D. Stancliffe. 1999. Field testing of a personal size-selective bioaerosol sampler. *Ann. Occup. Hyg.* 43:393–404.

Kotimaa, M. 1990. Occupational exposure to spores in the handling of wood chips. *Grana* 29:153–156.

Lacey, J. and J. Dutkiewicz. 1994. Bioaerosols and occupational lung disease. *J. Aerosol Sci.* 25:1371–1404.

Lach, V. 1985. Performance of the Surface Air System air samplers. *J. Hosp. Inf.* 6:102–107.

Li, C.-S., M.-L. Hao, W.-H. Lin, C.-W. Chang, and C.-S. Wang. 1999. Evaluation of microbial samplers for bacterial microorganisms. *Aerosol Sci. Technol.* 30:100–108.

Lighthart, B. and B. T. Shaffer. 1997. Increased airborne bacterial survival as a function of particle content and size. *Aerosol Sci. Technol.* 27:439–446.

Lin, W.-H. and C.-S. Li. 1999. Collection efficiency and culturability of impingement into a liquid for bioaerosols of fungal spores and yeast cells. *Aerosol Sci. Technol.* 30:109–118.

Lin, X., T. A. Reponen, K. Willeke, Z. Wang, S. A. Grinshpun, and M. Trunov. 2000. Survival of airborne microorganisms during swirling aerosol collection. *Aerosol Sci. Technol.* 32:184–196.

Lin, X., T. A. Reponen, K. Willeke, S. A. Grinshpun, K. Foarde, and D. S. Ensor. 1999. Long-term sampling of airborne bacteria and fungi into a non-evaporative liquid. *Atmos. Environ.* 33:4291–4298.

Lin, K., K. Willeke, V. Ulevicius, and S. A. Grinshpun. 1997. Effect of sampling time on the collection efficiency of all-glass impingers. *Am. Ind. Hyg. Assoc. J.* 58:480–488.

Lundholm, M. 1982. Comparison of methods for quantitative determinations of airborne bacteria and evaluation of total viable counts. *Appl. Environ. Microbiol.* 44:179–183.

Macher, J. M. 1989. Positive-hole correction of multiple-jet impactors for collecting viable microorganisms. *Am. Ind. Hyg. Assoc. J.* 50:561–568.

Macher, J. M., F.-Y. Huang, and M. Flores. 1991. A two-year study of microbiological indoor air quality in a new apartment. *Arch. Environ. Health* 46:25–29.

Madelin, T. M. and H. E. Johnson. 1992. Fungal and actinomycete spore aerosol measured at different humidities with an aerodynamic particle sizer. *J. Appl. Bacteriol.* 72:400–409.

Mainelis, G., S. A. Grinshpun, K. Willeke, T. Reponen, V. Ulevicius, and P. J. Hintz. 1999. Collection of airborne microorganisms by electrostatic precipitation. *Aerosol Sci. Technol.* 30:127–144.

Martinez, F. K., J. W. Sheehy, J. H. Jones, and L. B. Cusick. 1988. Microbial containment in conventional fermentation processes. *Appl. Ind. Hyg.* 3:177–181.

Matthias-Maser, S. and R. Jaenicke. 1995. The size distribution of primary biological aerosol particles with radii >0.2 μm in an urban/rural influenced region. *Atmos. Res.* 39:279–286.

May, K. R. 1966. Multistage liquid impinger. *Bacteriol. Rev.* 30:559–570.

Milton, D. K., R. J. Gere, H. A. Feldman, and I. A. Greaves. 1990. Endotoxin measurement: Aerosol sampling and application of a new Limulus method. *Am. Ind. Hyg. Assoc. J.* 51:331–337.

Morris, K. J. 1995. Modern microscopic methods of bioaerosol analysis. In *Bioaerosol Handbook*, eds. C. S. Cox and C. M. Wathes. Boca Raton, FL: Lewis Publishers.

Näsman, Å., G. Blomquist, and J.-O. Levin. 1999. Air sampling of fungal spores on filters. An investigation on passive sampling and viability. *J. Environ. Monit.* 1:361–365.

Nevalainen, A. 1989. *Bacterial Aerosols in Indoor Air*. Publications of the National Public Health Institute of Finland A3/1989. Academic Dissertation.

Nevalainen, A., J. Pastuszka, F. Liebhaber, and K. Willeke. 1992. Performance of bioaerosol samplers: Collection characteristics and sampler design considerations. *Atmos. Environ.* 26(A):531–540.

Noble, W. C. 1975. Dispersal of skin microorganisms. *Br. J. Dermatol.* 93:477–485.

Orr, C., Jr., and M. T. Gordon. 1956. The density and size of airborne *Serratia marcescens*. *J. Bacteriol.* 71:315–317.

Palmgren, U., G. Ström, G. Blomquist, and P. Malmberg. 1986. Collection of airborne microorganisms on Nuclepore filters, estimation and analysis—CAMNEA method. *J. Appl. Bacteriol.* 61:401–406.

Pasanen, A.-L., P. Pasanen, M. J. Jantunen, and P. Kalliokoski. 1991. Significance of air humidity and air velocity for fungal spore release into the air. *Atmos. Environ.* 25A:459–462.

Qian, Y., K. Willeke, V. Ulevicius, S. A. Grinshpun, and J. Donnelly. 1995. Dynamic size spectrometry of airborne microorganisms: Laboratory evaluation and calibration. *Atmos. Environ.* 29:1123–1129.

Qian, Y., K. Willeke, S. A. Grinshpun, and J. Donnelly. 1997. Performance of N95 Respirators: Reaerosolization of bacteria and solid particles. *Am. Ind. Hyg. Assoc. J.* 58:876–880.

Rahkonen, P., M. Ettala, M. Laukkanen, and M. Salkinoja-Salonen. 1990. Airborne microbes and endotoxins in the work environment of two sanitary landfills in Finland. *Aerosol Sci. Technol.* 13: 505–513.

Rantio-Lehtimäki, A. 1995. Aerobiology of pollen and pollen antigens. In *Bioaerosol Handbook*, eds. C. S. Cox and C. M. Wathes. Boca Raton, FL: Lewis Publishers, pp. 387–406.

Reponen, T. A., S. V. Gazenko, S. A. Grinshpun, K. Willeke, and E. C. Cole. 1998. Characteristics of airborne actinomycete spores. *Appl. Environ. Microbiol.* 64:3807–3812.

Reponen, T., A. Hyvärinen, J. Ruuskanen, T. Raunemaa, and A. Nevalainen. 1994. Comparison of concentrations and size distributions of fungal spores in buildings with and without mold problems. *J. Aerosol Sci.* 25:1595–1603.

Reponen, T., A. Nevalainen, M. Jantunen, M. Pellikka, and P. Kalliokoski. 1992. Normal range criteria for indoor air bacteria and fungal spores in a subarctic climate. *Indoor Air* 2:26–31.

Reponen, T., K. Willeke, V. Ulevicius, S. A. Grinshpun, and J. Donnelly. 1997. Techniques for dispersion of microorganisms into air. *Aerosol Sci. Technol.* 27:405–421.

Reponen, T., K. Willeke, V. Ulevicius, A. Reponen, and S. Grinshpun. 1996. Effect of relative humidity on aerodynamic size and respiratory deposition of fungal spores. *Atmos. Environ.* 30:3967–3974.

Riley, R. L., G. Murphy, and R. L. Riley. 1978. Airborne spread of measles in a suburban elementary school. *Am. J. Epidemiol.* 107:421–432.

Roszak, D. B. and R. R. Colwell. 1987. Survival strategies of bacteria in the natural environment. *Microbiol. Rev.* 51:365–379.

Rylander, R. and P. Haglind. 1984. Airborne endotoxins and humidifier disease. *Clin. Allergy* 14:109–112.

Rylander, R. and R. R. Jacobs. 1994. *Organic Dusts, Exposure, Effects, and Prevention*. Boca Raton, FL: CRC Press, Inc.

Schafer, M. P., J. E. Fernback, and P. A. Jensen. 1998. Sampling and analytical method development for qualitative assessment of airborne mycobacterial species of the *Mycobacterium tuberculosis* complex. *Am. Ind. Hyg. Assoc. J.* 59:540–546.

Sorenson, W. G., D. G. Frazer, B. B. Jarvis, J. Simpson, and V. A. Robinson. 1987. Trichothecene mycotoxins in aerosolized conidia of *Stachybotrys atra*. *Appl. Environ. Microbiol.* 53:1370–1375.

Stackebrandt, E., F. A. Rainey, and N. L. Ward-Rainey. 1997. Proposal for a new hierarchic classification system, *Actinobacteria* classis nov. *Int. J. Syst. Bacteriol.* 47:479–491.

Stewart, S. L., S. A. Grinshpun, K. Willeke, S. Terzieva, V. Ulevicius, and J. Donnelly. 1995. Effect of impact stress on microbial recovery on an agar surface. *Appl. Environ. Microbiol.* 61:1232–1239.

Terzieva, S., J. Donnelly, V. Ulevicius, S. A. Grinshpun, K. Willeke, G. Stelma, and K. Brenner. 1996. Comparison of methods for detection and enumeration of airborne microorganisms collected by liquid impingement. *Appl. Environ. Microbiol.* 62:2264–2272.

Trunov, M., S. Trakumas, K. Willeke, S. A. Grinshpun, and T. Reponen. 2001. Collection of bioaerosols by impaction: Effect of fungal spore agglomeration and bounce. *Aerosol Sci. Technol.* (in press).

Truper, H. G. and J. Kramer. 1981. Principles of characterization of prokaryotes. In *The Prokaryotes*, eds. M. P. Starr, H. Stolp, H. G. Truper, A. Balows, and H. Schlegel. Berlin: Springer-Verlag.

Verhoeff, A. P., J. H. van Wijnen, J. S. M. Boleij, B. Brunekreef, E. S. van Reenen-Hoekstra, and R. A. Samson. 1990. Enumeration and identification of airborne viable mould propagules in houses. *Allergy* 45:275–284.

von Bischoff, E. 1989. Sources of pollution of indoor air by mite allergen containing house dust. *Environ. Int.* 15:181–192.

Wang, Z., T. Reponen, S. A. Grinshpun, R. L. Górny, and K. Willeke. 2001. Effect of sampling time and air humidity on bioefficiency of filter samplers for bioaerosol collection. *J. Aerosol Sci.* 32:661–674.

Willeke, K., X. Lin, and S. A. Grinshpun. 1998. Improved aerosol collection by combined impaction and centrifugal motion. *Aerosol Sci. Technol.* 28:439–456.

Willeke, K. and J. M. Macher. 1999. Air sampling. In *Bioaerosols: Assessment and Control*, ed. J. M. Macher. Cincinnati, OH: American Conference of Governmental Industrial Hygienists, pp. 11-1–11-25.

Willeke, K., Y. Qian, J. Donnelly, S. Grinshpun, and V. Ulevicius, V. 1996. Penetration of airborne microorganisms through a surgical mask and a dust/mist respirator. *Am. Ind. Hyg. Assoc. J.* 57:348–355.

25 Aerosol Measurement in the Workplace

ANDREW D. MAYNARD

National Institute for Occupational Safety and Health, Centers for Disease Control and Prevention, Public Health Service, U.S. Department of Health and Human Services, Cincinnati, OH

PAUL A. JENSEN

National Institute for Occupational Safety and Health, Centers for Disease Control and Prevention, Public Health Service, U.S. Department of Health and Human Services, Morgantown, VA

INTRODUCTION

Industrial hygiene can be defined as the identification, evaluation, and control of occupational health hazards. It encompasses the complete process from a hazardous material being made accessible through use or lack of containment; to control of the exposure route; through exposure and dose; and finally to health effects resulting from received dose. Exposure may be through ingestion, inhalation, or dermal contact, although it is the inhalation route that is of prime importance when considering aerosols. Traditionally, workplace aerosols have been categorized as fumes (fine particles and agglomerates generated through combustion and vapor condensation), smokes (solid and liquid particles arising from incomplete combustion), dusts (solid particles generated through mechanical means), sprays (liquid aerosols with relatively large particle sizes, usually produced through mechanical means), and mists (liquid aerosols with finer particles, generally produced through condensation or atomization) (Vincent, 1995). Aerosols containing biological organisms, or bioaerosols, are also considered as a separate category, and are covered in more depth in Chapter 24. These definitions tend to be used as descriptors rather than as discrete classifications, and when considering sampling and health effects their use can be somewhat misleading. For example, a size selective sampler will not differentiate between a fume, smoke, or mist, and the distinction between health effects arising from a fume and a submicron dust can be somewhat blurred.

Historically there have been a number of different approaches to measuring aerosols generated in the workplace (Walton and Vincent, 1998). From the early 1900s through the 1950s and beyond, particle number was the dominant metric of exposure assessment, with

Aerosol Measurement: Principles, Techniques, and Applications, Second Edition, Edited by Paul A. Baron and Klaus Willeke.
ISBN 0-471-35636-0

devices such as the konimeter (Le Roux, 1970; Hewson, 1996), impinger (Greenburg and Smith, 1922), and thermal precipitator (Green and Watson, 1935; Hamilton, 1956) finding widespread use. Analysis was generally accomplished by using light microscopy, with electron microscopy finding increasing use as the instrument was developed. Although exposure to fibers is still assessed on a particle number basis, current sampling and analysis methods are dominated by the use of collection on filters and mass analysis (gravimetric or, for specific elements or compounds, by chemical analysis). Many of the methods employed are the same as or similar to those used in other areas of aerosol measurement. However, workplace aerosols, and the aims of applied measurement techniques, differ somewhat from those found in other circumstances. In most cases the bulk aerosol composition is known or can be deduced from the processes or products in use. The mass concentrations involved are typically an order of magnitude or so greater than those in the general environment. Finally, sampling is carried out specifically for assessing human exposure rather than characterizing the aerosol itself.

While philosophies and approaches may differ, there is a great deal of commonality between methods used in the workplace and those used in other areas of aerosol measurement. Thus, techniques and applications described elsewhere in this book will frequently be directly relevant to workplace sampling. Chapters 7, 9, 10, 12, and 15 are particularly pertinent, providing detailed information on approaches to aerosol monitoring; filter collection; inertial, gravitational, centrifugal, and thermal sampling; and direct-reading techniques using optical particle detection, respectively. Chapter 26 on measurements in mines covers a subfield of industrial hygiene, while Chapters 23, 24, and 27 on nonspherical particle measurement, bioaerosol measurement, ambient aerosol sampling and aerosol exposure measurement are all relevant to the workplace. In this chapter, the emphasis is therefore on the basic sampling philosophies and methods used on a daily basis in the workplace.

AEROSOL EXPOSURE MEASUREMENT IN THE WORKPLACE

Biologically Relevant Sampling

Aerosol sampling in the workplace is ultimately concerned with measuring that aspect of the aerosol that leads to specific health effects. Thus, the method and metric used aim to provide biologically relevant information. Aerosol particles can cause health problems when deposited on the skin and eyes, but generally the most sensitive route of entry into the body is through the respiratory system. The biological effects resulting from deposition of an aerosol in the *respiratory tract* will depend on the dose received and the body's response to the deposited particles. Physiological response to an aerosol depends on the chemical and physical nature of the particles and the location of the interaction (i.e., deposition region). The ultimate goal of industrial hygiene aerosol measurement is therefore to ascertain the dose of aerosol delivered to the body and to evaluate whether the dose or potential dose is sufficient to cause adverse health effects.

The respiratory system deposition region is primarily governed by particle size and shape. The health response may be a function of mass, chemical composition, or morphology and possibly particle size and surface chemistry. Ideally dose should be expressed in terms of the most appropriate metric. However, additional restraining factors on industrial hygiene aerosol measurements include the practical and economic application of measurement methods. In practice, it is simpler to measure penetration to the relevant areas of the respiratory system rather than dose, thus giving a measure of the *potential* dose. Mass and bulk chemical composition are easier to measure than parameters such as particle shape and surface area, and correlation between health effects and particle number and mass concentration (e.g., Bedford and Warner, 1943) indicates mass to be a suitable metric in many cases. Asbestiform fibers present an exceptional case where dose is best represented by particle number and shape, and accordingly a number and morphology-based metric is used (see Chapters 12 and 23).

Deposition Regions

The respiratory system is an effective size-selective aerosol sampler in its own right, and it is fallacious to assume that all airborne particles will enter it. Large particles are excluded from entering the nose and mouth (the nasopharyngeal region) through inertial separation. Aspiration is a function of a number of parameters, including particle size, external air speed, orientation to the prevailing air movement direction, and breathing rate and volume. However, for external wind speeds of a few m/s and lower, the probability of a particle entering the mouth or nose (termed *inhalable particles*) may be generalized as being around 100% for particles with aerodynamic diameters of a few micrometers and below, reducing to around 50% at 100 μm aerodynamic diameter (Vincent et al., 1990).

Aerosol deposition in the nasopharyngeal region is dominated by inertial impaction, although interception and (for particles in the nanometer size range) diffusion also contribute. Further inertial separation and interception occurs as the particles pass into the trachea and the upper lungs (tracheobronchial region). Although population variance is high (Lippmann, 1977), penetration into the tracheobronchial region may be typified by particles smaller than approximately 10 μm aerodynamic diameter (Lippmann, 1977; ISO, 1995). As the airways bifurcate to ever finer branches toward the alveolar region, aerosol particles are predominantly removed from the flow through a combination of impaction, interception, charge effects, and diffusion. In the preceding regions, deposited particles are cleared primarily by the action of cilia transporting them along to the upper airways. Particles depositing in the alveolar, or gas exchange, region are cleared either through the action of alveolar macrophages engulfing them and transporting them to ciliated airways (phagocytosis) or by dissolution in the lung fluid. Particle deposition is through impaction and diffusion, and penetration to the alveolar region is restricted to particles around 5 μm and less aerodynamic diameter (Lippmann, 1977; ISO, 1995). The clearance mechanism employed in the alveolar region, together with the close proximity of the bloodstream, leads to a number of health effects specific to particle deposition within this region.

Particle Characteristics and Biological Response

Although particle aerodynamic diameter dominates deposition within the respiratory tract, the subsequent effect on health is a combination of physical particle characteristics and biological response. On deposition, the body may react to the chemical substances contained within the particle, interact with the particle surface, or be influenced by physical parameters such as size and morphology.

Highly soluble particles and droplets will be rapidly assimilated by the body, particularly in the alveolar region. Local effects, such as irritation and inflammation, and systemic responses may become manifest over very short time periods. The gradual release of agents from low-solubility particles will have a much longer response time. However, low-solubility particles may also act as vectors for the transport of high-solubility solids, liquids, and gases present as thin surface layers, thus leading to a response not indicated by the bulk aerosol particle properties alone. For example, adsorption of nitrogen oxides and sulfur dioxide onto particles can lead to health effects at levels normally considered safe.

Very low-solubility particles are more likely to have health effects associated with their physical characteristics. Lung overload phenomena are associated with the physical limitations of the lungs' clearance mechanisms as opposed to chemical interactions with the deposited particles (Morrow, 1994). Particle shape is a factor for fibrous aerosols (Blake et al., 1998) (see Chapter 23). It also influences available surface area, which may be related to toxicity through surface interactions (Lison et al., 1997) or increased solubility. Where open agglomerates of particles exist, including those resulting from combustion (such as diesel exhaust particulates), metal processing, welding, or fine powder production, the aerosol may have a very high specific surface area and be formed from particles able to penetrate to the alveolar region. In some fine powders, including ultrafine titanium dioxide, carbon blacks, and fumed silicas, specific surface areas in excess of $2 \times 10^5\,m^2/kg$ [$200\,m^2/g$] are achieved among particles with aerodynamic diameters less than 4 µm. In comparison, an aerosol of spherical particles 4 µm in diameter and with unit density would have a specific surface area of $1.5 \times 10^3\,m^2/kg$ [$1.5\,m^2/g$]. There is evidence that for some low-solubility materials toxic response may be associated with surface area or even particle number (Oberdörster et al., 1994; Lison et al., 1997; Donaldson et al., 1998). However, little is known of the role of what may be termed *available surface area*, which will be influenced by particle surface structure and biological mechanisms.

Some of the responses observed on inhaling aerosols are reversible; some may be chronic. Some effects are cumulative; others are not. For some substances, there may be an exposure level below which no effects are observed (a "no-effect" level). For others, notably carcinogens, there may be no identifiable no-effect level. For a class of substances known as *sensitizers*, relatively high exposure levels may be experienced without obvious effect until a person becomes "sensitized" to the substance. Following sensitization, exposure to very low levels may result in a significant biological response.

Biologically relevant exposure monitoring requires the range of interactions and responses, together with aerosol dose and particle form, to be taken into account. It can be seen that in principle there are a number of particle characteristics that will influence the toxicity of inhaled particles. Although characteristics such as size, morphology, surface area, and structure may be influential, current technology lacks the means to characterize workplace aerosols as completely as may be desirable. Fortunately, the specificity of many workplace aerosols enables successful exposure monitoring to be carried out by linking a related metric (such as mass concentration) to empirical dose–response data. The extent to which this approach is tenable where toxicity data are sparse is questionable, however.

Sampling Conventions

The accurate measurement of aerosol exposure via inhalation requires sampling devices that match particle deposition to the relevant areas of the respiratory system. However, aerosol deposition is highly dependent on the individual (Lippmann, 1977) and not trivial to replicate in a sampling device. Broad standards have therefore been developed describing representative penetration characteristics of aerosol particles through the respiratory system as a function of aerodynamic diameter. These standards provide a basis for estimating the aerosol concentration potentially available to cause harm within specific areas of the respiratory system and underlie many industrial hygiene aerosol sampling methods.

Early estimates of penetration into what was considered the most vulnerable part of the system—the alveolar region—were proposed in the 1950s and 1960s, resulting in the British Mines Research Council (BMRC) and the American Conference of Government Industrial Hygienists (ACGIH) conventions describing respirable aerosols (BMRC, 1952; ACGIH, 1968). More recently, the International Standards Organization (ISO, 1995) and the ACGIH (1998) arrived at convergent conventions describing the probability of particles penetrating to the nasopharyngeal, tracheobronchial, and alveolar regions. However, it was not until the early 1990s that international consensus was reached on particle penetration standards between the ISO, ACGIH, and the European Committee for Standardization (CEN). The resulting conventions describe penetration as a function of particle aerodynamic diameter into the respiratory system (*inhalable* aerosol), into the tracheobronchial region (*thoracic* aerosol), and into the alveolar region (*respirable* aerosol), with thoracic and respirable aerosol as subfractions of the inhalable aerosol. These particle-size–dependent fractions shown in Figure 25–1 are now widely used as the standards to which industrial hygiene aerosol samplers should conform (ISO, 1995).

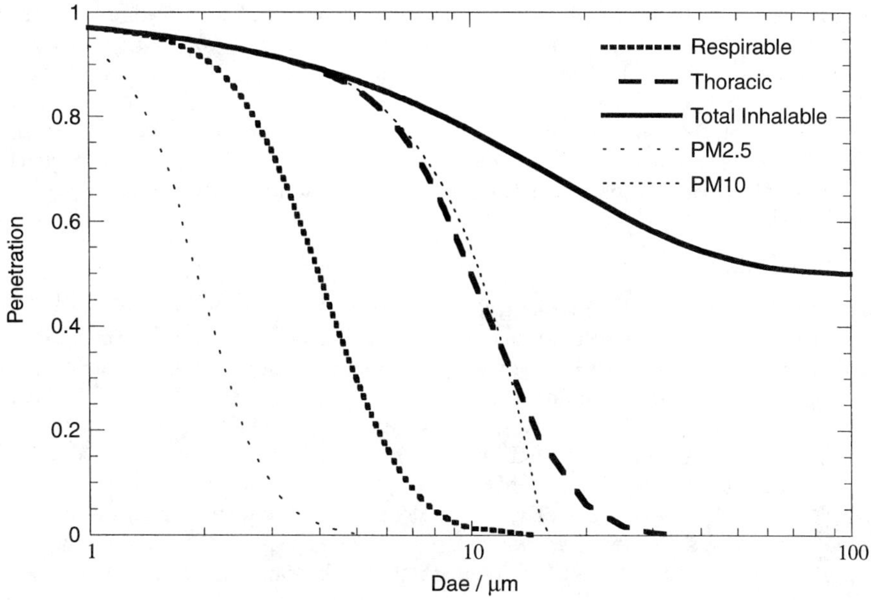

Fig. 25–1. International workplace sampling conventions (ISO 1995). Environmental conventions are also shown for comparison (chapter 27).

The inhalable convention is based on particle penetration through the mouth and nose of a breathing mannequin over a range of wind speeds and orientations with respect to the wind and is defined as

$$SI(d_{ae}) = 0.5 \times \left(1 + e^{-0.06 d_{ae}}\right) \qquad (25\text{–}1)$$

for $0 < d_{ae} < 100\,\mu m$. $SI(d_{ae})$ is the fraction of particle entering the system as a function of aerodynamic diameter d_{ae}.

Both the thoracic and respirable conventions are expressed as subfractions of the inhalable convention and are based on lung penetration measurements. The thoracic convention is given as

$$ST(d_{ae}) = SI(d_{ae}) \times [1 - F(x)]$$
$$x = \frac{\ln(d_{ae}/\Gamma)}{\ln(\Sigma)} \qquad (25\text{–}2)$$

$ST(d_{ae})$ is the fraction of particles penetrating beyond the larynx as a function of aerodynamic diameter. $F(x)$ is a cumulative lognormal distribution, with a mass median aerodynamic diameter (MMAD) Γ of $11.64\,\mu m$ and a geometric standard deviation (GSD) Σ of 1.5.

The respirable convention $SI(d_{ae})$ is similarly given as

$$SR(d_{ae}) = SI(d_{ae}) \times [1 - F(x)]$$
$$x = \frac{\ln(d_{ae}/\Gamma)}{\ln(\Sigma)} \qquad (25\text{–}3)$$

where the cumulative lognormal distribution has an MMAD Γ of 4.25 μm and a GSD Σ of 1.5. A respirable convention for susceptible groups is also defined, with $\Gamma = 2.5$ μm, although this has not been implemented in any exposure standards as yet. Standards relating to penetration to the tracheobronchial and extrathoracic regions are defined by the difference between the respirable and thoracic conventions (tracheobronchial) and between the thoracic and inhalable conventions (extrathoracic). Further information on particle size-selective sampling for workplace contaminants may be found in ACGIH (1998).

Occupational Exposure Limits

Health-based aerosol exposure limits follow country-specific systems, but in the majority of cases follow a similar philosophy (Vincent, 1998). In the United States, the primary sources of occupational exposure limits for the workplace are (1) National Institute for Occupational Safety and Health (NIOSH) recommended exposure limits (RELs); (2) the U.S. Department of Labor (OSHA and MSHA) permissible exposure limits (PELs), and (3) the American Conference of Government Industrial Hygienists' (ACGIH) threshold limit values (TLVs). NIOSH RELs are time-weighted average (TWA) concentrations for up to a 10 h work day during a 40 h work week. OSHA PELs are TWA concentrations that must not be exceeded during any 8 h work shift of a 40 work week. The ACGIH TLVs are 8 h TWA concentrations for a normal 8 h workday and a 40 h work week, to which nearly all workers may be exposed, day after day, without adverse effects.

In the United Kingdom, a two-tier system of occupational exposure standards (OES) and maximum exposure limits (MELs) is employed (HSE, 1999). Each represents an 8 h TWA exposure limit. An OES is set where a no-effect level can be identified for a substance, thus giving an exposure limit below which adverse effects are not expected (as for the ACGIH TLVs). MELs are employed where there is no clear no-effect level. As there will be a degree of resultant health effects manifest whatever exposure limit is chosen (above zero), the choice of limit is in essence a political decision. Reflecting the nature of substances having MELs, there is an obligation on U.K. industries to keep exposures as low as reasonably practicable, even when this results in a target exposure significantly below the limit.

Even below these various exposure limits, a small percentage of workers may experience adverse health effects due to individual susceptibility, a pre-existing medical condition, and/or a hypersensitivity (allergy). In addition, some materials may act in synergy with other substances to produce undesirable health effects, even if the occupational exposures to individual contaminants are controlled at the level set by the evaluation criteria. For example, gases such as oxides of nitrogen and sulfur dioxide may adsorb on dust particles and produce health effects at levels normally considered safe. Furthermore, some substances are absorbed by direct contact with the skin and mucous membranes and thus potentially increase the overall exposure. For substances that may potentially lead to health effects following short exposures, or high peak exposures, short-term exposure limits (STELs) are generally set to complement the 8 to 10 TWA limits. These are generally sampled over shorter time periods—typically 15 min—and are collected during periods when the concentration of contaminant is likely to be highest.

SAMPLING AGAINST EXPOSURE CONVENTIONS

The accuracy and relevance of aerosol samples taken within the workplace predominantly rely on selection of an appropriate sampling device. However, filter selection, pump selection and use, sampling strategy, and sample handling also play a role in determining the accuracy and suitability of samples. Useful sources of information include the ACGIH (1995, 1998).

Matching the Sampler to Sampling Requirements

A number of the industrial hygiene aerosol samplers introduced to the market in recent years have been developed and tested against international sampling conventions (ISO, 1995). However, many devices are still available that were brought into use before acceptance of the current conventions. Some of these agree reasonably well with the relevant convention, and others have been brought into line by altering the sampling flow rate (e.g., the SIMPEDS respirable cyclone; Bartley et al., 1994; Maynard and Kenny, 1995). Others, such as the closed-face 37 mm filter cassette, show poor agreement with the current conventions (Kenny et al., 1997).

The analytical development of inhalable samplers has been hampered by the complexities of how external conditions affect aspiration, together with the difficulties of making penetration measurements with particles up to 100 μm aerodynamic diameter. The IOM personal inhalable sampler was the first sampler built to match the inhalable convention and was developed following aspiration measurements with a breathing mannequin (Mark and Vincent, 1986). Although the sampler has shortcomings (e.g., it is very accessible to sample tampering, and there is evidence for significant projectile entry in some environments), it is still regarded as a benchmark sampler. More recent samplers such as the CIP10-I (***ARE***)* address some of the problems inherent in the IOM inhalable sampler, but still fall short of the ideal. Samplers such as the button sampler (***SKC***) have been developed specifically to reduce intersampler variability and wind speed dependence common to a number of inhalable samplers (Aizenberg et al., 2000). Samplers following the thoracic and respirable conventions have been easier to engineer. The development of an empirical understanding of particle penetration through cyclones and polyurethane foams in particular has led to sampling devices that match the respirable and thoracic conventions reasonably well (Vincent et al., 1993; Kenny and Gussman, 1997; Chen et al., 1998; Maynard, 1999).

In recognition that no sampler will agree with the current workplace sampling conventions at all times, performance criteria are under development to set acceptable bounds on how well a device performs (CEN, 1998). The mass fraction of a lognormal aerosol characterized by its MMAD and GSD that would be sampled by a device may be compared with the mass that would be sampled by an ideal sampler (i.e., one following the convention perfectly). The comparison gives the sampler's bias as a function of aerosol size distribution (Bartley and Breuer, 1982; Lidén and Kenny, 1992; Maynard and Kenny, 1995). Incorporating errors inherent in sampler performance measurements and typical usage into calculations of bias allows the sampler's accuracy as a function of the aerosol size distribution to be estimated. Ensuring that sampler accuracy and bias lie within acceptable bounds then gives a basis for determining good and poor sampler performance.

From the available samplers that lie within acceptable performance criteria, the choice of device will depend largely on the sampling requirements. Two general types of sampling are used in the workplace, fixed location sampling (also called *static* or *area sampling*) or personal sampling, where the sampler is placed on the worker. Static and personal samplers should not be interchanged, except where otherwise indicated. High flow-rate samplers should be used to increase the aerosol detection limit, for instance, during short-term sampling or when the sampled material has a low exposure limit (although the detection limit will also depend on the filter used and the analysis method). Where high air velocities are expected, samplers with a sampling efficiency that are not as prone to wind speed should be selected. Other considerations should include whether the aerosol charge is likely to affect sampling (e.g., Baron and Deye, 1990; Puskar et al., 1991), whether projectiles are likely to enter the sampling orifice and be included in the sample, and whether there is a possibility of significant sample loss during transport (see Chapter 7). Table 25–1 summarizes many of

* See Appendix I for full manufacturer addresses referenced to the italicized three-letter codes.

TABLE 25–1. Summary of Industrial Hygiene Aerosol Samplers

Sampler	Flow Rate (10^{-5} m^3/s [L/min])	Deployment	Manufacturer or Distributor (see Appendix I)	Agreement with Convention	Notes	References
Inhalable samplers						
IOM Inhalable	3.3 [2]	Personal	***SKC***	Good	Uses filter cassette. Susceptible to large projectiles. Wind speed dependent	Mark and Vincent (1986), Kenny et al. (1997, 1999a)
IOM Inhalable static	5 [3]	Static	***CAS***	Good		Mark et al. (1985)
CIP-10I	17 [10]	Personal	***ARE***		Rotating porous foam acts as an air mover and collection medium	Courbon et al. (1988), Kenny et al. (1997)
GSP Inhalable	5.8 [3.5]	Personal	***STR***	Good		Kenny et al. (1997)
Conical Inhalable Sampler	5.8 [3.5]	Personal	***CAS***, ***BGI***	Good	Based on the GSP sampler	Kenny et al. (1997)
Seven Hole	3.3 [2]	Personal	Various	Fair	Also known as the multiorifice or UKAEA sampler	Kenny et al. (1997)
Single Hole	3.3 [2]	Personal	Various	Poor	Used for lead aerosol sampling in the United Kingdom	Kenny et al. (1997)
PAS-6	3.3 [2]	Personal	***KOE***	Fair		Kenny et al. (1997)
Button sampler	6.7 [4]	Personal	***SKC***	Good	Perforated inlet reduces wind speed dependence and intersampler variability and leads to a uniform filterdeposit	Hauck et al. (1997), Aizenberg et al. (2000)
Thoracic samplers						
Elutriator	12.3 [7.4]	Static	***GMW***	Poor	Specific to cotton dust	Robert (1979)
CIP-10T	12 [7]	Personal	***ARE***	Fair	CIP-10I with a thoracic separation stage	Fabriès et al. (1998)
CATHIA	12 [7]	Static	***ARE***	Fair	Static version of the CIP-10T	Fabriès et al. (1998)
IOM Thoracic	3.3 [2]	Personal	***IOM***	Fair	Separation based on polyurethane foam	Maynard (1999)
GK 2.69 Cyclone	2.7 [1.6]	Personal	***BGI***	Good	Can also be used as a respirable sampler—see below	Maynard (1999)
Modified SIMPEDS cyclone	1.3 [0.8]	Personal	Not commercially available	Good	Developmental modification to the SIMPEDS cyclone	Maynard (1999)
IOM Inhalable + thoracic foam	3.3 [2]	Personal	***SKC***	Fair	IOM inhalable sampler with a size-selective polyurethane foam insert	Maynard (1999)

Respirable samplers						
CIP-10R	17 [10]	Personal	***ARE***	Good	CIP-10I with a respirable separation stage	Courbon et al. (1988)
SIMPEDS Cyclone	3.7 [2.2]	Personal	Various	Good	Also known as the Higgins and Dewell (HD) cyclone	Lidén and Kenny (1993), Bartley et al. (1994), Maynard and Kenny (1995)
SKC Cyclone		Personal	***SKC***	Good	Variant of the SIMPEDS cyclone	Lidén (1993)
GK 2.69 Cyclone	7 [4.2]	Personal	***BGI***	Good	Can also be used as a thoracic sampler—see above	Maynard (1999)
Dorr-Oliver (10 mm) Cyclone	2.8 [1.7]	Personal	Various	Good	Sampler constructed from nonconducting nylon	Bartley et al. (1994)
MRE 113A (Gravimetric Dust Sampler)	4.2 [2.5]	Static	***CAS***	Fair	Use limited to U.K. mines	Dunmore et al. (1964)
IOM Inalable + respirable foam	3.3 [2]	Personal	***SKC***	Good	IOM inhalable sampler with a size-selective polyur ethane foam insert	Kenny et al. (1999b)
Foam respirable sampler	3.3 [2]	Personal	Not commercially available	Good	Cowled sampler with size-selective polyurethane foam plugs	Chen et al. (1998)
Virtual Cyclone		Personal	Not commercially available	Good	Provides a good match with the respirable convention slope	Chen et al. (1999)
Spiral sampler	4.2 [2.5]	Personal	***SKC***		Uses centrifugal particle separation	John and Kreisberg (1999) ($PM_{2.5}$ operation)
Miscellaneous samplers						
37 mm Cassette (open)	3.3 [2]	Personal/static	Various		Standard filter cassette, worn facing down at 45° to the body. Conducting versions available	Kenny et al. (1997)
37 mm Cassette (closed)	3.3 [2]	Personal/static	Various		Standard filter cassette with a cap containing a 2 mm diameter inlet	Kenny et al. (1997)
Static sampler for "total" aerosol	Variable	Static	***CAS***		Open-faced filter. Widely used in the United Kingdom	Mark et al. (1986)
Passive sampler	—	Personal/static	HSE, UK		Electret-based sampler relying on aerosol charge and naturally occurring air movements. Correlation is good with some size-selective samplers	Brown et al. (1994, 1995)
Cowled sampler	3.3 [2] (typical)	Personal/static	Various		Used in the main for fiber sampling. Size selectivity not quantified	

the workplace sampling devices currently available or in use and gives some indication as to their application.

Filter and Substrate Selection

Industrial hygiene aerosol samples are generally collected onto a filter, within a polyurethane foam, or onto an impenetrable impaction substrate such as Mylar (which is usually coated with a layer of grease or oil to prevent particle bounce). Filters may be held in a cartridge within the sampler, as is the case with the IOM inhalable sampler, or may be mounted directly into the sampling head. Selection of a suitable collection substrate is governed by the sampling equipment used and by the subsequent sample analysis. Low-power lightweight pumps require filters with relatively low pressure drops at the operating flow rate. Gravimetric analysis requires a high degree of weight stability in changing environmental conditions. Chemical analysis requires that the collected material can be released from the substrate and/or that background levels of the analyte are low. Sample analysis by microscopy requires deposited particles to lie on the surface of the substrate. Chapter 9 gives further details of filter properties and selection. Table 25–2 summarizes the properties of filters, collection substrates, and filter holders commonly used within the workplace.

The accuracy of gravimetric samples may be affected by water adsorption onto substrates and filter holders and by losses or gains in material during transit (see Chapter 7) (van Tongeren et al., 1994; Awan and Burgess, 1996). In particular, cellulose ester membrane filters, polyurethane foams, and conducting plastic filter cassettes are particularly prone to weight changes following water uptake (Vaughan et al., 1989; Smith et al., 1998). To combat bias from such sources, it is common practice to weigh a number of control, or blank, filters with each

TABLE 25–2. Filter Selection for Industrial Hygiene Aerosol Sampling[a]

Substrate or Cassette	Typical Application	Weight Stability	Pressure drop
Cellulose fiber filter	General collection Gravimetric analysis	**	**
Cellulose nitrate filter	General collection Gravimetric analysis		
Glass fiber filter	General collection Gravimetric analysis	***	**
Quartz fiber filter	Chemical analysis	**	**
Cellulose ester membrane filter	Imaging, fiber sampling	***	***
PVC membrane filter	Chemical analysis	*	*
Teflon membrane filter	Gravimetric analysis Chemical analysis	**	*
Polycarbonate filter	Particle imaging	**	****
Silver membrane filter	Chemical analysis	**	****
Polyurethane foam	Various samplers	****	*_****
Mylar impaction substrate	Impaction substrate	**	N/A
Aluminum foil impaction substrate	Impaction substrate	**	N/A
Conducting plastic cassette	IOM inhalable sampler, conical inhalable sampler	****	N/A
Aluminum cassette	IOM inhalable sampler	*	N/A
Stainless steel cassette	IOM inhalable sampler	*	N/A

[a] A higher star rating indicates better weight stability or lower pressure drop.

set of sample filters (typically one blank per 10 samples, with a minimum of three blanks). It is advisable to condition filters in the weighing area (preferably in a temperature- and humidity-controlled environment) for up to 24 h before weighing to allow them to reach an equilibrium weight. It is generally not advisable to desiccate the filters before weighing, as weight changes after removal of the filter can be sufficiently rapid to lead to significant weight change during weighing (Smith et al., 1998). Where possible, blank filters should be transported with the sample substrates and exposed to the same conditions in order to minimize bias resulting from handling, transport, and changes in environment.

Other sources of bias include electrostatic attraction, where substrates are highly charged, and buoyancy effects. Electrostatic charge build-up may be significant for substrate materials such as PVC and PTFE, particularly when working at low relative humidity. In all instances, samples should be electrically neutralized using a source of bipolar ions. A common approach is to place samples close to a radioactive antistatic source before weighing. Buoyancy corrections only become necessary when the volume of the sample exceeds around 10^{-7} m^3 [0.1 cm^3]. For most substrates this is not a problem, although it may be significant when using large integral filter holders or substrate supports.

Pump Selection

Present-day personal sampling devices usually rely on either diaphragm or piston-type pumps to draw air through them. The pump is connected to a direct current (dc) motor, supplied by a battery pack of rechargeable nickel-cadmium cells. The achievable flow rates of pumps vary among manufacturers, but most will provide flows of 1.67×10^{-5} to 5×10^{-5} m^3/s [1 to 3 L/min] against a pressure drop of 6.25 kPA [25 inches of H_2O] for periods of up to 8 h. Personal pumps are available that will achieve flow rates of up to 1.67×10^{-4} m^3/s [10 L/min], but with current technology there is always a trade-off between sampling flow rate, sampling time, sustainable pressure drop, and pump weight. Most currently available pumps regulate the selected flow to minimize the impact of changes in temperature, pressure, and filter loading on the flow rate and the total volume of air sampled. Regulation is achieved in a number of ways, including using feedback from pressure drop across the filter, atmospheric temperature and pressure, pump rotational rate, and power usage. As the performance of some size-selective samplers is adversely affected by pulsations in the sampling flow (e.g., Bartley et al., 1984), some pumps incorporate flow dampers. Wood (1977) presents a useful review of personal sampling pumps, carried out in 1977, and, apart from limited advances in control technology, it still reflects much of the hardware available today.

The volumetric flow rate of sampling pumps needs to be set with the sampling device attached (including filter) and under the same conditions of temperature and humidity as sampling will be carried out under. Although many pumps incorporate a visual indication of flow rate such as a rotameter, this should be used for indication purposes only and the sampling flow measured and set using a primary standard such as a bubble flowmeter. Typically, the set flow rate is expected to be within 5% of the target flow rate, although the most recent guidelines on sampling in the United Kingdom specify flows to be set to $\pm 1.67 \times 10^{-6}$ m^3/s [±0.1 L/min] in all cases (HSE, 1997).

Sampling Strategy

While "static" or "area" sampling with fixed point samplers is still used in many situations, it is now widely accepted that representative aerosol sampling in the workplace should be carried out in the breathing zone—frequently defined as a region of the body not more than 0.3 m from the mouth and nose (Vincent, 1995). Breathing zone measurements generally give a better representation of worker exposure. However, Vincent (1995) notes that placement of sampling devices in this region does not guarantee representative sampling, and large

EXAMPLE 25–1: CALCULATION OF AN 8H TWA EXPOSURE

Three consecutive air samples for lead are collected at $3.3 \times 10^{-5}\,m^3/s$ [2 L/min] onto filters in the breathing zone of a worker in a brass foundry, with the results shown in Table 25–3.

The shift started at 08:00 and finished at 18:00. Breaks were taken between 09:30 and 10:00, 12:00 and 12:30, and 15:00 and 15:30. The work pattern was split into different tasks in the morning and the afternoon. Using Eq. 25–4, calculate the 8 h TWA exposure level over the total duration of the shift (600 min).

The assumption is made that during breaks exposure is zero. During the afternoon period, when no sampling was carried out, it is assumed that exposure is similar to that measured by sample 3. Table 25–4 therefore gives a complete account of the day's exposure.

The 8 h TWA mass concentration is therefore given as

$$c_m = \left[\frac{(111\times 90)+(0\times 30)+(104\times 120)+(0\times 30)+(17\times 150)+(0\times 30)+(17\times 150)}{8\times 60}\right]$$
$$= 57\,\mu g/m^3$$

using Eq. 25–4.

TABLE 25–3. Example Gravimetric Sample Data for a Worker in a Brass Factory

Sample No.	Time On	Time Off	Flow rate ($10^{-5}\,m^3/s$ [L/min])	Sample Duration (min)	Sample Volume (L)	Mass Collected (μg)	Mass Concentration (μg/m³)
1	08:00	09:30	3.3 [2]	90		20	
2	10:00	12:00	3.3 [2]	120		25	
3	12:30	15:00	3.3 [2]	150		5	

TABLE 25–4. Complete Account of a Worker's Exposure to Lead in a Brass Factory (from Table 25–3)

Sample No.	Time On	Time Off	Flow rate ($10^{-5}\,m^3/s$ [L/min])	Sample Duration (min)	Sample Volume (L)	Mass Collected (μg)	Mass Concentration (μg/m³)
1	08:00	09:30	3.3 [2]	90	180	20	111
1a (break)	09:30	10:00		30	—	0	0
2	10:00	12:00	3.3 [2]	120	240	25	104
2a (break)	12:00	12:30		30	—	0	0
3	12:30	15:00	3.3 [2]	150	300	5	17
3a (break)	15:00	15:30		30	—	0	0
4 (Est. from #3)	15:30	18:00		150	—	5 (est.)	17
Total				600	75		

variations in sampled aerosol concentration can be seen across the front of the body, depending on worker orientation, placement of the aerosol source, and local air movements (Raynor et al., 1975).

As a matter of convention, exposure measurements for chronic hazards are usually taken for the duration of a single work shift. An 8h TWA mass concentration (c_m) relates to the process whereby exposure occurring within a 24h period is treated as being equivalent to a single uniform exposure over 8h. A TWA mass concentration can be determined from a single full-shift sample, or it can be calculated from a series of consecutive samples (Leidel et al., 1977). Where sampling gaps occur over a shift, exposures during these periods should be estimated from adjacent measurements or from additional information (see Example 25–1). The TWA for a given time period (e.g., 8h, or 15min for a STEL) is calculated by

$$c_m = \frac{\sum_{i=1}^{n} c_{mi} \times t_i}{T}, \quad \sum_{i=1}^{n} t_i = \text{full shift duration} \tag{25–4}$$

where T is the given reference period (in minutes), t_i is the duration of sample i in minutes, and c_{mi} is the mass concentration of sample i.

For purposes of determination of compliance with occupational exposure limits, it is generally desirable to sample the workers assumed to be at maximum risk. When the maximum-risk employees cannot be ascertained, employees should be selected at random. Leidel et al. (1977) recommend calculating the 95% one-sided lower confidence limit (LCL) and the 95% one-sided upper confidence limit (UCL). These are calculated as follows:

$$\text{LCL}(95\%) = \chi - t_\alpha \times CV_T$$
$$\text{LCL}(95\%) = \chi + t_\alpha \times CV_T$$
$$\chi = \frac{c_m}{\text{OEL}} \tag{25–5}$$

where $t_\alpha = 1.645$ when $\alpha = 0.95$, CV_T is the coefficient of variation for the sampling/analytical method, and OEL is the exposure limit. If LCL and χ are above unity, then the exposure is classified as noncompliant. If UCL and χ are below unity, then the exposure is classified as compliant. Finally, if unity lies between LCL and χ, or between UCL and χ, the exposure is classified as possible overexposure.

MEASUREMENT OF SIZE DISTRIBUTION

Full characterization of the size distribution of an aerosol may be carried out during non-routine investigations using a range of available methods described in previous chapters (see Chapters 10, 13, and 15–19). Although many instrument types have been used in the workplace (Mark et al., 1984), cascade impactors (see Chapter 10) are often the instrument of choice, giving an indication of the mass-weighted size distribution of an aerosol. Impactors are generally capable of giving the size distribution of an aerosol between around 0.1 and 15µm aerodynamic diameter and above. Static cascade impactors such as the Andersen eight-stage impactor (***AND***) and the Micro Orifice Uniform Deposit Impactor (MOUDI) (***MSP***) have found relatively widespread use in the workplace. The Andersen consists of eight multiorifice stages with cut points between 10 and 0.4µm when operated at $4.72 \times 10^{-4}\,m^3/s$ [28.3L/min]. Collection is usually onto aluminum foils, although other substrates are available. The use of multiorifices in the Andersen impactor allows deposits to be distributed with relative evenness onto substrates. This is taken further within the MOUDI, where many orifices per stage, together with rotating substrates, lead to highly uniform deposits. The MOUDI is available in an 8- or 10-stage version and is capable of making aerosol size distribution measurements down to 0.056 μm at $5 \times 10^{-4}\,m^3/s$ [30 L/min].

Aerosol size distributions within the breathing zone are generally of greater relevance to health than static samples, and two cascade impactors have been developed to enable personal aerosol size distribution measurements to be made. The Marple personal cascade impactor (***AND***) (Rubow et al., 1987) is configurable with up to eight stages and will provide information on particle size distribution down to 0.5 μm at a flow rate of $3.33 \times 10^{-5}\,m^3/s$ [2 L/min]. The Personal Inhalable Dust Spectrometer (PIDS) is similar in concept to the Marple impactor, although the slot-shaped impactor jets of the Marple device are replaced by circular jets (Gibson et al., 1987). Cut points in the eight stages of the PIDS range from 0.9 to 19 μm at $3.33 \times 10^{-5}\,m^3/s$ [2 L/min].

Cascade impactors are of limited use for measuring aerosol size distributions up to the limit of the inhalable convention (100 μm aerodynamic diameter) due to the relatively low cut point of the upper stage in most cases. Extrapolation of measured size distributions above this cut point depends on assumptions about the sampled aerosol and the aspiration efficiency of the device and is generally not reliable. However, the PIDS was designed with an inlet designed to follow the inhalable convention (Gibson et al., 1987). It may be assumed that summing all deposits within the PIDS impactor gives a measure of the inhalable aerosol mass, and subsequent analysis of the deposits gives the size distribution as a function of the inhalable aerosol. Such an approach is advantageous to industrial hygiene measurements, where ultimately measurements need to be related to the mass of particles inhaled.

When the specific health-related fractions of the aerosol are of more concern than a detailed analysis of particle size distribution, a number of samplers allow simultaneous measurement of all three fractions. The IOM personal multifraction sampler uses aerosol separation within polyurethane foams to achieve this (Vincent et al., 1993). Aerosol is sampled through a 15 mm diameter inhalable inlet at $3.33 \times 10^{-5}\,m^3/s$ [2 L/min]. Two polyurethane foam selectors of different grades placed in series then separate the thoracic and respirable subfractions. The sampler enables the inhalable fraction to be measured by weighing deposits in both foams and the backup filter. The combined deposits on the filter and adjacent foam give the thoracic fraction, and the filter alone gives the respirable aerosol fraction. A similar approach using polyurethane foams has been developed for use with the conventional IOM inhalable sampling head (Kenny et al., 1999b). An alternative approach is used by the Personal Spectrometer (PERSPEC) (Prodi et al., 1988, 1989). The inhalable aerosol fraction is introduced to a highly divergent flow of clean air and deposited onto a 47 mm filter. Deposition position depends on particle size; thus by weighing the complete filter the inhalable fraction can be determined, or by weighing specific areas of the filter (after cutting them out) different subfractions can be measured (Kenny and Bradley, 1994). The Respicon sampler (***TSI***) achieves separation of the three aerosol size fractions using a series of virtual impactors. A modified version has been developed (Respicon TM, ***HUN***) that allows real-time monitoring of each fraction using light scattering (Koch et al., 1998).

USE OF DIRECT-READING INSTRUMENTS

Instruments giving a near-instantaneous, or rapid, measure of aerosol properties (commonly referred to as *real-time measurement instruments*) are widely used in the workplace. Vincent (1995) and Walton and Vincent (1998) provide a broad summary of techniques commonly used in the workplace. However, it is possible to find examples of most devices described in earlier chapters being applied in the workplace.

For routine measurements, aerosol photometers are widely used and available from an increasing number of manufacturers. Their use covers checking short-term, task-specific, or instantaneous exposure levels and identifying exposure hot spots. Systems have also been developed that combine photometer measurements with simultaneous video filming of workers, allowing direct comparison between work tasks and exposure levels (Gray et al., 1992; Gressel et al., 1993; Heitbrink et al., 1993; Unwin et al., 1993). The implementation of the measurement method has various guises, from passive instruments relying on convection to bring particles into the sensing zone (as with the Mini-RAM, and the later personal data-RAM, ***MIE***), to pumped devices such as the Microdust Pro (***CAS***), to instruments incorporating data loggers (e.g., the DustTrack [***TSI***] and Data-RAM [***MIE***]). Most devices are compact, with most being portable and a number of them being suitable for personal sampling.

Over a relatively narrow size range (approximating to the upper end of respirable size fraction) the light scattered from an aerosol is roughly proportional to the scattering volume (see Chapter 15; Baron, 1994). Thus, after correcting for density, scattered light may be used as an indirect measure of mass concentration. The method is relatively good for measuring respirable aerosol concentration, but becomes tenuous when used for the thoracic subfraction and potentially misleading when used to measure the inhalable aerosol mass concentration (the sensitivity to equivalent aerosol masses represented by 20 μm particles is approximately a factor of 10^2 lower than the sensitivity to 2 μm particles). Instruments such as the Respicon TM (***HUN***) go some way to overcoming this size dependence of photometry by selectively concentrating larger particles through the use of virtual impaction (Koch et al., 1998). In some situations it is feasible to calibrate a photometer to the inhalable mass concentration, but only when the fine particles detected form a constant fraction of the inhalable aerosol. Optical single-particle detection and sizing instruments such as the Grimm "Work-check" (***GRI***) overcome some of the limitations of photometers, but their sensitivity is still restricted to a similar range of particle sizes.

In all cases it is advisable to calibrate photometers before using them with different aerosols, as particle size distribution, shape, and refractive index will affect measurements (see Chapter 15). Calibration is usually performed by carrying out parallel gravimetric sampling and applying an adjustment factor to the photometer to ensure that results agree. Many photometers have the facility to collect aerosol passing through the sensing zone on a filter, thus simplifying calibration. Zero offset checks are also recommended before use by placing the photometer in a clean environment. Deposits on the optics and surfaces of the sensing zone can lead to a change in the instrument calibration.

Recent developments in condensation particle counter (CPC) technology have led to a commercially available portable device with logging capabilities, suitable for semiquantitative particle number measurements. The P-Trak (***TSI***) is designed to provide near-instantaneous measurements of particle concentration between 20 nm and approximately 1 μm. Although it is primarily aimed at investigating aerosol number concentration levels and variations and tracking contamination sources in indoor environments, it is also being applied to measuring real-time particle number concentration measurements in the workplace.

Respirator Fit Testing

Both photometers and CPCs are used extensively for measuring the fit factor of respirators. The fit of a respirator can either be measured using a qualitative fit test (QLFT) or quantitative fit test (QNFT) (OSHA, 1998). Both approaches rely on the respirator filter removing the majority of a test agent, leaving gaps in the respirator–face seal as the main route for the agent to penetrate the mask. QLFT methods generally rely on the wearer's perception of a test agent through odor or taste, the most common agents being isoamyl acetate, sodium saccharin, and irritant fume. QNFT methods, on the other hand, use quantitative measurement of the leakage rate around the mask. The first QNFT methods suitable for routine use exposed the wearer to a controlled atmosphere of dioctyl phthalate (DOP) aerosol and measured the aerosol inside and outside the mask using forward light-scattering photometry (Burgess et al., 1961; Hyatt et al., 1972). HEPA filters were used on the mask to ensure that most of the particles detected within the mask penetrated due to leakage. This method is still used, although alternative aerosol materials such as corn oil and sodium chloride have replaced the use of DOP.

The use of a specific aerosol in an enclosed system is somewhat restrictive, and in 1981 Willeke et al., investigated easily generated aerosols, including cigarette smoke, fine carbon particles from high-volume sampling pumps, and fine metal/metal oxide particles generated from the filament of a hair dryer (Willeke et al., 1981). However, the most significant aerosol investigated was the ambient aerosol found in the workplace. Using a CPC to compare the ambient particle number concentration to that inside a respirator being worn, they were able to show that this formed a basis for rapid quantitative fit tests. The ambient air–CPC method is now widely applied, using the PortaCount Respirator Fit Tester (***TSI***). With HEPA filters, penetration is lower than 0.03% for 0.3 μm particles (representing the particle size region of highest penetration). Thus, given a sufficiently high challenge aerosol concentration, fit factors of over 3000 are measurable. Ambient concentrations of submicrometer particles are rarely lower than 10^9 to 10^{10} particles/m^2 [10^3 to 10^4 particles/cm^3] unless the air is highly filtered, allowing fit factors greater than 1000 to be measured under most circumstances. A similar approach has been proposed for measuring leakage around filters in filter cassettes used for workplace sampling (Baron et al., 2001).

Alternative QNFT methods have been proposed and are currently used, including dynamic pressure measurements inside the respirator, large-particle penetration tests, and small-particle penetration tests. A comprehensive review of current and proposed methods may be found in Han et al. (1997).

FUTURE TRENDS

Perhaps the most significant change in industrial hygiene aerosol measurement over the past two decades has been the development and gradual adoption of size-selective sampling conventions. These now enable measurements that have greater biological relevance to be made in the workplace. Although the next few years are likely to see the current position being consolidated, there is scope for the present sampling conventions to be revised. The inhalable convention is limited in its scope and applicability. Its abrupt termination at 100 μm brings into question whether the ingression of large particles and projectiles into more open samplers is acceptable or leads to inaccurate measures of aerosol concentration (Aitken and Donaldson, 1996). In addition, the inability to develop inhalable samplers thus far that follow the convention over a wide range of wind speeds raises the question of whether the standard is unattainable or inappropriate.

At the other end of the size spectrum, toxicological information on responses to nanometer-sized low-solubility particles are challenging the applicability of current sampling conventions and philosophies. Recent toxicology on low-toxicity insoluble materials such as titanium dioxide has indicated that a more appropriate dose metric for depositing in the alveolar region may be particle number or surface area (Oberdörster et al., 1994; Donaldson et al., 1998). These studies appear to support some epidemiological investigations of the general population, indicating correlation between inhalation of fine particles and health effects (Dockery et al., 1993). The extent to which such findings are applicable to exposure within the workplace is not apparent at present. However, to begin to understand the relevance of exposure to very fine particles and the appropriate metric to use for dose, developments in the measurement of exposure in terms of particle number and surface area are necessary. As the debate on the appropriate dose and exposure metric develops, there will no doubt be further extensions to the manner in which aerosol exposure in the workplace is measured in the subrespirable size range.

Current trends in workplace sampling practice indicate a desire within the industrial hygiene community to adopt methods that provide measurements with greater rapidity and with less effort. In particular there is a growing interest in relating exposure to specific tasks and operations, thus requiring highly sensitive or rapid-response aerosol measurement methods. Such trends are perhaps most obvious in the increased use (and abuse) of direct-reading photometer-based instruments. These provide a rapid indication of exposure, both allowing a more rapid response to problem situations than filter collection and analysis allows, together with a means of avoiding the expense of sample collection and analysis where it is not necessary. However, their attraction has seen their increasing but erroneous use to estimate exposure to inhalable aerosol. This clear need for direct-reading instruments that extend to the inhalable fraction is likely to lead to the development of new devices. Although it is not at all clear at present whether workable technologies will present themselves, there are a number of possible contenders: The application of optical methods may be further increased to large particle sizes through the size-selective concentration of particles and the detection of individual large particles. The development of handheld oscillating microbalance methods is underway, although it is unclear whether there will be collection and sensing problems associated with particles approaching 100 μm in diameter. Aerosol mass sensing methods based on filter pressure drop are being developed for fine particles (Dobroski et al., 1995; Sioutas et al., 1999; Volkwein et al., 1998). At present, the indications are that size-dependent effects will hinder their extension to 100 μm, although size-selective concentration and aerosol-specific calibration may lead to successful applications.

Passive samplers provide another route to simplified exposure measurement and have been under development for some time. They offer the simplicity of a discrete lightweight badge-type sampler with no need for a sampling pump. The passive sampler developed by Brown et al., relies on electrostatic deposition onto a pre-prepared electret material. Aerosol is carried to the deposition zone through convection; thus no pump is required (Brown et al., 1994). Although the device is not designed along size-selective lines, good correlation has been seen with size-selective samplers in some cases, and it is likely that such samplers could be developed into indicative screening devices (Brown et al., 1995).

A different approach to making size-selective sampling more accessible and less expensive is seen in the increasing utilization of porous foam pre-separators. The use of foam allows relatively inexpensive size-separation devices to be constructed and provides the possibility of modifying existing samplers to different applications. Chen et al. (1998) have proposed a porous foam respirable sampler based on the cowled sampler used for asbestos sampling, and foam plugs to convert the IOM inhalable sampler to either a respirable or a thoracic sampler have been investigated (Kenny et al., 1999b). At the same time, new methods of creating size-selective samplers that are better suited to occupational aerosol sampling (e.g., by operating at higher flow rates, giving better agreement with the sampling conventions, providing a more compact, lighter sampler, and operating at lower pressure drops) are under constant development. Recent work includes the investigation of virtual, axial-flow, and tangential-flow cyclones (Chen et al., 1999; Maynard, 2000; Kenny and Gussman, 1997), the development of centrifugal personal samplers (John and Kreisberg, 1999), and the development of inhalable samplers with inlet screens (Kalatoor et al., 1995; Hauck et al., 1997; Aizenberg et al., 2000), reducing the adverse effects of wind speed and large-particle projectiles on samples.

The desire for simplification also extends to standards against which exposure is measured. While the current emphasis is on monitoring worker exposure, the concept of controlling emissions at the source is gaining ground. The application of such thinking to the workplace will possibly result in the use of exposure modeling to estimate exposure risk, accounting for materials used, generation processes involved, emission control measures applied, and dispersion in the workplace (Maidment, 1998). The logical end point is the estimation of exposures from materials and processes within the workplace and the relegation of aerosol exposure measurement to a supportive role. However, sufficient questions surround the classification of materials in terms of their ability to form an aerosol during specific processes, together with the containment or release and transport of generated aerosols, to ensure that developments in aerosol measurement methods in the workplace will continue for a number of years to come.

REFERENCES

ACGIH. 1968. *Threshold Limit Values of Airborne Contaminants.* Cincinnati, OH: American Conference of Government Industrial Hygienists.

ACGIH. 1995. *Air Sampling Instruments*, 8th Ed. Cincinnati: ACGIH.

ACGIH. 1998. *Particle Size-Selective Sampling for Particulate Air Contaminants.* Cincinnati, OH: American Conderence of Government Industrial Hygienists.

Aitken, R. J. and R. Donaldson. 1996. *Large Particle and Wall Deposition Effects in Inhalable Samplers.* Report Number 117/1996, ISBN 0 7176 1270 8. Health and Safety Executive, UK.

Aizenberg, V., S. A. Grinshpun, K. Willeke, J. Smith, and P. A. Baron. 2000. Performance characteristics of the button personal inhalable sampler. *AIHAJ* 61:398–404.

Awan, S. and G. Burgess. 1996. The effect of storage, handling and transport traumas on filter-mounted dusts. *Ann. Occup. Hyg.* 40:525–530.

Baron, P. A. 1994. Direct-reading instruments for aerosols. A review. *Analyst* 119:35–40.

Baron, P. A. and G. J. Deye. 1990. Electrostatic effects in asbestos sampling I: Experimental measurements. *Am. Ind. Hyg. Assoc. J.* 51:51–62.

Baron, P. A., A. Khanina, A. B. Martinez, and S. A. Grinshpun. 2001. Investigation of filter bypass leakage and a test for aerosol sampling cassettes. *Aerosol Sci. Technol.* In submission.

Bartley, D. L. and G. M. Breuer. 1982. Analysis and optimisation of the performance of the 10 mm cyclone. *Am. Ind. Hyg. Assoc. J.* 43:520–528.

Bartley, D. L., G. M. Breuer, and P. A. Baron. 1984. Pump fluctuations and their effect on cyclone performance. *Am. Ind. Hyg. Assoc. J.* 45:10–18.

Bartley, D. L., C. C. Chen, R. Song, and T. J. Fischbach. 1994. Respirable aerosol sampler performance testing. *Am. Ind. Hyg. Assoc. J.* 55:1036–1046.

Bedford, T. and C. Warner. 1943. Physical studies of the dust hazard and thermal environment in certain coalmines. In *Chronic pulmonary disease in South Wales coalminers. II. Environmental studies.* London: British Medical Research Council, HMSO. Special report series no. 244, pp. 1–78.

Blake, T., V. Castranova, D. Schwegler-Berry, P. Baron, G. J. Deye, C. H. Li, and W. Jones. 1998. Effect of fiber length on glass microfiber cytotoxicity. *J. Toxicol. Environ. Health Part A* 54:243–259.

BMRC. 1952. *Recommendations of the BMRC Panels Relating to Selective Sampling.* From the minutes of a joint meeting of Panels 1, 2 and 3, held on March 4th, British Medical Research Council.

Brown, R. C., M. A. Hemingway, D. Wake, and J. Thompson. 1995. Field trials of an electret-based passive dust sampler in metal-processing industries. *Ann. Occup. Hyg.* 39:603–622.

Brown, R. C., D. Wake, A. Thorpe, M. A. Hemingway, and M. W. Roff. 1994. Preliminary assessment of a device for passive sampling of airborne particulate. *Ann. Occup. Hyg.* 38:303.

Burgess, W. A., L. Silverman, and F. Stein. 1961. A new technique for evaluating respirator performance. *Am. Ind. Hyg. Assoc. J.* 22:422–429.

CEN. 1998. *Workplace Atmospheres: Assessment of Performance of Instruments for Measurement of Airborne Particle Concentrations.* Comité Européen de Normalisation, CEN prEN 13205.

Chen, C.-C., S.-H. Huang, W. Lin, T. Shih, and F. Jeng. 1999. The virtual cyclone as a personal respirable sampler. *Aerosol Sci. Technol.* 31:422–432.

Chen, C.-C., C.-Y. Lai, T.-S. Shih, and W.-Y. Yeh. 1998. Development of respirable aerosol samplers using porous foam. *Am. Ind. Hyg. Assoc. J.* 59:766–773.

Courbon, P., R. Wrober, and J.-F. Fabriès. 1988. A new individual respirable dust sampler: The CIP-10. *Ann. Occup. Hyg.* 32:129–143.

Dobroski, H., D. P. Tuchman, and R. P. Vinson. 1995. *Differential pressure as a means of estimating respirable dust mass on collection filters*. Presented at American Industrial Hygiene Association Conference and Exposition.

Dockery, D. W., C. A. Pope, X. Xu, J. D. Spengler, J. H. Ware, M. E. Fay, B. G. Ferris, and F. E. Speizer. 1993. An association between air pollution and mortality in six U.S. cities. *N. Engl. J. Med.* 329:1753–1759.

Donaldson, K., X. Y. Li, and W. MacNee. 1998. Ultrafine (nanometre) particle mediated lung injury. *J. Aerosol Sci.* 30:553–560.

Dunmore, J. H., R. J. Hamilton, and D. S. G. Smith. 1964. An instrument for the sampling of respirable dust for subsequent gravimetric assessment. *J. Sci. Instrum.* 41:669–672.

Fabriès, J. F., P. Görner, E. Kauffer, R. Wrobel, and J. C. Vigneron. 1998. Personal thoracic CIP10-T sampler and its static version CATHIA-T. *Ann. Occup. Hyg.* 42:453–465.

Gibson, H., J. H. Vincent, and D. Mark. 1987. A personal inspirable aerosol spectrometer for applications in occupational hygiene research. *Ann. Occup. Hyg.* 31(4A):463–479.

Gray, M. I., J. Unwin, P. T. Walsh, and N. Worsell. 1992. Factors influencing personal exposure to gas and dust in workplace air—Application of a visualization technique. *Safety Sci.* 15:273–282.

Green, H. L. and H. H. Watson. 1935. *Physical methods for the estimation of the dust hazard in industry.* London: Medical Research Council Special Report Series 119.

Greenburg, L. and G. W. Smith. 1922. *A New Instrument for Sampling Aerial Dust.* U.S Bureau of Mines, No. 2392.

Gressel, M. G., W. A. Heitbrink, and P. A. Jensen. 1993. Video exposure monitoring—A means of studying sources of occupational air contaminant exposure, Part 1: Video exposure monitoring techniques. *Appl. Occup. Environ. Hyg.* 8:334–338.

Hamilton, R. J. 1956. A portable instrument for respirable dust sampling. *J. Sci. Instrum.* 33:395–399.

Han, D.-H., K. Willeke, and C. E. Colton. 1997. Quantitative fit testing techniques and regulations for tight-fitting respirators: Current methods measuring aerosol or air leakage, and new developments. *Am. Ind. Hyg. Assoc. J.* 58:219–228.

Hauck, B. C., S. A. Grinshpun, A. Reponen, T. Reponen, K. Willeke, and R. L. Bornschein. 1997. Field testing of new aerosol sampling method with a porous curved surface as inlet. *Am. Ind. Hyg. Assoc. J.* 58:713–719.

Heitbrink, W. A., M. G. Gressel, T. C. Cooper, T. Fischbach, D. M. O'Brien, and P. A. Jensen. 1993. Video exposure monitoring—A means of studying sources of occupational air contaminant exposure, Part 2: Data interpretation. *Appl. Occup. Environ, Hyg.* 8:339–343.

Hewson, G. S. 1996. Estimates of silica exposure among metaliferous miners in Southern Australia. *Appl. Occup. Environ. Hyg.* 11:868–877.

HSE. 1997. *General Methods for the Gravimetric Determination of Respirable and Total Inhalable Dust.* London: HSE Books.

HSE. 1999. *EH40/99 Occupational Exposure Limits 1999*. London: HSE Books.

Hyatt, E. C., J. A. Pritchard, and C. P. Richards. 1972. Respirator efficiency measurement using quantitative DOP man tests. *Ann. Ind. Hyg. Assoc. J.* 33:635–643.

ISO. 1995. *Air Quality—Particle Size Fraction Definitions for Health-Related Sampling.* Geneva: International Standards Organisation, ISO Standard 7708.

John, W. and N. Kreisberg. 1999. Calibration and testing of samplers with dry, polydisperse latex. *Aerosol Sci. Technol.* 31:221–225.

Kalatoor, S., S. A. Grinshpun, K. Willeke, and P. A. Baron. 1995. New aerosol sampler with low wind sensitivity and good filter collection uniformity. *Atmos. Environ.* 29:1105–1112.

Kenny, L. C., R. J. Aitken, P. E. J. Baldwin, G. Beaumont, and A. D. Maynard. 1999a. The sampling efficiency of personal inhalable samplers in low air movement environments. *J. Aerosol Sci.* 30:627–638.

Kenny, L. C., R. Aitken, C. Chalmers, J. F. Fabriès, E. Gonzalez-Fernandez, H. Kromhout, G. Lidén, D. Mark, G. Riediger, and V. Prodi. 1997. A collaborative European study of personal inhalable aerosol sampler performance. *Ann. Occup. Hyg.* 41:135–153.

Kenny, L. C., A. Bowry, B. Crook, and J. D. Stancliffe. 1999b. Field testing of a personal size-selective bioaerosol sampler. *Ann. Occup. Hyg.* 43:393–404.

Kenny, L. C. and D. R. Bradley. 1994. Optimization of the Perspec multifraction aerosol sampler to new sampling conventions. *Ann. Occup. Hyg.* 38:23–35.

Kenny, L. C. and R. A. Gussman. 1997. Characterization and modelling of a family of cyclone aerosol preseparators. *J. Aerosol Sci.* 28:677–688.

Koch, W., W. Dunkhorst, and H. Lodding. 1998. RESPICON TM-3 F: A new personal measuring system for size segregated dust measurement at workplaces. *Occup. Health Ind. Med.* 38:161.

Leidel, N. A., K. A. Busch, and J. R. Lynch. 1977. *Occupational Exposure Sampling Strategy Manual.* DHEW (NIOSH) Publication No. 77–173. National Institute for Occupational Safety and Health.

Le Roux, W. L. 1970. *Recorded Dust Conditions and Possible New Sampling Strategies in South African Gold Mines.* Johannesburg International Pneumoconiosis Conference. New York: Oxford University Press.

Lidén, G. 1993. Evaluation of the SKC Personal Respirable Dust Sampling Cyclone. *Appl. Occup. Environ. Hyg.* 8:178–190.

Lidén, G. and L. C. Kenny. 1992. The performance of respirable dust samplers—Sampler bias, precision and inaccuracy. *Ann. Occup. Hyg.* 36:1–22.

Lidén, G. and L. C. Kenny. 1993. Optimisation of the performance of existing respirable dust samplers. *Appl. Occup. Environ. Hyg.* 8:386–391.

Lippmann, M. 1977. Regional deposition of particles in the human respiratory tract, eds. D. H. K. Lee and S. Murphy. In *Handbook of Physiology; Section IV, Environmental Physiology*. Philadelphia: Williams & Wilkins, pp. 213–232.

Lison, D., C. Lardot, F. Huaux, G. Zanetti, and B. Fubini. 1997. Influence of particle surface area on the toxicity of insoluble manganese dioxide dusts. *Arch. Toxicol.* 71:725–729.

Maidment, S. C. 1998. Occupational hygiene considerations in the development of a structured approach to select chemical control strategies. *Ann. Occup. Hyg.* 42:391–400.

Mark, D. and J. H. Vincent. 1986. A new personal sampler for airborne total dust in workplaces. *Ann. Occup. Hyg.* 30:89–102.

Mark, D., J. H. Vincent, H. Gibson, R. J. Aitken, and G. Lynch. 1984. The development of an inhalable dust spectrometer. *Ann. Occup. Hyg.* 28:125–143.

Mark, D., J. H. Vincent, H. Gibson, and G. Lynch. 1985. A new static sampler for airborne total dust in workplaces. *Am. Ind. Hyg. Assoc. J.* 46:27–133.

Mark, D., J. H. Vincent, D. C. Stevens, and M. Marshall. 1986. Investigation of the entry characteristics of dust samplers of the type used in the British nuclear industry. *Atmos. Environ.* 20:2389–2396.

Maynard, A. D. 1999. Measurement of aerosol penetration through six personal thoracic samplers under calm air conditions. *J. Aerosol Sci.* 30:1227–1242.

Maynard, A. D. 2000. A simple model of axial flow cyclone performance under laminar flow conditions. *J. Aerosol Sci.* 31:151–167.

Maynard, A. D. and L. C. Kenny. 1995. Performance assessment of three personal cyclone models, using an aerodynamic particle sizer. *J. Aerosol Sci.* 26:671–684.

Morrow, P. E. 1994. Mechanisms and significance of "particle overload," eds. D. L. Dungworth, J. L. Mauderly, and G. Oberdorster. In *Toxic and Carcinogenic Effects of Solid Particles in the Respiratory Tract.* Washington, DC: ILSI Press, pp. 17–25.

Oberdörster, G., J. Ferin, and B. E. Lehnert. 1994. Correlation between particle-size, in-vivo particle persistence, and lung injury. *Environ. Health Perspect.* 102(S5):173–179.

OSHA. 1998. *Respiratory Protection Standard 29CFR1910.134.* Washington DC.: Occupational Safety and Health Administration.

Prodi, V., F. Belosi, A. Mularoni, and P. Lucialli. 1988. Perspec—A personal sampler with size characterization capabilities. *Am. Ind. Hyg. Assoc. J.* 49:75–80.

Prodi, V., C. Sala, F. Belosi, S. Agostini, G. Bettazzi, and A. Biliotti. 1989. Perspec, personal size separating sampler—Operational experience and comparison with other field devices. *J. Aerosol Sci.* 20: 1565–1568.

Puskar, M. A., J. M. Harkins, J. D. Moomey, and L. H. Hecker. 1991. Internal wall losses of pharmaceutical dusts during closed-face, 37-mm polystyrene cassette sampling. *Am. Ind. Hyg. Assoc. J.* 52:280–286.

Raynor, G. S., E. C. Ogden, and J. V. Hayes. 1975. Spatial variation in airborne pollen concentrations. *J. Allergy Clin. Immunol.* 55:195–202.

Robert, K. Q. 1979. Cotton dust sampling efficiency of the vertical elutriator. *Am. Ind. Hyg. Assoc. J.* 40:535–545.

Rubow, K. L., V. A. Marple, J. Olin, and M. A. McCawley. 1987. A personal cascade impactor: Design, evaluation and calibration. *Am. Ind. Hyg. Assoc. J.* 48:532–538.

Sioutas, C., S. Kim, and M. Chang. 1999. Development and evaluation of a prototype ultrafine particle concentrator. *J. Aerosol Sci.* 30:1001–1017.

Smith, J. P., D. L. Bartley, and E. R. Kennedy. 1998. Laboratory investigation of the mass stability of sampling cassettes from inhalable aerosol samplers. *Am. Ind. Hyg. Assoc. J.* 59:582–585.

Unwin, J., P. T. Walsh, and N. Worsell. 1993. Visualization of personal exposure to gases and dust using fast-response monitors and video filming. *Appl. Occup. Environ. Hyg.* 8:348–350.

van Tongeren, M. J. A., K. Gardiner, and I. A. Calvert. 1994. An assessment of the weight-loss in transit of filters loaded with carbon black. *Ann. Occup. Hyg.* 38:319–323.

Vaughan, N. P., B. D. Milligan, and T. L. Ogden. 1989. Filter weighing reproducibility and the gravimetric detection limit. *Ann. Occup. Hyg.* 33:331–337.

Vincent, J. H. 1995. *Aerosol Science for Industrial Hygienists.* Bath, UK: Pergamon.

Vincent, J. H. 1998. International occupational exposure standards: A review and commentary. *Am. Ind. Hyg. Assoc. J.* 59:729–742.

Vincent, J. H., R. A. Aitken, and D. Mark. 1993. Porous plastic foam filtration media: Penetration characteristics and applications in particle size-selective sampling. *J. Aerosol Sci.* 24:929.

Vincent, J. H., D. Mark, B. G. Miller, L. Armbruster, and T. L. Ogden. 1990. Aerosol inhalability at higher windspeeds. *J. Aerosol Sci.* 21:577–586.

Volkwein, J. C., A. L. Schoeneman, and S. J. Page. 1998. *Laboratory Evaluation of Pressure Differential Dust Detector Tube.* AIHA 1998 Conference. Washington, DC: AIHA.

Walton, H. W., and J. H. Vincent. 1998. Aerosol instrumentation in occupational hygiene: An historical perspective. *Aerosol Sci. Technol.* 28:417–438.

Willeke, K., H. E. Ayer, and J. D. Blanchard. 1981. New methods for quantitative respirator fit testing with aerosols. *Am. Ind. Hyg. Assoc. J.* 42:121–125.

Wood, J. D. 1977. A review of personal sampling pumps. *Ann. Occup. Hyg.* 20:3–17.

26 Mine Aerosol Measurement

BRUCE K. CANTRELL and JON C. VOLKWEIN
National Institute for Occupational Safety and Health, Pittsburgh Research Laboratory, Pittsburgh, PA

INTRODUCTION

Exposure to mineral aerosol is an occupational health hazard in mining and mineral processing industries because of the risk of developing pneumoconiosis. Agricola (1556) described this hazard for metal mining in Carpathia. He described shortness of breath and consumption, conditions that are now associated with asthma and emphysema. By the end of the nineteenth century, several respiratory diseases were known to affect miners, including silicosis and coal workers' pneumoconiosis (Fletcher, 1948; Seaton et al., 1981). For example, silicosis, not tuberculosis, was clearly recognized as the hazard in the Vermont granite quarries and stone-cutting sheds (McFarland, 1927). The Gauly Bridge Tunnel disaster in the mid-1930s caused the deaths of an estimated 700 workers by acute silicosis and focused the nation's attention on this disease (Cherniack, 1986). By the mid-twentieth century, it was clear that risk of simple pneumoconiosis is associated with a miner's cumulative exposure to mine aerosol in the respirable size range and that prevention lies in reducing that exposure through regulation (Seaton, 1986).

In the United States, regulation of mine worker exposure to airborne mine dust is the responsibility of the Mine Safety and Health Administration (MSHA) in the U.S. Department of Labor. The key legislation relating to improved mine worker health was the Federal Coal Mine Health and Safety Act, enacted in 1969 (U.S. Congress, 1969) and subsequently amended as the Federal Mine Safety and Health Act of 1977 (U.S. Congress, 1977). MSHA receives technical assistance from the National Institute for Occupational Safety and Health (NIOSH). NIOSH also provides recommended exposure limits (RELs) for contaminants to MSHA. MSHA's regulation of mines includes the establishment of a dust standard administratively based on the these RELs. Some mine operators must submit a dust control plan designed to meet the dust standard. Periodic mine inspections determine if mines are complying with both the dust standard and the dust control plans (MSHA, 1991).

The most common dust measurements are those for respirable dust (according to the definition given in Fig. 26–7), that is, the fraction of aerosol particles able to reach the gas exchange region of the lungs. The other health-related size fractions (inhalable and thoracic—see Fig. 25–1, and "total" dust) are also occasionally measured. A "total" dust measurement is defined by the common practice of the last 30 years or so, using 25 mm or 37 mm plastic cassettes (e.g., from ***PAL***, ***MIL***, ***SKC****). The types of mines have been categorized

* See Appendix I for full manufacturer addresses referenced to the italicized three-letter codes.

Aerosol Measurement: Principles, Techniques, and Applications, Second Edition, Edited by Paul A. Baron and Klaus Willeke.
ISBN 0-471-35636-0

for regulatory purposes as either coal mines or other types of mines (indicated as metal and nonmetal mines).

This chapter summarizes aerosol measurement technology currently used in the U.S. mining industry as it relates to regulation, research, and personal exposure monitoring.

MINE AEROSOL SOURCES

The choice of measurement techniques and the performance of various aerosol monitors may be affected by the type of aerosol to be sampled. Mine aerosols typically originate from comminution, re-entrainment, and combustion sources.

Comminution

Mechanical crushing of the mined material by drill bits, crushers, and mining machine picks generate significant concentrations of respirable aerosol mass in the mine. In metal and nonmetal mines, a typical mining process begins with the cutting or drilling and blasting of the ore from the parent rock. The ore is then scooped or gathered and loaded into trucks or onto conveyers for transport to the surface. In some cases the ore is crushed underground to facilitate transportation. The aerosol composition generated from these activities is generally similar to the host ore; however, some selective fractionation of the composition may occur based on the cleavage and strength patterns of the host ore. After the ore is removed, underground construction activities then begin to support the roof, direct ventilation air, and extend the mine's infrastructure.

Continuous mining methods are the most prevalent underground coal extraction processes, accounting for most of the underground coal mined (Organiscak, 1989). In continuous mining, a mining machine, pictured in Figure 26–1, cuts coal from the face and loads it directly into shuttle cars. The shuttle cars transport the coal from the face area to the material transfer system, usually a crusher that loads onto a conveyor belt.

Fig. 26–1. Continuous miner. (From U.S. Bureau of Mines.)

Re-entrainment

Aerosols that result from comminution may settle on roadways or in return air entries and be subject to re-entrainment through the activities of mobile equipment such as shuttle cars, ventilation, or in-mine construction activities (Jankowski and Hake, 1989). In coal mining, an inert limestone dust is spread on the entryway roof, floor, and ribs to prevent coal dust explosions. This limestone dust is coarse and contributes only a small fraction to the overall mine aerosol concentrations.

Combustion

Combustion by-products are the other major source of aerosols in mines. Combustion aerosol particles are smaller in size and currently can contribute significantly to the overall mass of respirable aerosols. They have higher number concentrations than the coarser comminution aerosol and a greater interaction potential with the lungs. Typical mining sources include blasting agent fumes, welding fumes, and diesel engine particles and condensates.

Concentrations of blasting agent fumes and welding aerosols in the mine environment are primarily controlled by dilution with the mine ventilation air. Monitoring the aerosol concentration of these agents must focus on the transient nature of the aerosol and short-term exposure monitoring. Due to the intermittent nature of the sources and the substantial ventilation volumes needed for mining, the overall concentrations of these aerosols are generally low with some exceptions in areas with poor ventilation.

With the exception of coal mines, diesel engines provide the primary power source for mining equipment. In addition to being exposed to dust, a miner working in an underground mine with diesel-powered equipment is exposed to a wide array of pollutants from diesel exhaust. These include CO, CO_2, NO, NO_2, SO_2, diesel exhaust aerosol, and a variety of aerosol-associated and gas-phase hydrocarbon compounds (Cantrell and Watts, 1997). To date, ventilation rates in diesel-equipped mines have been dictated by gas concentrations. In the future, new standards may dictate that diesel exhaust aerosol concentration is the factor that will control required ventilation volumes (U.S. Department of Labor, 1998; MSHA, 2001a,b).

Of the combustion aerosols in the mining environment, diesel exhaust aerosol is of particular concern because it is almost entirely respirable in size, with more than 90% of the particles, by mass, having an aerodynamic diameter less than 1.0 μm (Cantrell, 1987). This means that the aerosol can penetrate to the deepest regions of the lungs and, if retained, cause or contribute to the development of obstructive or restrictive lung disease (Watts, 1987).

PHYSICAL CHARACTERISTICS OF MINE AEROSOL

Aerosols in mines sources have particle size distributions that are determined by both the mining method and the source of the aerosol (Cantrell and Rubow, 1992a,b). A size distribution summarizing the physical characteristics of mine aerosols is shown in Figure 26–2. The shape of the aerosol size distribution is influenced by the different sources contributing to the aerosol. Figure 26–2 displays some of these sources and the physical mechanisms, such as condensation and coagulation, which transfer aerosol mass from one size to another. It should be noted that these mechanisms and the general shape of the distribution are not unique to mine aerosols.

There are three distinct aerosol size ranges identifiable by features in measured mine aerosol size distributions. The smallest of these, from 0.001 to 0.08 μm, is the *Aitken nuclei range*, which contains primary aerosol from combustion sources, such as diesel engines, and

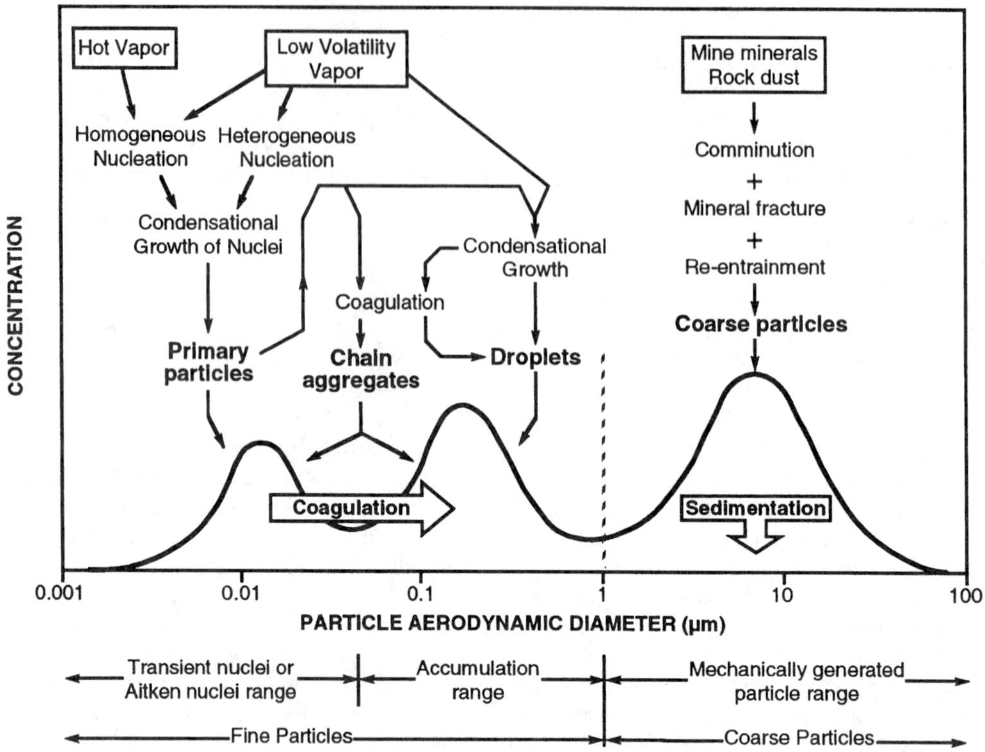

Fig. 26–2. Size distribution summarizing the general physical characteristics of mine aerosol. (From Cantrell et al., 1987.)

secondary aerosols or chain aggregates, formed by coagulation of primary aerosols. The next size range, from 0.08 to approximately 1.0 μm, is termed the *accumulation range*. This size range contains direct aerosol emissions plus aerosols that grow from the Aitken nuclei range by accumulating mass through coagulation and condensation processes. The last range, 1.0 μm to approximately 40 μm, is termed the *coarse particle range*. Aerosols within this size fraction generally result from mechanical processes such as rock fracture and bulk material handling. Mineral dust aerosol re-entrained by mine haulage vehicles during the load–haul–dump cycle is an example of an in-mine emission that will contribute aerosol to this size range.

For convenience, the Aitken nuclei and the accumulation ranges are combined in a single "fine" particle range. A division is usually made between this range and the "coarse" particle range at 1.0 μm. This distinction is possible because sources of aerosol in the two ranges are usually different, and the coarse particle range contains very little mass transferred from the accumulation range by coagulation.

In each of the ranges mentioned, the size distribution of mine aerosol can exhibit a maximum, or mode, which takes its name from the size range in which it occurs. Hence, the maximum in the accumulation range is termed the *accumulation mode*. Figure 26–3 presents a typical size distribution of aerosol mass concentration measured in a haulage entry of a diesel-equipped coal mine (Cantrell and Rubow, 1990). Here the modal character of the size distribution is discernible even though the nuclei mode is suppressed compared with the accumulation mode. In contrast, Figure 26–4 shows a mass size distribution measured in the

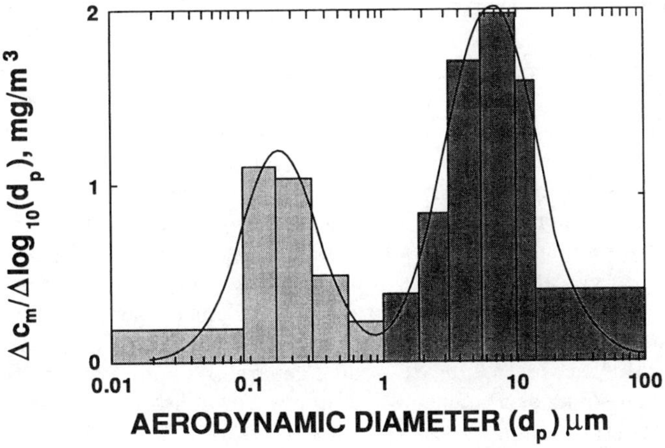

Fig. 26–3. Mass size distribution of mine aerosol in diesel-equipped mine. (From Cantrell and Rubow, 1990.)

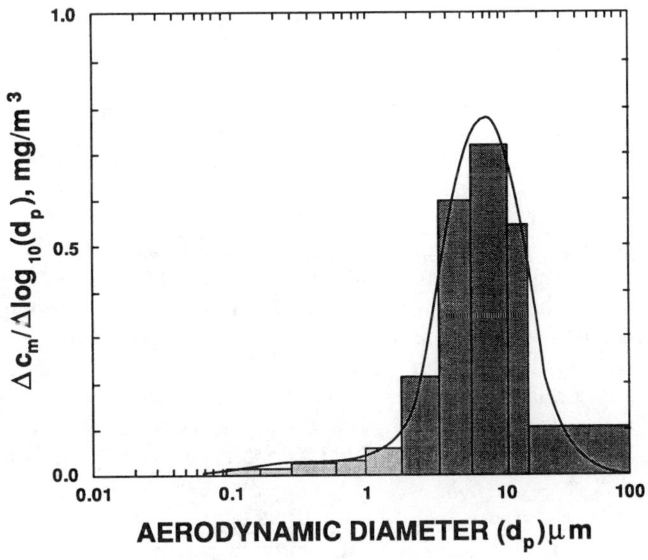

Fig. 26–4. Mass size distribution of mine aerosol in an all-electric–equipped mine. (From Cantrell and Rubow, 1990.)

haulage way of an all-electric-motor–equipped coal mine. Here the accumulation mode is much smaller than the coarse particle mode. Taken together, the figures indicate that diesel aerosols can make a strong contribution to accumulation mode aerosols in a diesel engine–equipped mine.

MEASUREMENT TECHNOLOGY

Aerosol measurement technology used by and for the mining industry can be conveniently separated into compliance measurements in support of regulation and research measurements that aid in the development of new compliance measurement techniques, determine control effectiveness, and assist in determining the fundamental properties of mine aerosols.

Compliance measurements are primarily intended for use in enforcement of regulations established by MSHA. Regulatory requirements determine the sampling strategy and instrumentation used for such measurements. Research aerosol measurements are used to evaluate engineering dust control techniques, to develop new aerosol instrumentation, to define new occupational health hazards, and to expand knowledge regarding mine aerosols. They draw on all of the technology available to the aerosol community. Consequently, compliance and research aerosol measurements are treated separately here. Also, the following discussion focuses on aerosol measurements in U.S. mines. For a discussion of measurement practices in the European Community, see Vincent (1991).

Compliance Measurements and the Regulatory Environment

Regulatory Requirements

Metal and Nonmetal. MSHA regulates practices affecting health and safety in metal and nonmetal mines and mills under the authority of the Federal Mine Safety and Health Act of 1977 (U.S. Congress, 1977). The specific regulations are found in the Code of Federal Regulations, Title 30 (MSHA, 1991). MSHA continues to use the 1973 recommended threshold limit values of the American Conference of Governmental Industrial Hygienists (ACGIH, 1973). Compliance with these regulations is determined by the collection of environmental samples by MSHA inspectors. Aerosol-related contaminants that are regulated include total dust, respirable dust, quartz, asbestos, silicates, radionuclides in air, diesel exhaust aerosol, and welding fumes (MSHA, 1990; 2001a,b).

The following example illustrates the type of sampling and analysis procedure for a respirable dust containing more than 1% quartz. A sample is collected using the personal respirable dust sampler shown in Figure 26–5. Sample air is first passed through a Dorr-Oliver nylon cyclone pre-classifier (***MSA***) at a flow rate of 1.7 L/min to remove the nonrespirable fraction of sampled dust. Respirable dust is then collected on a filter that is analyzed gravimetrically to determine mass concentration. The filter deposit is also analyzed for quartz content using X-ray diffraction (MESA, 1975). The measured mass concentration is compared with the PEL determined from the quartz content of the respirable dust by

$$\mathrm{PEL}_{\mathrm{quartz}} = \frac{10\,\mathrm{mg/m^3}}{\text{percent respirable quartz} + 2} \qquad (26\text{–}1)$$

For a given exposure level, the magnitude of the toxicity is proportional to the quartz content (ACGIH, 1980). The factor 2 in the denominator of the PEL formula ensures that dust exposures will not be excessively high when the quartz content is less than 5%. When quartz levels are less than 1%, nuisance particles listed by the 1973 ACGIH standard are regulated to 10 mg/m^3 of total dust. MSHA has proposed a revision of many of the existing health regulations (MSHA, 1989a). Included in these revisions is a proposed change in the PEL for respirable quartz. The current PEL is 100 μg/m^3 of respirable quartz (MSHA, 1971).

Coal. Respirable coal mine dust measurements are made to determine compliance with MSHA-established dust standards (MSHA, 1989b). In 1970, a mandatory total respirable dust

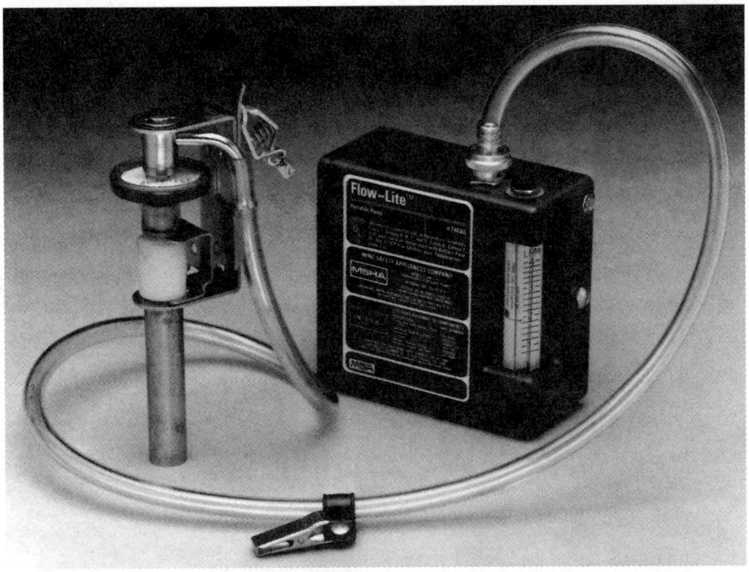

Fig. 26–5. Personal Respirable Dust Monitor. (Courtesy of MSA Co.)

standard of 3.0 mg/m^3 was established for underground coal mines in the Federal Coal Mine Health and Safety Act of 1969 (U.S. Congress, 1969). The respirable dust standard was subsequently lowered in 1972 to 2.0 mg/m^3. Mandatory dust standards for surface work areas of underground coal mines and surface mines also became effective in 1972. These regulations were continued under the Federal Mine Safety and Health Act of 1977 (U.S. Congress, 1977), which amended the 1969 Coal Act and merged coal and noncoal regulations into one law. In the 1969 Act, "concentration of respirable dust" was defined as that measured using a Mining Research Establishment (MRE) parallel plate elutriator (Casella, model 113A, United Kingdom) sampling instrument (Fig. 26–6) or such equivalent concentration measured with another device. This instrument was designed to have a sampling efficiency equivalent to the respirable response curve specified by the British Medical Research Council (BMRC) (Lippman, 1989). The 1977 Act changed the definition of "concentration of respirable dust" to be the "average concentration of respirable dust measured with a device approved by the Secretary (of Labor) and the Secretary of Health Education and Welfare."

The personal respirable dust sampler illustrated in Figure 26–5 is also approved for measuring respirable coal mine dust (MSHA, 1989b). The sampling rate used for coal, however, is 2 L/min (Tomb and Raymond, 1970). Sample analysis for total respirable dust is gravimetric (Raymond et al., 1987). Analysis for quartz is by Fourier transform infrared spectrometry (Ainsworth et al., 1989). Measurements are converted to equivalent MRE concentrations by multiplying the measured concentrations by an accommodation factor of 1.38 (Treaftis et al., 1984). The difference between the Dorr-Oliver and BMRC sampling efficiency curves is evident from Figure 26–7 (Caplan et al., 1977a,b; Lippman, 1989). Specific regulations detailing the methods for collecting respirable dust samples are found in the Code of Federal Regulations, Title 30 (MSHA, 1991).

A mine is in noncompliance with its dust standard if the arithmetic average concentration of five consecutive respirable dust samples is in excess of the applicable standard (MSHA, 1989b). If the percent quartz is less than 5%, the standard is 2.0 mg/m^3, calculated using Eq. 27–1. In underground coal mines, samples are collected on workers with a designated occupation, usually a mining machine operator. A mine may also not be in compliance

Fig. 26–6. Casella, Model 113a, MRE Respirable Dust Monitor. (From Cantrell et al., 1987.)

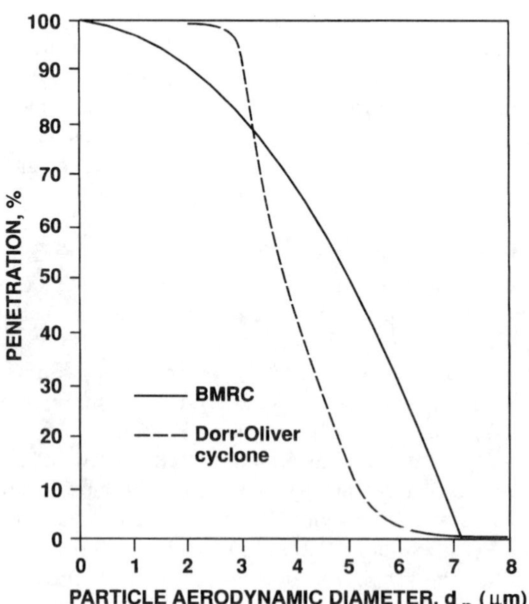

Fig. 26–7. Respirable aerosol sampling criteria; BMRC, Dorr-Oliver Cyclone. (From Caplan et al., 1977a,b; Lippman, 1989.)

with the law if the dust control plan is not being used. MSHA is required to inspect all underground coal mines four times a year, and the mine operators are required to sample bimonthly.

Diesel Exhaust Aerosol. NIOSH (1988) has recommended that "whole diesel exhaust be regarded as a 'potential occupational carcinogen,' as defined in the Cancer Policy of the Occupational Safety and Health Administration." NIOSH further stated that "though the excess risk of cancer in diesel-exhaust–exposed workers has not been quantitatively estimated, it is logical to assume that reduction in exposure to diesel exhaust in the workplace would reduce the excess risk." The International Agency for Research on Cancer (1989) has also classified diesel exhaust as "probably carcinogenic to humans." Additionally, the Mine Safety and Health Administration (MSHA, 1988) received a recommendation from an advisory committee to establish a diesel exhaust aerosol standard and to establish regulations to minimize exposure to all diesel pollutants in underground coal mines. The ACGIH has proposed a threshold limit value (TLV) for diesel exhaust particles of 50 $\mu g/m^3$ (ACGIH, 2000).

MSHA has proceeded to establish a diesel exhaust aerosol emission standard for diesel powered equipment in underground coal mines (MSHA, 2001a) and a PEL for diesel exhaust aerosol in metal and non-metal mines (MSHA, 2001b). The regulation for coal mines specifies a limit of 2.5 grams per hour for permissible heavy duty equipment. The regulation also specifies an interim aerosol emissions limit of 5.0 grams per hour and, after the year 2005, a limit of 2.5 grams per hour for non permissible heavy duty equipment. Light duty diesel powered equipment exhaust aerosol emissions are limited to 5.0 grams per hour. For metal/non-metal mines MSHA has proposed an interim total diesel aerosol carbon (TC) PEL of 400 $\mu g/m^3$ based on an 8 hour sample and, after the year 2005, a final PEL of 160 $\mu g/m^3$.

Research Aerosol Measurements

Research measurements are used to characterize the mine aerosol's mass and specific components such as quartz, trace elements, and carbon as a function of aerosol size. Recent emphasis on the potential health hazard associated with exposure to diesel exhaust aerosols has focused attention on specific techniques to measure these aerosols. The primary technique used to collect samples in coal and noncoal mines for all of these measurements is size-selective sampling using inertial impaction. In addition, several aerosol sensor techniques are being developed for continuous monitoring of respirable mine aerosols. These include both light-scattering and direct mass measurement.

Size-Selective Sampling. A size-selective sampling technique that has been useful for in-mine measurement of both diesel and mineral dust aerosol employs the Marple personal impactor (series 290, ***AND***) discussed in Chapter 10. The series 290 sampler was originally designed for NIOSH as a wood dust sampler by Rubow et al. (1987). More recently, it has been used in surveys of diesel-equipped mines by the U.S. Bureau of Mines (BOM) and NIOSH to measure the size distribution of mine aerosol (NIOSH, 1987). Estimates were made of average concentration levels of respirable diesel aerosol for the working shift using the sub-1.0 μm portion of each sample under the assumption that this accounted for most of the diesel exhaust in the mine atmosphere. The average concentration for submicrometer aerosol generated in the mining sections was $0.7 \pm 0.3\, mg/m^3$.

The micro-orifice, uniform deposit impactor (MOUDI) (model 100, ***MSP***), also discussed in Chapter 10, has also been used to measure the size distribution of mine aerosol over the size ranges in which respirable coal dust and diesel aerosols are expected to predominate (Marple et al., 1991). In addition to laboratory studies of these aerosols (Marple et al., 1986), the MOUDI has been used during field experiments in underground coal mines to evaluate its ability to separate diesel aerosol from coal dust aerosol on the basis of their size distributions (Rubow et al., 1990a). The field evaluations were conducted in underground mines that used only electric-powered haulage equipment and in other mines that used diesel-powered haulage equipment.

Typical mass size distributions of aerosols measured in the haulage entry of the diesel-equipped mines, shown in Figure 26–3, exhibit two distinct maxima: one submicrometer and the other greater than 1 µm. These measurements indicate that more than 90% of diesel exhaust aerosols in the diesel-equipped coal mines studied were submicrometer in size (Cantrell, 1987). The diesel-associated submicrometer aerosol accounted for approximately 40% to 60% of the respirable aerosol mass concentration. In contrast, aerosol size measurements in the all-electric coal mines, typified by Figure 26–4, exhibited a very small submicrometer maximum. Less than 10% of the measured respirable aerosol mass was in the submicrometer size range.

Diesel Aerosol Sampling and Analysis. Three analytical methods have been used for measurement of diesel exhaust aerosol in underground mines. These methods are (1) its mass concentration, (2) its carbon content, and (3) the combustible fraction of the sample. The first is measured using typical filter gravimetric techniques and the second by direct analysis of the organic carbon (OC) and elemental carbon (EC) content of the aerosol through thermal–optical analysis (see Chapter 11). The third is the respirable combustible dust (RCD) method that uses gravimetric techniques combined with ashing to separate mineral from combustible matter. In all cases, size-selective sampling can be used to provide a sample that contains most of the diesel-associated portions of the sampled respirable aerosol.

A personal diesel exhaust aerosol sampler based on size-selective sampling, developed for use in underground coal mines (Rubow et al., 1990b; McCartney and Cantrell, 1992), is pictured in Figure 26–8. It has three stages and employs inertial impaction for separating and collecting the diesel and mineral dust fractions of the sampled respirable aerosol. The first stage is an inertial preclassifier that separates and collects the larger, nonrespirable aerosol. The pre-classifier used in this design is a 10 mm Dorr-Oliver cyclone. Its second stage is a four-nozzle impactor with a 50% cut point of 0.8 µm aerodynamic diameter. The third stage, which is a filter, collects the remaining aerosol particles with less than 0.8 µm aerodynamic diameter. The sampler components are used with an ***MSA*** dust monitor sampling frame and operate at a sampling flow rate of $3.33 \times 10^{-5}\,m^3/s$ [2 L/min]. It is designed to be compatible with commercial personal sampling pumps.

Gravimetric Analysis. Coupled with gravimetric analysis, the personal diesel exhaust aerosol sampler can provide measurements of diesel exhaust aerosol concentrations in coal mines under worst-case sampling conditions, which are accurate to within 25%, with a confidence of 95%, for concentration levels greater than $0.3\,mg/m^3$ (Cantrell and Rubow, 1991). During field evaluation tests, the sampler was used to make numerous aerosol concentration measurements in underground coal mines that use diesel haulage equipment. Figure 26–9 summarizes respirable aerosol concentrations in these mines, determined from area samples collected in the haulage entries, on coal shuttle cars, and in the ventilation return entries (Cantrell et al., 1992). Figure 26–10 summarizes the diesel exhaust aerosol concentrations measured using the same samples.

The haulage, shuttle car, and return locations have similar distributions for the diesel exhaust aerosol portion of the respirable total. This implies that diesel exhaust aerosol concentrations are uniform regardless of where they are measured in the section. Total respirable aerosol concentration levels are different depending on where the samples are taken. The highest is in the ventilation return, and the lowest is in the haulage way. These results imply that exposure to respirable mineral dust is location and hence, is occupation dependent.

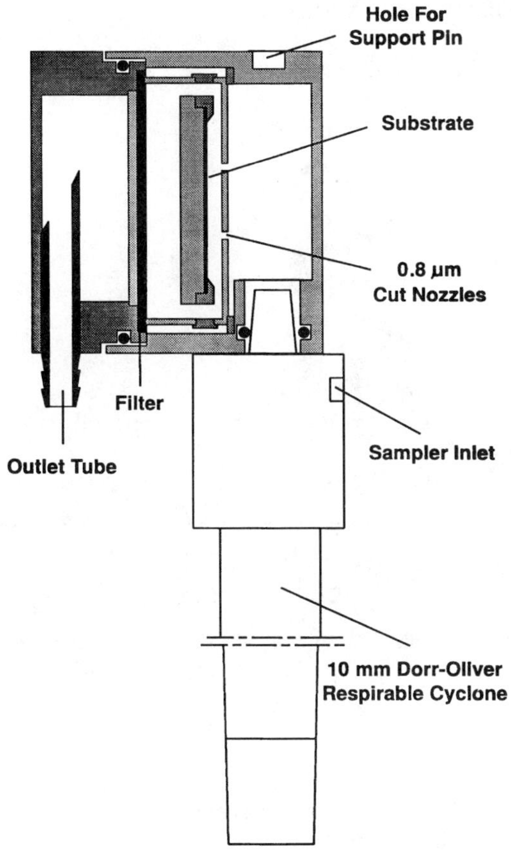

Fig. 26–8. Personal diesel exhaust aerosol sampler. (From McCartney and Cantrell, 1993.)

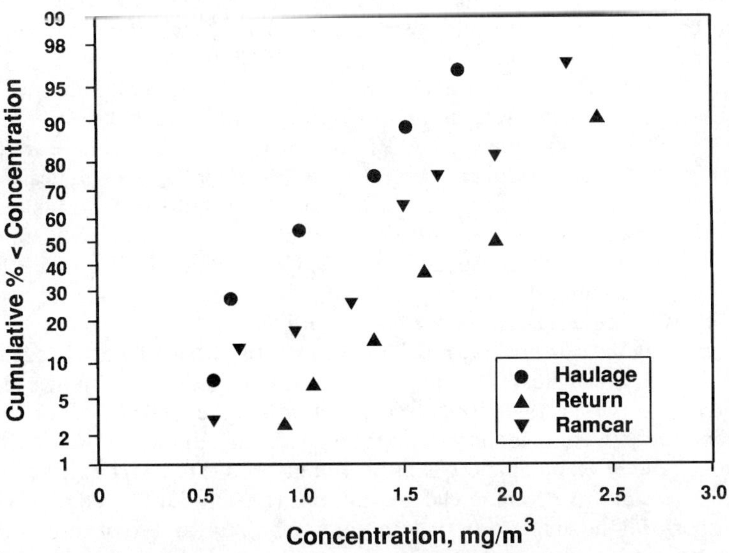

Fig. 26–9. Cumulative frequency plot of respirable aerosol concentrations in continuous mining operations using diesel-equipped haulage vehicles. (From Cantrell et al., 1991.)

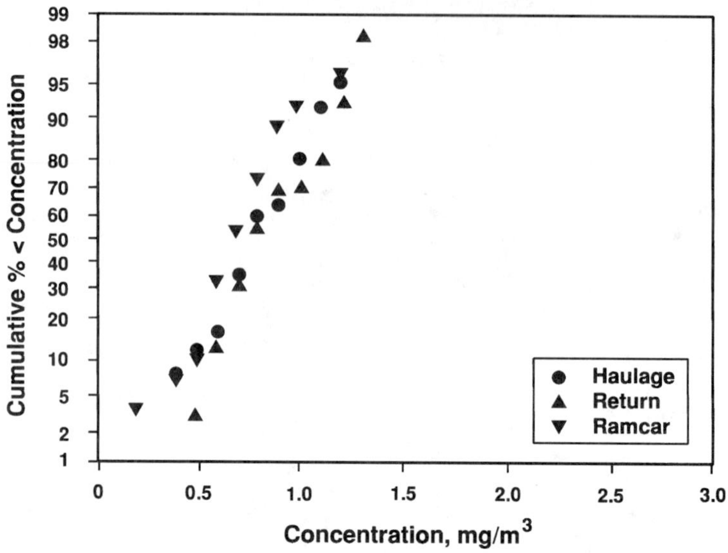

Fig. 26–10. Cumulative frequency plot of respirable diesel aerosol concentrations in continuous mining operations using diesel-equipped haulage vehicles. (From Cantrell et al., 1991.)

The median diesel exhaust aerosol concentration determined with the personal sampler at the haulage location for the three mines surveyed was 0.8 ± 0.2 mg/m^3. Diesel aerosol contributed 65% of the respirable aerosol at this location.

Direct Carbon Analysis. The use of size-selective sampling with gravimetric analysis is only intended for measuring diesel exhaust aerosol concentrations at the levels displayed in Figure 26–10 and cannot be used to support a standard below approximately 300 μg/m^3. Because control technology is currently capable of reducing diesel exhaust aerosol concentrations to levels well below 300 μg/m^3, there is a need for an analytical technique that can be used to measure diesel exhaust aerosol at this level (Ambs et al., 1991).

One approach is the thermal–optical method (Birch and Cary, 1996) published as Method 5040 by NIOSH. The method was initially described in 1996 and was recently updated (NIOSH, 1999). In the analysis, the OC and EC in a quartz fiber filter sample are determined through temperature and atmosphere control. The instrument has an optical feature that corrects for sample charring. The analytical technique on which the method is based was originally developed for atmospheric aerosols (Johnson et al., 1981; Cadel and Groblicki, 1980; Hering et al., 1990). NIOSH proposed its use for occupational monitoring of DPM. Although OC and EC are determined by the method, NIOSH researchers proposed use of EC as an exposure index because it is a selective marker of DPM in many occupational settings. In contrast, OC can originate from many nondiesel sources (e.g., drill oil mist, hydraulic oil mist, hydraulic fluids, and carbonates). The method evaluation included field studies and interlaboratory comparisons (Birch, 1998; Birch et al., 1999). Improvements in the original method and instrument design were made in collaboration with the laboratory that first offered the analysis and instrument on a commercial basis. The method can detect as little as 1 μg/m^3 of EC. Currently, there are five commercial laboratories (four in the United States and one in Canada) that perform the analysis. In underground mines, EC and total carbon (TC = EC + OC) account for about 50% and 80% respectively, of the submicrometer particulate mass, but these percentages vary depending on the engine duty cycle, fuel quality, after-treatment devices, and other factors. Thermal analysis techniques for OC and EC are also being used in several European countries.

The simplest sampling train for EC analysis, and the one used by MSHA to obtain field samples, consists of a 10 mm Dorr-Oliver cyclone followed by a 37 mm precombusted ultra-pure tissue–quartz fiber filter (PALL 2500QAT, ***GEL***) mounted in a 37 mm plastic cassette. Alternatively, the Dorr-Oliver cyclone can be followed by the 0.8 μm size-selective impactor described above. In this situation, diesel exhaust aerosol particulates, which are primarily smaller than 0.8 μm, are collected on the filter. This is advantageous because large, mechanically generated aerosols are not collected on the filter and cannot interfere with the carbon analysis.

Respirable Combustible Dust Analysis. In the RCD method, total respirable dust (TRD) is collected from sample air on a 25 or 37 mm, 0.8 μm pore size silver membrane (SM) filter after passing through a 10 mm Dorr-Oliver cyclone pre-classifier at a flow rate of $2.83 \times 10^{-5}\,m^3/s$ [1.7 L/min]. Flow is controlled using a personal sampling pump. At this flow rate, the cyclone is a respirable dust pre-classifier with a 50% cut point of 4 μm. TRD is determined gravimetrically by weighing the SM filter before and after the sample is collected. The RCD fraction of TRD is determined gravimetrically from the amount of material removed from the SM by controlled combustion at 673 K [400°C] for 1 to 2 h. A correction is made for loss of mass from the SM due to combustion (Gangal et al., 1990; Gangal and Dainty, 1993).

Elemental and Mineral Analysis. Another method for analyzing diesel aerosol is source apportionment. This technique has been used to evaluate measurements made in underground mines using size-selective sampling (Cantrell, 1987; Rubow et al., 1990a; Cantrell and Rubow, 1990). It relates elemental, mineral, or chemical components in an aerosol sample to those same components in the sources of the aerosol. Using this relationship, the contribution of each source to the sampled aerosol can be determined. For diesel-equipped coal mines the primary sources of aerosol are coal, rock dust used as a fire retardant on the mine walls, and diesel exhaust aerosol emissions. One specific analytical technique used to apportion collected aerosol among these sources is chemical mass balance model analysis (Watson, 1984; Henry et al., 1984; Davis, 1984).

The model is expressed as

$$c_{ei} = \sum_{j=1}^{p} a_{ij} S_j \qquad (26\text{–}2)$$

Here, c_{ei} is the mass concentration of the ith elemental, mineral, or chemical component of the sample in $\mu g/m^3$; a_{ij} is the fractional amount of component i in emissions from source j; S_j is the total contribution of source j to the sample; and p is the number of sources. Apportionment of the source is achieved by first characterizing the aerosol sources, obtaining values for a_{ij}, then analyzing the aerosol in the sample for the same components, and finally solving for the S_j. A least-squares regression analysis is used to determine the S_j of the overdetermined system of equations expressed by Eq. 26–2.

Real-Time Measurement

Real-time measurement of mine dust remains an objective for aerosol research. A report of the Secretary of Labor's Advisory Committee on the Elimination of Pneumoconiosis Among Coal Mine Workers listed several recommendations dealing with the need for continuous respirable dust monitors to help protect workers' health (U.S. Department of Labor, 1996). In addition, a NIOSH Criteria Document lists improved sampling devices as a research need pertinent to coal miner respiratory health and prevention of disease (NIOSH, 1995). In mining, personal exposure monitoring is preferable to area monitoring (Leidel et al., 1977). Dust concentrations have been shown to change dramatically over times of a few seconds and distances of just 0.6 m in underground coal mines (Kost and Saltsman, 1977). Therefore, to meaningfully approximate a worker's dust exposure, personal real-time monitoring is the preferred strategy.

Several approaches are being taken to address these needs. The more promising technologies include light-scattering dust monitors, a person-wearable tapered-element oscillating microbalance, and the Dust Dosimeter (Tsao et al., 1996; Lehocky and Williams, 1996; Volkwein et al., 2000). The principal goals of each of these efforts has been to develop an instrument that will give short-term or real-time measurements of worker dust exposure.

Light Scattering. Often called *photometers* or *nephelometers*, light-scattering instruments use a light source to illuminate sample aerosol and a light sensor to measure the scattered light that can be related to the mass concentration of the dust. The theory of these monitors is summarized in Chapter 15. Several of the monitors have been characterized in the laboratory for different dusts (Kuusisto, 1983; Marple and Rubow, 1981, 1984; Keeton, 1979; Rubow and Marple, 1983). The relationship of the instruments' responses to dust concentration is not simple but depends on particle size, particle composition, and instrument design and manufacturing differences. Significant changes in dust particle characteristics such as shape, size, surface properties, and density, can affect the instrument's correlation with mass concentration and require calibration of the instrument for each type of dust measured (Williams and Timko, 1984). These factors have limited the use of photometers to dust source identification and control technology evaluation. They are not useful for monitoring compliance with dust standards.

Most of the mining community's experience with real-time, photometric instruments is with the RAM-1 and MINIRAM devices (***MIE***), which were initially designed for use in the mine environment. Intrinsically safe versions of these instruments were certified by MSHA for use in the potentially explosive atmospheres of underground coal and gassy noncoal mines. They are normally operated with the Dorr-Oliver 10 mm cyclone to remove water droplets and ensure a respirable dust sample (Treaftis et al., 1984; Tomb et al., 1981). The RAM, pictured in Figure 26–11, is a real-time light-scattering aerosol monitor with a sampling pump that pulls aerosol through the sensing zone. The MINIRAM employs similar technology but typically operates in the passive sampling mode wherein size classification is based solely on light-scattering properties. An optional chamber and pump may be used to allow a cyclone classifier to physically determine the respirable fraction. McCawley and Cocalis (1986) obtained good correlations ($r^2 = 0.92$) between in-mine MINIRAM measurements and those obtained using a gravimetric sampler fitted with an intake impactor to restrict penetration of particles larger than 1.5 µm. Analysis of the collected sample was accomplished using RCD measurements. This successful comparison between diesel aerosol and the response of a photometric instrument holds promise that it might be used to monitor diesel aerosol.

Tapered Element Oscillating Microbalance. The theory of the Tapered Element Oscillating Microbalance (TEOM, ***R&P***) is summarized in Chapter 14. Despite limited application in the mining industry, this technology offers one notable advantage: direct measurement of dust mass. Because dust exposure standards are based on dust mass, this attribute of the TEOM is significant. In 1983, the BOM and NIOSH funded the development of a prototype TEOM personal dust monitor (Patashnick and Rupprecht, 1983). The prototype monitor developed is a system configured for end-of-shift measurements. It is not a real-time monitor,

Fig. 26–11. MIE RAM-1 Real-Time Respirable Aerosol Monitor equipped with a Dorr-Oliver Cyclone Sample Inlet. (From Olson and Veith, 1987.)

but uses oscillating microbalance technology to "weigh" the collection filter before and after dust sampling.

The BOM has evaluated this prototype system in the laboratory for both end-of-shift and near-real-time applications (Williams and Vinson, 1986). The effective standard deviation of repeated measurements was 1.6 μg. Tests at controlled temperature and humidity showed less than 20 μg of drift during an 8 h shift. These early attempts to construct a person-wearable form of the TEOM required a substantial mass in the base of the element to dampen the vibrations, reducing the concept's "wearability." Recent proprietary developments by ***R&P*** have solved the need for a substantial base mass electronically, and development of a person-wearable TEOM-based personal dust sampler is nearing completion under a NIOSH contract.

Dust Dosimeter. An alternative to highly accurate mine dust measurement devices is a new instrument that is based on a dust detector tube (Volkwein et al., 2000). The Dust Dosimeter, shown in Figure 26–12, is an inexpensive device that consists of a small pump with integral pressure transducer and a disposable detector tube, much like a gas detector tube, that correlates pressure drop across the filter with mass accumulation. It has recently been tested and shown to provide qualitative mid-shift and end-of-shift determinations of cumulative dust exposure. Figure 26–13 shows the laboratory responses for a variety of coal types. The Dust Dosimeter is an example of a new type of tool to help improve workers' health by enabling frequent and timely knowledge of current dust levels. Field testing of the device has shown that it gives a higher response to diesel particulate than to coal dust.

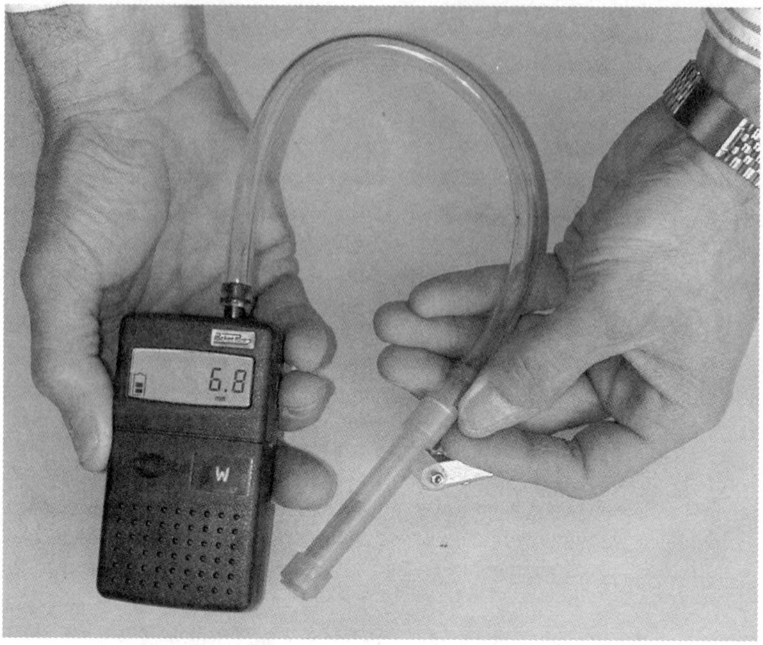

Fig. 26–12. Dust Dosimeter. (From NIOSH, 2000.)

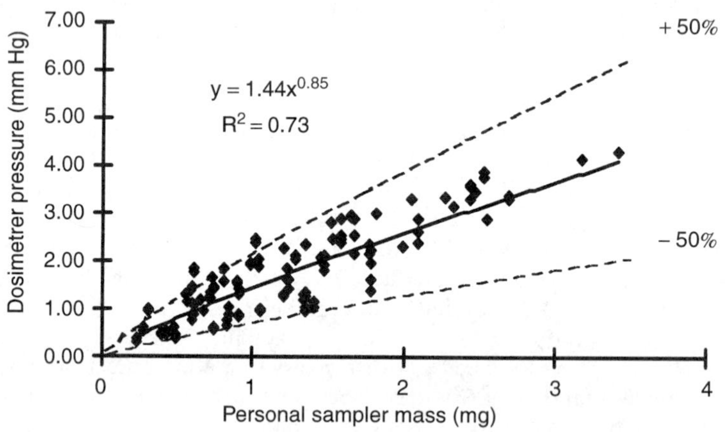

Fig. 26–13. Differential pressure response of Dust Dosimeter to six different laboratory coal dusts. (From Volkwein et al., 2000.)

REFERENCES

ACGIH. 1973. *TLVs—Threshold Limit Values for Chemical Substances in Workroom Air*. Cincinnati, OH: ACGIH.

ACGIH. 1980. *Documentation of the Threshold Limit Values*, 4th ed. Cincinnati, OH: ACGIH, pp. 364–365.

ACGIH. 2000. *TLVs—Threshold Limit Values for Chemical Substances and Physical Agents. Notice of Intended Changes for 2000*. Cincinnati, OH: ACGIH, p. 73.

Agricola, G. 1556. *De re Metallica*. Translated by M. C. Hoover and L. H. Hoover. New York: Dover Publications.

Ainsworth, S. M., P. S. Parobeck, and T. F. Tomb. 1989. *Determining the Quartz Content of Respirable Dust by FTIR*. MSHA IR 169, Arlington, VA.

Ambs, J. L., B. K. Cantrell, and A. D. Klein. 1991. Evaluation of a disposable diesel exhaust filter. *Am. Fil. Soc.* 4:287–291.

Birch, M. E. 1998. Analysis of carbonaceous aerosols: Inter-laboratory comparison. *Analyst* 123:851–857.

Birch, M. E. and R. A. Carey. 1996. Elemental carbon-based method for occupational monitoring of particulate diesel exhaust: Methodology and exposure issues. *Analyst* 121:1183–1190.

Birch, M. E., D. Dahmann, and H. Hennig-Fricke. 1999. Comparison for two carbon analysis methods for monitoring diesel particulate levels in mines. *J. Environ. Monitor.* 1:541–544.

Cadle, S. H. and J. Groblicki. 1980. *An Evaluation of Methods for Determination of Organic and Elemental Carbon in Particulate Samples*. General Motors Report GMR-3452, ENV #86. Detroit: General Motors.

Cantrell, B. K. 1987. *Source Apportionment Analysis Applied to Mine Dust Aerosols: Coal Dust and Diesel Emissions Aerosol Measurement*. Proceedings of the 3rd Mine Ventilation Symposium. Society for Mining, Metallurgy, and Exploration, Inc., Littleton, CO: Society for Mining Metallurgy and Exploration, pp. 495–501.

Cantrell, B. K. and K. L. Rubow. 1990. *Mineral Dust and Diesel Exhaust Aerosol Measurements in Underground Metal and Nonmetal Mines*. Proceedings of the VIIth International Pneumoconioses Conference. NIOSH Pub. No. 90-108, Cincinnati, NIOSH, pp. 651–655.

Cantrell, B. K. and K. L. Rubow. 1991. Development of personal diesel aerosol sampler design and performance criteria. *Mining Eng.* February:232–236.

Cantrell, B. K., K. L. Rubow, W. F. Watts, Jr., and D. E. Carlson. 1991. Pollutant levels in underground coal mines using diesel equipment. In *Proceedings of the SMME Annual Meeting and Exhibit*, Denver, CO, Feb. 25–28, 1991. Society for Mining Metallurgy and Exploration (SME), Littleton, CO, Preprint No. 91-35, p. 8.

Cantrell, B. K. and K. L. Rubow. 1992a. *Measurement of Diesel Exhaust Aerosol in Underground Coal Mines*. U.S. Bureau of Mines IC 9324, Washington, DC: Bureau of Mines, pp. 11–17.

Cantrell, B. K. and K. L. Rubow. 1992b. *Diesel Exhaust Aerosol Measurement in Underground Metal and Nonmetal Mines*. U.S. Bureau of Mines IC 9324, pp. 18–23.

Cantrell, B. K., K. L. Rubow, W. F. Watts, Jr., and D. E. Carlson. 1992. *Pollutant Levels in Underground Coal Mines Using Diesel Equipment*. Transactions of the Society for Mining, Metallurgy, and Exploration, Inc., v. 290, Littleton, CO: Society for Mining, Metallurgy and Exploration, pp. 1901–1907.

Cantrell, B. K. and W. F. Watts Jr. 1997. Diesel exhaust aerosol, review of occupational exposure. 2nd International Conference of the Health of Miners, November 11–13, 1995, Pittsburgh, PA. *Appl. Occup. Environ. Hyg.* 12:1019–1027.

Caplan, K. J., L. J. Doemeny, and S. D. Sorenson. 1977a. Performance characteristics of the 10-mm respirable mass sampler: Part I—Monodisperse studies. *Am. Ind. Hyg. Assoc. J.* 38:83–95.

Caplan, K. J., L. J. Doemeny, and S. D. Sorenson. 1977b. Performance characteristics of the 10-mm respirable mass sampler: Part II—Coal dust studies. *Am. Ind. Hyg. Assoc. J.* 38:162–173.

Cherniack, M. 1986. *The Hawk's Nest Incident, America's Worst Industrial Disaster*. New Haven: Yale University Press.

Davis, B. L. 1984. X-ray diffraction analysis and source apportionment of Denver aerosol. *Atmos. Environ.* 18:2197–2208.

Fletcher, C. M. 1948. Pneumoconiosis of coal-miners. *BMJ* 1:1015–1065.

Gangal, M. K. and E. D. Dainty. 1993. *Ambient Measurement of Diesel Particulate Matter and Respirable Combustible Dust in Canadian Mines*. Proceedings of the 6th U.S. Mine Ventilation Symposium, R. Bhaskar, ed. Littleton, CO: Society for Mining, Metallurgy, and Exploration, Inc., pp. 83–89.

Gangal, M., J. Ebersole, J. Vallieres, and D. Dainty. 1990. *Laboratory Study of Current (1990/91) SOOT/RCD Sampling Methodology for the Mine Environment*. Mining Research Laboratory, Canada Centre for Mineral and Energy Technology (CANMET). Ottawa, Canada.

Henry, R. C., C. W. Lewis, P. K. Hopke, and H. J. Williamson. 1984. Review of receptor model fundamentals. *Atmos. Environ.* 18:1507–1515.

Hering, S. V., et al. 1990. Comparison of sampling methods for carbonaceous aerosols in ambient air. *Aerosol Sci. Technol.* 12:200–213.

International Agency for Research on Cancer. 1989. *IARC Monographs on the Evaluation of Carcinogenic Risks to Humans—Diesel and Gasoline Engine Exhausts and Some Nitroarenes.* IARC, Lyon, France, 46:458.

Jankowski, R. A. and J. Hake. 1989. *Dust Sources and Controls for High Production Longwall Faces.* Proceedings of Longwall USA, Pittsburgh, PA.

Johnson, R. L., J. J. Shaw, R. A. Cary, and J. J. Huntzicker. 1981. An automated thermal-optical method for the analysis of carbonaceous aerosol. In *Atmospheric Aerosol: Source/Air Quality Relationships*, eds. W. S. Macias and P. K. Hopke. Washington, DC: ACS Symposium Series, No. 167, pp. 223–233.

Keeton, S. C. 1979. *Carbon Particulate Measurements in a Diesel Engine.* Albuquerque: Sandia Laboratories, Publication SAND 79-8210.

Kost, J. A. and R. D. Saltsman. 1977. *Evaluation of the Respirable Dust Area Sampling Concept as Related to the Continuous Miner Operator.* Bituminous Coal Research, Inc., Report L-792, January. Monroeville, PA: Bituminous Coal Research, Inc.

Kuusisto, P. 1983. Evaluation of the direct reading instruments for the measurement of aerosols. *Am. Ind. Hyg. Assoc. J.* 44:863–874.

Lehocky, A. H. and P. L. Williams. 1996. Comparison of respirable samplers to direct-reading real-time aerosol monitors for measuring coal dust. *AIHAJ* 57:1013–1018.

Leidel, N. A., K. A. Busch, and J. R. Lynch. 1977. The inadequacy of general (area) monitoring for measuring employee exposures. Technical Appendix C, *Occupational Exposure Sampling Strategy Manual*, NIOSH, NTIS #PB 77-274752. Washington, DC: Government Printing Office.

Lippman, M. 1989. Size-selective health hazard sampling. In *Air Sampling Instruments*, 7th Ed., S. V. Hering. Cincinnati: ACGIH, pp. 163–198.

Marple, V. A., D. B. Kittleson, K. L. Rubow, and C. Fang. 1986. *Methods for the Selective Sampling of Diesel Particulates in Mine Dust Aerosol.* U.S. Bureau of Mines OFR 44-87, NTIS PB 88-130810. Washington, DC: U.S. Bureau of Mines.

Marple, V. A. and K. L. Rubow. 1981. *Instruments and Techniques for Dynamic Particle Size Measurement of Coal Dust.* U.S. Bureau of Mines OFR 173-83; NTIS PB 83-262360. Washington, DC: U.S. Bureau of Mines.

Marple, V. A. and K. L. Rubow. 1984. *Respirable Dust Measurement.* U.S. Bureau of Mines OFR 92-85, NTIS PB 85-245843. Washington, DC: U.S. Bureau of Mines.

Marple, V. A., K. L. Rubow, and S. M. Behm. 1991. A micro-orifice uniform deposit impactor (MOUDI): Description, calibration, and use. *Aerosol Sci. Technol.* 14:434–446.

McCartney, T. C. and B. K. Cantrell. 1992. *A Cost-Effective Personal Diesel Exhaust Aerosol Sampler.* U.S. Bureau of Mines IC 9324, pp. 24–30. Washington, DC: U.S. Bureau of Mines.

McCawley, M. and J. Cocalis. 1986. *Diesel Particulate Measurement Techniques for Use with Ventilation Control Strategies in Underground Coal Mines.* Cincinnati, OH: ACGIH, Vol 14, pp. 271–281.

McFarland, W. 1927. Silicosis and tuburculosis as seen in the granite workers in Barre. *Br. J. Ind. Hyg.* 9:315–330.

MESA. 1975. *The Determination of Free Silica in Airborne Dust Collected on Membrane Filters.* U.S. Dept. of the Interior, Mining Enforcement and Safety Administration, IR No. 1021. Arlington, VA: Mine Safety and Health Administration.

MSHA. 1971. Code of Federal Regulations (CFR), Title 30—Mineral Resources, Chapter 1. Mine Safety and Health Administration, Department of Labor, Revision of 10 March 1971.

MSHA. 1988. *Report of the Mine Safety and Health Administration Advisory Committee on Standards and Regulations for Diesel-Powered Equipment in Underground Coal Mines.* Report to the Secretary of Labor, July 1988. Arlington, VA: MSHA.

MSHA. 1989a. Air quality, chemical substances, and respiratory protection standards; proposed rule. *Federal Register* 54(166):35760–35785.

MSHA. 1989b. Coal mine health inspection procedures. In *MSHA Handbook Series, Handbook No. 89-V-1*. Arlington, VA: MSHA.

MSHA. 1990. Metal and nonmetal mine health inspection procedures. *MSHA Handbook Series, Handbook No. PH 90-IV-4*. Arlington, VA: MSHA.

MSHA. 1991. Code of Federal Regulations (CFR), Title 30—Mineral Resources, Chapter 1. Mine Safety and Health Administration, Department of Labor, Revision of 1 July 1999, pp. 3–705.

MSHA. 1996. 30 CFR Parts 7, Exhaust Gas Monitoring, and Safety Requirements for the Use of Diesel-Powered Equipment in Underground Coal Mines; Final Rule. *Federal Register*, October 25, 1996, pp. 55411–55534. Washington, DC: U.S. Department of Labor, Mine Safety and Health Administration.

MSHA. 1998. 30 CFR Part 57, Diesel Particulate Matter Exposure of Underground Metal and Nonmetal Miners; Proposed Rule. *Federal Register*, October 29, 1998, pp. 58104–58270. Washington, DC: U.S. Department of Labor, Mine Safety and Health Administration.

MSHA. 2001a. Diesel Particulate Matter Exposure of Underground Coal Miners. *Federal Register* 66(13):5526–5706.

MSHA. 2001b. Diesel Particulate Matter Exposure of Underground Metal and Nonmetal Miners. *Federal Register* 66(13):5706–5910.

NIOSH. 1987. Diesel Particulate Measurement Techniques Applied to Ventilation Control Strategies in Underground Coal Mines. *BuMines contract No. JO 145006*. For information contact Mr. M. McCawley NIOSH, Morgantown, WV.

NIOSH. 1988. Carcinogenic effects of expsosure to diesel exhaust. *Current Intelligence Bulletin 50*, NIOSH Publ. 88-116. Cincinnati, OH: NIOSH.

NIOSH. 1995. Criteria for a recommended standard—Occupational exposure to respirable coal mine dust. Washington, DC: National Institute for Occupational Safety and Health, DHHS(NIOSH) Publication No. 95-107.

NIOSH. 1999. Elemental carbon (diesel exhaust). Method 5040. *NIOSH Manual of Analytical Methods* (NMAM), 5th Edition, January 1, 1999. Cincinnati, OH: NIOSH.

Olson, K. S. and D. L. Veith. 1987. Fugitive dust control for haulage roads and tailing basins. *BuMines RI 9069*, Washington, DC: U.S. Bureau of Mines.

Organiscak, J. A. 1989. *Respirable Dust Generation, Comparison of Continuous and Conventional Mining Methods when Excavating Rock*. RI 9233. Washington, DC: U.S. Bureau of Mines.

Patashnick, H. and G. Rupprecht. 1983. *Personal Dust Exposure Monitor Based on the Tapered Element Oscillating Microbalance*. U.S. Bureau of Mines OFR 56-84; NTIS PB 84-173749 Washington, DC: U.S. Bureau of Mines.

Raymond, L., T. F. Tomb, and P. S. Parobeck. 1987. *Respirable Coal Mine Dust Sample Processing*. Washington, DC: Mine Safety and Health Administration, IR 1156.

Rubow, K. L., B. K. Cantrell, and V. A. Marple. 1990a. *Measurement of Coal Dust and Diesel Exhaust Aerosols in Underground Mines*. Proceedings of the VIIth International Pneumoconioses Conference, NIOSH Publ. No. 90-108. Cincinnati: NIOSH, pp. 645–650.

Rubow, K. L. and V. A. Marple. 1983. An instrument evaluation chamber: Calibration of commercial photometers. In *Aerosols in the Mining and Industrial Work Environment*, V. A. Marple and B. Y. H. Liu, eds. Ann Arbor, MI: Ann Arbor Science Publishers, Inc., III:777–798.

Rubow, K. L., V. A. Marple, J. Olin, and M. A. McCawley. 1987. A personal cascade impactor: Design, evaluation and calibration. *Am. Ind. Hyg. Assoc. J.* 48:532–538.

Rubow, K. L., V. A. Marple, Y. Tao, and D. Liu. 1990b. *Design and Evaluation of a Personal Diesel Aerosol Sampler for Underground Coal Mines*. Preprint No. 90-132. Littleton, CO: Society for Mining, Metallurgy, and Exploration, Inc.

Seaton, A. 1986. Respiratory disease in miners. *Proc. Ann. Am. Conf. Gov. Ind. Hyg.* 14:21–25.

Seaton, A., J. A. Dick, J. Dodgson, and M. Jacobsen. 1981. Quartz and pneumoconiosis in coal miners. *Lancet* 1:1272.

Tomb, T. F. and L. D. Raymond. 1970. *Evaluation of the Penetration Characteristics of a Horizontal Plate Elutriator and of a 10-mm Nylon Cyclone Elutriator*. U.S. Bureau of Mines, RI 7367. Washington, DC: U.S. Bureau of Mines.

Tomb, T. F., H. N. Treaftis, and A. J. Gero. 1981. Instantaneous dust exposure monitors. *Environ. Int.* 5:85–96.

Treaftis, H. N., A. J. Gero, P. M. Kacsmar, and T. F. Tomb. 1984. Comparison of mass concentrations determined with personal respirable coal mine dust samplers operating at 1.2 liters per minute and the Casella 113a gravimetric sampler (MRE). *Am. Ind. Hyg. Assoc. J.* 45:826–832.

Tsao. C. J., T. S. Shih, and J. D. Lin. 1996. Laboratory testing of three direct reading dust monitors. *AIHA J.* 57:577–563.

U.S. Congress. 1969. Federal Coal Mine Heath and Safety Act of 1969, 30 U.S.C. Stat. 801.

U.S. Congress. 1977. Federal Mine Safety and Health Act of 1977, Public Law No. 95-164, 91 Stat. 1290 (1977), Codified at 30 U.S.C. Stat. 801.

U.S. Department of Labor. 1996. *Report of the Secretary of Labor's Advisory Committee on the Elimination of Pneumoconiosis Among Coal Mine Workers*; recommendation 8 and 17. Washington, DC: U.S. Department of Labor.

Vincent, J. H. 1991. Joint Investigations of New Generations of Sampler for Airborne Dust in Mines. Synthesis Report on a joint project carried out by six laboratories in five European member states on behalf of the Commission of European Communities during 1985–1988. *Report No. EUR 13414EN*. Luxembourg: Commission of European Communities.

Volkwein, J. C., A. L. Schoeneman, and S. J. Page. 2000. Laboratory evaluation of a pressure differential-based respirable dust detector tube. *Appl. Occup. Environ. Hyg.* 15:158–164.

Watson, J. G. 1984. Overview of receptor model principles. *JAPCA* 34(6):619–623.

Watts, W. F., Jr. 1987. Industrial Hygiene Issues Arising from the Use of Diesel Equipment in Underground Mines. *BuMines IC 9141*. Washington, DC: U.S. Bureau of Mines. pp. 4–8.

Williams, K. and R. J. Timko. 1984. *Performance Evaluation of a Real-Time Aerosol Monitor*. Washington, DC: U.S. Bureau of Mines, IC 8968.

Williams, K. L. and R. P. Vinson. 1986. *Evaluation of the TEOM Dust Monitor*. Washington, DC: U.S. Bureau of Mines, IC 9119.

27 Ambient Air Sampling

JOHN G. WATSON and JUDITH C. CHOW

Desert Research Institute, University and Community College System of Nevada, Reno, NV

INTRODUCTION

Goals for ambient sampling for suspended particulate matter include (1) determining compliance with air quality standards; (2) evaluating the extent and causes of high mass concentrations, deposition, and visibility impairment; (3) enhancing understanding of the chemical and physical properties of atmospheric pollution; (4) apportioning the chemical constituents of suspended particulate matter to their emitting sources; and (5) evaluating adverse health effects. A sampling system that serves one objective does not necessarily meet the needs of other objectives. Most notably, filter samplers designed to determine compliance with mass-based air quality standards have limited applicability to sampling for chemical characterization, hourly and daily sequential sampling, quantification of volatile aerosols, and particle sizing.

Most of the world's data on suspended particle mass concentrations are obtained by drawing a measurable volume of air through a particle-efficient filter over a 24h period. The filter is weighed in a temperature- and humidity-controlled laboratory before and after sampling to determine the particle deposit. The deposit weight is then divided by the sample volume to determine the particle mass concentration. The U.S. Environmental Protection Agency (EPA) has designated federal reference methods (FRMs) for total suspended particulate (TSP), PM-10, and PM-2.5 (mass of particles with aerodynamic diameters less than approximately 40 μm, <10 μm, and <2.5 μm, respectively), applying this method to determine compliance with national ambient air quality standards (NAAQS).

Current U.S. NAAQS (U.S. EPA, 1997a, 1998a) for particulate matter are (1) annual average PM-10 less than 50 μg/m^3, averaged over 3 years; (2) 99th percentile 24h average PM-10 less than 150 μg/m^3, averaged over 3 years; (3) annual average PM-2.5 less than 15 μg/m^3, averaged over 3 years; and (4) 98th percentile 24h average PM-2.5 less than 65 μg/m^3, averaged over 3 years. The use of percentiles is intended to provide a more stable measure of high concentrations than annual maxima that were the basis for prior NAAQS (U.S. EPA, 1987). Three year averaging was adopted to moderate the effects of uncommon emissions or meteorological events that might bias the indicator during a single year. FRMs have been optimized for accurate mass measurements over 24 hour periods with sixth-day, third-day, and everyday sampling frequencies.

Although these FRMs are necessary for the compliance monitoring objective, they are not sufficient for other research objectives. Many of the components developed for FRMs

Aerosol Measurement: Principles, Techniques, and Applications, Second Edition, Edited by Paul A. Baron and Klaus Willeke.
ISBN 0-471-35636-0

and other sampling systems can be used to assemble filter samplers tailored to a specific need, and the new NAAQS have stimulated substantial innovation in this area. This chapter specifies the generic requirements of ambient aerosol sampling systems, tabulates available components that can meet these requirements, and describes particle sampling systems that can be purchased or assembled to address noncompliance research objectives. Citations and references direct the reader to more detailed information on each of these topics.

SAMPLING SYSTEM COMPONENTS

Figure 27–1 illustrates the components of a generic particulate sampling system. Air enters the inlet where particles larger than the desired size are removed. The air stream containing the remaining particles passes through a transfer tube to a denuder that transmits more than 95% of the particles while removing gases that are particle precursors or are in equilibrium with particulate ammonium nitrate and organic compounds that have high vapor pressures. Denuder surfaces naturally retain these gases, or they are treated with gas-absorbing chemicals. Inlet and transfer tubes also remove gases to varying degrees, depending on the materials of which they are made and the gases of interest. The air stream is directed through a holder containing a collection filter that transmits the gases and retains more than 99% of the particles in an even deposit across its surface. Flat membrane or fibrous filters are selected to be compatible with the intended analysis methods. One or more backup filters that naturally absorb, or that are treated with absorbing chemicals that react with specific gases, are located behind the particle filter to capture gases in the air stream or those that volatilize from particles collected on the first filter. A flow measurement or control device behind the filter monitors the volume of air sampled, and a pump of sufficient capacity to obtain the desired flow rates draws air through the system.

Although Figure 27–1 is simple in concept, its practical implementation requires a careful integration of the components specific to the sampling objectives. Denuders and backup filters are commonly omitted owing to their high maintenance and cost, as is the case for FRMs. In many situations the omission of denuders and backup filters results in minimal bias to mass measurements, but there are other conditions under which deviations from actual ambient concentrations are large. The nature of the aerosol being sampled, environmental sampling conditions (e.g., temperature and relative humidity), and the types of chemical analyses applied to the filter deposit must be evaluated before the first sample is taken.

Size-Selective Inlets

Inertial classifiers (see Chapter 10) are used as size-selective inlets to remove particles exceeding a specified aerodynamic diameter. Inlets are characterized by sampling effectiveness curves, showing the fraction of particles that pass through as a function of aerodynamic diameter. Sampling effectiveness is summarized by a 50% cut point (d_{50}), the diameter at which half of the particles pass through the inlet, and a slope (or geometric standard deviation), which is the square root of the ratio of the diameter of particles with 84% removal (d_{84}) to the diameter with a 16% removal (d_{16}). A slope of 1 indicates a step function that is impossible to obtain in practice. Many inlets have slopes of up to 1.5; the lower the slope, the "sharper" the cut point.

Sampling efficiency quantifies the mass fraction of suspended particles passing through the inlet and depends on the particle size distribution as well as the inlet sampling effectiveness. Sampling efficiency is obtained by integrating the product of sampling effectiveness and aerosol mass–size distribution across all expected particle sizes (Watson et al., 1983; Wedding and Carney, 1983). Because many ambient mass–size distributions peak near 10 µm, but are

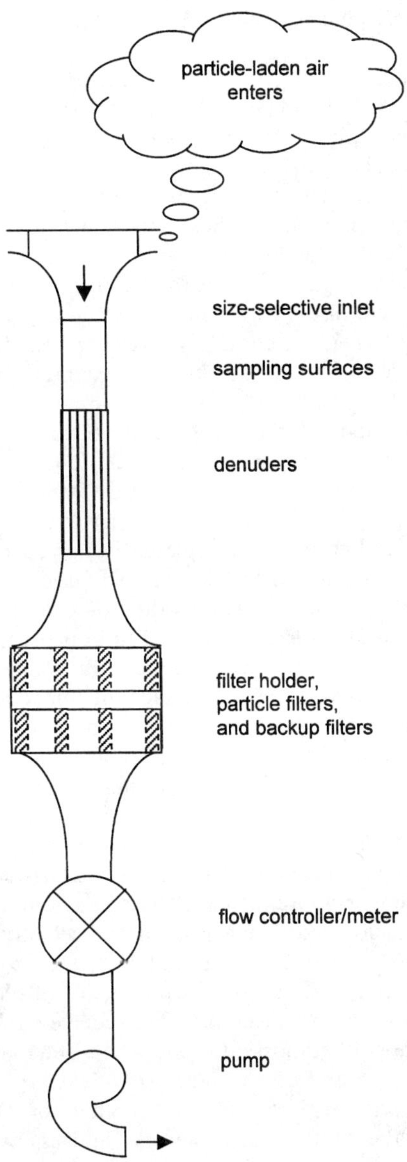

Fig. 27–1. Components of an aerosol sampling system.

at a minimum near 2.5 μm, changes in PM-10 inlet effectiveness curves have a larger effect on mass concentrations than do changes in PM-2.5 inlet effectiveness curves (Lundgren and Burton, 1995). PM-10 inlets with sharper cut points collect less mass than those with broader cut points. The dependence of mass on the sampling effectiveness slope is much smaller for PM-2.5 inlets.

Inlet transmission characteristics must be independent of wind speed and wind direction. The sampling efficiency of the rectangular peaked-roof inlet of the high-volume (hi-vol) sampler for TSP had a variable sampling effectiveness in response to wind direction and speed (Wedding et al., 1977; McFarland et al., 1980). PM-10 inlets have been symmetrically constructed to minimize these changes. Wind speeds and directions do not appreciably affect PM-2.5 cut points.

Chow (1995a), Hering (2001), and Chapter 10 describe a large number of size-selective inlets used for ambient, workplace, and personal exposure sampling. Table 27–1 identifies and describes several PM-10 and PM-2.5 inlets currently in use for ambient air sampling. Some of these are later models of those listed by Chow (1995a), although they are based on the same principles. Inlets of the same design are offered by different vendors under different names, and Table 27–1 attempts to classify these as a single design. Cut points and slopes of effectiveness curves are those given by the manufacturer, although different results are reported by independent tests from several of the citations. These differences result from a variety of test methods, the condition of the inlet at the time of the test, and experimental uncertainty. For PM-10 and larger size-cut inlets, large wind tunnels (McFarland et al., 1977; Wedding et al., 1977; Ranade et al., 1990) are used to examine changes over wind speeds of ~2 to 24 m/s.

Table 27–1 entries are grouped by the size separation principles (see Chapters 9 and 10) of direct impaction, impaction/elutriation, virtual impaction, cyclonic flow, and selective filtration. Direct impaction systems consist of one or more jets positioned above an impaction plate. The impactor dimensions are selected to allow particles with diameters exceeding the desired cut point to strike and adhere to the plate. Elutriator/impactors employ a vertical barrier before or after the impactor to further remove large particles by gravitational settling. When the particle settling velocity exceeds the upward air velocity, the particle is not transmitted through the inlet. The ***URG**** vertical elutriator precedes the impaction surface, while the other inlets in this category contain vertical transport pathways and sharp bends after the impaction surface. Virtual impactors replace the impaction surface with an opening that directs the larger particles elsewhere, sometimes to another filter; these impactors eliminate re-entrainment and separate PM-10 into PM-2.5 and coarse (PM-10 minus PM-2.5) fractions. Cyclones impart a circular motion to air entering the inlet with tangential vanes. This air enters a cylindrical or conical tube, and the centripetal force of the circular motion moves particles toward the walls of this tube and then to a collection hopper or "grit pot." Selective filtration uses filters or other porous materials that have consistent and measurable particle transmission properties. The Nuclepore etched polycarbonate filters (***CCO***) with 8 μm pores collect particles by interception and impaction in the vicinity of the pores to provide 50% cut points for particles between 2 and 3 μm at flow rates of ~1.7×10^{-4} m^3/s [10 L/min]. A ~0.4 μm pore size filter is place behind the 8 μm filter to collect the transmitted particles.

Mathai et al. (1990), in evaluating differences between several samplers used in western U.S. visibility studies, concluded that most differences between mass measurements from collocated sampling systems are caused by differences in inlet characteristics. These differences caused major controversies with respect to the development of PM-10 reference methods in the early 1980s (several of the citations in Table 27–1 address these controversies). Design theories for impactors (Marple and Willeke, 1976), virtual impactors (Loo and Cork, 1988; Sioutas and Koutrakis, 1998), and cyclones (Kenny and Gussman, 1997; Kenny et al., 2000) are in good agreement with empirical tests. However, as inlets become loaded with particles, their sampling effectiveness can change.

Impaction inlets require frequent cleaning and oiling or greasing to prevent impacted particles from disaggregating or becoming re-entrained in the air flow (e.g., John et al., 1991a,b). Even with the highly tactile surface used in the Well Impactor Ninety Six (WINS) PM-2.5 impactor/elutriator required in FRM samplers, Pitchford et al. (1997) observed a cone-like buildup that extended above the oil surface after 3 to 4 days of sampling. The tip of this could easily break off and become entrained in the air flow to the filter. Kenny et al. (2000) found that the WINS cut point shifted from 2.5 to 2.15 μm after its well was loaded with Aloxite

* Appendix I for full maunfacturer addresses referenced to the italieiced three-letter codes.

TABLE 27–1. Size-Selective Inlets and Characteristics for Ambient Aerosol Sampling

Name, Manufacturer, References[a]	Inlet ID: d_{50} (μm), Slope[b], Flow (L/min)	Description and Comments
Impactors		
Airmetrics Minivol Impactor (***ARM***) (Turner, 1998; Wiener and Vanderpool, 1992)	MV10: ~10, NA, 5 MV2.5: ~2.5, NA, 5	Machined polymeric propylene plastic or machined aluminum. PM-10 and PM-2.5 inlets are used in series for PM-2.5 sampling. Apiezon vacuum grease dissolved in hexane is pipetted onto impaction surfaces before each sample to minimize re-entrainment
Harvard Sharp Cut Impactors (***ADE***) (Marple et al., 1987; Turner et al., 2000)	MST123: 1, 1.22, 23 MST24: 2.5, 1.02, 4 MST210: 2.5, 1.06, 10 MST220: 2.5, 1.25, 20 MST104: 10, 1.11, 4 MST1010: 10, 1.09, 10 MST1020: 10, 1.06, 20	Machined aluminum that can also be coated with FEP Teflon
Impactor/elutriators		
Andersen (***GRA***) hi-vol PM-10 (John and Wang, 1991; Kashdan et al., 1986; McFarland et al., 1984; Ranade et al., 1990; Wedding et al., 1985)	G1200: 9.7, 1.4, 1, 133	Anodized spun aluminum with a single stage of opposing jets. The body is hinged to facilitate cleaning and re-greasing of the removable impaction plate that is sprayed with an aerosol adhesive after cleaning. The G1200 was preceded by the SA-320 single-stage PM-15 inlet and the SA321A and SA321B dual stage and SA321C single stage PM-10 inlets that are no longer sold but may still be in use. It is not entirely clear which sampling effectiveness tests apply to each of these inlets
Andersen (***GRA***) med-vol PM-10 (Olin and Bohn 1983)	SA254I: 10, 1.6, 113	Spun aluminum with 10 impactor jets and a central elutriation tube. The inlet can be disassembled for cleaning. The SA254I was preceded by the SA254, or "Blue Head" owing to the color of its enamel coating, that was nearly impossible to disassemble for cleaning
Andersen (***BGI***, ***GRA***, ***R&P***, ***URG***) Flat Top PM-10 (McFarland et al., 1978; Van Osdell and Chen, 1990; Wedding et al., 1980; Lai and Chen, 2000)	246B Flat Top: 10.2, 1.41, 16.7	Machined aluminum with one impactor tube and three vertical elutriator tubes. Rain drops are blown into the inlet beneath the flat top and accumulate on the impaction surface. Water exits through a small drain attached to a bottle on the outside of the inlet. The top unscrews for cleaning the impactor surface
FRM (***BGI***, ***GRA***, ***R&P***, ***URG***) Louvered PM-10 (U.S. EPA, 1997a)	Louvered PM-10: 10, NA, 16.7	Same materials and design as the SA246B but with a top that curves over the inlet bug screen to minimize the entry of windblown raindrops. Also available with a PTFE Teflon® coating
URG, Inc. (***URG***) Elutriator Impactors	25A: 1–2.5 & 10, NA, 4 22D: 2.5, NA, 1–20 30KN: 2.5, NA, 1–20	FEP Teflon coated glass with an elutriator tube preceding the impactor. The impactor plate is after the jet at the top of the elutriator, facing the ground
EPA (***BGI***, ***GRA***, ***R&P***, ***URG***) Well Impactor Ninety Six (U.S. EPA, 1997a; Kenny et al., 2000)	WINS: 2.48, 1.18, 16.7	Machined aluminum well with a detachable impactor jet. The impaction surface consists of a 37 mm quartz fiber filter immersed in 1 mL of vacuum pump oil to minimize particle re-entrainment over multiple sampling days between cleaning

(continued)

TABLE 27–1. *Continued*

Name, Manufacturer, References[a]	Inlet ID: d_{50} (μm), Slope[b], Flow (L/min)	Description and Comments
Virtual Impactors		
Andersen (***GRA***) Dichotomous Virtual Impactor (McFarland et al., 1978)	SA241: 2.5, NA, 16.7	Anodized machined aluminum integrated into the dichotomous sampler body. 2.8×10^{-5} m^3/s [1.7 L/min] goes through the virtual impactor to the coarse particle filter, with the remaining flow directed to the PM-2.5 filter. The unit can be disassembled for cleaning, but care must be taken to reassemble it such that the PM-10 inlet tube is over the virtual impaction tube rather than over the bypass zone to the PM-2.5 filter
VAPS (***URG***) Virtual Impactor	VAPSVI: 2.5, NA, 32	FEP Teflon coated aluminum. 3.3×10^{-5} m^3/s [2 L/min] goes through the virtual impactor to the coarse particle filter with two streams of 2.5×10^{-4} m^3/s [15 L/min] each directed to PM-2.5 denuders and filters
Harvard (***R&P***) Particle Concentrator (Sioutas et al., 1994a,b, 1997, 1998)	H1000: 2.5, NA, 1000	Particles are accelerated through a 0.32 × 280 mm rectangular nozzle and impact on an oil-soaked filter. Flow rates can be adjusted with different slit lengths. Particles less than 0.1 μm are diverted from the flow and not collected
Cyclones		
Wedding (***GRA***, ***TEI***) IP_{10} (Wedding and Weigand, 1985)	IP10: 9.6, 1.37, 1133	RFPS-1087–062. Spun aluminum inlet cleaning port on top of inlet
Andersen (***GRA***)/California Air Industrial Hygiene Laboratory Cyclone (AIHL) (John and Reischl, 1980)	SA3.68: 2.7, 1.16, 24 SA3.68: 2.3, 1.18, 28	Machined aluminum used on Andersen RAAS and IMPROVE speciation samplers
Sharp Cut Cyclones (***BGI***, ***GRT***, ***R&P***) (Kenny et al., 2000)	BGI1.062: 1.0, 1.21, 3.5 BGI2.229: 1, 1.17, 16.7 BGISCCA: 2.5, 1.19, 16.7 GRT1.118: 2.5, NA, 2 GRT2.141: 2.5, 1.24, 6.8 R&P1.829: 2.5, 1.23, 5	Nickel-plated aluminum. The cyclone has a horizontal design, with the hopper at one end. Particles are collected over the lower surface of the vortex cone and in the grit pot. A cap at the end facilitates regular cleaning
Bendix/Sensidyne 240 Medium Volume Cyclone (Chan and Lippmann, 1977; Mueller et al., 1983)	B240: 2.5, 1.7, 113	Welded steel that can be coated with PFA Teflon. Operates in a vertical position with a hopper on the bottom
URG, Inc. (***URG***) Cyclones (Moore and McFarland, 1993; Kenny et al., 2000)	30EHB: 1, 1.34, 16.7 30EN: 2.5, 1.32, 10 30EH: 2.5, 1.35, 16.7 30ED: 2.5, NA, 3 30EC: 3.5, NA, 28 30ENB: 10, NA, 16.7 30EA: 10, NA, 28.3	FEP Teflon coated aluminum. Cyclones can be operated vertically and horizontally

TABLE 27–1. *Continued*

Name, Manufacturer, References[a]	Inlet ID: d_{50} (μm), Slope[b], Flow (L/min)	Description and Comments
Stacked Filters		
Nuclepore Filters (***CCO***) (Spurny et al., 1969; Cahill et al., 1977; Parker et al., 1977; John et al., 1983; Droppo et al., 1995)	N8: ~2, NA, 10	Large pore (8 μm) polycarbonate filters remove particles >2 to 3 μm by interception and impaction. Small particles pass through the pores to another filter. Extensively used in U.S. visibility networks during the 1970s and early 1980s
CIS Foam (***BGI***) (Vincent et al., 1993)	CIS10: 10, NA, 3.5 CIS4: 4, NA, 3.5 CIS25: 2.5, NA, 3.5	A polyurethane foam with different thickness transmits particles with varying degrees. Used for personal sampling

[a] Refer to Appendix I for manufacturer codes and information. References cited provide more complete descriptions of the inlet and tests of its collection and transmission properties. This is a relatively complete list of inlets that are commercially available for ambient air sampling and are currently in use. It complements, rather than replaces, lists published elsewhere.

[b] Inlet IDs have been assigned for this table to facilitate later reference. These are contractions of the manufacturer's part number, where possible. "d_{50}" is the aerodynamic diameter at which half of the particles pass through the inlet and the other half deposit in the inlet, as determined by presentation and detection of known particle sizes. Slope = $\sqrt{d_{84}/d_{16}}$, the square root of the particle diameter ratios for inlet penetration at 84% and 16%. Values given are those provided by the vendor for the specified flow rate, and these may differ from those reported in some of the citations owing to different test and inlet conditions. "NA" in the slope position indicates that this value was not available.

dust. The U.S. EPA (1998b) recommends cleaning the WINS after five 24 h samples. Kenny et al. (2000) determined that the Sharp-Cut Cyclone had a sampling effectiveness curve similar to that of the WINS with a much larger collection capacity; no shift in cut point was found with increased loadings.

Table 27–1 shows that different PM-10 and PM-2.5 inlets have different cut points. The U.S. EPA (1987) required a PM-10 cut point of 10 ± 0.5 μm for FRMs. The Wedding IP-10 inlet had a 9.6 μm cut point, while the original SA-321A inlet had a 10.2 μm cut point. This difference made the Wedding inlet more attractive to some users because sampling with a lower cut point decreased the PM-10 measured and lowered the probability of exceeding the NAAQS. The Andersen SA-321A (***AND***) was replaced by the SA-321B with a 9.7 μm cut point to meet this competition; new impactor jets were also provided for the SA-321A inlet. Although both of these were replaced with the G1200 inlet described in Table 27–1, many of the original inlets are still in use, and it is difficult to distinguish them from each other by their appearance.

Inlet flow rates fall into ranges appropriate for high-volume (~0.017 m^3/s [1000 L/min]), medium-volume (~0.0017 m^3/s [100 L/min]), low-volume (~1.7×10^{-4} to 3.3×10^{-4} m^3/s [10 to 20 L/min]), and mini-volume (<8.3×10^{-5} m^3/s [5 L/min]) sampling systems. A massive volume (>0.17 m^3/s [10,000 L/min]) inlet and sampler has also been developed (Fitz et al., 1983) to obtain several grams of suspended particles for health studies. The medium- and high-volume inlets are especially useful when samples are taken in parallel on several substrates because flow rates can be kept high enough to obtain an adequate deposit for analysis.

Sampling Surfaces

Anodized aluminum is the most commonly used material for inlets and transfer tubes. Copper, stainless steel, conducting plastic, and glass are also used as transfer tubes in some systems. Nonconducting plastic surfaces and glass can acquire an electrical charge that might attract suspended particles to them, though the dimensions of most ambient sampling systems are sufficiently large that this attraction is negligible (Rogers et al., 1989).

Certain materials absorb or react with gases and particles, thereby removing them from the air stream (John et al., 1986; Hering et al., 1988). This is especially the case for nitric acid vapor, which sticks to nearly everything. Removal of nitric acid in an inlet or sampling duct can change the gas–particle equilibrium of particulate ammonium nitrate, causing this substance to dissociate into ammonia and nitric acid gases (Stelson and Seinfeld, 1982). This is also true for some volatile organic species.

John et al. (1986) tested different materials with respect to their affinity for nitric acid and found that surfaces coated with perfluoro alkoxy (PFA) Teflon can pass nitric acid with 80% to 100% efficiency. These results were confirmed by Neuman et al. (1999), who found that aluminum, stainless steel, nylon, glass, fused silica, silane-coated glass, silica-coated steel, and stainless steel tubes removed more than 80% of nitric acid over a 0.3 m tube length. PFA Teflon surfaces should be washed with a dilute solution of nitric acid to season them before sampling. Several of the Table 27–1 inlets and their associated transfer tubes are available with a Teflon coating to minimize removal of reactive gases from the air stream.

Denuders

Denuders (Kitto and Colbeck, 1999) are placed in the air stream to remove more than 95% of selected gases while transmitting more than 95% of the particles (see Chapter 19). Molecules in the gas phase diffuse rapidly to the surfaces of the denuder, while particles having lower diffusion coefficients pass through the denuder to the filter. The denuder surface can be coated with chemicals that retain the gas molecules. When properly treated and handled, the denuders can be extracted in a solvent for laboratory analysis to determine average gas concentrations over the sampling period. The denuder difference method (Shaw et al., 1982) operates two samplers in parallel, one with and one without a denuder. Gases and particles are collected on filters that naturally adsorb or are impregnated with gas-absorbing chemicals and are subsequently analyzed for the desired chemical compounds; the gas concentration is the difference between the nondenuded and denuded concentrations.

Tubular, parallel plate, annular, honeycomb, and cloth denuders have been used, the ***URG*** glass and Teflon-coated annular denuders being the most common. A tubular denuder (Gormley and Kennedy, 1949) is a single tube or a bundle of tubes. A bundle of tubes >0.3 m long removes gases at typical flow rates; a single tube is only practical for low (less than approximately 8.3×10^{-5} [5 L/min]) flow rates, depending on the diffusion coefficient of the gases. A parallel plate denuder (Eatough et al., 1993) is formed from rectangular panels with a small gap between them. An annular denuder (Possanzini et al., 1983; Allegrini et al., 1987; Koutrakis et al., 1988) consists of a solid center inside a tube of slightly larger diameter. Particles pass through the annulus between the two surfaces while the gases diffuse to them. ***URG*** annular denuders create several annuli by positioning tubes of slightly larger diameters inside one another. The honeycomb denuder (Koutrakis et al., 1993; Sioutas et al., 1994c, 1996) is a thick piece of glass or aluminum through which many tiny holes have been drilled or etched. These denuders can be as short as a few centimeters, depending on flow rate, and can be treated as part of a filter stack. Fitz and Motallebi (2000) evaluated loosely woven rayon fabrics that can be soaked in different chemicals to remove gases. Their tests show that PM-2.5 is collected with >95% efficiency through the fabric while nitric acid is quantitatively recovered from extraction and analysis of the fabric. These fabric denuders can be handled as if they were another component of a filter pack.

Kitto and Colbeck (1999) describe many of the coatings applied to denuders to retain specific gases. Magnesium oxide, sodium fluoride, sodium chloride, and sodium carbonate have been used to retain nitric acid. Ammonia has been removed with oxalic acid, tungstic acid, phosphoric acid, phosphorous acid, and citric acid. Organic gases have been removed by quartz filter strips (Fitz, 1990), carbon-impregnated filters (Eatough et al., 1993), finely ground

XAD resin (Gundel et al., 1995), and potassium hydroxide (Lawrence and Koutrakis, 1994).

The most common use of denuders is to obtain an accurate measurement of particulate ammonium nitrate, which often volatilizes during sampling as temperatures exceed 288 to 293 K [15° to 20°C] (Hering and Cass, 1999; Stelson and Seinfeld, 1982; Watson et al., 1994). Accurate nitrate measurements are obtained by removing nitric acid from the sample stream, collecting nitrate particles on a filter, and recovering volatilized nitric acid on an absorbing backup filter. John et al. (1988), while examining methods to estimate nitric acid dry deposition, found that an anodized aluminum denuder had a large capacity for nitric acid removal and developed one for the dichotomous sampler. Fitz and Hering (1996) determined that an anodized aluminum annular denuder (Chow et al., 1993) became saturated after 3 years of monitoring at a photochemically active southern California location; re-anodizing the surfaces was sufficient to restore its capacity.

Filter Holders

A filter holder consists of a frame that seals the edges of the filter so that air is forced to pass through its porous center. A porous support grid downstream of the filter keeps it from being sucked through the holder by the pump vacuum. Lippmann (2001) and Chow (1995a) describe several different filter holders. Filter holders should mate to the sampler and to the flow system without leaks, be composed of inert materials that do not absorb reactive gases, create a uniform particle deposit on the filter surface, have a low pressure drop across the empty holder, accommodate the sizes of commonly available air sampling filters (e.g., 37 or 47 mm), and be durable and reasonably priced. They should also allow for filter transport to and from the field sampler without contamination, minimize adherence or damage to the filter, and minimize removal of the substances being measured.

Filter holders are configured as in line or open faced. In-line holders concentrate particles in the center of the substrate, and this biases the results when analyses are performed on portions of the filter (Chow et al., 1994; Chow, 1995b). Inhomogeneous support grids, or grids with a low porosity (<50%), may also result in nonuniform deposits. Filter holders should be open faced or preceded by a diffusion zone that disperses the particles before collection.

The PM-2.5 FRM (U.S. EPA, 1997a) specifies the dimensions and materials for 47 mm diameter filter holder rings made of Delrin plastic with a stainless steel grid into which 100 μm diameter holes are etched 100 μm apart. Nuclepore (***CCO***) polycarbonate plastic filter holders accommodate 25, 37, and 47 mm diameter filters and are adaptable to many sampling systems. Each one costs a few tens of dollars, making it possible to purchase many of them for laboratory loading and unloading. Nuclepore filter holders are modified by drilling out the outlet hole to reduce flow resistance. Multiple extender sections can be used to stack several filters in series without having them touch each other. The rubber O-ring in the Nuclepore holder has been found to raise the carbon blank levels on quartz fiber filters, an artifact that can be eliminated by replacing the rubber with a Viton O-ring (Tombach et al., 1987). Little information is available on the extent to which the filter holder material affects the quantification of reactive gases. Filter holders made of or coated with Teflon are used to minimize losses of nitric acid. Several stainless steel filter holders are available, many of them with in-line connectors. These are usually expensive and absorb volatilized nitrate (Neuman et al., 1999). The speciation samplers described below usually have specially constructed filter holders that are specific to those samplers.

Filters

Filter media are judged for specific applications based on their mechanical stability, chemical stability, particle or gas sampling efficiency, flow resistance, loading capacity, blank values, artifact formation, compatibility with analysis methods, cost, and availability. Chow (1995a), Lippmann (2001), and Chapter 9 describe various types of filters, while Chapter 11 shows how filter characteristics relate to desired chemical analyses. Some filters are also used for optical transmission or reflection to measure particle absorption (Horvath, 1997) or to distinguish organic from elemental carbon (Birch and Cary, 1996; Chow et al., 2001), filter color and internal scattering properties important for these purposes.

U.S. EPA (1987) filter specifications for PM-10 compliance sampling include 0.3 µm dioctyl phthalate (DOP) sampling efficiency in excess of 99.9%, weight losses or gains due to mechanical or chemical instability of less than a 5 µg/m^3 equivalent, and alkalinity of less than 25 µEq/g to minimize absorption of sulfur dioxide (SO_2) and nitrogen oxides (NO_x). These are only the minimal requirements for samples that require chemical analyses. A sample from each batch (50 to 100 filters) must always be tested for contamination before sampling with that batch; high and variable blank levels typically invalidate subsequent quantification on particle deposits.

The most commonly used filter media for atmospheric particle and gas sampling are Teflon membrane, quartz fiber, nylon membrane, cellulose fiber, Teflon-coated glass fiber, etched polycarbonate membrane, and glass fiber. None of these materials is perfect for all purposes.

Ringed-Teflon membrane filters (***GEL***, ***WHA***) consist of a thin, porous polytetrafluoroethylene (PTFE) Teflon sheet stretched across a polymethylpentane ring; the thin membrane collapses without the ring, and the filter cannot be accurately sectioned into smaller pieces. The white membrane is nearly transparent and has been used to estimate light absorption (Campbell et al., 1995). PTFE Teflon is very stable, absorbing negligible water or gases. It has inherently low contamination levels, but chemicals have been found in some batches by acceptance testing. This filter is commonly used for mass and elemental analyses. The thin membrane is especially appropriate for X-ray and proton-induced fluorescence (XRF and PIXE) analyses that obtain elemental concentrations while leaving the filter intact. Teflon is hydrophobic, and emulsifiers (such as ethanol) need to be added for water-extraction methods for ion analysis. Carbon cannot be measured on Teflon membranes because of its high carbon content, although aerosol carbon has been inferred from hydrogen measurements (Kusko et al., 1989). Because of their high flow resistance and need for a ring support, high-volume flow rates are not attainable, though it is possible to obtain flow rates required for low-volume and medium-volume inlets. A variation on this filter consists of a PTFE Teflon membrane mounted on a woven PTFE mat instead of a support ring. This filter has a higher density, but it can be obtained in larger sizes for higher volume samples. The membrane and the mat sides are similar in appearance, and care must be taken to prevent mounting it upside down, particles being drawn through the mat rather than onto the surface of the membrane in this case.

Quartz fiber filters (***PAL***, ***WHA***) consist of a tightly woven mat of quartz filaments. These filters meet requirements in most categories. Quartz fiber filters adsorb hydrocarbon gases during sampling (McDow and Huntzicker, 1990), as determined by placing a clean quartz filter behind a Teflon filter (Turpin et al., 1994). The extent to which this is a positive bias due to gas adsorption or a negative bias due to the collection of volatilized particles is uncertain (Cui et al., 1997, 1998). Quartz filters are baked at ~117 °K before sampling to remove adsorbed organic vapors. Blank levels are high and variable for several elements (especially aluminum and silicon), though newer formulations are cleaner than earlier formulations. These filters are widely used for ion and carbon analyses. The greatest drawback of quartz fiber filters is their fragility; they require extremely careful handling for accurate mass measurements. The Whatman QM/A quartz fiber filter contains a 5% borosilicate glass binder that minimizes its friability. This filter is often used in high-volume PM-10 samplers for mass measurements. The Pallflex 2500 QAT-UP filter is pure quartz and undergoes a distilled water washing (thus the "UP," or "ultrapure," designation).

Nylon membrane filters consist of thin sheets of porous nylon and are used almost exclusively for the collection of nitric acid. Nylon filters have high flow resistance that increases with filter loading. These filters passively absorb nitric acid, depending on how long they have been exposed to an acid-rich environment, and they should be washed in a sodium bicarbonate solution followed by distilled water before use in the field. They also absorb small amounts of ammonia (Masia et al., 1994) and sulfur dioxide (Sickles and Hodson, 1999), thereby limiting their use in front of filters intended to collect these gases. Gelman Nylasorb (***GEL***) is most commonly used for ambient air sampling.

Cellulose fiber filters (***WHA***) consist of a tightly woven paper mat. These filters meet requirements in most categories with the exception of sampling efficiency and water vapor artifacts. Particle penetration for submicrometer particles have been observed, but these are highly dependent on the filter weave (Biles and Ellison, 1975). Cellulose fiber is hygroscopic and requires precise relative humidity control in the filter processing environment to obtain accurate mass measurements (Demuynck, 1975). These filters have low chemical blanks except for carbon, and they can be used for elemental and ionic analyses of the deposit. Cellulose fiber filters can be impregnated with gas-absorbing compounds, similar to those described above for denuders, and located behind more efficient particle-collecting filters. Nitric acid, ammonia, sulfur dioxide, and nitrogen dioxide gases are determined by this means. The most commonly used cellulose fiber filter is Whatman 41.

Teflon-coated glass fiber filters (***PAL***) imbed a Teflon slurry onto a loosely woven glass fiber mat. These filters meet requirements in all categories except blank element and carbon levels. Though a small amount of nitric acid absorption has been observed (Mueller et al., 1983), it is tolerable in most situations. These filters are used for ion analyses and for specific organic compounds, but not total carbon analyses owing to their Teflon coating. The most commonly used Teflon-coated glass fiber filters are Pallflex TX40HI20 and T60A20 (***PAL***).

Etched polycarbonate membrane filters (***CCO***) are constructed from a thin polycarbonate sheet through which pores of uniform diameter have been produced by radioactive particle penetration and chemical etching. Depending on pore size, these filters have variable particle collection properties, as noted in Table 27–1, that are used for size-specific measurements. These filters are nearly transparent and have been used to estimate light absorption (Horvath, 1993). Polycarbonate membrane filters have low elemental blank levels and are appropriate for elemental and ion analysis. They are the best filter media for single-particle analysis by electron microscopy because particles are easily distinguished from the flat filter surface. These filters are not used for thermal evolution carbon analysis owing to their carbon composition. The filters hold an electrostatic charge that influences mass measurements unless substantial effort is invested in discharging them with a small radioactive source (Engelbrecht et al., 1980). Electrostatic discharging is good practice for all filter media, even though others do not retain as much charge as the polycarbonate membranes. The Nuclepore 8.0 and 0.4 μm filters are most commonly used in ambient aerosol sampling. While the 0.2 μm pore size filter provides a higher sampling effectiveness, its higher flow resistance requires a large pressure drop across the filter for $>1.7 \times 10^{-4}\,m^3/s$ [10 L/min] flow rates.

Glass fiber filters (***WHA***) are obsolete and should not be considered for use in particle sampling. These filters consist of a mat of borosilicate glass filaments. The high alkalinity of these substrates causes sulfur dioxide, nitrogen oxides, and gaseous nitric acid to be adsorbed (Coutant, 1977; Spicer and Schumacher, 1979), thereby biasing mass and chemical measurements. Blank levels for most elements of interest are extremely high and variable (Witz et al., 1983). As with quartz filters, glass fiber filters adsorb organic carbon vapors that are measured as particulate carbon during analysis. Unlike quartz filters, the sodium in glass fiber filters catalyzes the combustion of elemental carbon at lower temperatures, causing biases in the quantification of the light-absorbing carbon fraction (Lin and Friedlander, 1988). Although their cost per filter is lower than that of the other filters, the savings are not worth the expense of compromised data.

Flow Measurement and Control

The quantity of air per unit time must be precisely measured and controlled to determine particle concentrations and to maintain the size-selective properties of the inlet. Manual volumetric, automatic mass, differential pressure volumetric, pump speed volumetric, and critical orifice volumetric flow control principles have been applied to aerosol samplers.

Manual control is accomplished when the operator initializes a setting, such as a valve adjustment, and then relies on the known and constant functioning of sampler components, such as pumps and tubing, to maintain flows within specifications. Flow rates that are set manually change as the collection substrate loads up and presents a higher flow resistance. For most concentrations ($<200\,\mu g/m^3$), the flow will not change by more than 10% during sampling, and the average of flow rates taken before and after sampling provides an accurate estimate of the actual flow.

Mass flow controllers use temperature sensors that measure the heat transfer between two points in the gas stream. To a first approximation, the heat transfer is proportional to the number of gas molecules passing between the two points, and hence the mass flow controller is able to sense the mass flux. An electronic feedback loop compensates for temperature and pressure variations of the gas and the sensing probe. Wedding (1985) estimated potential differences in excess of 10% between mass and volumetric measurements of the same flow rates, depending on temperature and pressure variations.

Differential pressure volumetric flow control maintains constant pressure across an orifice (usually a valve that can be adjusted for a specified flow rate) by a control valve located between the filter and the orifice. When the pressure between the filter and the valve increases, as it does when filters load up, the valve control opens to allow more air to pass.

The pump volumetric method changes the dc current to a variable speed pump in response to a flow sensor at the pump inlet or exhaust. As the flow rate decreases with filter loading, more current is supplied to the pump, thereby increasing its speed and throughput.

A critical orifice consists of a small circular opening between the filter and the pump. When the downstream pressure at the minimum flow area downstream of the orifice is less than 53% of the upstream pressure, the air velocity attains the speed of sound and it will remain constant, regardless of increased flow resistance. Critical orifices provide stable flow rates, but they require large pumps and low flow rates (typically less than $3.3 \times 10^{-4}\,m^3/s$ [20 L/min] with commonly available pumps) to maintain the high pressure difference. Wedding et al. (1987) developed a "critical throat" that uses a diffuser arrangement to recover much of the energy expended in back pressure behind a critical orifice. This design allows higher flow rates to be obtained with a given pump.

Flow Movers

Air is passed through the sampling substrates by means of a vacuum created by a pump. Monteith and Rubow (2001) and Chow (1995a) describe commercially available air pumps and their capacities and operating principles. High-volume samplers typically use a radial flow fan that rotates curved blades at a high velocity to move substantial quantities of air. The original fans for the TSP hi-vol were manufactured for household vacuum cleaners and do not tolerate large pressure drops caused by complex inlets, tightly woven filters, and stacks of filters and denuders. Hinds (1999) shows that a 20 kPa [15 cm Hg] resistance across the hi-vol fan reduces the flow to zero. Diaphragm and piston pumps convert the rotary motion of a driveshaft into the movement of a flexible membrane or piston that draws air through an opening on the backside of the filter and pushes it out through another opening at the pump exit. Hinds (1999) shows diaphragm pumps capable of 80 kPa [60 cm Hg] vacuum when flow is reduced to zero. The rotary pump turns carbon vanes within a circular chamber with an offset driveshaft. The centripetal force of the vanes seals their ends against the chamber surface. Air enters when the volume between vanes is large and exits when the volume decreases where the driveshaft is closest to the chamber walls. These pumps can create a vacuum of up to 93 kPa [70 cm Hg] (Hinds, 1999) and can pull more than 0.0017 m^3/s [100 L/min] through a stack of filters with high flow resistance. Carbon vane pumps are noisy, and their exhaust must be filtered as the carbon vanes are continually abrading into small particles.

SAMPLING SYSTEMS

Monteith and Rubow (2001), Hering (2001), and Chow (1995a) describe a large number of systems for ambient aerosol sampling. Chow (1995a) cites many researchers who have assembled different components (Fig. 27–1) to achieve research objectives beyond those of compliance monitoring. Typical research needs are

- Simultaneous sampling on multiple filters for different analyses: Parallel samples on Teflon membrane and quartz fiber filters are needed for simultaneous mass, elemental, ionic, and carbon analyses.
- Higher sample volumes for shorter sampling intervals: Samples of 3 to 6 h duration are needed to evaluate the diurnal evolution of chemical components, and higher flow rates are often needed to obtain enough sample for laboratory quantification.
- Gas and particle phase measurements of volatile species: Accurate measurements of ammonium nitrate and some volatile organic compounds require denuders and backup filters. Gas phase precursors collected by denuders are also needed when equilibrium models are used to determine which precursors limit particle formation.
- Multiple sequential samples without operator intervention: It is not practical or cost-effective for an operator to change samples every few hours for diurnal sampling or at midnight for everyday sampling, thereby requiring an automated sampler changer.
- Several particle size fractions: Multiple inlets, in series or in parallel, are needed to separate particles into different size fractions, with PM-10, PM-2.5, and coarse (PM-10 – PM-2.5) being the most widely used fractions.

Many of these systems are based on FRMs developed for PM-10 or PM-2.5 compliance monitoring. Table 27–2 specifies sampling systems that have attained FRM or federal equivalent method (FEM) status with their EPA designation numbers and *Federal Register* announcements.

High-volume samplers designated for PM-10 compliance monitoring are being replaced by low-volume samplers that are similar for PM-10 or PM-2.5; the PM-2.5 include a WINS inlet behind the PM-10 inlet. Most of the newer units control the volume throughout sampling by measuring the mass flow and then adjusting this for ambient temperature and pressure conditions. These units are controlled by a microprocessor that also logs flow rate, temperature, and pressure over periods as short as 1 min throughout the sampling period.

Chow et al. (1996) outline various strategies for applying FRMs to chemical speciation, including duplicate samplers with different filter media, collocated high volume and dichotomous samplers with quartz and Teflon filters, and supplementing a permanent FRM sampler with temporary "mini-vol" samplers using complementary filter packs and size-selective inlets.

The Oregon Department of Environmental Quality PM-10 Sequential Filter Sampler (SFS) has proved itself very adaptable for a variety of purposes. The transfer tube has been filled with aluminum denuder tubes, and quartz filters have been backed with sodium chloride or nylon filters to measure nitrate accurately. Two filter packs can be sampled in parallel with Teflon membrane and quartz fiber filters for different analyses. Flow rates from

TABLE 27–2. PM-10 and PM-2.5 Federal Reference Methods and Federal Equivalent Methods[a]

Sampler, EPA FRM Designation, Federal Register Notice	Description
Federal Reference Methods	
Andersen (***GRA***) RAAS10-100 PM-10 Single Channel PM-10 Sampler, RFPS-0699-13064, FR 33481, 06/23/99 RAAS2.5-100 PM-2.5 Sampler, RFPS-0598-119, 63 FR 31991, 06/11/98 RAAS10-200 PM-10 Audit Sampler, RFPS-0699-131, 64 FR 33481, 06/23/99 RAAS2.5-200 PM-2.5 Audit Sampler, RFPS-0299-128, 64 FR 12167, 03/11/99	Louvered PM-10 $2.78 \times 10^{-4} m^3/s$ [16.7 L/min] inlet with anodized aluminum sampling surfaces. FRM Delrin 47 mm ringed filter holder. Variable speed diaphragm pump. Volumetric flow measured by dry test meter at pump outlet that modulates pump speed. PM-2.5 samplers include a WINS impactor in the sample line after the PM-10 inlet. Audit samplers are of similar design but more compactly packaged for periodic collocated sampling with a permanent unit
Andersen (***GRA***) RAAS10-300 PM-10 Sequential Sampler, RFPS-0699-132, 64 FR 33481, 06/23/99 RAAS2.5-300 PM-2.5 Sequential Sampler, RFPS-0598-120, 63 FR 31991, 06/11/98	Similar to RAAS10-100 and RAAS2.5-300, but with a circular tray into which six 47 mm ringed filter holders can be located. A timer rotates the tray at preset intervals, and a plunger compresses the holder into the sample flow stream
BGI, Inc. (***BGI***) PQ100 FRM PM-10 Sampler, RFPS-1298-124, 63 FR 69625, 12/17/98 PQ200 or PA200A PM-10 FRM Sampler, RFPS-1298-125, 63 FR 69625, 12/17/98 PQ200 or PQ200A PM-2.5 FRM Sampler, RFPS-0498-116, 63 FR 18911, 04/16/98	Louvered PM-10 $2.78 \times 10^{-4} m^3/s$ [16.7 L/min] inlet with anodized aluminum sampling surfaces. FRM Delrin 47 mm ringed filter holder. Variable-speed double diaphragm pump controlled by mass flowmeter adjusted every minute to ambient temperature and pressure for volumetric equivalent. The PM-2.5 unit is identical to the PM-10 sampler with addition of a WINS impactor after the PM-10 inlet. The audit sampler (PQ200A) is of similar design but more compactly packaged for periodic collocated sampling with a permanent unit. These units can operate for 24 h on a battery
Rupprecht & Patashnick (***R&P***) Partisol FRM 2000 PM-10 Sampler, RFPS-1298-126, 63 FR 69625, 12/17/98 Partisol FRM 2000 PM-2.5 Sampler, RFPS-0498-117, 63 FR 18911, 04/16/98 Partisol 2000 Audit Sampler, RFPS-0499-129, 64 FR 19153, 04/19/99	Louvered PM-10 $2.78 \times 10^{-4} m^3/s$ [16.7 L/min] inlet with anodized aluminum sampling surfaces. Delrin 47 mm ringed filter holder with a tapered edge for sequential version. Differential pressure flow control using a servo-controlled valve controlled by a mass flowmeter adjusted every minute to ambient temperature and pressure for volumetric equivalent. WINS inlet added for the PM-2.5 model. An optional insulating jacket is available for cold weather. The audit sampler is packaged for portability
Rupprecht & Patashnick (***R&P***) Partisol Plus 2025 PM-10 Sequential Sampler, RFPS-1298-127, 63 FR 69625, 12/17/98 Partisol-Plus 2025 Sequential Sampler, RFPS-0498-118, 63 FR 18911, 04/16/98	Same as R&P Partisol FRM 2000 samplers but with a pneumatically controlled sampler insertion mechanism. Up to 18 47-mm filter holders can be stacked in a tube and inserted into the sample stream at preset intervals
Rupprecht & Patashnick (***R&P***) Partisol 2000 PM-10 Sampler, RFPS-0694-098, 59 FR 35338, 7/1/94	Similar to *R&P* Partisol FRM 2000 samplers, but not meeting all PM-2.5 FRM specifications. The sampler consists of a hub unit with up to three satellite units for sequential sampling

TABLE 27–2. *Continued*

Sampler, EPA FRM Designation, Federal Register Notice	Description
Thermo Environmental Instruments, Inc. (***TEI***) 605 "CAPS" PM-2.5 Sampler, RFPS-1098-123, 63, FR 58036, 10/29/98	Similar to the above FRM PM-2.5 samplers, but never made commercially available. Included here for completeness
URG Corp. (***URG***) URG-MASS100 Single PM-2.5 Sampler, RFPS-0400-135, 65 FR 26603, 5/8/2000 URG-MASS300 Sequential PM-2.5 Sampler, RFPF-0400-136, 65 FR 26603, 5/8/2000	Louvred PM-10 $2.78 \times 10^{-4}\,m^3/s$ [16.7 L/min] inlet with anodized aluminum sampling surfaces followed by WINS impactor with FRM Delrin 47 mm filter holders. Variable-speed diaphragm pump with volumetric flow measured by a dry test meter at the pump outlet that modulates pump speed. The sequential system contains a circular tray holding six filters. Filters are rotated into the sampling section at preset intervals and compressed into the flow stream with a mechanical plunger
Andersen/GMW (***GRA***) 1200 High-Volume PM-10 Sampler, RFPS-1287-063, 52, FR 45684, 12/01/87, 53 FR 1062, 01/15/88 321-B High-Volume PM-10 Sampler, RFPS-1287-064, 52 FR 45684, 12/01/87, 53 FR 1062, 01/15/88 321-C High-Volume PM-10 Sampler, RFPS-1287-065, 52 FR 45684, 12/01/87, 53 FR 1062, 01/15/88	Andersen G1200 $1.9 \times 10^{-2}\,m^3/s$ [1130 L/min] single-stage greased PM-10 inlet and 8 × 10 inch framed filter holder with steel screen support. Spun and sheet aluminum surfaces. Flow controlled by mass flowmeter that varies the speed of a blower motor. Also available with a critical throat, thus two designation numbers. The 321B and 321C were identical, but equipped with earlier versions of the G1200 that did not open for cleaning. These are no longer sold, but many are still in service. When the inlet is replaced with a peaked roof, this sampler is identical to the hi-vol TSP sampler that is widely used throughout the world
Andersen/GMW ***(GRA)*** SA241 and SA241M Dichotomous PM-10/PM-2.5 Sampler, RFPS-0789-073, 54 FR 31247, 07/27/89	Flat top SA246B PM-10 inlet and dichotomous virtual impactor for $2.78 \times 10^{-4}\,m^3/s$ [16.7 L/min] flow rates. Made from machined aluminum. Flow rates are controlled by a differential pressure regulator and a single-speed diaphragm pump. Ten percent of the fine particles are sampled on the coarse particle filters, and corrections must be made to the coarse particle measurements to compensate
Oregon DEQ PM-10 Sequential Filter Sampler, RFPS-0389-071, 54 FR 12273, 03/24/89	SA245I PM-10 inlet at $1.9 \times 10^{-3}\,m^3/s$ [113 L/min] on top of a ~0.4 m aluminum transfer tube that leads to a large removable plenum. 47 mm Nuclepore polycarbonate filter holders are located in quick disconnect fittings under the plenum. Solenoid valves beneath the plenum are switched by a timer for up to 12 sequential samples without operator intervention. Flow rates are controlled by a differential pressure regulator and a GAST 1022 or 1023 carbon vane vacuum pump with filtered exhaust
Wedding 600 ***(TEI)*** High-Volume PM-10 Sampler RFPS-1087-062, 52 FR 37366, 10/06/88	Wedding IP-10 cyclonic flow inlet followed by an 8 × 10 inch filter frame with flow controlled by a critical throat before the blower fan. Larger capacity blowers are used owing to the larger pressure drop across the throat

(continued)

TABLE 27–2. *Continued*

Sampler, EPA FRM Designation, Federal Register Notice	Description
Federal Equivalent Methods (FEM)	
Met One ***(PAC)*** BAM 1020, GBAM 1020, BAM 1020-1, and GBAM 1020-1 PM-10 Beta Attenuation Monitors, EQPM-0798-122, 63 FR 41253, 08/03/98	Flat-top 246B style $2.78 \times 10^{-4}\,m^3/s$ [16.7 L/min] PM-10 inlet, aluminum transfer tubes. Particles are collected on a quartz fiber filter tape each hour. The attenuation of electrons from a radioactive source is related to mass loading by calibration to athin metal sheet. Hourly PM-10 measurements are possible
Andersen ***(GRA)*** FH621-N PM-10 Beta Attenuation Monitor, EQPM-0990-076, 55 FR 38387, 09/18/90	Similar in principle to Met One BAM, but with different radioactive electron source, tape advance unit, and data-processing system
Wedding ***(TEI)*** 650 PM-10 Beta Attenuation Monitor, EQPM-0391-081, 56 FR 9216, 03/05/91	Similar in principle to the Met One unit, but using a Teflon membrane tape. No longer available but included for completeness
Rupprecht & Patashnick ***(R&P)*** TEOM Series 1400/1400a PM-10 Monitors, EQPM-1090-079, 55 FR 43406, 10/29/90	Flat-top $2.78 \times 10^{-4}\,m^3/s$ [16.7 L/min] PM-10 inlet. Aluminum inlet, copper transfer tube, and a heater (323 K [50°C]) for PM-10 equivalence, but adjustable to other temperatures. $5 \times 10^{-5}\,m^3/s$ [3 L/min] of air is drawn through a 12.7 mm diameter quartz filter on a hollow glass tube that vibrates. The vibration frequency changes as the filter mass increases with aerosol deposit, and this is related to mass concentration. Five minute averages are attainable in ambient air. A bypass is available to collect the remaining $2.28 \times 10^{-4}\,m^3/s$ [13.7 L/min] of makeup air on a filter for other analyses

[a] Reference and equivalent method requirements are specified in 40 CFR Part 50 and by the U.S. EPA (1987, 1997b). FEMs are designated after collocated sampling with FRMs in several different environments.

3.3×10^{-4} to $1.9 \times 10^{-3}\,m^3/s$ [20 to 113 L/min] have been drawn through the filter packs, with a makeup flow supplying the difference to retain the inlet cut point. A Bendix 240 PM-2.5 replaces the SA254I inlet for PM-2.5 particle measurements. Chow et al. (1993) demonstrated how a variation of this sampler could be coated with PFA Teflon for sampling of reactive species.

The *I*nteragency *M*onitoring of *Pr*otected *V*isual *E*nvironments (IMPROVE) network has operated at U.S. National Parks and Monuments since 1987 to provide a long-term database of chemical concentrations in support of regional visibility assessment. The IMPROVE sampler (Eldred et al., 1988) consists of different modules controlled by a common sample switching system. Each module can be tailored to a specific type of measurement. The standard configuration acquires one PM-10 sample for gravimetric and proton-induced X-ray emission (PIXE) elemental analyses and three PM-2.5 samples, one on a Teflon membrane for gravimetric and PIXE analyses, one on a nylon membrane for ion analysis, and one on a quartz fiber filter for carbon analysis. Up to four sequential samples can be taken without operator intervention by using a timer to switch solenoid valves. Flow rates are controlled by critical orifices.

The initiation of a PM-2.5 speciation monitoring network by the U.S. EPA has resulted in several commercially available monitoring systems. Figure 27–2 shows a typical configuration

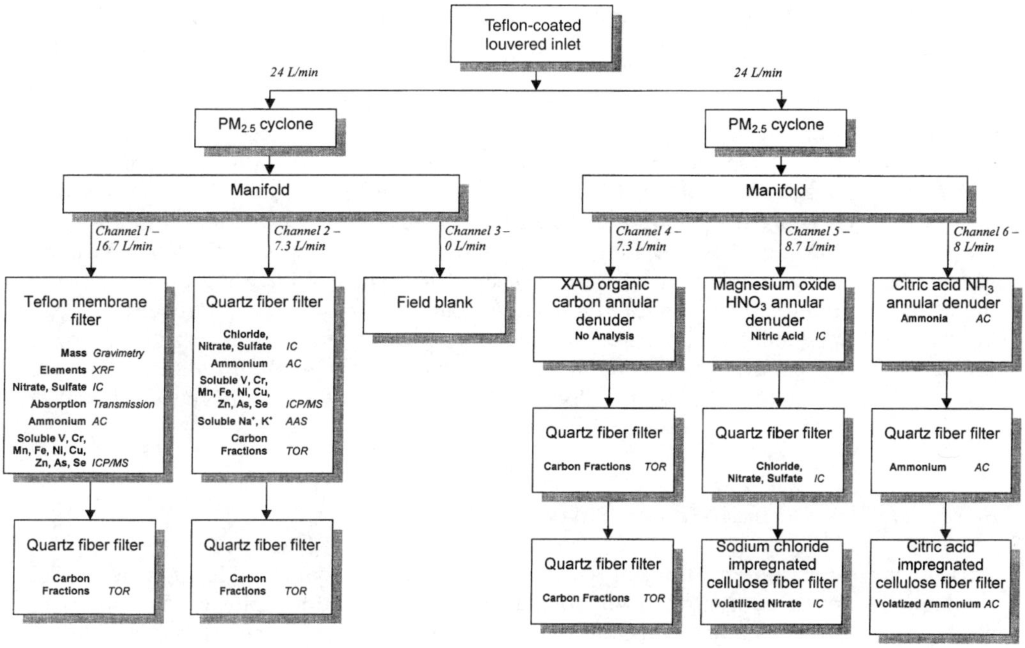

Fig. 27–2. Example sample configuration and analyses for the ***AND*** PM-2.5 RAAS. AAS, atomic absorption spectrometry; AC, automated colorimetry; IC, ion chromatography; ICP/MS, inductively coupled plasma/mass spectrometry; TOR, thermal/optical reflectance; Transmission, light transmission; and XRF, X-ray fluorescence.

based on the Reference Ambient Air Sampler (RAAS) speciation monitor (***GRA***). The sampler pulls $8 \times 10^{-4}\,m^3/s$ [48 L/min] through a louvered PM-10 inlet without impaction jets that removes very coarse particles to minimize maintenance of the PM-2.5 inlets. The flow is split and directed through two AIHL cyclone inlets into manifolds to which custom filter holders are attached. The filter holders are configured to contain up to three filters in series. Flow is controlled by critical orifices behind each filter with a vacuum created by a large rotary vane pump. The six available channels can be configured in different ways with flow rates varied by changing the orifices. All of the aluminum and stainless steel parts are coated with PTFE Teflon. Similar speciation monitors include the MASS and VAPS units from URG, Inc. (***URG***), the SASS from MetOne (***PAC***) and the Partisol 2300 and 2025 from Rupprecht & Patashnick, Inc. (***R&P***).

Good examples of aerosol samplers assembled from existing parts for specific purposes are represented by the "BOSS" (Brigham Young University Organic Sampling System) series of samplers (Eatough et al., 1993, 1995, 1996; Tang et al., 1994; Cui et al., 1997, 1998). These samplers allow filters to be located before and after denuders to evaluate the difference between artifact vapor adsorption and adsorption of volatilized particles. The PC-BOSS (Cui et al., 1998; Pang et al., 1998) uses the Harvard virtual impactor to concentrate particles for organic compound speciation. Annular denuders are used for inorganic species, while parallel plate denuders lined with carbon-impregnated filters are used for organic measurements.

Several cascade impactors have been developed to partition PM-10 into smaller size ranges on substrates suitable for chemical analysis. Each of these impaction methods uses low pressure stages to attain the lower particle cut points. The Low Pressure Impactor (LPI) (Hering et al., 1979a,b), the Davis Rotating-drum Unit for Monitoring (DRUM) (Raabe et al., 1988), the Berner impactor (Berner et al., 1979), and the Micro Orifice Uniform Deposit Impactor (MOUDI) (Marple et al., 1981) have been applied in ambient aerosol chemistry and visibility studies.

The low flow rate ($1.7 \times 10^{-5}\,m^3/s$ [1 L/min]) and nonuniform deposits, as well as substantial handling of the substrates, make the LPI impractical for many studies, though it was a precursor to the later impactors and provided valuable information for its time. The DRUM is a cascade impactor sampling at $5 \times 10^{-4}\,m^3/s$ [30 L/min] with low-pressure stages that acquires particle deposits in nine size ranges from 0.07 to 8.5 μm. Mylar impaction substrates are mounted on cylinders that rotate below the impaction jet so that particle deposits can be monitored as a function of time and analyzed by PIXE. The MOUDI and Berner cascade impactors are different in design but similar in function. These impactors operate at $5 \times 10^{-4}\,m^3/s$ [30 L/min] and have been used to acquire carbon and ion deposits on substrates that are amenable to chemical analysis by thermal evolution carbon analysis on aluminum foils and sulfate and nitrate analyses on Teflon substrates in at least eight size ranges from 0.03 to 15 μm.

SELECTING A SAMPLING SYSTEM

This chapter is not intended to identify every existing or every conceivable ambient aerosol sampling system. It concentrated on sampler components and sampler configurations in common use for which a body of knowledge and confidence have been established. Nevertheless, the practitioner is presented with many choices when confronted with the challenge of designing, installing, operating, and using the data from a sampling network to accomplish a specific purpose. The following steps can be used to select and implement a particle sampling system.

The first step is to clearly define monitoring objectives (see Chapter 7). Five types of monitoring goals were given as examples in the first part of this chapter. The specific objectives should be as comprehensive as possible. For example, networks of PM-10 and PM-2.5 FRMs have been established for the immediate objective of determining compliance with NAAQS. If one or more of these samples exceeds a standard, then the objectives broaden to include source apportionment. Sampling that meets the first objective is typically inadequate for the source apportionment objective that requires a speciation monitor and chemical analysis.

The second step is to determine the particle size fractions, chemical analyses, sampling frequencies, and sample durations needed to address the objectives. More frequent samples, or samples taken at remote locations, may require a sequential sampling feature. Shorter sample durations may require a larger flow rate to obtain an adequate sample deposit for analysis.

The third step is to calculate the expected amount of deposit on the filter for each chemical species and compare it with typical detection limits for the analyses being considered. The types of analyses and size fractions desired affect the number of sampling ports and different filter media needed. Chow (1995a) specifies lower quantifiable limits for many analysis methods that can be translated into the needed aerosol deposit. Urban samples acquire adequate deposits for analysis with flow rates as low as $\sim 3.3 \times 10^{-4}\,m^3/s$ [20 L/min] for as low as 4 h durations, while samples at nonurban sites may require larger flows and sample durations to obtain an adequate deposit. The analytical laboratory should be involved at the sampler design stage to ensure compatibility between sampling methods, analysis methods, filter media, and detection levels for the analytical methods.

The fourth step is to create, adapt, or purchase a sampling system that provides the most cost-effective and reliable means of meeting the monitoring needs. It is also worthwhile to contact the authors of cited references for the noncommercial sampling systems because these investigators may be in a position to loan, rent, sell, or build one of these units to meet a special need. If no existing unit can fulfill all of the requirements, then it will be necessary to assemble different sampling components into a new configuration that will meet those needs.

The final step is to create or adapt an operating procedure that specifies methods and schedules for inlet cleaning, filter transport and handling, calibration and performance tests, and record keeping. Sampling systems described here have such written procedures that can serve as guides for specific procedures.

CONCLUSIONS

Ambient aerosol sampling systems have evolved in order to meet different monitoring needs. Sampler inlets, monitoring surfaces, denuders, filter media, filter holders, flow monitors, flow movers, and operating procedures have been developed and tested. These components have been integrated into various sampler configurations for application in dozens of studies. Using these components and the lessons learned from previous studies, it is possible to develop sampling systems customized to meet specific objectives without extensive original development and testing.

ACKNOWLEDGMENTS

The authors are grateful to Barbara Hinsvark and Steve Kohl of DRI for compiling data on different inlet and aerosol samplers. Fruitful discussions with Mr. Robert Gussman of BGI, Inc., on inlet testing and FRM development history and with Ms. Julie Morris of URG, Inc., on sampling surfaces and filter holders also contributed to this chapter. Norman Mankim of DRI assisted with the manuscript and references.

REFERENCES

Allegrini, I., F. de Santis, V. Di Palo, A. Febo, C. Perrino, M. Possanzini, and A. Liberti. 1987. Annular denuder method for sampling reactive gases and aerosols in the atmosphere. *Sci. Total Environ.* 67: 1–11.

Berner, A., C. H. Lurzer, L. Pohl, O. Preining, and P. E. Wagner. 1979. The size distribution of the urban aerosol in Vienna. *Sci. Total Environ*. 13:245–261.

Biles, B. and J. M. Ellison. 1975. The efficiency of cellulose fiber filters with respect to lead and black smoke in urban aerosol. *Atmos. Environ.* 9:1030–1032.

Birch, M. E. and R. A. Cary. 1996. Elemental carbon-based method for monitoring occupational exposures to particulate diesel exhaust. *Aerosol Sci. Technol.* 25:221–241.

Cahill, T. A., L. L. Ashbaugh, J. B. Barone, R. A. Eldred, P. J. Feeney, R. G. Flocchini, C. Goodart, D. J. Shadoan, and G. Wolfe. 1977. Analysis of respirable fractions in atmospheric particulates via sequential filtration. *JAPCA* 27:675–675.

Campbell, D. S., S. Copeland, and T. A. Cahill. 1995. Measurement of aerosol absorption coefficient from Teflon filters using integrating plate and integrating sphere techniques. *Aerosol Sci. Technol.* 22: 287–292.

Chan, T. and M. Lippmann. 1977. Particle collection efficiencies of sampling cyclones: An empirical theory. *Environ. Sci. Technol.* 11:377–386.

Chow, J. C. 1995a. Critical review: Measurement methods to determine compliance with ambient air quality standards for suspended particles. *JAWMA* 45:320–382.

Chow, J. C. 1995b. Summary of the 1995 A&WMA critical review: Measurement methods to determine compliance with ambient air quality standards for suspended particles. *EM* 1:12–15.

Chow, J. C., E. M. Fujita, J. G. Watson, Z. Lu, D. R. Lawson, and L. L. Ashbaugh. 1994. Evaluation of filter-based aerosol measurements during the 1987 Southern California Air Quality Study. *Environ. Monit. Assess.* 30:49–80.

Chow, J. C., J. G. Watson, J. L. Bowen, C. A. Frazier, A. W. Gertler, K. K. Fung, D. Landis, and L. L. Ashbaugh. 1993. A sampling system for reactive species in the western U.S. In *Sampling and Analysis of Airborne Pollutants*, eds. E. D. Winegar and L. H. Keith. Ann Arbor, MI: Lewis Publishers, pp. 209–228.

Chow, J. C., J. G. Watson, D. Crow, D. H. Lowenthal, and T. Merrifield. 2001. Comparison of IMPROVE and NIOSH carbon measurements. *Aerosol Sci. Technol.* 34:23–34.

Chow, J. C., J. G. Watson, and F. Divita, Jr. 1996. Particulate matter with aerodynamic diameters smaller than 10 μm—Measurement methods and sampling strategies. In *Principles of Environmental Sampling*, ed. L. H. Keith. Washington, DC: American Chemical Society, pp. 539–573.

Coutant, R. W. 1977. Effect of environmental variables on collection of atmospheric sulfate. *Environ. Sci. Technol.* 11:873–878.

Cui, W., D. J. Eatough, and N. L. Eatough. 1998. Fine particulate organic material in the Los Angeles Basin I. Assessment of the high-volume Brigham Young University Organic Sampling System, BIG BOSS. *JAWMA* 48:1024–1037.

Cui, W., J. Machir, L. Lewis, D. J. Eatough, and N. L. Eatough. 1997. Fine particulate organic material at Meadview during the Project MOHAVE Summer Intensive Study. *JAWMA* 47:357–369.

Demuynck, M. 1975. Determination of irreversible absorption of water by cellulose filters. *Atmos. Environ.* 9:523–528.

Droppo, J. G., B. G. Krishnappan, S. S. Rao, and E. D. Ongley. 1995. Investigation of a sequential filtration technique for particle fractionation. *Environ. Sci. Technol.* 29:546–550.

Eatough, D. J., D. A. Eatough, L. Lewis, and E. A. Lewis. 1996. Fine particulate chemical composition and light extinction at Canyonlands National Park using organic particulate material concentrations obtained with a multisystem, multichannel diffusion denuder sampler. *J. Geophys. Res.* 101: 19515–19532.

Eatough, D. J., H. Tang, W. Cui, and J. Machir. 1995. Determination of the size distribution and chemical composition of fine particulate semi-volatile organic material in urban environments using diffusion denuder technology. *Inhal. Toxicol.* 7:691–710.

Eatough, D. J., A. Wadsworth, D. A. Eatough, J. W. Crawford, L. D. Hansen, and E. A. Lewis. 1993. A multiple-system, multi-channel diffusion denuder sampler for the determination of fine-particulate organic material in the atmosphere. *Atmos. Environ.* 27A:1213–1219.

Eldred, R. A., T. A. Cahill, M. L. Pitchford, and W. C. Malm. 1988. IMPROVE—A new remote area particulate monitoring system for visibility studies. *JAPCA* 54:3–12.

Engelbrecht, D. R., T. A. Cahill, and P. J. Feeney. 1980. Electrostatic effects on gravimetric analysis of membrane filters. *JAPCA* 30:391–392.

Fitz, D. R. 1990. Reduction of the positive organic artifact on quartz filters. *Aerosol Sci. Technol.* 12: 142–148.

Fitz, D. R., G. J. Doyle, and J. N. Pitts, Jr. 1983. An ultrahigh volume sampler for the multiple filter collection of respirable particulate matter. *JAPCA* 33:877–879.

Fitz, D. R. and S. V. Hering. 1996. *Study to evaluate the CADMP Sampler. Final report.* Report No. 93-333. Prepared for California Air Resources Board, Sacramento, CA, by CE-CERT, University of California, Riverside, CA.

Fitz, D. R. and N. Motallebi. 2000. A fabric denuder for sampling semi-volatile species. *J. Air Waste Manage.* 50:981–992.

Gormley, P. G. and M. Kennedy. 1949. Diffusion from a stream flowing through a cylindrical tube. *Proc. R. Ir. Acad. A* 52:163.

Gundel, L. A., R. K. Stevens, J. M. Daisey, V. C. Lee, K. R. R. Mahanama, and H. G. Cancel-Velez. 1995. Direct determination of the phase distributions of semi-volatile polycyclic aromatic hydrocarbons using annular denuders. *Atmos. Environ.* 29:1719–1733.

Hering, S. V. 2001. Impactors, cyclones, and other inertial and gravitational collectors. In *Air Sampling Instruments for Evaluation of Atmospheric Contaminants*, eds. B. S. Cohen and C. S. McCammon, Jr. Cincinnati, OH: Am. Conf. of Governmental Industrial Hygienists, pp. 315–376.

Hering, S. V. and G. R. Cass. 1999. The magnitude of bias in the measurement of $PM_{2.5}$ arising from volatilization of particulate nitrate from Teflon filters. *J. Air Waste Manage.* 49:725–733.

Hering, S. V., R. C. Flagan, and S. K. Friedlander. 1979a. Design and evaluation of new low-pressure impactor. I. *Environ. Sci. Technol.* 13:667–673.

Hering, S. V., S. K. Friedlander, J. J. Collins, and L. W. Richards. 1979b. Design and evaluation of a new low-pressure impactor. II. *Environ. Sci. Technol.* 13:184–188.

Hering, S. V., D. R. Lawson, I. Allegrini, A. Febo, C. Perrino, M. Possanzini, J. E. Sickles, II, K. G. Anlauf, A. Wiebe, B. R. Appel, W. John, J. L. Ondov, W. Wall, R. S. Braman, R. Sutton, G. R. Cass, P. A. Solomon, D. J. Eatough, N. L. Eatough, E. C. Ellis, D. Grosjean, B. B. Hicks, J. D. Womack, J. Horrocks, K. T. Knapp, T. G. Ellestad, R. J. Paur, W. J. Mitchell, M. Pleasant, E. Peake, A. MacLean, W. R. Pierson, W. W. Brachaczek, H. I. Schiff, G. I. Mackay, C. W. Spicer, D. H. Stedman, A. M. Winer, H. W. Biermann, and E. C. Tuazon. 1988. The nitric acid shootout: Field comparison of measurement methods. *Atmos. Environ.* 22:1519–1539.

Hinds, W. C. 1999. *Aerosol Technology: Properties, Bahavior, and Measurement of Airborne Particles*, 2nd ed. New York: John Wiley & Sons, Inc.

Horvath, H. 1993. Comparison of measurements of aerosol optical absorption by filter collection and a transmissometric method. *Atmos. Environ.* 27A:319–325.

Horvath, H. 1997. Experimental calibration for aerosol light absorption measurements using the integrating plate method—Summary of the data. *J. Aerosol Sci.* 28:1149–1161.

John, W., D. N. Fritter, and W. Winklmayr. 1991a. Resuspension induced by impacting particles. *J. Aerosol Sci.* 22:723–736.

John, W., S. V. Hering, G. Reischl, and G. V. Sasaki. 1983. Characteristics of Nuclepore filters with large pore size—II. Filtration properties. *Atmos. Environ.* 17:373–382.

John, W. and G. Reischl. 1980. A cyclone for size-selective sampling of ambient air. *JAPCA* 30:872–876.

John, W., S. M. Wall, J. L. Ondov, and H. C. Wang. 1986. *Dry Deposition of Acidic Gases and Particles.* Prepared for California Air Resources Board, Sacramento, CA, by Air and Industrial Hygiene Laboratory, California Department of Health, Berkeley, CA.

John, W., S. M. Wall, and J. L. Ondov. 1988. A new method for nitric acid and nitrate aerosol measurement using the dichotomous sampler. *Atmos. Environ.* 22:1627–1636.

John, W. and H. C. Wang. 1991. Laboratory testing method for PM_{10} samplers: Lowered effectiveness from particle loading. *Aerosol Sci. Technol.* 14:93–101.

John, W., W. Winklmayr, and H. C. Wang. 1991b. Particle deagglomeration and reentrainment in a PM_{10} sampler. *Aerosol Sci. Technol.* 14:165–176.

Kashdan, E. R., M. B. Ranade, L. J. Purdue, and K. A. Rehme. 1986. Interlaboratory evaluation of two inlets for sampling particles less than 10 μm. *Environ. Sci. Technol.* 20:911–916.

Kenny, L. C. and R. A. Gussman. 1997. Characterization and modelling of a family of cyclone aerosol preseparators. *J. Aerosol Sci.* 28(4):677–688.

Kenny, L. C., R. A. Gussman, and M. B. Meyer. 2000. Development of a sharp-cut cyclone for ambient aerosol monitoring applications. *Aerosol Sci. Technol.* 32:339–358.

Kitto, A. M. and I. Colbeck. 1999. Filtration and denuder sampling techniques. In *Analytical Chemistry of Aerosols*, ed. K. R. Spurny. Boca Raton, FL: Lewis Publishers, pp. 103–132.

Koutrakis, P., C. Sioutas, S. T. Ferguson, J. M. Wolfson, J. D. Mulik, and R. M. Burton. 1993. Development and evaluation of a glass honeycomb denuder/filter pack system to collect atmospheric gases and particles. *Environ. Sci. Technol.* 27:2497–2501.

Koutrakis, P., J. M. Wolfson, J. L. Slater, M. Brauer, and J. D. Spengler. 1988. Evaluation of an annular denuder/filter pack system to collect acidic aerosols and gases. *Environ. Sci. Technol.* 22:1463–1468.

Kusko, B. H., T. A. Cahill, R. A. Eldred, Y. Matsuda, and H. Miyake. 1989. Nondestructive analysis of total nonvolatile carbon by Forward Alpha Scattering Technique (FAST). *Aerosol Sci. Technol.* 10:390–396.

Lai, C. Y. and C. C. Chen. 2000. Performance characteristics of PM_{10} samplers under calm air conditions. *J. Air Waste Manage.* 50:578–587.

Lawrence, J. E. and P. Koutrakis. 1994. Measurements of atmospheric formic and acetic acids: Methods evaluation and results from field studies. *Environ. Sci. Technol.* 28:957–964.

Lin, C. I. and S. K. Friedlander. 1988. A note on the use of glass fiber filters in the thermal analysis of carbon containing aerosols. *Atmos. Environ*. 22:605–607.

Lippmann, M. 2001. Filters and filter holders. In *Air Sampling Instruments for Evaluation of Atmospheric Contaminants*, eds. B. S. Cohen and C. S. McCammon, Jr. Cincinnati, OH: American Conference of Government Industrial Hygienists, pp. 281–314.

Loo, B. W. and C. P. Cork. 1988. Development of high efficiency virtual impactors. *Aerosol Sci. Technol.* 9:167–176.

Lundgren, D. A. and R. M. Burton. 1995. Effect of particle size distribution on the cut point between fine and coarse ambient mass fractions. *Inhal. Toxicol*. 7(1):131–148.

Marple, V. A., B. Y. H. Liu, and G. A. Kuhlmey. 1981. A uniform deposit impactor. *J. Aerosol Sci.* 12: 333–337.

Marple, V. A., K. L. Rubow, W. A. Turner, and J. D. Spengler. 1987. Low flow rate sharp cut impactors for indoor air sampling: Design and calibration. *JAPCA* 37:1303–1307.

Marple, V. A. and K. Willeke. 1976. Inertial impactors: Theory, design and use. In *Fine Particles*, ed. B. Y. H. Liu. New York: Academic Press, pp. 411–466.

Masia, P., V. Di Palo, and M. Possanzini. 1994. Uptake of ammonia by nylon filters in filter pack systems. *Atmos. Environ*. 28:365–366.

Mathai, C. V., J. G. Watson, C. F. Rogers, J. C. Chow, I. H. Tombach, J. O. Zwicker, T. A. Cahill, P. J. Feeney, R. A. Eldred, M. L. Pitchford, and P. K. Mueller. 1990. Intercomparison of ambient aerosol samplers used in western visibility and air quality studies. *Environ. Sci. Technol*. 24:1090–1099.

McDow, S. R. and J. J. Huntzicker. 1990. Vapor adsorption artifact in the sampling of organic aerosol: Face velocity effects. *Atmos. Environ*. 24A:2563–2571.

McFarland, A. R., J. B. Wedding, and J. E. Cermak. 1977. Wind tunnel evaluation of a modified Andersen impactor and an all weather sampler inlet. *Atmos. Environ*. 11:535–539.

McFarland, A. R., C. A. Ortiz, and R. W. Bertch, Jr. 1984. A 10 μm cutpoint size selective inlet for Hi-Vol samplers. *JAPCA* 34:544–547.

McFarland, A. R., C. A. Ortiz, and C. E. Rodes. 1980. Characterization of sampling systems. In *Proceedings, the Technical Basis for a Size Specific Particulate Standard, Parts I & II*, ed. C. Cowherd. Pittsburgh: Air Pollution Control Association. pp. 59–76.

McFarland, A. R., C. A. Ortiz, and R. W. Bertch. 1978. Particle collection characteristics of a single-stage dichotomous sampler. *Environ. Sci. Technol*. 12:679–682.

Monteith, L. E. and K. L. Rubow. 2001. Air movers and samplers. In *Air Sampling Instruments for Evaluation of Atmospheric Contaminants*, eds. B. S. Cohen and C. S. McCammon, Jr. Cincinnati, OH: American Conference of Governmental Industrial Hygienists, pp. 233–280.

Moore, M. E. and A. R. McFarland. 1993. Performance modeling of single-inlet aerosol sampling cyclones. *Environ. Sci. Technol*. 27:1842–1848.

Mueller, P. K., G. M. Hidy, R. L. Baskett, K. K. Fung, R. C. Henry, T. F. Lavery, N. J. Nordi, A. C. Lloyd, J. W. Thrasher, K. K. Warren, and J. G. Watson. 1983. *Sulfate Regional Experiment (SURE): Report of Findings.* Report No. EA-1901. Prepared by Electric Power Research Institute, Palo Alto, CA.

Neuman, J. A., L. G. Huey, T. B. Ryerson, and D. W. Fahey. 1999. Study of inlet materials for sampling atmospheric nitric acid. *Environ. Sci. Technol*. 33:1133–1136.

Olin, J. G. and R. R. Bohn. 1983. A new PM_{10} medium flow sampler. In *76th Annual Meeting of the Air Pollution Control Association*. Pittsburgh: Air Pollution Control Association.

Pang, Y., D. J. Eatough, and N. L. Eatough. 1998. Determination of $PM_{2.5}$ sulfate and nitrate with a PC-BOSS designed for routine sampling for semi-volatile particulate matter. In *Proceedings, $PM_{2.5}$: A Fine Particle Standard*, eds. J. C. Chow and P. Koutrakis. Pittsburgh: Air & Waste Management Association, pp. 375–393.

Parker, R. D., G. H. Buzzard, T. G. Dzubay, and J. P. Bell. 1977. A two stage respirable aerosol sampler using nuclepore filters in series. *Atmos. Environ*. 11:617–621.

Pitchford, M. L., J. C. Chow, J. G. Watson, C. T. Moore, D. S. Campbell, R. A. Eldred, R. W. Vanderpool, P. Ouchida, S. V. Hering, and N. H. Frank. 1997. *Prototype $PM_{2.5}$ Federal Reference Method Field Studies Report—An EPA Staff Report.* U.S. Environmental Protection Agency, Las Vegas, NV.

Possanzini, M., A. Febo, and A. Liberti. 1983. New design of a high-performance denuder for the sampling of atmospheric pollutants. *Atmos. Environ.* 17:2605–2610.

Raabe, O. G., D. A. Braaten, R. L. Axelbaum, S. V. Teague, and T. A. Cahill. 1988. Calibration studies of the DRUM impactor. *J. Aerosol Sci.* 19(2):183–195.

Ranade, M. B., M. C. Woods, F. L. Chen, L. J. Purdue, and K. A. Rehme. 1990. Wind tunnel evaluation of PM_{10} samplers. *Aerosol Sci. Technol.* 13:54–71.

Rogers, C. F., J. G. Watson, and C. V. Mathai. 1989. Design and testing of a new size classifying isokinetic sequential aerosol sampler. *JAPCA* 39:1569–1576.

Shaw, R. W., R. K. Stevens, J. W. Bowermaster, J. W. Tesch, and E. Tew. 1982. Measurements of atmospheric nitrate and nitric acid: The denuder difference experiment. *Atmos. Environ.* 16:845–853.

Sickles, J. E. and L. L. Hodson. 1999. Short communication: Retention of sulfur dioxide by nylon filters. *Atmos. Environ.* 33:2423–2426.

Sioutas, C., S. T. Ferguson, J. M. Wolfson, H. Ozkaynak, and P. Koutrakis. 1997. Inertial collection of fine particles using a high-volume rectangular geometry conventional impactor. *J. Aerosol Sci.* 28: 1015–1028.

Sioutas, C. and P. Koutrakis. 1998. Development of small-cutpoint virtual impactors and applications in environmental health. In *Advances in Aerosol Filtration*, ed. K. R. Spurny. Boca Raton, FL: Lewis Publishers, pp. 453–470.

Sioutas, C., P. Koutrakis, and R. M. Burton. 1994a. Development of a low cutpoint slit virtual impactor for sampling ambient fine particles. *J. Aerosol Sci.* 25:1321–1330.

Sioutas, C., P. Koutrakis, S. Ferguson, J. M. Wolfson, and H. Ozkaynak. 1998. Inertial collection of fine particles using a high-volume rectangular geometry conventional impactor with a moving substrate. In *Proceedings, $PM_{2.5}$: A Fine Particle Standard*, eds. J. C. Chow and P. Koutrakis. Pittsburgh: Air & Waste Management Association, pp. 283–293.

Sioutas, C., P. Koutrakis, and B. A. Olson. 1994b. Development and evaluation of a low cutpoint virtual impactor. *Aerosol Sci. Technol.* 21:223–235.

Sioutas, C., P. Koutrakis, and J. M. Wolfson. 1994c. Particle losses in glass honeycomb denuder samplers. *Aerosol Sci. Technol.* 21:137–148.

Sioutas, C., P. Y. Wang, S. T. Ferguson, P. Koutrakis, and J. D. Mulik. 1996. Laboratory and field evaluation of an improved glass honeycomb denuder/filter pack sampler. *Atmos. Environ.* 30:885–896.

Spicer, C. W. and P. M. Schumacher. 1979. Particulate nitrate: Laboratory and field studies of major sampling interferences. *Atmos. Environ.* 13:543–552.

Spurny, K. R., J. P. Lodge, E. R. Frank, and D. C. Sheesley. 1969. Aerosol filtration by means of Nuclepore filters: Aerosol sampling and measurement. *Environ. Sci. Technol.* 3:453–468.

Stelson, A. W. and J. H. Seinfeld. 1982. Relative humidity and temperature dependence of the ammonium nitrate dissociation constant. *Atmos. Environ.* 16:983–992.

Tang, H., E. A. Lewis, D. J. Eatough, R. M. Burton, and R. J. Farber. 1994. Determination of the particle size distribution and chemical composition of semi-volatile organic compounds in atmospheric fine particles with a diffusion denuder system. *Atmos. Environ.* 28:939–950.

Tombach, I. H., D. W. Allard, R. L. Drake, and R. C. Lewis. 1987. *Western Regional Air Quality Studies; Visibility and Air Quality Measurements: 1981–1982.* Report No. EA-4903. Electric Power Research Institute, Palo Alto, CA.

Turner, J. R. 1998. *Laboratory and Field Evaluation of the Minivol $PM_{2.5}$ Sampler.* Report No. 97A-DIR-1. Prepared for U.S. Environmental Protection Agency, Research Triangle Park, NC, by Washington University, St. Louis, MO.

Turner, W. A., B. A. Olson, and G. A. Allen. 2000. Calibration of sharp cut impactors for indoor and outdoor particle sampling. *J. Air Waste Manage.* 50:484–487.

Turpin, B. J., J. J. Huntzicker, and S. V. Hering. 1994. Investigation of organic aerosol sampling artifacts in the Los Angeles Basin. *Atmos. Environ.* 28:3061–3071.

U.S. EPA. 1987. Revisions to the national ambient air quality standards for particulate matter: 40 CFR parts 51 and 52. *Federal Register* 52(July 1):24634–24669.

U.S. EPA. 1997a. National ambient air quality standards for particulate matter: Final rule. *Federal Register* 62:38651–38701.

U.S. EPA. 1997b. Revised requirements for designation of reference and equivalent methods for $PM_{2.5}$ and ambient air surveillance for particulate matter: Final rule. *Federal Register* 62:38763–38854.

U.S. EPA. 1998a. National ambient air quality standard for particulate matter: Final rule. *Federal Register* 63:6032–6037.

U.S. EPA. 1998b. *Quality Assurance Guidance Document 2.12: Monitoring $PM_{2.5}$ in Ambient Air Using Designated Reference or Class I Equivalent Methods.* U.S. Environmental Protection Agency, Research Triangle Park, NC.

Van Osdell, D. W. and F. L. Chen. 1990. *Wind Tunnel Test Report No. 28. Test of the Sierra–Andersen 246b Dichotomous Sampler Inlet at 2, 8, and 24 km/hr.* Prepared for Atmospheric Research and Exposure Assessment Laboratory, U.S. Environmental Protection Agency, by Research Triangle Institute, Research Triangle Park, NC.

Vincent, J. H., R. J. Aitken, and D. Mark. 1993. Porous plastic foam filtration media: Penetration characteristics and applications in particle size-selective sampling. *J. Aerosol Sci.* 24:929–944.

Watson, J. G., J. C. Chow, F. W. Lurmann, and S. Musarra. 1994. Ammonium nitrate, nitric acid, and ammonia equilibrium in wintertime Phoenix, Arizona. *J. Air Waste Manage.* 44:405–412.

Watson, J. G., J. C. Chow, J. J. Shah, and T. G. Pace. 1983. The effect of sampling inlets on the PM_{10} and PM_{15} to TSP concentration ratios. *JAPCA* 33:114–119.

Wedding, J. B. 1985. Errors in sampling ambient concentrations employing setpoint temperature compensated mass flow transducers. *Atmos. Environ.* 19:1219–1222.

Wedding, J. B. and T. C. Carney. 1983. A quantitative technique for determining the impact of non-ideal ambient sampler inlets on the collected mass. *Atmos. Environ.* 17:873–882.

Wedding, J. B., A. R. McFarland, and J. E. Cermak. 1977. Large particle collection characteristics of ambient aerosol samplers. *Environ. Sci. Technol.* 11:387–398.

Wedding, J. B. and M. A. Weigand. 1985. The Wedding ambient aerosol sampling inlet ($D_{50} = 10\,\mu m$) for the high volume sampler. *Atmos. Environ.* 19:535–538.

Wedding, J. B., M. A. Weigand, W. John, and S. M. Wall. 1980. Sampling effectiveness of the inlet to the dichotomous sample. *Environ. Sci. Technol.* 14:1367–1370.

Wedding, J. B., M. A. Weigand, and Y. J. Kim. 1985. Evaluation of the Sierra–Anderson 10-μm inlet for the high-volume sampler. *Atmos. Environ.* 19:539–542.

Wedding, J. B., M. A. Weigand, Y. J. Kim, D. L. Swift, and J. P. Lodge. 1987. A critical flow device for accurate PM_{10} sampling and correct indication of PM_{10} dosage to the thoracic region of the respiratory tract. *JAPCA* 37:254–258.

Wiener, R. W. and R. W. Vanderpool. 1992. *Evaluation of the Lane Regional PRO-1A and PRO-2 Saturation Monitor.* U.S. Environmental Protection Agency, Research Triangle Park, NC.

Witz, S., M. M. Smith, and A. B. Moore, Jr. 1983. Comparative performance of glass fiber hi-vol filters. *JAPCA* 33:988–991.

28 Fugitive Dust Emissions

CHATTEN COWHERD, JR.
Midwest Research Institute, Kansas City, MO

INTRODUCTION

Most of the more than 70 areas of the United States that have been unable to attain the national ambient air quality standards (NAAQS, 1990) for PM-10 (particles smaller than 10 µm in aerodynamic diameter) are in western states with significant emission contributions from fugitive dust sources.

Fugitive dust sources may be separated into two broad categories: process sources and open dust sources. Process sources of fugitive emissions are those associated with industrial operations such as rock crushing that alter the characteristics of a feed material. Open dust sources are those that generate nonducted emissions of solid particles by the forces of wind or machinery acting on exposed material.

Open dust sources include industrial sources of particulate emissions associated with the open transport, storage, and transfer of raw, intermediate, and waste aggregate materials and nonindustrial sources such as unpaved roads and parking lots, paved streets and highways, heavy construction activities, and agricultural tilling. Generic categories of open dust sources are listed in Table 28–1 (Cowherd et al., 1986).

On a nationwide basis, fugitive dust consists mostly of soil and other crustal materials. However, fugitive dust may also be emitted from other powdered or aggregate materials that have been placed in open storage or deposited on the ground or roadway surfaces by spillage or vehicle tracking. Toxic fugitive dust may originate from hazardous-waste treatment, storage, and disposal facilities (TSDFs) and abandoned waste dumps or industrial deposition areas where contaminated soils are subject to disturbance by wind or machinery. Dust emissions from paved roadways contain tire and brake wear particles and allergens (pollen, dander, biofragments) in addition to resuspended road surface dust composed mostly of crustal particles.

This chapter addresses the techniques used to determine particulate emission rates from fugitive dust sources before and after control application. Specific topics include the characteristics of fugitive dust sources, field techniques for emission quantification, and available models for emission estimation and control performance estimation.

FACTORS AFFECTING DUST EMISSIONS

Mechanical Dust Generation

Mechanically generated emissions from open dust sources exhibit a high degree of variability from one site to another, and emissions at any one site tend to fluctuate widely. The

Aerosol Measurement: Principles, Techniques, and Applications, Second Edition, Edited by Paul A. Baron and Klaus Willeke.
ISBN 0-471-35636-0

TABLE 28–1. Generic Categories of Open Dust Sources

Unpaved travel surfaces
Roads
Parking lots and staging areas
Storage piles
Paved travel surfaces
Streets and highways
Parking lots and staging areas
Exposed areas (wind erosion)
Storage piles
Bare ground areas
Materials handling
Batch drop (dumping)
Continuous drop (conveyor transfer, stacking)
Pushing (dozing, grading, scraping)
Tilling

site characteristics that cause these variations may be grouped into (1) properties of the exposed surface material from which the dust originates and (2) measures of energy expended by machinery interacting with the surface. These site characteristics are further discussed below.

Surface Material Texture and Moisture. The dry-particle size distribution of the exposed soil or surface material determines its susceptibility to mechanical entrainment. The upper size limit for particles that can become suspended has been estimated at about 75 μm in aerodynamic diameter (Cowherd et al., 1974). Conveniently, 75 μm in physical diameter is also the smallest particle for which size analysis by dry sieving is practical (ASTM, 1984). Particles passing a 200-mesh screen (74 μm opening) on dry sieving are termed *silt* by highway officials. Note that for fugitive dust particles, the physical diameter and aerodynamic diameter are roughly equivalent because of the offsetting effects of higher density and irregular shape.

Dust emissions are known to be strongly dependent on the moisture level of the mechanically disturbed material (Cowherd et al., 1974). Water acts as a dust suppressant by forming cohesive moisture films among the discrete grains of surface material. In turn, the moisture level depends on the moisture added by natural precipitation, the moisture removed by evaporation, and moisture movement beneath the surface. The evaporation rate depends on the degree of air movement over the surface, material texture and mineralogy, and the degree of compaction or crusting. The moisture-holding capacity of the air is also important, and it correlates strongly with the surface temperature. Vehicle traffic intensifies the drying process primarily by increasing air movement over the surface.

Mechanical Equipment Characteristics. In addition to the material properties discussed above, it is clear that the physical and mechanical characteristics of materials handling and transport equipment also affect dust emission levels. For example, visual observation suggests (and field studies have confirmed) that vehicle emissions per unit of unpaved road length increase with increasing vehicle speed (Cowherd et al., 1974). For traffic on unpaved roads, studies have also shown positive correlations between emissions and (1) vehicle weight and (2) number of wheels per vehicle (Cowherd et al., 1979). Similarly, dust emissions from materials-handling operations have been found to increase with increasing wind speed and drop distance.

Windblown Dust Generation

Wind-generated emissions from open dust sources also exhibit a high degree of variability from one site to another, and emissions at any one site tend to fluctuate widely. The site characteristics that cause these variations may be grouped into (1) properties of the exposed surface material from which the dust originates and (2) measures of energy expended by wind interacting with the erodible surface. These site characteristics are further discussed below.

Surface Material Texture and Moisture. As in the case of mechanical entrainment, the dry-particle size distribution of the exposed soil or surface material determines its susceptibility to wind erosion. Wind forces move soil particles by three transport modes: saltation, surface creep, and suspension. Saltation describes particles, ranging in diameter from about 75 to 500 μm, that are readily lifted from the surface and jump or bounce within a layer close to the air–surface interface. Particles transported by surface creep range in diameter from about 500 to 1000 μm. These large particles move very close to the ground, propelled by wind stress and by the impact of small particles transported by saltation. Particles smaller than about 75 μm in diameter move by suspension and tend to follow air motions. As stated above, the upper size limit of silt particles (75 μm in physical diameter) is roughly the smallest particle for which size analysis by dry sieving is practical. The threshold wind speed for the onset of saltation, which drives the wind erosion process, is also dependent on soil texture, with 100 to 150 μm particles having the lowest threshold speed. Saltation provides energy for the release of particles in the PM-10 size range that typically are bound by surface forces to larger clusters.

Dust emissions from wind erosion are known to be strongly dependent on the moisture level of the erodible material (Woodruff and Siddoway, 1965). The mechanism of moisture mitigation is the same as that described above for mechanical entrainment.

Nonerodible Elements. Nonerodible elements, such as clumps of grass or stones (larger than about 1 cm in diameter) on the surface, consume part of the shear stress of the wind that otherwise would be transferred to erodible soil. Surfaces impregnated with a large density of nonerodible elements behave as having a "limited reservoir" of erodible particles, even if the material protected by nonerodible elements is itself highly erodible. Wind-generated emissions from such surfaces decay sharply with time, as the particle reservoir is depleted. Surfaces covered by unbroken grass are virtually nonerodible.

Crust Formation. Following the wetting of a soil or other surface material, fine particles will move to form a surface crust. The surface crust acts to hold in soil moisture and resist erosion. The degree of protection that is afforded by a soil crust to the underlying soil may be measured by the modulus of rupture and thickness of the crust (Cowherd et al., 1985). This modulus of rupture is roughly a measure of the hardness of the crust. Exposed soil that lacks a surface crust (e.g., a disturbed soil or a very sandy soil) is much more susceptible to wind erosion.

Frequency of Mechanical Disturbance. Emissions generated by wind erosion are also dependent on the frequency of disturbance of the erodible surface. A disturbance is defined as an action that results in the exposure of fresh surface material. This would occur whenever a layer of aggregate material is either added to or removed from the surface. The disturbance of an exposed area may also result from the turning of surface material to a depth exceeding the size of the largest material present. Each time that a surface is disturbed, its erosion potential is increased by destroying the mitigative effects of crusts, vegetation, and friable nonerodible elements and by exposing new surface fines.

Wind Speed. Under high wind conditions that trigger wind erosion by exceeding the threshold velocity, the wind speed profile near the erodible surface is found to follow a logarithmic distribution (Gillette, 1978b):

$$u(z) = \frac{u^*}{0.4} \ln \frac{z}{z_0} \quad (z > z_0) \tag{28–1}$$

where u is wind speed (cm/s), u^* is friction velocity (cm/s), z is height above test surface (cm), z_0 is roughness height (cm), and 0.4 = von Karman's constant (dimensionless).

The friction velocity (u^*) is a measure of wind shear stress on the erodible surface, as determined from the slope of the logarithmic velocity profile. The roughness height (z_0) is a measure of the roughness of the exposed surface as determined from the y-intercept of the velocity profile (i.e., the height at which the wind speed is zero) on a logarithmic-linear graph.

Agricultural scientists have established that total soil loss by continuous wind erosion of highly erodible fields is dependent roughly on the cube of wind speed above the threshold velocity (Woodruff and Siddoway, 1965). More recent work has shown that the loss of particles in suspension mode follows a similar dependence. Soils protected by nonerodible elements or crusts exhibit a weaker dependence of suspended particulate emissions on wind speed (Cowherd, 1988).

Wind Gusts. Although mean atmospheric wind speeds may not be sufficient to initiate wind erosion from a particular "limited-reservoir" surface, wind gusts may quickly deplete a substantial portion of its erosion potential. In addition, because the erosion potential (mass of particles constituting the "limited reservoir") increases with increasing wind speed above the threshold velocity, estimated emissions should be related to the gusts of highest magnitude.

The current meteorological variable that appropriately reflects the magnitude of wind gusts is the fastest 2 min wind speed from the "First Order Summary of the Day," published by the U.S. Weather Service for first order meteorological stations (NEDS, 1992). The quantity represents the wind speed corresponding to the largest linear passage of wind movement during a 2 min period. The duration of the fastest 2 min wind speed matches well with the half-life (i.e., the time required to remove one-half the erodible particles on the surface) of the erosion process. It should be noted that instantaneous peak wind speeds can significantly exceed the fastest 2 min wind speed.

Because the threshold wind speed must be exceeded to trigger the possibility of substantial wind erosion, the dependence of erosion potential on wind speed cannot be represented by any simple linear function. For this reason, the use of an average wind speed to calculate an average emission rate is inappropriate.

Wind Accessibility. If the erodible material lies on an exposed area with little penetration into the surface wind layer, then the material is uniformly accessible to the wind. If this is not the case, it is necessary to divide the erodible area into subareas representing different degrees of exposure to wind. For example, the results of physical modeling show that the frontal face of an elevated pile is exposed to surface wind speeds of the same order as the approach wind speed upwind of the pile at a height matching the top of the pile (Billings-Stunder and Arya, 1988); on the other hand, the leeward face of the pile is exposed to much lower wind speeds.

EMISSION CALCULATION PROCEDURE

A calculation of the estimated emission rate for a given source requires data on source extent, uncontrolled emission factor, and control efficiency. The mathematical expression for this calculation is as follows:

$$R = Me(1 - c) \tag{28–2}$$

where R is estimated mass emission rate in the specified particle size range; M is source extent; e is uncontrolled emission factor in the specified particle size range, that is, mass of uncontrolled emissions per unit of source extent; and c is fractional efficiency of control.

The source extent (or activity level) is the appropriate measure of source size or the level of activity that is used to scale the uncontrolled emission factor to the particular source in question. For process sources of fugitive particulate emissions, the source extent is usually the production rate (i.e., the mass of product per unit time). Similarly, the source extent of an open dust source entailing a batch or continuous drop operation is the rate of mass throughput.

For other categories of open dust sources, the source extent is related to the area of the exposed surface that is disturbed by either wind or mechanical forces. In the case of wind erosion, the source extent is simply the area of erodible surface. For emissions generated by mechanical disturbance, the source extent is also the surface area (or volume) of the material from which the emissions emanate. For vehicle travel, the disturbed surface area is the travel length times the average daily traffic (ADT) count, with each vehicle having a disturbance width equal to the width of a travel lane.

If an anthropogenic control measure (e.g., treating the surface with a chemical binder that forms an artificial crust) is applied to the source, the uncontrolled emission factor in Eq. 28–2 must be multiplied by an additional term to reflect the resulting fractional control.

In broad terms, anthropogenic control measures can be considered as either continuous or periodic, as the following examples illustrate:

Continuous Controls	Periodic Controls
Wet suppression at conveyor transfer points	Watering or chemical treatment of unpaved roads
Enclosures/wind fences around storage piles	Sweeping of paved travel surfaces
Continuous vegetation of exposed areas	Chemical stabilization of exposed areas

The major difference between the two types of controls is related to the time dependency of performance. For continuous controls, efficiency is essentially constant with respect to time. On the other hand, the efficiency associated with periodic controls tends to decrease (decay) with time after application until the next application, at which time the cycle repeats but often with some residual effects from the previous application.

To quantify the performance of a specific periodic control, two measures of control efficiency are required. The first is "instantaneous" control efficiency and is defined by

$$c(t) = \left[1 - \frac{e_c(t)}{e_u}\right] \times 100 \tag{28–3}$$

where $c(t)$ is instantaneous control efficiency (percent), $e_c(t)$ is instantaneous emission factor for the controlled source, e_u is uncontrolled emission factor, and t is time after control application.

The other important measure of periodic control performance is average efficiency,

$$C(T) = \frac{1}{T}\int_0^T c(t)dt \tag{28–4}$$

where $c(t)$ is instantaneous control efficiency at time t after application (percent) and T is time period over which the average control efficiency is referenced. The average control efficiency is needed to estimate the emission reductions due to periodic applications.

To estimate the emissions of a specific contaminant that is present within the disturbed surface material, the emission rate should be adjusted by the fraction of contaminant in the specified airborne-particle size range (Cowherd et al., 1985). However, as an approximation, the fraction of contaminant in the silt portion of the disturbed surface material may be used to adjust the uncontrolled emission factor and, in turn, the calculated emission rate, so that it applies to the contaminant of interest.

EMISSION QUANTIFICATION TECHNIQUES

Fugitive dust emission rates and particle size distributions are difficult to quantify because of the diffuse and variable nature of such sources and the wide range of particle sizes, including particles which deposit immediately adjacent to the source. Standard source testing methods, which are designed for application to confined flows under steady-state, forced-flow conditions, are not suitable for the measurement of fugitive emissions unless the plume can be drawn into a forced-flow system. The available source testing methods for fugitive dust sources are described in the following paragraphs.

Mechanical Entrainment Processes

Because it is usually impractical to enclose open dust sources or to capture the entire emissions plume, only two methods are suitable for the measurement of particulate emissions from most open dust sources:

1. The upwind–downwind method involves the measurement of upwind and downwind particulate concentrations, utilizing ground-based samplers under known meteorological conditions, followed by a calculation of the source strength (mass emission rate) with atmospheric dispersion equations (Kolnsberg, 1976).
2. The exposure-profiling method involves simultaneous, multipoint measurements of particulate concentration and wind speed over the effective cross section of the plume, followed by a calculation of the net particulate mass flux through integration of the plume profiles (Cowherd et al., 1974).

In both cases it is customary to use high-volume air samplers so that quantifiable sample mass can be accumulated in sampling periods no longer than about 6 hours.

Upwind–Downwind Method. The upwind–downwind method involves the measurement of airborne particulate concentrations both upwind and downwind of the pollutant source. The number of upwind sampling instruments depends on the degree of isolation of the source operation of concern (i.e., the absence of interference from other sources upwind). Increasing the number of downwind instruments improves the reliability in determining the emission rate by providing better plume definition. To reasonably define the plume emanating from a point source, instruments need to be located at a minimum of two downwind distances and three crosswind distances. The same sampling requirements pertain to line sources except that measurements need not be made at multiple crosswind distances.

Net downwind (i.e., downwind minus upwind) concentrations are used as input to atmospheric dispersion equations (normally of the Gaussian type) to back-calculate the particulate emission rate (i.e., source strength) required to generate the pollutant concentrations measured. Emission factors are obtained by dividing the calculated emission rate by the source extent. A number of meteorological parameters must be concurrently recorded for input to this dispersion equation. As a minimum, the wind direction and speed must be recorded on site.

While the upwind–downwind method is applicable to virtually all types of sources, it has significant limitations with regard to the development of source-specific emission factors. Because of the impracticality of adjusting the locations of the sampling array for shifts in wind direction during sampling, it may be questionable to assume that the plume position is fixed in the application of the dispersion model. In addition, the usual assumption that a line or area source is uniformly emitting may not allow for a realistic representation of spatial variation in source activity.

Exposure-Profiling Method. As an alternative to conventional upwind–downwind sampling, the exposure-profiling technique utilizes the emission profiling concept, which is the basis for conventional ducted source testing (Method 5; U.S. EPA, 1977), except that, in the case of exposure-profiling, the ambient wind directs the plume to the sampling array. The passage of airborne particulate matter immediately downwind of the source is measured directly by means of a simultaneous, multipoint sampling of particulate concentration and wind velocity over the effective cross section of the fugitive emissions plume.

For the measurement of nonbuoyant fugitive emissions using exposure profiling, sampling heads are distributed over a vertical network positioned just downwind (usually about 5 m) from the source. Particulate sampling heads should be symmetrically distributed over the concentrated portion of the plume containing at least 80% of the total mass flux. A vertical line grid of at least three samplers is sufficient for the measurement of emissions from line or moving point sources (Fig. 28–1), while a two-dimensional array of at least five samplers is required for quantification of the fixed virtual point source of emissions. For quantifying emissions of particles larger than about 10 μm, the particulate samplers should have directional intakes, as discussed below. At least one upwind sampler must be operated to measure the background concentration, and wind speed and direction must be measured concurrently on site.

The particulate emission rate is obtained by a spatial integration of the distributed measurements of exposure (accumulated mass flux), which is the product of mass concentration and wind speed:

$$R = \int_A C(h, w)u(h, w)\mathrm{d}h\mathrm{d}w \qquad (28\text{–}5)$$

where R is emission rate (g/s), C is net particulate concentration (g/m^3), u is wind speed (m/s), h is vertical distance coordinate (m), w is lateral distance coordinate (m), and A is effective cross-sectional area of plume (m^2).

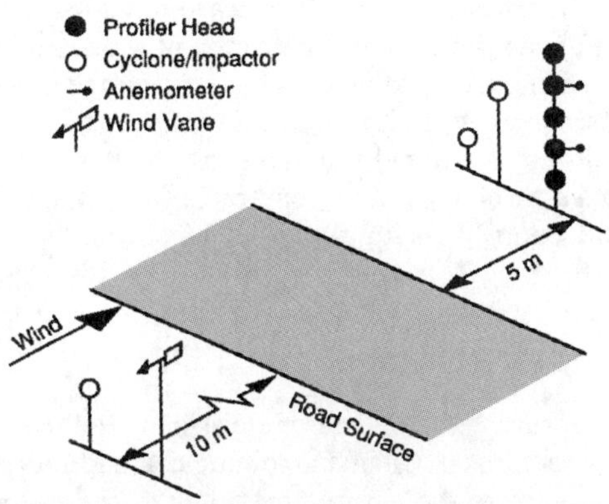

Fig. 28–1. Exposure profiling method—roadway.

Usually, a numerical integration scheme is used to calculate the emission rate. This mass-balance calculation scheme requires no assumptions about plume dispersion phenomena.

Isokinetic Sampling. Regardless of which method is used, isokinetic sampling is required for a representative collection of particles larger than about 10 μm in aerodynamic diameter. The directional sampling intakes are pointed into the mean wind direction, and the intake velocity of each sampler is periodically adjusted (usually with intake nozzles) to closely match the mean wind velocity approaching the sampling intake. Because of natural fluctuations in wind speed and direction, some anisokinetic sampling effects will always be encountered. If the angle α between the mean wind direction and the direction of the sampling axis equals 30°, the sampling error is about 10% (Watson, 1954). For an isokinetic flow ratio of sampling intake speed to approach wind speed between 0.8 and 1.2 (Watson, 1954), the sampling error is about 5%.

Wind Erosion

The two wind erosion source testing methods of interest are the upwind–downwind method as described above and the portable wind tunnel method. The wind tunnel method involves the use of a portable open-floored wind tunnel for in situ measurement of emissions from representative surfaces under predetermined wind conditions (Cuscino et al., 1983).

Upwind–Downwind Method. The upwind–downwind method is burdened with practical difficulties for the study of wind erosion in that the onset of erosion and its intensity are beyond the control of the investigator. In addition, background (upwind) particulate concentrations tend to be high during erosion events, making source isolation very difficult.

Wind Tunnel Method. The most common version of the wind tunnel method utilizes a pull-through wind tunnel with an open-floored test section placed directly over the surface to be tested. A device of this type is shown in Figure 28–2. Air is drawn through the tunnel at controlled velocities. The exit air stream from the test section passes through a circular duct fitted with a directional sampling probe at the downstream end. Air is drawn isokinetically through the probe by a high-volume sampling train.

The wind tunnel method incorporates the essential features of Method 5 stack sampling (U.S. EPA, 1977). The one prime difference, the use of single-point sampling, is justified by the high turbulence levels in the sampling module. The measurement uncertainty inherent in this method is of the same order as that in Method 5, which has been subjected to extensive collaborative testing by the EPA.

The wind tunnel method relies on a straightforward mass-balance technique for the calculation of emission rate. By sampling under light ambient wind conditions, background interferences from upwind erosion sources can be avoided. Although a portable wind tunnel does not generate the larger scales of turbulent motion found in the atmosphere, the turbulent boundary layer formed within the tunnel simulates the smaller scales of atmospheric turbulence. It is the smaller scale turbulence that penetrates the wind flow in direct contact with the erodible surface and contributes to the particle entrainment mechanisms (Gillette, 1978a,b).

Particle Sizing

Concurrent with the measurement of mass emissions, the aerodynamic particle size distribution should be characterized. Chemical, biological, and morphological analyses may also be performed to characterize the nature and origin of the particles. For particle sizing, a high-volume cyclone/cascade impactor featuring isokinetic sample collection has been used

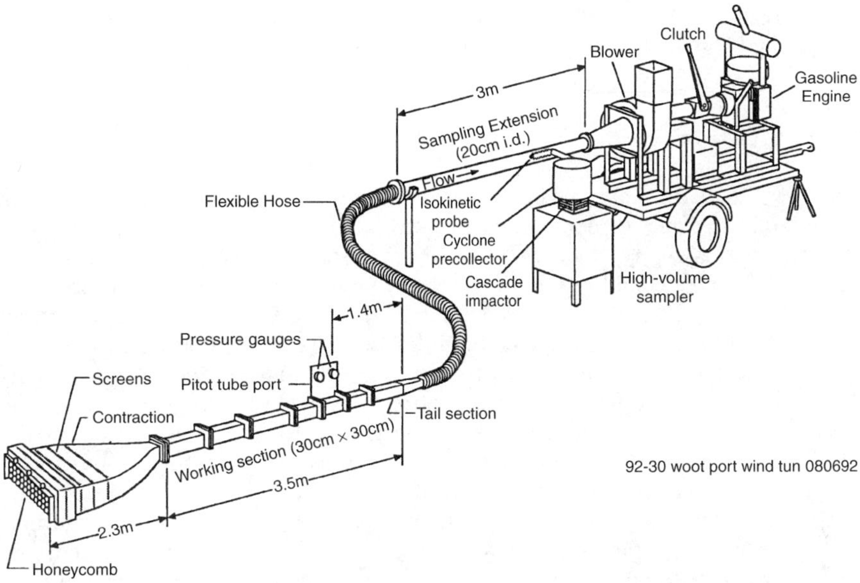

Fig. 28–2. Portable open-floored wind tunnel equipment.

(Cowherd et al., 1986). A cyclone preseparator (or other device) is needed to remove the coarse particles, which otherwise would bounce off the greased substrate stages within the impactor, causing fine-particle bias. Once again, the sampling intake is pointed into the wind and the sampling velocity adjusted to the mean local wind speed by fitting the intake with a nozzle of appropriate size. This system offers the advantage of a direct determination of aerodynamic particle size.

Another particle-sizing option includes an analysis of the particulate deposit by optical or electron microscopy. Disadvantages include (1) potential artificial disaggregation of particle clusters during sample preparation and (2) uncertainties in converting physical size data to equivalent aerodynamic diameters. In a collaborative field test of the exposure-profiling method, the cyclone/impactor method was judged to be more suitable than microscopy for the particle sizing of fugitive dust emissions (McCain et al., 1985).

Control Efficiency Estimation

Field evaluation of the control efficiency requires that the study design include not only adequate emission measurement techniques but also a proven "control application plan." In the past, two major types of plans have been used. Under the Type 1 plan, controlled and uncontrolled emission measurements are obtained simultaneously. Under Type 2, uncontrolled tests are performed initially, followed by controlled tests.

To ensure comparability between the operating characteristics of the controlled and uncontrolled sources, many evaluations are forced to employ Type 2 plans. An example would be a wet suppression system used on a primary crusher. One important exception to this, however, is unpaved-road dust control. In this instance, testing under a Type 1 plan may be conducted on two or more contiguous road segments. One segment is left untreated, and the others are treated with the dust suppressant. Under a Type 2 plan, a normalization of emissions may be required to allow for potential differences in source characteristics during the uncontrolled and controlled tests because they do not occur simultaneously.

EMISSION MODELS

Early in the U.S. EPA field testing program to develop emission factors for fugitive dust sources, it became evident that uncontrolled emissions within a single generic source category may vary over two (or more) orders of magnitude as a result of variations in source conditions (equipment characteristics, material properties, and climatic parameters). Therefore, it would not be feasible to represent an entire generic source category in terms of a single-valued emission factor, as traditionally used by the U.S. EPA to describe average emissions from a narrowly defined ducted source operation. In other words, it would take a large matrix of single-valued factors to adequately represent an entire generic fugitive dust source category. To account for emissions variability, therefore, the approach was taken that fugitive dust emission factors be constructed as mathematical equations for sources grouped by the dust generation mechanisms. The emission factor equation for each source category would contain multiplicative correction parameter terms that explain much of the variance in observed emission factor values on the basis of variances in specific source parameters. Such factors would be applicable to a wide range of source conditions, limited only by the extent of experimental verification.

For example, the use of the silt content as a measure of the dust generation potential of a material acted on by the forces of wind or machinery proved to be an important step in extending the applicability of the emission factor equations to a wide variety of aggregate materials of industrial importance.

A compendium of emission factors (referred to as AP-42) is maintained on a CD-ROM (Air CHIEF, 1998) by the U.S. EPA. Chapter 13 of AP-42 contains the predictive emission factor equations for fugitive dust sources.

Also with each equation is provided a set of particle size multipliers for adjusting the calculated emission factors to specific particle size fractions. The ratios of PM-2.5 to PM-10 typically range from 0.15 to 0.25.

Example: Vehicle Traffic on Unpaved Roads

For the purpose of estimating uncontrolled emissions, the AP-42 emission factor equation applicable to vehicle traffic on unpaved roads takes source characteristics into consideration (MRI, 1998):

$$e = \frac{0.73(s/12)^{0.8}(W/2.7)^{0.4}}{(M/0.2)^{0.3}}(kg/VKT) \qquad e = \frac{2.6(s/12)^{0.8}(W/3)^{0.4}}{(M/0.2)^{0.3}}(lb/VMT) \qquad (28\text{–}6)$$

where s is surface material silt content (%), W is mean vehicle weight, Mg (ton), and M is surface material moisture content (%).

The denominators in each of the multiplicative terms of the equation constitute normalizing default values, in case no site-specific correction parameter data are available. The default moisture content represents dry (worst-case) road conditions. Extrapolation to annual average uncontrolled emission estimates (including natural mitigation) is accomplished by assuming that emissions are occurring at the estimated rate on days without measurable precipitation and, conversely, are absent on days with measurable precipitation.

EMISSION CONTROL OPTIONS

Typically, there are several options for the control of fugitive particulate emissions from any given source. This is clear from Eq. 28–2, used to calculate the emission rate. Because the uncontrolled-emission rate is the product of the source extent and the uncontrolled-emission factor, a reduction in either of these two variables produces a proportional reduction in the uncontrolled-emission rate.

In the case of open sources, the reduction in the uncontrolled-emission factor may be achieved by adjusted "work practices." The degree of the reduction of the uncontrolled-emission factor can be estimated from the known dependence of the factor on source conditions that are subject to alteration. For open dust sources, this information is embodied in the predictive emission factor equations for fugitive dust sources as presented in Section 13 of AP-42. The reduction of source extent and the incorporation of adjusted work practices that reduce the amount of exposed dust-producing material are preventive measures for the control of fugitive dust emissions.

Add-on controls can also be applied to reduce emissions by reducing the amount (areal extent) of dust-producing material, other than by clean-up operations. For example, the elimination of mud/dirt carryout onto paved roads at construction and demolition sites is a cost-effective preventive measure. On the other hand, mitigative measures involve the periodic removal of dust-producing material. Examples of mitigative measures include clean up of spillage on travel surfaces (paved and unpaved) and clean up of material spillage at conveyor transfer points. Mitigative measures tend to be less favorable from a cost-effectiveness standpoint.

Periodically applied control techniques for open dust sources begin to decay in efficiency almost immediately after implementation. The most extreme example of this is the watering of unpaved roads, where the efficiency decays from nearly 100% to 0% in a matter of hours (or minutes). On the other hand, the effects of chemical dust suppressants applied to unpaved roads may last for several months.

TABLE 28–2. Controls for Fugitive Dust Sources

Source Category	Control Action
Paved roads	Water flushing/sweeping
	Improvements in sanding/salting applications and materials
	Truck covering
	Prevention of track-on/wash-on
	Construction site measures
	Curb installation
	Shoulder stabilization
	Storm water drainage
Unpaved roads	Paving
	Chemical stabilization
	Surface improvement (graveling)
	Vehicle speed reduction
Storage piles (transfer operations)	Wet suppression
Construction/demolition	Paving permanent roads early in project
	Truck covering
	Access apron construction and cleaning
	Watering of graveled travel surfaces
Open area wind erosion	Revegetation
	Limitation of off-road vehicle traffic

Consequently, to describe the performance of most intermittent-control techniques for open dust sources, the "time-weighted average" control efficiency must be reported along with the time period over which the value applies. For continuous-control systems (e.g., wet suppression for continuous-drop materials transfer), a single control efficiency is usually appropriate.

Table 28–2 lists fugitive dust control measures that have been judged to be generally cost-effective for application to metropolitan areas unable to meet PM-10 standards. The most highly developed performance models available apply to application of chemical suppressants on unpaved roads. These models relate the expected instantaneous control efficiency to the application parameters (application intensity and dilution ratio) and to the number of vehicle passes (rather than time) following the application. More details on available dust control measure performance and cost are presented by Cowherd et al. (1988) and Cowherd (1991).

REFERENCES

Air Chief. 1998. CD-ROM, Version 6.0, U.S. EPA. Research Triangle Park, NC, September 1998.

ASTM. 1984. American Society of Testing and Materials, Subcommittee C09.03.05. *Standard Method for Sieve Analysis of Fine and Coarse Aggregates*. Method C-136, 84a. Philadelphia: ASTM.

Billings-Stunder, B. J. and S. P. S. Arya. 1988. Windbreak effectiveness for storage pile fugitive dust control: A wind tunnel study. *JAPCA* 38:135–143.

Cowherd, C. Jr. 1988. A refined scheme for calculation of wind generated PM-10 emissions from storage piles. In *Proceedings: APCA/EPA Conference on PM-10: Implementation of Standards*. San Francisco: U.S. EPA.

Cowherd, C. Jr. 1991. *Best Available Control Measures (BACM) for Fugitive Dust Sources*. Revised Draft Guidance Document submitted to Air Quality Management Division. Research Triangle Park, NC: U.S. EPA.

Cowherd, C., Jr., K. Axetell Jr., C. M. (Guenther) Maxwell, and G. A. Jutze. 1974. *Development of Emission. Factors for Fugitive Dust Sources*, EPA Publication No. EPA-450/3-74/037, NTIS Publication No. PB-238 262.

Cowherd, C. Jr., R. Bohn, and T. Cuscino Jr. 1979. *Iron and Steel Plant Open Source Fugitive Emission Evaluation*, EPA-600/2-79/103, NTIS Publication No. PB-299 385.

Cowherd, C. Jr., J. S. Kinsey, D. D. Wallace, M. A. Grelinger, T. A. Cuscino, and R. M. Neulicht. 1986. *Identification, Assessment, and Control of Fugitive Particulate Emissions*, EPA-600/8-86-023, NTIS Publication No. PB86-230083.

Cowherd, C., Jr., G. E. Muleski, P. J. Englehart, and D. A. Gillette. 1985. *Rapid Assessment of Exposure to Particulate Emissions from Surface Contamination Sites*, EPA/600/8-85/002. Washington, DC: U.S. EPA.

Cowherd, C., Jr., G. E. Muleski, and J. S. Kinsey. 1988. *Control of Open Fugitive Dust Sources*, EPA 450/3-88-008. Research Triangle Park, NC: U.S. EPA.

Cuscino, T., Jr., G. E. Muleski, and C. Cowherd Jr. 1983. *Iron and Steel Plant Open Source Fugitive Emission Evaluation*. EPA-600/2-83-110, NTIS Publication No. PB84-110568.

Gillette, D. A. 1978a. A wind tunnel simulation of the erosion of the soil. *Atmos. Environ.* (Great Britain) 12:1735–1743.

Gillete, D. A. 1978b. Tests with a portable wind tunnel for determining wind erosion threshold velocities. *Atmos. Environ.* 12:2309.

Kolnsberg, H. J. 1976. *Technical Manual for the Measurement of Fugitive Emissions: Upwind/Downwind Sampling Method for Industrial Fugitive Emissions*. EPA-600/2-76-089a, NTIS Publication No. PB253092.

McCain, J. D., B. E. Pyle, and R. C. McCrillis. 1985. Comparative study of open source particulate emission measurement techniques. In *Proceedings of the Air Pollution Control Association Annual Meeting*. Pittsburgh, PA: APCA.

MRI. 1998. Midwest Research Institute. *Emission Factor Documentation for AP-42, Section 13.2.2, Unpaved Roads, Final Report*. Kansas City, MO.

NAAQS. 1990. CAA 109(b), 42 U.S.C. 7409(b) (1990), Public Law 101-549, 15 November 1990.

NEDS. 1992. National Environmental Data Service. *Local Climatological Data Annual and Monthly Summaries*. Asheville, NC: National Climatic Center.

U.S. EPA. 1977. Standards of performance for new stationary sources, revision to reference methods 1–8. *Federal Register*. 18 August 1977, Part II.

Watson, H. H. 1954. Errors due to anisokinetic sampling of aerosols. *Am. Ind. Hyg. Assoc. Q*. 15:21.

Woodruff, N. P. and F. H. Siddoway. 1965. A wind erosion equation. *Soil Sci. Soc. Am. Proc*. 29:602–608.

29 Indoor Aerosols and Exposure Assessment

CHARLES E. RODES

Center for Engineering and Environmental Technology, Research Triangle Institute, Research Triangle Park, NC

RUSSELL W. WIENER

National Exposure Research Laboratory,* U.S. Environmental Protection Agency, Research Triangle Park, NC

INTRODUCTION

This chapter provides an overview of both indoor aerosol concentration measurements and the considerations for assessment of exposure to aerosols in nonoccupational settings. The fixed-location measurements of concentration at an outdoor location, while commuting inside an automobile, or indoors in a residential living area represent only single microenvironments.† By comparison, personal exposure assessment reflects a time-weighted average across all microenvironments encountered. The relative emphasis on aerosol exposure in this chapter reflects the increasing importance of exposures assessments compared with making simple local concentration measurements in selected microenvironments. In reviewing research priorities for airborne particulate matter, the National Research Council (NRC) observed that "actual exposure to humans is determined by ambient air concentrations, contributions from indoor sources, and human time-activity patterns" (NRC, 1999). An issue highlighted by the NRC was an understanding of the potential "misclassification" of exposure, where concentration measurements from a single microenvironment (e.g., outdoors) are assumed to represent the integrated exposure to a specific aerosol when in fact the majority of the breathing zone exposure was contributed by another microenvironment.

* **Disclaimer**: This work has been funded in part by the U.S. Environmental Protection Agency. It has been subjected to Agency review and approved for publication. Mention of trade names or commercial products does not constitute endorsement or recommendation for use.

† *Microenvironment* is used in this chapter to represent a compartmentalized space that is represented by a single concentration for any of several reasons, including (1) the space is known to be suitably well mixed (e.g., a single room having no strong point sources) or (2) the space is often represented (correctly or incorrectly) by a single measurement location (e.g., an outdoor metropolitan area). The outdoors would be considered a *macroenvironment*.

Aerosol Measurement: Principles, Techniques, and Applications, Second Edition, Edited by Paul A. Baron and Klaus Willeke.
ISBN 0-471-35636-0

The focus here is primarily limited to assessments in nonoccupational settings. Occupational indoor microenvironments impose special requirements on aerosol characterization, which are discussed separately in Chapter 25. Outdoor aerosol sampling is important on its own as a regulated macroenvironment and is described in Chapter 27. The reader is cautioned to be aware of inadvertent biasing of nonoccupational air samples by occupational exposures (e.g., metal dusts resuspended into the breathing zone from work clothing worn while at home, as reported by Cohen et al., 1984). The ubiquitous contributions of ambient aerosol to indoor spaces and to integrated exposure assessments, however, also merits limited discussion in this chapter. The ability of outdoor aerosols to penetrate into indoor spaces is well documented (e.g., Alzona et al., 1979; Dockery and Spengler, 1989; Lioy et al., 1990; Wallace, 1996; Pellizzari et al., 1999a). A recent paper by Thornburg et al. (2001) demonstrated that the penetration process is complex and dependent on a number of factors, and it is a function of particle size. These latter considerations are especially important when attempting to understand the relationships between concentrations and exposures or between exposures and doses.

The methodologies and equipment needed to measure both fixed-location indoor concentrations and personal exposures to aerosols are described in modest detail. A representative review of the relevant literature is provided, and specific references should be reviewed for additional details. The limited scale and empirical nature of current indoor air and research studies has not prompted a wide range of commercially available measurement options for nonoccupational exposure assessments. Although occupational personal exposure monitors (PEMs) are readily available and have occasionally been pressed into service, their weight and obtrusiveness are not always suitable for use in nonoccupational studies. This is especially true for exposure studies where the selected participants exhibit a wide range of ages and physical sizes and may be reluctant to carry burdensome samplers for extended periods.

CONCENTRATIONS VERSUS EXPOSURES

The study of aerosols in nonoccupational indoor air, and especially the assessment of personal exposures, are relatively recent activities. Concentrations in outdoor and occupational microenvironments were studied extensively as important exposure predictors until the recognition of the relative importance of the time spent indoors (Ott, 1995) and the contributions of indoor aerosol sources to personal exposures (Wallace, 1996). A comprehensive summary of the key findings of recent nonoccupational indoor air and personal exposure studies, plus discussions of the factors most likely to contribute to elevated exposures, is provided by Wallace (1996). The observation that close human proximity to localized indoor sources (e.g., Sherwood, 1966; Rodes et al., 1991) could significantly bias exposure estimates made from fixed-location, residential concentration measurements by median factors as high as 3.3 also supported the move toward personal exposure assessments.

The control of outdoor and occupational aerosol source emissions, and their resulting concentrations, have historically (e.g. Buckley et al., 1991; Ogden et al., 1993) been used to minimize adverse health effects. A risk paradigm can be used to illustrate the intervening steps in the relationship between toxic emissions from sources and adverse health effects (NRC, 1999):

$$\text{sources} \rightarrow \text{emissions} \rightarrow \text{concentrations} \rightarrow \text{exposures} \rightarrow \text{doses} \rightarrow \text{effects} \qquad (29\text{–}1)$$

While concentrations can be linked statistically to effects through epidemiology, the sequential, direct linkage across the paradigm in Eq. 29–1 requires an understanding of the intervening relationships, including those between concentrations and human exposures. The transition from concentration measurements to the assessment of human exposure requires a shift in the strategies used to characterize aerosol, along with implicit requirements in the hardware technology. Relocating the sampler from a fixed location to a movable reference frame that follows the breathing zone location has proved to be a difficult transition for aerosol assessments. Significant improvements in the technology to provide acceptable data quality with relatively unobtrusive samplers have greatly facilitated occupational and indoor aerosol studies. Personal exposure assessments for the general population, however, require technologies with an even more stringent level of unobtrusiveness (quieter, smaller, lighter) while still maintaining the robust sampling volumes needed for subsequent sample analysis.

Although many occupational settings are "indoors," the aerosol concentration levels and chemical compositions can be substantially different from those in residential settings. They can also pose conflicting study design constraints. The simplest approach is to avoid overlapping the occupational and nonoccupational assessments, employing the respective measurement strategies that are appropriate. When overlapping activity patterns cannot be avoided, separate integrated aerosol collections for each microenvironment or time interval can be considered if the individual sampled aerosol loadings are sufficient to provide the necessary analytical data quality.

The constraints placed on aerosol exposure samplers, when considering children and the elderly, are more restrictive than those required for the general nonoccupational population. Children and the elderly have increasingly become the focus of aerosol exposure studies because of their identification as sensitive subpopulations with greater health risks (e.g., Liao et al., 1999). Compromises may be needed to study these populations, such as accepting the representativeness of area concentration measurements, in lieu of using unacceptably obtrusive personal monitors. The risks inherent in attempting to use obtrusive PEMs are the potential for substantial modifications of normal personal activity patterns and excessive drop-out rates during the study by the participants. The latter concern can affect the validity of population exposure studies in which participants are identified using a probability-based sample selection.

The time-weighted contribution of indoor air to total personal exposure is substantial, given the typical preponderance of time spent in all indoor microenvironments. While the substantially larger collection volumes of outdoor samplers readily lend themselves to complex sampling and analysis strategies, unobtrusive personal samplers with very small sampled volumes are still technologically limited to either relatively simple approaches or modest minimum detection limits.

The information base concerning aerosol character and concentrations in indoor air has grown steadily in the past two decades. Until 10 to 15 years ago, it was commonly believed that the general quality of indoor air was superior to that of the outdoor (ambient) air. However, even though ambient air quality may have improved for many pollutants, the indoor air quality for some compounds generated indoors (especially volatile organics) has not. The general decline in smoking during the past decade has significantly reduced the presence and mean concentrations of aerosols found indoors. Studies relating indoor aerosol concentrations and personal exposures to ambient concentrations have consistently shown that while indoor residential concentrations are generally lower than outdoors, personal aerosol exposures are significantly higher than fixed location concentrations of either (e.g., Alzona et al., 1979; Hayes, 1989; Rodes et al., 1991; Pellizzari et al., 1999b). This is especially true for particles $<2.5\,\mu m$ (PM-2.5), but less so for PM-10, due to the self-generation of particles during personal activities, including the resuspension of particles greater than about $1\,\mu m$ while walking on carpeting (Thatcher and Layton, 1998).

Changes in people's life styles have introduced new contaminants into the indoor environment, including synthetic fibers, residues from spray propellants, deodorants, pesticides, and combustion particles from kerosene space heaters. The average adult now spends 61%

to 78% of the day indoors in their private residence (U.S. EPA, 1989; Jenkins et al., 1990) and at least 87% to 89% indoors including occupational microenvironments (Ott, 1995). It could be surmised that less active individuals, and especially the elderly and those with pre-existing health problems, spend an even greater fraction of time indoors. Rodes et al. (2000) showed that the elderly in retirement centers spent as much as 95% of their time indoors.

MEASUREMENT STRATEGIES

Several questions are provided in Table 29–1 that can be reviewed to assist in developing the indoor or exposure aerosol study design and selecting the appropriate methodologies. Defining a clear understanding of the objectives of the study, the required data quality objectives, and the use of the collected data are essential in the selection of the sampling and analysis approaches and the development of the measurement strategy.

The simplest indoor aerosol measurements are made using a sampling device in a fixed location. These devices are often referred to as *microenvironmental monitors* (MEMs) and can provide acceptable data for well-mixed spaces. After a location is chosen that is expected to represent a microenvironment, sampling is conducted for a defined integration period. If a sufficient number of indoor (and outdoor) microenvironments are also characterized, integrated exposures can be estimated using a simple time-weighted-average model. Wallace (1996) noted, however, that it was often difficult to reconcile the differences between such simple model estimates and the consistently higher aerosol concentrations found using personal samplers carried by subjects in the same microenvironments. It was concluded that personal proximity to localized sources was the most likely contributor to elevated personal exposure mass concentration.

PEMs are small, self-contained, battery-powered sampling systems that can be carried by an individual to mimic the proximity of the breathing zone to local sources or spatial concentration gradients (Fig. 29–1). These relatively unobtrusive samplers are used to assess the

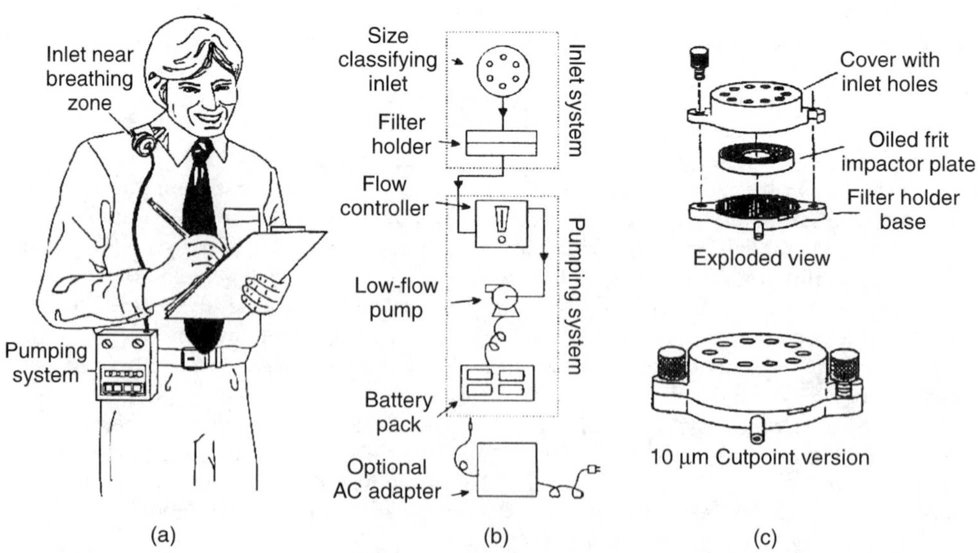

Fig. 29–1. Schematic diagram of a personal aerosol exposure monitor, showing (A) individual wearing the monitoring system, (B) block diagram of personal exposure system components, and (C) Diagram of ***MSP*** Model 200 Personal Environmental Monitoring Impactor. (Adapted from U.S. Environmental Protection Agency, 1990.)

TABLE 29–1. Design Considerations for Indoor Air and Exposure Aerosol Sampling Studies

No.	Design Consideration	Section Heading
1.	How are fixed-location, indoor and outdoor aerosol concentration measurements related to personal exposures?	Concentrations versus Exposures; Relating Indoor, Outdoor, and Personal Concentrations
2.	What specific aerosol characteristics are expected (e.g., size distributions, presence of volatile components, etc.), and how will they affect the collection and analysis methodologies?	Measurement Strategies; Sampling and Analysis Methods
3.	Are specific cut point size ranges or sampling efficiency curve shapes required by regulations or conventions?	Measurement Strategies
4.	What are the expected aerosol concentration levels, and are high and low concentrations equally important?	Sampling and Analysis Methods
5.	What are the expected influences of spatial and temporal gradients in aerosol concentration and size distribution across the important microenvironments?	Introduction
6.	What are the aerosol sources, their strengths, and their locations relative to the proposed sampling locations?	Measurement Strategies
7.	Is identification of the aerosol source categories contributing to collected samples important?	Measurement Strategies
8.	What are the ventilation characteristics of the indoor microenvironments or meteorology in the outdoors that may adversely affect aerosol sampling (e.g., turbulence levels, frequency of high winds, etc.)?	Indoor Air Studies
9.	What supplemental measurements may be useful to collect indoors to support IAQ aerosol modeling and to assist in linking concentrations to sources?	Indoor Air Studies
10.	Will speciation beyond mass be needed, will the sampler cut point influence the collected samples, and will multiple substrate collections be required to match the analysis methodology?	Sampling and Analysis Methods
11.	Is the majority of the aerosol collection expected to occur in occupational microenvironments?	See Chapter 25
12.	Are viable biological aerosols a concern?	See Chapter 24
13.	Are there suitable locations within the microenvironments (fixed locations) to accommodate the monitors to minimize safety concerns and alteration of normal activities?	Exposure Studies
14.	Are suitably precise and unobtrusive personal monitors available for the desired measurement, especially for special populations such as the elderly or children?	Exposure Studies
15.	What are the biases that may affect the MEM and PEM measurements?	Sampling and Analysis Methods; Personal Aerosol Sampling Biases
16.	Is there a sensitive focus population, and is it important to define the most exposed in the population?	Exposure Studies; Focused Panel Studies

time-integrated exposure of an individual to aerosols, with hopefully minimal influence on normal activity patterns. Although sampling near the breathing zone is usually desirable, alternative locations (e.g., at the waist) may be acceptable for a chosen particle size range. An intercomparison study should be considered before the exposure study to demonstrate the absence of a bias (relative to the breathing zone) from using an alternative location. Although breathing zone sampling is considered ideal, certain situations may require resorting to alternative locations (e.g., sampling adults with a waistpack or children using a backpack sampler; Figs. 29–2 and 29–3).

Fig. 29–2. Personal exposure monitoring system with aerosol inlet mounted on the front of a waistpack (Reproduced by permission; Research Triangle Institute, Research Triangle Park, NC 27709).

Fig. 29–3. Personal Exposure Monitoring System with the Aerosol Inlet Mounted on a Backpack for a Study of Children (Reproduced by permission; Research Triangle Institute, Research Triangle Park, NC 27709).

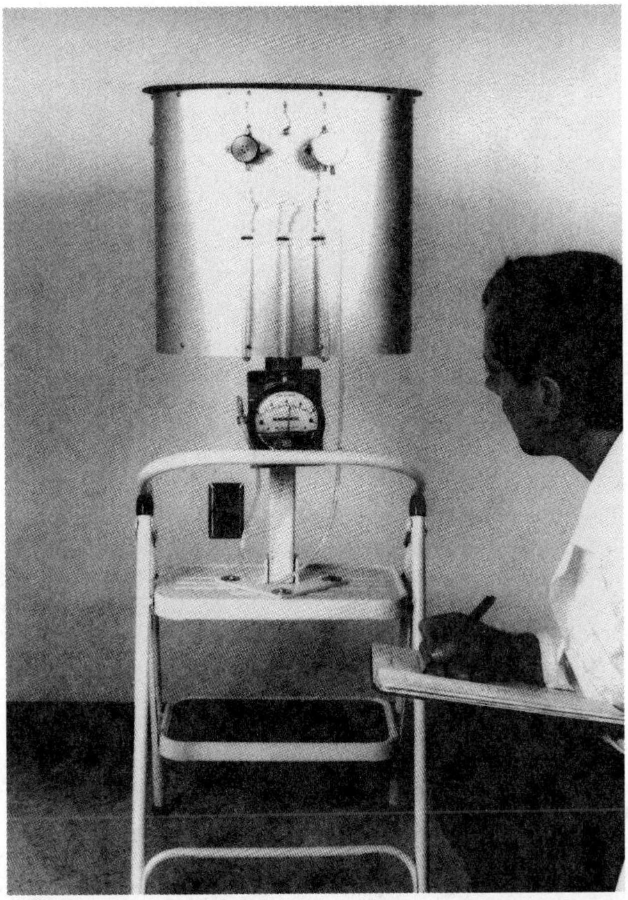

Fig. 29–4. Microenvironmental exposure monitoring system used indoors with the aerosol inlet mounted on the front of a "bluff" body (Reproduced by permission; Research Triangle Institute, Research Triangle Park, NC 27709).

Because air exposure is defined to occur when an aerosol particle reaches the entrance to a person's nose or mouth (Lioy, 1990), the best measurements for the assessment of exposure to aerosols typically require a sampler inlet be placed within the breathing zone. When personal exposure assessments are not possible or cost effective, a microenvironment (e.g., a room, work area) can be selected and monitored using a MEM or fixed-location sampler as shown in Figure 29–4. These stationary samplers operate at varying distances from the subject's breathing zone and provide correlated but biased measurements compared with personal exposures.

The aerosol dose that an individual receives in both industrial and residential settings is a combination of the exposure and specific elements of the breathing process (e.g., inhalation rate, tidal volume, respiratory physiology). The key exposure elements, include such factors as the aerosol size distribution, morphology, bulk or surface chemistry, and so forth. The aerodynamic size of aerosol particles provides a direct relationship to the deposition of such particles in the human respiratory system. This forms the most logical basis for selecting inlet sampling size cut points for health-based studies. Another important factor to consider is the distinct chemical compositions that may exist between aerosol size distribution modes (Wilson and Suh, 1997). Selecting an inlet cut point that clearly separates distributional modes may be more important to the study objectives than mimicking deposition in portions of the human respiratory system.

The respiratory system's particle deposition processes of impaction, settling, interception, and diffusion are best defined in terms of aerodynamic diameter (see Chapter 25). One or more of these removal processes may operate simultaneously. Depending on the location of particles in the individual's respiratory system, the action of the removal processes may produce different aerosol deposition patterns. Particles greater than about 50 μm in diameter are considered inhalable (see Chapter 25) and capable of entering the respiratory system by crossing the oral/nasal entry plane. Particles greater than 10 μm in diameter are deposited on the ventilation pathway surfaces above the trachea, while fine particles less than ~2 μm in aerodynamic diameter are removed primarily in the gas-exchange (alveolar) region of the lungs. Integrated sampling with either PEM or MEM samplers can use inlets with selected cut points, such as those just described, that approximate exposures in specific size fractions.

Individuals are exposed to ambient aerosols when they spend time outdoors; however, they may be exposed indoors due to the infiltration of ambient aerosols inside (Colome et al., 1982; Hayes, 1989). Vu Duc and Favez (1981) found lead (Pb) particles indoors, for which 60% of the total mass was submicrometer-sized particles, and attributed it to automobile exhaust. Polycyclic aromatic hydrocarbons (PAHs) were also found on submicrometer-sized particles by these researchers and were attributed to the infiltration of combustion aerosols. In fact, PAH concentrations are often higher indoors than outdoors (Pellizzari et al., 1983; Wallace et al., 1986; Wallace, 1987; Chuang et al., 1987).

The identification of an aerosol of ambient source origin for indoor and exposure samples has been studied for years (e.g., Alzona et al., 1979), but improvements in the methodologies are still considered an important research priority (NRC, 1999), especially for personal exposures. The commonality of significant components of both indoor- and outdoor-generated aerosol sometimes makes accurate source identification difficult, if not impossible. A potential assessment methodology is the application of ambient source apportionment to estimate the relative contributions of indoor and outdoor sources (Sexton and Hayward, 1987; Koutrakis et al., 1992). This powerful tool requires the establishment of singular measurement "fingerprints" for the potential indoor and outdoor source categories (a task requiring creative sampling and analysis schemes). The characteristic receptor sample chemistries are then related to these fingerprints through an apportionment model to estimate the relative contributions by source category. While ambient air samplers are sufficiently robust—a common flow rate is $2.78 \times 10^{-4}\,m^3/s$ [16.7 L/min]—to collect the sample volumes and parallel collection substrates to define these fingerprints and receptor samples, indoor air and especially personal exposure samplers often operate at flow rates that are two to eight times lower. Space limitations (even using indoor MEM samplers) can also preclude the luxury of parallel sampling channels to collect supplemental aerosol size ranges and substrates (e.g., Teflon filters for mass, ions, metals, and pre-fired quartz filters for elemental and volatile carbon).

SAMPLING AND ANALYSIS METHODS

Aerosol exposure sampling methods can be generally separated into three groups: physical, chemical, and biological. Physical sampling methods provide information regarding particle size, number, morphology, and mass concentration. Chemical sampling methods provide information on elemental and chemical constituents, as well as phase distributions. These two groups can be further subdivided into integrated sample assessments and those that are made as near-real-time measurements on a continuous basis. Biological exposure sampling is discussed in Chapter 24. For indoor and personal exposure measurements, the significant sampling and analysis factors must be overlaid on the physical size and flow rate limitation imposed by miniature personal aerosol exposure devices. Personal exposure devices also require consideration of the energy required (batteries) to operate the subcomponents, as well as the burden associated with their weight, bulk, and noise.

Particle Sizing and Characterization

The types of sampling methods selected for indoor exposure measurements depend initially on the physical characteristics of the aerosol being studied, as described in Chapter 8. Compromises are often required for exposure measurements, however, due to inherent limitations, including allowable size, noise, and weight. Integrated sample collection typically utilizes a simple fractionating inlet to provide an upper cut point for the collected sample. Inlets used in indoor aerosol sampling are often similar to those for ambient air sampling (Chapter 27). Personal exposure sampling inlets, however, are severely limited in size and weight, which strongly influences their design. Similarly, continuous measurements of aerosol optical properties indoors use the same devices for monitoring ambient air (Chapter 14). Continuous personal exposure measures are typically limited to miniature single-channel nephelometers.

Personal and microenvironmental inertial impactors, virtual impactors, and cyclones have been developed that are suitable for use in both indoor air and exposure studies. Some specific samplers were described by Turner et al. (1979), Rubow et al. (1985), Spengler et al. (1989), Dockery and Spengler (1981), McKenzie et al. (1982), Buckley et al. (1991), and Marple et al. (1988). A representative list of commercially available devices as of August 2000 are listed in Table 29–2.

Inlets with specific cut points, such as 2.5 μm, have been developed for indoor sampling (Marple and McCormack, 1983; Marple et al., 1987, 1989; Buckley et al., 1991) that use impaction on oil-soaked or greased, sintered metal surfaces. Cascade impactors can also be used to develop the mass size distribution from which the contribution of specific size fractions can be estimated (Martonen et al., 1984).

The measurement of respirable particles that would be deposited in the alveolar region of the lungs can be approximated by using an aerosol inlet with an impactor with a 2.5 μm cut point. This size range is the PM-2.5 fraction described by the U.S. EPA (1997) and is often measured in outdoor settings. Different types of inlets may be required to sample various micro- and macroenvironments, especially the outdoor macroenvironment with its wide range of wind speeds and precipitation types. While outdoor ambient samplers may be required to meet regulated performance standards, they are often unacceptably large and noisy for indoor usage. The sampling performance of the ***MSP**** Model 200 PM-2.5 personal aerosol inlet operating at $3.33 \times 10^{-5} m^3/s$ [2 L/min] outdoors is compared with the EPA ambient cut point requirements in Figure 29–5. This intercomparison suggests that the MSP inlet can be used for indoor and outdoor sampling, as well as PM-2.5 personal exposure assessment.

Limited data are available on the size distribution of residential indoor aerosols or on the distribution of particles to which people are commonly exposed (Kamens et al., 1988; Owen et al., 1990). Generally, indoors one would expect a bimodal (fine, coarse) size distribution, similar to that found outdoors. Indoor fine particles (<2.5 μm) typically originate from condensation and nucleation processes, such as smoking, heating, cooking, and infiltration of ambient particles. Coarse particles (>2.5 μm) can be inhalable and are generated from a wide variety of sources in the indoor environment. Mechanical (as opposed to chemical)

* See Appendix I for full manufacturer addressess referenced to the italicized three-letter codes.

TABLE 29–2. Representative Personal Exposure Monitor (PEM) and MicroEnvironmental Monitor (MEM) Components that are Commercially Available[a]

Type	Name	Vendor Code[b]	Remarks
MEM indoor sampling systems	AD&E Indoor Area Monitor (Harvard Impactor)	***ADE***	Up to 20 L/min with 1.0, 2.5, or 10 μm
	MEM system	***SKC***	2.5 or 10 μm cutpoints
	MEM system	***MSP***	10 L/min dual flow 2.5 and 10 μm
	MEM system	***RTI***	Up to 3 parallel channels; 2.5 μm, 10 μm, and inhalable; 2 or 4 L/min
Ambient samplers			See Chapter 27
Biological samplers			See Chapter 24
PEM personal sampling inlets	Series 200 Impactors	***MSP***	2, 4, or 10 L/min for 1.0, 2.5, or 10 μm
	Cyclones	***BGI***	1.7 to 2.2 lpm
	Respicon	***TSI***	3.1 L/min with simultaneous 2.5 μm, 10 μm, and inhalable
	Spiral Inlet	***SKC***	2 L/min for 2.5 μm
	IOM	***SKC***	Total inhalable
	Personal Samplers	***URG***	4 L/min with 1.0, 2.5, and 10 μm
PEM cascade impactor	Model 290	***MSP***	2 L/min, 8 stages, from 0.6 to 20 μm
Occupational PEMs			See Chapter 25
Personal sampling pumps	AFC123	***BGI***	0.5 to 4.5 L/min
	Air Check	***SKC***	Various ranges
	Escort	***MSA***	Various ranges
	Gilian-Sensidyne	***GIL***	Various ranges
	PEM system	***RTI***[c]	1 to 4 L/min; incorporates activity sensor
Denuders	Annular Denuder Indoor Sampling System	***URG***	2, 4 L/min flow, 2.5 μm cut size
Ultrafine aerosols	P-Trak	***TSI***	Hand-held; 0.01 to 0.10 μm range
Aerosol nephelometers	RAM	***MIE***	Active sampling
	MINI-RAM	***MIE***	Passive sampling (no pump)
	Dust-Trak	***TSI***	1.4 to 2.4 L/min
	Aerosol spectrometer	***GRI***	Inlets for 2.5 and 10 μm
Supplemental indoor environmental measurements	Q-Trak IAQ monitor	***TSI***	CO, CO_2, Temp., RH
	MPM 500e	***SOL***	Temp., RH, air velocity
	Thermal Comfort Logger	***INN***	Temp., RH, air velocity
	Air Tech	***INN***	SF_6; air exchange rate
	Compuflow	***ALR***	Temp., RH, air velocity

[a] Table entries represent only selected models; check with vendors for more complete information. RH, relative humidity.
[b] Appendix I contains vendor information indexed to codes.
[c] The RTI system was not yet commercially available as of August 2000.

generation and resuspension of dust are constantly occurring because of all human activities. Penetration of outdoor coarse particles to the indoor environment occurs infrequently because these particles are easily lost to surfaces. Inlet testing may be necessary to measure the sampling efficiencies for the inlets in still air, at typical indoor air or walking velocities, and at various wind speeds, if used outdoors.

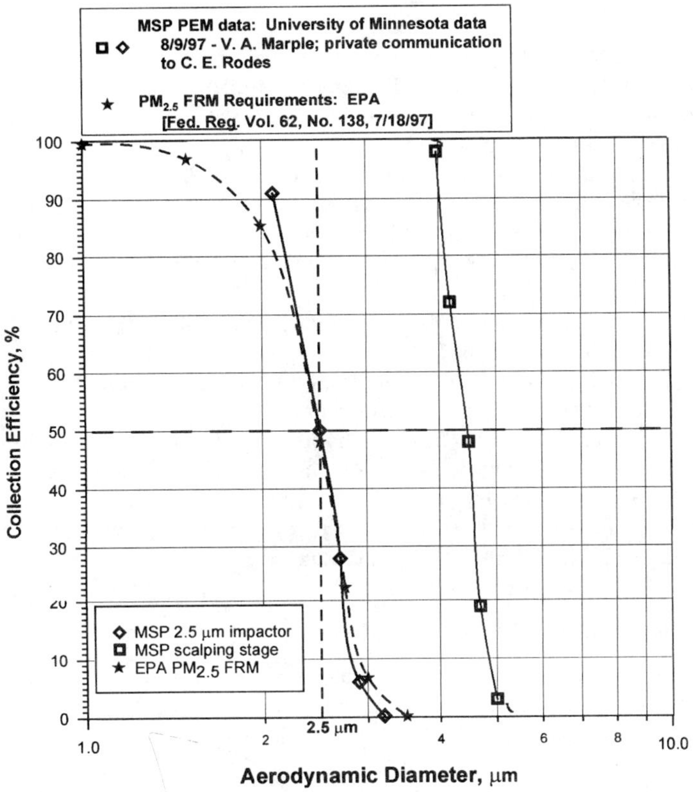

Fig. 29–5. Comparison of the sampling efficiences for the ***MSP*** Model 200 2 lpm, 2.5 μm inlet, compared with the EPA Federal Reference Method requirements for $PM_{2.5}$ (Sources: Rodes, 1997; Rodes, 1999).

In evaluating the performance of personal sampler inlets, it is important to consider bluff body effects (Vincent and Gibson, 1981; Wiener and VanOsdell, 1990; Ingham and Yan, 1996). This effect occurs as a result of the changes in aerosol particle trajectories induced by the alterations in the flow stream trajectories around the human body immediately adjacent to the PEM inlet (also see Chapter 8). A potential bias in collection performance for a PEM can occur if the PEM is used as a fixed-location sampler without a form (simulating the body shape) immediately behind the inlet. The interaction of human activity and the flow field particle trajectories may further bias the measurement. The microenvironmental indoor sampler shown in Figure 29–4 illustrates the use of an aluminum, elliptical backing body behind the inlet to minimize this effect.

For integrated methods, the characteristics of the filter media or impaction substrate become important. Chapter 9 provides a detailed discussion of the characteristics of filters and their particle capture mechanisms. Improper selection of the filter substrate can cause substantial biases in low-flow-rate indoor and personal exposure sampling systems. Teflon filters of 2 or 3 μm porosity are perhaps the most widely used substrates for indoor and personal exposure sampling and can be used for subsequent ion and XRF analyses. Lower porosity filters used for personal exposures are prone to clogging if the sampler collects heavy loadings of environmental tobacco smoke. While Teflon filters are generically susceptible to nitrate losses (Hering and Cass, 1999), the typically lower face velocities of personal exposure samplers may provide reduced losses compared with ambient samplers.

Continuous Monitoring

Only a limited number of miniature, portable aerosol spectrometers using optical nephelometry are currently commercially available (e.g., MiniRam [***MIE***] or the Dust Trak [***TSI***]; see Table 29–2) to provide estimates of the particle concentrations on a nongravimetric basis. The conversion of photometer response to estimated mass concentration is provided by the manufacturer's internal calibrations against test aerosols, such as Arizona test dust (***PTI***). The ability of these calibrations to represent the aerosols across a variety of microenvironments, however, is not clear. A portable ultrafine aerosol monitor (P-Trak ***TSI***; based on total condensation nuclei counting, see Chapter 19) is also available for making real-time indoor surveys.

To characterize indoor air aerosols in even greater detail (e.g., count, surface area, and volume distributions), the measurement devices described in Chapter 13 can be used indoors, most of which are limited to fixed-location sampling because of their physical size, vibration sensitivity, or dependence on AC power. Optical particle counters (e.g., LAS-X, ***PMS***), aerodynamic particle sizers (e.g., APS ***TSI***; see Chapter 17), and ultrafine monitors (e.g., SMPS system, ***TSI***; see Chapter 18) have been used in tandem in indoor studies to define the number distribution over a broad size. Merging and presentation of composite size distributions produced by different measurement techniques are complex tasks, however, and the reader is referred to Chapter 22 for additional information.

Gravimetric, Elemental, and Chemical Analyses

An extensive discussion of aerosol sample analysis methods are provided in Chapter 11 and need not be repeated here. Most of the same sampling considerations are used by indoor and personal exposure samplers (e.g., filter types, archive concerns) as their larger ambient sampler relatives. Sampling residential indoor and personal exposure aerosols poses special analysis problems because of the relatively low concentrations of the pollutants sometimes present (often less than 10 to 20 $\mu g/m^3$) and the resulting small collections for low-flow-rate samplers. This is especially true of gravimetric analysis, which is limited to the minimum detection levels of currently available electronic balances. For example, applying gravimetric analyses to samples of less than ~50 μg can result in poor precision because of the 2 to 5 μg composite error imposed by the weighing process (Lawless and Rodes, 1999).* Rather than attempt collection of larger sample volumes, Lawless and Rodes (1999) addressed improvements in the gravimetric analysis process for personal exposure samples in order to improve the minimum detection levels.

Serial inlet sampling systems have been developed in reduced scale versions for lower flow rates that sequentially combine inertial impactors with annular denuders and filter holders (sometimes stacked) to separate particles and gases (Koutrakis et al., 1989; Vossler et al., 1988). The phase distribution of species that are present as both particles and gases may be assessed with these devices. This can be especially important to accurately quantify nitrates (Hering and Cass, 1999) or the lighter semivolatile organic compounds (McDow and Huntzicker, 1990). Inlet denuder systems can also protect collected aerosols by removing potentially interactive vapors (e.g., removing ammonia to protect acid aerosol already deposited). The denuders can be used with either a MEM or, with some burden penalty, a PEM. A pre-separator, such as a cyclone or inertial impactor, is used to remove particles larger than 2.5 μm. The denuder removes gases of interest, and then an after-filter removes fine particles. The large particles, adsorbed gases, and fine particles can then be analyzed separately.

*It is encouraging to note that the estimated gravimetric error inherent in gravimetric analyses reported in the 1993 edition of this text were twice as large as are now possible—the result of significant improvements in the methodologies.

Personal Aerosol Sampling Biases

A number of important factors have been identified that can significantly influence the precision and accuracy of personal exposure measurements. Lawless and Rodes (1999) discuss in detail the gravimetric analysis errors that can result in significant precision and accuracy problems for lightly loaded, personal exposure filters. Key suggestions were the application of computer control with an electronic balance to provide paired average weighings for each filter (pre- and post-weight) and minimal operator intervention that may inadvertently alter the balance zero. Baron (1998) discusses a number of issues biasing personal exposure sampling, including sealing problems and the advantages of conductive construction materials to minimize static charge problems. Willeke and Baron (1990) also discussed various aspects that might bias sampling results and included subsequent treatment options and pitfalls to minimize misinterpretation of collected data.

Substrate handling and storage may cause loss of large particles from filter surfaces. The sampling media should remain in an airtight enclosure during shipping and handling to prevent nonsampled exposures. If the presence of semi-volatile organic carbon (SVOCs) is suspected, it is necessary to enhance their stability on the substrate through cold (<273 K [<0°C]), dark storage because of chemical and photochemical reactivity and artifact production.

Flow measurement and control are key factors in integrated sampling accuracy and precision. Changes in flow rate will alter the effective cut diameter, as well as change the amount of mass collected and the computed concentration. A personal sampler pumping system with no means of recording the flow continuously during the sampling period may experience periods of significant flow reduction, resulting from inlet obstruction (e.g., from overcoats). A robust flow control system may, however, consume as much power as the pumping system, significantly shortening battery life. Compromises may be required if longer sampling durations are necessary.

Other Sampling Considerations

In addition to meeting the study objectives, the sampling integration period must be selected based on other specific factors, including the battery life (for personal samplers), the expected temporal patterns of the aerosol source emissions and human activity patterns, and the time required to acquire enough material for analysis without overloading the filters. Levels of particulate matter and SVOCs in indoor environments can vary substantially because of local, strong sources such as environmental tobacco smoke (ETS). If analytical sensitivity is sufficient, it is sometimes desirable to make separate measurements for sampling by daytime and nighttime periods to highlight the influence of personal activity sources and outdoor diurnal chemistry.

The accuracy and precision of the measurements determine the level of researchers' confidence in the data collected. Collocated placement of instruments (for replicate samples) are used to estimate precision. Coefficients of variation (CVs) for ambient aerosol monitors meeting outdoor equivalency requirements (Code of Federal Regulations, 1990) range from 3% to 10% and those for indoor MEMs range from 3% to 15%. PEMs have poorer precision, primarily resulting from smaller sample collection and poorer flow controls, with CVs ranging from 5% to 25%. The accuracy of aerosol measurements can only be estimated by calibrating the instruments under controlled test atmosphere conditions, such as in wind tunnels. Accuracy is inferred from comparison with isokinetic nozzle measurements.

There are no mandatory federal indoor air quality or exposure standards for nonoccupational aerosols in the United States. The current outdoor standards apply to PM-2.5 and PM-10 aerosols in ambient air. The US EPA's national ambient air quality standard (NAAQS) for PM-10 is set at 50 $\mu g/m^3$ for a 24 h annual average and at 150 $\mu g/m^3$ for an individual 24 h daily limit.

When present, ETS is one of the most significant indoor aerosols to be addressed. Personal samplers must be capable of handling the elevated pressure drops that may occur on the filters, caused by elevated ETS aerosol deposits. Additionally, the complex mixture of vapor and aerosol phase compounds in ETS may interfere with some analytical schemes and result in volatilization losses after collection if not properly stored. ETS exposure studies have been performed by a number of investigators (e.g., Vaughan and Hammond, 1990). Repace and Lowrey (1983) estimated a mean exposure of 1.43 mg/day, with a range of 0 to 14 mg/day for a series of public buildings not randomly selected. The Harvard Six-Cities Study (Spengler et al., 1981; Samfield, 1985) found a mean increase in indoor PM-2.5 concentration of 13 $\mu g/m^3$ when one smoker was present. Several methods are available to apportion ETS from the bulk respirable suspended particles in indoor aerosol samples. The tests rely on marker species and metabolites to determine the degree of exposure to ETS. Nicotine is the most commonly used marker for ETS and is found almost exclusively in the vapor phase in air samples of ETS (Eatough et al., 1986; Hammond et al., 1987). Nicotine has a short half-life (30 min), but is stable once frozen. A number of specific tests for nicotine in indoor air are available (Caka et al., 1990). Cotinine, a metabolite of nicotine with a long half-life (20 h), can be measured in urine or saliva as a surrogate for aerosol exposure. Cotinine analyses are expensive, but specific, because cotinine is endogenously produced.

INDOOR AIR ASSESSMENTS

Focused Studies

Focused indoor aerosol studies are designed to better understand and evaluate selected physical parameters governing specific aspects of the generation of, distribution of, sampling of, and exposure to aerosols. The studies can be performed under controlled conditions in the laboratory or in field environments. Laboratory studies may include special studies to address aerosol source characteristics, sampler performance factors, aerosol distribution, and transport. Quality assurance studies may be required for sampler inlet validation and method standardization. Precision can be obtained from sampler intercomparison and collocational testing of inlets and sampling systems (Lioy et al., 1988). Estimation of aerosol sampling accuracy can only be made in controlled laboratory wind tunnel conditions. The physical differences between the measurements from PEM and MEM sampling can be studied using aerosols in a test chamber. While PEMs provide the best estimates of aerosol exposure, they often have a number of drawbacks, including poorer precision and sensitivity, anisokinetic sampling biases, and higher study costs. MEMs may provide a cost-effective substitute for personal exposures, especially in situations where the microenvironments are few and well mixed.

Field studies are needed to assess aerosol concentrations and exposures in complex indoor, microenvironmental, and macroenvironmental conditions and to determine source contributions (source/receptor analyses). Although respiratory deposition of specific aerosols can be measured in the laboratory, actual human exposure to aerosols cannot be studied directly. PEMs and MEMs are used as surrogate samplers to approximate the level of particulate matter to which subjects are exposed. Each microenvironment in which the subject might be significantly exposed must be considered. Most early residential indoor air quality (IAQ) studies used only fixed-location microenvironmental measurements.

Exposure studies also require the use of questionnaires, often referred to as *survey instruments*. Survey instruments are used to obtain information from the study participants, including information on activity patterns, source activities, ventilation system settings, and health responses. These questionnaires and activity logs should be pre-tested using focus groups in laboratory situations and using pilot field tests, with participants from the population intended for study. In contrast with most laboratory or field studies, the success of human exposure assessment studies depends on the degree of cooperation and the performance of the subjects in the study. In standard industrial hygiene sampling, the hygienist closely monitors the participant. Typically, the participant in an occupational setting is engaged in repetitive, well-defined tasks. When studying exposure in residential settings, the investigator must be as unobtrusive as possible to minimize biases in participant activity patterns.

Building Studies

The sick building syndrome (SBS) has been a key factor in promoting studies of IAQ and human exposure. SBS is specifically defined by the World Health Organization (1983) as symptoms from which building occupants suffer during the time they spend in the building that diminish when they leave the building. Some symptoms may have delayed onset and may occur after the occupant leaves the building; however, the symptoms diminish after he or she spends time away from the building. These conditions cannot be readily traced to specific pollutants within the building. Typical health responses include the following: irritation of eye, nose, and throat; dry mucous membranes and skin; erythema; mental fatigue; headache; airway infections; cough; hoarseness; wheezing; unspecific hypersensitivity reactions; nausea; and dizziness (Molhave, 1987).

Tight building syndrome (TBS) has symptoms similar to SBS and is due to poor ventilation practices, causing thermal discomfort or extremes in humidity, often resulting from energy conservation measures. The National Institute for Occupational Safety and Health (NIOSH) Hazard Evaluation Program concluded that about one-third to one-half of sick buildings have this problem. At least 5% of the SBS and illnesses are due to biological aerosol contamination problems (Wallingford and Carpenter, 1986).

Building-related illness (BRI) is defined as a discrete, identifiable disease or illness that can be traced to a specific pollutant or source within a building. Aerosols (especially bioaerosols) are undoubtedly a leading cause of SBS, in combination with ventilation deficiencies (Brief and Bernath, 1988; LaForce, 1984).

Initial preparation for a sick building study is important. Discussions before sampling must be conducted with the occupants regarding the complaints and symptoms reported. An assessment must also be made of (1) the mitigation steps already taken to improve the IAQ and (2) the heating, ventilation, and air conditioning (HVAC) system operation and maintenance schedules. A thorough inspection of the building must be conducted inside and out, including the HVAC system. The following determinations must be made: the type of system; the configuration, location, and effectiveness of air handlers, chillers, cooling tower, filtration or particle-removal devices, air supply and air return ducts, and air circulation path; location of outside air intake vents (check for cross-contaminants) and thermostats; and the presence of moisture, standing water, or microbial contaminants in the HVAC system.

The building must be surveyed for sources and strengths of suspected contaminants (e.g., tobacco smoke, combustion products), build up of pollutants from poor air ventilation/high occupant density (carbon dioxide monitoring can be used as a surrogate), insulation materials (fiberglass, urea-formaldehyde foam, asbestos), age and cleanliness of carpets and furnishings, radon decay products from soil or concrete blocks, construction materials or activities, aerosol cleaning products, pesticides, moisture (condensation on windows, walls), biological species (fungi or mold on walls or in closets or enclosed spaces), and the presence of unvented combustion heaters (kerosene heaters, unvented gas ranges). It is also desirable to determine if sociological or psychological factors have been introduced that would make personnel more receptive or responsive to sick building problems.

Relating Indoor, Outdoor, and Personal Concentrations

Rodes et al. (1991) suggested that strong spatial gradients, either indoors or outdoors, can cause substantial biases between MEM and PEM measurements. They also reported that there are lognormal relationships between microenvironmental and personal exposures when results are plotted as cumulative distributions (Fig. 29–6). The slopes of these relationships are hypothesized to be affected by a combination of source/receptor proximity and the spatial distribution of the sources. MEMs tend to underrepresent exposure when the aerosol being measured is being generated by personal activities. If the outside aerosol that infiltrates indoors is homogeneous, the indoor gradients are most likely due to the generation of particles indoors.

A number of researchers have recently reported methodologies for relating integrated personal exposures with indoor and outdoor aerosol concentrations (e.g., Rojas-Bracho et al., 2000; Janssen et al., 2000; Williams et al., 1999). Both PM-2.5 and PM-10, personal samples were collected by Rojas-Bracho et al. (2000) using reduced flow rate, impactor-based inlets and a single pump (with a flow splitter) operating at $5.33 \times 10^5\,m^3/s$ (3.2 L/min). A 0.1 m elutriator channel was added to the inlets to reduce the unwanted collection of clothing fibers. Janssen et al. (2000) utilized miniature personal cyclones to provide PM-2.5 cut points operating at $6.66 \times 10^5\,m^3/s$ (4 L/min), weighing 1.5 kg. They reported a sample failure rate of 38%, attributed to battery failures using the portable pumps. This highlights a current problem in personal aerosol exposure measurements because the available miniature pump technologies require compromises among the factors of battery life, noise level, and system weight. Williams et al. (1999) described using a single-channel, impactor-based PM-2.5 inlet system mounted on a vest with a total weight just under 1 kg. Battery problems were eliminated by

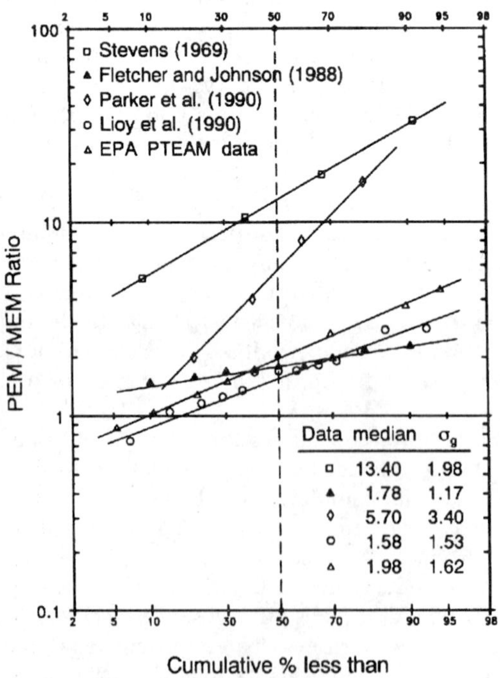

Fig. 29–6. Distributions of Personal exposure monitor (PEM) to microenvironmental monitor (MEM) ratio from literature data (Source: Rodes et al., 1991).

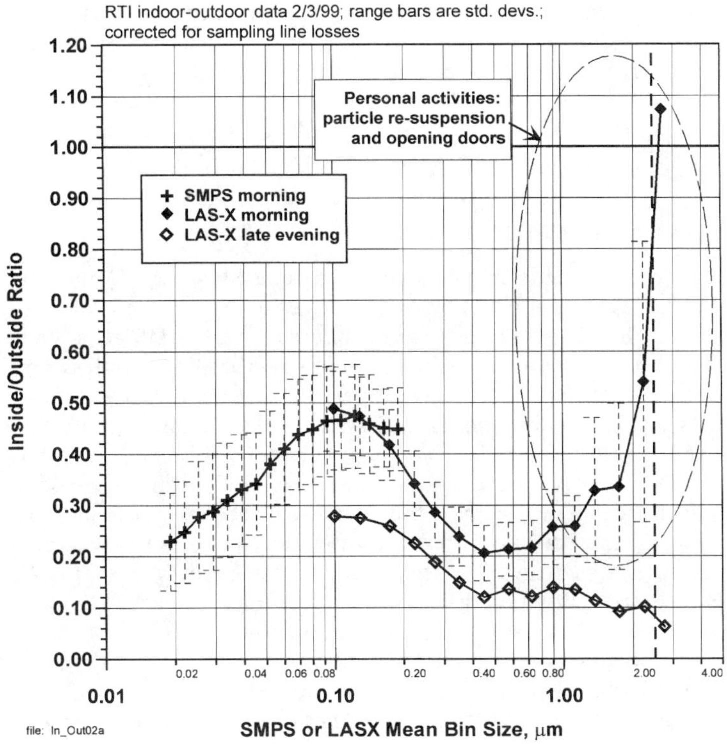

Fig. 29–7. Indoor-to-outdoor particle count ratios by size, using two different measurement devices (TSI SMPS and a PMS LAS-X); illustrating the influence of personal activities on particle resuspension from carpeting (Source: C. E. Rodes, 1999 unpulished data; Research Triangle Park, NC 27709).

using a low power drain pump/motor/flow controller combination and alkaline, rather than rechargeable, batteries.

A more robust (but nongravimetric) approach to studying indoor–outdoor relationships involves simultaneous collections of particle count distributions in both microenvironments. This permits assessing the time history of ratios as a function of particle size, allowing direct comparison of periods with and without indoor sources (or participants) present. An example of this approach is shown in Figure 29–7 in which a ***TSI*** SMPS and a ***PMS*** LAS-X aerosol spectrometer were used in tandem to cover the particle size range from 0.01 to 2.5 μm (data from Rodes, 1999). The data illustrate the influence of personal activities on the larger particle sizes. To eliminate the need to have duplicate pairs of monitors collecting indoor and outdoor data simultaneously, a low-loss manifold system was used, operating on a 5 min cycle. Note that the losses in this sampling manifold were of sufficient magnitude (and size dependent) to require correction.

Indoor Source Characterization

With the exception of ETS, only limited data are available characterizing aerosol emissions from indoor sources. There are few standardized test protocols for determining indoor source emission factors. Studies have characterized unvented combustion sources and particles generated by home humidifiers, leakage rates of particles from vented combustion sources, and the deposition rates of particles on indoor surfaces (Highsmith et al., 1988; Girman et al., 1982; Leaderer et al., 1990; Traynor et al., 1987; Traynor and Nitschke, 1984; Tu and Hinchliffe, 1983). Thatcher and Layton (1995) reported that resuspension of particles above 1 µm occurred readily by walking on carpeting, with the predominant larger sizes have a significant affect on the PM-10 fraction. Lioy et al. (1999) characterized the aerosol emissions as a function of particle size for several commercially available vacuum cleaners and found emissions in both the ultrafine and fine particle ranges.

Associated Indoor Air Quality Measures

Indoor ventilation parameters are important for understanding (1) mixing and dilution of aerosol sources, (2) aerosol transport from sources to receptors, (3) the loss mechanisms that occur in both the physical structure and internally in the HVAC system, and (4) the penetration of ambient aerosol indoors. Whole-house ventilation rate is measured as air exchange rate (AER) in units of air changes per unit of time. It is governed by factors such as the tightness of the building, the HVAC system's design, and the use of windows and doors. Aerosols can be transported throughout the microenvironment via the HVAC system or by natural ventilation pathways. Outdoor aerosols are lost during penetration through the building "shell" to the indoors as a function of particle size. Once indoors, aerosols are lost by a variety of mechanisms to indoor surfaces. The collective loss process is often referred to as *deposition* losses. Particles are also removed by the HVAC system as a function of particle size, filtration efficiency, and operational duty cycle. Supplemental make-up air introduces outside aerosols to the indoor environment in many commercial HVAC systems and is an obvious source of ambient aerosols. Thornburg et al. (2001) suggest that variables associated with the HVAC system must be added to IAQ models to appropriately account for all sources and sinks indoors. Note that filtered air emitted from an HVAC vent will gradually reduce the indoor aerosol concentrations if no other contributing sources are present. Even with no personal activity sources, this removal is balanced against the continuous penetration of outdoor aerosols. Portable high-efficiency particulate air (HEPA) filtration devices are commonly used indoors to provide supplemental air cleaning. Offermann et al. (1985) tested the removal efficiencies as a function of particle size for several commercially available units and showed a substantial performance range, especially for ETS. The presence of supplemental filtration devices can significantly alter the expected indoor concentration levels, and their presence and operational schedules should be documented by questionnaires.

Methods and instrumentation used to measure building tightness include tracer gases such as sulfur hexafluoride (SF_6) and perfluorocarbons, blower doors (whole-house pressurization or component pressurization), and thermography (U.S. EPA, 1990). Measurement of the effectiveness of mechanical ventilation equipment and characterization of fluid flow can be made with hot wire/film anemometry, tracers, neutral-density bubbles, and smoke pencil testing (flow visualization).

EXPOSURE STUDIES

The critical elements in designing a personal aerosol exposure study are a thorough understanding of the purpose and objectives and the establishment of experimental hypotheses. The overall expense of most exposure studies may be minimally impacted by collecting supplemental measures that might greatly enhance the subsequent interpretation of the data. What population is being studied, and are there special burden considerations? Are especially susceptible populations, such as asthmatic children, of interest? What is the specific aerosol feature to be measured, and for what period of time? How many samples will be needed to define the upper 95th percentile?

If the objective is to examine exposure to combustion aerosols, using PM-2.5 inlets would gather samples that would be more representative of that exposure than using PM-10 or total suspended particulate. If the purpose of the exposure study is for subsequent risk analyses, will exposure—dose relationships be needed? Will an IAQ model be used to interrelate the study data? If so, what additional measurements (e.g., AER) are required to apply the model?

When designing aerosol exposure studies, questions regarding whether to include or exclude smokers are especially important. Including smoking participants confounds the relationship between personal and microenvironmental data. The most significant personal exposure to aerosol particles of the smoker is undoubtedly the mainstream tobacco smoke. On the other hand, excluding smokers jeopardizes the representativeness of the study population because of possible cluster effects. Smokers are not uniformly distributed in the population, and other demographic biases are incorporated into the study by the exclusion of smokers.

An exposure survey study may be used to assess the distribution of exposure of a population. Inferences to a larger population (or frame) can be made by using statistical probability sampling methods to select the study population. The survey and sampling instruments, methodologies, and protocols used in defined-focus/objective studies can be adapted for this type of study. The ingredients of exposure assessment include taking a representative probability sample, measuring pollutant concentrations, measuring body burden (not possible for generic aerosols without a regular and identifiable chemical constituency), and recording daily personal activities. The objectives of population exposure studies are to produce representative frequency distributions of the exposure of human beings to aerosols, to determine how exposures compare with existing regulatory standards or mitigation guidelines, to establish indoor/outdoor/personal relationships for aerosol exposure, and to determine the significant sources of aerosol exposure (Ott, 1985, 1990; Wallace et al., 1986; Wallace, 1987).

Each microenvironment encountered in daily exposures must be defined. Microenvironments typically consist of the home (with one or more cells), the workplace, the transportation vehicle, and the ambient environment near the home. Relationships between personal and microenvironmental sampling, spatial and temporal variance in aerosol concentrations, exposure and activity, and the frequency distribution of exposure must be determined for the test population (Ju and Spengler, 1981).

Once exposure information has been collected, it can be used for risk assessment. Risk assessment requires determining the number of persons exposed, the sources and transport factors relating to the pollutant from its source to receptor (person or sampler), the exposure-related significance of sampling data, and the health effects of exposure. Risk assessment also requires an estimate of the population at risk. The risk assessment can then be used to help define the relationship of exposure to outdoor standards.

Focused Panel Studies

Exposure studies are designed to meet many objectives. Some are collecting data simply to expand the exposure knowledge base, while others are conducted concurrently with health studies attempting to associate various health indicators with aerosol features. A number of studies are reported in the literature that collect data to better define nonoccupational aerosol exposures. Only recently have "panel" studies been reported, where an exposure study is conducted in parallel with collection of health data on a small selected population of subjects. Williams et al. (1999) described the design of an aerosol exposure panel study in a retirement center. The aerosol measurements included a combination of integrated 24 h sampling (personal, indoors, and outdoors) for PM-2.5, as well as several continuous monitors and size distribution analyzers to supplement the collected data. Bahadori et al. (1999) described the design of a multifaceted aerosol study focusing on data collection for inhalation epidemiology, including a concurrent exposure assessment component. Lippmann et al. (1999) discussed a planned exposure study to define the contributions of aerosols originating both outdoors and indoors for asthmatic and chronic obstructive pulmonary disease study subjects. Rodes et al. (2001) summarized the personal aerosol exposure assessments across three panel studies of retirement center residents and provided estimates of the "personal cloud"* contributions that increased the personal exposures, and the suggested aerosol sources.

Population Exposure Studies

Exposure studies that provide the ability to extrapolate the results to the general population are highly desirable but costly as the result of the participant selection process and typically large number of participants. The probability-based selection of participants is complex, hypothesis driven, and dependent on a robust participation rate that is partly a function of the obtrusiveness of the samplers and the survey instruments. Only a limited number of aerosol population studies have been reported in the literature, including those of Ozkaynak et al. (1990), Pellizzari et al. (1999b), Clayton et al. (1999), and Rotko et al. (2000).

Participants in this type of study are selected from a known population, or frame, through a random draw. The frame must have a complete demographic evaluation. Random selection from the frame provides a probability sample. By this method, the participant data can later be used to make statistically valid inferences to the frame. Particular factors of interest in the population (e.g., smoking habits, vocation) can be used to stratify the sample. Weighting factors can then be applied so that specific subpopulations of interest can be represented in sufficient number to provide statistically significant approximations of exposure.

The response rate during selection and the subsequent participation rate during sampling must be kept high so that the test population is randomly represented. Low respondent rates suggest that possible biases may have been introduced to the study and are most often caused by excessive study burden on the participants. The potential burdens of both the personal exposure samplers and the survey instruments must be carefully considered as part of the total burden on the study population. Personal aerosol sampling systems should have minimal weight and noise levels, while indoor samplers should also be unobtrusive (see next section). Survey materials—such as diaries, logs, and questionnaires—must be easy to understand and simple to complete to reduce their burden level. Collection of survey data is critical because the information provides the ability to interpret the exposure measurement data relative to sources and/or activity patterns. Thus, the combined burden of the sampling systems and the survey instruments on the study population must be considered carefully.

Sampler Obtrusiveness and Burden

Indoor and personal exposure samplers can be obtrusive to individuals whose environments are being studied. The degree of allowable obtrusiveness and the participant burden are important factors in avoiding biasing normal activity patterns and in obtaining sustained participation by the study subjects. Obtrusiveness depends on a number of factors. For an indoor fixed-location sampler, this includes primarily the relative size of the sampler, relative to that of the microenvironment, and its noise level; sampler configuration (packaging, size, and weight); how the sampler is to be employed (fixed location or mobile); the environment in which it will be used (i.e., children or pets present); noise produced by the system relative to background noise level (for homes or offices, total instrument noise should be less than

*The aerosol *personal cloud* is generally defined as the portion of the measured concentration that cannot be reconciled with a time-weighted concentration calculation that uses "% of time spent" data for each microenvironment taken from activity diaries and microenvironmental, fixed-location concentration measurements.

40 dbA at 1 m to avoid nuisance; in bedrooms, instruments may need to be even quieter); conspicuousness (prefer dull color so as not to be interest-provoking); safety of operation (no sharp edges and tamperproof to prevent injury, sample bias, or destruction and to reduce liability); power requirements (battery operation is the least obtrusive); and operational independence (does not require technical maintenance during sampling period).

For personal aerosol exposure samplers, it is important that the sampling package be comfortable to carry for extended periods and not obstruct routine activities. The personal exposure sampling system shown in Figure 29–2 is contained in a waistpack to minimize obtrusiveness and simplify its application for extended exposure periods (e.g., up to 7 days). The proper use of equipment depends on the participant's behavior. Minimizing requirements for the participant's time and minimizing interruptions in his or her daily activities (by judicious scheduling of field visits) further reduces participant burden and its resulting bias. It is imperative that the researcher compromise his or her desire to constantly monitor the participant's activities and the need to not significantly alter the participant's behavior.

Outdoor, ambient aerosol samplers have often been used to characterize indoor microenvironments when less obtrusive samplers are not available with the desired capabilities. Although smaller MEM samplers used indoors can often be suitably designed to operate outdoors, using ambient monitors indoors imposes some additional considerations. In addition to the obvious obtrusiveness concerns, the following should be considered when using high-flow-rate ambient samplers indoors: (1) Provide an outside exhaust for pumps to reduce either the contributions of filtered air or pump emissions (e.g., carbon, copper, oil vapor) into the space; (2) account for the possible influence of the sampler on the AER if simultaneous sampling is conducted with several monitors; and (3) place pumps outdoors if possible to reduce the unnatural heat loads they may pose on the operation (and filtration) of the HVAC system.

MODELING

Exposure Modeling

Personal exposure monitoring studies to estimate population exposure levels can be expensive and, in some cases, impractical. An alternative approach combines limited-scale measurement studies and existing data with an appropriate exposure model. The most widely used exposure model (e.g., Letz et al., 1984; Spengler et al., 1985) is the simple summation model of the measurements of interest (e.g., mass concentration for a specific aerosol size fraction) in each microenvironment, weighted by the time spent in each microenvironment. These relationships can be expressed as the simple summation of concentrations:

$$E = \sum^{N} f_i C_i \tag{29–2}$$

where E is mean exposure over N microenvironments, f_i is fraction of time spent in microenvironment i, and c_i is mean concentration in microenvironment i.

A study participant would be required to keep a time-activity diary that would provide information on the fraction of time spent in each compartment. The mean concentration measurement would be obtained either from a PEM carried by the participant while he or she was in the microenvironment or, less desirably, from a fixed-location monitor positioned to gather readings considered representative of the microenvironment.

As noted by Letz et al. (1984), the variance about the mean exposure can be estimated from the variances of the mean concentrations and exposure times. If these individual parameter variances are small relative to their means and the parameters are uncorrelated, Gauss' law of error propagation (see Bevington, 1969; Chapter 22) can be applied. The mean exposures and variance estimates representing specific microenvironments can be catalogued and used subsequently to predict distributions of exposures for other combinations of activity patterns. Wilson et al. (2000) describe a more complex version of Eq. 29–1 that accounts the contributions of outdoor aerosol to personal exposures for each microenvironment.

Compartmental Indoor Air Quality Modeling

The mean value of the aerosol mass concentration for a microenvironment can be estimated using compartmental models. These models are based on differential mass balances and assume that the contaminant accumulation in a room or compartment is equal to the sum of material entering or being formed, less that which is being lost. Analytical solutions are possible for the least complicated differential equations, although more complex situations typically use numerical methods to estimate the time-dependent or equilibrium concentration. Compartmental models compute an integrated average value that may not exist at any single point in the real microenvironment. Because aerosol properties typically vary by particle size, the accuracy of the compartmental averages is improved by application to specific size fractions.

Several general compartmental models have been proposed for aerosols in indoor environments, including those of Offermann et al. (1989), Raunemaa et al. (1989), and Nazaroff and Cass (1989). These models account for a variety of situations, including polydisperse size distributions, transport between multiple microenvironments, coagulation, gas-phase reaction kinetics, and wall losses. The model of Offermann et al. (1989) was used to describe the relationship of the aerosol removal performance of air cleaners, as a function of particle diameter, and the mean room concentration. Raunemaa et al. (1989) noted that a remission term for indoor aerosols released after loss to surfaces should be included in compartmental models. They also noted that the size distributions of these resuspended particles differ from those during deposition. Nazaroff et al. (1989) applied the aerosol transport and fate model of Nazaroff and Cass (1990) to an estimate of aerosol soiling in museums from surface deposition.

More recent compartmental models (Thatcher and Layton, 1995; Thornburg et al., 2001) have been applied to estimate the penetration of ambient particles into indoor microenvironments. The Thornburg et al. (2001) model is the most comprehensive and includes the loss mechanisms added by the HVAC system. The model was applied to calculate particle penetrations and the resultant indoor/outdoor (I/O) ratios for a range of sizes for structures with and without HVAC systems. It was determined that I/O ratios are a maximum (~40% to 60%) for ambient particles in the 0.3 to 0.5 μm size range for all structure types. The I/O ratios for particles above about 5 μm were significantly lower, ranging from 0 to 40%, with the lowest values resulting from the removal mechanisms in the HVAC systems.

REFERENCES

Alzona, J., B. L. Cohen, H. Rudolph, H. N. Jow, and J. O. Frohliger. 1979. Indoor–outdoor relationships for airborne particulate matter of outdoor origin. *Atmos. Environ.* 11:55–60.

Bahadori, T., M. Van Loy, and P. Saxena. 1999. Paper 295P. In *Proceedings of the Joint ISEE/ISEA Annual Conference*, Athens, Greece. Boston: International Society for Exposure Analysis. p. S107.

Baron, P. A. 1998. Personal Aerosol Sampler design: A review. *Appl. Occup. Environ. Hyg.* 13:313–320.

Bevington, P. R. 1969. *Data Reduction and Error Analysis for the Physical Sciences*. New York: McGraw-Hill.

Brief, R. S. and T. Bernath. 1988. Indoor pollution: Guidelines for prevention and control of microbiological respiratory hazards associated with air conditioning systems. *Appl. Ind. Hyg.* 3:5–10.

Buckley, T. J., J. M. Waldman, N. C. G. Freeman, V. A. Marple, W. A. Turner, and P. J. Lioy. 1991. Calibration, intersampler comparison, and field application of a new PM-10 personal air sampling impactor. *Aerosol Sci. Technol.* 14:380–387.

Caka, F. M., D. J. Eatough, E. A. Lewis, H. Tang, S. K. Hammond, B. P. Leaderer, P. Koutrakis, J. D. Spengler, A. Fasano, J. McCarthy, M. W. Ogden, and J. Lewtas. 1990. An intercomparison of sampling techniques for nicotine in indoor environments. *Environ. Sci. Technol.* 24:1196–1203.

Chuang, J. C., S. W. Hannan, and N. K. Wilson. 1987. Field comparison of polyurethane foam and XAD-2 resin for air sampling of polynuclear aromatic hydrocarbons. *Environ. Sci. Technol.* 21:798–804.

Clayton, C. A., E. D. Pellizzari, C. E. Rodes, R. E. Mason, and L. L. Piper. 1999. Estimating distributions of long-term particulate matter and manganese exposures for residents of Toronto, Canada. *Atmos. Environ.* 33:2515–2526.

Code of Federal Regulations. 1990. *EPA Ambient Air Monitoring Reference and Equivalent Methods.* 40 C.F.R. Part 53.43(c). Washington, DC: EPA.

Cohen, B. S., N. H. Harley, and M. Lippmann. 1984. Bias in air sampling techniques used to measure inhalation exposure. *Am. Ind. Hyg. Assoc. J.* 45:187–192.

Colome, S. D., J. D. Spengler, and S. McCarthy. 1982. Comparisons of elements and inorganic compounds inside and outside of residences. *Environ. Int.* 8:197–212.

Dockery, D. W. and J. D. Spengler. 1981. Indoor–outdoor relationships of respirable sulfates and particles. *Atmos. Environ.* 15:335–343.

Eatough, D. J., C. Benner, R. Mooney, D. Bartholomew, D. S. Steiner, L. D. Hansen, J. D. Lamb, E. A. Lewis, and N. L. Eatough. 1986. *Gas and Particle Phase Nicotine in Environmental Tobacco Smoke.* Paper No. 86-68.5, presented at the 79th Annual Meeting of the Air Pollution Control Association, 22–27 June 1986, Minneapolis, MN.

Girman, J. R., M. G. Apte, G. W. Traynor, J. R. Allen, and C. D. Hollowell. 1982. Pollutant emission rates from indoor combustion appliances and sidestream cigarette smoke. *Environ. Int.* 8:213–221.

Hammond, S. K., B. P. Leaderer, A. C. Roche, and M. Schenker. 1987. Collection and analysis of nicotine as a marker for environmental tobacco smoke. *Atmos. Environ.* 21:457–462.

Hayes, S. R. 1989. Estimating the effect of being indoors on total personal exposure to outdoor air pollution. *J. Air Waste Manage.* 39:1453–1461.

Hering, S. and G. Cass. 1999. The magnitude of bias in the measurement of $PM_{2.5}$ arising from volatilization of particulate nitrate from Teflon filters. *J. Air Waste Manage.* 49:725–733.

Highsmith, V. R., C. E. Rodes, and R. J. Hardy. 1988a. Indoor particle concentrations associated with use of tap water in portable humidifiers. *Environ. Sci. Technol.* 22:1109–1112.

Ingham, D. B. and B. Yan. 1994. The effect of a cylindrical backing body on the sampling efficiency of a cylindrical sampler. *J. Aerosol Sci.* 25:535–541.

Janssen, N., J. deHartog, G. Hoek, B. Brunekreef, T. Lanki, K. Timonen, and J. Pekkanen. 2000. Personal exposure to fine particulate matter in elderly subjects: Relation between personal, indoor and outdoor concentrations. *J. Air Waste Manage.* 50:1133–1143.

Jenkins, P. L., T. J. Phillips, and E. J. Mulberg. 1990. Activity patterns of Californians: Use of and proximity to indoor pollutant sources. In *Proceedings of the 5th International Conference on Indoor Air Quality and Climate*, ed. D. S. Walkinshaw, Vol. 2. Ottawa: Canada Mortgage and Housing Corporation, pp. 465–70.

Ju, C. and J. D. Spengler. 1981. Room-to-room variations in concentration of respirable particles in residences. *Environ. Sci. Technol.* 15:592–596.

Kamens, R., R. Wiener, C. Lee, and D. Leith. 1988. The characterization of aerosols in residential environments. In *Proceedings of the 1988 EPA/APCA [Air Pollution Control Association] Symposium on Measurement of Toxic and Related Air Pollutants*, eds. R. K. M. Jayanty and S. Hochheiser. Pittsburgh: APCA, pp. 89–97.

Koutrakis, P., S. L. K. Briggs, and B. P. Leader. 1992. Source apportionment of indoor aerosols in Suffolk and Onondaga Counties, New York. *Environ. Sci. Technol.* 26:521–527.

Koutrakis, P., A. M. Fasano, J. L. Slater, J. D. Spengler, J. F. McCarthy, and B. P. Leaderer. 1989. Design of a personal annular denuder sampler to measure atmospheric aerosols and gases. *Atmos. Environ.* 23:2767–2773.

LaForce, F. M. 1984. Airborne infections and modern building technology. In *Indoor Air: Recent Advances in Health Sciences and Technology.* World Health Organization et al., Vol. 1. Stockholm: Swedish Council for Building Research, pp. 109–119.

Lawless, P. A. and C. E. Rodes. 1999. Maximizing data quality in the gravimetric analysis of personal exposure sample filters. *J. Air Waste Manage.* 49:1039–1049.

Leaderer, B. P., P. M. Boone, and S. K. Hammond. 1990. Total particle, sulfate, and acidic aerosol emissions from kerosene space heaters. *Environ. Sci. Technol.* 24:908–912.

Letz, R., P. B. Ryan, and J. D. Spengler. 1984. Estimated distributions of personal exposure to respirable particles. *Environ. Monit. Assess.* 4:351–359.

Liao, D., J. Creason, C. Shy, R. Williams, R. Watts, and R. Zweidinger. 1999. Daily variations of particulate air pollution and poor cardiac autonomic control in the elderly. *Environ. Health Perspect.* 107:521–525.

Lioy, P. J. 1990. Assessing total human exposure to contaminants. *Environ. Sci. Technol.* 24:938–945.

Lioy, P. J., T. Wainman, W. Turner, and V. A. Marple. 1988. An intercomparison of the indoor air sampling impactor and the dichotomous sampler for a 10 m cut-size. *JAPCA* 38:668–670.

Lioy, P. J., T. Wainman, J. Zhang, and S. Goldsmith. 1999. Typical household vacuum cleaners: The collection efficiency and emissions characteristics for fine particles. *J. Air Waste Manage.* 49: 200–206.

Lioy, P. J., J. M. Waldman, T. Buckley, J. Butler, and C. Pietarinen. 1990. The personal, indoor and outdoor concentrations of PM-10 measured in an industrial community during the winter. *Atmos. Environ.* 24B:57–66.

Lippmann, M., G. D. Thurston, K. Ito, J. Reibman, N. Xue, and M. Heikkinen. 1999. *Personal Exposure to PM of Outdoor and Indoor Origin.* Paper 140O, in Proceedings of the Joint ISEE/ISEA Annual Conference, Athens, Greece, International Society for Exposure Analysis, Boston, MA.

Marple, V., B. Liu, S. Behm, B. Olson, and R. Wiener. 1988. A new personal impactor sampler inlet. Presented at the American Industrial Hygiene Conference, 15–20 May 1988, San Francisco.

Marple, V., B. Liu, S. Behm, B. Olson, and R. Wiener. 1989. *A personal environmental monitor (PEM).* Presented at the American Association for Aerosol Research Annual Meeting, 10–13 October 1989, Reno, Nevada.

Marple, V. A. and J. E. McCormack. 1983. Personal sampling impactor with respirable aerosol penetration characteristics. *Am. Ind. Hyg. Assoc. J.* 44:916–922.

Marple, V. A., K. L. Rubow, W. Turner, and J. D. Spengler. 1987. Low flow rate sharp cut impactors for indoor air sampling: Design and calibration. *JAPCA* 37:1303–1307.

Martonen, T., M. Clark, D. Nelson, D. Willard, and E. Rossignal. 1984. Evaluation of a minicascade impactor for sampling exposure chamber atmospheres. *Fund. Appl. Toxicol.* 2:149–152.

McDow, S. R. and J. J. Huntzicker. 1990. Vapor adsorption artifact in the sampling of organic aerosol: Face velocity effects. *Atmos. Environ.* 24A: 2563–2571.

McKenzie, R. L., D. S. Bright, R. A. Fletcher, and J. A. Hodgeson. 1982. Development of a personal exposure monitor for two sizes of inhalable particulates. *Environ. Int.* 8:229–233.

Molhave, L. 1987. The sick buildings—A subpopulation among the problem buildings. In *Indoor Air: '87 Proceedings of the 4th International Conference on Indoor Air Quality and Climate*, eds. B. S. Seifert et al., Vol. 2. Berlin: Oraniendruck, GmbH Publishers, pp. 469–473.

National Research Council. Committee on Geosciences, Environment, and Resources. 1999. *Research Priorities for Airborne Particulate Matter II.* Washington, DC: National Academy Press.

Nazaroff, W. W. and G. R. Cass. 1989. Mathematical modeling of indoor aerosol dynamics. *Environ. Sci. Technol.* 23:157–166.

Nazaroff, W. W., L. G. Salmon, and G. R. Cass. 1990. Concentration and fate of airborne particles in museums. *Environ. Sci. Technol.* 24:66–77.

Offermann, F. J., R. G. Sextro, W. J. Fisk, D. T. Grimsrud, T. Raunemaa, M. Kulmala, H. Saari, M. Olin, and M. H. Kulmala. 1989. Indoor air aerosol model: Transport indoors and deposition of fine and coarse particles. *Aerosol Sci. Technol.* 11:11–25.

Ogden, T. L., I. W. Bartlett, C. J. Purnell, C. J. Wells, F. Armitage, and H. Wolfson. 1993. Dust from cotton manufacture: Changing from static to personal sampling. *Ann. Occup. Hyg.* 37(3):271–285.

Owen, M. K., D. S. Ensor, L. S. Hovis, W. G. Tucker, and L. E. Sparks. 1990. Particle size distributions for an office aerosol. *Aerosol Sci. Technol.* 13:486–492.

Ott, W. R. 1985. Total human exposure. *Environ. Sci. Technol.* 19:880–886.

Ott, W. R. 1995. Human exposure assessment: The birth of a new science. *JEAEE* 5:449–472.

Ozkaynak, H., J. D. Spengler, C. A. Clayton, E. Pellizzari, and R. W. Wiener. 1990. Personal exposure to particulate matter: Findings from the particle total exposure assessment methodology (PTEAM) prepilot study. In *Proceedings of the 5th International Conference on Indoor Air Quality and Climate*, Vol. 2, ed. D. S. Walkinshaw. Ottawa: Canada Mortgage and Housing Corporation, pp. 571–576.

Pellizzari, E. D., C. A. Clayton, C. E. Rodes, R. E. Mason, L. L. Piper, B. Fort, G. Pfeifer, and D. Lynam. 1999a. Particulate matter and manganese exposues in Toronto, Canada. *Atmos. Environ.* 33:721–734.

Pellizzari, E. D., T. D. Hartwell, C. Leininger, H. Zelon, S. Williams, J. Breen, and L. Wallace. 1983. Human exposure to vapor-phase halogenated hydrocarbons: Fixed-site vs. personal exposure. In *Proceedings from the National Symposium on Recent Advances in Pollutant Monitoring of Ambient Air and Stationary Sources*. Pub. No. EPA/600/9-83/007. Research Triangle Park, NC: U.S. Environmental Protection Agency, pp. 264–288.

Pellizzari, E. D., R. L. Perritt, and C. A. Clayton. 1999b. National Human Exposure Assessment survey: Exploratory survey of exposure among population subgroups in EPA region V. *JEAEE* 9(1): 49–55.

Raunemaa, T., M. Kulmala, H. Saari, M. Olin, and M. H. Kulmala. 1989. Indoor air aerosol model: Transport indoors and deposition of fine and coarse particles. *Aerosol Sci. Technol.* 22:11–25.

Repace, J. L. and A. H. Lowrey. 1980. Indoor air pollution, tobacco smoke and public health. *Science* 208:464–472.

Rodes, C. E. 1997. Unpublished data, Research Triangle Institute, Research Triangle Park, NC 27709; from private communication with V. A. Marple, University of Minnesota, Minneapolis, MN.

Rodes, C. E. 1999. Unpublished data, Research Triangle Institute, Research Triangle Park, NC 27709; partially supported by U. S. Environmental Agency Contract 68-D5-0040, WA041, Research Triangle Park, NC 27711.

Rodes, C. E., R. M. Kamens, and R. W. Wiener. 1991. The significance and characteristics of the personal activity cloud on exposure assessment measurements for indoor contaminants. *Indoor Air* 2: 123–145.

Rodes, C. E., P. A. Lawless, G. F. Evans, L. S. Sheldon, R. W. Williams, A. F. Vette, J. P. Creason, and D. Walsh. 2001. The relationships between personal PM exposures for elderly populations and indoor and outdoor concentrations for three retirement center scenarios. *JEAEE* 11:103–105.

Rojas-Bracho, L., H. Suh, and P. Koutrakis. 2000. Relationships among personal, indoor, and outdoor fine and coarse particle concentrations for individuals with COPD. *JEAEE* 10:294–306.

Rotko, T., L. Oglesby, N. Kunzli, and M. Jantunen. 2000. Population sampling in European air pollution exposure study, EXPOLIS: Comparisons between the cities and representativeness of the samples. *JEAEE* 10:355–364.

Rubow, K. L., V. A. Marple, J. Olin, and M. A. McCawley. 1985. *A Personal Impactor: Design, Evaluation, and Calibration*. Pub. No. 469. Minneapolis: University of Minnesota, Particle Technology Laboratory.

Samfield, M. 1985. Importance of cigarette sidestream smoke—Further aspects. *Tabak J. Int.* 6:448.

Sexton, K. and S. B. Hayward. 1987. Source apportionment of indoor air pollution. *Atmos. Environ.* 21(2):407–418.

Sherwood, R. J. 1966. On the interpretation of air sampling for radioactive particles. *Am. Ind. Hyg. Assoc. J.* 27:98–109.

Spengler, J. D., D. W. Dockery, W. A. Turner, J. M. Wolfson, and B. G. Ferris, Jr. 1981. Long-term measurements of respirable sulfates and particulates inside and outside homes. *Atmos. Environ.* 15:23–30.

Spengler, J., H. Ozkaynak, J. Ludwig, G. Allen, E. Pellizzari, and R. Wiener. 1989. Personal exposures to particulate matter: Instrumentation and methodologies (P-TEAM), In *Proceedings of the 1989 EPA/AWMA [American Waste Management Association] Symposium on Measurement of Toxic and Related Air Pollutants*, eds. R. K. M. Jayanty and S. Hochheiser. Pub. No. VIP-13. Pittsburgh: AWMA, pp. 449–463.

Spengler, J. D., R. D. Treitman, T. D. Tosteson, D. T. Mage, and M. L. Soczek. 1985. Personal exposures to respirable particulates and implications for air pollution epidemiology. *Environ. Sci. Technol.* 19: 700–707.

Thatcher, T. L. and D. W. Layton. 1995. Deposition, resuspension, and penetration of particles within a residence. *Atmos. Environ.* 29:1487–1497.

Thornburg, J., D. S. Ensor, C. E. Rodes, P. A. Lawless, L. E. Sparks, and R. B. Mosley. 2001. Penetration of particles into buildings and associated physical factors, part I: Model development and computer simulations. *Aerosol Sci. Technol.* 34:284–296.

Traynor, G. W., M. G. Apte, H. A. Sokol, J. C. Chuang, W. G. Tucker, and J. L. Mumford. 1990. Selected organic pollutant emissions from unvented kerosene space heaters. *Environ. Sci. Technol.* 24: 1265–1270.

Traynor, G. W. and I. A. Nitschke. 1984. Field survey of indoor air pollution in residences with suspected combustion-related sources (chemical characterization and personal exposure). In *Proceedings of the 3rd International Indoor Air Quality and Climate Conference*, Vol. 14. World Health Organization. Stockholm: WHO, pp. 343–348.

Tu, K. W. and L. E. Hinchliffe. 1983. A study of particulate emissions from portable space heaters. *Am. Ind. Hyg. Assoc. J.* 44:857–862.

Turner, W. A., J. D. Spengler, D. W. Dockery, and S. D. Colome. 1979. Design and performance of a reliable personal monitoring system for respirable particulates. *JAPCA* 29(7):747–774.

U.S. Environmental Protection Agency. 1989. *Report to Congress on Indoor Air Quality, Volume II: Assessment and Control of Indoor Air Pollution*. Pub. No. EPA/400/1-89/000c. Research Triangle Park, NC: EPA.

U.S. Environmental Protection Agency. 1990. *Compendium of Methods for the Determination of Air Pollutants in Indoor Air*, Pub. No. EPA/600/S4-90/010. Research Triangle Park, NC: EPA.

U.S. Environmental Protection Agency. 1997. *Federal Register* 62(138):1–102.

Vaughan, W. M. and S. K. Hammond. 1990. Impact of "designated smoking area" policy on nicotine vapor and particle concentrations in a modern office building. *J. Air Waste Manage. Assoc.* 40: 1012–1017.

Vincent, J. H. and H. Gibson. 1981. Sampling errors in blunt dust samplers arising from external wall loss effects. *Atmos. Environ.* 15:703–712.

Vossler, T. L., R. K. Stevens, R. J. Paur, R. E. Baumgardner, and J. P. Bell. 1988. Evaluation of improved inlets and annular denuder systems to measure inorganic air pollutants. *Atmos. Environ.* 22: 1729–1736.

Vu Duc, T. and C. M. P. Favez. 1981. Characteristics of motor exhausts in an underground car park: Mass size distribution and concentration levels of particles. *J. Environ. Sci. Health* A16:647–660.

Wallace, L. A. 1987. *The Total Exposure Assessment Methodology (TEAM) Study: Summary and Analysis*, Vol. 1. Pub. No. EPA/600/6-87/002a. Washington, DC: U.S. Environmental Protection Agency.

Wallace, L. A. 1996. Indoor particles: A review. *J. Air Waste Manage.* 46:98–126.

Wallace, L. A., E. D. Pellizzari, T. D. Hartwell, R. Whitmore, C. M. Sparacino, and H. S. Zelon. 1986. Total Exposure Assessment Methodology (TEAM) study: Personal exposures, indoor–outdoor relationships, and breath levels of volatile organic compounds in New Jersey. *Environ. Int.* 12:369–387.

Wallingford, K. and J. Carpenter. 1986. NIOSH field experience overview: Investigating sources of IAQ problems in office buildings. In *IAQ '86: Managing Indoor Air for Health and Energy Conservation*. Atlanta: American Society for Heating, Refrigeration, and Air Conditioning Engineering, pp. 448–453.

Wiener, R. W. and D. VanOsdell. 1990. *Validation of the Personal Cloud Phenomenon.* Presented at the 21st Annual Meeting of the Fine Particle Society, 21–25 August 1990. San Diego, CA.

Willeke, K. and P. A. Baron. 1990. Sampling and interpretation errors in aerosol monitoring. *Am. Ind. Hyg. Assoc. J.* 51:160–168.

Williams, R., J. P. Creason, R. Zweidinger, J. Suggs, R. Kwok, and L. Sheldon. 1999. *Personal Exposure and Ambient Air Sampling Related to an Elderly Population Living in a Baltimore Retirement Center: Preliminary Findings of the 1998 Baltimore Study*. Paper 144O, Proceedings of the Joint ISEE/ISEA Annual Conference, Athens, Greece, International Society for Exposure Analysis, Boston, MA.

Wilson, W. E., D. T. Mage, and L. D. Grant. 2000. Estimating separately personal exposure to ambient and nonambient particulate matter for epidemiology and risk assessment: Why and how. *J. Air Waste Manage.* 50:1167–1183.

Wilson, W. E. and H. H. Suh. 1997. Fine particles and coarse particles: Concentration relationships relevant to epidemiologic studies. *J. Air Waste Manage.* 47:1238–1249.

World Health Organization (1983). Indoor air pollutants: exposure and health effects. EURO Reports and Studies 78, Geneva: World Health Organization.

30 Measurement of Aerosol from Aircraft

JAMES CHARLES WILSON and W. RUSSELL SEEBAUGH
Department of Engineering, University of Denver, Denver, CO

INTRODUCTION

Aircraft platforms are used for measurements of aerosol properties from the boundary layer to altitudes in excess of 20 km. These measurements are made at velocities ranging from tens of meters per second to 250 m/s. These platform velocities raise a number of difficulties that are described in this chapter, along with some techniques for overcoming these difficulties. The chapter emphasizes aerosol inlets used to extract samples from the air stream and deliver them to instruments inside the aircraft and discusses features of instruments used on aircraft. Cloud water, cloud condensation nuclei measurements, and remote sensing with instruments like LIDAR are not discussed.

RESEARCH UTILIZING AEROSOL MEASUREMENT FROM AIRCRAFT

Aircraft are useful for measurement of aerosol over spatial scales ranging from tens of meters to thousands of kilometers. The particles of interest range in diameter from a few nanometers to tens of micrometers. Instruments and samplers have been deployed on aircraft to determine properties such as concentrations, size distributions, chemical composition, elemental composition, optical properties, radioactivity, biological nature, and aerosol charge. Airborne measurements are particularly useful for process studies. Data can be acquired on one flight for rapidly occurring events. Several flights can be timed to permit study of the evolution of multiday or seasonal phenomena. Aircraft have also been involved in characterization of relatively unknown aerosol and in monitoring.

Aircraft studies of air pollution and atmospheric chemistry have addressed the origins and impacts of pollution aerosol, emissions from aircraft and rockets, and characteristics and formation of natural aerosol. Quantitative characterizations of aerosol processes were important in understanding ozone depletion (World Meteorological Organization, 1998). Measurements from aircraft contributed significantly to those characterizations. Airborne aerosol measurement capabilities must again be improved because uncertainties about large-scale aerosol processes and effects impede understanding of climate change (Intergovern-

Aerosol Measurement: Principles, Techniques, and Applications, Second Edition, Edited by Paul A. Baron and Klaus Willeke.
ISBN 0-471-35636-0

mental Panel on Climate Change, 1994). Satellite validation studies have helped determined the accuracy of satellite measurements and data retrieval processes.

OBJECTIVES TO BE ACHIEVED IN AIRBORNE AEROSOL SAMPLING AND MEASUREMENT

Airborne instruments are used to characterize all the properties of aerosol that are studied in ground-based measurements. The investigators' objectives for the airborne measurements are often the same as those motivating ground-based measurements.

Most instruments or samplers accept air streams that are moving at much smaller velocities than the aircraft platform. The pressure at which the samplers or instruments operate may nearly equal the stagnation pressure rather than the ambient static pressure. Therefore, it is convenient to characterize the aerosol in terms of mixing ratios rather than concentrations. Mixing ratios are defined as the ratio of a given aerosol property to the mass of air sampled. The mixing ratios of aerosol properties will remain unchanged as air density changes as long as the particles follow the streamlines and the size of the sampled particles remains unchanged. A sample is representative of the ambient aerosol if its aerosol properties expressed in terms of mixing ratios are the same as those found in the free air stream.

The systems that extract a sample from the air stream and deliver it to instruments are judged by different criteria depending on the objectives of the measurements. In some investigations, inlets are expected to provide a representative sample of the ambient aerosol for characterization by procedures that determine integral properties of the size distribution without performing size-discriminated selection on the sampled aerosol. Examples include determination of aerosol mass by weighing of filter samples and measurement of light-scattering coefficients by use of a nephelometer. In such cases, the integral property is determined without determining the details of the size distributions, and size-dependent corrections are impossible.

It is not always possible to achieve representative sampling. The sampling and transmission efficiencies of the inlet and transport systems are a function particle of size, and the sampled particles may suffer significant size-dependent losses or enhancements. If the efficiencies are known and the size distribution is measured, or if size-segregated samples are acquired, then the ambient properties describing each size range can be calculated by dividing the measured results by the corresponding sampling-system efficiencies.

Not all sampling systems are designed to achieve representative sampling. For example, virtual impactors can be incorporated into the sampling system to improve the counting statistics for large particles or to provide sufficient mass to permit chemical analysis of size-segregated samples (Porter et al., 1992), and some inlets intentionally enhance particles. The ambient mixing ratios must again be calculated from the measured ones using the quantitative description of the impactor or inlet performance.

Inlets have also been used to separate particles from the flow in order to deliver nearly particle-free air to gas analyzers. The objective is to measure mixing ratios of gases such as water or nitric acid in the gas phase. If the particles were not removed from the air in the inlet, they would vaporize and confound the determination of gas phase mixing ratios. The separation of the phases depends on particle inertia, and small particles may follow the sampled streamlines and not be removed.

To study the composition of particles, particle inertia has been used to remove particles from the ambient gas stream and to deposit them in another gas of known composition. Inlets that accomplish this separation perform like virtual impactors. To avoid partial volatilization and loss of particle mass, the inertial separation must occur before the particles are heated significantly by ram heating or by heat transfer from the instrument. Following that separa-

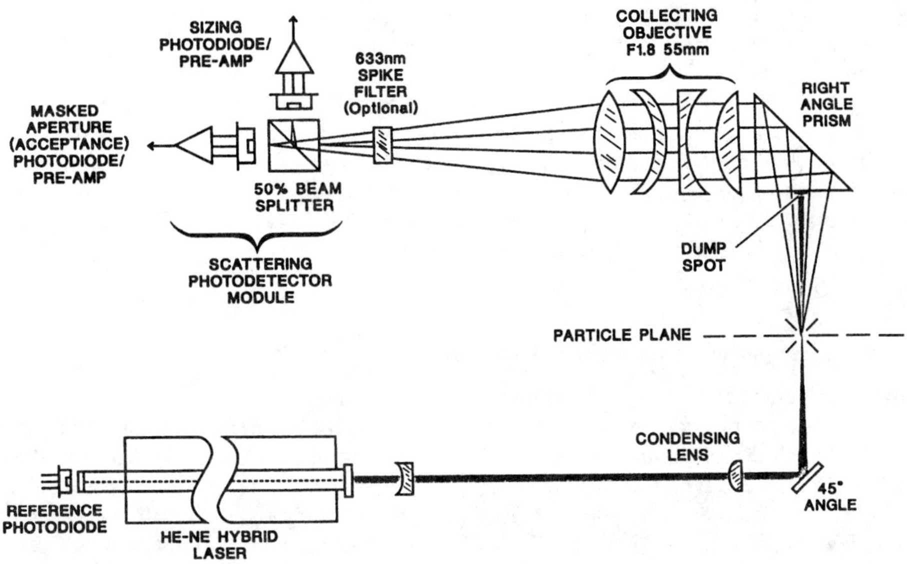

Fig. 30–1. Optical diagram of the forward-scattering spectrometer probe (FSSP). The particle plane is in the free stream. (Courtesy of ***PMS***.)

tion, the particles may be analyzed using aerosol techniques or vaporized to permit analysis with gas samplers. In either case, the composition of the gas phase in the free stream is not permitted to contaminate the sampled particles. Small particles may escape the separation and not be sampled.

AIRBORNE AEROSOL MEASUREMENT TECHNIQUES

Aerosol Measurements Made Outside the Aircraft

Bringing particles into the aircraft for measurement frequently causes significant heating and often alters their size and composition. Therefore, it is often desirable to measure the particles in the air stream to the extent possible. Three manufacturers market instruments capable of measuring particles without bringing a sample into the aircraft (***DMT***, ***PMI***, and ***SPE***).* PMI markets some probes that were formerly sold by ***PMS***. These measurements are easier to make on large particles than small ones, and most of these instruments are most effective on supermicrometer aerosol and even larger cloud particles.

The ***PMI*** forward-scattering spectrometer probe (FSSP; Fig. 30–1) measures the optical diameter of single aerosol particles having diameters larger than about 0.4 μm in the air stream in real time. The instrument case is constructed so that the measuring volume is outside the aircraft and is surrounded with a cylindrical housing aligned with the air flow. The objective of this design is to perturb the air stream and particles as little as possible. Particles pass through an HeNe laser beam and the forward-scattered light is measured to determine particle size. The concentration of particles is determined from the air speed and the size of the viewing volume. Aerosol spectrometers of this type collect light scattered in the near-forward direction. They often have a multivalued response over a portion of the size range that they measure. Discussions of aerosol and cloud measurements with the FSSP may

* See Appendix I for full manufacturer addresses referenced to the italicized three-letter codes.

be found in Knollenberg (1981), Pinnick et al. (1981), Baumgardner (1984), Cerni (1983), Dye and Baumgardner (1984), Baumgardner et al. (1985), Brenguier and Amodei (1989), and Brenguier (1989).

An instrument capable of measuring scattering into multiple angles by particles in the free stream has been developed. Information about the angular scattering can be used to determine the refractive index of the particles (Baumgardner et al., 1996). It is also possible to acquire images of larger particles in the free stream. These images provide information on shape and size and are of particular interest in cloud and precipitation studies. Imaging probes are available from the three companies listed above.

Instruments Requiring Sampling and Transport of Aerosols

Many instruments used on the ground have been used or adapted for use aboard aircraft and require the user to provide an inlet. These instruments include condensation nuclei counters (CNCs), optical particle counters (OPCs), differential mobility analyzers (DMAs), diffusion batteries (DBs), aerodynamic particle sizers (APSs), electrostatic precipitators, nephelometers, optical extinction cells, aethelometers, impactors, quartz crystal microbalances, and filtration devices (Woods and Chuan, 1983; Hering, 1987; Dreiling and Jaenicke, 1988; Leaitch et al., 1989; Brock et al., 1990). The ***PMI*** PCASP is an OPC with its own inlet and capable of sensing submicrometer particles. It is available in an aircraft-mountable package (Strapp et al., 1992; Liu et al., 1992).

When aerosol instruments designed for use at 1 atm pressure are operated at reduced pressures, the assumptions and approximations used in the design of the instruments may no longer be valid. For example, Heintzenberg and Ogren (1985) recommend that the ***TSI*** Model 3020 CNC not be used at pressures less than ~250 mb. Noone and Hansson (1990) have calibrated the ***TSI*** Model 3760 CNC at various pressures and find that it also functions at pressures down to 250 mb. At higher altitudes and lower pressures, other designs are required (Wilson et al., 1983). Impactor performance depends on Reynolds number as well as Stokes number, and impactor performance can degrade with altitude. Aerosol jets formed by inlets in optical counters may change shape as pressure falls, causing particles to miss the beam. The precision of counting depends on the number counted, and small aerosol concentrations are usually encountered at higher altitudes. Thus, normal, terrestrial sample flow rates may provide inadequate counting statistics in the free troposphere or stratosphere. Investigators are strongly cautioned to calibrate their instruments at the pressures to be encountered and to evaluate the response of instruments to the concentrations to be sampled.

The U.S. National Aeronautics and Space Administration (NASA), the U.S. Office of Naval Research (ONR), and the National Center for Atmospheric Research (NCAR) have programs that encourage development or use of instruments capable of operating on conventional or uninhabited aircraft (UAVs). The Center for Interdisciplinary Remotely Piloted Aircraft Studies (CIRPAS) is operated by the ONR and located at the Naval Post Graduate School. The Environmental Research Aircraft and Sensor Technology (ERAST) program is operated by NASA. The High Performance Instrumented Airborne Platform for Environmental Research (HIAPER) initiative is being conducted by NCAR. Because of the increasing complexity of research payloads, programs like these encourage development of autonomous aerosol instruments able to operate on aircraft and to meet space, weight, and strict airworthiness requirements. A sampling of such instruments includes scanning, radial differential mobility analyzers (Russell et al., 1996), aerosol mass spectrometers able to provide composition information on individual particles (Murphy et al., 1998), and aerosol spectrometers able to determine aerosol size distributions at stratospheric concentrations in the diameter ranges of 4 to 50 nm (Brock et al., 2000) and 90 to 2000 nm (Jonsson et al., 1995).

EFFECTS COMPLICATING AEROSOL SAMPLING FROM AIRCRAFT

Alteration of the Size Distribution by Disturbance of Air due to the Aircraft

Several experimental and theoretical studies have examined the effects of the air flow distortion on airborne measurements (Beard, 1983; Baumgardner, 1984; King, 1984, 1986a,b; Drummond, 1984; Drummond and MacPherson, 1984, 1985; MacPherson and Baumgardner, 1988; Norment, 1988; Twohy and Rogers, 1993). Convergence and divergence of streamlines around the fuselage, wings, and instrument pods can lead to areas of enhanced particle concentrations as well as zones where concentrations are greatly reduced. These effects result from the fact that particles have significant inertia and may not be able to follow the flow streamlines. The effects are more significant for particles with large Stokes numbers. Thus, although samples are drawn from beyond the aircraft's boundary layer, they may not contain a sample that is representative of the particles in the free stream. Rather, if the stream lines are bent upstream of the sample point due to the aircraft or probe, the sampled concentration may differ from that in the free stream. The effects vary substantially from aircraft to aircraft depending on fuselage shape and diameter, wing thickness and aspect ratio, placement of the engines and propellers, and the location of appendages and of the sample point.

Sampling locations thought to minimize effects of air flow distortion include below the center of each wing, on the lower aircraft fuselage, on the upper fuselage ahead of the trailing edge of the wings but well behind the cockpit, and extending into the air stream ahead of the aircraft nose. These locations minimize the influences of wing tip vortices and the distorted flow created by the nose and windshield areas on the fuselage. Less preferable locations include wing tip, near-wing tip, forward, dorsal fuselage mounts, and regions of propwash and engine exhaust. The area aft of the wing along the side of and over the top of the fuselage is a region that may be swept by horseshoe vortices produced at the juncture of the wing root and fuselage. This area may also present unexpected flows and sampling difficulties.

Aircraft are often electrified with respect to the air. Laboratory studies suggest that sharp-edged inlets should not be located at points on the airframe where electrical discharges are likely (Romay et al., 1996).

Following studies of sampling on propeller aircraft, it has been argued that the effects of airflow distortion are limited to particles larger than approximately 10 μm. However, aircraft such as the NASA ER-2, the NASA DC-8, the proposed NCAR HIAPER, and passenger jets reach speeds in excess of 200 m/s and often operate at higher altitudes. Under these conditions, aerodynamic distortion of streamlines upstream of sample points can affect sampling efficiencies of particles having diameters smaller than 10 μm. Anecdotal information suggests that upwind interactions with the fuselage or protuberances on the surface of the aircraft can remove submicrometer particles from the sample.

Wings bend upstream streamlines and induce an upwash velocity and angle of attack. These effects can be estimated and taken into account in designing orientation of probes located on the wings or wing pods (Soderman et al., 1991). Some published studies involve detailed flow modeling for specific aircraft. Some models use supercomputers, and such models may not be available for every installation. Studies of the impact of the flow around the fuselage or wing tanks can be done with simplified geometries and commercial programs such as FLUENT (***FLT***) that run on personal computers. Such models can provide an indication of what size particles are likely to be affected by the passage of the fuselage or of a wing tank through the sampled air.

Effects of Sensor Housings on Airflow

The sensor housing may also distort the airflow. The airflow effects of the ***PMI*** FSSP and OAP cloud and precipitation probes have been studied in some detail (MacPherson, 1985;

King, 1986b; MacPherson and Baumgardner, 1988; Norment, 1988). As an example, the FSSP sensing volume is housed within a cylindrical tube. Based on detailed numerical simulation of the air flow around this tube, its support arms, and the FSSP nacelle itself (Norment, 1988), the instrument was found to undercount particles with $d_p > 20\,\mu m$ by ~10%. Norment (1988) also found that approximately half of the flux and speed distortions at the point of measurement were caused by the presence of the sensor housing. Additionally, Norment pointed out that distortions induced by the aircraft and those caused by the sensor interact synergistically and thus should not be evaluated independently.

Anisokinetic Sampling

When the velocity of the air entering the inlet is different from the velocity upstream of the inlet, streamlines bend upwind of the sampling inlet, and particle mixing ratios entering the inlet can differ from ambient mixing ratios. This condition is referred to as *anisokinetic sampling* (see Chapter 8). Anisokinetic sampling results in size-dependent biases in the sample. Investigators routinely correct for the effects of anisokinetic sampling. Such corrections require knowledge of the free stream velocity, the flows entering the inlet, and the size and density of particles whose mixing ratio is being corrected. Claims of isokinetic sampling can be verified if the sample flow and free stream velocity are known.

A number of the inlets described in this chapter are intentionally anisokinetic. Slowing the flow outside of the inlet with shrouds or restricted ducts causes streamlines to bend upstream of the inlet and leads to enhancements for large particles. These inlets may not provide a representative sample for all particle sizes.

Inertial Effects and Turbulent Deposition in the Inlet

Some inlets use diffusers whose role is to slow the flow and the particles. The relaxation time for a unit density $10\,\mu m$ particle is approximately 3×10^{-4} s at the surface of the earth. If an aircraft is traveling at 100 m/s, that particle will have a stopping distance of 3 cm (see Chapter 4). This dimension is significant compared with the distances over which streamlines diverge in practical diffusing inlets. Thus, concentrations of particles in this size range are likely to be enhanced near the center of flows in diffusing inlets. The stopping distance for particles with diameters near $1\,\mu m$ is smaller by a factor of ~100 at 1 atm pressure, and these particles will not suffer as much enhancement near the center of the flow. This inertial redistribution of the aerosol may cause the samples to be nonrepresentative in cases where the diffuser flow is not ingested by a single instrument. If only part of the diffuser flow is delivered to the instrument or if the flow is split among several instruments, larger particles may not be accurately sampled by any or all of the instruments.

Diffusers generate turbulence as well as slow the flow. The turbulence will act to redistribute the particles in the flow and to deposit particles on the wall of the diffuser (Huebert et al., 1990; Sheridan and Norton, 1998). These studies have showen that as much as 50% of the mass can end up on the wall and that the effect is particle size dependent, with supermicrometer particles suffering larger losses than submicrometer particles.

Standard turbulence models and particle motion models may fail to accurately predict both the inertial effects on the spatial distribution of particles at the outlet of the diffuser and turbulent losses in the diffuser. The equations that are usually solved in turbulence calculations represent the mean flow, and all the scales of the turbulence are modeled. The user can vary five parameters in standard models, and the calculation of particle motion is likewise modeled (Fluent manual, ***FLT***). Parameters may be chosen that produce desired results in a given situation, but, because the fundamental physics of fluid and particle motion are not captured in the models, use of the these parameters in other situations is not guaranteed to produce accurate calculations. Therefore, even if a model produces accurate particle deposition rates for a given flow and geometry, the same choice of parameters may fail in another geometry or flow regime.

The situation is different for calculations of laminar flow and particle trajectories in laminar flow. The accuracy of the laminar calculations may be limited by availability of computing resources, but the theory that is applied does not suffer any fundamental limitations. Calculations of particle motion in laminar flow have been successful in explaining the performance of inertial impactors (see Chapter 10), the losses in transport around bends and effects of anisokinetic sampling (see Chapter 8), and the performance of the ***TSI*** Aerodynamic Particle Sizer (see Chapter 17).

Misalignment of the Inlet with the Free-Stream Flow

Misalignment of the inlet with the air stream of a few degrees can lead to separation and turbulence generation near the tip of sharp-edged inlets (Soderman et al., 1991). Turbulence can lead to particle deposition and alteration of particle mixing ratios. Misaligned inlets also cause the streamlines to bend before entering the inlet. If particles fail to follow the streamlines, then the sample will be altered (see Chapter 8).

It is often difficult to know the precise direction of the local wind vector at the sampling location. Numerical modeling, wind tunnel testing, and in-flight measurements can be used to examine the air flow field around specific inlet locations on research aircraft. The angle of attack of the aircraft may vary depending on altitude, rate of descent or climb, aircraft weight, true air speed, and the presence of turbulence, while aircraft yaw may vary with wind speed, turn coordination, and so forth. It is possible to measure the wind vector at a particular location on the aircraft with an instrumented probe such as the 858 Flow Angle Sensor (***ROS***). The wind vector should be characterized for all anticipated sampling conditions.

A number of investigators have devised and tested shrouded inlets to align the flow with the inlets (see below). The shrouds have been shown to straighten the flow before it impinges on the inlet. Proper shroud design takes the inertia of particles into account. There is an upper limit on the diameter of particles whose trajectories can be straightened by a given shroud (Twohy, 1998).

Inlets with an elliptical leading edge profile have been shown to be more resistant to turbulence generation in the inlet as the angle between the inlet and the flow changed (Soderman et al., 1991). A blunt leading edge increases the surface presented to the flow, and some area inside of the forwardmost contour is occupied by solid material. The definition of isokinetic sampling is less clear in this instance. If the flow velocity in the throat of the inlet is set equal to the free stream velocity, then streamlines are forced to curve around the blunt leading edge in order to flow around the inlet. Then particles will certainly strike the inlet and may shatter (Webber et al., 1998). If the flow is set so that the stagnation point is poised on the forwardmost contour, then air entering the inlet will have been forced to flow around the curve and particles may be lost by impaction and may also shatter. Rader and Marple (1988) calculate the enhancements for a tube with a wall of finite thickness presented to the flow and show that representative sampling is possible for some flow ratios.

Solid-walled, diffusing inlets generate turbulence. There is no evidence suggesting that this turbulence can be eliminated by careful alignment of the inlet axis with the free stream or by use of a shroud.

Losses in Transport to the Instruments

Flows carrying aerosol samples from external inlets to the instruments inside aircraft often experience bends and can be turbulent. Estimates of particle losses in transport can be made using the techniques in Chapter 8. Care should be taken in this analysis because the assumptions underlying the techniques in Chapter 8 may not correspond with the actual sampling situation. For example, the spatial distribution of particles entering the bend may not be uniform, or the flow may not be fully developed in the sampling situation. However, in the studies used to produce the relationships in Chapter 8, the flows were usually well developed and the particles were usually distributed uniformly. Despite this, the analysis may still provide important guidance concerning the size range of particles that experience significant sampling errors (see Chapter 2: Aerosol Calculator).

The most prudent application of this type of analysis is in the design of transport systems and not in the correction of measurements. Inertial effects often result in a rather steep curve of transmission efficiency versus aerodynamic diameter. It is preferable to design the transport system so that the transition from high transmission efficiency to low efficiency occurs for particle diameters somewhat larger than those of interest. Gravitational sedimentation must also be considered in studies of supermicrometer particles.

As in the case of losses in diffusers, computational fluid mechanics (CFD) codes can be used to calculate particle deposition in transport to the instruments. However, the cautions stated above concerning calculation of turbulent deposition apply here as well.

Alteration of Particles by Thermodynamic Changes in Sampling or Transport to the Instruments

Particles are often heated on the way from the inlet to the instrument. The heating due to slowing of the flow is unavoidable and can be estimated from the Mach number of the free-stream flow relative to the aircraft. Heating of the transported aerosol by conduction from the warm cabin may be avoided in some instances. However, the instrument often heats the aerosol sample before the measurement. The heating can volatilize part or all of sampled particles, and changes can occur within fractions of a second. In some cases, it is possible to reconstruct the ambient size distribution from the measured one if the composition and evaporation rates of the particles are understood (Wilson et al., 1992; Hermann et al., 2000). In other cases, it may be impossible to determine the volatile components from the ones that survived to the measurement.

REVIEW OF INLETS

Table 30–1 lists several inlets and selected characteristics including references. A number of investigators operate inlets that are not described in the literature. Inlets 1, 2, and 3 employ diffusers to slow the flow. Inlets 2 and 3 have blunt leading edges. The inlets are designed to achieve isokinetic sampling. Inlet 1 provides near isokinetic sample flows for Mach numbers between 0.4 and 0.8 without use of a pump or flow control system. A tube near the center line of the main diffuser extracts the sample and transports it to the instruments. A second diffuser on the sample extraction tube reduces the sample velocity to ~10 m/s. All flows are measured, and corrections are made for anisokinetic sampling at both the main and sample extraction diffuser. Although the flow in the diffuser of inlet 1 is turbulent, comparisons with LIDAR and SAGE II satellite extinctions show that the inlet provides accurate samples for submicrometer particles in cases where the number distribution is dominated by particles smaller than 0.3 μm and larger than 0.7 μm in diameter. Transmission of aerosol particles was studied in inlet 3 in the laboratory using Reynolds, Stokes, and Schmidt similarity. Inlet 3 provides representative samples of submicrometer aerosol. Approximately 50% of the 1.3 μm particles are delivered to the sampling device.

Inlets 4 and 5 use shrouded diffusers. Figure 30–2 shows inlet 4, which is similar to inlet 5. The flow out the rear of the shroud is restricted, and the velocity of the air impinging on the entrance of the diffuser is much less than the free-stream velocity. Thus the streamlines upstream of the intake diverge, and these inlets are designed to be anisokinetic. Wind tunnel

TABLE 30–1. Inlet Characteristics

Inlet No.	Classification	Characteristics	True Air Speed, Altitude	References
1	Sharp-edged diffuser	A passive, instrumented, near-isokinetic inlet	200 m/s 20 km	Wilson et al. (1992), Jonsson et al. (1995)
2	Blunt-edged diffuser	A near-isokinetic inlet with a velocity reduced by a factor of 16	60 m/s 3.5 km	Pena et al. (1977)
3	Blunt-edged diffuser	Intended to be isokinetic	235 m/s 12 km	Hermann et al. (2000)
4	Shrouded diffuser	Tolerates variations in angle of attack. Anisokinetic	15 m/s	McFarland et al. (1989)
5	Shrouded diffuser	Tolerates variations in angle of attack. Anisokinetic	100 m/s 6.1 km	Ram et al. (1995), Cain and Ram (1998), Cain et al. (1998)
6	Counterflow virtual impactor (CVI)	Separates large particles and droplets from the atmosphere and deposits them in a gas of known composition for analysis	100 m/s	Noone et al. (1988), Laucks and Twohy (1998)
7	Shrouded CVI	Shroud straightens flow up stream of the CVI. Shroud does not slow flow	250 m/s	Twohy (1998)
8	Shrouded capillary	Shroud straightens flow and provides high-speed sample to the capillary	200 m/s 20 km	Murphy and Schein (1998)
9	Nonseparating, double duct	Anisokinetic inlet. Turbulence is prevented from disturbing the slowed sample flow	200 m/s 20 km	Soderman et al. (1991)
10	Boundary-layer suction diffuser	Low turbulence in the diffuser. Elliptical leading edge and isokinetic operation	100 m/s 6 km	Lafleur et al. (2000)
11	Inertial aerosol separator	Large particles are unable to follow curved streamlines upstream of a sampling inlet	200 m/s 20 km	Fahey et al. (1989)
12	Gas/particle separating inlet	Variable geometry permits the inlet to operate either as a CVI or an inertial aerosol separator	200 m/s 20 km	Dhaniyala et al. (2000)
13	Trans-sonic and supersonic inlets	Solid wall diffusers		Martone et al. (1980), Ivie et al. (1990)

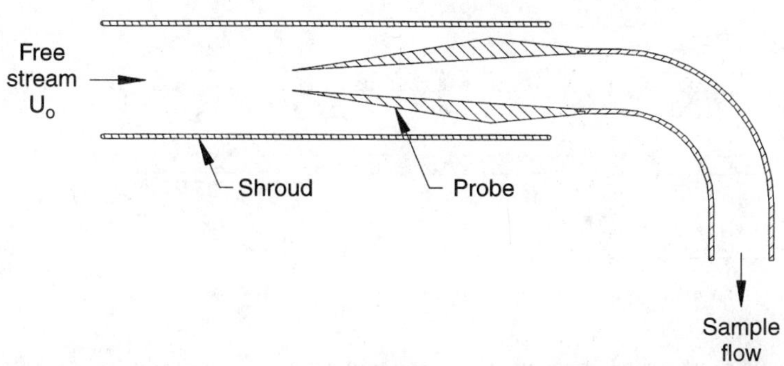

Fig. 30–2. Inlet number 4 is a shrouded, diffusing inlet. The sample flow is brought into the aircraft with a pump. The flow through the shroud is slowed by the restriction caused by the probe, and the velocity entering the probe is smaller than the free stream velocity. (After McFarland et al., 1989.)

studies of inlet 4 demonstrated transmission ratios between 0.93 and 1.11 for wind speeds of less than 15 m/s and for particles as large as 10 μm. Wind tunnel tests demonstrated the ability of the shroud in inlet 5 to straighten the flow. The investigators initially proposed this inlet for study of insoluble aerosol because the suggested collection method involves washing the inlet after sampling to recover particles deposited by turbulence on the diffuser wall. Application of corrections for enhancements due to anisokinetic sampling requires knowledge of particle size. Calculations of transmission efficiency of particles through the turbulent diffuser are presented for users who do not wish to wash down the inlet. These results are subject to the limitations of such calculations discussed above.

The counterflow virtual impactor (CVI), inlets 6 and 7, separates large particles from the air stream and deposits them in a flow of controlled composition. Gas is forced through the porous material near the front of the probe and is pulled back to the sampling instruments (Fig. 30–3). The counter flow gas and the free stream flow create a stagnation zone, and particles having sufficient inertia penetrate this zone and are deposited in the counterflow gas and carried to the sampling instruments. Wind tunnel tests showed that the cut point for separation could be varied from 9 to 30 μm by controlling the flow to and from the probe tip. Wind tunnel tests and numerical predictions were used to design a shroud for the CVI that reduced the impact of variations in the angle of attack.

Inlet 9 employs a double duct. The sample flow is slowed to approximately 20 m/s from 200 m/s and is laminar upon delivery to the inner duct. The inlet is mounted on a wing pod. The main duct has a rounded lip and is oriented into the incoming flow to prevent flow separation. Analysis of the impact of anisokinetic flow on particle sampling is not presented.

Turbulence normally generated in a diffuser is reduced to negligible levels by applying boundary layer suction through a porous wall in inlet 10 (Fig. 30–4). Velocity reductions by factors exceeding 20 are achieved, and an elliptical leading edge reduces the impact of variations in angle of attack. Large particles are enhanced in the laminar flow delivered to the rear of the inlet. The magnitude of the enhancement is determined by calculations of laminar flow and particle trajectories (Fig. 30–5). The resulting sample is not representative for supermicrometer particles, and measured concentrations of these particles must be corrected for inertial enhancements that depend on particle size. Flight tests have demonstrated the ability of this inlet to maintain laminar, isokinetic flow.

An obstacle in the form of a double ellipse of revolution resembling an American football is mounted below the fuselage in inlet 11. Streamlines reaching the sampling inlet near the rear of the football are forced to curve around the obstacle, and large particles are unable to follow the flow. At 10 kPa [100 mb] pressure and 200 m/s, particles larger than 5 μm in diam-

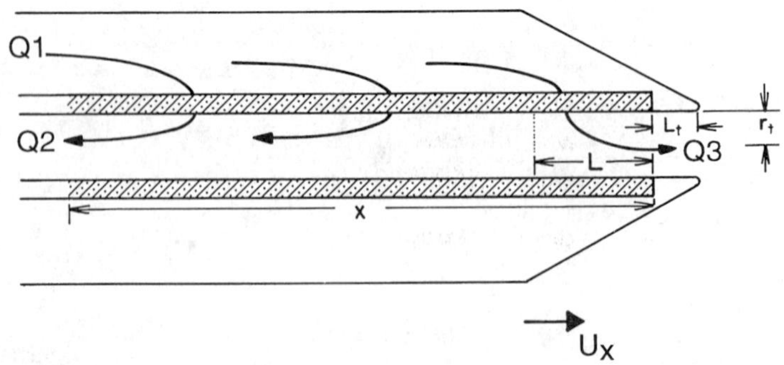

Fig. 30–3. Counterflow virtual impactor (CVI). The probe moves through the air at a velocity U_x. Flow *Q1* is filtered air and passes through the porous section of length *x*. Flow *Q3* goes out the end. Particles with sufficient inertia to traverse distance $L_t + L$ are entrained in flow *Q2* and carried to instruments. (Courtesy of C. Twohy, National Center for Atmospheric Research, Boulder, CO.)

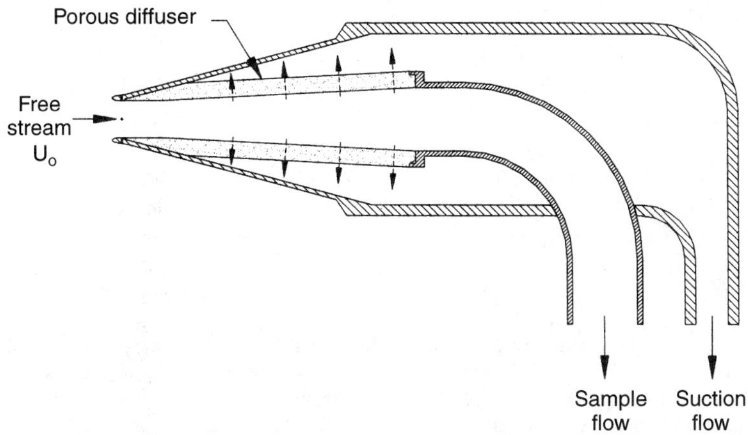

Fig. 30–4. Inlet 10 reduces the turbulence generated in the diffuser by boundary layer suction. The sample flow and suction flow are brought into the aircraft by pumps. The flows are controlled to maintain isokinetic sampling and laminar flow.

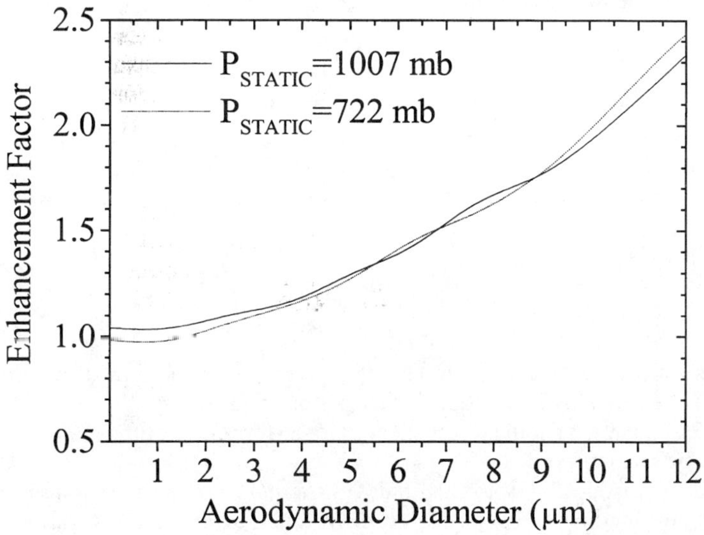

Fig. 30–5. Calculations of laminar flow and particle trajectories in inlet 10 show enhancements for large particles. In this case, 27% of the flow is delivered as sample flow, and the rest is suction flow. Thus, the maximum enhancement possible is 3.7 for large particles. The ambient mixing ratio equals the mixing ratio in the sample flow divided by the enhancement. Results are shown for two static pressures.

eter are excluded from the sample. This excludes most precipitation and type two polar stratospheric cloud particles.

Inlet 12 is designed for sampling vapors and semivolatile aerosol. Figure 30–6 shows the aerosol separating/enhancing geometry. Flow channels that are not shown in the figure deliver the sample stream to the rectangular slit intake. In the particle mode, particles larger than approximately 0.2 μm are separated from the 200 m/s free stream and deposited in the

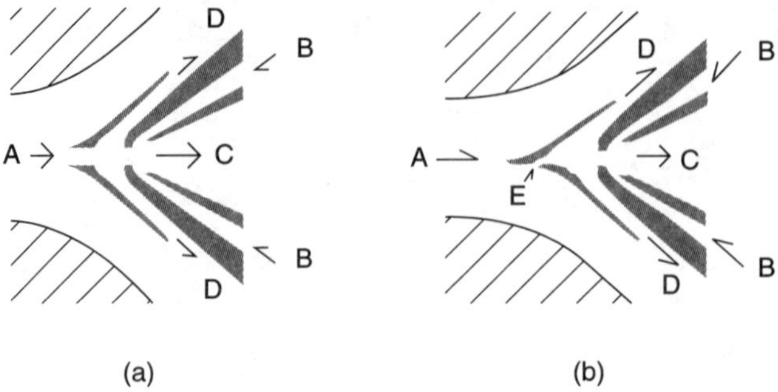

Fig. 30–6. **a**, Inlet 12 in the aerosol mode. Sampled air is delivered to point A by a flow system not shown here. Gas of known composition is injected at point B. A smaller flow goes to the instrument at C so some of the supplied gas flows out of the nozzle and away at D. The sampled air does not enter the sample nozzle and exits at D. Particles larger than about 0.2 μm in diameter have sufficient inertia to cross into flow C and are transported to the instrument in the injected gas. (Courtesy of S. Dhaniyala, California Institute of Technology, Pasadena, CA.) **b**, Inlet 12 in the gas mode. Sampled air is delivered to point A by a flow system not shown here. Gas of known composition is injected at point B, but this flow is less than the flow to the instrument at C. Some of the sampled air enters at E, and particles larger than about 0.2 um in diameter fail to follow the streamlines at E and are removed from the air that enters the sample nozzle. Air that has contacted the inlet is removed through D. Sampled air is transported to the instrument with the gas injected at B.

gas stream forming the counterflow in the inlet. The inlet is upstream of a chemical ionization mass spectrometer, and the particles are volatilized and are analyzed in a flow of known composition. In the gas mode, particles larger than approximately 0.2 μm are separated from the sample flow that is analyzed. The inlet shifts back and forth from particle to gas mode during a flight.

CONCLUSIONS

Inlet designers should consider the impact of the aircraft on the flow upstream of the inlet, the impact of the inlet itself, and the impact of the transport system from the inlet to the instruments. Investigators should also validate the performance of the instruments under the conditions to be encountered.

Improvements in our ability to manipulate flows and predict the motion of particles in laminar flows are leading to inlets with improved characteristics. Several inlets appear to provide representative samples of submicrometer aerosol. Supermicrometer particles and high speeds present difficult problems that appear to be yielding to engineering advances and may be ameliorated by choosing slower platforms.

REFERENCES

Baumgardner, D. 1984. The effects of airflow distortion on aircraft measurement: A workshop summary. *Bull. Am. Meteorol. Soc.* 65:1212–1213.

Baumgardner, D., J. E. Dye, B. Gandrud, K. Barr, K. Kelly, K. R. Chan. 1996. Refractive indices of aerosols in the upper troposphere and lower stratosphere. *Geophys. Res. Lett.* 23:749–752.

Baumgardner, D., W. Strapp, and J. E. Dye. 1985. Evaluation of the forward scattering spectrometer probe. Part II: Corrections of coincidence and dead-time losses. *J. Atmos. Ocean. Technol.* 2:626–632.

Beard, K. V. 1983. Reorientation of hydrometeors in aircraft accelerated flow. *J. Clim. Appl. Meteorol.* 22:1961–1963.

Brenguier, J. L. 1989. Coincidence and dead-time corrections for particle counters. Part II: High concentration measurements with and FSSP. *J. Atmos. Ocean. Technol.* 6:585–598.

Brenguier, J. L. and L. Amodei. 1989. Coincidence and dead-time corrections for particle counters. Part I: A general mathematical formalism. *J. Atmos. Ocean Technol.* 6:575–584.

Brock, C. A., L. F. Radke, and P. V. Hobbs. 1990. Sulfur in particles in arctic hazes derived from airborne in situ and lidar measurements. *J. Geophys. Res.* 95:22369–22387.

Brock, C. A., F. Schröder, B. Kärcher, A. Petzold, R. Busen, and M. Fiebig. 2000. Ultrafine particle size distributions measured in aircraft exhaust plumes. *J. Geophys. Res.* 105:26555–26567.

Cain, S. A. and M. Ram. 1998. Numerical simulation studies the turbulent airflow through a shrouded airborne aerosol sampling probe and estimation of the minimum sampler transmission efficiency. 1998. *J. Aerosol Sci.* 29:1145–1156.

Cain, S. A., M. Ram, and S. Woodward. 1998. Qualitative and quantitative wind tunnel measurements of the airflow through a shrouded airborne aerosol sampling probe. 1998. *J. Aerosol Sci.* 29:1157–1169.

Cerni, T. A. 1983. Determination of the size and concentration of cloud drops with an FSSP. *J. Appl. Meteorol.* 22:1346–1355.

Dhaniyala, S., R. Flagan, and P. O. Wennberg. 2000. A gas/particle separating inlet for sampling vapors and semi-volatile aerosols. American Association for Aerosol Research 2000 Conference, St. Louis, MO, November 6–10.

Dreiling, V. and R. Jaenicke. 1988. Aircraft measurement with condensation nuclei counter and optical particle counter. *J. Aerosol Sci.* 19:1045–1050.

Drummond, A. M. 1984. Aircraft flow effects on cloud droplet images and concentrations. Aeronautical Note NAE-AN-21 (NRC No. 23508). National Research Council Canada, 30 pp.

Drummond, A. M. and J. E. MacPherson. 1984. Theoretical and measured airflow about the Twin Otter wing. Aeronautical Note NAE-AN-19 (NRC No. 33184), National Research Council, Canada, 33 pp.

Drummond, A. M. and J. E. MacPherson. 1985. Aircraft flow effects on cloud drop images and concentrations measured by the NAE Twin Otter. *J. Atmos. Ocean. Technol.* 2:633–643.

Dye, J. E. and D. Baumgardner. 1984. Evaluation of the forward scattering spectrometer probe. Part I: Electronic and optical studies. *J. Atmos. Ocean. Technol.* 1:329–344.

Fahey, D. W., K. K. Kelly, G. V. Ferry, L. R. Poole, J. C. Wilson, D. M. Murphy, M. Loewenstein, and K. R. Chan. 1989. In situ measurements of total reactive nitrogen, total water and aerosol in a polar stratospheric cloud in the Antarctic. *J. Geophys. Res.* 94:11299–11315.

Heintzenberg, J. and J. A. Ogren. 1985. On the operation of the TSI-3020 condensation nuclei counter at altitudes up to 10 km. *Atmos. Environ.* 19:1385–1387.

Hering, S. V. 1987. Calibration of the QCM impactor for stratospheric sampling. *Aerosol Sci. Technol.* 7:257–274.

Hermann, M., F. Stratmann, M. Wilck, and A. Wiedensohler. 2001. Sampling characteristics of an aircraft-borne aerosol inlet system. *J. Atmos. Ocean. Technol.* 18:7–19.

Huebert, B. J., G. L. Lee, and W. L. Warren. 1990. Airborne aerosol inlet passing efficiency measurement. *J. Geophys. Res.* 95:16369–16381.

Intergovernmental Panel on Climate Change. 1994. *Radiative Forcing of Climate Change.* Cambridge University Press, Cambridge, Great Britain.

Ivie, J. J., L. J. Forney, and R. L. Roach. 1990. Supersonic particle probes: Measurement of internal wall losses. *Aerosol Sci. Technol.* 13:368–385.

Jonsson, H. H., J. C. Wilson, C. A. Brock, R. G. Knollenberg, R. Newton, J. E. Dye, D. Baumgardner, S. Borrmann, G. V. Ferry, R. Pueschel, D. C. Woods, and M. C. Pitts. 1995. Performance of a focused cavity aerosol spectrometer for measurements in the stratosphere of particle size in the 0.06–2.0 μm diameter range. *J. Ocean. Atmos. Technol.* 12:115–129.

King, W. D. 1984. Airflow and particle trajectories around aircraft fuselages. I: Theory. *J. Atmos. Ocean. Technol.* 1:5–13.

King, W. D. 1986a. Airflow and particle trajectories around aircraft fuselages. IV: Orientation of ice crystals. *J. Atmos. Ocean. Technol.* 3:439.

King, W. D. 1986b. Airflow around PMS canisters. *J. Atmos. Ocean. Technol.* 3:197–198.

Knollenberg, R. G. 1981. Techniques for probing cloud microstructure. In *Clouds, Their Formation, Optical Properties, and Effects*, eds. P. V. Hobbs and A. Deepak. New York: Academic Press, pp 15–91.

Lafleur, B. G., J. C. Wilson, and W. R. Seebaugh. 2000. Flight testing of low turbulence aerosol inlets. American Association For Aerosol Research 2000 Conference, St. Louis, MO, November 1–6.

Laucks, M. L. and C. H. Twohy. 1998. Size dependent collection efficiency of an airborne counterflow virtual impactor. *Aerosol Sci. Technol.* 28:40–61.

Leaitch, W. R., R. M. Hoff, and J. I. MacPherson. 1989. Airborne and lidar measurements of aerosol and cloud particles in the troposphere over Alert, Canada in April 1986. *J. Atmos. Chem.* 9:187–211.

Liu, P. S. K., W. R. Leaitch, J. W. Strapp, and M. A. Wasey. 1992. Response of Particle Measuring Systems Airborne ASASP and PCASP to NaCl and Latex Particles. *Aerosol Sci. Technol.* 16:83–96.

MacPherson, J. I. 1985. Wind tunnel calibration of a PMS canister instrumented for airflow measurement. Aeronautical Note NAE-AN-32 (NRC No. 24922). National Research Council, Canada.

MacPherson, J. E. and D. Baumgardner. 1988. Airflow about King Air wingtip-mounted cloud particle measurement probes. *J. Atmos. Ocean. Technol.* 5:259–273.

Martone, J. A., P. S. Daley, and R. W. Boubel. 1980. Sampling submicrometer particles suspended in near sonic and supersonic free jets. *JAPCA* 30:898–903.

McFarland, A. R., C. A. Ortiz, M. E. Moore, R. E. DeOtte, Jr., and S. Somasundaram. 1989. A shrouded aerosol sampling probe. *Environ. Sci. Technol.* 23:1487–1492.

Murphy, D. M. and M. E. Schein. 1998. Wind tunnel tests of a shrouded aircraft inlet. *Aerosol Sci. Technol.* 28:33–39.

Murphy, D. M., D. S. Thomson, and M. J. Mahoney. 1998. In situ measurements of organics, meteoritic material, mercury and other elements in aerosols at 5 to 19 kilometers. *Science* 282:1664–1669.

Noone, K. J., R. J. Charlson, D. S. Covert, J. A. Ogren, and J. Heintzenberg. 1988. Design and calibration of a counterflow, virtual impactor for sampling of atmospheric fog and cloud droplets. *Aerosol. Sci. Technol.* 8:235–244.

Noone, K. J. and H.-C. Hansson. 1990. Calibration of the TSI 3760 condensation nucleus counter for non-standard operating conditions. *Aerosol Sci. Technol.* 13:478–485.

Norment, H. G. 1988. Three-dimensional trajectory analysis of two drop sizing instruments: PMS OAP and PMS FSSP. *J. Atmos. Ocean. Technol.* 5:743–756.

Pena, J. A., J. M. Norman, and D. W. Thomson. 1977. Isokinetic sampler for continuous airborne aerosol measurements. *JAPCA* 27:337–341.

Pinnick, R. G., D. M. Garvey, and L. D. Duncan. 1981. Calibration of Knollenberg FSSP light-scattering counters for measurements of cloud droplets. *J. Appl. Meteorol.* 20:1049–1057.

Porter, J. N., A. D. Clarke, G. Ferry, and R. F. Pueschel. 1992. Studies of size dependent aerosol sampling through inlets. *J. Geophys. Res.* 97:3815–3824.

Rader, D. J. and V. A. Marple. 1988. A study of the effects of anisokinetic sampling. *Aerosol. Sci. Technol.* 8:293–299.

Ram, M., S. A. Cain, and D. B. Taulbee. 1995. Design of a shrouded probe for airborne aerosol sampling in a high velocity air stream. *J. Aerosol Sci.* 26:945–962.

Reagan, J. A., J. D. Spinhirne, D. M. Byrne, D. W. Thomson, R. G. de Pena, and Y. Mamane. 1977. Atmospheric particulate properties inferred from lidar and solar radiometer observations compared with simultaneous in situ aircraft measurements: A case study. *J. Appl. Meteorol.* 16:911–928.

Romay, F. J., D. Y. H. Pui, T. J. Smith, N. D. Ngo, and J. H. Vincent. 1996. Corona discharge effects on aerosol sampling efficiency. *Atmos. Environ.* 30:2607–2613.

Russell, L. M., M. R. Stolzenburg, S. H. Zhand, R. Caldow, R. C. Flagan, and J. H. Seinfeld. 1996. Radially classified aerosol detector for aircraft-based submicron aerosol measurements. *J. Atmos. Ocean. Technol.* 13:568.

Sheridan, P. J. and R. B. Norton. 1998. Determination of the passing efficiency for aerosol chemical species through a typical aircraft-mounted, diffuser-type aerosol inlet system. *J. Geophys. Res.* 103:8215–8225.

Soderman, P. T., N. L. Hazan, and W. H. Brune. 1991. Aerodynamic design of gas and aerosol samplers for aircraft. NASA Technical Memorandum 103854. National Aeronautics and Space Administration, Ames Research Center, Moffatt Field, CA.

Strapp, W. J., W. R. Leaitch, and P. S. K. Liu. 1992. Hydrated and dired aerosol-size-distribution measurements from the Particle Measuring Systems FSSP-300 Probe and Deiced PCASP-100x probe. *J. Atmos. Ocean. Technol.* 9:548–555.

Torgeson, W. L. and S. C. Stern. 1966. An aircraft impactor for determining the size distributions of tropospheric aerosols. *J. Appl. Meteorol.* 5:205–210.

Twohy, C. H. 1998. Model calculations and wind tunnel testing of an isokinetic shroud for high-speed sampling. *Aerosol Sci. Technol.* 29:261–280.

Twohy, C. H. and D. Rogers. 1993. Airflow and water-drop trajectories at instrument sampling points around the Beechcraft King Air and Lockheed Electra. *J. Atmos. Ocean. Technol.* 10:566–578.

Webber, R. J., A. D. Clarke, M. Litchy, J. Li, G. Kok, R. D. Schillawski, and P. H. McMurry. 1998. Spurious aerosol measurements when sampling from aircraft in the vicinity of clouds. *J. Geophys. Res.* 103:28337–28346.

Wilson, J. C., J. H. Hyun, and E. D. Blackshear. 1983. The function and response of an improved stratospheric condensation nucleus counter. *J. Geophys. Res.* 88:6781–6785.

Wilson, J. C., M. R. Stolzenberg, W. E. Clark, M. Loewenstein, G. V. Ferry, K. R. Chan, S. R. Kawa, and K. K. Kelly. 1992. Stratospheric sulfate aerosol in and near The northern hemisphere polar vortex: The morphology of the sulfate layer, multimodal size distributions and the effect of denitrification. *J. Geophys. Res.* 97:8:7997–8013.

Woods, D. C. and R. L. Chuan. 1983. Size-specific composition of aerosols in the El Chichón volcanic cloud. *Geophys. Res. Lett.* 10:1041–1044.

World Meteorological Organization. 1988. *Scientific Assessment of Ozone Depletion: 1988.* Geneva: World Meteorological Organization, pp. 3.1–3.40, 7.6–7.16.

31 Measurement of High-Concentration and High-Temperature Aerosols

PRATIM BISWAS
Environmental Engineering Science Program, Department of Chemical Engineering, Washington University in St. Louis, St. Louis, MO

INTRODUCTION

High-concentration aerosols are encountered in many industrial systems utilizing aerosol reactors. Such reactors are used in industry to make a wide variety of particulate commodities such as carbon black, pigments, and materials for high technology applications such as optical waveguides and powders for advanced ceramics (Stamatakis et al., 1991). A similar scenario is encountered in many other systems where a large quantity of the so-called undesirable aerosols is produced. Municipal waste incinerators, hazardous waste incinerators, welding systems, automobile, diesel engine and aircraft exhausts, coke ovens, smelters, nuclear reactor accidents, and utility boilers are some examples. Despite the inherent difference in the use of the particles, a number of common features stand out. The aerosols are produced in high concentrations, and in many cases are the result of a high-temperature process. In both cases, the particles need to be collected or deposited onto some substrate: In the first case, the particles might be used to fabricate a part for some application; in the latter case, the particles could be potentially very toxic, and emission to the ambient atmosphere needs to be prevented. With the recent findings that fine particles may be associated with enhanced deleterious health effects, it will be necessary to control the emissions of such particles from combustion sources. For efficient design of control devices, it is, therefore, essential to know the size distribution of the aerosol. Frequently, in the industrial aerosol processes, the focus is also on producing monodisperse particles or particles in very narrow size ranges. Measurements may, therefore, need to be carried out to control the process parameters. Also, to control and understand particle formation and growth rates, there is a need to measure the evolution of the aerosol size distribution.

There are certain characteristics that must be considered in the measurement of high-temperature, high-concentration aerosols. A number of dynamic physicochemical processes are in progress in the system. If a sampling system is used, it must be designed to quench all aerosol dynamics (nucleation, coagulation, condensation to avoid alteration of the aerosol size distribution in the sampling system) and chemistry to obtain a representative

Aerosol Measurement: Principles, Techniques, and Applications, Second Edition, Edited by Paul A. Baron and Klaus Willeke.
ISBN 0-471-35636-0

measurement. If the real-time measuring instruments described earlier in this book are to be used, the number concentration and temperature must be reduced to within the operational limits of the instrument. To alleviate some of these problems, in situ techniques can be used. The measurement of high-temperature and high-concentration aerosols using these two methods is described. Impactors used in high-temperature applications are also discussed. This is followed by a discussion of the in situ techniques for measurement of high-temperature aerosols in combustion environments.

DILUTION SYSTEMS

The hot gases with the particles must be diluted and cooled in the sampling system. The aerosol dynamics must be rapidly quenched, and so must changes in chemistry. High dilution ratios are often required to reduce the number concentration to the operational range of the aerosol instruments. Such high dilution ratios often require that multistage dilution systems be used. Dilution systems used by various researchers are summarized in Table 31–1.

Primary Dilution Systems

A simple design of a primary dilution probe used by Linak and Peterson (1984) is shown in Figure 31–1. Dry, filtered compressed air is used to entrain a constant flow rate sample by

TABLE 31–1. Summary of Dilution Systems

Researchers	Dilution Ratio			Dilution Gas Used	Sampled Source
	Primary	Secondary	Overall		
Newton et al. (1980)	5	N.A.	5	Air	Fluidized bed combustor
Pedersen et al. (1980)	N.A.	20	20	Air	Organic aerosols, automobiles
Ulrich and Riehl (1982)	N.R.	N.A.	N.R.	Nitrogen	Silica flame reactor
Houck et al. (1982)	N.A.	30	30	Air	Organic aerosols, stack
Sousa et al. (1987), Linak and Peterson (1984)	21.2	39.5	837.4	Air	Pulverized coal combustion
Du and Kittleson (1984)	10	33	330	Nitrogen	Organic aerosols, diesel engines
Bonfanti and Cioni (1986)	6	N.A.	6	Nitrogen, air	Organic aerosols, stack
Wu and Flagan (1987)	N.R.	2000	2000	Nitrogen, air	Silicon, tubular reactor
Biswas et al. (1989)	31	101	3131	Argon, air	Silica, tubular reactor
Hildemann et al. (1989)	N.A.	25–100	25–100	Air	Organic aerosols, stack
Sethi and Biswas (1990)	19.5	25	487.5	Argon, air	Lead, flame incinerator
Pratsinis et al. (1990)	5	200	1000	Argon, air	Titania, tubular reactor
Zimmer (2000)	10	100	1000	Air	Arc welding process

N.A., not applicable; N.R., not recorded.

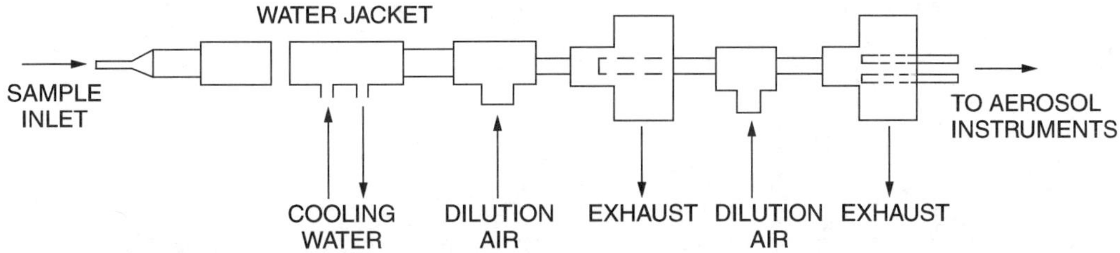

Fig. 31–1. A two-stage dilution system consisting of a particle probe as the primary dilution system. (Adapted from Linak and Peterson, 1984.)

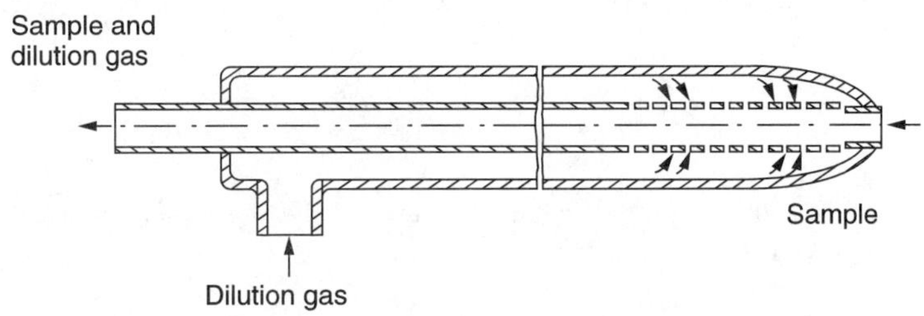

Fig. 31–2. Schematic diagram of a primary dilution probe. The dilution gas enters through the perforations in the central tube and mixes with the sample.

aspiration. Constant-pressure dilution air is supplied through an outer annulus, and, as it enters the inner tube, a sample stream is drawn through the orifice due to the lower pressure region that is created. This sample then mixes with the filtered and compressed dilution air. The dilution ratio is determined based on the capillary dimension, clearance between outer and inner annuli, and diluent gas pressure. The dilution ratio is not controlled directly (by flow control), and there may be associated fluctuations in the sample flow. Isokinetic conditions may also be difficult to maintain in such systems.

An aspirated isokinetic probe consisting of a water-jacketed air delivery system has been used by Peterson and co-workers (see Scotto et al., 1988; Gallagher et al., 1990) for sampling aerosols for studying alkali metal partitioning in ash from pulverized coal combustors. Sampling conditions were isokinetic, thus allowing sampling of larger particles. The aerosol sample was rapidly quenched (cooled and diluted) by a free turbulent jet that provided a high mixing rate.

The problem of not having direct control of the dilution ratio is alleviated by using a direct flow-controlled primary sampling probe. Newton et al. (1980) describe a radially injected dilution probe for sampling in fluidized bed coal combustion systems. The inlet surfaces of the probe are chamfered to reduce edge effects. Dilution air enters through a porous stainless steel cylinder perpendicular to the direction of the sample flow. The dilution air flow is independently controlled and is particle free. The system tends to compress the aerosol sample, allowing cooling and mixing to occur away from the walls, thus reducing losses. A sample flow of $4.17 \times 10^{-4}\,m^3/s$ [25 L/min] was diluted with a total air flow of $1.67 \times 10^{-3}\,m^3/s$ [100 L/min], leading to a dilution ratio of 5. While thermophoretic losses are shown to be minimized, alteration of the aerosol by condensation and coagulation is not necessarily eliminated.

A design of a flow-controlled primary dilution probe is shown in Figure 31–2 (Biswas et al., 1989). The probe is made of quartz and is constructed using two concentric tubes. Small 1/6 inch (4 mm) diameter holes are drilled on the walls of the inside tube. The dilution

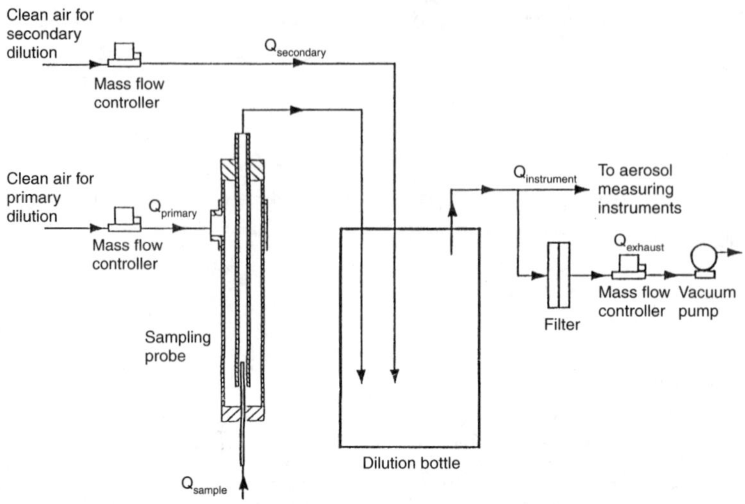

Fig. 31–3. Schematic of a dilution probe for flame measurements. Also shown is the secondary dilution system.

gas can thus enter the inside tube and mix and dilute the sample aerosol. The probe is easily movable to sample at different locations in the system. In systems where chemical reactions are rapid, an inert gas such as argon is used for dilution as it participates in no chemical reactions.

A more rugged version of a probe designed by Sethi and Biswas (1990) for measurement in flame environments is shown in Figure 31–3. The following criteria are adopted in this modified design: The cross section of the probe tip should be small so as not to disturb the flow in the system at the sampling location; cleaning and fabrication should be easy; and material should withstand flame environments. A stainless steel probe with a small outer diameter (1/8 inch, 3.2 mm) is used for the sampling end of the probe and is housed in a slightly larger diameter tube. The outer tube also serves as the outer jacket for the dilution gases. The sample is drawn out of the probe from a tube housed inside the outer tube. Thermophoretic and diffusional particle losses in the probe are negligible. The residence time in the sampling tube is short. Sampling losses due to anisokinetic sampling are estimated using the criteria for probe alignment and sampling to mainstream velocity ratio. The probe is aligned along the flame axis; hence there are no losses due to misalignment. The Stokes number of particles sampled in the flame is very much smaller than 0.01; thus, the losses due to velocity differences are unimportant (Hinds, 1999). A similar design probe has been used by Zimmer and Biswas (2001) for effectively monitoring the evolution of the aerosol size distribution during an arc welding process. Very high dilution ratios had to be used due to the high number concentrations and high temperatures in the zone of the plasma arc. A nano-differential mobility analyzer was successfully used in conjunction with the sampling system to map out the evolution of the ultrafine particle size distribution, both spatially and temporally.

Ulrich and Riehl (1982) used a nitrogen-quenched, sonic expansion probe for collection of flame-generated silica particles for surface area analysis. The probe is constructed using two concentric tubes with nitrogen gas flowing in the region between the inner and outer tubes. A vacuum is applied at the inner tube that pulls some sample from the mainstream. The nitrogen gas not only cools and dilutes the sample but also prevents clogging of the probe tip.

In the above designs, a probe-type configuration is used to draw a small sample from the mainstream flow and dilute it by mixing with some particle-free gas. This is done so that higher dilution ratios can be attained. Designs where the entire reactor flow is diluted have been used by Alam and Flagan (1986) and then modified and used by Wu and Flagan (1987). A schematic diagram is shown in Figure 31–4. As the aerosol comes out of the reactor, a coaxial flow is introduced from the sides of a sintered tube. This prevents the aerosol from coming into contact with the cooler walls and depositing due to thermophoresis. Downstream, a large volume of diluent gas is mixed with the aerosol. The flow becomes turbulent, and good mixing is obtained with the gases, causing both dilution and cooling simultaneously. A similar set up used by Pratsinis et al. (1990) is shown in Figure 31–5. The reactor effluent is rapidly mixed and cooled with argon in a perforated quartz dilutor. The perforated quartz dilutor is used in place of the sintered tube used by Alam and Flagan (1986) for ease of fabrication and cleaning.

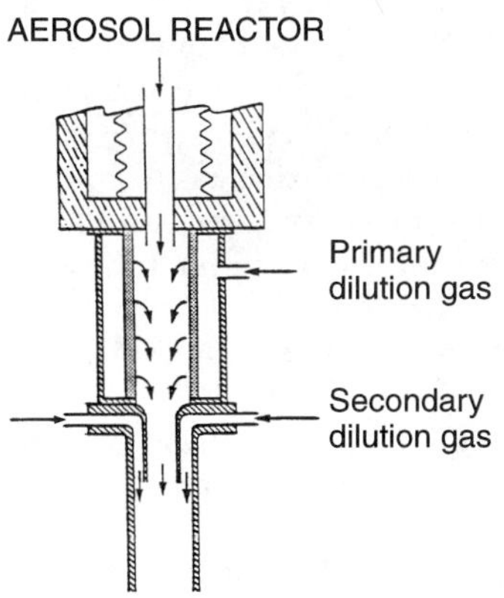

Fig. 31–4. Dilution system (primary and secondary) wherein the entire reactor flow is diluted. (Adapted from Alam, 1984.)

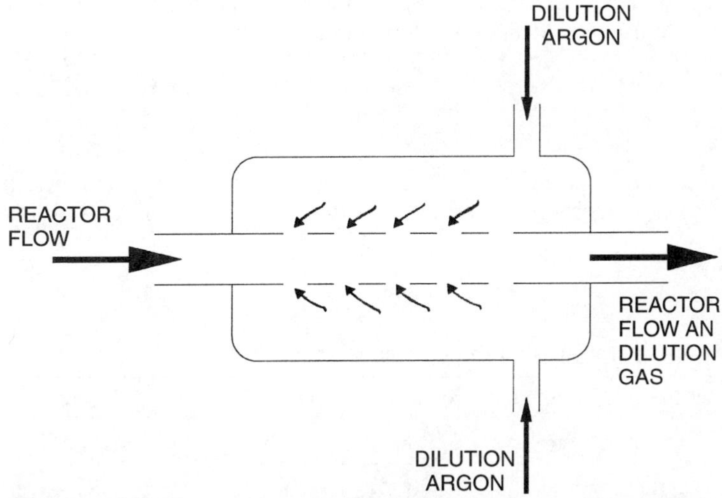

Fig. 31–5. Details of a primary dilution system wherein the entire reactor flow is diluted with a volume of inert diluent gas to quench chemical reactions. (Adapted from Pratsinis et al., 1990.)

The following conditions must be met in the design of primary dilution probes:

1. Sampling under as close to isokinetic conditions as possible (see Chapter 8), matching the sample velocity to the face stream velocity
2. Choosing the dilution ratio to ensure minimal deposition losses due to thermophoresis and diffusion and minimal alteration of the size distribution by various dynamic aerosol phenomena such as nucleation, coagulation, and condensation

Secondary Dilution

The primary dilution probes should rapidly cool and dilute a sample in order to quench aerosol growth and chemical reaction. As described earlier, in the case of high mainstream aerosol concentrations (10^{16} to 10^{21} particles/m^3 [10^{10} to 10^{15} particles/cm^3]), a secondary dilution is often necessary as real-time aerosol instruments are limited to measuring a maximum of about 10^{13} particles/m^3 [10^7 particles/cm^3]. Secondary dilution can be done in configurations similar to the primary dilution system. The flow with the primary dilution system mixes with a large volume of particle-free air, and a part of this flow is pulled into the aerosol sampling instrument (Linak and Peterson, 1984; Fig. 31–1). As the residence times are relatively short, proper mixing has to be ensured to obtain a representative sample. In the dilution system developed by Alam and Flagan (1986), a coaxial flow is used to obtain secondary dilution (see Fig. 31–4). In both these systems, as the entire flow from the primary dilution system is being diluted, a relatively high flow rate of diluent gas must be chosen, depending on the number concentration of the aerosol in the mainstream (see Example 31–1).

An alternative secondary dilution system developed by Biswas (1985) and used for particle sampling in high-temperature systems (Biswas et al., 1989; Sethi and Biswas, 1990) is shown in Figure 31–6. The dilution takes place in a 50 L bottle, with a small flow from the dilution system being drawn in and mixed with a large volume of particle-free air. Very high dilution ratios can be obtained as precise flow control is possible by using mass flow controllers. To minimize uncertainties in the dilution ratio, the dilution flow is often cooled, filtered, and recirculated (Biswas, 1985:125; Biswas and Flagan, 1988; Wu and Flagan, 1987). A schematic diagram of the set up indicating the various flows is shown in Figure 31–6. The residence time is sufficiently long to allow good mixing. The expression developed by Crump et al. (1983) can be used to calculate the losses in the dilution vessel

$$\frac{c_n}{c_{n0}} = \exp(-\beta t) \qquad (31\text{–}1)$$

where c_n is the number concentration of the aerosol after being resident in the dilution vessel for a time t, c_{n0} is the number concentration at time $t = 0$, and β is the wall loss coefficient (Table 31–2).

TABLE 31–2. Fraction of Particle Loss in the Dilution Bottle

Particle Diameter (μm)	β (s^{-1}) Loss Coefficient (Crump et al., 1983)	Fraction of Particle Loss
0.024	1.8×10^{-4}	0.05
0.042	8.4×10^{-5}	0.02
0.34	1.5×10^{-5}	0.01
0.51	4.8×10^{-5}	0.02
0.79	1.26×10^{-4}	0.04

EXAMPLE 31–1

(1) For the dilution system shown in Figure 31–6, if the sample flow rate is $Q_{\text{sample,in}}$ and the primary dilution gas flow rate is $Q_{\text{primary,in}}$, compute the dilution ratio in the probe. If the flow rate of clean air (particle free) mixed with a flow of $Q_{\text{sample,out}}$ coming from the probe is $Q_{\text{secondary,in}}$, compute the overall dilution ratio.

The following are a set of flow rates used in the experiments described by Biswas et al. (1989):

$$Q_{\text{primary,in}} = Q_{\text{primary,out}} = 5\times10^{-5}\,\text{m}^3/\text{s}[3\,L/\text{min}]$$

$$Q_{\text{secondary,in}} = 1.67\times10^{-4}\,\text{m}^3/\text{s}[10\,L/\text{min}], Q_{\text{secondary,out}} = 1.63\times10^{-4}\,\text{m}^3/\text{s}[9.8\,L/\text{min}]$$

$$Q_{\text{sample,in}} = Q_{\text{sample,out}} = 1.67\times10^{-6}\,\text{m}^3/\text{s}[0.1\,L/\text{min}]$$

$$Q_{\text{instrument,in}} = 5\times10^{-6}\,\text{m}^3/\text{s}[0.3\,L/\text{min}]$$

Answer: Primary dilution: Assuming that the number concentration of the aerosol in the sample is c_n and that in the diluted sample is c_{n1}, a mass (or number, particle properties unchanged) balance yields

$$c_n Q_{\text{sample,in}} + 0\,Q_{\text{primary,in}} = c_{n1}\left(Q_{\text{sample,in}} + Q_{\text{primary,in}}\right)$$

where it is assumed that the diluent gas is particle free (filtered). The dilution ratio is defined as the ratio by which the number concentration must be multiplied to obtain the number concentration of the aerosol in the system. Therefore, the dilution ratio in the sampling probe, DR_1, is

$$DR_1 = \left(Q_{\text{sample,in}} + Q_{\text{primary,in}}\right)/Q_{\text{sample,in}}$$

Secondary dilution: In a similar manner, the dilution ratio for the secondary system, DR_2, is

$$DR_2 = \left(Q_{\text{secondary,in}} + Q_{\text{sample,in}}\right)/Q_{\text{sample,out}}$$

The overall dilution ratio, DR is thus given by

$$\begin{aligned} DR &= DR_1 \times DR_2 \\ &= \left(Q_{\text{sample,in}} + Q_{\text{primary,in}}\right)\times\left(Q_{\text{secondary,in}} + Q_{\text{sample,out}}\right)/\left(Q_{\text{sample,in}} + Q_{\text{sample,out}}\right) \end{aligned}$$

The number concentration of the aerosol in the system is then determined by multiplying the measured size distribution of the aerosol by DR. For the given flow rates, using the formulae above, one computes $DR = 3131$.

(2) If the number concentration in a system is 10^{16} particles/m^3 [10^{10} particles/cm^3], what flow rates are to be used in the dilution system illustrated in Figure 31–6 for measurement with commonly used real time aerosol instruments?

Answer: As most instruments have a measurement upper limit of around 10^{13} particles/m^3 [10^7 particles/cm^3], and using a margin of safety, we choose an overall dilution ratio, $DR = 10^4$. Hence, for a number concentration in the system of 10^{16} particles/m^3 [10^{10} particles/cm^3], the instrument would see a concentration of $10^{16}/10^4 = 10^{12}$ particles/m^3. Let $Q_{\text{primary,in}} = Q_{\text{primary,out}} = 5\times10^{-5}\,\text{m}^3/\text{s}$ [3 L/min], $Q_{\text{sample,in}} = 1.67\times10^{-6}\,\text{m}^3/\text{s}$ [0.1 L/min] as in Example 31–1a. Therefore, $DR_1 = 5.17\times10^{-5}/1.67\times10^{-6}$ [3.1/0.1] = 31. Hence, the required $DR_2 = DR/DR_1 = 10^4/31 = 322.6$. If $Q_{\text{sample,out}} = 1.67\times10^{-6}\,\text{m}^3/\text{s}$ [0.1 L/min], then we obtain $Q_{\text{secondary,in}} = DR_2\,Q_{\text{sample,out}} - Q_{\text{sample,out}} = 5.38\times10^{-4} - 0.017\times10^{-4}$ [32.26 − 0.1] $= 5.36\times10^{-4}\,\text{m}^3/\text{s}$ [32.16 L/min].

The required degree of accuracy of flow control should be appreciated.

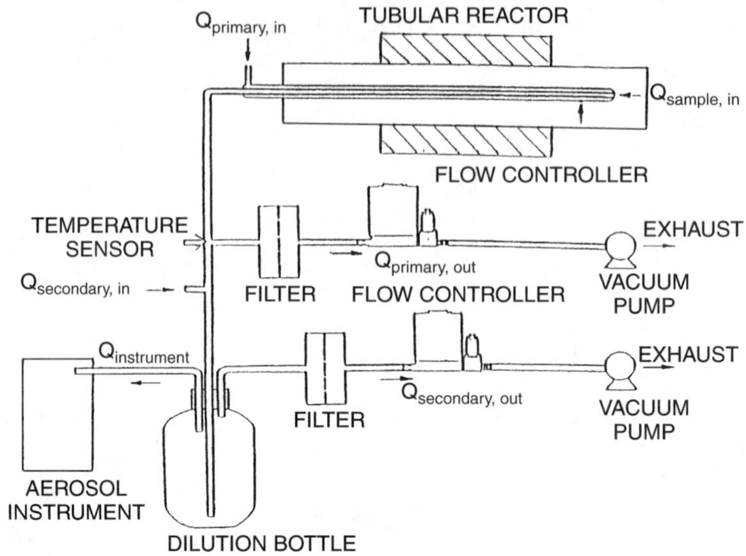

Fig. 31–6. Entire dilution system used by Biswas et al. (1989) showing primary dilution probe and secondary dilution system. Also indicated are the various flow rates through the different sections.

EXAMPLE 31–2

Consider the secondary dilution bottle system to be spherical with a volume of $0.05\,m^3$ [50L]. Assuming a total flow rate of $1.67 \times 10^{-4}\,m^3/s$ [10L/min], estimate the fraction of particle loss in the bottle for 0.024, 0.042, 0.34, 0.51, and 0.79 μm particles.

Answer: A simplistic calculation will be carried out, using Eq. 31–1 and the measured values of β reported by Crump et al. (1983). The residence time in the dilution bottle is 50/10 = 5 min = 300 s. The results, computed assuming that β remains constant for this time period for particles in a certain size range, are listed in Table 31–2.

Dilution Sampling of Organic Aerosols

Due to the toxic nature of organic compounds, it is of interest to determine the contribution of emission sources to ambient organic aerosol concentrations (Daisey et al., 1986). The ratio of the vapor phase to particulate organics is a strong function of the temperature. Direct filtration of hot gases may not lead to the capture of the organic aerosols that exist primarily as vapors at these temperatures, leading to underestimation of source organic aerosol contributions. Alternatively, the use of cooled impingers or cryogenic traps results in an overestimation of organic aerosol concentrations due to the condensation of organics that normally would not condense. Thus, the sampling of organic aerosols from combustion sources requires that dilution systems be used. Hildemann et al. (1989) have reviewed the various systems for organic aerosol sampling.

Large, nonportable dilution tunnels have been used for automotive emissions measurements. These systems were designed to simulate the dilution and cooling that take place in the atmosphere when exhaust gases leave the exhaust pipe of a vehicle (Pedersen et al., 1980). The exhaust was partially cooled (to 433 K [160°C] to prevent the condensation of moisture) and mixed with dilution air at a dilution ratio of 20 to further cool the sample to 308 K [35°C]. Du and Kittleson (1984) used a sampling system capable of removing, quenching, and diluting the entire contents of a cylinder during actual engine combustion. The extracted gas was made to go through an adiabatic expansion into the sampling apparatus whereby it was diluted with nitrogen at a ratio of about 10. These diluted combustion chamber contents were further diluted in a polyethylene sampling bag with nitrogen to give an overall dilution ratio of about 330.

Portable dilution systems for stack sampling of organic aerosols have been described by Houck et al. (1982), Hyunh et al. (1984), Bonfanti and Cioni (1986), Sousa et al. (1987), and Hildemann et al. (1989) with dilution ratios in the range of 6 to 100. The important design issues for such systems have been summarized by Hildemann et al. (1989):

1. Simulation of atmospheric dilution as closely as possible to collect a sample that includes all organics that would condense into the aerosol phase in the atmosphere. This is achieved by a high degree of dilution and cooling to ambient temperatures. The system should allow a long residence time for mixing.
2. Sampler surfaces should be inert and withstand rigorous cleaning and heat treatment.
3. The sampler should be configured and operated to minimize particle and vapor losses.

EPA STACK SAMPLING METHODS

There are two U.S. EPA proposed methodologies for obtaining particle mass concentrations in stacks: an external filtration method (Method 5) and an in-stack filtration method (Method 17). The U.S. EPA Method 5 Sampling Train is used to determine total particulate emissions on a mass basis from stationary sources such as utility plants, coal-fired boilers, incinerators, and other stacks (U.S. EPA, 1990). It is widely used for stationary-source sampling for air pollution control equipment performance evaluation for determining the effect of operating variables on mass emissions and for regulatory compliance tests.

The sampling train (Fig. 31–7) consists of a nozzle, probe, filter housing, impingers, pitot tube, and metering system. The nozzle is designed to operate under isokinetic conditions, and its size is chosen accordingly. It has a button hook design to create a smooth change in the flow direction by 90°. Typical nozzle sizes range from 1/6 inch (1.6 mm) to 1.2 inches (12.7 mm) diameter and are constructed of 316 stainless steel. The function of the probe is to support the nozzle and connect it to the filter housing. The probe length varies from 2 feet (0.61 m) to 12 feet (3.67 m) and is typically made of borosilicate glass. When long lengths are required (>15 feet, 4.6 m), they are made of stainless steel. The probe and the filter assembly is maintained at 393 K [120°C] to prevent moisture condensation. Glass fiber filters are used as the sampling media.

The recommended point of sample extraction should be 10 duct diameters away from the nearest upstream disturbance (bends, dampers, duct outlets) and two duct diameters away from the nearest downstream disturbance. The number of sampling or traverse points ranges from 8 to 24, depending on the stack diameter and the downstream distance from the nearest flow disturbance. The sampling time should be long enough to collect $0.6\,m^3$ at a rate no greater than $0.021\,m^3/min$. Leak tests also need to be performed to ensure that the correct volume of gas is sampled. The particulate sample is recovered from all the components, including the nozzle, probe, cyclone, filter holder, filter, rubber gasket, and impingers, by washing with water or acetone. This extract is then dried and the amount of particulate matter weighed to determine the mass concentration.

The EPA Method 5 uses a flexible connecting hose (in contrast to a rigid connection) linking the probe to the impingers, thus making it simpler to operate. The other difference is that the filter is placed in the stack. Thus, the filter material has to be able to withstand the

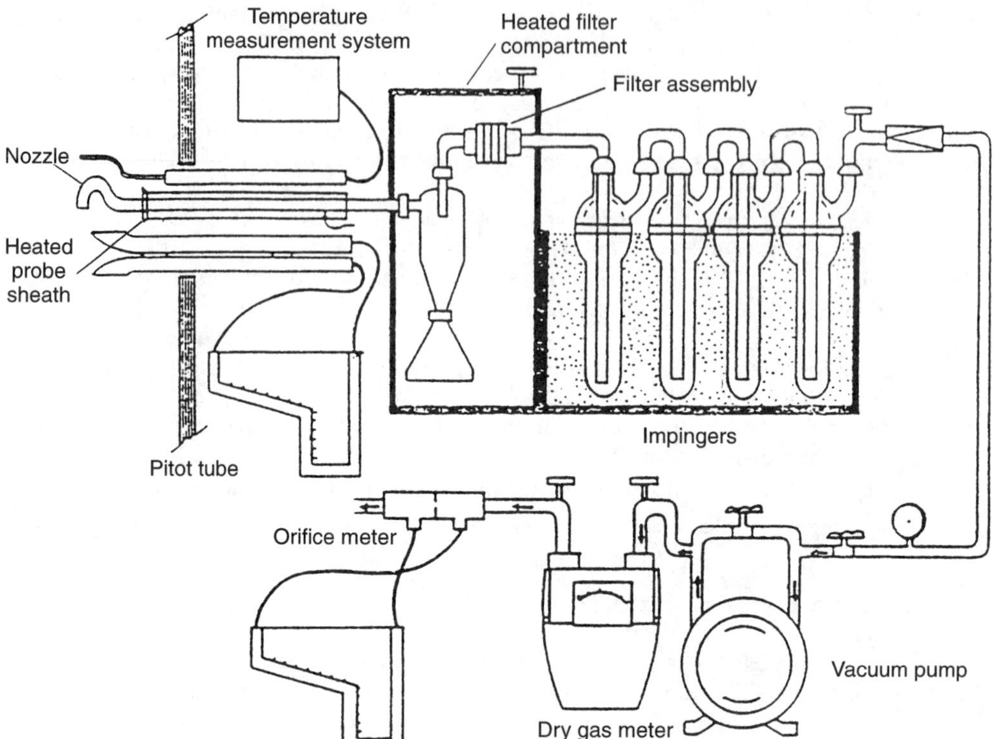

Fig. 31–7. The EPA Method 5 Sampling Train showing the various components. (Adapted from U.S. EPA, 1990.)

local stack temperatures. The measurements taken by EPA Method 5 and 17 may be vastly different, especially if there are condensibles present in the stack gases.

HIGH-TEMPERATURE IMPACTORS

Inertial impactors have been discussed in Chapter 10. To prevent bounce and re-entrainmnet of particles, the collection surfaces are often coated with a thin layer of grease or oil. These coatings are not practical for use in sampling of high-temperature gas streams. Other substrates such as ceramic fibers have been used at temperatures as high as 870 K, with no apparent shift in the 50% cut-off size; however, the size cuts were considerably less sharp when compared to greased plate impaction at room temperature (Parker et al., 1981). Newton et al. (1982) used ungreased stainless steel substrates for high-temperature and high-pressure measurements; however, the impaction velocities were limited to prevent particle bounce. Pilat et al. (1970) designed a source stack cascade impactor for measuring particle size distributions in stacks and ducts of emission sources. By operating the impactor inside the duct, true isokinetic sampling could be achieved with minimum wall losses and condensation problems. However, the problem of particle bounce would remain in these impactors at temperatures greater than 423 K [150°C] as the collection surfaces cannot be coated with a sticky medium. Pilat and Steig (1983) used a version of this impactor with filter media as the collection surface for sampling in a pressurized fluidized bed coal combustor. Although filter media reduce particle bounce in comparison with uncoated collection surfaces, impactor collection efficiency characteristics were found to be significantly different when compared to greased collection surfaces (Rao and Whitby, 1977).

Virtual impactors are a good alternative as the particles are collected off line and greased plates do not need to be used (Schott and Ranz, 1976; Marple and Chien, 1980). A disadvantage of the virtual impactor is that secondary flows need to be controlled precisely at every stage, and a multistage device becomes cumbersome and difficult to operate.

A particle trap impactor was developed by Biswas and Flagan (1988) for use in high-temperature systems. The particles were collected in a cavity instead of having a secondary flow stream as in a virtual impactor. Extensive calibration at room temperatures was carried out, and the device yielded sharp efficiency characteristics. The impactor could also be used under high jet velocity conditions for classification of submicrometer-sized aerosols. The impactor was also demonstrated to be functional up to 770 K [500°C] with sharp cut-off characteristics (Biswas and Flagan, 1988).

Distortions may also occur in the size measurements due to condensation of vapors and other species. The distortion of the particle size distributions in both conventional and low-pressure cascade impactors was determined for aerosols at different humidities by Biswas et al. (1987). Differential impaction of aerosols within the impactor may also cause shifts in the measured distributions as high number concentrations are encountered in these systems. The allowable upper limit in the number concentration for sampling of such aerosols has been determined by Biswas (1988).

IN SITU MEASUREMENTS

In addition to the probe techniques described earlier, a variety of diagnostic techniques have been used to measure particle sizes in flames. Thermophoretic sampling techniques have been used by Megaridis and Dobbins (1989) to deposit particles on a cold probe and then examine them by electron microscopy. Many optical techniques have been used as they offer advantages of in situ, nonintrusive measurements and greater optical resolution. The principles of light scattering for particulate measurements have been outlined in Chapters 15 and 16. Instruments that are commercially available for in situ measurements using optical techniques have been discussed in Chapter 17. Two techniques, elastic light scattering (ELS) and dynamic light scattering (DLS), used recently by various researchers for measurement in high-concentration aerosol systems are discussed here (see Table 31–3).

Elastic Light Scattering

An elastic light scattering process is one in which there is no energy exchange between the incident photons of light and target particles, resulting in no shift in the incident light frequency. These techniques have been used by many researchers in various ways such as combination of scattering and extinction, angular dissymetry, polarization ratio measurements, and two-color techniques. Lights of different wavelengths (two colors) provide a small range of measurements and are costly to implement (Zachariah et al., 1989); they are not discussed in this chapter.

Combination of Scattering and Extinction. Santoro et al. (1983, 1987) and Presser et al. (1990) summarized the measurement procedure for light-absorbing aerosols using a combination of scattering and extinction measurements. The technique was demonstrated by measurements of soot particles formed by combustion of a gaseous fuel in a diffusion flame. The scattered light detection system allowed an absolute determination of the differential scattering cross section per unit volume. Light scattering and extinction measurements were

TABLE 31–3. Light Scattering Techniques for Particle Measurement

Elastic light scattering	
Scattering/extinction	Mean particle size and number concentration for absorbing aerosols. No size limitation
Angular dissymetry	Mean size, number concentration and spread in distribution
Polarization ratio	Measurable particle size greater than 0.1 λ. Good for nonabsorbing aerosols with known refractive index
Nonelastic light scattering	
Dynamic light scattering (also called photon correlation spectroscopy)	No size limitation; sensitivity decreases with particle size. Determines diffusion coefficient; hence requires system temperature to compute particle size. Refractive index of particles not needed

obtained simultaneously at different radial and axial positions. The scattering cross section applicable for the vertically polarized light, $Q_{1,1}$ and the extinction coefficient, k_{ext}, are determined from the measurements.

Using the Rayleigh theory (Chapter 15), expressions for $Q_{1,1}$ and k_{ext} are obtained, and the ratio of the two is proportional to an average particle diameter, $d_{p6,3}$ defined by

$$d_{p6,3} = \frac{\int c_n f(d_p) d_p^6 \mathrm{d}d_p}{\int c_n f(d_p) d_p^3 \mathrm{d}d_p} = \frac{4\, E(m) Q_{1,1}}{\pi^2 F(m) k_{ext}} \tag{31–2}$$

where $f(d_p)$ is the size distribution function, m is the complex refractive index of the particles, $E(m) = -\mathrm{Im}\,[(m^2-1)/(m^2+2)]$, and $F(m) = |(m^2-1)/(m^2+2)|$. If a certain shape of the size distribution is assumed, $d_{p6,3}$ can be related to the volume mean diameter, $d_{p3,0}$. Knowing $d_{p3,0}$, the particle number concentration, c_n, can be determined (Santoro et al., 1983):

$$c_n = \frac{\lambda k_{ext}}{\pi^2 E(m) d_{p3,0}^3} \tag{31–3}$$

Santoro et al. (1983) suggested that, due to high number concentrations encountered in flames, the size distribution can be assumed to be self-preserving. They assumed $(d_{p6,3}/d_{p3,0})^3 = 2$ for the self-preserving size distribution and, using the experimentally measured k_{ext}, determined the number concentration using Eqs. 31–2 and 31–3.

Angular Dissymetry Measurements. The scattering to extinction ratio technique described above is not applicable to the study of nonabsorbing aerosols such as silica. The dissymmetry technique utilizes the angular dependence of the scattered-light intensity for particles of known refractive indices in the Mie regime. The intensity of the scattered light from the particles is a function of the measured angle, the refractive index of the particle, and particle size in the Mie regime. The well-established relation between particle size and the scattering intensity is described by van de Hulst (1981).

A description of the angular dissymmetry technique to determine the mean particle size and number concentration is provided by Zachariah et al. (1989). The vertically polarized scattering intensity, $I_{1,1}$ can be measured and is given by

$$I_{1,1}(\alpha,\theta_1) = C_{1,1}(\alpha,m,\theta_1) c_n S(\theta) \tag{31–4}$$

where $\alpha = \pi d_p/\lambda$, λ is the wavelength of the incident light beam, $C_{1,1}$ is the Mie scattering cross section for spherical particles and the subscript (1,1) denotes state of vertical polarization, c_n

is the number concentration, and S is the optical system parameter. Zachariah et al. (1989) assumed the aerosol size to be described by a mean size (monodisperse) and determined the dissymmetry ratio from their measurements:

$$\frac{I_{1,1}(\alpha,\theta_1)}{I_{1,1}(\alpha,\theta_2)} = \frac{C_{1,1}(\alpha,m,\theta_1)}{C_{1,1}(\alpha,m,\theta_2)} \tag{31–5}$$

With the Mie scattering cross section (Dave 1968), the above equation can be solved to determine α and, therefore, the mean particle size. Using the value of the mean particle size thus determined, Eq. 31–4 can be used to determine the number concentration, c_n, with $S(\theta)$ determined from calibration experiments. Thus, intensity measurements at two angles could provide information on the mean particle size and number concentration.

The angular dissymmetry technique has been extended by Chang and Biswas (1992) to determine the spread of the particle size distributions. The dissymmetry ratio for a size distribution of particles is given by

$$\frac{I_{1,1}(\alpha,\theta_1)}{I_{1,1}(\alpha,\theta_2)} = \frac{\int C_{1,1}(\alpha,m,\theta_1)f(d_p)dd_p}{\int C_{1,1}(\alpha,m,\theta_2)f(d_p)dd_p} \tag{31–6}$$

As the above equation has an unknown $f(d_p)$, the size distribution function, it cannot be solved directly as the monodisperse case described earlier. Chang and Biswas (1992) represented a flame-generated silica aerosol with a lognormal distribution, and Eq. 31–6 can then be written with two unknowns, the geometric mean size and the geometric standard deviation. Thus, two dissymmetry ratios were determined by scattering-intensity measurements at three angles and then used to determine the geometric mean diameter and the geometric standard deviation. Chang and Biswas (1995) further developed a statistical regression procedure for determining the best-fit parameters of the size distribution function using measurements at a number of angles. They also applied this technique to the measurement of multicomponent aerosols where the refractive index is unknown and is used as a fitting parameter.

Polarization Ratio Measurements. The ratio of the intensity of the parallel to perpendicular component of the scattered light for unpolarized incident light is termed the *polarization ratio*. Kerker and La Mer (1950) suggested that the polarization ratio at any angle could be used to determine the particle size. This technique has been used by several researchers and has been reviewed by Bonczyk (1979). The polarization ratio ($I_{1,1}/I_{1,2}$) is a function of the refractive index, the size distribution of the scatterer, and the angle of the measurement (Bonczyk and Sangiovanni, 1984):

$$\frac{I_{1,1}}{I_{1,2}} = \frac{\int C_{1,1}(\alpha,m,\theta)f(d_p)dd_p}{\int C_{1,1}(\alpha,m,\theta)f(d_p)dd_p} \tag{31–7}$$

where $I_{1,1}$ is the vertically polarized scattered-light intensity, $I_{1,2}$ is the horizontally polarized scattered-light intensity (see Chapter 15), and C is the Mie scattering cross section with subscripts (1,1) and (1,2) denoting states of vertical and horizontal polarization, respectively. If the aerosol is assumed to be monodisperse, the mean particle size can be determined from Eq. 31–7 using the expressions for the scattering cross section (Dave, 1968) and the measured polarization ratio.

The advantage of the technique is that the ratio is independent of geometrical and system parameters. The disadvantage is that the polarization ratio measurements may not be very reliable for high-concentration aerosols due to the depolarization effects resulting from agglomerated nonspherical scatterers (Maron and Elder, 1963).

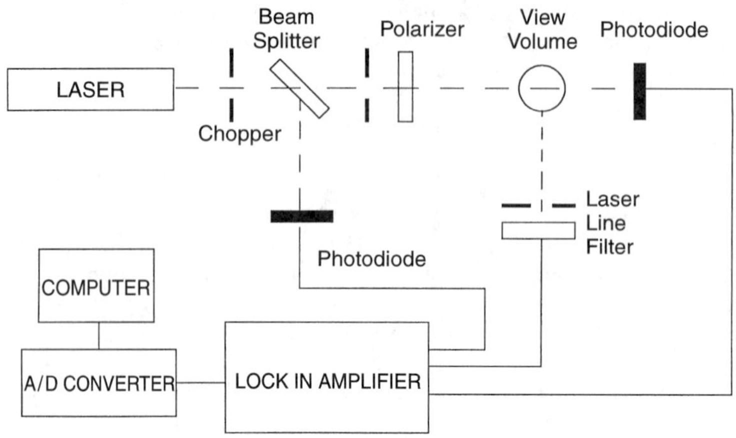

Fig. 31–8. An optical system for light scattering and extinction measurements. The detection system is rotated for measurement at different angles.

Elastic Light-Scattering Measurement System. This section is concluded with a description of the components of an optical system used to carry out light-scattering and transmission measurements. A schematic diagram of such a system is shown in Figure 31–8. It consists typically of a laser light source with a mechanically chopped beam divided into two parts by a beam splitter. One beam is directed through the region of the flame for the light-scattering and extinction measurements. The other beam is directed to a photodiode that provides a reference signal to a lock-in amplifier, thus distinguishing the scattered light intensity signal from the unchopped background noise. If the polarization ratio is to be measured, an unpolarized light source is used. The scattered signal from the probe volume is measured using a photomultiplier tube (PMT) connected to a lock-in amplifier for signal conditioning. Two slits on the PMT allow a very small scattering volume to be defined, thus enhancing spatial resolution. The laser line filter centered at the light source wavelength with a $\Delta\lambda$ of around 1 nm is also mounted on the PMT to reject the extraneous signals from flame emissions. A polarizer is used before the PMT to ensure that the scattered light corresponds to the incident beam if the source is also polarized. The PMT is typically installed on a rotator so that it can measure the scattered-light intensities at various angles. The optical system or the burner can be installed on a moveable system that can be traversed for spatial measurements. The output signal from the lock-in amplifier is digitized with an A/D converter and sent to a microcomputer for data storage and further processing.

Dynamic Light Scattering

Dynamic Light Scattering (DLS), also known as *photon correlation spectroscopy*, has been used in the measurement of soot formation (Flower 1983a). Monochromatic light incident on particles in a defined probe volume results in a fluctuating scattering intensity due to the Brownian motion of the particles. These fluctuations are correlated to the mean speed or the diffusion coefficient of the particles, both of which are related to the particle diameter. DLS is discussed in Chapter 15. The procedure to obtain the particle size is discussed herewith (Flower and Hurd, 1987; Zachariah et al., 1989).

Flower (1983a,b) used DLS techniques for the measurement of soot particles in flames. The advantage of the DLS technique is that the measurement is independent of the particle refractive index. However, the technique provides a value for the particle diffusion coefficient, which is used to compute the particle size. If the mean motion of the particles is small, the power spectra $P(\omega)$ of the scattered light exhibits a Lorentzian function (Cummins and Swinney, 1970)

$$P(\omega) = \frac{I^2 q^2 D/\pi}{(q^2 D)^2 + \omega^2} \tag{31–8}$$

where I is the intensity of the scattered light, ω is the angular frequency, q is the magnitude of the scattering wave vector [$= 4\pi\sin(\theta/2)/\lambda_o$], λ_o is the wavelength of the incident light, D is the diffusion coefficient, and θ is the scattering angle measured in the forward direction. When the scattered light is focused on the PMT, self beating of the scattered light results in photocurrent fluctuations. From the above equation, the frequency at half-width is given by

$$\Delta f = \frac{\Delta\omega}{2\pi} = \frac{q^2 D}{2\pi} \tag{31–9}$$

The frequency at half-width, Δf, can be determined from the spectral plot described above, and the diffusion coefficient can be computed using Eq. 31–9. Knowing the relationship of D to particle size ($D = kTC_c/3\pi\,\eta\, d_p$), the particle size, d_p, can be determined. However, an accurate estimate of the temperature and the properties of the medium are necessary.

Nonspherical Scatterers

Due to high number concentrations encountered in combustion environments, particle agglomerates may be formed. Light scattering has been used for examining such aerosol agglomerates. However, the relation of the scattered intensity to particle size from even a small agglomerate is rather complicated. The fractal dimension can characterize statistically similar but geometrically irregular objects. Light scattering has been used to determine the fractal dimension of clusters in liquid media (Schaefer and Martin, 1984) and in gas media (Martin et al., 1986; Gangopadhyay et al., 1991; Yang and Biswas, 1997). Details are provided in Chapter 23, and only a brief description is provided here.

Mountain and Mulholland (1988) established light-scattering relationships from simulated smoke agglomerates. The measured scattering intensity and its angular dependence were related to the fractal dimension, the radius of gyration of the agglomerate, and the total number of primary particles in the agglomerate. They also reported the resulting error in the volume mean diameter between an effective sphere model and their proposed agglomerate model. The error increased with the number of primary particles in the agglomerate, with a 26% difference being obtained for a cluster with 119 primary particles and an 8% difference for a cluster with 32 primary particles.

Hurd and Flower (1988) studied flame-generated silica particle aggregates by in situ dynamic and static light scattering and reported fractal dimensions. A morphological description of flame-generated fractal particles is provided by Megaridis and Dobbins (1990), and the absorption and scattering of light by polydisperse aggregates is discussed by Dobbins and Megaridis (1991).

Assuming Rayleigh-Debye scattering (Kerker, 1969), the differential scattering coefficient, $Q_{1,1}$, for an aggregate of particles can be expressed as

$$Q_{1,1}(\theta) = c_N C_{1,1,\mathrm{Rayleigh}} \sum_{j=1}^{N_p} e^{j\delta_j} = c_N C_{1,1,\mathrm{Rayleigh}} N_p^2 S(q) \tag{31–10}$$

where c_N is the number concentration of agglomerates, N_p is the number of primary particles in an agglomerate, and $C_{1,1,\mathrm{Rayleigh}}$ is the Rayleigh scattering cross section of a primary particle in an agglomerate. For a single agglomerate, the structure factor, $S(q)$ is defined as (Berne and Pecora, 1976)

$$S(q) = \frac{1}{N_p^2} \sum_{j=1}^{N_p} e^{i\delta_j} \tag{31–11}$$

where δ is the phase lag of the scattered beam caused by the interference between two distinct primary particles in an agglomerate. Expanding the structure factor for small values of q leads to the relation

$$S(q) = 1 - \frac{R_G^2 q^2}{3} + \ldots \tag{31–12}$$

where q is $[= 4\pi \sin(\theta/2)/\lambda_o]$; R_G is the radius of gyration and is defined as

$$R_G = \sqrt{\frac{\sum_{j=1}^{N_p} r_j^2}{N_p}} \tag{31–13}$$

where r_j is the distance from the center of mass of the agglomerate to the jth primary particle. For small values of q (so that $R_G q \ll 1$), Eq. 31–11 can be rewritten as

$$S(q) = 1 - \frac{R_G^2 q^2}{3} \tag{31–14}$$

For larger values of q ($R_G q > 1$), the behavior of $S(q)$ becomes (Mountain and Mullholland, 1988)

$$S(q) \propto q^{-D_f} \tag{31–15}$$

D_f is the fractal dimension and defines the mass distribution in an agglomerate by the relation

$$N_p = k_0 \left(\frac{R_G}{a} \right)^{D_f} \tag{31–16}$$

where a is the radius of a primary particle and k_0 is a constant, different values of which have been established both experimentally and computationally (details are provided in Chapter 23). For a fixed position, the intensity $Q_{1,1}(\theta)$ can be measured at different angles (or q). As $c_N\ C_{1,1,\text{Rayleigh}}\ N_p^2$ is a constant at a fixed location, the intensity measurement is proportional to $S(q)$. With Eq. 31–15, a plot of $S(q)$ with respect to q on a logarithmic scale yields a slope that is equal to $-D_f$, the fractal dimension. At small q, $S(q)$ has a parabolic dependence on q (Eq. 31–14), and fitting the data yields a value of the radius of gyration. If the primary particle size is known (determined from microscopy and assumed to be monodisperse), Eq. 31–16 can be used to determine N_p, the number of primary particles in an aggregate. These parameters can then be substituted in Eq. 31–10 to determine c_N, the number concentration of the agglomerate particles.

Ex situ small-angle X-ray scattering and small-angle neutron-scattering measurements have been used to elucidate the morphological characteristics of fumed aggregates (Schaefer et al., 1991, 1994). In situ light-scattering measurements have also been extensively used to establish the fractal dimensions of flame-generated particles (Hurd and Flower, 1988; Gangopadhyay et al., 1991; Yang and Biswas, 1997). A schematic diagram of a system to perform multiangle measurements is shown in Figure 31–9. Yang and Biswas (1997) have used in situ light-scattering measurements in conjunction with TEM observations to study the growth, aggregation, and evolution of the aggregate structure of titania particles in flame

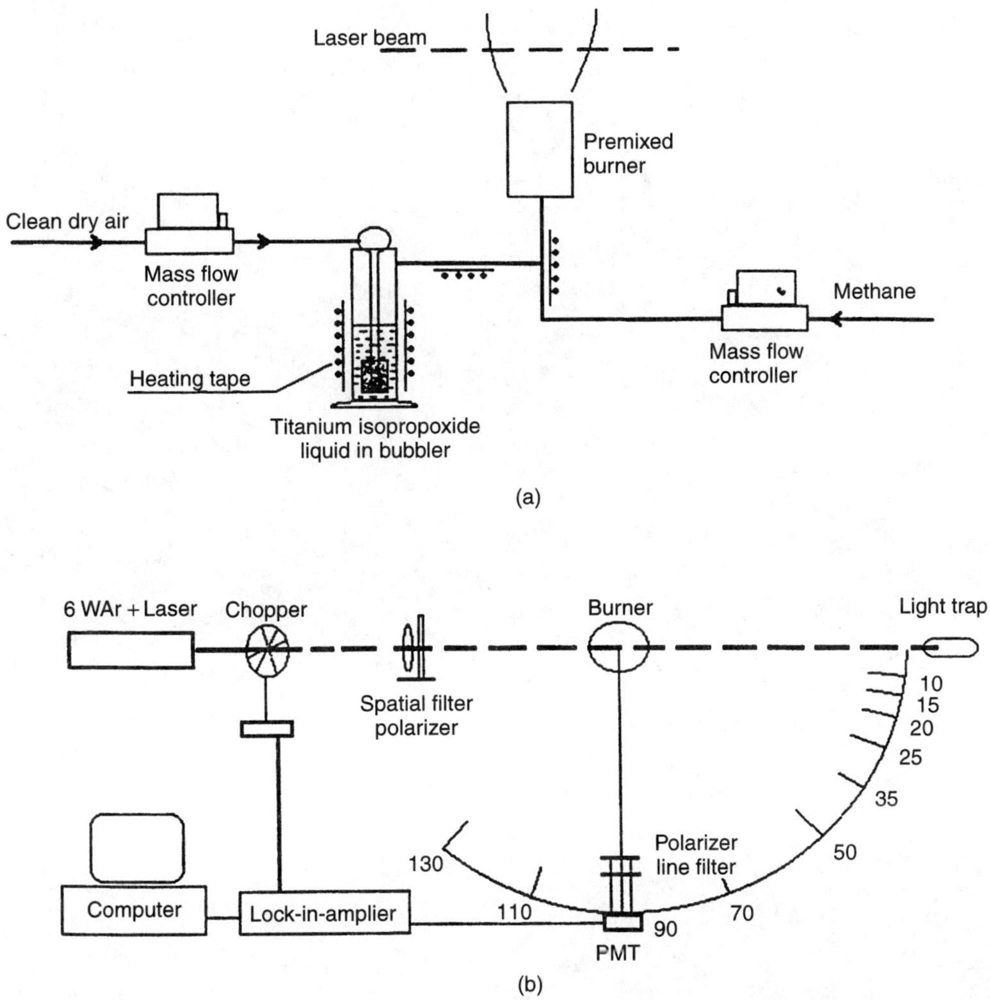

Fig. 31–9. **a**, Schematic diagram of the premixed flame aerosol reactor system. **b**, Schematic illustrating multiangle light-scattering measurement system.

reactors. Using their experimental light-scattering measurements, they determined (using Eq. 31–15) the fractal dimension of the aggregates as a function of position along the burner axis. The fractal dimension of these small aggregates was found to vary (unlike the scale invariance observed in large agglomerates) and was attributed to sintering-type processes. They also used a numerical simulation (Yang and Biswas 1999) to relate the change in fractal dimension to the change in the normalized surface area. Using this relationship, they were able to establish the characteristic sintering time of the agglomerates. Results of the experiments are illustrated in Figure 31–10.

Laser-Induced Fluorescence

To gain a detailed understanding of the particle formation mechanisms, it is essential to map out the vapor phase precursor concentrations. Moreover, there are several engineered

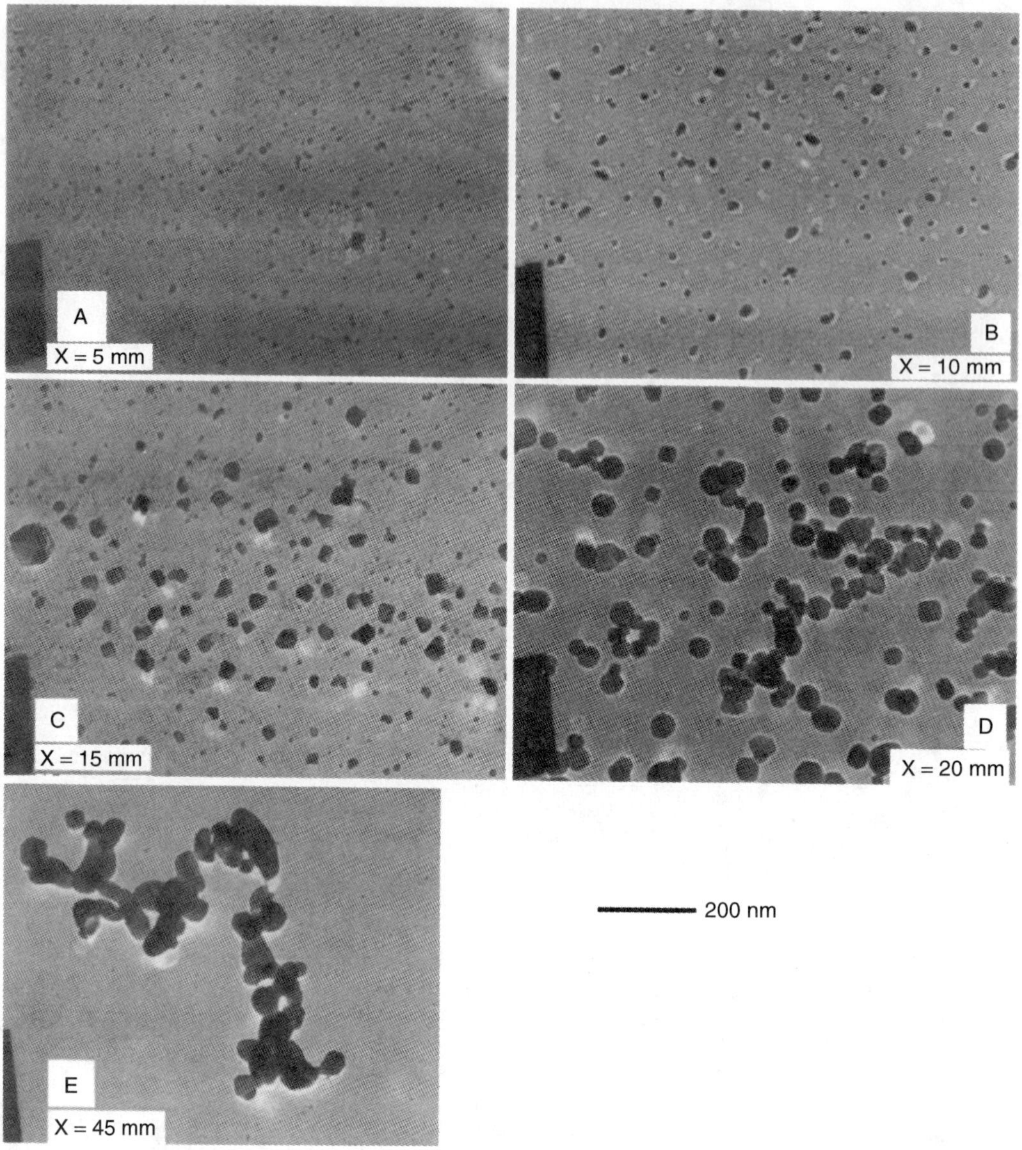

a

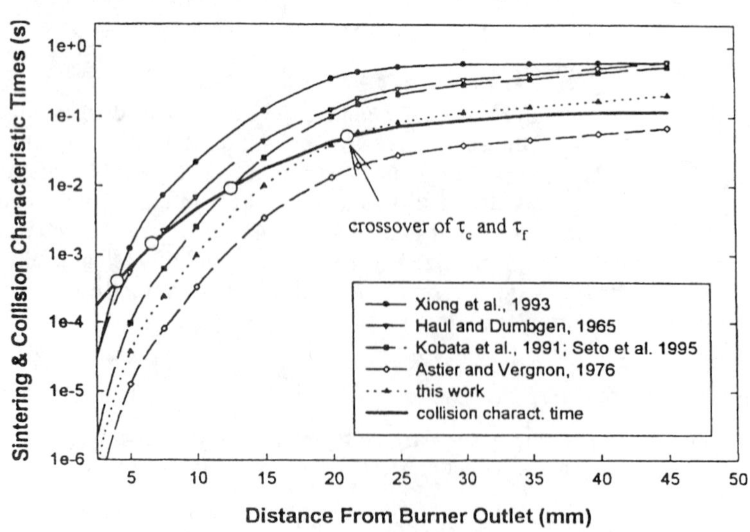

b

Fig. 31–10. **a**, Transmission electron microscopy scans of particles in the premixed flame at different axial distances (X) above the burner outlet. **b**, Characteristic sintering time as a function of the center line axial distance.

sorbent processes that rely on gas to particle conversion for control of fine particle emissions (Biswas and Wu, 1998), and the vapor phase concentration profiles are important at establishing the gas to particle conversion rates. Laser-induced fluorescence is an in situ measurement technique that allows the mapping of the gas phase precursors to particle formation. The theory and application of laser-induced fluorescence spectroscopy has been described in detail elsewhere (Eckbreth, 1988). Planar laser-induced fluorescence (PLIF) is a well-established spectroscopic flow diagnostic that is based on the spontaneous radiative emission (fluorescence) following absorption of laser photons by a specific molecular species. In a typical PLIF measurement, a thin sheet of frequency-tuned pulsed-laser light is used as the excitation source and directed through the flow region of interest (Fig. 31–11). The resulting broadband fluorescence from the illuminated region is imaged with a lens onto an intensified charge-coupled device (CCD) camera. Through various excitation and detection strategies, the pulse integrated fluorescence signal can be related to the absorbing species concentration (McMillin et al., 1996; Biswas and Zachariah, 1997). The absorbing species concentration generally requires knowledge of the local collisional quenching rate coefficient and temperature. One strategy that is used is to assume a constant quenching rate (Zachariah and Burgess, 1994; McMillin et al., 1996) and an absorption transition that is insensitive to temperature (discussed later). With these approximations, the fluorescence signal is a direct representation of the relative vapor phase precursor concentration in the flame.

It is important to know the partial fluorescence excitation–detection scheme and associated energy level diagram, and an example for PbO is shown in Figure 31–12a (Biswas et al., 1998b). Represented schematically is the $A^3\Pi(0^+) \leftarrow X^1\Sigma^+(0^+)(0,3)$ excitation and the ground level vibration states, $v'' = 0, 1, 2, 3, 4, 5$, and 6. The upward-pointing arrow indicates the laser excitation frequency that was tuned to excite transitions from a particular rotational level within the $v'' = 3$ ground state to the $v' = 0$ level in the excited A state. The downward-pointing arrows depict the radiative (fluorescent) or nonradiative (quenched) decay from the excited state (A) to various lower states. The partial excitation frequency spectrum obtained by scanning the laser from 567.5 to 571.5 nm in steps of 0.001 nm is shown in Figure 31–12b. With the spectroscopic constants reported by Oldenberg et al. (1973), the synthetic spectra were calculated for the above-mentioned transition, and rotational line assignments were made (Fig. 31–12b).

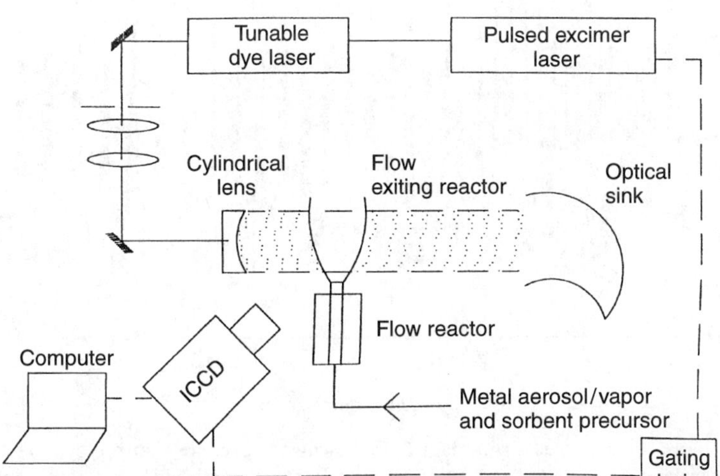

Fig. 31–11. Schematic diagram of the system for performing laser-induced fluorescence measurements.

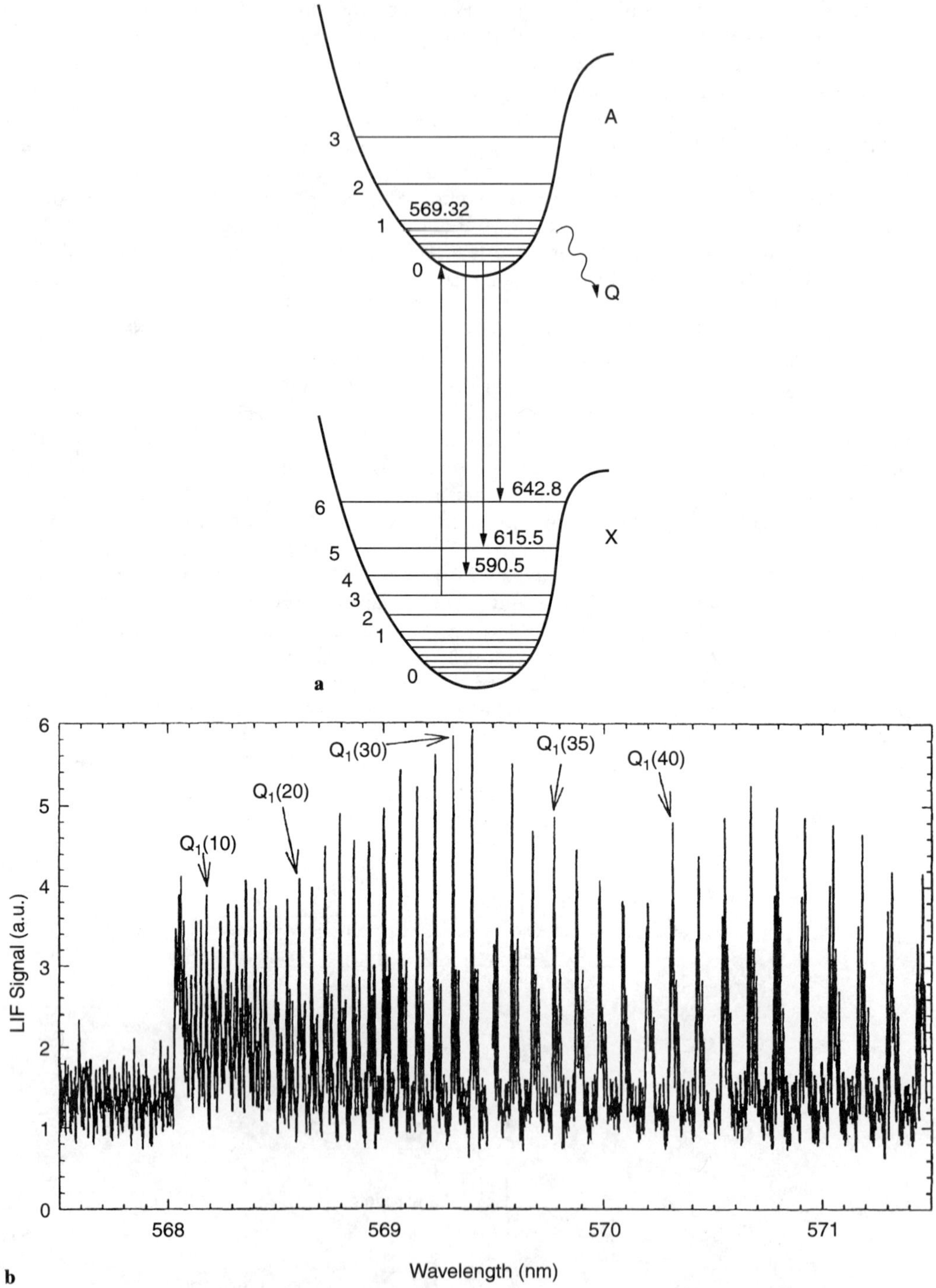

Fig. 31–12. **a**, Partial energy level diagram for PbO showing the excitation and detection scheme. **b**, Laser excitation spectrum of lead oxide between 567.5 and 571.5 nm established with a resolution of 0.001 nm. Also shown are rotational states for a few lines.

While transition probabilities are relatively independent of rotational state, the population in each rotational state is a function of temperature. Hence the absorption and fluorescence are also a function of temperature, this being related to the fractional population (F_B) of the jth rotational state (Banwell, 1983) being excited

$$F_B = \frac{(2j+1)\exp\left\{-\frac{hc}{kT}[T_e + G(v) + F(j)]\right\}}{Q_e Q_n Q_r} \tag{31–17}$$

where h is the Planck's constant, c is the velocity of light, k is the Boltzmann constant, T is the temperature, T_e is the electronic energy level, $G(v)$ is the vibrational energy level, $F(j)$ is the rotational energy level, Q_e (= 1) is the electronic partition function, Q_v (= $1/[1 - \exp(-hc\omega_e/kT)]$) is the vibrational partition function, ω_e is the vibrational frequency, Q_r (= kT/hcB_v) is the rotational partition function, and B_v is a rotational constant. It is important to choose the appropriate frequencies that have a minimal dependence on the temperature (Biswas et al., 1998b).

CHARACTERIZATION OF COMBUSTION AEROSOLS

Several characterization techniques can be used in the analysis of combustion-generated particulate matter. These include electron microscopy and X-ray diffraction, which are routinely used and are discussed in Chapter 12. Two additional spectroscopic techniques—Mössbauer and Raman—are briefly discussed here.

Raman Spectroscopy

Due to quantum absorption and re-emission, light can be scattered at different wavelengths from the incident beam; this is known as *inelastic scattering*. Raman scattering is an inelastic scattering process that provides information on the different vibrational states of the molecule. The intensities of the Raman scattering signal are orders of magnitude smaller than elastic scattering signals, and limited number of in situ measurements have been made, especially in combustion environments. However, ex situ Raman measurements have been extensively used for the chemical analysis of particles.

Biswas et al. (1998a) have used Raman scattering to characterize the optical phonons and crystallinity of flame-processed titania powders and films. The E_g mode in nanophase titania (20 nm) was broadened and blue shifted, similar to trends observed by Gonzalez and Zallen (1997). The mode frequency shift is due to a relaxation of the $k = 0$ selection rule, leading to blue or red shifts depending on the $w(k)$ dispersion of phonon modes in the Brillouin zone near $k = 0$. These shifts can be quantified and related to particle size. Wang et al. (2000) have used this to study iron-doped titania samples and showed elegantly the decrease in grain size with increasing iron dopant concentration. Thus, in addition to the chemical analysis of the combustion-generated particles, Raman spectroscopy can be used to establish the grain size of particles and other characteristics of particles such as oxygen deficiency (Parker and Siegel, 1990).

Mössbauer Spectroscopy

Mössbauer spectroscopy provides microscopic information on the chemical environment of the emitting or absorbing nuclei of particles. It involves the measurement of resonant absorption occurring in a solid as a result of the resonant emission taking place from another solid. γ-Rays are emitted (or absorbed) by energy level transitions in the nucleus without recoil. The γ-ray energies associated with resonant emission and absorption are sensitive to changes in the solid-state environment via hyperfine interactions. When the resonant absorption is analyzed by monitoring the number of γ-rays passing through the absorber, the measurement is said to be carried out in the transmission mode. One obtains a spectrum of transmitted rays versus velocity (emitted by the source and passed through the absorber) as a function of the velocity of the source relative to the absorber. This has been extensively used to study the characteristics of aerosol reactor synthesized superconducting powders and to establish the role of oxygen deficiency (Boolchand et al., 1990). The Mössbauer effect has been successfully used to obtain the kinetics of formation of the γ-phase iron oxide during aerosol synthesis. Not only was the Mössbauer effect used as an atomic scale probe of the iron rich phases, but could also be readily used to discriminate small (nano) sized grains of γ-Fe_2O_3 grains from the "large-sized" ones (Lin et al., 1996).

REFERENCES

Alam, M. K. 1984. *Nucleation and Condensational Growth of Aerosols: Application to Silicon Production*. Ph.D. thesis, California Institute of Technology, Pasadena, CA.

Alam, M. K. and R. C. Flagan. 1986. Controlled nucleation aerosol reactors. Production of bulk silicon. *Aerosol Sci. Technol.* 5:237–248.

Banwell, C. N. 1983. *Introduction to Molecular Spectroscopy*. New York: McGraw Hill.

Berne, B. R. and R. Pecora. 1976. *Dynamics of Light Scattering*. New York: John Wiley & Sons.

Biswas, P. 1985. *Impactors for Aerosol Measurements: Developments and Sampling Biases*. Ph.D. thesis, California Institute of Technology, Pasadena, CA.

Biswas, P. 1988. Differential impaction of aerosols. *J. Aerosol Sci.* 19:603–610.

Biswas, P. and R. C. Flagan. 1988. The particle trap impactor. *J. Aerosol Sci.* 19:113–121.

Biswas, P., C. L. Jones, and R. C. Flagan. 1987. Distortion of size distributions by condensation and evaporation in aerosol instruments. *Aerosol Sci. Technol.* 7:231–246.

Biswas, P., X. Li, and S. E. Pratsinis. 1989. Optical waveguide preform fabrication: Silica formation and growth in a high temperature aerosol reactor. *J. Appl. Phys.* 65:2445–2450.

Biswas, P. and C. Y. Wu. 1998. Control of toxic metal emission from combustors using sorbents: A review. *J. Air Waste Manage.* 48:113–127.

Biswas, P., G. Yang, W. Bresser, and P. Boolchand. 1998a. An aerosol process for depositing multifunctional titania films. *Adv. Technol. Part. Process*. New York: AICHE Publications. vol. I, pp. 239–245.

Biswas, P., G. Yang, and M. R. Zachariah. 1998b. In situ processing of ferroelectric materials from lead waste streams by injection of gas phase titanium precursors: Laser induced fluorescence and X-ray diffraction measurements. *Combust. Sci. Technol.* 134:183–200.

Biswas, P. and M. R. Zachariah. 1997. In situ immobilization of lead species in combustion environments by injection of gas phase silica sorbent precursors. *Environ. Sci. Technol.* 31:2455–2463.

Bonczyk, P. A. 1979. Measurement of particulate size by in situ laser-optical methods: A critical evaluation applied to fuel-pyrolyzed carbon. *Combust. Flame* 35:191–206.

Bonczyk, P. A. and J. J. Sangiovanni. 1984. Optical and probe measurements of soot in a burning fuel droplet stream. *Combust. Sci. Technol.* 36:135–147.

Bonfanti, L. and M. Cioni. 1986. Sampling of polynuclear aromatic hydrocarbons at stack with a dilution sampler. In *Aerosols: Formation and Reactivity*. 2nd International Aerosol Conference, Berlin, pp. 952–955.

Boolchand, P., C. Blue, K. Eglaid, W. Huff, A. Kilinic, D. McDaniel, P. Biswas, D. Zhou, and J. Oostens. 1990. Signature of Twin Planes in Mossbauer Spectroscopy of Cuprate Superconductors", *Hyperfine Interactions*, 62: 73–88.

Chang H. and P. Biswas. 1992. In situ light scattering dissymmetry measurements of the evolution of the aerosol size distribution in flames. *J. Colloid Interface Sci.* 153:157–166.

Chang, H. and P. Biswas. 1995. Multiangle light scattering dissymmetry measurements to determine evolution of silica particle size distributions in flames. *Aerosol Sci. Technol.* 22:24–32.

Crump, J., R. C. Flagan, and J. H. Seinfeld. 1983. Particle wall loss rates in vessels. *Aerosol Sci. Technol.* 2:303–309.

Cummins, H. Z. and H. L. Swinney. 1970. Light beating spectroscopy. *Prog. Opt.* 8:135–138.

Daisey, J. M., J. L. Cheney, and P. J. Lioy. 1986. Profiles of organic particulate emissions from air pollution sources: Status and needs for receptor source apportionment models. *JAPCA* 36:17–33.

Dave, J. V. 1968. *Subroutine for Computing the Parameters of the Electromagnetic Radiation Scattered by a Sphere*. IBM Scientific Center Report No. 320–3237. New York: IBM Watson Research Center.

Dobbins, R. A. and C. M. Megaridis. 1991. Absorption and scattering of light by polydisperse aggregates. *Appl. Opt.* 30:4747–4754.

Du, C. J. and D. B. Kittleson. 1984. In-cylinder measurements of diesel particle size distributions. In *Aerosols* eds. B. Y. H. Liu, D. Y. H. Pui, and H. Fissan. New York: Elserier. 744–748.

Eckbreth A. C. 1988. *Laser Diagnostics for Combustion Temperature and Species*. Cambridge, MA: Abacus Press.

Flower, W. L. 1983a. Optical measurements of soot formation in premixed flames. *Combust. Sci. Technol.* 33:17–33.

Flower, W. L. 1983b. Measurements of the diffusion coefficient for soot particles in flames. *Phys. Rev. Lett.* 51:2287–2290.

Flower, W. L. and A. J. Hurd. 1987. In situ measurement of flame formed silica particles using dynamic light scattering. *Appl. Opt.* 26:2236–2239.

Gallagher, N. B., L. E. Bool, J. O. L. Wendt, and T. W. Peterson. 1990. Alkali metal partitioning in ash from pulverized coal combustion. *Combust. Sci. Technol.* 74:211–221.

Gangopadhyay, S., I. Elminyawi, and C. M. Sorensen. 1991. Optical structure factor measurements of soot particles in a premixed flame. *Appl. Opt.* 30:4859–4864.

Gonzalez, R. J. and R. Zallen. 1997. Optical studies of nanophase titania. In *Amphorous Insulators and Semiconductors*, eds. M. F. Thorpe and M. I. Mitkova. Dordrecht: Kluwer Academic Publishers, pp. 395–403.

Hildemann, L. M., G. R. Cass, and G. R. Markowski. 1989. A dilution stack sampler for collection of organic aerosol emissions: Design, characterization and field tests. *Aerosol Sci. Technol.* 10:193–204.

Hinds, W. C. 1999. *Aerosol Technology*. New York: John Wiley & Sons.

Houck, J. E., J. A. Cooper, and E. R. Larson. 1982. *Dilution Sampling for Chemical Receptor Source Finger-Printing*. Paper read at the 75th Annual Meeting of the Air Pollution Control Association, New Orleans, LA, 82-61M.2.

Hurd, A. J. and W. L. Flower. 1988. In situ growth and structure of fractal silica aggregates in aflame. *J. Colloid Interface Sci.* 122:178–192.

Huynh, C. K., T. Vu Duc, C. Schwab, and H. Rollier. 1984. In-stack dilution technique for the sampling of polycyclic organic compounds. Application to effluents of a domestic waste incineration plant. *Atmos. Environ.* 18:255–259.

Kerker, M. 1969. *The Scattering of Light and Other Electromagnetic Radiation*. New York: Academic Press.

Kerker, M. and V. K. La Mer. 1950. Particle size distribution in sulfur hydrosols. *J. Am. Chem. Soc.* 72: 3516–3525.

Lin, S. Y., J. Ferg, P. Biswas, R. Enzweiler, and P. Boolchand. 1996. Characterization of maghemite ferric oxide crystals processed by an aerosol technique. *J. Magn. Magn. Mater.* 159:147–158.

Linak, W. P. and T. W. Peterson. 1984. Effect of coal type and residence time on the submicron aerosol distribution from pulverized coal combustion. *Aerosol Sci. Technol.* 3:77–96.

Maron, S. H. and M. E. Elder. 1963. Determination of latex particle size by light scattering. *J. Colloid Sci.* 18:107–118.

Marple, V. A. and C. M. Chien. 1980. Virtual impactors: A theoretical study. *Environ. Sci. Technol.* 14: 796–805.

Martin, I. E., D. W. Schaefer, and A. I. Hurd. 1986. Fractal geometry of vapor-phase aggregates. *Phys. Rev. A* 33:3540–4353.

McMillin, B., P. Biswas, and M. R. Zachariah. 1996. In situ characterization of vapor phase growth of iron oxide-silica nanocomposites, Part I. 2-D planar laser induced fluorescence and imaging *J. Mater. Res.* 11:1552–1561.

Megaridis, C. M. and R. A. Dobbins. 1989. Comparison of soot growth and oxidation in smoking and non-smoking ethylene diffusion flames. *Combust. Sci. Technol.* 66:1–16.

Megaridis, C. M. and R. A. Dobbins. 1990. Morphological description of flame generated materials. *Combust. Sci. Technol.* 71:95–109.

Mountain, R. D. and G. W. Mullholland. 1988. Light scattering from simulated smoke agglomerates. *Langmuir* 4:1321–1326.

Newton, G. I., R. L. Carpenter, Y. S. Cheng, E. B. Barr, and H. C. Yeh. 1982. High temperature–high pressure: cascade impactor design, performance and data analysis methods. *J. Colloid Interface Sci.* 87: 279–290.

Newton, G. I., R. L. Carpenter, H. C. Yeh, and E. R. Peele. 1980. Respirable aerosols from fluidized bed coal combustion. 1. Sampling methodology for an 18-inch experimental fluidized bed coal combustor. *Environ. Sci. Technol.* 14:849–853.

Oldenborg R. C., C. R. Dickson, and R. N. Zare. 1973. Oldenborg et al. Title: Spectroscopic Bands of PbO in the 450–850 nm range. *J. Mol. Spectrosc.* 58:283.

Parker, R., S. Calvert, D. Drehmel, and I. Abbott. 1981. Inertial impaction of fine particles at high temperature and high pressure. *J. Aerosol Sci.* 12:297–306.

Parker and Siegel. 1990. Calibration of the Raman spectrum to oxygen stoichiometry. *Appl. Phys. Lett.* 57:943–945.

Pedersen, P. S., I. Ingwersen, T. Nielsen, and E. Larsen. 1980. Effects of fuel, lubricant, and engine operating parameters on the emission of polycyclic aromatic hydrocarbons. *Environ. Sci. Technol.* 14: 71–79.

Pilat, M. I., D. S. Ensor, and I. C. Bosch. 1970. Source test cascade impactor. *Atmos. Environ.* 4:671–679.

Pilat, M. I. and T. W. Steig. 1983. Size distribution of particulate emissions from a pressurized fluidized bed coal combustion facility. *Atmos. Environ.* 17:2429–2433.

Pratsinis, S. E., H. Bai, P. Biswas, M. Frenklach, and S. V. R. Mastrangelo. 1990. Kinetics of titanium (IV) chloride oxidation. *J. Am. Ceram. Soc.* 73:2158–2162.

Presser, C., A. K. Gupta, H. G. Semerjian, and R. I. Santoro. 1990. Application of laser diagnostic techniques for the examination of liquid fuel spray structure. *Chem. Eng. Commun.* 90:75–102.

Rao, A. K. and K. T. Whitby. 1977. Nonideal collection characteristics of single stage and cascade impactors. *Am. Ind. Hyg. Assoc. J.* 38:174–179.

Santoro, R. I., H. G. Semerjian, and R. A. Dobbins. 1983. Soot particle measurements in diffusion flames. *Combust. Flame* 51:203–218.

Santoro, R. I., T. T. Yeh, I. I. Horvath, and H. G. Semerjian. 1987. The transport and growth of soot particles in laminar diffusion flames. *Combust. Sci. Technol.* 53:89–115.

Schaefer, D. W. and I. E. Martin. 1984. Fractal geometry of colloidal aggregates. *Phys. Rev. Lett.* 52(26): 2371–2374.

Schaefer, D. W., B. J. Oliver, A. J. Hurd, G. Beaucage, J. J. Ivie, and C. R. Herd. 1991. Structure of combustion aerosols. *J. Aerosol Sci.* 22:S447–S450.

Schaefer, D. W., B. J. Oliver, C. Ashley, G. Beaucage, D. Richter, B. Fargo, B. Frick, and A. Fischer. 1994. Structure and Topology of Silica Aerogels During Densification. *J. Non-Cryst. Solids* 172–174:647–655.

Schott, I. H. and W. E. Ranz. 1976. Jet-cone impactors as aerosol separators. *Environ. Sci. Technol.* 10: 1250–1256.

Scotto, M. V., E. A. Bassham, J. O. L. Wendt, and T. W. Peterson. 1988. Quench induced nucleation of ash constituents during combustion of pulverized coal in a laboratory furnace. In *Proceedings of 22nd International Symposium on Combustion*, pp. 239–247. The Combustion Institute.

Sethi, V. and P. Biswas. 1990. Fundamental studies on particulate emissions form hazardous waste incinerators. In *Proceedings of the 16th Annual RREL Hazardous Waste Research Symposium*. EPA/600/9-90073, pp. 59–67. Cincinnati: USEPA.

Sousa, I. A., J. E. Houck, I. A. Cooper, and I. M. Daisey. 1987. The mutagenic activity of particulate organic matter collected with a dilution sampler at coal fired power plants. *JAPCA* 37:1439–1444.

Stamatakis, P., C. A. Natalie, B. R. Palmer, and W. A. Yuill. 1991. Research needs in aerosol processing. *Aerosol Sci. Technol.* 14:316–321.

Ulrich, G. D. and J. W. Riehl. 1982. Aggregation and growth of submicron oxide particles in flames. *J. Colloid Interface Sci.* 87:257–265.

U.S. EPA. 1990. *Code of Federal Register*, CFR 42.9.

Van de Hulst, H. C. 1981. *Light Scattering by Small Particles*. New York: Dover.

Wang, Z. M., G. Yang, P. Biswas, W. Bresser, and P. Boolchand. 2001. Processing of iron-doped titania powders in flame aerosol reactors. *Powder Technol.* 114:197–204.

Wu, J. J. and R. C. Flagan. 1987. Onset of run away nucleation in aerosol reactors. *J. Appl. Phys.* 61: 1365–1371.

Yang, G. and P. Biswas. 1997. Study of the sintering of nanosized titania agglomerates in flames using in situ light scattering measurements. *Aerosol Sci. Technol.* 27:507–521.

Yang, G. and P. Biswas. 1999. Computer simulation of the aggregation and sintering restructuring of fractal-like clusters containing limited number of primary particles. *J. Colloid Interface Sci.* 211: 142–150.

Zachariah, M. R. and D. R. F. Burgess. 1994. Strategies for laser-excited fluorescence spectroscopy: Measurements of gas-phase species during particle formation. *J. Aerosol Sci.* 25:487–497.

Zachariah, M. R., D. Chin, H. G. Semerjian, and J. L. Katz. 1989. Dynamic light scattering and angular dissymmetry for the in situ measurement of silicon dioxide particle synthesis in flames. *Appl. Opt.* 28:530–536.

Zimmer, A. 2000. *Aerosol Formation Mechanisms, Metallurgical Aspects and Engineering Control of Fumes Generated During Arc Welding Operations*. Ph.D. thesis, Aerosol and Air Quality Research Laboratory, University of Cincinnati, Cincinnati, OH.

Zimmer, A. and P. Biswas. 2001. Characterization of arc welding aerosols as a function of distance from welding zone. *J. Aerosol Sci.* Submitted.

32 Manufacturing of Materials by Aerosol Processes

SOTIRIS E. PRATSINIS and GEORGIOS SKILLAS
Institut für Verfahrenstechnik, CH-8092 ETH Zürich, Switzerland

TOIVO T. KODAS
Superior Powder Technologies, Albuquerque, NM

MATERIALS

Flame aerosol processes are by far the most widely used ones for commercial quantity material manufacture, the most important of which are carbon blacks (e.g., ***CAB***, ***DEG***, and ***COL****), fumed silica (***CAB***, ***DEG***) and pigmentary titania (***DUP***, ***MLN***, ***KEM***, ***KER***, ***ISH***) and optical fibers (***LUC***, ***SIC***, ***SUM***). The annual production volume of the flame industry is several million metric tons (Pratsinis, 1998). Hot-wall aerosol reactors are used on an industrial scale for synthesis of filamentary (nanostructured) nickel and iron powders from decomposition of the corresponding metal carbonyls (***BAS***, ***INC***). In addition these reactors have been used for commercial synthesis of nanostructured carbides, nitrides, borides and other nonoxide ceramics (***DOW***, ***BAY***). Rather recently, the inert gas condensation technique has been scaled up by small (start-up) companies (***NAN***) for manufacture of rather costly (about $100/kg) nanostructured metals and ceramics powders. With respect to droplet-to-particle conversion processes, spray pyrolysis technologies are used primarily by small (start-up) companies (***PTI***, ***NAC***) for manufacture of precious metal, ceramic and especially nanostructured composite ceramic powders from nitrate, organic and other solutions.

Carbon blacks are the oldest manufactured aerosols. Today, they are made primarily in flame reactors. An aromatic ("feedstock") residual oil is sprayed into a hot gas-air flame in a furnace where carbon black is formed as soot. The flue gas is rapidly quenched and the carbon black is collected by electrostatic precipitators, cyclones, and baghouse filters (Medalia and Rivin, 1976). The product carbon black is made of aggregates each containing several elementary spherical particles. Depending on the commercial application, different grades of carbon black are manufactured by empirically controlling the primary particle size and extent of aggregation. Fine particles, for example, can be made at short residence times and high temperatures by adjusting the air flow rate and the length-to-diameter ratio of the furnace

* See Appendix I for full manufacturer addresses referenced to the italicized three-letter codes.

Aerosol Measurement: Principles, Techniques, and Applications, Second Edition, Edited by Paul A. Baron and Klaus Willeke.
ISBN 0-471-35636-0

(Mezey, 1966). The aggregation of particles is usually controlled by addition of traces of potassium salts in the flames (Dannenberg, 1971). Particle formation proceeds by hydrocarbon oxidation in the diffusion flame which is followed by soot particle formation by cyclization reactions at the fuel rich side of the flame (Santoro and Miller, 1987). The soot particles then grow by coagulation and surface reactions (material addition from the gas phase). Measurement problems that arise in this case include determination of particle size and morphology.

The other major application of aerosol reactors is the generation of pigmentary titania and fumed silica powders in flame reactors. These reactors are similar to those used in manufacture of carbon blacks. Precursor vapor ($TiCl_4$ or $SiCl_4$) with small amounts of hydrocarbon fuel are injected with nearly stoichiometric amounts of oxygen into the reactor. Rapid, exothermic oxidation of the vapor takes place forming oxide powders and chlorine gas (Clark, 1977). These powders are loose aggregates of submicrometer particles. The crystalline structure of the particles can be controlled by adding dopant vapors in the feed stream and by tailoring the temperature history of the reactor (Mezey, 1966).

A number of other examples of ceramic powder generation by aerosol routes exist. Spray pyrolysis is used on an industrial scale to produce simple and complex metal oxides (Ruthner, 1983). These ceramic powders are used in a wide variety of applications for manufacture of substrates for catalysts and integrated circuits, structural ceramics and ceramics with specific optical, magnetic and electrical properties. In most of these cases, manufacture of submicrometer powders with precisely controlled physical and chemical properties is of utmost importance in making ceramic parts with the minimal number of imperfections (microcracks and flaws). Currently, ceramic powders are made primarily in spray pyrolysis and flame reactors. Other types of reactors such as furnace, plasma, and laser reactors are used on a much smaller scale. In all of these cases, measurement problems include determination of particle size distributions and morphology, chemical and phase content, and microstructural information. Examples of different types of materials and the determination of their physical and chemical properties are presented in later sections.

Generation of powders using flame processes relies on introducing species into the flame that react to form gaseous species with low vapor pressures (Table 32–1). The reactants that have been used most often are metal chlorides because of their low cost and high volatility. Some processes have relied on using precursors in the aqueous phase. The solution is formed into droplets that are sent into the flame where evaporation takes place to provide gaseous reactants (Zachariah and Huzarewicz, 1991).

Aerosol routes also allow direct production of thick and thin films of a variety of materials via aerosol deposition onto surfaces. A variety of aerosol processes for film formation exist. Films can be formed by either deposition of droplets or by deposition of solid particles onto surfaces. Examples of deposition of droplets onto surfaces include spray pyrolysis (see reviews by Tomar and Garcia, 1981; Mooney and Redding, 1982). This route begins with a solution of molecular precursors which is converted into droplets. These droplets are directed onto a heated surface where the solvent evaporates and the precursors decompose resulting

TABLE 32–1. Powders Made by Flame Reactors

Product	Reactant	Author
Al_2O_3	Al acetylacetonate	Sokolowski et al. (1977)
SiO_2	$SiCl_4$	Ulrich (1984)
TiO_2	$TiCl_4$	George et al. (1973)
Carbon black	Alkanes	Dannenberg (1971)
Al_2O_3, Cr_2O_3, Fe_2O_3, GeO_2, SiO_2, SnO_2, TiO_2, V_2O_5, ZrO_2	Chlorides	Formenti et al. (1972)

in film formation. A related process involves delivery of volatile species into a reactor using an aerosol (Koukitu et al., 1989; Viguie and Spitz, 1975; Siefert, 1984; Salazar et al., 1992). An example of solid particle deposition for film formation is the cluster deposition process (Takagi et al., 1980; Andres et al., 1989). Many other examples also exist including formation of thick films of ceramics (Kodas et al., 1988; Baker et al., 1989; Komiyama et al., 1987; Adachi et al., 1988; Koguchi et al., 1990). However, the most well known and successful example of a solid particle deposition process is optical fiber generation in which the unique capacity of aerosol processes for manufacture of high-purity particles has been exploited (MacChesney et al., 1974).

Optical fibers are made by fabrication of a preform glass rod, rod sintering and fiber drawing and coating (Nagel et al., 1982). The key process with respect to the light transmission and mechanical properties of the fiber is the fabrication of the preform rod. The manufacturing goal of the latter process is to make a preform with a prescribed radial refractive index profile at the maximum yield. Preforms are made by silica and dopant particle deposition onto thin substrate rods (external processes) or into hollow, substrate glass tubes (internal processes).

In the external processes that were invented by Siecor, Inc. (***SIC***), the particles are generated by a flame reactor and deposit primarily by thermophoresis onto the substrate. The flame-generated particles are too large for rapid Brownian diffusion and too small for inertial impaction to be important for particle deposition.

In the internal processes that were invented at Lucent Technologies (***LUC***), halide vapors and oxygen flow through a rotating, hollow, quartz tube that is externally heated by a slowly, axially traversing torch (or plasma). Inside the tube, the halide gases are oxidized forming particles that either deposit to the tube wall by thermophoresis or exit the tube with the process gases. The preform is completed when the quartz tube is almost filled with glassy layers of particles that were fused by the torch. Currently, preforms are made on an industrial scale at rather low yields [50% for silica and much lower for costly dopants such as germania (Bohrer et al., 1985)].

A major challenge in manufacture of optical fibers is to understand the relationship between deposit properties, glass particle size and composition and process conditions. This is important since it has been observed that during sintering of multicomponent particulate deposits, small particles tend to be depleted of dopants much faster than large particles and bubbles can be present in deposits. This relationship is also important for improvement of the current low process yields. As the fiber optics market becomes more competitive, the low yields become a critical issue from both economic and environmental viewpoints (toxic halide vapors can be released when they are not converted to particles). Thus, there is a need for in situ measurements of particle size and composition.

AEROSOL PROCESSES

In general, aerosol processes are attractive for their simplicity since they do not involve large volumes of liquid byproducts and can be used for production of high-purity materials (Pratsinis and Mastrangelo, 1989). Most aerosol measurement techniques have been developed for aerosols in environmental systems. Industrial aerosols, however, are quite different from their environmental counterparts. Typically, high concentrations (10^{14} to 10^{20} particles/m^3) of irregular, fine particles suspended in a variety of gases are encountered in industrial processes. These aerosols are made at high temperatures (1300 K) and massive production rates (10 tons/h) in closed systems and experience a spectrum of physicochemical phenomena within a few seconds. Product yields vary widely. The yield, for example, for pigmentary titania is usually over 90% (Xiong and Pratsinis, 1991), while that of optical fibers is under 50% (Bohrer et al., 1985).

Despite the economic significance of industrial aerosol processes and the importance of aerosol characterization in operation and control of these processes, very little has been done for the development of real-time instruments for aerosol characterization in manufacturing processes. There are three reasons for this oddity. First, industrial aerosols have notoriously high concentrations, and furthermore, particle formation, growth and transport take place simultaneously over very short time scales. Removing an industrial aerosol from its environment for sampling and analysis, may drastically change its characteristics. For example, an aerosol with 10^{16} particles/m^3 of average diameter 0.1 μm reduces its concentration by coagulation by 90% within 1 s (Friedlander, 1977). Second, powders made by aerosol processes tend to be nonspherical aggregate structures making difficult their rapid characterization by conventional aerosol instruments. Third, the development of industrial aerosol processes has historically been Edisonian rather than guided by the principles of aerosol science. In fact, the development of some of these processes was well ahead of aerosol science!

Aerosol routes in material synthesis can be classified as gas-to-particle and droplet-to-particle conversion depending on the starting material. In gas-to-particle conversion, precursor gases or vapors react forming product particles that grow further by coagulation and/or surface reactions. Powders made by gas-to-particle conversion have relatively narrow size distributions and consist of aggregates of nonporous primary particles. Production of chemical commodities such as carbon black, silica and titania involves gas-to-particle conversion at high temperatures in the so-called flame reactors (Ulrich, 1984).

In the droplet-to-particle route, solution or slurry droplets are suspended in gases by liquid atomization. These droplets are converted to powders by direct pyrolysis or by in-situ reactions with another gas. The product powder distribution is determined primarily by the droplet distribution and in some cases by particle break-up during pyrolysis or drying. Powders made by this route are rarely agglomerated but can be porous depending on precursor solute concentration and drying rate (Zhang et al., 1990; Charlesworth and Marshall, 1969; Leong, 1981, 1987). Spray drying (Lukasiewicz, 1989) and spray pyrolysis (Kodas, 1989) are typical industrial processes employing droplet to powder conversion.

Solids can also be used as starting materials in aerosol processes involving either complete or partial vaporization or melting of the solid and subsequent formation of the product (Bolsaitis et al., 1987; Weimer et al., 1991).

In this chapter, processes and materials employed in industrial scale manufacture of aerosols are reviewed and the most common techniques for particle characterization are outlined.

Gas-to-Powder Conversion

The gas-to-powder conversion route refers to "building" particles from individual molecules in the gas phase (Fig. 32–1). The particle formation process is driven by the generation of molecules by chemical reaction from precursor gases or by the rapid cooling of a superheated vapor. Cooling of the vapor can be accomplished by contact with a cooler inert gas or by expansion through a nozzle even of a supercritical fluid (Petersen et al., 1986; Matson et al., 1987). The process of evaporation–condensation is carried out either under atmospheric pressure in inert gases or in a vacuum chamber that can be evacuated to pressures of 10^{-6} to 10^{-10} Torr in order to reduce contamination levels. High temperatures are usually required to accomplish the reaction or to bring the vapor to the superheated state. Evaporation followed by condensation of a species is a convenient method for the formation of nanometer-size particles. Depending on the thermodynamics of the process (Ulrich, 1971; Rao and McMurry, 1989; Xiong and Pratsinis, 1991), the product molecules can form particles either by uninhibited collisions (collision-controlled nucleation) or by balanced condensation and evaporation to and from molecular clusters (condensation-evaporation–controlled nucleation). The newly formed particles grow further by collision with product molecules

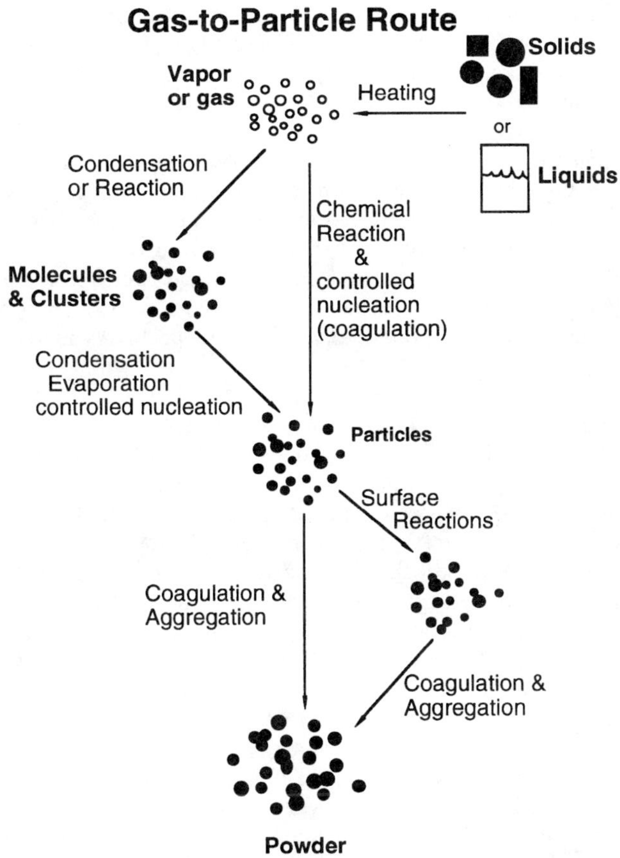

Fig. 32–1. Physicochemical processes occurring during powder production from gases (gas-to-particle conversion). (Adapted from Pratsinis, 1990.)

(condensation) and/or particles (coagulation). When the rate of particle collision is faster than that of particle coalescence (fusion), aggregates of spherical (primary) particles are formed. The latter ones are termed hard or soft aggregates (or agglomerates) depending on how easy it is to break the bonds connecting the primary particles. Rapid cooling favors the formation of aggregates (Schaefer and Hurd, 1990; Dobbins and Megaridis, 1987; Hurd and Flower, 1988; Ulrich and Riehl, 1982).

Various energy sources are used to create the high temperatures required during gas-to-particle conversion. The name of these sources is frequently used to signify a specific aerosol reactor in which gas-to-particle conversion takes place. Thus, flame reactors use combustion of hydrocarbons in premixed or diffusion flames. Plasma reactors utilize the high energy of an ionized gas (plasma). Laser reactors employ the high energy and precision of a laser beam. Furnace reactors electrically heat ceramic tubes through which precursor gases flow.

The gas-to-particle conversion route is used industrially for production of pigmentary titania, fumed silica and carbon blacks and on a laboratory scale for production of metals, semiconductors, and oxide and nonoxide ceramics in the nanometer range. As a result, this route has been studied the most and mathematical models describing particle formation in various gas-to-particle conversion processes have been developed. The effects of various process variables on average powder properties (polydispersity and average diameter) are easily described by moment models (Dobbins and Mulholland, 1984; Pratsinis, 1988). The evolution of the detailed powder size distribution in gas-to-particle processes has been also modeled (Wu and Flagan, 1988; Landgrebe and Pratsinis, 1990). More recently these models have been used to create design diagrams facilitating the understanding of the effect of process variables on powder size characteristics (Pratsinis et al., 1989; Landgrebe et al., 1990).

The gas-to-particle coversion route is most useful for generation of single-component powders. It is not as convenient, however, for multicomponent materials where differences in the vapor pressures and subsequently in the nucleation and growth rates of the various species may lead to nonuniform product composition from particle to particle and even within particles. The primary advantages of gas-to-particle routes are: small particle size, narrow particle size distribution, solid particles, and high purity. Disadvantages include difficulties in producing multicomponent materials, and problems in handling hazardous gases.

Flame Reactors. These reactors involve formation of spherical or aggregate particles from gaseous precursors in diffusion or premixed flames (Ulrich, 1984; Pratsinis, 1998). Aromatic hydrocarbons are used as precursors of carbon blacks in fuel rich flames (Medalia and Rivin, 1976; Ivie and Forney, 1988). A variety of metal oxides, most notably silica and titania, has been produced by oxidation of the respective metal halide vapors in hydrocarbon supported flames (Ulrich and Riehl, 1982; Kriechbaum and Kleinschmidt, 1989; Formenti et al., 1972; Hurd and Flower, 1988; Sokolowski et al., 1977; George et al., 1973). Additives are introduced in the process to control the phase, shape and size distribution of the product particles (Mezey, 1966; Dannenberg, 1971; Haynes et al., 1979). Electrical charges offer another tool for control of the characteristics of flame-made powders. Hardesty and Weinberg (1973) found that the primary particle size of product silica was reduced up to a factor of 3 with increasing applied electric potential. This was attributed to the rapid particle deposition on the electrodes as a result of reduced residence time for particle sintering. Katz and Hung (1990), found that the number density of the aggregates decreased and the size of titania and germania particles made in an H_2/O_2 counterflow diffusion flame reactor increased by 3 to 10 times upon the application of an electrical field across the flame. In premixed and diffusion flames (Vemury and Pratsinis, 1995, 1996; Vemury et al., 1997) making titania and silica, the aggregate and primary particle size decrease proportionally to the applied electric field as they were measured by transmission electron microscopy (TEM) and specific surface area analysis. Flame reactors provide an attractive method for particle generation since they are simple in construction and operation relative to other systems.

Particle formation in flame reactors closely follows the standard description of the gas-to-particle route. Flame-generated aggregates contain from a few up to many thousand primary particles. The typical size of the primary particles ranges from 1 to 500 nm. Maximum flame temperatures are usually on the order of 2500 K. In contrast to other gas to particle conversion systems, temperatures are lower in flame reactors than in plasmas but much higher than in furnace, evaporation–condensation and laser reactors. The residence time and temperature determine whether the product powder is amorphous or crystalline and determines the degree of agglomeration and coalescence.

Advantages of flame reactors include: Scale up has been demonstrated for carbon blacks, silica, titania and other oxides; simplicity; ability to use volatile and nonvolatile precursors; high product purity; and large range of diameters for particles in aggregates. The disadvantages of flame reactors are that hard agglomerates are formed under most conditions and that broad particle size distributions (coagulation controlled) are usually obtained. A detailed review of this technology can be found in Pratsinis (1998)

Furnace, Laser and Plasma Reactors. Furnace reactors involve driving chemical reactions of powder precursor vapors in a tubular, furnace-heated reactor (Masdiyasni et al., 1965;

Kanapilly et al., 1970; Suyama and Kato, 1976; Prochaska and Greskovich, 1978; Alam and Flagan, 1986; Kruis et al., 1998a; Okuyama et al., 1986; Kim, 1997). The reactants are introduced into the reactor either as a gas or as an aerosol of a solution. In the case of solution droplets, the solution and precursors in the droplet evaporate in the reactor before gas phase reaction occurs. Thus, the aerosol facilitates introduction of low volatility precursors into the reactor (Kagawa et al., 1983).

Particle formation in furnace reactors is reasonably well understood through detailed models for specific chemical systems such as silicon particle formation by silane decomposition (Wu et al., 1988), silica by $SiCl_4$ oxidation (Kim and Pratsinis, 1988, 1989), titania by oxidation of $TiCl_4$ or titanium isopropoxide (Landgrebe et al., 1990; Okuyama et al., 1986, 1989). Furnace reactors provide excellent control of temperature and flow velocity so they are frequently used for measurement of global chemical reaction rates of particle forming systems (SiO_2: Powers, 1978; French et al., 1978; TiO_2: Pratsinis et al., 1990).

Lasers can be used to drive chemical reactions that result in particle formation by photothermal and photochemical processes (Cannon et al., 1982; Casey and Haggerty, 1987). In the photothermal process, the laser energy is absorbed by the carrier gas and raises its temperature, thus, driving the chemical reactions. In the photochemical process, the reactant molecules absorb the laser light and then dissociate to form condensable species. Thus, temperatures can be near ambient in contrast to the photothermal case where temperatures are near 1300 K or more (Kruis et al., 1998b). The photochemical process presently requires expensive reactants and lasers. In a typical photothermal process, a CO_2 laser is used to heat gas molecules by light absorption (Alexandrescu et al., 1998). The laser acts as a heat source since molecules are in thermal equilibrium with the gas. High energy lasers can provide sufficient energy to force coalescence of agglomerated particles. This can be used to produce spherical, nonagglomerated particles in the nanometer range (Lee and Choi, 2000). The absence of walls leads to high-purity products. The advantage of using a laser is that its narrow spectral width allows coupling between the light source and the molecular precursor. The coupling requires near coincidence between laser light wavelength and wavelength of absorption. Efficiencies for use of the laser power are about 15%. Models for particle formation in laser reactors have been developed indicating that particle coagulation determines the product powder size distribution (Flint and Haggerty, 1990).

Plasma reactors use an ionized gas (plasma) to provide the energy required to drive reactions that result in particle formation (Phillips and Vogt, 1987; Pickles and McLean, 1983; Gani and McPherson, 1980; Girshick and Chiu, 1989, 1990). A plasma is a system with a high energy content in which a significant fraction of the species are ionized. Two types of plasma systems are of interest: high temperature equilibrium thermal plasmas and low temperature nonequilibrium plasmas. In the thermal plasmas, the temperatures of the electrons and ions are equal. In low temperature plasmas such as glow discharges, the temperature of the electrons is much greater than that of the ions. Thermal plasmas are most common for powder generation. Applications of thermal plasma for processing of materials include plasma synthesis, spraying, and consolidation (Peng et al., 1998). Glow discharges are not used extensively for powder generation or processing (Singh and Doherty, 1990) though they are used in film coating.

Numerous variations of plasma reactors exist depending on how the reactants are introduced into the system. Temperatures (600 to 25,000 K) are higher than in all other aerosol reactors and complete destruction of reactants is common. This allows use of molecular and solid feed streams so, in principle, any material can be processed. Common processes for all of these methods are formation of product species that nucleate to form particles during cooling while exiting the plasma, following the standard formation pathways of gas-to-particle conversion. Two broad classes of thermal plasma reactors are used: DC arc jet and high frequency (microwave or radio frequency) induction systems. In the case of the DC arc jet, current is supplied to the ionized gas (plasma) by physical contact with a metallic electrode surface. This system is relatively simple and inexpensive. However, the electrodes are consumed and end up in the product, resulting in contamination. In the case of a high frequency induction plasma (microwave or radio frequency plasma reactor), there is no contact between the plasma and its power source. The induction coil lies outside the reactor walls like the electrical thermal elements in furnace reactors. Energy transfer takes place through the electromagnetic field of the induction coil, so there is no contamination of the product. An overview of microwave plasmas can be found in Vollath et al. (1997). Variables that can be controlled are plasma composition and frequency. The most common case is an argon plasma operated at 200 kHz to 20 MHz with typical temperatures about 15,000 K. As a result, plasma reactors can handle high melting point materials and solid powder feeds.

Evaporation–Condensation. Evaporation followed by condensation of a species is a convenient method for the formation of nanometer-size particles that is used for the synthesis of metal nanoparticles. The technique is called inert gas condensation in that field. Metals and nanocrystalline ceramics with improved mechanical properties can be produced by these methods (Siegel, 1990; Ramsey and Avery, 1974; Granqvist and Buhrman, 1976). A metal is vaporized into a host gas where the vapor is cooled resulting in particle formation by nucleation. Cooling of the vapor can be accomplished by contact with a cooler inert gas or by expansion through a nozzle. Expansion of a supercritical fluid through a nozzle can also be used for particle formation (Petersen et al., 1986; Matson et al., 1987).

The process of evaporation–condensation is often carried out below atmospheric pressure in inert gases. A typical system involves a vacuum chamber that can be evacuated to pressures of 10^{-6} to 10^{-10} Torr in order to reduce contamination levels. Often a boat is used to hold a metal heated to vaporize it into the gas phase. A natural plume is established by free or forced convection from the metal containing boat. This boat can be heated continously or in pulses. Temperatures are usually lower than those in plasma and flame reactors.

In evaporation–condensation/reaction processes, reactions take place at the newly formed particle surface altering the chemical and/or phase composition of the particles. A precursor compound is vaporized into a carrier gas, which is then cooled resulting in droplet formation by nucleation (Fig. 32–1) (Matijevic, 1986, 1987; Ingebrethsen et al., 1983; Visca and Matijevic, 1979; Ingebrethsen and Matijevic, 1980; Kodas et al., 1987). After the droplet formation a reaction step particle formation takes place resulting in solid.

Evaporation–condensation processes are reasonably well understood especially when precursors with known physicochemical properties are employed and well-defined flow patterns exist such as in tube flows (Tsantilis et al., 1999).

Aerosol processes are used industrially to produce a variety of products. Each of the gas to particle conversion processes discussed in this chapter has its own advantages and disadvantages. Table 32–2 presents a comparison of these different processes.

Droplet-to-Particle Conversion

The most common processes for generation of powders with liquid phase precursors are spray pyrolysis, spray drying and freeze drying (Fig. 32–2). The common step in all of the processes is formation of droplets containing molecular or particulate precursors using some type of droplet generator.

In spray drying processes a variety of physical and chemical processes can occur while the particles are in the gas phase. These processes include: droplet generation, evaporation of solvent from droplets, initial crystallization of the solute in the droplet and further evaporation from the droplet consisting of solution/solid along with further solute crystallization to form a dried particle.

Spray and freeze drying is similar to spray pyrolysis since it only involves heating the droplets to lower temperatures so solid phase reactions do not occur (Masters, 1972; Johnson,

TABLE 32–2. Comparison of Aerosol Processes for Powder Production

	Flame	Evap.–Cond. Reaction	Laser	Plasma	Hot Wall	Spray Pyrolysis
Maximum size	1 μm	0.1–10 μm	1 μm	1 μm	1.0 μm	0.10–100 μm
Spread	Broad	Narrow	Narrow	Broad	Narrow	Broad
Morphology	Agglomerates solid	Solid	Solid	Agglomerates solid	Spherical solid	Spherical solid, porous hollow
Maximum temperature (K)	2,500	<2,000	1,000	25,000	1,000	1,600
Material	Oxides	Metals and oxides	Nonoxides oxides	Nonoxides oxides	Nonoxides oxides semiconductors	Nonoxides oxides
Complexity	Low	Low	Medium	High	Low	Low

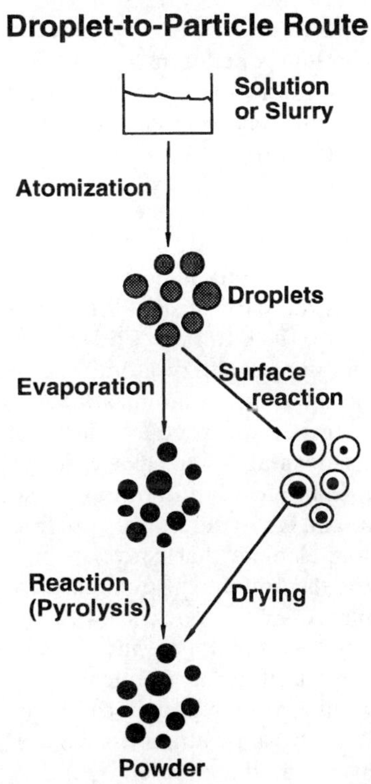

Fig. 32–2. Physicochemical processes occurring during powder production from liquid droplets (droplet-to-particle conversion). (Adapted from Pratsinis, 1990.)

1981). An additional difference is that much larger droplets are used in spray drying. This is a consequence of the goal of spray drying, which is usually to form powder consisting of relatively large free-flowing granules (100 μm). Charlesworth and Marshall (1969) have discussed some of the possible particle morphologies and how they are produced in spray drying.

In the case of spray pyrolysis, these steps are followed by: reaction of the precursors in the dried particle to form the product powder and gaseous products, intraparticle transport processes resulting in changes of the particle morphology, evaporation of volatile metals, metal oxides, and other species followed by condensation on reactor walls and on particles and by new particle formation and coagulation.

During freeze drying, once the droplets are formed and collected in the cold liquid (hexane or nitrogen), the characteristics of the product powder are determined by the nature of the cold liquid, solvent, precursors, drying rate, and subsequent processing conditions. Freeze drying processes are similar to spray drying as far as droplet generation, but the droplets are frozen and then dried once in the solid state (Johnson et al., 1987). The solution containing dissolved compounds is sprayed into liquid nitrogen or some other cold liquid where the droplets freeze. The liquid is removed by sublimation and the dry product is converted to the desired product by further heating.

In addition to the processes discussed above, coagulation, diffusion, sedimentation, impaction, and thermophoresis in the process vessels and manifold can modify the particle size distribution and influence yields. Further operations may be required to form the final powder product when droplet-to-solid processes are involved. These processes are typically operated at atmospheric pressure and involve particles in the 0.1 to 100 μm range.

In general, advantages of all of these processes are the ability to process organic and inorganic materials; the ability to form a variety of multicomponent materials; simplicity; cost effectiveness (only a few unit operations); many choices for inexpensive liquid phase precursors; doping is possible; scale-up has been already demonstrated to ton quantities; uniform chemical composition; relatively safe process since volatile precursors not required. The disadvantages for these processes are: porous or hollow particles can be formed at certain conditions and the spread of particle sizes is limited by the spread of the generated droplets.

Droplet Generation. The droplet generation process is critical to all droplet-to-particle processes since it largely controls the particle size distribution of the product. A variety of generators are available (Lefebvre, 1989; Kerker, 1975): pressure, rotary, air-assist, air-blast, ultrasonic, electrostatic, and vibrating capillary atomizers. Other atomizers such as sonic, windmill, flashing liquid jets and effervescent atomization are also used.

Each of these types of droplet generators produces different particle size distributions and has different advantages and disadvantages. Typical average droplet diameters are 1–100 μm. Pressure atomizers work by discharging a liquid through a small aperture under high pressure into a slow-moving gas stream. Rotary atomizers operate by centrifugal atomization. A liquid is directed onto a rotating element that converts the liquid stream to droplets. Air-assist atomizers involve exposing the liquid to a stream of air or steam flowing at high velocity. Air-blast atomizers are similar to air-assist nozzles with the exception that the former use much higher quantities of air or steam and at lower velocities. Ultrasonic atomizers use a transducer or horn which vibrates at ultrasonic frequencies to produce the short wavelength required for atomization. Flow rates are lower than can be obtained for air blast, air assist, pressure and rotary atomizers. Electrostatic atomizers expose a liquid jet or film to a strong electrical pressure. The expansion in the droplet is opposed by surface tension forces. Although this method is capable of producing submicrometer droplets, it does not allow high liquid flow rates and is commonly used on a laboratory scale. Vibrating capillary atomizers produce droplets down to 30 μm. Again, only low liquid flow rates are possible and the latter atomizers are mostly used on a laboratory scale.

Measurements of droplet size distributions are routinely made by optical techniques. The prediction of droplet size distributions is accomplished through empirical correlations accounting for liquid properties, atomizer design and fluid flows (Lefebvre, 1989). The most relevant properties of a liquid are surface tension, viscosity, and density.

Spray Pyrolysis. Spray pyrolysis involves passing suspended precursor droplets through a high temperature region (furnace or a flame). There, rapid evaporation of the volatile components of the droplet takes place first. Further residence in this region allows chemical reactions to occur in the solid phase that form the product powder. Depending on the process conditions and the material properties solid or porous powders are formed and collected on a filter. This process has been called evaporative decomposition, spray roasting, spray calcining, Aman process, atomizing burner technique, decomposition of misted solutions, aerosol or mist decomposition method and others (Sokolowski et al., 1977; Pebler and Charles, 1989; Chadda et al., 1991; Wenckus and Leavitt, 1957; Vollath, 1990; Epstein, 1976; Imai and Takami, 1985; Kodas et al., 1988, 1989a,b; Biswas et al., 1989b, 1990; Zhang et al., 1990; Gardner et al., 1984; Zachariah and Huzarewicz, 1991; Messing et al., 1993).

Spray pyrolysis allows generation of particles with a uniform chemical composition from particle to particle since the precursors are present in the correct stoichiometry in each droplet. Volatile precursors are not required; thus, a wide variety of reactants can be used while avoiding the need for carbon-containing precursors. This allows use of inexpensive reactants such as metal nitrates, chlorides, and fluorides. The generation of multicomponent ceramic powders can be carried out on an industrial scale using only a few unit operations; the ability to scale up these systems for production of ton quantities of conventional ceramic oxide powders has already been demonstrated (Ruthner, 1979). Unagglomerated particles are produced since particle growth rarely involves collisions between particles. High purities are also obtained since milling is not required.

Key challenges include determining the morphology of the particles, their chemical composition, their phase composition, and their behavior in liquids during ceramic powder processing operations. Recently, Brewster and Kodas (1997) produced nonaggregated, dense $BaTiO_3$ particles by flame-spray pyrolysis of aqueous solutions of barium acetate and titanium lactate. Low flame temperatures resulted in hollow particles while high temperatures or long residence times resulted in dense and homogeneous particles. Xiong and Kodas (1993) and Messing et al. (1993) have nicely summarised some of the current quantitative understanding on particle formation by spray drying, along with applications for synthesis of solid or hollow particles and even nanoparticles. Superconducting powders can be made by flame processes also. Merkle et al. (1990) produced $YBa_2Cu_3O_x$ superconducting oxide by reacting an atomized nitrate solution containing yttrium:barium:copper in the atomic ratio 1:2:3, respectively, in an oxy-hydrogen diffusion flame. The primary particle size was about 100 to 300 nm. X-ray diffraction measurements of oxygen-annealed particles revealed a crystal structure identical with that of the conventionally produced superconducting oxide. Magnetic susceptibility measurements demonstrated that the material is superconducting with a T_c of about 93 K. Zachariah and Huzarewicz (1991) also made submicrometer $YBa_2Cu_3O_{7-\delta}$ particles with T_c = 92 K in an aerosol flame reactor by a flame spray pyrolysis method in which an aerosol composed of an aqueous solution of Y, Ba, and Cu nitrate salts was introduced into a $H_2/O_2/Ar$ flame (1000 to 1400 K) as was pointed out earlier in this review. Lewis (1991) has produced mullite and other mixed oxides (SiO_2, Al_2O_3) by spray combustion of triethylaluminum with various organosilicon compounds and solvents. Recently, Bickmore et al. (1996), have further developed a similar flame spray pyrolysis process for synthesis of spinel powders from Mg-Al alkoxides with diameters 10 to 100 nm and specific surface areas of 40 to 60 m^2/g at a production rate of 50 to 100 g/h.

MEASUREMENT TECHNIQUES

Physical and chemical characterization of materials produced by aerosol processes involves many techniques that are summarized in Tables 32–3 and 32–4. Niessner (1990) has reviewed in

TABLE 32–3. Summary of Characterization Techniques for Collected Particles

Physical	Chemical
Size	Phase composition
Sedimentation velocity	XRD
Electrical mobility measurements	Analytical TEM
Image analysis and microscopy	
Light scattering	
Sieving	
Chromatography and field flow fractionation	
Electronic methods	
Shape	Elemental composition
SEM	Bulk
TEM	X-ray fluorescence
Optical microscopy	Emission and absorption spectroscopies
STM/AFM	FTIR
Image analysis and microscopy	PIXE
	NRA
	GC/MS
	NMR
	Surface sensitive
	AES
	EDS
	XPS
	Electron photoemission spectroscopy
	Mass spectrometry
Charge	
Electrometer	
Microstructure/Porosity	Thermochemical and thermophysical
Mercury porosimetry	DTA/DSC
Surface area	TGA
He density	TGA/MS
Pellet green density	
Microstructure/grain size	Other
TEM	SAXS
SEM	SANS
STM/AFM	ESR
	EPR
	Raman spectroscopy
	Magnetic and electrical properties
	Optical properties (pigments)

situ aerosol measurement techniques, most of which are included in Chapters 12 and 13 in this book.

Physical Properties

Size. A wide variety of methods exist for measurement of particle size distributions in the gas and liquid phases. Measurements of particle size distributions in the gas phase are discussed in Chapters 15, 16, and 18. Recent advances in particle size measurements for aerosol particles that have been collected and placed in liquids have been summarized by Miller and Lines (1988). Table 32–4 lists techniques for measurement of particle size. A critical problem

TABLE 32–4. Particle Size Analysis Techniques

Method	Particle Size (μm)	Sample Size
Sedimentation velocity		
Gravity	0.2–100	<1 g
Centrifuge	0.02–100	<1 g
Electrical	0.003–1	<1 μg
SEM	0.01–50	mg level
TEM	0.001–1	
Optical microscope	0.2–400	
Sieving	30–5000	5–20 g
Electronic methods	0.3–400	<1 g
Chromatography and field flow fractionation	0.001 to 200 (depending on technique)	<1 g

that arises when using several techniques for measurement of size is that each technique relies on different physical characteristics of the particles such as optical properties, electrical properties and sedimentation velocity. As a result, particle sizes determined from different techniques can vary significantly. Examples of this will be given in later sections.

Light extinction by aerosols is a powerful method for determination of particle size distributions. The method can be applied in-situ in the gas phase and can also be used for particles that have been collected and suspended in liquids. Details of light absorption and scattering processes (static and dynamic) are discussed in Chapters 15 and 16.

The most basic particle size analysis involves a microscope. In recent years scanning and transmission electron microscopes (SEM and TEM) are used for direct size characterization of fine particles. These methods involve the examination of particle images followed by collection of the images by a computer and presentation to an image analyzer for automatic characterization. In addition to particle size distributions, particle shape is also best determined by SEM and TEM examination of particles.

The most common image analysis systems consist of a microscope, a video camera and an image analyzing computer. The images of the particles can be collected with optical, scanning electron and transmission electron microscopes, depending on the size range of interest. For example, TEM allows sizing of particles from the nanometer up to the micrometer range. Only small amounts of materials are needed. For example, a filter can be inserted into a flow path to collect particles. A good dispersion of particles is required so only single particles are present on the filter surface. The filter surface can then be examined by a microscope. Figure 32–3 shows a TEM photograph of TiO_2 particles produced by a furnace aerosol reactor. A comparison of results from SEM/image analysis and a differential mobility particle size analyzer are shown in Figure 32–4.

Sedimentation velocity measurements of particle size rely on Stokes law (Reed, 1988). Particles are collected and then placed in a fluid. The settling velocity of the particles is detected by measuring light or X-ray transmission. Calculation of particle size distributions assumes that the particles are solid and spherical since Stokes law is used to relate velocity to diameter. In practice, particles greater than 0.1 μm can be sized by this method while particles below 0.1 μm are influenced too strongly by Brownian motion to allow measurements unless centrifuges are used. Particles larger than 50 μm do not follow Stokes law, defining an upper limit for this technique. However, extending or converting the Sedigraph (***MCM***) distributions for example to sieve distributions and vice versa is reasonably well successful, allowing easy and robust comparisons (Austin and Shah, 1983). Figure 32–5 shows a particle size distribution obtained by this technique for $Ba_{0.86}Ca_{0.14}TiO_3$

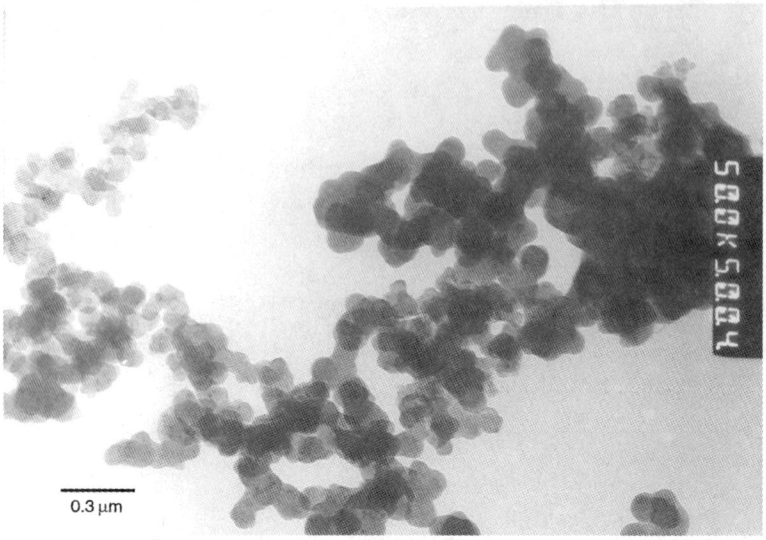

Fig. 32–3. TEM of titania particles produced at 1300 K by $TiCl_4$ oxidation in a furnace aerosol reactor. (Courtesy of M. K. Akhtar.)

particles produced by spray pyrolysis using a mixture of metal acetates, lactates and nitrates (Ortega et al., 1991).

Sieving is a simple and easy way to fractionate particles according to size (Reed, 1988). This method is limited to particles that are much larger than 1 μm, usually greater than about 50 μm. Agglomeration becomes a problem below this size. Smaller particle sizes can be sized using liquids, but these methods are often difficult to carry out reproducibly.

Chromatography and Field Flow Fractionation. These are separation methods whereby a liquid suspension of particles is classified into size fractions that must later be quantified in order to produce a particle size distribution (Miller and Lines, 1988). The chromatographic techniques involve separation of particles according to size within a flow channel as in conventional chromatography. In field flow fractionation, the particle retention is induced by an electrical, centrifugal or thermal gradient.

Hydrodynamic chromatography (HDC) involves passing particles through a packed bed of non-porous column material. The rate of transport of particles depends on the size of the particles, the size of the spheres in the bed, and the flow rate and ionic strength of the eluting solution. This technique is useful for particles from 20 nm to 2 μm. Size exclusion chromatography utilizes a porous bed. The flow through the porous material imposes a separation effect by steric exclusion. The upper limit of this technique is roughly 500 nm with a lower limit of essentially molecular size. Capillary particle chromatography is an extension of HDC. This method used a long capillary tube instead of a packed bed to provide the separation. Particles from 20 nm to 200 μm can be separated.

Field flow fractionation methods can be used for particles from 10 nm to 1 μm. Fractionation in the centrifugal sedimentation version of this process occurs within closely spaced parallel plates. The particles are caused to roll or tumble along the outer wall to produce the segregation.

These methods have not been used extensively for characterization of particles produced by aerosol processes, primarily because they require resuspension of collected particles in a liquid. Resuspension of submicrometer particles into a liquid is not simple and can lead to significant errors in measurements of size distributions.

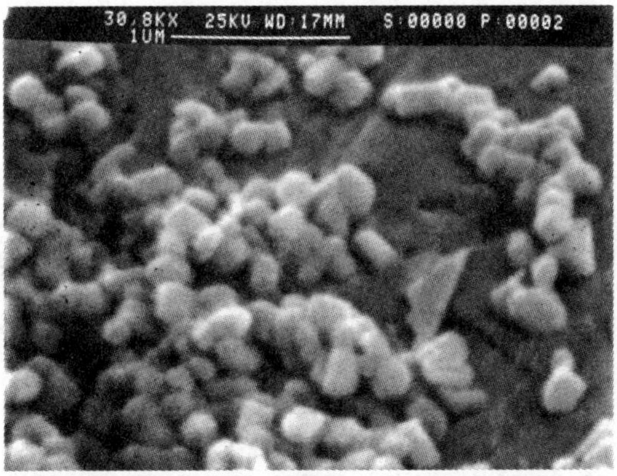

(a)

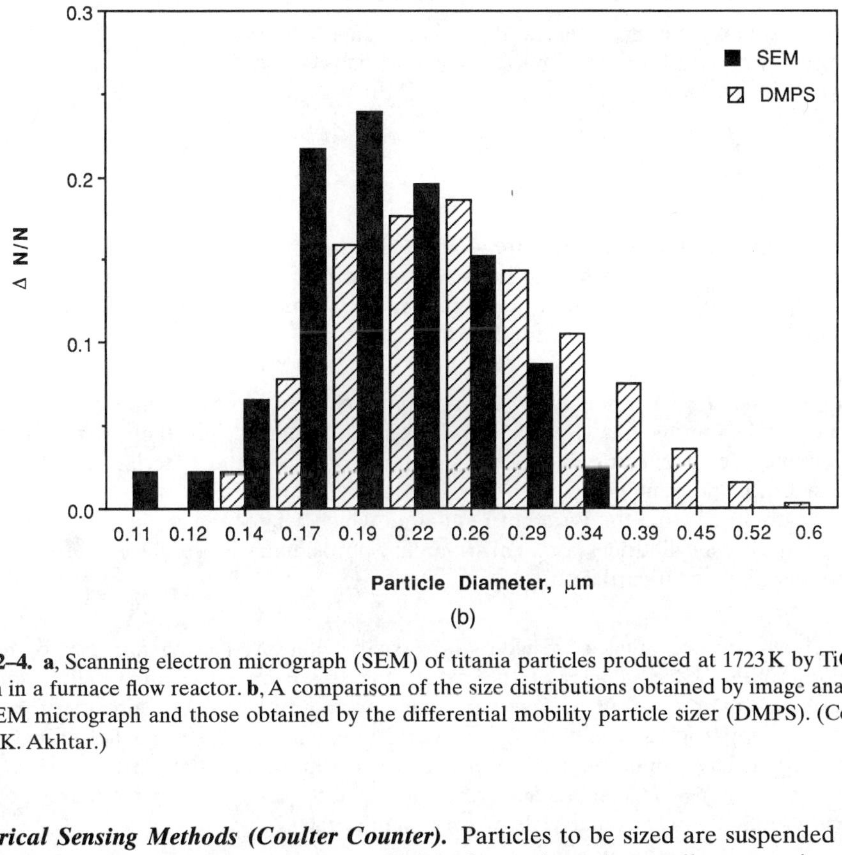

(b)

Fig. 32–4. **a**, Scanning electron micrograph (SEM) of titania particles produced at 1723 K by $TiCl_4$ oxidation in a furnace flow reactor. **b**, A comparison of the size distributions obtained by image analysis of the SEM micrograph and those obtained by the differential mobility particle sizer (DMPS). (Courtesy of M. K. Akhtar.)

Electrical Sensing Methods (Coulter Counter). Particles to be sized are suspended at low concentration in an electrolyte solution, which is drawn through a small aperture in an insulating wall, across which a current also flows (Miller and Lines, 1988). Each particle that passes through the sensing zone displaces a certain amount of electrolyte solution that results in a momentary change in the electrical impedance across the aperture. The volume of solution displaced by the particle is determined by the amplitude of the electrical pulse produced as the particle passes the sensing zone. Extreme dilution is required to avoid coincidence errors. Particles 100 nm to roughly 400 μm can be sized using different size apertures.

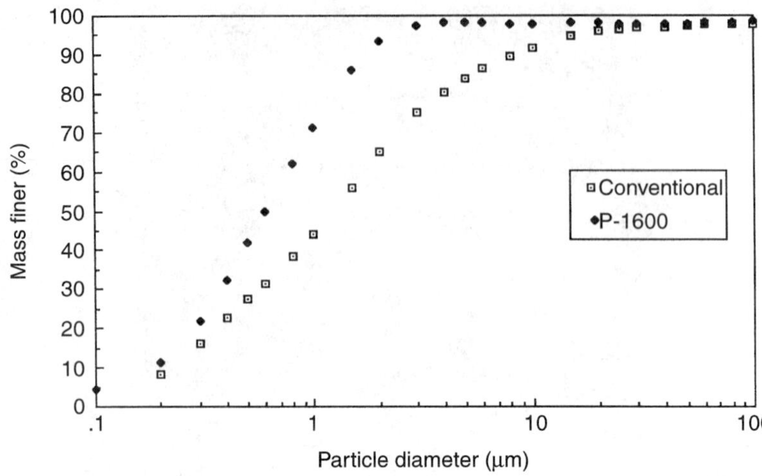

Fig. 32–5. Particle size distribution obtained by sedimentation velocity for $Ba_{0.86}Ca_{0.14}TiO_3$ particle produced by spray pyrolysis (P-1600) and by conventional milling of ceramic oxide mixtures. (Adapted from Ortega et al., 1991.)

Microstructure

The description of the microstructure of particles involves characterization of porosity, surface area, density, and grain size. The microstructure of particles plays a critical role in determining rates of densification of ceramic particles. Similarly, particles used for catalytic applications require knowledge of total surface areas.

Mercury Porosimetry. This technique utilizes mercury intrusion to obtain pore size distributions in powders and solids (Reed, 1988). A sample is placed into the instrument where mercury under a controlled pressure is forced into the pores of the material. As the pressure is increased, more mercury enters the pores. The accessible pore size corresponding to a given pressure is obtained from the Washburn equation which relates the applied pressure to the intruded radius of a cylindrical pore. Thus, mercury porosimetry is useful for determining the pore size distribution in particles.

Surface Area. Surface areas of powders are usually measured using nitrogen or helium adsorption followed by analysis of the data using the Brunauer-Emmet-Teller isotherm (Reed, 1988). Estimates of surface areas provide information about the porosity of particles and the primary particle size in particles that consist of agglomerates of smaller particles. The primary particle size can be related to surface area assuming that the particles are smooth, monodisperse spheres. Specific surface areas can range from a few m^2/g to hundreds of m^2/g in the case of highly porous materials or very fine particles.

Density. In many cases, it is useful to know the density of a material in particles. Helium density measurements provide the density of a material assuming that no closed porosity is present (Reed, 1988). A sample is weighed and placed in a vessel into which helium is introduced. The volume of the sample is obtained from the volume of gas displaced by the powder. The density is obtained from the volume and weight of the sample. For this reason, hollow particles with no porosity can lead to erroneously low densities. In contrast, if the density is known, He density measurements become a method for the determination of the amount of closed porosity in a sample.

Pellet Green Density. An indirect but practical measure of the quality of a powder is the green density that can be obtained by dry pressing or other powder processing techniques. Typically about 100 mg of powder is needed. Hollow particles usually lead to low green densities, 20% to 40%. In contrast, solid particles can provide green densities of 50% to 60%. It must be noted that this is a crude measure since the type of processing can result in large variations in green density (Fig. 32–6). A related measurement is the tap density. This technique, which involves measuring the apparent density of powders, also provides a way of quantifying the packing characteristics of powders from an applications point of view.

Tranmission Electron Microscopy. Processing of ceramic or metallic powders involves particle sintering. Sintering rates and the grain size distribution of densified ceramics depend on the initial grain size distribution within each particle and the size distribution of the particles. For example, Edelson and Glaser (1988) have shown that sintering of spherical particles consisting of agglomerates of smaller particles can result in broad grain size distributions. Broad grain size distributions are usually undesirable (Barringer et al., 1984). Thus, it is useful to know the size of grains in particles. In addition, particles with smaller grain sizes have larger grain surface areas which provides a large driving force for increase of grain size (Kingery et al., 1976). Thus, particles consisting of nanocrystalline grains are usually more active than single-crystal particles. TEM provides a powerful method for determination of grain sizes. Crystallites with dimensions ranging from nanometer to micrometer can be observed. Figure 32–7 shows how TEM can be used to distinguish if particles are nanocrystalline or single crystals (Feldman and Mayer, 1986). The separate crystallites in the particles are visible within each particle. Special 3D TEM techniques can enhance our understanding of the morphological properties of particles being examined (Hoppe et al., 1968; Frank, 1995). To extract volume information the probe is rotated and different projections of the sample are imaged. At present these methods are time intensive and therefore not widespread. However, advances in computing power, electro-optical imaging and microscope design promise much faster and versatile acquisition of 3D TEM pictures.

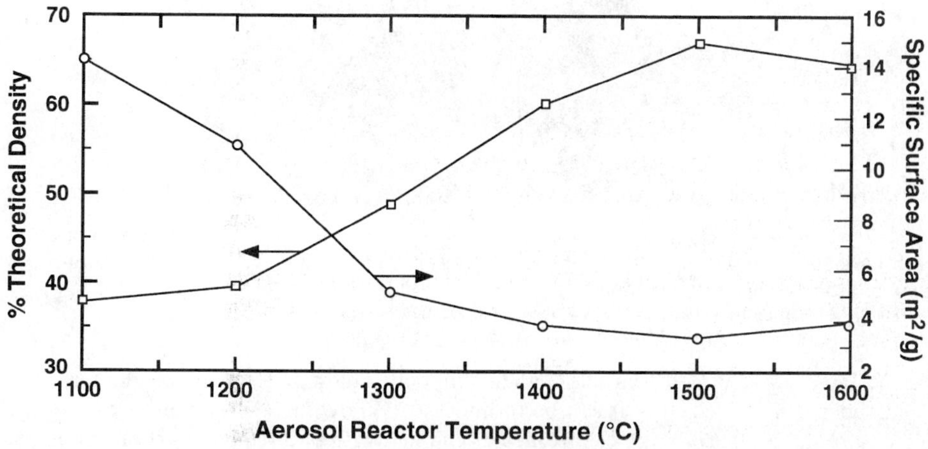

Fig. 32–6. Green density (left-hand scale) and surface area (right-hand scale) of $Ba_{0.86}Ca_{0.14}TiO_3$ particles produced by spray pyrolysis as a function of production temperature. Particles produced at lower temperatures are hollow and lead to low green densities. (Adapted from Ortega et al., 1991.)

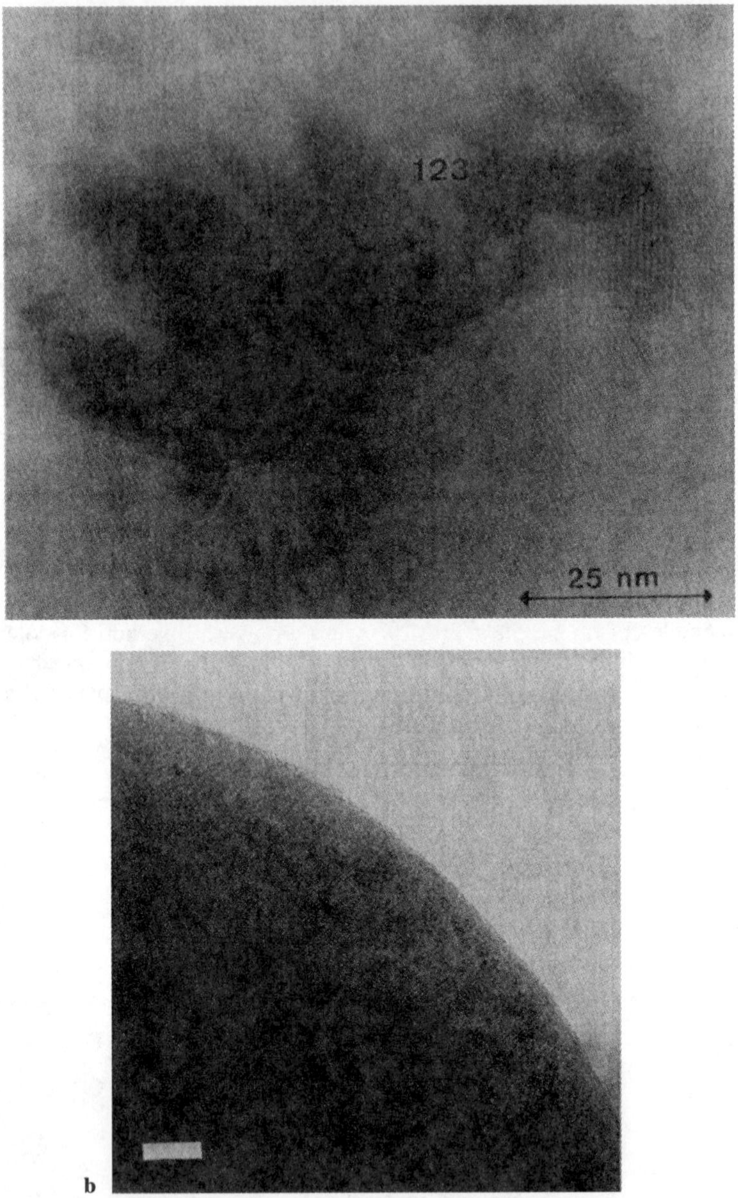

Fig. 32–7. **a**, Superconducting (so-called 123 superconductors) ($YBa_2Cu_3O_{7-x}$ particles with nanometer grains. (Adopted from Carim et al., 1989.) **b**, Single-crystal $YBa_2Cu_3O_{7-x}$ particles (bar = 10 nm). Particles were produced by spray pyrolysis.

Charge. Multiply charged particles with unipolar charges tend to agglomerate less (Friedlander, 1977). Therefore it can be of interest to know the charge per aerosol particle. To measure the particle charge electrometers coupled with particle counters are often used. The electrometer is a well insulated total particle filter. The charge accumulating on the filter is responsible for a small potential difference between ground and the particle filter. By amplifying this signal, typically by a factor of 10^{10} to 10^{12} the amount of charge being deposited on the filter can be measured. If the particle count is known the average particle charge can be easily obtained.

Phase Composition

X-Ray Diffraction. X-ray diffraction (XRD) is a standard tool for identification of crystalline phases in powder samples (Feldman and Mayer, 1986). The crystalline phases in the sample diffract X-rays according to Bragg's equation which relates lattice spacing to the wavelength of the X-rays used as a probe. The amount of sample needed in practice is roughly 100 mg or more. Crystalline phases present at levels of about 1 percent or greater can be detected. The unit cell dimensions (a, b, c) of the crystal can also be determined and provide more information for complex structure materials (Biswas et al., 1990). Amorphous materials are not observed. In routine work, XRD relies on availability of standards that allow identification of peaks in the diffraction pattern. Figure 32–8 shows phases of aerosol made titania and aluminum nitride identified by XRD (Akhtar et al., 1991).

TEM/Electron Diffraction. Microdiffraction with transmission electron microscopy (TEM) allows determination of the crystalline structure of phases present in particles (Feldman and Mayer, 1986). Since single particles can be examined, only very small amounts of a sample are needed. The electron diffraction pattern allows determination of whether the particles is polycrystalline (ring patterns), amorphous, or a single crystal (spot patterns). Figure 32–9

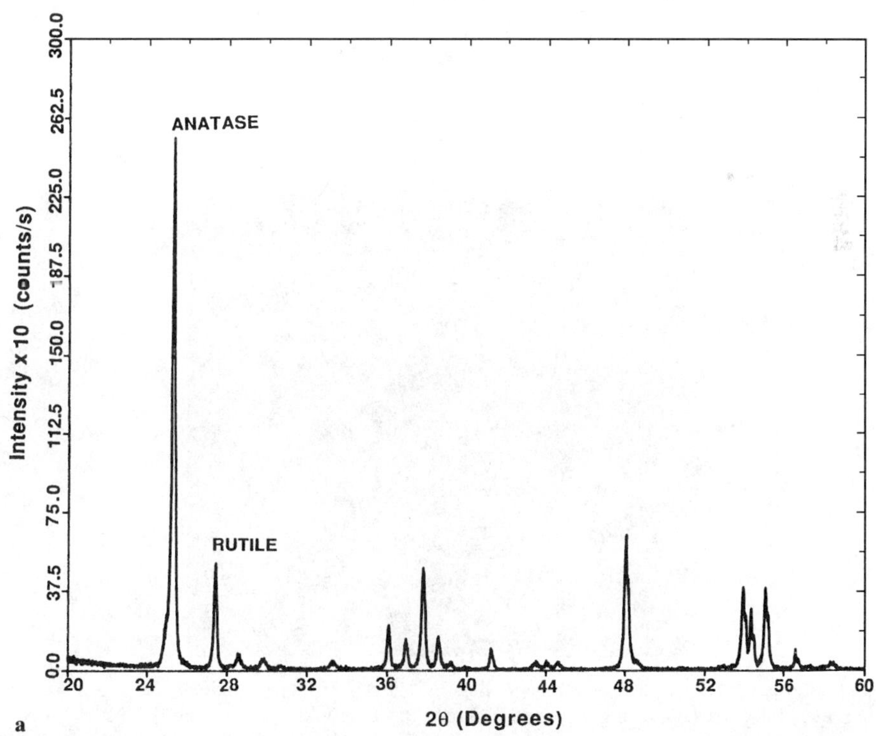

a

Fig. 32–8. X-ray diffraction (XRD) patterns for **a**, titania powder produced in a furnace aerosol reactor. The peaks resulting from the two different crystal phases (anatase and rutile) in which titania can be found are clearly visible; and for **b**, aluminum nitride powder produced in a furnace aerosol (Courtesy of M. K. Akhtar.)

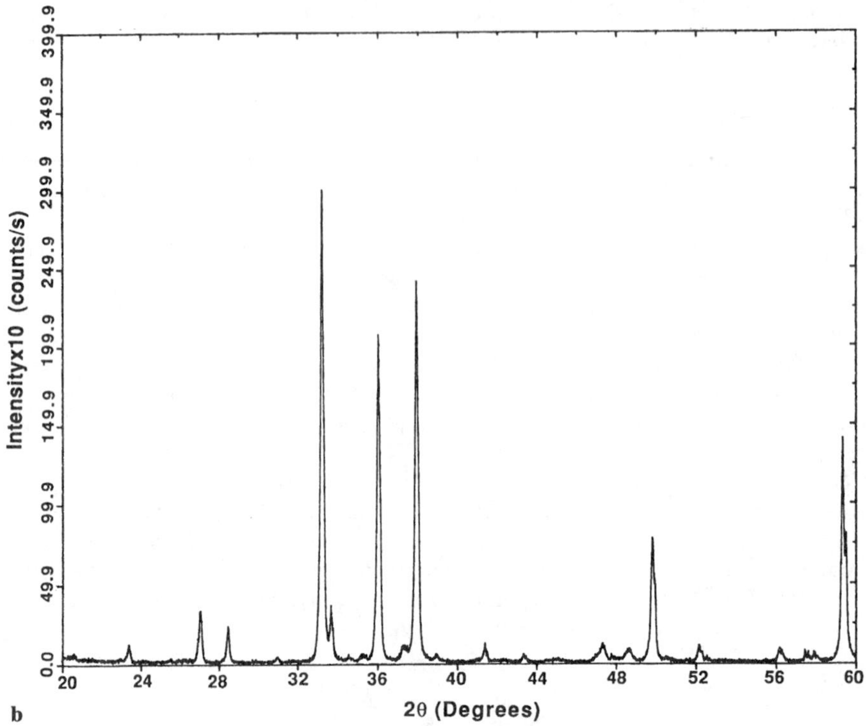

b

Fig. 32–8. *Continued*

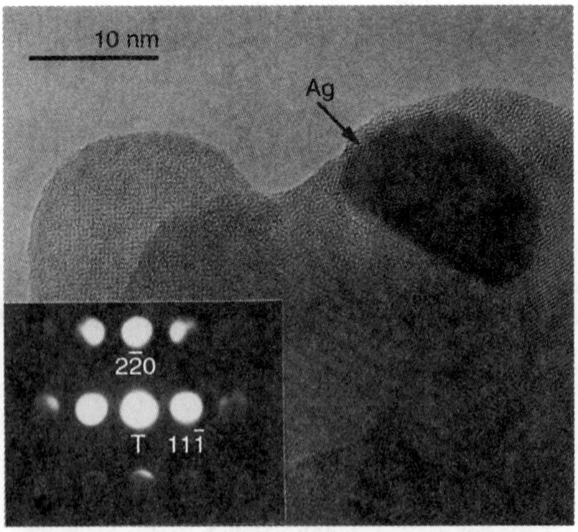

Fig. 32–9. Microdiffraction pattern for crystallite in a $YBa_2Cu_3O_{7\text{-}x}$ particle. Spot diffraction pattern demonstrates that particle is a single crystal. (Courtesy of Larry Allard and Abhaya Datye.)

shows how different phases in a submicrometer sized particle can be identified by electron diffraction by TEM (Carim et al., 1989).

Chemical Composition

A variety of techniques exist for determining the chemical composition of aerosol particles. A more extensive discussion of X-ray fluorescence, proton-induced X-ray emission, nuclear reaction analysis, and emission and absorption spectroscopies is presented in Chapter 13. This chapter outlines these methods and provides extensive reference materials in the context of material synthesis.

Methods for Bulk Materials

X-Ray Fluorescence. X-ray fluorescence (XRF) is a method for determination of the chemical composition of liquid and solid samples (Feldman and Mayer, 1986). The sample is irradiated with X-rays, which results in fluorescence. The wavelength of the emitted light is characteristic of the elements in the sample. This technique can be used for examination of multicomponent solid samples where it is desirable to determine the ratio of the elements present in the sample. Elements with $Z > 11$ are detectable down to 10 ppm. Only 10 to 100 ng levels of particles are required on a filter in order to determine chemical composition.

Fourier Transform Infrared Spectroscopy. Fourier transform infrared spectroscopy (FTIR) measures the amount of infrared radiation absorbed by a sample as a function of frequency (Reed, 1988). This technique provides information about the different types of bonds present in a sample. The amount of sample required for obtaining a spectrum depends on the nature of the sample, but 100 mg amounts are often adequate.

If the FTIR spectra are obtained in-situ, the technique permits measurement of the aerosol temperature. A black body temperature can be obtained both for the gas and the particles as demonstrated by Morrison et al. (1997). Also by analyzing the IR spectra the gas as well as particle chemical composition can be obtained.

Proton-Induced X-Ray Emission Spectroscopy. Proton-induced X-ray emission (PIXE) spectroscopy involves the use of ions to induce emission of x-rays from a sample (Feldman and Mayer, 1986). Elements from Na to U in the periodic table can be detected. PIXE allows the use of 10 μm beams for high sensitivity detection of trace elements. PIXE also allows identification of a number of elements simultaneously in complex samples.

Nuclear Reaction Analysis. Nuclear reaction analysis (NRA) (also known as *instrumental neutron activation analysis*) involves probing a sample using an ion beam and examining the light ion reaction products (Feldman and Mayer, 1986). The probe beam induces radioactivity in the sample, which is examined to determine the nature of the elements in the sample. Light elements such as H, Be, B, C, O, F can be detected. The detection limit for thermal neutron activation analysis is 10^{-8} to 10^{-10} g. Because this method has poor spatial resolution, it is commonly used for samples of particles collected on filters of impactors. A drawback of this technique is that it requires a nuclear reactor and specialized expertise. However, several laboratories provide this service on a routine basis.

Emission and Absorption Spectroscopies. Emission and absorption spectroscopies are extremely sensitive methods for detection of trace amounts of elements in samples (Reed, 1988). Emission spectroscopy provides analyses to the ppm level with 5 mg samples. The powder is usually excited in an electric arc or by a laser flash. Chemical information is provided by the wavelength and intensities of the spectra of the emitted light. This technique is often used for quick qualitative surveys of samples.

Flame Emission Spectroscopy provides quantitative analyses of alkali elements and boron to the ppm level, with ppb detectability for some elements. This technique is commonly used for liquid samples that can be sprayed into flames. Inductively coupled plasma emission spectroscopy (ICP-ES) also utilizes liquid samples and is a reliable, fast and multielement method.

Atomic absorption spectroscopy (AAS) can analyze 30 to 40 elements present at concentrations of <0.1%. This technique is the industry standard for quantitative analysis to ppm levels for solution samples though it analyzes one element at a time whereas ICP can analyze about 40. The sample in solution form is sprayed into a flame where the radiation from a lamp is passed through the flame. Dissociated atoms in the flame absorb the radiation and reduce the transmitted intensity. The identity of the species present is determined from the absorption wavelengths.

Surface Sensitive Methods

Auger Electron Spectroscopy. Auger electron spectroscopy (AES) is a surface sensitive method for determination of the chemical composition of materials (Feldman and Mayer, 1986). This method requires an ultrahigh vacuum chamber and involves directing an electron beam at a sample which results in emission of electrons (the so-called Auger electrons) with energies that are characteristic of the elements in the sample. Depth profiling can be obtained in combination with sputtering for solid samples. This method is surface sensitive with a penetration depth of 0.5 to 2 nm and can detect elements from Li through U with up to 0.3% sensitivity. Instruments with 0.1 μm spatial resolution are available. Since even single particles can be examined, the amount of material required is very small.

Due to the special properties of nanoparticles, it is possible to extract electrons from aerosol particles under ultraviolet light irradiation (Schleicher et al., 1993). The photons are absorbed by surface atoms and the deposited energy results in the subsequent emission of a photoelectron (Greber et al., 1995). The probability of the emitted photoelectron returning to the surface is small, and therefore the particles can be efficiently charged. After removing the highly mobile negative ions from the gas stream the accumulated charge can be counted by an electrometer. The method is very sensitive allowing for the detection of submonolayer coating of particles. The photoemission depends on the photon energy being equal or bigger than the energy an electron needs to escape from the atom. Therefore this technique is sensitive to the particle surface composition. For example, most metal and most carbonaceous particles give a strong signal (Hüglin et al., 1997; Skillas et al., 1999) for wavelengths around 200 nm. Due to space charge effects there is an upper limit on the particle concentration of about 10^{12} particles/m^3 (Mohr et al., 1993). A detailed description can be found in Chapter 14. For the PAS to be used effectively in an industrial environment it has to be combined with an effective dilution system (Hueglin et al., 1997).

Energy Dispersive Spectroscopy. Energy dispersive spectroscopy (EDS) involves the probing of a sample with electrons resulting in emission of X-rays with energies characteristic of the elements present in the sample (Feldman and Mayer, 1986). Because an electron beam is used to probe the sample, the composition of particles in the nm range can be examined. Single particles can be examined by this technique. The probing depth is on the order of 1 μm. EDS has 0.1% detectability for $Z > 11$. Thus, elements with atomic weights greater than sodium can be detected in most instruments (Fig. 32–10). Resolution down to 1 μm is available with SEM and down to 50 nm is available with TEM instruments. X-ray maps of a particle can be used to examine the chemical homogeneity of multicomponent particles. These maps show regions of high or low concentrations of a given element. The technique can also be used to examine a large number of particles and to look for variations in stoichiometry

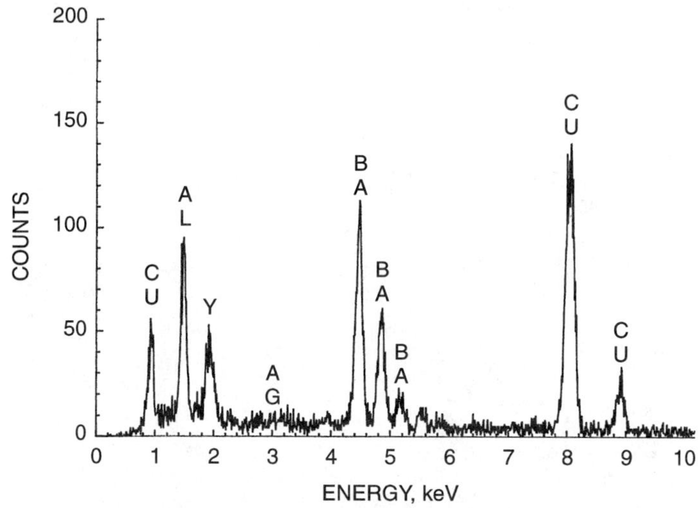

Fig. 32–10. Energy dispersive spectrometer results for an Ag-$YBa_2Cu_3O_{7-x}$ particle. The location of the peaks and their intensities determine the element and the relative amount of the element present in the probed portion of the sample. (Adapted from Carim et al., 1989.)

from particle to particle. Quantitative measurements usually require flat surfaces, but semi-quantitative information can be obtained for particles.

X-Ray Photoelectron Spectroscopy. X-ray photoelectron spectroscopy (XPS) (also known as *electron spectroscopy for chemical analysis*) can detect elements from Li—U (Feldman and Mayer, 1986). This method irradiates the particles with x-rays resulting in the emission of photoelectrons with energies characteristic of the elements in the sample and their chemical bonding. The effective probing depth is 3 nm and the method has 0.1 monolayer sensitivity. Spatial resolution of 10 μm can be obtained in state-of-the-art instruments. This limits this technique to relatively large particles. Because monolayer amounts can be detected, the amount of sample needed is small. The primary advantage of XPS over AES and EDS is that the chemical bonding of the elements can be determined.

Mass Spectrometric Methods. Mass spectrometers (MS) are extremely sensitive instruments for detection of elements from H to U and can detect ppm levels (Feldman and Mayer, 1986). The various mass spectrometric methods differ by the method used to introduce the sample into the MS. Particles on surfaces can be vaporized using pulsed lasers or can be introduced directly from the gas phase into ionization regions. These methods are usually not used on a routine basis because of their complexity.

Thermophysical and Thermochemical Analysis

Thermogravimetric Analysis. Thermogravimetric analysis (TGA) measures the change in the weight of a sample as it is heated at a known rate (Hench and Gould, 1971). The primary use of TGA for materials produced by aerosol processes is for determination of the extent of reaction and for detection of species such as water vapor. For determination of the identity of the evolved species, TGA/mass spectrometry can be used. This variation determines the identity of the species released at a certain temperature. Using rate of weight loss, kinetic parameters can also be determined (Biswas et al., 1989a,b). TGA and differential thermal

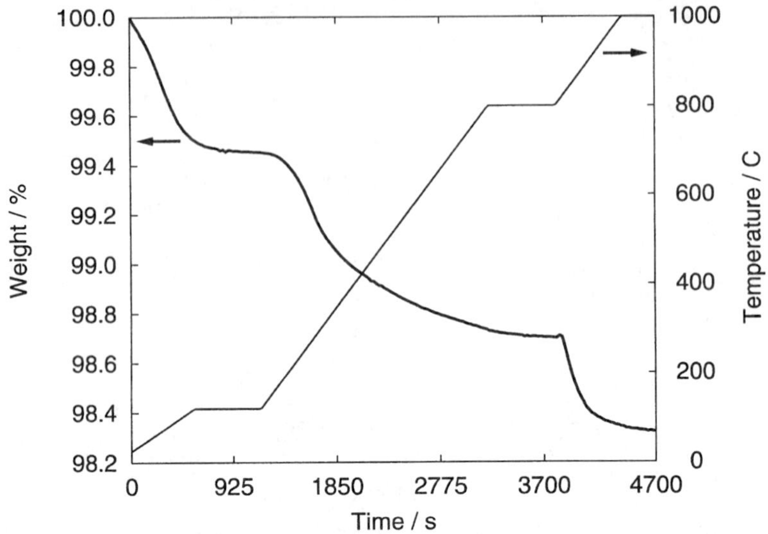

Fig. 32–11. Weight loss as a function of temperature for a silica-carbon nanocomposite. The first and second weight loss stem from the removal of physically adsorbed and chemically bound water. The third step stems from the oxidation of the carbon (Courtesy of R. Müller).

analysis (DTA) (below) typically require 50 mg samples. Figure 32–11 shows TGA for a silica-carbon nanocomposite produced in a flame reactor.

Differential Thermal Analysis. This technique (DTA) measures the amount of heat released or absorbed by a sample as it is heated at a known rate (Hench and Gould, 1971). When the enthalpy change is determined, the method is called differential scanning calorimetry (DSC). The presence of exothermic or endothermic processes at certain temperatures provides information about the nature of phase changes and chemical reactions occurring in the material as it is heated. DTA can often be used as a sensitive method for establishing the presence or absence of secondary phases in samples if these phases undergo phase transformations at known temperatures.

Surface Properties

The surface properties of powders suspended in liquids can play a critical role in powder processing. For this reason, it is often important to understand the behavior of the particles in solution. A variety of measurements are available to characterize surface properties in solution: electrophoresis, sedimentation potential, electro-osmosis, streaming potential, Zeta potential. These measurements are related and can provide information about the nature of ionized functional groups on the surfaces of the particles.

Other Methods

A variety of other methods exist for characterization of particles produced by aerosol processes. These methods include Raman spectroscopy, electron spin resonance (ESR), small angle X-ray scattering (SAXS), small-angle neutron scattering (SANS), and other processes. These are relatively specialized techniques that are expensive and are not usually used on a routine basis.

A class of techniques that is being used on an increasing basis is scanning tunneling microscopy (STM) and atomic force microscopy (AFM). These techniques provide information about structure down to the atomic level (Neddermeyer, 1990) for both conductors and insulators. This resolution is important in the case of nanometer-size clusters on surfaces that may be difficult to image by techniques such as TEM.

REFERENCES

Adachi, A., K. Okuyama, Y. Kousaka, and H. Tanaka. 1988. Preparation of gas sensor films by deposition of ultrafine tin oxide particles. *J. Aerosol. Sci.* 19:253–263.

Akhtar, M. K., Y. Xiong, and S. E. Pratsinis. 1991. Vapor synthesis of titania powder by oxidation of titanium tetrachloride. *AIChE J.* 37:1561–1570.

Alam, M. K. and R. C. Flagan. 1986. Controlled nucleation aerosol reactors: Production of bulk silicon. *Aerosol Sci. Technol.* 5:237–248.

Alexandrescu, R., I. Morjan, A. Crunteanu, S. Cojocaru, S. Petcu, V. Theodorescu, F. Huisken, B. Kohn, and M. Ehbrecht. 1998. Iron-oxide–based nanoparticles produced by pulsed infrared laser pyrolysis of $Fe(CO)_5$. *Mater. Chem. Phys.* 55:115–121.

Andres, R. P., R. S. Averback, W. L. Brown, L. E. Brus, W. A. Goddard III, A. Kaldor, S. G. Louie, M. Moscovits, P. S. Peercy, S. J. Riley, R. W. Siegel, F. Spaepen, and Y. Wang. 1989. Research opportunities on clusters and cluster assembled materials. *J. Mater. Res.* 4:704–736.

Austin, L. G. and I. Shah. 1983. A method for the inter-conversion of Microtrac and sieve size distributions. *Powder Technol.* 35:271–278.

Baker, R., W. Hurng, and H. Steinfink. 1988. Oriented high T_c superconductive layers on silver by devitrification of glasses formed in the Bi-Sr-Ca-Cu-O system. *Appl. Phys. Lett.* 54:371–373.

Barringer, E., N. Jubb, B. Gegley, R. L., Pober, and H. K. Bowen. 1984. Processing of monosized powders. In *Ultrastructure Processing of Ceramics, Glasses and Composites*, L. L. Hench and D. R. Ulrich, eds. New York: John Wiley & Sons., pp. 315–333.

Bickmore, C. R., K. F. Waldner, D. R. Treadwell, and R. M. Laine. 1996. Ultrafine spinel powders by flame spray pyrolysis of a magnesium aluminum double alkoxide. *J. Am. Ceram. Soc.* 79:1419–1423.

Biswas, P., X. Li, and S. E. Pratsinis. 1989a. Optical waveguide preform fabrication: Silica formation and growth in a high-temperature aerosol reactor. *J. Appl. Phys.* 65:2445–2450.

Biswas, P., D. Zhou, J. Grothaus, P. Boolchand, and D. McDaniel. 1990. Oxygen evolution from $YBa_2Cu_3O_{7-\delta}$ superconducting powders generated by aerosol routes. *Mater. Res. Symp. Proc.* 169:197–200.

Biswas, P., D. Zhou, I. Zitkovsky, C. Blue, and P. Boolchand. 1989b. Superconducting powders generated by an aerosol process. *Mater. Lett.* 8:233–237.

Bohrer, M. P., J. A. Amelse, P. L. Narasimham, B. K. Tariyal, J. M. Turnipseed, R. F. Gill, W. J. Moebuis, and J. L. Bodeker. 1985. A process for recovering germanium from effluents of optical fiber manufacturing. *J. Lightwave Technol.* LT-3:699–705.

Bolsaitis, P. P., J. F. McCarthy, and G. Mohiuddin. 1987. Formation of metal oxide aerosols for conditions of high supersaturation. *Aerosol Sci. Technol.* 6:225–246.

Brewster, J. and T. T. Kodas. 1997. Generation of unagglomerated, dense, BaTiO3 particles by flame-spray pyrolysis. *AIChE J.* 43:2665–2669.

Cannon, W. R., S. C. Danforth, J. H. Flint, J. S. Haggerty, and R. A. Marra. 1982. Sinterable ceramic powders from laser-driven reactions. *J. Am. Ceram. Soc.* 65:324–335.

Carim, A., P. Doherty, and T. T. Kodas. 1989. Nanocrystalline $Ba_2YCu_30_7$/Ag composite particles produced by aerosol decomposition. *Mater. Lett.* 8:335–339.

Casey J. and J. Haggerty. 1987. Laser-induced vapor phase synthesis of boron and titanium diboride powders. *J. Mater. Sci.* 22:737–744.

Chadda, S., T. T. Kodas, T. L. Ward, A. Carim, D. Kroeger, and K. C. Ott. 1991. Synthesis of $YBa_2Cu_3O_{7-x}$ and $YBa_2Cu_4O_8$ by aerosol decomposition. *J. Aerosol Sci.* 22:601–616.

Charlesworth, D. and W. Marshall. 1969. Evaporation from droplets containing disolved solids. *AIChE J.* 6:9–23.

Clark, H. B. 1977. Titanium dioxide pigments. In *Treatise on Coatings: Pigments, 3*, P. R. Myers and J. S. Long, eds. New York: Marcel Dekker, pp. 479–532.

Dannenberg, E. M. 1971. Progress in carbon black technology. *J. IRI* Oct:190–195.

Dobbins, R. A. and C. M. Megaridis. 1987. Morphology of flame-generated soot as determined by thermophoretic sampling. *Langmuir* 3:254–259.

Dobbins, R. A. and G. W. Mulholland. 1984. Interpretation of optical measurements of flame generated particles. *Combust. Sci. Technol.* 40:175–191.

Edelson L. and A. Glaser. 1988. Role of particle substructure in the sintering of monosized titania. *J. Am. Ceram. Soc.* 71:225–235.

Epstein, J. 1976. Utilization of the dead sea minerals. *Hydrometallurgy* 2:1–10.

Feldman L. and J. Mayer. 1986. *Fundamentals of Surface and Thin Film Analysis*. New York: North-Holland.

Flint, J. H. and J. S. Haggerty. 1990. A model for growth of silicon particles from laser-heated gases. *Aerosol Sci. Technol.* 13:72–84.

Formenti, M., F. Juillet, P. Mereaudeau, S. Techner, and P. Vergnon. 1972. Preparation in a hydrogen-oxygen flame of ultrafine metal oxide particles, oxidative properties toward hydrocarbons in the presence of ultraviolet radiation. In *Aerosols and Atmospheric Chemistry*, G. Hidy, ed. New York: Academic, pp. 45–55.

Frank, J. 1995. Approaches to large-scale structures. *Curr. Opin. Struct. Biol.* 5:194–201.

French, W. G., L. J. Pace, and V. A. Foertmeyer. 1978. Chemical kinetics of the reactions of $SiCl_4$, $SiBr_4$, $GeCl_4$, $POCl_3$ and BCl_3 with oxygen. *J. Phys. Chem.* 82:2191–2194.

Friedlander, S. K. 1977. *Smoke, Dust and Haze.* New York: Wiley Interscience.

Gani, M. and R. McPherson. 1980. The structure of plasma-prepared Al_2O_3 and TiO_2 powders. *J. Mater. Sci.* 15:1915–1925.

Gardner, T., D. Sproson, and G. Messing. 1984. Precursor effects on development of particle morphology during evaporative decomposition of solutions. *Mater. Res. Soc. Symp. Proc.* 32:227–232.

George, A. P., R. D. Murley, and E. R. Place. 1973. Formation of TiO_2 aerosol from the combustion supported reaction of $TiCl_4$ and O_2. *Faraday Symp. Chem. Soc.* 7:63–77.

Girshick, S. L. and C.-P. Chiu. 1989. Homogeneous nucleation of particles from the vapor phase in thermal plasma synthesis. *Plasma Chem. Plasma P.* 9:355–369.

Girshick, S. L. and C.-P. Chiu. 1990. Numerical study of MgO powder synthesis by thermal plasma. *J. Aerosol Sci.* 21:641–650.

Granqvist, C. and R. Buhrman. 1976. Ultrafine metal particles. *J. Appl. Phys.* 47:2200–2219.

Greber, T., T. Giessel, C. Pettenkofer, H. C. Siegmann, and G. Ertl. 1995. Substrate mediated autoionisation of benzene on graphite. *Surf. Sci.* 343: L1187–L1191.

Hardesty, D. R. and F. J. Weinberg. 1973. In *Fourteenth Symposium (International) on Combustion*. Pittsburgh: The Combustion Institute, pp. 907–918.

Haynes, B. S., H. Jander, and H. G. Wagner. 1979. In *Seventeenth Symposium (International) on Combustion*. Pittsburgh: The Combustion Institute, pp. 1365–1374.

Hench, L. L. and R. W. Gould. 1971. *Characterization of Ceramics.* New York: Marcel Dekker.

Hinds, W. C. 1999. *Aerosol Technology.* New York: John Wiley & Sons.

Hoppe, W., R. Langer, G. Knesch, and C. Poppe. 1968. Proteinkristallstrukturanalyse mit Elektronenstrahlen. *Naturwissenschaften* 55:333–336.

Hüglin, C., L. Scherrer, and H. Burtscher. 1997. An accurate, continuously adjustable dilution system (1 : 10 to 1 : 10^4) for submicron aerosols. *J. Aerosol Sci.* 28:1049–1055.

Hurd, A. J. and W. L. Flower. 1988. In-situ growth and structure of fractal SiO_2 aggregates. *J. Colloid Interface Sci.* 122:178–192.

Imai H. and K. Takami. 1985. Preparation of fine particles of carnegieite by a mist decomposition method. *J. Mater. Sci.* 20:1823–1827.

Ingebrethsen, B. J. and E. Matijevic. 1980. Preparation of uniform colloidal dispersions by chemical reactions in aerosols—2. Spherical particles of aluminum hydrous oxide. *J. Aerosol Sci.* 11:271–280.

Ingebrethsen, B. J., E. Matijevic, and R. Partch. 1983. Preparation of uniform colloidal dispersions by chemical reactions in aerosols iii. mixed titania alumina colloidal spheres. *J. Colloid Interface Sci.* 95:228–239.

Ivie, J. J. and L. J. Forney. 1988. A numerical model of the synthesis of carbon black by benzene pyrolysis. *AIChE J.* 34:1813–1820.

Johnson, D. 1981 Non-conventional powder preparation techniques. *Ceram. Bull.* 60:21–29.

Johnson, S. M., M. I. Gusman, and D. J. Rowcliffe. 1987. Freeze drying. *Adv. Ceram. Mater.* 2:237–241.

Kagawa, M., F. Honda, H. Onodera, and T. Nagae. 1983. The Formation of ultrafine Al_2O_3, ZrO_2, and Fe_2O_3 by the spray ICP technique. *Mater. Res. Bull.* 18:1081–1087.

Kanapilly, G., O. Raabe, and G. Newton. 1970. A new method for the generation of aerosols of insoluble particles. *J. Aerosol Sci.* 1:313–323.

Katz, J. L. and C. H. Hung. 1990. *Twenty-Third Symposium (International) on Combustion.* Pittsburgh: The Combustion Institute, pp. 1733–1738.

Kerker, M. 1975. Laboratory generation of aerosols. *Adv. Colloid Interface* 5:105–172.

Kim, K. S. 1997. Analysis on SiO_2 particle generation and deposition using furnace reactor. *AIChE J.* 43:2679–2687.

Kim, K.-S. and S. E. Pratsinis. 1988. Manufacture of optical waveguide preforms by modified chemical vapor deposition. *AIChE J.* 34:912–921.

Kim, K.-S. and S. E. Pratsinis. 1989. Modeling and analysis of modified chemical vapor deposition of optical fiber preforms. *Chem. Eng. Sci.* 44:2475–2482.

Kingery, W. D., H. K. Bowen, and D. R. Uhlmann. 1976. *Introduction to Ceramics.* New York: Wiley Interscience.

Kodas, T. T. 1989. Generation of complex metal oxides by aerosol processes: Superconducting ceramic particles and films. *Angew. Chem. Int. Ed.* 28:794–807.

Kodas, T. T., A. Datye, V. Lee, and E. Engler. 1989a. Single-crystal $YBa_2Cu_30_7$ particle formation by aerosol decomposition. *J. Appl. Phys.* 65:2149–2151.

Kodas, T. T., E. Engler, and V. Lee. 1989b. Generation of thick $YBa_2Cu_30_7$ films by aerosol deposition. *Appl. Phys. Lett.* 54:1923–1925.

Kodas, T. T., E. M. Engler, V. Lee, R. Jacowitz, T. H. Baum, K. Roche, S. S. P. Parkin, W. S. Young, S. Hughes, J. Kleder, and W. Auser. 1988. Aerosol flow reactor production of fine $YBa_2Cu_30_7$ powder: Fabrication of superconducting ceramics. *Appl. Phys. Lett.* 52:1622–1624.

Kodas, T. T., S. E. Pratsinis, and A. Sood. 1987. Submicron alumina powder production by a turbulent flow aerosol process. *Powder Technol.* 50:47–53.

Koguchi, M., Y. Matsuda, E. Kinoshita, and K. Hirabayashi. 1990. Preparation of $YBa_2Cu_3O_{7-x}$ films by flame pyrolysis. *J. Appl. Phys. Jpn.* 29: L33–35.

Komiyama, A., T. Osawa, H. Kazi, and T. Konno. 1987. Rapid growth of AlN films by particle precipitation aided CVD. In *High Tech Ceramics*, P. Vincenzini, Ed. Amsterdam: Elsevier, pp. 667–676.

Koukitu, A., Y. Hasegawa, H. Seki, H. Komijama. I. Tanaka, and Y. Kamioka. 1989. Preparation of YBaCuO superconducting thin films by the mist microwave plasma decomposition method. *J. Appl. Phys. Jpn.* 28:L1212–1213.

Kriechbaum, G. and P. Kleinschmidt. 1989. Superfine oxide powders: Flame hydrolysis and hydrothermal synthesis. *Adv. Mater. Angew. Chem.* 10:330–337.

Kruis, F. E., K. Nielsch, H. Fissan, B. Rellinghaus, and E. F. Wassermann, 1998a. Preparation of size-classified PbS particles in the gas phase. *Appl. Phys. Lett.* 73:547–549.

Kruis, F. E., W. Oostra, J. Marijnissen, J. Schoonman, and B. Scarlett. 1998b. Particle formation paths in the synthesis of silicon nitride powder in a laser-heated aerosol reactor. *J. Eur. Ceram. Soc.* 18: 1037–1043.

Landgrebe, J. D. and S. E. Pratsinis. 1990. A discrete-sectional model for powder production by gas-phase chemical reaction and aerosol coagulation in the free molecular regime. *J. Colloid Interface Sci.* 139:63–86.

Landgrebe, J. D., S. E. Pratsinis, and S. V. R. Mastrangelo. 1990. Nomographs for vapor synthesis of ceramic powders. *Chem. Eng. Sci.* 45:2931–2941.

Lee, D. and M. Choi. 2000. Control of size, morphology and phase of nanoparticles using CO_2 laser during flame synthesis. *J. Aerosol Sci.* 31:1145–1163.

Lefebvre, A. H. 1989. *Atomization and Sprays.* New York: Hemisphere Publishing.

Leong, K. 1981. Morphology of aerosol particles generated from the evaporation of solution drops. *J. Aerosol Sci.* 12:417–435.

Leong, K. 1987. Morphology control of particle generated from the evaporation of solution drops. *J. Aerosol Sci.* 18:525–552.

Lewis, D. J. 1991. Technique for producing mullite and other mixed-oxide systems. *J. Am. Ceram. Soc.* 74:2410–2413.

Lukasiewicz, S. 1989. Spray drying of ceramic powders. *J. Am. Ceram. Soc.* 72:617–624.

MacChesney, J. B., P. B. O'Connor, and H. M. Presby. 1974. A new technique for the preparation of low-loss graded-index optical fibers. *Proc. IEEE.* 62:1280–1281.

Masdiyasni, K., C. Lynch, and J. Smith. 1965. Preparation of ultra-high-purity submicron refractory oxides. *J. Am. Ceram. Soc.* 48:372–375.

Masters, K. 1972. *Spray Drying.* New York: John Wiley & Sons.

Matijevic, E. 1986. Monodispersed colloids: Art and science. *Langmuir* 2:12–20.

Matijevic, E. 1987. Colloid science in ceramic powder preparation. In *High Tech Ceramics*, P. Vincenzini, ed. Amsterdam: Elsevier, pp. 441–458.

Matson, D. W., R. Peterson, and R. Smith. 1987. Production of powders and films by the rapid expansion of supercritical solutions. *J. Mater. Sci.* 22:1919–1928.

Medalia, A. I. and D. Rivin. 1976. Carbon blacks. In *Characterization of Powder Surfaces*, G. D. Parfitt and K. S. W. Sing, eds. New York: Academic Press, pp. 280–351.

Merkle, B. D., R. N. Kniseley, F. A. Schmidt, and I. E. Anderson. 1990. Superconducting $YBa_2Cu_3O_x$ particulate produced by total consumption burner processing. *Mater. Sci. Eng. A* 124:31–38.

Messing, G. L. S. Zhang, and G. Jayanthi. 1993. Ceramic powder synthesis by spray pyrolysis. *J. Am. Ceram. Soc.* 76:2707–2726.

Mezey, E. J. 1966. Pigments and reinforcing agents. In *Vapor Deposition*, C. F. Powell, J. H. Oxley, and J. M. Blocher Jr., eds. New York: John Wiley & Sons. pp. 423–451.

Miller, B. and R. Lines. 1988. Recent advances in particle size measurements: A critical review. *CRC Crit. Rev. Anal. Chem.* 20:75–116.

Mooney, J. and S. Redding. 1982. Spray pyrolysis. *Annu. Rev. Mater. Sci.* 12:81–90.

Mohr, M., B. A. Kwetkus, and H. Burtscher. 1993. Improvement of electrostatic precipitation by UV-charging of submicron particles. *J. Aerosol Sci.* 24(Suppl. 1):S247–S248.

Morrison, P. W., R. Raghavan, A. J. Timpone, C. P. Artelt, and S. E. Pratsinis. 1997. In situ Fourier transformed infrared characterisation of the effect of electrical fields on the flame synthesis of TiO_2 particles. *Chem. Mater.* 9:2702–2708.

Nagel, S. R., J. B. MacChesney, and K. L. Walker. 1982. An overview of the modified chemical vapor. *IEEE J. Quantum Electron.* QE-18:459–476.

Neddermeyer, H. 1990. STM studies of nucleation and the initial stages of film growth. *Crit. Rev. Solid State* 16:309–335.

Niessner, R. 1990. Chemical characterization of aerosols. *Fresenius J. Anal. Chem.* 337:565–576.

Okuyama, K., J.-T. Jeung, Y. Kousaka, H. V. Nguyen, J. J. Wu, and R. C. Flagan. 1989. Experimental control of ultrafine TiO_2 particle generation from thermal decomposition of titanium tetraisopropoxide vapor. *Chem. Eng. Sci.* 44:1369–1375.

Okuyama, K., Y. Kousaka, N. Tohge, S. Yamamoto, J. J. Wu, R. C. Flagan, and J. H. Seinfeld. 1986. Production of ultrafine metal oxide aerosol particles by thermal decomposition of metal alkoxide vapors. *AIChE J.* 32:2010–2019.

Ortega, J., T. T. Kodas, S. Chadda, D. M. Smith, M. Ciftcioglu, and J. Brennan. 1991. Generation of dense barium calcium titanate particles by aerosol decomposition. *Chem. Mater.* 3:746–751.

Pebler, A. and R. Charles. 1989. Synthesis of superconducting oxides by aerosol pyrolysis of metal EDTA solutions. *Mater. Res. Bull.* 24:1069–1076.

Peng, J. H., P. J. Hong, S. S. Dai, D. Vollath, and D. V. Szabo. 1998. Microwave plasma sintering of nanocrystalline alumina. *J. Mater. Sci. Technol.* 14:173–175.

Petersen, R., D. Matson, and R. Smith. 1986. Rapid precipitation of low vapor pressure solids from supercritical fluid solutions: The formation of thin films and powders. *J. Am. Chem. Soc.* 108:2100–2102.

Phillips D. and G. Vogt. 1987. Plasma synthesis of ceramic powders. *MRS Bull.* Oct:54–58.

Pickles C. and A. McLean. 1983. Production of fused refractory oxide spheres and ultrafine oxide particles in an extended arc. *Am. Ceram. Soc. Bull.* 62:1004–1009.

Powers, D. R. 1978. Kinetics of $TiCl_4$ oxidation. *J. Am. Ceram. Soc.* 61:295–297.

Pratsinis, S. E. 1988. Simultaneous nucleation, condensation and coagulation in aerosol reactors. *J. Colloid Interface Sci.* 124:416–427.

Pratsinis, S. E. 1990. Vapor synthesis of ceramic powders. In *Ceramic Powder Science—III*, G. L. Messing, H. Hausner, and S.-C. Hirano, eds. Columbus, OH: American Ceramic Society, pp. 227–238.

Pratsinis, S. E. 1998. Flame aerosol synthesis of ceramic powders. *Prog. Energy Combust. Sci.* 24:197–219.

Pratsinis, S. E., H. Bai, P. Biswas, M. Frenklach, and S. V. R. Mastrangelo. 1990. Kinetics of $TiCl_4$ oxidation. *J. Am. Ceram. Soc.* 73:2158–2162.

Pratsinis, S. E., J. D. Landgrebe, and S. V. R. Mastrangelo. 1989. Design correlations for gas phase manufacture of ceramic powders. *J. Aerosol Sci.* 20:1457–1460.

Pratsinis, S. E. and S. V. R. Mastrangelo. 1989. Material synthesis in aerosol reactors. *Chem. Eng. Prog.* 85(5):62–66.

Prochaska, S. and C. Greskovich. 1978. Synthesis and characterization of silicon nitride powder. *Ceram. Bull.* 57:579–586.

Ramsey, D. and R. Avery. 1974. Ultrafine oxide particles prepared by electron beam evaporation. *J. Mater. Sci.* 9:1681 1688.

Rao, N. R. and P. H. McMurry. 1989. Nucleation and growth of aerosol in chemically reacting systems: A theoretical study of the near collision-controlled regime. *Aerosol Sci. Technol.* 11:120–132.

Reed, J. S. 1988. *Introduction to the Principles of Ceramic Processing*. New York: John Wiley & Sons.

Ruthner, M. I. 1979. Preparation and sintering characteristics of Mgo, $MgO-Cr_2O_3$ and $MgO-Al_2O_3$. *Sci. Sintering* 11:203–208.

Ruthner, M. I. 1983. Industrial production of multicomponent ceramic powders by means of the spray roasting method. In *Ceramic Powders*, P. Vincenzini, ed. Amsterdam: Elsevier, pp. 515–531.

Salazar, K., K. C. Ott, R. C. Dye, K. M. Hubbard, E. J. Peterson, J. Y. Coulter, and T. T. Kodas. 1992. Aerosol assisted chemical vapor deposition of superconducting $YBa_2Cu_3O_{7-x}$. *Physica C.* 198:303–308.

Santoro, R. J. and J. H. Miller. 1987. Soot particle formation in diffusion flames. *Langmuir* 3:244–254.

Schaefer, D. and A. J. Hurd. 1990. Growth and structure of combustion aerosols. *Aerosol Sci. Technol.* 12:876–890.

Schleicher, B., H. Burtscher, and H. C. Siegmann. 1993. Photoelectric quantum yield of nanometer metal particles. *Appl. Phys. Lett.* 63:1191–1193.

Siefert, W. 1984. Properties of thin In_2O_3 and SnO_2 films by corona spray pyrolysis and a discussion of the spray pyrolysis process. *Thin Solid Films* 121:275–282.

Siegel, R. 1990. Nanophase materials assembled from atomic clusters. *MRS Bull.* Oct:60–67.

Singh, R. and R. Doherty. 1990. Synthesis of TiN powders under glow discharge plasma. *Mater. Lett.* 9: 87–89.

Skillas, G., C. Hüglin, and H. C. Siegmann. 1999. Determination of air exchange rates of rooms and deposition factors for fine particles by means of photoelectric aerosol sensors. *Indoor Built Environ.* 8: 246–254.

Sokolowski, M., A. Sokolowska, A. Michalski, and B. Gokieli. 1977. The in-flame reaction method for Al_2O_3 aerosol formation. *J. Aerosol Sci.* 8:219–230.

Suyama Y. and A. Kato. 1976. TiO_2 produced by vapor-phase oxygenolysis of $TiCl_4$. *J. Am. Ceram. Soc.* 59:146–149.

Takagi, T., K. Matsubara, and H. Takaoka. 1980. Optical and thermal properties of BeO thin films prepared by reactive ionized cluster beam technique. *J. Appl. Phys.* 51:5419–5429.

Tomar, M. and F. Garcia. 1981. Spray pyrolysis in solar cells and gas sensors. *Prog. Cryst. Growth Charact.* 4:221–247.

Tsantilis, S., S. E. Pratisinis, and V. Haas. 1999. Simulation of synthesis of palladium nanoparticles in a jet aerosol flow condenser. *J. Aerosol Sci.* 30:785–803.

Ulrich, G. D. 1971. Theory of particle formation and growth in oxide synthesis flames. *Combust. Sci. Technol.* 4:47–57.

Ulrich, G. D. 1984. Flame synthesis of fine particles. *Chem. Eng. News* Aug 6:22–29.

Ulrich, G. D. and J. W. Riehl. 1982. Aggregation and growth of submicron oxide particles in flames. *J. Colloid Interface Sci.* 87:257–265.

Vemury, S. and S. E. Pratsinis. 1995. Corona-assisted flame synthesis of ultrafine titania particles. *Appl. Phys. Lett.* 66:3275–3277.

Vemury, S. and S. E. Pratsinis. 1996. Charging and coagulation during flame synthesis of silica. *J. Aerosol Sci.* 27:951–966.

Vemury, S., S. E. Pratsinis, and L. Kibbey. 1997. Electrically controlled flame synthesis of nanophase TiO_2, SiO_2 and SnO_2 powders. *J. Mater. Res.* 12:1031.

Viguie, J. C. and J. Spitz. 1975. Chemical vapor deposition at low temperature. *J. Electrochem. Soc.* 122: 585–588.

Visca, M. and E. Matijevic. 1979. Preparation of uniform colloidal dispersions by chemical reactions in aerosols. i. spherical particles of titanium dioxide. *J. Colloid Interface Sci.* 68:308–319.

Vollath, D. 1990. Pyrolytic preparation of ceramic powders by a spray calcination technique. *J. Mater. Sci.* 25:2227–2232.

Vollath, D., D. V. Szabó, and J. Hausselt. 1997. Synthesis and properties of ceramic nanoparticles and nanocomposites. *J. Eur. Ceram. Soc.* 17:1317–1324.

Weimer, A. W., W. G. Moore, R. P. Roach, C. N. Haney, and W. Rafaniello. 1991. Rapid carbothermal reduction of boron oxide in a graphite reactor. *AIChE J.* 37:759–768.

Wenckus, J. and W. Leavitt. 1957. Preparation of ferrites by the atomizing burner technique. *Magnetism and Magnetic Materials Conference.* Boston, MA, 1956, Amer. Inst. Elect. Engs, pp. 526–530.

Wu, J. J. and R. C. Flagan. 1988. A discrete-sectional solution to the aerosol dynamic equation. *J. Colloid Interface Sci.* 123:339–352.

Wu, J. J., H. V. Nguyen, R. C. Flagan, K. Okuyama, and Y. Kousaka. 1988. Evaluation and control of particle properties in aerosol reactors. *AIChE J.* 34:1249–1256.

Xiong Y. and T. T. Kodas. 1993. Gas phase production of particles in reactive turbulent flows. *J. Aerosol Sci.* 24:893–908.

Xiong Y. and S. E. Pratsinis. 1991. Gas phase production of particles in reactive turbulent flows. *J. Aerosol Sci.* 22:637–656.

Zachariah, M. R. and S. Huzarewicz. 1991. Aerosol processing of YBaCuO superconductors in a flame reactor. *J. Mater. Res.* 6:264–269.

Zhang, S. C., G. L. Messing, and M. Borden. 1990. Synthesis of solid spherical zirconia particles by spray pyrolysis. *J. Am. Ceram. Soc.* 73:61–67.

33 Aerosol Measurements in Cleanrooms

ROBERT P. DONOVAN
L & M Technologies, Inc., Albuquerque, NM

INTRODUCTION

The concentration of aerosol particles in a cleanroom is one of the primary metrics used to specify the performance of that cleanroom, and in many cleanroom applications it is the only performance specification of concern. It is the most common metric by which cleanrooms are bought and sold. Measurement of aerosol particle concentration in a cleanroom and measurement of aerosol particle emission from cleanroom apparatus are the measurements discussed in this chapter.

Background

The term *cleanroom* (one word; it is not just a *clean room*, a term appropriate for, say, a bedroom after being swept, dusted, and straightened up) describes a specialized work area built expressly for the purpose of carrying out activities whose success can be adversely affected by the presence of any of the many extraneous airborne contaminants typically found in indoor atmospheres. The term implies the use of active measures to eliminate or reduce the concentration of contaminants in the work area atmosphere. It also implies some sort of isolation between the atmosphere of the work area and that of the atmosphere external to the work area.

While the need to create special, controlled work spaces for certain activities has long been recognized and implemented (e.g., hospital operating rooms), the first modern cleanroom, employing unidirectional air flow, was built at Sandia National Laboratories in the early 1960s to improve reliability of parts manufactured for the weapons program. The concept was to construct a room isolated from its surroundings, to filter all air entering that room, and to locate the inlet and outlet ducts so that the air flow was more or less unidirectional between them. This design establishes a convective flow that sweeps contaminated air, especially that generated by the activities going on inside the room, out of the room and replaces it with freshly filtered air. A room so designed and built became the prototype "cleanroom" for what has now become an indispensable work environment for manufacturing many of today's precision products.

Contemporary cleanrooms are simply more sophisticated versions of the early Sandia concept. In today's state-of-the-art cleanrooms, such as those used in manufacturing semi-

Aerosol Measurement: Principles, Techniques, and Applications, Second Edition, Edited by Paul A. Baron and Klaus Willeke.
ISBN 0-471-35636-0

conductor chips, magnetic head assemblies, and other precision products, vertical laminar flow (VLF) is the rule. The entire ceiling of the cleanroom is covered with HEPA or ultra low penetrating air (ULPA) filters (see Chapter 9), which feed air into the work space beneath at flow velocities typically in the 0.25 to 0.50 m/s range. In like fashion, the entire floor of the work space is the return duct so that in an empty cleanroom the air flow is essentially laminar and directed vertically downward. This arrangement means that the air reaching the work space level is freshly filtered and that any contaminant emissions generated at the work level by the ongoing activity are convected away from the work space and into the floor return ducts. The exhaust air recirculates through conditioning filters and re-enters the cleanroom through the ceiling ULPA filters along with any makeup air needed to replace air losses (hoods venting to the outside, leaks, and so forth). Figure 33–1 is a photograph of the interior of a cleanroom used for manufacturing state-of-the-art semiconductor chips. The ceiling surface facing the work space is composed of 100% filters except for the light fixtures. The perforated squares making up the floor are the entry ports to the return ducts for recirculating the cleanroom air.

The primary impetus for cleanroom development has come from those industries most vulnerable to particulate contamination. The concentration of aerosol particles in the ambient air can be greater than $10^9\,m^{-3}$. The concentration of aerosol particles in a cleanroom is often less than $10\,m^{-3}$. This dramatic reduction in concentration is the primary value contributed by cleanroom suppliers of which there are many with many differing designs. As a result, standards for cleanroom performance were initiated as part of the pioneering cleanroom work at Sandia and since then have been periodically revised and updated by broadbased industrial/professional groups having vital interests in the use and performance of cleanrooms. The most recent versions of these standards have been developed by international committees and are now internationally accepted.

The International Cleanroom Standards

According to ISO 14644-1 (ISO, 1999), a cleanroom is a "room in which the concentration of airborne particles is controlled and which is constructed and used in a manner to minimize the introduction, generation and retention of particles inside the room and in which other relevant parameters, e. g. temperature, humidity and pressure, are controlled as necessary." No mention of contaminants other than aerosol particles appears in this definition. The concentration of nonparticulate contaminants ("molecular contaminants") in a cleanroom is not part of the definition and has no bearing or influence in the system used by ISO 14644-1 to classify cleanrooms. While ignoring the concentration of molecular contaminants in a cleanroom may be a serious oversight in many applications, this chapter, following the lead of the standards and in accordance with the subject of this handbook, discusses only the measurement of the concentration and size distribution of aerosol particles in a cleanroom.

The aerosol particles found in a state-of-the-art cleanroom represent one extreme of the aerosol particle size/concentration spectrum. The particles of interest in the highest quality cleanrooms are those at the smallest end of the aerosol particle size range, and the concentrations of interest are among the lowest detectable. A state-of-the-art cleanroom is a space in which particles have not only been controlled but in many cases eliminated insofar as possible. Measuring aerosol particle concentration in such a cleanroom thus means measuring small-sized aerosol particles at low concentrations.

This chapter consists of three major topics:

1. A detailed review of the newly published ISO 14644-1 standard and its companion standard, ISO 14644-2 (ISO, 2000). ISO 14644-1 defines an international system for classifying the quality of cleanroom air, defined in terms of aerosol particle

Fig. 33–1. Contemporary VLF cleanroom used in semiconductor manufacturing. Particle-free air continuously enters through the ceiling filters and exits through the perforated floor panels. (Courtesy of Semiconductor International Magazine.)

concentration, and spells out procedures for verifying that the air quality in a given cleanroom meets the requirements of its claimed classification. This verification procedure requires measuring the concentration and size distribution of aerosol particles in that cleanroom. ISO 14644-2 offers monitoring guidelines that provide some assurance that the performance of a clean room, previously certified by the methods of ISO 14644-1, has been maintained.

2. A brief description of the commercially available instrumentation suitable for making measurements of aerosol particle concentration and size distribution in a cleanroom. This list of instrumentation properties is based on literature from U.S. instrument suppliers only. However, the performance specifications given here for just U.S. manufacturers can be assumed to be representative of instrumentation that is commercially available from manufacturers worldwide, as reported in Chapter 15.
3. A test method proposed for measuring the emission rate of aerosol particles from specific particle sources inside a cleanroom, such as apparatus, equipment, or even personnel. This test method provides a basis for ranking candidate cleanroom apparatus according to their propensity to introduce particulate contamination into a cleanroom. It is taken from an Institute of Environmental Sciences and Technology (IEST) recommended practice (IEST, 1992).

Thus the primary goals of the chapter are to provide a contemporary summary of (1) the measurements of aerosol particle concentrations that are required in today's cleanrooms in order to satisfy business contracts and to maintain control of the environment for manufacturing precision products; and (2) the performance and properties of the instrumentation that is commercially available for making those required measurements of aerosol particle concentration in a cleanroom. A secondary goal is to emphasize that anything brought into a cleanroom is a potential source of airborne particulate contamination and to present a test method for measuring the emission rate of aerosol particles from the tools and other objects so introduced.

INTERNATIONAL STANDARDS FOR CLASSIFYING, VERIFYING, AND MONITORING CLEANROOMS: ISO 14644-1 AND -2

Cleanroom standards classify cleanroom air quality by code names that specify a maximum allowable concentration of aerosol particles equal to or greater than a specified reference particle size. ISO 14644-1 defines particle size as the "the diameter of a sphere that produces a response by a given particle-sizing instrument that is equivalent to the response produced by the particle being measured." The composition of the reference sphere is not specified by ISO 14644-1, but optical particle counters are most often calibrated using polystyrene latex spheres.

Typically, cleanroom standards use a code name composed of numbers from which the maximum allowable aerosol particle concentration permitted under that class code name can be directly read for the reference particle size. For example, Class 10 air quality, according to Federal Standard 209E (1992), means that the concentration of aerosol particles, 0.5 μm and larger, is no greater than 10 particles/ft^3 (350 particles/m^3). Class 1 implies a concentration of aerosol particles, 0.5 μm and larger, that is no greater than 1 particle/ft^3 (35 particles/m^3), and so on. In this specific U.S. federal standard, the reference particle size is always 0.5 μm and larger. Other cleanroom standards specify other reference particle sizes and use other units of aerosol particle concentration. The discussion in this section is of one classification system only—the classification system defined by the newly published ISO 14644-1 (ISO, 1999) standard and its companion standard, ISO 14644-2 (ISO, 2000).

ISO Standards 14644-1 and -2

On May 1, 1999, the International Organization for Standardization (ISO), Geneva, Switzerland, issued ISO 14644-1: *Cleanrooms and Associated Controlled Environments—Part 1: Classification of Air Cleanliness* (ISO, 1999). Its companion standard was issued on September 15, 2000: ISO 14644-2: *Cleanrooms and Associated Controlled Environments—Part 2: Specifications for Testing and Monitoring to Prove Continued Compliance with ISO 14644-1* (ISO, 2000).

These documents were prepared by an international committee with the goal of replacing the many differing national standards previously used for cleanroom classification. While the general scope and purpose of all the preceding national standards have always been basically the same, each country developed its own definitions so that a given class code name in one country's standard invariably had a different name than the same cleanroom air quality would have in another country's standard. When, by coincidence perhaps, the same class code name was used in two different national standards, it invariably referred to a different concentration of aerosol particles in one standard than in the other. The particle size to which the class code name referred varied among the various national standards. Except for the most recent version (FED-STD-209E, which used both English and its own unique, metric class designations), the U.S. 209 standards expressed particle concentrations exclusively in ft^{-3}; all other national standards used m^{-3}, and so it went. Table 34–2 of Fissan et al. (1993) presents a short summary comparing the designations used by five of the predecessor national standards. No easy conversion from the usage of one national standard to that of another existed. This confusing jumble of conflicting names and unique designations has long made obvious the need for a single, universal standard for cleanroom classification. The ISO has now provided just such a consensus international standard, which is the topic of this section.

Class Codes as Defined by ISO 14644-1

Table 33–1 (a reproduction of Table 1 from ISO 14644-1), and Figure 33–2 (a modified reproduction of Fig. A.1 from ISO 14644-1) summarize the classifications of this new international standard. The class designations are expressed as "ISO Class *N*" where *N* can be any number between 1.0 and 9.0 in increments of 0.1. As in predecessor standards, N describes the maximum allowable concentration for its class. The maximum allowed aerosol particle concentration for ISO Class *N* is 10^N particles/m^3 that are 0.1 μm in size or larger. While the reference particle size in the international standard is 0.1 μm, a maximum allowable concentration can be calculated for any particle size between 0.1 and 5.0 μm by the following equation (Eq. 1 from ISO, 1999):

$$C_{\max} = 10^N \times (0.1/d)^{2.08} \tag{33–1}$$

where $C_{\max}$ (particles/m^3) is the maximum allowed concentration of aerosol particles of size *d* and larger; *N* is the class code ($9.0 \geq N \geq 0.1$); and *d* (μm) is the lower bound on the particle size at which the aerosol particle concentration is measured.

Only for $d = 0.1\,\mu m$ is $C_{\max} = 10^N$. At all other values of *d*, a factor less than 1, as given in Eq. 33–1, reduces the maximum allowed aerosol particle concentration. The standard specifies that *d* must always be in the range $5.0\,\mu m \geq d \geq 0.1\,\mu m$. For most values of *N*, allowable values of *d* are further restricted as indicated in both Table 33–1 and Figure 33–2. The standard does not allow $C_{\max}$ to be measured at particle sizes outside the range of the filled blocks in the row corresponding to each integer class code in Table 33–1 or beyond the solid circles on the end of the curves plotted in Figure 33–2 for each integer class. For example, ISO classes 7, 8, and 9 can be verified by measuring the concentration of aerosol particles only in the size

TABLE 33–1. Cleanroom Classification Codes as Defined in ISO 14644-1

ISO Classification Number (N)	Maximum Concentration Limits (Particles/m³ of Air) for Particles Equal to and Larger Than the Considered Sizes Shown Below (Concentration Limits are Calculated in Accordance with Equation [1] in 3.2)					
	0.1 µm	0.2 µm	0.3 µm	0.5 µm	1 µm	5 µm
ISO Class 1	10	2				
ISO Class 2	100	24	10	4		
ISO Class 3	1,000	237	102	35	8	
ISO Class 4	10,000	2,370	1,020	352	83	
ISO Class 5	100,000	23,700	10,200	3,520	832	29
ISO Class 6	1,000,000	237,000	102,000	35,200	8,320	293
ISO Class 7				352,000	83,200	2,930
ISO Class 8				3,520,000	832,000	29,300
ISO Class 9				35,200,000	8,320,000	293,000

Note: Uncertainties related to the measurement process require that concentration data with no more than three significant figures be used in determining the classification level.

Source: From ISO (1999), Table 1.

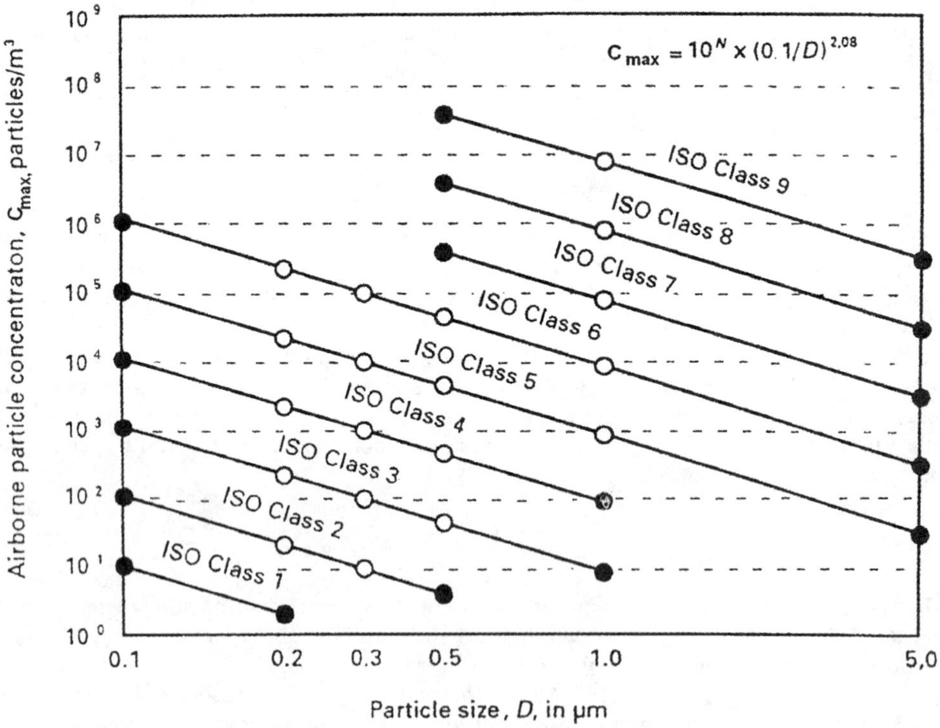

Fig. 33–2. Maximum allowable aerosol particle concentrations for the Class codes of ISO 14644-1. C_{max} represents the maximum permitted concentration (in particles per cubic metre of air) of airbome particles equal to and larger than the considered particle size. N represents the specified ISO class number. (ISO 1991.)

range $5.0\,\mu m \geq d \geq 0.5\,\mu m$. Similarly, showing compliance with the requirements for ISO Class 1 air quality requires measurements in the size range $0.2\,\mu m \geq d \geq 0.1\,\mu m$. The range of d allowed for non-integer values of N is not explicitly defined in the standard. Because one intent of limiting the ranges of d is to not allow verification of cleanrooms of high quality (low values of N) at large particle sizes where aerosol particle concentrations are very low, the safe procedure is to assume that the allowable particle size range for non-integer values of N less than 5 is that of the next lowest integer range. For $N = 4.5$, for example, the allowed range of d should be assumed to be the same as that for $N = 4$ or $1.0\,\mu m \geq N \geq 0.1\,\mu m$. Similarly, at large values of N, the d ranges specified in Table 33–1 and Figure 33–2 are those that avoid measuring excessively large concentrations of small particles, vulnerable to errors associated with coincidence counting (see Chapter 15), so the reasonable approach for non-integer values of $N > 6$ is to assume the d range of the next larger integral value of N. For $N = 6.2$, for example, the allowed value of d should be those of the $N = 7$ curve or $5.0\,\mu m \geq d \geq 0.5\,\mu m$.

Equation 33–1 assumes an inverse power-law relationship between concentration and particle size with an exponent of 2.08. This size distribution is unlikely to be an accurate description of most (or even any) size distributions encountered in actual operating cleanrooms. Conceivably, the air quality in a given cleanroom could meet the requirements for its claimed classification at one of the allowed particle sizes but not at another or any other. Thus the particle size at which the measurement of aerosol particle concentration is made must be specified in advance and be included in the test report. When measurements at multiple particle sizes are part of the test protocol, the different particle diameters at which the measurement of aerosol particle concentration are made must differ by a factor of at least 1.5:

$$\ldots d_3 \geq 1.5\, d_2 \geq 1.5\, d_1 \ldots$$

Because all d_i must be within the allowed ranges of d, this additional restriction means that ISO Class 1 can be measured at no more than two different particle sizes in the 0.1 to $0.2\,\mu m$ range.

"U" and "M" Descriptors

While any measurement of aerosol particle concentration that will be used to verify air quality class must be made in the particle size range $5.0\,\mu m \geq d \geq 0.1\,\mu m$, the standard provides guidance for reporting the concentrations of particles outside this size range. A "*U* descriptor" is the concentration (m^{-3}) of all measurable aerosol particles including the ultrafine particles, where ultrafine particles are defined as aerosol particles smaller than $0.1\,\mu m$. It is specified and reported as $U\,(x; y)$, where

x is the maximum permitted or, in a test report, the actually measured concentration (m^{-3}) of aerosol particles, including the ultrafine particles, of size greater than y (μm)

y is the lower cut-off limit (μm at 50% counting efficiency) of the particle counter used to measure the U descriptor.

A maximum permitted U descriptor of $140{,}000\,m^{-3}$ in the particle size range $\geq 0.01\,\mu m$, for example, would be written "U (140,000; 0.01)."

Similarly, an "*M* descriptor" is the measured concentration of macroaerosol particles, defined as aerosol particles larger than $5.0\,\mu m$. It is specified and reported as $M\,(a; b); c$, where

a is the maximum permitted or, in a test report, the actually measured concentration of macroparticles (m^{-3})

b is the lower bound on the particle size of the measuring instrument (μm)
c identifies the measurement method

A maximum permitted *M* descriptor of 10,000 aerosol particles ≥5 μm measured by a time-of-flight aerodynamic particle counter would be written "*M* (10,000; 5); time of flight aerodynamic particle counter."

Both the *U* descriptor and the *M* descriptor simply provide supplementary information regarding the air quality. In principle, these descriptors can have any value whatsoever without affecting the ISO classification codes of the cleanroom air quality.

Procedures for Verifying a Cleanroom Classification per ISO 14644-1

As in previous cleanroom standards, ISO 14644-1 spells out, in its Annex B, explicit procedures for verifying that a given cleanroom does in fact meet the requirements of its claimed classification. Simply making a measurement of aerosol concentration at a random location or two within the cleanroom does not constitute adequate verification. This section summarizes the procedures given in ISO 14644-1 as a reference test method for demonstrating compliance. However, alternative test methods "having comparable accuracy" may be used if specified and agreed on by both the certifier and his customer.

Number of Sampling Locations; Number of Separate Measurements/Location. The minimum number of sampling locations required to certify a given cleanroom is the square root of the cleanroom area in m^2:

$$N_L = [A]^{1/2} \qquad (33\text{–}2)$$

where N_L is the minimum number of required sampling locations (rounded up to the next whole number) and *A* is the area of the cleanroom or clean zone in m^2.

These sampling locations "are evenly distributed throughout the area of the cleanroom or clean zone and positioned at the height of the work activity." For small area clean zones the minimum number of locations can be one. However, the minimum number of individual sample measurements required for the verification of any cleanroom or clean zone is three. If only one location is required and chosen, at least three separate, individual sample measurements must be made at that one location. If more than one location is required, only one sample measurement per location is permissible provided that the total number of separate sample measurements made at all locations in the cleanroom is still at least three.

Sample Volume. The minimum air volume required for any one individual measurement is the larger of (1) 2 L or (2) that air volume that would be required to contain 20 particles, assuming that the particles are uniformly distributed in the air at the maximum allowed concentration of the designated ISO class:

$$V_S = 20/C_{max,dmax} \times 1{,}000 \qquad (33\text{–}3)$$

where

V_S (liters) is the minimum air volume required to be sampled in each individual measurement
20 is an arbitrarily selected minimum number of particles assumed to represent an acceptable balance between sampling practicality and statistical significance

$C_{max,dmax}$ (m^{-3}) is the maximum allowed particle concentration at the largest particle size, dmax, to be measured

1,000 converts m^3 to liters

Regardless of Eq. 33–3 and particle counter flow rates, the sample collection time at each location must be at least 1 min.

Sample Probe Size and Orientation. While sizing a sampling probe to sample air flow isokinetically would seem to be good measurement practice (see Chapter 8), ISO 14644-1 does not require it, recognizing that the errors introduced by anisokinetic sampling of cleanroom air are negligible for particles smaller than 0.5 µm and likely to be minor for aerosol particles smaller than 5 µm (Appendix C, FED-STD-209E).

In cleanrooms with unidirectional air flow, the center line of the probe should be aligned so as to face into the direction of the air flow streamlines. In those cleanrooms characterized by nonunidirectional air flow, the probe should be oriented vertically, 180° to the direction of the gravitational force.

Acceptance Criteria for Meeting the Class Requirement. Two conditions must be satisfied in order to verify a claimed cleanroom classification.

Condition 1. The *average* aerosol particle concentration calculated from all the individual measurements made at one location cannot exceed the maximum concentration allowed for the claimed cleanroom classification, C_{max}, as calculated by Eq. 33–1. While an individual measurement at a given location can exceed the target C_{max} concentration, the average concentration, calculated by combining all individual measurements made at that same location, cannot.

$$A_i = (c_{i,1} + c_{i,2} + \ldots + c_{i,n})/n \tag{33–4}$$

where A_i is the average particle concentration at location i, $c_{i,n}$ are the particle concentrations of the individual measurements at location i, and n is the number of measurements made at location i. Individual values of $c_{i,n}$ can exceed C_{max} without causing the cleanroom to fail the verification as long as no value of A_i does.

Condition 2. The mean of all the location averages must also meet the target C_{max} concentration with a 95% upper confidence limit as calculated by the Student t statistic. Demonstration of compliance with this condition is required only when fewer than 10 locations are sampled ($i < 10$), because the probability of failing this requirement is vanishingly small when all 10 or more locations satisfy condition 1.

$$M = (A_1 + A_2 + \ldots + A_L)/L \tag{33–5}$$

where M is the mean of the location averages, A_i are the individual location averages (Eq. 33–4), and L is the total number of locations.

The 95% upper confidence limit (UCL) of M is calculated from

$$95\%\ \text{UCL} = M + t_{0.95}(s/[L])^{1/2} \tag{33–6}$$

where $t_{0.95}$ is the Student t distribution factor for 95% UCL (see Table 33–2) and s is the standard deviation of M:

$$s = \left([A_1 - M]^2 + [A_2 - M]^2 + \ldots + [A_L - M]^2/[L-1]\right)^{1/2} \tag{33–7}$$

TABLE 33–2. Student *t* Distribution Factor for the 95% UCL

Number of locations, L	2	3	4	5	6	7–9	>9
$t_{0.95}$	6.3	2.9	2.4	2.1	2.0	1.9	Not applicable

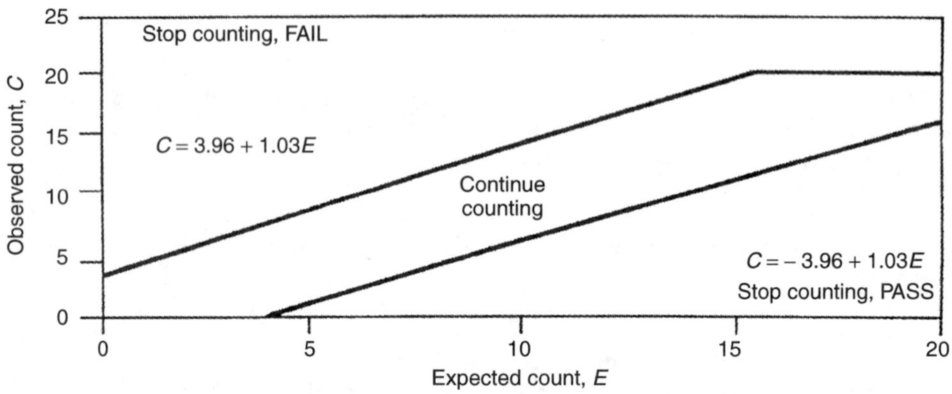

Fig. 33–3. Graph for making pass/fail decisions per the sequential sampling procedure of ISO 14644-1. (Adapted from ISO 1991.)

Sequential Sampling. ISO 14644-1 recognizes that Eq. 33–3 can call for impracticably long sampling times in state-of-the-art cleanrooms. For example, when certifying ISO Class 1 air quality at $d_{max} = 0.2\,\mu m$, Table 33–1 shows that $C_{max,dmax} = 2\,m^{-3}$ and hence, from Eq. 33–3, $V_S = 10{,}000\,L$. With a particle counter sampling air at a rate of $4.67 \times 10^{-4}\,m^3/s$ [28 L/min or 1 cfm, a typical flow rate for a particle counter], collecting a single sample of the required volume would take nearly 6 h! The standard provides potential relief in such instances by allowing the use of a sequential sampling procedure. This procedure can lead to early pass/fail decisions based on observed cumulative particle count versus cumulative sampling volume rather than the concentration determined only after measuring the entire volume prescribed by Eq. 33–3. When the aerosol particle concentration in the air being sampled is either significantly greater or significantly less than the maximum allowed particle concentration of the claimed class code, sequential sampling spots these trends early in the sampling period, dramatically reducing the time required to verify class compliance. At the same time, the pass/fail probability of the sequential sampling procedure is essentially the same as for the single sampling procedure (Cooper and Milholland, 1990).

Unlike the single, fixed-volume sampling procedure, sequential sampling requires continuous data monitoring and reduction in order to make the pass/fail decision as soon as the data indicate sufficiently large departures from the observed concentration versus sampling time (or volume) curve expected for air at the maximum particle concentration allowed by the claimed class code. Software for automating the required data logging and real-time analyses is readily available.

Figure 33–3 illustrates the procedure by plotting the boundaries of the observed particle count, C, against the expected particle count, E. The expected particle count in a valid sample of air at the maximum allowed aerosol particle concentration of any class code is 20. Thus the abscissa in Figure 33–3 starts at 0 and ends at 20. The upper and lower boundaries of C plotted in Figure 33–3 are derived from published formulas for sequential sampling, assuming a 0 to 20 range for the expected particle counts (Cooper and Milholland, 1990). As soon as the cumulative particle count exceeds the upper bound ($C = 3.96 + 1.03\,E$), counting stops

TABLE 33–3. Pass/Fail Criteria per the Sequential Sampling Plan in ISO 14644-1

Fails if Count, *C*, Comes Earlier Than Expected		Passes if Count, *C*, Comes Later Than Expected	
Fractional Time (*t*)	Observed Count	Fractional Time (*t*)	Observed Count
0.0019	4	0.1922	0
0.0505	5	0.2407	1
0.0992	6	0.2893	2
0.1476	7	0.3378	3
0.1961	8	0.3864	4
0.2447	9	0.4349	5
0.2932	10	0.4834	6
0.3417	11	0.5320	7
0.3902	12	0.5805	8
0.4388	13	0.6291	9
0.4873	14	0.6676	10
0.5359	15	0.7262	11
0.5844	16	0.7747	12
0.6330	17	0.8233	13
0.6815	18	0.8718	14
0.7300	19	0.9203	15
0.7786	20	0.9689	16
1.0000	21	1.0000	17

Note: Fractional times are given as the fraction of total times ($t = 1.0000$ at the class limit).
Source: ISO (1999).

and the sample is declared to have failed to meet the class code requirement. Conversely, as soon as the cumulative particle count falls below the lower bound ($C = -3.96 + 1.03$), counting again stops with the verdict that the sample has passed the class code requirement. Counting continues as long as the cumulative counts remain between the two bounds. If the cumulative counts remain between the two bounds throughout the sampling period, the sequential sample provides a concentration measurement that meets the class code requirement and that can serve as a single measurement of the concentration average for that location.

Table 33–3 is an alternative presentation of the sequential sampling process. Here the upper and lower bounds are expressed in terms of the fractional time at which these bounds equal the observed counts. An expected count of 20 equals a normalized time of unity; the real time is that time required by the selected particle counter to sample the required volume, V_S (Eq. 33–3). The left-hand side of Table 33–3 describes the upper bound. When the observed count reaches the listed value before the corresponding fractional time, the sample fails the class requirement. Conversely, the right-hand side of Table 33–3 represents the lower bound. When the fractional sampling time reaches the listed value before the corresponding number of counts is observed, the sample passes the class requirement. When neither of these conditions is met, the sampling continues. The software merely has to note the time at which each integral count is reached and determine if that time is earlier than the upper bound (a failure), longer than the lower bound (a pass), or in between the two (no decision, so counting continues).

The number of required sampling locations and the acceptance criteria for sequential sampling are similar to those of single-volume sampling. What changes is the procedure for making the pass/fail decision. Sequential sampling allows an early decision to be made on the basis of one trial per location with potentially a smaller sample volume. A fail decision rendered by that trial can be overturned only by reverting to the single-volume sampling procedure with the data from the failed sequential sample included in the calculation of the average particle concentration for that location. A second sequential sampling at a location is not part of the procedure. When the number of locations sequentially sampled is less than 10, the 95% UCL calculation is carried out by assuming each location average to be the cumulative particle count at the time the pass decision was reached divided by the cumulative sample volume at the same time.

For the ISO Class 1 verification example cited previously (certifying ISO Class 1 air quality at $d_{max} = 0.2\,\mu m$), sequential sampling would allow the observation of five counts in less than 18 min to render a fail decision and stop the counting of that sample at that time. Conversely, sequential sampling would allow the absence of any observed counts in 68 min to render a pass decision and stop the counting of that sample, a significant reduction in sampling time from the nearly 6 h needed by the standard's single sampling procedure. While 68 min is still a long sampling period, this example represents a worse case in that the $C_{max,dmax}$ value of $2\,m^{-3}$ is the smallest listed in Table 33–1 and represents the largest required sampling volume (hence the longest sampling time) to be calculated from Eq. 33–3.

Test Report

ISO 14644-1 requires a "comprehensive" report, including an explicit statement of compliance or noncompliance with a specific classification code of cleanroom air cleanliness. The report must include all details of the test method used and present the concentration data measured at all sampling locations, identified by coordinates within the cleanroom. Any special conditions or departures from the test method need to be noted. The occupancy state of the clean room—as-built, at-rest, or operational*—is part of the report. The report must name the particle counter used and provide a copy of its current calibration certificate. Any measurements of *U* descriptors or *M* descriptors also become part of the report even though they do not impact the verdict of compliance with a class code.

Maintaining Continued Compliance

ISO 14644-2 specifies requirements for demonstrating, by monitoring procedures, that a cleanroom continues to comply with its class code as designated by ISO 14644-1. The document considers two cases: (1) a cleanroom having no monitoring in place following its initial verification and (2) a cleanroom operating with "continuous or frequent monitoring" of aerosol concentration following its initial verification. Monitoring is defined as "observations made in accordance with a defined method and plan to provide evidence of the performance of an installation." The monitoring plan must be written, and the standard suggests that it be based on risk assessment. At a minimum it must specify the

1. Predetermined sample locations
2. Minimum air volume and time per sample
3. Required replicates
4. Time interval between measurements
5. Measured particle size(s)

* Cleanroom occupancy states:

- *As-built*: an empty cleanroom without any equipment or personnel inside but with filter fans operating
- *At-rest*: a cleanroom with filter fans operating, equipment in place, and idling power at each work station but no processing underway or personnel present (the state of a cleanroom between shifts; the overnight state of a cleanroom that runs just two shifts)
- *Operational*: a cleanroom performing its usual functions; equipment is processing product, and operating personnel are present.

TABLE 33–4. Schedule of Testing to Demonstrate Continued Particle Count Compliance

Class Code	Maximum Time Interval	Test Method
≤ISO Class 5	6 mo	ISO 14644-1, Annex B
>ISO Class 5	12 mo	ISO 14644-1, Annex B

Source: ISO (2000), Table 1.

6. Acceptance, excursion, and alert limits
7. Types of response actions required and when

Placing a particle counter at a critical location or two within a cleanroom for continuous logging of particle concentration, while not a satisfactory verification procedure, could be part of an adequate monitoring plan. Specific details, however, of how to monitor a given cleanroom are left to the cleanroom owner.

Without such a monitoring plan, Table 33–4 (Table 1 of ISO 14644-2) lists the maximum time interval allowed between repeating the test method of ISO 14644-1. With a monitoring plan in place, the maximum time interval between requalification can "be extended, provided the results of continuous or frequent monitoring remain within the specified limit(s)." No explicit time interval is specified for a monitored cleanroom. Monitoring simply exempts a cleanroom from the maximum requalification intervals given in Table 33–4.

The following conditions also trigger a requalification (repeating the test procedures of ISO 14644-1):

1. Remedial action to correct an out-of-compliance condition
2. Significant change in the operational use of the cleanroom or its performance specification
3. Interruption of air movement that significantly affects the cleanroom operation
4. Major maintenance such as a filter changeout.

ISO 14644-2 also discusses the monitoring of other variables not directly related to the measurement of aerosol particle concentration, such as air pressure, air flow velocity and volume, temperature, and humidity. Annex B of ISO 14644-2 lists a number of useful optional tests, including installed filter leakage and air flow visualization, that are based on test methods to be specified in ISO 14644-3 (ISO, 2002), yet to be issued.

COMMERCIALLY AVAILABLE AEROSOL PARTICLE COUNTERS FOR CLEANROOM CLASSIFICATION AND MONITORING

While a large number of particle-counting techniques and commercial versions exist, as discussed in Chapter 15, not all of them match up well with the cleanroom application. To be useful in a cleanroom, a particle counter must be able to count individual particles. Instruments that measure the properties of particle ensembles, while important in some aerosol measurement roles, are of limited value in cleanroom measurements.

The primary properties of interest for a cleanroom particle counter are *size sensitivity* (the smallest particle size at which the counter detects just 50% of the particles of that size actually passing through the counter); *the background counting rate* (generally expressed as the particle concentration that corresponds to the number of counted particles/unit time when measuring "zero aerosol" air [the absence of any "real" particles]); and *the sampling volume flow rate*. (See Chapter 15 for a fuller discussion of the properties of optical particle counters and their calibration). Size resolution is typically of less interest in cleanroom measurements because the primary measurements required by the classification standards have always been expressed in terms of the concentrations of cumulative size distributions. However, being able to measure cumulative concentration at any of the allowed particle size cuts affords flexibility in sampling strategy that can be advantageous.

Selecting the optimum particle counter for a cleanroom depends on all three of the primary properties. First, the size sensitivity of the particle counter must cover the particle size range of interest. For certifying cleanroom air quality according to ISO 14644-1, particle counters that can count particles as small as 0.1 μm are those best suited for state-of-the-art cleanrooms. Table 33–1 and Figure 33–3 show that in the size distribution assumed by ISO 14644-1, the allowed particle concentrations are about an order of magnitude larger at 0.1 μm than at 0.3 μm and about four times larger than at 0.2 μm. This greater particle concentration at 0.1 μm means that sampling volumes required for certification can be smaller (Eq. 33–3) and hence be less time consuming than at the larger particle sizes. Regardless of the sample flow rate through the counter, counting 20 particles when the particle concentration is high requires less sample volume and takes less sampling time than when the particle concentrations are lower.

Similarly, higher volume air flow rates through a particle counter also mean less time to sample a required volume of air, regardless of the value of the required sampling volume. The most common sampling air flow rates of today's cleanroom particle counters are either $4.67 \times 10^{-5}\,m^3/s$ or $4.67 \times 10^{-4}\,m^3/s$ [2.8 L/min or 28 L/min]. The product of particle concentration × air flow rate yields the 100% particle-counting rate—the maximum number of particles passing through the counter per unit time. This value increases with both increasing particle concentration and sampling flow rate.

Counting 0.3 μm particles at the maximum allowed concentration of ISO Class 3 air with a particle counter whose sampling flow rate is $0.028\,m^3/min$ yields the same 100% particle counting rate as counting 0.1 μm particles at $0.0028\,m^3/min$:

$$
\begin{aligned}
&102\ \text{particles/m}^3 \times 0.028\,\text{m}^3\text{/min} = 2.8\ \text{particles/min} \\
&\quad = 1000\ \text{particles/m}^3 \times 0.0028\,\text{m}^3\text{/min} = 2.8\ \text{particles/min}
\end{aligned}
$$

Reaching 20 particles with either of these configurations requires about 7 min. However, measuring particles ≥0.1 μm at $0.028\,m^3/min$ reduces the required sampling time to about 0.7 min (remember, however, in all cases ISO 14644-1 requires a minimum sampling time of 1 min). For verifying ISO Class 3 and cleaner air, counting particles ≥0.1 μ at $0.028\,m^3/min$ makes even the ISO fixed-volume sampling procedure practical. A $4.67 \times 10^{-4}\,m^3/s$ [28 L/min] counter also accelerates the sequential sampling procedure, which can be used to verify these very clean air qualities.

The third property of importance in selecting a particle counter for verifying air quality in a cleanroom is the background count rate. This background count rate is that which is characteristic of the counter response when measuring air having no particles. It can be determined by placing a ULPA filter on the inlet of the counter or by measuring the particle counts with no air flowing through the counter (pump Off). Light scattering from air molecules contributes to this count as well as cosmic ray strikes on the detector and electronic noise. Chapter 15 presents data showing that air flow anomalies often dominate the background count, which is significantly lower when measured with the pump Off than with the pump running and a filter on the inlet. Ideally, the background count rate should be less than 10% of the particle count rate expected at the maximum allowed concentration of the class code (the 100% counting rate for that particle size).

Table 33–5 lists the particle counter properties suitable for measuring particle concentration at the classifications and particle sizes explicitly called out in Table 33–1. This table

TABLE 33–5. Preferred Particle Counter Properties for Verifying the ISO 14644-1 Cleanroom Classifications

ISO Class Code	Detection Limit of Particle Counter Used (Smallest Detectable Particle Size [μm])	Recommended Volume Air Flow Rate (L/min)	Preferred Background Concentration at Measured Particle Size (m^{-3})
1	0.1	28	<1
	0.2	28	<<1
2	0.1	28	<10
	0.2	28	<2
	0.3	28	<1
	0.5	28	<<1
3	0.1	2.8; 28	<100
	0.2	28	<30
	0.3	28	<10
	0.5	28	<4
	1.0	28	<1
4	0.1	2.8; 28	<1000
	0.2	2.8; 28	<300
	0.3	2.8; 28	<100
	0.5	28	<400
	1.0	28	<10
5	0.1	2.8; 28	<1000
	0.2	2.8; 28	<1000
	0.3	2.8; 28	<1000
	0.5	2.8; 28	<400
	1.0	2.8; 28	<100
	5.0	28	<3
6	0.1	2.8; 28	<1000
	0.2	2.8; 28	<1000
	0.3	2.8; 28	<1000
	0.5	2.8; 28	<1000
	1.0	2.8; 28	<1000
	5.0	28	<300
7	0.5	2.8	<1000
	1.0	2.8; 28	<1000
	5.0	2.8; 28	<1000
8,9	0.5	2.8	<1000
	1.0	2.8	<1000
	5.0	2.8	<1000

assumes that the only available particle counters suitable for cleanroom measurements are those having flow rates of either $4.67 \times 10^{-5} m^3/s$ or $4.67 \times 10^{-4} m^3/s$ [2.8 or 28 L/min]. Lower flow rates are not suitable, except for measuring particle concentration in the lowest quality air (ISO Class 7 and above). A $9.33 \times 10^{-4} m^3/s$ [56 L/min or 2 cfm] particle counter has been advertised but seems to be not readily available or in widespread use. Its higher flow rate would be highly suitable for the highest quality air (ISO Classes 1 to 3).

The criteria for choosing the particular set of particle counter properties given in Table 33–5 are

1. Background concentration <10% of the maximum allowed concentration, C_{max}, for the classification and in the size bin(s) at which the measurement(s) is (are) made
2. Maximum particle concentration $<35 \times 10^5/m^3$ [$10^5/ft^3$] to keep coincident counting errors less than 10% for all counters

3. Sampling time per individual measurement <10 min as an arbitrarily selected, practical sampling time
4. Minimum sampling time of 1 min as required by ISO 14644-1. A $4.67 \times 10^{-4}\,m^3/s$ [28 L/min] counting rate offers no time advantage when certifying cleanrooms of air quality having C_{max} values exceeding $7000\,m^{-3}$. However, higher total counts imply higher quality measurements.

The first three of these targets are arbitrary, not always achievable, and, for some particle counters, unduly restrictive. Particle counters having background concentrations less than 10% of C_{max} will not always be available; some particle counters can operate with less than 10% coincidence error at aerosol particle concentrations exceeding $35 \times 10^5\,m^{-3}$; and, obviously, limiting sampling time to 10 minutes per sample point is important primarily for certifying cleanrooms of large areas.

A typical background concentration claimed by manufacturers of optical particle counters is on the order of $35/m^3$ [$1/ft^3$] at the sensitivity limits of the counter. This background concentration exceeds that of some of the values listed in column 4 of Table 3–5. In such instances, counter selection becomes a trade-off to achieve the best compromise. The reduced background concentration typically exhibited by a particle counter in a larger particle size bin may be worth the longer sampling time required by certifying at that larger particle size.

Condensation particle counters (see Chapter 19) usually have background concentrations on the order of $0.35/m^3$ [$0.01/ft^3$]—one to two orders of magnitude lower than that of an OPC. They are highly suitable for measuring *U* descriptors, but their lack of sizing discrimination without additional upstream conditioning stages and their typical flow rate limitation of $0.0028\,m^3/min$ make them less useful for verifying classifications by measuring particle concentrations in the size ranges ≥0.1 μm.

Observations

ISO Class 1 can be very time consuming to certify (the C_{max} of ISO Class 1 at 0.5 μm, if allowed for certification, which it is not, would be that corresponding to FED-STD-209E Class 0.01). Even at a flow rate of $4.67 \times 10^{-4}\,m^3/s$ [28 L/min] using a particle counter sampling 0.1 μm particles, the sampling time per individual measurement is 71 min. As the area of the cleanroom in which air of this quality must be certified increases, the required number of sampling locations also increases so that the sequential sampling plan becomes increasingly attractive. Without a particle counter having a flow rate of $4.67 \times 10^{-4}\,m^3/s$ [28 L/min] and 0.1 μm sensitivity, the certification of ISO Class 1 air quality could become onerous indeed. On the other hand, lower quality air can be reasonably certified with less sensitive particle counters having lower flow rates, as indicated in Table 33–5.

MEASURING PARTICULATE EMISSIONS FROM CLEANROOM EQUIPMENT

In most contemporary cleanrooms the major sources of particulate contamination are within the cleanroom itself rather than the outside. The air filters used in today's cleanrooms can reduce particle penetration in both the make-up air, introduced from the outside ambient, and the exhaust air being recirculated from the cleanroom to very low concentrations. Aerosol particle concentration in many cleanrooms, when measured in the *as-built* state (earlier definition), is often at the background concentration of the particle counter making the measurement. This conclusion holds especially true inside small clean zones that have their own filters, air flow controls, and enclosures to isolate a single processing unit within a larger cleanroom from the environment of that larger "ballroom" cleanroom (a small, "super clean" cleanroom within a standard cleanroom—the so-called minienvironment). In its *operational* state, cleanroom ambient air is not as contaminant-free as that in either the *as-built* or the *at-rest* state because of the particulate emissions from the equipment itself. Thus the measurement of particulate emissions from operating equipment is important and can provide an additional property for specifying and selecting process equipment. Neither ISO 14644-1 or -2 specifies how to measure this potentially important property (although aggressive manufacturers sometimes claim "Class 1" performance or use similar phrasing in touting the merits of their equipment or product). Methods for making such measurements are given in the literature, however (Donovan et al., 1987; Donovan, 1990) and have been incorporated into a Recommended Practice prepared by the IEST (1992). The discussion presented here follows the methods described in these publications.

Relating measurements of aerosol particle concentration, made by a particle counter in the vicinity of a source, to the particle emission rates characterizing that source requires that the volume into which the particles are emitted be known and also that the air flow through that volume, if any, be controlled and known. In the *closed box method*, the source is completely isolated inside an enclosure that confines the particle emissions. A particle counter monitors the particulate concentration within the enclosure as a function of time following source activation. There is no air flow except for what is drawn through a sampling line leading to the particle counter located outside the enclosure. Make-up air for this sampling loss is through an orifice containing an air filter that removes all aerosol particles from the make-up air. In the *flowing duct method*, the source is located inside a duct open at one end and having filters at the opposite end through which air continuously enters. Particle concentration is measured downstream of the source placed inside the duct and after a steady state exists following source activation. In both methods, complete mixing of the aerosol particles throughout the closed box or across the flowing duct is assumed and must be promoted either by the insertion of mixing fans or, in the flowing duct, by flow conditions that create turbulent mixing. The following section presents the analyses appropriate for calculating a particle emission rate with either of these configurations.

Closed Box Method

The closed box configuration allows two independent determinations of particle emission rate, both based on the following rate equation:

$$V\,\mathrm{d}C/\mathrm{d}t = S - K_r C - Q_x C \tag{33–8}$$

where S is the number of particles emitted/unit time $[\mathrm{t}^{-1}]$*, V is the volume of the enclosure $[\mathrm{l}^3]$, C is the measured particle number concentration $[\mathrm{l}^{-3}]$, t is time measured starting with the source activation [t], K_r is the wall loss rate $[\mathrm{l}^3\mathrm{t}^{-1}]$, and Q_x is the sampling flow rate of the particle counter $[\mathrm{l}^3\mathrm{t}^{-1}]$.

Equation 33–8 relates the rate of change of the particle concentration within the enclosure to the particle generation rate of the source less the particle loss rate to the enclosure walls and less the particle loss rate attributable to the air sample drawn by the particle counter.

Determination 1. Assuming that the starting particulate concentration within the enclosure is zero (this condition can be approximated by allowing the particle counter, or other external pump, to replace the initial air within the enclosure with filtered make-up air), the particle emission rate is related to the initial time rate of increase in particle concentration by

*Symbols in brackets give the dimensions of the variables being defined: t = time; l = length.

$$S = V\,\mathrm{d}C/\mathrm{d}t \tag{33–9}$$

Equation 33–9 is just Eq. 33–8 with $C \cong 0$, as established by the above starting conditions. The volume of the enclosure is presumed to be known, and the initial slope of the concentration versus time plot provides a value of dC/dt. Thus, Eq. 33–9 provides a measure of the particle emission rate, S, of the source within the enclosure.

Equations 33–8 and 33–9 assume that the particles emitted by the source are uniformly distributed throughout the volume. A mixing fan within the enclosure helps create this condition, although such a fan is also a source of particulate emissions. The fan's particle emission rate must be measured first and subsequently subtracted from any value of S determined when both it and the source under evaluation are operating.

Determination 2. Under continued source operation and consequent build up of particle concentration within the enclosure, C eventually reaches a value at which the sum of particle loss terms in Eq. 33–8, both of which depend linearly on C, equals the particle emission rate of the source (S). At this particle concentration, C remains constant and the dC/dt term of Eq. 33–8 goes to 0. Under these conditions, Eq. 33–8 becomes

$$S = K_r C_s + Q_x C_s \tag{33–10}$$

C_s is the steady-state concentration measured by the particle counter sampling at a known volume flow rate, Q_x. Both of these factors are thus known. K_r, however, is not known and must be measured.

K_r can be measured by deactivating the source once C_s has been reached and recorded. Deactivating the source is equivalent to setting $S = 0$ in Eq. 33–8, which then becomes

$$V\,\mathrm{d}C/\mathrm{d}t = -C(K_r + Q_x) \tag{33–11}$$

or

$$K_r = -Q_x - V\,\mathrm{d}(\ln C)/\mathrm{d}t \tag{33–12}$$

The slope of a plot of ln$C$ versus t, following source deactivation, then provides a value of d(ln$C$)/dt to use in Eq. 33–12 to solve for K_r. This K_r value inserted in Eq. 33–10 then yields a second independent measure of S, which can be compared with that obtained from the conditions of Eq. 33–9.

Flowing Duct Method

The flowing duct method consists of inserting the source to be evaluated inside a duct through which filtered ("particle-free") air continuously flows. Once the source is activated, the emitted particles increase the particle concentration measured by a particle counter sampling the air downstream of the source. The particle emission rate of the source is given by:

$$S = CQ \tag{33–13}$$

where S is the particle emission rate [t^{-1}], C is the downstream concentration measured after activating the source [l^{-3}], and Q is the volume air flow rate through the duct [l^3/t].

Particle concentration upstream of the source is assumed to be 0 so that all particles measured downstream of the source are assumed to originate from the source under test. The air flow dilutes the particle concentration in the downstream air, making this method less appropriate than the closed box method for measuring sources of low emission rates. Wall losses can be ignored—any $K_r C$ term, analogous to that in Eq. 3–8, will usually be small compared with the CQ term of Eq. 33–13. The dominant particle removal mechanism in the flowing duct method is the convective flow of particles out the end of the duct.

Turbulent mixing of the air flow can be achieved by making the Q/A of the duct (A is the cross-sectional area of the duct) that needed to produce a Reynold's number in excess of, say, 2000 to 4000. Sampling under these conditions lessens the cross-sectional position dependency of the sampling probe. Ideally the probe should be 10 or so duct widths downstream of the source. Isokinetic sampling is preferred but not critical for the particle sizes likely to be of most importance in a cleanroom.

Limitations

These methods for measuring particle emissions rates of sources work well with sources that can be easily moved in and out of enclosures or ducts and that can be activated or not by the simple flip of a switch. Large, permanently anchored sources that are always powered and activated present additional but not insurmountable problems. Enclosures or ducts can be built around such sources. The vertical laminar flow in a cleanroom represents a source of particle-free air that can feed air into a temporary duct, perhaps assembled out of flexible plastic sheets, that surrounds a stationary source. The floor vents already in place then serve as the open exhaust end of the duct. If powering a source On and Off is not practical, particle concentration upstream and downstream of the source can be used to determine a ΔC that can be used in Eq. 33–13 in place of C. The closed box method, however, depends on being able to transition from source On to source Off. It is thus more appropriate for measuring the emission rates of equipment undergoing acceptance checks at the manufacturer's site where startup and shutdown do not yet interfere with production schedules. The closed box configuration can be simulated by simply dropping a tent constructed with flexible walls over the equipment to be tested. Sealed entry ports for the sampling line and the power cords are easily installed. The analyses presented in the preceding paragraphs then apply to these large, immobile sources.

CONCLUSIONS

The aerosol technology developed over the past half century proves very useful in both the design of today's cleanrooms and in assessing their performance. Contract specifications between suppliers and buyers of cleanrooms rely heavily on highly developed, sophisticated particle counters as the means of verifying performance and winning customer acceptance. Conversely, the needs of the cleanroom user community with its significant market impact have motivated manufacturers of particle counters to focus particular attention on this segment of the particle counter spectrum. About 75% of all aerosol particle-counting instruments sold today are for cleanroom applications. Without this driving force, the impressive performance of today's particle counters, as presented here and in Chapter 15, most likely would not be available.

Indeed, instrumentation for measuring the concentration and size of aerosol particles in a cleanroom and the general aerosol technology base developed over the years have become key elements in preventing the specter of contamination limitations creating a road block to the manufacture of new products. Thus far, continually improving aerosol particle measuring instrumentation, a prerequisite for improving cleanroom design and performance, has maintained the pace of development needed to create the more demanding cleanroom work spaces required by each new generation of precision products.

Aerosol particle counters can be used for more than just verifying and monitoring the particulate concentration in the ambient cleanroom air. They can also be used to measure particulate emissions from sources within a cleanroom and to discriminate between more cleanroom-compatible and less cleanroom-compatible hardware and processing equipment. The methods for measuring particle emissions presented here are for determining the contribution of particulate sources within a cleanroom to the cleanroom ambient. Other researchers have adapted sampling probes to measure particulate concentrations inside process equipment, including equipment operating at subatmospheric pressures. This use of an aerosol particle counter is an industry-specialized application not reviewed here but is noted as simply another example of the versatility and importance of aerosol measurements in cleanrooms. Future industrial needs will likely lead to more capable and versatile instrumentation for measuring particles of interest within a cleanroom, wherever they are deemed to be important and whatever their size and concentration.

REFERENCES

Cooper, D. W. and D. C. Milholland. 1990. Sequential sampling for Federal Standard 209 for cleanrooms. *J. Instrum Environ. Sci.* 33:28–32.

Donovan, R. P. 1990. Measurement of particle emission rates from equipment. In *Particle Control for Semiconductor Manufacturing*, R. P. Donovan, ed. New York: Marcel Dekker, pp. 243–253.

Donovan, R. P., B. R. Locke, and D. S. Ensor. 1987. Measuring particle emissions from cleanroom equipment. *Microcontamination* 5(10):36–39, 60–63.

Federal Standard 209E. 1992. *Airborne Particulate Cleanliness Classes in Cleanrooms and Clean Zones.* September 11, 1992. Mount Prospect, IL: IEST.

Fissan, H., W. Schmitz, and A. Trampe. 1993. Clean-room measurements. In *Aerosol Measurement: Principles, Techniques and Applications*, K. Willeke and P. Baron, eds. New York: Van Nostrand Reinhold. pp. 747–767.

IEST. 1992. IEST Recommended Practice CC018.1, *Cleanroom Housekeeping—Operating and Monitoring Procedures*. Mount Prospect, IL: IEST.

ISO. 1999. ISO 14644-1. *Cleanrooms and Associated Controlled Environments—Part 1: Classification of Air Cleanliness*. May 1, 1999. Mount Prospect, IL: IEST.

ISO. 2000. ISO 14644-2. *Cleanrooms and Associated Controlled Environments—Part 2: Specifications for Testing and Monitoring to Prove Continued Compliance with ISO 14644-1*, September 15, 2000. Mount Prospect, IL: IEST.

ISO. 2002. ISO 14644-3. *Cleanrooms and Associated Controlled Environments—Part 3: Metrology and Test Methods.* In press.

34 Radioactive Aerosols

MARK D. HOOVER and GEORGE J. NEWTON
Lovelace Respiratory Research Institute, Lovelace Biomedical and Environmental Research Institute, Albuquerque, NM

INTRODUCTION

Measurement of radioactive aerosols involves most of the standard tools of aerosol science and technology, as well as a number of specialized techniques that take advantage of the unique physical properties of radioactive materials. Instruments and techniques for characterizing radioactive aerosols have been described extensively in numerous manuscripts, books, and standards, including those by the International Electrotechnical Commission (IEC, 1997, 2000a–e), the International Commission on Radiological Protection (ICRP, 1965, 1968, 1982, 1985, 1994a–c), the American National Standards Institute (ANSI, 1969, 1978, 1979, 1980a,b, 1983, 1989a–c, 1999), the International Commission on Radiation Units and Measurements (ICRU, 1971), Raabe (1972), the International Atomic Energy Agency (IAEA, 1978), the National Council on Radiation Protection (NCRP, 1978a,b), Hoover and Newton (1993), the American Conference of Governmental Industrial Hygienists Air Sampling Instruments books (e.g., Cohen, 1995), the U.S. Nuclear Regulatory Commission (NRC) regulatory guide 8.25 on calibration and error limits of air sampling instruments (U.S. NRC, 1992) and the associated NUREG-1400 document on air sampling in the workplace (Hickey et al., 1991), the U.S. Department of Energy *Operational Health Physics Training Manual* (U.S. DOE, 1988) and the DOE *Implementation Guide for Air Monitoring* (U.S. DOE, 1999), and numerous nuclear radiation and health physics text books (e.g., Price, 1965; Eicholz and Poston, 1979; Shapiro, 1981; Eisenbud, 1987; Turner et al., 1988; Turner, 1996; Cember, 1996). Requirements for proper use of these instruments have been further described in documents such as the *American National Standard for Radiation Protection Instrumentation Test and Calibration* (ANSI, 1978), the standard for *Performance Specifications for Reactor Emergency Radiological Monitoring Instrumentation* (ANSI, 1979), the standard for *Specification and Performance of On-Site Instrumentation for Continuously Monitoring Radioactivity in Effluents* (ANSI, 1980a), the standard for *Performance Criteria for Instrumentation Used for Inplant Plutonium Monitoring* (ANSI, 1980b), the *Performance Specifications for Health Physics Instrumentation* (ANSI, 1989a–c), and the *American National Standard for Sampling and Monitoring for Releases of Airborne Radioactive Substances from the Stacks and Ducts of Nuclear Facilities* (ANSI, 1999). Published proceedings of many scientific meetings have been devoted to describing the measurement of radioactive aerosols (e.g., the Proceedings of the Committee on the Safety of Nuclear Installations, Meeting on Nuclear Aerosols in

Aerosol Measurement: Principles, Techniques, and Applications, Second Edition, Edited by Paul A. Baron and Klaus Willeke.
ISBN 0-471-35636-0

Reactor Safety [U.S. NRC, 1980]). Detailed review and application of this information, as well as the development of new techniques and applications, continue to occupy the careers of many aerosol scientists and health protection professionals.

This chapter provides an overview of the principles, techniques, and applications of measuring radioactive aerosols. Information needed for a basic understanding is presented, and several newer measuring techniques are described.

RADIATION AND RADIOACTIVE DECAY

Aerosols of radioactive materials have all the physical and chemical forms of nonradioactive aerosols. They can range from ultrafine metal fumes to large liquid droplets. Their physical form has the usual influence on their aerodynamic behavior and on the choice of measurement technique. Their chemical form has a similarly important influence on their biological or environmental behavior and on the technique that will be used to confirm their chemical form after collection. It is their radioactive properties that can make them easier to detect, yet, in many cases, more hazardous to handle.

Types of Radiation

Three types of radiation are of concern in studies of airborne radioactive materials: alpha, beta, and gamma radiation (Fig. 34–1). Neutrons and positrons are also of concern in some special circumstances. Details of the origin and characteristics of all types of radiation as well as the radioactive decay schemes for radionuclides are presented in references such as the *Table of Isotopes* (Lederer, 1978) and the *Handbook of Health Physics and Radiological Health* (Schleien et al., 1998).

Alpha Radiation. Alpha radiation is the least penetrating, but most highly ionizing type of radiation. Alpha particles consist of two neutrons and two protons, carry two positive charges, are identical to a helium nucleus, and are created spontaneously during the radioactive decay of high atomic number elements such as radium, uranium, and plutonium. Alpha-emitting radionuclides can be identified by the characteristic energy of their emissions. They are of special concern when they are inhaled and deposited in the respiratory tract. If inhaled in large quantities, damage to cells in the lung or in other organs to which the material is translo-

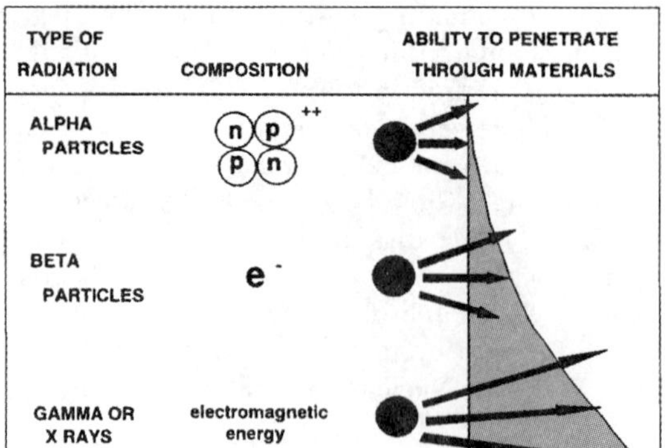

Fig. 34–1. Characteristics of the three major forms of radiation of concern for measurements of radioactive aerosols: Alpha particles, beta particles, and gamma radiation.

cated can cause acute health effects such as fibrosis and long-term health effects such as cancer (e.g., Hobbs and McClellan, 1986).

Beta Radiation. Beta radiation is more penetrating than alpha radiation and consists of negatively charged particles that are identical to electrons. Beta radiation can penetrate the skin and, like alpha radiation, is of concern when large amounts of beta-emitting radionuclides are deposited in the body. The energy spectrum of beta emissions covers a broad range up to a characteristic maximum. Iodine-131, cesium-137, and strontium-90 are beta-emitting radionuclides that are of special concern for accidents involving nuclear reactors.

Gamma Radiation. Gamma radiation and X-rays are penetrating quanta of electromagnetic energy. They are not charged particles, but they can cause ionization in materials and biological damage in tissues. Gamma radiation originates in the nucleus of atoms during radioactive decay. Gamma radiation is emitted at discrete, characteristic energies for each radionuclide. Many alpha- and beta-emitting radionuclides also emit gamma radiation. X-rays originate in disruptions of electrons from their orbits around the nucleus. X-rays can also arise as a secondary form of radiation called *bremsstrahlung* (braking radiation). Bremsstrahlung is created when the path of a negatively charged beta particle is altered by the coulombic attraction to positively charged nuclei of the absorbing material. This alteration of direction causes a radial acceleration of the beta particle, which is accompanied, in accordance with classical theory, by a loss of electromagnetic energy at a rate proportional to the square of the acceleration (e.g., Cember, 1996).

Half-Life for Radioactive Decay

The rate at which a radioactive material decays is described by its half-life: the time for half of the atoms present to undergo a spontaneous nuclear transformation. Half-life is an important consideration in determining how to collect, handle, and quantify samples of a radioactive material. For example, samples of short-lived materials may need to be analyzed immediately after collection, or even in real time. Conversely, samples of longlived radionuclides may remain radioactive for years, centuries, or even millennia. Experimental determination of half-life is also one way of helping to identify a given radionuclide.

The equation describing radioactive decay is

$$A(t) = A_0\, e^{-0.693t/t_{1/2}} \tag{34–1}$$

where $t_{1/2}$ is the decay half-life for the radionuclide (in time units), t is the elapsed time (in the same time units as $t_{1/2}$), e is the base of the naperian logarithm system (2.718), 0.693 is the natural logarithm of 2, A_0 is the activity of the sample at time $t = 0$ (in disintegrations per second), and $A(t)$ is the activity of the sample after an elapsed time of t.

Another way of writing the relationship involves the decay constant, λ, where

$$\lambda = 0.693/t_{1/2} \text{ and } A(t) = A_0 e^{-\lambda t} \tag{34–2}$$

In terms of the number of elapsed half-lives, n, the amount of activity remaining can be determined as

$$A = A_0(\tfrac{1}{2})^n \tag{34–3}$$

Thus, after two half-lives, only one-fourth (25%) of the original activity is present, and after seven half-lives only 1/128 (0.8%) of the original activity remains.

Specific Activity

The specific activity of a radioactive material is the rate of decay per unit mass. Materials with a short half-life have a high specific activity. The historical unit for specific activity, the Curie (2.22×10^{12} disintegrations per minute), was defined in terms of the specific activity of radium (1 Ci/g). The system international (SI) unit for radioactive decay is the Becquerel, which is 1 disintegration per second. Thus, 1 nC is 37 Bq. Specific activity is of biological concern for radioactive materials in the body because it determines the rate at which energy will be deposited and damage will occur in tissues. Specific activity is of practical concern for aerosol measurement because it determines the amount of radioactivity that will be present in a sample of a given mass and the mass or number of particles that will be associated with a sample of a given activity.

The dependence of airborne mass concentration, c_m (g/m^3), on activity concentration, c_a (Bq/m^3), involves the specific activity of the material, A (Bq/g), in the following way:

$$C_m = C_a / A \qquad (34\text{–}4)$$

The number of particles per cubic meter, c_n (particles/m^3), required to provide a given activity air concentration, c_a, involves a similar relationship of specific activity, A, particle density, ρ(g/m^3), and the volume of the airborne particles. For a simple example in which particles are assumed to be monodisperse in size and spherical in shape (volume = $\pi d_p^3/6$, with d_p in m), this relationship is

$$C_n = C_a / \left[A\rho\left(\pi d_p^3/6\right)\right] \qquad (34\text{–}5)$$

Inhalation Exposure Limits

Statutory limitations on exposures of humans to radioactive materials are provided in regulations such as the U.S. Nuclear Regulatory Commission *Standards for Protection Against Radiation* (U.S. NRC, 1991) and the U.S. Department of Energy *Standards for Protection Against Radiation* (U.S. DOE, 1993). Specific activity and biological behavior of materials are considered in determining acceptable annual limits on intake (ALIs) for workers exposed to radioactive aerosols (see ICRP [1979] and its addendums). A related concept is the derived air concentration (DAC). A DAC is the calculated air concentration to which a worker could be exposed 8 h per day, 5 days per week, for 50 weeks (an entire work year of 2000 h), without exceeding the ALI for the radionuclide. The worker is assumed to breathe at the rate of 20 L/min as given for the ICRP Reference Man (ICRP, 1975). Thus, the DAC is the ALI divided by the volume of air inhaled during the working year (2400 m^3).

Particle number concentrations for several materials at their DAC are illustrated in Table 34–1 as a function of particle size. These materials include three that are radioactive ($^{238}PuO_2$, $^{239}PuO_2$, and enriched uranium) and one that is nonradioactive (beryllium, a toxic metal of concern in many nuclear facilities). Insoluble ^{238}Pu has a specific activity of $6.44 \cdot 10^{11}$ Bq/g and a DAC of 0.3 Bq/m^3. Insoluble ^{239}Pu has a specific activity of $2.26 \cdot 10^9$ Bq/g and a DAC of 0.2 Bq/m^3. For enriched uranium, the specific activity is $2.35 \cdot 10^6$ Bq/g (dominated by the contribution from ^{234}U, which is present at 1% by mass), and the DAC for the insoluble material is 0.6 Bq/m^3. The occupational exposure limit for the nonradioactive beryllium is 2 μg/m^3. This example demonstrates that the number of particles of concern for different radionuclides covers a broad range.

TABLE 34–1. Number of Particles Per Cubic Meter as a Function of Monodisperse Particle Size for Selected Toxic Materials at Their Derived Air Concentration (DAC)

Particle Diameter (μm)	$^{238}PuO_2$	$^{239}PuO_2$	Enriched Uranium	Beryllium Metal
10	0.0001	0.02	54	2,065
5	0.0008	0.15	433	16,518
3	0.004	0.7	2,007	76,471
1	0.1	19	54,180	2,064,715
0.5	0.8	150	433,443	16,517,716

Note the following:

Insoluble ^{238}Pu has a specific activity of $6.44 \cdot 10^{11}$ Bq/g and a DAC of 0.3 Bq/m^3.

Insoluble ^{239}Pu has a specific activity of $2.26 \cdot 10^{9}$ Bq/g and a DAC of 0.2 Bq/m^3.

For enriched uranium, the specific activity is $2.35 \cdot 10^{6}$ Bq/g (dominated by the contribution from ^{234}U, which is present at 1% by mass), and the DAC is 0.6 Bq/m^3.

The effective density of beryllium metal aerosol particles with a slight oxide coating is 2 g/cm^3 (Hoover et al., 1989) and the occupational exposure limit for beryllium is 2 μg/m^3.

RADIATION DETECTION

Radioactive materials can be detected in a number of ways. Most of them depend on the ionization process that occurs when gamma rays or charged particles pass through a gas, liquid, or solid and form ion pairs by disrupting atoms and electrons in the material. Radiation detection methods can be used as direct tools for quantifying radioactive aerosols and as support tools for the safe handling of radioactive materials. Measurement of alpha-emitting radionuclides generally requires direct, or nearly direct, contact between the alpha-emitting material and the detection device. Proximity requirements are less severe for the more penetrating beta and gamma forms of radiation. Investigators must have a thorough understanding of the decay schemes of the radionuclides of interest. In many cases the radioactive emissions of progeny radionuclides are of greater biological concern than the emissions of the parent radionuclide. For example, although strontium-90 has a relatively long half-life (28.8 years) and emits beta radiation at a relatively low energy (0.54 MeV), its progeny radionuclide yttrium-90 has a short half-life (64.2 h) and is a high-energy beta emitter (2.28 MeV). An overview of the measuring processes for all three types of radiation is given below. Additional details can be found in resources such as Evans (1955), the chapter on health physics instrumentation in Cember (1996), or the section on radiation detection principles in the *Operational Health Physics Training* manual (U.S. DOE, 1988).

The Cloud Chamber

Condensation tracks are visible in a cloud chamber as charged particles pass through a supercooled vapor (e.g., carbon dioxide from dry ice). Photographs of the particle tracks observed in such devices readily reveal the manner in which charged particles travel through a medium and deposit their energy (e.g., Rutherford et al., 1930; Rasetti, 1947). Cloud chambers provide a useful insight into the physical processes that make radiation detection devices work. Improved versions of the cloud chamber are still used in high-energy physics experiments.

A popular, simple, historical experiment involved placing a camping lantern mantle (Welsbach mantle) in a cloud chamber and observing the tracks from the naturally occurring uranium and thorium progeny in the phosphors of the mantle. In the mid-1990s, Coleman Outdoor Products (Wichita, KS) stopped using uranium and thorium in their lantern mantles. Before that, concerns for inhalation hazards from aerosols of the older mantles led to the warnings that lantern mantles should be installed and lighted in well-ventilated areas and that users should avoid breathing the vapors.

Scintillation Counting

Phosphors are materials that absorb energy during the ionization process and re-emit a fraction of it as light flashes (scintillations). Light emission occurs when electrons are elevated to a higher energy state and then make the transition back to the ground state. Scintillation bursts can be detected with a photodiode or photomultiplier device and counted to determine the number of charged particles or gamma rays that passed through the material. The intensity and duration of the scintillation can also be analyzed to determine the energy of the radiation being detected.

A number of different phosphors are suitable for radiation detection. For gamma or low-level beta detection, the most notable are solid phosphors, such as thallium-doped sodium iodide (NaI[Tl]), cesium iodide (CsI[Tl]), or potassium iodide (KI[Tl]). The thallium activator serves as an impurity in the crystal structure and converts the energy absorbed in the crystal to light. Zinc sulfide (usually doped with silver), ZnS(Ag), is an excellent detector for alpha radiation. Scintillations from the interactions of alpha particles with ZnS are bright enough to be viewed with the human eye in a darkened room. ZnS must be used as a thin layer (usually a fine crystalline coating on a clear plastic film) because it has a low transmission factor for visible light. Plastic or liquid scintillators such as trans-stilbene are also available. Samples containing radioactive materials can be dissolved in a scintillation "cocktail" involving a solvent such as toluene to improve the contact between the radionuclides being detected and the phosphors. There is a wide range of commercially available cocktail media, many with optimized features for dissolving the test material or functioning at special conditions of pH or salt content.

Ionization Chamber Devices

Gas proportional counters and other ionization chamber devices can detect the number and, in some cases, the energy of radioactive emissions. When gamma rays or charged particles pass through the chamber, ion pairs are formed in the gas. These bursts of ion pairs are drawn by an electric field to a charged sensor, and the number and energy of the emissions can be recorded. Counting efficiency, fidelity of energy determination, and counting rate limits are available for standard ionization devices.

Solid-State Detection

More recently, solid-state detectors have been developed that use special layers of semiconducting materials to provide a site for ion-pair production. These high-efficiency devices provide excellent energy resolution and can be used with a multichannel analyzer to identify the energy spectrum of the detected radiation. In detectors such as germanium-lithium (GeLi) or high-purity germanium (HPGe), the sensitive region extends far enough into the surface that they are highly effective for gamma rays. Diffused-junction or surface-barrier detectors have a thin, sensitive layer that is especially useful for alpha and beta radiation detection. When these detectors are used for alpha spectroscopy, the energy resolution is highest when the alpha-emitting material is placed directly on the detector surface or when the air in the gap between the alpha source and the detector is evacuated. New developments in surface coatings for solid-state detectors are making them highly resistant to damage from liquids, acids, bases, and abrasion.

Autoradiography, Track-Etch, and Other Detection Systems

Radiation-induced ionization can cause chemical changes in materials like photographic film. The film can be developed to reveal the location of the damage tracks. This process is known as *autoradiography*; the radioactive material provides its own exposure of the film. Solid-state track recorders can also be made from materials such as polycarbonate or cellulose nitrate. Following exposure to radiation, the surface of the material can be chemically etched to reveal the location, length, and diameter of the damage tracks. Other detection systems use materials that respond to ionizing radiation by undergoing optical density changes, radiophotoluminescence, thermoluminescence, or conductivity changes (e.g., U.S. DOE, 1988, Chapter 13).

Calibration Considerations

Calibration is a critical part of the detection process. Background samples are analyzed, followed by counting of calibration standards of known activity. The normal calibration process involves the use of sealed sources or electroplated sources with radioactivity levels that are traceable to the National Institute of Standards and Technology. Quantities of radioactivity are generally chosen to correspond to those expected in the samples to be analyzed. This allows appropriate corrections to be made for nonlinear phenomena such as instrument dead time (the inability of a system to detect a new event until it has finished reporting or recovering from a previous event). At high count rates (several million counts per minute or more, depending on the instrument), effects such as dead time can be substantial, but at very low count rates such errors are negligible. When instruments are to be used at very low count rates, higher activity standards are used to ensure that statistically valid counting efficiencies are determined for the instruments. Note that detection uncertainty is generally calculated from Poisson statistics as being proportional to the square root of the number of events detected. Thus, the relative uncertainty is high for low-activity sources or for short counting times. The influence of background interference must also be taken into account. For errors of less than 1%, net counts of 10,000 or more are generally required (Price, 1965).

SAFE HANDLING OF RADIOACTIVE AEROSOLS

The handling techniques for radioactive aerosols are usually very stringent. Personal protection and contamination control are major concerns. Contamination of people, equipment, other samples, the workplace, and the environment must be avoided.

Facility Design and Licensing

Radioactive materials are generally handled in special facilities with access control, HEPA filtration, and ventilation systems that provide successively more negative air pressures from the outside environment to general office areas, to laboratory or work areas, and finally to any special enclosures in which actual handling occurs. Figure 34–2 illustrates a laboratory design with graded pressure zones and access control.

Operation of facilities that handle radioactive materials may require federal, state, and local approvals. Laboratories involved in measuring radioactive aerosols must meet the same standards for worker protection, posting of warning signs, effluent monitoring, and waste disposal as the production or reactor facilities to which they provide measuring services.

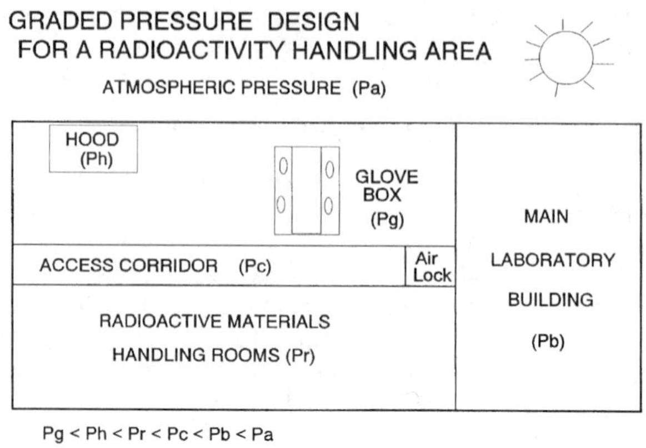

Fig. 34–2. Illustration of the successively decreasing pressure conditions in a facility designed for handling radioactive aerosols.

Use of Special Enclosures

Special laminar flow hoods, glovebox enclosures, and handling rooms are generally needed for work involving radioactive materials. Flow hoods for handling radioactive materials draw room air into the handling area and exhaust it through HEPA filtration. Temporary, special-use enclosures may be required for nonroutine activities such as maintenance or decommissioning that are being performed (e.g., Newton et al., 1987; Fig. 34–3). Such systems may have to be built when existing radioactivity handling areas are being fitted with special features for aerosol sampling or when special sampling efforts are underway. Portable aerosol sampling systems may include local ventilation and filtration around the sampling instruments (e.g., Hoover et al., 1983). Ultraviolet and chemically resistant glove materials such as Hypalon® and specially designed modular glovebox links may used to minimize the spread of contamination and exposures of workers (Hoover et al., 1999).

Contamination Control for Radioactive Aerosol Samples

Samples of radioactive aerosols collected for radiation detection can be stabilized for sample preservation and contamination control by several techniques. For example, filters containing gamma-emitting samples can be stored in sealed containers or sandwiched between tape without degrading the radioactivity counting results. For low-energy, beta-emitting materials or for alpha-emitting materials, samples may be sandwiched between tape and a thin (e.g., 1.5 μm thick) piece of Mylar® film. A slight energy correction must be made to account for energy loss in the Mylar®. In a further variation of this technique, alpha-emitting particles collected on substrates such as a filter can be prepared for counting by sandwiching the substrate between a piece of tape and a piece of ZnS(Ag)-coated clear plastic with the scintillator side facing the particles. This stabilizes the sample and prevents contamination of the counting equipment or laboratory.

Requirements for Radiation Shielding

Workers and radioactivity-counting equipment must also be appropriately shielded from radiation. As shown in the radiation penetration illustration in Figure 34–1, shielding requirements for alpha-emitting radionuclides are minimal; normally the packaging used for contamination control is adequate. Conversely, shielding requirements for gamma-emitting radionuclides may involve several centimeters of lead, steel, or other high atomic number

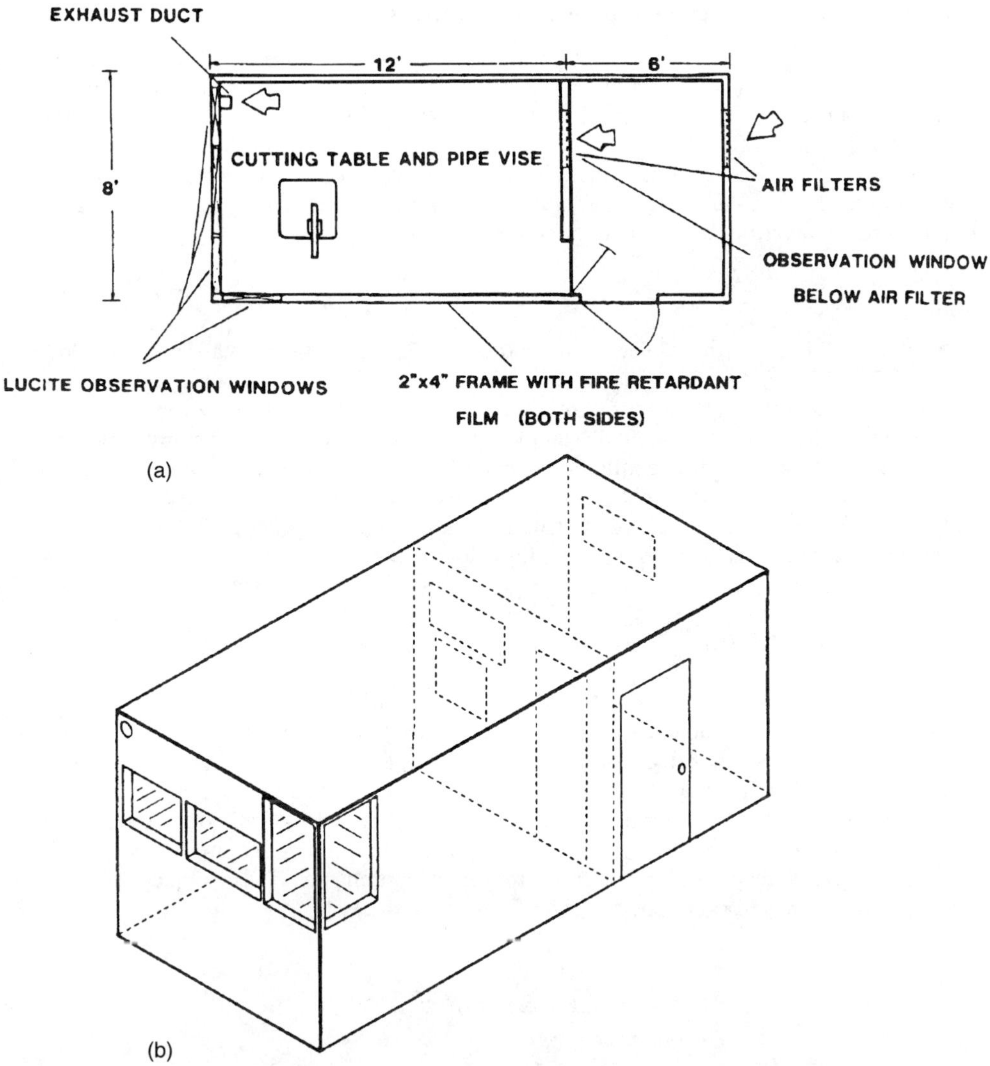

Fig. 34–3. Sketches of the temporary containment structure ("tent") built to enable controlled cutting of radioactively contaminated pipe. Sampling probes were placed at breathing zone levels for a worker at the cutting table.

(high-Z) material. Shielding for beta radiation involves two considerations: attenuation of the beta particles themselves and attenuation and minimization of secondary X-rays from bremsstrahlung. Because bremsstrahlung production increases with the atomic number of the absorber, low-Z materials (less than atomic number 13, aluminum) are normally used to shield beta-emitting radionuclides. Thus, a typical storage unit for aerosol samples of beta-emitting radionuclides involves an inner layer of several millimeters or centimeters of methylmethacrylate acrylic plastic or graphite for beta attenuation, supplemented with an outer layer of high-Z material for absorption of any bremsstrahlung photons. Failure to provide adequate shielding for radioactive samples can cause unnecessary radiation exposures of people and may interfere with proper radioactivity counting.

Proper Handling and Disposal of Radioactive Materials

Federal, state, and local laws must be met in the packaging, labeling, and transportation of radioactive materials (U.S. DOT, 1990). Federal regulations for handling and disposal of radioactive wastes are contained in Title 10 of the Code of Federal Regulations Part 20 and related documents (U.S. NRC, 1991). State and local laws must also be consulted. Even when radioactive particles from air streams that have concentrations below legal limits for environmental release are sampled, the process of concentrating the particles may create samples that eventually must be treated as radioactive waste.

Minimization of Mixed Wastes

"Mixed waste" is a concept defined under the U.S. Resource Conservation and Recovery Act (RCRA) (U.S. EPA, 1990). Wastes that are mixtures of chemically toxic and radioactive materials cannot, by law, be disposed of in either chemical disposal sites or radioactive disposal sites. The lead used for shielding samples of radioactive aerosols and many of the chemicals such as toluene used for scintillation cocktails to count radioactive samples are listed as hazardous wastes under RCRA. Production of mixed wastes should be minimized because they must be handled at special processing facilities, usually at great cost. Mismanagement of mixed wastes can result in fines and/or imprisonment.

OBJECTIVES FOR MEASURING RADIOACTIVE AEROSOLS

As is true for all air sampling and air monitoring, the selection of methods for measuring radioactive aerosols should be determined by the underlying objectives of the sampling effort. "What information is really needed?" This concept seems trivial, but it must be emphasized so that needed data will not be missed and time and fiscal resources will not be wasted. As shown in Figure 34–4, sampling objectives can be grouped into six areas: basic characterization and toxicological testing, process control, health protection, environmental monitoring, emergency response, and demonstration of compliance. These objectives are not mutually exclusive, and other special objectives may also exist.

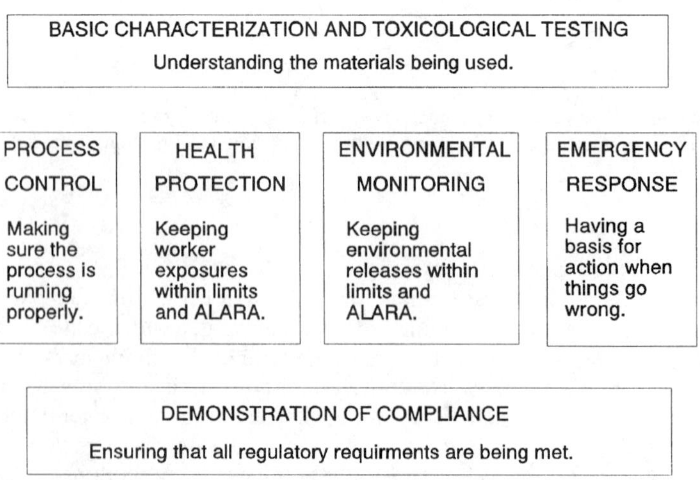

Fig. 34–4. Illustration of six major objectives for sampling radioactive aerosols. The objectives are not mutually exclusive, and other special objectives may also exist.

Basic Aerosol Characterization and Toxicological Testing

Basic aerosol characterization and toxicological testing are ideally done before initiation of any process involving radioactive materials. They involve both collecting and characterizing relevant samples of selected materials and creating laboratory surrogate aerosols that have well-controlled and well-characterized properties. The full spectrum of aerosol sampling tools is usually applied to measure parameters such as particle size, concentration, morphology, chemical composition, and solubility. This approach provides a technical basis for selecting appropriate requirements for designing and managing the entire process in order to keep exposures of people within limits and "as low as reasonably achievable" (ALARA). It also provides a technical basis for balancing the relative risks of internal radiation exposures from inhalation or ingestion and external radiation exposures from working or living in a radiation area. Because both internal and external radiation doses pose risks to people, choices must be made. Historically, the choice has been to accept larger than necessary external exposures in an effort to avoid inhalation exposures. This is not reasonable in light of the underlying risks, and an equal treatment of both types of exposures is recommended by bodies such as the International Commission on Radiological Protection (ICRP, 1973).

Dorrian and Bailey (1995) have summarized the particle size distribution of radioactive aerosols that have been measured in a wide range of industrial operations. The typical particle size distribution had an activity median aerodynamic diameter of 5 μm, with a geometric standard deviation of 2, although smaller size distributions were observed in operations involving high temperatures or fumes, and larger particle size distributions were observed in operations such as coarse powder handling. The typical size distribution values reported by Dorrian and Bailey have now been accepted as the default values for use in ICRP Report 66 on the new *Human Respiratory Tract Model for Radiological Protection* (ICRP, 1994b). The previous default assumption for aerosol particle size in the workplace had been an activity median aerodynamic diameter of 1 μm (ICRP, 1979). The 1 μm aerodynamic diameter is still the default value for the particle size of radioactive aerosols in the environment (ICRP, 1994b). The recommendations of the ICRP are internationally accepted as a coherent and consistent approach to radiation protection. An alternate model developed by the National Council on Radiation Protection and Measurements (NCRP, 1997) focuses on fundamental considerations of human respiratory tract structure and function in deriving an alternate mathematical model to describe the deposition, clearance, and dosimetry of inhaled radioactive substances.

The characteristics of aerosols from a wide range of activities such as powder handling, spills, and fires have also been summarized in a handbook on airborne release fractions/rates and respirable fractions for nonreactor nuclear facilities (U.S. DOE, 1994).

Process Control

Routine sampling or monitoring for aerosols in processes such as nuclear fuel fabrication, reactor operations, or radioactive waste disposal ensures that the processes are working properly or provides an early warning that conditions are changing. *Sampling* refers to methods that involve off-line analysis, and *monitoring* refers to methods that provide a real-time response. Such measurements are usually made within containment areas and generally focus on concentration or size measurements. Other parameters may be measured periodically.

Health Protection

Measurements for health protection purposes can be taken in the breathing zone of workers using fixed or personal samplers or can be taken in the general work area. Results of breathing zone and area sampling can indicate the need for bioassay, the need for improved containment or work practices, and the need for air monitoring and personal protective equipment. Neither breathing zone nor area sampling data should be used to assign radiation dose to workers except in cases where no bioassay method is available and where the sampling method can reasonably be expected to provide a representative sample of the airborne radioactivity (U.S. DOE, 1998). Problems with correlations between air sampling results and bioassay can be especially severe for high specific activity materials such as plutonium-238 because a small number of particles can be significant from a dose standpoint (Scott et al., 1997; Scott and Fencl, 1999). Stochastic concerns for whether or not an individual particle was inhaled do not exist for isotopes with lower specific activity (e.g., tritiated particulate materials) where the aerosol cloud can be expected to be more homogenous at concentrations of concern.

Basic characterization and process control measurements provide the guidance for determining the appropriate level of health protection measurements; processes involving potentially small source terms of low-toxicity materials do not require as extensive a measurement program as would be appropriate for potentially large source terms of high-toxicity materials. In all cases, the objective is to keep worker exposures within limits and ALARA. Concerns for particle solubility are generally high because of the influence it has on the biological behavior on inhaled material and because knowledge of solubility is needed to correctly apply biokinetic models and interpret bioassay information from urine, fecal, or blood samples obtained from workers. Other measurement techniques, such as radiation monitoring for hand and foot contamination or for contamination of work place surfaces, are also part of a total health protection program. They often signal increased airborne concentrations before worker exposures become excessive.

Environmental Monitoring

Because the Clean Air Act sets limits on allowable releases of radioactive materials to the environment (U.S. EPA, 1991), effluent monitoring or stack sampling is required. This must be done routinely to ensure that environmental releases are within limits and ALARA. Measurements are often made both on site and off site. Because concentration limits for environmental releases are generally lower than for the workplace, greater sensitivity is usually required. This is achieved by using higher sampling flow rates, sampling for longer periods of time, or more sensitive analytical techniques.

Emergency Response

Good emergency response involves graded levels of reaction, depending on the severity of any releases. This implies a greater reliance on real-time or near-real-time information than is necessary during normal operations. In addition, instruments may need to remain operational and provide useable results at aerosol concentrations much higher than are seen in routine process, health protection, or environmental monitoring. It is sometimes necessary to deploy portable or mobile instrumentation. Instruments may also need more frequent cleaning or replacement.

Demonstration of Compliance

Throughout all the measuring regimes, it is necessary to demonstrate that the instruments are working properly and to document actual releases of airborne radionuclides in comparison to statutory release limits. In a sense, this effort is a subset of all other measuring objectives. Once the operator is convinced of what has occurred, the regulators and other interested parties must be convinced.

APPLICATION OF STANDARD MEASURING TECHNIQUES

Nearly all standard aerosol sampling techniques can be applied to the measurement of radioactive aerosols. Some limitations are related to the amount of material that may be significant or available. For example, the mass amounts of concern for radionuclides may be below the limits of detection of piezoelectric mass monitoring systems or optical monitoring devices. On the other hand, radioactivity levels associated with particles collected for electron microscopy may require use of designated and restricted equipment. Considerations such as these are discussed below for a number of standard measuring techniques.

Optical Particle Counting

Optical particle counters have not been widely used for radioactive aerosols, but there are circumstances under which they might provide useful information to help protect workers from exposures to radioactive aerosols, especially to warn of unusual particle releases. Optical particle counters have two main advantages: They sample air streams continuously, and they provide real-time information. Their main disadvantages are that radioactive particles cannot be differentiated from inert particles on the basis of light-scattering behavior and that light scattering generally provides an estimate of the physical size distribution of the observed particles rather than an estimate of the aerodynamic size distribution, which is more relevant for assessing inhalation risks. A number of commercially available instruments are being used to detect nonradioactive aerosols in clean rooms, to monitor work area dust levels in industries such as mining and textiles, and to provide quality control monitoring for processes such as paint pigment preparation that fabricate or use fine particles. Optical monitoring has also been used in systems for generating and characterizing inhalation exposure atmospheres of radioactive aerosols (Hoover et al., 1988a). Issues related to their application for radioactive aerosols include (1) level of detection compared with allowed air concentrations for radioactive aerosols, (2) level of detection compared with background levels of nonradioactive aerosols in the workplace, (3) aerosol characterization requirements to determine relationships between radioactive and nonradioactive aerosols in the workplace, (4) calibration requirements to quantify instrument response to specific aerosols, and (5) health protection management strategies for using optical monitoring information in a total program for workplace control and worker protection.

Figure 34–5 illustrates a graphical scheme that we recently developed (Hoover and Newton, 1991) to evaluate whether a given optical particle counter can meet useful requirements for a level of detection. The abscissa of the two-dimensional field describes the level of an airborne radionuclide concentration in relevant units such as Bq/m^3. This permits placement of a vertical line for any concentration level of interest, such 1 DAC for a specific radionuclide. For convenience, we refer to this level-of-interest line as the "alarm limit." This line divides the field in half. The ordinate of the two-dimensional field describes the particle number concentration (particles/m^3) or mass concentration (mg/m^3) that corresponds to a given radionuclide concentration. The ordinate uses the units in which the optical counter measures or reports information, and it permits placement of two horizontal lines corresponding to the lower limit of detection and the upper limit of detection for the instrument. The lower limit of detection depends intrinsically on the sensitivity of the instrument (a function of flow rate, sensing volume dimensions, and internal signal-to-noise ratio), and it also depends in practice on the concentration of background dust (external signal to noise ratio). Except in very clean environments, the lower limit of detection is due to the background aerosol concentration.

The upper limit of detection is caused by saturation of the counter (more than one particle in the sensing volume at once). In very dusty environments, the concentration of background aerosols may actually be *above* the upper limit of detection due to coincidence (the

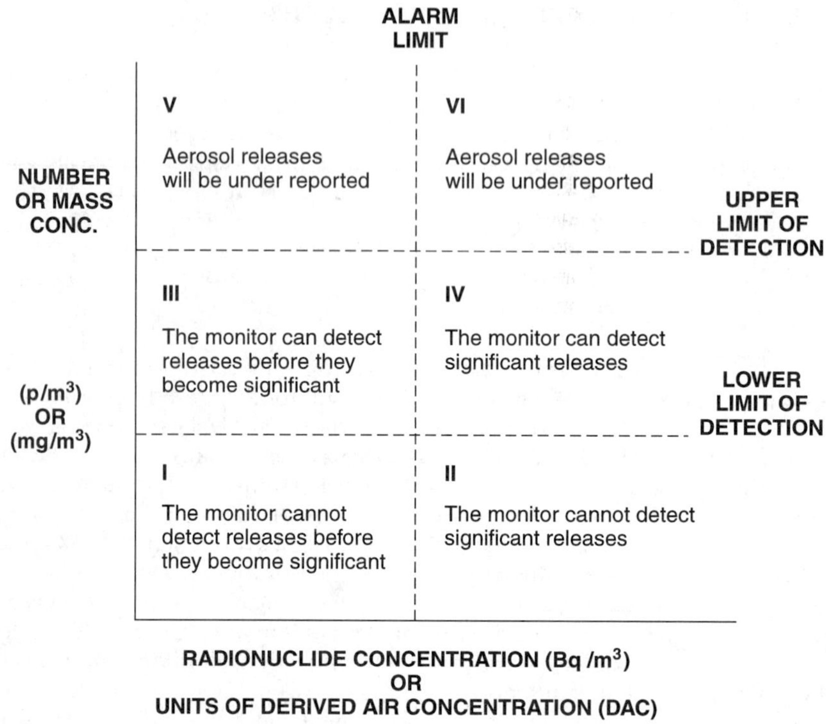

Fig. 34–5. Format of graphical scheme for evaluating if an optical monitor could be useful for detecting radioactive or other toxic aerosols.

monitor is saturated by background aerosols). The combination of the horizontal limit of detection lines and the vertical concentration of interest line divides the field into six sectors. When a curve of mass versus activity or particle number versus activity (assuming a particle size) is added to this plot, it reveals the usefulness of the instrument for detecting a given radionuclide.

The ideal situation would involve a monitor whose effectiveness spans sectors III and IV. This would involve a robust instrument that tracks routine airborne levels at concentrations well below the alarm limit, allows detection of increased airborne particle levels (thus providing opportunity for preventive action or progressive responses), provides a reliable alarm when airborne concentrations exceed an action level, and continues to track airborne concentrations well above the alarm limit (thus providing reliable information for estimating human exposures or environmental releases if accidents have occurred).

Figure 34–6 illustrates a practical application of the evaluation scheme for applying an optical particle monitor to the three radioactive materials ($^{238}PuO_2$, $^{239}PuO_2$, and enriched uranium) and one nonradioactive material (beryllium) presented in Table 34–1. For convenience, the abscissa units of Figure 34–6 are DAC. That equalizes the scale for all radioactive materials and also allows treatment of nonradioactive materials. The particle size in the Figure 34–6 example is assumed to be monodisperse 3 μm diameter spheres. Optical particle counters typically give information as "number greater than a given size." The saturation limit is assumed to be $10^8/m^3$. The lower limit of detection depends on the flow rate and sampling interval. For a flow rate of $0.03\, m^3/min$ (nominal 0.1 cfm, typical of small, portable units) and a sampling interval of 1 min, detection of a single particle during the sample interval would correspond to a concentration of 347 particles/m^3. Figure 34–6 indicates that optical particle

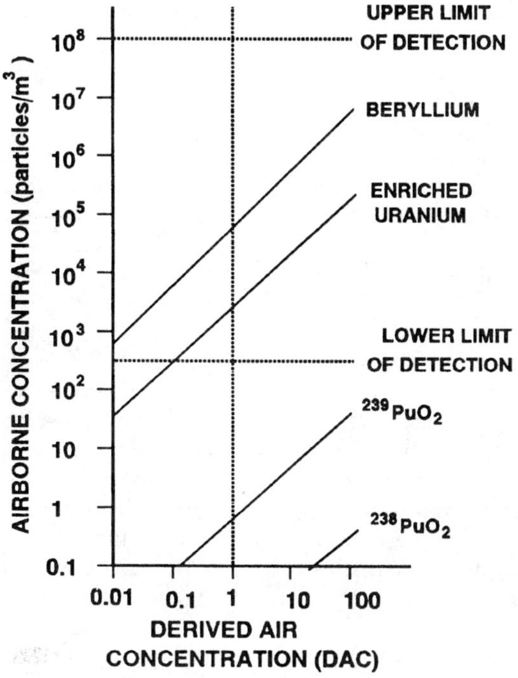

Fig. 34–6. Illustration of the evaluation scheme for usefulness of optical particle monitors. Particles are assumed to be monodisperse with diameters of 3 μm. Sampling flow rate is $5 \times 10^{-5}\,m^3/s$ [3 L/min].

counting will not be effective for ^{238}Pu and ^{239}Pu because number concentrations of concern are below normal limits of detection. However, optical particle counting may be useful for enriched uranium and beryllium metal (assuming that background levels of other dusts are not excessive).

Parameters that still need to be evaluated include the unique response of light-scattering devices to each material. This is of special concern because the particles are unlikely to be spherical. Correlations of optical diameter or real diameter with aerodynamic diameter are also important because movement of the aerosol through the workplace and inhalation of the aerosol by workers will depend on aerodynamic, not optical, diameter. Strategies also need to be developed on precisely how optical monitors would be integrated into a total monitoring program (placement, worker response to alarm, record keeping, and other considerations). The use of optical monitors as early warning devices to detect radioactive releases appears promising, especially for special applications, such as during maintenance operations to detect unusual leaks and to guide in the placement of local ventilation. Early warning could speed reactions to protect workers and regain control of the workplace.

Particle Collection for Microscopy

Collection of radioactive particles for morphological examination by transmission or scanning electron microscopy can easily be done using standard instruments such as the point-to-plane electrostatic precipitator (Morrow and Mercer, 1964). A small hood is normally sufficient for sample handling. The air flow rate through the hood opening should be maintained at the lowest effective level (approximately 75 linear feet per minute) to avoid problems in handling the small, fragile electron microscope grids. Standard Formvar-coated copper grids can be used for transmission electron microscopy. Degradation of the Formvar by radiation damage is usually only a problem for high specific activity radionuclides such as plutonium-238 (half-life = 87.7 years).

Care should always be taken to use contamination control features, such as liquid nitrogen cold fingers, to minimize the spread of contamination within the microscope. High-specific-activity alpha-emitting radionuclides such as plutonium-238 are prone to migration from the collection grid by radioactive decay-induced recoil and spallation. The need for respiratory protection and radiation monitoring should be considered before servicing any microscope used for radioactive materials.

Filtration

Filtration is the most widely used method for collecting samples of radioactive aerosols. The methods and equipment range from high-volume samplers (sampling rates up to about $60\,m^3/h$) for environmental or short-term workplace sampling to low-volume, miniature lapel samplers (1 L/min or less) for collecting aerosols in the breathing zone of individual workers (U.S. DOE, 1988). Low-pressure-drop, cellulose filters are commonly used, and samples can be easily reduced to ash or dissolved for analysis by analytical chemistry or radiochemistry.

Concerns for penetration of particles into the filter matrix are a function of the type of filter, the type of radiation, and the radiation counting method being used. Membrane filters with their superior front-surface collecting characteristics are preferred over fibertype filters when alpha-particle spectroscopy is applied. Shielding by the filter media is seldom a concern for detection of gamma radiation. Although energy degradation concerns are greatest for alpha-emitting radionuclides, we have found that even glass microfiber filters such as the Gelman A/E glass (***GEL***)* collect particles near enough to the filter surface that radioactivity counting results from the ZnS method are as accurate as radiochemical results.

Long-term storage of filter samples for archival purposes is not always feasible for high-specific-activity alpha-emitting radionuclides such as plutonium-238. Radiation damage to the filter, packaging tape, or plastic container may allow release of radioactivity. Care should be taken when handling samples that have aged.

Inertial Sampling

Inertial sampling using cascade impactors, spiral duct centrifuges, and cyclones has been the major approach for characterizing the aerodynamic particle size distribution of radioactive aerosols. A number of specialized versions of these instruments have been developed specifically for use with radioactive aerosols (Mercer et al., 1970; Kotrappa and Light, 1972). Special requirements for instruments used in handling radionuclides generally include being compact and easy to assemble and disassemble while wearing protective gloves in a confined space such as a glove box enclosure. They also need to be easy to clean. Low sample collection rates are usually adequate for collection of small sample masses. Analytical methods for collected samples are straightforward by radioactive counting. The spiral duct centrifuge has been widely used to estimate the density or shape factor of individual particles. This works equally well for radioactive and nonradioactive particles. Because the aerodynamic diameter associated with each particle deposition location is known, electron microscopy can be used to determine the physical size and shape of particles found at those locations, and density or shape factor can be calculated (Stöber and Flachsbart, 1969).

Real-time inertial techniques, such as time-of-flight measurements of particles accelerated through a nozzle, are also useful for radioactive aerosols. This assumes a willingness to

* See Appendix I for full manufacturer addresses referenced to the italicized three-letter codes.

purchase dedicated instruments for use with radioactive aerosols because decontamination of equipment for return to unrestricted use is not always easy.

Measurement of Electrical Properties

Measurement of aerosol electrostatic charge distribution can be done using a standard aerosol charge spectrometer (Yeh et al., 1976). The theory of Yeh et al. (1976, 1978) predicts that self-charging due to alpha or beta emission in radioactive aerosols will occur in addition to friction charging due to comminution. Even when aerosols are created by highly charging processes like grinding, radioactive decay processes such as alpha or beta decay may quickly result in a charge distribution that is near Boltzmann equilibrium. Measurements of particle charge distribution on a plutonium-uranium aerosol obtained by Yeh et al. (1978) with and without use of a krypton-85 discharge unit were identical. At high alpha radioactivity concentrations (>25 nCi/L), it is likely that sufficient ion pairs are present to reduce the charge on the aerosols to near Boltzmann equilibrium.

At lower radioactivity concentrations, this equilibrium condition may not be reached. Raabe et al. (1978) reported the anomalous results of two cascade impactor samples taken without the use of a ^{85}Kr discharge unit in the blending step of a mixed plutonium-uranium oxide fuel preparation process. The alpha radioactivity concentration at the time those samples were taken was only 1 to 2 nCi/L. Because the activity median aerodynamic diameter of these samples was larger than observed in samples taken with a discharger, it is likely that anomalous deposition on the upper stages of the impactors occurred as a result of electrostatic charge effects. It is therefore recommended, without advance information on the radioactivity concentration being sampled, to include an in-line ^{85}Kr discharge unit as a standard procedure, even though it may not be needed to reduce the charge distribution to Boltzmann equilibrium. Fabrication and use of ^{85}Kr discharge units are described by Teague et al. (1978).

Volumetric Grab Samples, Impingers, Cold Traps, and Adsorbers

Sampling techniques such as evacuated volumes, liquid bath impingers, cold traps, and activated charcoal adsorbers work equally well for capturing radioactive and nonradioactive vapors and particles. Standard radioactivity counting techniques can be applied, depending on sample geometry. The Lucas cell is an interesting variation of a grab sampling approach (see Fig. 35–3). Radioactive particles or gases are drawn into a chamber whose interior walls are coated with a layer of crystalline ZnS(Ag) or other scintillator. Nearly 100% of the alpha rays reaching the scintillator will result in a flash of light. Any flashes of light occurring within the chamber are observed by a detector. This method can be applied to radon gas, radon progeny, or other alpha-emitting radionuclides that can be drawn into the chamber. The half-life of the radionuclides being sampled influences the delay time or cleaning requirements before the cells can be reused.

Analytical Chemical Techniques

Traditional analytical chemistry methods such as infrared spectrometry, flame or furnace atomic absorption spectrometry, energy dispersive X-ray analysis, electron or neutron diffraction, and inductively coupled plasma (with either emission spectroscopy or mass spectroscopy) have sensitivities that are compatible with the small sample sizes usually associated with radioactive aerosols. Dedicated equipment is usually required for handling radionuclides, and appropriate controls must be applied. Techniques requiring tens or hundreds of milligrams, such as X-ray diffraction, have much more limited application for radioactive aerosols.

SPECIAL TECHNIQUES FOR RADIOACTIVE AEROSOLS

Detection of Individual Particles by Autoradiography

The interaction of ionizing radiation with photographic film or nuclear track detector foils such as CR-39 creates tracks or material defects than can be photographically developed or chemically etched for observation by scanning electron microscopy or light microscopy. The location, number, and length of the tracks can be used to determine the position and estimate the radioactivity of individual radioactive particles. In one example, Voigts et al. (1986) identified single aerosol particles as the alpha sources from industrial plume samples with particle number concentrations of 2000 particles/mm^2. Cohen et al. (1980) used cellulose nitrate track etch film to measure the alpha radioactivity on human autopsy specimens of the bronchial epithelium. In combination with microscopy, autoradiography can be used to determine how cells and organelles are irradiated by particles that are inhaled and deposited in the body. Figure 34–7 shows an example of autoradiography used with histological slides of lung tissue to determine the microdosimetry of an inhaled uranium-plutonium oxide aerosol.

Measurement of Particle Solubility and Biological Behavior

Particle solubility and biokinetic studies can be done on all classes of aerosol particles, but detection of dissolved material is especially straightforward for radioactive aerosols (Kanapilly et al., 1973). Samples can be sandwiched between filters and subjected to continuous solvent flow (dynamic systems) or placed sequentially in fresh containers of solvent (static systems). Particles can also be placed in a tube with the solvent and periodically centrifuged to concentrate the particles in the bottom of the tube and to allow sampling of dissolved material from the supernatant. Radioactivity counting provides the very low limits of detection needed for accurate determination of particle dissolution, especially for highly insoluble materials, and radioactive aerosols have played a unique role in many unusual

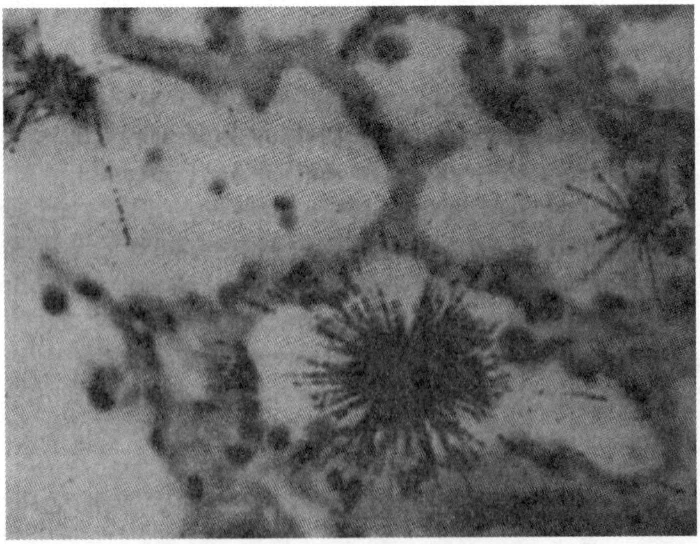

Fig. 34–7. Autoradiograph of alpha-emitting particles of varied specific activity in the lungs of a rat following inhalation exposure to mixed uranium plutonium oxide aerosols. (Adapted from Mewhinney, 1978.)

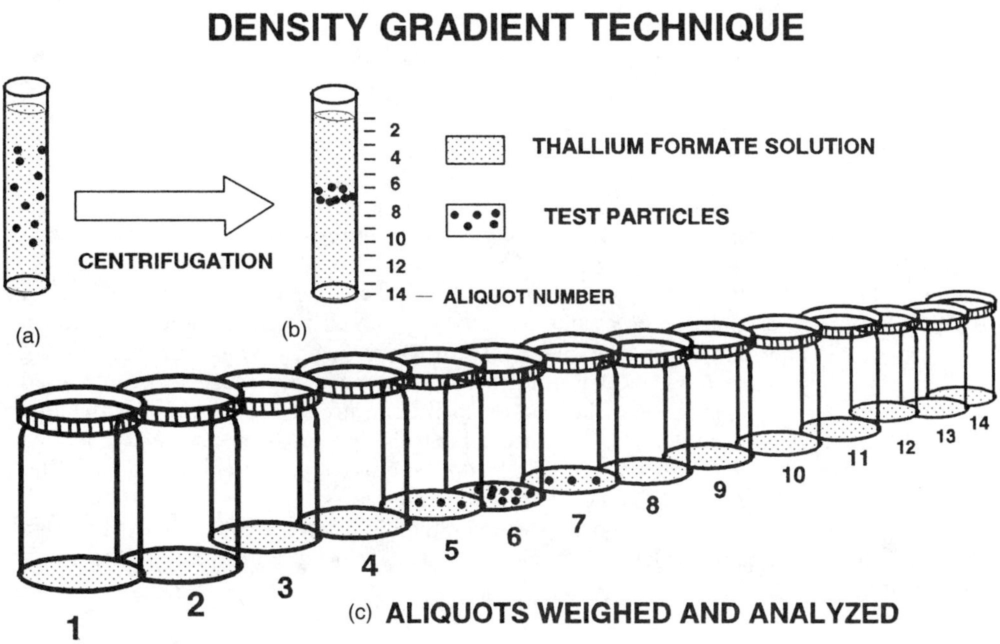

Fig. 34–8. Illustration of the density gradient ultracentrifugation technique for determining the density of individual particles. (Adapted from Finch et al., 1989.)

discoveries about aerosol behavior. The high degree of measurement sensitivity was a major factor in the studies of Mewhinney et al. (1987a) that showed a rapid initial release of material whenever particles were reintroduced into a solvent. That work provided insight into the possible environmental effects of wet and dry weathering on particles released to the biosphere.

Density Measurement by Isopycnic Gradient Ultracentrifugation

Isopycnic density gradient ultracentrifugation has been shown to be a useful technique for measuring the density of small quantities (0.1 to 5 mg) of a variety of particles (Allen and Raabe, 1985; Finch et al., 1989). Normal density measurement techniques, such as air or gas pycnometry or liquid displacement, are not suitable for the small sample volumes normally associated with radioactive aerosols. Thallium formate has been the usual heavy-metal solution for this technique, but sodium metatungstate has been shown to be an economical, nontoxic alternative (Hoover et al., 1991). Figure 34–8 illustrates the technique. Particles are added to a centrifuge tube containing the heavy liquid. The tube is subjected to centrifugation to form a gradient of density from top to bottom. Density near the top of the tube normally approaches that of water, and higher density near the bottom of the tube can be higher than $3.0\,g/cm^3$. Particles move to the location within the tube where their density equals that of the surrounding liquid. Successive samples of known volume are then removed, weighed to confirm the density of the liquid in the sample, and then analyzed by radioactivity counting or other suitable methods to determine the fraction of the particle material in the sample. Refractive index measurements of the liquid samples can also be used to determine the density of the liquid samples, but that requires a dedicated refractometer and is usually more time-consuming than simple weighing. The density gradient ultracentrifugation technique provides information about the density distribution of particles within an aerosol sample.

Surface Area Measurement by Krypton-85 Adsorption

Particle-specific surface area (m^2/g) influences the rate of surface phenomena such as dissolution. When adequate sample masses are available (10 to 50 mg or more), measurement of specific surface area is reliable and straightforward. The most widely used approach involves the Brunauer, Emmett, and Teller method of calculating nitrogen adsorption onto the surface of a sample of known mass. A number of commercial instruments are available. Rothenberg et al. (1982, 1987) have focused attention on the special problems of surface area measurement when sample size is less than 10 mg, which is a typical restriction when characterizing radioactive particles. They have described and evaluated a method for adsorbing ^{85}Kr gas onto the sample surface (Rothenberg et al., 1987). Radioactive decay of ^{85}Kr emits a 0.514 MeV gamma ray that can be readily detected by a standard scintillation method such as NaI(Tl). They note that a 1 cm^2 monolayer of ^{85}Kr gas having a specific activity of 10 Ci/g of gas will give approximately 10,000 disintegrations per minute that can easily be measured with an uncertainty of less than 1%. The major statistical uncertainty is associated with blank correction for the sample holder. The ^{85}Kr-adsorption technique has been successfully applied to the characterization of small samples of mixed uranium and plutonium dioxide particles (Mewhinney et al., 1987b), and the technique can be used for samples as small as 1 mg, with specific surface areas greater than 1 m^2/g. However, the disadvantage of the method rests in the lack of a commercially available instrument and the significant effort required for a user to set up the method.

Real-Time Monitoring for Airborne Radionuclides

The International Electrotechnical Commission has developed standards for radiation protection instrumentation for continuously monitoring radioactivity in gaseous effluents, including (1) general requirements and (2) specific requirements for (a) radioactive aerosol monitors including transuranic aerosols, (b) radioactive noble gas monitors, (c) iodine monitors, and (d) tritium monitors (IEC, 2000a–e). Many instruments for real-time monitoring of airborne radioactivity are commercially available. The typical instrument configuration involves a radiation detector in close proximity to a filter that is collecting aerosols. Radioactivity concentration information can be accessed at the instrument itself, or instruments can be networked to central monitoring stations. Radiation shielding and background correction for external radiation sources are required for beta and gamma radiation detection systems, and background correction, calibration, and geometry considerations come into play for alpha radiation detection systems. Some of the important considerations for design, calibration, and operation of continuous air monitors (CAMs) for alpha-emitting radionuclides are given below.

Mitigation of Interference from Radon Progeny. The alpha emissions of naturally occurring radon progeny such as polonium-218 and bismuth-212 (with alpha energies of 6.0 and 6.08 MeV, respectively) are similar enough in energy to the alpha emissions of plutonium-239 (alpha energy 5.2 MeV) and plutonium-238 (alpha energy 5.5 MeV) to cause interference or false-positive reports of plutonium air concentrations. Figure 34–9 shows a typical radon progeny spectrum as it is detected by an alpha spectrometer. Early alpha CAM designs did not include spectrometry, but used a single-channel analyzer to detect radioactivity in a plutonium energy region of interest (ROI) and a second analyzer to both detect radon progeny activity in a second energy region and allow a simple background correction for the counts seen in the plutonium ROI. The correction method was crude and the limit of detection high, but it provided a useful, real-time means for detecting relatively large airborne releases of plutonium.

An alternate approach to handling radon progeny background interference was developed at the Department of Energy Savannah River Site (Tait, 1956; Alexander, 1966). It used

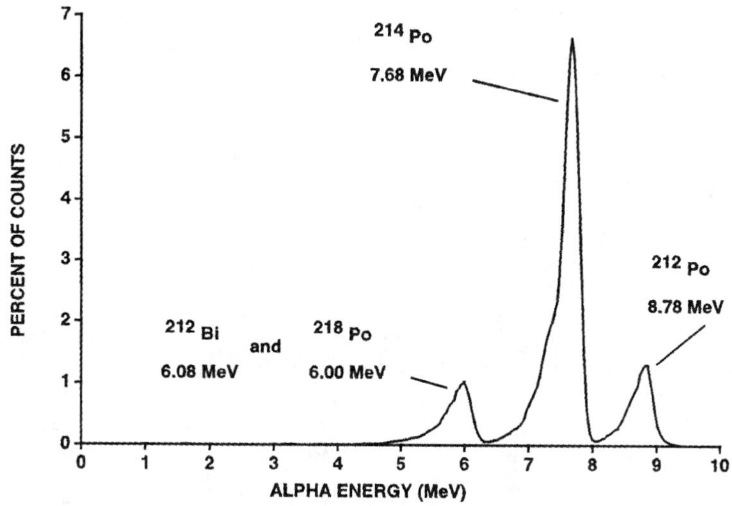

Fig. 34–9. Typical radon progeny spectrum collected in an alpha continuous air monitor. The energy region for ^{234}Pu (5.5 MeV) is in the tail of the ^{218}Po peak.

an impactor jet to deposit aerosol particles directly onto a photomultiplier tube that was coated with a thin layer of ZnS(Ag). This eliminates most of the radon progeny, which are usually attached to the small particle fraction of ambient aerosol (diameter less than 0.3 μm) and are smaller than the effective cut-off diameter of the impactor, and provides a real-time capability for detecting plutonium particles deposited on the collection substrate (Chen et al., 1999). Smaller plutonium particles are also undetected, but most aerosol releases in the workplace have an activity median aerodynamic diameter of about 5 μm with a geometric standard deviation of about 2, and therefore most of the radioactivity is associated with larger particles (Dorrian and Bailey, 1995). Radioactivity counting efficiency for the collected plutonium was approximately 50% (nearly 100% of the emissions occurred in the direction of the ZnS layer) because the light emissions from the ZnS(Ag) are emitted isotropically, and all emissions can be seen with equal likelihood by a photomultiplier tube or photodiode detector. A recent variation of the Savannah River approach uses an impactor jet to deposit the aerosol directly onto the surface of a solid-state detector (Model 8300, ***KUR***). Detection efficiency is excellent, and direct detection of the alpha particles causes good separation of the plutonium and radon progeny peaks. Particle bounce off the detector surface degrades the collection efficiency of large particles (we found that more than 90% of the 10 μm aerodynamic diameter particles are lost to bounce), but this might be solved by use of a virtual impactor or a particle trap approach (for a description of the particle trap, see Biswas and Flagan, 1988).

Recent advances in microcomputers have enabled a major development effort to include alpha spectroscopy in solid-state detector CAMs for real-time monitoring of alpha-emitting radionuclides. Again, a solid-state detector is placed just above the surface of a filter onto which alpha-emitting aerosols are drawn, but the detector output goes to an embedded spectrometer in the CAM rather than to single-channel analyzers. In a batch operation, the air gap between the filter and the detector can be evacuated. This prevents attenuation of the alpha-particle energies as they pass through the air gap and minimizes any spectral overlap of the plutonium and radon progeny alpha energy peaks. For real-time sampling of the aerosol, however, the region between the filter and the detector cannot be evacuated, and attenuation of the alpha particle energies occurs in the air gap.

The first successful algorithm for subtraction of radon progeny interference in the plutonium energy region was developed by Unruh (1986) at Los Alamos National Laboratory. Four ROIs are established: ROI-1 on the lower energy tail of the ^{218}Po peak, where plutonium is expected; ROI-2 on the upper energy tail of the ^{218}Po peak; ROI-3 on the lower energy tail of the ^{214}Po peak; and ROI-4 on the upper energy tail of the ^{214}Po peak. Basically, the ^{218}Po and ^{214}Po peaks are assumed to be congruent in shape, and the radon progeny activity in the plutonium ROI (ROI-1) can be estimated by taking a ratio of the radon progeny counts in the other three ROIs:

$$(\text{ROI-1}) = k \cdot (\text{ROI-2}) \cdot (\text{ROI-3})/(\text{ROI-4}) \tag{35–6}$$

This approach is used in the Eberline Alpha 6 (***EBE***) and in the Victoreen Model 758 (Victoreen Inc., Cleveland, OH; note that the Model 758 is still in use in some nuclear facilities, but it is no longer sold, and no replacement is available). The Radeco Model 452 (SAIC/Radeco, San Diego, CA) uses a similar detector and filter arrangement with a triple window background subtraction algorithm (counts in regions above and below the plutonium region are used to make the background subtraction) or a peak-shape subtraction algorithm that uses peak stripping to remove interfering counts from the plutonium ROI.

The Canberra Alpha Sentry (Canberra Industries, Inc., Meriden, CT; McFarland et al., 1992) and the Eberline Alpha 7 (***EBE***) use peak-shape algorithms for correction of radon progeny background. The Canberra Alpha Sentry also includes a fine screen on the aerosol inlet to prevent unattached radon progeny from penetrating to the aerosol collection region of the instrument (McFarland et al., 1992).

The Merlin-Gerin alpha–beta CAM (***MGP***) uses a very large detector-to-filter separation distance (more than 1 cm) and a stainless steel baffle assembly (installed between the filter and the detector) to collimate the "beam" of alpha emissions reaching the detector. Although the overall detection efficiency is severely reduced because of the small solid angle of view from the detector to the filter, the collimation results in a very narrow full-width-at-half-maximum for the alpha energy peaks. That obviates the need for a correction for radon progeny interference in the plutonium ROI.

Other instruments have used coincidence-counting measurements of beta or gamma emissions from radon progeny to correct for alpha emission interference. However, these indirect measurements are susceptible to errors when the composition or concentration of the radon progeny background is altered.

Performance testing of alpha continuous air monitors is addressed in an international standard promulgated by the International Electrotechnical Commission (IEC, 1997). Test facilities meeting the requirements of that standard have been described by Hoover and Newton (1998) and Grivaud et al. (1998). Concerns relate to particle collection efficiency and stability of the radon correction algorithms over a range of radon progeny concentrations.

Filter Requirements for Continuous Alpha Air Monitoring. An active area of recent work has related to the selection of appropriate filter media for alpha CAMs. The quality of the alpha energy spectrum reported by a solid-state detector system is strongly dependent on the type of filter substrate used for particle collection. Membrane filters have long been accepted as having both the excellent particle collection efficiency and the excellent surface collection characteristics needed for alpha spectroscopy. Lindeken et al. (1964) demonstrated that, compared with submicrometer filters, large-pore membrane filters show no serious sacrifice in collection efficiency until the pore diameters exceed 5 μm. Thus, larger pore filters, with their lower pressure drop, are preferred. The ANSI N13.1 (1999) standard states that if a filter with a collection efficiency lower than 95% is required to meet the overall sampling objectives, then a correction for efficiency shall be made.

The filter that has been widely recommended by CAM manufacturers is the 5 μm pore size, membrane filter (mixed cellulose ester, type SMWP; ***MIL***). However, we found the fragile nature of that filter to be unacceptable. Particle collection efficiency and gravimetric confirmation of particle mass collection were unreliable because of filter breakage under field conditions.

Results of evaluations to find a more rugged membrane filter that would have a reasonably low pressure drop and excellent surface collection efficiency are summarized in Annex D of ANSI N13.1 (1999). The Versapor-3000 (an acrylic co-polymer on a nonwoven nylon fiber support, from Gelman Sciences, now Pall-Gelman [***PAL***]) performs similarly to that of the SMWP (***MIL***). The Durapore 5 μm pore size, polyvinylidene fluoride membrane filter (***MIL***) is acceptable but provides a poorer alpha energy spectrum than the SMWP. The AW-19 filter (***MIL***) is a reinforced mixed cellulose ester filter that is rugged and provides a better alpha energy spectrum than the SMWP. It is also easy to dissolve for radiochemical analysis. We found the performance of the Fluoropore FSLW filter (***MIL***) to be superior to all the others because it has a very low pressure drop and it is an excellent front-surface-collecting filter. It is a polypropylene-backed, polytetrafluoroethylene (PTFE) filter that is extremely rugged and provides excellent alpha energy spectral separation of the collected radionuclides. The qualities of radon progeny spectra collected with the Fluoropore FSLW (***MIL***) and SMWP (***MIL***) membrane filters are compared in Figure 34–10a. Note the much lower tail of the ^{218}Po peak in the plutonium ROI (around 5.2 MeV). As a warning, Figure 34–10b shows the extremely poor spectral quality obtained when a conventional fiber filter such as the Whatman 41 (***WHA***) is used in an alpha CAM. Thus, the use of fiber-type filters must be avoided. These and other factors that affect alpha particle detection in continuous air monitor applications are presented by Moore et al. (1993).

A useful innovation in filtration technique involves mounting each filter in a thin cardboard carrier. The card protects the filter during handling and can be labeled to readily identify the sample. Bar code labeling and computer data logging can be used for both the filter cards and the sampling instruments. This improves chain-of-custody control and facilitates long-term archiving of samples. If samples are to be destructively analyzed instead of

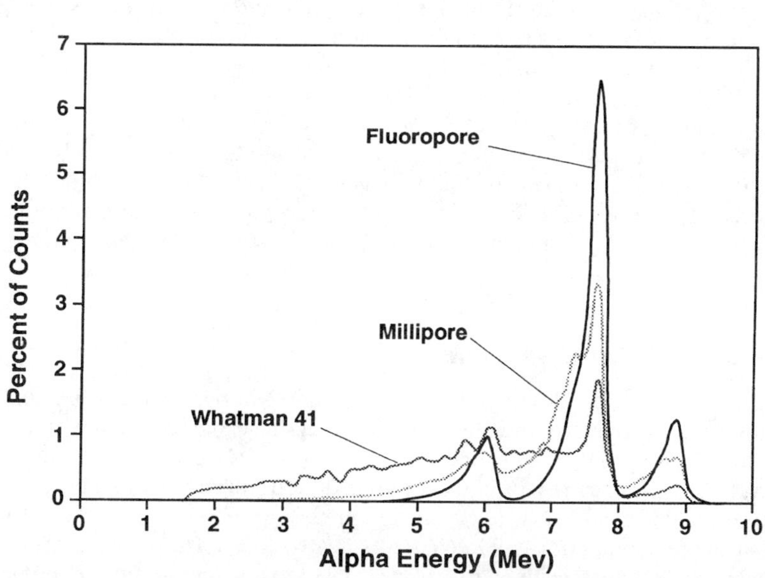

Fig. 34–10. Illustration of the influence of filter type on the quality of the radon progeny energy spectrum in an alpha continuous air monitor. The Fluoropore filter provides superior resolution.

archived, they can be easily removed from the cards. The choice of filter for use in the cards depends on the quality of the energy spectrum that is required and on compatibility requirements of the filter for the analytical technique to be used.

Mitigation of Interference from Airborne Dust. The accumulation of ambient dust on the collection filter of an alpha CAM leads to attenuation of alpha energy, just as the air gap above the filter degrades the alpha energy. Such burial of plutonium leads to underreporting of air concentrations, ranging from 10% to 100% when airborne dust concentrations are greater than 1 mg/m^3 (Hoover et al., 1988b, 1990). Alpha particles from plutonium that is buried by 20 μg/mm^2 of salt on a filter are prevented from reaching the detector. This does not prevent the CAM from responding to large puff releases of radioactivity, but it does raise the limit of detection for slow, continuous releases. Dust concerns are primarily associated with decommissioning activities where metal piping and structures are being cut (see Newton et al., 1987), with environmental restoration activities where soil is being disturbed, and with special activities such as in the underground salt mine environment of the Department of Energy Waste Isolation Pilot Plant near Carlsbad, New Mexico.

Efficiency Considerations for Filter/Detector Geometry. Concerns for dust loading make it necessary to consider some trade-offs in filter and detector size. Many CAMs use a 25 mm diameter detector and a 25 mm diameter filter. This arrangement has an overall detection efficiency of approximately 20%. Retaining the same detector size but increasing the diameter of the filter collection surface to 43 mm cuts the detection efficiency in half (down to 10%), but increases the collection surface area by nearly a factor of 4. At first, this appears to be a reasonable trade-off.

However, a closer examination of detection efficiency as a function of filter diameter reveals that material collected at the filter edges contributes very little to overall efficiency. This is because of solid angle considerations that reduce both the efficiency at which alphas are intercepted by the detector and the energy at which they are detected. With a larger filter, alphas from the filter edge lose a significant amount of energy as they travel to the detector. Alphas traveling from one edge of the filter to the opposite edge of the detector have the longest path through air and thus suffer the greatest energy loss, often enough to remove them from the plutonium energy region. Plutonium collected at the center of the filter is detected in the plutonium ROI at an efficiency of 30%, with little energy degradation. At a diameter of 25 mm, detection efficiency has dropped to 15%, but only marginal energy degradation has occurred. At a diameter of 43 mm, only 0.04% of the emitted alpha particles reach the detector and have energies in the plutonium ROI. Thus, the larger filter has marginal utility, especially in the presence of dust, where additional energy degradation will degrade any alphas from the filter edge to energies below the plutonium region.

Remote Detection of Radioactive Particles

A new technique has been developed at Los Alamos National Laboratory for remotely detecting alpha radioactivity contamination on internal surfaces of equipment or in areas that cannot be directly surveyed by radiation instrumentation (MacArthur and Allander, 1991). It uses the fact that ion pairs persist in the air after they are formed by the movement of positively charged alpha particles through air. This persistence was a surprise to many who believed that the recombination of ion pairs was nearly instantaneous. By drawing clean air across a contaminated surface, ion pairs can be carried to an electrometer and detected. The new technique should receive wide use in facilities having airborne alpha-emitting aerosols and makes it possible to identify contamination with a greater degree of certainty than was previously possible.

Sampling from Stacks and Ducts

The original version of the American National Standard *Guide to Sampling Airborne Radioactive Materials in Nuclear Facilities* (ANSI, 1969) has been revised and reissued as the American National Standard for *Sampling and Monitoring for Releases of Airborne Radioactive Substances from the Stacks and Ducts of Nuclear Facilities* (ANSI, 1999). Both versions of the standard emphasize the importance of obtaining a representative sample. However, the original standard included faulty assumptions about aerosol mixing in exhaust ducts and faulty technical guidance on the use of multipoint sampling to obtain a representative sample. The new standard minimizes reliance on simplistic concepts such as isokinetic sampling (considered by many to be broadly misapplied in turbulent sampling conditions) and emphasizes "qualification" of a well-mixed sampling location, which enables the more efficient collection of samples from a single point. Compact and cost-effective mixer designs have been developed to create a well-mixed condition in an exhaust stack (McFarland et al., 1999a,b). The revised standard also contains information about using the new and more efficient shrouded probes for sample extraction (McFarland et al., 1989; Rodgers et al., 1996). Significant improvements have also been made in modeling the deposition of radioactive aerosols in sampling transport lines. Modern codes such as DEPOSITION 4.0 (Anand et al., 1996; Riehl et al., 1996) can now be used to predict particle deposition as a function of transport line geometry. An overview of the methodology for sampling effluent air from stacks and ducts of the nuclear industry is available (McFarland, 1998).

PRACTICAL OPTIONS FOR DATA TRANSMISSION AND NETWORKING

For many years, radiation monitoring instruments were considered stand-alone devices that provided information locally. A new emphasis on demonstrating "control of the workplace" has lead many health protection professionals to consider networking and central monitoring capabilities as essential. Applications in nuclear power plants have long demonstrated the advantages of networking, but those applications have largely involved fixed physical situations with hardwired cables installed during original construction. Development of new networking capabilities is especially difficult when cables must be installed in radioactive areas or when work tasks change and new monitoring locations are needed. The cost of installing the cables for data transmission can be prohibitive.

Some of the most notable breakthroughs in recent years are in new radiation monitoring devices that provide continuous dose rate information by radio signal to a remotely located individual. To accomplish this, the most beneficial devices use spread spectrum radio transmission. This is basically the same high-technology approach used for cellular telephones and remote microphones. For the first time, it is possible to cut radiation doses in half by removing the health physics technician from direct proximity to the work site. Someone totally outside a contaminated area can now monitor air concentrations and worker whole-body exposures and provide instructions and guidance for minimizing dose. This is an area of rapid progress.

ADEQUACY OF THE EXISTING AEROSOL SCIENCE DATA BASE

An ever-increasing emphasis is being placed on having a "technically defensible basis" for measuring radioactive aerosols in the workplace and environment. New terms such as *conduct of operations*, *total quality management*, and *integrated safety management* have been coined to provide new paradigms for management and worker attitudes and responsibilities, record keeping and quality assurance requirements, demonstration of compliance, and proper integration of concerns for industrial hygiene, health physics, and traditional safety hazards.

Some of the notable deficiencies in the technical basis for radioactive aerosol measurement are quality and handling of radioactive calibration sources, selection of appropriate instrumentation for specific tasks, placement of instrumentation in the workplace or environment, and criteria for collection of an adequate sample. Improvements in the limit of detection for workplace and environmental releases are also needed. This might come through wider application and improvements in virtual impaction techniques to concentrate airborne particles before collection.

Pressure for improved quality and capabilities can be expected to be the greatest in four classes of nuclear applications: environmental restoration of radioactively contaminated facilities such as those located throughout the Department of Energy weapons production complex; design, construction, and operation of facilities for safe disposal of low-level and high-level radioactive waste; monitoring and handling of radioactive materials in medical settings; and continued operation or design of new nuclear facilities for generation of electricity, including those that would use mixed uranium-plutonium oxide fuels from disposition of nuclear weapons.

Opportunities are available for improving the quality of regulations and standards and for preparing guidance documents for state-of-the-art sampling and monitoring. The *American National Standard for Radiation Protection Instrumentation Test and Calibration* (ANSI, 1983) is being revised to include a specific new standard to be issued as ANSI-N323C for air monitoring instruments. International standards for air monitoring and radiation protection instrumentation are under continuous development and improvement. However, continued work is needed to fully understand the basic behavior of radioactive aerosols and to synthesize total measurement approaches that are technically defensible and cost effective.

CONCLUSIONS

The measurement of radioactive aerosols is a challenging and specialized subset of aerosol science. The basic physics of radiation and radioactivity provides some unique technical advantages and disadvantages for accomplishing the task. Overall, the advantages outweigh the disadvantages. At the same time, a range of political, regulatory, and emotional obstacles arise from both the justifiable and the unreasonable health and environmental concerns that surround radioactivity and the use of radioactive materials. Because the health, economic, psychological, and political costs to society can be high, the tools of aerosol science must be used in wise and technically defensible ways when dealing with radioactivity.

ACKNOWLEDGMENTS

Preparation of this manuscript was supported by the U.S. Department of Energy under Cooperative agreement No. DE-FC04–96AL76406.

REFERENCES

Alexander, J. M. 1966. A continuous monitor for prompt detection of airborne plutonium. *Health Phys.* 12:553–556.

Allen, M. D. and O. G. Raabe. 1985. Slip correction measurements of spherical solid particles in an improved Millikan apparatus. *Aerosol Sci. Technol.* 4:269–286.

Anand, N. K., A. R. McFarland, V. R. Dileep, and J. D. Riehl. 1996. *Deposition: Software to Calculate Particle Penetration Through Aerosol Transport Lines*. NUREG/GR-0006. Washington, DC: U.S. Nuclear Regulatory Commission.

ANSI. 1969. *American National Standard Guide to Sampling Airborne Radioactive Materials in Nuclear Facilities*. ANSI N13.1-1969. New York: American National Standards Institute.

ANSI. 1978. *American National Standard for Radiation Protection Instrumentation Test and Calibration.* ANSI N323-1978. New York: American National Standards Institute.

ANSI. 1979. *American National Standard for Performance Specifications for Reactor Emergency Radiological Monitoring Instrumentation.* ANSI N320-1979. New York: American National Standards Institute.

ANSI. 1980a. *American National Standard for Specification and Performance of On-Site Instrumentation for Continuously Monitoring Radioactivity in Effluents.* ANSI N42.18-1980. New York: American National Standards Institute.

ANSI. 1980b. *American National Standard for Performance Criteria for Instrumentation Used for Inplant Plutonium Monitoring.* ANSI N317-1980. New York: American National Standards Institute.

ANSI. 1983. *Radiation Protection Instrumentation Test and Calibration.* ANSI-N323-1978, reaffirmed 1983. New York: American National Standards Institute.

ANSI. 1989a. *American National Standard on Performance Specifications for Health Physics Instrumentation—Portable Instrumentation for Use in Normal Environmental Conditions.* ANSI N42.17A-1989. New York: American National Standards Institute.

ANSI. 1989b. *American National Standard on Performance Specifications for Health Physics Instrumentation—Occupational Airborne Radioactivity Monitoring Instrumentation.* ANSI N42.17B-1989. New York: American National Standards Institute.

ANSI. 1989c. *American National Standard on Performance Specifications for Health Physics Instrumentation—Portable Instrumentation for Use in Extreme Environmental Conditions.* ANSI N42.17C-1989. New York: American National Standards Institute.

ANSI. 1999. *American National Standard for Sampling and Monitoring for Releases of Airborne Radioactive Substances from the Stacks and Ducts of Nuclear Facilities.* ANSI N13.1-1999. New York: American National Standards Institute.

Biswas, P. and R. C. Flagan. 1988. The particle trap impactor. *J. Aerosol Sci.* 19:113–121.

Cember, H. 1996. *Introduction to Health Physics*, 3rd Ed. New York: McGraw-Hill.

Chen, B. T., M. D. Hoover, G. J. Newton, S. J. Montano, and D. S. Gregory. 1999. Performance evaluation of the sampling head and annular kinetic impactor in the Savannah River site alpha continuous air monitor. *Aerosol Sci. Technol.* 31:24–38.

Cohen, B. 1995. Sampling airborne radioactivity. In *Air Sampling Instruments for Evaluation of Atmospheric Contaminants*, 8th Ed., B. S. Cohen and S. V. Hering, eds. Cincinnati: American Conference of Governmental Industrial Hygienists.

Cohen, B. S., M. Eisenbud, and N. H. Harley. 1980. Measurement of the α-radioactivity on the mucosal surface of the human bronchial tree. *Health Phys.* 39:619–632.

Dorrian, M. D. and M. R. Bailey. 1995. Particle size distributions of radioactive aerosols measured in workplaces. *Radiat. Protect. Dosim.* 60:119–133.

Eicholz, C. G. and J. S. Poston. 1979. *Principles of Nuclear Radiation Detection.* Ann Arbor, MI: Ann Arbor Science.

Eisenbud, M. 1987. *Environmental Radioactivity from Natural, Industrial, and Military Sources.* San Diego: Academic Press.

Evans, R. D. 1955. *The Atomic Nucleus.* New York: McGraw-Hill.

Finch, G. L., M. D. Hoover, J. A. Mewhinney, and A. F. Eidson. 1989. Respirable particle density measurements using isopycnic density gradient ultracentrifugation. *J. Aerosol Sci.* 20:29–36.

Grivaud, L., S. Fauvel, and M. Chemtob. 1998. Measurement of performances of aerosol type radioactive contamination monitors. *Radiat. Protect. Dosim.* 79(1–4):495–498.

Hickey, E. E., G. A. Stoetzel, and P. C. Olsen. 1991. *Air Sampling in the Workplace.* NUREG-1400. Washington, DC: U.S. Nuclear Regulatory Commission.

Hobbs, C. H. and R. O. McClellan. 1986. Toxic effects of radiation and radioactive materials. In *Casarette and Doull's Toxicology: The Basic Science of Poisons*, 3rd Ed., C. D. Klaassen, M. O. Amdur, and J. Doull, eds. New York: MacMillan Publishing Co.

Hoover, M. D., B. T. Castorina, G. L. Finch, and S. J. Rothenberg. 1989. Determination of the oxide layer thickness on beryllium metal particles. *Am. Ind. Hyg. Assoc. J.* 50:550–553.

Hoover, M. D., A. F. Eidson, J. A. Mewhinney, G. L. Finch, B. J. Greenspan, and C. A. Cornell. 1988a. Generation and characterization of respirable beryllium oxide aerosols for toxicity studies. *Aerosol Sci. Technol.* 9:83–92.

Hoover, M. D., G. L. Finch, and B. T. Castorina. 1991. Sodium metatungstate as a medium for measuring particle density using isopycnic density gradient ultracentrifugation. *J. Aerosol Sci.* 22:215–221.

Hoover, M. D., C. J. Mewhinney, and G. J. Newton. 1999. Modular glovebox connector and associated good practices for control of radioactive and chemically toxic materials. *Health Phys.* 76:66–72.

Hoover, M. D. and G. J. Newton. 1991. *Preliminary Evaluation of Optical Monitoring for Real-Time Detection of Radioactive Aerosol Releases*. Albuquerque, NM: Inhalation Toxicology Research Institute.

Hoover, M. D. and G. J. Newton. 1993. Radioactive Aerosols. In *Aerosol Measurement: Principles, Techniques, and Applications*, 1st Ed., K. Willeke and P. A. Baron, eds. New York: Van Nostrand Reinhold.

Hoover, M. D. and G. J. Newton. 1998. Performance testing of continuous air monitors for alpha-emitting radionuclides. *Radiat. Protect. Dosim.* 79:499–504.

Hoover, M. D., G. J. Newton, R. A. Guilmette, R. J. Howard, R. N. Ortiz, J. M. Thomas, S. M. Trotter, and E. Ansoborlo. 1998. Characterisation of enriched uranium dioxide particles from a uranium handling facility. *Radiat. Protect. Dosim.* 79:57–62.

Hoover, M. D., G. J. Newton, H. C. Yeh, and A. F. Eidson. 1983. Characterization of aerosols from industrial fabrication of mixed-oxide nuclear reactor fuels. In *Aerosols in the Mining and Industrial Work Environments*, V. A. Marple and B. Y. H. Liu, eds. Ann Arbor: Ann Arbor Science.

Hoover, M. D., G. J. Newton, H. C. Yeh, F. A. Seiler, and B. B. Boecker. 1988b. *Evaluation of the Eberline Alpha-6 Continuous Air Monitor for Use in the Waste Isolation Pilot Plant: Phase I Report*. December 21, 1988. Albuquerque, NM: Inhalation Toxicology Research Institute.

Hoover, M. D., G. J. Newton, H. C. Yeh, F. A. Seiler, and B. B. Boecker. 1990. *Evaluation of the Eberline Alpha-6 Continuous Air Monitor for Use in the Waste Isolation Pilot Plant: Report for Phase II*. January 31, 1990. Albuquerque, NM: Inhalation Toxicology Research Institute.

IAEA. 1978. *Particle Size Analysis in Estimating the Significance of Airborne Contamination*. Technical Report Series No. 179. Vienna: International Atomic Energy Agency.

ICRP. 1965. *Principles of Environmental Monitoring Related to the Handling of Radioactive Materials*. International Commission on Radiological Protection Publication 7. Oxford: Pergamon Press.

ICRP. 1968/1982. *General Principles of Monitoring for Radiation Protection of Workers*. International Commission on Radiological Protection Publications 12 and 35. Oxford: Pergamon Press.

ICRP. 1973. *Implications of Commission Recommendations that Doses be Kept as Low as Reasonably Achievable*. International Commission on Radiological Protection Publication 22. Oxford: Pergamon Press.

ICRP. 1975. *Reference Man: Anatomical, Physiological and Metabolic Characteristics*. International Commission on Radiological Protection Publication 23. Oxford: Pergamon Press.

ICRP. 1979. *Limits on Intakes of Radionuclides by Workers*. International Commission on Radiological Protection Publication 30 and addendums. Oxford: Pergamon Press.

ICRP. 1985. *Principles of Monitoring for the Radiation Protection of the Population*. International Commission on Radiological Protection Publication 43. Oxford: Pergamon Press.

ICRP. 1994a. *1990 Recommendations of the International Commission on Radiation Protection*. International Commission on Radiological Protection Publication 60. Ann. ICRP 21 (1/3). Oxford: Elsevier Science, Ltd.

ICRP. 1994b. *Human Respiratory Tract Model for Radiological Protection*. International Commission on Radiological Protection Publication 66. Ann. ICRP 24 (4). Oxford: Elsevier Science, Ltd.

ICRP. 1994c. *Dose Coefficients for Intakes of Radionuclides by Workers*. International Commission on Radiological Protection Publication 68. Ann. ICRP 24 (4). Oxford: Elsevier Science, Ltd.

ICRU. 1971. *Radiation Protection Instrumentation and its Application*. ICRU Report 20. Washington, DC: International Commission on Radiation Units and Measurement.

IEC. 1997. *Radiation Protection Instrumentation. Calibration and Verification of the Effectiveness of Radon Compensation for Alpha and/or Beta Measuring Instruments*. IEC Standard 61578. Geneva, Switzerland: International Electrotechnical Commission.

IEC. 2000a. *Radiation Protection Instrumentation. Equipment for Continuously Monitoring Radioactivity in Gaseous Effluents, Part 1: General Requirements*. IEC Standard 61761-1. Geneva, Switzerland: International Electrotechnical Commission.

IEC. 2000b. *Radiation Protection Instrumentation. Equipment for Continuously Monitoring Radioactivity in Gaseous Effluents, Part 2: Specific Requirements for Radioactive Aerosol Monitors Including Transuranic Aerosols*. IEC Standard 61761-2. Geneva, Switzerland: International Electrotechnical Commission.

IEC. 2000c. *Radiation Protection Instrumentation. Equipment for Continuously Monitoring Radioactivity in Gaseous Effluents, Part 3: Specific Requirements for Radioactive Noble Gas Monitors*. IEC Standard 61761-3. Geneva, Switzerland: International Electrotechnical Commission.

IEC. 2000d. *Radiation Protection Instrumentation. Equipment for Continuously Monitoring Radioactivity in Gaseous Effluents, Part 4: Specific Requirements for Radioactive Iodine Monitors*. IEC Standard 61761-4. Geneva, Switzerland: International Electrotechnical Commission.

IEC. 2000e. *Radiation Protection Instrumentation. Equipment for Continuously Monitoring Radioactivity in Gaseous Effluents, Part 5: Specific Requirements for Tritium Effluent Monitors*. IEC Standard 61761-5. Geneva, Switzerland: International Electrotechnical Commission.

Kanapilly, G. M., O. G. Raabe, C. H. T. Goh, and R. A. Chimenti. 1973. Measurement of the in vitro dissolution of aerosol particles for comparison to in vivo dissolution in the respiratory tract after inhalation. *Health Phys.* 24:497–507.

Kotrappa, P. and M. E. Light. 1972. Design and performance of the Lovelace aerosol particle separator. *Rev. Sci. Instrum.* 43:1106–1112.

Lederer, C. M., J. M. Hollander, and I. Perlman, editors. 1978. *Table of Isotopes*, 7th Ed. New York: John Wiley & Sons.

Lindekin, C. L., F. K. Petrock, W. A. Phillips, and R. D. Taylor. 1964. Surface collection efficiency of largepore membrane filters. *Health Phys.* 10:495–499.

MacArthur, D. W. and K. S. Allander. 1991. *Long-Range Alpha Detectors*. LA-12073-MS. Los Alamos, NM: Los Alamos National Laboratory.

McFarland, A. R. 1998. *Methodology for Sampling Effluent Air from Stacks and Ducts of the Nuclear Industry*. LA-UR-96-2958. Revised July 1998. Los Alamos, NM: Los Alamos National Laboratory.

McFarland, A. R., N. K. Anand, C. A. Ortiz, R. Gupta, S. Chandra, and A. P. McManigle. 1999a. A generic mixing system for achieving conditions suitable for single point representative effluent air sampling. *Health Phys.* 76:17–26.

McFarland, A. R., R. Gupta, and N. K. Anand. 1999b. Suitability of air sampling locations downstream of bends and static mixing elements. *Health Phys.* 77:703–712.

McFarland, A. R., C. A. Ortiz, M. E. Moore, R. E. DeOtte, Jr., and A. Somasundaram. 1989. A shrouded aerosol sampling probe. *Environ. Sci. Technol.* 23:1487–1492.

McFarland, A. R., J. C. Rodgers, C. A. Ortiz, M. E. Moore. 1992. A continuous sampler with background suppression for monitoring alpha-emitting aerosol-particles. *Health Phys*. 62:400–406.

Mercer, T. T., M. I. Tillery, and G. J. Newton. 1970. A multi-stage low flow rate cascade impactor. *J. Aerosol Sci.* 1:9–15.

Mewhinney, J. A. 1978. *Radiation Exposure and Risk Estimates for Inhaled Airborne Radioactive Pollutants Including Hot Particles, Annual Progress Report July 1, 1976–June 30, 1977*. NUREG/CR-0010. Albuquerque, NM: Inhalation Toxicology Research Institute.

Mewhinney, J. A., A. F. Eidson, and B. A. Wong. 1987a. Effect of wet and dry cycles on dissolution of relatively insoluble particles containing Pu. *Health Phys.* 53:337–384.

Mewhinney, J. A., S. J. Rothenberg, A. F. Eidson, G. J. Newton, and R. Scripsick. 1987b. Specific surface area determination of U and Pu particles. *J. Colloid Interface Sci.* 116:555–562.

Moore, M. E., A. R. McFarland, and J. C. Rodgers. 1993. Factors that affect alpha particle detection in continuous air monitor applications. *Health Phys.* 65:69–81.

Morrow, P. E. and T. T. Mercer. 1964. A point-to-plane electrostatic precipitator for particle size sampling. *Am. Ind. Hyg. Assoc. J.* 25:8–14.

NCRP. 1978a. *Instrumentation and Monitoring Methods for Radiation Protection*. NCRP Report No. 57. Bethesda, MD: National Council on Radiation Protection and Measurements.

NCRP. 1978b. *A Handbook of Radiation Protection Measurements Procedures*. NCRP Report No. 58. Bethesda, MD: National Council on Radiation Protection and Measurements.

NCRP. 1997. *Deposition, Retention and Dosimetry of Inhaled Radioactive Substances*. NCRP Report No. 125. Bethesda, MD: National Council on Radiation Protection and Measurements.

Newton, G. J., M. D. Hoover, E. B. Barr, B. A. Wong, and P. D. Ritter. 1987. Collection and characterization of aerosols from metal cutting techniques typically used in decommissioning nuclear facilities. *Am. Ind. Hyg. Assoc. J.* 48:922–932.

Price, W. S. 1965. *Nuclear Radiation Detection*. New York: McGraw-Hill.

Raabe, O. G. 1972. Instruments and methods for characterizing radioactive aerosols. *IEEE Trans. Nuclear Sci.* NS-19(1):64–75.

Raabe, O. G., G. J. Newton, C. J. Wilkenson, and S. V. Teague. 1978. Plutonium aerosol characterization inside safety enclosures at a demonstration mixed-oxide fuel fabrication facility. *Health Phys.* 35:649–661.

Rasetti, F. 1947. *Elements of Nuclear Physics*. New York: Prentice-Hall.

Riehl, J. R., V. R. Dileep, N. K. Anand, and A. R. McFarland. 1996. *DEPOSITION 4.0: An Illustrated User's Guide*. Aerosol Technology Laboratory Report 8838/7/96. Department of Mechanical Engineering. College Station: Texas A&M University.

Rodgers, J. C., C. I. Fairchild, G. O. Wood, C. A. Ortiz, A. Muyshondt, and A. R. McFarland. 1996. Single point aerosol sampling: Evaluation of mixing and probe performance in a nuclear stack. *Health Phys.* 70:25–35.

Rothenberg, S. J., P. B. Denee, Y. S. Cheng, R. L. Hanson, H. C. Yeh, and A. F. Eidson. 1982. Methods for the measurement of surface areas of aerosols by adsorption. *Adv. Colloid Interf. Sci.* 15:223–249.

Rothenberg, S. J., D. K. Flynn, A. F. Eidson, J. A. Mewhinney, and G. J. Newton. 1987. Determination of specific surface area by krypton adsorption, comparison of three different methods of determining surface area, and evaluation of different specific surface area standards. *J. Colloid Interface Sci.* 116:541–554.

Rutherford, E., J. Chadwick, and C. D. Ellis. 1930. *Radiation from Radioactive Substances*. New York: Macmillan.

Schleien, B. S., L. A. Slabeck, Jr., and B. K. Kent, eds. 1998. *Handbook of Health Physics and Radiological Health*, 3rd Ed. Baltimore, MD: Williams & Wilkens.

Scott, B. R. and A. F. Fencl. 1999. Variability in PuO_2 intake by inhalation: Implications for worker protection at the US Department of Energy. *Radiat. Protect Dosim.* 83:221–232.

Scott, B. R., M. D. Hoover, and G. J. Newton. 1997. On evaluating respiratory tract intake of high-specific activity emitting particles for brief occupational exposure. *Radiat. Protect Dosim.* 69:43–50.

Shapiro, J. 1981. *Radiation Protection, a Guide for Scientists and Physicians*. Cambridge: Harvard University Press.

Stöber, W. and H. Flachsbart. 1969. Size-separating precipitation of aerosols in a spinning spiral duct. *Environ. Sci. Techol.* 3:1280–1296.

Tait, G. W. C. 1956. Determining concentration of airborne plutonium dust. *Nucleonics* 14:53–55.

Teague, S. V., H. C. Yeh, and G. J. Newton. 1978. Fabrication and use of krypton-85 aerosol discharge devices. *Health Phys.* 35:392–395.

Turner, J. E. 1996. *Atoms, Radiation, and Radiation Protection*, 2nd Ed. New York: John Wiley & Sons.

Turner, J. E., J. S. Bogard, J. B. Hunt, and T. A. Rhea. 1988. *Problems and Solutions in Radiation Protection*. New York: Pergamon.

Unruh, W. P. 1986. *Development of a Prototype Plutonium CAM at Los Alamos*. LA-UR-90-2281, December 15, 1986. Los Alamos, NM: Los Alamos National Laboratory.

U.S. DOE. 1988. *Operational Health Physics Training*. ANL-88–26. Prepared by Argonne National Laboratory, Argonne, IL, for the U.S. Department of Energy Assistant Secretary for Environment, Safety, and Health. Oak Ridge, TN: National Technical Information Service.

U.S. DOE. 1993. *Occupational Radiation Protection*, Title 10, Code of Federal Regulation, Part 835 (10 CFR 835), December 14, 1993. Washington, DC: U.S. Government Printing Office.

U.S. DOE. 1994. *Airborne Release Fractions/Rates and Respirable Fractions for Nonreactor Nuclear Facilities*. DOE-HDBK-3010-94, Vols. 1 and 2, December 1994. Washington, DC: U.S. Department of Energy.

U.S. DOE. 1998. *The Department of Energy Laboratory Accreditation Program for Radiobioassay*. DOE STD 1112-98, December 1998. Washington, DC: U.S. Department of Energy.

U.S. DOE. 1999. *Implementation Guide for Air Monitoring for use with Title 10 Code of Federal Regulations Part 835 Occupational Radiation Protection*. DOE G 441.1-8. March 17, 1999. Washington, DC: U.S. Government Printing Office.

U.S. DOT. 1990. *Regulations for Transportation of Hazardous and Radioactive Materials*. Title 49, Code of Federal Regulations, Parts 171–177. October 1, 1990. Washington, DC: U.S. Government Printing Office.

U.S. EPA. 1990. *Hazardous Waste Management Regulations*. Title 40, Code of Federal Regulations, Parts 260 through 268, July 1, 1990. Washington, DC: U.S. Government Printing Office.

U.S. EPA. 1991. *National Emission Standards for Hazardous Air Pollutants*. Title 40, Code of Federal Regulations, Part 61, July 1, 1991. Washington, DC: U.S. Government Printing Office.

U.S. NRC. 1980. *Proceedings of the CSNI Specialists Meeting on Nuclear Aerosols in Reactor Safety*. NUREG/CR-1724. Washington, DC: U.S. Nuclear Regulatory Commission.

U.S. NRC. 1991. *Standards for Protection Against Radiation*. Title 10, Code of Federal Regulations, Part 20 et al., May 21, 1991. Washington, DC: U.S. Government Printing Office.

U.S. NRC. 1992. *Calibration and Error Limits of Air Sampling Instruments for Total Volume of Air Sampled*. Regulatory Guide 8.25. Washington, DC: U.S. Nuclear Regulatory Commission.

Voigts, C., G. Siegmon, M. Berndt, and W. Enge. 1986. Single alpha-emitting aerosol particles. In *AEROSOLS, Formation and Reactivity*, 2nd Int. Aerosol Conf. Berlin. Oxford: Pergamon Press, p. 1153.

Yeh, H. C., G. J. Newton, O. G. Raabe, and D. R. Boor. 1976. Self-charging of ^{198}Au-labeled monodisperse gold aerosols studied with a miniature electrical mobility spectrometer. *J. Aerosol Sci.* 7:245–253.

Yeh, H. C., G. J. Newton, and S. V. Teague. 1978. Charge distribution on plutonium-containing aerosols produced in mixed-oxide reactor fuel fabrication and the laboratory. *Health Phys.* 35:500–503.

35 Radon and Its Short-Lived Decay Product Aerosols

BEVERLY S. COHEN
New York University School of Medicine, Nelson Institute of Environmental Medicine, Tuxedo, NY

INTRODUCTION

Radon, a radioactive noble gas, with its short-lived progeny, or "daughters," are known to cause cancer in uranium and other underground miners. Radon is present at elevated concentrations in many homes and workplaces, and concentration measurements are required for evaluation and control. Recent measurements have demonstrated radon levels in some homes high enough to deliver a radiation dose to the bronchial epithelium equal to that known to produce cancer in underground miners. Exposure in most homes is substantially lower, but experience with ionizing radiation demonstrates that any dose, however small, confers some risk. The large number of people potentially exposed to elevated radon concentrations has led to a recognition of the need for measurement of indoor as well as outdoor concentrations.

Measurement of radon and its daughters presents unique problems because there occurs a serial transformation of the radionuclides. The airborne concentrations that are biologically significant are extremely low and difficult to measure. In addition, radon is a gas and the progeny are particles so that sampling methods differ. This chapter presents the background information needed to understand the reason for radon sampling; briefly defines the relationships between radon, its progeny, and radiation dose because these interactions shape the sampling strategy; and describes methods for sampling and measuring both radon and the short-lived decay products.

RADON IN THE ENVIRONMENT

Radon is formed in the decay chains that begin with the naturally occurring primordial radionuclides uranium-238, thorium-232, and uranium-235 present in the earth's crust. When radon is formed, the gas may diffuse from the rocks and soils to enter the atmosphere. The amount of radon formed depends only on the concentration of the parent nuclide. The fraction that emanates to the atmosphere depends on the rock or soil matrix, water content, atmospheric pressure, and other geological and climatic factors. Also important is the half-

Aerosol Measurement: Principles, Techniques, and Applications, Second Edition, Edited by Paul A. Baron and Klaus Willeke.
ISBN 0-471-35636-0

TABLE 35–1. Parent Nuclide and Radon Isotope of the Natural Radioactive Series

	Long-Lived Parent					Radon Isotope			
			Crustal Abundance						
Series	Isotope	Half-life ($\times 10^9$ years)	ppm	Bq/kg	pCi/g	Isotope	Common Name	Half-life	α-Particle Energy (MeV)
Uranium	^{238}U	4.5	2.7	33	0.89	^{222}Rn	Radon	3.82 days	5.49
Thorium	^{232}Th	14.1	8.5	34	0.92	^{220}Rn	Thoron	55.6 s	6.29
Actinium	^{235}U	0.7	0.02	1.5	0.04	^{219}Rn	Actinon	3.96 s	6.82

Source: Adapted from U.S. NCRP (1988).

life of the radon, that is, the time needed for half of a given quantity to decay. The half-life is different for each radon isotope (Table 35–1). The concentration of a radioactive element may be designated by mass, that is, mg/kg or ppm, or by activity (i.e., Bq/kg or pCi/g).

The unit of activity is the becquerel (Bq):

$$1\,\text{Bq} = 1\,\text{s}^{-1} \tag{35–1}$$

Thus, 1 Bq represents one transformation or disintegration per second. The conventional unit of activity is the curie (Ci):

$$\begin{aligned} 1\,\text{Ci} &= 3.7 \times 10^{10}\,\text{s}^{-1} \\ &= 3.7 \times 10^{10}\,\text{Bq} \end{aligned} \tag{35–2}$$

Further conversions are shown in Table 35–2.

Both ^{238}U and ^{232}Th are present in all rocks and soils at concentrations of roughly 1 to 5 and 2 to 12 ppm, respectively, or about 7 to 60 Bq/kg of soil for both nuclides. The ^{232}Th concentration of some rocks is as high as 20 ppm (U.S. NCRP, 1988). The crustal abundance (by mass) of ^{232}Th averages about three times that of ^{238}U, but the activity concentrations are about equal because of the much longer half-life of ^{232}Th (Table 35–1). The content of ^{235}U in crustal matter is less than 1% that of ^{238}U. These nuclides and their progeny will be in radioactive equilibrium unless separated by a physical process, such as diffusion of the gas radon away from its site of origin.

The most significant isotope with respect to exposure of the population to background radiation is ^{222}Rn, of the ^{238}U decay series. Its half-life of 3.82 days allows time for substantial diffusion to soil gas and to the atmosphere. ^{220}Rn (thoron), with a half-life of 55 s, contributes less significantly to the airborne inventory of radon. ^{219}Rn (actinon) is the least abundant isotope both because of its short half-life (3.96 s) and because of the low content of ^{235}U in crustal matter. The following discussion centers on sampling and evaluation of airborne ^{222}Rn because (1) the atmospheric concentration far exceeds that of ^{220}Rn and (2) the radiation dose that results from exposure to ^{220}Rn is only about one-third as great as that from an equal activity concentration of ^{222}Rn. For special circumstances (e.g., geological formations with highly elevated thorium concentrations or manufacturing processes utilizing thorium), thoron exposure may be significant and must be recognized if radon sampling is required. A few methods are available for sampling thoron progeny (U.S. NCRP, 1988), but the need is rare and is not considered further in this chapter. Detailed information on concentrations of the radon isotopes, sources, emanation rates, atmospheric mixing and transport, the radiation dose resulting from exposure and the measurement of radon and radon daughters in air can be found in a series of reports issued by the National Council on Radiation Protection and Measurements (U.S. NCRP, 1984a,b, 1987, 1988, 1989). These authoritative reports have been freely consulted in the preparation of this chapter. Primary sources for material referenced to U.S. NCRP documents can be found in those reports.

TABLE 35–2. Units and Conversion Factors for Radon and Radon

	Activity
Becquerel (SI unit)	$1\,Bq = 1\,s^{-1}$
Curie (conventional unit)	$1\,Ci = 3.7 \times 10^{10}\,s^{-1}$
Useful conversions	
$1\,Ci = 3.7 \times 10^{10}\,Bq$	$1\,Bq = 27\,pCi$
$1\,pCi = 3.7 \times 10^{-2}\,Bq$	$100\,Bq = 2.7\,pCi\,l^{-1}$
$1\,pCi\,l^{-1} = 37\,Bq\,m^{-3}$	
	Special Units
SI Units	
^{222}Rn	$WL = 2.8 \times 10^{-5}\,(A) + 1.4 \times 10^{-4}\,(B) + 1 \times 10^{-4}\,(C)$ EEC in $Bq\,m^{-3} = 0.105\,(A) + 0.516\,(B) + 0.379\,(C)$ where A, B, C = concentrations of ^{218}Po, ^{214}Pb, and ^{214}Bi in $Bq\,m^{-3}$ EEC in $Bq\,m^{-3} = (3700)\,(WL)$
^{220}Rn	$WL = 3.3 \times 10^{-5}\,(B) + 3.1 \times 10^{-4}\,(C)$ EEC in $Bq\,m^{-3} = 0.91\,(B) + 0.09\,(C)$ where B, C = concentrations of ^{212}Pb and ^{212}Bi in $Bq\,m^{-3}$. The contribution of ^{216}Po to WL and EEC is negligible EEC in $Bq\,m^{-3} = 275\,(WL)$
Either	$J\,m^{-3} = 5.6 \times 10^{-9}$ (EEC in $Bq\,m^{-3}$) $J\,h\,m^{-3} = (J\,m^{-3})$ (hours exposed)
Conventional Units	
^{222}Rn	$WL = 1.05 \times 10^{-3}\,(A) + 5.16 \times 10^{-3}\,(B) + 3.79 \times 10^{-3}\,(C)$ EEC in $pCi\,l^{-1} = 0.105\,(A) + 0.516\,(B) + 0.379\,(C)$ where A, B, C = concentrations of ^{218}Po, ^{214}Pb, and ^{214}Bi in $pCi\,l^{-1}$ EEC in $pCi\,l^{-1} = (100)\,(WL)$

EEC, equilibrium equivalent concentration; WL, working level.

TABLE 35–3. Major Sources of Worldwide Atmospheric Radon-222

Source	Ci/year
Emanation from soil	2×10^9
Ground water (potential)	5×10^8
Emanation from oceans	3×10^7
Phosphate residues	3×10^6
Uranium tailings piles	2×10^6

Source: Adapted from U.S. NCRP (1984b).

Sources

Direct emanation to the atmosphere from soils and rocks is the major source of atmospheric radon (Table 35–3). Ground water is the second most significant source. Radium, the immediate predecessor of radon, is dissolved by ground water and carried into streams and rivers; the decay of the radium provides a reservoir of potential atmospheric radon. Lesser amounts are contributed from the ocean and other radium-containing crustal materials. The major source of indoor radon is the infiltration of soil gas into a structure drawn by pressure differentials between the indoor air and the soil gas. Small pressure differentials caused by temperature differences between indoor and outdoor air, or surface winds, draw soil gas from large areas into a dwelling (U.S. NCRP, 1989).

Airborne Concentrations

Typical atmospheric concentrations of radon in the United States range from about 4 to 15 Bq m^{-3} [0.1 to 0.4 pCi l^{-1}] (U.S. NCRP, 1988), with an average of about 7 Bq m^{-3} (U.S. NCRP, 1989). Concentrations are much lower over the oceans. Indoor concentrations are reported to be lognormally distributed with a geometric mean of about 48 Bq m^{-3} [1.3 pCi l^{-1}] (U.S. EPA, 1992a) and a geometric standard deviation (GSD) of about 2 to 2.5 (Harley, 2000). This implies that about 5% of homes exceed 1500 Bq m^{-3} [4 pCi l^{-1}], the level at which the U.S. EPA (1992) suggests remedial action be considered. These concentrations represent a few thousand atoms of radon and only a few atoms of the short-lived progeny per liter so that identification by chemical methods is not possible. The radiometric properties of the nuclides are used for detection; these radiometric properties also confer the potential for induction of biological effects.

RADIOMETRIC PROPERTIES OF RADON AND DAUGHTERS

Radon, an alpha-particle emitter, transforms to a polonium isotope that further decays through a series of short-lived isotopes of bismuth, lead, and polonium (also thallium for ^{220}Rn and ^{219}Rn) (Figs. 35–1 to 35–3). The short-lived progeny of ^{222}Rn include the series from ^{218}Po to ^{214}Po (RaA to RaC′). This portion of the decay chain has an overall half-life of about 30 min. Various alpha, beta, and gamma rays emitted from these nuclides are detectable (Table 35–4), and specific radiations may be used to quantitate the air concentration of individual progeny. The next nuclide in the series, ^{210}Pb, has a half-life of 21 years, is largely

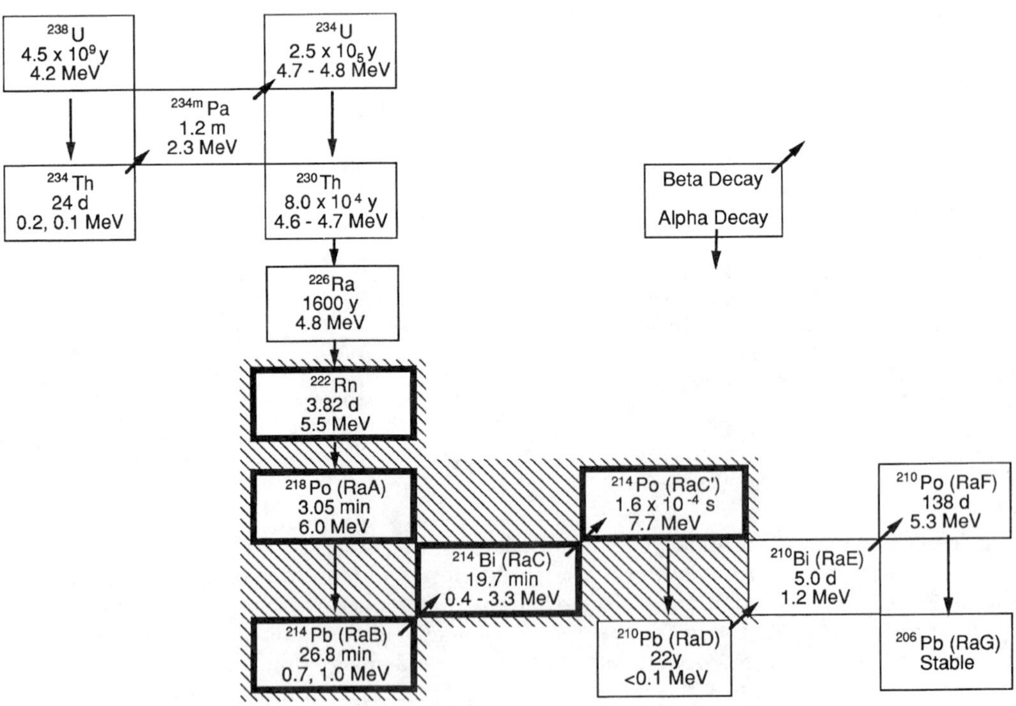

Fig. 35–1. Decay scheme of the uranium-238 series. The hatched area highlights radon and its short-lived progeny.

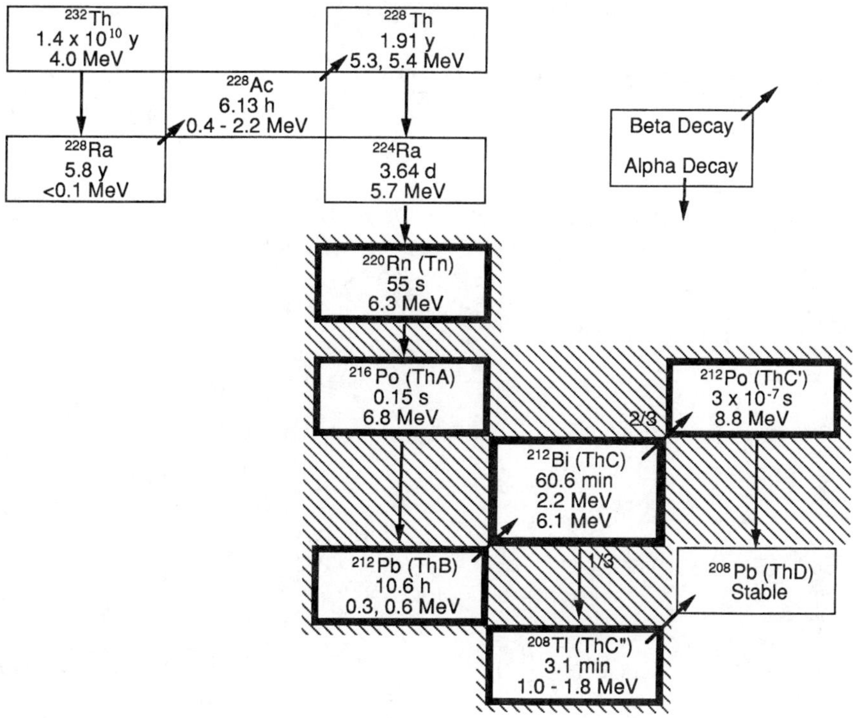

Fig. 35–2. Decay scheme of the thorium-232 series. The hatched area highlights thoron and its short lived progeny.

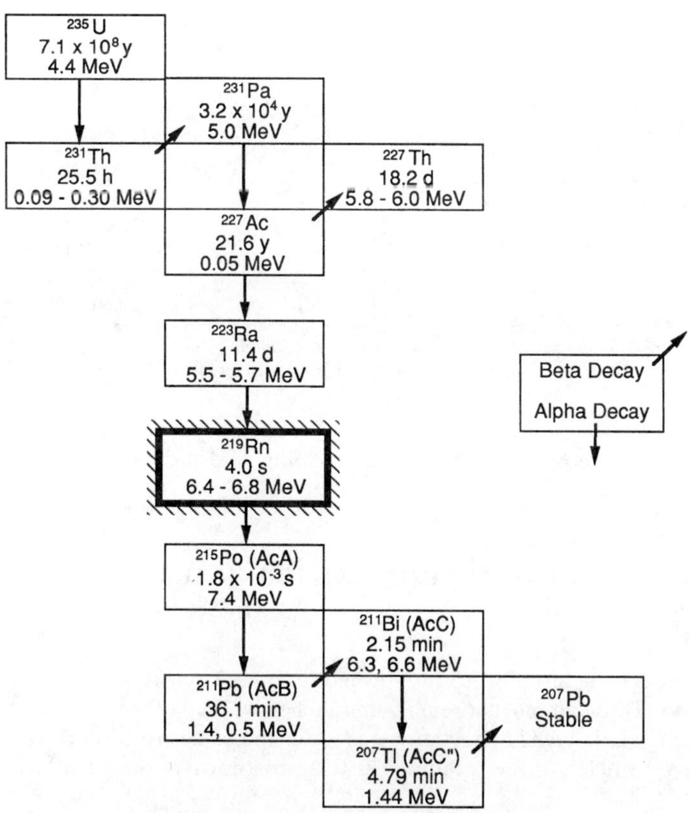

Fig. 35–3. Decay scheme of the uranium-235 series.

TABLE 35–4. Uranium Series from Rn-222 to Pb-210

Nuclide	Historical Name	Half-life	Major Radiation Energies (Mev) and Intensities		
			α	β	γ
$^{222}_{86}$Rn	Emanation Radon (Rn)	3.823 days	5.49 (99.92%)	—	—
$^{218}_{84}$Po	Radium A	3.05 min	6.00 (99.98%)		
99.98% | 0.02%					
$^{214}_{82}$Pb	Radium B	26.8 min	—	0.178 (2.4%) 0.665 (46.1%) 0.722 (40.8%) 1.02 (9.5%)	 0.295 (18.4%) 0.352 (35.4%) 0.768 (1.04%)
$^{216}_{85}$At	Astatine	~2 s	6.65 (6.4%) 6.99 (90%) 6.76 (3.6%)	—	—
$^{214}_{83}$Bi	Radium C	19.9 min	—	1.06 (5.56%) 1.15 (4.25%) 1.41 (8.15%) 1.50 (16.9%) 1.54 (17.5%) 1.89 (7.56%) 3.27 (19.8%)	0.609 (44.8%) 0.768 (4.76%) 1.12 (14.8%) 1.24 (5.83%) 1.76 (15.3%) 2.20 (4.98%)
99.98% | 0.02%					
$^{214}_{84}$Po	Radium C′	164 μs	7.69 (100%)	—	—
$^{210}_{81}$Tl	Radium C″	1.30 min		1.86 (24%) 2.02 (10%) 2.41 (10%) 4.20 (30%) 4.38 (20%)	0.298 (79.1%) 0.800 (99.0%) 1.07 (12%) 1.21 (17%) 1.36 (21%)
$^{210}_{82}$Pb	Radium D	22.3 years	—	0.0165 (87%) 0.063 (18%)	0.046 (4.18%)

Source: From Cohen, B. Sampling airborne radioactivity. In *Air Sampling Instruments for Evaluation of Atmospheric Contaminants*, 7th Ed. Ed. S. Hering, ed. Cincinnati, OH: American Conference of Governmental Industrial Hygienists, Inc., pp. 73–109.

removed from the atmosphere before disintegrating, and ultimately decays to stable lead (Evans, 1969).

AEROSOL PROPERTIES OF RADON AND DAUGHTERS

Unattached Fraction

When ^{222}Rn decays, the atom of freshly formed ^{218}Po may quickly attach to a particle of the ambient aerosol. Those atoms that remain unattached will rapidly grow by reacting with trace gases, or if charged, they will attract ambient water molecules to form clusters. These small, highly diffusive particles, or the resulting clusters, are known as the "unattached fraction." A

TABLE 35–5. Characteristics of Radon Progeny Aerosols

Atmosphere	Number Concentration (No. cm^{-3})	AMD (nm)	Equilibrium Factor
Outdoor	10^3–10^5	30–500	0.7[a], 0.8[b]
Indoor	10^3–10^5	5–150	0.4[a,b]
Mines	10^4–10^5	90–300	0.3[b]

AMD, activity median diameter.
[a] U.S. NCRP (1987).
[b] UNSCEAR (1988).
Source: U.S. NCRP (1988).

review of the diffusion coefficients measured for this size fraction can be found in U.S. NCRP (1988) and Hopke (1989). They range from 0.0025 to 0.08 $cm^2 s^{-1}$. This corresponds to a size range of about 0.5 to 5.0 nm. These highly mobile particles diffuse rapidly and deposit on larger atmospheric particles or plate out to the walls and other surfaces, but because new RaA is constantly formed, there is always a small percentage of unattached ^{218}Po. Similarly, a few ^{214}Pb or ^{214}Bi (RaB or RaC,C′) remain unattached, but the fraction is even lower than that of the RaA.

It is estimated that, after formation, about 90% of the RaA atoms have a single positive charge at the end of the recoil path, the rest being neutral (Hopke, 1989; U.S. NCRP, 1988). Although the ^{218}Po is rapidly neutralized, there remains some charged fraction. U.S. NCRP (1988) reports that a positive charge is regained during growth or attachment to ambient particles.

Attached Fraction

Most of the radon progeny attach to ambient aerosol particles, the size of the fraction depending on the number of airborne particles available. Their subsequent physical behavior tracks that of the stable airborne particles except for radioactive decay. The latter is the primary mode of removal from the atmosphere because the overall radiological half-life, 30 min, is faster than removal by deposition or plateout.

The important measurable quantities for the attached fraction are the concentration and particle size. The size distribution of the aerosol determines its physical behavior in the environment, including particle transport and deposition. Particle size is the most important factor in determining whether the particles are inhaled and where they will deposit in the respiratory tract. For these reasons, particle size is a critical determinant of the radiation dose to the respiratory airways. The size distribution and, thus, the radiation dose differ for mine, outdoor, and indoor air (Table 35–5). For radioactive aerosols, the diameter of interest is the activity median diameter (AMD), that is, the geometric mean diameter determined by measuring the radioactivity of the size fractionated sample, assuming a lognormal distribution. The AMD reported for radon progeny varies with ambient conditions, as seen from the representative ranges shown in Table 35–5, but it is always less than 1 μm. AMD is one of the most significant variables when comparing the dose per unit exposure for underground miners with that of the general population.

HUMAN EXPOSURE PARAMETERS

When atmospheric radon decays, both attached and unattached short-lived progeny are inhaled and deposit on the bronchial epithelium, where they decay before they can be removed by natural clearance processes. The dose delivered by the alpha particles emitted after deposition is the significant quantity in determining the carcinogenic potential of a particular radon atmosphere. For ^{222}Rn the alpha emitters are ^{218}Po (RaA) and ^{214}Po (RaC′) (Table 35–4). The ^{214}Pb (RaB) and ^{214}Bi (RaC), both beta-particle emitters, are not significant dosimetrically, but their behavior determines the ultimate concentration of ^{214}Po.

Units of Exposure

The historical unit of exposure is the working level (WL). This is defined as equivalent to an exposure atmosphere containing 100 pCi l^{-1} of ^{222}Rn in equilibrium with all of the short-lived progeny. The alpha-particle energy that would eventually be emitted into this air volume is 1.3×10^5 Mev l^{-1} (Table 35–6). In practice, one WL is any combination of short-lived daughters in 1 L of air that will result in the emission of 1.3×10^5 MeV of potential alpha energy. An exposure atmosphere with this quantity of potential alpha-particle energy is called a WL regardless of whether equilibrium exists. Note that WL is a measure of the air concentration of short-lived decay products, not of the radon gas.

The unit of cumulative exposure is the working level month (WLM) and is equal to exposure in WL multiplied by exposure duration in multiples of the occupational month (170 h). To calculate the cumulative exposure of a member of the population in terms of WLM, the hours must be adjusted because exposure times exceed 170 h/month. Then

$$\text{WLM} = \text{WL}[(\text{hours exposed})/170] \qquad (35\text{–}3)$$

Atmospheric clearance processes, such as deposition to surfaces, usually result in disequilibrium between the concentration of ^{222}Rn and its offspring. The airborne activity concentration of ^{218}Po is usually less than that of radon and so on for each successive decay product, except that the extremely short half-life of ^{214}Po (RaC′) ensures that it remains in equilibrium with ^{214}Bi (RaC). The resulting airborne radioactivity concentration may be described as an equilibrium equivalent concentration (EEC). The EEC is the concentration of radon that in equilibrium with each of the daughters would have the same potential alpha energy per unit volume as the actual mixture. By calculating the fractional alpha energy shown in the last column of Table 35–6, it can be seen that

$$\text{EEC} = 0.105\,[\text{A}] + 0.516\,[\text{B}] + 0.379\,[\text{C}] \qquad (35\text{–}4)$$

where, [A], [B], and [C] are the concentrations of ^{218}Po, ^{214}Pb, and ^{214}Bi, respectively, in either Bq m^{-3} or pCi l^{-1}. EEC and WL are related as follows:

$$\text{WL} = \text{EEC}\,(\text{Bq m}^{-3})/3700 \qquad (35\text{–}5)$$

TABLE 35–6. ^{222}Rn Short-Lived Decay Products

Isotope	Alpha Ray Energy (MeV)[a]	Half-Life	Number of Atoms/100 pCi	Ultimate Alpha Ray Energy (MeV)	Total Alpha Ray Energy (MeV)
^{222}Rn	5.49	3.82 days	1.76×10^6	—	—
^{238}Po	6.00	3.05 min	977	6.00 + 7.68	1.34×10^4
^{214}Pb	0	26.8 min	8580	7.68	6.50×10^4
^{214}Bi	0	19.7 min	6310	7.68	4.85×10^4
^{214}Po	7.68	2.7×10^{-6} min	8.5×10^{-4}	7.68	0.007
				12.7×10^4	

[a] The total amount of alpha ray energy that will ultimately be emitted by the short-lived decay products of the 100 pCi of ^{222}Rn = 1.3×10^5 MeV.

$$[\mathrm{WL} = \mathrm{EEC}\ (\mathrm{pCi\,l^{-1}})/100] \quad (35\text{–}6)$$

The state of equilibrium is also described by the equilibrium factor, F. F is the ratio of the potential alpha-energy concentration that exists in the mixture to that which would exist if all the decay products were in equilibrium with the radon present. Then F is the ratio of EEC to the radon concentration:

$$F = \mathrm{EEC}/[\mathrm{Rn}] \quad (35\text{–}7)$$

where [Rn] is the concentration of radon gas. If F is about 0.4, a level commonly found in homes in the United States (Table 35–5; U.S. NCRP, 1987), in an atmosphere with 100 pCi l^{-1} Rn, the EEC is 40 pCi l^{-1} and the exposure is 0.4 WL; in an atmosphere with 4 pCi l^{-1} of Rn, the EEC is 1.6 for an exposure of about 0.02 WL.

Dose Conversion Factor

The dose conversion factor (DCF) gives the relationship between the radon progeny exposure and the radiation dose to the critical target cells of the lung. It is derived from lung model calculations taking into account ambient aerosol conditions.

When calculating the dose to the lungs from inhaled short-lived radon progeny, particle size is very important. Particle number concentration is important because of the relationship between particle number and the unattached fraction (Porstendorfer and Reineking, 1999).

Pörstendorfer and Reineking (1999) calculated average dose conversion factors (DCF) for a variety of well-characterized atmospheres. They took into account the AMD, σ_g, and number concentration of the unattached fraction and of each mode of the attached fraction of the ambient decay products, and they allowed for hygroscopicity of the particles. Average DCFs calculated for the unattached fraction ranged from 0.3 to 32 mSv WLM^{-1} and 4 to 10 mSv WLM^{-1} for the attached fraction.

AIR SAMPLING FOR RADON AND ITS SHORT-LIVED DECAY PRODUCTS

Sampling Strategy

When air sampling is planned, the reason for making a measurement will determine the type of measurement that is needed. Additional considerations are the estimated amount of the activity present and whether suitable equipment is available. For any measurement, it is necessary to have adequate sensitivity and adequately calibrated instruments.

The purpose of the measurement, in order of complexity, may range from simply determining the indoor or outdoor air concentration, through quantitation of sources, to a health effects evaluation. If a determination is to be made as to whether there is a potential for people to be significantly exposed, then decision-making criteria will be required. Various national and international regulatory bodies have recommended limits for air concentrations of radon and radon daughters in buildings. These range from 74 to 400 Bq m^{-3} [0.02 to 0.11 WL]. *The Indoor Radon Abatement Act* of 1988, which is an amendment to the *Toxic Substances Control Act*, established a long-term national goal of reducing radon in U.S. buildings to ambient outdoor levels. There is some controversy about the appropriate limit for initiating remedial action, but there is general agreement that remedial action be taken if radon concentrations are greater than about 800 Bq m^{-3}, or 0.1 WL (20 pCi l^{-1} Rn, EEC = 10 pCi l^{-1}).

Activity concentrations can be determined by measuring radon, the daughters, or both. Measurement of Rn is easier and will give an upper limit to the amount of potential alpha energy (WL) in the atmosphere. In addition, if an average equilibrium factor for homes of about 0.4 is assumed, the radiation dose that will result from inhalation exposure can be estimated. Determining the sources of radon is more complex. This may require a variety of measures to assess the emanation rate from surfaces, soil, building materials, water, and so forth. To perform a complete health effects evaluation requires the measurement of both the concentration and size distribution of the radon progeny.

Sampling Methods

Sampling for either radon or progeny may be accomplished by grab sampling, continuous sampling, or collecting an integrated sample. Grab samples are either instant or very short period samples. Either radon or daughter samplers may be used. Grab samples are useful for screening on a small scale. Samples can be collected periodically over several seasons, and seasonal average concentrations can be assessed from a few measurements taken in each season.

Continuous sampling with an instrument that produces sequential on-site readout is more descriptive because there are spatial and temporal variations in radon and progeny concentrations. It is necessary for any in-depth assessment of exposure, but requires skilled personnel, a high level of quality control, and relatively complex instrumentation. Continuous sampling is essential, however, for remedial work so that sources and ventilation effects may be correlated with the observed concentrations. Some of the modern real-time air monitors for alpha-emitting radionuclides are described in Chapter 34.

Integrated sampling will result in a single value for the average concentration over the duration of the sampling. Average sampling times may be days, weeks, or months. Three important advantages for radon sampling are that (1) integrated samples enable the measurement of concentrations that are too low for grab sampling, (2) many long-duration samples can be collected at the same time, and (3) the samples can be analyzed at a convenient time and place. Both passive and active integrated samplers for radon are available. These samplers are generally less expensive than continuous samplers.

Most measurements are undertaken to measure the air concentration of radon. Separation of radon from its daughters is feasible because radon is a gas and the progeny are particles. To sample radon only, the ambient daughters are removed by filtration. The progeny that are subsequently formed by the decay of the isolated radon are either counted or quantitatively removed from the sample volume. Counting the progeny increases the sensitivity of the sampling system because it provides three alpha particles for each ^{222}Rn decay (see Fig. 35–1 or Table 35–4). It is only necessary to wait for radioactive equilibrium to be established. This will occur in about six half-lives, or approximately 3 h for the ^{222}Rn short-lived decay products.

Grab samplers and integrating samplers have been developed for both radon and the short-lived decay products. Continuous monitors are available for radon, but only semicontinuous monitors are available for the short-lived progeny. The operation of these samplers is based on a few collection and detection principles, but there are multiple variations of many of the basic types of monitors. This chapter presents at least one example of each type of monitoring system (grab, continuous, and integrating) and some of the more common variations. Details are given where the information is instructive of the sampling principles. Standard methods for indoor air measurements are presented; however, because new monitors are developed frequently, it is advisable to review the recent literature before selecting a sampling method for any extensive radon detection program. Commercially available samplers and monitoring equipment are described and discussed in the current edition of the air sampling instruments text published by the American Conference of Governmental (Cohen and Heikkinen, 2001).

Detection Principles. Most detection systems rely on counting of the alpha particles emitted from radon, ^{218}Po (RaA), and ^{214}Po (RaC′). The methods are limited by the range of the alpha particles, which can penetrate only a few centimeters in air and less than 0.1 mm in unit-density material. However, this limited range permits the development of very low background detection systems for alpha-particle counting. When counting the gamma rays emitted from ^{214}Pb (RaB) and ^{214}Bi (RaC), it is more difficult to reduce the always-present background radiation and, thus, more difficult to detect low levels of activity. To calculate WL, it is necessary to know the concentrations of ^{218}Po, ^{214}Pb, ^{214}Bi, and ^{214}Po (RaA, RaB, RaC, and RaC′). Either alpha, beta, or gamma spectrometers may be used to identify the radiation emitted at a specific energy for the identification of individual decay products. The different half-lives of the radon progeny provide another means of identification, and several sampling systems utilize the temporal pattern of the decay of total alpha activity to separate out ^{218}Po, ^{214}Pb, ^{214}Bi, and ^{214}Po.

Lower Limits of Detection. An important measure of system performance is the smallest quantity of radon or progeny that can be detected. This will generally depend on both the volume sampled and the duration of the sampling. For example, if radon gas is concentrated from an air volume, increasing the air volume will allow adequate detection of a lower initial concentration. Similarly, increasing the exposure time of an integrating sampler will generally permit quantitation of lower concentrations. The determinants of the lowest detectable amount are the sensitivity with which the instrument responds to radiation (i.e., counts/min/unit activity) and the background count rate. A good measure is the lower limit of detection (LLD) as defined by Pasternack and Harley (1971). For a 5% risk of concluding falsely that activity is present and for a 95% predetermined degree of confidence for detecting its presence, a simplified expression is

$$\mathrm{LLD} = 3.29\ \gamma\sigma_n \tag{35–8}$$

where γ is a calibration constant to convert counts into activity and σ_n is the standard deviation of the net count rate.

Grab Sampling for Radon

A common and popular device for grab sampling of radon is the scintillation cell or flask (Fig. 35–4). It consists of a 100 to 1000 mL glass, plastic, or metal chamber with a flat transparent bottom and either one or two valves. Either the flask is evacuated at the laboratory and filled by opening the valve in the field or an air sample may be drawn through the cell. The inside surfaces, except for the base, are coated with ZnS phosphor. The phosphor emits a flash of light when struck by an alpha particle. The air sample is held for six half-lives to allow the decay of any progeny originally present and for build up of equilibrium with the trapped radon. Counting is done by viewing the scintillations with a photomultiplier tube through the optically transparent base. Detection limits are about 10 pCi l^{-1}, but can be as low as 0.1 pCi l^{-1} with longer counting periods. The system is simple and reliable as long as the containers are thoroughly leak tested. The drawbacks are that the flasks are fragile and that time must be allowed for decay and to establish radioactive equilibrium.

A second grab-sampling method for radon is the "two-filter method" (Fig. 35–5). Air is drawn through a tube with filters at both inlet and outlet. The first filter removes all airborne decay products. As the radon traverses the tube, some of it decays and the resulting progeny are collected on the second filter. Knowing the time for flow through the tube, the tube volume, and the deposition loss to the walls as a function of the time of formation of the ^{218}Po

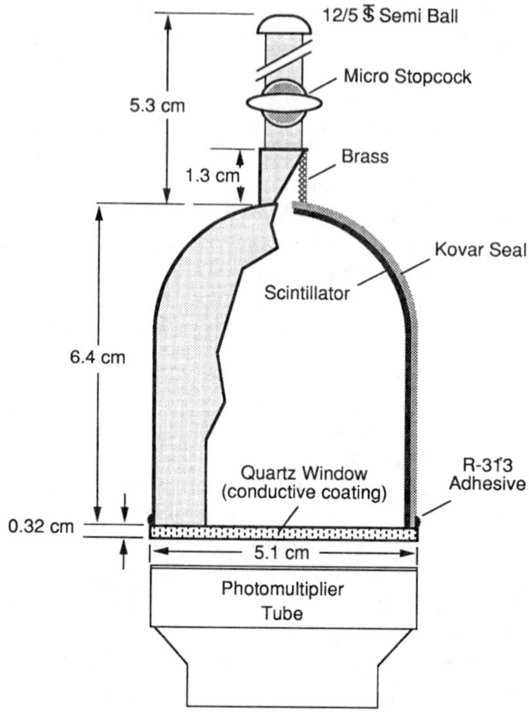

Fig. 35–4. Radon grab sampler. Alpha particles emitted by ^{222}Rn, ^{218}Po, and ^{214}Po in the sampled air volume interact with phosphor coated on the inner walls to produce scintillations. The light pulses are transmitted through the optically clear base to a photomultiplier tube and converted to electrical pulses. The signal is then amplified and counted. (Adapted from Lucas, 1977.)

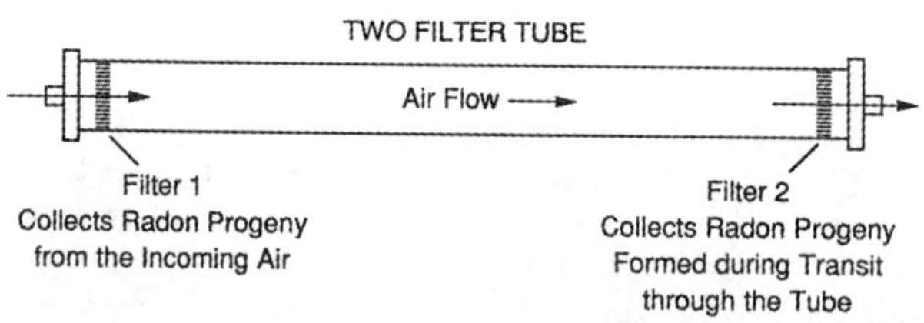

Fig. 35–5. Schematic of a two-filter tube for measurement of radon in air. A holding tank may be placed in front of the first filter to allow time for decay of any thoron that may be present.

(RaA), assay of the second filter permits a calculation of the radon concentration. The sensitivity depends on the tube dimensions, but can be as low as 0.1 pCi l^{-1}.

Grab Sampling for Radon Daughters

Determination of EEC or WL is made by collecting the short-lived progeny from a known air volume onto a filter and measuring the amount of each decay product collected on the filter. The collected progeny continue to decay during sampling; this complicates the analysis. In addition, the amount collected is often close to the LLD of the detection system, so counting errors are relatively large. A method still in use for mine atmospheres was developed by Kusnetz (1956). He used the simple approach of alpha counting the filter sample at 60 min after sampling. The basis of the method is that if an atmosphere of 100 pCi l^{-1} of radon is in equilibrium with the short-lived decay products (1 WL), an air filter will collect 100 pCi l^{-1} of each daughter. At 60 min after collection there will be 112.6 disintegrations per minute (dpm) on the filter, all due to ^{214}Po (RaC′). The number of multiples of 112.6 dpm l^{-1} is a measure of the number of WL. The complications are that the equilibrium ratios in the air vary, the samples may not be counted at precisely 60 min, and build up and decay occur on the filter during sampling. Kusnetz showed that for a sample collected for 5 or 10 min, the error due to build up and decay on the filter will be <12% under normal mine conditions, and he provided correction factors for counting times from 40 to 90 min. The method has been modified and improved by several investigators (Harley and Pasternack, 1969; Rolle, 1972) and can measure down to 0.0005 WL with a reproducibility of 35%. At 0.04 WL the reproducibility is about 4%. If a substantial amount of thoron is present the filter may be counted again after about 4 h have elapsed (to permit decay of the short-lived ^{222}Rn daughters) to obtain a correction factor. The later count will result from remaining ^{212}Bi (ThC) and ^{212}Po (ThC′) that are formed after decay of ^{212}Pb (ThB) (10.6 h half-life).

Another grab-sampling method, the Thomas method (Thomas, 1972), is a modification of an older method (Tsivoglou et al., 1953). A filter sample is collected for exactly 5 min, and the total number of alpha counts is determined for the periods 2 to 5, 6 to 20, and 21 to 30 min. These counts are entered into three simultaneous equations, together with the air sampling rate in 1 min^{-1} and the counting efficiency of the system, from which the concentration in air of ^{218}Po, ^{214}Pb, and ^{214}Bi, (RaA, RaB, and RaC) may be individually determined. This method does not make any assumptions about equilibrium, but it is slow. Scott (1981) has adapted the method for a reduced counting time.

Alpha spectrometry can directly measure the concentration of each of the progeny on a filter sample. It requires the use of solid-state detectors and a spectrometry system that is more expensive than a simple alpha-counting system. Counting at two separate times is required in order to calculate the initial concentration of each nuclide.

Continuous Sampling for Radon

Continuous radon samplers count continuously, but totals are recorded for specified time periods (e.g., 30 or 40 min counts). The continuous scintillation cell is based on the same principle as the Lucas flask. Air passes through the cell and the counts are recorded for 30 min counting periods, and then another count cycle begins. Calculation of the radon concentration in these samplers is complicated by the "memory effect," that is, the presence of radon progeny formed and deposited on cell walls during an earlier sampling period. Corrections needed for this effect are determined by sampling from a test radon atmosphere with a clean sampler (Thomas and Countess, 1979).

Another continuous sampler for radon is an automated version of the two-filter method. Air flows continuously through a two-filter tube, and an alpha particle detector is placed facing the second filter. Alternatively, the second filter may be replaced by a moving tape so that the sample is collected for a specified time period, after which the tape moves the collected sample under an alpha-particle detector and is replaced by a clean filter.

A useful class of radon samplers operate by allowing radon to diffuse into a detection volume from which the progeny are excluded. The atmospheric particles are filtered by a diffusion barrier of foam or other material. The progeny that subsequently form in the volume are collected by a charged surface. The shapes and sizes of these monitors vary, as do the means for maintaining charge on the particle collector. Alpha particles emitted from the progeny collected on the charged surface are counted with a surface barrier detector, or the surface may be coated with ZnS and coupled to a photomultiplier tube for scintillation counting. The response of these detectors must be determined in a calibration atmosphere.

These monitors also suffer from the "memory effect," which can be troublesome in a changing radon atmosphere. The same principle is applied in several integrating radon monitors that use film or thermoluminescent dosimeters as alpha detectors. This popular method is sensitive to relative humidity, which alters the collection efficiency of the progeny by the charged surface, thus changing the calibration of the system. The details of individual monitors based on this method can be found in U.S. NCRP (1988).

An alternate approach used by Chittaporn et al. (1981) is to collect the progeny on a surface that keeps the alpha particles out of the range of the detector. The radon progeny are collected by an electret that can maintain a field of $100\,V\,cm^{-1}$ in the sensitive volume. The sensitive volume, coated with scintillation phosphor, detects only the alpha particles from the decay of the parent radon. Removing the daughters from the counting volume avoids any changes in efficiency due to humidity. The monitor can detect $0.03\,pCi\,l^{-1}$ for a 1 h counting time.

To avoid counting thoron in continuous radon monitors, a decay chamber may be placed in front of the detection volume. The volume needed will depend on the sampling flow rate and must be large enough to allow the decay of the 55 s half-life thoron. Passive radon samplers, that is, those that depend on diffusion of the gas into the sensitive volume, can reduce thoron detection by use of a diffusion barrier thick enough to delay the thoron beyond its average life.

Continuous Samplers for Radon Daughters

Radon daughters are sampled by collecting them onto a filter and then counting the alpha particles emitted as they decay. The samplers can only be semicontinuous, but sampling can be done for 2 to 5 min and counting for 2 to 3 min so that a measurement can be made every 10 min. Membrane filters with pore sizes $<1\,\mu m$ are usually used because they are highly efficient for all particle sizes and because the counting efficiency of the samples thus collected is not too variable. Some other filters exhibit a variability in counting efficiency caused by self-absorption of alpha particles in the filter material. As described for the grab samplers, the filter may be counted in place, or it may move to a counting position and be replaced by the next filter collector. Counting the filter in place avoids mechanical problems that are troublesome in harsh environments. Such instruments frequently incorporate computer capability to directly calculate the concentrations of each of the progeny and the WL from the measured ^{218}Po (RaA) and RaC′ alpha particles or decay of the beta count rate. Several commercial continuous "working level monitors" are available (Cohen and Heikkinen, 2001).

Integrated Sampling for Radon

Integrating samplers for radon may be either passive or active. Passive samplers that require no power supply are extremely useful field instruments. A significant advantage is that they are quiet and, thus, acceptable to occupants when measurements are needed in homes. There are currently four types of passive radon monitors. As with other radon samplers, the progeny particles are excluded by diffusion barriers. The two most common are activated-charcoal-containing collectors and monitors containing etched-track detectors. Activated-charcoal-containing monitors are perforated cannisters 2.5 to 5 cm [1 to 2 in] high and 5 to 10 cm in diameter. The container is sealed at the laboratory and then opened to the test atmosphere for a fixed time period, usually 4 to 7 days for a cannister with >50 g of charcoal, less for smaller quantities. Atmospheric radon diffuses onto the charcoal where it is adsorbed; the container is resealed and returned to the laboratory for counting. Counting is usually done by gamma-ray spectrometry of the entire cannister. Charcoal adsorbs water vapor as well as radon so that for open-faced carbon containers an experimentally derived correction factor is needed when the humidity is high. Each sampler must be weighed before and after exposure; any difference is attributed to adsorbed water vapor, and an experimentally derived correction factor is applied. These detectors can be reused after regeneration by heating. The heating drives off the water vapor before the sampler is resealed. George and Weber (1990) report that if a sintered metal filter is used as a diffusion barrier over the inlet, no correction is needed for adsorbed water vapor under most conditions. Activated charcoal adsorbant is satisfactory for most sampling environments, but if the concentration of radon varies dramatically, sample loss may occur by desorption when air concentration declines. The sintered filter cannister is less sensitive to changing Rn concentrations than other open-faced or diffusion barrier cannisters (George and Weber, 1990).

Etched-track detectors are plastic films, usually cellulose nitrate (Kodak LR115, Eastman Kodak, Rochester, NY) or allyl diglycol carbonate (CR-39, Tech/OPs Landauer, Inc., Glenwood, IL) that develop tracks when traversed by an alpha particle. Chemical etching enlarges the tracks so they can be optically counted. These detectors do not respond to beta or gamma radiation. The detectors are placed in a container into which the radon diffuses. The alpha particles from the subsequent progeny expose the detector. The results are in terms of tracks $cm^{-2} day^{-1}$. Calibration must be done in a test atmosphere because the geometry of the detector and container will determine the detection efficiency. They are usually left in place for several months or a year, thus giving a seasonal or annual average exposure.

Another passive monitor is the passive environmental radon monitor (PERM). This unit, developed by Breslin and George (1978), is similar to the continuous passive diffusion radon monitors described above. Radon diffuses through a silica gel and filter barrier, the progeny are collected by electrostatic attraction to a rod on which a voltage (900 V) is maintained by batteries, and the radiation dose is monitored by a thermoluminescent dosimeter (TLD). The PERM sampler is usually deployed for a few weeks, and then the TLD is collected and read out on a TLD analyzer.

A small passive monitor, the environmental gamma and radon detector (EGARD) (Maiello and Harley, 1987), also utilizes TLDs to record exposure. The detector consists of a small aluminum container providing a $0.001 m^3$ [1 L] volume into which the radon diffuses. There are three sets of TLDs. One set is covered with aluminized mylar to prevent alpha particles from reaching the TLDs and records background gamma radiation. The second set is placed on an electret to which the progeny are attracted and records both background gamma and the alpha radiation. The third set is for quality control. This small detector is easy to mail and has an LLD of $100 pCi l^{-1}$ days.

Active sampling into gas sampling bags or sampling tanks may be used for radon as for other gases. The sample is transported to the laboratory, where the gas is transferred to a detection system.

Integrated Sampling for Radon Daughters

There are no passive samplers for radon daughters. Because radon progeny aerosols must be sampled onto filters, power is required. Once collected, the resulting radiation may be measured by an integrating detector such as a film, a TLD, or an etched-track detector. One such monitor, the RPISU (for radon progeny integrating sampling unit), collects daughters on a filter near a TLD (Schiager, 1974).

Radon and radon progeny measurement devices and instruments are rapidly becoming more portable and more often incorporate microprocessors and data-recording devices. However, all operate within the basic framework for measurement discussed above. A recent review of available instruments and sources will be found in Cohen and Heikkinen (2001). Additional information on instrument suppliers can be found at www.epa.gov/radonpro/manufact.htm.

Other Monitors

Several active and passive samplers have been developed that are small enough, sensitive enough, and reliable enough to be useful as personal monitors, while others are under development. The operating principle of these monitors is basically the same as for those described above.

A personal radon monitor that can also be used for thoron has recently been developed (Chittaporn and Harley, 1999). The monitor is made of lightweight conducting plastic and has three chambers. Radon diffuses into each chamber through a different diffusion barrier for signal differentiation. CR-39 plastic is the radon detector. The new monitor is based on miniaturization of an earlier version (Harley et al., 1991b). These monitors were worn by subjects to make "personal" exposure measurements, and more of the monitors were simultaneously placed in the individual's home to measure ^{222}Rn on different levels (Harley et al., 1991a). The personal exposure correlated well with exposure on the first floor of the homes tested. The reported ratio of personal to first floor exposure was 0.71.

A system for measuring the fraction of airborne radon decay products that will deposit in the nasal cavity or on the bronchial tree has been suggested (Hopke et al., 1990). Wire screen collectors are used preceding a filter. The wire screen sizes and flow rates are selected based on lung deposition models to mimic respiratory deposition. A parallel filter sampler measures the total airborne activity. The activity deposited is then obtained from the difference between the amount of alpha activity measured on the two filters.

Special techniques have been developed for measuring the flux of radon from soils and building materials. It is difficult to measure radon flux without perturbing the system under study. Various methods are described in U.S. NCRP (1988). Detection of the radon and progeny for this special class of samplers is also based on the principles described above.

CALIBRATION

All sampling and counting systems must be calibrated to ensure accurate measurement. Flow calibration is required for active sampling systems (see Lippmann, 2001). Calibration of counting systems may be accomplished with standard reference sources traceable to the National Institute of Standards and Technology (NIST). Determination of the counting efficiency of any detector, that is, the counts per minute per unit activity of radon or daughters, is an essential part of any quality control program. Two reports present detailed information on calibration procedures (U.S. NCRP, 1988; Beckman, 1972). The counting efficiency of radon samplers may be determined by emanating a known amount of radon into the sampler from a standard solution of ^{226}Ra obtainable from NIST. A few radon chambers exist in which carefully controlled atmospheres are maintained. These atmospheres are standardized by using samplers that have been standardized with reference sources. Among these are a chamber at the Environmental Measurements Laboratory of the U.S. Department of Energy (New York, NY) and a few that are commercially operated. Arrangements can usually be made to calibrate sampling equipment at one of these chambers. For radon progeny there is no standard atmosphere available. Thus, alpha-emitting standard sources are used for comparison with laboratory counters of known efficiency; alternatively, beta or gamma standards of appropriate energy are used. If counters and detectors are properly calibrated, it is then assumed that the progeny count rates are correct. In addition to counting efficiency it may be necessary to determine the collection efficiency of a sampling system (e.g., by passive monitors containing etched-track detectors). This can only be done in a chamber with a known radon atmosphere. If any changes are made in the detection system, the calibration may no longer be valid and should be reevaluated.

PROTOCOLS FOR INDOOR MEASUREMENT

The U.S. Environmental Protection Agency (U.S. EPA, 1992b, 1993) has prepared a set of protocols to be used for obtaining standardized indoor measurements. The protocols require sampling to be done in a closed house with minimum ventilation in order to maintain a stable environment. The locations at which measurements are to be made are also specified. The purpose is to permit an intercomparison of the measurements in different houses. The values of airborne radon and decay product concentrations obtained using these protocols do not necessarily reflect the concentrations to which occupants would be exposed. The EPA gives specifications for measurements of radon with continuous radon monitors, charcoal cannisters, alpha track detectors, and grab samplers. For radon daughters, procedures have been specified for the continuous working-level monitors, the RPISU, and for grab samplers. For each of these, requirements are listed for the location of the measurement device and the house conditioning; a set of minimum requirements for quality control are established for each procedure.

CONCLUSIONS

The study of radon and its progeny is challenging. There are few atoms of either radon or its daughters in air, even at concentrations that are considered high in terms of the potential for causing harm to lung tissue. The size distribution of the progeny aerosol particles is of special significance because it is an important determinant of the efficiency with which they will deposit in the lung. This, in turn, determines the radiation dose that results from a particular exposure, and it is the radiation dose that confers the health risk. A variety of sensitive methods for radon (and progeny) sampling have been developed, many of which have been noted in this chapter. These methods, and those for sample analyses, must take into account the formation and decay of the short-lived progeny. The time frame (minutes to a few hours) for the build up and decay of these nuclides complicates the measurements. It is not possible to simply allow decay until only one nuclide is present or to assay samples quickly before the build up of interferences. New sampling instruments are rapidly being developed and marketed because of the current intense interest in determining the exposure of the general population. Thus, the reader who plans to sample for these airborne materials is advised to investigate the current literature as well as sources that compile and update a list of available instruments (e.g., www.epa.gov/iaq/radon).

REFERENCES

Beckman, R. T. 1972. *Calibration Procedures for Radon and Radon-Daughter Measurement Equipment.* Informational Report 1005. Denver, CO: United State Department of the Interior, Denver Technical Support Center.

Breslin, A. J. and A. C. George. 1978. *An Improved Time Integrating Radon Monitor*. Presented at the NEA Specialist Meeting on Personal Dosimetry and Area Monitoring Suitable for Radon and Daughter Products, November 20–22, 1978, Paris.

Chittaporn, P., M. Eisenbud, and N. H. Harley. 1981. A continuous monitor for the measurement of environmental radon. *Health Phys.* 41:405–410.

Chittaporn, P. and N. H. Harley. 1999. A new personal ^{222}Rn and ^{220}Rn (RnTn) monitor. *Health Phys.* 76: S163–S164.

Cohen, B. S. and M. L. Heikkinen. 2001. Sampling airborne radioactivity. In *Air Sampling Instruments for Evaluation of Atmospheric Contaminants*, eds. B. S. Cohen and C. S. McCammon, 9th Ed. Cincinnati, OH: American Conference of Governmental Industrial Hygienists, pp. 221–240.

Evans, R. D. 1969. Engineers' guide to the elementary behavior of radon daughters. *Health Phys.* 17: 229–252.

George, A. C. and T. Weber. 1990. An improved passive activated C collector for measuring environmental ^{222}Rn in indoor air. *Health Phys.* 58:583–589.

Harley, N. H. 2000. Radon and daughters. In *Environmental Toxicants*, 2nd Ed., ed. M. Lippmann. New York: Wiley Interscience.

Harley, N. H., P. Chittaporn, M. H. Roman, and J. Sylvester. 1991a. Personal and home ^{222}Rn and gamma-ray exposure measured in 52 dwellings. *Health Phys.* 61:737–744.

Harley, N. H., P. Chittaporn, and S. C. Scarpitta. 1991b. *The Influence of Time–Activity Patterns and Lifestyle on Human Exposure to Radon in Air. Part 1: Development of a Personal Radon Monitor.* Report to New Jersey Department of Environmental Protection, Trenton, NJ.

Harley, N. H. and B. S. Pasternack. 1969. The rapid estimation of radon daughter working levels when daughter equilibrium is unknown. *Health Phys.* 17:109–114.

Hopke, P. K. 1989. Initial behavior of ^{218}Po in indoor air. *Environ. Int.* 15:299–308.

Hopke, P. K., M. Ramamurthi, and E. V. Knutson. 1990. A measurement system for Rn decay product lung deposition based on respiratory models. *Health Phys.* 58:291–295.

Kusnetz, H. L. 1956. Radon daughters in mine atmospheres—A field method for determining concentrations. *Ind. Hyg. Q.* 3:85–88.

Lippmann, M. 2001. Calibration of air sampling instruments. In *Air Sampling Instruments for Evaluation of Atmospheric Contaminants*, 9th Ed. eds. C. S. McCammon and B. S. Cohen. Cincinnati, OH: American Conference of Governmental Industrial Hygienists, pp. 73–109.

Lucas, H. F. 1977. Alpha scintillation counting. In *Atomic Industrial Forum Workshop on Methods for Measuring Radiation In and Around Uranium Mills*, Vol. 3, No. 9, ed. E. D. Harwood. Washington, DC: Atomic Industrial Forum.

Maiello, M. L. and N. H. Harley. 1987. EGARD: An environmental gamma-ray and radon detector. *Health Phys.* 53:301–305.

Pasternack, B. S. and N. H. Harley. 1971. Detection limits for radionuclides in the analyses of multi-component gamma-ray spectrometer data. *Nucl. Instrum. Methods* 91:533–540.

Porstendorfer, J. and A. Reineking. 1999. Radon: Characteristics in air and dose conversion factors. *Health Phys.* 76:300–305.

Rolle, R. 1972. Rapid working level monitoring. *Health Phys*. 22:233–238.

Schiager, K. J. 1974. Integrating radon progeny air sampler. *Am. Ind. Hyg. Assoc. J.* 35:165–174.

Scott, A. G. 1981. A field method for measurement of radon daughters in air. *Health Phys.* 41:403–405.

Thomas, J. W. 1972. Measurement of radon daughters in air. *Health Phys.* 23:783–789.

Thomas, J. W. and R. J. Countess. 1979. Continuous radon monitor. *Health Phys.* 36:734–738.

Tsivoglou, E. E., H. E. Ayers, and D. A. Holaday. 1953. Occurrence of non-equilibrium atmospheric mixtures of radon and its daughters. *Nucleonics* 11:40–45.

UNSCEAR. 1988. *Sources, Effects and Risks Of Ionizing Radiation*. Report to the General Assembly, United Nations Scientific Committee on the Effects of Atomic Radiation.

U.S. EPA. 1992a. *A Citizens Guide to Radon*. EPA 402-K-92-001. Washington, DC: Environmental Protection Agency, Office of Air and Radiation.

U.S. EPA. 1992b. *Indoor Radon and Radon Decay Product Measurement Device Protocols*. EPA 402-R-92-004. Washington, DC: Environmental Protection Agency, Office of Air and Radiation.

U.S. EPA. 1993. *Protocols for Radon and Radon Decay Product Measurements in Homes*. EPA 402-R-92-003. Washington, DC: Environmental Protection Agency, Office of Air and Radiation.

U.S. NCRP. 1984a. *Exposures From the Uranium Series with Emphasis on Radon and its Daughters*. NCRP Report No. 77. Bethesda, MD: National Council on Radiation Protection and Measurements.

U.S. NCRP. 1984b. *Evaluation of Occupational and Environmental Exposures to Radon and Radon Daughters in the United States*. NCRP Report No. 78. Bethesda, MD: National Council on Radiation Protection and Measurements.

U.S. NCRP. 1987. *Exposure of the Population in the United States and Canada from Natural Background Radiation*. NCRP Report No. 94. Bethesda, MD: National Council on Radiation Protection and Measurements.

U.S. NCRP. 1988. *Measurement of Radon and Radon Daughters in air*. NCRP Report No. 97. Bethesda, MD: National Council on Radiation Protection and Measurements.

U.S. NCRP. 1989. *Control of Radon in Houses*. NCRP Report No. 103. Bethesda, MD: National Council on Radiation Protection and Measurements.

36 Measurement of Pharmaceutical and Diagnostic Aerosols

ANTHONY J. HICKEY
School of Pharmacy, University of North Carolina, Chapel Hill, NC
DAVID SWIFT*
Division of Environmental Health Engineering, Johns Hopkins University, Baltimore, MD

INTRODUCTION

Aerosols find applications in a number of health-related settings, and their measurement plays an important role in their optimization, development, and monitoring. These applications may conveniently be grouped into three categories: (1) the use of pharmaceutical aerosols in therapy by various routes of administration; (2) the use of aerosols in diagnostic procedures; and (3) the inadvertent exposure to aerosols and its control in various areas of health care.

Aerosols employed for health care purposes share many characteristics with aerosols in other fields. Their origin and fate consist of generation, maturation with inherent physicochemical changes, and ultimate demise due to deposition or some other physical process. It is unnecessary to reiterate the techniques of measurement that have been described for other applications, but the unique aspects of aerosols employed in health care can be described, emphasizing the rationale for devising, adopting, or adapting aerosol measurement techniques to these special circumstances. Some special aerosol measurement problems in health care applications requiring future development are also discussed.

Some aerosol measurement applications in industrial hygiene, indoor and outdoor air pollution, and radioactive particulates are closely related to aerosol measurement techniques in the health care industry. All these applications involve potential human exposure to aerosols, whether for efficacy or toxicity. Consequently, it is not uncommon to see aerosol techniques in one field being adapted for another. The study of aerosols in the health care field is a recent development. However, applications of aerosols in health care stretch back in modern times at least as far as the pioneering work of Lord Lister, who employed phenol aerosols in his original disinfection studies. Classically, there are documented examples of therapeutic agents

* Deceased.

Aerosol Measurement: Principles, Techniques, and Applications, Second Edition, Edited by Paul A. Baron and Klaus Willeke.
ISBN 0-471-35636-0

derived from natural products administered as smokes as far back as 2000 to 4000 BCE. in China and India (Sciarra, 1970). In recent times, pharmaceutical aerosol products were developed for inhalation in the mid-1950s (Sciarra and Cutie, 1990).

Routes of Administration

While inhalation aerosol products represent the most sophisticated delivery system for therapeutic agents, another major application of aerosol technology involves topical administration for local activity. The physicochemical characteristics of droplets delivered from topical aerosol delivery systems are different from those required for delivery by inhalation and are considered later in the text.

Types of Pharmaceutical and Diagnostic Aerosols

Pharmaceutical aerosol delivery systems may be divided in five general categories of drug formulation/device combinations. These categories are topical aerosol products, intranasal pumps, propellant-driven metered dose inhalers, dry powder inhalers, and nebulizer systems. Each of these categories of delivery technology allows for presentation of the aerosol as an aqueous solution or suspension, a nonaqueous solution or suspension, or a dry powder. Each of these approaches has merit, and the purpose for which the drug formulation/device is being selected must be considered to allow optimal use of the available techniques (Dalby et al., 1996).

Diagnostic aerosols may be delivered from more traditional aerosol generators. Because these devices do not have to be portable, as the subject will be studied in a clinical setting, more sophisticated and fundamental approaches can be taken to aerosol generation (Gonda, 1990).

Physicochemical Properties of Pharmaceutical and Diagnostic Aerosols

The physicochemical properties of pharmaceutical and diagnostic aerosols play a major role in the ease of dispersion, site of deposition, and rate and route of clearance from the lungs.

Among aerosol measurement applications, aerosols in the health care field have certain unique features that bear on the appropriate measurement methods. One important feature is the presence in many (but not all) such aerosols of substances, which are extremely biologically potent, including macromolecules of various configurations and weights. Some of these substances have the capacity to initiate marked biological changes at very small concentration and are able to be detected and quantified with extremely sensitive and specific tests (Byron and Patton, 1994). Such macromolecules and other biologically active substances are often physicochemically fragile, and their existence in the aerosol state may alter their biological activity.

Other aerosol substances in the health care field carry living organisms such as fungi, bacteria, or viruses. Such aerosols can be detected and quantified with sensitive and selective methods of bacteriology and virology. They are similarly often physiologically fragile in certain environments, and these features are the concern of a significant historical interest in airborne infection. The introduction of vaccines and other preventive agents has markedly reduced the concern about many classic diseases spread by aerosols. However, the concern with newly discovered pathogens has renewed the interest in airborne infection, and the continuing inability to prevent infection from the common cold or influenza provides impetus to understand the airborne route. Despite a lack of any positive evidence to date, there is continuing interest in the possibility that human immunodeficiency virus infection may have an airborne component, and many diseases, such as tuberculosis and measles, are known to be transmitted in air.

PHARMACEUTICAL AEROSOLS BY ROUTE OF ADMINISTRATION

Topically Applied Aerosols

Types of Topically Applied Aerosols. A number of therapeutic aerosols are applied topically. This may be for antiseptic applications or for antimicrobial activity, for example, in burn treatment. These products operate on the principle of top pressurization with gases or by the use of liquefied propellants. Where the head space in the canister is filled with gas above atmospheric pressure, the ideal gas law may be used to predict the loss of pressure over the lifespan of the product. For liquefied propellant systems Raoult's and Dalton's laws may be used to predict the vapor pressure over the lifetime of the product.

Principles of Delivery and Delivery Systems. Figure 36–1 depicts a typical topical aerosol product. The product is displaced through the valve by depressing the valve stem using the actuator as can be seen in the schematic Figure 36–1b. The product consists of a number of components. Typically the container is composed of stainless steel or tin-plate stainless steel. The valve contains a number of components including, potentially, delrin and/or buna or neoprene rubber gaskets, stainless steel springs, a variety of possible metal mounting cups, and a polypropylene or polyethylene dip tube. Each of these components must be compatible with the product.

These aerosols, while having a broad consumer application, are not a major focus for evaluation in this chapter as the droplet size of interest is quite large, generally >50 μm.

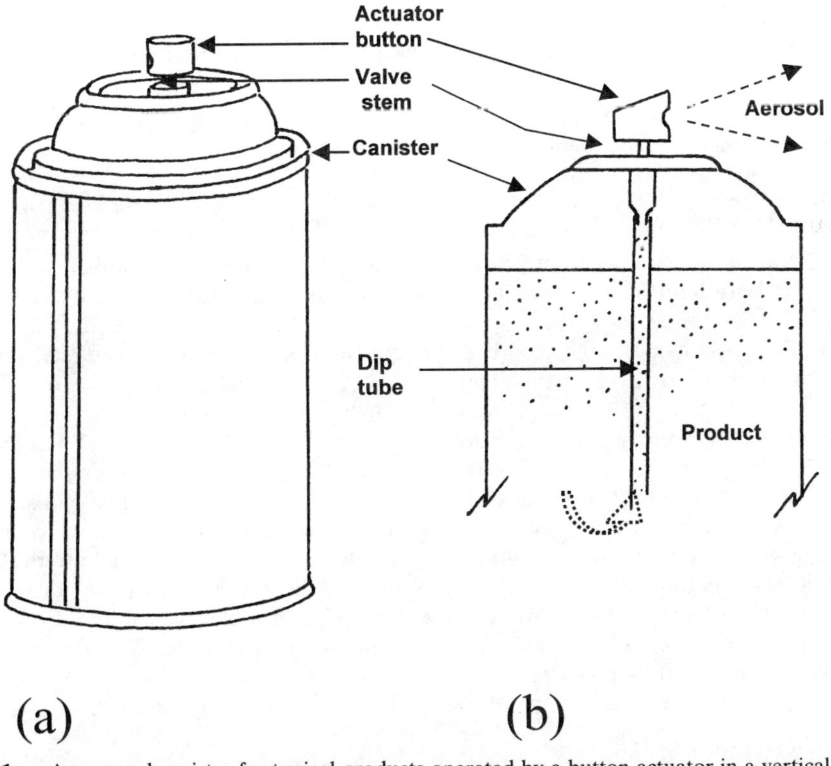

Fig. 36–1. **a**, An aerosol canister for topical products operated by a button actuator in a vertical orientation. **b**, A vertical section showing the dip tube, continuous valve, gas headspace and product. Dotted lines indicate direction of product into dip tube and aerosol from actuator.

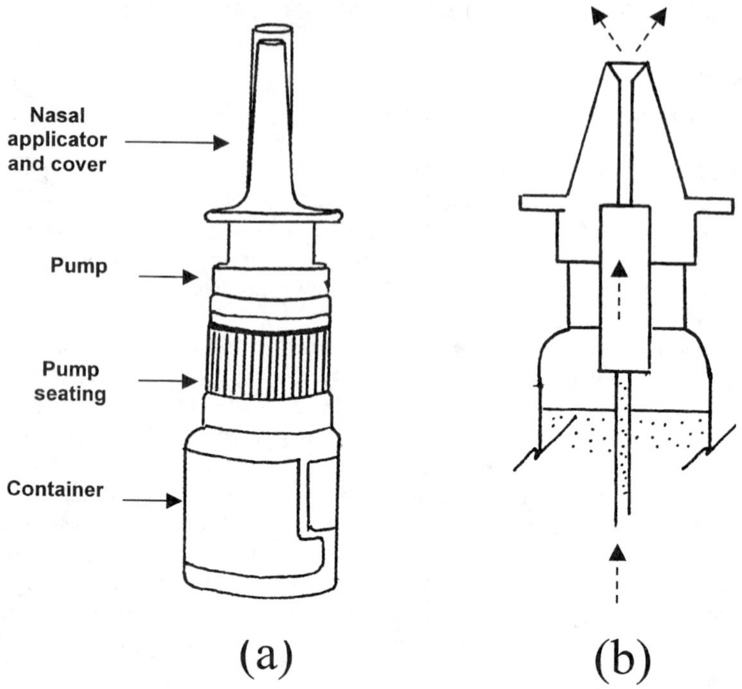

Fig. 36–2. **a**, A nasal spray product. **b**, A schematic vertical cross section through the product indicating the metering chamber, dip tube, and swirl nozzle. Dotted lines indicate direction of flow.

Large droplets are required as the intention is to use this product to meter the therapeutic substance to a local site and deposit it by inertial impaction.

Intranasal Aerosols

Types of Intranasal Aerosols. Intranasal aerosols are delivered for a number of purposes. Saline sprays are applied to rehydrate the nasal mucosa. Steroidal and nonsteroidal anti-inflammatory agents are delivered to alleviate allergic rhinitis. More recently, calcitonin, a systemically active agent that is absorbed through the nasal mucosa, has been delivered for the treatment of osteoporosis in the elderly.

Principles of Delivery and Delivery Systems. Figure 36–2 illustrates a typical pharmaceutical nasal spray product. Figure 36–2b shows that the principle of delivery is similar to that of a topical product except the force for emission of the product is now supplied by the patient. As the base of the product is pushed the container moves upward with respect to the actuator finger grip, and the contents of the metering valve are propelled throughout the nozzle, which, based on its geometry, can swirl the product or deliver it directly thereby dictating the droplet size and plume geometry of the emitted spray. The droplet size emitted from a nasal product is generally >20 μm and is intended for direct application to the nasal mucosa by inserting the nozzle into the nostril.

Inhalation Aerosols

Types of Inhalation Aerosols. A therapeutic aerosol for inhalation is intended to deliver a substance to the respiratory tract that will bring about some desired physiological change

(Newman, 1984). The respiratory tract may be the locus of activity, or it may be the portal of entry for systemic distribution of a therapeutic drug. Several different types of therapeutic agents may be delivered to the respiratory tract. These include bioactive molecules, which act locally or systemically at the cellular level, substances that physically alter the properties of respiratory secretions, and substances that treat infection. They range in chemical complexity from macromolecular solutions to water or saline aerosols.

For systemic drugs, it is acceptable to deliver the substance to any site of the respiratory tract that will permit the desired absorption characteristics (usually rapid absorption). Therapeutic agent aerosols that act locally usually require a more specific site of delivery, depending on the desired site of action. For example, aerosol bronchodilator drugs that are acting locally must be delivered to the bronchial airways in sufficient quantity to achieve the desired result. If the aerosol is entirely deposited in the airways above the trachea or predominantly delivered to the alveolar spaces, the local bronchodilation may not occur.

The considerable body of knowledge that now exists concerning the deposition of "ideal" aerosols in the respiratory tract (Heyder and Rudolph, 1984) has been drawn on to predict and design aerosol delivery systems for therapeutic substances. However, as is detailed below, there are a number of "nonidealities" concerning many therapeutic aerosol systems that hinder reliable prediction, even though some general statements about deposition can be made. Measurement methods for therapeutic aerosols must take these nonidealities into account in linking aerosol properties to deposition sites.

Principles of Delivery and Delivery Systems. A number of different techniques for aerosol generation are employed for therapeutic aerosols, each of which bears on the appropriate measurement methods for these aerosols. The initial physical state of the substance to be aerosolized is a bulk dry powder, a solution, an emulsion, or a solid-particle suspension. Depending on the method of generation, the liquid vehicle may be either water or a liquid propellant. Properties of aerosols differ according to the device utilized to generate the aerosol. Some of the properties are listed in Table 36–1.

Several methods exist for aerosolizing bulk powders of therapeutic agents (Dunbar et al., 1998). The Spinhaler was originally described by Bell et al. (1971). Bulk powder in a gelatin capsule is mounted in a rotor holder inside the inhalation tube. Two holes for powder dissemination are punched just before use. The rapid inspiratory flow of the patient causes the rotor to spin and vibrate on the undersized shaft, causing the powder to be discharged out of the capsule into the air stream. In its original form, the bulk powder was a mixture of the micronized drug and powdered lactose, the latter intended to enhance aerosolization by preventing excessive cohesion of the drug particles. Currently, the micronized drug is loaded into the capsule without lactose or other diluent.

The Rotahaler (Fig. 36–3) is a similar device for delivery of a dry powdered drug in which the powder is discharged into a small cylindrical chamber before inspiration and is aerosolized during rapid inspiration (causing high air velocity and turbulence in the chamber) and transported through a mouth tube to the respiratory tract (Kjellman and Wirenstrand, 1981).

TABLE 36–1. Characteristics of Aerosols (Velocity, Particle Size, and Concentration) Actively or Passively Delivered from a Variety of Devices

Device	Velocity	Size (μm)	Concentration	Emission/Sampling
pMDI	High	1–10	Variable	Active
DPI	Variable	1–10	High	Passive or active
Nebulizer	Low	1–5	Low	Passive or active

DPI, dry powder inhaler; pMDI, propellant-driven metered dose inhaler.

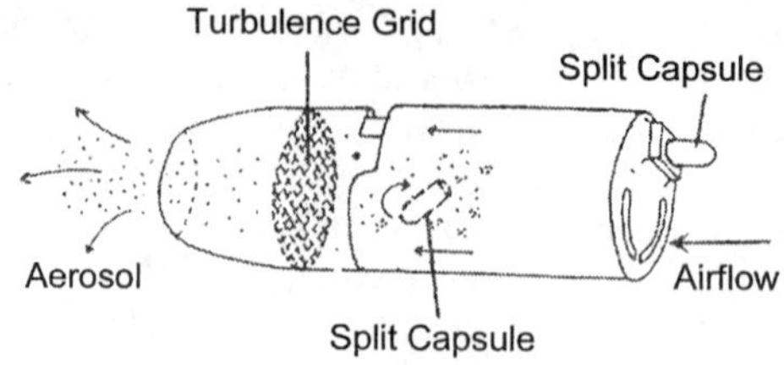

Fig. 36–3. Schematic of a Rotahaler indicating the two sections and groove by means of which rotation is controlled and showing the position of the capsule and the grid through which particles pass on the patient's inspiratory flow. The broken capsule moves on the inspiratory flow and disperses aerosol particles.

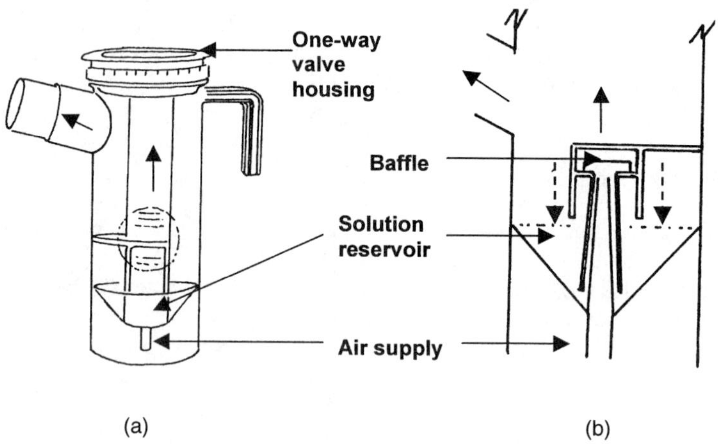

Fig. 36–4. **a**, Schematic depiction of a Pari LC-Star nebulizer. **b**, A section through the device indicating the coaxial liquid feed and air supply orifices and the blade baffle upon which the atomized spray impinges. Solid directional lines indicate aerosol flow; dotted lines indicate large impact and return to reservoir.

Both of these devices require a loading step before inspiration. In the Turbohaler, a disk containing many powder charges (up to several hundred) can be rotated for each application to locate a fresh powder charge in the path of inspired air. Entry of the inspired air through the device suspends the powder charge. The turbulent flow through the flow channel to the mouth tube is intended to disperse and mix the powder so that primarily individual powder particles enter the respiratory tract (Jaegfeldt et al., 1987).

Many therapeutic aerosols are generated by nebulizers (Niven, 1996; Mercer, 1981). Jet nebulizers employ high-velocity gas (usually compressed air) to shear bulk liquid into fine particles (for example, Fig. 36–4). This principle was originally applied to generate liquid aerosols at least 150 years ago, and all such devices are linearly descended from the original. It is well known that for a given liquid the median particle size produced decreases with increasing air velocity, reaching a practical limit at sonic velocity. For a given air velocity, the median-size particle decreases with decreasing liquid viscosity and surface tension. Modern nebulizers achieve smaller median particle sizes by inertial removal of larger particles, thus "recycling" a large fraction (usually 99%) of the liquid mass atomized and, therefore, reducing the airborne droplet mass concentration. All aerosols produced by jet nebulizers are polydisperse such that the majority of the aerosol mass (even after inertial removal) is usually found in a relatively small number of large particles whose size is much greater than the count median diameter.

The other major type of nebulizer employed for therapeutic aerosols is the ultrasonic nebulizer, in which ultrasonic energy originating from an electric transducer is focused within a bulk liquid, producing airborne particles whose median size is related to the liquid properties and the ultrasonic frequency. The amount of aerosol produced is determined by the ultrasonic energy level, and the aerosol mass concentration is determined by the generation rate and air flow dilution rate, which can be independently set. This is the primary practical difference between jet and ultrasonic nebulizers, in that the aerosol mass concentration of the jet is limited by the jet flow for atomization.

The propellant-driven metered dose inhaler (pMDI) is a widely used method of generating therapeutic aerosols. In the pMDI, drug is aerosolized by the release of a mixture of drug, propellant, and (in some instances) surfactant to maintain a stable dispersion of the drug in the liquid propellant within the canister before release (Purewal and Grant, 1998). This is the well-known "aerosol can" concept (used to dispense numerous commercial products) miniaturized and fitted with a dosing valve that dispenses a given liquid volume on each depression of the valve cap (Fig. 36–5). In most instances, the drug is maintained as a suspension of micronized powder within the propellant. Some drugs are liquid emulsions or solutes. The propellant normally makes up most of the material within the canister; a surfactant to prevent particle agglomeration is sometimes a minor constituent.

The MDI enjoys wide popularity as a therapeutic drug generator because it is self-contained (requires no air pressure, battery, or other energy source), pocket sized, generates a relatively constant mass of drug, and is relatively easy to use. The phase-out of widely used chlorofluoromethane propellants for environmental reasons has clouded the future use of MDIs. Alternative propellants are now in development that should have similar generative properties provided they do not demonstrate undesirable physicochemical interactions with therapeutic agents.

Some aerosols for localized delivery in the nasal or oral airways are generated by liquid spray devices. These generation devices are known technically as *hydraulic atomizers* as opposed to the jet nebulizer; in the former a liquid is mechanically forced through a small orifice and breaks up to form large droplets (50 to 200 μm) while, as described above, jet nebulizers use air streams to break up liquids, producing much smaller particles. Liquid spray devices differ from MDIs in that no explosive evaporation of the droplets takes place to disperse the particles and reduce particle size.

Because therapeutic aerosols are meant to be breathed, a system is needed to couple the aerosol generator to the breathing pattern of the user (Swift, 1989). Account must be taken of both the generative process of the aerosol and the breathing pattern of the user. A mismatch between these may result in nonoptimal conditions under which either aerosol is generated but not breathed or breathing takes place without sufficient aerosol being provided. Because the object of aerosol therapy is to deliver an optimum amount of agent to a specified respiratory tract location, nonoptimal conditions can result in the potentially undesirable conditions of underdose or overdose.

In the case of dry-powder aerosol generators, the delivery system is simply a suitably sized tube to be inserted into the oral or nasal passage. Ideal coupling between breathing and aerosol generation is ensured by making generation of the aerosolized powder breath actuated. The only condition of failure occurs when the user cannot achieve a high enough inspiratory flow rate to aerosolize the dry powder. Exhaling through the system before inhalation could humidify the powder and cause particle adhesion, reducing the ability to aerosolize individual particles; therefore, such devices usually include some means to reduce or obviate such conditions.

Jet and ultrasonic nebulizers normally run at a constant flow rate of 1.3 to $1.7 \times 10^{-4}\,m^3/s$ [8 to 10 L/min], thus the necessity to provide a path for exhaled air and aerosol generated

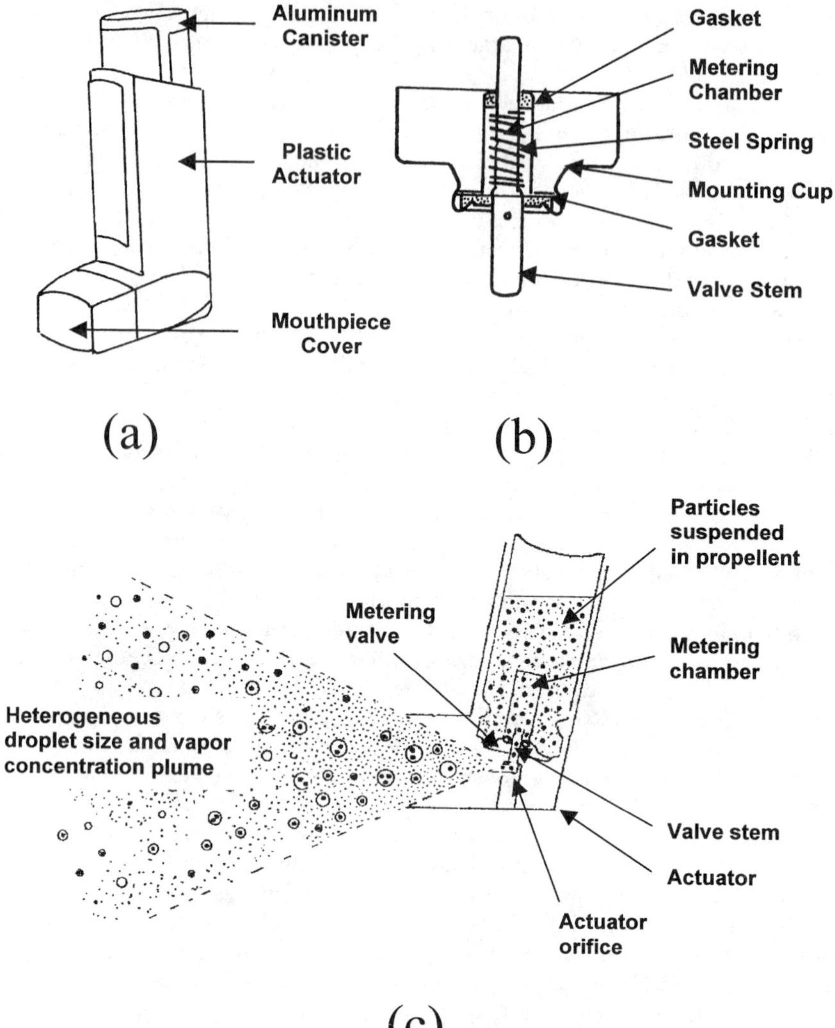

Fig. 36–5. **a**, A pressure-packaged metered dose inhaler shown in its operational orientation with canister inverted and valve down. **b**, Metering Valve. **c**, Vertical cross section through device showing aerosol generation.

during the exhalation phase. Because inspiratory flow rate varies, it is also necessary to provide for extra "make-up" air when the inspiratory rate exceeds the generation flow rate (and vice versa). In some jet nebulizer systems, an air pressure bypass is provided to permit generation only during inspiration; the user simply covers the bypass by finger pressure during the inspiratory phase to produce aerosol. This does eliminate aerosol generation during exhalation, but does not provide for flow matching during inspiration. Because ultrasonic generation does not require a constant flow for aerosol generation, the flow through the generator can be inspiratory-flow-driven, resulting in some temporal variation in aerosol mass concentration.

The temporal generation pattern of MDIs is quite different from the above types; when the actuator is pressed, an aerosol is formed during a very short period of time (~0.1 s). This aerosol generation phase is accompanied by a very high air flow rate resulting from the rapid evaporation of the propellant and the induction of air. It is impossible to match the temporal flow pattern of the MDI to the inspiratory flow of the user. Thus, if the canister is fired directly into the oral cavity through a short mouth tube, the particles, which consist of the evaporating propellant and agent, have a very high velocity and a strong tendency to preferentially impact on the posterior pharyngeal wall. In studies of deposition of MDI aerosols in the human respiratory tract, it has been shown (Newman et al., 1982) that only 6% to 14% of the aerosol reaches the thoracic region of the respiratory tract distal to the trachea.

Efforts to reduce high pharyngeal deposition and improve the matching between aerosol generation and inspiratory breathing have centered on "spacers." These are conduits of various geometries and volumes that are placed between the MDI and the mouth into which the MDI is discharged. The aerosol is then inspired from the spacer (Moren, 1985). The spacer is intended to allow the aerosol particles to both slow down and evaporate before they are inhaled, resulting in greater deep-lung delivery. It is observed that spacers do not result in greater thoracic deposition but do reduce pharyngeal deposition; larger particles impact or settle in the spacer, and the remaining aerosol is inhaled with less resulting oral-cavity deposition. This delivery system allows generation and breathing to be temporally separated.

A special issue associated with delivery systems for therapeutic aerosols is product utilization. Many currently used drug substances are relatively expensive, and there is considerable motivation to reduce the cost by utilizing as much of the drug as possible. The ultimate measurement of utilization is how much of the provided therapeutic substance ends up at the desired site in the respiratory tract during the breathing periods. Often, the utilization efficiency is a very small percent, even in the case of MDIs, in which practically all of the agent in the container is available for aerosolization.

Other therapeutic aerosol systems have lower utilization for a variety of reasons. Optimal design of generation-delivery systems should strive for high utilization as well as proper particle size (Swift, 1989).

DIAGNOSTIC AEROSOLS

Types of Diagnostic Aerosols

Diagnostic aerosols differ from therapeutic aerosols in that the intention for their use is to learn something about the state of the respiratory tract rather than the delivery of an agent to produce a specific local or systemic effect (Wagner, 1976). It is usually important to characterize properly the properties of such a diagnostic aerosol because the information to be determined may be critically dependent on the quantity and site of deposition of the aerosol.

There are two widely used diagnostic procedures involving aerosols: the measurements of pulmonary ventilation and those of airway reactivity. The remainder of the diagnostic methods employing aerosols are rather more specialized and, thus, limited to research laboratory settings where special equipment and expertise is available.

Ventilation measurements using aerosols may have several purposes, including the diagnosis of pulmonary embolism (PE) and the identification of regions of either acute or chronic airway constriction. Ventilation is defined as the actual quantity of air in various compartments of the lungs compared to what would be expected if the lung were operating normally with respect to the inspired air. The measurement of ventilation for diagnosis of PE is carried out in tandem with a lung perfusion measurement. Lung perfusion is likewise defined as the quantity of blood delivered to various vascular compartments of the lung compared to the expected quantity. A specific region of the lung that has both a ventilation and a perfusion defect, as evidenced by a significantly reduced amount of radioactivity of both aerosol and vascular injected labeled microspheres, is considered a positive indication for PE. The use of aerosols for ventilation measurements is not universal; many clinicians avoid using aerosols (despite their superiority in several respects) and prefer to measure ventilation with ^{133}Xe or ^{82}Kr gas. While these gas methods do not require aerosol generation equipment, the radiation dose required to obtain several views of the thoracic region for thorough analysis is greater than that for aerosols. In the case of aerosol use, one administration of the radiolabeled aerosol is adequate for all views that can be taken sequentially. It is common to use a jet nebulizer (see below) for such aerosols.

The other common diagnostic aerosol procedure is the measurement of airway hyperreactivity, often referred to as *bronchoprovocation*. In this method aerosol containing a specific activating substance, usually methacholine or histamine, is administered to the subject in progressively concentrated solutions. Measurement of pulmonary mechanical behavior is made after each administration. When the mechanical behavior has changed by a significant amount, the test is stopped, noting the total dose required to achieve the change. It is found that allergic and asthmatic individuals have a markedly greater sensitivity to these agents than do individuals who have no such clinical history, and the test is a convenient means to quantify the degree of sensitivity. Other environmental agents may modulate the response of sensitive individuals. This provides a convenient means for investigating the interaction between common environmental airborne agents (indoor or outdoor air pollutants) and specific reactivity producing agents or the conditions to which individuals may be exposed.

Aerosols are also employed in diagnostic procedures for measuring the permeability of the alveolar membrane, the degree of constriction of medium and small bronchi, the mucociliary system, and the mixing of tidal and residual air in the lungs (Agnew, 1984). All these methods are rather more specialized, requiring a greater degree of understanding of the aerosol characteristics and the means for delivery. Aerosols used for alveolar permeability must not only have the proper biological properties to differentiate high from low permeability, but must be of such size and nature to achieve predominantly alveolar deposition. High degrees of bronchoconstriction measured by aerosol penetrance, the ratio of the quantity of aerosol in the lung periphery to the quantity in the central hilar zone, are difficult to make without well-defined aerosols and techniques such as gamma-camera scintiography. Measurements of lung mucociliary clearance are usually made with radioisotope-labeled particles, which are not readily dissolved in the lung fluid or otherwise transported away from the airway epithelium by means other than mucus transport. The particles must deposit predominantly on ciliated airway surfaces. Similarly, measurements of mucociliary clearance in the nasal airway require particles to deposit or be transported to ciliated epithelium. Aerosol mixing experiments require that the aerosols suffer minimal respiratory deposition, a condition that can be realized with a highly monodisperse aerosol of a specific size range.

Principles of Delivery and Delivery Systems

For the commonly used aerosol diagnostic methods, it is usual for the aerosol to be generated by a jet or ultrasonic nebulizer, with the liquid containing the radioisotope in solution or colloidal suspension (Newman, 1984). Although this may not be the ideal method, the equipment and expertise to generate stable monodisperse aerosols for specific procedures are not usually available. It is likely that the aerosol generated for bronchoprovocation tests has a large mass fraction, which deposits in the alveolar region; even so, the test is reasonably reproducible for a single subject. Bronchoconstriction aerosol measurements are also made with polydisperse radiolabeled aerosols generated from jet or ultrasonic nebulizers. If the mucociliary clearance of large bronchi is to be measured, large-diameter particles ($d_a >$ 5 μm) must predominate in the aerosol, which must be breathed at high inspiratory rate to achieve significant deposition at bifurcations of large airways. Under such conditions, a significant fraction of the aerosol is deposited in the posterior oral cavity, reducing effective delivery. Nebulization of suspended large particles (3 to 10 μm diameter) can be employed for such procedures. For mixing studies, it is necessary to measure aerosol concentration profiles during inspiration and expiration. This is normally accomplished by photometric light-scattering measurement of a monodisperse aerosol generated by an evaporation condensation (Sinclair-LaMer) generator. This method of generation requires expertise in determining the degree of aerosol monodispersity.

Most delivery systems for diagnostic aerosols are similar to certain delivery systems for therapeutic aerosols. The primary difference in most cases is that the material limitations on substances used for diagnosis are usually less stringent. The exception is that the utilization of radioisotopic aerosols used in diagnosis is very low. Thus, most of the isotope remains behind after the procedure. This is a situation, which may be costly or involve high activity, that must be shielded during handling and use.

In some diagnostic procedures it is important to measure the breathing characteristics during delivery to estimate aerosol deposition mass. Because diagnosis is usually done in a clinical laboratory, the space and portability features of delivery systems, which are important for some therapeutic aerosols, are not a factor (Heyder, 1988). The MDI is rarely, if ever, used in diagnosis employing aerosols. In bronchoprovocation aerosol tests the usual method of aerosol delivery is by nebulized solution.

CHARACTERIZATION OF PHARMACEUTICAL AND DIAGNOSTIC AEROSOLS

Measurement of Physicochemical Properties

Topically Applied Noninhaled Aerosols. Some therapeutic agents appear in aerosol form, although they are not intended for inhalation. Usually, these aerosols are to be transported to some surface where their therapeutic action occurs. This was the case with Lister's aerosol generation of phenol, which presumably had its antiseptic action by depositing on biological and other surfaces to render harmful pathogens impotent. It is not proposed that inhalation of the phenol had any therapeutic benefit; in fact, it was ultimately abandoned because of the irritant effect of the aerosol on the eyes and mucous membranes.

Such agents, which may be used to deliver therapeutic agents by aerosol to the skin and other body surfaces (such as during surgery), must have desired characteristics. One key feature is a high probability of being directly deposited on the intended surface, with a minimum of "overspray" that might lead to other undesirable effects or (at minimum) less than optimal utilization of the agent.

Agents could be in the form of dry dusts, solutions, or suspensions, but in order to achieve even coverage of a surface the use of propellants as "drivers" and dispersers of the agent from pressure-packaged canisters ("aerosol" cans) seems the most ideal approach. As has been detailed above, the aerosol emerging from such a device is composed primarily of large propellant drops (~25 μm) containing the active substance and travelling at high velocity (~40 ms^{-1}). According to the theory of inertial deposition, these are the ideal conditions for a high percentage of the droplets to deposit on a surface perpendicular to the spray axis and near enough to prevent total evaporation and deceleration by viscous drag.

The major disadvantage of this method of administration of aerosols to biological surfaces is the cooling effect on the surface as a result of the propellant evaporation. However, this is not ordinarily a significant problem if a distance is maintained to allow some evaporation to take place in the gas phase before deposition. In fact, some applications of propellant-driven droplets are actually intended to retard inflammation and pain by "freezing" the tissues, similar to the effect of cold water or other cooling upon burns.

Little information exists in the scientific literature on the optimal conditions for therapeutic aerosol delivery to surfaces, but manufacturers of these and other consumer products applied to surfaces by propellant atomization (e.g., waxes, cleansers, polishes, and hair sprays) have determined by experiment the conditions maximizing even-spray application and minimizing overspray. Some of these products are irritants to some individuals when inhaled. Notable among these products are nonstick spray containing lecithin and certain hair sprays.

Therapeutic skin sprays contain agents such as bactericides, steroids for control of inflammation and itch, antibiotics, and antifungal agents. In their normal use, the lifetime of the aerosol between generation and surface deposition is short enough to render standard methods of aerosol sampling and measurement ineffective. Aerosol overspray of such applications can be characterized in a standard form using area cascade impactors, filter cassettes, or personal aerosol samplers with cyclones to determine the respirable fraction. However, the irritant effect of such aerosols is likely to be observed for any region of the respiratory tract.

In summary, no aerosol measurement techniques unique to the noninhalation aerosol area have been developed despite the importance of proper minimization of overspray for many of the agents used.

Inhaled Aerosols. The properties of therapeutic aerosols necessary to describe their behavior are similar to the properties of other aerosols, that is, size distribution (number, surface, and volume), number and mass concentration, electric charge, hygroscopicity, and the distribution of the active ingredient among the particles. Because inhaled particle transport and deposition depend on particle aerodynamic behavior in a flowing gas, particles in the inertial regime are properly characterized by their aerodynamic equivalent diameter. The conditions of generation, inhalation, and aerodynamic behavior in the respiratory tract are such that for isometric particles with $d_a \leq 5.0\,\mu m$, this is the appropriate size descriptor.

There are three major problems with therapeutic aerosols that complicate the measurement process. The first is that, in many instances, the distance between the generator and inhalation site of the aerosol is very short, and there is not an opportunity to place aerosol-measuring devices in this location. It is conceivable to sample (remove) the aerosol being generated before it is inhaled, but the influence of sampling (for subsequent measurement) on inhalation of the aerosol is not known in general.

A second related issue is the matching of flows between generation, inhalation, and measurement involving a sampling flow. As discussed above, generation of therapeutic aerosols may be either at a constant flow rate (e.g., nebulizers) or under pulse flow conditions (e.g., MDIs). In the first case, the matching between the generation and sampling can be achieved, but the cyclical breathing pattern is not matched to either generation or sampling. In the case of pulse generation, neither sampling nor breathing can be properly flow matched to the pattern of generation, and this leads to potential errors in aerosol characterization.

In the case of dry-powder aerosol generation, the generation is forced to match the inspiratory flow, but if sampling is envisioned for aerosol characterization, the sampling flow cannot readily be matched to the inspiratory flow. Thus, under any circumstance, it is not feasible to match generation, sampling, and breathing flows simultaneously.

The third problem of therapeutic aerosols that make aerosol characterization difficult is the unstable nature of the liquid aerosols produced by nebulizers and MDIs. Aqueous aerosols produced by jet or ultrasonic nebulizers are subject to rapid changes in particle diameter and other properties due to evaporation or growth of aqueous particles as they seek equilibrium at the changing conditions of temperature and humidity within the delivery system. This is particularly true when the delivery systems encounter both inspiratory and expiratory flows; the expiratory flow consists primarily of air at near to the body core temperature and fully saturated, while aerosol generated by nebulizers is often at 5 to 7 K below the ambient temperature (Mercer, 1981).

In the case of MDIs the aerosol emerging from the canister consists of propellant particles containing the active ingredient and other substances (e.g., surfactant). These particles are travelling at high velocity and are rapidly evaporating, with attendant cooling. The particle size distribution at the front end of the distribution tube is likely not the same as the distribution at the distal end of the oral passage.

Because of these experimental difficulties in the measurement of aerosols entering the respiratory tract during actual breathing, the usual practice has been to connect the delivery system output directly to an appropriate instrument for aerosol measurement. The most common instrument used for nebulizer-generated aerosols is the cascade impactor (see Chapter 10). This size separation device is operated at a constant flow rate, which can be set according to one of the several mutually exclusive criteria. The first is that of exactly matching the output flow rate of the delivery system so that all the aerosol generated enters the impactor, but this requires that the impactor be capable of such matching, a situation not possible with all commercially available impactors. Next, the flow rate into the impactor can be less than the output flow of the nebulizer, with the excess allowed to escape. Third, if the impactor flow rate exceeds the output from the delivery system, additional air must be drawn in, simulating the possible real situation during the breathing cycle when inspiratory flow exceeds nebulizer output flow.

As alluded to above, no one of these criteria is totally satisfactory when the inspiratory breathing is cyclical and the nebulizer output is constant, especially if the size and concentration of the aerosol depend strongly on the make-up air temperature and humidity. No impactor presently exists that can be operated predictably under cyclical flow conditions to match the breathing profiles.

A further difficulty with using cascade impactors to measure aerosols from nebulizers is the apparent evaporation of the particles taking place as they traverse the stages of the impactor. It is observed that when aqueous aerosols containing a solute such as 0.9% NaCl (isotonic with respect to extracellular fluids) are passed through a cascade impactor, the upper-stage deposition is that of liquid particles while the final stage(s) collect solid salt particles. The evaporation may be primarily due to the pressure drop through the impactor in conjunction with the increase in the vapor pressure of submicrometer-sized particles (Kelvin effect) (Hinds, 1999).

The situation when dry particle aerosols are generated by breath actuation presents many of the same difficulties when the cascade impactor is employed to measure the particle size distribution during actual or simulated inspiration. A sample can be drawn from the delivery system to an impactor during this period of generation and delivery, but it would be normally done at constant flow, whereas the flow, particle concentration, size distribution, and static pressure are likely to vary markedly during the inspiratory period. The equal weighting of all time periods during inspiratory flow by the cascade impactor will not likely reflect the highly time-dependent character of the aerosol.

Aerosols produced from MDIs have similar problems when sampled by instruments such as cascade impactors during their passage from the canister to the respiratory tract. In this case, the very rapid evaporation of the propellant and the pulsatile velocity of the generated particles, as described above, are not consistent with conditions ideal for cascade impactor sampling.

To avoid the practical and theoretical difficulties and sampling complications associated with therapeutic aerosols outside the respiratory tract, many investigators have elected to deliver the aerosol into a large-volume chamber. The residual aerosol from this chamber is sampled into a cascade impactor for size and mass analysis (Mercer et al., 1968), after a short period of evaporation or mixing. If the mixed aerosol is sampled, integration over the time of generation has already been performed, and the aerosol sampled represents the mixed average throughout the entire period of generation. While this method offers a means of comparison for various methods and conditions of aerosol generation and delivery, the aerosol that is eventually sampled is the "fossil remains" of the aerosol, which is delivered to the respiratory tract in the actual therapeutic application. It remains, therefore, for the investigator to calculate or estimate backward in time and space to the aerosol condition at the point of delivery in order to determine what aerosol actually entered the respiratory tract at a particular time during the inspiratory maneuver.

The use of size-selective instruments such as cascade impactors faces similar interpretive difficulties when other occupational or environmental aerosols are sampled due to their labile nature and spatial and temporal variability. However, these difficulties seem to be more severe in the case of therapeutic aerosols because less time is available between their generation and inhalation compared with other situations. A number of particle sizing techniques and their application to pharmaceutical aerosols have been reviewed in a series of articles in the pharmaceutical literature. These techniques include not only microscopic (Evans, 1993) and inertial sampling methods (Atkins, 1992; Milosovich, 1992), but also laser diffraction (Ranucci, 1992), laser time-of-flight (Niven, 1993), laser holography (Gorman and Carroll, 1993), right-angled light scattering (Jager et al., 1993), and phase Doppler anemometry (Ranucci and Chen, 1993).

Respiratory Tract Deposition

The difficulties of measuring therapeutic aerosols outside the respiratory tract and the objective of delivering a specified quantity of an agent to some respiratory tract region make it desirable to conduct practical measurements of aerosol deposition within the respiratory tract. This has led to the development of an expanding technology of aerosol deposition measurement. It should be noted that mathematical models have been used successfully to describe lung deposition and clearance (Swift, 1996). These models are currently being validated using imaging techniques to allow prediction of lung disposition.

In almost all cases, this is accomplished by means of gamma-emitting radioisotopes, which are either physically or chemically attached to the aerosol particles inhaled. The isotopes emit gamma rays, which are detected outside the body by a radiation detector (Taplin and Chopra, 1978).

The type of technology used depends on the level of spatial resolution required to describe the location of the aerosol particle. In the simplest example, one might wish simply to estimate the quantity of aerosol deposited in the entire respiratory tract. Even then, it is usually necessary to separate the respiratory tract into the thoracic and extrathoracic compartments, the division line usually taken to be at the mid-trachea. In this division, the extrathoracic airways include the upper trachea, larynx, pharynx, oral, and nasal passages. These are alternatively known as the head airways or (taking the division slightly more proximal) the nasal-oralpharyngeal-laryngeal (NOPL) region. The simplest measurement of deposition in this extrathoracic region is obtained with a single scintillation detector, with some collimation to exclude thoracic deposition oriented either to view from lateral, anterior, or posterior positions. Similarly, an estimate of the thoracic deposition can be made with a single scintillation detector aimed at the lungs and situated far enough from the chest to accept gamma rays from the lung periphery. Such measurements provide an estimate of the total amount of deposition in one or the other major region, without any further spatial information.

With this degree of spatial information, the amount of activity required to exceed background radiation is not great; subjects can inhale aerosol containing a radioisotope and receive a dose within the annual limit considered safe for the general population. When the isotope used is ^{99m}Tc, several breaths can be taken by the same subject to achieve a total dose well below the annual exposure limit of 0.5 rem.

A gamma scintillation camera can detect the deposited aerosol if a greater degree of spatial resolution is desired. The amount of radioactivity required for a detailed image of the thoracic or extrathoracic region is much greater than with an uncollimated detector because of the collimation required to attain good spatial resolution (~5 mm). The dose received for a single study limits the number of studies that can be performed with the same subject to keep within the annual dose requirement. Even so, the information from such a study can only be grouped into about three concentric regions for each lung or three regions in the head airways. A planar (two-dimensional) image of activity in the lung does not allow a good degree of separation between the peripheral (alveolar) region and the medium and small bronchial airways because these regions overlap in such a planar view.

An even more detailed spatial look at the respiratory tract deposition of aerosol can be obtained using the technique of SPECT (single-photon emission computed tomography) (Phipps et al., 1989). This technique allows the activity in the lung or extrathoracic region to be assigned to three-dimensional volume elements, "voxels," the equivalent of two-dimensional "pixels," so that activity distributions in various planes can be visualized and quantified. This added spatial information is obtained at the cost of additional radioactivity, a measurement time for a single image of ~20 min (compared with ~5 min for a gamma-camera scintigram), and additional computer capacity to process all of the information from the multiposition scan about the region. The scan time of 20 min may be unacceptably long because rapid upper bronchial or extrathoracic airway clearance is taking place. For example, aerosol deposition onto the mucociliary surface of the nasal passage is normally cleared to the pharynx and swallowed within about 15 min. The SPECT procedure is suitable for aerosol deposition and clearance measurement only when the temporal scales of clearance are much greater than the scanning time.

In some cases of therapeutic aerosol delivery to the respiratory tract the spatial distribution is not necessary, and the uptake of the agent can be monitored by serial measurements of blood or urine concentration of the agent or its metabolites (Walter et al., 1972). Likewise, in the extrathoracic region, the aerosol deposition can sometimes be measured by washing out the deposited material for quantitative analysis of an appropriate tracer. For example, it is well known that MDIs deliver large fractions of drug to the posterior pharyngeal wall, which can readily be removed by oral rinse and gargle for analysis.

Nonideal Behavior of Therapeutic Aerosols

The quantitative measurement of aerosol deposition of therapeutic agents is important because in many instances it is difficult to predict how therapeutic aerosol will behave based on well-established models of respiratory deposition. This is because such models are based on a number of simplifying assumptions defining an "ideal" aerosol. This means that the particles are of spherical shape, are solid, nonhygroscopic, nonevaporating, noninteracting, are present in the air at moderate concentration, and are travelling through the airways essentially at the velocity of the inspired or expired air. In most therapeutic aerosols, one or more of these conditions does not hold (Newman, 1984).

For example, the aerosol from an MDI begins as a very high velocity jet of large, rapidly evaporating propellant particles at high concentrations. After losing most of the propellant, the particles are often hygroscopic and, thus, subject to growth by accretion of water at the temperature and humidity of the central airways. The jet begins in the center of the inhaled air and may not be spatially uniform during part of its path. Because of its short duration, the inspired dose is an aerosol bolus, not a continuous volume of aerosol, subject to dispersion within the respiratory tract. Other therapeutic aerosols consisting of liquid droplets or dispersed solid particles are likewise "nonideal" to a degree as a result of their physicochemical properties and/or their mode of generation and exposure.

An additional factor that makes the prediction of therapeutic aerosol deposition difficult is the nonideal nature of the respiratory tract in many people using such aerosols to treat respiratory diseases. Even in "normal" individuals, aerosol deposition exhibits significant biological variability, but, in many individuals using therapeutic aerosols, lung morphology and/or breathing pattern is markedly abnormal. For example, individuals who have cystic fibrosis inhale bland saline aerosols to "liquify" the secretions. However, observations suggest that the preferential sites of deposition of these aerosols are in the "healthy" regions of the lung while the diseased regions receive little aerosol because they are poorly ventilated (Alderson et al., 1974). Much work remains to be done to clarify this situation.

Respiratory Tract Disposition (Pharmacokinetics)

Disposition of airborne particles from the lungs occurs by absorption, mucociliary transport, and cell-mediated (macrophage) disposition. The pharmacokinetics of disposition depend on the site of deposition and the residence time of the particle; the latter is related to the dissolution rate of the particle. Particles deposited in the upper airways are subject to rapid mucocilary transport through the trachea to the throat where they are swallowed. Particles deposited in the periphery of the lungs may be absorbed as a function of their dissolution rate, solubility, and partition coefficient. Sparingly soluble particles will be phagocytosed and subjected to intracellular mechanisms of dissolution and potentially degradation including continued hydrolysis and enzyme activity. Particles exhibiting very long dissolution times will be translocated by the macrophages from the lungs to the lymphatics and ultimately to the systemic blood circulation. Various models have been proposed for the general disposition of drugs from the lungs (Byron, 1986; Gonda, 1988) and for the disposition of specific drugs from the lungs (Hochhaus et al., 1998; Edsbacker et al., 1998). The pharmacokinetics of disposition can be conducted utilizing radiolabeled materials, which can be detected by imaging techniques or by scintillation counting in blood samples. Chromatographic separation methods such as high-performance liquid chromatography with drug-specific detection methods such as ultraviolet, fluorescence, or mass spectroscopy may be employed to detect drug in plasma, serum, or other biological fluids (e.g., urine, bronchoalveolar lavage fluid) to monitor the appearance/disappearance of drug. Model independent methods may be employed to analyze the data, or mathematical models can be employed to predict the disposition of drug.

Efficacy and Toxicity

Pharmaceutical Aerosols. Pharmaceutical aerosols exhibit a dose-dependent efficacy. For most aerosols delivered to the lungs the therapeutic ratio is large, which is to say that the toxic dose is much larger than the therapeutic dose. Targeted delivery in itself allows small doses to be delivered, thereby circumventing both local and systemic toxicity. The use of specific adrenergic agonists and anticholinergics has long been regarded as both safe and efficacious. The local delivery of glucocorticosteroids is considered much less hazardous than systemic administration of these drugs. However, the potential for local immunosuppression, particularly in the oropharynx and upper respiratory tract, where significant deposition of aerosols occurs, does present the increased risk of local infection most frequently in the form of candidiasis. The safety concerns regarding drug delivery will become increasingly important as the lungs are considered as a route of administration for systemically acting agents. The chronic delivery of materials, some of which are of biological origin, to the periphery of the lungs may require unique immunological and toxicological considerations. Some studies endeavor to relate the lung dose and distribution as measured by gamma scintigraphy with clinical response to therapeutic aerosols (Dolovich, 1993). This integration of aerosol measurements, lung imaging, pharmacokinetics/pharmacodynamics, and efficacy are a desirable goal to strive for in product development.

Diagnostic Aerosols. The major risk associated with the inhalation of diagnostic aerosols relates to the use of radioactive isotopes for imaging. These risks are minimized by the use of short half-life radiation sources such as technetium-99m and by advances in the sensitivity of detection systems. Nevertheless, all studies involving radioactivity are subject to oversight by radiation safety committees and Institutional Review Boards on the protection of the rights and safety of human subjects.

Inadvertent Exposure to Aerosols in Health Care. There are a number of potential and actual aerosol exposure situations in the health care field that are unintended and can result in a risk of ill effect. In such cases, aerosol measurement provides the estimates of potential exposure and expected benefit of exposure control methodology. Some of these instances involve chemical therapeutic agents that have potent biological effects, while other instances involve living organisms or their bioactive remains that may result in infections, transmission of disease, or other ill effects.

The anecdotal observations of airborne (aerosol) transmission of disease have an illustrious history dating back at least to the Greeks. Despite their lack of knowledge of the nature of the agents (usually attributing such transmission to "bad airs"), modern biochemistry, physiology, bacteriology, virology, and parasitology have provided the tools to understand the nature of the airborne transmission of many diseases, ranging from the common cold to tuberculosis. The understanding of immunologic defense mechanisms has led to the development of vaccines and other strategies to prevent or attenuate the harmful spread of these diseases.

Despite these advances, the appearance of "new" diseases such as AIDS, the proliferation of potent therapeutic agents such as antitumorals, and the increased processing of human and other biological tissues and fluids in diagnostic and research laboratories has made an understanding of airborne exposure and transmission as important now as in the days when tuberculosis, influenza, and pneumonia were responsible for pandemics.

Human–Human Aerosol Transmission of Disease. It is a well-established observation that aerosols are produced from the respiratory tract by a number of means, including sneezing, coughing, conversation, and singing (Wells, 1955). Particles thus emitted have a wide size range, the larger of which ($d > 10\,\mu m$) settle rapidly to horizontal surfaces. Smaller particles emitted as such or resulting from evaporation remain airborne for longer time periods. These "droplet nuclei," which may contain any of the numerous biological organisms, are the means by which disease is transmitted over distance by air between humans. Depending on their environmental "hardiness," such droplets may even be transmitted through ventilation systems and infect individuals quite remote from the source individual (Riley, 1974).

Two approaches toward aerosol characterization have been used for such agents: (1) the number concentration of organisms collected as a suitable medium and determined by colony (or equivalent) counts and (2) the empirical number of "infectivity units" (each of which contains the minimum number of organisms needed to produce an active disease case). In some cases this may be only a single viable organism, while in other cases many organisms are needed. Control of transmission by air dilution, killing of organisms, or physical removal (filtration) can be measured by the reduction of organisms or disease units by standard measurement methods. Various schemes for killing organisms have been proposed. For example, reduction of viable aerosol particles by ultraviolet light has been quantified with liquid impingers or agar-plate cascade impactors (see Chapter 24).

With extensive use of vaccines and effective chemotherapy for tuberculosis, it was thought, not long ago, that airborne transmission of disease in developed countries was limited to the common cold, influenza, and other treatable illnesses. However, the recurrence of tuberculosis as a side effect of AIDS and other immune deficiency conditions has revived the concern, particularly among health care personnel, that more serious effects of airborne transmission may still occur. Although there is no evidence that AIDS itself is transmitted by the airborne route, there are other opportunistic pathogens that deserve attention with respect to airborne spread. Thus, viability techniques to assess the magnitude of the problem and the effect of control technology employing aerosol measurement need to be developed (Riley, 1972).

Inadvertent Exposure to Therapeutic-Agent Aerosols. There is a growing concern that the production, handling, and administration of therapeutic agents may pose health hazards for health care workers and others who routinely or infrequently are exposed by several routes, including the formation of aerosols. Measurement techniques for bioaerosols, which result during bioprocessing, are discussed in Chapter 24. However, it is important to emphasize that the rapid development of new therapeutic agents (many of which result from developments in genetic engineering and molecular biology) should be accompanied by an examination of the consequences resulting from exposure of pharmaceutical workers and health care providers to these agents. These include not only the agents for treating human disease, but an increasingly expanding array of agents used in animal husbandry and veterinary medical practice. Means must be developed to ensure that exposure, including that by airborne particles, is controlled to minimize or eliminate ill effects.

The detection and quantification of airborne concentrations of such substances can be carried out with standard aerosol collection techniques as long as the collection substrates do not interfere with the sensitive and selective bioassay methods available for many of these substances. One of the advantages of environmental sampling for such agents is that their biological potency permits highly sensitive and selective detection tests to be used; these detection capabilities are already in place in many laboratories where the agents are employed.

Inadvertent Exposure to Bioaerosols from Tissue, Cell, and Fluid Tests. The increasing number of routine tests being carried out with tissues, cells, and body fluids (including blood) of living organisms (mainly human) and the possibility that many routes of infection may contribute to disease transmission to health care workers (Zimmerman et al., 1981) is a concern similar to that referred to above. Although high on the list of transmission routes is the introduction of substances directly into the blood via inadvertent needle sticks, and so forth, the possibility of aerosol formation and transmission in such settings has received considerable attention. Numerous processes such as centrifugation, stirring, and pipetting can lead to aerosol formation as well as poor housekeeping procedures such as spilling liquids that may dry and become aerosolized. Even procedures that are intended to minimize other transmission routes, such as hypodermic needle clipping, can produce aerosols directly or result in fluid spills on surfaces (Binley et al., 1984).

These possibilities have achieved heightened public interest with the advent of AIDS. Even without the fear of AIDS transmission, the possible transmission of much more robust organisms such as hepatitis virus ought to be reason enough to understand and develop control measures to eliminate such pathways. The health literature contains several occupational epidemiology studies demonstrating that transmission takes place in such settings, but little environmental sampling including the airborne route has been done. As with therapeutic agents, good microbiological techniques for the detection of most organisms are available and can be combined with appropriate aerosol sampling technology when indicated.

Inadvertent Exposure to Bioaerosols in Dental Procedure. It has been known for some time that many dental procedures such as tooth drilling, abrasive cleaning, and certain oral surgical procedures can produce airborne particles containing irritant dust and pathogenic flora from the mouth (Macdonald, 1987). The newer technologies being introduced into dentistry, such as laser cutting, water abrasive cleaning, and dry powder abrasive cutting, are all sources of aerosols that could pose hazards for the dental patient, the dentist and staff, and other patients being treated in the same area (Pagniano et al., 1986). Greater drilling efficiency with very high-speed drills produces finer sized dust, and the escape of this dust from the mouth due to its high velocity leads to concern for aerosol exposure. Aerosol samples taken in dental suites have demonstrated the existence of viable and pathogenic airborne organisms. This has led to the introduction of exhaust ventilation and other methods to eliminate or minimize exposure. A practical problem in developing such exhaust systems is the difficulty of getting good capture efficiency without interfering with access to the work area.

With new techniques and materials, it is important to evaluate the degree of aerosol exposure (as well as other routes) in dental practice and minimize the risk of infection. The amount of inhaled aerosol producing an effect may be very small for some highly infectious and pathogenic organisms.

Inadvertent Exposure to Aerosols in Surgery. Advances in surgical techniques, as in dentistry, are sometimes accompanied by new modes of transmission, affecting not only the patient but also health care workers. This has been demonstrated in the case of the increasing use of lasers and electrocautery in many surgical procedures. Both these techniques, as well as bone and tissue cutting procedures, are capable of producing airborne particles that can transmit infections and pathogenic organisms both from one region of the patient to another and from the patient to other individuals (Hoye et al., 1968).

This is a reversal of the traditional concept of infection transmission, where concern was (and still is) centered on preventing infection agents arising from surgeons or other operating-room personnel from gaining access to the patient via an open surgical wound. This was attempted (not always successfully) by using clean clothing, skin cleansing and disinfection, surgical masks, and (more recently) filtered laminar flow of air over the site of surgery as well as the application of disinfecting materials directly on the wound. This becomes a difficult task, especially in large-area surgical procedures such as hip replacement or open-heart surgery.

Aerosol characterization of smokes and other aerosols produced during surgery is very limited and mostly anecdotal at present, but it has been demonstrated that viable organisms can be aerosolized by these procedures (Merritt and Myers, 1991). Because both lasers and electrocautery produce high temperatures at the site of cutting, it might be questioned how aerosol generated by such processes could indeed remain viable. It appears that boiling of tissues adjacent to the cutting site results in the emission of cells and tissue fragments into the air. No studies linking particle size to the nature of the viable organisms have been performed, although such investigations do not appear to be beyond the present capability of aerosol sampling and analysis by one of the several biological tests for viable species.

Inadvertent Exposure to Aerosol in Aerosol Therapy. There are several therapeutic aerosol procedures currently used in which aerosol production, either from the therapeutic agent source or from the patient, poses a concern for health workers (Arnold and Buchan, 1991). One of these that has achieved some attention is the increasing use of aerosol pentamidine in the treatment of AIDS-related *Pneumocystis carinii* pneumonia (PCP). This is presently carried out in hospitals where several patients are normally placed in a room, each with an aerosol delivery system containing an aqueous solution of pentamidine. Although it is the intent of the systems to deliver all the pentamidine to the patient's alveolar compartment, it is often the case that pentamidine aerosol escapes the delivery system and is transported throughout the room. Health care workers in the room and nearby are exposed to this aerosol, which is designed to be accessible to the deep lung, and there are anecdotal reports of cough and irritation among health care workers (Boulard, 1991). The possible long-term effects of this exposure are unknown.

Aerosol measurements of the pentamidine have been performed in a few instances employing fairly crude means for area aerosol sampling, and these confirm that pentamidine aerosol of inhalable size is present. Several methods to reduce exposure have been proposed, but none has been evaluated; these include increased room ventilation, isolation of the patients in booths containing a separate air supply and filtered exhaust, and respiratory protection devices for the health care personnel. The disadvantages of all these proposed control methods have been recognized, but ideal solutions, which do not entail high cost or questionable benefits, have not been proposed.

Another widely used aerosol treatment method is the administration of saline aerosol to induce sputum production for the collection of bronchial cells and fluid and cytological examination. It is well known that the coughing that accompanies the aerosol administration and aids in bringing the bronchial secretions upward also produces aerosols, which may contain pathogenic organisms. As above, little quantitative aerosol characterization has been performed to test the control methods to reduce exposure.

It is likely that in the future many more drug substances will be administered to patients in similar settings. It is important that aerosol measurement technology be employed to evaluate both the effectiveness of the treatment and the possibility that health care workers may be inadvertantly exposed at harmful levels.

CURRENT ISSUES IN PHARMACEUTICAL AND DIAGNOSTIC AEROSOL MEASUREMENT

It is clear that advances in many of the health care areas alluded to above carry with them both the possibility of increased effectiveness of treatment and diagnosis employing aerosolized substances and the possibility of undesirable exposure of health care workers to biologically potent aerosol agents. Aerosol measurement technology has an important part to play in these developments both to guide in the most effective and efficient use of therapeutic and diagnostic aerosols and in the evaluation of the degree of exposure and the effectiveness of exposure control strategies.

It is not surprising therefore that there are ongoing regulatory and pharmacopeial considerations regarding the control and characterization of pharmaceutical aerosols. The regulatory (FDA) and pharmacopeial (USP) agencies of the United States are currently collaborating with similar agencies worldwide to harmonize (ICH) the requirements for evaluation and release of products thereby standardizing approaches to health and safety internationally and facilitating commerce.

CONCLUSIONS

There is a long history of combining effective aerosol measurement and biological assay techniques to characterize aerosols of importance to health. Some of this effort has been directed toward health care worker protection, but much more has been directed toward workers in the "dusty trades." The design of methods to deliver therapeutic and diagnostic aerosols has borrowed heavily from the knowledge base developed by industrial hygienists, health physicists, and air pollution researchers.

These interactions ought to be fostered by continued professional contacts through conferences, journals, and collaborative research projects and ought to be supported by both public and private funding. The results of these efforts will be beneficial to health care consumers and health care providers, both in cost and life quality. These developments give further strength to the widening role of aerosol science and technology, including aerosol measurement, as a scientific discipline and practice that touches many important fields of endeavor.

REFERENCES

Agnew, J. E. 1984. Aerosol contributions to the investigations of lung structure and ventilation function. In *Aerosols and the Lung*, eds. S. W. Clarke and D. Pavia. London: Butterworths.

Alderson, P. O., R. H. Secher-Walker, D. B. Strominger, J. Markham, and R. L. Hill. 1974. Pulmonary deposition of aerosols in childern with cystic fibrosis. *Pediatrics* 84:479–489.

Arnold, S. D. and R. M. Buchan. 1991. Exposure to ribavirin aerosol. *Appl. Occup. Environ. Hyg.* 6:271–279.

Atkins, P. J. 1992. Aerodynamic particle-size testing—Impinger methods. *Pharm. Tech.* 16(8):26–32.

Bell, J. H., P. S. Hartley, and J. S. G. Cox. 1971. Dry power aerosols 1: A new powder inhalation device. *J. Pharm. Sci.* 60:1159–1164.

Binley, R. J., D. O. Fleming, D. L. Swift, and B. S. Tepper. 1984. Release of residual material during needle cutting. *Am. J. Injection Control.* 12:202–288.

Boulard, M. 1991. Evaluation of health care worker's exposure to aerosolized pentarnidine in the treatment of HIV-positive individuals. Presented at the 1991 Am. Ind. Hyg. Conf., Salt Lake City, UT, May 18–24.

Byron, P. R. 1986. Prediction of drug residence times in regions of the human respiratory tract following aerosols inhalation. *J. Pharm. Sci.* 75:433–436.

Byron, P. R. and J. S. Patton. 1994. Drug delivery via the respiratory tract. *J. Aerosol Med.* 7:49–75.

Dalby R. N., S. Tiano, and A. J. Hickey. 1996. Medical devices for aerosol delivery. In *Inhalation Aerosols, Physical and Biological Basis for Therapy*, ed. A. J. Hickey. New York: Marcel Dekker, pp. 441–473.

Dolovich, M. 1993. Lung dose, distribution and clinical response to therapeutic aerosols. *Aerosol Sci. Technol.* 18:230–240.

Dunbar, C. A., A. J. Hickey, and P. Holzner. 1998. Dispersion and characterization of pharmaceutical dry powder aerosols. *KONA* 16:7–45.

Edsbacker, S. and M. Jendbro. 1998. Modes to achieve topical selectivity of inhaled glucocorticosteroids—Focus on budesonide. In *Respiratory Drug Delivery VI*, eds. R. N. Dalby, P. R. Byron, and S. J. Farr. Buffalo Grove, IL: Interpharm Press, pp. 71–82.

Evans, R. 1993. Determination of drug particle size and morphology using optical microscopy. *Pharm. Tech.* 17(3):146–152.

Gonda, I. 1988. Drugs administered directly into the respiratory tract modeling of the duration of effective drug levels. *J. Pharm. Sci.* 77:340–346.

Gonda, I. 1990. Aerosols for delivery of therapeutic and diagnostic agents to the respiratory tract. *Crit. Rev. Ther. Drug* 6:273–313.

Gorman, W. G. and F. A. Carroll. 1993. Aerosol particle-size determination using laser holography. *Pharm. Tech.* 17(2):34–37.

Heyder, J. 1988. Assessment of airway geometry with inert aerosols. *J. Aerosol Med.* 1:167–171.

Heyder, J. and G. Rundolph. 1984. Mathematical models of particle deposition in the human respiratory tract. *J. Aerosol Sci.* 15:697–707.

Hinds, W. C. 1999. *Aerosol Technology: Properties, Behavior, and Measurement of Airborne Particles.* New York: John Wiley & Sons.

Hochhaus, G., S. Suarez, R. J. Gonzalez-Rothi, and H. Schreier. 1998. Pulmonary targeting of inhaled glucocorticoids: How is it influenced by formulation. In *Respiratory Drug Delivery VI*, eds. R. N. Dalby, P. R. Byron, and S. J. Farr. Buffalo Grove, IL: Interpharm Press, pp. 45–52.

Hoye, R. C., G. C. Riggle, and A. S. Ketcham. 1968. Laser destruction of experimental tumors: State of the art and protection of personnel. *Am. Ind. Hyg. Assoc. J.* 29:173–180.

Jaegfeldt, A., J. Anderson, E. Trofast, and K. Welterlin. 1987. A new concept in inhalation therapy. In *Proc. Int. Workshop on a New Inhaler*, eds. S. Newman, F. Moren, and G. Kropmton. London: Medicom.

Jager, P. D., G. A. DeStefano, and D. P. McNamara. 1993. Particle-size measurement using right-angle light scattering. *Pharm. Technol.* 17:102–120.

Kjellman, N. and B. Wirenstrand. 1981. Letter to the editor. *Allergy* 36:437–438.

Macdonald, G. 1987. Hazards in the dental workplace. *Dent. Hyg.* 61:212–218.

Mercer, T. T. 1981. Production of therapeutic aerosols: Principles and techniques. *Chest* 80(Suppl 1): 813–818.

Mercer, T. T., M. L. Tillery, and H. Y. Chow. 1968. Operating characteristics of some compressed air nebulizers. *Am. Ind. Hyg. Assoc. J.* 29:66–78.

Merritt, W. H. and W. R. Myers. 1991. Real time aerosol monitoring during laser surger. Paper presented at the 1991 Am. Ind. Hyg. Conference, Salt Lake City, UT, May 18–24.

Milosovich, S. M. 1992. Particle size determination via cascade impaction. *Pharm. Tech.* 16:82–86.

Moren, F. 1985. Aerosol dosage forms and formulations. In *Aerosols in Medicine*, eds. F. Moren, M. T. Newhouse, and M. B. Dolovich. Amsterdam: Elsevier.

Newman, S. 1984. Therapeutic aerosols. In *Aerosols and the Lung*, eds. S. W. Clarke and D. Pavia. London: Butterworths.

Newman, S. P., F. Moren, D. Pavia, O. Corrado, and S. W. Clarke. 1982. The effects of changes in metered volume and propellant vapor pressure on the deposition of pressurized inhalation aerosols. *Int. J. Pharm.* 11:337–344.

Niven, R. W. 1993. Aerodynamic particle size testing using a time-of-flight aerosol beam spectrometer. *Pharm. Tech.* 17:72–78.

Niven, R. W. 1996. *Atomization and Nebulizers in Inhalation Aerosols, Physical and Biological Basis for Therapy*, ed. A. J. Hickey. New York: Marcel Dekker, pp. 273–312.

Pagniano, R. P., R. C. Scheid, S. Rosen, and F. M. Beck. 1986. Reducing airborne microbes in the preclinical dental laboratory. *J. Dent. Educ.* 50:234–235.

Phipps, P., I. Gonda, D. Bailey, P. Borham, G. Bautovich, and S. Anderson. 1989. Comparison of planar and tomographic gamma scintigraphy to measure the penetration index of inhaled aerosols. *Am. Rev. Respir. Dis.* 139:1516–1523.

Purewal, T. S. and D. J. W. Grant. 1998. *Metered Dose Inhaler Technology*. Buffalo Grove, IL: Interpharm Press.

Ranucci, J. 1992. Dynamic plume-particle size analysis using laser diffraction. *Pharm. Tech.* 16(10):109–114.

Ranucci, J. and F.-C. Chen. 1993. Phase Doppler anemometry: A technique for determining aerosol plume-particle size and velocity. *Pharm. Tech.* 17:62–74.

Riley, R. L. 1972. The ecology of indoor atmospheres: Airborne infections in hospitals. *J. Chronic Dis.* 25:421–430.

Riley, R. L. 1974. Airborne infection. *Am. J. Med.* 57:466–475.

Sciarra, J. J. 1970. Pharmaceutical aerosols. In *The Theory and Practice of Industrial Pharmacy*, eds. L. Lachman, H. A. Lieberman, and J. L. Kanig. Philadelphia: Lea & Febiger, pp. 605–638.

Sciarra, J. J. and A. J. Cutie. 1990. Pharmaceutical aerosols. In *Modern Pharmaceutics*, eds. G. Banker and C. T. Rhodes. New York: Marcel Dekker, pp. 605–634.

Swift, D. L. 1989. Design of aerosol delivery systems to optimize regional deposition and agent utilization. *J. Aerosol Med.* 2:211–220.

Swift, D. L. 1996. Use of mathematical aerosol deposition models in predicting the distribution of inhaled therapeutic aerosols. In *Inhalation Aerosols, Physical and Biological Basis for Therapy*, ed. A. J. Hickey. New York: Marcel Dekker, pp. 51–81.

Taplin, G. V. and S. K. Chopra. 1978. Inhalation lung imaging with radioactive aerosols and gases. *Prog. Nuclear Med.* 5:119–120.

Wagner, H. N. 1976. The use of radioisotopes techniques for the evaluation of patients with pulmonary disease. *Am. Rev. Respir. Dis.* 113:203–218.

Walter, S. R., M. Evans, M. E. Richards, and J. W. Paterson. 1972. The fate of (″C) disodium cromoglycate in man. *J. Pharm. Pharmacol.* 24:525–531.

Wells, W. F. 1955. *Airborne Contagion and Air Hygiene*. Cambridge: Harvard University Press.

Zimmerman, P. F., R. K. Larsen, E. W. Barkley, and J. F. Gallelli. 1981. Recommendations for the safe handling of injectable antineoplastic drug products. *Am. J. Hosp. Pharm.* 38:1693–1695.

37 Inhalation Toxicology: Sampling Strategies Related to Control of Exposure Atmospheres

OWEN R. MOSS
CIIT Centers for Health Research, Research Triangle Park, NC

INTRODUCTION

In contrast to environmental and industrial monitoring, a critical consideration for aerosol, gas, and vapor measurement in support of inhalation toxicology research is that the measurement process must not significantly change the concentration or air flow pattern through the system. Control of exposure conditions must be made to ensure an accurate estimation of inhaled dose and, subsequently, the best possible correlation between the concentration of material presented and the biological response at the tissue, cellular, and molecular levels.

The preceding chapters in this book have provided the technical basis for accurate and effective sampling. These aerosol sampling devices are generally used in the sampling of exposure atmospheres during the conduct of inhalation toxicology research. Although some exposure atmospheres may have particle concentrations sufficiently high to overwhelm most of the aerosol detection methods, these problems can be overcome in practical application because the aerosol is known and highly controlled. Insight into these solutions can be found in the chapter conclusions within Part II of this book.

In this chapter the basic structure and properties of the exposure systems used in inhalation toxicology are described. Generic sampling strategies are then presented for use with a variety of sampling instrumentation. Finally, the sampling techniques are incorporated into basic sampling strategies that meet the goals of no change in concentration and a minimum change in flow during the measurement of exposure concentration of aerosols and/or gases. For a broader background in the measurement and control of exposure atmospheres related to inhalation toxicology, the reader is referred to the latest editions of texts by Phalen (1984), McClellan and Henderson (1995), and the American Conference of Governmental Industrial Hygienists (Cohen and Hering, 1995), and to the Materials and Methods sections in journals such as *Inhalation Toxicology*, *Journal of Aerosol Science*, *Aerosol Science and Technology*, *Journal of Aerosols in Medicine*, *Toxicology and Applied Pharmacology*, and *Toxicological Sciences*.

Aerosol Measurement: Principles, Techniques, and Applications, Second Edition, Edited by Paul A. Baron and Klaus Willeke.
ISBN 0-471-35636-0

BASIC ATMOSPHERE GENERATION AND CONTROL SYSTEMS

The types of exposure systems used in inhalation toxicology are whole-body exposure chambers, nose-only and head-only exposure systems, or single or multiple animal, closed-loop, metabolism chambers. The construction and operation of these devices have been described elsewhere (Phalen, 1984; Hinner et al., 1968; Moss, 1995; McClellan and Henderson, 1995; Leavens et al., 1996). In this section the basic properties of air flow in each of these systems is discussed and related to the concepts of air changes per hour and system half-time.

Whole-Body Exposure Chamber

Whole-body exposure chambers are either well-mixed systems or laminar or plug flow systems (Fig. 37–1). In a well-mixed system, it is assumed that any incremental volume of air entering the chamber will be instantaneously and completely mixed with all other volume elements present. With this assumption, the concentration build up and decay curves as described by Silver (1946) can be calculated. The equations are based on the exponential relations used to predict reaction rates in continuously stirred tank reactors and in predicting heat transfer and mixing rates in heating and air-conditioning systems.

Airflow through a well-mixed or completely mixed whole-body exposure chamber is produced in either a "push," a "pull," or a "push–pull" mode. Energy (to create compression) is used to push exposure air into the inlet of the chamber. Energy (to create suction) is used to pull exposure air from the outlet of the chamber. In the "push–pull" mode of operation two driving systems, one on the inlet side and the other on the exhaust side, are adjusted to maintain a proper atmospheric pressure inside the chamber. In a "pull" system, energy is used to pull exposure atmosphere from the chamber, while the total input into the chamber is partially supplied from a pressure-driven exposure atmosphere generation system and partially from passive suction of air from the room or supply duct into the chamber. Although seldom used, a "push-only" system is operated by pushing exposure atmosphere into the chamber and letting the increased pressure drive air out of the chamber, through filtration systems and into an exhaust area. The latter approach is not commonly used because it would result in the chamber being operated above atmosphere pressures, increasing the possibility for contamination of the surrounding room and personnel. The "pull" system naturally operates at lower than ambient pressure, and the "push–pull" system is generally adjusted to ensure that the chamber is at a pressure lower than ambient. Thus, if there are leaks in this system, room air will be sucked into the chamber, preventing the exposure atmosphere from

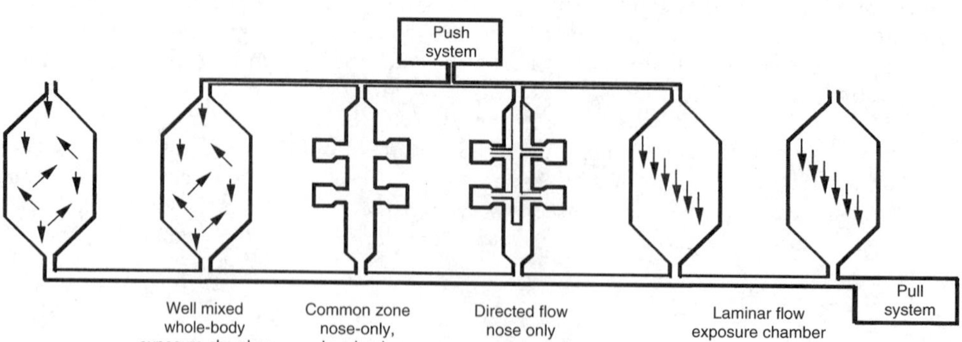

Fig. 37–1. Configurations of four basic exposure systems.

escaping. Such an inward leak will, however, change the exposure atmosphere disposition within the chamber.

Laminar (or Plug) Flow Exposure Chamber

In a laminar flow or plug flow exposure chamber, an incremental element of exposure air volume, upon entering the chamber, maintains its uniqueness throughout its passage through the chamber to the exhaust side. It is assumed that there is no mixing, horizontally or vertically, for any element of volume and that the flow is simple plug flow from entrance to exhaust. A laminar flow exposure chamber is normally operated as a push–pull system (exposure air is pushed into the horizontal or vertical laminar flow portion of the exposure chamber and then pulled from the chamber by equipment on the exhaust side). The pressure of the chamber and total flow through the chamber are controlled. A laminar flow whole-body exposure chamber can also be operated using a pull system (whereby exposure atmosphere is pulled from the chamber by equipment in the exhaust side and exposure air is introduced into the laminar flow system, in part by a pressure-driven aerosol generation system and in part by passive suction of material from the surrounding room or open supplied air duct [Fig. 37–1]).

Nose-Only and Head-Only Exposure Systems

Nose-only and head-only exposure systems have important applications in inhalation toxicology, especially when only the head or nose of the animal is to be exposed to eliminate surface contamination or when very little research chemical is available for study. These systems have been described previously by Phalen (1984), McClellan and Henderson (1995), Cannon et al. (1983), and Hemenway and Jakab (1990). Nose-only exposure systems, by their very design, are push–pull systems for they all include some point where compressed air flow is balanced or exceeded by suction. There are two basic types of nose-only exposure systems: the ones with a common distribution zone and the ones that incorporate a directed flow system (Fig. 37–1). Nose-only exposure systems with common distribution zones are designed and operated essentially around a small cavity that is several times larger than the head of the animal. The nose, or the head of the animal, is placed through the wall of this cavity, and air is directed through the cavity. The exhaled breath from the animal is captured in this region and may be carried past other animals before passing into the exhaust side of the nose-only exposure system. Usually re-breathing of air from other animals is avoided by staggering the position of the animals in the common distribution zone. The assumption is made that there is very little mixing of the exposure atmosphere with the cleaner air exhaled by the animal and thus very little dilution of atmosphere delivered to the subsequent levels of animals.

Directed flow, nose-only exposure systems were designed to eliminate the potential dilution of exposure air by the exhaled air of other animals and to reduce the consumption of material used in the exposure. Such systems have a separate, small-volume exposure air distribution line that feeds short delivery tubes positioned and sized so that the flow through them is directed at the breathing zone of the animal. The nose of the animal is positioned in a side port of an exhaust tube or duct. The exhaled breath of the animal is directed into the exhaust tube or duct and carried away. Each animal is exposed to a unique, well-controlled exposure atmosphere, undiluted by its own breath or the breath of other animals present on the system. Such systems as described by Cannon et al. (1983), Baumgartner and Coggins (1980), and Hemenway and Jakab (1990) operate on the premise that the directed exposure air has, at least, the minimum velocity and mass flow to dominate the air space in front of the nose of the animal at the beginning of each breath. In general, such systems must still be operated at least with air flows above 1.5 times and preferably above 2 times the minute ventilation of the rodent (Moss and Asgharian, 1994).

Closed-Loop and Partial Closed-Loop Exposure Systems

In single or multiple animal exposure systems used in metabolism studies, the animal is placed in a closed vessel that is attached to a closed-loop air circulation system. All air withdrawn from the system must be replaced in order to preserve the closed loop during sampling and analysis of the atmosphere. Such systems are predominantly used for gases and aerosols that have very low diffusion and sedimentation loss to the fur of the animal, to the sides of the tubing, or to the walls of the animal-holding chamber. Partial recirculation has successfully been used in whole-body systems to reduce the impact adsorption on exposure concentration (National Toxicology Program, 1999).

PROPERTIES OF EXPOSURE SYSTEMS

An operating exposure system has two basic measured properties: volume (L) and flow (L/min). In discussing the operation of an exposure system, the flow through the system is normally described in terms of air changes per hour. The flow, Q (L/min), is equal to n, the number of air changes per minute, times the volume, v, of the system (Fig. 37–2). The number of air changes per hour is equal to (60 min/h) $\times n$, where $n = Q/v$.

The definition of volume, v, is the key in this normalization. For example, the volume of a single-animal closed-loop exposure system is the entire volume of the animal containment vessel and all tubing and lines associated with the closed recirculating air system minus the volume of the animals. The volume of a directed flow, nose-only exposure system is taken to be the volume of the system contained between the end of the delivery tube and the nose of the rodent. The key factors affecting exposure in such nose-only systems are the animal's minute ventilation, the peak flow rate that occurs during each breath, the velocity of the exposure air as it exits the end of the delivery tube, and the air-exchange rate in the volume. Directed flow nose-only exposure systems must be set so that the flow rate through the delivery tube is greater than the maximum inspiratory flow rate of the animal. This is necessary in order to minimize dilution of the delivered exposure air by exhaled air. In practice a flow of two times the animals' minute ventilation is sufficient to keep the inhaled concentration greater than 90% of the concentration of the exposure atmosphere at the end of the delivery tube.

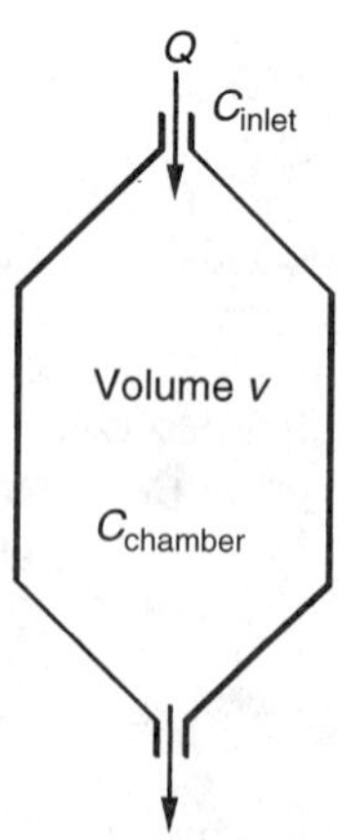

Air change per hour

$$n = \frac{Q}{V}$$

Chamber concentration buildup

$$C_{chamber} = C_{inlet}\,(1 - e^{-nt})$$

Chamber concentration decrease for $C_{inlet} = 0$ and $C_{chamber,\,t=0} = C_0$

$$C_{chamber} = C_0\, e^{-nt}$$

Fig. 37–2. Air flow and concentration in a well-mixed exposure chamber. Q is the flow through the system (L/min); v is the relevant chamber volume (L); and n is the number of air changes per minute.

In common distribution zone nose-only exposure systems the volume of the system is the volume of the common distribution zone in which the nose of the animal or the head of the animal is placed. In a laminar flow whole-body exposure system, the volume of the system is taken to be the volume in which each level of animals are housed. Finally, in a well-mixed whole-body exposure chamber, the chamber volume is defined as that volume above and below the location of the animals where there is contribution (mixing) to the inhaled exposure atmosphere. This latter definition is essential in order to meet the basic assumption of complete and thorough mixing made in the operation of these chambers. This assumption also leads to subtle differences in the definitions of volumes for whole-body exposure chambers.

For example, in exposures where air is drawn from and mixed in both the top and bottom of the chamber, the entire volume of the chamber and its transition pieces must be taken to be the volume of the chamber. In exposure systems where there is no back flux of exposure air, once the housing level of the animal is passed, the exposure system volume is the volume of the top transition piece plus the volume of the region housing the animals.

The term *half-time* comes from the equations used to describe concentration change in a well-mixed exposure system and is a function of the air-exchange rate (such as "air changes per hour or minute"). In a well-mixed system, the rate of change in concentration is directly proportional to its current value. Under such an ideal system, concentration build up in the chamber is equal to the concentration of the air entering the chamber times the quantity $1 - e^{-nt}$ (where n is the air change rate). When the incoming concentration is set to zero, the concentration in such a chamber at any subsequent time is equal to $C_o e^{-nt}$, where C_o is the concentration in the chamber at the start of the time period. In both build up and decay of concentration, the time it takes to move from any starting concentration to 50% closer to the equilibrium concentration (i.e., the target concentration for build up and zero for decay) is given as the half-time, $t_{1/2}$, and is equal to $(\ln 2)/n$ (i.e., $0.693/n$). For example, if it takes 3 min for one chamber volume of air to be pulled into the chamber inlet (i.e., $n = Q/v = (1/3)\ \text{min}^{-1}$), then $t_{1/2} = 0.693 \times 3 = 2.1$ min. Thus, after 2.1 min from any starting concentration, the chamber concentration becomes 50% closer to equilibrium concentration. Two half-times, or 4.2 min from the start of the concentration, would be 75% closer to the equilibrium concentration. The relation can be generalized. The time it takes to change in concentration from the start to some percentage, P, of the difference between the starting concentration and the equilibrium concentration can be calculated as $t_{(P/100)} = -(1/n)\ln(1 - P/100)$. For example, $t_{(1/2)} = 0.693/n$; and similarly $t_{(0.75)} = 1.386/n$; $t_{(0.90)} = 2.3/n$; $t_{(0.95)} = 3/n$; and $t_{(0.99)} = 4.61/n$.

The term *half-time* does not accurately apply to a true laminar flow or plug flow system. In a perfect laminar flow or plug flow system the concentration changes instantaneously from its current value to the new equilibrium value as the "plug" moves past the levels of the animals and the sampling probe. In such an ideal system the characteristic time of the system is taken to be the time for an increment of air to reach the animals after it leaves the point where the generator air is first diluted. In such cases, the reciprocal of this time is n, the air-change rate for the system. However, there is usually some level of mixing in such systems, and there is a build up or decay of concentration near the breathing zone of the animal that can be discussed on the basis of a half-time, as described above for the well-mixed system where the volume in this case is taken to be the volume of the housing around the animals.

BASIC SAMPLING TECHNIQUES AND STRATEGIES

The sampling and analyses of aerosol and gas concentrations in support of inhalation toxicology must be carried out so that the measurement process does not significantly alter the stability of the sampling device or change concentration or total air flow through the system.

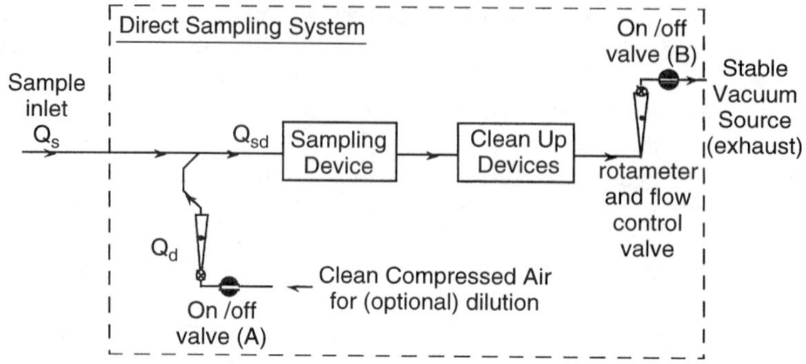

Fig. 37–3. Direct sample system for measuring concentrations from inhalation exposure systems. Q_{sd} is the total flow through the sampling device; Q_d is the flow of dilution air (if required); and Q_s is the sample flow.

The latter is important in dealing with well-mixed systems because changing the total air flow can significantly change the half-time of the system and concentration level. Nearly all exposure systems are constant-flow systems except for the case of exposure of a single animal by head-only or intratracheal intubation through a non-rebreathing valve. In such cases the flow in the system fluctuates with the breathing cycle of the animal.

Sampling System for Constant Flow Exposure Systems

In the commonly used direct sample system (Fig. 37–3), a stable vacuum source or vacuum pump internal to the sampling device is used to pull a sample at flow rate Q_s from the exposure chamber. Exhaust from the sampling device is passed through a clean-up system and then through a flow control valve to a stable vacuum source and then into the facility exhaust system. An optional source of clean compressed air at flow rate Q_d may be used to continuously dilute the sample before entry of the combined flow ($Q_{sd} = Q_s + Q_d$) into the sampling device. Such direct sampling systems are turned on or off by closing the air and vacuum valves, shutting off the sampling device, or unplugging the sampling line from the exposure system. The operation of a direct sampling system may change the concentration and air flow rate in the exposure system. However, in most cases Q_s is less than 2% of the total air flow through the chamber, and the effect on concentrations can be ignored.

The goal of stabilizing the sampling device by leaving it on at all times can be met by incorporating an additional clean air line of flow Q_s in the direct sample system (Figs. 37–3 and 37–4). Sampling of exposure air in such a system is turned off by diverting a flow of clean sample-replacement air into the sampling device. The total clean air flow rate matches the air flow rate through the sampling device and shuts off flow from the sampling inlet, eliminating the need for a valve in the sampling line. Clean air is sampled at rate Q_s directly into the sampling device after passing a Y connector. Clean air is provided to one or two rotameters at the front of the system as shown. Sampling is initiated when the flow, Q_s, of clean replacement air through the first rotameter (just below three-way valve C in Fig. 37–4a) is diverted to the chamber exhaust. The total flow, Q_{sd}, being pumped by the sampling device is maintained by removing an equal flow rate, Q_s, from the exposure chamber (Fig. 37–4).

The flows in Figure 37–4a are "balanced" when the flow through the first rotameter is diverted to the sampling device and set equal to the sampling flow, Q_s and $Q_s + Q_d = Q_{sd}$. One way to set this rotameter is to place a soap bubble volume meter at the sample inlet. Flow is balanced when the soap bubble does not move.

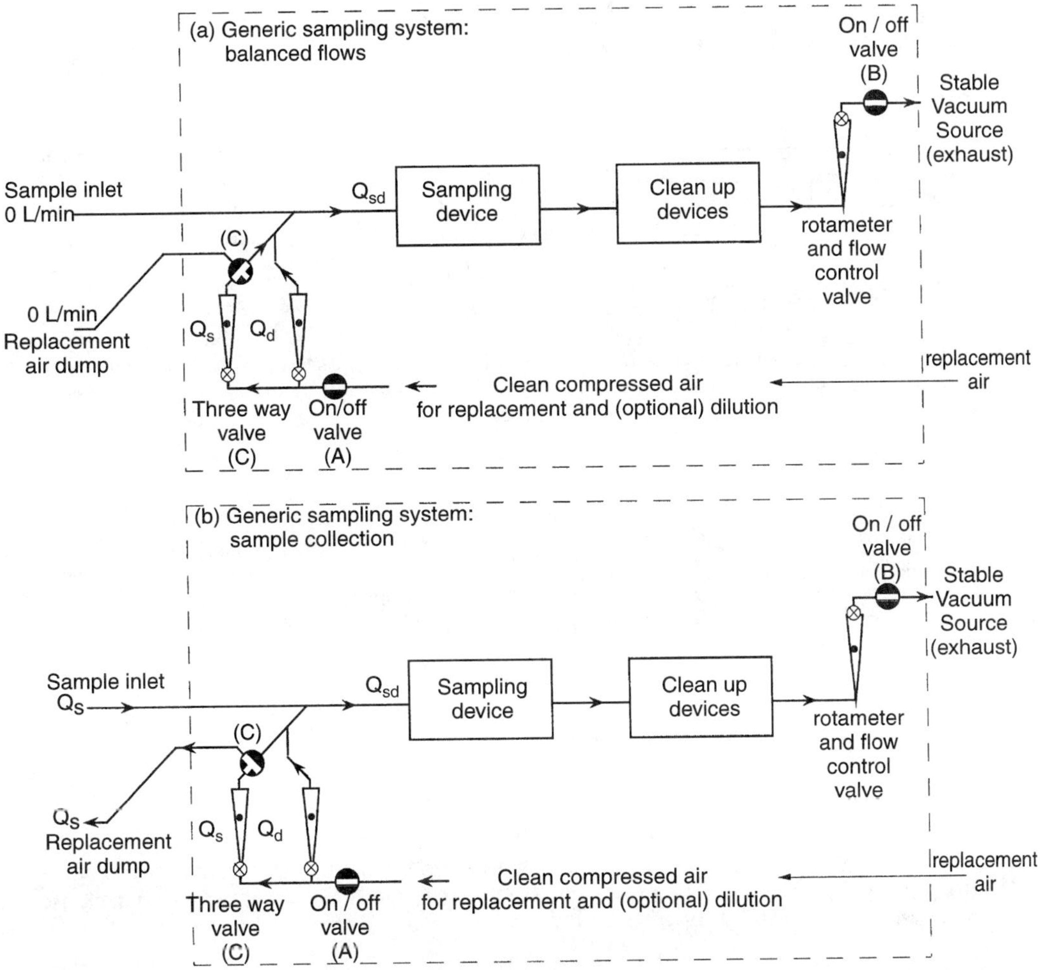

Fig. 37–4. Generic sampling system. **a**, Diversion of Q_s to the sample device with valve (C) allows continuous operation of the sampling device without pulling a sample from the exposure system. **b**, Diversion of Q_s to the sample exhaust with valve (C) allows chamber air to enter the sampling device with the same flow rate Q_s. Valves (A) and (B) together can remain on or off and will not affect chamber concentration. Q_{sd} is the total flow through the sampling device; Q_d is the flow of dilution air (if required); and Q_s is the sample flow.

Sampling System for Constant Flow: Push–Pull Exposure Systems

A direct sampling system (Fig. 37–3) should not be used when the sample flow rate, Q_s, is greater than 2% to 5% of the total flow through the chamber. The generic sampling system, shown in Figures 37–4 and 37–5, can be hooked to an exposure chamber with the sample being pulled from the breathing zone of the animals and system flow balance maintained by attaching the "clean replacement air dump" to the chamber exhaust (Fig. 37–5). Sampling commences by turning the three-way valve so that "clean replacement air," Q_s, flows from the rotameter to the chamber exhaust. With such a generic system, both the total flow and the concentration within the exposure chamber remain constant. The concentration in the exhaust line decreases downstream of the entry of clean air at flow rate Q_s. On the other hand, if the direct sample system shown in Figure 37–3 is attached directly to a pull-only

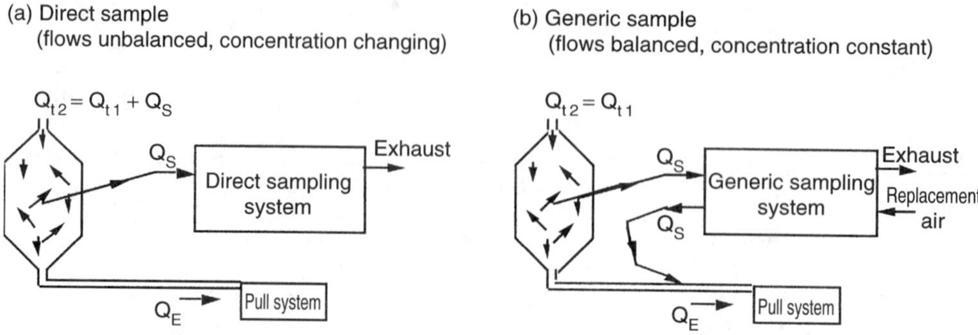

Fig. 37–5. Direct (**a**) and generic (**b**) sampling systems attached to a well-mixed exposure chamber operated in the "pull-only" mode. Q_s is the sample flow; Q_t is the total flow before (Q_{t1}) and during (Q_{t2}) sampling; and Q_E is the exhaust flow from the exposure chamber.

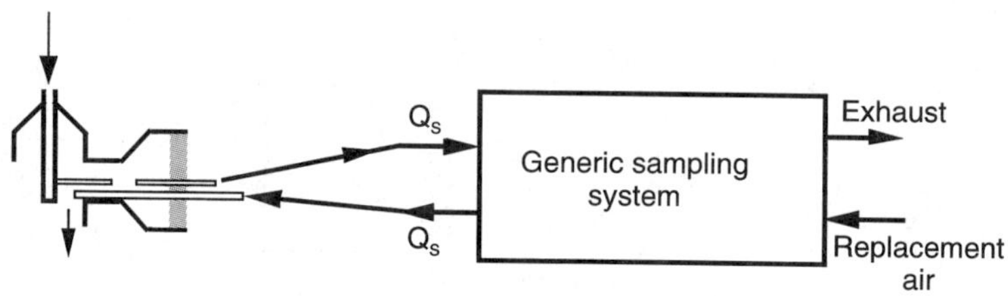

Fig. 37–6. Generic sampling system attached to a directed flow nose-only exposure system. Q_s is both the sample flow and the replacement air exhaust flow.

exposure system (Figs. 37–1 and 37–5), then during the sampling cycle the flow through the system is increased by the amount Q_s. The new total flow, Q_{t2}, through the chamber is equal to the original flow, Q_{t1}, plus the sampling flow: $Q_{t2} = Q_{t1} + Q_s$. The concentration will decrease by the factor $[1 - Q_s/(Q_{t1} + Q_s)]$.

Sampling System for Directed Flow Nose-Only Exposure Systems

In sampling directed flow nose-only exposure systems, special care must be taken to maintain the flow balance. The generic sampling system must be attached as shown in Figure 37–6. The clean replacement air exhaust is so attached that it vents into the exhaust duct near the sampling site. The inlet sampling point must be placed so that it is in the same relative position as would be the nose of the animal.

Sampling System for Pulsed Flow Exposure Systems

Pulsed flows are produced when an exposure system is attached to a single-direction non-rebreathing valve, being used by a subject or an animal. Such flows can also be produced when a constant flow sampling system (such as a cascade impactor) is being used to collect the entire output of a pulsed generator (such as a metered dose inhaler). If sampling can be accomplished upstream from the "T" used to divert flow to the non-rebreathing valve, then the sampling techniques for whole-body or nose-only chambers apply as discussed above. If, on the other hand, sampling of concentration in the delivery air must take place down stream from the "T" or "Y" used to divert flow to the non-rebreathing valve, then the flow through the system will fluctuate while the concentration remains constant. In such a case, for a sampling system that operates

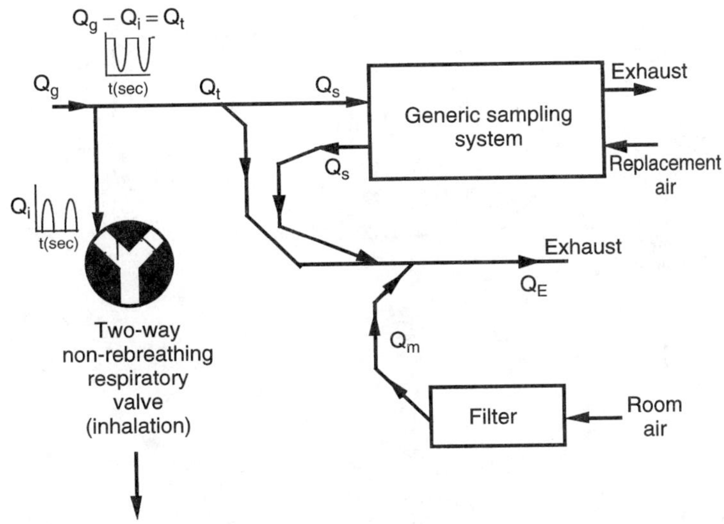

Fig. 37–7. Sampling for concentration of a fluctuating air stream where the total flow in the air stream is always greater than the sampling flow. Q_g is the total flow entering the system from the generator; Q_i is the pulsating suction from the non-rebreathing valve; Q_s is the constant sample flow; Q_m is the flow of make-up air from the room; and Q_E is the exhaust flow.

under constant sampling flow, Q_s, the generic sampling system shown in Figure 37–7 should be used. In this system, the fluctuating total flow, Q_t, is the result of the difference in the constant flow from the generator, Q_g, minus the fluctuating inhalation flow, Q_i: $Q_t = Q_g - Q_i$. The total flow, Q_t, is divided between the sampling device and the exhaust system. The sample flow, Q_s, is kept less than the lowest value of Q_t. The loss of flow caused by the inspiration of the animal, Q_i, is passively replaced by suction of clean air from the room through a low-pressure clean-up device, $Q_m = -Q_i$. In this way the flows in the system can be fully controlled for accurate sampling. The amount of exposure atmosphere sampled at flow Q_s is a true measure of the concentration inhaled through the non-rebreathing valve.

To measure the total amount of material exhaled from the exhaust side of a non-rebreathing valve, or from the output of a pulsed generator, a different sampling train must be used (Fig. 37–8). Constant total flow, Q_t, in this sampling system is maintained by using a flow of clean compressed air, Q_a, that is either vented into the room (or hood) or sucked into the sampling system. In the schematic shown in Figure 37–8, the flow of exposure air, Q_i, is from the exhaust side of the non-rebreathing valve. The flow of Q_a is set to be equal to or greater than the peak flow rate coming from Q_i, the non-rebreathing valve ($Q_a \geq Q_i$). As the animal exhales, an additional portion of the clean air flow Q_a, is diverted through the filter, flow Q_f, and into the room or into the exhaust system ($Q_i \leq Q_f$). As the animal stops exhaling and Q_i decreases to zero, the total flow through the system is made up by Q_a. This dynamic air flow system allows a constant flow, Q_t, to be established at the middle of the system dividing the sampling side from the non-breathing valve. The flow through the generic sampling system, Q_s, is less than or equal to the maximum value of Q_t.

CONCLUSIONS

Even though the sampling device itself may be very accurate, the measurement may be inaccurate because the process of sampling changes the equilibrium of the exposure system. The

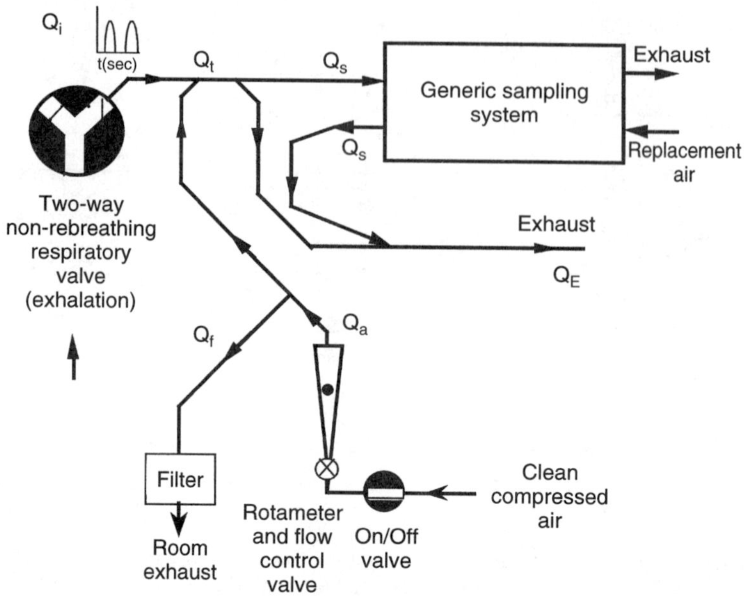

Fig. 37–8. Sampling for total output from a fluctuating air stream where the total flow periodically drops to zero. Q_i is the flow entering the system from the non-rebreathing valve; Q_a is the flow of clean air into the system; Q_s is the constant sample flow; Q_f is the flow of clean air out to the room; and Q_E is the exhaust flow.

basic sampling systems shown meet the general requirement of obtaining a concentration measurement from a variety of exposure systems without any significant change in concentration or total air flow. For large flow systems, where the sampling flow, Q_s, is small compared with the total flow through the system, the use of the direct sample system will not change the exposure concentration. However, for lower flow systems, or for systems requiring special air flow balancing, such as directed flow nose-only and closed-loop exposure systems, care must be taken to allow sampling to proceed in an accurate and efficient manner.

REFERENCES

Baumgartner, H. and C. R. E. Coggins. 1980. Description of a continuous-smoking inhalation machine for exposing small animals to tobacco smoke. *Beitrage Tabakforschung Int.* 10:169–174.

Cannon, W. C., E. F. Blanton, and K. E. McDonald. 1983. The flow-past chamber: An improved nose-only exposure system for rodents. *Am. Ind. Hyg. Assoc. J.* 44:923–928.

Cohen, B. S. and S. V. Hering. 1995. *Air Sampling Instruments for Evaluation of Atmospheric Contaminants*, 8th Ed. Cincinnati, OH: American Conference of Governmental Industrial Hygienists, Inc.

Hemenway, D. R. R. and G. J. Jakab. 1990. Nose-only inhalation system using the fluidized-bed generation system for coexposures to carbon black and formaldehyde. *Inhal. Toxicol.* 2:69–89.

Hinner, R. G., J. K. Burkart, and E. L. Punte. 1968. Animal inhalation exposure chambers. *Arch. Environ. Health* 16:194–206.

Leavens, T. L., O. R. Moss, and J. A. Bond. 1996. A dynamic inhalation system for individual whole-body exposure of mice to volatile organic chemicals. *Inhal. Toxicol.* 8:655–677.

McClellan, R. O. and R. F. Henderson, eds. 1995. *Concepts in Inhalation Toxicology*, 2nd Ed. Washington, DC: Taylor and Francis.

Moss, O. R. 1995. Calibration of gas and vapor samplers. In *Air Sampling Instruments for Evaluation of Atmospheric Contaminants*, 8th Ed., eds. B. S. Cohen and S. V. Hering. Cincinnati, OH: American Conference of Governmental Industrial Hygienists, Inc., pp. 151–164.

Moss, O. R. and B. Asgharian. 1994. Precise inhalation dosimetry with minimum consumption of product: The challenge of operating inhalation exposure systems at their design limits. In *Respiratory Drug Delivery IV*, eds. P. R. Byron, R. N. Dalby, and S. J. Farr. Richmond, VA: Virginia Commonwealth University, pp. 197–201.

National Toxicology Program. 1999. *Toxicology and Carcinogenesis Studies of Glutaraldehyde (CAS No. 111-30-8) in F344/N Rats and B6C3F1 Mice (Inhalation Studies)*. Technical Report No. 490, Public Health Service, National Institutes of Health, NIH Publication No. 99-3980, Research Triangle Park, NC.

Phalen, R. F. 1984. *Inhalation Studies: Foundations and Techniques.* Boca Raton, FL: CRC Press.

Silver, S. D. 1946. Constant flow gassing chambers: Principles influencing design and operation. *J. Lab. Clin. Med.* 31:1153–1161.

APPENDIX A
Glossary of Terms

Term	Explanation
ablation	transfer of material directly from the solid phase to the vapor phase
absorption	process whereby gas or vapor molecules are transferred to the liquid phase
accumulation mode	a mode in the atmospheric particle size distribution, formed primarily by coagulation of smaller particles
accuracy	an indication of correctness of a measurement
actinomycetes	group of bacteria with a morphology resembling that of fungi
active sampling	aerosol measurement using an air-moving device to draw the air into the collector or detector, as opposed to passive sampling
activity	measurable property of a particle population; for example, number, surface, mass, especially radioactivity
activity coefficient	a measure of the deviation from ideal solution behavior
activity concentration	amount of radioactive material in a unit of air
activity median diameter	particle diameter with 50% of airborne activity above and 50% below this size
A/D convertor	electronic device used to change an analog signal to a digital one
adsorption	transfer of gas or vapor molecules from the surrounding gas to a solid surface
adsorption isotherm	function relating the volume of vapor adsorbed on a surface to the pressure of the vapor in the gas phase at a fixed temperature
aerodynamic (equivalent) diameter	diameter of a unit-density sphere having the same gravitational settling velocity as the particle in question
aerodynamic lens	an orifice that can focus particles onto the axis of flow exiting the orifice
aerodynamic particle sizer	particle spectrometer that uses an acceleration system to differentiate particles by aerodynamic diameter and a laser velocimeter to detect particles
aerosol	assembly of liquid or solid particles suspended in a gaseous medium long enough to be observed and measured; particle size is generally about 0.001 to 100 μm
aethalometer	instrument used to measure the optical absorption of collected aerosol samples

Term	Explanation
agglomerate	group of particles held together by van der Waals forces or surface tension
aggregate	heterogeneous particle in which the various components are not easily broken apart
air monitoring	sampling and analysis of air to determine the quantity of pollutants present
air pollution	condition of materials present in the air at levels detrimental to the health and/or welfare of human beings
air quality standards	level beyond which air pollutants can cause damage to humans, animals, plants, or materials
Airy disk	disk-like image of a small point produced by an optical system with diffraction-limited resolution
Aitken nuclei	atmospheric particles in the approximate size range of 0.01 to 0.1 μm
algae	microscopic plant
alveolar	part of the respiratory system in which gas exchange occurs; alveoli are small sacs at the ends of the bronchioles
ambient air	surrounding air
anisoaxial sampling	sampling condition in which the air flowing into an inlet has a different direction from the ambient air flow
area sample	sample taken in a fixed location assumed to be representative of the area being investigated
aspiration efficiency	fraction of particles entering an inlet from the ambient environment
atomizer	device used to produce droplets by mechanical disruption of a bulk liquid
autocorrelation	relationship of measured values to previously measured ones
bacteria	single-celled microorganisms; some genera produce endospores
beta gauge	method of mass measurement that relies on the attenuation of a beam of beta particles
beta particle	energetic electron emitted in certain nuclear decay processes
BET method	Brunauer-Emmett-Teller method; procedure using the adsorption isotherm of a material to measure its surface area
bias	consistent difference between a measurement and a true or accepted value
bimodal size distribution	particle size distribution with two distinct maxima
bioaerosol	suspension of particles of biological origin; viable or dead cells; spores or pollen grains; fragments, products, or residues of organisms
bipolar ion field	region in which ions of both polarities exist
Boltzmann charge distribution	residual or minimum charge distribution on particles after exposure to a bipolar ion field
boundary layer	region of flow near bounding surface where the flow is dominated by friction forces resulting in reduced flow velocity relative to the free stream

Term	Explanation
breathing zone sample	sample taken as close as possible to the point at which the subject inhales air, usually within 0.3 m of the nose and mouth; represents a subject's inhaled air
Brownian motion	random motion of particles due to collisions with gas molecules
bubble meter	tube with a defined volume into which bubbles are injected to measure flow rate
bulk analysis	analysis of a sample in its entirety versus analysis of individual particles
capillary pore filter	filter consisting of solid membrane with an array of cylindrical holes of uniform size penetrating the membrane
carcinogen	agent that causes cancer
cascade impactor	device that uses a series of impaction stages with decreasing particle cut size so that particles can be separated into relatively narrow intervals of aerodynamic diameter; used for measuring aerodynamic size distribution of an aerosol
centrifuge	device in which particles are removed by centrifugal forces from an aerosol flowing in a helical path; device is usually characterized by high-resolution particle size separation
closed-face sampler	filter cassette sampler with the inlet smaller than the filter, as used in in-line liquid filtration
colony-forming units	number of biologically active entities (e.g., bacteria, fungi) that can form a colony present in a unit volume of air
cloud	assembly of particles with an aerosol density that is more than about 1% higher than the density of the gas alone
coagulation	aerosol growth process resulting from the collision of aerosol particles with each other
coarse particle mode	largest particle mode (>2 μm) in atmospheric particle size distributions, consisting primarily of particles generated by mechanical processes
coincidence	simultaneous presence of two or more particles in the sensing volume of a particle counter
comminution	breakup of particles by mechanical action
condensation	process with more vapor molecules arriving at a particle's surface than leaving the surface, resulting in a net growth of the particle
condensation nuclei counter	device in which submicrometer-sized particles are grown by vapor supersaturation to a larger size and are detected by light scattering
confidence limits	values defining a range around a sample statistic
continuum flow	flow governed by the macroscopic properties of the gas or fluid such as viscosity and density
corona	region of intense ionization, often surrounding an electrode at high voltage
Coulter counter	instrument that measures individual particle volume in a liquid by measuring the change in resistivity of the liquid as it passes through an orifice

Term	Explanation
cowl	cylindrical tube used in front of filter cassette to prevent direct impaction or contamination of samples; used primarily for asbestos fiber sampling
critical orifice	an orifice through which there is a constant air flow when a sufficient pressure drop across the orifice causes sonic flow
Cunningham slip correction factor	same as slip correction factor
cut-off particle diameter	diameter of a particle that has 50% probability of being removed by the device or stage and 50% probability of passing through; also called 50% cut point, d_{50}, or the effective cut-off diameter
cyclone	device in which particles are removed by centrifugal forces in a cyclonic path
Dean number	ratio of square root of the product of inertial and centrifugal forces to viscous forces
dichotomous impactor	virtual impactor with two emerging aerosol flows
differential mobility classifier	device that passes only particles within a narrow range of electrical mobilities
diffraction	change in direction and amplitude of radiation after passing near an object or through an orifice
diffusion	net movement of particles or gas from a higher to a lower concentration
diffusion battery	aerosol spectrometer used for submicrometer-sized aerosols in which size is measured by the diffusive loss of particles in an arrangement of ducts (e.g., tubes, filters, screens)
diffusion charging	process by which airborne particles acquire charge from ions undergoing Brownian motion
diffusion denuder	device that passes particles (low diffusivity) and removes gases (high diffusivity)
diffusion (equivalent) diameter	diameter of a unit-density sphere with the same rate of diffusion as the particle in question
diffusiophoresis	particle motion under the influence of a gas concentration gradient
dilution ratio	factor by which measured concentration is multiplied to obtain mainstream concentration
dilution system	system wherein aerosol is mixed with particle-free dilution gas in known volumetric ratio to reduce concentration
disinfection	destruction of the majority of microorganisms, not necessarily of all the spores
dispersion	system consisting of particles suspended in a fluid
drag coefficient	coefficient that relates the particle's drag force to the velocity pressure
drag force	resistance experienced by a particle when moving in a fluid
dust	solid particles formed by erosion or other mechanical breakage of a parent material; generally consists of particles of irregular shape and larger than about 0.5 μm

Term	Explanation
dust generator	device used to disperse dry particles in the air in a controlled fashion
dynamic shape factor	ratio of the drag force on a particle to that on a sphere of equivalent diameter
eddy diffusion	aerosol transport due to gas turbulence; theoretical description is similar to molecular diffusion
effective density	density of a particle with voids, in contrast to a compact particle's bulk material density
elastic light scattering	process in which there is no energy exchange between incident photons of light and target particles
electrical aerosol analyzer	aerosol size spectrometer in which the particles are separated by removing those with an electrical mobility greater than a selected value
electrical aerosol classifier	aerosol size spectrometer in which the particles are separated by selecting those within a narrow range of electrical mobilities
electrical mobility	parameter that indicates a particle's ability to move in an externally applied flow field
electrical mobility (equivalent) diameter	diameter of a unit-density spherical particle moving at the same velocity in an electric field as the particle in question
electrodynamic balance	device that uses superimposed ac and dc fields to levitate particles
electrophoresis	charged particle motion induced by an electric field
electrostatic balance	device that uses a dc field to levitate particles (e.g., the Millikan condenser)
electrostatic precipitator	device in which airborne particles are charged in a unipolar ion field and deposited with a high-voltage electric field
elutriator	device used to separate particles by aerodynamic diameter by allowing them to settle in a moving air stream
emission	material being discharged into the outdoor atmosphere
endotoxin	toxic cell wall component of gram-negative bacteria
envelope (equivalent) diameter	diameter of a sphere composed of the particle bulk material and included voids that has the same mass as the particle in question
epiphaniometer	instrument that measures the surface area of aerosol particles
equivalent diameter	diameter of a sphere having the same value of a specific physical property (activity) as the particle in question
evaporation	process with more vapor molecules leaving a particle's surface than arriving at the surface, resulting in shrinkage of the particle
extinction coefficient	measured parameter given by the amount of light scattered and absorbed by a particle divided by the amount incident upon it
extrathoracic	region of the respiratory system above the larynx containing the nose and mouth
fabric filter	filter consisting of a woven or felted fabric
fastest 2 minute wind speed	wind speed corresponding to the largest linear passage of wind movement during a 2 minute period

Term	Explanation
Feret's diameter	particle dimension determined by the projection of the particle's silhouette onto a selected axis
fibrous filter	filter consisting of a mat of individual fibers
field charging	process by which particles are charged by ions moving in a strong electric field
filter	porous membrane or mat of fibers used to collect particles from the air
fine particle	particle less than about 2 µm in size, consisting of particles in the nuclei and accumulation modes; term used in describing atmospheric aerosols
flocculate	group of particles very loosely held together, often by electrostatic forces; flocculates can easily be broken apart by shear forces within the air
fluidized bed generator	device using an agitation force, e.g. air pressure, to fluidize a powder to release dust particles
fly ash	particles of ash entrained in flue gas produced by fossil fuel combustion
fog	liquid particle aerosol, typically formed by condensation of supersaturated vapors
fractal dimension	measure of complexity of a particle's shape
free molecular flow	flow governed by discrete impacts of gas molecules
Froude number	proportional to inertial force/gravitational force
fume	small particles that are usually the result of condensed vapor (often from combustion) with subsequent agglomeration
fungi	multicellular organisms that produce spores
Gaussian curve	profile of distribution or curve similar to that observed for the normal distribution
Geiger-Müller tube	radiation-sensing instrument that relies on an avalanche process initiated by the production of electron-ion pairs
geometric	refers to a size parameter on a logarithmic size scale where a given ratio of two sizes appears as the same linear distance
geometric standard deviation	measure of dispersion in a lognormal distribution (always 1)
graticule	transparent disk with calibrated scale placed in the focal plane of an optical system (e.g., a microscope) used for measurement of particles or other objects
gravitational deposition parameter	ratio of particle settling distance during transport in the sampling inlet region to the diameter of the inlet
gravitational settling velocity	particle velocity in a gravitational field after equilibrium between gravity and aerodynamic drag forces has been reached
half-life	time interval required to reduce the rate of emission of a radioisotope by a factor of two
Hatch-Choate equations	expressions that, given a characteristic diameter and geometric standard deviation of a distribution, allow the calculation of any other characteristic diameter of the distribution
heterogeneous	consisting of individual components that may differ from each other in size, shape, and chemical composition

Term	Explanation
heterogeneous nucleation	formation of droplets on condensation nuclei (existing submicrometer particles)
homogeneous nucleation	formation of droplets in the absence of condensation nuclei; also called *self-nucleation*
horizontal elutriator	horizontal channel through which aerosol flows and particles above a given size or size range are removed by gravitational settling
hot wire anemometer	device used to measure air velocity by measuring the change in resistance of a heated wire
house-dust mites	common insects living in mattresses and carpets; excreta are common allergens
hydraulic diameter	hypothetical diameter of an object equal to four times the object's cross-sectional area divided by the perimeter of that area
hydrosol	suspension of particles in a liquid
hygroscopicity	property of a chemical that indicates its tendency to absorb water from the air
hyphae	chain of fungal cells
ideal fluid	hypothetical fluid having no viscosity
impactor	device in which aerosol particles with sufficiently high inertia in a deflected air stream are impacted onto a surface
impinger	device in which particles are removed by impacting the aerosol particles into a liquid
inhalable	fraction of an aerosol that can enter the human respiratory system (defined for sampling purposes)
inlet efficiency	fraction of ambient particles that is delivered to the aerosol transport section of a sampling system by the inlet; it is the product of the aspiration and transmission efficiencies
inspirable	same as inhalable; *inhalable* is the currently preferred term
interception	collision with and deposition of a particle on an object when the particle passes within one particle radius of the object
ionization chamber	radiation-sensing instrument that relies on the detection of free electron-ion pairs
isoaxial sampling	sampling condition in which the air flowing into an inlet has the same direction as the ambient air flow
isokinetic sampling	sampling condition in which the air flowing into an inlet has the same velocity and direction as the ambient air flow
jet nebulizer	a nebulizer employing air pressure to aerosolize a bulk liquid
Kelvin effect	increase in partial vapor pressure for a particle's curved surface required to maintain mass equilibrium relative to the vapor pressure above a flat liquid surface
Knudsen number	ratio of gas molecular mean free path to the physical dimension of the particle; indicator of free molecular flow versus continuum gas flow
Kuwabara flow	solution of the two-dimensional viscous flow field for a system of cylinders perpendicular to the flow, taking into account the interference effects of neighboring fibers (has also been applied to spheres); used to model flow in fibrous filters

Term	Explanation
laminar flow	gas flow with a smooth, nonturbulent pattern of streamlines, with no streamline looping back on itself; usually occurs at very low Reynolds numbers
light scattering	change in direction of light radiation due to reflection, diffraction, and refraction from a particle
lognormal size distribution	particle size distribution characterized by a bell-shaped or Gaussian distribution shape when plotted on a logarithmic size scale
lung model	representation of the respiratory system used to make quantitative estimates of particle deposition
Mach number	ratio of gas to acoustic velocity; indicator of compressibility
manometer	device used to measure pressure differences
Martin's diameter	length of horizontal line bisecting particle cross section into equal areas
mass (equivalent) diameter	diameter of a sphere composed of the particle bulk material with no voids that has the same mass as the particle in question
mass median size	size with an equal mass of particles above and below this value (see also *median size*)
mean free path	mean distance a molecule in a gas travels before colliding with another molecule
mean size	average of all sizes, that is, the sum of all sizes divided by the number of particles
mechanical mobility	aerosol parameter that indicates a particle's ability to move in a suspending medium; see *mobility*
median size	size with an equal number of particles above and below this value (see also *mass median size*)
membrane filter	filter that is formed as a gel from a colloidal suspension; characterized by tortuous air passages
micronize	process by which coarse powders are mechanically reduced to a particle size suitable for redispersion as an aerosol from a solvent or propellant
microparticles	particles with sizes of the order of micrometers
Mie scattering theory	general theory describing light scattered by spherical particles
mildew	visible fungal growth on surfaces
minute volume	volume of air passing in and out of the respiratory system in 1 min
mist	liquid particle aerosol, typically formed by physical shearing of liquids, such as in nebulization, spraying, or bubbling
mobility	ratio of particle velocity to the force producing that velocity
mobility (equivalent) diameter	diameter of a spherical particle with the same dynamic mobility as the particle in question
mode	value occurring most often in a distribution of values; peak value of a distribution
mold	visible fungal growth on surfaces
monodisperse	composed of particles with a single size or a small range of sizes
mycelium	mass of fungal hyphae
mycotoxin	toxic chemical produced by fungi

Term	Explanation
nanoparticles	general term indicating particles with sizes on the order of nanometers, usually limited to less than 100 nm
nasopharyngeal compartment	portion of the respiratory tract between the epiglottis and the anterior nares
nebulizer	device in which droplet aerosols are produced by dispersion of a bulk liquid
nephelometer	instrument that measures the amount of light scattered from an aerosol; also called a *photometer*
neutralizing	reduction in electronic charge on particles by exposure of the aerosol to ion clouds (often produced by radioactive sources)
normal size distribution	particle size distribution characterized by a bell-shaped or Gaussian distribution shape when plotted on a linear size scale
nucleation	process of initial formation of particles from a vapor
nuclei mode	smallest mode in atmospheric particle size distributions, formed by condensation of atmospheric gases or emissions from hot processes, typically containing particles <0.1 μm in size
opacity	degree to which an aerosol obscures an observer's view
open face sampler	filter cassette sampler with the inlet approximately the same size as the filter
optical (equivalent) diameter*	diameter of a calibration particle that scatters as much light in a specific instrument as the particle being measured
optical (single) particle counter	aerosol size spectrometer that differentiates particles by the amount of light scattered by each particle
orifice meter	device used to measure flow rate in a duct by measuring the pressure drop across a calibrated constriction
owl	device used to measure diameter of monodisperse aerosol particles by illumination with white light and detection of higher order Tyndall spectra
packing density	ratio of fiber or membrane volume of a filter to its total volume; also solidity
partial pressure	pressure that a vapor would exert if it were the only component present in a volume of gas
particle	small discrete object, often having a density approaching the intrinsic density of the bulk material; it may be chemically homogeneous or contain a variety of chemical species; it may consist of solid or liquid materials or both
particle bounce	rebound of particles that fail to adhere after impacting on a collecting surface
particle size distribution	relationship expressing the quantity of a particle property (activity) associated with particles in a given size range
particulate	an adjective indicating that the material in question has particle-like characteristics; this term is also used colloquially as a noun to describe particles or particle-containing material

* For glossary of terms related to optics, see the ***MIE*** internet site.

Term	Explanation
passive sampling	aerosol measurement using natural convection or diffusion to draw the air into the measurement device, as opposed to active sampling
pathogen	microorganism that causes disease
Peclet number	ratio of a particle's convective to diffusive transport
personal sampler	device attached to a person in order to sample air in the person's immediate vicinity
phantom particles	particles that appear in a measured distribution that are due to coincidence or other nonideal aspects of the measurement process and not due to real particles
photometer	instrument that measures the amount of light scattered from a particle cloud; also called a *nephelometer*
photophoresis	particle motion under the influence of asymmetrical light absorption within a particle
Pitot tube	device used to measure velocity pressure in a flow stream
plug flow	gas flow through a tube characterized by a uniform velocity across the entire cross section
plume	flow of visible effluent from an outlet (e.g., a stack or vent)
PM	particulate mass; usually used in relation to regulatory standards having a specific particle aerodynamic diameter cut-off (e.g. PM-10)
point-to-plane precipitator	electrostatic precipitator using a corona from a single point to deposit particles onto a flat plane
Poiseuille flow	laminar flow with a parabolic velocity profile occurring in a circular duct; the gas velocity in the center of the tube equals twice the average velocity in the tube
Poisson distribution	mathematical function relating the number of particles in a given volume element to the average concentration of randomly distributed particles in the entire volume
polydisperse	composed of particles with a range of sizes
porosity	(1 – packing density)
precision	an indication of the degree of variation in the results of repeated measurements of a variable
pre-classifier	device that removes particles ahead of an aerosol sensor, often in a manner similar to the particle removal occurring ahead of the respiratory region of interest; also called a *pre-separator* or a *pre-cutter*
primary particle	particle introduced into the air in solid or liquid form
projected area (equivalent) diameter	diameter of a circle that has the same area as the projected area of a particle seen under a microscope
pulmonary compartment	portion of the respiratory tract in which gas exchange occurs (includes alveoli and respiratory bronchioles)
pycnometer	device for measuring density of particles
radiometric force	force produced by light pressure
Rayleigh scattering	scattering of radiation occurring when the size of the scattering object is much smaller than the radiation wavelength

Term	Explanation
re-entrainment	return of particles to an air stream after deposition on a surface
refraction	change in speed and direction of radiation passing from one medium into another
refractive index	ratio of the speed of light in a vacuum to that in a material in question
relative settling velocity	ratio of the terminal gravitational settling velocity to sampling air velocity in an inlet
relative standard deviation	ratio of standard deviation to the mean; also described as coefficient of variation (CV)
relaxation time	time for a particle to reach 1/e of its final velocity from an initial velocity or from rest when subjected to an external force; an indicator of a particle's ability to adjust to changes in flow velocity
respirable fraction	fraction of an aerosol that can reach the gas exchange region of the human respiratory system (defined for sampling purposes)
reticle	transparent disk with lines or other marks placed in the focal plane of optical systems for calibration or alignment
Reynolds number	flow similitude parameter, expressed as the ratio of the inertial force of the gas to the friction force of the gas moving over the surface of an object; flow Reynolds number is the gas flow in a tube, and particle Reynolds number describes the gas flow around a particle
rotameter	device used to measure flow rate as indicated by the height of a float centered in a vertical tapered tube
sampling probe	device to withdraw aerosol from a system
sampling ratio	ratio of the ambient air velocity to the air velocity in an inlet
saturation ratio	ratio of the partial pressure of a vapor to its saturation vapor pressure
saturation vapor pressure	partial pressure of a liquid's vapor required to maintain the vapor in equilibrium with the condensed liquid or solid; also referred to as *vapor pressure*
Sauter mean diameter	diameter of a droplet whose surface to volume ratio is equal to the mean of all the surface-to-volume ratios of the droplets in a spray distribution; also referred to as the *surface area mean diameter* or the *mean volume-surface diameter*
Schmidt number	ratio of Peclet number to Reynolds number, or the ratio of kinematic viscosity to diffusion coefficient
scintillation spectrometer	radiation-sensing instrument that relies on the excitation of optical emission followed by detection with a photomultiplier tube
secondary particle	particle formed in the air, usually by gas to particle conversion; also sometimes used to describe agglomerated or redispersed particles
sedimentation	movement of particles by the influence of gravity
semiconductor detector	radiation-sensing instrument that relies on the generation of free carriers in a semiconductor material

Term	Explanation
shape factor	factor that relates the drag force on a particle to that on an equivalent sphere
Sherwood number	dimensionless mass transfer coefficient that relates the particle's diffusive deposition velocity to the particle's diffusion coefficient
Sinclair-LaMer generator	device that produces monodisperse aerosols by condensation of vapor onto nuclei
slip correction factor	factor that allows slip flow behavior to be calculated using continuum gas flow equations
slip flow regime	transition between free molecular flow and continuum gas flow
smog	an aerosol consisting of solid and liquid particles, created, at least in part, by the action of sunlight on vapors; the term is a combination of the words *smoke* and *fog* and often refers to the entire range of such pollutants, including the gaseous constituents
smoke	solid or liquid aerosol, the result of incomplete combustion or condensation of supersaturated vapor; most smoke particles are submicrometer in size
Snell's law	fundamental principle in optics that the sines of the angles of incidence and refraction are in a constant ratio to one another
solidity	see *packing density*
soot	conglomeration of particles formed by incomplete combustion of carbonaceous material
source apportionment	analysis of an aerosol sample so that fractions of the aerosol can be assigned to specific sources
source sampling	collection of materials emitted from an air pollutant–generating source
specific surface	particle surface area per unit mass or volume of particles
spinning disk atomizer	device that produces monodisperse droplets from the break up of a thin film of liquid ejected from the surface of a spinning disk
spirometer	device used to measure gas volume (or flow rate with a timer) using an expandable can sealed with a liquid
spores	dormant cells of microorganisms
standard	maximum allowable level of an air contaminant established by law
Stephan flow	special case of diffusiophoresis with particle motion toward or away from evaporating or condensing surfaces; also written as *Stefan flow*
sterilization	complete destruction of microorganisms and their spores
Stokes diameter	diameter of a spherical particle with the same density and settling velocity as the particle in question
Stokes law flow	flow around a body under the influence of viscous, but not inertial forces (nonturbulent flow)
Stokes number	ratio of a particle's stopping distance to a characteristic dimension; generally used as an indicator of similitude in particle behavior in a given aerosol flow configuration
Stokes regime	condition for which Stokes law applies

Term	Explanation
stopping distance	product of relaxation time and the initial particle velocity; an indicator of a particle's ability to adjust to directional changes in aerosol flow
subisokinetic sampling	sampling condition in which the air flowing into an inlet has a lower velocity than the ambient air flow
superisokinetic sampling	sampling condition in which the air flowing into an inlet has a higher velocity than the ambient air flow
surface barrier detector	a type of semiconductor detector used primarily for charged particle emissions
terminal settling velocity	equilibrium velocity of a particle approached when falling under the opposing influences of gravity and fluid drag
thermal precipitator	device that deposits particles using a temperature gradient
thermophoresis	particle's motion in a temperature gradient (i.e., from a hotter to a colder region)
thoracic	region of the respiratory tract from the larynx down
thoracic fraction	fraction of an aerosol reaching the respiratory region below the larynx system (defined for sampling purposes)
tidal volume	volume of gases inhaled or exhaled during each breath
total lung capacity	volume of air contained in the lung at maximum inspiration
tracheobronchial compartment	region of the respiratory tract from the larynx to the terminal bronchioles
transfer function	the change of one function into another; used to describe the change in size distribution or spatial distribution as an aerosol passes through a classifier
transmission efficiency	fraction of aspirated particles that is transmitted through an inlet to the rest of the sampling system
turbulent flow	chaotic flow with streamlines looping back on themselves; less "well behaved" than laminar flow
ultrasonic nebulizer	a nebulizer employing focused sound waves to aerosolize a liquid into droplets
ultra-Stokesian	condition in which flow velocity relative to an object is high enough to be outside the Stokes regime
unipolar ion field	region containing ions of only one polarity
vapor pressure	partial pressure of a liquid's vapor required to maintain the vapor in equilibrium with the condensed liquid or solid; also referred to as *saturation vapor pressure*
variability	measure of spread of repeated measurements of a parameter
variance	square of the standard deviation; a measure of variability
vena contracta	flow contraction with flow separation from the wall, usually occurring after constriction of a flow channel or just downstream of the entry point of an inlet
Venturi meter	device used to measure flow rate in a duct by measuring the pressure drop across a calibrated streamlined constriction
vertical elutriator	vertical channel that gravitationally retains or removes particles above a given size or size range and emits the remaining airborne particles

Term	Explanation
virtual impactor	device in which particles are removed by impacting them through a virtual surface into a stagnant volume, or a volume with a slowly moving air flow, so that large particles remain in this volume, while smaller particles are deflected with the bulk of the original air flow; the dichotomous impactor is a frequently used virtual impactor
virus	microorganism that needs a complete cell to reproduce
vital capacity	maximum volume of gas that can be exhaled from the lung after maximum inhalation
wall loss	deposition of particles in a sampler on surfaces other than those designed for particle collection
Weber number	indicates the ratio of the air pressure force to the surface tension force for a liquid droplet undergoing acceleration in a gas
weighting	application of a factor to one particle activity to obtain another activity (e.g., count, surface, volume, or mass)
yeast	unicellular fungus

APPENDIX B
Conversion Factors

Length

1 micrometer (µm) = 10^{-6} m = 10^{-4} cm = 10^{-3} mm = 10^{3} nm = 10^{4} Å = 3.937×10^{-5} in = 3.281×10^{-6} ft
1 nanometer (nm) = 10^{-3} µm = 10^{-9} m
1 Ångstrom (Å) = 10^{-4} µm = 10^{-10} m
1 inch (in) = 2.540 cm
1 foot (ft) = 12 in = 0.3048 m

Volume

1 m^3 = 10^{18} µm^3 = 10^{3} L = 6.102×10^{4} in^3 = 35.31 ft^3
1 µm^3 = 10^{-15} L = 10^{-18} m^3 = 6.102×10^{-14} in^3 = 3.531×10^{-17} ft^3
1 liter (L) = 10^{15} µm^3 = 10^{-3} m^3 = 61.02 in^3 = 3.531×10^{-3} ft^3
1 in^3 = 5.787×10^{4} ft^3 = 1.639×10^{3} µm^3 = 1.639×10^{-2} L = 1.639×10^{-5} m^3
1 ft^3 = 1.728×10^{3} in^3 = 2.832×10^{16} µm^3 = 28.32 L = 2.832×10^{-2} m^3

Force

1 Newton (N) = 10^{5} dyne = 0.2248 lb = 102.1 g
1 dyne = 10^{-5} N = 2.248×10^{-6} lb = 1.021×10^{-3} g
1 pound (lb) = 4.448×10^{5} dynes = 4.448 N = 453.6 g
1 gram (g) force = 980.7 dyne = 9.807×10^{-3} N = 2.205×10^{-3} lb
1 grain (gr) = 63.55 dynes
1 poundal = 1.383×10^{4} dynes

Temperature

degrees Kelvin (K) = T°C + 273.15 = 5/9 (T°F + 459.67)
degrees Celsius (°C) = T K − 273.15 = 5/9 (T°F −32)
degrees Fahrenheit (°F) = 1.8T°C + 32 = 1.8T K − 459.67
degrees Rankine (R) = T°F + 459.67
where T is temperature in the indicated units

Pressure

1 Pascal (Pa) = 1 N/m^2 = 10 dyne cm^2
= 9.869×10^{-6} atm = 1.450×10^{-4} lb/in^2 = 7.501×10^{-3} mm Hg = $4.015 \cdot 10^{-3}$ in H_2O
1 atmosphere (atm) = 1.013×10^{5} N/m^2 = 1.013×10^{5} Pa = 101.3 kPa = 1.013×10^{6} dyne/cm^2
= 14.70 lb/in^2 = 760 mm Hg = 406.8 in H_2O

1 inch of water (in H_2O) (at 4°C) = 2.458×10^{-3} atm = 2491 dyne/cm^2 = 249.1 N/m^2 = 3.613×10^{-3} lb/in^2 = 1.868 mm Hg

1 mm of mercury (mm Hg) (at 0°C) = 1.316×10^{-3} atm = 1.333×10^{3} dyne/cm^2 = 1.333×10^{2} N/m^2 = 0.535 in H_2O = 1.934×10^{-2} lb/in^2

1 torr = 1 mm Hg

Viscosity

1 Pa·s = 10 poise (P) = 10 g/cm·s = 10 dyne·s/cm^2

Electrical Units

electronic charge = 1.6022×10^{-19} C
1 ampere (amp) = 2.998×10^{9} statamp
1 statampere (statamp) = 3.336×10^{-10} amp
1 volt (V) = 3.336×10^{-3} statV
1 statvolt (statV) = 299.8 V
1 farad (F) = 10^{6} μF = 8.987 statF
1 statfarad (statF) = 1.113×10^{-12} F
1 ohm = 1.113×10^{-12} statohm
1 statohm = 8.987×10^{11} ohm

APPENDIX C
Commonly Used Constants

Boltzmann constant	k	1.381×10^{-23} N·m/K [1.381×10^{-16} dyne·cm/K]
Avogadro's number	N_a	6.022×10^{23} molecules/mole
gas constant	R	8.314 J/(mole·K) [8.314×10^{7} dyne·cm/mole·K]
Stephan-Boltzmann constant	s	5.670×10^{-8} N/(m^2·s·K^4) [5.670×10^{-5} dyne/(cm^2·s·K^4)]
elementary charge	e	1.602×10^{-19} C [4.803×10^{-10} statC]
permittivity of free space	ε_o	8.854×10^{-12} F/m [1 electrostatic unit]
speed of light in vacuum	c	2.998×10^{8} m/s [2.998×10^{10} cm/s]
gravitational acceleration	g	9.807 m/s^2 [980.7 cm/s^2]

APPENDIX D
Some Properties of Air and Water

Air at 293.13 K [20°C] and 101.3 kPa [1 atm] (NTP)

density (ρ_g)	1.205 kg/m³ [1.205×10^{-3} g/cm³ = 1.205 g/L = 0.075 lb/ft³]
viscosity (ν)	1.832×10^{-4} P [1.832×10^{-5} Pa·s]
mean free path (λ)	0.0665 μm
average molecular weight ($\overline{M}$)	28.96 g/mole
specific heat ratio (γ)	1.40
diffusion coefficient (D)	1.9×10^{-5} m²/s [0.19 cm²/s]

For some properties of other gases, see Table 4–1.

Composition of Dry Air by Volume

Gas	Content (Volume %)	Molecular Weight (g/mole)
N_2	78.08	28.01
O_2	20.95	32.00
A	0.934	39.95
CO_2	0.033[a]	44.01
Other	<0.003	

[a] CO_2 concentration may vary from place to place.

Water at 293 K [20°C]

viscosity	1.002×10^{-3} N·s/m² [0.01002 dyne·s/cm²]
surface tension	0.07275 N/m [72.75 dyne/cm]
vapor pressure	2.338 kPa [17.54 mm Hg]

Water Vapor at 293 K [20°C]

Diffusion coefficient	2.4×10^{-3} m²/s [0.24 cm²/s]
Density	0.75 kg/m³ [0.75×10^{-3} g/cm³]

APPENDIX E
Major Dimensionless Numbers

Froude (Fr)	$= \dfrac{U^2}{gL}$	inertial force/gravitational force
Knudsen (Kn)	$= \dfrac{\lambda}{d_p}$	mean free path of molecules/particle diameter
Mach (Ma)	$= \dfrac{U}{U_{sonic}}$	flow velocity/sonic velocity
Peclet (Pe)	$= \dfrac{LU}{D}$	bulk mass transfer/diffusive mass transfer
Prandtl (Pr)	$= \dfrac{\nu}{\alpha}$	momentum diffusivity/thermal diffusivity
Reynolds (Re)	$= \dfrac{d_p U}{\nu}$	inertia force/viscous force
Schmidt (Sc)	$= \dfrac{\nu}{D}$	momentum diffusivity/mass diffusivity
Stokes (Stk)	$= \dfrac{\tau U}{L}$	stopping distance/characteristic flow dimension

APPENDIX F
Properties of Particles

Standard Density (1000 kg/m³) Spheres at NTP—293.15 K, 101.3 kPa [1 atm]

Particle Diameter (μm)	Slip Correction Factor	Settling Velocity (m/s)	Diffusion Coefficient (m²/s)	Mobility m/(N·s)	rms Brownian Displacement (m)[a]
0.00037[b]	611.5	2.490×10^{-9}	3.930×10^{-5}	9.700×10^{15}	26.58
0.001	226.5	6.737×10^{-9}	5.381×10^{-6}	1.330×10^{15}	10.37
0.002	109.9	1.326×10^{-8}	1.305×10^{-6}	3.227×10^{14}	5.110
0.005	44.31	3.340×10^{-8}	2.105×10^{-7}	5.203×10^{13}	2.052
0.01	22.45	6.768×10^{-8}	5.332×10^{-8}	1.318×10^{13}	1.033
0.02	11.53	1.390×10^{-7}	1.369×10^{-8}	3.385×10^{12}	0.5233
0.05	5.014	3.779×10^{-7}	2.382×10^{-9}	5.888×10^{11}	0.2182
0.1	2.888	8.708×10^{-7}	6.861×10^{-10}	1.696×10^{11}	0.1171
0.2	1.879	2.267×10^{-6}	2.232×10^{-10}	5.517×10^{10}	0.06681
0.5	1.333	1.005×10^{-5}	6.335×10^{-11}	1.566×10^{10}	0.03559
1	1.166	3.515×10^{-5}	2.769×10^{-11}	6.846×10^{9}	0.02353
2	1.083	0.0001306	1.286×10^{-11}	3.179×10^{9}	0.01604
5	1.033	0.0007787	4.908×10^{-12}	1.213×10^{9}	0.009908
10	1.017	0.003065	2.415×10^{-12}	5.969×10^{8}	0.006949
20	1.008	0.01216	1.198×10^{-12}	2.960×10^{8}	0.004894
50	1.003	0.07395	4.766×10^{-13}	1.178×10^{8}	0.003087
100	1.002	0.2605	2.379×10^{-13}	5.881×10^{7}	0.002181

[a] Displacement in 10 s.
[b] Effective diameter of air molecule.

APPENDIX G
Geometric Formulas

Circle	
circumference	$= \pi d$
area	$= \pi r^2 = \dfrac{\pi d^2}{4}$
Ellipse	
circumference (approx.)	$= 2\pi \sqrt{\dfrac{a^2 + b^2}{2}}$
area	$= \pi ab$
	where a and b are the major and minor semi-axes, respectively
Sphere	
surface area	$= 4\pi r^2 = \pi d^2$
volume	$= \dfrac{4\pi r^3}{3} = \dfrac{\pi d^3}{6}$
Ellipsoids: prolate	
surface area	$= 2\pi b^2 + \dfrac{2\pi ab}{\varepsilon}\sin^{-1}\varepsilon$
	where $\varepsilon = \dfrac{\sqrt{a^2 - b^2}}{a}$
volume	$= \dfrac{4\pi ab^2}{3}$
Ellipsoids: oblate	
surface area	$= 2\pi a^2 + \dfrac{\pi b^2}{\varepsilon}\ln\left(\dfrac{1+\varepsilon}{1-\varepsilon}\right)$
volume	$= \dfrac{4\pi a^2 b}{3}$
Right cylinder	
surface area	$= 2\pi rL + 2\pi r^2 = \pi dL + \dfrac{\pi d^2}{2}$
	where L is the length
volume	$= \pi r^2 L = \dfrac{\pi d^2 L}{4}$

APPENDIX H
Bulk Densities of Some Common Aerosol Materials*

Material	Density (×1000) (kg/m^3)
Solids	
Aluminum	2.70
Corundum (Al_2O_3)	4.0
Ammonium sulfate	1.77
Asbestos	2.4–3.3
Calcite ($CaCO_3$)	2.7–2.9
Coal	1.2–1.8
Coal fly ash	ca. 2.0
Glass	2.4–2.8
Granite	2.4–2.7
Iron	7.86
Iron oxide	5.2–5.7
Limestone	2.1–2.9
Lead	11.3
Lead oxide	8.0–9.5
Methylene blue	1.26
Mineral wool	ca. 2.7
Plant particles	1.1–1.5
Paraffin	0.9
Pollen	ca. 1.4
Polystyrene	1.05
Polyvinyl toluene	1.03
Portland cement	3.2
Potassium biphthalate	1.64
Rock wool	ca. 2.5
Quartz	2.64–2.66
Sodium chloride	2.17
Sulfur	2.07
Starch	1.5
Talc	2.6–2.8
Titanium dioxide	4.26
Uranine dye	1.53
Wood	ca. 1.5
Zinc oxide	5.61

Continued

Material	Density (×1000) (kg/m³)
Liquids	
Isopropyl alcohol	0.7855
Dibutyl phthalate	1.043
Dioctyl phthalate (DOP)	0.981
Dioctyl sebecate	0.915
Hydrochloric acid	1.19
Mercury	13.6
Oils	0.88–0.94
Oleic acid	0.894
Polyethylene glycol	1.13
Sulfuric acid	1.84

* Additional references for particle densities and other properties: McCrone, W. C. and J. G. Delly. 1973. *The Particle Atlas*, Vol. IV. Ann Arbor: Ann Arbor Science Publishers Inc.; Weast, R. C. 1994. *Handbook of Chemistry and Physics*. Cleveland, OH: CRC Press.

APPENDIX I
Manufacturers and Suppliers

Note that addresses and internet sites may not be up to date. Companies have been included that are not mentioned elsewhere in the book, but may have useful aerosol-related products.

AGI
Ace Glass, Inc.
P.O. Box 688
1430 Northwest Blvd.
Vineland, NJ 08360-0688
Tel: 609-692-3333 or 800-223-4524
Fax: 800-543-6752
www.aceglass.com
Chapters 10, 24

ADT
Adept Scientific, Inc.
257 Great Rd.
Acton, MA 01720
Tel: 978-635-5360
Fax: 978-635-5330
www.adeptscience.com
Chapter 2

ADE
Air Diagnostics and Engineering, Inc.
110 Alpine Village Rd.
Harrison, ME 04040
Tel: 207-583-4834
www.airdiagnostics.com
Chapters 10, 29

AIR
Air Techniques, Inc.
11403 Cronridge Dr.
Owings Mills, MD 21117
Tel: 410-363-9696
Fax: 410-363-9695
www.atitest.com

AER
Aerosol Dynamics, Inc.
2329 Fourth St.
Berkeley, CA 94710
Tel: 510-649-9360
Fax: 510-649-9260
Chapter 16

ALL
Allergen LLC
Allergenco/Blewstone Press
P.O. Box 8571
Wainwright Station
San Antonio, TX 78208-0571
Tel: 210-822-4116
Fax: 210-829-1883
www.txdirect.net/corp/allergen
Chapter 24

ALR
Alnor Instrument Company (part of ***TSI***)
7555 N Linder Ave.
Skokie, IL 60077-3223
Tel: 847-677-3500
Fax: 847-677-3539
www.alnor.com
Chapter 29

AND
(Andersen Instruments)
Thermo Andersen
500 Technology Court
Smyrna, GA 30082-5211
Tel: 770-319-9999 or 800-241-6898
Fax: 770-319-0336
www.anderseninstruments.com
Chapters 10, 14, 24, 25, 26, 27

API
Apiezon Products
M&I Materials, Ltd.
P.O. Box 136
Manchester
M60 1AN, UK
Tel: 44-161-875-4444
www.apiezon.com
Chapter 10

ARE
ARELCO
2 Avenue Ernest Renan
F-94120 Fontenay-sous-Bois
France
Tel: 33-1-48-75-82-82
Fax: 33-1-43-94-07-21
www.arelco-arc.com
Chapter 25

BEI
Baldwin Environmental, Inc.
895 E. Patriot Blvd., Suite 107
Reno, NV 89511
Tel: 775-828-1300 or 888-234-7366
Fax: 775-828-1305
www.bei-reno.com

BAN
Bangs Laboratories, Inc.
9025 Technology Dr.
Fishers, IN 46038-2886
Tel: 317-570-7020 or 800-387-0672
Fax: 317-570-7034
www.bangslabs.com
Chapters 16, 21

BAR
Barramundi Corp.
P.O. Drawer 4259
Homosassa Springs, FL 34447
Tel: 352-628-0200
Fax: 352-628-0203
www.mattson-garvin.com
Chapter 24

BAS
BASF Aktiengesellschaft
Carl-Bosch-Strasse 38
D-67056 Ludwigshafen
Germany
Tel: 49-621-600
Fax: 49-621-604-25-25
www.basf.de
Chapters 24, 32

BAY
Bayer AG
Werk Leverkusen
D-51368 Leverkusen
Germany
Tel: 49-214-301
www.bayer.de
Chapter 32

BEC
Beckman Coulter, Inc.
4300 N. Harbor Blvd.
P.O. Box 3100
Fullerton, CA 92834-3100
Tel: 714-871-4848
Fax: 714-773-8283
www.beckmancoulter.com
Chapter 16

BEL
Belfort Instrument Co.
727 S. Wolfe St.
Baltimore, MD 21231
Tel: 410-342-2626
Fax: 410-342-7028
www.belfortinstrument.com

BGI
BGI Incorporated
58 Guinan St.
Waltham, MA 02154
Tel: 781-891-9380
Fax: 781-891-8151
www.bgiusa.com
Chapters 2, 10, 25

BSI
Bioscience International
11607 Magruder Lane
Rockville, MD 20852-4635
Tel: 301-230-1418
www.biosci-intl.com
Chapter 24

BII
BIOS International
10 Park Place
Butler, NJ 07405
Tel: 973-492-8400 or 800-663-4977
Fax: 973-492-8270
www.biosint.com
Chapters 21, 24

BIO
Biotest Diagnostics Corporation
66 Ford Rd., Suite 131
Denville, NJ 07834
Tel: 973-625-1300 or 800-522-0090
Fax: 973-625-9454
www.biotest.com
Chapter 24

BKR
Booker Systems, Ltd.
P.O. Box 5894
Towcester, Northants
NN12 8ZX U.K.
Tel: 44-(0)870-241-1557
Fax: 44-(0)870-241-1558
e-mail: sales@bookersystems.co.uk
Chapter 14

BIR
Bristol Industrial and Research Associates, Ltd.
P.O. Box 2, 1 Beach Road West
Portishead, Bristol
BS20 7JB U.K.
Tel: 44-(0) 1275 847787
Fax: 44-(0) 1275 847303
www.biral.com
Chapters 15, 23

BRK
Brookhaven Instruments Corp.
750 Blue Point Rd.
Holtsville, NY 11742-1896
Tel: 631-758-3200
Fax: 631-758-3255
www.bic.com
Chapter 16

BUC
A.P. Buck, Inc.
7101 Presidents Dr., Suite 110
Orlando, FL 32809
Tel: 407-851-8602
Fax: 407-851-8910
www.apbuck.com
Chapter 21

BUR
Burkard Manufacturing Co., Ltd.
Woodcock Hill Industrial Estate
Richmansworth, Hertfordshire
WD3 1PJ England
Tel: 44-(0)923-773134
Fax: 44-(0)923-774790
www.burkard.co.uk
Chapter 24

CAB
Cabot Corp.
700 E. U.S. Highway 36
Tuscola, IL 61953-9643
Tel: 217-253-3370
www.cabot-corp.com/cabot
Chapter 32

CAN
Canberra Industries
800 Research Parkway
Meriden, CT 06450
Tel: 203-238-2351 or 800-243-3955
Fax: 203-235-1347
www.canberra.com

CMI
California Measurements, Inc.
150 E. Montecito Ave.
Sierra Madre, CA 91024
Tel: 626-355-3361
Fax: 626-355-5320
www.californiameasurements.com
Chapters 10, 14

CAS
Casella Group, Ltd.
Regent House, Wolseley Rd.
Kempston, Bedford
MK42 7JY U.K.
Tel: 44-(0)1234-841441
Fax: 44-(0)1234-841490
www.casella.co.uk
Chapters 23, 24, 25

CLI
Climet Instruments Co.
1320 W. Colton Ave.
Redlands, CA 92374
Tel: 909-793-2788
Fax: 909-793-1738
www.climet.com
Chapters 7, 15

CIL
(CILAS) Compagnie Industrielle des Lasers
Route de Nozay
BP 27
F-91460 Marcoussis
France
Phone: 33-1-64-54-48-00
Fax: 33-1-69-01-37-39
www.cilas.com
Chapter 16

COL
Columbian Chemicals Company
1800 West Oak Commons Ct.
Marietta, GA 30062-2253
Tel: 770-792-9400
Fax: 770-792-9625
www.columbianchemicals.com
Chapter 32

COP
Copley Scientific, Ltd.
Private Road #7, Cowlick Industrial Estate
Cowlick, Nottingham
NG 2ER U.K.
Tel: 44-115-96-16229
Chapter 10

CRL
Corel Corp.
1600 Carling Ave.
Ottawa, Ontario
K1Z 8R7 Canada
Tel: 353-1-213-3912
www.corel.com
Chapter 2

CCO
Corning Costar
One Alewife Center
Cambridge, MA 02140
Tel: 617-868-6200 or 800-492-1110
Fax: 617-868-2076
www.corningcostar.com
Chapters 9, 27

COR
Corning, Inc.
45 Nagog Park
Acton, MA 01720-3413
Tel: 978-635-2200 or 800-492-1110
Fax: 978-635-2476
www.scienceproducts.corning.com
Chapter 24

DAI
DAN Industry Co., Ltd.
Japan
Tel: 03-488-1111
Fax: 03-488-1118

DAN
Dantec Dynamics, Inc.
777 Corporate Dr.
Mahway, NJ 07430
Tel: 201-512-0037
Fax: 201-512-0120
www.dantecmt.dk
Chapters 16, 21

DAT
Datatest
6850 Hibbs Lane
Levittown, PA 19057
Tel: 215-943-0668
Fax: 215-547-7973
www.datatest-inc.com

DEG
Degussa-Hüls AG
Hauptverwaltung
Weissfrauenstrasse 9
D-60311 Frankfurt am Main
Germany
Tel: 49-69-218-01
Fax: 49-69-218-3218
www.degussa.de
Chapter 32

DEK
Dekati, Ltd.
Osuusmyllynkatu 13
FIN-33700 Tampere
Finland
Tel: 358-3-3578-100
Fax: 358-3-3578-140
www.dekati.fi
Chapters 10, 14

DEV
The DeVilbiss Co.
P.O. Box 635
Somerset, PA 15501
Tel: 814-443-4881
www.sunrisemedical.com
Chapter 21

DOW
The Dow Chemical Company
Midland, MI 48667
www.dow.com
Chapter 32

DMT
Droplet Measurement Technologies, Inc.
P.O. Box 20293
Boulder, CO 80308
Tel: 303-440-5576
Fax: 303-440-1965
www.dropletmeasurement.com
Chapter 30

DUK
Duke Scientific Co.
2463 Faber Pl.
Palo Alto, CA 94303
Tel: 650-424-1177 or 800-334-3883
Fax: 650-424-1158
www.dukescientific.com
Chapters 16, 21

DUP
DuPont Co.
1007 Market St.
Wilmington, DE 19898
Tel: 302-774-1000 or 800-441-3515
www.dupont.com
Chapter 32

DWY
Dwyer Instruments, Inc.
P.O. Box 373
Michigan City, MI 46361
Tel: 219-879-8000
Fax: 219-872-9057
www.dwyer-inst.com
Chapter 21

DYN
Dynal Particles AS
Svellevn. 29
P.O. Box 160
N-2001 Lillestrøm, Norway
Tel: 47-63-89-71-00
Fax: 47-63-89-74-72
www.dynalbiotech.com
Chapter 21

EBE
Eberline Instrument Corp.
P.O. Box 2108
504 Airport Rd.
Santa Fe, NM 87504
Tel: 505-471-3232
Fax: 505-473-9221
www.eberlineinst.com
Chapter 34

ECO
EcoChem Technologies, Inc.
22605 Valerio St.
West Hills, CA 91307
Tel: 818-347-4369
Fax: 818-347-5639
www.ecochem-analytics.com
Chapter 14

EDC
Environmental Devices Corporation
88 Essex St.
Haverhill, MA 01832-5675
Tel: 978-521-1514
Fax: 978-521-1628
www.hazdust.com

ESM
ESM-Andersen Instruments GmbH
Frauenauracher Strasse 96
D-91056 Erlangen
Germany
Tel: 49-9131-909-0
Fax: 49-9131-909-156
http://www.esm-andersen.de
Chapter 14

FSI
Fisher Scientific, Inc.
585 Alpha Dr.
Pittsburgh, PA 15283
Tel: 800-766-7000
Fax: 800-926-1166
www.fisherscientific.com
Chapter 9

FLU
Fluid Energy Aljet
P.O. Box 428-T
Plumsteadville, PA 18949
Tel: 215-766-0300
Fax: 215-766-0555

FLT
FLUENT, Inc.
10 Cavendish Court,
Centerra Park
Lebanon, NH 03766
Tel: 603-643-2600
Fax: 603-643-3967
www.fluent.de
Chapters 2, 30

GEL
Gelman Sciences (Pall-Gelman)
600 South Wagner Rd.
Ann Arbor, MI 48103-9019
Tel: 734-665-0651 or 800-521-1520
Fax: 734-913-6114
www.pall.com/gelman
Chapters 9, 19, 26, 27, 34

GEN
Genitron Instruments GMBH
Heerstrasse 149
D-60488 Frankfurt am Main-90
Germany
Tel: 49-69-976-514-0
Fax: 49-69-765-327
www.genitron.de

GIL
Gilian Instrument Corp. (part of ***SEN***)
Chapters 21, 29

GMW
Graseby-GMW (part of ***AND***)
Chapters 10, 25

GRA
Graseby-Andersen (part of ***AND***)
Chapter 27

GRE
Greenfield Instruments
P.O. Box 971
Amherst, MA 01004
Tel: 413-253-4216
Fax: 413-253-1397
www.greenfieldinst.com
Chapter 27

GRT
GreenTek, Inc.
295 NW Hillcrest Drive
Grants Pass, OR 97526
Tel: 541-955-5386
Fax: 541-479-4285
www.greentekusa.com
Chapter 27

GRI
Grimm Technologies, Inc.
9110 Charlton Place
Douglasville, GA 30135
Tel: 877-474-6872
Fax: 770-577-0955
www.dustmonitor.com
Chapters 25, 29

HAA
Haan and Wittmer GmbH
Birkenstrasse 31
D-71292 Friolzheim
Germany
Tel: 49(0)-7044-4064
Fax: 49(0)-7044-4040

HAM
Hampshire Glassware
77-79 Dukes Rd.
Portswood, Hampshire, Southampton
S014 OST U.K.
Tel: 44-02380-553755
Fax: 44-02380-553020
www.hgl-uk.com
Chapter 24

HAU
Hauke GmbH KG
P.O. Box 63
A-4810 Gmunden
Austria
Tel: 43-(0)76-12-63758
Tel: 43-(0)76-12-64133
members.aon.at/meissl/hauke.html
Chapters 10, 18

HIA
Hiac-Royco (part of ***PAC***)
141 Jefferson Dr.
Menlo Park, CA 94025
Chapter 15

HIQ
Hi-Q Environmental Products Co.
7386 Trade St.
San Diego, CA 92121
Tel: 858-549-2820
Fax: 858-549-9657
www.hi-q.net

HOR
Horiba
1080A East Duane Ave.
Sunnyvale, CA 94086
Tel: 408-730-4772
Fax: 408-730-8975
www.horibastec.com
Chapter 16

HOS
Hosokawa Micron Powder Systems
10 Chatham Rd.
Summit, NJ 07901
Tel: 908-273-6360
Fax: 908-273-7432
www.hosokawamicron.com
Chapter 17

HUN
Helmut Hund GmbH
Wilhelm-Will-Str. 7
35580 Wetzlar
Germany
Tel: 49-6441-20040
Fax: 49-6441-200444
www.hund.de
Chapters 7, 15, 25

INC
Inco, Ltd.
145 King Street West, Suite 1500
Toronto, Ontario
M5H 4B7 Canada
Tel: 416-361-7511
Fax: 416-361-7781
www.inco.com
Chapter 32

INN
Innova AirTech Instruments A/S
Energivej 30
2750 Ballerup
Denmark
Tel: 45-44-20-01-00
Fax: 45-44-20-01-01
www.innova.dk
Chapter 29

INO
Inovision Radiation Measurements (includes Victoreen)
6045 Cochran Rd.
Cleveland, OH 44139
Tel: 440-248-9300
Fax: 440-349-2307 or 800-850-4608
www.victoreen.com
Chapter 34

IDC
Interfacial Dynamics Corp.
Suite 120
17300 SW Upper Boones Ferry Rd.
Portland, OR 97224
Tel: 503-684-8008 or 800-323-4810
Fax: 503-684-9559
www.idclatex.com
Chapter 21

INS
Insitec Measurement Syst. (see ***MAL*** and ***PRM***)
2110 Omega Rd., Suite D
San Ramon, CA 94583
Tel: 510-837-1330
Fax: 510-837-3864
www.insitec.com

IOM
Institute of Occupational Medicine Ltd.
8 Roxburgh Place
Edinburgh EH8 9SU
UK
Tel: 44 1 31 667 5131
Fax: 44 31 667 0136
Chapter 25

INT
In-Tox Products
P.O. Box 2070
Moriarty, NM 87035
Tel: 505-832-5107
Fax: 505-832-5092
www.intoxproducts.com
Chapter 10

ISH
Ishihara Sangyo Kaisha, Ltd.
3–15, Edobori 1-chome
Nishi-ku, Osaka 550
Japan
Tel: 06-444-1451
Fax: 06-445-7798
Chapter 32

JSR
Japan Synthetic Rubber
11–24, Tsukiji 2-chome, Chuo-ku
Tokyo 104, Japan
Tel: 03-5565-6521
Fax: 03-5565-6645
Chapter 21

KAN
Kanomax USA, Inc.
250 West 57th St., Suite 816
New York, NY 10107
Tel: 212-489-3755
Fax: 212-489-4104
www.kanomax-usa.com
Chapter 14

KEM
Kemira Pigments B.V.
P.O. Box 1013
NL-3180 AA Rozenburg
The Netherlands
Tel: 31-10-295-2540
Fax: 31-295-2536
www.kemira.com
Chapter 32

KER
Kerr-McGee
P.O. Box 25861
Oklahoma City, OK 73125
Tel: 405-270-1313
www.kerr-mcgee.com
Chapter 32

KIM
Kimoto Electric Co., Ltd.
3-1 Funahashi-cho, Tennoji-ku,
Osaka 543-0024, Japan
Tel: 81-6-6-768-3401
Fax: 81-6-6-764-7040
www.kimoto-electric.co.jp
Chapter 14

KOE
Koenders Instruments
Postbus 1189
1300 BD Almere
The Netherlands
Tel: 31-36-5480101
Fax: 31-36-5480102
www.koendersgroup.com
Chapter 25

KRA
Kratel SA
CH-1222 Geneve-Vesenaz
64 Ch. De St. Maurice
Switzerland

KUR
Kurz Instruments, Inc.
2411 Garden Rd.
Monterey, CA 93940
Tel: 831-646-5911 or 800-424-7356
Fax: 831-646-8901
www.kurz-instruments.com
Chapter 34

LAN
Lanzoni, S. R. L.
Via Michelino 93/B
40127 Bologna, Italy
Tel: 39-(0)51-504810
Fax: 39-(0)51-6331892
www.lanzoni.it
Chapter 24

LAS
Laser Holography
Mammoth Lakes, CA 93546
Chapter 17

LEA
Monitor Labs, Inc.
(supports Lear Siegler Corp. equipment)
74 Inverness Dr. East
Englewood, CO 80112-5189
Tel: 303-792-3300 or 800-422-1499
Fax: 303-799-4853

LOT
LOT—ORIEL GmbH
Im Tiefen See 58
6100 Darmstadt, Germany
Tel: 49(0)6151-88060
Fax: 49(0)6151-896667
www.lot-oriel.com

LUC
Lucent Technologies
600 Mountain Ave.
Murray Hill, NJ 07974
www.lucent.com
Chapter 32

LUD
Ludlum Measurements, Inc.
P.O. Box 810
501 Oak Street
Sweetwater, TX 79556
Tel: 915-235-5494 or 800-622-0828
Fax: 915-235-4672
www.ludlums.com
Chapter 34

MMM
3M
3M Center Building
St Paul, MN 55144
Tel: 612-737-6501
www.3m.com
Chapter 21

MAG
Magee Scientific
1829 Francisco Street
Berkeley, CA 94703
Tel: 510-845-2801
Fax: 510-845-7137
www.mageesci.com
Chapter 11

MAL
Malvern Instruments, Inc.
10 Southville Rd.
Southborough, MA 01772
Tel: 508-480-0200
Fax: 508-460-9692
www.malvern.co.uk
Chapter 16

MAS
MathSoft Engineering & Education, Inc.
101 Main Street
Cambridge, MA 02142-1521
Tel: 617-577-1017
Fax: 617-577-8829
www.mathsoft.com
Chapter 2

MAT
Matter Engineering
Bremgarterstrasse 62
CH-5610 Wohlen
Switzerland
Tel: 41-56-618-66-30
Fax: 41-56-618-66-39
www.matter-engineering.com
Chapter 14

MET (see ***PAC***)
Met One, Inc.
481 California Ave.
Grants Pass, OR 97526
Tel: 503-479-1248 or 800-866-7889
Fax: 503-479-3057
www.pacsciinst.com

MGP
MGP Instruments,
5000 Highlands Parkway, Suite 150
Smyrna, GA 30082
Tel: 770-432-2744
Fax: 770-432-9176
www.mgpi.com
Chapter 34

MEL
MetroLaser, Inc.
18010 Skypark Circle, Suite 100
Irvine, CA 92614
Tel: 949-553-0688
Fax: 949-553-0495
www.metrolaserinc.com
Chapters 15, 16

MFS
Advantec MFS, Inc.
(Formerly Micro Filtration Systems, Inc.)
6723 Sierra Ct., Suite A
Dublin, CA 94568
Tel: 925-479-0625 or 800-334-7132
Fax: 925-479-0630
www.advantecmfs.com
Chapter 9

MCM
Micromeritics Instrument Corp.
One Micromeritics Dr.
Norcross, GA 30093-1877
Tel: 770-662-3620
www.micromeritics.com
Chapter 32

MSI
Micron Separations, Inc. (see ***OSM***)
5951 Clearwater Dr.
Minnetonka, MN 55343
www.osmolabstore.com
Chapter 9

MIC
Microsoft, Inc.
Redmond, WA 98052
Tel: 425-882-8080
www.microsoft.com
Chapter 2

MCR
Microtrac, Inc.
12501-A 62nd Street North
Largo, FL 33773
Tel: 727-507-9770
Fax: 727-507-9774
www.microtrac.com
Chapter 16

MIE
MIE, Inc.
7 Oak Park
Bedford, MA 01730
Tel: 781-275-1919
Fax: 781-275-2121
www.mieinc.com
Chapters 15, 17, 23, 25, 26, 29

MLN
Millennium Inorganic Chemicals
3901 Fort Armistead Rd.
Baltimore, MD 21226
Tel: 410-355-3600
Fax: 410-229-5003
www.millenniumchem.com
Chapter 32

MIL
Millipore Corp.
80 Ashby Rd.
Bedford, MA 01730
Tel: 781-533-6000
Fax: 781-533-3110
www.millipore.com
Chapters 9, 24, 26, 34

MSA
Mine Safety Appliances Co.
RIDC Industrial Park
Pittsburgh, PA 15238-2919
Or P.O. Box 426
Pittsburgh, PA 15230
Tel: 412-967-3000 or 800-672-2222
Fax: 412-967-3451
www.msanet.com
Chapters 10, 26, 29

MSP
MSP Corp.
1313 Fifth Street S.E., Suite 206
P.O. Box 549
Minneapolis, MN 55414
Tel: 612-379-3963
Fax: 612-379-3965
www.mspcorp.com
Chapters 10, 25, 26, 29

MUL
Multidata LLC, Operations
4838 Park Glen Rd.
St. Louis Park, MN 55416
Tel: 952-285-9890 or 800-264-1338
Fax: 952-285-9902
www.multidata.com
Chapter 10

NAC
Nanochem, Inc.
2901 Maximillian
Albuquerque, NM 87104-1817
Chapter 32

NAL
Nalge Nunc International
75 Panorama Creek Dr.
Rochester, NY 14602-0365
Tel: 716-586-8800
www.nalgenunc.com
Chapter 9

NAN
Nanophase Technologies Corp.
453 Commerce Street
Hinsdale, IL 60521-0750
www.nanophase.com
Chapter 32

NAT
National Instruments Corp.
11500 N. Mopac Expwy.
Austin, TX 78759-3504
Tel: 512-794-0100
Fax: 512-683-8411
www.ni.com
Chapter 2

NBS
New Brunswick Scientific Co., Inc.
P.O. Box 4005
44 Talmadge Rd.
Edison, NJ 08818-4005
Tel: 732-287-1200 or 800-631-5417
Fax: 732-287-4222
www.nbsc.com
Chapters 10, 24

NOV
Novelec, North America, Inc.
113 W. Outer Dr.
P.O. Box 6621
Oak Ridge, TN 37831
Tel: 423-482-9287
Fax: 423-483-0305

OSI
Omega Specialty Instrument Co.
4 Kidder Rd., Unit 5
Chelmsford, MA 01842
Tel: 978-256-5450 or 800-343-8253
Fax: 978-256-8015
www.omegaspec.com

OPS
Opsis AB
Box 244
SE-244 02 Furulund, Sweden
Tel: 46-46-722500
Fax: 46-46-722501
www.opsis.se
Chapter 14

OSM
Osmonics
5951 Clearwater Dr.
Minnetonka, MN 55343-8995
Tel: 612-933-7979 or 800-848-1750
www.osmonics.com
Chapter 9

OXF
Oxford Lasers, Inc.
29 King Street
Littleton, MA 01460-1528
Tel: 978-742-9000
Fax: 978-742-9100
www.oxfordlasers.com
Chapter 21

PAC
Pacific Scientific Instruments
481 California Ave.
Grants Pass, OR 97526
Tel: 541-479-1248 or 800-866-7889
Fax: 541-479-3057
www.pacsciinst.com
Chapter 15

PAS
Palas, GmbH
Greschbachstrasse 3B
D-76229 Karlsruhe, Germany
Tel: 49-721-96213-0
Fax: 49-721-96213-33
www.palas.de
Chapters 15, 16, 21

PAL
Pall Corp.
25 Harbor Park Dr.
East Hills, NY 11548
Tel: 516-484-5400 or 800-645-6532
Fax: 516-484-5228
www.pall.com
Chapters 9, 11, 26, 27, 34

PAR
PARI GmbH
Moosstrasse 9
D-82319 Starnberg
Germany
Tel: 49(0)8150-2790
Fax: 49(0)8151-279101
www.pari.de
Chapter 15

PMS
Particle Measuring Systems, Inc.
5475 Airport Blvd.
Boulder, CO 80301-2339
Tel: 303-443-7100 or 800-238-1801
Fax: 303-449-6870
www.pmeasuring.com
Chapters 7, 15, 16, 29

PMI
Particle Metrics, Inc.
6315 Monarch Park Place, Suite B
Longmont, CO 80503
Tel: 303-247-0411
Fax: 303-247-1318
Chapter 30

PSS
Particle Sizing Systems
75 Aero Camino, Suite B
Santa Barbara, CA 93117
Tel: 805-968-1497
Fax: 805-968-0361
www.pssnicomp.com
Chapter 16

PTI
Particle Technologies, LLC
14554 Lee Rd.
Chantilly, VA 20151-1632
Tel: 703-378-6200 or 301-896-0257
Fax: 703-378-7274
Chapters 29, 32

PBI
International PBI
Via Novara, 89
20153 Milan
Italy
Tel: 39(0)2-487791
Fax: 39(0)2-40090010
www.wheatonsci.com
Chapter 24

PIX
PIXE International, Inc.
2306 Domingo Dr.
Tallahassee, FL 32303
Tel: 850-574-6469

PCS
Pollution Control Systems Corp.
P.O. Box 15770
Seattle, WA 98115
Tel: 206-523-7220
Fax: 206-523-7221
www.cascadeimpactor.com
Chapter 10

POL
Polysciences, Inc.
400 Valley Rd.
Warrington, PA 18976
Tel: 215-343-6484 or 800-523-2575
Fax: 215-343-0214
www.polysciences.com
Chapter 21

PLY
Polytec Pl, GmbH
Polytec Platz 5–7
P.O. Box 1140
W-7517 Waldbronn, Germany
Tel: 49(0)-7243-60-0
Fax: 49(0)-72436994
www.polytecpi.com
Chapters 15, 16

POR
Poretics, see ***OSM***

PTI
Powder Technology, Inc.
P.O. Box 1464
Burnsville, MN 55337
Tel: 952-894-8737 or 800-718-8737
Fax: 952-894-0734
www.powdertechnologyinc.com
Chapter 29

PPM
ppm Inc.
9737 Cogdill Rd.
Suite 215
Knoxville TN 37932
Tel: 865-966-8796
Fax: 865-966-2750
www.ppmcorp.com

PRM
Process Metrix (Formerly Insitec; see also ***MAL***)
2110 Omega Rd., Suite D
San Ramon, CA 94583-1295
Tel: 925-837-1330
Fax: 925-837-3864
www.processmetrix.com
Chapter 16

PYL
Pylon Electronics, Inc.
147 Colonnade Rd.
Nepean, Ontario
Canada K2E 7L9
Tel: 613-226-7920 or 800-896-4439
Fax: 613-226-8195
www.pylonelectronics.com

QCM
QCM Research
2825 Laguna Canyon Rd.
Laguna Beach, CA 92651
Or P.O. Box 277
Laguna Beach, CA 92652
Tel: 949-497-5748
Fax: 949-497-9828
www.qcmresearch.com
Chapters 10, 14

RAD
Rad Elec, Inc.
5714-C Industry Lane
Frederick, MD 21704
Tel: 301-694-0011 or 800-526-5482
Fax: 301-694-0013
www.radelec.com

RIO
RION Co., Ltd.
Ikeda Bldg.
7-7 Yoyogi 2-chome
151 Tokyo
Japan
Tel: 33-79-23-52
Fax: 33-70-48-28
www.rion.co.jp

RTI
Research Triangle Institute
P.O. Box 12194
Research Triangle Park, NC 27709
Tel: 919-541-6000
www.rti.org
Chapter 29

ROS
Rosemount Aerospace
(changed to Goodrich Aircraft Sensors)
14300 Judicial Rd.
Burnsville, MN 55306-4898
Tel: 952-892-4000
Fax: 952-892-4800
www.bfg-sensors.com
Chapter 30

R&P
Rupprecht & Patashnick Co.
25 Corporate Circle
Albany, NY 12203
Tel: 518-452-0065
Fax: 518-452-0067
www.rpco.com
Chapters 7, 11, 14, 26, 27

SAM
(Sampling Technologies, Inc.)
Multidata LLC
4838 Park Glen Rd.
St. Louis Park, MN 55416
Tel: 952-285-9890 or 800-264-1338
Fax: 952-285-9902
www.multidata.com
Chapter 24

SAR
Sartorius North America, Inc.
131 Heartland Blvd.
Edgewood, NY 11717
Tel: 631-254-4249 or 800-635-2906
Fax: 631-254-4253
www.sartoriuscorp.com
Chapter 9

SAS
SAS Institute, Inc.
SAS Campus Dr.
Cary, NC 27513-2414
Tel: 919-677-8000
Fax: 919-677-4444
www.sas.com
Chapter 22

S&S
Schleicher and Schuell, Inc.
10 Optical Ave.
Keene, NH 03431
Tel: 603-352-3810
Fax: 603-357-3627
www.s-and-s.com
Chapter 9

SCI
Scientific Computing & Instrumentation Magazine
301 Gibraltar Dr.
Box 650
Morris Plains, NJ 07950-0650
Tel: 973-292-5100
Fax: 973-539-3476
www.scimag.com
Chapter 2

STS
(SciTech Science)
Cole-Parmer
625 E. Bunker Ct.
Vernon Hills IL 60061-1844
Tel: 847-549-7600 or 800-323-4340
Fax: 847-247-2929
www.coleparmer.com
Chapter 2

SEN
Sensidyne, Inc.
16333 Bay Vista Dr.
Clearwater, FL 33760
Tel: 727-530-3602 or 800-451-9444
Fax: 727-539-0550
www.sensidyne.com
Chapter 10

SRD
Seradyn Particle Technology
7998 Georgetown Rd., Suite 1000
Indianapolis, IN 46268
Tel: 317-610-3800 or 800-428-4072
Fax: 317-610-3888
www.seradyn.com
Chapter 16

SHI
Shimadzu Scientific Instruments, Inc.
7102 Riverwood Dr.
Columbia, MD 21046
Tel: 410-381-1227 or 800-477-1227
Fax: 410-381-1222
www.ssi.shimadzu.com
Chapter 16

SIB
Sibata Scientific Technology, Ltd.
1-25, Ikenohata 3-Chome, Taito-Ku,
Tokyo 110-8701
Japan
Tel: 81-3-3822-2112
Fax: 81-3-5685-1394
www.sibata.co.jp

SIC
(Siecor, Inc.)
Corning Cable Systems
800 17th St. NW
Hickory, NC 28601
Tel: 828-327-5000 or 800-743-2671
Fax: 828-327-5973
www.corningcablesystems.com
Chapter 32

SIE
Siemens AG Energie- und Automatisierungstechnik
Balanstrasse 73
8000 Muunchen 80
Germany
Tel: 49(0)89-4144-0
Fax: 49(0)89-4144-8002
www.siemens.com

SKC
SKC, Inc.
863 Valley View Rd.
Eighty Four, PA 15330-9614
Tel: 412-941-9701 or 800-752-8472
Fax: 412-941-1369
www.skcinc.com
Chapters 10, 24, 25, 26, 29

SOL
Zellweger Analytics, Inc.
Neotronics/Solomat Division,
4331 Thurmond Tanner Rd.
P.O. Box 2100,
Flowery Branch, GA 30542
Tel: 770-967-2196 or 800-535-0606
www.zelana.com
Chapter 29

SON
Sono-Tek Corp.
2012 Route 9W, Building 3
Milton, NY 12547
Tel: 845-795-2020
Fax: 845-795-2720
www.sono-tek.com
Chapter 21

SPI
Spiral Biotech, Inc.
7830 Old Georgetown Rd.
Bethesda, MD 20814
Tel: 877-657-2030
www.spiralbiotech.com
Chapter 10, 24

STA
Staplex Co., Air Sampler Div.
777 Fifth Ave.
Brooklyn NY 11232-1695
Tel: 718-768-3333 or 800-221-0822
Fax: 718-965-0750
www.staplex.com

SPE
Stratton Park Engineering Company (SPEC), Inc.
5401 Western
Boulder, CO 80301
Tel: 303-449-1105
Fax: 303-449-0132
www.specinc.com
Chapter 30

STR
Ströhlein GmbH & Co.
Girmeskreuzstrasse 55
P.O. Box 14 63
D-41564 Kaarst 1
Germany
Tel: 49-21-31-606-0
Fax: 49-21-31-606-167
Chapter 25

SUM
Sumitomo Electric USA, Inc.
Park Avenue Tower,
65 East 55th St., 16th Floor
New York, NY 10022
Tel: 212-308-6444
Fax: 212-308-6575
www.sumitomo.com
Chapter 32

SUN
Sun Nuclear Corp.
425-A Pineda Court
Melbourne, FL 32940-7508
Tel: 321-259-6862
Fax: 321-259-7979
www.sunnuclear.com
Chapter 34

SYM
Sympatec Inc., System-Partikel-Technik
3490 U.S. Route 1
Princeton, NJ 08540-5706
Tel: 609-734-0404
Fax: 609-734-0777
www.sympatec.com
Chapter 16

TEC
Technical Associates
7051 Eton Ave.
Canoga Park, CA 91303-2197
Tel: 818-883-7043
Fax: 818-883-6103
www.tech-associates.com

TOP
Topas GmbH
Wilischstrasse 1
D-01279 Dresden
Tel: 49-351-2-54-10-08
Fax: 49-351-2-54-10-13
www.topas-gmbh.de

TSI
TSI Inc.
500 Cardigan Rd.
P.O. Box 64394
St. Paul, MN 55164
Tel: 651-483-0900
Fax: 651-490-2748
www.tsi.com
Chapters 2, 6, 7, 8, 13, 14, 15, 16, 17, 18, 19, 21, 24, 25, 29, 30

UNI
(United Sciences Inc.)
Monitor Labs
5310 N. Pioneer Rd.
Gibsonia, PA 15044
Tel: 724-443-8610
Fax: 724-443-4025
www.monitorlabs.com

URG
URG Corp.
116 South Merrit Mill Rd.
Chapel Hill, NC 27516
Tel: 919-942-2753
Fax: 919-942-3522
www.urgcorp.com
Chapters 10, 27, 29

VER
Verewa Umwelt- und Prozeßmeßtechnik GmbH
Kollaustrasse 105
D-22453 Hamburg, Germany
Tel: 49-40-55-42-18-0
Fax: 49-40-58-41-54
www.durag.de
Chapter 14

VIR
The VirTis Co.
815 Route 208
Gardiner, NY 12525-9989
Tel: 914-255-5000 or 800-765-6198
Fax: 914-255-5338
www.virtis.com

WED
Wedding & Associates, Inc. (Part of ***AND***)
www.andersoninstruments.com
Chapter 10

WHA
Whatman International Ltd.
Whatman House
St. Leonard's Rd.
20/20 Maidstone
Kent ME16 OLS, U.K.
Tel: 44(0)-1622-676670 or
(USA) 800-441-6555
Fax: 44(0)-1622-677011
www.whatman.com
Chapters 9, 27, 34

WOL
Wolfram Research, Inc.
100 Trade Center Dr.
Champaign, IL 61820-7237
Tel: 217-398-0700 or 800-965-3726
Fax: 217-398-0747
www.wolfram.com
Chapter 2

WYA
Wyatt Technologies
30 South La Patera Lane, B-7
Santa Barbara, CA 93227-3253
Tel: 805-681-9009
Fax: 805-681-0123
www.wyatt.com
Chapter 16

ZAA
Zefon Analytical Instruments
2860 23rd Avenue North
St. Petersburg, FL 33713-4211
Tel: 727-327-5449 or 800-282-0073
Fax: 727-323-6965
www.zefon.com
Chapter 24

INDEX

Note that many terms are listed primarily in terms of their acronyms. This has been done to include definitions and references for acronyms in the index rather than in a separate list. See also Appendix A for the glossary of terms. *Indicates commercial designations of products